P9-DGJ-371

Not to be
taken from
the Library

ENCYCLOPEDIA OF

ENERGY

TECHNOLOGY

— AND THE —

ENVIRONMENT

VOLUME 2

WILEY ENCYCLOPEDIA SERIES
IN ENVIRONMENTAL SCIENCE

ATTILIO BISIO AND SHARON BOOTS
ENCYCLOPEDIA OF ENERGY TECHNOLOGY AND THE ENVIRONMENT

ROBERT MEYERS
ENCYCLOPEDIA OF ENVIRONMENTAL ANALYSIS AND REMEDIATION

EDITORIAL BOARD

ENCYCLOPEDIA OF

ENERGY

TECHNOLOGY

—AND THE—

ENVIRONMENT

VOLUME 2

Attilio Bisio
Sharon Boots

Editors

A Wiley-Interscience Publication
John Wiley & Sons, Inc.

New York / Chichester / Brisbane / Toronto / Singapore

Copyright© 1995 by John Wiley & Sons, Inc.

Library of Congress Cataloging in Publication Data:

Encyclopedia of energy technology and the environment / Attilio Bisio,
    Sharon Boots, editors.
        p.   cm.
    Includes bibliographical references (p.) and index.
    ISBN 0-471-54458-2
    1. Power resources—Handbooks, manuals, etc.   2. Environmental
protection—Handbooks, manuals, etc.   I. Bisio, Attilio.
II. Boots, Sharon.
TJ163.235.E53   1995
333.79'03—dc20   94-44119

Printed in the United States of America

10 9 8 7 6 5 4 3 2 1

ENCYCLOPEDIA OF

ENERGY

TECHNOLOGY

— AND THE —

ENVIRONMENT

VOLUME 2

COAL GASIFICATION

L. Douglas Smoot
Brigham Young University
Provo, Utah

Coal gasification is the partial combustion of coal or other carbonaceous fuels in order to produce a fuel-rich gas. Fuels can be coal, char, liquefaction residue, or other solid or liquid hydrocarbons, while oxidizing gases include air, oxygen, steam, and carbon dioxide. Hydrogen may be added to promote the formation of certain products. Desired gaseous products are typically carbon monoxide, hydrogen, and methane. Some gasification technologies also produce a solid fuel or char and aromatic, liquid hydrocarbons. However, the large majority of current processes attempt to maximize the production of carbon monoxide and hydrogen. Potential uses of this product gas are as a:

1. Feedstock in the production of substitute natural gas (SNG)
2. Synthesis gas for subsequent production of alcohols, fertilizers, gasoline, or plastics
3. Gaseous fuel for generation of electrical power
4. Gaseous fuel for production of industrial steam or heat
5. Liquid and solid fuels from mild gasification

Coal gasification is thus a very flexible process, capable of producing gaseous, liquid, and solid fuels for varied uses in power generation, transportation, space heating, or chemical feedstocks.

See Combustion modeling; Commercial availability of energy technology; Fuel, synthetic; Lignite and brown coal; Underground gasification.

History

Coal gasification originated near the end of the 18th century, when there was an incentive to produce tar from coal as a substitute for wood tar. The manufacture of gas from coal was pioneered in 1792 by William Murdoch, who used the fuel to light a number of factories. The first commercial company to provide gas from coal was founded in 1812 by Friedrich Albert Winsor to supply gas to London (1). This coal gas was produced by the thermal decomposition of coal in a closed, cast-iron retort and was called town-gas and had a heating value of about 17 MJ/m^3, compared to 39 MJ/m^3 for a typical natural gas (2). This process not only produced the town-gas, which was composed of a mixture of hydrogen, methane, and carbon monoxide, but also produced large quantities of residual carbon, called gas coke, which could be used as a smokeless fuel. The operating temperature of the retorts was limited to about 900 to 1200 K and their lifetime was measured in terms of months. By the middle of the 19th century, the use of batch, horizontal, and firebrick retorts became widespread. These retorts had a life expectancy of several years rather than months and allowed operating temperatures above 1200 K, which increased the total gas yield. Horizontal retorts were replaced in the first decade of the 20th century by continuously-fed, vertical retorts. Continuous operation allowed for higher efficiency than the batch operation of the horizontal retorts. About this same time, offgas from slot-type metallurgical coke ovens was being collected and distributed as domestic gas. Coal gasification in horizontal or vertical retorts or coke ovens not only yielded gas and gas coke but also a tar by-product that could be utilized as a fuel or chemical feedstock. Only about 25% of the coal's original energy content was delivered to the gas with the gas coke and tar receiving about 50% and 5%, respectively; the remaining energy was used in gasification reactions or lost in the system. The economics of the process were driven by the high proportion of gas coke yields (1).

Attempts were made in the 1920s and 1930s to increase the total gas yield and reduce the coke by-product. The development of the cyclic water-gas process allowed the coke to be completely gasified. A fixed bed of hot coke was alternately reacted with air (air-blown) and steam (steamed). During the blowing portion of the cycle, the coke bed temperature increased and a low heating-value gas, called producer gas, was given off. Producer gas contained high amounts of nitrogen and was not distributed to consumers but rather used for heating the retorts. During the steaming part of the cycle, the bed was cooled and a nitrogen-free, medium heating-value gas called water gas was produced which was blended with the coal carbonization gases for distribution. A few integrated gasification processes were established, particularly in the United States, which carbonized coal in a retort and then transferred hot gas coke to a cyclic gas plant. However, in the mid-20th century, production of gas from coal gradually decreased as natural gas obtained prominence (2).

Major technological advances in gasification were accomplished by Germany just prior to and during World War II. The first Winkler fluidized bed gasifier was tested in 1926. It was the first use of fluidized-bed technology in any commercial application. Development and use of Lurgi fixed-bed gasifiers occurred in the mid-thirties. The demand for aviation and motor fuels in the war preparations and war efforts increased the attention these technologies received; however, the synthetic fuels industry, which supplied 70% of Germany's fuel, (a total of nearly 4 million metric tons per year) operated for the main part with the older style, water-gas reactors and coking ovens. Experimental testing on the Koppers process also took place during the war with the first commercial applications occurring in 1952. After the war, these technologies were commercialized. In 1955, SASOL (South African Coal, Oil, and Gas Corporation, Ltd.) started up its first gasification plant, combining Ruhrchemie-Lurgi fixed-bed gasifiers with the Arge Fischer-Tropsch reactors, (both of which are discussed later in the chapter) in order to produce liquid fuels and chemical feedstocks from their abundant reserves of coal. Other Lurgi, Winkler, and Koppers-Totzek installations were built in Europe, India, and Asia to provide gas for the production of ammonia, methanol, and synthetic fuels. A recent installation to produce synthetic natural gas using Lurgi fixed-bed reactors was built at Beulah, N. Dak. and started operation in 1984.

With the advent of the oil crisis in the United States in the early 1970s, interest in the use of coal increased dramatically. Major gasification development efforts were sponsored by the Federal government during that decade.

Gasification processes such as Bi-Gas, Hygas, and Sythane were developed, but none were commercialized, in part due to the increased availability of oil world-wide and corresponding decline in oil prices (3). However, industrial gasification process development continued into the 1980s. Of particular note were the industrial developments of entrained gasification processes by Texaco, Dow, and Shell. In the late 1980s, spurred by environmental concerns, the U.S. government initiated the Clean Coal Technologies program funded jointly by industry. Gasification processes that have been included in this new initiative are discussed elsewhere in this chapter. This demonstration and commercialization of technologies will continue through the end of this century. These government and industry cooperative efforts have been focused on the clean and efficient production of power from coal rather than synthetic fuels production.

Recently, interest has been generated for "mild" gasification processes which are designed to remove a fraction of the total volatile matter from the coal. The resulting product slate includes fuel gas and condensed liquids as well as a mildly treated char containing a sufficient fraction of volatile matter to maintain adequate combustion properties for use as a utility boiler fuel. This type of process shows promise in upgrading coals and lignites for utility use as well as providing gas and liquid products for fuels and chemical feedstocks. Earlier processes of this type have tended to treat the coal too severely, which produced a char with almost no volatile material. However, the difficulty in finding markets for large quantities of severely treated chars, which had poor combustion properties, has hampered the commercialization of these processes (4).

Basic Features

Many of the fundamental physical and chemical processes in coal gasification are similar to those in coal combustion (see COAL COMBUSTION). The coal is mixed with the gas, either in a fixed bed, a fluidized bed, or an entrained-flow gasifier. Moisture in the coal is released as the temperature of the coal begins to rise. Volatile matter, composed of gases and tarry substances, is evolved from the coal as the temperature continues to increase. This release of volatile material is called devolatilization. The volatiles react with the oxidizing gas to produce a variety of products including carbon dioxide, gaseous water, and carbon monoxide. The char, which is the solid that remains after devolatilization, is then oxidized by the carbon dioxide and steam, if temperatures are sufficiently high. Depending on gasification operating conditions, these fundamental processes can occur sequentially or simultaneously.

Comparison with Combustion. Coal combustion, which is the exothermic reaction of coal with oxygen or air to produce carbon dioxide and water, is a fundamental part of coal gasification. However, in coal gasification, typically only 20 and 40% of the oxygen or air needed for complete combustion is provided (5). The purpose of this partial combustion is to supply the energy necessary to complete the gasification of the coal particles. The solid carbonaceous material that is not combusted by oxygen reacts endothermically with carbon dioxide and steam, producing carbon monoxide, hydrogen, and methane.

Gasification of coal can be viewed as fuel-rich combustion. There are many similarities between gasification and combustion technologies, including use of the same types of furnaces or reactors, coal preparation and grinding, and use of varieties of coal. There are, also, many differences between the technologies of direct coal combustion and coal gasification, which are summarized in Table 1. Methods of modeling coal gasification processes (see COMBUSTION MODELING) are essentially identical to those for coal combustion (6).

Physical and Chemical Processes. While essentially the same fundamental physical and chemical processes occur during gasification and direct combustion, these basic processes interact in different ways with different results. Significant differences among the interactions of these basic reactions also occur in the different types of gasifiers: fixed bed, fluidized bed, and entrained bed. The following physical and chemical processes control both combustion and gasification of coal:

1. Turbulent mixing of reactants (except fixed beds)
2. Turbulent dispersion of particles (except fixed beds)
3. Convective coal particle heatup
4. Radiative coal particle heatup
5. Coal devolatilization

Table 1. Differences in Direct Coal Combustion and Coal Gasification

	Direct Coal Combustion	Coal Gasification
Operating temperature	Lower	Higher
Operating pressure	Usually atmospheric	Often high pressure
Ash condition	Often dry	Often slagging
Feed gases	Air	Oxygen, steam
Product gases	CO_2, H_2O	CO, H_2, CH_4
Gas cleanup	Post-scrubbing	Intermediate scrubbing
Pollutants	SO_x, NO_x	H_2S, HCN, NH_3
Char reaction rate	Fast (wtih O_2)	Slow (with CO_2, H_2O)
Oxidizer	In excess	Deficient
Tar production	None	Sometimes
Purpose	High temperature gas	Fuel-rich gas

[a] With permission from Ref. 7.

6. Gaseous reaction of volatiles products
7. Heterogeneous reaction of char
8. Formation of ash/slag residue

Features of basic gasification processes, compared to combustion include the following, particularly in oxygen-blown, fluidized and entrained beds:

1. Peak gasification temperatures are often higher in gasification due to absence of diluent nitrogen. Thus, the extent to which the coal devolatilizes, which is a strong function of peak temperature (see Fig. 1) (8), can be greater during gasification, reaching as high as 80% of the total dry, ash free (daf) coal (9).

2. These volatile products quickly mix with available oxygen in the gas phase, completely depleting oxygen, and producing the very high temperatures that cause the increased extent of devolatilization. High concentrations of CO_2 and increased concentrations of H_2O are produced through these gaseous reactions.

3. The residual char is reacted relatively slowly through heterogeneous attack by the CO_2 and

steam. This process is faster at higher temperatures, but is still slow compared to the oxygen-char reaction that occurs in direct combustion.

In direct combustion, oxygen is usually present in excess and dominates the consumption of residual char.

Process Types

Key process types for gasification include, as illustrated in Fig. 2 (10): fixed (or moving) bed, entrained flow, fluidized bed, and other process types (eg, molten bath). Further division is made between those containing nitrogen (ie, air-blown) in the product gas and those without nitrogen (ie, oxygen-blown), those that are pressurized and those that are not, and those that are slagging and those that are not.

Conventional Gasification. Coal gasification in fixed beds is the oldest and most common method. In fixed-bed gasification, coal is crushed to centimeter-sized pieces and fed to the top of a bed. Fixed-bed gasifiers generally use counter-current flow systems as shown in Figure 2 (10). The coal is heated by the upward flow of hot gases as the solid bed of particles moves slowly downward. As the coal moves downward, it passes through a devolatilization zone, then a char gasification zone, finally reaching a combustion zone at the bottom of the bed where the reactant gases are injected and heated by the ash. Because of the counter-current flow nature of fixed-bed gasifiers, volatile matter is released from the coal and removed from the gasifier without passing through a high-temperature zone, minimizing thermal decomposition of the hydrocarbon products. Therefore the product gas for fixed-bed gasifiers is relatively high in tar vapor and methane. The solids residence time for fixed beds is the longest of the flow type reactions, being several hours. Gas velocities in fixed beds are limited to prevent elutriation of small particles by the gas stream (11).

Modifications have been made to traditional fixed-bed gasifiers to promote uniform reaction of the coal and the oxidizer. The distributor or grate can be rotated to more uniformly distribute the coal. The bed may be stirred or agitated in order to remove voids, prevent channeling of the gases, and create a more uniform bed. Kiln-type gasifiers are logical extensions of this concept where the gasifier is placed at a near-horizontal angle and rotated, like a cement kiln. Rotating grates are also used where the coal is placed in a shallow layer on the grate, gases passed up through the grate and the coal, and the reacted coal removed at the end of a single cycle of the grate.

In an entrained-bed gasifier, the coal is pulverized to a fine powder and injected into the gasifier with the reactant gases where the gasification reactions take place in a reaction zone similar to a pulverized coal combustion system (see Fig. 2). Gas residence times are on the order of seconds for all gasification flow systems, with entrained systems having characteristically the shortest gas residence times. In entrained flow systems, solids residence times are similar to that of the gas. Because of the short residence time of the solid particles, entrained flow sys-

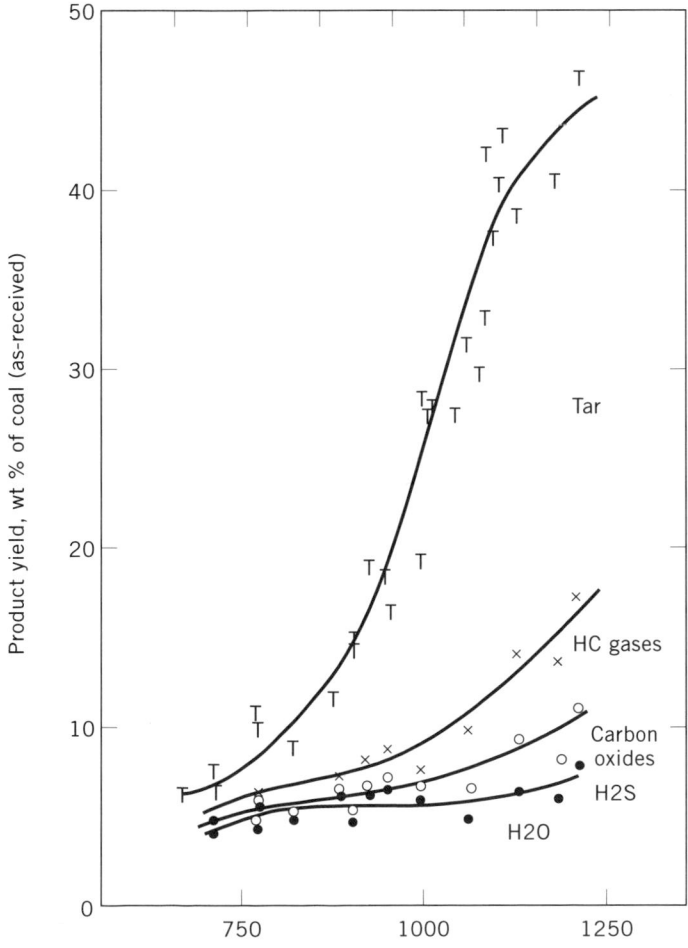

Figure 1. Devolatilization product distribution for coal heated to various temperatures (with permission from ref. 8).

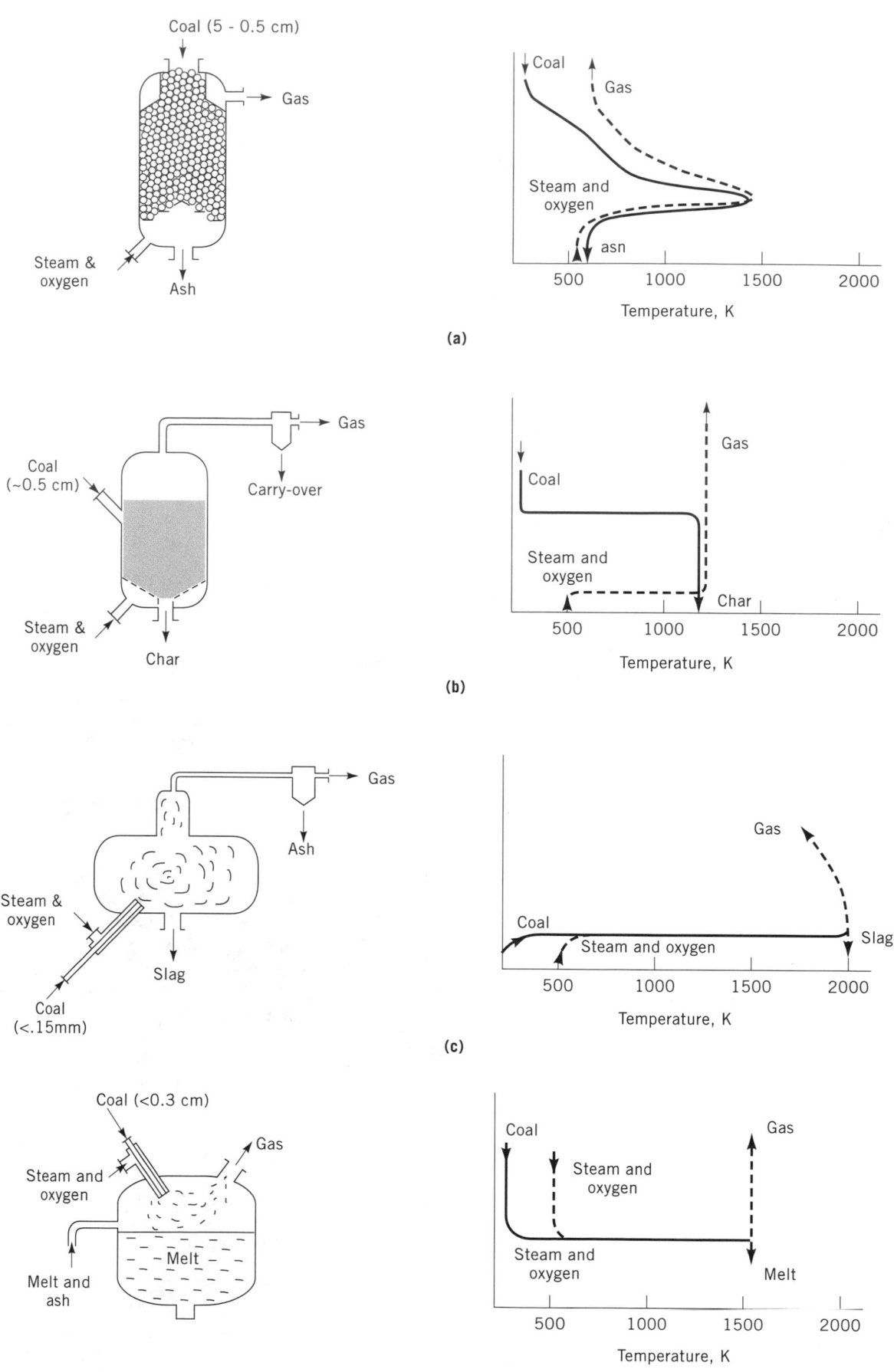

Figure 2. Classification and characteristics of principal gasification systems (with permission from ref. 10). **(a)** Fixed bed, nonslagging, **(b)** fluidized bed, **(c)** entrained flow, **(d)** molten bath.

tems generally operate under high temperature, slagging conditions in order to obtain reaction rates fast enough for high carbon conversion. Tars and methane are not produced at these high temperatures. Nonslagging operation in an entrained-flow gasifier may only be attractive for hydrogasification where partial conversion of the coal is acceptable (11). Slag has been shown to be less environmentally hazardous than ash.

A fluidized bed consists of a bed of particles set in vigorous, turbulent motion by the reactant gases blowing upward through the bed (Fig. 2). The particles are mostly inert materials such as coal ash, sand, or sulfur sorbents such as limestone or dolomite. Coal particles make up only a small percentage of the bed. To prevent the ash from sintering and subsequently losing fluidization, fluidized-bed gasifiers are restricted to operating at non-slagging temperatures. Nevertheless, fluidization temperatures can be high enough to cause thermal decomposition of tars and methane. This process results in relatively low concentrations of these constituents in the product gas which may or may not be advantageous, depending on the application. The solids residence time in a fluidized bed is about one hour. Because of the shorter residence times and lower gasification temperatures, a lower carbon conversion can be expected in fluidized beds than for a fixed-bed system. In some instances, unconverted char from a fluidized bed is collected and fed to a separate gasifier or combustor (1,11).

In a molten-bath system, coal is gasified in a bath of molten slag, metal, or salt. Both slagging and non-slagging operations are possible, depending on the temperature of the bath. Molten slag and metal baths operate at slagging temperatures (1700 to 2000 K) whereas molten salts can be operated under non-slagging temperatures as low as 1300 K. The reactant gases may either be introduced from above through jets which penetrate the bath (Fig. 2) or may be fed to the bottom of the bath (1).

Underground Coal Gasification. Underground coal gasification (UCG) processes are similar to the coal gasification processes in the fixed-bed gasifiers. Boreholes for blast injection and gas removal are drilled through the overburden of a coal field. Linkings between the blast injection and the gas removal boreholes are provided within the coal seam. Blast air or oxygen, with steam, are injected in one borehole and product gas is removed from the other. Three overlapping zones are established between the blast and the gas boreholes: combustion, gasification, and devolatilization zones. In the combustion zone, carbon dioxide and steam are generated. Additional steam comes from the blast and from the coal moisture. Carbon dioxide and steam are reduced in the gasification zone to carbon monoxide and hydrogen. More carbon monoxide and hydrogen, as well as light hydrocarbons and tar, are evolved in the devolatilization zone. Air injection yields a low heating value gas; a medium heating value gas is produced with oxygen.

UCG combines extraction and conversion of coal into a single step. Thus, the need for mining equipment and gasification reactors is eliminated. It is a promising technology for coal deposits that are not economically or technically feasible to recover by conventional mining technologies, particularly by shaft mining. UCG can also eliminate some of the health, safety, and environmental problems associated with conventional shaft mining of coal. UCG is expected to have much smaller environmental impacts than surface mining or shaft mining of coal, but it is not without problems. Particularly important is the potential for ground water contamination (11).

Comparison and Selection of Process Type. A brief summary of the various features of major coal gasification process types is given in Table 2. Key differences among the various processes relate to coal particle size, operating temperature, and coal and gas mixing patterns. Specific

Table 2. Comparison of Coal-Gasification-Process Types (Air- and Oxygen-Blown)[a]

	Fixed Bed	Fluidized Bed	Entrained Bed
Residence time	1–3 h[b]	20–150 min[b]	0.4–12 s
Coal size	6–50 mm	500–2400 μm	10–150 μm
O_2/coal	0.14–0.81	0.25–0.97	0.28–1.17
Steam/coal	0.28–3.09	0.11–1.93	0.1–1.20
Coal type	Most types; no fines	Noncaking coals	All types
Temperature range, K	1150–1300	600–1470	1150–2500
Pressure range MPa (atm)	0.1–2	0.1–10	0.1–30
	(1–20)	(1–100)	(1–100)
Product gases, mol %			
$CO + H_2$	39–66	2–80[b]	35–91
CH_4	2–15	3–68	0.1–17
HHV, kJ/std m^3	9,000–12,000	11,000–30,000	4,000–20,000
Commercial operations	Extensive;	Slight	Moderate;
	Lurgi	Winkler	Koppers-Totzek
	Wellman-Galusha		Texaco, others
Principal advantages	High turndown ratio	Lower temperature	Smaller, simple design
	Mature technology	Reduced thermal losses	All coal types
	Low thermal losses	Variety of coal sizes	Highest capacity/volume
		Moderate residence time	

[a] From Ref. 11 with permission.
[b] Information from ref. 12. All other information from ref. 13.

gasification processes are designed to accept a limited coal size range. Acceptable particle sizes decrease from fixed-bed to fluidized-bed to entrained-flow systems. Allowable particle sizes are determined mainly by particle residence time in the gasifier but also, in the case of fluidized-beds, the size is limited by the requirements for proper bed fluidization. Fixed-bed gasifiers have a difficult time handling a large amount of fine particles which affect bed temperature as well as overload cyclonic recycle systems. Developments in fixed-bed gasification are moving toward the acceptance of a greater proportion of fine material in the feed. Entrained-flow gasifiers have very short particle residence times and so require pulverized particles to ensure high carbon conversion efficiencies (10).

Large coal sizes (6–50 mm) and moderate operating temperatures lead to very long residence times (hours) in fixed or moving beds. Highly pulverized coal (1–150 μm) is used in entrained systems with very high operating temperatures (up to 2200 K in oxygen-blown gasifiers) that result in very short (<1 s) residence times. Fluidized-bed gasifiers operate with residence times of minutes and with coal size (500–1000 μm) but at low temperatures, (1100–1200 K) which result from cooling by in-bed tubes. Rates and the nature of coal reaction differ dramatically among these process types. Other key process variables include operating pressure and reactant composition (ie, steam/oxygen/coal ratio). High-pressure operation is common in all types of gasifiers and is particularly attractive for combined cycle operation when generating electrical power (7).

The choice of a gasification process is determined mainly by the requirements of the products. If the production of a low heating value gas is desired, then the presence of nitrogen is no great disadvantage, and air-blown gasifiers are usually chosen. If, however, the gas is to be used for the production of hydrogen or chemicals, nitrogen dilution is undesirable and an oxygen-blown gasifier is required. If methane is an undesirable constituent, then it may either be separated and used as a fuel or reformed at some stage, or high temperature gasifier operation can be employed which minimizes methane production. The requirements for methane processing add to production costs and a low methane concentration in the initial product might be preferred (2). If liquid and solid products are sought, then mild gasification with fixed or rotary beds

operating at lower temperatures is chosen. Table 3 shows typical product compositions of a number of gasifiers, some of which are described later. While product gas composition is a strong factor influencing the choice of a gasification process, other factors must also be included such as efficiency, capital cost and reliability. These factors will also naturally affect the choice of a particular gasification process (2).

SCIENCE OF COAL GASIFICATION

Chemical Reactions

Chemical reactions can either occur between gaseous species and solid coal or char (heterogeneous) or between gaseous reactants and the products of devolatilization and gasification (homogeneous). A principal heterogeneous reaction during the gasification of coal and other carbonaceous material is:

$$C(s) + \tfrac{1}{2} O_2 \rightarrow CO$$
$$\Delta H^\circ_R = -1110.5 \text{ mJ/kmol} \tag{1}$$

ΔH°_R is the standard heat of reaction at 298 K and atmospheric pressure. This partial-combustion reaction is exothermic; that is, it liberates heat, as signified by the negative sign. Further, the reaction of carbon does not stop at CO, but any remaining oxygen rapidly reacts with CO in the gas phase to produce CO_2:

$$CO + \tfrac{1}{2} O_2 \rightarrow CO_2$$
$$\Delta H^\circ_R = -283.1 \text{ mJ/kmol} \tag{2}$$

Thus, for a fuel-rich system, in order to consume the remaining solid carbon, the much slower endothermic (heat consuming) reaction must occur:

$$C(s) + CO_2 \rightarrow 2 \, CO$$
$$\Delta H^\circ_R = +172.0 \text{ mJ/kmol} \tag{3}$$

In order to control high temperatures resulting from $C(s) - O_2$ reactions, and to increase the heating value of the product gas through production of hydrogen, steam is

Table 3. Typical Product Compositions of Various Gasification Processes[a]

Gasification Process	Raw Gas Composition (Mole %)				
	CH_4	CO	H_2	CO_2	H_2/CO
Hydrane	73.2	3.9	22.0		5.9
CO_2—Acceptor	20.9	17.0	53.8	6.6	3.2
Lurgi	9.4	18.5	40.4	29.5	2.2
Synthane	24.5	16.7	27.8	28.9	1.7
Molten Salt	7.5	33.6	45.0	13.3	1.3
Hygas	18.7	23.8	30.2	24.5	1.3
Winkler	3.1	33.4	41.9	20.5	1.3
Koppers-Totzek		55.8	36.6	6.0	0.7
Bi-gas	15.6	55.8	36.6	6.0	0.7
Atgas	20.0	69.7	9.6		0.1

[a] From ref. 2 with permission.

often added as a reactant:

$$C(s) + H_2O(g) \rightarrow CO + H_2$$
$$\Delta H_R^{\circ} = +131.4 \text{ mJ/kmol} \tag{4}$$

This reaction is also endothermic and must rely on the heat release from the $C-O_2$ reactions for energy requirements. Further, the rates of reactions 3 and 4 are very slow compared to the rates of 1 and 2 (3). However, the resulting product gas has an increased heating value.

The results of a series of computations for various mixtures of a typical, high volatile (HV) bituminous coal with various proportions of steam and oxygen, assuming thermodynamic and chemical equilibrium can be found in ref. 14. Temperature increases with increasing oxygen content and decreases with increasing steam content. Also shown is the peak cold gas efficiency, which is the ratio of heating value of the product gas at ambient temperature to the heating value of the coal (dry, ash free–daf). The cold gas efficiency increases with increasing steam and decreases with increasing oxygen. Thus, for development of a given gasification process, a balance between sufficiently high reacting temperature for flame stability and carbon conversion, and acceptably high cold gas efficiency, must be achieved.

In a coal–steam–oxygen system, the homogeneous, water-gas shift reaction is also important:

$$CO + H_2O(g) \rightarrow CO_2 + H_2$$
$$\Delta H_R^{\circ} = -41.0 \text{ mJ/kmol} \tag{5}$$

This slightly exothermic shift reaction produces CO_2 from CO and tends to control the final product distribution. In some gasification processes, hydrogen is added as a reactant in order to increase the quantity of methane as a product according to the hydrogasification reaction:

$$C(s) + 2 H_2 \rightarrow CH_4 \tag{6}$$

and the methanation reaction:

$$CO + 3 H_2 \rightarrow CH_4 + H_2O \tag{7}$$

In fuel-rich combustion, the sulfur in the coal is released mainly as hydrogen sulfide with a small amount of carbonyl sulfide and the fuel-bound nitrogen is released as elemental nitrogen, ammonia, and hydrogen cyanide. In order to capture the sulfur, limestone or dolomite may be fed to the gasifier:

$$CaO + H_2S \rightarrow CaS + H_2O \tag{8}$$

If the limestone is calcined prior to feeding to the gasifier, it may also capture carbon dioxide (1):

$$CaO + CO_2 \rightarrow CaCO_3 \tag{9}$$

Reactant Gases

Oxygen, steam, and hydrogen compose the three basic feed reactant gases for coal gasification and are introduced into the gasifier independently or in combination.

Oxygen can be supplied either by air injection or as pure oxygen. Steam is generally used to control temperature but also can increase the hydrogen content of the product gas through the water-gas shift reaction. Hydrogen addition can be utilized to increase product gas methane yields. If large amounts of nitrogen in the product gas are acceptable, air can be used as the oxidizing gas. Sensible heat is required to increase the temperature of large quantities of nitrogen. Slagging temperatures can be reached using air alone as the gasifying agent since the cooling potential of the steam is not present. However, if the air is preheated or if oxygen-enriched air is used, some steam addition may be required to control the temperature to maintain a heat balance (1). If nonslagging operation with air alone is desired, then additional heat removal capabilities other than the endothermic reactions may be required such as heat transfer to water walls to generate steam.

The product gas requirements dictate which reactant gases are suitable. If a product gas with medium heating value is required, then oxygen-steam, steam alone, or hydrogasification is required. If a low heating value gas with diluent nitrogen is acceptable, then air or air-steam gasification is suitable. The choice of reactant gas also affects the system throughput. Gasification with oxygen typically allows for two to three times the throughput as air gasification for an equivalent system. Hydrogasification typically has high throughputs since it is operated at elevated pressures (1).

Reaction Processes and Times

The time required for the different gasification reactions varies with reaction type. Therefore, gasifier design must allow for adequate residence time for desired reactions to occur. The rate-determining process for heterogeneous reactions is more difficult to determine than for homogeneous reactions. For heterogeneous reactions to take place, the gaseous reactant must diffuse from the bulk gas to the particle and, perhaps, into the particle. This diffusion causes the overall heterogeneous reaction often to be much slower than the devolatilization process, requiring seconds for small particles to several minutes for larger particles. These rates vary with coal type, temperature, pressure, char characteristics, and oxidizer concentration (see COAL PYROLYSIS). Reactions of char with CO_2, H_2, and steam are considerably slower than with oxygen. The rates of carbon oxidation by steam and CO_2 are of the same order of magnitude while the hydrogasification reaction (ie, H_2 + C) is several orders of magnitude slower than the steam-char and CO_2-char reactions (15).

Devolatilization and char oxidation may take place simultaneously, especially at very high heating rates. During both of these processes, particularly devolatilization, product gases leaving the particle create a flow against which the oxidizer must diffuse. Therefore, heterogeneous reaction rates are determined by competing processes that include diffusion of the gaseous reactants and products to and from the particle surface and within the particle, and chemical reaction at the particle surface. Depending on gasification conditions and particle properties, either of these two processes can be rate-controlling.

The reactivity of the chars to the gasifying agent is important to gasifier operation and product gas yields. Chars vary in their chemical reactivity to hydrogen to a greater extent than they do to steam. Highly reactive chars decompose steam more readily and remain reactive at a lower temperature than do less reactive coals. Reactivity can be measured in the laboratory, but care must be taken in the interpretation of laboratory measurements since it is difficult to achieve the temperature and the reaction histories in the laboratory that exist in the gasifier. High char reactivity influences the gasification process by improving methane formation which is an advantage to SNG processes. It also reduces oxygen consumption by allowing steam decomposition to occur at a lower gasification temperature as well as reducing the demand for steam used for bed temperature control (10). Different coals and chars have different properties such as porosity, surface area, and chemical and mineral matter composition which can greatly affect reactivity. Also pretreatment of coals and chars can strongly affect reaction rates (16).

Effects of Temperature

Reaction temperature can significantly affect gasifier operation by determining chemical reaction rates, equilibrium conditions, and ash properties. Many gasifier designs are limited in their operating temperature due to the physical state of the ash. Gasifiers can be divided into categories based on the physical state of the ash, ie, slagging and non-slagging reactors. Non-slagging reactors include both dry ash and agglomerating ash reactors. For most coals, the ash from a gasifier operating at temperatures below about 1300 K can be removed dry without sintering or slagging. Ash-agglomeration, or sticky ash, generally occurs in the temperature range 1300 to 1500 K depending on the chemical composition of the ash. Operation above approximately 1500 K generally results in the ash forming a molten slag (1).

The equilibria and kinetics of the gasification reactions and hence product gas composition are also dependent on gasification temperature. High temperatures favor production of carbon monoxide and hydrogen and discourage methane production. Therefore, the product gas from slagging gasifiers generally has relatively low concentrations of carbon dioxide and steam and relatively high concentrations of carbon monoxide and hydrogen. At slagging temperatures, quantities of methane and tars are negligible since they undergo thermal decomposition. Countercurrent gasifiers allow for coal devolatilization to occur at low temperature with oxidation of the residual char at slagging temperatures. In this case, methane and tar may be present in significant quantities in the product gas. Since methane decomposes at high temperatures, gasifiers designed for high methane production are operated under non-slagging conditions.

Gasifiers operating under slagging conditions generally have higher throughputs because high temperatures allow increased reaction rates requiring shorter gas and solids residence times. However, slagging gasifiers also require more oxygen, resulting in greater energy use for air separation. Slagging gasifiers are limited to coals with an ash type capable of creating a sufficiently mobile slag at the operating temperature. Therefore, ashes with a high fusion temperature are generally unfavorable for slagging operation although slag properties can be modified by the addition of a fluxing agent such as limestone. Coals with ash that create a slag that is too mobile may cause excessive attack of the refractory lining of the gasifier (1).

Low-rank coals are usually preferred when gasifying under non-slagging conditions due to their higher reactivity. In order to increase reaction kinetics and improve equilibrium yields, non-slagging gasifiers using steam as a reactant are generally operated at the highest temperature possible. Therefore, coals with a high ash fusion temperature are preferred. Successful operation at ash-agglomerating temperatures depends on control of the rate of agglomeration, the size of the agglomerates, and the rate of removal. Gasification at high temperatures has a number of advantages such as higher reaction rates and the ability to utilize higher-rank, less reactive coals. However, the technology is generally regarded as more difficult than gasification at low temperatures (1).

Effects of Pressure

Gasifiers are designed to operate at either atmospheric or elevated pressure. The major advantages of elevated pressure are the increased throughputs of coal and the opportunity to extract energy more efficiently. Increased pressure has the effect of increasing the overall gasification reaction rate somewhat. However, the increase is not proportional to pressure increase since only reactions which are chemically rate-controlled are significantly influenced. Reactions that are diffusion-controlled, such as char oxidation at higher temperatures, are not affected significantly by pressure. High pressure favors the formation of methane. The effect on equilibrium concentrations, however, is small for pressures up to about 3 MPa, compared with reaction temperature.

Many units located downstream of the gasifier for emissions control or gas processing are designed to operate at elevated pressure. In such cases the option is available to either operate the gasifier at atmospheric pressure and compress the product gases or to compress the reactants and operate the gasifier at elevated pressure. Since the product gas volume can be 50 to 250% greater than the reactant gases, depending on gasifier operation, additional energy is required to compress the product gases which results in a significant decrease in efficiency for atmospheric pressure gasification (1). High pressure operation allows for increased throughput per unit reactor volume. However, the increase in throughput per unit volume of the gasifier has been shown to be only proportional to the square root of the pressure. At high pressures, process efficiency and throughput are generally increased for large-scale operations. For this reason, much of the current interest in coal gasification is directed towards elevated pressure systems even though elevated pressure gasification is a more difficult technology than atmospheric pressure gasification (1).

Effects of Coal

The major coal properties that must be considered when choosing a gasification process include coal composition (ie, fixed carbon, volatile matter, moisture, and ash), caking tendency, reactivity, and particle size (10). The COAL CHARACTERISTICS section describes the ASTM method for proximate analysis of coal wherein moisture, fixed carbon, proximate volatiles and ash are defined. The coals used and product gas desired influence the choice of gasifier type and gasifying agent, as well as gasifier design and operation. Large-scale gasification plants may be installed near a dedicated source of coal. With such a dedicated fuel supply, gasification process type must be tailored to accept the fuel available and the choice of a particular process becomes important, with regard to both the gasifier design and operation, and the effluent treatment facilities. For a relatively small-scale operation where coal is shipped to the gasification site, the coal supply characteristics may be optimized in relation to a desired gasification process chosen for other reasons. The ability to choose coal characteristics allows purchasing flexibility and optimization of fuel cost and facility operation with the added ability to select fuel supplies to achieve an economic advantage. The advantage of having a dedicated fuel supply is guaranteed fuel availability and minimum fluctuation in raw material costs. In either case, coal properties must be known to operate the gasification process efficiently.

The moisture content of coals varies significantly with coal type from typically less than 1–2% with anthracites to more than 40% with some low-rank subbituminous coals and lignites. Gasifier design may dictate acceptable coal moisture levels. Moisture has the effect of increasing the hydrogen content of the product gas and cold gas efficiency as well as decreasing gasification temperature. In instances where high coal moisture contents are detrimental, some drying may be required. Fixed-bed gasifiers can generally accept coal moisture levels up to about 35%, providing the ash content is less than about 10%. For entrained-flow and fluidized-bed gasifiers, where the coal is pulverized or ground to a fine granular composition, high coal moisture content may adversely affect coal handling ability during grinding, crushing, and feeding to the gasifier. In such situations, it may be required to reduce the moisture content to less than 5% by drying. In the fully entrained system, the residual moisture contributes to the gasification steam but requires process heat to evaporate it.

Ash is the inert mineral matter present in all coals and varies in amount from a few percent up to around 50%. Although in some gasifier designs, such as fixed-beds, ash can provide beneficial preheating of the gasifying agent, it is generally considered best to minimize ash content in gasification systems. Ash can vary in its properties such as acidity, fusion temperature, and corrosiveness. Therefore, ash composition must be considered when selecting materials for gasifier construction, especially in slagging gasifiers. When operating at slagging temperatures, it may be desirable to add reagents such as fluxes to improve the flow characteristics of the slag or to adjust its chemical behavior.

Provisions must be made for handling ash throughout the gasifier system while solutions to problems associated with fouling and slagging in high-temperature systems increase the complexity and cost. These problems are often more challenging for elevated pressure systems. Ash-handling techniques are dictated mainly by the gasification temperature which determines the properties of the ash. Ash is either removed as a liquid slag from gasifiers operating under high-temperature slagging conditions or as a fine dust from non-slagging gasifiers. Ash content of coal can be reduced by physical and chemical cleaning processes prior to gasification. In fluidized beds, the ash tends to accumulate in the fuel bed which presents a special problem. Ash removal from these gasifiers can be accomplished by either causing ash sintering and agglomeration to allow it to separate from the bed, or external circulation of ash particles through an entrained flow combustor to melt and separate the ash as a liquid slag.

The composition of the ash determines the temperature at which the ash changes between a solid or liquid phase called the ash fusion temperature. Since gasifiers are designed to handle ash in a particular phase, the ash fusion temperature dictates the maximum temperature at which a fuel bed can be operated without causing fusion or liquefaction of the ash. Alternatively, it determines the approximate temperature that must be exceeded if slagging operation is desired to remove the slag as a liquid. Ash can be diluted significantly with carbon which can allow a fluidized bed to operate at temperatures well in excess of the fusion point without incurring ash fusion. Also, ash fusion can occur below the fusion point if ash concentrations increase above a certain threshold limit.

In fixed-bed gasifiers, ash fusion is prevented by controlling bed temperature through the addition of steam. This limits the maximum temperature generated in the combustion zone to below the ash fusion temperature. For slagging gasifier operation, not only must the ash be above the fusion point but the viscosity of the molten slag must lead to good flow characteristics for proper slag removal and gasifier operation. Ash can have a low fusion temperature but poor slag flow characteristics which may require the addition of fluxes for satisfactory slag removal. Slag viscosity-temperature correlations can be determined by direct measurement. Extensive information in the form of phase diagrams and composition-viscosity-temperature correlations currently exists so that the necessary information can be calculated from an ash analysis (10).

Caking is the tendency of coals to stick together as they soften during heating. Caking occurs with certain coals as the volatile matter is driven off during devolatilization. This phenomenon is most prevalent in higher rank bituminous coals which not only become sticky but also swell. Non-caking coals may either retain their original shape during heating or may fracture, creating a large number of small pieces. The caking behavior of coals must be taken into account when specifying a gasification process design. While some gasifiers may be specifically designed to handle caking and swelling coals, others require caking coals to be pretreated to eliminate or reduce their caking tendencies. Typically, fixed-bed and fluidized-bed gasifiers tend to be sensitive to strongly caking and swelling coals while entrained systems can more readily accept coals with these properties (10).

Pollutants from Coal Gasification

Pollutant formation processes in coal gasification differ from those in coal combustion. The main difference is that under fuel-rich conditions, sulfur from coal is converted to H_2S and small amounts of COS rather than to SO_2. The sulfur species, H_2S and COS, can be removed from the product gas by a variety of commercially available processes more easily than can the SO_2 from combustors. Another difference is that under reducing conditions, almost no nitrogen oxide pollutants (NO_x) are formed. Nitrogen from coal is converted to NH_3 and HCN. However, when the product gas is used as a fuel, NO_x is still formed due to gas combustion (11).

Environmental impacts of some secondary pollutants generated by coal gasification may be of greater concern than from coal combustion. These pollutants are mainly solid residues including: (1) spent catalysts, (2) inorganic solids and sludges from acid gas removal and water pollution control, and (3) tar and oil sludge (17). The types and amounts of these pollutants generated depend on the particular gasification processes and waste treatment methods used. Liquid wastes may also be generated in larger quantities during coal gasification compared to direct coal combustion since water is used as a coolant and as a means of quenching and washing the crude gas. In these processes water can collect organic, inorganic, and trace element contaminants. Both solid and liquid wastes may contain trace elements and organics such as lead and mercury which may require special processing (17,18).

Coal ash contains a wide variety of trace elements depending on the origin of the coal. Many of these elements, such as mercury, arsenic, lead, cadmium, and beryllium are toxic and some may be in vapor form during the high temperature conditions present during gasification. Metals present in coal ash may be in reduced state in the gasifier due to the highly reducing environment present. Many trace elements such as mercury, arsenic, cadmium, selenium, and zinc have low boiling points and may be carried out of the gasifier with the raw gas. Metal halides are also fairly volatile. Other sources of volatile toxic compounds are reactions of metals with carbon monoxide to form carbonyls of iron, nickel and cobalt, and formation of arsenic, antimony, and selenium hydrides. Also, most alkali metal salts have significant vapor pressures at gasification conditions. Table 4 shows the estimated volatility of trace elements present in coal and also the estimated amount of trace elements emitted in the various waste streams for a 11,000 metric ton/day gasification plant. Studies have indicated that over 90% of the chlorine in the coal is ultimately released in gas streams as hydrogen chloride. However, the chemical form and process stream in which most of the other components are found is unknown (19).

APPLICATIONS

Synthesis Gas Production

Hydrogen and carbon monoxide produced by coal gasification are widely used commercially to produce ammonia and some methanol. Typically, the raw product gas requires purification by removing sulfur compounds and carbon dioxide, and its composition is adjusted as required by reforming and shifting, that is creating more hydrogen through the water-gas shift reaction (see SYN-

Table 4. Estimated Volatility of Trace Elements[a]

	Typical Coal ppm	% Volatile[b]	kg/day[c]
Cl	1,500	90+	14,700
Hg	0.3	90+	3
Se	1.7	74	14
As	9.6	65	68
Pb	5.9	63	5
Cd	0.8	62	5
Sb	0.2	33	1
V	33	30	108
Ni	12	24	31
Be	0.9	24	31
Zn	44	eg, 10	0.7
B	165	eg, 10	180
F	85	eg, 10	93
Cr	15	nil	nil

[a] From Ref. 19 with permission.
[b] Volatility is based on gasification experiments, but chlorine volatility is taken from combustion tests, while zinc, boron, and fluorine taken at 10% for illustration in absence or data.
[c] Estimated amount volatile for 11,000 mt/day of coal.

THESIS GAS). The production of ammonia requires hydrogen. To produce additional hydrogen from coal gasification product gas, carbon monoxide is converted by the shift reaction with steam, to hydrogen and carbon dioxide. A cobalt-molybdenum catalyst has been used to enhance this reaction. Removal of the carbon monoxide from the raw gas enables the sulfur compounds to be removed with the carbon dioxide in a downstream stage. Methane is readily separated from hydrogen through cryogenic processing since the boiling points of methane (109 K) and hydrogen (21 K) are significantly different. The methane-rich stream may then be used as a fuel or recycled after passing through a reforming stage. Methane reforming to hydrogen can be carried out using steam at high temperature and pressure in the presence of a nickel-based catalyst. Using these procedures, H_2/CO ratios up to 5:1 can be obtained. The hydrogen produced by this type of process is sufficiently pure for many industrial purposes. However, hydrogen used in ammonia production requires the removal of all oxygen-containing compounds such as carbon monoxide and carbon dioxide (by adsorption or washing with liquid nitrogen) (2). Ammonia is important in fertilizers and as a feedstock for other chemical syntheses.

Carbon monoxide may be separated as a pure gas cryogenically or by adsorption techniques such as the CO-SORB process reported by Haase and Walker (20). In this technique, carbon monoxide is separated from the gas stream by formation of a molecular complex by using a solution of cuprous aluminum chloride in an aromatic base. The off-gas is composed of a mixture of methane and hydrogen which may either be separated or used as a fuel gas. The carbon monoxide is recovered as a pure (approximately 99%) gas which may then be used as a chemical feedstock. Use of the purified carbon monoxide gas includes the production of esters, acrylic acid, hydroxyl acids, formamides, and metal carbonyls. Some of these products are further processed into higher esters, ketones, aldehydes, glycols, and plasticizers. The metal carbonyls, of which nickel is an example, can be converted into powdered metals of high purity (2).

Methanol may be synthesized from gasification product gas which contains carbon monoxide, hydrogen, and carbon dioxide. Methanol synthesis generates a large amount of heat and uses carbon dioxide to help control the reaction temperature since its heat of reaction with hydrogen is only about half that of carbon monoxide. In addition, the water produced aids in cooling the reaction temperature. Surplus carbon dioxide can then be removed in the stage following the shift reaction. Processes aimed at methanol synthesis are carried out under pressure at slightly elevated temperatures (about 500 to 550 K) and in the presence of a catalyst, typically copper-based. The crude methanol is condensed and distilled while unreacted gases are recycled in the synthesis loop. A purge stream of methane-containing gas is either used as fuel or reformed (2). Methanol is used in the production of formaldehyde and acetic acid. It is also a very desirable fuel, since it is easily stored and burns with low NO_x, particulates, and sulfur emissions. However, cost estimates could not justify the production of methanol as a fuel (5).

Substitute Natural Gas

The production of substitute natural gas (SNG) from coal gasification requires the manufacture of a synthesis gas in a gasifier which is oxygen-blown or which uses indirect heating. Since methane is the major component of natural gas, it is desirable to maximize the production of methane from the gasifier. However, it may be argued that the choice of gasifier in a commercial system should depend more on process reliability than on the proportion of methane in the off-gas. Nevertheless, the methane content of product gas for oxygen-blown gasifiers rarely exceeds 15%. Therefore, a key objective is to convert a mixture of hydrogen and carbon oxides to methane.

The reactions required to produce methane from a mixture of hydrogen and carbon oxides can be promoted by catalysts, usually nickel-based. Reaction efficiency is generally low. The removal of residual carbon oxides from the product gas is required, which is relatively easy with carbon dioxide but more difficult with carbon monoxide. Also, sulfur compounds resulting from the gasification process which are present in the feed gas tend to poison the catalyst. A typical SNG synthesis method first reduces the amount of carbon monoxide and increases the hydrogen content of the raw gas from the gasifier through the shift reaction. The gas is then cleaned to remove carbon dioxide and sulfur compounds. This processed gas is then introduced to the stage where methanation is promoted by the nickel-based catalyst. Temperature control is a major problem in the methanation stage. It is achieved by performing methanation in a number of stages, and also by recycle of cold product gas (2).

Mild Gasification

Mild gasification processes pyrolyze coal under conditions which permit only a portion on the volatiles content of the coal to be released. As a result, a gas product, a liquid product, and solid char are produced. The liquids are highly aromatic and may be used as a chemical feedstock or heavy fuel oil replacement. The solid char product retains sufficient volatile matter, and hence sufficient reactivity, to be used as a feedstock for utility boilers. Application of mild gasification processes are currently aimed at upgrading low quality coals and lignites.

Integrated Gasification Combined Cycle (IGCC) Electric Power Generation

Integrated Gasification Combined Cycle (IGCC) is a promising use of gasification recently commercialized. The flow chart for a generic IGCC plant is shown in Figure 3 (21). The key components of the system are the gasifier, the gas turbine, and the steam cycle. The IGCC plant can be built in a phased fashion (21). The gas turbine is put online first to take advantage of its low capital cost, its ability to respond to changes in load demands, and the present relatively low cost of natural gas and oil. As the demand for power increases, then a steam cycle is added, which increases the power output from the system through increased efficiency. Finally, as demand increases even more, a gasification plant is built to supply the combined cycle with a steady, low-cost supply of fuel gas and some additional steam. Figure 4 (21) shows projections as to where each phase is the most economical. Increased efficiencies in gasification cleanup systems and gas turbines have recently been achieved.

Other advantages that IGCC plants have other traditional coal combustion power plants are much lower quantities of gaseous and solid emissions, the ability to handle high sulfur coals, potentially higher availability due to parallel construction of gasifiers, the possibility of co-products, and higher possible efficiencies, of over 42%. As will be discussed further, the sulfur species emissions from gasification are easily controlled since they are usually in a reduced state. Further, the sulfur is easily recovered and can be marketed. Lower levels of nitrogen oxide pollutants (NO_x) are also produced. The ability of IGCC to remove sulfur at low cost, permits the use of cheaper high sulfur coals without substantial increase in sulfur emissions. As well as producing fewer gaseous emissions, gasification produces less solid waste than conventional coal power plants. The solids from gasification are often environmentally benign slags compared to sludges from flue gas desulfurization processes.

Gasifiers are also usually designed to be built in parallel trains, where one gasifier may be taken off-line and the rest of the plant may still stay in operation. Even if all the gasifiers in an IGCC plant were off-line, the combined cycle could still be fueled by natural gas or oil. Gasification technology also permits the plant to produce synthesis gas in its off-load mode which may then be sold or used on-site to produce ammonia, methanol, or other chemical products as discussed above. Finally, conventional power plants are limited by the Rankine cycle to efficiencies, usually below 37%. Current technology promises IGCC efficiencies of 42% with possible improvements above that (22,23). These efficiencies are obtained through combined cycle technology together with high temperature, high performance turbines, and novel hot-gas cleanup systems.

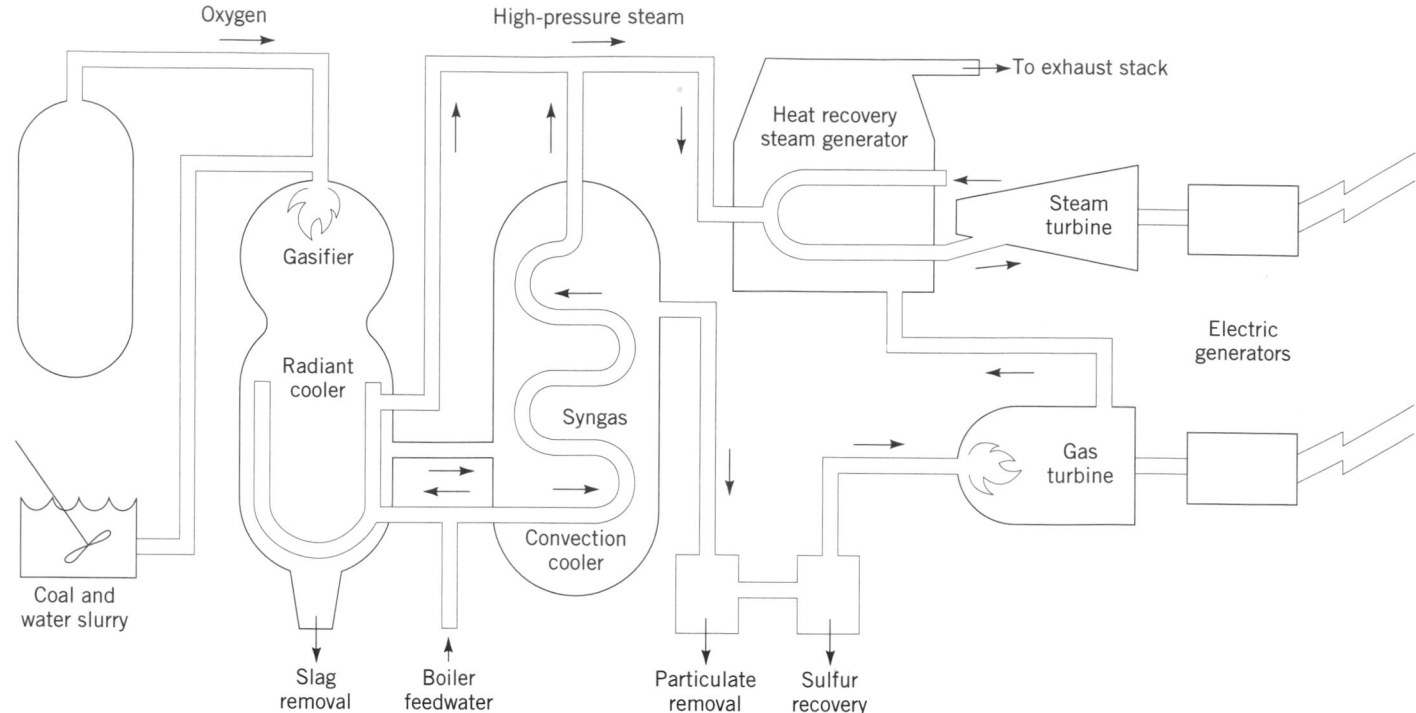

Figure 3. Schematic diagram of integrated gasification combined cycle facility (with permission from ref. 21).

Gasifiers can also be paired with various types of fuel cells in order to produce electricity. Molten carbonate fuel cells have been proposed that could have efficiencies of 45%. Improvements such as recycle of the hydrogen to the gasifier could increase efficiencies to 60%. However, unlike IGCC plants, these fuel cells have not been demonstrated on commercial scales.

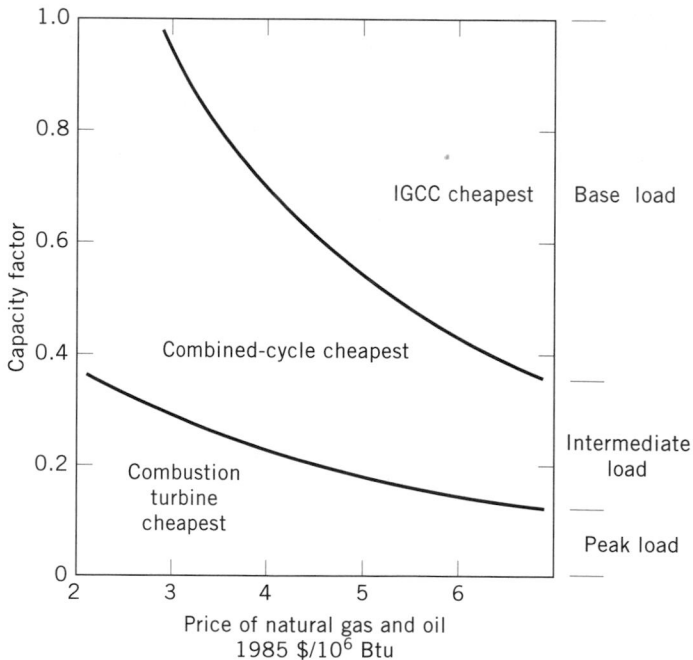

Figure 4. Economics of phased construction of an IGCC facility (with permission from ref. 21).

PROCESSES AND TECHNOLOGIES

There is a limited number of technologies that are commercially distributed or have been demonstrated on a commercial scale. Some other technologies that have been demonstrated on a pilot-plant scale are also briefly presented.

Lurgi Fixed Bed Process

The Lurgi gasifier, shown in Figure 5 (24), is a pressurized, dry-ash, oxygen-blown, fixed bed gasifier. The feed coal, typically sized from 5 to 50 mm, enters the top of the bed through a lock hopper and moves downward under gravity. The coal movement is controlled by a distributor or a stirrer and a rotary grate. The ash falls from the grate and is removed through another lock hopper. Steam and oxygen enter the gasifier below the grate, move upward, and react with the coal. The dry-ash Lurgi gasifier requires a large quantity of steam to reduce the combustion zone temperature below the ash fusion temperature. Some of this steam is generated in a water jacket around the gasifier. Only a small part of steam reacts with the coal. The temperatures are very non-uniform because of the countercurrent flow and near the bottom, in the combustion zone, the temperatures are around 1400 K. Near the top, after leaving the devolatilization and drying zone, the temperatures are around 800 K. For a high moisture coal, the gasifier exit temperature may be as low as 600 K (5). Because of the low temperature and the lack of oxygen in the devolatilization zone, the product gas shows a high content of hydrocarbon liquids such as tars, oils, and phenols. The product gas is water-quenched to condense and remove hydrocarbon liquids and cooled to generate

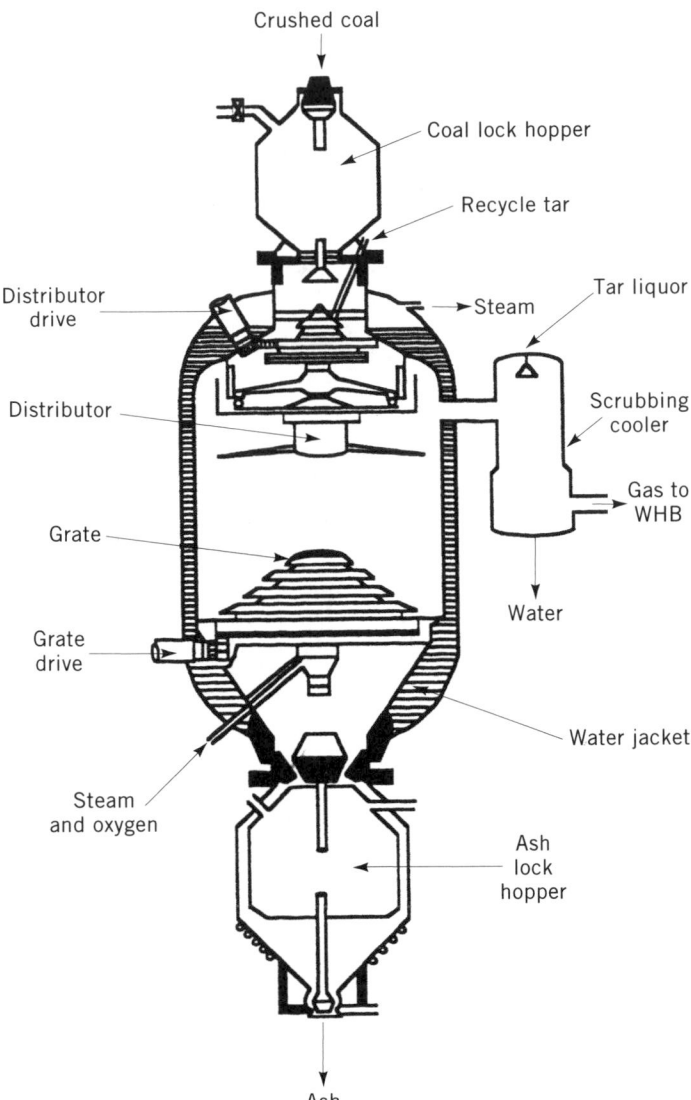

Figure 5. Dry-ash Lurgi fixed-bed gasifier (with permission from ref. 24).

additional steam. The cleaned and cooled gas is then desulfurized in a sulfur removal system. A typical Lurgi gasifier is 4 m in diameter. It has a nominal dry gas capacity of 55,000 m^3/h at standard conditions. This size is equivalent to approximately 535 mt/d of dry, ash-free coal. A larger Lurgi gasifier is 5 m in diameter. Its nominal dry gas capacity is 85,000 m^3/h at standard conditions, which is equivalent to approximately 830 mt/d of daf coal (5).

The most important application of the Lurgi gasification technology is in the SASOL plants in Sasolburg and Secunda, South Africa. The three SASOL plants, SASOL I, II, and III, produce nearly 90% of the total world production of gas from coal. The synthesis gas is used primarily as a feedstock to produce liquid transportation fuel by the Fisher-Tropsch synthesis. However, a broad product slate includes gasoline, distillate fuel oil, light olefins, light alcohols, waxes, phenols, tars, and town gas. The three SASOL plants have a total of more than 90 Lurgi gasifiers and consume approximately 82,000 mt/d of sub-

bituminous coal (25,26). The dry-ash Lurgi gasifier, originally designed for sized, noncaking coals, it cannot readily handle caking coals and coals with fines. In addition, the operating temperature is kept low to prevent ash melting and agglomeration. The low gas temperature results in high content of tars, oils, and phenols in the product gas.

British Gas/Lurgi Slagging Gasifier (BGC/Lurgi)

In order to overcome the limitations of the Lurgi gasifier, a slagging Lurgi gasifier has been developed by British Gas Corporation and Lurgi (BGC/Lurgi). The BGC/Lurgi gasifier is a slagging, oxygen-blown, pressurized, fixed-bed gasifier. This gasifier operates at higher temperatures than the dry ash Lurgi so that the ash melts and forms liquid slag. As in a conventional Lurgi gasifier, the coal is fed to a gasifier through a lock hopper system and a distributor. A fluxing agent is added to some coals to reduce slag viscosity. The coal reacts while moving downward through the gasifier. The coal ash melts and passes through a slag tap hole. The slag is then water-quenched and removed through a slag lock hopper. Steam and oxygen are injected through tuyeres at the bottom of the bed. The temperatures in the raceway zone are above the ash melting temperature. The steam requirement is reduced to only about 15% of that for the dry-ash Lurgi gasifier when gasifying bituminous coal. The product gas passes through a water-quench scrubber into a waste heat boiler. The particulates and the condensed tars, oils, and phenols are recycled to the top of the gasifier or reinjected through the tuyeres. Coal fines can also be fed through the tuyeres. The cleaned and cooled product gas is then desulfurized in a sulfur removal system (5,24,27).

Koppers Totzek Entrained Flow Gasifier

The Koppers-Totzek System is a commercially proven entrained-flow gasification system. As of 1983, over 26 GKT gasifiers have been operating world-wide, with a total product gas rate estimated to be at 9.3 million standard cubic meters per day. Some of these gasifiers have been operating for over 30 years. The Koppers-Totzek gasifier, shown in Figure 6, (24) is an atmospheric pressure, entrained flow gasifier. The feed coal is dried and pulverized to fine powder in a pulverizer. Typically 70 to 80% of the pulverized coal particles pass through a 200 mesh sieve of 74 μm apertures. The dry, pulverized coal is fed by screw feeders to two or four opposing burners. The coal reacts with oxygen and steam to produce a temperature of approximately 2200 K in the flame zone. The temperature of the gas mixture is reduced to approximately 1760 K by heat losses to the refractory-lined and steam-jacketed walls of the gasifier. At this temperature the coal is converted primarily to CO, H_2, and CO_2 and much of the ash is converted to molten slag. Most of this molten slag flows down the gasifier walls into a quench tank where it solidifies before being removed. A portion of the particulate matter leaves the gasifier as fly ash. The product gas leaving the gasifier is water-quenched to approximately 1200 K to solidify entrained molten ash before entering a waste heat boiler. The fly ash is removed after the waste heat boiler by wet scrubbers and electrostatic precipitators. The scrubbed and cooled product gas is then processed in a sulfur removal system. The resulting product gas has a

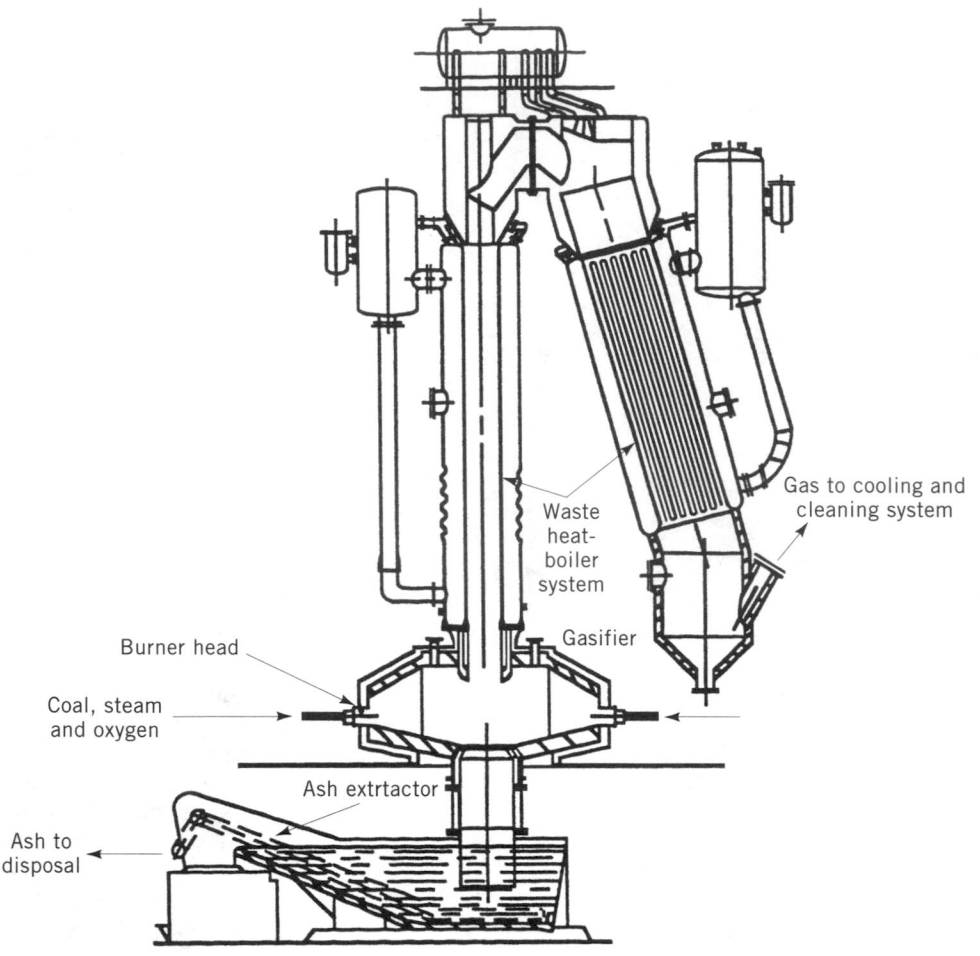

Figure 6. Koppers-Totzek two-headed entrained-flow gasifier (with permission from ref. 24).

heating value of about 11 MJ/m³ compared to 37 to 41 MJ/m³ for natural gas. Most of the Koppers-Totzek gasifiers are with two opposed burners. These two-headed gasifiers have a nominal dry gas capacity of 17,000 m³/hr at standard conditions. This capacity is equivalent to approximately 190 mt/d of daf coal. Four-headed gasifiers with the burner heads at 90 degrees have been the more common configuration recently. Their nominal dry gas capacity is 35,000 m³/hr at standard conditions which is equivalent to approximately 390 mt/d of daf coal. The most common application of the Koppers-Totzek gasifiers has been in ammonia and methanol production (5,24).

Texaco Entrained Flow Gasifier

The Texaco gasifier, shown with auxiliary units in Figure 7 (5), is a coal-water slurry-fed, pressurized, entrained-flow gasifier. The feed coal is pulverized and slurried in a wet rod mill. The slurry water consists of recycled condensate and make-up water. The coal-water slurry, together with oxygen or air, is pumped through a burner into the top of a refractory-lined gasifier. Gasifier pressure is typically around 4 MPa. The presence of slurry water moderates the temperatures of the coal-oxygen reaction and gasification takes place at temperatures between 1530 K and 1760 K. At these conditions, the coal is converted primarily to CO, H₂, and CO₂ and small amounts of CH₄; no

tars or oils are present. Fuel nitrogen is converted to N₂ and NH₃. Fuel sulfur is reduced to H₂S and small amounts of COS. The ash leaves the gasifier as molten slag. The product gas is either water-quenched or cooled in radiative and convective waste heat boilers before wet scrubbing. The solidified slag, with carbon content of less than 1%, is removed through a lockhopper for disposal. The product gas of the oxygen-blown Texaco gasifiers has a medium heating value like that of the Koppers-Totzek oxygen-blown system (5,24,28,29).

An important demonstration of the Texaco gasification technology has been the 120 MW "Cool Water" plant, an integrated-gasification, combined-cycle power plant located in the Mojave Desert in California, that began power production in 1989. The plant used an oxygen-blown, entrained-flow Texaco gasifier to convert approximately 820 metric tons of coal per day to a medium heating value gas. The product gas was cooled to ambient temperature before removing sulfur species and particulates by conventional methods. The gas, after particulate and sulfur removal, was burned in a gas turbine to produce electricity. Additional electricity was produced in a steam turbine. Steam for the steam turbine was generated from the hot product gas in the waste heat boilers and from the gas turbine exhaust gas in the heat recovery steam generator. The emissions of SO₂, NOₓ, and particulates were only one tenth of the United States New

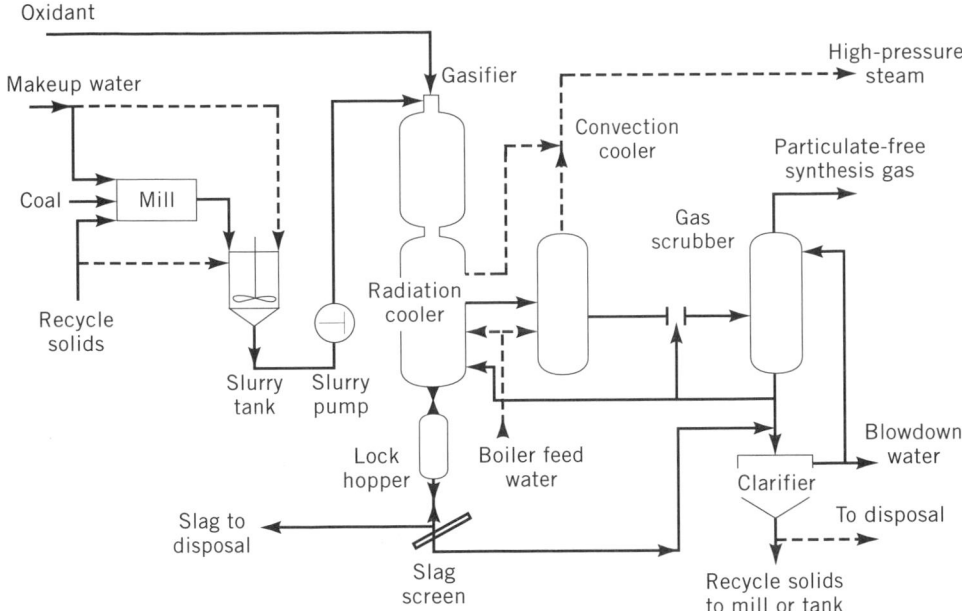

Figure 7. Texaco gasification process (with permission from ref. 5).

Source Performance Standards for coal-fired plants (28,29). This demonstration program was used to test many aspects of gasification of particular interest to the power generating community: effective environmental control, use of a wide variety of feedstocks, reliability and ease of operation and control, the integration of gasification with combined cycle electric production, and the economics and efficiency of the overall system. However, the Cool Water Project was the first to demonstrate viability of gasification for energy production. Analysis of the Cool Water results has also led to added impetus in the development of hot-gas cleanup systems, high-temperature turbines, and the integration of turbine compressors with the air separation stage.

Dow Entrained Flow Gasifier

Dow operates what is currently the world's largest single train gasifier at Planquemine, La. The 2200 metric ton/day gasifier/combined cycle, constructed in 1986, is in a phased fashion to provide power to Dow's electrolysis and chemical process plants. The Dow gasifier is a slurry-fed, pressurized, slagging, two-stage design. The slurry is fed with oxygen to opposite ends of a horizontal, refractory-lined, cylindrical reaction chamber. The majority of mineral matter in the coal is removed as slag from the bottom of this first stage. The hot gases and remaining particles exit the first stage into a vertical second stage where additional slurry is added to complete the gasification reactions and cool the gases. A cyclone removes the unreacted char particles, which are water-quenched and added to the feed slurry. Hydrogen sulfide and carbon dioxide are removed through a proprietary, acid-gas cleaning system, and the hydrogen sulfide is converted to elemental sulfur. The cleaned product gas is sent to the gas turbines for power production. These turbines have been designed for natural gas operation but have been modified to use the product gas alone or a mixture of the product gas with natural gas (30).

Shell Entrained Flow Gasifier

The Shell system has been operated in a 230 to 360 metric ton/day pilot plant. The development of Shell technology has benefited from a partnership with Krupp-Koppers. A 150 metric ton/day pilot plant first operated in 1978 has successfully tested a wide variety of coals. The 230 to 360 metric ton/day pilot plant at Deer Park, Tex., completed in 1987, has operated successfully at cold gas efficiencies of up to 83%. The Shell gasifier is a dry-feed, entrained-flow, pressurized, slagging gasifier. A dense phase feed system is used to convey the coal into the gasifier. The coal reacts with oxygen at very high temperatures which are moderated by the addition of steam. Slag is formed and runs down the sides of the reactor into a water bath. The process leads to high carbon conversion and a product gas that is mostly carbon monoxide and hydrogen. Ceramic candles, which are used to control particulate emissions in hot gases, have also been demonstrated.

Winkler Fluidized Bed Process

The Winkler gasifier, shown in Figure 8 (24), is an atmospheric pressure, fluidized bed gasifier. The gasifier is also available at pressures of up to four atmospheres. The feed coal is crushed to less than 9.5 mm size and fed through a variable speed screw feeder to the fluidized bed in the lower portion of the gasifier. The gasifier is a refractory-lined, steel-shell cylinder. The fluidized bed occupies around one third of the gasifier volume; the remainder is a freeboard zone. Steam and oxygen or air are blown through a grate at the bottom of the gasifier, fluidizing the bed and reacting with the coal.

The coal bed temperature is usually kept between 1070 and 1370 K by use of steam and heat transfer to avoid ash fusion. At these temperatures, the coal is converted primarily to CO, H_2, and CO_2 and small amounts of CH_4; no tars or other heavy hydrocarbons are present. These

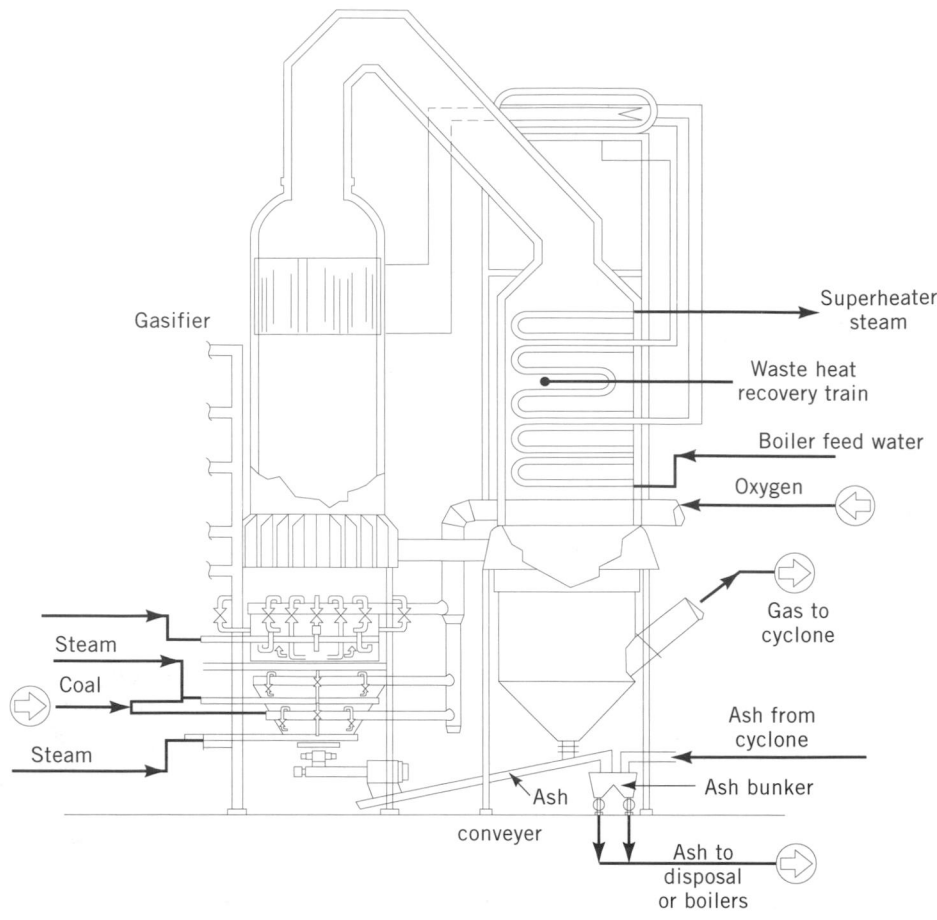

Figure 8. Winkler fluidized bed gasifier (with permission from ref. 24).

relatively low temperatures restrict the Winkler gasifier to more reactive lignites and subbituminous coals. The reactivity of high rank coal is, as a rule, insufficient at these temperatures. The heavier ash particles are removed from the bottom of the gasifier by a water-cooled screw conveyor. The lighter particles are entrained in the gas and carried over to the freeboard. The unconverted carbon from these particles reacts with steam and oxygen or air, blown in the freeboard; some of the ash particles melt. To solidify these molten ash particles, the gas is cooled approximately 200 K in a radiant steam boiler installed inside the gasifier at the top. The product is further cooled in a waste-heat boiler and cleaned in a cyclone, a wet scrubber, and an electrostatic precipitator. The cooled and cleaned gas is then desulfurized in a sulfur removal system. A typical commercial Winkler gasifier is 5.5 m in diameter and 23 m in height. It can gasify approximately 100 mt/d of coal at atmospheric pressure and 1640 mt/d of coal at four atmospheres (24,31). Recently, work has been done to pressurize the Winkler process and use ash agglomeration to reduce ash recirculation and disposal problems. This new process is called the High Temperature Winkler (HTW) process (32).

Developments in Gasification Technology

Much developmental work is continuing in gasification technologies. Fixed-bed technologies being developed in Europe and in the Clean Coal Technology (CCT) Program, were listed by Radulovic and Smoot (25). The "Air-Blown/Integrated Gasification Combined-Cycle Project," Figure 9, at Tampa Electric Company's Polk Power Station in Lakeland, Florida is described as a representative of these developments (33). The project objectives are to demonstrate an air-blown, fixed-bed, integrated gasification combined-cycle technology and to assess reliability, availability, and maintainability at commercial scale. The coal is gasified in a pressurized, air-blown, fixed-bed gasifier. The low heating value gas leaves the gasifier at approximately 810 K and goes to a hot-gas cleanup system. The removal of sulfur compounds is accomplished in a moving bed of solid sorbent. Approximately 1,150 mt/d of high sulfur Illinois bituminous coal are converted to a low heating value gas. The cleaned gas is used in a combined-cycle system, a gas turbine and a steam turbine, to produce 120 MW of electricity.

New entrained flow gasification technologies being developed in Europe and the U.S. have also been listed by Radulovic and Smoot (25). The Combustion Engineering Co. IGCC Repowering Project at the City Water, Light, and Power's Lakeside Station in Springfield, Illinois is a Clean Coal Technology Demonstration Program to demonstrate an advanced, dry-feed, air-blown, two-stage, pressurized, entrained-flow coal gasifier with limestone injection and moving bed, zinc ferrite, hot-gas cleanup system. The dry, pulverized coal is fed by a pressurizing feeder (kinetic extruder) and a fluidizing feeder to the gasifier. The gasifier consists of a bottom combustion section and a top reduction section. The coal is fed into both sections. The molten slag flows into a water-quench tank.

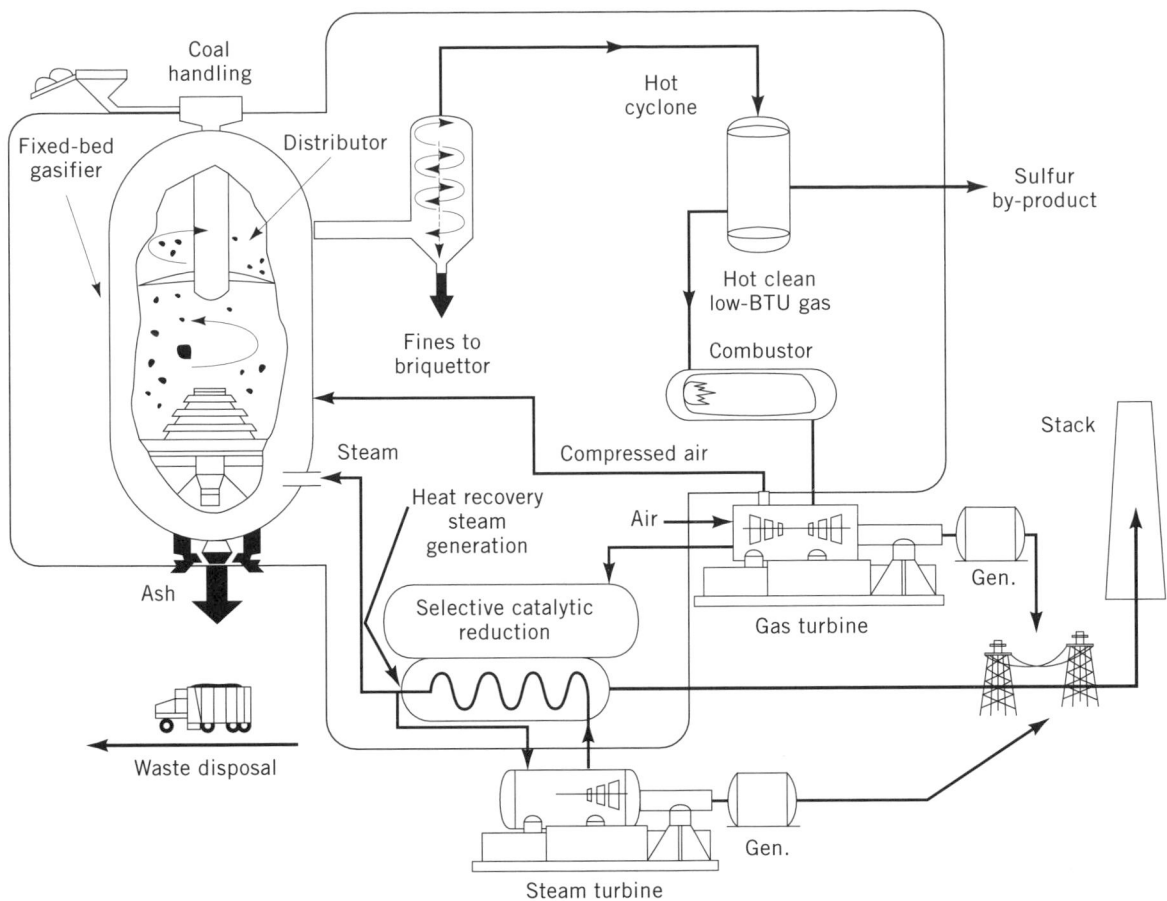

Figure 9. Air-blown/integrated gasification combined-cycle project (with permission from ref. 33).

Sulfur is removed from the hot gas by two processes: in situ and final desulfurization. In situ desulfurization is achieved by limestone injection. Final desulfurization is performed by a moving bed, zinc ferrite system downstream of the gasifier. The cleaned, low heating value gas (about 3700 kJ/scm) is used in a combined cycle system to generate 65 MW of electricity. Construction started near the end of 1992.

Fluidized bed gasification technologies being developed in Europe and the U.S. CCT program were also listed by Radulovic and Smoot (25). The Clean Coal Technology Pinon Pine IGCC Power Project at Sierra Pacific Power's Tracy Station in Reno, Nevada, seeks to demonstrate an air-blown, fluidized bed gasification technology incorporating hot gas cleanup. The dried, crushed coal is fed into a KRW pressurized, air-blown, fluidized bed gasifier through a lock hopper system. The coal bed is fluidized by blowing air and steam through special nozzles into the bed. Crushed limestone is also fed into the gasifier to absorb sulfur and to inhibit conversion of fuel nitrogen into ammonia. The product gas passes through cyclones to remove particulates and recycle fines. A hot-gas cleanup system, a fixed bed of zinc ferrite sorbent, is used to remove remaining sulfur. The cleaned, low-heating value gas is used in a combined cycle system, a gas turbine and steam turbine, to produce 80 MW of electricity (25).

Mild gasification is a modification of conventional coal gasification where a slate of gaseous, solid, and liquid products are produced rather than solely producing a fuel-rich gas product. This process is achieved by removing only a fraction of the volatile hydrocarbons from the parent coal, leaving a volatile-rich char. Mild gasification processes generally limit the peak operation temperature to about 800–1000 K. With the removal of coal moisture and a fraction of the volatile matter, the resulting heating-value of the char product is significantly increased from that of the parent coal. This upgraded char product may retain sufficient volatile matter (10–20%) to be useful as a fuel for utility boilers. The liquid products are potentially usable as a replacement for heavy fuel oil, diesel fuel, or diesel fuel diluents (34). Mild gasification processes are particularly well-suited for high volatile subbituminous coals and lignites since these low-rank coal fields are generally remote from population centers, and the char product has a higher heating value than the parent coal and the liquid and gas products are valuable fuels and chemical feedstocks.

Few mild gasification projects have progressed past the laboratory-scale. Notable exceptions have included the Occidental and the ENCOAL processes. The Occidental process was developed in the early 1970s by the Garrett Research and Development Company, a subsidiary of the Occidental Petroleum Corporation (35). This entrained-flow process operated at atmospheric pressure where Wyoming subbituminous coal was pulverized and dried similar to that used for utility boilers (70% minus 200 mesh).

The coal was introduced to the entrained flow reactor where it was heated to about 1000 K in about 0.1 sec. This rapid heating and short residence time allowed a high volatiles yield. The resulting char contained between 1–4% moisture and 12–15% volatile matter (36). The char performed well in boiler tests. No further developmental work has been done on the Occidental process since the early 1980s.

The ENCOAL project is one of the Clean Coal Technology Demonstration projects (33). It uses a mild gasification process based on the liquid-from-coal (LFC) process developed by SGI International and the Shell Mining Company (see Figure 10). The moving bed process utilizes rotary grates in two stages for coal drying and subsequent partial devolatilization followed by char cooling. Coal is fed onto the first rotary grate where it is heated by a hot-gas stream to reduce its moisture content. The dried coal is then transferred to the second rotary grate pyrolyzer where the temperature and residence time are controlled to achieve the desired pyrolysis products. Solids leave the pyrolyzer and are cooled. The 910 mt/d commercial demonstration plant is located at the mouth of the Triton Coal Company's Buckskin Mine near Gillette, Wyo. (33). Plant operation began in 1992.

Examples of technologies incorporating molten bath gasification are the Atgas or molten iron process, and the Atomics International process designed to produce sulfur-free gas to fire gas turbines and boilers. Coal and air are introduced into a bath of molten sodium carbonate which acts as a solvent and heat reservoir for the reactants. The alkali salt bath also catalyzes gasification reactions and captures sulfur that evolves from the coal. The gasifier operates at about 1250 K which allows coal to be gasified to a product gas consisting mostly of carbon monoxide and hydrogen. The ash and sulfur from the coal are retained in the molten carbonate bath, requiring the bath to be continuously regenerated. To accomplish this task, the bath is first quenched and dissolved in water, and then filtered to remove the ash. Hydrogen sulfide is stripped from the filtrate using flue gas from the boiler. The filtrate is then carbonated, crystallized, centrifuged and dried before recycling to the gasifier. It is claimed that the process can handle all kinds of coal (caking and non-caking) and that gasification rates are very high (2). This technology was developed in the 1970s and is not currently commercially active.

To successfully accomplish underground coal gasification (UCG), the permeability of the coal seam must by enhanced. The natural permeability of coal seams is not sufficient for the UCG processes. Permeability must be sufficiently high to permit gasification and to allow for loss in permeability due to the condensation of tars and

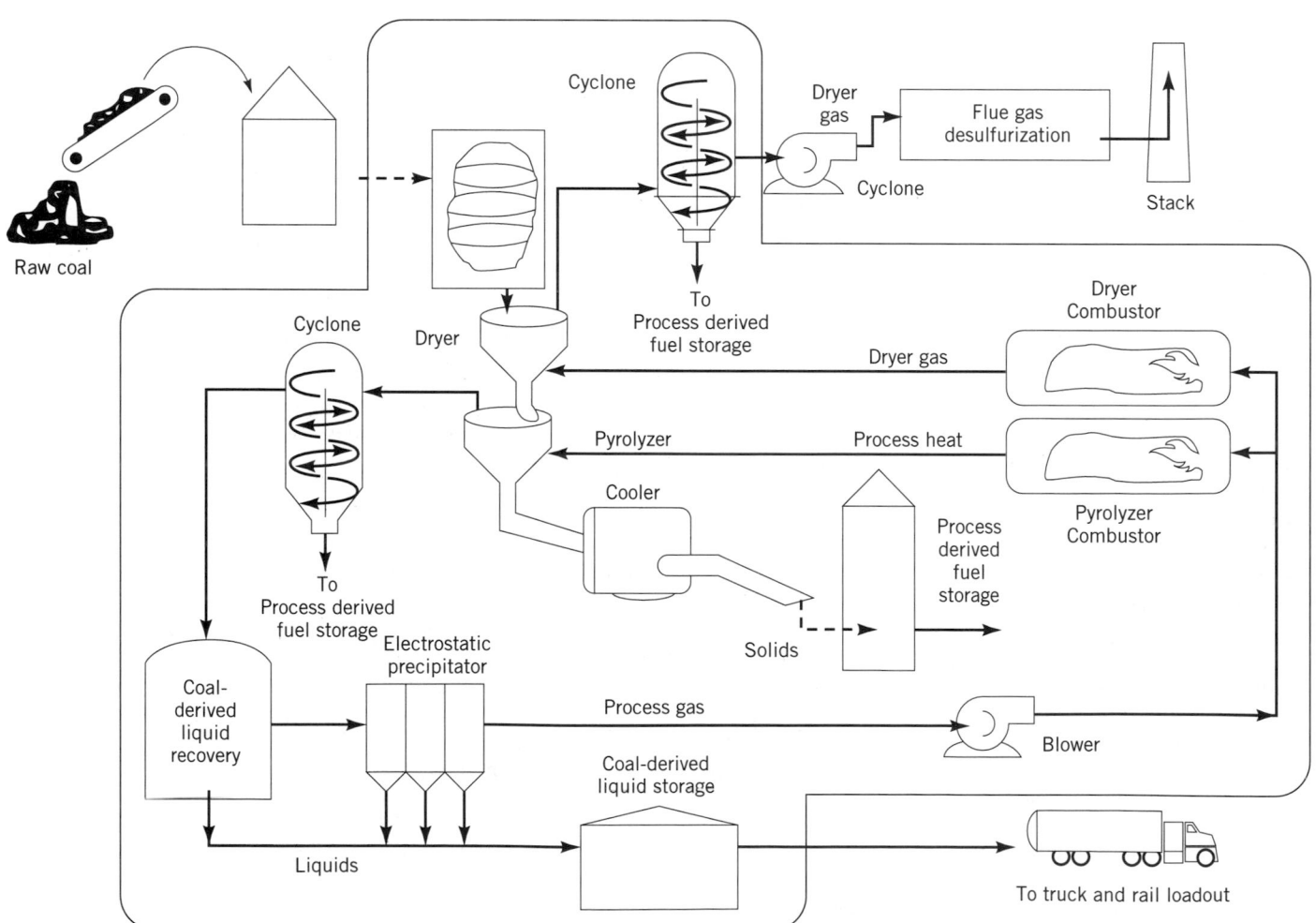

Figure 10. ENCOAL mild gasification project (with permission from ref. 33).

other heavier hydrocarbons. This action of permeability is referred to as "linking." Methods include directional drilling, countercurrent combustion, electrolinking, and hydraulic fracturing. These methods form narrow, highly permeable, nearly-cylindrical channels between the blast injection and the gas removal boreholes. Countercurrent combustion linking has been the method most extensively tested (25). In the United States, several field test studies have been conducted: the Hanna field test study (37), the Hoe Creek field test study (38), and the large block underground coal gasification study (39). The state-of-the-art in UCG has also been reviewed (40,41). At present, there is no commercial utilization of the UCG processes in the United States. UCG research and development in the USSR has been extensive. Several relatively large-scale installations operated successfully for many years (42).

ENVIRONMENTAL CONTROL TECHNOLOGIES

Gasifiers and gasification systems have been primarily used as sources of chemical feedstocks. Therefore, technologies have been developed that remove undesired species from the product gas. Further, these undesired species or by-products are usually converted to some saleable form to aid the overall profitability of the plant. These same technologies are available for gasification when it is used in power generation. Gaseous emissions, liquid effluents, and solid waste are still generated, but in much lesser quantities than by traditional coal power plants. Sources of atmospheric pollutants from gasification processes are coal dryers, coal gasifiers, desulfurization plants, cooling towers, process furnaces, and carbon dioxide stripping plants. Liquid effluents may be composed of gas liquors from process condensates, ash and slag quench water, and rain runoff from coal storage areas. Gas liquors may contain fairly high levels of tars, oils, phenols, naphthas, sulfur and nitrogen compounds, particulate matter, and dissolved solids. Solid wastes are generated primarily from coal cleaning processes, ash or slag disposal, and sludge generated from wastewater treatment facilities. Trace element emissions are present in each of these three streams resulting from gas cleanup volatiles, oil products, and solid waste leaching (19).

Emissions

Gaseous Emissions. An example of the amounts of major gaseous pollutants emitted from a coal gasification and process plant is shown in Table 5 for a typical SNG facility. Assuming there are no vagrant emissions due to leaks in the gasifier, gasifier emissions composed of gases such as carbon monoxide, carbon dioxide, hydrogen sulfide, and nitrogen compounds may occur when coal is charged to the reactor or when slag or ash is withdrawn (19).

Emissions from cooling towers present the same problems in gasification systems as in other industries. Process cooling water which passes through the towers also circulates through other process equipment, such as heat exchangers, gasifier cooling coils, and gas clean-up reboilers, where leakage can contaminate the cooling water with significant amounts of sulfur compounds, oil, particulates, and trace elements. During the cooling operation

Table 5. Gas Emissions from a Typical SNG Plant[a]

Stream	Million std, m^3/day	Relative Volume
Synthetic natural gas	7	1.0
Coal dryer vent gas	2.8	0.4
CO_2 from acid gas removal	8	1.1
N_2 from oxygen plant	12	1.7
Plant tail gas	2.4	0.3
Flue gas from furnace	22	3.1
Air from cooling tower	1570	220.0
H_2O from cooling tower	36	5.1

[a] From Ref. 19, with permission.

many of these contaminants may be stripped out of the cooling water and emitted as airborn pollutants. The problem of leakage into the cooling water is enhanced in coal gasification processes which operate at elevated pressures. Another problem arises from the practice of utilizing partially cleaned waste waters as make-up water for cooling towers. This practice may result in evaporation of residual contaminants present in those wastewater streams. Further, water vapor emitted from cooling towers often forms mists or plumes of water droplets. These mists can condense and cause fog or icing of roads during winter months and can also form corrosive deposits on equipment, buildings, roads, and vehicles.

Liquid Effluents. The major sources of liquid effluent streams from coal gasification plants are from the quenching of the product gas and ash, and from the blowdown of process condensates. As the temperature of the raw gas stream from the gasifier is lowered, either by direct quenching or heat exchange, certain components in the vapor phase begin to condense. This gas quench liquor is collected and removed between successive stages of cooling. Generally, larger volumes of the liquor are generated from fixed-bed gas quenching operations as compared to fluidized-bed and entrained flow gasifiers, since the latter two processes convert most of these components. If a direct water quench is used, the condensate collected during the cooling process will consist mainly of water. The amount of water produced will be determined by the process, the coal moisture content, and the operation of the cooling stage. Besides water, other components such as tars, ammonia, oils, hydrogen sulfide, naphtha, organic sulfur compounds, phenols, hydrogen cyanide, particulates, and trace elements may be found in this stream.

The concentrations of these components in the waste stream depend on their raw gas concentration and the type and degree of quenching utilized. Particulate matter in the waste stream is composed mainly of ash which is washed out of the raw gas stream. Although the ash is generally separated from the quench liquor in a settling tank, this type of separation technique is not completely effective. A wastewater treatment system is generally required to remove these contaminants and produce discharge water to meet minimum allowable standards. Of the components of the gas quench liquor listed above, phenols require the most complete processing. Extensive treatment systems to remove phenols may need to be used in large gasification plants. Biological oxidation treatment (Biox) has been used in many cases for dephenolization, the effect of which is shown in Table 6 (data taken from pilot plant and commercial coal gasification units) (19).

Recycled gas quench liquor is generally used in the ash quenching process for ash cooling and transport. Streams from other process units may also be used. Therefore, the effluent water from the ash quench water contains ash particles as well as any water leachable substances dissolved during the ash quenching process, in addition to the contaminants present in the water before its use for ash quenching. These contaminants can be concentrated by evaporation. However, some of the components are volatile, which may result in air emissions. Settling tanks may also be used to separate a concentrated ash slurry from the liquor. The liquor recovered from the ash settlers is generally suitable to be recycled to process condensate or gas quench liquor streams. The ash slurry can be dewatered, which produces a solid waste product requiring disposal (43).

Solid Waste Disposal. The primary solid wastes generated during coal gasification, which require disposal, are from coal-cleaning facilities, ash or slag from the gasifier, wastewater sludge, and solids generated from treatment of cooling tower liquors or other wastewater disposal facilities. A coal cleaning plant can produce large volumes of solid refuse (up to 1200 m³/year for a large gasification plant). This solid waste is generally landfilled which may also require revegetation. Similar disposal methods are required for the large quantities of coal ash or slag produced from gasifier operation. Wastewater sludge generated in a biological oxidation unit may require special treatment prior to being landfilled due to potential odor problems. Char from the gasification unit itself may be useful in treating this odor problem since its high surface area makes it effective for use as activated carbon. Possibly 90% of the phenols present in the wastewater can be thus removed. However, disposal of the char is also required.

The leaching behavior of salts and trace elements from ash and slag from low-rank coals is considerably different from that for higher rank coals. Solid waste disposal dumps must be evaluated for each proposed gasification

site based on local geology and coal type used in the gasification process. Monitoring of minerals such as magnesium sulfate, sodium salts, calcium chloride, sulfur compounds, and trace metals may be required to insure that they do not contaminate ground water sources (44).

Trace Elements. Trace quantities of various elements enter the gasification process in the ash portion of the feed coal with a small amount produced by refractory degradation or flux addition. These elements exit the gasifier through three streams: gasifier slag, particulates, and flue gases (18). The partitioning of trace elements in the gasifier depends largely on their boling point and chemical behavior (45) and must be properly handled in the effluent stream in which it is emitted.

Some gasification processes have utilized particulate recycling systems to attempt to deliver more of the trace elements to the slag where it is mostly inert. Figure 11 (46,47) shows the distribution of trace elements in an entrained-flow gasifier effluent streams with, and without, particulate recycle. Elements such as Co, Cr, Cu, and Ni become enriched in the slag with recycling whereas Pb, Sn, and Zn become concentrated in the filter cake and As, Cd, Hg, and Sb become volatilized and are removed in the raw gas (45).

The BGL slagging gasifier incorporates a high degree of particulate recycle and most trace elements are captured in the slag including many volatile elements. Analysis of BGL slags from various feed coals shows that most trace elements are retained in the slag including about 95% of the Co, Cr, Mn, Se, Sr, Ti, and Th (48). Conventional gas cleaning processes such as electrostatic precipitators, wet scrubbers, and fabric filters may be used to remove material carried over from the gasifier, reducing air-borne emissions of trace elements (49).

UCG Emissions. Underground coal gasification eliminates many of the environmental problems associated with coal mining. The fuel-rich environment found in the gasifier causes the sulfur in the coal to be converted mainly to hydrogen sulfide, which is readily removed from the product gas stream. Solid waste disposal problems are reduced, since inert ash material is left underground, and large scale excavation of the overburdening or mine-dump problems are avoided. However, underground gasification does present potential water pollution problems due to leaching into local aquifers.

The pollutants generated as part of the raw gas stream from underground gasification are similar to those found in a standard, air-blown gasification process. Therefore, the phenol disposal problems encountered with other gasification techniques also apply to underground gasification. The metals and organics which are not readily soluble in water can be expected to be transferred by leaching into ground water (10). Underground hydraulic connections can present unique environmental problems with underground gasification. For example, one study reported that vapors from the gasification process were noted at some distance (100 m away) from the gasification site. Deviations in ground water pH also can occur during gasification.

Table 6. Some Contaminants in Gasifier Sour Water, Parts/Million[a]

Item	Gas Liquor	To Biox[b]	From Biox[b]
Phenol	2,000–4,000	100–500	0.1–0.3
Fatty acids		500–1,500	9
Ammonia	8,000–11,000	200–1,000	5–10
Thiocyanates	5–10	1	0.1
Fluoride		56	6
BOD[c]		2,500	75
COD[d]	10,000–20,000	1,100	82

[a] From Ref. 19, with permission.
[b] Biox = biological oxidation treatment.
[c] BOD = biochemical oxygen demand.
[d] COD = chemical oxygen demand.

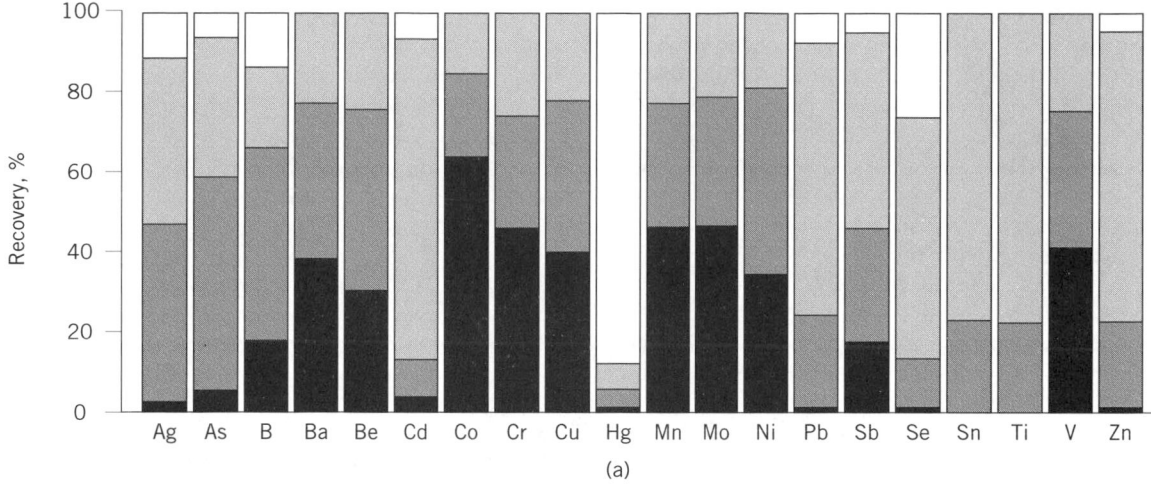

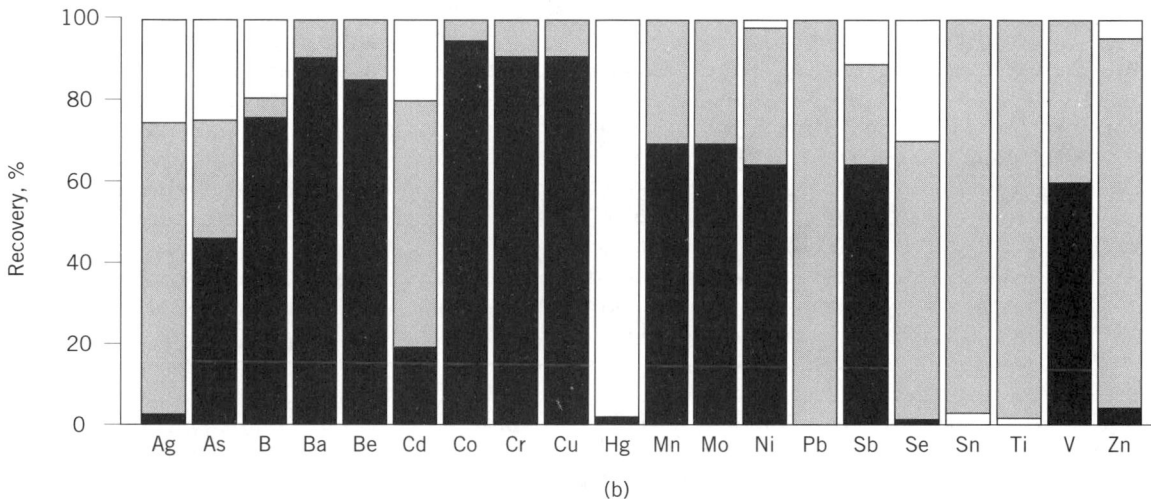

Figure 11. Relative distribution of trace elements in a gasification system (with permission from refs. 46 and 47).

Water located in the vicinity of the gasification area may contain high concentrations of dissolved gases, including H_2, CH_4, CO_2, H_2S, and NH_3 immediately after gasification is completed. However, due to natural ground water movement, the gas and mineral content of the water soon returns to its natural level. In areas where there is no appreciable ground water in the gasification area, the formation remains hot, at around 360 K, up to one year after gasification. Underground gasification studies have shown the presence of organic species in ground water after gasification is short-lived, but the released mineral and hydrogen sulfide may persist for up to a year. Water characteristics returned to their normal state about two years following gasification. The ability of ground water to naturally renovate itself is likely due to the high sorption capacity of the formation for organic pollutants such as phenol. However, the overburden material may also play an important role in natural renovation since certain gaseous species such as ammonia and hydrogen sulfide may be adsorbed by clays. Since the underground pollution is extremely localized, pollution control could be carried out by pumping out the contaminated water. The collected water could then be treated in typical wastewater treatment facilities developed for industrial plants such as coking operations (10).

Cleanup Systems

Coal gasification systems tend to be cleaner from an environmental standpoint than combustion, since sulfur, nitrogen, and ash are removed during the production of fuel gas. Acid gas (H_2S and CO_2) removal systems are capable of removing more that 99% of sulfur species produced during gasification. Since IGCC systems are more efficient than traditional coal power plants, less CO_2 is produced per kilowatt. Further, oxygen-blown gasifiers have no nitrogen in the flue gas, only CO_2 and H_2O. The CO_2 is easily separated from the water and may be used in such applications as tertiary oil recovery. Since nitrogen oxides are not formed in a significant amount during coal gasification, NO_x remediation is not a concern. Ammonia and hydrogen cyanide are easily stripped from the crude gas and recovered. Slagging gasifiers remove the bulk of the ash as environmentally inert slag. Other particulates can be controlled to levels well below New Source Performance Standards (NSPS) (17). Environmental characteristics for typical coal gasification systems are outlined in Table 7.

Sulfur Removal Through Dolomite/Limestone Injection. While sorbent injection is a relatively recent technology, it has benefited from much interest and application on the part of the power industry. The electric power industry, forced by the Clean Air Act to install expensive and relatively unproved gas scrubbing systems on its power generation facilities, has energetically sought other alternatives. One with great promise has been the injection of limestone or dolomite sorbent either directly into the reaction zone or into the flue. Six of the eight projects funded in the first round of Clean Coal Technology awards included some type of sorbent injection, either as part of a fluidized bed or in staged combustion or as part of flue-gas cleanup. Three of these projects have been completed as of 1991: Advanced Cyclone Combustor with Internal Sulfur, Nitrogen and Ash Control; LIMB Demonstration Project Extension and Coolside Demonstration; and NUCLA CFB Demonstration Project (33). These three projects have demonstrated the use of sorbents in combustion systems rather than in gasifiers. However, while the demonstration of sorbents in gasification systems has not been as widely published, similar efforts have been made. The CE Springfield Repowering Project uses sorbent injection in an entrained-flow gasifier, the Piñon Pine IGCC Power Project includes the addition of sorbent to a Winkler, fluid-bed gasifier. Similarly, the Toms Creek IGCC Demonstration Project purposes the use of sorbent in a Tampella U-Gas fluid-bed gasifier. Laboratory-scale experiments have shown the effectiveness of sorbent injection in reducing atmospheres as well as combustion atmospheres.

Conventional Acid-Gas Cleanup. Many acid-gas removal systems have been developed and are currently used in the oil, gas, and chemicals industries worldwide. Two of these, Rectisol and Selexol, are used currently by the majority of gasification systems with excellent results, removing more than 99.9% of the sulfur in the off-gas. Another process, Sulfinol, is used by a plant in Turkey and achieves 99.7% removal. The Rectisol and Selexol processes use physical adsorption and are good for high-pressure sulfur removal. The Sulfinol process uses a combination of chemical and physical adsorption. Once H_2S and COS are removed from the gas stream, the absorbent is regenerated, and the H_2S and COS are converted to sulfur using the Claus process with SCOT Tail Gas Treating, or the Stretford Process, or some similar process. Once again sulfur recovery is very high. All of these processes have been proven commercially.

Hot-Gas Sulfur Removal. Sulfur systems that operate at elevated temperatures have been and are being investigated in order to improve the efficiency of IGCC operations. A zinc oxide adsorption system is perhaps the prototype of these systems. Zinc oxide systems are primarily used as a sulfur guard to protect against incursions of H_2S in process gas streams into sulfur-sensitive catalysts. They can operate at temperatures in the range of 620–720 K and at pressures up to 5 MPa. There are over 100 commercial installations. Research has been undertaken to find and test sorbents with better selectivity, sulfur breakthrough, and sorbent regeneration. Bulk metal oxides, such as zinc ferrite and zinc titanate have been explored as well as supported metal oxides containing copper, molybdenum and manganese. Critical factors in sorbent performance are time until breakthrough, sorbent sulfur loading, fines generation, and fragmentation and deterioration of sorbent with cycling. Current physical set-up of hot-gas clean-up systems have the sorbent situated in fixed or moving beds, with sorbent recovery accomplished either through the removal of the reactor from the process stream or the mechanical conveyance of the sorbent into the regeneration bed. A fluid-bed design common to the petroleum industry to this sulfur removal process has also been tested. None of these hot-gas cleanup systems has been applied to a commercial or demonstration-scale gasifier. Significant research, engineering, and testing remains to prove the durability and performance of these sorbents in actual gasification systems.

Char Recycle. Gasifiers tend to have once-through carbon conversions lower than combustors with similar residence times. Gasifier design usually provides for a small surplus of carbon in the once-through mass balance. The remaining carbon is removed either through a recycle or through the separate combustion of the remaining char. Gasifier systems, as opposed to most pulverized coal combustion systems, are usually designed with a char recycle. The common physical element in these systems is a cyclone. Cyclones in these systems are designed to operate at high efficiencies at high temperatures. A key factor in the design is the selection of materials. As with all parts of the char recycle system, erosion is a major consideration. Corrosion and erosion are compounded by the high temperatures required for efficient operation of the gasifier as well as the highly reducing nature of the gases. Experimentation and testing has been undertaken to de-

Table 7. Typical Environmental Performance Characteristics for Coal Gasification[a]

SO$_x$		NO$_x$		Particulates		Secondary[b]	
Emissions % Removal	Rating	Emissions % Removal	Rating	Emissions % Removal	Rating	Emissions kg/kg coal	Rating
0.1 >97%	Very Good	<0.1 >97%	Very Good	0.0012 >99.5%	Very Good	0.45	Poor

[a] From Ref. 17, with permission.
[b] Includes slags and ashes, spent catalysts, and sludges.
* Emissions of primary pollutants are given in kg pollutant/kJ.

termine materials most suitable for application in these systems. Much practical experience has been gained through the operation of existing commercial gasifiers as well as the demonstration phases of several gasifier projects.

Particulate Cleanup. In IGCC systems, in order to retain the efficiency gained through use of a hot-gas sulfur removal system, final particulate removal before introduction of the gases into the gas turbine must also be accomplished at high temperatures and pressures. Traditional methods for particulate cleanup, such as fabric filters, are not suitable at these temperatures and pressures. Possible candidates for such systems are ceramic candles and ceramic filters. Results are still confined to laboratory and pilot-scale apparatus. Areas of particular concern are durability of components, and thermal and chemical degradation. Filters have been shown to perform well, attaining and surpassing desired performance levels. Upscaling to demonstration scale and resolution of durability concerns are still required.

Acknowledgments

Dr. Craig Eatough and Dr. Stephen Kramer, post-doctoral associates in the Advanced Combustion Research Center, have contributed substantially to this section. Also, the contribution of Dr. Predrag Radulovic, whose recently published work on coal gasification technologies provides a foundation to part of this section, is acknowledged. Support of the Advanced Combustion Engineering Research Center at Brigham Young University in preparing this manuscript is appreciated.

BIBLIOGRAPHY

1. D. Merrick, *Coal Combustion and Conversion Technology,* Elsevier Applied Science Publishing Co., Inc., New York, 1984.
2. G. J. Pitt, and G. R. Millward, *Coal and Modern Coal Processing, An Introduction,* Academic Press, Inc., New York, 1979.
3. L. D. Smoot, "Coal and Char Combustion," in W. Bartok, and A. F. Sarofim, eds. chapter 10, *Fossil Fuel Combustion: A Science Source Book,* John Wiley & Sons, Inc., New York, 1991.
4. J. B. Howard, "Fundamentals of Coal Pyrolysis and Hydropyrolysis," in M. A. Elliott, ed., chapter 12 in *Chemistry of Coal Utilization,* 2nd suppl. vol., Wiley-Interscience, New York, 1981.
5. D. R. Simbeck, R. L. Dickenson, and E. D. Oliver, *Coal Gasification Systems: A Guide to Status, Applications, and Economics,* Final Report, Electric Power Research Institute, Palo Alto, Calif., Synthetic Fuels Associates, Mountain View, Calif., 1983, Project 2207, EPRI AP-3109.
6. L. D. Smoot, and D. T. Pratt, eds., *Pulverized Coal Combustion and Gasification,* Plenum Publishing Corp., New York, 1979.
7. L. D. Smoot and P. J. Smith, *Coal Combustion and Gasification,* Plenum Press, New York, 1985.
8. J. M. Wooten, M. Nawaz, R. G. Duthie, R. A. Knight, M. Onischak, S. P. Babu, and W. G. Bair, *Literature Survey of Mild Gasification Processes, Co-Products Upgrading and Utilization, and Market Assessment,* U.S. Department of Energy, Morgantown Energy Technology Center, Report No., 1988, DOE/MC/24266-2666, Morgantown, W. Va.
9. N. R. Soelberg, L. D. Smoot, and P. O. Hedman, "Entrained Flow Gasification of Coal Part 1. Evaluation of Mixing and Reaction Processes from Local Measurements," *Fuel* **64,** 776–781(1985).
10. M. A. Elliott, *Chemistry of Coal Utilization,* 2nd suppl. vol., Wiley-Interscience Inc., New York, 1981.
11. L. D. Smoot, ed., *Fundamentals of Coal Combustion: for Clean and Efficient Use,* Elsevier, New York, 1993.
12. H. Perry, "The Gasification of Coal," *Scientific American* **19,** 230(1974).
13. D. Hebden and H. F. J. Stroud, "Coal Gasification Processes," in M. A. Elliott, ed., *Chemistry of Coal Utilization,* 2nd suppl. vol., Wiley-Interscience, Inc., New York, 1981, p. 1599.
14. F. D. Skinner, L. D. Smoot, and P. O. Hedman, "Mixing and Gasification of Coal in an Entrained Flow Gasifier," *ASME Conf.,* Chicago, Ill., Nov. 1980. Paper No., 80-WA/HT-30.
15. P. L. Walker, M. Shelef, and R. T. Anderson, "Catalysis of Carbon Gasification," in P. L. Walker, ed. *Chemistry and Physics of Carbon,* vol. **4,** Marcell Dekker, New York, 1986.
16. K. L. Smith, L. D. Smoot, T. H. Fletcher, and R. J. Pugmire, *The Structure and Reaction Processes of Coal,* Plenum Press, New York, 1994.
17. Engineering and Economics Research, Inc., Hagler, Bailly and Company, Inc., PEI Associates, Inc., and Paul W. Spaite, *Emerging Clean Coal Technologies,* Noyes Data Corporation, Park Ridge, N.J., 1986.
18. L. B. Clarke and L. L. Sloss, "Trace Elements-Emissions from Coal Combustion and Gasification," *IEA Coal Research,* London, 1992, IEACR/49.
19. G. H. Gronhovd, E. A. Sondreal, J. Kotowski, and G. Wiltsee, *Low-Rank Coal Technology, Lignite and Subbituminous,* Noyes Data Corporation, Park Ridge, N.J., 1982.
20. D. J. Haase, and D. G. Walker, "The COSORB Process," *Chemical Engineering Progress* **70**(5), 74–77(1974).
21. J. Douglas, "IGCC Phased Construction for Flexible Growth," *EPRI Journal,* Electric Power Research Institute, Palo Alto, Calif., Sept., 1986, pp. 4–11.
22. J. Douglas, "Beyond Steam," *EPRI Journal,* Electric Power Research Institute, Palo Alto, Calif., Dec. 1990, pp. 4–11.
23. N. Holt, "Highly Efficient Advance Cycles," *EPRI Journal,* Electric Power Research Institute, Palo Alto, Calif., April/May, 1991, pp. 40–43.
24. R. D. Parekh, Handbook of Gasifiers and Gas Treatment Systems, U.S. Department of Energy Report, UOP/SDC, McLean, Va., 1982, Contract No. DE-AC01-78ET10159, WD-TR-82/008-010.
25. P. T. Radulovic and L. D. Smoot, "Coal Processes and Technologies," in L. D. Smoot, ed., *Fundamentals of Coal Combustion: for Clean and Efficient Use,* Elsevier, New York, 1993, pp. 1–77.
26. H. Jüntgen, J. Klein, K. Knoblauch, H. J. Schröter, and J. Schulze, "Conversion of Coal and Gases Produced from Coal Into Fuels, Chemicals, and Other Products," in M. A. Elliott, ed., chapter 30 in *Chemistry of Coal Utilization,* Wiley-Interscience Inc., New York, 1981.
27. J. G. Speight, *The Chemistry and Technology of Coal,* Marcel Dekker, New York, 1983.
28. W. N. Clark and V. R. Shorter, "Cool Water Coal Gasification: A Mid-term Performance Assessment," *Mining Engineering* **39,** 97–102(1987).
29. W. N. Clark, R. L. Litsinger, and T. R. Straughan, "Cool Water: Performance and Economics," *Energy Progress* **7,** 99–104(1987).

30. J. P. Henley and D. G. Sundstrom, "Operational and Performance Data of the DOW Coal Gasification Plant-First One and a Half Years," *5th Annual Pittsburgh Coal Conference,* Pittsburgh Pa., Sept., 1988.

31. P. Nowacki, ed., *Coal Gasification Processes,* Energy Noyes Data Corporation, Park Ridge, N.J., 1981, Technology Review No. 70.

32. U. Femmer, J. Lambertz, and K. A., Theis, "Gasification of Brown Coal for the Generation of Synthesis Gas," *Energy Progress* **7,** 130–133(1987).

33. *Clean Coal Technology Demonstration Program: Program Update 1991* (as of Dec. 31, 1991), U.S. Department of Energy Report, Washington D.C., 1992, DOE/FE-0219P.

34. R. L. Graves, S. S. Lestz, S. S. Trevitz, and M. D. Gurney, "Screening Tests of Coal Pyrolysis Liquids as Diesel Fuel Extreders," SAE Technical Paper Series, Aug. 1984, Paper No., 841002.

35. J. R. Longanbach and A. Sass, "Ignition Characteristics of Chars Made in the G R&D Coal-Gasification and Coal-to-Liquids Processes," *Fuel* **54,** 29–33(1975).

36. W. F. Wells, S. K. Kramer, L. D. Smoot, and A. U. Blackham, "Reactivity and Combustion of Coal Chars," *20th Symposium (International) on Combustion,* The Combustion Institute, Pittsburgh, Pa., 1984, pp. 1539–1546.

37. T. C. Barthe and R. D. Gunn, "The Hanna Wyoming Underground Coal Gasification Field Test Series," in W. B. Krantz and R. D. Gunn, eds. *Underground Coal Gasification: The State of the Art,* American Institute of Chemical Engineers, New York, *AIChE Symposium Series,* **79**(226), 4–14(1983).

38. C. B. Thorsness and J. R. Creighton, "Review of Underground Coal Gasification Field Experiments at Hoe Creek," in W. B. Krantz and R. D. Gunn, eds. *Underground Coal Gasification: The State of the Art,* American Institute of Chemical Engineers, New York, *AIChE Symposium Series,* **79**(226) 15–43(1983).

39. R. W. Hill and C. B. Thorsness, "The Large Block Experiments in Underground Coal Gasification," in W. B. Krantz and R. D. Gunn, eds. *Underground Coal Gasification: The State of the Art,* American Institute of Chemical Engineers, New York, NY, *AIChE Symposium Series,* **79**(226), 57–65(1983).

40. G. J. Bell, C. F. Brandenburg, and D. W. Bailey, "ARCO's Research and Development Efforts in Underground Coal Gasification," in W. B. Krantz and R. D. Gunn, eds. *Underground Coal Gasification: The State of the Art,* American Institute of Chemical Engineers, New York, NY, *AIChE Symposium Series,* **79**(226), 44–56(1983).

41. T. F. Edgar, "Research and Development on Underground Gasification of Texas Lignite," in W. B. Krantz and R. D. Gunn, eds. *Underground Coal Gasification:* American Institute of Chemical Engineers, New York, NY, *AIChE Symposium Series,* **79**(226), 66–76(1983).

42. D. Olness and D. W. Gregg, "The Historical Development of Underground Coal Gasification." U.S. Energy Research and Development Administration Report, University of California, Lawrence Livermore Laboratory, Livermore, Calif., 1977, Contract No. W-7405-Eng-48, UCRL-52283.

43. D. E. Roe, A. W. Karr, K. R. Lemmerman, and J. E. Sinor, *Solid Fuels for U.S. Industry,* Vol. 2, Cameron Engineers, Mar., 1979.

44. A. Attari, *The Fate of Trace Constituents of Coal During Gasification,* U.S. Environmental Protection Agency, Research Triangle Park, N.C., Aug., 1973 Report EPA 650/2-73-004.

45. L. B. Clarke, "The Behavior of Trace Elements During Coal Combustion and Gasification: An Overview," Paper presented at the *EPRI Conference: Managing Hazardous Air Pollutants: State of the Art,* Washington D.C., Nov., 1991.

46. K. Hufan, "Origin and Properties of Slag and Fly Ash Obtained in the PRENFLO Process," *IEA Expert Meeting on Coal Gasification Slag,* Arnhem, Netherlands, May, 1990.

47. L. B. Clarke, "Management of By-Products from IGCC Power Generation," *IEA Coal Research,* IEACR/38, London, May, 1991, p. 73.

48. D. S. Beishon, J. Hood, and H. E. Vierrath, "The Fate of Trace Elements in the BGL Gasifier," *Sixth Annual International Pittsburgh Coal Conference,* Pittsburgh, Pa., 1989, pp. 539–547.

49. J. S. Klingspor and J. L. Vernon, "Particulate Control for Coal Combustion," *IEA Coal Research,* London, 1988, IEACR/05.

COAL LIQUEFACTION

LARRY L. ANDERSON
University of Utah
Salt Lake City, Utah

Coal liquefaction describes the conversion of solid coal to fuel liquids (1,2). Coal, although composed mostly of carbon and hydrogen like petroleum, unlike petroleum, is a solid basically due to its relatively lower hydrogen content. Therefore, the proceess of coal liquefaction consists of producing compounds that are liquids containing hydrogen at levels of approximately 10 to 15% by weight (3). Typical compositions for coals, liquid fuels, and some hydrocarbons are given in Table 1. One can observe from these and similar data that coal and hydrocarbon liquids cover a considerable range of composition and H/C values (4–8). The liquefaction of coal has always consisted of either adding hydrogen to that part of the coal produced as liquids or rejecting carbon from that same entity. Methods for accomplishing coal liquefaction are complex and numerous but can be classified as either direct, (9,10) where coal is converted to liquids without complete breakup of the coal molecular structure, or indirect, (11,12) where coal is first gasified to carbon monoxide and hydrogen (synthesis gas) and afterwards, liquid compounds are fabricated from the synthesis gas in one or more steps.

See also COMMERCIAL AVAILABILITY OF ENERGY TECHNOLOGY; FUELS, SYNTHETIC; LIGNITE AND BROWN COAL.

Coal liquefaction has been practiced only sporadically at commercial scale because the price of coal liquefaction products has usually been significantly higher than comparable liquids obtained from petroleum. The significant commercial application of coal liquefaction has been in unusual circumstances where price has been a less critical factor than secure or strategic supply. During World War II the German production of a wide range of transportation fuels was used to supply the war machine since Germany had an abundance of coal but little or no access to petroleum resources. The Germans used both direct and indirect methods to produce liquid fuels but the direct method was more emphasized (13,14). At the end of the war the German plants had mostly been destroyed by the Allied bombings. Both the U.S. and the former USSR took documents and personnel from the German operations to

Table 1. Compositions and Hydrogen/Carbon Ratios for Selected Fuels and Hydrocarbons (H/C ratios are atoms of H/atom of C)

Fuel	Carbon	Hydrogen	Oxygen	Sulfur	Nitrogen	H/C
Typical crude oil	86.0	11.0	0.7	1.5	0.5	1.6
Fuel oil	86.0	13.4	0.2	0.3	0.1	1.8
Anthracite coal	93.0	3.0	2.5	0.7	0.8	0.4
Bituminous coal	78.0	5.7	11.6	3.3	1.0	0.9
Subbituminous coal	71.9	6.1	20.2	0.6	1.0	1.0
Lignite	72.3	5.1	19.9	1.6	1.1	0.84
Gasoline	85.0	15.0	0.0	0.1	0.0	2.1
Benzene	92.3	7.7	0.0	0.0	0.0	1.0
Naphthalene	93.7	6.3	0.0	0.0	0.0	0.8
Methane	74.9	25.1	0.0	0.0	0.0	4.0

Table header (spanning Carbon–Nitrogen columns): Element, Weight % (moisture and ash-free)

evaluate the feasibility of commercial coal liquefaction in their nations (15–26).

More recently (since 1955) South Africa, due to a lack of petroleum and natural gas and an abundance of coal reserves, has carried out at large commercial scale the indirect liquefaction of coal (27).

Other efforts at commercial production of liquid fuels from coal have taken place in the U.S., the UK, and Japan without actual full scale development (28–30). Reasons for not achieving commercial operation in these nations include the availability of inexpensive liquid fuels from petroleum supplies, both domestic and imported. Because both industrialized and developing nations are increasing their demand for and use of petroleum and natural gas, world petroleum supplies are expected to be either unreliable or inadequate in the near future (31). The nonuniform distribution of petroleum in the world has resulted in several crises and supply interruptions, especially since 1973 when the Organization of Petroleum Exporting Countries (OPEC) first achieved sufficient power to have a major impact on world oil markets (32).

Many scientists and engineers believe the future will require the liquefaction of coal to meet the demand for liquid transportation fuels. The basis for such a viewpoint can be seen in Figure 1 where the U.S. demonstrated coal reserve base is compared with world petroleum and natural gas reserves. World fossil fuel reserves mostly consist of coal whereas consumption of fossil fuels since the 1960s has been mostly petroleum and natural gas (33). For example, since 1970 in the U.S., where coal comprises about 88% of the fossil fuel resources, approximately 75% of the U.S. energy supply has been in the form of petroleum and natural gas. Further world development of coal liquefaction will depend on both economics and the reliability of petroleum and natural gas supplies from the Middle East and other main exporting areas.

HISTORICAL BACKGROUND

Although many events could be listed as significant in the invention, development, or commercialization of coal liquefaction, the following are some of the principal milestones. These milestones have occurred in different parts of the world as a result of different pressures, incentives, and economic or political conditions.

1869 Chemical reduction of coal to produce liquids is accomplished by Berthelot (34);

1911 Hydrogenation of coal to liquids by Bergius;

1926 First large scale production of liquids from coal is begun at the Leuna plant in Germany (36);

1935 Liquefaction plant put into operation in Billingham, England;

1927–1945 Full-scale production of German coal liquefaction plants (both direct hydrogenation and indirect gasification and Fischer-Tropsch synthesis) (38);

1949–1954 Evaluation in the U.S. and USSR of both direct and indirect liquefaction technologies; In the U.S., studies were made using U.S. coals at Louisiana, Missouri for direct liquefaction and at Brownsville, Texas for indirect liquefaction (39);

1955 Commercial operation began for indirect liquefaction at Sasolburg in South Africa

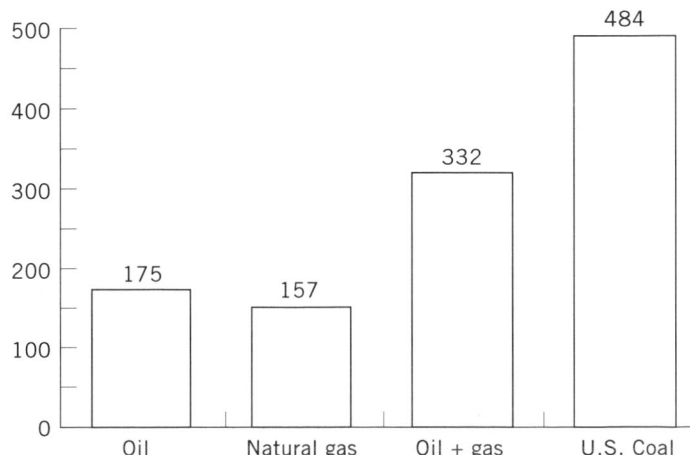

Figure 1. Comparison of World Oil and Natural Gas reserves and U.S. Demonstrated Coal reserves.

(SASOL I); Lurgi gasifiers and both ARGE (fixed-bed) and synthol (entrained-bed) synthesis reactors were utilized (40);

1974   Operation of the Solvent Refined Coal (SRC) plant in Fort Lewis Washington; Later, the process was modified to produce more valuable (distillate) products (SRC II) (41);

1976   Development of the Mobil Methanol-to-Gasoline (MTG) process is established, using ZSM-5, a shape selective zeolite catalyst (42);

1980   Operation began for SASOL II at Secunda, South Africa, an indirect liquefaction plant with a capacity of approximately 32,000 tons of coal per day and daily producing 4800 tons of liquids, 450 tons of ethylene and 280 tons of ammonia (43);

1980   The New Energy Development Organization (NEDO, Later changed to NEDOL) is begun in Japan. This organization was to centralize the research and development efforts of the Japanese government and industrial concerns, especially in the implementation of "Project Sunshine." Projects planned included a pilot plant to test lignite (brown coal) liquefaction and one to evaluate the feasibility of bituminous coal liquefaction (44);

1980   Operation of the 200–600 ton/day H-Coal plant at Catlettsburg, Kentucky to test the feasibility of the ebullated-bed H-Coal process (45);

1980   Operation of the Exxon donor solvent (EDS) process at Baytown, Texas and evaluation of this process in the 250 ton/day plant (46);

1981   Plant initiated to produce methanol from natural gas in New Zealand; The methanol was converted to gasoline using the Mobile MTG process. Plant design capacity: 17,000 barrels/day of gasoline (47);

1986   Operations begin for the NEDO pilot plant to test the liquefaction of Victorian (Australian Brown) coal by the BCL Process. The plant operated until October of 1990 (48);

1991   Construction of the Bituminous coal liquefaction plant to test the NEDOL process began at Kashima, Japan; Operations have been planned to commence in 1996, processing 150 tons/day of coal (Wyoming, U.S., Wandoan, Australian, and Illinois #6, U.S.) (49);

1992   Wilsonville Plant in Alabama concludes operation and is shut down bringing to a close the testing of many U.S. coals and catalysts as well as single and two-stage liquefaction.

There are many other significant milestones one could list as important in the history of coal liquefaction. In addition to the large pilot and demonstration plants operated, many attendant process development units (or processing supporting units) and bench scale units have been built and operated. The date of some plants being shut down are sometimes more important than the date of initial operation, since the former indicated a change in government or industrial policy with regard to the feasibility of coal liquefaction.

COAL STRUCTURE, COMPOSITION, AND CHEMISTRY

To understand the processes and chemistry of coal liquefaction some basic knowledge of the chemical structure of coals is necessary. Here one must go back to the origin and even the definition of coal. The definition by Francis given in 1954 (50) still seems appropriate:

> "Coal may be defined as a compact, stratified mass of mummified plants which have been modified chemically in varying degree, interspersed with smaller amounts of inorganic matter. The chemical properties of a coal depend upon the proportions of the different constituents present in the parent vegetable mass, the nature and extent of the changes which these constituents have undergone since their deposition and the nature and quantity of the inorganic matter present."

Although the exact nature of the chemical structures in coals are not known, a great deal is known about both the physical and chemical nature of coals. Extensive analytical and characterization studies have been done on many coals and their fractions. Some of the greatest impediments to precise knowledge about the molecular makeup of coal are its heterogeneity and its insolubility. Since the 1960s several coal structure models or schematic representations have been proposed to assist scientists and engineers with the formidable problem of determining the most efficient way to convert coal to valuable products. One model (Fig. 2) shows the variety

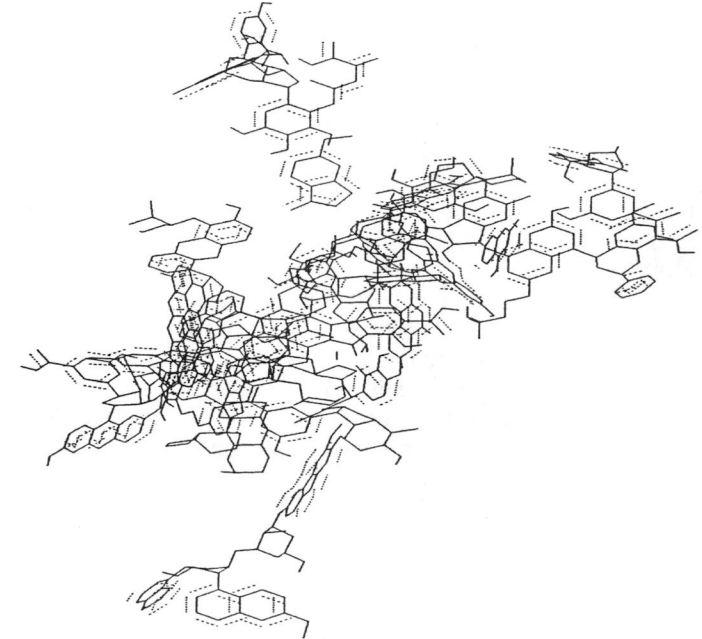

Figure 2. Oblad model of carbon structure showing a variety of chemical structures, ie, aromatics as well as cyclic and open-chain aliphatics (75).

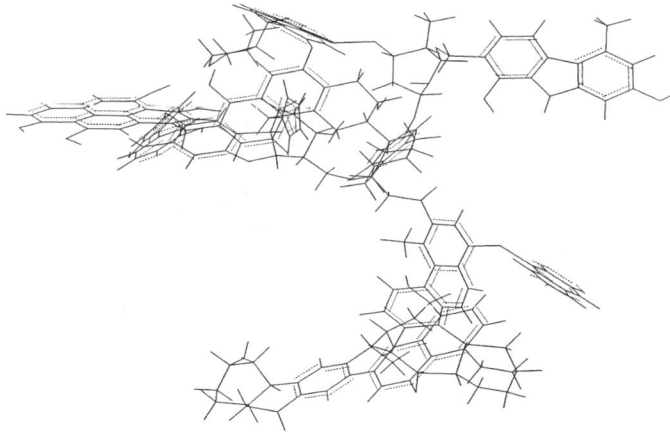

Figure 3. Given model of carbon structure showing a greater predominance of aromatic and other cyclic structures (52).

of chemical structures present, ie, aromatics, cyclic and open-chain aliphatics, both saturated and unsaturated. Most other models, eg, Figure 3, show a greater predominance of aromatic and other cyclic structures. The myriad of products possible include phenols, waxes, and liquid and gaseous fuels. Information on other structures is available (52–55). The most difficult challenge in trying to represent the molecular structures in coal arises from the fact that coals are three dimensional solid materials which may be bonded together in different ways in different directions. One of the most useful descriptions of coal concerning its heterogeneity is that given by Neavel (56) who described coal as "an aggregate of microscopically distinguishable, physically distinctive and chemically dif-ferent macerals and minerals. Coal is analogous to a fruitcake, formed initially as a mixture of diverse ingredients, then "baked" to a product that is visibly heterogeneous.

The analogy by Neavel is appealing since like a fruitcake the material which became coal was highly heterogeneous and was "cooked" but still retains some of the character of its original ingredients. Also like a fruitcake, coal has parts which can visually be distinguished from each other, although for coal, the distinction may require microscopic examination. Solubility studies on coals show another complicating factor: some parts of a coal are not tied to the main molecular framework of that coal. These entities can often be dissolved from a coal with ordinary organic solvents and their amounts vary with the coal being studied (57,58). The main framework of the molecular coal structure is considered by most coal scientists to consist of aromatic and related cyclical hydrocarbon structures of one to three or four rings linked together by carbon-carbon, carbon–oxygen, carbon–nitrogen, or carbon–sulfur bridges or bonds (59–61).

From the above definitions and descriptions it is easy to see that the term "coal" can mean different things. Coal is indeed classified into different classes, designated as ranks. The rank names, lignite, subbituminous, bituminous, and anthracite have resulted from the application of coal as a combustible fuel (62). Peat is considered as a precursor to coal and not a separate rank. Several different nations and areas have established criteria for rank designation. Table 2 gives the most recent standards by the ASTM (ASTM D388) (63).

More appropriate classifications for coal which relate to the important properties for coal liquefaction are those

Table 2. American Society for Testing and Materials Classification of Coals by Rank (ASTM D388)[a]

| Coal Rank | Subgroup | Fixed Carbon[a] Value Limitations[b] | | Gross Calorific Value, Btu[c] | | Agglomerating Character |
		Equal or greater than	less than	Equal or greater than	less than	
Anthracite	1. Meta-anthracite	98	—			
	2. Anthracite	92	98	—		Nonagglomerating
	3. Semianthracite[d]	86	92			
Bituminous	1. Low vol bituminous	78	86			
	2. Med. vol. bituminous	69	78			
	3. High vol A bituminous		69	14,000[d]		Commonly agglomerating
	4. High vol B bituminous		69	13,000	14,000	
	5. High vol C bituminous		69	11,500	13,000	
				10,500	11,500	Agglomerating
Subbituminous	1. Subbituminous A			10,500	11,500	
	2. Subbituminous B			9,500	10,500	
	3. Subbituminous C			8,300	9,500	
						Nonagglomerating
Lignite	1. Lignite A			6,300	8,300	
	2. Lignite B				6,300	

[a] This classification only applies to coals composed mostly of vitrinite. Liptinite- or inertinite-rich coals do not fit into the classification because rank determining parameters (calorific value, fixed carbon, and agglomerating propensity) differ greatly for these macerals compared to vitrinite.

[b] Dry, mineral matter-free fixed carbon $= \dfrac{Fixed\ carbon}{[100 - M + 1.1A + 0.15]} \times 100.$

[c] Moist refers to coals containing their natural inherent moisture. Moist Mm-free Btu $= \dfrac{(Btu\text{-}505)}{[100 - (1.08A + 0.55S)]} \times 100.$

[d] If agglomerating classify as low volatile bituminous.

[e] Coals with fixed carbon greater than 69% shall be classified only by fixed carbon.

Table 3. Type Classification of Coal

Coal Type	Transmitted Light Observation (Thiessen)	Reflected Light Observation (Stopes-Hearlen)
Banded Coals		
Bright coal	>5% Anthraxylon <20% Opaque attritus	Vitrite Clarite
Semisplint coal	>5% Anthraxylon 20–30% Opaque attritus	Duroclarite Vitrinertite
Splint coal	>5% Anthraxylon >30% Opaque attritus	Clarodurite
Nonbanded Coals		
Cannel coal	Attritus with little or no algae	Durite
Baghead coal	Attritus with predominant oil algae	

which classify coal according to "type"; that is, by the variation in the proportion of the different microscopically identifiable components. Such components in coal are called macerals and coal classification by type based upon type distribution has been done by Stopes (64) in England in 1919 and by Thiessen (65) in the U.S. in 1920 and later. The different methods of microscopic examination by these two coal petrographers have led to different terminology. Stopes had used polished samples or particles trapped in resinous materials and had examined these samples with reflected light. Thiessen had based his observations on microscopic examination of thin sections of coal or its components in transmitted light. A comparison of the two systems and their nomenclature are shown in Table 3. In recent years the terminology used by Stopes, now called the Stopes–Heerlen system has generally prevailed (66).

Because of the difficulty of studying whole coal with its heterogeneity, methods have been developed to separate and study the individual macerals. Dyrkacz and coworkers at Argonne National Laboratory in the U.S. pioneered such studies and some indication of the differences of the various macerals are shown in Table 4 (67,68). However, a complete treatment of the chemistry and molecular structure of coals is beyond the scope of this discussion of coal liquefaction and more detailed information can be found elsewhere (69,70).

DIRECT COAL LIQUEFACTION

Direct coal liquefaction involves the conversion of solid coal to liquids without the production of synthesis gas, (carbon monoxide and hydrogen) as an intermediate step. Direct coal liquefaction should be the most energetically efficient method of producing liquids. If specific reactions could be selectively carried out to break just the linking bonds between clusters of hydrocarbon moieties in the coal and thereby produce molecules with five to approximately twenty three carbon atoms per molecule (light gasoline to heavy fuel oil), all of the products would be liquids. So far such an ideal direct liquefaction process has not been found because coals are not homogeneous. There may be several linking bonds between clusters, but there are steric hindrances and other obstacles to liquefaction reactions. In all direct liquefaction processes solids and gases are produced along with liquids.

In reactions that break up the coal structure, free radicals and reactive fragments are produced that can react with each other and produce species that are high in molecular weight and less amenable to liquefaction reactions than the original coal. Direct liquefaction is observed to take place in two steps: The first step is the thermal or catalytic breakup of the coal macromolecular structure. This step is often referred to as "depolymerization" although coal is not strictly a polymer since it does not have uniform repeating units. The "depolymerization" reaction is followed by stabilization and/or upgrading to the final liquefaction products. Reactions occurring in this second step are those that add hydrogen to the reactive fragments or free radicals produced by the thermal or other bond-breaking (depolymerization) reactions of the first stage.

The main processes that have been used to directly liquefy coal are (71,72)

1. Pyrolysis
2. Hydrogenation
3. Solvent extraction or solvent refining
4. Coprocessing (liquefaction of coal and residual oil mixtures)

Table 4. Physical Properties and Compositions of Macerals in a Typical High Volatile Bituminous Coal[a]

Maceral	Element, weight %					Density, g/cc	Oil Reflectance	H/C
	C	H	O	S	N			
Vitrinite	84.1	5.3	8.0	1.0	1.6	1.29	0.91	0.75
Exinite	86.2	6.5	5.5	0.7	1.1	1.21	0.32	0.904
Micrinite	85.9	4.8	8.0	0.6	0.7	1.32	1.61	0.67
Fusinite	91.5	3.2	4.3	0.4	0.6	1.48	2.65	0.41

[a] H/C ratios are atoms of H/atom of C. Values given are only typical and emphasize the differences between the main maceral groups. As pointed out by Dyrkacz, relatively pure macerals can be prepared by using density gradient centrifugation (DGC) with only cross contamination of a few percent. Even then macerals exhibit a range of values for density and other properties.

If no catalysts are employed in these reactions the processes are thermal and indiscriminate, resulting in products that are less valuable than can be produced when catalysis is used to control the results. All commercial, pilot plant, and demonstration plant coal liquefaction projects have employed catalysis since well designed catalytic reactions can result in higher conversion to liquids and improved selectivity. German scientists and engineers tested thousands of catalyst combinations (51,73,74). The most often tested and used catalysts were iron-based since these are usually inexpensive and have minimal environmental consequences (75).

The catalysis of coal liquefaction is a complex and extensive subject and only a brief summary is appropriate here. Each of the main processes have employed catalysts, the functions of which are to facilitate desired reactions and to prevent regressive reactions which result in increased yields of coke, char, and/or heavy residual products. Both hydrogenation and solvent refining are hydrogenation processes where the differences are mostly in the way hydrogen is reacted with the coal material or the products of thermal breakup of the coal. Pyrolysis, often termed carbonization, involves only the application of heat to thermally break down the solid coal structure and produce liquids, although in recent years hydrogen has also been used (hydropyrolysis) for the same purpose. Usually pyrolysis results in minimal liquid production, about 20% of the mineral and ash-free (maf) coal material. Pyrolysis has mostly been applied when the main product desired was a solid (metallurgical coke). Yields of both total liquids and distillates can be considerably increased by hydropyrolysis (76).

In catalytic hydrogenation the hydrogen is usually present as a gas at high pressure. Early German processes and many processes recently developed have used two-stage liquefaction schemes (77). In the first stage, coal and usually a solvent or vehicle oil, are reacted to break the linking bonds in coal, producing reactive fragments or free radicals. The hydrogen functions to limit the recombination of these reactive fragments or free radicals thus producing liquid products (called middle oil) rather than solids. In the second stage the middle oil is upgraded to produce lighter liquids, (distillate fuels). In order to have a catalyst involved in the first stage dispersed catalysts were used. In the second stage, since the reactant is a liquid (middle oil), supported catalysts can be applied (78).

As indicated earlier, direct liquefaction of coal includes those processes which produce liquid fuels by only breaking up the coal structure enough to obtain liquid range molecules. In theory, direct liquefaction should be more efficient than indirect, since less thermal or catalytic degradation is necessary. Such reactions must break some of the C–C, C–H, C–O, C–S or C–N bonds. For each of these simple bond representations, there are various bond energies involved. For example C–C bonds may be aliphatic, olefinic, aromatic, or (rarely) alkynic. The other bonds likewise may have different energies, depending on the particular structures. In addition to breaking up the coal structure (which may have molecular weights in the order of 10^4–10^6), the resulting fragments must be stabilized to materials that can be stored, transported, etc, as usable liquid fuels. Both of these types of reactions are carried out in different ways in the different, direct coal liquefaction methods (79).

Pyrolysis

Pyrolysis or carbonization is probably the oldest and most practiced coal liquefaction method (80). Coal is heated without gas or air present so that the coal is thermally decomposed by destructive distillation. Carbonization has been used for the production of metallurgical coke for many years but is limited to "coking", "caking" or agglomerating coals, usually high- and medium-volatile bituminous ranks. In coking and all pyrolysis processes, thermal decomposition cracks the polycyclic coal structure into smaller, lighter molecules, either gases or liquids. At the same time many cracking reaction products polymerize or combine to produce high carbon coke. By rejecting a large portion of the carbon as coke, the liquids produced in pyrolysis are higher in hydrogen content than the starting coal. Liquids produced during the coking process are of poor quality and still contain the heteroatoms, S, N, and O. The greatest shortcoming of pyrolysis is that only a minor fraction of coal is converted to liquids. The amount produced by pyrolysis has been studied (81). Dryden's equation (82) was used for volatile yield in pyrolysis and results were obtained from 130 coals. Approximate relations found were:

$$\text{Volatile matter (mass \%)} = 97.3\,\text{H/C} - 40.4$$

$$\text{Tar + Light oils (mass \%)} = 29.1\,\text{H/C} - 12.1$$

Since the atomic H/C ratio for many coals is 0.8–1.0, the liquid yield from such coals, as found by Probstein and Hicks falls in the range of 11–17% of the coal mass. Other liquefaction processes typically yield more than 60% of the coal weight as liquids.

Despite low liquefaction yields in pyrolysis or carbonization processes, they have been the most consistently operated industrial coal conversion processes. Since the main product from carbonization is solid coke, the processes have not been optimized for liquid product yields or quality. However, significant information about coal chemistry and other coal properties has been obtained from carbonization. Carbonization processes are arbitrarily divided into low-temperature (300–800°C) or high temperature (900–1100°C) (83,84). Low temperature carbonization products are solid char, liquids, and gases. Typical liquids from low-temperature carbonization and hydrocarbonization processes are shown in Table 5. Tar, light-oil, and phenol compositions are a function of both the coals reacted and the conditions.

PROCESS YIELDS FROM LOW TEMPERATURE CARBONIZATION AND HYDROCARBONIZATION

High temperature pyrolysis or carbonization where final temperatures are above 900°C, is referred to as coking. Coking usually takes place in coke ovens which are built

Table 5. Process Yields from Low-Temperature
Carbonization and Hydrocarbonization

Yields	Carbonization, $N_2{}^a$	Hydrocarbonization, $H_2{}^b$
Char	67	49
Tar	12	25
Gas	9	14
Water	12	14
Light oil[c]	4	10
Phenols[d]	2	5.2
Phenol	0.5	1.9

[a] Temperature = 567°C.
[b] Temperature = 567°C. Hydrogen partial pressure = 648 1kPa (940 psi), 1.6% H_2 reaction.
[c] bp ≤ 260°C.
[d] bp ≤ 230°C.

in batteries of 24 to 80 ovens. Coals used for coking are fines from medium-volatile bituminous coals or, more common now, blends of high- and medium-, and/or low-volatile bituminous coals. The feed coal or blend of coals is selected to produce coke with the desired properties.

Tars produced from high-temperature coking are extremely complex mixtures but are less diverse than low-temperature tars and are less related to the properties of the coal from which they are produced (85–87). The main components are cyclic aromatic hydrocarbons and derivatives which range in molecular size from benzene and alkyl benzenes to compounds having 20 or more rings. There are many heteroatom-containing compounds in high temperature tars but they usually contain only one S, N, or O atom per molecule. High temperature tars also contain small amounts of aliphatic and partially saturated aromatic compounds.

Hydrogenation

The main methods of altering the composition of coal to convert a large fraction into liquids involve either direct hydrogenation by gaseous hydrogen or addition of hydrogen from a solvent or solvent mixture. Both of these routes have been investigated in Germany as well as other countries early in the twentieth century (88).

In direct hydrogenation using gaseous hydrogen, coal is usually fed as a slurry composed of powdered coal and recycled coal-derived liquids from the process. Successful processes have used catalysts and the technologies have been derived from the Bergius hydrogenation techniques patented in 1913 (35,89). Catalysts used are usually hydrocracking/hydrogenation (bifunctional) catalysts and thousands have been tested and evaluated by the Germans, by the U.S. Bureau of Mines and Department of Energy, Japanese scientists and engineers, and others (90). Most hydrogenation processes and developments have used rather severe reaction conditions with temperatures at 450–480°C and hydrogen pressures of 10–70 MPa (100–700 atmospheres). Times required for these reactions are typically near one hour. Although carbon conversions are about 90–95 mass % of the dry, ash-free (daf) coal, only 50–60 weight % of the coal is actually obtained as distillable liquids (91). Older German processes carried out at the high end of the pressure range (near 70 MPa or 700 atmospheres) have given mostly dis-

tillate fuel liquids. The composition of hydrogenation liquids depends on the degree of hydrogenation with 4–7% hydrogen added (as per cent of the coal weight) being typical for processes which result in high distillate liquid yields (92).

Solvent Extraction or Solvent Refining

In the 1920s Pott and Broche pioneered the processing of coal using the anthracene oil fraction from coal carbonization to extract liquids from coal. Even at low pressures, 35 to 40% of the coal weight could be extracted from a bituminous coal. At first this process was used to produce low-cost newspaper inks (93). Later solvents such as tetrahydronaphthalene (tetralin) were found to extract more of the coal as liquids. Solvents which could transfer hydrogen to the extracts from coal were used to produce a main fraction of the coal as liquids. The process that became the Pott-Broche process (94) used a mixture of cresol and tetralin to dissolve about 75% of German brown coals (lignite) at 427°C and 13.8 MPa. Similar processes were developed in Japan (95) and later in the U.S. (96). In the U.S., Spencer Chemical Company and the U.S. Office of Coal Research developed the Solvent Refined Coal (SRC) process which was tested at pilot plant scale. Gulf Oil Company and later Southern Company Services Inc. operated pilot plants to improve the solvent refining of coal (97,98).

One solvent extraction process which is different in several aspects than those previously discussed is supercritical gas extraction. This procedure can be considered as an extraction process or, since no chemical conversion is involved, as a pyrolysis process. Since there is a hydrocarbon solvent present it seems logical to include supercritical gas extraction in the solvent extraction category. This treatment gives higher yields than pyrolysis and takes advantage of the fact that coal volatilizes more readily in the presence of a compressed gaseous material. The volatilization degree depends on the gas density, which is greatest near the critical point. In supercritical gas extraction of coal a solvent such as toluene (C_7H_8) with a critical temperature of 319°C has been used at 400°C and 10 MPa to give over 30% extraction of coal to liquids (99). The extract, while higher in hydrogen than the feed coal, is usually still solid at room temperature but can be easily hydrocracked to distillable liquid fuels.

Coprocessing of Coal and Residual Oil

With advanced technologies available for processing heavy oil fractions in the petroleum industry the use of such materials has extended the usable fraction of crude oil. Commercialized technologies such as LC Fining, H-Oil Processing, and Veba Combicracking were developed in the U.S. and Germany (100). The potential to combine petroleum residues, the so-called "bottom of the barrel," with coal was investigated to use secure domestic resources as an alternative to importing greater quantities of crude or refined products. This coprocessing was intended to produce light liquids by coal conversion and hydrocracking of the 800+ petroleum fraction residues at the same time. Because of the experience in the petroleum industry with coking to produce solids-free oil from heavy fractions and the Bergius technology for coal conversion, this method of coal liquefaction was visualized by many as likely to be the least expensive method for coal liquefaction. It is significant that most of the German coal liquefaction plants operated during World War II used mixtures of coal tar, petroleum pitch and coal depending on availability. In other words most of the German plants were coprocessing plants, since only three of the twelve coal conversion plants used only coal as the feed (101). The U.S. government policy regarding coprocessing was demonstrated in 1993 by the awarding of an $18.6 million, three year contract to operate the Hydrocarbon Research Inc. (HRI) coal-to-oil test plant. HRI's experience with the H-Oil process (and the closely related H-Coal Process) was scheduled to continue to scale up coprocessing, as well as other liquefaction concepts.

Coal Liquefaction Processes

Interest in coal liquefaction has fluctuated since World War II with the level of interest a function of the price and availability of crude petroleum and the strategic climate between the major military nations. Many processes have been proposed at different times with the peak in activity in the U.S. around 1981. At that time there were approximately 69 coal liquefaction facilities and projects either in operation, under construction, being planned, or proposed. Another 12 projects had been previously operated, 8 small process development or pilot units, 2 pilot plants, and 2 demonstration plants (102). By 1993 essentially all of these plants and projects had been terminated. The chemistry of coal liquefaction was established but the economics were unfavorable compared with the price of liquids from petroleum.

Many research projects and a few test plants remain active to scale up new liquefaction concepts which might lower the cost of coal-derived liquid fuels. Direct liquefaction processes have also been developed in Germany, Japan, and elsewhere. Only representative direct processes will be described here including the more successful and developed processes. Table 6 summarizes the main features of the processes still active or being developed after about 1990 (103).

DIRECT LIQUEFACTION PROCESSES

Recent direct liquefaction processes that have been developed to the point of large pilot plant operation are:

- Nedol Process, bituminous coal liquefaction, Japan
- H-Coal Process, high sulfur coal liquefaction, U.S.
- SRC-II and CTSL Processes, U.S.
- Exxon Donor Solvent (EDS) Process, U.S.
- Brown Coal Liquefaction, Japan
- DMT Coal Hydrogenation (IGOR), Germany

Several other processes have had extensive bench and PDU (Process Development Unit) testing and evaluation. All direct liquefaction processes have some common characteristics, such as: (1) processing closely related to the earlier German high hydrogenation processes, and (2) synthetic crude products containing high levels of oxygen, nitrogen, and high-molecular weight asphaltols or asphaltenes. In recent years new generation processes have improved their product characteristics by employment of novel reactor configurations and catalysts which permit production of cleaner and more volatile fuel liquids. Emphasis has shifted from conversion level to selectivity. Liquefaction processes now include not only conversion to distillates but also upgrading to clean burning gasoline, diesel, and other fuels.

Nedol Process

This process was designed and developed to efficiently produce high quality transportation fuels under the Sunshine Program promoted by AIST/MITI. A pilot plant, in the process of construction during 1993–1995, has had a capacity of 150 tons of coal per day. The coal is fed in a slurry containing 40 weight % coal (with almost 1.2% pyrite), catalyst, and solvent. The essential steps of the process are shown in Figure 4. In addition to the basic process, other accompanying processes include hydrogen production, solvent hydrogenation and liquified oil distillation.

In the NEDOL process the slurry is heated in a preheater vessel to 400°C along with recycle gas containing

Table 6. Direct Coal Liquefaction Processes

Developer	Process	Products	Catalyst	Catalytic Upgrading
PAMCO, Gulf Oil Corp	SRC-II	Middle, heavy distillates, SRC	Mineral matter	
Southern Service Co		C$_1$–C$_4$ gas, naphtha		
Exxon Research & Engineering	EDS	Distillates, gas	Yes	Yes
Hydrocarbon Research Inc.	H-Coal	Gas, light and heavy distillates, fuel oil	Yes	No
DMT Forschung and DMT-Gesellschaft für Forschung und Prüfung mbH	IGOR	Gases, distillates residuum	Yes	Yes
New Energy Development organization. (NEDO) (Japan)	Brown	Light and medium coal liquefac-oils	Yes	Yes
Ministry of Internatl Trade & Industry	NEDOL	Light, middle and heavy oils, gases	Yes	Yes
SGI International	LCF	Solid process-derived fuel (PDF) gases and fuel oil	No	Yes

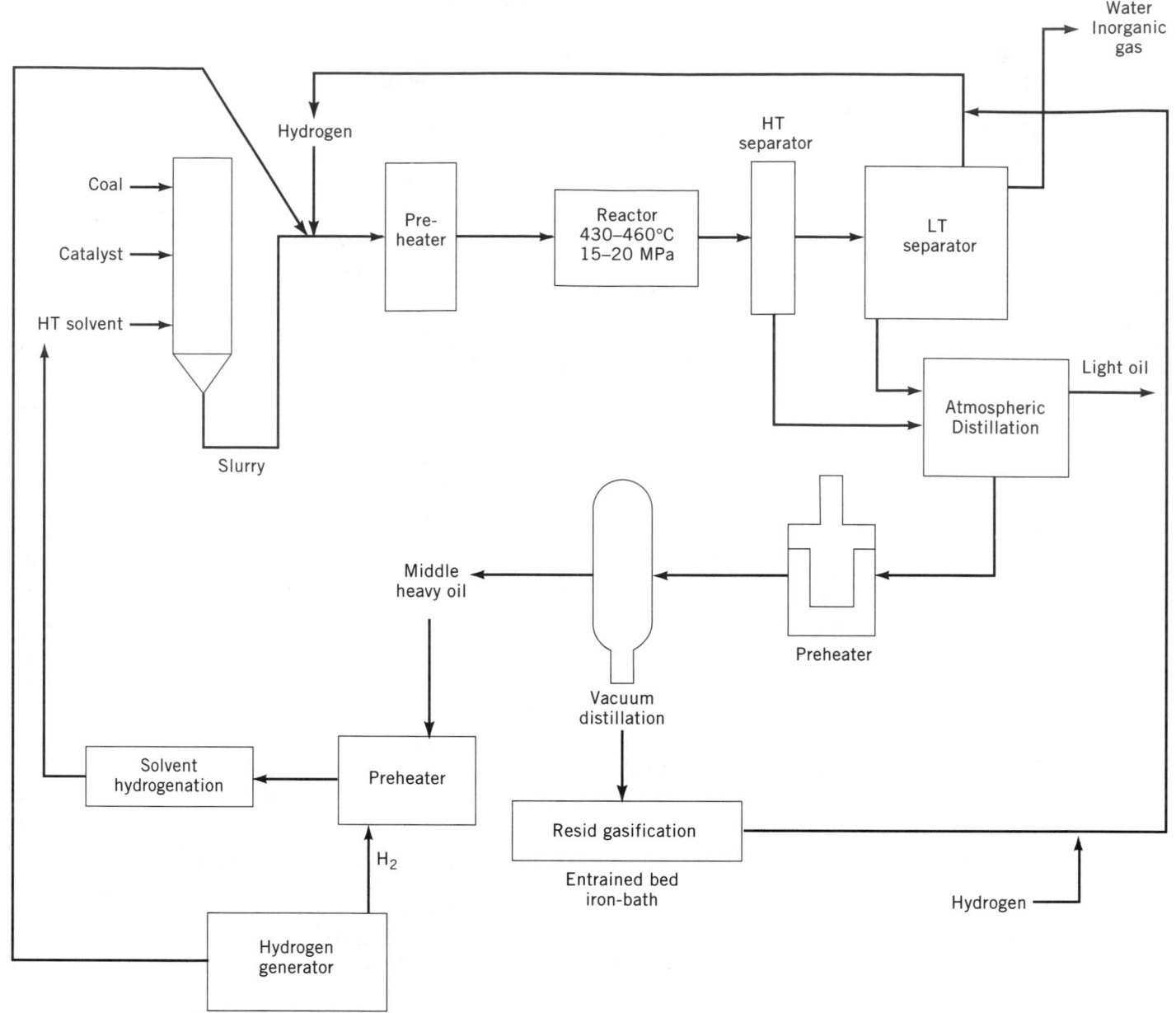

Figure 4. Nedol Bituminous Coal Liquefaction process for direct liquefaction.

approximately 85% hydrogen. Three suspended bubble-type reactors follow the preheater and operate at 450°C and 16.7 MPa. Slurry residence time in the reactors is one hour and the gas/liquid ratio is 700 Nm³/t. Products from the reactor are separated into light gas, light oil, medium/heavy oil, and residue. Typical results from small pilot-plant studies are, in weight %: light gas (up to C_5): 13; oils (C_5, bp = 538°C): 52; water and inorganic gases: 17; and residue, 24. Hydrogen consumption for the products listed is approximately 6% of the maf coal fed into the system. The solvent used to slurry the feed solids is part of the medium/heavy product oil and is mildly hydrogenated in a fixed bed reactor over a Ni–Mo/Al₂O₃ catalyst at 320–360°C under hydrogen at 9.8 MPa. At least four scenarios have been considered for the NEDOL process including the following combinations:

- NEDOL with conventional coal-fired power generation and coal-based hydrogen production
- NEDOL with integrated gasification, combined cycle (IGCC) power generation and hydrogen production by the HYCOL process
- NEDOL with hydraulic power generation and natural gas steam reforming
- NEDOL with power generation and hydrogen production from renewable energy sources

Exxon Donor Solvent Process (EDS)

Exxon Research and Engineering Company during the 1960s and later, developed the EDS process through bench scale and pilot plant (0.5 tons of coal per day) operation and testing. After 1976 a fully integrated liquefaction program tested EDS in a 250 ton/day pilot plant. By the 1990s Exxon had developed propriety second generation technology to produced premium transportation fuels using integrated process configurations only partly based on EDS.

The EDS process was a catalytic liquefaction technique where the hydrogen used to produce light liquids was contained in the solvent which transported the coal to the liquefaction reactor. As shown in Figure 5 the rehydrogenation of the solvent and hydrogen production took place in separate reactors from the main liquefaction path. Unique to the EDS process in addition to the solvent preparation was the flexicoking step whereby the residual primary product was converted to liquids and fuel gas leaving only an ash residue as solid product. Liquefaction conditions were 370–480°C 2–17 MPa, and residence times of 15 to 240 minutes. Solvent/coal ratios of 1.2–4.0 were used and relatively high hydrogen consumption (up to 17 weight % of the maf coal). The solvent was catalytically regenerated at 260–450°C by reaction with hydrogen at 8 to 21 MPa. Typical yields from high volatile coals using EDS were the following (weight % of maf coal):

YIELD	WT %
C_1–C_3 gases	5.3
Chemical gases (H_2S, NH_3, CO, CO_2)	4.7
C_4, 204°C (Naphtha)	17.9
204–371°C Liquids	7.6
371–538°C Liquids	7.6
+538°C (Vacuum bottoms)	52.0
Water	8.6

Research and development by Exxon in the late 1980s and after 1990 resulted in liquefaction to high yields of distillate fuels. Although only partially based on the EDS process emphases were similar; that is, maximizing transportation fuels such as diesel, jet fuel, or reformate. For example, in the diesel option liquid distributions were: diesel, 50% (cetane number = 42–53); reformate, 30%; and light naphtha, 11%.

The diesel produced contained no sulfur (0 ppm) and no aromatics (0 ppm). In the diesel/jet option as much as 63% of the liquids were high density jet fuel with less than 5% aromatics and less than 50 ppm sulfur.

H-Coal Process

The H-Coal process, developed by HRI Incorporated, utilized an ebullated bed reactor and introduced a catalyst for hydrogenation. The reactor operated at 20 MPa and 450°C and utilized cobalt–molybdenum catalyst in the form of millimeter-sized pellets. H-Coal technology was based on the H-Oil process. Products could be a synthetic crude refinery feedstock, a low-sulfur fuel oil, or a slate of distillate fuels.

In the H-Coal process crushed coal is slurried with recycle oil (2–3 : 1, oil: coal) and fed with hydrogen at 24 MPa to an ebullated bed reactor which operated at temperatures up to 455°C (851°F). Essential features of H-Coal are shown in Figure 6. The ebullating bed reactor brings the coal into contact with the Co/Mo catalyst where cracking of the coal molecular structure and hydrogenation take place. Heteroatoms, S, N, and O, are converted to hydrogen sulfide, ammonia, and water. The reactor is unique for direct liquefaction processes and uses an up-

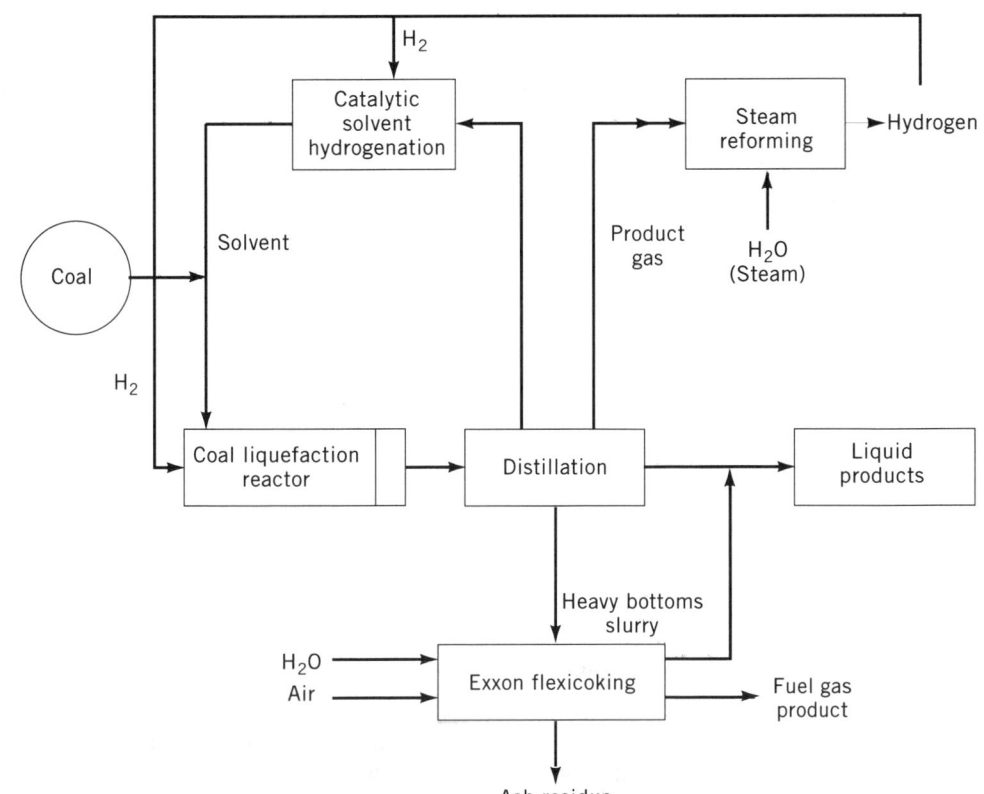

Figure 5. Exxon Donor Solvent (EDS) Process for direct coal liquefaction.

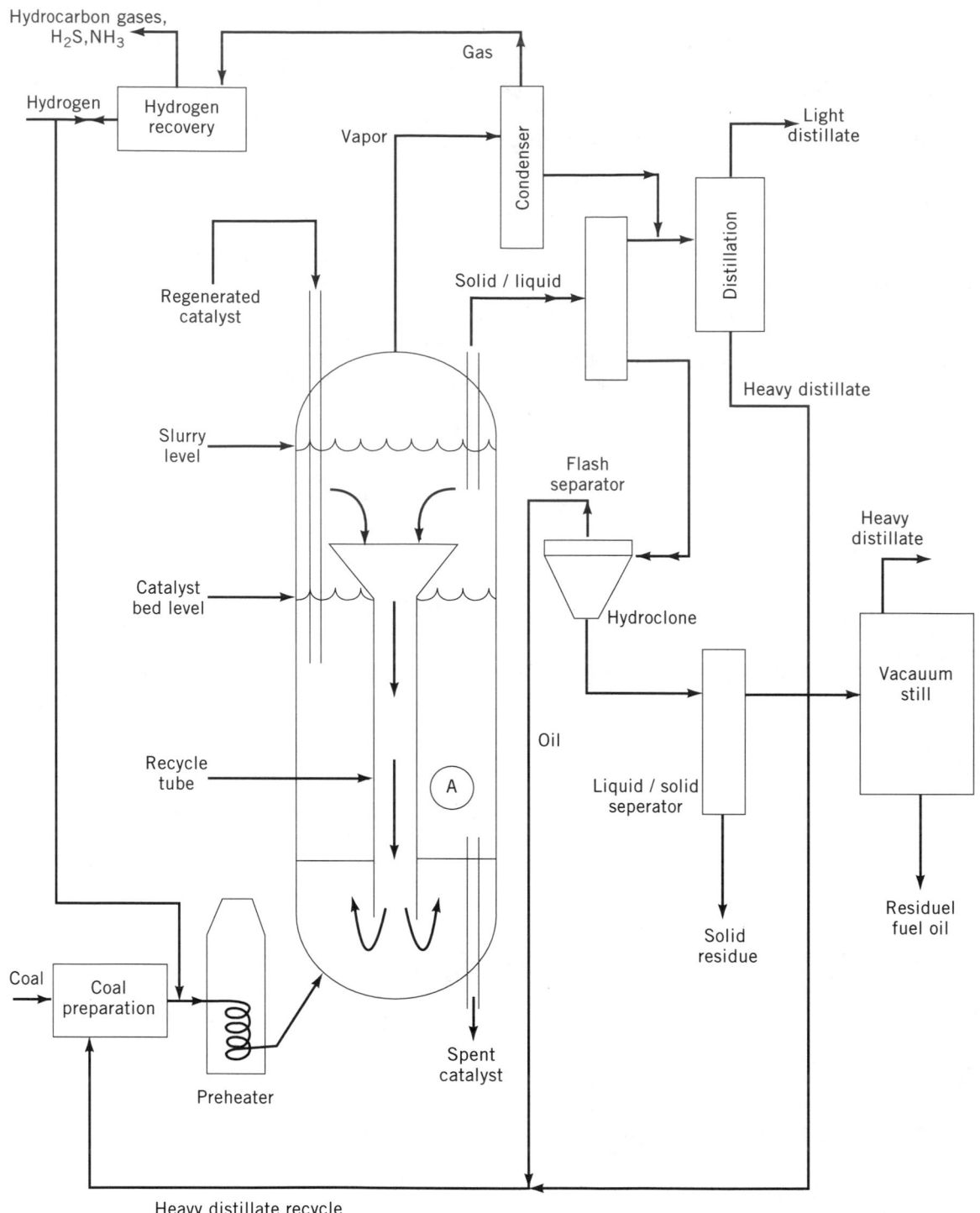

Figure 6. Process Scheme For the H-Coal Process for direct coal liquefaction.

ward flow of liquid and hydrogen which expands the catalyst bed and permits accumulation of solids without plugging. Since hydrogenation reactors are exothermic, temperature control is maintained by regulation of the circulation of gases, liquids, and solids. Advantages of H-Coal include technology based on proven processes from the oil industry which have been tested at commercial scale. Conversion to liquids of over 90% of the maf coal

have been attained with the H-Coal process at pilot-plant scale.

DMT-Integrated Gross Oil Refining Process (IGOR)

This process, developed by DMT Forschung und Prufung, Essen, Germany, employs a hydrorefining reactor into the hydrogenation scheme. This configuration has been done

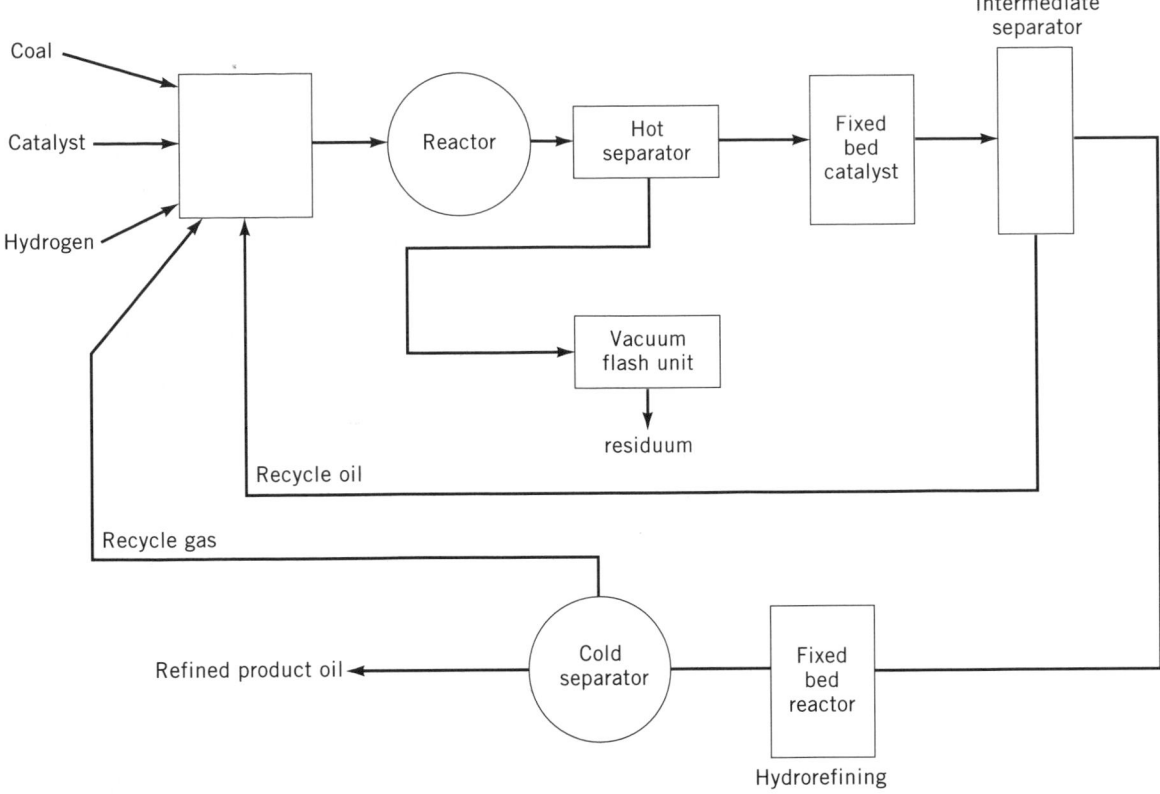

Figure 7. Integrated Gross Oil Refining (IGOR) Process for direct coal liquefaction.

to reduce the levels of oxygen and nitrogen in the product liquids. As shown in Figure 7, two fixed bed reactors are used in the process. For the feed to the IGOR process coal, catalyst and a distillate-type recycle oil are fed through heat exchangers where the mixture reaches 420°C. The upflow tubular reactor is operated at 475°C. From this reactor the products are separated into vapors and gases (product syncrude), high-boiling overhead oil (recycle), and an underflow liquid containing about 50% solids. The underflow stream is fed to a gasifier for hydrogen production. The product vapors are hydrotreated then cooled and recovered as a liquid along with water from the process. After scrubbing, the hydrogen and other gases are recycled to the hydrogenation unit.

The two fixed bed catalytic units are employed to ensure that both the recycle oil (gross oil) and the product liquids (net oil) are sufficiently treated. Even with relatively unreactive German bituminous coals the DMT parent process has yielded C_4+ refined oils at 54% of daf coal. With the IGOR 2-fixed-bed reactor configuration oil yields were increased to 60%. With more reactive bituminous coals, yields as high as 66% have been obtained. At the same time total heteroatom contents were kept below 5 ppm.

The IGOR process has been tested at coal feed rates of up to 0.2 tons per day (Process Development Unit) and 200 tons per day (Pilot Plant). Pressures used were 30 MPa with red mud catalyst used in the coal conversion reactor and commercial Ni–Mo–alumina catalysts in the fixed-bed reactors.

Indirect Liquefaction

Indirect coal liquefaction is defined as any liquefaction process which requires gasification as an initial step. Early German technology for coal liquefaction had included processes which began with gasification. Coal gasification had been used for many years and many processes were available (104). Indirect coal liquefaction completely avoids the problems of selective bond breaking in the coal structure and essentially breaks all C-H and C-C bonds in the coal. All coal gasification processes are not conducted as part of indirect liquefaction. Coal gasification can be used to produce a low calorific value (CV), fuel gas, a medium CV synthesis gas, or a high CV (pipeline quality) fuel gas. Most gasification processes involve reaction of the coal with steam and oxygen and produce carbon monoxide, hydrogen, carbon dioxide, methane, and other gases in small concentrations (105). Low CV gases are produced when the source of oxygen for gasification is air, thus introducing nitrogen as a major component of the product gas. When oxygen is introduced without nitrogen the gasification product is mostly a mixture of carbon monoxide and hydrogen (synthesis gas) although some carbon dioxide and methane are present. High CV gas is usually referred to as substitute natural gas (SNG) and its production requires a methanation step in addition to the initial gasification process. As shown in Figure 8, a simplified schematic of the general gasification process, coal and steam/oxygen (or steam/air) reactants move in opposite directions through the gasifier.

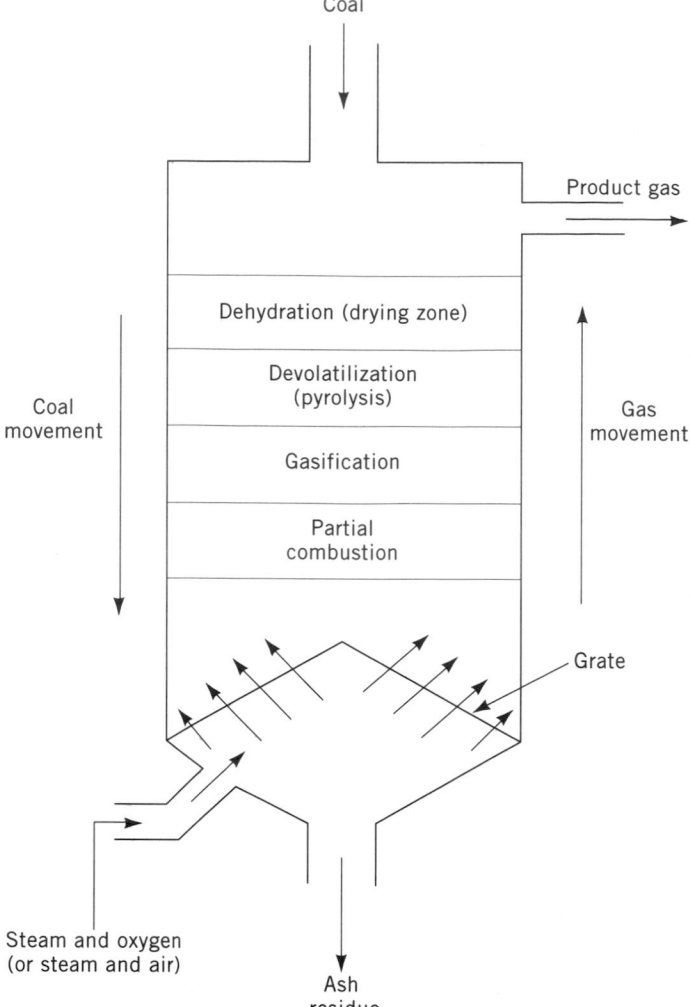

Figure 8. Coal Gasification (Fixed-Bed Gasifier).

The coal enters the gasification process where it contacts hot product gases. The coal is dried and then in the pyrolysis section of the gasifier any volatile components are driven off and become part of the product gases (106). At even higher temperatures in the gasification section, char (carbon) is gasified and hydrolysis takes place. The following reactions produce gases in the gasifier:

Oxidation

$$C + O_2 \rightarrow CO_2 \text{ (exothermic)}$$

$$C + CO_2 \rightleftharpoons 2\,CO \text{ (endothermic)}$$

Hydrogenation

$$C + 2\,H_2 \rightarrow CH_4 \text{ (exothermic)}$$

Hydrolysis

$$C + H_2O \rightleftharpoons CO + H_2 \text{ (endothermic)}$$

At gasifier conditions, steam can also react in two other endothermic reactions:

$$C + H_2O \rightleftharpoons \tfrac{1}{2} CO_2 + \tfrac{1}{2} CH_4$$

or

$$C + 2\,H_2O \rightleftharpoons CO_2 + 2\,H_2$$

With carbon, hydrogen, and oxygen present in the gasifier in the form of CO_2, CO, H_2, H_2, O_2, and CH_4 there are many possible reactions. The oxidation and hydrogenation reactions must provide the heat to carry out the other gasification reactions. Oxidation reactions such as reaction of carbon to produce carbon monoxide or carbon dioxide are not reversible but in the gasification section reactions are considered to be roughly in equilibrium. The product composition from the gasifier is controlled by the operating temperature, typically between 500 and 1200°C. In indirect liquefaction processes the gasification step is designed to maximize the production of CO and H_2. Synthesis of liquid products from $CO/H_2/CO_2$ mixtures can also be improved if the ratio of H_2/CO in the gases is optimized. This ratio can be somewhat controlled by the gasification conditions but usually requires additional adjustment after the gasification step. This adjustment is accomplished by "shifting" the composition by the reaction:

$$CO + H_2O \rightleftharpoons CO_2 + H_2 \text{ (water gas shift reaction)}$$

Since the water gas shift reaction is an equilibrium reaction it can be controlled by the temperature at which it is carried out. High H_2/CO ratios are especially important for synthesis of hydrocarbons in the Fischer-Tropsch reaction (101).

Indirect liquefaction includes processes that synthesize liquid hydrocarbons from synthesis gas when this gas is produced from coal. Also included are processes where liquid fuels are synthesized from other liquids produced from coal-derived gas.

Indirect Liquefaction Processes

Once coal is gasified the synthesis gas is reacted over a catalyst to liquid fuel products. Figure 9 shows the indirect liquefaction pathways for different liquid fuels. Traditional fuels such as gasoline and diesel fuel as well as oxygenated fuels may be produced. Advanced technology for indirect liquefaction has been enhanced by the great number of catalysts developed. Each catalyst and process mode imparts selectivity so that many fuel and non-fuel products can be produced from synthesis gas. The selectivity range can be illustrated by realizing that nickel catalysts allow almost 100% of CO in a synthesis gas mixture to be converted to methane, while using copper-based catalysts results in nearly 100% conversion to methanol. Table 7 gives many of the catalysts and reactions possible for synthesis gas (108,109). Related but not included in the table are the following; the water gas shift reaction, the fermentation of sugars and starch to ethanol, and the nickel catalyzed-production of synthesis gas from methanol.

Since German, U.S., Japanese, South African, and other scientists and engineers have worked on the development of catalysts for syntheses of products from synthesis gas, many are available. Catalyst structure/functional

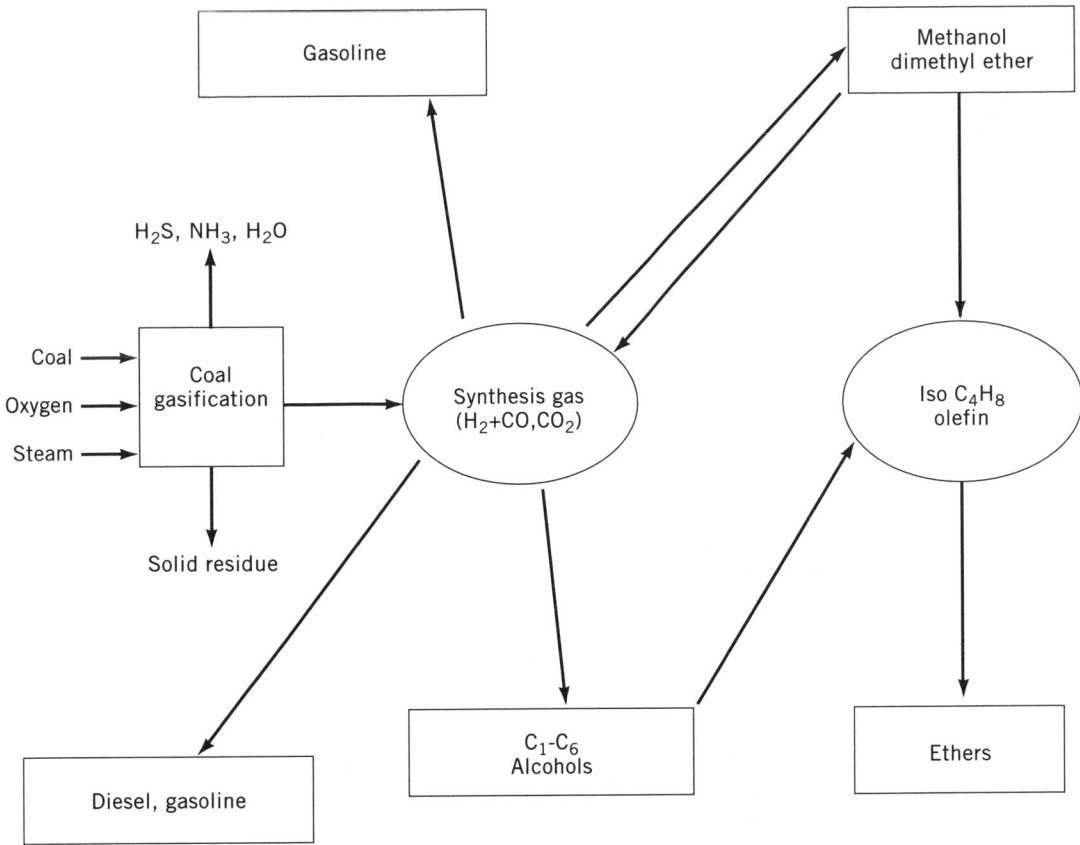

Figure 9. Indirect Coal Liquefaction Pathways

relationships have also been advanced and new catalysts continue to be discovered, along with new reactor systems for synthesis.

Methanol Production. The reaction of synthesis gas to produce methanol is straightforward:

$$CO + 2\,H_2 \rightarrow CH_3OH$$

However, thermodynamic limitations make multiple passes through the reactor necessary with only about 10% of synthesis gas converted to methanol per pass. Costly recycling makes methanol manufacture relatively expensive. The other big problem with the production of methanol is the removal of heat from the catalytic reaction zones in the synthesis reactors.

Modern plants for the production of methanol often use natural gas as a feed rather than synthesis gas. Where abundant supplies of natural gas are available this greatly simplifies the process.

Methanol has been used in blends with gasoline (110). M85, 85% methanol and 15% gasoline, is an excellent fuel that is being marketed in several U.S. locations. Use of high percentage methanol blends requires vehicles specifically designed for such fuels.

Table 7. Reactions and Catalysts of Synthesis Gas

Product	Reaction	Catalyst
Methanol	Synthesis gas → CH_3OH	$Cu–ZnO/Al_2O_3$
Higher alcohols	Synthesis gas → $C_nH_{2n+1}OH$	$Cu–ZnO–K/Al_2O_3$
Gasoline	Fischer-Tropsch Synthesis	
	($H_2 + CO \rightarrow CnH_{2n+2}$)	Fe, K
Hydrogen	$H_2 + CO + H_2O \rightarrow H_2 + CO_2$	$Fe_2O_3–Cr_2O_{3j}$ Cu–Zn oxides
ISA, SBA[a]	$CnH_{2n} + H_2O \rightarrow ROH$	H_3PO_4/SiO_2
MTBE, ETBE, TAME[b]	CnH_{2n} + alcohol → ether	acid resin
Isobutylene	TBA → iso-C_4H_8 + H_2O	H_3PO_4/SiO_2
Isobutylene	$C_4H_{10} \rightarrow$ iso-C_4H_8 + H_2	Cr_2O_3/Al_2O_3
Gasoline	(MTG)	
	$CH_3OH \rightarrow CnHm$	ZSM-5 zeolite

[a] IPA = isopropyl alcohol, SBA = *sec*-butyl alcohol
[b] MTBE = methyl-*tert*-butyl ether, ETBE = ethyl-*tert*-butyl ether, TAME = *tert*-amyl methyl ether, TBA = *tert*-butyl alcohol, MTG = Mobil's methanol-to-gasoline process.

Table 8. Properties of Oxygenated Fuels and Fuel Additives[a]

Oxygenated Fuel or Compound	Octane No. (R + M)/2	Heating Value		Boiling Point, °C
		MJ/L	Btu/gal	
Methanol (CH₃OH)		18.8	67,200	65
Ethanol (C₂H₅OH)	115	21.3	76,000	79
Isopropyl alcohol (C₃H₇OH)		29	103,500	82
Gasoline (U.S.)	90–92	33–35	118–125,000	
(Europe)	95			
Methyl *tert*-butyl ether (MTBE)	109	26.3	94,000	55
Ethyl *tert*-butyl ether (ETBE)	112	27.1	96,900	72
tert-Amyl methyl ether (TAME)	104.5	28.2	100,600	86
Di-isopropyl ether	105	28	100,000	68
tert-Amyl ethyl ether (TAEE)	100			101
Isopropyl *tert*-butyl ether (IPTBE)	113			87

[a] Some of these materials can be produced by indirect coal liquefaction.

Diesel and Gasoline Production. As shown by Figure 9, more than one process can be used to produce gasoline from synthesis gas or its products. The classical Fischer-Tropsch synthesis could produce hydrocarbon compounds of various chain lengths including distillate fuels such as diesel and gasoline (111). A recently developed synthesis, the Shell Middle Distillate Synthesis (SMDS), consists of three main steps, two of which would fit into indirect liquefaction of coal (112). The three steps being used in the Shell plant which began operation in 1993 in Sawawak, Malaysia are:

1. Conversion of natural gas to synthesis gas (H₂/CO = 2).
2. Conversion of synthesis gas to high molecular weight waxes.

3. Hydrocracking and hydrogenation of waxes to middle distillates (diesel and kerosene).

Exxon Process. Exxon began the demonstration of a gas to liquids process similar to SMDS, (10) but with the three steps:

1. Conversion of natural gas to synthesis gas.
2. Conversion of synthesis gas to liquid hydrocarbons.
3. Hydroprocessing of the liquids to clean distillate fuels (transportation fuels).

In this case the synthesis takes place in a fluidized-bed reactor system (advanced reactor and catalyst system).

Mobil Methanol-to-Gasoline (MTG) Process

Mobil scientists invented the MTG process about 1975 and developed it further with funding from the U.S. Department of Energy (DOE) (114). The first plant to utilize the MTG technology (1985) in New Zealand used a fixed-bed catalyst mode although fluidized beds have since been shown to give higher yields and higher quality gasoline. Figure 10 shows the two-stage MTG process.

Mobil has also developed several MTG-related processes such as MTO (Methanol-to-Olefins) and MOGD (Mobile-Olefins-to-Gasoline and Distillates) (109). In addition to methanol there are several other oxygenated products of synthesis gas that are recognized as high quality fuel additives or fuels. Some of these are listed in Table 8 along with some of their important fuel properties.

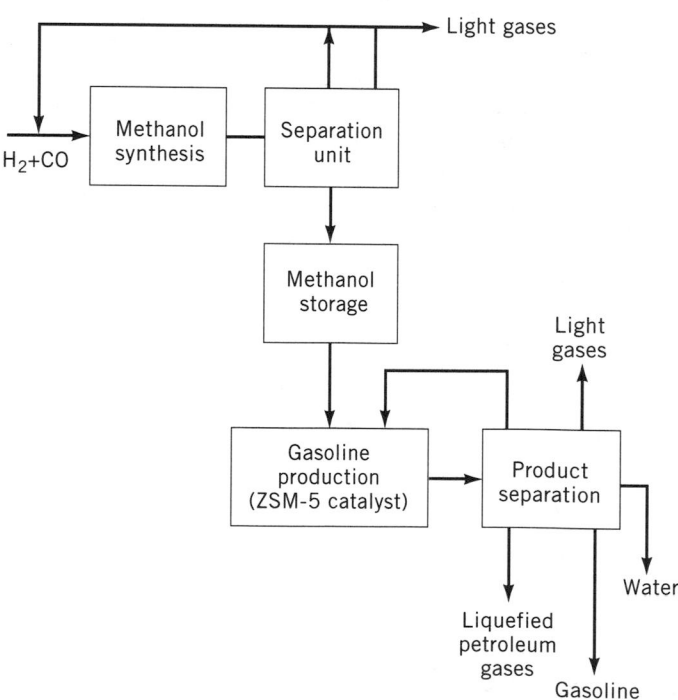

Figure 10. Two-Stage Mobile Methanol-To-Gasoline (MTG) Process.

BIBLIOGRAPHY

1. E. Gorin, "Fundamentals of Coal Liquefaction", in M. A. Elliott, ed., *Chemistry of Coal Utilization*, Sec. Supp. Vol., John Wiley & Sons, Inc., New York, 1981, pp. 1846–1847.
2. S. B. Alpert, "Liquefaction Processes", in Ref. 1, pp. 1923–1926.
3. W. R. K. Wu and H. H. Storch, *BuMines Bull.* **633**, 104 (1968).
4. E. J. Hoffman, *Coal Conversion*, Energon Co., Laramie, Wyo., 1978, p. 5.

5. K. S. Vorres, *Users Handbook for the Argonne Premium Coal Sample Program*, Div. of Chem. Sci., U.S. DOE, Contract No. W-31-109-ENG-38, 1989, p. 3.

6. F. W. Richardson, *Oil from Coal*, Noyes Data Corp., New Jersey, 1975, pp. 18–20.

7. J. B. O'Hara, "Liquid Fuels from Coal", R. A. Meyers, ed., *Coal Handbook*, Marcel Dekker, Inc., 1981, 732.

8. R. F. Probstein and R. E. Hicks, *Synthetic Fuels*, pH Press, Cambridge, Mass., 1990, pp. 13–14.

9. *Ibid.*, pp. 260–262.

10. Ref. 7, pp. 722–724.

11. Ref. 8, pp. 263–266.

12. Ref. 7, pp. 724–726.

13. E. S. Lee, "Coal Liquefaction", in C. Y Wen and E. S. Lee, eds., *Coal Conversion Technology*, Addison-Wesley Publ. Co., London, 1979, p. 436.

14. W. R. K. Wu and H. H. Storch, *BuMines Bull.* **633**, 1–4 (1968).

15. R. Holroyd, *Report on the Petroleum and Synthetic Oil Industry of Germany*, BIOS Overall Rept., No. 1, 1947.

16. W. F. Farragher and W. A. Horn, *Manufacture and Regeneration of Catalysts at I.G. Ludwigshafen-Oppau, FIAT Final Report*, No. 422, 1945.

17. R. Holroyd and I. G. Leuna, *CIOS XXVIII-107* (1945).

18. J. A. Oriel, I. H. Jones, and H. H. Weir, The High Pressure Hydrogenation Plant, Brown Coal, Wesseling, *CIOSXXVIII*-40, 1945.

19. C. Cockram, Scholven, *Combined Intelligence Objectives Subcommittee Rept., Off. of the Publication Board*, Dept of Commerce, *XXVIII-102*, Washington, D.C., 1945.

20. R. Holroyd and I. G. Werke Ludwigshafen-Oppau, *Combined Intelligence Objectives Subcommittee Rept., Office of the Publication Board*, Dept of Commerce, 103, Washington, D.C., 1945.

21. C. Cockram and Welheim, *Combined Intelligence Objectives Subcommittee Rept., Office of the Publication Board*, Dept of Commerce, Washington, D.C., *XXX-104*, 1945.

22. C. Cockram and Gelsenberg, *Combined Intelligence Objectives Subcommittee Rept., Office of the Publication Board*, Dept of Commerce, XXX-105, Washington, D.C., 1945.

23. J. F. Ellis and R. G. Morley, Zeitz, *Combined Intelligence Objectives Subcommittee Rept., Office of the Publication Board*, Dept of Commerce, Washington, D.C., XXXIII-24, 1945.

24. H. Hollings, Wintershall A. G. Lutzkendirf, *Combined Intelligence Objectives Subcommittee Rept., Office of the Publication Board*, Dept of Commerce, Washington, D.C., XXXII-90, 1945.

25. J. F. Ellis and R. G. Morley, Brabag Bohlen, *Combined Intelligence Objectives Subcommittee Rept., Office of the Publication Board*, Dept of Commerce, Washington, D.C., XXXII-92, 1945.

26. P. W. Sherwood, *FIAT Final Report*, (*952*) (1947).

27. F. C. Thyrion, "Indirect Liquefaction" in I. Romey, P. F. M. Paul and G. Imarisio, eds., *Synthetic Fuels from Coal*, Graham and Trotman, Ltd., London, 1987, pp. 8–19.

28. Ref. 7, pp. 750–769.

29. I. Romey, P. F. M. Paul and G. Imarisio, eds., *Synthetic Fuels from Coal*, Graham & Trotman, London, 1987, pp. 226–230.

30. P. Nowacki, *Coal Liquefaction Processes*, Noyes Data Corp., New Jersey, 1979.

31. *Coal Liquefaction—A Technology Review*, The Organization for Econ. Coop. and Dev., Internat. Energy Agency, Paris, 1982, p. 7.

32. C. L. Wilson, *Coal—Bridge to the Future*, Ballinger Publ. Co., Cambridge, Mass., 1980, pp. 64–66.

33. *Ibid.*, p. 99.

34. M. Berthelot, *Bull. Soc. Chim., Ser. 2*, **11**, 278–286(1869).

35. Ger. Pat. 301, 231 (Nov. 16, 1919) (Applied, Aug. 8, 1913) to F. Bergius and J. Billwiller; *Chem. Abstr.* **7**, 3219(1913).

36. R. Holroyd and I. G. Leuna, *Combined Intelligence Objectives Subcommittee Rept., Office of the Publication Board*, Dept of Commerce, Washington, D.C., XXXII-107, 1945.

37. I.C.I., *FUEL* **10**(11), 481–486 (1931).

38. E. E. Donath, in H. H. Lowry, ed., *Chemistry of Coal Utilization*, Supp. Vol., John Wiley & Sons, Inc., New York, 1963, pp. 1041–1048.

39. R. B. Anderson, in P. H. Emmett, ed., *Catalysis*, **V,** IV, CH 2, Ed., P. H. Emmett, 1961, 34.

40. F. C Thyrion in Ref. 29, pp. 9–14.

41. Ref. 29, pp. 153–165.

42. E. C. Mangold, M. A. Muradaz, R. P. Ouellette, O. G. Farah, and P. N. Chermisinoff, *Coal Liquefaction and Gasification Technologies*, Ann Arbor Sci. Publ., Ann Arbor, Mich., 1982, pp. 79–93.

43. Ref. 29, pp. 14–19.

44. Ref. 29, pp. 226–230.

45. Ref. 29, pp. 137–152.

46. Ref. 29, pp. 122–136.

47. Ref. 29, pp. 70–106.

48. Ref. 29, pp. 226–230.

49. Ref. 29, pp. 193–197.

50. W. Francis, *Coal—Its Formation and Composition*, Edward Arnold, Ltd., London, 1954, 1.

51. A. G. Oblad, *Catal. Rev.–Sci. Eng.* **14**(1), 83–95 (1976).

52. P. H. Given, *FUEL* **39,** 147 (1960).

53. W. H. Wiser, *Preprints, Fuel Chem. Div., Amer. Chem. Soc.* **20**, 122(1975); W. H. Wiser, L. L. Anderson, S. A. Qader, and G. R. Hill, *J. Appl. Chem. Biotechnol.* **21**(81)(1971).

54. L. A. Heredy and I. Winder, *Preprints, Fuel Chem. Div., Amer. Chem. Soc.* **25**(4), 38 (1980).

55. P. R. Solomon, in B. D. Blaustein, B. C. Bockrath and S. Friedman, eds, *New Approaches in Coal Chemistry, ACS Symp. Ser. 169*, Washington, D.C., 1981, p. 61.

56. R. C. Neavel, *Preprints, Fuel Chem. Div., Amer. Chem. Soc.*, **24**(1), 72–83.

57. L. L. Anderson and W. H. Yuen, "Solubilization of Coal", in Y. Yurum, ed., *Clean Utilization of Coal*, Kluwer Academic Publ., Dordrecht, The Netherlands, 1991, pp. 33–38.

58. W. S. Wise, *Solvent Treatment of Coal*, Mills & Boon, Ltd., London, 1971, pp. 1–55.

59. N. C. Deno, *FUEL* **59**(10), 694 (1980).

60. J. W. Larsen and J. Kovac, *Organic Chemistry of Coal, ACS Symp. Ser., 71*, Amer. Chem. Soc., Washington, D.C., 1978, pp. 36–49.

61. J. G. Speight "The Application of Spectroscopic Techniques to the Structural Analysis of Coal and Petroleum", *Appl. Spec. Rev.* **5**, 211(1971).

62. Ref. 18, pp. 11–16.

63. *Gaseous Fuels; Coal and Coke Annual Book of ASTM Standards*, Sect. 5, Vol. 05.05, American Society for Testing Materials, Philadelphia, Rev. annually.

64. M. C. Stopes, *Proc. Ray. Soc.* **90B,** 470(1991).

65. R. Thiessen, *Coal Age* **18,** 1183(1920); *J. Geol.* **28,** 185 (1920).

66. W. Francis, *Coal—Its Formation and Composition*, Edward Arnold, Ltd., London, 1954, 276–280.

67. G. D. Dyrkacz and C. A. Bloomquist, "The Efficient Large-Scale Separation of Coal Macerals", *Preprints, Fuel Chem. Div., Amer. Chem. Soc.* **33**(3), 128–135(1988).

68. G. R. Dyrkacz and E. P. Horwitz, *FUEL* **61,** 3–12(1982).

69. I. Romey, P. F. M. Paul and G. Imarisio, eds., *Synthetic Fuels from Coal*, London, 1987.

70. R. M. Davidson, *Molecular Structure of Coal*, IEA Coal Research, London, 1980.

71. B. Alpert and R. Wolk, "Liquefaction Processes", in M. A Elliott, ed., *Chemistry of Coal Utilization*, Sec. Supp. Vol., John Wiley & Sons, Inc., New York, 1981, pp. 1923–1926.

72. Ref. 29, pp. 259–338.

73. D. G. Skinner, *FUEL* **10**(3), 109–137 (1931).

74. W. Kronig, *Katalytische Druckhydrierung*, Springer, Berlin, 1950, p. 266.

75. Ref. 29, p. 342–344.

76. J. B. Howard, "Fundamentals of Coal Pyrolysis and Hydropyrolysis", in Ref. 71, pp. 688–788.

77. S. N. Rao, H. D. Schindler, and G. V. McGurl, "Advances and New Directions in Direct Liquefaction" *Preprints, Fuel Chem. Div., Amer. Chem. Soc.* **33**(3), 145–156 (1988).

78. F. J. Derbyshire, *Catalysis in Coal Liquefaction*, IEA Coal Research, London, 1988, pp. 16–28 and 45–50.

79. *Coal Liquefaction—A Technology Review*, The Organization for Econ. Coop. and Dev., Internat. Energy Agency, Paris, 1982, p. 32.

80. R. Wigginton, *Coal Carbonisation*, Bailliere, Tindall & Cox, London, 1930, pp. 1–49.

81. Ref. 8, pp. 100–107.

82. I. G. C. Dryden, *U. Instit. of Fuel* **30,** 193–214(1957).

83. L. Eglin and S. A. Bresler, "Low-Temperature Pyrolysis Technology", in Ref. 71, pp. 785–846.

84. J. B. Howard, "Fundamentals of Coal Pyrolysis and Hydropyrolysis", in Ref. 71, pp. 848–917.

85. C. Karr, "Low-Temperature Tar", in Ref. 38, pp. 539–579.

86. J. F. Weiler, "High-Temperature Tar", in Ref. 38, pp. 580–628.

87. N. Berkowitz, *The Chemistry of Coal*, Elsevier Sci. Publ. Co., Inc., New York, 1985, pp. 213–308, 341–352.

88. R. K. Hessley, J. W. Reasoner, and J. T. Riley, *Coal Science*, John Wiley & Sons, Inc., New York, 1986, pp. 135–140.

89. N. Berkowitz, *The Chemistry of Coal*, Elsevier Sci. Publ. Co., Inc., New York, 1985, pp. 405–444.

90. H. Charcosset and A. Genard, "Catalysis in Direct Coal Liquefaction" in Ref. 29, pp. 340–361.

91. J. B. O'Hara, "Liquid Fuels from Coal", R. A. Meyers, ed., *Coal Handbook*, Marcel Dekker, Inc., New York, 1981, pp. 775–776.

92. Ref. 1, p. 1856.

93. J. G. C. Dryden, "Chemical Constitution and Reactions of Coal", in Ref. 38, pp. 249–252.

94. H. H. Lowry and J. J. Rose, "Pott Borche Coal-Extraction Process and Plant of Ruhrol", *BuMines I.C. 7420; CIOS XXX7-27*, Item 30.

95. J. B. O'Hara, "Liquid Fuels from Coal", in Ref. 91, pp. 766–768.

96. J. B. O'Hara, "Liquid Fuels from Coal", in Ref. 91, pp. 728–735.

97. I. F. W. Romey, "Single-stage Direct Coal Liquefaction", in Ref. 29, pp. 201–219.

98. S. A. Moore, G. M. Kimber and J. C. Whitehead, "Direct Liquefaction (Two Stage Process)", in Ref. 29, pp. 201–219.

99. Ref. 8, pp. 288–291.

100. H. M. Oelert, "Coprocessing", in Ref. 29, pp. 265, 281–283, 288–291.

101. M. Hoering and E. E. Donath, *Ullmanns Encyklopadie der Technischen Chemie*, 3rd ed., Urban and Schwarzenberg, Munchen, 1958, Vol. 10, pp. 483–569.

102. NCA, *Coal Synfuel Facility Survey*, National Coal Association, Washington, D.C., 1981, pp. 9–22.

103. H. D. Schindler, *Coal Liquefaction—A Research & Development Needs Assessment*, Final Rept. Vol. 2, 1989, DOE Contract DE-AC01-87ER30110.

104. E. C. Mangold, M. A. Muradaz, R. P. Ouellette, O. G. Farah, and P. N. Cheremisinoff, *Coal Liquefaction and Gasification Technologies*, Ann Arbor Sci. Publ., Ann Arbor, Mich., 1982, pp. 121–123, 133–260.

105. R. K. Hessley, J. W. Reasoner, and J. T. Riley, *Coal Science*, John Wiley & Sons, Inc., New York, 1986, pp. 163–178.

106. Ref. 8, pp. 144–256.

107. Ref. 8, pp. 128, 222–223.

108. F. C. Thyrion, "Indirect Liquefaction", in Ref. 29, pp. 29–31.

109. G. Alex Mills, *Status and Future Opportunities for Conversion of Synthesis Gas to Liquid Energy Fuels*, National Renewable Energy Lab, U.S. DOE, 1992.

110. K. Owen and T. Coley, *Automotive Fuels Handbook*, Society of Automotive Engineering, Inc., Warrendale, Pa., 1990, pp. 265–266, 464–469.

111. R. B. Anderson, *The Fischer-Tropsch Synthesis*, Academic Press, Inc., Orlando, Fla., 1984, pp. 9–17.

112. F. C. Thyrion, "Indirect Liquefaction", in Ref. 29, pp. 58–61, 103–105.

113. F. C. Thyrion, "Indirect Liquefaction", in Ref. 29, pp. 87–93, 102–103.

114. F. C. Thyrion, "Indirect Liquefaction", in Ref. 29, pp. 93–99.

COAL PYROLYSIS

ETUAN ZHANG
PATRICK G. HATCHER
The Pennsylvania State University
University Park, Pennsylvania

Pyrolysis is a process of thermal decomposition in the absence of oxygen. The overall intent of pyrolysis is concerned with providing sufficient thermal energy to break bonds, more specifically carbon–carbon bonds that link organic molecules. The application was specifically devised to cleave large macromolecules into smaller fragment molecules so that it would be possible to analyze the fragment molecules by conventional spectroscopic and chemical methods not amenable to large macromolecules. Although the intent is for carbon–carbon sp^3 bond cleavage, it is important to consider whether sufficient energy is available to cleave carbon–oxygen, carbon–sulfur, carbon–nitrogen, and any other sp^3 bonds whose activation energy is on the order of the energy required for carbon–carbon bond cleavage. There is usually insufficient energy to cleave sp^2 bonds or carbon–hydrogen bonds.

This methodology is particularly well suited to macromolecular polymeric systems in which the structure of the polymers can be deduced by fragmentation of the macromolecules into monomeric, dimeric, and oligomeric units more easily analyzed than the polymers themselves. Presumably, pyrolysis cleaves at select sites, and analysis of the monomers provides information on the structural makeup of the polymers. The pyrolytic methods have been used extensively in the field of polymer science, and the technique has provided some valuable insights on the structures of polymers.

As applied to coal, the technique has been used for both characterization of macromolecular structures and measurements of coal reactivity, a more applied use. In regard to structural characterization, pyrolysis affords the same advantages as does its use for structural characterization of polymers, namely the cleavage of the macromolecules into lower molecular weight fragments that are more amenable to chemical analysis. Unfortunately, coal is not a polymer in that it is composed of nonrepeating structural units. Consequently, it cannot be assumed that monomers formed by pyrolysis are part of a repeating network defining the entire structure. Nonetheless, it is possible to deduce that the monomers comprise part of a complex macromolecular structure of largely undefined proportions. The assumption is made that pyrolysis cleaves one or two bonds that link monomers, which are identified to a macromolecular network. In its other use, pyrolysis of coal has a more applied slant, because it is possible to monitor the rate, chemistry, and efficiency of pyrolytic breakdown and deduce information regarding the propensity of coal to form commercial chars or cokes or to undergo pyrolytic gas and oil formation. This aspect is called applied pyrolysis, and this article focuses on both the applied and the analytical pyrolysis of coal.

See also PYROLYSIS; COKING, DELAYED.

APPLIED PYROLYSIS

Applied pyrolysis, though not specifically meant to describe only a commercial application, embodies processes which are important on a large scale. Accordingly, much research has been conducted to understand these processes as they relate to commercial operations such as coal combustion and conversion to liquid fuels. In both of these operations, coal is either initially or *in toto* converted to a gaseous or liquid product formed primarily by pyrolytic destruction of covalent linkages which maintain the coal structure as a solid. Thus, an understanding of coal pyrolysis is essential to knowing more about the preparative-scale or commercial operations.

Research devoted to understanding the pyrolytic degradation process for coal has involved numerous approaches of varying degrees of complexity. The simplest means of studying pyrolysis is by studying mass loss during the process. This can be accomplished by heating in crucibles (ASTM procedures) or by more sensitive mass loss measurement devices such as thermal gravimetric analyzers (TGA) (1–3). Mass loss can be measured under the influence of high heating rates, not possible with the above mentioned methods, using heated grid reactors (4–6) or entrained flow reactors (7–9). If one is interested in examining the kinetics of coal pyrolysis, this can be accomplished by use of heated grids (10–12), entrained flow reactors (13,14) fluidized bed reactors (15), TGAs (16–18), and very sensitive mass evolution devices, mass spectrometers (19,20).

In all these kinetic studies, it is important to have exact measurements of temperature. On the macro scale such measurements are accomplished by thermocouples in close proximity to the heating zones. However, the measurement of temperature in cases where small coal particles are being tested is not trivial. For example, laser pyrolysis of coal particles on a heated grid employs two-color pyrometry devices (21). Temperatures are measured by Fourier transform infrared spectrometry (FTIR) (22–24) combined with two-color pyrometry. A recent review by Solomon and co-workers (25) of the various methods for temperature measurement and pyrolysis reactors provides a comprehensive evaluation of the instrumentation available.

The development of a true chemical understanding of the pyrolysis process such that one may eventually derive predictive capabilities requires study of a complex set of processes involving both thermal and mass transfer processes. In addition, coal's heterogeneity does not lend itself well to systematic study in that it is often difficult to perform definitive experiments on one coal that can be extrapolated to other coals (26). Also, numerous investigators studying coals of different composition and rank have not been able to provide concordant information on pyrolysis rates and mechanisms. Fortunately, much progress has recently been made in both experimental and theoretical approaches and these developments have shed new light on the subject. Much of the current knowledge of applied coal pyrolysis has been reviewed in several articles (25,27,28). The reader is referred to these articles for more detail than can be provided in this article. Perhaps most of the recent advances in coal pyrolysis have been in the assessment of factors which govern the rate of evolution of molecularly distinct gaseous products from coal and on the rate of production of more complex molecules such as tar which is initially pyrolytically formed but distills from the coal during heating (4,14,29). This article focuses on these recent developments, in particular on the kinetic rated of pyrolysis and on the generation of volatile products.

Generation of Volatile Products

Pyrolysis of coal, which by definition requires an inert atmosphere, leads to the production of three types of products: gases, tar, and a residue called char or coke. The gases are primarily CH_4, CO, CO_2, H_2O, and the C_2–C_5 hydrocarbons. In addition, many sulfur- and nitrogen-containing gases are present. It is generally agreed that one might be capable of deducing the mechanism of pyrolysis if one could determine the origin of these gases. Thus, much emphasis has been placed on measuring the evolution of certain gas species over the course of pyrolysis (4,30). For example, in studies of the evolution of CH_4 from low rank coal, several distinct maxima are observed over the course of heating. This suggests that CH_4 derives from several types of structures: methoxyl groups of lignin precursors, methyl substituents on aromatic rings, and terminal methyl groups an paraffinic structures (13). In an attempt to pin down the mechanism for methane formation during pyrolysis, a series of pyrolysis experiments were conducted on chemically modified coals (31). The results showed that CH_4 was primarily derived from demethylation reactions of methoxyl groups, especially in low rank lignite and subbituminous coals. The higher rank coals do not generate CH_4 in this fashion because they contain few methoxyl groups. Also, paraffinic structures are not particularly responsible for CH_4 generation because they primarily produce C_2 and C_3 hydrocarbons. Methyl groups on aromatic rings also do not appear to produce CH_4. The most likely mechanism for CH_4 production during pyrolysis of medium to high rank coals is one involving hydrogen transfer reactions that convert hydroaromatic structures to aromatic structures where opportunities exist for demethylation from benzylic or reactive aliphatic positions (31).

Other gaseous products, CO_2, CO, and light hydrocarbon gases, obviously derive from specific structures in coal. Delineating their source, however, is not as straightforward as it is for CH_4. Figures 1 and 2 show the typical gaseous product yields from a low rank coal and a high rank coal, respectively, as a function of increasing pyrolysis temperature. These data have been used as input for models (32,33) whose purpose is to predict pyrolysis behavior of coal.

Another major product obtained from pyrolysis of coal is tar, a term used as a generic description of materials other than water which condense to liquids or solids at room temperature from the vapor state. Solomon and co-workers (26) have stated that "tar formation involves the following steps: (1) depolymerization by rupture of weaker bridges in the coal macromolecule to release smaller fragments that make up the "metaplast"; (2) repolymerization (cross-linking) of metaplast molecules; (3) transport of lighter molecules away from the surface of the coal particles by combined vaporization, convection, gasphase diffusion; and (4) internal transport of molecules to the surface of the coal particles by convection and diffusion in the pores of non-softening coals".

The yields of tar formed from coal during pyrolysis are, in large part, influenced by mass transfer as are other volatile species. This conclusion was obtained by varying gas pressures external to coal particles during pyrolysis (27,30,32,34–40). Figures 3 and 4 illustrate this fact well. As pressure increases the yields of tar decrease dramatically for the higher rank coals and not so significantly for

low rank coals (32,40). An inverse trend is observed for light gases and char, the residue not volatilized.

The inverse trends, as a function of external pressure between tar yields and char or light gases, can be explained by mass transport-limited escape of tar into the gas phase during heating (26). At higher pressures escape of tar from the coal is reduced and, thus, the tar experiences a longer residence time in contact with a nonvolatile char. This increased contact at high temperature promotes crosslinking or covalent bonding of the tar with char such that the tar is rendered nonvolatile. Consequently, additional heating promotes only the formation of light gases as pyrolysis fragments of the tar bind to char.

Another explanation for the dependence of tar yield on pressure has been proposed (41). In this scenario, the tars present in the gas phase permeate the pores of the coal and are released external to the coal matrix by transport along with the gas phase. Since the gas phase transport is pressure dependent, tar transport is pressure dependent because the tar yield is proportional to the volume of escaping gas.

Suuberg and co-workers (32) proposed, on the basis of modified boiling points of aromatic hydrocarbons under

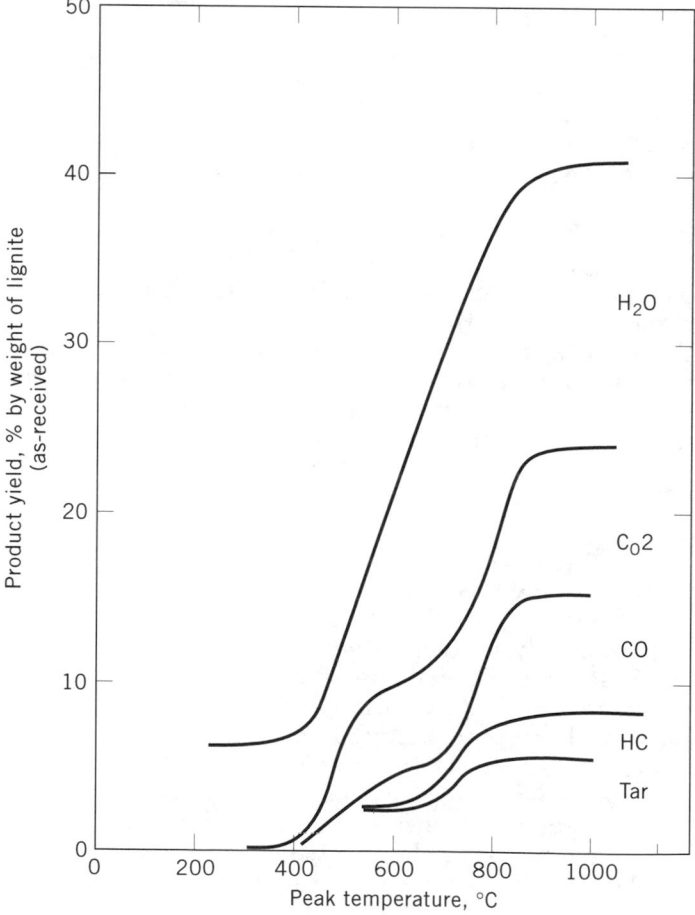

Figure 1. Distribution of pyrolysis products from lignite heated to different peak temperatures. From Ref. 32.

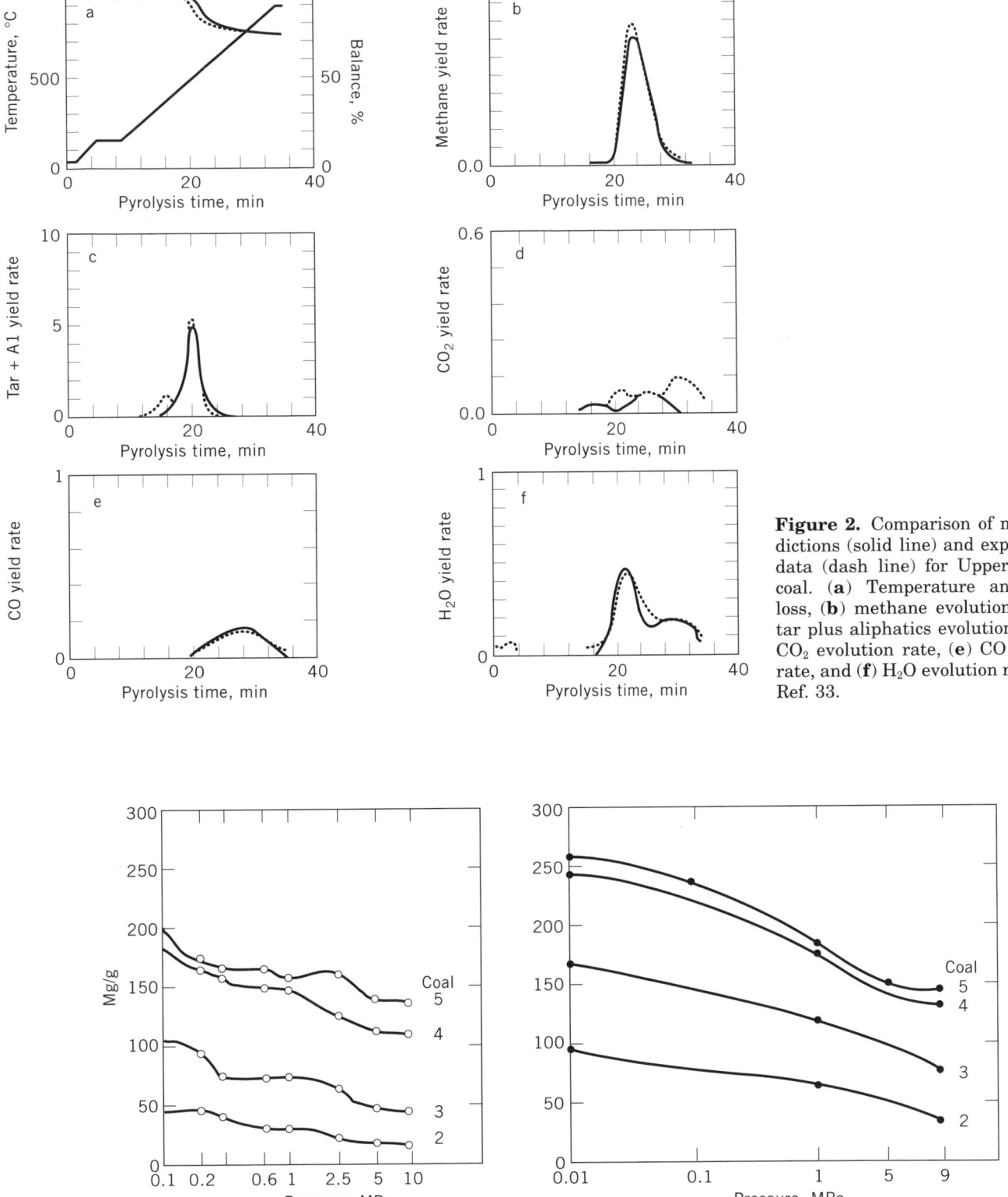

Figure 2. Comparison of model predictions (solid line) and experimental data (dash line) for Upper Freeport coal. (**a**) Temperature and weight loss, (**b**) methane evolution rate, (**c**) tar plus aliphatics evolution rate, (**d**) CO_2 evolution rate, (**e**) CO evolution rate, and (**f**) H_2O evolution rate. From Ref. 33.

Figure 3. Tar yields under inert gas. ○, 3 K/min; ●, 210 K/s, 2, low volatile coal; 3, medium volatile coal; 4 and 5, high volatile bituminous coals. From Ref. 40.

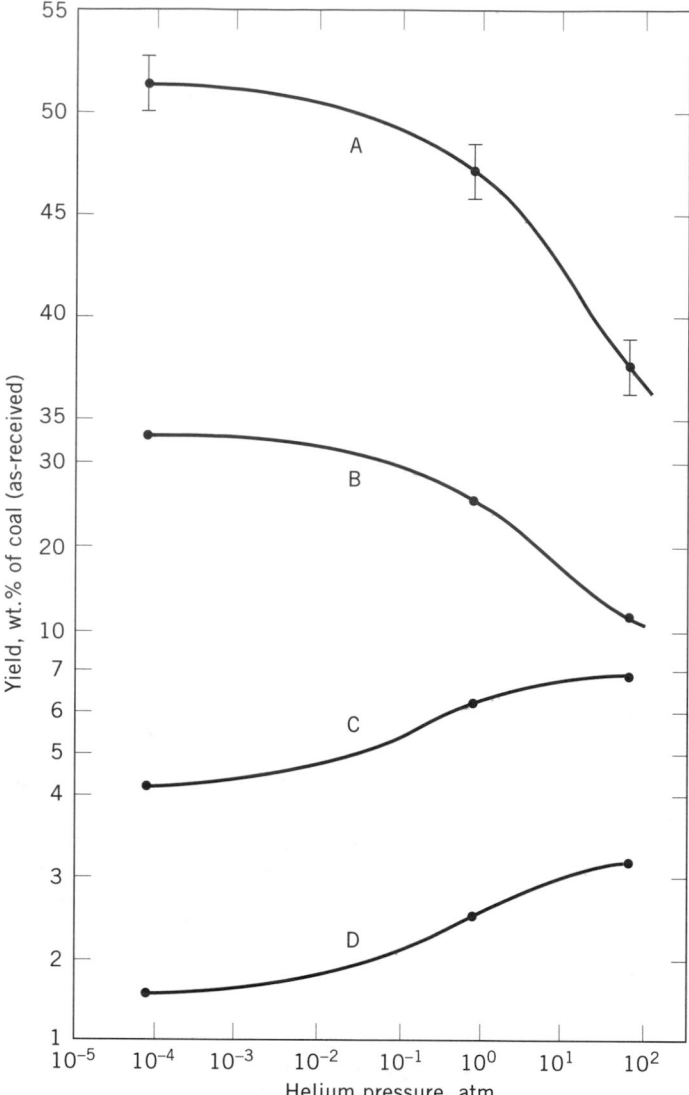

Figure 4. Effects of pressure on yields of (**a**) total volatiles, (**b**) tar plus hydrocarbon liquids, (**c**) all hydrocarbons gases, and (**d**) methane from bituminous coal pyrolysis. Heating rate = 1000°C/s, temperature = 1000°C, isothermal holding time = 2–10 s, particle diameter = 74 μm. From Ref. 32.

vacuum (42), that the pressure dependence of tar yield was due to vaporization of tar components from the coal particle surface. A correlation of vapor pressure with the molecular weight of the aromatic hydrocarbons (32) was used. More appropriate vapor pressure–molecular weight correlations were obtained (43) based on twelve narrow-boiling fractions of coal liquids with molecular weights ranging from 110 to 315 Da and vapor pressures ranging from 3Pa to 3.5 MPa. The Fletcher-Grant pressure (FGP) correlation agrees well with the measured vapor pressures of the different fractions (Fig. 5a). Such good agreement between data and model was not possible using the model proposed earlier, the Unger-Suuberg correlation (32,42). The FGP correlation also holds for pure

model compounds of the kind which might be observed in tars (Fig. 5**b**). The basic difference between the Unger-Suuberg correlation and the FGP correlation is that, at high temperatures, the FGP correlation predicts higher vapor pressures and attendant lower boiling points than the Unger-Suuberg correlation and this is dependent on molecular weights. Thus, one might overestimate the boiling point of a compound of a certain molecular weight using the Unger-Suuberg correlation.

In summary, it is clear that much is yet to be learned about devolatilization of coal during pyrolysis. The implementation of models to describe the evolution of products is in its infancy and the application of come new models such as the chemical chemical percolation devolatilization

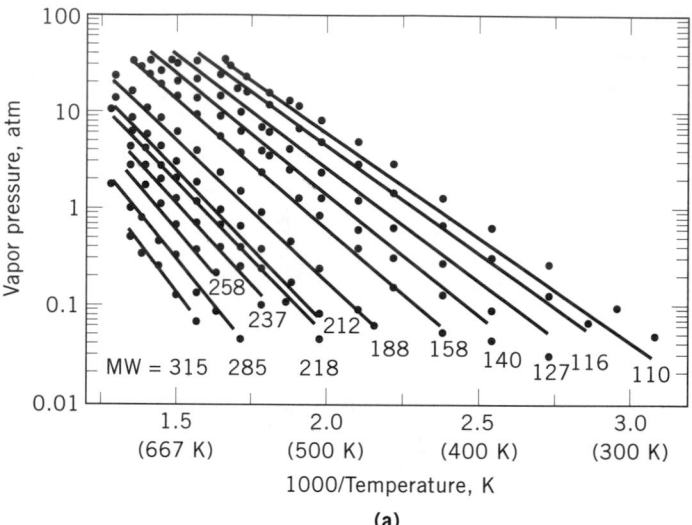

(a)

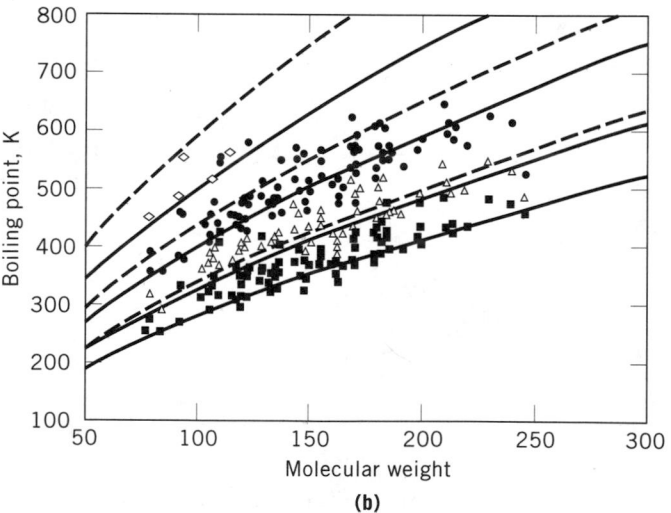

(b)

Figure 5. (a) Comparison of the FGP vapor pressure correlation (43) with data for 12 narrow-boiling fractions of coal liquids from Pittsburgh Seam coal. (b) Comparison of the FGP (42,43) vapor pressure correlations with boiling point data for 111 organic compounds at various pressures. From Ref. 26.

(CPD) model (44,45) promises to lead us into a better understanding of the product yields.

Kinetic Rates of Pyrolysis

Understanding kinetic rates of pyrolysis is of fundamental importance to the field of applied coal pyrolysis because the kinetics of thermal degradation of coal will determine most process variables in commercial applications. Unfortunately, there is a general lack of agreement on rate constants describing weight loss during pyrolysis, as tar and volatiles are removed. Figure 6 illustrates well this lack of agreement as there are at least three orders of magnitude variations in pyrolysis rate constants at any given temperature (13,25). These variations are, in part, due to the heterogeneity of coal. Samples may contain a variable mixture of macerals and minerals and even pure macerals themselves may be mixtures of heterogeneous components from apparently different origins. Another explanation may be that different laboratories using the same coal may are in fact using different samples from the same seam but different locations in the mine. No standardization existed in the past. Now with the implementation of Argonne Premium samples, we can expect better agreement among laboratories. This variability is illustrated in Figure 7 which shows the weight loss from Pittsburgh seam coal particles on a heated grid where temperature rise times are 10³K/s. Variations of up to 400 K for the same weight loss typify the problem. The same coal analyzed by various different groups exhibited a wide range of pyrolysis weight loss.

In principle, the variations can be explained as variations in coal chemistry due to the fact that different samples from the same seam were being used and that in-seam variations are responsible. This explanation seems unlikely to account for a wide range of variability due to the well known uniformity of the Pittsburgh seam. Another possible explanation is variation in mass transfer rates of volatiles based upon particle size variations. Large tar molecules might have difficulty migrating from

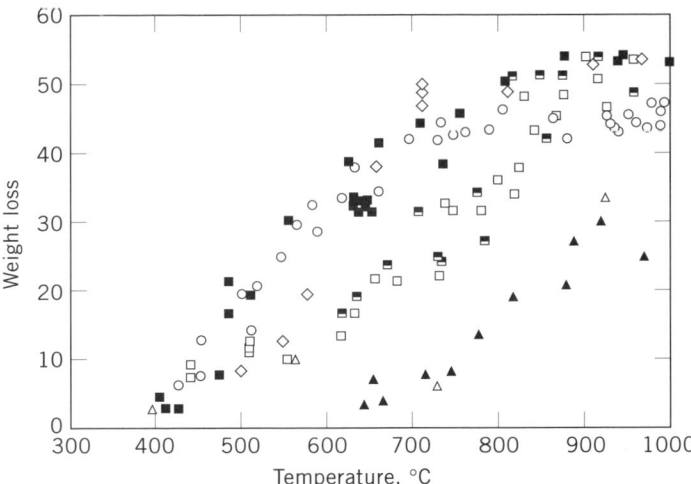

Figure 7. Comparison of experimental weight loss data for non-isothermal pyrolysis of Pittsburgh seam coal in a heated grid (1000 K/s, 0.1 MPa, no holding time). From Ref. 26.

within large particles, whereas, it would be facilitated in smaller particles. However, as pointed out above, this would enhance formation of smaller gaseous molecules due to the longer residence time of the tar within the particles and these low molecular weight species would not be mass transport limited.

Solomon and co-workers (26) attribute the variations in the pyrolysis rates to inaccuracies in measuring temperatures of coal particles. This is especially true for heated grid experiments (12) and entrained flow reactor studies (46). Others have also suggested that variations in reaction rates are related to inaccurate temperature measurements (13,14,47). The question of temperature measurement is of such great importance that countless studies have been devoted to obtaining information on temperature of coal particles during pyrolysis (26). This is especially true for heated grid experiments where additional uncertainties in rate measurements derive from assumptions used for calculations (48).

Coal particle temperatures have been measured directly by use of FTIR emission-transmission spectroscopy (13,14,22,23,49) and two- and three-color pyrometry (21,24). These techniques effectively measure surface particle temperatures but fail to measure the temperature within coal particles. Because pyrolysis proceeds from the surface inward, significant temperature gradients could exist, undetected by surface measurements. This is particularly true for pyrolysis methods such as laser pyrolysis methods (21) where heating rates are on the order of 10⁶K/s and large temperature gradients ensue. These gradients may be as large as 1500 K for 65-mm diameter particles whose surface temperature is approximately 2000 K. When such large gradients exist, especially toward the direction of laser heating, coal particles may undergo fracture or may be propelled at high velocities.

Measuring particle temperature with FTIR-emission-transmission spectroscopy is not a trivial feat in itself. The technique relies on measuring temperature by its radiative emission during heating, but the coal particle's

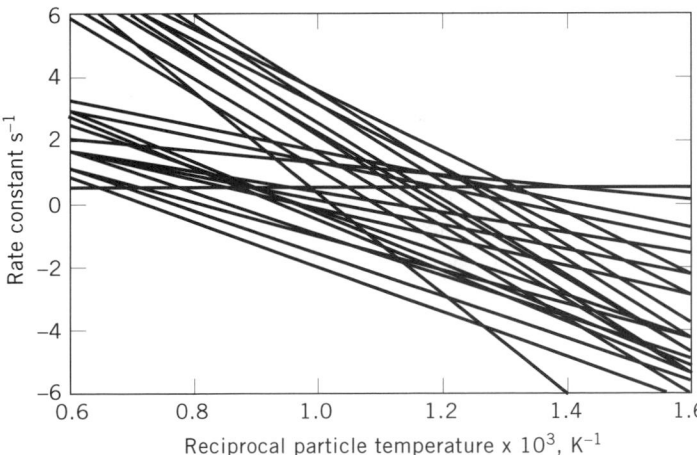

Figure 6. Arrhenius plot of mean kinetic rates for coal pyrolysis weight loss or tar evolution. From Ref. 25.

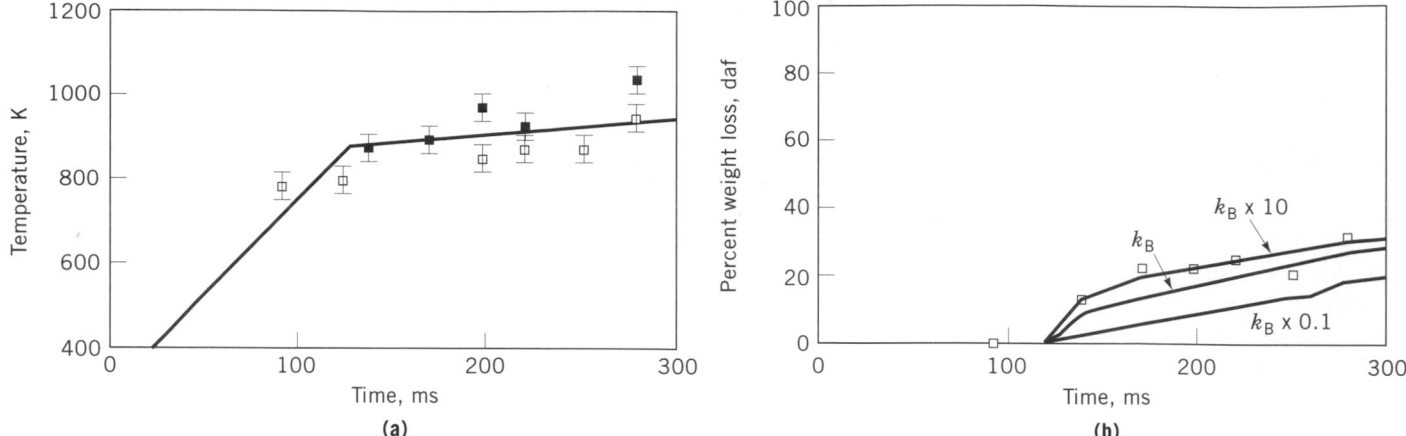

Figure 8. Pyrolysis of bituminous coal in a flow reactor. (**a**) Temperature for Pittsburgh seam coal, (**b**) mass loss for conditions given in **a**. From Ref. 49.

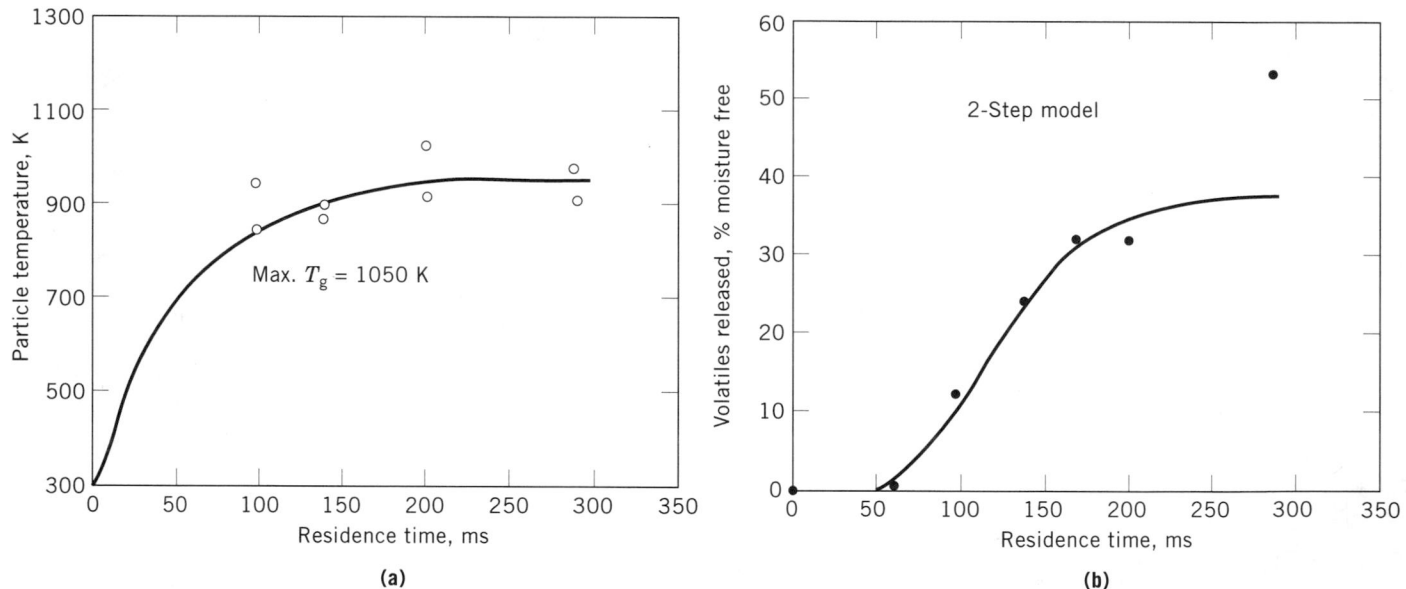

Figure 9. Pyrolysis of bituminous coal in a flow reactor. (**a**) Comparison of measured (o) and calculated (solid line) particle temperatures as a function of residence time for 69-μm-dia particles (PSOC 1415 d) in gas at 1050 K, (**b**) mass loss for conditions in **a**. From Ref. 26.

spectral emittance must be known. In one-color pyrometry, one can measure the radiance of the particle at a single frequency and use that as a measure of particle temperature. In two- or three-color pyrometry, two or three frequencies are used.

In an attempt to obtain more accurate temperature measurements Solomon and co-workers (49) recently compared temperature measurements by two techniques at high heating rates. Using bituminous coal as the feedstock and comparing measured coal particle temperatures in an entrained-flow reactor and heated tube reactor, they provided ample evidence that wide variations in kinetic rates of pyrolysis can be eliminated by appropriately modeling particle temperatures using FTIR-emission-transmission spectroscopy. Figure 8 shows the data obtained for particle temperature (open squares) and CO_2 (solid squares) in a transparent-wall reactor with flowing nitrogen gas. In this experiment solids and gases were collected to measure mass loss of coal particles (shown in Fig. 8b). Applying first order kinetic analysis to the data and modeling particle temperature history, the best correlation was found (solid lines in Figure 8) with a rate ten times the previously determined rate ($k_B \times 10$). Thus the particle rate defining particle heating in this experiment must be at least as high or higher than previously measured rates.

Data from another experiment by Fletcher (24), where two-color FTIR pyrometry was used to measure temperature, size, and particle velocities in an inert-gas flow reactor, is depicted in Figure 9. Heating rates of approximately 10^4K/s were determined and experiments were conducted at two different gas temperatures (1050 K and 1250 K). Details of the experiment are given by Fletcher (24) but it is important to mention that the tar and char were separated using a helium quench probe and the mass loss during volatilization was measured by accounting for the aluminum, silicon, titanium and ash contents. Figure 9 shows the calculated particle temperatures match well with the data. Using first order kinetics the Arrhenius pre-exponential factor for these data is $A = 2.3 \times 10^4 \text{s}^{-1}$ and an activation energy of $E = 230 \text{ kJmol}^{-1}$ is calculated. This data for the pre-exponential factor agrees well with previous work (14,22,46,49), at least it is within errors of particle temperature measurements.

While many uncertainties have recently been nearly eliminated by use of accurate temperature measurement systems and more appropriate modeling of kinetic heating rates (50,51), much needs to be done to describe a system which is inherently complex. Gross, first order approximations can be made with current technology, but second order variables introduced by the heterogeneous nature of coal need to be understood. Such second order variables may become more important in the future when predictions of coal devolatilization behavior will be required to evaluate with greater accuracy the emission of pollutant gases.

ANALYTICAL PYROLYSIS

Coal is a complex assemblage of fossilized plant remains which have been modified over extended periods of geologic time to be fused into a black, lustrous material of predominantly organic macromolecules. Petrographically, one can recognize discreet physical structures at both the macro-scale and micro-scale (52). The macroscopic entities are called lithotypes whereas the microscopic components are called macerals. Structural characterization of this physically and chemically heterogeneous material is a uniquely challenging task being pursued intensively (53).

While it is physically heterogeneous, coal is also chemically heterogeneous. Large numbers of different macromolecular structures can be expected to be associated with the different macerals. If we could unravel the nature of these structures we could learn much about coal's reactivity (eg, coal processing, conversion, and combustion) (54,55) and could begin to understand the chemical processes which transformed the plant materials to the macerals observed now. Unfortunately, there are few methods available for detailed chemical examination of coal macromolecules directly.

One technique which has recently made great strides in its application to coal science is analytical pyrolysis. The prefix analytical is used to denote that the products of pyrolysis of coal are quantitatively and qualitatively analyzed. The reason analytical pyrolysis has become popular is that it degrades the macromolecular structures in coal and associated macerals to smaller molecular fragments which are more amenable to study by gas chromatography (56,57). This is an important aspect of analytical pyrolysis which has contributed greatly to its increasing popularity as a characterization tool for coal.

Traditional means of examining coal components, prior to the advent of analytical pyrolysis, involved extraction of coal by organic solvents and analysis of the extracted material by gas chromatography–mass spectrometry. Generally less than 10% of the coal could be extracted and only a small fraction of this extract could be identified and quantified. Many new sophisticated techniques have been applied to the study of coal structures (56,58–75). Many of these techniques are not only capable of analyzing extractable components but the macromolecular components are being characterized as well. This has allowed for the development of macromolecular structural models to depict coal structure (76). Perhaps the most important technique addressing macromolecular components of coal has been analytical pyrolysis and the subject of the current review focuses of this method as applied to coal. There are other methods of pyrolysis which are conducted in a batch mode (eg, hydrous pyrolysis and gold-tube pyrolysis), but these will not be discussed here because they have not been extensively applied to coal and generally yield similar products.

Analytically pyrolysis reduces the coal macromolecules to simpler fragments which are then amenable to study by this gas chromatography–mass spectrometric technique. In essence, small molecules are clipped from the macromolecular structure by breaking of one or two bonds via pyrolysis. These molecules immediately enter the gas phase of the gas chromatograph where they are subsequently analyzed before they can recombine with other pyrolysis products to yield less nolatile residues.

There are several aspects of this type of pyrolysis

which distinguish it as being different than applied pyrolysis (57,65). First, the pyrolysis can be conducted on a small scale in an inert atmosphere, usually the inlet of a gas chromatograph. Second, small amounts of sample, usually 0.1 to 5 mg of coal, are more than sufficient for product formation and analysis by the very sensitive technique of mass spectrometry. Third, the temperature is usually increased to 600 to 800°C at rise times on the order of more than 10^3K/s but maintained only for a short period of time (seconds). These combinations of conditions allow sufficient thermal energy to cleave one or two bonds and to produce volatile substructures of coal which might be representative of the macromolecular structure (65–67). Figure 10 shows typical gas chromatograms from high and low rank coals obtained by analytical pyrolysis. Note that in low rank coals the peaks obtained are mostly methoxyphenolic and phenolic structures characteristic of macromolecular structures derived from wood or other vascular plant remains which have not been altered extensively. In addition, one usually observes an homologous series of n-alkane/n-alkene doublets representing pyrolysis products from aliphatic biopolymers in the coal (71). In the pyrogram of higher rank coal, bituminous coal, the product distribution is somewhat different, indicative of vastly different macromolecular structures. Benzenes, alkylbenzenes, phenols, naphthalenes, alkylnaphthalenes, and numerous other three- to four-ring condensed aromatic compounds are observed.

It is clear from these two chromatograms that significant differences in macromolecular composition of the two samples are resulting in different product distributions. This is what makes analytical pyrolysis an ideal method of characterization. What is perhaps another advantageous feature of analytical pyrolysis is that the pyrograms in Figure 10 can be reproduced time-and-again. So, the technique is quite reproducible.

The clear drawback of the technique is its unknown quantitative accuracy in representing macromolecular structural features. The deficiency has, to some extent, been partially relieved by addition of polymeric internal standards (70) from which one obtains peaks that can be quantified. Such an approach does not address the problem that one does not have a good measure of the absolute yields from the different macromolecular components of the coal. For example, some biopolymers (eg, aliphatic biopolymers) in coal may efficiently produce a certain suite of products with high yields. If these exist with other macromolecules which yield only modest or few volatiles, the resulting chromatograms will be biased toward those products most efficiently produced even though the relative proportions of precursor macromolecules are minor components. Thus, the pyrogram in Figure 10 shows significant total amounts of n-alkanes and n-alkenes, these may derive from a macromolecular component contributing only a small weight fraction of the coal. There is no simple method of evaluating the relative response of volatile product yield from small coal samples in analytical pyrolysis systems. It suffices to say that in most cases responses are approximately representative of the bulk structure and we know this from studies in which other independent means, such as NMR or FTIR, are used to measure product quality.

Chemical Composition of Coal Based on Pyrolysis

The product distribution obtained from analytical pyrolysis is inferred to represent the macromolecular composition. Thus, as the macromolecular coal fragments change in response to coalification or variations in plant sources contributing to the coal, the pyrolysis data should reflect this change. While it has generally been recognized qualitatively that highly aromatic coals generally yield aromatic pyrolysis products and aliphatic coals yield predominantly aliphatic products, the absolute yields of each type of product may provide a different picture. From a study (70) of coals and related kerogens, in which absolute product yields were measured, the yields of aliphatic products showed a positive correlation with yields of aromatic products, rather than the inverse correlation expected. The explanation for these results is that absolute product yields depend on efficiency of pyrolysis and in highly aliphatic coals the efficiency is greater for both aromatic and aliphatic products than in aromatic coals and kerogens, where only small amounts of volatiles are obtained in comparison.

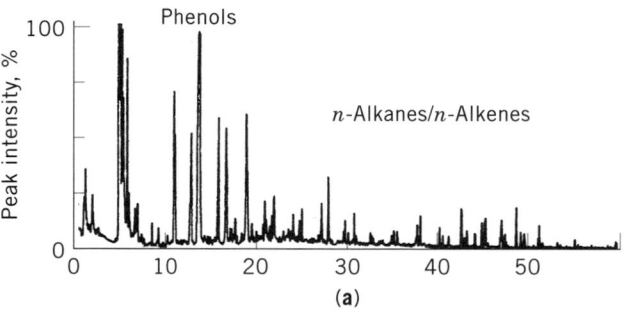

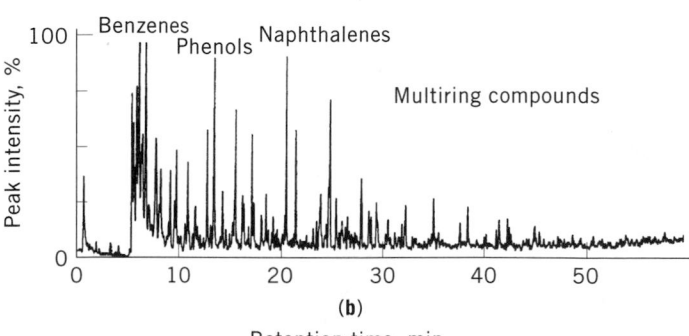

Figure 10. Pyrolysis chromatograms for a low rank coal from the Wyodak-Anderson Coal in Wyoming (upper) and a high-volatile A bituminous coal (lower) from the Lower Kittanning Seam in Western Pennsylvania.

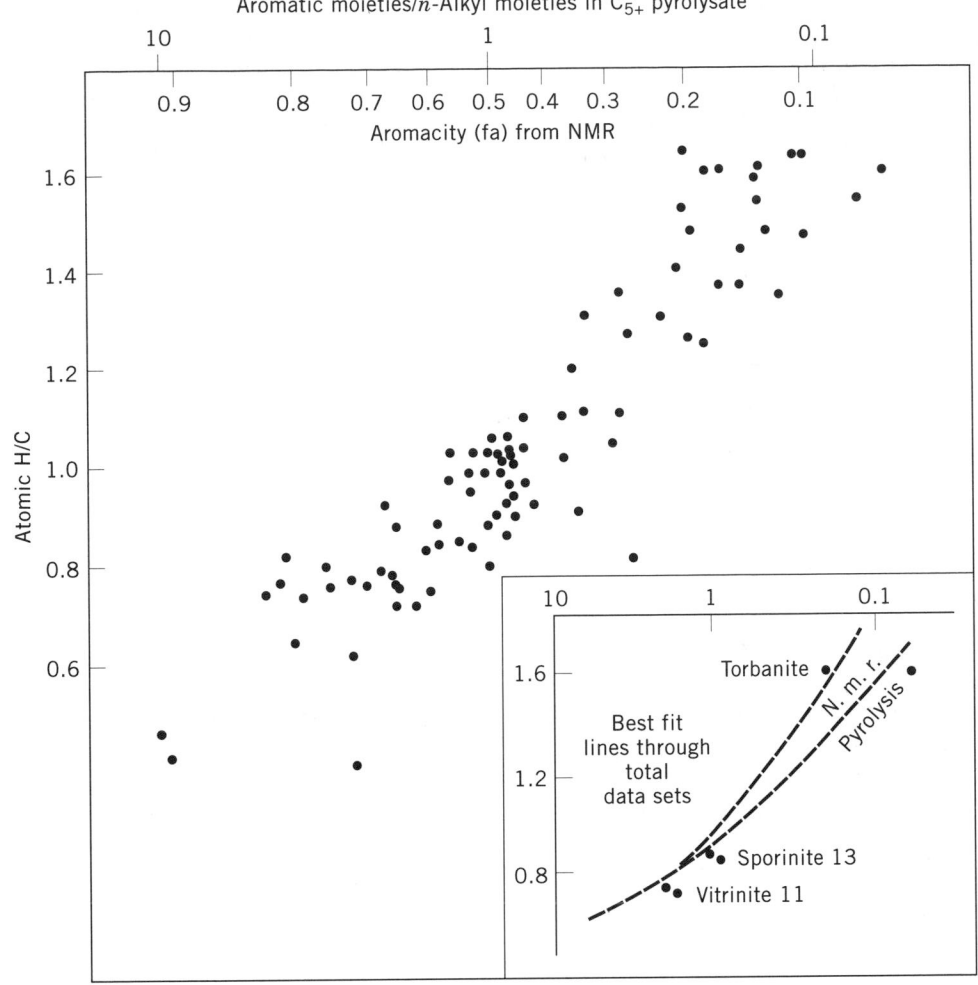

Figure 11. The aromaticity of coals and kerogens indicated by both Py-gc ratios and ^{13}C-nmr measurements. Modified from Ref. 69.

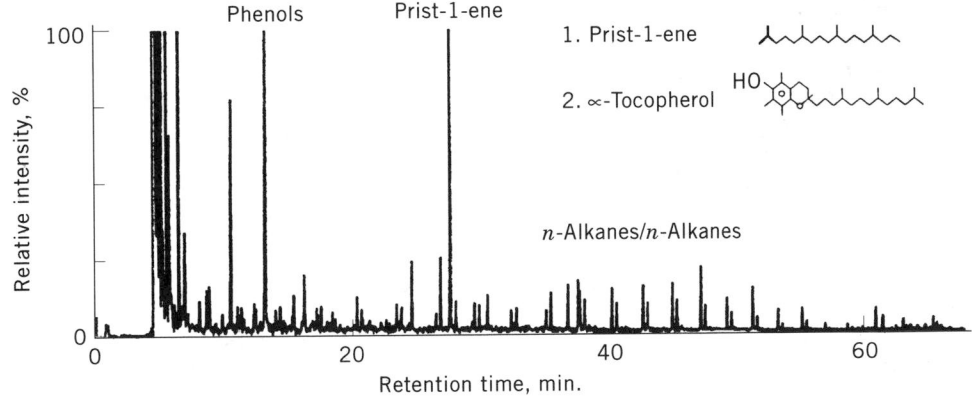

Figure 12. Pyrolysis chromatogram for an inertinite-rich coal from the Wyodak-Anderson Seam in Wyoming.

From these results, it is clear that absolute yields may not be appropriate indicators of compositional differences among coals. The relative yields, though they also may be misrepresentative, often provide a better measure of composition. Thus, analytical pyrolysis is viewed mainly as a qualitative technique; however, when combined with other quantitative methods such as ^{13}C NMR (72,73), a powerful means exists to assess the chemical composition of coal (67,76).

As mentioned above and shown in Figure 10, the dominant pyrolysis fragments observed are aromatic and aliphatic hydrocarbons with aromatic heterocompounds such as phenols, methoxyphenols and sulfur containing species. Of course these are the volatile species produced and also those which can be chromatographed. If tars and very polar compounds were formed upon pyrolysis, it is unlikely that they would be analyzed by this technique.

Aromatic Compounds. The aromatic hydrocarbons typically observed in coal (and kerogen) pyrolyzates are alkylbenzenes with alkyl chains which may be short or long (77–84). If long chains are attached, they often exist in homologous series analogous to the n-alkane series. It is well known that pyrolysis mimics the processes responsible for petroleum formation in the subsurface (85). Solli and co-workers (86) have observed a close correspondence between alkylbenzene distributions in kerogens and their associated crude oils and it is likely that a similar correspondence exists between oil-forming coals and their associated oils. Undoubtedly, the distributions of alkylbenzenes are related to the structural entities from which they were pyrolyzed.

While single ring aromatics are the dominant aromatic compounds, two- and three-ring aromatics are often observed in coal pyrolyzates (65,74,76,87). While ring condensation occurring during coalification probably explains the occurrence of these structures (76), aromatization of hydroaromatic structures induced by the pyrolysis could also account for their presence (87,88). Alkylindanes, alkylindenes, alkylbenzofurans, and sulfur-containing hydroaromatic compounds have all been detected in pyrolyzates (89–91). The aromatic hydrocarbons are most likely derived from defunctionalized precursors in lignin (67,71) and amino acids (92), whereas, the sulfur-containing aromatics are probably formed by sulfidization reactions during early diagenesis (91). Also, many aromatic hydrocarbons could derive from cyclization of condensed lipids (93) which aromatize over the course of coalification (94).

The yields of aromatic compounds produced from analytical pyrolysis have been shown to agree well with other means of assessing structural composition (69). Figure 11 shows pyrolysis data for coals and kerogens which have also been analyzed by elemental analysis and ^{13}C NMR. The ratio of aromatic to aliphatic moieties in pyrolyzates was directly correlated to NMR-measured aromaticities and both of these parameters were found to agree well with atomic H/C measurements. The only divergence between carbon aromaticity and pyrolyzate aromaticity was observed for very aliphatic Torbanites, the higher yield of aromatic pyrolysis products being explained as artifacts caused by aromatization of aliphatic hydrocarbons.

Perhaps the most abundant aromatic pyrolysis products in coal are the phenols (methoxyphenols, phenol, and alkylphenols). This is readily observed from the pyrograms in Figure 10. These compounds owe their presence to defunctionalized lignin structures (75). Even the substitution patterns of alkylphenols (eg, 2,4-dimethylphenol) can be rationalized as being related to lignin which has undergone demethylation and dehydroxylation reactions at specific sites (75). In low rank coals, the me-

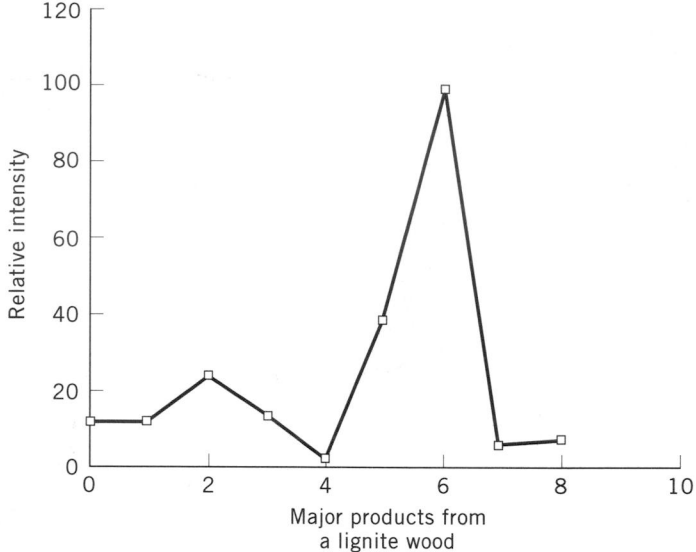

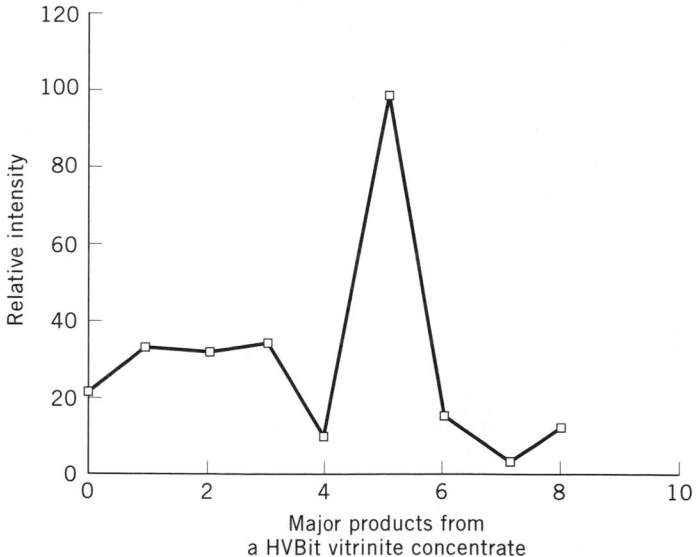

Figure 13. The relative distributions of different pyrolysis products produced by summing the appropriate molecular ion series in pyrolysis mass spectra (610°C/10 s) of a lignite wood and a high volatile bituminous vitrinite concentrate. 0 = alkanes; 1 = alkenes; 2 = dienes; 3 = benzenes; 4 = naphthalenes; 5 = phenols; 6 = methoxyphenols; 7 = sulfur compounds; 8 = carboxylated compounds. Modified from Refs. 57 and 65.

thoxyphenols predominate (75) whereas the higher rank coals the C_2 and C_3 phenols become more dominant (76), due in part to crosslinking reactions of the macromolecular structures.

Aliphatic Compounds. A wide variety of aliphatic hydrocarbons have been identified in coal pyrolyzates using the technique of analytical pyrolysis (56,57,65,71,77,82,95–98). The major aliphatic compounds are straight chain n-alkanes, n-alkenes, and n-alkadienes, with carbon numbers extending from n-C_5 to n-C_{33} for the three types of compounds. In addition, one commonly observes acyclic isoprenoids and cyclic terpenoids and steroids. The n-alkanes, n-alkenes, and n-alkadienes are common in low rank coals and often exist in nearly all maceral fractions (71). They derive from pyrolysis of large macromolecular structures commonly associated with liptinites. Alginite- and cutinite-rich coals are abundant sources of these structures formed by pyrolysis (65,71,98). Resistant aliphatic biopolymers in plant cuticles which survive coalification are thought to be the primary sources (99,100). There are variations in the distribution patterns of these aliphatic homologous series indicating that variations exist in the nature of the macromolecular species responsible for generating these hydrocarbons, even though they may all have a generic or common source from lipid-derived components of the coal (101).

Though land-plant waxes and associated macromolecular cuticular materials are responsible for the evolution of most alkane/alkene distributions, macromolecular resistant aliphatic biopolymers from algae are also possible. These could impart a differing distribution pattern as in the case of a specific Ordovician fossil algae named *Gleocapsomorpha* which shows n-alkanes and n-alkenes only in the region of n-C_9 to n-C_{18} (102).

Another group of significant aliphatic hydrocarbons which contribute to the pyrolyzates of most coals are the acyclic isoprenoids, prist-1-ene being the most notable (103). This compound is easily released in pyrolyzates (100) and usually dominates coal pyrograms (104). Figure 12 shows a pyrogram of a specific lithotype from the Wyodak-Anderson coal seam in Wyoming. Aside from the homologous series of n-alkanes and n-alkenes, prist-1-ene is the most obvious feature of the pyrogram. The chemical formula for prist-1-ene is shown above the peak. Also shown is the structure of what is believed to be its precursor molecule in plant remains which contributed to the coal, tocopherol (105). It is also possible that prist-1-ene originates from archaebacterial contributions (103) or from the phytol side-chain of chlorophyll. Decreasing abundances of prist-1-ene in coals of higher rank (104) is an indication that the prist-1-ene derives from a similar structure bound to the coal by a fairly weak bond, possibly a C—O bond which is more readily cleaved during coali-

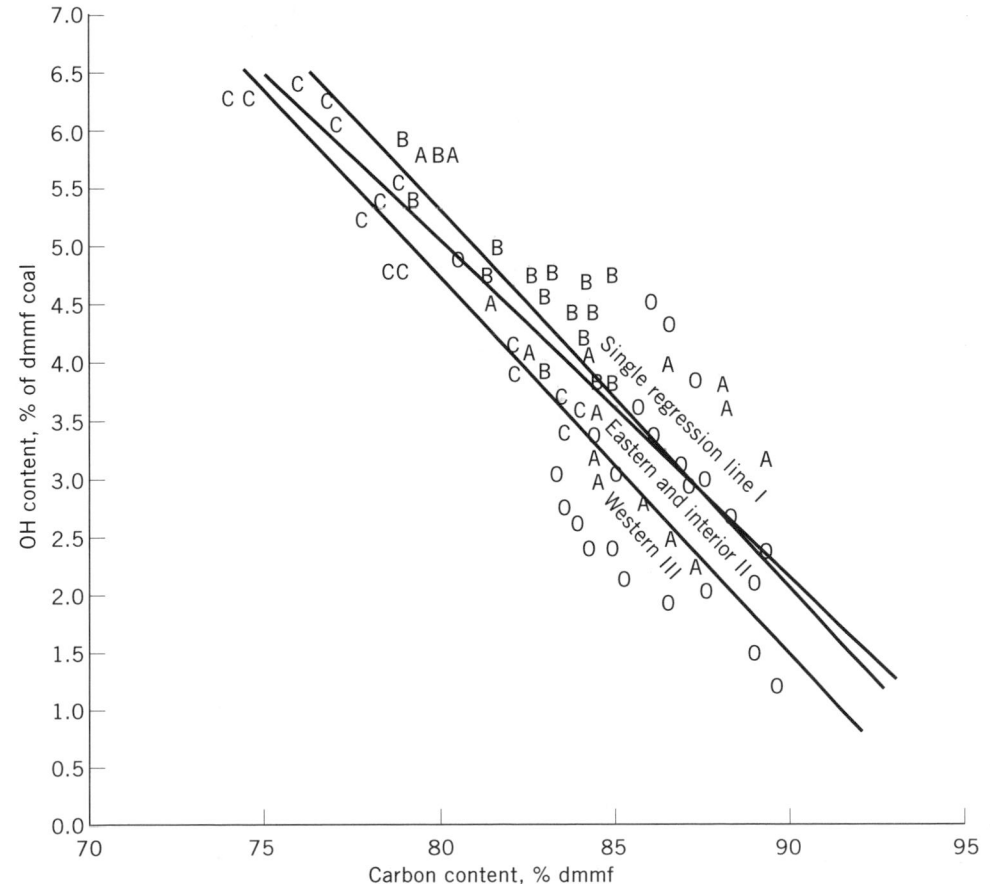

Figure 14. Concentrations of hydroxyl oxygen content for U.S. coals and pyrolytic phenol yields for 800°C/20 s flash pyrolyzates of eastern U.S. vitrinite concentrates. A = eastern coals; B = interior coals; C = western coals; O = pyrolysis data converted to OH content. Modified from Refs. 128 and 129.

fication. Finally, the last group of aliphatic compounds observed in coal pyrolyzates are steroidal and terpenoidal alkanes and alkenes (65,106–109). The alkenes are probably produced from pyrolytic cleavage of C—O and C—S bonds which bind these biomarkers to the coal macromolecules. Isomeric pairs are observed in abundances which have been suggested to be related to coal rank (90,110,111). The steroidal isomers are probably derived from the vascular and nonvascular vegetation contributing to the coal-forming peat whereas the terpenoidal isomers have various sources including bacteria and higher plant resins.

Numerous other aliphatic moieties have been identified in coal pyrolyzates. These include furan derivatives, short-chain aliphatic alcohols, aldehydes, and ketones (112,113) which are typically found in low-rank coals where contributions from carbohydrates and proteins are still significant. Their contributions are only significant in

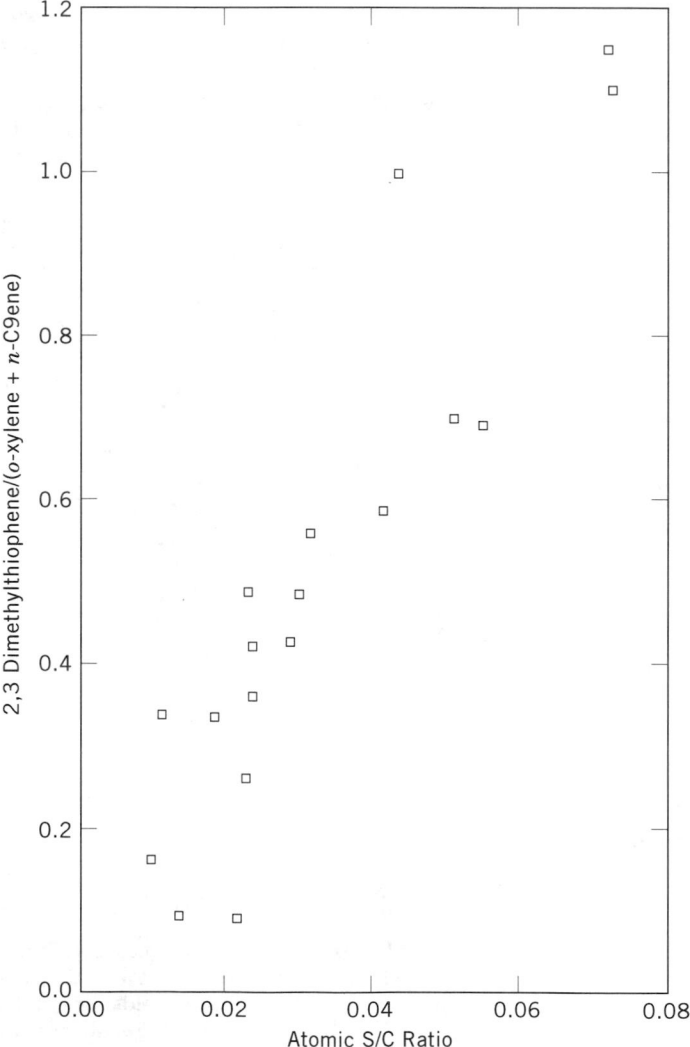

Figure 15. The ratio of (2,3-dimethyl-thiophene:(1,2-dimethyl-benzene + n-nonene)) from 800°C flash pyrolysis of oil asphaltenes and coals plotted versus chemically determined S:C atomic ratio. Modified from ref. 131.

low rank coals and peat, and they are normally present in trace quantities.

Use of Analytical Pyrolysis for Structural Determination

Analytical pyrolysis yields individual compounds cleaved from the macromolecular backbone of coal by rupture of one or two bonds. Thus, their relative abundance gives us clues as to the nature of the macromolecular backbone from which a structural reconstruction could be made. However, some caution should be employed in this approach to defining coal structure. The quantitative and qualitative representation can be compromised by numerous factors, mentioned above, leading to uncertainties about the use of analytical pyrolysis as a structural tool if used alone. If used in conjunction with established chemical degradation methods (58,60,79,114) or spectroscopic techniques (73,76,115–117) such as FTIR or ^{13}C NMR, a great deal of structural detail can be obtained. It is even possible to construct large macromolecular models (in 3-D space on a computer) based on the combined data from pyrolysis and NMR (116). ^{13}C NMR and elemental analysis provide the basic global structural constraints and the pyrolysis data provide clues with regard to the molecular units giving rise to the global characteristics. Using a program called the SIGNATURE program, Faulon and co-workers (116) have used this approach to construct three-dimensional macromolecular models for high volatile bituminous coal.

The major criticism of such an approach is that analytical pyrolysis products represent only a small fraction of the coal macromolecules, calling to question their degree of representation. The outlook is not as bad as it may seem in this regard because, as shown in Figure 11, there appears to be a good correspondence between aromaticities calculated from pyrolysis data (118–121) and the more quantitative aromaticities from ^{13}C NMR (69,72,122–125) for both coal and kerogen samples. Thus, direct comparison of pyrolysis data with its inherent problems to a more quantitative but bulk determinant technique, ^{13}C NMR, indicates that the pyrolysis data are representative of the overall chemical structure. There are, no doubt, exceptions where due caution is in order, especially in higher rank coals where the presence of condensed ring systems are significant and pyrolysis conditions are insufficient to produce representative volatile products.

For most low and medium rank coals, analytical pyrolysis provides quantitatively representative structural information of the individual fragments comprising the average macromolecular structure. Larter and Horsfield (126) have assembled data showing the relative concentrations of various pyrolysis product classes for low- and high-rank coals (Fig. 13). The low-rank coals generally contain methoxylated and alkylated phenols as the dominant pyrolysis products (75,117,127) with minor amounts of alkanes and alkenes, benzenes, and naphthalenes. Coals of higher rank, subbituminous and higher, are dominated by alkylphenols and increasing amounts of benzenes and aliphatic hydrocarbons. The methoxyphenols have diminished to very low levels. This is mostly due to the fact that the lignin-derived methoxyphenols which contribute heavily to low rank coals have been altered by coalification reactions, first undergoing demethylation to catechol and later dehydroxylation to form phenols, the dominant structural entities in the higher rank coals (75,117).

As rank coal increases, the phenols also diminish in relative abundance (76,128). Figure 14 shows the trend in phenol, reported as OH content (129), pyrolysis products in coals from the eastern and interior regions of the U.S. as well as the western coals. The decreasing trend of phenols is clearly evident, and it reflects the fact that increasing coalification is modifying the nature of phenolic units in coal macromolecules. Recently, Hatcher and coworkers (76) have suggested that such a trend reflects condensation of phenolic units to diaryl ethers and benzofurans as the hydroxyl content diminishes. Such a structural change is consistent with increased yields of benzenes and naphthalenes which are likely products from pyrolysis of diaryl ethers or dibenzofurans.

Even though the pyrolysis mechanisms leading to the formation of phenolic products may be complex (130), there are many trends in pyrolysis data of coals which make sense conceptually. This is an indication that analytical pyrolysis is a satisfying tool for structural characterization, especially with regard to phenolic entities in coal structure.

Similar observations have been made with regard to sulfur-containing molecules in pyrolyzates (131). Close correspondence between S:C ratios in coals and asphaltenes and the ratios of various alkylthiophenes suggest that the pyrolysis data are representative of coal-bound sulfur compounds and that the pyrolysis technique is providing a representative picture of coal structure.

SUMMARY

This article has focused on two currently important aspects of coal pyrolysis, applied pyrolysis and analytical pyrolysis. It is clear that these two areas have, in their own way, made a major impact on our knowledge of coal structure and its reactivity. A great deal of research is continuing at the present time in these two areas, and this review will likely become outdated in a few years. Nonetheless, we hope that the reader has assimilated a flavor for the nature of research on coal pyrolysis while obtaining an appreciation for the complexity of problems inherent in the approach.

BIBLIOGRAPHY

1. D. W. van Krevelen, C. van Heerden, and F. J. Huntjens, *Fuel* **30,** 748 (1951).
2. P. M. J. Wolfs, D. W. van Krevelen, and H. I. Waterman, *Fuel* **39,** 25 (1968).
3. V. T. Ciuryla, R. F. Weimer, A. Bivans, and S. A. Motika, *Fuel* **58,** 748 (1979).
4. H. Juntgen, *Fuel* **63,** 731 (1984).
5. J. R. Bautista, W. B. Russel, and D. A. Saville, *Ind. Eng. Chem. Proc. Des. Dev.* **25,** 988 (1986).
6. J. Gibbins-Matham and R. Kandiyoti, *Energy Fuels* **2**(4), 505(1988).
7. W. Wanzl, P. Kassler, K. H. van Heek, and H. Juntgen, *ACS Div. Fuel Chem. Prepr.* **32**(3), 125 (1987).
8. L. D. Timothy, D. Froelich, A. F. Sarofim, and J. M. Beer in the *21st Symposium (International) on Combustion*, Combustion Institute, Pittsburgh, Pa., 1988, p. 1141.
9. G. J. Hausmann and C. H. Kruger, in the *22nd Symposium (International) on Combustion*, Combustion Institute, Pittsburgh, Pa., 1989, p. 233.
10. P. C. Stangeby and P. L. Sears, *Fuel* **60**(2), 131 (1982).
11. J. D. Freihaut, M. F. Zabielski, and D. J. Seery in the *19th Symposium (International) on Combustion*, Combustion Institute, Pittsburgh, Pa., 1983, p. 1159.
12. J. D. Freihaut and W. M. Proscia, *Energy Fuels* **3**(5), 625 (1989).
13. P. R. Solomon and D. G. Hamblen in R. H. Schlosberg, ed., *Chemistry of Coal Conversion*, Plenum Press, New York, 1985, p. 121.
14. M. A. Serio, D. G. Hamblen, J. R. Markham, and P. R. Solomon, *Energy Fuels* **1,** 138 (1987).
15. J. P. Morris and D. L. Keairns, *Fuel* **58,** 465 (1979).
16. R. M. Carangelo, P. R. Solomon, and D. G. Gerson, *Fuel* **66,** 960 (1987).
17. A. K. Burnham, M. S. Oh, R. W. Crawford, and A. M. Samoun, *Energy Fuels* **3,** 42 (1989).
18. P. R. Solomon, D. G. Hamblen, M. A. Serio, Z. Z. Yu, and S. Charpenay, *ACS Div. Fuel Chem. Prepr.* **36,** 267 (1991).
19. V. Koch, K. H. van Heek, and H. Juntgen, *Dyn. Mass. Spectrom.* **1,** 15 (1970).
20. N. Simmleit and H. Schulten in H. Meuzelaar, ed., *Advances in Coal Spectroscopy*, Plenum Press, New York, 1990.
21. A. B. Witte and N. Gat, paper presented at the DOE Direct Utilization AR and TD Contractors' Meeting, Pittsburgh, Pa., 1983.
22. P. R. Solomon, M. A. Serio, R. M. Carangelo, and J. R. Markham, *Fuel* **65,** 182 (1986).
23. M. A. Serio, P. R. Solomon, Z. Z. Yu, and R. Bassilakis, *ACS Div. Fuel Chem. Prepr.*, **34**(4), 1324 (1989).
24. T. H. Fletcher, *Combust. Flame* **78,** 223 (1989).
25. P. R. Solomon, M. A. Serio, and E. R. Suuberg, *Prog. Energy Combust. Sci.* **18,** 133 (1992).
26. P. R. Solomon, T. H. Fletcher, and R. J. Pugmire, *Fuel* **72**(5), 587 (1993).
27. J. B. Howard in M. A. Elliott, ed., *The Chemistry of Coal Utilization*, Wiley-Interscience, New York, 1981, Chapt. 12.
28. S. D. Brandes, *Coal Pyrolysis: Literature Survey*, Office of Scientific and Technical Information, U.S. Department of Energy, Oak Ridge, Tenn. 1988.
29. W. H. Calkins, E. Hagaman, and H. Zeldes, *Fuel* **63,** 1113 (1984).
30. E. M. Suuberg, W. A. Peters, and J. B. Howard, *Ind. Eng. Chem. Process Des. Dev.* **17,** 3729 (1978).
31. L. M. Stock, *Accounts Chem. Res.* **22,** 427 (1989).
32. E. M. Suuberg, W. A. Peters, and J. B. Howard in the *17th Symposium (International) on Combustion*, Combustion Institute, Pittsburgh, Pa., 1979, p. 117.
33. M. A. Serio, P. R. Solomon, S. Charpenay, Z. Z. Yu, and R. Bassilakis, *ACS Div. Fuel Chem. Prepr.* **35,** 808 (1990).
34. G. R. Gavalas, *Coal Pyrolysis*, Elsevier, Science Publishing Co., New York, 1980.
35. S. Niksa, W. B. Russel, and D. B. Saville in Ref. 11, p. 1151.

36. T. Hirajima, E. W. Chan, and S. G. Whiteway, *Fuel* **65**, 844 (1986).

37. M. Fatemi, A. W. Scaroni, C. W. Lee, and R. G. Jenkins, *ACS Div. Fuel Chem. Prepr.* **32**(3), 117 1987).

38. T. Chakravarty, W. Windig, G. R. Hill, and H. L. C. Meuzelaar, *Energy Fuels* **2**(4), 400 (1988).

39. M. R. Khan, *Fuel* **68**, 1522 (1989).

40. P. Arendt and K. H. van Heek, *Fuel* **60**, 779 (1981).

41. P. R. Solomon, D. G. Hamblen, R. M. Carangelo, M. A. Serio, and G. V. Deshpande, *Energy Fuels* **2**, 405 (1988).

42. P. E. Unger and E. M. Suuberg, *ACS Div. Fuel Chem. Prepr.* **28**(4), 278 (1983).

43. T. H. Fletcher, D. M. Grant, and R. J. Pugmire, *ACS Div. Fuel Chem. Prepr.* **36**, 250 (1991).

44. D. M. Grant, R. J. Pugmire, T. H. Fletcher, and A. R. Kerstein, *Energy Fuels* **3**, 175 (1989).

45. T. H. Fletcher, A. R. Kerstein, R. J. Pugmire, and D. M. Grant, *Energy Fuels* **4**, 54 (1990).

46. P. R. Solomon, and M. A. Serio in J. Lahaye and G. Prado, eds., *Fundamentals of the Physical-Chemistry of Pulverized Coal Combustion* Martinus Nijhoff, The Netherlands, 1987, p. 121.

47. D. J. Maloney and R. G. Jenkins in the *20th Symposium (International) on Combustion*, Combustion Institute, Pittsburgh, Pa., 1985, p. 85.

48. A. S. Jamaluddin, J. S. Truelove, and T. F. Wall, *Combst. Flame* **62**, 85 (1985).

49. P. R. Solomon, M. A. Serio, and J. R. Markham in the *Proceedings of the International Conference on Coal Science, Tokyo*, 1989, p. 575.

50. A. K. Burnham, *Geochim. Cosmochim. Acta* **53**, 1693 (1989).

51. M. S. Oh, W. A. Peters, and J. C. Howard, *AIChE J.* **35**, 775 (1989).

52. E. Stach, M.-Th. Mackowsky, M. Teichmüller, G. H. Taylor, D. Chandra, and R. Teichmüller, *Stach's Textbook of Coal Petrology*, 3rd ed., Gebruder Borntraegaere, Stuttgart, 1982.

53. H. H. Schobert, *Structure and Properties of Coal*, The Pennsylvania State University, University Park, PA, unpublished course notes, 1992.

54. F. J. Vastola and L. J. McGahan, Report No. GRI-82-0070, Gas Research Institute, 1983.

55. P. H. Given in M. L. Gorbaty and co-workers, eds., *Coal Science*, Vol. 3, 1984, p. 63.

56. H. L. C. Meuzelaar, J. Haverkamp, and P. D. Hileman, *Py-MS of Recent and Fossil Biomaterials: Compendium and Atlas*, Elsevier, Amsterdam, The Netherlands, 1982.

57. B. Horsfield, in J. Brooks and D. H. Welte, eds., *Advances in Petroleum Geochemistry*, Vol. 1, Academic Press, Inc., New York, 1984, p. 247.

58. L. M. Stock in M. L. Gorbaty, J. W. Larsen, and I. Wender, eds., *Coal Science*, Vol. 1, Academic Press, Inc., New York, 1982, p. 161.

59. T. E. Hammond, D. G. Cory, W. M. Ritchey, and H. Morita, *Fuel* **64**, 1687 (1985).

60. K. Ouchi, Y. Yokoyama, T. Katoh, and H. Itoh, *Fuel* **66**, 1115 (1987).

61. P. C. Painter, M. Sobkowiak, and J. Youtcheff, *Fuel* **66**, 973 (1987).

62. P. G. Hatcher, H. E. Lerch, R. K. Kotra, and T. V. Herheyen, *Fuel* **67**, 1069 (1988).

63. P. R. Solomon, and R. M. Carangelo, *Fuel* **67**, 949 (1988).

64. H. C. K. Chang, K. D. Bartle, K. E. Markids, and M. L. Lee in L. C. Meuzelaar, ed., *Advances in Coal Spectroscopy*, Plenum Press, New York, 1992, p. 141.

65. S. R. Larter in K. Voorhees, ed., *Analytical Pyrolysis Methods and Applications*, Butterworth & Co. (Publishers) Ltd., Kent, UK, 1984, p. 212.

66. W. J. Irwin, *Analytical Pyrolysis*, Marcel Dekker, Inc., New York, 1982.

67. P. G. Hatcher, *Earth Mineral Sci.* **61**, 3 (1992).

68. W. L. Maters, D. Van De Meent, P. J. W. Schuyl, J. W. De Leeuw, and P. A. Schenck in Jones and Cramer, eds., *Analytical Pyrolysis*, Elsevier, Amsterdam, 1977, 203.

69. B. Horsfield, *Geochim. Cosmochim. Acta* **53**, 891 (1989).

70. S. R. Larter and J. Senftle, *Nature* **318**, 277 (1985).

71. M. Nip, J. W. de Leeuw, and J. C. Crelling, *Energy Fuels* **6**, 125 (1992).

72. K. W. Zilm, R. J. Pugmire, S. R. Larter, J. Allan, and D. M. Grant, *Fuel* **60**, 717 (1981).

73. F. P. Miknis, J. W. Smith, E. K. Maughan, and G. E. Maciel, *Bull. AAPG* **66**, 1396 (1982).

74. M. Nip, J. W. de Leeuw, and P. A. Schenck, *Geochim. Cosmochim. Acta* **52**, 637 (1988).

75. P. G. Hatcher, H. E. Lerch, and T. V. Verheyen, *Int. J. Coal Geol.* **13**, 65 (1989).

76. P. G. Hatcher, J. L. Faulon, K. A. Wenzel, and G. D. Cody, *Energy Fuels* **6**, 813 (1992).

77. S. R. Larter and A. G. Douglas in W. Krumbein, ed., *Environmental Biogeochemistry and Geomicrobiology*, Vol. 1, Ann Arbor Science Publ., Woburn, MA, 1978, p. 373.

78. H. Solli, S. R. Larter, and A. G. Douglas in A. G. Douglas and J. R. Maxwell, eds., *Advances in Organic Geochemistry 1979*, Pergamon Press, Oxford, UK, 1980, p. 591.

79. J. Allan, M. Bjoroy, and A. G. Douglas, in Ref. 78, p. 239.

80. Z. H. Baset, R. J. Pancirov, and T. R. Ashe, in Ref. 78, p. 619.

81. W. H. Calkins, *Fuel* **63**, 1125 (1984).

82. C. E. Snape, W. R. Ladner, and K. D. Bartle, *Fuel* **64**, 1394 (1985).

83. P. F. Nelson, *Fuel* **66**, 1264 (1987).

84. J. Z. Dong and K. Ouchi, *Fuel* **68**, 710 (1989).

85. G. P. Cooles, A. S. Mackenzie, and T. M. Quigley in D. Leythaeuser and J. Rullkotter, eds., *Advances in Organic Geochemistry 1985*, Pergamon Press, Oxford, UK, 1986, p. 235.

86. H. Solli, S. R. Larter, and A. G. Douglas, *J. Anal. Appl. Pyrolysis* **1**, 231 (1980).

87. G. Van Graas, J. W. De Leeuw, and P. A. Schenck in Ref. 78, p. 485.

88. J. Allan and S. R. Larter in M. Bjoroy and co-workers, eds., *Advances in Organic Geochemistry 1981,* Wiley Heyden, 1983, p. 534.

89. G. Van Graas, J. W. De Leeuw, P. A. Schenck, and J. Haverkamp, *Geochim. Cosmochim. Acta* **45**, 2465 (1981).

90. T. I. Eglinton, R. P. Philp, and S. J. Rowland, *Org. Geochem.* **12**, 33 (1988).

91. J. S. Sinninghe-Damste, T. I. Eglinton, J. W. De Leeuw, and P. A. Schenck, *Geochim. Cosmochim. Acta* **53**, 873 (1989).

92. H. H. Schobert, *The Chemistry of Hydrocarbon Fuels*, Butterworth & Co. (Publishers) Ltd., Kent UK, 1990.

93. S. R. Larter, H. Solli, and A. G. Douglas in Ref. 88, p. 513.

94. B. P. Tissot and D. H. Welte, *Petroleum Formation and Occurrence*, 2nd ed., Springer-Verlag, New York, 1984.

95. J. Allan and A. G. Douglas in B. Tissot and F. Binner, eds., *Advances in Organic Geochemistry 1973*, Editions Technip, Paris, 1974, p. 203.

96. D. Van de Meent, S. C. Brown, R. P. Philp, and B. R. T. Simoneit, *Geochim. Cosmochim. Acta* **44,** 999 (1980).

97. B. Horsfield, H. Dembicki, and T. T. Y. Ho, *J. Geol. Soc. Lond.* **140,** 431 (1983).

98. R. Narayan, *Fuel* **68,** 1101 (1989).

99. M. Nip, E. W. Tegelaar, J. W. De Leeuw, P. A. Schenck, and P. J. Holloway, *Naturwissenschaften* **73,** 579 (1986).

100. M. Nip, E. W. Tegelaar, H. Brinkhuis, J. W. De Leeuw, P. A. Schenck, and P. J. Holloway in Ref. 85, p. 769.

101. K. Øygard, S. R. Larter, and J. T. Senftle in L. Mattavelli and L. Novelli, eds., *Advances in Organic Geochemistry 1987,* Pergamon Press, Oxford, UK, 1989, p. 1153.

102. J. D. Reed, H. A. Illich, and B. Horsfield in Ref. 85, p. 347.

103. S. R. Larter, H. Solli, A. G. Douglas, F. De Lange, and J. W. De Leeuw, *Nature* **279,** 405 (1979).

104. D. J. Curry and T. K. Simpler in Ref. 101, p. 995.

105. H. Goossens, J. W. De Leeuw, P. A. Schenck, and S. C. Brassell, *Nature* **312,** 440 (1984).

106. E. J. Gallegos, *Anal. Chem.* **47,** 1524 (1975).

107. E. J. Gallegos in P. C. Uden, S. Siggia, and H. G. Jensen, eds., *ACS Advances in Chemistry Series*, Vol. 170, 1978, p. 236.

108. R. P. Philp, *Trends Anal. Chem.* **1**(10), 237 (1982).

109. R. P. Philp, T. D. Gilbert, and N. J. Russell, *Fuel* **61,** 221 (1982).

110. R. P. Philp and T. D. Gilbert, *Geochim. Cosmochim. Acta* **49,** 1421 (1985).

111. F. Behar and R. Pelet, *J. Anal. Appl. Pyrolysis* **8,** 173 (1985).

112. D. Van de Meent, J. W. De Leeuw, and P. A. Schenck, *J. Anal. Appl. Pyrolysis* **2,** 249 (1980).

113. H. R. Schulten and W. Gortz, *Anal. Chem.* **50,** 428 (1978).

114. R. Liotta, *Fuel* **58,** 734 (1979).

115. P. G. Rouxhet and P. L. Robin, *Fuel* **57,** 533 (1978).

116. J.-L. Faulou, P. G. Hatcher, G. A. Carlson, and K. A. Wenzel, *Fuel Proc. Technol.* **34,** 277 (1993)

117. P. G. Hatcher, *Org. Geochem.* **16,** 959 (1990).

118. P. Leplat, *J. Gas Chromatog.*, **5,** 128 (1967).

119. J. Romovacek and J. Kubat, *Anal. Chem.* **40,** 1119 (1986).

120. A. Giraud, *Bull. AAPG* **54,** 439 (1970).

121. S. R. Larter and A. G. Douglas, *Geochim. Cosmochim. Acta* **44,** 2087 (1980).

122. J. M. Dereppe, J. P. Boudou, C. Moreaux, and B. Durand, *Fuel* **62,** 575 (1983).

123. A. J. G. Barwise, A. L. Mann, G. Eglinton, A. P. Gowar, A. M. K. Wardroper, and C. S. Gutteridge in P. A. Schenck and co-workers, eds., *Advances in Organic Geochemistry 1983,* Pergamon Press, Oxford, UK, 1984, p. 343.

124. J. D. Saxby and L. C. Stephenson, *Chem. Geol.* **63,** 116 (1987).

125. S. Derenne, C. Largeau, E. Casadevall, and F. Laupretre, *Fuel* **66,** 1084 (1987).

126. S. R. Larter and B. Horsfield in M. H. Engel and S. A. Macko, eds., *Organic Geochemistry,* Plenum Press, New York, 1993, p. 271.

127. R. P. Philp, N. J. Russell, T. D. Gilbert, and J. M. Friedrich, *J. Anal. Appl. Pyrolysis* **4,** 143 (1982).

128. J. T. Senftle, S. R. Larter, B. W. Bromley, and J. H. Brown, *Org. Geochem.* **9,** 345 (1986).

129. R. F. Yarzab, Z. Abdel-Basset, and P. H. Given, *Geochim. Cosmochim. Acta* **43,** 281 (1979).

130. S. R. Larter, *Geol. Rund.* **78**(1), 349 (1989).

131. T. I. Eglinton, J. S. Sinninghe Damste, M. E. L. Kohnen, J. W. De Leeuw, S. R. Larter, M. Thatcher, and R. S. Patience in W. L. Orr and C. M. White, eds., *Geochemistry of Sulphur in Fossil Fuels, ACS Symposium Series*, American Chemistry Society, Washington, D.C., 1990, p. 471.

COALBED GAS

DUDLEY D. RICE
BEN E. LAW
JERRY L. CLAYTON
U.S. Geological Survey
Denver, Colorado

Coal is the most abundant energy source in the world with resources estimated to be as much as 25 trillion metric tons (1). Although most of this resource (>90%) is concentrated in four countries—Canada, China, Russia, and the United States—coal resources are widely distributed and commonly are the only abundant source of fossil energy in many countries, such as China, Czech Republic, Poland, and Ukraine (Table 1) (1,2). With the expansion of global economies, particularly in underdeveloped countries, world coal production has been increasing rapidly on a yearly basis with total production totaling more than 5 billion metric tons in 1990; China alone produced more than 1 billion metric tons (see COAL).

In addition to minable reserves, coal is a major source of hydrocarbons, particularly natural gas (see NATURAL GAS). The gas is a product of the coalification process whereby plant material is progressively converted to coal and large quantities of gas are generated, some of which is stored within the coal matrix. The rank sequence from peat to anthracite and other commonly used indices of rank and thermal maturity are shown in Figure 1. The presence of methane-rich gas in coal has been recognized

Table 1. Estimates of Principal Coal and Coalbed Resources by Country[a]

Country	Coal Resources, 10^9 mt	Coalbed Gas Resources, 10^{12} m^3
Russia	6,500	17–113
China	4,000	30–35
United States	3,950	11
Canada	7,000	6–76
Australia	1,700	8–14
Germany	326	3
United Kingdom	190	2
Kazakhstan	170	1
Poland	160	3
India	160	1
Southern Africa[b]	150	1
Ukraine	140	2
Total	*24,446*	*85–262*

[a] Ref. 1.
[b] Includes Botswana, South Africa, and Zimbabwe.

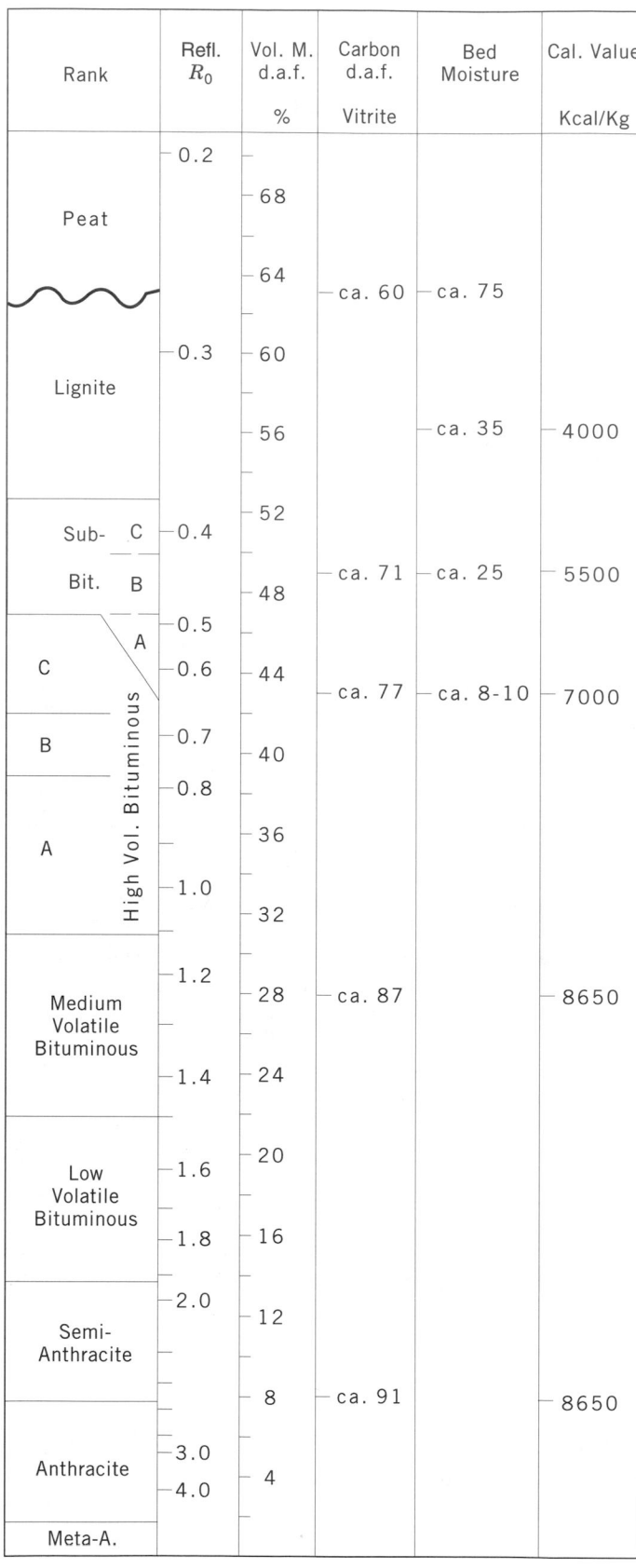

Figure 1. Coal ranks, from lowest (peat) to highest (meta-anthracite), and common properties of measurement. Daf, dry and ash free; Vol. M., volatile matter; Refl R_0, vitrinite reflectance in percent; Cal. value, caloric value; meta-A., Meta-anthracite (10).

for a long time because of dangerous explosions and outbursts associated with gassy underground mining. Only recently has coal been recognized as a reservoir rock in addition to being a source rock for gas, thus representing an enormous undeveloped "unconventional" energy resource.

Major resources of coalbed gas are associated with the immense amounts of coal. Worldwide estimates of coalbed gas are estimated to range from 85 to $262 \times 10^{12} \mathrm{m}^3$ (Table 1); the range of values indicates the scarcity of basic coal data in many coal-bearing areas of the world (1). Although the resource is widely distributed, most is concentrated in 12 countries where coal is commonly an important source of energy (1,2). The principal U.S. coal-bearing basins are estimated to have coalbed gas resources of about $11 \times 10^{12} \mathrm{m}^3$ of which $3 \times 10^{12} \mathrm{m}^3$ is expected to be recoverable (Fig. 2) (3,4). These coal-bearing areas are generally in cratonic and foreland basins without major structural deformation. The largest part of the resource is in Rocky Mountain basins where the coal beds are of Cretaceous and Tertiary age; lesser amounts are in the eastern and central United States where the coal beds are of Paleozoic age.

Some of the first attempts to produce gas from coal were undertaken in Europe in the late 19th century in order to reduce mining hazards (5). It has been uncertain when the first wells were drilled specifically for coalbed gas; however, in the late 1970s some commercial production of coalbed gas was established in the Black Warrior and San Juan basins in the United States (Fig. 2). Significant exploration for and production of coalbed gas in the United States began in the mid-1980s because of a federal tax credit given for the production of coalbed gas. Gas has been commercially produced from coal beds that range in thickness from 1 to 30 meters and occur at depths of 45 to 2730 meters. By the end of 1991, about 5000 wells, located mainly in the Black Warrior (2800) and San Juan basins (1800), were producing about $28 \times 10^6 \mathrm{m}^3$ of gas per day. The tax credit not only increased activity, but also was responsible for the development of new drilling and completion technology.

The knowledge and technology gained from exploration and production in the United States is currently being transferred to other countries where there has been no significant coalbed gas production because of economic, logistical, political, and geological reasons. However, considerable exploration activity is currently taking place in Australia, Canada, China, Czech Republic, the former Soviet Union, Poland, South Africa, Spain, and the United Kingdom as they become aware of the enormous resource potential.

GENERATION OF COALBED GAS

Generation of natural gas in coal beds takes place by two distinct processes: biogenic and thermogenic. Biogenic gas is composed primarily of methane and is the end result of the degradation of organic matter resulting from a complex series of processes by a diverse population of microorganisms. The major requirements for the generation of significant amounts of biogenic gas by methanogenic mi-

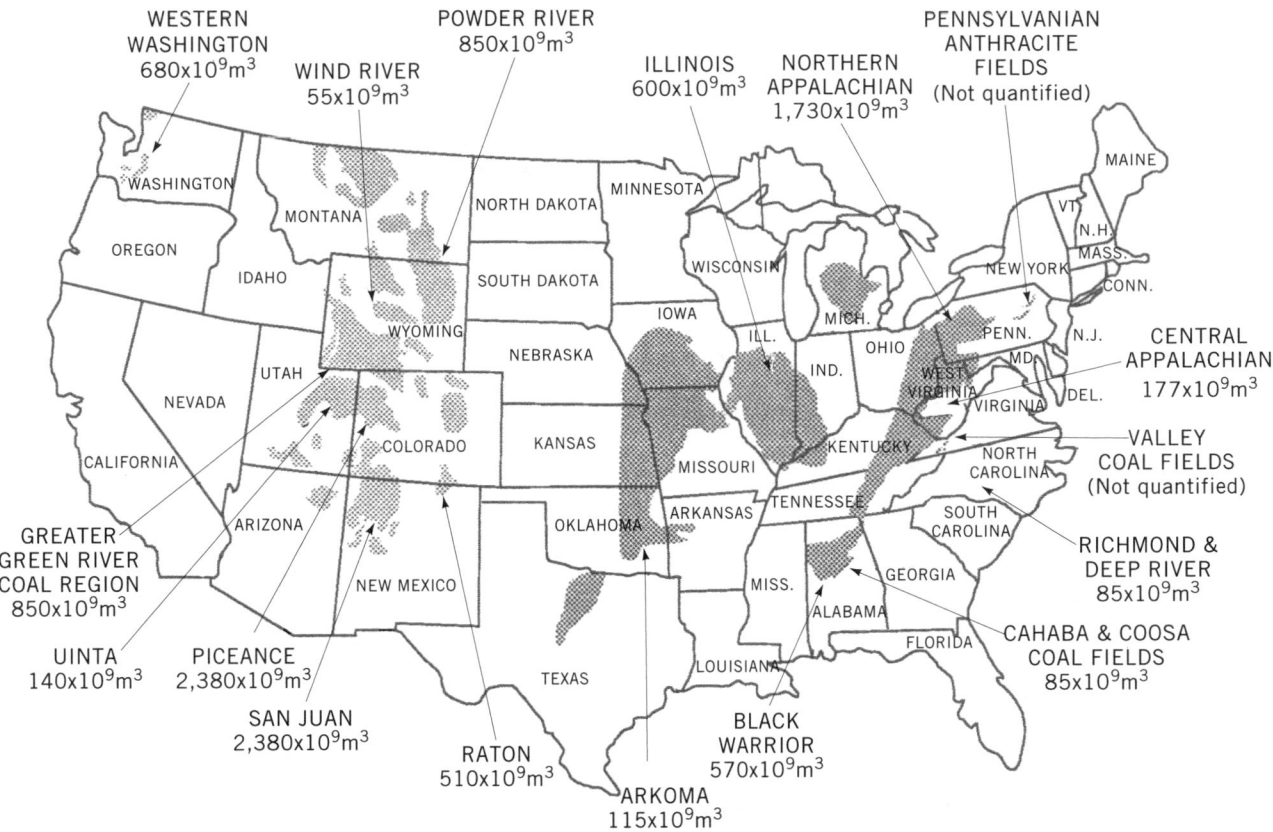

Figure 2. Map showing principal coal-bearing basins of the United States and estimates of in-place coalbed gas resources. Coals in basins of central and eastern United States are of Paleozoic age and those in western U.S. basins are of Cretaceous and Tertiary age (4).

croorganisms, which are assigned to the Archaea domain (6), are: anoxic environment, low sulfate concentrations, low temperatures, abundant organic matter, high pH values, and adequate space (7,8). If these conditions are met, economically significant amounts of biogenic gas can be generated over a period of tens of thousands of years after burial. Although biogenic gas can be generated by two pathways—carbon dioxide reduction and methyl-type fermentation—most ancient accumulations have probably resulted from the former as indicated by molecular and isotopic composition (8).

Biogenic gas can be generated during two different stages in coal beds (9). Biogenic gas is formed early in the burial history of low rank coals (peat to subbituminous rank—vitrinite reflectance (R_O) values <0.5%). This biogenic gas is referred to as early stage and its generation and accumulation are favored by rapid sedimentation. In addition, biogenic gas can be generated in coal beds in recent geologic time (tens of thousands to a few million years ago) where there is groundwater flow in an aquifer creating a favorable environment for microbial activity. This type of biogenic gas is referred to as late stage and its generation can take place in coal beds of any rank provided that the requirements outlined previously have been met.

With increasing degree of coalification resulting from higher temperatures and pressures, coals become enriched in carbon as large amounts of volatile matter rich in hydrogen and oxygen are released. Methane and carbon dioxide (Fig. 3), and water are the most important products of this devolatilization (10). This methane is thermogenic and its generation begins at a rank of high-volatile bituminous (R_O values >0.6%). The quantities of these main volatile products can be estimated using the major elemental composition (C-H-O) of the coal which will vary for the different types of kerogen (11). Four types of kerogen (I-IV) are recognized using principal elemental analysis and a van Krevelen diagram (Fig. 4). These four types of kerogen generally correlate with the major maceral groups identified by petrography—liptinite, vitrinite, and inertinite. Liptinite-rich coals generally correspond to types I and II kerogen, vitrinite-rich coals to type III kerogen, and inertinite-rich coals to type IV kerogen.

Major amounts of methane are generated from coals during the coalification process. The estimates vary from about 100 to 300 cm³/g depending on the elemental values used, starting rank, and the assumption made about the products (11–13). Under natural conditions, the actual yield is probably in the range of 150 to 200 cm³/g.

In addition to methane, some coals are capable of generating wet gas and oil. Generation of these heavier hydrocarbons takes place in coals having significant amounts of hydrogen-rich components, such as liptinite or certain components of the vitrinite maceral, such as humic gel (14–17).

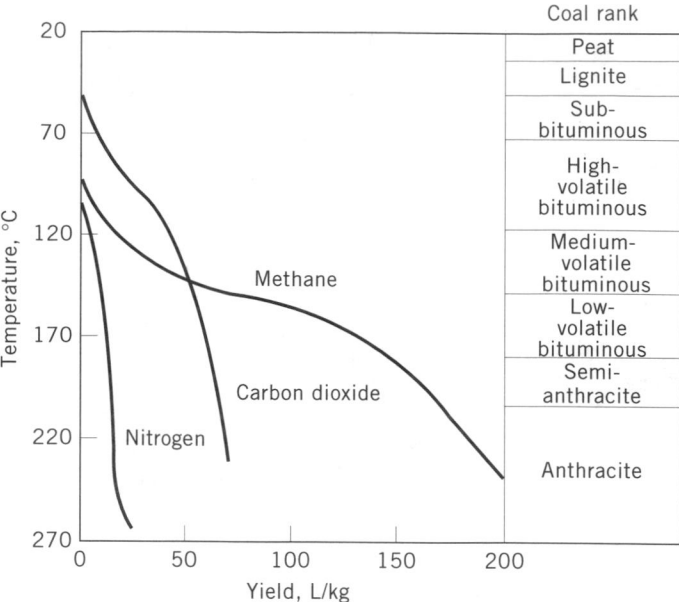

Figure 3. Calculated amounts of gases generated from coal during coalification (12).

CHARACTERISTICS OF COALBED GAS

Compositional data for coalbed gas is obtained from desorption tests of coal samples from underground mines and exploration boreholes or from production of coalbed reservoirs. Compositional data presented in this article are for gas samples collected from coal beds and are not for gas samples from adjacent reservoirs that have been interpreted to be coal derived. Analyses of coalbed gas samples are available from Australia, Canada, China, Germany, Poland, and the United States. The associated coals range in age from Pennsylvanian (Upper Carboniferous) to Tertiary and in rank from lignite to anthracite (R_O values of 0.3 to 4.9%) (9).

The molecular and isotopic composition of coalbed gases are highly variable (9). Methane is usually the major component with other hydrocarbon gases and carbon dioxide occurring in lesser amounts. The hydrocarbon composition of the gases, which is expressed as gas wetness (C_{2+} C_2–C_5/C_1–C_5 = C_2—C_5/C_1—C_5) ranges from 0 to 70.5%. The wettest gases are from coal samples that were desorbed and crushed. Carbon dioxide is the other major component of the coalbed gases and varies from less than 1 to 99%. Because of the variable molecular composition, gases from coal beds should be referred to as coalbed gas, and not coalbed methane. Coalbed gases are also variable in their isotopic composition; methane $\partial^{13}C$ values range from −80 to −16.8‰ and carbon dioxide $\partial^{13}C$ values range from −26.6 to +18.6‰.

The relation between coalbed gas composition (C_{2+} and methane $\partial^{13}C$ values) and rank of associated coal, expressed as R_O, is illustrated by coalbed gas samples from Canada and the United States (Fig. 5) (9). There is a tendency for all of the gases to be methane rich at low and high ranks, and for some to be wet at intermediate rank (R_O values of about 0.6 to 0.8%); however, many samples

are methane rich at intermediate ranks (Fig. 5a). Methane $\partial^{13}C$ values generally are isotopically lighter at lower ranks and isotopically heavier at higher rank. However, gases from coal of a given rank display a wide range of methane $\partial^{13}C$ values and the values commonly fall above (isotopically lighter) the regression lines shown on Figure **5b** which were determined for gases interpreted to have been generated from coal and dispersed types II/III kerogen. Similar patterns are exhibited by coalbed gases from other countries (9).

Although wetness and methane $\partial^{13}C$ values of coalbed gases are scattered when plotted against rank of the associated coal, they illustrate a more systematic trend when plotted against present-day depth of burial (9). Shallow coalbed gases are composed of isotopically light methane, as compared with deeper coalbed gases, regardless of rank. The methane-rich nature of shallow coalbed gases has been known for years because of the compositional data provided by underground mines. This information resulted in the expression "coalbed methane". The depth-related change in molecular and isotopic composition of coalbed gas can take place over a transition zone, can be abrupt, and can take place at varying depths, but usually in the range of a kilometer of the surface.

The reported coalbed gases are interpreted to be both biogenic and thermogenic based on molecular and isotopic

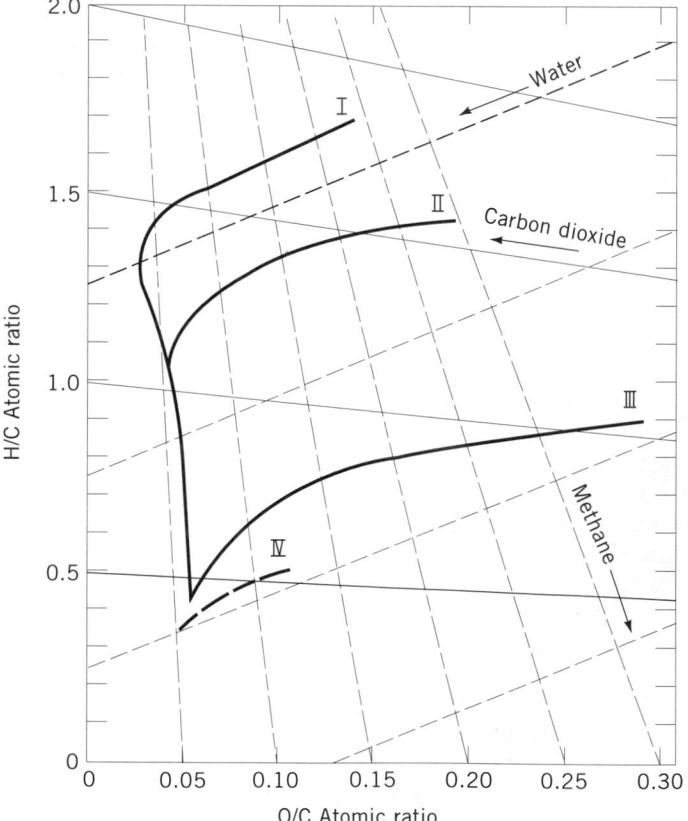

Figure 4. Van Krevelen diagram showing types of kerogen (I, II, III, IV) and pathways (solid and dashed lines) for generation of methane and elimination of carbon dioxide and water during coalification. Liptinite corresponds to types I and II kerogen, vitrinite to type III, and inertinite to type IV (13).

composition (9). The primary controls on the composition of the gas are considered to be rank, composition, and depth/temperature of associated coal. Secondary controls are also reflected in the coalbed gas composition and they will be discussed later.

Biogenic gas consists mainly of methane; the presence of significant amounts of heavier hydrocarbon gases with biogenic gas indicates mixing of late-stage biogenic methane in mature coals that have already generated heavier thermogenic hydrocarbons. Biogenic gas generation is restricted to shallow depths and low temperatures, and can occur in coals of all ranks. Biogenic gas can generally be distinguished by its isotopic composition; methane $\partial^{13}C$ values are generally in the range of -55 to $-90‰$.

In comparison, thermogenic coalbed gases are characterized by (1) common presence of heavier hydrocarbons at intermediate ranks of high-volatile and medium-volatile bituminous coal ($R_O > 0.6\%$), and (2) enrichment of heavy isotope ^{13}C in methane with increasing rank (methane $\partial^{13}C$ values commonly more positive than $-55‰$). At intermediate ranks, high-volatile and medium-volatile bi-

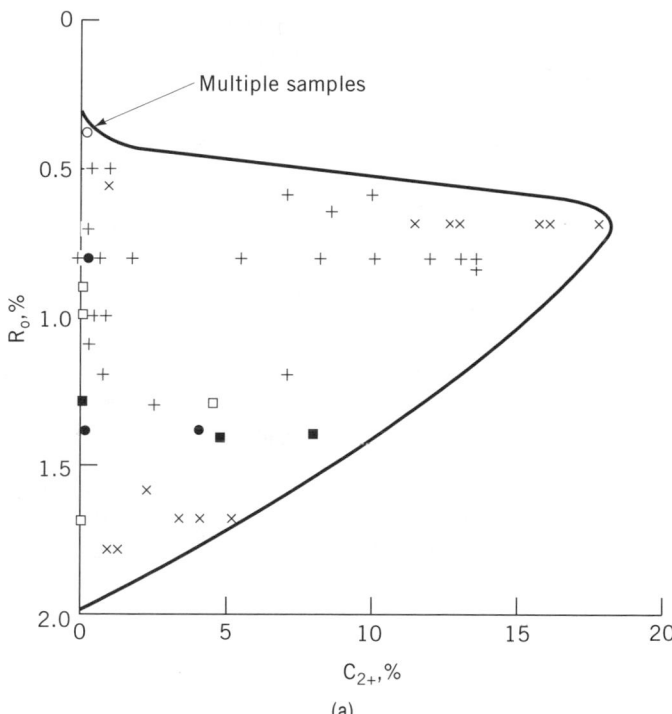

(a)

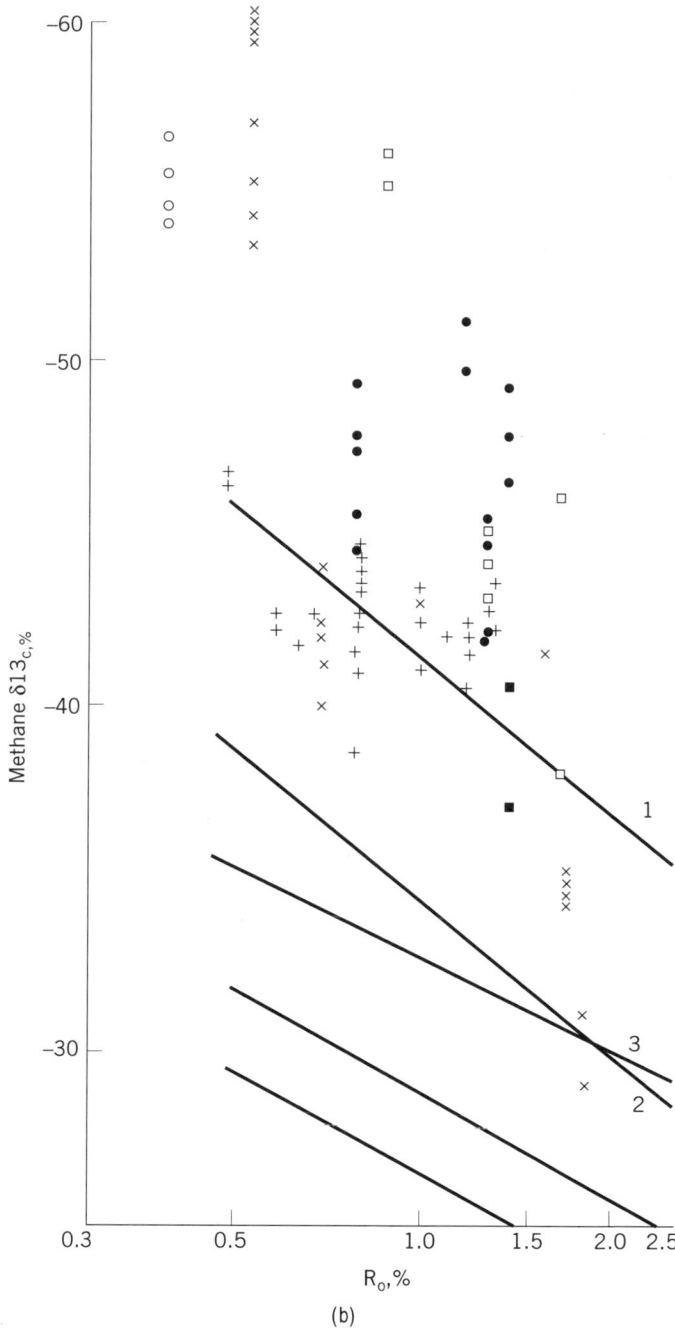

(b)

Figure 5. (*Continued*).

Figure 5. (**a**) Gas wetness (C_{2+}) versus vitrinite reflectance (R_O) for coalbed gases and coals, respectively, in Canada and the United States. (**b**) Methane $\partial^{13}C$ versus vitrinite reflectance (R_O) for coalbed gases and coals, respectively, in Canada and the United States. Regression line 1 is for gases generated from types II/III kerogen, line 2 is for coalbed and coal pyrolysis gases in China, line 3 is for coal-derived gases in China, and stippled area is for coal-derived and coal pyrolysis gases from several authors (9).

○   Powder River basin, USA
□   Arkoma basin, USA
●   Black Warrior basin, USA
×   Piceance basin, USA
+   San Juan basin, USA
■   Deep basin, Alberta, Canada

tuminous, sapropelic coals (liptinite and hydrogen-rich vitrinite) contain both wet gas and liquids, whereas humic coals (mostly vitrinite) contain a drier gas. These liquids contain high pristane : phytane ratios, a feature that is typical for coal-derived oils, and a slight predominance of odd-carbon numbered n-alkanes and in the C_{25+} fraction (18). However, the saturated fraction consists predominantly of low-molecular-weight components ($C_4–C_{10}$), which results in the liquids having high API gravities ($\approx 50°$). In addition, the coal-derived liquids are commonly characterized by high pour points because they contain abundant long-chain n-alkanes derived from terrestrial plant waxes. At high ranks, coalbed gases are mostly methane generated from residual kerogen and cracking of previously formed heavier hydrocarbons.

Thermogenic methane is isotopically more positive than biogenic methane because of the smaller kinetic isotope effects associated with thermal cracking (19,20). In general, methane $\partial^{13}C$ values for thermogenic methane become isotopically heavier with increasing rank because residual gas-producing carbon becomes enriched in ^{13}C due to $^{12}C-^{12}C$ bonds being broken more frequently by thermal processes than $^{12}C-^{13}C$ bonds. Also, coal composition, in addition to rank, influences the isotopic composition of the coalbed gas, whereby gases generated from coals with oxygen-rich kerogen (mostly vitrinite) (1) have more positive methane $\partial^{13}C$ values than those generated from hydrogen-rich kerogen (liptinite and hydrogen-rich vitrinite) at similar levels of thermal maturity, and (2) exhibit less spread in methane and ethane $\partial^{13}C$ values (20,21). These differences result from isotopically lighter methane being produced by thermal cracking of aliphatic-type structures, which are prevalent in hydrogen-rich kerogen, whereas isotopically heavier methane results from cracking of aromatic structures which are predominant in oxygen-rich kerogen. This effect of kerogen composition on isotopic composition is illustrated by regression lines shown on Figure 5**b**.

Secondary processes also affect the composition of coalbed gases, particularly at shallow depths, such that isotopically light methane is dominant in shallow coalbed gases, regardless of rank (9). The depth interval where these secondary processes are active and affect the composition of the gases is referred to as the zone of alteration. Unaltered gas generally occurs in the deeper parts of basins and is referred to as original gas.

At shallow depths, coal beds are commonly aquifers where microorganisms can thrive. Microbial activity can affect the observed composition of coalbed gases in two ways. First, significant amounts of late-stage, isotopically light biogenic methane generated by anaerobic microorganisms can be mixed with previously generated thermogenic gas or fill degassed coal beds (9). Second, aerobic bacteria can preferentially destroy most of the wet gas components resulting in a methane-rich gas (22).

The zone of alteration, in which mixing and oxidation affect the coalbed gas composition, can extend from depths of a hundred meters to a kilometer below the surface. The zone is usually restricted to the margins of basins, however it can extend into and throughout a basin under certain conditions. The primary controls of the depth and lateral extent of the zone of alteration are the physical nature of the coal beds, burial history, and hydrology. Thick, laterally continuous, permeable coal beds are conducive to the development of an aquifer system. Uplift and erosion result in the depressurization of the coal beds and subsequent degassing of the original gas. In addition, uplift and erosion result in reduced stress and increased permeability, thereby aiding both aquifer development and degassing. Finally, microorganisms become widely distributed in active aquifer systems leading to widespread microbial activity, including aerobic oxidation and anaerobic methanogenesis.

In addition to hydrocarbon gases, carbon dioxide is the other significant component of coalbed gases. As previously stated, major quantities of carbon dioxide are generated by thermal decarboxylation or devolatilization of coal, particularly prior to the main stage of thermogenic methane production (Fig. 3). However, carbon dioxide is highly soluble in water and very reactive, and can have different origins (12). As a consequence, the carbon dioxide presently in a coalbed gas may have an origin other than devolitilization. In addition, the concentration and isotopic composition ($\partial^{13}C$) of carbon dioxide in coalbed gases is commonly different from those in gases of adjacent reservoirs, suggesting different origins. The wide range in concentrations and isotopic composition ($\partial^{13}C$) of carbon dioxide suggests that significant amounts of carbon dioxide in coalbed gas are not commonly a product of the coalification process, but rather of other processes active in localized areas. Documented sources of large amounts of carbon dioxide in coalbed gas are (1) thermal destruction of carbonates, (2) bacterial degradation of organic matter, (3) bacterial hydrocarbon oxidation, and (4) migration from magma chambers or deep crust (9).

GAS STORAGE

One of the most important characteristics of coalbed gas reservoirs that set them apart from more conventional gas reservoirs is the amount of and manner in which gas is stored. In conventional reservoirs, such as sandstones or carbonates, gas occurs as either a free or dissolved phase. While some free and dissolved gas may occur in the coal, as much as 98% of the gas is sorbed in the coal (23). The free and dissolved gas in the coal occurs within the fractures (cleats) and pores, whereas the sorbed gas occurs as a monomolecular layer on the internal surfaces of coal (24,25). Because coal is a microporous solid with large internal surface areas (10's to 100's of m^2/gm), it has the ability to sorb very large amounts of gas and can hold much more gas than the same rock volume of a conventional reservoir. In addition, the microstructure of coal has been described as a molecular sieve or cage (26) within which the methane molecule can be stored.

In general, gas content increases with increasing rank, although there is a wide range of gas contents within each coal rank (Fig. 6). Low-rank coals contain as much as 2.5 cc/g, whereas higher rank coals contain as much as 31

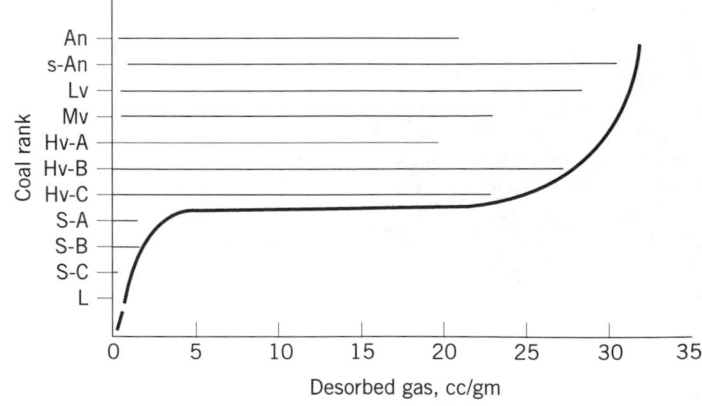

Figure 6. Cross-plot showing range of gas contents versus coal rank. Data from measurements of desorbed coalbed gas in the United States. From R. R. Charpentier and B. E. Law, 1992.

cc/g. The large increase in gas content between the ranks of sub-bituminous A and high-volatile C bituminous is mainly due to the larger amounts of methane generated by thermal processes associated with the coalification of high-rank coals relative to the smaller volumes of methane generated by only microbial processes at lower ranks. At higher ranks (medium-volatile bituminous and higher), coals may have generated more methane than they can store resulting in possible expulsion into adjacent reservoirs (11,27).

Pressure and temperature also play an important role in the gas content of coal. For a given rank, sorbed gas content increases with increasing pressure and decreases with increasing temperature (11). Because high pressures and temperatures are more commonly associated with high-rank coals than low-rank coals, relatively high gas contents are generally expected in high rank coals. In some cases, however, natural desorption of gas facilitated by uplift and erosion in the subsurface environment may result in unexpected low gas contents, thereby accounting for some of the wide variation of gas content within a given coal rank (Fig. 6).

Regardless of rank, nearly all coals contain at least some water. For practical purposes, moisture content of coal is defined as the water present as H_2O and released at temperatures of 104° to 110°C (28). Water in coal occurs as free water, water of decomposition, and water of hydration. Free water occurs in fractures and macropores, water of decomposition is bound to oxygen-containing functional groups (—OH, —COOH, —C, =O) by hydrogen bonding (29), and water of hydration is found in accessory minerals, such as gypsum and clay (30). Some of this water competes with methane for sites of sorption. The amount of water sorbed in a coal is referred to as the critical moisture content and is dependent on the oxygen content of the coal (31). Moisture contents above the critical moisture content do not affect the sorption of methane. Low-rank coals have larger amounts of oxygen than high-rank coal and therefore can sorb more water than high-rank coal.

The amount of inorganic material and the maceral composition of the coal are additional factors affecting the gas sorption capacity of coal. There is no affinity for gas to sorb onto the surfaces of inorganic matter. Inorganic matter occupies space that otherwise would be filled with organic matter, thereby reducing the surface area available for gas sorption (32). As a consequence, coals of equal rank with relatively high ash content will not contain as much gas as low-ash coals. The gas storage capacity of hydrogen-rich, high-volatile bituminous coals is reduced because of the generation of high-molecular weight hydrocarbons (33). These hydrocarbons have the effect of plugging the microporosity, thus reducing the storage capacity. The sorption capacity is increased again at higher ranks (low-volatile bituminous and higher) when these heavier hydrocarbons are thermally cracked.

DETERMINATION OF GAS CONTENT

The most common method of determining the volume of gas contained in coal is by direct measurement from core

or drill-cutting samples retrieved from a well during drilling. The direct method has three components of measurement—desorbed gas, residual gas, and lost gas. The desorbed gas component is derived by placing core or drill-cuttings samples in a sealed canister immediately after the samples arrive at the surface. Periodic measurements of the desorbed gas are recorded over a span of several weeks to months until there is virtually no more gas desorbing from the coal. The coal is then removed from the canister, crushed to a very small size, and the amount of evolved or residual gas is measured. The lost gas component represents the amount of gas lost from the sample from the time the sample was penetrated by the drill bit to the time the sample was placed in a canister. It is calculated graphically by the linear extrapolation of the square root of time versus the cumulative volume of desorbed gas. The sum of the three components then constitute the calculated amount of gas contained in the coal sample. In general, the measured values tend to be less than the actual gas contents as revealed by pressure cores (34). Although there are several sources of potential error in measuring the gas content, particularly in calculating the lost gas amount, the results are usually representative of the relative volume of gas contained in a coal. Detailed procedures for measuring gas content are provided by several workers (35–37).

The sorption capacity of a coal is commonly determined from a sorption isotherm (Fig. 7), which provides the sorbed gas content as a function of the free gas pressure at a constant temperature (38). The sorption isotherm is a measure of the maximum amount of gas that a coal can sorb. However, in many cases, coals are undersaturated with respect to gas and the sorption isotherm, in conjunction with direct measurement of gas content, can be used to determine if a coal is gas saturated. Common reasons for the undersaturation of gas are miscalculation of gas content, laboratory errors in measuring sorption isotherm, degassing of coal beds resulting from uplift and

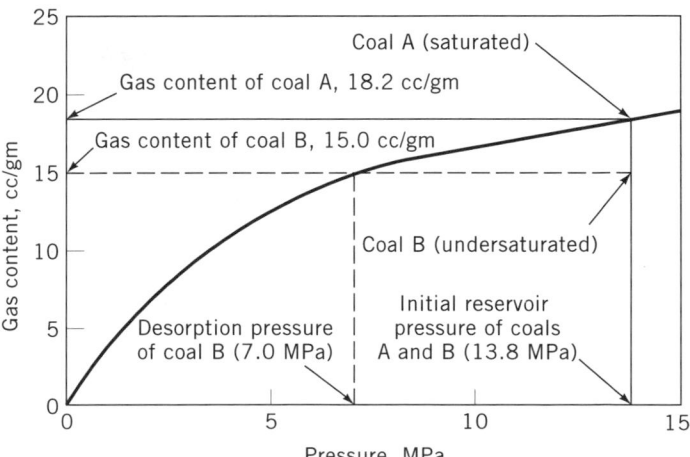

Figure 7. Idealized coalbed gas sorption isotherm showing relations between reservoir pressure and gas content for a saturated (A) and undersaturated (B) coal with respect to gas. Heavy solid line indicates the maximum amount of gas that can be sorbed at reservoir temperature and pressure (39).

erosion, and differential generation of coalbed hydrocarbons resulting from different coal composition.

GAS PRODUCTION

Economic quantities of gas can be produced from coalbed reservoirs. The diagrammatic sketch in Figure 8 shows the path that methane molecules, and other gases, must take in order to be produced. From the sites of gas sorption within the matrix, the gas must desorb and migrate by diffusion out of the matrix into the cleats where it then flows to the wellbore. The coalbed gas is held in the microstructure of the coal by pore pressure and gas migration takes place when the pressure is reduced. The sorption isotherm, in conjunction with direct measurement of gas content, can provide information on the production potential of a coalbed reservoir. For example, coal A, as shown in Figure 7, contains 18.2 cc/g at a reservoir pressure of 13.8 MPa. Because this coal is saturated with methane, gas will immediately begin desorbing from the matrix when the coal is penetrated by the drill bit and the pore pressure begins to drop. In contrast, coal B is undersaturated with methane with only a measured content of 15.0 cc/g, and methane will not desorb at the present reservoir pressure of 13.8 MPa. Significant gas desorption will only begin after reducing the reservoir pressure by about 6.8 MPa to 7.0 MPa. Because most coals are characterized by high water saturations, depressurization results from dewatering of the coal beds.

The development of permeability is a critical aspect of gas production from coalbed reservoirs. The absolute permeability of coal beds ranges from less than 0.1 millidarcy (md) to more than 100.0 md. Commonly, permeability in coalbed reservoirs is in the range of 1.0 to 10.0 md. For practical purposes, there is essentially no permeability

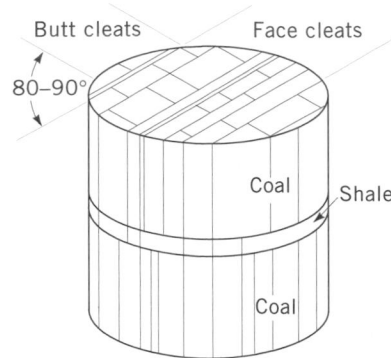

Figure 9. Diagram showing relation between face and butt cleats in coal and carbonaceous shale. Note that the frequency of cleats in carbonaceous shale is less than in coal. Modified from ref. 41.

within the coal matrix. Virtually all of the permeability in coal occurs within the fracture system, commonly referred to as the cleat system. Without a well-developed cleat system, commercial gas production from coal beds is not possible. Cleats constitute a roughly orthogonal set of fractures referred to as face and butt cleats (Fig. 9). Face cleats are the dominant, more extensive set and butt cleats are the subordinate set. Due to the better development of face cleats, permeability in coal beds commonly exhibits varying degrees of anisotropy, with the better permeability developed parallel to the face cleat direction (42). The factors that affect the permeability of cleats include cleat frequency, connectivity, and aperture width. These factors are, in turn, affected by bed thickness, coal quality, rank, tectonic deformation, and stress (43–45). The preservation of cleats is largely dependent on the severity of the structural history. Interestingly, current coalbed gas production is from relatively structurally undeformed areas where the cleat system is preserved. In areas of severe structural deformation, the cleat system is commonly obliterated (46). Thus, the economic production of gas in structurally complex regions may be severely hindered, even though the coals may contain large amounts of gas.

One of the largest obstacles to the economic recovery of coalbed gas is water. The presence of water in coal inhibits desorption of gas from the coal matrix and flow to the wellbore, particularly in coals undersaturated with gas (Figures 7 and 8). Coal beds are commonly dewatered by drilling several wells and producing water by pumping to a point at which gas begins to desorb from the matrix. The period of time necessary to accomplish sufficient dewatering is highly variable (days to months) and in some cases is never reached. The success of dewatering efforts is dependent on the source of the water and the number of wells involved in the dewatering process.

Sources of water in coal may include inherent moisture, water of meteoric origin, and water from adjacent aquifers. First, nearly all coals contain some inherent moisture which decreases with increasing degree of coalification; lignite commonly contains as much as 70% water which decreases to less than 5% in medium-volatile bituminous and higher rank coals (26). Second, in areas

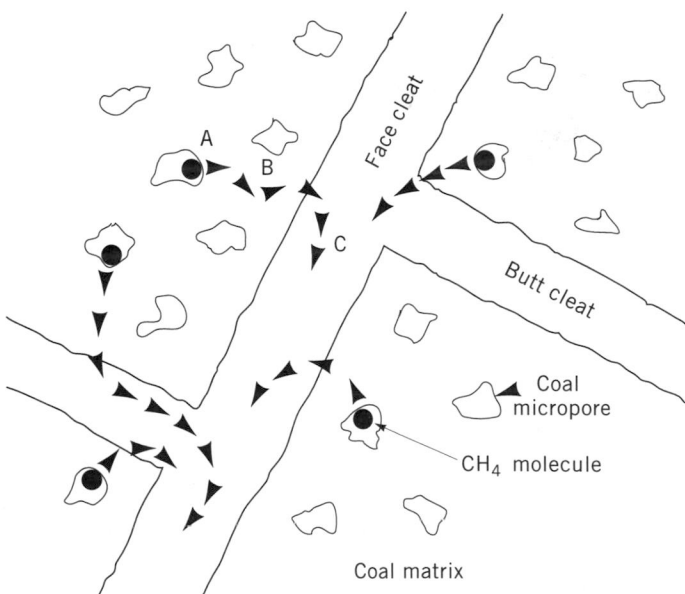

Figure 8. Diagram showing (A) desorption of methane from micropore in coal matrix, (B) diffusion path of methane through coal matrix, and (C) flow of methane in cleats (40).

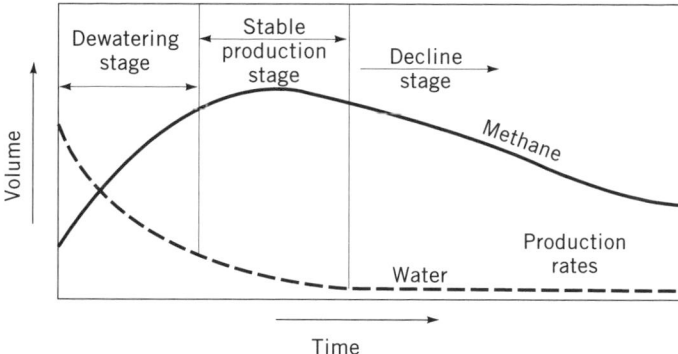

Figure 10. Generalized production history showing volumes of methane and water produced over time for a typical coalbed gas well (48).

where coal beds are relatively permeable and continuous, and precipitation is abundant, atmospheric water may enter the coal beds at the surface and travel great distances into the subsurface. Differences in elevation between the surface and subsurface may cause abnormally high formation pressures (overpressured). This artesian overpressuring is developed in the northern part of the San Juan basin which is characterized by high rates of both water and gas production from the coalbed reservoirs (47). Finally, communication may exist between a coal bed and adjacent aquifers because of natural or man-related causes.

In areas where water is the pressuring phase, dewatering simultaneously reduces the reservoir pressure and the water content, thereby allowing gas to desorb from the coal. In areas where the coalbed reservoir is discontinuous and/or has low permeability and the supply of water is limited, the probability of conducting successful dewatering programs is favorable.

As a consequence of the gas storage characteristics of coal and the relation between water and gas production, coalbed gas wells exhibit a distinctive production history (Fig. 10). In the early dewatering stage of production, large amounts of water are produced along with small amounts of gas until sufficient water is produced to allow for larger amounts of gas to desorb from the matrix and flow through the cleats to the wellbore. During the stable production stage, the quantities of gas increase as the quantities of water decrease. This coalbed gas production during the stable production stage is referred to as "negative decline" and distinguishes it from production from other types of reservoirs. Finally, in the declining stage of production, the amount of gas gradually declines and water production remains low.

In nearly all cases, however, coalbed gas wells require special drilling or completion techniques to effectively connect the reservoir to the wellbore. Both vertical and horizontal wells have been utilized, although vertically drilled wells are far more common than horizontally drilled wells, except in some of the coal mining areas. The completion techniques range from hydraulic fracturing of cased holes to open holes, some with enlarged cavities (49). Although open-hole cavity completions in the main production fairway of the San Juan basin have yielded

the best results (as much as $179 \times 10^3 m^3$ per day), the completion technique in each area must be tailored to the localized coalbed reservoir characteristics, such as rank, pressure, permeability, and gas content.

ENVIRONMENTAL CONCERNS

Methane Emissions From Coal Mining

Methane is a strong infrared absorber and important greenhouse gas. According to recent estimates, atmospheric methane accounts for approximately 15% of the "radiative forcing" added to the atmosphere during the past decade (50). On an atomic basis, methane contributes 25 times more radiative forcing than carbon dioxide, and 70 times more on a weight basis (50). Methane has a short residence time in the atmosphere of 8 to 12 years compared to more than 200 years for carbon dioxide. This means that warming associated with methane emissions is realized in the first few decades after its emissions, whereas the warming from carbon dioxide takes place gradually over centuries. The short residence time indicates that reductions in methane emissions will have noticeable short-term impacts on atmospheric concentrations and that smaller reductions will be needed to stabilize atmospheric concentrations.

Currently, the concentration of methane in the atmosphere is about 1,700 parts per billion (ppb) (51). Evidence from analysis of gases trapped in ice cores indicates that atmospheric methane has more than doubled over the past 200–300 years from a preindustrial range of 600–700 ppb and has increased at a rate of about 1% per year over the past 15 years or about 28 to 45 teragrams (Tg) per year (52).

This increase in atmospheric methane is correlative with human population growth and human activities. The principal anthropogenic sources of methane are rice paddies, domestic livestock, landfills, biomass burning, venting of natural gas, losses of natural gas during transmission of gas and petroleum, and coal mining (Fig. 11) (50). These human activities account for about 60% of the total global emissions that are estimated to be in the range of

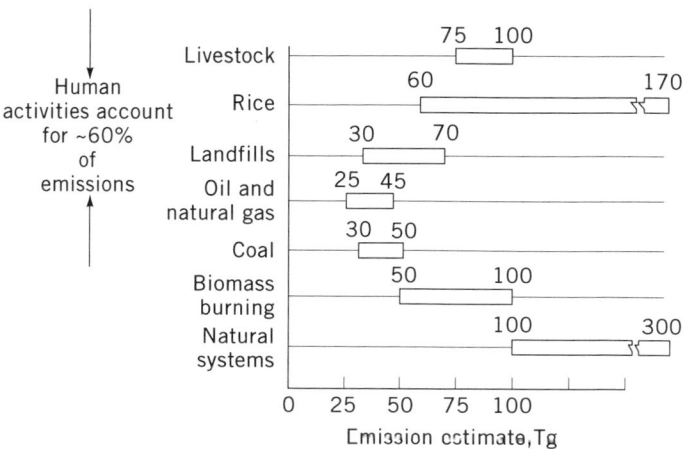

Figure 11. Principal sources of methane emissions, in teragrams (Tg) per year (52).

440 to 640 Tg per year. The methane reductions required to stabilize atmospheric concentrations by the year 2000 are on the order of 25 to 50 Tg (50). These reductions are much lower than those required to stabilize other greenhouse gases, such as carbon dioxide and nitrous oxide. Coal mining probably accounts for about 10% of the total human-related emissions; however, actions resulting in reductions of about 10% are predicted to be economically recovered and utilized (52).

Pressure reduction during coal mining results from the removal of overburden and dewatering. This pressure reduction leads to the desorption of methane, dominant in coalbed gas at shallow depths, and possible emissions to the atmosphere. In general, more methane is released from underground mines than surface mines because of the relation between methane storage, depth, and pressure. In underground mines, emission rates are variable depending on factors, such as the mining method (room and pillar versus longwall) and age of the mine (53). Methane emissions also continue during the transport, processing, and storage of the coal. In 1987, underground mining accounted for about 40% of the coal production in the United States and about 45% of the world-wide production (53). This methane release is not only an environmental concern, but also a serious safety hazard because methane is highly explosive at relatively low concentrations of 5 to 15%. In the United States and some other countries, methane levels in the mines are kept at very low levels by installation of high-grade ventilation systems.

Annual emissions of methane from coal mining have been estimated to be about 30 to 50 Tg (Fig. 11) (50). However, currently available data are insufficient to satisfactorily assess the relative source strength of coalbed methane; current global estimates contain only "order of magnitude" levels of reliability. Even though methane emissions from mining ventilation systems are measured in the United States, estimates of the total methane flux from coal mines include broad assumptions and contain a high degree of uncertainty. Estimates of methane emissions from coal mines in countries other than the United States contain greater uncertainty because even less data are available. In some cases, global estimates of methane flux from coal mines are based solely on data from United States mines which have been extrapolated to a global scale (54).

Based on 1987 coal production data and using U.S. mines as an analog, methane emissions from worldwide coal mines are estimated to be in the range of 33 to 64 Tg (54). More than 90% of the total emissions are from the 10 primary coal producing countries shown in Table 2. About 75% of the total estimated methane emissions from coal mining are from four countries—China, the former Soviet Union, United States, and Poland; China, by itself, accounts for about 34% of the world's annual emissions (Table 2) (53). Because of increasing energy demand, particularly from the developing countries and from the growth in the world's population, coal production is forecast to grow substantially over the next decade and coal mining could become an even greater source of atmospheric methane.

Table 2. Estimated Methane Emissions in Teragrams from Principal Coal-Producing Countries[a]

Primary Coal Producing Countries	Estimated Methane Emissions, Tg
China	16.1
Former U.S.S.R.	8.3
United States	7.0
Poland	3.4
Germany	2.1
South Africa	2.1
India	1.7
United Kingdom	1.6
Australia	1.1
Czech Republic	0.6
Total	*44*

[a] Ref. 54.

Coalbed gas can be recovered before, during, or after mining and the quality and quantity of the recovered methane will depend on the method used. Ventilation is the primary method of controlling methane levels in most mines and fresh air is typically circulated across the actively mined coal face (53). Sometimes, methane emissions can be controlled using ventilation alone, but commonly this method has to be supplemented. Other recovery methods include: vertical wells; gob wells which remove gas released from rocks above and below the coal bed during and after mining; and horizontal and cross-measure boreholes (54). Because large amounts of air are mixed with methane during ventilation, this gas mixture is useful only for combustion air in a gas turbine or coal-fired boiler. Typical uses for gases recovered by wells and boreholes, which generally have higher percentages of methane, are pipeline injection and power generation.

An excellent example of the economic recovery of methane from coal mining operations, resulting in reduced emissions, comes from the Black Warrior basin of southeastern Alabama (Fig. 2) (55). A coal mining company that has captured mining-related methane, operates gassy mines at depths of 396 to 640 m; the methane content of the coal beds range from 15.6 to 18.7 cm^3/g and presents a hazard to miners. Initially, the methane had been vented to the surface and later, the methane had been recovered and sold successfully, improving both mine safety and overall productivity. Daily production of methane averages about $1 \times 10^6 m^3$ per day which represents about 35% of the methane released during the mining process. About 83% of the gas is recovered from gob wells; the remaining 17% is from hydraulically stimulated vertical wells and horizontal drainage wells which recover methane before mining.

Methane release related to coal mining can also cause safety hazards in nearby areas as documented in Powder River basin, Wyoming, in the United States (Fig. 2) (56). On the east flank of the Powder River basin, thick, subbituminous Tertiary coal beds are surface mined from some of the most economically important deposits in the United States. Due to dewatering operations, the ground water levels have fallen as much as 24 meters in wells near the active mines and up to 3 meters as much as 5 kilometers away from the mines (57). This lowering of the ground-water level has resulted in desorption of the gas from the coal matrix in the area surrounding the surface coal mine. The desorbed gas has migrated within the coal to a later-ally continuous anticline near the mine to form a commer-cial coalbed gas accumulation with only minor amounts of water. In addition, stress fractures and faults on the anticline have formed conduits for the gas to migrate out of the coal beds and upward to the surface and into base-ments of nearby houses, causing an explosion danger.

Water Disposal

Water, commonly in large quantities, is produced from coalbed gas wells, especially during the early stages of gas production (Fig. 10). The disposal of this produced water cannot only affect the economics of the development of the coalbed gas, but also pose serious environmental con-cerns. In the United States, water disposal is controlled by Federal, State, and/or local agencies. The disposal of produced waters will also be an environmental and eco-nomic concern in other countries as they develop their coalbed gas resources.

The most economic methods of water disposal from coalbed gas reservoirs are discharge into streams, direct use for agricultural, wildlife, or industrial purposes, and evaporation. In addition, produced coalbed waters can be directly used in oil and gas fields as hydraulic fracture fluids and for enhanced oil recovery projects. More costly alternatives of disposal are underground injection and surface discharge after treatment, by methods such as distillation, dilution, ion-exchange, reverse osmosis, and electrodialysis. The specific method of disposal will de-pend on the quality and quantity of the water. Problems and methods of water disposal are discussed in refs. 58 and 59.

The quantity and quality of water in coal-bearing ba-sins vary considerably and reflect the water's residence time and rock-water interactions in the subsurface and groundwater flow patterns. Total dissolved solids (TDS) of produced coalbed water range from less than 200 mg/L to more than 90,000 mg/L with most values less than 30,000 mg/L (47,60). TDS and salinity tend to increase with the water's residence time in the subsurface and its increased time for rock-water interactions. As a conse-quence, TDS and salinity are generally higher in the cen-tral, deeper parts of basins as compared with the shal-lower parts of basins near recharge areas of relatively fresh meteoric water. Typically produced water is neutral to slightly alkaline with pH values ranging from 7.0 to 8.5 and contains the ions carbonate, bicarbonate, chloride, sulfate, sodium, potassium, calcium, and magnesium in variable amounts. If enough chemical data are available, chlorinity maps and different water types can be used to map groundwater flow patterns (47,60). For example, con-centrations of chloride, which are unaffected by rock-wa-ter interactions, indicate directions of groundwater flow with low chloride waters occurring in recharge areas of relatively fresh water circulation.

Production rates of water from a well can be as high as 950 m³ per day. However, most commercial coalbed gas wells produce less than 40 m³ per day after the initial dewatering phase. By water-well standards, most coalbed gas wells are low-yield wells. Higher rates of production are generally near areas of recharge where the waters are relatively fresh (low TDS); low rates of water production are in deeper parts of basin and/or areas of low perme-ability where the water is commonly very saline. Al-though the volume of water from an individual coalbed gas well is generally low as compared to a water well, the cumulative volume of produced water can be enormous when a large number of wells are drilled to develop the resource. For example, about $15 \times 10^6 m^3$ of water have been produced from more than 2,800 wells in the south-eastern part of the Black Warrior basin in 1991 (Fig. 2). Because of the humid climate, most of this water has been discharged into streams without noticeable environmen-tal consequences. In comparison, about $6 \times 10^6 m^3$ of water have been produced from more than 1700 wells in the San Juan basin during the same time. Because of the quality and quantity of the water and the dry climate of this re-gion, most of this water has been disposed in under-ground injection wells, a more costly method which was offset by higher gas production rates from individual wells than those in the Black Warrior basin.

Aquifer Contamination by Coalbed Gas

In the San Juan basin (Fig. 2), the upsurge of coalbed gas development and production has been partially contempo-raneous with the contamination of an alluvial aquifer by natural gas in a populated area along a river valley (61). Based on water analyses (measurements of organic va-pors), pressure testing of gas-producing wells, and molec-ular and isotopic composition of gas samples, some of the near-surface gas was determined to be coalbed gas which migrated vertically from a deeper interval (about 1000 meters). The coalbed gas has been interpreted to have been released by depressurization after the drilling of gas-producing wells, some completed in reservoirs below the coal beds, and the dewatering of coalbed-gas wells. The coalbed gas then migrated upward from the coal bed into the overlying strata via the uncemented portions of producing gas wells that penetrate the gas-bearing coal beds. The gas also escaped through leaks in the casing. Once inside the alluvial aquifer, the gas sometimes in-vaded the cathodic protection wells and the domestic wa-ter wells.

The State regulatory agency discovered that many gas wells were not cemented through the shallow alluvial aq-uifer and/or opposite the gas-bearing coal beds. To pre-vent further migration, the State and Federal regulatory agencies ordered operators to apply casing cement to all existing and new gas wells that penetrate the coal beds. In addition, they have continued to conduct pressure tests on producing wells in areas where gas wells or cathodic protection wells are blowing out. The San Juan basin is the first basin where large-scale development of coalbed gas has taken place subsequent to gas production from other types of reservoirs. Similar aquifer contamination problems might happen in other areas under similar cir-cumstances.

Environmental Impact Studies

As demonstrated by experiences in the United States, mainly in the Black Warrior and San Juan basins (Fig. 2), large-scale development of coalbed gas resources can strain a government and its regulatory agencies to adequately address the full range of environmental concerns. For example, rapid development of coalbed gas resources in the United States has taken place because of a production tax credit which had a time limit. Many environmental concerns, such as the disposal of water and aquifer contamination, have not been considered and/or accounted for until problems arose. Multdisciplinary studies of all anticipated environmental concerns should be incorporated into a development plan prior to any development of the coalbed gas resource, and monitored and modified throughout the various stages of development of the resource. These studies should consider factors such as the siting of well pads, access roads, pipelines for both water and gas, and facilities for disposal of water. Finally, operators and their planners should be aware of the environmental regulations and concerns and incorporate the use of sound environmental practices into their project plans. An example of environmental planning is given from the Black Warrior basin in ref. 62.

ENVIRONMENTAL IMPACTS OF COAL MINING AND COMBUSTION VERSUS COALBED GAS DEVELOPMENT AND UTILIZATION

The environmental impacts of the mining and combustion of coal, as documented in Poland (63), appear to be more serious than those associated with the development and utilization of coalbed gas. First of all, coal mining results in large amounts of solid waste, only a small part of which is commonly utilized. In addition, the following water resource problems are associated with coal mining: (1) subsidence above and near mine areas can change local hydraulic gradients and drainage basin limits, and create numerous ponds; (2) drainage of fresh water aquifers can lower the water table and result in dry water wells; (3) leaching of coal waste piles can contribute to chlorides, sulfates, and heavy metals; and (4) highly mineralized mine waters can be discharged to streams. The combustion of coal commonly results in emissions of sulfur dioxide, nitrogen oxides, volatile organic compounds, such as benzene and toluene, carbon dioxide, particulate matter, and heavy metals.

In comparison, the main environmental concerns about the development of coalbed gas are the disposal of large amounts of production water. When burned, coalbed gas emits none of the sulfur dioxide or particulates associated with coal burning, and much less nitrogen oxides, carbon dioxide, and volatile organic compounds (64).

CONCLUSIONS

Large amounts of methane-rich gas are generated and stored in coal beds. Recently, commercial production of coalbed gas has been established in the United States with development to be extended to other countries as they become aware of the resource potential and as technology is transferred. Coalbed gas can be utilized as an energy source which (1) is environmentally more acceptable than the mining and combustion of coal, (2) can partly replace coal as a fossil energy source, and (3) sometimes occurs where other conventional hydrocarbon resources are not present. Environmental concerns about coalbed gas are related mainly to methane emissions from mining and the disposal of water produced along with the gas. In reference to emissions, which result in atmospheric warming and also create a hazard to miners, significant amounts of the methane-rich gas can be captured prior to and during mining and utilized. Although sometimes costly, produced water can be disposed of in an environmentally acceptable manner, such as discharge to streams or injection wells.

BIBLIOGRAPHY

1. V. A. Kuuskraa, C. M. Boyer II, and J. A. Kelafant, *Oil and Gas Journal* **90**(40), 49–54 (1992).

2. J. R. Kelafant, S. H. Stevens, and C. M. Boyer, II, *Oil and Gas Journal* **90**(44), 80–85 (1992).

3. C. T. Rightmire, G. E. Eddy, and J. N. Kirr, eds., *Coalbed Methane Resources of the United States*, American Association of Petroleum Geologists Studies in Geology No. 17, Tulsa, Okla.,1984.

4. ICF Resources, *Quarterly Review of Methane from Coal Seams Technology* **7**, 10–28 (1990).

5. J. G. Bromilow and J. M. Jones, *Colliery Engineering* **32**, 222–232(1955).

6. C. R. Woese, O. Kandler, and M. L. Wheelis, *Proceedings of the National Academy of Science (U.S.)* **87**, 4576–4579 (1990).

7. D. D. Rice and G. E. Claypool, *American Association of Petroleum Geologists Bulletin* **65**, 5–25 (1981).

8. D. D. Rice, "Controls, Habitat, and Resource Potential of Ancient Biogenic Gas," in R. Vially, ed., *Bacterial Gas*, Editions Technip, Paris, 1992, pp. 91–118.

9. D. D. Rice, "Composition and Origins of Coalbed Gas," in B. E. Law and D. D. Rice, eds., *Hydrocarbons from Coal*, American Association of Petroleum Geologists *Studies in Geology*, no. 38, Tulsa, Okla., 1993, pp. 159–184.

10. E. Stach, M.-Th., Mackowsky, M. Teichmuller, G. H. Taylor, D. Chandra, and R. Teichmuller, *Stach's Textbook of Coal Petrology*, 3rd ed., Gebruder Borntraeger, Berlin, 1982.

11. H. Juntgen and J. Karwell, *Erdol Kohle, Erdgas, Petrochem* **19**, 251–258, 339–344 (1966).

12. J. M. Hunt, *Petroleum Geochemistry and Geology*, W. H. Freeman, San Francisco, Calif., 1979.

13. J. R. Levine, "Influence of Coal Composition on the Generation and Retention of Coalbed Natural Gas" in *The 1987 Coalbed Methane Symposium Proceedings*, 1987, pp. 15–18.

14. P. Bertrand, F. Behar, and B. Durand, *Organic Geochemistry* **1–3**, 601–608 (1986).

15. J. D. Saxby and M. Shibaoka, *Applied Geochemistry* **1**, 25–36 (1986).

16. G. K. Khorasani, "Oil-Prone Coals in the Walloon Coal Measures, Surat Basin, Australia," in A. C. Scott, ed., *Coal and Coal-Bearing Sequences-Recent Advances*, Geological Society Special Publication 32, London, 1987.

17. R. W. Stanton, D. D. Rice, J. L. Clayton, and R. M. Flores, "Matrix-Gel Vitrinite Types and Rock-Eval Analysis of Coal Samples, Cretaceous age, from the San Juan and Piceance

Basins," in *Ninth Annual Meeting of the Society for Organic Petrography*, Abstracts and Agenda, 1992, pp. 57–58.

18. J. L. Clayton, "Composition of Crude Oils Generated and Expelled from Coal and Coaly Organic Matter Dispersed in Shales," in Ref. 9, pp. 185–201.

19. H. M. Chung, J. R. Gormly, and R. M. Squires, *Chemical Geology* **71**, 97–103 (1988).

20. C. Clayton, *Marine and Petroleum Geology* **8**, 232–240 (1991).

21. A. T. James, *American Association of Petroleum Geologists Bulletin* **74**, 1441–1458 (1990).

22. A. T. James and B. J. Burns, *American Association of Petroleum Geologists Bulletin* **68**, 957–960 (1984).

23. I. Gray, "Reservoir Engineering in Coal Seams," in *Society of Petroleum Engineers Preprints*, SPE Paper 12514, 1987.

24. R. F. Selden, *The Occurrence of Gases in Coal*, U.S. Bureau of Mines Report of Investigations 3233, 1934.

25. A. G. Kim, *Estimating Methane Content of Bituminous Coal from Adsorption Data*, U.S. Bureau of Mines Report of Investigations 8245, 1977.

26. D. W. van Krevelan, *Coal (Typlogy, Chemistry, Physics, and Constitution)*, Elsevier, Amsterdam, 1981.

27. F. F. Meissner, "Cretaceous and lower Tertiary Coals as Sources for Gas Accumulations in the Rocky Mountain Area," in J. Woodward, F. F. Meissner, and J. L. Clayton, eds., *Hydrocarbon Source Rocks of the Greater Rocky Mountain Region*, Rocky Mountain Association of Geologists, Denver, 1984, pp. 401–431.

28. American Society for Testing Material, "Moisture in the Analysis Sample of Coal and Coke", *ASTM Standards*, Part 26, 1973.

29. M. S. Iyengar and A. Lahiri, *Fuel* **36**, 286–297 (1957).

30. D. J. Allardice and D. G. Evans, "Moisture in Coal," in C. Karr, Jr., ed., *Analytical Methods for Coal and Coal Products*, Vol. 1., Academic Press, Inc., New York, 1978.

31. J. I. Joubert, C. T. Grein, and D. Beinstock, *Fuel* **52**, 181–185 (1973).

32. J. Gunther, *Revue de l'Industrie Minerale* **47**, 693–708 (1965).

33. J. R. Levine, "The Impact of Oil Formed During Coalification on Generation and Storage of Natural Gas in Coalbed Reservoir Systems," in *The 1991 Coalbed Methane Symposium Proceedings*, 1991, pp. 307–315.

34. D. Yee, J. P. Seidle, and W. B. Hanson, "Gas Sorption on Coal and Measurement of Gas Content," in Ref. 9, pp. 203–218.

35. F. N. Kissell, C. M. McCulloch, and C. H. Elder, *The Direct Method of Determining Methane Content of Coalbeds for Ventilation Design*, U.S. Bureau of Mines Report of Investigations 7767, 1973.

36. D. M. Smith and F. L. Williams, "A New Method for Determining the Methane Content of Coal," in *Proceedings of the 16th Intersociety Energy Conversion Engineering Conference*, 1981, pp. 1276–1272.

37. J. P. Ulery and D. M. Hyman, "The Modified Direct Method of Gas Content Determination: Applications and Results," in *The 1991 Coalbed Methane Symposium Proceedings*, 1991, pp. 489–500.

38. M. J. Mavor, L. B. Owen, and T. J. Pratt, "Measurement and Evaluation of Coal Sorption Isotherm Data," in *Society of Petroleum Engineers Preprints*, 1990, SPE Paper 20728.

39. J. C. McElhiney, G. W. Paul, G. B. C. Young, and J. A. McCartney, in Ref. 9, pp. 361–372.

40. W. Diamond in Ref. 9, pp. 237–267.

41. A. H. Jones, U. Ahmed, D. Bush, M. Holland, S. Kellar, K. Rakop, K. C. Bowman, and G. J. Bell, "Methane production characteristics for a deeply buried coalbed reservoir in the San Juan basin," *Quarterly Review of Methane from Coal Seams Technology*, **2**(1), figure 7, 23 (1984).

42. C. M. McCulloch, M. Deul, and P. W. Jeran, *Cleat in Bituminous Coal*, U.S. Bureau of Mines Report of Investigations 7910, 1974.

43. J. C. Macrae and W. Lawson, *Transactions of Leeds Geological Association*, **VI**, 1954, pp. 227–242.

44. I. I. Ammosov and I. V. Eremin, *Fracturing in Coal*, IZDAT Publishers, Moscow 1963. Translated from Russian by Israel Program for Scientific Translations, Jerusalem.

45. C. R. McKee, A. C. Bumb, and R. A. Koenig, "Stress-Dependent Permeability and Porosity of Coal," in *The 1987 Coalbed Methane Symposium Proceedings*, 1987, pp. 195–205.

46. B. E. Law, "The Relationship Between Coal Rank and Cleat Spacing: Implications for the Prediction of Permeability in Coal," in *The 1993 International Coalbed Methane Symposium Proceedings*, 1993, pp. 435–441.

47. W. R. Kaiser, T. E. Swartz, and G. J. Hawkins, "Hydrology of the Fruitland Formation, San Juan Basin," in W. B. Ayers, Jr. and co-workers, *Geologic and Hydrologic Controls on the Occurrence and Producibility of Coalbed Methane, Fruitland Formation, San Juan Basin*, Gas Research Institute, Chicago, Ill., 1991, Topical Report GRI-91/0072.

48. V. A. Kruuskraa and C. F. Brandenberg, "Coalbed methane sparks a new energy industry," *Oil & Gas Journal*, **87**(41), fig. 5, 53 (1989).

49. I. D. Palmer, S. W. Lambert, and J. L. Spitler, "Coalbed Methane Well Completions and Stimulations," in Ref. 9, pp. 303–339.

50. Intergovernmental Panel on Climate Control, *Methane Emissions and Opportunities for Control*, EPA/400/9-90/007, Japan Environmental Agency and U.S. Environmental Protection Agency, Washington, D.C., 1990.

51. D. R. Blake and F. S. Rowland, *Science* **259**, 1129–1131 (1988).

52. D. W. Kruger, "Coalbed Methane: Environmental Protection at a Profit," in *The 1991 Coalbed Methane Symposium Proceedings*, 1991, pp. 100, 193–202.

53. M. Duel and A. G. Kim, eds., *Methane Control Research: Summary of Results, 1964–1980*, U.S. Bureau of Mines Bulletin 687, 1988.

54. C. M. Boyer II, J. R. Kelafant, V. A. Kuuskraa, K. C. Manger, and D. Kruger, *Methane Emissions from Coal Mining: Issues and Opportunities for Reduction*, U.S. Environmental Protection Agency, Washington, D.C., 1990, EPA/400/9-90/008.

55. R. A. Mills and J. W. Stevenson, "History of Methane Drainage at Jim Walter Resources, Inc.," in *The 1991 Coalbed Methane Symposium Proceedings*, 1991, pp. 143–151.

56. B. E. Law, D. D. Rice, and R. M. Flores, "Coalbed Gas Accumulations in the Paleocene Fort Union Formation, Powder River Basin, Wyoming," in S. D. Schwochow, D. K. Murray, and M. F. Fahy, eds., *Coalbed Methane of Western North America*, Rocky Mountain Association of Geologists, Denver, Colo., 1991, pp. 179–190.

57. L. J. Martin, D. L. Naftz, H. W. Lowham, and J. G. Ranki, *Cumulative Potential Hydrologic Impacts of Surface Coal Mining in the Eastern Powder River Structural Basin, Northeustern Wyoming*, 1988, U.S. Geological Survey Water-Resources Investigations Report 88-4046.

58. G. L. Zimpfer, E. J. Harmon, and B. C. Boyce, "Disposal of Production Waters from Oil and Gas Wells in the Northern

San Juan Basin, Colorado," in J. E. Fassett, ed., *Geology and Coal-Bed Methane Resources of the Northern San Juan Basin, Colorado and New Mexico*, Rocky Mountain Association of Geologists, Denver, Colo., 1988.

59. W. C. Burkett, R. McDaniel, and W. L. Hall, "The Evaluation and Implementation of a Comprehensive Production Water Management Plan," in *The 1991 Coalbed Methane Symposium Proceedings*, 1991, pp. 43–56.

60. J. C. Pashin, *Regional Analysis of the Black Creek-Cobb Coalbed-Methane Target Interval, Black Warrior Basin, Alabama*, Geological Survey of Alabama Bulletin 145, 1991.

61. C. Shuey, "Policy and Regulatory Implications of Coal-Bed Methane Development in the San Juan Basin, New Mexico and Colorado," in *Proceedings of the First International Symposium on Oil and Gas Exploration and Production Waste Management Practices*, 1990, pp. 757–769.

62. B. J. Luckianow, W. C. Burkett, and C. Bertram, "Overview of Environmental Concerns for Siting of Coalbed Methane Facilities," in *The 1991 Coalbed Methane Symposium Proceedings*, 1991, pp. 1–11.

63. R. C. Pilcher, C. J. Bibler, R. Glickert, L. Machesky, J. M. Williams, D. W. Kruger, and S. Schweitzer, *Assessment of the Potential for Economic Development and Utilization of Coalbed Methane in Poland*, U.S. Environmental Protection Agency, Washington, D.C., 1991, EPA/400/1-91/032.

64. *Oil & Gas Journal* **89**(29), 31–32(1991).

COGENERATION

CLEMENS M. THOENNES
General Electric Company
Schenectady, New York

Cogeneration is a technology used by many industries since the beginning of this century as an economic means of providing plant energy requirements. Before facilitating initiatives on the part of various governments (eg, privitization in the UK and the National Energy Act (NEA) in the United States), most systems were developed based on site-specific heat and power demands. The resultant facilities were commonly referred to as by-product power generation, in-plant generation, and in some cases, total energy systems. Presently, *cogeneration* is a term frequently used to describe projects' selling power to electric utility entities. However, power sale contracts are not a prerequisite for cogeneration.

See also ELECTRIC POWER GENERATION; ENERGY EFFICIENCY; ENERGY MANAGEMENT; GAS TURBINES; WASTE-TO-ENERGY.

These initiatives have expanded the potential for cogeneration by providing a mechanism by which cogenerators can sell power to utilities. This permits the user a greater degree of freedom in developing the cogeneration system, because the industrial plant electric power requirement is not a constraint on system design. Furthermore, it has spawned an industry in which entrepreneurial firms actively pursue utility power sales opportunities, with the steam host ensuring the thermal demand necessary for system qualification for the benefits of the initiatives. Frequently large power generation systems are developed with relatively small steam demands.

Before 1960, most applications were developed based on steam turbine cogeneration systems. However, more recently, the economic benefits resulting from high power:heat ratios, the wide range of system integration options, and attractive cogeneration system performance levels have made gas turbines highly desirable prime movers in applications for which suitable fuels are economically available. These characteristics also have been instrumental in the development of many large systems involving power sales to electric utility companies.

COGENERATION

Cogeneration is frequently defined as the sequential production of necessary heat and power (electrical or mechanical) or the recovery of low level energy for power production. The sequential energy production yields fuel savings relative to separate production facilities. With the rapid changes in energy prices in the 1970s, the fuel-conserving aspect of cogeneration became a major driver for the increased interest in this technology.

Power can be cogenerated in topping or bottoming cycles. In a topping cycle, power is generated before the delivery of thermal energy to the process. Typical examples are noncondensing steam turbine cycles, gas turbine heat recovery cycles, and combined cycles in which the gas turbine exhaust energy is ultimately used for process requirements. In bottoming cycles, power is produced from the recovery of process thermal energy that would normally be rejected to the heat sink. Bottoming cycle examples include power generation resulting from recovery of excess thermal energy from exothermic process reactions and heat recovery from kilns, process heaters, and furnaces.

The fuel use effectiveness for a modern coal-fired utility plant and an industrial facility using a noncondensing cogenerating steam turbine generator is illustrated in Figure 1. The figure suggests that the power generation cycle energy losses can be reduced from 65% to 16% of the fuel input through use of cogeneration. In reality, the process becomes the heat sink for the cogeneration cycle, thus minimizing the power cycle energy losses.

Similar performance benefits are also available in gas turbine cogeneration systems. For example, a natural gas–fired MS7001EA gas turbine generator in a combined cycle providing 103.4×10^3 Pa process steam will yield an overall energy effectiveness of about 75% on a higher heating value basis (see Table 1 for GE model designations). This cogeneration system performance is significantly better than typical steam turbine or gas turbine combined cycles that produce power only. Few applications have been developed for which all the thermal energy available from a power cycle has been used in process. Thus the question becomes: How closely can a cycle approach a straight power generation system and still be categorized as cogeneration?

The influence of decreasing the thermal energy to a process from a steam turbine cycle is illustrated in Figure 2. As less steam is delivered to process, the electrical output ratio relative to the electrical output at 100% steam to process increases, becoming a maximum of about 2.0 for the steam conditions noted if no steam is delivered to process. The overall efficiency decreases from 84% to 35% as process steam delivery is eliminated. Similar evaluations for gas turbine cogeneration systems with unfired heat recovery generally yield overall efficiencies in excess

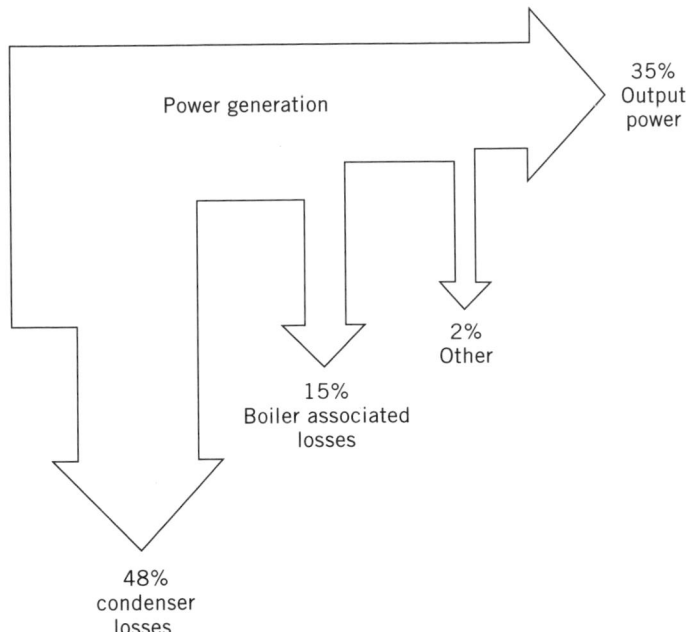

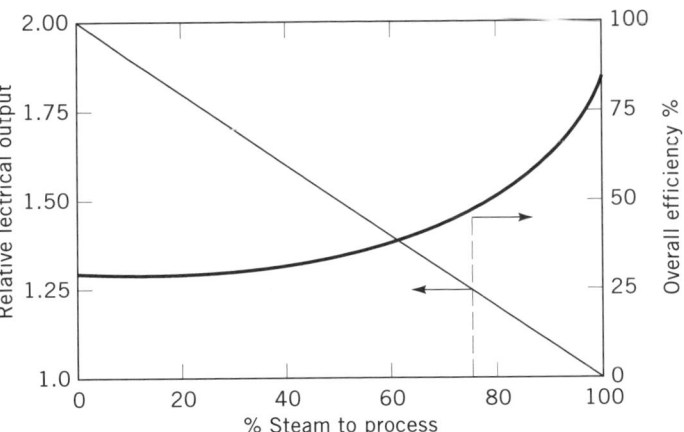

Figure 2. Steam turbine cycle performance at various process steam demands. (1) steam conditions 9998 × 10³ Pa, 510°C, 1034 × 10³ Pa process, 6.35 × 10⁻²m HgA condenser pressure, (2) three stages of feed-water heating, (3) boiler efficiency of 85%.

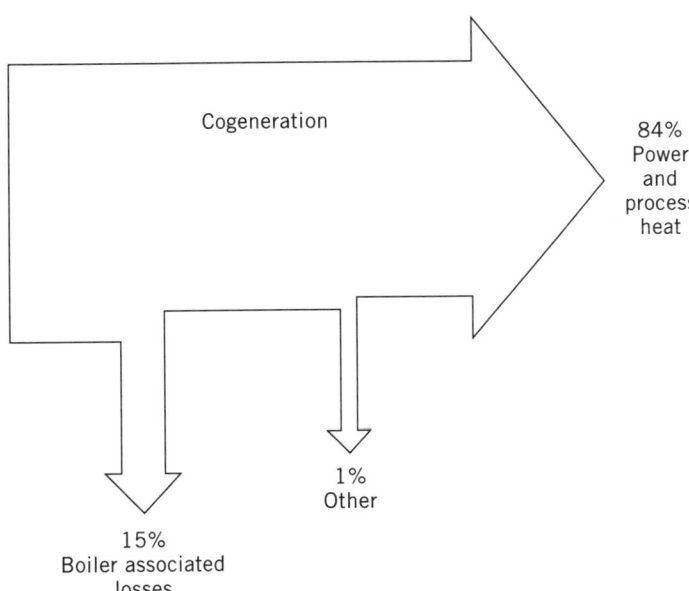

Figure 1. Fuel utilization effectiveness. Basis: (1) modern coal-fired systems, (2) effectiveness on higher heating value of coal.

condensing units operated at minimum flow to the condenser for cooling purposes. Furthermore, gas turbine cycles without steam turbines or including steam turbines with minimum condenser flow are included.

FUEL CHARGEABLE TO POWER—DEFINITION

A parameter used to define the thermal performance of a topping cogeneration system is the fuel-chargeable-to-power (FCP). The FCP is the incremental fuel : power ratio for the cogeneration system relative to the case with which it is being compared (usually a no-cogeneration alternative). For a plant generating electric power only, FCP and net plant heat rate are interchangeable terms. Net plant heat rate in But/kWh is the more commonly used term for plants generating electric power only. The FCP concept is illustrated in Figure 3. Stated in simple terms, the FCP is the total fuel burned in the cogeneration system minus the fuel that would have been required to produce the process heat in the absence of cogeneration (process fuel credit) divided by the gross power generated minus the difference in powerhouse auxiliaries.

of 70% if all thermal energy generated is delivered to process. This is in contrast to about 45% thermal efficiency on a higher heating value basis for a natural gas–fired MS7001EA combined-cycle system.

Because of the wide range of allowable steam flows that fall into the category "cogeneration," it is difficult to establish general guidelines suitable for all applications. Large power generation systems that deliver minimum thermal energy will be similar to straight power generation cycles. Other systems that deliver large quantities of process heat will usually be developed to optimize electrical production, yielding high thermal efficiencies and favorably impacting the economics.

For purposes of the following discussion, *thermally optimized* cogeneration systems are defined as those developed using noncondensing steam turbine generators or

STEAM TURBINE CYCLES

In thermally optimized steam turbine cogeneration cycles, steam is expanded in noncondensing or automatic-extraction, noncondensing steam turbine generators that extract or exhaust into the process steam header(s). The FCP for these systems is typically 4000 to 4500 But/kWh higher heating value (HHV). Steam turbine generators have been produced for industrial and cogeneration applications since the early 1900s, and to date, the leading global supplier has placed more than 5000 units into service around the world. Throughout this period, the performance, reliability, and cost-effectiveness of these turbines have been continuously improved through product and packaging innovations.

To achieve the cost and reliability benefits of standard-

Table 1. GE Gas Turbine Performance Characteristics of Generator Drive Units

Model Number	Fuel[a]	Output (kW)	Heat Rate Btu/kWh (LHV)	Exhaust Flow (kg/h)	Exhaust Temperature (°C)	Frequency (Hz)
PG5271(RA)	G	20260	12820	354,262	520	
	D	19940	12920	355,179	521	
PG5371(PA)	G	26300	11990	446,796	487	50 and 60
	D	25800	12070	448,157	488	
PG6541(B)	G	38340	10860	500,774	539	50 and 60
	D	37520	10970	502,135	539	
PG7111(EA)	G	83500	10480	1,066,414	530	60
	D	82100	10560	1,609,589	530	
PG7221(FA)	G	159000	9500	1,536,343	589	60
	D	144800	9580	1,540,879	591	
PG9171(E)	G	123400	10100	1,476,922	538	50
	D	121300	10170	1,481,004	539	
PG9311(FA)	G	226500	9570	2,212,207	589	50
	D	222000	9650	2,219,011	591	
LM6000(PA)	G	39970	8790	445,571	449	60
	D	39920	8850	445,481	458	
LM6000(PA)	G	39170	8960	445,571	449	50
	D	39120	9030	445,481	458	

[a] G, gas; D, distillate.

ization without compromising turbine performance, contemporary suppliers have adopted a modular product structure for their line of industrial turbines. This structure enables optimization of the turbine configuration for a customer's specific operating conditions. Preengineered and field-proven components are selected from an array of alternative offerings, and a custom steam path is designed that satisfies the unique requirements of the application. Component modules making up the "building blocks" of the product line include front bearing standards, inlet sections, valve gear, extractions, and exhaust sections. These modules are shown for a typical turbine in Figure 4. Note that the barrel section of the turbine is custom designed for each unit based on the user's specific operating conditions. By using this flexible modular structure, maximum reliability and performance are achieved while product cost and delivery cycle are minimized. Development efforts associated with this product line center on the development of new and improved component modules to replace or augment the existing component modules.

Figure 5 illustrates the type of turbines most frequently used in industrial and cogeneration applications. Figures 5a through 5d show noncondensing designs that exhaust to a header from which the steam is used for process or supply to a lower pressure turbine. Figures 5e through 5h represent condensing units that exhaust at the lowest pressure obtainable using water or air as a heat sink. Figures 5a and 5e illustrate straight noncondensing and straight condensing turbines, which are sim-

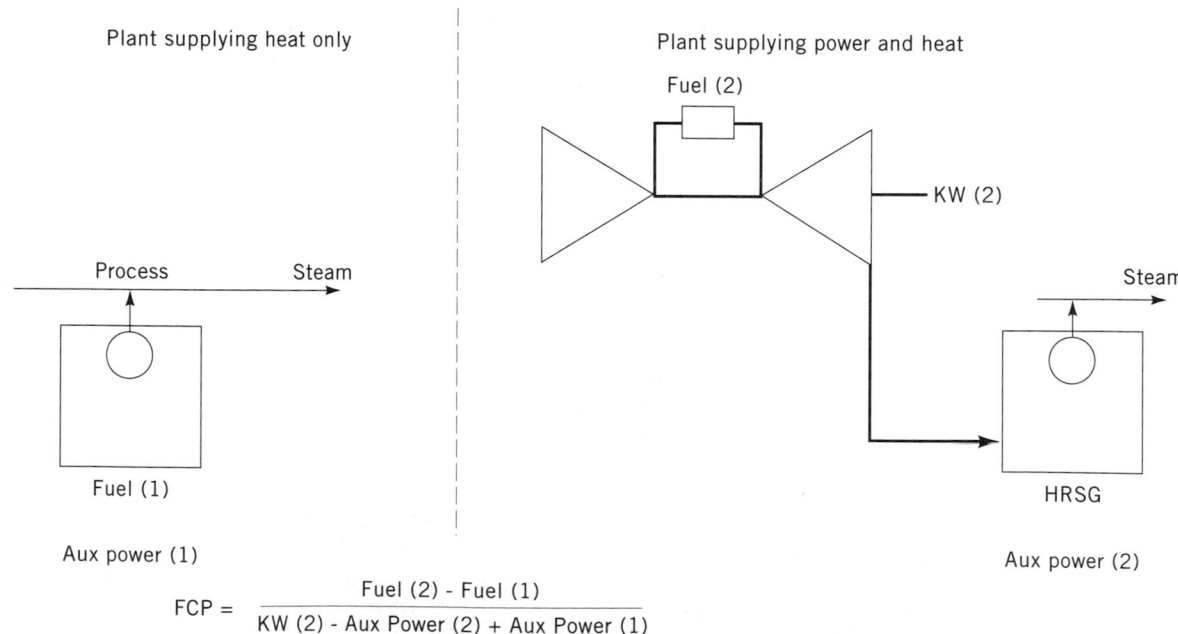

$$FCP = \frac{Fuel\ (2) - Fuel\ (1)}{KW\ (2) - Aux\ Power\ (2) + Aux\ Power\ (1)}$$

Figure 3. Fuel chargeable to power.

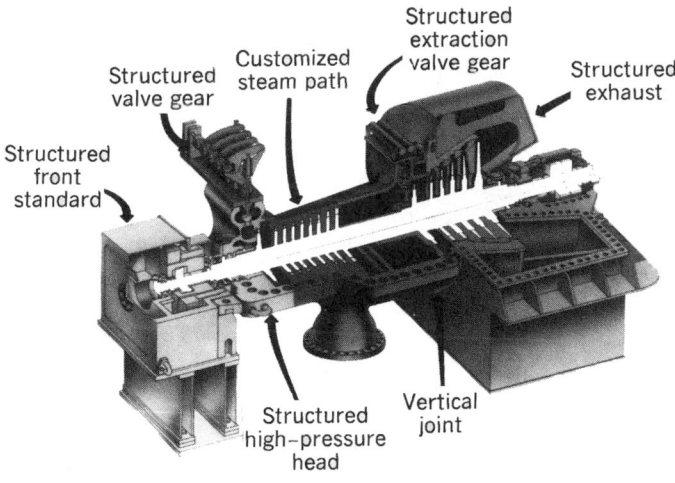

Figure 4. Steam turbine modular "building block" concept.

ple types in which no flow is removed from the turbine between its inlet and exhaust. Figures 5b and 5f show the next simplest variations, in which steam is made available for process from an uncontrolled, or nonautomatic, extraction. The extraction pressure is proportional to the flow passing beyond the extraction through the unit to its exhaust and is thus related to the inlet steam flow and the extraction itself. Variations may include two or more such uncontrolled extractions.

Figures 5c and 5g illustrate automatic extraction units, providing process steam at one controlled pressure. The extraction-control valves regulate the flow to the exhaust section of the turbine. Should an increase in process demand cause the extraction pressure to fall below the set value, the valve closes, reducing exhaust section flow, raising the extraction pressure, and diverting additional flow to the extraction. Figures 5d and 5h show double-

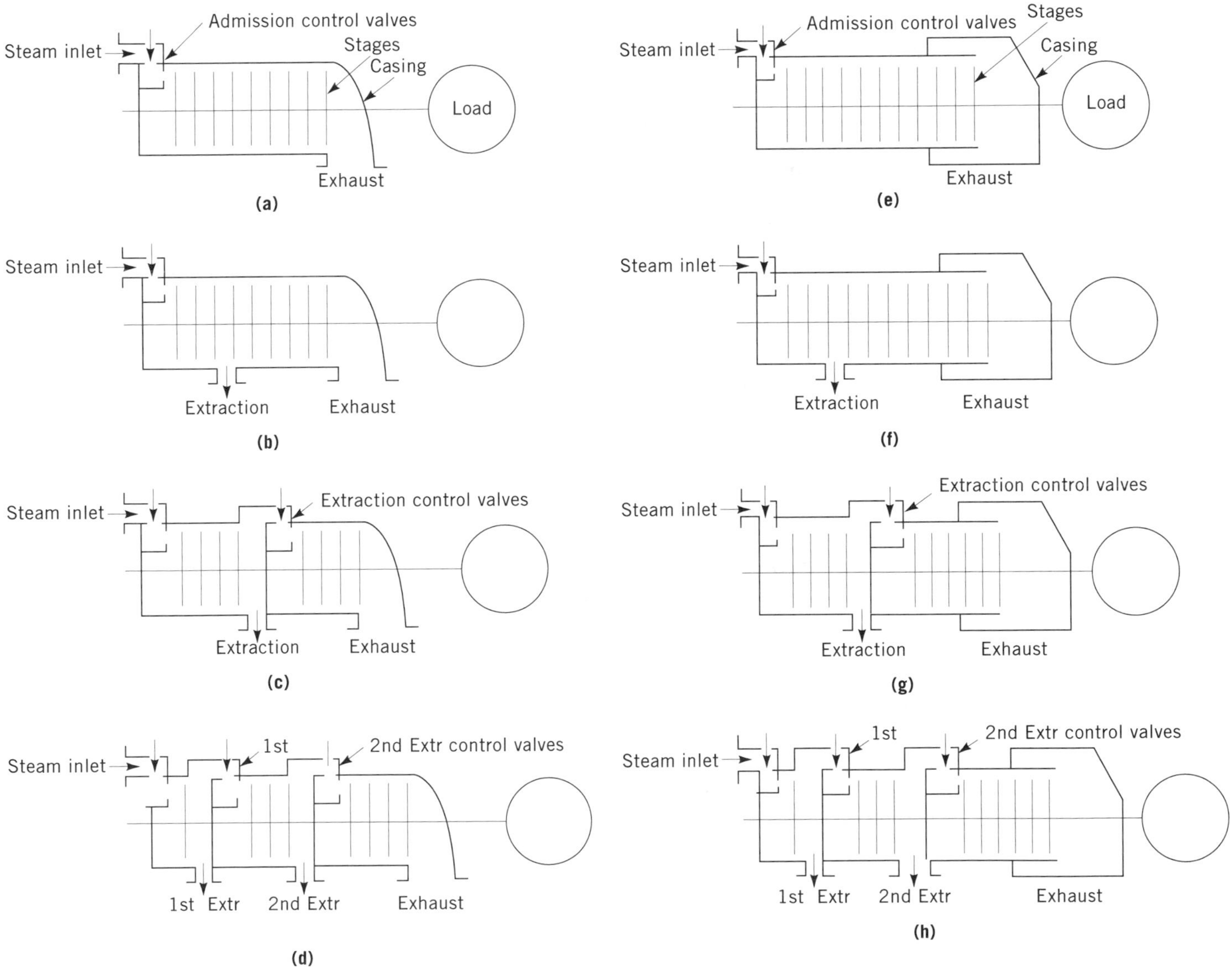

Figure 5. Basic types of industrial and cogeneration steam turbines. (**a**) Straight noncondensing (SNC), (**b**) single-nonautomatic extraction noncondensing (SNAXNC), (**c**) single-automatic extraction noncondensing (SAXNC), (**d**) double-automatic extraction noncondensing (DAXNC), (**e**) straight condensing, (**f**) single-nonautomatic extraction condensing (SNAC), (**g**) single-automatic extraction condensing (SAXC), (**h**) double-automatic extraction condensing (DAXC).

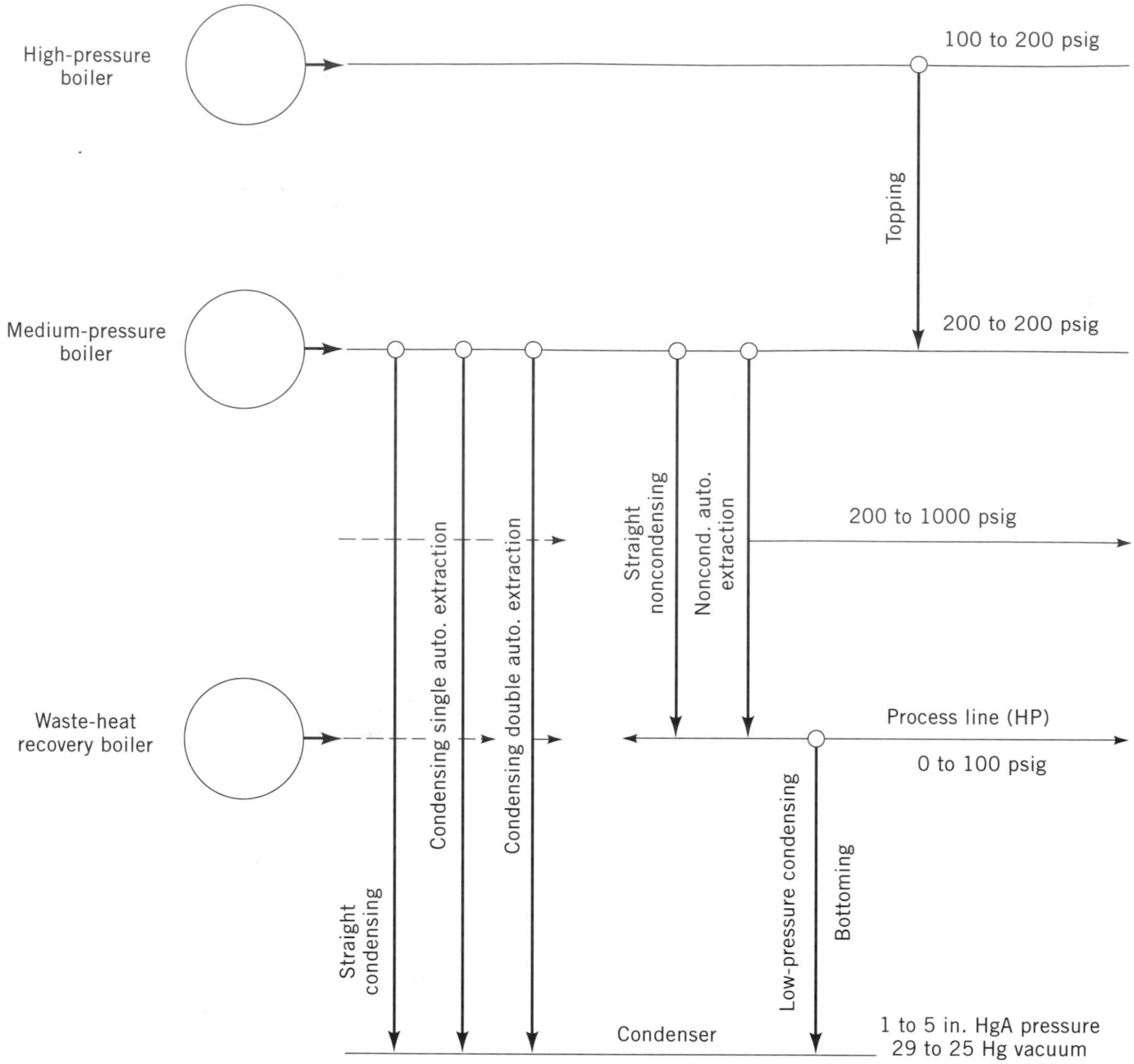

Figure 6. Regimes of application of various types of steam turbines.

automatic extraction units in which a second set of internal extraction-control valves allows controlled extraction at two pressures. Triple-automatic extraction is a further variation.

Figure 6 is a schematic map indicating where these various turbine types are applied. The pressure ranges selected for the boilers and the process lines are chosen for illustration. Topping turbines, high pressure (HP) boiler to medium pressure (MP) boiler, are used where an existing boiler is replaced with one at higher pressure to gain additional generation and improved efficiency. A bottoming turbine, from a process pressure to the condenser, is used to recover energy when process heat needs are reduced either permanently or seasonally.

The representative units of Figure 5 are shown taking steam from the MP boiler and exhausting to process in the noncondensing cases and to the condenser in the condensing designs. Any of these types can be designed for use with HP boilers when needed. Automatic extraction units can be designed to accept steam from a process line (admission or mixed pressure) when the steam available

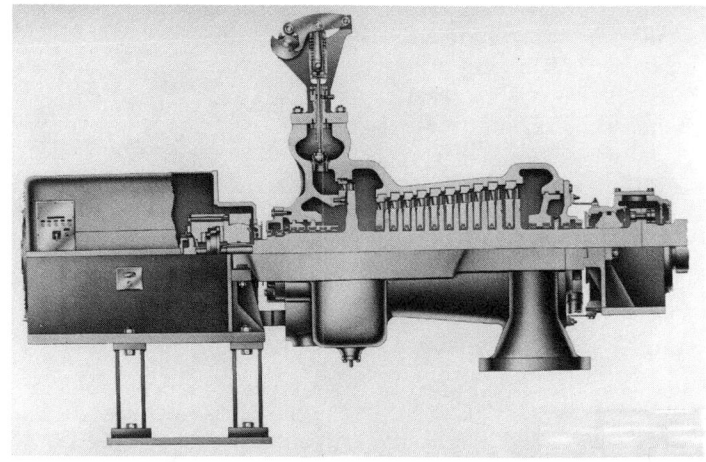

Figure 7. Noncondensing steam turbine.

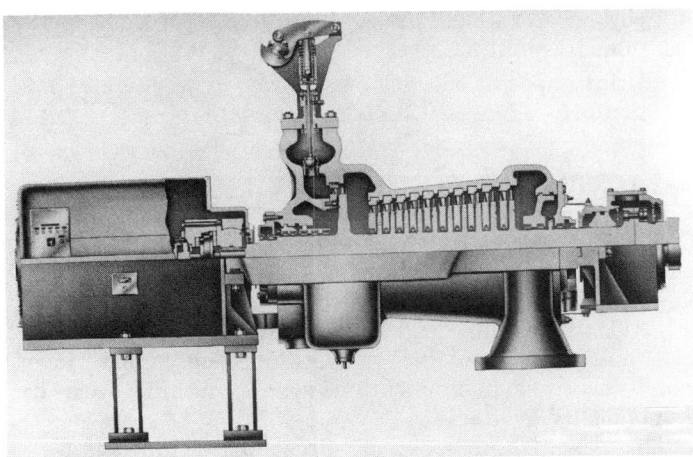

Figure 8. Double-automatic extraction condensing steam turbine.

from other sources exceeds the process needs. Two turbine types from Figure 5 are illustrated with cross-section in Figures 7 and 8.

Straight Noncondensing

The simplest steam turbine configuration is the straight noncondensing design. The output of the turbine is a function of the initial steam conditions, the turbine exhaust pressure, and the process steam demand. The power production of this unit type is limited by the process demand, unless an artificial demand is created by the use of a steam vent at the exhaust.

A cross-section of a typical noncondensing steam turbine is shown in Figure 7. The plug-type inlet valves are shown on the upper casing. The casing is in two halves, each made from a single steel casting. The front standard, shown to the left, contains the thrust bearing, first journal bearing, and control devices. The turbine is anchored at its exhaust end. Thermal expansion of the casing is accommodated by the flexible support under the front standard. The steam path is of the impulse, wheel-and-diaphragm type, in which the moving buckets are carried on the periphery of wheels machined from a solid forging. The packing diameters between the wheels are made small to minimize the interstage packing leakage. A small-diameter shaft acts to minimize transient thermal stresses, optimizing starting and loading characteristics. Most of the stage pressure drop is taken across the nozzles in the stationary diaphragms. Spring-backed packing rings seal against the rotating shaft. The solid coupling to the driven load is shown to the right.

Automatic Extraction Noncondensing

Industrial plants having steam demands at two or more pressure levels can benefit from the use of automatic extraction, noncondensing turbines. They provide the flexibility to respond automatically to variations in steam demands at the extraction and the exhaust. In recent years, typical noncondensing turbines have been rated 10 to 40 MW with inlet steam conditions of 4137×10^3 Pa 399°C to 9999×10^3 Pa 510°C. Extraction pressures of

1034×10^3 to 4482×10^3 Pa for the high pressure and 25 to 200 psig for the low pressure are typical of the broad spectrum of steam conditions used in industrial plants where these units are applied.

Automatic Extraction Condensing

Automatic extraction, condensing units provide additional operating flexibility and the ability to control power generation as well as process header pressures. They are well suited to third-party cogeneration systems because of their ability to handle variations in the steam host's steam requirements while maintaining electric power delivery to the utility. They can be sized for electrical generation considerably in excess of that associated with the extraction steam flows.

A double-automatic extraction condensing steam turbine is shown in Figure 8. It is a single-casing, single-flow machine with two bearings. The casing consists of cast-steel shells down to the condensing section, joined to a fabricated-steel exhaust hood using a vertical joint. The upper and lower casing components are bolted together with a horizontal joint. The inlet valves are plug type, similar to those of Figure 7. The extraction valves illustrated are of the internal spool type. (Poppet-type and axial-flow extraction valves are also used, for appropriate pressure and flow duty.) With spool valves, a horizontal, external camshaft lifts four vertical stems. Each stem positions two internal spool valves, one each in the upper and lower halves of the turbine. The valves are designed to open sequentially, providing the efficiency advantage of multiple partial-arc admissions to the immediate downstream stage group. Steam volume increases rapidly as the steam expands to condenser pressure. Thus the length of the buckets increases rapidly between the inlet of the LP section and the last-stage buckets.

Effect of Steam Conditions

The influence of initial steam conditions and process steam pressure on the amount of cogenerated power per 1054×10^8 J/h net-heat-to-process (NHP) is shown in Figure 9. The increase in cogenerated power through use of higher initial steam conditions as well as lower process pressures is readily apparent. Studies have shown that the higher steam conditions can be economically justified more easily in industrial plants having relatively large process steam demands. Data given in Figure 10 provide guidance in regard to the initial steam conditions that are normally considered for industrial cogeneration applications. Higher energy costs experienced since the mid-1970s are favoring the upper portion of the bands shown in Figure 10.

Even through the use of the most effective steam turbine cogeneration systems, the amount of power that can be cogenerated per unit of heat energy delivered to process will usually hot exceed about 85 kW/1054×10^6 J net heat supplied. This is generally less power than that required to satisfy most industrial plant electrical energy needs. Thus with thermally optimized steam turbine cogeneration systems, a purchased power tie or condensing power generation is necessary to provide the balance of the plant power needs. Condensing steam turbine power

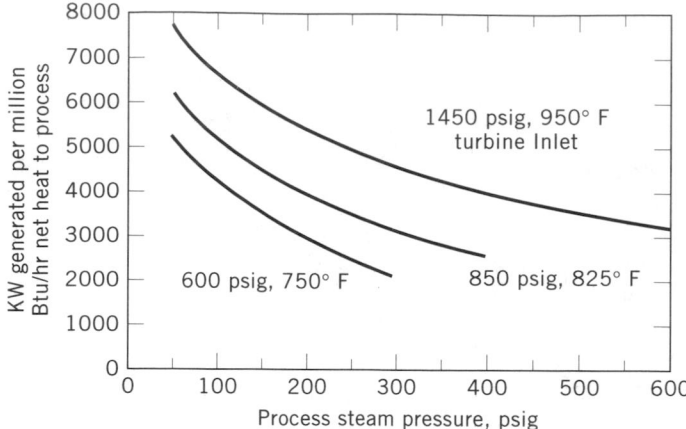

Figure 9. Cogeneration power with steam turbines. Basis: (*1*) average temperature of process returns and makeup is 73.9°C; (*2*) power cycle credited for feed-water heating is 235°C for 9998 × 10³ Pa, 204°C for 5861 × 10³ Pa, and 188°C for 4137 × 10³ Pa; (*3*) turbine efficiency is 75%.

generation, although not necessarily energy efficient, has proven economic in many industrial applications. Favorable economics are often associated with systems in which:

- Condensing power is used to control purchased power demand.
- Low cost fuels or process by-product fuels are available.
- Adequate low level process energy is available for a bottoming cogeneration system.
- Condensing provides the continuity of service in critical plant operations in which loss of the electric

power can cause a major disruption in process operations or plant safety.
- Utility specific situations favoring power sales, particularly if low cost fuels are available.

GAS TURBINE AND COMBINED CYCLES

Gas turbine cycles provide the opportunity to generate a larger power output per unit of heat required in process than noncondensing steam turbine cogeneration systems (Fig. 11). This characteristic, combined with a favorable FCP and proven reliability, has made this prime mover widely accepted in applications where suitable fuels are economically available.

The gas turbine is an air-breathing engine that responds to the mass flow entering its compressor. For constant speed units, the gas turbine output will generally vary in proportion to the inlet air temperature (density) as shown for the MS6001B in Figure 12. For gas turbine designs in which mechanical limitations exist, the characteristic may be similar to that shown for the LM6000 in Figure 12. A schematic diagram for a simple-cycle, single-shaft gas turbine is shown in Figure 13. Air enters the axial flow compressor at point 1 at ambient conditions. Because these conditions vary from day to day and from location to location, it is convenient to consider some standard conditions for comparative purposes. The standard conditions used by the gas turbine industry are 15°C (59°F), 1.013 × 10⁵ Pa and 60% relative humidity, which are established by the International Standards Organization (ISO). These conditions are frequently referred to as ISO conditions.

Air entering the compressor at point 1 is compressed to some higher pressure. No heat is added; however, the temperature of the air rises due to compression, so that

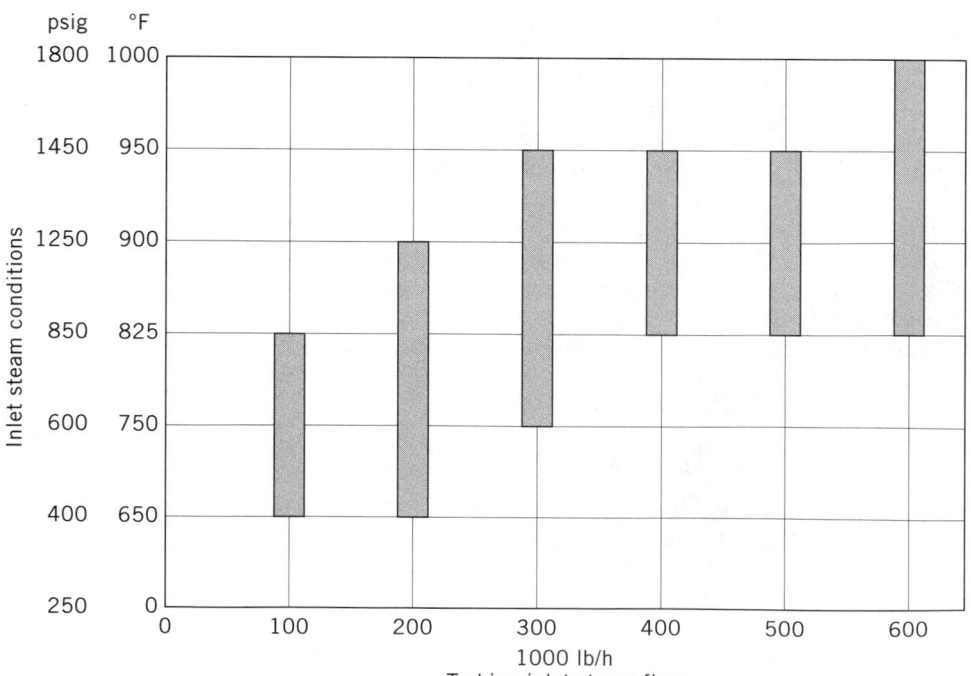

Figure 10. Range of initial steam conditions normally selected for industrial steam turbines.

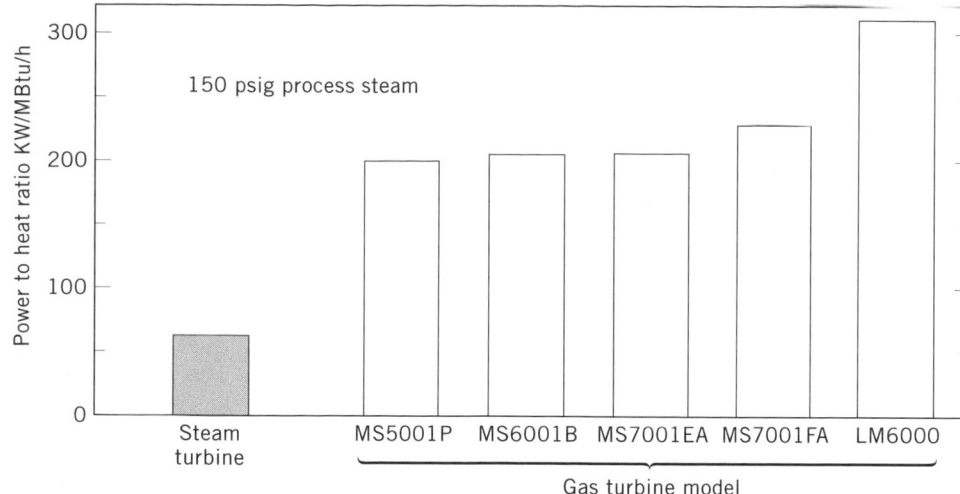

Figure 11. Power:heat ratio. Basis: gas turbine with unfired LP HRSG.

the air at the discharge of the compressor is at a higher temperature and pressure. Upon leaving the compressor, air enters the combustion system at point 2, where fuel is injected and combustion takes place. The combustion process occurs at essentially constant pressure. Although high local temperatures are reached within the primary combustion zone (approaching stoichiometric conditions), the combustion system is designed to provide mixing, dilution, and cooling. Thus by the time the combustion mixture leaves the combustion system and enters the turbine at point 3, it is a mixed average temperature.

In the turbine section of the gas turbine, the energy of the hot gases is converted into work. This conversion actually takes place in two steps. In the nozzle section of the turbine, the hot gases are expanded and a portion of the thermal energy is converted into kinetic energy. In the subsequent bucket section of the turbine, a portion of the kinetic energy is transferred to the rotating buckets and converted to work. Some of the work developed by the turbine is used to drive the compressor, and the remainder is available for useful work at the output flange of the gas turbine. Typically, more than 50% of the work

developed by the turbine sections is used to power the axial flow compressor.

As shown in Figure 2, single-shaft gas turbines are configured in one continuous shaft, and therefore, all stages operate at the same speed. These units are typically used for generator-drive applications in which significant speed variation is not required. A schematic diagram for a simple-cycle, two-shaft gas turbine is shown in Figure 14. The low pressure or power turbine rotor is mechanically separate from the high-pressure turbine and compressor rotor. This unique feature allows the power turbine to be operated at a wide range of speed and makes two-shaft gas turbines ideally suited for variable-speed applications.

All of the work developed by the power turbine is available to drive the load equipment, because the work developed by the high-pressure turbine supplies all the necessary energy to drive the compressor. Furthermore, the starting requirements for the gas turbine load train are reduced because the load equipment is mechanically separate from the high pressure turbine.

The thermodynamic cycle on which all gas turbines operate is called the Brayton cycle. Figure 15 shows the classical pressure–volume (PV) and temperature–entropy (TS) diagrams for this cycle. The numbers on this diagram correspond to the numbers used in Figure 13. Path 1 to 2 represents the compression occurring in the com-

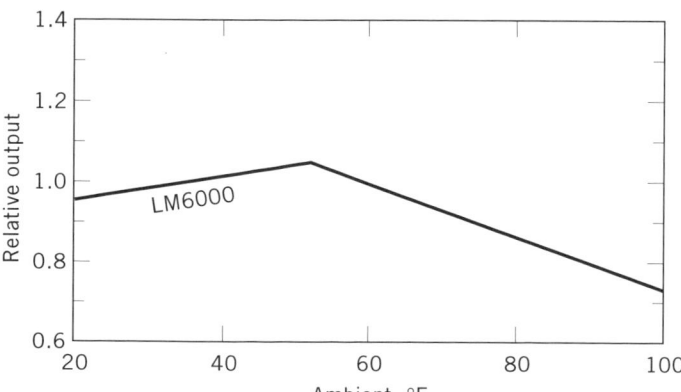

Figure 12. Gas turbine ambient output characteristics. Basis: (1) sea level site; (2) natural gas fuel; (3) standard inlet–exhaust losses; (4) NO_x at 25 ppmvd, water injection; (5) simple cycle operation.

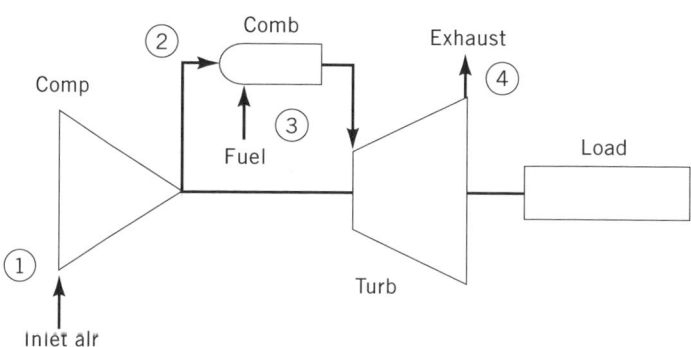

Figure 13. Simple cycle, single-shaft gas turbine.

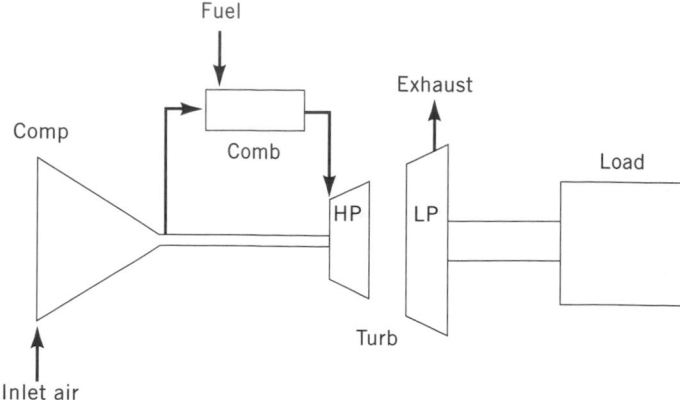

Figure 14. Simple cycle, double-shaft gas turbine.

pressor, path 2 to 3 represents the constant pressure addition of heat in the combustion systems, and path 3 to 4 represents the expansion occurring in the turbine.

The path from 4 back to 1 on the cycle diagrams is indicative of a constant pressure cooling process. In the gas turbine, this cooling is done by the atmosphere, which provides fresh, cool air at point 1 on a continuous basis in exchange for the hot gases exhausted to the atmosphere at point 4. The actual cycle is an open rather than closed cycle, as indicated. In power generation applications, gas turbines are available from approximately 1000 to 226,000 kW. Generally, specific models are offered with unique performance characteristics. Table 1 illustrates a typical range of models available.

Gas Turbine Power Enhancement

The gas turbine output for any unit may be enhanced at high ambient temperatures and low humidity levels by application of an evaporative cooler. This system decreases the compressor inlet temperature by the chilling effect of evaporating water introduced into the area upstream of the compressor. This approach frequently can

be economically justified in both base load and peaking applications. Output increases from 10% to 20% can be experienced at a 32.2°C ambient temperature at a relative humidity of 20%. The greater the output change (loss) with increasing ambient temperature, the larger the economic potential associated with various power enhancement alternatives.

Gas Turbine Exhaust Heat Recovery

The economics of gas turbines in process applications usually depend on effective use of the exhaust energy. The increase in overall system efficiency as the exhaust temperature is decreased through use of effective heat recovery is illustrated in Figure 16. The most common use of this energy is for steam generation in heat recovery steam generators (HRSGs), with unfired and fired designs. However, the gas turbine exhaust gases can also be used as a source of energy for unfired and fired process fluid heaters as well as preheated combustion air for power boilers.

Heat Recovery Steam Generators

The overall FCP in a gas turbine HRSG system is a function of the amount of energy recovered from the turbine exhaust gas. The greater the amount of energy recovered, the lower the HRSG stack temperature and the better the FCP. Thus gas turbine HRSG cycles should use the lowest practical feed-water temperature to the economizing section of the HRSG, within constraints imposed due to gas-side corrosion considerations. The typical feed-water temperature is 110°C if corrosion is not a problem, and with an integral deaerating section, the inlet water temperatures can be much lower. For applications using sulfur-bearing fuels, a feed-water temperature of about 140°C should be used to ensure that metal temperatures remain above the condensation temperature of the sulfurous products of combustion. These feed-water temperatures are in contrast to steam turbine cycles, which provide increased cogenerated power as more regenerative feed-

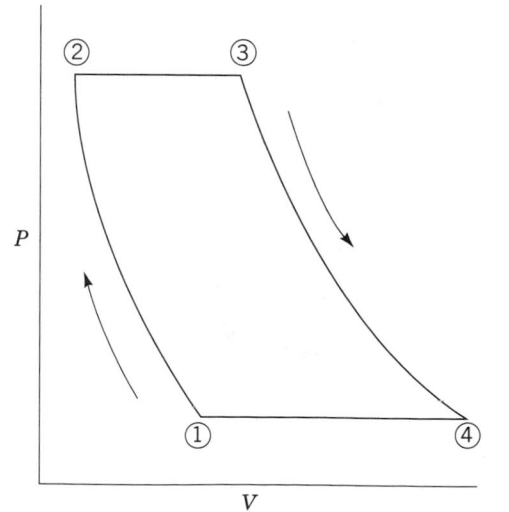

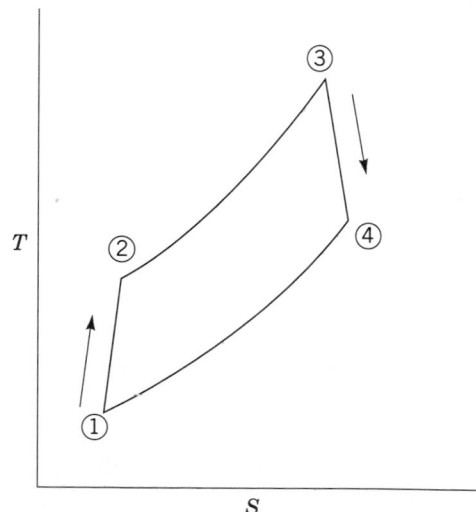

Figure 15. Brayton cycle.

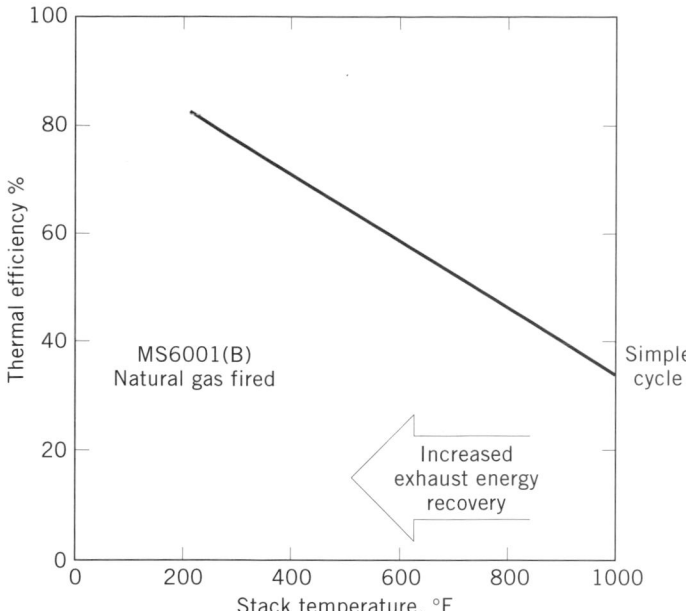

Figure 16. Thermal efficiency versus stack temperature.

wate heating (higher feed-water temperature to the boiler) is incorporated into the cycle.

HRSG units are available in unfired, supplementary fired, and fully fired designs. The appropriate selection is established through economic evaluations of various potential configurations for the application.

Unfired HRSG. An unfired unit is the simplest configuration. Characteristically, steam conditions range from 150 psig saturated to approximately $10,618 \times 10^3$ Pa 513°C. Steam temperatures are usually 23.9°C or more below the turbine exhaust gas temperature. With the introduction of F technology gas turbines, exhaust conditions will permit superheated steam temperatures of 537.8°C and reheat steam cycles, if the application warrants that approach. Generally speaking, unfired units can be economically designed to recover approximately 92% of the energy in the turbine exhaust gas available for steam generation. Higher performance levels are possible; however, the increased cost of the heat transfer surface must be evaluated against the additional energy recovered to establish whether the higher costs are warranted. When unfired units are designed with higher steam conditions for a combined cycle, multiple pressure units are usually applied to increase exhaust heat recovery and enhance system performance. The intermediate level may be that required for a process level. In applications using natural gas, a third pressure level, the integral deaerator, will further enhance overall system performance. Unfired HRSGs are convective heat exchangers that respond to the exhaust conditions of the gas turbine. Thus unfired HRSG units are slaves of the gas turbine operating mode and cannot be easily controlled.

Supplementary Fired HRSG. The oxygen content of the gas turbine exhaust generally permits supplementary fuel

firing ahead of the HRSG to increase steam production rates relative to an unfired unit. A supplementary fired unit is defined as an HRSG fired to an average temperature not exceeding 926.7°C. Because the turbine exhaust gas is essentially preheated combustion air, the supplementary fired HRSG fuel consumption is less than that required for a power boiler, providing the same incremental increase in steam generation. Characteristically, the incremental steam production from supplementary firing over an unfired HRSG will be achieved at 100% efficiency based on the lower heating value of the fuel fired. The incremental fuel will be about 10% to 20% less than for a natural gas–fired power boiler providing the same incremental increase in steam produced. A supplementary fired HRSG is basically a convective unit whose design is quite similar to an unfired HRSG. However, the firing capability provides the ability to control the HRSG steam production, within the capability of the burner system, independent of the normal gas turbine operating mode.

Fully Fired HRSG. A few industrials have used the exhaust of the gas turbine as preheated combustion air for a fully fired HRSG. A fully fired HRSG is defined as a unit having the same amount of oxygen in its stack gases as an ambient air–fired power boiler. The HRSG is essentially a power boiler with the gas turbine exhaust as its air supply. Steam production from fully fired HRSGs (10% excess air) may range up to six or seven times the unfired HRSG steam production rate. The actual increase is a function of the oxygen remaining for combustion and the gas turbine exhaust temperature. Because of the use of preheated combustion air, fuel requirements for fully fired units will usually range between 7.5% and 8% less than those of an ambient air–fired boiler providing the same incremental steam-generating capacity. With the more efficient gas turbines (higher firing temperatures resulting in lower oxygen content in the exhaust gases), the ability to ignite and maintain stable combustion in the HRSG should be confirmed with the HRSG manufacturer.

Even though fully fired units can provide a significant amount of steam, few applications of this nature can be found in industry. Evaluations show that the higher power : heat ratio available using unfired or supplementary fired HRSGs is usually economically preferred over fully fired HRSGs using small gas turbine sizes.

HRSG Steam Production Rates. The amount of steam that can be generated using the exhaust gas from various gas turbine generators frequently considered in industrial cogeneration systems is given in Table 2. These data are useful in performing gas turbine cogeneration feasibility studies.

Cycle Configurations

The simplest gas turbine cogeneration cycle is one in which the exhaust energy is used to generate steam at conditions suitable for the process steam header (Fig. 17). Generation of steam at higher initial steam conditions than those required in process will allow use of a steam turbine in addition to the gas turbine in the cogeneration

Table 2. Steam Generation and FCP with Gas Turbine and Exhaust Heat Boilers[a]

Generator Drives, Natural Gas Fuel, Dry Performance

Gas turbine type	MS5001(PA)	MS6001(B)	MS7001(EA)	MS7001(FA)	MS9001(FA)	LM2500(PE)	LM5000(PC)	LM6000(PA)
Gas turbine model	PG5371(PA)	PG6541(B)	PG7111(EA)	PG7221(FA)	PG9331(FA)	PGLM2500(PE)	PGLM5000(PC)	PGLM6000(PA)
ISO Base Rating (kW)	26300	38340	83500	159000	226500	22020	33480	39970
Performance at 15°C, sea level, natural gas fuel output, kW								
Unfired	25890	38000	82680	157100	223800	21800	33050	39700
Supplementary fired	25710	37820	82330	156300	222600	21680	32820	39550
Fully fired	25430	37530	81780	155000	220800	21510	32480	39320
Speed, rpm	5100	5100	3600	3600	3000	3600	3600	3600
Fuel, MBtu/h (HHV)	344.6	462.3	967.8	1677.3	2406.6	239.5	356.4	390.0
Exhaust flow, kg/h	440,627	491,249	1,062,785	1,536,343	2,212,207	245,806	430,058	445,571
Exhaust temp, °C								
Unfired	486	542	531	594	594	534	439	451
Supplementary fired	487	543	532	596	596	535	441	452
Fully fired	489	546	534	598	598	537	443	454
HRSG performance Fuel, MBtu/h (HHV)								
Supplementary fired	221.5	214.2	475.7	564.5	820.8	109.8	242.6	244.4
Fully fired	878.7	904.4	1989.6	2592.3	3742.6	444.2	841.1	850.0

Steam conditions, (Pa/°C)	HRSG Steam K lb/h	FCP GT Btu/kWh	HRSG Steam K lb/h	FCP GT Btu/kWh	HRSG Steam K lb/h	FCP GT Btu/kWh	HRSG Steam K lb/h	FCP GT Btu/kWh	HRSG Steam K lb/h	FCP GT Btu/kWh	HRSG Steam K lb/h	FCP GT Btu/kWh	HRSG Steam K lb/h	FCP GT Btu/kWh	HRSG Steam K lb/h	FCP GT Btu/kWh
Unfired																
$1,103 \times 10^3/188$	143.4	6650	193.2	6060	403.5	5840	704.0	5320	1010.0	5330	94.4	5790	117.2	6520	127.8	5960
$2,895 \times 10^3/346$	114.4	7240	159.0	6420	331.5	6200	592.9	5530	850.0	5550	77.6	6110	90.4	7030	99.8	6370
$4,344 \times 10^3/402$	104.4	7560	148.0	6610	307.0	6410	559.2	5630	800.0	5660	72.0	6280	80.5	7310	89.5	6610
$6,171 \times 10^3/443$	96.2	7870	139.4	6800	288.5	6600	533.0	5740	765.0	5760	67.7	6440	—	—	—	—
$6,171 \times 10^3/443$	96.2	6360	139.4	5920	288.5	5680	533.0	5270	765.0	5330	67.7	5690	—	—	—	—
$1,103 \times 10^3/188$	32.3	—	27.6	—	63.0	—	61.4	—	79.6	—	13.7	—	—	—	—	—
$9,067 \times 10^3/485$	—	—	130.8	5920	269.5	5680	510.0	5270	732.0	5330	—	—	—	—	—	—
$1,103 \times 10^3/188$	—	—	34.3	—	78.7	—	74.9	—	99.6	—	—	—	—	—	—	—
$10,515 \times 10^3/513$	—	—	125.8	5920	258.5	5670	495.0	5270	709.0	5330	—	—	—	—	—	—
$1,103 \times 10^3/188$	—	—	37.8	—	86.9	—	82.0	—	111.6	—	—	—	—	—	—	—
Supplementary fired																
$2,895 \times 10^3/346$	301.0	5960	338.0	5630	730.0	5380	1059.0	5080	1525.0	5110	169.4	5410	296.5	5870	307.5	5380
$4,344 \times 10^3/402$	289.5	5980	324.5	5670	701.0	5410	1017.0	5100	1465.0	5120	162.6	5430	285.0	5880	295.5	5400
$6,171 \times 10^3/443$	281.0	6030	315.0	5710	681.0	5440	988.0	5130	1423.0	5160	158.0	5460	276.5	5940	287.0	5430
$9,067 \times 10^3/485$	273.5	6100	306.5	5770	663.0	5490	962.0	5170	1385.0	5200	153.8	5500	269.5	5960	279.5	5470
$10,515 \times 10^3/513$	269.0	6080	301.5	5740	652.0	5470	946.0	5150	1362.0	5180	151.2	5490	265.0	5950	274.5	5470
Fully fired																
$4,344 \times 10^3/402$	777.0	4610	836.0	4710	1826.0	4390	2526.0	4380	3630.0	4460	410.5	4610	729.0	4930	745.0	4560
$6,171 \times 10^3/443$	757.0	4610	815.0	4710	1779.0	4390	2460.0	4380	3537.0	4460	400.0	4610	710.0	4930	726.0	4560
$9,067 \times 10^3/485$	740.0	4610	796.0	4710	1739.0	4390	2405.0	4380	3457.0	4460	391.0	4610	694.0	4930	710.0	4560
$10,515 \times 10^3/513$	726.0	4610	782.0	4710	1708.0	4390	2362.0	4380	3395.0	4460	384.0	4610	681.0	4930	697.0	4560

[a] Gas turbines and boilers fueled with natural gas and all fuel data based on higher heating value (HHV). Unfired single-pressure boilers 92% effectiveness for SH and evaporator; supplementary fired to 871°C, 86.8% to 90.5% effectiveness; fully fired to 10% excess air with 149°C stack temperature. For two-pressure boilers, criteria of minimum 149°C stack temperature may require less than 92% low-pressure boiler effectiveness. Assumes 0% exhaust bypass stack damper leakage, 3% blowdown, 1.5% radiation and unaccounted losses, and 109°C feed water for all cases. Standard gas turbine and inlet losses; exhaust 25.4×10^{-2} m H_2O for unfired, 35.56×10^{-2} m H_2O for supplementary fired, and 50.8×10^{-2} m H_2O for fully fired. LM2500, LM5000, and LM6000 values based on guarantee, not average, engine performance. Fuel chargeable to gas turbine power assumes GT credit with PH auxiliaries and equivalent 84% boiler fuel required to generate steam. Lower heating value (LHV) − 21515 Btu/lb, HHV = LHV × 1.11.

cycle (Fig. 18). This configuration derives the benefits of both gas and steam turbine cogeneration and yields a higher power:heat ratio than the arrangement given in Figure 17.

A multipressure HRSG system is shown in Figure 18. This arrangement is common for unfired and moderately fired (about 648.9°C) HRSG systems. The multipressure HRSGs provide increased recovery of the gas turbine exhaust energy and thus contribute to the favorable FCP associated with these cycles. For example, an unfired multipressure HRSG used in conjunction with an MS7001EA combined cycle supplying steam to process at 1034×10^3 Pa will yield about 5700 Btu/kWh HHV FCP, whereas a low pressure unfired HRSG used with the same gas turbine would have an FCP of about 6100 Btu/kWh HHV.

The steam turbine design shown in Figure 18 provides considerable cycle flexibility in cogeneration applications. The condenser provides a heat sink for HRSG steam-generating capability in excess of that extracted from the turbine for process use. Furthermore, the admission capability will permit the introduction of lower pressure steam

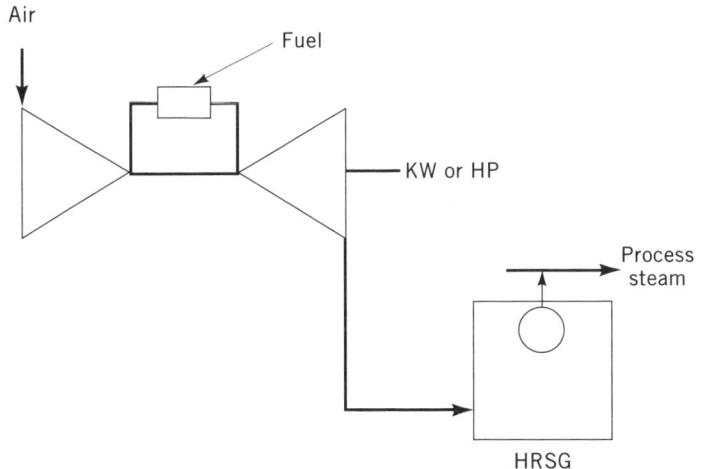

Figure 17. Gas turbine with LP HRSG.

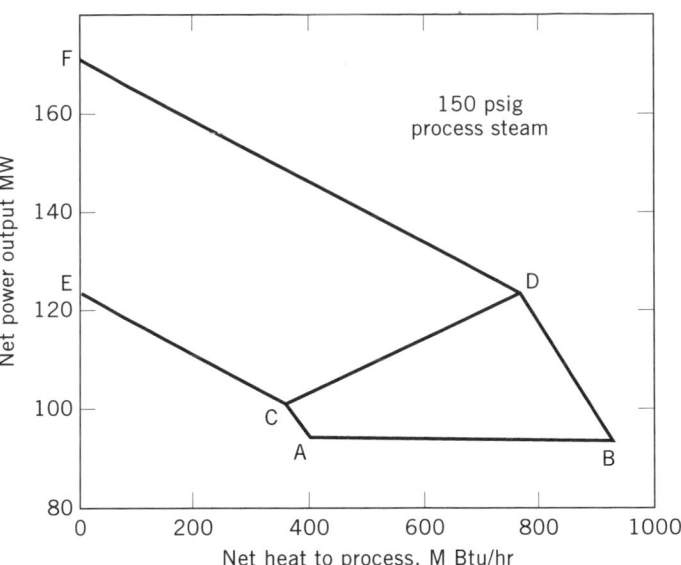

Figure 19. Performance envelope for gas turbine cogeneration system. Basis: (*1*) gas turbine operating at its 16°C capability, sea level site, natural gas fuel, steam injection for 25 ppmvd NO$_x$ emission level; (*2*) cycle A: unfired HRSG, cycle B: supplementary fired (871°C) HRSG, LP process steam, cycle C: combined cycle, unfired, two-pressure level HRSG, HP at 8619 × 10³ Pa, 482°C, 1034 × 10³ Pa saturated, noncondensing steam turbine generator, cycle D: combined cycle, supplementary fired HRSG, steam at 8619 × 10³ Pa, 482°C, noncondensing steam turbine generator, cycle E: same as cycle C, but with extraction–admission condensing steam turbine generator, cycle F: same as cycle D, but with extraction–admission condensing steam turbine generator; (*3*) process returns and makeup enter the integral 34 × 10³ Pa deaerating heater at a mixed temperature of 43°C.

Cycle Design Flexibility

One method of displaying the many options available using a gas turbine in a cogeneration application is shown in Figure 19. The figure was developed for the MS7001EA gas turbine generator (83,500 kW ISO, natural gas fired). A summary of the performance used to develop the envelope is given in Table 3. Point A represents the MS7001EA gas turbine generator exhausting into an unfired low pressure HRSG. Point C is a combined cycle configuration based on use of a two pressure level unfired HRSG. The steam turbine in the C cycle is a noncondensing unit expanding the HP HRSG steam to the 1034 × 10³ Pa process steam header.

Points B and D in Figure 19 represent operation of the HRSG with supplementary firing to a 871°C average exhaust gas temperature entering the heat-transfer surface. The temperature used for the HRSG firing has been arbitrarily limited to 871°C, even though higher firing tem-

peratures (and thus higher steam production rates) are possible in the exhaust of this unit. The envelope defined by A, B, C, D represents the most thermally optimized use of a gas turbine in a cogeneration application (ie, provides the lowest FCP). Operation along the line CE, DF, or any intermediate point to the left of line CD represents use of condensing steam turbine power generation with points E and F applicable for combined cycle operation without any heat supplied to process. Thus the cycles along EF are combined cycles providing power alone.

Performance envelopes for many of the gas turbines included in Table 2 are presented in Figure 20. These data are on the same basis as Figure 19 except for point C. Point C for all units (except the various MS7001 models) is based on 5861 × 10³ Pa 441°C initial steam temperature to the noncondensing steam turbine. Furthermore, the only condensing power illustrated is based on unfired, two pressure level HRSG designs.

Cogeneration Opportunities and Evaluation

Circumstances under which cogeneration should be considered include the following:

- Development of new industrial facilities.
- Major expansions to existing facilities.
- Replacement of aging steam generation equipment.

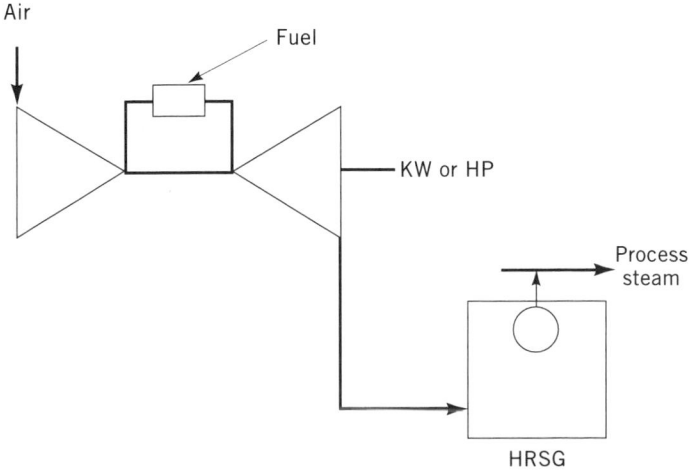

Figure 18. Typical industrial gas turbine cycle.

Table 3. Performance of MS7001EA Gas Turbine Cycles[a]

Cycle	A	B	C	D	E	F
Net output, MW	89.5	88.7	104.0	125.2	126.5	172.5
NHP, J/h	410×10^9	894×10^9	366×10^9	780×10^9	0	0
FCP-Btu/kWh, HHV	6090	5730	5730	5090	7980	8800

[a] Basis: (1) cycle definitions as given in Figure 19, (2) net output is the total power credited to the cogeneration cycle, (3) net fuel incudes credit for the NHP at an 84% process boiler efficiency.

- Significant changes in energy costs (fuel and power).
- Power sales opportunities.

New industrial plants or major expansions to existing facilities having large process heat demands and continuous process operations provide ideal opportunities to evaluate cogeneration. In these instances, cogeneration is compard with a base case, in which process heat is produced on site with power requirements purchased from the utility. Cogeneration represents an incremental investment relative to the base case, and thus the capital cost on a dollar per kilowatt basis is less than for an in place base case facility without this "investment credit." For example, assuming a new facility requires 208,656 kg/h of gas-fired boiler capacity at 1034×10^3 Pa and 90 MW, the incremental investment for a MS7001EA with an unfired HRSG providing a portion of the required steam may be about \$325/kW. Whereas, installation of a separate facility with the MS7001EA and supplementary fired HRSG system may approach \$560/kW making a potential project more difficult to justify economically (Table 4).

Replacement of old low pressure process steam boilers or even boilers with higher steam conditions used to support a steam turbine cogeneration system often provides an attractive cogeneration opportunity. Boiler steam capacity can be replaced by a gas turbine HRSG system, significantly increasing the system power:heat ratio at an attractive FCP. In addition, the investment credit for the replacement boiler generally ensures that the per kilowatt cost is reasonable.

When a facility anticipates a significant change in energy costs, the economic potential of cogeneration should be examined. This is particularly true in locations where purchased power costs may be increasing disproportion-

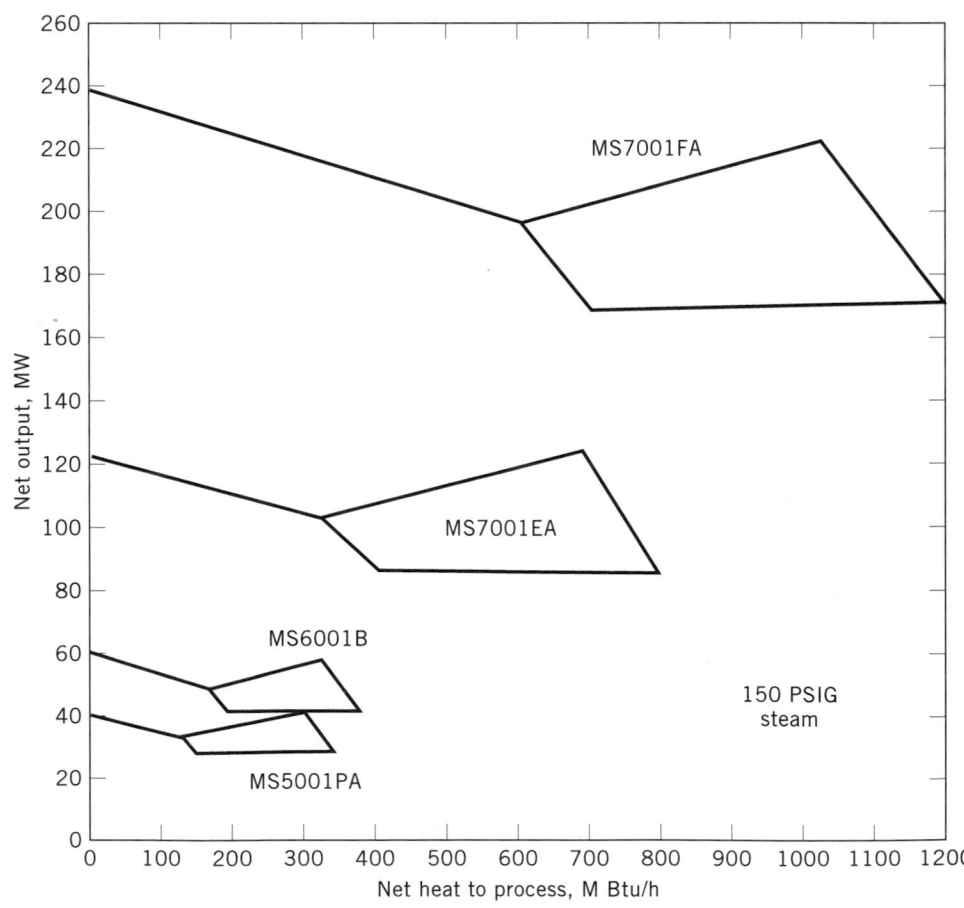

Figure 20. Gas turbine cogeneration systems MS options, 60 Hz.

Table 4. Feasibility Grade Installed Cost Comparisons for a Grass-roots Facility[a]

Alternative	Base	MS7001EA
Generation, MW	NA	89.5
Estimated total installed, cost, $ millions, 1992	21	50
Incremental investment, $ millions, 1992	Base	29
Unit Cost, $/kW	NA	559
Incremental unit cost, $/kW	Base	324

[a] Basis: (1) plant requires 208,656 kg/h of $1,034 \times 10^3$ Pa saturated steam, 90 MW electric power, (2) gas turbine performance based on sea level site, 16°C ambient, 60% relative humidity, natural gas fuel, steam injection for NO_x control to 25 ppmvd NO_x, (3) costs are feasibility grade values that do not include escalation, interest during construction, spares, or project soft costs.

ately relative to fuel. A cogeneration evaluation may suggest attractive economics even if there are no offsetting investments. Furthermore, if the cogeneration system results in an attractive FCP, the profitability may increase as fuel costs increase.

Many projects have been developed from favorable power sales opportunities. The size of some projects allows simple displaced power purchases. Others are based on circumstances in which large process heat demands permit generation of electric power significantly in excess of plant power needs, such as the enhanced oil recovery projects using steam injection.

Acknowledgments
The author acknowledges the significant support and contribution provided by the technical staff of General Electric's Power Generation Business.

BIBLIOGRAPHY

1. W. I. Rowen, paper presented at the Gas, Turbine and Aeroengine Congress, Amsterdam, The Netherlands, June 6–9, 1988.
2. H. E. Miller, paper presented to the American Cogeneration Association, San Francisco, March 1988.
3. L. B. David and R. M. Washam, *Development of a Dry Low NO_x Combustor*, Paper No. 89-GT-255, ASME, June 1989.
4. A. Morson, *Steam Turbine Long Bucket Developments*, GE Power Generation Paper No. GER-3647, General Electric Co., 1990.
5. C. V. Dinh, W. W. Kellyhouse, M. F. O'Connor, S. G. Ruggles, and J. C. Williams, GE Power Generation Paper No. GER-3577, General Electric Co., 1988.
6. J. A. Booth, B. L. Morrison, and P. Schofield, paper presented at the EPRI Heat Rate Improvement Conference, Knoxville, Tenn., 1989.
7. P. Schofield, paper presented at the Missouri Valley Electric Association Engineering Conference, Kansas City, 1981.
8. *Steam Turbines*, ASME PTC 6-1976, ASME.
9. *Guidance for Evaluation of Measurement Uncertainty in Performance Tests of Steam Turbines*, ASME PTC 6 Report-1985, ASME.
10. *Interim Test Code for an Alternative Procedure for Testing Steam Turbines*, ASME PTC 6.1-1984, ASME.
11. *Procedures for Routine Performance Tests of Steam Turbines*, ASME PTC 6S Report-1988, ASME.
12. D. Johnson, R. W. Miller, and W. I. Rowen, *SPEEDTRONIC Mark V Gas Turbine Control System*, GE Power Generation Paper No. GER-3658A, General Electric Co., 1991.
13. J. Dombrosky, T. Drummond, and J. Kure-Jensen, *Turbine Digital Control and Monitoring (DCM) System*, 88-JPGC/Pwr-33, ASME.
14. R. Hanisch and J. Kure-Jensen, *Integration of Steam Turbine Controls into Power Plants*, '89 JPGC 863-2 EC, 1989.
15. J. K. Salisbury, *Steam Turbines and Their Cycles*, John Wiley & Sons, Inc., New York.
16. R. S. Couchman, K. E. Robbins, and P. Schofield, *GE Steam Turbine Design Philosophy and Technology Programs*, GE Power Generation Paper No. GER-3705, General Electric Co., 1991.
17. J. D. Peterson, P. G. Stephens, and H. G. Stoll, paper presented at the Power-Gen '91, Tampa, Fla., Dec. 1991.
18. H. G. Stoll, *Least Cost Electric Utility Planning*, John Wiley & Sons, Inc., New York, 1989.
19. M. M. Schorr, *The Impact of Energy Legislation on Electric Generation Choices*, GE Power Generation Paper No. GER-3739, General Electric Co., 1992.
20. D. E. Brandt, *The Design and Development of an Advanced Heavy-Duty Gas Turbine*, Paper 87-GT-14, ASME.
21. J. H. Moore, paper presented at the Joint ASME/IEEE Power Generation Conference, Oct. 1989.
22. *Power Magazine* (Apr. 1991).
23. J. S. Anderson and J. M. Kovacik, paper presented at the Industrial Energy Technology Conference, Texas A&M University, Houston, June 1990.
24. M. M. Schorr and R. L. Yates, paper presented at Power Gen '91, Orlando, Fla., 1991.
25. M. M. Schorr and R. L. Yates, paper presented at Power Gen '90, Orlando, Fla., 1990.
26. M. M. Schorr, paper presented at the AWMA Annual Meeting, Kansas City, 1991.
27. M. M. Schorr, *A 1992 Update on Legislation and Regulations Affecting Power Generation*, Turbomachinery International, 1992.
28. M. M. Schorr, paper presented at the 1992 ASME Cogen Turbo Power Congress, 1992.
29. *DRI* (Winter 1991).
30. *DRI* (Summer 1992).
31. *TAG Technical Assessment Guide*, EPRI P-6587-L, Electric Power Research Institute, Sept. 1989.
32. D. E. Brandt and R. R. Wesorick, *Gas Turbine Design Philosophy*, GE Power Generation Paper No. GER-3434B, General Electric Co., 1992.
33. F. J. Brooks, *Gas Turbine Performance Characteristics*, GE Power Generation Paper No. GER-3567C, General Electric Co., 1992.
34. M. W. Horner, *Aeroderivative Gas Turbines—Design and Operating Features*, GE Power Generation Paper No. GER-3965A, General Electric Co., 1992.
35. D. Johnson, R. W. Miller, and W. I. Rowen, *SPEEDTRONIC Mark V Gas Turbine Control System*, GE Power Generation Paper No. GER-3658B, General Electric Co., 1992.
36. L. B. Davis, *Low NO_x Combustion Systems for Heavy-Duty Gas Turbines*, GE Power Generation Paper No. GER-3568C, General Electric Co., 1992.

37. A. M. Beltran, A. D. Foster, J. J. Pepe, and P. W. Schilke, *Advanced Gas Turbine Materials and Coatings*, GE Power Generation Paper No. GER-3569C, General Electric Co., 1992.

38. A. A. Slaterpryce, *Gas Turbine Support Systems*, GE Power Generation Paper No. GER-3452B, General Electric Co., 1992.

39. W. I. Rowen, *Design Considerations for Gas Turbine Fuel Systems*, GE Power Generation Paper No. GER-3648A, General Electric Co., 1992.

40. R. W. Fisk and J. M. Kovacik, *Cogeneration Application Considerations*, GE Power Generation Paper No. GER-3430C, General Electric Co., 1992.

41. M. A. Freeman and R. Hoeft, *Heavy-Duty Gas Turbine Operating and Maintenance Considerations*, GE Power Generation Paper No. GER-3620A, General Electric Co., 1992.

42. J. M. Davidson, *Aeroderivative Operation and Maintenance Considerations*, GE Power Generation Paper No. GER-3694A, General Electric Co., 1992.

43. J. I. Confer IV, J. K. Reinker, and W. J. Sumner, *Advances in Steam Path Technology*, GE Power Generation Paper No. GER-3713A, General Electric Co., 1992.

44. R. W. Bjorge, W. T. Parry, and J. S. Wright, *Steam Turbine Cycle Optimization, Evaluation and Performance Testing Considerations*, GE Power Generation Paper No. GER-3642A, General Electric Co., 1992.

45. D. R. Leger and W. A. Ruegger, *Recent Advances in Steam Turbines for Industrial and Cogeneration Applications*, GE Power Generation Paper No. GER-3706, General Electric Co., 1992.

46. W. Barker and J. Kure-Jensen, *SPEEDTRONIC Mark V Steam Turbine Control System*, GE Power Generation Paper No. GER-3687A, General Electric Co., 1992.

47. P. Schofield, *Steam Turbine Sustained Efficiency*, GE Power Generation Paper No. GER-3750, General Electric Co., 1992.

48. R. T. Bievenue, W. A. Ruegger, and H. G. Stoll, *Features Enhancing Reliability and Maintainability of Steam Turbines*, GE Power Generation Paper No. GER-3741, General Electric Co., 1992.

49. J. J. Gibney III, *Generators—An Overview*, GE Power Generation Paper No. GER-3688A, General Electric Co., 1992.

COKING, DELAYED

A. K. BHARGAVA
W. S. LOUIE
A. N. STEFANI
ABB Lummus Crest Inc.
Bloomfield, New Jersey

Delayed coking is a thermal cracking-type process used to convert petroleum residue and other feedstocks to coke, gas, and distillates. This process uses multiple coking chambers to permit continuous feed processing wherein one drum is making coke and one drum is being decoked.

Delayed coking is the most widely used process to upgrade petroleum residue. Coking technology is mature but changes are being made mainly in the areas of equipment safety, environmental, compliance, decoking and coke recovery (1). (See also PETROLEUM REFINING; PYROLYSIS.)

Today, delayed cokers are built not only to upgrade the bottom of the barrel, but also to produce high quality coke

for the metallurgical industry. Thus, the value of the coke can almost be equal to the liquid/vapor products when charging the proper feedstock.

Feedstocks charged to delayed coking units may be divided into two categories; namely, (1) Petroleum Derived Feedstocks and (2) Non-Petroleum Derived Feedstocks.

In the first group there are virgin feedstocks, such as heavy whole crudes, atmospheric and vacuum residues which, depending on feed properties, produce coke for fuel or electrodes in aluminum manufacture, depending on feed properties. Nonvirgin feedstocks derived from petroleum are principally thermal tars, pyrolysis heavy oils, and catalyst cracker decant oils. These materials produce premium anode coke or needle-type cokes for graphitized electrodes (2).

The non-petroleum derived feedstocks are coal tar pitch which produces premium anode coke or needle coke, hydroliquefied coal derived liquids (SRC-solvent refined coal) and coal gasified liquids which potentially can produce premium anode coke (3).

CHEMICAL REACTION

The kinds of chemical reactions which occur during the early stages of coking can be delineated by studying the carbonization of several aromatic hydrocarbons. Their initial reactions lead to the intermediates which control the subsequent course of carbonization. There are three types of thermal reaction processes that are important: (1) dehydrogenation, (2) rearrangement, and (3) polymerization. These reactions do not proceed in distinct steps, but occur simultaneously throughout the coking process.

The thermal reactions taking place are both endothermic and exothermic. Heat is applied to the material in the coking heater to initiate the dehydrogenation and rearrangement reactions. As polymerization proceeds in the coke drum, the exothermic reactions provide the heat to continue the dehydrogenation process.

The formation of synthetic carbon and graphite involves thermal dehydrogenation and polymerization of aromatic hydrocarbons. The thermal reactivity and the course of carbonization are controlled by the structure of the starting aromatic molecule. More rapid reaction during the early stages of carbonization usually leads to a more disordered graphite structure (sponge coke) (4).

FEED AND PRODUCTS

Feeds

There are distinctive differences between feedstocks such as vacuum residues and the aromatic oils. Vacuum residues have low aromaticity, and high coking value. They retain most of the sulfur and metals contained in the crude. The residues contain asphaltic compounds which are large heterocyclic molecules. (Ring compounds of different elements, such as quinoline). On the other hand, the aromatic oils, such as decant oil and pyrolysis residues have a high concentration of polynuclear aromatics. They contain single structures mainly of six-carbon aromatic rings. These are the precursors of the well-known hexagonal lattice structure which give graphite its unique physical properties, especially electrical conductivity.

PRODUCTS

Coke

Sponge Coke. Coke produced from virgin petroleum feedstocks, such as atmospheric and vacuum residues will exhibit quite different characteristics from cracked feedstocks. Feedstocks from virgin petroleum residues are non-uniform having a large number of cross-linkages or growth in all directions and resist graphitization because they contain less than six carbon atoms. Thus, they are termed isotropic or amorphous, and often called sponge coke because of such appearance. Generally, this is sold as anode (aluminum anodes) or fuel grade coke, depending on the sulfur and metal content (5).

Needle Coke. The starting point for the production of graphite electrodes is delayed coking of the highly aromatic molecular weight fractions from refinery cracking or petrochemical operations. The coke thus produced, is highly structured and crystalline in nature with a minimum number of cross-linkages. The reacting materials remain plastic for a longer time during carbonization allowing the crystals to be oriented in an elongated fashion or needle-like structure by the upward flow of gases in the coke drum. A convenient tool for classification of coke is the coefficient of thermal expansion (CTE) on graphites. As high temperatures and rapid temperature changes are encountered in the operation of electric steel furnaces, a low CTE is important in order to maintain high current densities and to avoid excessive thermal stresses. Indeed, the electrical conductivity of large diameter graphite electrodes is on the order of five times those fabricated from amorphous carbon fillers (6).

Shot Coke. The term "shot coke" is used to describe an abnormal type of delayed coke (resembling small balls or buckshot-like material) which is normally unintentionally made. Upon decoking, the shot coke pours loosely from the drum. The size of these agglomerates is 0.4–10 cm size. Although, the mechanism of shot coke formation is not yet known, it appears that residue, high in asphaltenes, are more likely to produce shot coke. Shot coke has been produced when high asphaltenic crudes (such as Maya crude) is charged to the unit. There is a high correlation between low API and high asphaltenic crude and shot coke formation.

Shot coke is considered low quality (even if low in sulfur) by aluminum anode producers and is to be avoided. It is difficult to grind and size and it does not give a baked anode of good strength and density.

Shot coke can be eliminated by blending aromatic material with the coker feedstock and/or increasing the coker recycle ratio.

Liquid Products. Olefinic liquid products are formed when coking petroleum residues (Tables 1 and 2). Thermal cracking produces products with a similar olefin content, octane and cetane number regardless of feedstock.

Table 1. Composition of Petroleum Medium Used as Feedstock

Feedstock	Light Arabian Vacuum Residue
TBP cut point, °C	565
°API	6.5
S.G.	1.025
Sulfur, wt %	4.4
Nitrogen, wt %	0.34
Conradson carbon, wt %	22.0
Metals, wt ppm	
V	69
Ni	21

These properties can be correlated vs gravity or boiling range with sufficient accuracy to define the product for downstream processing (hydrotreating, hydrocracking, etc) or blending. However, sulfur and nitrogen content in products is feedstock dependent and has a major effect upon downstream processing requirements.

PROCESS VARIABLES

Coke Yield

The key process variables that affect coke yield and quality are

- Conradson carbon
- Coker recycle ratio
- Coke drum pressure
- Coke drum temperature

See Figure 1 for the relationship between pressure, temperature, and recycle ratio.

The Conradson carbon residue (CCR) value determined in the fresh feedstock is a reliable property for predicting coke yield. The coke yield and Conradson carbon in the feedstock can be correlated by way of a coke factor, which is defined as coke yield (wt %) divided by the CCR (wt %) in the feed.

The second process variable, recycle ratio, is defined as the ratio of equivalent feed above the gas oil end-point to fresh feed. An increase in the recycle ratio will increase the coke yield, and at the same time will reduce the distil-

Table 2. Estimated Yields for Feedstock Described in Table 1, in Maximum Liquid Product Mode

Component	Wt %	S, Wt %	°API
H₂S	1.2		
C₂-(Fuel Gas)	4.1		
C₃/C₄ LPG	3.5		
C₅–350°F (Naphtha)	12.1	0.9	57.9
350–650°F (Light gasoil)	22.7	2.4	35.0
650°F+ (Heavy gasoil)	25.8	3.8	22.6
Green coke[a]	30.6	5.4	
Total	*100.00*		

[a] Green coke: VCM, wt % = 10 max, moisture, wt % = 15 max.

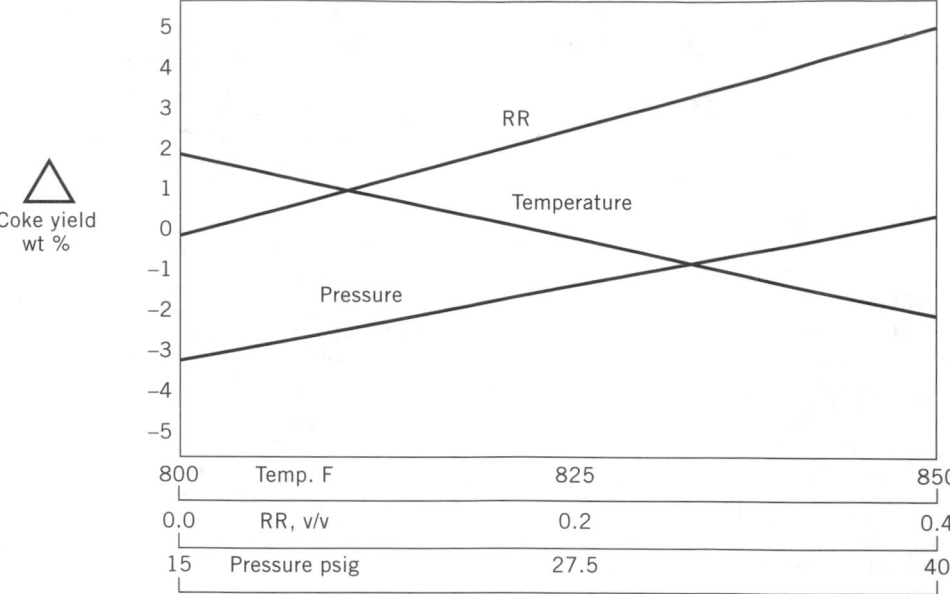

Figure 1. Differential coke yield versus operating conditions.

late yield. This formula occurs since a lower end point of the heavy gas oil must be taken to make more recycle. Recycle ratio is usually set by the product specifications of the heavy coker gas oil downstream processing units. Such product specifications include endpoint, metals and conradson carbon. The increase in recycle is obtained by increasing the bottom reflux and, therefore, lowering the vapor temperature passing through the heavy gas oil total drawoff tray in the fractionator. Since more recycle passing through the coking heater is cracked, there is an accompanying small increase in yields of coke and gas.

The third process variable, coke drum pressure, affects coke yield. A higher pressure favors higher coke yield by increasing the concentration of liquid in the coke drum at the same temperature and, therefore, higher conversion.

Typically the coke drums would operate between 15 and 100 psig measured at the top of the coke drum. Thus, for minimum coke, they would be at the lowest pressure possible, consistent with the pressure necessary for material to pass through the coke drum and fractionation system.

Lastly, the temperature inside the coke drum influences coke yield, hardness (VCM) and quality. Higher temperatures tend to decrease coke yields since more liquid is vaporized and thereby more vapor is carried out of the drum uncracked. Similarly, a higher temperature will reduce the coke's residual oil content called VCM (volatile combustible matter). The temperature of the vapor leaving the coke drum is a measure of coke drum operation, and typically is about 426–471°C. This temperature is controlled only by changes made to the coking heater outlet temperature, other things being equal. These changes are limited by the heater design to prevent coke laydown in the tubes.

A counter effect, however, to higher temperatures in the coke drum in order to produce lower VCM, is an increase in coke hardness. The harder the coke, the more time that will be required to cut the coke out of the coke

drum with the hydraulic jet water cutting system. Thus, the coke drum schedule may be jeopardized and more coke fines could be produced (7).

PROCESS DESCRIPTION

The following is a description of the processing scheme which is shown in a simplified manner on Figure 2.

Coking and Primary Fractionation

The feed enters the unit from battery limits, is heated passing through preheat exchangers and then flows into the bottom of the fractionator. There it mixes with the recycle liquid condensed from the coke drum effluent. Introduction of relatively cool feed to the bottom of the fractionator reduces coke formation there and helps to wash out any coke fines. This combined feed and recycle is pumped from the bottom of the fractionator through the coking heater where each pass is flow controlled.

A controlled quantity of boiler feedwater or high pressure steam is injected into each heater pass to assure satisfactory velocity in order to minimize coking in the heater tubes. The prime function of the coking heater is to quickly heat the feed to the required coking temperature to initiate the cracking reaction without premature coke formation in the heater tubes.

The effluent from the coking heater at a temperature between 493 to 510°C flows through a switch valve into the bottom of one of the two coke drums where further cracking and then polymerization takes place to form coke. Each drum is designed to be filled to a safe operating level with coke produced in a nominal 24-hour period. Antiform is injected into the coke drum during the latter part of the filling cycle to minimize foam carryover into the fractionator. The coke drums are operated in cycles to maintain continuity of operation. One drum is always in service to receive the coking heater effluent.

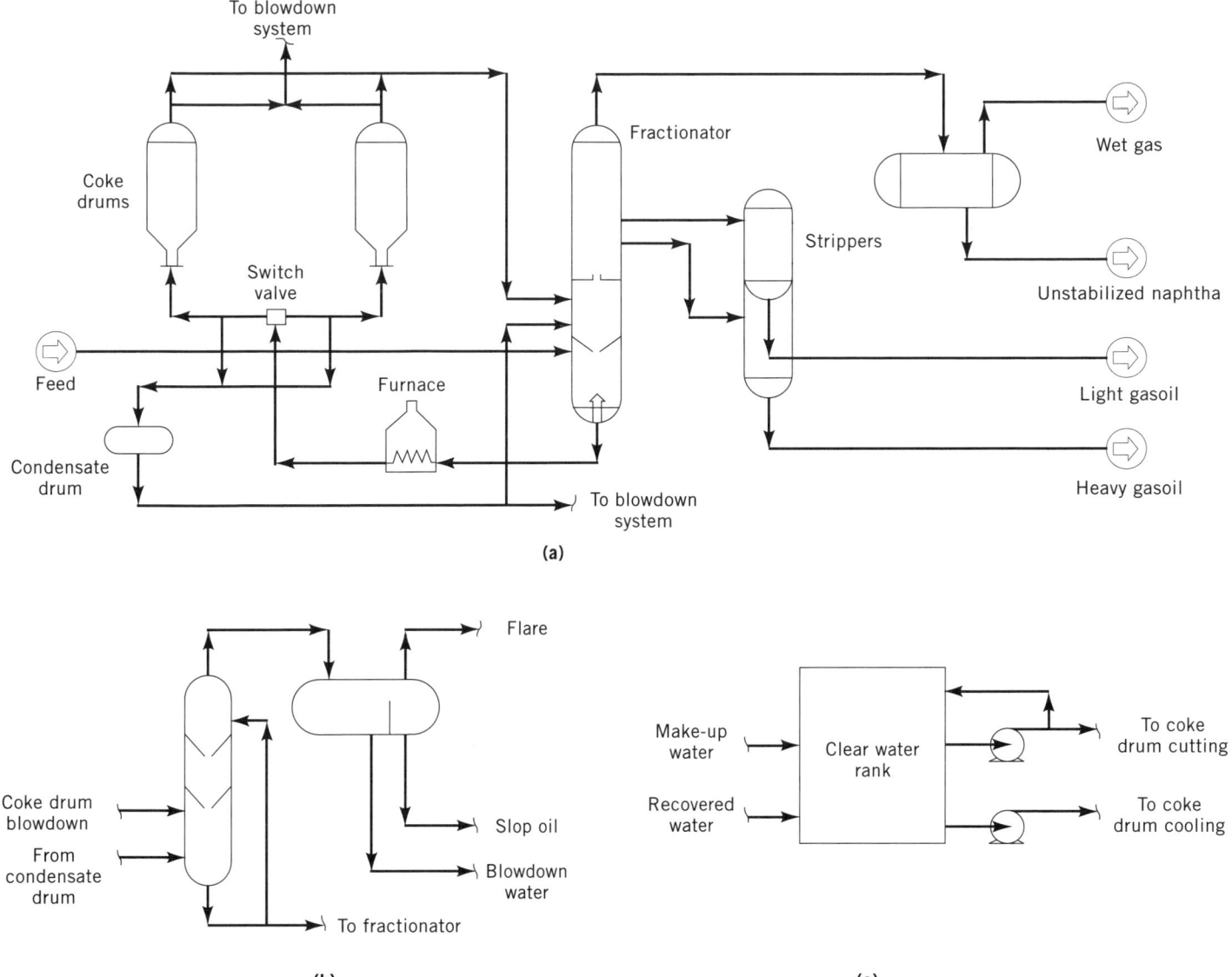

Figure 2. Delayed coking flow diagram.

The coke drum overhead vapor is quenched to approximately 421°C by heavy gasoil to stop the cracking/polymerization reactions and thereby reduce coke formation in the transfer line from the coke drums to the fractionator.

The fractionator is divided into two sections by the heavy gasoil drawoff pan. The upper section consists of valve-type trays; the lower section contains two bubble cap trays and several baffle trays. The coke drum vapor enters the fractionator below the bottom baffle tray.

The vapor flows upward through the trays in the lower section and is cooled by contacting the downflowing reflux liquid. The recycle stream thus condensed is collected at the bottom of the tower where it is mixed with the feed charge.

Vapor leaving the lower section of the tower flows to the upper section through the risers in the heavy gasoil drawoff pan. This vapor consists of light hydrocarbons,

naphtha, light and heavy gasoil, vaporized reflux, and steam. It is fractionated in the upper section of the tower.

The heavy gasoil stream collected in the total liquid drawoff pan is used as circulating reflux and gasoil product. Part of it is pumped without cooling to reflux the bottom section. The quantity of this reflux is controlled to set the heavy gasoil end-point and recycle quantity. The remainder of the reflux is used for feed preheat, reboilers and steam generation before it is returned to the fractionator. Another part is used as quench for coke drum vapors.

The heavy gasoil product portion, after steam stripping in the heavy gasoil stripper, may be used for feed preheat, BFW preheat and generation before it is cooled and sent to the battery limit. To improve heat recovery from the tower, a second light gasoil circulating reflux stream is withdrawn from a partial drawoff pan. This stream may be used for feed preheat, reboilers and steam generation before it is returned to the fractionator. A part of this stream is used as lean sponge oil, in the vapor recovery section.

The light gasoil product is steam stripped in the LCGO stripper and may be used for reboilers and then cooled and sent to storage. Top reflux to the tower is provided by a circulating reflux stream. The quantity is set to control the naphtha end point. The vapor leaving the top tray is partially condensed in the fractionator overhead condenser and sent to the fractionator overhead receiver for vapor-liquid separation. The liquid is pumped to the vapor recovery section, while the reflux is returned to the tower. The wet-gas is also sent to the vapor recovery section after it is compressed. Instrumentation at the compressor maintains a constant pressure in the overhead receiver (8).

Blowdown

Coking is a batch operation, normally on a 48-hour cycle, with 24 hours of online coke deposition and 24 hours of off-line decoking. A typical 48-hour drum cycle is shown below:

Operation	Hours
Coking	24.0
Switch drums	0.5
Steamout to coker fractionator	0.5
Steamout to blowdown system	1.0
Slow water cooling	1.0
Fast water cooling	4.0
Drain coke drum	3.0
Remove top and bottom heads	1.0
Hydraulic coke boring/cutting	4.0
Reheading/pressure testing	2.0
Drum heatup	7.0
	48.0

At the end of a coking period, the coking heater effluent is switched to the empty drum of the pair. Steam is injected into the base of the coke-filled drum for 1.5 hours to remove volatile hydrocarbons. During the first 30 minutes of steamout, the overhead vapors are sent to the coker fractionator. During the last hour of steamout, the overhead vapors are sent to the blowdown system. The function of the blowdown system is to condense the steam-and-oil vapor mixture from the coke drum during the coke drum steaming and water cooling operation. In the blowdown tower, these vapors are cooled and washed with an oil quench to remove heavy hydrocarbons. The overhead vapor from the blowdown tower is condensed in the blowdown overhead condenser. The water and oil are separated at the settling drum (7).

Wet and dry slops produced from the blowdown system are sent to slop and the coker fractionator, respectively. Condensed steam, is sent to the clear water tank and noncondensible gases are sent to the flare.

Prior to completion of the steamout, water is slowly injected into the coke-filled drum to cool the coke. The cooling is done slowly at first to avoid excessive thermal stresses in the drum shell. During slow and fast cooling,

water is vaporized and this steam and any stripped oil are sent to the blowdown system. After water cooling, the drum is drained and the heads removed at the top and bottom. Drained water flows to the pit sump and from there it is returned to the clear water tank.

Decoking

Coke is cut from the drum using high pressure water. The cut coke mixed with water falls out of the drum (see section on hydraulic decoking equipment and coke handling for additional details).

After all the coke is removed, the drum is reheated, deaerated, and pressure tested with steam. The drum is then heated with vapors from the coke drum which is being filled. The vapor/liquid mixture, which forms as the gas is cooled in the drum, flows to the condensate drum where the vapor and the liquid are separated. The vapor is sent to the coker fractionator and the condensate is pumped to the blowdown system until it is "dry" (ie, at a temperature of 148°C (300°F)); then it is pumped to the coker fractionator. When the coke drum is adequately heated, it is ready to go on-stream for the coking period (7).

DESIGN CONSIDERATIONS

More information on design can be found in refs. 7 and 8.

Coking Cycles

Large cokers are typically designed for a preferred nominal 48 hour drum cycle, ie, 24 hours of coking service and 24 hours offline for cooling, decoking and preheating.

It is possible to shorten the cycle to 16–22 hours principally by reducing the time for cooling and preheating. The minimum cycle depends upon the coke drum number and size. Many refineries have increased coker capacity by shortening the cycle from the original 24-hour design basis.

It is also possible to decrease the cost of a new unit by designing for a short-cycle since the coke drums, coke cutting equipment, pit and blowdown system can be specified for a smaller batch of coke.

However, the increased thermal stresses and number of cycles per year will shorten coke drum life. Short cycles are generally provided for cokers with up to four coke drums. Shortening the cycle for a six-drum coker would require the decoking operations to be accomplished smoothly and as planned. Failure to do so could result in operating problems.

Coke Drum Size

The largest commercial size coke drum is 27'–0" (8 m) diameter and 114'–0" (34.7 m) flange to flange.

The diameter of the coke drum is usually determined by the allowable velocity in the coke drum (vapor controlled design).

The length of the coke drum is limited by the design of the coke cutting equipment.

Certain feedstocks produce a large quantity of coke and a relatively small quantity of liquid products. Drums in this application are not controlled by the vapor velocity but by the quantity of coke produced (coke-controlled design).

Vapor Controlled Design

The maximum allowable superficial vapor velocity is derived from commercial experience. New delayed cokers are generally designed for lower velocities which result in minimum carry-over of coke fines, pitch and foam to the fractionator and requires minimum antifoam usage. Also, since the coke drum is not designed for maximum velocity and high antifoam use, it is possible to increase capacity in the future by reducing the cycle time.

The coke drums may also be designed for operation at high superficial velocity. This will require a greater use of antifoam injection. This high velocity design is usually provided in a revamp situation where the throughput is increased by a simultaneous reduction in cycle time. Operating at these high velocities presents certain risks and operating disadvantages such as antifoam carry-over, pitch and coke fines entrainment, foamover and shot coke formation.

Coke drums controlled by vapor velocity offer greater design flexibility. The operating pressure could be easily increased to reduce the velocity to either operate with lower velocity and/or reduce the number of coke drums required. The green coke yield increases when the coke drum operating pressure is increased.

Coke Controlled Design

Coke drums in this category should be operated at as low a pressure as possible. Any increase in pressure would increase the amount of coke produced. The selection of the drum size will be based on minimizing the number of coke drums required. Since the design is not vapor controlled, it is easy to increase the plant capacity by reducing the cycle time.

Energy Recovery

Energy recovery is carefully considered during the design of any unit, the delayed coker being no exception. However, energy recovery in a coker differs from that of other units because of the cyclic nature of the process. Each 24-hour period, the process is upset by heating up a cold, empty coke drum and by drum switching.

Significant upsets during these times can cause fractionation products to go off specification. To minimize such process upsets, heat integration must be limited to a certain extent. Coker design allows for a drop off in liquid products during this part of the coking cycle and steam generators are provided on the fractionator pumparounds to absorb the swings in duty. The energy is efficiently recovered during normal operation, and the drop off in steam production during switching has no noticeable effect on the overall steam system.

Coking Heater

Recently a double-fired heater design has been adapted for the delayed coking heaters. The advantage of the double-fired heater design is lower peak temperature and flux rate but a higher average flux rate which results in longer heater run length and overall less radiant surface area.

On-line Decoking

The coking of any of the heater process coils would normally require heater shutdown for decoking. This would result in the loss of unit throughput. The on-line decoking procedure permits the decoking of one process coil while the other process coils remain in normal process service. This procedure allows the unit to continue to operate at a reduced throughput and increases the heater run length.

The on-line decoking procedure involves the blocking off of the hydrocarbon flow to the coil and replacing it with steam or condensate at a controlled rate. The temperature of the coil to be decoked is modulated while injecting steam or condensate. The coke deposits in the coil are spalled off the tube surface and swept out of the tubes into the transfer line to the coke drums.

Coke Drums

The cyclic nature of the coke drum operation introduces considerable stresses at the shell/skirt joint. A finite element analysis should be undertaken for the coke drum/skirt attachment design. Other design features for stress minimization include hot box insulation, self-reinforced nozzles with full penetration welds and elimination of stiffening rings.

Coke Drum Level Indication

Neutron backscatter instruments are used to check the level in the coke drums. This device will indicate the vapor/foam interface and foam/liquid interface during the filling cycle and the water level during drum cooling.

Hydraulic Decoking Equipment

Equipment provided for hydraulic removal of coke from the coke drums consists in general of the following items:

1. A high pressure multistage jet pump which supplies water to the combination boring and cutting tool is provided with a special bypass valve system to permit periodic diversion of water from the decoking equipment to the clear water tank, to avoid stopping the pump.

2. A cutting head assembly for each drum consists of the combination boring and cutting tool, drill stems, air motor and swivel joint for rotating the drill stem, connecting water and air hoses, hoist for lifting the drilling assembly and guide plates for the drill stem. The combination cutting tool has nozzles at the bottom for boring and nozzles at the side for cutting. A single combination cutting tool is provided for each pair of drums.

3. A remote operated automatic coke drum top head unheading system is provided to remove the top manway cover.

4. A coke drum bottom unheading system is provided to remove the bottom manway cover and attach the telescopic chute. The system consists of a hydrau-

lically operated flange cover, unheading cart, telescopic chute, feed pipe disconnect system and hydraulic power unit.

5. The telescopic chute is attached to the bottom flange of the coke drum during decoking to direct the flow of falling coke and water thus preventing splashing and scattering of coke about the operating platform.

The drilling equipment is supported by a derrick structure, mounted over the coke drums, permitting the drill assembly to be raised or lowered. Each coke drum has its individual cutting head assembly. The drill stem and cutting tools are rotated by means of an air motor driven swivel joint. Each drilling assembly includes a drill stem which is long enough to reach the entire length of the coke drum, and to which the cutting tool can be attached. The water line from the jet pump branches at the top coke drum platform to serve the drilling assembly over each drum. When one branch is in service, the other is disconnected, using a swing joint, and then plugged for safety. A flexible hose, designed for high pressure service, connects the water line to the drill stem (through the swivel joint) thus permitting free vertical travel of the drilling assembly.

The coke is removed from the drum over a period of 4 to 5 hours. The drilling operation is performed in two steps: boring and cutting. During the boring operation, the high velocity water stream is directed through the downward pointing nozzles to cut a hole through the coke bed. The drill stem is lowered quite rapidly through the drum to keep the amount of water in the drum to a minimum before the hole is cut through. When the cutting tool breaks through, accumulated water and coke flow from the coke drum bottom manway. During the cutting operation, the high velocity water is directed through the side nozzles to cut the coke bed radially. The cutting tool is rotated slowly and is lowered in steps from the top of the coke drum.

When the cutting of coke is completed, the water to the cutting tool is stopped.

Coke Handling

Coke and water discharged from the coke drums during coke cutting need to be separated prior to coke reclamation and re-use of the water. There are a variety of schemes and equipment employed for coke dewatering and handling. The selection of a scheme or equipment is based on several considerations including environmental issues, capital investment, equipment maintenance, available plot-area and means of coke transport from the unit.

Coke handling systems can be generally classified into four types (Fig. 3):

1. Direct loading into railroad cars beneath the coke drums

2. Hydrobin system, utilizing a coke-water slurry for initial transport of coke

3. Discharging into a coke drop-out pit and reclaiming coke utilizing a crane with a clamshell bucket

4. Discharging into a coke drop-out pad and reclaiming coke utilizing a front-end loader

Direct Loading Into Railroad Cars

In this system, railroad tracks are laid under the coke drum structure so that coke and the water used to cut the coke, fall through the bottom head of the coke drum into a railroad hopper or gondola cars. As one car is filled, the line of cars is pulled along for another to fill. The number of cars required depends on the amount of coke produced and the size of the cars. A crusher can be utilized after the coke drum if sized coke is required.

Water separates from the coke by overflowing or draining through holes in the car. It is collected in a trench below the car and pumped from a sump for clarification and reuse.

Hydrobin System

The hydrobin system is a slurry handling method for dewatering coke. Coke and cutting water, as it drops from the coke drum, fall into a rail mounted portable coke feeder-crusher car which reduces large lumps to a size suitable for pumping (up to 15 cm). It also controls the flow of coke to the sluiceway located below the car. As the coke falls into the sluice it is conveyed to the sump by the water jet.

The coke and water are pumped away from the sump by a coke slurry pump. This mixture is discharged over a suitable sizing screen or directly into a hydrobin. The hydrobin is a separating vessel for the coke and water. The overflow water flows by gravity from the hydrobin to the clarifier from which the clear water is pumped to the clear water tank for re-use in the jet pump. When all the coke has been removed from the coke drum, the decanting valves are opened on the hydrobin and the remaining water drained to the clarifier. After the water has drained from the coke, the coke is emptied from the bin by opening a large piston-operated coke gate at the bottom of the bin. The coke can drop directly into railroad cars positioned under the hydrobin, or it can be chuted to mechanical conveyors for disposal elsewhere. Often two hydrobins are used, one for filling with coke slurry while the other is draining water.

Coke Drop-out Pit

Coke and water from the coke drum are discharged via a chute onto a concrete pit and form a pile. The concrete pit is located below grade and is large enough to typically hold two days of coke production and allow adequate space for water drainage. The coke pile is allowed to remain on the pit to allow the water to drain from the coke pile. The pit is sloped to a side which has multiple screened ports packed with coke. The coke helps in retaining the fines while the water drains across the ports into a maze. The maze is typically a 3- or 4-compartment settling basin which directs the water to flow from one compartment to the other across an overflow weir. A sump is located at the end of the last maze compartment.

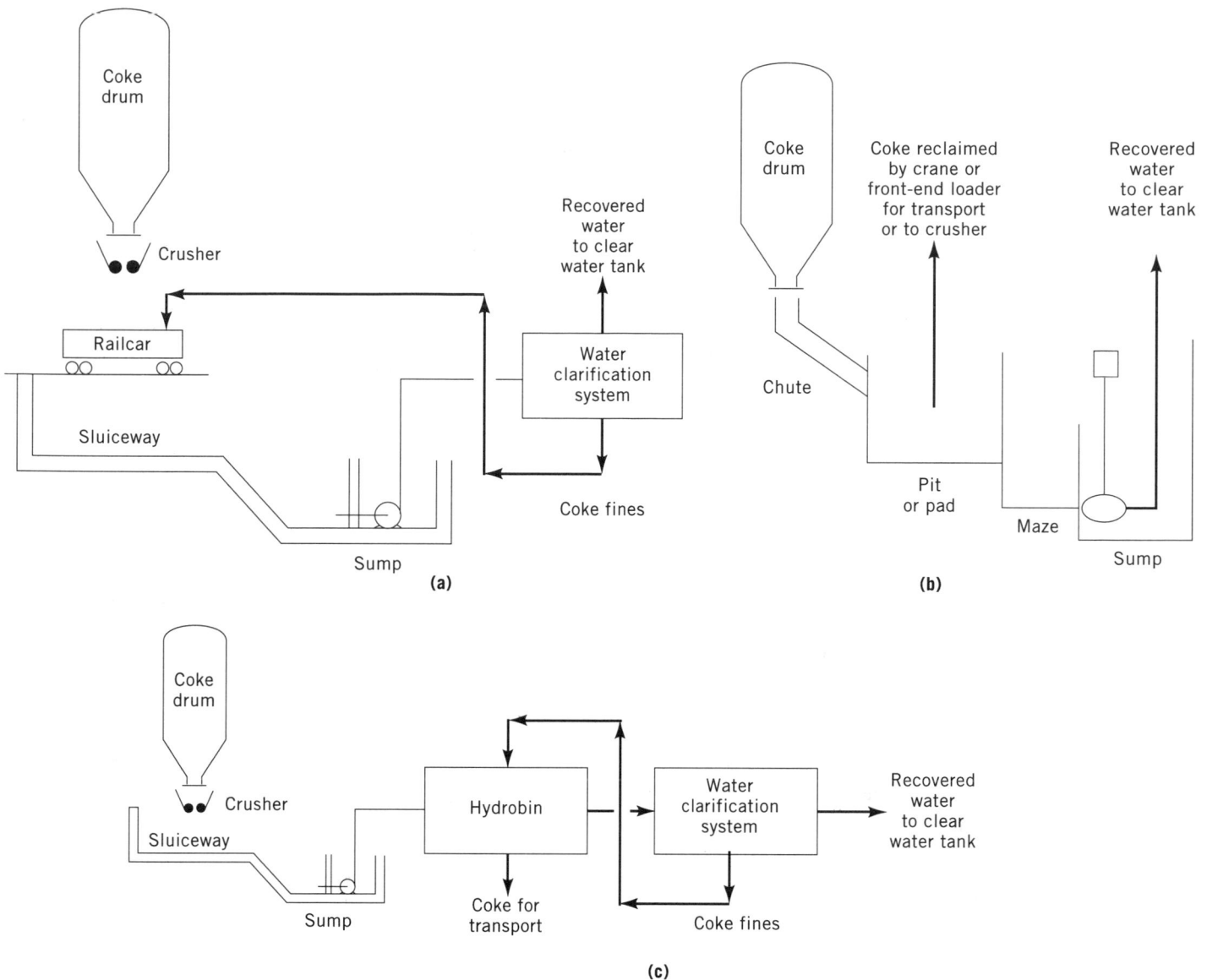

Figure 3. Coke handling systems.

The clarified water is pumped from the sump to the clear water tank for reuse.

The pit is spanned by a mobile crane with a clam-shell bucket. After the coke has had sufficient time to drain, the crane picks up the coke and transports it to a crusher, railcar, truck or conveyor for shipment. The mobile crane is also used to remove the coke fines from the maze.

COKE DROP-OUT PAD

The coke drop-out pad functions similar to a pit except the bottom of the pad is typically at grade level or slightly lower than grade. The water clarification is also accomplished via a maze. However, the coke is typically removed by a front-end loader fitted with a large bucket. The front-end loader is also used to remove coke fines from the maze.

SAFETY

Several safety features for the protection of unit personnel and equipment are incorporated into the design of the delayed coker unit (7–9). Some of these safety features are systems typical for a unit handling flammable liquids and gases. However, there are certain safety issues which are typical for a coker unit and they are described below:

Hydraulic Decoking System

The hydraulic decoking system is equipped with an elaborate and extensive interlock system to prevent injury to personnel from the high pressure water jet. The interlock system is designed to accomplish the following primary objectives:

- Pump can be started only from the pump location and when the flow is directed to the clear water tank.

- Water flow to the cutting tool is possible only when the tool is inside the drum and only the corresponding isolation valve is open.
- Permissive start and automatic shutdown of pump when certain equipment safe operating requirements are within acceptable levels eg, adequate lube oil pressure.
- Equipment to prevent free fall of drill stem.
- Interlocks to stop water flow in case of hose rupture.

Coke Drum Bottom Flange Unheading System

Prior to coke cutting, the coke drum bottom flange must be removed. Sometimes, the water used for coke cooling remains trapped in the coke or free flowing shot coke is formed during coking. The manual opening and removal of the bottom flange and the manual attachment of the telescopic chute to the coke drum bottom flange exposes the operator to significant safety hazard.

Several semi-automatic or fully automated systems have been developed and used to enhance safety during the bottom flange removal and telescopic chute attachment. The design features of these systems vary considerably but in general include the following:

- Remote operated and automated system.
- Reduction in time allocated for unheading and reheading.
- Ability to handle shot coke.

Coke Drum Top Flange Unheading System

The opening and removal of the top flange may release harmful hydrocarbon vapor which remains trapped in the drum and, thereby presents a safety hazard to the operator. An automatic hydraulic system typically is provided for safe operation. This system permits the top flange to be opened and removed by the operator from a remote location.

ENVIRONMENTAL

Delayed coking units should be environmentally safe addressing all the four major categories of air, liquid, solid and noise pollution occurring within the unit battery limit (8). Typical delayed cokers are designed to meet the US–EPA New Source Performance Standards (40CFR60) and Refinery Effluent Guidelines (40CFR419).

Gaseous Emissions

The gaseous emissions generated with the delayed coking unit can be grouped as follows:

- Combustion flue gases from the coking heater.
- Safety valve and pressure control valve discharges.
- Fugitive emissions.

The pollutants from combustion, ie, nitrogen oxides (NO_x), carbon monoxide (CO), sulfur dioxide (SO_2) and particulars are controlled by the application of the Best Available Control Technology (BACT) such as Low-NO_x burners, selective catalytic reduction, combustion air control and using desulfurized fuel.

BACT is employed to control fugitive emissions in accordance with the New Source Performance Standards for control of fugitive emissions from refineries.

Liquid Effluents

The liquid effluent generated within the delayed coking unit can be grouped into the following source categories:

- Sour water
- Slop oil
- Process vessel drains

Sour water is sent outside the unit for further processing and treatment. The stripped sour water can be recycled to the coking unit. Slop oil and process vessel drain flow is recovered and recycled to the coking unit.

Solid Effluents

There are no major solid wastes generated within the delayed coking unit except coke fines. The BACT for the control of coke fines consists of a combination of one or more of the following engineering design and operating practices to minimize the emissions of fugitive dust in material storage and handling systems:

- Reducing the number of material transfer points.
- Use of enclosed transfer points and conveyors.
- Use of dry dust collection systems.
- Use of wet dust suppression systems.

Noise

Delayed coking units are designed to meet the federal, state or local regulations. Typical noise sources are rotating equipment, control valves, heater coils, heater burners, and coke drum vents.

OPERATING REQUIREMENTS

Utilities

For a typical 20,000 BPSD fuel grade coker, with an integrated vapor recovery unit, estimated utilities and chemicals requirements are as follows:

Utilities (continuous): net steam, produced, 8618 kg/h; BFW, 11,340 kg/h; power, 3,500 kWh/h; cooling water, 2,700 gpm; fuel, 137×10^9 J/H.

Utilities (intermittent): steam, 13,608 kg/day for 2 h/day; power, 8,000 kWh/day for 5 h/day; and industrial water, 302,832 L/d (max).

Chemicals (continuous): antifoam, 15.8 kg/day; corrosion inhibitor, 27 kg/day; deemulsifier, 0 (max 4.5) kd/day.

On Stream Factor

The typical on-stream factor for a delayed coker unit is 0.9 to 0.97. Heater decoking takes approximately 2–3 days and the normal turnaround time is 10 days.

Typical Operating Labor Per Shift (Does Not Include Supervision)

Operators, 3; decoking crew (shift including swing), 4; and coke handling, 2.

Maintenance Costs

The delayed coker maintenance costs typically run 3–4% of the total installed cost of the plant.

FUTURE DEVELOPMENTS

Areas where significant improvement is expected in the delayed coking technology include:

- Disposal of API sludge with coker products.
- Disposal of plastics with coker products.
- Pollution abatement such as lining of pit/pad to prevent ground water pollution.
- Safety enhancements in the areas of coke drum top and bottom head unheading.
- Advanced process control.
- Alternatives to hydraulic decoking.
- Coke cutting technology.

BIBLIOGRAPHY

1. V. Mekler and M. Brooks, "When is Delayed Coking Worthwhile?" *Petroleum Refiner* (June 1959).
2. D. B. Meyer and J. C. Webb, "Coker Can Handle High Carbon Stock," *Petroleum Refiner* (Feb. 1960).
3. D. T. Shedd, "Technical Status and Position-Delayed Coking," ABB Lummus Crest internal document.
4. D. T. Shedd, "Highlights of Delayed Coking," ABB Lummus Crest internal document.
5. A. Rhoe, "Meeting the Refiner's Upgrading Needs," Presented at *NPRA*, San Francisco, Calif., Mar. 20–22, 1983.
6. "Feedstocks Key to High-Quality Needle Coke," *Oil Gas Journal*, **73**(21), 108–111 (May 26, 1975).
7. D. T. Shedd and J. F. S. Frith, "Process Design Policy Manual—Delayed Coking Units," ABB Lummus Crest internal document.
8. "Delayed Coking Technology Brochure," ABB Lummus Crest document.
9. G. H. Stockman, "Assure Delayed Coker Safety," *Hydrocarbon Processing*, 109–110 (Mar. 1992).

COLD FUSION

R. M. Shaubach
N. J. Gernert
Thermacore, Inc.
Lancaster, Pennsylvania

Two scientists at the University of Utah, Stanley Pons and Martin Fleischmann, announced on March 23, 1989, that they had carried out fusion through the electrolysis of water. The process was quickly dubbed "cold fusion" (1).

The announcement set off a flurry of activity directed at explaining and harnessing this apparent new source of energy. Over the next year, the world's scientific community expended significant funds trying to confirm the discovery at Utah. Cold fusion became the most controversial scientific event in decades. Work in the area has almost ceased, but a few groups continue to study the technology; at best, it remains a laboratory curiosity.

A theory consistent with the limited experimental observations is presented here. If the technology is proven to be feasible, the potential for generating large quantities of heat would exist. However, conversion of this heat to useful energy may present difficulties; its economic potential is not at all clear (ca. 1994).

See also FUSION ENERGY; MAGNETOHYDRODYNAMICS.

THEORY

Many theories have been offered to explain the physics and chemistry behind the excess energy. Some, like Pons and Fleischmann's theory, include a nuclear fusion reaction, others include a chemical reaction. Experimental results to date show nuclear by-products orders of magnitude below those required to explain the excess energy. As a result, the authors believe a chemical rather than a nuclear explanation best fits existing observations in nature and experiments. A theory developed by Mills (2) provides a focal point for an explanation. However, this theory is controversial and is not widely accepted by the world's scientific community.

The theory starts with the hydrogen atom. Hydrogen is selected for ease of discussion; however, the theory applies to deuterium as well. The hydrogen atom has a single proton surrounded by an electron. The electron shell around the proton behaves in analogous terms much like an elastic balloon.

Energy is required to inflate a balloon; energy is stored in the expanded balloon. Similarly, the electron shell of a normal hydrogen atom can be made to expand by the absorption of energy. This is a common phenomenon known to scientists as photon absorption. When the stored energy of a balloon is released, the balloon contracts. Similarly, the electron of a normal hydrogen atom can be induced to contract below its normal ground state and release some of its stored energy. This is not a common phenomenon, and it runs counter to the usual model of the atom. The excess heat observed during electrolysis is derived from the energy released as hydrogen atoms contract.

If deuterium is used in place of normal hydrogen, the electron shell of a few deuterium atoms may be reduced to the point where fusion occurs, producing small amounts of tritium and other nuclear particles as reported by many research groups reporting "cold fusion." However, the small number of nuclear particles emitted can only account for a billionth of the heat released, and the fundamental nature of the phenomenon is physical and chemical, not nuclear.

The process has been stated by Mills as follows: "The predominant source of heat of the phenomenon denoted Cold Fusion is the electrocatalytically induced reaction whereby hydrogen atoms undergo transitions to quantized energy levels of lower energy than the conventional 'ground state.' These lower energy states correspond to fractional quantum numbers. The hydrogen electron transition requires the presence of an energy hole of approximately 27.21 eV provided by electrocatalytic reactant(s) (such as Pd^{2+}/Li^+, Ti^{2+}, or K^+/K^+), and results in 'shrunken atoms' analogous to muonic atoms. In the case of deuterium, fusion reactions of shrunken atoms yielding predominantly tritium are possible."

According to quantum mechanics, hydrogen can only have energies given by the following Rydberg formula:

$$E = \frac{-13.6\,eV}{n^2} \tag{1}$$

where n is an integer. Mills predicts that in addition to the energy states of hydrogen given by Eq. 1 with n as an integer, new lower energy states are possible given Eq. 1 with n as a fraction.

The hydrogen atoms that have achieved energy levels below the ground state are smaller in diameter than normal hydrogen. Mills has named these shrunken atoms hydrinos, Latin for baby hydrogen.

In summary, excess energy can be extracted from a hydrogen atom when it chemically interacts with another reactant, providing an energy hole of 27.21 eV. For example, hydrogen atoms produced at a nickel cathode of an electrolytic cell have unpaired electrons binding them to the surface of the nickel. These atoms react with the K_2CO_3 electrolyte, which provides the energy hole of 27.2 eV. The K^+/K^+ interaction is catalytic; K_2CO_3 is not consumed. The same reaction can be created without electrolysis by diffusing hydrogen atoms through a nickel membrane contacting K_2CO_3. The product of the reaction, a hydrino, has less energy than normal hydrogen and is the ash of the process. The hydrino atom will join with a second atom to form a dihydrino molecule, much as hydrogen atoms join to form hydrogen molecules.

PRODUCT OF THE REACTION

Several methods are available to detect the presence of hydrinos. They appear to have been detected in interstellar space as well as in the laboratory.

Labov and Bowyer (3) instrumented a sounding rocket to measure the emissions from dark matter in the extreme ultraviolet background of interstellar space. These results shown in Table 1 indicate 15 peaks in the spectrum at wavelengths ranging from 84.8Å to 633Å.

These peaks and associated energy levels correlate with the predicted values for the formation of hydrinos. This implies that hydrinos are being formed from hydrogen in outer space and may be the constituents of dark matter.

Dihydrino and dideutrino molecules may have been identified by mass spectroscopy. For example, Miles (4–7) and Bockris (8) report production of mass four as identified by mass spectroscopy of the cryofiltered gases evolved from an electrolysis cell comprising a palladium cathode and a LiOD/D_2O electrolyte. According to Miles (6), the intensity of the mass four peak maintained an approximate correspondence to the amount of excess power or heat observed in electrochemical calorimetric cells. At the time, Miles believed the mass four peak to be helium as a result of fusion reactions. According to Miles, "Ignoring the helium/heat relationship [Table 1 of reference 6], the

Table 1. Observed Emission Data[a,b]

Peak	Observed Wavelength, Å	Observed Energy, eV	Peak Assignment	Predicted Energy[a], eV	Predicted Wavelength, Å
1	84.8	146.2	$\frac{1}{5} \to \frac{1}{6}$ H transition	149.6	82.9
2	101.5	122.2	$\theta \to \frac{1}{3}$ and $\frac{1}{4} \to \frac{1}{5}$ H transition	122.4	101.3
3	116.8	106.2	$1 \to \frac{1}{3}$ H transition	108.8	114.0
4	129.6	95.6	$\frac{1}{3} \to \frac{1}{4}$ H transition	95.2	130.2
5	139.6	88.8	He scattered peak #3	87.6	139.6
6	163.2	75.9	Second order of peak #1	74.8	163.2
7	181.7	68.3	$\frac{1}{2} \to \frac{1}{3}$ H transition	68.0	182.3
8	200.6	61.8	Second order of peak #2	61.2	202.6
9*	233.8	53.0	$\theta \frac{1}{2}$ H transition	54.4	227.9
10	261.2	47.5	He scattered peak #7	46.8	265.0
11	302.5	41.0	$1 \to \frac{1}{2}$ H transition	40.8	303.9
12	459.1	27.0	Second order of peak #9	27.2	455.8
13	584	21.2	He resonance scattered emission	21.2	584
14	607.5	20.4	Second order of peak #11	20.4	607.8
15	633.0	19.7	He scattered peak #11	19.6	633.0

[a] Ref. 3. Bowyer and Labov used three monochrometers for maximal sensitivity in each energy range: 80–230Å, 230–430Å. The monochrometer change at 230Å resulted in the 6Å discrepancy between the calculated and observed lines.

[b] For hydrogen transitions, $\frac{1}{n_1} \to \frac{1}{n_2}$; $n_1 = 0$ for $r = \theta$

$$E = \left[\frac{1}{n_2^2} - \frac{1}{n_1^2} \right] \times 13.6\ eV$$

For helium Compton scattered peaks of hydrogen transitions, $\frac{1}{n_1} \to \frac{1}{n_2}$; $n_1 = 0$ for $r = \theta$

$$E = \left[\frac{1}{n_2^2} - \frac{1}{n_1^2} \right] \times 13.6\ eV - 21.21$$

simple yes/no detection of helium in 7/7 experiments producing excess heat and the absence of helium in 6/6 experiments not producing excess heat (1 in D_2O, 5 in H_2O) implies a chance probability of $[\frac{1}{2}]^{13} = 1/8192$ or 0.012%" (6).

The data may not be consistent with a fusion reaction as the source of the excess heat or the mass four peak. The correct assignment is D_2^*, the dideutrino molecule. These molecules form from deutrino atoms on the surface of the palladium cathode. The deutrino atoms form the same way hydrino atoms form, which is the source of the observed excess heat.

The dideutrino molecule can also be identified by high-resolution mass spectroscopy. Yamaguchi (9) reported high-resolution (.001 AMU) mass spectroscopic data of the gases released from deuterium or hydrogen-loaded palladium sheets coated on one side with a hydrogen-impermeable gold layer and coated on the other surface with an oxide coat (MnO_x, AlO_x, SiO_x). Heat was observed from light and heavy hydrogen only when the mixed oxide coat was present. The mass spectroscopic data of the gases released when current was applied to a deuterium (99.9%)-loaded MnO_x coated palladium sheet indicate the presence of a large shoulder on the D_2 peak. The authors assert that the data are consistent with the production of D_2^*, the dideutrino molecule. These molecules form from deutrino atoms on the surface of the palladium sheet.

The dihydrino molecule (H_2^*) is predicted to be spin paired, to be smaller than the hydrogen molecule, to have a higher ionization energy than H_2, and to have a lower liquefaction temperature than H_2. Dihydrino molecules present in the gases from an electrolytic cell having an electrolyte of the electrocatalytic couple, K^+/K^+, can be separated from normal hydrogen by cryofiltration (10). Following cryofiltration, the dihydrino molecule is distinguished from normal molecular hydrogen using mass spectroscopy. Mass spectroscopy distinguishes a sample containing dihydrino molecules versus a sample containing H_2 by showing both a different ion production efficiency as a function of ionization potential and a different ion production efficiency at a given ionization potential.

Typical results for hydrogen and dihydrino gas samples are shown in Table 2.

EXPERIMENTAL PROTOCOL

This section describes a procedure to make an electrolytic cell using regular light water and potassium carbonate as the electrolyte. Some experimenters claim to have reproduced the excess heat if care is taken in preparing the apparatus. As usual in electrochemistry, measures must be taken to avoid impurities in the system, especially organic substances. *Electrolysis of water-based electrolytes produces hydrogen and oxygen gas. A dangerous explosion will occur if a spark ignites the gas.*

1. The overall assembly is shown in Figure 1. It consists of a vacuum dewar, cathode, anode, heater, thermocouple, electrolyte, and stirrer. Each of these parts and their fabrication is discussed below.

2. Use a 350-ml silvered vacuum jacketed dewar (Cole Palmer #8600) with a 7-cm opening covered with a 19.05-mm thick styrofoam stopper lined with Parafilm. Clean with Alconox and rinse with distilled water, then use 0.1 M nitric acid and rinse thoroughly with distilled water to remove all organic contaminants.

3. The cathode can be made from a 7.15 cm wide by 5 cm long by 0.125-mm thick nickel foil (Aldrich 99.9+%, cold roll, clean Ni) spiral of 9 mm diameter and 2 mm pitch with a nickel lead strip. The cathode material should be removed from its container with rubber gloves and cut and folded in such a way that no organic substances are transferred to the nickel surface.

   The nickel cathode can be prepared by tightly rolling the nickel foil about a 9-mm diameter rod. Remove the rod. The spiral should be formed by partially uncoiling the foil. The lead should be inserted into Teflon tubes to ensure that no recombination of the evolving gases occurs. Clean the cathode by inserting it into a solution of 0.57 molar K_2CO_2/3% H_2O_2 for 30 minutes. Remove the cathode and immediately rinse with distilled water and allow to dry.

4. The anode can be made from a 10 cm long by 1 mm diameter spiraled platinum wire (Johnson Matthey) with a 0.127-mm Pt lead wire. The lead should be inserted into a Teflon tube to prevent recombination, if any, of the evolving gases. The Pt anode should be mechanically scoured with steel wool, soaked overnight in concentrated HNO_3, and rinsed with distilled water.

5. A 10-ohm resister housed in a 2-mm outer diameter Teflon tube can be used to calibrate the cell. An AC or DC source can be used to power the resister.

6. Make the electrolyte from 40 g of K_2CO_3 powder mixed into 500 cc of distilled water. This will provide a 0.6 molar solution.

7. A Teflon-jacketed type J thermocouple can be used to measure electrolyte temperature.

8. Magnetic stirring should be used to avoid mechanical penetrations. Flexamix Model 76, Fisher or equivalent.

The apparatus should be assembled as shown in Figure 1. The cathode should be added to the cell as the last step of the assembly. Electrical power to the cathode should be initiated immediately after inserting the cathode in the electrolyte. Maintain about 2.5 to 3.0 VDC across the cathode at all times while the cathode is inserted in the

Table 2. Mass Spectroscopy Intensity Cryofiltered DiHydrino and Hydrogen Molecules at Nominal Mass 2[a]

Species	Ionization Potential, eV		
	15	20	35
Hydrogen, H_2	0.000	0.005	0.005
Dihydrino, H_2^*	0.025	0.020	0.240

[a] These results show that H_2^* has an ionization potential above 20 eV; H_2 has an ionization potential below 20 eV.

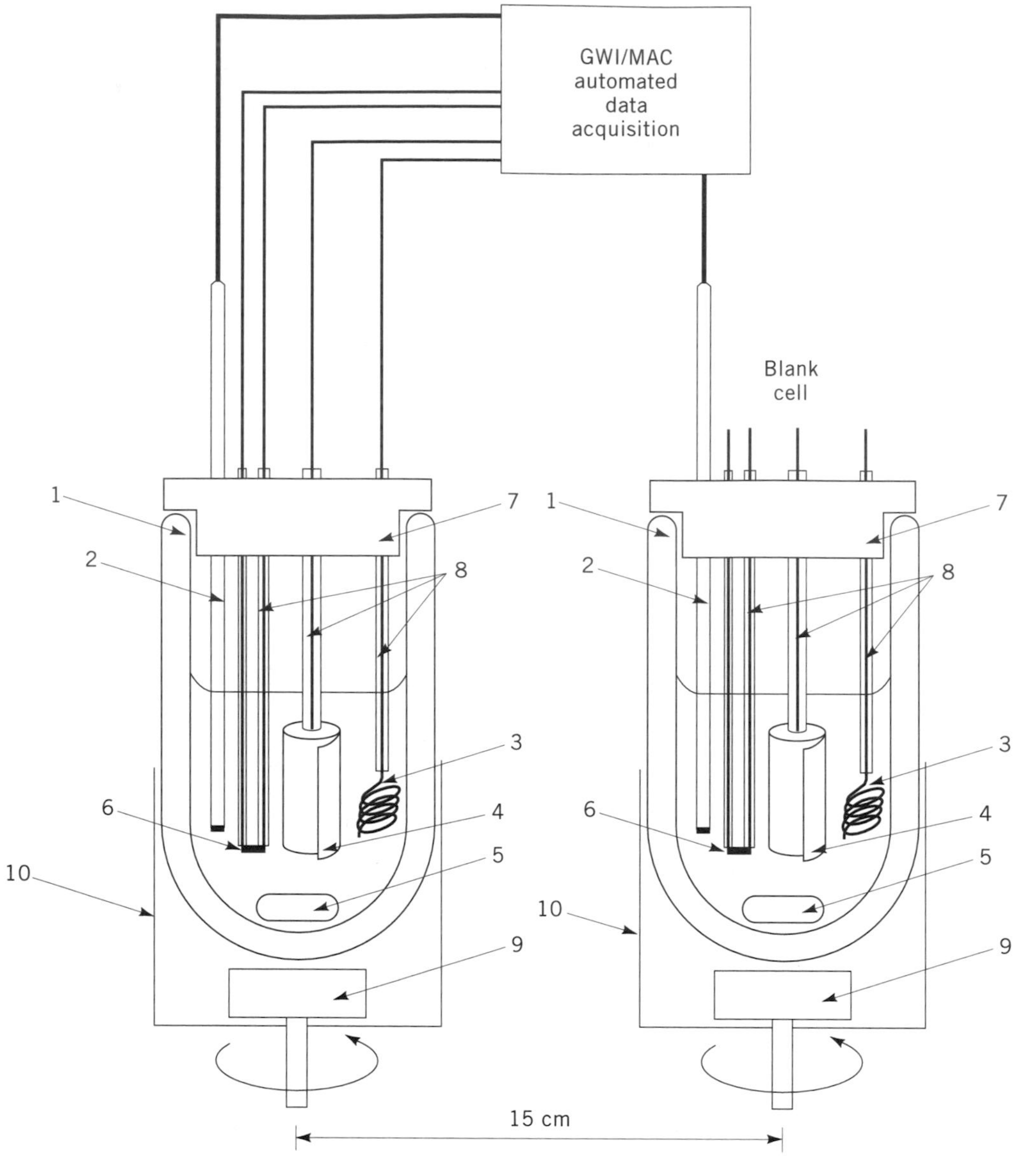

Figure 1. Experimental calorimeter set-up 1, vacuum jacketed dewar; 2, thermistor; 3, Pt anode; 4, Ni cathode; 5, magnetic stirring bar; 6, resistor heater; 7, rubber stopper; 8, Teflon tubing; 9, magnetic stirrer; 10, aluminum cylinder.

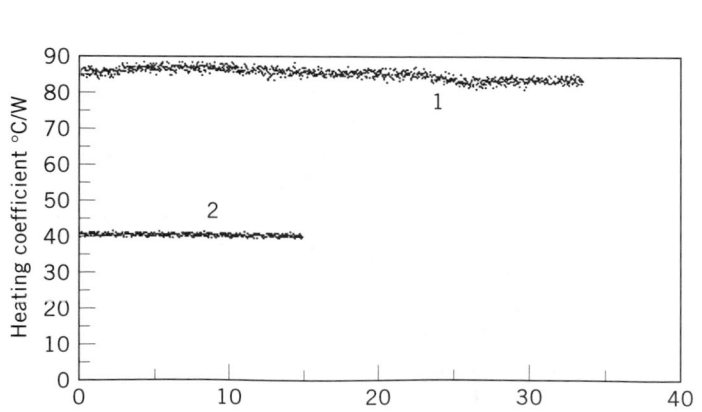

Figure 2. Plot of heating coefficients versus time, 1, electrolysis with a nickel cathode at 0.083A in K_2CO_3; 2, resistor working in K_2CO_3.

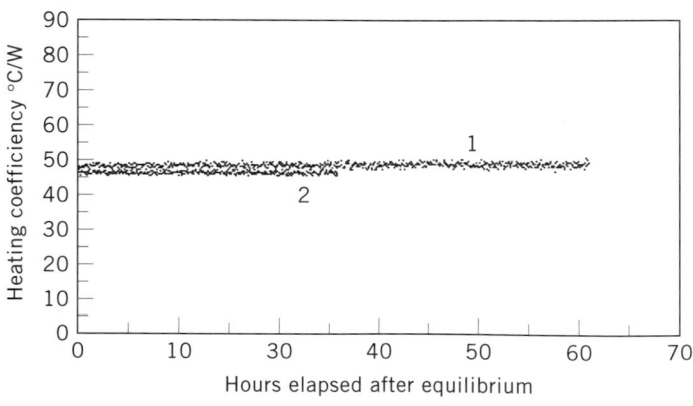

Figure 3. Plot of the heating coefficents versus time 1, electrolysis at 0.018 A in Na_2CO_3; 2, resistor working in Na_2CO_3. Almost all electrolysis experiments will be similar to the case of Na_2CO_3. Only a few combinations of electrolytes/electrodes such as the K_2CO_3 case above, will yield excess heat.

electrolyte. Reclean the cathode each time it is removed from the electrolyte.

The presence of excess energy can be established by comparing the electrolyte temperature rise above ambient for two nominally identical cells—one using K_2CO_3 as an electrolyte, the other using $NaCO_3$ as an electrolyte. The temperature rise above ambient for K_2CO_3 will be larger than that for Na_2CO_3 because the K^+/K^+ combination provides the required 27.21-eV energy hole whereas the Na^+/Na^+ does not.

Representative data are plotted in Figure 2 for the K_2CO_3 cell and in Figure 3 for the Na_2CO_3 cell. These figures also contain data from a third cell with equivalent I^2R heat input without electrolysis. Notice that the cell using K_2CO_3 as an electrolyte runs about 40°C/watt hotter than the Na_2CO_3 cell.

BIBLIOGRAPHY

1. B. Stanley Pons and W. P. Pons, Mortin Press Conference, University of Utah, March 23, 1989. (As reported in Taubes, G., Bad Science, Random House, New York, 1993).

2. R. Mills, *Unification of Spacetime, the Forces, Matter, and Energy*, Science Press, Ephrata, Pa., 1992.

3. S. Labov and S. Bowyer, "Spectral Observations of the Extreme Ultraviolet Background," *Astrophysical J.* **771**, 810–819 (1991).

4. M. H. Miles, B. F. Bush, G. S. Ostrom, and J. J. Lagowski, *Electroanal. Chem.*, **304**, 271 (1991).

5. R. Daganl, *Chemical and Engineering News*, 31–33, (April 1, 1991).

6. M. H. Miles, B. F. Bush, G. Ostrom, and S. Lagowski, "The Science of Cold Fusion," *J. J. Conference Proceedings*, Vol. 33. T. Bressani, E. Del Giudice, and G. Preparata, eds., SIF, Bologna, 1992, pp. 363–372.

7. M. H. Miles, R. A. Hollins, B. F. Bush, J. J. Lagowski, and R. E. Miles, *J. Electroanal. Chem.*, **346**, 99 (1993).

8. C. C. Chien, D. Hodko, Z. Minevski, and J. Bockris, *J. J. Electroanal. Chem.* **338**, 189–212 (1992).

9. E. Yamaguchi and T. Nishioka, "Direct Evidence for Nuclear Fusion Reactions in Deuterated Palladium," *Proceedings of the Third International Conference on Cold Fusion*, Nagoya, Japan, 21–25, October 1992.

10. R. Mills, W. Good, and R. Shaubach, "DiHydrino Molecule Purification and DiHydrino Molecule and Hydrino Atom Identification," *Fusion Technology*, **25**(1), 103–119 (1994).

COMBUSTION MODELING

L. Douglas Smoot
Stephen K. Kramer
College of Engineering and Technology
Brigham Young University
Provo, Utah

HISTORY AND CLASSIFICATION

Combustion models are mathematical expressions or formulations that describe various aspects of combustion processes. They are based on the laws of physics and chemistry (such as the conservation of mass, momentum, and energy) and on experimental observations. They can be simple or complex in form. Combustion models can be generally categorized as follows:

1. Limited analytical or asymptotic models that focus on a few key aspects of combustion and that are commonly used to test combustion theories.

2. Models of a specific combustion process such as particle devolatilization, radiation, or chemical kinetics that are often used as submodels in comprehensive codes but that can sometimes model simplified systems.

3. Models based primarily on correlations of experimental data; development of these correlative models may be guided by insights from theories or dimensional analysis.

4. Comprehensive models that include many of the governing aspects of a complex combustion process such as a coal gasifier or a gas turbine; such models typically require numerical solution of a large number of differential equations, include several essential submodels, and are not "universal" in their application but are designed for specific classes and types of combustion systems and processes.

This article discusses each of these modeling types but focuses on comprehensive combustion models. See also Coal combustion; Combustion systems-measurements.

History

Models of combustion are based on the fundamentals of chemistry and physics that have been developed during the last four centuries. Areas of chemistry and physics that are especially important to the understanding of combustion are chemical reactions and kinetics, thermodynamics, molecular motion and diffusion, heat transfer, fluid flow, and turbulence. The historical developments of these disciplines have been detailed (1). Of particular note are the efforts of researchers who have demonstrated that molecular systems can be understood and the properties calculated using thermodynamics. This line of investigation led to the determination of adiabatic flame temperatures and calculations of chemical compositions. The efforts of many molecular physicists have been summarized (2). Understanding of the molecular behavior of gases has been essential for understanding combustion behavior. Equally important are the conservation equations (3). In the field of chemical kinetics, Svante Arrhenius in 1889 postulated that the temperature dependence of the reaction rate is related to the Boltzman factor, $e^{(-E_A/RT)}$. Mechanisms for first-, second-, and third-order reactions have been postulated (4), which have aided the development of rate equations.

The earliest combustion problem that combined transport phenomena with chemical kinetics was the propagation of a laminar combustion wave (5). In 1883, Mallard and Le Chatelier considered the heat transfer from the flame to the incoming gas to be the dominant mechanism (6,7). They divided the temperature variation across a laminar flame into two regions: a preheat zone and a reaction zone (Fig. 1). An energy balance, coupled with an overall mass balance, leads to the significant result:

$$S_L \propto \sqrt{\frac{\lambda}{\rho_0 c_p} w} \tag{1}$$

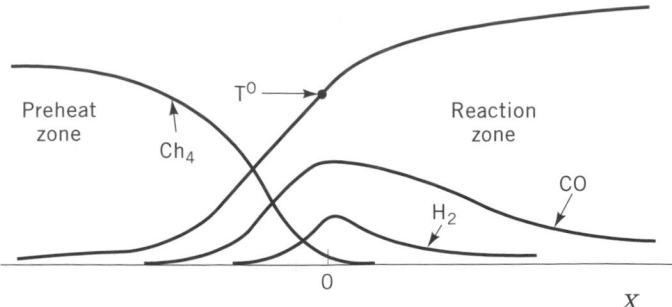

Figure 1. The structure of a simple, laminar, premixed methane-air flame. From Ref. 8.

where S_L is laminar flame velocity, λ is thermal conductivity of the mixture, ρ_0 is the initial density of the mixture, c_p is the heat capacity of the mixture, and w is the reaction rate. The temperature dependence of flame velocity is found by replacing the reaction rate with the Arrhenius relationship. These relationships were proved and expanded on soon thereafter (5).

The international nature of the advances in combustion modeling is a characteristic that has continued. Significant additions to fundamental combustion theory were made in the 1930s (6). Researchers coupled the species conservation equation with the energy equation, assumed a constant pressure, adiabatic flame, and a Lewis number (the ratio of conductive heat transfer to molecular diffusivity) of 1. Simplifications such as these and the use of dimensionless relationships and asymptotic analyses became standards in flame modeling. These analytical and asymptotic models have been reviewed in depth (9).

Possibly the first numerical calculations on a electronic computer of laminar flame properties were performed by Hirschfelder, Curtiss, and Campbell (10) in 1953 (5,7). Other efforts followed soon thereafter with the majority of these models being set up in a time-dependent formulation (11). The success of the computerized modeling of premixed, laminar flames is such that the various current models generally arrive at nearly identical answers that agree well with experimental results (12).

Semenov is credited with the identification of critical conditions for self-ignition for a homogeneous mass, followed by Frank-Kamenetskii's extension of this idea to burning carbon and Zel'dovich's application of it to the continuous-flow stirred-tank reactor (13). The key concept in these theoretical developments is that for an exothermic reaction to be self-propagating, the heat flux out of the reacting volume must be less than the heat generated by the reactions. Semenov is also credited with the identification of chain theory as being key to the understanding of many combustion phenomena such as the rapidity of low temperature hydrocarbon combustion, the variation of ignition with temperature and pressure, and "cool" flames (14).

The study and modeling of turbulent flames may have begun with Damköhler in 1940. He related the turbulent burning velocity to the square root laminar flame velocity law (Eq. 1) by introducing a turbulent transport term. The turbulent flame was basically seen as a wrinkled laminar flame. This was the understanding of turbulent combustion until the middle of the 1960s (5). Turbulent stresses were represented as functions of the mean velocity gradient. The currently popular k-ε model for describing turbulent shear flows (15) calculates turbulent properties such as eddy viscosity through modeled conservation equations fitted to experimental data. The direct numerical solution of Reynolds turbulent stresses and numerical simulations of turbulence remains a major field of investigation.

Comprehensive models of combustion are relatively recent developments. Comprehensive models couple fluid dynamics with chemical reactions to simulate practical combustion systems. Computational fluid dynamics (CFD) is a large and important area of research in its own right and is commonly used in such scientific and engineering fields as hydrodynamics, heat transfer, machine and chemical process design, and analysis. Given the complexity of combustion phenomena in practical combustion systems, such models were not possible until computer technology and numerical solution methods reached a certain level of sophistication and speed. Among the earliest reports of comprehensive models were those for a gas turbine (16) and for anthracite coal flames in a concrete kiln furnace (17). Spalding was credited for originating the method used by Richter and Quack (17). Interestingly, Spalding reported that his program required less than 20 kilobytes of memory with total storage requirements varying from 20 to 150 kilobytes, depending on the grid size. Current comprehensive codes require program memory on the order of 1000 kilobytes with total storage requirements to operate the code up to several orders of magnitude higher. Comprehensive code development in the past 20 years has included substantial advances in the understanding of combustion reactions of particles, droplets, and gases, sophisticated numerical schemes, and graphical interfaces made possible by advances in computer technology, increased insights into turbulence and radiation, and considerable refinements of computer codes by comparison with experimental data.

No mention has been made of the considerable work that has been done on detonation waves. These phenomena have been discussed in detail (5–7,18,19), and regions of detonation have been defined on the Hugoniot curve (20,21). Understanding of detonation waves is useful for the comparison of deflagration waves and for the analysis of hazards in various environments.

Purposes

Modeling has two broad purposes: (*1*) to aid in the understanding of combustion processes and (*2*) to serve as an engineering tool in the design, operation, and evaluation of combustion equipment. Models of combustion can improve understanding of combustion processes in many ways:

1. They relate the fundamentals of chemistry and physics to observed phenomena.
2. They illuminate the interaction of complex or rapid processes that are difficult to examine experimentally.
3. They help determine the relative significance of the various parameters that influence combustion.
4. They can predict phenomena that have not been ob-

served experimentally, thereby indicating areas for further experimental work.

5. They can be used to improve the design and operation of experiments.

Combustion modeling has been used to further the understanding of combustion phenomena such as burning rates of various mixtures under a variety of conditions, cool flames, counterflow diffusion flames, ignition, deflagration-to-detonation transition, details of flame structure and chemical kinetics, flammability limits, flame stretch, the role of turbulence in flames, particulate and droplet combustion, soot formation, and microgravity flames.

Combustion models can be used as engineering tools in the following ways:

1. To scale up apparatus and equipment.
2. To predict the operation of established combustion apparatus given different conditions.
3. To help in the diagnoses and remediation of hazards, operating difficulties, and sources of emission.
4. To complement experiments that are costly to conduct.
5. To explore novel designs.

The engineering use of models is applicable to a variety of combustion processes and equipment, including the burning of natural gas, oil, or coal to generate electricity, heat, or power for home or commercial use; the use of liquid fuels in automobiles, trains, aircraft, and ships; the use of solid or liquid propellants in rockets and missiles; the combustion of gaseous, liquid, and solid fuels in the production of steel, other metals, cement, glass, and ceramics; the prevention and handling of fire and explosions in mines, mills, industry, homes, farms, and wildlands; the use of incinerators and flares to burn hazardous wastes and unwanted fuels and gases; and the emission of pollutants and exhaust gases such as unburned hydrocarbons, oxides of nitrogen and sulfur, carbon monoxide and dioxide, soot, and particulates from any of the above listed sources.

Classification

Combustion models can be classified according to a variety of characteristics (Table 1). Their geometry can be nondimensional or one-, two-, or three-dimensional. The equations can include transient (ie, time dependent) effects or be restricted to steady-state conditions. The reactants can be either premixed or initially separate (ie, diffusion flame), have only one or several phases, and may react homogeneously (ie, in one phase) or heterogeneously among phases. The flow may be quiescent, laminar, or turbulent, with natural or forced convection, and compressible effects may or may not be important. The reactions may occur so rapidly that only equilibrium chemistry need be considered or so slowly that the kinetics of the reactions must be included. The combustion waves themselves may propagate at either supersonic speeds (detonations) or at subsonic rates (deflagrations).

Given the multitude of purposes and the variety of classifications, many different types of models exist. More in-depth treatments of this subject are available (5,6,23).

Table 1. Classification of Comprehensive Combustion Models[a]

Criteria	Classification
Mathematical complexity	Nondimensional
	One-dimensional
	Multidimensional
	Steady-state
	Transient
Flame type	Premixed
	Diffusion
Flow type	Quiescent
	Laminar
	Plug flow
	Well stirred
	Recirculating
Process type	Entrained flow
	Fixed (moving) bed
	Fluid bed
	Other (MHD, etc)

[a] From Ref. 23.

Furthermore, several journals consider combustion models, including *Combustion and Flame*, *Combustion Science and Technology*, *Energy and Fuels*, *Fuel*, and *Progress in Energy and Combustion Science*; the Combustion Institute sponsors international symposia and on combustion.

FOUNDATIONS

This section discusses the key fundamental relations necessary for comprehensive modeling of combustion phenomena. Only a few equations are shown; detailed derivations of the relations discussed here may be found in the references.

Overview of Combustion Processes

A wide variety of combustion processes can be included in a combustion model (Fig. 2) (*see* COAL COMBUSTION). Similar processes also occur in fixed-bed and fluid-bed furnaces, gas turbines, and internal combustion engines. In such furnaces, the primary air–coal stream is fed into the furnace, where the dust is dispersed (point *1* in Fig. 2) as the primary stream mixes with the secondary air and the gaseous reactants and products already in the furnace (*2*). These processes are described by the conservation equations of mass, momentum, and energy, with adaptations for turbulent flows and multiple phases. The coal particles are heated by radiation and convection (*3*), and devolatilize, releasing hydrocarbon volatiles (*4*) that react in the gas phase (*5*). Particle heat up and devolatilization are usually modeled separately from the fluid dynamics of the gas flow and are included in comprehensive models as submodels. Gas phase reactions may be handled with equilibrium models or with simplified kinetics models, as is needed for the formation of some pollutants. The particles and the gas can be trapped in a recirculation zone (*6*), where the particles can continue to react heterogeneously (*7*). The hot char particles, along with soot, radiate energy back to the incoming particles and to the walls and other heat transfer surfaces (*8*). Radiative heat transfer involves temperatures to the fourth power (T^4) and thus requires special consideration when modeling com-

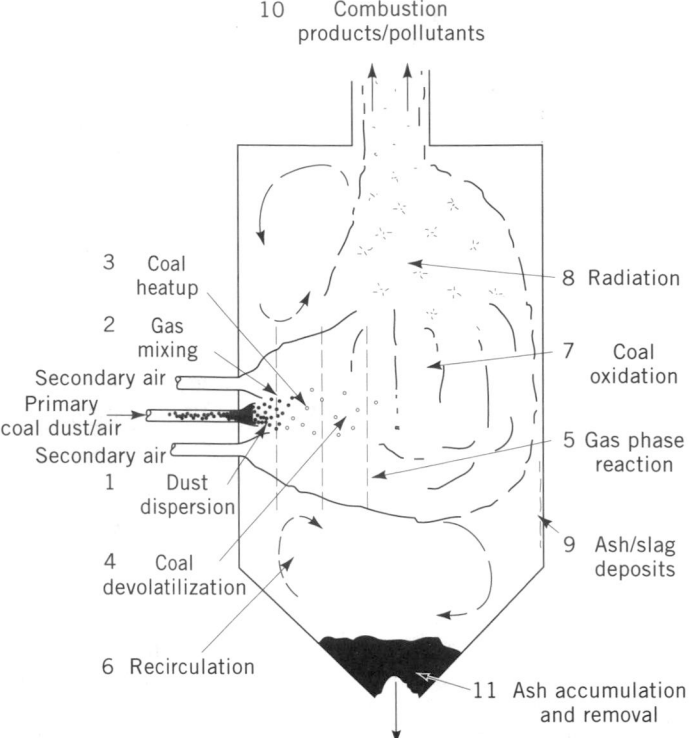

10   Combustion
products/pollutants

3   Coal
heatup

2   Gas
mixing

Secondary air

Primary
coal dust/air

Secondary air

1   Dust
dispersion

4   Coal
devolatilization

6   Recirculation

8   Radiation

7   Coal
oxidation

5   Gas phase
reaction

9   Ash/slag
deposits

11   Ash accumulation
and removal

Figure 2. Physical and chemical processes of pulverized coal flames. From Ref. 24.

bustion. The mineral matter in the combusted particles can adhere to the walls and other surfaces as ash or slag (*9*), which can strongly influence the operation of a furnace. Like other particle reactions, ash and mineral matter deposition are handled in comprehensive codes through submodels. The gaseous products of combustion, including such pollutants as NO_x and SO_x (*see* COAL COMBUSTION), leave the furnace (*10*) and are treated to remove sulfur species, while the ash and slag must also be collected and disposed of (*11*). Models that attempt to describe these combustion systems must include the key physical and chemical processes described above in a numerical method that coordinates and converges these various aspects and submodels with some type of interactive software interface to simplify model formulation and help interpret the results.

Conservation Equations and Fluxes

Three defining conservation equations constitute the foundations on which comprehensive combustion models are based: (*1*) mass conservation, sometimes termed continuity; (*2*) momentum conservation, which is a statement of Newton's law of motion for fluids; and (*3*) the conservation of energy, which is a statement of the first law of thermodynamics for open, fluid systems. These equations have been given for homogeneous systems (3) and for multiphase systems (25). These equations are shown in their overall form for a single phase in Figure 3. Many combustion processes can be modeled as steady state, where the time dependence is removed. This is true even for those that are, at first glance, transient, such as a

flame progressing along a pipe. In such a case, if the flame is moving at a relatively constant velocity, the flame can be modeled as stationary with the gases moving into the flame at a steady rate. Special consideration must be given to the modeling of turbulent and multiple phase flows.

Turbulence. For many engineering designs, values obtained from the conservation equations are sufficient. However, turbulence can substantially affect the behavior of the flow as well as introduce fluctuations that are important to the chemistry of combustion. The most widely used techniques for the treatment of turbulent combustion follow a method pioneered by Reynolds (26). These techniques separate or "decompose" the variables that fluctuate chaotically due to turbulence, such as velocity, pressure, chemical composition, and temperature, into mean and fluctuating components:

$$P = \bar{P} + P' \tag{2}$$

where P is the property. These decomposed variables are substituted into the conservation equations and the various terms expanded. The equations are then averaged, using either Reynolds averaging (ie, time averaging), Favre averaging (ie, using a density-weighted time average), or a combination of the two averages. Other methods of averaging, such as volume averaging and ensemble averaging, have also been used. Favre averaging is currently favored for variable-density flows, such as are found in combustion applications in which there are large temperature differences. A problem with this Reynolds decomposition technique is that additional variables have been incorporated (ie, the fluctuating components) without any additional equations. The total set of equations is no longer a closed or sufficient set and cannot be solved without a method of "closure."

Several closure methods have been proposed. When the equation for momentum is decomposed and Favre averaged, a term called the Reynolds stress tensor appears. One of the earliest suggestions was to represent Reynolds stress as a function of the mean velocity gradients. Given that a major characteristic of turbulence is its faster mixing, this is done by defining a coefficient to the gradient, called eddy diffusivity or eddy viscosity, that is a function of the mean velocity, somewhat analogous to laminar diffusivities and viscosities. Various methods exist for estimating these terms. One is the Prandtl mixing length model for which eddy viscosity is considered to be proportional to a characteristic velocity and some length scale. Another method is to relate the eddy viscosity to the square root of the turbulent kinetic energy. With these two methods, additional assumptions and experimental measurements, often of some turbulent length scale, are needed to close the equations (27). One way of obtaining the length scale is to assume that it is related to turbulent transport properties, such as turbulent convection, vortex stretching, or viscous dissipation. The most commonly used correlation, the k-ε model, relates length scale to the 3/2 power of the turbulent kinetic energy k and inversely to the dissipation of turbulent energy ε. Still,

Mass Continuity

$$\frac{\partial \rho}{\partial t} = -(\nabla \cdot \rho \mathbf{u})$$

Rate of density Net rate of mass
change   influx

Momentum Conservation

$$\frac{\partial (\rho \boldsymbol{v})}{\partial t} = -(\nabla \cdot \rho \, \boldsymbol{vv}) - \nabla p - \nabla \cdot \boldsymbol{\tau} + \rho \boldsymbol{g}$$

Rate of momentum Rate of momentum Pressure force on Rate of momentum gravitational force
increase   gain by convection element gain by viscous on element
                   transfer

Total Energy Conservation

$$\frac{\partial \left[\rho \left(i + \frac{v^2}{2} \right) \right]}{\partial t} = -\left(\nabla \cdot \rho \boldsymbol{v} \left(h + \frac{v^2}{2} \right) \right) - \nabla q + q_r - \nabla \cdot (\boldsymbol{\tau} \cdot \boldsymbol{v}) + \rho(\boldsymbol{v} \cdot \boldsymbol{g})$$

Rate of gain of Rate of energy increase Energy Energy Work done Work done
energy   by convection and increase by increase by on fluid by on fluid by
     pressure forces   conduction radiation viscous forces gravitational
                          forces

Figure 3. Overall conservation equations for a single phase. Symbols in boldface type are vector quantities. ∂, partial derivative; ∇, vector differentiation operator; ρ, density; τ, viscous stress tensor; t, time; v, velocity vector; p, pressure; g, gravitational force vector; i, internal energy; h, enthalpy; q, thermal conduction vector; q_r, radiative flux (Ref. 3).

experimentally determined correlations are needed to obtain values for k and ε (28).

While fundamental difficulties exist with the k-ε turbulence model, it is the most commonly used model in comprehensive combustion models (29). In some refinements, a nonlinear k-ε model (30) has been adopted to account for nonisotropic eddy viscosity. Laminarization terms (15) have also been included to model more accurately the dead zones in industrial-scale furnaces. These extensions enable the k-ε model to predict more accurately negligible eddy diffusivity at low Reynolds numbers and then revert to normal k-ε predictions at high Reynolds numbers (31).

Many of the difficulties with the eddy viscosity models can be avoided by solving a transport equation for the Reynolds stress term directly (32). Higher order approaches such as direct numerical simulation (DNS) and large eddy simulation (LES) have not yet been applied in comprehensive modeling of combustion systems (27).

Multiple Phases (Sprays and Particles). Some modification to the conservation equations, including additional terms, are required if multiple phases or species are to be considered (25). For instance, the mass continuity equation is the mathematical formulation of the principle of conservation of mass. In combustion processes, while the overall mass remains constant, the amounts of the various species and phases, reactants and products are rapidly changing. Use of the continuity equation for individual species entails rigorous bookkeeping of these conversions. The conservation equations for momentum and energy are affected in similar fashions. Additional phases, such as particles or droplets, can exchange momentum with

the gas phase or can affect the viscosity of the gas phase. Energy from chemical reactions is also partitioned between the phases and transferred from one phase to another by radiation, conduction, and convection.

Practical furnaces often have large recirculation zones (see Fig. 2). Swirling, incoming flows also contribute to the complexity of the flow patterns. Particles may or may not collide significantly, depending on the combustion process. Many comprehensive, entrained-flow coal combustion models use the particle-source-in-cell method (PSICM) (33) for particle–gas interactions. This well-known method ignores particle–particle interactions and thus is not readily applicable to highly loaded, dense-phase flows (eg, fluidized beds). The trajectories or paths of representative particles through the gas phase (continuum) field are modeled in a Lagrangian, or moving, reference frame. Mean particle velocity, position, temperature, and composition are obtained along the trajectory by integrating Lagrangian equations of motion, energy, and continuity for an ensemble of particles. Total momentum, energy, and mass of the particle cloud are recorded when the cloud crosses cell boundaries. The difference in particle properties between entering and leaving any given computational grid cell provides a contribution to the particle source term for the gas equations for that cell.

In turbulent systems, the effect of turbulent fluctuations on particle motion must be taken into account. Some comprehensive models use a stochastic approach to simulate these effects (34–36). In the less rigorous diffusion velocity approach, total particle velocity is decomposed into convective and diffusive components. The convective component is defined as the velocity that would arise in the absence of turbulent fluctuations or the ballistic velocity based on the mean gas velocity. An empirical diffusive component is then added to account for the turbulence effects. The diffusive velocity is modeled using the mean particle concentration gradient.

The temperature of the particle is obtained implicitly from particle enthalpy. Heat transfer between the particles and gas is calculated using classic methods. Vaporization of liquid (if present) is controlled by either heat or mass transfer. The vaporization rate below the boiling point is the sum of diffusive and convective components.

Thermodynamic Relationships

Commencing with the first three laws of thermodynamics, which concern the conservation of energy, the increase of entropy, and the definition of absolute zero, many important and useful equations have been derived, which can be found in standard textbooks of thermodynamics and physical chemistry (37,38). Of particular importance to combustion modeling are the gaseous equations of state, heats of reactions, and determination of equilibrium concentrations.

Ideal Gas Law and Heats of Reaction. Most combustion occurs at high temperatures and at near atmospheric pressures. Therefore, the ideal gas law (Fig. 4) is appropriate for use. It is derived from the chemical potential

for a perfect gas or from experimental observation (38). Use of the ideal gas law in combustion modeling provides simple relationships among density, pressure, and volume and permits straightforward formulation of many other thermodynamic and transport relationships.

The heat of combustion and adiabatic flame temperature can also be calculated from thermodynamic relationships without necessary reference to the path of the reactions. Simply, the enthalpy of the products of combustion is equal to the enthalpy of the reactants, given no heat transfer or mechanical work to or from the system, and without any consideration as to how those products were formed. These values can be obtained from tabulated standard heats of formation and heat capacities (39). If the substances are ideal gases, then heat capacities are functions of temperature only, often expressed as polynomials, which simplifies the integration.

Equilibrium Compositions. There are two methods commonly used to determine the composition of products of combustion (40): direct calculation of equilibrium constants and minimization of a state function such as free energy. Both methods are straightforward, yield the same results, and require computer solutions for realistic combustion systems. The two techniques can be reduced to the same number of iterative equations; however, the minimization of energy method requires less computation, has fewer numerical difficulties, is better able to test for condensed species, and is easier to extend to nonideal

Ideal Gas Law

$$pV = nRT$$

Pressure times volume　　Moles times gas law constant times temperature

Enthalpy Balance for Adiabatic Combustion

$$\sum_i^R \Delta H_R^\circ + \sum_i^R \int_{T=\text{ref}}^{T_a} C_p\, dT = \sum_i^P \Delta H_P^\circ + \sum_i^P \int_{T=\text{ref}}^{T_a} C_p\, dT$$

The sum of the standard heat of formation of the reactants　　The sum of the integrals of the heat capacities of the reactants　　The sum of the standard heat of formation of the products　　The sum of the integrals of the heat capacities of the products

Partial Molar Gibbs Free Energy

$$g_i = h_i + T s_i$$

Partial molar Gibbs free energy　　Partial molar enthalpy　　Temperature times the partial molar entropy

Minimization Equations (must be solved for each species)

$$g_i + \sum_j^E \lambda_j a_j = 0$$

Partial molar Gibbs free energy　　The sum of the number of atoms j per mole of species i times a Lagrange multiplier　　Criteria for minimization

Figure 4. Thermodynamic relationships often used in combustion modeling. λ, Lagrange multiplier; Δ, difference; P, pressure and products; V, volume; n, number of moles; R, gas law constant and reactants; T, absolute temperature; i, individual species; a, adiabatic and number of atoms per mole of species i; ref, reference; C_p, heat capacity at constant pressure; g, molar Gibbs free energy; S, molar entropy; E, number of elements.

equations of state. When temperature and pressure are the desired characteristics of the final state of the products, then formulation with Gibbs free energy is most natural. The key equations to this solution technique are also shown in Figure 4. The first is the definition of Gibbs free energy per unit mass of the mixture. The second is arrived at after the Gibbs free energy for the system is formulated in terms of the mass balance, then the Lagrange method of undetermined multipliers is used, and the resulting equation differentiated and set to zero. A Newton-Raphson numerical iteration technique is commonly used to perform the calculations. Several widely available programs use this technique, including chemical equilibrium calculations 1976 (CEC76) (40), commonly called NASA-Lewis; the Edwards Air Force Base program (41), and combustion reaction equilibrium and kinetics (CREK) (42). Given a list of reactants, such codes will calculate the equilibrium compositions for various thermodynamic states (such as temperature and pressure, enthalpy and pressure, and temperature and volume), and some will also calculate theoretical rocket performance, Chapman-Jouguet detonations, and shock tube parameters.

Equilibrium calculations provide a theoretical end state and often a reasonable approximation for many combustion processes. Due to the high activation energies of the reactions, combustion reactions do not commonly occur at room temperature; rather the vast majority of the reactions take place near the flame temperature, and the products, once formed, are not easily converted back to the reactants. Complex combustion systems that may not attain equilibrium conditions at a given point, often approach equilibrium composition when the overall process is considered. For instance, it has been demonstrated that the equilibrium values are good first approximations for the maximum temperature and exit gas composition of fixed-bed combustors and gasifiers (43). This also is true for many entrained flows.

Equilibrium calculations also are important submodels of comprehensive combustion codes that endeavor to de-scribe the kinetic details of complex combustion processes. Chemical reactions for the major species are often assumed to be limited by mixing rates. With this assumption, the local instantaneous gas properties are determined from equilibrium, without regard to chemical kinetics. Mixing is modeled with two or more mixture fractions, and instantaneous properties are averaged over variations in time due to turbulence by formal techniques (29).

Chemical Reactions

The reactions of most combustion systems are inherently complex and, once initiated, proceed rapidly. Even simple combustion reactions, such as the combustion of hydrogen with oxygen, has many elementary steps. The rates of elementary reactions follow the law of mass action (Fig. 5). Elementary reactions are usually first or second order, by simple bimolecular collisions (3), although third-order reactions are possible. The reaction rate constants commonly follow the form developed by Arrhenius (Fig. 5). The rate of the forward reaction is related to the reverse by the equilibrium constant. In some reactions, the temperature term in the preexponential factor is important. Activation energies of combustion reactions are usually large. There are many excellent texts on chemical kinetics that treat these processes in a general fashion (5,6,44–46).

Complete Combustion Kinetics. Models of many simple gas flames are dominated by chemical kinetics. The flow of premixed reactants is usually one-dimensional and laminar. The flames are gaseous and often nonluminous. Molecular diffusion and heat transfer by conductance are well characterized. The reactions are usually formulated as shown in Figure 5. The number of reactions needed to describe completely the combustion of methane (CH_4) with air is 40 or more, depending on the degree of detail sought (46). Numerical methods (46) have permitted the

Elementary Bi-Molecular Reaction

$$A \quad + \quad B \quad \leftrightarrow \quad C \quad + \quad D$$

Species A   Species B   Forward and reverse reaction   Species C   Species D

Law of Mass Action

$$r_f \quad = \quad k_f[A]\,[B]$$

forward reaction rate   Reaction rate constant times the concentration of reactants

Modified Arrhenius Rate Expression

$$k_f \quad = \quad A\,T^n\,\exp(E_a/RT)$$

Reaction rate constant   Pre-exponential factor times temperature correction factor time Arrhenius factor

Figure 5. Formulations of chemical kinetics often used in combustion models. A, B, C, D, chemical species (A, also preexponential fitting factor); $[A]$, $[B]$, concentration of those species; r, reaction rate; f, forward reaction; k, reaction rate constant; T, absolute temperature; n, fitting exponent; E_a, activation energy; R, gas constant.

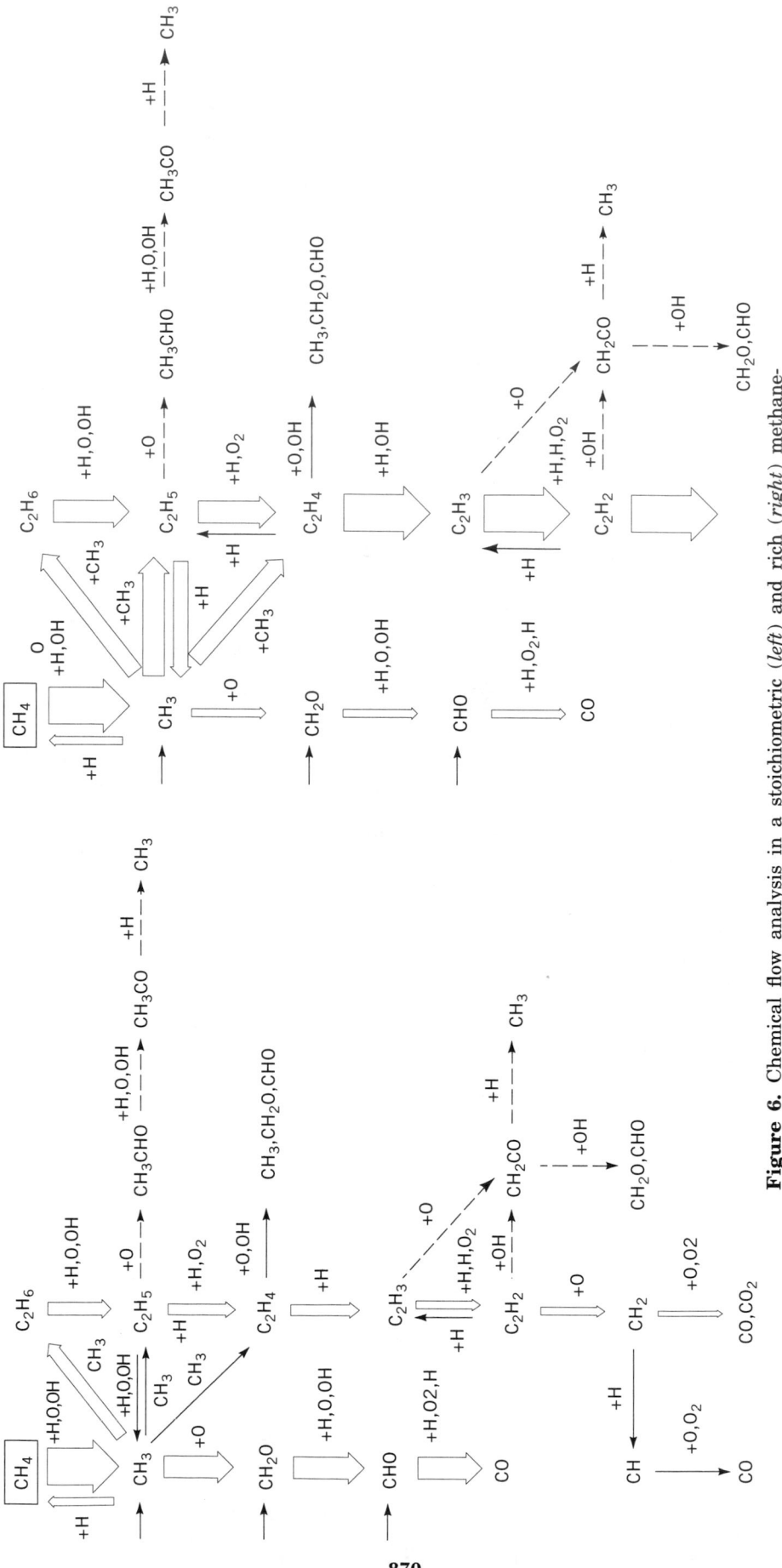

Figure 6. Chemical flow analysis in a stoichiometric (*left*) and rich (*right*) methane-flame (pressure = 10^5 P, unburned gas temperature = 298 K). From Ref. 46.

analysis of these reactions to determine the relative magnitude of each in fuel-lean and fuel-rich combustion with the result shown in Figure 6. Dominant reactions in gaseous hydrocarbon combustion are always (47):

$$H + O_2 \leftrightarrow OH + O \qquad (3)$$

$$OH + CO \leftrightarrow CO_2 + H \qquad (4)$$

The importance of these reactions has enabled reduced mechanisms for methane combustion to be formulated, which involve three to five reactions that describe important flame characteristics.

Most reactions in hydrocarbon combustion are of a type commonly called chain reactions and usually involve chain branching. In chain reactions, there is an initiation step in which intermediate reactants, almost always free radicals, are produced. These intermediates react with the initial reactants and other intermediates to form products and more intermediates, thereby propagating the overall reaction. There are also chain-termination steps, which are reactions that consume the intermediates without the production of more intermediates. When a chain-propagation step produces more intermediates than it consumes, it is called chain branching. Chain-branching reactions can produce chain carriers exponentially, causing the overall reaction to accelerate in an explosive fashion (5). Chain branching of highly exothermic reactions with high activation energies is responsible for the rapid combustion of fuel and oxidizer with substantial heat release. These characteristics cause asymptotic models of combustion to be fairly accurate and create difficulties in the numerical solution of complete sets of chemical reactions.

Limited Combustion Kinetics for Use in Comprehensive Models. Comprehensive combustion models often assume that the major gas species are in instantaneous, local equilibrium throughout the reactor. This assumption allows the incorporation of turbulence effects on chemistry through averaging of equilibrium gas properties (composition, temperature, etc) with respect to the statistics of the turbulence. While this procedure has been successfully used in the prediction of the compositions of major gas species in various practical combustors, it has not been as successful for pollutants such as NO_x, and possibly in some cases for such major species as carbon monoxide and carbon dioxide.

The kinetics of nitrogen oxide formation in gas and liquid flames are fairly well known and are discussed elsewhere (see NO_x; COAL COMBUSTION). Comprehensive kinetics models of the type described above exist for the C/H/N/O system, which contains more than 200 elementary-step reactions (48). Depending on stoichiometric conditions, certain reactions become more or less significant, which has made it difficult to create a 3- to 5-step reduced reaction mechanism. Therefore, efforts have centered on producing overall or "global" mechanisms that do not predict intermediate species but that are simple enough to be included in comprehensive combustion codes. One of these models (49–52) assumes that HCN, NH_3, NO, and N_2 are adequate for describing the conversion of coal nitrogen to pollutant and inert products.

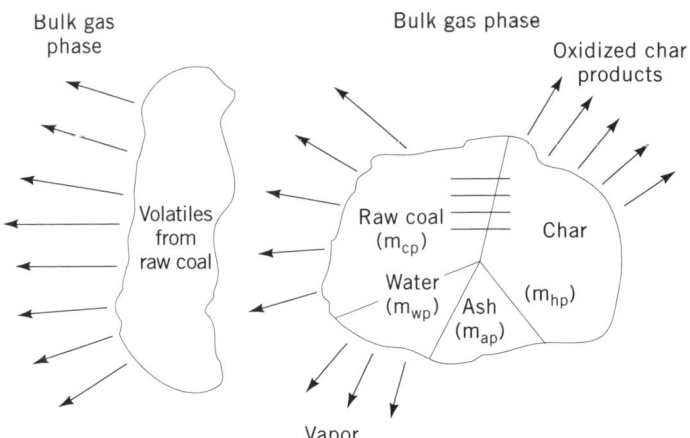

Figure 7. Schematic of coal particle and reaction processes. From Ref. 25.

Solid and Liquid Reactions

The modeling of solid and liquid fuel combustion includes elements not present in homogeneous gas phase models (Fig. 7). There can be moisture in these fuels that vaporizes on heating. As heating continues, the organic constituents of the solid or liquid fuels vaporize or pyrolyze (ie, chemically decompose due to the action of heat), with the lighter products being released. The released volatile matter mixes and reacts with the gaseous oxidizer. The solid carbonaceous residue, termed char or coke, reacts heterogeneously with the oxidizer. The included mineral matter can vaporize, liquefy, or remain solid and can form slag, foul heat transfer surfaces, or leave the combustor with the gas.

Vaporization and Devolatilization. Vaporization and devolatilization are among the most critical and difficult combustion processes to model. For combustors fueled only with solid or liquid fuels, the vaporization and devolatilization processes determine the type and amount of volatiles that are available to react in the gas phase. The nature and amount of residual coke or char also depend on the vaporization and devolatilization processes. Furthermore, in practical combustors, the liquid droplets and solid particles vaporize and devolatilize under conditions that are difficult to duplicate experimentally.

The vaporization/devolatilization process involves many steps that are particle and fuel-specific. First, heat is transferred into the particles. The included moisture evaporates. While some of the moisture may be at the surface of the droplet or particle, there also is moisture in the structure of the liquid or solid. This imbedded moisture vaporizes, then is transported either through bubble transport or pore diffusion, where it is released to the bulk gas. As the droplet or particle continues to heat up, other bonds start to break and light molecular weight hydrocarbons are evolved. Like moisture, these volatiles must diffuse or flow to the surface, then into the bulk gas. As the bonds continue to break, the gases leave, and a carbon skeleton is formed. In fuel oils, this residual solid is termed coke and usually accounts for less than 10% of the original droplet weight. In coals, this solid is termed char and usually accounts for between 25 and 75% of the original particle weight. There are some coals (eg, anthracite) and processed fuels (also called coke and chars and derived from oil and coal) that have nearly no volatile portion, while others (eg, gilsonite) can be almost completely devolatilized.

Moisture loss from solid fuels is mass and heat transport limited. Heat is transferred to the surface by conduction, convection, and radiation. Heat is transferred to the interior of solid particles by conduction and to the interior of liquid droplets by conduction and convection. The rate at which the water vapor leaves the particle is calculated by adding diffusional resistances to the bulk diffusion rate of the vapor in the gas. In liquids, the rate at which the vapor leaves the droplet is calculable first by interphase transport; then, as the particle heats up, by interphase transport with a high blowing factor; and finally, as the droplet boils, the transport rate is dominated by bubble formation and transport. The approach for evaporating moisture in a droplet also is useful for vaporizing fuel–oil droplets. Many of the procedures for calculating liquid spray combustion have been reviewed (53). A common assumption used in the calculations is the *thin-skin* approximation, which is that all the heat reaching the droplet preheats and vaporizes the liquid surface and that there is no bulk heating of the liquid.

As the particle continues to heat up, simple vaporization ceases and the pyrolysis of hydrocarbon molecules in the fuel particle or droplet begins. Pyrolysis refers to the breaking of chemical bonds by heat, while devolatilization includes transport of the released species to the bulk phase and the formation of the carbon skeleton. These processes are fuel specific and complex, with heat and mass transport and chemical reaction rates all playing significant roles in the overall rate determination. Several models have been proposed for the rate of devolatilization of coal (54,55). Devolatilization has been modeled as a first-order reaction process (56); as two parallel, first-order, irreversible reactions (57,58); and as an infinite series of parallel reactions (59). The two-reaction model has been favored for inclusion in comprehensive codes, because of its simplicity and its ability to predict variable composition of the volatiles with changing temperatures and heat-transfer rates. More advanced models of devolatilization have been proposed that treat the coal as a collection of functional groups (60) and as a macromolecular network. The network models have received the most attention recently; three of the most prominent are the functional group-depolymerization, vaporization and cross-linking (FG-DVC) model (54); the FLASHCHAIN model (61); and the chemical percolation devolatilization (CPD) model (62). These network models employ statistics, such as Monte Carlo methods and percolation theory, to describe the macromolecular breakup and cross-linking. The network models are particularly adept at predicting the volatiles composition and include the possibility of calculating the residual char structure. Furthermore, these models include the effect of rank, and results from these models agree quite well with a wide range of experimental results for different coal types. These models are more complex than the two-reaction model and have only recently been included in comprehensive combustion codes.

Heterogeneous Combustion. Heterogeneous char and coke oxidation can proceed simultaneously with or after vaporization and devolatilization, depending on reaction conditions. The time required for char and coke combustion is usually substantially greater than that required for devolatilization and often is a determining factor in furnace design. There is also much emphasis on reducing the amount of carbon in the ash, which means that the burnout of the char or coke needs to be more effective. The physical structure of the char or coke, such as pore structure, surface area, particle size, and inorganic content, controls the reaction processes of the char. However, the structure of char is determined by the devolatilization process as well as by the parent fuel. Models of char combustion can include the following processes: (1) diffusion of gaseous reactants and products through the boundary layer surrounding the particle, (2) diffusion of reactants through the porous structure of the particle, (3) conduction of heat through the particle, (4) the chemical reaction of the oxidizer with the particle, (5) the homogeneous reaction of carbon monoxide with oxidizer within the pores and the boundary layer, and (6) the evolution of particle and pores during the reactions.

Global models of heterogeneous reaction typically include only processes 1, 2, and 4 described above as resistances in the reaction rate expression. The resistance due to film diffusion is usually calculated through the use of diffusional coefficients corrected for the mass leaving the particle surface, ie, corrected for the "blowing factor." The effective diffusivity of the gases into the particle and the portion of the particle that participates in the reaction can be estimated by using methods that have been published (63). The char or coke reacts with oxygen to form CO, primarily, and CO_2, depending on conditions (64). The rate of oxidation is measured and correlated to provide reaction rate expressions (65). Heat capacity of the solid fuels, which changes with burnout due to the changing elemental composition of the particle, can be calculated (66). Changes of the particle structure are included by modeling the particle as a shrinking sphere or as having constant diameter but changing density (67).

Better descriptions of pore development and structure require microscopic models of the particle. These models include intrinsic kinetics and pore structural changes during burnoff. Three of the most popular microscopic models are a random capillary pore model, one in which the pores are considered spherical vesicles connected by cylindrical micropores, and one in which the pores have a treelike structure. These models allow for pore growth and coalescence in their respective fashions and provide estimates of reactive surface area. Parameters required for these models are obtained from experimental measurements of the various chars. Other microscopic models have been developed that are based on percolation theory and could be tied to devolatilization models with the same bases. Due to their computational requirements, these microscopic models have not been included in comprehensive combustion codes (55).

Modeling of Mineral Matter Effects. While gaseous and liquid fuels contain little or no inorganic material, ie, less than 0.05%, coals typically contain between 5 and 15% inorganic components or ash, with the amount sometimes reaching 50% in commercially used, low quality coals used in various parts of the world. The interaction of this mineral matter with various heat-transfer surfaces in furnaces and boilers is important to utility and industry users. Furthermore, the resulting ash must be disposed of in an environmentally acceptable manner. Models of fouling and slagging by mineral matter ideally begin with the amount and chemical composition of the mineral matter in the coal; then track the transformations of this inorganic material in the combustion process; and then model the transport of the inorganic component to the surfaces, its attachment to the surface, and the growth of deposits (68). Such models must predict the thermal and physical properties of the deposits so that the heat transfer through the deposit may be estimated and the effectiveness of deposit removal methods such as soot blowing predicted (see COAL COMBUSTION).

Computer-controlled scanning electron microscopy (ccsem) provides detailed characterization of coal and coal minerals. The distribution of the minerals in the coal particles is not homogeneous but is best represented by statistical methods, such as probability distribution functions and Monte Carlo approaches. As the coal is consumed, the ash can coalesce into larger ash particles. Partial coalescence and no coalescence also occur. Some inorganic elements vaporize and are commonly modeled with equilibrium calculations. The impingement of inorganic materials is modeled through the use of detailed particle trajectories or generalized correlations. Models have been devised to describe the flow in and around boiler tube banks. Furthermore, particle deposition from flows parallel to the surfaces have been modeled using turbulence models. The probability that the particle will attach or stick is influenced by particle velocity, viscosity, surface tension, temperature, size, impact angle, and condition of the substrate. Detailed models of all these factors are computer intensive, so it is more common to apply a sticking probability. However, the value of this sticking probability is currently under much discussion. Models of deposit thermal conductivity await additional information concerning the nature of the deposits and more accurate models of deposition. Current models are empirical correlations that highly depend on the nature of the deposit. Models of deposit strength are at roughly the same stage of development (68).

Radiative Heat Transfer

Radiation is the dominant mode of heat transfer in pulverized coal combustors and an important mode of heat transfer in gaseous flames and liquid spray flames, particularly when soot is formed (69). Radiative heat transfer has been extensively studied in various disciplines, with many of the results applicable to combustion modeling. The radiative properties of combustion gases are particularly well characterized. The radiative properties of dispersed particles also can be calculated from the concentration, shape, size distribution, chemical composition, radiative properties of the material, and the temperature of a cloud of particles. However, in comprehensive combustion codes, these properties are coupled to the incident radiative flux. Calculation of radiative properties also is computationally intense and not favored for comprehensive combustion codes; simplifications have been used.

The dominant gases in gas phase radiation during combustion are water, carbon dioxide, and carbon monoxide. Other gases participate, but their concentrations are usually small enough so their contributions can be neglected or added to the contributions of the major species. Gases are transparent to radiation at certain wavelengths and absorb and emit energy at other wavelengths. Wavelengths can be divided into discrete wavelength bands. Only certain bands need be considered for the dominant gaseous species, and these can be collected into a few wide bands. The level of detail required is determined by the needs of the model. Even simpler formulations such as polynomial expressions or the weighted sum of gray gases methods are used and are well suited to some comprehensive codes (69).

Commonly used models of radiative transport in particle-laden flows are simplifications of the radiative transport equation and include particle radiative properties. One of the most widely used methods is the long-standing zonal approach, by which the volume is divided into many small volumetric elements and the contribution of each element to the radiative flux distribution on every surface and every other volume element is considered (70). Because of the computational demands of zonal methods, simpler approximations have been developed. In the multiflux method, radiative fluxes are computed along the axes of the coordinate system. The discrete ordinates approximation is a multiflux model in which a quadrature scheme is used to integrate the incoming radiation, and the number of angular subdomains is determined by the order of the quadrature. In moment methods, the angular dependence of the radiation can be expressed as a power series or with spherical harmonics. Statistical methods have also been employed that consider photon bundles distributed in all directions in a Monte Carlo fashion and the absorption and scattering of the photons calculated. Hybrid techniques, which combine methods have also been tried. All these methods have been recently discussed (69). Radiative heat transfer models for gas- or oil-fired furnaces (71), and pulverized coal-fired furnaces (72) have been reported using zonal, multiflux, differential or moment, and statistical methods, with simple empirical assumptions for flow characteristics, temperatures, heat release rates, and gas composition. Recent furnace modeling has focused on the use of the more efficient multiflux (especially discrete ordinates) methods.

Pollutant Formation

Several pollutants are generated and emitted from combustion processes (see COAL COMBUSTION; COAL GASIFICATION), including carbon oxides, nitrogen oxides, sulfur oxides, particulates, trace metals, and unburned hydrocarbons. Modeling of these emissions has focused on the acid rain precursors (nitrogen and sulfur oxides) driven by current emission regulations. Particular focus has been on NO_x because of the demonstrated success in reducing NO_x through combustion control. Modeling work for these pollutants has been reviewed (73).

Numerous investigations into the chemistry of gas and liquid spray flames have resulted in a good understanding of NO_x formation reactions in simple hydrocarbon flames; however, only a partial understanding of nitrogen chemistry in coal flames is currently available. An important goal of research has been to determine the dominant reaction sequences that lead to the formation and destruction of nitrogen oxides in coal flames. One challenge is in identifying those reactions that are indispensable over all stoichiometries. Other challenges include accounting for the effects of turbulence on the chemical kinetics and mathematically describing the fuel oxidation process for liquid and solid fuels.

The current level of knowledge of comprehensive mechanisms and rate parameters for the gas phase reactions of nitrogen compounds that are relevant to gaseous, combustion-generated nitrogen oxides has been given (48), which is a compilation of the work by several investigators to model simple hydrocarbon flames in premixed, one-dimensional, or perfectly stirred reactors. The combined C/H/N/O mechanisms describing simple hydrocarbon combustion comprises more than 200 elementary-step reactions.

A general, computerized pollutant model for predicting NO formation during turbulent pulverized-coal combustion (49–52) predicts nitric oxide formation and destruction for gaseous and pulverized coal combustors. Other groups have developed similar nitrogen oxide models based on this model (74,75). The NO_x pollutant submodel requires integration with generalized combustion codes (eg, PCGC-3) that provide theoretical predictions for the temperature, velocity, particle burnout, and major species throughout the combusting flow field. This and other NO_x pollutant models that have been successfully applied to a variety of combustion cases have been summarized (73).

Sulfur oxides are formed in stationary combustors from the sulfur entering with the fuel. Natural gas is sulfur free with the exception of trace amounts of odor additives such as mercaptans. However, both fuel oil and coal commonly contain significant amounts of sulfur. Sulfur oxide emissions can be reduced substantially by simply converting from high to low sulfur coal. For most coals, the sulfur content can be reduced by cleaning the coal before burning it. Sulfur-containing pollutants formed by burning fossil fuels include SO_2, SO_3, H_2S, COS, and CS_2. Under normal boiler operating conditions, with overall excess oxygen being used, virtually all of the sulfur is oxidized to SO_2 and small quantities of SO_3 (76). When coal is gasified, the sulfur predominantly remains as reduced forms of H_2S, COS, and CS_2 (77). Even though the reactions of sulfur species in coal flames are numerous and complex, they are usually not included in comprehensive combustion models, since in combustion processes, essentially all of the sulfur is converted to sulfur oxides.

During combustion, injection of solid, finely pulverized, calcium-based sorbents ($CaCO_3$, CaO, $Ca(OH)_2$) has been effective in capturing sulfur oxides from the gas for easier disposal. Significant sulfur capture occurs only in the 1140–1500 K window (78). Above these temperatures, sulfation is inhibited by particle sintering and growth. Below this range, the sulfation rate is negligible. Thus the sorbent particles need to be injected into the combustor at the point where the longest residence time can be obtained at optimum temperature conditions.

A number of theoretical sulfur capture models have been developed for interpreting laboratory sorbent capture data. Most of the mathematical models assume that the solid particles are porous spheres composed of spherical grains or cylindrical pores. The more sophisticated models incorporate a distribution of grains or pores and account for void reduction as the sulfation reactions proceed (73). As with other particulate submodels, a sorbent capture submodel must be integrated into a comprehensive combustion model.

Modern computers have made mathematical modeling of practical combusting systems possible. A comprehensive combustion code that accurately predicts pollutant concentrations throughout the reacting domain of combustion facilities can be a significant aid in understanding and optimizing processes that minimize emissions. A reliable model must adequately describe the governing chemistry and physical processes. When only an empirical model can be formulated, then the model must be verified with experimental data and extrapolated to new conditions with caution.

Solution Methods

Only for simple theories and models can the equations be solved explicitly to yield an algebraic solution. More generally, the complete set of equations that describe common combustion processes cannot be solved exactly in the sense that the equations can be manipulated to arrive at an unambiguous equality. Instead, a solution to these equations is obtained through the use of simplifying assumptions and approximations, dimensional analysis, asymptotic techniques, and numerical methods. The solution to any particular combustion problem may make use of one or several of these techniques in combination.

Assumptions and Approximations. Solutions to combustion problems employ simplifying assumptions and approximations. These assumptions and approximations are specific to the particular problem in relation to the requirements of the solution method. Historically, certain assumptions have been commonly applied to the majority of combustion situations. Furthermore, a limited number of key assumptions have been applied to specialized situations, such as particle combustion and turbulence, which have been mentioned in passing but deserve specific attention.

Common Assumptions and Approximations. The majority of combustion problems that have been historically considered in modeling have been steady-state deflagrations (ie, flames that propagate at velocities much below the speed of sound). This has led to the adoption of a large number of common, classic assumptions. First, the conservation equations outlined above have been shown accurately to describe combustion phenomena. If the gas density was sufficiently small to allow the molecules to act as individual particles or if there were severe discontinuities that occurred on the molecular scale, this would not be true. The ideal gas law has also been shown accurately to approximate the behavior of high temperature gases. Given these as bases, the following quantities are commonly neglected in many applications (5): (1) body forces (those due to gravity), (2) the Soret and Dufour effects (diffusion of mass and species caused by thermal gradients), (3) pressure-gradient diffusion, (4) bulk viscosity, and (5) radiant heat flux. Body forces are important and cannot be neglected when natural convection, ie, buoyant forces, is significant. Radiant heat flux cannot be neglected in systems with particles or soot.

To simplify the equations even further, additional assumptions have commonly been made (6). These include neglecting viscous effects altogether, assuming that the binary diffusion coefficients of the different species are equal and that the Lewis number (which is the rate of energy transport divided by the rate of mass transport) is equal to 1. To simplify the chemical reaction process, it has often been assumed that the reactions are simple, with one or two steps, and irreversible. These chemical reactions are often assumed to be infinitely fast. Finally, the heat capacity of the mixture has often been assumed to be constant.

In more recent work, many of these assumptions have been avoided with advances in solution techniques, numerical methods, and computers. Others have been retained, such as the ideal gas law and incompressible flow, because they greatly simplify the equations without introducing noticeable error, in the large majority of current situations.

Assumptions and Approximations for Specific Systems. Many other assumptions are made for specific combustion systems and submodels, some of which have already been discussed. However, a few are key to understanding some major systems and so are considered here.

The combustion of sprays and entrained particles involves the interactions among the droplets or particles and the continuum gas. A key assumption is that the droplets or particles do not interact, because they are sufficiently small and dispersed. This assumption is a reasonable estimate in a majority of practical situations but is clearly not valid in such situations as fluidized beds or fixed beds, dense phase flows, or the atomizing regions of jets. Another common assumption used for particles and droplets is that the temperature throughout the droplet or particle is uniform. This appears to be true for small particles, usually considered to be less than 100 μm in diameter, depending on heating rate. This is not true for large particles or particles with a large amount of exothermic reaction at the surface. Other assumptions that are often made are that the particles are spherical, that their sizes may be accurately represented by a distribution function, that they are initially uniformly dispersed, that body forces can be neglected, and that the presence of the particles does not affect the movement or behavior of the gas.

With respect to turbulence, a common key assumption is that the turbulence is isotropic (ie, the fluid has the same turbulence properties in all directions), that the length and time scales of the turbulence are of the same order of magnitude as the motion of the mean flow field, and that the turbulence adjusts instantly to local changes in the mean flow field. Large-eddy simulations do not require these assumptions except on a small scale (27). The interactions of the particles and chemistry with turbulence also have many assumptions that are made in the

various models. These include the shapes of the distribution functions that describe the interactions, the time and length scales of the turbulence, and the noneffect of chemical reaction and particle motion on the turbulence. These assumptions can be significant limitations in current comprehensive models, and much research is being conducted to provide improved relationships.

Dimensional Analysis. Several dimensionless quantities such as the Lewis, Prandtl, and Schmidt numbers are commonly used in combustion modeling. These quantities automatically appear as the conservation equations are cast into dimensionless form (3). Dimensional analysis can be used in two broad fashions. The first is to correlate data and results in some physically significant way when a model is absent. The second is to simplify models through the elimination of various parameters and variables. These two uses of dimensional analysis have been described for a wide range of phenomena (79).

The well-known correlation of pipe friction factors during turbulent flow with the dimensionless quantity known as the Reynolds number is an example of using dimensional analysis to correlate data. In combustion processes, this type of dimensional analysis is commonly used in two ways. The first is to provide relationships for combustion phenomena that appear to be simple overall but difficult to model fundamentally, such as explosion bomb data or the entrainment of air into free jets and plumes. The second use of correlative dimensional analysis is to provide submodels from experimental data for inclusion into comprehensive models. For example, heat loss from a flame burning in a pipe may be calculated through the use of the well-characterized Nusselt correlations for the heat-transfer coefficient.

The use of dimensional analysis to simplify equations and sets of equations is widely used in the field of combustion modeling, as it is in the general fields of heat and mass transport. Several important dimensionless quantities are common to many models. One is the mass fraction of fuel and oxidizer in a mixture, termed the mixture fraction. Another is the Zel'dovich number, which is the product of the nondimensional activation energy (the activation energy divided by the gas constant and the flame temperature) with the temperature rise divided by the final temperature.

Asymptotic Techniques. Asymptotic techniques simplify the set of equations by taking advantage of the mathematical characteristics of the large activation energies and rapid reaction rates common to combustion processes. Or, in other words, this method uses limiting conditions to solve the conservation equations. These techniques have been formalized (9) and have been used to examine a wide variety of combustion processes (8).

The conservation equations are arranged to be strongly dependent on reaction rate. The activation energy asymptotic technique of von Kármán has been described in detail (5). This involves performing series expansions of the appropriate terms in the equations. These expansions are performed at the limit at which the reaction rate is negligible and at the limit at which the reaction rate dominates. A few of these terms are substituted back into the systems of equations and the equations are solved either analytically or numerically. The solutions for negligible reaction-rate and the reaction-rate–dominant zones are matched, possibly through the use of the parametric limits of some intermediate variable. Only the first term or the first few terms are needed to provide reasonable accuracy to the solutions, due to the asymptotic nature of the activation energy.

This asymptotic technique can be easily based on other characteristics of the combustion process such as the reaction rates. Reaction rate asymptotics examine the ratio of two important reactions, the ratio of the flow rate to a key reaction, or other such ratios in a critical zone where the mechanisms shift (8). Asymptotic expansions in these zones can illuminate the importance of various reactions and lead to reduced rate expressions of chemical reactions that can be used in comprehensive codes as well as providing accurate approximations by themselves. Asymptotic techniques are not restricted to laminar, planar flames but can also be used to examine flame instabilities, flame resonance, and turbulent flows.

Numerical Methods. Advances in the modeling of combustion have been closely linked to the development of sophisticated numerical methods that make use of high speed computers. A large number of such methods have been developed. However, the majority of combustion problems are solved using the same general procedures, such as finite differencing, diagonal matrices, staggered and adaptive grids, and pressure-linked solution algorithm. Both Cartesian and body-fitted coordinates have been used successfully.

In numerical methods, there are two widely used techniques used to solve field equations: finite difference and finite element. Finite-element methods are more suited to complex geometries but fairly simple equations. Finite-difference methods do not handle the boundaries as well but simplify the solution of complex sets of differential equations. Finite-difference methods are almost universally used for comprehensive combustion codes. Indeed, one algorithm, teaching elliptic axisymmetric calculations heuristically (TEACH) (80) is used for many industrially available codes.

In finite difference methods (81,82), the combustor or furnace is divided into a lattice or "grid" with many connecting points or "nodes" (Fig. 8). Finite-difference algorithms set up the differential equations as difference-quotient approximations, often using Taylor series expansions. The difference-quotient approximation improves as the distance between node points decreases. The difference approximations can be arranged to give equal weight to the values of all the surrounding nodes in approximating the value of the central node (central differencing), more weight to the nodes that are solved previously in the solution algorithm (upwind differencing), or some combination of the two (hybrid schemes). The resulting polynomial expressions are arranged as diagonal matrices. Sweeping techniques are favored for the solution of these matrices, due to their large size. The Thomas algorithm is a popular method to solve tridiagonal matrices. Often the equations are highly coupled (ie, interrelated), which can cause "stiffness" or difficulties in arriving at a solution. To obtain stable convergence, the

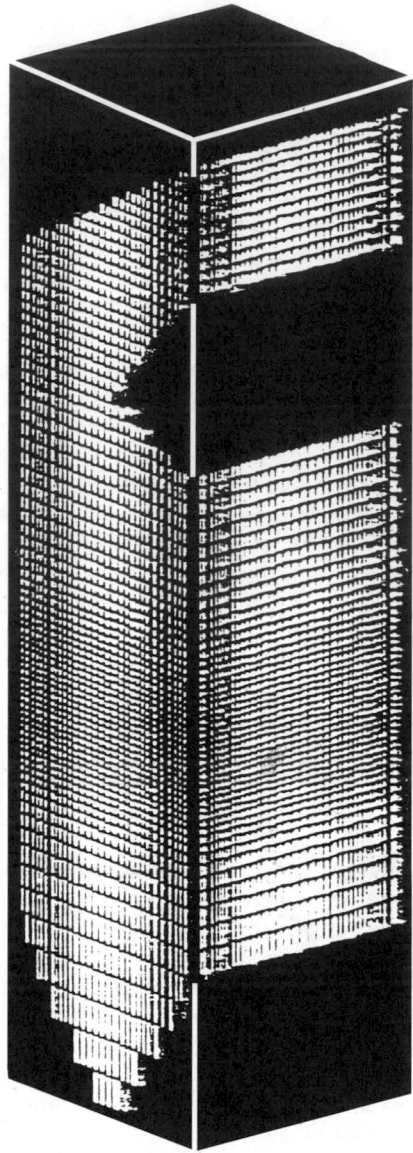

Figure 8. Computational mesh ($52 \times 52 \times 84 = 227{,}136$ cells) for an industrial furnace. From Ref. 29.

amount of change between successive iterations may need to be underrelaxed (ie, changed less than suggested by the solution method). Proper alignment or fineness of the grids increases solution accuracy as well as provides better resolution in areas of interest. For large utility furnaces, several hundred thousand nodes are required for acceptable accuracy (Fig. 8). Adaptive gridding techniques allow the computer to be programmed to perform the necessary grid changes automatically. However, the complexity of combustion problems often make adaptive gridding techniques difficult to apply.

Boundary conditions are required at lines and planes of symmetry, wall or intrusion faces, inlets, and outlets. Boundary conditions are implemented by appropriately modifying the coefficients in the finite-difference equations for the boundary cells. Symmetry can be used to reduce the computations and should be used to advantage whenever possible. In axisymmetric geometry, symmetry at the centerline reduces the dimensionality of the problem to two dimensions. In three dimensions, nonaxisymmetric cylindrical or Cartesian geometries, symmetry often exists and should be used to reduce the size of the computational domain. At lines and planes of symmetry, all normal gradients should be constrained to be zero to satisfy continuity.

If particles or droplets are present and have been treated in a LaGrangian fashion, then a solution procedure that couples the Eulerian gas-phase grid and the Lagrangian particles must be established. Commonly, geometric parameters are calculated and the flow field variables are initialized, including particle source terms. Then the differential equations for the variables shown in Table 2 are solved successively. This procedure is repeated until the gas flow field is converged for the assumed particle source terms. After gas phase convergence, gas properties are updated, Eulerian particle number density is calculated, and the radiation flux field is calculated. This completes the Eulerian calculations. Lagrangian particle trajectories are then calculated one

by one. Particle source terms for mass, momentum, and enthalpy are determined from the changes in the particle properties while passing through a computational cell, multiplied by the particle number flow rate for the trajectory and divided by the cell volume. After solving all particle trajectories, the new source terms are compared with the old values. New values of the source terms are assumed and the entire procedure is repeated until the calculated source terms at the end of the Lagrangian calculations are approximately equal to the assumed values at the beginning of the Eulerian calculations. Overall convergence is enhanced when all of the coupled partial differential equations converge at similar rates. An overly tight convergence criterion for a single equation can slow the convergence of the coupled equation set.

The CPU time on an advanced, high end engineering workstation required to converge a solution for combustion in a coal-fired furnace (see COAL COMBUSTION) may be a few days of continuous computer operation. Examination of the solution status as it progresses toward convergence is usually required. The declining costs of computing as well as the increasing availability of computers has permitted more exact and detailed combustion models coupled with finer grids to be formulated and solved economically. Therefore, the CPU time required to converge ever more sophisticated combustion models has either remained the same or decreased as the capacity and speed of computers has increased.

Table 2. Key Comprehensive Model Variables[a]

Reactor Parameters

Configuration (eg, shape, upflow, downflow)
Inlet configurations (eg, location, presence of quarl)
Inlet locations
Dimensions
Wall properties
Wall thickness

Input Data for Each Inlet Stream

Gas velocity
Gas composition
Gas temperature
Gas turbulence intensity
Gas mass flow rate
Pressure
Particle or droplet slip velocity
Particle or droplet composition
Particle or droplet temperature
Particle or droplet size distribution
Particle or droplet loading
Particle or droplet bulk density

Independent Variables

Physical coordinates (x,y,z) (x, r, θ)
Time (t)

Dependent Variables (results)

Gas species composition
Gas temperature
Gas velocity
Pressure
Turbulent kinetic energy
Turbulent energy dissipation rate
Mixture fraction (mean and variance)
Particle or droplet composition
Particle or droplet temperature
Particle or droplet velocity
Particle or droplet diameter
Gas density
Gas viscosity

[a] From Ref. 29.

Model Evaluation and Reliability

Model predictions for combustion are often viewed with uncertainty because of the numerous engineering and mathematical approximations that are required. The model may have subprocess descriptions that are inexact, with unknown or neglected coupling effects between subprocesses. For example, in many comprehensive models, the effect of turbulence on chemistry is included, but the reciprocal effect of chemistry on turbulence is neglected. Continual evaluation and reevaluation are, therefore, needed to establish confidence in model predictions, establish ranges of model applicability, and identify submodels and approximations needing improvement. The following methods have been used to evaluate comprehensive models (24): numerical evaluation, sensitivity analysis, trend analysis, and comparison of model predictions with profile data.

Numerical Evaluation. Numerical evaluation examines numerical accuracy, robustness, and efficiency. A numerically accurate model minimizes numerical error and accurately approximates the solution to the model equations. A numerically robust model reliably approximates the solution for a wide range of input conditions with minimal user intervention. A numerically efficient model finds the approximate solution with a reasonable amount of computer time. Numerical evaluation does not address the model's ability accurately to predict physical reality, provide information about which model input parameters or submodels must be improved to increase model accuracy, or determine the level of technical expertise required to use the model. These questions are addressed by other methods of evaluation.

Numerical accuracy is best evaluated by comparing model predictions with analytical solutions. This can be done for limiting cases, such as nonreacting, gaseous flow in simple geometries (24) or for practical equation sets that have been altered to be consistent with an assumed analytical solution. Other methods have also been used to evaluate model numerics, including investigating the effects of the differencing scheme on numerical diffusion, investigating the effect of convergence criteria for each individual equation on overall convergence speed, investigating the appropriateness of the overall convergence criteria, and investigating the effect of numerical solution algorithm on convergence efficiency and robustness. Numerical evaluation is the first method of model evaluation. Without it, results from other evaluation methods are difficult to interpret.

Sensitivity Analysis. The second method of evaluating comprehensive combustion models addresses the sensitivity of model predictions to uncertainty in specific submodel and model parameters. Attention can then be focused on improving the precision of those submodels and parameters that are making the greatest contribution to uncertainty in the predictions.

A method for rigorously investigating sensitivity was recently demonstrated (83). This particular procedure is suited to large, complex models having numerous input parameters and requiring substantial computer time. The method has two steps: (1) identification of primary effects through use of a linear screening design and (2) evaluation of the higher order effects of the identified parameters with global, nonlinear analysis. The method was demonstrated using 19 potentially significant input parameters in a two-dimensional combustion code. A normal distribution about the mean was assumed for each parameter, and standard deviations were identified from the literature or estimated from experience. A total of 20 reactor simulations were performed for a coal combustion case, and seven output functions were examined: (1) total carbon conversion, (2) maximum centerline temperature, (3) maximum centerline temperature at the flame front, (4) flame front location, (5) carbon conversion at the flame front, (6) fraction of total carbon conversion occurring before the flame front, and (7) NO_x concentration at the reactor exit. The initial screening showed that uncertainty in devolatilization and turbulent particle dispersion pa-

rameters had a dominant impact on the predictions, while uncertainty in the radiation coefficient, char oxidation parameters, and gas turbulence parameters had a lesser impact. Based on the results of Step 1, 9 of the original 19 parameters were retained for Step 2. The global nonlinear analysis (Step 2) was performed with a technique known as the Fourier amplitude sensitivity test (FAST) (84). FAST statistically varies all parameters simultaneously. The most significant model parameter governing coal burnout was the activation energy for the high temperature coal devolatilization reaction. This parameter dominates the uncertainty in the postignition region of the reactor and illustrates the dominant effect that overall coal volatiles yield has on burnout predictions. The second most important parameter was the activation energy for char oxidation.

Trend Analysis. Trend analysis is the third method for model evaluation and involves comparing model predictions with experimental data in which a key variable was varied systematically. Examples of this method have been given (73) for variation of predicted nitric oxide concentration with stoichiometric ratio and secondary swirl number, and for variation of SO_x sorbent particle conversion with particle residence time and variation of SO_2 concentration with the Ca:S ratio. Trend analysis confirms the phenomenological basis of the model and increases confidence in predictions made at conditions different from those at which the model was evaluated.

Profile Data Comparisons. In the fourth method, experimental measurements of temperature, gas composition, and the like from inside combustors are used to validate models (85). These detailed reactor profiles are commonly organized into categories of increasing complexity for modeling (Table 3). Categories A and B allow model evaluation without the complicating effects of chemistry, while categories A and C allow model evaluation without the complicating effects of particles. Category E is the most difficult to simulate because of the fuel-rich chemistry and steeper gradients.

Predictions from a generalized two-dimensional combustion code were compared with experimental data for 33 cases from a comprehensive data book (86,87). The comparisons included gas velocity, inlet gas mixture frac-

Table 3. Categories of Data for Model Evaluation and the Submodels That They Test[a]

Category	Description	Submodels Tested
A	Nonreacting gaseous flow	Gas fluid mechanics
B	Nonreacting, particle-laden flow	Same as A plus particle dispersion, particle–turbulence interaction
C	Gaseous combustion	Same as A plus homogeneous chemistry, gas radiation, thermal NO_x
D	Coal combustion	Same as B plus C plus devolatilization, heterogeneous chemistry, particle phase heat transfer, fuel NO_x
E	Coal gasification	Same as D plus fuel-rich chemistry, steeper temperature gradients

[a] From Ref. 29.

tion, coal–gas mixture fraction, particle flux, gas species mole fractions, and gas temperature.

An extensive evaluation of various turbulence submodels applied to swirling flows was conducted (88). Swirling flow results when tangential momentum is imparted to incoming coaxial gas streams. Swirl is often used in pulverized coal combustors, particularly wall-fired units, to promote mixing and particle heating, stabilize the flame, and control pollutant formation. Based on their results, the authors identified uncertainties in turbulence models to predict accurately strongly swirling flows. However, they stated that predicted species concentrations may be considered acceptable in the recovery or near-exit regions of coal flames, because these values are controlled primarily by char and volatile reaction rates and particle dispersion. Furthermore, the k-ε turbulence model performed competitively with several other turbulence models in that region.

A comparison of predictions with a generalized three-dimensional combustion model with gas velocity data in the absence of combustion from two furnace test facilities was performed (31). The experimental facility was 5.4 m high, 1.9 m wide, and 1.6 m deep. It was corner fired with a global Reynolds number of 200,000. Ambient air was blown into the furnace at equal velocities in both the fuel and air ports of all four burners. Experimental velocity data were obtained using a calibrated, five-hole pitot tube coupled to a computer-controlled, traversing, data-acquisition system. This furnace was simulated with computational grids of various sizes from several thousand to about 120,000 cells to determine the effect of grid density on model predictions. Agreement between the predicted and measured values was poor for the coarsest grid, but improved as the number of computational cells increased. Around and below the furnace nose, the finest grid simulation demonstrated good agreement with the measured values. None of the simulations totally resolved the velocities near the top of the reactor where the flow was dominated by interactions with the pendant panels, and even the finest grid probably lacked sufficient resolution in this region.

LIMITED MODELS

Many combustion models are purposefully limited. In contrast to the comprehensive models discussed in the next section, these models do not attempt to include, or have the potential to include, all the various combustion processes discussed above. Instead, the models illustrated in this section emphasize key controlling processes, simplify geometrical considerations, or limit the type of flows that can be considered to focus on a particular aspect of combustion at often greatly reduced computational time and difficulty. These models have been classified for the purposes of discussion as: (1) correlative models that relate input factors to results in specific circumstances, (2) models that model complex combustion processes but in simple geometries, including equilibrium models (3) kinetic models that detail the chemical reactions that occur in a flame completely for laminar flows, and (4) asymptotic models that apply the solution procedures discussed

above to examine particular aspects of combustion such as combustion instability.

Correlative Models

Dimensional analysis and statistical regression methods can be applied to combustion processes and subprocesses and various correlations obtained. These correlations can be used as stand-alone analysis tools or incorporated into larger process or comprehensive models. Examples of simple models of this type include the following:

1. The cubic law of dust and gas explosions by which explosion violence is related to the cube root of enclosure volume by an experimentally derived constant (89).
2. Models in which the height of freely convecting flames from pool fires have been correlated to the diameter of the pool bed through the use of various dimensionlesss parameters such as the Grashof and Froude numbers (90).
3. The correlation of the burning rate of wood cribs to the "porosity" of the bed (90).
4. The correlation of such chemical aspects of coal combustion as slagging, fouling, slag viscosity, and incomplete combustion to the composition and absorptivity of the ash (91).

More sophisticated models of this type have been used by utility operators to assess the impact of coal quality on overall power station performance. For example, data from coal-fired power stations were statistically examined for meaningful relationships. Much data were found to be lacking and many of the key variables were confounded (92). Other correlative models of power plant performance with more of a technological basis include CIVEC, COALBUY, coal quality adviser (CQA), coal quality engineering analysis (CQEA) model, coal quality impact model (CQIM), and coal quality expert (CQE). These models have been developed by industry, industry research groups, and the government. These models are mainly concerned with the cost of power, pollution-control measures, and operational characteristics. The modeling of the combustion portion of the process is usually quite limited, perhaps being restricted to heat and mass balances. Boiler efficiency and potential slagging and fouling are often estimated.

For example, CQIM, from The Electric Power Research Institute and Black and Vetch (92) includes models for coal handling, coal preparation, boilers, bottom ash system, air heaters, fans, particulate removal, fly ash handling, fluid-gas desulfurization system, scrubber additive, waste disposal, and gas reheat. Input required includes coal characteristics such as heating value; ultimate analysis; moisture, ash, chlorine, and sodium content; ash fusion temperatures; and ash analysis. Generating unit input includes unit size, capacity factor, net power level, hours of operation, excess air level, boiler losses, boiler dimensions, soot blowing details, and tub bank configurations. CQIM calculates such quantities as discount rate and cost of power, limestone, fuel, and transport and performance data such as slagging, fouling, and erosion po-

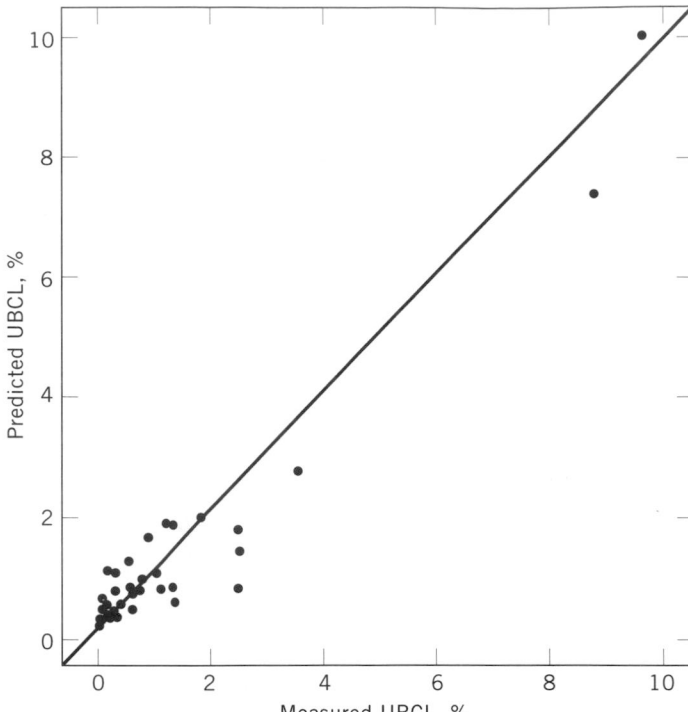

Figure 9. Comparison of measured and predicted unburned carbon loss (UBCL) in 39 test cases from industrial, utility, and test furnaces, with four different burners, using coals that ranged from anthracite to lignite in rank and operating conditions that included single burner, single wall fired, opposed wall fired, and staged. From Ref. 95.

tentials; equipment performance and de-rate information; maintenance availability data; and generation cost. It also has been used to help evaluate strategies to reduce emissions (93) and acid rain (94).

Simplified Geometries

Much of the computational effort in comprehensive codes is caused by the multidimensionality of most combustion problems. The adoption of simplified geometries often provides reasonable approximations of such characteristics as heat flux and exit temperatures and compositions at a fraction of the time and cost of a comprehensive model. Often these limited models take a zonal approach, by which the properties of the flow are passed en masse from one essentially well-stirred section to the next. The behavior in each zone or volume can be modeled with detailed conservation equations or by simpler correlative equations. The solution of the equations can be done through a rigorous finite-difference scheme with various matrix solvers, or in a marching fashion.

An example of such a model capable of rapid convergence on a personal computer has been presented (95). It models an industrial or utility boiler with several volumes arranged in a one-dimensional fashion. Particle and gas velocities are taken to be equal. The k-ε equation is used to obtained turbulence intensity. A simplified devolatilization model is used, and the volatiles are assumed to be a single gaseous fuel. The equation set is tridiagonal and solved using a tridiagonal matrix solver. The solution

takes about 5 min on a desktop PC and a few seconds on an engineering workstation. Figure 9 shows the comparison of the estimates of unburned carbon to data obtained for high and low volatile coals, for small-scale and full-scale furnaces, for various burner types, and for staged and unstaged combustion. Given the range of input variables and the simplified geometry of the model, the comparison of data to prediction is reasonable.

Equilibrium models represent a form of simplified geometry modeling. From thermodynamic equilibrium principles discussed above, effluent composition and temperature can be readily calculated for a variety of combusting fuels such as gasoline, natural gas, coal, and oil. The development of such a model for calculating effluent compositions and temperatures in a fixed-bed gasification of coal has been reported (43).

Kinetic Models

Kinetic models are currently limited only by the nature of the flows they are capable of handling. These models contain detailed reaction schemes for simple to more complex molecules, with accurate calculation of molecular diffusion and heat transfer. The good agreement between these models and flame data is shown in Figure 10, which compares methane-air flame velocities as calculated by various groups with measurements. However, the complexity of the reaction schemes, a lack of precise under-

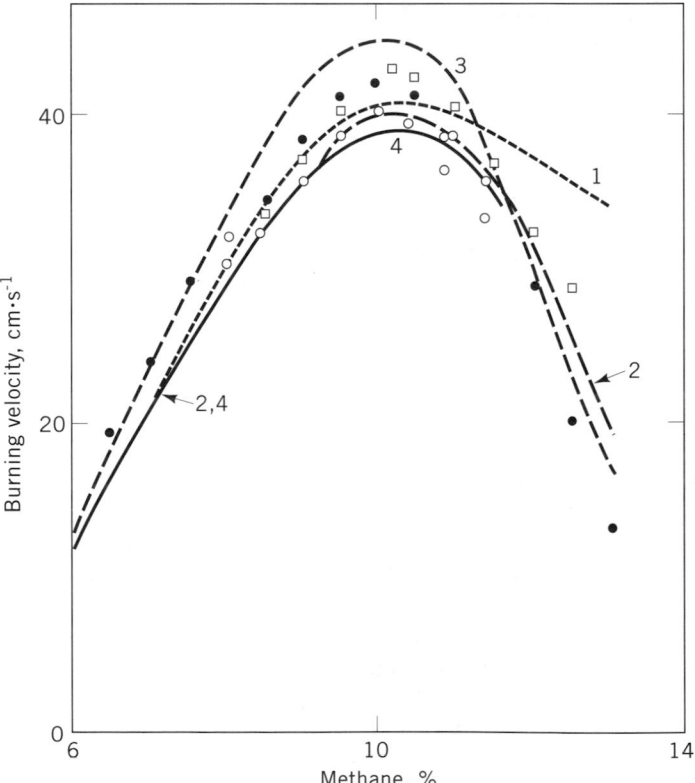

Figure 10. Comparison of measured and computed methane-air flame velocities at atmospheric pressure and an unburned gas temperature of 298 K. Measurements from three investigators (1972–1983) and predictions from four investigators (1976–1981). From Ref. 96.

standing of how turbulence interacts with this chemistry, and the limits of current computing capability have restricted the use of these codes in comprehensive combustion models and complex geometries. How detailed reaction mechanisms can be developed for simple hydrocarbons, extended to large hydrocarbons, and how these complex reaction mechanisms with multiple elementary reactions can be simplified for possible use in comprehensive codes by such methods as lumping techniques and equilibration of fast reactions coupled with sensitivity analyses have been discussed (46).

Much work has been done on various aspects of these codes (12,46,47). CHEMKIN is an example of a well-recognized model of combustion kinetics. Most models use implicit finite difference methods, by which the solution is the steady-state limit of the problem posed in transient form. CHEMKIN combines the transient approach with a hybrid-Newton, steady-state finite difference technique (97). It can describe pressure-dependent reactions as well as those that fit the traditional Arrhenius expression (98). CHEMKIN can be linked with various modules to predict well-stirred reactors, chemistry behind shocks, and surface chemistry as well as to perform sensitivity analyses.

Aymptotic Models

Asymptotic techniques were briefly described above. They have been used to examine such combustion phenomena as laminar diffusion flames, autoignition, detonation, explosion, deflagration of solids and liquids, flame spread through condensed fuels, flame instabilities, flames in nonuniform flows, and even turbulent combustion (8). Each of these combustion processes have an aspect that is dominated either by activation energy or by some reaction or diffusion rate and, therefore, can be treated with asymptotic techniques. Many of these solutions are quite successful in approximating the modeled phenomena. Figure 11 compares the asymptotic solution with the burning velocity of methane-air flames, experimental data, and a solution obtained by a kinetic technique, as described above. The asymptotic analysis agrees quite well in the fuel-lean region, because the asymptotic expansion was about the reactions in the fuel-lean region. As was shown in Figure 6, the dominant reactions in the fuel-rich region are different from those in the fuel-lean region and, therefore, would require a different expansion.

COMPREHENSIVE MODELS

Background

Comprehensive modeling of combustion must consider a number of complex processes. Typical components of a combustion system are illustrated in Figure 2. As shown, the model must account for fluid flow, turbulence, particle motion, heat transfer, and chemical reactions of particles and gas. In addition, the model may consider pollutant formation and mineral behavior as well as other processes of interest (eg, sorbent injection and sulfur capture). Numeric solution, graphic illustration, model evaluation, and technology transfer also play significant roles. Such models are a combination of many submodels (Fig. 12).

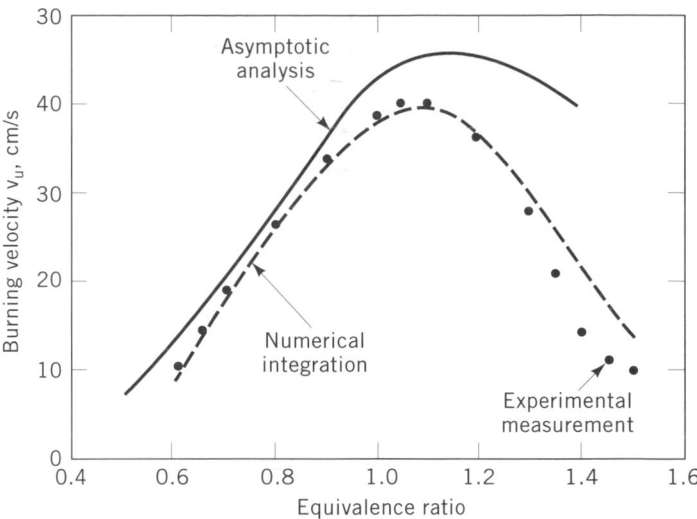

Figure 11. The burning velocity of methane-air flames at atmospheric pressure and an initial temperature of 300 K, according to asymptotic analysis with reduced chemistry, numerical integration with full chemistry, and experiment. From Ref. 8.

Comprehensive, computerized combustion models can be used to help design, test, and analyze combustors and gasifiers for clean and efficient use of fossil fuels. Such models provide a framework for effectively integrating combustion-related technology from a wide array of disciplines and are a method for transferring this technology to industry. However, to be useful, these models must satisfy at least three criteria: (1) the input and output must be easily accessible (user-friendly graphical capabilities are important), (2) the computer algorithms must be robust and computationally efficient, and (3) the models must be thoroughly evaluated to demonstrate applicability to industrial processes and to justify confidence in their predictions. Developing and implementing user-friendly, robust, efficient, applicable, accurate models requires a significant, on-going technical commitment.

Model Applications. Potential applications of comprehensive models have been identified (99,100). They include

1. Identifying reactor characteristics.
2. Interpreting and extrapolatng test data.
3. Evaluating technical options or advanced concepts.
4. Predicting consequences of fuel changes.
5. Predicting effects of process changes.
6. Identifying and evaluating pollution control methods.
7. Identifying sensitive process variables.
8. Identifying controlling (limiting) processes.
9. Identifying areas requiring further investigation.
10. Assisting in scaleup.
11. Assisting in design and optimization.
12. Assisting in process control.

Performing such tasks experimentally is costly, both in time and in resources. The attractiveness of modeling,

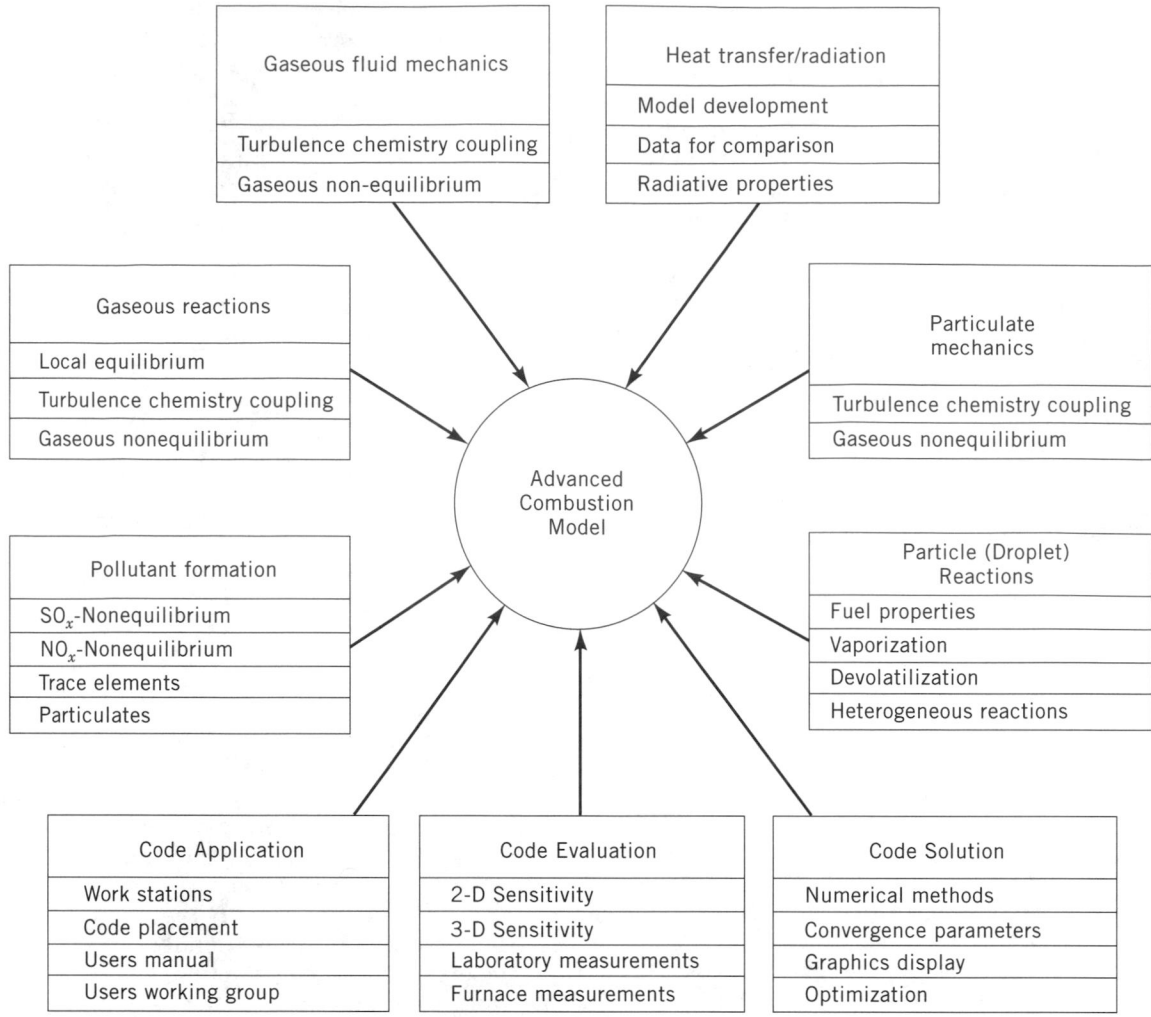

Figure 12. Elements required for the development of comprehensive combustion models. From Ref. 29.

which is becoming more sophisticated, more reliable, and less costly, is increasing. Recent developments are helping improve the accuracy, generality, and reliability of comprehensive models and the ease of interpreting results. Improved diagnostic methods are providing more accurate and more complete data for model evaluation. More powerful and less costly computers are providing quicker turnaround at lower cost. Sophisticated graphics programs are providing increased capability for visualization of data in three dimensions. With these and other advancements, comprehensive modeling is becoming accepted as an important tool in combustor design.

Use of Comprehensive Models by Industry. Table 4 is a list of typical applications for combustion modeling technology, and Table 5 is a list of the kinds of properties that such models can compute. Most models of this type provide spatial variation of the listed quantities, and some give temporal dependence.

Recently, a survey was conducted to determine the status of the worldwide use of advanced comprehensive combustion models by industry (22). A total of 12 organizations in seven countries responded to two key questions.

The first question was, "In the 1990s [1990–1992], for how many cases have you applied comprehensive combustion modeling at the specific request of industry?" All 12 respondents indicated that they had made such applications. Details are provided in Table 6. For example, 11 groups noted that they had modeled utility boilers. The number of completed applications for utility boilers ranged from 1 to 20 per respondent. Comprehensive combustion modeling has been applied to 13 energy-related industries (Table 6). Combustion modeling has most commonly been used for combustion efficiency, flow patterns, general features, temperature–heat transfer; gas composition, and trends analysis.

The survey also asked, "Have your calculations with your comprehensive combustion model, in any of these cases, been a significant factor in an industry having made a substantial commitment of resources?" A total of 9 of the 12 respondents said yes, which indicates that industry supports combustion modeling.

Premises and Variables for Comprehensive Models. A number of key premises as a foundation for the development of comprehensive combustion models have been sug-

Table 4. Some Typical Applications of Combustion Model Technology[a]

Reduce carbon loss
Optimize low NO_x burner swirl number
Find regions of particle impact on wall
Test effects of feedstock change
Examine effects of high pressure
Determine importance of CO_2, H_2O, and O_2 oxidizers
Evaluate changes in burner configuration
Calculate impact of larger or smaller coal size
Optimize furnace size
Check peak temperatures for materials impact
Identify peak radiative heat flux zones
Compute effects of load variation
Find regions of maximum SO_x capture
Estimate effectiveness of flue gas recirculation on NO_x
Evaluate impact of high moisture coals

[a] From Ref. 22.

gested (100). These premises are summarized below with illustrative examples.

Premise 1. Controlling physical and chemical mechanisms must be accounted for, while noncontrolling mechanisms can be ignored or treated in a limited fashion with only minor losses to the overall accuracy of the code. Example: in turbulent diffusion flames, turbulent mixing is assumed to be rate limiting and is modeled carefully, and homogeneous chemical reactions are often considered to be in equilibrium.

Premise 2. Model components (submodels) can be independently investigated and modeled with competing processes suppressed or absent. Example: the rate of particle weight loss due to coal devolatilization can be measured in inert atmospheres where the resulting char does not oxidize. Thus all of the weight loss can be attributed to devolatilization, and the kinetics for devolatilization in inert atmospheres is assumed to apply in oxidative atmospheres as well.

Premise 3. Similarly, basic rate measurements made in laminar systems can be applied to turbulent flows when the effects of turbulence on kinetic processes are taken into account. Example: kinetic measure-

Table 5. Typical Properties That Combustion Models Can Calculate[a]

Temperature distribution
Gas composition
Velocities
Particle trajectories
Extent of burnout
NO_x formation and reduction
SO_x formation and capture
Pressure distribution
Particle size distribution
Ash–slag accumulation
Time variations
Local variations

[a] From Ref. 22.

Table 6. Multidimensional Combustion Models: Recent Applications to Industry[a,b]

Use of Combustion Model for Industry

Application	Number of Respondents Who Modeled Application	Range of Completed Applications
Utility boiler	11	1–20
Gas turbine	4	2–5
Industrial furnace	7	1–12
Gasifier	4	1–5
Reactor, combustor	7	1–12
Recovery boiler	1	10
Fire	1	2
Incinerator	1	5
Blast furnace	1	3

Types of Fuels

Black liquor	Oil
Coal	Solid wastes
Coke	Wood
Gas	

Types of Industries

Aerospace	Oil refineries
Boiler manufacturers	Power generating
Concrete kilns, gypsum	Pulp, paper
Environmental	Small energy
Glass	Steel
Incinerator	Utilities
Metallurgical	

Indicated Purposes

Most Frequent (>10)	Frequent (5–9)	Less Frequent (1–4)
combustion efficiency	minerals effects	SO_x
flow patterns	No_x	other pollutants
general features	reactor size	ignition stability
temperature, heat transfer	fuel change design	pressure distributions flame shape
gas composition	test planning	residence times
trends analysis	configuration change	

[a] From Ref. 22.
[b] A total of 12 respondents from seven countries: Australia (2), Canada (1), China (1), the UK (2), Germany (1), The Netherlands (1), and the United States (4).

ments made in laminar systems for nitric oxide formation are applied to entrained-flow combustion by convolving instantaneous reaction rates over the statistics (eg, mixture fraction probabilities) of the turbulence (24).

Premise 4. Fundamental, steady-state relationships can be used to predict transient or pseudo–steady-state behavior. Example: devolatilizing and reacting coal particles represent a pseudo–steady-state situation in which the particle mass and composition are changing. However, steady-state correlations for external heat, mass, and momentum transfer are used to predict reaction rate, temperature, and particle motion.

Premise 5. The behavior of individual particles or droplets can be used to predict the behavior of clusters or clouds of particles or droplets in entrained flows. Example: the Lagrangian approach is often used for particles and droplets in entrained combustion. Individual trajectories are independently calculated without any particle–particle effects. Particle and droplet cloud properties are currently a subject of research (101) and have not yet been incorporated into comprehensive combustion models.

Premise 6. The combination of individual model components (submodels) gives a total description of the overall process. Example: separately derived models for turbulence, turbulent chemistry, particle or droplet reactions, and radiation are combined in comprehensive codes that are thought to predict accurately many of the phenomena in complex combustion processes.

Premise 7. Numerical methods validated for simple systems can be applied to complex systems that defy rigorous formal verification. Example: the governing equations of a comprehensive combustion model, for which there is no known analytical solution, can be modified to be exactly satisfied by a simple, analytical solution (102), which can then be used to validate the finite difference approximations used in the model.

These premises help make the development of comprehensive models for complex combustion processes feasible with today's technology.

Key variables in comprehensive models include independent and dependent variables, input data for each inlet stream, and reactor parameters. Reactor parameters and input data for each inlet stream are specified by the user. Dependent variables are calculated by the model as functions of the independent variables. Physical coordinates are typically Cartesian (x, y, z) or cylindrical (x, r, θ) (Table 2).

Model Classifications. Comprehensive models of combustion can be classified according to process type, flow type, mathematical complexity, and flame type (*see* COAL COMBUSTION; COAL GASIFICATION). The model types that will be considered here are gas only entrained flow (such as might be used for gas turbines), entrained flow with particles or droplets (such as might be used for furnaces), transient entrained flows (typically applied to internal combustion engines), fluid-bed reactors, and fixed-bed reactors. Entrained-flow models typically describe recirculating, steady-state flows. Fixed-bed models are typically plug flow and steady state, while fluid-bed models are usually well stirred or recirculating and transient. Flame type is a special criterion that applies to entrained flows. In a premixed system, the fuel and oxidizer are fed to the furnace in a single stream. In a diffusion flame, the fuel and oxidizer are fed separately and mixed within the furnace. Most practical combustors are of the diffusion type, for reasons of safety.

User Friendliness. User friendliness ensures that models developed by experienced comprehensive modelers in universities, government laboratories, and industry can be reliably and efficiently applied and used by technical professionals in industry. User friendliness and technology transfer pose significant challenges in today's technical environment because of the ever-increasing level of sophistication of the models and the need for technical specialization by researchers and model developers. Significant effort is, therefore, required to develop sophisticated models that can be used safely and efficiently by combustor design engineers or furnace operators who may not understand much of the model technology and numerics. Model user friendliness is best evaluated by model users rather than model developers. Developers must, therefore, establish and maintain close ties with model users, and improving user friendliness must be a conscious, ongoing effort. Graphics should play a significant role.

Entrained-flow Models

Model Elements. Entrained-flow or suspended-flow combustion models treat a variety of systems. These include gas only combustion such as takes place in gas-fired turbines, gas-fired furnaces, flares, and home heating systems. Particle and droplet submodels have been included in several models to treat spray combustion in turbines and furnaces and particulate combustion and gasification that occur in industrial and utility furnaces and gasifiers (*see* COAL COMBUSTION; COAL GASIFICATION). Special formulations allow for the treatment of transient combustion like that in internal combustion engines. These systems can have complex geometries such as those found in transportation turbines or more simple geometries as found in cylindrical combustors. Flows range from highly swirling, highly turbulent as found in internal combustion engines or near the burners in furnaces and turbines to near-laminar directed as found in flares and the higher zones of furnaces. There can be multiple entry ports for fuel, air, and diluent gas (as found in turbines and furnaces) or a single burner. Particles and droplets are dispersed by the turbulence of the gas and heated by radiation and convection, causing vaporization, devolatilization, and heterogeneous reaction. Gaseous reaction rate can be dominated by mixing or controlled by reaction rates. Pollutants are formed from fuel-included species as well as through thermal formation pathways. Particle and droplet reaction rates are influenced by internal and external diffusion, heat transfer processes, and intrinsic kinetics. Fuel volatiles and oxidizers can continue to react in the gas phase, while mineral matter can collect on walls and heat transfer surfaces.

Historically, many different types of submodels and model components have been applied to various physical and chemical processes in entrained-flow combustor design and simulation. CFD models have been used to predict the fluid flow field for nonreacting conditions (103). CFD models have also been used with specified local heat input to predict the effects of chemical reaction (104). Models for gaseous combustion have been applied to predict the reacting flow field in the absence of particles. Heat-transfer models have been used to calculate the effects of radiation on wall heat flux, based on a user-

Table 7. Some Groups That Provide Comprehensive Combustion Codes and Modeling Services for Commercial Users

Advanced Combustion Engineering Research Center (PCGC)[a]
Advanced Scientific Computing (TASCflow)
Computational Fluid Dynamics Services (Harwell-FLOW3D)
Fluent (FLUENT)
Innovative Research (COMPACT and customized codes)
Phoenics North America (PHOENICS)
Reaction Engineering International (JASPER and others)

[a] Names of codes are in parentheses.

supplied, turbulent gas flow field. Some of these calculations have been extended to include particle burnout and ash deposition (103).

Existing Comprehensive Models

Entrained-flow Models Available to Industry. Several models are available commercially that are capable of comprehensive simulation of gaseous, entrained-flow combustion (Table 7). Several of these also have capabilities to handle particulate and droplet combustion. The majority of these models are available from commercial concerns, such FLUENT, FLOW-3D, and PHOENICS. Others are available from university-located research centers (ie, PCGC). These models have similar bases and ways of treating aspects of the overall combustion processes. They also have various levels of customer support.

The models are based on computational fluid dynamics codes that solve the momentum, mass, and energy equations in two or three dimensions. The equations are commonly pressure linked and solved using methods such as SIMPLE (80), or versions thereof. FLUENT and FLOW-3D use body-fitted coordinates, while PCGC employs a staggered grid rectangular computational mesh. FLOW-3D also has multiblock gridding, which is desirable for complex flow geometries. Commonly, a discrete ordinate radiative heat-transfer model is used. A variety of turbulence models are available. The standard k-ε, two-equation model is the most common option; nonlinear k-ε as well as algebraic and differential Reynolds stress models are sometimes available. FLUENT also has a "renormalization group theory" turbulence model for use, which is well suited for flow separation and associated heat transfer, streamline curvature and rapid distortion, and regions with mixed high and low Reynolds numbers. Combustion is usually modeled by assuming that the rate-limiting step is the intimate mixing of fuel and oxidizer and that all reactions proceed to equilibrium.

Although many such codes are available for implementation by industry users, the codes remain "specialist usable" in that a qualified professional who has required experience and a substantial time commitment to the effort is necessary to operate the codes and interpret the results. Grid generation can be complex, and the analyst must understand the assumptions. It is typically a challenge to obtain converged solutions that demand a high level of computational skill and code-specific expertise. Many combustion research questions remain to be answered, such as turbulent mixing and its interactions with chemical reactions. Interactive graphics preprocessors and postprocessors are generally included to increase ease of application and understanding. The commercially available codes do not commonly allow access to the source code for user adaptation. However, the codes are quite flexible and are continually being upgraded.

Models with Particulate Combustion. A summary of multidimensional entrained-flow combustion and gasification models capable of handling particulate combustion that have appeared in the literature since 1970 is shown in Table 8 (29). Three of the models appeared during the first half of the 1980s and continue to evolve. Researchers at Sheffield University and the International Flame Research Foundation extended a CFD model for application to coal combustion (108). Other models were developed

Table 8. Selected Multidimensional Entrained-bed Coal Reactor Models[a]

Name	Classification	Turbulence	Gaseous Combustion	Particle Dispersion	Devolatilization	Char Oxidation	References
PCGC-2 and PCGC-3	two-dimensional steady state and three-dimensional steady state	k-ε and nonlinear k-ε with laminarization	probability density function (pdf) with fast chemistry; constant coal off gas	Lagrangian-Eulerian; empirical diffusion velocity	two competing, first-order rxns for weight loss; empirical swelling; submodel based on coal structure under development	diffusion–surface rxn of O_2, CO_2, H_2, H_2O with blowing (simultaneous devolatilization); constant dia or density	18, 105, 106
Imperial College	two-dimensional steady state	k-ε	Eddy breakup (EBU) with fast chemistry	Lagrangian; empirical diffusion velocity	multiple, parallel rxns with distributed activation energy	diffusion–surface rxn of O_2	109
University of Newcastle	two-dimensional steady state	k-ε	EBU with fast chemistry	Lagrangian; stochastic dispersion	competing rxns; empirical swelling	diffusion–surface reaction of O_2	34
Modified FLUENT	two-dimensional steady state	Reynolds stress; k-ε	EBU with fast chemistry	Lagrangian; stochastic dispersion	competing rxns	diffusion–surface rxn of O_2	35

[a] From Ref. 29.

specifically for coal combustion with simpler application to gaseous flames.

Many similarities exist among the models. All of the models are steady state. Of the models listed in one report (29), 8 are two-dimensional and 6 are three-dimensional. The FLUENT model is capable of three-dimensional transient CFD calculations, but only two-dimensional steady-state results have been reported for coal combustion. Most of the models use a k-ε submodel for turbulence; only one has incorporated a Reynolds stress closure. All 14 assume fast chemistry (24) for the gas phase, and 1 incorporates kinetics for carbon monoxide oxidation. A total of 9 use the less rigorous eddy breakup method for modeling the chemistry/turbulence interaction, 4 use the statistical probability density function (pdf) method, and 1 neglects the interaction altogether and calculates time-mean gas properties based on local equilibrium.

A total of 13 of the models perform Lagrangian calculations for particles; only 1 uses a strictly Eulerian approach. There were 4 models that ignored turbulent particle dispersion. Most of the models use a competing-reactions submodel for devolatilization, which predicts weight loss as a function of heating rate. A single model uses a distributed activation energy submodel, and 3 models use a single-reaction devolatilization submodel with specified products. Several models include an empirical swelling factor. Most models account for external diffusion of oxidizing species to the particle surface. Only 2 (PCGC-2 and PCGC-3) are reported to account for the blowing effect of volatiles on mass transfer. All of the models neglect the effect of turbulence on the solid phase and heterogeneous reactions. Multiflux radiation models (including discrete ordinates) are popular because of their reasonably accurate predictions at moderate computational effort. A few models use the more accurate and more computationally burdensome zone and Monte Carlo methods. Most models do not currently predict fouling, slagging, and pollutant formation, although interest in these areas is high and model development is ongoing. Most models solve the fluid mechanical equations using methods based on the well-known SIMPLE algorithm (80). PCGC-2 has been widely applied to both combustion and gasification.

One major difference in the models is the method of modeling turbulent particle dispersion. Several investigators incorporate the effects of gas turbulence on particle motion through an empirical turbulent diffusion velocity, which is expressed as a gradient of either the mean velocity (109) or particle bulk density (105). The use of empirical constants makes it possible to perform the particle phase calculations with relatively few particle trajectories (eg, 50–1200), each presenting not a single particle, but an ensemble of particles with a particle number flow rate. A second approach uses the more rigorous Monte Carlo method (110), by which a statistically significant number (eg, 5–10 times the number required for the deterministic approach) of individual particle trajectories are calculated. Particle velocity and motion are calculated using the instantaneous gas velocity, including turbulent fluctuations. During each particle's flight, the fluctuating velocity is randomly sampled at appropriate time intervals and added to the mean velocity. The difference in compu-

tational burden between the two methods is substantial, given the difference in number of required trajectories and the need to compute fluctuating velocity at appropriate intervals in the stochastic method. The empirical and stochastic approaches have been compared in pulverized coal combustion simulations (111). The results show the stochastic approach is more realistic, and an important consideration is that of correctly specifying the instantaneous particle conditions at the inlet.

The three methods shown in Table 8 of treating the gas chemistry–turbulence interaction (ie, local equilibrium, eddy dissipation, and full pdf) were compaerd (112). Ignoring the effects of turbulence resulted in a localized flame with steep temperature and concentration gradients.

Transient Combustion Models. Transient (ie, time-dependent) combustion poses particular difficulties in entrained-flow modeling. For example, in the special situation of internal combustion engines, the enclosure geometry is changing with respect to time as well as pressure, and it affects the combustion significantly. KIVA-II (113) is a code that was developed specifically for internal combustion engines, yet has been applied to many other combustion situations as well. KIVA-II solves the conservation equations including the transient portion of the equation. The position of the grid nodes can be specified as a function of time, thus allowing the overall grid to be Lagrangian as well as Eulerian, or any a mixture of the two. The mesh need not be orthogonal. The gas phase is solved using finite volumes, with the basic equations used in integral form and with the divergence terms transformed to surface integrals. The velocity vector is calculated at the node points; however, during a portion of the solution procedure, staggered gridding is also used. The solutions are carried out in consecutive or "marching" time steps with the grid being remapped after each time step. The size of the time steps is determined by the accuracy of the solution. The solution of each time step uses coupled equations in a fashion similar to that employed for the majority of steady-state codes. The mass transfer among the cells that occurs when the grid is remapped between time steps is calculated explicitly.

The transient nature of KIVA-II allows the chemical reactions to be slow or fast. Fast reactions are assumed to achieve equilibrium in a time step. The reaction rates for the slow reactions are solved using a partially implicit procedure. Turbulence is modeled either with the standard k-ε model modified for volume expansion or a KIVA-II–specific subgrid scale turbulence model. Droplet distributions are represented by several particle size classes. Particle–fluid interchange of mass, momentum, and energy are treated with coupling procedures, with some reduction of time steps occurring when exchange rates are large.

Example Model Applications. As shown in Table 6, entrained-flow combustion models have been applied to many different combustion situations. The most popular are utility boilers, gas turbines, and industrial furnaces. Examples of the work being done in gas turbines, utility boilers, and internal combustion engines are discussed below.

Gas Turbines. There are two basic types of gas turbines; those for transportation and those for power generation. Because of the needs of national security, substantial effort has gone into the development of advanced turbines for aircraft applications. The modeling of gas turbines has emphasized these engines, with little attention, until recently, given to the modeling of turbines used for power generation.

The combustors used in transportation turbines have complex geometries. Figure 13 is a cross-section of a typical turbine and shows the complex shape with the multiple entry points for air and fuel and the grid system needed to model it. Simulations of this particular turbofan-engine combustor have been performed (114) and compared with turbine near-effluent measurements. Similar calculations have been made by the major turbine manufacturers and have been reported in the combustion literature (115,116). The simulated turbine combustor was annular with 20 burners (114). Five dilution holes were placed on the top and bottom surfaces in each burner sector. There are also seven film-cooling slots on the top and six on the bottom. For the case presented here, 18% of the flow was assumed to enter through the burner, 13% through the splash plate, 39% through the film-cooling slots, and 30% through the dilution holes. This example did not model the spray; instead the grid domain inlet was positioned slightly downstream of the burner where the flow had not yet ignited but the fuel was fully vaporized. The flow pattern at the inlet was taken from measured data at atmospheric pressure with the temperature at 755 K. The mixture fraction was set at 0.1375.

Figure 14 shows the calculated temperature at the exit plane of the combustor compared with temperature contours measured with thermocouples. Many of the same features are seen in both figures, such as the wavy pattern along the top wall in the temperature contours. The range of calculated temperatures is more extreme than that measured, which may be attributed to factors such as averaging and flow disturbance introduced by the thermocouple probes, not enough datum points, and the use of equilibrium chemistry along with the k-ε model in the code as well as the neglect of radiation heat transfer.

Utility Furnaces. Simulations of an 80-mW electric power coal-fired utility boiler were performed with a generalized three-dimensional, combustion code (PCGC-3) and compared with *in situ* measurements from the same boiler (117,118). Measured quantities included spatially resolved gas velocity, temperature, and concentration; particle size distribution; velocity; number density; and radiative heat flux. Test variables included coal particle grind, furnace load, burner tilt, and percent excess air. The measurements were made on the Goudey Station power plant in Johnson City, New York. The furnace was tangentially fired, with forced recirculation. The boiler had seven access levels, six of which had ports available for data acquisition. There were 16 burners, 4 at each corner. A medium volatile bituminous coal was used in the experimental tests. The mass mean particle size ranged from 25 to 50 μm.

Simulation of the Goudey furnace required a large computational mesh to describe the flow field inside the furnace, particularly in the near burner region. The computational mesh was $52 \times 52 \times 84$ (227,136 cells) (Fig. 8). Two of the corner burner banks could be seen on the outside edges near the middle of the furnace. One study (31) of similar tangential- and wall-fired boilers showed that in excess of 200,000 cells were required for grid independence with gaseous combustion. The model calculations used variable grid spacing, with the computational nodes concentrated in the near burner region. The simulation was run on a super–mini computer with 256 MB of main memory; it required approximately 50 h of CPU time.

Figure 15**a** shows a volume rendering of the heat loss, with an exterior corner block of the plot cut away to show the interior of the furnace. Figure 15**b** shows side and top views of the particle trajectories inside the furnace. A total of 1440 particle trajectories were used to represent the particle field, and 288 trajectories are shown in Figure 15. The color along the particle trajectory represents the temperature of the particle at that location. Temperature, oxygen, carbon dioxide, and nitric oxide measurements inside the furnace were compared with the combustion model predictions with reasonable agreement outside the burner zones (106,119). Data and simulations for temperature and nitric oxide concentrations at two locations and oxygen and carbon dioxide concentrations at one location were compared (Fig. 16). Port 7 was located between two burners in a difficult zone to model, and agreement between model and experiment is only fair. These burner regions are highly swirled, with large turbulent fluctuations and large temperature and concentration gradients. Obtaining sufficiently fine spatial and time scale resolution is difficult for both experiment and model. Port 5 is halfway between the top row of burners and the top of the furnace, where the properties of the flow are more uniform and stable. There is good agreement between the experiment and model here (Fig. 16).

Comprehensive modeling of gaseous and entrained-flow coal combustion processes has made significant progress during the last decade. Reasonable success has been achieved in developing and combining independent submodels to simulate the overall process. The significant subprocesses on the fire side of the process are reasonably well represented, although important questions remain. Today's computers and workstations are rapidly gaining the ability to perform comprehensive modeling computations on closely-spaced, three-dimensional grids to resolve the flow structure around burners and inlets in full-scale boilers where the mixing takes place and particle reaction and pollutant formation occur.

Fluid-bed Models

Fluid-bed combustion (FBC) differs markedly from entrained-flow combustion in that interparticle contact dominates heat transfer and fluid flow. Development of these models has been pursued for the past 15 to 20 yr, but the foundations were established by still earlier research in fluid beds. Reviews of developments of one-dimensional FBC models are available (120,121). Work on two-dimensional FBC models is limited and no general reviews of these models are known.

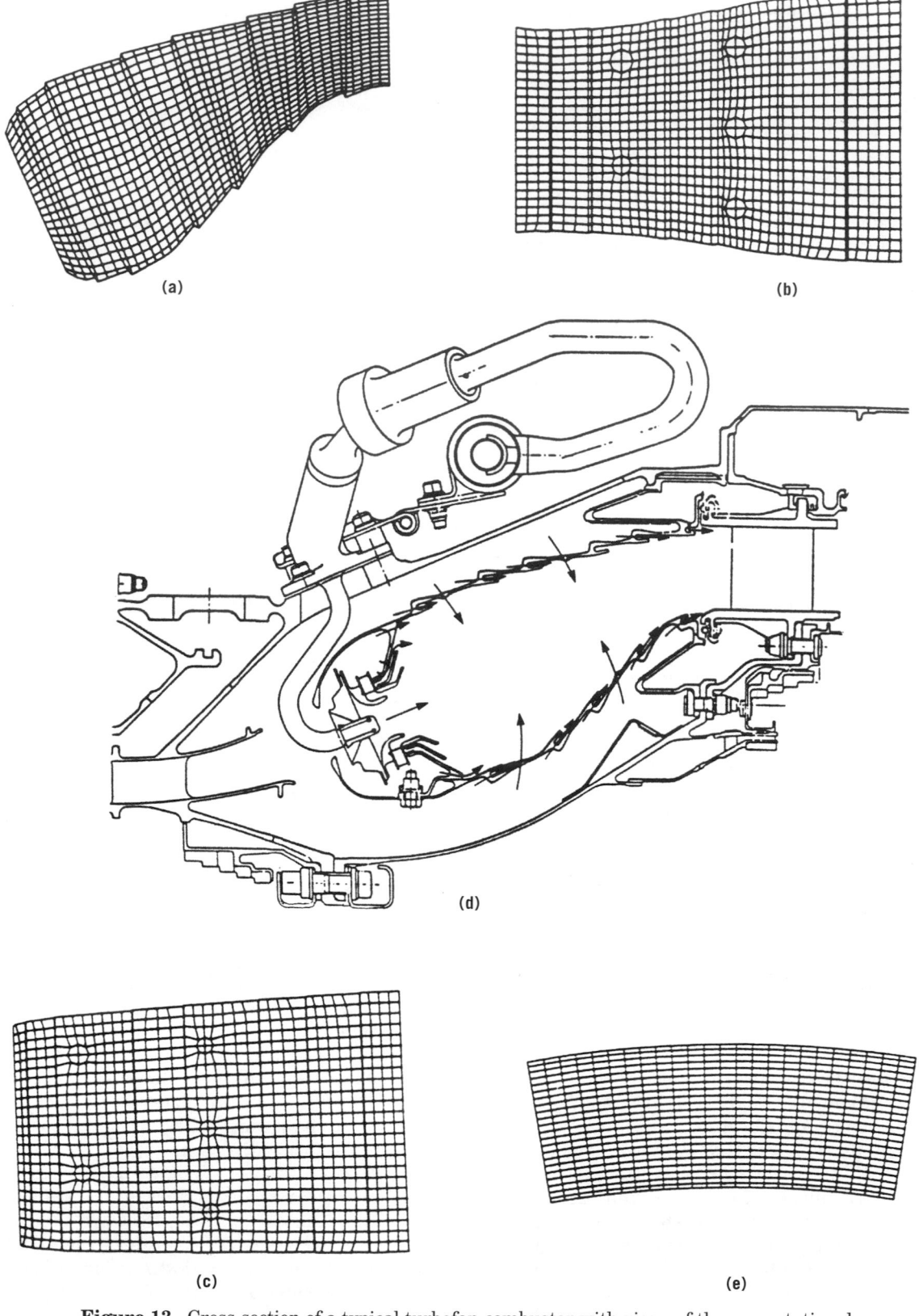

Figure 13. Cross-section of a typical turbofan combustor with views of the computational grid used to model it. (**a**), Midplane of side view; (**b**) bottom wall; (**c**) top wall; (**d**) combustor; (**e**) exit plane of a single-cup sector. From Ref. 114.

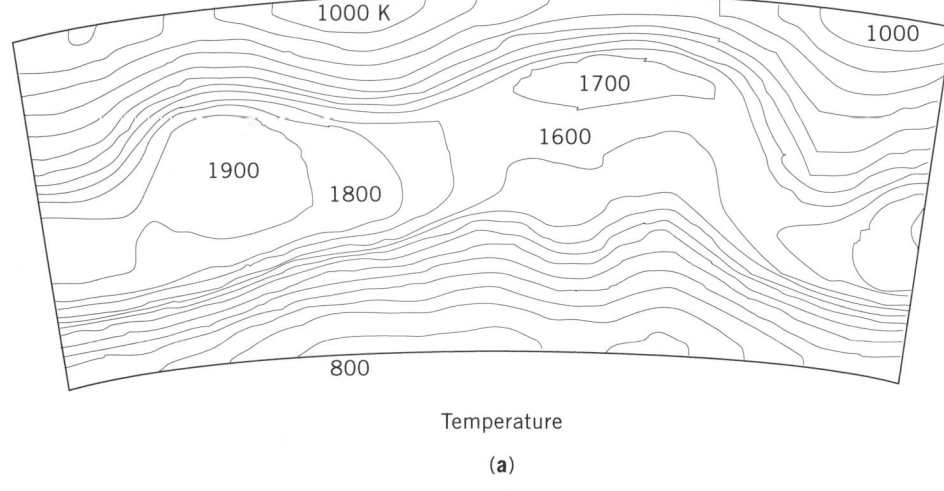

Temperature

(a)

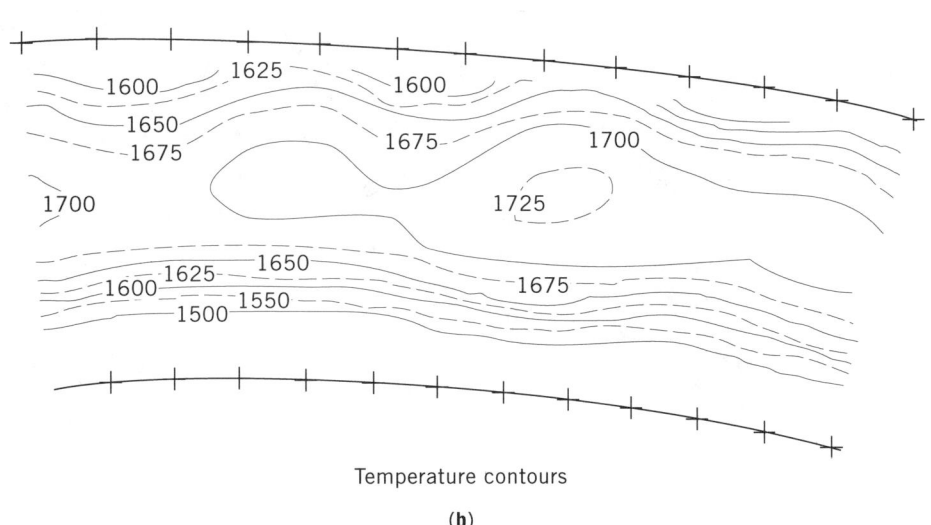

Temperature contours

(b)

Figure 14. Calculated and measured exit temperature for a typical turbofan combustor. (**a**) Calculated; (**b**) measured. From Ref. 114.

Model Elements. Fluid-bed combustion and gasification is described elsewhere (*see* COAL COMBUSTION; COAL GASIFICATION). Fluid-bed combustors are distinguished by low operating temperatures (1100 K), high levels of excess air (30%), intermediate-size particles (1–3 mm), long particle residence times (several minutes), and vigorous particulate motion that dominates heat-transfer processes. Comprehensive modeling of this process is greatly complicated by the complex fluid motion of the interacting gas and particles. Therefore, much of the modeling reported to date for FBCs has been one-dimensional and has been based heavily on correlations of laboratory data for various controlling aspects of the process.

A comprehensive review of comminution of carbons during fluid-bed has been combustion published (122). The authors identified four phenomena: primary and secondary percolation, fragmentation, and attrition of residual materials (chars, cokes, etc). Such comminution is shown to control size reduction and, therefore, govern elu-

triation of carbon fines, particle emission, and combustion efficiency.

Development of one-dimensional FBC models has proceeded under some common premises and assumptions that have been accepted by several independent investigators. Most of the treatments have been for direct atmospheric combustion. However, applications to FB gasification also exist (123). Simplifying assumptions have been made to avoid complex fluid motion. The predictions rely heavily on correlations of laboratory data for various model elements, together with mass and heat balances. The models are usually one-dimensional, because the properties (eg, oxygen concentration) do not vary significantly with radial position in the bed; at most, they vary with axial position or are uniform. Such models usually consider two distinct phases: bubbles and emulsion of air and coal. The bed-particle phase is considered to be well mixed. The particles are reduced in size in the bed (attrition) and are lost to the free board region (elutriation).

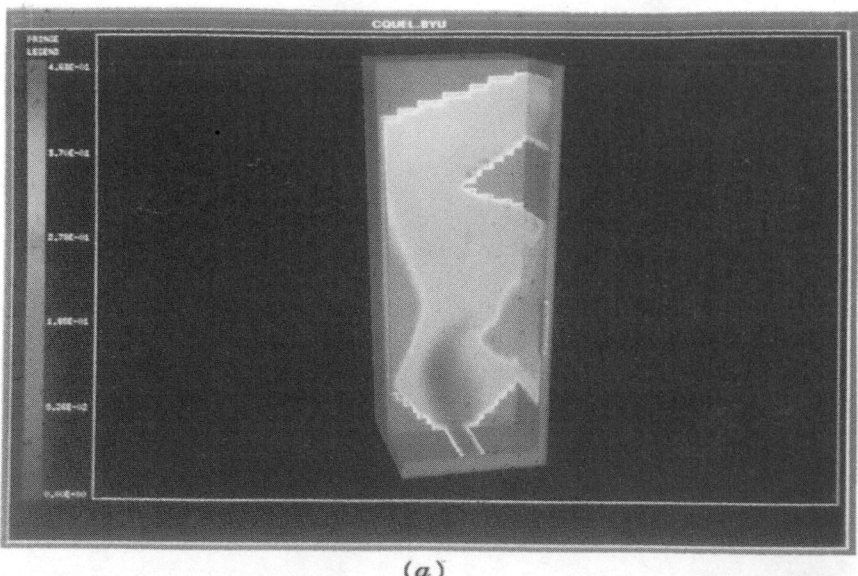

(a)

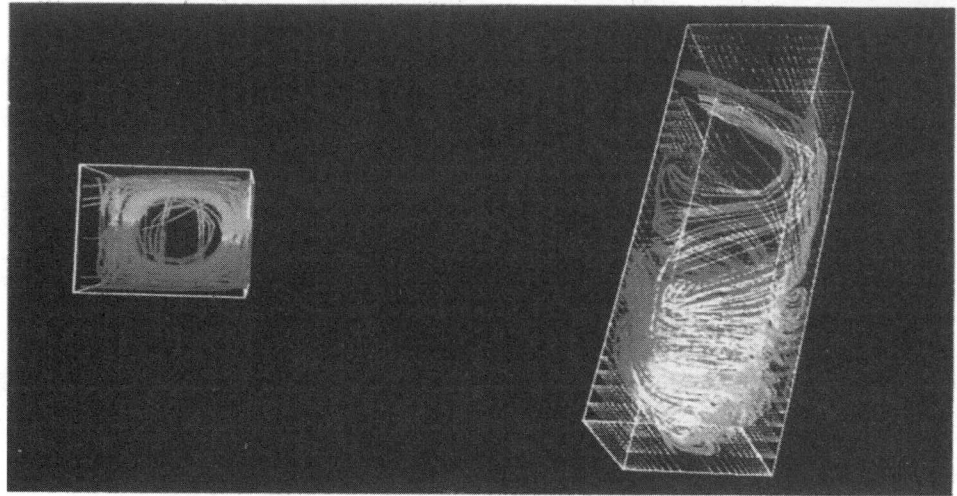

(b)

Figure 15. Simulation results for coal combustion in a tangentially fired utility furnace. (**a**) Volume rendering of heat loss in the furnace. (**b**) Particle trajectories showing their swirling motion. From Ref. 29.

Devolatilization is taken to be fast in most treatments. Recent treatments consider both rates for selected coals (124). Char oxidation is treated by coupled diffusion and surface reaction. Gas reaction rates such as $CO + \frac{1}{2} O_2 \rightarrow CO_2$ are modeled to account for carbon monoxide emissions. Early treatments neglected gaseous reaction. Subsequent treatments included this effect (124).

One-dimensional Fluid-bed Models. Typical components for a comprehensive, one-dimensional FBC model are

Fluid dynamics.
Bed and free board heat transfer.
Gas and coal combustion.
Particle dynamics.
Sulfur retention.
NO_x emissions.

Work is continuing on the modeling of fluid beds on an international level as evidenced by recent publications (109,125). This effort is in keeping with the rapid world-wide expansion in the use of fluid-bed combustors for generating power from coal in small- and intermediate-scale units.

Several one-dimensional FBC models were documented between 1973 and 1989 (Table 9). A consistent approach has been used in the progressive development of one-dimensional FBC models, with new components being added by subsequent investigators. The complex motion of particles has been neglected, while gas behavior has been divided into two or three categories. One-dimensional FBC models include the following aspects:

Gas fluid dynamics.
Particle fluid dynamics.
Coal devolatilization.
Char oxidation.
Volatiles combustion.
Particle attrition (size reduction).
Particle elutriation (loss from bed).

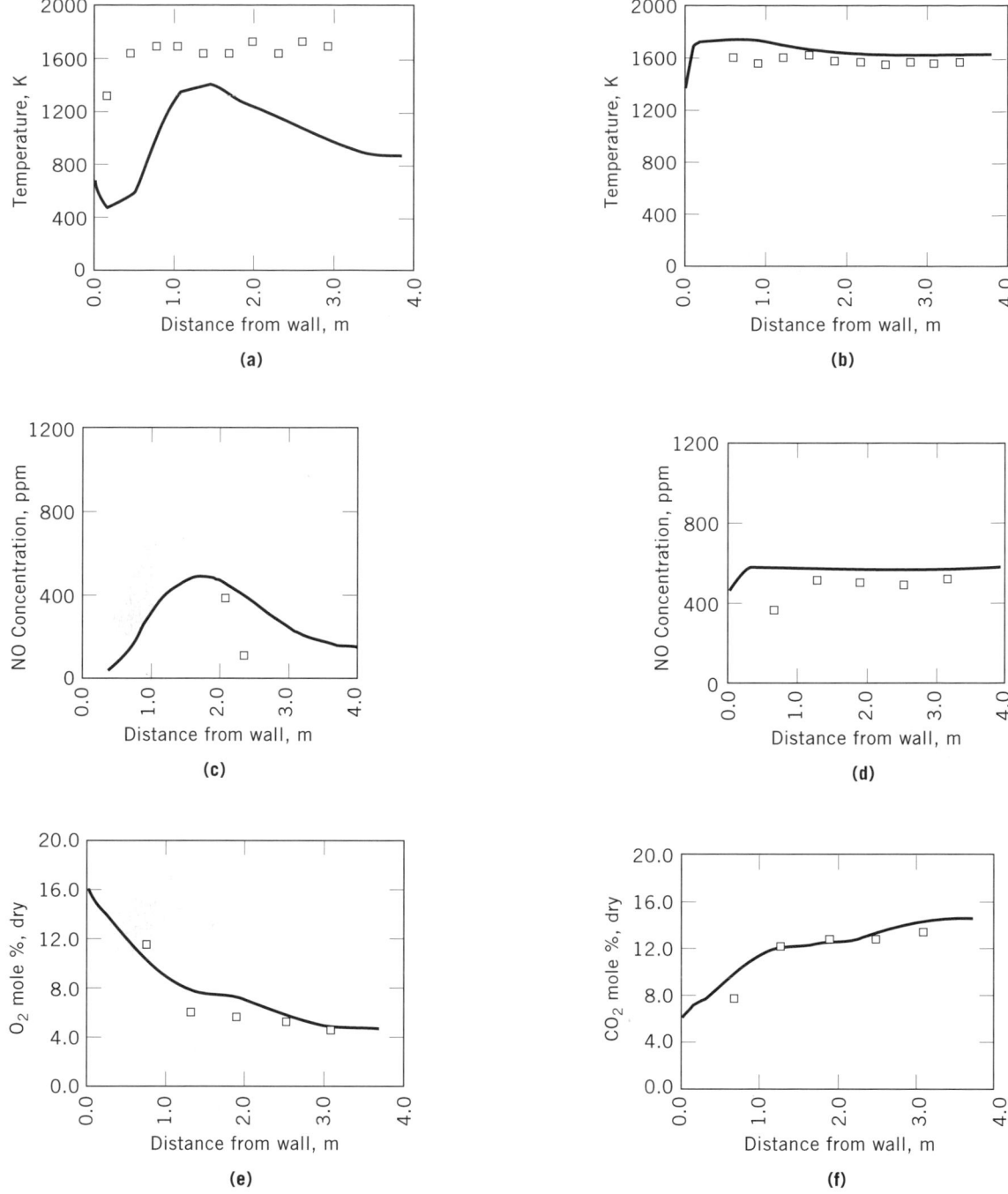

Figure 16. Comparison of measured and predicted gas temperatures, and NO, oxygen, and CO_2 concentrations at two locations in a tangentially fired utility furance. From Refs. 106,119. (**a**) and (**c**) are Port 7, between upper two burners; (**b**), (**d**), (**e**), (**f**) are port 5, about half-way between top row of burners and furnace top.

Elevated pressures.

Gasification reactions.

Polydispersed particle sizes.

Vertical and horizontal bed tubes.

Convective and radiative heat transfer.

Formation of oxides of nitrogen.

Sorption of sulfur by acceptor particles (eg, limestone or dolomite).

Transient systems.

Particle drying.

One-dimensional FBC models that consider all of these issues have evolved to the point of application. Two developed and documented one-dimensional FBC models, with user's manuals, are available (124,132). These models are largely algebraic, with ordinary differential equations to integrate along the length of the bed. Foundations for

Table 9. Summary of Selected One-dimensional Fluid-bed Combustor Models[a]

Year	Comments	Reference
1973	early FBC model for spherical carbon-batch system	126
1975	extended model to coal and continuous feed; diffusion-controlled char reaction	127
1977	treated sulfur-capture and NO formation and added kinetic term to char consumption, polydispersed sizes, and heat-transfer tubes	128
1978	treated gas phase reaction to bubbles, cooling tubes, carbon fuel	129
1978	three-phase bed with bubbles, wake and dispersion; ash restriction to char burnout	130
1978	considered NO_x formation; CO formation and reaction.	120
1979	considered particle elutriation and attrition, CO_2-C gasification, volatiles release and combustion for different coals, heat-transfer surfaces in free board region	124
1980	considers char and limestone elutriation, particle size distribution in bed and entrained materials, bed temperature profile, SO_2 retention in bed, SO_2 and NO_x emissions; allows variable bed area cross-section	131
1982	added transient and high-pressure treatment	132
1984	allowed wide particle size distribution in a nonisothermal bed.	134
1989	combustion and gasification model; predicts composition and temperature profiles for all gas and solid phases in bed and freeboard; predicts dynamics of drying processes for various solids added to the reactor; user selectable shrinking-core or exposed-core model capability	

[a] From Ref. 29.

both models are similar. Both models can be applied to high pressure and to transient flow. Two recent studies (134,135) made use of existing FBC models to investigate overall characteristics of fluid-bed power plants.

Some comparisons of one-dimensional model predictions with laboratory and pilot-scale FBC effluent data have been reported. However, the extent of evaluation by direct comparison with in-bed profile data is limited. Submodels have often been evaluated quite extensively, compared with limited evaluation of the comprehensive one-dimensional models. Models have also been applied to practical FBC systems (123,124,136). The one-dimensional models are somewhat restrictive for generalized application. Because model formulation relies significantly on empirical correlations, often for limited regions of operation, extrapolation of the model predictions to a variety of coals, pressures, or scales may not give reliable results. Specific research needs for one-dimensional FBC models have been discussed in more detail (121). Models for two-dimensional systems are far less developed. Two-

dimensional models show significant potential for generalizing FBC computations, but they have not reached the point at which the accuracy of the models has been demonstrated.

Fixed-bed Models

Existing Models. Fixed-bed combustors and gasifiers are described elsewhere (see COAL COMBUSTION; COAL GASIFICATION). The term *fixed bed* is in common use, although the bed is often slowly moving. *Fixed bed* and *moving bed* are used here interchangeably. Fixed-bed coal combustion and gasification models have been reviewed, with an emphasis on one-dimensional models (137). A total of 21 models were identified: 2 zero-dimensional, 16 one-dimensional, and 3 two-dimensional models (Table 10). Of the models identified, 12 considered different solid and gas temperatures. No heterogeneous two-dimensional model was found. Modeling of biomass gasification, including fixed-bed gasification has been reviewed (143). Fixed-bed models were classified as thermodynamic models, kinetic-free or zone models, steady-state rate models, and transient models. The models identified were all developed for coal gasification. Mathematical modeling of cocurrent, fixed-bed gasifiers were considered (144,145) but only one cocurrent model ((146) was identified. Similarly, a consideration of modeling of cross-current fixed-bed gasifiers (147) found only countercurrent model. A related review of packed-bed reactor modeling, emphasizing dynamic modeling, was presented (148). Modeling of iron ore packed-bed industrial processes was discussed (149), and fuel combustion in the blast furnace raceway zone was reviewed (150). The models reviewed were similar to fixed-bed combustion and gasification models. Modeling of smoldering combustion propagation was also related to modeling of fixed-bed combustion and gasification (151).

Model Applications. *Countercurrent Processes.* A recently published one-dimensional, steady, heterogeneous fixed-bed model (43) considered axially variable solid and gas flow rates and bed void fraction. Accounting for axially variable bed void fraction was found to be necessary to predict realistic axial temperature, concentration, and pressure profiles. The solid–gas heat transfer coefficient (152) was corrected to account for chemical reaction. The pressure drop in the reactor was calculated from Ergun's equation (153), because Reynolds numbers are typically less than 500 in fixed-bed coal gasifiers. The small pressure drop in the reactor is sensitive to the bed void fraction. Void fractions of the feed coal at the bed top and of the product ash at the bed bottom are estimated based on coal and ash bulk and apparent density measurements. The devolatilization phenomena were treated in detail: the gas evolution rates were determined by a functional group model (154), and the tar evolution rate was determined by a semiempirical correlation (155). Three gasification agents were considered: steam, carbon dioxide, and hydrogen. The gas phase species were assumed to be in partial equilibrium by holding tar out of chemical equilibrium. All gas phase species were in thermal equilibrium.

Table 10. Summary of Selected Fixed-bed Coal Gasification Models[a,b]

Scope	Devolatilization	Oxidation, Gasification	Gas Reactions	References
One-dimensional, steady, heterogeneous, countercurrent	none reported	SP model, diffusion and kinetics for O_2 reaction; kinetics for H_2O and H_2 reactions; CO_2 reaction through WGS	WGS equilibrium, instantaneous H_2 and CO oxidation reactions in combustion zone	138
One-dimensional, steady, homogeneous, countercurrent	yields by coal devolatilization, tar cracking, and tar deposition kinetics; product distribution assumed	SP model, diffusion for O_2 reaction; kinetics for H_2O, H_2, and CO_2 reactions	WGS kinetics; H_2 oxidation kinetics	139
One- or two-dimensional transient, homogeneous, countercurrent, true fixed bed	none reported	SP or AS model; film, ash, and core diffusion and kinetics	Catalyzed WGS kinetics with SP or AS model; diffusion limited CO, H_2, and CH_4 oxidation kinetics	140
One-dimensional steady, heterogeneous, countercurrent	yields of 19 functional groups and tar by functional group model and empirical tar correlation	SP or AS model; film, ash, and core diffusion and kinetics for O_2, CO_2, H_2O, and H_2 reactions	Partial or full equilibrium for tar, full equilibrium for gases, Gibbs free energy minimization	43, 141
Two-dimensional transient, heterogeneous, countercurrent, cocurrent, true fixed bed	yields by coal devolatilization and tar cracking kinetics; product distribution assumed	SP model, film and ash diffusion for O_2 reaction; kinetics for H_2O, CO_2, and H_2 reactions	WGS kinetics	142

[a] From Ref. 29.
[b] SP, shell progressive; WAS, watergas shift; AS, ash segregation.

The split boundary value problem was reduced to an initial value problem and solved by a shooting method. Sensitivity analysis considered eight model parameters and either operational variables. The model results were compared with experimental data obtained from commercial-scale gasifiers. The predicted and measured temperature and pressure profiles for the gasification of eight coals, ranging from lignite to bituminous, were found to be in a reasonable agreement (141).

Cross-current Processes. Cross-current coal combustion and gasification processes are also common. The traveling grate stokers and the rotary kiln gasifiers are the best known examples. Only one model of cross-current coal gasification was identified (147). The model was based on reducing a section of the steady two-dimensional, cross-current moving-bed process. The heat, mass, and momentum transport processes along the cross-current moving bed as well as free board processes above the moving bed were neglected. The gas and solid phases were assumed to be at different temperatures. Drying and devolatilization were not considered. The combustion reaction (oxygen) was assumed to be irreversible, whereas the gasification reactions (steam, carbon dioxide, hydrogen) as well as the water–gas shift reaction were assumed reversible. The water–gas shift reaction was also assumed to be in equilibrium. The heat of the water–gas shift reaction was assigned to the gas phase. All heterogeneous reactions were modeled by the shrinking-core model. Gas film and ash diffusion were taken into account, but pore diffusion was not. The char–steam, char–carbon dioxide, and char–hydrogen gasification reactions were modeled after Wen and Chaung (156). Reactivity factors were introduced for the steam and carbon dioxide reactions to account for the reactivity of the specific char used; 0.1 was assumed for both to match measurements in a laboratory pot gasifier. The solid–gas heat-transfer coefficient was adjusted to account for chemical reactions in the bed.

The governing equations were simplified by neglecting the accumulation of mass and energy in the gas phase. The water–gas shift equilibrium was taken into account by correcting the species concentrations, and the temperatures were calculated by the governing rate equations. Integration was performed from the top, and the iteration proceeded until the carbon in the ash reached 1% throughout the bed. The predicted variations in time of the product gas composition, the maximum solid temperature, and the gas and solid temperatures at a selected bed location were compared with the measured values from a laboratory pot gasifier. The product gas composition and the composition profiles at selected times compared well with measured values from the literature.

A heat-transfer model of rotary kiln incineration is related to cross-current coal combustion and gasification modeling. The model predictions were compared with the available bed temperature measurements, and the agreement was found to be excellent (157), but additional experimental data are needed to validate the mass transfer portion of the code.

SUMMARY

Comprehensive modeling of combustion is increasingly used by industry. Comprehensive models have become more sophisticated and are supported by increasingly powerful computer hardware and software, while experimental approaches to industrial combustor design have become more costly. In addition, design constraints, such as allowable pollutant levels and required efficiency, are becoming more stringent, and process optimization is becoming more critical. Comprehensive modeling is an important tool needed by the combustion industry to remain competitive into the 21st century.

Acknowledgments

Dr. Craig Eatough, postdoctoral associate in the Advanced Combustion Engineering Research Center, has contributed substantially to this section. The contribution of Scott Brewster and coauthors (29), whose recently published work on combustion modeling provides a foundation to this article is acknowledged. Support of the Advanced Combustion Engineering Research Center at Brigham Young University in preparing this manuscript is appreciated, as is the editorial and format work of Elaine Alger.

REFERENCES

1. A. J. Berry, *From Classical to Modern Chemistry*, Dover Publications, Inc., New York, 1954.

2. J. O. Hirschfelder, C. F. Curtiss, and R. B. Bird, *Molecular Theory of Gases and Liquids*, John Wiley & Sons, Inc., New York, 1964.

3. R. B. Bird, W. E. Stewart, and E. N. Lightfoot, *Transport Phenomena*, John Wiley & Sons, Inc., New York, 1960.

4. F. A. Lindemann, *Trans. Faraday Soc.* **17**, 598 (1922).

5. F. A. Williams, *Combustion Theory*, The Benjamin/Cummings Publishing Co., Inc., Menlo Park, Calif., 1985.

6. K. K. Kuo, *Principles of Combustion*, John Wiley & Sons, Inc., New York, 1986.

7. B. Lewis and G. von Elbe, *Combustion, Flames and Explosions of Gases*, Harcourt Brace Jovanovich, Orlando, Fla., 1987.

8. F. A. Williams in *24th Symposium (International) on Combustion*, Combustion Institute, Pittsburgh, Pa., 1992, pp. 1–17.

9. Ya. B. Zel'dovich, G. I. Barenblatt, V. B. Librovich, and G. M. Makhviladze, *The Mathematical Theory of Combustion and Explosions*, Consultants Bureau, Plenum Publishing Corp., New York, 1985.

10. J. O. Hirschfelder, C. F. Curtiss, and D. E. Campbell in *4th Symposium (International) on Combustion*, Williams & Wilkins, Baltimore, Md., 1953.

11. D. B. Spalding, *Phil. Trans. R. Soc. London* **249A**, 1 (1957).

12. N. Peters and J. Warnatz, eds., *Numerical Methods in Laminar Flame Propagation*, Friedr, Vieweg, and Sohn Verlagsgesellschaft mbH, Braunschweig, 1982.

13. P. Gray in *23rd Symposium (International) on Combustion*, Combustion Institute, Pittsburgh, Pa., 1990, pp. 1–19.

14. F. L. Dryer in W. Bartok and A. F. Sarofim, eds., *Fossil Fuel Combustion: A Source Book*, John Wiley & Sons, Inc., New York, 1991, pp. 121–213.

15. W. P. Jones and B. E. Launder, *Int. J. Heat Mass Transfer* **15**, 301–314 (1972).

16. S. V. Patankar and D. B. Spalding in N. H. Afghan and J.

M. Beer, eds., *Heat Transfer in Flames*, Scripta Book Co., Washington, D.C., 1974, pp. 73–94.

17. W. Richter and R. Quack in Ref. 16, pp. 73–94.

18. I. Glassman, *Combustion*, Harcourt Brace Jovanovich, Orlando, Fla., 1977.

19. R. A. Strehlow, *Fundamentals of Combustion*, International Textbook Co., Scranton, Pa., 1968.

20. D. L. Chapman, *Phil. Mag.*, **47**, 90–103 (1899).

21. E. Jouguet, *J. Mathematique*, 347 (1905).

22. L. D. Smoot in Ref. 14, pp. 653–781.

23. L. D. Smoot, ed., *Fundamentals of Coal Combustion: For Clean and Efficient Use*, Elsevier Amsterdam, The Netherlands, 1993.

24. L. D. Smoot and P. J. Smith, *Coal Combustion and Gasification*, Plenum Publishing Corp., New York, 1985.

25. L. D. Smoot and D. T. Pratt, eds., *Pulverized-Coal Combustion and Gasification*, Plenum Press, New York, 1979.

26. O. Reynolds, *Phil. Trans. R. Soc. London*, 174 (1883).

27. P. A. McMurtry and M. Queiroz in Ref. 23, chap. 7, pp. 511–566.

28. B. E. Launder and D. B. Spalding, *Mathematical Models of Turbulence*, Academic Press, London, 1972.

29. B. S. Brewster, S. C. Hill, P. R. Radulovic, and L. D. Smoot in Ref. 23, chap. 8, pp 567–706.

30. C. G. Speziale, *J. Fluid Mech.* **178**, 459–475 (1987).

31. P. A. Gillis and P. J. Smith in Ref. 13, pp. 981–991.

32. W. P. Jones in W. Kollman, ed., *Prediction Methods for Turbulent Flows*, Hemisphere Publishing Co., New York, 1980.

33. C. T. Crowe, M. P. Sharma, and D. E. Stock, *J. Fluids Eng. Trans. ASME* **99**, 325–332 (1977).

34. J. S. Truelove in *20th Symposium (International) on Combustion*, Combustion Institute, Pittsburgh, Pa., 1984, pp. 523–530.

35. J. Swithenbank, E. S. Garbett, F. Boysan, and W. H. Ayers in J. Feng, ed., *Coal Combustion: Science and Technology of Industrial and Utility Application*, Hemisphere Publishing Co., New York, 1988, pp. 39–63.

36. R. K. Boyd and J. H. Kent in *21st Symposium (International) on Combustion*, Combustion Institute, Pittsburgh, Pa., 1986, pp. 265–274.

37. P. W. Atkins, *Physical Chemistry*, W. H. Freeman and Co., San Francisco, 1978.

38. K. Denbigh, *The Principles of Chemical Equilibrium*, Cambridge University Press, Cambridge, UK, 1981.

39. D. R. Stull and H. Prophet, *JANAF Thermochemical Tables*, NSRDS-NBS **37**, Office of Standard Reference Data, National Bureau of Standards, Washington, D.C., 1971.

40. S. Gordon and B. J. McBride, *Computer Program for Calculation of Complex Chemical Equilibrium Compositions, Rocket Performance, Incident and Reflected Shocks, and Chapman-Jouguet Detonations*, NASA SP-273, NASA, Washington, D.C., Mar. 1976.

41. C. Selph, *Generalized Thermochemical Equilibrium Program for Complex Mixtures*, Rocket Propulsion Laboratory, Edwards Air Force Base, Calif., 1965.

42. D. T. Pratt and J. Wormeck, *CREK—A Computer Program for Calculation of Combustion Reaction Equilibrium and Kinetics in Laminar or Turbulent Flows*, Report WSU-TEL-76-1, Thermal Energy Laboratory, Washington State University, Pullman, May 1976.

43. M. L. Hobbs, P. T. Radulovic, and L. D. Smoot, *Fuel* **71**, 1177–1194 (1992).

44. O. Levenspiel, *Chemical Reaction Engineering*, John Wiley & Sons, Inc., New York, 1972.

45. J. M. Smith, *Chemical Engineering Kinetics*, McGraw-Hill Inc., New York, 1981.
46. J. Warnatz in Ref. 8, pp. 553–579.
47. G. Dixon-Lewis in Ref. 13, pp. 305–324.
48. J. A. Miller and C. T. Bowman, *Prog. Energy Combust. Sci.* **15,** 287–338 (1989).
49. P. J. Smith, S. C. Hill, and L. D. Smoot in *19th Symposium (International) on Combustion*, Combustion Institute, Pittsburgh, Pa., 1982, p. 1263.
50. S. C. Hill, L. D. Smoot, and P. J. Smith in Ref. 34, p. 1391.
51. R. D. Boardman and L. D. Smoot, *AIChE J.* **34,** 1573 (1988).
52. R. D. Boardman, *Development and Evaluation of a Combined Thermal and Fuel Nitric Oxide Predictive Model*, Ph.D. dissertation, Brigham Young University, Provo, Utah, 1990.
53. G. M. Faeth, *Prog. Energy Combust. Sci.* **13,** 293–345 (1987).
54. P. R. Solomon, M. A. Serio, and E. M. Suuberg, *Prog. Energy Combust. Sci.*, in press.
55. K. L. Smith, L. D. Smoot, T. H. Fletcher, and R. J. Pugmire, *The Structure and Reaction Processes of Coal*, Plenum Publishing Corp., New York, 1994.
56. S. Badzioch and P. G. W. Hawksley, *Ind. Eng. Chem. Process Des. Dev.* **9,** 521 (1970).
57. H. Kobayashi, J. B. Howard, and A. F. Sarofim in *16th Symposium (International) on Combustion*, Combustion Institute, Pittsburgh, Pa., 1977, pp. 411–425.
58. S. K. Ubhayakar, D. B. Stickler, C. W. von Rosenberg, and R. E. Gannon in Ref. 57, pp. 427–436.
59. D. B. Anthony, J. B. Howard, H. C. Hottel, and H. P. Meissner in *15th Symposium (International) on Combustion*, Combustion Institute, Pittsburgh, Pa., 1975, pp. 1303–1317.
60. G. R. Gavalas, *Coal Pyrolysis*, Elsevier Amsterdam, The Netherlands, 1982.
61. S. Niksa and A. R. Kerstein, *Energy Fuels* **5,** 647–665 (1991).
62. T. H. Fletcher, A. R. Kerstein, R. J. Pugmire, and D. M. Grant, *Energy Fuels* **4,** 54–60 (1990).
63. I. W. Smith in Ref. 49, pp. 1045–1065.
64. R. H. Hurt and R. E. Mitchell in Ref. 8, pp. 1243–1250.
65. R. E. Mitchell, R. H. Hurt, L. L. Baxter, and D. R. Hardesty, *Compilation of Sandia Coal Char Combustion Data and Kinetic Analyses—Milestone Report*, SAND92-8208, UC-361, Sandia National Laboratory, Livermore, Calif., 1992.
66. D. Merrick, *Fuel* **62,** 540–546 (1983).
67. C. Y. Wen, *Ind. Eng. Chem.* **60,** 34 (1968).
68. S. A. Benson, M. L. Jones, and J. N. Harb in Ref. 23, chap. 4, pp. 299–373.
69. M. P. Mengüçand B. W. Webb in Ref. 23, chap. 5, pp. 375–431.
70. H. C. Hottel and A. F. Sarofim, *Radiative Transfer*, McGraw-Hill Inc., New York, 1967.
71. K. A. Bueters, J. G. Cogoli, and W. E. Habelt in Ref. 59, pp. 1245–1260.
72. A. Lowe, T. F. Wall, and I. Stewart in Ref. 59, pp. 1261–1270.
73. R. D. Boardman and L. D. Smoot in Ref. 23, chap. 6, pp. 433–509.
74. W. Zinser and U. Schnell in J. Lahaye, ed., *Fundamentals of Physical Chemistry of Pulverized Coal*, NATO ASI Series, Matinus Nijhoff Publishers, 1987.
75. W. A. Fiveland and R. A. Wessel, *J. Eng. Gas Turbines Power* **110,** 117 (1988).
76. J. C. Kramlich, P. C. Malte, and W. L. Grosshandler in *18th Symposium (International) on Combustion*, Combustion Institute, Pittsburgh, Pa., 1981, p. 151.
77. K. M. Nichols, P. O. Hedman, L. D. Smoot, and A. U. Blackham, *Fuel* **74,** 255 (1989).
78. W. Bartok, B. A. Folsom, and F. R. Kurzynske, paper presented at the EPRI/EPA Joint Symposium on Stationary Combustion NO$_x$ Control, New Orleans, LA, 1987.
79. S. W. Churchill, *The Interpretation and Use of Rate Data: The Rate Concept*, Scripta Publishing Co., Washington, D.C., 1974.
80. S. V. Patankar, *Numerical Heat Transfer and Fluid Flow*, Hemisphere Publishing Co., Washington, D.C., 1980.
81. P. J. Roache, *Computational Fluid Dynamics*, Hermosa Publishers, Albuquerque, N.Mex., 1985.
82. E. S. Oran and J. P. Boris, *Numerical Simulation of Reactive Flow*, Elsevier Amsterdam, The Netherlands, 1987.
83. J. D. Smith, *Foundations of a Three-Dimensional Model for Predicting Coal Combustion Characteristics in Industrial Power Generation Plants*, Ph.D. dissertation, Brigham Young University, Provo, Utah, 1990.
84. R. I. Cukier, H. B. Levine, and K. E. Shuler, *J. Comp. Phys.* **26,** 1–42 (1978).
85. S. D. Philips and L. D. Smoot, *Data Book for Evaluation of Three-Dimensional Combustion Models, Vol. III, Detailed Model for Practical Pulverized Coal Furnaces and Gasifiers*, final report, Brigham Young University, Provo, Utah, 1989.
86. M. W. Rasband, *PCGC-2 and the Data Book: A Concurrent Analysis of Data Reliability and Code Performance*, M.S. thesis, Brigham Young University, Provo, Utah, 1988.
87. K. R. Christensen, M. W. Rasband, and L. D. Smoot, *Revised Data Book: For Evaluation of Combustion and Gasification Models, Vol. III, Prediction and Measurement of Entrained Flow Coal Gasification Processes*, DOE-METC Contract No. DE-AC21-85MC22059, final report, Brigham Young University, Provo, Utah, 1987.
88. D. G. Sloan, P. J. Smith, and L. D. Smoot, *Prog. Energy Comb. Sci.* **12,** 63–250 (1986).
89. W. Bartknecht, *Explosions, Course, Prevention, Protection*, Springer-Verlag, Berlin, 1981.
90. D. Drysdale, *An Introduction to Fire Dynamics*, John Wiley & Sons, Inc., New York, 1985.
91. E. J. Badin, *Coal Combustion Chemistry-Correlation Aspects*, Elsevier Amsterdam, The Netherlands, 1984.
92. N. M. Skorupska, *Coal Specifications—Impact on Power Station Performance*, IEACR/52, IEA Coal Research, London, Jan. 1993.
93. R. J. Evans, *PETC Rev.* 4, 24–28, 1991.
94. G. S. Stallard and A. Mehta in *Effects of Coal Quality on Power Plants*, EPRI report no. **GS-7361,** Palo Alto, Calif., July 1991, pp. 2.57–2.75.
95. W. A. Fiveland and A. S. Jamaluddin, *Combust. Sci. Tech.* **81,** 147–167 (1992).
96. G. Dixon-Lewis in W. C. Gardiner, ed., *Combustion Chemistry*, Springer-Verlag, New York, 1984, pp. 21–126.
97. R. J. Kee, J. F. Grcar, M. D. Smooke, and J. A. Miller, *A Fortran Program for Modeling Steady Laminar One-Dimensional Premixed Flames*, SAND85-8240, Sandia National Laboratory, Albuquerque, N.Mex., 1985
98. R. J. Kee, F. M. Rupley, and J. A. Miller, *Chemkin-II: A Fortran Chemical Kinetics Package for the Analysis of Gas Phase Chemical Kinetics*, SAND89-8009, Sandia National Laboratory, Albuquerque, N.Mex., 1989.

99. L. D. Smoot, *Prog. Energy Combust. Sci.* **10**, 229–272 (1984).

100. L. D. Smoot and P. J. Smith, paper presented at the 1987 ASME/JSME Thermal Engineering Joint Conference, Honolulu, Hawaii, 1987.

101. K. Annamalai and S. C. Ramalingam, *Comb. Flame* **70**, 307–332 (1987).

102. P. A. Gillis, *Three-dimensional Computational Fluid Dynamics Modeling in Industrial Furnaces*, Ph.D. dissertation, Brigham Young University, Provo, Utah, 1989.

103. M. R. Modarres-Razavi and A. K. Gupta, in *Heat Transfer in Furnaces*, American Society of Mechanical Engineers, New York, 1987, pp. 29–38.

104. H. Shida, *Nensho Kenkyu*, 71 (1986).

105. P. J. Smith, T. H. Fletcher, and L. D. Smoot in Ref. 76, pp. 1285–1293.

106. S. C. Hill and L. D. Smoot, *Energy Fuels*, **7**, 874–883 (1993).

107. F. C. Lockwood, S. M. A. Rizvi, G. K. Lee, and H. Whaley in Ref. 34, pp. 513–522.

108. B. M. Visser, J. P. Smart, W. L. Van de Kamp, and R. Weber in Ref. 13, pp. 949–955.

109. Z. Lou, K. Cen, and M. Ni, paper presented at the 2nd International Conference on Combustion, Beijing, China, Oct. 1991.

110. A. A. Mostafa and H. C. Mongia, *Int. J. Heat Mass Transfer* **31**, 2063–2075 (1988).

111. D. S. Jang and S. Acharya, *AIChE J.* **34**, 514–518 (1988).

112. P. J. Smith and T. H. Fletcher, *Comb. Sci. Tech.* **58**, 59–76 (1988).

113. A. A. Amsden, P. J. O'Rourke, and T. D. Butler, *KIVA-II: A Computer Program for Chemically Reactive Flows with Sprays*, LA-11560-MS, UC-96, Los Alamos National Laboratory, N.Mex., May 1989.

114. W. Shyy, S. M. Correa, and M. E. Braaten, *Combust. Sci. Tech.* **58**, 97–117 (1988).

115. H. C. Mongia, R. S. Reynolds, and R. Srinivasan, *AIAA J.* **24**(6), 890–904 (1986).

116. C. H. Priddin and J. Coupland, *Combust. Sci. Tech.* **58**, 119–133 (1988).

117. M. P. Bonin and M. Queiroz, *Comb. Flame* **85**, 121–133 (1990).

118. B. W. Butler and B. W. Webb, *Heat Transfer Combust. Sys.* **142**, 49 (1990).

119. L. D. Smoot, R. D. Boardman, B. S. Brewster, S. C. Hill, and A. K. Foli, *Energy Fuels*, **7**, 786–745 (1993).

120. A. F. Sarofim and J. M. Beér in *17th Symposium (International) on Combustion*, Combustion Institute, Pittsburgh, Pa., 1979, pp. 189–204.

121. J. Olofsson, *Mathematical Modelling of Fluidized Bed Combustors*, No. ICT is/RT 14, IEA Coal Research, London, 1980.

122. R. Chirone, L. Massimilla, and P. Salantino, *Prog. Energy Combust. Sci.* **17**, 297–326 (1991).

123. M. L. de Souza-Santos, *Fuel* **68**, 1507–1520 (1989).

124. J. R. Wells and R. P. Krishnan, *Interim Annual Report for 1979*, ORNL/TM-7498, Oak Ridge National Laboratory, Oak Ridge, Tenn., Oct. 1980.

125. F. N. Fett, K. J. Burkle, J. Dersch, J. Scholer, H. Edelmann, and I. Heinbockel, paper presented at the 2nd International Conference on Coal Combustion, Beijing, China, Oct. 1991.

126. M. M. Avedesian and J. R. Davidson, *Trans. Inst. Chem. Eng.* **51**, 121–131 (1973).

127. B. M. Gibbs, *Inst. Fuel (Symp. Ser. No. 1)* **1**, A5-1 (1975).

128. M. Horio, P. Rengarajan, R. Krishnan, and C. Y. Wen, *Fluidized Bed Combustor Modeling*, report NAS3-19725, West Virginia University, Morgantown, 1977.

129. A. L. Gordon, H. S. Caram, and N. R. Amundson, *Chem. Eng. Sci.* **33**, 713–722 (1978).

130. S. C. Saxena, N. S. Grewal, and M. Benhatoramana, *Three Phase Fluidized Bed Model*, U.S. DOE Report FE-1787-10, Argonne National Laboratories, Ill., 1978.

131. R. R. Rajan and C. Y. Wen, *AIChE J.* **26**, 642–655 (1980).

132. J. R. Lewis and S. E. Tung, *Modeling of Fluidized Bed Combustion of Coals*, DOE/MC/16000-1294, Final Report, MIT, Cambridge, Mass., May 1982.

133. N. Selcuk and A. Pekyilmaz in Ref. 36, pp. 585–592.

134. P. L. Douglas and B. E. Young, *Fuel* **70**, 145–154 (1991).

135. F. N. Fett, K. J. Burkle, U. Jungst, J. Scholer, and V. Weiss in Ref. 35, pp. 307–316.

136. J. P. Meyer, J. W. Wells, J. R. Cox, J. P. Belk, G. C. Grazier, and R. M. Wham, *Mathematical Model of the HYGAS Pilot Plant Reactor*, ORNL-5475, Oak Ridge National Laboratory, Oak Ridge, Tenn., Nov. 1980.

137. M. L. Hobbs, P. T. Radulovic, and L. D. Smoot, *Prog. Energy Combust. Sci.*, **19**, 505–586 (1993).

138. N. R. Amundson and L. E. Arri, *AIChE J.* **24**, 87–101 (1978).

139. C. Y. Wen, H. Chen, and M. Onozaki, *User's Manual for Computer Simulation and Design of the Moving Bed Coal Gasifier*, Contract No. DE-AT21-79MC16474, DOE/ME/16474-1390 (DE 83009533), Final Report for the U.S. Department of Energy, Morgantown Energy Technology Center, West Virginia University, Morgantown, 1982.

140. C. B. Thorsness and S.-W. Kang, *A General-Purpose, Packed-Bed Model for Analysis of Underground Coal Gasification Processes*, UCID-20731, Lawrence Livermore National Laboratory, University of California, Livermore, 1986.

141. M. L. Hobbs, P. T. Radulovic, and L. D. Smoot, *AIChE J.* **38**(5), 681–702 (1992).

142. M. Syamlal, *METC Gasifier Advanced Simulation (MGAS) Model*, Contract No. DE-AC21-90MC26328, Report No. 93AA-R91-001, Topical Report for the U.S. Department of Energy, Morgantown Energy Technology Center, Morgantown, W.Va., 1991.

143. A. Buekens and J. G. Schoeters in A. V. Bridgewater, ed., *Thermochemical Processes in Biomass*, Butterworth & Co. (Publishers) Ltd., Kent, UK, 1984, pp. 177–200.

144. A. Bliek, *Mathematical Modeling of a Cocurrent Fixed Bed Coal-Gasifier*, Ph.D. dissertation, Twente University of Technology, Enschede, The Netherlands, 1984.

145. J.-S. Chen, *Kinetic Engineering Modeling of Co-current Moving Bed Gasification Reactors for Carbonaceous Material*, Ph.D. dissertation, Cornell University, Ithaca, N.Y., 1986.

146. M. J. Groenveld, *The Cocurrent Moving Bed Gasifier*, Ph.D. dissertation, Twente University of Technology, Enschede, The Netherlands, 1980.

147. E. R. Monazam, *A Theoretical and Experimental Investigation of Coke Gasification in a Batch Reactor*, Ph.D. dissertation, West Virginia University, Morgantown, 1986.

148. L. Windes, *Modeling and Control of a Packed Bed Reactor*, Ph.D. dissertation, University of Wisconsin, Madison, 1986.

149. F. Hoiselbauer and C. Jaquemar in R. W. Lewis, K. Morgan, and B. A. Schrefler, eds., *Numerical Methods in Heat Transfer*, John Wiley & Sons, Inc., New York, 1983, Vol. 2, pp. 485–510.

150. J. M. Burgess, *Prog. Energy Combust. Sci.* **11**, 61–82 (1985).

151. T. J. Ohlemiller, *Prog. Energy Combust. Sci.* **11**, 277–310 (1985).

152. A. Sen Gupta and G. Thodos, *AIChE J.* **9**, 751–754 (1963).

153. S. Ergun, *Chem. Eng. Prog.* **48**, 89–94 (1952).

154. P. R. Solomon, D. G. Hamblen, R. M. Carangelo, M. A. Serio, and G. V. Deshpande, *Energy Fuels* **2**, 405–422 (1988).

155. G. H. Ko, D. M. Sanchez, W. A. Peters, and J. B. Howard in *22nd Symposium (International) on Combustion*, Combustion Institute, Pittsburgh, Pa., 1988, pp. 115–124.

156. C. Y. Wen and T. Z. Chaung, *Ind. Eng. Chem. Process Des. Dev.* **18**, 684–695 (1979).

157. W. D. Owens, G. D. Silcox, J. S. Lighty, X. X. Deng, D. W. Pershing, V. A. Cundy, C. B. Leger, and A. L. Jakway, *Combust. Flame* **86**, 101–114 (1991).

COMBUSTION SYSTEMS, MEASUREMENTS

NORMAN CHIGIER
Carnegie Mellon University
Pittsburgh, Pennsylvania

INTRODUCTION

Before the advent of the laser, measurements in flames were made by inserting probes, such as pitot tubes for velocity, suction pyrometers for temperature, and suction probes for gas concentration measurements. Particles were collected on filters inside suction probes for subsequent removal and size analysis. In the past two decades, important developments have taken place in combustion diagnostic techniques. There has been a large-scale increase in the quantity of research conducted by university, government, and industrial laboratories, resulting in a greater understanding of fundamental processes in turbulent, reacting, high temperature systems. Instead of relying on global measurements, instruments have been developed that probe into flames and combustion environments, allowing measurements to be made as a function of both space and time.

See also AIR QUALITY MODELING; COAL COMBUSTION; CARBON DIOXIDE EMISSIONS; ENERGY EFFICIENCY, CALCULATIONS.

The presence of particles of solid fuel, liquid drops, and soot created special difficulties in making measurements of temperature and concentration, and interest developed in obtaining more information about these particles. This led to a number of new concepts for particle size and concentration measurements. At the same time, alternative techniques were developed, allowing measurements to be made of temperature in heavily laden, radiating, particle flowfields.

Computer models that predict turbulent mixing and combustion flowfields are being developed and are becoming more sophisticated and reliable. Computational fluid dynamics uses the fundamental equations of conservation of mass, momentum, energy, and individual species, but models are required in order to obtain closure and solution to these equations. Detailed measurements of velocity, temperature, species concentration, particle size and velocity, and number density are required for formulation and verification of computer models. There is particular interest in accurate determination of initial and boundary conditions, and providing information on distributions as a function of both space and time. The new developments in measurement techniques and the increasing sophistication and quantity of these measurements is resulting in improved understanding of fundamental mechanisms as well as more accurate predictions.

VELOCITY

Measurements of the distributions of gas velocity provide the data needed to determine the flowfield. Streamlines determined from velocity measurements show recirculation and reverse flow zones, as well as regions of acceleration and deceleration. Velocity and streamline distributions are the foundation for computational schemes. Information on flow patterns can help designers to identify locations for excessive erosion and particle deposition. Changes in design geometry can frequently be assessed by comparing flow patterns. Information on turbulence is obtained almost exclusively from velocity measurements. Three components of turbulence intensity, shear stresses, and kinetic energy of turbulence can be measured. Turbulent length scales and eddy sizes can be derived. Conditional sampling is used to identify coherent structures and periodic events. Separate measurements are made of particle and gas velocities in high temperature flows.

Laser anemometry is used for measurement of local instantaneous velocity of particles suspended in the flow (1). In the dual-beam fringe anemometer, two light beams of equal intensity are focused to intersect, producing a pattern of successive light and dark fringes. When a particle crosses these fringes, a periodic variation in intensity of the scattered light is generated as a Doppler burst. The measured frequency difference is directly proportional to the instantaneous component of velocity of the particle lying in the plane of the beams. For measurements in recirculating and high turbulence intensity flows, Bragg cells or rotating diffraction gratings are used for frequency shifting.

The measured intensity of scattered light is a complex function of scattering angle, beam angle, particle size and shape, particle material, and ratio of fringe spacing to particle diameter. Frequency counters use high frequency clocks to measure the time durations for a number of zero crossings, and transient recorders are fast analog-to-digital converters able to digitize a Doppler burst fully using fast Fourier transform instrumentation. Aluminum oxide particles, 0.5 mm in diameter, are used for seeding with particle densities on the order of 10^{10} particles/m^3. Data rates are on the order of 10,000/s. Power spectral and autocorrelations are obtained from continuous records of velocity versus time using high seeding levels.

Three-color Laser Doppler Velocimeter (LDV) systems allow simultaneous measurement of three velocity components. Independent selectable frequency shifting is provided for each component, so that three-dimensional flows with swirl, recirculation, large fluctuations, and high turbulence intensities can be readily measured. Receiving optics modules may be placed in on-axis forward scatter or in off-axis backscatter. Fiber optics are used for measurements in small or restricted spaces and in complex geometric configurations. Fiber links separate the optics from the laser and the system electronics, which can be protected from the harsh environment.

Menu-driven software packages process raw data to provide individual velocity components and their projections in three orthogonal planes with statistical properties including mean, turbulence intensity, skewness, and flatness of velocity components. Correlations, Reynolds stresses, velocity, and flow angle histograms and vectors expressed in terms of magnitude and angle are displayed or printed in tabular or graphic form. Software packages are also available to control the automatic traversing of the flow or instrument system with three axes of translation. These software packages are greatly increasing the accuracy of measurements, and the ease and convenience of making LDV measurements.

Comparison of measurements made in noncombusting and combusting jet flows shows significant differences in mean velocity and kinetic energy of turbulence. Detailed measurements of axial and radial distribution in combusting and noncombusting flows are made of mean axial velocity, turbulent kinetic energy, temperature, normal turbulent intensities, shear stress, relative axial turbulent intensity, probability density functions, power spectra, and turbulent macroscale.

Detailed measurements of axial and azimuthal mean and root-mean-square (rms) velocity have been made with a two-color LDV in a swirl-stabilized model combustor, which has the important features of practical gas turbine combustors. Comparison of measurements made in reacting and nonreacting flows show that, in the case of reaction, the recirculation zone is stronger, more compact, and radially wider than for nonreacting flows. Rms axial velocities and normalized Reynolds stresses are substantially higher as a result of combustion.

Heat release that accompanies combustion results in acceleration of the flow. In shear flows with combustion, the mean shear is enhanced by the heat release. Reynolds stresses, turbulence levels, and production of turbulence kinetic energy by the mean shear are all increased.

Simultaneous measurement of velocity, temperature, and species concentration allows direct determination of correlation terms in the energy and species concentration conservation equations. Scalar measurements can be made jointly with LDV using Mie scattering, laser Rayleigh scattering, laser-induced fluorescence, spontaneous Raman scattering, and coherent anti-Stokes Raman scattering.

Measurements of velocity, drop size distribution, drop velocity distribution, drop number density, and liquid flux are made in air-assisted spray flames with the phase Doppler laser anemometer. Comparisons between burning and nonburning sprays show reduction in drop size and number density as a result of evaporation and changes in shape of the drop size distribution. The presence of the flame did not affect the shape of the mean axial velocity distribution.

HOLOGRAPHY

Holography, using pulsed lasers with a pulse duration of several nanoseconds, allows the freezing of events in three dimensions. Reconstruction of the hologram allows the recorded events to be analyzed. Measurements of particle size and shape can be readily made at a large number of planes from a single hologram. Velocities and trajectories of particles can be measured by using multiple-exposure holography. New developments in automatic scanning image analyzers are allowing more rapid data acquisition and analysis.

In-line holograms are made by passing a plane or spherical wavefront through the field and recording the interference pattern of the scattered and unscattered waves (2). The diffraction process can be caused by amplitude transmission variations, phase shifting, or path-length difference on reflection. The process can occur in a single plane, in a volume, on transmission, or on reflection. Holography is essentially an information buffer memory between the experiment and data analysis, and is used in dynamic cases where the requirement is to record high resolution optical information over a large volume in a short time. A hologram retains all optical information, including phase, and stores wavefront information. In experiments where time is short or extremely expensive, data reduction can be done after the experiment. Holocameras are designed to use pulsed ruby, YAG, and dye lasers for recording and a collinear HeNe CW (continuous wave) laser for alignment and reconstruction.

In holographic interferometry, two mutually coherent waves produce interference fringes that are used to determine the phase distribution of one of the waves relative to the other. Several special cases of holographic interferometry are considered: double exposure, double reference wave, double plate, tilt in both plates during reconstruction, real time, single-exposure live reference wave, local reference wave, and glint.

In particle field holography, dynamic fields are optically frozen to provide a three-dimensional (3D) high resolution image that can be studied microscopically. Multiple-exposure holography allows study of the dynamics of the field. By using holographic subtraction, the signal-to-noise ratio in a reconstructed image is improved. Image analyses augment data extraction by scanning the image volume while counting and sizing individual particles. The Fourier transform of a scattered light field yields instantaneous extraction of the size distribution. Image analyzers can be programmed to average many data frames to improve picture quality and signal-to-noise ratio. Phase shift interferometry and heterodyne interferometry offer promise for making automated data reduction easier.

TEMPERATURE

Temperature measurements provide important information on heat release as a result of chemical reaction. Heat transfer by radiation convection and conduction is temperature-dependent; knowledge of temperature is required to avoid burning and damage to surfaces. Calculations of thermodynamic processes, development of thrust, and transfer of heat require information on the variation of temperature with space and time. Computer predictions are based on mean temperature for calculation of density and the energy equation. Determination of correlation of temperature with velocity and concentration is needed to complete the turbulence information. Previous measurements by thermocouples and suction pyrometers were mainly restricted to time-averaged "point" measurements of temperature.

Temperature measurements are now made by light-scattering methods using spontaneous vibrational and rotational Raman scattering, Rayleigh scattering, and laser-induced fluorescence (3). These methods provide capability for nonintrusive, *in situ* measurements with spatial resolution of less than 0.1 mm^3 and temporal resolution on the order of 10 ns. Precise temperatures are measured in laminar flows by averaging over many laser shots, and accurate probability density functions are obtained in turbulent flows by compiling single-shot data at one spatial location. Raman and Rayleigh scattering measurements require high species concentration and clean laboratory conditions, whereas fluorescence provides intensity and spectral selectivity in flows moderately laden with particles where there is interference due to blackbody radiation, flame emission, or laser-induced incandescence.

For laminar flows, temperatures can be measured to within 2% of thermocouple measurements. For turbulent systems, the Stokes–Anti-Stokes method with a pulsed laser and gated detection, yields temporally resolved temperatures with a precision of 4%. For laminar systems, slow excitation or fluorescence scans can be used to determine the rotational temperature of molecular species. For excitation scans, the spectral bandwidth of the detection is fixed, and the excitation wavelength is varied. For fluorescence scans, the excitation frequency is fixed, and the grating of the monochromer is rotated to determine the spectral distribution of the emissive signal. In two-line fluorescence thermometry, a pair of excitation wavelengths is used to generate two broadband fluorescence signals. Two-line molecular fluorescence is employed to monitor OH radical flame temperatures and to measure ambient temperatures by single- and two-photon-excited fluorescence from NO. The dynamic range and sensitivity of two-line fluorescence methods can be enhanced by employing two-line, laser-saturated fluorescence for OH thermometry.

In thermally assisted fluorescence, the laser-induced populations of those energy levels higher than the laser-excited level are presumed to be collisionally equilibrated so that an electronic, vibrational, or rotational temperature can be extracted from the resulting fluorescent spectrum. Planar thermometry is achieved by sheet illumination with right-angle detection of scattered light using a two-dimensional array detector with a microchannel plate image intensifier. Combination of two-dimensional Rayleigh thermometry with planar laser-induced fluorescence measurements of the C_2 radical are used to identify the location of the flame front in turbulent non-premixed flames.

JOINT MEASUREMENTS OF VELOCITY, TEMPERATURE, AND SPECIES CONCENTRATION

In turbulent reacting flows, velocity, temperature, and species concentration fluctuate, related to time. The basic flowfields, movements of eddies, and large flow structures result in variations of each quantity, but the variations of each quantity can differ. The local fluxes of momentum, heat, and species are directly related to the time-averaged correlation of instantaneous values of velocity, temperature, and concentration. By using a pulsed laser, it has recently become possible to make simultaneous measurements of velocity–temperature, velocity–concentration correlations, and joint probability density functions in turbulent combustion flows.

Joint measurements of velocity, temperature, and species concentration are being made in flames using joint laser Raman spectroscopy and anemometry (4). For the understanding of turbulent combustion, time-resolved measurements of velocity, temperature, and concentration are necessary. Simultaneous measurements are required of velocity–scalar correlations such as $\overline{uc}$, $\overline{vc}$, $\overline{uT}$, $\overline{vT}$, and the joint probability density function $P(u, \phi, x)$. To perform such measurements in turbulent combustion flows, both Raman spectroscopy and laser anemometry systems have to be spatially and temporally integrated.

Single-pulse vibrational Raman spectroscopy (vrs) has been used for time-resolved measurements of scalar probability density functions. Concentrations of individual major species (H_2, N_2, O_2, H_2O) are determined from their respective Stokes vibrational Raman intensities. Temperature is determined from the anti-Stokes–Stokes N_2 vibrational Raman scattering intensities ratio. Due to the inherent weakness of the observed signal, the vrs does not permit measurement of temporal power spectral density functions or autocorrelation functions required to compute temperature or concentration scales in turbulent reactive flows.

In rotational Raman spectroscopy (rrs), a high power CW laser and a multipass cell optics configuration are used to detect pure rotational Raman transitions. This combination provides signals 1,000 times stronger than those observed from the N_2 Q-branch using a 1W CW laser. Simultaneous measurement is made of Rayleigh intensity to determine either the concentration of two gases in cold flow or the temperature and concentration fluctuations of a single gas species, such as N_2, in a flame. The LDA system used for integration with the rrs system is a two-component, real fringe system. The two LDA measurement channels are separated by polarization, and the optical train provides three parallel beams. The Raman interface unit (RIU) provides a variety of joint measurement schemes to detect CH–stretch Raman vibrational bands of the fuel species or an oxygen Raman rotational line. It also allows laser-induced fluorescence measurements and background signal measurements to be made.

Mass conservation balances carried out by integrating velocity and mass function profiles across the flowfield show agreement to within 5%, and conservation of momentum is verified to within 5%. Raman measurements have been found accurate to 7% at 1,727°C and 1 mol % for species mole fraction.

Combined Raman–LDA determinations of correlations, joint probability density functions (pdf)s, and spectra in combusting flows are providing acceptable spatial and temporal resolutions, as well as accuracy. Both Rayleigh and Raman techniques are better understood and simpler to use than the more sophisticated coherent anti-Stokes Raman scattering (cars), inverse Raman scattering, or planar imaging techniques. Sources of error, such as Raman signal contamination due to background flame luminosity, interference from LDA seed, and soot particle Mie scattering, uncertainties due to Poisson statistics, beam steering, beam defocusing, and LDA particle-bias errors, have been carefully analyzed and corrected.

Measurements of joint pdfs, velocity–scalar correlations, scalar power spectra, and scalar length scales have been made in variable-density, nonreactive round jets; turbulent premixed flames; and turbulent diffusion flames. Measurements reveal the mixing process between jet fluid and outer airstream. Combustion measurements in turbulent premixed flames demonstrate the wrinkled, wavy nature of the flame front and the relatively high scalar dissipation rates associated with practical thick flames compared to laboratory thin flames. Measurements have been made of conserved scalar fluxes and mass fluxes in turbulent diffusion flames. Measured data demonstrate the interaction between mean gradients of the scalar and fluctuating velocity field, effects of buoyancy, countergradient diffusion at selected locations in a flame, and gradient transport at other locations.

PARTICLE SIZING

Liquid fuel is atomized during injection into combustion chambers. Droplets in the sprays penetrate into the gas flowfields, and the largest drops can have lifetimes sufficient to result in impingement on surfaces. The trajectories of droplets and their rates of evaporation determine the local air–fuel mixture ratios, which in turn determine ignition and local heat release. Atomizers are designed to provide a spectrum of drop sizes and specific spray angles. Several different techniques have been developed for measurement of particle size (5).

Photographic and holographic methods are used to determine particle size distributions by counting and sizing the diameter of particles from photographs and holograms. Image analysis is used for automatic scanning of negatives or video images. Resolution for drop size is approximately 5 microns. Holograms provide 3D pictures, which are subsequently analyzed similar to photography. Q-switched pulse lasers with light pulse durations of several nanoseconds succeed in freezing the motion of particles. Velocities are determined by using double-pulse holography.

Particle sizing interferometry is an extension of laser Doppler anemometry. Using a two-color LDA for validation, visibility from Doppler signals is used for particle size measurement. Only particles that pass through the central portion of the beam, where light intensity is uniform, are measured. Alternatively, all particles crossing the measurement volume are recorded, and subsequent corrections are made based on the distribution of particle trajectories and corresponding incident intensities, assuming that all possible trajectories through the volume are equally probable.

For measurement of particle size distribution and concentration in low particle density systems, a portable Malvern particle sizer has been developed for measurements in the range of 1–30 μm at concentrations as low as 20 mg/m^3. A 20-mW solid-state laser increases the de-

tector–amplifier sensitivity. A light baffle is used to block outside light from entering the measurement volume.

The conventional Malvern particle sizer can be used for the measurement of concentration as well as the particle size distribution for obscurations between 5 and 50%. Beer-Lambert's law is used, based on measurements of light intensity measured by the central diode, light path length, mean drop diameter, extinction cross-section efficiency, and the drop volume distribution.

For dense particle fields with high obscuration, a mathematical model has been constructed to simulate the effect of multiple scattering. Parametric studies show that the multiple-scattering correction factor is both a function of obscuration and the actual particle dispersion. The two-parameter size distribution correction for multiple scattering has been extended to the 15-parameter, model-independent size distribution model.

A new technique has been developed for on-line measurement of both particle size and velocity distribution. To obtain drop velocity information, a dual-collimated laser beam system is employed. An acousto-optic Bragg cell is used as a beam splitter and a beam switch for rapid switching between the diffracted and undiffracted beams. Separation between the two parallel collimated beams is varied, and the transit time of particles crossing the two beams is measured. Velocities of drops of different size are obtained by analyzing signals received from different detector rings.

Tomographic techniques are used to measure particle size distribution and concentration in volume elements within axisymmetric sprays by scanning a cross-section of the spray. Abel transformation of the series of line-of-sight measurements yields the profiles of particle size and concentrations distributions. Analysis of diffraction patterns for cylindrical and rectangular particles indicates that information can be obtained from laser diffraction instruments of average dimensions of noncircular particles.

Modelling and computational fluid dynamic procedures for two-phase reacting flows are based on the FLUENT code using an interactive menu-driven interface. A finite difference method is used to solve the nonlinear differential Navier-Stokes equations using additional equations for solving turbulence models, chemical species, radiation fluxes, and enthalpy. The equations of motion and trajectories of drops and particles are solved in a Lagrangian frame of reference, with the path through each finite difference cell broken into a number of time steps. Allowance is made for heating of drops and evaporation. Initial size distributions are represented by the Rosin Rammler continuous distribution. Equations of motion and auxiliary equations are solved for ensembles of drops of differing sizes and initial conditions. Account is taken of the interactions that particles and drops can have when encountering boundary surfaces. The importance of measuring quantities to provide accurate information on spray characteristics is emphasized for computation and for design of spray combustion systems.

Simultaneous Particle Size and Velocity

The trajectories of particles are governed by initial particle size, particle velocity, injection angle, and subsequent interaction with the gas flowfield. Collision, coalescence, deflection, drag, acceleration, deceleration, and evaporation all influence particle trajectories. Momentum–drag ratios will determine the extent that particles follow or deviate from gas flow streamlines. Asymmetries in the spray result in asymmetries in the flame, which can result in "hot spots" where flame elements can melt or burn wall surfaces. The development of the phase Doppler analyzer, which measures simultaneously and instantaneously the velocity and diameter of single particles in sprays, has been a major breakthrough in instrument technology (6).

The liquid fuel sprays, for which the phase Doppler analyzer has been developed, influences air–fuel mixing, ignition, flame stability, combustion efficiency, combustor durability, and pollutant emissions. Drop spacing, spatial oxygen distribution, ambient temperature, relative velocity between drops and air, drop size, and drop interactions have been found to influence the production of pollutants. Drop number density and mass flux also contribute to the processes, leading to soot formation. Drop number densities are usually high in fuel sprays, which hinders obtaining coincident occurrences in sample volumes; beam extinction also reduces the accuracy of measurements. Steep density gradients under burning conditions cause laser beam steering and spreading, which induce gradients in the index of refraction. The phase Doppler analyzer developed by Bachalo and Houser uses light-scattering interferometry to acquire simultaneously the size and velocity of spherical particles (6). Measurements depend on the wavelength of the scattered light, which is unaffected by attenuation of the spray or intervening optics.

Bachalo derived a theory for drop sizing, utilizing the phase shift of the light transmitted through or reflected from spherical particles (6). The phase shift is obtained by using light-scattering interferometry produced with a standard dual-beam laser Doppler velocimeter. Drop size measurement is obtained from accurate measurement of the spatial frequency of the interference fringe pattern. The temporal frequency of the fringe pattern is the Doppler difference frequency, which is directly related to the velocity of the particle. The interference pattern is measured directly by using pairs of detectors located at known angles to the laser beam and separated by fixed spacings. Doppler burst signals are produced by each detector with a phase shift between them. The phase difference between the detectors is determined by measuring the time between zero crossings of the signals. Measurements of the phase shift are directly related to particle size.

The on-line signal processing and data management computer stores data packets, which include drop size, velocity, and time of arrival, for each drop measured. Data is stored by direct memory access and processed at a continuous rate of approximately 2,000 samples per second. Data is processed to form velocity probability density functions for each size class. Typically, 10,000 particle measurements are acquired at each point in the flowfield. Frequency shifting is used to measure the small transverse velocity components and the resolution of the directional ambiguity in recirculating flows. Information is provided on size, velocity, number density, volume flux, angle of trajectory, and time-resolved data.

To provide comparative measurements of number density and mass flux, a light extinction system is used to acquire a line-of-sight measurement of the beam attenuation on the optical path. Determination of the transmittance is made using Beer's law relating the extinction cross-section to the measured drop area mean diameter. The phase Doppler instrument provides mean diameters at points along the optical path, which are integrated to estimate the transmittance, which is used as a consistency check on the number density determined by the phase Doppler instrument. Comparisons between measurements made with sprays in quiescent air and with sprays injected into the recirculation zone downstream of a bluffbody show drastic differences in drop size–drop velocity distribution and in size–velocity correlations.

The phase Doppler particle analyzer (PDPA) tracks particle arrival times as they sweep through the probe volume at a particular location. New software provides a time analysis of particle arrivals at the measuring probe volume for specified drop size and drop velocity ranges. Distance between two particles is obtained from the time elapsed between particle arrival times and the instantaneous velocity of the particles. Events are timed to an accuracy of 1 μs with a mean data acquisition rate of 100,000 particles per second. Processing time is 3 μs per particle, allowing an equivalent rate of 300,000 particles per second in short duration bursts. Results indicate that drops tend to form clusters, which vary widely in number and concentration. Local number density within individual clusters can be determined.

Particle Sizing in Large Furnaces

Measurements in large furnaces are usually made with water-cooled probes that are inserted through furnace doors. Suction probes collect particles on filters, which are then withdrawn and analyzed. Attempts to make optical measurements in furnaces using conventional instrumentation have encountered difficulties in preventing damage to and deposition on lenses, and problems associated with high temperature radiating particles in the flow. Probes have now been developed, using fiber optics, that overcome many of these problems.

Instruments have been developed for particle size measurements in large-scale combustion environments (7). Measurements of particle size, concentration, and velocity distributions inside furnaces provide information that leads to reduction of slagging, fouling, and erosion of surfaces. A light-scattering technique is used for making in-line, real-time measurements of particle laden flows. Problems due to thermal gradients, which cause changes in the refractive index and result in deflection of laser beams (beam steering), arise in large-scale combustion applications. (Aerosol opacity refers to the additional problem of transmitting light signals through gas with high particle concentrations over long distances in industrial systems.) It is difficult to maintain clean windows for optical access in particle-laden streams, and to avoid degradation of light signals through secondary scattering by other particles in the flow. Finally, the measuring instrument must be designed to cause the minimum of interference to the flow.

The Insitec probe has been designed to solve these problems; it measures liquid, solid, or slurry particles that may vary in shape and may exhibit wide ranges of refractive index and surface reflectivity (7). Spatially resolved measurements are made at "points" in space. Count rates are up to 500,000 particles/second; data analysis requires less than one second. Absorbing particles with irregular shapes, such as coal or flyash, can be measured as well as nonabsorbing particles, such as water drops or latex spheres. The probe is water-cooled for operation up to temperatures of 1,400°C and high pressures; windows in the probe are gas-purged. Fiber optic signal transmission and a computer-driven motor system allow instruments to be remotely operated at distances up to 61 m.

The Insitec probe is capable of measuring size in the range of 0.2–200 μm, with particle concentrations of $10^7/$cm^3 for the submicron range, and up to 10 g/m^3 for the supermicron range at particle speeds in the range of 0.1–400 m/s. Continuous in situ alignment is maintained under changing thermal conditions. A large-diameter, multimode, fiber optic cable is used to transfer the scattered light signals from the optical head to the signal processor for detection and analysis. Computer keyboard controls allow the fiber optic system to be translated within the optical head to ensure proper alignment of the optical system during measurements in hostile environments. A narrow band pass interference filter, centered at 632.8 nm, screens out background radiation from the system.

The principle of operation is based on single-particle counting. The peak intensity of scattered light is measured together with the width of scattered light signals produced by single particles moving through the sample volume of a single, focused laser beam. The signal processor determines particle size from the measurement of peak intensity, and particle velocity from the pulse width of each scattered light signal. Since the laser light intensity varies across the measurement volume, a particle trajectory through the center of the measurement volume results in a much higher signal intensity than does a particle trajectory near the boundary. The light intensity distribution throughout the sample volume must be known in order to solve for particle size and concentration. However, the sample volume also varies with particle size; larger particles experience larger sample volumes than do smaller particles. Thus, the probability of counting a larger particle is greater than that of counting small particles.

The intensity deconvolution algorithm is based on the statistical analysis of a large number of scattered light signals from single particles passing through the measurement volume. Use of the intensity deconvolution algorithm allows the absolute particle concentration and particle size distribution to be obtained directly from the experimental data. A near-forward light-scattering configuration is selected for minimum sensitivity to particle shape and refractive index. The sample volume size must be made small enough so that only one particle is in the sample volume at a time. In order to cover the wide dynamic range in concentration and size, two beams with different beam diameters and different sample volume sizes are used.

Optical Properties of Soot and Droplets

Soot particles are in the submicron range and cannot be detected by most particle sizing instruments. Thus, measurements of light scattering and extinction are made through clouds of particles to yield number concentration and average size, as well as the structural properties of particles. Use of interferometry allows simultaneous measurement of velocity and size, whereas polarization ratio measurements allow a clear discrimination between soot and droplets. This set of optical techniques provides information of particle size, velocity, and optical properties of submicron solid and liquid particle clouds (8).

Condensed phases are found in combustion systems in the form of (1) liquid fuel drops, (2) pulverized solid fuel, (3) polycyclic aromatic hydrocarbons, (4) carbon clusters, and (5) submicron soot particles. The theories of light scattering for particles much smaller than the wavelength of the incident radiation and also for particles much larger than the wavelength of the incident radiation are summarized (8).

Characterization of the optical properties of soot formed in flames is obtained from measurements of the angular patterns of the scattering coefficients. When the complex refractive index is known, combined measurement of the scattering and extinction coefficients yields the number concentration and average size of soot particles present in a cloud. Scattering and extinction measurements at different wavelengths reveal the structural properties of soot particles. When the average size and number concentration of particles or the optical properties at reference wavelength are known by independent measurements, the complex refractive index of soot can be quantitatively evaluated.

The experimental apparatus is based on interferometry, which allows the simultaneous determination of velocity and size from which the joint distribution can be evaluated. The same apparatus and technique is used for measuring the scattering cross-sections and the polarization ratio of partially absorbing drops. Measurements of average drop size are significantly smaller than those obtained using photographic techniques, as the ensemble scattering method gives the lower moments of the size distribution function, which automatically accounts for the very large number of small drops that are not detectable by photography.

Fuel droplets in combustion systems undergo chemical reactions resulting in soot formation; spectrophotometric measurements conducted on samples of liquid removed from flames show blackening of the drops. Taking into account that initially transparent drops become partially absorbing, a combined set of measurements in both forward- and side-scattering regions simultaneously yields the average size and the imaginary part of the refractive index of the drops.

The structure of spray flames has been investigated by measuring radial and axial distributions of the scattering coefficients and the polarization ratios. For measurements in transient diesel sprays, quantitative imaging techniques are used. A Nd-YAG laser tuned to its second harmonic, is shaped into a thin light sheet through two cylindrical lenses. The collection optics, perpendicular to the laser beam, focuses the light-scattered images using macro-objective lenses. Images are split and focused on two microchannel plate intensifiers gated on the laser pulse. Photoelectrons from the intensifier are directed through bundles of fiber optics to an interfaced video camera. Video signals are sent to an A–D converter and stored in a buffer memory. Selected digital images are processed by a host computer, where image enhancement, noise reduction, stray light elimination, intensity calibration, and comparison between images are performed numerically.

The simultaneous use of the two arrays allows two-dimensional (2D) frozen light-scattering measurements at two different polarization states, different scattering angles, or different wavelengths. When the laser is operated in a double-pulse mode, instantaneous temporal gradients can be measured to determine 2D velocity fields of drop and spatial correlations of other combustion quantities.

GENERALIZED LORENZ-MIE THEORY FOR LIGHT SCATTERING

For many years, the Lorenz-Mie theory has been the basis for computation of the properties of light scattered by particles. Classical Lorenz-Mie theory was found to be inappropriate for most cases where laser sources are used in combustion systems. Because of the limitation of the classical theory, Gouesbet and colleagues have spent several years in developing a more general theory (9). The classical theory is based on plane–wave scattering, whereas the new generalized theory describes the interaction between a laser beam and a sphere. When the particle diameter is too large with respect to the beam width, the classical theory can be misleading. The new theory is considered to be a breakthrough in light scattering, allowing more rigorous theories to be developed for visibility and phase Doppler instruments, and influencing the design of instruments and interpretation of measurement data.

The generalized Lorenz-Mie theory allows for computation of the properties of light scattered by a Mie scatter center illuminated by a nonplane wave. The Maxwell equations are solved, accounting for the specific boundary conditions using the Bromwich formulation, which allows special solutions satisfying the boundary conditions in special curvilinear coordinate systems. The properties of scattered light observed at a point are computed together with associated integral quantities. Separate equations are used for external waves scattered by the particle and for spherical waves inside the particle. Scattering coefficients of the external wave are determined from the tangential continuity of the electric and magnetic fields at the sphere surface. The field components of the scattered wave are obtained from the Bromwich scalar potentials. Scattered intensities are computed with the aid of the Poynting theorem. Radiative balances are carried out on spheres surrounding the scatterer. Balances are made of

the incident field, and scattering cross-sections are computed. The radiation pressure force exerted by the beam on the scatterer is proportional to the net momentum removed from the incident beam.

Numerical computations of the mathematical functions involved in the theory are made, and the required scattered properties are determined. Numerical results are compared for three methods: quadratures, finite series, and localized approximations.

The optical levitation technique is used to study the interaction between a laser beam and a particle. Scattered light from the levitating scatter center is collected by an optical fiber and fed to a detector, and the experimental scattering diagram is compared with the computations' results. Accurate predictions will be possible for the scattering of Gaussian beams by spheres and for two crossing incident Gaussian beams. Doppler signals for particles larger than the beam diameter can be computed. These new computational techniques will improve the design of instruments and data-processing software packages, and will also lead to the development of new instruments for two-phase flow studies. These calculations will initiate the wave of the future in Mie scattering measurement analysis.

FOURIER TRANSFORM INFRARED SPECTROSCOPY

Fourier transform infrared (ftir) spectroscopy provides measurements of temperature and concentration of individual gas species and particles in flames. Line-of-sight measurements are made over an ensemble of particles, from which point measurements are derived by tomography. Separate radiation contributions can be determined for soot and char. The shape of the transmittance is used to determine particle size and composition, whereas the amplitude of the transmittance yields the concentration of particles.

Fourier transform infrared (ftir) spectroscopy is used for measurement of gas, particle, and soot temperatures and concentrations in flames (10). The advantages of the ftir emission and transmission (et) technique include the capability (1) to determine separate temperature and concentration for individual gas species as well as for solid particles, including soot, by employing different regions of the infrared spectrum; (2) to determine temperatures as low as 100°C and consequently the ability to follow particle or droplet temperatures prior to ignition; (3) to make measurements in densely loaded streams to study cloud effects; (4) to separate the radiative contribution from soot and from char particles; and (5) to measure particle sizes.

Ftir et spectroscopy makes line-of-sight measurements of an ensemble of particles. Tomographic techniques must be used to obtain spatially resolved data. Spectra acquisition times are on the order of 0.1 s to measure time-averaged properties. Successful measurements have been made in coal, ethylene, hexane spray, and coal water fuel spray flames. The Fourier transform technique processes all wavelengths of a spectrum simultaneously and can be used to measure spectral properties of particulate streams with varying flow rate. Radiation passing through the interferometer is amplitude-modulated, whereas particle emission passes directly to the detector and does not interfere with the measurements of scattering or transmission.

Transmission tomography is used to construct three-dimensional images from two-dimensional slices of the image. For objects with axial symmetry, Abel's radial inversion (onion peeling) equations are used. In spectral regions for which Beer's Law applies, the two-dimensional image reconstructed from transmittance leads to the spatial dependence of absorbances and hence concentration. A Fourier reconstruction of two-dimensional images was found to be consistent with the measured transmittance. Gas temperatures determined using an automated least-squares fitting routine in a homogeneous sample provided good agreement with average thermocouple measurements, with an accuracy of 10 K. Particle temperatures, obtained by fitting theoretical blackbody curves to the experimental normalized radiance, was in good agreement with calculated temperatures and temperatures measured with a thermocouple. Particle size information is obtained from extinction spectra for particles less than 80 μm in diameter, based on calculations using Rayleigh theory for large particles or Mie theory for small particles.

In a laminar diffusion flame, soot particle concentration and temperature were measured at several axial positions. Different temperatures were recorded for soot, CO_2, and HC gases, which were concentrated at different radial positions in the flame. Local soot concentrations were in good agreement with measurements made by laser extinction–scattering. Excellent agreement is also found between ftir and cars temperature measurements.

The shape of the transmittance provides information on the particle size and composition, whereas the amplitude is proportional to concentration. Experiments were performed on unignited sprays and spray flames of hexane and coal water fuels. From Mie calculations for each particle size distribution, best-fit theoretical spectra were overlaid on the measured spectra. The extinction measurement was found to be a remarkably sensitive gauge of particle size distribution. For spray combustion, both emission and transmission measurements were obtained to determine concentrations of pyrolysis and combustion products and their temperatures.

In coal water fuel spray flames, the following quantities were determined: (1) gas concentrations and temperatures for CO_2, H_2O, and CH_4; (2) concentrations of particles and soot; (3) particle temperatures; (4) percent of particles ignited; and (5) flame radiation intensity from individual components (particles, soot, gases).

COAL AND CHAR PARTICLES

A wide range of laboratory instruments are used for measurements of single particles or small samples of burning coal and char particles. Measurements are made of reactivity, weight loss, particle temperature, pore volume, and particle structure. Particles are analyzed in thermogravimetric analyzers, electrodynamic balances, electrodynamic thermogravimetric analyzers, and laminar-flow, drop-tube reactors. Pyrolysis and char oxidation are ob-

served as samples are continuously weighed, and programmable temperature controllers are used for heating. Single particles are held in suspension by electric fields in the electrodynamic balance while lasers heat the particle. Particle temperatures are measured by radiant emissions from hot particles at selected wavelengths, brightness pyrometry, or by disappearing filament pyrometer. The volume of pores and pore structure is determined by gas adsorption isotherms and mercury porosimetry.

Measurement techniques used in the study of coal and char combustion are discussed by Sahu and Flagan, with emphasis on reactivity measurements, mineral matter transformations, the measurement of particle temperatures, and particle structure characterization (11). Experimental reactors using electrical, radiant, flame, plasma, laser, and shock tubes for heating have been used to achieve temperatures as high as those attained in pulverized coal combustors, ie, 1,727°C. Laminar-flow reactors with a dilute stream of coal particles introduced on the reactor centerline are used to study rate processes in entrained-flow reactors where the reaction time is the same for all particles. Kinetic studies require that systems be large enough to permit sample extraction from the reaction zone or have optical access to facilitate noninvasive measurements of particle size, particle temperature, and gas composition. Sampling probes introduce a heat load that can alter reactor temperature profiles and possibly introduce sampling biases. Kinetic measurements examine the evolution of an ensemble of particles, eg, bulk property measurements made on extracted samples, optical measurements of a cloud of particles, or single-particle measurements made at a fixed point along a reactor length. Measurements of the combustion history of a single particle can be made by viewing along the axis of a drop-tube furnace and measuring temperature histories with a two-color pyrometer. To simulate coal flames in practical combustion systems, large reactors are needed with turbulence and flow recirculation.

The thermogravimetric analyzer (tga) is used extensively in low temperature studies of pyrolysis and char oxidation reactions. The tga consists of a sensitive balance that continuously weighs a sample as it is heated and reacts. A programmable temperature controller is used to heat the sample at a prescribed rate. A gas flow continuously supplies reactants and removes reaction products from samples ranging in mass from 1 to 100 mg. Temperatures of particles can be as high as 1,427°C, but the tga is most useful for reaction studies at temperatures below 727°C.

The electrodynamic balance (EDB) is used to suspend and weigh a single charged particle in an electric field. Although high temperature experiments are usually limited to particles larger than 100 μm, particles ranging in size from a fraction of a micron to several hundred microns can be held in suspension in an EDB. A charged particle is suspended in the electric field created by a potential across the top and bottom electrodes. An alternating potential on the ring electrode creates a time-varying force on the charged particle. That force increases with displacement from the center of the cell, causing an uncentered particle to oscillate. Due to the particle's inertia and aerodynamic drag, the motion of the particle lags the field, leading to a time-averaged force that tends to push the particle back toward the center of the chamber. This dynamic focusing makes possible the study of particles undergoing rapid change. The edb has a sensitivity in the range of 10^{-9}–10^{-14} g.

An electrodynamic thermogravimetric analyzer (edtga) uses laser heating to raise a particle trapped in an electrodynamic balance to temperatures at which reaction occurs. The particle temperature is measured optically, and its mass is determined from the field required to hold it at a null point of the cell. Laser heating of a particle produces photophoretic forces that can push a particle from the view volume unless the forces are carefully balanced. A hot particle in a cold gas generates a buoyant flow, which tends to lift the particle. At temperatures above 1,127°C, the particle begins to lose charge, making mass-loss measurements impossible. Thus, edtga is the only technique that allows detailed characterization of a burning coal particle. Photographs are taken, and the porous microstructure of the particle is studied by gravimetric gas adsorption. A monolayer of adsorbed gas such as carbon dioxide is weighed. Reactivity and ignition characteristics of levitated char particles have been determined by edtga.

Measurements of reactivity at high temperatures have been made in laminar-flow, drop-tube reactors with heat supplied by a flame, plasma, or electric heating. In vitiated combustion, the temperature depends on the fuel–oxidant mixture ratios. In an entrained-flow reactor, an electrically heated air preheater tube is followed by a reactor tube. The pulverized sample is entrained in the reactant gas and introduced into the coflowing hot primary stream. Optical measurements of particle size and temperature are made through transparent reactor walls. A moveable particle collection probe is inserted from the exhaust end of the reactor to collect partially burned samples or combustion products for characterization and analysis. Gas or liquid quench is used in the probes to stop reaction of the particles immediately on entering the collector. Samples can be obtained at different particle residence times. Feed rates of particulate material range from 100 mg/h to several grams per hour.

For measurements of the temperatures in a cloud of burning particles with a range of particle size, the gas environment varies greatly with position in the flame, resulting in considerable uncertainty in the temperature measurement. Particle temperature measurements are based on radiant emissions from the hot particles at selected wavelengths. In brightness pyrometry, the absolute emission intensity is measured at a single wavelength. The disappearing filament pyrometer matches the particle emission to that from a calibrated filament in the red. More recent brightness pyrometers make electronic measurement of the absolute emittance from the particle. Taking advantage of the wavelength dependence of the spectral emittance and assuming that the spectral emissivity does not vary with wavelength over the region of interest, two-color pyrometers compare intensity measurements at two wavelengths.

Spectrally resolved measurements are made using a monochrometer or band-pass filters to select wavelength intervals viewed by separate detectors. If the difference between the two wavelengths selected for measurement is small, the variation in the emissivity is assumed not to vary. The temperature of single-burning bituminous char particles has been measured in one instance to be 1,527°C over a period of 30 ms. Optical methods have been used for simultaneous measurement of particle size, velocity, and temperature in a combustion environment. Temperature measurements were made using two-color pyrometry, whereas particle size and velocity were measured by imaging the particle onto a coded aperture. The signal from one of the two photodetectors in the two-color pyrometer is used to determine the particle size. For a uniformly heated, spherical particle of uniform emissivity, the ratio of the peak signals determined by the photodetector is a monotonic function of the particle diameter.

Noninvasive optical measurements of particle size are made by light scattering or imaging. Typical particle concentrations are between 10^8 and 10^{11}/cm^3, hence, *in situ* measurements of fine particles are generally based on measurements of light scattered from a large number of particles. Conditional sampling ensures that larger particles are not present in the sample volume when measurements are made. *In situ* measurements are generally limited to small or dilute systems in which the region of interest is optically thin. Extractive sampling must be employed when the density of the aerosol is such that only the perimeter regions can be probed optically.

Fume particles produced by condensation of refractory vapors are generally agglomerates of small spherules. These aggregate particles have remarkably similar structures, which have been characterized as fractal structures in which particle mass scales as radius. The value of the fractal dimension conveys information about the mechanisms of particle formation and growth. The dynamics of the aggregation process depends on the aerodynamics of the particles and on particle structure. Small-angle x-ray scattering and small-angle neutron scattering are used to measure mass correlation functions for particles in the range of 1–100 nm.

Extraction of samples by probes can result in sampling biases caused by probe and particle aerodynamics and deposition of particles within the probe, particularly where hot particle-laden gas is exposed to cool surfaces and changes in aerosol properties. Coagulation can be minimized by diluting and quenching the sample. The temperature gradient between the hot particle-laden gas and the cool surface of the probe can result in particle deposition on the surface due to thermophoresis. Deposition can be reduced by using a transpiration-cooled sampling probe with diluent injection. Isokinetic sampling reduces aerodynamic biases. Particles can also be lost to walls of sampling systems by convective diffusion, inertial impaction at bends in the flow, sedimentation, and electrostatic deposition. Compared to metal tubing, many plastics create localized electric fields that result in increased deposition of charged particles.

Particles larger than 0.3 μm can be classified aerodynamically in inertial impactors. Cyclone separators are used for measurements of particles larger than a few microns in diameter. Particles collected in size-classified samples are chemically analyzed. Measurement of the size distributions of particles smaller than 100 nm is most commonly performed using electrical mobility analysis in which the particles are given an electrical charge and then classified in an electrical field. Gas adsorption isotherms are used to determine the volume of pores in char particles. Mercury porosimetry is widely used to study the pore structure properties of porous solids. Mercury, which is nonwetting with most substances, is forced under pressure to penetrate pores, openings, and voids in materials. The volume of mercury penetrated is recorded as a function of pressure. Pore surface area distributions are calculated from the pore volume.

CONCLUSIONS

Laser technology has changed the field of combustion measurements. In the past, scientists deduced combustion characteristics on the basis of pressure and temperature measurements made at the periphery of the system. Global calculations were made of heat release and flow rates. The insertion of water-cooled probes into flames enabled samples of gas and particles to be withdrawn for subsequent physical and chemical analysis. However, these probes cause disturbances to the temperature and flowfields, and withdrawn samples are not fully representative of the flame conditions. All of these measurement techniques were laborious and cumbersome.

Laser anemometry has become the established technique for measuring velocity and turbulence characteristics in flames. With two- or three-component LDA systems, normal and shear stresses are measured to determine kinetic energy of turbulence. Distributions of these quantities provide the fundamental basis for Computational Fluid Dynamic models and calculations.

Imaging techniques include high speed photography, holography, and cinematography. The use of seeding and color film reveal detailed dynamic structures within flames. Laser sheet lighting shows two-dimensional thin sections of the flowfield, from which dynamic movements of large eddy structures can be clearly seen and analyzed. Holography allows the freezing of events in three dimensions by using pulsed lasers with pulse durations of several nanoseconds. Automatic image analysis provides more rapid data acquisition and analysis of particle size, shape, and velocity.

Noninvasive laser techniques for simultaneous gas temperature and concentration measurement include spontaneous vibrational and rotational Raman scattering, Rayleigh scattering, coherent anti-Stokes Raman spectroscopy, and laser-induced fluorescence. The accuracy of temperature and concentration measurements is well-established, and laser techniques are superior to methods using physical probes. These optical methods are providing details of the chemical processes as they are taking place in the flame, showing the wide range of chemical reactions that occur even with simple fuels. This information is leading to formulation of complex kinetic models, which, when coupled with the fluid dynamic models, are leading to more accurate predictions and analyses.

Instead of requiring a series of sequential measurements, instruments are being coupled, using the same laser beams and the same measurement control volumes, but with different detection systems. Velocity, gas temperature, and gas concentration may now be measured simultaneously, as well as velocity–temperature, velocity–concentration correlations, and joint probability density functions in turbulent combustion flows. Eventually, several instruments will be incorporated into a single hybrid instrument for making simultaneous instantaneous measurements of several quantities in flames. Earlier difficulties in making measurements in the presence of irradiating particles appear to be overcome.

Important advances have also been made in the methods used to measure the size of liquid and solid fuel, char, soot, and dust particles in flames. Liquid fuel sprays with atomization, break-up, collision, coalescence, acceleration, deceleration, and evaporation are being analyzed. Laser diffraction with line-of-sight measurements yields average size and size distribution of particles. The phase Doppler analyzer simultaneously measures velocity and size of single spherical particles from which number density, volume flux, and size–velocity correlations are determined, as well as angle of trajectory and time-resolved data. Originally, measurements, based on refraction, were only made on transparent droplets. More recently, measurements, based on reflection using backscatter, have also been made on opaque spherical particles. For large furnaces, light scattering is used to measure velocity and concentration of liquid, solid, and slurry particles. The probes are water-cooled, and fiber optics is used for signal transmission. This instrument is becoming increasingly useful for reduction of slagging, fouling, and erosion of surfaces and reduction of emission of particulates. Polarization ratio measurements are particularly useful for determining particle size and concentration in dense particle clouds, such as soot or diesel sprays.

Fourier transform infrared spectroscopy, which provides temperature and concentration of individual gas species, particles, and soot, is a line-of-sight measurement of an ensemble of particles. Particle size information is obtained from extinction spectra. Interesting experiments have recently been performed on spray flames of hexane and coal water slurries.

Observing the changes in characteristics of burning coal and char particles has become possible with the use of electrodynamic balances and thermogravimetric analyzers, where single coal particles, heated by laser, are held in suspension by electrodynamic fields. Particle temperatures are measured by radiant emissions, whereas pore volume and structure are determined by gas adsorption and mercury porosimetry.

In solid propellant rocket research, attempts are being made to utilize the wide array of measurement techniques that have been proven in laboratory conditions for gaseous and liquid fuels. There are special difficulties, associated with the high rates of burning, high temperatures, and other unique dangers, in making measurements in solid propellant combustion chambers. Direct rates of burning are determined from surface regression rates, and the structure of combustion waves is studied by radiation pyrometers, spectral absorption–emission, and line reversal methods. However, some measurements, such as impact, shock, friction, and cook-off, are special to solid propellant systems.

Overall, significant developments have occurred in combustion instruments during the past two decades, which allow much more sophisticated and detailed research in a wide variety of combustion systems. Combustion environments were considered to be hostile because of the high temperatures, gas flows, and concentration of particles. Each one of these problems has been resolved. Laser beams that penetrate through the flames, without interference, have made it possible to measure velocities, particle size, temperature, and concentrations of many species. These combustion instruments are also being applied to other industrial processes with high temperature and particle-laden flows; the measurements are leading to greater understanding, increased control, and better predictions of the processes.

BIBLIOGRAPHY

1. N. Chigier, "Velocity Measurements by Laser Anemometry," *Combustion Measurements,* (1991).

2. J. D. Trolinger, "Particle and Flow Field Holography," *Combustion Measurements* (1991).

3. N. M. Laurendeau, "Temperature Measurements by Light-Scattering Methods," *Combustion Measurements* (1991).

4. D. R. Ballal, "Joint Laser Raman Spectroscopy and Anemometry," *Combustion Measurements* (1991).

5. J. Swithenbank, J. Cao and A. A. Hamidi, "Spray Diagnostics by Laser Diffraction," *Combustion Measurements* (1991).

6. W. D. Bachalo, A. Brena de la Rosa, and S. V. Sankar, "Diagnostics for Fuel Spray Characterization," *Combustion Measurements* (1991).

7. D. J. Holve and P. L. Meyer, "*In Situ* Particle Measurements in Combustion Environments," *Combustion Measurements* (1991).

8. A. D'Alessio, F. Beretta, and A. Cavaliere, "Optical Characterization of Condensed Phases in Combustion Systems," *Combustion Measurements* (1991).

9. G. Gouesbet, G. Grehan, and B. Maheu, "Generalized Lorenz-Mie Theory and Applications to Optical Sizing," *Combustion Measurements* (1991).

10. P. R. Solomon and P. E. Best, "Fourier Transform Infrared Emission/Transmission Spectroscopy in Flames," *Combustion Measurements* (1991).

11. R. Sahu and R. D. Flagan, "Particle Measurements in Coal Char Combustion," *Combustion Measurements* (1991).

COMMERCIAL AVAILABILITY OF ENERGY TECHNOLOGY

SAM RASKIN
Research and Development Office MS 43
Sacramento, California

California's vulnerability during the energy crises of the 1970s demonstrated the importance of establishing energy programs and policies to provide the state with a secure mix of energy technologies. The California Energy Commission's *Energy Development Report (EDR)* has addressed this issue directly by establishing policy recommendations for California energy technology development based on evaluation criteria such as cost, rate impacts, diversity, environmental impacts, operating flexibility, planning flexibility and reliability. The *ETSR* is intended to serve as an important resource for this planning effort by supporting the *EDR* with a continually updated assessment of more than 230 energy technology options for electric generation and end-use efficiency. This assessment includes identifying the status of commercial availability, areas where research is needed most for further development and issues constraining deployment.

Four levels of energy technology status information are provided: general conclusions, summary matrices, fact sheets and detailed technology evaluations. Each of these information levels covers the three evaluation components: commercial status, research and development goals, and deployment issues.

Conclusions provide an overview of evaluations for all fuel cycle, electric generation and end-use energy technologies and a list of possible state actions for supporting further energy technology development. This information provides a quick assessment of technology status and developments and also can be used to understand the broad range of options available to stimulate a preferred mix of technologies in California.

Summary Matrices provide the results of detailed analyses included in separate appendices on a single set of matrices for all *ETSR* technologies. In addition, this chapter includes results of detailed cash-flow analyses used to determine cost competitiveness. This summary section can be used to quickly identify analysis results or compare results for a number of different technologies.

Fact Sheets are developed by compressing the most important information from the detailed evaluations into an easy-to-use graphic format. As a result, the fact sheets can be used as an energy technology reference that describes each technology and the basis for all results found on the matrices.

Detailed Technology Evaluations include unabridged evaluations for electric generation and end-use technologies, respectively. The level of detail varies substantially among different technologies based on the availability of the latest technical information and each author's expertise.

See also COAL; ELECTRIC POWER GENERATION; ENERGY CONSERVATION, ENERGY EFFICIENCY; FUELS FROM BIOMASS; FUELS FROM WASTE; FUELS, SYNTHETIC; NUCLEAR POWER; PETROLEUM; RECYCLING; RENEWABLE RESOURCES.

COMMERCIAL STATUS

Commercial status assessment for all *ETSR* technologies indicates energy options now available *in California,* or expected to be available within a 20-year planning period. The base year (first year of operation) for these assessments is 1997 for electric generation technologies and 1992 for end-use technologies. Three criteria are used for commercial status determinations: 1) technology maturity, 2) existence of supplier(s), and 3) competitive cost. All three criteria must be satisfied for a technology to be considered "commercially available" for operation (not order) in the base year. Technologies where any one criterion is not satisfied are automatically assessed as "not commercially available." The basis for analyzing each technology according to the criteria is further discussed.

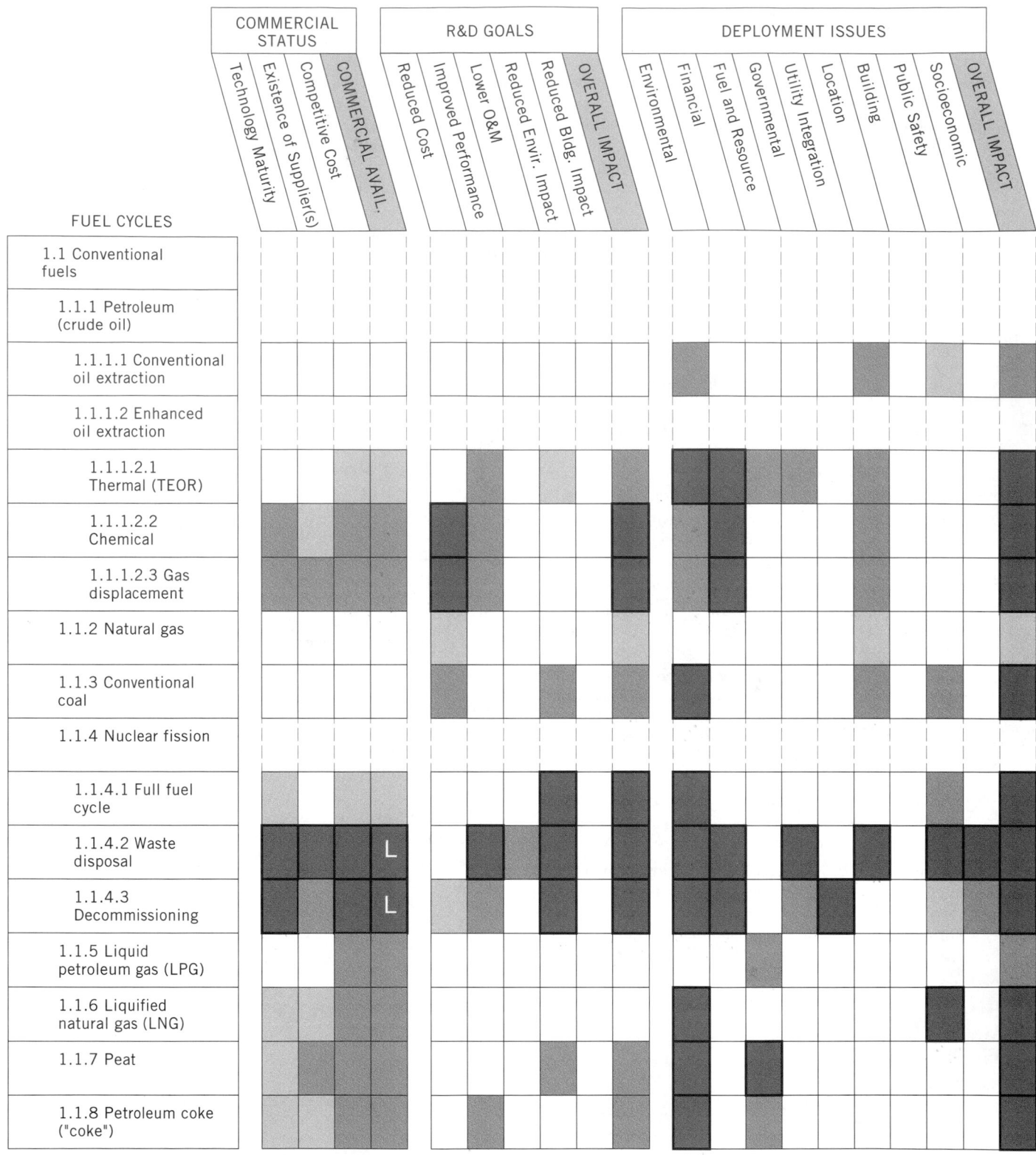

Figure 1. Technology Evaluation Matrix

Criteria for Analyzing Technology

Technology Maturity. This criterion verifies that each technology has reached a level of technical development demonstrating its readiness for market introduction. Four general phases of technology development have been identified:

- Scientific feasibility
- Technical feasibility
- Engineering feasibility
- Commercial demonstration

Figure 1. (Continued)

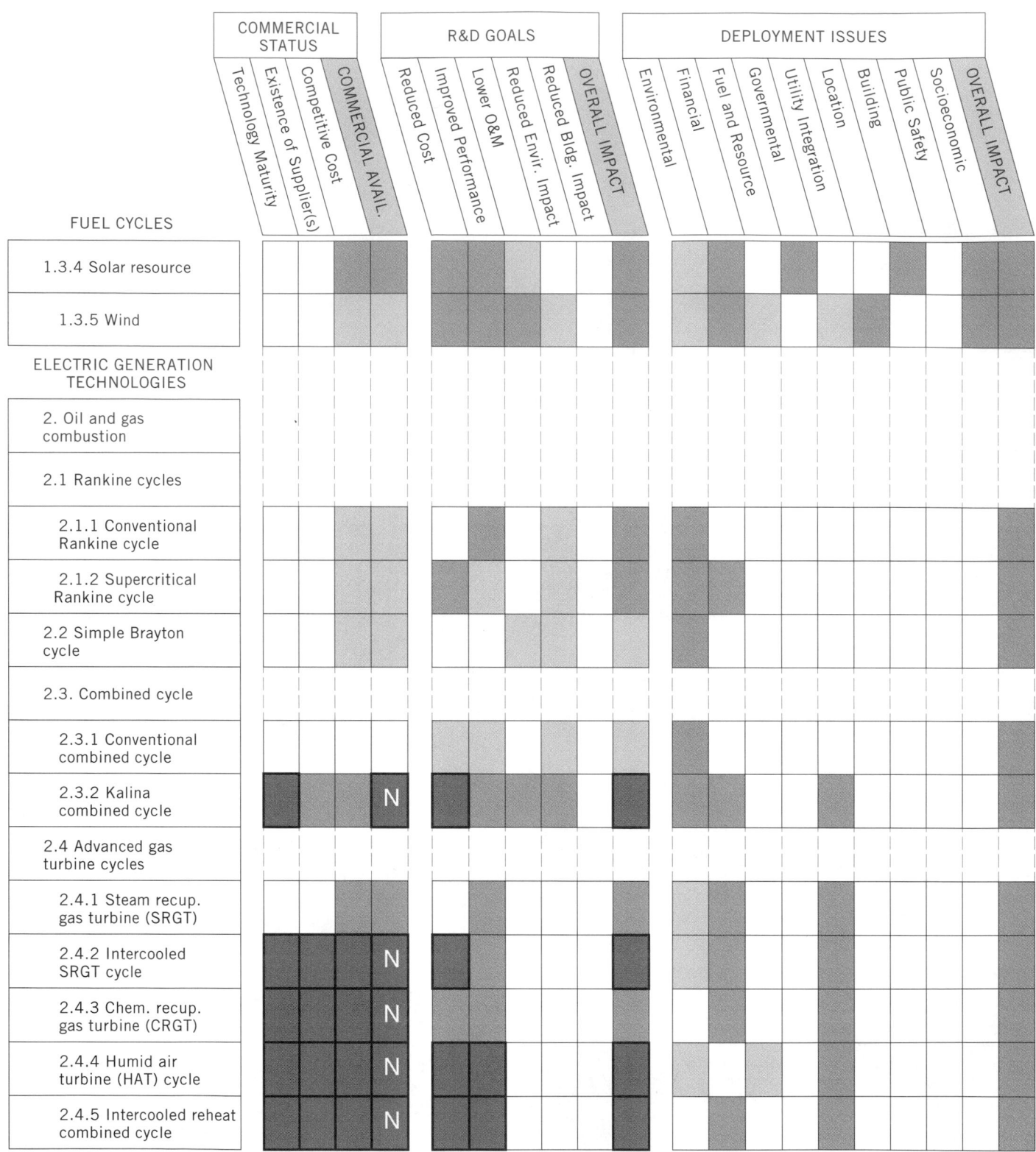

Figure 1. (Continued)

Each phase is explained in terms of its major objectives. The purpose of the scientific feasibility phase is to demonstrate at an intellectual level the feasibility of a given concept and compliance with known physical laws. The second phase, technical feasibility, seeks to confirm the scientific principles underlying a particular concept. This phase typically involves laboratory-scale experiments that test the general design concept and show necessary materials exist (or might through appropriate research and development) to permit application or operation of the concept. The engineering feasibility phase, if successful, verifies that adequate engineering design methods ex-

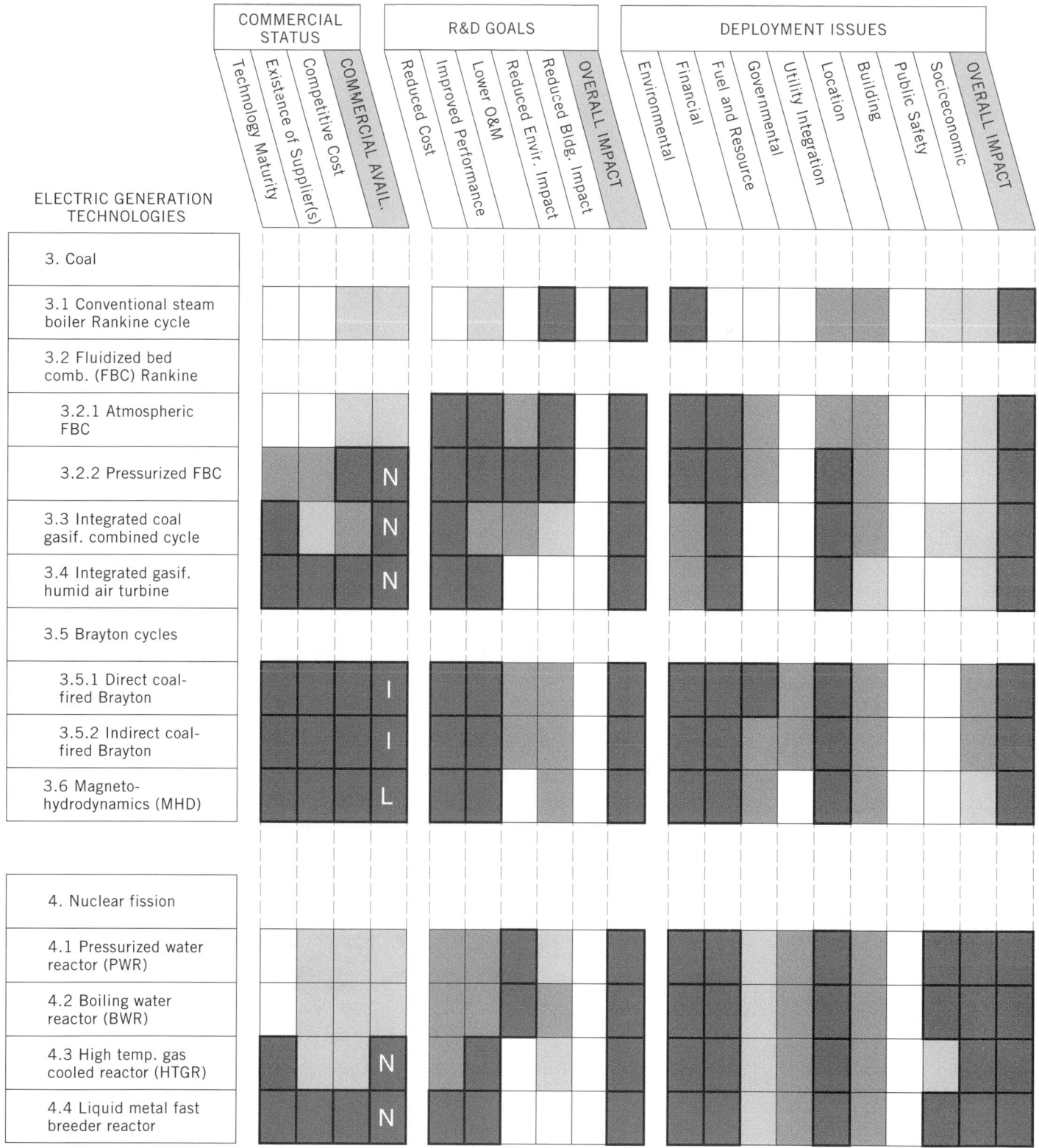

Figure 1. (Continued)

ist to produce a facility or product which may be operated and maintained. This phase is often accomplished through a reduced-scale demonstration project typically ranging from one-tenth to one-half full system size. The last phase, commercial demonstration, is necessary to establish full-scale operation of the technology to verify that it is reliable, durable, safe and has predictable performance characteristics under actual commercial conditions. Considerable overlap could exist between the various development phases. To move ahead, continuous feedback is necessary as designs evolve, are tested and data are acquired. The delineation of development phases

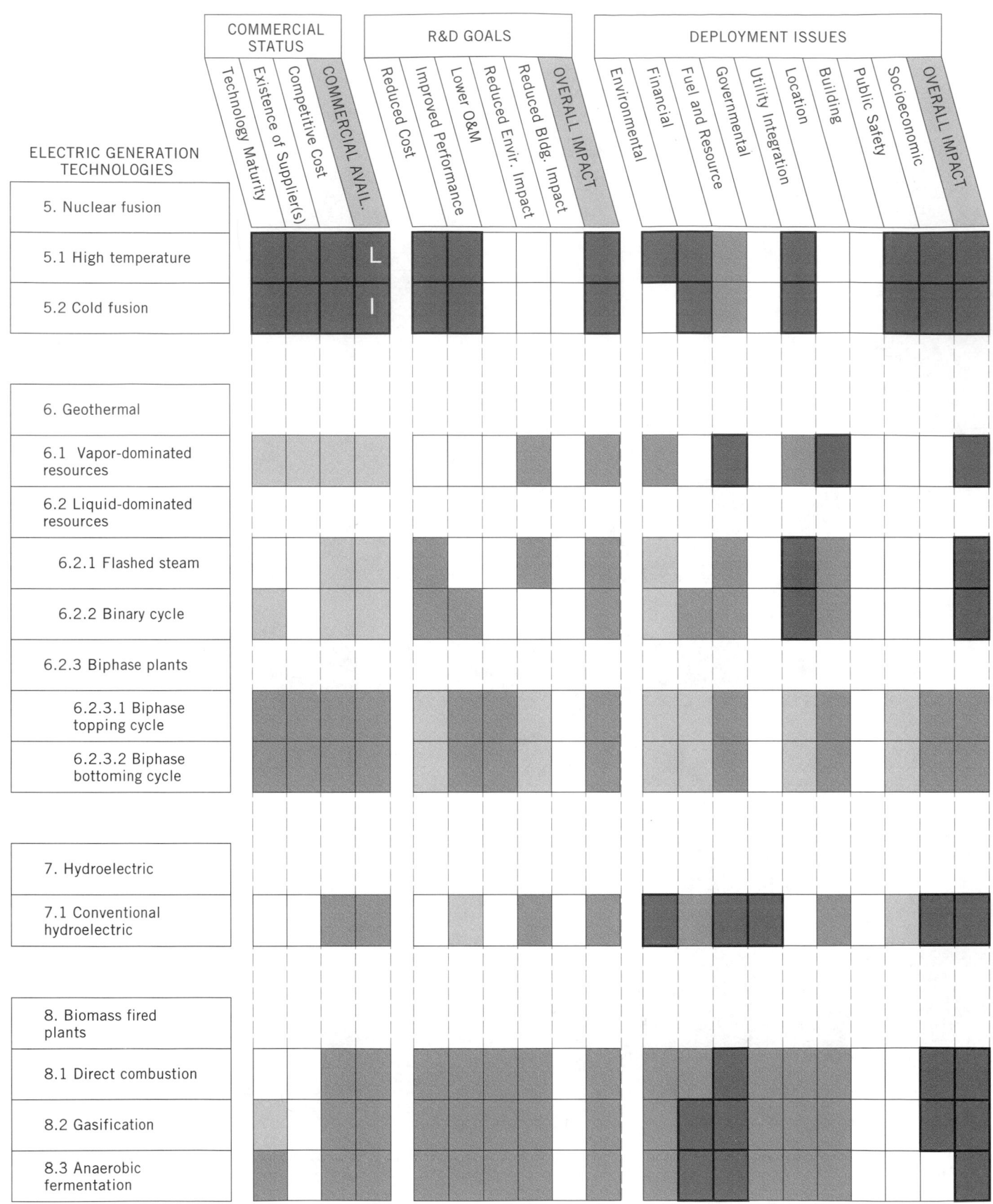

Figure 1. (Continued)

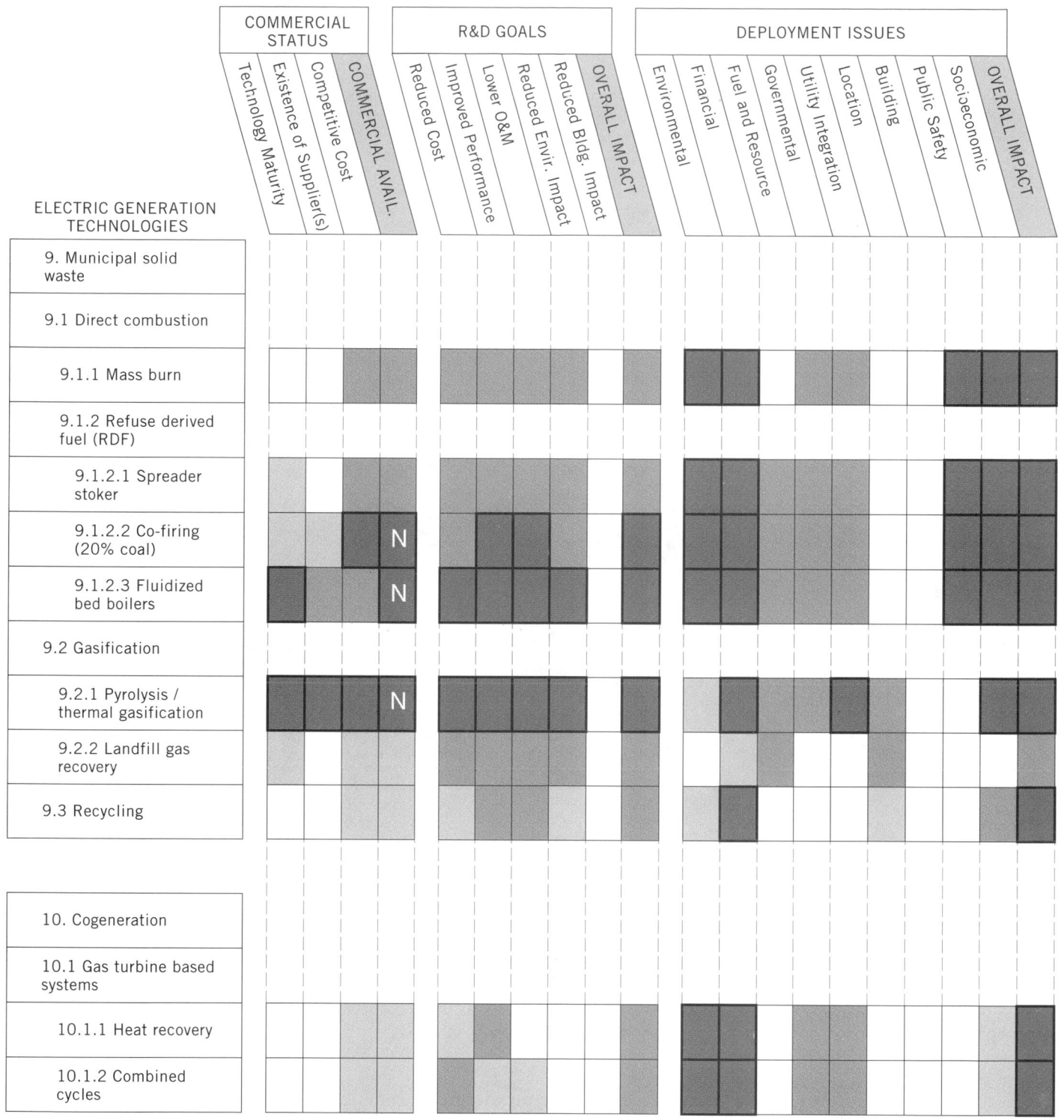

Figure 1. (Continued)

permits a technology to be classified as to its "technical maturity."

For a technology to have achieved technology maturity, it must have evolved through the technology development process and reached the point of a successful commercial demonstration. At a minimum, this typically will involve a full-scale demonstration for electric generation techno-

logies and market introduction of a final product for end-use technologies.

Existence of Supplier(s). This criterion is intended to establish whether adequate means exist to make each technology available for wide-scale use. For this criterion to be satisfied for generation technologies, either a recognized

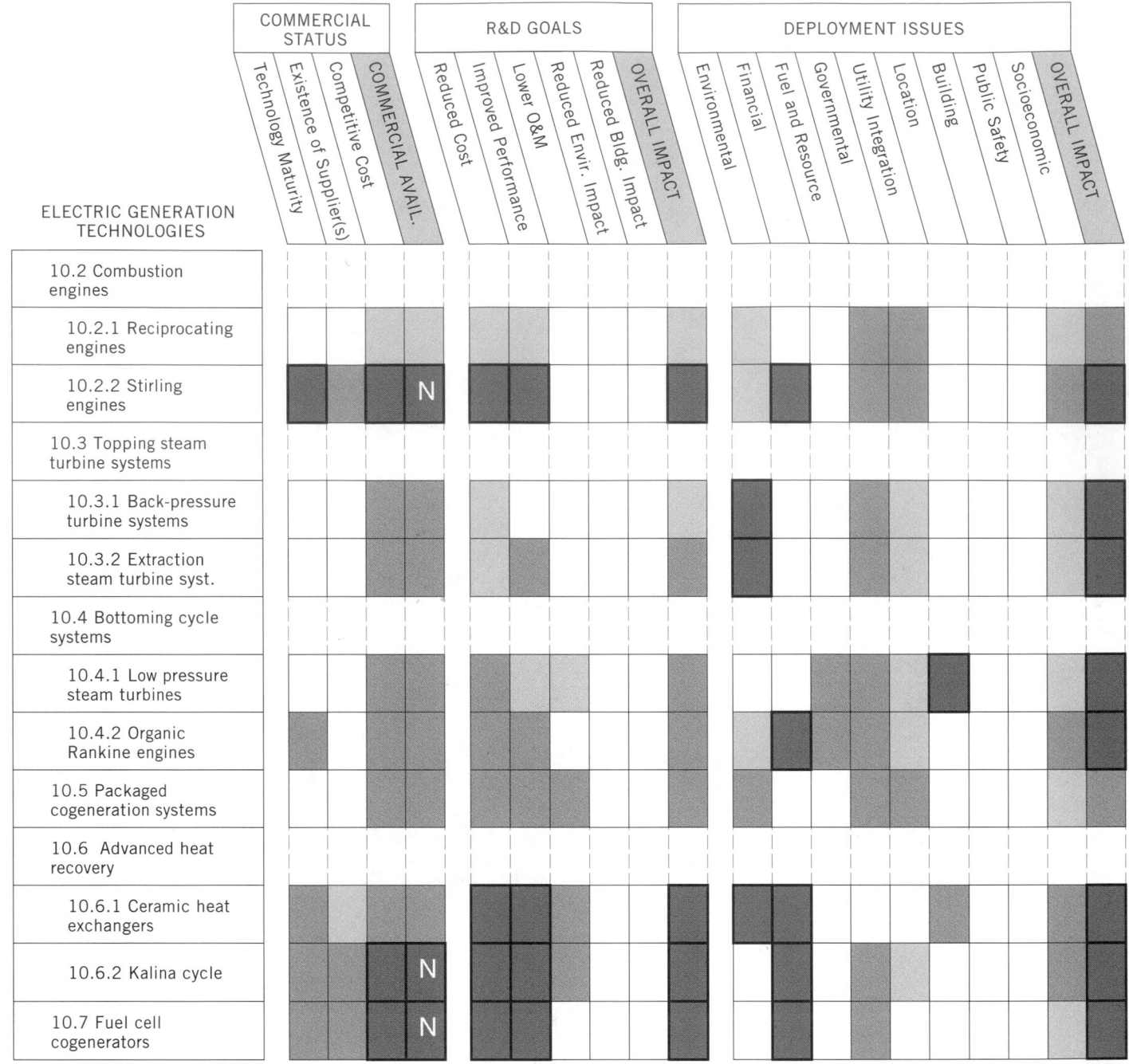

Figure 1. (Continued)

supplier for required hardware, design expertise and installation skills must exist as demonstrated by the existence of the technology in operation or under construction. Or a potential supplier must have indicated an interest in supplying the technology and must have the technical and engineering expertise and the manufacturing infrastructure to supply the technology in commercial quantities.

For most end-use technologies, specialized hardware and design and installation skills are required to satisfy this criterion (eg, solar systems). However, some end-use technologies require only hardware availability because installation procedures are identical to conventional technologies (eg, energy efficient refrigerators). Passive solar

heating and cooling are a unique exception where only special design skills are required to satisfy the criterion because its availability typically relies only on the ability to rearrange standard building materials.

Competitive Costs. This criterion establishes whether the life-cycle cost of using an energy technology is reasonably competitive with other currently available options. This criterion has been analyzed by comparing the costs of each energy technology with "business as usual" energy costs for six major ownership sectors involved in decisions to use energy technologies: investor owned utility, government owned utility, non-utility generator, industrial, commercial and residential. The degree each energy technol-

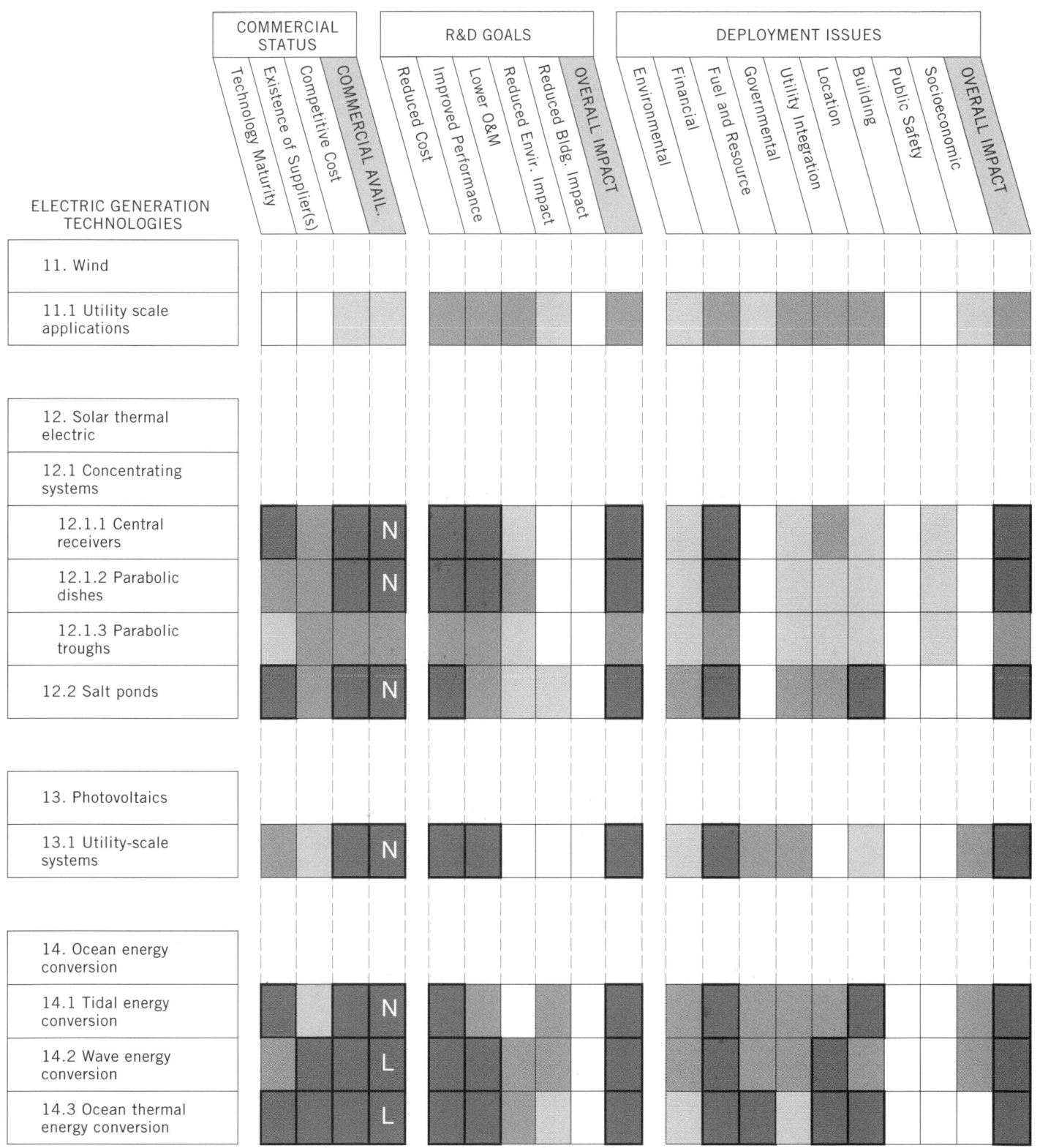

Figure 1. (Continued)

ogy satisfies the "business as usual" cost for at least one of the ownership sectors appropriate to that technology is used to assess cost competitiveness.

The competitive cost analysis is based only on monetary costs that include capital, fuel, and operation and maintenance expenses, and excludes all external costs (environmental, health and safety) and government subsidies. Social cost analyses, including all of these technology cost components, currently are the subject of extensive studies by many government and research organizations and will be considered in more detail in subsequent versions of the *ETSR*.

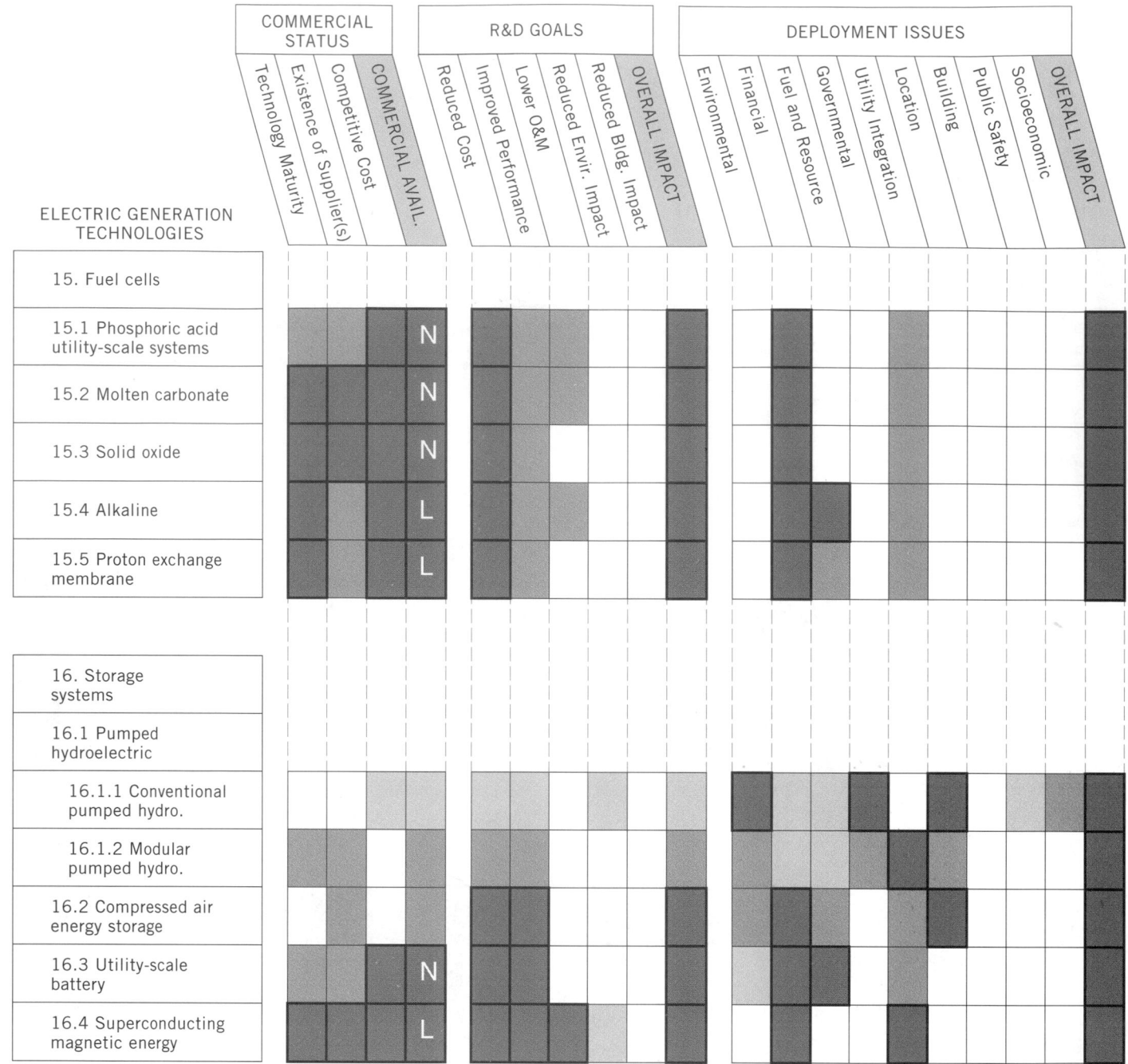

Figure 1. (Continued)

RESEARCH AND DEVELOPMENT GOALS

Energy technologies typically use mechanical equipment, electric devices and specialized materials that can be improved through research targeted at alternative materials, designs, manufacturing techniques and volume production. The ultimate objective of research is to make energy technologies more acceptable to users and society. To achieve this objective, individual research and development activities frequently focus on a narrow aspect of technology development. Technological advancements usually are the result of an incremental process involving many small improvements that together lead to a commercially viable product or refinement.

Energy policymakers are most concerned with the end results of successful research and development rather than the details associated with highly specialized incremental developments. This "big picture" point of view allows policymakers to decide on broader issues such as the timeframe for commercial availability; necessity and suitability of research and development funding; and applicability of the technology in California's energy future.

To facilitate this type of policy analysis, each specific technology evaluation in this article categorizes research and development goals according to five identified generic research goals based on research activities and programs currently underway at government and private sector research facilities: reduced cost, improved performance,

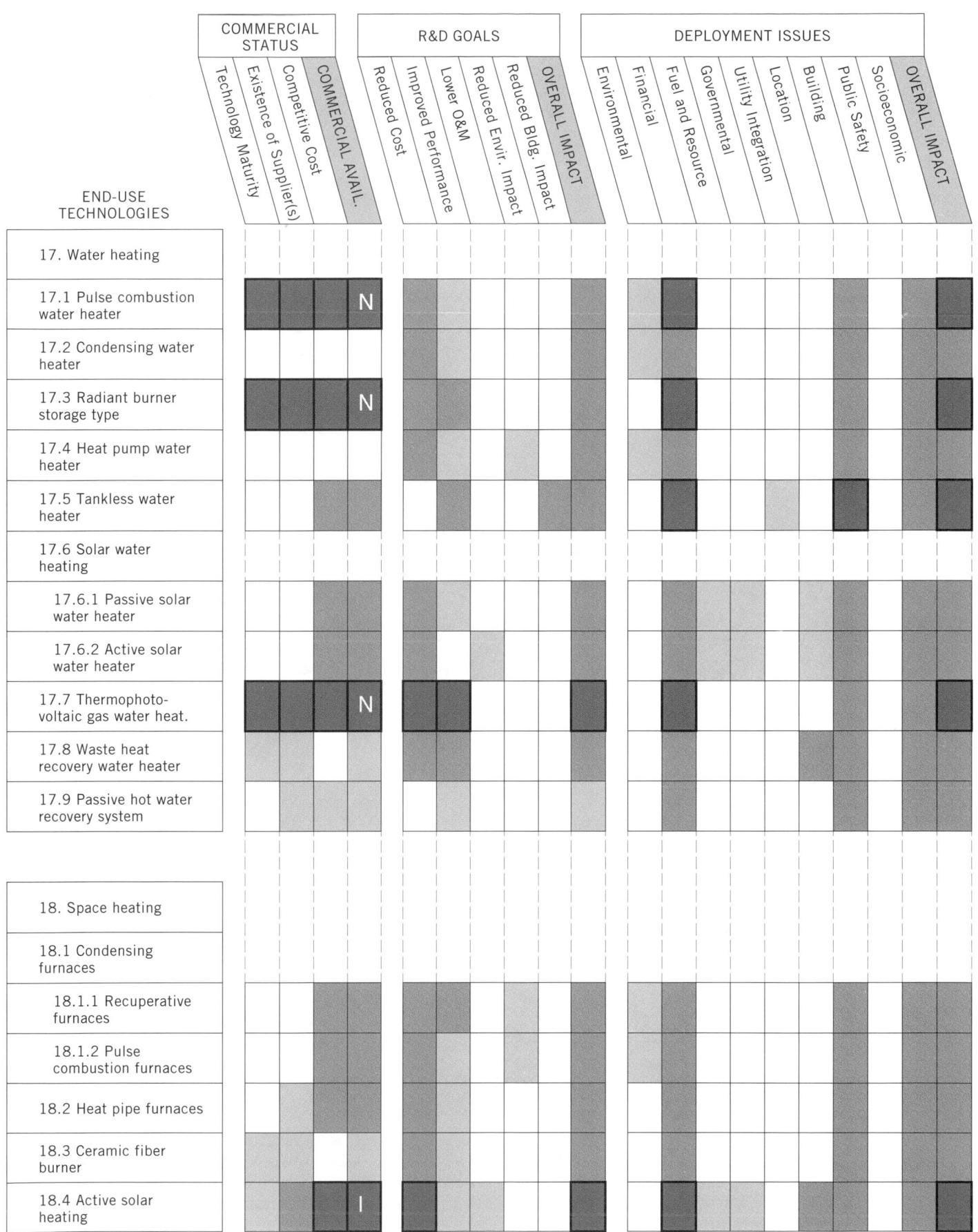

Figure 1. (Continued)

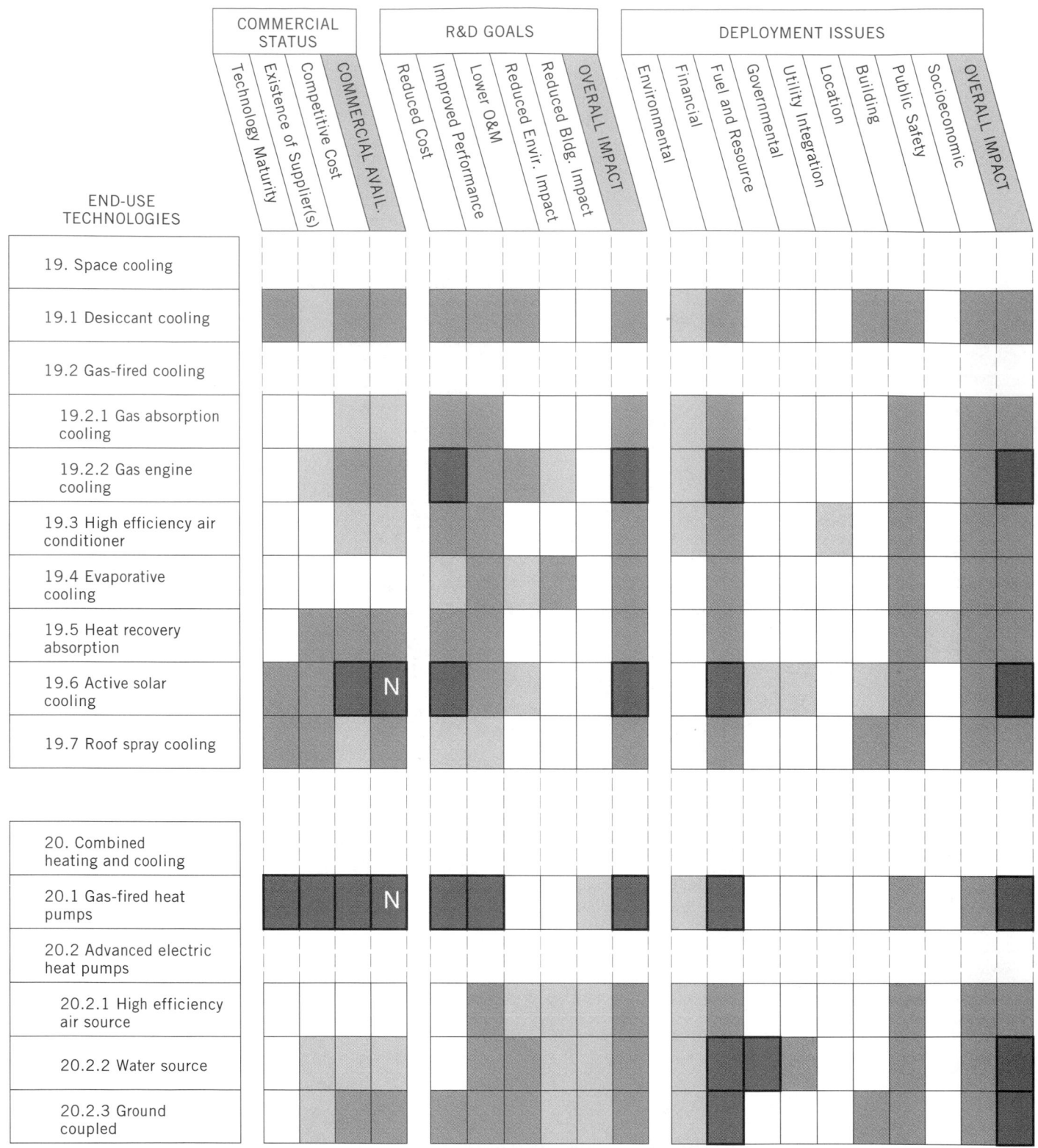

Figure 1. (Continued)

lower operation and maintenance costs, reduced environmental impacts and reduced building impacts. These generic issues are listed with a full range of subordinate issues in the following paragraph.

The following list includes five goals for technology development along with generic research activities that address these goals.

Reduced Capital Cost. Capital cost reductions are needed for many energy technologies before they can effectively compete in the marketplace.

- Materials
- Manufacturing
- Installation

Figure 1. (Continued)

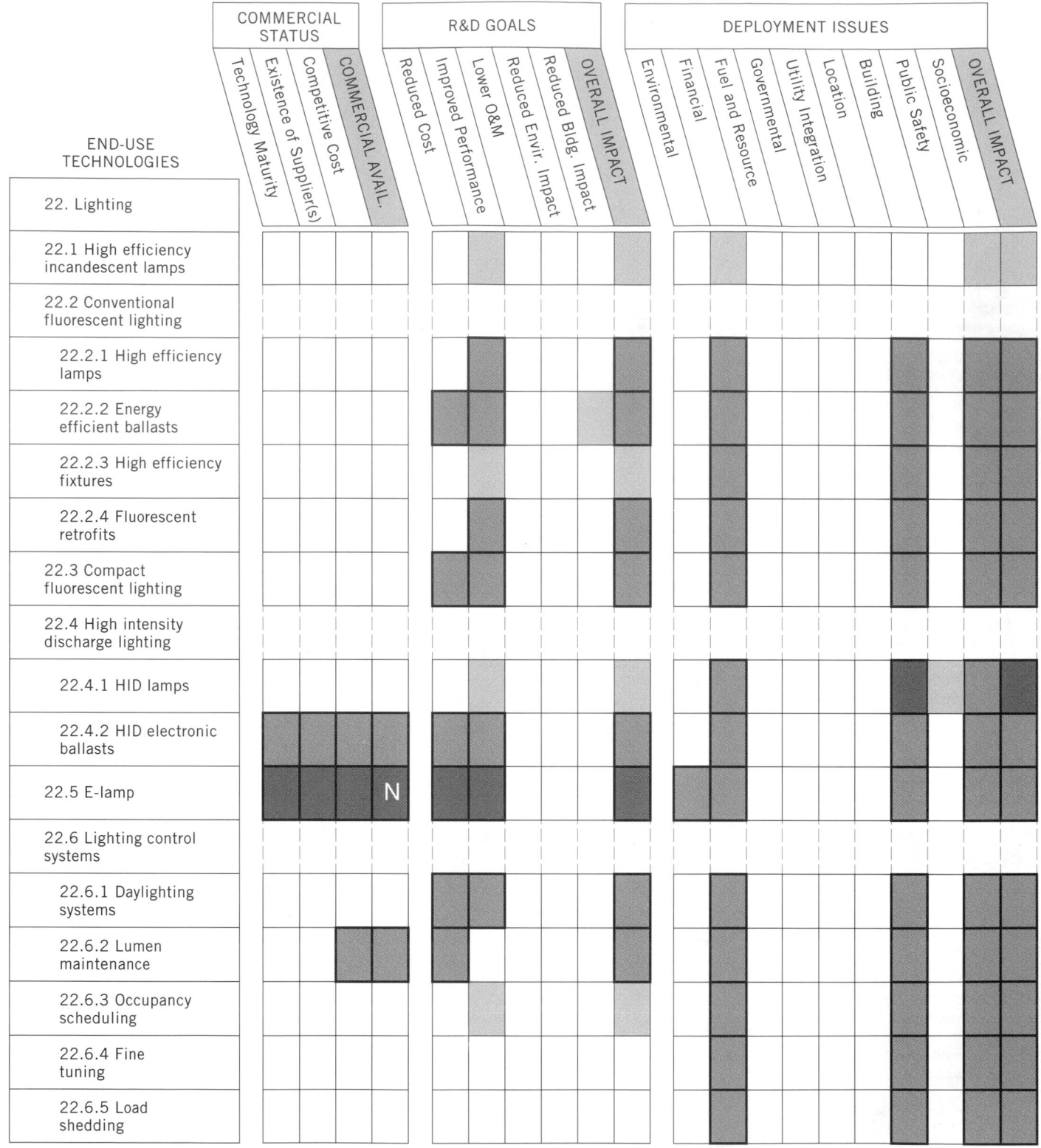

Figure 1. (Continued)

- Systems/Design Development
- Resource Development

Improved Performance. Unless one or more performance goals are achieved for some technologies, their ultimate economic viability may be limited or fall short of

their technological potential. Frequently there is a trade off between capital cost and performance.

- Efficiency
- Availability/Reliability/Durability
- Match Output to Time of Use (Demand)

Figure 1. (Continued)

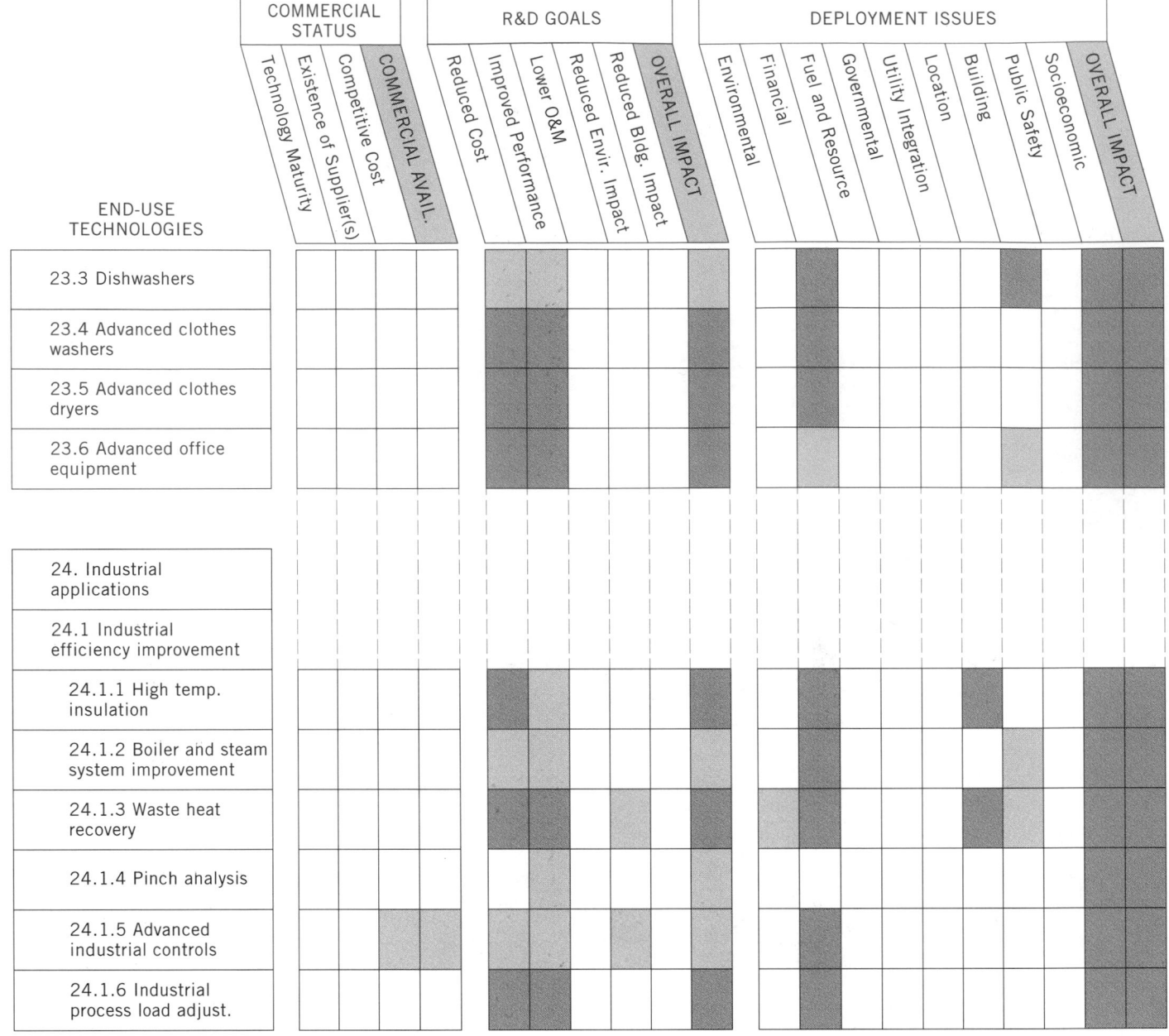

Figure 1. (Continued)

- Health and Safety
- New Applications

Lower Operation and Maintenance Costs. Operation and maintenance costs, including non-recurring replacement of components and major overhauls, can decrease the economic viability of a technology.

- Systems Integration
- Instrumentation and Controls
- Fuel Modification

Reduced Environmental Impacts. Air pollution impacts are of major concern in much of California. However, water usage, water discharges, and solid waste disposal are of concern also.

- Pre-event Cleanup
- Process Modifications
- Post-Event Cleanup
- Adverse Site Impacts

Reduced Building Impacts. Market acceptance will not be possible or limited for many end-use technologies unless they can be incorporated into new and existing building stock without significant problems.

- Structural
- Appearance
- Occupant Comfort/Health/Safety

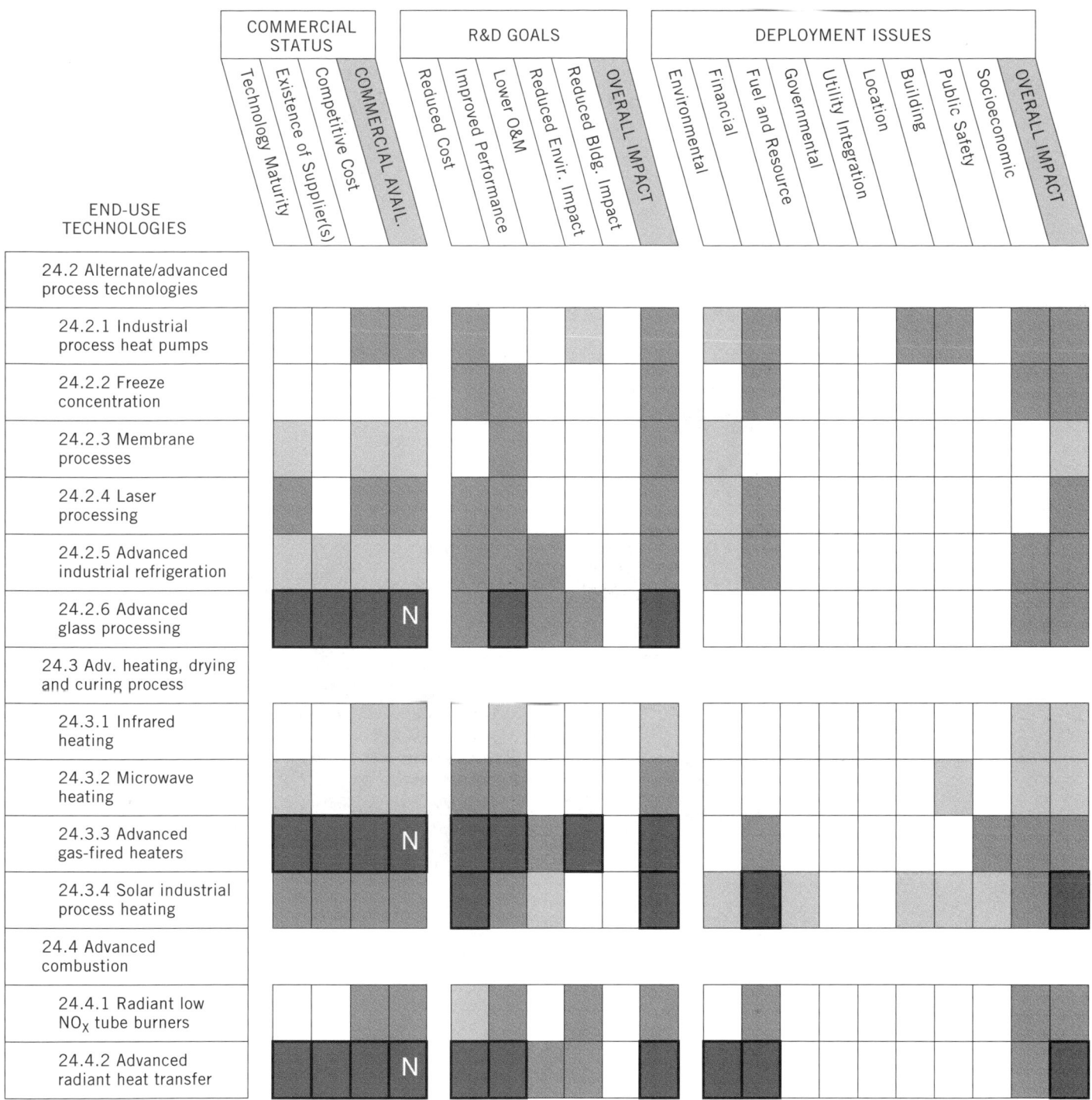

Figure 1. (Continued)

DEPLOYMENT ISSUES

A technology can become commercially available through research and development efforts yet require many years to achieve widespread market adoption due to constraints that preclude or limit a technology from consideration as a viable alternative. Energy policymakers must evaluate these issues when identifying programs and activities to support preferred energy options. To meet this need, a master list of generic deployment issues, included in the next paragraph, has been developed by identifying constraints to energy technology use. Each technology evaluation in this article details deployment issues evaluated according to these generic issues. These issues are most fully defined for commercially available technologies and less well defined for non-commercially available technologies. The absence of an identified deployment issue for a non-commercially available technology does not mean that a certain deployment issue will not exist. Some deployment issues become important only when a technol-

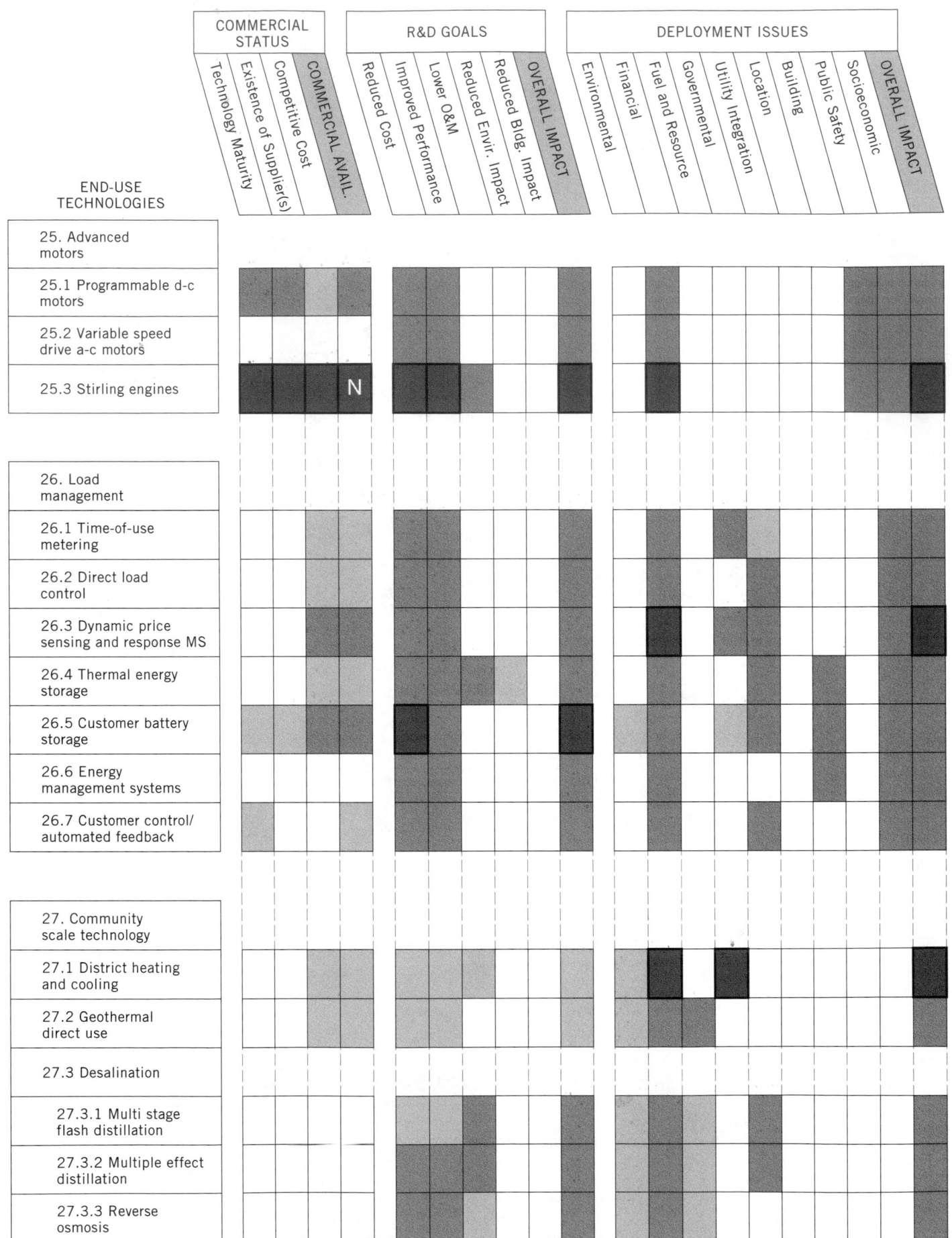

Figure 1. (Continued)

924

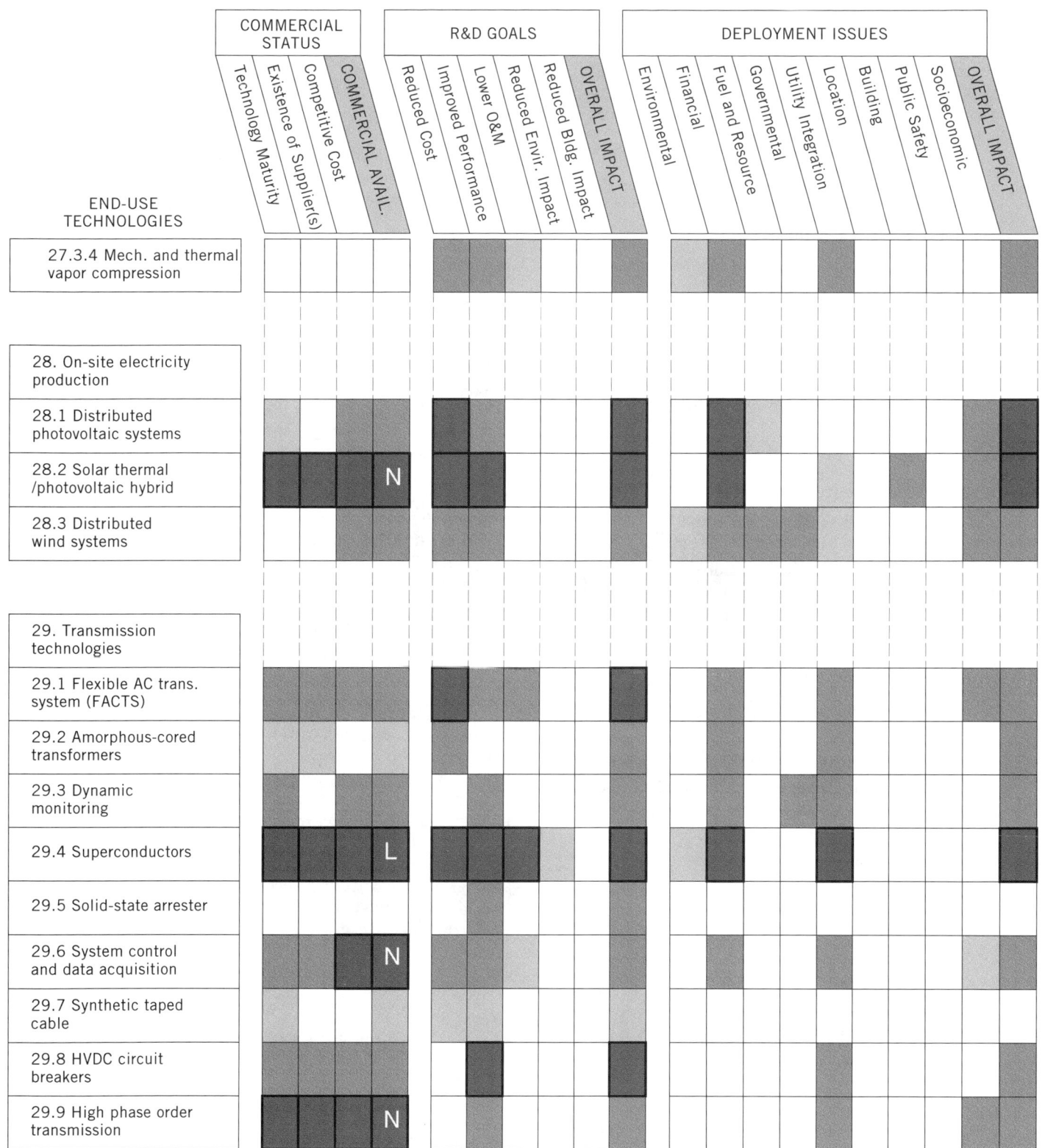

Figure 1. (Continued)

ogy has penetrated the marketplace above a certain level. All of these issues are discussed in more detail.

The following list includes nine issues constraining the deployment of energy technologies along with sub-issues that address more detailed aspects associated with each issue.

Environmental Constraints

- Air Pollution Impacts
- Water Pollution
- Waste Disposal
- Noise Pollution

- Radio/TV Signal Interference
- Thermal Discharge
- Destruction/Disturbance of Habitat
- Scenic Resource Impacts

Financial Constraints

- Availability of Financing
- High Capital Costs
- High Operation and Maintenance Costs
- Availability of Tax Incentives

Fuel and Resource Constraints

- Availability of Fuel or Resource
- High Cost of Fuel
- Variation in Fuel or Resource Quality

Governmental Constraints

- Agency-Government Coordination
- Building Code/Planning Restrictions
- Undependable Avoided Cost Contracts
- Regulatory/Legislative Restrictions
- Permit Restrictions
- Intergovernmental Coordination

Utility Integration Constraints

- Control of Intermittent Sources
- Need Conformance
- Lack of Demonstrated Reliability/Performance
- Conformance with Interconnection Requirements
- Lack of Incentive for Utility Companies

Location Constraints

- Fuel Delivery Restriction/Cost
- Lack of Suitable Sites
- Adverse Subsidence Impacts
- Availability of Transmission Lines
- Availability of Water
- Risk of Seismic Damage

Building Constraints

- Adverse Structural Impacts
- Adverse Appearance Impacts
- Adverse Occupant Impacts
- Minimal Industry Acceptance
- Lack of Incentive for Building Owners/Developers

Public Safety Constraints

- Catastrophic Risks
- Fire Hazards

- Toxic Gas Hazards
- Health Risks

Socioeconomic Constraints

- Poor Public Opinion
- Low End User Awareness
- Complexity of Operation
- Adverse Agricultural Impacts

Environmental Constraints

Air Pollution. Currently, oxides of sulfur (SO_x), oxides of nitrogen (NO_x), unburned hydrocarbon (UHC), carbon monoxide (CO), particulates and lead are regulated by the State of California and the federal Environmental Protection Agency. Future regulations (federal and state) concentrate on hazardous air pollutants (National Emission Standard for Hazardous Pollutants [NESHAP]) and airborne toxics. Under local New Source Review (NSR) regulations, electric generating facilities in nonattainment areas must use best available control technology (BACT) and obtain offsets or tradeoffs for pollutants which will be emitted. Carbon dioxide (CO_2) also is developing as a global air quality issue.

Water Pollution. Water emissions are regulated by the Clean Water Act, usually administered in California by local water quality districts. The emissions regulated by the Clean Water Act probably will be tightened in the future.

Waste Disposal. Solids, such as fly-ash from a coal-fired power plant, are generated by many processes and disposal must be off-site. Disposal of the wastes from each process must be reviewed on a case-by-case basis. In addition, many energy technologies create hazardous wastes in the form of chemicals that exhibit generic characteristics such as corrosivity, reactivity, toxicity or combustibility. Certain organic compounds that are listed in the Resource Conservation Recovery Act–Appendix 8, are classified as hazardous waste. Chemicals determined to be hazardous are difficult and, in some cases, expensive to discard.

Noise Pollution. Noise is not federally regulated, except within plant boundaries where the Occupational Safety and Health Administration (OSHA) imposes noise limitations. Outside of plant boundaries, noise is only regulated by some local governments. In general, noise constraints are easily met, except in urban areas. However, public pressure can be brought against appropriate agencies to influence plant locations or, in some cases, deny approval of plant construction.

Radio/Television Signal Inteference. VORTAC stations provide directional and/or range information for air navigation. The Federal Aviation Administration has guidelines for siting of large objects near VORTAC stations. In addition, where new energy facilities impact radio and television signal reception for nearby residents, restrictions may be imposed.

Thermal Discharge. The two methods of thermal energy discharge are emitting heat directly into the atmosphere (such as gas turbine exhaust) and emitting heat into a body of water or the atmosphere by means of a once-through cooling or a cooling tower. Significant temperature changes that occur in the vicinity of thermal discharge can have a serious impact on the surrounding environment.

Destruction/Disturbance of Habitat. Several technologies cause essentially permanent change to land or water use over relatively large areas. Other technologies such as strip mining of coal and harvesting of wood cause at least temporary disturbance of relatively large natural areas. Effects on indigenous wildlife and vegetation must be considered.

Scenic Resource Impact. Certain technologies establish substantial changes to preexisting vistas and natural settings. Local opposition to these changes can lead to regulations restricting development. Siting of windmills or hydroelectric dams are examples of such impacts.

Financial Constraints

Availability of Financing. Financing of some technologies can be difficult to obtain for reasons including: long payback period; high initial cost; perceived risk and uncertainties relative to revenue, cost and the value of the product.

High Capital Cost. Some technologies have high or uncertain capital costs that reduce financing options and make them more risky to purchase or develop.

High Operation and Maintenance Costs. Some technologies require labor intensive resources, costly fuel handling procedures or expensive materials to maintain performance and reliability.

Availability of Tax Incentives. The uncertainty of continued availability of a tax credit can limit a developer's ability to obtain financing. In addition, some energy technologies qualify for more tax incentives and other forms of government support (oil, gas and nuclear) than others (conservation, end-use and renewable technologies).

Fuel and Resource Constraints

Availability of Fuel or Resource. Most of the conventional fuels used in California are natural gas, petroleum or nuclear. Availability of these fuels is related to the worldwide energy picture. Intermittent energy availability affects other technologies such as solar and wind energy. Long-term availability and affordability of fuels is a concern for other technologies such as biomass electric generating plants.

Variation in Fuel or Resource Quality. Where the resource or fuel is not of consistent quality, energy projects are more difficult to plan and finance.

High Cost of Fuel. Uncertainty about the long-term relative cost of a particular fuel can limit commitment of its use.

Governmental Constraints

Agency-Government Coordination. Requirements that a technology installation obtain multiple use permits increases installation development time, cost and uncertainty about eventual approval.

Building Code/Planning Restrictions. Building code and planning restrictions may affect the on-site use of some technologies. Building code and planning requirements also can be used as a barrier to impede construction of new electric generating plants in areas where they are not desired.

Undependable Avoided Cost Contracts. The price paid for electricity is largely regulated by the California Public Utilities Commission. Buy-back rates and electricity sales provisions have substantial impact on the economic competitiveness of some technologies. The economic viability of third-party financed generating plants hinges on the ability of plant owners to obtain desirable electricity purchase contracts with utilities. Because the final long-term, fixed-price contracts have not been instituted, third-party developers are limited to either short-term "as available" avoided cost contracts or negotiated contracts. As a result, new projects cannot count on a dependable revenue stream. It may be difficult to obtain power purchase contracts in areas where excess capacity or energy exists or transmission line constraints exist.

Regulatory/Legislative Restrictions. Government regulations and legislation can increase the cost of energy production. Health, safety and emissions are just a few of the regulations requiring power plant compliance. Complying with regulations often requires additional equipment and technical expertise that increase project costs.

Permit Restrictions. Permitting for new electric generating plants is highly dependent on the type of plant and site specific factors. Substantial delays can affect project feasibility.

Intergovernmental Coordination. Multi-state cooperation is needed for out-of-state projects involving a sensitive blend of many perspectives.

Utility Integration Constraints

Control of Intermittent Resources. Energy technologies such as solar and wind that cannot control their resource availability will be more difficult to coordinate with utility power needs.

Need Conformance. The California Energy Commission makes an integrated assessment of need in its *Electricity Report*. This assessment requires the Commission's judgment of how resources will satisfy growth in electricity demand while balancing the factors of environmental quality, conservation, economic health, and public safety and health. Issues may include fuel dependency levels, uncertainty in basic plannng assumptions and assuring flexibility in the electricity system. These strategic issues then determine the Commission's findings regarding quantity, timing and characteristics of needed resource additions and other generating system modifications. Resources selected for addition should provide at least the same level of benefit at comparable or lower cost.

Lack of Demonstrated Reliability/Performance. Because of prudency review requirements and the utility obligation to serve, utilities require demonstrated performance in any resource addition.

Conformance with Interconnection Requirements. Utility requirements for tying into the electricity grid can sometimes impose barriers to new projects. In particular, power quality, safety concerns and equipment specifications must all be satisfied.

Lack of Incentive for Utility Companies. Historically, electric utility companies had little incentive to upgrade the average heat rate of their systems since the California Public Utilities Commission allows most fuel costs to be passed on to ratepayers. Future competitive bidding processes may increase the incentive for better energy efficiency.

Location Constraints

Fuel Delivery Restrictions/Cost. For systems using chemicals or hazardous fuels, the site is restricted to locations where delivery is practical. Also, if fuel delivery requires transportation, the extra costs involved may affect the economic viability of the technology.

Lack of Suitable Sites. Many technologies depend on specific geographic, resource availability and climatic conditions that limit their consideration.

Adverse Subsidence Impacts. Subsidence involves the settling of land areas where underground water resources are depleted. This impact particularly affects some geothermal projects which draw hot water from underground reservoirs without injecting them back into the reservoir. Agricultural industries can be affected by subsidence impacts.

Availability of Transmission Lines. Power plant projects can only be developed where transmission interconnections are reasonably accessible because the cost of extending transmission lines can impose a burden to a project's cost-effectiveness.

Availability of Water. Many technologies depend on large amounts of water for cooling. Utilities using ocean water to reject heat can have problems with pipe corrosion and scaling. Thus, it is optimal to find a fresh water source. In particular, nuclear, oil/gas and geothermal technologies can be dependent on large sources of water.

Risk of Seismic Damage. The degree of exposure to seismic activity varies within California. Some technologies may be excluded from a geographic area due to seismic risk. For example, it is not acceptable to locate nuclear power plants in areas with high seismic activity.

Building Constraints

Adverse Structural Impacts. Some technologies cannot be incorporated into existing buildings without structural modifications. These extra costs can sometimes preclude the use of the technology in retrofit situations. For new construction, most energy technologies' structural requirements can be planned for without cost premiums.

Adverse Appearance Impacts. The use of energy technologies may be substantially reduced where they have a negative effect on exterior building appearance.

Adverse Occupant Impacts. The use of energy technologies affecting the comfort of building occupants will be severely limited. In particular, any substantial impact to space conditioning, lighting, water quality, noise level or air quality severely limits technology acceptability.

Minimal Industry Acceptance. Unless the building industry accepts a new or improved energy technology, it will not be available to most new home buyers because the predominant share of new buildings are built by subdivision developers and "spec" builders. Unless the building industry accepts a technology, market applications will often be limited to replacement of old equipment.

Lack of Incentive for Building Owners. For commercial buildings constructed for rental income, incentive is lacking to provide cost-effective energy options as tenants, rather than building owners, usually pay the energy bills. In addition, most tenants do not shop for lower energy costs because energy costs usually do not represent a meaningful portion of total business expenses. Similarly, most residential buildings are built by developers who pass on the cost of energy to new home buyers or renters.

Public Safety Constraints

Catastrophic Risks. The possibility of catastrophic failure is more of a factor with some technologies than with others. For example, nuclear, coal and oil/gas technologies rely on high temperatures and pressures which are inherently more risky than technologies such as photovoltaics and wind. In addition, some technologies such as hydroelectric facilities are more susceptible to catastrophic failure during seismic loading than others.

Fire Hazards. Certain technologies that are candidates for siting in metropolitan areas must be concerned with potential fire hazards.

Toxic Gas Hazards. Certain technologies that are candidates for siting in metropolitan areas must be concerned with potential safety hazards involved with gas leakage.

Health Risks. Technologies that impose potential health risks both on-site or as the result of fuel or waste-product transport may face intensive public scrutiny.

Socioeconomic Constraints

Poor Public Opinion. Public opinion can restrict the use of energy technologies. In many cases, poor public opinion has a greater influence over technology use than more substantial economic and environmental constraints.

Low End-User Awareness. If potential technology users are not aware of specific energy options, those options will not be deployed. It is critical that decision-makers involved in selecting energy technologies are provided with adequate performance and reliability information.

Complexity of Operation. For some utility applications, the complexity of use may impose a barrier to technology use. For residential and commercial applications, the complexity of energy technologies could be an issue that may require specialized maintenance services or training. Failure to address this complexity issue may make customers reluctant to use them.

Adverse Agricultural Impacts. Technologies that release certain metals or reactive chemical compounds may be limited in their siting in agricultural regions.

COMMERCIAL AVAILABILITY

This section provides an overview of all specific energy technologies assessed in terms of their commercial availability and unavailability and general conclusions concerning energy technology status, research and development goals and deployment issues. The conclusions are organized according to fuel cycle, electric generation and end-use energy technologies.

Fuel Cycle Technologies

Fuel Cycle Technologies Commercially Available. The following fuel cycle technologies have been determined to be commercially available. They are listed as "F" where all three criteria are "fully" satisfied, "MC" where one or more criteria are satisfied under "most conditions," and "LC" where one or more criteria are satisfied under rare or "limited conditions."

Conventional Fuels:
Conventional Oil Extraction (F)
Thermally Enhanced Oil Extraction (MC)
Chemical Enhanced Oil Extraction (LC)
Gas Displacement Enhanced Oil Extraction (LC)
Natural Gas (F)
Conventional Coal (F)
Full Fuel Nuclear Fission Cycle (MC)
Liquefied Petroleum Gas (LC)
Liquefied Natural Gas (LC)
Peat (LC)
Petroleum Coke (MC)

Alternative Fuels:
Tar Sands (LC)
Ethanol (LC)
Methanol (Non-Coal) (LC)

Renewable Fuels:
Hydrothermal Geothermal Resource (F)
Biomass Fuel (MC)
Municipal Solid Waste (LC)
Solar (LC)
Wind (MC)

Fuel Cycle Technologies Not Commercially Available. The following fuel cycle technologies have been determined to be not commercially available because they do not satisfy one or more of the commercial availability criteria. Dates of expected commercial availability are indicated in parentheses as either "near-term" (within 12 years), "long-term" (beyond 12 years) or "indeterminate" where there are a number of unresolved R&D issues or a low likelihood of commercialization in the foreseeable future.

Alternative Fuels:
Waste Disposal Nuclear Fission (long-term)
Decommissioning Nuclear Fission (long-term)
Oil Shale (long-term)
Nuclear Fusion (long-term)
Coal Gasification (near-term)
Coal-Based Methanol (near-term)
Methanol/Electricity Coproduction From Coal (near-term)
Coal-Based Synthetic Oil (near-term)
Hydrogen (long-term)

Renewable Fuels:
Hot Dry Rock Geothermal Resource (near-term)
Geopressured Geothermal Resource (near-term
Magma Geothermal Resource (long-term)

Fuel Cycle Technologies General Conclusions

- Although more than 85% of all energy consumed in California is provided by only two fuels, natural gas and petroleum, many other alternative and renewable fuel technologies are commercially available.
- Several technologies for processing coal as a fuel are being pursued at the national level.
- The primary research goals for alternative and renewable fuel options are to reduce cost and improve performance.
- Financial constraints are the primary barrier to alternative and renewable fuel use. This barrier follows directly from the research goal of reducing cost and improving performance. Environmental and resource issues need to be resolved for renewable energy sources such as geothermal, biomass and municipal solid waste.

Electric Generation Technologies

Electric Generation Technologies Commercially Available. The following electric generation technologies are determined to be commercially available. They are listed as "F" where all three criteria are "fully" satisfied, "MC" where one or more criteria are satisfied under "most conditions," and "LC" where one or more criteria are satisfied under rare or "limited conditions." The distinction between satisfaction of the commercial availability criterion under "most conditions" or "limited conditions" may depend on subjective judgments made in the scenario analysis formulated.

Oil and Gas Combustion:
Conventional Rankine Cycle (MC)
Supercritical Rankine Cycle (MC)
Simple Brayton Cycle (MC)
Conventional Combined Cycle (F)
Steam Recuperated Gas Turbines (LC)

Coal:
Conventional Steam Boiler (MC)
Atmospheric Fluidized Bed Combustion (LC)

Nuclear Fission:
Pressurized Water Reactor (MC)
Boiling Water Reactor (MC)

Geothermal:
Vapor-Dominated Resources (MC)
Liquid-Dominated Resource Flashed Steam (MC)
Liquid-Dominated Resource Binary Cycle (MC)
Biphase Topping Cycle (LC)
Biphase Bottoming Cycle (LC)

Hydroelectric:
Conventional (LC)

Biomass-Fired Plants:
Direct Combustion (LC)
Gasification (LC)
Anaerobic Fermentation (LC)

Municipal Solid Waste:
Mass Burn (LC)
Refuse-Derived Fuel Spreader Stoker (LC)
Landfill Gas Recovery (MC)
Recycling (MC)

Cogeneration:
Gas Turbine Based Systems with Heat Recovery (MC)
Gas Turbine Based Systems with Combined Cycles (MC)
Reciprocating Engines (MC)
Back-Pressure Topping Steam Turbine Systems (LC)
Extraction Topping Steam Turbine Systems (LC)

Low Pressure Steam Turbine Bottoming Cycle Systems (LC)
Organic Rankine Engine Bottoming Cycle Systems (LC)
Packaged Cogeneration Systems (LC)
Ceramic Heat Exchangers (LC)

Wind:
Utility-Scale Applications (MC)

Solar Thermal Electric:
Parabolic Troughs (LC)

Storage Systems:
Conventional Pumped Hydroelectric (MC)
Modular Pumped Hydroelectric (LC)
Compressed Air Energy Storage (LC)

Electric Generation Technologies Not Commercially Available. The following generation technologies have been determined not commercially available because they do not satisfy one or more of the commercial status criteria. Dates of expected commercial availability are indicated in parentheses as either "near-term" (within 12 years); "long-term" (beyond 12 years) or "indeterminate" if having excessive unresolved R&D issues or if there is a low likelihood of commercialization in the foreseeable future.

Oil and Gas:
Kalina Combined Cycle (near-term)
Intercooled Steam Recuperated Gas Turbine (near-term)
Chemically Recuperated Gas Turbine (near-term)
Humid Air Turbine (HAT) Cycle (near-term)
Intercooled Reheat Combined-Cycle (near-term)

Coal:
Pressurized Fluidized Bed Combustion (PFBC) (near-term)
Integrated Coal Gasification Combined-Cycle (near-term)
Integrated Gasification Humid Air Turbine (near-term)
Direct Coal-Fired Brayton (indeterminate)
Indirect Coal-Fired Brayton (indeterminate)
Magnetohydrodynamics (MHD) (long-term)

Nuclear Fission:
High-Temperature Gas-Cooled Reactor (HTGR) (near-term)
Liquid Metal Fast Breeder Reactor (near-term)

Nuclear Fusion:
High-Temperature Nuclear Fusion (long-term)
Cold Nuclear Fusion (indeterminate)

Municipal Solid Waste:
Refuse-Derived Fuel Co-Firing (20% Coal) (near-term)
Refuse-Derived Fuel Fluidized Bed Boilers (near-term)
Pyrolysis/Thermal Gasification (near-term)

Cogeneration:
Stirling Engines (near-term)
Kalina Cycle Advanced Heat Recovery (near-term)
Fuel Cell Cogenerators (near-term)

Solar Thermal Electric:
Central Receivers (near-term)
Parabolic Dishes (near-term)
Salt Ponds (near-term)

Photovoltaics:
Utility-Scale Systems (near-term)

Ocean Energy Conversion:
Tidal Energy (near-term)
Wave Energy Conversion (long-term)
Ocean Thermal Energy Conversion (long-term)

Fuel Cells:
Phosphoric Acid Utility-Scale Applications (near-term)
Molten Carbonate (near-term)
Solid Oxide (near-term)
Alkaline (long-term)
Proton Exchange Membrane (long-term)

Storage Systems:
Utility-Scale Battery (near-term)
Superconducting Magnetic Energy Storage (long-term)

Electric Generation Technologies General Conclusions

- For electric generation technologies which are not commercially available, the research and development goals of Reduced Cost and Improved Performance and the Financial Deployment Issue are frequently listed as "Show Stoppers." The R&D Goals, the Deployment Issues and the Non-Commercial Status follow one from the other. High first cost and high operational cost (poor performance) are common reasons that a technology is not commercially available and has financial constraints on use. While a technology is in the R&D phase, production costs are high because of high engineering and fabrication costs. Once a technology has been successfully demonstrated, better and lower cost designs, equipment and materials can be chosen.
- For modular technologies, the high first cost results from high cost in tooling and labor. The development of a manufacturing infrastructure is an important R&D goal. Once production has started, cost reductions can be expected to follow a "learning curve" with perhaps a 13% reduction in production cost for every doubling of production volume.

- Although not always identified in this article, materials issues tend to impact the cost and performance of many technologies which must overcome problems of high materials costs, and corrosion, erosion, fatigue and thermal stress resistance.
- The Utility Integration (Lack of Demonstrated Reliability/Performance) and Socioeconomic (Poor End-User Awareness) Deployment Issues are frequently associated with Non-Commercially Available Technologies. The Deployment Issues follow directly from the commercial status: if a technology is not commercially available because it is immature (not demonstrated), utilities will not use it and they (or other end-users) may not be fully aware of the technology's potential.
- Deployment Issues associated with commercially available technologies are immediate opportunities for government or other organizations or individuals to affect the use of a technology, perhaps at a relatively low cost.
- Although gas technologies are substantially commercialized, research and development support to reduce costs is needed to design and demonstrate advanced gas turbine cycles. The advanced gas turbine cycles have the potential for low capital cost and high efficiency.
- Compared to other generation technologies, gas combustion technologies have fewer constraints—focused primarily on environmental and utility integration issues.
- Most coal technologies have significant research and development needs for cost reduction, improved performance, lower operations and maintenance costs, and environmental impact mitigation.
- As a group, coal technologies face multiple deployment issues associated with environmental, financial, resource, governmental, utility integration and location constraints. Environmental constraints must be considered because the resource extraction and some combustion processes are inherently dirty. Resource constraints become a factor for California because of limited indigenous resources and transportation costs for out-of-state resources to California.
- Among all coal technologies, integrated coal gasification combined cycle and integrated gasification humid air turbine may offer California the most environmentally acceptable options.
- Over the past few years, coal-fluidized bed technologies have developed rapidly. Applicability to utility operation remains uncertain.
- Nuclear energy technologies are faced with major research and development issues relative to cost reduction, improved performance, lower operations and maintenance costs, and reduced environmental impacts. Environmental impacts are particularly critical because future deployment of nuclear energy technologies in California depends on resolving difficult high-level waste disposal issues. The future of nuclear power may depend on the developments and acceptance of standardized designs.

- Many deployment issues confront nuclear technologies including environmental (high-level waste disposal), financial, public safety and socioeconomic (poor public opinion).
- Nuclear fusion technologies are long-term prospects that depend on meeting cost and performance research goals.
- Many advanced cogeneration technologies face cost and performance research goals before becoming commercially available.
- Most cogeneration technologies face varying degrees of environmental and financial constraints.
- Although fuel-cell technologies are not currently available, phosphoric acid, molten carbonate and solid oxide systems are expected in the near-term.
- All fuel-cell technologies must address substantial research goals for reduced cost and improved performance.
- All fuel-cell technologies must overcome financial and utility integration constraints before being effectively deployed.
- Storage technology options are lacking due to environmental, governmental and location constraints for pumped hydroelectric facilities, and the noncommercial status of compressed air, utility-scale battery and superconducting magnetic energy storage technologies.
- High capital costs and the need for improved performance are major research and development issues associated with most renewable energy technologies, including geothermal, biomass, municipal solid waste, wind, solar thermal electric, photovoltaics and ocean energy.
- Renewable and alternative energy technologies often are confronted with financial and utility integration constraints. Qualifying facility development is expected to be constrained because the low short-term avoided cost rate does not economically support technology development. In addition, short-term contracts inherently include too many uncertainties to risk intensive capital investments. A new long-term standard offer contract is available in locations that need baseload power, with the price set by competitive bidding.
- Resource constraints are a potential "show stopper" limiting the extensive use of many renewable energy technologies in California including vapor-dominated and advanced geothermal resources, hydroelectric, biomass and ocean energy conversion.
- Because utilities pass fuel costs on to ratepayers, they lack incentive to develop new technologies that could improve the energy efficiency of new and existing power plant facilities. The future bidding process for electric power generation resources may increase the incentive for utilities to improve energy efficiency.

End-Use Technologies

End-Use Technologies Commercially Available. The following end-use generation technologies are determined to be commercially available. They are listed as "F" where all three criteria are "fully" satisfied, "MC" where one or more criteria are satisfied under "most conditions" and "LC" where one or more criteria are satisfied under rare or "limited conditions."

Water Heating:
Condensing Commercial Water Heater (F)
Heat Pump Water Heater (F)
Tankless Water Heater (LC)
Passive Solar Water Heater (LC)
Active Solar Water Heater (LC)
Waste Heat Recovery Water Heater (MC)
Passive Hot Water Recovery System (MC)

Space Heating:
Recuperative Furnaces (LC)
Pulse Combustion Furnaces (LC)
Heat Pipe Furnaces (LC)
Ceramic Fiber Burner (MC)

Space Cooling:
Desiccant Cooling (LC)
Gas Absorption Cooling (MC)
Gas Engine Cooling (LC)
High-Efficiency Air Conditioner (MC)
Evaporative Cooling (F)
Heat Recovery Absorption Cooling (LC)
Roof Spray Cooling (LC)

Combined Heating and Cooling:
High-Efficiency Air Source Heat Pumps (F)
Water Source Heat Pumps (MC)
Ground Coupled Heat Pumps (LC)
Heat Pump Setback Thermostats (LC)
Integrated Appliances (LC)
Heat Pipe Assisted Air Conditioning (LC)
Passive Solar Heating and Cooling (MC)

Building Envelope Technologies:
Advanced Glazing Films and Coatings (MC)
Gas Filled Glazings (MC)
Fenestration Control High R-Value Systems (F)
Advanced Insulation (F)
Radiant Barriers (LC)

Lighting:
High-Efficiency Incandescent Lamps (F)
High-Efficiency Lamps (F)
Energy Efficient Ballasts (F)
High-Efficiency Fluorescent Fixtures (F)
Fluorescent Lighting Retrofit (F)
Compact Fluorescent Lighting (F)
HID Lamps (F)
HID Electronic Ballasts (LC)
Daylighting Systems (F)

Lumen Maintenance Lighting Control (LC)
Occupancy Scheduling Lighting Control (F)
Fine Tuning Lighting Control (F)
Load Shedding Lighting Control (F)

Appliances:
High-Efficiency Refrigerators (LC)
Advanced Residential Electric Cooktops (LC)
Residential Solar Cooker (MC)
Advanced Commercial Fryers (MC)
Advanced Commercial Griddles (LC)
Advanced Commercial Ovens (LC)
Advanced Dishwashers (F)
Advanced Clothes Washers (F)
Advanced Clothes Dryers (F)
Advanced Office Equipment (F)

Industrial Applications:
High-Temperature Insulation (F)
Boiler and Steam System Improvements (F)
Waste Heat Recovery (F)
Pinch Analysis (F)
Advanced Industrial Controls (MC)
Industrial Process Load Adjustment (F)
Industrial Process Heat Pumps (LC)
Freeze Concentration (F)
Membrane Processes (MC)
Laser Processing (LC)
Advanced Industrial Refrigeration (MC)
Infrared Heating (MC)
Microwave heating
Solar Industrial Process Heating (LC)
Radiant Low NO_x Tube Burners (LC)

Advanced Motors:
Programmable DC Motors (LC)
Variable Speed Drive AC Motors (F)

Load Management:
Time-of-Use Metering (MC)
Direct Load Control (MC)
Direct Price Sensing and Response Meter System (LC)
Thermal Energy Storage (MC)
Customer Battery Storage (LC)
Energy Management Systems (F)
Customer Control/Automated Feedback (MC)

Community-Scale Technology:
District Heating and Cooling (MC)
Geothermal Direct Use (MC)
Multistage Flash Distillation (F)
Multiple Effect Distillation (F)
Reverse Osmosis (F)
Mechanical and Thermal Vapor Compression (F)

On Site Electricity Production:
Distributed Photovoltaic Systems (LC)
Distributed Wind Systems (LC)

Transmission Technology:
Flexible AC Transmission (LC)
Amorphous-Cored Transformers (MC)
Dynamic Monitoring (LC)
Solid State Arrester (F)
Synthetic Taped Cable (MC)
HVDC Circuit Breaker (LC)

End-Use Technologies Not Commercially Available. The following end-use technologies have been determined not commercially available because they do not satisfy one or more of the commercial status criteria. Dates of expected commercial availability are indicated in parentheses as either "near-term" (within 12 years), "long-term" (beyond 12 years) or "indeterminate" if having excessive unresolved R&D issues or low likelihood of commercialization in the foreseeable future.

Water Heating:
Pulse Combustion Heaters (near-term)
Radiant Burner Storage Type Heater (near-term)
Thermophotovoltaic Gas Water Heater (near-term)

Space Heating:
Active Solar Heating (indeterminate)

Space Cooling:
Active Solar Cooling (near-term)

Combined Heating and Cooling:
Gas-Fired Heat Pumps (near-term)
Bivalent Heat Pumps (near-term)

Building Envelope Technologies:
Aerogel High R-Value Insulation (near-term)
Evacuated Glazing High R-Value Insulation (near-term)
Switchable Windows (near-term)

Lighting:
E-lamps (near-term)
Light Pipe (near-term)
Fiber Optic Systems (near-term)

Appliances:
Advanced Insulation for Refrigerators (near-term)
Advanced Residential Ovens (near-term)
Advanced Commercial Burners (near-term)

Industrial Applications:
Advanced Glass Processing (near-term)
Advanced Gas-Fired Heaters (near-term)
Advanced Radiant Heat Transfer (near-term)

Advanced Motors:
Stirling Engines (near-term)

On-Site Electricity Production:
Solar Thermal/Photovoltaic Hybrid (near-term)

Transmission Technologies:
Superconductors (long-term)
System Control and Data Acquisition (near-term)
High Phase Order Transmission (near-term)

End-Use Technologies General Conclusions

- The lack of incentive for building owners and developers to incorporate energy efficiency measures and minimal industry acceptance of cost-effective energy technologies limit the deployment of most end-use technologies provided as original equipment in new buildings. This situation is because building owners and developers have no incentive to improve energy efficiency where tenants or new owners pay their own energy bills. Until builders include more efficient technologies in new buildings, the only major market for energy efficient end-use technologies will be for replacing old equipment. Without government or utility actions, these constraints will be mitigated only when demands of tenants and building buyers create a market advantage to include cost-effective energy technologies.

- Low end-user awareness is a particularly critical deployment issue for new end-use technologies. Unlike electric generation technologies, end-users are widely dispersed, have fewer resources and lack familiarity with technical concepts.

- Substantial energy savings can be derived from many commercially available end-use technologies that have very limited use though they are highly cost competitive. These technologies include: condensing and waste heat recovery commercial water heating, heat pump water heaters, ceramic fiber burner heating, evaporative cooling, passive solar heating and cooling, high efficiency lighting and control systems, high efficiency refrigerators, advanced clothes washers and dryers, advanced office equipment, advanced industrial technologies, variable speed drive AC motors, energy management systems, customer controlled automated feedback transmission systems, amorphous cored transformers, and synthetic taped transmission cables. Thus, government or utility programs targeted at the "lack of incentive for building owners and developers" and "low end-user awareness" issues identified in the previous two items could yield reduced energy consumption.

- Most advanced energy conservation technologies have favorable levelized costs compared to electric generation technologies.

- A broad range in the levelized cost of heating and cooling technologies exists because their economic viability is dependent on load requirements. Typically, high-cost heating and cooling technologies are not cost competitive in many of California's moderate climate areas.

- Most new lighting technologies are cost-effective energy options for new commercial buildings.

TECHNOLOGY EVALUATION MATRIX

The Technology Evaluation Matrix in Figure 1 summarizes the results of assessments of each energy technology for the three evaluation factors: 1) Commercial Status, 2) Research and Development Goals, and 3) Deployment Issues.

The Commercial Status section of the matrix provides assessments of each technology according to three criteria: 1) technology maturity, 2) existence of supplier(s), and 3) cost competitiveness and an overall commercial availability determination. Since a technology must meet the minimum requirements of all criteria to be judged commercial, the overall determination has been labeled "not commercially available" if any one criterion was not satisfied.

The matrix uses a graphic system to indicate the evaluation results. The darker the box under each criterion, the less a technology satisfies that criterion: a white box indicates a criterion is fully satisfied; a light gray box means that a criterion is satisfied under most conditions; a medium gray box indicates that the criterion is satisfied only under very limited or rare conditions; and a dark gray box means the criterion is not satisfied under any conditions.

The Commercial Availability column uses the same graphic code: a white box indicates a technology is fully commercialized; a light gray box means commercial availability under most conditions; a medium gray box indicates commercial availability only under limited or rare conditions; and a dark gray box indicates a technology is not commercially available.

Technologies determined to be "not commercially available" also include staff's best estimates of expected future availability in the "Commercial Availability" column. These estimates are based on information from experts and literature on specific technology research development and commercialization activities. However, due to the uncertainty involved in forecasting energy demand and technological breakthroughs, these estimates are limited to broad designations as either "N" for near-term (within 12 years), "L" for long-term (beyond 12 years) or "I" for "indeterminate" where noncommercial technologies have too many unresolved research and development deployment issues or limited activity to further develop them. Thus, technologies noted as "near-term" are expected to be commercially available—at least to a limited degree—before the year 2004; technologies noted as "long-term" are not expected to be commercially available to any degree until after the year 2004; and technologies noted as "indeterminate" have too many uncertainties to project availability or are unlikely ever to be commercially available.

The Research and Development Goals section of the matrix includes results of evaluating each technology according to the five major categories in the generic list of research and development goals discussed earlier in this chapter. The Deployment Issues section of the matrix includes results of evaluating each technology according to the nine major categories in the generic list of deployment issues also discussed earlier. Both of these matrix sections provide an assessment of each major category as well as the overall impact.

These sections of the matrix use a graphic system similar to the Commercial Status Matrix. The darker a color, the greater the impact: a white box indicates no impact; a light gray box indicates a minor impact; a medium gray box indicates a significant impact; and a dark gray box indicates a potential "show stopper" unless resolved. The overall assessment is based on the highest impact assessed for any generic goal or issue. Thus, technologies with only one high impact generic goal or issue would have a higher overall impact than technologies with many less serious goals or issues.

The Deployment Issues section of the matrix should be reviewed after the Research and Development Goals section. If a technology has a research and development goal identified as a potential "show stopper," then that goal may be far more of a barrier to commercial use of the technology than any of the deployment issues. The deployment issues will be less well defined and may increase in importance as research and development goals are achieved.

COMPUTER APPLICATIONS FOR ENERGY-EFFICIENT SYSTEMS

VIKAS R. DHOLE*
ROBIN SMITH
BODO LINNHOFF
University of Manchester Institute of Science and Technology

Energy is used whenever anything is made and traditional energy use means polluting emissions. This article deals with techniques that can, largely thanks to the development of computers, minimize that energy for industrial applications. The techniques, known collectively as process integration, were initially directed toward the design of chemical plants. But because they are based on fundamental principles, they may generally be applied to any process that uses networks of heating, cooling, and power to drive it. That includes industries as wide ranging as food, pulp and paper, cement, brewing, and dairy produce. See also ENERGY EFFICIENCY; ENERGY MANAGEMENT.

So how can one design the perfect plant from the point of view of energy consumption? How can one be sure the least energy possible is being used, when you have a complicated setup of units, all working at different temperatures and pressures? Setting aside small items such as pumps and valves, there may be as many as 50 major plant items to consider in, eg, a modern petrochemical plant. If there are just three possible variations for each of these, that gives 3^{50} possible process designs. Of course,

many of these alternatives will be too impractical for realistic consideration, but this simple calculation gives an idea of the immensity of the task of finding the best design (1,2).

The situation is not helped by the traditional approach to design. Figure 1 shows the overall process as an onion, with the hierarchy of conventional design represented by concentric layers (2). Designers would start at the core of the onion to design for the path from feed to products: the reactors, where chemical changes take place, and the separators, which change the cocktail produced in the reactors into salable products. The designer would choose the pressures and temperatures of the various units at this stage and draw up a heat and material balance for the process.

The heat-transfer network forms the next layer of the onion. This could include heat exchangers to transfer energy either from one part of the process to another (process–process heat exchangers) or either to or from a utility such as steam and cooling water (process–utility heat exchangers). Only then would the utilities be defined, and it is the demand of the process for heating and cooling from outside utilities that defines its energy requirement.

All the layers of the onion interact with each other. So if the process is changed, the heat-transfer network and the utility demands will also change. Similarly, changes in the heat exchanger network are likely to force changes in both the process and its energy requirement (Fig. 2).

Before the use of computers, designing a plant was an extremely laborious process. If any adjustment to completed designs was made, it took a lot of engineering time to recalculate the effect that this might have on energy consumption. So the advent of computer-aided engineering (CAE) (in this context building a mathematical model of the process inside the computer) allowed designers to get a step closer to the ultimate in energy efficiency by simulating the results of many possible changes to the process while it was still at the conceptual stage. This obviously made investigations much cheaper, quicker,

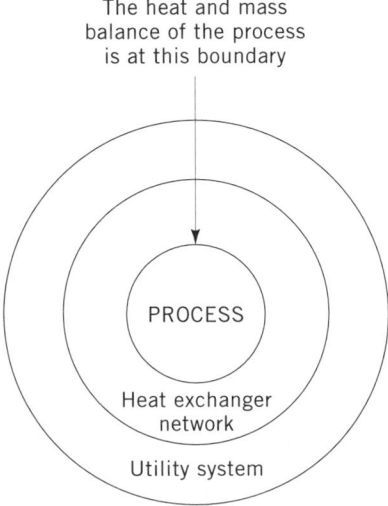

Figure 1. Onion diagram: traditional hierarchy in chemical process design.

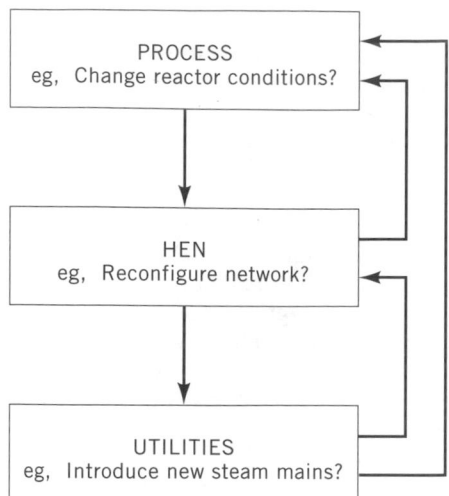

Figure 2. Design interactions between the layers of the onion diagram.

and more wide ranging than when they were carried out by hand.

There is an important aspect of design, however, that this number crunching approach does not address. That is the problem of synthesis: knowing where exactly the process configuration should be changed and targeted and what the possible energy savings should be. Without proper synthesis and targeting, designers were effectively still working in the dark. While they might be able to find out the consequences of proposed changes more quickly, they still had to rely on common sense, rules of thumb, and experience to come up with suggestions as to which changes were the right ones. A lot of expensive computer time could be wasted trying to achieve the impossible, or the designer might (and this is the most likely outcome) end up with a process that uses more energy than it really needs to.

Toward the end of the 1970s, the initial steps were taken that would eventually lead to the family of tech-

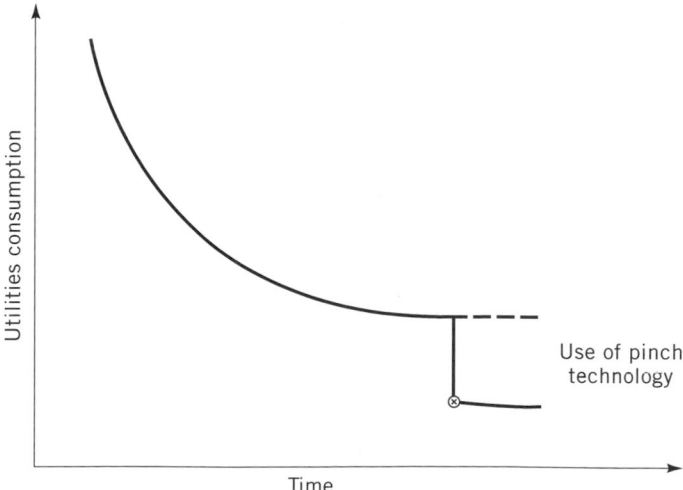

Figure 3. The use of pinch technology results in a step change in energy savings.

niques known today as process integration. The first big breakthrough was known as pinch technology and was centered around a graphic and thermodynamic approach. Figure 3 shows the effect that the use of pinch technology can have on the learning curve toward reduced energy consumption. Rather than building plants and thinking that the next one will be better, pinch technology allowed designers to take an entirely fresh and objective view of the energy demands of the process. The result was a "step change" in energy savings.

The use of pinch technology yielded typical energy savings of 20% to 30% at a sweep, coupled with capital cost savings related to optimizing the number and size of the heat exchangers needed to service the process (3). Pinch technology takes a two-stage approach to the problem. First, it defines realistic targets so that designers know where they are headed. Second, from the basic thermodynamic data generated in the targeting stage, it takes a logical, step-by-step approach to lead the designer to the right design to achieve the targets. The following two sections look at these stages in more detail.

BASIC ENERGY TARGETING

Energy targeting using pinch technology starts from a basis of complete heat and material balance for the process. In other words, the design has progressed through the central layer of the onion and has determined how much material and how much energy each process stream contains (Fig. 1). This initial stream information can be obtained from the plant data or from a converged computer simulation of the process.

Data Extraction

There may be several process streams, at a variety of different temperatures, and a majority of them must be heated or cooled before they can progress to the next stage of the process. But they can be simply divided into two groups: heat sources and heat sinks. Heat sources (also known as "hot streams," regardless of their temperature) need to be cooled down, whereas heat sinks (also known as "cold streams") need heat input.

Composite Curves

Each stream can be represented on a graph of heat content (enthalpy) versus temperature (2,5). Figure 4a shows two hot streams drawn in this way. The mass flow rates and heat capacities (CP) (how much energy is needed to heat the streams for each degree Celsius of temperature rise) of the streams are known. Stream 1 is cooled from 200° to 100°C. It has a total heat capacity (the mass flow rate times the specific heat capacity) of 1 kW/°C so it loses 100 kW of heat energy. Stream 2 is cooled from 150° to 50°C. It has a CP of 2 and loses 200 kW of heat.

The hot composite curve is produced by simply adding the heat content of the two streams over the different temperature ranges (Fig. 4b). Between 200° and 150°C, only stream 1 is present with its CP of 1 kW/°C, so 50 kW of heat is lost over that temperature range. Between 150° and 100°C, both streams are present. Their combined

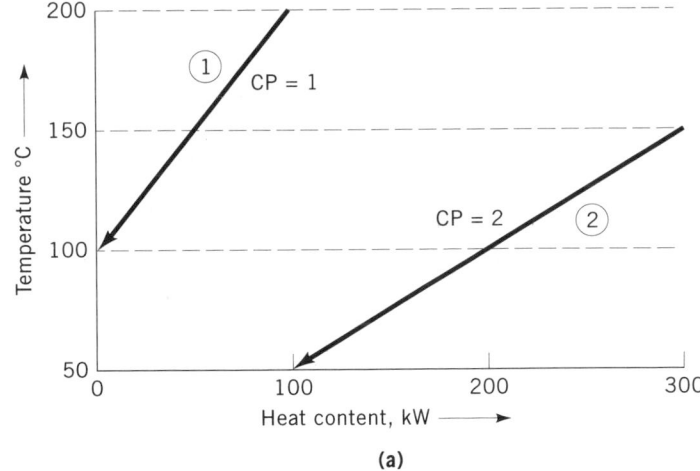

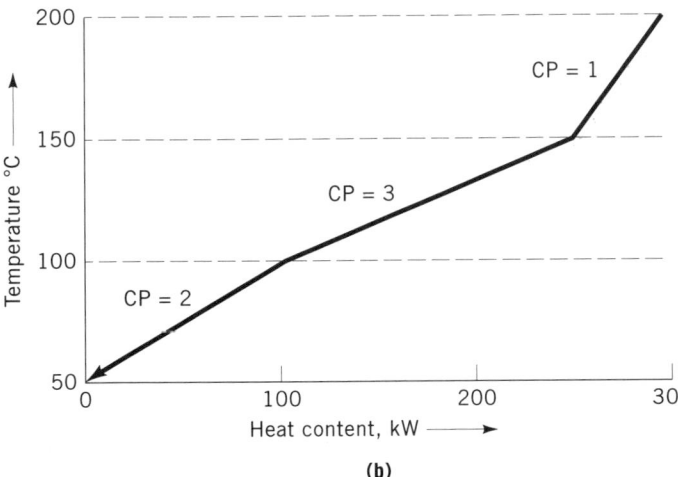

Figure 4. Composite curve construction.

heat capacity is 3 kW/°C, so the total heat loss over the temperature range is 150 kW. Because the heat loss in this section is greater than that of the 200° to 150°C temperature range, this section of the composite curve is flatter. Only stream 2, with its CP of 2 kW/°C is present between 100° and 50°C, giving a heat loss over this interval of 100 kW. The cold composite curve is constructed in exactly the same way. In practice, the number of streams is likely to be much greater than in the example used here, but the process of constructing the curves is exactly the same.

Figure 5 shows hot and cold composites plotted on the same temperature–enthalpy graph. The two curves overlap along the enthalpy axis for most of their length. In this region, it should be possible to heat the cold streams using the heat removed from the hot ones without the need to use utilities to provide external energy. But at each end of the curves there is an overhang where heat from within the process itself cannot be used. The overhang at the top of the cold composite curve represents the minimum amount of external heating that will be needed to drive the process Q_H. Similarly, the overhang at the bottom of the hot composite represents the minimum

amount of external cooling Q_C. The minimum hot and cold utility requirements are targets. They are set before design, purely on the basis of heat and material balance information.

At one point along their length, the composite curves reach their closest temperature approach. This is known as the "pinch," and recognizing its implications allows energy targets to be realized in practice. It will start designers on the road to designing a heat exchanger network that conserves energy.

Once the pinch has been identified, it is possible to consider the process as two systems: one above and one below the pinch. The system above the pinch requires a heat input and is, therefore, an overall heat sink. Below the pinch the system rejects heat and so is an overall heat source. This gives three golden rules that the designer must stick to in achieving the minimum energy targets (3):

1. Heat must not be transferred across the pinch.
2. There must be no outside cooling above the pinch.
3. There must be no outside heating below the pinch.

These rules are illustrated in Figure 6. Consider rule 1 (Fig. 6a): if heat is traveling across the pinch from a higher to a lower temperature, heat is being removed from above the pinch (the heat sink) so additional hot utility must be supplied to make up the difference. Similarly, such movement adds heat to the region below the pinch (the heat source), so more cold utility will also be needed. Rules 2 and 3 follow a similar logic (Figs. 6b and 6c). Breaking rule 2 would involve external cooling of the heat sink, which must already be heated from the outside, and so would put extra demands on the hot utilities. Rule 3 ensures that extra heat is not added to a part of the process that is a net heat source, which would put extra demands on the cold utilities.

These golden rules play a crucial role in creating a heat exchanger network that meets the minimum energy targets set by the composite curves.

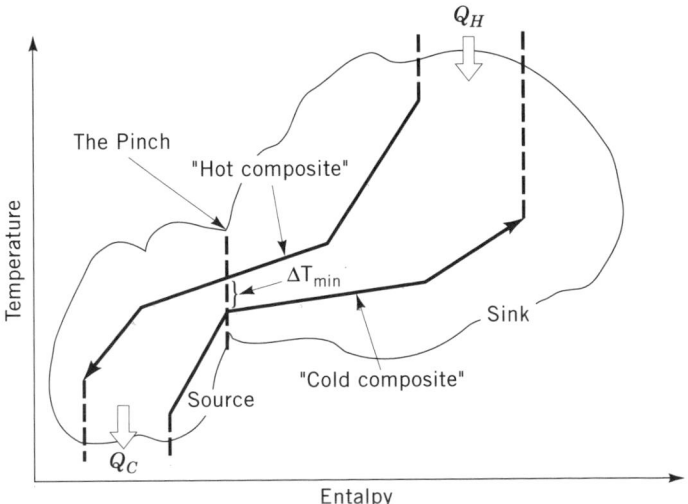

Figure 5. Using composite curves to set minimum energy targets.

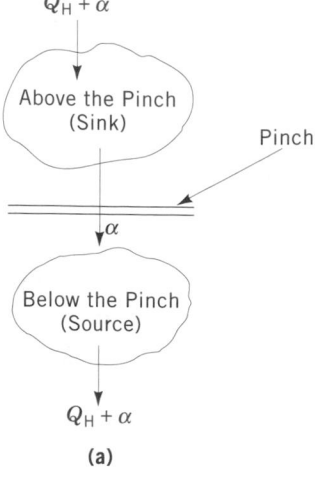

$Q_H + \alpha$

Above the Pinch (Sink)

Pinch

α

Below the Pinch (Source)

$Q_H + \alpha$

(a)

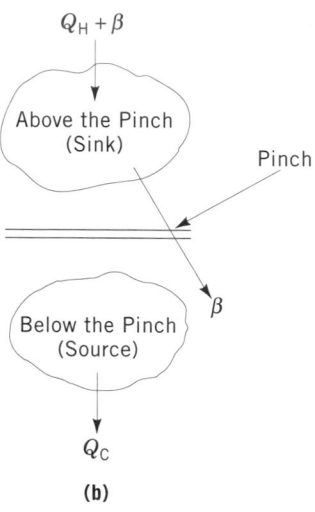

$Q_H + \beta$

Above the Pinch (Sink)

Pinch

β

Below the Pinch (Source)

Q_C

(b)

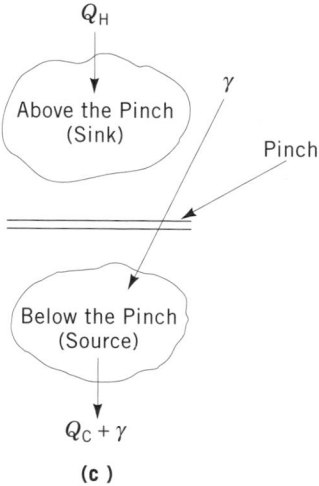

Q_H

γ

Above the Pinch (Sink)

Pinch

α

Below the Pinch (Source)

$Q_C + \gamma$

(c)

Figure 6. Three key design rules that achieve minimum energy targets.

Targeting for Multiple Utilities: The Grand Composite Curve

The vast majority of industrial sites use steam to transfer heat to the process and cooling water for removing the excess heat. However, for high temperature heating hot oil circuits or furnace heating is usually used, while refrigeration is used for the low temperature cooling duties. Steam is normally provided by an on-site boiler or combined heat and power system using steam turbines. This design gives operators the flexibility to provide steam at several different temperatures and pressures.

It is usually preferable to use the cheapest utilities wherever possible, eg, steam at the lowest feasible temperature and pressures and cooling water instead of refrigeration. The composite curves give the designer overall energy targets but do not indicate how much energy needs to be supplied by different utilities. The tool that provides this information is called the grand composite curve (2,6).

The construction of the grand composite curve is shown in Figure 7. The designer starts with the hot and cold composite curves (Fig. 7**a**). It is assumed that the temperature difference between the composite curves at the pinch is the minimum permissible temperature difference (ΔT_{min}). This also must be taken into account when transferring heat between the process streams and the utilities. The first step is to shift all the streams in the hot composite curve colder by $\frac{1}{2}\Delta T_{min}$. Similarly, all the streams in the cold composite are shifted to a temperature that is higher by $\frac{1}{2}\Delta T_{min}$ (Fig. 7**b**). The reasons for this will become clear shortly. The grand composite curve can then be constructed from the enthalpy (horizontal) differences between the shifted composite curves at different temperatures (Fig. 7**c**). Because of the temperature shifts made to the composite curves, the curves will actually touch each other at the pinch. Thus the point on the grand composite curve that touches the temperature axis corresponds to the pinch, and the extremes of the curve correspond directly to the overall energy requirements. The section of the curve above the pinch is the overall heat sink of the process, and the heat source is represented below the pinch.

Figure 8 illustrates the use of grand composite curve in targets for multiple utilities servicing a process. In Figure 8**a** the process is serviced by high pressure (HP) steam and refrigeration (Ref.). Using the grand composite it is possible to identify opportunities for minimizing the utilities cost. The grand composite in Figure 8**b** suggests that a part of the high pressure steam consumption can be replaced by medium pressure (MP) steam and that a part of the refrigeration load can be replaced by cooling water (CW). Medium pressure steam and cooling water are cheaper utilities than high pressure steam and refrigeration. Using the grand composite curve it is also possible to set the energy targets for furnace heating alongside steam heating (Fig. 8**c**). The flue gas profile extends from the flame temperature to the pinch temperature. In conclusion, the grand composite curve is used for selection of the appropriate utilities and for setting the targets for the minimum utilities cost.

Remember that the cold process streams making up the grand composite curve have already been raised by

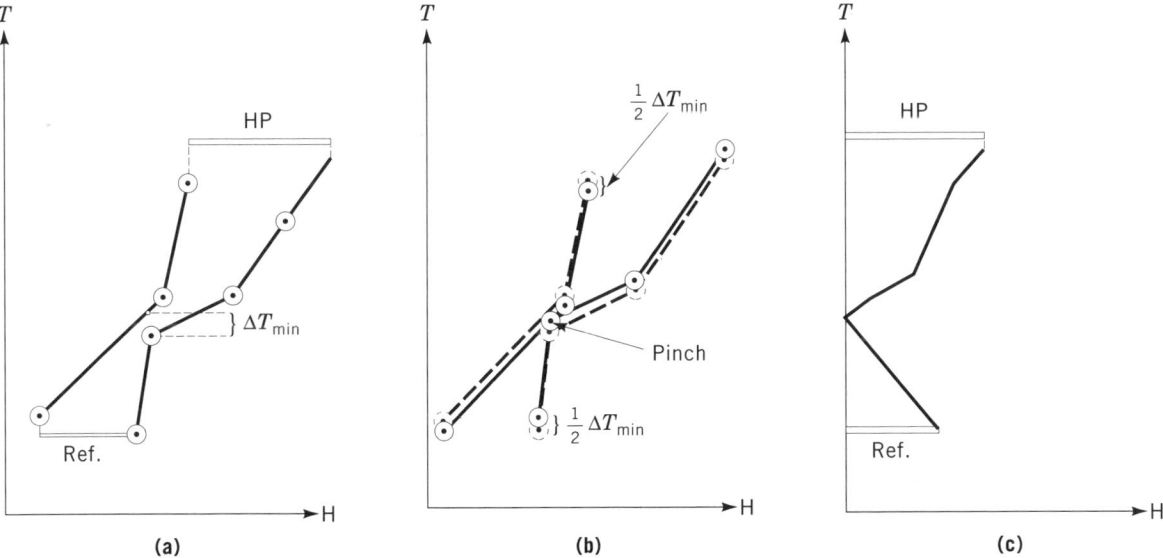

Figure 7. Construction of the grand composite curve.

half the minimum temperature difference for heat transfer (Fig. 7**b**). The temperatures of the hot utilities (such as steam levels) on the grand composite curve are $\frac{1}{2}\Delta T_{min}$ cooler than they actually are. This means that where the temperatures of the utilities and the process streams apparently coincide they will actually be separated by the minimum temperature difference (ΔT_{min}) for heat recovery. This is the reason for shifting the stream temperatures; it simply makes the construction of the diagram and the utility requirements more easy to visualize. The temperatures of the process streams for the cold utilities in the heat source section have been adjusted down. So the cold utility temperatures (eg of cooling water and refrigeration) are adjusted up by $\frac{1}{2}\Delta T_{min}$ for the same reasons given for adjusting the hot utility temperatures down; it will make visualization of the problem easier.

Capital Energy Trade-offs

Capital cost considerations are vital in determining whether an energy-saving project will go ahead, so they must be considered at an early stage. For heat-transfer networks, the overall capital cost is made up of two principal components: the minimum overall surface area for

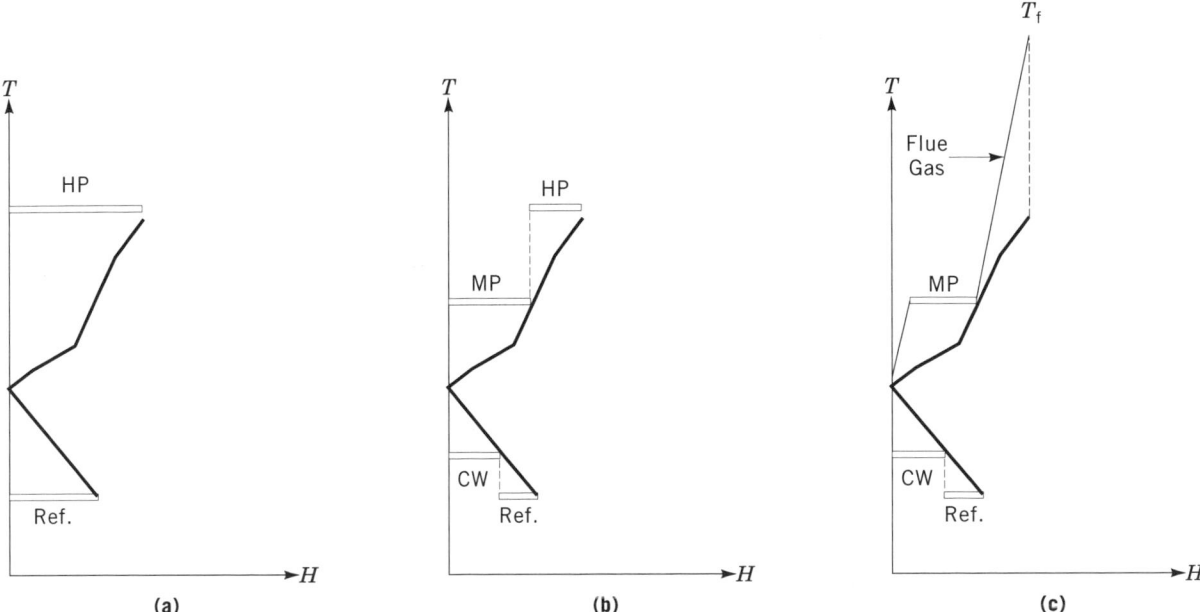

Figure 8. Using the grand composite curve for multiple utilities targeting.

heat transfer (in effect the size of the heat exchangers) and the minimum number of units. The number of units will be discussed first. Using more units than the minimum implies building more foundations, more pipework, more instrumentation, and carrying out extra maintenance. The minimum number of units in a network $Umin$ depends on the number of process streams and utilities N (2,7). It can be expressed as:

$$Umin = N - 1$$

This has been explained as a special case of a mathematical theorem called Euler's general network theorem (8).

Independent studies have established that the minimum number of units and the minimum utility use are often incompatible (9). The first of the three golden rules is to blame for the incompatibility of the minimum number of units and minimum energy consumption. This is because, without heat transfer across the pinch, the network is divided into two thermodynamically independent systems (Fig. 5). If the system on each side of the pinch is considered separately, any of the process streams that cross the pinch must be counted twice. Thus the minimum number of units for the maximum energy recovery, $Umin_{MER}$ can be expressed as:

$$Umin_{MER} = Umin \text{ (above the pinch)} \\ + Umin \text{ (below the pinch)}$$

Surface area targets are discussed next. The composite curves are used as a basis for surface area targets (Fig. 9). The composite curves provide a countercurrent representation of the overall heat transfer in the process (as indicated by the vertical arrows in Fig. 9). It can be shown that if the process streams in a network exchange heat in such a way that the stream matches are "vertical" between the composite curves then the overall heat transfer area is minimized. This arrangement is equivalent to pure countercurrent flow in a single heat exchanger. Based on the vertical model for heat transfer, the designer can set surface area targets for the network (7,10). First, the composite curves must be divided into enthalpy intervals (Fig. 9). Within each interval i a stream j has an individual stream duty q_j and film transfer coefficient h_j. The minimum overall surface area is then approximated by:

$$A_{target} = \sum_i^{intervals} (1/\Delta T_{LM_i}) \sum_j^{streams} (q_j/h_j)$$

Coupled with the target for the minimum number of units, this simple formula allows approximate capital cost targets to be set for the whole heat exchanger network before design. It does not yield exact results, because of several simplifying assumption that will not be discussed here (7,10). But experience has shown that predicted overall costs are typically accurate to within 5%.

Based on the capital cost targets and the energy targets, it is now possible to evaluate the trade-off between capital cost and energy recovery explicitly (Fig. 10). For the various values of ΔT_{min} (obtained by shifting the composite curves) the energy target and the associated capital

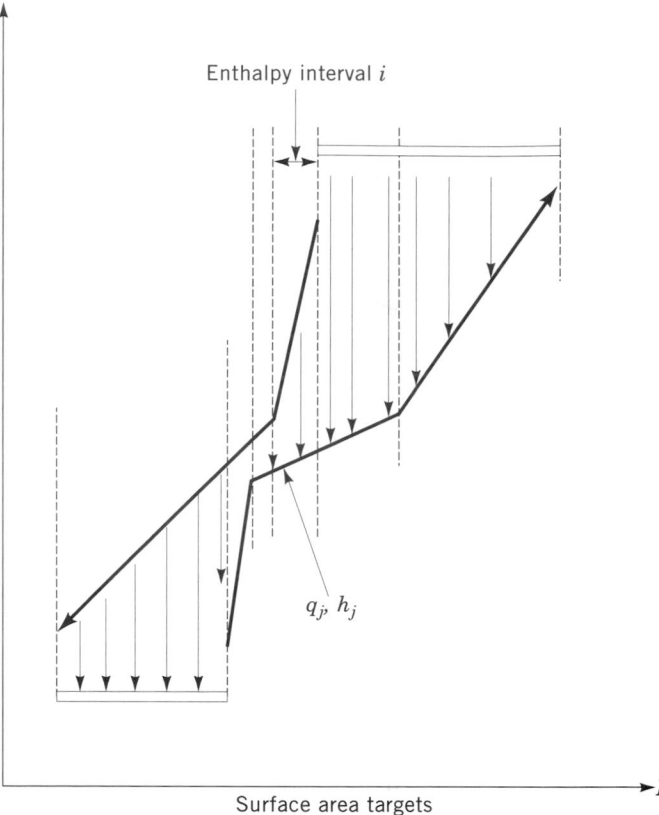

Figure 9. Vertical heat transfer model for overall surface area targets.

cost target are determined. The larger the ΔT_{min}, the larger the energy target and the lower the surface area. Thus the lowest possible overall cost is clearly established.

Note that the procedure yields the optimum ΔT_{min} before design. In other words, the designer obtains an optimized starting point for design. Using tools for network design, it is possible to obtain a network that achieves the predicted minimum cost, in line with the predicted component costs for energy and capital. The network is already optimized as it is put together.

HEAT EXCHANGER NETWORK DESIGN

Targeting is only the first stage in creating a finished design. When it comes actually to deciding how to set up the network of heat exchangers, the temperature–enthalpy representations used so far are not helpful. The grid diagram (2) representation of the network is much clearer.

The Grid Diagram

A basic grid diagram is shown in Figure 11. The hot streams are shown at the top of the figure, running left to right. Cold streams run across the bottom, from right to left. A heat exchanger transferring heat between the process streams is shown by a vertical line joining circles on the two matched streams. A heater is shown as a circle with an H and a cooler is a circle with a C.

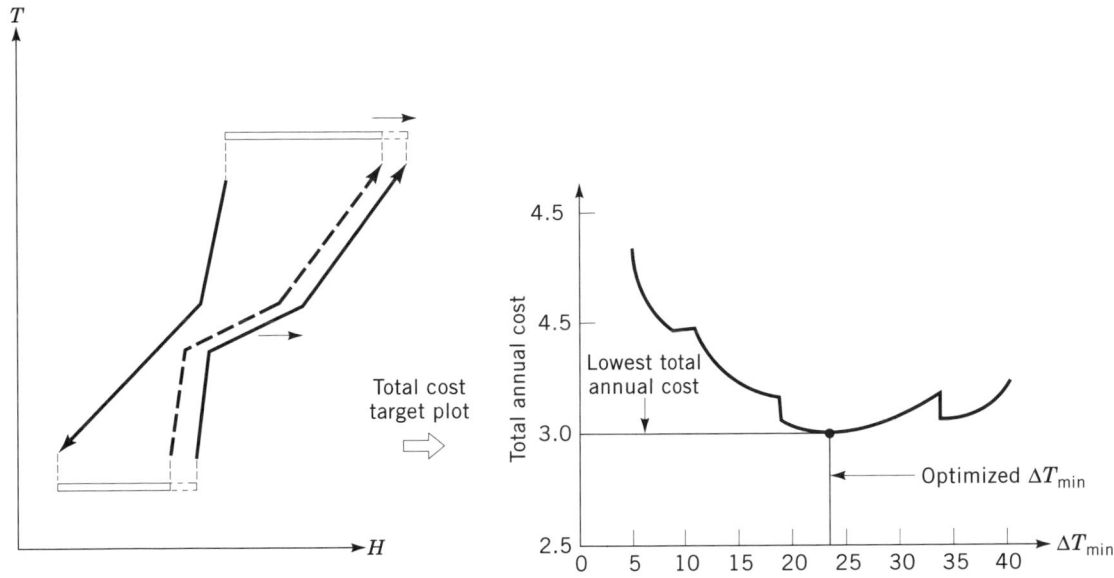

Figure 10. Using capital and energy targets to determine optimum ΔT_{min} before designing.

The pinch is shown as a vertical line cutting the process into two parts. From the composite curves and optimum ΔT_{min}, it is known at what temperatures the hot and cold streams encounter the pinch. The heat sink section (above the pinch on the T–H plots) is shown to the left and the heat source section is to the right.

Pinch Design Rules

The three golden rules, discussed earlier, are central to the way a designer goes about designing the network. There must be no external cooling used above the pinch (on the left) so hot streams on this side must be brought

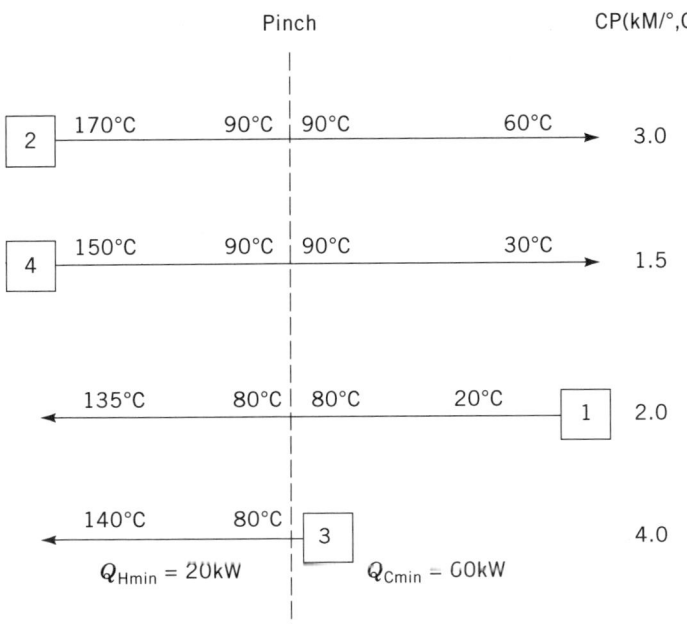

Figure 11. Grid diagram for the example problem.

to pinch temperature by heat transfer with cold streams on the same side, ie, on the left. Similarly, cold streams on the right must be brought up to the pinch temperature using hot streams on the right rather than utility heating. The ΔT_{min} puts another constraint on the design because it has been defined as the minimum temperature difference for heat transfer anywhere in the system.

Some ground rules for setting up the heat exchanger matches between the hot and the cold streams on each side of the pinch are now considered (Fig. 12). A proposed match can be drawn between streams 1 and 2 in Figure 12**a**. A temperature–enthalpy diagram of the proposed match is shown in the inset. Note that the stream directions have been reversed so that they can be more easily related to the grid diagram. Because the CP of stream 1 is *less* than stream 2, as soon as a heat load is applied, the temperatures of the two streams converge to closer than ΔT_{min}. This means that the proposed heat exchanger is not feasible and another match must be found.

In Figure 12**b**, streams 2 and 3 are matched. In this case the CP of stream 3 is *greater* than CP of stream 2. The relative gradients on the temperature–enthalpy plot show that the streams diverge from ΔT_{min}; the match is feasible. For the temperature feasibility of the matches close to the pinch, the CP of the streams going out of the pinch needs to be greater than the CP of the stream coming into the pinch. Thus for temperature feasibility there is a "CP rule" (2): $CP_{out} \geqslant CP_{in}$.

Stream 4 also must be brought to the pinch temperature. This is achieved using a match with stream 1. Since CP1 (the stream going out) is greater that CP4 (the stream going in), the temperature–enthalpy plot is again divergent, and the match is feasible. Now that both the hot streams on the left side of the pinch are completely matched with the cold streams, the possibility of using a cooling utility above the pinch has been eliminated. The designer is now free to provide the remaining heating load on the cold streams to the utilities.

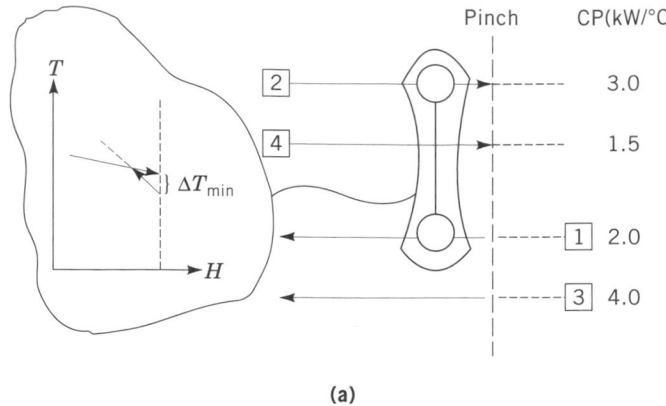

(a)

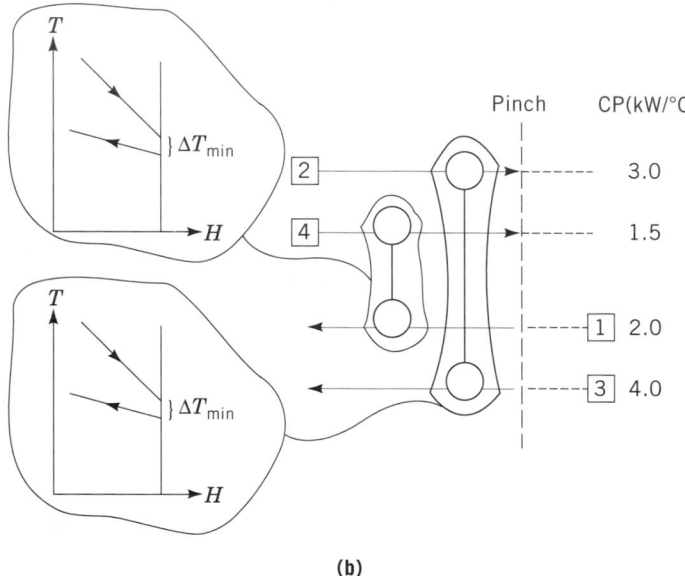

(b)

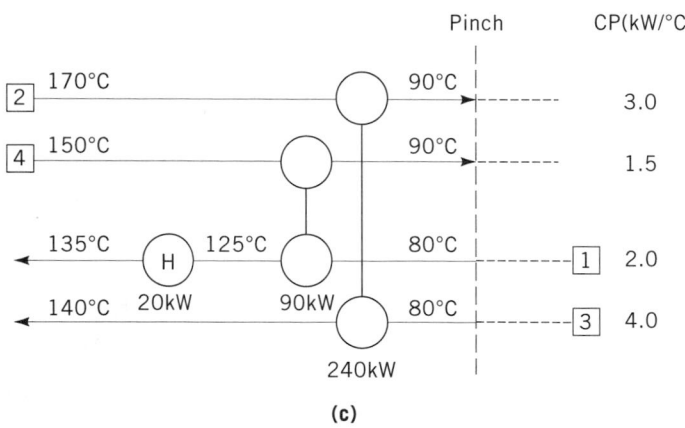

(c)

Figure 12. Criteria for temperature feasibility at the pinch.

Because stream 2 requires 240 kW of cooling and stream 3 needs 240 kW of heating on the left side of the pinch, the 2–3 match satisfies the requirements of both streams. But the 4–1 match can only satisfy stream 4, because stream 4 can only give up 90 kW of heat before it reaches the pinch temperature. This will heat stream 1 to 125°C, but the target temperature is 135°C. An extra

20 kW must be provided by utilities to satisfy the heating requirements of stream 1. This is indeed the minimum utility requirement for the network. Figure 12**c** shows the completed design for the section to the left of the pinch.

Figure 13 shows the equivalent design process below the pinch (on the right of the grid). The golden rule below the pinch is not to use utility heating, so streams 1 and 2

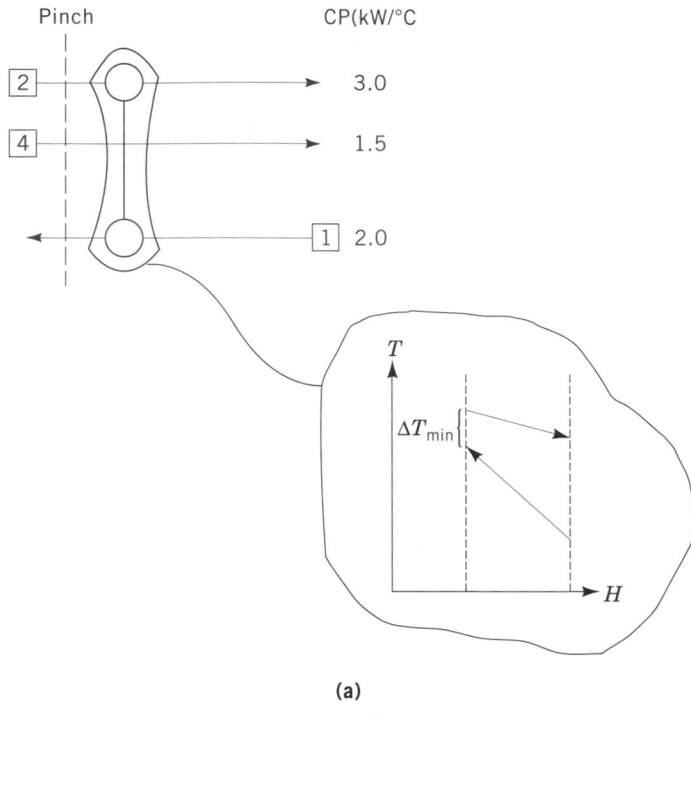

(a)

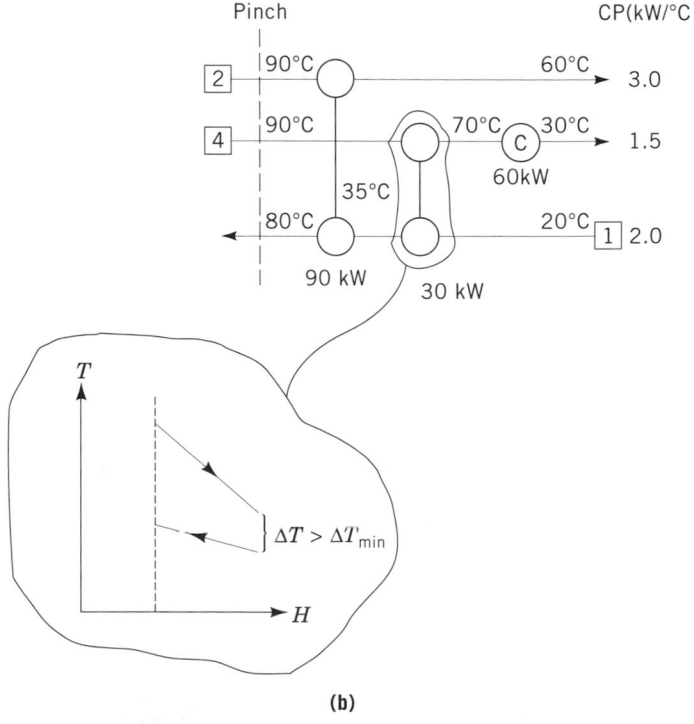

(b)

Figure 13. Network design below the pinch.

are matched. The inset shows that the match is feasible, because the CP of stream 2 (the stream going out of the pinch) is greater than the CP of stream 1 (the stream going into the pinch). As expected, the temperature–enthalpy plots diverge as they leave the pinch.

It is important to realize that the CP rule for temperature feasibility is rigid only for units directly adjacent to the pinch. Figure 13b shows that the proposed match between streams 4 and 1 displays converging temperatures on the temperature–enthalpy plot. However, since the match is not adjacent to the pinch, it does not have to bring the streams to the pinch temperature difference. This means that even though the temperatures converge they do not necessarily end up at a temperature difference smaller that the minimum. In this instance, the final ΔT is greater than ΔT_{min}, and the match is still feasible.

Figure 12b shows the completed design to the right of the pinch. Putting a heat load of 30 kW on the 1–4 match (all the cooling that stream 1 can provide before it reaches its target temperature) means that residual cooling of 60 kW must be provided to stream 4 by utilities. The resulting overall network is shown in Figure 14 and is known as the minimum energy requirement (MER) design.

Special Cases

For the constrained region around the pinch, there is a problem if there are more streams going into the pinch than there are coming out. According to the golden rules, each stream going into the pinch needs to be matched with an outgoing stream to bring it to the pinch temperature. Therefore, it is necessary to split the outgoing streams to equal the number of streams going into the pinch. Consider, for example, the grid representation on the left side of the pinch in Figure 15a. In this case, the number of streams going into the pinch is three whereas the number of outgoing streams is two. It is necessary to split an outgoing stream, as shown in Figure 15b. The splitting should also ensure that the CP rule is not violated.

There is also a problem if there are matches at the pinch that do not comply with the $CP_{out} \geq CP_{in}$ rule. In this case, an incoming stream could be split to reduce its CP. Consider the example shown in Figure 16a. One of the solutions is to split the incoming stream that has a CP of 4 (Fig. 16b). Now the number of outgoing streams is less than that of the incoming streams. Therefore, an

outgoing stream can be split as shown. However, in practice, it is possible to find simpler solutions (Fig. 16c). This network design requires only one split and it achieves the minimum energy requirement as well as the temperature feasibility. One of objectives in network design is to minimize the number of stream splits. Stream splitting has been examined in more detail (11).

Network Optimization

After the network has been designed according to the pinch rules it can be further subjected to capital energy optimization. Remember that the network is subjected to optimization before design (Fig. 10). By choosing the optimum ΔT_{min} for the network design the designer addresses capital energy trade-offs. Optimization after network design, therefore, will only fine-tune the network costs.

Network optimization is mainly carried out by using standard optimization routines. Optimization plays on extra degrees of freedom within the network. The first of the other two degrees of freedom is illustrated in Figure 17a and is known as a heat load loop (2). The loop is shown by the dotted line that marks a circuit between matches two and four. Heat duties can be shifted around a loop without affecting the heat duties imposed on other units in the network. What the designer will aim to do here is to shift the duties such that the overall network cost is reduced. This may result in reducing the heat load on a unit to zero; a zero-duty unit means no exchanger at all, saving the cost of one unit. The other degrees of freedom the computer looks for are heat load paths (2) (Fig. 17b). A path allows the transfer of heat loads between heat exchanger units and utilities. By exploiting these degrees of freedom the designer fine-tunes the network for capital energy trade-off.

RETROFIT MODIFICATION OF HEAT EXCHANGER NETWORKS

The designs examined so far started from scratch, but the majority of process integration projects are aimed at achieving energy savings on existing plants. Because of environmental problems, it is necessary to restrict emissions from industry; thus it could prove vital that energy consumption on existing installations is checked, helping industry to meet energy targets without having to cut production.

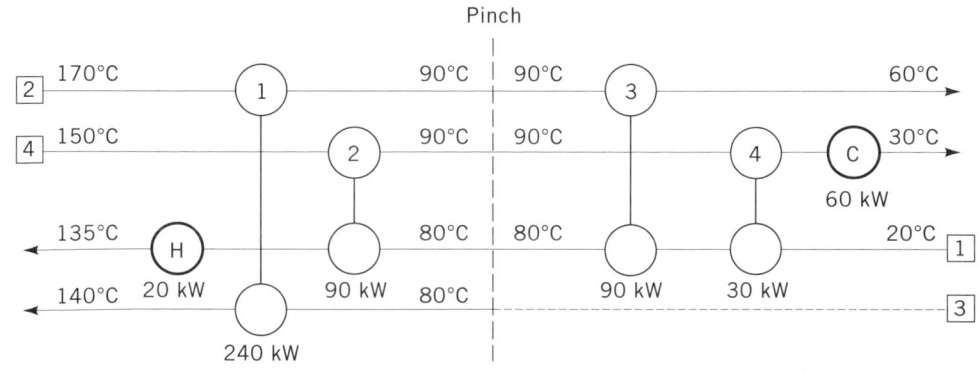

Figure 14. Completed design based on pinch design method.

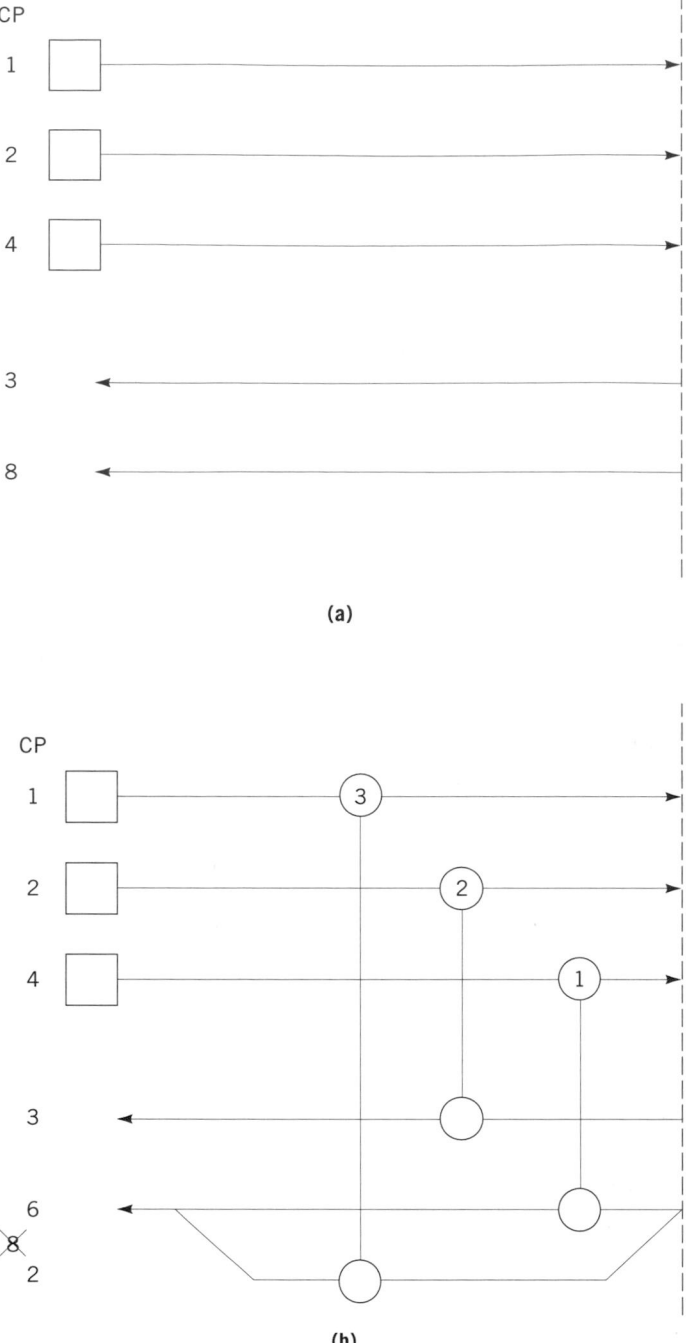

Figure 15. Criteria for stream splitting at the pinch.

maining doubts. Because it does not necessarily involve a true understanding of all the factors at work, there might always be a better answer. The second approach uses computers purely as number crunchers. When simulation packages are available, some designers think that churning out many alternative network designs could be the answer. But, as with design by inspection, this approach lacks a true understanding of the process. It can take up a lot of computing time, prove expensive, and, most important, it can miss some key design options for energy savings.

Targeting for Retrofits

Like grass-roots designs, proposed retrofit projects will always have a trade-off between energy and capital costs. In industry, this is usually expressed as a payback time on capital investment. It is common, for example, to expect to get the money that the company has invested at the start of a project back in energy savings over 2 years.

The basic difference between the cost considerations when designing for retrofits and designing from scratch is that a significant amount of hardware is already in place. So minimizing the overall heat-transfer area for heat exchangers is not necessarily a primary consideration. After all, if the equipment is available, there is no cost benefit in rendering it obsolete. The aim is rather to take up any slack in the system by making sure that existing units are used as effectively as possible. It then follows that any extra capital investment will also be minimized.

Figure 18 illustrates the point. From an existing design at point X, the designer wants to reduce energy consumption. But there is no reason to move toward point B on the capital energy curve, because the designer would not be making use of all the existing heat exchangers, even if point B is where it would end up on a grass-roots design. The aim is rather to head for point A, where the maximum energy recovery would be possible without incurring capital costs. But in real situations, designers are unlikely to reach this ideal; they will be unable to reuse existing exchangers in an ideal way. So the path of a retrofit tends to steer upward to a higher capital cost.

The pinch technology targeting procedure, like grass-roots design, begins with the composite curves for the process. The designer uses the heat-transfer area algorithm, discussed above (10), to get a plot of energy versus heat-transfer area for varying $\Delta T_{\min}$. This is, in fact, the area–energy curve shown in Figure 18.

It is known that the path of the retrofit project is going to tend up, but by how much? Choosing the best path would be a difficult task, so instead look at the least the designer can expect. Assume that the new design will be at least as efficient as the existing one in the way the capital is used. Define an area efficiency α equal to the ratio of the target area for the existing energy consumption to the area used by the existing design (12).

$$\alpha = (A_{\text{target}}/A_{\text{existing}})$$

for existing energy consumption. Assume a constant value of α across the full energy span (the new design is only as efficient as the old one) to obtain the curve shown in Figure 19a. The experience of applying the techniques to in-

It is a common assumption that a good retrofit design will make the existing process similar to a good grass-roots design. This is not always the case (12). An existing process is likely to have special constraints built in, but it is also likely to offer up some good opportunities. Exploiting those opportunities may make the retrofit look quite different from a design built up from scratch.

Before process integration, there were two approaches to retrofit design, both of which are still used to some extent. The first is known as design by inspection. The designer looks at the plant and chooses a project intuitively. The problem with this is that it will always leave re-

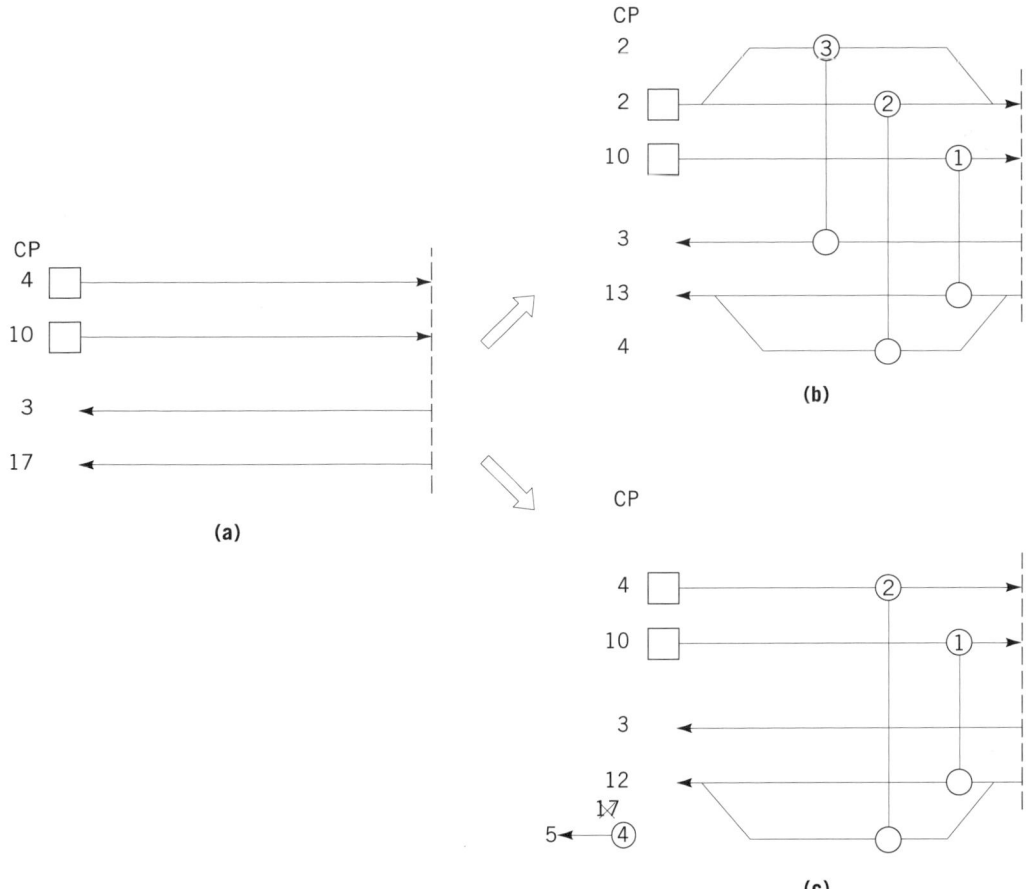

Figure 16. Number of stream splits should be minimized to reduce complexity.

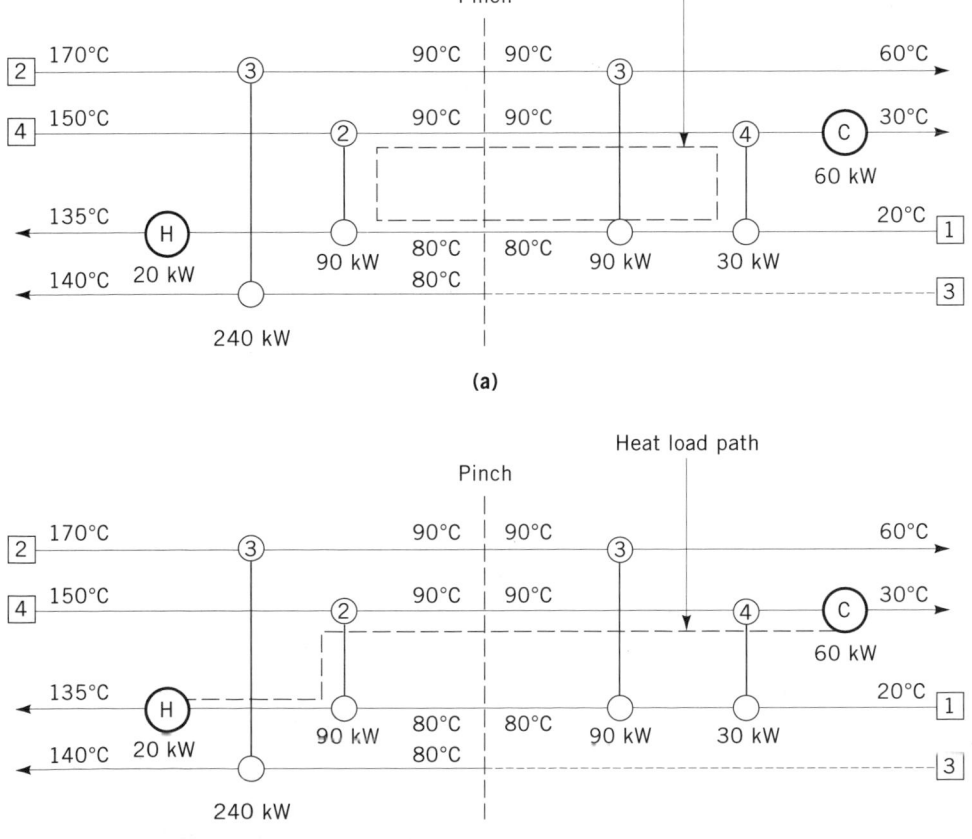

Figure 17. Heat load loops and paths provide additional degrees of freedom for network optimization.

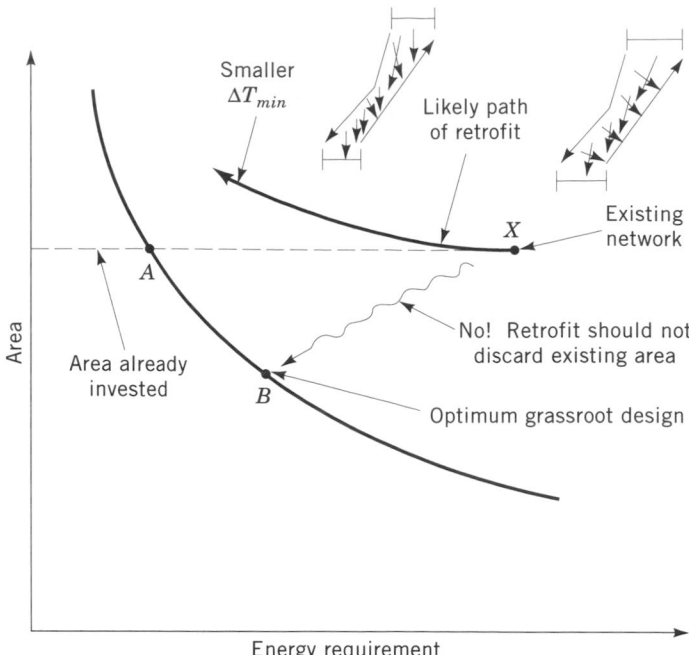

Figure 18. A retrofit design. The designer should try to reach point A, not point B, to take full advantage of the existing area.

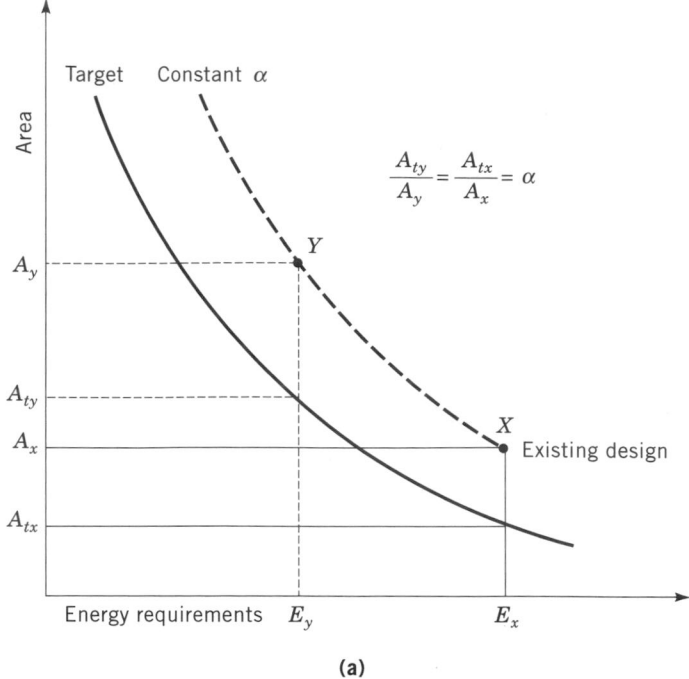

(a)

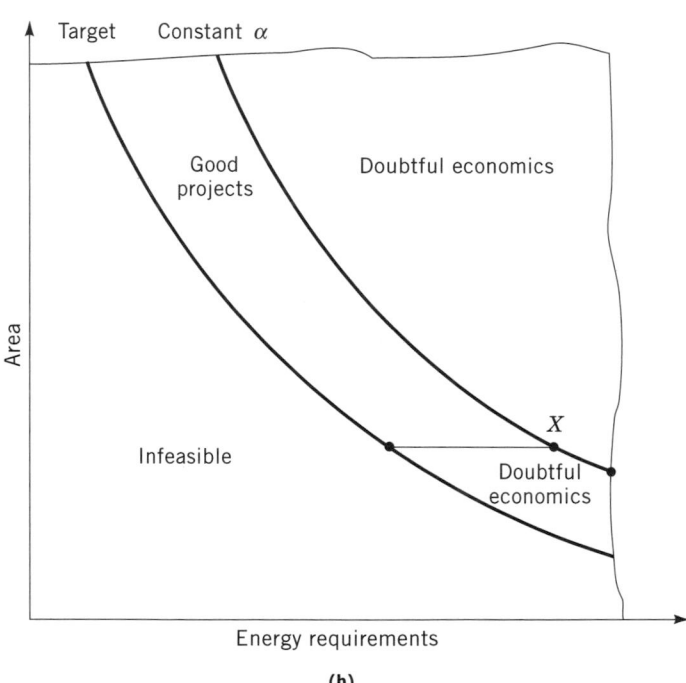

(b)

Figure 19. The new, retrofitted, design. (**a**) The designer assumes a constant value of α across the energy span, which yields a curve that serves as a boundary for design. (**b**) The best results appear in a distinct region on the area–energy plot.

dustrial processes has shown that using the constant α curve as a conservative target (Fig. 19**b**) works well in practice (13).

Because the heat-transfer area is directly related to capital cost and the energy savings can be easily related to operating cost savings, it is possible to translate the area–energy targeting curve from Figure 19 into a savings–investment targeting curve (Fig. 20**a**). The savings and investments required for the various payback periods can be easily identified. Figure 20**b** shows the targeting curve for a specific example. The payback constraint of 2 yr is used as a target for energy savings. These in turn will give the optimized value for ΔT_{min}, which in this case equals 19°C. So rather than having to work intuitively, the designer is provided, through pinch technology, with targets for energy savings, capital cost, and the minimum temperature difference for heat transfer in the process.

Designing for Retrofits

The targets indicate what the designer should be able to achieve. To get a clear picture of the situation, the existing heat-exchanger network is drawn up on a grid diagram. The position of the pinch and ΔT_{min} were identified during the targeting stage. For this example, the target ΔT_{min} of 19°C was identified. The corresponding grid representation is shown in Figure 21**a**. This allows the designer to see which heat exchangers are working across the pinch (cross-pinch exchangers). These units must be removed from the design if heat transfer across the pinch is to be eliminated and hence the energy requirements of the process. This leaves the partial network shown in Figure 21**b**.

It is then necessary to complete the network using the same procedures as used for new heat exchanger networks. In this case, however, the designer will obviously save on capital costs if he or she uses the existing units that were removed in the first step wherever possible. Figure 22 shows the resulting network for the example helping to make old exchangers fit new duties. The resulting optimized network is shown in Figure 23.

PROCESS MODIFICATIONS

It has been accepted that heat-exchanger network design requires, as a starting point, the underlying process heat and material balance, and that much could be gained by

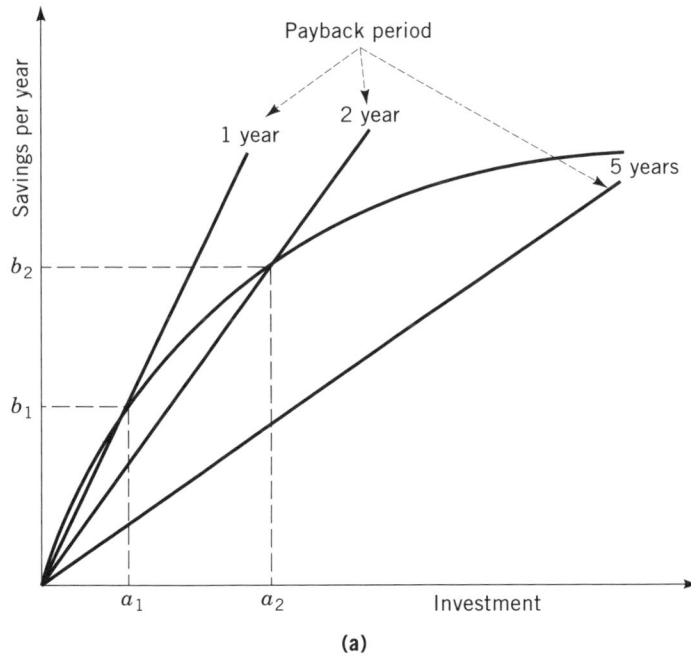

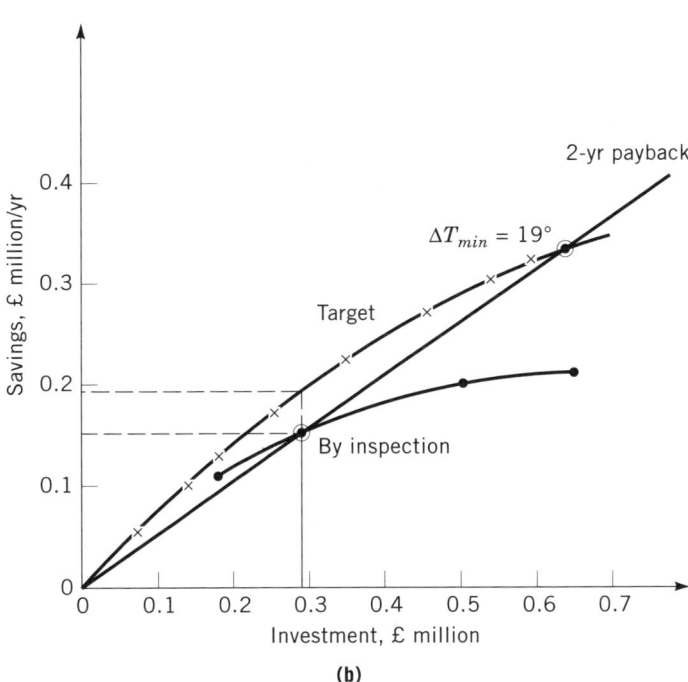

Figure 20. Savings–investment targeting curves. (**a**) The best area–energy targeting curve can be translated into a savings–investment curve. (**b**) The pinch method yielded a more economical design than the inspection method.

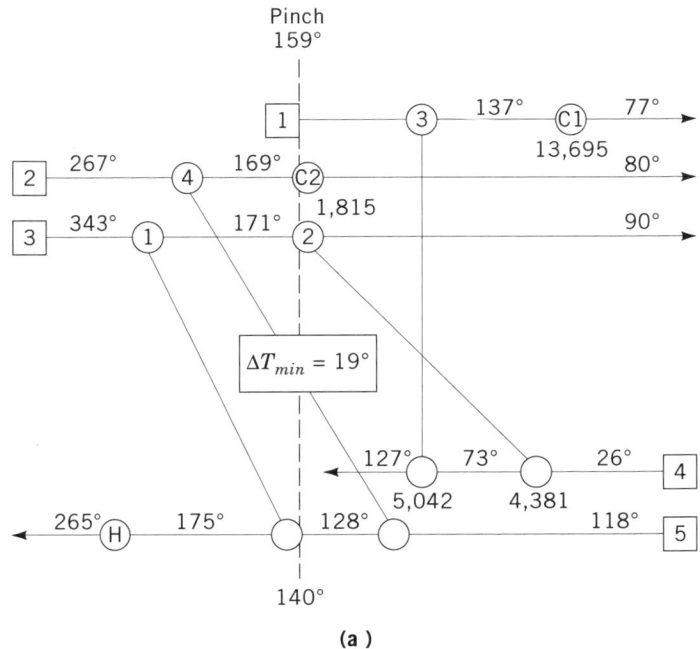

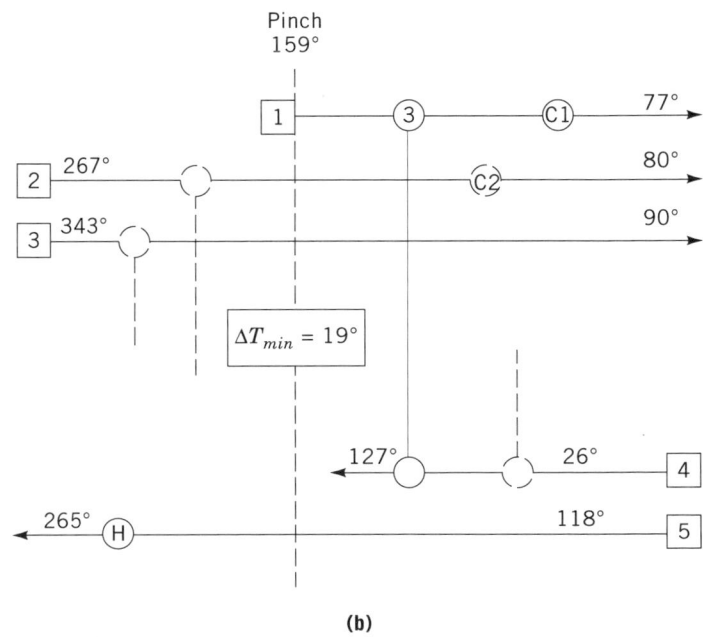

Figure 21. Designing the retrofit. (**a**) The network, initialized for retrofit, highlights exchangers that work across the pinch. (**b**) The cross-pinch exchangers must be eliminated before the network design is developed.

problem. The utility heater, exchanger 1, and exchanger 4 are all reused above the pinch. Below the pinch, exchangers 2 and 3 are reused. The cooler C2 has a reduced duty and exchanger A is new. A computerized optimization program can be used at this stage (eg, Supertarget). Loops and paths can be used to give added flexibility by

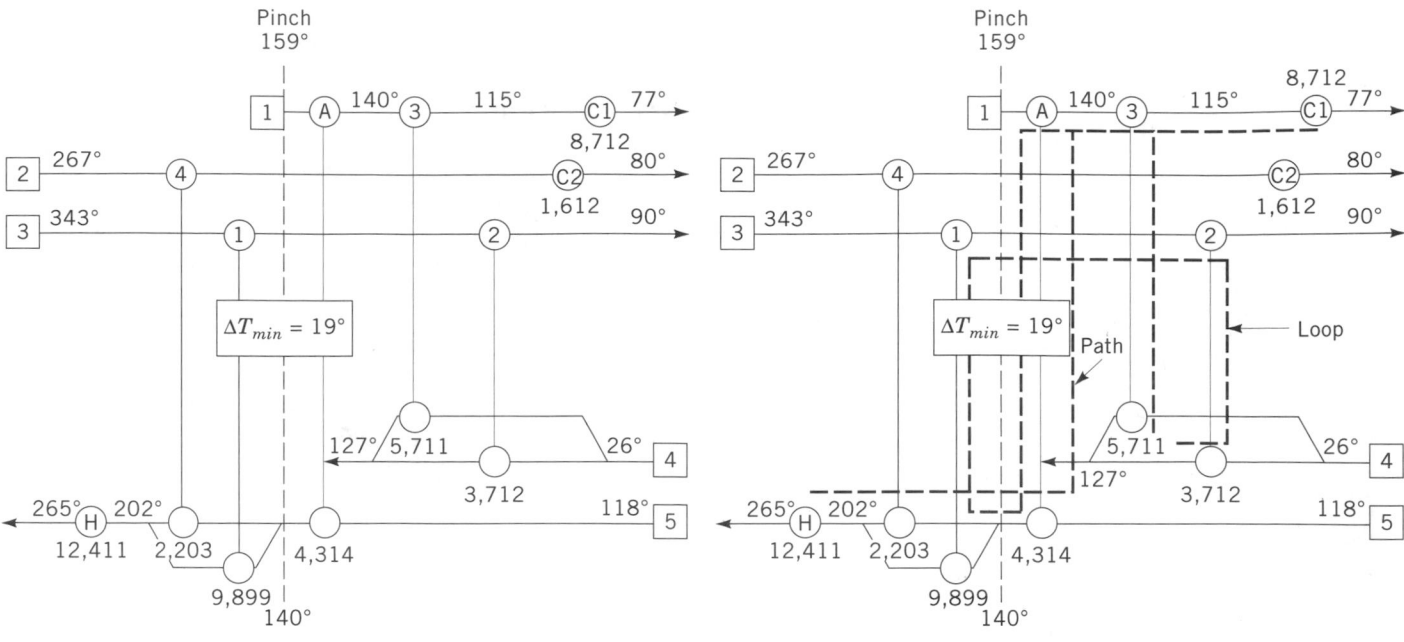

(a)

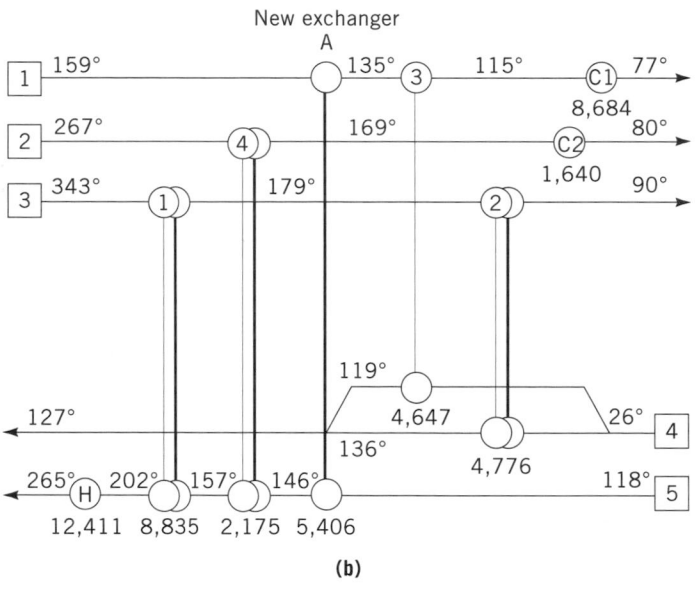

(b)

Figure 22. The completed network. (**a**) The preliminary design involves redeploying existing exchangers and adding new units. (**b**) The improved design employs all existing exchangers and offers a 1.9-yr payback.

identifying process changes to complement network design changes. However, it has also been accepted that opening up the question of process changes is like opening up Pandora's box. There are an infinite number of settings for reactor conversion, evaporator stages, distillation column pressures and reflux ratios, feed vaporization pressures, pump-around flow rates, etc. The number of choices is so large that it seems an impossible goal to predict confidently which three or four such parameters could be changed to advantage in the overall context.

Figure 23. Loops and paths enhance design flexibility, permitting reuse of existing exchangers.

Against this background, the designer now has simple results that allow him or her to discuss with great confidence many different process parameters from the point of view of their ultimate impact on the overall design.

Consider the composite curves in Figure 24**a**. What are the changes to the underlying process heat and material balance that will change the hot and/or cold energy target beneficially? If the targets are given by the process enthalpy balance above and below the pinch, then the designer clearly must try to

Increase hot stream duty above the pinch.
Decrease cold stream duty above the pinch.
Decrease hot stream duty below the pinch.
Increase cold stream duty below the pinch.

These simple guidelines provide a definite reference for the adjustment of single heat duties, such as vaporization of a recycle and pump-around condensing duty, and indicate which modifications would be beneficial and which would be detrimental.

Often it is possible to change temperatures rather than heat duties. It is clear from Figure 24**a** that temperature changes that are confined to either side of the pinch can have no effect on the energy targets. It is also clear that temperature changes across the pinch will shift heat duties from one part of the process to the other. Thus the pattern in Figure 24**b** emerges:

Shift hot streams from below the pinch to above.
Shift cold streams from above the pinch to below.

For example, if the pressure for a feed vaporizer can be chosen, it should be set to fall below the pinch. This prin-

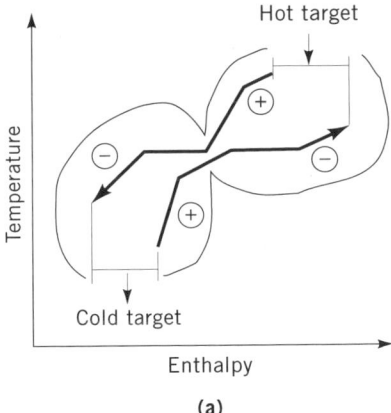

(a)

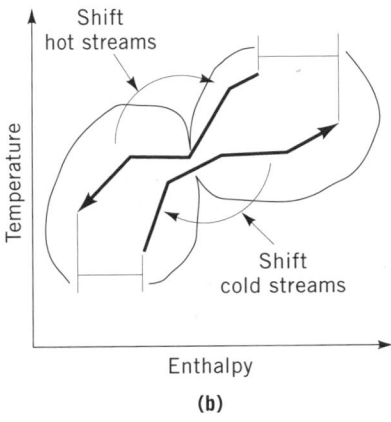

(b)

Figure 24. Modifying the design. (**a**) The energy targets can be modified only if the process stream enthalpy changes are modified. (**b**) Temperature changes can affect the energy targets only if streams are shifted through the pinch.

ciple is in line with the general idea that it ought to be beneficial to increase the temperature of hot streams (this must make it easier to extract heat from them) and that likewise cold streams should be cold. Changing the temperature of streams in this fashion will improve driving forces but cannot possibly decrease the energy targets *un-*

less the temperature changes extend across the pinch. Again, a definite reference is obtained. The designer can predict which modifications would be beneficial, detrimental, and inconsequential. These observations, shown in Figure 24, are summarized as the plus–minus principle (3).

So far the discussion has concentrated on the effect of process modifications on energy costs. Evidently, if capital cost changes are not considered, the conclusions are likely to be a little tenuous. There are two relevant elements of capital cost. First, capital costs will change in the unit operations of the process (eg, the distillation columns and reactors). Second, capital costs will change in the heat exchanger network, which has as yet to be designed. Capital cost changes in the unit operation must be estimated by shortcut calculations. Capital cost changes in the network can be predicted by using network capital cost targets. With a little practice, the principles enable the engineer to screen quickly from among, say, 50 possible modifications 3 or 4 that will lead to beneficial overall cost effects.

A Process Modification Example

Consider the outline flowsheet and the composite curves of the acetic anhydride process (6) shown in Figure 25. The composite curves do not indicate that there is much scope for energy recovery. However, the question that should be asked is, "Are there any modifications to the 'inner' process that can be undertaken?" In the case of the acetic anhydride plant, the plus–minus principle brings into focus the pressure of the feed vaporizer and the distillation columns. This, in turn, suggests a small number of promising process changes (14). One is shown in Figure 26. The pressure in distillation column 1 has been increased to allow feed vaporization by heat recovery (from the distillation column condenser).

The effects of such a design change on distillation column sizing and capital cost is readily determined for different pressure levels. However, this would have been premature in this case, because inspection of the new curves (Fig. 26) immediately raised further possibilities. With the proposed modification, the overheads from the acetone recycle column would be heat exchanged with the feed vaporizer (ie, acetone vaporized) and a simple further modification becomes self-apparent: only condense the overheads from the column partially, feeding the rest, as

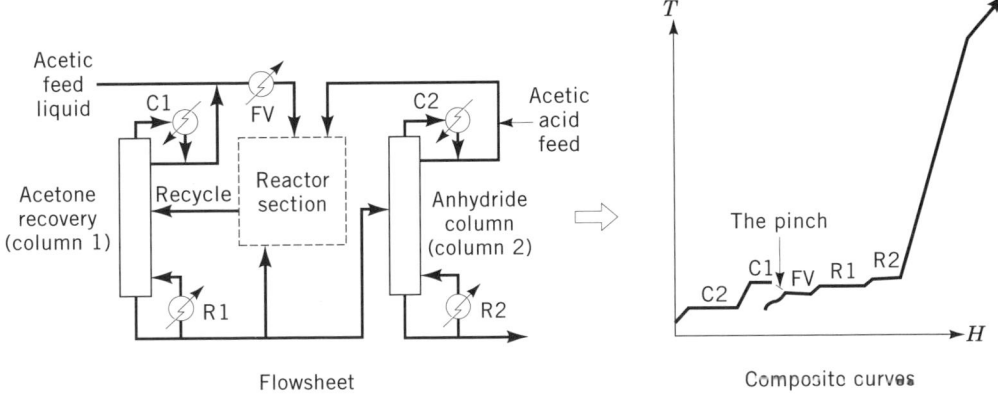

Figure 25. Original flowsheet and corresponding composite curves for the acetic anhydride process.

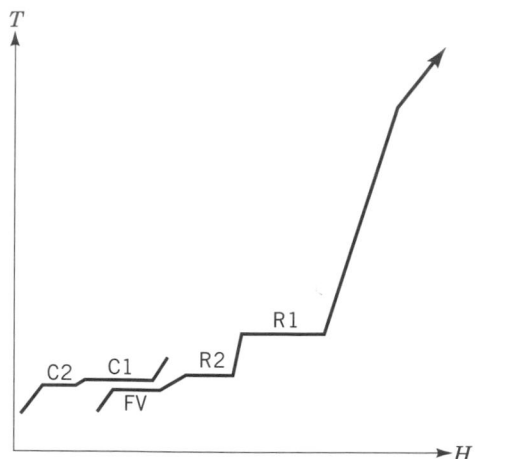

Figure 26. Composite curves obtained after raising the acetone column pressure.

vapor, to the reaction furnace. Then use the fresh process feed to provide the remaining reflux to the column. The resultant flowsheet and composite curves are shown in Figure 27.

The Energy Savings

This simple process modification not only requires less capital investment (no feed vaporizer, smaller overhead condenser) but also saves around 18% of the process's energy consumption. It appears obvious through the systematic application of pinch technology. The composite curves in Figure 27 show, if anything, less scope for heat integration than those in Figures 25 and 26. This is well in line with the main premise, namely that pinch technology has developed from a heat integration tool into an overall process design methodology. Frequently, process changes are the main outcome.

PLACEMENT OF HEAT AND POWER UTILITY SYSTEMS

It would be a rare process that did not use mechanical energy, or power. Pumps, fans, process compressors, and refrigeration systems all use power, not heat, to drive them. Because mechanical energy is generated by heat in the first place, eg, by burning fossil fuels to produce the electricity to drive motors, and because no conversion of energy can ever take place without incurring losses, power is more expensive than heat. Typically, the price ratio between power and heat (or fuel) varies between 3 and 4.

So far, the discussion has only examined heat exchanger networks and their interaction with process conditions, but integrating power or combined heat and power systems correctly into the process is vital in regard to overall energy efficiency. First placing heat engines will be examined. Then extend the principle of heat engine placement will be expanded to distillation columns that act as separation engines. Then the placement of heat pumps will be considered as will shaftwork targeting for refrigeration systems.

Heat Engines: Combined Heat and Power

A heat engine takes in heat from a higher temperature and rejects it at a lower temperature. The difference between the two heat loads goes into mechanical work or power. In practical terms, this is usually turbine systems. There are two main types of turbine systems: stream and gas. A turbine generates power as a hot gas stream at high pressure flows through it. The gas stream at the turbine outlet is at reduced pressure and temperature. The steam turbines use steam as the motive fluid, and gas turbines use combustion gases at high temperature and pressure.

Usually, processes use the heat engines to provide for the process heat as well as power demand. Steam turbine systems are used more often in the process industry than gas turbine systems. This is mainly because steam is a more convenient heating medium than gas. This section considers the use of pinch technology in setting up key design parameters for the heat engines so that the heat engine is appropriately integrated with the process.

Figure 28 shows three possibilities for the placement of heat engines with a process. The process is represented by two regions: one above and other below the pinch. If a heat engine falls across the pinch, it wastes energy (15). The use of the heat engine increases the hot and the cold utility requirements (Figure 28a). The energy is wasted

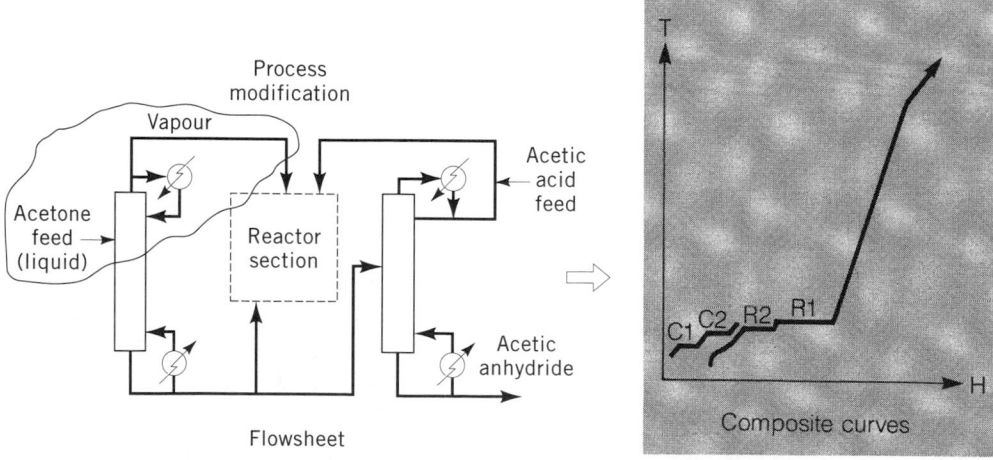

Figure 27. The modified acetic anhydride process.

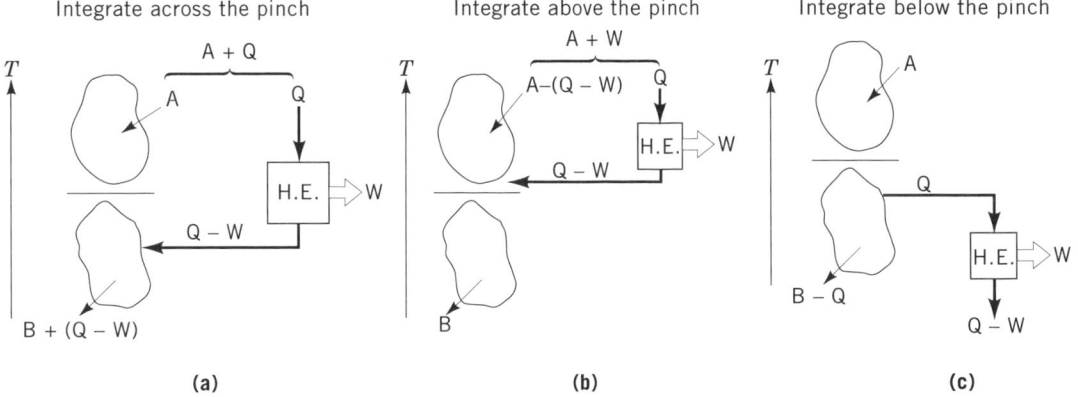

Figure 28. Placement of the heat engines.

because the heat engine would remove heat from the overall heat sink and transfer it to the overall heat source, therefore placing extra demands on both the hot and cold utilities.

If the heat engine is placed so that it rejects the heat into the process above the pinch temperature (Fig. 28**b**), it transfers that heat to the process heat sink, thereby reducing the hot utility demand. Due to the heat engine, the overall hot utility requirement is only increased by W (ie, shaft work). This implies a 100% efficient heat engine. The heat engine is, therefore, appropriately placed.

If the heat engine is placed so that it takes in energy from the process below the pinch temperature (Fig. 28**c**), it can take that energy from the overall process heat source and reject it directly to utilities. Here the engine effectively runs on process heat free of fuel cost and reduces the overall cold utility requirement by W. The heat engine is again appropriately placed.

To summarize, the appropriate placement of heat engines is either above the pinch or below the pinch but not across the pinch. When integrating the heat engines with the process, designers must consider the grand composite curve of the process. Figure 29 shows the use of the grand composite curve in setting the design parameters for steam and gas turbines. The grand composite curve determines the appropriate temperature levels and the heat loads for heat rejection from the steam and the gas turbines.

Placement of Distillation Columns as Separator Engines

Distillation is one of the most commonly used techniques for separation in the process industry. A distillation column separates a mixture into a low boiling fraction and a high boiling fraction. The column accepts heat at higher temperature (ie, reboiler temperature) and rejects heat at a lower temperature (ie, condenser temperature). Thus a distillation column can be treated as a separation engine. The column can be represented by a temperature–enthalpy "box" as shown in Figure 30.

Similar to the heat engines, a distillation column is inappropriately placed if it runs across the pinch as shown in Figure 30**a** (16). There is no benefit in integrating a distillation column with the background process if it runs across the pinch. The hot and the cold utility consump-

tions are both increased. The designer must avoid the temperature overlap between the distillation column and the grand composite curve of the background process. Figure 30**b** shows an appropriate placement of the distillation column achieved by increasing the column pressure. The appropriate placement for a distillation column when the column lies entirely above the pinch or below the pinch and provided that the grand composite curve of the background process can accommodate the column. The integration possibilities between the column and the background process can be improved by introducing column modifications such as side condensing–reboiling and feed preheating–cooling. Recent developments in this area have been discussed (17).

Heat Pumps in the Process Context

A heat pump accepts heat at a lower temperature and, by using mechanical power, rejects the heat at a higher temperature (Fig. 31). Heat pumps, therefore, operate in an opposite manner from heat engines. There are three arrangements that a heat pump may take relative to the pinch (Fig. 31). First, heat can be taken in above the pinch and rejected at a higher temperature, also above the pinch. This saves on hot utilities but only at the expense of an equal outlay in power. Because power is more costly than heat, this is not helpful. Second, the heat pump can take in heat below the pinch and reject it at a higher temperature that is also below the pinch. This is even more obviously unhelpful than the first arrangement. Because the section of the process below the pinch is an overall heat source, the designer is simply changing power into extra waste heat. The third option is to take in heat below the pinch and reject it at a temperature above the pinch. This provides savings in both hot and cold utilities because it is pumping heat from a source (below) to a sink (above) (2,15). So the best place to integrate heat pumps is across the pinch.

Shaft Work Targeting for Low-Temperature Processes

A low temperature process is serviced by a refrigeration system. A refrigeration system is in principle a heat pump that accepts heat at below ambient temperatures and rejects it to the ambient. In doing so it requires shaft work. In the process industry a refrigeration system usually op-

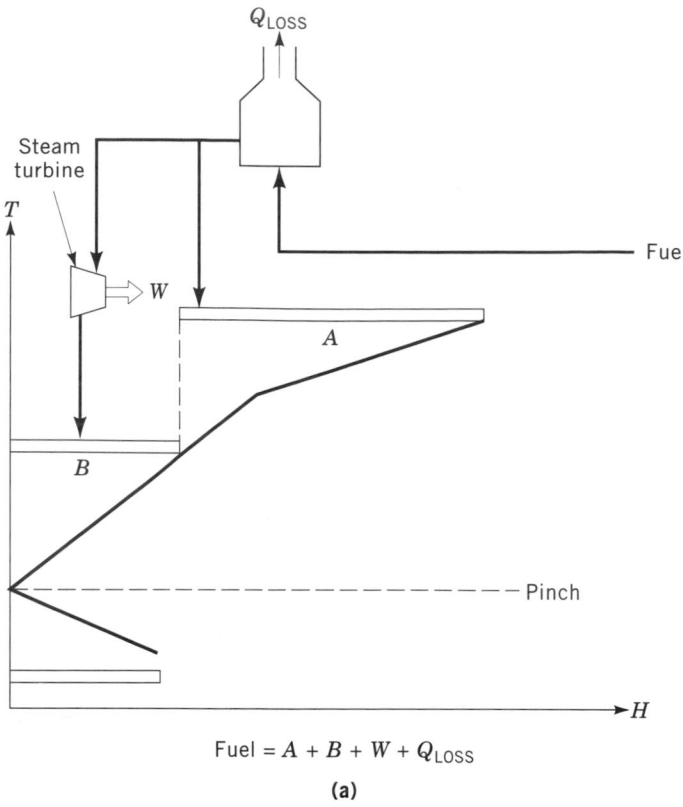

Fuel $= A + B + W + Q_{LOSS}$

(a)

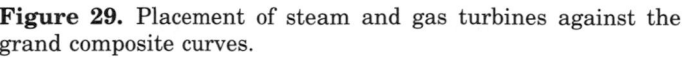

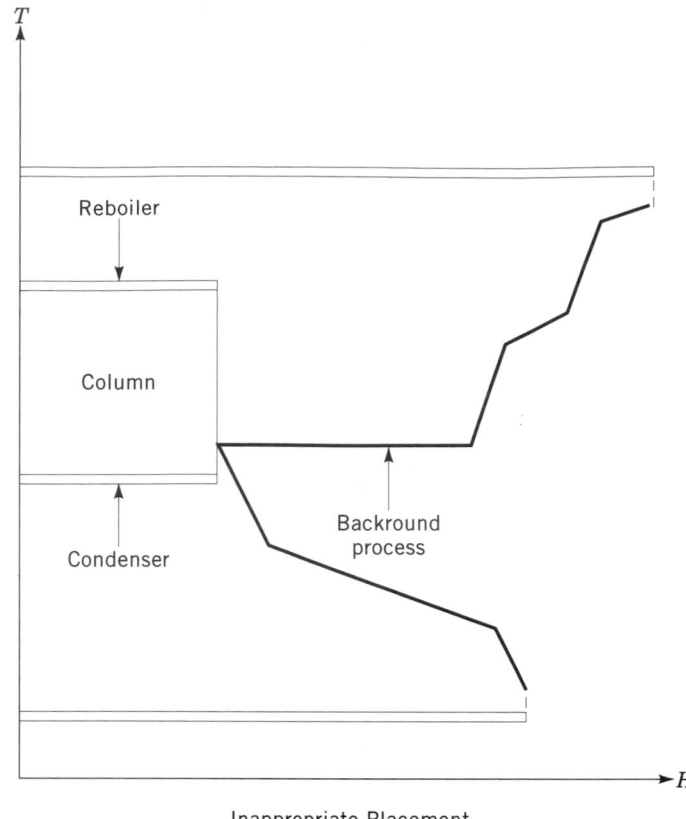

Fuel $= A + B + W + Q_{LOSS}$

(b)

Figure 29. Placement of steam and gas turbines against the grand composite curves.

Inappropriate Placement

(a)

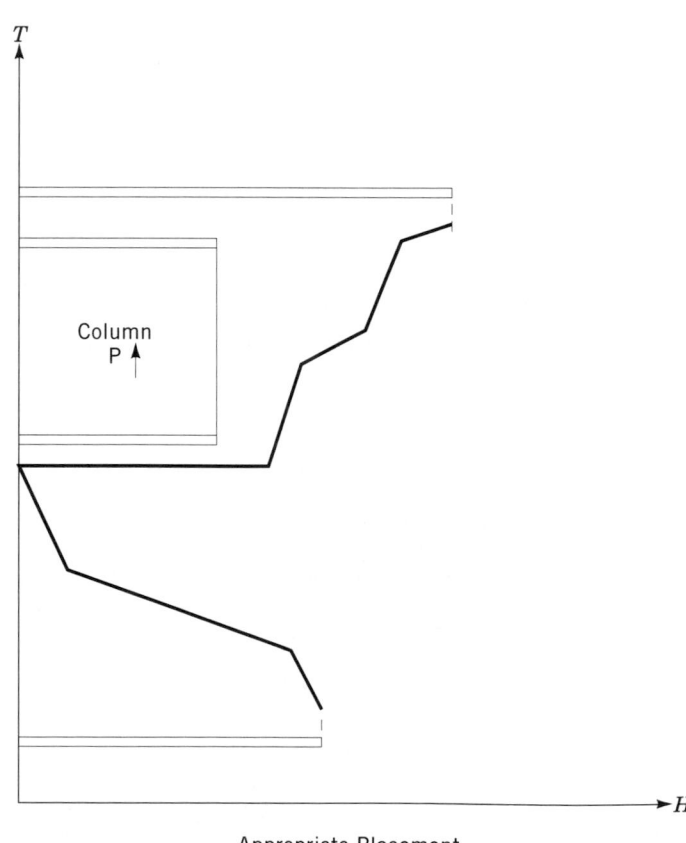

Appropriate Placement

(b)

Figure 30. Appropriate placement of a distillation column against the background process.

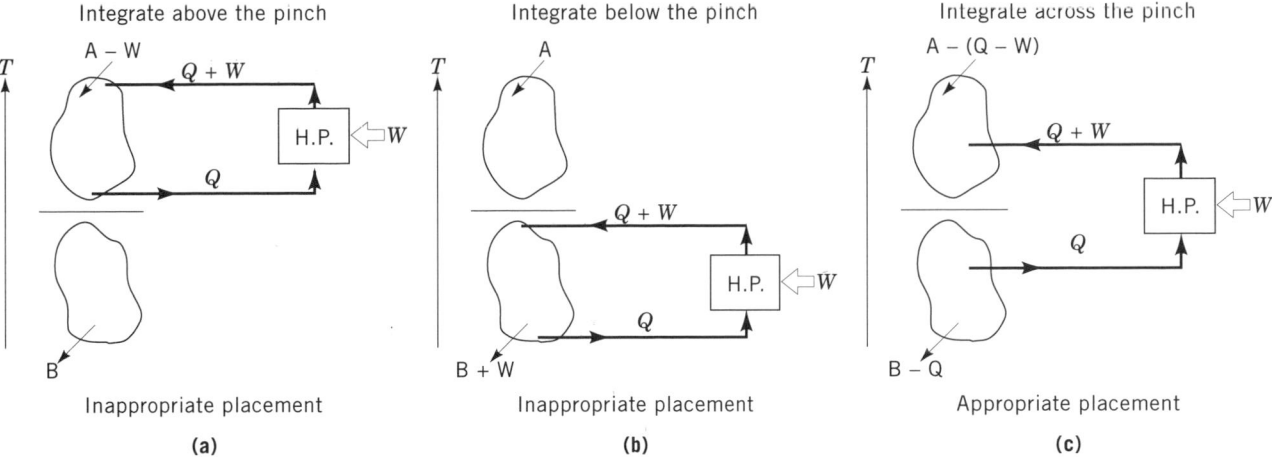

Figure 31. Placement of the heat pumps.

erates through multiple levels of refrigeration and heat rejection. Using the grand composite curve for a low temperature process it is possible to determine appropriate heat loads for refrigeration and heat rejection. Based on these heat loads, the shaft work requirement for the refrigeration system can be determined through simulation.

A recent development in pinch technology provides a technique that bypasses the effort of refrigeration system simulation to predict its shaft work requirement (18). It directly sets shaft work targets for the refrigeration system (Fig. 32). When the temperature axis of the grand composite curve is replaced by the Carnot factor ($\eta_c = (1 - T_o/T)$), the new grand composite curve is termed the exergy grand composite curve. The area between the exergy grand composite curve and the refrigeration levels (shown as shaded areas) is directly proportional to "exergy loss," or ideal shaft work loss in heat transfer. As the number of refrigeration levels is increased, the exergy loss (shaded areas) will reduce (Fig. 32). It can be shown that the change in the shaded area is directly proportional to the change in the refrigeration shaft work (18). Therefore, targets can be set for the change in the refrigeration shaft work consumption for the various options by simply evaluating the change in the shaded areas. This concept can be further extended to minimize the overall refrigeration shaft work requirement of a low temperature process (19).

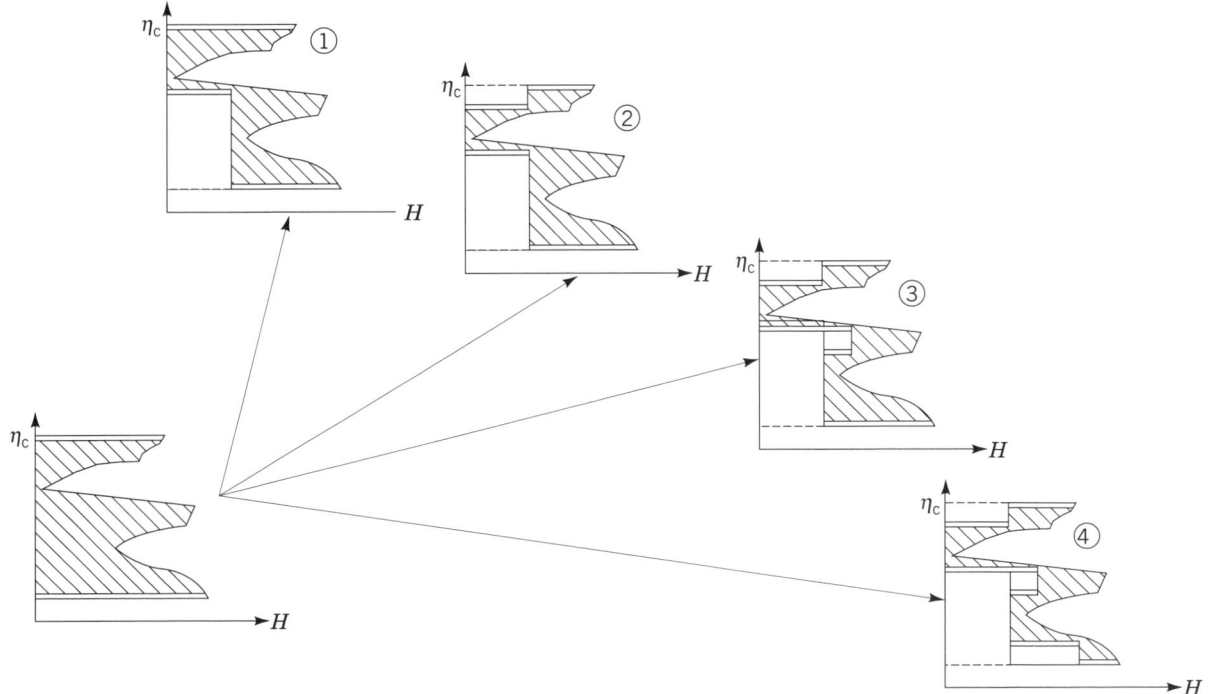

Figure 32. Shaft work targeting for refrigeration systems. The changes in the shaded areas are directly proportional to the shaft work.

TOTAL SITE TARGETING

The discussion so far has concentrated on heat and power targeting for a single process. Typically refinery and petrochemical processes operate as parts of large sites or factories. These sites have several processes serviced by a centralized utility system involved in steam and power generation. Figure 33 shows a schematic of a total site, involving several processes and a central utility system. There is both consumption and recovery of process steam via the steam mains. The utility system consumes fuel (eg, gas and lignite), generates power, and supplies the necessary steam through several steam mains. The site imports or exports power to balance the power generated.

The process's steam heating and cooling demands dictate the sitewide fuel demand and co-generation potential via the utility system. The heating and cooling requirements of the individual processes are represented by the respective grand composite curves. The grand composite curve lays open the process–utility interface for a single process. However, for a site involving several processes, the grand composite curve of each process will suggest different steam levels and loads. How can the correct compromise in steam levels on a sitewide basis be identified?

Total Site Profiles: Sitewide Utilities Targets

A recent development in pinch technology enables the designer to set utilities targets for total sites. This development is based on thermal profiles for the entire site called total site profiles (20). Total site profiles are constructed from the grand composite curves of the processes in the site. Figure 34 illustrates the construction of the total site profiles, using two processes. The construction starts with the grand composite curves of the individual processes. The grand composite curves are then modified in two steps. (1) The nonmonotonic parts, or the "pockets," in the individual grand composite curves are sealed off through vertical lines. (2) The source and sink elements of the resulting grand composite curves are shifted by $1/2\Delta T_{min}$ (Fig. 34). Source element temperatures are reduced by $1/2\Delta T_{min}$ while the sink element temperatures are increased by $1/2\Delta T_{min}$.

The construction of the total site profiles from the modified grand composite curves is shown in Figure 34. The construction simply involves the generation of composite curves from the source and sink elements of the modified grand composite curves. The composite of the source elements is called the site source profile and that of the sink elements is called the site sink profile. The individual sink elements (A, B, D, E, and F) and the source elements (C and G) have been highlighted in Figure 34.

Figure 35 illustrates an approach for setting the co-generation target using a plot of Carnot factor versus enthalpy. To demonstrate the method, a simple VHP to HP turbine system is considered. The exergy (ideal shaft work equivalent) of the VHP steam equals the shaded area 1-2-3-4, while the exergy of the HP steam equals the shaded area 4-5-6-7. The net ideal shaft work from VHP to HP equals the exergy difference, ie, the shaded area 1-2-3-7-6-5. The real shaft work is proportional to the ideal shaft work, taking into account the exergetic efficiency of the turbine system, n_{ex} (18). Thus the target for co-generation is shaded area 1-2-3-7-6-5, multiplied by n_{ex}.

Figure 36 illustrates sitewide utilities targeting based on total site profiles. The figure also shows the corresponding utilities schematic. Note that the vertical axis has been changed from temperature to the Carnot

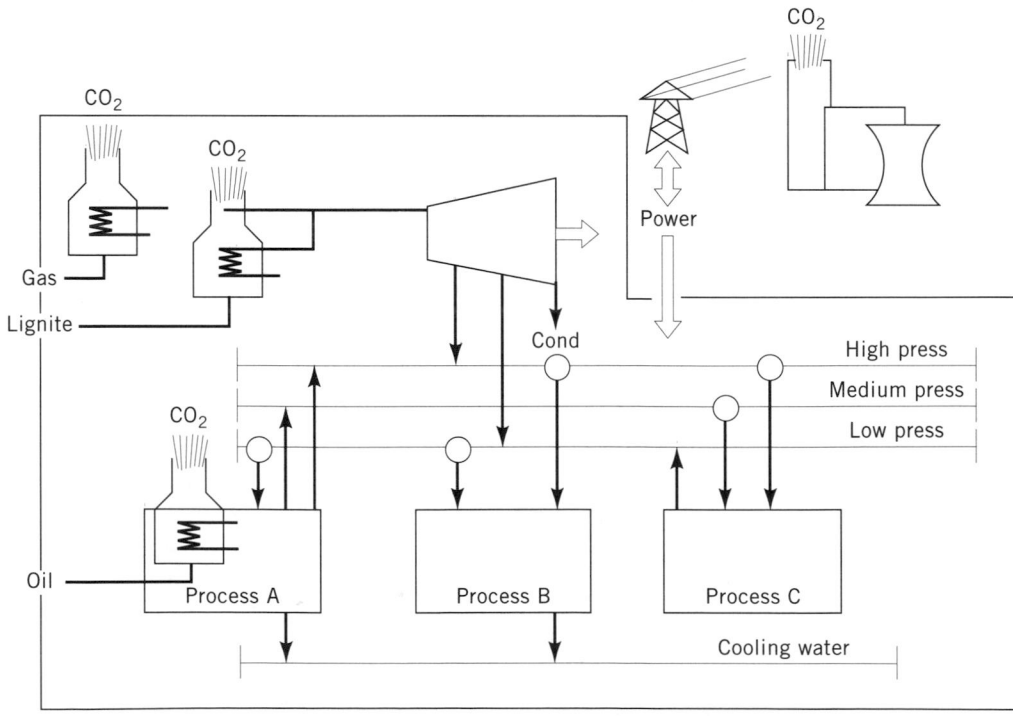

Figure 33. Schematic of a typical total site.

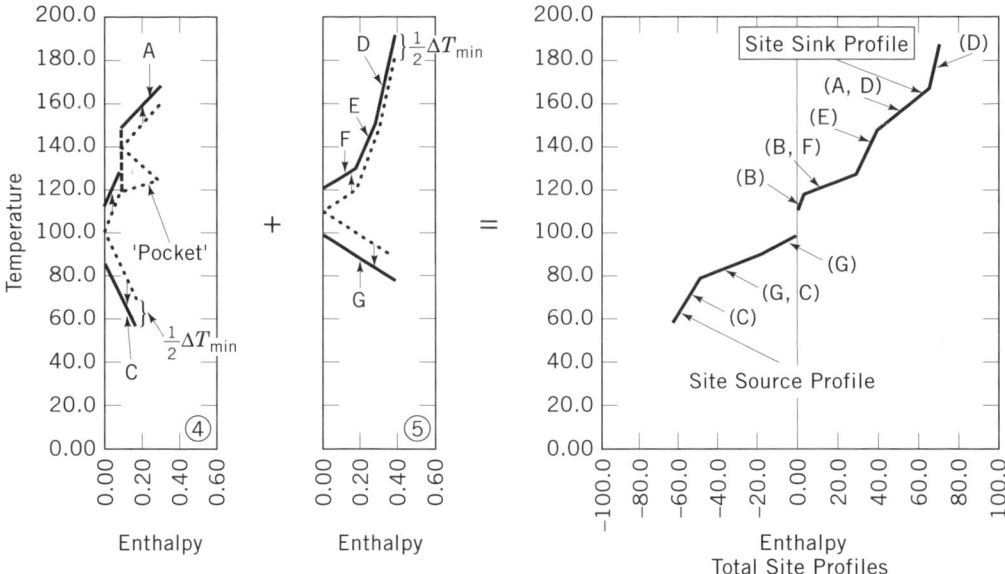

Figure 34. Constructing total site profiles from the process grand composite curves.

factor. This makes it possible to set direct targets for co-generation. The target for MP steam generation from all the processes is D. The remaining targeted MP demand (C) is provided by the turbine system. The targeted HP steam demand for all the processes is B. Thus the turbine system must satisfy the MP (C) and HP steam (B) demands. Because of the vertical axis factor, the shaded area is proportional to co-generation (W). Thus it is possible to identify the generating steam requirement (A) at the VHP level. The VHP load sets the target for fuel requirement of the boiler (F) as a horizontal projection of the fuel profile on the enthalpy axis. Knowing the fuel composition, the designer set fuel-related emissions targets. The cooling demand for the total site as identified by the total site profiles is E. Thus starting from individual processes grand composite curves the designer can set targets for site fuel, co-generation, site emissions, and cooling (20).

Case Study: Total Site Integration

The approach developed above can be used to identify design modifications in processes or in the utility system

based on a coherent approach (20,21). Consider a case study to illustrate the use of total site profiles in a total site project. Figure 37 shows the existing utility placement on the total site profiles for the example site. There is a high pressure steam main that receives steam from the central turbine system. The medium pressure main is not connected to the turbine system; rather steam generation from processes and consumption by processes at MP level are in balance. The construction reveals an overall cross-pinch heat transfer (in all processes and the utility system) of 40 MW.

A site expansion is planned that involves adding a new process to the existing site. Figure 38 describes the proposed expansion plan with total site profiles and a utility sketch. Both the source and the sink profiles are enlarged as a result of the expansion. In the case study the new process was subjected to a careful individual optimization, resulting in a minimum energy requirement of 70 MW of MP steam. That steam is to be supplied via a new turbine, which links into the existing VHP and MP mains. To generate the additional VHP steam required, a new boiler is planned. An expansion of the cooling water system also is necessary (from 240 to 280 MW).

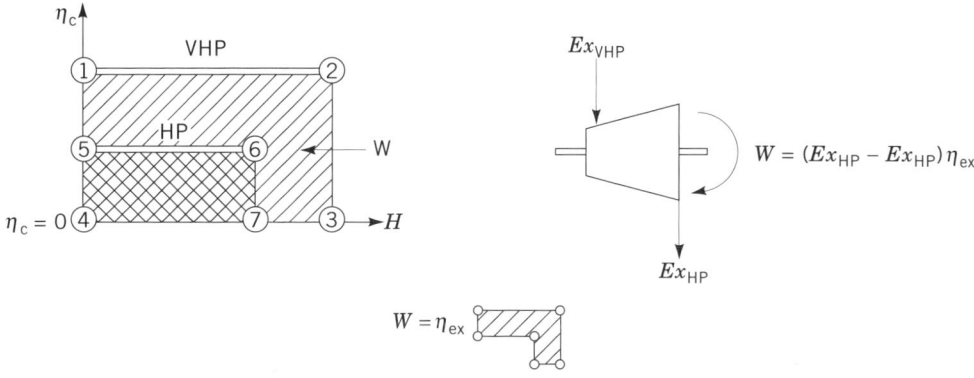

Figure 35. Shaft work targeting for co-generation.

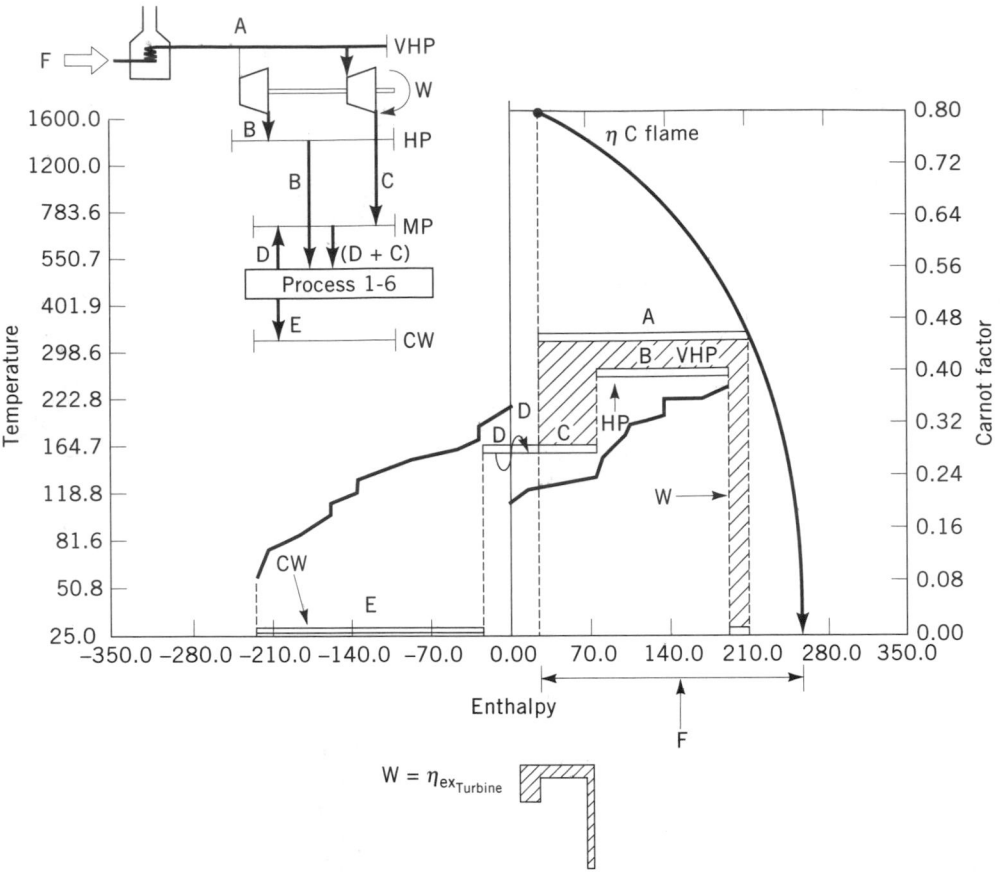

Figure 36. Total site targeting for fuel, co-generation, emissions, and cooling.

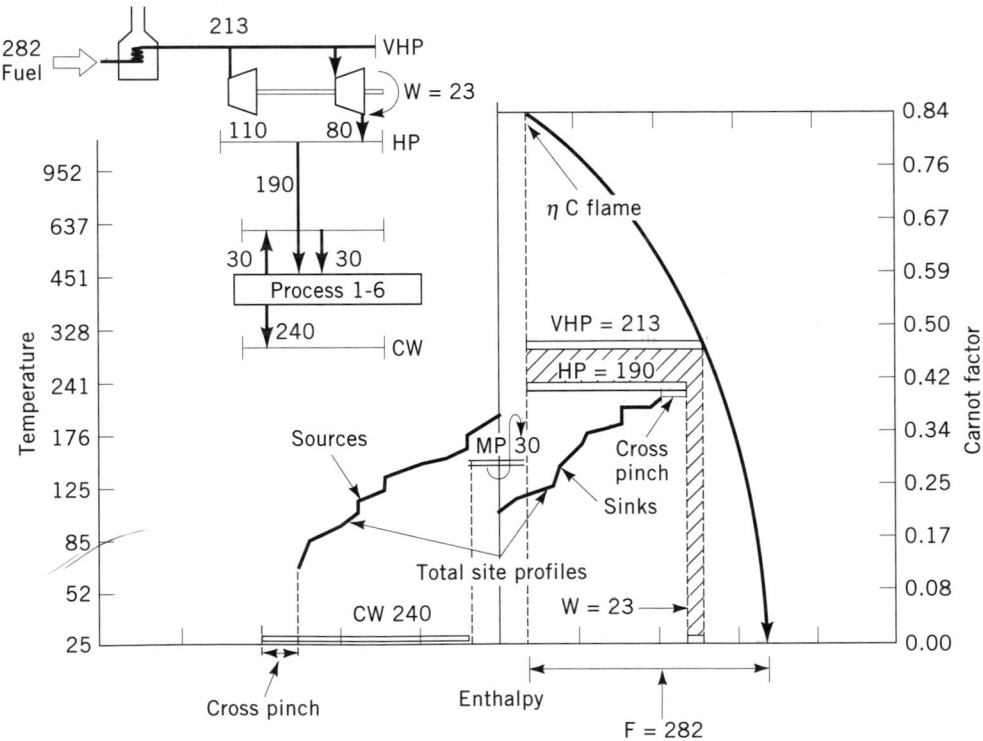

Figure 37. Total site profiles for the example site.

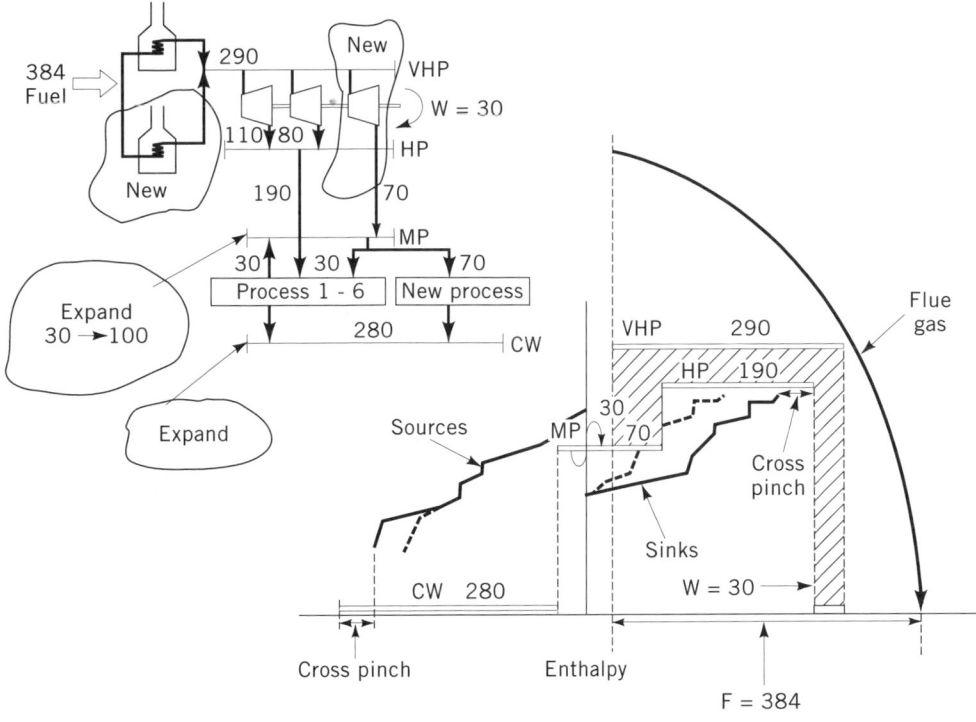

Figure 38. Proposed expansion of the site, involving the addition of a new process.

The total site profiles, however, reveal additional information. Because the profiles have changed significantly as a result of the expansion, the pressure of the MP steam is no longer appropriate. It is apparent that MP steam does not exploit the driving forces available between the steam levels and the site sink profile. The user of total site pinch analysis will clearly recognize that MP steam pressure must be reduced. This is implemented in Figure 39. The reduction in the MP steam pressure has been selected so that MP steam generation is in balance with its use *and its load is maximized.* Only a small pressure shift was found to be necessary. The VHP steam demand is reduced significantly.

Comparing the proposed expansion plan (Fig. 38) with the alternate expansion plan a significant reduction in energy as well as capital costs can be seen. The alternative

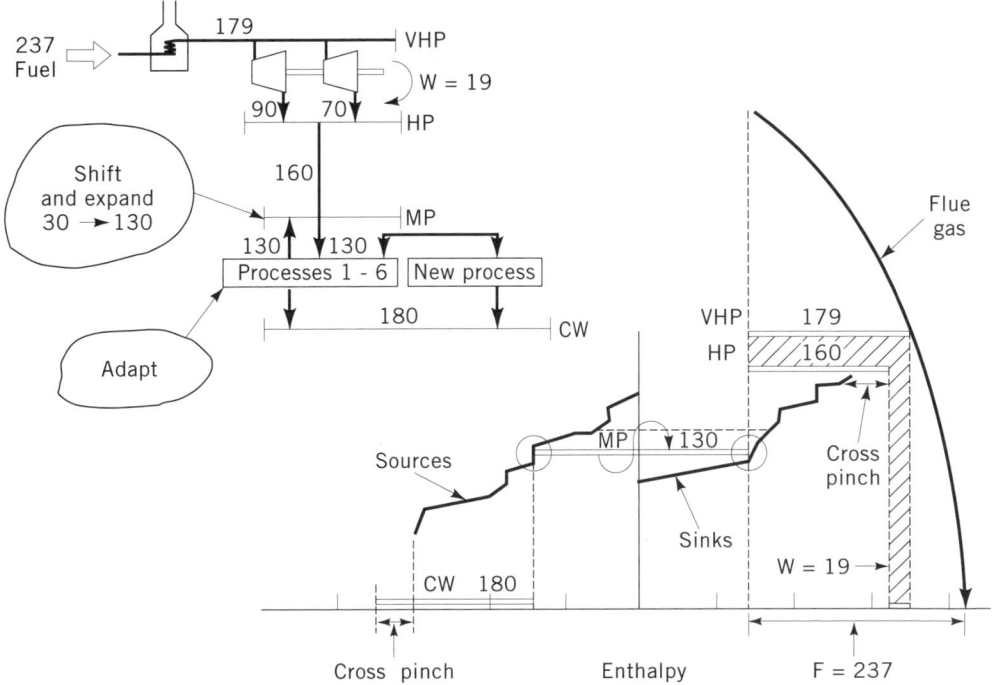

Figure 39. An alternative option based on the total site profiles.

plan does not require investment into a new turbine or a new boiler. It also does not require any expansion in the cooling capacity. Table 1 compares the operating and capital costs for the proposed expansion and the alternative plan.

Total site integration has been a major breakthrough in pinch technology. Within the short time since its development, it has been used successfully in several applications (22,23). The new approach enables the designer to set sitewide targets for fuel, co-generation, emissions, and cooling ahead of design. The targets allow the designer to analyze before design various site-related proposals with reference to site expansion, process and utilities modification, etc, both in grassroots and in retrofit situations.

TARGETING FOR EMISSIONS

This article has summarized the techniques used to target the amount of energy a plant or a site needs. Remember that the principal aim is to use the energy-efficient designs to improve the environmental performance of plants; thus how can the designer target for the specific pollutants that relate to energy consumption?

Defining Emissions Targets

Historically, air pollution control was established to tackle local problems like smog in the Los Angeles basin or in London. It is now recognized that problems such as global warming and acid rain must be tackled on a worldwide basis, with absolute limits placed on emissions. It is preferable to define the emissions of specific pollutants in terms of mass flow rate rather than concentration, which can be misleading (24). It is also important to look at the total energy-related emissions arising from the operations at a site on a global basis, ie, taking into account emissions from the power station supplying the site, rather than just emissions at the site itself.

Modeling for Specific Pollutants

Four pollutants will be considered:

1. Carbon dioxide (implicated as a major contributor to global warming).
2. Sulfur oxides (a major contributor to acid rain).
3. Nitrogen oxides (a factor responsible for acid rain).
4. Particulates (lead to smog and health hazards).

Table 1. Comparison of Operating and Capital Costs

Costs	Proposed Model	Alternative Model
Total utilities operating cost	100%	81%
Capital cost implications	· Expand MP	· Shift MP pressure and expand
	· New turbine	
	· New boiler	· Adapt processes to suit
	· Expand cooling	

Carbon dioxide emissions are directly related to the fuel burned via a straightforward stoichiometric model:

$$C_nH_m + (n + m/4)O_2 \rightarrow n\,CO_2 + (m/2)H_2O$$

For predicting the amount of CO_2 it is usually accurate enough to assume that the combustion system is an efficient one with enough excess air added to the fuel to ensure that carbon monoxide emissions are negligible. In other words, all the carbon in the fuel results in CO_2.

The same type of model can be used for sulfur oxides:

$$C_nH_mS_p + (n + m/4 + p)O_2 \rightarrow n\,CO_2 + m/2\,H_2O + p\,SO_2$$

Some of the sulfur dioxide will also react to form SO_3, but because it forms roughly only 10% of the total sulfur oxide emissions, to predict the amount of SO_2, it is usually good enough to assume that SO_2 is the only product.

The situation with nitrogen oxides is more complex. Nitrogen oxides (NO_x) form via two main reaction paths (25). Thermal NO_x is formed from nitrogen in the combustion air, particularly at high temperatures:

$$N_2 + O_2 \rightleftharpoons 2\,NO$$
$$NO + \tfrac{1}{2}O_2 \rightleftharpoons NO_2$$

Fuel-bound NO_x is formed from nitrogen in the fuel at low as well as high temperatures. But part of the fuel nitrogen reacts directly to form N_2. To complicate things even more, N_2O and N_2O_4 are also formed in various reactions. The mixture of NO_x in stack emissions depends on kinetics, mixing, mass transfer, and thermodynamics.

It is virtually impossible to calculate a precise value for the NO_x for a real device. Modelers tend to use estimates based on experimental observations for different combustion technologies such as boilers, gas turbines, and fired heaters (26). At best, NO_x predictions are usually good enough only to indicate trends rather than precise figures.

Particulates derive from two sources. First, metals may be present in the fuel, which oxidize during combustion. Second, unburned hydrocarbons and carbon may exist. As with NO_x, kinetics, mixing, and thermodynamics will influence the amount of fuel that may remain unburned. It is usually sufficient to consider only the metals, assuming that all metals present in the fuel and inorganic ash are emitted after combustion.

Analyzing the Utility Options

It is possible to relate the emissions to the amount of fuel used. Thanks to pinch technology, it is also possible to calculate the amount of energy a utility system must supply. This still leaves a gap: it is necessary to model the utility systems, such as furnaces or turbines, to relate the amount of energy they product to the amount of fuel they consume. There are simple models available in the literature for furnaces and steam boilers and gas and steam turbines (2,24).

The method of relating utilities to the process using the grand composite curve and total site profiles has been discussed. By adding the models for utility fuel consumption and the emissions from the given fuel, the designer can use already established techniques to target directly for on-site emissions. Computers obviously provide the simplest means of putting this into practice.

A more global view can be taken if there is on-site power generation from the utility turbine system. While the turbine is creating emissions on site by placing extra demands on the furnace or boiler, it is reducing the emissions off site by cutting down on the amount of power that must be imported from the external supply grid (24,27–31). And because the big power generators may be operating less efficient plants (or using a different, dirtier fuel such as coal instead of gas) the global emissions from an on-site power device can even turn out to be negative. The use of total site profiles provides a direct interrelation between the process changes, fuel switching alternatives, and heat exchanger network modifications on local and global CO_2 (20,21).

THE ROLE OF SOFTWARE

The sheer size and complexity of most practical problems demands the use of software of some kind. In the implementation of an energy integration study it is possible to identify two broad activities that call for the use of software: targeting and network design.

Targeting

Targeting uses physical principles such as thermodynamics to define key parameters for the energy system before design. The parameters that can be targeted before design include the following:

- Minimum energy consumption.
- Minimum heat-exchanger surface area.
- Minimum number of heat exchangers.
- Minimum total cost of the heat-exchanger network (capital and energy).
- Minimum shaft work required by refrigeration systems.
- Maximum shaft work generated by co-generation systems.
- Minimum flue gas emissions (CO_2, SO_x, and NO_x).

The actual design task can be lengthy and it is usually impractical to study many design alternatives by carrying designs through in detail. The power of targeting is that it can be used to screen many design alternatives without having to resort to the design task itself. Thus, during the course of a study, targeting will be carried out many times. Targeting will be used to screen different choices of utility systems, co-generation systems, process changes, etc. Targeting will also be used to explore the sensitivity of options to future changes in energy prices, environmental regulations, etc.

Targeting clearly calls for software to explore these alternatives and sensitivities However, because effective targeting also calls for creativity on the part of the designer to obtain the best results, it is important that the targeting software should be interactive with a graphical user interface.

Network Design

Once the major design parameters have been initialized during targeting, the task turns to network design. In network design, the topology of the heat-exchanger network and utility system is created. From the targeting, the designer knows what to expect in terms of the major design parameters such as energy consumption. Knowledge of the targets allows the validity of the design to be immediately assessed. More important, the targets allow the design to be initialized at the correct settings.

Two types of software are used at the network design stage. The designer can interact with a graphical presentation of the design, adding heat exchangers one at a time, while following certain rules and heuristics. The role of the software in this interactive mode is to carry out an evaluation of the design as pieces of equipment are added. Alternatively, the designer can use software that attempts to carry out an automatic network design. With large complex networks automated network design is attractive. However, the designer is removed from the decision-making processes, which can lead to designs that have some undesirable features.

Two different approaches can be used for targeting and network design.

Pinch Technology

Pinch technology as reviewed here has a philosophy of analysis and design of energy systems that is based entirely on physical principles. Targeting uses thermodynamics, graph theory, basic heat transfer theory, and economics to predict the key design parameters. Design relies on the decomposition of the problem into subproblems at the pinch (or pinches). In addition to the decomposition, thermodynamic rules are applied as the network is built up. A key feature of the approach is that it relies on the intervention of the designer. This has advantages and disadvantages. The principal advantage is that the designer can address the many intangibles, such as safety and layout, as the design evolves, maximizing the practicality of the final design. The major disadvantage of the intervention is that the designer must be an expert in pinch technology to obtain the best results.

Mathematical Programming

When mathematical programming is used, the intervention of the designer is replaced by an optimization algorithm that makes discrete decisions. This approach starts with the construction of a superstructure. The superstructure is a design that has all feasible alternatives embedded within it. The superstructure is then subjected to optimization. Should a structural parameter be optimized to zero, that feature is deleted from the design. The major disadvantage of this approach is that the designer is removed from the decision-making process, and so the many intangibles of process design cannot be accounted for.

The two types of software used in targeting and network design have advantages and disadvantages. It is likely, therefore, that each has a role to play. Ideally, the software should allow the user to use both approaches. There are likely to be parts of the design in which the designer would wish to intervene and parts where this would not be required. Finally, once a topology has been defined for the heat-exchanger network and utility system, whether this be by pinch technology or mathematical programming, then design detail must be added using simulation software. The simulation software should be based on detailed models of individual units to allow a proper evaluation of the topology that has been created. The detail incorporated within the simulation software allows more detailed sizing and costing of the unit operations. The degrees of freedom in the network can then be varied and the design subjected to detailed optimization.

BIBLIOGRAPHY

1. B. Linnhoff, *Chem. Eng. Res. Des.* **61** (July 1983); latest version in *Chem. Eng. Prog.* (Aug. 1994).
2. B. Linnhoff and co-workers, *A User Guide on Process Integration for the Efficient Use of Energy,* The Institution of Chemical Engineers, Rugby, UK, 1982.
3. B. Linnhoff and D. Vredeveld, *Chem. Eng. Prog.,* 33–40 (July 1984).
4. B. Linnhoff and V. Sahdev in *Ullmann's Encyclopedia of Industrial Chemistry,* 1987, Vol. B3, pp. 13-1–13-6.
5. B. Linnhoff, G. Polley and V. Sahdev, *Chem. Eng. Prog.,* 51–58 (July 1988).
6. B. Linnhoff and G. Polley, *Chem. Eng.* 25–32 (Feb. 1988).
7. S. Ahmed and B. Linnhoff, *Comput. Chem. Eng.* **14**(7), 751–767 (1990).
8. F. Harary, *Graph Theory,* Addison-Wesley Publishing Co., Inc., Reading, Mass, 1972.
9. L. Grimes, *The Synthesis and Evolution of Networks of Heat Exchangers That Feature the Minimum Number of Units,* Master's Thesis, Carnegie-Mellon University, Pittsburgh, Pa., 1980.
10. D. Townsend and B. Linnhoff, paper presented at the IChemE annual research meeting, Bath, UK, 1988.
11. B. Linnhoff and E. Hindmarsh, *Chem. Eng. Sci.* **38,** 745–763 (1983).
12. T. Tjoe and B. Linnhoff, *Chem. Eng.* 47–60 (Apr. 1986).
13. T. Tjoe, Ph.D. dissertation, University of Manchester Institute of Science and Technology, UK, 1986.
14. B. Linnhoff and S. Parker, paper presented at the IChemE annual research meeting, Bath, UK, 1984.
15. D. Townsend and B. Linnhoff, *AIChE J.* **29**(5), 742–748 (1983).
16. R. Smith and B. Linnhoff, *Chem. Eng. Des. Res.* **66,** 195–228 (May 1988).
17. V. Dhole and B. Linnhoff, *Comput. Chem. Eng.* **17**(5–6), 549–560 (1993).
18. B. Linnhoff and V. Dhole, *Chem. Eng. Sci.* **47**(8), 2081–2091 (1992).
19. V. R. Dhole and B. Linnhoff, *Comput. Chem. Eng.* **18**(Suppl.), S105–S11 (1994).
20. V. Dhole and B. Linnhoff, *Comput. Chem. Eng.* **17**(Suppl.), S101–S109 (1993).
21. B. Linnhoff and V. R. Dhole, *Chem. Eng. Tech.* **16,** 252–259 (1993).
22. R. Davidson, *Eur. Chem. News Special Report* (July 1992).
23. Y. Natori, paper presented at the IEA Workshop on Process Integration, Gothenburg, Sweden, 1992.
24. R. Smith and O. Delaby, *Trans IChemE* **69** (Part A), 495–505 (1991).
25. J. Glassman, *Combustion,* 2nd ed., Academic Press, Inc., Orlando, Fla., 1987.
26. Mills and co-workers, A Summary of Data on Air Pollution by Oxides of Nitrogen Vented from Stationary Sources. *Final Report, Report No. 4,* Emissions of Oxides of Nitrogen from Stationary Sources in Los Angeles County, California, 1961.
27. R. Smith and E. Petela, *Chem. Eng.* (Oct. 1991).
28. R. Smith and E. Petela, *Chem. Eng.* (Dec. 1991).
29. R. Smith and E. Petela, *Chem. Eng.* (Feb. 1992).
30. R. Smith and E. Petela, *Chem. Eng.* (Apr. 1992).
31. R. Smith and E. Petela, *Chem. Eng.* (July 1992).

COOLING TOWERS

ROBERT BURGER
Burger Associates
Dallas, Texas

Cooling towers are used in industry worldwide for chemical plants, power-generating stations, refrigeration/air-conditioning systems, and manufacturing processes, primarily to reject waste heat generated by this equipment. Electric-generating plants, whether they be a small 100-megawatt (100-MW) installation for an apartment house complex that burns fossil fuel or a 1,000-MW nuclear steam–electric facility for a large public utility, possess one common cooling tower denominator. Chemical plants throughout the country yield a limitless number and variety of utilized and necessary products. In these chemical manufacturing operations, there is that one cooling tower common denominator. In a commercial office building the central air-conditioning/refrigeration system can be rated at 3,000 t circulating 34,065 liters per minute (LPM) (9,000 gallons per minute, GPM) of cooling water, whereas a small-town shoe store cools its premises with a 10 t air-conditioning system circulating 114 LPM (30 GPM). However, again, there is one cooling tower common denominator. This common denominator thread that runs through all cooling towers is the removal of waste heat.

See also ELECTRIC POWER GENERATION; ENERGY MANAGEMENT; AIR CONDITIONING; BUILDING SYSTEMS.

In the above examples, waste heat is absorbed by water having a temperature colder than the process. This hot water must now be either discharged into a body of water, or cooled and recycled. The alternative similar to all systems is that the hot water temperature is lowered in a mechanical device called a cooling tower. The hot water is cooled by the exchange of sensible heat and the release of latent heat of vaporization in the cooling tower for its return trip to the process. The waste heat is then rejected into the atmosphere, whereas the cold water is then recirculated, thereby conserving water resources.

The problem is enormous, as industry uses approximately 1,900 billion liters of water a day, which if not recirculated, would seriously deplete the available water supply. The unwanted heat from the water must be restored to the environment without creating ecological havoc or adding unnecessary expenses to the inflationary spiral (1). The development of cooling towers solved the problem of possible ecological damage of returning hotter water to the environment and enabled industry to recycle the greater proportion of the water it requires at reasonable capital investment costs.

Operating procedures, costs, and upgrading potential for greater efficiencies must be constantly investigated to provide economic returns for the installation. To achieve this intelligently, knowledge of the fundamental principles of how the cooling tower performs, and identification and understanding of the various elements, are necessary to be able to upgrade them. A comprehensive maintenance program must be instituted to obtain maximum utilization of the equipment.

OPERATIONAL RESPONSIBILITY

Power plant operators understand the impact that poorly performing water cooling towers can have on overall plant operations, both in efficiency and cost. However, often the evaluated operating cost is not considered when purchasing the low bidder's equipment. It is therefore the operating engineers' function to utilize equipment as efficiently as possible to produce profits for management (2). Inefficient towers can place an enormous cost burden on plant operation; upgrading can change this potential loss picture to one of profit.

Whereas the cooling tower uses energy to generate cooling air, the main area of profit to analyze is the utilization of the cold water. In the refrigeration/air-conditioning system, the colder the water, the less energy is required to power the equipment. In chemical product manufacturing plants, the colder the water, the more efficient the condensation and the greater the volume of saleable condensate product yielded for the facility at lower costs. In power plants, colder water reduces the electrical generation penalty (the cost of electricity used in the process of producing saleable or usable power) (3).

The second area of profit to consider is the electrical and/or steam dollar expenditure required to power the systems. The cooling tower is a high energy utilization machine as well as a mass and heat transfer device.

After the cooling tower has been erected, proper operation and maintenance is required to yield the design conditions of cooling. By not maintaining towers adequately, they can lose appreciable performance.

IMPORTANCE OF WET BULB TEMPERATURE

Specifying and installing a cooling tower can be a mysterious undertaking. The uninitiated user is often at the mercy of the cooling tower salesperson, and the owner may end up with a tower that does not meet plant needs. A little knowledge, however, about design conditions and tower terminology will help to protect the user from purchasing an inadequate heat transfer system. The following points are intended to provide the user with some basic information about building or retrofitting a cooling tower and to help assure that the equipment performs as expected.

Design Conditions

Design conditions of a cooling tower reflect the amount of water circulating per minute, at the thermal level the tower is required to cool. For example, in Figure 1, 22,710 LPM (6,000 GPM) of water enters the tower at a hot water temperature (HWT) of 41°C and leaves at a cold water temperature (CWT) of 29°C. The ambient wet bulb temperature (WBT) for that particular geographic area is 26°C. The cooling range (ΔT) equals the HWT minus the CWT (in this case, 12°C). The approach to the WBT is the CWT minus the WBT, in this case, 3°C (4).

Importance of Design Conditions

Design conditions are the criteria for specifying the level of thermal operation of the tower; in other words, how much work it does. When the tower operates at 100% of its design conditions, the user obtains the total performance paid for, particularly when the WBT equals the design level. Performance at anything less than 100% at design conditions indicates the user was short-changed.

However, design conditions are often difficult to determine. One difficulty is the inability to measure the level of heat rejection, which is a complicated procedure. A tower may achieve the design range and approach to the wet bulb temperature, but if the WBT is less than the design temperature, the tower is not operating at 100%. Constantly fluctuating design variables and interchanging relationships make it difficult to compare an operating result to the design conditions. A thermal example of this concept is shown in Figure 1. It is unlikely that 22,710 LPM (6,000 GPM) flow through the tower constantly; the amount can actually fluctuate by ± 1,892 LPM (500 GPM). If only one pump is operating, 1,135 LPM (3,000 GPM) may be circulating. Also, the incoming temperature is rarely exactly 41°C and can vary.

The temperature varies several degrees if the process is not up to full load. Because the CWT depends on the volume of water circulating and the entering HWT, the

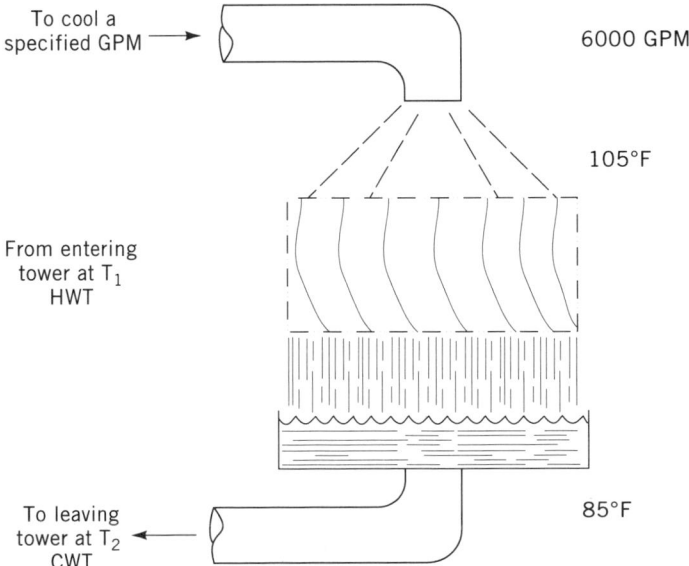

Figure 1. Components of design conditions.

CWT varies as these criteria change. The WBT is critical and can change fairly significantly within an hour; the WBT is a combination of the ambient temperature and either the relative humidity or the dew point, all of which constantly fluctuate (5).

Engineers set design conditions after studying the cooling water requirements of the process. They determine the temperature of the water leaving the particular process (HWT) and the optimum cold water temperature leaving the tower (CWT) on the basis of the area WBT. For example, 15 LPM (4 GPM) of cooling water is required to produce 1 liter of high octane gasoline from an overhead catalytic cracker. The temperature of the process water leaving the cracker (HWT) is 49°C. Cold water at 32°C (CWT) is needed to produce the optimum product. If 3,785 LPM (1,000 GPM) of gasoline are produced, design conditions for the cooling tower are to cool 15,140 LPM (4,000 GPM) from 49°C to 32°C at the geographic area WBT. For example, if the plant is located on the Texas Gulf Coast, the design WBT is 27°C; if the tower is in Wyoming, the WBT is 17°C. Because of the 10°C difference in WBT, each location requires a different sized tower to achieve the same performance.

In another example, a tower is used for refrigeration to cool condensers and compressors. Mechanical refrigeration uses water at a rate of 11 LPM/ton (3 GPM/ton) of refrigeration. For every 1°F colder water to the compressors, $2\frac{1}{4}\%$ electric energy is required (6). Conversely, for every 1°F colder water, $2\frac{1}{4}\%$ energy is saved. If the cooling tower can be upgraded and retrofitted to produce 4°F colder water, 10% of the utility cost is saved. For a 1,000-ton system utilizing $500,000 of electricity a year for the compressors, this can amount to a savings of $50,000 per year.

If a building is equipped with a 1,000-ton air-conditioning system operating at 12°C ΔT, design conditions specify a 11,355 LPM (3,000 GPM) tower with water entering at 35°C and leaving at 29°C at the area WBT. In Connecticut, that temperature is 24°C; in Florida, it is 27°C. Again, different sized towers are required.

Cooling tower designs vary with the amount of heat rejected or refrigeration requirements, and are greatly affected by the wet bulb temperature. For example, a Texas Gulf Coast chemical plant has a design condition of 37,850 LPM (10,000 GPM) entering the tower at 74°C HWT and leaving at 32°C CWT in an ambient WBT of 27°C. The process determines the design conditions for the cooling tower, and the WBT determines the size of the tower. (In this case, the high HWT requires special construction and internal materials.) The higher the WBT, the larger the tower must be and the more mechanical equipment and depth of fill are needed. The changes in size are seen by comparing two identical systems with different wet bulb temperatures.

In one instance, a 75,700 LPM (20,000 GPM) counterflow cooling tower with a 46°C inlet HWT and 32°C CWT is designed for an area WBT of 23°C. In the second, the WBT is elevated at 27°C. The different requirements generated by this increase in WBT are shown in Table 1. A main difference in the two towers is the closer approach temperature required by the 27°C WBT. Horsepower and fan diameters are larger, depth of fill greater, and shaft

Table 1. Size–Cost Relationship as the Wet Bulb Fluctuates

	Area 1	Area 2
Flow rate, liters per minute	75,700	75,700
Wet bulb temperature, °C	23	27
Hot water temperature, °C	46	46
Cold water temperature, °C	32	32
Motor horse power	100	100
Fan diameter, meters	8	9
Cooling range, °C	−4	−4
Approach to wet bulb, °C	−9	−12
Airflow, ft³/min	825,350	979,400
Box size, meters	11 × 22	11 × 22
Erected cost, $	350,000	425,000

and gear reduction units bigger to achieve that closer 5°C approach with the same sized tower for both conditions. In addition, the air inlet must be larger to accommodate the greater volume of cooling airflow. These greater requirements increase the cost of the tower by approximately $75,000.

WBT for an area is based on average readings compiled over more than 25 years by the Armed Services; the published meteorological information is available (7). (This datum is reproduced and expanded by the American Society of Heating, Refrigerating, and Air-Conditioning Engineers.) Table 2 is a sample from the Engineering Weather Data Book that illustrates the WBT fluctuations, in this case, within Texas. In Table 2, the design wet bulb category has three entries: 1%, 2.5%, and 5%. The percentages indicate the number of hours in the air-conditioning season that the wet bulb is expected to equal or exceed the listed temperature. The industry defines the air-conditioning season as June–September, or 2,930 hours. The accepted normal area wet bulb criteria for most equipment requirements is 2.5%. Therefore, the 2.5% column indicates that the wet bulb temperature equals or exceeds the listed figure, 73.25 hours (2.5%), out of the 2,930 hours in the season.

In addition, the wet bulb temperature can also be obtained using a sling psychrometer. Or, the plant engineer determines the WBT by obtaining the ambient temperature and either the relative humidity or dew point for the day of the test from the Weather Bureau and consulting the psychrometric chart (8).

Too often, the WBT is neglected and abused. For example, an engineer may complain that a cooling tower is rated for a −7°C ΔT, but a recent test showed only a −8°C range. Another engineer may announce that although a cooling tower is also rated for a −7°C spread, a recent test indicated an outstanding −6°C cooling range. Both statements are incorrect because neither checked the WBT. Perhaps the tower cited for poor performance was checked with the WBT at 27°C, although it was rated for 26°C. When the WBT falls to the design WBT of 26°C, the ΔT is −7°C. The tower whose performance appears outstanding may have been checked when the WBT was low. When the WBT rises to the area design of 26°C, the range may be a deficient −9°C.

Circulating water volume (LPM) and airflow (CFM) also have a significant impact on performance. The op-

Table 2. Wet Bulb Variation Within One State

State (Texas) Station	Elev., feet	Winter 99%, °F	Winter 97.5%, °F	Winter Pvig. Wind Dir.	Mean Speed, knots	Degree Days Heating, Annual	Summer DB 1%, °F	MCWB, °F	Summer DB 2.5%, °F	MCWB, °F	Summer DB 5%, °F	MCWB, °F	Mean Daily Range, °F	Summer Pvig Wind Dir.	WB 1%, °F	WB 2.5%, °F	WB 5%, °F	DB =93°F, hrs	DB =80°F, hrs	WB =73°F, hrs	WB =67°F, hrs
Beaumont Army Hospital	4,185	20	24	N	7	2,678	100	64	98	64	96	64	25	S	69	68	68	370	1,917	0	376
Beeville/Chase Field NAS	190	30	33	N	9	1,189	99	78	97	77	95	77	21	SSE	82	81	79	301	2,144	2,776	3,736
Bergstrom AFB/Austin	541	24	28	N	10	1,712	99	74	97	75	96	75	22	S	79	78	77	363	1,966	1,844	3,384
Brooke Army Medical Center	785	25	30	N	8	1,570	99	72	97	73	96	73	22	SSE	77	76	76	397	2,049	1,699	3,426
Brooks AFB	598	28	32	N	10	1,272	100	74	98	77	97	74	22	SSE	78	77	77	460	2,189	2,185	3,551
Brownsville IAP	19	35	39	NNW	13	650	94	77	93	77	92	77	16	SE	80	79	79	103	2,456	3,391	4,090
Brownwood	1,386	18	22	N	9	2,437	101	73	99	73	96	73	23	S	77	76	75	297	1,922	1,045	3,043
Camp Bullis	1,400	23	28	N	8	1,952	100	74	97	74	94	74	23	S	78	77	76	220	1,626	1,481	3,548
Carswell AFB/Fort Worth	650	18	23	N	10	2,301	101	75	99	78	97	78	22	SSE	78	77	76	415	1,969	1,182	2,928
Corpus Christi IAP	41	31	35	N	12	930	95	78	94	78	92	78	17	SSE	80	80	79	134	2,507	3,238	3,989
Corpus Christi NAS	19	34	38	N	12	899	92	79	91	79	90	79	12	SE	82	81	80	15	2,845	3,329	4,036
Dallas/Love Field	481	18	22	N	11	2,290	102	75	100	76	97	75	21	S	78	78	77	474	2,208	1,675	3,103
Dallas NAS/Hensley Field	495	20	25	NNW	9	2,308	102	76	100	76	98	76	22	S	79	78	77	497	2,229	1,764	3,092
Del Rio IAP	1,026	26	31	NNW	9	1,523	100	73	98	73	97	73	23	SE	79	77	76	509	2,349	1,260	3,367
Dyess AFB/Abilene	1,789	15	19	N	9	2,682	100	71	98	71	96	71	23	S	75	74	73	342	1,700	209	2,089
Eagle Pass AFS	884	27	32	NNW	9	1,423	101	73	99	73	98	73	24	ESE	78	78	77	605	2,517	1,426	3,515
Ellington AFB/Houston	40	28	31	N	8	1,384	95	78	94	78	92	78	19	S	81	80	80	114	1,763	2,373	3,616
El Paso IAP	3,918	20	24	N	7	2,678	100	64	98	64	96	64	25	S	69	68	68	370	1,917	0	376
Fort Bliss/Biggs AAF	3,947	19	23	N	5	2,432	100	65	97	65	95	65	25	W	70	69	68	325	1,813	5	373
Fort Hood/Hood AAF	923	20	25	N	9	1,959	99	73	97	73	95	73	23	S	77	76	75	295	1,791	1,045	3,043
Fort Hood/Robert Gray AAF	1,015	20	25	N	9	1,959	99	73	97	73	95	73	23	S	77	76	75	295	1,791	1,045	3,043
Fort Sam Houston	760	25	30	N	8	1,570	99	73	97	73	96	73	22	SSE	77	76	76	397	2,049	1,699	3,426
Fort Wolters	900	17	21	N	6	2,432	102	75	100	74	98	74	24	S	78	77	76	489	1,921	1,136	2,880
Fort Worth IAP	537	17	22	NW	11	2,382	101	74	99	74	97	74	22	S	78	77	76	469	2,095	1,415	3,087
Galveston	7	31	36	NW	15	1,224	90	79	89	79	88	78	9	S	81	80	80	4	2,603	2,998	3,932
Garland ANG Station	558	18	22	NNW	10	2,290	102	75	100	75	97	75	21	S	78	78	77	474	2,208	1,675	3,103
Goodfellow AFB/San Angelo	1,877	18	22	NNE	10	2,240	101	71	99	71	97	70	24	SSE	75	74	73	465	1,978	245	2,424
Harlingen	35	35	39	NNW	10	693	96	77	94	77	93	77	18	SSE	80	79	79	223	2,442	3,294	4,059
Hondo MAP	901	26	29	N	8	1,596	99	74	97	74	95	74	22	SSE	77	77	76	480	2,159	1,703	3,374
Houston IAP	96	27	32	N	11	1,434	96	77	94	77	92	77	19	S	80	79	79	132	1,888	2,694	3,695
Kelly AFB/San Antonio	690	26	29	N	8	1,520	99	74	97	74	96	74	22	SSE	78	77	76	352	1,920	1,774	3,444
Kingsville NAS	50	32	35	N	9	970	97	78	95	78	94	78	18	SE	81	80	80	258	2,422	3,154	3,935
Lackland AFB	670	26	29	N	8	1,520	99	74	97	74	96	74	22	SSE	78	77	76	352	1,920	1,774	3,444
La Porte ANG Station	24	29	32	N	9	1,284	94	78	93	78	91	78	18	S	81	80	80	66	1,568	2,347	3,782

erating efficiency of all cooling towers should be tested on installation to determine performance and to establish a basis for comparison. A tower may be tested according to one of two accepted industry standards: ASME *PTC-23-58* or the Cooling Tower Institute Acceptance Test Code *(ATC)-105*.

WBT is critical; as it falls, the temperature of the process water or compressor cooling water also drops. As the WBT fluctuates, the corresponding CWT can be deter-

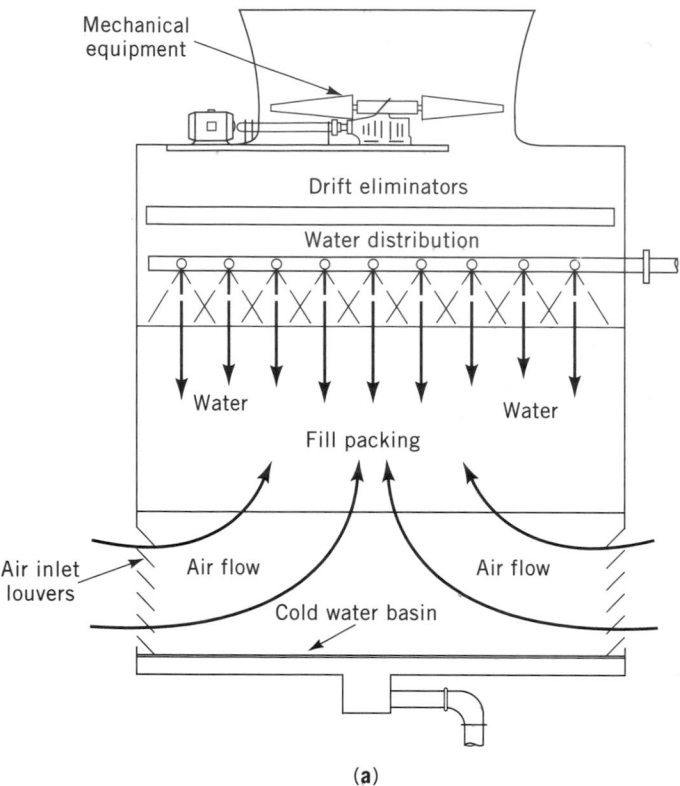

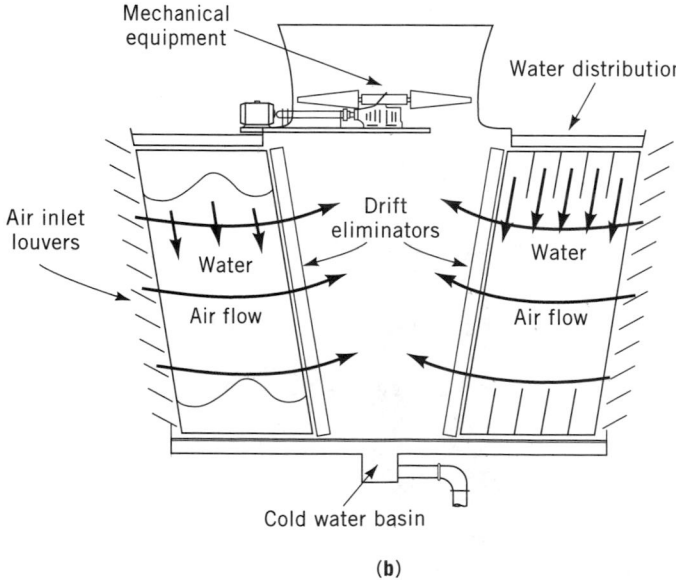

Figure 2. (**a**) Counterflow cooling tower and (**b**) crossflow cooling tower.

mined. A tower built to operate at 100% of capability generates cold water economically, whereas a deficient tower costs more to operate and is never 100% efficient.

Most plant engineers rely on sales representatives when purchasing a cooling tower. All too often, manufacturers feel pressured economically to cut corners on thermal performance. A recent survey by a government agency showed that 65% of the cooling towers reported on either failed to meet design specifications or caused decreased plant efficiency or capacity, costing industry millions of dollars in lost production a year due to higher energy costs needed to make up the deficient discharge cooling tower water temperature. One area in particular to view with caution is the nominal cooling tower. According to Webster's Collegiate Dictionary, nominal means "that which is not what it seems to be," eg, a nominal lumber 2 × 4, which measures 20% short. If a cooling tower is nominally rated, will it operate at 100% or is it designed to operate below the area WBT? The deficiency may not be large enough for the novice to notice but can make a significant difference in overall performance and operating costs.

Lack of understanding concerning design WBT and design conditions can lead to the specification and installation of "short" towers (those whose performance falls short of requirements). Too many plant engineers state that their towers work fine most of the year but cannot produce enough cold water in the summer. This deficient performance is routinely accepted as inevitable. However, when a tower is purchased, 100% capability should be expected at all times, and the user should insist on getting this performance. To ensure optimum performance, a cooling tower should be inspected by an independent cooling tower consultant who will report on the physical, structural, mechanical, and thermal conditions of the installation. Thermal upgrading potential should be outlined, together with specifications and budget for suggested maintenance and upgrading. Further, by paying close attention to the quality of cold water from the cooling tower, a considerable difference can be made in the company's profit and loss statement.

Cooling Tower Technology

Before proceeding further, the fundamentals of cooling tower technology should be considered. The basic principle of cooling tower operation is that of evaporative cooling and exchange of sensible heat. The air and water mixture releases latent heat of vaporization. Water exposed to the airstream evaporates; as a portion of the water changes to vapor, heat is consumed, which amounts to approximately 1,000 BTU/lb of water evaporated. The heat is taken from the water that remains, thereby lowering its temperature.

However, there is a penalty involved, that is, the loss of water that goes up the cooling tower and is discharged into the atmosphere as hot moist water vapor. The evaporation rate of typical design conditions is approximately 1% of the water flow rate for each 7°C of cooling range (5). Sensible heat that changes temperature is also responsible for part of the cooling tower's operation. When water is hotter than the air, there is a tendency for the air to cool the water. The air then gets hotter as it gains the sensible heat of the water, and the water is cooled as its sensible heat is transferred to the air.

Cooling due to the evaporative effect of the release of latent heat of vaporization amounts to approximately 75%, whereas 25% of the heat exchange in the cooling tower is sensible heat transfer. The cooling tower, like any other device, process, or operation, does not escape the law of conservation of matter. A cooling tower is merely a machine that takes a mass of heat from one area and moves it to another area. In more technical terms, it is referred to as "the heat rejection solution of the chemical process" or "correction of the heat penalty generation of compression equipment." The cooling tower is a machine that moves heat from point A to point B and ultimately discharges this heat into the atmosphere, which thermal engineers euphemistically call a "heat sink."

All cooling towers, except the small-packaged specialty models, are custom-made; the manufacturer varies the shape, size, configuration and input to meet the particular set of thermal parameters required by the customer. Because cooling towers are available in two basic operational designs, counterflow and crossflow, it is useful to list a few terms and their definitions to aid in investigat-

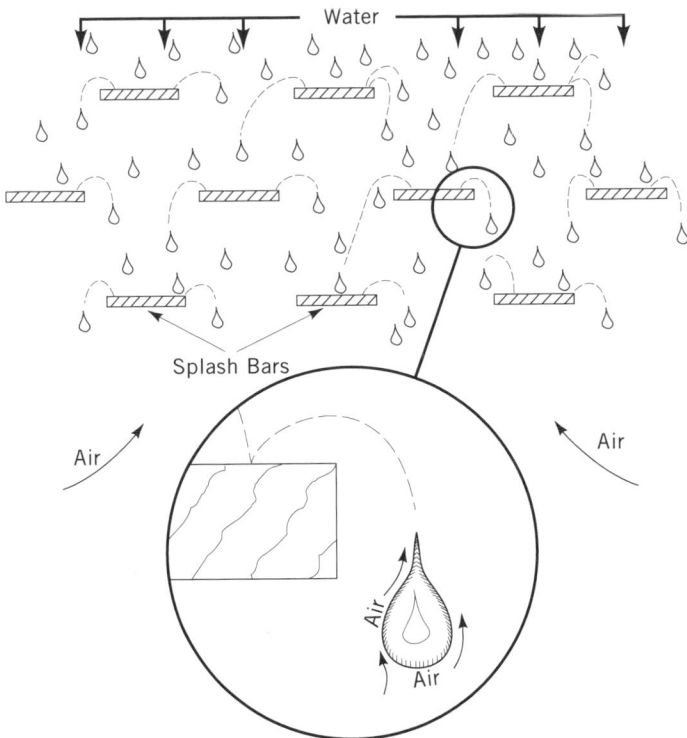

Figure 5. Old-fashioned splash bar system only cools exterior of droplet.

ing these differences. The letters in the circles on the schematic diagrams, Figures 3 and 4 show commercial counterflow and slant flow cooling towers.

To purchase a new state-of-the-art cooling tower, or upgrade the performance of an existing cooling tower, the three major areas to investigate are:

1. Wet decking fill
2. Water distribution systems
3. Drift eliminators

Wet Decking Fill. Generally, the most significant improvements can be made simply by changing the wet decking fill to state-of-the-art. This, however, is not done capriciously. The heat transfer must be investigated from a thermal engineering point of view in conjunction with the fill characteristics, as determined by the manufacturers' performance curves (which are developed painstakingly by trial, error, and experimentation, and expressed as KaV/L or heat transfer characteristics) (9). In wood slat splash bars (see Fig. 5), droplets of water bounce from one layer of wood to the other, and the moving airstream cools the exterior of each water droplet. However, cellular fill takes the same droplet of water and spreads it out in a very thin film where the air can now affect the entire surface of the film, as in Figure 6. Considerably more surface is then available to the flowing air for vaporization and sensible heat exchange to take place. The film pack contains more surface area than splash bars and as the design of cellular fill permits air to flow through it with less static pressure, it is extremely efficient compared to the old-fashioned splash bar mixture system for the same

Figure 3. Chemical plant counterflow cooling tower.

Figure 4. Electric generating plant crossflow cooling tower.

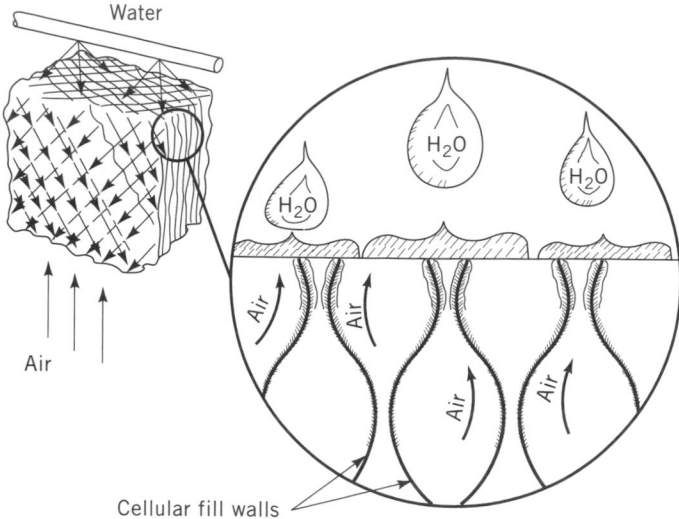

Water

Air

Cellular fill walls

Figure 6. State-of-the-art cellular film fill stretches droplet into thin film for optimum and faster cooling.

parameters of cooling. Plus, the cellular fill has a lower static pressure of operation, which further enhances the heat transfer by more efficient utilization of the existing air.

It is extremely important to investigate the proper selection of cellular film fill, as many factors do affect its operation:

1. Temperature: standard PVC is viable to about 52°C; up to 60°C, C–PVC can be selected (the extra chlorine atom elevates the temperature deformation level); above 60°C, high temperature polypropylene is recommended.

2. Water treatment is extremely essential and important, as cellular fill will clog up if maintenance and chemical treatment are poor.

3. If "dirty" water (with possible bacteriological buildup) is being used, larger port openings of the fill should be selected.

4. Possible chemical upsets such as solvents will deteriorate PVC cellular fill; a resistant cellular fill material can be utilized.

5. In some cases where makeup water and the system's water quality is poor, cellular fill might not be the material of choice.

Water Distribution Systems. For maximum performance, the water distribution system must provide a uniform pattern over the fill for optimum air–water interface. The older type trough has an uneven splash distribution, based on columns of water falling vertically and hitting splash plates which, when accurately placed, distribute the water throughout the fill. Over time, this delicate balance is destroyed as the tower deteriorates, and the end result is a vertical column of water, which in many places, drops up to 122 cm (4 ft) through the fill before it is broken up, losing entire areas of efficiency. When a nozzle is clogged, it leaves a dry spot on the fill, and the air follows the path of least resistance, rushing up to this dry spot and thereby wasting energy and cooling potential.

Modern or retrofitted cooling towers utilizing film fill are very sensitive to water distribution. Because the surface of the film fill represents nearly all of the heat and mass transfer surfaces in the tower, it is important that the water be spread evenly over all of the fill. Reduction of overall capacity of the tower will occur when areas of the fill are not wetted, which results in lost heat transfer area (10). Towers employing splash bars or grids do not require too sophisticated a distribution system, but here also, uniformity is essential.

The preferred distribution systems for crossflow towers utilize target nozzles, which consist of an adjustable orifice that can be removed and replaced with a different opening and a "target" that is suspended about 7–14 cm (3–4 in) below the orifice. The water sprays uniformly over the fill as the target breaks it up; the nozzles are typically spaced about 30 cm (12 in) on center containing a fair amount of overlap spray pattern to ensure adequate coverage.

Counterflow towers have greater flexibility with water distribution. Low pressure drop nozzles from 21 to 103 × 10³ Pa (3–15 psi) provide solid cone patterns with little or no overlap. A spray nozzle developed specifically for counterflow towers produces a square pattern and can be placed on up to 120 cm (48 in) centers providing a wide selection of water flows. Large industrial or power plant applications employ low pressure nozzles of less than 28 × 10³ Pa (4 psi) and produce a hollow cone pattern. A considerable amount of overlap is generated so that there will be no "umbrella" effect creating dry spots. The overlap approach, however, is mostly engineered for large towers with little or no internal bracing and support columns. Generally, this system is used in concrete structures as redwood towers have too many internal vertical and horizontal structural members that create dry areas blocked off by the beams.

Whether it be crossflow or counterflow, uniformity is the key to success: Uniformity of water distribution, uniformity of static pressure drop, uniformity of fill configuration, and uniformity of air velocity (even though in a crossflow tower, the pressure differential will vary with the height).

Drift Eliminators. Drift eliminators are mechanical devices utilized to prevent solid droplets of water from being drawn or blown out of the tower. Mist or fog, which is the product of the evaporation process, cannot be eliminated from the tower, but solid droplets can be removed from the discharge airstream with a good drift eliminator system. A baffle, or drift eliminator, is placed between the water distribution system and the air discharge point to minimize dispersal of entrained water droplets into the surrounding atmosphere.

The performance of cooling towers, ie, the level of heat removal, is a balance between water flow and air volume; therefore, drift eliminators are normally designed to be efficient through a calculated range of airflow velocity. Too high an airflow can result in excessive drifting of water from the tower, whereas poorly designed eliminators will increase static pressure through the tower, slowing the air and lowering the cooling capabilities and efficiency of the unit. Proper drift eliminators can add up to 0.3°C additional performance to a cooling tower.

The need for an efficient mechanical system to retain the water is obvious since the spray or water distribution system turns the heavy stream of condenser or process water into light droplets preparatory to being cooled by the airstream. The evaporative cooling is generated by the mixture of air and water, plus loss of sensible heat. As the air moves counter or crossflow to the water, it will pick up much of the mist and droplets and carry them with the airflow out of the tower.

The schematic operation of a three-pass drift eliminator is shown in Figure 6. A newer, more efficient drift eliminator fabricated from cellular PVC has been developed. It causes the air to change its direction approximately six times, thereby obtaining greater surface contact to release water droplets. Figure 7 illustrates the schematic mechanics of the six-pass cooling.

One key to the rebuilding and upgrading procedure is that the considerably lower static pressure resistance of the cellular drift eliminators permit the available air to flow at a higher velocity at no additional cost. The versatility of cellular drift eliminator design can be seen in its applications to varying conditions. One common method of trying to improve cooling tower performance is to increase the air velocity by rotating the fan with more power and higher pitch angle to the blades. This often causes increased drift loss, because the eliminators were not originally designed for this type of higher air velocity service. Cellular drift eliminators, which do not increase the static pressure but will increase the drift collecting capability, can minimize such an eventuality. In fact, reduced static pressure resistance permits more air through the tower at the same horsepower expenditure, but it will permit more work, ie, colder water at the same cost as before the conversion.

Complete drift elimination is possible in theory, but impractical in application. An acceptable level found to be generally satisfactory is "not in excess of 0.002%" of the circulating water.

Crossflow towers are more susceptible to drift problems

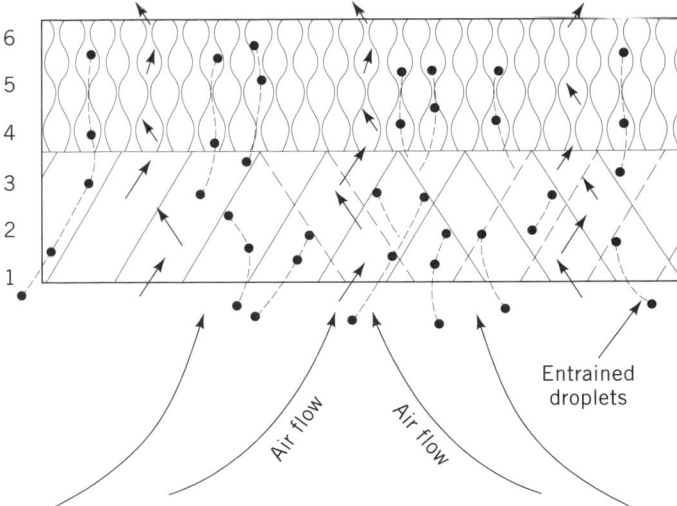

Figure 8. Six-pass low pressure drop cellular drift eliminators.

than counterflow, because of a tendency toward higher airflow velocities in this type of design. Because the drift requirements are more stringent, a different type of eliminator system is used with a very delicate balance design and performance. If one or two eliminator blades are out of position, drift will occur quite readily. Several manufacturers have developed cellular eliminators to overcome the problem. Here again, the lower resistance of these new types of eliminators, plus their drift elimination effectiveness, make them a desirable conversion from the less efficient wood.

1. *Capital Cost.* Lowest price should not be the purchasing criteria, but "lowest evaluated cost," consisting of a comparison of fan motor horsepower, pumping head horsepower, maintenance requirements, and life cycle costs.
2. *Thermal Performance.* The savings in first cost which could result in poor thermal performance can cost the facility a hundred times the alleged savings of the low bid.

A classic example of the importance of upgrading is that of a Mississippi Power Co. Generating Station. The power plant paid millions of dollars to have their natural draft tower fitted with ceramic tile fill blocks, PVC drift eliminators, and a water spray distribution system. The tower was tested by the Cooling Tower Institute (CTI) in accordance with the CTI Acceptance Test Code, *ATC-105*, which indicated substandard operation of at least 60% of design capability, or 40% short thermally. The power plant could not operate at that low level of "hot" cooling water economically because of high turbine back pressure, and the facility put out a contract to upgrade the tower. The retrofit of state-of-the-art PVC cellular film fill blocks installation and new cellular drift eliminators were then retested, again by the Cooling Tower Institute, and the retrofitted performance was measured to perform at 100% of design capability. The manufacturer who supplied the 500,000 ft³ of cellular film fill circulated these facts in a

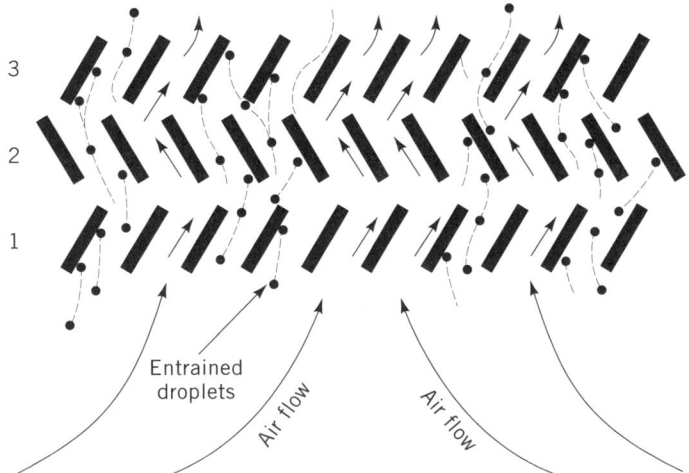

Figure 7. Three-pass slat type high pressure drop drift eliminators.

newsletter to the industry, together with the consulting engineer's report (11). It is alleged that the first installer paid a $250,000 penalty for substandard performance and then walked away from the job in accordance with contract terms. The retrofit, to bring the generating station up to design, cost in the area of $4,000,000. This additional high retrofit cost does not consider the millions of dollars lost by the facility in excessive generating fuel costs or loss of saleable electric power from the time the problem was recognized until the fix was performed.

It is thus contended that for new cooling towers or retrofits, there should be no financial penalty for poor performance, as that is a one-time payment from the manufacturer to the owner. However, the facility then suffers financially for the life of the installation. Retention must be specified, of a sufficient sum to ensure that the manufacturer or installer will perform the job in accordance to design contract conditions.

Counterflow vs Crossflow Towers

1. Counterflow vs crossflow—with the advent of cellular film fill over 25 years ago, counterflow towers are more energy-efficient and cost-effective over the life of the equipment than crossflow towers with splash bar fill.

2. Counterflow (CF) towers have a larger average driving force for heat and mass transfer as the coldest water contacts the coldest air at the bottom of the tower, and warmest water contacts the warmest air.

3. To accomplish the same given cooling duty, a crossflow (XF) tower will always require either more airflow, more fill, or both, plus a larger footprint.

4. Crossflow cooling towers have a tendency to recirculate hot air at the top of the tower as the fan velocity is greater in that area, creating negative pressure at the top of the air intake louvers. Prevailing winds will blow the hot discharge into the top air intake louvers degrading the performance of the induced draft XF tower by ingesting this higher warmer wet bulb air.

5. Inspections and maintenance are easier in large CF towers as the fill and drift eliminators are more accessible than in the large XF towers.

6. Pumping head requirements for large XF towers, to raise the water to the hot water distribution basins, can be as much as twice as high as similar capacity CF designs, costing hundreds of thousands of dollars and, in cases of larger installations, millions of dollars in extra energy costs over the life of the equipment.

7. Windage losses through XF towers are greater as they have a lower air inlet velocity than CF towers.

8. Air distribution in CF towers is more uniform. Large air velocity gradients tend to reduce the theoretical efficiency of XF towers even further.

9. Icing in CF towers has more controllable options such as auxiliary spray system beneath wet decking fill, better heat retention in reverse fan mode, and operation closer to freezing point without ice formation; if icing does occur, CF towers typically sustain less damage than XF towers operated in a similar matter.

10. CF towers can be built less expensively and more efficiently without air intake louvers, whereas XF towers must have pressure drop inducing air intake louvers to restrain water splash out.

11. Whereas many XF towers are still built up to 18 m (60 ft) high with wood or plastic splash bars, cellular fill has overtaken the industry as the heat exchange surface of choice due to the cellular fill efficiency and low profile requirements of 50% less pumping head.

THE IMPORTANCE OF INSPECTION

About every five years, large industrial, chemical plant, and power-station cooling towers should be professionally inspected by a cooling tower engineer. The physical, structural, and mechanical conditions should be investigated, plus the thermal upgrading potential, with a comprehensive report submitted. Photographs should be submitted, highlighting the inspector's findings. Further, specifications for repairs and upgrading should be included, together with budgetary estimates of the work proposed. Whereas operating personnel are quite competent to perform inspections, it is advisable to call in a specialist who can give the facility the benefit of new industry advances.

Before spending large amounts of capital for a new cooling tower when faced with plant expansion or additional heat rejection requirements, the upgrading potential of the existing equipment needs to be investigated. More often than not, the capacity is available and can be examined by the cooling tower professional and analyzed by computer expertise.

CONCLUSION

When considering the purchase of a large volume of circulating water cooling tower, investigate the operating costs (pumping head, fan horsepower, estimated maintenance costs) of a crossflow and counterflow tower to determine that, in fact, over a 20-year life of the cooling tower, the counterflow configuration will be far and away the least expensive overall purchase. Do not only consider the importance of the first cost difference.

The state-of-the-art cellular film fill should be specified in a new cooling tower together with high efficient cellular drift eliminators and modern fiberglass air foil type propeller fan blades. To upgrade existing cooling towers to cool greater volumes of circulating water and/or colder discharge water, look at the operations of state-of-the-art wet decking fill and drift eliminators. In all cases, it would be a prudent investment to have a cooling tower consultant review the specifications and recommend to the facility which purchase and procedure would be in the best interest of the owner.

The bottom line of a cooling tower is that colder water makes money by reducing back pressures of turbines, low ering operating energy requirements for compressors, and assisting a chemical facility in producing more saleable product at lower costs. Good maintenance procedures and state-of-the-art rebuilding on cooling towers can make a significant difference in the company's bottom line profit and loss picture.

BIBLIOGRAPHY

1. M. LeFevre, *Cooling Tower Institute Paper*, Jan. 1989.
2. C. Weiss, *National Engineer*, 22 (Mar. 19, 1985).
3. R. C. Rittenhouse, *Power Engineering*, 12 (July 1987).
4. "Operations," *Cooling Tower Institute Handbook*, 1977, Ch. 1.
5. "Equipment," *ASHRAE Handbook*, 1983, p. 21.2.
6. *The Pressure Enthalpy Diagram, Its Construction, Use, and Values*. Allied Chemical Corp.
7. *Army, Navy, Air Force Engineering Weather Data*, U.S. Government Printing Office.
8. "Fundamentals," *ASHRAE Handbook*, 1983, Ch. 21.
9. R. Burger, "Thermal Evaluation," *Cooling Tower Technology*, Ch. 6.
10. M. Smith, *Understanding Cooling Tower*, AlChE National Meeting, Apr. 1987.
11. "Cooling Tower Upgrade," *Newsletter*, Munters Corp., Aug. 26, 1987.

CRYOGENICS

RANDALL BARRON
Louisiana Tech University
Ruston, Louisiana

Cryogenics is the branch of science and technology concerned with the design and utilization of very low temperature processes and equipment. The field of cryogenics (1) generally involves systems that operate at temperatures below $-150°C$. Although this dividing point is somewhat arbitrary, it is logical because normal boiling points of commonly used cryogens, such as oxygen, hydrogen, and helium, are $\leq -150°C$; whereas normal boiling points of conventional refrigerants, such as Freon compounds, ammonia, etc, are $\geq -150°C$.

See also REFRIGERATION; THERMODYNAMICS.

PHYSICAL PRINCIPLES

Joule-Thomson Effect

In 1853 an experiment to determine temperature changes resulting from pressure drops caused by flow through a constriction was conducted (2,3). The physical phenomenon corresponding to the change in temperature when a fluid flows through a constriction, such as a valve or porous plug, is called the Joule-Thomson effect, and the expansion valve is often called a Joule-Thomson (J-T) valve. Many cryogenic refrigeration systems utilize a J-T valve to achieve necessary low temperatures in the system.

For flow of a fluid through a J-T valve, heat transfer and changes in kinetic energy are negligible and there is no external work transfer; ie, enthalpy remains constant (isenthalpic). The constant enthalpy curves are shown in Figure 1, and the Joule-Thomson coefficient (eq. 1) is the slope of these curves.

$$\mu_{JT} = \left(\frac{\partial T}{\partial p}\right)_h \tag{1}$$

In the region where $\mu_{JT} < 0$ (negative), expansion through the J-T valve (decrease in pressure) results in an increase of the fluid temperature. On the other hand, expansion through the J-T valve where $\mu_{JT} > 0$ (positive) results in a decrease in the fluid temperature. The curve separating these two regions is called the inversion curve; here $\mu_{JT} = 0$ because the slope of the isenthalpic line is zero.

Thermodynamic analysis (4) may be used to relate the J-T coefficient to other thermodynamic properties, as follows:

$$\mu_{JT} = \frac{1}{C_p}\left[T\left(\frac{\partial v}{\partial T}\right)_p - v\right] \tag{2}$$

where v = specific volume of the material and C_p = specific heat at constant pressure.

For an ideal gas, $v = RT/p$ and

$$\left(\frac{\partial v}{\partial T}\right)_p = \frac{R}{p} = \frac{v}{T}$$

where R = gas constant and p = pressure. Substitution of this expression into equation 2 yields $\mu_{JT} = 0$ for an

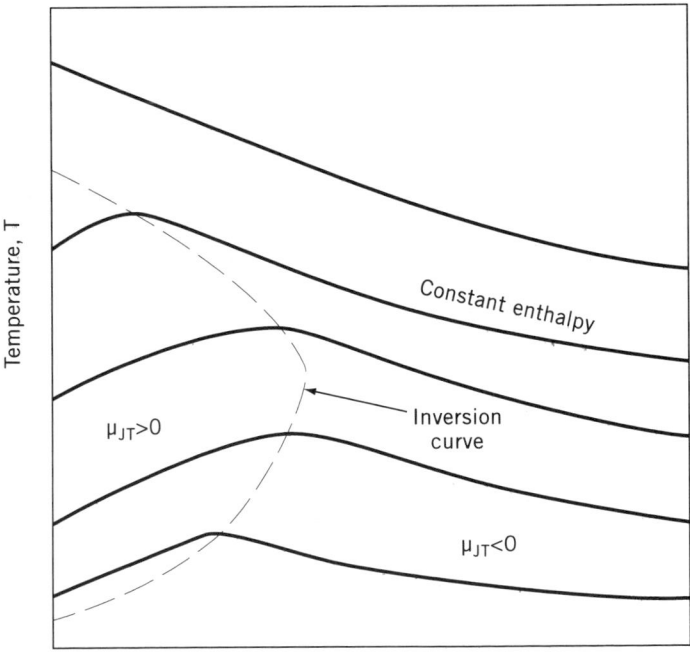

Figure 1. Temperature–pressure diagram for a real gas, where (—) is constant enthalpy and (---) is the inversion curve. See equation 1.

ideal gas. For a real gas, the J-T coefficient may be negative or positive, depending on the temperature and pressure of the gas. It may be observed that nonideal gas behavior is responsible for the J-T coefficient being nonzero for a real gas.

The van der Waals equation of state illustrates the behavior of real gases, although the numerical accuracy of the equation is not always satisfactory. Equation 3 considers intermolecular forces of attraction, represented through the constant a, and finite molecular volumes, represented through the constant b.

$$\left(p + \frac{a}{v^2}\right)(v - b) = RT \tag{3}$$

The J-T coefficient for a van der Waals gas may be found from equations 2 and 3:

$$\mu_{JT} = \frac{(2a/RT)(1 - b/v)^2 - b}{C_p[1 - (2a/vRT)(1 - b/v)^2]} \tag{4}$$

For large values of the specific volume, equation 4 can be approximated by

$$\mu_{JT} = \frac{1}{C_p}\left(\frac{2a}{RT} - b\right) \tag{5}$$

This expression illustrates the two regions of behavior for the J-T coefficient for real gases. According to equation 5, the J-T coefficient is positive when $T < 2a/bR$ and negative when $T > 2a/bR$.

The inversion curve for the van der Waals gas is found by setting $\mu_{JT} = 0$ in equation 4.

$$T_i = \frac{2a}{bR}(1 - b/v)^2 \tag{6}$$

The maximum inversion temperature for a van der Waals gas corresponds to the temperature on the inversion curve where $p = 0$ (or $b/v = 0$). From equation 6 the following is found:

$$T_{max} = \frac{2a}{bR} \tag{7}$$

Measured values of maximum inversion temperatures for several gases are found in Table 1.

REFRIGERATION SYSTEMS

Cryogenic refrigeration systems (cryocoolers) are used in cooling of advanced electronic systems, superconducting equipment, and space environmental simulation chambers. Several cryocoolers utilize the J-T effect to liquefy the working fluid, which is used as the refrigerant. Other cryocoolers, eg, Stirling, Gifford-McMahon, and Vuilleumier refrigerators, employ other physical principles in maintaining low temperatures.

Table 1. Maximum Inversion Temperatures for Selected Gases

Gas	Maximum Inversion Temperature	
	K	°R
Helium-4	45	81
Hydrogen	205	369
Neon	250	450
Nitrogen	621	1118
Air	603	1085
Carbon monoxide	652	1174
Argon	794	1429
Fluorine	709	1276
Oxygen	761	1370
Methane	939	1690

J-T Refrigerator

The basic J-T refrigerator is shown in Figure 2. The gas is compressed from ambient conditions to a pressure of approximately 20 MPa (200 atm). Warm incoming high pressure gas is cooled as it flows through a heat exchanger. Cold high pressure gas leaving the heat exchanger is expanded through a J-T valve to a near ambient pressure. At the J-T valve exit, the fluid becomes two-phase (liquid plus vapor), and in the evaporator heat is added to the liquid from the space to be cooled. The liquid is evaporated, and the resulting vapor flows back through the heat exchanger to complete the cycle.

The refrigeration effect, or heat transfer per unit mass in the evaporator, is given by

$$Q_a/m = h_1 - h_2 = (h_1' - h_2) - (1 - e)(h_1' - h_g) \tag{8}$$

where h_1 = enthalpy at the outlet at the warm end of the exchanger, h_2 = enthalpy at the inlet at the warm end of the exchanger, h_1' = enthalpy of the low pressure stream at a pressure p_1 and a temperature $T_1' = T_1$, h_g = enthalpy of the saturated vapor at the exit of the evaporator, and e = heat exchanger effectiveness.

The work requirement per unit mass flow of the gas is shown in equation 9.

$$-W/m = \frac{T_2(s_1 - s_2) - (h_1 - h_2)}{\eta_{CO}} \tag{9}$$

where η_{co} = compressor overall efficiency. The coefficient of performance (COP) for the refrigerator is shown in equation 10.

$$COP = \frac{Q_a/m}{-W/m} = \frac{\eta_{CO}[(h_1' - h_2) - (1 - e)(h_1' - h_g)]}{T_2(s_1 - s_2) - (h_1 - h_2)} \tag{10}$$

The effectiveness of the heat exchanger must be near unity for the J-T refrigeration system to produce a practical refrigeration effect. In fact, the system produces no cooling if the effectiveness of the heat exchanger is less

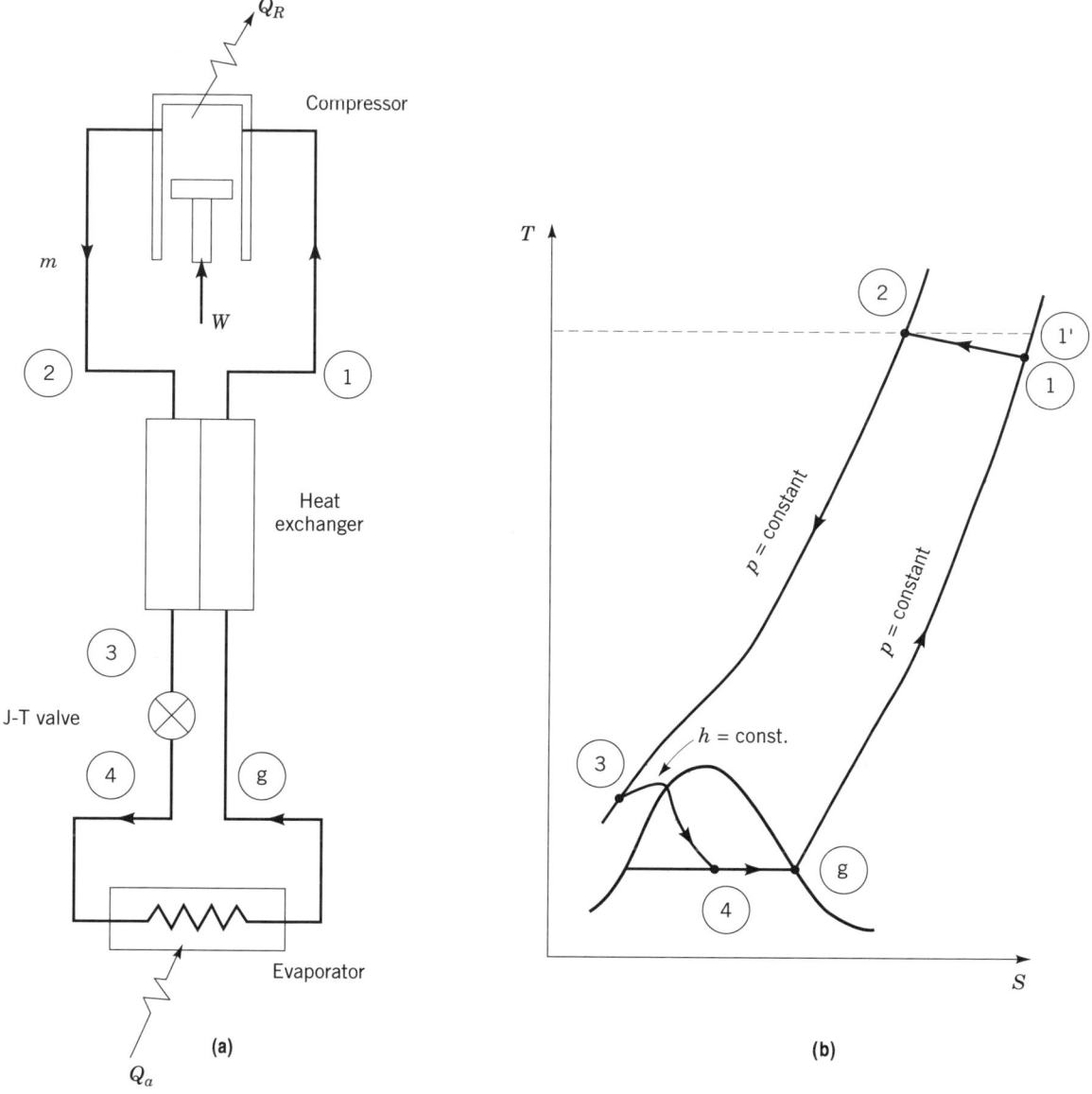

Figure 2. (**a**) Basic Joule-Thomson (J-T) refrigerator. The numbers 1, 2, etc. represent similarly designated points on the T-s diagram (**b**).

than e_{min}, where

$$e_{min} = \frac{h_1' - h_2}{h_1' - h_g} \quad (11)$$

A typical value of the limiting heat exchanger effectiveness for a J-T cryocooler operating between ambient pressure and temperature (101.3 kPa and 300 K) and 20.3 MPa is approximately 0.86 (5).

Because the cooling effect in the J-T refrigerator is provided solely by the J-T valve, the maximum inversion temperature of the refrigerator working fluid must be above ambient temperature. Without precooling, helium, hydrogen, and neon cannot be used as working fluids in a J-T cryocooler operating between room temperature and cryogenic temperatures. The use of an expansion engine

may also be utilized to achieve cooling with these gases (see Table 1).

Claude Refrigerator

Expansion through a J-T valve is a thermodynamically irreversible process; therefore, the performance of the refrigeration system can be improved by using an expansion engine to remove energy from the working fluid. The Claude (expander) refrigerator, shown in Figure 3, uses an expander to provide the bulk of the cooling effect.

The refrigeration effect for the Claude refrigerator is given by

$$Q_a/m = (h_1 - h_2) + (m_e/m)\eta_{ad}(h_3 - h_e') \quad (12)$$

where η_{ad} = expander adiabatic efficiency, m_e = mass flow rate through the expander, m = mass flow rate through

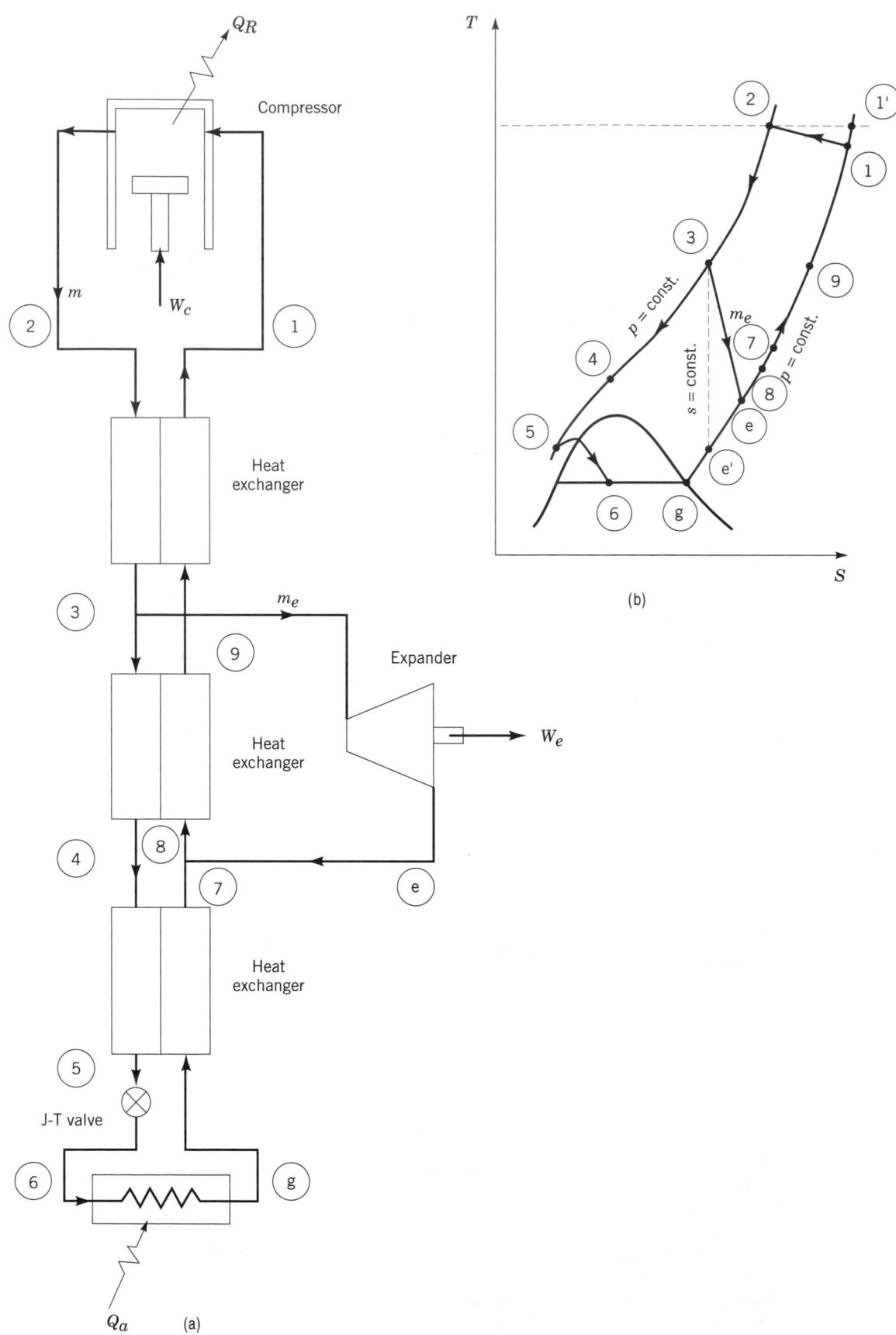

Figure 3. (**a**) Claude (expander) refrigerator. The numbers 1, 2, etc. represent similarly designated points on the T-s diagram (**b**).

the compressor, and h_e' = enthalpy of gas at the expander outlet pressure for reversible and adiabatic (isentropic) expansion.

If the expander work is utilized to aid in compression of the gas, the work requirement per unit mass compressed is given by

$$-W/m = [T_2(s_1 - s_2) - (h_1' - h_2)]/\eta_{CO}$$
$$- (m_e/m)\eta_m\eta_{ad}(h_3 - h_e') \qquad (13)$$

where η_m = expander mechanical efficiency.

Stirling Refrigerator

The Stirling cycle was originally developed in 1816 for use as a hot-air engine (6). The first Stirling refrigerator was described in 1864 (7). In 1953 workers at the Philips Co. (Eindhoven, the Netherlands) led in the development of the Stirling refrigerator as a commercially viable cryogenic refrigeration system. Since that time several other organizations have designed and constructed Stirling refrigerators for a variety of applications, including cooling infrared thermal imaging systems, spacecraft instrument systems, and superconducting elements.

The operating cycle for the Stirling refrigerator is shown in Figure 4. The refrigerator consists of a cylinder

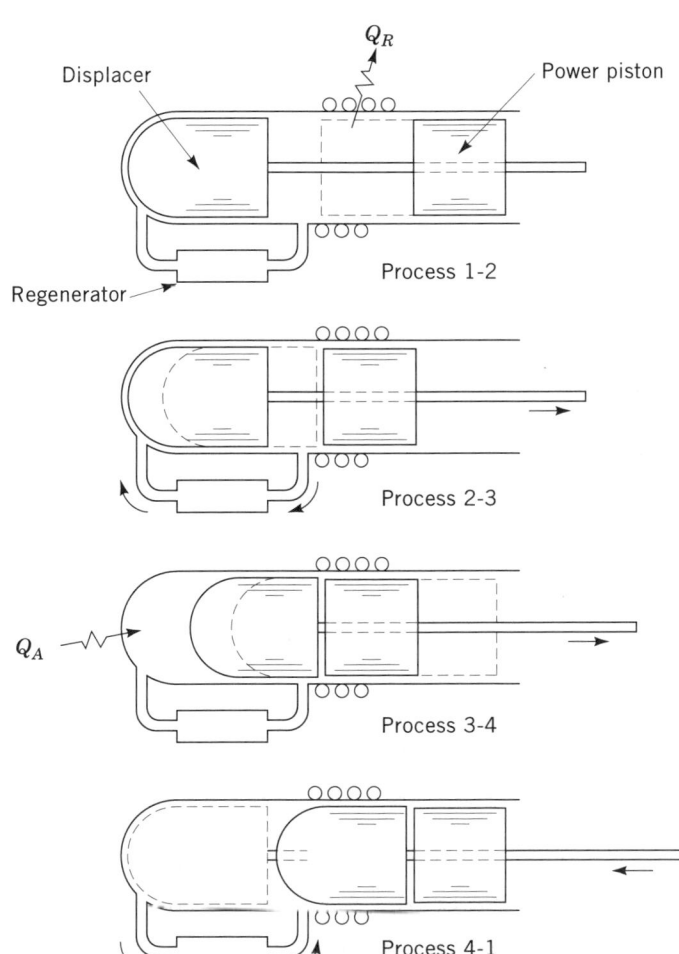

Figure 4. Operating cycle for the Stirling refrigerator. See text.

enclosing a power piston and a displacer piston. The two chambers of the refrigerator are connected through a regenerator.

The working fluid (usually helium gas, although hydrogen gas has also been used) is compressed by the power piston while heat is removed from the gas by cooling coils at the ambient-temperature end of the cylinder (Fig. 4a). Ideally, the compression process is isothermal. With the power piston approximately stationary, the gas is forced through the regenerator by moving the displacer piston and is cooled by heat exchange with the regenerator matrix (Fig. 4b). Ideally, this process is a constant-volume cooling process. The gas is expanded by moving both the power piston and the displacer piston, while heat is added to the gas from the space to be refrigerated at the cold end of the cylinder (Fig. 4c). Ideally, the expansion process is isothermal. The cold gas is forced back through the regenerator by motion of the displacer piston, while the power piston remains approximately stationary. The gas is warmed to near ambient temperature by heat exchange with the regenerator matrix (Fig. 4d). Ideally, this process is a constant-volume heating process.

The COP for the ideal Stirling refrigerator (eq. 14) is the same as the thermodynamically ideal system of the Carnot refrigerator.

$$COP = \frac{T_3}{T_1 - T_3} \qquad (14)$$

Irreversible effects, such as frictional energy dissipation and regenerator ineffectiveness, all interact to reduce the COP of the actual Stirling refrigerator to a value ~36% of the ideal value, depending on the temperature at the cold end of the cylinder. An excellent analysis of the operation of the Stirling refrigerator is given in Reference 8, and a more exact numerical analysis is presented in Reference 9.

Vuilleumier Refrigerator

The Vuilleumier (VM) refrigerator was first patented in 1918 (10). Though similar in principle to the Stirling cryocooler, the VM refrigerator uses a thermal rather than mechanical process to supply the energy requirement needed for production of refrigeration (11). The VM refrigerator is shown in Figure 5.

Heat is added from a high temperature source to the gas in the hot cylinder, and the displacer piston is moved downward to maintain the temperature of the gas constant at T_h. At the same time, near ambient temperature gas flows from the intermediate temperature space through the warm regenerator to the hot space. The displacer moves upward to displace the high temperature gas from the hot space to the intermediate space. Heat is rejected from the intermediate space to maintain the gas temperature constant at T_0.

The motion of the cold piston is approximately 90° out of phase with the motion of the hot piston. As the cold piston moves to the left, heat is added to the cold space from the low temperature volume to be refrigerated to maintain the gas temperature constant at T_c. At the same time, gas flows from the intermediate temperature space

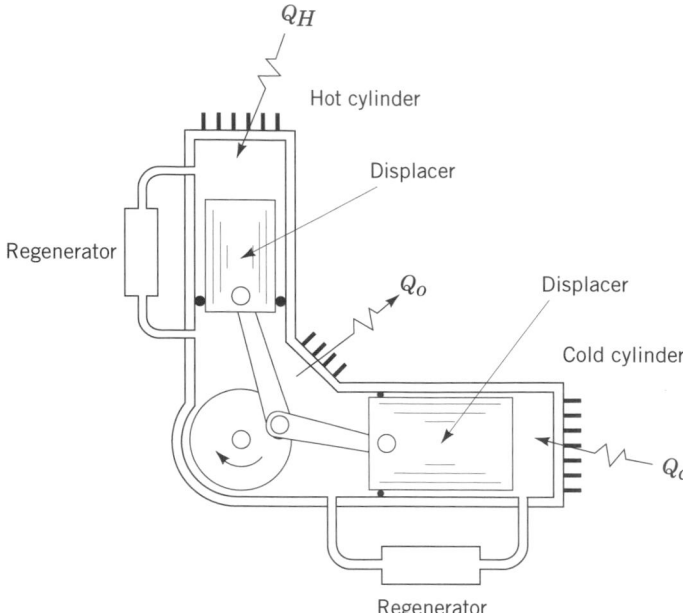

Figure 5. The Vuilleumier (VM) refrigerator. See text.

through the cold regenerator to the cold space. The cold displacer moves back to the right, and the gas is moved from the cold space back to the intermediate space.

The ideal COP for the VM refrigerator is given by

$$\text{COP} = \frac{Q_c}{Q_h} = \frac{T_c(T_h - T_0)}{T_h(T_0 - T_c)} \qquad (15)$$

Several factors contribute to the reduction of the COP below the value given by equation 15 in the actual system, including shuttle loss, caused by the mismatch of thermal gradients between the displacer and the cylinder wall; pumping loss, resulting from pumping of gas into and out of the clearance volume between the displacer and cylinder wall; heat transfer through the displacers from the warmer to the cooler end; heat transfer through the cylinder wall from the warmer to the cooler end; frictional energy dissipaton between the cylinder wall and piston seals; and regenerator ineffectiveness effects.

One significant advantage of the VM cryocooler is that the thermal input at the hot space may be provided by a variety of sources, including solar and isotope energy. This feature makes the VM cryocooler attractive for cryogenic cooling on long-term space missions and other applications where mechanical vibrations of a drive engine must be avoided (12).

Gifford-McMahon Refrigerator

The Gifford-McMahon cryocooler, a new cryogenic refrigeration cycle proposed in 1959 (13), has been extensively used in miniature cryocooler applications (Fig. 6).

The first process of operation involves pressure buildup. With the displacer at the bottom of the cylinder, the high pressure inlet is opened and the pressure within the upper expansion space is increased from the low pres-

sure p_1 to the high pressure p_2. The volume of the lower expansion space is practically zero during this process. The displacer piston is moved from the bottom to the top position while the high pressure inlet valve remains open. The gas that was originally in the upper (warm) expansion space is displaced through the regenerator to the lower (cold) expansion space by the intake stroke. Because the gas is cooled as it flows through the regenerator, the gas volume tends to decrease and additional gas is drawn into the system through the high pressure intake valve to maintain a constant pressure within the system. With the displacer at the top of the cylinder, the high pressure intake valve is closed and the low pressure outlet valve is opened to allow the gas in the lower expansion space to expand from the high pressure to the low pressure of the cycle. Ideally, this expansion takes place isentropically. The temperature of the gas remaining in the cylinder drops during the expansion process. The low temperature gas is forced from the lower expansion space through the heat exchanger by moving the displacer from the top to the bottom position in the cylinder. As the cold gas flows through the heat exchanger, the gas absorbs energy from the low temperature source. This is called the refrigeration stroke. Finally, the gas flows back through the regenerator, where it is warmed to ambient temperature and exhausted through the low pressure outlet valve.

The COP of the ideal Gifford-McMahon system is given by

$$\text{COP} = \frac{(\gamma - 1)(p_2/p_1 - 1)}{\gamma \left(\dfrac{p_2 T_h}{p_1 T_c} - 1 \right) \left[\left(\dfrac{p_2}{p_1} \right)^{(\gamma-1)/\gamma} - 1 \right]} \qquad (16)$$

where p_1 and p_2 are the low and high pressures, respectively, in the cycle; T_h is ambient temperature; T_c is the temperature at which refrigeration is delivered; and γ = specific heat ratio, c_p/c_v.

The Gifford-McMahon cryocooler has several advantages over the Claude-type refrigerators, which also use an expander. All valves and seals in the Gifford-McMahon refrigerator operate at ambient temperature; therefore, problems with low temperature seals and valves are eliminated. Leakage around the piston seals is reduced because the viscosity of the gas near ambient temperatures is higher than the gas viscosity at cryogenic temperatures. The expanders in the Claude refrigerators require special dry-sealing rings, whereas the Gifford-McMahon refrigerators can use seals constructed of conventional materials.

A regenerator is used in the Gifford-McMahon cryocooler instead of the heat exchanger employed in the J-T and Claude refrigerators. The regenerator can be constructed with a larger heat-transfer area per unit volume than a heat exchanger (recuperator), which is an important feature for miniature systems. In addition, the regenerator is self-cleaning and impurities deposited during the intake process are swept out of the system during the exhaust stroke.

The Gifford-McMahon system is well-suited for multistaging (14). Refrigeration may be required at two differ-

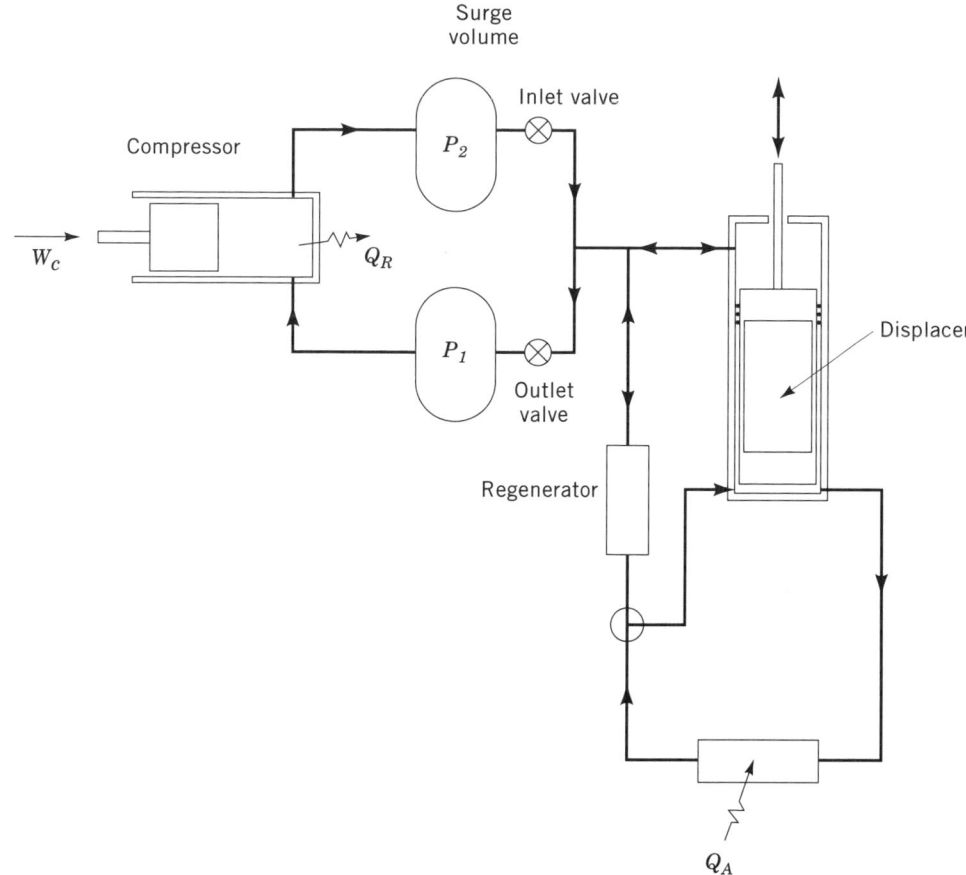

Figure 6. The Gifford-McMahon refrigerator. See text.

ent temperature levels for some cryogenic systems. For example, cooling of thermal radiation shields may be required at 80 K while the main refrigeration load occurs at 20 K.

Other Cryogenic Refrigerators

Two other refrigerators are used to attain temperatures below 2 K: magnetic refrigerators and dilution refrigerators (15).

The magnetic refrigerator uses the magnetocaloric effect with a paramagnetic salt to produce temperatures as low as 10 mK. This refrigerator has the advantage that the presence of a gravitational force is not required for effective operation of the system; therefore the magnetic refrigerator can operate in a microgravity environment.

The dilution refrigerator utilizes a special property of mixtures of liquid helium-3 and liquid helium-4. At temperatures below 0.86 K, the mixture separates into a dilute phase and a phase concentrated in He-3. The concentrated phase is less dense than the dilute phase and, therefore, the concentrated phase floats on top of the dilute phase. The heat of mixing absorbed when the two phases separate provides the refrigeration effect for the He-3/He-4 dilution refrigerator. These refrigerators can attain temperatures as low as 5 mK.

A comparison of the performance of several cryocooler systems is given in Table 2. The figure of merit (FOM) values in Table 2 give a comparison of the thermodynamic performance of the cryocooler with that of the thermodynamically ideal Carnot refrigerator.

CRYOGENIC HEAT EXCHANGERS

The heat exchanger is one of the most critical components in any cryogenic refrigeration system. The regenerative gas turbine power system can operate satisfactorily with a heat exchanger having an effectiveness as low as 50%. On the other hand, cryocoolers generally require a heat exchanger with an effectiveness exceeding 90%; otherwise, the system produces no refrigeration.

Cryogenic heat exchangers may be broadly classified as either recuperators or regenerators.

Recuperators

Recuperators used in cryogenic systems may be classified as tubular, plate-fin, and perforated-plate heat exchangers. A recuperator has separate flow passages for each fluid stream and generally operates in steady flow.

Tubular Exchangers. Several types of tubular exchangers have been used in smaller cryogenic liquefaction and refrigeration systems (16); however, the classic heat exchanger used for medium- to large-scale cryogenic systems is the Giauque-Hampson type (Fig. 7). Other types of tubular cryogenic heat exchangers include the double-

Table 2. Comparison of Cryogenic Refrigeration Systems[a]

Refrigerator	Coefficient of Performance, COP = Q_a/W	Figure of Merit, FOM = COP/COP_{id}
Ideal Carnot	0.3636	1.000
Joule-Thomson		
Ideal: $p_1 = 136.7$ kPa[b]	0.0648	0.1783
$p_2 = 20.3$ MPa[c]		
Actual: $p_1 = 136.7$ kPa[b]	0.328	0.0903
$p_2 = 20.3$ MPa[c]		
$\eta_{CO} = 70\%$; $e = 0.965$		
Claude		
	0.2776	0.7635
Ideal: I$p_1 = 136.7$ kPa[b]		
$p_2 = 4.05$ MPa[c]		
$m_e/m = 0.60$;		
$T_3 = 220$ K		
Actual: $p_1 = 136.7$ kPa[b]	0.1435	0.3947
$p_2 = 4.05$ MPa[c]		
$m_e/m = 0.60$;		
$T_3 = 220$ K		
$\eta_{CO} = 70\%$; $e = 0.965$		
$\eta_{ad} = 85\%$; $\eta_m = 95\%$		
Actual Stirling[d]		
	0.1309	0.360
Gifford-McMahon[d]		
Ideal: $p_2/p_1 = 2$	0.1926	0.530
Actual: $p_2/p_1 = 2$	0.115	0.315

[a] The working fluid for each system is nitrogen unless otherwise noted. The high temperature for the refrigerator is 300 K (26.8°C) and the low temperature is 80 K (−193.3°C).
[b] To convert kPa to atm, divide by 101.
[c] To convert MPa to atm, divide by 0.101.
[d] The working fluid is helium.

pipe or concentric-tube exchanger, the concentric-tube exchanger with wire spacer, the multiple-tube exchanger, the bundle-tube exchanger, and the Collins heat exchanger.

The construction of the Giauque-Hampson heat exchanger involves winding one or more coils of stainless steel, aluminum, or copper tubes onto a low conductivity mandrel, such as a stainless steel cylinder. A close-fitting sheath of the same low conductivity material is placed around the coil, and thermal insulation is added on the outer surface. The high pressure stream flows inside the small tubes while the low pressure stream flows across the outside of the helical coil of small tubes between the inner mandrel and the outer sheath.

Achieving a uniform spacing between the tubes is an important design consideration for the Hampson exchanger; otherwise, flow maldistribution occurs, and the thermal performance of the heat exchanger is degraded. An important contribution to solving the tube spacing problem was introduced through the use of punched brass spacers (15). A thin strip of acetate was placed between the spacer and the tube as the coil was being wound. After winding was completed, the acetate was dissolved by acetone, and a uniform spacing was achieved between each coil layer. To minimize the pressure drop, several layers of tubes were used to provide multiple parallel paths for the high pressure stream. The length of each path of the high pressure fluid was made uniform by adjusting the pitch of the tube helix from layer to layer, thereby alleviating the problem of flow maldistribution.

Plate-Fin Exchangers. The plate-fin heat exchanger (Fig. 8) consists of stacks of alternating layers of corrugated metal sheets (fins) between thin metal sheets or plates (17). The edges of each layer are sealed with bars to provide both sealing and mechanical support at the edges. The flow passages within the corrugations are terminated with headers at each end of the heat exchanger. This type of heat exchanger may be constructed in the counterflow, crossflow, or multipass arrangements.

The plate-fin heat exchanger has the advantage that a compact system (high surface area-to-volume ratio) can be achieved. For example, as much as 1475 m² of heat-transfer surface area per cubic meter of exchanger volume (450 ft²/ft³) can be attained with a plate-fin heat exchanger.

Perforated-Plate Exchanger. The perforated-plate heat exchanger (Fig. 9) consists of several parallel perforated plates or screens with spacers between each plate (18).

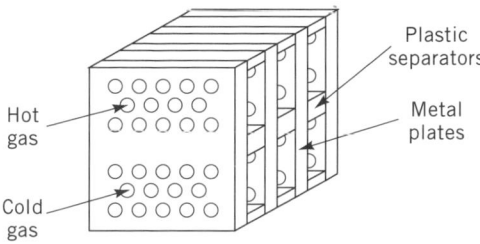

Figure 9. The perforated-plate heat exchanger.

The plates are made of high conductivity material, such as copper or aluminum, and the spacers are made of a low conductivity plastic. The spacers are used to provide sealing between the plates and to reduce longitudinal conduction in the heat exchanger. The two fluid streams generally flow in counterflow, and heat exchange between the streams occurs by lateral conduction through the copper or aluminum plates.

The diameter of the perforations can be made very small to attain a large heat-transfer surface area per unit volume. The gaps between the plates allow a uniform flow pattern to be achieved between each plate. Because the length-to-diameter ratio for the perforations is relatively small, both the hydrodynamic and thermal boundary layers are disrupted at the exit of the perforation before the flow within the perforation becomes fully established. This effect results in both high convective heat-transfer coefficients and high friction factors.

Regenerators. In an ordinary heat exchanger (recuperator), the two fluids exchanging energy are separated by a solid surface, such as a tube wall or a plate. In a valved-type regenerator (Fig. 10), the two fluids occupy the same space at alternate times in the regenerator, while energy is stored and released from the porous regenerator matrix material. Several types of configurations have been used for this matrix, including small metal spheres, woven wire screens, punched copper disks, and corrugated metal strips. The temperature of the fluid within the regenerator and the temperature of the regenerator matrix are functions of both position and time of operation. In addition to the conventional thermal parameters, the regenerator effectiveness is also dependent on the mass of the matrix, the specific heat of the matrix material, and the switching frequency of the fluid flow through the regenerator (19).

There are several advantages of regenerators over recuperators. (1) A larger heat-transfer surface area per unit volume, as much as 4100 m² of surface area per cubic meter of volume (1250 ft²/ft³), can be obtained in a regenerator, compared with 1475 m²/m³ (450 ft²/ft³) for a compact plate-fin heat exchanger. (2) Regenerators are generally more simple to fabricate, thus costs are lower than those for a heat exchanger having the same heat-transfer capability. (3) The regenerator tends to be self-cleaning because of periodic flow reversal; therefore impurity buildup, such as frozen CO_2 or other solids, is not as severe.

The primary disadvantage of a regenerator is that some mixing or carryover of the warm and cold streams occurs during the switching operation.

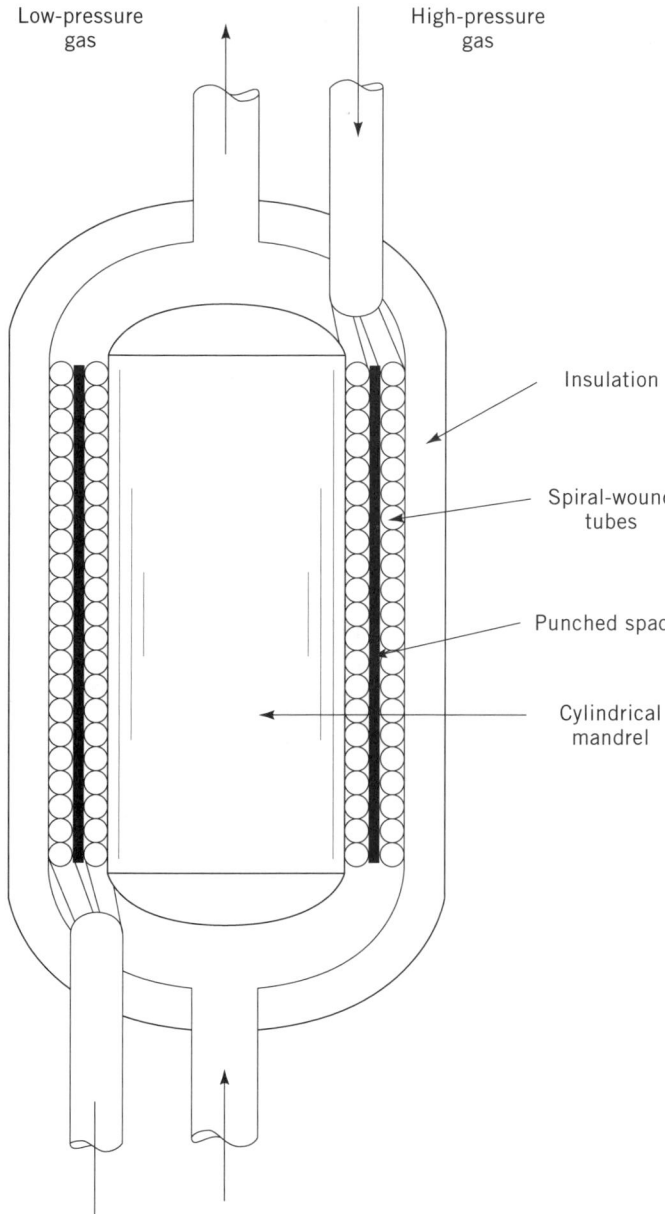

Figure 7. The Giauque-Hampson heat exchanger.

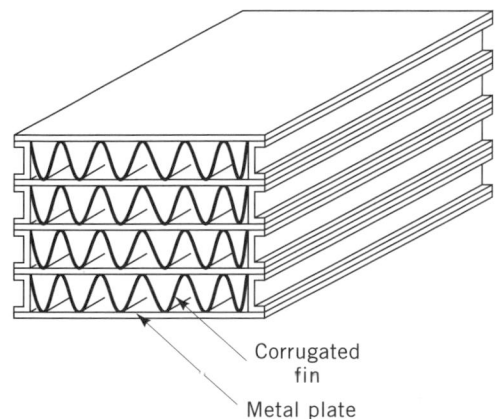

Figure 8. The plate-fin heat exchanger.

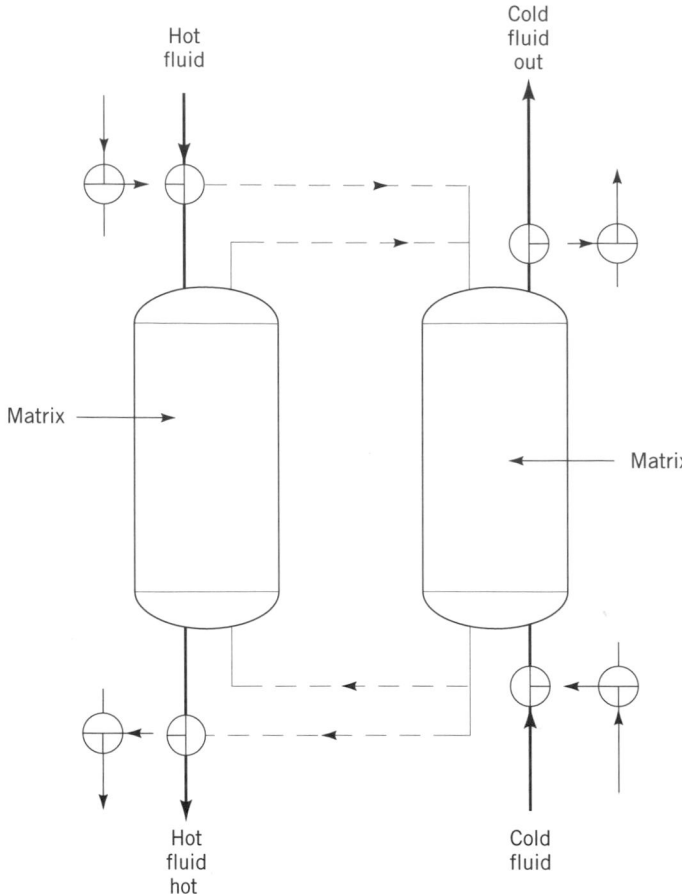

Figure 10. Valved-type regenerator where (**a**) shows heating and (**b**) shows cooling of the matrix. (---) Corresponds to the switched condition.

Heat Exchanger Design Problems. Four effects present particular challenges in the design of heat exchangers and regenerators for cryogenic service: (*1*) the effect of variable fluid properties, in particular the variation of specific heat; (*2*) the effect of longitudinal conduction through the solid material in the heat exchanger; (*3*) the effect of uneven distribution of flow across the heat-exchanger cross section; and (*4*) for regenerators, the effect of the T^3 variation of the specific heat of the solid matrix at very low temperatures.

In many heat exchanger design situations, the assumption of constant specific heats of the two streams is adequate; however, for gaseous hydrogen in the cryogenic temperature range, the specific heat varies considerably, ie, by about 40% in the temperature interval between 80 K and ambient temperature, 300 K. This specific heat variation can result in a thermal pinch effect within the heat exchanger. A thermal pinch effect is the phenomenon in which the hot and cold fluid temperatures approach a common value at some point within the heat exchanger.

The problem of thermal pinch may be alleviated by increasing the flow rate of the hydrogen gas, ie, thermally unbalancing the heat exchanger. An alternative solution is to decrease the flow rate of the other gas stream, eg, helium, by bleeding off a portion of the flow at some intermediate point in the exchanger.

Longitudinal heat conduction along the separating surfaces of conventional heat exchangers usually does not present a serious problem because the length of the heat exchanger is quite large. However, miniature cryogenic refrigerators use heat exchangers that may be as short as 150 mm. In the perforated-plate heat exchanger, for example, the effect of longitudinal conduction is minimized by the use of low conductivity spacers between the perforated plates.

Many cryogenic heat exchangers have several parallel paths for one or both fluid streams, instead of a single path for each stream. Ideally, the flow divides equally between each flow passage, but it is not always possible to assure that the flow is evenly distributed because of manufacturing tolerances. As a result, the flow rate is larger in the flow passage that has the smaller hydraulic resistance. The performance degradation that results from flow maldistribution is most severe for heat exchangers that have a very high effectiveness, such as those in cryogenic service. The effect of uneven flow distribution in cryogenic heat exchangers has been analyzed (20) where it was shown that uneven flow distribution could reduce the heat exchanger effectiveness from 0.995 to as low as 0.95.

The volumetric specific heat of the matrix material in a regenerator is ordinarily much larger than that of the gas flowing through the regenerator. This characteristic allows the solid material to store energy during the heating or cooling cycle without experiencing large temperature swings. At temperatures below about one-tenth of the Debye characteristic temperature of the material, the specific heat begins to follow a T^3 variation, which means that the specific heat of the solid matrix decreases rapidly as the temperature is lowered. As a result, storing energy effectiveness is rapidly degraded. For example, the specific heat of helium gas at 405 kPa (4 atm) and 20 K is approximately the same as the specific heat for copper at the same temperature (21).

One solution for this problem is to use regenerator matrix materials that have very low Debye temperatures (eg, lead shot). Another approach (22) is to utilize certain rare-earth materials, such as europium sulfide, that have large peaks in the specific heat curve at very low temperatures (eg, 16 K for EuS).

COST RELATIONSHIPS

The costs of cryogenic refrigerators and components of refrigerators (transfer lines, storage dewars, etc) have been reported (23–26). The cost of cryogenic refrigerators as a function of the required input power is shown in Figure 11. The input power requirement is observed to be a more important parameter in determining the cost of a cryogenic refrigerator than is the refrigeration temperature (23).

For refrigerators having a cooling capacity between 30 W and 15 kW, the following expression is given (26) for the cost of the refrigerator as a function of the refrigeration capacity:

$$C(M\$) = 4.33(Q_a/T_c)^{0.7} \qquad (17)$$

where Q_a = refrigeration capacity (kW) and T_c = refrigeration temperature (K). The cost C is given in $\times 10^6$ (ca

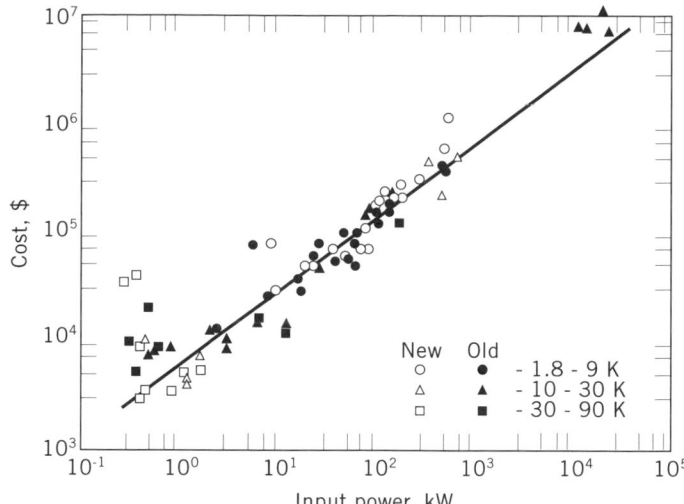

Figure 11. Correlation of cryogenic refrigerator cost as a function of the input power, where $\circ$, $\triangle$, $\square$ (new) and $\bullet$, $\blacktriangle$, $\blacksquare$ (old) are 1.8–9, 10–30, and 30–90 K, respectively (23).

1991). Very small refrigerators (capacity <30 W) have costs that exceed the values predicted by equation 17.

The following expression for estimation of the annual cost of refrigeration is also given (26).

$$C_A(M\$/\text{yr}) = 8.79(Q_a/T_c)^{0.78}(C_E)^{0.56} \quad (18)$$

where C_E = energy costs ($/kW · h). The annual cost of refrigeration includes amortization of the refrigerator, depreciation of the system, operation and maintenance labor, electric power, and auxiliary cooling costs. If amortization and depreciation costs are not incurred, the annual cost of refrigeration is approximately two-thirds of that given by equation 18.

The emphasis in refrigerator development and design has been directed toward minimizing equipment costs and maximizing the system reliability. As a result, the coefficient of performance of contemporary cryogenic refrigerators has not been significantly increased over that for refrigerators produced in the 1970s.

The cost for vacuum-jacketed (V-J) cryogenic pipe has been correlated (25). For "off the shelf" or catalog standard products, the cost per unit length ($/m) is given by

$$C(\$/m) = 120 + 57.5\,D \quad (19)$$

where D = nominal inner pipe diameter (in.). This expression applies for V-J lines in the size range between 0.5 and 6 in. If the lines have liquid nitrogen tracing (for helium transfer lines), the cost is considerably higher:

$$C(\$/m) = 660 + 130\,D \quad (20)$$

Transfer lines that require long lengths and are field erected follow a cost curve given by

$$C(\$/m) = 250 + 140\,D \quad (21)$$

The cost of dewars for storage of cryogenic liquids has been examined (25). For liquid helium dewars, and ap-

proximately for liquid hydrogen dewars, the cost for standard sized dewars may be estimated from

$$C(\$) = 70\,V^{0.8} \quad (22)$$

where V = nominal capacity of the dewar (L). This expression applies for vessels in the size range between 200 and 40,000 L (50 and 10,000 U.S. gallons, respectively). For liquid nitrogen, and also for liquid oxygen and liquid argon dewars, the cost for standard units may be estimated from

$$C(\$) = 120\,V^{0.6} \quad (23)$$

The costs for dewars estimated from equations 22 and 23 correspond to 1980 U.S. dollars and have generally followed the general cost index.

APPLICATIONS

Superconductivity. There are many engineering applications of superconductivity. Superconductivity is a phenomenon involving the simultaneous disappearance of all electrical resistance and the appearance of near perfect magnetic insulation (diamagnetism). Superconducting magnets (solenoids) may be used as levitation systems for commuter trains, developed by the Japanese National Railways, that operate at speeds up to 500 km/h (27). The levitation system consists of a normal track conductor coil on the ground and a superconducting coil on the train. The superconducting coils are maintained at liquid helium temperature (4.2 K) by means of a dewar filled with liquid helium. The propulsion unit for the train is a linear induction motor with the primary coil located on the ground and the secondary coil placed on the train.

The energy density (energy per unit volume) for a 10-tesla magnetic field is $E/V = B^2/2\mu_0 = 40$ MJ/m^3 (0.3 kW · h/ft^3). This value is considerably higher than the capacity of capacitor banks, about the same order of magnitude as that for mechanical flywheels, and somewhat lower than the energy storage capacity for lead-acid batteries. Superconducting magnetic energy storage (SMES) has been proposed (28) for smoothing out the daily load demand in large central electric generating plants. The super-conducting coils, approximately 100 m in diameter, will have a storage capacity of 30,000 kW · h of electric energy. SMES also has been considered for energy storage requirements in communication systems and advanced weapon systems.

Superconducting electric motors have been used in applications such as ship propulsion systems (29). The superconducting d-c motor is smaller and more efficient than conventional motors, which use large quantities of iron in the armatures and field cores. Superconducting motors as large as 3250 hp capacity have been placed into service.

There are numerous medical applications including the use of superconducting magnets in magnetic resonance imaging (MRI) systems. Because these magnets require less space than conventional magnets and can produce an absolutely constant magnetic field, they meet the stringent requirements of MRI equipment.

Tiny superconducting electronic elements, Superconducting QUantum Interference Devices (SQUIDs), are used in extremely sensitive digital magnetometers and voltmeters (30). These devices are so sensitive that they can be used effectively in detecting excessive iron concentrations in patients having hemochromatosis, a condition in which the liver has lost the ability to regulate iron. Superconducting sensors are used in magnetoencephalography (MEG) to detect epileptic centers deep within the brain by producing a three-dimensional picture of brain activity.

Space Vehicle Propulsion Systems

The first flight test of a liquid fuel rocket engine using liquid oxygen was made in 1926. Since that time, cryogenics has been used extensively in rocket propulsion systems (31). The NASA Saturn V Apollo vehicle used in the 1960s used liquid oxygen (normal bp, $-183°C$) as the oxidant for all stages and liquid hydrogen (normal bp, $-253°C$) as the fuel for the upper stages. The Space Shuttle propulsion system uses liquid oxygen as the oxidant and liquid hydrogen as the fuel.

Recycling of Rubber

Automobile or truck tires are one of the more difficult materials to recycle. Even year 160 to 200 million tires, which contain about 1.5 million metric tons of rubber, are scrapped in the United States alone (32). The conventional reclamation method involves chemical devulcanization of the rubber and mechanical grinding of the resulting material, a technique that is uneconomical when applied to steel belted tires.

When rubber is cooled to a temperature below about -60 to $-100°C$, it passes through a glass transition and becomes brittle. In this brittle condition, which can be achieved by spraying liquid nitrogen onto the tire, the tire can be crushed easily into small particles. The cord and metal components in the original tire can be separated economically by magnetic and air-blast techniques.

Food Freezing

Foods are prepared in the cryogenic food freezing process by moving the food packages on a conveyor belt through a liquid nitrogen spray in a specially insulated tunnel (33). Initial contact with the liquid nitrogen freezes the exposed surfaces and seals in both the flavor and aroma of the food product. The frozen food packages are then passed through an equilibrium zone in the tunnel that is cooled by nitrogen vapor. Very little (0.2 to 0.3% by weight) moisture is lost from the food during the cryogenic freezing process, resulting in better flavor retention, no degradation of texture, and improved appearance of the frozen food when compared with the corresponding product frozen by conventional techniques.

Medical Applications

There are several applications of cryogenics in the field of medicine, in addition to those involving superconducting elements. Whole blood, tissue, bone marrow, cornea, and animal semen have been preserved for extended periods by freezing the material in containers immersed in liquid nitrogen (34). Cryosurgery has been used successfully in a variety of surgical procedures, including eye, brain, and tumor surgery, as well as for tonsilectomies and in gynecology (35). Cryogenic surgery has several advantages over conventional surgery, including reduced problems with bleeding, greater control of the affected area, and less trauma to the patient's system.

FUTURE DEVELOPMENTS

The future of cryogenics is characterized by the diversity of its applications. During the first half of the twentieth century, problems facing engineers and scientists working in the low temperature arena included gas liquefaction, and cryogenic liquid transfer and storage. Storage losses were relatively high, because high performance multilayer insulations were not developed. Liquid helium was only available in small quantities and from relatively few laboratories. Superconductivity was known only at temperatures below 20 K.

During the last half of the twentieth century the production and sale of liquified gases has become a multibillion dollar business in the United States. Advanced multilayer insulations are being used to reduce cryogen boil-off in large containers to less than 0.05% per day. Large-scale liquid helium liquefaction plants have been brought on-line, and the development of the Collins helium liquefier has brought liquid helium production capability to almost all low temperature laboratories.

One of the most exciting developments in cryogenics was the discovery of the high Tc ceramic superconductors (HTSC) in 1986 (36). The compound $YBa_2Cu_3O_{7-d}$ has a transition temperature in zero magnetic field of 95 K ($-178°C$), which is above the normal boiling point of liquid nitrogen (77 K). This means that the inexpensive refrigerant liquid nitrogen can be used to cool the superconductor, instead of the more expensive liquid helium required with conventional superconductors. When the technology of production of HTSC materials is developed to the point that commercial equipment containing high performance HTSCs can be constructed, cryogenics will make a significant impact. Small superconducting motors may be used in many drive systems. Superconducting solenoids cooled by liquid nitrogen may replace the conventional lead–acid battery, thereby eliminating environmental problems of battery disposal. Magnetic levitation using HTSC magnets may become an economic reality for the transportation system in the United States.

Another cryogenic development for the future is the use of liquid hydrogen as a fuel for automobiles and trucks (37). More than a dozen vehicles have been modified and tested with liquid hydrogen fuel. A pumping station for liquid hydrogen has been built and tested by workers at Los Alamos National Lab in 1974 (38). Liquid hydrogen is a nonpolluting fuel, because the products of combustion include water vapor, having a minor quantity of oxides of nitrogen. When the cost of gasoline becomes excessive, or gasoline is not available at any cost, liquid hydrogen may fill the fuel needs of the world's transportation system.

NOMENCLATURE

a	constant in van der Waals equation, $Pa \cdot m^6/kg^2$
B	magnetic induction, T
b	constant in van der Waals equation, m^3/kg
C	cost, U.S. \$ or \$/m
COP	Q_a/W, coefficient of performance
C_p	specific heat at constant pressure, $J/kg \cdot K$
D	diameter, m
E	energy
e	heat exchanger effectiveness
h	enthalpy, J/kg
m	mass flow rate, kg/s
p	pressure, Pa
Q_a	heat-transfer rate (refrigeration), W
R	gas constant, $J/kg \cdot K$
s	entropy, $J/kg \cdot K$
T	absolute temperature, K
T_c	low temperature in the cycle, K
T_i	inversion temperature, K
T_n	high temperature in the cycle, K
v	specific volume, m^3/kg
V	storage vessel capacity, m^3
W	power requirement, W

Greek symbols

γ	specific heat ratio, C_p/C_v
η_{ad}	expander adiabatic efficiency
η_{CO}	compressor overall isothermal efficiency
η_m	expander mechanical efficiency
μ_{JT}	Joule-Thomson coefficient, K/Pa
μ_0	permeability of vacuum, $\mu_0 = 4\pi \times 10^{-7} \ T \cdot m/A$

BIBLIOGRAPHY

1. R. B. Scott, *Cryogenic Engineering,* D. van Nostrand Co., Princeton, N.J., 1959, p. 1.
2. J. P. Joule and W. Thomson, *Philosoph. Trans. Roy. Soc. London* **143**(I), 357 (1853).
3. J. P. Joule and W. Thomson, *Philosoph. Trans. Roy. Soc. London* **144**(I), 321 (1854).
4. M. L. Moran and H. N. Shapiro, *Fundamentals of Engineering Thermodynamics,* 2nd ed., John Wiley & Sons, Inc., New York, 1992, p. 498.
5. R. F. Barron, *Cryogenic Systems,* 2nd ed., Oxford University Press, New York, 1985, p. 129.
6. G. Walker, *Stirling Engines,* Oxford University Press, London, 1980.
7. A. Kirk, *Proc. Inst. Civil Eng. (London)* **37,** 244–315 (1874).
8. G. Walker, *Cryocoolers, Part 1,* Plenum Press, New York, 1983, pp. 280–282.
9. T. Finkelstein, in K. D. Timmerhaus, ed., *Advances in Cryogenic Engineering,* Vol. 20, Plenum Press, New York, 1975, pp. 269–282.
10. U.S. Pat. 1,275,507 (1918), R. Vuilleumier.
11. R. White, "Vuilleumier Cycle Cryogenic Refrigeration," *AFFDL-TR-76-17,* (AD/A-)27-055, Wright-Patterson Air Force Base, Dayton, Ohio, 1976.
12. C. K. Chan, E. Tward, and W. W. Burt, *Advances in Cryogenic Engineering,* Vol. 35, Plenum Press, New York, 1990, pp. 1239–1250.
13. W. E. Gifford, *Advances in Cryogenic Engineering,* Vol. 11, Plenum Press, New York, 1961, pp. 152–159.
14. W. E. Gifford and T. E. Hoffman, *Advances in Cryogenic Engineering,* Vol. 6, Plenum Press, New York, 1961, pp. 82–94.
15. G. Walker, *Cryocoolers, Part 2,* Plenum Press, New York, 1983, pp. 177–251.
16. R. F. Barron, in Ref. 5, pp. 109–113.
17. R. B. Scott, in Ref. 1, p. 19.
18. R. B. Fleming, in K. D. Timmerhaus, ed., *Advances in Cryogenic Engineering,* Vol. 14, Plenum Press, New York, 1969, pp. 197–204.
19. W. M. Kays and A. L. London, *Compact Heat Exchangers,* 3rd ed., McGraw-Hill Book Co., New York, 1984, pp. 29–33.
20. R. B. Fleming, *Advances in Cryogenic Engineering,* Vol. 12, Plenum Press, New York, 1967, pp. 240–251.
21. A. Daniels and F. K. du Pre, *Advances in Cryogenic Engineering,* Vol. 16, Plenum Press, New York, 1971, pp. 178–184.
22. H. Hashimoto and co-workers, *Advances in Cryogenic Engineering,* Vol. 37, Plenum Press, New York, 1991, pp. 859–865.
23. T. R. Strobridge, "Cryogenic Refrigerators–An Updated Survey," NBS Technical Note 655, U.S. Government Printing Office, Washington, D.C., 1974.
24. C. Trepp, *Low-Temperatures and Electric Power,* Pergamon Press, New York, 1970, pp. 31–41.
25. G. Y. Robinson, *Advances in Cryogenic Engineering,* Vol. 25, Plenum Press, Inc., New York, 1981, pp. 342–349.
26. M. A. Green, R. A. Byrns, and S. J. St. Lorant, *Advances in Cryogenic Engineering,* Vol. 37, Plenum Press, Inc., New York, 1991, pp. 637–643.
27. K. Oshima and Y. Kyotani, *Advances in Cryogenic Engineering,* Vol. 19, Plenum Press, New York, 1974, pp. 127–136.
28. R. W. Boom and H. A. Peterson, *IEEE Trans. Mag.* **8,** 701 (1972).
29. A. D. Appleton, *Advances in Cryogenic Engineering,* Vol. 16, Plenum Press, New York, 1971, pp. 11–18.
30. J. E. Zimmerman and R. Radebaugh, "Operation of a SQUID in a Very Low-Power Cryocooler," *NBS Special Publication 508,* U.S. Government Printing Office, Washington, D.C., 1978.
31. C. A. Bailey, *Advanced Cryogenics,* Plenum Press, New York, 1971, pp. 355–371.
32. N. R. Brayton, *Cryogenic Recycling and Processing,* CRC Press, Boca Raton, Fla., 1980, pp. 211–219.
33. D. C. Brown, *Advances in Cryogenic Engineering,* Vol. 12, Plenum Press, New York, 1967, pp. 11–22.
34. A. U. Smith, *Current Trends in Cryobiology,* Plenum Press, New York, 1970, pp. 153–180.
35. I. S. Cooper, in H. von Leden and W. G. Cahan, eds., *Cryogenics in Surgery,* Medical Examination Publishing Co., Flushing, N.Y., 1971, pp. 31–38.
36. C. P. Poole, T. Datta, and H. A. Farach, *Copper Oxide Superconductors,* Wiley-Interscience, New York, 1988, pp. 3–5.
37. W. F. Stewart, *Advances in Cryogenic Engineering,* Vol. 25, Plenum Press, New York, 1980, pp. 822–830.
38. W. F. Stewart and F. J. Edeskuty, *Mech. Eng.* **96,** 22 (1974).

Reading List

Advances in Cryogenic Engineering, Vols. 1–40, Plenum Press, New York, 1960–1994.

A. Arkharov, I. Marfenina, and Ye. Mikulin, *Theory and Design of Cryogenic Systems,* Mir Publishers, Moscow, 1981.

R. F. Barron, *Cryogenic Systems,* 2nd ed., Oxford University Press, London, 1985.

B. A. Hands, *Cryogenic Engineering,* Academic Press, Inc., New York, 1986.

V. Z. Kresin and S. A. Wolf, *Fundamentals of Superconductivity,* Plenum Publishing Corp., New York, 1990.

P. V. E. McClintock, D. J. Meridith, and J. K. Wigmore, *Matter at Low Temperatures,* Wiley-Interscience, New York, 1984.

K. D. Timmerhaus and T. M. Flynn, *Cryogenic Process Engineering,* Plenum Publishing Corp., New York, 1991.

S. W. Van Sciver, *Helium Cryogenics,* Plenum Publishing Corp., New York, 1986.

G. Walker, *Cryocoolers, Part 1: Fundamentals; Part 2, Applications,* Plenum Publishing Corp., New York, 1983.

K. D. Williamson, Jr., and F. J. Edeskuty, *Liquid Cryogens,* Vol. 1, *Theory and Equipment;* Vol. 2, *Properties and Applications,* CRC Press, Cleveland, Ohio, 1983.

M. N. Wilson, *Superconducting Magnets,* Oxford University Press, London, 1983.

D

DEWAXING OF PETROLEUM

James Speight
Western Reserve Institute
Laramie, Wyoming

Dewaxing is the process of separating hydrocarbons which solidify readily (waxes) from petroleum fractions (1). Dewaxing processes were originally designed, and are still used, to remove wax from lubricating oils to give the product good fluidity characteristics at low temperatures (eg, low pour points). Thus, dewaxing is an important step in the manufacture of lubricating oils. The wax removed in the process may need to be purified further to produce paraffin or microcrystalline waxes for commercial use.

See also Petroleum refining; Lubricants.

In the 1930s two types of stocks, naphthenic and paraffinic, were used to make motor oils. Both types were solvent extracted to improve their quality, but in the high-temperature conditions encountered in service the naphthenic type could not stand up as well as the paraffinic type. Nevertheless, the naphthenic type was the preferred oil, particularly in cold weather, because of its fluidity at low temperatures. Such products could be further treated with phenol to yield products having pour points of −40°C to −7°C, depending on the viscosity of the oil. Paraffinic oils were also available and could be treated with phenol to produce a higher quality oil, but their wax content was so high that the "oil" products were often solid at ambient temperature.

The lowest viscosity paraffinic oils were dewaxed by the *cold press method* to produce oils with a pour point of 2°C. The light paraffin distillate oils contained a paraffin wax that crystallized into large crystals when chilled and could be readily separated from the oil by the cold press filtration method. The more viscous paraffinic oils (intermediate and heavy paraffin distillates) contained amorphous or microcrystalline waxes, which formed small crystals that plugged the filter cloths in the cold press and prevented filtration. Because the wax could not be removed from intermediate and heavy paraffin distillates, the high-quality, high-viscosity lubricating oils in them could not be used except as cracking stock.

Methods were therefore developed to dewax the high viscosity paraffinic oils (2–4). The methods were essentially alike in that the waxy oil was dissolved in a solvent that would keep oil in solution, whereas crystalline wax separated when the temperature was lowered.

Most commercial dewaxing processes utilize *solvent dilution,* chilling to crystallize the wax, and filtration. The *MEK process* (methyl ethyl ketone–toluene solvent) is widely used. Wax crystals are formed by chilling through the walls of scraped surface chillers, and wax is separated from the resultant wax–oil–solvent slurry by using fully enclosed rotary vacuum filters. In one process modification, most of the chilling is accomplished by multistage injection of very cold solvent into the waxy oil with vigorous agitation, resulting in more uniform and compact wax crystals which filter faster. The processes differed chiefly in the use of the solvent. Commercially used solvents were naphtha, propane, sulfur dioxide, acetone–benzene, trichloroethylene, ethylene dichloride–benzene (Barisol), methyl ethyl ketone–benzene (benzol), methyl-*n*-butyl ketone, and methyl-*n*-propyl ketone. Other solvents in commercial use for dewaxing include MEK–MIBK (methyl isobutyl ketone), acetone–benzene, dichloroethane–methylene dichloride, and propylene–acetone.

The mechanism of *solvent dewaxing* can involve either the separation of wax as a solid that has been crystallized from the oil solution at low temperature or the separation of wax as a liquid that has been extracted at temperatures above the melting point of the wax by preferential selectivity of the solvent.

In the first solvent dewaxing process (developed in 1924), the waxy oil was mixed with naphtha and filter aid (fuller's earth or diatomaceous earth). The mixture was chilled and filtered, and the filter aid assisted in building a wax cake on the filter cloth. This process is now obsolete, and most of the modern dewaxing processes use a mixture of methyl ethyl ketone and benzene. Other ketones may be substituted for dewaxing but, regardless of which ketone is used, the process is generally known as *ketone dewaxing.*

The process is carried out by mixing waxy oil with one to four times its volume of ketone and heating the mixture until the oil is in solution (Fig. 1). The solution is then chilled at a slow, controlled rate in double-pipe scraped-surface exchangers. Cold solvent, such as filtrate from the filters, passes through the 2-in. annular space between the inner and outer pipes and chills the waxy oil solution flowing through the inner 6-in. pipe. To prevent wax from depositing on the walls of the inner pipe, blades or scrapers (extending the length of the pipe and fattened to a central rotating shaft) scrape off the wax. Slow chilling reduces the temperature of the waxy oil solution to 2°C and, then, faster chilling reduces the temperature to the approximate pour point required in the dewaxed oil. The waxy mixture is pumped to a filter case into which the bottom half of the drum of a rotary vacuum filter dips. The drum (2.4 m-dia, 4.3 m long), covered with filter cloth, rotates continuously in the filter case. Vacuum within the drum sucks the solvent and the oil dissolved in the solvent through the filter cloth and into the drum. Wax crystals collect on the outside of the drum to form a wax cake and, as the drum rotates, the cake is brought above the surface of the liquid in the filter case and under sprays of ketone that wash oil out of the cake and into the drum. A knife edge scrapes off the wax, the cake falls into the conveyor and is moved from the filter by the rotating scroll.

The recovered wax is actually a mixture of wax crystals with a little ketone and oil and the filtrate consists of the dewaxed oil dissolved in a large amount of ketone. Ketone is removed from both by distillation but, before the wax is distilled, it is deoiled, mixed with more cold ketone, and pumped to a pair of rotary filters in series where further washing with cold ketone produces a wax cake that con-

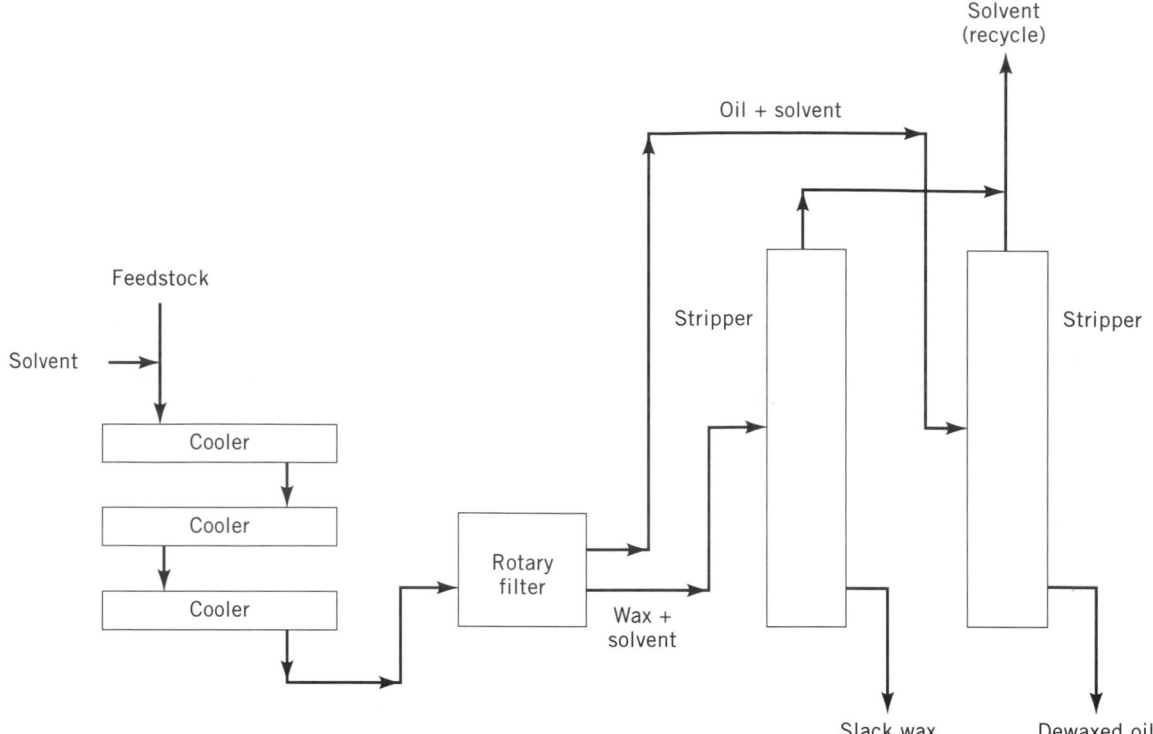

Figure 1. Solvent dewaxing.

tains very little oil. The deoiled wax is melted in heat exchangers and pumped to a distillation tower operated under vacuum where a large part of the ketone is evaporated or flashed from the wax. The rest of the ketone is removed by heating the wax and passing it into a fractional distillation tower operated at atmospheric pressure and then into a stripper where steam removes the last traces of ketone.

An almost identical system of distillation is used to separate the filtrate into dewaxed oil and ketone. The ketone from both the filtrate and wax slurry is reused. The dewaxed oil is finished by clay treatment or by hydrotreatment. The wax product, also known as slack wax, even though it usually contains no oil (as compared to 50% in the slack wax obtained by cold pressing), is the raw material for either wax sweating or wax recrystallization, which subdivides the wax into a number of wax fractions with different melting points.

Solvent dewaxing can be applied to light, intermediate, and heavy lubricating oil distillates but each distillate produces a different kind of wax. Each of these waxes is actually a mixture of a number of waxes. For example, the wax obtained from light paraffin distillate consists of a series of paraffin waxes that have melting points in the range of 30–70°C and are characterized by a tendency to harden into large crystals. However, heavy paraffin distillate yields a wax composed of a series of waxes with melting points in the range of 60–90°C that harden into small crystals from which they derive the name of microcrystalline waxes, or microwaxes. On the other hand, intermediate paraffin distillates contain paraffin waxes and waxes intermediate in properties between paraffin and microwaxes. Thus, the solvent dewaxing process produces three

different slack waxes (also known as crude or raw waxes) depending on whether light, intermediate, or heavy paraffin distillate is processed. The slack wax from heavy paraffin distillate may be sold as dark raw wax, the wax from intermediate paraffin distillate as pale raw wax. The latter is treated with lye and clay to remove odor and improve color.

In the propane process, part of the propane diluent is allowed to evaporate by reducing pressure, so as to chill the slurry to the desired filtration temperature, and rotary pressure filters are employed.

Complex dewaxing requires no refrigeration, but depends upon the formation of a solid urea/n-paraffin complex which is separated by filtration and then decomposed. This process is used to make low-viscosity lubricants which must remain fluid at very low temperatures (refrigeration, transformer, and hydraulic oils) (5). Similar use is anticipated for catalytic dewaxing processes, which are based on selective hydrocracking of the normal paraffins; it uses a molecular sieve-based catalyst in which the active hydrocracking sites are accessible only to the paraffin molecules.

Another method of separating petrolatum from reduced crude is centrifuge dewaxing. In this process, the reduced crude is dissolved in naphtha and chilled to −18°C or lower, which causes the wax to separate. The mixture is then fed to a battery of centrifuges where the wax is separated from the liquid. However, the centrifuge method has now been largely displaced by solvent dewaxing methods and by more modern methods of wax removal.

Catalytic dewaxing is a hydrocracking process (Figure 2) and is operated at elevated temperatures (280–400°C) and pressures of 2.1–10.3 MPa (300–1500 psi) (6). How-

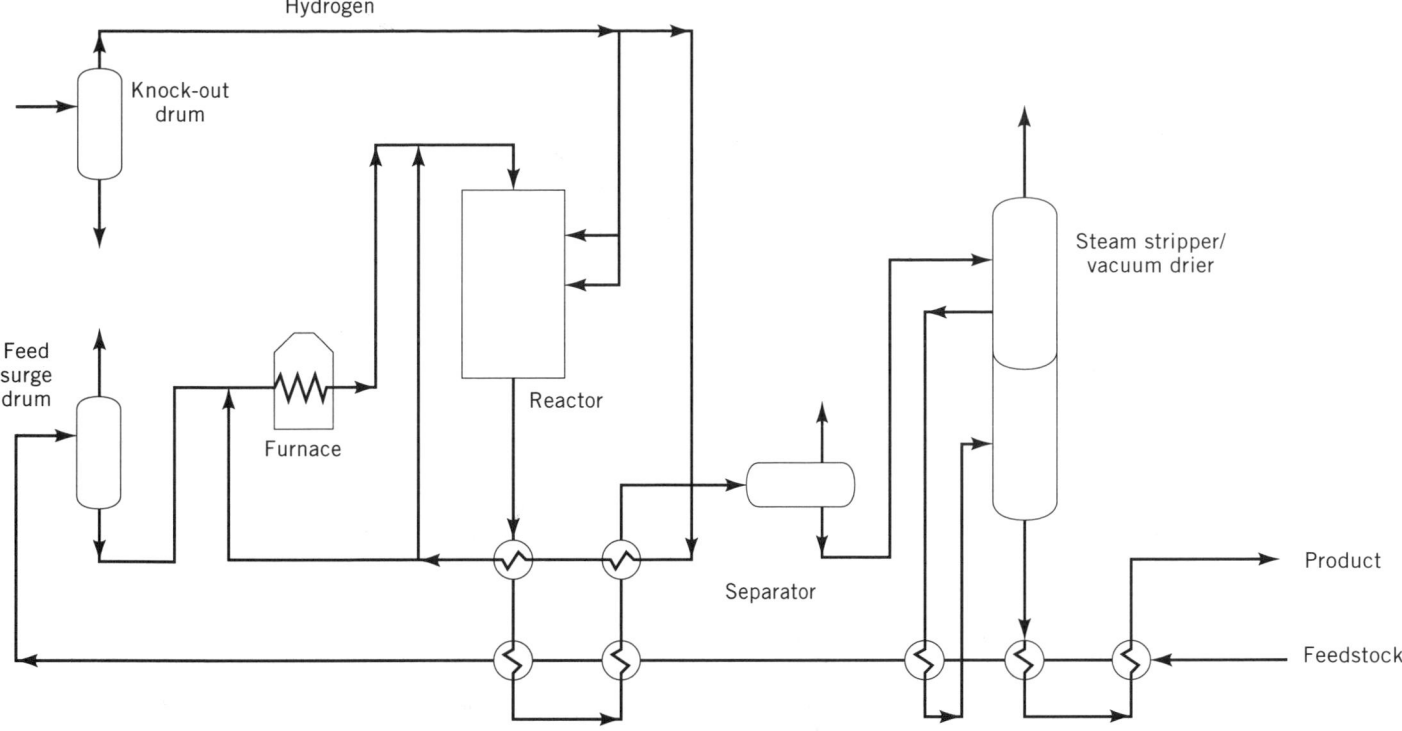

Figure 2. Catalytic dewaxing.

ever, the conditions for a specific dewaxing operation depend upon the nature of the feedstock and the product pour point required. The catalyst employed for the process is a mordenite-type catalyst that has the correct pore structure to be selective for normal paraffin cracking. Platinum on the catalyst serves to hydrogenate the reactive intermediates so that further paraffin degradation is limited to the initial thermal reactions. The process has been employed to successfully dewax a wide range of naphthenic feedstocks (7), but it may not be suitable to replace solvent dewaxing in all cases. The process has the flexibility to fit into normal refinery operations and can be adapted for prolonged periods on-stream.

Another catalytic dewaxing process (Fig. 3) also in-

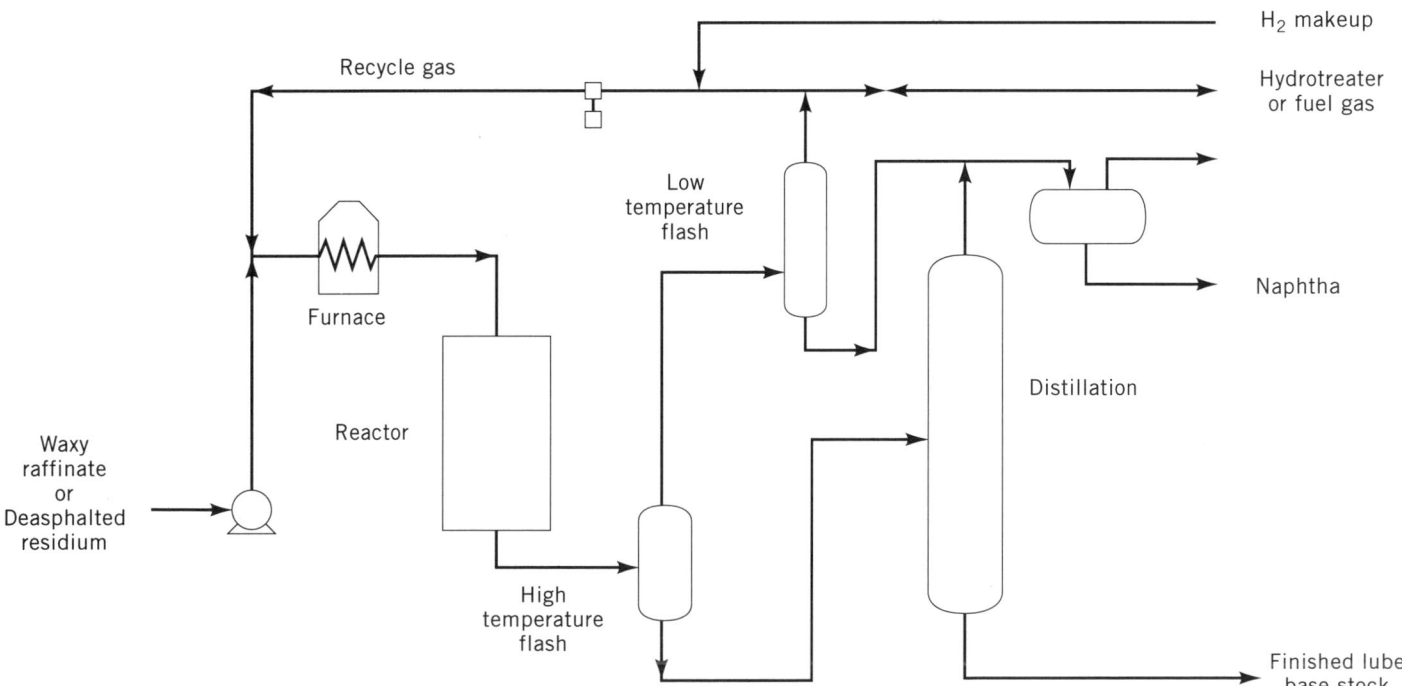

Figure 3. Catalytic dewaxing.

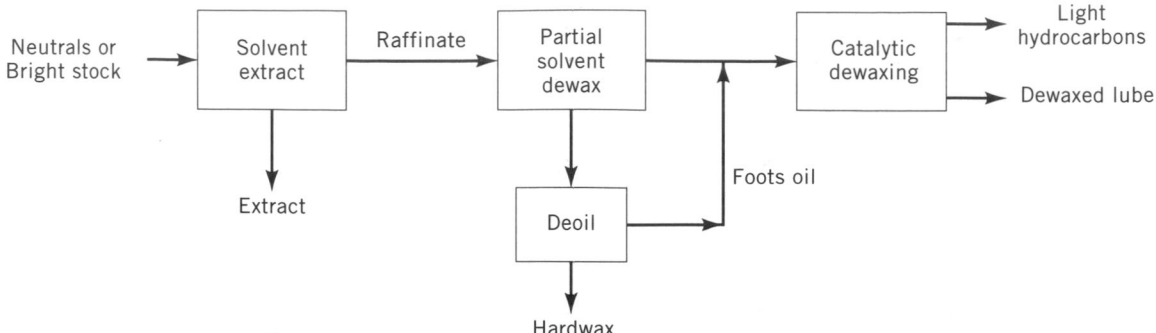

Figure 4. Catalytic dewaxing used in conjunction with solvent dewaxing.

volves selective cracking of normal paraffins and those paraffins that may have minor branching in the chain (8). In the process, the catalyst can be reactivated to fresh activity by relatively mild nonoxidative treatment. Of course, the time allowed between reactivations is a function of the feedstock, but after numerous reactivations it is possible that there will be coke buildup on the catalyst.

The catalytic dewaxing process(es) can be used to dewax a variety of lubricating base stocks and, as such, the processes have the potential to completely replace solvent dewaxing or can even be used in combination with solvent dewaxing (Fig. 4) as a means of relieving the bottlenecks which can, and often do, occur in solvent dewaxing facilities.

BIBLIOGRAPHY

1. J. G. Speight, *The Chemistry and Technology of Petroleum, Second Edition,* Marcel Dekker Inc., New York, 1991, Chapt. 18.

2. S. J. Marple and L. J. Landry, *Advances in Petroleum Chemistry and Refining,* **10,** 190 (1965).

3. G. H. Unzelman and C. J. Wolf, in W. F. Bland and R. F. Davidson, eds., in *Petroleum Processing Handbook,* McGraw-Hill Book Co. Inc., New York, 1967, pp. 3–92.

4. G. G. Scholten, in J. J. McKetta, ed., in *Petroleum Processing Handbook,* Marcel Dekker Inc., New York, 1992, p. 565.

5. *Ibid,* p. 583.

6. J. D. Hargrove, in J. J. McKetta, ed., in *Petroleum Processing Handbook,* Marcel Dekker Inc., New York, 1992, p. 558.

7. J. D. Hargrove, G. J. Elkes, and A. H. Richardson, *Oil and Gas Journal,* **77**(3), 103 (1979).

8. K. W. Smith, W. C. Starr, and N. Y. Chen, *Oil and Gas Journal* **78**(21), 75 (1980).

DISTILLATION

JAMES R. FAIR
The University of Texas at Austin
Austin, Texas

Distillation is a method of separation that is based on the difference in composition between a liquid mixture and the vapor formed from it. The composition difference is due to differing effective vapor pressures, or volatilities, of the components of the liquid mixture. When such a difference does not exist, (eg, at an azeotropic point), separation by distillation is not possible. Distillation normally involves condensation of the vaporized material, usually in multiple vaporization–condensation operations, and thus differs from evaporation, which is usually applied to separation of a liquid from a solid but which can be applied to simple liquid concentration operations. In the chemical and petroleum industries, distillation is one of the largest consumers of energy among the several processing operations employed. (See also PETROLEUM REFINING.)

Distillation is the most widely used industrial method of separating liquid mixtures and is at the heart of the separation processes in many chemical and petroleum plants. The most elementary form of the method is simple distillation, in which the liquid is brought to boiling and the vapor formed is separated and condensed to form a product. In some cases, a normally gaseous mixture is condensed by using refrigeration to obtain a liquid that is amenable to distillation separation, for example, the separation of air. If the process is continuous with respect to feed and product flows, it is called flash distillation. If the feed mixture is available as an isolated batch of material, the process is a form of batch distillation, and the compositions of the collected vapor and residual liquid are thus time-dependent. The term fractional distillation (also referred to as fractionation) was originally applied to the collection of separate fractions of condensed vapor; currently, the term is applied to distillation processes in general. When the vapors are enriched by contact with counterflowing liquid reflux, the process is called rectification. When fractional distillation is accomplished with a continuous feed of material and with a continuous removal of product fractions, the process is called continuous distillation. When steam is added to the vapors to reduce the partial pressures of the components to be separated, the term steam distillation is used.

Most distillations conducted commercially operate continuously, with a more volatile fraction recovered as distillate and a less volatile fraction recovered as bottoms (or residue); these are the large energy consumers. If a portion of the distillate is condensed and returned to the process to enrich the vapors, the liquid is called reflux. The apparatus in which the enrichment occurs is usually a vertical, cylindrical vessel called a still or distillation column. This apparatus normally contains internal devices for effecting vapor–liquid contact; the devices may be categorized as plates or packings.

Distillation has been practiced in one form or another for centuries and was of fundamental importance to the alchemists. Because it is a process involving vaporization of a liquid, energy must be supplied if it is to function.

VAPOR–LIQUID EQUILIBRIA

The equilibrium distributions of mixture component compositions in the vapor and liquid phases must be different if separation is to be made by distillation. The compositions at thermodynamic equilibrium are termed *vapor–liquid equilibria* (VLE) and may be correlated or predicted with the aid of thermodynamic relationships. The driving force for any distillation is a favorable vapor–liquid equilibrium, which provides the needed composition differences. Reliable VLE are essential for distillation column design and for most other operations involving liquid–vapor phase contacting, such as absorption and stripping. Many VLE have been measured and reported in the literature, and compilations of such data are available (1,2). Also, bibliographic guides have been published, providing source references for thousands of publications presenting VLE (3–5). VLE may also be measured, or estimated by generalized methods (6–8), with some sacrifice in reliability. Even if carefully measured data are available, thermodynamic models are usually required to extrapolate or interpolate them for conditions not represented by the experiments. Whatever the source and extent of the VLE, some evaluation should be made with regard to accuracy.

The equilibria for the distillation system may be simple and easily represented by an equation or, in some systems, may be so complex that they cannot be adequately measured or represented. Excellent treatises are available for assistance with the selection and implementation of vapor–liquid equilibrium studies (9–12).

Typical VLE are shown in Figure 1. The diagrams are for binary (two-component) systems. Figure 1(**a**) is a representative boiling point diagram; it shows equilibrium compositions as functions of temperature at a constant pressure. The lower line is the liquid bubble point line, the locus of points at which a liquid on heating forms the first bubble of vapor. The upper line is the vapor dew point line, representing points at which a vapor on cooling forms the first drop of condensed liquid. The liquid and vapor compositions are conventionally plotted in terms of the lower boiling (more volatile) component of the mixture; it is also customary to use mole fractions rather than weight fractions when defining compositions. In this discussion of the binary (two-component) system L–H, the low boiler is L. The system point A has a vapor composition of y_L^A in equilibrium with a liquid composition of x_L^A at a temperature of T^A. Figure 1(b) is a typical isobaric phase diagram (also called a y–x diagram). For further discussion, see References 13 and 14.

Thermodynamic Relationships

A closed container with vapor and liquid phases at thermodynamic equilibrium may be depicted as in Figure 2, where at least two mixture components are present in

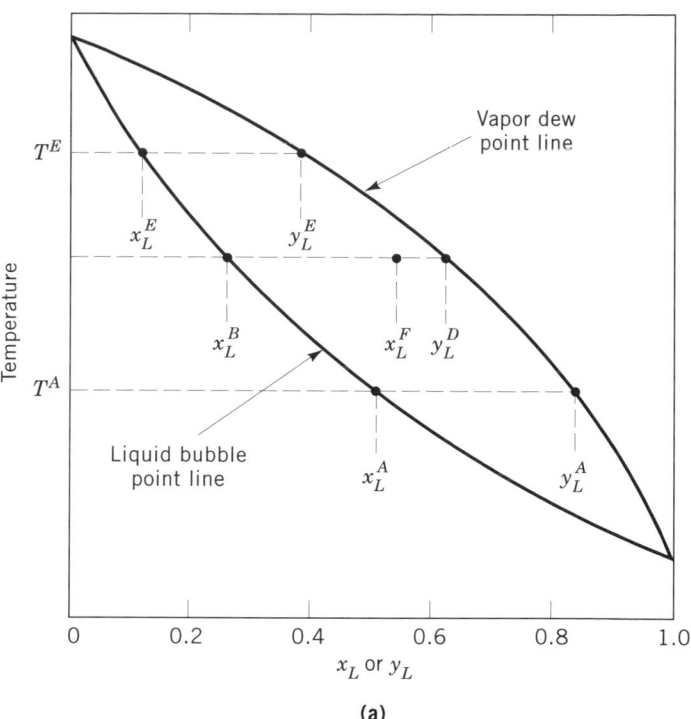

(a)

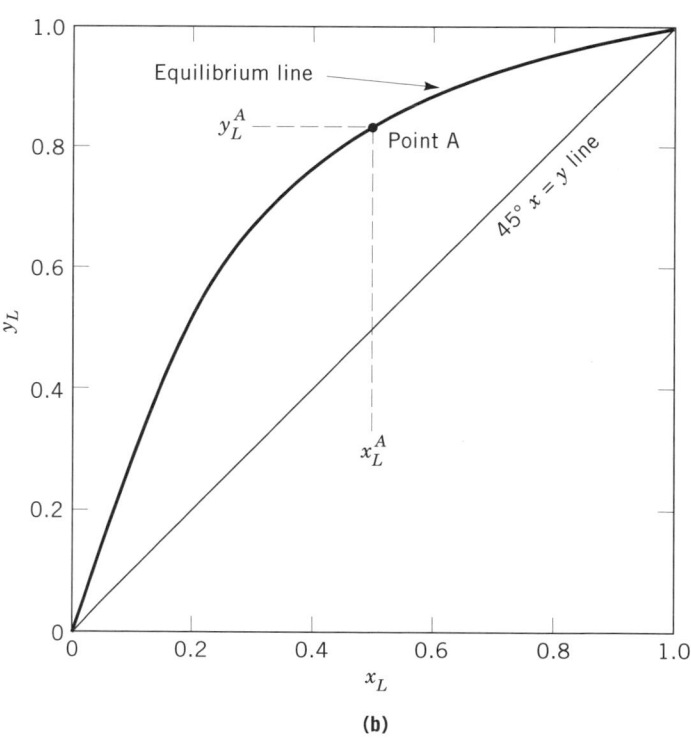

(b)

Figure 1. Isobaric VLE diagrams: (**a**) dew and bubble point; (**b**) vapor–liquid (y–x) equilibrium.

each phase. The components distribute themselves between the phases according to their relative volatilities. A distribution ratio for mixture component i may be defined using mole fractions:

$$K_i = y_i^* / x_i$$

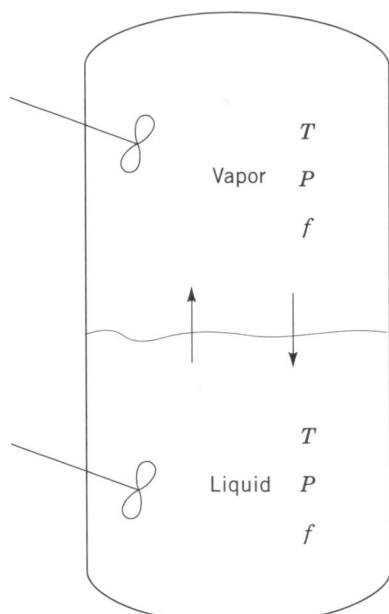

Figure 2. Equilibrium between vapor and liquid. The conditions for equilibrium are $T^v = T^L$ and $P^V = P^L$. For a given T and P, phase fugacities are equal, ie, $f^v = f^L$ and $f_i^v = f_i^L$.

where the asterisk denotes an equilibrium condition. The K value, also known as the vapor–liquid equilibrium ratio, is widely used, especially in the petroleum and petrochemical industries. For any two mixture components i and j, their relative volatility (often called the alpha value) is defined as

$$\alpha_{ij} = \frac{K_i}{K_j} = \frac{y_i x_j}{x_i y_j} = \frac{y_i(1 - x_i)}{x_i(1 - y_i)}$$

This equation may be rearranged to form an expression for all or a part of the equilibrium curve in Figure 1(**b**):

$$y_i = \frac{\alpha_{ij} x_i}{1 + (\alpha_{ij} - 1)x_i}$$

The relative volatility α is a direct measure of the ease of separation by distillation. If α equals 1, then component separation is impossible, as the liquid and vapor phase compositions are identical. Separation by distillation becomes easier as the value of the relative volatility becomes increasingly greater than unity. Distillation separations with α values less than 1.2 are relatively difficult, whereas those with values above 2 are relatively easy.

When both phases form ideal thermodynamic solutions (no heat of mixing, no volume change on mixing, etc), Raoult's law applies:

$$p_i^v = x_i P_i^0$$

where P_i^0 is the vapor pressure of i at equilibrium temperature. Combining this expression with Dalton's law of partial pressures, K values and relative volatilities may be obtained:

$$K_i = P_i^0 / P$$

$$\alpha_{ij} = P_i^0 / P_j^0$$

Examples of ideal binary systems are benzene–toluene and ethylbenzene–styrene; the molecules are similar and within the same chemical families. (Thermodynamics references should be consulted before assuming that a chosen binary or multicomponent system is ideal.) When pressures are low and temperatures are at ambient or above, but the solutions are not ideal (dissimilar molecules), corrections to the above equations may be made:

$$K_i = \gamma_i^L P_i^0 / P$$

$$\alpha_{ij} = \gamma_i^L P_i^0 / (\gamma_j^L P_j^0)$$

where the Raoult's law correction factor γ^L is a thermodynamically important liquid-phase activity coefficient.

The liquid-phase coefficients are strong functions of liquid composition and temperature, and, to a lesser extent, of pressure. Typical activity coefficient plots are shown in Figure 3. A system with positive deviation, ie, having activity coefficients greater than one (logarithm of the coefficient is positive), is shown in Figure 3(**a**); a system with negative deviation (coefficients less than unity, logarithms negative), is shown in Figure 3(**b**). In a few cases, one component of a binary mixture has a positive deviation and the other a negative deviation. Most commonly, both coefficients have positive deviations, and as indicated in Figure 3, the nonideality of a component (as measured by the value of γ) is greatest when at its most dilute condition.

Terminal activity coefficients (γ_i^∞) (often called infinite dilution coefficients) are noted in Figure 3 and for some, representative systems are listed in Table 1. The hexane–heptane mixture is included as an example of an ideal system. As the molecular species become more dissimilar, they are prone to repel each other, tend toward liquid immiscibility, and have large positive activity coefficients, as in the case of hexane–water.

If the molecular species in the liquid tend to form complexes, the system will have negative deviations and activity coefficients less than unity, eg, the system chloroform–ethyl acetate. In azeotropic and extractive distillation and in liquid–liquid extraction, nonideal liquid behavior is used to enhance component separation. An extensive discussion on the selection of nonideal addition agents is available (15).

A great deal of study and research has gone into the development of working equations that can represent the curves of Figure 3. One of the simplest and most often used equations is that of Van Laar (16). For a binary system of components 1 and 2, the Van Laar equations are

$$\ln \gamma_1 = \frac{A_{12}}{\left(1 + \dfrac{A_{12} x_1}{A_{21} x_2}\right)}$$

$$\ln \gamma_2 = \frac{A_{21}}{\left(1 + \dfrac{A_{21} x_2}{A_{12} x_1}\right)}$$

Only the two parameters directly related to the terminal activity coefficients are involved:

$$\ln \gamma_1^\infty = A_{12}$$

$$\ln \gamma_2^\infty = A_{21}$$

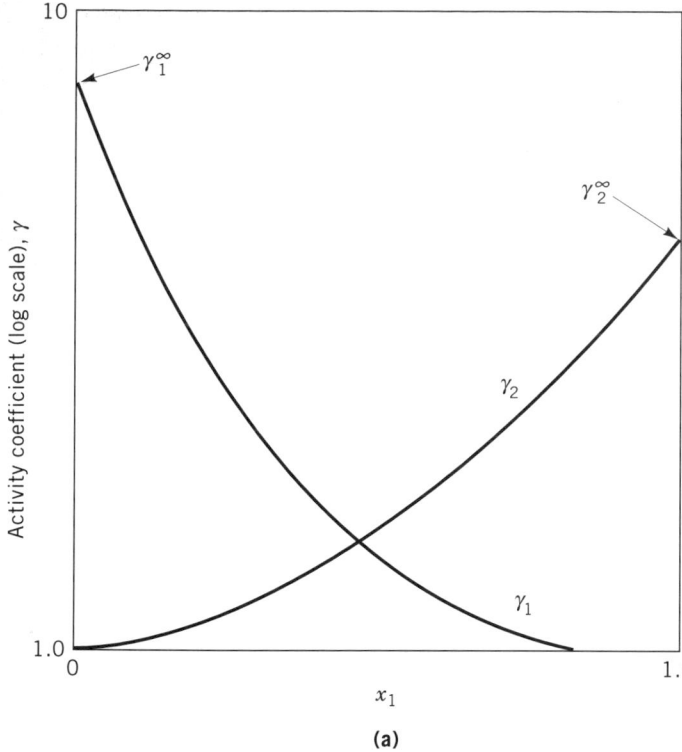

(a)

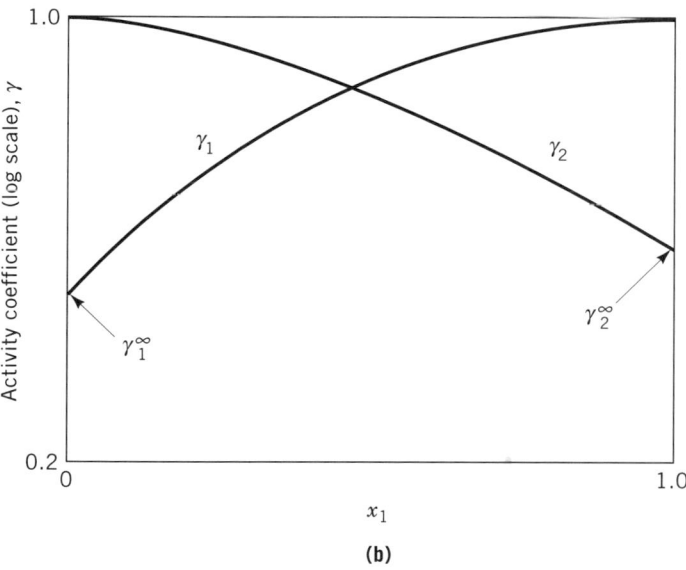

(b)

Figure 3. Binary activity coefficients for two component systems having (**a**) positive and (**b**) negative deviations from Raoult's law. Conditions are either constant pressure or constant temperature, and terminal coefficients, γ_i^{∞}, are noted.

The Van Laar model is only one of many that have been proposed for the correlation of VLE. Others include the Wilson (17) and the nonrandom two-liquid (NRTL) phase (18) models. The latter can represent immiscible systems but requires a third parameter. The most recently developed relationship is called UNIQUAC (19). The series of volumes by Gmehling and co-workers (1) in-

Table 1. Approximate Terminal Activity Coefficients at Atmospheric Pressure[a]

System			
Component 1	Component 2	γ_1	γ_2
Chloroform	Ethyl acetate	0.3	0.3
Chloroform	Benzene	0.9	0.7
n-Hexane	n-Heptane	1.0	1.0
Ethyl acetate	Ethanol	2.5	2.5
Ethanol	Toluene	6.0	6.0
Benzene	Methanol	9.0	9.0
Ethanol	Isooctane	11.0	8.0
Methyl acetate	Water	20.0	7.0
Ethyl acetate	Water	100.0	15.0
Water	Water	> 100.0	> 100.0

[a] Equilibrium data from various sources.

cludes comparisons against measured VLE of predictions by the Van Laar, Wilson, NRTL, and UNIQUAC models, as well as the older model of Margules (20). Thousands of comparisons have been made in this reference, which covers the Dortmund Data Base, available for purchase and use with standard computers. The predictive VLE models can be accommodated to multicomponent systems through the use of certain combining rules. These rules require the determination of parameters for all possible binary pairs in the multicomponent mixture. It is possible to use more than one model in determining binary pair data for a given mixture (21).

To estimate VLE when no experimental data or model parameters are available, and the cost of special measurements cannot be justified, a group contribution method called UNIFAC (26) has been developed and is being continuously improved as more confirmatory data become available. The method makes use of contributions by the molecular structure groups involved. The UNIFAC method, as well as the other models mentioned above, are critically important in extending limited data to conditions in distillation columns that cover wide ranges of temperatures, pressures, and compositions. Handling of all the models by computer solution has been described in some detail (22).

The above discussion concerns nonidealities in the liquid phase, under conditions where the vapor phase mixes ideally and where pressure–temperature effects do not result in deviations from the ideal gas law. Such conditions are by far the most common in commercial distillation practice. However, when high pressures and very low temperatures are encountered, more rigorous forms of the corrected K value expression should be used.

Vapor–liquid equilibria for dilute solutions are often expressed in terms of Henry's law:

$$p_i^v = H_i^* x_i$$

where the Henry's law coefficient is

$$H_i^* = \gamma_i^L P_i^0$$

This coefficient is widely used in environmental engineering work with contaminated wastewater or groundwater. Concentrations of the dissolved species are normally quite low, and the terminal value of the activity coefficient in the previous equation applies. A recent tabulation of measured and predicted terminal activity coefficients for aqueous systems is available (23). The Henry's law expression is also useful for the dilute gas concentrations involved in air pollution abatement by absorption or scrubbing.

Compendia listing equilibrium data, or references to such data, have been mentioned earlier. Popular journals for the publication of VLE are *Fluid Phase Equilibria* and *Journal of Chemical and Engineering Data*. A comprehensive tabulation of azeotropic data has been authored (24); if the composition and temperature of the azeotrope are known (at a given pressure), then such information may be used to calculate activity coefficients using expressions such as those of Van Laar. At the azeotropic point, by definition $y_i = x_i$; thus,

$$\gamma_i^L = P / P_i^0$$

with the vapor pressure P_i^0 being obtained from one of many reference sources.

The measurement of VLE can be carried out in several ways. A common procedure is to use a recycle still that is designed to ensure equilibrium between the phases; samples are then taken and analyzed by suitable methods. It is possible in some cases to extract equilibrium data from chromatographic procedures (3,9). For the more challenging measurements, eg, under conditions where one or more components in the mixture can decompose or polymerize, commercial laboratories can be used.

Azeotropic Systems

An azeotropic mixture is one that will vaporize without any change in composition. Figure 4 represents a homogeneous minimum boiling azeotropic system, 1-propanol–toluene. The calculated points were obtained from the known azeotrope composition plus the application of the Van Laar expressions; the measured points were obtained by Lu (25). The relative volatility of the components reverses at the azeotropic point, as shown in Figure 4(b), which limits the separation that can be made between the components. Maximum boiling azeotropes, where the azeotrope temperature is greater than the boiling points of the pure components, may also be found. The compilation by Horsley (24) helps to determine whether the azeotrope is maximum or minimum boiling. The azeotropic composition is also identified as a constant boiling mixture (CBM). Positive activity coefficients tend to produce minimum boiling azeotropes, and negative coefficients tend to produce maximum boiling azeotropes.

Heterogeneous azeotropes are formed when the positive activity coefficients are sufficiently large to produce two liquid phases: at the boiling point, and a constant boiling mixture that is formed at some composition, generally within the liquid immiscibility composition range. An example of a heterogeneous azeotropic system is the water–1-butanol system shown in Figure 5. The abscissa values M and N represent the liquid miscibility limits. Within the immiscible range M–N, the equilibrium vapor is the heterogeneous azeotrope Z of constant composition,

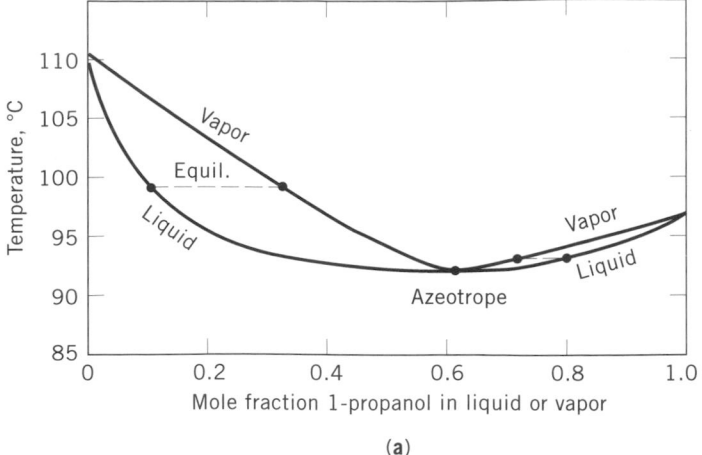

(a)

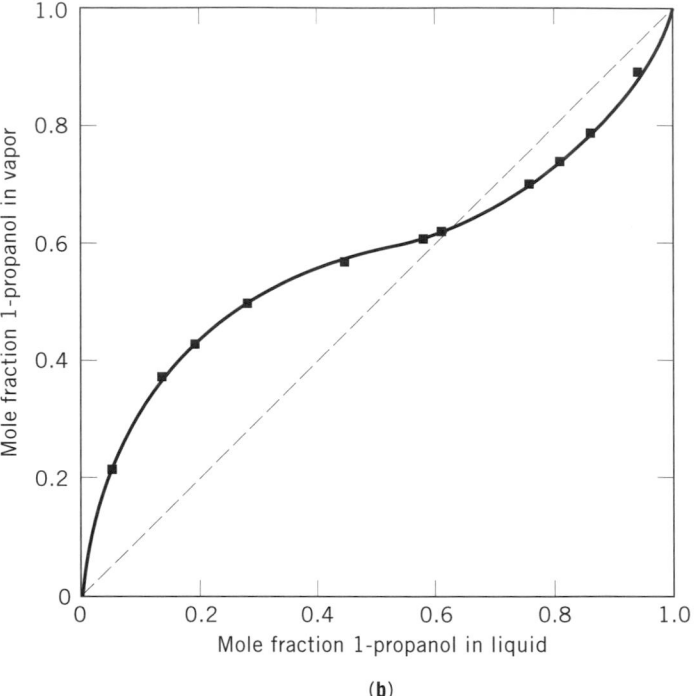

(b)

Figure 4. Boiling point (**a**) and phase diagram (**b**) for the minimum boiling binary azeotropic system 1-propanol–toluene at atmospheric pressure. The curves are calculated on the basis of Van Laar coefficients $A_{12} = 1.466$ and $A_{21} = 1.268$. The points shown in (**b**) are measured.

and the equilibrium temperature is constant. For the water–1-butanol mixture shown at liquid compositions lower in water content than for the azeotrope, the relative volatility is greater than one; at liquid compositions higher in water than for the azeotrope, the relative volatility of water–1-butanol is less than one.

DISTILLATION PROCESSES

Basic distillation involves application of heat to a liquid mixture, vaporization of part of the mixture, and removal

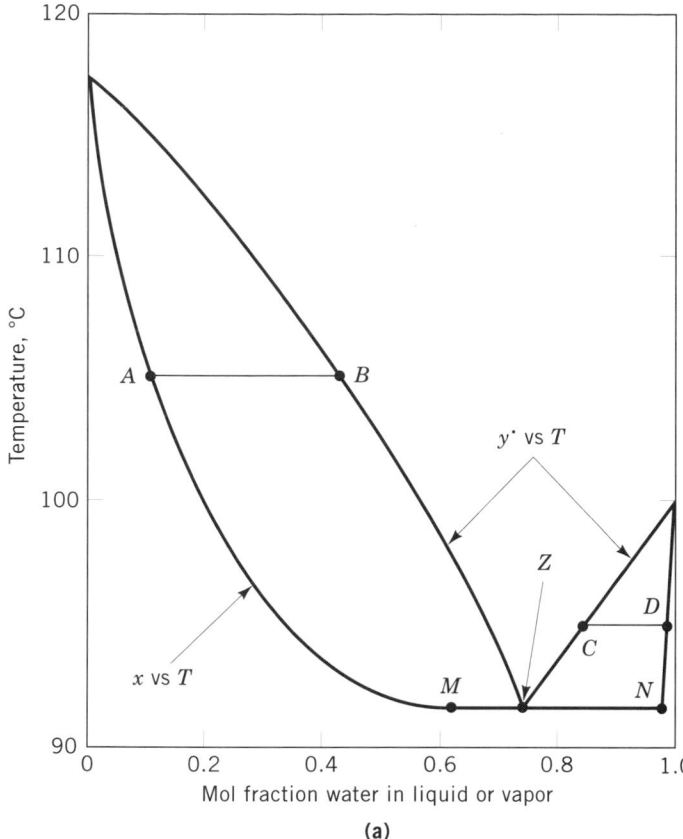

(a)

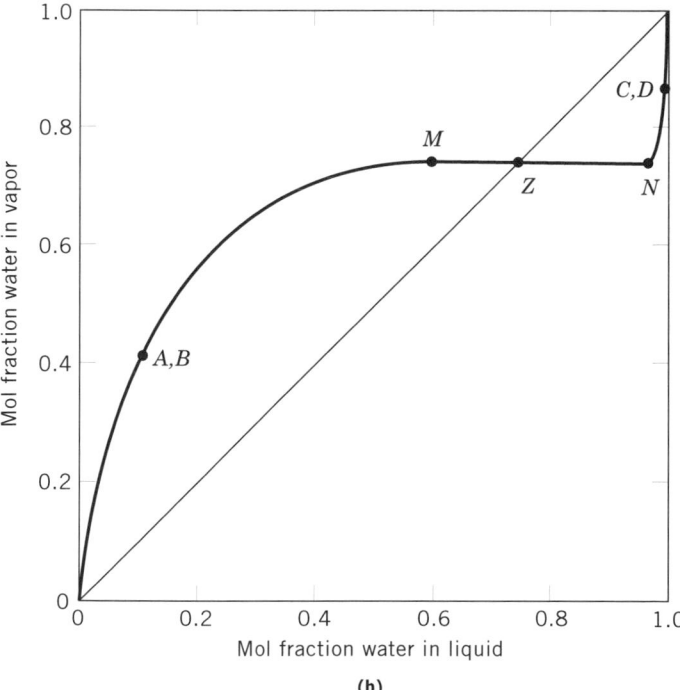

(b)

Figure 5. Boiling point (**a**) and phase diagram (**b**) for the heterogeneous azeotropic system, water–1-butanol at atmospheric pressure. A, B and C, D are representative equilibrium points; Z is the azeotropic point, M and N are liquid miscibility limits.

of the heat from the vaporized portion. The resultant condensed liquid, the distillate, is richer in the more volatile components, and the residual (unvaporized) bottoms are richer in the less volatile components. Most commercial distillations involve some form of multiple staging in order to obtain a greater enrichment than is possible by a single vaporization and condensation.

For ease of presentation and understanding, the initial discussion of distillation processes is based on binary (two-component) systems. With reference to the boiling point diagram and phase diagrams shown in Figures 1(**a**) and 1(**b**), the enrichment from liquid composition x_L to vapor composition y_L represents a theoretical step, or equilibrium stage.

Simple Distillations

Simple distillations utilize a single equilibrium stage to obtain separation. Simple distillations may be either batch or continuous. Simple batch distillation (also called differential distillation) may be represented on boiling point or phase diagrams. In Figure 1(**a**), if the batch distillation begins with a liquid of composition x_L^A, the initial distillate vapor composition will be y_L^A. As the distillate is removed, the remaining liquid becomes less rich in L (low boiler), and the boiling liquid composition moves to the left along the bubble point line. If the distillation is continued until the liquid has a composition of x_L^E, the last vapor distillate will have had a composition of y_L^E. Simple batch distillation is only minimally used in industry, eg, for the processing of high valued chemicals in small production quantities or for distillations requiring regular sanitization. Calculation methods are found in most standard distillation references and computer programs, such as BATCHFRAC® of Aspen Technology (26), are available for handling the more complex, multicomponent batch distillations.

Simple continuous distillation (also called flash distillation) has a continuous feed to a single equilibrium stage; the liquid and vapor leaving the stage are considered to be in phase equilibrium (see Fig. 6). On the boiling point diagram in Figure 1(**a**), the feed is represented by X_L^F, the bottoms liquid by X_L^B, and the equilibrium vapor distillate by y_L^D. The mass balances are

$$F = D + B \qquad \text{(overall balance)}$$
$$x_L^F F = y_L^D D + x_L^B B \qquad \text{(component } L \text{ balance)}$$

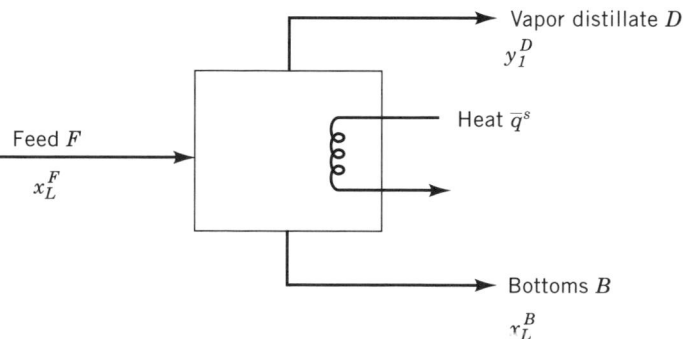

Figure 6. Simple continuous distillation with single equilibrium stage.

Flash distillations are widely used where a crude separation is adequate. They may be adiabatic or isothermal; in the latter case, heat must be supplied to compensate for the cooling effect of partial vaporization. Examples of flash multicomponent calculations are given in standard distillation references (eg, see Ref. 27).

Multiple Equilibrium Staging

The component separation in simple distillation is limited to the composition difference between liquid and vapor in phase equilibrium. To overcome this limitation, multiple equilibrium staging is used to increase the component separation. Figure 7 represents a continuous distillation that employs multiple equilibrium stages, stacked one upon another. This arrangement results in the familiar vertical distillation column found in all oil refineries and most chemical plants. The feed F enters the column at equilibrium stage f. The heat q^s required for vaporization is added at the base of the column in a reboiler (or calandria). The vapors V^T from the top of the column flow to a condenser, from which heat q^C is removed. The liquid condensate from the condenser is divided into two streams: the first, a distillate D, which is the overhead product, is withdrawn from the system. The second stream, reflux R, is returned to the top of the column. A bottoms stream B is withdrawn from the reboiler or directly from the bottom of the column. The overall separation is represented by feed F separating into a distillate D and a bottoms B.

Above the feed, a typical equilibrium stage is designated as n; the stage above n is $n + 1$, and the stage below is $n - 1$. The section of column above the feed is called the rectification section, and the section below the feed is referred to as the stripping section.

The mass balance across stage n is based on (1) vapor (V^{n-1}) from the stage below, (2) liquid (L^{n+1}) from the stage above, (3) the vapor V^n leaving the stage, and (4) the liquid L^n leaving the stage. Thus,

$$V^{n-1} + L^{n+1} = V^n + L^n$$

According to the equilibrium stage concept, V_n and L_n are in equilibrium, because all of their components are in equilibrium:

$$y_i^n = K_i^n x_i^n = K_j^n x_j^n \ldots .$$

The vapors moving up the column from one equilibrium stage to another are increasingly enriched in the more volatile components. Similarly, the liquid streams moving down the column are increasingly diminished in the more volatile components.

The overall column mass balances are

$$F = D + B$$

and for any component i,

$$Fx_i^F = Dx_i^D + Bx_i^B$$

The overall enthalpy balance is

$$H^F F + H^S = H^D D + H^B B + H^C$$

A mass balance around plate n and the top of the column gives

$$V^{n-1} = L^n + D$$

And for any component:

$$V^{n-1} y_i^{n-1} = L^n x_i^n + D x_i^D$$

therefore,

$$y_i^{n-1} = \left(\frac{L^n}{V^{n-1}} \right) x_i^n + \left(\frac{D}{V^{n-1}} \right) x_i^D$$

Below the feed, a similar balance around plate m and the bottom of the column results in

$$y_i^{m-1} = \left(\frac{L^m}{V^{m-1}} \right) x_i^m + \left(\frac{B}{V^{m-1}} \right) x_i^B$$

The equation for y_i^{n-1} represents the upper (or rectifying) operating line equation and the equation for y_i^{m-1} represents the lower (or stripping) operating line equation. The slopes L^n/V^{n-1} and L^m/V^{m-1} can vary, depending on heat effects.

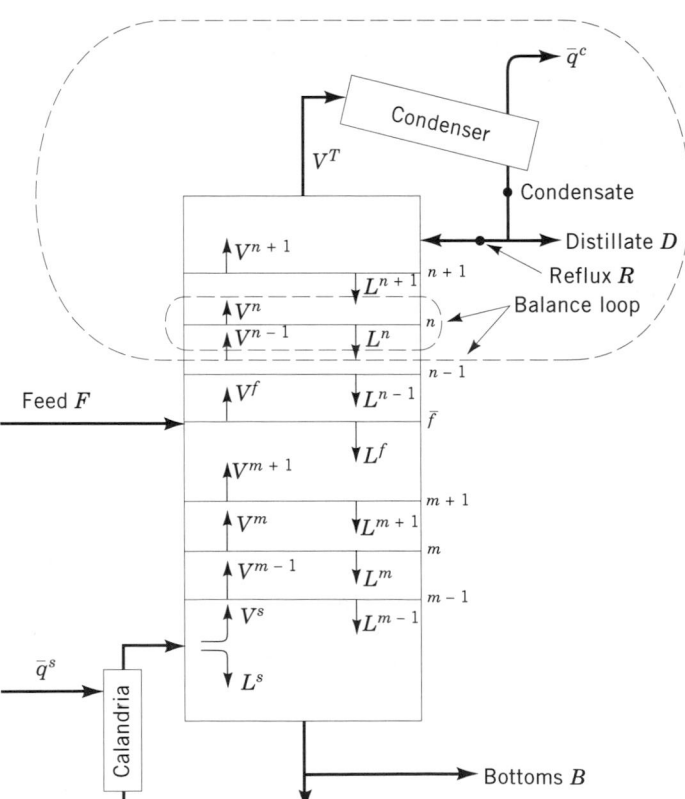

Figure 7. Distillation column with stacked multiple equilibrium stages.

McCabe-Thiele Method. The graphical McCabe-Thiele design method (28) facilitates a visualization of distillation principles and provides a solution to the material balance and equilibrium relationships. Details of the method are available (13,27) but for orientation purposes, a McCabe-Thiele general diagram is given in Figure 8. In the figure, it is necessary for the upper and lower operating lines to intersect at a point on the q line, which has a slope of $q/(q-1)$ where,

$$q = \frac{\text{Heat required to take the feed mixture to a saturated vapor}}{\text{Heat of vaporization of the feed mixture}}$$

Thus, for a saturated liquid feed (as is often used) q equals 1, the slope of the q line is infinite, and the line itself is vertical.

The McCabe-Thiele method employs the simplifying assumption that the molal overflows in the stripping and the rectification sections are constant, which reduces the rectifying and stripping operating line equations to

$$y^{n-1} = \left(\frac{\overline{L}}{\overline{V}}\right)_R x^n + \left(\frac{D}{\overline{V}_R}\right) x^D$$

$$y^{m-1} = \left(\frac{\overline{L}}{\overline{V}}\right)_S x^m \left(\frac{B}{\overline{V}_S}\right) x^B$$

The constant molal flows in each section are designated by $\overline{L}$ and $\overline{V}$. The McCabe-Thiele assumption of constant molal overflow implies that the molal latent heats of the two components are identical, the sensible heat effects are negligible, and the heat of mixing and the heat losses are negligible. This simplified situation is closely approximated for many distillations. The equilibrium curve gives

the vapor–liquid relationships of y^n and x^n above the feed and of y^m and x^m below the feed.

Reflux and Reflux Ratio. The liquid returned to the top of the column is called reflux. The molar R–D ratio is the external reflux ratio. The ratio $(L/V)_R$, which is the slope of the rectifying operating line, is the rectifying internal reflux ratio. Similarly, the ratio $(L/V)_S$, which is the slope of the stripping operating line, is the stripping internal reflux ratio. As the R–D ratio increases, the rectifying internal reflux ratio increases and numerically approaches unity; similarly, the stripping internal reflux ratio decreases and numerically approaches unity. When unity is approached by these slopes, the external reflux ratio approaches infinity and very little, if any, distillate product can be removed. Thus, little, if any, feed can be introduced to the column, if a material balance is to be maintained. This condition is known as total reflux and is a method of operation often used during the start-up of a distillation system. For a McCabe-Thiele type of plot, the minimum stages are as shown in Figure 9(**a**). Similarly, the slope of the upper operating line can be reduced, with an accompanying increase in stages, until the line intersects with the equilibrium curve (before it intersects with the q line, or when the q line, operating line, and equilibrium curve have a common intersection). This is the minimum possible slope of the operating line and corresponds to the minimum reflux ratio (see Fig. 9(**b**)). On these plots, each "step" represents a theoretical stage.

For a given separation specification, there is an infinite number of combinations of reflux ratio and theoretical stages that can be used. This idea can be represented as a stages–reflux plot as shown in Figure 10. Clearly, there is some optimum combination of these two parameters. The results of numerous studies indicate that the R–R_m ratio is usually in the range of 1.10 to 1.25.

Simple analytical methods are available for determining minimum stages and minimum reflux ratio. Although developed for binary mixtures, they can often be applied to multicomponent mixtures if the two key components (the components between which the specification separation must be made) are used. Frequently, the heavy key will be the component with a maximum allowable composition in the distillate, and the light key will be the component with a maximum allowable specification in the bottoms. On this basis, minimum stages may be calculated by means of the Fenske (29) relationship:

$$N_{\min} = \frac{\ln[(y_i/y_j)^D(x_j/x_i)^B]}{\ln \alpha_{ij}}$$

where i and j are the light and heavy components of a binary mixture, or the light key and heavy keys in a multicomponent mixture. The value of α_{ij} is often taken as the geometric average between the value at the top of the column and that at the bottom of the column, ie, $\alpha_{ij} = (\alpha_i \alpha_j)^{0.5}$. As an example of the use of the above equation, let a 50-50 molar mixture of benzene and toluene be distilled into fractions with 95 mol % benzene in the distillate and 5 mol % benzene in the bottoms. Let the average relative volatility between benzene and toluene be 2.5.

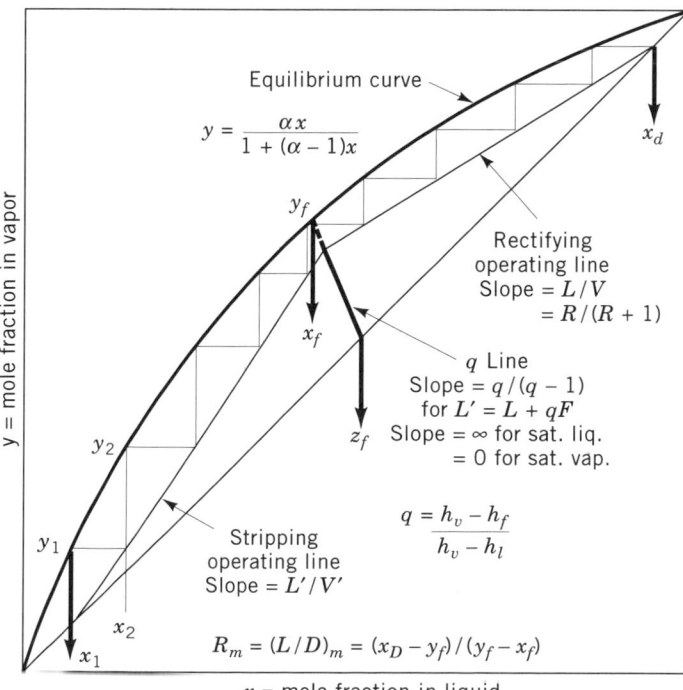

Figure 8. Summary of McCabe-Thiele graphical method for binary distillation.

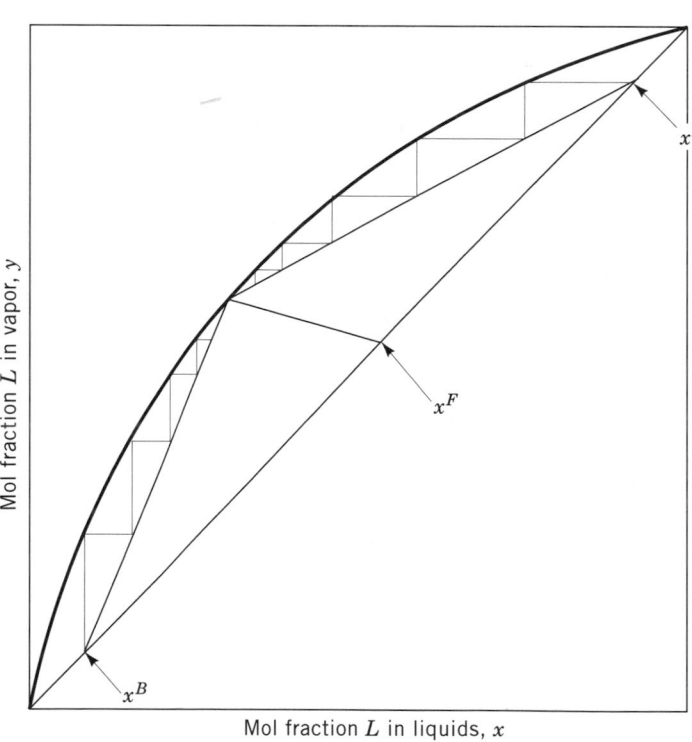

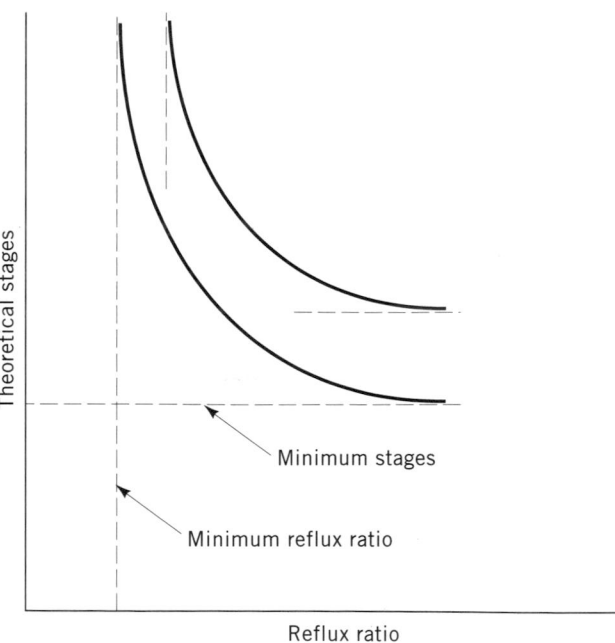

Then at total reflux,

$$N_{\min} = \frac{\ln[(0.95/0.05)(0.95/0.05)]}{\ln 2.5} = 6.43 \text{ stages}$$

For minimum reflux ratio, the equations of Underwood (30) may be used:

$$\sum_i \frac{\alpha_i x_{if}}{\alpha_i - \phi} = 1 - q$$

$$\sum_i \frac{\alpha_i (x_{id})}{\alpha_i - \phi} = R_{\min} + 1$$

where the value of q is determined as demonstrated previously. There are two possible values of root ϕ that can be obtained; the value to use must lie between 1.0 and the light key volatility.
Thus,

$$\frac{2.5\,(0.5)}{2.5 - \phi} + \frac{1.0\,(0.5)}{1.0 - \phi} = 1 - q = 0 \quad (q = 1)$$

Root $\phi = 1.43$ and thus,

$$\frac{2.5\,(0.95)}{2.5 - 1.43} + \frac{1.0\,(0.05)}{1.0 - 1.43} = R_{\min} + 1$$

from which, $R_{\min} = 1.10$.

Both of these limits, the minimum number of stages and the minimum reflux ratio, are impractical for useful operation, but they are valuable limits as shown in Figure 11. The operating, fixed, and total costs of a distillation

Figure 9. Limiting conditions in binary distillation. (**a**) Minimum stages at total reflux; (**b**) minimum reflux at infinite stages.

Figure 10. Representative plot of theoretical stages vs reflux ratio for a given separation. Each curve is the locus of points for a given separation. Note the limiting conditions of minimum reflux and minimum stages.

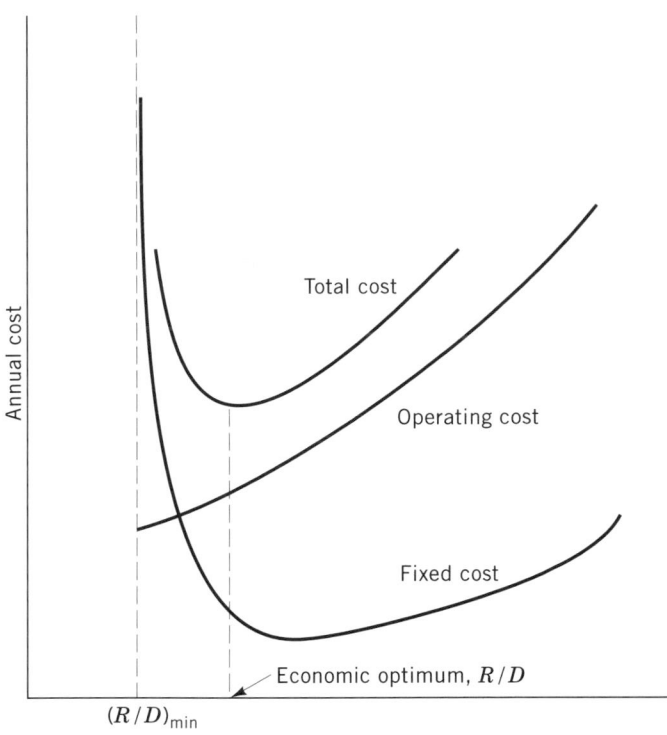

Figure 11. Fixed, operating, and total costs of a typical distillation, as a function of reflux ratio.

system are functions of the relation of operating reflux ratio to minimum reflux ratio. Figure 10 shows a typical plot of costs; as the operating to minimum reflux ratio increases, the operating cost (principally, the energy cost for the boil-up) increases almost linearly. Similarly, the fixed costs at first decrease from the infinite number of stages, pass through a minimum, and then increase again as the diameter of column increases with increased reflux ratio. These costs for typical distillations have been calculated (30); the ratio of the economic optimum reflux to the minimum reflux is often 1.2 or less.

Multicomponent Calculations. The calculations that determine the reflux and stage requirements are more difficult to make for multicomponent systems than for binary systems. When the concentration of a component in the distillate and in the bottoms is specified for the overall solution of a binary distillation, the component balance around the column is also completely specified. In the multicomponent case, only a single high boiling key component can be specified in the distillate and a single low boiling key component in the bottoms; the split of other components can be determined only by detailed calculations. Thus, a series of trial and error computations is required to obtain the solution at any given reflux ratio and number of stages. As the number of components and number of stages become large, the mathematical problem becomes formidable. Two approaches may be followed: use of approximate ("shortcut") methods, or use of a suitable computer program that provides rigorous solutions. The former are used when approximate solutions are adequate or when a computer program is not available. For

the latter, numerous commercial programs are available and may be used with personal computers.

Most shortcut methods involve: (1) calculating the minimum number of stages, (2) calculating the minimum reflux ratio, and (3) estimating, from empirical correlations, the actual number of theoretical stages at an operating reflux. For minimum stages, the Fenske relationship is used, whereas for minimum reflux ratio, the Underwood relationships are used (31). The relationship between operating and minimum reflux ratio, and between operating and minimum number of plates, is then estimated from the Gilliland correlation (32), or from a more recent correlation such as that of Erbar and Maddox (33).

The Gilliland correlation as published was in graphical form. The curve of Gilliland has been fitted by several workers, eg, Eduljee (34):

$$\frac{N_t - N_{\min}}{N_t + 1} = 0.75 - 0.75 \left(\frac{R - R_{\min}}{R + 1}\right)^{0.5668}$$

For the example given above, and using the Eduljee equation for a selected operating reflux ratio 30% greater than the minimum, ie, 1.10 (1.30) = 1.43,

$$\frac{N_t - 6.43}{N_t + 1} = 0.75 - 0.75 \left(\frac{1.43 - 1.10}{1.43 + 1}\right)^{0.5668}$$

from which, $N_t = 14.1$ stages.

A stages–reflux plot for the example problem is shown in Figure 12. The asymptotic approach to the minimum reflux ratio is typically sharper than that to minimum stages. The example deals only with a binary mixture, but the method may be applied to multicomponent mixtures. If the example had included additional components in

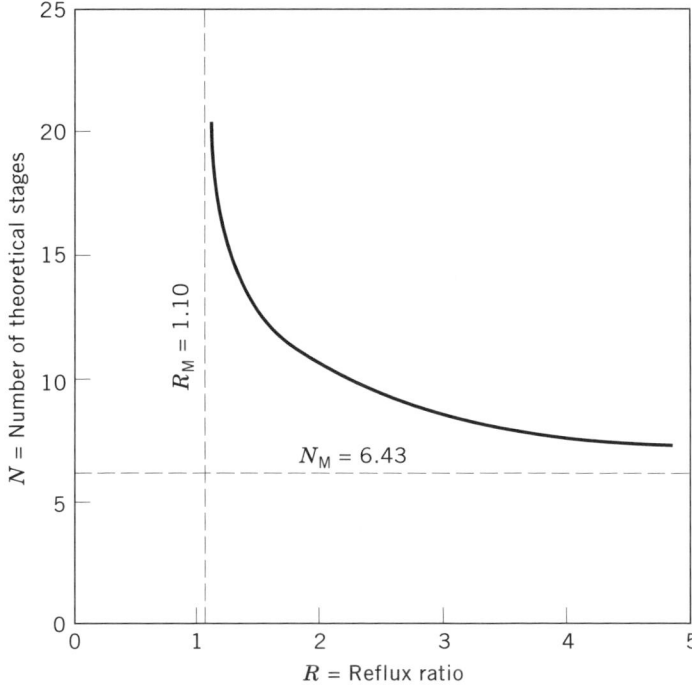

Figure 12. Stages–reflux relationship for the example benzene–toluene separation.

the feed, eg, xylenes, the methodology would be the same, but concentrations and possibly relative volatiles would change.

Rigorous computer solutions are used for complex distillations involving multiple stages, multiple components, nonideal phase equilibria, multiple feeds and drawoffs, and heat addition or removal at intermediate stages. With the availability of computer programs able to handle the complex distillation problems, most calculations today are made by computer. The algorithms are generally based on the Thiele-Geddes model (35), which rates a given number of stages and reflux ratio for separation capability. A detailed discussion of computer solutions, including the handling of convergence problems, has been provided (36).

Computer-generated solutions entail setting up component equilibrium and component mass and enthalpy balances around each theoretical stage, and specifying the required design variables as well as solving the large number of simultaneous equations required for the final solution. The explicit solution to these equations remains too complex for present methods. Studies to solve the mathematical problem by algorithm or iterative methods have been successful and, with few exceptions, the most complex distillation problems can be solved.

Multiple Products

If each component of a multicomponent distillation is to be essentially pure when recovered, the number of columns required for the distillation system is $N^* - 1$, where N^* is the number of components. Thus, in a five-component system, recovery of all five components as essentially pure products requires four separate columns. However, those four columns can be arranged in 14 different ways (37).

The number of columns in a multicomponent train can be reduced from the $N^* - 1$ relationship if sidestream draw-offs are used for some of the component cuts. The feasibility of multicomponent separation by such draw-offs depends on sidestream purity requirements, feed compositions, and equilibrium relationships. In most cases, sidestream draw-off distillations are economically feasible only if component specifications for the sidestream are not tight. If a single component is to be recovered in an essentially pure state from a mixture containing both lower and higher boiling components, a minimum of two columns is required, one column to separate the lower boilers from the desired component and another column to separate the component from the higher boilers.

The economics of the various methods that are employed to sequence multicomponent columns have been studied, eg, Freshwater and Henry (38) considered the separation of three-, four-, and five-component mixtures. They examined the heuristics (rules of thumb) developed by earlier investigators and made an economic analysis of various methods of sequencing the columns. The study of sequencing of multicomponent columns is part of a broader new field, process synthesis, which attempts to formalize and develop strategies for the optimum overall process (39).

Batch Distillation

Although most commercial distillations are run continuously, there are certain applications where batch distillation is the method of choice. For this method, a charge, or batch, of the initial mixture is placed in a vessel where it can be heated and distilled over a period of time. Compositions of the charge and product thus vary with time, which is not the case for continuous distillation. Particular cases where batch distillation may be preferred are (1) semiworks operations producing interim amounts of product in equipment that is used for multiple purposes; (2) distillations of specialty chemicals where contamination can be a problem, and the batch equipment be cleaned or sterilized between batch runs; (3) operations involving wide swings in feed compositions and product specifications, where batch operating conditions can be adjusted to meet the varying needs; and (4) laboratory distillations where separability is being investigated without concern over the scale-up to commercial continuous operations.

Batch distillations are generally more expensive than their continuous counterparts, in terms of cost per unit of product. Close supervision and/or computer control are required, the equipment is more complex if several products are to be recovered, and total throughput is limited by the needs for changing operating modes and for recharging the system with feed material.

The simplest batch distillation method is the straight takeover approach, without reflux, often called a Rayleigh distillation. As indicated in Figure 13, a batch of material is charged to a stillpot and brought to boiling. The most volatile constituent of the charge mixture distills over first, followed the other components in order of descending volatility. The distilled products are condensed and placed in appropriate receivers. Invariably, there are mixtures condensed that do not meet specifications, and these must be sent back to the stillpot as part of the next charge.

The analytical method for handling a Rayleigh distillation may be summarized as follows. Let S equal total moles in stillpot with light key mole fraction x. A material balance for a differential amount of vaporization is

$$Sx_s = (S - dS)(x_s - dx_s) + x_D\, dS$$

(Amount before vaporization) = (Remaining amount)

+ (Vaporized amount)

Rearranging and integrating between initial stillpot moles S_o and final stillpot moles S_t:

$$\ln \frac{S_o}{S_t} = -\int_{x_{so}}^{x_{st}} \frac{dx_s}{x_D - x_s}$$

Whereas this equation may be evaluated easily by graphical integration, if the vapor is assumed to be in equilibrium with the liquid at any time and if the relative volatility is constant, the following analytical expression may be developed:

$$\ln \frac{S_o}{S_t} = \frac{1}{\alpha - 1} \ln \frac{x_{so}}{x_{st}} + a \ln \frac{1 - x_{st}}{1 - x_{so}}$$

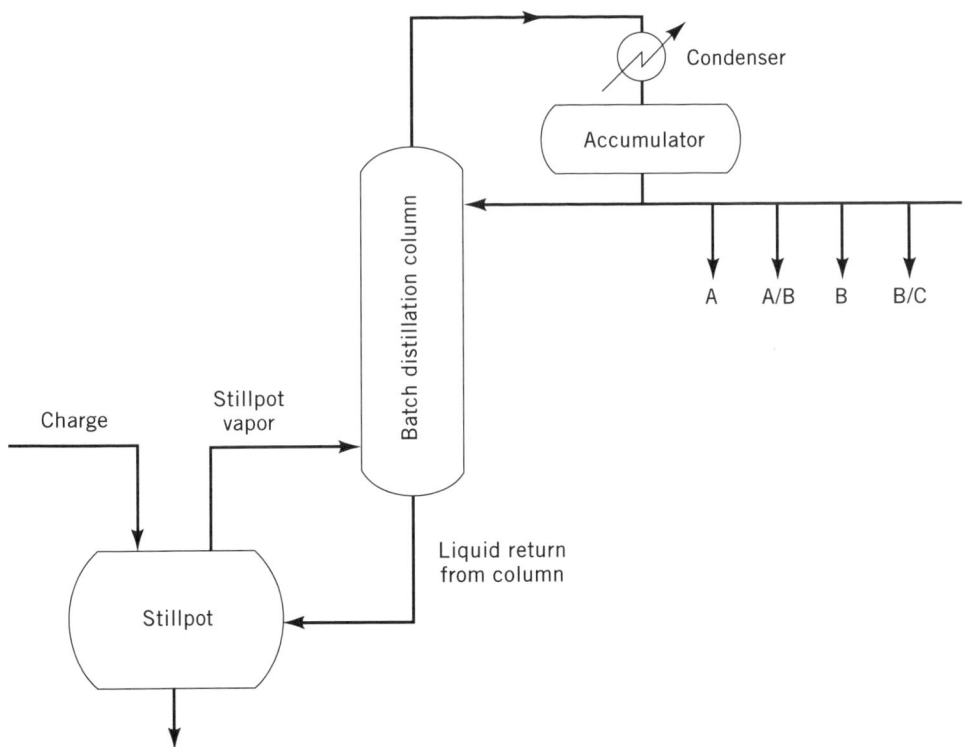

Figure 13. Simple batch distillation flow diagram.

As an example of the use of the previous equation, if 100 moles of a 50-50 mixture of A (light key) and B (heavy key) are distilled in a simple takeover process, and the relative volatility between A and B is 2.0, then the total moles remaining in the stillpot, if the distillation is allowed to proceed until the mole fraction of A in the stillpot is 0.20, is

$$\ln\frac{S_o}{S_t} = \frac{1}{2.0-1}\ln\frac{0.5}{0.2} + 2.0\ln\frac{1-0.2}{1-0.5} = 6.40$$

which gives $S_t = 100/6.40 = 15.6$ moles left.

Batch distillations may be carried out in plate or packed columns with the capability of a number of theoretical stages, and with reflux returned to the column. If the distillate product composition is to be maintained constant during a portion of the run, then reflux must be varied. There are many possible operating modes for batch distillation columns. Calculation methods are given in Holland and Liapis (40) and, to a limited extent in standard texts. Computer programs are available for rigorous simulations of multicomponent, multistage batch distillations, one example of which is BATCHFRAC of Aspen Technology (26).

DISTILLATION COLUMNS

Distillation columns are vertical, cylindrical vessels which contain devices that provide the needed contacting of the rising vapor with the descending liquid. This contacting provides the opportunity for the two streams to achieve some approach to thermodynamic equilibrium. Depending on the type of internal devices used, the contacting may occur in discrete steps, called plates or trays, or in a continuous differential manner on the surface of a packing material. The fundamental requirement of the column is to provide efficient and economic contacting at a required mass transfer rate. Individual column requirements vary from high vacuum to high pressure, from low to high liquid rates, from clean to dirty systems, and so on. As a result, a variety of internal devices have been developed to fill these needs. The column devices discussed in this section are used for both absorption and stripping, as well as for distillation. The principal operational difference is that in absorption or stripping, the gas flowing up the column is primarily a noncondensable phase at column conditions, whereas in distillation, the gas phase is a condensable vapor.

Plate Columns

Two general types of plates are used: crossflow and counterflow. The names refer to the direction of the liquid flow relative to the rising vapor flow. On the crossflow plate, the liquid flows across the plate and from plate to plate via downcomers (see Fig. 14). On the counterflow plate, liquid flows downward through the same orifices used by the rising vapor.

Crossflow Plates. For this type of device, liquid enters the plate from the bottom of the downcomer of the plate above and flows across the active area (or "bubbling area") where it is aerated by the vapors flowing through orifices from the plate below. It is in this aerated zone where most of the vapor–liquid mass transfer occurs. The aerated mixture flows over the exit weir into a downcomer, where vapor–liquid disengagement occurs and where most of

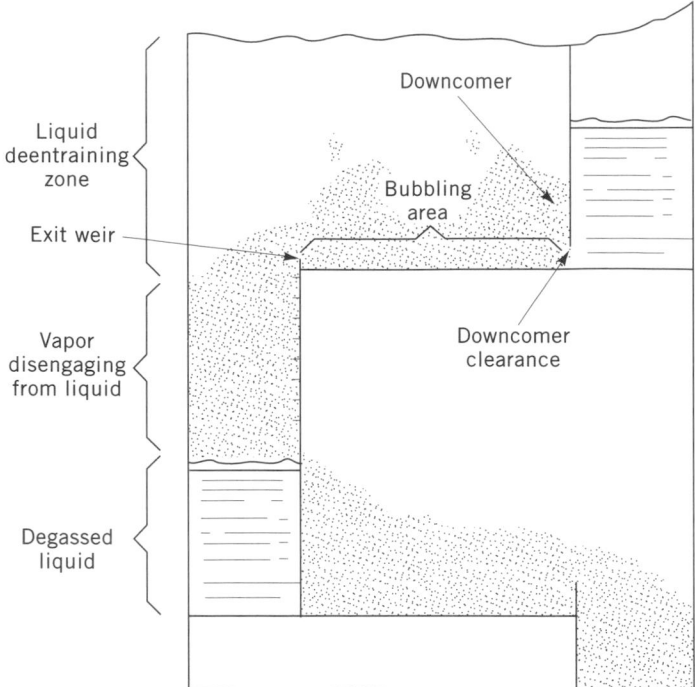

Figure 14. Flow pattern in a crossflow plate distillation column.

(a)

(b)

Figure 15. Views of (**a**) sieve plate and (**b**) Koch Flexitray, a representative valve plate.

the trapped vapor escapes from the liquid and flows back to the interplate vapor space. The liquid, essentially free of entrapped vapor, leaves the plate by flowing under the downcomer to the inlet side of the next lower plate. The vapor, disengaging from the aerated mass on the plate, rises to the next plate above.

The pressure drop incurred by the vapor as it passes through the orifices of the plate is fundamental to plate operation. In most plate designs, the pressure drop prevents the crossflowing liquid from falling through the plate. It also results from the energy consumed to disperse the vapor–liquid mixture, eg, to atomize a portion of the liquid to provide increased interfacial area for mass transfer. Diameters of commercial crossflow plate columns range from 0.3 to 15 m and plate spacings range from 0.15 to 1.2 m. The total pressure drop per plate is often in the range of 0.25–1.6 kPa (2–12 mm Hg).

Two principal vapor–liquid contacting devices are used in current crossflow plate design: the sieve plate and the valve plate. These devices provide the needed intimate contacting of vapor and liquid to maximize transfer of mass across the interfacial boundary. A formerly popular crossflow device, the bubble-cap plate, is rarely used today for new designs.

Sieve Plates. The conventional sieve or perforated plate is inexpensive and the simplest of the devices normally used. The contacting orifices in the conventional sieve plate, as shown in Figure 15(**a**), are holes which measure 1–12 mm diameter and which exhibit ratios of open area to active area ranging from 1:20 to 1:7. If the open area is too small, the pressure drop across the plate is excessive; if the open area is too large, the liquid will "weep" or "dump" through the holes.

Valve Plates. Valve plates are categorized as proprietary, and details of their design vary from one vendor to another. They represent a variation of the sieve plate in which the holes are larger and are fitted with liftable valve units (see Fig. 15(**b**)). Their principal advantage over sieve plates is their ability to maintain efficient operation over a wider operating range through the use of variable orifices ("valves") which open or close depending on vapor rate. The most common valve units consist of flat disks with attached legs, which allow the valve to open or close (see Fig. 16). Sometimes, two weights of valves are used on a single plate to extend operating range and improve vapor distribution. The valve units usually have a tab or indentation that provides a minimum open area of vapor flow, even when the valve is closed, and also prevents the valve from sticking under corrosive or fouling conditions. Details on valve plate geometry, along with methods for valve plate design, may be found in literature from the vendors (41–43).

Bubble-Cap Plates. Until the early 1950s, bubble caps were industry's standard design. Today, their use in new installations is limited to very low liquid flow rate applications, or to those cases where the widest possible operating range is desired. Detailed descriptions and design methods for bubble-cap plates are available (44). Bubble-cap trays are more expensive and have lower capacity

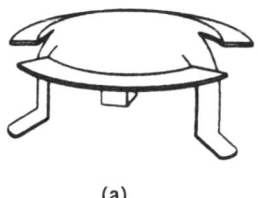 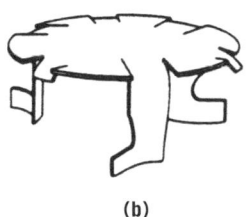 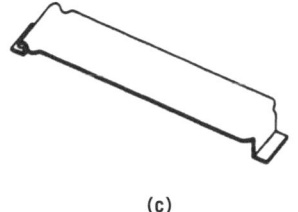

(a)   (b)   (c)

Figure 16. Representative individual valve units: (**a**) Koch Flexitray valve (courtesy of Koch Engineering Co.); (**b**) Glitsch Ballast valve (courtesy of Glitsch, Inc.); (**c**) Nutter Float Valve (courtesy of Nutter Engineering Co.).

than sieve or valve plates; therefore, their use has dropped to a very small percentage of new column designs.

Multiple Liquid-Path Plates. As the liquid flow rate increases in large diameter crossflow plates (4 m or larger), the crest heads on the overflow weirs and the hydraulic gradient of the liquid flowing across the plate become excessive. To obtain improved overall plate performance, multiple liquid-flow-path plates may be used, with multiple downcomers. These designs are illustrated and discussed in detail by Bolles (45).

Counterflow Plates. Counterflow plates are used less frequently than crossflow plates. The liquid flows downward and the vapor upward through the same orifices in a counterflow plate, and the plate does not have downcomers. The openings are round holes ("dualflow tray") or slots ("Turbogrid tray"); A variation of the dualflow tray is the Ripple tray in which the tray floor is shaped in a corrugated fashion (46). Counterflow plates are used advantageously in fouling services since for each hole, vapor and liquid flow alternately; this provides a self-cleaning action that is quite effective. The dualflow and Turbogrid trays have similar operating characteristics, and typical operating data have been published (47).

Another important plate that has characteristics similar to a counterflow plate is the Multiple Downcomer (MD) plate (48). This is a plate where the active area occupies the full column cross-section but with a plurality of small downcomers interspersed among the perforations. The downcomers are specially sealed to prevent upflow of vapor through them. The plate has been used successfully in many high liquid flow cases.

Vapor Capacity Parameters. The diameter of a distillation column is determined by the capacity of the column to handle the required flows of vapor and liquid. The vapor capacity parameter is

$$C_{sb} = V^* \left(\frac{\rho_g}{\rho_L - \rho_g} \right)^{0.5}$$

and its simplification:

$$F^* = V^* (\rho_g)^{0.5}$$

The term C_{sb} is called a Souders-Brown capacity parameter and is based on the tendency of the upflowing vapor to entrain liquid with it to the plate above. The term F^* is called an F factor, and for C_{sb} and F^* to be meaningful, the cross-sectional area to which they apply must be spec-

ified. The capacity parameter is usually based on the total column cross-section, minus the area blocked for vapor flow by the downcomer(s). For the F factor, typical operating ranges for sieve plate columns are

	Area Basis	$kg^{0.5}/(m^{0.5}\ s)$ $[lb^{0.5}/(ft^{0.5}\ s)]$
F_S^*	Total cross-section	0.6–3.0 (0.5–2.5)
F_A^*	Active area	0.85–4.3 (0.7–3.5)
F_H^*	Hole area	8.5–30 (7–25)

Entrainment Flooding. The vapor capacity of a column is limited by excessive entrainment, usually called flooding. A flooding condition can be observed when the holdup of liquid becomes excessive, the pressure drop increases dramatically, and the mass transfer efficiency falls precipitously. Estimates of the vapor velocity for a flooding condition may be made from the chart in Figure 17. The abscissa term $L/G(\rho_g/\rho_L)^{0.5}$ is called a flow parameter, and its value can indicate several things about the character of the aerated mass on the plate. For example, a very low value can indicate a phase inversion in which the vapor flow is continuous ("spray flow"), whereas a high value can indicate a bubbly mass ("emulsion flow"). The value of the flow parameter is easily determined from the stage calculations (reflux and boilup ratios) and densities of the phases. The ordinate value in Figure 17 leads to a value of the flooding velocity, and prudent design calls for limiting actual flows to 70–80% of this velocity.

Downcomer Flooding. For cases of very high liquid–vapor flow ratios, the limiting capacity of the column is based on the ability of the downcomers to move the deaerated liquid from a plate to the plate next below. It is clear that there can be constrictions in the downcomer design or that even with no constrictions there is simply not enough flow area to accommodate the high volume of liquid. Thus, the downcomer can flood, or "choke," when it becomes completely filled with liquid or aerated mass. Typical design heuristics include limiting the downcomer velocity (clear liquid basis) to no more than 0.12 m/s. Also, to allow for complete disengagement of vapor from liquid in the downcomer, a minimum residence time of 4 s is often used. The actual limiting values of these parameters will vary somewhat with the properties of the fluids and the exact dimensions of the plate components.

Stable Operating Range. All plates have a stable operating envelope bounded by a range of liquid and vapor flow rates, as shown in Figure 18. The size and shape of the stable area depends on the plate design and on the

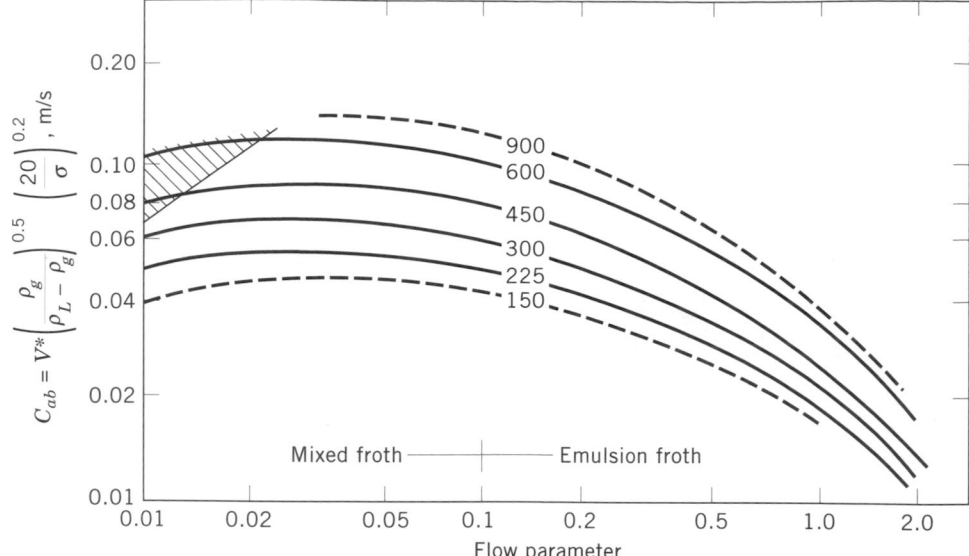

Figure 17. Flooding correlation for crossflow plates (sieve, valve, bubble-cap) where the numbers represent the plate spacing, in mm. Also shown are approximate boundaries of spray zone ▨, and mixed froth and emulsion flow regimes (49).

system properties. The line AD represents the minimum operable vapor flow rate at various liquid flow rates. Below AD, the vapor rate is too low to maintain the liquid on the plate and, as a result, the liquid will weep excessively or dump through the plate orifices. Above line BC, the column will flood by entrainment. To the right of CD, the high liquid rate will cause downcomer flooding. The area to the left of AB represents high entrainment at low liquid flow rates, with vapor jets at the orifices.

Design procedures for bubble-cap plates (44) and sieve plates (49,50) have been published. A comprehensive treatment of crossflow plate fundamentals has been pub-

lished by Lockett (51). As noted earlier, vendors of valve plates routinely make available their design methods.

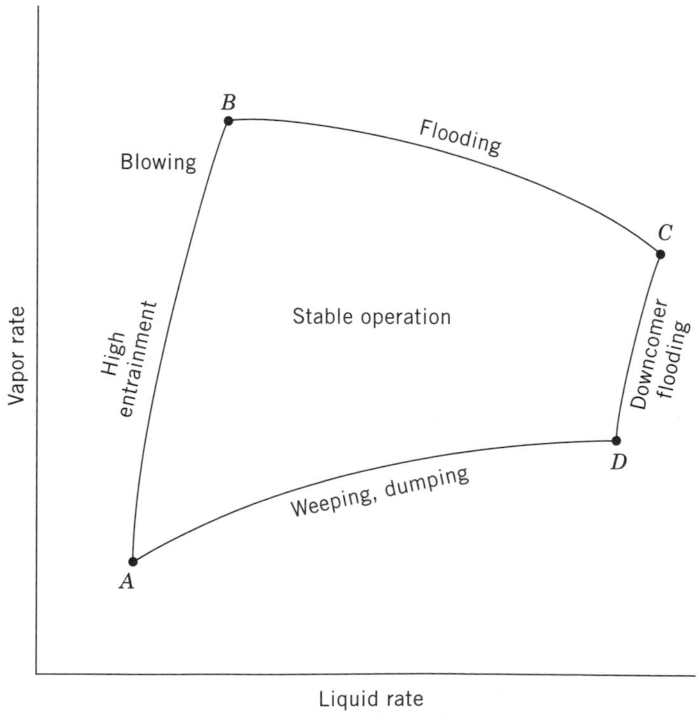

Plate Efficiencies. Column requirements are calculated in terms of theoretical stages or plates; the designer, however, must specify actual plates. To do, the designer must predict how effective the plate must perform in approaching the equilibrium condition. This approach, called the plate efficiency, measures the rate of mass transfer on the actual plate. This efficiency, expressed either as a fraction or as a percentage, depends on three main factors:

1. The geometry of the plate (hole arrangement, valve design, etc).
2. The loading of vapor and liquid traffic on the plate.
3. The diffusional properties of the fluids.

The simplest efficiency is the overall column efficiency, which is the number of theoretical plates in a column divided by the number of actual plates:

$$E_o = \frac{\text{Number of theoretical stages calculated}}{\text{Number of actual plates or stages required}} = \frac{N_t}{N_a}$$

Thus, the overall efficiency is an averaged efficiency of all the individual plates. A more useful plate efficiency for theoretical prediction is the Murphree plate efficiency (52):

$$E_{mv} = \frac{y^n - y^{n-1}}{y^{n*} - y^{n-1}}$$

where y^n and y^{n-1} are the vapor compositions from plate n and $n - 1$ (the plate below n), and y^{n*} is the vapor composition that would be in equilibrium with the liquid composition leaving plate n. Thus, for a given plate, E_{mv} is the ratio of the actual vapor composition change to the change that would occur if the plate were effective enough to

Figure 18. Stable operating range for crossflow plates.

bring the vapor and liquid to thermodynamic equilibrium. Note that the definition is based on the outlet liquid composition; it says nothing about the average liquid composition on the plate. In cases where a significant concentration gradient exists in the liquid composition across the plate, it is possible for E_{mv} to have a value greater than 1.0 (100%). The Murphree plate efficiency equation is written in terms of vapor compositions; a similar equation can be written in terms of the liquid compositions and is denoted as E_{mL}.

Of still more theoretical importance is the efficiency at some point on the plate:

$$E_{og} = \left(\frac{y^n - y^{n-1}}{y^{n*} - y^{n-1}}\right)_{\text{point}}$$

This parameter is called the point efficiency (or local efficiency). It cannot have a value greater than 1.0, and it has a counterpart term for liquid compositions.

Prediction of Plate Efficiency. The most comprehensive study of plate efficiency to date was made in the mid-1950s by the AIChE Research Committee (53) and was based on the then still popular bubble-cap plates. Unfortunately, the predictive model developed by the AIChE effort has been shown inadequate for many industrial distillations. There has been continuing research directed toward a better understanding of the mechanisms that occur in the rather complex aerated mass on a typical plate; later models have been discussed extensively (51,54). A complicating factor is the lack of uniform liquid flow across the plate, and situations have been found where the liquid actually stagnates in certain zones of larger diameter plates. As mentioned earlier, for large columns it is possible for the observed Murphree efficiency to exceed 100%. A completely satisfactory method for predicting plate efficiency does not exist, and probably will not exist for many years. Recently, there have been studies of the various types of flow regimes that occur on operating plates and of the effect of these regimes on tray performance, including plate efficiency. Pursuit of the flow regime studies (55–58) may lead to improved plate efficiency prediction methods. For example, a new model (59) accounts for the regime as well as the vapor bubble (froth flow) or liquid drop (spray flow) characteristics in determining mass transfer coefficients in the aerated zone on the plate.

Empirical Efficiency Prediction Methods. Numerous empirical methods for predicting plate efficiency have been proposed. Probably the most widely used method correlates overall column efficiency as a function of feed viscosity and relative volatility (60); this model, along with its companion for absorption and stripping, is shown in Figure 19.

General Comments on Plate Efficiency. The plate efficiencies of well-designed commercial bubble-cap, sieve, and valve plates are approximately the same when the plates are operated within their normal design range. The plate efficiency decreases both at the low end of the plate's operating range, where the liquid tends to leak through the plate, and at the high end of the operating range, where liquid entrainment becomes substantial.

Most distillation systems in commercial columns have Murphree plate efficiencies of 70% or higher. Lower efficiencies are found under system conditions of a high slope of the equilibrium curve (see Fig. 1(**b**)), of high liquid viscosity, and of large molecules with characteristically low diffusion coefficients. Finally, most measured efficiencies have been for binary systems, where by definition the efficiency of one component is equal to that of the other component. For multicomponent systems, it is possible for each component to have a different efficiency. Practice has been to use a pseudobinary approach involving the two key components. However, the theory for multicomponent efficiency prediction has been developed by Toor and Burchard (61) and by Krishna and Standart (62), and is amenable to computational analysis.

Packed Columns

In these columns, the vapor–liquid contacting takes place in continuous beds of solid packing elements rather than in discrete individual plates. The contacting can be visualized as occurring in differential increments across the height of the packing; thus, packings are known as counterflow devices rather than stagewise devices. Mechanically, the packed column is a relatively simple structure; in its simplest form, it comprises a vertical shell with dumped or carefully arranged packing elements on an open-type support, together with a suitable liquid distribution device above the packed bed. A packed column with two packed beds and a midcolumn feed is shown in Figure 20. The vapor enters the column below the bottom bed and flows upward through the column. The liquid (reflux or other liquid stream) enters at the top through the liquid distributor and flows downward through the packing countercurrently to the rising vapor. The height of the individual packed beds is limited to 2–9 m by the mechanical strength of the packing or by the need to redistribute the liquid so that good mass transfer efficiency can be maintained.

Packings. For many years, random packings were used almost exclusively, with occasional applications of regularly stacked packings or pads of woven or knitted wire. In the late 1960s, a partial trend away from random packings began when a special structured packing made of wire gauze was introduced (63). The indicated advantages of these structured packings were high mass transfer efficiency and very low pressure drop; the devices appeared to be ideal for high vacuum distillations. However, their cost of fabrication was very high and they were considered mainly for the vacuum distillation of specialty chemicals. In 1977, a lower cost sheetmetal version was introduced (64), and since that time, business in structured sheetmetal packings increased for a number of packing manufacturers. At the same time, improved random packings have been developed and a comprehensive discussion of their characteristics has been published (65). The Raschig ring, one of the oldest of packings, is an open cylinder of equal height and diameter. The Berl saddle and the ceramic Intalox saddle (Norton Co.) have a higher capacity and efficiency than the Raschig ring. The Pall ring is a modification of the Raschig ring which allows

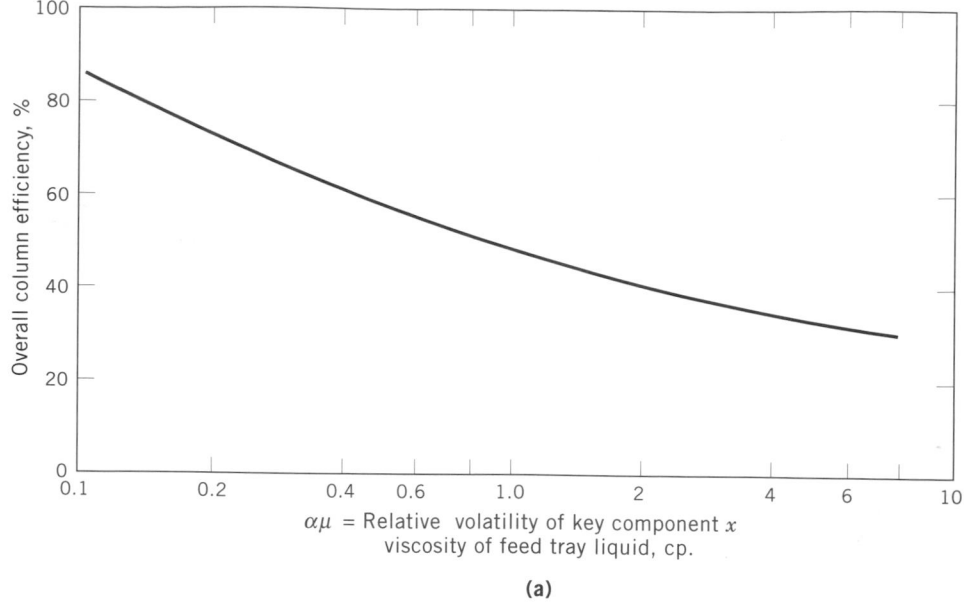

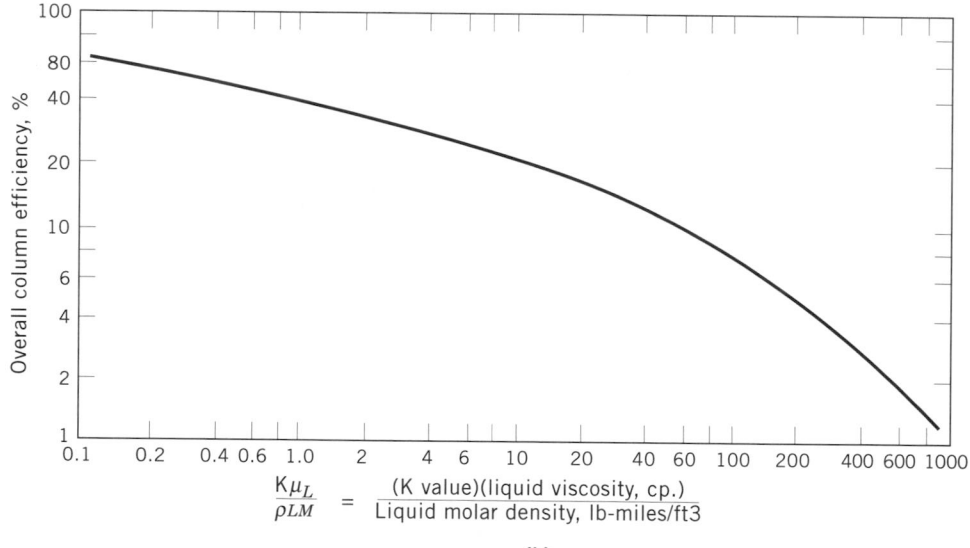

Figure 19. Charts for making approximate estimates of column efficiency of (**a**) distillation columns and (**b**) absorbers or strippers.

throughflow of liquid and vapor, with consequent lower pressure drop and better efficiency. The newer Intalox metal saddle (IMTP) is an example of a random packing with a very high void fraction and low resistance to the flowing phases. Other newer random packings (not shown in Fig. 21) include the CMR ring (Glitsch, Inc.) and the Nutter ring (Nutter Engineering Co.). The random-type packings can generally be made from metal, plastic, or ceramic materials; the approximate nominal size range for the individual elements is 12–75 mm.

Common structured packing geometries are shown in Figure 22. Flat plates of gauze or sheetmetal are perforated (or embossed or lanced) and corrugated. Corrugated sheets are then stacked together with adjacent sheets having opposite directions of the corrugations. The corrugations have angles with the horizontal of 45–60°. Vapor and liquid contact each other in wetted-wall fashion, and

the perforations plus other surface enhancements (eg, texturing) serve to promote liquid spreading into thin films. Dimensions, performance characteristics, and design procedures for the structured packings have been summarized (66).

Packed Column Internals. In order to ensure good packed-column mass transfer efficiency, the liquid must be distributed uniformly over the surface of the packing. Generally, there should be at least 100 pour points/m² (10 points/ft²), although fewer points may be used for random packings of the bluff-body type (Raschig rings, Berl saddles). Although they have capacity and pressure drop limitations, the bluff-body packing elements are able to divide the downflowing liquid and thus improve on an initially marginal distribution. On the other hand, the throughflow type random packings (eg, Pall rings, Intalox

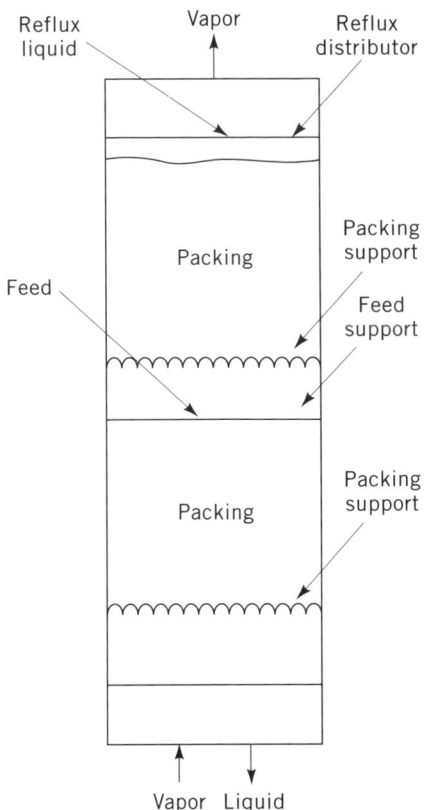

Figure 20. Packed column shell and internals. Column shown has single packed beds above and below the feed. For separations requiring a large number of stages, additional beds, separated by redistribution devices, are likely to be needed.

metal tower packings), as well as the structured packings, are not able to correct the initial distribution and in fact may allow some deterioration of distribution if the bed heights are greater than about 5 m.

Considerable research continues on methods for ensuring good liquid distribution in large diameter columns, and the packing manufacturers maintain large test stands where a particular design of distributor can be tested with water before being installed in the column. The distributor design problem becomes more severe at low liquid rates (< 700 cm^3/(s·m^2)) [1 gal/(min·ft^2)] or in large (> 3 m) diameter towers. An example of a more fundamental study of liquid distribution has been described (67), as well as typical liquid distributor designs and packing supports (65).

Packed Column Operation. In the packed column, liquid flows downward in opposition to the upward flow of vapor; both phases flow through the same open space or interstices between the packing elements. At low liquid and gas flow rates, the descending liquid occupies only a small fraction of the interstices and, therefore, offers little hindrance to the rising vapor flow. Figure 23 shows a schematic plot of pressure drop per unit of height as a function of the gas rate at low and high liquid flow rates. At a low rate of gas flow, the log slope of each curve is approximately 2. As the gas flow rate increases, there is an increasing tendency for the liquid to be held up in the void space, thereby decreasing the space available for the gas flow. As the gas flow rate increases further, more liquid is held up until at some high gas rate the packing floods. At this point, the liquid is essentially filling the interstices and can no longer flow downward. At flooding, the log slope is practically infinite. The pressure drop at the inception of flooding ranges from 1.6 to 3.3 kPa/m (2–4 in water/ft) of packing. More comprehensive discussions of packed column hydraulics may be found (13,65,68–71).

Capacity of Packed Columns. Packed columns are usually designed to operate at some percentage approach to flooding, eg, 60–70%, or at some specified pressure drop per unit height of packing, eg, 0.8 kPa/m (1 in water/ft) of packing. The most commonly used flooding correlation was initially proposed by Sherwood and co-workers (72) and it has been revised by several authors. One revision introduced constant pressure drop lines (73); the most recent revision is shown in Figure 24. The idea of "flooding" has been eliminated from the chart with the stipulation that the topmost curve represents the maximum capacity. Experimentally determined packing factors F_p from Table 2 should be used in the ordinate group; these factors distinguish between the various shapes and sizes of the available packings. The curves are for constant pressure

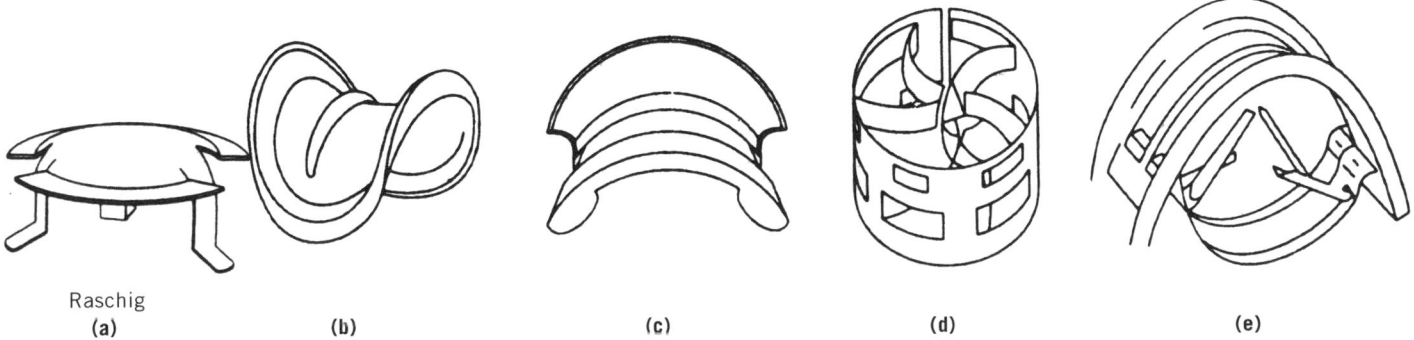

Raschig
(a)   **(b)**   **(c)**   **(d)**   **(e)**

Figure 21. Random packing elements for distillation columns: (**a**) Raschig ring (metal); (**b**) Berl saddle (ceramic); (**c**) Intalox saddle (ceramic); (**d**) Pall ring (metal); and (**e**) Intalox saddle (metal).

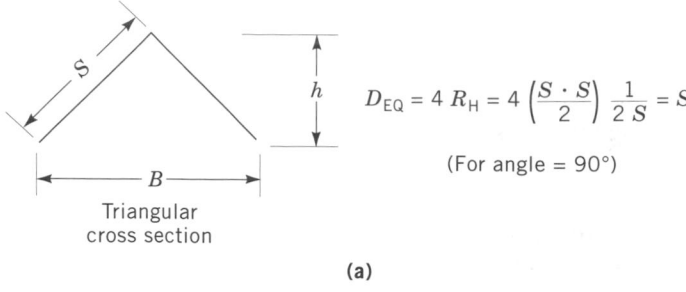

Triangular
cross section

$$D_{EQ} = 4\,R_H = 4\left(\frac{S \cdot S}{2}\right)\frac{1}{2\,S} = S$$

(For angle = 90°)

(a)

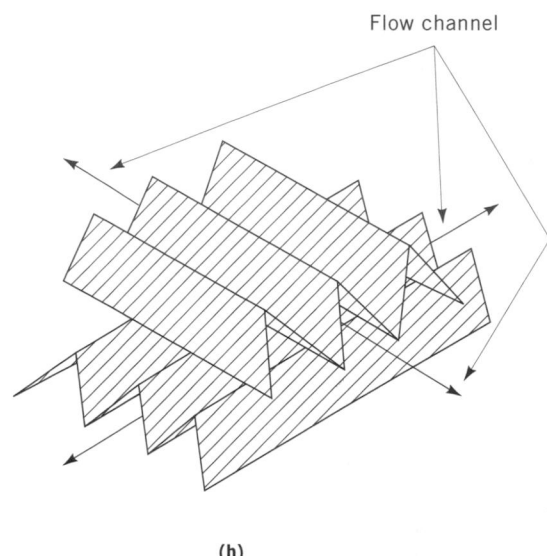

Flow channel

(b)

Figure 22. Geometric variables for structured packings: (**a**) flow channel cross-section; (**b**) flow channel arrangement.

drop and thus the chart enables estimation of both capacity and pressure drop.

Packing Mass Transfer Characteristics. The contacting for mass transfer in a packed column occurs differentially along the length of the column. The separation calculations can thus be made on a differential basis along this length, using mass transfer coefficients or heights of transfer units. The calculations are somewhat imprecise because of the uncertainty in the fundamental mass transfer mechanisms in larger scale columns. Useful models for predicting the mass transfer efficiency of randomly packed columns have been published (74,75), using the same commercial-scale performance data. These models cover the better known packings (metal and ceramic Raschig rings, ceramic Berl saddles, and metal Pall rings), in nominal sizes in the range of 12–50 mm. To avoid excessive maldistribution of liquid near the wall, a ratio of column diameter to packing element size of at least 8 should be maintained. Thus, to conduct pilot-scale packed column tests, a minimum column diameter of about 100 mm would be used, together with 12-mm packing elements. The models would then permit scale-up to large columns containing 50-mm size elements of the same type (eg, Pall rings).

These models provide values of the height of a transfer unit for the liquid phase H_L and the vapor phase H_v.

These values are combined to form the height of an overall transfer unit H_{ov}:

$$H_{ov} = H_v + (m'V/L)H_L$$

where V and L are molar flow rates of vapor and liquid, and m' is the average slope of the y–x equilibrium curve (Fig. 1(**b**)) in the concentration range of interest. The required total height of the packed section is then obtained from the simple relationship,

$$Z_p = (N_{ov})(H_{ov})$$

In order to determine the packed height Z_p, it is necessary to obtain a value of the overall number of transfer units N_{ov}; methods for doing this are available for binary systems and, in a more complex way, multicomponent systems (76). However, it is simpler to calculate the number of required theoretical stages and make the conversion:

$$N_{ov} = N_t\,(\ln m'V/L)/(m'V/L - 1)$$

An alternative approach to determining packed height is through the use of an empirical term, height equivalent to a theoretical plate (HETP). This term can be measured in a fashion similar to that used for the overall plate efficiency of a column:

$$\text{HETP} = \frac{\text{Total packed height}}{\text{No. of theoretical plates}} = \frac{Z_p}{N_t}$$

Typical experimental values of HETP for a random packing (50-mm Pall rings) and a structured packing (Intalox 2T, Norton Co.), and for the same system conditions, are shown in Figure 25. Many designers of packed columns prefer the use of HETP instead of H_{ov}, but the latter

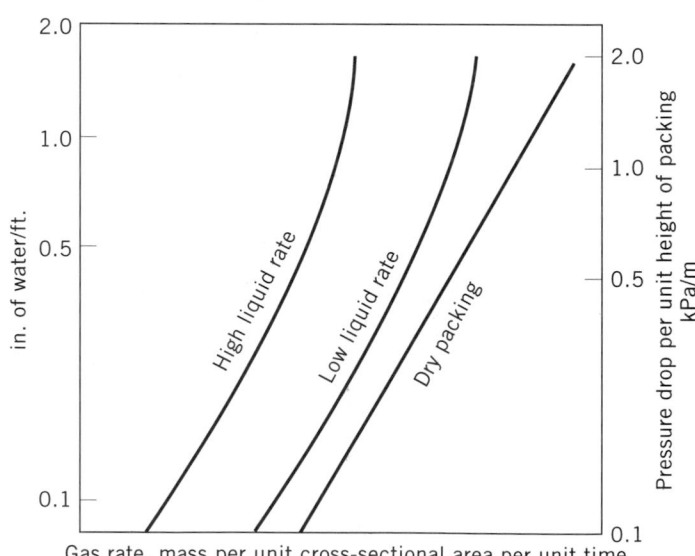

Figure 23. Pressure drop per unit of height as a function of the gas rate at low and high liquid flow rates.

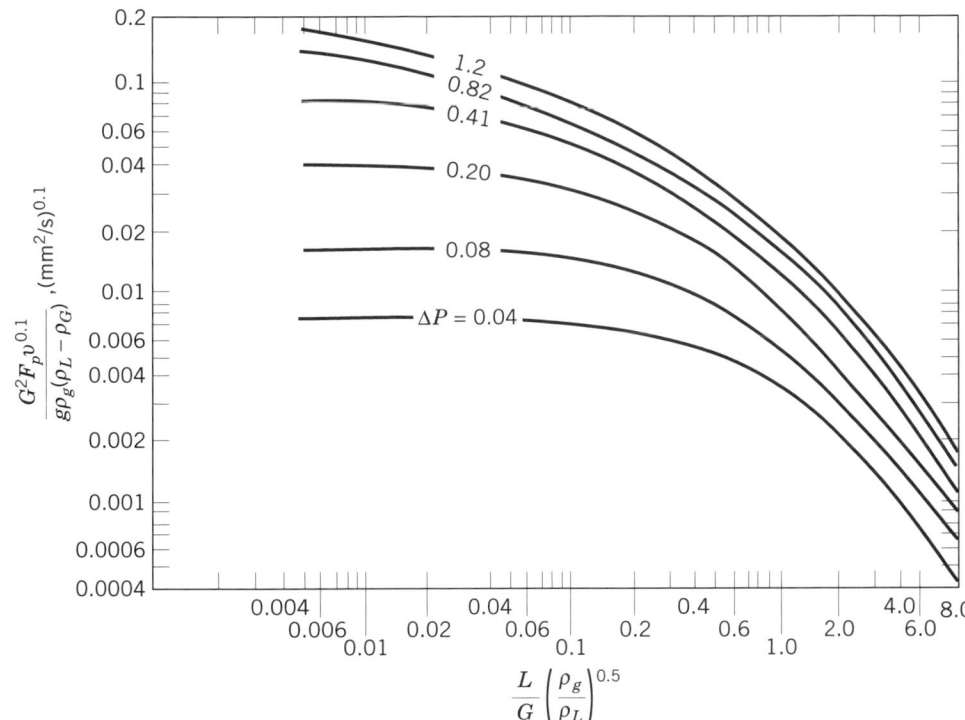

Figure 24. Generalized method using log scales for estimating packed column flooding and pressure drop ΔP in kPa/m; g = gravitational constant, 9.81 m/s²; ν = kinematic viscosity in mm²/s (= cSt); L, G have units of kg/(m²s); ρ_L, ρ_g are in kg/m₃; and the packing factor F_p in m⁻¹ can be found in Table 2. To conver kPa/m to mm Hg/m, multiply by 7.5 (65).

is more fundamental and discriminates between liquid and vapor phase resistances. (Terms such as H_{ov} and N_{ov} are based on vapor phase concentrations; equivalent terms based on liquid concentrations could also be used.)

For structured packings, methods for predicting H_v and H_L are somewhat more reliable than those for random packings. Perhaps the most frequently used efficiency correlation for these packings is that of Bravo and co-workers (77); a slightly different model that covers a broader packing size range is that of Spiegel and Meier (78). Methods for predicting pressure drop and flooding in beds of structured packings have been reviewed (66).

Packed vs Plate Columns

Relative to plate towers, packed towers are more useful for multipurpose distillations, usually in small towers (under 0.5 m), or for the following specific applications: severe corrosion environment—some corrosion-resistant materials (plastics, ceramics, and certain metallics) can easily be fabricated into packing but may be difficult to fabricate into plates; vacuum operation, where a low pressure drop per theoretical plate is a critical requirement; high liquid rates (eg, above 49,000 kg/(h·m² [~10,000 lb/(h·ft²)]; foaming systems; or debottlenecking plate towers with plate spacings that are relatively close (under 0.3 m).

Plate columns have the advantage of lower fabrication cost, less dependence on good liquid and gas distribution, and protection against vapor bypassing the liquid in critical zones (eg, regions of extremely low impurities). Further, methods for the design on plate columns are somewhat more reliable than those for many of the packings, especially those packings of a proprietary nature.

There are notable cases where plate columns have been converted to packed columns to gain advantage of the low

Table 2. Characteristics of Packings[a]

Packing	Nom. Size, mm	Surface Area, m²/m³	Void Fraction	Packing Factor, m⁻¹
Intalox ceramic saddles	D[b] 13	625	0.78	660
	25	255	0.77	300
	50	118	0.79	130
Intalox metal saddles (IMTP)	D 25		0.97	135
	40		0.97	79
	50		0.98	59
Pall rings, metal	D 16		0.92	265
	25	205	0.94	183
	50	115	0.96	88
Raschig rings, ceramic	D 13	370	0.64	1902
	25	190	0.74	587
	50	92	0.74	215
Berl saddles, ceramic	D 13	465	0.62	790
	25	250	0.68	360
	50	105	0.72	150
Intalox saddles, plastic	D 25	206	0.91	130
	50	108	0.93	92
	75	88	0.94	59
Pall rings, plastic	D 16	341	0.87	310
	25	207	0.90	180
	50	100	0.92	85
FLEXIPAC				
1	S 6[c]	558	0.91	108
2	12[c]	246	0.93	72
3	37[c]	134	0.96	52
4	50[c]	69	0.98	30
Sulzer—BX	S 6[c]	490	> 0.90	66

[a] Ref. 65.
[b] D = dumped (random) packing; S = structured packing.
[c] Crimp height, mm.

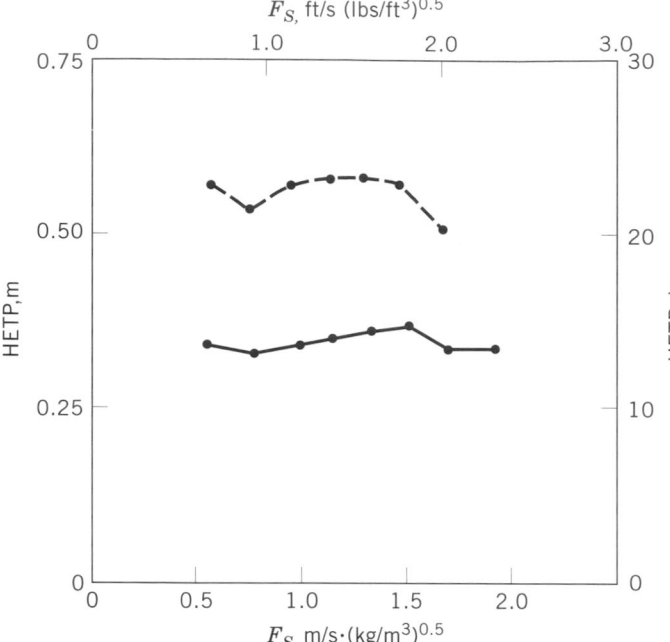

Figure 25. Values of height equivalent to a theoretical plate (HETP) as a function of throughput for (−−) 50-mm metal Pall rings and (−·−) No. 2 structured packing at 12-mm crimp height. Conditions are cyclohexane–*n*-heptane system, 165 kPa (24 psia) operating pressure, total reflux, 0.43-m diameter column. Courtesy of The University of Texas at Austin.

pressure drop exacted from the vapor stream. In recent years, the packings have been largely of the structured type, as illustrated by the trend toward the use of structured packing in ethylbenzene–styrene fractionators, some of which have diameters of 10 m or higher.

STEAM DISTILLATION

Steam distillation is used to lower the distillation temperatures of high boiling organic compounds that are essentially immiscible with water. If an organic compound is immiscible with water, both it and the water will exert their full vapor pressure upon vaporization from the immiscible two-component liquid. At a system pressure of P, the partial pressures will be

$$P = P_{\text{water}} + P_{\text{organic}}$$

And since the water and organic compound are immiscible:

$$P = P^0_{\text{water}} + P^0_{\text{organic}}$$

One example of the steam distillation is of *N*-ethyl-aniline at atmospheric pressure (68). The vapor pressures at 99.15°C of water and *N*-ethylaniline are 98.27 and 3.04 kPa (737 and 22.8 mm Hg), respectively. Thus,

$$P = 98.27 + 3.04 = 101.3 \text{ kPa (760 mm Hg)}$$

The concentration of the *N*-ethylaniline in the vapor is

$$y = 3.04/101.3 = 0.030 \text{ mole fraction}$$

The normal boiling point of *N*-ethylaniline is 204°C. Therefore, steam distillation makes possible the distillation of *N*-ethylaniline at atmospheric pressure at 99.15°C instead of its normal boiling point of 204°C. Commercial applications of steam distillation include the fractionation of crude tall oil (79), the distilling of turpentine, and certain essential oils. A detailed calculation of steam distillation of turpentine has been reported (80).

DISTILLATION AS A SEPARATION METHOD

Distillation is the most important industrial method of separation and purification of liquid components. Liquid separation methods less commonly used include liquid–liquid extraction, membrane permeation, ion exchange, and adsorption. Unlike these alternative methods, distillation does not require a mass separating agent (solvent, adsorbent, membrane); it utilizes energy in a convenient heating medium (often steam). Also, the wealth of experience with design and operations makes distillation column performance prediction more reliable than equivalent predictions for other methods. At times, distillation also competes indirectly with methods involving solid–liquid separations such as crystallization. A discussion of distillation in context with other commercial separation methods is available (81).

The suitability and economics of a distillation separation depend on many factors, including favorable vapor–liquid equilibria, feed composition, number of components to be separated, product purity requirements, the absolute pressure of the distillation, heat sensitivity, corrosivity, and continuous vs batch requirements. Distillation is somewhat energy-inefficient, as usually, heat added at the base of the column is largely rejected overhead to an ambient sink. However, the source of energy for distillations is often low pressure steam, which characteristically is in abundant supply and thus relatively inexpensive. Also, schemes have been devised for lowering the energy requirements of distillation and are described in many publications (eg, see Ref. 82).

Favorable Vapor–Liquid Equilibria

The suitability of distillation as a separation method is strongly dependent on favorable vapor–liquid equilibria. The absolute value of the key relative volatilities directly determines the ease and economics of a distillation. The energy requirements and the number of plates required for any given separation increase rapidly as the relative volatility becomes lower and approaches unity. For example, given an ideal binary mixture with a 50 mol % feed and a distillate and bottoms requirement of 99.8% purity each, the minimum reflux and minimum number of theoretical plates are shown below for assumed relative volatilities of 1.1, 1.5, and 4.

Relative Volatility	Minimum Reflux Ratio	Minimum No. of Theoretical Plates
1.1	20	130
1.5	4.0	31
4	0.66	9

The minimum reflux ratio and minimum number of theoretical plates decreased 14- to 33-fold, respectively, when the relative volatility increased from 1.1 to 4. Other distillation systems would have different specific reflux ratios and numbers of theoretical plates, but the trend would be the same. As the relative volatility approaches unity, distillation separations rapidly become more costly in terms of both capital and operating costs. The relative volatility can sometimes be improved through the use of an extraneous solvent that modifies the VLE. Binary azeotropic systems are impossible to separate into pure components in a single column, but the azeotrope can often be broken by an extraneous "entrainer."

Feed Composition

Feed composition has a substantial effect on the economics of a distillation. Distillations tend to become uneconomical as the feed becomes dilute. There are two types of dilute feed cases: one in which the valuable recovered component is a low boiler and the second when it is a high boiler. When the recovered component is the low boiler, the absolute distillate rate will be low, but the reflux ratio and the number of plates will be high, eg, the recovery of methanol from a dilute solution in water. When the valuable recovered component is a high boiler, the distillate rate, the reflux relative to the high boiler, and the number of plates all are high. An example for this case is the recovery of acetic acid from a dilute solution in water. For the general case of dilute feeds, alternative recovery methods are usually more economical than distillation.

Product Purity

Product purity requirements influence the choice between separation methods. With favorable equilibria, distillation energy requirements do not increase significantly as purity specifications become tighter. For example, in an ideal binary distillation of 60 mol % of A in the feed, the minimum and operating reflux ratios would be essentially the same whether the required purity of A was 99 or 99.9999%. The number of plates would increase substantially, however, as the purity requirements became more stringent. The shortcut methods of calculating minimum reflux ratio, minimum number of plates, operating reflux ratio, and number of operating plates (as discussed previously) allow a rapid evaluation of the effect of changes in purity requirements on the key economic factors in distillation.

Operating Pressure

The absolute pressure of the distillation may have substantial economic impact. The temperatures at which heat is supplied to the reboiler and removed from the condenser determines the unit cost of the energy. The cost of removing heat in the condenser increases rapidly as the condensing temperature drops below the range of air or water cooling capability, eg, the cost of removing a unit quantity of heat at −25°C may be 100 times as high as removing it at 100°C. Similarly, the cost of the energy required for the reboiler increases rapidly as the boiler temperature increases above some level determined by local conditions. For example, at a particular site, low pressure waste steam at 110°C may be essentially without cost but if a temperature level of 200°C is required, the unit cost of the heat will be much higher. The relative cost of the heat being removed and supplied is the controlling factor determining the design of some distillations. The use of multiple interstage reboilers and condensers at different energy levels, as well as the use of other operational modes to optimize the overall economics, has been discussed (83).

The absolute pressure may have a significant effect on the vapor–liquid equilibrium. Generally, the lower the absolute pressure, the more favorable the equilibrium. This effect has been discussed for the styrene–ethyl benzene system (84). In a given column, increasing the pressure can increase the column capacity by increasing the capacity parameter. Selection of the economic pressure can be facilitated by guidelines (85) that consider the pressure effects on capacity and relative volatility. Low pressures are required for distillation involving heat-sensitive material.

Heat Sensitivity

The heat sensitivity or polymerization tendencies of the materials being distilled influence the economics of distillation. Many materials cannot be distilled at their atmospheric boiling points because of high thermal degradation, polymerization, or other unfavorable reaction effects that are functions of temperature. These systems are distilled under vacuum in order to lower operating temperatures. For such systems, the pressure drop per theoretical plate is frequently the controlling factor in contactor selection. An excellent discussion of equipment requirements and characteristics of vacuum distillation may be found (86).

Corrosivity

Corrosivity is an important factor in the economics of distillation. Corrosion rates increase rapidly with temperature; in distillation, the separation is made at boiling temperatures. Such temperatures may require expensive construction materials for the distillation equipment, and some of them may be difficult to use for fabricating the column components. For some materials, eg, ceramics, random packings may be specified, and the application of such packings to highly corrosive services has been practiced widely. On the other hand, the extensive surface areas of metal packings may make them more susceptible than plates to corrosion. Again, cost may be the final arbiter.

DISTILLATION EQUIPMENT COSTS

A recently published compilation of costs of distillation and related equipment is available (87). Some of the commercial computer-aided process design packages also contain equipment cost information. For specialized internals (distributors, support plates, packings, crossflow plates, and so on), it is usually necessary to obtain cost information directly from the equipment vendors. The cost of a distillation system includes many components in addition to the column itself, for example, an expensive packing may be justified on the basis that it can reduce the cost of the column shell, foundations, piping, and so on. Discussions of economics of distillation systems are contained in the two recent publications (88,89).

EXPERIMENTAL TESTING FOR NEW DESIGNS

In the process of scale-up from bench experiments to commercial design, it is usually desirable to obtain some small-scale operating experience with the intended distillation separation. It is possible to emulate plate column design in laboratory columns as small as 28 mm diameter; it has been shown that the use of a small laboratory device, the Oldershaw column, can lead to reliable prediction of large-size plate column efficiency for the sample system (90). Similar methods have been proposed for columns containing structured packing (91). Scale-up procedures for columns with random packing are not as straightforward and usually require testing in a minimum column diameter of 100 mm.

For testing new designs of devices (packings, plates) the facilities of Fractionation Research, Inc. (FRI) may be available. This company is a nonprofit, industry-sponsored, research corporation with laboratories in Stillwater, Oklahoma. Much of the current research on commercial-size distillation equipment is being done by FRI. The industrial sponsors are fabricators, designers, and constructors, or users of distillation equipment. For the most part, the research is confidential to the members or to the developers of the devices for testing.

DISTILLATION LITERATURE REVIEWS

For several years, annual reviews of distillation (along with other unit operations) were published in *Industrial and Engineering Chemistry*. Since the mid-1970s, there have been no exhaustive literature reviews, but the publications by Ray (92,93) provide general coverage of the period 1967–1990.

BIBLIOGRAPHY

1. J. Gmehling, U. Onken, and W. Arlt, *Vapor-Liquid Equilibrium Collection,* DECHEMA, Frankfurt, 1979.

2. M. Hirata, S. Ohe, and K. Nagahama, *Computer Aided Data Book of Vapor–Liquid Equilibria,* Elsevier, Amsterdam, 1975.

3. E. Hala, J. Pick, V. Fried, and O. Vilim, *Vapor–Liquid Equilibrium,* 2nd ed., Pergamon Press, Oxford, 1967.

4. E. Hala, I. Wichterle, J. Polak, and T. Boublik, *Vapor–Liquid Equilibrium at Normal Pressures,* Pergamon Press, Oxford, 1968.

5. I. Wichterle, J. Linek, and E. Hala, *Vapor–Liquid Equilibrium Data Bibliography,* Elsevier, Amsterdam, 1975.

6. A. Fredenslund, J. Gmehling, and P. Rasmussen, *Vapor–Liquid Equilibria Using UNIFAC,* Elsevier, Amsterdam, 1977.

7. J. H. Hildebrand, J. M. Prausnitz, and R. L. Scott, *Regular and Related Solutions,* Van Nostrand Reinhold, New York, 1970.

8. E. L. Derr and C. H. Deal, *I. Chem. E. Symp. Ser.* **32,** 3:40 (1969).

9. D. A. Palmer, *Handbook of Applied Thermodynamics,* CRC Press, Boca Raton, Fla., 1987.

10. J. M. Prausnitz, R. N. Lichtenthaler, and E. G. Azeredo, *Molecular Thermodynamics of Fluid-Phase Equilibria,* 2nd ed., Prentice-Hall, Englewood Cliffs, N.J., 1986.

11. R. C. Reid, J. M. Prausnitz, and B. Pohling, *The Properties of Gases and Liquids,* 4th ed., McGraw-Hill, New York, 1987.

12. S. M. Walas, *Phase Equilibria in Chemical Engineering,* Butterworths, Reading, Mass., 1985.

13. E. J. Henley and J. D. Seader, *Equilibrium-Stage Separation Operations in Chemical Engineering,* John Wiley & Sons, Inc., New York, 1981.

14. P. Wankat, *Equilibrium-Staged Separations,* Elsevier, New York, 1988.

15. M. Van Winkle, *Distillation,* McGraw-Hill, New York, 1967.

16. J. J. Van Laar, *Z. Physik. Chem.* **72,** 723 (1910); **83,** 599 (1913).

17. G. M. Wilson, *J. Am. Chem. Soc.* **86,** 127 (1964).

18. H. Renon and J. M. Prausnitz, *AIChE J.* **14,** 135 (1968).

19. D. S. Abrams and J. M. Prausnitz. *AIChE J.* **21,** 116 (1975).

20. Margules, *Sitzber. Math. Naturw. Kl. Kaiserlichen Akad. Wiss.* **104,** 1243 (1895).

21. H. H. Chien and H. R. Null, *AIChE J.* **18,** 1177 (1972).

22. J. M. Prausnitz, T. F. Anderson, E. A. Grens, C. A. Eckert, R. Hsieh, and J. P. O'Connell, *Computer Calculations for Multicomponent Vapor–Liquid and Liquid–Liquid Equilibria,* Prentice-Hall, Englewood Cliffs, N.J., 1980.

23. Y.-L. Hwang, J. D. Olson, and G. E. Keller, *Ind. Eng. Chem. Res.* **31,** 1759 (1992).

24. L. Horsley, "Azeotropic Data–III," *Adv. Chem. Ser.* **116,** American Chemical Society, Washington, D.C., 1973.

25. B. Lu, *Can. J. Tech.* **34,** 468 (1957).

26. J. F. Boston, H. I. Britt, S. Jiraphongphan, and V. B. Shah, "An Advanced System for the Simulation of Batch Distillation Operations," in *Foundations of Computer-Aided Chemical Process Design,* Vol. 2, American Institute of Chemical Engineers, New York, 1981.

27. W. L. McCabe, J. C. Smith, and P. Harriott, *Unit Operations of Chemical Engineering,* 5th ed. McGraw-Hill, New York, 1993.

28. W. L. McCabe and E. W. Thiele, *Ind. Eng. Chem.* **17,** 605 (1925).

29. M. R. Fenske, *Ind. Eng. Chem.* **24,** 482 (1932).

30. J. R. Fair and W. L. Bolles, *Chem. Eng.* **75** (9), 156 (Apr. 22, 1968).

31. A. J. V. Underwood, *Chem. Eng. Progr.* **44,** 603 (1948).

32. E. R. Gilliland, *Ind. Eng. Chem.* **32,** 918 (1940).

33. J. H. Erbar and R. N. Maddox, *Petrol. Ref.* **40** (5), 183 (1961).

34. H. E. Eduljee, *Hydrocarbon Proc.* **54** (9), 120 (1975).

35. E. W. Thiele and R. L. Geddes, *Ind. Eng. Chem.* **25,** 290 (1933).

36. C. D. Holland, *Fundamentals of Multicomponent Distillation,* McGraw-Hill, New York, 1981.

37. R. N. S. Rathore, K. A. Van Wormer, and G. J. Powers, *AIChE J.* **20,** 491 (1974).

38. D. C. Freshwater and B. D. Henry, *Chem. Eng. London* (301), 533 (1975).

39. J. E. Hendry, D. F. Rudd, and J. D. Seader, *AIChE J.* **19,** 1 (1973).

40. C. D. Holland and A. Liapis, *Computer Methods for Solving Dynamic Separation Problems,* McGraw-Hill, New York, 1983.

41. "Ballast Tray Design Manual," *Bulletin 4900,* Glitsch, Inc., Dallas, Tex., 1974.

42. "Flexitray Design Manual," *Bulletin 960,* Koch Engineering Co., Wichita, Kan., 1960.

43. *Float Valve Tray Design Manual,* Nutter Engineering Co., Tulsa, Okla., 1976.

44. W. L. Bolles in B. O. Smith, ed., *Design of Equilibrium Stage Processes,* McGraw-Hill, New York, 1963, ch. 14.

45. W. L. Bolles, *AIChE J.* **22,** 153 (1976).

46. M. H. Hutchinson and R. F. Baddour, *Chem. Eng. Progr.* **52** (12), 503 (1956).

47. F. Kastanek, M. V. Huml, and V. Braun, *I. Chem. E. Symp. Ser.* **32,** 5:100 (1969).

48. W. V. Delnicki and J. L. Wagner, *Chem. Eng. Progr.* **52** (1), 28 (1956).

49. J. R. Fair in B. D. Smith, ed., *Design of Equilibrium Stage Processes,* McGraw-Hill, New York, 1963, ch. 15.

50. J. R. Fair in R. H. Perry and D. Green, eds., *Perry's Chemical Engineers' Handbook,* 6th ed., McGraw-Hill, New York, 1984, sect. 18.

51. M. J. Lockett, *Distillation Tray Fundamentals,* Cambridge University Press, New York, 1986.

52. E. V. Murphree, *Ind. Eng. Chem.* **17,** 747, 960 (1925).

53. American Institute of Chemical Engineers, *Bubble-Tray Design Manual,* AIChE, New York, 1958.

54. M. M. Dribika and M. W. Biddulph, *Trans. I. Chem. E.* **70** (Part A), 142 (1992).

55. K. E. Porter, M. J. Lockett, and C. T. Lim, *Trans. I. Chem. E.* **50,** 91 (1972).

56. W. V. Pinczezewski, N. D. Benke, and C. J. D. Fell, *AIChE J.* **21,** 1210 (1975).

57. K. E. Porter, A. Safekouri, and M. J. Lockett, *Trans. I. Chem. E.* **51,** 265 (1973).

58. M. Prado, K. L. Johnson, and J. R. Fair, *Chem. Eng. Progr.* **83** (3), 32 (1987).

59. M. Prado and J. R. Fair, *Ind. Eng. Chem. Res.* **29,** 1031 (1990).

60. H. E. O'Connell, *Trans. AIChE* **42,** 741 (1946).

61. H. L. Toor and J. K. Burchard, *AIChE J.* **6,** 202 (1960).

62. R. Krishna, H. F. Martinez, R. Sreedhar, and G. L. Standart *Trans. I. Chem. E.* **55,** 178 (1977).

63. A. Sperandio, M. Richard, and M. Huber, *Chem. Ing. Tech.* **37,** 22 (1965).

64. W. D. Stoecker and B. Weinstein, *Chem. Eng. Progr.* **73** (11), 71 (1977).

65. R. F. Strigle, *Random Packings and Packed Tower Design,* Gulf Publications, Houston, Tex., 1987.

66. J. R. Fair and J. L. Bravo, *Chem. Eng. Progr.* **86** (1), 19 (1990).

67. P. J. Hoek, J. A. Wesselingh, and F. J. Zuiderweg, *Chem. Eng. Res. Des.* **64,** 431 (1986).

68. R. E. Treybal, *Mass Transfer Operations,* 3rd ed., McGraw-Hill, New York, 1980.

69. W. S. Norman, *Absorption, Distillation and Cooling Towers,* John Wiley & Sons, Inc., New York, 1961.

70. P. A. Schweitzer, ed., *Handbook of Separation Techniques for Chemical Engineers,* 2nd. ed., McGraw-Hill, New York, 1988, chs. 1.1–1.8.

71. R. W. Rousseau, ed., *Handbook of Separation/Process Technology,* John Wiley & Sons, Inc., New York, 1987, ch. 5.

72. T. K. Sherwood, G. H. Shipley, and F. A. L. Holloway, *Ind. Eng. Chem.* **30,** 765 (1938).

73. M. Leva, *Chem. Eng. Progr. Symp. Ser.* **10,** 50, 51 (1954).

74. W. L. Bolles and J. R. Fair, *Chem. Eng.* **89** (14), 109 (July 12, 1982).

75. J. L. Bravo and J. R. Fair, *Ind. Eng. Chem. Proc. Des. Dev.* **21,** 162 (1982).

76. R. Krishnamurthy and R. Taylor, *AIChE J.* **31,** 449, 456 (1985).

77. J. L. Bravo, J. A. Rocha, and J. R. Fair, *Hydrocarbon Proc.* **64** (1), 91 (1985).

78. L. Spiegel and W. Meier, *I. Chem. E. Symp. Ser.* **104,** A203 (1987).

79. J. Drew and M. Propst, eds., *Tall Oil,* Pulp Chemicals Assn., New York, 1981.

80. W. L. McCabe and J. C. Smith, *Unit Operations of Chemical Engineering,* 3rd ed., McGraw-Hill, New York, 1976, ch. 19.

81. G. E. Keller, "Separations: New Directions for an Old Field," *AIChE Monograph Series* **17,** American Institute of Chemical Engineers, New York, 1987.

82. J. R. Fair in Y. A. Liu, H. A. McGee, and W. R. Epperly, eds., *Recent Developments in Chemical Process and Plant Design,* John Wiley & Sons, Inc., New York, 1987, ch. 3.

83. W. C. Petterson and T. A. Wells, *Chem. Eng.* **84** (20), 79 (1977).

84. C. J. King, *Separation Processes,* 2nd ed., McGraw-Hill, New York, 1980.

85. H. Z. Kister and I. D. Doig, *Hydrocarbon Proc.* **56** (7), 132 (1977).

86. P. G. Nygren and G. K. S. Connolly, *Chem. Eng. Progr.* **67** (3), 49 (1971).

87. M. Peters and K. D. Timmerhaus, *Plant Design and Economics for Chemical Engineers,* 4th ed., McGraw-Hill, New York, 1991.

88. H. Z. Kister, *Distillation–Operation,* McGraw-Hill, New York, 1990.

89. H. Z. Kister, *Distillation–Design,* McGraw-Hill, New York, 1992.

90. J. R. Fair, H. R. Null, and W. L. Bolles, *Ind. Eng. Chem. Proc. Des. Devel.* **22,** 53 (1983).

91. J. R. Hufton, J. L. Bravo, and J. R. Fair, *Ind. Eng. Chem. Res.* **27,** 2096 (1988).

92. M. S. Ray, *Chemical Engineering Bibliography, 1967–1988,* Noyes Publications, Park Ridge, N.J., 1990; M. S. Ray, *Sepn. Sci. Techn.* **27,** 105 (1992).

93. M. S. Ray, *Sepn. Sci. Technol.* **27,** 106 (1992).

DISTRICT HEATING

ATTILIO L. BISIO
Atro Associates
Mountainside, New Jersey

A range of technology is used to provide space heating in commercial buildings: residential-style oil and natural gas furnaces in smaller buildings, oil and natural gas boilers, heat pumps, and electric boilers. Gas boilers and furnaces are used to heat almost half of the commercial space in the United States (Table 1).

Table 1. Space Heating Technologies in Commercial Buildings[a]

Technology	Percent[b]
Gas furnace/boiler	48
Oil furnace/boiler	25
Electric boiler	22
Electric heat pump	2
Other	3

[a] Ref. 1.
[b] The percent of all commercial square footage heated with the technology in 1988.

District heating involves the production of heat in the form of hot water or steam in a central plant. The steam or hot water is then distributed directly through underground pipes to buildings. District heating supplies the heat for over 10% of commercial building space (2).

Most district heating systems in the United States rely on dedicated steam generation facilities close to the buildings being served. In some cities, particularly New York, high-pressure steam is generated and used to drive turbines for a multitude of pumps and refrigerant compressors.

There is only a limited use of district heating in the United States to service detached residences. The cost to install the underground piping and the small quantities of heat required makes this kind of service generally uneconomical.

District heating is particularly appropriate in colder climates that have large space heating needs. Many European countries apply these systems more widely. In Denmark, for example, almost half of all building space heating needs are met with district heating systems (3).

The efficiency of district heating depends on the method used to produce the heat. If a cogeneration system is used to produce both heat and electricity, the overall efficiency can be quite high (see COGENERATION). However, one must have a demand for hot water large enough to justify the cogeneration system. (See also AIR CONDITIONING; BUILDING SYSTEMS.)

BIBLIOGRAPHY

1. *Baseline Projection Data Book,* Gas Research Institute, Washington, D.C., 1991, p. 127.
2. U.S. Department of Energy, Energy Information Administration, *Commercial Building Characteristics,* 1989, DOE/EIA-024689, Washington, D.C., June 1991, p. 128.
3. P. Kunjeer, "District Heating and Cooling: Solution for the Year 2000," *Proceedings from the International Symposium,* *Energy Options for the Year 2000,* Center for Energy and Urban Policy Research, University of Delaware, Newark, Del., 1988, pp. 1–109.

DRILLING FLUIDS

MARIO ZAMORA
M-I Drilling Fluids L.L.C.
Houston, Texas

The vast quantity of petroleum energy that has accumulated beneath the Earth's surface over geologic time is recovered by drilling wells to the rock formations containing the crude oil and/or natural gas. Drilling is a complex, multidisciplinary activity fraught with uncertainty and involving great effort and expense. By its very nature, the drilling process is a disturbance to the natural environment, even if only for a short period of time. However, there are no other practical means to win hydrocarbons from the ground.

Today's wells use the rotary drilling method, characterized by a rotating bit, to drill the hole, and a circulating drilling fluid (or drilling mud) to transport cuttings to the surface and to preserve the drilled hole. Whereas most drilling fluids are basically a suspension of solids in a liquid phase, some can be chemically complex. Drilling-fluids technology, with emphasis on environmental concerns, is the subject of this article.

Drilling-fluid technology is dominated by three factors: performance, economics, and environmental concerns. Drilling fluids must contend with an extensive range of subsurface conditions, including extreme pressures and temperatures, unstable formations, and harsh contaminants. Clearly, their principal goal is to contribute to the successful drilling of the well. However, drilling fluids must also be affordable and have a minimal impact on the natural environment. Finding the right balance among the three factors is a significant challenge for the drilling-fluids companies who provide this technology and for the drillers who use it.

HISTORICAL PERSPECTIVE

Growth of technology in the petroleum industry has been a more evolutionary, rather than revolutionary, process. Nonetheless, there are three distinct dates of importance to the drilling-fluids segment. In 1859, the celebrated well drilled by "Colonel" Edwin Drake near Titusville, Pennsylvania, struck oil at a depth of 21.2 m (69.5 ft) and signaled the start of the modern petroleum industry. In 1901, the discovery well of the Spindletop field on the Texas Gulf Coast became the first notable well to use the rotary drilling method successfully (1). A quicksand problem encountered at a shallow depth was solved by driving a herd of cattle through the slush pit to create muddy water. For many, this was the birth of drilling muds and the embryonic seed for the drilling-fluids industry which would blossom a few years later. Finally in 1978, landmark environmental regulations aimed at controlling discharge of mud and cuttings into the sea forever changed the face of the drilling-fluids industry.

Use of the rotary drilling method spread rapidly after the Spindletop discovery well, especially in areas where formations were soft and unconsolidated (2). Hole stability was achieved by plastering the hole walls with clay materials, contained in formation cuttings or added intentionally to the mud at the surface. Generally overlooked until years later, however, was the contribution of mud weight (density) to hole stability and, more importantly, to control of subsurface pressures. In the early 1920s, barite (barium sulfate ore) was selected from a group of high specific-gravity minerals as the best mud-weight additive to handle pressured formations. Other important mud additives followed. Bentonite clay, mined in Wyoming, was recognized for its superior mud-making properties and became the primary additive for viscosity, solids suspension, and filtration control.

Increasing demands for petroleum promoted new drilling challenges; drilling-fluids technology had to keep pace. Considerable resources were allocated to improve the understanding of mud chemistry, apply new testing methods, refine field procedures, and develop special additives to minimize and correct mud-related problems. Innovative mud systems, some radically different, emerged to satisfy expanding technical requirements and to lower costs. Air and natural gas were exploited to increase drilling rates in hard, dry formations. Muds based on oil instead of water, originally introduced to minimize damage to producing formations, evolved into high performance systems suitable for handling the most hostile of drilling conditions. Natural and synthetic polymers became the foundation for entire families of drilling fluids. Technology had matured to the point where cost-effective mud systems could be formulated for almost any application.

By the mid-1970s, the petroleum industry was in the midst of an extraordinary drilling boom that eventually would peak in 1981–1982; drilling-fluids technology had mushroomed alongside. Although some attention had been given to safety issues related to the storage and handling of corrosive and flammable materials, health and environmental concerns were not high priorities at this time. However, U.S. federal and state agencies, in separate moves in 1978, demonstrated their resolve to minimize the environmental impact of discharging whole mud and drilling-mud contaminated cuttings into the sea (3). These actions crystallized environmental issues and initiated the growth of awareness worldwide; environmental concerns thus became a dominant force in drilling-fluid decisions.

SCOPE OF DRILLING TECHNOLOGY

Drilling for oil and natural gas is unique among all of the industrial processes for generating an energy source. Indeed, wells are drilled only where commercial quantities of hydrocarbons are believed or known to exist. Drillsite locations are nearly always temporary and oftentimes remote, inhospitable, and environmentally sensitive. Subsurface drilling conditions can vary significantly from well to well, and even within the same well. Moreover, it is neither easy nor inexpensive to control a process occurring up to 8 km (5 miles) below the Earth's surface.

The scope of drilling technology ranges between the extremes of shallow wells in mature producing fields, to deep exploration wells drilled from rigs floating on the high seas. Challenges and risks often are greatest when drilling wildcat and exploration wells with limited geological information, especially when severe drilling hazards are possible. Optimization usually is the primary focus when development wells are drilled to exploit commercial fields. Onshore wells are drilled from land locations (including populated areas), the open plains, deserts, jungles, mountainous areas, and Arctic regions. Offshore wells are drilled in marine locations, including rivers, bays, lakes, and oceans as deep as 2.4 km (1.5 miles). Because of the great expense to establish and equip an offshore drillsite, it is not uncommon for multiple wells to be drilled from a single location.

Whereas wells are normally drilled vertically, directional wells allow the drilling of multiple wells from a single location to reach a distant target, avoid problem zones, or maximize production. Sometimes wells are drilled directionally from a land location to a distant offshore target to minimize environmental impact. If the lateral extent of a directional well is significant, it may be classified as an extended-reach well. Wells drilled horizontally through a productive formation to expose a longer productive interval or to improve production are aptly named horizontal wells. Horizontal-well technology, the benefactor of considerable research and development efforts, has rejuvenated drilling in many parts of the world and promises to contribute to the industry for a long time to come. Horizontal-drilling techniques are also being applied to environmental wells to characterize, monitor, and remediate contaminated areas. Vertically drilled slimhole wells represent emerging exploration technology designed to reduce costs and environmental impact by drilling small-diameter holes with special, small-footprint rigs.

Many drilled wells do not require advanced technology. However, hostile conditions encountered in deep, high pressure, high temperature wells require the combined efforts of numerous drilling technologies. The deepest well drilled in the United States reached 9,583 m (31,441 ft) in 1974. Formation temperatures in oil and gas wells have been recorded above 288°C at 7,315 m (24,000 ft). Some wells must progress through troublesome intervals, including "gumbo" shales, unconsolidated sands, massive salt zones, and formations containing contaminants such as deadly hydrogen sulfide (H_2S). Drilling conditions are most difficult when multiple hazards occur in the same well, such as the deep wells of the Mississippi Salt Basin. High pressure gas in this basin can contain up to 75% H_2S; bottomhole temperatures can exceed 204°C below 6,096 m (20,000 ft).

ROLE OF DRILLING FLUIDS

Oil and gas wells simply cannot be drilled without drilling fluids. The drilling fluid is the lifeblood of the modern drilling operation and must be continually circulated, as shown in Figure 1, in order to perform its functions. Whereas drilling-fluid expenditures average less than 6%

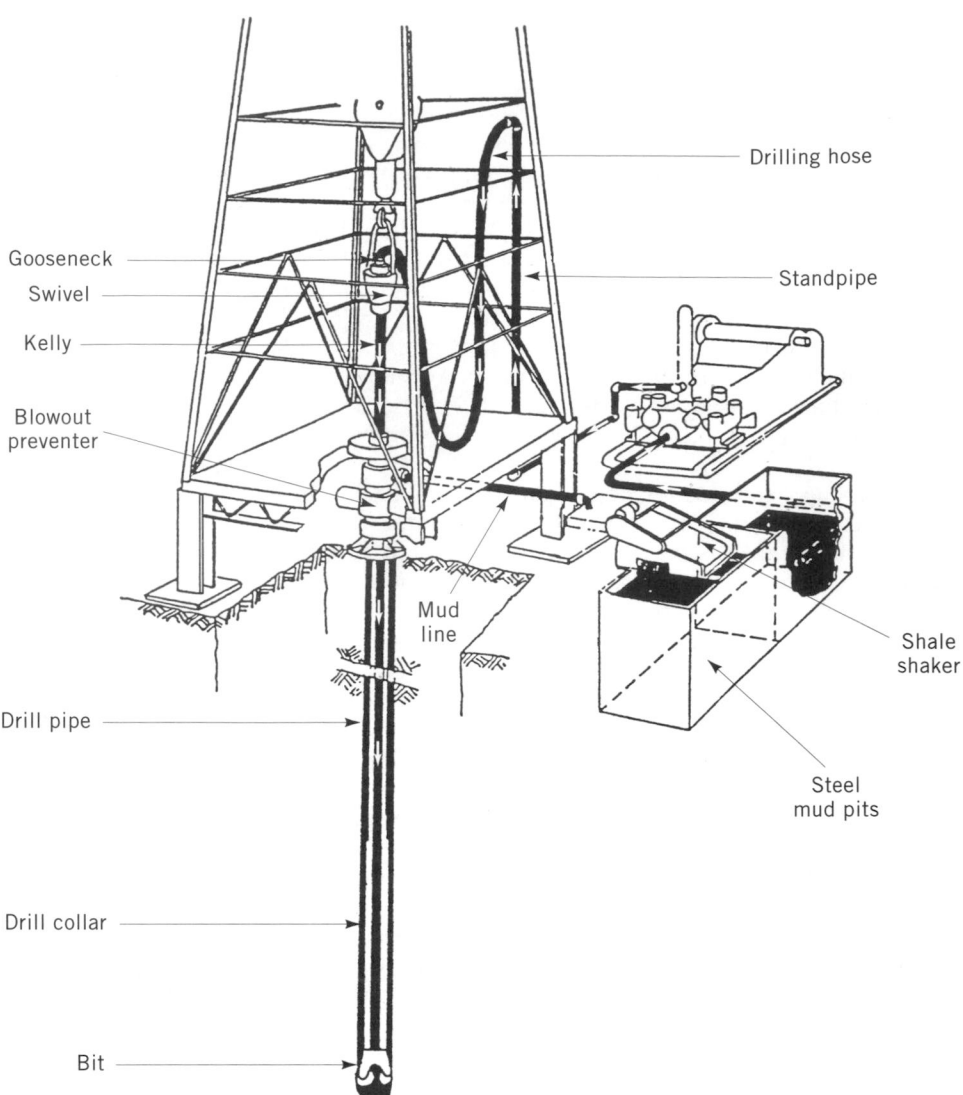

Gooseneck

Swivel

Kelly

Blowout
preventer

Drill pipe

Drill collar

Bit

Drilling hose

Standpipe

Mud
line

Shale
shaker

Steel
mud pits

Figure 1. Schematic of the circulatory system of a rotary rig (4). Reprinted with permission of Gulf Publishing Company.

of the tangible costs to drill and case a well, the drilling fluid plays a key role in the success of any drilling project. Mud cost in itself is important; however, the mud should perform such that overall drilling, completion, cleanup, and disposal expenses are minimized.

The two basic drilling-fluid functions are to (1) transport drilled cuttings and cavings to the surface and (2) control subsurface pressures. Figure 2 illustrates the spectrum of mud densities that can be theoretically achieved with different fluids and additives. Until casing is run to preserve the drilled hole, the drilling fluid must support and protect the walls of the wellbore. In addition, the mud should promote high penetration rates, cool and lubricate the bit and drill string, buoy the weight of the drill string and casing, and help to obtain information on subsurface formations.

In the process of providing the above functions, the drilling fluid should not create side effects that could compromise the drilling-equipment integrity, damage productive formations, endanger the health and safety of personnel, or contaminate the environment. It also should minimize mud-related problems that could delay, suspend, or even cancel the drilling project. Stuck pipe, lost

circulation, excessive surge or swab pressures, poor cement jobs, and wireline logging difficulties are among the most common problems. The drilling mud must remain fluid, stable, and usable even though it may be continually exposed to subsurface contaminants and hostile conditions. Highly reactive drill solids (especially colloidal-sized particles), corrosive acid gases, saltwater flows, evaporites, and cement are common contaminants; hostile conditions include high temperatures and pressures.

The drilling process relies on successively protecting and isolating troublesome intervals by casing strings run in the hole and cemented in place using a telescopic scheme. The required number of casing strings depends on the well depth and the severity and diversity of drilling conditions. For most wells, drilling conditions change and, consequently, drilling-fluid requirements must also change, sometimes significantly, for each casing interval. Drilling efficiency largely depends on matching the drilling fluid to the formations being drilled. In some instances, the drilling fluid is completely displaced by another fluid specially formulated to handle different drilling conditions expected in the next interval; one example of this is shown in Figure 3.

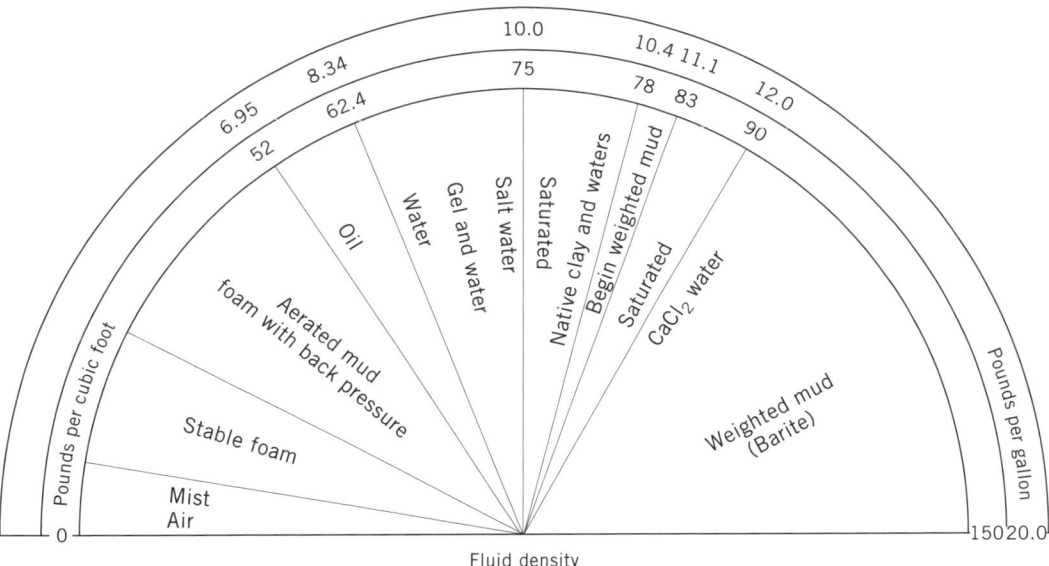

Figure 2. Density ranges for a variety of drilling fluids and additives (5). Reprinted with permission of Gulf Publishing Company.

PHYSICAL AND CHEMICAL PROPERTIES

Drilling-fluid properties are continually measured and adjusted at the wellsite to satisfy efficiency and economic requirements. Despite the complexity of drilling-fluid chemistry, basic tests in concert with extensive field experience serve to monitor the drilling-fluid condition and guide its optimization. The American Petroleum Institute (API) issues recommended practices which specify equipment and procedures for testing and monitoring water-based (6) and oil-based drilling fluids (7), as follows:

STANDARD API DRILLING-FLUID FIELD TESTS

Water- and Oil-Based

Mud weight (density)
Viscosity and gel
  strength
  Marsh funnel
  Direct-indicating
  viscometer
Filtration
Water, oil, and solids
Shear strength

Oil-Based (RP 13B-2)

Chemical analysis
  Whole mud alkalinity
  Whole mud chlorides
  Whole mud calcium
Electrical stability
Oil and water content
  from cuttings
Aqueous-phase activity
  by:
  Electrohygrometer
  Aniline point

Water-Based (RP 13B-1)

Sand
Methylene blue capacity
pH
Chemical analysis
  Alkalinity and lime
    content
  Chloride
  Total hardness as calcium
  Calcium
  Magnesium
  Calcium sulfate
  Formaldehyde
  Sulfide
  Carbonate
  Potassium
Resistivity
Drill-pipe corrosion ring
  coupon

Additional tests are sometimes run in the field to evaluate special properties. Still others, many of which require sophisticated analytical equipment and/or carefully controlled environments, are conducted in laboratories by chemists, scientists, and technicians.

The mud balance (for density) and the Marsh funnel (for relative viscosity) are two basic pieces of mud-testing equipment found on every rig. A typical field "mud lab" contains, in a single cabinet, the other equipment needed to run daily mud checks. This unit includes a viscometer, low pressure filter press, pH paper, sand-content set, stopwatch, thermometer, chemicals for titration tests, and retort. The filter press and a high temperature, high pressure version force mud through a small filter. The residual solids deposition is called the filter cake; the liquid passing through is the filtrate. The retort is a distillation unit used to determine the mud's water, oil, and solids content (percent by volume).

The rotational viscometer measures rheological properties (viscosity) at different shear rates (velocities). Most drilling muds are non-Newtonian fluids; many are rheologically complex. Rheological parameters such as plastic viscosity and yield point are used to calculate frictional pressure losses (8), optimize drilling hydraulics, and serve as mud-condition indicators to guide mud treatments. The viscometer is also used to measure gel strengths, the tendency for muds to develop a semirigid gel structure for suspension of drill solids and weight material under static conditions.

COMPOSITION OF DRILLING FLUIDS

Drilling fluids are composed of chemicals, minerals, and fluids that are dissolved, suspended, and/or emulsified in water or oil. Some are based on synthetic liquids, air, or natural gas. To satisfy the requirements discussed previously, both efficiently and economically, drilling fluids

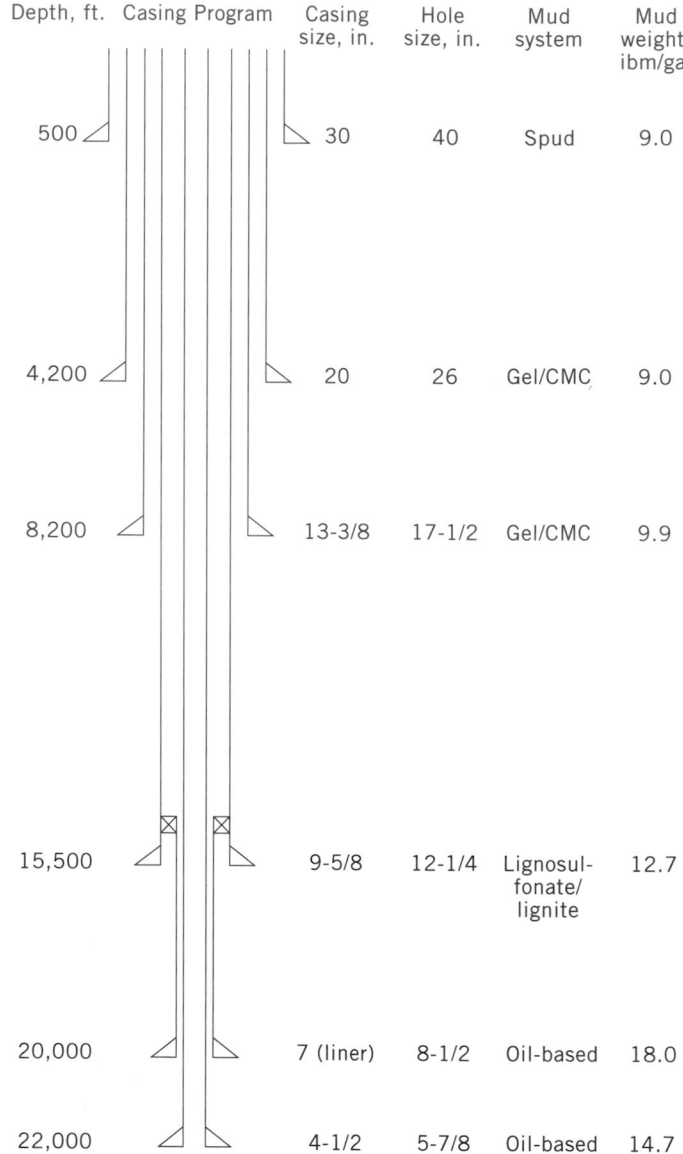

Figure 3. Typical six-string casing program, used at a deep on-shore well in Abu Dhabi, U.A.E.

Depth, ft.	Casing Program	Casing size, in.	Hole size, in.	Mud system	Mud weight, lbm/gal
500		30	40	Spud	9.0
4,200		20	26	Gel/CMC	9.0
8,200		13-3/8	17-1/2	Gel/CMC	9.9
15,500		9-5/8	12-1/4	Lignosul-fonate/lignite	12.7
20,000		7 (liner)	8-1/2	Oil-based	18.0
22,000		4-1/2	5-7/8	Oil-based	14.7

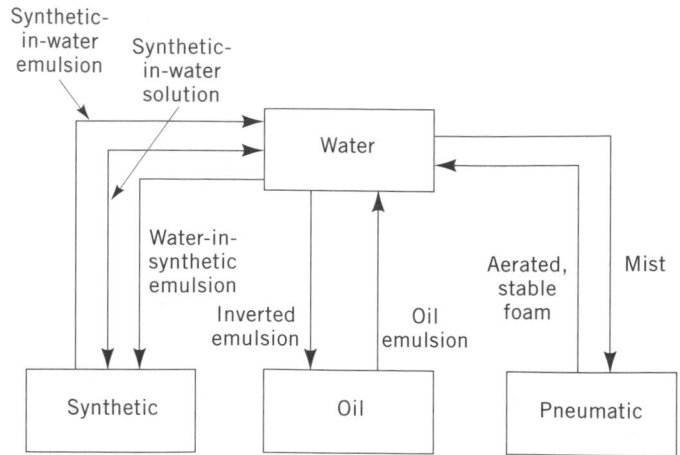

Figure 4. Drilling-fluid classification according to the principal component or base fluid.

must be individually formulated for specific applications; no single formulation is suitable for all situations. The drilling fluid of choice depends on the specific well requirements.

Drilling-fluid types are usually grouped according to their principle component. As depicted in Figure 4, the four basic types of drilling fluids are: water-based, oil-based, synthetic-based (or pseudo-oil-based), and pneumatic (which include air and natural gas). It is not uncommon for different base fluids to be present in the same drilling fluid; various combinations are also shown in Figure 4. For example, water-based muds containing emulsified oil drops are called oil-emulsion muds; oil-based muds containing emulsified drops of brine are called invert-emulsion muds. Drilling-fluid names are traditionally formed by concatenating terms identifying the base fluid, distinguishing additives, and important physical and chemical characteristics.

The large number of drilling fluids and specialty additives that exist today results from the wide variety of drilling conditions and problems encountered during drilling. Trade journals regularly catalog drilling-fluid additives according to their primary use in major mud systems; recent editions also include environmental information on products (9). Details on how to formulate, mix, test, and run various generic and proprietary mud systems are generally provided by the drilling-fluids companies in their technical manuals (10).

Most drilling-fluid components are associated with one of three phases. The continuous phase is the base fluid in which chemicals, minerals, formation solids, and other fluids are dissolved, suspended, and/or emulsified. The continuous phase is nearly always the principle component of the drilling-fluid system and filtrate. The solids phase consists of particles suspended in the continuous phase. Solids are typically categorized by their specific gravity (low or high) and/or by their reactivity with water (active or inert). Fluids emulsified in the continuous phase are the discontinuous phase. The solids and discontinuous phases are the main sources of viscosity and filter cake in a drilling mud.

Water-Based Muds

The vast majority of drilling fluids are water-based. Water-based muds range from clear water to muds highly treated with chemicals. The base liquid can be fresh water, seawater, saltwater, or saturated saltwater, depending on the availability of make-up water and desired mud properties. Common water-based mud additives are listed below.

Common Water-Based Mud Additives

Weight

Materials. Barite, hematite, calcium carbonate

Viscosifiers. Bentonite, attapulgite, sepiolite, beneficiated bentonite, biopolymer (xanthan and welan gum), guar gum, hydroxyethyl cellulose, mixed-metal hydroxide

Dispersants and Deflocculants. Lignite (standard, causticized, potassium, chrome), lignosulfonate (chrome, chrome-free, calcium), tannin (extract blend, chrome, chrome-free), polyacrylate, sodium tetraphosphate, sodium acid pyrophosphate

Filtration-Control Agents. Starch, cellulose polysaccharide, cellulose (carboxymethyl, polyanionic), polyacrylate, resinated organic polymer, lignite

Shale Stabilizers. Polyacrylamide (PHPA), potassium acetate, asphalt (blown, sulfonated), cationic PHPA, polyamino acid, quaternary ammonium compound, polyvinyl alcohol

Lubricants. Petroleum—distillate based, polyglycol, fatty acid blend, diesel oil, mineral oil, graphite, silicon, solid beads (plastic, glass)

Defoamers. Alcohol blend, mineral-oil—alcohol blend, aluminum stearate with diesel

Corrosion Inhibitors. Water and brine-dispersible blended amines, persistent filming amine, phosphate organic, zinc-based sulfide scavenger, organic biocide, ammonium bisulfite

Commercial Chemicals. Sodium chloride, potassium chloride, calcium chloride, caustic soda, potassium hydroxide, magnesium oxide, sodium bicarbonate, soda ash, lime, gyp, sodium bichromate, diesel, mineral oil, citric acid

Lost Circulation Materials. Nut shells, mica, shredded wood fiber, cane fiber, cottonseed hulls, diatomaceous earth, shredded paper, plastic chips, cellophane flakes

API specifications are available for barite, hematite, bentonite, attapulgite, sepiolite, carboxymethyl cellulose (CMC), and starch (11).

The large and diverse class of water-based muds is difficult to classify without overlap. Clear water and brines, the simplest water-based muds, are suitable for many competent, nonreactive, low pressure formations. Lightly treated, noninhibitive muds are inexpensive, low toxicity muds that provide hole-cleaning capability and some filtration control to drill routine top-hole sections. More footage has been drilled with these muds than any other type. Common lightly treated muds include muds viscosified with native formation clays or commercial bentonite (gel), clay-based muds dispersed with phosphates or low concentrations of organic thinners, and seawater and saltwater muds viscosified with attapulgite clay. Saltwater muds are often the result of using seawater, brackish water, or field brines for make-up. Salt can be added to inhibit bentonitic shales. High salt levels (up to saturation) are used to prevent washout of massive salt formations and to prevent formation of gas hydrates, which can plug flowlines and interfere with rig safety equipment in deepwater offshore wells. Gas hydrates are icelike compounds that can form when natural gas in the mud is exposed to low seafloor temperatures, pressure, time, and water.

Inhibitive muds reduce the chemical interaction between the mud and water-sensitive formations. Sodium, potassium, and calcium ions added to the muds minimize hydration and swelling of clays and reactive shales. Lime muds are calcium-based and able to maintain low viscosities and gel strengths in a high solids environment. Additionally, their high alkalinity neutralizes corrosive acid gases effectively. Gyp muds, originally developed to drill massive anhydrite or gyp formations, also rely on the calcium ion to provide shale inhibition. The viscosity and gel strengths of lime and gyp muds are controlled by strong deflocculants, which disperse fine solids in the mud. When the most common deflocculant, chrome lignosulfonate, is used in freshwater and saltwater, the muds are known as lignosulfonate muds. High concentrations of lignosulfonate and lignite improve inhibition, solids tolerance (more dispersion), filtration control, temperature stability, and resistance to salt, gyp, and cement contamination. Oil-emulsion muds contain up to 10% emulsified oil to improve penetration rates, lubricity, filtration control, and shale inhibition. To reduce mud toxicity, however, the oil has been systematically replaced by oil-like synthetic materials and glycol–glycerol additives.

Polymers, both natural and synthetic, are routinely used in many water-based drilling fluids for viscosity, filtration control, shale inhibition, high temperature stability, total or selective flocculation, and/or deflocculation. Most polymers are effective at very low concentrations and can be formulated for low toxicity. Polymer muds depend heavily on these materials to achieve desired properties. Polymers in low solids, nondispersed muds lower solids levels by enhancing the performance of bentonite and by providing filtration and rheology characteristics without using conventional dispersants. PHPA muds use a high molecular-weight, partially hydrolyzed polyacrylamide to adsorb on shale surfaces to minimize dispersion. PHPA muds are particularly effective in combination with a potassium source for additional inhibition and a biopolymer for improved suspension and viscosity. The biopolymer either enhances or completely replaces commercial clays. Cationic muds, which use cationic polymers to achieve high inhibition levels, are considered among the most inhibitive water-based muds. Inorganic, cationic compounds used to flocculate bentonite and form extremely shear-thinning and fragile gels give mixed-metal–hydroxide muds excellent hole-cleaning and suspension for horizontal wells. "Rheologically engineered" biopolymer muds also achieve superior hole-cleaning capabilities when polymer levels are maintained at optimum levels. In addition, these systems are environmentally friendly and nondamaging.

Oil-Based Muds

The unique performance characteristics of oil-based drilling fluids serve a wide range of applications, some of which currently cannot be adequately handled by any other mud type. Oil-based muds are highly inhibitive, resistant to contaminants, stable at high temperatures and pressures, lubricious, noncorrosive, and flexible. As such, they are particularly effective for drilling highly reactive shale and evaporite formations; extended-reach wells; and deep, high pressure, high temperature, H_2S wells. They are also used for special applications such as freeing stuck pipe. Economics of oil-based muds can be attractive when considering high drilling rates achieved with modern bits, reduction and elimination of mud-related problems, and lowering of overall drilling costs. Unfortunately, oil-based muds are considered highly toxic, and their use is restricted. Disposal costs must always be included in economic evaluations.

Petroleum oils are used for the continuous phase of oil-based muds. The crude oil initially used has long since been replaced by diesel oil and, more recently, by low toxicity mineral oils. Unlike the early muds that were water-intolerant, most oil-based muds today contain up to 40% emulsified brine. The volume percentages of oil and water are expressed as an oil–water ratio that can range from 100:0 to 50:50. Because the term "oil emulsion" already had been chosen for oil-in-water emulsion muds, "invert emulsion" was coined to describe water-in-oil emulsion muds. The emulsified water droplets help to suspend weight material and lower fluid loss. Additionally, high salinity levels in the water phase improve wellbore stability by creating osmotic pressures, which dehydrate and harden reactive shales.

The basic components of an invert-emulsion mud include oil, brine (usually calcium chloride), primary and supplementary emulsifiers, oil-wetting agent, viscosifier–gellant (oil-dispersible bentonite), filtration-control additives, and slaked lime. A rheology modifier is sometimes added to improve hole-cleaning in directional and large-diameter wells. Fatty acid soaps are the most common emulsifiers used in oil-based muds. Polyamines, polyamides, imidazolines, and other cationic emulsifiers are also effective. Not all modern oil-based muds contain emulsified brine; some all-oil muds have demonstrated exceptional performance and have been suitable for special disposal methods (12).

Toxicity is the most serious and perhaps insurmountable drawback of oil-based muds. Environmental regulations strictly control and even forbid their use and disposal in some areas. Whereas great strides have been made to reduce toxicity, the very characteristics that give oil-based muds their superior performance also can contaminate the environment. A wide variety of refined mineral oils have been developed for use in low toxicity oil-based muds (LTOBMs) to reduce environmental problems. These oils have lower aromatic contents than diesel and reduced acute toxicity to various organisms. However, even the cleanest of these oils is considered toxic, and disposal of mud and contaminated cuttings must be handled in accordance with local environmental regulations.

Synthetic-Based Muds

Synthetic-based systems represent the latest technology for providing high performance with minimal environmental impact. The goal of synthetic-based drilling fluids is to provide the performance and inhibitive properties of oil-based muds without the objectionable toxicity. Synthetics are manmade, nonaqueous liquids whose molecules are wholly dissimilar to the raw materials used for manufacture. The continuous phase of a synthetic-based drilling fluid is at least 99% synthetic material.

To date, three different compounds have been used in the field as the base liquid for synthetic-based drilling fluids: polyalphaolefins (PAOs), esters, and ethers. Table 1 compares the typical properties of these synthetics with diesel and mineral oil, including aromatic content, a key toxicity indicator. Conceptually, synthetic-based drilling fluids are mixed and run in the field much like conventional oil-based muds. They have been primarily used offshore, where permits have allowed discharge of cuttings and mud directly into the sea. Performance has been exceptional, even to the point of outperforming muds based on diesel and mineral oil. Whereas material costs are still high, synthetic-based systems have proven economical because of exceptional drilling performance and reduction in disposal costs.

Pneumatic Drilling Fluids

Pneumatic drilling fluids are used in special applications, primarily to minimize damage to productive formations, prevent loss of circulation, and achieve high penetration rates. These fluids function adequately, except for cuttings suspension, filter-cake deposition, and control of subsurface pressures. As such, their use is limited to competent, low permeability formations such as limestones and dolomites.

The principal component of a pneumatic drilling fluid is air or natural gas, although use of either is commonly referred to as air drilling. Air or gas must be circulated just like conventional liquid drilling muds. The circulation pressure for air is provided by large air compressors installed as part of the surface drilling equipment; gas pressure is usually obtained from a high pressure field source. A surface rotating seal around the annulus directs the flow, with cuttings a safe distance away from the rig. Extra safety precautions are required because of the inherent dangers of explosions and fires on the surface or downhole. Risks of a downhole explosion are greater when

Table 1. Comparison of Base Fluids Used for Nonaqueous Drilling Fluids[a,b]

Typical Properties	Polyalphaolefin	Ester	Ether	Diesel Oil #2	Mineral Oil
Density, sg	0.79–0.85	0.85–0.92	0.81–0.83	0.865	0.790
Viscosity, cSt, at 40°C	3.9–9.6	5.0–10.0	3.9–6.0	2.8–3.4	1.6
Flash point, °C	156–172	150–195	144–166	57–63	79
Pour point, °C	< −101	−66 to 5	−76 to −44	−18	−54
Aniline point, °C	104–155	20–30	35–45	54–60	76
Aromatic content, %	None	None	None	30–60	< 0.25

[a] Ref. 13.
[b] Reprinted with permission of Society of Petroleum Engineers.

using air, since the combination of air and formation hydrocarbons can readily form an explosive mixture.

The basic forms of air drilling are dusting, mist, stable foam, and aerated mud. Dusting, which involves circulation of dry air or gas at very high velocities, can only be used when formations are nearly dry. Small amounts of water may be tolerated (mist), but injection of dilute mixtures of foaming agents and polymers in water may be required to keep cuttings from sticking together. Superior foaming agents are used to form a stable foam with the consistency of shaving cream. Stable foams can provide excellent hole-cleaning and still maintain the benefits of reduced-density drilling. Some of the advantages of air drilling can be obtained by aerating conventional drilling muds. Aerated muds are used for severe loss circulation problems when other forms of air drilling are not possible. Nitrogen generated at the wellsite has been used instead of air to reduce high corrosion rates associated with aerated muds.

ENVIRONMENTAL CONCERNS

Drilling fluids, like many industrial chemicals, can be hazardous to humans if not handled safely. Also, contamination of the natural environment could occur if fluids and formation cuttings are not disposed of properly. Regulations vary by country and by jurisdictional area within a country. For example, in the Norwegian sector of the North Sea, cuttings generated with a LTOBM can be discharged into the sea if the oil-on-cuttings is less than 10 g/kg; discharge of any oil-based mud cuttings is prohibited in the Gulf of Mexico; regulations in Mobile Bay strictly prohibit all discharges (zero discharge).

Hydrocarbons, chlorides, and heavy metals are the principle sources of toxicity in drilling fluids; products containing these contaminants have been used in drilling muds at one time or another. However, they also occur naturally and are sometimes incorporated into the mud during the drilling operation.

Examples include crude oil found in productive formations, salt from massive salt formations, and trace metals contained in organically rich shales drilled offshore. Heavy metals include chrome, lead, zinc, arsenic, barium, mercury, and cadmium. Common sources include drilling-fluid additives, corrosion inhibitors, pipe dope, and subsurface formations.

The key to reducing environmental impact is to address the combined drilling-fluid issues of performance, economics, composition, disposal, and environmental impact. Although environmental problems and regulations have been a source of consternation to the drilling industry, they have stimulated and significantly advanced drilling-fluid developments (3). These developments have involved toxicity testing, waste-minimization techniques, and treatment–disposal options.

Toxicity Testing

The toxicity of a drilling fluid is determined by its composition and the test protocol (bioassay). Bioassays measure the acute toxicity (lethality) on a test population of organisms, and results are used to determine if mud and cuttings can be discharged in offshore waters. Mysid shrimp (*Mysidopsis bahia*) is the species specified by the U.S. Environmental Protection Agency for use in drilling-fluid toxicity tests (14). The shrimp are exposed for 96 h to several concentrations of the suspended particulate phase of the effluent (the effluent is the test additive mixed in one of eight generic muds and diluted 1:9 with seawater). The 96-h, LC_{50} value (lethal to 50% of the population) determines the toxicity. Bioassays are required on a monthly, change-of-mud, and end-of-well basis. An LC_{50} value less than 30,000 ppm prevents discharge of mud and cuttings in the Gulf of Mexico. Other countries use similar tests with different species and discharge limits.

Most drilling areas now require all used or discharged chemicals to be tested for toxicity and health effects. In addition to acute toxicity for organisms representing different trophic levels, chemicals may undergo testing for bioaccumulation, biodegradation, by-products and effects of degradation, duration of impact, and rate of seabed recovery.

Waste Minimization

The most desirable method of controlling pollution is to minimize or eliminate it at the source. The two most common ways to achieve this goal are product substitution and a change in operating practices. In the case of drilling-fluids technology, both techniques are being heavily researched. Regarding product substitution, new products are formulated to meet performance requirements without the objectionable materials. Substituting chrome-free lignosulfonate for chrome lignosulfonate is an example of product substitution that minimizes heavy metal contamination. Low toxicity synthetics have replaced petroleum-based lubricants, and polyglycols are now used instead of alcohol and petroleum-based defoamers. Synthetic-based muds have been developed to replace oil-based muds eventually. This type of product research is continuing with each new generation of products to offer higher performance to meet drilling objectives, and at the same time to lessen the impact on the environment.

Improvements in drilling fluids and drilling technology have also achieved source reduction through operational practices. Use of inhibitive mud systems and excellent solids-control equipment have led to dramatic decreases in the volume of solids and liquids that need to be discharged into the environment. Systematic segregation of contaminated waste from uncontaminated waste is important. In addition, waste volume can be reduced by using a closed-loop, solids-control system. Finally, new packaging methods for drilling fluids products have also led to waste reduction.

Recycling is an alternative to control pollution for situations where source reduction is limited or not possible. After the well is completed, the remaining drilling fluid may be stored in tanks until needed on the next well; this is particularly advantageous for oil-based and synthetic-based muds. Another type of recycling involves conversion of drilling fluids for other uses at the wellsite. Mud can be treated with blast furnace slag (15), converted to cement, and used to hold casing in place. Cuttings generated from the wellbore are difficult to reuse, because they are detrimental from a performance standpoint to the drilling fluid. However, some drilled cuttings can be used as landfill cover or construction materials.

Treatment–Disposal Options

Drilling mud and formation cuttings must be disposed of in a manner that is economical and still limits long-term liability. Onshore, the concerns are oil and grease, chlorides, and heavy metals. Some of the available disposal processes are

1. *Onsite Separation.* Waste-material dewatering using water-treatment methods utilized at some municipal and industrial plants yield water and a dry mudcake as the two end products.
2. *Solidification.* Cementing compounds such as fly ash, kiln dust, and portland cement are mixed with waste fluids in the reserve pit and allowed to dry.
3. *Bacterial Degradation.* Hydrocarbons are removed from the drilling-fluid waste (commonly oil-based mud) by special bacteria that "eat" the hydrocarbons off the solids. The remaining solids are then solidified.
4. *Dewatering / Backfilling.* Water in the reserve pit is allowed to evaporate, or the water is pumped out after solids are flocculated chemically and allowed to settle. Top soil from the reserve pit dike is then backfilled to complete the reserve-pit closure.
5. *Landfarming.* An even distribution of mud and cuttings is mechanically mixed with the soil and tilled to accelerate biodegradation of hydrocarbons (12).
6. *Washing.* Free oil on cuttings is removed using a solvent or washing solution and mixer.
7. *Landfill Disposal.* Mud and formation cuttings are first treated to remove free liquids and then buried in a secure landfill.
8. *Annular Injection.* Cuttings are slurried and injected through a casing annulus into downhole formations protected by casing (16).
9. *Experimental Techniques.* Experimental waste-disposal techniques include incineration, distillation, and critical fluids extraction.

Disposal options for offshore drilling are limited. If allowed by local regulations, the mud and cuttings can be discharged directly into the sea, returned to the formation using annular injection, or converted to cement and used in place of conventional oilfield cement. Otherwise, the waste must be hauled onshore for treatment and ultimate disposal, using one of the previously mentioned methods.

FUTURE DEVELOPMENTS

Unless current trends reverse dramatically, the pressure on the petroleum industry to find and produce more oil and natural gas will intensify well into the next century. Unfortunately, drilling for oil and gas is not easy, inexpensive, or sanitary. Clearly, the interactions among performance, economics, and environmental concerns will continue to dominate. Historically, drilling-fluids technology has always responded to new challenges. Real progress has been made in all three areas—more is needed.

The highest research priority in drilling fluids continues to be the elusive, nontoxic, water-based mud that performs like an oil-based mud. Advancements in polymer technology continue to lead the way. Or, perhaps the torch will be passed to the new synthetic-based muds that already surpass some oil-based mud performance levels, but are still costly and not yet fully embraced by the environmental community worldwide. Then again, maybe the final answer will be less revolutionary and found in the collection of individual products that are being developed as a matter of course to replace lesser performing, more costly, and/or environmentally objectionable counterparts.

BIBLIOGRAPHY

1. J. A. Clark and M. T. Halbouty, *Spindletop,* Gulf Publications Co., Houston, Tex., 1980.
2. H. C. H. Darley and G. R. Gray, *Composition and Properties of Drilling and Completion Fluids,* 5th ed., Gulf Publications Co., Houston, Tex., 1988.
3. R. D. Bleier, A. J. J. Leuterman, and C. L. Stark, "Drilling Fluids: Making Peace with the Environment," *J. Petr. Tech.,* 6–10 (Jan. 1993).
4. R. A. Bobo, R. S. Hoch, G. S. Boudreaux, and R. R. Angel, *Keys to Successful Competitive Drilling,* Gulf Publishing Co., Houston, Tex., 1958, p. 67.
5. S. O. Hutchison and G. W. Anderson, "What to Consider When Selecting Drilling Fluids," *World Oil,* 83–94 (Oct. 1974).
6. *Recommended Practice Standard Procedure for Field Testing Water-Based Drilling Fluids, API RP 13B-1,* 1st ed., American Petroleum Institute, June 1, 1990.
7. *Recommended Practice Standard Procedure for Field Testing Oil-Based Drilling Fluids, API RP 13B-2,* 2nd ed., American Petroleum Institute, Dec. 1, 1991.
8. *Bulletin on the Rheology of Oil-Well Drilling Fluids, API 13D,* 2nd ed., American Petroleum Institute, May 15, 1985.
9. "1992–93 Environmental Drilling and Completion Fluids Directory," *Offshore / Oilman,* 41–56 (Sept. 1992).
10. *Drilling Fluid Engineering Manual,* M-I Drilling Fluids Co., Houston, Tex., 1991.
11. *Specification for Drilling-Fluid Materials, API Spec 13A,* 14th ed., American Petroleum Institute, Aug. 1, 1991.
12. T. S. Carter and G. L. Faul, "Successful Application of the AOBM System in a Deep West Texas Well," *Proceedings of the SPE Annual Technical Conference,* Washington, D.C., Oct. 4–7, 1992.
13. J. E. Candler, J. H. Rushing, and A. J. J. Leuterman, "Synthetic-Based Mud Systems Offer Environmental Benefits Over Traditional Mud Systems," *Proceedings of the SPE / EPA Expl. & Prod. Env. Conference,* San Antonio, Tex., Mar. 7–10, 1993.
14. F. V. Jones, C. M. Moffitt, and A. J. J. Leuterman, "Drilling

Fluids Disposal Regulations: A Critical Review," *Drilling,* 21–24 (Mar.–Apr. 1987).

15. J. J. Nahm, K. Javanmardi, K. M. Cowan, and A. H. Hale, "Slag Mix Mud Conversion Cementing Technology: Reduction of Mud Disposal Volumes and Management of Rig-Site Drilling Wastes," *Proceedings of the SPE/EPA Expl. & Prod. Env. Conference,* San Antonio, Tex., Mar. 7–10, 1993.

16. R. C. Minton, A. Meader, and S. M. Wilson, "Downhole Cuttings Injection Allows Use of Oil-Base Muds," *World Oil,* 47–52 (Oct. 1992).

E

ELECTRIC HEATING

KARL S. KUNZ
The Pennsylvania State University
University Park, Pennsylvania

Electric heating can occur via a number of processes and over a wide frequency range and may be intended or not. When intended, it may serve a host of useful purposes (Table 1). The electromagnetic phenomenon of current flow induced within a material body results in random particle motion in the material or heat, which may be used directly, ie, drying, curing, cooking, welding, etc., or indirectly, as in the production of light, as from an incandescent light bulb or the flow of electrons from a heated filament in a cathode ray tube (CRT). (See also ELECTRIC POWER; ENERGY EFFICIENCY.)

These intended effects cover a wide range of important technologies and demonstrate the importance of electrical heating in current technology. The treatment of electrical heating presented here starts with ohmic heating, the basis of electrical heating, and demonstrates how this effect is described by the Maxwell equations with the inclusion of a conduction current $J_c = \sigma E$. It then addresses the definition of heat as random atomic motions induced by collisions and goes on to treat the sources of electrical heating. Sources are distinguished according to whether or not they are intentional or unintentional, how the conduction current was produced, and the frequency range of operation. The intended sources are presented in order of increasing frequency of operation from power line frequencies to radio and microwave frequencies to infrared and visible light frequencies. Their applications, including the advantages and disadvantages, are cited. Unintentional electrical heating is discussed only for the two relatively important areas of waveguide heating and the heating of solid-state devices. The last area includes how the computational modeling of solid-state device heating can be performed with the theory and techniques presented here.

DEFINITION OF ELECTRICAL HEATING

Current passing through a conductor obeys Ohm's law, $V = IR$, where V is the voltage across the conductor, I the current flowing through the conductor, and R a bulk or extensive property characterizing the resistance to the current flow. Here R is a macroscopic quantity based on atomic and electron interactions at the microscopic level that relates V and I in the conductor, which can be virtually any material. The physical process underlying the resistance R involves charge carriers, generally electrons, colliding with the atoms of the material. The atoms may be coupled to form a lattice structure within the material. Charge carrier motion and the attendant energy are gradually lost to motion of the material in the form of atomic or lattice vibrations. Energy imparted to the electrons, eg, from the electric field, is thereby converted via atomic col-

lisions into mechanical motion of the atoms. This motion is random because of the random nature of the collisions and represents the electrical heating of the material.

Formally the electric power flow in the material P is given by $P = VI$. The heating is equal to this power flow for the case of a bulk material with voltage V across it and current I flowing through it when a lossless external circuit is assumed. Using Ohm's law, the power flow producing heating, P_{heat}, is either $P_{heat} = I^2 R$ or $P_{heat} = V^2/R$. For a fixed current source, then, the heating rate increases as R does. Of course, as R increases for a fixed current source, the voltage $V = IR$ also increases and some practical limit to the permissible voltage will be reached. Similarly, for a fixed voltage source the heating rate increases as $1/R$. In this case, as R decreases, $1/R$ increases. A decreasing R for a fixed voltage implies an increasing current. Once again some practical limit is reached. For example, a short or a zero resistance across the outputs of an audio amplifier in the form of a copper bar will very briefly result in a very large current flow, often followed by amplifier failure.

GOVERNING EQUATIONS

Electrical heating is a phenomenon that is subject to the Maxwell equations (1). The Maxwell equations may be expressed as

$$\nabla \times E = -\frac{\partial B}{\partial t}$$

$$\nabla \times H = \frac{\partial D}{\partial t} + J$$

with

$$B = \mu H \qquad D = \varepsilon E$$

These last two equations are the constitutive relations relating fields E and H to fluxes D and B. In addition

$$\nabla \cdot J + \frac{\partial \rho}{\partial t} = 0$$

This is the continuity equation relating current density flow J from an infinitesimal volume to the change in time of the charge ρ residing within the infinitesimal volume. The continuity equation is just the expression of the conservation of charge. Combined with the preceding two Maxwell equations, often referred to as the Maxwell curl equations, it yields the following two divergence equations:

$$\nabla \cdot D = \rho \qquad \nabla \cdot B = 0$$

These are ancillary equations contained in the curl and continuity equations that are often included in the set of Maxwell equations.

Table 1. Typical Electrical Heating Applications

Techniques	Frequency	Application	Advantage	Disadvantage
Resistive heating	50–60 Hz	Heating and lightning	Simple inexpensive	Only filaments heated to high temperatures
Induction heating	Kilo- to megahertz	Welding	Heat melts large metal objects	Heating over large area not for precision welding
Radio frequency and microwave	~100 kHz to gigahertz	Cooking, curing, drying	Efficient	Effects vary with material, especially depth of penetration
Infrared	~10^{-10} Hz	Similar to microwave	Robust, inexpensive	Broad area coverage only
Laser welding	1.06 and 10.6 μm	Welding	Precise, efficient	Cannot work Al and Cu or their alloys; expensive
Electron beam	—	Welding	Precise for vacuum implementation efficient	X-ray shielding required; expensive

Maxwell's contribution was the inclusion of $\partial D/\partial t$ in the second curl equation. This term is referred to as the displacement current, and its inclusion results in coupled first-order differential equations that can be recast as wave equations for E and H that permit time harmonic, ie, oscillatory solutions. The J term in the second curl equation describes the motion of charge carriers as opposed to the time behavior of electric flux in the $\partial D/\partial t$ term.

The current density J is composed of two parts: a conduction current density $J_c = \sigma E$ and a convection current density $J_{conv} = \sum_i n_i q_i v_i\,(t)$, where $v_i(t)$ is not proportional to the electric field. The relationship given for the conduction current, $J_c = \sigma E$, is just a restatement of Ohm's law written as $I = (1/R)V$ when one accounts for density, electric field E as a voltage V over a path length d, and σ, the conductivity of the material, an intensive property characterizing the material and given by $\sigma = 1/\rho$, where ρ is the resistivity. The resistivity ρ is also an intensive property of the material. It can be related to the extensive property R, the resistance of a "slab" of material A in cross section and l in length by the formula

$$\rho = RA/l$$

The conduction current represents charge carriers moving in response to the applied electric field with a velocity proportional to the electric field. This situation pertains when the charge carrier is acted on by fields that change slowly with respect to a collision time τ_c. In this case many collisions occur while the field remains much the same. The multiple collisions ensure a randomization in the charge motion and through the collisions in the motion of the atoms or lattice of the material.

Imposed on the random motion of the charge carriers is a drift velocity μ that represents a net average velocity in the direction of the applied electric field of the charge carriers. This drift velocity arises from the collective motion of the charge carriers in the direction of E for the times between collisions. Thus a charge experiences a force along the direction of E for time τ_c, collides, and is

sent into a new direction, but once again experiences the same force along the direction of E and hence, over time, while fluctuating in direction and velocity, maintains an average or drift velocity in the E direction. Thus we write

$$V_{\text{charge } i} = \mu_i E$$

where μ_i is the average or drift velocity of the ith charge species in the material where the collisions occur. Since current density is, by definition,

$$J_i = n_i q_i v_i$$

one can write

$$J_i = n_i q_i \mu_i E$$

or

$$J_i = \sigma_i E$$

where $\sigma_i = n_i q_i \mu_i$

The force equation for a charge carrier q_i is given by the Lorentz force law:

$$\mathbf{F} = q_i\,(\mathbf{E} + \mathbf{v} \times \mathbf{B})$$

Since $\mathbf{v} \times \mathbf{B}$ is always perpendicular to $\mathbf{v}$, no energy is added to the charge by the $\mathbf{B}$ field; it merely changes the direction of motion from that given by $\mathbf{E}$ alone. For a large enough $\mathbf{v} \times \mathbf{B}$ compared to $\mathbf{E}$ the motion is no longer along $\mathbf{E}$ and the above analysis would need modification. However, typically, $\mathbf{v}$ would need to have a magnitude approaching the speed of light c or B would have to be impractically large. In effect, for most practical situations the $\mathbf{v} \times \mathbf{B}$ term is ignored and the approximation $\mathbf{F} = q_i \mathbf{E}$ is made.

When a charge moves in free space, this approximation is often invalid when assessing motion over distances that are large compared to distances between collisions in a material and B must be considered. This motion is not the motion of charge associated with conduction current; rather it is the motion associated with convection current. More generally convection current is charge moving in response to the fields with no collisions occurring. Neglecting B, one can still write $\mathbf{F} = q_i\mathbf{E}$; however, without collisions there is no resulting drift velocity and in fact the approximate force equation can be written $\mathbf{a}_i$ $(q_i/m_i)\mathbf{E}$. Thus the acceleration, not the velocity, is proportional to $\mathbf{E}$. This was the reason for the earlier distinction on $v_i(t)$ for convection current, namely, that it is not proportional to $\mathbf{E}$.

It is possible to get an appreciation of where the energy that goes with heating comes from by considering the following two cases: an electric field applied to a cloud of electrons without any material present and the same cloud and electric field only with the material present. In the first case the electrons continue to accelerate and gain more and more kinetic energy. In the second case they reach their drift velocity on average and the kinetic energy is much less. Energy in the form of heat imparted to the material is what makes up the difference.

From the preceding discussion it is evident that only the conduction current involves collisions, and therefore only the conduction current produces electric heating. At very low frequencies, eg, power line frequencies, and for applications such as a resistive coil in, say, a stove top burner, there are no currents in the form of electrons transversing free space and the displacement current, $\partial D/\partial t$, is negligibly small so that the current is only conduction current. As a counterexample at microwave frequencies $\partial D/\partial t$ need not be negligible compared to conduction current. As another counterexample for CRTs convection currents are sizable. Thus for modern technological applications one must note that Ohm's law, $V = IR$, is best viewed as $J_c = \sigma E$ and that J_c appears along with J_{conv} and $\partial D/\partial t$ in the Maxwell equations, which must be carefully solved to obtain the correct answer at higher frequencies and for situations involving electron and/or ion beams.

In keeping with this field formulation of current flow in a material and the identification of conduction current with heating, the earlier expression for heating in terms of V, I, and R can be reexpressed as

$$U_{\text{heat}} = J_{\text{conduction}}E$$
$$= \sigma E^2$$

where U_{heat} is the heat produced in a unit volume of material with conductivity σ and a field strength $E = |\mathbf{E}|$ in the volume.

The Maxwell equations can be solved directly in the time domain using finite differencing (2). Only the two curl equations are required. Conduction currents in the form of $J_c = \sigma E$ and the displacement current, $\partial D/\partial t$, are a part of these equations and their time behavior is part of the solution. The solution requires a specification of the material geometry and electrical properties, namely permittivity ε, permeability μ, and conductivity σ. A source current, voltage, or incident electromagnetic field must be prescribed. At low frequencies $\partial D/\partial t$ is often negligible and can be ignored, allowing for large time steps. At

higher frequencies where wavelengths approach the material object size the $\partial D/\partial t$ term must be retained and the time step is set by roughly the transit time across the "cells" into which the problem space has been discretized.

All the heating problems discussed here can be treated in this very straightforward fashion with the advent of powerful computers. When convection currents are present, a particle distribution function describing the particles and subject to the Lorentz force law must be added, as will as shown.

DEFINITION OF HEAT

Heat is a form of mechanical energy that is random in nature (3). Any material body can be heated either mechanically or electrically. The material can be in any of the states of nature: gas, liquid, solid, or plasma. In three of the four states the mechanical motion is assigned to individual particles, atoms, molecules, or electrons and ions. For the solid this picture is superceded by a phonon description when the atoms or molecules of the solid are arrayed in a lattice and so coupled together so as to support collective vibrations. A completely rigorous description of the heat in a crystalline solid where the lattice is well established is not discussed in detail here as it is quite complex and is more a matter of solid-state theory.

The electrical heating of any material can be described as the collision or scattering of an electron from atoms, molecules, electrons, ions, or phonons, in short any material body. The simplest material system to consider is a neutral gas. The interaction between the electrons and the gas atoms or molecules is short range and the scattering is binary, one electron interacting with one gas atom or molecule with a well-defined scattering angle. An electron interacting with a number of ions and other electrons in a plasma simultaneously is in sharp contrast.

If the gas is at equilibrium at temperature T, a Maxwellian velocity distribution will hold. That is, the velocity distribution function for the gas atoms or molecules, $f(v)$, is given by

$$f(v) = (m/2\pi KT)^{3/2}\exp(-v^2/v_{\text{th}}^2)$$

where $K = 1.38 \times 10^{-23}$ J/K is the Boltzmann constant and $v_{\text{th}} = (2KT/m)^{1/2}$ ($\sim 0.95 \times 10^5$ m/s for electrons at 300 K). As T increases and assuming equilibrium conditions still hold, the particles are distributed more broadly over velocity, as might be intuitively expected.

More generally the particle distribution can be written as a function $f(\mathbf{r}, \mathbf{v}, t)$ of space, velocity, and time. For an arbitrary distribution function the following definitions hold. The number density of particles $n(\mathbf{r}, t)$ at $\mathbf{r}$ in space is given by

$$n(r, t) = n_0 \int f(\mathbf{r}, \mathbf{v}, t)\, d\mathbf{v}$$

where

$$n_0 = N/V$$

is the average particle density of N total particles distributed throughout a volume V [if $n(\mathbf{r}, t) = $ constant for a homogeneous space distribution, then $\int f(\mathbf{r}, \mathbf{v}, t) \, d\mathbf{v} = 1$]. The "fluid" velocity $\mathbf{u}(\mathbf{r}, t)$ is calculated as

$$\mathbf{u}(\mathbf{r}, t) = \frac{n_0}{n(\mathbf{r}, t)} \int_v f(\mathbf{r}, \mathbf{v}, t) \mathbf{v} \, d\mathbf{v}$$

The random component of the velocity of a particle moving with velocity $\mathbf{v}$, ie, one particle in the distribution $f(\mathbf{r}, \mathbf{v}, t)$ at velocity $\mathbf{v}$ and position $\mathbf{r}$, is $w(\mathbf{r}, t)$, given by

$$\mathbf{w}(\mathbf{r}, t) = \mathbf{v}(\mathbf{r}, t) - \mathbf{u}(\mathbf{r}, t)$$

The kinetic temperature T, or simply temperature, is given by

$$T = \frac{n_0}{3Kn(\mathbf{r}, t)} \int m[\mathbf{w}(\mathbf{r}, t) \cdot \mathbf{w}(\mathbf{r}, t)] f(\mathbf{r}, \mathbf{v}, t) \, d\mathbf{v}$$

and the pressure tensor P by

$$P = mn_0 \int \mathbf{w}(\mathbf{r}, t) \mathbf{w}(r, t) f(\mathbf{r}, \mathbf{v}, t) \, d\mathbf{v}$$

Finally heat flow q is calculated as

$$q = \frac{mn_0}{2} \int \mathbf{w}(\mathbf{r}, t)[\mathbf{w}(\mathbf{r}, t) \cdot \mathbf{w}(\mathbf{r}, t)] f(\mathbf{r}, \mathbf{v}, t) \, d\mathbf{v}$$

For a Maxwellian distribution in a homogeneous media it is easily shown that

$$n(\mathbf{r}, t) = n_0$$
$$\mathbf{u}(\mathbf{r}, t) = 0$$
$$\mathbf{w}(r, t) = v(\mathbf{r}, t)$$
$$T = \frac{|w^2|}{3K}$$

where

$$|w^2| = \int \mathbf{w}(r, t) \cdot \mathbf{w}(r, t) f(\mathbf{r}, \mathbf{v}, t) \, d\mathbf{v}$$

and $|w^2|^{1/2}$ is the root-mean-square random velocity, $q = 0$.

Electrons flowing through the gas increase $\mathbf{w}(\mathbf{r}, t)$ via the collisions, and with increasing $\mathbf{w}(\mathbf{r}, t)$, the temperature T goes up, and with increasing T the heat in the gas increases. The relationship between T and heat Q can be described in terms of heat capacity; ie,

$$\Delta Q = C_v \, \Delta T$$

specifies the heat added to a system kept at constant volume when the temperature is increased by ΔT. Alternately

$$\Delta Q = C_p \, \Delta T$$

where the pressure is kept constant.

It should be evident that obtaining the heating effects of conduction current flow in a material from first princi-

ples is quite complex. It is certainly not always accurate and may not even be feasible. Experiment is often the only recourse. Nonetheless, it is important to see how heating is a phenomenon associated with materials that do not require any specific source other than one that increases $\mathbf{w}(\mathbf{r}, t)$, the random motion of the material.

SOURCES OF ELECTRICAL HEATING: TYPES

Electrical heating is produced in materials by the passage of charge carriers, typically electrons, through the material. The electrons colliding with the material produce random motion in the atoms, molecules, or ions and electrons in the material. This random motion represents the heat in the material. Electrical heating then must concern itself with how the electron flow is generated and how collisions transfer energy of motion in the electron to random thermal motion of the material.

The generation of electron flow in a material, the source of the electrical heating, is a fundamental issue requiring its own treatment. The treatment presented here distinguishes between different sources or categories of electron or current flow.

The first distinction is between processes that require current flow for their operation, eg, the heating of a hot water heating coil, and processes for which current flow was not sought after or was unintended or a secondary effect, eg, heating of the walls of a waveguide.

The second distinction is between the forces that can produce charge carrier motion. A charge carrier (an electron for ease of discussion) can be set in motion by a number of forces including electromagnetic fields, short-range atomic forces in collisions with material particles, and gravitational effects. This last force has no technological utility for electrical heating and is mentioned only for completeness. Quantization effects, on the other hand, are not just a curiosity. High-energy photons in the form of gamma rays are an important form of ionizing radiation that can produce charge carriers from neutral materials and set these carriers in motion.

The third distinction that is useful to make is frequency regime. There are a number of distinctive frequency regimes in which electrical heating processes display unique characteristics.

This last distinction requires somewhat greater elaboration than the preceding two. Electrical heating effects can occur over a very broad range of frequencies. The frequencies can be as low as tens of hertz for low-frequency ionospheric wave effects, ELF signal generation and propagation effects, and power line heating effects, which can be of the lines themselves, of the nearby earth, or somewhat controversially, of biological material, such as human bodies, in close proximity to the power lines. The frequencies can be significantly higher at radio wave or microwave frequencies. For microwaves thermal motion can be efficiently produced by the internal vibrations of water molecules that result from the strong coupling of the microwave field energy to this mode of vibration. At even higher frequencies, such as infrared, visible, and ultraviolet, a different set of processes occur, such as the photoelectric effect. Finally for sufficiently high frequencies (eg, for gamma rays) any number of atomic scattering processes can take place, one example being Compton scattering of electrons.

As frequency increases, the simple picture of electron flow throughout the material characterized by a drift velocity gives way to a description in which individual scattering events and the details of the charge carrier motion and location are considered. When material size is very large compared to wavelength and conductivity is large, the effects can be mostly at the surface or more precisely over a depth of a few skin depths. Conversely, when the material size is small compared to a wavelength, as occurs in the structures making up a very large scale integrated (VLSI) chip, the effects are not noticeably attenuated. Thus material size and material properties, most importantly those related to the materials' conductivity, play an important role in how electrical heating may be produced at different frequencies.

A brief categorization of electrical conductivity for a limited set of different materials is given in Table 2.

INTENTIONAL PRODUCTION OF ELECTRICAL HEATING

Resistance Heating

The most commonly encountered form of intentional electrical heating occurs in poor conductors, metals treated for low conductivities, at low frequencies, namely power line frequencies. For a poorly conducting metal ring 1 cm in cross section and 20 cm in diameter a conductivity of 10^{-5} mho/m or equivalently a resistivity of 10^5 Ω-m yields a bulk resistance of around 10 Ω. The applied household voltage is 120 V peak so that the instantaneous peak power delivered to the coil and hence the heating is just $P_{\text{heat}} = V^2/R = 1440$ W. Integrating over time yields the average or root-mean-square (RMS) power $P_{<\text{heat}>}$, which is $\frac{1}{2}P_{\text{heat}}$, or about 700 W. Applications would include immersion water heating coils. The use of smaller wires produces high resistances even when the material has a higher conductivity. Such a configuration may be used for a toaster or, with higher conductivity and hence less resistance, for an electric blanket. Material properties of interest outside of conductivity are the ability to survive thermal cycling and, for the electric blanket, flexibility.

Table 2. Representative Conductivity Values

Good conductors	
Silver	6.17×10^7 mhos/m
Copper	5.8×10^7 mhos/m
Gold	4.1×10^7 mhos/m
Aluminum	3.82×10^7 mhos/m
Iron	1.0×10^7 mhos/m
Poor conductors	
Salt water	4 mhos/m
Average human tissue	0.95 mhos/m
Silicon	1.56×10^{-3} mho/m
Earth	10^{-3}–3×10^{-2} mho/m
Water	10^{-3} mho/m
Insulators	
Glass	10^{-10}–10^{-14} mho/m
Polystyrene	10^{-16} mho/m
Quartz	1.3×10^{-18} mho/m

Other examples of desired heating are incandescent light filaments and electron emission from resistively heated cathodes in CRTs.

The incandescent filament uses a metal filament made of tungsten capable of being heated to over 3000 K. The thinness of the filament is essential to achieving the high resistivity needed to obtain the high temperature. The filament radiates electromagnetic energy because of the acceleration imparted to the electrons in the metal in their collisions with the atoms of the metal. The radiation is described by the Planck blackbody radiation law,

$$U(\omega) = \frac{\hbar}{\pi c^3} \frac{\omega^3 \, d\omega}{e^{\hbar\omega/KT} - 1}$$

where $U(\omega)$ is the spectral energy density of the emitted radiation, Planck's constant $\hbar = 1.054 \times 10^{-34}$ J-s, Boltzmann's constant $K = 1.38 \times 10^{-23}$ J/K, T is in degrees Kelvin, and ω is angular frequency. Only when T is as large as around 3000 K is there appreciable radiation is the visible spectrum between 300 and 500 nm. Even then, for typical incandescent light bulbs efficiency is only 1 or 2%.

The CRT relies on thermionic emission (4) for the emission of electrons from the cathode surface. In this process thermal energy from electrical heating produces a Maxwellian distribution in the motion of the atoms and a Fermi–Dirac distribution in the motion of the electrons. The atoms and electrons are in thermal equilibrium and have the same temperature. The high-velocity tail end of the Fermi–Dirac distribution has the same distribution as a Maxwellian distribution. For thermionic emission the work potential ϕ an electron must overcome to escape from the surface of the cathode represents an equivalent velocity v_{eq}, where $\phi = \frac{1}{2}m_e v_{\text{eq}}^2$ that is less than the velocities of the electrons in the "tail" of the approximately Maxwellian distribution of the electrons at these high velocities. If enough of the electrons in the "tail" of this distribution are above this velocity, significant numbers of electrons are emitted.

Accounting for the Fermi-Dirac distribution of the electrons yields the Richardson-Dushman equation for thermionic emission,

$$J = AT^2 e^{-\phi/KT}$$

where

$$A = 4\pi m e K^2 h^{-3} = 120 \text{ A/cm}^2\text{-deg}^2$$

Experimentally for tungsten it is found that $A = \sim 75$ A/cm^2-deg^2 and $\phi = 4.5$ eV for clean uniform surfaces.

Induction Heating

Induction heating (5) works via magnetic induction. The Maxwell curl equation with a time-varying magnetic field is based on the Faraday law of induction. It can be converted to an integral equation via Stokes' theorem, which shows that a time-varying magnetic field linking a loop induces a voltage around the loop. A metal work piece placed in a time-varying field with less than perfect conductivity has electric fields induced on its surface and roughly over a depth given by the skin depth (see the section on Unintentional Heating for a definition). These fields induce currents in the metals called eddy currents that produce heating via the same ohmic collision processes associated with direct resistance heating.

The advantage is that the work piece is not materially connected to the heat source. The connection is via the fields, and large pieces can be heated to very high temperatures without damaging the sources. Frequencies range from power line frequencies to radio frequencies with power levels as high as 100 MW at the lower frequencies to between 10 kW and 1 MW at radio frequencies. Applications include preheating and melting metals and surface hardening of steels to improve strength, wear, and fatigue properties. These applications are often in a continuous-line operation. Another application is sintering of carbide at high (2550°C) temperature. Induction heating provides on-demand heating because of its quick starting and quick heating. Induction heating does not lend itself to precision work, such as fine welds.

Microwave Heating

At higher frequencies in the radio frequency to microwave regime heating effects may also be desired, as in the case of microwave heating of food.

Table 11.1 of Ref. 6 gives an exhaustive list of industrial RF and microwave applications including Pasteurization, rendering, sterilization, and vulcanization. The important features here are uniformity of the field incident to the object to be heated and complete or at least adequate penetration of the field into the object being heated, which requires a skin depth comparable to the object's characteristic dimensions. This in turn may place limits on the frequency employed as the skin depth decreases with increasing frequency.

A medical application of microwave heating is the microwave generation of hyperthermia in tumors where as little as a 5°C increase in temperature exceeds certain tumor thermoregulatory capabilities and the tumor dies. Here field uniformity is not sought; rather the goal is a highly focused field on the tumor for rapid heating and minimal fields and minimal heating outside the tumor in the surrounding healthy tissue.

Field uniformity in a microwave oven is difficult to achieve because the oven is effectively a cavity and strong standing-wave patterns are generated within the oven. The only mitigating factors are losses to the walls of the oven in ohmic heating, radiation out of the oven and ohmic heating of the food. The first two factors are kept low, the first because the oven should not itself be heated and this can be ensured by the use of highly conducting metals to form the cavity, the second because microwave radiation illuminating a microwave oven user is considered harmful at levels exceeding 10 μW/cm^2 according to current standards. The use of solid cavity walls or a fine mesh ($\sim$0.5–1.0 mm^2) screen and tightly fitting door seams ensures these standards are met. The last factor then is the main loss mechanism. The food as a terminating load for the microwave circuit representing the oven is not a matched load, and the standing wave pattern in the oven, while moderated, is not eliminated. Illumination is therefore not uniform, and the reflected wave portion of the field, which combined with an equal measure of the incident wave forms the standing wave, is energy not delivered to the load. Thus only a fraction of the rated power of the microwave oven appears as heating of the food.

It is perhaps fortunate that the process works this way. A microwave oven capable of delivering full rated power to the cavity with or without food and with no losses to the walls or to radiation would be very dangerous indeed. Such an oven with presumably very sturdy doors and a 1000-W capability with no food present would in 1000 s or roughly 17 min store 1 MJ of energy in the cavity. It is from simple considerations such as this that it becomes very clear why a microwave oven operates near its rated power only when a considerable amount of food is present in the cavity and even then not all the power is delivered and what is delivered is not entirely uniform.

Simple considerations also are important to the issue of why the walls of the cavity do not heat significantly. At the end of the section on governing equations it was noted that heating in a unit volume of material is given by $U_{\text{heat}} = \sigma E^2$. Since the walls are good metal conductors, σ is extremely high. Why then do the walls not become extremely hot? The answer lies in the fact that only a very small electric field actually penetrates the metal at the 915-MHz operating frequency. Most is reflected back into the cavity. Thus the high σ is more than offset by a very small E inside the metal.

It is clear from these discussions that determining the electrical heating of an object requires an understanding of not just the object's material properties, namely σ and the heating process, namely ohmic, but also the fields within the object. This determination of the fields is often a challenging task and can only be performed using relatively sophisticated computational tools.

Infrared Heating

Midway in frequency between microwave heating and visible and near visible light laser heating is infrared (IR) heating (7). Infrared can be produced from a bulb much like an incandescent bulb only with the filament heated to a lower temperature. This shifts the Planck blackbody radiation spectrum first presented for the incandescent bulb filament in the resistive heating section to a lower spectral peak. An industrial lamp with a filament at 2750 K has its spectrum centered at 10,500 Å. Alternatively, vapor arc lamps can be employed. Here the arc produces atomic collisions that excite one or more spectral lines in the atoms. Mercury vapor arc lamps, eg, have strong spectral lines at 1.01, 1.13, 1.37, 3.94, and 4.02 μm. The lamp bulb itself is heated and is a secondary source of radiation with a continuous spectrum as it radiates a Planck blackbody spectrum.

Finally, IR can be produced by heating in air filaments of various alloys such as Ni (80%) and Cr (20%) or Ni (60%) and Fe (25%) and Cr (15%), called 80/20 and 60/15 alloy, with resistivities of 106 and 110 $\mu\Omega$/cm^3, respectively. The temperature is in the range of 1000°C. This approach is ideally suited to space heaters as used in homes. Other IR applications include cooking, drying, and paint baking.

Laser Heating

At even higher frequencies there are fewer examples of intentional electrical heating. One example is laser metal cutting and welding (8). The requirement is to reach electric field strengths high enough to transfer enough energy into the metal's surface over a skin depth to heat that portion of the metal to melting, in fact beyond, so that it vaporizes and the heating process can continue at the layer below the one that was vaporized. Assume quite simplistically that adiabatic conditions hold, namely that no heat flows out of the heated volume, an approximation good only for times less than the thermal transit time across the heated spot (9). Then the heating must raise the metal to above its melting point in a time less than or equal to the thermal transit time. For 500 nm light illuminating mild steel the skin depth $\delta = 6 \times 10^{-9}$ m. The thermal transit time across the skin depth is on the order of $(6 \times 10^{-9}$ m)$/(1/2 \times 10^{-4}$ m/s) $\cong 10^{-12}$ s. The specific heat per kilogram c is approximately 0.46×10^3 J/K-kg and the temperature must increase roughly 1000 K. Thus

$$(\sigma E^2)10^{-12} \text{ s} = c \cdot \text{density} \cdot \Delta T$$

and the required E field is found to be on the order of 2×10^7 V/m.

Laser cutting and welding heating is strictly a surface phenomenon requiring very large power levels, as may be seen by converting the calculated E field levels into average power flow per square meters, using

$$P = Uc = (\tfrac{1}{2}\varepsilon E^2)c = \tfrac{1}{2} \times 10^{12} \text{ W/m}^2$$

For a spot size of 0.1 mm^2 this corresponds to 50 kW.

In practice, first the laser light is largely reflected; then the small amount absorbed produces a metal vapor that results in most of the energy being absorbed. A keyhole, ie, a small-diameter cylindrical hole, appears in the metal and is held open by the metal vapor pressure. The heating process is no longer confined to the unvaporized metal skin depth and power levels of 10^{10} W/m^2 or 10^4 W/mm^2 produces excellent welds. Common sources are pulsed Nd–YAG lasers operating at 1.06 μm and continuous CO_2 lasers at 10.6 μm. Power levels are over 1 kW for the pulsed lasers and 25 kW for the continuous gas laser.

Commercial applications cover the same range as other fusion welding techniques with the exception of aluminum and copper and their alloys. Applications run from spot welding of pacemakers, CRT electron guns and automobile doors, as well as steel plates over 25 mm thick. Little collateral heating occurs, and unlike the somewhat similar process of vacuum electron beam heating, no vacuum is required. In comparison to nonvacuum electron beam welding, the source and the work can be widely separated, as opposed to a spacing of inches with the nonvacuum electron beam welders. Precision pieces that fit together tightly are required for laser welding and the bulk of the equipment is large and the costs are high. Table 2.1 of Ref. 9 provides a more extensive discussion of the advantages and disadvantages.

Electron Beam Welding

Electron beam welding (10) is in some respects similar to laser beam welding. It relies on a stream of electrons from a high-power electron gun to weld the metal instead of a laser beam. The advantage of an electron beam over a laser beam is that the electron beam can weld any metal or combination of dissimilar metals. The disadvantage is the need for a vacuum between the beam source and the work piece when the weld must be of narrow width. Nonvacuum electron welding produces a wider weld and more of the material is heated to where material properties may be adversely affected. Just as for laser welding, the weld can be very narrow.

Electron beam welders can be in the range of 100 kW for nonvacuum welders using electron energies in the hundreds-of-kilovolt range for beam currents of several hundred milliamperes. Vacuum electron beam welders can go even higher. Lower powers are used for processes requiring less welding depth than the more than 5-cm depth possible with nonvacuum machines in the 60-kW power range. Higher powers in nonvacuum machines have other advantages than just faster speed of welding. When the power is obtained by increasing the voltage of the electron gun, the scattering of the electrons in the air is reduced as the scattering cross section goes down at higher electron velocities. This results in a narrower weld.

Typically, the electron gun is a triode system similar to a TV tube. It operates at around 10^{-4} torr, and in vacuum welding the vacuum in the welding chamber is on the order of 10^{-1}–10^{-2} torr. For nonvacuum welding a semivacuum is maintained along the beam transmission path to reduce scattering. Overpressurized air or helium at the end of the beam exit nozzle can be used to keep metal vapor out of the transmission path and provides 1.2- and 1.8-in. distances between the exit nozzle and the work piece, respectively. Heating is produced by the transfer of kinetic energy from the electrons to the work piece resulting from the abrupt deceleration of the electrons. X-rays are also produced by the electron collisions, so the equipment must be shielded.

UNINTENTIONAL ELECTRIC HEATING

The simplest example of unintentional electric heating is that occurring with overhead power lines. Even though the picture is one of very large currents flowing on very low resistance wires at very low frequencies, fields are still present. This is required by the Maxwell equations, where

$$\nabla \times H = \frac{\partial D}{\partial t} + J$$

can be recast via the Stokes theorem into the integral form

$$\oint H \, ds = \oint \left(\frac{\partial D}{\partial t} + J \right) dA \approx I \quad \text{for the wire}$$

Thus whenever there is a conduction current I flowing through the wire, there is an H field of

$$H = I/2\pi r$$

in free space (wire removed from the ground). Note that only the conduction current is considered; the displacement current $\partial D/\partial t$ is vanishingly small in comparison at power line frequencies.

The conservation of change requires a charge distribution on the line arising from the current. A radial E field about the wire as the result of this charge therefore arises. This radial E field is modified by the presence of a conducting earth beneath the wire in all real applications. The wire, its currents, and charges are imaged in the ground with electric field lines now bent between the wire and its image.

An exact solution includes the imaging and the attenuation in the fields arising from the finite conductivity of the earth. For a perfect earth, one with infinite conductivity, the image is an exact opposite, $I_{image} = -I_{wire}$ and $\rho_{image} = -\rho_{wire}$, but no fields penetrate the earth. For a real earth with nominal conductivity σ of 10^{-3}–30×10^{-3} mho/m and with $\varepsilon = \varepsilon_r \varepsilon_0$, where ε_r is on the order of 10 and $\varepsilon_0 = 8.85 \times 10^{-12}$ F/m, the permittivity of free space, the fields penetrate the earth. These fields, in particular the electric field, heat the earth at a rate given by $\sigma_{earth} E_{earth}^2$ over an infinitesimal volume element. Integrating over all of the conducting space gives the power losses to the ground. The key to the solution is finding the field distribution in the presence of the earth, a problem solved in a number of texts for simplified power line geometries.

Another loss mechanism is the ohmic heating of the wire itself. Since the wire is highly conductive and the frequency is very low, so that $\sigma/\omega\varepsilon \gg 1$, the skin depth δ is given by

$$\delta = (2/\omega\mu\sigma)^{1/2}$$

and for the copper with $\sigma = 5.7 \times 10^7$ mhos/m and $\mu = \mu_0$, $\delta = 0.85$ cm at 60 Hz. The current can be approximated as flowing uniformly throughout the wire for a solid wire 0.5 cm in diameter. A larger wire provides little improvement in resistance per unit length of wire, given by

$$R_0 = \frac{1m}{\sigma A} = \frac{1}{(5.7 \times 10^7)(\pi/4)(10^{-2})^2} = 2.23 \times 10^{-4} \ \Omega$$

as the fields remain primarily at the surface for a larger wire.

In comparison the earth and the resistance associated with it behave differently. The earth is very much thicker than its skin depth. For low frequencies in the earth $\sigma \gg \omega\varepsilon$ so it is a good conductor. In this case the earth has a resistance found from integrating over the exponentially decaying electric fields. The same result is obtained by assuming a constant electric field over one skin depth. Thus the surface resistance of the earth R_s is calculated as

$$R_s - 1/\sigma\delta$$

and for a nominal earth conductivity $\sigma = 10^{-2}$ mho/m and $\delta = 650$ m, $R_s = 0.15 \ \Omega$. For a strip of earth 1 m long (the

same as a 1-m wire length) and roughly 100 m wide to encompass the majority of the field distribution, the total resistance is $1.5 \times 10^{-3} \ \Omega$. Losses to the earth can therefore exceed wire losses.

Similar considerations hold at higher frequencies, namely in the microwave regime, for transmission lines supporting TEM waves predominately and for waveguides, generally either rectangular or coaxial supporting any number of modes, but principally TE_{10} for rectangular waveguides and TE_{11} for circular waveguides. The walls of a rectangular waveguide, eg, are treated as surfaces with surface impedance Z_s, which is obtained by relating the E and H propagating into the metal as governed by the wave equation for a conducting media with $\sigma \gg \omega\varepsilon$. This surface impedance for a good conductor is given by

$$Z_s = \sqrt{\frac{\omega\mu}{2\sigma}} (1 + j)$$

The real part of Z_R is the surface resistance R_s, where

$$R_s = \sqrt{\frac{\omega\mu}{2\sigma}}$$

The surface resistance can be used to find the power dissipated in the walls of the waveguide according to

$$P_c = \frac{R_s}{2} \iint_{\text{wall area}} J_s \cdot J_s^* \, dS$$

where the current densities J_s are well approximated by the current density for the limit of lossless walls

$$J_S \simeq n \times H_{\text{lossless}}$$

Explicitly performing these operations for the TE_{10} mode of a rectangular waveguide yields

$$\frac{(P_c)_{TE_{10}}}{l} = \frac{aR_s}{2\eta^2} E_0^2 \left[1 + \frac{b}{a}\left(\frac{f_c}{f}\right)^2\right]$$

where η is the intrinsic impedance (377 Ω for air-filled waveguides), a and b the guide dimensions $(a > b)$, E_0 the maximum electric field strength, f_c the cutoff frequency, R_s the surface resistance, and l the length of the waveguides.

The signal attenuation in the waveguide is found from this power loss. For the TE_{10} mode it is

$$\alpha_{TE_{10}} = \frac{R_s}{\eta b} \frac{[1 + (2b/a)(f_c/f)^2]}{\sqrt{1 - (f_c/f)^2}} \ m^{-1}$$

The TE_{10} mode for rectangular waveguides, along with being the principal mode, is also less lossy than the higher modes, such as TE_{20}, TE_{11}, and TM_{11}. The TE_{11} mode of the circular waveguide is less lossy than all but the TE_{on} modes where attenuation theoretically decreases indefinitely with frequency. For real geometries with slight irregularities this advantage of the TE_{on} modes is lost.

SOLID-STATE DEVICE HEATING

A final example of unintentional heating occurs in solids used for solid-state devices. Until now the ohmic heating processes underlying electrical heating have been described using the macroscopic constitutive parameter σ or conductivity. In a solid at high frequencies a steady-state description based on drift diffusion solutions of the Boltzmann transport equation (BTE) may be needed (11). Such a solution is based on a probabilistic description of the collisions experienced by the electrons in the solid.

Transit Solid-State Device Heating Computational Modeling

If transient effects are desired, then even this BTE solution may prove inadequate. With large computational resources it is now possible to solve directly the Boltzmann–Vlasov equation describing charged particle motion in the presence of external and self-induced fields while accounting for collisions at the microscopic or atomic level in a self-consistent treatment.

Conceptually this is a very direct method of modeling. Charge motion in response to external fields produces a change in the particle distribution, from equilibrium typically, that can be appropriately integrated to obtain currents throughout the space occupied by the moving charges. These currents then act as sources in a Maxwell equation solver such as the finite difference time domain (FDTD) technique. From the FDTD field solver the fields produced by the currents are found. These fields modify the motion of charges as determined by the application of the Lorentz force equation using the external and self-induced fields. These solutions are stepped in time at a time step given by the Courant stability condition. Periodically a scattering term is evaluated to determine the number of charge carriers scattered by collision processes in the material. This scattering determines the transfer of energy from the external fields into charge carrier motion, which via the scattering may be turned into random motion in the material or more simply the heating of the material.

Note that for such a complete treatment a detailed description of the charge carrier behavior in the form of a particle distribution is required along with a computational tool for determining the exact field distributions as they evolve in time and a detailed description of the scattering process.

The scattering processes of interest are associated with solids having a crystalline lattice structure of the type found in silicon and gallium arsenide. The processes that can occur include acoustic deformation potential scattering (ADP), optical deformation potential (ODP) scattering, polar acoustic phonon (PAP) scattering, polar optical phonon scattering (POP), scattering from defects, and carrier–carrier scattering.

Of these only ADP will be described to illustrate how heating is produced by the scattering process. The scattering term that describes how an electron scatters off an acoustic phonon is given by (11)

$$S(p, p') = \frac{\pi D_A^2 \beta}{\rho \omega_s \Omega} \left(N_{AP} + \frac{1}{2} \mp \frac{1}{2} \right)$$

$$\times \delta(\mathbf{p'} - \mathbf{p} \mp \hbar\boldsymbol{\beta})\delta(E' - E \mp \hbar\omega_s)$$

Here $S(p, p')$ describes the rate at which an electron in a state with momentum $\mathbf{p}$ and energy E scatters into a state $\mathbf{p'}$ and E'. The difference between E' and E is the energy transferred to the lattice that produces the heating of the lattice and therefore the material. The constants in the above expression are $\hbar\omega_s$, the energy of the acoustic phonon; D_A, the acoustic deformation potential (9.5 eV in silicon); β, the wave number of the phonon; $N_A = [\exp(\hbar\omega_s/kT - 1)^{-1}$, the number of acoustic phonons given by the Bose–Einstein distribution; ρ, the mass density; and Ω, the normalization volume.

BIBLIOGRAPHY

1. J. D. Jackson, *Classical Electrodynamics,* 2nd ed., Wiley, New York, 1975.
2. K. S. Kunz and R. J. Luebbers, *The Finite Difference Time Domain Method for Electromagnetics,* CRC, Boca Raton, Fla., 1993.
3. P. M. Morse, *Thermal Physics,* W. A. Benjamin, 1965.
4. C. Kittel, *Introduction to Solid State Physics,* 3rd ed., Wiley, New York, 1968.
5. S. Zinn and co-workers, *Elements of Induction Heating,* Electric Power Research Institute, 1988.
6. A. C. Metaxas and R. J. Meredith, *Industrial Microwave Heating,* IEE Power Engineering Series 4, Peter Peregrinies LTD, 1983.
7. W. Summer, *Ultraviolet and Infrared Engineering,* Interscience, New York, 1962.
8. C. Dawes, *Laser Welding,* McGraw-Hill, New York, 1992.
9. H. S. Carslaw and J. C. Jaeger, *Conduction of Heat in Solids,* 2nd ed., Oxford University Press, Oxford, 1971.
10. M. M. Schwartz, *Source Book on Electron Beam and Laser Welding,* American Society for Metals, Engineering Bookshelf, 1981.
11. M. Lundstrom, *Modular Series on Solid State Devices,* Vol 10: *Fundamentals of Carrier Transport,* Addison-Wesley, Reading, Mass., 1990.

ELECTRIC POWER DISTRIBUTION

J. DUNCAN GLOVER
Failure Analysis Associates
Boston, Massachusetts

Major components of an electric power system are generation, transmission, subtransmission, distribution, and utilization. Distribution, including primary and secondary distribution, is that portion of a power system that runs from distribution substations to customers' service-entrance equipment. (See also ELECTROMAGNETIC FIELDS; LIGHTING.) As an introduction to electric power distribution, the major components shown in Figure 1 are briefly discussed in the following paragraphs (1–7).

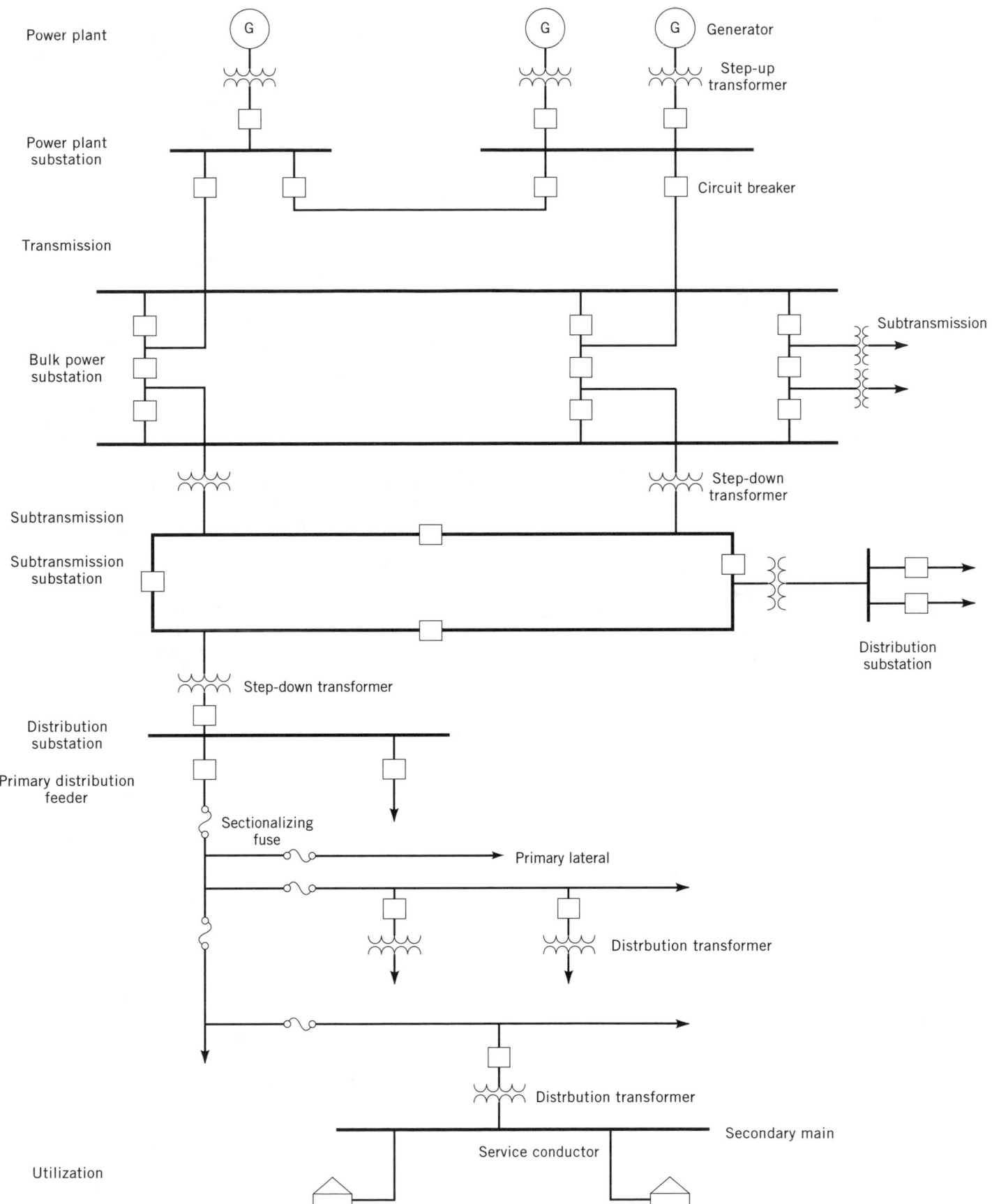

Figure 1. Principal components of an electric power system.

Power plants convert energy from fuel (coal, oil, gas, nuclear, etc) and from water flow, wind, or other forms into electric energy. Power plant generators, with ratings varying from 50 to 1300 MVA, are of three-phase construction, with three-phase armature windings embedded in the slots of stationary armatures. Generator terminal voltages, which are limited by material and insulation capabilities, range from a few kilovolts for older and smaller units up to 20 kV (line-to-line rms) for newer and larger units.

To reduce transmission energy losses, step-up transformers at power plant substations increase voltage and decrease current. Both the transformers and the buses in these substations are protected by circuit breakers, surge protectors, and other protection equipment.

The transmission system serves three basic functions: it delivers energy from generators to the system, provides for energy interchange with other utilities, and supplies energy to the subtransmission and distribution systems. The transmission system consists of a network of three-phase transmission lines and transmission substations, also called bulk power substations. Typical transmission voltages range from 230 up to 765 kV. Single-circuit, three-phase line ratings vary from 400 MVA at 230 kV up to 4000 MVA at 765 kV.

The subtransmission system consists of step-down transformers, subtransmission lines, and substations that connect bulk power substations to distribution substations. In some cases, a subtransmission line may be tapped, usually through a circuit breaker, to supply a single-customer distribution load such as a large industrial plant. Typical subtransmission voltages range from 69 to 138 kV.

Distribution substations include step-down transformers that decrease subtransmission voltages to primary distribution voltages in the 2.2- to 46-kV range for local distribution. These transformers connect through associated circuit breaker and surge arrester protection to substation buses, which in turn connect through circuit breakers to three-phase primary distribution lines called distribution circuits or feeders. Each substation bus usually supplies several feeders. Typical distribution substation ratings vary from 15 MVA for older substations to 200 MVA or higher for newer installations. Distribution substations may also include equipment for regulating the primary voltage.

Primary distribution feeder ratings include 4 MVA for 4.16-kV, 12 MVA for 13.8 kV, 20 MVA for 22.9-kV, and 30 MVA for 34.5-kV feeders. Feeders are usually segregated into several three-phase sections connected through sectionalizing fuses or switches. Each feeder section may have several single-phase laterals connected to it through fuses. Three-phase laterals may also be connected to the feeders through fuses or reclosers. Separate, dedicated primary feeders supply industrial or large commercial loads.

Feeders and laterals run along streets, as either overhead lines or underground cables, and supply distribution transformers that step the voltage down to the secondary distribution level (120 to 480 V). Distribution transformers, typically rated 5 to 2500 kVA, are installed on poles for overhead lines, and on pads at ground level or in vaults for underground cables. The transformers are protected from overloads and faults by fuses or circuit breakers on the primary and/or the secondary side. From these transformers, energy flows through secondary mains and service conductors to supply single- or three-phase power directly to customer loads (residential, commercial, and light industrial).

Service conductors connect through meters, which determine kilowatt-hour consumption for customer billing purposes as well as other data for planning and operating purposes, to service panels located on customers' premises. Customers' service panels contain circuit breakers or fuses that connect to wiring that in turn supplies energy to utilization devices (lighting, appliances, motors, heating-ventilation-air-conditioning, etc).

Distribution of electric energy from distribution substations to meters at customers' premises has two parts: (a) primary distribution, which distributes energy at voltages in the 2.2- to 46-kV range from distribution substations to distribution transformers, where the voltage is stepped down to customer utilization levels; and (b) secondary distribution, which distributes energy at customer utilization voltages of 120 to 480 V to meters at customers' premises.

PRIMARY DISTRIBUTION

Table 1 shows typical primary distribution voltages in the United States (1–7). Primary voltages in the "15-kV class" predominate among U.S. utilities. The 2.5- and 5-kV classes are older primary voltages that are gradually being replaced by 15-kV class primaries. In some cases, higher 25- to 50-kV classes are used in new, high-density-load areas.

The three-phase, four-wire primary system is the most widely used. Under balanced operating conditions, the voltage of each phase is equal in magnitude and 120° out of phase with each of the other two phases. The fourth wire in these Y-connected systems is used as a neutral for the primaries, or as a common neutral when both primaries and secondaries are present. Usually, the windings of transformers at distribution substations are Y-

Table 1. Typical Primary Distribution Voltages in the United States

Class, kV	Voltage, kV
2.5	2.4
5	4.16
8.66	7.2
15	12.47
	13.2
	13.8
25	22.9
	24.94
34.5	34.5
50	46

connected, with the neutral point grounded and connected to the common neutral wire. The common neutral is also grounded at frequent intervals along the primary, at distribution transformers, and at customers' service entrances. California is one state that uses three-phase, three-wire primary systems.

Rural areas with low-density loads are usually served by overhead primary lines, with distribution transformers, fuses, switches, and other equipment mounted on poles. Urban areas with high-density loads are served by underground cable systems, with distribution transformers and switchgear installed in underground vaults or in ground-level cabinets. There is also an increasing trend toward underground residential distribution (URD), particularly single-phase primaries serving residential areas. Underground cable systems are highly reliable and unaffected by weather. But the installation costs of underground distribution are significantly higher than overhead.

Primary distribution includes three basic types: radial, loop, and primary network systems.

Primary Radial Systems

The primary radial system, as illustrated in Figure 2, is a widely used, economical system often found in low-load-density areas (2,3,8). It consists of separate, three-phase feeder mains (or feeders) emanating from a distribution substation in a radial fashion, with each feeder serving a given geographical area. Single-phase laterals (or branches) are usually connected to feeders through fuses, so that a fault on a branch can be cleared without interrupting the feeder. Single-phase laterals are connected to different phases of the feeder, so as to balance the loading on the three phases.

To reduce the duration of interruptions, overhead feeders can be protected by automatic reclosing devices located at the substation or at the first overhead pole (9). Studies have shown that the large majority of faults on overhead primaries are temporary, caused by lightning flashover of line insulators, momentary contact of two conductors, momentary bird or animal contact, or momentary tree limb contact. The recloser or circuit breaker with reclosing relays opens the circuit either "instantaneously" or with intentional time delay when a fault occurs, and then recloses after a short period of time. The recloser can repeat this open and reclose operation if the fault is still on the feeder. A popular reclosing sequence is two instantaneous openings (to clear temporary faults), followed by two delayed openings (allowing time for fuses to clear persistent downstream faults), followed by opening and locking out for persistent faults between the recloser and fuses. Reclosing is not used on circuits that are primarily underground.

To further reduce the duration and extent of customer interruptions, sectionalizing fuses are installed at selected intervals along radial feeders. In the case of a fault, one or more fuses blow to isolate the fault, and the unfaulted section upstream remains energized. In addition, normally open tie switches to adjacent feeders are incorporated, so that during emergencies, unfaulted sections of a feeder can be tied to the adjacent feeder. Spare capacity is often allocated to feeders to prevent overloads during such emergencies, or there may be enough diversity between loads on adjacent feeders to eliminate the need for spare capacity. Some utilities have also installed automatic fault locating equipment and remote controlled sectionalizers (automatic switches) at intervals along radial lines, so that faulted sections of a feeder can be isolated and unfaulted sections reenergized rapidly from a dispatch center, before the repair crew is sent out.

Shunt capacitor banks including fixed and switched banks are used on primary feeders to reduce voltage drop, reduce power loss, and improve power factor. Capacitors are switched off at night during light loads and switched on in the morning during heavy loads. Computer programs are available to determine the number, size, and location of capacitor banks that optimize voltage profile, power factor, and installation and operating costs.

One or more additional feeders along separate routes may be provided for critical loads, such as hospitals that cannot tolerate long interruptions. Switching from the normal feeder to an alternate feeder can be done manually or automatically with circuit breakers and electrical interlocks to prevent the connection of a good feeder to a faulted feeder. Figure 3 shows a primary selective system, often used to supply concentrated loads over 300 kVA

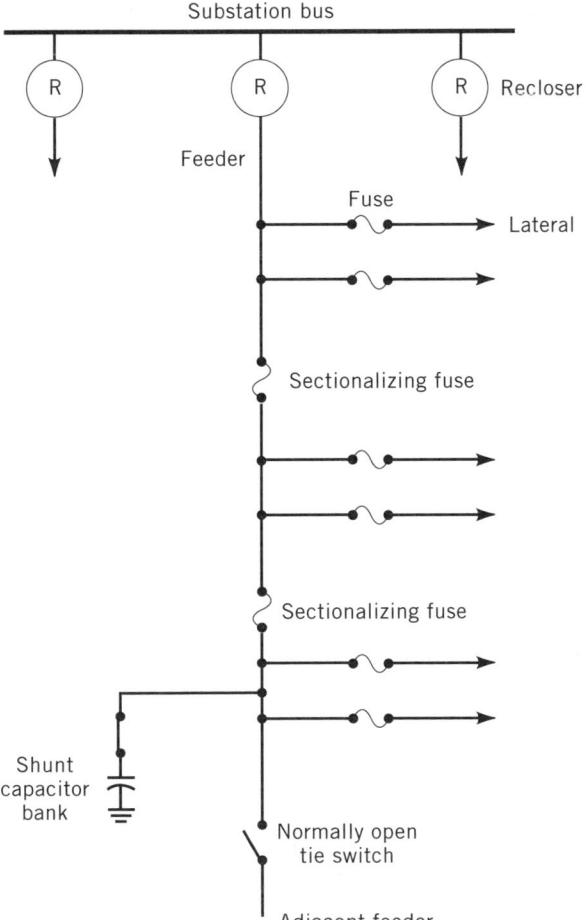

Figure 2. Primary radial system.

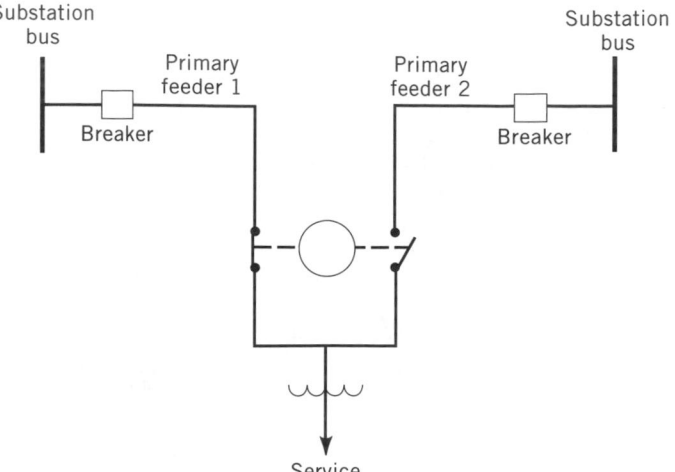

Figure 3. Primary selective system.

(2,3,8). There are two primary feeders with automatic switching in front of the distribution transformer. In case of feeder loss, automatic transfer to the other feeder is rapid and does not require fault locating before transfer.

Primary Loop Systems

The primary loop system, as illustrated in Figure 4 for overhead, is used where high service reliability is important (2,3,8). The feeder loops around a load area and

returns to the distribution substation, especially providing two-way feed from the substation. The size of the feeder conductors, which are kept the same throughout the loop, is usually selected to carry the entire load connected to the loop, including future load growth. Reclosers and sectionalizers are used to reduce customer interruptions and isolate faulted sections of the loop. The loop is normally operated with the tie switch (or tie recloser) open. Power to a customer at any one time is supplied through a single path from the substation, depending on the open/close status of the reclosers/sectionalizers. Each of the circuit breakers at the substation can be connected to separate bus sections and fed from separate transformers.

Figure 5 shows a typical primary loop for underground residential distribution (URD). The size of the cable, which is kept the same throughout the loop, is selected to carry the entire loop load, including future load growth. Underground primary feeder faults occur far less frequently than in overhead primaries, but are generally permanent. The duration of outages caused by primary feeder faults is the time to locate the fault and perform switching to isolate the fault and restore service. Fault locators at each transformer help to reduce fault locating times.

Primary Network Systems

Although the primary network system, as illustrated in Figure 6, provides higher service reliability and quality than a radial or loop system, only a few primary networks

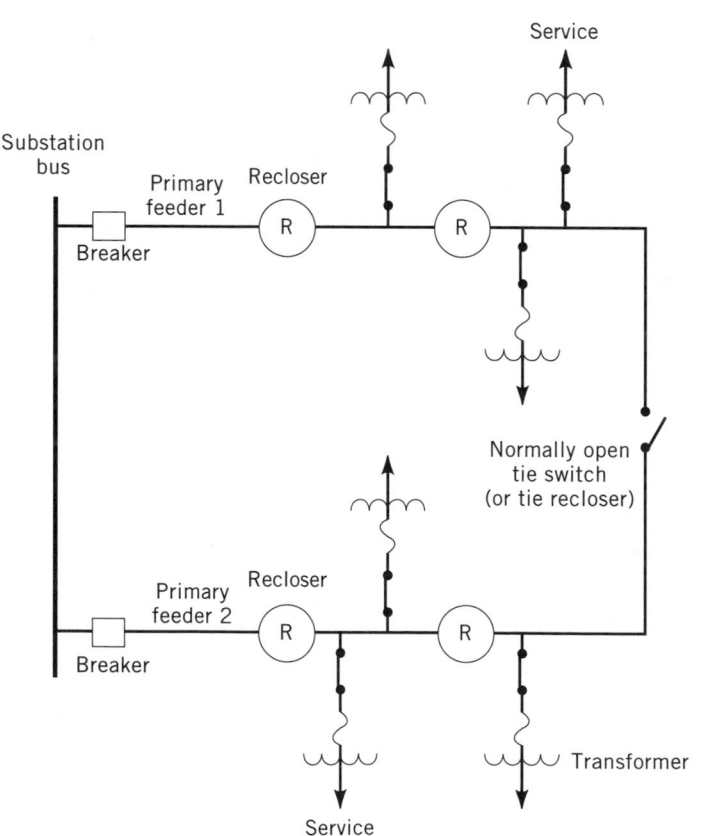

Figure 4. Overhead primary loop.

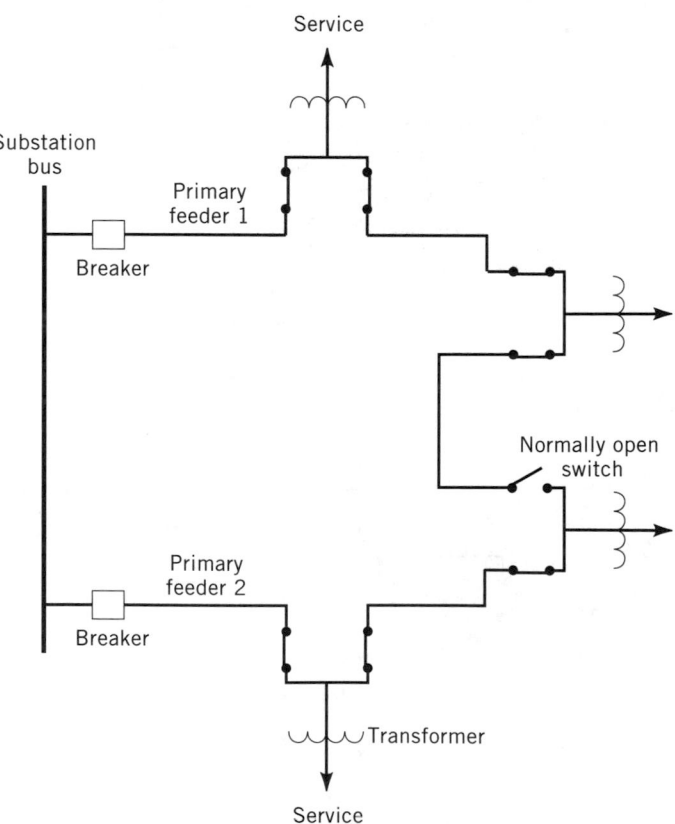

Figure 5. Underground primary loop.

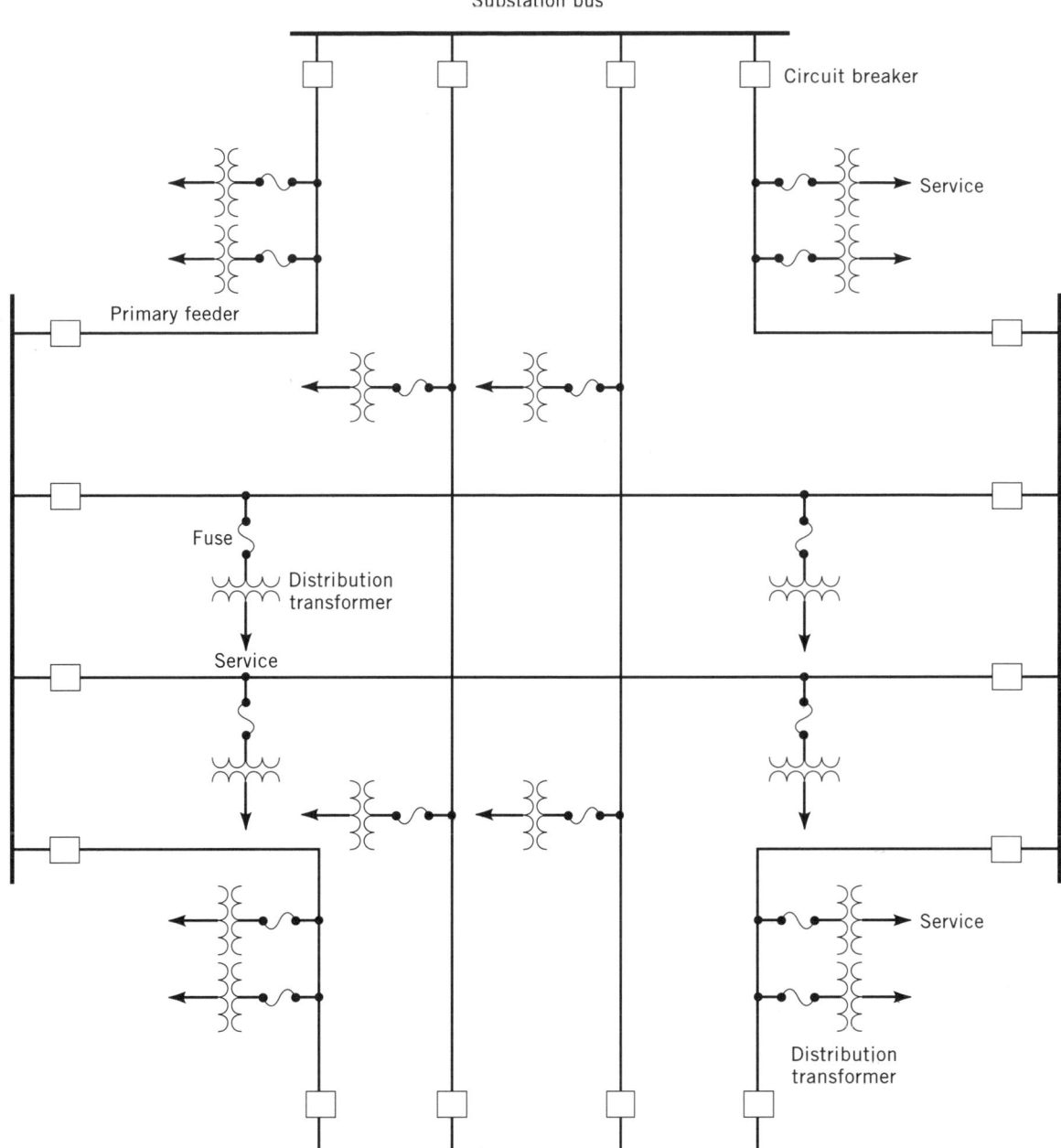

Figure 6. Primary network.

remain in operation today (2,3,8). They are typically found in downtown areas of large cities with high load densities. The primary network consists of a grid of interconnected feeders supplied from a number of substations. Conventional substations can be replaced by smaller, self-contained unit substations at selected network locations. Adequate voltage is maintained at utilization points by voltage regulators at substations and by locating distribution transformers close to major load centers on the grid. However, it is difficult to maintain adequate voltage everywhere on the grid under various operating conditions. Faults on interconnected grid feeders are cleared by circuit breakers at substations and, in some cases, by fuses on the grid. Radial primary feeders protected by circuit breakers or fuses can be tapped off the grid or connected directly at a substation.

SECONDARY DISTRIBUTION

Secondary distribution distributes energy at customer utilization voltages from distribution transformers up to meters at customers' premises. Table 2 shows typical secondary voltages and applications in the United States (1–7). In residential areas, 120/240-V, single-phase, three-wire service is the most common, where lighting loads and outlets are supplied by 120-V, single-phase connections, and large household appliances such as ranges,

Table 2. Typical Secondary Distribution Voltages in the United States

Voltage	# Phases	# Wires	Application
120/240 V	Single-phase	Three	Residential
208Y/120 V	Three-phase	Four	Residential/Commercial
480Y/277 V	Three-phase	Four	Commercial/Industrial/High Rise

clothes dryers, water heaters, and electric space heating are supplied by 240-V, single-phase connections. In urban areas serving high density residential and commercial loads, 208Y/120-V, three-phase, four-wire secondary networks are common, where lighting and small motor loads are supplied by 120-V, single-phase connections and larger motor loads are supplied by 208-V, three-phase connections. In areas with very high-density commercial and industrial loads as well as high-rise buildings, 480Y/277-V, three-phase, four-wire service is common, with fluorescent lighting supplied by 277-V, single-phase connections and motor loads supplied by 480-V, three-phase connections. A separate 120-V radial system fed by small transformers from the 480-V system supplies outlets in various rooms.

There are four general types of secondary systems:

1. Individual distribution transformer per customer
2. Common secondary main
3. Secondary network
4. Spot network

Individual Distribution Transformer Per Customer

Figure 7 shows an individual distribution transformer with a single service supplying one customer, which is common in rural areas where distances between customers are large and long secondary mains are impractical (2,3). This type of system may also be used for a customer that has an unusually large load or for a customer that would otherwise have a low-voltage problem with a common secondary main. Although transformer installation costs and operating costs due to no-load losses are higher than those of other types of secondary systems, the installation and operating costs of secondary mains are avoided.

Common Secondary Main

Figure 8 shows a primary feeder connected through one or more distribution transformers to a common secondary main with multiple services to a group of customers (2,3).

This type of system takes advantage of diversity among customer demands that allows a smaller capacity of the transformer supplying a group compared to the sum of the capacities of individual transformers for each customer in the group. Also, the large transformer supplying a group can handle motor starting currents and other, abrupt load changes without severe voltage drops.

In most cases, the common secondary main is divided into sections, where each section is fed by one distribution transformer and is also isolated from adjacent sections by insulators. In some cases, fuses are installed along a continuous secondary main, which results in banking of distribution transformers, also called banked secondaries.

Secondary Network

Figure 9 shows a secondary network or secondary grid, which may be used to supply high-density load areas in downtown sections of cities, where the highest degree of reliability is required and revenues justify grid costs (2,3,8). The underground secondary network is supplied simultaneously by two or more primary feeders through network transformers. Most networks are supplied by three or more feeders with transformers that have spare capacity, so that the network can operate with two feeders out of service. New York City has secondary networks fed by as many as 24 primary feeders operating in parallel (8).

Network transformers are protected by network protectors between the transformers and the secondary mains. A network protector is a circuit breaker with relays and auxiliary devices that opens to disconnect the transformer from the network when the transformer or the primary feeder is faulted, or when there is a power-flow reversal. Fuses may also be used for backup of network protectors. Also, special fuses called cable limiters are commonly used at tie points in the network to isolate faulted secondary cables.

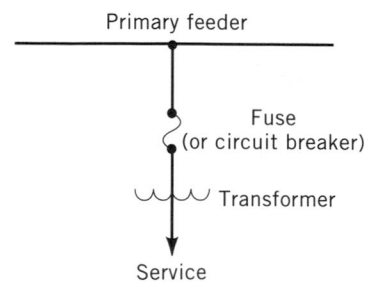

Figure 7. Individual distribution transformer supplying single-service secondary.

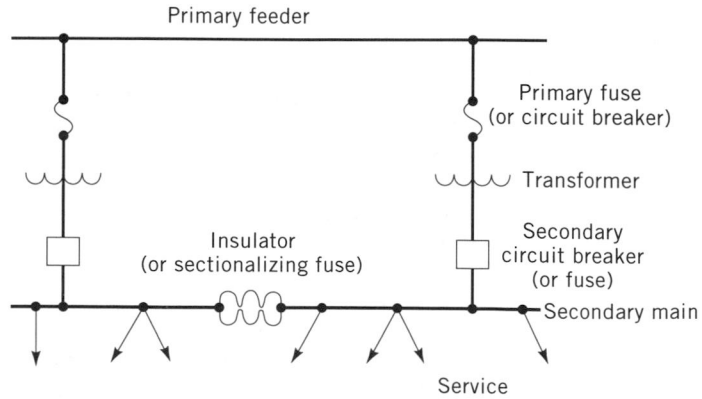

Figure 8. Common secondary main.

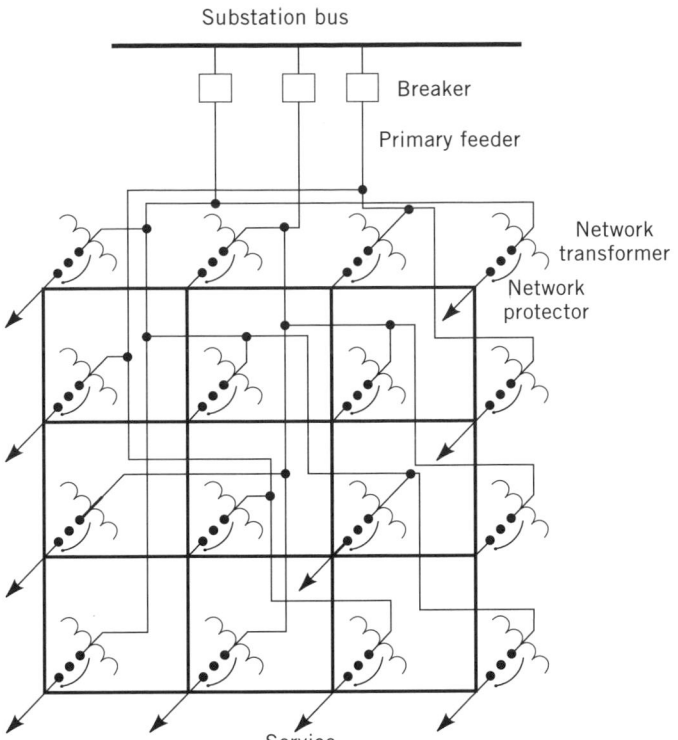

Figure 9. Secondary network.

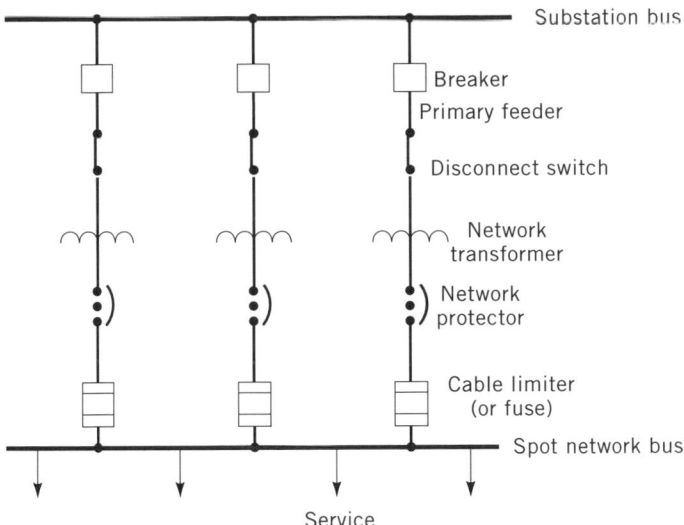

Figure 10. Secondary spot network.

In secondary network systems, a forced or scheduled outage of a primary feeder does not result in customer outages. Because the secondary mains provide parallel paths to customer loads, secondary cable failures usually do not result in customer outages either. Also, the network is designed to share the load equally among the transformers and to handle large motor starting and other abrupt load changes without severe voltage drops. In New York City, which has the largest network system in the United States with approximately 23,000 transformers feeding various secondary networks, an on-line monitoring system continuously monitors transformer loadings (8).

Spot Network

Figure 10 shows a spot network consisting of a secondary network supplying a single, concentrated load such as a high-rise building or a shopping center, where a high degree of reliability is required (2,3,8). The secondary spot network bus is supplied simultaneously by two or more primary feeders through network transformers. In some cases, a spot network load as large as 25 MVA may be fed by up to six primary feeders. Network protectors are used to disconnect transformers from the spot network bus for transformer/feeder faults or for power-flow reversal, and cable limiters are used to protect against overloads and faults on secondary cables.

DISTRIBUTION SOFTWARE

Computer programs are available for the planning, design, and operation of electric power distribution systems.

Program functions include

1. Capacitor placement optimization
2. Circuit breaker duty
3. Conductor and conduit sizing—ampacity and temperature computations
4. Database management
5. Distribution reliability evaluation
6. Distribution short circuit computations
7. Graphics for single-line diagrams and mapping systems
8. Harmonics analysis
9. Motor starting
10. Power factor correction
11. Power flow/voltage drop computations
12. Power loss computations and costs of losses
13. Protective (overcurrent) device coordination
14. Tie capacity optimization
15. Transformer sizing—load profile and life expectancy.

Some of the vendors that offer distribution software packages are given as follows:

- ABB Network Control Ltd., Switzerland
- Cooper Power Systems, Pittsburgh, Pennsylvania
- Cyme International, Burlington, Massachusetts
- EDSA Corporation, Bloomfield, Michigan
- Electrocon International Inc., Ann Arbor, Michigan
- Power Technologies Inc., Schenectady, New York
- Operation Technology, Irvine, California

DISTRIBUTION RELIABILITY

Reliability in engineering applications, as defined in the *IEEE Standard Dictionary*, is the probability that a device

Table 3. Example—1994 Reported Outage Data

Item	# Customers	kVA Connected	kVA Out	kVA-hours Out
Circuit	696	12,692	25,266	33,702
Town	6,984	95,926	144,081	210,078
District	147,565	1,625,068	2,812,993	2,502,605
System	630,261	6,902,756	11,065,118	11,617,338

will function without failure over a specified time period or amount of usage. In the case of electric power distribution, reliability concerns have come from customers who want uninterrupted continuous power supplied to their facilities at minimum cost (10–16).

A typical goal for an electric utility is to have an overall average of one interruption of no more than two hours' duration per customer per year. Given 8760 hours in a year, this goal corresponds to an Average Service Availability Index (ASAI) of:

$$\text{ASAI} \geq \frac{8758 \text{ service hours}}{8760} = 0.999772 = 99.9772\%$$

At present, there is no industrywide, uniform standard for computing reliability indices. The following formulas are typically used for reliability computations based on a given time period (12,13):

(1) Interruption Frequency (IF)

$$\text{IF} = \frac{\text{Total Interrupted Power}}{\text{Total Installed Transformer Capacity}}$$
$$= \frac{\text{Total kVA Out}}{\text{kVA Connected}}$$

(2) Interruption Duration (ID)

$$\text{ID} = \frac{\text{Total kVA-hours Out}}{\text{kVA Connected}}$$

(3) Average Service Availability Index (ASAI)

$$\text{ASAI} = \frac{(H - \text{ID})}{H} \times 100\%$$

Where H is the total number of hours in a given time period ($H = 8760$ hours for one year). In the preceding formulas, add the kVA out and add the kVA-hours out for all reported outages during the given time period to obtain the totals. Outages of duration of less than one minute are usually not reported. The formulas use distribu-

tion transformer kVA out and kVA-hours out data, which are often readily available to utilities. Alternatively, customers out and customer-hours out data may be used to compute reliability indices. The following example illustrates the use of the preceding formulas to compute reliability indices.

Example: The power system of a large electric utility is divided into districts, where each district is composed of several towns (municipalities), and each town has several circuits. Reported outage data during 1994 are given in Table 3 for one circuit, for one town, for one district, and for the entire system. Determine the IF, ID, and ASAI values for the circuit, the town, the district, and the system.

Solution: Using the circuit outage data in Table 3 and the preceding formulas:

$$\text{IF}_{\text{circuit}} = \frac{25266}{12692} = 1.991 \text{ outages/customer}$$

$$\text{ID}_{\text{circuit}} = \frac{33702}{12692} = 2.665 \text{ hours/customer}$$

$$\text{ASAI}_{\text{circuit}} = \frac{(8760 - 2.655)}{8760} \times 100 = 99.970\%$$

Similar calculations are made for the town, the district, and the system. The results are shown in Table 4. Assuming that the utility's 1994 goal is $\text{ASAI}_{\text{min}} = 99.9772\%$, the results show that the system and the district meet the goal, but the town and the circuit do not.

There are presently no industrywide, uniform standards for reporting outages. Table 5 lists basic outage reporting information recommended by an IEEE committee (14). Table 6 lists generic and specific causes of outages, based on a U.S. Department of Energy study (15). Electric utilities routinely prepare distribution outage reports monthly, quarterly, and annually by town (municipality) or by district. The purposes of the reports are to monitor and evaluate distribution reliability, uncover weaknesses and potential reliability problems, and make recommen-

Table 4. Example—Reliability Index Calculations

Item	IF, outages/customer	ID, hours/customer	ASAI, %
Circuit	1.991	2.655	99.970
Town	1.502	2.190	99.975
District	1.731	1.540	99.982
System	1.603	1.683	99.981

Table 5. Basic Outage Reporting Information[a]

1 Type, design, manufacturer, and other descriptions for classification purposes.
2 Date of installation, location on system, length in case of a line.
3 Mode of failure (short circuit, false operation, etc.)
4 Cause of failure (lightning, tree, etc.)
5 Times (both out of service and back in service, rather than outage duration alone), date, meteorological conditions when the failure occurred.
6 Type of outage, forced or scheduled, transient or permanent.

[a] Ref. 14.

dations for improving reliability. These reports may include

1. Frequency and duration reports, which provide data for the number of interruptions on distribution circuits together with power interrupted, average interruption duration, and causes.
2. Analysis reports that sort outages according to cause of failure and according to circuit classifications (for example, sort for each primary voltage; or sort for each conductor type including overhead open wire, overhead spacer cable, underground direct-burial cable, and cable in conduit).
3. 5- or 10-year trends for reliability indices, and outage trends for specific causes such as tree-contact outages for overhead distribution or dig-in outages for underground distribution.
4. Lists of problem circuits such as the 20 "worst" (lowest ASAI) circuits in a district, or all circuits with repeated outages during the reporting period.

Methods for improving distribution reliability include replacement of older distribution equipment, upgrades of problem circuits, crew staffing and training for fast response to outages and rapid restoration of service, formal maintenance programs, and public awareness programs to reduce hazards in the vicinity of distribution equipment such as contractor dig-ins. In addition, utilities have developed computer databases to track distribution equipment failure rates and failure causes. Reliability evaluation has also become an important component of bid selections to procure new distribution equipment. Recently, however, great strides in distribution reliability have come through distribution automation.

DISTRIBUTION AUTOMATION

Distribution automation is defined in a study by the Electric Power Research Institute (EPRI) as "an integrated systems concept for the digital automation of distribution substation, feeder and user functions. Distribution automation features include control, monitoring, and protection of distribution systems as well as load management and remote metering" (16).

Benefits of distribution automation include

- Improved distribution reliability
- Reduced customer outages and outage durations by automatically locating and isolating faulted sections of distribution circuits and automatically restoring service to unfaulted sections
- Reduced customer complaints
- Reduced power losses for substation transformers, distribution feeders, and distribution transformers

Table 6. Generic and Specific Causes of Outages[a]

Weather	Miscellaneous	System Components	System Operation
Blizzard/snow	Airplane/helicopter	Electric and mechanical:	System conditions:
Cold	Animal/bird/snake	Fuel supply	Stability
Flood	Vehicle:	Generating unit failure	High/low voltage
Heat	Automobile/truck	Transformer failure	High/low frequency
Hurricane	Crane	Switchgear failure	Line overload/transformer
Ice	Dig-in	Conductor failure	overload
Lightning	Fire/explosion	Tower, pole attachment	Unbalanced load
Rain	Sabotage/vandalism	Insulation failure:	Neighboring power system
Tornado	Tree	Transmission line	Public appeal:
Wind	Unknown	Substation	Commercial and industrial
Other	Other	Surge arrestor	All customers
		Cable failure	Voltage reduction:
		Voltage control equipment:	0–2% voltage reduction
		Voltage regulator	Greater than 2–8% voltage
		Automatic tap changer	reduction
		Capacitor	Rotating blackout
		Reactor	Utility personnel:
		Protection and control:	System operator error
		Relay failure	Power plant operator error
		Communication signal error	Field operator error
		Supervisory control error	Maintenance error
			Other

[a] Ref. 15.

- More effective use of distribution through automatic voltage control, load management, load shedding, and other automatic control functions
- Improved methods for logging, storing, and displaying distribution data
- Improved engineering, planning, operating, and maintenance of distribution.

The installation and operating costs of distribution automation equipment are justified by increased revenues from faster service restoration and by lower operating costs from reduced power losses and corresponding generator fuel savings. Lower operating costs are also achieved from reduced personnel time for repair and maintenance, meter reading, and handling customer complaints. In addition, investment savings come from deferral of additions or upgrades to generation, transmission, and substation facilities. Distribution automation can accomplish all these cost savings through more effective utilization of distribution (16,17).

Candidate distribution automation functions are listed below (16).

Automatic Control
   Switching and Sectionalizing
      Overload detection
      Fault location
      Fault isolation
      Service restoration
      Circuit reconfiguration
   Switching and Sectionalizing
      Voltage regulators
      Transformer load tap changers
      Switched shunt capacitors
   Substation Transformer Load Balancing
   Cold Load Pickup
Data Acquisition and Processing
   Data Monitoring
      Bus voltages
      Equipment loadings
      Circuit configurations
      Sensors (CTs, VTs, equipment operating temperatures and pressures, liquid levels)
      Alarms
      Relay settings
   Data Logging
      Sequence of events
      Relay targets
Load Management
   Load control
   Remote service disconnect/reconnect
   Load shedding
Remote Metering
   Load survey
   Peak demand metering
   Remote meter reading
   Remote meter programming
   Tamper detection
Computer Databases and Graphics
   Database management

Automated mapping/facilities management (AM/FM)
Global positioning system (GPS)
Reports
Single line diagrams

These functions involve actions such as opening and closing circuit switching devices, changing transformer taps, switching capacitor banks, monitoring voltages and equipment loadings, and reading customers' meters.

The main components of a distribution automation system are computer hardware and software, communications systems, remote terminal units, and distribution devices that are monitored or controlled. These components are briefly discussed as follows (16,17).

Computers at distribution dispatch centers receive and process data to execute the automated functions given earlier. Supervisory control and data acquisition (SCADA) may be interfaced with automated mapping and facilities management (AM/FM), giving dispatchers the ability to locate equipment outages geographically with the real-time monitoring system. Computers at distribution dispatch centers may also interface with computers at a utility's main offices, so that when a customer calls the main office to report an outage, a service representative can use AM/FM to associate the customer's address with a corresponding feeder and inform the customer of restoration status.

A number of communication options are available for distribution automation, including distribution line carrier, radio (UHF, VHF, FM, and AM), telephone, fiber optics, satellite, and cable TV. Distribution automation requires two-way communication for both data acquisition and control. Functional requirements including communication reliability, security, speed, and cost are factors included in selection of the optimum communication system for each automated distribution function.

Remote terminal units (RTUs) provide a communications interface between master stations and devices that are monitored or controlled. RTUs are installed at substations, on poles, and at customers' premises. The RTU communications interface interprets messages from the master and formats messages to be sent to the master, in order to supply data or execute a control command. In addition to communications interface, RTUs also contain digital input modules to monitor contact positions of switches, breakers, and relays, and analog input modules with A/D converters to monitor transducers (eg, CTs and VTs). Also, digital output modules send TRIP and CLOSE commands to switches and relays. RTUs also contain data processing and control processing modules, and in some cases modules for self diagnostics, user interface testing, and database management. Large RTUs at distribution substations are also used to control a number of small RTUs along distribution circuits.

Distribution automation monitoring and control devices are listed below (16).

Alarms
Fault detectors
Meters
Reclosers

Recorders
Relays
  Auxiliary relays
  Circuit breaker relays
Switches
  Capacitor bank switches
  Load break switches
  Load control switches
  Sectionalizer switches
  Transformer load tap changer (LTC) switches
Transducers
  Ambient temperature transducers
  Current transducers
  Current transformers (CTs)
  LTC position indicators
  Transformer top oil temperature transducers
  Var transducers
  Voltage transducers
  Voltage transformers (VTs)
  Watt transducers

These devices include primary and secondary devices such as fault detectors, reclosers, relays, switches, transducers, and end-use customer devices such as meters and load-control switches.

Design of the distribution automation system requires a flexible or "open" architecture that accommodates various types of equipment from different suppliers. The products encompassing the computer hardware and software, communications, and monitored/controlled devices and RTUs are too numerous for any one vendor to supply.

BIBLIOGRAPHY

1. D. G. Fink and H. W. Beaty, *Standard Handbook for Electrical Engineers*, 11th ed., McGraw-Hill, New York, 1978, Sec. 18.

2. T. Gonen, *Electric Power Distribution System Engineering*, Wiley, New York, 1986.

3. A. J. Pansini, *Electrical Distribution Engineering*, 2nd ed., The Fairmont Press, Liburn, Ga., 1992.

4. Various co-workers, *Electric Distribution Systems Engineering Handbook, Ebasco Services Inc.*, 2nd ed., McGraw-Hill, New York, 1987.

5. Various co-workers, *Electrical Transmission & Distribution Reference Book*, Westinghouse Electric Corporation, Pittsburgh, 1964.

6. Various co-workers, *Distribution Systems Electric Utility Engineering Reference Book*, Vol. 3, Westinghouse Electric Corporation, Pittsburgh, 1965.

7. Various co-workers, *Underground Systems Reference Book*, Edison Electric Institute, New York, 1957.

8. R. Settembrini, R. Fisher, and N. Hudak, "Seven Distribution Systems: How Reliabilities Compare," *Electrical World* **206:5**, 41–45 (May 1992).

9. J. L. Blackburn, *Protective Relaying*, Marcel Dekker, New York, 1987.

10. R. Billinton, *Power System Reliability Evaluation*, Gordon and Breach, New York, 1988.

11. R. Billinton, R. N. Allan, and L. Salvaderi, *Applied Reliability Assessment in Electric Power Systems*, Institute of Electrical and Electronic Engineers, New York, 1991.

12. R. Billinton and J. E. Billinton, "Distribution Reliability Indices," *IEEE Transactions on Power Delivery*, **4:1**, 561–568 (January 1989).

13. D. O. Koval and R. Billinton, "Evaluation of Distribution Circuit Reliability," Paper F77 067-2, *IEEE Winter Power Meeting*, New York, NY, (January/February 1977).

14. IEEE Committee Report, "List of Transmission and Distribution Components for Use in Outage Reporting and Reliability Calculations," *IEEE Trans. on Power Apparatus and Systems* **PAS 95:4**, 1210–15 (July/August 1976).

15. U.S. Department of Energy, *The National Electric Reliability Study: Technical Study Reports*, DOE/EP-0003, April 1981.

16. J. B. Bunch and co-workers, *Guidelines for Evaluating Distribution Automation*, EPRI-EL-3728, Project 2021-1, Electric Power Research Institute, Palo Alto, CA, November, 1984.

17. T. Desmond, "Distribution Automation: What is it?, What does it do?," *Electrical World* **206:2**, 56–57 (February 1992).

ELECTRIC POWER GENERATION

Sheppard Salon
Rensselaer Polytechnic Institute
Troy, New York

When we switch on the lights at home we are the end users of electric power. This power was generated in a power plant by converting mechanical power from a turbine into electricity and then transmitted from the power plant to our homes. The electricity was generated by a synchronous machine which was driven by a turbine. On a global basis about 20% of the worlds electricity supply comes from turbines driven by falling water (hydro power), about 17% from steam turbines in which the steam is produced by nuclear fission (nuclear power), and almost all the rest from steam or gas turbines driven by heat produced by burning fossil fuels (coal, oil, and natural gas) (1). If the power system is operating in steady state, the power being consumed is equal to the power being generated. Due to our mode of living and working there are typical load curves which have daily, weekly, and annual cycles. These curves illustrate the use of electricity or the load versus time.

In the United States it is estimated that 40% of all energy use goes into the production of electricity and this figure continues to rise. The widespread use of electricity for transportation, either automobiles or mass transit, could cause this percentage to rise dramatically.

See also Coal combustion; Commercial availability of energy technology; Gas turbines; Geothermal energy; Hydropower; Renewable resources.

ELECTROMECHANICAL ENERGY CONVERSION

The conversion of mechanical energy from the prime mover (usually a turbine) to electrical energy takes place in the electric generator. To see how this change of energy forms takes place consider the two coils of Figure 1.

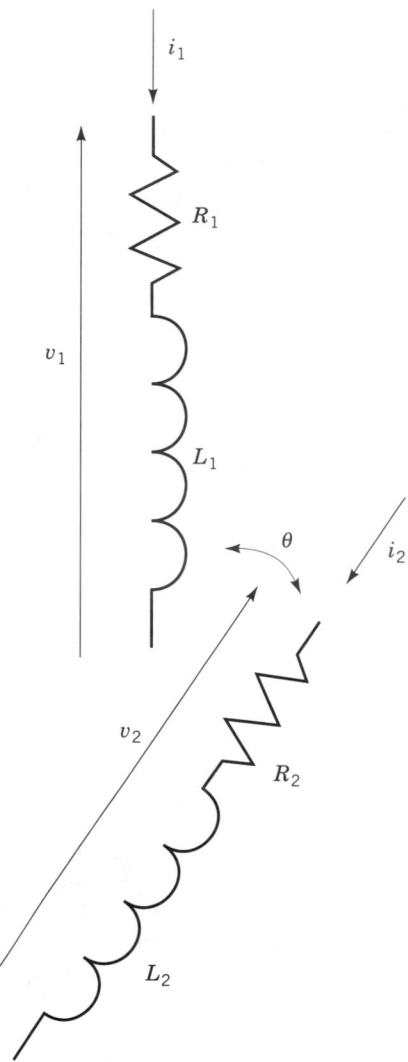

Figure 1. Two coils in relative motion.

Coil 1 has resistance R_1 and self-inductance L_1 and coil 2 has resistance R_2 and self-inductance L_2. The coils have mutual inductance M which depends on the relative position of the coils.

The magnetic flux linkages of the coils can be expressed as

$$\psi_1 = L_1 i_1 + M i_2$$
$$\psi_2 = L_2 i_2 + M i_1 \tag{1}$$

The voltage across the coils is then

$$v_1 = i_1 R_1 + \frac{d\psi_1}{dt}$$
$$v_2 = i_2 R_2 + \frac{d\psi_2}{dt} \tag{2}$$

The energy stored in the magnetic field is

$$W = \tfrac{1}{2} L_1 i_1^2 + \tfrac{1}{2} L_2 i_2^2 + M i_1 i_2 \tag{3}$$

In terms of the flux linkages

$$W = \tfrac{1}{2} (\psi_1 i_1 + \psi_2 i_2) \tag{4}$$

If the stored energy changes from one value to another, the time rate of change of that stored energy is the instantaneous power required to change energy states.

$$P = \frac{dW}{dt}$$

$$P = L_1 i_1 \frac{di_1}{dt} + L_2 i_2 \frac{di_2}{dt} + M\left(i_2 \frac{di_1}{dt} + i_1 \frac{di_2}{dt} \right)$$
$$+ \frac{1}{2} i_1^2 \frac{dL_1}{dt} + \frac{1}{2} i_2^2 \frac{dL_2}{dt} + i_1 i_2 \frac{dM}{dt}$$

The total electrical power is the product of voltage and current for each of the circuits.

$$P_e = R_1 i_1^2 + R_2 i_2^2 + L_1 i_1 \frac{di_1}{dt} + L_2 i_2 \frac{di_2}{dt}$$
$$+ M\left(i_1 \frac{di_1}{dt} + i_2 \frac{di_2}{dt} \right) + i_1^2 \frac{dL_1}{dt}$$
$$+ i_2^2 \frac{dL_2}{dt} + 2 i_1 i_2 \frac{dM}{dt}$$

The first two terms represent the power dissipated in the resistances R_1 and R_2. The remaining terms do not equal the power required to change the energy stored in the inductances. In other words there is some power left over. The difference in these powers is due to the mechanical work done on or by the system. The total mechanical power required to produce a change in the inductances is therefore,

$$P_{\text{mech}} = -\frac{1}{2}\left(i_1^2 \frac{dL_1}{dt} + i_2^2 \frac{dL_2}{dt} + 2 i_1 i_2 \frac{dM}{dt} \right) \tag{5}$$

If the self-inductances are constant, then the mechanical power is

$$P_{\text{mech}} = -i_1 i_2 \frac{dM}{dt} \tag{6}$$

The change in mutual inductance must be due to a change in position for electromechanical energy conversion to take place. A change in inductance due to saturation, for example, will not result in work being done by or on the circuit.

SYNCHRONOUS GENERATORS

Essentially all of the world's electric power is generated by synchronous machines. The synchronous generator has proven to be a reliable and efficient device for converting mechanical power to electric power. Since the typical power system uses alternating current (60 Hz in the U.S.), the chief requirement of such a device is that it produces a constant and controllable voltage at a constant frequency. Some early power systems used direct current. The development of devices such as the power transformer, which made high voltage transmission possible, made a-c generation much more attractive. In recent times dc has been used in large transmission projects but the power is still generated as ac and then converted to dc for transmission.

Many configurations of a synchronous machine are possible but the most common synchronous machine consists of a rotor with a d-c winding and a stator with a three-phase a-c winding. The rotor has a d-c power supply and the stator is connected to the power system through a generator step up transformer.

The d-c winding on the rotor is distributed in such a way as to produce a magnetic field that is approximately sinusoidally distributed around the air gap. In a round rotor machine or turbine generator, the rotor conductors are placed in slots that are machined in a solid magnetic steel forging. These conductors are held in place by metallic wedges at the top of the slot which also help form a damper cage. Turbine generator rotors are cooled with atmospheric air or with high pressure hydrogen. Depending on the design, the rotor conductors can be conventionally cooled, in which case the conductor is not in direct contact with the cooling gas. The heat generated in the winding must pass through the insulation and into the tooth where it enters the cooling gas.

For large machines, the rotor winding may be inner cooled. In this case the cooling gas (high pressure hydrogen) passes through channels in the copper winding. By avoiding the high thermal resistance of the insulation, the temperature rise between cooling fluid and copper is much smaller and the allowable current density is much higher.

Experimental turbine generators have been constructed with superconducting rotor windings. As of 1993 there had been no commercial plants operating with superconducting generators.

The stator consists of a core made of laminated magnetic steel, with slots cut out to accept a three-phase winding. The three phases, A, B, and C are offset by 120° electrical. Electrical degrees are measured relative to the pole pitch in the machine. Each two poles consist of 360° electrical. For a two-pole machine electric degrees and mechanical degrees are the same. For a four-pole machine the circumference of the machine consists of 720° electrical. The stator conductors may also be conventionally or indirectly cooled, or inner cooled with hydrogen or water. The stator conductors are normally thin strands to reduce eddy current losses caused by the alternating magnetic fields.

To see how the machine operates, we first assume that the stator winding is open circuited. As the rotor turns, the magnetic field produced by the rotor induces a voltage in the stator windings. This voltage is approximately sinusoidal in time and due to the relative positions of the phases, the voltage in phase B lags the voltage in phase A by 120° and the voltage in phase C lags the voltage in phase A by 240°. This voltage, due to the rotor dc current alone, is called the internal voltage. Thus the turbine generator produces a sinusoidal three-phase voltage.

If the stator windings are connected to a load, current flows through the windings and the load. The magnetic field due to this load current produces a sinusoidal field in the air gap similar in spatial distribution to the rotor produced field. This magnetic field distribution rotates at the same speed as the rotor. For a given operating point (load and power factor), the stator-produced field and rotor-produced field are therefore always in the same relative position. If the machine is delivering load to the system (ie, normal operation) the magnetic field produced by the stator produces a torque on the rotor to oppose the rotation. As the electrical load increases, the prime mover (turbine) must expend more mechanical energy to keep the rotor turning at a constant speed. As the rotor turns there is a change in mutual inductance with position between the stator and rotor windings. Thus mechanical energy input by the turbine is being transformed into electrical energy.

Generators in hydro-plants are also synchronous machines, but due to the lower speed of the hydro-turbines these machines typically have a large radius and large number of poles. Mainly because of the low speed, the rotors of hydro-generators are salient pole instead of the round rotor construction. The salient pole is constructed of magnetic steel laminations. The d-c field winding is wound around the salient pole and is held in place by the overhang of the pole. The rotor is usually air cooled. The stator is similar in construction to the round rotor machines with a laminated steel core and three-phase winding. The stator is either air or water cooled.

Regulation of Power

The electrical power produced by a synchronous generator is almost equal to the mechanical power input, the efficiency being in the range of 98%. The division of electric load among a number of generators is determined by a number of factors, including economy. At a given operating point each turbine generator has an incremental fuel cost, which is the cost per kWh to generate an additional small amount of power. When this is included with other factors, such as the transmission losses, any increase in load to the system is allocated by an algorithm which minimizes the cost of producing the extra energy.

The control of the real power and regulation of the speed (which must be held constant to provide a constant frequency) is done with the speed governor. A simple configuration is shown in Figure 2. There are two basic classifications of governors: isochronous and droop. An isochronous governor holds the speed constant regardless of the load. In a droop governor the speed changes (slightly) as the load changes. Droop governors are used in electric power plants since loads are more easily shared among a number of generators with a droop characteristics. The simple Watt governor of Figure 2 operates as follows. Assume that the turbine is operating at steady state speed and load. If the speed drops, the fly-balls move inward. This causes point A to move upward. With the linkage arrangement shown, point B moves downward, opening the steam valve. This eventually increases the speed. A

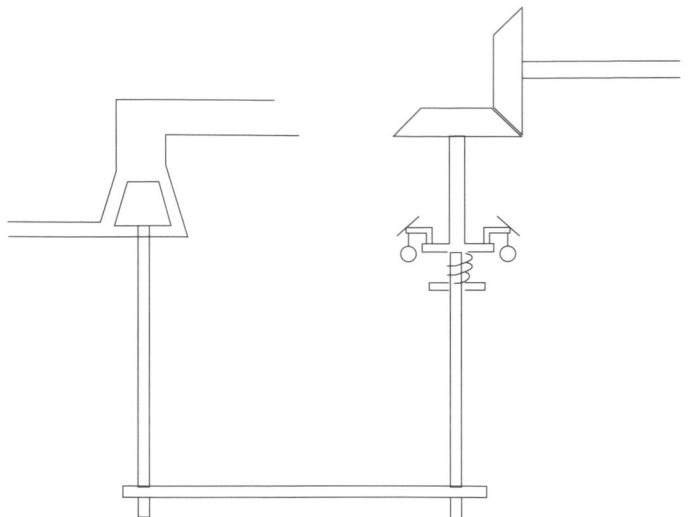

Figure 2. Simple watt governor.

similar argument shows that as the speed increases the valve closes. A modern governor, usually electro-hydraulic, is more complicated, including hydraulic amplifiers, servo systems, damping, and possibly a number of auxiliary signal inputs for stability considerations.

Regulation of Reactive Power

Real and Reactive Power. There are two distinct types of power produced by the generator. Real power, measured in watts, is available to do work or produce heat and comes directly from the conversion of mechanical power into electrical power. Real power is sometimes called active power. Another form of power is the reactive power measured in volt-amperes. Although reactive power has no work equivalent, it is needed to maintain the voltage in the power system at desired levels and to run devices such as motors, which make up most of the power system load. The synchronous generator is capable of being a producer or an absorber of this reactive power. To see where this concept comes from consider the basic definition of instantaneous power:

$$p(t) = v(t)i(t) \qquad (7)$$

Assume that both the voltage and current are sinusoidal in time:

$$v(t) = \sqrt{2}V \cos(\omega t)$$
$$i(t) = \sqrt{2}I \cos(\omega t + \theta) \qquad (8)$$

where θ is the angle by which the current lags the voltage. Then the instantaneous power is

$$p(t) = 2VI \cos(\omega t) \cos(\omega t + \theta) \qquad (9)$$

Using the trigonometric identity

$$\cos \alpha \cos \beta = \frac{1}{2}(\cos(\alpha + \beta)) + \cos(\alpha - \beta)) \qquad (10)$$

we obtain

$$p(t) = VI \cos\theta + VI \cos(2\omega t + \theta) \qquad (11)$$

The average power, P, is then

$$P = VI \cos \theta \qquad (12)$$

This is what is usually meant by real power. The apparent power, S, is the product of the RMS voltage and current.

$$S = VI \qquad (13)$$

The ratio of the real power to the apparent power is called the power factor and is equal to $\cos \theta$.

The instantaneous power as the sum of two sinusoidal quantities can be written

$$p(t) = VI \cos \theta(1 + \cos(2\omega t)) + VI \sin \theta \sin(2\omega t) \qquad (14)$$

The second term is the instantaneous reactive power. The peak value, $VI \sin \theta$, is called the reactive power Q. This power has no average value and physically represents the time rate of change of energy stored in the magnetic field (inductive) or in the electric field (capacitive).

Steady-State Generator Performance

The control of reactive power is accomplished by means of the rotor field winding current. For a given terminal voltage and real power output, the stator current can be represented by a set of phasors whose tips lie on the straight line perpendicular to the terminal voltage.

In the phasor diagram of the synchronous machine using the angle of the terminal voltage of the machine as the reference angle, we see the current I_a lagging the voltage by the power factor angle θ. In the convention of generators, this is the case in which the generator is delivering reactive power to the system. The internal voltage E_i is generated by the d-c field winding. This is the voltage which would appear at the generator terminals if the stator current ceased to flow and the field current remained the same. The voltage difference between the internal voltage and the terminal voltage is due to the load current. There is a resistance drop, I_aR, in phase with the stator current, and a reactance drop, I_aX_s, perpendicular to the stator current. The reactance X_s is called the synchronous reactance and is the equivalent inductive reactance of each phase of the generator when it is operating in steady state. As is shown, for constant terminal voltage, if the real power (input from the turbine) remains constant, the stator current magnitude and angle can be changed by adjusting the internal voltage, which is proportional to the field current. Say we increase the field current. The stator current magnitude and the power factor angle increase in such a way that the real power ($VI \cos \theta$) remains the same. The reactive power ($VI \sin \theta$) delivered to the system goes up. By a similar argument if the rotor current decreases the reactive power decreases. The rotor current is controlled by a d-c voltage source (excitation system) and the reactive power generated or absorbed can be varied by field voltage adjustment.

Capability Curve. The limits within which the turbine generator can supply real and reactive power are illustrated in the capability curve. The curve is constructed for the case of constant terminal voltage. It is made of three different zones. The region where the generator is exporting reactive power is sometime called the overexcited or lagging power factor region. The reactive power is increased as the field current is increased. At some point the thermal limit of the rotor winding is reached. One section of the capability curve is a locus of constant field current, as the stator current and power factor change. The next section of points represents constant stator current. At some point on this curve all the power is real (unity power factor). In the region below the real axis the generator is absorbing reactive power. As the power factor changes to leading certain components of the end region, magnetic fields increase and stray losses in the end region thermally limit the operation.

STEAM GENERATION

The steam generation cycle begins when water is pumped into the boiler. The water absorbs heat from the furnace forming wet steam. Normally this liquid-vapor mixture flows through a series of tubes to increase the surface area for the heat exchange (watertube design). It is also possible to have the hot combustion gases flow through tubes to heat the vapor (firetube design). If this wet steam were used to drive the turbine, condensation would take place in the last stages. Further, the overall efficiency of the cycle would be rather low. Therefore, the steam is then heated again in a superheater which raises the temperature of the steam above it saturation point.

After the steam passes through the turbine it is changed back into water in a condenser and reused. The heat removed from the vapor is discharged to the atmosphere. To increase the overall cycle efficiency a feedwater heater is used to preheat the water before it goes back into the pump and the boiler.

STEAM TURBINES

There are two basic classifications of steam turbines, impulse turbines and reaction turbines. In an impulse turbine the steam flows through stationary nozzles against buckets mounted on the periphery of a rotating disk. When the nozzles are located on the rotating disk this is called a reaction turbine. In a modern power plant the turbines may be a combination of the two having impulse stages and reaction stages. In an impulse stage, the steam enters and passes through the stationary nozzles. Moving blades or buckets mounted on the rim of a rotating disk absorb the kinetic energy of the steam converting it to shaft rotation. The steam, now flowing in the axial direction, may enter a number of impulse stages where its energy is absorbed in a number of smaller pressure drops. This is called velocity compounding.

In the reaction stage, the steam is diverted through a row of stationary blades and its velocity with respect to a moving set of blades is increased. This creates the force which turns the turbine. The high temperature and enormous forces necessitate that the mechanical design be very exacting. High strength materials have been specially developed for these applications. The dovetail mounting ensures a strong mechanical coupling between the rotor and the turbine blade. In large power plants many stages of turbines, sometimes with reheating between stages is used.

STEAM PLANTS

In a steam turbine generating plant, a fossil fuel is burned and the heat is used to produce steam. This steam drives a turbine which in turn drives a turbine generator. Steam turbines are heat engines and as such are limited in efficiency by the Carnot efficiency, which is the highest efficiency possible for a system such as this which turns heat energy into mechanical energy. The Carnot efficiency, E, is given by

$$E = \frac{T_{in} - T_{out}}{T_{in}} \quad (15)$$

where in our case T_{in} is the temperature of the steam input to the turbine in degrees Kelvin and T_{out} is the temperature of the exhaust. Taking typical numbers for a utility steam turbine, the steam entering the turbine may be 811 K (1000°F). The exhaust may be 311 K (100°F). This gives an efficiency of 61%. Although this is the theoretical limit, in practice the turbine efficiency is usually somewhat less than 50% (2). To find the overall efficiency of a power plant this efficiency is multiplied by the efficiency of the boiler and by the efficiency of the generator. Again using typical numbers a boiler efficiency of 88% and a generator efficiency of 98% are used to find an overall plant efficiently in the range of 40%. A range of between 30 and 35% has been given (3). It is common to classify these thermal plants in terms of heat rate. The heat rate is defined as the heat input from fuel divided by the electrical output of the plant. These are commonly expressed in units of Btu/KW·h (one kilowatt hour of energy is equivalent to 3412 Btu). Therefore high efficiency plants have lower heat rates than low efficiency plants. A typical number is 10,500 Btu/kW·h. In nuclear plants the steam temperature is somewhat lower than in a fossil fuel plant, eg, 623 K (2). The overall efficiency of the nuclear plant is in the range of 30%.

A steam power plant has the following basic structure which is illustrated in Figure 3. Fossil fuel and air are mixed in a furnace where combustion takes place. The heat produced in the furnace is used to boil water in the boiler converting it into steam. This steam then flows through a turbine where the energy in the steam is converted to mechanical energy. The turbine drives the generator (synchronous machine) which converts the mechanical energy to electrical energy. The steam, after leaving the turbine, goes through a condenser stage which cools the steam and therefore produces a greater pressure

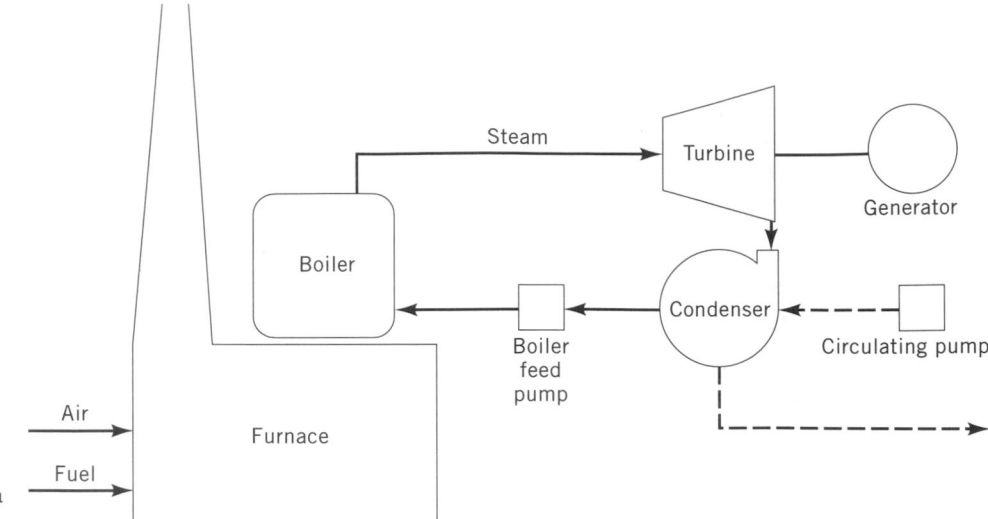

Figure 3. Basic configuration of a steam plant.

drop across the turbine. The condensed water is then pumped back into the boiler.

The steam turbine used in power plants operates on a Rankine cycle shown in Figure 4. Liquid water at low pressure leaves the condenser where it flows through a pump (path 1–2) where its pressure is raised to the turbine inlet pressure. The water is then heated at constant pressure in the boiler to its saturation temperature and then at constant temperature until it is a 100% saturated vapor (path 2–3). The steam can then either pass through the turbine (path 3–4) or through a superheater (path 3–3a) and then through the turbine path (3a–4a). After expanding through the turbine the steam passes into a condenser and emerges as a saturated liquid (path 4 or 4a–1). One problem that arises is that as the saturated vapor travels through the turbine, its quality is reduced (larger percentage of liquid). If the liquid exceeds more than a few per cent of the weight, severe erosion of the turbine blades can result. One solution is to extract the steam after it passes through the high pressure turbine and reheat it, then expand it in the low pressure stages of the turbine.

GAS TURBINES

An alternative and complement to steam turbines, gas turbines have enjoyed a resurgence in recent years (Fig. 5). Whereas a gas turbine normally has a higher cost than a steam turbine, a number of factors make them attractive.

Gas turbines are often used as peaking plants and have the following advantages:

1. The cost of installing a gas turbine plant is lower than installing a steam turbine plant.
2. The units have fast starting capability.

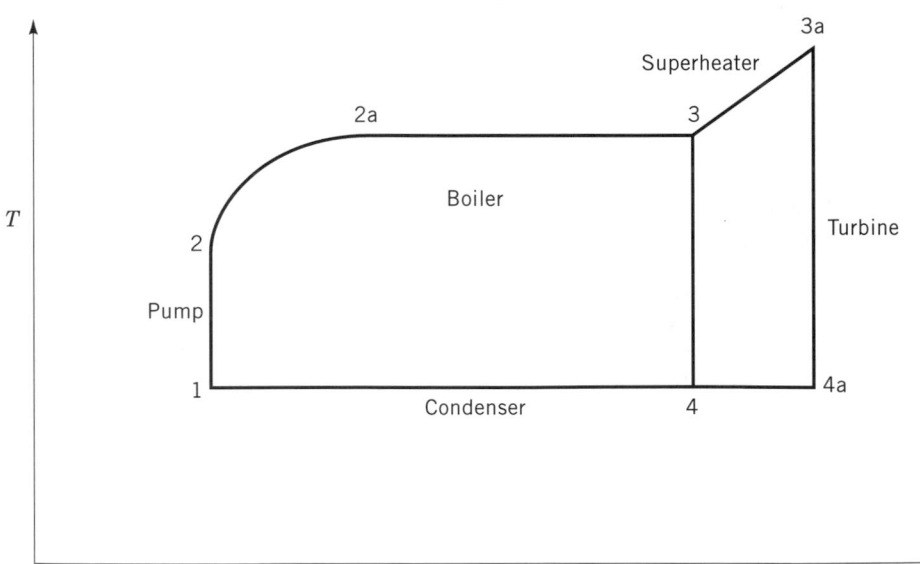

Figure 4. Rankine cycle.

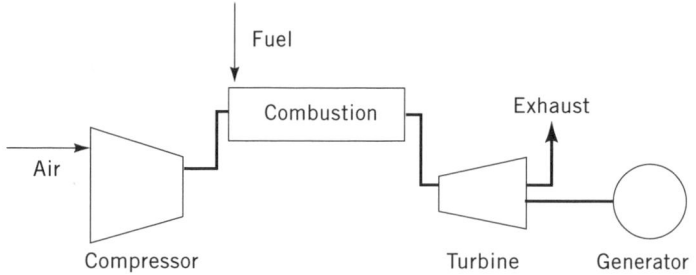

Figure 5. Simple cycle gas turbine.

3. The units have a short delivery and installation time.
4. The units can pick up load very quickly.
5. The units have a black start capability.

Although there are many implementations of the gas turbine, in its basic configuration atmospheric air is drawn into a rotary compressor. The compressor increases the pressure and temperature of the air. The air then enters the combustor where it mixes with fuel and burns at approximately constant pressure. The fuel may be gas or liquid or a combination of the two. The combustion gases then enter the turbine. The combustion gases expand in the nozzle increasing kinetic energy. This gas then delivers its kinetic energy to the blades or buckets of the turbine which drive the generator. The gases then exit at atmospheric pressure. Modern gas turbines have efficiencies competitive with steam turbines.

The gas turbine operates on the Brayton cycle illustrated in Figure 6. Line 1–2 is the compression of air. Line 2–3 is the constant pressure combustion. Line 3–4 is the expansion of the gas through the turbine. Line 4–1

is the constant pressure discharge to atmospheric air. This is actually an open system.

Practical considerations such as noise abatement, air filtering, and exhaust heat recovery add pressure drops to the system. These result in a drop in the power output or, correspondingly, an increase in the heat rate.

The gas turbine is also used in a combined cycle plant along with a heat recovery system and a steam turbine. These plants have excellent heat rates. These gases still have a lot of heat energy and much of this energy can be recovered as illustrated below in the combined cycle plant (see Fig. 7).

Combined cycle plants use a simple cycle gas turbine. Instead of exhausting the high temperature gas output of the gas turbine to the atmosphere, the combined cycle plant uses the exhaust in heat recovery steam generators (HRSG) which drives a separate steam turbine generator. These plants can achieve an efficiency of approximately 50%.

NUCLEAR POWER

Nuclear power provides an alternative to the furnace described above under fossil fuel. In nuclear power plants the heat is generated by the nuclear reaction (fission) and used to heat water and produce steam. The fuel used is enriched uranium containing between 2–4% of U_{235}. In the United States the nuclear power plants use light (normal) water reactors. Heavy water is an isotope in which the hydrogen atoms in the water have a proton and a neutron. The water is used both to slow down the fast neutrons produced by the nuclear reaction and to cool the reactor and produce the steam necessary for the turbine.

There are two principal types of reactors used in the United States today: the pressurized water reactor and the boiling water reactor. Figure 8 gives a schematic of the pressurized water reactor.

In the pressurized water reactor the water is pumped through the reactor by a reactor coolant pump. This forms a closed loop referred to as the primary loop. In the secondary loop, the feedwater pump circulates water through a heat exchanger where it is turned to steam by heat from the primary loop. This steam then flows through the turbine driving a synchronous generator.

In the boiling water reactor, there is only one loop. The boiling water reactor generally has a higher overall efficiency then the pressurized water reactor, gained at the expense of having the turbine become radioactive.

HYDRO-GENERATORS

Hydro-generators use the potential energy stored in water to produce electricity. Water falls through a turbine and turns a synchronous generator. The arrangement of most hydro-generators is with vertical shafts as opposed to steam or gas turbine generators which have horizontal shafts. Some exceptions are so called low head hydro-generators usually defined as less than 66 feet of head (4). The energy in the water is proportional to the weight of

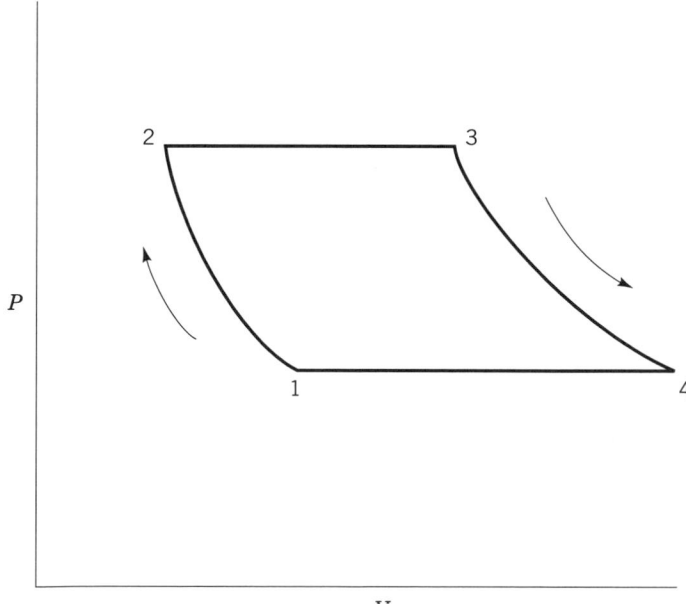

Figure 6. Brayton cycle.

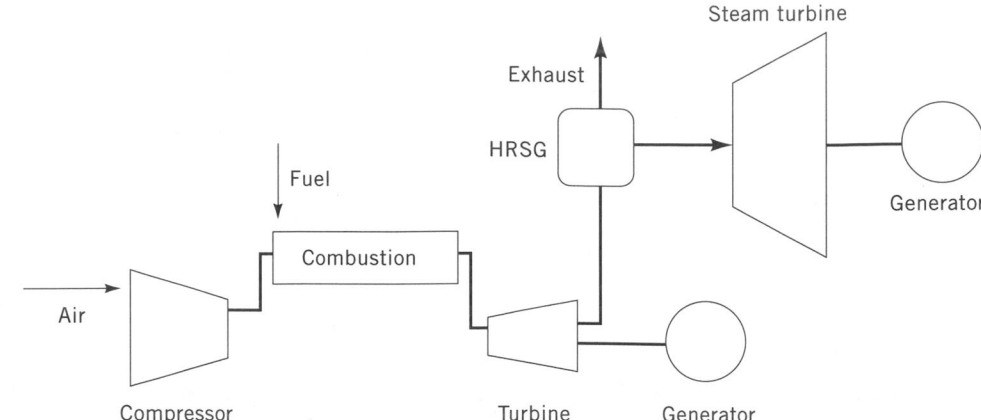

Figure 7. Configuration of a combined cycle plant.

the water through the turbine times the head or vertical distance the water travels to the turbine. The power is proportional to the head times the volumetric flow rate (m³/s) through the turbine.

Pumped Storage

Pumped storage plants are designed so that during times of light load and inexpensive power from the rest of the system, water can be pumped back into a reservoir to be later run through the turbine and produce power at times of greater load. For example, during the evening when the system load is light it may be economical to use electricity produced by nuclear power to pump water into a reservoir. That water is then used during the high load period during the day instead of burning, for example, oil or natural gas in peaking units. The energy lost in pumping the water up only to let it flow through the turbine later is typically in the range of 30%. In early pumped storage plants, separate motor driven pumps were used when the plant was in pumping mode. In recent times reversible hydraulic pump turbines are used. Some of these are now equipped with adjustable speed drives used to increase the pumping efficiency.

WIND POWER

The interest in renewable energy sources has also spurred the development of wind energy generated electricity. The wind is used to drive a turbine which drives a generator. Because wind speed and direction are not controllable, substantial research has recently gone into turbine configurations of high efficiency and energy capture.

Due to the variable nature of the energy source, the speed of the wind turbine is not constant and machines other than synchronous generators have been applied. These include induction generators, d-c generators, and variable reluctance generators. The further development of power electronic controls will doubtless result in new configurations for wind power generating systems (5).

The installed cost of wind energy in 1980 was above $2000/kW. In 1993 this cost decreased by more than 50% with quotes as low as $735 (6). Wind energy is already cost competitive on a $/kW basis with oil and natural gas.

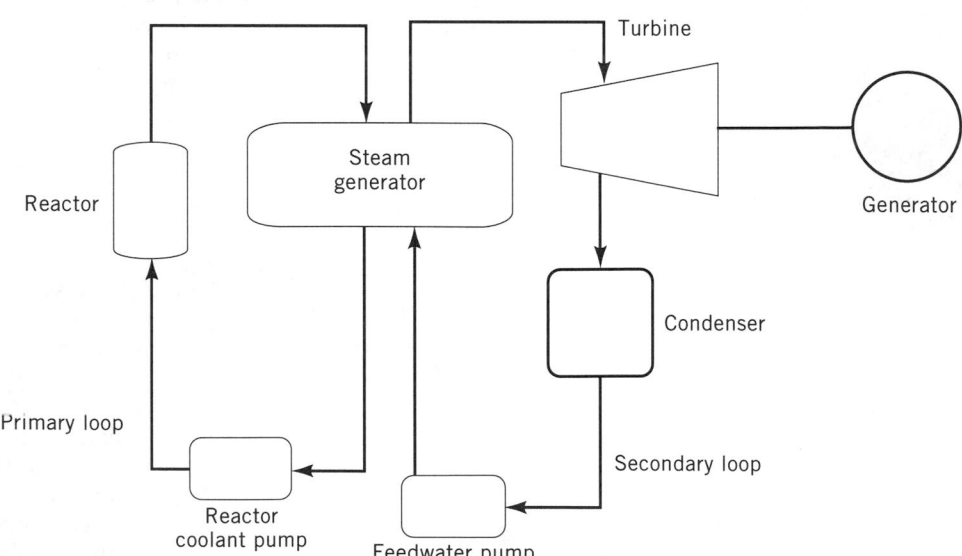

Figure 8. Pressurized water reactor.

The U.S. Department of Energy has projected that the cost of wind-produced electricity will be competitive with coal energy by the year 2000 in sites with good wind resources (averaging 7.2 m/s at 10 m height) (7). In the United States, California leads the nation in wind energy production. In 1992, 2.7 billion kWh, or more than 1% of the state's electricity requirement, was generated by wind energy (8).

SOLAR POWER

Producing power directly from solar energy is not currently a significant factor in electricity generation. However, this technology has attracted the attention of electric utilities and industry as a viable alternative for future energy production. Photovoltaics are semiconductor devices that convert solar radiation (sunlight) directly into electricity. While the electricity is essentially free in that there are no fuel costs, photovoltaics have not been widely used even in attractive climates due to the high initial investment required for the devices. Since 1973 the price of photovoltaic electricity has dropped from $15 to $0.30/kW·h with costs projected to drop to $0.15/kW·h by the mid-1990s (9). Collector types include thin-film technologies common in solar-powered calculators and crystalline silicon plates, the latter being somewhat more efficient (23%). Also of interest are concentrators or mirrors that focus the sunlight on a smaller area of cells thus reducing the number and cost of cells. The efficiency of these can be 35%. Also of interest is the solar–thermal concept in which solar energy is concentrated by a series of reflectors to boil water producing steam which is expanded through a turbine to produce electricity in the conventional manner. This technology is already in place in the California system.

BIBLIOGRAPHY

1. C. Starr, M. Searl, and S. Alpert, "Energy Source: A Realistic Outlook," *Science*, **256** (May 1992).

2. C. Summers, "The Conversion of Energy," *Sci. Am.*.

3. A. Wood and B. Wollenberg, *Power Generation Operation and Control*.

4. G. Friedlander, *Elect. World*, 57–63 (July 1980).

5. D. Torrey and S. Childs, "Development of Variable Reluctance Wind Turbine Generators," *Windpower '93 Conference Proceedings*, American Wind Energy Association, San Francisco, Calif., July 12–16, 1993, pp. 258–265.

6. S. Rashkin, "Is the Sustained Orderly Development of Wind Technology a US Success Story?" in Ref. 5, pp. 118–123.

7. R. Thresher, S. Hock, H. Dodd, and J. Cohen, "Advanced Technology for the Year 2000: Utility Applications," in Ref. 5, pp. 61–73.

8. S. Rakow, "Wind Energy in California: A Decade of Progress," in Ref. 5, pp. 9–15.

9. J. R. Rannels and J. E. Bradley, "Progress in Photovoltaic Power Production," in D. A. Jones, ed., *Power Generation Technology 1994* Sterling Publications Ltd., London.

ELECTRIC POWER SYSTEMS AND TRANSMISSION

J. A. Casazza
D. J. LeKang
CSA Energy Consultants
Arlington, Virginia

ELECTRIC POWER TRANSMISSION

The transmission system provides the means by which large amounts of power are transported from the generating stations that produce this power to other parts of the utility system or to other utilities. In terms of the three primary functions of the electric utility industry, transmission can be viewed as the link between the production and distribution functions (1). (see also Commercial availability of technology.)

The most common form of transmission in use in the United States today is three-phase, alternating current (ac) having a frequency of 60 Hz (cycles per second). In recent years, the use of direct current (dc) transmission, while a small portion of total transmission, has increased for a number of special applications.

The most common ac voltages in use in the United States today for transmission purposes are between 115,000 and 765,000 V. A standard voltage has been agreed upon for industry use.

Purpose

The purposes of the transmission system (2) are as follows:

- To deliver electric power from generating plants to large consumers and distribution systems.
- To interconnect systems and generating plants to reduce overall required generating capacity requirements by taking advantage of:

    The diversity of generator outages, ie, when outages of units occur in one plant, units in another plant can provide an alternative supply.

    The diversity of peak loads since peak loads occur at different times in different systems.

- To minimize fuel costs in the production of electricity by allowing its production at all times at the available sources having the lowest incremental production costs.
- To facilitate the location and use of the lowest cost additional generating units available.
- To make possible the buying and selling of electric energy and capacity in the marketplace.

In recognition of these functions, transmission systems have been designed to achieve acceptable reliability at minimum cost. This includes provision and maintenance of service without loss of load for "normal" generation and transmission contingencies under various load conditions. It also requires that transmission systems be able to withstand major contingencies, such as the unexpected loss of an entire power plant or a transmission right of way, without a total system shutdown.

History

During the early stages of the electric power industry, generating stations were located within the cities very close to their loads and were directly connected to the distribution system. As customer demand grew and in-city generating sites became scarce, new generating stations were built farther away from their loads. This led to the development of a single transmission system capable of delivering electric energy short distances from generating stations to load centers in the cities (2).

By 1925 it was clear that larger generating stations were needed to achieve greater economies of operation. These larger generating stations of one city were linked (through transmission) to stations in adjoining cities to provide more reliable service. This led to the present-day transmission systems that link to most cities across the country.

Elements of Electric Power Systems (1)

The basic elements of an electric power system are shown in Figure 1. The illustration is divided into three sections: generation, transmission, and distribution. Function and to some extent voltage level are used to distinguish between these sections (1).

The power supply components are located at the left side of the diagram. These include the generators used to produce the power, along with their prime movers and the auxiliary equipment needed for the functioning of the generators. Another source of power supply consists of transmission ties between other systems used for purchases and sales.

The bulk power transmission system is at the highest voltage levels and is used to transmit generated power over long distances. These bulk transmission lines terminate at transmission substations that contain transformers and switchgear used to transform the electric power to lower voltages. As shown in Figure 1, some transmission substations are used to facilitate power flow in the network by interconnecting several transmission lines at the same voltage but do not include transformers. Other transmission substations provide both a connecting point for two or more lines and transformers that connect the voltage to higher or lower voltages. The right side of Figure 1 shows distribution substations that receive power from the transmission system, transform it to a lower voltage, and supply the distribution system. The distribution system then delivers power to customers for their respective end uses.

Components of Transmission System

Transmission Lines. The primary components of overhead transmission lines are conductors, insulators, support structures, and rights-of-way.

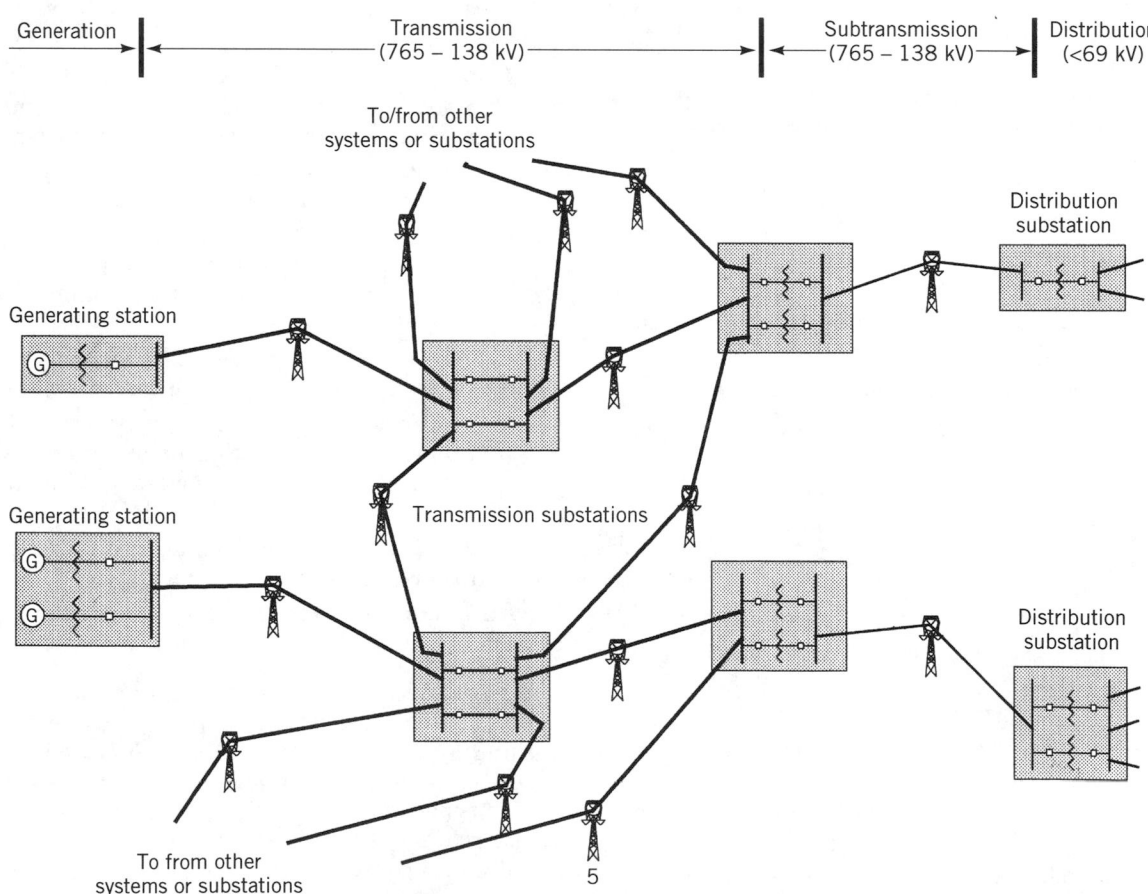

Figure 1. Key elements of electric power transmission.

Conductors are the wires through which the electricity passes. They are usually made of aluminum with steel reinforcements. Insulators are made of materials (such as porcelain) that do not permit the flow of electricity. They are used to support the conductor while separating it from the supporting structure.

The most common form of support for transmission lines is a steel, lattice tower. Sometimes wood poles are used for lower transmission voltages. The primary purpose of the support structure is to maintain the electricity-carrying conductor at a safe distance above the ground.

Rights-of-way are strips of land along which the transmission line is built. In many cases rights-of-way are wide enough to accommodate two support towers with two transmission lines built on each tower.

In some areas of the country, population densities and environmental considerations preclude the possibility of obtaining a right-of-way for a transmission line. In such a case transmission is often built underground by using shielded cables located in large protective pipes. The pipes are filled with oil under pressure to provide insulation (3). This type of transmission is many times more costly to construct than overhead transmission and presents technical and accessibility problems.

Transmission Substations. Transmission substations occur wherever transmission lines terminate or connect to one another. Various types of equipment are installed in substations in order to transform the electrical energy to a different voltage, transfer electrical energy from one transmission line to another, or provide protection for the transmission lines under emergency or fault conditions.

Busbars are used within substations to connect circuits of the same voltage level with one another and with transformers.

Transformers are used to transfer power between electrical circuits and buses operated at different voltages. Tap-changing equipment is used on transformers to vary the amount of voltage change in the transformer and to help compensate for voltage variations in the system.

Phase shifters, also called phase angle regulators, are devices similar in construction to transformers that induce a power flow into a circuit, in order to increase or decrease the power loading of that circuit by inserting a phase angle difference.

Switchgear is a general name given to equipment that can open connections and interrupt the flow of electricity between circuits, busbars, and transformers under normal and emergency conditions. The main types of switchgear are circuit breakers and disconnect switches.

Circuit breakers are designed to interrupt the flow of electricity in power system circuits and devices. When a fault or a short circuit occurs, the amount of current flowing through the transmission line and all of its substation equipment increases to many times the current under normal conditions. The circuit breaker is capable of interrupting the flow of electricity under these abnormal levels of current. In fact, the capability of these circuit breakers is given in terms of the maximum current they can interrupt and in terms of their current-carrying capability under normal continuous operation. An important characteristic of circuit breakers is the time that is required to interrupt the short circuit current, typically a very small fraction of a second.

Disconnect switches are used to keep circuits open once a circuit breaker had tripped. They generally cannot interrupt load or fault currents and are usually used in conjunction with circuit breakers.

Surge arresters are devices used in substations to protect equipment from overvoltages caused by lightning and circuit switching. They are designed to conduct high current to ground and away from substation equipment when voltages reach an undesirable level, thus limiting maximum voltages.

Protective relays are used, often in connection with communications equipment, to detect faults or abnormal conditions and signal circuit breakers or other equipment to take corrective action. They receive system information in the form of voltages and currents from potential transformers (PTs) and current transformers (CTs) and, when there is a disturbance on the system, use a combination of measurements and logic to determine whether a specific piece of the system should be disconnected in order to prevent the spread of the disturbance and to protect the equipment.

System Protection. Because faults and short circuits occur from time to time on electric transmission systems, system protection techniques have been used since the earliest days to disconnect the faulted equipment, thus allowing the remainder of the network to continue operation. This has been accomplished through the use of relays, which sense the existence of such faults and send an appropriate signal to circuit breakers, which disconnect the faulted line or apparatus. Relays have also been used to stabilize system conditions resulting from various types of perturbations.

Since transmission lines are predominantly carried overhead on towers, they are subject to lightning and other external forces. To protect these lines from damage and to ensure their reliable operation required significant improvements in protective devices for lightning strokes. New lightning arresters and new methods of grounding lines to reduce the magnitude of "traveling waves" were developed to make possible the technology required by the systems (4).

To protect against faults or short circuits, new types of circuit breakers and relaying protection were developed. This reflected recognition that the speed with which the disturbance could be removed from the system was of vital importance. As the protection system became more important, it also became apparent that a failure in the protection system was a contingency one had to consider in developing the system. System layouts and designs were developed so that a circuit breaker could fail or a relay could fail but the system would go on working.

Reactive Power Equipment Reactive power is required to supply the magnetic fields of electrical equipment (1, 3). It uses no energy and produces no useful work in itself.

An undersupply of reactive power in a part of the transmission system results in excessively low voltages in that area, and an oversupply produces excessively high voltages. In order to prevent these occurrences, reactive power must be supplied or absorbed as required.

The unit of reactive power is the VAR (volt-ampere reactive), but the practical unit used in transmission system operation and design is the megavar (MVAR), equal to 1×10^6 VAR. Almost all types of load require the supply of some reactive power along with the active power producing useful work. Reactive power can be produced or absorbed by several types of equipment:

- Generators produce reactive power along with producing real power. Each generator has a "field" supplied by an "exciter." The amount of reactive power produced depends upon the field current. Generator reactive output is required to maintain proper transmission system voltage conditions.
- Shunt capacitors are devices connected between a conductor and ground that produce reactive power. They are located in the distribution system and the transmission system. Shunt capacitors are generally used to provide the required reactive power not supplied by generators.
- Shunt reactors absorb reactive power. They are sometimes needed on power systems to prevent increased voltages under some conditions. These are also usually switched for reactive supply control.
- Static VAR compensators (SVCs) utilize reactors, capacitors, and high-speed switching devices to provide a source of controlled reactive power. The high-speed switching devices allow SVC equipment to respond very quickly to changing system conditions.

Capacitors are sometimes also connected in series with lines. While a series capacitor produces some reactive power, its main function is to reduce the effective reactance of the line and cause it to deliver more current than other parallel lines. Similarly, series reactors can be used to increase the reactance of a line in order to reduce the load currents, but this application occurs infrequently.

Voltage Control. Voltage control on a transmission system is achieved by a combination of the settings of transformer fixed taps (which can only be changed when a transformer is out of service), transformer tap changing under load, and adjustment of the previously discussed reactive sources.

Voltage regulators monitor the voltage at the generator terminals and automatically adjust the excitation system for voltage fluctuations to obtain the voltage that the generating station operator has indicated or the voltage that is required under some emergency condition.

Because the transmission of reactive power over transmission lines causes large voltage drops, transmission system operation usually tends to minimize the flow of reactive power from one area to another and encourage the production of reactive power as close as possible to the place where it is needed. To minimize the transfer of reactive power over the transmission system, reactive power supply is often provided in the distribution system to supplement transmission operations.

Direct Current Transmission

Direct current transmission, frequently referred to as high-voltage direct current (HVDC), is used in a growing number of instances where its characteristics are particularly beneficial (1,5).

HVDC Components. The dc transmission lines generally use two conductors, as compared with the three used in three-phase ac transmission. Both terminals of a dc line consist of dc conversion stations that transform ac power into dc (rectifiers) or dc into ac (inverters). The main components of a dc conversion station are

- Banks of thyristors (controllable solid-state diodes) arranged in "bridges" for the conversion of ac into dc or vice versa.
- Transformers.
- Capacitors used to filter out harmonics on the ac side and to provide reactive power to support the thyristor bridges.
- Switchgear on the ac side.
- Control systems to regulate the amount and direction of power flow.

Operating Characteristics of dc Transmission. By not being subject to the physical laws governing ac transmission, dc transmission offers some characteristics that are valuable in some situations.

- The amount and direction of the power converted from ac to dc and back is directly controllable and is independent of the characteristics and power flows on other ac lines in parallel that join the same areas.
- Reactive power, being inherently associated with ac transmission, is not carried by dc transmission. Conversion stations at both the sending and receiving terminals, however, require substantial reactive power support from their respective ac systems. Some of this support is generally supplied by capacitors at the conversion stations. These capacitors also help absorb some of the harmonics produced by the conversion stations.
- dc lines do not transmit short-circuit currents from adjoining ac systems and do not spread transient instability disturbances.

Application of dc Transmission. The physical characteristics of dc transmission make it particularly valuable for the following applications:

- Linking two large synchronous systems with relatively weak interties, which if linked by ac would be subject to instability or to an excessive inrush of power in case of severe disturbances.
- Providing links in parallel with ac transmission under conditions in which power would be transferred too much on some links and too little on others.

- Linking systems by submarine cables, which, if ac were used, would carry too much reactive power and severely limit the remaining capacity for real power.

- Linking areas that, if linked by ac circuits, would cause excessive currents to flow in the event of short circuits exceeding the capability of circuit breakers to interrupt them.

- Controlling the amount of power flowing from one area to another, allowing the link to be treated in a manner similar to a controllable generation source.

DEVELOPMENT AND OPERATION OF TRANSMISSION SYSTEMS

Economic Dispatch

The incremental production cost of a generating unit is the additional cost per kilowatt-hour of generating a small additional quantity of energy from that unit or the cost reduction per kilowatt-hour due to generating a lesser quantity of energy (1,6,7). Generally, this incremental cost consists mostly of additional fuel costs, but it also includes any other operating costs that vary with the level of power production. The incremental cost of a generator is almost entirely the cost of its fuel and by how efficiently it converts this fuel into electric power. For a typical thermal unit, the incremental cost increases as production *increases* from the minimum to the maximum rated output. The incremental cost of a unit is not the same as its average cost, which generally decreases as its output increases (1–7).

Economic dispatch consists of supplying the system's total power requirements at any time by lading each available generating unit to the point where its incremental cost is the same as that of all other generations. Units whose highest incremental cost is lower than this common incremental cost are, of course, loaded to their maximum capability.

Today's electric power systems are being dispatched and operated based on the incremental costs of production. This general practice often results in one utility producing electricity for use by some other utility's customers when the first utility can produce the incremental power at a lower cost. *This leads to maximum economic efficiency in the use of the investment and in the cost of operating these systems.* No other industry operates in this way, whereby all suppliers work together to produce a common product, using plants that have the lowest incremental cost to produce the additional units of energy needed regardless of plant ownership.

The only exception to this production based on incremental costs has resulted from the recent addition of co-generators and qualifying facilities. The proposed addition of independent power producers could increase further the amount of generation that is not economically operated based on incremental production costs.

The principles of economic dispatch are applied to a system once it has already been determined which units will operate. The process leading to the operation of certain generating units rather than others is called "unit commitment." Like economic dispatch, unit commitment

methods are intended to produce the maximum economy while satisfying reliability requirements.

Transmission Losses

There are two basic types of power losses in transmission systems (3):

- *Core losses* are dissipated in the steel cores of transformers (8). These losses typically vary with at least the third, and often higher, powers of the voltage at their terminals. If transmission voltages are fairly constant, core losses are also constant. They depend on the transformer's design voltage tap settings and its rating. They are not affected by how much power is actually flowing through the transformer. An increase in the power carried by a transmission system does not substantially affect the core losses; a change in transmission voltage can.

- *Conductor losses* are dissipated in transmission lines and cables and transformer windings. As these losses vary with the square of the current, and therefore approximately with the square of the power carried by each component, they vary greatly between light-load and heavy-load times and are also affected by increases in the power carried by a transmission system (9). It should be noted that because of the "square" effect mentioned above, a given percent increase in power flow results in a much higher percent increase in power losses on the system.

When electrical energy is transported across large distances through the transmission system, a portion of the energy is "lost." For a given amount of electric power, the higher the operating voltage of the transmission line, the lower the current flowing through the line. Therefore, use of higher transmission voltages permits the transmission of electric power with lower currents and a resulting reduction in energy losses (9).

The losses in the transmission system at a specific moment, typically at the time of system peak, are measured in megawatts and are referred to as "capacity losses." The energy expended in transmission losses over a given period is given in mega-watt-hours and is referred to as "energy loss." Capacity losses require the installation of additional generation and transmission equipment; energy losses require the consumption of fuel or equivalent energy sources. The increase in a system's total losses due to a specific action is referred to as "incremental loss."

In addition to real power losses, significant reactive power losses occur in the transmission systems, and these reactive losses are typically about 4 or 5 times the real power capacity losses.

Transmission lines and cables generate a certain amount of reactive power (their inherent capacitance produces a "charging" current) that varies with the square of the voltage (3). Currents flowing through them also cause reactive power losses proportional to the square of the currents. The net effect is that the lines and cables tend to generate net reactive power when their loading is light and to absorb retroactive power when their loading is heavy.

Interconnection and Pooling

Benefits of Interconnections between Systems. With interconnections, a utility may be in a position to call upon its neighbors in the case of severe outages on its own system (2). This is similar to a series of mountain climbers that rope themselves together; if one stumbles, the others can hold that person up. Once neighboring utilities are interconnected, it is possible for them to exchange energy or buy and sell energy to each other in order to minimize their total expenses for energy. For example, if a utility has capacity available that is not required by its own load and that capacity has an associated fuel cost less than its neighbors, it makes economic sense, as well as satisfying the public interest, for one utility to make a momentary sale of energy to its neighbor instead of having the neighbor utility operating higher cost generating facilities. The sale generally will be made at a price that is halfway between the cost to the producing utility and the cost avoided by not operating the generation of the purchasing utility. The customers of both utilities benefit through lower costs for electricity (10).

Electric utilities in the United States have been coordinating the development and use of their transmission systems and generating capacity for many years (11). As a result of this coordination, considerable savings have accrued. These savings have resulted from various causes:

- Reduced investment in installed generating capacity from sharing of installed generation reserves, load diversity and equipment outage diversity, optimum use of available generation sites, use of larger unit sizes, coordination of maintenance schedules, and long-term firm capacity purchases.
- Reduced operating costs from economic energy exchanges, use of regional economic dispatch and unit commitment, optimum use of hydro and pumped-hydro facilities, coordination of maintenance schedules, short-term capacity purchases and sales, and reduced spinning reserves.
- Reduced transmission investment from coordinated regional transmission planning, supply to other systems' loads, and back-up to other systems' substations.

Estimates show that the ties between regions, coupled with regional coordination procedures, produced annual savings on the order of $15 billion in 1986 and about $20 billion in 1989 (12). About two-thirds of these savings were due to reduced capital investment for generating capacity; the remainder resulted from reduced fuel costs and fuel use.

Synchronous Operation. Because of the huge benefits, the addition of interconnections between systems and between regions in North America evolved rapidly, as they did elsewhere in the world. Figure 2 shows the five synchronous areas that currently exist in North America. All the systems within each area operate in synchronism. Four of these areas are interconnected by dc ties.

Because of the synchronous operation of all its generation, an interconnected electric power system functions as

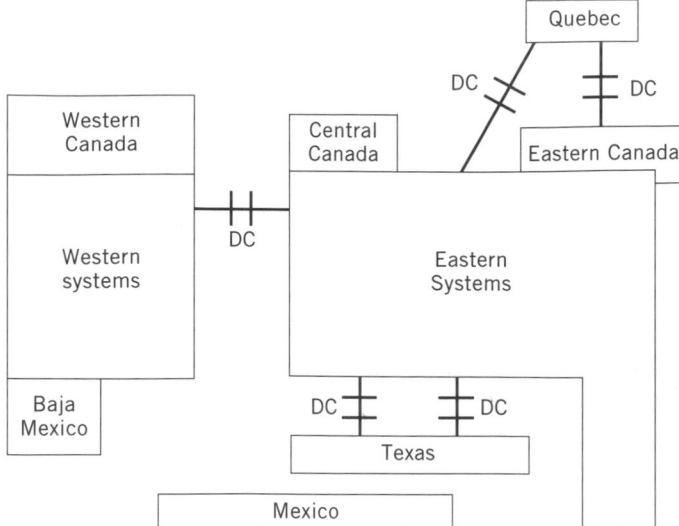

Figure 2. Fire synchronous systems of North America.

a single large machine (13–16) that can extend over thousands of miles. One of this machine's characteristics is that changes in any single portion instantly affect the operation of all other portions. Future plans for any single part can affect other parts. The obvious questions are to what degree does this affect planning the addition of a new generator and a new transmission line and will it have a significant effect on the development of systems' plans 500 and 1000 miles away.

In system operation the effects of developments in one system can be felt throughout a large geographic area. For example, if a large generating unit is lost in New York City, the interconnected power system (the machine) suddenly has less power input than power output. Any machine in which this occurs will slow down. This is also what happens in electric power systems. As individual rotors of every generator in this system slow down, each of the rotors gives up a certain amount of its rotating energy ($\omega^2 R$) to compensate for the lost input from the unit that has tripped.

Instantaneously, with the loss of a large unit, there is an inrush of power from units all over the synchronous network, feeding into the system or region that has lost the generator unit. While these inrushes of power from individual units long distances away are not large, they accumulate and build up like water flowing from creeks into a river, approaching the system that has lost a large unit as a flood.

The reason that systems have been tied together and operate in synchronism causes this effect. By having the various generator units throughout the region assist with the loss of a large generator unit in a specific system, the total amount of spare or reserve-generating capacity required can be reduced. This is similar to the insurance business. The larger the number of policy holders, the better able the insurance company is to cope with any specific major disaster and the less percentage reserves it requires. The great strength of operating in synchronism and being tied together in an integrated system is the ability of one system to be helped by others. Its greatest weakness, however, is that the inrush of power into any one system can cause transmission system overloads.

Because of these characteristics of modern electric power systems, the design and operation of the key elements in the synchronous network must be coordinated. Business decisions, government legislation and regulations, and other institutional processes must be compatible with the technical characteristics. Many problems can be solved by technical solutions or institutional solutions and, in some cases, the use of both.

Transmission System Capacity

The capacity of an ac transmission system (17) to deliver power is also a complex matter and depends on the network arrangement, size and type of conductors, lengths of lines, location and size of generation sources and loads, etc. The network capacity may be limited by transient or steady-state stability conditions for longer lines, voltage instability, or thermal heating of conductors. The capability of the circuit breakers and protective systems must be carefully checked as new generators are added and system changes are made. These factors require that many highly technical analyses be made by skilled engineers to determine the capacity of a transmission network (1).

Equipment Ratings. When electricity flows through the conductors of a transmission line, the conductors and terminal equipment are heated. The amount of heat flowing through it increases. Each transmission line, transformer, and piece of switchgear is capable of withstanding a given amount of heat based on certain electrical and mechanical design factors. This amount of heat is usually defined in terms of the amount of current a line can carry. This current limit determines the maximum amount of power the equipment can deliver. The limiting factor for a line may be determined by conductor limitations, line sag limitations, current breaker limitations, wave trap limitations, etc., with the element having the lowest rating determining the overall line rating.

The rating depends upon the period for which the loading will exist and also what conditions (weather and line loading) existed prior to that period. Ratings are also dependent upon the loss of conductor life that the utility is willing to tolerate per high loading incident. Occasionally, transmission line loading is limited by the amount of conductor expansion and resulting increase in sag caused by heating the transmission line. In some instances, excessive sag could cause the conductor to contact underlying lines or structures.

Because heating effects also depend on ambient conditions, different line ratings are usually used by utilities for summer and winter conditions. Hearing effects also depend on duration of the loading; ratings are also frequently used for various time periods, such as 15 min, 4 h, 24 h, and continuous.

Voltage Drop. The passage of current through a wire results in a reduction in the voltage along that wire. Since utilization equipment is sensitive to voltage, there are limits beyond which the voltage cannot be permitted to decrease.

Power and Energy Losses. The passage of current through a line results in a loss of energy due to the internal heating of the conductor. At a certain point, the losses would render the operation uneconomic.

Stability of Operations. Stability refers to the property of the system to be able to maintain synchronism after a severe perturbation of some kind, such as a short circuit or the sudden outage of a transmission line. Such disturbances cause dynamic oscillations on the system with large power savings on various generators. This savings cause loss of synchronism and tripping of generators.

Voltage drop, power and energy losses, and stability limitations improve in proportion to the square of the voltage at which the line operates. A 230-kV transmission line, therefore, would tend to have the power-carrying capability of four 115-kV transmission lines. Thus, transmission line voltage is related to the amount of power that has to be transmitted. Further, all three considerations are directly influenced by the distance over which power is to be transmitted. Thus, the longer is the transmission distance involved, the more likely is the voltage to be higher in order to avoid voltage drop, power and energy loss, and stability problems.

Division of Flow Among Transmission Routes

The flow of power over the specific circuits in an ac transmission network cannot be controlled. Kirchhoff's laws determine how the power will divide. Legislation, regulations, and contracts cannot specify the load to be earned by specific transmission circuits. This sometimes results in one transmission circuit being excessively loaded while other circuits in a network are lightly loaded.

Loop Flows. When various transmission systems are interconnected in a network, power from one system will flow over the lines of every other system even when each system is supplying its own load from its own generating source, as shown in Figure 3. This is caused by the physics of power flow. With this condition there is a circulating power flow around a closed loop. This flow is in addition to the flow that would exist on the lines of each system if there were no interconnections to other systems. Such flows are called "loop flows." They are the result of the network design and the distribution of generation and loads in each system. They are not the result of any purchases or sales over the transmission system. These loop flows can change considerably at various times and system loading levels. Loop flows are caused by the interconnection of systems to achieve generation capital cost savings and fuel cost reductions. These flows can be reduced or corrected through use of expensive phase-shifting transformers.

Parallel-Path Flow. When one utility sends power to another utility, a change in power flow occurs along all "parallel" electrical paths leading from the point (or points) of the seller's increased generation to the point(s) of the buyer's increased load or decreased generation. This occurs whether the seller or the buyer is a private entity or util-

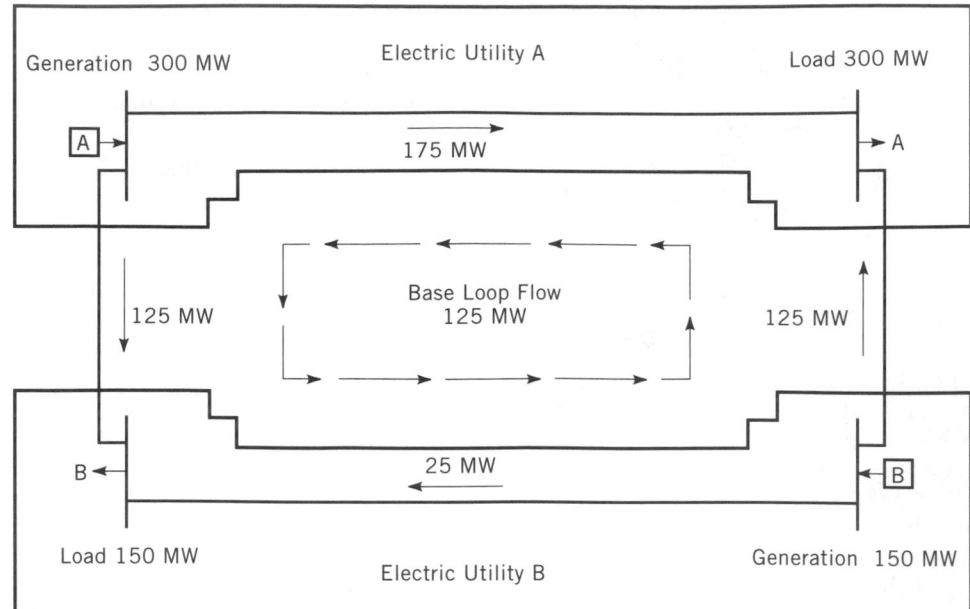

Figure 3. Example of loop flow.

ity. Even if the two utilities are directly connected, some of the transacted power will flow over any and all available paths, including those through third-party utilities. In this case as well, however, some of the transacted power may flow in other utilities' systems. These flows through the systems of other utilities that are not involved in the transmission service transaction are generally referred to as "parallel-path flows." Figure 4 provides an excellent illustration, showing that more than 30% of the power being shipped from the Pacific Northwest and Utah will flow through southern California. Such flows can be reduced through the use of phase-shifting transformers.

Virtually every transmission service transaction will lead to parallel-path flows in transmission systems that are not directly a party to the transaction. These may be large or insignificant, depending on the quantities transmitted and the topography of all the transmission systems involved. They will generally increase the power flows on some of the lines and transformers of the third-party systems and reduce the power flows on others.

Control Areas

As the individual electric utility systems were being interconnected in the United States, it became quite apparent that new control technologies and procedures had to be developed so that operating decisions and problems in one system did not unnecessarily affect another (14–16). This led to the development of what is now known as control areas. A control area is a bounded geographical area within which all generation is controlled by a single dispatch center. All transmission lines entering or leaving this bounded geographic area are metered; the total flow in or out of the area is determined. The control center has a dispatch computer that controls generators to meet optimum economic dispatch and the scheduled amount of power flowing in or out on all of the tie lines (7,18,19).

In addition to the automatic computer control features of the control center, it monitors voltages. There are other manually performed duties such as scheduling of maintenance, handling of emergencies, agreeing on schedules for power interchanges, developing data for incremental cost dispatch, etc. The control area concept has continued to evolve and will grow in the future.

Transmission System Reliability Criteria

Transmission system capacities sufficient to meet needs are determined on the basis of specified planning and operating criteria, such as those agreed to by the North American Electric Reliability Council (NERC) and members of the NERC Regional Electric Reliability Councils (20,21). Where justified by local requirements, individual companies often use criteria that exceed those specified by the reliability councils. The transmission capacities provided recognize both initial needs and future needs so the system can be expanded in the most economic manner.

The reliability criteria define the permissible range of deviations from the normal transmission conditions that may be tolerated for various contingencies. Typically, a minimal deviation (or none) is permitted for the types of disturbances that are the most frequent, while greater deviations (and possibly loss of service to a limited number of customers for a limited time) are accepted for very severe and unusual contingencies. Similarly, a higher thermal rating, called "emergency rating," may be applied for a limited amount of time during specified types of line, transformer, or generation outages.

Reliability standards are generally based on both technical and economic considerations, often balancing increased costs against lower service reliability. The NERC sets these standards for North America. Regional Electric Reliability Councils may set more stringent standards for their specific regions, but the NERC standards must be adhered to by all regions.

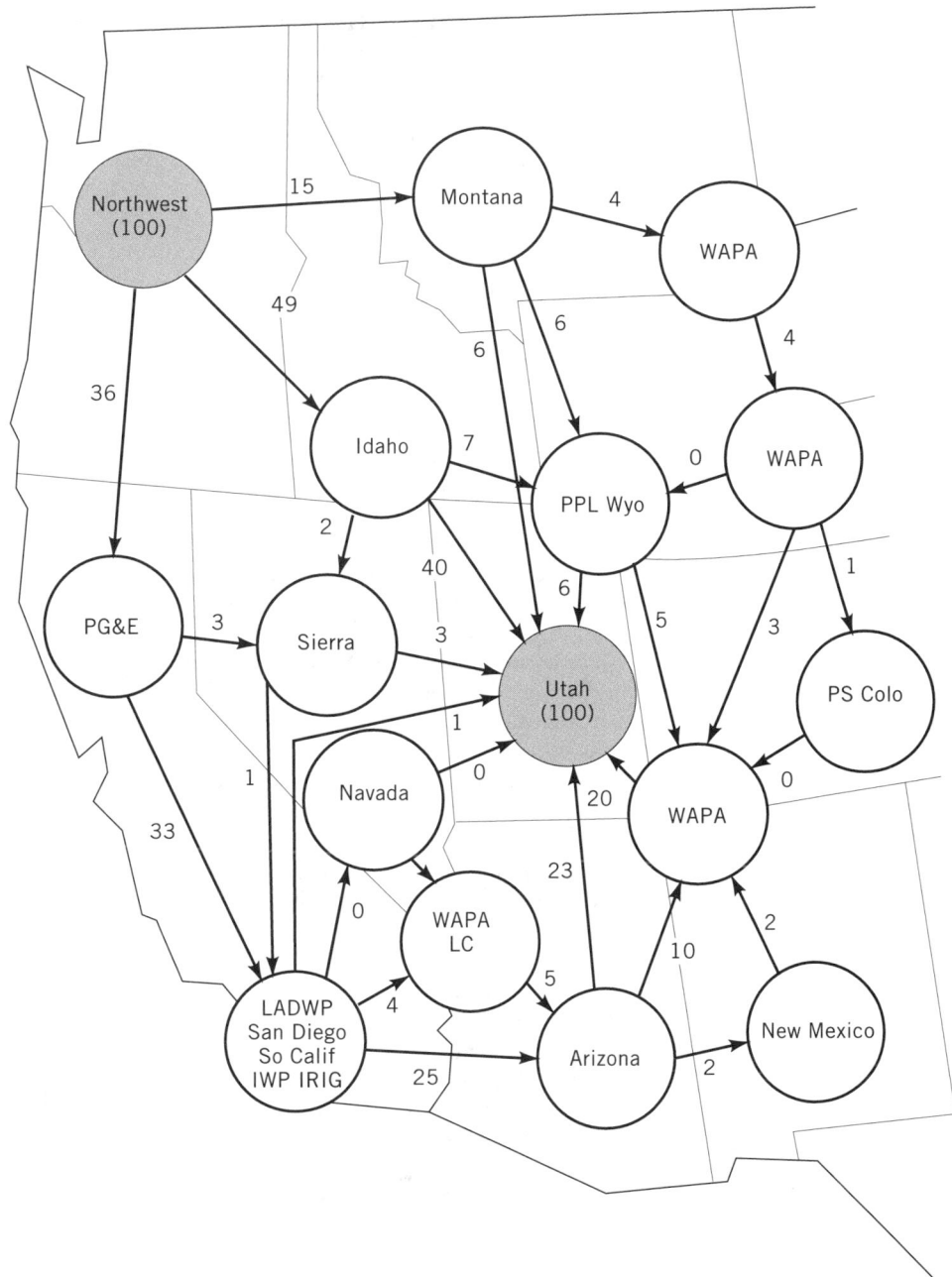

Figure 4. Distribution of flow from a 100-MW transfer (2).

Single-Contingency Criteria. One of the basic reliability criteria applied in system operation on most U.S. transmission systems consists of "single-contingency operation." It provides that the loadings on a transmission system must be limited to levels that will permit any single contingency (such as the loss of any single line or any generating unit) to occur without exceeding acceptable circuit or system limits, as discussed in the next two sections. Some systems include as single contingencies disturbances affecting more than one component, such as both circuits on one tower line or the failure of a single circuit breaker to properly operate.

Multiple-Outage Criteria. The possibility of a combination of disturbances and other conditions must also be

considered, such as multiple transmission outages during lightning storms; transmission line outages occurring while other equipment such as generators, cables, or transformers are being maintained; or the outage of lines sharing a common right-of-way. The criteria involving such contingencies may (but are not always) able to accept more severe consequences to system conditions than for single contingencies.

Transmission Reserves. One immediate effect of operation taking into account single or multiple contingencies is that a significant fraction of each system's transmission capacity must be held in reserve so that any likely contingency can be sustained. The practice of defining the transfer capability of a transmission system including consideration of the contingencies specified in the reliability criteria may also be viewed as "setting aside" transmission reserves for reliable operation.

Transmission Systems Operation

Operation of a transmission system is concerned with two principal goals:

- The reliability of supply to the customers.
- The production of power in the most economical way possible.

In pursuing these goals, the usual approach is for the operator to seek the most economic pattern of generation, while complying with all limitations imposed by the need for reliable operation (22).

The practice of limiting transfers to what can be safely transmitted even if the worst single contingency occurs is recommended by the NERC for general use by all systems. This limit is called the network transfer capability (23). It is based on what is sometimes called *preventive* operation. In some situations, operation based on *corrective* operation is used. In these situations, fast automatic controls are used to quickly reduce the transmission loading to safe limits if a contingency should occur. Corrective operation results in higher average transmission system loadings with some potential reliability risks.

The safe and reliable operation of the transmission system is based on three basic requirements:

- All components operating within their thermal ratings.
- Voltages at all points remaining within acceptable lower and upper limits. (This imposes both static and dynamic constraints.)
- Synchronous operation of all the generators connected to the system.

These conditions must be satisfied in system operation under a variety of conditions, including the effect of:

- Disturbances of varying severities, including short circuits, the sudden loss of transmission system components, or the sudden loss of generating units.
- Temporal changes in the loads and resulting changes in generation dispatch.
- Changing weather conditions.
- Different external power sales and purchases and transmission service transactions.
- Maintenance outages of transmission equipment and generation.

Transmission System Planning

Transmission system planning has been defined as the process of determining *what* facilities should be provided and *when* and *where* they should be provided to assure adequate transmission service at minimum present and future costs on a present-worth basis, consistent with environmental standards.

The planning of the transmission system depends on future load projections, generation plans, and expected capacity and energy transfers. The transmission planning process must ensure that the transmission system will be possible to operate the system in accordance with the applicable reliability and service quality criteria under a reasonable combination of future conditions (24).

The design of the transmission system and the generating system must take into account the need to provide for customer loads both under normal conditions and under contingencies in the transmission system (transmission line failures) and in the generation system (generator failures). The transmission system must be designed so that even under peak-load conditions and peak-transfer conditions, the operator can always operate a system in such a way that any reasonable contingency will not cause service failures and unusually heavy contingencies will cause limited failures as provided in the NERC, regional, and utility standards. This may require the installation of additional transmission facilities including lines, cables, protective devices and reactive sources, etc., to allow this reliable operation at all times.

The conditions that must be considered in system planning include any actions that the system may be required to take due to commercial, legal, contractual, and regulatory requirements. This includes any transmission service that the transmission system may be required to provide because of legally contracted transmission service or because of legal or regulatory requirements.

Time Horizons. A transmission system must be planned many years in advance in order to satisfy a number of needs:

- To make decisions on the construction of transmission facilities far enough in advance that all the time needed to obtain the required governmental authorizations, obtain rights-of-way, make detailed designs, obtain construction bids, and construct the facilities be available before the facilities are needed.
- To determine whether planned generation facilities at given locations can be connected to the system and at what cost.
- To determine the best long-term transmission additions to meet near-term needs.
- To determine the additional facilities required, if any, to provide the transmission service requested by prospective transmission customers.

Since future expected circumstances, including load forecasts, available generation sources, fuel prices, and other conditions, change constantly, transmission planning is essentially a continuing process, requiring frequent reviews of the effect of changed circumstances and projections on the long-range plan. These reviews may cover the entire planning period, or only a fraction of it, depending on what circumstances are changed. For example, a transmission planning review for a given requested transmission service transaction needs only to be made for the period covered by the proposed transaction if it is found that the transaction does not require a change in the plan for that period; if, however, a change in plans is required to accommodate the transaction, the study may need to be carried out for more than 10 years to determine any further consequences of the charge in plan and to determine which of a number of alternative solutions may be the best.

Uncertainties. In vertically integrated utilities, generation additions and transmission additions are generally planned on a coordinated basis. While some uncertainties regarding future generation additions always exist, the planning of the transmission system is based on reasonably reliable information. As new electric power institutional mechanisms develop, uncertainties are increasing (25).

Developing Transmission Technology

Research is in progress on new transmission technologies that may have long-range effects both on the types of transmission facilities installed and the capacity of the various networks. These technologies fall into the general classification of "network improvement" technologies and include devices such as solid-state phase shifters and fast control over series capacitors (26). They are typically "power electronic devices" that utilize thyristor and other solid-state devices to provide high-speed response. The SVC, which is in occasional use, was the first of such power electronic device to be developed and implemented. The research program that is in progress in the United States and other in countries is an attempt to develop other useful power electronic devices.

TRANSMISSION SYSTEM COMPUTATIONAL METHODS

Types of Transmission Studies

The ac transmission networks are complex and must be carefully designed and operated to be able to withstand various generator and transmission contingencies under continually changing load conditions. Unlike telephone systems, gas systems, and water systems, the power *cannot be directed* or controlled to flow over prescribed transmission paths. The power will divide at all times among the various paths based on the law of engineering known as Kirchhoff's laws. In addition, as discussed above, the electricity required must be produced and delivered the instant it is needed. The product cannot be stored in a "warehouse" for future potential use (27).

Transmission studies are performed to determine if existing transmission lines will become overstressed under future loads, where and when new transmission facilities might be required, under what circumstances transmission facilities would be inadequate, and how existing and future transmission facilities will respond to emergency conditions.

In performing these studies, complex digital computer programs are used. These computer programs require detailed descriptions of the design and operating characteristics of the transmission and generation systems. Information such as transmission line ratings, circuit breaker ratings, impedances and length of each transmission line,

impedances of transformers, turns ratios of transformers, estimated customer demand on each transmission and subtransmission substations, as well as estimated generating unit output levels on an existing and projected basis are typically recorded and maintained by the electric utility for use in these transmission studies.

Load Flow Studies

Load flow studies are used to simulate the operation of the electrical network under normal operation and various equipment outages. They determine what the voltage, current, power, and reactive power flows would be on various lines and at various points in the system. Load flow studies are an essential tool in the planning, development, and operation of the system. Reliable operation of the system depends on knowing the effects of various outages and new generation, new load, and new transmission lines before they are installed.

Short Circuit Studies

Every addition to a network increases the power that in the case of a short circuit is to be conducted or interrupted by switching equipment. Studies of potential fault currents are needed to ensure proper relaying and circuit breakers are installed.

Voltage Security Studies

These studies aim at detecting risks of voltage collapse due to severe contingencies, which may occur at high level of exchanges between the utilities.

Transient Stability Studies

All generators within an interconnected network should operate at a given speed accurately determined by the common network frequency. Sudden disturbances may create electrical and mechanical forces that may cause loss of synchronism for specific generators or portions of a system.

To analyze these different possibilities, stability studies have to be made that check the behavior of generators in different contingencies.

WHEELING AND TRANSMISSION ACCESS

Transmission Access

Interutility sales have long comprised a form of competition in generation as utilities with excess capacity competed to sell power to those that needed additional power. Typically these interutility transactions benefitted all parties and encouraged efficient operation.

The Energy Policy Act of 1992 provided expanded access to the transmission system. This was intended to increase competition by permitting (when possible) generating companies to deliver power to customers other than the local utility. The widened market is expected to increase opportunities available to independents and also to utilities with excess power to sell.

As the number of players in the electric power industry increases, demand for wheeling increased. Competing generators will want to sell to whomever will pay the most, whether that is the local utility or a distance customer, and some consumers will want to shop for supplies. In either case, they will require transmission services.

The subject of compulsory wheeling is a contentious one with many believing that it will result in overall increases in the use of electricity and adverse environmental effects because of increases in fuel consumption. In its report on Electric Power Wheeling and Dealing (29) U.S. Congress' Office of Technology Assessment stated, "separating a system's mutually dependent areas of decision making may introduce a different kind of inefficiency that could be costlier than that intended to be addressed by competition" (p. 7).

The technical and economic challenges of increased transmission access are significant. The available capacity on transmission systems is difficult to determine. It depends on the specific conditions at the time transmission is desired, the reliability and longevity levels selected by the utility, and the parallel-path flows that will result. Therefore disputes over the feasibility and cost of wheeling are difficult to resolve.

Wheeling Services

Wheeling services with many different conditions and arrangements are used: firm service and interruptible service.

Firm service, in almost all cases, refers to the firmness of the commitment to wheel only after the wheeling utility has agreed to provide the service and a rate schedule or service agreement has been signed stating the amount of power to be wheeled. Firm service generally means that once the wheeling utility has made a commitment to wheel, it will not interrupt the service except under certain specified conditions. Some utilities further subdivide their firm service into short-term and long-term service. For example, short-term service may be for less than a year and long-term service for a year or more.

Interruptible service means that the wheeling service can be curtailed or interrupted for reasons other than those specified for firm service.

A very important characteristic of a transmission service is the priority of the particular transaction should curtailment be necessary. In general, firm transactions will have a much higher priority than nonfirm transmission.

Transmission Pricing

The provision of transmission service requires the use of facilities in generating plants and in the distribution systems. Costs for use of these facilities are being included in transmission costs. This becomes quite complicated since, in many cases, the generation and distribution companies are separately owned.

Also, power will flow over the various transmission circuits in an ac network based on Kirchhoff's laws, not on contracts or legislation. Transactions frequently result in the use of transmission systems other than that of the transmitting company. Procedures for handling this problem are proving difficult to develop.

SITING, ENVIRONMENTAL, AND HEALTH ISSUES

The continued need for additional transmission systems raises some public policy issues. Three of the most significant and contentious of these issues are transmission line siting, environmental impacts, and potential public health effects of electric and magnetic fields.

Siting

The siting of new electric transmission lines has become more difficult because of the obstacles encountered in the process of regulatory review and approval. The process of gaining approval for transmission line construction has become more formalized as opportunities have been provided for public involvement and greater scrutiny of potential environmental and social impacts of proposed projects. Competition for land to route transmission lines has become more intense and right-of-way costs are increasing (30).

Planned investment in new transmission lines has been declining. Eventually, however, new and expanded transmission systems will have to be built to provide an adequate and reliable power supply. Improved procedures for resolving siting disputes are needed. If these cannot be achieved, the only alternative remaining will be the very expensive installation of transmission cables.

Environmental Impacts

Overhead transmission lines are generally considered unsightly and undesirable. Transmission systems do provide, however, environmental benefits by reducing the total amount of generating capacity needed and by reducing the total amount of fuel burned to supply national electric energy needs.

Health Effects

Several recent epidemiologic studies have suggested an association between exposure to electric and magnetic fields and cancer. While these epidemiologic studies are controversial and incomplete, they do provide a basis for concern about effects from exposure.

The research results to date are complex and inconclusive (31,32). Many experiments have found no difference in biological systems that have been exposed to fields and those that have not (33,34). It still is not possible to demonstrate that such risks exist, and they may not. However, the emerging evidence no longer allows one to conclude that there are no risks. Health effects are the most prominent concerns raised by people living near existing or proposed transmission lines. However, it is important to recognize that exposure from high-voltage transmission lines is only one, perhaps minor, source. Exposure to local electric distribution lines, appliances, lighting fixtures, and all wall wiring is more common and could play a more significant role in any public health risks (35).

BIBLIOGRAPHY

1. Rustebakke, *Electric Utility Systems & Practices*, 4th ed., Wiley, New York.

2. J. A. Casazza, *The Development of Electric Power Transmission*, The Institute of Electrical and Electronics Engineers, New York, 1993.

3. *Electrical Transmission and Distribution Reference Book*, Westinghouse Electric Corp., 1950.

4. L. V. Bewley, *Traveling Waves on Transmission Systems*, 2nd ed., Wiley, New York, 1961.

5. *Standard Handbook for Electrical Engineers*, McGraw-Hill, New York, 1993.

6. M. J. Steinberg, *Economic Loading of Steam Power Plants and Electric Systems*, 2nd ed., John Wiley & Sons, Inc., New York, 1947.

7. L. K. Kirchmayer, *Economic Control of Interconnected Systems*, Wiley, New York, 1959.

8. G. R. Slemon and A. Straughen, *Electric Machines*, Addison-Wesley, Reading, Mass., 1982.

9. L. K. Kirchmayer and G. W. Stagg, "Analysis of Total and Incremental Losses in Transmission Systems," *AIEE Transactions* **70**, 1197–1204 (1951).

10. J. A. Casazza, "Coordinated Regional EHV Planning in the Middle Atlantic States—U.S.A.," CIGRE Paper No. 315, Paris, France, 1964.

11. Federal Power Commission, *National Power Survey—A Report by the Federal Power Commission*, U.S. Government Printing Office, Washington D.C., October 1964.

12. J. A. Casazza, P. J. Palermo, J. Lucas, and F. Branca, "Generation Planning and Transmission Systems," paper presented at the CIGRE International Conference on Large High Voltage Electric Systems, Paris, France, August 1988.

13. T. P. Hughes, "Systems Builders Technology's Master Craftsmen," *CIGRE Electra*, **109**, 21–30 (Sept. 1986).

14. N. Cohn, *Control of Generation and Power Flow on Interconnected Systems*, Wiley, New York, 1966.

15. C. Concordia and L. K. Kirchmayer, "Tie-Line Power and Frequency Control of Electric Power Systems," *AIEE Transactions*, **72**, 562–572 (1953).

16. C. Concordia and L. K. Kirchmayer, "Tie-Line Power and Frequency Control of Electric Power Systems, Part II," *AIEE Transactions*, **73**(pt. III-A), 133–141 (1954).

17. National Electric Reliability Council (NERC), "Transfer Capability—A Reference Document," NERC, October 1980, pp. 6–7.

18. R. H. Travers, "Load Control and Telemetering," *AIEE Transactions* **73**(pt. III-B), 522–527 (1954).

19. "Control Area Concepts and Obligations," North American Electric Reliability Council, 1992.

20. "Overview of Planning Reliability Criteria of the Regional Reliability Councils of NERC," North American Electric Reliability Council.

21. "Discussion of Regional Council Planning Reliability Criteria and Assessment Procedures," NERC.

22. P. J. Palermo, R. A. Bolley, and T. R. Woodward, *The Effects of Coordinated Operation on Energy Exchanges, System Operation and Data Exchange Requirements: A Comparison of Methods Used in the USA*, 1992.

23. "Transfer Capability—A Reference Document," NERC, 1980.

24. G. S. Vassell and R. M. Maliszewski," AEP 765-kv System: System Planning Considerations," *IEEE Transactions on Power Apparatus and Systems*, **PAS-88**(9), 1320–1328 (1969).

25. T. J. Nagel and G. S. Vassell, "The American Electric Power System's Transmission Grid: A Major Asset in an Uncertain World," CIGRE Paper No. 31-01, Paris, France, 1982.

26. "The Future in High-Voltage Transmission: Flexible AC Transmission Systems (FACTS)," *Proceedings from the EPRI Workshop*, 1990.

27. J. A. Casazza, "Understanding the Transmission Access and Wheeling Problem," *Public Utilities Fortnightly*, October 31, 1985.

28. "The Great 'Retail Wheeling' Illusion—And More Productive Energy Futures," National Resources Defense Council.

29. "Electric Power Wheeling and Dealing: Technological Considerations for Increasing Competition," U.S. Congress Office of Technology Assessment, 1989.

30. *Non-Technical Impediments to Power Transfers*, National Regulatory Research Institute, 1987.

31. "Biological Effects of Power Frequency Electric and Magnetic Fields," U.S. Congress Office of Technology Assessment, 1989.

32. *Electric and Magnetic Field Cases*, Edison Electric Institute, 1989.

33. P. Nicolini and R. Conti, "ENEL Research on Biological Effects of 50-H7 EMF Show No Harmful Impact on Animal and Vegetation Life," *Transmission and Distribution International*, June 1990, pp. 100–103.

34. C. Kowalczuk, "Health Effects of Low-Frequency Electromagnetic Fields," *Power Technology International 1993*, 1993, pp. 27–31.

35. *Your Guide to Understanding EMC and Your Home*, Culver Company, 1993.

ELECTRICAL ENERGY EFFICIENCY, COMMERCIAL APPLICATIONS

KAREN LARSEN
Office of Technology Assessment
Washington, D.C.

Commercially available energy-efficient technologies offer abundant opportunities to cut electricity consumption in the residential, commercial, and industrial sectors. The major electricity uses across all sectors are lighting, space conditioning, water heating, motors, drives, and appliances. Studies of energy efficiency opportunities have identified a variety of technologies for each of these applications that offer cost-effective savings and rapid paybacks. Still other energy-saving technologies are not currently cost-effective in most applications, but could prove more financially attractive if economies of scale cut costs, if energy prices rise, or if policy interventions provide additional incentives to install them.

This article briefly examines some of the energy efficiency opportunities in the residential, commercial and industrial sectors, including a profile of electricity use in each sector, examples of electricity-saving technologies, estimates of potential savings, and major factors influencing technology adoption. See also LIGHTING, ENERGY EFFICIENCY; ENERGY EFFICIENCY; COMMERCIAL AVAILABILITY OF ENERGY TECHNOLOGY; AIR CONDITIONING; REFRIGERATION; ENERGY AUDITING; ENERGY CONSERVATION.

HOW MUCH ELECTRICITY CAN BE SAVED?

Estimates of how much energy can be saved through more efficient electric technologies, vary. Some of the differences in the estimates are attributable to what measure of energy efficiency is used–maximum technical potential, cost-effective potential, or achievable or likely savings potential. Estimates of potential energy savings from efficient technologies vary considerably. At least part of the difference in estimates can be attributed to what is being estimated. Most published estimates use one of the following measures:

Maximum technical potential, or MTP, is a measure of the most energy that could be saved if all possible efficiency improvements were made with the most efficient technologies adopted in all new and existing applications (ie, 100% penetration reached). Achieving MTP savings assumes aggressive government and private efforts and implementation of policies designed to make efficient alternatives attractive to everyone. Supporting policies might include, for example, increased R&D to lower costs, information programs, and rebates and other financial incentives.

Cost-effective potential is an estimate of the energy savings that could be obtained if efficient technologies are installed in new and replacement applications whenever they are cost-effective. Cost-effective potential is lower than MTP and depends on projections of future marginal electricity costs and rates. Several cost-effectiveness tests are in common use in utility planning and rate regulation.

Likely energy efficiency savings estimates are used in utility planning and reflect judgments about the savings from efficient technologies adopted in response to utility programs. Likely impacts are lower than cost-effective potential because of the influence of various factors including, for example: lack of customer awareness of potential savings or utility programs; customer reluctance to convert with new or different technologies; and constraints on the supply or deliverability of the technology.

Natural occurring energy efficiency savings estimates reflect estimates about the penetration of energy efficient technologies in response to normal marketplace conditions and existing standards in the absence of new utility or other programs to encourage their adoption. The savings arise from installation of newer, more efficient technologies—but not necessarily the most efficient technologies commercially available—in new and replacement applications. Estimates of naturally occurring savings are used by utilities to evaluate the effectiveness of efficiency programs. Actual electricity use is compared to what consumption would have been if efficiency levels were frozen at a base year's level and then the effects of naturally occurring savings are subtracted to yield the savings attributable to the utility program.

Figure 1 shows a conceptual comparison of the relative magnitudes of different estimates of energy efficiency potential.

The Electric Power Research Institute (EPRI) has estimated that if the existing stock of equipment and appliances were replaced with the most efficient commercially available technologies, projected U.S. electricity use in the year 2000 could be cut by 27 to 44% without any diminution of services (1). (Table 1.) The EPRI analysis presents its best-case estimates of the most energy that could be saved through efficient technologies, further improvements in existing technologies, and policy initiatives such

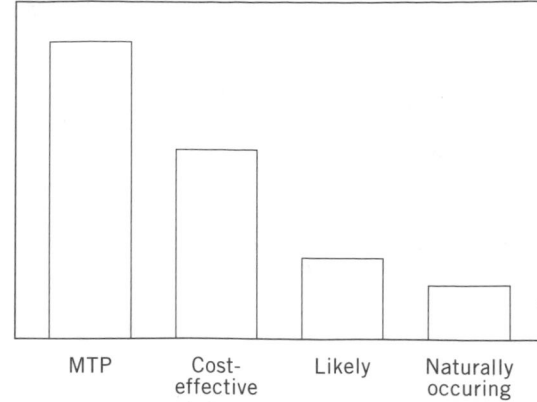

Savings in 2000

Figure 1. EPRI base case usage and maximum technical potential (MTP) from electricity-savings technologies.

as information programs, rebates and other incentives such as information programs, rebates and other incentives that make the alternatives attractive to everybody. The range in their estimates from "conservative low impacts" to best-case, "high" impacts reflects uncertainties in technology applicability, manufacturing capabilities, and performance characteristics.

The EPRI MTP estimates of savings from efficient electric technologies in the year 2000, included savings from: (1) using the most efficient electricity-saving technologies available for new installations and replacement of all the existing stock of installed electric equipment; and (2) replacing less-efficient fossil-fired equipment with more efficient electrotechnologies in industrial processes. EPRI's MTP estimates compared with current and projected electricity use by sector are shown in Table 1.

The estimates of savings were developed using a baseline projection of electricity demand in the year 2000, which includes naturally occurring improvements in efficiency, the effects of mandatory standards and a best case scenario in which all applicable technologies are replaced instantaneously with the most efficient commercially available electric equipment.

The MTP estimates are subject to a great deal of uncertainty including:

- The efficiency levels of new and existing equipment.
- The unknown impacts from naturally occurring efficiency improvements.
- Physical constraints that limit the applicability, compatibility, or deliverability of efficient equipment.

To account for these uncertainties, the EPRI report used two scenarios reflecting a range of impacts from technology adoption: an "optimistic" or high impact scenario assuming adoption of all commercially available technologies (ie, no prototypes, demonstration models, or lab bench-scale technologies), and a "conservative" or low impact scenario reflecting possible constraints on the penetration rates due to technology applicability and manufacturer capacity. Neither estimate reflects considerations of cost-effectiveness, the economic trade-offs between efficiency improvements and equipment cost.

Table 1. EPRI Base Case Usage and Maximum Technical Potential (MTP) From Electricity-Savings Technologies (gigawatt-hours)

Uses	Electricity Consumption		Electricity Savings			
	1987 Base GWh	2000 Base GWh	Low Case GWh	% of Base	High Case GWh	% of Base
Residential Uses Sector						
Space heating	159,824	223,024	71,915	32.2%	122,285	54.8%
Water heating	103,499	134,509	43,481	32.3	88,995	66.2
Central air conditioning	78,127	90,134	26,265	29.1	30,996	34.4
Room air conditioning	15,254	13,063	2,421	18.5	4,222	32.3
Dishwashers	15,308	23,707	1,240	5.2	6,233	26.3
Cooking	30,390	39,271	3,115	7.9	7,132	18.2
Refrigeration	146,572	139,255	30,716	22.1	66,896	48.0
Freezer	59,779	48,073	11,534	24.0	15,594	32.4
Residual appliances	240,861	353,620	98,242	27.8	141,552	40.0
Total residential	849,613	1,064,656	288,929	27.1%	483,904	45.5%
Industrial Uses						
Motor drives	570,934	780,422	222,226	28.5%	351,040	45.0%
Electrolytics	98,193	138,273	25,950	18.8	41,124	29.7
Process heating	83,008	125,274	9,928	7.9	16,606	13.3
Lighting	84,527	114,097	19,016	16.7	38,032	33.3
Other	8,453	9,192	0	0.0	0	0.0
Total industrial[a]	845,266	1,167,413	277,119	23.7	446,802	38.3%
Commercial Uses						
Heating	77,245	128,322	16,335	12.7%	30,333	23.6%
Cooling	154,299	208,106	62,432	30.0	145,674	70.0
Ventilation	76,959	96,094	28,828	30.0	48,047	50.0
Water heating	24,068	39,794	15,917	40.0	23,876	60.0
Cooking	16,172	26,381	5,276	20.0	7,914	30.0
Refrigeration	60,883	81,652	9,925	12.2	27,857	34.1
Lighting	238,488	283,124	62,916	22.2	157,291	55.6
Miscellaneous	108,447	177,254	32,228	18.2	64,456	36.4
Total commercial	756,561	1,040,726	233,858	22.5%	505,448	48.6%
Total	2,451,440	3,272,795	799,905	24.4%	1,436,154	43.9%

[a] Ref. 1.
[b] Sum of end uses may not add to total due to rounding.

The analysis did not include assessments of the cost-effectiveness of the technologies in particular applications or projections of future electricity costs and rates that would strongly influence cost-effectiveness determinations. Considerations of cost, practicality, and capital availability may preclude attainment of the maximum savings potential, but nevertheless EPRI believes that many opportunities remain for substantial gains.

OTA's own analysis concluded that cost effective, energy-efficiency measures could yield savings of one-third in total energy use in the residential and commercial sectors by 2015 over a business as usual scenario. In fact total energy use in these sectors would decline somewhat under an aggressive efficiency strategy. These two sectors combined are often dubbed "the buildings sector" because energy use for building systems (space heating and conditioning, ventilation, lighting, and water heating) has made up the overwhelming bulk of energy consumption in these two sectors. Reported energy use for the buildings sector includes building systems, appliances, office systems, and other electrical equipment.

Amory Lovins and others at the Rocky Mountain Institute have estimated the maximum technical potential of efficiency savings as high as 75% by the year 2010 (3). Other studies have included considerations of cost-effectiveness in their estimates.

OTA's report, Energy Technology Choices: Shaping Our Future (4) a moderate-efficiency scenario, assumes adoption of all cost-effective efficiency measures (defined as measures that repay their added incremental costs with energy savings over their lifetimes). The scenario also assumes adoption of a variety of government policy initiatives to overcome significant market, institutional, and behavioral barriers that have hampered full use of cost-effective, energy-savings opportunities. Under the moderate-efficiency scenario, electricity demand in 2015 would be 25% less than the baseline demand (which assumes some naturally occurring efficiency improvements, but no significant policy initiatives).

The 1991 National Energy Strategy, projects that electricity consumption in 2010 will be about 12% less than the current policy baseline due to cost-effective energy savings from proposed initiatives to promote utility integrated resource planning (and associated demand-side management programs), building and appliance efficiency standards, and industrial conservation research and development (5). Other studies on energy efficiency opportunities in specific sectors or regions have yielded similar estimates of cost-effective savings potential.

There is considerable agreement among the various energy efficiency potential studies about the most promising strategies for achieving more efficient use of electricity. They include:

- Improvements in the thermal integrity of building shells and envelopes.
- Improvements in the efficiency of electric equipment.
- Lighting improvements.
- Net efficiency gains from shifting energy sources from fossil fuels to electricity (electrification).
- Optimization of electricity use through better energy management control systems, shifts in time of use, and consumer behavior and preference changes.

ENERGY-EFFICIENCY OPPORTUNITIES FOR RESIDENTIAL CUSTOMERS

The residential sector essentially consists of all private residences including single and multifamily homes, apartments, and mobile homes. Institutional residences, such as dormitories, military barracks, nursing homes, and hospitals are included in the commercial sector. About 22% of total primary energy consumption in the United States can be attributed to residential sector energy demand. Total energy expenditures by the residential sector in 1990 were $110.5 billion (6).

Figure 2 shows direct on-site energy consumption in the residential sector. Historical energy use statistics of the Energy Information Administration do not separate residential and commercial property energy use. Residential energy use share is based on Gas Research Institute estimates (7,8). Electricity at present supplies about 30% of the residential energy needs and this share is expected to grow as electric heating and appliance loads grow. Nat-

ural gas supplies 47% of residential energy use mostly for space and water heating. The remaining residential energy consumption consists of oil (15%), coal (1%), and other energy sources (7.6%), predominantly firewood.

If the residential sector's share of direct primary energy consumption is augmented by its pro-rate share of primary energy consumed by electric utilities in the generation, transmission and distribution of electricity for residential customers, electricity accounts for some 60% of primary energy consumption attributable to the residential sector. The existence of these sizable conversion and delivery losses associated with end-use electricity consumption means that energy savings at the point of use are magnified in their impacts on utilities and overall primary energy use.

The residential sector accounts for about 34% of all U.S. electricity sales. In 1990, total residential electricity sales (exclusive of conversion and transmission losses) were 924 billion kilowatt-hours (kWh) at a cost of $72 billion (9). Residential electricity demand growth is driven by population, climate, number of households, the number of persons per household, regional population growth patterns, increased demand for electricity-intensive services (eg, air-conditioning, clothes-dryers) and size of residences.

Among factors that tend to limit growth are the decline in population growth, the increased efficiency of new housing stock and appliances, and retrofits of existing housing units. Various forecasts peg expected growth in residential electricity demand at from 1 to 2% per year (10).

Figure 3 shows household electricity use by application. Within each of the categories shown there are a number of attractive and cost-effective options for cutting household electricity use, without diminishing the services provided.

EPRI's analysis of maximum technical potential estimated that residential electricity use in 2000 could be reduced by from 27 to 45% if the most efficient end-use technologies currently available commercially were used to replace the existing stock of electric appliances in homes. The EPRI study did not include estimates of total costs for achieving this maximum technical potential, nor any

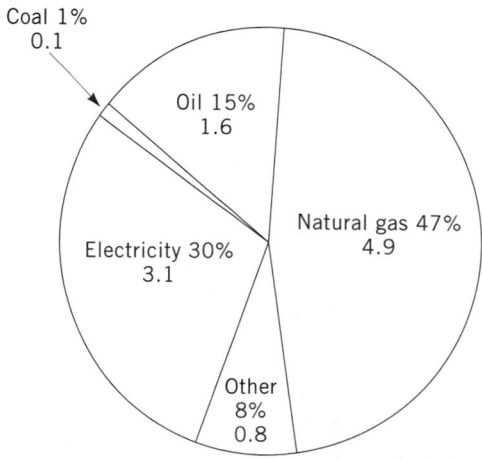

Figure 2. Residential on-site energy consumption by source, 1990 (quadrillion Btus).

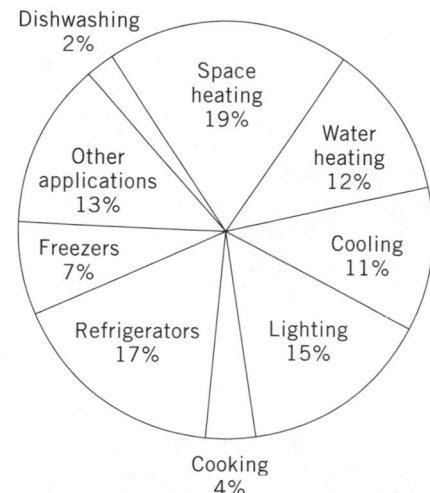

Figure 3. Residential electricity use by application, 1987.

analysis of the cost-effectiveness of replacing working appliances with more efficient models. Other studies have included cost-effectiveness considerations in their analyses and generally found considerable opportunities for electricity savings in the residential sector at a cost less than that of supplying electricity.

A study of electricity use in U.S. residences by researchers at Lawrence Berkeley Laboratories estimated that residential electricity demand in 2010 could be cut by 37% from a "frozen" efficiency baseline (ie, excluding "naturally" occurring efficiency gains over the period) by aggressive use of commercially available technologies with a cost of conserved energy below 7.6 cents/kWh, using a discount rate of 7% (11). Another analysis of possible electricity savings in Michigan found achievable savings of 29% in residential electricity use by 2005 at reasonable cost over a business-as-usual baseline with aggressive conservation programs and commercially available technologies (12). Researchers estimated that current residential electricity use in New York State could be cut 34% at a cost below that of supplying electricity-less than 7 cents/kWh, assuming a 6% discount rate (13).

Residential Energy Efficiency Technologies

There are a variety of technologies available to cut residential energy use without diminishing the services provided. Some of these technologies are listed in Table 2. The basic strategies for cutting electricity use in the residential sector are

- Improving residential building shell efficiency through better insulation by cutting conductive heat losses and gains through ceilings, walls, and floors; in-stalling storm doors and windows; and cutting air infiltration by caulking gaps and weatherstripping around doors, windows, joints and other spaces;
- Choosing more efficient appliances for new installations and accelerating the retirement of older less efficient appliances;
- Improving the management of residential energy use through better maintenance, as energy management controls, load shifting, and changes in occupant behavior.

Improving the energy efficiency of existing buildings is one of the most promising and vital areas for energy savings. Space heating and cooling account for 30% of residential electricity use. Improved thermal integrity in new and existing residential buildings can reduce heating and cooling loads and save electricity.

Replacement of existing buildings by energy-efficient new buildings is slow and expensive; most of the existing

Table 2. Selected Energy Efficiency Technology Options for the Residential Sector[a]

Building envelope improvements	Air-conditioners
Cut conductive heat losses/gains; control infiltration	Central air-conditioners
Weatherstripping and caulking	More efficient units
Insulation improvements	Frequent cleaning of filters and coils
Storm windows and doors	
Design and siting of new structures	Room air-conditioners
	More efficient units
Space heating	Frequent cleaning of filters and coils
Use heat pumps instead of resistance heat	
Air source heat pumps	Refrigerators and freezers
More efficient models	Efficient motors and controls
Improved technology	Improved gaskets and seals
Ground-source heat pumps	Improved insulation
Solar heating	Improved maintenance
	Clean coils often
Energy management controls and systems	
Set-back thermostats	Lighting
Smart house/smart systems	Replace incandescents with fluorescents and compact fluorescents
Zoned heat systems	Reduced wattage incandescents
	Dimmers, controls, and sensors
Air distribution systems	Reflective fixtures
Improved insulation	
Reduced duct leakage	Cooking
	More efficient ovens and stoves
Water heating	Alternative cooking devices
Blanket wrap of existing tanks	Microwave ovens
More efficient tanks	Convection ovens
Increased insulation for tanks and pipes	Induction cooktops
Low-flow devices	
Thermal traps	Dishwashers
Set-back thermostats	Energy-saver cycles
Heat-pump water heaters	No-heat drying
Alternative water heating systems	Reduced hot water usage
Heat recovery water heaters	
Solar water heat systems	
Reduced thermostat settings	

[a] From Office of Technology Assessment, 1993.

housing stock will continue in use for the next 30 to 40 years or more. There are over 90 million residential units in the United States, and we are adding between 1 and 2 million units per year. Although by the year 2000 there will be 10 to 15 million new units, about 90% of the units existing in 2000 have already been built, and by the year 2010 it is estimated that about 70% of homes will consist of housing stock built before 1985 (14).

The most cost-effective time to incorporate energy-saving measures into buildings is when they are built, re-modeled or rehabilitated. In fact, failure to make accom-modation for energy-saving technology in material and design choices at this stage causes lost energy savings op-portunities—for example, using the standard 2-by-4 di-mension lumber in exterior walls instead of 2-by-6 con-struction that allows for more insulation, or not selecting the most energy efficient windows.

Careful attention to energy efficiency features in the design, setting, and construction of residential housing can save electricity. Over the past two decades, because of high energy prices, building code requirements, and greater attention to energy efficiency, newer residential buildings make greater use of energy-efficient features. In fact, new houses built in 1985 were much more energy efficient than those built in 1973 and were better insu-lated and had more energy-efficient windows. Design fea-tures to take advantage of passive solar heating and day-lighting can also be incorporated into new units for additional savings.

The rate of replacement of major appliances with newer, more efficient models has been slow and will con-tinue to be so in the absence of policy initiatives or large changes in energy prices. Major electric appliances such as furnaces, heat pumps, central air-conditioners, water heaters, and refrigerators often are in use for 10 to 20 years or more and are unlikely to be replaced unless they fail. It could take as long as 20 years to realize potential savings from currently available efficient new equipment. Not installing the most energy-efficient model initially creates lost efficiency opportunities for a decade or more. Assuring the installation of the most efficient appliances and accelerating the replacement of older inefficient ap-pliances offer prospects for reaping energy savings.

Building shell improvements in existing buildings are effective means of cutting heating and cooling costs and increasing occupant comfort. The most common weatheri-zation retrofits include: installing more insulation in ceil-ings, walls, and floors; adding storm windows and doors, and weatherstripping and caulking windows and doors. One study of home retrofits found variations in savings attributable to climate and differences in individual build-ing characteristics. Average savings of 12 to 21% in heat-ing energy demand and payback periods of about 6 years were found for ceiling and wall insulation. Another inten-sive experiment in weatherization cut space heating elec-tricity use by two-thirds (2). According to DOE surveys, many Americans have already taken some steps to im-prove the energy efficiency of their homes. Even where some weatherization measures have been reported, it is likely that additional efficiency upgrades are possible.

Various measures are used to indicate the energy effi-ciency of electrical devices. The following are among the most common measures for residential and commercial equipment:

The energy efficiency ratio (EER) is used to measure the cooling performance of heat pumps and air-condition-ers. EER is expressed as the number of Btus of heat re-moved from the conditioned space per watt-hour of elec-tricity consumed (ie, the cooling output divided by the power consumption). Typical EERs for room air-condition-ers are 8.0 to 12.0 Btus per watt-hour. The higher the EER the more efficient the air conditioner.

The seasonal energy efficiency ratio (SEER) is used to measure the seasonal cooling efficiency of heat pumps. SEER is expressed as the number of Btus of heat removed from the conditioned space per watt-hour of electricity consumed under average U.S. climate conditions. Unlike the EER, the SEER incorporates seasonal performances under varying outdoor temperatures and losses due to cy-cling. Typical SEERs are 9.0 to 12.0 Btus per watt-hour.

The heating seasonal performance factor (HSPF) is a measure of the seasonal heating efficiency of heat pumps under varying outdoor temperatures, losses due to cy-cling, defrosting, and backup resistance heat for average U.S. climate conditions. HSPF is expressed as the number of Btus of heat added to the conditioned space per watt-hour of electricity consumed. Typical values are 7.0 to 12.0 Btus per watt-hour.

The efficiency factor (EF) is a measure of the energy efficiency of water heaters based on the energy used to provide 64 gallons of hot water per day.

The annual energy cost (AEC), required by Federal ap-pliance labeling regulations, reflects the cost of energy (usually electricity) needed to operate a labeled appliance for 1 year at a specified level of use. The AEC label pro-vides information on the costs of operating the labeled ap-pliance and similar models over a range of energy prices (eg, cents per kilowatthour) to account for variations in local rates.

Space Heating

About one-quarter of American homes (22 million units) depend on electric heat and each year more and more electrically heated units are added (15). Electric space heating accounts for 19% of residential electricity con-sumption. There are two basic categories of electric space heating systems: electric resistance heat systems (includ-ing electric furnaces, baseboard heaters, and portable heaters) and electric heat pumps (including air-source heat pumps, and ground-source heat pumps). Electric re-sistance heating systems are virtually 100% efficient; that is, 100% of the energy delivered to the system is con-verted to heat, so that there are few technical opportuni-ties to improve on their performance.

Electric heat pumps use a reversible vapor compres-sion refrigeration cycle to transfer heat from an outside source to warm indoor spaces in the winter; in summer, the cycle reverses to cool indoor spaces by removing heat from inside and discharging it outdoors. The most com-monly used heat pump is the air-source heat pump that uses the ambient air as its heat source. On average heat pumps are twice as energy efficient as electric resistance systems. However, the performance of heat pumps is highly variable and dependent on sizing, climate, and the rated performance of the heat pump. At about $-5°C$ (23°F), heat pumps begin to lose their heating capacity and in moderate to cold climates they must have a backup heat source, usually an electric rhesustance heater. There is a considerable range in the performance of residential heat pumps currently on the market. The typical heat pump has a heating efficiency (heating season performance factor, or HSPF) of about 6.9 Btus per watt-hour and a cooling efficiency (seasonal energy efficiency ratio, or SEER) of about 9.1 Btus per watt-hour. The best units currently on the market have efficiencies of 9.2 HSPF and 16.4 SEER. Federal mini~mum efficiency standards for heat pumps sold after 1992 specify 6.8 HSPF and 10.0 SEER.

Another variant of the heat pump, the ground-source heat pump uses groundwater, or the ground itself as the heat source. This technology offers an advantage over air-source heat pumps, in that ground temperatures seldom drop below freezing, thus there is no loss of heating capacity or resultant need for supplemental resistance heat.

For both heat pump and electric rhesustance heat systems, improving the thermal integrity of the building shell or envelope and insulating and plugging leaks in air distribution ducts can also cut heat losses and reduce the heating loads.

EPRI estimated that a combination of envelope improvements, a shift to electric heat pumps, and improvements in heat pump efficiencies could result in savings of 40 to 60% in space heating electricity demand in 2000 over 1987 stock.

Space Cooling

Air-conditioning accounts for about 11% of the residential energy consumption and this demand is projected to grow as more homes are air-conditioned. Over two-thirds of U.S. homes are now air-conditioned; 40% have central air-conditioning and 29% have room units. Over three-quarters of new housing units have central air-conditioning, but this growth in air-conditioning demand has been offset by increases in the efficiency of both central and room air-conditioning units.

The most efficient central air units on the market today have a SEER of 16.9 Btus per watt-hour (16) and new Federal appliance standards in effect in 1992 will require a mini~mum SEER of 10 Btus per watt-hour. Just 10 years ago, the average efficiency for new central air systems was 7.8 Btus per watt-hour. These gains were due to more efficient fan motors and compressors, larger evaporator coils and condensers, and reduced air~flow resistance. EPRI has estimated that as of 1987, the stock of central air units in use had an average SEER of 7 Btus per watt-hour—considerably below the most efficient systems on the market. New installations and replacement of existing units with higher-efficiency central air units could cut electricity use by central air-conditioners in 2000 by 29 to 34% or more according to EPRI.

Room or "window" air-conditioners have also improved with the addition of more efficient motors for fans and compressors, better fan blade design, larger heat exchangers, reduced airflow path resistance and better low-

temperature refrigerant line insulation (17). Efficiencies vary according to model sizes and features, but nevertheless new units today use about 30% less electricity to operate than units sold in 1972. The most efficient units available today, with SEERs of 12 consume half the electricity of 1972 models. EPRI estimated that the 1987 stock of room air-conditioners had average SEER of 6.5 Btus per watt-hour. Using the most efficient room units for new and replacement installations could cut room air-conditioner electricity use by 19 to 32% by 2000 according to EPRI's analysis.

Better maintenance of air-conditioners can also boost efficiency. A dirty filter can cut efficiency by 10%. Cleaning air-conditioner coils and cleaning or replacing dirty filters can preserve efficiency.

Heat pumps are also used for space cooling. Today's typical heat pump has a SEER of 9, but commercially available high-efficiency models have SEERs up to 16.4. New Federal standards effective in 1992 will set minimum cooling efficiency for new heat pumps at 10. Careful selection and sizing of heat pumps to match cooling loads, especially in hot climates, can increase efficiency.

Water Heating

Electric water heating is used in about 37% of American homes and makes up about 12% of residential electricity consumption. Electric resistance water heaters are the most common type of electric water heater in use today and new units incorporating better tank insulation and improved heat transfer surfaces, use 10 to 15% less electricity than models of 10 years ago. (On average, larger size hot water tanks are less efficient.) Other electricity-saving measures including wrapping the out~side of the hot water tank with an insulating blanket, insulating hot water pipes, and installing devices such as low-flow shower~heads, aerators, and self-closing hot water faucets. EPRI has estimated that use of these energy-saving measures could cut water heating power needs by 20 to 30% in 2000.

Shifting to alternative electric water heating systems, such as heat-pump water heaters, heat-recovery water heaters, and solar hot water systems can achieve efficiencies of up to 70%. Overall, EPRI has estimated that the range of efficient electric water heating technologies offered savings of from 40 to 70%.

Refrigerators and Freezers

Together, refrigerators and freezers make up about 24% of residential electricity demand. Both technologies have seen substantial increases in efficiency over the past 20 years, but opportunities for significant improvements in performance remain.

The typical refrigerator on the market today uses just 45% of the electricity needed to run the average 1972 model. ≪ceo~27≫ A combination of technological gains has produced these savings, including: more efficient fans, motors, and compressors; better and more compact insulation; improved door seals and gaskets; and dual compressors. DOE researchers believe that it is technically feasible to cut electricity needed to run today's average new model almost 50% further. EPRI's analysis estimates that more efficient refrigerators could cut energy use about 22 to 48% in 2000 over the 1987 stock. Even more efficient refrigerators are available today than those assumed in the EPRI report, so that the maximum potential savings probably under~state the potential.

Freezers account for 7% of residential electricity use and are found in about 34% of U.S. households. Stand alone freezers also have seen significant efficiency gains over the past 20 years as a result of advances in refrigeration technology. The typical model sold today uses half the electricity of the average 1972 model and as with refrigerators, additional efficiency gains are probable. More efficient freezers could save 24 to 32% over energy required for the 1987 stock according to EPRI analyses.

Complicating the drive for more efficient refrigerators and freezers is the need to find replacements for the chlorofluorocarbons (CFCs) used as refrigerants and in insulation that offer equivalent or improved performance. In an innovative effort to overcome market barriers that have slowed the commercialization of more energy-efficient consumer appliances, 25 U.S. utilities joined to offer a "Golden Carrot"™ in the form of a $30-million award to the winner of a design competition for an advanced, energy-saving refrigerator that is free of ozone-depleting chlorofluorocarbons (CFCs). The consortium, the Super-Effficient Refrigerator Program, Inc. (SERP), was formed in collaboration with the U.S. EPA Global Change Division's green programs, EPRI, and others. Its member utilities provide electric service to more than one quarter of the Nation's households. The aware will provide the winning manufacturer with a subsidy of over $100 per refrigerator. In return, the new super-efficient refrigerator will be delivered in participating utilities' service areas before it is available to other distributors.

Among the disincentives that the SERP program and possible future "Golden Carrot" competitions are intended to counter are consumer reluctance to try new products and the higher first cost of more energy-efficient or green products. By offering a subsidy for development of the winning design and guaranteed orders for a sizable initial manufacturing run, SERP hopes to create a market pull for the energy-saving product, lower product development risks, and allow the manufacturer to achieve economies of scale in production. This strategy should accelerate commercialization and result in a lower market price for the product than in the absence of the incentive. It will also help speed the commercialization of replacements of CFCs that are to be phased out of production by 1995.

The competition challenged manufacturers to commit to producing a CFC-free refrigerator at least 25% more efficient than the 1993 Federal energy efficiency standards require and to deliver them to participating utilities' service areas in 1994–1997. The manufacturer must agree to assemble the refrigerators in North America. Additional points in the competition could be awarded for achieving greater efficiency levels.

Bids were received in October 1992 and all but 1 of the 15 major U.S. manufacturers entered the competition. Submittals were reviewed based on a number of key factors including proposed design, efficiency levels, incentive requested, marketing plans, and technological experience.

In December 1992, Whirlpool Corp. and Frigidaire Co. were selected as finalists to design the new refrigerator. The winner announced in June 1993 was Whirlpool Corp., which will deliver about 250,000 SERP refrigerators in various models between 1994 and 1997. SERP refrigerators will be priced the same as other models with similar features.

EPA estimates that a super-efficient refrigerator has the potential to save 300 to 400 kWh/year over its lifetime and save its owners a total of about $500 on utility bills. It also is expected to eliminate 9,000 pounds of carbon dioxide emissions compared with current models.

As with air-conditioning, maintenance practices can affect the efficient operation of refrigerators and freezers. Cleaning refrigerator coils two to three times per year can save about 3% of annual refrigerator electricity use at little or no cost.

Lighting

About 15% of household electricity load is lighting. As in other sectors, use of more energy-efficient lighting products can save electricity for residential customers. OTA's report Building Energy Efficiency estimated that efficient lighting could cut residential lighting electricity use by one-third if one-half of all residential incandescent lights were replaced by compact fluorescents (2). Assuming the light is used 6 hours per day, OTA calculated a payback period of 1.7 years for a $20 compact fluorescent bulb. Compact fluorescents also last 10 to 13 times longer than standard incandescent bulbs. EPRI estimated maximum potential lighting related savings at from 20 to 40% in 2000.

Depending on applications, compact fluorescent bulbs can cut energy use per bulb by two-thirds over standard fluorescents. Even standard fluorescents offer energy savings over incandescent bulbs for equivalent lighting output. But consumers often find fluorescent lighting unacceptable or unattractive for some purposes. The extent to which energy-efficient lighting can cut electricity demand in the residential sector is highly uncertain and depends on consumer preferences and applications. Manufacturers of compact fluorescents continue to make progress on adapting these lamps for more common residential fixtures and to improve the quality of light provided, which may hasten acceptance by residential customers.

Other options such as lower-wattage "energy-saver" incandescents, reflector fixtures, task lighting, dimmers, and automatic lighting controls can also shave lighting energy use. Increased use of daylighting through windows, skylights, and clerestories can also reduce the need for interior lighting.

Cooking

Electric ranges and ovens account for 4% of household electricity demand. Newer models, particularly self-cleaning ovens are more efficient than current stock owing to a number of changes: more insulation, better seals, improved heating elements and reflective pans, reduced thermal mass, reduced contact resistance, and better controls. The penetration of microwave ovens, convection ovens, and induction cooktops also offer energy savings. It is uncertain whether microwave ovens, which cook food with one-third the electricity required for standard electric ranges and ovens, will actually result in reduced cooking loads as consumers may tend to use them more as an adjunct to conventional appliances. EPRI estimates that replacement of the 1987 stock of ranges and ovens with more efficient models could produce savings of 10 to 20% in electricity demand for cooking in 2000.

Dishwashers

Dishwashers account for about 2% of household electricity use and are found in 43% of households. Energy-saving features such as better insulation, water temperature boosters, water saver cycles, and air drying cycles can cut electricity consumption. Total savings are dependent on the customer's use of energy-saving cycles. EPRI estimates that improved dishwashers could cut dishwasher electricity demand in 2000 by 10 to 30% over 1987 stock.

Other Appliances

The remaining household electric appliances, such as clothes washers and dryers, televisions, stereos and other electronic equipment, vacuums, small household appliances, power tools, and home computers account for about 13% of present residential electricity use. This portion of household electricity demand is expected to grow with greater saturation of clothes washers and electronic equipment. Newer models will be more energy efficient, and EPRI estimates that this trend is expected to result in electricity consumption that is 10 to 20% less than equivalent 1987 models by 2000.

Estimating net efficiency gains from more efficient appliances is difficult; however, because energy services are growing, and households may use the energy savings to buy larger appliances or increase the utilization of the equipment.

OBSTACLES TO RESIDENTIAL ENERGY EFFICIENCY

Total residential energy use in 1990 was over 1 quad less than it was in 1978, even as the number of households grew from 77 million to 94 million, reflecting a steady improvement in residential energy efficiency (19). Over this period the energy intensity of new living space has decreased and many older units were retrofitted with a variety of energy-saving measures. Major household appliances use significantly less electricity to operate than comparable models of 20 years ago.

Household electricity use also has grown from 24% of residential energy use in 1978 to 30% in 1990, but growth in residential electricity demand has been less than it might have been without energy efficiency gains. These gains are attributable to several factors in addition to evolutionary efficiency gains: higher energy prices during the 1970s and early 1980s; energy efficiency requirements in building codes; appliance labeling and efficiency standards; government and utility energy education efforts; utility conservation programs; and more awareness of awn energy efficiency by consumers, equipment vendors, and building professionals and tradespeople.

Even with the admirable gains that have been made in energy efficiency since the 1970s, there remains a sizable gap between the most energy-efficient products on the market today and the products in use in American homes. More efficient options exist for almost all of the major electricity uses at home. The potential energy and cost savings from residential energy-efficiency investments are significant according to many efficiency proponents. For many measures the energy savings over the lifetime of the investment would exceed the initial cost, in some cases offering payback periods of 2 years or less.

If energy efficiency investments are such attractive investments, why then haven't they been enthusiastically embraced by American consumers? Analysts commonly cite a host of disincentives that have tended to dampen the pace and extent of efficiency savings. These include a number of institutional, economic, behavioral, and practical matters.

OTA's report, Building Energy Efficiency, has found a confluence of factors that resulted in underinvestment in residential energy efficiency. Decision-making affecting household energy efficiency is fragmented among: residents (homeowners and renters); architects; developers; builders; equipment manufacturers and vendors; and a host of federal, state, and local government agencies. For all of these decision-makers, energy efficiency is only one of many attributes considered in making choices that affect home energy use and it competes against such characteristics as lower first cost, appearance, convenience, features, and hassle-avoidance. For most decision-makers, awn-energy efficiency has not been a high priority. In all too many instances, residential consumers are effectively precluded from energy efficiency opportunities because design and major equipment choices are made by others–by architects, builders, and developers for new housing, and by landlords for the one-third of residential units that are rented.

Although energy-efficient residences and high-efficiency appliances offer electricity savings and lower life-cycle costs over less efficient versions, these potential cost savings provide only weak financial incentives for several reasons.

First, residential electricity prices seem to have only a weak influence on energy choices for most ratepayers, and almost no influence on third-party decision-makers (developers, builders, equipment vendors and manufacturers, and landlords and tenants who do not pay monthly electric bills). Residential electricity prices have declined steadily in real terms over the past decade. Moreover, residential rates usually do not reflect the higher costs of using electricity at times of peak demand, nor the social and environmental costs (externalities) of generating electricity.

Future savings from energy-efficiency investments are heavily discounted. Studies have found that residential consumers demand a short pay-back period for efficiency investments–2 years or less for home appliances, for example.

Many decision-makers are driven by the desire to keep first-costs low; few pursue the goal of minimizing life-cycle costs (the sum of capital and operating costs over the life of the equipment–or eg, the initial purchase cost of an appliance plus the cost of annual electric bills, maintenance and repairs). This so-called first-cost bias is especially strong when energy-efficient equipment costs more and others (home purchasers or tenants) will reap the benefits of lower electric bills. First-cost bias is also strong for low-income consumers who lack either the cash or access to credit to pay for the more efficient and expensive equipment.

Reliable, understandable information on energy use and costs is often lacking or hard to use. Consumers that would like to give greater weight to energy efficiency in their decisions—whether motivated by lower life-cycle costs, environmental concern, technological fascination—have few alternatives. Government and private programs for energy-efficiency ratings of homes and apartments are only just beginning. The effectiveness of federally-required labeling for major appliances is uncertain and has not been adequately assessed (2,20).

Energy efficiency is often misperceived as requiring discomfort or sacrifice, rather than as providing equivalent services with less energy. The poor popular image of home energy efficiency as meaning cold showers, dark rooms, and warm beers hampers consumer acceptance and diminishes incentives for housing developers and equipment manufacturers to make efficiency a selling point for their products. Without a market pull for efficiency, equipment manufacturers and building suppliers give less emphasis to efficiency in product design and research.

The typical low turnover rate in the housing stock and slow rate of replacement of major appliances mean that efficiency improvements in the residential sector will significantly lag behind technical potential. Without aggressive efforts in response to government policy and/or an energy crisis, this lagging response will continue.

From a somewhat different analytical perspective, the Bush Administration also found progress in residential energy efficiency unacceptably slow. President Bush's National Energy Strategy (NES) found that "a number of institutional and market barriers" limited consumer responses to the higher energy prices of the 1970s and early 1980s. Strongly reflecting the economic policy framework of its analysis, the NES concluded that "our stock of housing and appliances is still far less energy efficient than would be economically optimal" (5).

Among the "significant market barriers" in the residential sector identified by the NES were:

- Traditional energy price regulation and ratesetting do not reflect the full costs to society of energy use, and thus cause individual consumers to undervalue energy-efficiency investments and renewable resources.
- Failure of market mechanisms to induce adoption of economical energy-saving measures by residential customers, particularly in situations where those who must pay for such devices cannot expect any economic benefits.
- First-cost bias tendency of buyers (especially builders and homebuyers) to minimize upfront costs of residential property and major appliances.

- Mortgage lending practices that fail to consider the lower total cost of energy-saving homes in calculating mortgage eligibility.
- Low incomes of some energy users that often make them unable to finance energy-efficiency improvements no matter what the payback period is.
- Absence of credible data on reliability and cost of energy-saving technologies for builders, architects, utility programs, mortgage lenders, and individual consumers.
- Fragmented and cyclical nature of homebuilding industry that contributes to a reluctance to try innovative energy-saving designs, products, and construction techniques and makes concerted industry-led efficiency initiatives unlikely.
- Inadequate implementation and enforcement of energy building codes because of lack of resources to check actual plans and construction sites and to educate builders.
- Inadequate energy-efficiency investment in public sector housing because many local housing authorities lack funds and management incentives to improve efficiency.
- Slow turnover of residential structures and long lifetimes of heating and cooling systems.

The premise of institutional and market barriers to energy efficiency has wide acceptance among energy analysts, government policymakers, state regulators and utility executives. There are others, generally economists of the classical and neoclassical persuasions, who reject this conclusion of market failure, however. They adhere to a belief that present energy efficiency characteristics represent the informed decisions of knowledgeable consumers who have compared alternative investment opportunities and selected energy conservation that offers equal or better returns (21).

ENERGY EFFICIENCY OPPORTUNITIES IN THE COMMERCIAL SECTOR

The commercial sector has consisted of all businesses that are not engaged in transportation or industrial activity and includes, for example, offices; retail stores; wholesalers; warehouses; hotels; restaurants; religious, social, educational and healthcare institutions; and federal, state, and local governments. In 1990 the commercial sector accounted for about 14% of total primary energy use (8). Figure 4 shows energy consumption (excluding electricity conversion and transmission losses) in the commercial sector. Electricity and natural gas each supply about 42% of commercial sector energy needs, with oil (15%) and coal (1%) supplying the remainder. Adjusting for conversion and distribution losses of utilities for serving commercial loads, electricity has accounted for 69% of total primary energy consumption by the commercial sector.

In 1990 the commercial sector consumed about 751 billion kWh of electricity at a cost of $55 billion (9). Commercial establishments made up about 28% of the total electric utility retail sales in 1990. In addition to purchased

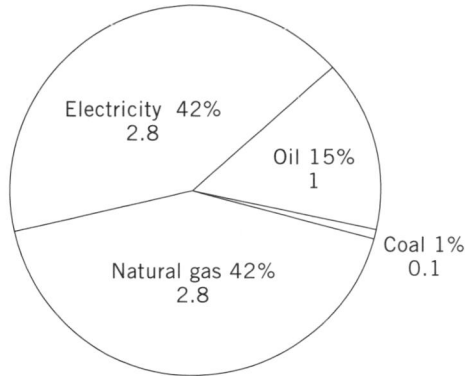

Figure 4. Commercial sector on-site energy consumption, by source, 1990 (quadrillion Btus). From Office of Technology Assessment, 1993, based on data from the U.S. Department of Energy, Energy Information Administration, and the Gas Research Institute.

electricity, a growing number of commercial facilities have resorted to cogeneration or self-generation to meet some or all of their electricity demand; this output is not included in commercial sector electricity consumption estimates, but fuels used to produce this power are included in overall commercial energy consumption. Many commercial facilities are cogenerators—with natural gas the most common fuel. Opportunities to combine heating and or cooling plants with power generation abound in large institutions and concentrated urban commercial areas. Cogeneration can add to overall efficiency of energy use in the sector, but in part means a shift of primary energy consumption from the electric utility sector.

Figure 5 shows commercial electricity use by application. Heating, ventilating, and air-conditioning (HVAC) dominates, comprising 37% of commercial electricity use (space heating, percent; cooling, percent; and ventilation,

percent). Water heating accounts for an additional 3%. Lighting accounts for an estimated 29% of commercial load. Estimates of commercial lighting electricity use vary, some estimates place lighting at 40% of commercial load reflecting the high percentage of lighting loads in office buildings. For purposes of this analysis we have adopted the estimates used in EPRI's analysis.

Refrigeration (7%); cooking (2%), and miscellaneous equipment including elevators, escalators, office computers, printers, telephone systems, and other commercial equipment (21%). Sixty percent of electricity use in commercial establishments is for nonspace heating purposes. These nonspace conditioning applications are projected to grow faster than commercial square footage to over 65% of electric load by the year 2010. The heat generated by miscellaneous equipment add to demands for cooling, but lowers space heating loads.

Electricity demand in the commercial sector is driven by the growth in square footage in commercial buildings and the intensity of service demand—for space cooling, lighting, and office equipment, for example (2). On average, office, health care, and food service establishments are the most energy-intensive commercial buildings. Between 1970 and 1989, the amount of commercial square footage and electricity use each grew by 45% (19). Even so, newer commercial buildings have tended to be more energy efficient incorporating more insulation, better windows, lighting and more efficient space-conditioning equipment, thus tempering the growth in electricity demand.

Commercial building energy intensity (ie, energy use per square foot) has remained flat for the past two decades, even as demand for air-conditioning, computers and other equipment grew. Complicating this trend has been the growth in commercial electricity demand due to a shift from on-site use of primary fuels—oil, gas, and coal—to electricity. Thus primary fuel use has transferred from the commercial sector to the utility sector, and may even have resulted in a net increase in primary energy consumption, because of the losses involved in electricity generation and delivery.

At present there are over 4.5 million commercial buildings in the United States with a total of over 61 billion square feet (22). Each year about 1 billion square feet of new commercial space is added—10 to 15 billion total square feet will be added this decade. There is great diversity in the size and energy using characteristics of these commercial buildings. Smaller commercial building energy systems are similar to those in houses and small apartment buildings. Larger buildings, however, have complex HVAC systems and activities inside the building—lighting, occupancy, electric and other equipment—can add to energy demand and determine equipment choices. Buildings larger than 10,000 square feet make up almost 80% of building square footage and offer many opportunities for electricity savings.

Energy Efficiency Technologies for the Commercial Sector

Space-conditioning, lighting, and building shell weatherization are primary targets for improving energy efficiency and saving electricity in the commercial sector. In addi-

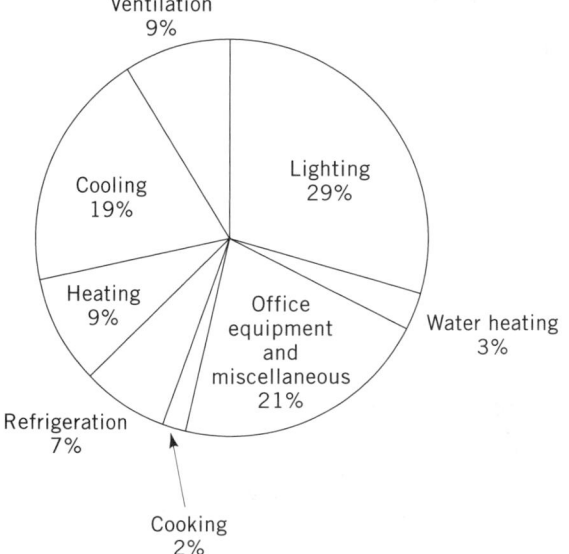

Figure 5. Commercial sector electricity use by application, 1987. From Office of Technology Assessment, 1993, based on data from the Electric Power Research Institute and U.S. Department of Energy.

tion, large commercial buildings are suitable targets for utility load management programs designed to shift energy use away from peak hours, but not necessarily resulting in lower overall energy demand, through installation of technologies such as storage heating and cooling systems. There are also potential energy savings in other commercial applications, as shown in Table 3.

Analysis of potential efficiency opportunities by EPRI found that commercially available electric equipment could reduce commercial electricity in year 2000 by 22 to 49% from what consumption would be without the use of these technologies if efficiency were frozen at 1987 levels. Commercial applications with the most significant savings potential in the EPRI analysis were lighting, cooling, and miscellaneous electric equipment.

IMPROVEMENTS IN COMMERCIAL BUILDING EFFICIENCY

Turnover of commercial building space is more rapid than residential, but it is evident that a large portion of commercial space in use for the next few decades is already in place. Analysts estimate that one-half of the commercial space in 2010 has already been built, and 80% of the existing stock of commercial buildings will still be in use for the next 30 years (14).

The pace of new commercial construction provides opportunities for efficiency gains in both building shell, equipment, and appliances. Measures to increase the efficiency of commercial buildings include improved design, setting, and construction techniques, better insulation, and more efficient equipment choices.

The remodeling and rehabilitation of commercial space offers additional opportunities. There is considerable potential for energy-efficiency improvements in existing commercial buildings. According to DOE surveys, while 84% of buildings are reported to have installed building shell conservation features, there remains a considerable pool of buildings that have not installed basic measures. The most frequently reported measure is ceiling insulation, 67.5%, weatherstripping or caulking, 61%, and wall insulation, 47%. Storm windows and multiple-glazing have been reported in 32% of buildings, and shades and

---

Table 3. Selected Energy Efficiency Technology Options for the Commercial Sector[a]

Heating, ventilation, and air-conditioning (HVAC) systems	Water heating
Building envelope efficiency improvements	Blanket wrap for water tanks
Weatherstripping and caulking	Commercial heat pump water heaters
Insulation	Integrated heating and hot water systems
Storm windows and doors	Heat recovery water heat systems
Window treatments	Increased insulation of tanks and pipes
Space heating	Flow restrictors
Improved commercial heat pumps	Service/point of use water heaters
Air-source heat pumps	Commercial lighting
More efficient models	Delamping
Improved technology	Lighting fixture retrofits
Ground-source heat pumps	Electronic ballasts for fluorescents
	High-efficiency lamps
Heat recovery systems	Reflectors
Energy management controls and systems	Increased use of of daylighting
Set-back thermostats	High-intensity lighting applications
Smart buildings and smart systems	Increased use of task lighting
Zoned heat systems	Compact fluorescents
Thermal storage systems	(LED) signs
Cogeneration systems	Lighting control systems: timers, occupancy sensors,
District heating systems	photocells, dimmers
Space cooling	Commercial refrigerators and freezers
More efficient cooling systems	Efficient motors and controls
Cool storage systems	Improved insulation and seals
District cooling systems	Commercial cooking
Ventilation	Energy-efficient commercial electric ranges, stoves, fryers,
Air distribution systems	ovens and broilers
Improved insulation	Microwave cooking
Reduced duct and damper leakage	Convection cooking
Separate make up airflows for cooling exhaust systems	Induction cooking
Economizer controls	Miscellaneous electrical equipment and office machines
Improved HVAC maintenance	More efficient motors and drives for elevators, escalators,
	and other building systems
Integrated HVAC systems	Improved hardware and software for office equipment
	Integrated building energy management and control
	systems

[a] From Office of Technology Assessment, 1993.

awnings and reflective shading glass or films have been reported for 21% of buildings (22).

Heating, Ventilation, and Air-Conditioning

Space Heating. Just under one-quarter of commercial buildings rely on electric heating systems (22). Most of these buildings are located in the south and west.

Installation of more efficient electric heating equipment, such as heat pumps instead of rhesustance heat, coupled with a combination of measures such as building shell improvements, window treatments, heat recovery, and improved maintenance practices can cut electricity demand for space heating. Further savings are possible with integrated heat pump systems that provide heating, cooling and water heating. These potential savings are offset by the expected increase in heating load attributable to reduced internal heating gains from installation of energy-efficient lighting measures. Use of the best available energy efficiency measures could reduce space heating electricity demand in 2000 by 20 to 30% from what would be required from the 1987 stock of commercial buildings and equipment, according to EPRI.

District heat, in which a central plant provides heat, and often hot water for all buildings within a complex or downtown area, also offers efficiency opportunities, particularly if coupled with cogeneration (2).

Cooling. Commercial cooling loads are the biggest component of summer peak load for most utility systems. Over 70% of commercial buildings have cooling systems and 96% of these systems are electric. Common commercial cooling equipment includes a packaged cooling system, individual air-conditioners, central chillers, and heat pumps. Often these systems are integrated with the building ventilation and air transport systems. Commercial cooling load is driven by building size, external temperature, and internal heat gains from electric and other equipment and occupants. Over 6% of commercial buildings maintain separate cooling systems for computer areas (22).

Energy-efficient cooling options for commercial buildings include more efficient air-conditioners, heat pumps, high-efficiency chillers, chiller capacity modulation and downsizing, window treatments, radiant barriers, energy management control systems, and improved operation and maintenance. Reduced internal heat gain from installing efficient lighting systems also cuts cooling load. Excluding lighting-related savings, EPRI has estimated that cooling requirements can be reduced by 30% or more in commercial buildings. By including lighting efficiency packages with cooling system improvements total savings of over 80% could be provided according to EPRI estimates. However, the need to find replacements for CFCs now used in cooling systems could result in newer cooling technologies that may reduce some possible efficiency gains. EPRI therefore estimates maximum potential electricity savings in commercial space cooling in the year 2000 to be from 30 to 70% over 1987 performance levels (1).

Another energy efficiency strategy for commercial cooling that may not always result in a net reduction in electricity demand is the use of cool storage systems that shift all or part of a buildings' air-conditioning electricity demand from peak to off-peak hours. Typically, ice or chilled water is produced in a refrigeration system at night and used to meet some or all of the next day's air-conditioning needs. Cool storage systems offer financial savings for customers through lower off-peak rates and peak reduction for utilities (27).

Ventilation. Air transport and ventilation systems are a critical component of modern large commercial buildings. Improving the energy efficiency of ventilation and air transport systems can be attained through a variety of measures: viable air volume systems; low-friction air distribution designs; high-efficiency electric motors; variable speed drives; heating, cooling, and lighting improvements; and improved operation and maintenance practices. EPRI estimates that ventilation electricity use can be reduced by 30 to 50% through a comprehensive package of measures.

Lighting

About 29% of commercial electricity consumption is for lighting. Commercial lighting requirements are met with a combination of incandescent, fluorescent, and high-intensity discharge lamps and most commercial buildings have a mixture of these fixtures. Fluorescent lamps are already extensively used in the commercial sector. About 78% of commercial floorspace is lit with fluorescents and high-efficiency ballasts have been installed in about 40% of this space (22).

A range of cost-effective technologies is available to cut lighting loads. Ready savings can be achieved in many commercial buildings by delamping to lower lighting levels, using lower wattage fluorescents, and replacing incandescents with more efficient fluorescent or compact fluorescent lamps where appropriate. More advanced lighting system efficiency upgrades include installation of high-efficiency electronic ballasts, aluminum and silver film reflectors, daylight dimming, occupancy sensors, and use of high-pressure sodium lamps instead of mercury vapor lamps in high-intensity discharge fixtures. In new construction and remodeling, better lighting system design and greater use of daylighting can also cut lighting requirements.

Estimates of lighting savings involve interactions among package components and are not necessarily the sum total of individual measures. Building characteristics also influence potential savings. In addition, lighting upgrades can cut cooling costs by reducing internal heat gain, but add to heating loads. EPRI estimates potential electricity savings from more efficient commercial lighting in the year 2000 to range from 30 to 60% over 1987 stock (2).

Commercial Refrigeration Systems

Commercial refrigeration in retail stores, restaurants, and institutions can be a significant load. About 20% of commercial buildings are equipped with commercial refrigeration systems; about 16% have commercial freezers.

EPRI estimates that commercial refrigeration electricity use can be cut by 20 to 40% from 1987 performance levels by combining a variety of efficiency improvements. Examples include: more efficient fan motors and compressors, multiplex unequal parallel compressors, advanced compressor cycles, variable speed controls, evaporatively cooled condensers, floating head pressure systems, air barriers, food case enclosures, electronic controls, and improved maintenance practices. Electricity savings are highly site specific and depend on the previous saturation of these technologies.

Water Heating

About 48% of commercial buildings with hot water systems (22) use electricity as the sole or supplemental water heat source. Hot water heating accounts for about 3% of commercial electricity use.

There are a number of efficiency measures for commercial hot water systems on the market. These measures include many also used in residential applications, such as water heater wraps, low flow devices, hot water pipe insulation, and installation of valves that reduce convection loses. Commercial heat-pump water heaters and heat recovery systems can provide energy savings of one-third or more over conventional resistance systems. Integrated heat pumps can provide heating, cooling, and hot water for commercial buildings. Lowering the hot water thermostat can reduce electricity use while still providing adequate water temperatures for most uses. EPRI estimates potential savings in water heating electricity use in the year 2000 of 40 to 60% over 1987 stock (1).

Cooking

Commercial cooking equipment accounts for about 2% of the commercial sector electricity use. Microwaves, convection ovens, and magnetic induction cooktops can cook food with less time and energy than more conventional electric stoves and ovens and are seeing greater use in commercial establishments. A range of technological improvements are available to cut electricity use in commercial ranges, ovens, broilers, griddles, and fryers. Examples include: increased insulation, better heating elements, more precise temperature controls, reflective pans, reduced thermal mass, and less contact resistance. EPRI estimates that by incorporating a combination of efficiency measures, electricity use by commercial electric stoves and ovens in the year 2000 could be from 20 to 30% less than that required for 1987 stock (1).

Miscellaneous Commercial Systems and Equipment

Residual electric systems and equipment (eg, elevators, escalators, telephone systems, office machines, food preparation and other equipment) account for 21% of the commercial sector electricity use and will continue to grow. EPRI estimates that overall savings from expected efficiency advances in miscellaneous commercial sector equipment will range from 10 to 30%. Expected improvements in hardware, software, and system operations could offer maximum potential savings of up to 50% for office equipment in 2000. EPRI also calculates maximum

potential savings of up to 35% in 2000 from the use of high-efficiency motors and adjustable-speed drives in elevators and escalators (1).

The Federal Government, through the Environmental Protection Agency's green programs and Federal procurement policies, is seeking to overcome some of the market barriers to more energy-efficient computer equipment.

Computer equipment and other electric office machines are among the fastest growing components of commercial energy consumption. They now total about 5% and are expected to total 10% by 2000. Surveys have determined that most personal computers are left turned on when not in use during the day, overnight, and on weekends. Desktop computers typically have been designed with little consideration for energy efficiency, unlike portable or laptop models that incorporate a number of energy-saving measures to save battery power. If desktops were equipped with technologies that allowed them to "nap" or shut down when not in use and return quickly to full power capability when needed, EPA has estimated that such computers could save 50% of the energy used to run them. Green computers thus have become one of the first commercial consumer products targeted by DPA's pollution prevention programs to increase consumer and manufacturer awareness of energy efficiency benefits, and to create a new market for energy-efficient equipment.

Using a model similar to the Green Lights Program, EPA has entered into discussions with manufacturers of computers, peripherals, and microprocessors. Manufacturers agree to produce products that meet certain efficiency improvements and sign a memorandum of understanding with EPA. The manufacturers are then eligible to use the "Energy Star–EPA Pollution Preventer" logo in the marketing and displaying of the products. For example, personal computers with the capability of switching to a low power mode of 30 watts or less (about 75% less than current models) qualify for the EPA logo that identifies new high efficiency equipment. EPA is expanding the use of such voluntary agreements for related computer products including printers, monitors and other pieces of office equipment.

By May 1993, the EPA had reached agreement with an impressive array of companies producing personal computers and related products. Charter partners in EPA's Energy Star computer program represent 60% of the U.S. market for computers and monitors, and 80% of the laser printer market. An Energy Star allies program has been established enlisting agreements from components and software makers. Intel Corporation, one of the world's major microprocessor manufacturers, has committed to incorporating energy-saving technologies into all future microprocessors. The first products bearing the Energy Star logo have become available in 1993.

The widespread penetration of energy-saving computer technologies offers significant benefits to consumers, the economy, and the environment. The cost of operating a typical 150-watt personal computer 24 hours per day year round can be $105/year (assuming electricity costs at $0.08/kWh) and uses 1,314 kWh/yr. Turning the machine off at night reduces the operating cost to $35/year and cuts energy consumption to 433 kWh/year. Using technology that conserves power when the machine is not active during the day could cut costs to $17/year for 216 kWh/ year. EPA estimates that green computers could save a total of $1.5 to $2 billion in annual electricity bills and avoid emissions of 20 million tons of carbon dioxide, 140,000 tons of sulfur dioxide, and 75,000 tons of nitrogen dioxide by 2000.

BARRIERS TO ENERGY EFFICIENCY IN THE COMMERCIAL SECTOR

There remains a significant gap between the electricity-using characteristics of the present stock of commercial buildings and equipment and the energy-saving potential of the most efficient buildings and equipment marketed today. As with the residential sector, many economic, institutional, and behavioral influences hamper greater commercial sector investment in energy efficiency.

Some influences are shared with other sectors. The normally slow turnover in commercial buildings and major equipment, albeit more rapid than in the residential sector, means that actual efficiency savings lag considerably behind technical potential. Relatively low energy prices that do not reflect all societal and environmental costs of energy production and use, also lead to undervaluing of energy and underinvestment in efficiency by commercial consumers. (This problem persists even though commercial customers are in general more price-sensitive than residential customers, and utility bills for commercial establishments can be quite large). Choices affecting commercial energy demand are made by a large number of decision-makers—architects, designers, developers, building owners, tenants, equipment manufacturers and vendors, and local building authorities. The plethora of decision-makers and the absence of any direct economic benefit in efficiency for many of them, lessens the impact of existing weak financial incentives and fragments the potential constituency for efficiency improvements.

Several factors contribute to limited financial incentives to invest in efficiency. Energy costs of buildings can often be a small fraction of total business expenses and thus gain little management attention as a means of saving money. According to some estimates, for large office buildings and retail space, energy costs are less than 5% of total annual operating costs per square foot and are dwarfed by other business costs. Energy efficiency is only one consideration in decisions affecting energy use—first-cost, appearance, comfort, and other performance features may overshadow potential lifecycle cost savings from efficiency. Building owners and tenants tend to place greater emphasis on occupant comfort and productivity and may be reluctant to make any changes that might affect building operations. One-quarter of commercial space is leased and lower energy bills offer no incentives for building landlords when the tenants are responsible for paying electric bills. Where landlords pay utility bills and energy prices are included in rent, building occupants may have little financial incentive to choose high-efficiency equipment or to invest in energy-savings maintenance (2).

When efficiency investments are considered, commercial sector decision-makers also tend to require short pay-back periods of 1 to 3 years. Lack of resources or access to capital can discourage some possible commercial sector efficiency investments, particularly for nonprofit institutions and small businesses. Cost-effective, low-risk measures that could cut operating costs are often given low priority in government facility management. Even when government facility managers are aware of potential savings, budgetary and procurement constraints limit investments in efficiency for government owned or occupied facilities (23).

The energy efficiency industry is still in its infancy and the small pool of trained vendors, installers, and auditors available to serve commercial establishments and utility programs can limit achievable energy savings at least in the short term. The relative newness of the industry and absence of a proven track record of delivering savings may make many in the commercial sector reluctant to make significant investments in energy efficiency. Indeed, savings from early building retrofit investments have been less than expected on average, and unpredictable for individual buildings, adding to the perceived riskiness of the investment (14).

Nevertheless, the commercial sector remains a prime and potentially profitable target for utility, private sector and government efforts at improving energy efficiency.

ENERGY EFFICIENCY OPPORTUNITIES IN THE INDUSTRIAL SECTOR

The industrial sector includes both manufacturing enterprises (ie, businesses that convert raw materials into intermediate or finished products) and nonmanufacturing enterprises, such as agriculture, forestry, fishing, construction, mining, and oil and gas production. The industrial sector is characterized by the diversity of energy uses, equipment, and processes and is the largest energy sector, consuming 37% of U.S. total primary energy use in 1990. Patterns of industrial energy use are further complicated by the use of oil, gas, and coal as feedstocks and for cogeneration. Figure 6 shows industrial energy use for fuel and power only.

Industrial energy use is variable, reflecting economic conditions, structural changes, interfuel competition, and rate of investment. Patterns of industrial energy use and energy intensity of industry also vary significantly by region. Price is the major determinant in most industrial energy choices, and head-to-head competition among fossil fuels is intense. Price however is not the sole consideration–availability, reliability, and quality also drive industrial energy decisions. Another trend is the growth in industrial cogeneration, which is generally viewed as a positive development for efficiency, but, which in effect transfers demand and losses between industrial sector and utilities. Moreover there has been a general trend toward electrifying many process technologies and a shift in energy and electric intensity of manufacturing. The relationship of efficiency gains and structural changes in U.S. industry has been examined in detail in an OTA background paper, Energy Use and the U.S. Economy (24). A companion OTA report, Industrial Energy Efficiency, had been published in the summer of 1993 (25).

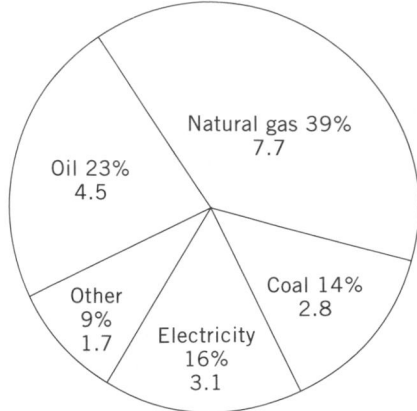

Figure 6. Industrial energy use for fuel and power, 1989 (quadrillion Btus). From Office of Technology Assessment, 1993, based on data from the Gas Research Institute.

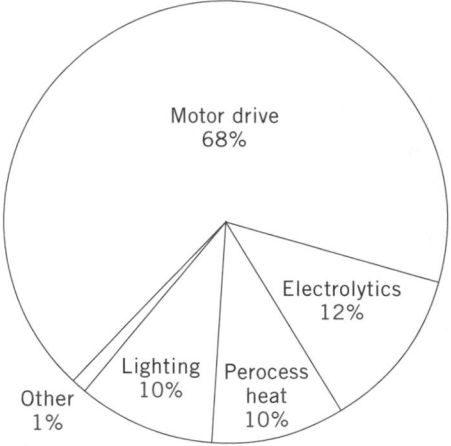

Figure 7. Industrial electricity use by application, 1987. From Office of Technology Assessment, 1993, based on data from the Electric Power Research Institute.

There are five major fuel and power demands in the industrial sector: process steam and power generation (36%), process heat (29%), machine drive (14%), electrical services (4%), and other (including off-highway transportation, lease and plant fuel use, and mining) (16%) (7). The industrial sector derives 40% of its fuel and power needs from natural gas, 25% from oil, 15% from purchased electricity, 9% from coal, and the remaining 9% from waste fuels and other sources. Electricity competes with other fuels, particularly natural gas, for direct heat applications (7). For other uses, purchased electricity competes with the options of self-generation or cogeneration. It is estimated that in 1989, the industrial sector produced about 153,270 gigawatt-hours of electricity onsite. Surplus electricity production had been sold to local utilities (7). To avoid doublecounting, fuel used for industrial self-generation or cogeneration is usually attributed to primary fuels.

In 1990, industrial consumers purchased 946 billion kWh from electric utilities at a cost of $45 billion (9). Sales to industrial users accounted for 35% of electric utility revenues from sales to end-users/ultimate customers. Electricity consumption in the industrial sector is divided among the manufacturing enterprises (87%); agriculture (5%) and construction and mining (8%).

The major industrial electricity uses are motor drive, electrolytics, process heat, and lighting (see Figure 7). Table 4 summarizes EPRI estimates of 1987 industrial energy consumption for these applications by industrial subsectors (SIC codes), manufacturing loads and nonmanufacturing loads.

The most electricity-intensive manufacturing activities (including on-site generation) are chemical products, primary metals, pulp and paper, food, and petroleum refining, together accounting for more than half of manufac-

Table 4. Industrial Electricity Use by Application and Industry, 1987 (gigawatt-hours)[a]

Category SIC Code	Total Electricity Consumption (GWh)	Motor Drive (GWh)	Percent	Electrolytics (GWh)	Percent	Process Heating (GWh)	Percent	Lighting (GWh)	Percent	Other (GWh)	Percent
Manufacturing, main users[b]											
28 Chemicals	141,191	90,250	63.9%	36,810	26.1%	668	0.5%	13,464	9.5%	0	0.0%
26 Paper	83,219	74,364	89.4	0	0.0	1,870	2.2	6,985	8.4	0	0.0
20 Food	47,213	40,544	85.9	0	0.0	1,202	2.5	5,466	11.6	0	0.0
33 Primary metals	146,410	54,482	37.2	58,956	40.3	25,785	17.6	7,187	4.9	0	0.0
29 Petroleum	41,444	8,108	91.9	0	0.0	401	1.0	2,936	7.1	0	0.0
32 Stone, clay, and glass	34,019	27,192	79.8	0	0.0	5,077	14.9	1,822	5.3	0	0.0
37 Transportation equipment	37,560	21,539	57.3	101	0.3	13,895	37.0	2,025	5.4	0	0.0
35 Industrial machinery	33,194	16,598	51.1	101	0.3	12,692	38.2	3,442	10.4	0	0.0
34 Fabricated metal prod.	31,045	13,937	44.9	2,225	7.2	267	41.7	1,923	6.2	0	0.0
36 Electronics	32,299	27,679	85.7	0	0.0	267	0.8	4,353	13.5	0	0.0
22 Textiles	25,509	20,760	81.4	0	0.0	802	3.1	3,948	15.5	0	0.0
30 Rubber and plastics	28,809	26,510	88.9	0	0.0	1,069	3.6	2,227	7.5	0	0.0
Total all manufacturing	736,950	495,012	67.2%	98,193	13.3%	78,959	10.7%	64,787	8.8%	0	0.0%
Nonmanufacturing											
Agriculture	44,541	23,283	52.3%	0	0.0%	4,049	9.1%	1,985	26.9%	5,132	11.5%
Mining	55,676	50,615	90.9	0	0.0	0	0.0	3,525	6.3	1,509	2.7
Construction	8,098	2,025	25.0	0	0.0	0	0.0	4,230	52.2	1,811	22.4
Total nonmanufacturing	108,315	75,923	70.1%	0	0.0%	4,049	3.7%	19,740	18.2%	8,452	7.8%
Total industrial	845,266	570,934	67.5%	98,193	11.6%	83,008	9.8%	84,527	10.0%	8,453	1.0%

[a] Ref. 1.
[b] Industries using more than 25,000 (GWh) annually.

turing electricity use. The pulp and paper and chemical products subsectors have significant cogeneration capacity—mostly fired by waste fuels.

Efficient Industrial Technologies

There are several strategies for improving energy efficiency in the industrial sector, including making existing electricity applications more efficient, shifting industrial processes from fossil fuel to electrotechnologies for net energy savings, and using more industrial cogeneration for net energy savings over purchased electricity.

EPRI estimates that application of more efficient industrial equipment and processes offer potential savings of from 24 to 38% of their projected base-case electricity use in 2000 (1). The most promising targets for potential efficiency gains are high efficiency electric motors and variable speed drives, improved electrolytic processes, industrial process waste heat recovery, and more efficient lighting technologies (see Table 5). All but electrolytic technologies have a wide and diverse range of potential applications across the industrial sector.

Table 5. Selected Energy Efficiency Technology Options for the Industrial Sector[a]

Electric motors and drives
    High-efficiency motors
    Variable speed drives
    Optimal sizing of motors and loads, serial motors
Waste heat recovery systems
    Industrial process heat pumps
    Industrial heat exchangers
    Vapor recompression systems
Electrolytic processing
    Chlor-alkali production
        Improved membrane and diaphram cells for chlor-alkali production
    Aluminum smelting
        Improved efficiency in Hall-Heroult smelting process
        Alternative aluminum reduction technologies
Industrial lighting
    Delamping
    Lighting fixture retrofits
    Electronic ballasts for fluorescents
    High-efficiency lamps
    Reflectors
    Increased use of daylighting
    High-intensity lighting applications
    Increased use of task lighting
    Compact fluorescents
    LED signs
    Lighting control systems—timers, occupancy sensors, photocells, dimmers
Industrial electro-technologies
    Plasma processing
    Electric arc furnaces
    Induction heating
    Industrial process heat pumps
    Freeze separation
    Ultraviolet processing/curing
Industrial cogeneration systems
    High-efficiency industrial boilers
    Integrated process heat/steam and power production

[a] From Office of Technology Assessment, 1993.

Energy-Efficient Electric Motors and Drives

There is great diversity in industrial applications of electric motors and drives: pumps, fans, compressors, conveyors, machine tools, and other industrial equipment. Motor drive end-uses account for an estimated 70% of electricity load in manufacturing. High-efficiency electric motors combined with adjustable-speed drives (ASDs) offer significant electricity savings potential.

Electric motors are available in standard and high-efficiency models and energy efficiency of both vary according to size. In general, larger motors are more efficient than smaller ones in both standard and high efficiency models. The high-efficiency models cost from 10 to 30% more than the standard versions (25), but have efficiency increases of 8% for smaller motors and 3% for larger motors (17). Energy-efficient motors typically have a longer operating life than standard motors. The initial capital costs of electric motors are usually only a fraction of their operating costs. For example, annual energy costs for an electric motor might run as much as 10 times its initial capital cost; increasing its efficiency from 90 to 95% could mean savings of 50 to 60% of its capital costs in a single year (26).

Many industrial motors are often run at less than maximum power because of varying loads. Electronic adjustable speed drives allow an electric motor to operate at reduced speed when maximum power is not needed, saving energy. ASDs are appropriate in applications with high operating hours where motors are often operated at less than full load.

There are three targets for displacing standard-efficiency motors with high-efficiency motors: selecting new or replacement motors, rewinding of existing motors, and retrofitting of existing motors that do not need repair or replacement.

High-efficiency variable-speed motors offer tremendous potential for efficiency. Various studies have yielded estimates of potential savings of 20 to 50% depending on circumstances for application of ASDs. Use of high-efficiency electric motors can provide savings of an additional 3 to 10%. Overall efficiency improvements in motor drive of 35 to 50% over 1987 equipment have been assumed in EPRI's analysis (1). Motor drive improvements have offered nearly 80% of estimated savings in their analysis, with over 90% of these savings in just a few industry categories.

Waste Heat Recovery

Waste heat recovery systems improve energy efficiency by using heat from fuel combustion or excess thermal energy from a process steam product. An estimated one quarter to one-half of the process heat used by industry is discharged as hot gases or liquids (17). There are various approaches to capturing energy from these sources of waste heat. The choice depends on characteristics of the heat source, process needs, and economics. Heat exchangers are used to transfer heat from a high-temperature waste exhaust source, such as combustion gases, to a cooler supply stream such as steam for lower temperature uses. Low-temperature waste heat streams can be upgraded to supply heat for higher temperature processes via industrial heat pumps or vapor recompression systems. Analyses for EPRI have found that installation of heat recovery devices can reduce a plant's overall energy requirements by at least 5% with paybacks of less than 2 years. The most cost-effective time to incorporate the systems is during new construction or modernization projects and most applications have been custom designed. Heat recovery devices displace conventional energy sources (such as purchased electricity) and are used in processes requiring a constant heat source. Hence they are attractive to utilities as a means to reduce base loads and peak loads.

Waste heat recovery in industrial process heat systems can provide electricity savings of 5 to 25% according to EPRI estimates. Very little waste heat recovery currently exists, so there is a potential for significant improvement with an average of 10 to 15% savings.

Electrolysis

An estimated 12% of industrial electricity use is used for electrolysis. Electrolysis is a method for separating and synthesizing chemicals or metals by using electricity to produce chemical reactions in aqueous solutions or molten salts. At present the two largest industrial applications of electrolysis are aluminum reduction in the primary metals processing industry and the production of chlorine and caustic soda from salt brines in the chemical products industry.

Electricity is the most costly material in aluminum production. In the century-old Hall-Heroult process alumina refined from bauxite ore is reduced via electrolysis to molten aluminum (27). The smelting process is continuous. Alumina is dissolved in a molten electrolytic bath in carbon lined steel cells or pots. In each pot a direct current is passed from an carbon anode suspended in the cell through the bath to the carbon lining of the cell producing a chemical reaction. Molten aluminum is siphoned from the bottom of the pots and is then formed into aluminum ingots or further refined and/or alloyed into fabricating ingot. A single potline can consist of from 50 to 200 cells with a total voltage of 1,000 volts at currents of 50,000 to 250,000 amperes. U.S. smelters use from 6 to 8 kWh to produce each pound of aluminum.

The efficiency of aluminum production has improved steadily. Following World War II about 12 kWh of electricity has been needed to produce one pound of aluminum; today, through greater economies of scale and process controls, the most efficient smelters use half that electricity per pound (27). Further efficiency gains are promised by advanced electrolytic reduction methods including bipolar cells, inert anodes, and wettable cathodes. None of these technologies, however, is currently installed, but EPRI estimates that they could potentially yield efficiency savings by year 2000 of some 30 to 50% over current methods. These improvements are highly attractive given the high electric intensity of aluminum production and are significant for regions where such production is concentrated, such as the Pacific Northwest.

Chlor-alkali production is second to aluminum in terms of electricity consumption and uses about 30% of electric power used for electrochemical production (17). Chlorine and caustic soda (sodium hydroxide) are produced from salt brine by electrolysis in either the diaphragm or mercury cell. Mercury cells account for about 20% of U.S. capacity. Throughout this century economies of scale have produced steady efficiency gains in chlor-alkali production as newer and larger cells have required less energy to drive the chemical reactions (27). In the membrane cell, different constituents of the solution are separated by selective diffusion through the membranes. EPRI analyses have estimated that the use of membrane cells to replace diaphragm cells could save 10% of electricity used in chlor-alkali production. Other analyses have estimated savings of up to 25% over current methods.

Adaptation and improvement of electrolytic separation methods, including electrodialysis which uses electric current to accelerate membrane separation, for other inorganic and organic processes also can yield efficiency gains over conventional methods.

Lighting

Lighting accounts for about 10% of electricity use in the industrial sector. As in the commercial and residential sectors, more efficient lighting technologies offer promises of electricity savings across the industrial sector too. Some industrial lighting efficiency upgrades strategies are: delamping, reduced wattage fluorescents, high-efficiency ballasts, reflective fixtures, occupancy sensors, replacing incandescent lamps with compact fluorescents, and greater use of daylighting. EPRI analyses estimate that lighting efficiency packages offer savings of from 36 to 49%. Lighting upgrades can also lower cooling loads, but increase heating loads.

Electrification of Industrial Processes

Electrification offers the potential for net savings in fossil fuel use even as it increases electricity demand in the industrial sector. There has been a continuing trend toward electrification of many industrial processes and end-uses. Cost has been a major factor, but increasingly, reliability, flexibility, and reduced environmental impacts on-site have made electrification an attractive option for improving industrial productivity. There are a variety of electrotechnologies that could boost industrial electricity use over the next several decades, while providing net savings in fossil fuel consumption. EPRI has looked at the possible net energy savings from five such technologies.

Freeze concentration uses refrigeration processes to separate and concentrate constituents from mixed dilute streams. Separation of constituents from process streams is a major energy use in the industrial sector and many techniques such as distillation rely on high temperatures produced by burning fossil fuels. It takes less energy (about 150 Btu) to freeze a pound of water than the 1,000 Btu needed to boil it (14). Shifting to freeze separation could cut overall energy consumption and displace industrial fossil fuel use. More energy-efficient refrigeration technologies add to the attractiveness of freeze concentration as an alternative separation technique. Currently used for treating hazardous wastes, concentrating fruit juices, and purifying organic chemicals, the technique is being investigated for broader industrial application.

Industrial process heat pumps can replace indirect resistance heating for certain low temperature applications below 138 to 149°C (below 280 to 300°F) in lumber, pulp and paper, food, chemical, and petroleum subsectors.

Electric arc furnaces allow direct melting of raw steel and uses less energy than fossil-fired furnaces. Electric arc furnaces have already gained a significant foothold in the steel industry accounting for an estimated 34% of steel produced in 1985. Continuation of this trend to 56% or more by 2000 has been projected. Electric arc furnace foundries are also used to produce steel castings and increased use of this technology also promises net fossil fuel savings.

Plasma processing uses a high intensity electric arc to generate ionized gases at temperatures up to 5538°C (10,000°F) and more, far exceeding the 1538°C (2,800°F) practical limit for fossil fuel combustion (17). The technology offers high energy density and temperature capability, controllability, and fuel flexibility compared with conventional combustion technologies. Plasma processing can be expanded in already established uses for cutting, welding, heat treating, and burning and into promising new applications in electric arc furnace dust processing, cupola refits with plasma torches, ferroalloy production, and ore reduction. Use for chemical production also is said to have future commercial potential.

Ultraviolet curing uses ultraviolet radiation produced by ionizing gases in an electrical arc or discharge, such as in a high-pressure mercury vapor lamp, to change the molecular structure of a coating to make it a solid. UV curing offers large energy and cost savings compared with thermal curing and is expected to gain increasing market penetration especially in quickcuring applications. An additional and significant environmental and health benefit is the elimination of solvents in the curing process.

Potential Savings. EPRI estimates that all these technologies offer strategic load growth to electric utilities, while resulting in net savings in fossil fuel use overall. Maximum application of these technologies could add 319 trillion Btu of fossil fuel in electric utility generation, but at the same time yield a net savings of 290 trillion Btus in these industrial processes.

Cogeneration

Cogeneration is the simultaneous or sequential production of both electrical or mechanical power and thermal energy from a single energy source (4). On-site industrial cogeneration has grown significantly since the late 1970s as a result of higher energy prices, volatile energy prices, and uncertainty over energy supplies. Implementation of the Public Utility Regulatory Policies Act of 1978 (PURPA), which required electric utilities to provide interconnections and backup power for qualifying cogeneration facilities and to purchase their excess power at the utilities' avoided cost, has reduced institutional barriers to the expansion of cogeneration. PURPA was intended to promote industrial cogeneration as a means of improving efficiency especially in the use of premium fossil fuels (gas and oil) and encouraging the use of waste fuels.

In most industrial cogeneration systems, fuel is burned first to produce steam that is then used to produce mechanical energy at the turbine shaft or to turn the shaft of a generator to produce electricity. The steam leaving the turbine is then used to provide process heat or drive machines throughout the host industrial plant and related facilities. From an energy policy perspective, the attraction of cogeneration is the ability to improve fuel efficiency. Cogeneration systems achieve overall fuel efficiencies 10 to 30% higher than if power and heat were provided by separate conventional energy conversion systems, ie, less energy than if the fossil fuel were burned in an industrial boiler to provide process heat and at an off-site utility power plant to generate electricity to be transmitted to the industrial site. This aspect of cogeneration efficiency depends on the fuel that is burned to produce electricity. Cogeneration can also be attractive as a means of quickly adding electric-generating capacity at sites where thermal energy is already being produced.

Industrial cogeneration is concentrated in the pulp and paper, chemicals, steel, and petroleum refining industries. Often the industrial cogenerators can take advantage of waste fuels to fire their boilers for heat and power. Natural gas has been the fuel of choice for many qualifying cogeneration plants under PURPA.

Cogeneration does not always provide significant efficiency advantages, however. Almost the entire output of newer combined-cycle, natural gas-fired cogeneration systems is electric power generation with little steam for process applications. In this case, there is a much smaller efficiency gain from cogeneration and a net shift in primary fuel demand from the utility sector to the industrial sector. Thermal conversion losses in electric utility and industrial combined cycle generating units are similar and there are some small savings in avoided transmission and distribution losses. If a significant portion of the cogenerated power is sold to the local electric utility, these transmission and distribution gains would largely disappear.

Industrial cogeneration has made up the overwhelming bulk of the explosive growth of so called independent power producers in the past decade. While cogeneration was initially viewed by many utilities as a threat to their market share, it is increasingly accepted as an alternative power source and has been integrated into some utilities load management and resource plans. In fact a number of utility companies have independent power subsidiaries or affiliates that are partners in industrial cogeneration projects.

In 1989, Edison Electric Institute estimated that cogeneration accounted for about 73% of the operating capacity of nonutility power plants (28).

Industrial cogeneration plants will benefit from many of the same efficiency improvements as utility generation because many use the identical technologies. In addition, better integration of industrial cogeneration and utility system operations through planning and dispatch offers net improvements to system efficiencies.

CONSTRAINTS ON EFFICIENCY GAINS IN THE INDUSTRIAL SECTOR

There have been significant energy efficiency gains in the industrial sector over the past two decades. Industrial energy use per unit of output (energy intensity) has been declining since 1970. At the same time, more and more industrial processes have been electrified. Even so, OTA has found that opportunities for further gains in energy efficiency have by no means been exhausted (4).

The industrial sector faces some of the same constraints as other sectors: low energy prices, failure of energy prices to reflect societal and environmental costs, multiplicity of decision-makers, and reluctance to adopt unproven new technologies. Energy efficiency choices tend to be made in new investments and when equipment must be repaired or replaced which creates a normal lag time between the development of new electricity-saving technologies and their dispersion throughout industry. But certain barriers are less applicable—for example, the disconnect between those who pay for energy-efficient improvements and those who benefit, is rarely present. Of all sectors, the industrial sector is probably the most responsive to price signals, so that the argument that there are market failures resulting in an underinvestment in energy efficiency here (from the perspective of myriad industrial consumers) is hardest to make. Nevertheless, certain characteristics of industrial decision-making about energy choices can result in lower adoption rates for energy-efficient equipment than might be desirable from a societal or utility perspective.

Economic considerations dominate investment decisions in the industrial sector. For most industries energy costs and electricity costs are only a small part of operating costs and thus may not enjoy a high priority. Industries that are highly energy and electricity intensive have a stronger incentive to invest in efficiency, while others do not even though there may be substantial and cost effective opportunities. Most firms regard energy efficiency in the context of larger strategic planning purposes. Investments are evaluated and ranked according to a variety of factors: product demand, competition, cost of capital, labor, and energy. Energy-related projects are not treated differently from other potential investments and must contribute to increased corporate profitability and enhanced competitive position. As a result incentives aimed at reducing energy demand growth or improving efficiency in the industrial sector must compete with other strategic factors and therefore have to be substantial to make a significant impact.

In addition to the lack of strong financial incentives and management indifference, industrial energy efficiency gains are also hampered by lack of information, and shortages of skilled designers, installers, and auditors. Highly specialized and plant- or application-specific analyses are often required to identify optimal and appropriate energy savings improvements because of the diversity of industrial processes, equipment, and energy applications. A report of the Bush administration, the National Energy Strategy report, had found that the industrial sector tended to underfund investment in energy efficiency R&D because of the belief that competitors could quickly adopt process or technology advances, thus minimizing any potential competitive advantage (5).

Overall, past studies have found that the best way to improve energy efficiency in the industrial sector is to promote general corporate investment in new plant and equipment because newer generally means more energy-efficient.

BIBLIOGRAPHY

1. Barakat and Chamberlin, *Efficient Electricity Use: Estimates of Maximum Energy Savings*, EPRI CU-6746, Electric Power Research Institute, Palo Alto, Calif., Mar., 1990.

2. U.S. Congress, Office of Technology Assessment, *Building Energy Efficiency*, OTA-E-518, U.S. Government Printing Office, Washington, D.C., May, 1992, p. 3.

3. A. P. Fickett, C. W. Gellings, and A. B. Lovins, "Efficient Use of Electricity," *Scientific American* 65–74 (Sept. 1990). (see eg, the estimates from Rocky Mountain Institute cited).

4. U.S. Congress, Office of Technology Assessment, *Energy Technology Choices: Shaping Our Future*, OTA-E-493, U.S. Government Printing Office, Washington, D.C., July, 1991.

5. *National Energy Strategy: Powerful Ideas for America, First Edition 1991/1992* U.S. Government Printing Office, Washington, D.C., Feb. 1991, app. C, pp. C25–26.

6. U.S. Department of Energy, Energy Information Administration, "Energy Preview: Residential Energy Consumption and Expenditures Preliminary Estimates, 1990," *Monthly Energy Review* (Apr. 1992), DOE/EIA-0035(92/04) U.S. Government Printing Office, Washington, D.C., April, 1992, p. 1.

7. P. D. Holtberg, T. J. Woods, M. L. Lihn, and A. B. Koklauner, *Gas Research Insights: 1992 Edition of the GRI Baseline Projection of U.S. Energy Supply and Demand to 2010* Gas Research Institute, Chicago, Ill., Apr., 1992.

8. U.S. Department of Energy, Energy Information Administration, *Annual Energy Review 1991*, DOE/EIA-0384(91), U.S. Government Printing Office, Washington, D.C., table 17, June, 1992.

9. U.S. Department of Energy, Energy Information Administration, *Electric Power Annual 1990*, DOE/EIA-0348(90), U.S. Government Printing Office, Washington, D.C., Jan., 1992, table 1, p. 16.

10. U.S. Department of Energy, Energy Information Administration, *Annual Energy Outlook 1993*, DOE/EIA-0383(93), U.S. Government Printing Office, Washington, D.C., Jan., 1993, table 21, p. 78.

11. J. Koomey and co-workers, *The Potential for Electricity Efficiency Improvements in the U.S. Residential Sector*, LBL-30477, Lawrence Berkeley Laboratory, Berkeley, Calif., July, 1991, pp. 35–36.

12. F. Krause and co-workers, *Final Report: Analysis of Michigan's Demand-Side Electricity Resources in the Residential Sector*, vol. 1, Executive Summary, LBL-23025, Lawrence Berkeley Laboratory, Berkeley, Calif., Apr., 1988.

13. American Council for an Energy Efficient Economy, *The Potential for Energy Conservation in New York State*, NYSERDA Report 89–12, New York State Energy Research and Development Authority, Albany, N.Y., Sept., 1989, pp. S-5–6.

14. Oak Ridge National Laboratory, *Energy Technology R&D: What Could Make a Difference?* vol. 2, part 1 of 3, ORNL-6541/v2/p1, Oak Ridge National Laboratory, Oak Ridge, Tenn., Dec., 1989, pp. 15, 45.

15. U.S. Department of Energy, Energy Information Administration, Housing Characteristics 1987 DOE/EIA-0314(87), U.S. Government Printing Office, Washington, D.C., May, 1989.

16. American Council for an Energy-Efficient Economy, *The Most Energy-Efficient Appliances 1989–1990* American Council for and Energy-Efficient Economy, Washington, D.C., 1989, pp. 18–19.

17. Battelle-Columbus Division and Enviro-Management & Research, Inc., *DSM Technology Alternatives*, EPRI-EM-5457,

Interim Report, Electric Power Research Institute, Palo Alto, Calif., Oct., 1987.

18. S. Cowell, S. Gag, and J. Kelly, *Home Energy* **9**(2), 27–33 (Mar./Apr. 1992).

19. U.S. Department of Energy, Energy Information Administration, *Annual Energy Review 1991*, DOE/EIA-0384(91), U.S. Government Printing Office, Washington, D.C., June, 1992, tables 17 and 21.

20. U.S. Congress, Office of Technology Assessment, *Changing by Degrees: Steps to Reduce Greenhouse Gases*, OTA-O-482, Government Printing Office, Washington, D.C., Feb., 1992, ch. 4.

21. F. Kraus and J. Eto, *Least-Cost Utility Planning: A Handbook for Public Utility Commissioners: Vol. 2, The Demand Side: Conceptual and Methodological Issues,* National Association of Regulatory Utility Commissioners, Washington, D.C., Dec., 1988.

22. U.S. Department of Energy, Energy Information Administration, *Commercial Building Characteristics 1989*, DOE/EIA-0246(89), U.S. Government Printing Office, Washington, D.C., Jan., 1991, table 61, p. 122.

23. U.S. Congress, Office of Technology Assessment, *Energy Efficiency in the Federal Government: Government by Good Example?* OTA-E-492, U.S. Government Printing Office, Washington, D.C., May, 1991.

24. U.S. Congress, Office of Technology Assessment, *Energy Use and the U.S. Economy*, OTA-BP-E-57, U.S. Government Printing Office, Washington, D.C., June, 1990.

25. American Council for an Energy-Efficient Economy and New York State Energy Office, *The Achievable Conservation Potential in New York State from Utility Demand-Side Management Programs*, final report, Energy Authority Report 90-18, New York State Energy Research and Development Authority, Albany, N.Y., Nov., 1990, p. 48.

26. U.S. Congress, Office of Technology Assessment, *Industrial Energy Use*, OTA-E-198, U.S. Government Printing Office, Washington, D.C., June, 1983, p. 50. (Available from the National Technical Information Service, Springfield, Va., NTIS Order #PB83-240606.)

27. U.S. Congress, Office of Technology Assessment, *Nonferrous Metals: Industry Structure: Background Paper*, OTA-BP-E-62, U.S. Government Printing Office, Washington, D.C., Sept. 1990, pp. 25–26.

28. Edison Electric Institute, *1989 Capacity and Generation of Non-Utility Sources of Energy*, Washington, D.C., Apr., 1991, p. 29.

ELECTRICAL ENERGY EFFICIENCY, UTILITY PROGRAMS

KAREN LARSEN
Office of Technology Assessment
Washington, D.C.

With ample untapped opportunities to save electricity, demand rising but long-term growth rates uncertain, and powerplant construction costs soaring, it is not surprising that energy efficiency has become the byword for cost-conscious consumers, regulators, and utilities seeking new ways to hedge future strategies. The potential of energy efficiency as a means to lessen the environmental impacts of energy use has also attracted the interest of conservationists. The prospective new business opportuni-ties have garnered the attention of energy service companies and equipment manufacturers and vendors, as well as utilities. See also LIGHTING, ENERGY EFFICIENCY; ENERGY EFFICIENCY; COMMERCIAL AVAILABILITY OF ENERGY TECHNOLOGY; AIR CONDITIONING; REFRIGERATION; ENERGY AUDITING; ENERGY CONSERVATION.

SCOPE OF UTILITY ENERGY EFFICIENCY PROGRAMS

U.S. utilities and state regulators have now had more than a decade's worth of experience with utility-sponsored energy efficiency programs. Broadly speaking, energy efficiency programs are aimed at reducing the energy used by specific end-use devices and systems without degrading the services provided. Such savings are generally achieved by substituting technically more advanced equipment to produce the same level of energy services (eg, lighting or warmth) with less electricity (1). Energy efficiency programs are sometimes referred to as energy conservation programs. However, because to some people the term *conservation* implies an overall reduction in electricity use and energy services, many industry analysts prefer to use the more neutral and inclusive term, *energy efficiency*. Energy conservation measures can be included in efficiency programs.

Demand-side management (DSM) programs are organized utility activities intended to affect the amount and timing of customer electricity use. In theory, successful DSM programs can reduce the need to build powerplants by controlling demand for electricity and thereby creating room for expansion without providing additional supply resources.

Utility load management programs are closely related to energy efficiency DSM programs. Load management programs refer to utility programs intended to influence customer demand through economic or technical measures, usually with the objectives of reducing demand during peak periods, and/or encouraging demand during off-peak periods (2). In pursuit of the first goal, load management programs can include many of the same technologies and measures used for overall reductions in electricity use. Load management programs usually employ a combination of load management incentives, metering to measure the time and quantity of customer electricity use, and load control equipment. Because of the timeshifting aspect of load management programs, they may be targeted at peak loads and not necessarily at an overall reduction in electricity consumption. Load management programs can also be directed at retaining load or customers, and expanding customer loads. Table 1 shows common utility load management strategies and their load shape objectives. These same load shape objectives are used for utility DSM programs.

Electric utilities have used load control measures for more than 50 years, but interest in these measures has increased significantly in the 1970s and 1980s. Over this period, interest in load control has been high among utilities that purchase most of their power from others (primarily municipal utilities and rural cooperatives) because load control offered an additional means to reduce wholesale power costs (3).

Table 1. Load Shape Objectives[a]

PEAK CLIPPING, or the reduction of the system peak loads, embodies one of the classic forms of load management. Peak classic is generally considered as the reduction of peak load by using direct load control. Direct load control is most commonly practiced by direct utility control of customers' applicances. While many utilities consider this as a means to reduce peaking capacity or capacity purchases and consider control only during the most probable days of system peak, direct load control can be used to reduce operating cost and dependence on critical fuels by economic dispatch.

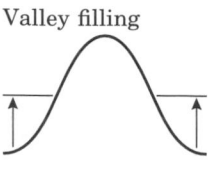

Peak clipping

VALLEY FILLING is the second classic form of load management. Valley filling encompasses building off-peak loads. This may be particularly desirable where the long-run incremental cost is less than the average price of electricity. Adding properly priced off-peak load under those circumstances decreases the average price. Valley filling can be accomplished in several ways, one of the most popular of which is new thermal energy storage (water heating and/or space heating) that displaces loads served by fossil fuels.

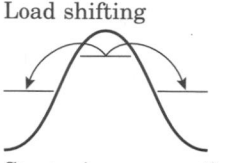

Valley filling

LOAD SHIFTING is the last classic form of load management. This involves shifting load from on-peak to off-peak periods. Popular applications include use of storage water heating, storage space heating, coolness storage, and customer load shifts. In this case, the load shift from storage devices involves displacing what would have been conventional appliances served by electricity.

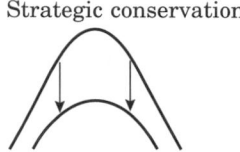

Load shifting

STRATEGIC CONSERVATION is the load shape change that results from utility-stimulated programs directed at end-use consumption. Not normally considred load management, the change reflects a modification of the load shape involving a reduction in sales as well as a change in the pattern of use. In employing energy conservation, the utility planner must consider what conservation actions would occur naturally and then evaluate the cost effectiveness of possible intended utility programs to accelerate or stimulate those actions. Examples include weatherization and appliance efficiency improvement.

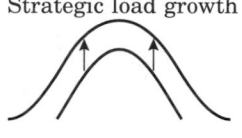

Strategic conservation

STRATEGIC LOAD GROWTH is the load shape change that refers to a general increase in sales beyond the valley filling described previously. Load growth may involve increased market share of loads that are, or can be, served by competing fuels, as well as area development. In the future, load growth may include electrification. Electrificiation is the term currently being employed to describe the new emerging electric technologies surrounding electric vehicles, industrial process heating, and automation. These have a potential for increasing the electric energy intensity of the U.S. industrial sector. This rise in intensity may be motivated by reduction in the use of fossil fuels and raw materials resulting in improved overall productivity.

Strategic load growth

FLEXIBLE LOAD SHAPE is a concept related to reliability, a planning constraint. Once the anticipated load shape, including demand-side activities, is forecast over the corporate planning horizon, the power supply planner studies the final optimum supply-side options. Among the many criteria used is reliability. Load shape can be flexible—if customers are presented with options as to the variations in quality of service that they are willing to allow in exchange for various incentives. The programs involved can be variations of interruptible or curtailable load; concepts of pooled, integrated energy management systems; or individual customer load control devices offering service constraints.

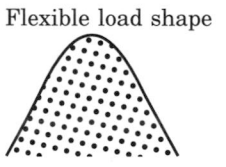

Flexible load shape

[a] From Ref. 2.

Utilities can have many goals for DSM and load management programs. Maximizing energy savings is one. Others, and perhaps more important to different utilities, are maximizing customer satisfaction, minimizing lost revenues (utility revenues lost when consumers reduce electricity use), minimizing free riders, or minimizing the cost per kilowatt (kW) or kilowatt-hour (kW·h) saved.

The development of utility energy efficiency programs coincides with the trend toward adoption of integrated resource planning (IRP) processes by electric utilities. IRP involves a comprehensive and open utility planning process that includes greater consideration of potential demand-side measures on a par with generation and other supply-side additions in order to meet projected loads. The prospect of greater reliance on demand-side measures to delay the need for new powerplant construction requires that potential energy savings be estimated with greater certainty and that actual savings be validated. Adoption of IRP has created new challenges for electric utilities planners and their regulators in incorporating rapidly expanding DSM programs into the resource mix.

INFLUENCING CUSTOMER BEHAVIOR

Electric utilities, with the approval and encouragement of state regulatory bodies, have adopted a variety of mechanisms to influence customer electricity use: load controls, differential or incentive rates, rebates, loans, grants, shared-savings agreements, energy audits, technical assistance, direct equipment installation and replacement, comprehensive energy management programs, and so forth. Many of these programs are of recent vintage and limited in scope, but overall the initial savings have been promising even though not as high as expected. Certain issues have recurred in the design, implementation, and evaluation of these programs, including: cost-effectiveness determinations, choice of effectiveness tests, free riders, measurement of savings, persistence of savings, customer participation rates, utility cost recovery, and financial incentives.

All utility DSM programs fit into one or both of the following programs: 1) those affecting the way energy-using equipment is operated, and 2) those that focus on the installation of efficient technologies. Utilities typically operate separate programs for commercial, residential, and industrial customers.

Load control measures differ based on the degree of control and input exercised by the utility and the customer. They range from programs in which the utility asks customers to reduce load and the customer individually decides which appliances to turn off, to direct load control systems that are highly automated and have little customer input.

Direct control systems are by far the most common form of load control. They typically consist of a communications system that links the customer's equipment with the utility and a decision logic system (ie, a computer program) that dispatches commands to the customer equipment in response to information on utility and/or customer loads. In a residential load management program, equipment might be installed to allow the utility to cycle participating home air-conditioners and water heaters on and off briefly during times of peak load with little or no disruption to the customer. With widespread participation, this system represents a critical strategic tool for utilities to shave peak load. Typically, the customer enters into an agreement with the utility that gives them either lower rates and/or a small monthly payment for participation in the program. For example, Potomac Electric Power Company (PEPCO) offers a credit of $110 to households that join its "Kilowatchers Plus Club" and allow the company to shut off their air conditioning for short periods of time to offset summer peak loads if needed. Some 100,000 members of PEPCO's "Kilowatchers Club" receive a $45 credit for allowing the utility to cycle their compressors off and on for brief periods. PEPCO estimates that by 1995, cycling will pare 170 megawatts from its summer loads (4).

Utilities and regulators have experimented with various incentive rates in an attempt to encourage greater efficiency in electricity use. They have instituted variations in rates by charging more for peak load and higher volume usage to reflect the increased costs of providing such service. There has been a great deal of activity involving time of use rates for large industrial and commercial customers, but only limited experience with time of use rates for residential customers. Participation in time of use rates generally requires installation of meters that allow measurement of both the quantity and time of customer electricity use.

Information programs are intended to alert customers about potential electricity savings measures. Examples include informational advertising campaigns, energy audits, and bill enclosures. According to an analysis of utility DSM programs prepared by the American Council for an Energy-Efficient Economy (ACEEE) for New York State, information-only programs that provide customers with general information about energy efficiency opportunities and/or combine information with energy audits have low participation rates and low energy savings (4). The most effective programs are the free energy audits coupled with post-audit followup. According to ACEEE, the programs can achieve high participation rates (60 to 90%) and energy savings among participating customers of 6 to 8% (4). Revamped information programs are reported to be achieving greater levels of participation.

Rebate programs provide money to customers, contractors, homebuilders, vendors, or others who make equipment choices to help defray some or all of the cost of DSM measures. Rebate programs are the most common utility program offered. The form of rebate mechanism can be cash, discount coupons, or bill credits. ACEEE found that the most successful rebate programs in their survey reached about 10% of eligible customers (and about 25% of the larger customers with peak demand of 100 to 500 kW) over a period of 3 to 7 years. The most successful programs cut electricity use by 5% at utility costs of $0.01/kWh saved. The most effective targets have been lighting and heating, ventilation, and air conditioning (HVAC) equipment improvements.

Rebate programs have not historically been very effective at promoting system improvements—those involving the interaction of many pieces of equipment. Generally, participation levels are moderate, as are energy savings. They effectively cut utility peak demand and electricity sales by about 1%/year in successful cases surveyed in the ACEEE study. Some analyses indicate that participation drops off after several years of aggressive program promotion; however, more analysis is needed of this possible pattern according to ACEEE's study.

Loan programs provide cash to finance energy-savings investments and are attractive for customers who lack cash. The program may allow the customer to repay energy efficiency investments on the monthly utility bill, often at a low interest rate. They are offered by only a few utilities. Studies of consumer loan programs have found that customers offered a choice of rebates or low-interest loans have generally opted for the rebates (5).

Increasingly, utility programs bundle various DSM approaches into a single package. For example, the City of Fort Collins Light and Power offers residents of Fort Collins the Energy Score Home Energy Rating Service that combines information, audit, building efficiency standards, rebates, loans, and eligibility for energy efficient mortgages. Homeowners, builders, sellers, or purchasers of new or existing homes can contact an independent utility-certified rater to inspect and report on a home. The rater provides a comprehensive home energy efficiency analysis covering the orientation of the lot, insulation, windows, doors, air leakage, heating and hot water systems, and other factors influencing energy use. The house is given a rating from 0 to 100, with 100 being the most efficient. The cost of the rating is $100 to $175 and the city utility picks up $50 of the cost. Homes with higher energy efficiency ratings (G-70 for gas-heated homes, and G-65 for electric heated homes) may be eligible for a 2% ratio increase on an energy efficient mortgage from participating lenders, increasing the purchaser's buying power. The rating also identifies opportunities for efficiency improvements and may qualify the homeowner for the utility's zero-interest "Zilch" home improvement loan to finance the upgrades.

Performance contracting programs offer payments based on the amount of energy saved as a result of efficiency improvements. They generally rely on energy service companies (ESCOs) or other vendors to recommend, install, and finance efficiency measures. Utilities can also contract directly with large customers. According to the ACEEE study, the most successful of these programs have included high incentives, but have achieved significant savings. On the whole these programs have been more costly than some other types of programs. Experience has indicated that ESCOs have tended to focus on the largest customers and the most lucrative measures (especially lighting and cogeneration) to achieve savings. ESCO contracts provide one mechanism for reaching some of the most cost-effective, energy-efficient opportunities with significant economies of scale. Other approaches can target and achieve these same savings opportunities. Initial experience with performance contracting and ESCOs has been mixed. Many utilities are substantially revising their performance contracting programs or are complementing them with other types of programs. Performance contracting with ESCOs or with large customers still remains an attractive alternative for financing and installing energy-saving technologies.

Comprehensive programs combine regular personal contacts with customers, comprehensive site-specific technical assistance, and financial incentives that pay the majority of the installation costs of efficiency measures. According to the ACEEE study, these programs were highly successful, but also tended to be among the most expensive at a typical cost of $0.03/kwh saved. There is little experience with large-scale programs over time. The analysis suggests that this type of program may be particularly appropriate for serving small customers and for new construction (where there is a one-time opportunity to capture substantial savings at only the marginal cost of efficient equipment over standard equipment).

Request for proposal (RFP) and bidding programs have been in operation for several years. Under these programs, the utility issues a request for proposals to provide demand-side resources and receives and evaluates proposals from ESCOs and customers. After evaluating the bids, the utility negotiates contracts with the winning bidders for specific energy savings and load reductions. Based on preliminary results analyzed for the ACEEE New York study, these programs offer the promise of significant savings (up to 1.5% of peak demand after 2 years). The success has been tied to reaching large customers directly or through ESCOs who participate in the process. The programs generally are less than utility avoided cost, but a tendency has been noted for bids to approach utility avoided costs. Much of the initial experience with DSM bidding programs has been in Maine and New York. As utility competitive resource procurements have expanded, so too have the number of bidders offering demand-side installations. Moreover, these demand-side bids are proving to have a higher success rate in winning bids than conventional supply options.

Fuel switching programs involve incentives to utility customers to reduce load by switching to an alternative energy source for all or some of the service provided by electricity. Examples include the installation of a gas-powered air conditioning system in a large commercial office complex or switching from an electric resistance to a gas water heater. Fuel switching and electrification measures generally involve complicated site-specific tradeoffs, and no generalizations can as yet be made about their overall performances and costs.

MEASURING ENERGY SAVINGS AND EVALUATING EFFECTIVENESS

There has been over a decade of experience with utility load management and DSM programs around the country. Substantial dollar and energy savings have been claimed, but much remains to be learned.

To be successful from the perspective of least-cost planning objectives, the program should achieve maximum long-term, cost-effective net energy and demand savings (net of what would be required in absence of the utility program). Generally this finding means a long-term strategy aimed at serving the most customers (including all but the very smallest). This is to assure maximum savings and to minimize equity issues of cross-subsidization. In addition, the strategy should promote efficiency/load management measures that customers are unlikely to install without utility efforts in the short term and for longer-term measures with long-lives or that have a high probability of replacement.

The ACEEE study found that utility DSM programs as a whole had not yet had a dramatic impact. The programs surveyed were reaching less than 5% of target customers on a cumulative basis and were reducing their energy use by less than 10%. As of 1989–1990, it was estimated that utility peak demand had been reduced by less than 1%. They did find a number of highly successful programs, however. A few reached 70% or more of eligible customers—with customer energy savings of 10 to 30%, and reductions in utility peak demand of up to 5%. Many of the most successful programs, however, have still been in pilot- or small-scale programs and had yet to be applied on a large-scale basis. The good news is that all of the energy savings reported came at a cost to utilities of less than $0.04/kWh saved even including free riders. These reported costs have been less than many utilities' avoided costs to generate new power, making it likely that the programs would prove cost-effective using the utility cost test.

Since the ACEEE study has been published, utility DSM programs have continued to grow, and many utilities are now projecting significant savings from their efforts. A recent analysis by Oak Ridge National Laboratory, based on reports to the Energy Information Administration, has found that existing utility programs are projected to offset 14% of the growth in electricity demand by the year 2000 (6).

Measuring Savings

The savings from a DSM program are estimated by comparing energy demand both before and after the program is implemented. Evaluating the success of utility DSM programs is difficult both on a local and a national basis. Most savings estimates reflect engineering estimates, and more sophisticated measured and validated estimates of savings are rare. Engineering estimates generally rely on simple rules of thumb calculations using manufacturers' data, or engineering simulations. Engineering estimates can be fairly accurate for some simple DSM actions (eg, domestic water heater wraps). However, in practice, engineering estimates have been found to overestimate actual electricity savings (7). As experience with DSM programs increases, and energy savings are subjected to more rigorous impact evaluation, it should be possible to develop other techniques, or at least more accurate engineering rules of thumb, to support reliance on this technique to estimate potential savings. Until then, such estimates should be viewed with caution.

In order to show the effectiveness of efficiency measures, it must be determined how much electricity use is actually reduced over what it would be in the absence of the measure. Depending on the goal of the program, monitoring usage by the time of use (on-peak vs. off-peak) will also be important to determine impacts on load shape.

There are several means of measuring (collecting data on) electricity use and the impacts of efficiency improvements: monthly customer bills, spot metering (either on a short-term before and after retrofit or permanent basis), whole building load research monitoring, and end-use load research monitoring (Table 2) (8).

Measuring exact savings is not necessary in all cases and could become prohibitively complex as the number and extent of DSM programs increase. For simple measures, where there is substantial experience (more efficient residential refrigerators, for example), past measurements and engineering estimates may suffice to calculate savings for the program. For more complex and site-specific DSM measures (eg, retrofitting and relamping a large commercial building), detailed site-specific measurements of specific load shapes may be needed to estimate savings.

Comparison of customer billing data is the most straightforward and least expensive method for many applications, but is not adequate for new construction or for

Table 2. Methods Used to Measure Electricity Consumption

Approach	Explanation	Advantages	Disadvantages
Monthly electricity bills	Obtain electricity bills for a year before and a year after participation, adjust annual electricity use for weather and other relevant factors, and compute the difference between pre-and post-participation use in kWh/year.	Measures actual changes in electricity use, permits adjustment for changes in weather and other factors, and requires little primary data collection.	Provides no estimate of demand (kW) reductions unless customers face demand changes. Analysis of monthly billing data can yield ambiguous results. Estimates of kWh savings affected by changes in facility use unrelated to devices installed.
Spot metering of electricity use	Monitor electricity use before and after participation for short times (eg, a few days); also measure other relevant factors (eg, operating hours for equipment and heating degree days) for a longer time (eg, up to a year).	Measures electricity savings (both kWh and kW) for well-defined, short time periods. Modest cost.	Could yield estimates of savings not realized if measurements taken incorrectly or at atypical times, or if building use changes. Difficult to apply to devices that are season- or weather-dependent.
Whole-building load-research monitoring	Monitor electricity use of facility to record kW demand before and after participation.	Measures actual electricity use and demand (kWh and kW). Can be combined with other data to adjust for changes in weather and other factors.	Expensive and time consuming. Large amounts of data produced. Results may be affected by changes in facility use unrelated to equipment installed.
End-use, load-research monitoring.	Monitor specific circuits affected by new systems to record kW demand before and after participation.	Measures actual electricity use and demand (kWh and kW) for specific end uses affected by program. Can be combined with other data to adjust for changes in weather and other factors.	Most expensive and time consuming method. Large amounts of data require sophisticated computer programs and analysts to interpret. Results may be affected by changes in facility use unrelated to equipment installed.

[a] From Ref. 6.

large and complex installations. In the former case, bill data will be absent, thus engineering calculations or comparisons with similar buildings for which data are available might be used. In the latter case, normal fluctuations in energy use could mask the effects of efficiency improvements and so specific end-use metering that tracks the time and quantity of electricity may be required.

Once total end-use savings have been determined, the impacts on utility load shape and supply must be calculated. In general, because of transmission and distribution losses, the actual kilowatts saved at the powerplant from customer efficiency measures are about 8% higher than that saved on site. Kilowatts saved by the customer may also reduce utility reserve margins, ie, customer savings plus the reserve margin percentage (allowance for powerplant downtime). Improved measurement and monitoring of end-use efficiency savings and documentation of actual reductions in utility-generating demand over time may contribute to less uncertainty about demand-side measures in utility resource planning.

Tracking the persistence of energy savings from efficiency measures is also important. Some measures may prove to be fairly reliable and long-lived (for example building insulation that the customer is unlikely to remove). But other measures may be affected by declines in the technical performance of equipment, the lifetime of the measure, user replacement of measures when they wear out, changes in operating conditions induced by the DSM program, or market-related changes in electricity use (1). The energy-savings benefits of a compact fluorescent light may disappear, if when it is worn out, the user replaces it with a standard incandescent lamp.

Similarly, an occupant might be induced to raise thermostat settings to take advantage of improved building insulation or an efficient space heating system, thus, at least partially offsetting the efficiency gain. This phenomenon of losses in efficiency gains because of customer behavior is often referred to as takeback.

Participation Rates

The success of DSM programs often hinges on the number of customers and/or trade allies (businesses that sell or influence choices of energy using equipment such as architects, designers, builders, appliance dealers) that participate in the program. Participation rates are the ratio of the number of participants to the number of eligible customers. In many cases, determination of the pool of potential participants is fairly straightforward and based on information a utility readily has at hand (eg, commercial office buildings, all residential customers). However, for more specialized programs, additional market research may be needed to identify potential participants, for example, homes with electric resistance heat, or industrial motor applications.

As a practical matter, most estimates of DSM program participation rates generally include free riders: customers who participate in a program, but would have undertaken the same conservation actions even if the program were not offered (1). (Some discussions also brand as free riders ratepayers who benefit from conservation pro-

grams, but do not participate; however in this article we include only program participants). The presence of free riders tends to overstate program results. Some economists maintain that free riders should not be eligible for program incentives and will drive up program costs and ratepayer impacts to an unacceptable degree. Conversely, another complication is the general exclusion from participation rates of free drivers. Free drivers are customers who take DSM program-recommended actions, but do not participate directly in the program (ie, claim rebates). The absence of free drivers will result in understating the program's effectiveness.

The presence of free riders, setting aside the issue of whether they should be eligible for financial incentives, complicates evaluations of the effectiveness of utility DSM programs. In determining whether the program has actually had an impact on customer energy use, the focus must be on net savings—calculated by determining the share of free ridership and excluding the associated savings.

But the presence of a high portion of free riders in a program is not necessarily an indication that the program is not effective for several reasons. First, one should expect a high degree of free riders early in the program and then as the program becomes more successful and participation increases, the free rider share should approach a floor defined by the penetration of the efficiency measure in the market place or the market share of efficient devices versus standard devices in absence of the program. Various utility programs have estimated free riders at 5 to 10% for replacement of working motors and 5 to 35% for new motors (5).

Second, many estimates of free ridership are based on self-identification by those who say they would have adopted the measure anyway, thus tending to overestimate actual free ridership. The bias problems with surveys are well documented and show a tendency of respondents to give the perceived "right" answer to the interviewer rather than the "true" answer. Additionally, while a participant might be favorably disposed to installation of the efficiency measure in the absence of the program, it is difficult to estimate with any accuracy how many of the self-identified free riders would actually have installed the measure without the program or the extent to which the existence of the program accelerated their actions.

Costs of DSM Programs

Monitoring and estimating the costs to utilities and customers is necessary to determine whether the costs of efficiency programs are outweighed by their benefits, and to provide for adequate cost recovery in regulatory proceedings.

For newly authorized programs, very little actual cost data may be available, but as experience increases, costs should be calculated with greater accuracy. The ACEEE review of 58 existing utility DSM programs has found that reported cost figures per kilowatt and kilowatt-hour for efficiency measures have only been approximate, often have ignored customer costs, and sometimes rely on rough estimates of indirect utility costs. The lack of accurate cost data is troubling when one considers that over $2 billion has been invested in utility energy efficiency programs in 1991 and that by the end of the decade some experts estimate that DSM could be a $30 billion/year industry (9). Moreover, more reliable and detailed cost data are needed for DSM resources to be more fully integrated into utility resource planning processes.

Determining Cost-Effectiveness

There is a wide variation in how different utilities and state regulators calculate the cost-effectiveness and costs of DSM programs. The cost-effectiveness of DSM measures is commonly estimated from either the utility, customer, or the societal perspective. The utility perspective considers the utility's costs and benefits for program, including rebate and other costs, avoiding energy and capacity benefits. It excludes customer costs and the value of revenues lost by the utility because of energy savings.

The total resource cost perspective (adopted in New York State) includes the money paid by program participants for materials, installation, and maintenance (including credits for reducing customer costs, such as reduced maintenance costs in addition to factors considered from the utility perspective). In practice, the total resource cost test suffers from the fact that extensive data on customer costs are not generally collected by utilities.

There are several alternative units used in estimating cost-effectiveness: Cost per kilowatt-hour saved simply uses program expenses divided by kilowatt-hours saved. Other measures calculate levelized cost per kWh saved (discounting the cost over time) to provide a long-term cost estimate. More rigorous approaches involve calculation of total levelized costs of the program and comparison with avoided total costs of the energy saved (avoided energy costs plus levelized value of annual capacity cost divided by 8,760 hours/year).

Evaluation of DSM Programs

Evaluation is the systematic measurement of the operations and performance of DSM programs and should rely to the extent possible on objective measurements and well-defined and executed research methods. Program and impact evaluations of DSM and load management programs are critical components of both utility and government assessments of the cost-effectiveness and success of efficiency measures. Program evaluation is a rapidly evolving specialty that relies on social-science research methods together with technical data to provide valid and reliable documentation and quantification of program results and costs in order to analyze program usefulness. Good impact evaluation efforts are not cheap or easy to perform, and yet are an indispensable element of any expansion of efficiency programs. Adequate funding of evaluation and monitoring can amount to 10% of the costs of utility programs. As the costs, extent, and expectations of utility energy efficiency programs grow, the resources devoted to monitoring and evaluation will have to expand and the evaluation techniques must also become more technically sophisticated and reliable.

MIXED RESULTS FROM UTILITY ENERGY EFFICIENCY PROGRAMS

While there are clearly successful utility energy efficiency programs with demonstrable energy savings, experience so far indicates that the energy savings achieved fall far short of the full technical potential that is cost-effective to end users.

The ACEEE review of 58 utility programs has found evidence for this conclusion in low participation rates and in actual savings well below cost-effective technical potential (5). Programs with the highest participation rates have reached only 10 to 70% of eligible customers. Among participating custoemrs, the programs with the highest energy savings have been found to yield only 20 to 60% of the cost-effective technical potential. Cost-effective technical potential has been defined as measures having equipment and installation costs less than $0.05 kWh saved, ie, less than the retail commercial and industrial electricity rates and/or utilities' long-run avoided costs. The gap between technical potential and actual savings has been large for the best programs and larger still for typical programs.

Low participation and savings rates are typical of the startup stage of most programs, and many programs have still been limited in scale and had only a few years experience. However, other utility programs have been operating for some time and it is reasonable to expect better performance.

No utility operates state-of-the-art programs in all areas. The largest commercial and industrial efficiency programs have been found to reduce kilowatt-hour sales by only 2 to 14%–far less than the estimated 35% cost-effective savings potential (from the consumer perspective) found in the ACEEE study of New York State potential (9). These performance shortfalls raise questions about the viability of ambitious state and utility efficiency goals.

Some economic analysts are challenging utilities and regulators cost-benefit equations and questioning the claimed successes of DSM programs. One controversial analysis performed for the U.S. Environmental Protection Agency has discarded the more commonly used cost-benefit tests and applied an alternative cost-benefit measure to DSM program savings calculated by two utilities. The report has examined a total of 16 separate DSM programs operated by the two utilities and has concluded that none of the programs passed its "conventional" economic cost-benefit test if environmental benefits were excluded from the calculations (11).

The New York State Energy Plan sets a goal for utility conservation and load management programs to reduce electricity use and demand by 15% by the year 2008. To do this, ACEEE estimates that DSM programs will have to reach 50 to 70% of customers and achieve savings among participants of 20 to 30%.

Nevertheless, many utilities have now enthusiastically embraced DSM programs. The New England Electric System (NEES) has been an early leader in utility DSM programs, spurred in part by financial incentives adopted by state regulators in Massachusetts, New Hampshire and

Rhode Island. For 1990, NEES have reported that potential system profits from DSM programs were $10 million. Estimated savings have been a total of 194,300 megawatt-hours saved and 116.5 megawatts of demand reduced (12).

NEES's third resource plan adopted in 1991 relies on DSM programs to displace a total of 850 megawatts by the year 2000, constituting more than 12% of the utility's capacity resources. NEES's resource plan will also achieve a 45% reduction in net air emissions by the year 2000 through its resource strategies including DSM, converting/repowering an existing plant to natural gas, accelerating environmental compliance, power purchases from nonutility generators and Canadian hydroelectric facilities, and various initiatives to offset greenhouse gas emissions (13).

Pacific Gas & Electric Company, the Nation's largest utility, also has long experience in DSM programs. PG&E plans to spend about $2 billion on customer energy efficiency in the 1990s and cut their energy growth by half and peak demand by 75% (2,500 megawatts). PG&E projects that these savings can be achieved at a cost of from $0.03 to $0.04/kilowatt-hour—less than half the cost of building new fossil generation (1).

Elements of Successful Programs

Even though no utility was found to perform at state-of-the-art levels in all of its efficiency programs, a number had demonstrated notable success. Reviews of utility efficiency and conservation programs have indicated that some utilities consistently do a better job than others in operating these programs. Among the most successful cited in the 1990 ACEEE study were: the City of Palo Alto, Calif.; Central Maine Power; New England Electric System; Pacific Gas and Electric: Southern California Edison; and Wisconsin Electric (1).

Certain program elements are believed to contribute to above-average participation and savings:

- Marketing strategies that use multiple approaches (direct mail, media, etc.) combined with personal contacts with the target audiences; particularly successful are those strategies that develop regular, person-to-person contacts and followup contacts after installation to assure that the measures are working properly and that they promote additional measures.

- Targeting of program approaches and marketing strategies to different audiences (customers, architects, equipment dealers, engineers, developers) and for different types of investment decisions (new construction, remodeling, replacement, retrofitting); targeting audiences in program design is especially successful in producing a program that meets consumer needs.

- Technical assistance to help targeted customers assess efficiency opportunities and identify and implement DSM measures; assistance might include energy audits, advice on equipment, contractors, computer modeling of possible savings alternatives, information on new state-of-the-art technologies. Detailed technical assistance is generally only cost-effective when coupled with incentive programs that induce high levels of customer participation and savings.

- Simple program procedures and materials that make it easier for the customer to understand program potential and to participate; examples are one-step application procedures, assistance in filling out forms, packaged rebate programs.

- Financial incentives that attract customer attention and reduce first costs of implementing DSM measures; analyses of the effects of varying incentive rates are scanty, but initial results indicate that offering free measures produces the highest participation rate. High incentives (50% or more of a measure's cost) generally appear to produce higher participation rates than moderate incentives (one-third of a measure's cost). Moderate incentives may not produce higher participation rates than low incentives, however.

- Multiple measures available for customers to choose from that increase the likelihood that customers will find a measure or program that is appropriate for their needs and/or to implement more than one measure and gain more savings; there are a plethora of programs limited solely to lamps and air conditioners including additional HVAC, efficient lighting, and motor measures; allowing customers to propose their own qualifying measures tends to boost participation rates and savings.

- Programs promoting new technologies not yet widely adopted in the marketplace; these programs for high-efficiency technologies tend to reduce free riders and achieve higher savings than available through first generation technologies. A high percentage of free riders (about 30%) have been found with some technologies, especially when rebates are provided for products that are already being purchased by many customers, such as, reduced wattage lamps and moderate efficiency air conditioners. Because customers may be unfamiliar with and wary of new, advanced energy-saving technologies, programs that focus on these technologies may require substantial marketing efforts to boost typically low initial participation rates.

Additional factors that have contributed to the success of utility DSM programs were: top management commitment to energy efficiency measures: staff and organizational commitment, skills, support, creativity, and flexibility. Personal contact marketing and follow-up by utility personnel are also key to successful programs. Lastly, the most successful utility programs have been those where utilities are offered incentives for successful programs.

Problem Areas

The ACEEE study identified several problem areas that must be addressed if DSM programs are to have a significant effect.

Most utility commercial and industrial DSM programs have had only a limited focus. The programs must expand beyond lighting and small HVAC improvements to include advanced lighting and motor technologies and comprehensive industrial system improvements. There is no one-size-fits-all comprehensive demand-side program. Regulators and utilities must develop packages of programs tailored for the utility, load, and customer characteristics if the initiative is to be a success. Many utilities in an effort to structure their services to enhance customer values are examining ways to provide more comprehensive energy efficiency services.

Participation rates have been low. Marketing efforts must be expanded to reach and persuade more customers to participate.

More data and research are needed to support DSM program development and evaluation. Program design and evaluation is hampered by the lack of credible data on energy use and target populations (building characteristics, motors and other equipment), and by the lack of accessible and useful documentation and evaluation of existing programs. Information on actual percentage reductions in energy use is rarely collected and yet would be of invaluable assistance to utilities, regulators, and consumers.

Additionally, because energy and load management efforts have been limited in scope and long-term experience is lacking, mistakes will be made. But utilities may fear to publicize mistakes and shortcomings for fear that regulators will punish them. There is, however, much to be learned from mistakes. Therefore unsuccessful program experiences should be investigated and the results publicized so that others might avoid these pitfalls.

Need for Complementary State and Federal Efforts

Even the best DSM programs cannot achieve all the cost-effective savings. Some customers will not participate, no matter what incentives utilities offer. Many will not adopt all cost-effective measures. Because of this tendency, utility programs need complementary approaches–eg, building codes and appliance and equipment efficiency standards–in order to maximize the overall adoption of energy-efficient technologies. The California Energy Commission analysis of the effectiveness of utility DSM measures in 1983 has found that the reduced peak demand of 2,718 megawatts was due 45% to utility programs, 37% to building code requirements, and 16% to various appliance efficiency standards (11). Federal and state efficiency initiatives can also boost the availability of energy efficiency products in the marketplace.

INCORPORATING ENERGY EFFICIENCY INTO UTILITY OPERATIONS–THE ROLE OF IRP

IRP has become the main process through which utilities incorporate DSM measures into their mid- and long-term resource planning. As a planning tool, IRP allows a utility to incorporate a variety of information about load, system characteristics, demand growth, resource options, and corporate goals into an analysis that. explicitly evaluates supply- and demand-side resources; performs this task in a consistent manner; and expressly confronts the uncertainties inherent in utility planning to produce a flexible resource plan for meeting customer needs at least-cost. IRP also generally includes opportunities for public involvement and regulatory review, as well as consideration of environmental and other social impacts of utility resource alternatives.

By mid-1993, utilities in at least 41 states were actively involved in some sort of IRP process. At least 33 states have required IRP or least-cost planning by their utilities. Under Federal law, utilities that purchase power from the Bonneville Power Administration, the Western Area Power Administration, and the Tennessee Valley Authority also must adopt IRP planning principles as a condition of their power contracts.

Resource planning is an integral part of utility operations and is driven by the three fundamental goals–serving customer load reliably, minimizing customer costs, and maintaining the financial stability of the utility. Today's IRP process has evolved from traditional utility planning, which focused narrowly on supply-side resource additions to meet ever-growing customer demand. With experience, IRP planning processes are continuously evolving in both theory and application. Each utility's IRP process is different reflecting its system characteristics and planning needs, corporate culture and organization, and regulatory environment. However, every IRP process follows a general framework in evaluating a broad range of resource options to develop a long-term resource plan typically covering 20 to 30 years, and an action plan covering from two to five years. New or revised integrated resource plans are prepared on average every two to five years. Figure 1 shows a simplified IRP process.

The process typically begins with preparation of long-term load forecasts projecting both energy sales (megawatt-hours) and peak demand (megawatts) over the planning period. The forecasts are based on historical con-

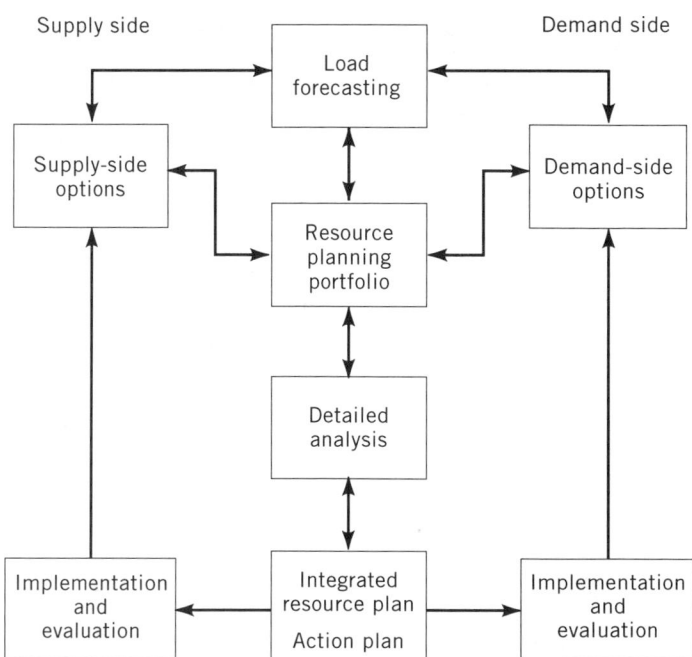

Figure 1. Simplified integrated resource planning framework. From Office of Technology Assessment, 1993.

sumption data, weather, population, economic data, and electric equipment use. The load forecasts must also take into account expected load growth and potential changes in energy consumption patterns due to new technologies, DSM programs, and other conservation measures. The detailed forecasts are used for financial and resource planning to identify an appropriate mix of generation, transmission, distribution, power purchase, and energy efficiency options to meet system needs under a range of alternative future scenarios.

Using the initial load forecasts, utility planners then survey potential demand- and supply-side resource options to identify appropriate measures for inclusion in the integrated resource planning portfolio.

For supply-side options, planners will consider existing generation, transmission and distribution resources, utility generating plant additions, life extensions and efficiency upgrades, plant retirements, power purchases, and improvements to transmission and distribution facilities. During this initial evaluation, planners will compare the resources on considerations of: load profiles, reliability and dispatch capabilities; capital, fuel, operating, and maintenance costs; environmental and siting requirements; and capital availability. The result will be a supply-side resource stack.

Demand-side options will be identified based on considerations of existing customer use patterns, availability of energy-efficient technologies, demographic data, and evaluations of existing utility DSM programs. Planners will estimate costs, load impacts, and participation rates to produce an initial stack of demand-side resources.

The IRP process then proceeds to detailed and iterative evaluation of the potential resource options to identify the best resource mix, taking into consideration the utility's strategic goals, load profile impacts, production and capital costs, revenue requirements, rate impacts, environmental and other regulatory requirements, and other planning uncertainties. Planning uncertainties related to demand-side resources include participation rates, and the costs, effectiveness, durability, and verification of efficiency measures. For supply resources, uncertainties include construction time and costs, regulatory approvals, fuel availability and costs, operating and maintenance costs, and public attitudes towards the technology and the specific facility proposed. Overall uncertainties complicating resource planning affecting load growth and costs, include impacts of inflation and interest rates, changing economic conditions, availability of purchased power, and changes in environmental and economic regulatory policies. During the process, resources may be added or removed from the portfolio based on the initial evaluation.

The typical IRP process includes opportunities for participation by the public and by regulators. The extent and type of participation vary. Some utilities have relied upon a collaborative consultation with interested parties over the entire course of plan preparation. Others may prepare a draft plan and then solicit public and regulatory comment before preparing a final plan.

The costs and benefits of alternative resource options are compared individually and in combination and then ranked according to the appropriate cost-effectiveness test and planning goals. This cost-benefit ranking may be

conducted under a number of separate scenarios with different assumptions about factors affecting energy demand, financial conditions, or regulatory requirements. In selecting a final integrated resource stack, planners must balance many different and sometimes competing goals and expectations about the future. Because resource planning involves many qualitative and strategic judgments, a least-cost plan will not always be the option that minimizes power production costs. Considerations of reliability, flexibility, resource diversity, and business strategy/policy may outweigh options that are the cheapest alternatives at the time the plan is developed.

The integrated resource plan lays out the utility's least-cost long-term strategy. It is coupled with a short- to mid-term action plan that details the specific actions and resource additions that the utility will take to carry out the plan objectives. Based on the plan, the utility, with any necessary regulatory approvals, will proceed to plan and acquire the preferred supply- and demand-side resources to meet customer loads. During implementation of the plan, adjustments can be made to reflect changing conditions using the plan as a guide. The utility's experience in implementing the action plan and evaluating its results are then used in the next round of the IRP process.

BIBLIOGRAPHY

1. E. Hirst and C. Sabo, *Electric Utility DSM Programs: Terminology and Reporting Formats*, ORNL/CON-337 Oak Ridge National Laboratory, Oak Ridge, Tenn., Oct., 1991.

2. Battelle-Columbus Div. and Synergic Resources Corp., *Demand-Side Management* vol 3, *Technology Alternatives and Market Implementation Methods* EPRI EA/EM 359, Electric Power Research Institute, Palo Alto, Calif., 1984.

3. U.S. Congress, Office of Technology Assessment, *New Electric Power Technologies: Problems and Prospects for the 1990s*, OTA-E-246, U.S. Government Printing Office, Washington, D.C., July, 1985, p. 142.

4. "Utilities Field Peak Power Demand with Incentives for Homeowners," *Wall Street Journal*, A1 (June 6, 1991).

5. American Council for an Energy Efficient Economy, *Lessons Learned: A Review of Utility Experience with Conservation and Load Management Programs for Commercial and Industrial Customers* Final Report, Energy Authority Report 90–8, New York State Energy Research and Development Administration, New York, Apr., 1990.

6. E. Hirst, *Electric Utility DSM-Program Costs and Effects: 1991–2001*, ORNL/CON-364 Oak Ridge National Laboratory, Oak Ridge, Tenn., May, 1993.

7. S. M. Nadel and K. M. Keating, "Engineering Estimates vs. Impact Evaluation Results: How Do They Compare and Why," *Energy Program Evaluation: Uses, Methods, and Results, Proceedings of the 1991 International Energy Program Evaluation Conference*, CONF-910807, Aug., 1991, cited in Hirst and Sabo, Ref. 1, note 1, pp. 24–33.

8. Ref. 1, p. 36.

9. E. Hirst and J. Reed, eds., *Handbook of Evaluation of Utility DSM Programs*, ORNL/CON-336, Oak Ridge National Laboratory, Oak Ridge, Tenn., Dec., 1991.

10. American Council for an Energy-Efficient Economy, and the New York State Energy Office, *The Achievable Conservation Potential in New York State from Utility Demand-Side Man-*

agement Programs, Energy Authority Report 90–18, New York State Energy Research and Development Administration, Albany, N.Y., Nov., 1990.

11. A. L. Nichols, *Estimating the Net Benefits of Demand-Side Management Programs Based on Limited Information*, National Economic Research Associates, Inc., Cambridge, Mass., Jan. 25, 1993, p. 34, cited in K. Maize and J. McCaughey, "DSM at Mid-Passage: A Discussion of the State of the Art and Science of Demand-Side Management in Electric Utilities," *The Quad Report*, Special Report, Spring, 1993.

12. Association of Demand-Side Management Professionals, "NEES Credits Regulatory Incentives in 'Overwhelming' 1990 DSM Success," *Strategies* **2**(2), (Spring 1991).

13. "New England Utility Outlines Plans to Cut Greenhouse Emissions by 45%," *Energy Conservation Digest* 1 (Dec. 23, 1991).

ELECTRIC AND MAGNETIC FIELDS IN THE ENVIRONMENT

Markus Zahn
Massachusetts Institute of Technology
Cambridge, Massachusetts

We live in natural electromagnetic fields due to atmospheric electricity and the earth's magnetism. In addition, living organisms have internal electric fields due to electrolytic processes for cell and nerve functions, and magnetic fields from the magnetic moments of molecules. Superposed onto these natural fields are such man-made fields as from electric power lines, appliances and wiring. In order to better understand how electromagnetic fields can interact with living systems, we review the fundamentals of electric and magnetic field interactions with materials.

SOURCES OF ELECTRIC AND MAGNETIC FIELDS

All matter is composed of atoms that are in turn composed of negatively charged particles called electrons and positively charged particles called protons. Throughout this article, SI units are generally used for electrical quantities for which the base units are taken from the rationalized MKSA system of units. The unit of charge is a coulomb = 1 A·s. The charge magnitude on an electron and proton is 1.6×10^{-19} C so that 1 C of electricity contains about 6.24×10^{18} elementary charges. Each of these charges have associated with them an electric field **E**, which has SI units of volts per meter (V/m). The electric field emanates radially from a point charge and is proportional to the force that the charge exerts on other charges as given by Coulomb's law, where opposite charges attract and like charges repel. Usual matter is charge neutral with an equal amount of protons and electrons so that there is no external electric field. An electric field arises when there is an excess of protons or electrons on the material so that it is not charge neutral. See also ELECTRIC POWER DISTRIBUTION; ELECTRIC POWER SYSTEMS AND TRANSMISSION.

Most people have experienced nuisance frictional electrification caused by charge separation when walking across a carpet or pulling clothes out of a dryer. Using a plastic comb can cause hair to become charged and stand up, as the like-charged hairs repel one another. In a dry environment, in which large amounts of charge can significantly accumulate, these effects often result in small sparks because the electric forces from large amounts of charge actually pull electrons from air molecules. These sparks occur when the electric fields are larger than the electrical breakdown strength of air, $\sim 3 \times 10^6$ V/m.

When charges move, they constitute a current, and they give rise to the magnetic flux density **B**, which has SI units of teslas (T). Because a 1-T magnetic field is uncommonly large, a more common unit for **B** is the gauss (G); 1 T = 10,000 G. A related quantity, the magnetic field intensity **H**, which has units of amperes per meter (A/m), is also often used to describe magnetic fields; $\mathbf{B} = \mu_0 \mathbf{H}$ in free space. The quantity $\mu_0 = 4\pi \times 10^{-7}$ H/m is called the magnetic permeability of free space. The magnetic field from a small element of current is given by the Biot-Savart law, is in the direction perpendicular to the current element, and is proportional in magnitude to the force that the current element exerts on other current elements as given by the Lorentz force law where opposite flowing currents repel and like flowing currents attract. The magnetic Lorentz force on a current element is in the direction perpendicular to both the direction of current flow and the direction of the magnetic field acting on the current element. Currents generally flow in conductors, such as metallic wires, where electrons can easily flow. Insulators do not allow easy electron flow.

Net positive charge often accumulates on clouds with a negatively charged earth giving rise to a "fair weather" vertical dc electric field at ground level, typically of order 130 V/m. However, in thunderstorms these charges can increase, and when the electric field in air anywhere exceeds 3×10^6 V/m, air breaks down with luminous spark discharges as the charge passes from the cloud to another cloud or toward the ground, which is seen as lightning and heard as thunder (caused by the pressure wave of heated air). A typical lightning stroke passes 10,000–100,000 A of current for about 100 μs for a total charge of 1–100 C.

The earth's core consists of iron. An iron molecule behaves as if its nucleus of protons were spinning. This moving charge creates a dc magnetic field distribution around the earth, which at ground level has values of about 0.25–0.6 G. Because motion through a magnetic field creates an electric field, walking through the earth's magnetic field can induce body currents that add to biologically generated body currents.

Human bodies are composed of electrolytic materials, and bodily functions depend on ion concentrations and membrane potentials that are of order of the thermal voltage (kT/q), where $k = 1.38 \times 10^{-23}$ J/K is Boltzmann's constant, T is the absolute temperature in degrees Kelvin, and $q \approx 1.6 \times 10^{-19}$ C is the charge on an electron. At room temperature ($T \approx 300$ K), the thermal voltage is about 25 mV. The brain communicates with the rest of the body by electrical impulses. A typical nerve cell fires at $\sim$ 50–100 mV. It is possible to monitor the health of the body by measuring these electrical signals using electrocardiography (ECG) of the heart and electroencepha-

Table 1. Typical Steady-state Electric and Magnetic Power Frequency Fields[a]

Field	E (V/m) rms[b]	B (mG) rms
Home wiring	1–10	1–5
At electrical appliances	30–300	5–3000
Under distribution lines serving homes	10–60	1–10
Inside railroad cars on electrified lines		10–200
Under high voltage transmission lines	1000–7000	25–100

[a] From Ref. 2.

[b] rms is the root mean square; it is equal to the square root of the average of the square of periodic voltage or current over a period. For a sine wave with peak amplitude A, the rms value is $A/\sqrt{2} \approx 0.707\, A$.

lography (EEG) of the brain. Heart pacemaker electrodes in contact with heart muscle provide current pulses 0.1–10 mA a few milliseconds long to synchronize the firing of heart cells. Magnetic resonance imaging (mri) can image internal organs from the magnetic moments of molecules in the body using d-c magnetic fields up to 2T.

Generated electricity adds additional electric and magnetic fields to the environment, which can differ from natural electromagnetic fields in amplitude, direction, and frequency (1). In the United States, electric power uses 60 Hz ac, while Europe and some other parts of the world use 50 Hz. Table 1 lists representative values of generated electromagnetic fields.

The typical household background at 60 Hz is 1–10 V/m for the electric field and 0.1–10 mG for the magnetic field (3). People can generally sense 60 Hz electric fields above ~ 20 kV/m through hair and skin sensations, although there is great variability among individuals (4). Some individuals can sense electric fields as low as 0.35 kV/m (4). People cannot generally sense magnetic fields even up to 20,000 G, although low frequency (~ 20 Hz) magnetic fields greater than 100 G produce luminous images known as "phosphenes," apparently due to induction of electric currents in the retina (3,4).

In the United States, maximum exposure standards for 60-Hz electromagnetic fields at transmission lines right of way are typically over the range of 1–11 kV/m rms for the electric field, and up to 250 mG in Florida for the magnetic field (3,5). Elsewhere there are no exposure standards for 50/60 Hz magnetic fields. Because these representative environmental fields are not greatly different from natural electromagnetic fields, it was generally thought that there were no health hazards (4–16).

ELECTROMAGNETIC FIELDS

The purpose of this article is to review fundamental electromagnetic field interactions with media so that people can try to minimize their exposure to fields and so that sound physical science can be used in the design and interpretation of continuing and future health studies to identify nonambiguously mechanisms of biological interaction between electromagnetic fields and living systems.

Quasistatics and Electrodynamics

There are three broad types of electromagnetic field interactions: (1) electroquasistatics, a low frequency range for which charges and voltages are the sources of the electric

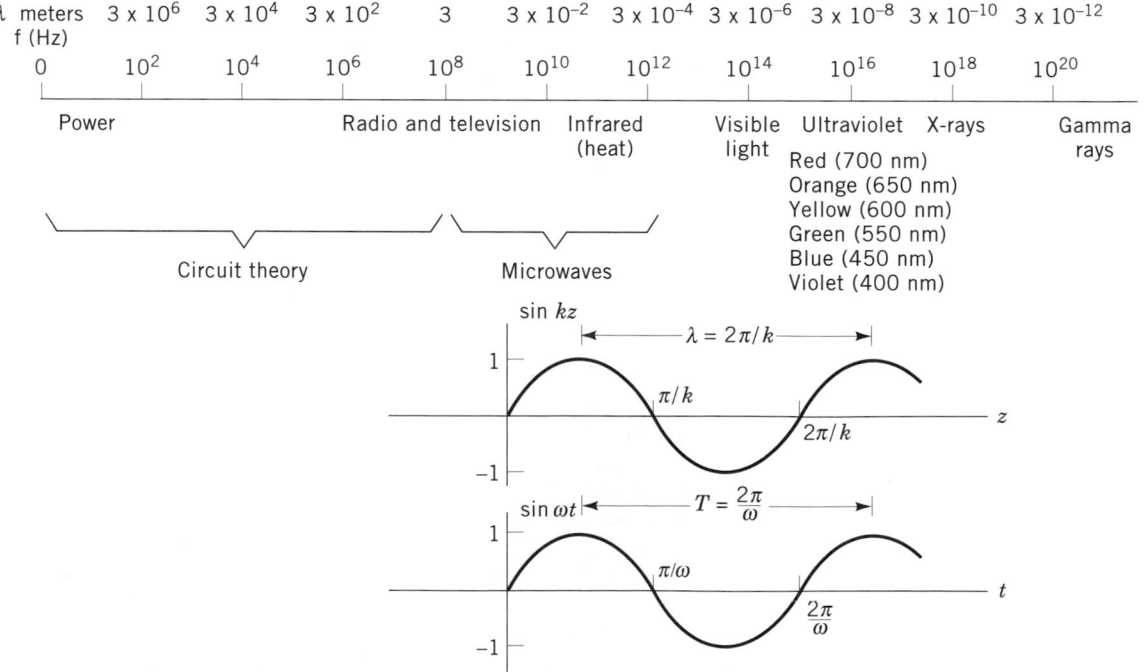

Figure 1. Time-varying electromagnetic phenomena differ by scaling of time (frequency) and size (wavelength). In free space, the frequency f in Hz (angular frequency $\omega = 2\pi f$ rad/s) and wavelength λ in m (wave number $k = 2\pi/\lambda$) are related as $f\lambda = c$ ($\omega = kc$) where $c = [\varepsilon_0 \mu_0]^{-1/2} \approx 3 \times 10^8$ m/s is the speed of light in free space.

field coupled to dielectric and conducting media with negligible magnetic field; (2) magnetoquasistatics, a low frequency range for which currents are the source of the magnetic field coupled to magnetizable and conducting media with electromagnetic induction generating an electric field; and (3) electrodynamics, a high frequency range for which the electric and magnetic fields are of equal importance, resulting in radiating waves that travel at the speed of light ($c \approx 3 \times 10^8$ m/s) in free space. Figure 1 summarizes the frequency ranges for many common applications. Quasistatic fields occur when the wavelength of electromagnetic waves at frequency f ($\lambda = c/f$) is much larger than the size of the system. For power frequency of $f = 60$ Hz in the United States, $\lambda \approx 5 \times 10^6$ m, so that the usual small electrical appliances can be considered quasistatic. This power frequency regime is often called the extra low frequency (ELF) range. Quasistatic fields are confined to the immediate vicinity of the electrical device. Radio frequency and higher frequencies have propa-

gating electromagnetic waves that travel at the speed of light away from a source.

The energy W in each photon, the fundamental equivalent particle of an electromagnetic field, is $W = hf$, where h is 6.6256×10^{-34} J·s is Planck's constant. Frequencies below the uv light region ($f < 10^{15}$ Hz) are called nonionizing radiation, because the energy W is too low to break chemical or molecular bonds. The energy of a 60-Hz photon ($W \approx 4 \times 10^{-32}$ J) is 11 orders of magnitude smaller than the thermal energy at room temperature ($T \approx 300$ K), $kT \approx 4 \times 10^{-21}$ J. Ultraviolet, x-ray, and γ-ray radiation have sufficiently high energy to break chemical and molecular bonds and can cause cancer and other health problems, depending on total exposure. Radio and microwaves do not have enough energy to break bonds, but they are absorbed by water in body tissues to cause heating. This is the principle of cooking with a microwave oven and of medical treatment by heating body tissue with diathermy machines.

Figure 2. Examples of electromagnetic fields in daily life.

Examples of electromagnetic field use in daily life are shown in Figure 2. Electroquasistatic applications include capacitors, vacuum tubes, and semiconductors in circuit applications; microphones; electrocardiography and electroencephalography; and electrostatic precipitators, printers, and copiers. Magnetoquasistatic applications include inductors and transformers in circuit and power system applications; audio speakers; magnetic resonance imaging devices; and motors and generators. The proposed new train system of magnetically levitated trains (Maglev) and perhaps electric cars provides further urgency to resolving adverse health issues in low level magnetic fields. Electrodynamic fields are present in radio, television, communications, optics, and microwave ovens.

Imposed charges and voltages generate electric fields and imposed currents generate magnetic fields. However, these fields induce further charges and currents in dielectric, conducting, and magnetizable materials that then in turn create additional electric and magnetic fields. Often, the presence of material concentrates the fields so that the fields at sharp points at the surface can be much larger than when the materials are absent. Fields inside materials are also often much less than the fields just outside, being shielded by induced surface charges and currents. To further complicate analysis and understanding, time-varying magnetic fields create electric fields (Faraday's law) and time-varying electric fields create magnetic fields (Ampere's law). Electric fields in conducting materials cause current flow, voltage differences, and heating.

Quasistatic Fields from Electrical Devices

Small Appliances: Point Dipole Fields. The fundamental source for the electric field is a point charge q, while the source for a magnetic field is a current i of vector differential length $\mathbf{d\ell}$. Opposite polarity charges attract through the electric force, whereas like-charged particles repel. Currents flowing in the same direction attract through the magnetic force, while currents flowing in opposite directions repel. The electric field is the Coulomb force per unit charge on a charge due to all other charges. The magnetic field causes a Lorentz force on a moving charge at right angles to its motion due to all other moving charges or currents. The electric field from a single point charge q is given by Coulomb's law, and the magnetic field from a single current element $i\mathbf{d\ell}$ is given by the Biot-Savart law:

$$\mathbf{E} = \frac{q}{4\pi\varepsilon_0 r^2}\mathbf{i_r}; \mathbf{B} = \frac{\mu_0 i\mathbf{d\ell}\mathbf{x}\mathbf{i_r}}{4\pi r^2} \qquad (1)$$

where $\varepsilon_0 \approx 10^{-9}/36\pi \approx 8.854 \times 10^{-12}$ F/m is the dielectric permittivity of free space. The quantities ε_0 and μ_0 are related by the speed of light in free space, $c = [\varepsilon_0\mu_0]^{-1/2} \approx 3 \times 10^8$ m/s. Both fields at any point a distance r from the sources are inversely proportional to the square of the distance r. The unit vector $\mathbf{i_r}$ is in the direction from the source to the field point at distance r. The electric field $\mathbf{E}$ is in the direction of $\mathbf{i_r}$ while the magnetic field $\mathbf{B}$ is perpendicular to both the direction of the current $\mathbf{d\ell}$ and $\mathbf{i_r}$. If there is a distribution of charges and currents throughout space, the total electric and magnetic field is

the vector superposition of equation 1 from all source charges and currents.

However, electrical devices must be charge neutral with at least two electrodes so that if one electrode has positive charge, the other electrode has an equal magnitude but negative charge. The charge on each electrode is proportional to the voltage difference between electrodes through the capacitance. The capacitance depends on electrode geometry and on the permittivity (dielectric constant) of the materials between the electrodes. From a distance away that is larger than the spacing d between opposite polarity charges ($r \gg d$), the system looks like a point electric dipole as in Figure 3a. Similarly, current must flow in closed loops as in the magnetic dipole in Figure 3b with small area S. For distances far from the dipoles compared with their size ($r \gg d$, $r \gg \sqrt{S}$), the electric and magnetic fields given by equation 1 approximately cancel, leaving weaker strength point dipole fields:

$$\mathbf{E} = \frac{qd}{4\pi\varepsilon_0 r^3}[2\cos\theta\mathbf{i_r} + \sin\theta\mathbf{i_\theta}];$$

$$\mathbf{B} = \frac{\mu_0 iS}{4\pi r^3}[2\cos\theta\mathbf{i_r} + \sin\theta\mathbf{i_\theta}] \qquad (2)$$

These fields depend in magnitude and direction on angle θ, but most significant, they increase linearly with the source strength (charge or voltage for the electric field and current for the magnetic field), and source size (spac-

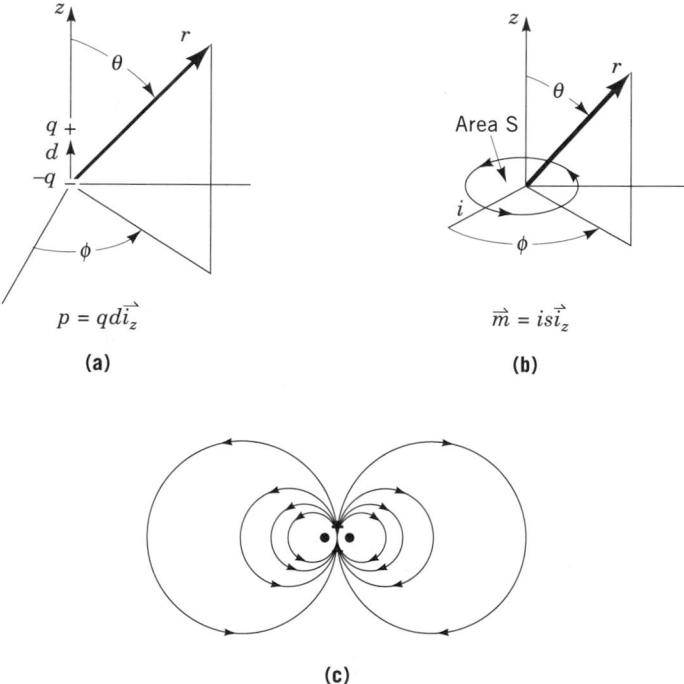

(a)

(b)

(c)

Figure 3. (a) Electrical devices must be charge neutral with positive and negative charges $\pm q$ a small distance d apart forming an electric dipole with dipole moment $p = qd$. **(b)** Current i must flow in closed loops with area S forming a magnetic dipole with moment $m = iS$. **(c)** The electric field from an electric dipole and the magnetic field from a magnetic dipole have the same shapes, which vary in magnitude and direction with radial distance r and angle θ. Shown are the field lines that are tangent to the electric and magnetic fields.

ing or area), and decrease with distance as $1/r^3$. Figure 3c shows the dipole field lines that are tangent to the fields given in equation 2. The strength of the field is proportional to the density of drawn lines, which decreases as $1/r^3$. Electric field lines start on the positive charge and terminate on the negative charge, whereas magnetic field lines always form closed loops encircling a current loop.

Small appliances and motors have approximate point dipole fields at large distances compared with their size (17,18). Thus the near fields depend directly on the sources and size, but a doubling of the distance from the device center would approximately decrease the fields by a factor of eight. This indicates that by moving electric clocks, lights, motors, and other small electrical devices as far as possible from the body, the electric and magnetic field exposure would greatly decrease (16,19,20).

Within a typical home, appliances generally have the electrical and magnetic field values summarized in Table 2. Immediately next to these appliances, the maximum magnetic field is typically about 2 G and the maximum electric field is about 100 V/m (19,20,22–25). A 325-W soldering gun and hair dryer have localized magnetic fields from 10 to 25 G, whereas a clothes dryer, toaster, vacuum cleaner, dishwasher, clothes washer, electric iron, and refrigerator have magnetic fields less than 1 G (21). Electric blankets are thought to cause high exposure to power-frequency electric and magnetic fields because they are close to a body for a long period of time (3,25). Older-style designs have magnetic fields from 5 to 100 mG, with whole body–averaged magnetic flux densities of ~22 mG and electric fields of 100–2000 V/m. Newer designs using partial magnetic field cancellation of parallel conductors with current flow in opposite directions have the magnetic field strength reduced by about 30 times while the electric field remains essentially unchanged (25). Twisting the two wires would further decrease the magnetic fields.

Home Wiring: Line Dipole Fields. The two wires connecting to an appliance are generally at an ac voltage difference of 120 V rms in the United States. At distances close to the wires compared with their length, the fields are essentially the same as if the wires were infinitely long.

The voltage on each line also imposes a line charge density $\pm \lambda$ C/m of opposite polarity on each wire. Each single cylindrical conductor alone with uniform line charge density λ and uniform current i have radial electric and encircling ϕ directed magnetic fields outside the conductor for radial distances r much less than the

length:

$$\mathbf{E} = \frac{\lambda}{2\pi\varepsilon_0 r}\mathbf{i_r}; \quad \mathbf{B} = \frac{\mu_0 i}{2\pi r}\mathbf{i_\phi} \qquad (3)$$

The amount of charge $\pm \lambda$ on each conductor is proportional to the voltage difference between conductors through the capacitance, which only depends on the geometry and dielectric constant of the insulation. Because the pair of conducting wires at a distance d apart carry opposite polarity line charges $\pm \lambda$ and oppositely flowing currents $\pm i$, the fields in the vicinity of the wires tend approximately to cancel so that the fields die off quicker than the $1/r$ dependence in equation 3. These line dipole fields decay as $1/r^2$ with distance, and the field direction depends on angle ϕ:

$$\mathbf{E} = \frac{\lambda d}{2\pi\varepsilon_0 r^2}[\sin\phi\mathbf{i_r} - \cos\phi\mathbf{i_\phi}];$$

$$\mathbf{B} = \frac{\mu_0 i d}{2\pi r^2}[\sin\phi\mathbf{i_\phi} + \cos\phi\mathbf{i_r}] \qquad (4)$$

The electric and magnetic fields can be even further decreased if the pair of wires is twisted, thereby greatly decreasing the effective spacing d. However, because for safety reasons additional grounds in the home wiring are often attached to water pipes, all the return current does not necessarily travel back via the second wire. Then the net current along the line pair is nonzero, and the magnetic field decays more slowly as $1/r$ as given by equation 3, greatly increasing a person's exposure to magnetic fields. Because the voltage difference across the line pair always imposes equal magnitude but opposite polarity line charges, the electric field dies off as $1/r^2$ as given by equation 4, even when the currents are unbalanced.

Europe and other parts of the world use 240 V residential power, which halves the current through the home wiring and to appliances. Then the resulting electric field is twice and magnetic field is half that in U.S. residences.

Fields from Power Lines

An electric power system consists of generating stations, transmission lines, and distribution systems (Fig. 4) (3,20,23,24). Transmission lines connect generators to the distribution systems, and the distribution systems connect to individual loads. A generator can typically produce on the order of 100–1000 MW power at 11–35 kV rms at typically 3,000–50,000 A rms. Because transmission line losses are primarily due to series resistance in the cables, to minimize these losses it is necessary to minimize the line current. Because the transmitted power is proportional to the product of line voltage and line current, the same power can be transmitted at reduced current by increasing the voltage. A voltage step-up transformer can increase the generator voltage at the primary winding up to the typical range of 69–765 kV rms at its secondary winding, thereby reducing the secondary current from the primary current by the inverse ratio of voltage step-up.

Typical power transmission systems use balanced four-wire, three-phase power, and the generator and voltage step-up transformer have voltages and currents of the fol-

Table 2. Typical Residential Magnetic and Electric Field Exposures[a]

Field	B (mG rms)	E (V/m rms)
Power lines	5	
Appliances at 25 cm		
Electric range	20	4
Color television	12	30
Refrigerator	5	60
Clock radio (analogue)	24	15
Fluorescent light	21	

[a] From Refs. 20 and 21.

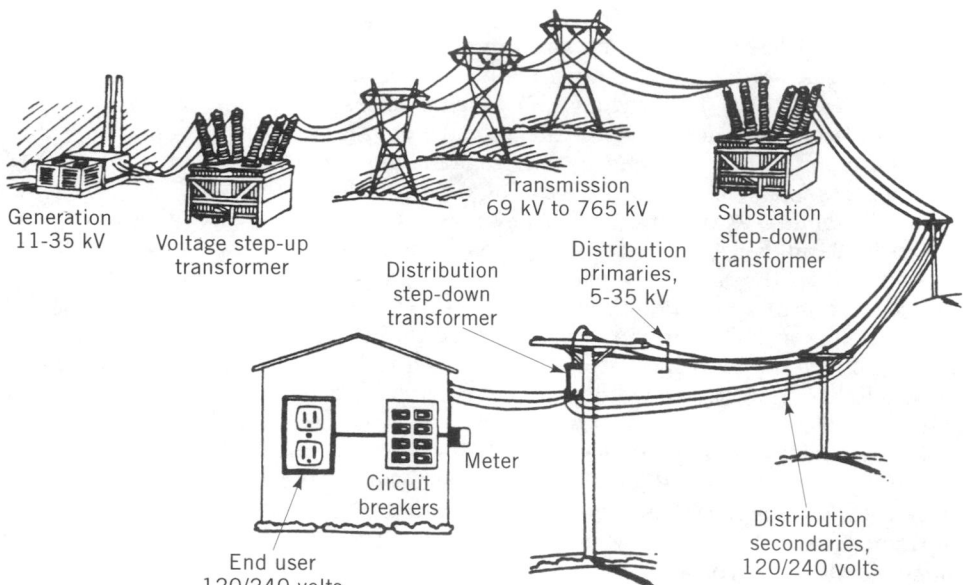

Figure 4. Three-phase electric power carried from the generator through the transmission and distribution systems to a house.

lowing form for each phase:

$$v_A = V \cos \omega t;$$

$$v_B = V \cos (\omega t - 2\pi/3);$$

$$v_C = V \cos (\omega t + 2\pi/3)$$

$$i_A = I \cos (\omega t + \phi);$$

$$i_B = I \cos (\omega t + \phi - 2\pi/3);$$

$$i_C = I \cos (\omega t + \phi + 2\pi/3) \tag{5}$$

where ϕ is the phase angle between voltage and current, and for 60 Hz power, $\omega = 120 \pi \approx 377$ rad/s. Each phase voltage is with respect to the neutral fourth wire, said to be at ground potential. Each phase has the same magnitude voltage and current, but each differs in phase angle from the other two phases by $\pm 120°$. Three-phase power is used because the total power delivered by the generator

$$p = v_A i_A + v_B i_B + v_C i_C = \frac{3}{2} VI \cos \phi \tag{6}$$

is constant in time and thus mechanically easier on the generator to deliver power at a constant rate.

Algebraically, the sum of voltages and the sum of currents for balanced three-phase lines are zero:

$$v_A + v_B + v_C = 0 \tag{7}$$
$$i_A + i_B + i_C = 0$$

so that the electric and magnetic fields die off as line dipoles, given in equation 4. Because the neutral wire must carry the algebraic sum of currents in the three phases, a balanced system as in equation 7 has zero current in the neutral wire.

A three-phase ac 525 kV rms transmission line with cylindrical conductors about 3 cm in diameter, spaced about 10 m apart and 10.6 m above ground have a peak ground level electric field of about 8 kV/m rms. If this

three-phase transmission line carries a representative current of 2000 A rms per phase, the peak ground level magnetic field is about 0.2 G rms.

Typically, each phase of a primary distribution line reaches a pole transformer to reduce the voltage to the secondary distribution line voltage of 120/240 V in the United States, which connects to residences by three wires. One wire is a neutral return and there are two hot wires at $\pm$ 120 V rms with respect to the neutral wire. Small appliances operate at 120 V between one hot line and neutral while large power appliances operate at 240 V between the two hot lines. The small pole transformer is like a small point dipole and so the fields die off as $1/r^3$ as given by equation 2. The power line to the house is designed to have no net current as all the current that flows in the two hot wires is supposed to return via the neutral. Such a balanced line again has fields that decrease as $1/r^2$ as given by equation 4. However, because for safety reasons electrical grounds are often attached to water pipes, cable television lines, telephone lines, and other connections to ground, all the return current does not necessarily travel back via the neutral wire. Also, each phase of the three-phase secondary distribution line is connected to different homes, and if the power requirements of each home differ, the current magnitudes flowing in each phase are not necessarily the same. Either of these cases leads to unbalanced line currents so that the net current along a set of distribution lines is not zero and the magnetic field decays more slowly as $1/r$ (eq. 3), greatly increasing a person's exposure to magnetic fields. However, because the line voltages in a three-phase system are always essentially balanced even if the currents are unbalanced, the power line electric fields die off as $1/r^2$, even when the magnetic field dies off as $1/r$ with unbalanced currents.

To lower electric and magnetic field exposure under power lines, the Electric Power Research Institute (EPRI) High Voltage Transmission Research Center (HVTRC) has been examining different conductor configurations such as multiple conductors per phase, twisted lines, shielding wires, and six-phase power (20,23). For a 115 kV rms line at 1000 A rms, at a person's waist level 50 feet from the line center, a conventional line configuration would give a magnetic field of 40 mG; the best experimental configuration would produce a much weaker magnetic field of 1.5 mG.

The step down of voltage from the transmission levels at a bulk-power substation typically reduces the voltage to the range of 35–138 kV rms. The next step down in voltage is at the distribution substation, reducing the voltage commonly to 5–35 kV rms. The first epidemiological study (10) noted a higher incidence of childhood leukemia for those homes near pole distribution transformers. When the voltage is stepped down from the 5–35 kV rms range to 120/240 V rms, about a factor of 40–300, the secondary distribution current from the transformer similarly increases by the same factor, thus increasing the magnetic field strength by the same factor. It was hypothesized that children living and sleeping near such a distribution transformer with high magnetic fields may be subject to adverse health effects (10,11).

Electric Railroads

Conventional electric railroads have typical steady-state magnetic fields within the cars of order 10–25 G (26,27), while the German Maglev System has a magnetic field mostly below 0.5 G.

Radiation Fields

Radio, television, communication, and radar systems operate at much higher frequencies f, 500 kHz (0.5×10^6 Hz) to 1000 GHz (10^{12} Hz). Measurements of FM radio and UHF and VHF broadcast signal field intensities indicate that typical population exposure is $\sim 50\ \mu W/m^2$, corresponding to $E \sim 0.14$ V/m and $B \sim 4.6\ \mu G$ (1). The simplest model for transmitting antennae is the point electric or magnetic dipole antennae (Fig. 5) excited by a sinusoidal current of the form $i(t) = I_0 \cos \omega t$. If the antenna length ℓ of the electric dipole antenna is small compared to the wavelength $\lambda = c/f$, ($\ell \ll \lambda$) the electric and magnetic fields are

$$\mathbf{E}(r, \theta, t) = \frac{I_0 \ell k^2}{4\pi} \sqrt{\mu_0/\varepsilon_0} \left[-\mathbf{i_r} \left[2 \cos \theta \left(\frac{\cos(\omega t - kr)}{(kr)^2} \right. \right. \right.$$
$$\left. \left. + \frac{\sin(\omega t - kr)}{(kr)^3} \right) \right] + \mathbf{i_\theta} \left[\sin \theta \left(\frac{\sin(\omega t - kr)}{kr} \right. \right.$$
$$\left. \left. \left. - \frac{\cos(\omega t - kr)}{(kr)^2} - \frac{\sin(\omega t - kr)}{(kr)^3} \right) \right] \right] \quad (8)$$

$$\mathbf{H}(r, \theta, t) = -\mathbf{i_\phi} \frac{I_0 \ell k^2}{4\pi} \sin \theta \left[\frac{\sin(\omega t - kr)}{kr} - \frac{\cos(\omega t - kr)}{(kr)^2} \right] \quad (9)$$

where $k = 2\pi/\lambda = \omega/c$ is called the wave number. At distances close to the antenna, the fields die off as $1/r^3$ like a point electric dipole. There is an intermediate field region that varies as $1/r^2$, but the most important terms are the far radiation field terms that decrease only as $1/r$ and thus dominate at distances far from the antenna. In this far field, electric power flows radially and $E_\theta = [\mu_0/\varepsilon_0]^{1/2} H_\phi$ where $\eta = [\mu_0/\varepsilon_0]^{1/2} \approx 120\pi \approx 377\ \Omega$ is known as the radi-

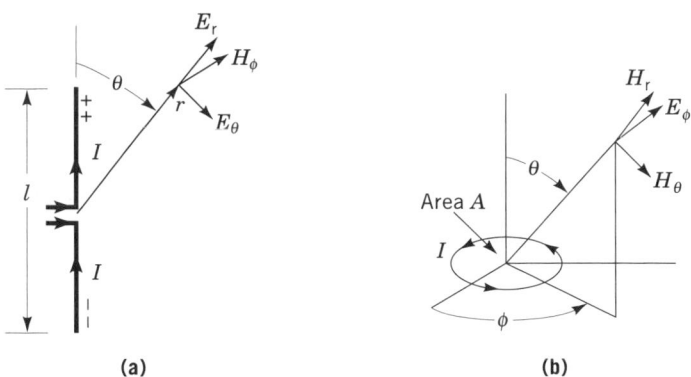

Figure 5. (a) Electric dipole antenna and (b) magnetic dipole antenna generate electromagnetic waves that radiate away from the antenna.

ation resistance of free space. In the far field, the electromagnetic power density in units of W/m² is $S_r = E_\theta H_\phi \approx \eta H_\phi^2 = E_\theta^2/\eta$. The magnetic dipole radiation fields are the dual to the electric dipole, where the electric and magnetic fields reverse roles ($E \to H$ and $H \to E$), replacing $I_0 \ell \cos \omega t/\omega \varepsilon_0$ by $-I_0 S \sin \omega t$, where S is the area of the small current loop.

If one uses transmitted radio or telephone communications, such as with CB radio or cellular phones, the electric and magnetic fields decrease by $1/r^3$ in the vicinity of the user. However, the electric and magnetic fields from large commercial transmitting and radar antennae die off as $1/r$ if one is more than a few wavelengths away. Low power radar is used for monitoring weather in aircraft, as navigational aids on small boats, and to determine vehicle speed by police. The maximum power density varies over the range of $S = 1 - 30$ W/m² so the fields are about $E = [S\eta]^{1/2} = 19 - 580$ V/m and $B = \mu_0 [S/\eta]^{1/2} = 0.6 - 20$ mG (1).

A microwave oven is a microwave resonator at about 2.45 GHz with typical power levels of ~ 500 W. The majority of microwave ovens have leakage below 50 W/m², which is the emission limit from ovens at a distance of 5 cm, as given by the U.S. Bureau of Radiological Health (1). This corresponds to microwave rms fields of $E \approx [S\eta]^{1/2} = 137$ V/m and $B = \mu_0[S/\eta]^{1/2} = 4.6$ mG. The 60 Hz magnetic field from a microwave oven at 25.4 cm is 1–120 mG (19).

Field Concentration

Dielectric and conducting objects placed into an electric field, change the electric field distribution in the vicinity of the object, often tending to enhance the electric field at selected positions along the medium surface. Similarly, conducting and magnetizable media tend to enhance magnetic fields at selected locations. For example, Figure 6 shows a perfectly conducting (ohmic conductivity $\sigma \to \infty$) sphere of radius R placed into uniform z directed electric and magnetic fields. The electric and magnetic fields inside the perfectly conducting sphere must be zero. To shield the interior of the sphere from the electric field, a dipolar surface charge forms on the sphere surface. A $-\phi$ directed surface current flows so that its self-field is in the opposite direction to the imposed magnetic field, as illustrated by the right-hand rule by which the thumb

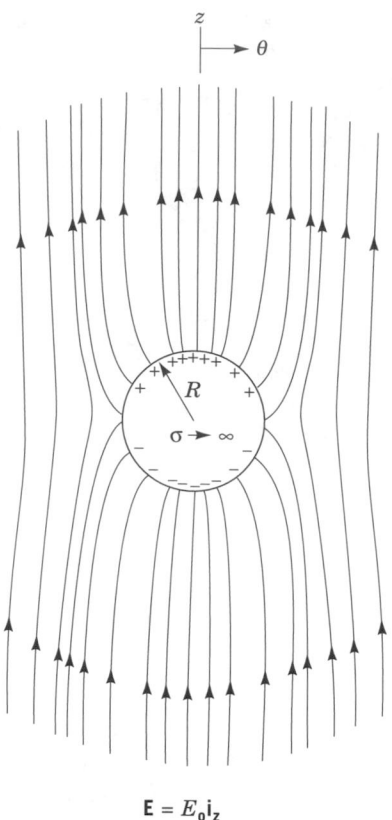

$$\mathbf{E} = E_0 \mathbf{i}_z$$

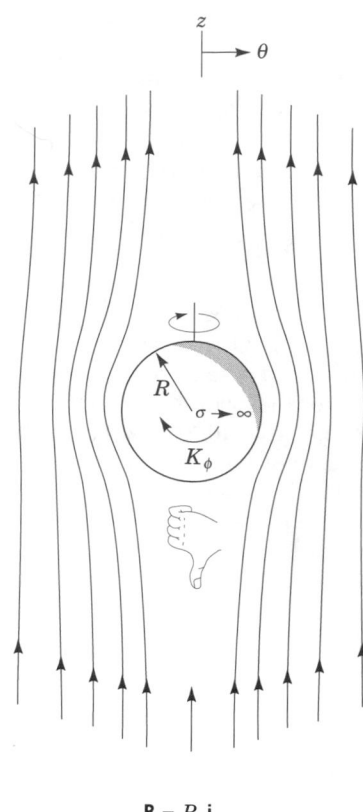

$$\mathbf{B} = B_0 \mathbf{i}_z$$

Figure 6. A perfectly conducting sphere is placed into uniform electric and magnetic fields. The induced surface charge causes the electric field to terminate perpendicularly and the induced surface current causes the magnetic field to pass tangentially to the sphere surface, with zero fields inside the sphere.

points in the direction of current-generated magnetic field when the fingers of the right hand are curled in the direction of current flow. The reaction fields due to induced surface charges and currents are point dipole fields so that the total electric and magnetic fields for $r > R$ are

$$\mathbf{E} = E_0 \left[\left(1 + \frac{2R^3}{r^3} \right) \cos \theta \mathbf{i_r} - \left(1 - \frac{R^3}{r^3} \right) \sin \theta \mathbf{i}_\theta \right]$$

$$\mathbf{B} = B_0 \left[\left(1 - \frac{R^3}{r^3} \right) \cos \theta \mathbf{i_r} - \left(1 + \frac{R^3}{2r^3} \right) \sin \theta \mathbf{i}_\theta \right] \quad (10)$$

Note that at the sphere surface, $r = R$, boundary conditions require that the electric field be purely perpendicular (radial) to the sphere and the magnetic field to be purely tangential (θ directed) to the sphere. The peak surface electric field is $3E_0$ at $\theta = 0$ and $\theta = \pi$, while the peak magnetic field is $3/2\, B_0$ at $\theta = \pi/2$. The sharper the geometry, the greater the field enhancements at the sharpest points.

Shielding

Electroquasistatic Fields. Most nonmagnetic materials can be described by their dielectric permittivity ε, or equivalently their relative dielectric constant $\varepsilon_r = \varepsilon/\varepsilon_0$, and their ohmic conductivity σ. When an electric field is first imposed in the vicinity of such a medium, the total electric field distribution depends only on the dielectric constants of all materials. However, the electric field in the conducting media causes a volume current to flow with current density $\mathbf{J} = \sigma\mathbf{E}$. As time progresses, the current carries charge to surfaces so that a new steady-state

electric field distribution results. If such an object is surrounded by free space in a dc electric field, the resulting steady-state surface charge distribution is such that the dc electric field inside the conducting medium is zero. The time constant for the system to evolve to the dc steady-state is called the dielectric relaxation time $\tau = \varepsilon/\sigma$. Taking the relative dielectric constant of the body to be that of water, $\varepsilon_r = 80$, and a typical body conductivity to be $\sigma \approx 0.2$ Siemen/m, the dielectric relaxation time is $\tau = \varepsilon_r\varepsilon_0/\sigma \approx 3.5 \times 10^{-9}$ s. When a dc electric field is first turned on, the electric field penetrates into the conductor, but after a few multiples of τ, the electric field inside the isolated conductor is essentially zero. For low sinusoidal frequencies with $f \ll 1/(2\pi\tau)$ ($f \ll 45$ MHz), the body acts like a good conductor, and the electric field within the body is small, being 10^{-7}–10^{-4} times less inside the body than just outside for 60 Hz electric fields (8). Because the body is a good electrical conductor, it significantly perturbs the 60 Hz electric field from the distribution present when the body is absent. The external time varying electric fields must terminate essentially perpendicularly on the highly conducting body, which induces a surface charge that also varies with time. A time-varying surface charge causes surface currents. However the magnitude of these surface currents is typically much less than normal biologically induced currents. Measured body currents in grounded humans approximately obey the relation:

$$i = 15 \times 10^{-8} f W^{2/3} E_0 \quad (11)$$

where i has units of mA, frequency f is in Hz, body weight W is in g, and E_0 is the ambient electric field in V/m (8).

Thus a 68-kg person in a 1000 V/m 60 Hz electric field would have ~ 15 mA current.

Magnetoquasistatic and Electrodynamic Fields. Time-varying magnetic fields generate electric fields as given by Faraday's law causing an ohmic current with density $\mathbf{J} = \sigma\mathbf{E}$. These induced "eddy" currents flow in the direction such as to generate self-magnetic fields in the opposite direction to the original magnetic field, tending to decrease the magnetic field strength in the conductor. The magnetic field then penetrates only about a skin depth distance, $\delta = [2/(\omega\mu_0\sigma)]^{1/2}$ into the conductor, where $\omega = 2\pi f$ is the angular frequency of the magnetic field. For a person, $\sigma \approx 0.2$ S/m, and at power frequency ($f = 60$ Hz) $\omega = 2\pi f = 120\pi \approx 377$ rad/s, giving $\delta \approx 145$ m. Because this value is so much greater than the size of a person, the magnetic field essentially completely passes through the body with essentially no difference in distribution due to the presence of the body. At $f = 10$ MHz, the skin depth in the body is $\delta \sim 0.36$ m, somewhat comparable to a person's thickness. Thus for frequencies less than 10 MHz, the body has essentially no effect in perturbing a magnetic field distribution.

A house or vehicle also typically shields the interior from power line electric fields but has essentially no effect on power line magnetic fields. If one really wanted to shield their vehicle or house from power line magnetic fields, the good electrical conductor copper, which has conductivity $\sigma = 5.8 \times 10^7$ S/m, could be used. The skin depth of copper at 60 Hz is 8.5 mm, and to greatly attenu-ate magnetic field strength, copper sheet a few skin depths thick would be necessary, but with great weight and cost penalties. Of course, there would still be magnetic fields inside the home from home wiring and electricity use, so such an approach seems noneffective.

Power Frequency Fields and the Body. At 60 Hz the body is a good conductor for electric fields but a poor conductor for magnetic fields. The introduction of a body into an electric and magnetic field will locally change the electric field distribution but will have negligible effect on the magnetic field distribution. The heating rate in an ohmic conductor is $Q = \sigma E^2$ W/m^3. If the body is in an external electric field $E_{ext} = 10$ kV/m, the internal electric field is less than $E_{int} = 1$ V/m. With a body conductivity of $\sigma = 0.2$ S/m, the heating rate at $E = E_{int}$ is $Q = 0.2$ W/m^3. This power frequency heating is much less than body heating from normal metabolic processes.

Figure 7 shows a person standing under a three-phase power line (28,29). The electric field must terminate essentially perpendicularly on the body and on the ground with a surface charge distribution, with essentially no electric field inside the body or below ground. The magnetic field distribution is essentially the same whether the person is there or not and is not significantly perturbed by the presence of the ground. The time-varying electric field thus induces a time-varying surface charge density on the body and on the ground. A typical person standing on the ground in a 60 Hz, 10 kV/m rms electric field in the absence of the person has a body-averaged

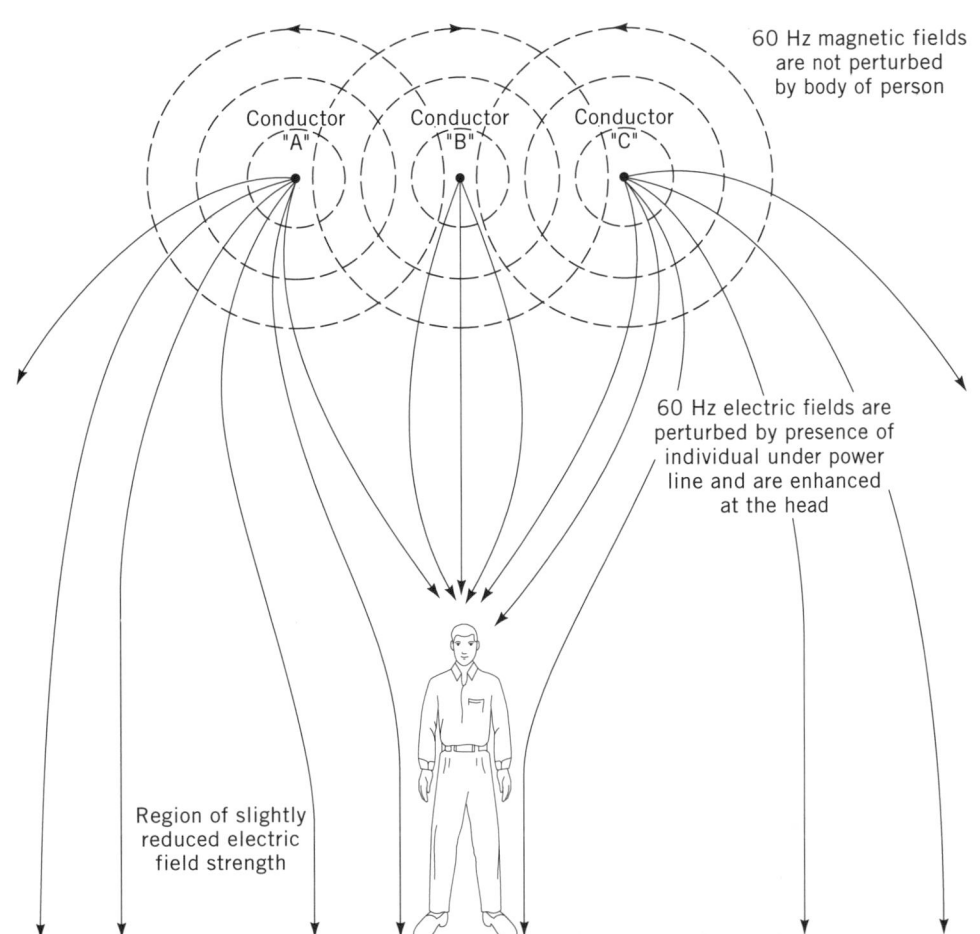

60 Hz magnetic fields are not perturbed by body of person

Conductor "A"   Conductor "B"   Conductor "C"

60 Hz electric fields are perturbed by presence of individual under power line and are enhanced at the head

Region of slightly reduced electric field strength

Figure 7. A person under a transmission line acts like a good conductor to the electric field, which terminates perpendicularly on the body but has negligible effect on the magnetic field (28,29).

electric field enhanced to about 27 kV/m, with peak electric field of 180 kV/m rms at the head and a head current density of 60 nA/cm² (8). The current density through a person's foot to ground increases to about 2000 nA/cm².

Moving at velocity **v** through magnetic field **B** generates an electric field $\mathbf{E} = \mathbf{v} \times \mathbf{B}$, which is perpendicular to both **v** and **B** with magnitude $|E| = |vB \sin \theta|$ where θ is the angle between the vectors **v** and **B**. If a person walks at $v = 1$ m/s perpendicular ($\theta = 90°$) to the earth's dc magnetic field, $B \sim 0.5 \times 10^{-4}$ T, the induced dc electric field is $E \sim 0.5 \times 10^{-4}$ V/m. With the body's current density of $\sigma \sim 0.2$ S/m, the resulting dc current density is $J = \sigma E \sim 10^{-5}$ A/m² = 1 nA/cm².

Under a distribution line, the electric field is about 10 V/m, 1000 times less than the example of a 10 kV/m electric field, appropriate for standing under a transmission line. The induced current would similarly be 1000 times less or about 2 nA/cm² through a person's foot. Thus the induced dc current density of walking through the earth's magnetic field is much less than the induced 60 Hz currents of standing under a transmission line, but comparable to that of standing under a distribution line. Under most situations, 60 Hz induced currents in the body are comparable in magnitude to dc currents induced by walking through the earth's magnetic field.

Figure 8 illustrates the same principles as in Figure 7 for a video display terminal (VDT) operator (28,29). VDTs have 60 Hz fields from the power supply, dc fields with a

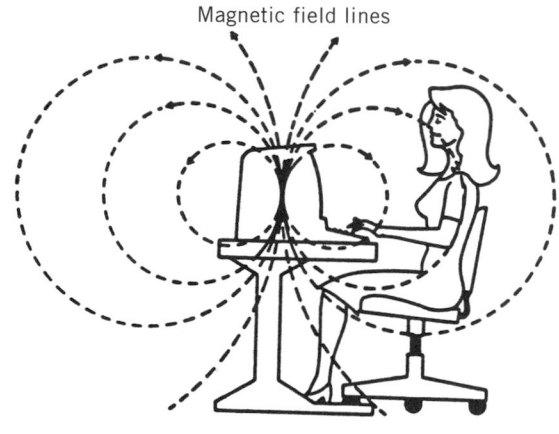

Magnetic field lines

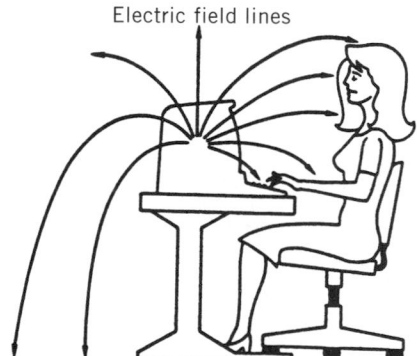

Electric field lines

Figure 8. A video display terminal has its external electric field terminate perpendicularly on the operator, while the magnetic field is negligibly perturbed by the presence of the operator because the skin depth of all frequencies present are much larger than the person's thickness (29).

60 Hz modulation from the high voltage dc power supply applied to accelerate the electron beam in the cathode ray tube (CRT), and radio frequency (RF) fields (~10–300 kHz) caused by the circuitry associated with the deflection of the electron beam (28,29). At 300 kHz, the skin depth in the body is about 2 m, so that the magnetic field from the currents in the VDT are approximately that of a magnetic dipole and are negligibly perturbed by the presence of an operator or of ground and other nearby objects being about $H = 150$ mA/m ($B = 1.9$ mG) 10 cm in front of the screen. In the absence of an operator, the 300 kHz electric field would be essentially that of an electric dipole over a ground plane with electric field about 24 V/m rms just in front of the screen. However, with the operator present, the electric field is perturbed to terminate essentially perpendicularly on the operator, with field strengths increased at sharper parts, such as the head.

REPRESENTATIVE STANDARDS

The first ANSI electromagnetic field standard established in 1966, limited exposure to 100 W/m² over the frequency range of 10 MHz to 100 GHz, which corresponds to an electric field limit of ~ 200 V/m rms and magnetic field limit of ~ 6.5 mG rms (30). In 1982, the ANSI standard was revised to make absorbed power dependent on frequency with a goal of limiting absorbed body average power density to 0.4 W/kg in the frequency range of 300 kHz–100 GHz. The International Radiation Protection Association (IRPA) formed the International Non-Ionizing Radiation Committee (INIRC) to formulate limits of exposure of 50/60 Hz and magnetic fields as given in Table 3 (31,32). The values in Table 3 were chosen so that the induced body current density is less than 10 mA/m², because this value does not exceed normal biological values.

IRPA and INIRC guidelines on radio frequency and microwaves (Table 4) assume no adverse health effects for energy deposition rates below 4W/kg. They further use a safety factor of 10 for long-term exposure and a further safety factor of 5 for general public limits.

PERSONAL MEASUREMENTS

Environmental electric and magnetic fields were measured in a house and community in Lexington, Massachusetts (33). A Holaday Industries, Inc. HI-3600-02 Power Frequency Field Strength Meter was used, which is designed to assist in the evaluation of electric and magnetic fields associated with 50/60 Hz electric power transmission and distribution lines and electrically operated equipment and appliances (28,29). A Field Star 1000 mag-

Table 3. IRPA/INIRC Limits of Exposure to 50/60 Hz Electric and Magnetic Fields[a]

Class	E (kV/m rms)	B (G rms)
Occupational workers		
Whole working day	10	5
Short-term	30	50
General public		
Whole day	5	1
Few hours per day	10	10

[a] From Refs. 31 and 32.

Table 4. IRPA/INIRC Guidelines for Exposure Limits at Radio and Microwave Frequencies[a]

f(MHz)	Occupational Exposures		General Public Exposures	
	E (V/m rms)	B (mG rms)	E (V/m rms)	B (mG rms)
0.1–1	614	$20/f$	87	$2.9/f^{1/2}$
1–10	$614/f$	$20/f$	$87/f^{1/2}$	$2.9/f^{1/2}$
10–400	61	2	27.5	0.9
400–2000	$3f^{1/2}$	0.1	$1.37f^{1/2}$	$0.465f^{1/2}$
2000–300,000	137	4.5	61	2

[a] Refs. 31 and 32.

netic field meter with memory was also used to record magnetic fields as a function of time.

Electric fields are detected using a capacitive current sensor, which consists of two thinly separated conductive discs that are electrically connected together. When placed in an alternating electric field, charge is redistributed on the two parallel discs so that the electric field between the two discs at the same potential is zero. In a sinusoidally varying electric field, the redistribution of charge also changes sinusoidally with time giving a measured sinusoidal current in the connecting wire between the discs whose measured amplitude is proportional to the local electric field.

The magnetic field in the Holaday instrument is measured using a coil consisting of several hundred turns of fine gage wire. When placed in an alternating magnetic field, a current is induced in the coil whose amplitude is proportional to the magnetic field strength perpendicular to the coil. The Field Star 1000 instrument has three mutually perpendicular coils to measure the three vector components of magnetic field. The magnetic field magnitude is also computed by the instrument by taking the square root of the sum of squares of the three components.

The measured ambient electric field through the house varied over the range of 3–1600 V/m, the highest value being in front of an operating television. Far from an appliance the ambient electric field was of order 10 V/m. The magnetic field far from appliances was about H = 90 mA/m ($B \sim 1$ mG), with the highest values at a stove heating element at the high setting with 3330 mA/m ($B \sim 42$ mG). In the course of a 0.80-km distance from the house the outdoor electric field varied over the range of 0.35–17.2 V/m and the magnetic field varied from H = 5.2 to 620 mA/m (B = 0.065–7.8 mG). Within 3 m of a substation transformer, the electric field was 0.8 V/m and magnetic field was H = 1083 mA/m (B = 13.6 mG). For a typical high school student, the Field Star 1000 recorded a school-day magnetic field exposure of 0–2 mG; this increased to 16 mG in the school lunchroom (33).

CONCLUSION

Despite epidemiological, laboratory, and human and animal studies that show possible biological effects from nonionizing electromagnetic fields, there are no well-established physical mechanisms or conclusive proof of adverse health effects. Nevertheless, it remains prudent

that when possible, people should minimize their exposure to electromagnetic fields. This article has shown that typical exposures from power frequency and radio frequency fields result in voltages, currents, fields, and heating that are comparable or less in magnitude than those naturally occurring in the environment or from normal metabolic processes in the body.

Because at frequencies below $\approx$ 10 MHz the body is a good conductor to electric fields but a poor conductor for magnetic fields, the presence of a body in an ambient electric and magnetic field will perturb the electric field distribution but will have essentially no effect on the magnetic field distribution. The electric field inside the body is greatly reduced from the electric field just outside the body, and the external electric field must terminate essentially perpendicularly to the body, resulting in a surface charge distribution and a surface current distribution if the external electric field is time varying. Those studies that show adverse health effects linked to the magnetic field but not the electric field may perhaps be due to the body conductivity that almost completely shields the body interior from the electric field but not the magnetic field.

Charges and voltages are the source of electric fields and currents are the source of magnetic fields. Appliances are charge neutral with current flow in a closed loop and thus approximately have electric fields like that of a point electric dipole and magnetic fields like that of a point magnetic dipole (fields that decrease with distances as $1/r^3$). Home wiring and balanced power lines have electric and magnetic fields like line dipoles that decrease with distances as $1/r^2$, but lines with unbalanced currents can have a larger magnetic field due to the net current flow on the lines that decreases with distance as $1/r$.

Because in the United States normal residential power is at 60 Hz, 120 V rms, whereas in Europe and elsewhere it is 50 Hz, 240 V, the normal household current is about half in Europe than in the United States. Health studies around the world need to recognize this difference in voltage and current magnitudes and perhaps frequency, although interpretation is difficult because health studies thus far have not shown the normal dose–response relationship that greater exposure is more harmful. While continuing health research tries to sort out physical mechanisms of possible health hazards, it is prudent for everyone to minimize unnecessary exposure to electric and magnetic fields.

Acknowledgments
Great appreciation is given to Donald L. Haes Jr., assistant radiation protection officer in the MIT Environmental Medical Service, and James C. Weaver of MIT for sharing of references on electromagnetic fields in the environment and for providing constructive comments on the manuscript. Appreciation is also given to Haes for loan of the Holaday Industries Power Frequency Field Strength Meter and the Field Star 1000.

BIBLIOGRAPHY

1. O. P. Gandhi, ed., *Biological Effects and Medical Applications of Electromagnetic Energy*, Prentice-Hall, Inc., Englewood Cliffs, N.J., 1990.

2. Oak Ridge Associated Universities Panel for the Committee

on Interagency Radiation Research and Policy Coordination, *Health Effects of Low Frequency Electric and Magnetic Fields*, June 1992.

3. M. G. Morgan, *Electric and Magnetic Fields from 60 Hertz Power: What Do We Know About Possible Health Risks?*, Department of Engineering and Public Policy, Carnegie-Mellon University, Pittsburgh, Pa., 1989.

4. L. E. Anderson in B. W. Wilson, R. G. Stevens, and L. E. Anderson, eds., *Extremely Low Frequency Electromagnetic Fields: The Question of Cancer*, Battelle Press, Columbus, Ohio, 1990.

5. C. Polk and E. Postow, eds., *CRC Handbook of Biological Effects of Electromagnetic Fields*, CRC Press, Boca Raton, Fla., 1986.

6. M. Blank, ed., *Electricity and Magnetism in Biology and Medicine*, San Francisco Press, 1993.

7. E. L. Carstensen, *Biological Effects of Transmission Line Fields*, Elsevier, Amsterdam, The Netherlands, 1987.

8. W. T. Kaune and L. E. Anderson, Chap. 3 in Ref. 4.

9. T. S. Tenforde, Ch. 12 in Ref. 4.

10. N. Wertheimer and E. Leeper, *Am. J. Epidemiol.* **109**, 273–284 (1979).

11. P. Brodeur, *Currents of Death: Power Lines, Computer Terminals, and the Attempt to Cover Up Their Threat to Your Health*, Simon & Schuster, New York, 1989.

12. Electric Power Research Institute, *EMF Health Effects Research: A Selected Bibliography*, EPRI, Mar. 1992.

13. Electric Power Research Institute, *Exposure Assessment Fundamentals*, EPRI, Nov. 1992.

14. Electric Power Research Institute, *Fundamentals of Epidemiology: Parts I and II*, EPRI, Oct. 1993.

15. I. Nair, M. G. Morgan, and H. K. Florig, *Biological Effects of Power Frequency Electrical Fields*, NTIS PB89-209985, U.S. Congress Office of Technology Assessment, Washington, D.C., May 1989.

16. K. Fitzgerald, I. Nair, and M. G. Morgan, *IEEE Spectrum*, 22–35 (Aug. 1990).

17. D. L. Mader and S. B. Peralta, *Bioelectromagnetics* **13**, 287–301 (1992).

18. J. R. Gauger, *IEEE Trans. Power Apparatus Sys.* **PAS-104**(9), 2436–2444 (1985).

19. J. Douglas, *EPRI J.*, 18–25 (Apr.–May 1993).

20. J. Douglas, *EPRI J.*, 6–13 (July–Aug. 1993).

21. R. L. Loftness, *Energy Handbook*, Van Nostrand Reinhold Co., Inc., New York, 1978.

22. T. Moore, *EPRI J.*, 4–17 (Jan.–Feb. 1990).

23. T. Moore, *EPRI J.*, 4–19 (Oct.–Nov. 1990).

24. T. Moore, *EPRI J.*, 4–13 (Mar. 1992).

25. H. K. Florig and J. F. Hoburg, *Health Phys.* **58**(4), 493–502 (1990).

26. R. B. Goldberg, W. A. Creasey, and K. R. Foster, pp. 248–250 in Ref. 6.

27. B. W. Wilson, R. J. Reiter, and A. A. Pilla pp. 251–254 in Ref. 6.

28. *User Manual, HI-3600-2 Power Frequency Field Strength Meter*, #600040, Holaday Industries, Inc., Sept. 1989.

29. *User Manual, HI-3600 VDT Radiation Survey Meter*, #600031 B, Holaday Industries, Inc., Jan. 1988.

30. K. R. Foster and A. W. Guy, *Sci. Am.* **255**(3), 32–39 (1986).

31. A. S. Duchene, J. R. A. Lakey, and M. H. Repacholi, eds., *IRPA (International Radiation Protection Association) Guidelines on Protection against Non-Ionizing Radiation*, Pergamon Press, Oxford, UK, 1991.

32. M. Grandolfo and M. H. Repacholi pp. 77–80 in Ref. 6.

33. A. E. Zahn, *Environmental Electric and Magnetic Fields in the Lexington, MA Community*, 9th Grade Science Project, Lexington High School, Lexington, Mass., April, 1994.

ELECTROMAGNETIC FIELDS, HEALTH EFFECTS

Stephen F. Cleary
Medical College of Virginia
Virginia Commonwealth University
Richmond, Virginia

Life on earth has evolved in an electromagnetic environment comprised of natural terrestrial sources such as quasi-steady-state fields due to thunderstorms, lightening, and low frequency Schumann resonances (1). The sun and other extraterrestrial sources also contribute to our electromagnetic environment (1). With the exception of short duration transients from sun spots, giant radio bursts or lightening, the intensities of natural environmental low frequency electromagnetic fields, are of very low intensity and have remained relatively constant during past milleniums. This status is in distinct contrast to environmental levels of man-made electromagnetic fields which have continued to increase steadily since the turn of the century such that present levels are more than a million times higher than naturally occurring electromagnetic fields. See also ELECTRIC POWER DISTRIBUTION; ELECTRIC POWER SYSTEMS AND TRANSPORTATION.

The term electromagnetic indicates the coexistence in time and space of electric and magnetic field components, the interrelationship of which depends upon the nature of the source and the frequency (ν) of variation of the field amplitudes. The electric and magnetic field components are coupled since a time-varying electric field will induce a magnetic field at the same frequency and a time varying magnetic field will in turn induce an electric field in conducting objects such as the human body. The extent of electric–magnetic coupling is directly proportional to the rate-of-change, or frequency, of the field components. At low frequencies, such as in the extremely low frequency (ELF) range (0 to 10^3 Hertz (Hz)) coupling is minimal and thus either the electric or magnetic component will predominate depending upon whether the source is a current or voltage source. Under these minimal coupling conditions the components are referred to as electric or magnetic fields. At higher frequencies there is significantly greater coupling or energy exchange between the field components which results in propagation of the field through space as an electromagnetic wave or electromagnetic radiation. The basic relationship between the frequency (ν) and the wavelength (λ) of electromagnetic radiation is $\lambda\nu = c$, where c is the velocity of light (3×10^8 m/s). In the context of this article, electromagnetic fields (EMF) refer to fields having frequencies of 10^3 Hz or less, whereas higher frequency fields will be referred to as electromagnetic radiation (EMR). Predominant man-made environmental EMF sources are associated with the generation, transmission, and usage of 60 Hz power frequency fields. Common environmental sources of man-made EMR include radio frequency (RF) and microwave (MW) radiation in the frequency range of approximately 10^5 to 10^{11} Hz used for communications and heating. As the frequency increases, localized EMR energy coupling increases such that at frequencies of 10^{15} Hz or greater (corresponding to ultraviolet (UV) radiation) molecules or atoms are ionized. EMR in this range of frequencies, which includes in addition to UV, X- and gamma-rays, is referred to as ionizing radiation. Adverse health effects of ionizing EMR, including somatic and genetic mutations, have been well characterized as a result of numerous studies over the past 50 years (1). The remainder of this article will focus on the much less well-defined current health concerns resulting from exposure to nonionizing electromagnetic radiation, specifically ELF EMFs and RF and MW EMR.

Until relatively recently, recognized adverse health effects due to ELF EMFs were primarily shocks and/or burns resulting from inadvertent human contact with electrical conductors. Adverse effects of exposure to high frequency EMR, such as microwaves, were commonly attributed to radiation energy absorption in tissue resulting in excessive heating and consequent thermal damage. In either case, the acute nature of the exposure and the immediacy of the injury resulted in obvious cause–effect relationships (2).

Recently, attention has been focused on health effects of chronic or long-term low-intensity EMF and EMR exposure. The principal reason for concern are reports indicating elevated cancer incidence in human populations exposed to such fields in the home or in the workplace. In contrast to acute, high intensity exposure effects, it has been more difficult to establish cause–effect relationships resulting from chronic low-intensity exposure. This difficulty may be attributed to several factors including the limited amount of data presently available and the lack of a theoretical basis to explain how such low-intensity fields alter living systems. In contrast to effects of high intensity EMFs that involve significant field/tissue energy exchange, low-intensity field interactions in many instances occur at levels well below those associated with energies involved in classical physicochemical processes in living systems. Additional difficulty has been introduced by the dependence of such effects on field frequency or modulation, especially in the ELF range. It should be noted that in addition to potentially adverse exposure effects beneficial aspects of the biological activity of electromagnetic fields have also been demonstrated. For example, the capability of RF and MW EMR to induce localized heating has been exploited in tumor therapy and low intensity ELF pulsed magnetic fields have been used clinically to enhance rates of tissue healing (2,3).

Considering the ubiquitous and ever increasing levels of human exposure to EMFs and EMR in our environment, there is an obvious need to more precisely define the potential health effects. The remainder of this article will attempt to summarize the present state of knowledge regarding the health effects of the most common sources of human exposure; namely, ELF electric and/or magnetic fields and high frequency RF and microwave radiation.

ELF MAGNETIC FIELD EXPOSURE EFFECTS ON HUMANS

Cancer Incidence and Residential Exposure

In 1979, the results of an epidemiological study were published indicating a positive association between residential 60 Hz magnetic field exposure and childhood cancer (4). In this case-control study, general levels of 60 Hz magnetic field exposure of cancer victims and control subjects were estimated by observing the number and size of electric power supply lines and their distance from homes. This method of estimating magnetic field exposure levels is referred to as wire code configuration estimation. Children who lived closest to high electric current wiring configurations, and hence the highest potential environmental levels of 60 Hz magnetic fields, experienced a 2- to 3-fold increased risk of cancer compared to children who lived in lower magnetic field environments. The unexpected and ominous implications of this study generated significant controversy and criticism which was primarily related to the use of wire code configurations to categorize magnetic field exposures, an approach potentially subject to inaccuracies and bias.

Subsequent studies of 60 Hz magnetic field exposure and childhood cancer involved measurements of magnetic fields in the homes as well as exposure assessments based upon wire code configurations (5,6). The results of these studies presented an apparent inconsistency with respect to the previous study of Wertheimer and Leeper (4). In these studies, the association of measured residential magnetic field intensities and childhood cancer risk was not as strong as the risk indicated by wire codes. In a study (5), magnetic field intensities were not associated with childhood leukemia when measurements were made under conditions of high electric current use in the home. Under conditions of low average home power usage, there was a moderate association of magnetic field exposure with childhood leukemia risk, whereas there was a more obvious association when exposure was estimated by the use of wire code configurations, as in Ref. 4. In the study reported by London and co-workers (6), measured magnetic field intensities were not related to childhood leukemia risk, but again when risk was estimated using wire code configurations, the risk factor was approximately two, similar to that found in previous studies (4,5). In the studies in Refs. 5 and 6, there was a positive trend between the intensity of 60 Hz magnetic field exposure and leukemia risk.

Wire Code Configuration Estimation Versus Measurements

Questions regarding the apparent inconsistency between childhood leukemia risk and measured magnetic field intensities versus estimates based on wire code configuration remain unanswered. There are, however, a number of possible explanations. It is possible, for example, that an unidentified risk factor may be responsible for the positive association between leukemia and residential wire code configurations. Considering the designs of these studies and specific attempts to detect confounders and/or unknown risk factors, it does not appear likely that the outcomes resulted from an unknown risk factor. This conclusion is supported by the results of a recent Swedish epidemiological study of the association of magnetic field exposure from high voltage power lines and cancer, as discussed below (7). It has been suggested that social class may represent a confounding element in the association of residential magnetic field exposure and cancer. However, based on available information, it does not appear likely that social class *per se* is a significant confounder (8).

An alternative explanation for the inconsistencies in the risk estimates may be that the measured magnetic fields in the studies in Refs. 5 and 6 may not have characterized the long-term residential magnetic field exposure history as adequately as estimates based on wire code configuration. This possibility is supported by the results reported in Ref. 7.

Magnetic Field Interaction Mechanisms

As previously noted, there is limited theoretical understanding of basic interaction mechanisms responsible for effects of low intensity ELF magnetic fields on living systems. For example, it is uncertain whether effects are due to electric fields and/or currents induced in cells or tissue by time varying magnetic fields, or if the effects are due, in whole, or in part, to direct magnetic field interactions. Until quite recently, it was believed that magnetic fields did not interact directly with tissue, an assumption that was recently challenged by detection of trace levels of biogenic magnitite in human tissue and other living systems (9). As a consequence of uncertainties surrounding interaction mechanisms, the magnetic field exposure parameter(s) of greatest relevance to leukemia risk, or other effects, are presently undefined. It is possible, for instance, that instead of short term magnetic field intensities measured at one or a few locations (referred to as "spot measurements"), the most significant parameter could be one of the following: (a) instantaneous or peak magnetic or electric field intensity, (b) frequency or rate of occurrence of field transients, (c) rate of change of electric or magnetic fields, or (d) long-term average exposure intensity. Lacking knowledge of relevant magnetic field exposure parameters, it is not surprising that there are apparent inconsistencies in the results of epidemiological studies related to differences in exposure metrics.

Assuming that exposure to 60 Hz residential magnetic fields is, in fact, associated with increased cancer incidence in human populations, one may ask what effect inadequacies in exposure estimation, such as the use of wire code configurations, would have on risk estimation. In general, if exposure estimators such as wire code configurations are equally inadequate estimators for cancer cases as well as controls (non-cancer cases), the true risk would be underestimated rather than overestimated.

Estimates of cancer risk from residential exposure to 60 Hz magnetic fields differed significantly depending upon the selection of cutpoints distinguishing exposed from unexposed subjects (10). Using a higher magnetic field exposure cutpoint than used in an earlier study (5), Wartenburg and Savitz using the same data, obtained larger odd ratios, two of which achieved statistical significance (10). In addition they found that by using probability plots, the data showed greater consistency with measures of magnetic fields in both low- and high-power use situations in contrast to results previously reported (5). They also reported a lack of concordance with results based upon measures of electric fields, suggesting that increased leukemia risk was more directly associated with

magnetic than electric field exposure. This study indicated statistically significant acute lymphocytic leukemia and lymphoma odds ratios for children exposed to residential 60 Hz magnetic field intensities of 0.3 micro Tesla (μT) [3 milliGauss (mG)] or greater (10). Magnetic field intensity units are Tesla (T) and Gauss (G) (1T = 10^4G).

A recent epidemiological study of exposure to 50 Hz magnetic fields and childhood leukemia involved both measured residential magnetic field intensities as well as long-term average intensities calculated on the basis of distance of homes from high voltage transmission lines and records of the electrical current carried by the lines during previous years of exposure (7). It is unfortunate that such records of line current are not available in the United States since they would permit studies of this type to be conducted in the U.S. Such studies could lead to a better understanding of the significance of long-term average magnetic field exposure as related to cancer incidence.

Children exposed in their homes to time-averaged magnetic fields of 0.3 μT (3mG) or more had a relative risk almost four times greater than children exposed to magnetic fields of less than 0.1 μT (1mG) (7). A significant aspect of this study was that cancer risk was directly proportional to average magnetic field exposure intensity. The detection of a statistically significant dose–response relationship is a criteria applied by epidemiologists to assess the internal consistency of study results. Children exposed to average magnetic field intensities of greater than 0.1 μT (1mG) had a doubling of leukemia incidence; exposure to more than 0.2 μT (2mG) was associated with a three-fold increase and, as indicated above, exposure to more than 0.3 μT (3mG) resulted in an approximate four-fold increase in leukemia risk. The childhood leukemia risk ratios, which were controlled for potential confounders such as socioeconomic status and air pollution, were statistically significant (7). There was a 70% increased incidence of both acute myeloid leukemia (AML) and chronic myeloid leukemia (CML) in adults exposed in their residences to magnetic field intensities of greater than 0.2 μT (2mG). Although these results suggested an association of adult cancer with 50 Hz magnetic fields, as had the study in Ref. 10, the risk ratios were not statistically significant. It was also noted that statistically significant elevation in childhood leukemias occurred in children who lived in single- as opposed to multifamily dwellings (7).

This study involved the most detailed assessment of residential magnetic field exposure to date. The assessment involved reconstruction of magnetic field exposure levels in homes for periods of up to 10 years prior to diagnosis, using records of transmission line electric current flow. This strategy enabled the investigations to take into account short-term as well as seasonal changes in magnetic field intensities. Comparison of the long-term magnetic field exposure assessments with results of short term "spot" measurements made in the homes indicated a lack of correlation. "Spot" measurements tended to overestimate past exposures, which suggested that studies using measured values could be biased. Since the use of wire code configurations may provide a better estimate of long-term residential magnetic field exposure than spot measurements, the result of the study, in Ref. 7 provide a possible explanation for the consistent statistically significant association of childhood cancer with magnetic field exposure when wire code configurations were used to estimate exposure levels (4–6).

Sources of Magnetic Field Exposure

The apparent association of residential ELF magnetic fields and cancer focussed attention on the types and magnitudes of exposure, as well as methods of exposure assessment. Principal sources of ELF residential magnetic fields included: high voltage transmission lines, distribution lines, electrical service wiring, and household electric wiring. In this context, it should be noted that since the vast majority of residences in the United States and other industrialized countries are electrified, magnetic field exposure is essentially ubiquitous. Considering the results of epidemiological studies associating magnetic field exposure and cancer, it must be kept in mind that there are no truly unexposed persons to serve as controls for those exposed to higher intensity magnetic fields. It has been estimated, for example, that power lines and the grounding circuits in houses, result in average residential fields of approximately 0.1 μT (1mG) (12). The findings of Wartenberg and Savitz (10) and Feychting and Ahlbom (7) that exposures of 0.3 μT (3mG) or greater may significantly increase childhood cancer risk imply an unexpectedly high sensitivity for this magnetic field exposure effect.

The actual sensitivity of humans to ELF magnetic field health effects is difficult to assess since, as noted above, the exposure parameters of primary concern have not been adequately described and there are multiple sources of exposure having different spatial and temporal characteristics. Controversy exists, for instance, regarding the contribution of electric and magnetic fields from appliances compared to other sources of exposure (13). In contrast to other residential sources of magnetic fields that vary relatively little over time or space, appliance fields are intermittent and highly dependent upon distance from the appliance. The magnetic field intensity of appliances such as dishwashers, toasters, and irons varies from about 10 μT (100mG) at 3 cm from the appliance to approximately 0.01 μT (0.1mG) at 1 meter (11). Significantly higher magnetic field intensities exist in close proximity to appliances such as electric hair dryers that can produce fields of 2,000 μT (20G) at 3 cm (13). The volume- and time-averaged magnetic fields emitted by home appliances have been estimated (12). Calculations based on measurements for 98 appliances indicated that appliance generated magnetic fields were not a significant source of whole-body exposure but that body extremities could be exposed intermittently to peak fields of up to 100 μT (1G) or higher (12). Table 1 lists the spatial average magnetic field intensities in close proximity (3 to 30 cm) to appliances together with time averaged intensity. The time averaged intensity of magnetic field exposure of extremities is calculated by multiplying the spatial average intensity by the fraction of time during the day the appliance is in use.

Table 1. Spatial and Time Average Exposure of Extremities to 60 Hz Magnetic Fields from Appliances[a]

Source	Magnetic Field Spatial Average	(3–30 cm) (μT)	Magnetic Field Time Average (μT)	
Can opener	60.6	(606)[b]	0.47	(4.7)
Electric saw	32.5	(325)	0.11	(1.1)
Electric shaver	16.9	(169)	0.06	(0.6)
Mixer	16.2	(162)	0.18	(0.2)
Electric drill	12.7	(127)	0.06	(0.1)
Hair dryer	12.0	(120)	0.07	(0.1)
Blender	5.6	(56)	0.02	(0.2)
Iron	0.7	(7)	0.01	(0.1)

[a] Adapted from data of Mader and Peratta (12).
[b] Magnetic field intensity in milliGauss units.

Cancer Incidence and Occupational Exposure

Epidemiological studies have also suggested an association of occupational exposure to electric or magnetic fields and cancer incidence. A meta analysis combining the results of occupational studies conducted prior to 1987 led Savitz and Calle to conclude that there was a consistent pattern of increased leukemia risk in workers exposed to electric or magnetic fields (15). Increased incidence of leukemia, acute leukemia (especially acute myeloid leukemia) were reported in post-1987 occupational studies as well (16–18). Workers at risk included powerline workers, electric utility workers, electricians and electronics workers (16).

A dose–response relationship for the association of cancer incidence and occupational exposure to ELF was reported (18). Based on measurements, workers were assigned to magnetic field exposure categories: Group 1, greater than 0.41 μT (4.1mG); Group 2, between 0.41 and 0.29 μT (4.1-2.9mG); Group 3, between 0.29 and 0.16 μT (2.9-1.6mG); and Group 4, less than 0.16 μT (1.6mG). There was a three-fold increase in chronic lymphocytic leukemia (CLL) and a 60% overall increase in leukemia in workers in Group 1. No association was found for acute myeloid leukemia. Among the most highly exposed Group 1 workers, there was a four-fold increased CLL risk. The study was based upon the occupational experiences of 850 cancer cases (approximately one-half leukemia cases and one-half brain tumors) and 1700 age-matched controls (non-cancer cases). Since the study group was not limited to any particular industries, the cases and controls were representative samples from the Swedish male working population. In addition to age, other potential confounders, including exposure to benzene, other organic solvents, and ionizing radiation, pesticides, and smoking, were taken into consideration (18).

The results of this study support the hypothesis that there is an association between ELF EMF occupational exposure and the development of CLL. Since the elevated risk was related to the job held longest during the 10 year period before diagnosis, the outcome was consistent with the possibility the EMF exposure acted as a cancer promoter.

The study also provided some evidence of elevated risk of brain tumors, but the dose–response relationship was less evident. Brain tumor relative risk was increased in the highest EMF exposure categories, based on median values for the job held longest during the decade before diagnosis. The EMF-brain tumor association was strongest for workers less than 40 years of age (18). Previous occupational studies suggested a positive association of brain cancer and EMF exposure (19,20).

Interaction Mechanisms

Insight regarding possible mechanisms involved in the elevation of cancer risk from low intensity EMF has been hindered by limited knowledge of cancer etiology in general. This limitation is particularly true in the case of specific types of cancer such as brain cancer and childhood leukemia. Some insight may, however, be provided by studies indicating that a rare form of cancer, male breast cancer, may be linked to occupational exposure to EMFs (21–23). If verified by future studies, the association of male breast cancer and EMFs suggests a possible mechanism related to the fact that breast cancer involves hormonal alterations.

Hormonal alterations have been reported to occur in laboratory animals exposed to ELF fields. Specifically, exposure affected biorhythms resulting in the suppression of the normal nocturnal increase in melatonin (24,25). This finding is potentially significant due to the interaction of melatonin with other hormones. Reduction in plasma melatonin concentrations causes increased levels of circulating steroid hormones such as estrogen and testosterone as well as increased prolactin release by the pituitary gland (26). Such hormonal alterations increase the rate of proliferation of breast tissue and suppress the immunological system, effects consistent with increased breast cancer risk (27). Animal experiments have indicated an association of melatonin levels with mammary tumorigenesis. Exposure of rats to 50 Hz magnetic fields of greater than 1 μT (10mG) resulted in statistically significant decrease in pineal and plasma melatonin levels (28). When the source of melatonin, the pineal gland, was removed, the incidence of breast tumors was increased significantly (29). The effect of ELF EMFs on melatonin levels in experimental animals thus suggests neuroendocrine involvement in the observed increase in breast cancer in occupationally exposed males. Obviously, this mechanism could be involved in exposure of females as well as males and could relate to other tumor types.

Epidemiological studies have also indicated an association of ELF EMF exposure and other physiological alterations including: altered biorhythms (30), behavioral and neurological disorders (31) and reproductive outcomes (32). The sensitivity of living systems to extremely weak electric fields is indicated by the alteration of normal biorhythms in humans exposed to fields as low 2.5 V/m, a field strength commonly encountered in the home or workplace (33). Exposure of human volunteers to combined electric and magnetic fields of higher intensity (9 kV/m, 20 μT (200mG) for 3 hr. periods resulted in statistically significant slowing of heart rate, changes in late components of event-related brain potentials, and decreased error rates on a choice reaction-time task (34).

Functional changes in the central nervous system (CNS), including alterations in brain wave potentials and learning were also reported following 45 min. exposures of human volunteers to 1.25 mT (12.5G) 45 Hz magnetic fields (35). A 45 min. exposure of rats to 60 Hz magnetic fields at intensities of 0.75 mT (7.5G) altered brain neurochemistry (36). The results of this study indicated that magnetic fields altered endogenous brain opioids (36).

Exposure to residential magnetic fields from electric blankets, heated water beds and ceiling cable heat that produce fields in the range of 0.4 to 1.5 μT (4–15mG) have been associated with seasonal variations in fetal growth and fetal loss (36,37). The results of a recent epidemiological study indicated an association of residential 50 Hz magnetic fields and early pregnancy loss (EPL). Magnetic field measurements in residences indicated that exposure to fields of 0.63 μT (6.3mG) or more resulted in an EPL odds ratio of 5.1 (38). Animal studies have also indicated that pulsed low-intensity magnetic fields may, in some instances, increase the incidence of malformations (39).

ELF Magnetic and Electric Field Effects on Experimental Animals

In a general sense the results of studies of the effects of ELF magnetic and electric fields on laboratory animals are consistent with effects reported in humans. However, the animal studies database is limited due to the rather recent awareness of the potentially adverse effects of low intensity ELF magnetic fields and outcomes such as cancer. Prior to this time, attention was focussed on health effects of exposure to high intensity electric fields such as those encountered near high-voltage transmission lines. Few studies have investigated the association of long-term exposure of experimental animals to low-intensity magnetic fields and carcinogenesis. The results of one such study reported that 60 Hz magnetic fields had a tumor-promoting effect on mouse skin cancer (40). Considering the paucity of relevant experimental data and the complex nature of carcinogenesis, no conclusions are possible regarding the mechanisms of effects of 60 Hz magnetic fields on this endpoint. It may be concluded tentatively, however, that it is more likely that low-intensity ELF electric or magnetic fields act as tumor promoters rather than initiators. Biophysical considerations, as well as experimental cell studies indicating that such fields do not appear to induce direct chromosomal alterations, support this conclusion.

Although limited in extent and applicability to questions such as association with cancer incidence, effects of ELF electric and/or magnetic fields on cats, dogs, swine and nonhuman primates have been reported. Endpoints investigated include: behavioral changes, reproductive outcomes, brain neurochemistry, hormone levels, cardiovascular responses, and hematopoietic changes. The effects of ELF EMFs on laboratory animals have been the subject of detailed reviews (41,42). While these data do not reveal well-defined dose-responses, they do indicate general sensitivity of mammalian systems to ELF EMF exposure that are not explainable by known interaction mechanisms.

Dosimetric Considerations

Difficulties are encountered in establishing cause-effect relationships between exposure to ELF EMFs and health effects due to dosimetric complications. For example, the electric fields induced in laboratory animals or human beings exposed to ELF magnetic fields are complex and nonuniform. Nonuniformities are due in part to the fact that there are large differences in the electrical conductivity (or resistance) of tissues such as a muscle (high conductivity) versus fat (low conductivity). These differences cause significant differences in induced current densities in tissue even when the body is exposed to a spatially uniform magnetic field. Additional complications are introduced by the dependence of the magnitude of induced currents on the body dimensions. In an ELF magnetic field of constant intensity, tissue induced electric fields will be significantly greater in a human body as compared to a rodent, for example. Such factors introduce significant uncertainty into the extrapolation of data derived from animal studies to the prediction of human health effects.

ELF EMF Effects on Mammalian Cells

To minimize ambiguity introduced by uncertainty about the distribution and/or intensity of tissue induced, electric or magnetic fields studies have been conducted using isolated mammalian cells. Such *in vitro* studies afford an opportunity to directly relate cellular alterations to well characterized electric or magnetic fields, thus providing data of use in determining basic interaction mechanisms. Whereas it is obvious that *in vitro* results cannot be directly extrapolated to *in vivo* systems, knowledge of basic interaction mechanisms obtained from such studies will be an essential element in understanding health effects of nonionizing electromagnetic fields.

In vitro studies of the effects of ELF EMFs on normal as well as transformed (cancer) mammalian cells have identified a number of sensitive physiological endpoints including: a) cell proliferation, b) membrane signal transduction, c) biomolecular synthesis, d) ion fluxes and binding, e) immune responses, and f) energy metabolism. Generally, alterations were induced by short-term (ie, minutes or hours) exposure to low intensity electric and/or magnetic fields at frequencies of 100 Hz or less. The results, which provide extensive evidence that ELF EMFs are biologically active, have been reviewed in detail (43).

The results of *in vitro* studies provide much needed insight about EMF effects on living systems such as indicating that the cell membrane is a likely field interaction site. To date, the results of such studies have not provided the detailed data necessary for the development of interaction mechanisms. *In vitro* data have, however, led to the generation of a number of interesting hypothetical mechanisms that may advance understanding of the biological effects of ELF EMFs (44).

Electric Power Consumption and Cancer Mortality

It has been suggested that trends in electric power consumption should be considered in assessing the relationship of ELF EMFs (such as 60 Hz magnetic fields from electric power distribution) to adverse health effects. If there is a direct correlation between electric power consumption and time averaged level of human exposure to 60 Hz magnetic fields, and if exposure is directly related to cancer incidence, trends in power consumption should be reflected in trends in cancer incidence. Since electric power consumption has increased steadily during the past 40 years or so, this should be reflected in an increased cancer incidence during this period. The validity of this assumption is based upon the following premises: a) increased overall electric power consumption is directly correlated with *per capita* time-averaged 60 Hz magnetic field exposure levels, and b) cancer incidence has increased during the period of concern.

The relationship between increased power consumption and *per capita* exposure to 60 Hz magnetic fields is uncertain due to a number of factors. During the period of increased power consumption, demographic changes occurred such as large population migration from urban to suburban areas. In addition, there have been changes in the electrical wiring methods, materials, and building codes that may well have affected average exposure levels. There is evidence, in fact, that such factors may have acted to limit or reduce *per capita* exposure to 60 Hz EMFs (45,46).

Whereas overall cancer mortality has not increased during the past 20 years for persons under 45 years of age, there have been significant increases in the incidence of specific types of cancer (47,48). The fact that there has not been a substantial reduction in the overall cancer death rate in industrial societies may be due to the fact that new therapies have not kept pace with increased cancer incidence. For example, during the period 1950 to 1986, there was a 20% decrease in childhood leukemia mortality but leukemia incidence increased by 60% over this period (49). Changes in age-specific tumor incidence were also detected. In the United States during the period 1969–1986, there was a 20% decrease in brain tumor mortality in white males and females ages 0–44, but brain tumor mortality in the age group 65–84 increased by more than 80% and there was a 200% increase in persons aged 75 to 84 (50). Female breast cancer incidence and mortality have increased at a constant rate for the past 50 years or more for unknown reasons (51). Although overall mortality from multiple myeloma has remained essentially constant in industrialized countries, there has been persistent increased mortality in people over the age of 70 (52). This increase could be attributed to an ubiquitous environmental risk factor common to industrialized nations that increased during the latter quarter of the past century and the first quarter of the 20th century (52).

Trends in electric power consumption and cancer incidence thus are not inconsistent with the possibility of an age-dependent cause-effect relationship, especially for specific cancer types. It is obvious that there are multiple cancer risk factors in industrial societies and that cancers are most likely multi-factor diseases. Thus, the extent to which ELF magnetic field exposure contributes to cancer risk is difficult to determine from currently available information. The results of animal experiments as well as cell studies *in vitro* provide ample evidence of the biological activity of low intensity ELF electric and magnetic fields thus adding plausibility to concern about exposure-related health effects. There is an obvious need to conduct further investigations of the biological effects of ELF EMFs to determine interaction mechanisms and to better define the nature and extent of adverse health effects.

RADIOFREQUENCY AND MICROWAVE HEALTH EFFECTS

Radiofrequency (RF) and microwave radiation have physical characteristics, and hence interaction mechanisms, quite distinct from ELF electromagnetic fields such as 60 Hz fields associated with electric power transmission. The frequency of RF radiation extends from approximately 10 KHz (10^4 Hz) to 300 MHz (3×10^8 Hz). EMR in the frequency range of 300 MHz to 300 GHz (3×10^{11} Hz) is referred to as microwave radiation. Typical sources of RF radiation include AM, television, and FM broadcast signals, whereas the most common environmental sources of microwave radiation are microwave ovens in the home that operate at a frequency of 2450 MHz. The contrast between ELF and RF/microwave EMR is most dramatically illustrated by considering differences in wavelength. The wavelength of a 60 Hz electric field is measured in thousands of miles compared to RF/microwave radiation having wavelengths of yards to fractions of an inch. Based on the knowledge that the way EMR energy is coupled to a living system, such as the human body, relates to the wavelength and the size of the absorber, it would be logical to expect different biological effects of ELF versus RF/microwave radiation. Surprisingly, in spite of differences in physical characteristics, there are some similarities in biological responses resulting from exposure to ELF EMFs and EMR.

EMR and Cancer

A possible association of RF and MW EMR exposure and human cancer incidence emerged recently. The use of hand-held cellular telephones operating in the frequency range of 825–890 MHz has been linked with brain cancer. Exposure of police officers to higher frequency microwave radiation in the frequency range of 10 to 24 GHz, emanating from traffic speed detecting radar units, has been associated with increases in testicular, brain and eye cancer. Exposure to hand-held cellular telephone radiation and police radar both involve localized radiation absorption. Tumors reportedly occurred at or near the apparent site of maximum microwave energy absorption in either case. For example, parietal lobe brain tumors were reported to occur in the brain region closest to the location of the hand-held cellular telephone antenna. Localized microwave exposure of testicular tissue of police officers presumably resulted from resting the radar gun in the groin region while the unit was emitting radiation. Neither in the case of hand-held cellular telephones or police radar have epidemiological studies been conducted to establish the validity of the association with cancer incidence. To date, few studies have been conducted of the

effects of these microwave frequencies on experimental animals or mammalian cells *in vitro*. Therefore, in the absence of a relevant scientific database, the possible association of microwave sources with cancer incidence must be viewed in terms of effects reported from exposure to other RF or microwave frequencies. The results of epidemiological studies, studies that involved exposure of laboratory animals, and investigations of the effects of RF or microwave radiation on mammalian cells, provide evidence for a possible association of such exposure with cancer incidence. The limited amount of relevant data precludes drawing firm conclusions. However, the results are most consistent with the hypothesis that under certain, presently not well-defined, exposure conditions, RF or microwave radiation may act as a cancer promoter, rather than as a cancer initiator. This hypothesis is consistent with the hypothesis discussed above that exposure to low intensity 60 Hz magnetic fields may also affect tumor promotion.

Epidemiological Studies

Until recently there has been limited concern about the relationship of RF or microwave exposure and cancer incidence. Consequently, few epidemiological studies have been conducted to investigate this possibility. This status is in distinct contrast to the number of epidemiological studies of the relationship of residential or occupational exposure to 60 Hz magnetic fields and cancer discussed previously in this chapter.

Milham (53) analyzed death certificates of 1,691 amateur radio operators from California and Washington states and found 24 leukemia cases. Compared to an expected number of such cases of 12.6, this was a statistically significant excess number of leukemias amongst amateur radio operators. In a subsequent study, Milham (54) used standardized mortality ratios (SMR) to characterize the cancer experience of amateur radio operators. The SMR for all cases of death of amateur radio operators was 71 and for combined cancers it was 89, indicating a better than average mortality history for the study group. This increase was expected based upon a comparison of this group with professional and academic cohorts. There was, however, significantly elevated leukemia mortality for amateur radio operators who had a SMR of 162 for cancers of lymphatic tissues including multiple myelomas and non-Hodgkins lymphomas. The SMR of 176 for chronic myeloid leukemia was statistically significant, whereas the elevated SMR of 124 for all leukemias was not significant. In addition to potential RF exposure from the operation of amateur radios, there was evidence of possible occupational RF exposure of some members of the study group from the state of Washington. It was noted that the elevated leukemia risk in the study group could have been due to other risk factors but that exposure to RF radiation should be considered as an etiological factor. In this study, it was not possible to determine the extent of exposure to RF radiation.

The association between EMR exposure and cancer including data from epidemiological as well as experimental studies has been investigated (55). An unexpectedly large number of cancer cases, including lymphomas, chronic myelocytic leukemia, acute myelocytic leukemia and pancreatic cancer was observed among military personnel exposed to microwave radiation. A retrospective epidemiological study of cancer morbidity in Polish military personnel aged 20–59 during the period 1971–1980 was reported (56). Cancer morbidity in personnel occupationally exposed to RF/microwave EMR was three times greater than the control group. Statistically significant increased cancer incidence was found for lymphatic and hematopoietic malignancies and for stomach, colorectal and skin neoplasms, including melanoma. The overall risk factor for lymphatic and hematopoietic malignancies in RF/microwave exposed workers was 6.7, which was statistically significant ($p < 0.01$) (56).

Evidence of an association of occupational EMR exposure and brain cancer was reported (57) in an epidemiological study of 951 cases of brain tumors among white male residents of Maryland during the period 1969–1982. Fifty cases of glioma and astrocytoma were observed among electricians, electrical engineers and high voltage transmission linemen. Compared to the expected number of 18 tumors, there was a statistically significant increase in brain tumors in EMR workers.

Epidemiological studies, although quite limited in number, provide evidence of an association of long term exposure to RF/microwave and lower frequency EMFs and cancer incidence. Although a number of different cancers have been reported to result from such exposure, leukemia and brain cancer appear to be the most prevalent.

Animal Studies

Effects of RF and microwave radiation on various psychophysiological responses in experimental animals such as rodents, dogs and non-human primates were reported. Effects on hematological and immunological systems, reproduction, nervous system, behavior, sensory systems and endocrine systems were the subject of extensive review articles (2,58). Numerous physiological alterations were reported to be induced by RF or microwave exposure, depending upon species and exposure conditions. Most studies involved acute or relatively short-term exposure to RF field intensities that induced varying degrees of temperature elevations in either the entire body or localized regions of the body. Whereas there were indications that certain physiological alterations, such as changes in behavior, might occur in the absence of tissue heating, it was generally assumed that such effects were of minor significance relative to assessments of potential effects of RF or microwave exposure on human health. Results of more recent studies involving long-term or chronic low-intensity exposure, studies of the relationship of microwave exposure to cancer promotion, and *in vitro* cell studies, have indicated the need to reconsider health effects issues from long-term low-intensity EMR exposure.

The potential tumor promoting effect of microwave exposure was investigated (56,60,61) and co-workers. Mice were exposed 2 h per day for 3 to 6 months to 2450 MHz microwave radiation at power densities of from 5 or 15 milliWatts per square centimeter (mW/cm^2) (specific ab-

sorption rate (SAR) 3–4 or 6–8 Watts per kilogram (W/kg), respectively). Microwave exposure accelerated the appearance and growth of skin neoplasms induced by benzopyrene, suggesting a tumor-promoting effect (56). EMR exposure decreased the natural antineoplastic resistance of mice to implanted tumor cells and also accelerated the growth of tumors induced by the carcinogens di-ethyl-nitrosoamine and methylchlorantrene. Szmigielski and co-workers (56) indicated that long-term exposure to low intensity (5 mW/cm^2) microwave radiation had an effect on tumor growth in mice similar to the effect of chronic stress.

Additional evidence of a possible association of chronic low-intensity microwave exposure and cancer in experimental animals was reported (62). Male rats were exposed, or sham-exposed, 21.5 h/day for 25 months to pulse-modulated 2450 MHz EMR at SARs that ranged from 0.15 to 0.4 W/kg. Physiological assays were routinely conducted throughout the experiment. Statistically significant alterations were reported in the following endpoints during the course of the experiment: behavior, serum corticosterone, lymphocyte (B and T cell) number, lymphocyte mitogenic stimulation, eosinophil and neutrophil counts, adrenal mass, O_2 consumption and CO_2 production, and benign adrenal neoplasia. Potentially of most significance with respect to the relationship of microwave exposure to cancer incidence, was the detection of 18 primary malignant neoplastic lesions in microwave exposed rats compared to 5 such lesions in the sham-exposed animals. This difference in malignant neoplasia was highly statistically significant (p = 0.006). There was also a statistically significant decrease in time of occurrence of primary tumors in microwave-exposed animals (62).

Results support the hypothesis that long-term or chronic low-intensity microwave exposure is associated with increased incidence and/or growth rate of cancer in laboratory animals (55,62). The significance of these observations with respect to human cancer incidence has yet to be determined. However, it should be noted that the microwave intensities used in these studies were in the range considered by certain regulatory and advisory bodies as being safe for long-term human exposure. It should also be noted that whole body RF or microwave absorption rates of 0.4 W/kg or less, as used in the study reported in Ref. 62, are assumed to be well below the level that causes tissue heating.

Animal studies have also provided evidence that low-intensity EMR alters brain neurochemistry under conditions not involving tissue heating. Effects of microwave exposure on: (*1*) actions of various psychoactive drugs, (*2*) the activity of the cholinergic systems in the brain, and (*3*) on neural mechanisms in the rat, were investigated by Lai (63). Neurological alterations were induced in specific parts of the rat brain by 45 minute exposures to 2450 MHz EMR at SARs of 0.6 W/kg. It was concluded that alteration of the levels of endogenous opiates were responsible for the observed EMR effects and that the effects depended upon the exposure parameters (63). The results of these studies are of interest since they provide evidence that low-intensity EMR can alter brain function which is consistent with numerous reports of behavioral, neurological and neuroendocrine alterations in humans due to EMR exposure.

In Vitro Studies

An impediment to assessing EMR effects on organisms is the fact that energy absorption in tissue is highly nonuniform and dependent upon the wavelength of the radiation and the size and orientation of the body and organs. Consequently, it is difficult to directly apply the results of studies using experimental animals to predict effects of RF or microwave exposure on human beings. In addition to dosimetric uncertainty, interspecies physiological differences and the interactive nature of various organ systems introduce additional complexities. In order to overcome such difficulties, studies have been conducted of the effects of EMR on mammalian cells *in vitro*. These studies can be conducted with precise knowledge and control of exposure conditions, thus providing data for the determination of cause-effect relationships. As in the case of ELF EMRs, effects of EMR on a variety of cellular endpoints have been investigated. Such effects include: a) membrane cation transport and binding, b) membrane structure, c) single ion channel kinetics, d) neuronal activity, d) energy metabolism, e) proliferation and activation and f) transformation. The results of these studies have been the subject of review articles (58,59,64). Due to their possible relevance to cancer incidence, effects of RF and microwave radiation on mammalian cell proliferation and transformation will be reviewed here.

To test the hypothesis that microwave radiation may interact synergistically with tumor promoters, mouse embryonic fibroblasts were exposed to low intensity 2450 MHz microwave radiation either in the presence or absence of the tumor promoter 12-0-tetradecanoylphorbol-13-acetate (TPA) (65). Cells were exposed to microwave radiation pulse modulated at 120 Hz at SARs of 0.1, 1, or 4.4 W/kg for 24 h. In the absence of TPA, EMR exposure did not affect fibroblast cell survival or the induction of neoplastic transformation. Neoplastic transformation was significantly enhanced, however, in cells treated with TPA. A synergistic TPA–EMR dose-response was detected for the frequency of neoplastic transformation in the SAR range of 0.1 to 4.4 W/kg. The neoplastic transformation frequency resulting from exposure at 4.4 W/kg, in the presence of TPA, was comparable to that induced by exposure to 1.5 Gy (1 Grey (Gy) = 100 rads) of X-radiation. It was noted that this dose of whole body X-radiation results in an approximate 6% risk of tumor induction. Comparison of the effects of X-radiation exposure, with or without 4.4 W/kg 2450 MHz microwave exposure, in the presence of TPA, indicated that the combined treatment induced two independent types of transformation damage, one due to microwave exposure and another to X rays, both promoted by TPA. This damage indicated separate sites of interaction for EMR versus X-radiation. Since genomic interactions of X-radiation are associated with neoplastic transformation, these data indicate the possibility of a non-genomic effect of microwave radiation leading to TPA promoted neoplastic transformation. Biophysical considerations, as well as results of

studies of microwave radiation effects on other cell physiological endpoints, suggest the cell plasma membrane as a probable site of interaction leading to neoplastic transformation.

In addition to evidence that low-intensity EMR may affect neoplastic transformation of mammalian cells, *in vitro* studies have also revealed that the EMR may alter the rate of proliferation of cells that are already transformed or malignant. Cleary and co-workers (66,67,68) exposed human or rat glioma cells (LN71, RT2, respectively) for 2 h to 27 or 2450 MHz CW or pulse modulated RF radiation under isothermal (37 ± 0.2°C) conditions. Cell proliferation was increased in a dose-rate dependent manner following exposures to either type of radiation at SARs of 0.5 to 25 W/kg. Maximum increased proliferation of 160% occurred following exposure to 25 W/kg 27- or 2450 MHz, 5 or 3 days after exposure, respectively. The effect on glioma proliferation was biphasic in that exposure at SARs of greater than 25 W/kg decreased the rate of cell proliferation. Statistically significant time-dependent effects were detected for up to 5 days postexposure which suggested a kinetic cellular response to EMR at these frequencies. The persistence of the RF or microwave radiation effect suggested the possibility of a cumulative effect on cell proliferation.

Evidence that EMR RF or microwave radiation may have a more general effect on cell proliferation was reported (66,68,69). Using the same experimental procedures used to study effects on glioma, human peripheral lymphocytes were exposed isothermally (37 ± 0.2°C) for 2 h to 27- or 2450 MHz radiation. Exposure to EMR at these frequencies had similar biphasic effects upon the rate of proliferation of mitogen (phytohemagglutinin (PHA)) stimulated lymphocytes. The maximum increase in proliferation, which occurred 3 days after exposure, was 40%, approximately one-fourth the magnitude of the proliferative effect of RF radiation on glioma (68). Cumulative effects were investigated by exposing lymphocyte cultures at the same SAR to two 1 h irradiations spaced 24 h apart. Such split-dose exposures caused similar effects but of reduced magnitude, indicating time dependent reversal of the exposure effect on lymphocyte proliferation with a time constant of somewhat greater than 24 h (69). Pulse modulation of 27- or 2450 MHz radiation caused generally similar effects on cell proliferation, at SARs of 5 W/kg or less, but there were differences in the dose responses. At higher SARs, pulse modulation resulted in significantly increased lymphocyte proliferation compared to CW exposures at the same SARs (69). The effects of 27- or 2450 MHz RF radiation on glioma or lymphocyte proliferation were attributed to a nonthermal, direct RF-induced alteration of the cell cycle (68).

In summary, *in vitro* studies of the effects of low-intensity RF and microwave radiation indicate dose-rate dependent increases in neoplastic transformation frequency and proliferation. In view of limitations on the extrapolation of *in vitro* results to *in vivo* responses, these results cannot be related directly to cancer incidence in human populations exposed to such radiation. However, these results are not inconsistent with the hypothesis that human exposure to RF- or microwave radiation, under presently not well-defined conditions, may affect cancer incidence.

CONCLUSIONS

There is increasing evidence of possible health effects of environmental exposures to EMFs and EMR in the home and in the work place. Epidemiological evidence indicates possible associations of long-term exposure and cancer incidence, adverse reproductive outcomes, and behavioral and neurological changes. Inherent limitations on exposure assessment, common to epidemiological studies, provide imprecise knowledge regarding time- or exposure intensity thresholds for these effects, thus making risk assessment difficult at this time. Whereas the results of animal experimentation and cellular studies of ELF EMFs and EMR effects are generally consistent with results of epidemiological studies, they provide insufficient data for meaningful risk assessment.

The greatest impediment to understanding the effects of EMFs and EMR on living systems is the limited knowledge of interaction mechanisms. One consequence is that research in this area has been treated with skepticism that has, together with other factors, resulted in serious limitations on research support. In view of the diverse nature of the physical properties of electromagnetic fields reviewed here, as well as the great variety of reported effects in living systems, the large gaps in our understanding are perhaps not surprising. The potential magnitude of exposure-related health effects in industrial societies indicates that these uncertainties must be resolved.

BIBLIOGRAPHY

1. *The Effects on Populations of Exposure to Low Levels of Ionizing Radiation,* Committee on Biological Effects of Ionizing Radiation (BEIR III), National Academy of Sciences, Washington, D.C., 1980.

2. S. F. Cleary, "Biological Effects of Nonionizing Electromagnetic Radiation," in J. G. Webster, ed. *Encyclopedia of Medical Devices and Instrumentation.* John Wiley & Sons, Inc., New York, 1988.

3. C. A. L. Bassett, S. N. Mitchell, and S. R. Gaston, *J. Am. Med. Assoc.* **247,** 623–628 (1982).

4. N. Wertheimer and E. Leeper, *Am. J. Epidemiol.* **109,** 273–284 (1979).

5. D. A. Savitz, H. Wachtel, F. A. Barnes, E. M. John, and J. G. Tvrdik, *Am. J. Epidemiol.* **128**(1), 21–38 (1988).

6. S. J. London, D. C. Thomas, J. D. Bowman, E. Sobel, L. Cheng, and J. M. Peters, *Am. J. Epidemiol.* **134,** 923–937 (1991).

7. M. Feychting, A. Ahlbom, "Magnetic fields and cancer in people residing near Swedish high voltage power lines," Report prepared for the *Swedish National Board for Industrial and Technical Development,* Karolinska Institute, Stockholm, Sweden, 1992.

8. M. R. Salzberg, S. J. Farish, and V. Delpizzo, *Bioelectromagnetics* **13,** 163–167 (1992).

9. J. L. Kirschvink, A. Kobayashi-Kirschvink, J. C. Diaz-Ricci, and S. J. Kirschvink, *Bioelectromagnetics* **Suppl 1,** 101–113 (1992).

10. D. Wartenberg and D. A. Savitz, *Bioelectromagnetics* **14,** 237–245 (1993).

11. N. Wertheimer and E. Leeper, *Int. J. Epidemiol.* **11,** 345–355 (1982).

12. D. L. Mader and S. B. Peratta, *Bioelectromagnetics* **13,** 287–301 (1992).

13. A. Leonard, R. Neutra, M. Yost, and G. Lee, "Electric and Magnetic Fields: Measurement and Possible Effects on Human Health," *Special Epidemiological Studies Program, California Dept. of Health Services,* Berkeley, Calif., 1990.

14. S. M. Harvey, *Evaluation of Residential Magnetic Field Sources,* 1988 Ontario Hydro Research Report No. E88-17-K.

15. D. A. Savitz, E. E. Calle, *J. Occup. Med.* **29,** 47–51 (1987).

16. G. Theriault, "Health Effects of Electromagnetic Radiation on Workers: Epidemiological Studies," in Peters Birenbaum eds., *Proceedings of the Scientific Workshop on the Health Effects of Electric and Magnetic Fields on Workers.* U.S. Dept. of Health and Human Services, Cincinnati, Ohio, 1991, Publ. No. 91-111.

17. G. Hutchinson, "The Epidemiology of EMF Exposures and Cancer Risks," presented at the *EPRI Conference on Future Epidemiologic Studies of Health Effects of EMF,* EPRI Workshop Agenda, Carmel, Calif., 1991.

18. B. Floderus and co-workers, "Occupational Exposure to Electromagnetic Fields in Relation to Leukemia and Brain Tumors. A Case-Control Study," *Report to the National Institute of Occupational Health,* Solna, Sweden, 1992.

19. M. A. Speers, J. G. Dobbins, and V. S. Miller, *Am. J. Ind. Med.* **13,** 629 38 (1988).

20. S. Preston-Martin, W. Mack, and B. E. Henderson, *Cancer Res.* **49,** 6137–6143 (1989).

21. G. M. Matanowski, P. N. Breysse, and E. A. Elliott, *Lancet* **1,** 737 (1991).

22. P. A. Demers and co-workers, *Am. J. Epidemiol.* **132,** 775–776 (1990).

23. T. Tynes and A. Andersen, *Lancet* **2,** 1596 (1990).

24. B. W. Wilson, L. E. Anderson, D. I. Hilton, and R. D. Phillips, *Bioelectromagnetics* **4,** 293 (1983).

25. B. W. Wilson, E. K. Chess, and L. E. Anderson, *Bioelectromagnetics* **7,** 239–242 (1986).

26. R. J. Reiter in B. W. Wilson, R. G. Stevens, L. E. Anderson, eds., *Extremely Low Frequency Electromagnetic Fields: The Question of Cancer,* Batelle Press, Columbus, Ohio, 87–107 (1990).

27. R. G. Stevens, *Am. J. Epidemiol.* **125,** 556–561 (1987).

28. L. Tamarkin, *Cancer Res.* **41,** 4432–4436 (1981).

29. M. Kato, K. Honma, T. Shigemitsu, and Y. Shiga, *Bioelectromagnetics* **14,** 97–106 (1993).

30. C. Martin, R. Moore-Ede, S. Scott, S. Campbell, R. J. Reiter, eds., *Electromagnetic Fields and Circadian Rhythmicity,* Birkhauser, Boston, Mass., 1992.

31. Z. Davanipour, *Neuroepidemiology* **10,** 308 (1991).

32. N. Wertheimer and E. Leeper, *Am. J. Epidemiol.* **129** (1), 220–224 (1989).

33. R. Wever, "ELF-Effects on Human Circadian Rhythms," in M. Persinger, ed. *ELF and VLF Electromagnetic Field Effects* Plenum Press, New York and London, 1974, pp. 101–144.

34. M. R. Cook, C. Graham, H. D. Cohen, and M. M. Gerkovich, *Bioelectromagnetics* **13,** 261–285 (1992).

35. E. B. Lyskov, J. Juutilainen, V. Jousmaki, J. Partanen, S. Medvedev, and O. Hanninen, *Bioelectromagnetics* **14**, 87–95 (1993).

36. H. Lai, M. A. Carino, A. Horita, and A. W. Guy, *Bioelectromagnetics* **14**, 5–15 (1993).

37. N. Wertheimer and E. Leeper, *Bioelectromagnetics* **7**, 13–22 (1986).

38. J. Juutilainen, P. Matilainen, S. Saarikoski, E. Laara, and S. Suonio, *Bioelectromagnetics* **14**, 229–236 (1993).

39. E. Berman and co-workers, *Bioelectromagnetics* **11**, 169–187 (1990).

40. M. A. Stuchly and co-workers, *Cancer Lett.* **65**, 1–7, 1992.

41. L. E. Anderson, "Biological Effects of Extremely Low-Frequency and 60 Hz Fields," in O. P. Gandhi, ed. *Biological Effects and Medical Applications of Electromagnetic Energy,* Prentice-Hall, Inc., Englewood Cliffs, N.J., 1990, pp. 196–235.

42. I. Nair, M. G. Morgan, and H. K. Florig, *Biological Effects of Power Frequency Electric and Magnetic Fields-Background Paper,* U.S. Congress, Office of Technology Assessment, U.S. Government Printing Office, Wash., DC, May, 1989, OTA-BP-E-53.

43. S. F. Cleary, *Am. Ind. Hyg. Assoc. J.,* **54**, 178–185 (1993).

44. S. F. Cleary, "Biophysical Mechanisms of Interaction," *URSI Review of Radio Science: 1990–1992* (in press).

45. J. P. Fulton, S. Cobb, L. Preble, and co-workers, *Am. J. Epidemiol.* **111**, 292–296 (1980).

46. N. Wertheimer and E. Leeper, *Am. J. Epidemiol.* **112**, 167–168 (1980).

47. Office of Concensus and Population Surveys, *Cancer Statistics Registrations, 1985,* 1990, Series MBI, No. 18.

48. *U.S. National Cancer Institute Cancer Statistics Review,* 1973–1987, NIH, Bethesda, Md., 1990.

49. D. Davis, Discussion comments, in D. E. Davis, D. Hoel, eds., *Trends in Cancer Mortality in Industrial Countries, Annals New York Acad. Science,* **609**, 56 (1990).

50. D. L. Davis, D. Hoel, D. Percy, A. Ahlbom, and J. Schwartz, in D. E. Davis, D. Hoel, eds., *Trends in Cancer Mortality in Industrial Countries, Annals New York Acad. Science,* **609**, 191–204 (1990).

51. L. Kohlmeier, J. Rehm, and H. Hoffmeister, "Lifestyle and Trends in Worldwide Breast Cancer Rates," in D. E. Davis, D. Hoel, eds., *Trends in Cancer Mortality in Industrial Countries, Annals New York Acad. Science,* **609** (1990).

52. J. Cuzick, "International Time Trends for Multiple Myeloma," in D. E. Davis, D. Hoel, eds., *Trends in Cancer Mortality in Industrial Countries, Annals New York Acad. Science,* **609**, 205–214 (1990).

53. S. Milham, "Silent Keys," *Lancet* **1**, 812 (1985).

54. S. Milham, *Am. J. Epidemiol.* **127**, 50–54 (1988).

55. S. Szmigielski and J. Gil, "Electromagnetic Fields and Neoplasms," in G. Franceschetti, O. P. Gandhi, M. Grandolfo, eds., *Electromagnetic Biointeraction,* Plenum Press, New York, 1989, pp. 81–98.

56. S. Szmigielski, M. Bielec, S. Lipski, and G. Sokolska, "Immunological and Cancer-Related Aspects of Exposure to Low-Level Microwave and Radiofrequency Fields," in A. Marino, ed., *Modern Bioelectricity,* Marcel Dekker, New York, Basal, 1988, pp. 861–925.

57. R. S. Lin, P. C. Dischinger, J. Conde, and K. P. Farrel, *J. Occup. Med.* **27**, 413–419 (1985).

58. J. A. Elder, D. F. Cahill, eds., *Biological Effects of Radiofrequency Radiation,* Health Effects Research Laboratory, Office of Research and Development, U.S. Environmental Protection Agency, Research Triangle Park, N.C., 1984, EPA-600/8-83-026F.

59. S. F. Cleary, in O. P. Gandhi, ed., "Cellular Effects of Radiofrequency Electromagnetic Radiation," *Biological Effects and Medical Applications of Electromagnetic Fields,* chapter 14, Prentice-Hall, Inc., Englewood Cliffs, N.J., 1990.

60. S. Szmigielski, A. Szudsnski, A. Pietroszek, M. Bielec, M. Janiak, and J. K. Wrembe, *Bioelectromagnetics* **3**, 179–188 (1982).

61. A. Szudzinski, A. Pietraszek, A. Roszkowski, and S. Szmigielski, *Arch Dermatol. Res.* **274**, 303–311 (1982).

62. C. K. Chou and co-workers, *Bioelectromagnetics* **13**, 460–496 (1992).

63. H. Lai, *Bioelectromagnetics* **13**, 513–526 (1992).

64. S. F. Cleary, *IEEE Eng. in Medicine and Biology EMB 3-06,* 1987, pp. 26–30.

65. E. K. Balcer-Kubiczek and G. H. Harrison, *Rad. res.* **126**, 65–72 (1991).

66. S. F. Cleary, L. M. Liu, and G. Cao, "Functional Alteration of Mammalian Cells by Direct High Frequency Electromagnetic Field Interactions," in M. J. Allen, S. F. Cleary, F. M. Hawkridge, eds., *Charge and Field Effects in Biosystems-2,* **2**, Plenum, New York, London, 1989, pp. 211–221.

67. S. F. Cleary, L. M. Liu, and R. Merchant, *Rad. Res.,* **121**, 38–45 (1990).

68. S. F. Cleary, L. M. Liu, and G. Cao, *Annals N.Y. Acad. Science,* **649**, 166–175 (1992).

69. S. F. Cleary, L. M. Liu, R. Merchant, *Bioelectromagnetics* **11**, 47–56 (1990).

ENERGY AUDITING

NARESH K. KHOSLA
SEMRA T. LOVE
Enviro-Management and Research, Inc.
Springfield, Virginia

An energy audit is the process that first identifies where, how, and why a building uses energy and then recommends energy conservation and savings opportunities that may exist. A general building audit examines building envelope, lighting, heating, air-conditioning, and ventilation requirements. A more detailed audit identifies specific system deficiencies and recommended solutions.

See also ENERGY EFFICIENCY; ENERGY MANAGEMENT.

TYPES OF AUDITS

There are three types of energy audits: walk-through, intermediate, and comprehensive. These audits are different, but they are not defined by sharp boundaries. The type of audit performed determines the level of analysis accorded to the various systems. The following sections provide a more detailed description of each type of audit.

Walk-Through Energy Audit

A walk-through audit is the least expensive energy audit. It evaluates a building's energy costs and efficiency and identifies preliminary energy savings that can be obtained through low-cost capital investments. It typically begins with a review of a building's energy consumption over the previous year. A visual walk-through inspection of the building is then performed to detect obvious opportunities for energy savings in operations and maintenance. The data collected during this audit becomes the foundation for a more detailed analysis in which capital-intensive investments are considered.

A team that includes a facility manager, an energy management consultant, and sometimes a building owner's representative performs the walk-through audits. The team conducts two tours of the building to collect two specific types of data: one during occupied hours and another during unoccupied hours.

The first type of information collected identifies the various energy systems installed, the energy consumption of each system, and technical data from equipment nameplates. This information forms the basis for recommending areas where specific energy savings can be found. The second type of information evaluates the existing condition of systems and subsystems. This information determines opportunities for energy savings based on these conditions. A typical notation might read:

Exterior North Facade: Window on the second floor does not close properly. Caulking around the window frame is cracked and some weatherstripping is missing.

Recommendations resulting from a walk-through audit include: reducing infiltration/exfiltration; improving the efficiency of heating, ventilating, and air-conditioning (HVAC) equipment and controls; modifying the lighting system; and optimizing tenant use practices.

Intermediate Energy Audit

An intermediate audit involves more detailed surveys and analyses of the building than a walk-through audit. Tests and measurements quantify energy uses and losses and estimate the economics of upgrades. The scope of this audit is determined by the building owner's concerns and economic criteria. Because of its adaptable format, an intermediate audit may provide detailed analysis of some systems but only limited information on others. This level of analysis is sufficient for most buildings and provides the following information:

Energy usage: identifies the areas of energy use within the building.

Savings and cost analysis: analyzes the savings and cost measures that would satisfy the owner's concerns and economic criteria.

Operations and maintenance procedures evaluation: examines the effect of operations and maintenance procedures on energy usage in the building.

Capital-intensive improvements list: lists potential capital-intensive improvements for consideration, recognizing that more thorough data collection and analysis is required. This list also includes an initial judgment of potential costs and savings.

Comprehensive Energy Audit

A comprehensive audit is a full-scale audit that evaluates the amount of energy used for each energy system function (such as lighting, HVAC, etc). It involves a comprehensive survey of all systems and identifies additional capital-intensive investments to provide maximum energy and cost savings. It requires highly detailed field data gathering, testing and measurements, engineering analysis, and model analysis (such as computer simulation). A comprehensive audit determines the energy usage of the building and its systems for an entire year and accurately predicts future energy usage. Unlike the walk-through and intermediate audits, a comprehensive audit provides detailed project cost and savings information for major capital investment modifications. This type of audit consists of the following elements:

Envelope Audit: surveys the building envelope for energy losses or gains due to leaks, building construction, doors, glass, lack of insulation, etc.

Functional Audit: determines the amount of energy required for a main system and identifies each system's energy savings potential. Main systems include HVAC, building envelope, lighting, domestic hot water, and air distribution.

Utility Audit: analyzes the daily, monthly, or yearly energy usage for each utility that services the building. This audit typically requires utility data from at least the three previous years.

The data collected from these elements is used to develop an energy-use profile that includes all end-use categories. When these energy-use profiles are analyzed, opportunities for energy savings are then developed and evaluated.

RECOMMENDED PRACTICES

To achieve the most accurate results from an energy audit, the building systems must be inspected thoroughly. Examining the end uses of a system as well as the system itself will provide more valuable information for developing opportunities for energy savings. For example, auditing a building with steam boilers would first observe all or most of the end uses of steam in the building before evaluating the efficiency of the boilers themselves. These end-use observations may uncover considerable quantities of steam being wasted by venting to the atmosphere, venting through defective steam traps, uninsulated lines, and unused heat exchangers. After end-use waste is eliminated, the boiler combustion efficiency can be measured and improvements recommended. However, discretion is required when evaluating end uses because tracking down every end use is not cost-effective.

It is also important to focus on operations and maintenance procedures during the audit. A judicious review of the operating and maintenance logs may suggest improvements to procedures that would provide energy savings in addition to the system savings.

It is always important to evaluate indoor air quality during an energy audit. Evaluating indoor air quality examines the ventilating system's ability to introduce and distribute air through the system, filter airborne contaminants, and maintain acceptable temperature and relative humidity levels. Manifestations of poor indoor air quality include an increasing number of health problems such as cough, eye irritation, headache, allergic reaction, and, in some extreme cases, life-threatening conditions (Legionnaire's disease, carbon monoxide poisoning); decreasing productivity due to discomfort; increasing absenteeism; and accelerating deterioration of furnishings and equipment. An indoor air quality profile should be developed to evaluate the effects of energy savings upgrades on the building's indoor air quality.

Four factors affect the development of indoor air quality problems:

Sources of contamination: There is a source of contamination or discomfort indoors, outdoors, or within the mechanical systems of the building.

HVAC systems: The HVAC system may not control existing air contaminants and ensure thermal comfort (temperature and relative humidity conditions comfortable for most occupants).

Pollutant pathways: One or more pollutant pathways connect the pollutant source to the occupants.

Building occupants: The building occupants are adversely affected by the pollutants.

It is important to ascertain how each of these factors affects the current indoor air quality conditions when investigating, resolving, and preventing indoor air quality problems.

PERFORMING THE ENERGY AUDIT

The exact parameters of an energy audit are based on the size of the building, the complexity of its systems, and budget constraints. The more comprehensive the audit, the more energy and cost saving opportunities it will identify. Guidelines for performing the audit are listed in the next section.

Assembling the Audit Team

Each member of the team will be responsible for completing different tasks in the audit. The team should include design professionals, qualified contractors, the building owner's representatives, operations and maintenance personnel, and utility representatives. Some, such as a representative of the electric utility, may provide off-site assistance.

Depending on the type and size of the audit, it may be necessary to include an architect on the team. For smaller buildings, an architect with mechanical and electrical engineering experience may be the only design professional required. If a significant percentage of the energy usage is from lighting, an illumination engineer should be on the team. If a significant part of the audit centers around the ventilation system, an indoor air quality professional should be included on the team or the consulting engineers should have sufficient experience in analyzing indoor air quality problems.

Design Professionals. For simple buildings, a qualified consultant or contractor with mechanical engineering expertise and electrical engineering experience (or vice-versa) will be the only design professional required. For larger, more complex buildings, a consultant(s) with expertise in both mechanical and electrical engineering will be required and should be designated as the team leader. The consulting engineer will conduct a comprehensive energy audit, develop a report indicating all savings opportunities, perform feasibility studies, provide recommendations and design guidelines for specific systems, and provide an economic analysis of the opportunities for energy savings.

Qualified Contractors. A reputable contractor with extensive experience in energy audits may substitute for the design professional. Contractors can provide information on the latest technology, products, and materials, including purchase and installation expenses, actual field performance evaluations, and the relative ease of operation and maintenance. Contractors should provide the same services as the consulting engineer.

Building Owner's Representatives. Building owner's representatives manage internal concerns. They help the data collection process by providing historical information on the building and its energy consumption. They are the critical liaisons that maintain cooperation between operations and maintenance personnel and provide access to the building for after-hours inspections. If required, the owner's representatives can elaborate on details about current lease agreements that affect entry into key areas, or that pertain to cost-sharing of modifications.

Operations and Maintenance Personnel. The facility manager or the chief operating engineer should escort members of the project team on most of their inspections. Staff members should be available to answer any questions. It is important to maintain amicable relations with members of the operations and maintenance staff.

Utility Representatives. Depending on the policies of the local electric utility, a utility representative may need to be involved with on-site inspections. All appropriate utility representatives should be contacted to obtain information on rate structures.

Obtaining Specific Information About the Building

The consultants performing the audit should have specific information about the building before actually visiting a building. Typically, this information includes the following:

Energy consumption data for previous years: Typically, energy consumption data for the three years prior to the base year is collected and available.

Weather data: Weather data should be obtained for each year that energy consumption data is available. Weather data usually are available from the local electric utility. It should include the daily minimum and maximum temperature and relative humidity readings.

Building data: The building owner/manger or the original consulting engineers or general contractor should provide original building plans and specifications, as-built plans, testing and balancing reports, and commissioning reports.

Operations and maintenance logs, manuals: Operations and maintenance (O&M) logs and equipment manuals provide information on equipment operation and maintenance and the manner in which equipment performance changes have been handled by modifications to O&M procedures. Complaint logs should also be included.

Utility rate schedules: Reviewing rate schedules will determine if modifications can result in better rates. The analysis should reflect future changes; eg, will utility policy changes concerning demand metering equipment or rate schedules affect restructuring? This information is important when formulating an energy savings strategy for the future.

Miscellaneous data: Other information available can assist in the auditing procedure. For example, details about energy conservation modifications that have already been completed should be obtained, along with plans for any future modifications. Of particular importance are any anticipated major tenant turnovers or scheduled physical modifications such as renovation, remodeling, or expansion. Other information pertains to codes that apply to the building and warranties and guarantees on equipment that may be modified.

Investigating On-Site Conditions

The main components of an on-site investigation are (1) testing and measurement and (2) inspection. These are addressed below.

Testing and Measurement. To accomplish the energy audit, qualify building energy uses and losses and perform extensive testing and measurements on each system. The instrumentation needed for this procedure is varied, and depends on the types of measurements required. It includes the following:

Cameras are used to take pictures of the condition of the equipment. The pictures highlight the importance of the modifications and are used in the report.

Infrared scanning devices detect the heat emitted by an object. The higher the temperature emitted by an object, the greater the infrared signature. These devices can locate building heating losses, inspect power transmission equipment, locate water leakage into building roof insulation, check for poor building insulation, inspect cooling coils for plugged tubes, inspect electronic circuits, detect plugged furnace tubes, and find leaks in buried steam lines.

Electrical Measurement Devices. These devices are used when measuring the electrical systems. These include the following:

Ammeters measure alternating current. For example, an ammeter determines the amount of current a motor is using under normal operating conditions. The amount of current used will establish if the motor is oversized or undersized.

Voltmeters measure the difference in electrical potential between two points in an electrical current. For example, a voltmeter would be used to determine the amount of incoming voltage a motor is receiving under normal operating conditions. If the voltage is too low or too high, the motor could fail prematurely.

Wattmeters measure electrical energy directly in watts. For example, a wattmeter uses a motor operating under normal operating conditions to determine how much energy it is using. This will verify if the motor should be replaced with a high efficiency motor that uses less energy.

Power factor meters measure the ratio of real power to apparent power. Utilities typically add a charge to the electric bills of buildings that have a low power factor. The power factor meter finds the sources of the low power factor, which can then be corrected.

Light intensity meters measure illumination in units of footcandles to determine lighting levels in office areas. This will establish where improvements in lighting can be made to reduce utility costs and occupant complaints.

Energy and demand meters measure electrical energy and electrical demand. These devices combine ammeters, voltmeters, and power factor meters into one meter to measure energy and demand. Measuring demand will locate equipment that incurs high demand during the peak demand period. This equipment can then become part of the peak-shaving or demand-limiting program.

Temperature Measurement Devices. These are an important part of any energy audit. These include the following:

Thermometers measure air, liquid, and surface temperatures. Glass-stem thermometers, electronic thermometers, and portable recording thermometers measure space temperature.

Surface pyrometers measure the temperature of various surfaces. They are typically used to measure heat loss through walls and to test steam traps.

Psychrometers measure relative humidity through a relationship between dry-bulb and wet-bulb temperatures. Sling psychrometers are typically used to measure the relative humidity in a space. Some electronic devices combine the thermometer and the psychrometer, taking both readings in the space with just one device.

Air Velocity Measurement Devices. These measure HVAC system performance. Smoke pellets, anemometers, pitot tubes, impact tubes, and heated thermocouples are all used to measure the air velocity on various types of systems and air movement in the space.

Pressure Measurement Devices. These measure pressure in HVAC systems. These include manometers, bourdon tubes, and draft gauges.

Flow Measurement Devices. These measure the flow of water, steam, oil, and gas. These instruments must be inserted into a pipe to take the measurements. Typically, three types of measurements are made: pressure differential, velocity, and positive displacement.

Combustion Efficiency Measurement Devices. They determine the composition of the flue-gas maximizing combustion efficiency. A boiler test kit that evaluates fireside boiler operation contains carbon dioxide, oxygen, and carbon monoxide gas analyzers. To measure stack conditions, an inclined manometer and/or a draft gauge measures pressure. A smoke detector measures the combustion completeness.

Portable PH Meters. PH meters measure the acidity or alkalinity of water treatment systems.

Tachometers and Stroboscopes. They measure the rotating speeds of fans and motors.

Indoor Air Quality Measurements. These measurements are made by thermometers and psychrometers or hygrothermographs to measure temperature and humidity; tracer gas or chemical smoke tubes to find pollutant pathways; a device to measure airborne particles; micromanometers to pressure differentials at fans, intakes, and ducts; flow hoods, anemometers, or velometers to measure the airflow at diffusers; and detector tubes or meters to measure carbon dioxide, carbon monoxide, and other contaminants.

Inspection. Because energy consumption in buildings is determined by the characteristics and interactions of all systems, information on these systems should aid in inspecting and monitoring operations. As equipment is inspected, information that identifies the specific type of equipment installed, the manufacturer, model number, nameplate ratings, current conditions, and other factors as required should be recorded.

Evaluations of maintenance operations and personnel are accomplished by examining equipment records and logs and inspecting the various fields of responsibility. If necessary, arrangements can be made for the team to witness a demonstration or explanation of how a given task is routinely performed by the personnel. Observing the office environment and general routine can provide insight into energy consumption. In the office environment, it is important to observe how people operate their equipment, lighting, appliances, drapes, blinds, etc. It is also important to observe the daily arrivals and departures, lunch breaks, and after-hours occupancy, including duration and equipment used to support the task performed.

The indoor air quality inspection consists of determining HVAC operating and maintenance schedules; inspecting the usage and storage of chemicals; looking for indoor air quality problem indicators, such as odors, unsanitary conditions, visible biological growth, poorly maintained filters, presence of hazardous substances, unusual noises from light fixtures or mechanical equipment, inadequate maintenance, signs of occupant discomfort and overcrowding, or blocked airflows; inspecting HVAC system condition and operation; performing an inventory of pollutant sources and pathways; collecting information on building occupancy; checking for indicators of adequate ventilation; determining whether the layout of air supplies, returns, and exhausts promotes efficient air distribution to all occupants and isolates or dilutes contaminants; and observing air movement direction.

Performing the Computer Simulation

In larger buildings, or those with more complex building systems, the consultant will use computer simulations or modeling to evaluate various opportunities for energy savings. Many computer programs are available and some, such as the DOE-II program, are more accurate than others.

The report from the simulation should provide an accurate estimate of monthly and annual energy consumption. If the monthly and annual data are not accurate, the cause must be determined. Once the program simulates the existing performance with a high degree of accuracy, several different modifications can be simulated to determine the energy savings impact of each modification or a combination of modifications.

Developing the Report

The end result of the audit is an extensive report detailing areas of energy consumption (problems) and listing the opportunities for energy savings (solutions) for each system. The report also contains information on how each energy saving opportunity affects other systems. It provides details on projected annual energy savings, annual dollar savings, the cost of modifications, payback periods, life-cycle costs, and return on investment. It establishes priorities specified by the owner's criteria, details proposed new procedures and types of equipment and controls to be added or modified, and may even provide specific guidelines for the modifications.

A feasibility study is provided whenever a modification involves a significant capital outlay or a main renovation that would result in closure of part of the building for a time (or an action of similar impact). This type of study provides the comprehensive information necessary for decision-making. The feasibility study is typically separate from the initial report and is prepared subsequently to review of the initial study, in full consideration of other alternatives that exist.

The report should also include an indoor air quality profile that can be referred to in planning for renovations, negotiating leases and contracts, or responding to future complaints. This profile should include information on how each opportunity for energy savings affects the indoor air quality.

BIBLIOGRAPHY

Reading List

Building Energy Audits, National Electrical Contractors Association, Washington, D.C., 1979.

Energy for the Year 2000: A Total Energy Management Handbook, National Electrical Contractors Association, Washington, D.C., 1993.

Energy Management and Control Systems, The Electrification Council, Washington, D.C., 1982.

ENERGY CONSERVATION

Dan Steinmeyer
Monsanto Co.
St. Louis, Missouri

The main driver for energy conservation is broadscale technological progress and continues in times of both rising and falling energy prices. It is responsible for the historical rise in energy efficiency of 1 to 3% per year achieved by the process industries. It includes a wide range of big and little steps such as:

- improved gas turbine efficiency
- structured packing in distillation
- computer control
- variable speed drives
- computer design tools
- improved catalysts and processes for a wide range of materials such as low density polyethylene, acrylonitrile, NH_3, and acetic acid.

The second component that has driven energy conservation increases is the trade of capital for energy. This is shown in Figures 1a,b,c,d,e. This trade is optimization within an existing technology and nets large increases when energy prices rise rapidly compared to capital price as it did in the 1975 to 1985 time period.

Many plants also offer opportunities for significant energy conservation by various cogeneration processes (see ENERGY EFFICIENCY; THERMODYNAMICS).

ENERGY BALANCE

An energy balance has historically been prepared for components of a process, primarily to assure that heat exchangers and utility supply are adequate. Often, an overall process energy balance was not developed, but today, the energy balance for the overall process has become a document almost as important as the material balance.

The energy balance should analyze the energy flows by type and amount (ie, electricity, fuel gas, steam level, heat rejected to cooling water, etc). It should include realistic loss values for turbine inefficiencies and heat losses through insulation.

Exergy, Lost Work, and Second-Law Analysis

When energy is critically important to process economics, the simple energy balance is sometimes carried into an analysis of lost work. This compares the actual design against the theoretical ideal at each step and defines where the true energy use (or lost work) is occurring (see Thermodynamics).

In the following discussions of reaction, separation, heat exchange, compression, refrigeration, and steam systems, the importance of this concept is illustrated. A few terms are defined below.

Exergy (E) is the potential to do work. Thermodynamically, this is the maximum work a stream can deliver by coming into equilibrium with its surroundings.

$$E = (H - H_0) - T_0(S - S_0)$$

where E = exergy, the maximum theoretical work potential; H, S = enthalpy and entropy of the stream at its original conditions; H_0, S_0 = enthalpy and entropy of the same stream at equilibrium with the surroundings; and T_0 = temperature of the surroudings (sink).

Exergy is also sometimes called availability or work potential.

Free energy (G) is a related thermodynamic property. It is most commonly used to define the condition for equilibrium in a processing step.

$$\Delta G = \Delta H - T\Delta S$$

It is identical to ΔE if the processing step occurs at T_0.

Lost work (LW) is the irreversible loss in exergy that occurs because a process operates with driving forces or mixes material at different temperature or composition.

$$LW = E_{in} - E_{out}$$

Second-Law Analysis looks at the individual components of an overall process to define the causes of lost work. Sometimes it focuses on the efficiency of a step and ratios the theoretical work needed to accomplish a change (eg, a separation) to that actually used. Sometimes it is more cost effective simply to compare design against a second-law-violation checklist covering items such as (2): mixing streams at different temperatures and compositions; high pressure drops in control valves; reactions run far from equilibrium, high temperature differentials; and pump-discharge recirculation.

REACTOR DESIGN FOR ENERGY CONSERVATION

How close a designer comes to minimum energy is heavily determined by the raw materials and catalyst system chosen. However, if the designer only has the freedom to choose reaction temperature, residence time, and diluent, he or she still has a tremendous opportunity to influence energy use via their effect on yield. But even given none of these, there is still wide freedom to optimize the heat interchange system.

Maximizing Yield. Often the greatest single contribution to reduced energy cost is increased yield. High yield reduces the amount of material to be pumped, heated, and cooled while also simplifying downstream separation. This says nothing about the indirect energy reduction achieved through reduced raw material use. On average, the chemical industry uses almost as much energy in its raw materials as it does in direct purchases of fuel.

Minimizing Diluent. The case concerning diluent is less clear. A careful balance needs to be made of the benefits

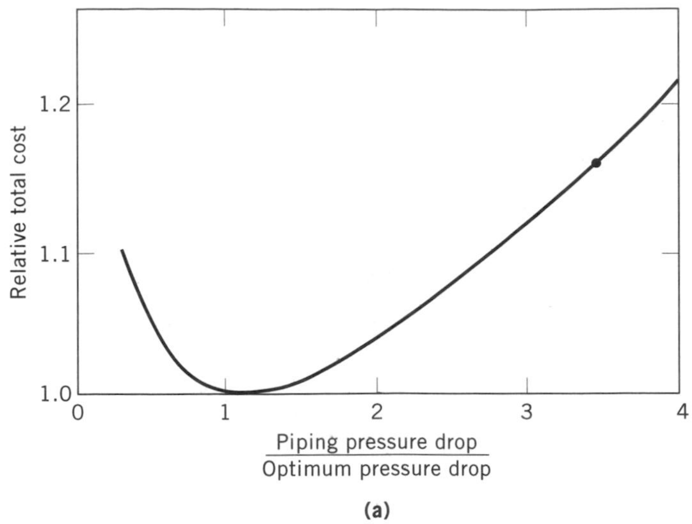

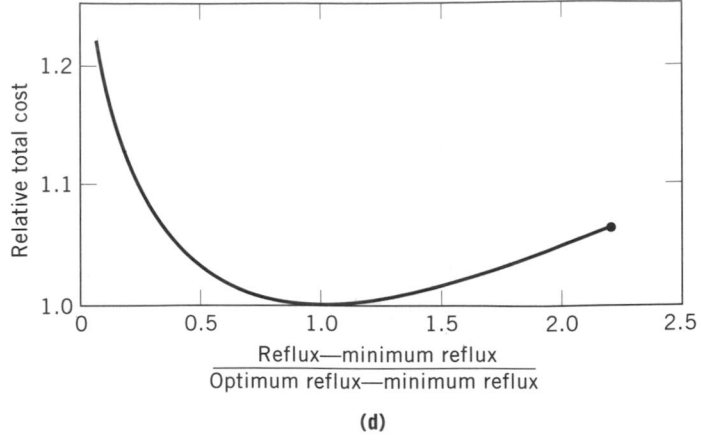

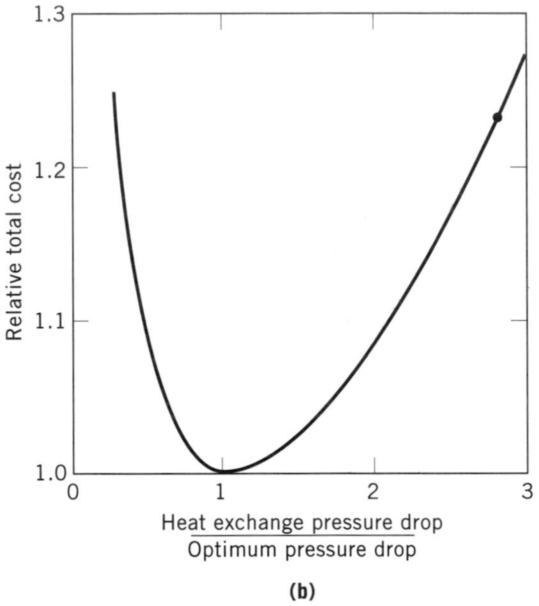

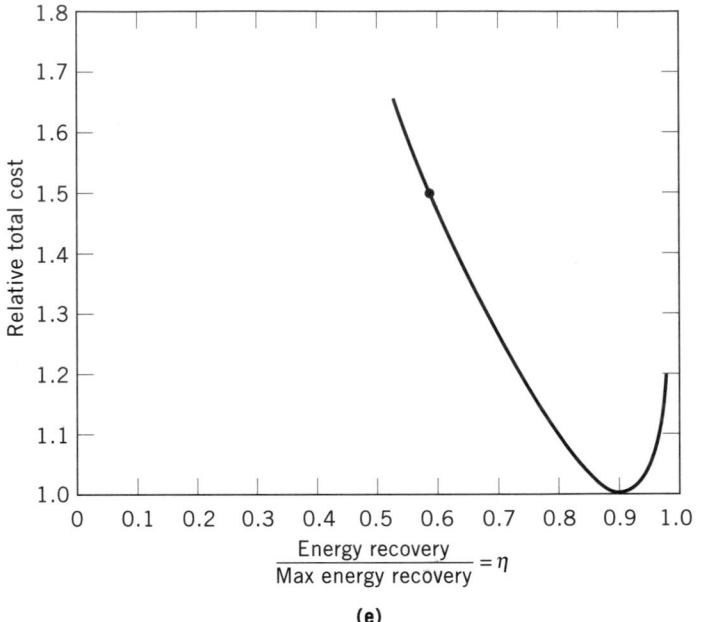

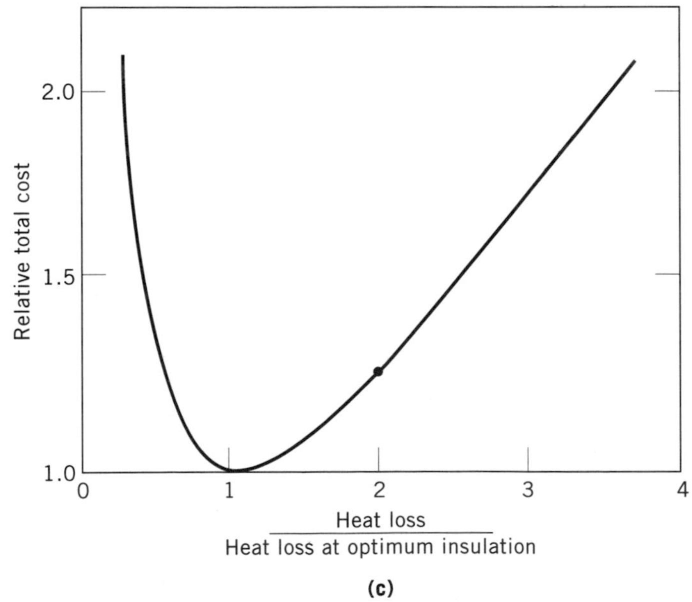

Figure 1. (**a**) Effect of pressure drop in piping on costs; (**b**) effect of pressure drop in exchangers on costs; (**c**) effect of heat loss through insulation on costs; (**d**) effect of reflux use on costs; and (**e**) effect of energy recovery through waste-heat boiler use on cost. Total cost is the sum of capital plus energy costs (for the lifetime of the plant, discounted to present value). Point marked on each graph is the design point if energy price is low by a factor of 4.

it gives in higher yield against the costs in mass handling and separation.

Optimizing Temperature. Temperature is usually dictated by yield considerations and often this opposes the simple dicta of energy: in an endothermic reaction, put the heat in at the lowest practical temperature; in an exothermic reaction, operate (and remove the heat) at the highest possible temperature.

In an exothermic reaction, there is an inevitable loss of work potential which is proportional to the free energy change (ΔG_{T_r}). With all reactants preheated to reaction temperature (T_r) and all heat recovered at T_r (3),

$$\text{lost work} = \Delta G_{T_r}^0 \left(\frac{T_0}{T_r} \right)$$

For highly exothermic reactions, eg, oxidations, this is typically the dominant loss in the entire process.

Heat Recovery and Feed Preheat. The objective, to bring the reactants to and from reaction temperature with the least utility cost and to recover maximum waste heat at maximum temperature is generally achieved by the criteria given below under Heat Exchange. Sometimes, control and safety conditions prevent these criteria from being completely followed. The impact of feed preheating merits a particularly careful look. In an exothermic reaction, it permits the reactor to act as a heat pump, ie, "to buy low and sell high." The most common example is combustion-air preheating for a furnace.

Batch vs Continuous Reactors. Usually, continuous reactors yield much lower energy use because of increased opportunities for heat interchange. Sometimes the savings are even greater in downstream separation units than in the reaction step itself.

Especially on batch reactors, the designer should critically review any use of refrigeration to remove heat. Batch processes often have evolved little from the laboratory-scale glassware where refrigeration was a convenience.

SEPARATION

About one third of the chemical industry's energy is used for separation. A correlation exists between selling price and feed concentration (Fig. 2) (4). This in turn can be traced to the minimum work of separation.

$$W = RT_0 \, \Sigma N_i \ln (x_i \gamma_i)$$

where T_0 is the sink temperature; N_i is the number of moles of a species present in the feed; $x_i = N_i/\Sigma N_i$; and γ_i is an activity coefficient.

This looks complicated, but actually, it provides a target that is easily calculated and approachable in practice. For example, work calculated from this expression closely approaches the performance of a real-world distillation after inefficiencies for driving forces are taken into account as illustrated below.

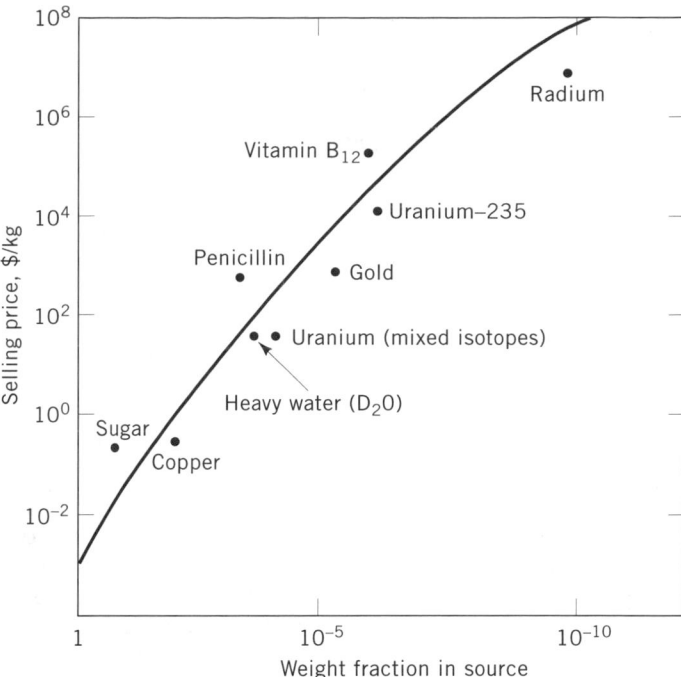

Figure 2. Commercial selling prices of some separated materials (6). Courtesy of McGraw-Hill Book Company.

For ideal solutions ($\gamma_i = 1$) of a binary mixture, this simplifies to:

$$W = RT_0[x_1 \ln x_1 + (1 - x_1) \ln (1 - x_1)]$$

This applies whether the separation is by distillation or any other technique.

When a separation is not completed, less work is required. For x_1 equal to 0.5:

Product purity, %	Relative work
100	1
99.9	0.99
99.0	0.92
90.0	0.53

Note that it takes only a little work to move from 99% separation to 100%, but a great deal to move from 90% to 100%. This is important to recognize when comparing separation techniques. Some leave much of the work undone, as for example, in crystallization involving an unseparated eutectic mixture.

Distillation

Distillation is overwhelmingly the most common separation technique because of its inherent advantages: phase separation is clean; it is relatively easy to build a multistage, countercurrent device; and equilibrium is closely approached in each stage.

Minimum work for an ideal separation at first glance appears unrelated to the slender vertical vessel with a condenser at the top and a reboiler at the bottom. The connection becomes evident when one calculates the work embedded in the heat flow which enters the reboiler and leaves at the condenser. An ideal engine could extract work from this heat.

$$\text{work potential} = QT_0 \left[\frac{1}{T_{\text{condenser}}} - \frac{1}{T_{\text{reboiler}}} \right]$$

Comparison of actual use of work potential against the minimum allows calculation of an efficiency relative to the best possible separation:

$$\eta = \frac{RT_0[x_1 \ln_1 + (1 - x_1) \ln (1 - x_1)]}{QT_0 \left[\dfrac{1}{T_{\text{condenser}}} - \dfrac{1}{T_{\text{reboiler}}} \right]}$$

There is still no obvious reason to believe that the efficiency of separating a mixture with an α (relative volatility) of 1.1 will be related to that for an α of 2; however, it is known that when α is small, the required reflux and Q are large, but ($T_{\text{condenser}} - T_{\text{reboiler}}$) is small (see Distillation).

The two effects almost cancel to yield an approximation for the minimum thermal work used in a distillation (4–5).

$$\text{min thermal work} = RT_0(1 + [\alpha - 1]x_1)$$

When this is combined with the definition of minimum separation work for an ideal binary, an approximation for distillation efficiency can be obtained:

$$\eta = \frac{x_1 \ln x_1 + (1 - x_1) \ln (1 - x_1)}{1 + (\alpha - 1)x_1}$$

This efficiency is high and shows only minor dependence on α over a broad range of α. For $x_1 = 0.5$,

$\alpha = 1.1$	$\alpha = 1.5$	$\alpha = 2$
$\eta = 0.66$	$\eta = 0.55$	$\eta = 0.46$

The dependence on x_1 is greater:

	$\alpha = 1.05$	$\alpha = 2$
η for $x_1 = 0.1$	0.32	0.20
η for $x_1 = 0.01$	0.056	0.053

This suggests that distillation should be the preferred method for feed concentrations of 10–90% and that it is probably a poor choice for feed concentrations of less than 1%. This matches experience. Techniques such as adsorption, chemical reaction, and ion exchange are chiefly used to remove impurity concentrations <1%.

The high η values above conflict with the common belief that distillation is always inherently inefficient. This belief arises mainly because past distillation practices utilized such high driving forces (pressure drop, reflux ratio, and temperature differentials in reboilers and condensers). A real example is instructive:

	C_2 splitter (relative numbers)
Theoretical work of separation	1.0
Min thermal work potential used	1.4

$$\eta = \frac{\text{Theoretical work}}{\text{Min thermal work}} = \frac{1.0}{1.4} = 0.7$$

Losses for driving forces:	0.1
Reflux above the minimum	0.1
Exchanger ΔT	2.1
ΔP in tower	0.5
ΔP in condenser and tower	0.8
Total losses	3.5

$$\eta_{\text{Including losses}} = \frac{1.0}{1.4 + 3.5} = 0.2$$

These numbers show first that the theoretical work can be closely approached by actual work after known efficiencies are identified, and second, that the dominant driving force losses are in pressure drop and temperature difference. This is a characteristic of towers with low relative volatilities.

What Does Optimum Design Look Like?

Condenser and Reboiler ΔT's. As shown by this example, the losses for ΔT typically are far greater than those for reflux beyond the minimum. As shown below under Heat Exchange, the economic optimum for temperature differential is typically under 15°C. This contrasts with the values over 75°C often used in the past. This is probably the biggest opportunity for improvement in the practice of distillation.

Adjusting the Process to Optimize ΔT. First glance may show only three or four utility levels (temperatures) to choose from. These might well be 100°C apart. Some ways to increase the options are to consider multieffect distillation, which spreads the ΔT across two or three towers; to use waste heat for reboil; and to recover energy from the condenser. Often to make some of these possible, the pressure in a column may have to be either raised or lowered.

Reflux Ratio. A number of studies have shown that the optimum reflux ratio is generally below 1.15 minimum. At this point, excess reflux is a minor contributor to column inefficiency. When designing to this tolerance, correct vapor–liquid equilibrium (VLE) and adequate controls are essential.

State-of-the-Art Control. Computer control with feedforward capability can save 5–20% of a unit's utilities. It does this by reducing the margin of safety. This sounds hard to believe, particularly for systems designed to operate within 10% of minimum reflux; however, unless the discipline of a controller forces this to happen, operators typically opt for increased safety. They are probably right to do so unless the proper set of analyzers and controllers is provided and maintained.

Right Feed Enthalpy. Often it is possible to heat the feed with a utility considerably less costly than that used for bottom reboiling. Sometimes the preheating can be directly integrated into the column-heat balance by exchange against the condensing overhead or against the net bottoms from the column. Simulation and a careful look at the overall process are required to assess the value of feed preheating accurately.

A vapor feed is favored when most of the column feed leaves the tower as overhead product. The use of a vapor feed was a key component in the high efficiency cited above for the C_2 splitter where most of the feed goes overhead.

Low Column-Pressure Drop. The penalty for column-pressure drop is an increase in temperature differential

$$\Delta T = \left(\frac{dT}{dP}\right)\Delta P$$

$$\frac{dT}{dP} = \frac{R}{\Delta H}\frac{T^2}{P}$$

As this suggests, the penalty becomes very large for low vapor pressure materials, ie, for components that are distilled at or below atmospheric pressure. The work penalty associated with this ΔT is approximately defined by the ratio:

$$\frac{\Delta T \text{ for pressure drop}}{T_{\text{reboiler}} - T_{\text{condenser}}} = \text{fraction of work potential for } \Delta P$$

This penalty is severest for close-boiling mixtures. The most powerful technique for cutting ΔP is the use of packing. Conventional packings such as 5-cm (2-in.) pall rings can achieve a factor of four reduction over trays, and structured packing can give a factor of 10.

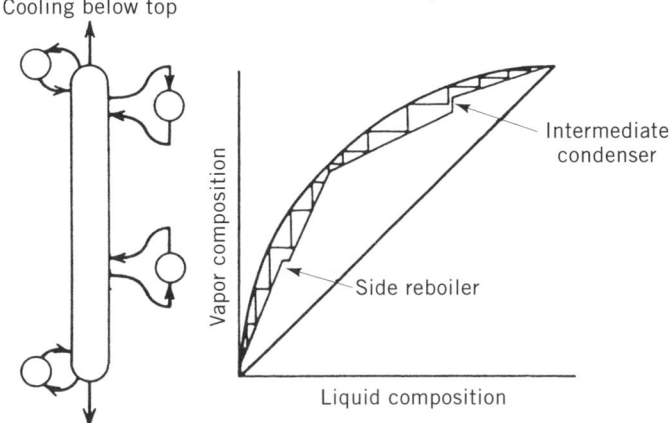

Figure 3. Intermediate condenser and reboiler.

In applying this analysis, one must consider the overhead line and condenser pressure drop as well. (Note the high loss in the C_2 splitter example.)

Intermediate Condenser. As shown by Figure 3, an intermediate condenser forces the operating line closer to the equilibrium line, thus reducing the inherent inefficiencies in the tower. With the use of intermediate condensers and reboilers, it is possible to exceed the efficiency given above, particularly when the feed composition is far from 50:50 in a binary mixture.

	Max efficiency 50% of heavy component in feed	Max efficiency 95% of heavy component in feed
1 condenser, 1 reboiler	67	20
2 condensers, 1 reboiler	73	47
3 condensers, 1 reboiler	77	62

The intermediate condenser is most effective when the feed leaves the column mainly as bottoms. This approach is economical if a less costly coolant can be substituted for refrigeration, or if it permits reuse of the heat of condensation.

Intermediate Reboiler. Inclusion of an intermediate reboiler moves the heat-input location up the column to a slightly colder point. It can permit use of waste heat for reboil when the bottoms temperature is too hot for the waste heat.

Heat Pumps. Because of added capital and complexity, heat pumps are rarely economical.

Lower Pressure. Usually, relative volatility increases as pressure drops. For some systems, a 1% drop in absolute pressure cuts required reflux by 0.5%. Again, if operating at reduced pressure looks promising, it can be evaluated by simulation (see Simulation and process design).

Steam Stripping. Steam (or other stripping gas) and vacuum are largely interchangeable. Steam stripping allows more tolerance for pressure drop, but this comes at the penalty of much higher energy use.

Other questions:

Is the separation necessary?

Is the purity necessary?

Are there any recycles that could be eliminated?

Can the products be sent directly to downstream units, thereby eliminating intermediate heating and cooling?

Other Separation Techniques

Under some circumstances, distillation is not the best method of separation. Among these instances are the following: when relative volatility is <1.05, as often happens with isomers; when ca 1% of a stream is removed, as in gas drying (adsorption or absorption) or C_2H_2 removal (reaction or absorption); when thermodynamic efficiency of distillation is <5% (for whatever reason); and when a high boiling point pushes thermal stability limits. A variety of other techniques may be more applicable in these cases.

Reaction. Purification by reaction is relatively common when concentrations are low (ppm) and a high energy but a low value molecule is present. Some examples are hydrogenation of acetylene and oxidation of waste hydrocarbons:

$$C_2H_2 + H_2 \rightarrow C_2H_4$$

$$\text{waste hydrocarbon} + O_2 \rightarrow H_2O + CO_2$$

Absorption (Extractive Distillation). As a separation technique, absorption starts with an energy deficit because it mixes in a pure material (solvent) and then separates it again. It is nevertheless quite common, because it shares most of the physical property advantages of distillation, and because it separates by molecular type it can be tailored to obtain a high α. The following ratios are suggested for equal costs (6):

$\alpha_{distillation}$	$\alpha_{extraction\ distillation}$	$\alpha_{extraction}$
1.2	1.4	2.5
1.4	1.9	5
1.6	2.3	8

In practice most of the applications have come where a small part (<5%) of the feed is removed. Examples include H_2S/CO_2 removal and gas drying with a glycol (see Azetropic and extractive distillation).

Extraction. The advantage of extraction is that it purifies a liquid rather than a vapor, allowing operation at lower temperatures and removal of a series of similar molecules at the same time, even though they differ widely in boiling point. An example is the extraction of aromatics from hydrocarbon streams.

The disadvantage of extraction relative to extractive distillation is the greater difficulty of getting high efficiency countercurrent processing.

Adsorption. Adsorbents can achieve even more finely tuned selectivity than extraction. The most common application is the fixed bed with thermal regeneration, which is simple, attains essentially 100% removal, and carries little penalty for low feed concentration. An example is gas drying. A variant is pressure-swing adsorption. Here, regeneration is attained by a drop in pressure. By use of multiple stages, high impurity rejection can be achieved, but at the expense of also losing part of the desired product.

Another adsorption approach is the simulated moving-bed system. It has large-volume applications in normal-paraffin separation and *para*-xylene separation. Since its introduction in 1970, it has largely displaced crystallization in xylene separations. The unique feature of the system is that the bed is fixed but the feed point shifts to simulate a moving bed.

Melt Crystallization. Crystallization from a melt is inherently attractive in competition to distillation because the heat of fusion is much lower than that of evaporation. It also benefits from lower operating temperature. In addition, organic crystals are virtually insoluble in each other so that a pure product is possible in a one-stage operation.

However, crystallization has a unique set of disadvantages that generally outweigh the virtues and have sharply limited its application. Industry practice suggests the use of a workable alternative, if one exists. The disadvantages of melt crystallization are as follows:

Physical separation is difficult. Impure liquid is trapped as occlusions as the crystals grow, and liquid wets all surfaces.

It requires a second separating process to separate the eutectic mixture. In some senses, it is akin to formation

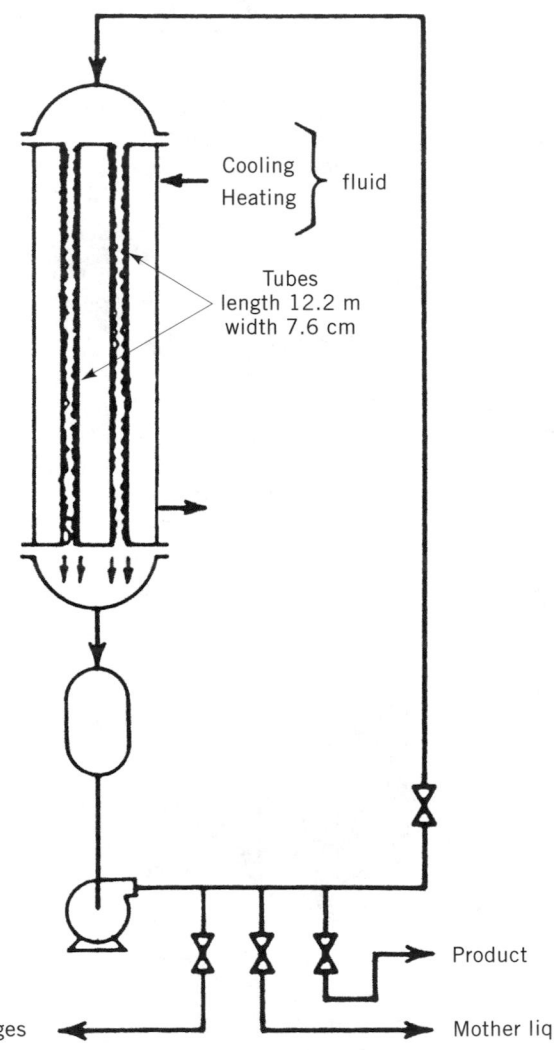

Figure 4. Falling-film crystallizer, semibatch.

of two liquid phases: little energy is required to get the two phases, but a great deal is required to finish the purification.

It is difficult to add or remove heat because of the thermal resistance of the crystal.

It is difficult to move the liquid countercurrent to the crystals.

Thermodynamic efficiency is hurt by the large ΔT between the temperatures of melting and freezing. In an analogy to distillation, the high α comes at the expense of a big spread in reboiler and condenser temperature. From a theoretical standpoint, this penalty is smallest when freezing a high concentration (ca 90%) material.

One process is shown in Figure 4. It is a semibatch operation in which liquid falls down the walls of long tubes. This permits both staged operation and sweating

of crystals. Typically, the sweating and staged operation require melting 5 kg of material for each kilogram of product (7).

Membranes. Membrane separators are used in purification of hydrogen and air. The working principle is a membrane that is chemically tuned to pass molecular type (see Fig. 5).

Liquid separation via membranes (reverse osmosis) is used in production of pure water from seawater. The chief limit to broader use of reverse osmosis is the high pressure required as the concentration of reject rises.

Mole fraction of reject	Min ΔP, MPa (psi)
0.05	7.6 (1100)
0.10	15.2 (2200)
0.20	31.8 (4600)

For most processes, the probability of finding a membrane with requisite strength, as well as selectivity and permeability, is low, and most systems are limited to achieving a mole fraction reject below 0.10.

HEAT EXCHANGE

Most processing is thermal. Reaction systems and separation systems are typically dominated by their associated heat exchange. Optimization of this heat exchange has a tremendous leverage on the ultimate process efficiency (see Heat-exchange technology).

Heat exchangers use energy two ways: as frictional pressure drop and as the loss in ability to do work when heat is degraded.

$$\text{lost work} = QT_{\text{sink}} \left(\frac{1}{T_{\text{cold}}} - \frac{1}{T_{\text{hot}}} \right) + \text{frictional work}$$
$$(\text{for } \Delta\text{P})$$

In an optimized system the lifetime value of the lost work associated with Δt typically equals the cost of the heat exchanger. The lifetime value of the ΔP lost work in an optimized system is $\frac{1}{3}$ as great as the heat exchanger capital.

The selection of design numbers for ΔP and ΔT is frequently the most important decision the process designer makes, but often this goes unrecognized.

The designer commonly becomes lost in the detail of tube length and baffle cut in an effort to optimize the hardware to meet a target and spends far too little time on choosing that target.

Heat-Exchange Networks

A basic theme of energy conservation is to look at a process broadly, ie, to look at how best to combine process elements. The heat-exchange network analysis can be a useful part of this optimization Figure 6 illustrates a simple example of the basic concept of what network analysis does; it builds cumulative heating and cooling curves and merges them until a minimum ΔT is reached.

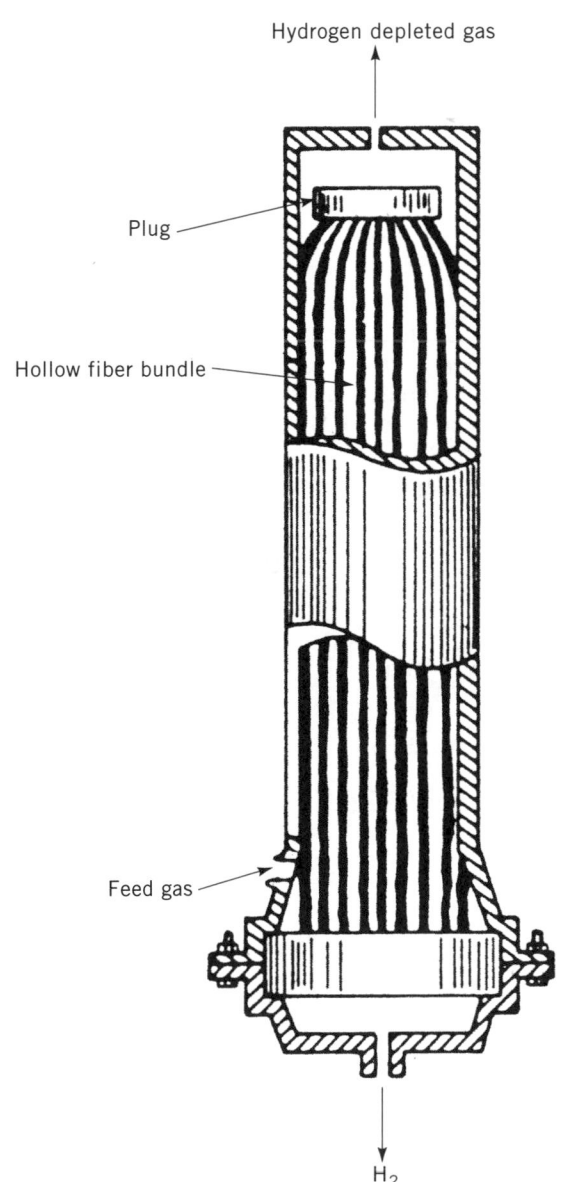

Hydrogen depleted gas

Plug

Hollow fiber bundle

Feed gas

H₂

Figure 5. Membrane hydrogen purifier.

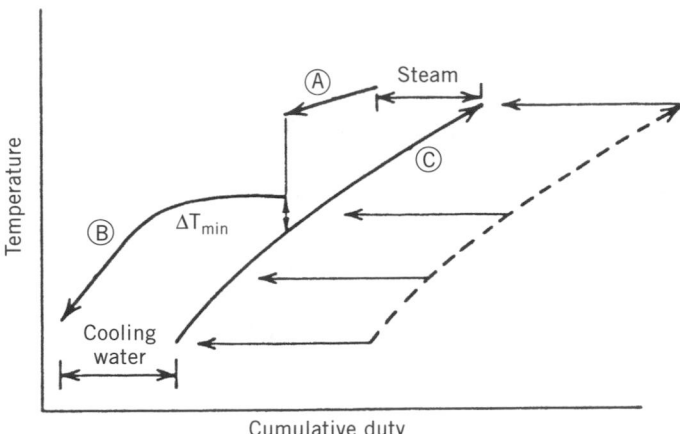

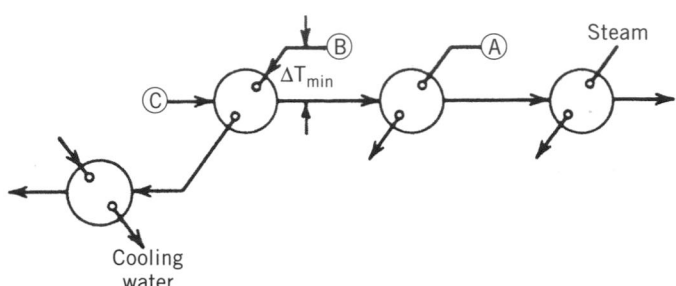

Figure 6. Simple heat-exchange network. Stream C is heated, and streams A and B are cooled.

This example also illustrates one of the targets of network analysis, ie, obtaining the minimum number of shells:

minimum shells = no. of streams − 1

= 3 process streams + 2 utilities − 1 = 4

Network analysis (or pinch technology) has become an increasingly powerful approach to process design that includes most of the virtues of 2nd law analysis (see ref. 8).

Overdesign

Overdesign also has a great impact on the cost of heat exchange and sometimes is confused with energy conservation. The best approach is to define clearly what the objective of overdesign is and then to explicitly specify it. If the main concern is a match to other units in the system, a multiplier is applied to flows. If the concern is with the heat balance or transfer correlation, the multiplier is applied to area. If the concern is fouling, a fouling factor is called for. But if low ΔT or ΔP is the principal concern, that should be specified. Adding extra surface saves energy only if the surface is configured to do so. Doubling the area may do nothing more than double the ΔP, unless it is configured properly.

ΔT and ΔP Optimization

Ideally, ΔT and ΔP are optimized by trying several values, making preliminary designs, and finding the point where savings in utility costs just balance incremental surface costs. Where the sums at stake are large, this should be done. However, for many cases the simple guidelines given below are adequate. The primary focus is the impact of surface and utility prices; a secondary focus is the impact of fluid properties on heat-transfer coefficient (9).

Optimum ΔT. There are three general cases of high importance: the waste-heat boiler, in which only one fluid involves sensible heat transfer, ie, a temperature change; the feed–effluent exchanger, in which both fluids involve sensible heat transfer and are roughly balanced, ie, undergo essentially the same temperature change; and the reboiler, in which neither fluid involves a temperature change, ie, one fluid condenses and the other boils.

Waste-Heat Boiler

In a waste-heat boiler (Fig. 7), the optimum ΔT occurs when

$$\Delta T_{\text{approach}} = \frac{K_1}{K_v} \frac{1.33}{U}$$

where K_1 = annualized cost per unit of surface, \$/(m^2·yr); K_v = annualized cost per unit of utility saved, \$/(W·yr); and U = heat transfer coefficient, W/(m^2·K). The factor 1.33 includes the value of the pressure drop for the added surface.

For example, the optimum $\Delta T_{\text{approach}}$ is computed

$$K_1 = \frac{\$215/\text{m}^2}{2 \text{ yr}} = \frac{\$107.5}{(\text{m}^2 \cdot \text{yr})}$$

$$K_v = \frac{0.017}{\text{kW} \cdot \text{h}} \cdot 8322 \text{ h/yr}$$

$$= \frac{\$142}{(\text{kW} \cdot \text{yr})} = \frac{\$0.142}{(\text{W} \cdot \text{yr})}$$

$$\Delta T_{\text{approach}} = \frac{107.5/(\text{m}^2 \cdot \text{yr})}{0.142/(\text{W} \cdot \text{yr})} \frac{1.33}{56.8 \text{ W}/(\text{m}^2 \cdot \text{K})} = 17.7 \text{ K}$$

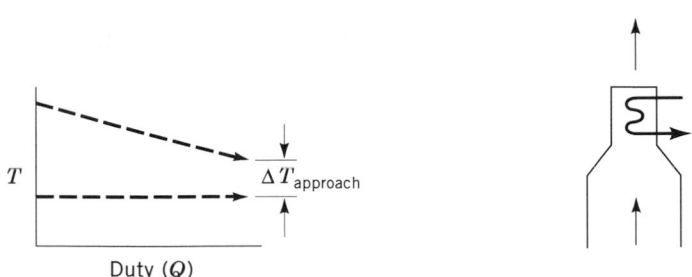

Figure 7. ΔT in a waste-heat boiler.

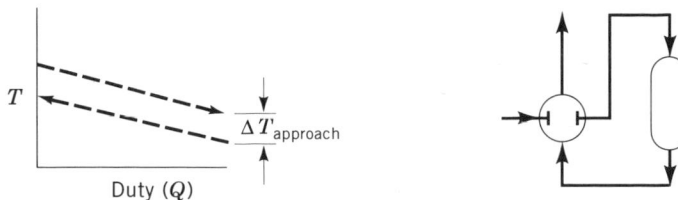

Figure 8. ΔT in a feed–effluent exchanger.

where U = 56.8 W/(m^2·K) (10 Btu/(h·ft^2·°F)); surface cost = \$215/m^2 (\$20/ft^2); payout time = 2 yr; energy price \$0.017/kW·h (\$5/10^6 Btu); and onstream time = 8322 h/yr. This case underlines a dramatic change in process design. Note that $\Delta T_{approach}$ varies to the first power of the ratio of surface price to energy price. The most visible result has been a change in typical heater design efficiency from 65–75% to 92–94%. A secondary result has been appearance of waste-heat recovery units in many processes at the point where air coolers were once used.

Feed–Effluent Exchanger

The detailed solution for the optimum ΔT in a feed–effluent exchanger (Fig. 8) involves a quadratic equation for $\Delta T_{approach}$, but within the restrictions

$$0.8 < \frac{\Delta T_{hot}}{\Delta T_{cold}} < 1.25$$

$$\frac{T_{hot_{in}} - T_{cold_{in}}}{\Delta T_{log\,mean}} < 10$$

An excellent approximation is given by

$$\Delta T_{log\,mean} = \left[\frac{K_1}{K_v} \frac{1.33}{U} (T_{hot_{in}} - T_{cold_{in}}) \right]^{1/2}$$

For example, the optimum $\Delta T_{log\,mean}$ for a feed–effluent exchanger is computed

$$K_1 = \frac{\$107.5/m^2}{2\,yr} = \$53.8/(m^2 \cdot yr)$$

$$K^v = \frac{\$0.027}{kW \cdot h} \cdot 8322 \frac{h}{yr} \cdot 1000 \frac{kW \cdot h}{W} = \frac{\$0.227}{W \cdot yr}$$

$$\Delta T_{log\,mean} = \left[\frac{53.8}{0.227} \frac{1.33}{284} (200 - 100) \right]^{1/2} = 10.5°C$$

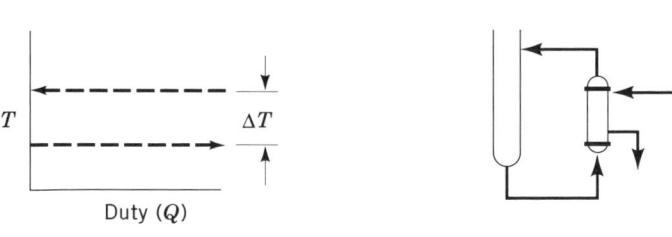

Figure 9. ΔT in the reboiler.

where $T_{hot_{in}}$ = 200°C; $T_{cold_{in}}$ = 100°C; U = 284 W/(m^2·K) (50 Btu/(h·ft^2·°F)); surface cost = \$107.5/m^2 (\$10/ft^2); payout time = 2 yr; energy price = \$0.027/kW·h (\$8/10^6 Btu); onstream time = 8322 h/yr; and $\Delta T_{hot}/\Delta T_{cold}$ = 1.20.

The Reboiler

The case shown in Figure 9 is common for reboilers and condensers on distillation towers. Typically, this ΔT has a greater impact on the excess energy use in distillation than does reflux beyond the minimum. The capital cost of the reboiler and condenser is often equivalent to the cost of the column they serve.

The concept of an optimum reboiler or condenser ΔT relates to the fact that the value of energy changes with temperature. As the gap between supply and rejection widens, the real work in a distillation increases. The optimum ΔT is found by balancing this work penalty against the capital costs of bigger heat exchangers.

If the Carnot cycle is used to calculate the work imbedded in the thermal flows with the assumption that the heat-transfer coefficient (U) is constant and that process temperature is much greater than ΔT,

$$\Delta T_{optimum} = T_p \left[\frac{K_1}{K_v U T_{sink}} \right]^{1/2}$$

where T_{sink} = temperature (absolute) at which heat is rejected; T_p = process temperature (absolute); K_1 = annualized cost per unit of surface; and K_v = annualized value of power.

For utilities above ambient temperature,

$$K_v = K_p \text{ (turbine efficiency)}$$

where K_p is the annualized cost of purchased power. The above relations will typically give ΔT values in the 10°C to 20°C range.

One strong caution is that the assumption of a constant U is usually inaccurate for boiling applications. Simulation is generally needed to fix ΔT accurately, particularly at ΔT values below 15°C.

Optimum Pressure Drop

For most heat exchangers, there is an optimum pressure drop. This results from the balance of capital costs against the pumping (or compression) costs. A common predjudice is that the power costs are trivial compared to the capital costs. The total cost curve is fairly flat within ± 50% of the optimum (see Fig. 1b), but the incremental costs of power are roughly one third of those for capital on an annualized basis. This simple relationship can be extremely useful in quick design checks.

The best approach is to have a computer program check a series of pressure drops and see how energy requirements decrease as surface increases. If this option is not available, the simple method shown below can be used to obtain specification sheet values.

Start with a pressure drop of 6.9 kPa (1 psi), and apply three correction factors ($F_{\Delta T}$, F_{cost}, F_{prop}):

$$\Delta P_{opt} = 6.9 \, (F_{\Delta T})(F_{cost})(F_{prop})$$

The correction for temperature difference is given by

$$F_{\Delta T} = \frac{T_{in} - T_{out}}{(T_{hot} - T_{cold})_{mean}}$$

This term is a measure of the unit's length. Sometimes it is referred to as the number of transfer units (see Mass Transfer).

The correction for costs is

$$F_{cost} = 0.017 \left(\frac{\$/(m^2 \cdot yr)}{\$/kW \cdot h} \right)^{0.75}$$

The correction for physical properties is

$$F_{prop} = \left(\frac{c}{c_w} \right)^{0.6} \left(\frac{k_w}{k} \right)^{0.6} \left(\frac{\mu}{\mu_w} \right) \left(\frac{\rho}{\rho_w} \right)^{0.5}$$

where c = specific heat; k = thermal conductivity; μ = viscosity; ρ = density; and c_w, k_w, μ_w, and ρ_w are the same properties for water at 25°C.

From these equations, the optimum ΔP for a feed–effluent exchanger where the fluid has the physical properties of water and:

$$T_{in} - T_{out} = 20°C$$

$$\Delta T = (T_{hot} - T_{cold})_{mean} = 10°C$$

$$\left. \begin{array}{l} \text{surface cost} = \$215/m^2 \\ \text{payout time} = 2 \text{ yr} \end{array} \right\} \$107.5/(m^2 \cdot yr)$$

$$\text{power cost} = \$0.03/kW \cdot h$$

is calculated:

$$F_{\Delta T} = \frac{20}{10} = 2$$

$$F_{cost} = 0.017 \left(\frac{107.5}{0.03} \right)^{0.75} = 7.8$$

$$F_{prop} = 1$$

$$\Delta P_{opt} = 6.9(2)(7.8)(1) = 107.67 \text{ kPa } (15.6 \text{ psi})$$

Table 1. Impact of Fluid Density on Optimum ΔP

	$\dfrac{\rho}{\rho_w}$	$\left(\dfrac{\rho}{\rho_w} \right)^{0.5}$	=	Relative Optimum ΔP
Water	1	$(1)^{0.5}$		= 1
Oil	0.8	$(0.8)^{0.5}$		= 0.9
High pressure gas	0.05	$(0.05)^{0.5}$		= 0.22
Atm pressure gas	0.001	$(0.001)^{0.5}$		= 0.03
Vacuum	0.0002	$(0.0002)^{0.5}$		= 0.014

If all else remains the same as above except a gas is obtained with the following properties: $\mu = 0.02 \, \mu_w$; $\rho = 0.00081 \, \rho_w$; $c = 0.25 \, c_w$; and $k = 0.066 \, k_w$; then,

$$F_{prop} = (0.25)^{0.6} \left(\frac{1}{0.066} \right)^{0.6} (0.02)^{0.3}(0.00081)^{0.5} = 0.0194$$

$$\Delta P_{opt} = 6.9(2)(7.8)(0.0194) = 2.1 \text{ kPa } (0.3 \text{ psi})$$

The great impact of density in this example and in Table 1 should be noted. Probably the most common specification error is to use the large ΔPs characteristic of liquids in low density gas systems.

Fired Heaters

The fired heater is first a reactor and second a heat exchanger. Often, in reality, it is a network of heat exchangers (see Burner technology; Furnaces, fuel-fired).

The Fired Heater as a Reactor. When viewed as a reactor, the fired heater adds a unique set of energy considerations.

Can the heater be designed to operate with less air by O_2 and CO analyzers?

How does air preheat affect fuel use and efficiency?

How could a lower cost fuel (coal) be used?

Can the high energy potential of the fuel be used upstream in a gas turbine?

CO Control. Control of excess air by carbon monoxide is one of the most significant new energy technologies. The key is that CO is highly sensitive to excess air as shown by Figure 10.

Coal and Low Btu Gas. The much lower cost of coal has caused a rapid resurgence of coal-firing for steam generation. Its direct use in process heaters has been negligible

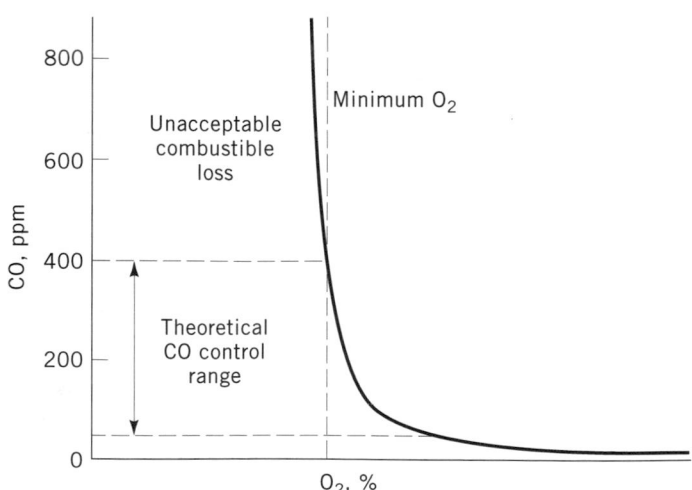

Figure 10. Relationship of CO concentration to O_2 concentration, in a fired heater.

Table 2. Comparison of Combustion Characteristics of Natural and Low Btu Gas

	Natural Gas	Low Btu Gas
Relative heating value per volume		
Fuel	1	0.17
Fuel and air mixture	1	0.78
Relative vol air:vol fuel	1	0.13
Relative flue-gas flow rate/unit of energy	1	1.11
Adiabatic flame temperature, °C	1927	1760

because of historical prejudice and the space and capital demands of coal-handling facilities. Concern about attack on high alloy tubing by ash also precludes its direct use in applications such as ethylene furnaces.

Low Btu gas appears to be an attractive way of utilizing coal in process heaters for a number of reasons: Its technology is old and well established; it keeps the ash out of the process heater; and it requires only mild retrofitting, chiefly of burners. The flame temperature and total flue-gas generation for low Btu gas are both tolerably close to natural gas as shown in Table 2.

Air Preheating. Use of unpreheated air in the combustion step is probably the biggest waste of thermodynamic potential in industry (see Table 3). It is not practical to preheat to the flame burst temperature, the optimum thermodynamic situation, but some preheating is invariably profitable. Air preheating has the unique benefit of giving a direct cut in fuel consumed. It also can increase the heat-input capability of the firebox because of the hotter flame temperature.

The most common type of air preheater on new units is the rotating wheel. On retrofits, heat pipes or hot-water loops are often more cost-effective because of duct-work costs or space limits.

Upstream Firing of Fuel, Gas Turbines. Limitations in the material of construction make it difficult to use the high temperature potential of fuel fully. This restriction has led to insertion of gas turbines into power-generation steam cycles and even to use of gas turbines in preheating air for ethylene-cracking furnaces.

The Fired Heater as a Heat-Exchange System

Improved efficiency in fired heaters has tended to focus on heat lost with the stack gases. When stack temperatures exceed 149°C, that attention is proper, but other losses

Table 3. Lost-Work Analysis for a Fired Heater

	Lost-Work Potential, %
Combustion step	54
Radiant section ΔT	7
Convection section ΔT	24
Stack losses	13
(Exit temp 225°C)	
Wall losses	2

can be much bigger when viewed from a lost-work perspective. For example, a reformer lost-work analysis gave the breakdown shown in Table 3.

The losses for ΔT in the convection section are almost twice those for the very hot exit flue gas. Furnace optimization is the clearest illustration of the benefits of lost-work analysis. If losses from a stack are nearly transparent, the losses imbedded in an excessive ΔT in a convection section are even harder to identify. They do not show up even on the energy balance that highlights the hot stack. These losses can be cut by adding surface to the convection section and shifting load from the radiant section, as well as by looking at the overall process (including steam generation) for streams to match the cooling curve of the flue gases.

Concern over corrosion from sulfuric acid when burning sulfur-bearing fuels often governs the temperature of the exit stack gas. However, the economics of heat recovery are so strong that flue gases are being designed into the condensing range of weak sulfuric acid. It is not a forbidden zone, but the designer should recognize that tube replacement is necessary.

Simple heat losses through the furnace walls are also significant. This follows from the high temperatures and large size of fired heaters, but these losses are not inevitable. In an optimized system, losses through insulation are roughly proportional to

$$\left(\frac{\text{refractory price}}{\text{energy price}}\right)^{1/2}$$

This means that if the price ratio has decreased by a factor of 9, then losses should be down by a factor of 3. If the optimum allowed a 3% loss in 1973, today's optimum would be closer to 1%.

Dryers

A drying (qv) operation needs to be viewed as both a separation and a heat-exchange step. When it is seen as a separation, the obvious perspective is to cut down the required work. This is accomplished by mechanically squeezing out the water. The objective is to cut the moisture in the feed to the thermal operation to less than 10%. In terms of hardware, this requires centrifuges and filters and may involve mechanical expression or a compressed air or superheated steam blow. In terms of process, it means big crystals.

When the dryer is seen as a heat exchanger, the obvious perspective is to cut down on the enthalpy of the air purged with the evaporated water. Minimum enthalpy is achieved by using the minimum amount of air and cooling as low as possible. A simple heat balance shows that for a given heat input, minimum air means a high inlet temperature. However, this often presents problems with heat-sensitive material and sometimes with materials of construction, heat source, or other process needs. All can be countered somewhat by exhaust-air recirculation.

Minimum exhaust-air enthalpy also means minimum temperature. If this cannot be attained by heat exchange

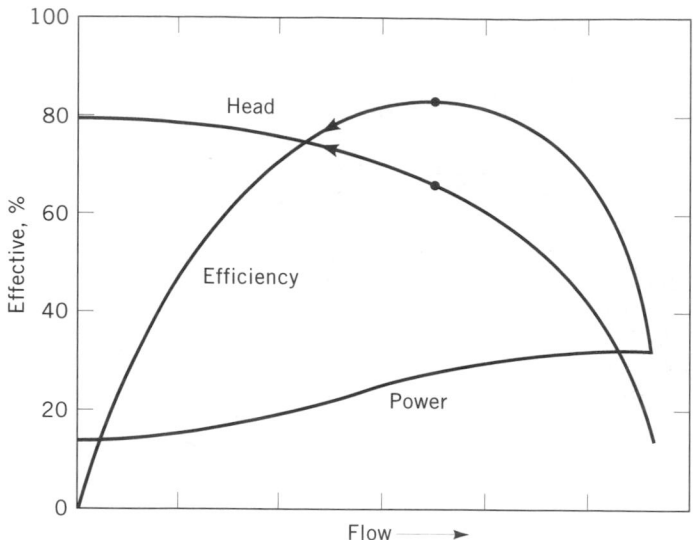

Figure 11. Impact of excess design capacity on pump energy use.

within the dryer, preheating the inlet air is an option. The temperature differential guidelines of the feed–effluent interchange apply.

Like the fired heater, the dryer is physically large, and proper insulation of the dryer and its allied ductwork is critical. It is not uncommon to find 10% of the energy input lost through the walls in old systems.

OPTIMUM DESIGN OF PUMPING, COMPRESSION, AND VACUUM SYSTEMS

Pumping

Is piping pressure drop optimal? Many companies have optimum-pipe-sizing programs, but in the absence of one, a good rule of thumb is that in an optimized system the annualized cost for pumping power should be one seventh the annualized cost of piping (1).

Is exchanger pressure drop optimal? Similarly, for an optimized heat exchanger the annualized cost for pumping should be one third the annualized cost of the surface for the thermal resistance connected with that stream.

Is the pump specified for the right flow? As Figure 11 shows, a 50% overdesign factor will increase power by 35% in a combination of higher head and lower efficiency.

Can the allowance for control be reduced? One option is the use of a variable-speed drive. This eliminates the control valve and its pressure drop and piping. Its best application is where a large share of the head is required for friction and where process demands cause the required flow to vary.

What can be done to get a more efficient pump? Sometimes a higher available net positive suction head (NPSH) permits a more efficient machine.

Compression

The work of compession is typically compared against the isentropic-adiabatic case.

$$\eta_{comp} = \frac{W_{min}}{E_{out} - E_{in}}$$

For an ideal gas, this can be expressed in terms of temperatures

$$\eta_{comp} = \frac{W_{min}}{W_{actual}} = \frac{T_{in}\left[\left(\dfrac{P_{out}}{P_{in}}\right)^{R/cp} - 1\right]}{T_{out} - T_{in}}$$

where R/cp is the ratio of gas constant to molar specific heat. Minimum work is directly proportional to suction temperature. This means that cooling-water systems should be run as cold as possible. Simply measuring temperature rise permits monitoring efficiency for a fixed pressure ratio and suction temperature.

Sometimes, W_{min} for compression is expressed for the isothermal case, in which it is always lower than for the adiabatic. The difference defines the maximum benefit from interstage cooling.

Efficiencies should always exceed 0.6, and 1.00 is approachable in reciprocating devices. Their better efficiency needs to be balanced against their greater cost, greater maintenance, and lower capacity.

The guidelines on pressure drop in piping and exchangers discussed above also apply here. The opportu-

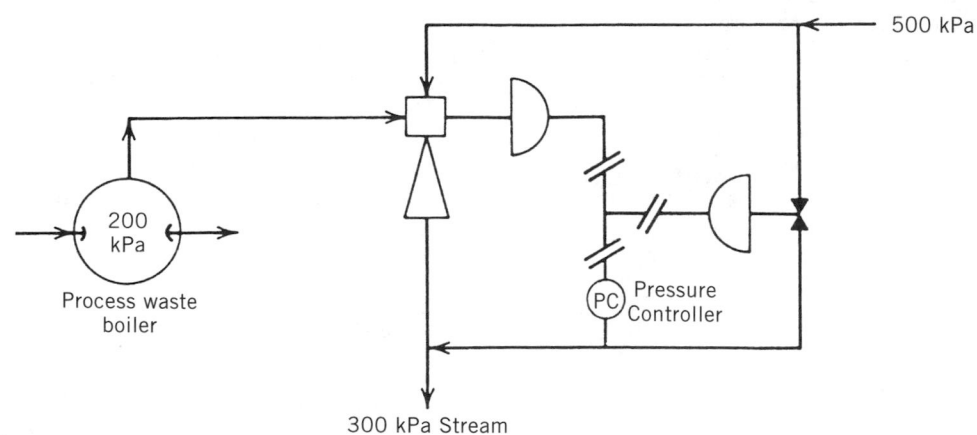

Figure 12. A thermocompressor.

nity for variable-speed drive is possibly even greater, as is the importance of tight control of minimum flows.

Thermocompressors

A thermocompressor is a single-stage jet using a high pressure gas stream to supply the work of compression. The commonest application is in boosting waste-heat-generated steam to a useful level. An example is shown in Figure 12. Thermocompressors can also be used to boost a waste combustible gas into a fuel system by use of high pressure natural gas. The mixing of the high energy motive stream with the low energy suction stream inherently involves lost work, but as long as the pressures are fairly close, the net efficiency for the device can be respectable as the pressures are fairly close, the net efficiency for the device can be respectable (25–30%). Here, efficiency is defined as the ratio of isentropic work done on the suction gas to the isentropic work of expansion that could have been obtained from the motive gas. The thermocompressor has the big advantage of no moving parts and low capital cost.

Vacuum Systems

The most common vacuum system uses the vacuum jet. Because of the higher ratio of motive pressure to suction pressure, the efficiency of vacuum systems is lower than thermocompressors. Generally, it is 10–20%. The optimum system often employs several stages with intercondensers. Steam use in this range varies roughly as $(1/P)_{0.3}$, where P is absolute suction pressure (see Vacuum technology).

Because of the low efficiency of steam–ejector vacuum systems, there is a range of vacuum above 13 kPa (100 mm Hg) where mechanical vacuum pumps are usually more economic (13). The capital cost of the vacuum pump

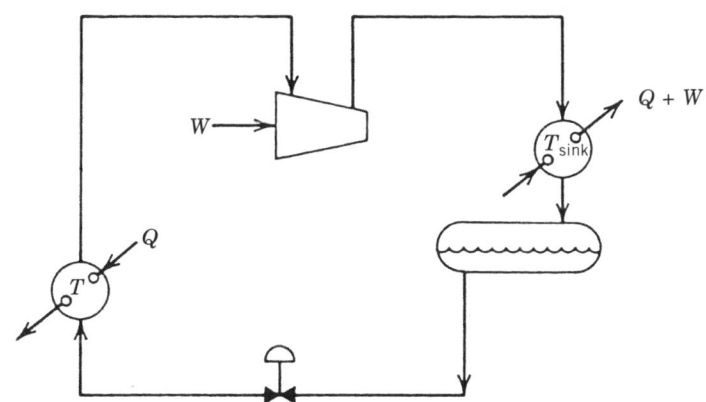

Figure 14. Compression refrigeration.

goes up roughly as (suction volume)$_{0.6}$ or $(1/P)_{0.6}$. This means that as pressure falls, the capital cost of the vacuum pump rises more swiftly than the energy cost of the steam ejector which increases as $(1/P)_{0.3}$. Usually below 1.3 kPa (10 mm Hg), the steam ejector is more cost-effective.

Other factors that favor the choice of the steam ejector are the presence of materials that could form solids and high alloy requirement. Factors that favor the vacuum pump are credits for pollution abatement and high cost steam. The mechanical systems require more maintenance and some form of backup vacuum system, but they can be designed with adequate reliability.

REFRIGERATION

Refrigeration (qv) is a very high value utility. The value of heat in a hot stream is the work it can surrender:

$$\frac{W}{Q} = \left(\frac{T - T_{\text{sink}}}{T}\right)\eta_{\text{turbine}}$$

And the value of refrigeration is the work required to heat pump it to the sink temperature:

$$\frac{W}{Q} = \left(\frac{T_{\text{sink}} - T}{T}\right)\frac{1}{\eta_{\text{compressor}}}\frac{1}{\eta_{\text{fluid}}}$$

The value of refrigeration is compared to heating in Figure 13 for $\eta_{\text{turbine}} = \eta_{\text{compressor}} = 0.7$ and for $\eta_{\text{fluid}} = 0.8$. Here, η_{fluid} accounts for cycle inefficiencies such as the letdown valve shown by Figure 14.

Because of its value, refrigeration justifies thicker insulation, lower ΔTs in heat exchange, and generally much more care in engineering (14). Some questions that the designer should ask are

Is refrigeration really necessary? Could river water or cooling-tower water be used directly? Could they be used for part of the year? Could they replace part of the refrigeration?

Can the refrigerant-condensing temperature be reduced? Could it be reduced during part of the year?

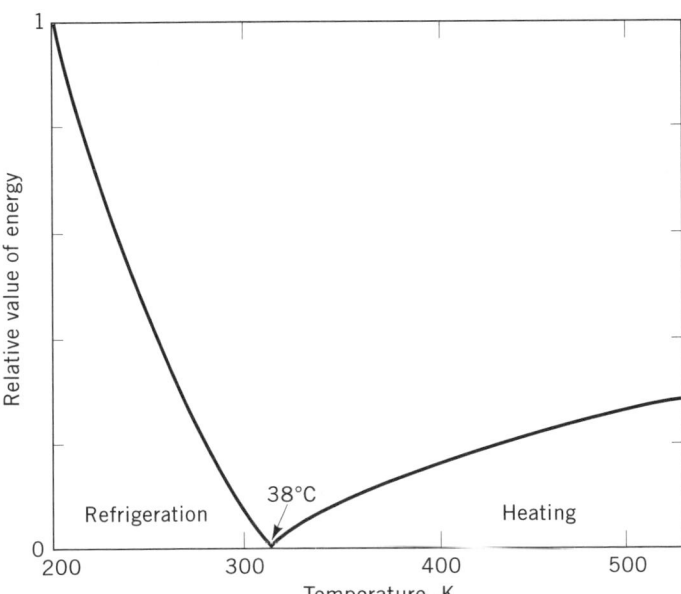

Figure 13. Relative value of energy at various temperatures.

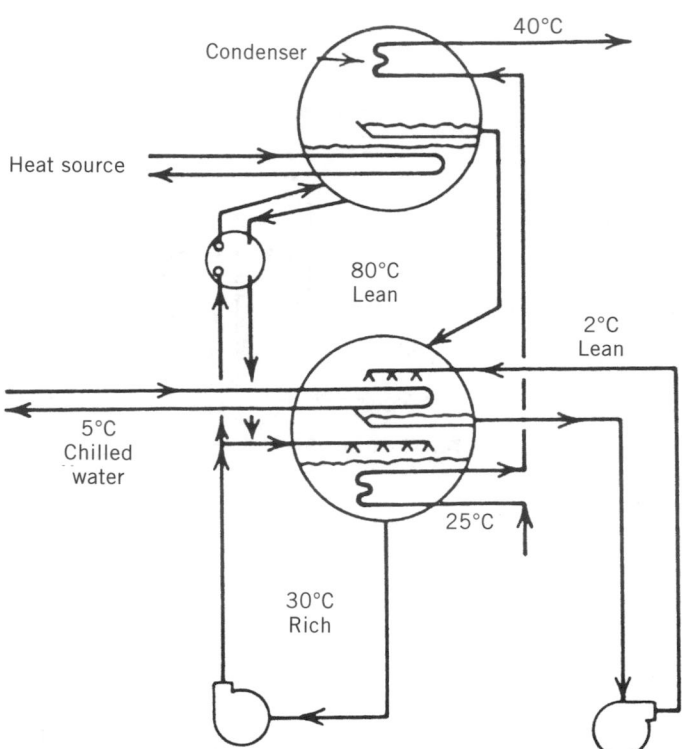

Figure 15. Absorption refrigeration.

Can the system be designed to operate without the compressor during the cold weather?

Is the heat-transfer surface the economic optimum?

Is a central system more efficient than scattered independent systems?

Does the control system cut required power for part-load operation? (Multiple units could yield increased reliability as well as efficiency.)

Are enough gauges and meters provided to monitor operation?

Is there an abundance of waste heat available from the plant (above 90°C)? If so, refrigeration could be supplied by an absorption system.

Absorption chiller units (Fig. 15) need 1.6–1.8 J of waste heat per joule of refrigeration. Commercially available LiBr absorption units are suitable for refrigeration down to 4.5°C.

For low level waste heat (90–120°C), absorption chillers utilize waste heat as efficiently as steam turbines powering mechanical refrigeration units. Absorption refrigeration using 120°C saturated steam delivers 4.5°C refrigeration with an efficiency of 35% where efficiency is referred to the work potential in the steam.

STEAM AND CONDENSATE SYSTEMS

In the process industry, steam serves much the same role as money does in an economy, ie, it is the medium of exchange. If its pricing fails to follow common sense or thermodynamics, strange design practices are reinforced. For example, many process plants employ accounting systems where all steam carries the same price regardless of temperature or pressure. This may be appropriate in a polymer or textile unit where there is no special use for the high temperature; but it is clearly wrong in a petrochemical plant.

Some results of the constant-value pricing system are typically the following: generation in a central unit at relatively low pressure, <4.24 MPa (600 psig); tremendous economic pressure to use turbines rather than motors for drives; lack of incentive for high efficiency turbines; excessively high temperature differentials in steam users; tremendous incentive to recover waste heat as low pressure steam; and a large plume of excess low pressure steam vented to the atmosphere.

A number of alternative pricing systems have been proposed that hinge on turbine efficiency and the relative pricing of fuel and electricity (15–16). Perhaps the best system relates the value of steam to that at the generation pressure by its work potential (exergy content).

$$\frac{\text{value at pressure}}{\text{value at highest generation pressure}} = \frac{\text{exergy at pressure}}{\text{exergy at highest generation pressure}}$$

Design of a central power–steam system is beyond the scope of this discussion, but the interaction between the steam system and the process must be considered at all stages of design. There is a long list of factors to consider in designing a steam-using system:

Can steam-use pressure be lowered? (If ΔT in the heater is above 20°C, the steam pressure is probably above the economic optimum.)

Are there any turbines under 65% efficiency? (Today, turbines are being limited to large sizes above 500 kW, where good efficiencies can be obtained; they are used only for those small drives essential to the safe shutdown of the unit.)

Can a gas turbine be utilized for power generation upstream of the boiler?

Are there any waste streams with unutilized fuel value?

Is there a program to monitor turbine efficiency by checking temperatures in and out?

Is condensate recovered?

Is the flash steam from condensate recovered?

Is feedwater heating optimized?

Is there any pressure letdown without power recovery?

Has enough flexibility been built into the overall condensing–turbine system? (The balance changes over the history of a unit as a process evolves, generally in the direction of less condensing demand.)

Is steam superheat maintained at the maximum level permitted by mechanical design?

Can a thermocompressor be used to increase steam pressures from waste heat?

Are all users metered?

Is low level process heat used to preheat deaerator makeup?

Are ambient sensing valves used to turn off steam tracer systems?

Are steam traps appropriate to the service?

COOLING-WATER SYSTEMS

Cooling water is a surprisingly costly utility. On the basis of price per unit energy removed, it can cost one fifth as much as the primary fuel. Roughly half of this cost is in delivery (pump, piping, and power). This fact has several important implications for design.

Heat exchangers should be designed to use the available pressure drop. A heat exchanger that is designed for 10 kPa when 250 kPa is available will have five times the design flow.

If an exchanger cannot be economically designed to use available ΔP, orifices should be provided to balance the system. This can be done without compromising the guidelines that no unit should be designed for less than 0.8 m/s on tubeside or less than 0.3 m/s on shellside.

If temperature requirements permit, the system will cost less to operate with exchangers in series.

An installed measuring element is usually justified.

If only part of the system requires a high head, this should be supplied by a booster pump. The whole system should not be designed to use or waste the high head.

Other energy considerations for cooling towers include use of two-speed or variable-speed drives on cooling-tower fans, and proper cooling-water chemistry to prevent fouling and excess ΔT in users.

Air coolers can be a cost-effective alternative to cooling towers at 50–90°C (just below the level where heat recovery is economic).

SPECIAL TECHNIQUES

Heat Pumps and Temperature Boosting

A heat pump is a refrigeration system that raises heat to a useful level. The most common application is the vapor recompression system for evaporation (Fig. 16). Its application hinges primarily on low cost power relative to the alternative heating media. If electricity price per unit energy is less than 1.5 times the cost of the heating medium,

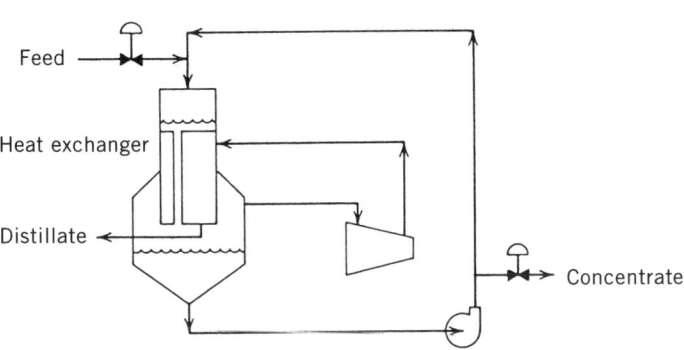

Figure 16. Vapor-recompression system.

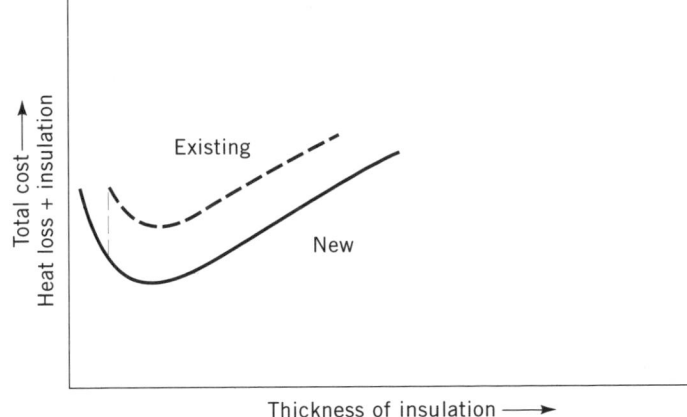

Figure 17. Tank-insulation costs, the existing vs the new.

it merits a close look. This tends to occur when electricity is generated from a cheaper fuel (coal) or when hydroelectric power is available.

Use in distillation systems is rare. The reason is a recognition that almost the same benefits can be achieved by integrating the reboiling–condensing via either steam system (above ambient) or refrigeration system (below ambient).

In an optimized system,

$$\frac{Q}{W} = \frac{T_{\text{hot}}\eta_{\text{compressor}}}{T_{\text{hot}} - T_{\text{cold}}}$$

where T_{hot} and T_{cold} are in absolute units, K.

This provides another criterion for testing whether a heat-pump system may be cost-effective. A power plant takes three units of Q to yield one unit of W; therefore, to provide any incentive for less overall energy use, Q/W must be far in excess of 3.

Energy-Management Systems

The considerable reduction in computing costs has made it possible to do a wide range of routine monitoring and controlling. One can, for example, continuously monitor a distillation system and compare energy use against an optimum and display the cost-per-hour deviation from optimum setpoint. The computer can also test specific actions to achieve the optimum. A computer can monitor a steam system, advise how best to load a set of boilers and choose which turbines to run.

THE EXISTING PLANT

How do existing plants differ from new plants? Good ideas for new plants are also good ideas for existing plants, but there are three basic differences: *1.* Because a plant already exists, the capital vs operating cost curve differs; usually, this makes it more difficult to reduce utility costs to as low a level as in a new plant. *2.* The real economic justification for changes is more likely to be obscured by the plant accounting system and other nontechnical inputs. *3.* The real process needs are measurable and better defined.

An example in support of the first of these is the case of optimum insulation thickness. Suppose a tank was optimally insulated ten years ago. If the value of heat quadrupled in the interim, this change would justify twice the old insulation thickness on a new tank. However, the old tank may have to function with its old insulation. The reason is that there are large costs associated with preparations to insulate. This means that the cost of an added increment of insulation is much greater than assumed in the optimum insulation thickness formulas (Fig. 17). An example of the second difference is that many things appear to be strongly justified by savings in low pressure steam if the steam is valued artificially high. A designer of a new plant has the advantage of focusing attention on savings in the primary budget items, ie, fuel and electricity at the plant gate, rather than on cost-sheet items such as steam at battery limits. The third difference is that many process details are relatively uncertain when a plant is designed. For example, inert loading for vacuum jets is rarely known to within 50%. Although the first two differences are negative, the third provides a unique opportunity to measure the true need and revise the system accordingly.

Energy Survey. The energy survey has seven components: *"as-it-is" balance; field survey; equipment tests; check against optimum design; idea-generation meeting; evaluation* and *followup.*

As-It-Is Balance. This is a mandatory first step for the energy survey. It permits targeting of principal potentials; check of use against design; check of use against optimum (how a new plant would be designed today); definition of possible hot or cold interchanges; definition of the unexpected uses (eg, the large steam purge to process or high pressure drop exchanger); and contribution from specialists not familiar with the unit.

Field Survey. This is often done by a team of two: one who knows what to look for and one who knows the process. On field surveys, it is as important to talk to the operators ("What runs when the unit is down?"; "What happens when you cut reflux?"; "Where are the guidelines for steam–feed ratio? How close do you usually run?") as it is to record items like air leaks, high pressure drops across values, frost on piping, lights of the wrong type or at the wrong time, steam plumes (a reason to climb to the top of the unit), or minimum-flow bypasses in use. The field survey should develop detailed "fix" lists for leaking traps, uninsulated metal, lighting, and steam leaks.

Equipment Tests. Procedures for rigorous, very detailed, efficiency determination are available (ASME Test Codes) but are rarely used. For the objective of defining conservation potentials, relatively simple measurements are adequate. For fired heaters, stack temperature and excess O_2 in stack should be measured; for turbines, pressures (in and out) and temperatures (in and out) are needed.

Check Against Optimum Design. This attempts to answer the question, "Need a balance be as it is?"

In the large view, the first thing to compare against is the literature claims for chemicals such as NH_3, HNO_3, CH_3OH, and ethylene.

The second thing to do is look for the obvious: stack temperatures $>149°C$; process streams $>121°C$, cooled by air or water; process streams $>65°C$, heated by steam; $\eta_{turbine} <65\%$; reflux ratio >1.15 times minimum; and excess air $>10\%$ on clean fuels.

Idea-Generation Meeting. The idea-generation meeting is most productive if three important guidelines are followed: *1.* Get the right people. A good guiding principle is to get "two wise old Turks" for each corporate expert. It is extremely important to ask, "Given free choice, whom would you choose to attend?," and then get them. *2.* Discuss the as-it-is balance for each area and record all ideas. *3.* Assign follow-up responsibility.

Evaluation. The evaluation of each idea should include a technical description and its economic impact and technical risk. The ideas should be ranked for implementation. A report should provide a five-year framework for energy projects.

Follow-up. If no savings result, the effort has been wasted. The survey leader needs to be sure that the potential of the good ideas is recognized by management and the project-generation channels of his company.

What Do You Not Do?

Often, what looks like negligence can be a tried and proven practice, and one should be cautious about experts bearing lists and offering huge savings. The process has to work, and present utility saving may or may not compensate for future repair bills or lost products. Some examples are the following: an idling turbine may be necessary to permit a safe plant shutdown if a power failure occurs; a cooling-water flow that is throttled to below 0.6 m/s (in winter) will likely assure a heat-exchanger cleaning in late spring; a furnace that runs too low on excess air may run into after-burning; and a column run too close to the minimum reflux ratio without adequate controls runs a risk of off-specification product.

One also does not trust the plant accounting system unquestioningly. All energy is not created equal. The energy that is recovered from flashed steam or that is shaved off a reboiler's duty may not be worth its cost-sheet value. The meters that matter are the primary meters at the plant gate. Only if the recovered energy reduces the meter settings does it save the plant money.

One does not accept the first solution to an energy-waste problem without seeing how this problem fits into the overall plant-energy balance. The sudden rise in energy prices has lifted many options into the justifiable range. There might be better alternatives available.

Acknowledgment

Revised from the *Kirk-Othmer Encyclopedia of Chemical Technology*, 3rd ed., Suppl. Vol. John Wiley & Sons, Inc., New York, 1984.

BIBLIOGRAPHY

1. D. E. Steinmeyer, *Chemtech,* 188 (Mar. 1982).
2. W. F. Kenney, *Proceedings 1981 Industrial Energy Conservation Technology Conference,* Texas Industrial Commission, p. 247.
3. K. G. Denbigh, *Chem. Eng. Sci.* **6,** 11 (1956).
4. C. J. King, *Separation Processes,* 2nd ed., McGraw-Hill, Inc., New York, 1980.
5. C. S. Robinson and E. R. Gilliland, *Elements of Fractional Distillation,* 4th ed., McGraw-Hill, Inc., New York, 1950.
6. M. Souders, *Chem. Eng. Prog.* **60**(2), 75 (1964).
7. D. Carter, personal communication, Monsanto Corp., St. Louis, Mo., 1982.
8. B. Linnhoff, *Chem. Eng. Prog.,* **90**(8), 33 (1994).
9. D. Steinmeyer, *Hydrocarbon Process.,* **53** (April 1992).
10. W. F. Furgerson, *Conserving Energy in Refrigeration,* Manual 12 of *Industrial Energy–Conservation,* MIT Press, Cambridge, Mass., 1982.

ENERGY CONSUMPTION IN THE UNITED STATES

PAUL KOMOR
Office of Technology Assessment, U.S. Congress
Washington, D.C.

ANDREW MOYAD
U.S. Environmental Protection Agency
Washington, D.C.

Energy is a crucial national and international concern for several reasons: it impacts the U.S. economy, it contributes to many environmental problems, and it is imported and therefore tied to issues of national security and international political stability. Further, its unequal global consumption raises questions of international resource equity and sustainability. The following facts highlight these points:

- The U.S. economy cannot function without energy, which is a principal component of Gross Domestic Product (GDP). In 1989, U.S. businesses, consumers, and government spent a total of $437 billon, about 8% of GDP, on energy (1).
- Energy consumption is closely tied to environmental concerns, notably global climate change and urban air quality. For example, virtually all U.S. carbon dioxide emissions stem from fossil fuel consumption; similarly, vehicles are responsible for much of the carbon monoxide and other harmful emissions found in urban areas (2).
- More than 40% of U.S. oil consumption is imported. This dependence is likely to climb in the future, contributing to the trade deficit and increasing U.S. vulnerability to economic shocks stemming from major shifts in international oil prices and availability.
- In 1992, total U.S. energy consumption was 82.4 quads of energy (3), more than any year in the past, and far more than any other country. In fact, the

United States accounts for almost one-fourth of global energy use (4), despite having less than 5% of the total world population (1).

Given this U.S. dominance in world energy use, as well as the important role energy plays in economic, environmental, and international political issues, an understanding of U.S. energy consumption (how energy is used, how this use has changed over time, what factors influence this use, and how energy use may change in the future) is essential to understanding energy's potential impact on these various domestic and international issues. This discussion provides an overview of energy consumption from several perspectives: aggregate national energy use and trends, consumption by individual end-use sectors, and consumption by fuel.

See also NATURAL GAS; PETROLEUM MARKETS; COAL; COMMERCIAL AVAILABILITY OF ENERGY TECHNOLOGY; ENERGY EFFICIENCY; TRANSPORTATION FUELS.

ENERGY USE: RECENT HISTORY

U.S. energy consumption increased steadily from 1950 to 1973. Since 1973, however, national energy consumption has moved erratically and has at times even decreased (see Fig. 1). A number of factors have contributed to these changes in U.S. energy use, including:

- Energy prices
- Population growth and migration
- Economic growth and structural economic shifts
- Technical advancements
- Lifestyle changes
- Major political events, notably the 1973 and 1979 oil crises
- Policy changes, at both the state and federal levels

The first four of these factors (price, population, economic growth and change, and technical advances) probably had the largest impact on national energy use, particularly since 1973. Energy prices in real terms, that is, adjusting for the effects of inflation, were generally stable until 1973 but shot up dramatically that year. In 1974 alone, energy prices increased 56%. As Figure 1 illustrates, real energy prices peaked in 1981 at more than 3.5 times the 1973 levels. Since 1981, though, real energy prices have generally decreased, dropping below the 1974 levels by 1991. Note that the prices discussed here are "composite," that is, a weighted average of the prices of the different types of energy. Trends in the prices of specific fuels are discussed below.

When confronted with rapidly rising prices, energy users generally act to reduce their energy costs (although the exact relationship between energy prices and energy consumption is not well-understood) by substituting fuels, choosing cheaper over more expensive fuels where possible; reducing consumption, through short-term behavioral changes (eg, lowering thermostats) and long-term struc-

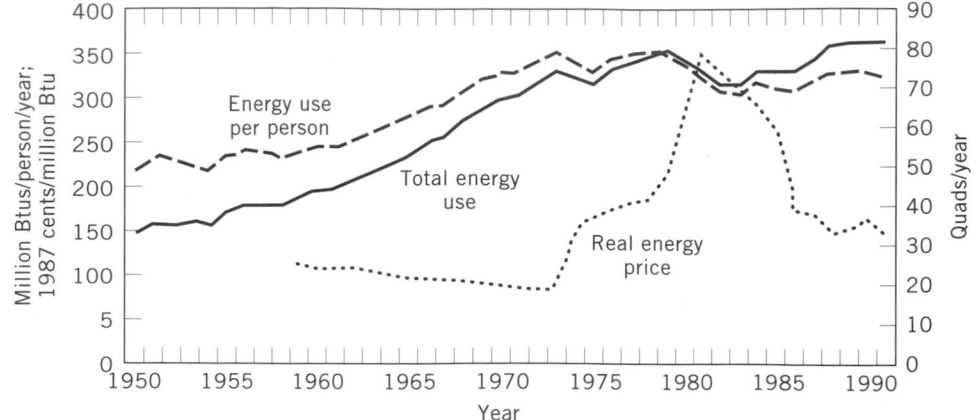

Figure 1. U.S. energy consumption, total and per person; and fossil fuel real energy price (1,3,5).

tural changes (eg, shifting production away from energy-intensive goods); and increasing energy efficiency through behavioral and technical changes, such as adding insulation to houses, installing process controls in industry, and switching to energy-efficient fuel injection instead of carburetors in automobiles.

Prices clearly influence energy use, but other factors are also important. Energy provides energy services: moving goods and people, running industrial motors, space heating for buildings, and so on. Clearly, the more people requiring these services, the greater the consumption of energy. Population has increased steadily in the United States, from 151 million in 1950 to 252 million in 1992 (5); however, trends in energy use per capita have matched trends in total energy use, indicating the importance of other factors in influencing consumption (see Fig. 1).

Recently, the basic outputs of the U.S. economy have undergone a slow but steady shift, with attendant impacts on national energy use. The service sector, such as hotels, restaurants, and legal and medical services, account for an increasing fraction of total economic output. Industry, notably manufacturing, has shown a corresponding decrease. In 1970, goods accounted for 46% of GDP, whereas services accounted for 43%. (The remaining 11% was for structures.) By 1991, these numbers had shifted to 39% and 53%, respectively (1). These structural economic shifts have had major impacts on U.S. en-

ergy use, which grew more slowly than did the economy over the last two decades.

From 1950 to 1973, the growth in national energy use mirrored that of the GDP, suggesting that increased energy use was required for economic growth. Since 1973, however, these two variables have diverged. GDP continued to climb, whereas energy use remained relatively unchanged (see Fig. 2). One important reason for this decoupling of energy use and economic growth was structural change. Simply put, less energy is required to produce, for example, $1,000 of computers than $1,000 of steel. By one estimate, about one-third of the divergence between energy use and GDP growth is attributed to structural economic shifts that have occurred over the last two decades (6).

The other two-thirds of this divergence is attributed to technical improvements in energy efficiency. In all sectors of the economy, numerous technologies have been introduced that wring more useful energy service (eg, heat, light, and motor drive) from each unit of energy consumed. Refrigerators, electric motors, automobiles—all have shown considerable gains in energy efficiency over the past 20 years, and considerable potential for further improvements remain.

Although debates continue over what caused these efficiency improvements, some argue that the energy price increases of the 1970s led to a search for ways to decrease energy use without reducing service. Others maintain

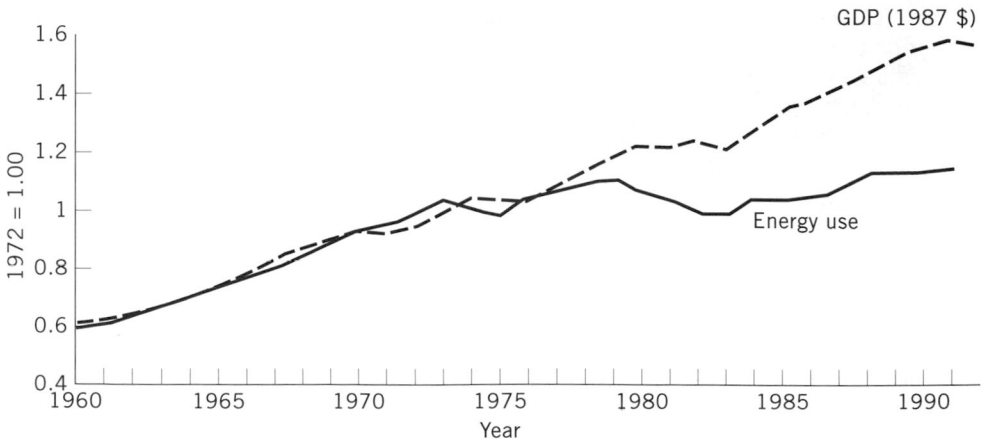

Figure 2. U.S. energy use and Gross Domestic Product (1,3,5).

that technical progress, including energy-efficiency improvements, has and will continue regardless of energy price changes. In any case, it is generally agreed that technical-efficiency improvements will continue, and that, all else being equal, higher prices lead to greater energy efficiency.

ENERGY USE BY SECTOR

Energy is used in three distinct end-use sectors: residential and commercial buildings, transportation, and industry.

Buildings

Energy use in buildings accounts for an increasing share of total U.S. energy consumption: from 27% in 1950, to 33% in 1970, to 36% in 1990. (See Reference 7 for a more complete discussion.) At present, more than 60% of national electricity consumption and almost 40% of national natural gas consumption occurs in the two basic building types: residential and commercial.

The Residential Sector. In 1989, residential buildings accounted for nearly 17 quads of energy (most as electricity, with significant contributions from natural gas and oil) at a total cost of $104 billion. As Figure 3 illustrates, space heating is responsible for almost half of this energy use, followed by water heating, refrigerators and freezers, space cooling, and lights. In the past 20 years, residential energy use increased at a modest average annual rate of about 1.2%.

Between 1970 and 1990, the combination of a growing population and a shrinking average household size (ie, fewer people per household) led to an almost 50% increase in the number of households. As each household requires space conditioning, hot water, and other energy services, these changes drove the growth in energy use in the residential sector. Increased demand for particular energy-intensive services also contributed to the growth in residential energy use, for example, central air conditioning is now routinely installed in over three-fourths of new single-family homes, and color televisions are found in almost all households.

Although total residential energy use increased from 1970 to 1989, annual energy consumption per household actually decreased by 15% in the same period. Whereas several factors contributed to this intensity drop, improved technology and better building practices were key: retrofits in older houses, greater use of energy-efficient building practices in new homes, and a dramatic improvement in the energy efficiency of household appliances and equipment.

Considerable efforts to improve the energy efficiency of U.S. buildings have been made. For example, by one estimate, about 26 million owner-occupied U.S. households added storm windows and/or storm doors, and 17 million added insulation, between 1983 and 1988. Careful evaluations of retrofits indicate that energy savings are often substantial. New houses have benefited from greater use of energy-efficient techniques; houses built in 1985, for ex-

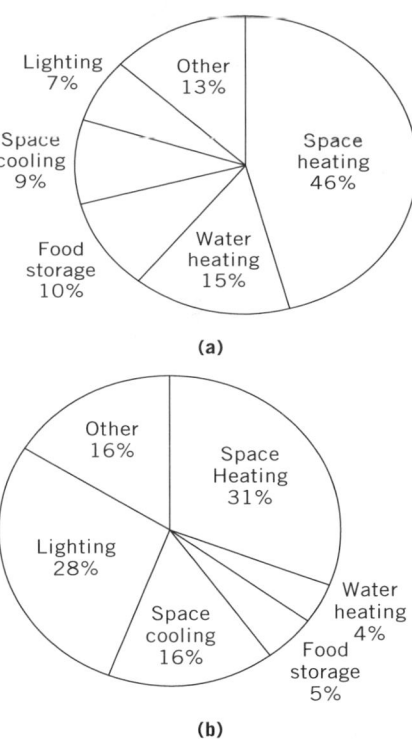

Figure 3. (**a**) Residential buildings and (**b**) commercial buildings; energy use (7).

ample, were better insulated and included more energy-efficient windows than those built in 1973. Residential energy-using equipment is now more energy-efficient as well. The typical new gas furnace sold in 1975 had an efficiency of 63%, and by 1988, this increased to 75%. The efficiency gains in appliances were even greater; the typical new refrigerator sold in 1990 uses less than half as much electricity as a comparable unit sold in 1972.

The fuel mix of residential energy use also has changed. Whereas oil use for space heating has dropped sharply, electricity has become increasingly prevalent for both space and water heating. In 1970, electricity supplied 41% of residential energy; by 1988, this had climbed to 61%. Yet this 20-year trend toward greater residential electrification may be changing, as electric space heating in new single-family homes has decreased dramatically in recent years (from 49% in 1985 to 33% in 1990).

The Commercial Sector. In 1989, commercial buildings accounted for about 13 quads of energy use, at a cost of $68 billion. Electricity represented about two-thirds of this energy. Space heating, lighting, and space cooling were the principal end uses, as shown in Figure 4. Energy use in commercial buildings has increased rapidly since 1970, at an average annual rate of 2.3%, or about twice the rate of energy increase realized in residential buildings in the same period. A number of factors contributed to this growth, the most significant being the rapid growth in new commercial buildings. As measured by total square footage, the commercial building stock has increased more than 50% since 1970. Heating, cooling, and lighting these new buildings has considerably increased

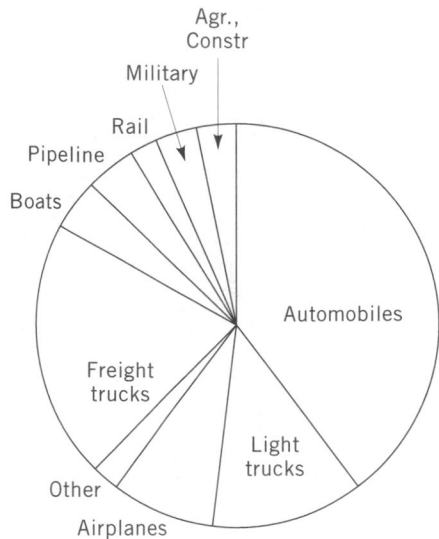

Figure 4. Commercial buildings energy use (7).

energy consumption. A greater demand for energy-intensive services, notably air conditioning and electronic office equipment, has further increased commercial building energy use. For example, in 1984, U.S. businesses used less than 2 million personal computers; their use in the commercial sector grew to 14 million, a sevenfold increase, only five years later.

Despite increases in the use of air conditioning and electronic office equipment, annual energy use per commercial square foot actually stayed flat between 1970 and 1990. As in the residential sector, improved technology helped to dampen the growth in commercial building energy use. New commercial buildings use improved windows and shells, more efficient space-conditioning equipment, and better lighting systems. For example, commercial buildings constructed in the 1980s contained more ceiling and wall insulation, multipane and reflective windows, and shadings or awnings compared to buildings constructed in earlier years. Computer advances have permitted greater use of computer-aided building design and analysis methods, and retrofits have improved the energy efficiency of previously constructed commercial buildings.

Buildings—Summary and Outlook. Energy use in both residential and commercial buildings has grown in the past 20 years. Sheer increases in numbers underlie much of this growth: more people, more households, and more offices. Increased service demand (more air conditioning, more computers, larger houses) has contributed as well. However, the application of improved technology in the areas of building shells (windows and insulation), appliances (furnaces, air conditioners, refrigerators), and building design have helped to moderate the growth in building energy use.

Several factors will influence future energy use in buildings. First, studies indicate that greater use of commercially available technologies could reduce building energy use up to one-third. Second, the service sector offices, restaurants, and other energy-intensive buildings grew

rapidly in the 1980s. If this growth resumes, then building energy use will also increase rapidly. Third, information technologies (eg, computers, printers, and copiers) are a small but rapidly growing energy user in commercial buildings. If their use continues to grow at current rates, these technologies will soon account for a significant share of commercial energy use. Fourth, electric and gas utility investments in energy efficiency, often called "demand-side management" (DSM), may substantially increase the use of energy-efficient technologies in buildings. Current utility DSM investments total about $2 billion per year. If these investments are as successful as planned, then significant energy savings will result.

Transportation

The movement of goods and people accounts for about one-fourth of U.S. energy use and almost two-thirds of U.S. oil. Although not the largest energy-consuming sector, transportation is often the most visible and controversial, due in part to its dependence on oil and to its impact on urban air quality. Transportation energy use is best understood by examining passenger and freight transport separately.

Passenger Transport Energy Use. The demand for passenger transport has grown rapidly in recent years, as the population has become more mobile and automobiles have become more widely available. Although vehicle energy efficiency has improved dramatically, this improvement has been outpaced by increased demand, resulting in significant increases in energy use. U.S. transport energy use grew faster between 1970 and 1990 than energy use in either the buildings or industrial sectors (8).

Private automobiles and light trucks used for personal transport account for over half of transport energy use (see Fig. 5) and over 85% of passenger-miles (8). In the past 20 years, the number of automobile and passenger trucks, and the miles driven per year, have both increased, as illustrated in Figure 6. The combination of these two factors would have resulted in very rapid growth in energy use; however, energy use by these vehicles actually increased quite slowly in recent years, at an annual average rate of 0.3% from 1970 to 1989. This large increase in transportation services with only a small increase in energy use was made possible largely through efficiency improvements.

The efficiency of the private automobile fleet increased over 50% between 1975 and 1989, as measured by miles per gallon (8). The main impetus behind this improvement was corporate average fuel economy (CAFE) requirements. Federal legislation passed in 1975 set minimum fleet fuel economy standards, starting at 18 miles per gallon (mpg) in 1978 and increasing to 27.5 mpg by 1985 (see Fig. 7). This efficiency increase was achieved by several vehicle changes, including decreased weight, reduced engine size, and increased use of fuel-injection and other efficient technologies. Contrary to some popular perceptions, however, these efficiency improvements did not reduce vehicle performance. From 1977 to 1993, passenger car interior volume remained essentially constant and performance (as measured by 0–60 mph acceleration) improved (9).

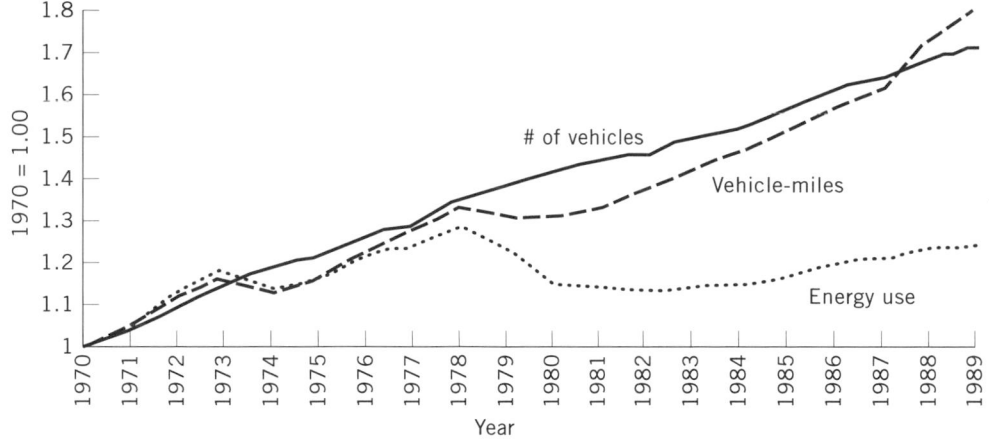

Figure 5. Automobile and passenger truck travel and energy use (8).

Light trucks are playing an increasingly important role in passenger transport and energy use. These trucks are often used exclusively for passenger transport and now account for almost one-third of the new sales of light duty vehicles (ie, automobiles and light trucks) (8). Light trucks are generally far less energy-efficient than automobiles; the 1991 CAFE standard for automobiles was 27.5 mpg but only 20.2 mpg for light trucks (see Fig. 7). As a result, the growing population of light passenger trucks is dragging down the energy efficiency of the light duty fleet.

Airplanes account for 9% of transport energy use. Airplane travel, as measured by passenger-miles, has grown at an average annual rate of 6.6% from 1970 to 1989 (8). As with the other modes of travel, energy use increased more slowly than passenger-miles due to technical and operational improvements. More seats per aircraft, higher load factors (ie, the ratio of occupied to total seats), and improved engine efficiencies and aerodynamics all contributed to greater efficiency in passenger air transport.

Freight Transport Energy Use. Freight transport accounts for one-third of transport energy use (see Fig. 5). Trucks account for the bulk of this energy use, followed by barges, pipelines, and trains. However, as measured by ton-miles of travel, a common yardstick in freight, a different energy pattern emerges: trains and barges both exceed trucks (see Fig. 8).

In the past 20 years, a gradual but steady shift in the U.S. economy occurred, away from basic materials and toward greater consumption of services and higher value-added goods. Although production of raw materials (such as coal and minerals) has grown, production of manufactured goods has grown much faster. These economic shifts directly impacted the freight transport system, which has changed to accommodate the altering mix of industrial production. Over the past 20 years, movements by trains and barges, which typically carry basic commodities (coal, farm products, chemicals), grew slowly, whereas truck and air freight movements, which carry value-added goods, grew more rapidly. Truck and air movements generally require more energy per ton-mile than do trains and barges; therefore, these economic shifts have resulted in relatively rapid growth in freight transport energy use.

The volume of freight moved by trucks increased rapidly in the past 20 years. The energy intensity (Btu/ton-mile) of trucks, however, stayed roughly constant since 1970. (This discussion excludes light trucks used primarily for personal transport.) During this time, there were

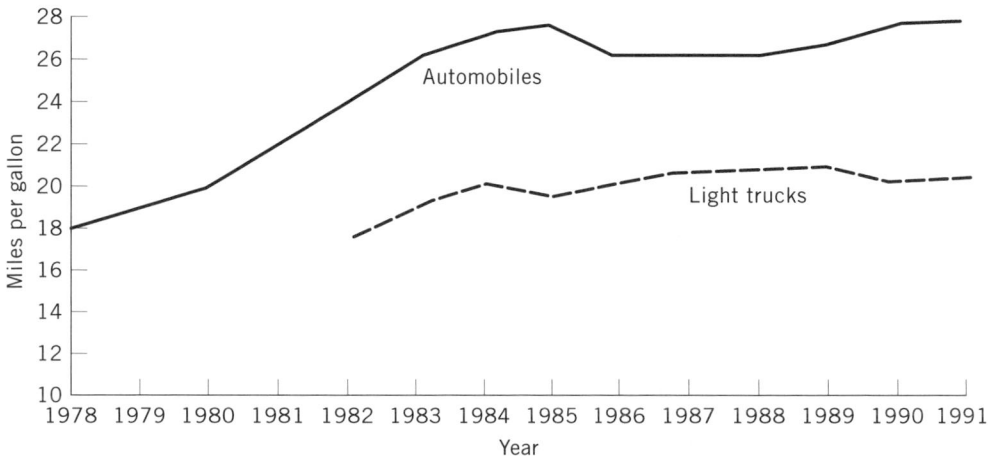

Figure 6. Corporate average fuel economy (CAFE) requirements (8).

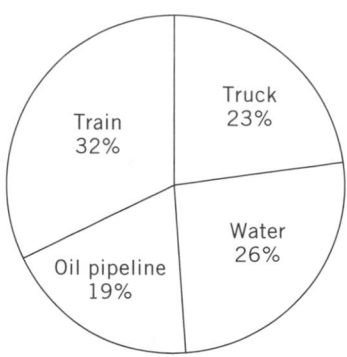

Figure 7. Freight ton-miles by mode, 1989 (8).

gines, fuel management computer systems, and the use of larger barges and tugs.

The growth of intermodalism has strongly shaped the freight transport system. Intermodalism usually refers to the carriage of trailers and containers by trains, with delivery to and from the train terminals by truck, but the term can also refer to the use of barges or open ocean ships to move containers, which are then moved by train or truck. Several innovative technologies have been implemented, including sealed containers that can be moved by train, barge, or truck; roadrailers (truck trailers that can ride directly on train tracks); piggybacking (putting truck trailers onto railcars); and double-stack containers (putting two levels of containers onto railcars). Intermodal loadings on freight trains grew at an average annual rate of 4.9% from 1970 to 1990 (10). The energy implications of this shift are not well-documented, but it is thought that trains use about one-half the energy of trucks for long-distance movements of high density goods (11). Therefore, to the extent that it has led to a shift from trucks to trains for long-distance movements, intermodalism represents a significant improvement in energy efficiency.

Transport—Future Issues. Several key factors will influence transport energy use in the future. First, alternative fuels (including electricity, ethanol, hydrogen, methanol, and natural gas) are currently being evaluated for both passenger and freight transport. Although alternative fuels in transportation are not widely used at present, their use may increase in the near future, particularly with growing concerns about urban air quality. For example, beginning in the year 1998, 2% of new vehicles available for sale in California are required to be zero tailpipe-emission vehicles. Second, if oil prices remain low, then there will be little economic incentive to invest in efficiency, but if they increase, investment in efficient vehicles is likely to increase. Third, regulatory changes, notably increases in the CAFE requirements, have been proposed. If such changes are signed into law, significant improvements in the energy efficiency of the transportation system may occur. With maximal use of existing technologies, the efficiency of new U.S. automobiles could approach 40 mpg in the next decade (12).

several technical improvements, notably electronic engine controls, demand-actuated cooling fans, aerodynamic improvements, and multiple trailers. The market penetration of these technologies varies, but some (eg, cab air deflectors) are now found in almost all trucks today. Despite these numerous technical improvements, however, truck energy intensity did not actually improve due to: (1) low turnover of the truck fleet, leading to slow adoption of new technologies, many of which cannot easily be retrofit to existing trucks; (2) increased highway speeds, which are less energy-efficient due to greater wind resistance; (3) changes in freight movements, eg, trucks may now be carrying less dense goods, causing trucks to fill up ("cube out") before they reach weight limits; and (4) other factors, including low load factors, poor driver training, and increased urban congestion.

For the train system, energy intensity has improved considerably. Between 1970 and 1990, train energy use actually decreased, despite an increase of over one-third in ton-miles carried (10). Several factors contributed to these gains: increases in average trip lengths, operations and communications improvements, and technical improvements such as reduced idling speeds for locomotives and greater use of flange lubricators. Water-based freight transport showed a moderate improvement in energy intensity. Both technical and operational factors contributed to this improvement, including improved diesel en-

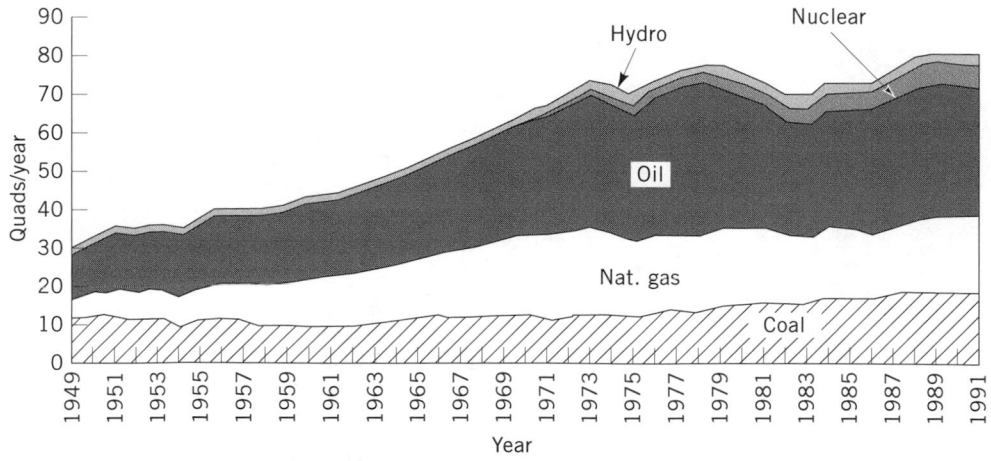

Figure 8. Energy consumption by fuel (3,5).

Industry

Since at least 1950, industry has been the largest energy-consuming sector in the U.S. economy, although its share of total energy use has decreased from 47% in 1950, to 37% in 1990 (5). In 1990, the U.S. industrial sector consumed about 30 quads of energy, about twice as much as it had in 1950 (5). The industrial sector consumes significant portions of all major energy sources in the United States: 44% of all natural gas, 34% of electricity, 25% of petroleum, and 12% of coal (5).

Several heavy manufacturing industries consume about three-fourths of the energy in this sector: ceramics and glass, chemicals, food, petroleum refining, primary metals, and pulp and paper. Nonmanufacturing industries, such as agriculture, construction, and mining, account for another 15% of sectoral energy use. Finally, lighter manufacturing industries, such as automobile manufacturers and textiles, consume the remaining 11% of industrial energy, but these industries consume a larger share of industrial electricity, in part due to their extensive use of electric motors (13). Energy's share of production costs varies widely by industry. In most industries, energy represents 5% or less of production costs, but the portion exceeds 20% for several large industries, including aluminum, cement, and certain chemical manufacturers (13).

Industrial activities are even more diverse than their fuel supplies, ranging from agriculture and construction to manufacturing and mining. Despite this diversity, however, there are four basic uses for industrial energy: steam production, direct process heat, electric motors, and feedstocks. The main use of industrial energy, steam production, is fueled mostly from natural gas burned in conventional boilers and cogenerating equipment. Industrial steam is commonly used in steel and pulp and paper production. Direct process heat is used for curing, drying, melting, and smelting. As with steam, most process heat derives from natural gas burning. Electric motors account for most of the electricity used in industry. Feedstocks include natural gas for chemical and fertilizer manufacturing and coal for steel production (13).

One important trend associated with industrial energy use is cogeneration, which involves the combined production of heat (typically as steam) and electricity. With the passage of the Public Utility Regulatory Policies Act of 1978, cogeneration in the industrial sector has increased substantially; the legislation requires electric utilities to purchase cogenerated power produced by a qualifying nonutility generator. This legislation has encouraged many industries to install cogeneration capacity, both to help meet their own power needs and to generate salable electricity. Today, cogeneration provides about 12% of the electricity used by manufacturing industries, with major contributions from the chemicals, food, paper, petroleum, and steel industries (13).

To simplify comparisons between the relative energy use of different industrial activities, analysts often compare the amount of energy consumed per unit of economic output, a comparison termed "energy intensity." The existing range of industrial energy intensities, as the diversity of industrial activities may suggest, varies by a factor of about 200. Printing, for example, requires about 800 Btu per dollar of output, whereas manufacturing nitrogen fertilizers consumes about 160,000 Btu per dollar of output (13). Between 1970 and 1990, the energy intensity of U.S. industry declined more than 20%, from more than 10,000 Btu per dollar to less than 8,000 Btu (in constant 1982 dollars). For the economy as a whole, this intensity decline was due both to structural changes (eg, less steel and more computers) and to technical efficiency improvements. Over the next two decades, that intensity is projected to decline further, to about 6000 Btu per dollar (14). In that same period, industrial energy use is projected to increase slightly more than 1% annually (14), the lowest rate of growth projected among all three sectors.

The increasing international exchange of goods and services points to an important limitation of energy intensity measures: they fail to reflect the embodied energy of products. Many industries use materials with substantially differing amounts of embodied energy. Although analyzing energy intensities often provides the first indication of where to focus government and other efficiency programs, the measure is an imperfect way to determine where optimal efficiency savings are possible; it simply reflects the relative energy use at one point in an often extended production chain.

Because energy represents a relatively minor portion of production costs for most industries, corporate attention to its use and efficiency has historically been low (13). However, at least three recent trends have and are likely to continue to improve industrial energy efficiency, which may slow the growth of energy use in this sector. First, increasingly stringent federal environmental regulations, particularly those pursuant to the Clean Air Act, have prompted many firms to examine the potential of improved energy efficiency as a means to limit stack and other emissions. Second, a growing number of utility DSM programs are targeting industrial consumers as a means to offset the need for additional generating capacity; as more money and attention are channeled to DSM programs, industrial energy efficiency may increase considerably. Finally, process changes and materials substitutions are being increasingly used to improve outputs and reduce production costs; an additional benefit of such changes is often reduced energy use.

ENERGY USE BY FUEL

Greater insight can be gained by looking at consumption patterns of the various fuels consumed in the United States, as illustrated in Figure 9. Three fuels account for the bulk of direct consumption: oil, natural gas, and electricity (strictly speaking, not a fuel but a carrier). Coal, nuclear power, and hydropower are not discussed here, as they are used largely for electricity generation.

Oil

Oil is the single largest U.S. energy source, accounting for 41% of national energy use and 48% of national energy spending. This fuel is also the most problematic and politically volatile form of energy. Oil has many attractive features: it has a high energy density, making it less expen-

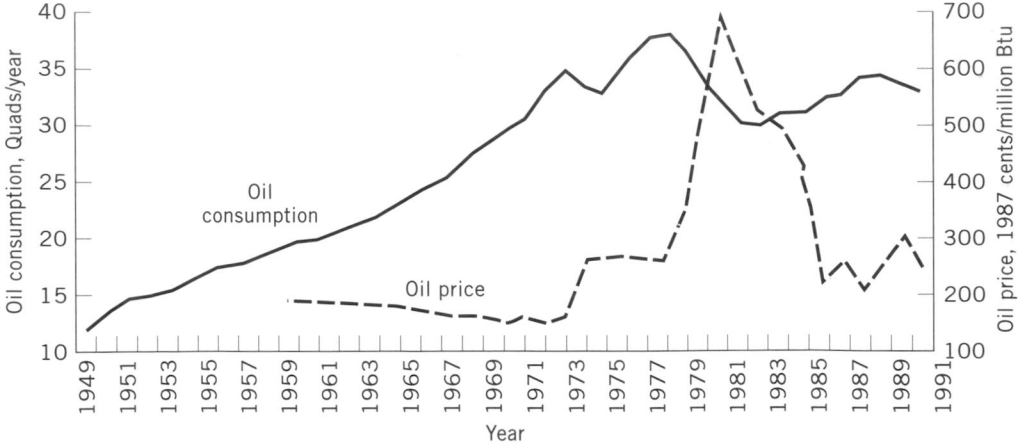

Figure 9. U.S. oil consumption and price (5).

sive to transport; it is a liquid and therefore relatively easy to produce and transport; and it can be used for a variety of purposes in different sectors, eg, transportation and large motor drive via internal combustion engines, space heating for buildings via oil-burning furnaces, industrial heat via oil-burning boilers, and so forth. Oil also has its problems. Its global distribution is uneven, leading to concentrations of wealth and power in those regions that have it and to disagreements and occasionally wars with those who do not. Further, like all fossil fuels, its production and use has detrimental environmental effects, including emissions of carbon dioxide (CO_2), volatile organic compounds (VOCs), and nitrogen oxides (NO_x), and occasionally widely publicized and locally disastrous oil spills.

Oil consumption in the United States increased rapidly from 1950 to 1973, due to growth in demand by all users—transportation, industry, buildings, and electric utilities. In 1973, however, Arab oil-producing countries, due in part to U.S. arms sales to Israel, briefly stopped selling oil to the United States. Oil prices climbed, and demand dipped slightly. In 1979, Iran briefly stopped exporting oil, and the market reacted much the same as it had in 1973, with large price increases and small demand decreases. Since then, both price and consumption have fluctuated (see Fig. 10).

At present, U.S. oil consumption is at about the same level it was in 1974, despite large increases in population, economic output, and the number of automobiles. This has been possible due in part to technical-efficiency improvements, and in part to shifts away from oil by those users who have other options. For example, industry and buildings both substituted other fuels, notably natural gas, when faced with increasing and volatile oil prices. The transport sector, however, has few nonoil technical alternatives; thus, the fraction of U.S. oil used for transport has increased from 53% in 1970, to 65% today. (Alternative fuels for transportation are being investigated and used in small numbers; but are currently used by only a small fraction of the total U.S. vehicle fleet.) Other oil users are industry (25%), buildings (7%), and electric utilities (3%) (5).

Natural Gas

Natural gas is an increasingly popular fuel, due to its reputation for environmental cleanliness, the relative abundance of domestic reserves (which makes it less susceptible to international price fluctuations), and to technical improvements that have allowed it to be used with impressive efficiency for electricity generation, space cooling, and other applications.

Historically, natural gas has generally been the second most-consumed U.S. energy source (behind oil) since 1958 (5). Low prices in the 1950s and 1960s (due in part to regulation) led to steady increases in demand for natural gas (5); demand peaked at 22 trillion cubic feet (tcf) in 1972. Since then, supply uncertainty, increasing prices, and restrictions in the Powerplant and Industrial Fuel Use Act of 1978 (repealed in 1987) slowed demand (5), although natural gas remains less expensive than oil or electricity on a per energy unit basis (5).

Industry consumes more natural gas than any other sector, accounting for more than 40% of total use. Over the last 15 years, greater use of cogeneration accounted for a significant share of the growth in industrial gas use (15). The relative use of natural gas in the residential and electric utility sectors has been fairly steady, representing about 25% and 15–20% of total demand since 1950, respectively. The commercial sector experienced the largest relative growth in natural gas demand over the last 40 years (from 7% to 14%), driven primarily by winter heating needs.

Over the next two decades, domestic natural gas use is expected to increase steadily, with growth averaging slightly more than 1% annually, in part due to its comparatively low cost and its environmental benefits relative to coal and oil. The greatest demand increases are expected from electric utilities and industry, where combined-cycle generation and cogeneration, respectively, are forecasted to increase quickly (14). By the year 2000, domestic natural gas consumption is projected to match its 1972 peak, and to increase to more than 24 trillion cubic feet by the year 2010 (14).

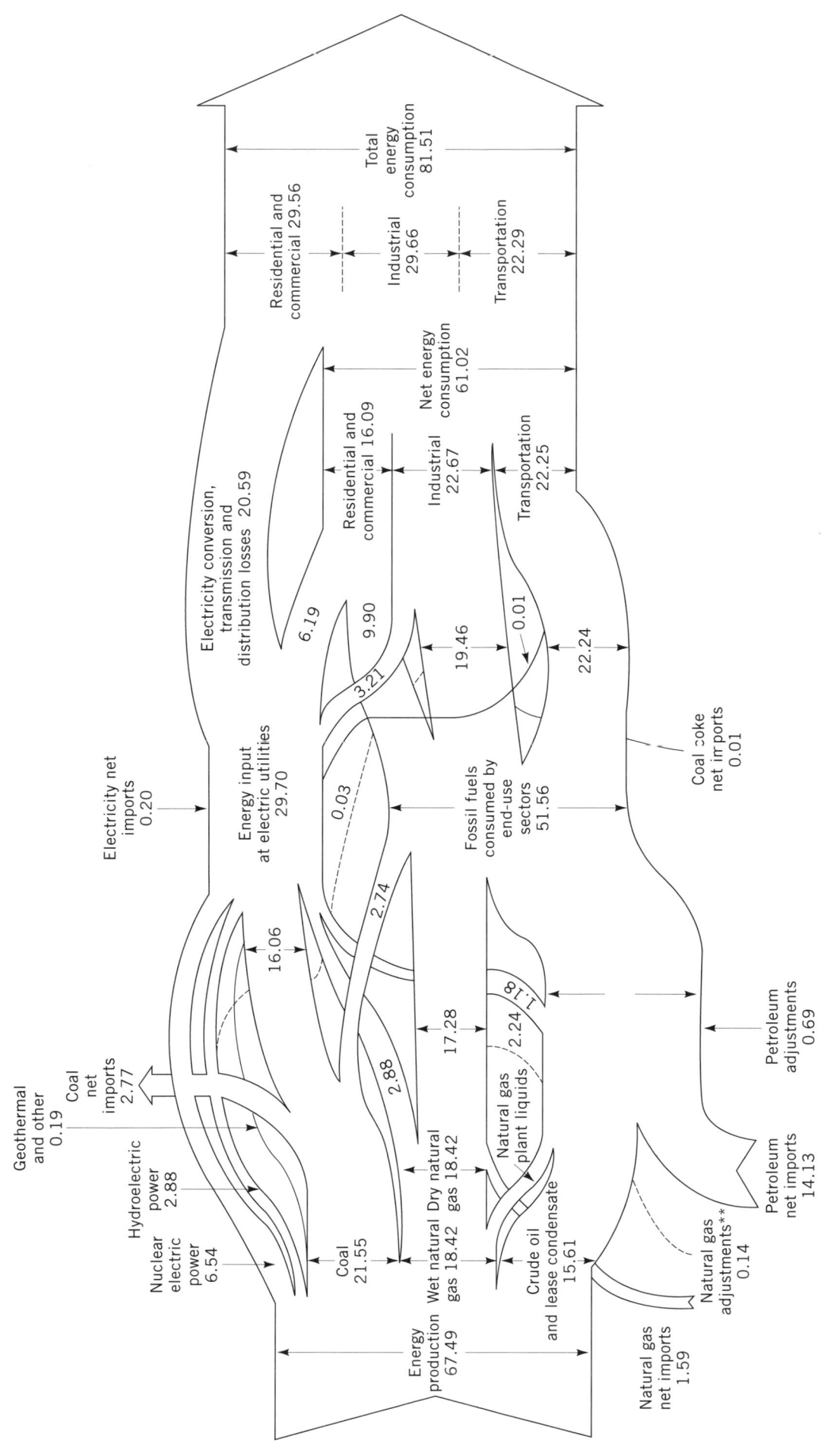

Figure 10. Total U.S. energy flow, 1991 (in quadrillion Btu) (5).

Electricity

U.S. electricity use has shown robust and steady growth since at least 1950. Over the past 40 years, annual electricity demand growth has averaged almost 6% (5), which translates to a doubling of total demand about every 12 years. However, demand growth during each decade has varied considerably. In the 1960s, annual electricity demand growth exceeded 7%, dropping to about 4% in the 1970s and less than 3% in the 1980s (14). Nonetheless, no other major energy source experienced as great of a demand growth in the past 40 years; since 1950, national electricity use has increased almost 10-fold (5). This dramatic growth in electricity use is largely explained by its convenience, end-use efficiency (which can approach or even exceed 100%), versatility, and cleanliness (for the end user), qualities important enough to overshadow the expense of this fuel (it is several times more expensive per unit of energy than other fuels).

In the 1950s, the industrial sector purchased about half of all U.S. electricity, but that relative share began dropping by the 1960s (5), as the residential and commercial sectors became more electrified, particularly with greater use of electric home appliances (notably space cooling) and electronic office equipment. Today, the residential, commercial, and industrial sectors each accounts for roughly one-third of electricity demand (5). Historically, coal has accounted for roughly half of all electricity generation; today, 55% of U.S. electricity is produced from coal, more than twice as much as the next largest source, nuclear power, at 22% (5).

In the next 20 years, national electricity demand is projected to increase between 1 and 2% per year, far lower than historic averages, in part due to the saturation of electric appliances and equipment, as well as to the proliferation of government and utility energy-efficiency programs. Most of the demand increase is expected in the commercial and industrial sectors, whereas demand growth in the residential sector will be more modest, due in part to low projected increases in U.S. population (14). If these projections are correct, annual U.S. electricity use will increase from about 2.7×10^{12} kWh in 1990 to 3.7×10^{12} by the year 2010 (14).

WHERE WE ARE TODAY

In 1991, the United States consumed about 81.5 quads of energy. One concise way of understanding where this energy comes from and where it goes is in the form of a "spaghetti chart." This chart shows all significant energy flows, starting with production on the far left, moving through transformation (eg, to electricity), and concluding with final consumption on the far right.

THE FUTURE OF ENERGY CONSUMPTION

Uncertainties over oil prices, technical advances, market penetration of energy-efficient technologies, economic growth, and other factors make it extremely difficult to forecast future energy consumption. It is possible, however, to identify likely trends and key uncertainties that will influence future levels of energy consumption.

Likely Future Trends

In the past 20 years, dramatic improvements in the energy efficiency of energy-using devices has occurred. New refrigerators, for example, use less than one-half the electricity used by units sold in the early 1970s. In addition, research by the Congressional Office of Technology Assessment and others suggests that dramatic further reductions in energy use are possible. Although the cost-effectiveness and market penetration rates of these technologies are unclear, technical advances will likely lead to continued improvements in energy efficiency.

There has been a gradual yet clear economic shift in the United States away from heavy manufacturing and toward greater demand for services. These trends, such as slow growth in the demand for transport of raw materials and minerals, rapid growth in computers and other information technologies, and increases in vehicle-miles per person per year, will likely continue in the near future.

Due largely to changes in state-level regulation, electric and gas utilities have invested considerable resources in energy efficiency. Current utility investments in energy efficiency exceeds $2 billion/year. Although the energy savings resulting from these investments are uncertain, they will likely result in increased market penetration of energy-efficient technologies.

Key Uncertainties

Considerable research effort focuses on alternative fuels for transportation. Currently, no fuel is a clear successor to gasoline; some argue that in the future, the light-duty vehicle fleet will use several different fuels, for example, electric vehicles for short urban trips and gasoline for longer trips. In any case, a shift to nonoil-based fuels by a significant fraction of the vehicle fleet could dramatically reduce U.S. oil consumption.

Oil price fluctuations, often driven by political factors, have strongly influenced energy use in the past. Despite these fluctuations, the United States still heavily depends on imported oil; in the future, oil prices will continue to affect U.S. energy consumption.

The possible effects of global climate change, due largely to CO_2 emitted from fossil fuel burning, is attracting increased scientific and political attention. Although the effects of increased global temperatures are uncertain, concern over possible detrimental effects may result in efforts to reduce CO_2 emissions. This could significantly change energy use (eg, large increases in energy efficiency, and a shift toward renewable and nuclear energy).

Natural gas is seen by many as the fuel of the 1990s, due to its environmental benefits, low price, and high availability. However if demand increases, then existing pipeline capacity may be exceeded, requiring expensive and time-consuming construction of new pipelines. Furthermore, deregulation of natural gas prices will likey affect prices and therefore demand. In general, changes in natural gas prices may influence future energy consumption levels.

Finally, the market penetration of energy-efficient technologies significantly influences consumption levels. For example, by one estimate, full-market penetration of cost-effective technologies in the buildings sector would lead to a savings of 14 quads/year by the year 2015. These market penetration rates depend on energy and technology prices, perceptions of comfort and service effects, changes in consumer preferences, and other factors, and are a source of considerable uncertainty in future energy consumption.

Disclaimer

The opinions expressed in this article are those of the authors, and are not necessarily those of the Office of Technology Assessment, the United States Congress, or the United States Government.

BIBLIOGRAPHY

1. U.S. Department of Commerce, Bureau of the Census, *Statistical Abstract of the United States 1992*, Washington, D.C., 1992.
2. U.S. Congress, Office of Technology Assessment, *Changing By Degrees: Steps to Reduce Greenhouse Gases, OTA-O-482* U.S. Government Printing Office, Washington, D.C., Feb. 1991.
3. U.S. Department of Energy, Energy Information Administration, *Monthly Energy Review, DOE/EIA-0035(93/06)*, Washington, D.C., June 1993.
4. U.S. Department of Energy, Energy Information Administration, *International Energy Annual 1991, DOE/EIA-0219(91)*, Washington, D.C., Dec. 1992.
5. U.S. Department of Energy, Energy Information Administration, *Annual Energy Review 1991, DOE/EIA-0384(91)*, Washington, D.C., June 1992.
6. U.S. Congress, Office of Technology Assessment, *Energy Use and the U.S. Economy, OTA-BP-E-57*, U.S. Government Printing Office, Washington, D.C., June 1990.
7. U.S. Congress, Office of Technology Assessment, *Building Energy Efficiency, OTA-E-518*, U.S. Government Printing Office, Washington, D.C., May 1992.
8. Oak Ridge National Laboratory, *Transportation Energy Data Book 12, ORNL-6710*, Oak Ridge, Tenn., Mar. 1992.
9. J. Murrell, K. Hellman, and R. Heavenrich, "Light-Duty Automotive Technology and Fuel Economy Trends Through 1993," *EPA/AA/TDG/93-01*, U.S. Environmental Protection Agency, Ann Arbor, Mich., May 1993.
10. *Railroad Facts 1992*, Association of American Railroads, Washington, D.C., Sept. 1992.
11. *Energy Use in Freight Transportation*, staff working paper, U.S. Congress, Congressional Budget Office, Feb. 1982.
12. U.S. Congress, Office of Technology Assessment, *Improving Automobile Fuel Economy: New Standards, New Approaches, OTA-E-504*, U.S. Government Printing Office, Washington, D.C., Oct. 1991.
13. U.S. Congress, Office of Technology Assessment, *Industrial Energy Efficiency, OTA-E-560*, U.S. Government Printing Office, Washington, D.C., Aug. 1993.
14. *Annual Energy Outlook 1993, DOE/EIA-0383(93)*, U.S. Department of Energy, Energy Information Administration, Washington, D.C., Jan. 1993.
15. *Natural Gas Annual 1991, DOE/EIA-0131(91)*, U.S. Department of Energy, Energy Information Administration, Washington, D.C., Oct. 1992.

ENERGY EFFICIENCY

MARC ROSS
University of Michigan
Ann Arbor, Michigan

Energy use by people provides enormous benefits but causes substantial harm; and the harm will grow as energy use continues to grow. Improved energy efficiency is the centerpiece of action to limit the harm while enabling continued growth in the services provided by energy. Not only does improved efficiency reduce energy requirements and related environmental impacts, but the reduced spacial concentration or density of energy requirements eases the task of introducing more benign alternative energy supplies.

This article begins with needed background, including explanations of energy use and its efficiency. Next analytical tools and concepts are discussed. Opportunities for improving energy efficiency in industry and transportation are also examined. The discussion ranges from improving energy conversion devices such as engines and furnaces to that of whole processes such as steelmaking and automobiles to that of the whole system. An understanding of these different levels of organization and change is needed if one is intelligently to consider future developments. It is also needed for intelligent policymaking. For both industry and transportation, a wide range of policies is also considered from the perspective of their influence on energy efficiency. See also THERMODYNAMICS; ENERGY CONSERVATION.

BACKGROUND

The Physics of Energy

The First Law. Energy is a quantitative property of all things (1). It has many forms, such as energy of bulk motion, gravitational energy, energy of light (or electromagnetic waves), chemical energy, and thermal energy. Transfer of energy from one form to another and from one place to another, or one set of matter to another, characterizes all happenings. The conversion of the chemical energy in gasoline to the energy of motion of a car plus considerable thermal energy as a by-product or waste, will be discussed.

An essential strength of energy as a concept is that energy is neither created nor destroyed. Its forms change, and it can move from one place to another, but for any region of space or system of matter the total energy is constant, after accounting for flows into and out of the system (the first law of thermodynamics). The constant numerical value for energy is a summary characteristic of any system. To determine it, all the different forms of energy must be expressed in the same units and added. This property of energy is similar to money accounting. Accounting for money flow and for changes in form makes it possible to conduct powerful analyses. There are many analogies between money and energy accounting, including the need to convert quantities of money into a common unit and to account for losses or diversions. Money can, however, be created and destroyed, whereas energy cannot.

The Second Law. Most energy forms are high quality, which means they can, in principle, be fully converted one into another. Thermal energy, the energy of random motion of atoms of matter, is of variable quality. Thermal energy can only be partly converted into a high quality form (the second law of thermodynamics).

The second law of thermodynamics is qualitatively shown in Figure 1. Two high quality forms such as chemical energy and electrical energy can, in principle, be wholly converted one into the other, and either of them can be converted into thermal energy, but the thermal energy cannot be wholly converted to a high quality form. Energy converted into a high quality form, or delivered from another place in a high quality form, is called work. Energy converted to the thermal form or delivered in that form is called heat.

Consider the hot gas in the cylinder of an engine, ie, the fuel–air mixture after combustion. As the gas pushes

the piston, this thermal energy is converted to energy of bulk motion, a high quality form; so the hot gas does work. But this thermal energy cannot all be converted to work. Most of the energy is carried away as lower quality thermal energy in the exhaust, and substantial "losses" of this kind cannot be avoided. Heat engines convert chemical fuel energy to thermal energy and then only partially into energy of bulk motion. They are fundamentally inefficient in this respect.

The quality of thermal energy depends on its temperature. High temperature thermal energy is of relatively high quality. At high temperatures most, but not all, of the thermal energy can, in principle, be converted to a high quality form. When the temperature is similar to the ambient temperature (that of the surroundings with which the system is in contact), then the thermal energy is of low quality. Little if any of it can be converted to a high quality form. The second law is illustrated in practical terms by heat pumps (or refrigerators) and heat engines: for a given amount of high quality energy input W, a heat pump delivers a larger amount of thermal energy Q_1 at a desired temperature higher than that of the ambient environment or the source from which thermal energy Q_0 is drawn (Fig. 2). The work primarily serves to in-

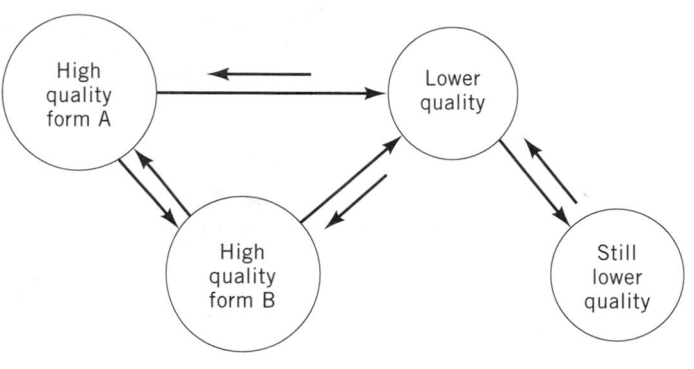

Although the total amount of energy does not change (first law), net transformations of energy increases the amount in low-quality forms (second law).

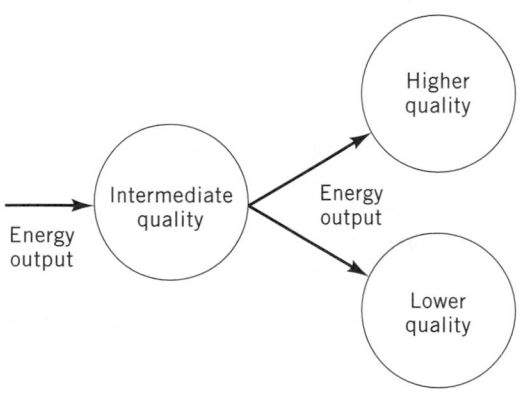

One cannot raise the quality of energy without either (1) reducing the quality of some of the energy shown, or (2) adding some higher quality energy (as in Fig. 2).

Figure 1. The second law of thermodynamics.

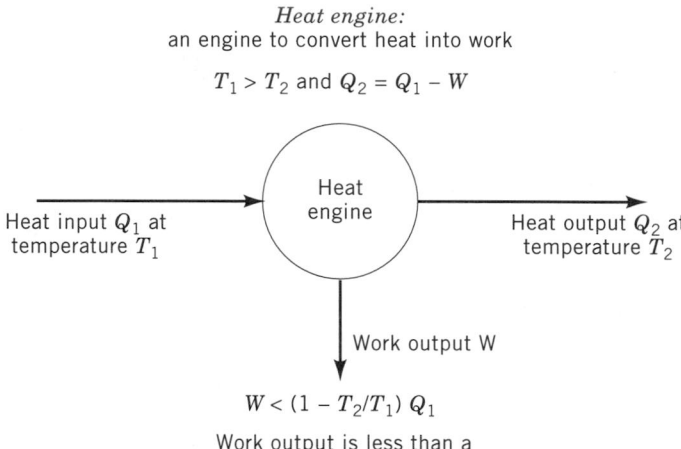

Heat engine:
an engine to convert heat into work

$$T_1 > T_2 \text{ and } Q_2 = Q_1 - W$$

Heat input Q_1 at temperature T_1

Heat engine

Heat output Q_2 at temperature T_2

Work output W

$$W < (1 - T_2/T_1) Q_1$$

Work output is less than a fraction of the heat input.

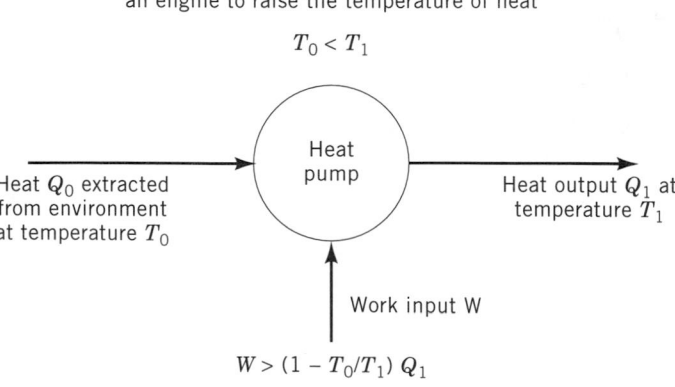

Heat pump:
an engine to raise the temperature of heat

$$T_0 < T_1$$

Heat Q_0 extracted from environment at temperature T_0

Heat pump

Heat output Q_1 at temperature T_1

Work input W

$$W > (1 - T_0/T_1) Q_1$$

Work input must be more than a certain fraction of the heat output.

Figure 2. Second-law limitations on heat engines and heat pumps. Temperatures are absolute temperatures.

crease the temperature of the thermal energy Q_0. The greater the quality (temperature) difference between the two forms of thermal energy, corresponding to Q_0 and Q_1, the more work must be done relative to the heat delivered. (A refrigerator operates in the same way except that the emphasis is on the thermal energy extracted at the lower temperature Q_0.) The first law is obeyed because the energy Q_0 is being extracted from an existing source, and $Q_1 = Q_0 + W$.

Turn this machine around, and it is a heat engine: with an input of thermal energy Q_1 at a high temperature T_1, work W can be done, but substantial thermal energy at lower temperature (ie, closer to ambient conditions) will have to be discharged. Thus $W < Q_1$. The greater the quality difference between the two forms of thermal energy, the greater the work output can be relative to the thermal energy input Q_1.

Consider the example of an automobile engine. Let 100 units of chemical (fuel) energy be introduced along with air into the cylinders. Almost all of this is converted into thermal energy by combustion. Some 40 units of work are done by the hot gases expanding against the pistons, while 60 units of lower quality energy are lost, primarily in the forms of exhaust and thermal energy transferred to the engine (eg, to the engine coolant). Thus $Q_1 = 100$, $W = 40$, and $Q_2 \lesssim 60$ energy units. Q_1 is slightly greater than $W + Q_2$ because some energy is lost in other forms, such as unburned fuel.

In contrast to converting one form of high quality energy into another using a heat engine, direct conversion between two high quality forms, such as chemical energy to energy of bulk motion, can be achieved without substantial losses of thermal energy. Efficient, direct conversion from chemical fuel to bulk motion (eg, using a fuel cell) has been technically difficult to achieve in practice, but it is a promising and active area for research and development.

Similar concepts apply, eg, to lighting. Light from a gas flame or incandescent electric bulbs is produced through the intermediary of thermal energy at high temperature. Only a modest fraction of the thermal energy can be converted to light, however. Fluorescent electric lights and other lamps not depending on the intermediate thermal stage are much more efficient.

Energy Use and Efficiency. What is meant then by energy use and efficiency of energy use? Energy use is the conversion of energy into a desired form at a desired time and place. For example, chemical fuel is converted into thermal energy at a temperature appropriate for industrial process heating, to operate an engine, or to keep a building warm. When energy is used it does not disappear, although much or all of it usually quickly degrades into thermal energy at near-ambient temperature, energy of low quality.

The thermal energy obtained by burning fuel in an automobile engine mostly goes out with the exhaust, where it quickly becomes useless, spreading into the surrounding air and moving toward ambient temperature. As mentioned, about 40% of the original energy remains as work done on the pistons. In average engine operations roughly 50% of this is lost to the thermal energy form as

a result of frictional processes inside the engine, and only some 15% of the original fuel energy typically reaches the drive wheels in the form of rotating machinery. There it does what is desired; it moves the vehicle on the road. After doing so, this energy too is converted to thermal energy of low quality, via the frictional mechanisms of air resistance to the vehicle's motion, resistance of the tires to the deformation that occurs as the wheels roll, and braking friction. Thus all the fuel energy is soon converted to low temperature thermal energy; but on the way some 15% of it does the desired work.

As the automotive example shows, two effects mitigate against the desired transformation and movement of energy. First, the second law prevents thermal energy from being wholly converted into the high quality forms that may be desired. Losses of roughly 50% are implied for the best high temperature engines, and much more severe limitations apply near ambient temperatures. The second and independent effect is dissipation, of which the primary kind is friction. Friction in an internal-combustion engine, for instance, involves the rubbing of solid parts, the fluid friction in pulling air through small orifices into the cylinders and pushing exhaust out, and the fluid friction of pushing coolant and lubricant through yet smaller orifices.

There are other forms of dissipation. For example, it has proven difficult to reduce losses to near zero in chemical to electrical (and electrical to chemical) transformations. Electrical resistances in charging and discharging a good car battery result in loss (to heat, of course) of roughly 33% of the energy. Careful, inventive improvements over a long period have resulted in reducing the dissipative losses in converting aluminum oxide to metal in commercial electrolytic cells to about 50%. Meanwhile, success has been easier in conversions between bulk motion and electricity. Large motors and generators achieve these transformations with little loss.

One measure of the energy efficiency of a car is the ratio of the desired energy reaching the drive wheels to the fuel energy input, which is about 15% for average driving. The efficiency of the vehicle's engine, considering only the work done by the hot gases on the pistons and neglecting all friction in the engine, is typically about 40%. These are strictly defined thermodynamic efficiencies, ratios of energy outputs to inputs.

These thermodynamic efficiencies are useful, but a broader efficiency concept is also needed, because the fuel energy required to run a car is affected by how large the car is, how streamlined it is, and how hard its tires are. The fraction of fuel energy reaching the wheels is only part of the story. In principle, there are no limits to reducing the impediments to moving the vehicle (air, tires, and brakes) once the energy reaches the drive wheels. A car can be made more streamlined, the tires can be made harder, the use of frictional brakes can be minimized. In the extreme, it is possible to imagine vehicles running in evacuated tubes with almost frictionless magnetic cushions. Thus no simple ratio of energies can be defined as the efficiency for moving the vehicle as a whole.

It is useful, therefore, to define an energy intensity (EI), such as fuel use per passenger mile, and give the word *efficiency* an additional meaning. Efficiency improvement is the increase in the ratio of old to new energy intensity EI_0/EI. Thus the efficiency of a vehicle has been improved 10% when the energy use per passenger mile, for a certain kind of trip, is reduced some 9%, ie, 1.00/0.91 = 1.10.

Although the car has been used as an example, the same arguments apply to other energy uses, such as heating buildings and producing industrial products. For heating a building a thermodynamic efficiency can be defined, which expresses the ratio of desired thermal energy output from the heating system to the high quality input energy. (Actually, there are several definitions with different applications, such as nominal furnace efficiency, seasonal furnace efficiency, and efficiency compared with an ideal heat pump.) Analogous to the car, this approach can describe the furnace or heating system but not the effectiveness of the building as a whole from a heating standpoint. The building could be designed so that it needs no energy input as such, eg, by storing ambient warmth, or zero quality energy from the previous summer. Again the term *efficiency* can also be used to compare the energy intensities of the heating of buildings.

Comparing a heating system to an ideal heat pump was mentioned parenthetically as one way to define a strict thermodynamic efficiency. This is the second law efficiency approach (2,3). The first law efficiency is the ratio of desired heat output to actual energy input. The second law efficiency is the ratio of energy input to an ideal heat pump system to the input to the actual system (both systems delivering the same desired output). Household heating systems have first law efficiencies of 60% and higher, ie, 60% of the input energy is delivered as desired heat and 40% is lost from the system before delivery, eg, up the flue. Their second law efficiencies are roughly 5%, depending on how cold it is. From this perspective, low quality thermal energy from the outside air or water can, in principle, be extracted, upgraded, and delivered at the desired temperature with the work input only 5% of the energy delivered, because the energy delivered is of such low thermodynamic quality. From this second law perspective, any low temperature heating system is highly inefficient, unless it is based on a heat pump or makes use of heat that is waste from another process. In a much broader attempt using these techniques, it has been concluded that the second law efficiency of the entire U.S. economy is roughly 2.5% (4).

Stocks, Flows, and Units. It is often necessary to discuss energy numerically as a stock or quantity: barrels of oil, kilowatt hours of electricity, kilocalories of chemical energy. It is also necessary to discuss the time rate of energy flows: barrels of oil per day, kilowatts of electrical work being done, British thermal units (Btu) of heating per hour. Energy flows are called power by physicists and engineers. The basic power unit, the Watt (W), expresses simultaneously the quantity of energy and time of flow (analogous to the velocity unit knot). If, eg, a household uses an average of 1 kilowatt (kw) of electrical power per hour, its average power is 24 kw·h per day, or 8760 kw·h per year. The Watt unit includes time in the denominator and is converted to a stock or energy unit by multiplying by time, as in kilowatt hour (the energy corresponding to

the flow of 1 kw for 1 h). This can again be converted to a power unit if one desires, as in kw·h/yr. The latter form of power unit is usually used to describe the average rate of flow, while the form kw can be used to describe the rate of flow at an instant, or an average rate.

Some common units for energy and power and conversion factors into Btu and Joules (the principal traditional and metric units, respectively) are given in the front of this volume. For example, 7.2 million barrels (bbl) of gasoline are used in the United States each day. Here are some conversions: (7.2×10^6) (5.25×10^6) 365 = 13.8 quads/yr = $13.8 \times 10^9 \times 33.4$ = 461×10^9 W = 461 GWs (average rate of use). The conversion factors are 5.25×10^6 Btu = 1 bbl gasoline (using the higher heating value on combustion), 10^{15} Btu = 1 quad, 33.4 W = 10^6 Btu/yr, and 10^9 W = 1 GW. The energy values associated with fuels are, of course, the heat release on combustion. A significant part of the definition is whether it is higher heating value, in which the heat released converting any water from gas to liquid is counted, or lower heating value in which it is not counted.

Power units, like Watts, Btu/h, horsepower (hp), and bbl/day, commonly characterize energy equipment and facilities. The oil well, power plant, automobile, home appliance, transmission line, and pipeline all have a nominal maximum capacity or rate of energy conversion or energy carriage. An electric power plant may have a rating of 1 GW, and the capital cost of such plants is often expressed in dollars per kilowatt. In this case, Watts refer to the capability to produce electric power. The rate of fuel use is roughly three times higher.

The Earth's Energy Balance. One face of the earth is always bathed in sunlight, at least at the top of the atmosphere. Sunlight is absorbed by the earth at an enormous rate in terms of power units. (Some energy also reaches the surface of the earth from the interior and from radioactivity, and some is released by burning fossil fuels.) The energy arriving must be balanced by energy leaving, or the earth's average temperature would change. The balance is achieved by direct reflection and by the outflow of "earthshine" (Fig. 3). Earthshine is infrared light, similar to sunshine but of lower frequency and invisible to the human eye.

The flux of sunshine in space at the earth's orbit is F_s = 1.4 kw/m². The total solar energy absorbed by the earth is $F_s \times$ (area of earth facing the sun) $\times$ (1 − albedo), where the albedo (≈ 0.3) is the fraction of sunlight reflected. Thus

$$\text{sunlight absorbed (rate)} = F_s \cdot \pi R_e^2 \cdot 0.7 = 1.25 \times 10^{17} \text{ W}$$

On the average, this flow is balanced by the outward flow of earthshine. The total earthshine emitted is

$$\text{earthshine emitted (rate)} = F_e \cdot 4\pi R_e^2$$

Setting these flows equal, the average flux of earthshine at the earth is F_e = $0.7 F_s/4$ = 240 W/m². (Much of the absorption and the emission into space is from the top of the atmosphere.) For comparison, the rate of energy use by people is about 1.2×10^{13} W, about 1 part in 10,000 of the average solar influx. However, energy use in U.S.

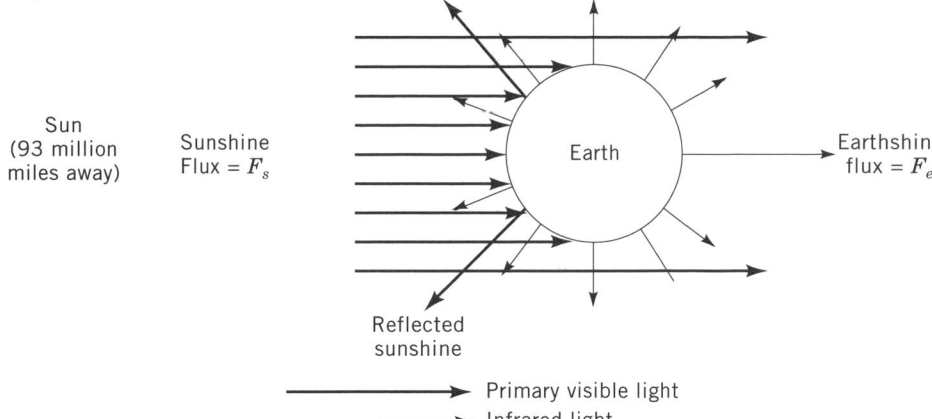

Primary visible light

Infrared light

Figure 3. Energy balance of the earth.

metropolitan areas is roughly 2% of the average solar influx.

The propensity to emit earthshine depends on the temperatures of the surface and upper atmosphere and on the transparency of the atmosphere to infrared light. Greenhouse gases in the atmosphere decrease this transparency. If the concentration of greenhouse gases is increased, eg, by people's activities, then the surface will get warmer to emit the same amount of earthshine.

To summarize, energy, as used by people, is initially extracted from the natural environment, usually then converted into an energy carrier (a form convenient for use), transported to where it is needed, and finally used. The use consists of transforming the energy into a desired form at a desired place and time. None of the energy disappears in this use; however, inefficiencies in each step result in energy being lost to the system, almost always in the form of low temperature heat. In addition, the useful energy, such as chemical change, bulk motion, or light, is also usually transformed by natural processes into low temperature heat. All this heat is eventually radiated into space as infrared light as part of the earth's thermal balance.

The Commercial Energy System

The commercial energy system (Fig. 4) comprises extraction of energy supplies, conversion to energy carriers, and final use in the various sectors of the economy. The principal supplies are fossil fuels (primarily coal, oil, and natural gas), nuclear fuel (uranium for fission), and solar-related energy (especially hydropower, biomass, wind, and direct sunlight). Typically, the extracted energy forms are converted into energy carriers appropriate for fuel use (5). Fossil fuel is used to generate the carrier electricity via the intermediary of thermal energy, with an average efficiency of about 32% (counting losses in transmission and distribution). That is, 32% of the fuel energy is delivered as electrical energy to customers, while 68% is lost as near ambient temperature heat. Crude oil is converted into products like gasoline with an overall efficiency of about 90%.

The principal energy carriers are electricity, dry natural gas, petroleum products, and washed and graded coal. In the future, gaseous hydrogen (eg, extracted from fossil

fuel or from water) might become a major carrier. The properties of an energy carrier (especially its storability, its transportability, the ease of its conversion at the point of end use, and its environmental acceptability) establish its relative value for a particular end use. There is a wide range in the storage characteristics of different energy carriers. The storage characteristics of solid and liquid carriers (high energy density and low capital cost) are hard to beat.

As a general rule, the storability of a form of energy is closely linked to its transportability. Fuels that are easily stored (coal, oil, and uranium) can be economically shipped in batch mode via barge, railroad, truck, etc. Those forms that are expensive to store (electricity and natural gas) are most efficiently transported by line, or continuous, mode. The systems required for distribution to, and conversion at, the site of end use are also important determinants of the appropriateness of a carrier. In small-scale applications (buildings, transportation, and small industry) the use of solid fuels is inhibited.

The storability of energy affects not only the physical structure of the delivery system but also the institutional

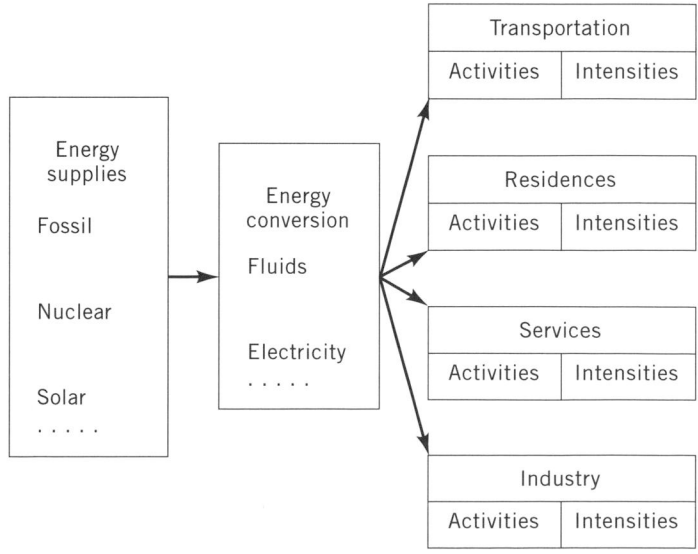

Figure 4. The commercial energy system.

structure. Because coal and oil can be brought to the retail market in batch mode, they can be sold competitively by multiple distributors. Since natural gas and electricity, on the other hand, entail the installation of an immobile distribution system, they are most efficiently distributed by a local monopoly. If methods were discovered to store these carriers in an inexpensive, compact form, they might also come to market in batch mode and be sold competitively.

The efficiency with which raw energy supply is converted to carriers and the efficiency of transportation of energy are issues of major concern. In the interests of coherence and brevity, they will not be dealt with here.

Societal Concerns with Energy Use

Societal concerns are only partly associated with today's conditions (6). They are largely focused on predictable future developments. In the United States, energy consumption used to grow about 2% faster than the population. For the past 20 yr, it has been growing more slowly than the population, ie, per capita consumption has declined slightly. (The population has been growing 1.0%/yr.) Two principal uses are growing substantially faster: electricity in buildings and petroleum for transportation. Moreover, in the industrializing world these two kinds of use are growing extremely rapidly. Thus when considering the adequacy of resources, cost of facilities and environmental damage, it is important to consider the increasing pressure caused by population growth and growth in physical consumption per capita in these two areas of use.

Energy Resources. The energy resources in use are primarily fossil fuels; they account for 75% of the world's total consumption. On the scale of a few decades, conventional petroleum will become scarce, while natural gas will become more prominent, although it also will begin to be in short supply (7–11). Natural gas is a convenient fuel once pipelines and distribution are established, and the ultimate resources of natural gas are somewhat larger than those for petroleum; greenhouse gas emissions are also lower. Supplies of coal are even larger. Alternative fossil fuels are also available (heavy oils, tar sands, and shale oil), but the environmentally related costs of coal and the alternatives are likely to be high (12).

Solar-related energy flows are already being tapped for some 20% of current world energy use, primarily biomass and hydropower. New biofuels and wind and photovoltaic technologies to produce electricity are highly promising (13). Wind-generated electricity is beginning to become competitive in good locations without major subsidies. Biofuels grown for the purpose, rather than being created as by-products of other agricultural and forestry activities, are not important today, but with selected species in marginal croplands and using new methods of cultivation and conversion into carriers, they appear promising at today's prices. Photovoltaics are competitive today in niche markets, and reasonable projections of manufacturing costs at large scale suggest that photovoltaics could become competitive in favorable large markets, such as in the U.S. Southwest, in another decade or two.

Although some are promising from a cost standpoint, solar-related sources are diffuse. Biomass can take up a lot of land, for example. So there are capacity limitations. There is also powerful geographical variation, eg, in Michigan, there is not much sunlight in the winter and relatively little persistent wind. There is water, however, so biomass can be grown, but clearly there are limitations. In general, most forms of solar-related energy are constrained in most geographic areas. In desert country, however, photovoltaic electricity could be produced on a large scale.

Nuclear fission, at 5% of the world's total use, is faltering in competition with natural gas and coal and may be left behind in the foreseeable future by cost reductions in the solar-related energies. However, countries that are affluent and have high population density, like Japan and France, may continue to develop nuclear power.

It must be added that because energy purchases now make up a substantial 7.5% of all final expenditures in the United States society will not embrace much higher costs simply to have the energy. There are too many other things people want to do with their incomes and so many ways to do with less energy, that people will not accept large price increases, at least after a period for adjustment. In other words, only a fraction of energy consumption is "essential" and there are many strings to the supply bow in the long run. When expenditures on energy rose to 14% of GNP in 1980–1981, many things were done on both the supply and the demand sides to bring the costs down. Costs are back down. What was needed was some time and investment to make supply and demand adjustments. The main point here is that potentially huge new kinds of energy supply, which some people are enthusiastic about, will not be of wide interest if they are costly. Nuclear fusion may be an example.

Given this perspective, there is a mixed conclusion. There are limitations to the cheap fuels used now. These limitations will be felt strongly in a few decades. Certainly not all the proposed energy supply panaceas are practical, but there are a variety of promising new energy supplies that provide options. With these options and the critical option of using energy more efficiently, society has choices for developing a future energy system that can meet its goals.

Disruptions and National Security. The concentration of petroleum resources in the Middle East combined with the dependence of nations on petroleum as the principal transportation fuel has created concerns about vulnerability to interruptions in the petroleum supply. Twice in the 1970s there were serious interruptions: the Arab oil embargo of 1973–1974, and the interruption associated with the Iranian revolution of 1979. To diminish the impact of any future interruption, policies were established to help create alternative supplies, especially synthetic fuels from coal and shale oil, and to reduce petroleum demand (efficiency improvement and fuel switching). In addition, the United States has a strong military presence in the Middle East. In terms of the country's overall energy situation, the supply-side policies were not successful, and the demand-side policies were rather successful. However, the United States has returned to a strong dependence on imported oil. Although most other industrial countries rely even more heavily on imported oil, concern is stronger in the United States, because, as a world power it is thought that the country should not be sensitive to the influence of the principal oil-producing nations.

Capital Costs. The economic problem society faces with energy supply, conversion, and use is that it costs a great deal up front. The use of energy by people is a mammoth activity. In the United States, some 23 kg of fuels are used per day per person. A huge capital investment is required to accomplish this: wells, mines, pipelines, refineries, power plants, transmission lines, motors, furnaces, etc. Roughly 20% of all capital investment is energy related (14). This fraction rose to about 30% in the late 1970s to early 1980s. Energy investments strain the U.S. economy, and they severely constrain development in industrializing countries.

The constraint on development is one of the most important concerns about energy. Developing countries need energy, especially electricity and transportation fuels, for new industries and for improving lifestyles. The scale of conventional energy supply and conversion investment would soak up much of their capital, domestic and foreign. There are two ways to reduce this capital cost and yet provide the desired services: low cost energy efficiency improvement and energy supply systems with relatively low capital cost. Both these kinds of measures need to be strongly pursued. Energy efficiency opportunities are discussed here at some length. Biomass and natural gas systems are two supply options that tend to have relatively low capital costs.

The key conclusion is that capital for energy supply systems is scarcer than energy resources, in the limited meaning of the latter term as a natural endowment. Most of what consumers spend on energy goes for capital expenditures already made and to provide for their replacement.

In its simplest form, the price of energy pays for fixed (capital) and variable (operating) costs:

$$PQ = K \cdot CRF + VQ$$

where P is the energy price (eg, dollars per kilowatt hour), Q is the quantity (eg, kw·h) sold in a year, K is the capital cost of the plant and equipment (eg, of power plant, transmission, and distribution facilities), CRF is the capital recovery factor (fraction per year of the initial cost, primarily to repay the lenders or investors with interest), and V is the variable cost per unit of energy (primarily labor for operations and maintenance and fuel to operate the power plant). Considering an electric utility, a new power plant plus new transmission and distribution facilities might have the following cost parameters: $CRF = 15\%$ per year, capital cost = \$2000/kw of capacity, average use = 70% of rated capacity over the year, and variable costs = \$0.03/kw·h (of which \$0.02/kw·h is for fuel). With these numbers

$$P = \frac{2000 \times 0.15}{0.70 \times 8760} + 0.03 = \$0.08/\text{kw} \cdot \text{h}$$

where 8760 is the number of hours in a year. The average price of electricity in 1992 was \$0.68/kw·h. Of course, this calculation is crude and it neglects taxes and regulatory requirements, but it does illustrate that electricity from new plants can be somewhat more expensive than at present (though this opinion is controversial).

Environmental Impact of Energy Supply and Use. Energy supplies and intensive energy uses cause most air pollution and much other environmental damage (although there are other important anthropogenic sources of environmental degradation such as in agriculture and forestry) (15). Air pollution is the most important (16), and it is conveniently categorized in terms of a geographical scale.

Metropolitan smog associated with ozone at low altitude, especially in warm weather, and carbon monoxide in cold weather, are problems in many metropolitan areas around the world. Low altitude ozone is produced by sunlight acting on nitrogen oxides (NO_x) and hydrocarbons (HC). Fuel combustion is a significant source of these pollutants, especially from motor vehicles. Carbon monoxide is also largely created by motor vehicles. Power plants and heavy industry (such as refineries and steel mills) are also sources of air pollution in some cities. In China, eg, combustion of coal results in soot and sulfur oxide as well as NO_x pollution, which can be severe, especially in winter.

Low altitude ozone is also a regional problem, eg, northeast of Los Angeles and in the north central United States (17). More highly publicized is SO_x and NO_x pollution, primarily from power plants and heavy industry, which results in acid rain (18). This causes acidification of lakes, making them sterile, and is a factor in forest dieoffs, eg, in Germany's Black Forest. The sulfur is emitted in the combustion of high sulfur coal and high sulfur oil (although sulfur can be, and to a large extent is, removed at refineries). Most nitrogen oxide is a direct by-product of combustion in air, ie, although nitrogen is relatively unreactive, the high temperatures created when oxygen in the air combines with fuel lead to some reaction of the nitrogen in the air with oxygen.

Carbon dioxide from combustion of fossil fuels and, to a lesser extent, methane leaked from natural gas systems, are the greenhouse gases responsible for most of the global warming potential (19–21). The consequences of global warming are expected to be severe: summer temperatures that may be high enough to be dangerous to health, shifting rainfall patterns likely to dry up the U.S. grain belt, and eventual flooding of extensive low lying areas. In addition to these forms of air pollution, energy systems are also responsible for oil spills on the oceans and shorelines, severe degradation in many coal mining areas, and dams that displace people from valuable river areas and degrade important environmental values.

The environmental problems associated with energy have resulted in extensive regulations and new technology aimed at vehicular, industrial, and power plant air pollution as well as oil spills and damage caused by coal mining. (U.S. regulations do not as yet address global warming.) In the face of growth, especially in vehicular travel, these environmental policies have only been able to stop the increase in environmental damage at the metropolitan and regional levels in the United States (although absolute progress has been made in some areas). In the absence of greater progress on metropolitan and regional impacts and in the face of global climate change, environmental problems are probably the main challenge for energy policy. A gradual shift in the forms of energy supplies and substantial progress in energy efficiency are tools that must be brought into play to supplement cleanup technologies. Another option, cleanup technologies for capturing and sequestering greenhouse gases like carbon dioxide, may be too difficult but should be studied.

Policies to Improve Energy Efficiency, An Overview

A preliminary look at the social institutional opportunities to improve the efficiency of energy use is in order (22–24). Policies will be discussed in some detail later. The first point to be made is that the United States and most other governments have strenuously encouraged creation of energy supply. (Often these policies created extremely profitable opportunities to induce the desired investment.) Now that efficient energy use is a critical tool for managing energy-related economic, environmental, and security problems, it is timely for societies to take measures to encourage efficiency improvement. Fortunately, promising measures are available because efficiencies are typically low, and there are many opportunities in technology and practice to improve efficiencies cost effectively.

The policies are justified because society's interest in improving energy efficiency is much greater than the individual's. For most individuals and firms, energy costs are a minor part of total costs and potential savings do not warrant the priority needed to improve efficiency. Moreover, the individual cannot do much as an independent energy user to help manage goods held in common, like environmental quality.

Policies to encourage individuals and firms to deal with technical issues associated with efficiency include performance regulations for mass-produced energy-intensive products like light bulbs, refrigerators, and cars, and thermal standards for buildings. Policies to help energy users gain access to capital (a serious difficulty for many individuals and firms and a serious hassle for many more) include demand-side management by utilities based on financial assistance for efficiency projects. The creation of efficient new technologies is encouraged by policies to "push and pull" technology: government support of research and development, and directed government procurement, respectively.

ANALYZING THE EFFICIENCY OF ENERGY USE

Effective analysis of energy use is important because energy supply is capital intensive and needs must be anticipated and because energy efficiency calls for innovative public policies. In both areas good planning calls for good analysis. In this section the focus is on the language and methods of analysis. Later sections will deal with actual levels of energy use in different sectors and actual opportunities to improve efficiency.

Activity-Intensity Analysis

Effective analysis of energy use requires disaggregation. It is important to study lighting, driving, steelmaking, and other individual energy-intensive activities. Energy consumption is the sum over the energy uses in the various activities:

$$\text{Consumption Rate} = \Sigma_i A_i \, (EI)_i \qquad (1)$$

where A_i is the level of activity and EI_i is the energy intensity in subsector i. For example, when the use of lighting is involved, the activity could be area in square meters lit to a certain standard; thus the energy intensity is average electrical watts per square meter. The product of A and EI is then in Watts, a rate of energy use. Consider driving as another example. The activity could be vehicle miles per year; then the energy intensity is gallons of a standard fuel per mile of average driving. For steelmaking, the activity could be annual tons of steel produced, and the energy intensity is energy use per ton of an average mix of steel products.

It is critical for policy analysis to account separately for the amount of each activity and the intensity associated with it. It is also important to avoid the trap of thinking too narrowly about efficiency improvement. When analyzing furnace and process efficiencies, eg, it is important to avoid taking the activities as givens, independent of social and technological change and of energy-related policymaking. Passenger transportation is another example. It is provided by several modes, including walking. Planning related to changing the mix of modes is important to energy analysis, because the different modes have rather different energy implications.

Divisia Analysis. Change in aggregate energy consumption depends on three factors: the overall growth in activity, the change in the mix or composition of activities, and the change in the individual energy intensities. Overall activity can be represented by a summary index, depending on one's needs. It could be gross domestic product. One way to define summary measures for changes in the other two factors is Divisia analysis (25). The energy-weighted average energy intensity defines the real energy intensity. The energy-weighted average activity represents the combined effect of overall growth and relative changes in activities.

It is possible to express aggregate energy use as the product of three indices:

$$I_E = I(Q)I(S)I(EI)$$

or, equivalently, in terms of three growth rates:

$$G(E) = G(Q) + G(S) + G(EI) \qquad (2)$$

where Q refers to the summary activity, such as GNP; S, to sectoral the mix of activities; and EI, to the real energy intensity. To obtain an expression like equation 2, define for the compound growth rate (in percent per year) of a variable X for the period t to $t + T$ years:

$$G(X) \equiv \frac{100}{T} \ln \left[\frac{X(t + T)}{X(t)} \right] \qquad (3)$$

and the energy-weighted-average growth rate:

$$\langle G(X) \rangle = \Sigma_i W_i G(X_i) \qquad (4)$$

where the sum is over the sectors or modes of equation 1 and the time-dependent energy weight for subsector, or mode, i is

$$W_i(t) = \frac{1}{2} \left(\frac{E_i(t + T)}{E(t + T)} + \frac{E_i(t)}{E(t)} \right) \qquad (5)$$

which is the fraction of total energy used in the subsector. It is found that

$$G(E) \approx \langle G(A) \rangle + \langle G(EI) \rangle \qquad (6)$$

To relate equation 6 to equation 2, $\langle G(A) \rangle = G(Q) + [\langle G(A) \rangle - G(Q)]$. The quantity in brackets is $G(S)$, the contribution of the change in the sectoral mix of activities. Equation 6 is not exactly what was wanted; it is an approximation because activity and intensity interact, but it is a good approximation.

Categorizing Energy Uses. One step in analyzing energy consumption is defining the categories of users. For example, should energy use by automobiles include the associated fuel consumption at petroleum refineries, the fuel consumption in maintaining and repairing automobiles, and that in manufacturing them? The user categories must be defined carefully or the same energy use will crop up in more than one place; there will be double counting. Economists have two ways of representing production (so that it can be summed without double counting) that provide good models for energy analysis. One is the final-product representation. In this picture, the values of all final products are added; these are products purchased by final consumers, not intermediate products purchased by businesses for further processing or resale. The categories that are summed are the different kinds of final products. The other method is the value-added representation. In this picture, the contributions of each category of production to the whole of production are added, eg, one term is the increase in the value of materials processed by the steel industry.

Theoretically, these two representations lead to quite different energy analyses. In the final product representation, automotive energy use is expressed as the combination of gasoline for driving and all the related energy uses for servicing cars, manufacturing cars, road building, manufacturing cement for the roads, and so on, in an endless sequence. This combination can be calculated with the help of input–output (IO) analysis, with its coefficients relating expenditures in one sector to expenditures in others (26). The advantage of the IO approach is that it expresses energy use in terms of more or less fundamental activities, tying the intermediate activities, like the part of steelmaking that is associated with cars, into the calculation of energy use for driving (27).

The value-added representation is simpler. It involves as separate items energy use in driving cars, servicing cars (and other equipment), assembling cars, steelmaking (for all purposes), etc. There is no doubt that, for a good energy analysis, analysts must in any case carefully consider developments in all such separate sectors that are energy intensive. That is not where the crux of the difference between the two representations lies. The weakness of the final-product analysis is that the time dependence of the IO coefficients may be inaccurately specified, yet may be critical. The relative changes in demand for final products, like driving or housing, may be less important than changes in the input–output coefficients that, eg, relate driving to a certain amount of steelmaking. The complex final-product analysis is not easily used to analyze technical change, such as the changing materials content of final products. It can, however, be effective for analysis of a particular product, where a few coefficients instead of the n^2 coefficients of the full input–output matrix are involved. IO analysis is also ideal for analysis of the implications of imports and exports in terms of overall content of energy in products (28).

In this article, energy consumption will be discussed using the value-added representation. This does not mean that connections, eg, between driving and manufacturing, can be ignored. Analysts have the obligation to consider important connections whether or not their formulations are designed to internalize those connections.

Energy Accounting Issues

Commercial Energy. Commercial energy involves the conversion of natural fuels into convenient energy carriers and the delivery of those carriers to customers. These carriers, like electricity, gasoline, and dry natural gas are essential for services like high quality lighting, high speed transportation, communications, most building appliances, and most kinds of manufacturing. Total current annual consumption of primary commercial energy in the United States is 82 quadrillion Btu (quads); world consumption is about 340 quads (29). So, with 4.6% of the world's population, the United States consumes 24% of the commercial energy.

Noncommercial Energy. The noncommercial energy is mainly biomass, either wood harvested for the purpose (both trees and brush), or by-products of other activities (paper- and lumber-mill residues, crop residues, dung, household trash, etc). Total current annual consumption in the United States is about 3 quads, primarily industrial by-products (in the paper, wood products, forestry, and food industries), and firewood used in residences. World consumption of noncommercial energy is 40 to 50 quads (30).

Primary or Resource Energy. Primary energy is the energy from natural sources that is converted for use by people. In the United States, some 92% of the primary energy is used for heat and power. The remaining use of fuels is for materials in making products like plastics. This feedstock use is often counted in energy consumption totals, even though the energy is not converted into other energy forms. Furthermore, hydroelectricity and nuclear electricity output is often expressed as primary energy by multiplying by the fossil fuel to electricity conversion factor of about 3.3.

End Use Energy. What is called end use energy is usually the content of the carriers converted at the point of use. It is poor practice routinely to sum up the Btu values of all end use energies, even though it is often done. Electricity should normally be accounted for separately from fuels, because electricity typically costs about three times as much as fuels, and involves about three times as much primary energy use. (For petroleum-based carriers like gasoline the losses in conversion and delivery are much smaller, so for them the issue is much less important.) In addition, substitution between electricity and fuel is usually more difficult than among fuels.

Because a quantity of electricity expressed as electrical energy differs greatly from the primary energy involved, it is essential to know which accounting convention is being used for electricity and whether it is used consistently. Unfortunately, many authors are careless on this point, especially when reporting total energy that combines consumption of electricity with that of other forms of energy.

Cost of Conserved Energy and Conservation Supply Curves

Let K be the up-front or capital cost of equipment whose purpose is to save energy in some energy-using facility (31). The cost of conserved energy (CCE) is the annualized payment for the conservation equipment divided by the annual energy savings:

$$\text{CCE} = \frac{K \cdot CRF}{S} \qquad (7)$$

where S is energy savings per year when the equipment is installed in the facility and used normally (the units for S might be kw·h/yr) and CRF is the capital recovery factor. Consider that K has been obtained by borrowing at interest d (annual interest rate). Then $K \cdot CRF$ is the annual payment that covers both interest and capital, paying off the loan in n years.

$$CRF = d/(1 - (1 + d)^{-n}) \qquad (8)$$

From a simple economic viewpoint, the behavior that should occur is that all investments will be made for which CCE is less than the price of energy (or the expected price of energy during the life of the equipment).

Example. Overhead cams are estimated to increase the retail cost of a typical automobile by $\Delta K = \$74$ (in 1987 dollars). The fuel savings in average driving are estimated to be 6%. Consider applying this technology to a base car with 22 mpg and driven 10,000 mi/yr. The savings rate is then

$$S = 0.06 \times \frac{10,000}{22} = 27.3 \text{ gal/yr}$$

Assume the car has a useful life of 10 yr and the inflation-corrected, or real, loan rate to the buyer is 10% per year ($d = 0.1$). Then

$$CRF = 0.1/(1 - 1.1^{-10}) = 0.163/\text{yr}$$

Then the cost of conserved energy is

$$\text{CCE} = 0.163 \times 74/27.3 = \$0.44/\text{gal}$$

If the price of gasoline is more than \$0.44/gal, as it of course is, this equipment should be adopted.

The interest rate in equation 8 is the decision maker's "discount rate." It is the overall cost to the decision maker of raising money (percent per year), including the costs of providing the down payment and other commitments that may be required. Proponents of energy conservation argue that a longer time horizon, ie, a lower discount rate, should be considered in designing energy efficiency policies. If, as they often suggest, a societal discount rate of 3% per year is adopted for the analysis, then $CRF = 0.117$ per year, and CCE = \$0.32/gal. Thus the use of this equipment is justified at a lower fuel price. Others, who oppose government intervention in new areas like energy conservation, argue that the correct calculation should only reflect the original buyer's perspective. If $n = 4$ yr is the original buyer's time of concern and is adopted for the analysis, and d is 10%, then $CRF = 0.315$ and CCE = \$0.86/gal. In this view, the equipment change is only justified at a higher fuel price.

Proponents of energy conservation also argue that for energy-policy analysis the price of energy to be compared with the CCE should include externalities or the net costs to society of the energy use beyond the costs included in the actual price. For example, energy suppliers and users do not pay for the air pollution damage caused by the energy system (which occurs even when all players satisfy environmental regulation). Unfortunately, the analysis of external costs is extremely uncertain.

The initial form of a conservation supply curve (CSC) to consider is shown in Figure 5. The y-axis, K/S, is the unit capital cost of each project in, eg, dollars per kilowatt saved. The x-axis shows the cumulative energy savings when the projects being analyzed are put in order of increasing unit capital cost. The CSC is analogous to an oil or gas supply curve: as money is spent to develop facilities, the capacity to deliver energy (eg, bbl/day) increases (32). With the CSC it is the capacity to save energy.

Alternatively, to facilitate comparison with the purchase of energy, analysts often annualize project costs to obtain a CCE on the y-axis. Then the CSC takes on the form shown in Figure 6. Unlike Figure 5, it now depends on the CRF, or the discount rate.

These are normative views; they describe the effect of energy conservation investments that should occur according to a simple financial criterion. One change is often

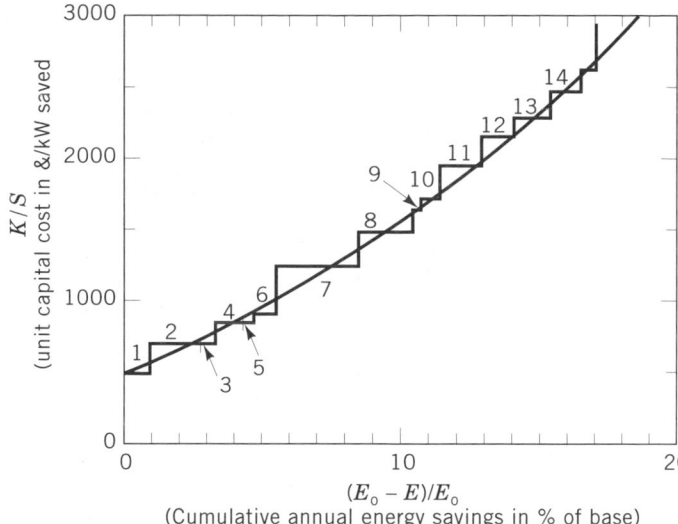

Figure 5. Conservation supply curve construction. Numbers on the curve correspond to hypothetical project numbers.

made in moving toward a description of actual behavior: the use of an implicit discount rate. After correcting for any investment inhibitions one may want explicitly to account for, one adjusts the discount rate d in equation 8 so that the actual investment behavior is described, or fit, by equation 7. Implicit discount rates are usually much higher than the apparent cost of raising money.

The Thermodynamic Potential for Improved Efficiency

The laws of physics, especially thermodynamics, determine the minimum conversion of energy that is needed to perform any task or "service." Many cases are straightforward: lifting a weight, creating a certain amount of light, raising the temperature of some material, and causing a certain chemical reaction in given input materials. For

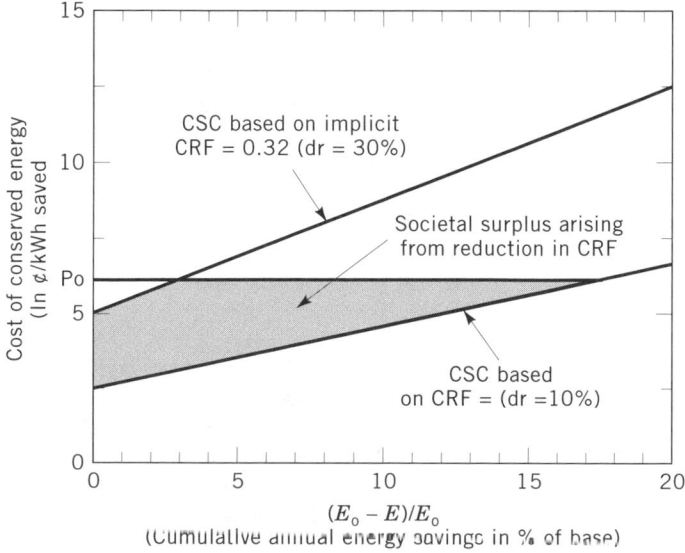

Figure 6. Dependence of conservation supply curve on capital recovery factor. CCE calculated assuming project life is equal to 10 yr.

these cases, it is possible to compare the ideal energy use to that typically used in the economy.

Typically, this ratio, the thermodynamic efficiency, is in the range 0–50%. There is great variability. The more narrowly one specifies the service to be provided the higher the efficiency of today's equipment tends to be. For example, the process typically used to convert iron ore to molten iron, the blast furnace, is some 80% efficient. For the entire steelmaking process from iron ore to rolled products, the efficiency is about 25%.

As indicated by this example, if the minimum energy use as required by physical laws is relatively high, today's thermodynamic efficiencies tend to be high. However, if little or no energy is required or if, in principle, energy could be extracted in performing the service, then the efficiencies tend to be low. Thus fabrication and assembly operations in manufacturing, where ideally the minimum required energy is negligible, have thermodynamic efficiencies of essentially 0%. In transportation and building heating, as was mentioned, most energy conversion devices have substantial thermodynamic efficiencies; but the entire vehicle or building has a low, although essentially undefined, efficiency, because the heating or the transportation could, in principle, be achieved with little or no energy inputs.

In the following two sections, the methods described in this section and the policies alluded to earlier will be developed in connection with the energy efficiency of industry and transportation, respectively. Discussions of energy efficiency in residences and services, often collectively referred to as the buildings sector, are available (33,34).

INDUSTRIAL ENERGY EFFICIENCY

Around the globe, industry consumes almost 50% of all the commercial energy used and is responsible for roughly similar shares of greenhouse gases, the emissions responsible for regional air pollution, and the waste solids and liquids that contaminate land and water. In this section, industrial production and energy consumption are considered, focusing on their ratio, energy intensity. The aggregate energy intensity of the industrial system as a whole (E/A) depends on the composition of production and on the energy intensities of each sector. These dependencies can be separately summarized: the rate of change in aggregate intensity caused by the shift in the sectoral composition of production and the rate of change in the energy-weighted average of the several energy intensities, or real energy intensity (see above). The development of the real energy intensity is discussed from the perspectives of economics and technology. The section concludes with a discussion of policies. An up-to-date reference on the U.S. industry (35) and a global perspective (36) are available.

Historical Statistics

Production. Aggregate industrial production in the United States grew at an average rate of about 3.3% per year from the late 1950s to 1970. It grew about 2.4% per year during the 1970s and 1980s. Industrial production grew rapidly in the 1960s and early 1970s. Growth then slowed but has still been substantial. Heavy industry, however, including the energy-intensive materials sectors, grew quite slowly in the 1970s and 1980s, as will be discussed.

Unfortunately, these production growth rates are rather uncertain, because different measures of production disagree. One prominent measure is value added, industry's contribution to the gross national product, with dollar values deflated or adjusted for price changes. Another measure is deflated gross output, the value of shipments collected on a census form at the factory or establishment, deflated by the Bureau of Labor Statistics, and then added for all establishments. Another is an index of production created by the Federal Reserve Bank (FRB), which is based on quickly available measures of activity. These measures disagree, often at the 0.5 to 1.0% per year level. All these measures have serious flaws, as seen by examining each carefully and by comparing each with production tonnage in the bulk-materials sectors.

Value added is hard to deflate accurately because information about what materials a factory buys may not be available in sufficient detail. In addition, the quantitative connection between value added and production is weak because, in this time period, major changes have occurred in two aspects of value added; a declining return to capital from industrial establishments and a declining fraction of services done in-house.

A principal purpose of the FRB industrial production indices is quick publication. The indices are determined from a short list of quickly available indicators, often quite inappropriate for long-term energy analysis. Although widely used, eg, in modeling of the economy, FRB indices are poor for industrial analysis.

Gross output has the virtue that it is much closer to the primary data, the census questionnaire. And it involves deflation only of the value of output, which is relatively well defined. Unfortunately, several energy-intensive sectors are parts of vertically integrated firms, and both the value of shipments and price deflators are substantially in question because prices internal to a firm are not meaningful. The difficulty is apparent when comparing this index with tonnage in the homogeneous materials subsectors of primary aluminum and cement and with various measures of petroleum refining, another vertically integrated sector. A second difficulty is the extensive double counting that occurs in aggregating. All in all, however, this is the best general measure of growth of production in industrial sectors.

This digression on production data is worthwhile because the analysis of industrial efficiency can be no better than the data; and production data, surprisingly perhaps, are a weaker link than energy data, even though the latter are collected less frequently, and in less detail, and are less important in the general scheme of national accounts. Perhaps the key is that energy is a simpler concept than money value.

Aggregate Energy Intensity. Use of primary energy by industry, including the fuel used to generate the electricity consumed by industry, grew from about 20 quads/yr in the late 1950s to 31 quads/yr in the early 1970s. The 1960s were an extraordinary age of material growth. Then in the early 1970s came the natural gas shortages and the beginnings of real (ie, inflation-corrected) price increases for electricity, a reversal of the historical decline that blessed that energy form. In the fall of 1973 came the first oil shock and in 1979, the second, with major increases in the price of oil and gas. Industrial energy consumption hesitated in the mid and late 1970s, fell to a low of 26 quads in 1982–1983 and climbed back to 30 quads in the early 1990s, as oil and gas prices declined.

The aggregate energy intensity, ie, the ratio of the energy consumed by all of industry to total production was steady in the period from 1958 to 1972 and then fell by more than 33% to 1985. (The quantitative remarks about energy intensities here and below refer to energy use for heat and power in manufacturing, which accounted for 63% of total industrial energy use in 1985. The remainder is feedstock uses of fuel, and energy use in agriculture, mining, and construction.)

The different forms of energy had quite different histories, however. From 1958 to 1985, the aggregate coal and oil intensities fell 70 and 60%, respectively. (When specific fuels are referred to, it is consumption at the site, not including consumption at power plants.) During this period, coal was being eliminated, except at large facilities and in heavy industries. Coal is an inconvenient fuel to handle without special skills and special equipment; more recently coal has also been hampered by environmental regulation. During this period, petroleum was gradually losing share, and after the second oil shock, its use was drastically curtailed. Most consumption of petroleum is now as a feedstock, at remote sites such as forest products, mining, and agriculture and at refineries where by-products of the process supply much of the energy. Between 1958 and 1971, the aggregate natural gas intensity rose 30%, but by 1985, it dropped from its peak by more than 50%. Natural gas had been widely introduced through pipeline construction by the beginning of this period, and it was favored for its low price and convenience until the shortages associated with federal price controls occurred in 1971, shortly before the first oil shock.

The combined result of these developments was that the aggregate fossil fuel intensity fell 15% from 1958 to 1971 because of fuel efficiency improvement in energy-intensive sectors, even though real fuel prices were low and falling. Beginning in 1971, the decline quickened and, by 1985, the aggregate fossil fuel intensity declined by another 50%. Part of the reason for this accelerated decline was a relative shift in production away from energy-intensive production (see below).

The pattern for electricity is similar, except that continuing electrification (new uses of electricity) overlies the other developments (Fig. 7). Thus the aggregate electricity intensity grew rapidly between 1958 and 1970, even though the efficiency of electricity-intensive processes, such as electrolysis of brine to produce chlorine and smelting of aluminum, was being improved. Real electricity prices fell during this period. Since 1970, electricity prices have mostly been rising and the aggregate electricity intensity has gradually declined. The two forces for

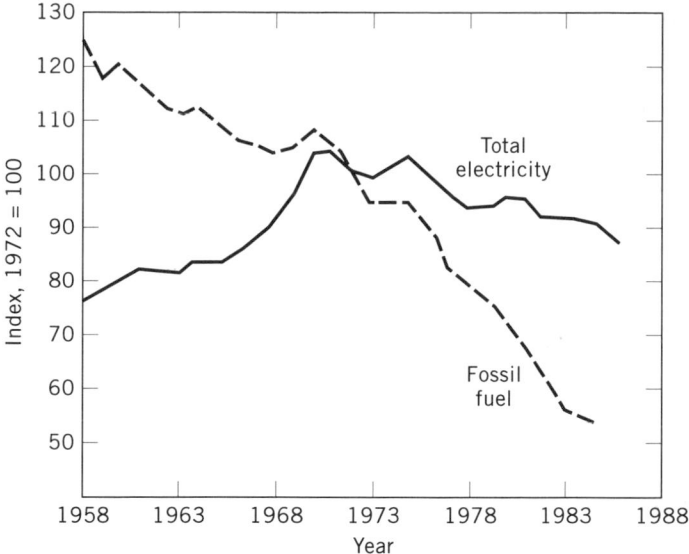

Figure 7. The aggregate electricity and fossil fuel intensities of U.S. manufacturing, relative to 1972.

and to 11.4% in 1990. These findings indicate that, up to 1981, electric utility sales to manufacturers were growing about 0.7% per year faster than total electricity use. Utility sales growth is now slower than the growth of total electricity use.

The Shifting Composition of Industrial Production

The relative roles of the different sectors in industrial production is important to energy efficiency because the energy intensities vary strongly from sector to sector (37) (Fig. 8). With a broad brush, Figure 8 shows roughly 10-times higher intensities in chemicals and primary metals than in general manufacturing. The ratio of the energy intensity of aluminum reduction to that of assembling computers or telephone equipment is about 300 to 1 (with value added as the denominator). Lumping these sectors together is similar to lumping cars with bicycles in the transportation sector.

decline, efficiency improvement and the relative decline in electricity-intensive production, have slightly outweighed ongoing electrification in the aggregate.

In 1958, more than 20% of the total electricity used was generated and used on-site by manufacturers. On-site generation had been high in the early days of electricity but had long been falling. It dropped to less than 8% of the total by 1981. With the Public Utilities Regulatory Policy Act of 1978, on-site generation has begun a comeback, rising to 10% of the manufacturing total in 1986

The Maturing of Bulk Materials. The manufacture of bulk materials is much more energy intensive than manufacturing in general, and the use of materials in tons per year is declining relative to total production in all affluent societies. The consumption of materials used in large quantities (paper, fertilizers, synthetic fibers, plastics, other industrial chemicals, cement, glass, pottery, and metals) is declining relative to total industrial production. This phenomenon, called dematerialization (38), has contributed a 0.5–1.0% per year decline in the aggregate energy intensity of industry Organization for Economic Cooperation and Development (OECD) countries (25,39). Note that dematerialization is not associated with increasing imports of some materials by some nations.

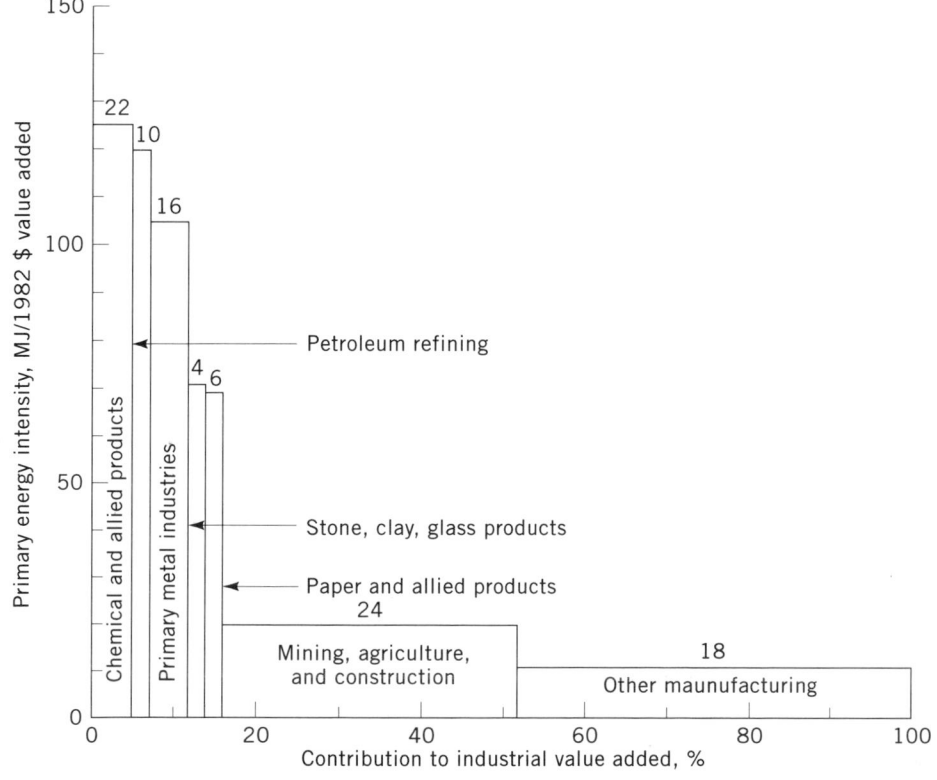

Figure 8. Primary energy intensities of U.S. industries versus value added by sector in 1981.

Dematerialization is somewhat controversial. Consider two kinds of evidence for it: general statistics and a case study of a particular material, plastics. The conclusion from general statistics is controversial because (1) the long-term decline in older materials like steel (in kilograms per dollars of GNP) could be caused by materials substitution, (2) the decline in overall materials (in kilograms per dollars of GNP) dates only from the early 1970s and might be a temporary phenomenon associated with the oil shocks, and (3) there are disagreements about how to measure and aggregate materials consumption. Upon examining the developments for particular materials, these objections do not appear to be critical (40).

The only bulk materials whose consumption still grows as fast or faster than the economy in the United States are in the chemical products family: plastics and industrial gases. But growth rates for these materials as for all the bulk-chemical groups have, nevertheless, been falling dramatically. What is happening to plastics production? Saturation effects apply even to plastics. Markets for heavy consumer products are saturating. For example, while the application of plastics to motor vehicles is increasing, unit sales of vehicles are no longer growing with the economy. Innovative consumer products tend to have a low materials:cost ratio (kilograms per dollar). For example, although electronic equipment typically has a plastic structure or body, the ratio of weight to cost is low. Materials are being used more efficiently. For example, linear low density polyethylene, introduced in the late 1970s, allows the use of thinner films and is taking over low density polyethylene markets. This increasing efficiency is being driven, in part, by the competition among materials, eg, the competition for materials for grocery

bags and beverage containers is fierce, putting a premium on efficient design and even beginning to bring in considerations of plastics recycling. Improved materials are increasing product durability. Plastic pipes, other uses of plastic in construction, and plastic auto parts that reduce rust and corrosion often contribute to longer product life.

In conclusion, materials manufacturing, which accounts for the lions' share of manufacturing fuel use and somewhat more than 50% of the electricity use, is likely to grow roughly with population rather than with the economy in industrialized countries.

The Fast-growing High Technology Sectors. Another structural development is taking place among some of the low energy-intensity sectors. Statistical information and case histories support the concept that high technology product sectors will play an important role in reducing the aggregate energy intensity of industry in the long term. During the 1970s and 1980s, the fastest growing sectors, especially electronic equipment, drugs, and instruments, were the main sources of growth in per capita industrial production. It seems likely that these and other research-intensive, innovative sectors will continue to propel growth in the U.S. economy as it matures.

In Figure 9, the growth rate in output is compared with energy intensity for the 71 manufacturing sectors in the National Energy Accounts (41). Total energy use in each sector is shown on the vertical axis (for 1985). The distribution is L-shaped with all the big energy users having low growth and high energy intensity (upper arm of the L), and all the high-growth sectors with real gross output growing roughly 4% per year and faster for 1971–1985 are furniture and fixtures (excluding household fur-

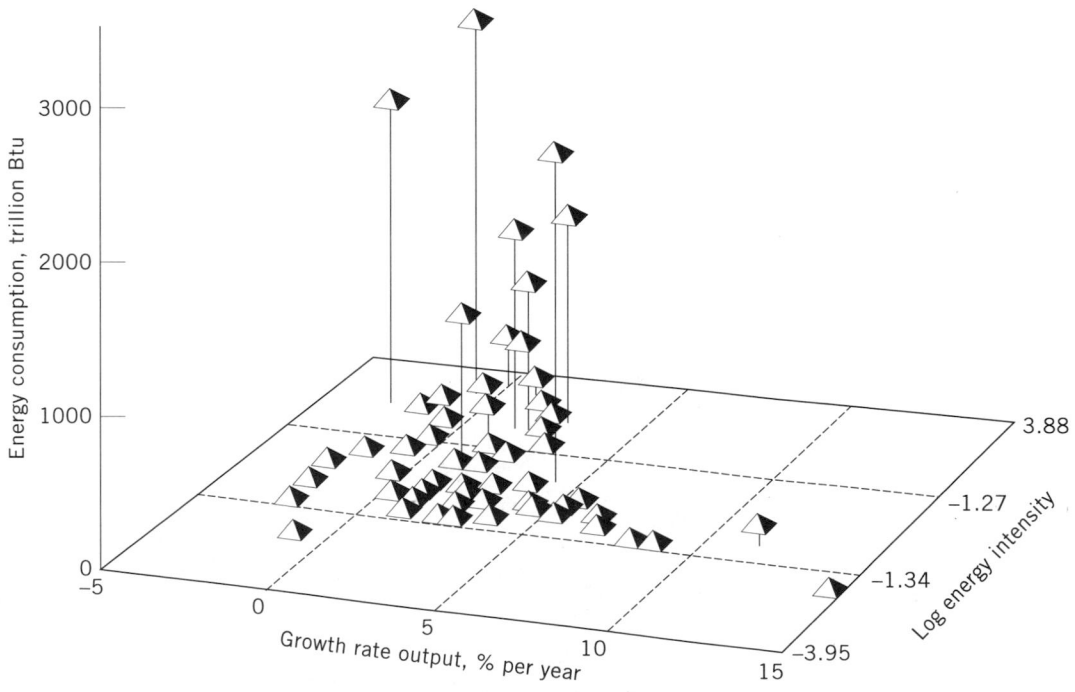

Figure 9. Growth rate of output (1971–1985), energy intensity, and total energy consumption (1985) for 71 sectors of U.S. manufacturing.

niture), printing and publishing, drugs and toiletries, rubber and miscellaneous plastics products, computer and office equipment and miscellaneous industrial machinery, electronic equipment and components, communication equipment and miscellaneous electrical equipment, selected military equipment, and instruments.

The key energy-related behavior is the low and declining energy intensity of these high growth sectors. (Here, energy intensity can be measured as energy per unit of value added in a base year. Relative changes in energy intensity are, however, much better measured in terms of the ratio of energy to deflated gross output.) The decline in the energy intensity, or equivalently, the increase in energy efficiency of these sectors is rapid. This decline is not due to efforts directed at improving the energy efficiency of production. Rather, it is due to continuing product innovation embodying design improvements that have principal impact on the product but relatively little on the materials and energy requirements. It is also due to the rapid creation of new, more modern, production facilities.

On the basis of statistics for 1975–1985, a judgment whether the rapidity of decline in energy intensity will match the rapid growth of these sectors is premature. During that decade there was an approximate balance, such that total energy use by these sectors grew with population rather than with the economy. Further experience and analysis are needed to form a judgment, but it is tempting to speculate that in the presence of substantial innovation, industrial energy consumption (as well as material tonnage consumption) will not grow with the economy, even in the absence of efforts to improve the real energy intensity.

Real Energy Intensities

In principle, most energy intensities could be reduced to zero or near zero (42). Thermodynamically, the difference between a collection of materials and the same materials shaped and assembled is nil. Substantial energy use is required for some endothermic chemical reactions, like reduction of metal ores (ie, removing the oxygen) certain organic reformations, and dissociating brines (such as

separating the chlorine from salt). These constitute a modest part of all industry.

Actual and absolute minimum energy intensities are shown for a few materials in Table 1. Shown is energy consumption within the corresponding manufacturing sector. This and other accounting conventions affect the analysis. It is seen that in terms of fossil and electrical energy arriving at the site, aluminum smelting is about 50% efficient (thermodynamic minimum divided by carrier-energy consumption). The steel industry is about 25% efficient; it would appear more efficient if the shaping and finishing steps were not included. The industries that are less focused on a major endothermic chemical reaction are less efficient in these simple physical terms. For example, industries that use fuel for low temperature heating, as for drying paper, are relatively inefficient in these simple physical terms.

In practice, energy intensities are substantial even when no energy is needed in principle. But, historically, established energy-intensive processes, like smelting aluminum or making cement, have shown substantial ongoing reductions in energy intensity, even in the period 1958–1970 when energy prices were low and falling. Energy intensities, of course, fell more rapidly in the period 1971–1985.

There are three sources of improved energy efficiency in making a particular product: operational improvements; autonomous change, investment projects to improve the production process that are not associated with energy prices; and energy conservation investments associated with differences in the cost of energy.

Operational Improvements. Much of the efficiency improvement of the mid to late 1970s was achieved through better management without substantial investment in equipment. Basically, it is a question of giving energy a high priority in management goals. Employee involvement through training, inspection, suggestion programs, and contests are important, as are audits to identify no-cost and low cost opportunities. Metering and reporting procedures and accounting practices to provide feedback

Table 1. Energy Intensities for Selected Basic Materials

Material	SIC	Energy Intensity (GJ/t)[a]		Thermodynamic Minimum
		U.S. Industry, 1988		
		Primary[b]	Carrier[c]	
Paper	26	24.1[d]	17.4[d]	—[e]
Petroleum refining[f]	2911	5.3	4.9	0.4
Cement	3241	5.4	4.4	0.8
Steel	3312	32.4[g]	28.1[g]	7.1
Primary aluminum	3334	207	69	29.3

[a] For millions of BTU per short ton, divide by 1.163.
[b] Electricity evaluated at 3.3 × electrical energy.
[c] Electricity evaluated at electrical energy.
[d] Excludes wood-derived fuels.
[e] The absolute value of the minimum is small, and its sign depends on accounting conventions and product.
[f] Per tonne of crude processed; thermodynamic minimum is nominal.
[g] Per tonne of rolled steel mill products.

on energy use to groups within the plant are powerful tools. More thorough and frequent maintenance and use of sophisticated inspection and maintenance equipment are important. Often what works to improve operations depends on the culture at the plant. Modifications introduced by technical people may not be effective if production workers are not involved in the planning. Moreover, many of the best opportunities are known to production workers and not to management and technical staff.

The responses to the natural gas shortages of 1970–1971 and the price shock of 1973–1974 were rapid and resulted in a 10 to 20% reduction in energy intensities during the 1970s (relative to existing trends). Until the late 1970s there was little energy-efficiency investment, so the improvement was mainly the result of operational changes.

Autonomous Efficiency Improvement. The adoption of a fundamental production process usually has important energy efficiency implications but is usually autonomous. That is, the decision is essentially independent of energy price differences. For example, steel firms do not choose electric arc furnace technology with a primary concern for electricity prices (30). For minimills, the scrap-based electric-arc process involves much lower capital costs than the ore-based process: scrap is available in the region, the scale of production can be kept small, and product markets are promising. In this context, energy price differences are unimportant. This conclusion is even stronger for less energy-intensive sectors than steel. It may not be valid in some cases, however, eg, in the primary aluminum sector, where the cost of energy is an extreme 25 to 33% of total costs (and greater than value added within the sector).

General progress in production technology tends to have certain implications for energy consumption. (1) Total energy costs associated with established types of processes are reduced along with other costs (43). There is the evidence just mentioned from energy-intensive industries in the 1950s and 1960s when energy prices, including electricity prices, were low and, often, falling. (2) New applications of electricity continue to occur, with surface treatment and specific heating (as contrasted with general volume heating in a typical oven) being two currently important areas (44). In addition, some electricity consumption is associated with added environmental controls. Thus most current electrification is based on markedly superior production technology rather than energy price effects. (3) Learning or experience curve studies of particular processes show fairly steady and remarkably large cost reductions over long periods (45). The reductions correlate better with cumulative production than time. Energy costs decline along with total costs, although less rapidly.

An example of an experience curve is shown in Figure 10 (46). The x-axis is cumulative production of low density polyethylene, using the high pressure production process. The y-axis is manufacturing cost per pound, which is seen to decline to a mere 3% of its initial value in just under 40 yr. Energy intensity declined to 8% of its initial value during the same period. Moreover, just when it might be

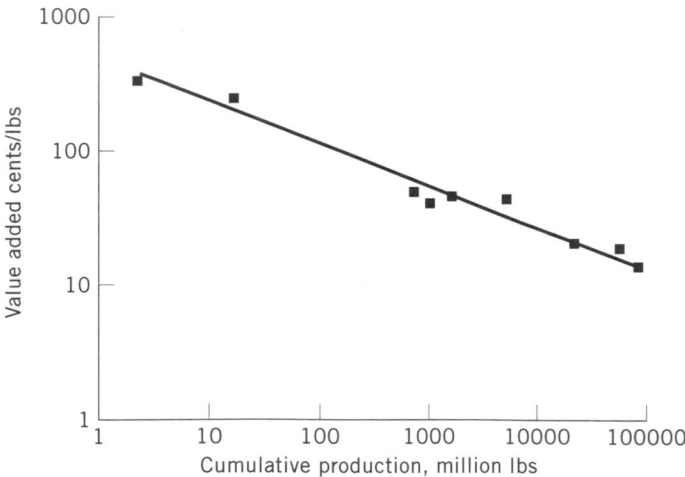

Figure 10. Experience curve for high pressure low density polyethylene. Based on constant dollar value added, obtained with 1981 factor prices. Energy use was reduced 88% (down by a factor of 0.125). Adapted from Ref. 46.

expected that opportunities for cost reduction were running out, a new low pressure process for making polyethylene was developed and is being adopted, enabling continuation of the downward trend in cost and energy intensity. With these developments, historical data indicate that there has been an autonomous trend for real fuel intensities to decline about 0.5% per year, and (excluding electricity-intensive sectors) for real electricity intensities to increase perhaps 2% per year (47).

Fundamental Process Technologies. Perhaps the most important recent examples of autonomous improvements, or fundamental process changes, are catalytic processes. In petroleum refining in the past half century, several processes to convert a larger fraction of the crude into gasoline have been created: catalytic cracking, catalytic reforming, alkylation, and isomerization. Molecular size, hydrogen:carbon ratio and molecular geometry are adjusted, and with enormous throughput and relatively modest cost. These technologies are now being turned to adjusting the chemical makeup to obtain reformulated gasoline that leads to reduced emissions, eg, to reduced evaporation.

Another industry that has been characterized by fundamental process change in recent decades is steel. While the initial step of ore reduction, or iron making in the blast furnace, has not fundamentally changed, steelmaking and shaping of products have. Refining of molten iron to make steel now takes place in a converting vessel (basic-oxygen furnace), in which oxygen is blown in, and the carbon already in solution with the molten iron is the fuel. This process greatly increases product throughput and saves energy compared with the previous standard, the open hearth furnace. Solidification, or casting, is now done continuously into a strand a few inches thick, instead of into thick ingots, which must then be heated and rough rolled to reach the stage achieved by the continuous caster. The continuous process greatly reduces the need to recycle unsatisfactory material from the beginning and end of each batch. In speeding up the process, these technologies have also helped the industry to bring the entire process under much better control, so that a much larger fraction of product meets specifications. Of course, in this way energy intensities are also substantially reduced.

The steel industry has also added a new subsector based on melting postconsumer scrap to make construction products like concrete reinforcing bars. The process technology is the electric arc furnace and continuous casting. Recent fundamental process changes in other industries and with major energy intensity implications are the float-glass process, induction heating and melting of metals, and surface heating of metals with electromagnetic beams.

Some examples of fundamental process change that may be achieved in the foreseeable future are listed below.

Chemicals, paper, and food processes.

Improved separation processes based on membranes, adsorbing surfaces, critical solvents, freeze concentrations, etc.

Ethylene chemistry based on natural gas feedstocks.

Waste reduction using closed systems.

New and improved catalysts for chemical processing.

Recycling paper and plastics into new material and product areas.

Bioprocessing with genetically engineered organisms.

Metals processes

Recycling postconsumer scrap into new product areas.

Near net shape casting, spray forming of products.

New surface treatments with electromagnetic beams.

Direct and continuous steelmaking.

Coal-based aluminum smelting, fuel-based scrap melting.

The improved separation techniques listed are all in areas where new processes have been developed and adopted in the last decade or two, and it is felt that further applications can be developed. Ethylene is the basic building block for most petrochemicals. It is now made from propane, butane, and naphtha feedstocks and might in the longer term have a more secure base in methane. The closing of systems by recycling, such as with water in bleaching of paper, is one way to avoid end-of-the-line treatment with discharge of harmful chemicals into the environment. The development of new catalysts, and new substrates on which to place the catalysts, is continuing to progress and will continue to lead to major innovations in manufacturing of chemicals, including creation of new and improved products. The challenge for recycling of paper and plastics is for manufacturers to make more valuable products from the postconsumer materials. This is an area of rapid change. Bioprocessing by organisms tailored to make the desired products and avoid undesirable by-products is a principal opportunity offered by genetic engineering. It is not known if these bioprocesses, which tend to have low throughput, can be competitive with petroleum-based processes, which have high throughput, in producing bulk products.

For the metals industry, as for plastics and paper, a principal opportunity is conversion of postconsumer scrap into more valuable products. The creation of sheet steel from scrap is one such area. High quality sheet requires low residual concentrations of copper and tin, so one approach would be to develop scrap processing techniques to separate these contaminants before melting the metal. Progress is being made in casting thinner slabs of metal at the steel mill, to reduce reheat and rolling requirements, and in casting closer to final shapes at foundries, to reduce machining requirements. Surface treatment, eg, with lasers and electron beams, is progressing as electromagnetic sources become stronger and more controlled. Direct and continuous steelmaking is under development in Germany, Sweden, Japan, and the United States. In the United States a joint government program with the old-line integrated steel firms was developed as the Steel Initiative in 1984. The concept is to replace four batch steps (agglomeration, coke ovens, blast furnace, and steel furnace) with a one- or two-step continuous process. Two motivations are the elimination of coke making with its air pollution and the reduction of capital cost per unit of production to help enable the industry to modernize. Replacing electricity-based aluminum smelting with fuel-based smelting would be partly motivated by the desire to retain the primary aluminum industry in industrialized regions. The primary aluminum industry is now in a gradual process of moving away to remote areas of the world where electricity is cheap and has few competing users.

If this were an analysis of the buildings sector, it would be necessary to discuss the percent energy savings offered by each technology; there are two reasons why this is not discussed here: (1) the technologies mentioned are only examples, and they represent only a few industrial sectors; and (2) these are fundamental process technologies. The decision to adopt such technologies, when and if they become available, will be based on broad business strategies, not on the price of energy or narrow energy-related considerations. In every example mentioned, issues involving product, environment, capital requirements, and/or location of production are more important than energy considerations as such. Still to be addressed are energy-efficiency technologies for which energy is a major component of the decision.

What can be concluded is that there are many opportunities to produce products differently. In terms of science and technology, this is a time of ferment in production process development, because of the new capabilities to design and evaluate systems through computation rather than trial and error, because of the new capabilities to operate systems accurately under demanding conditions using new sensors and controls, and because of the capabilities of new materials. There is every reason to believe that autonomous improvement in industrial processes, including energy efficiency, will continue at roughly the pace of the recent past, if investments in new facilities continue at roughly the same pace (see below).

Energy Efficiency Investments. Projects in industry that are largely motivated by energy cost reduction tend to be small or medium size (up to a few tens of million dollars but usually much smaller) and tend not to involve fundamental process change. In terms of technology, some of the projects are generic, like high efficiency motors and lighting systems. Others have names that are generic sounding, like variable-speed motor controls and automatic process controls, but these are mostly custom-designed systems, with a large cost for engineering the system (ie, in addition to the cost of equipment and installation as such). Other energy efficiency technologies, most of them in energy-intensive materials manufacturing, are process specific. For example, specific technologies from the grinding phase of cement manufacturing are use of roller mills instead of ball mills for grinding raw material, use of high efficiency classifiers for separating ground material that meets specifications from material that needs to be ground some more, and use of grinding media of increased hardness.

As these names suggest, energy efficiency investments are often not generic devices, like more insulation or more heat recovery. They are instead incremental production process improvements that, eg, reduce maintenance problems, improve product uniformity, increase throughput, and save energy. These multipurpose projects tend to be favored by decision makers over the single-purpose project whose only benefit is a simple reduction of costs. In other words, decision makers are attracted to projects that reduce uncertainty and hassle, and they prefer marketing advantages other than simple cost reduction. (Of course, everyone likes projects with really large cost savings.) The multipurpose project often is more easily sold to different players at the plant. It has potential advantages in a dynamic view of production, whereas the heat exchanger or insulation project carries the risk of having decreased value if the process changes. The multipurpose project can have the advantage of moving the state of the art at a plant, preparing the way for further technological advance.

In the Industrial Energy Efficiency Improvement Program (1978–1985), trade associations were asked to report examples of energy efficiency projects to the U.S. Department of Energy (DOE). Roughly 50 different kinds of projects, most of them process specific and few of them generic, can be found in each of these lists. Similarly, studies of five industries in the late 1980s (paper, steel, cement, glass, and textiles) list 50–100 different kinds of projects, mostly process specific, for each industry (48). The point is that each of the energy-intensive "process" industries is a world in itself, even if generic sounding terms like *controls* and *separation* are used. Energy conservation analysts have tended to select short lists of technologies, exaggerating their efficiency improvement potential, in the interest of simplification. Instead, achieving improved efficiency is not simple, even if some aspects

of it, like efficient mass-produced devices for households may be. Achieving high efficiency in industry requires technological sophistication with specialized knowledge of the particular industry, its production processes, products, and markets. And then there is the hard work of implementing change. There is little simple efficiency magic.

That said, it is time to switch hats and consider some simplistic economics, while not entirely forgetting the above message. Assume that a good energy audit for a plant has been carried out. (For all the multipurpose projects for which energy is important, the energy-related part of the capital cost has been identified, eg, by estimating the benefit streams associated with the different purposes and allocating the capital costs according to the present values of these benefits, although normally such elaborate procedures are not carried out, except for large projects.) The projects can be ordered according to their cost of conserved energy, and a conservation supply curve can be created.

An estimate of the electricity conservation opportunity in fabrication and assembly sectors of industry is shown in Figure 11 for purposes of illustration. (The presentation of this conservation supply curve is illustrative. Such curves are not widely available for industry, nor are they likely to be accurate because of the heterogeneity of industry and the lack of experience at high energy prices.) Figure 11 shows the cumulative electricity savings that will be achieved by conservation investments induced at a given electricity price. The curve represents typical behavior. The figure shows that with the financial criterion represented by a capital recovery factor (*CRF*) of 33%/yr, which roughly characterizes conservation investment by manufacturers, there is a relatively small opportunity for further conservation at the present price of about $0.05/kw·h. In particular, it shows that the electricity intensity is only about 5% higher than its equilibrium value, given the present price and investment behavior. (High implicit discount rates are discussed below.)

However, with a criterion more representative of capital markets, a *CRF* of 16%/yr, a 30–40% reduction of electricity intensity, would be justified at the present price and largely achieved after a delay. (With a 10% real dis-

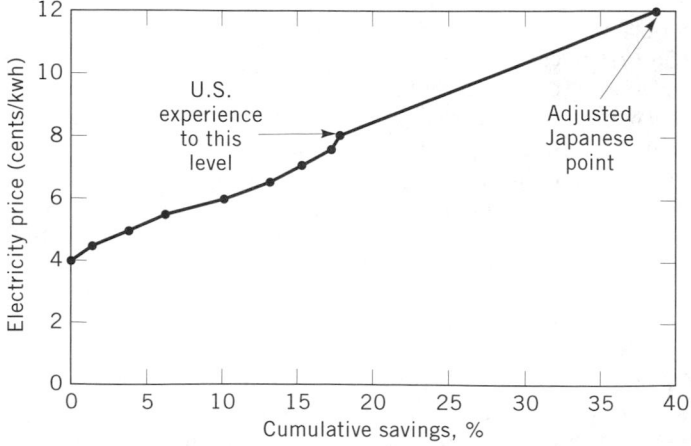

Figure 11. Supply curve: fabrication and assembly.

count rate and a 10-yr project life, the capital recovery factor is 16%/yr.) Figure 11 thus shows that there is substantial potential for efficiency improvement at negative overall cost, or a "free lunch" is to be had, if only there could be found an inexpensive way to change management practices and financing so that a longer time perspective would apply to conservation investments. There are schemes to accomplish just that. One that is beginning to be tried, but is highly controversial, is utility investment on the manufacturer's side of the meter, bringing into play the same interest rates that are used for financial decisions on energy supply projects. The barriers to investment in these projects and policies to reduce those barriers are discussed below.

Econometric Analyses of Energy Efficiency

The discussion continues to focus on real energy intensity, ie, the intensities of individual sectors, but shifts from a microeconomic view to statistical analysis of historical experience. In each sector of industry the energy intensity patterns for 1958–1985 were roughly as sketched in Figure 12 (47). The typical pattern for fuel intensity (Fig.

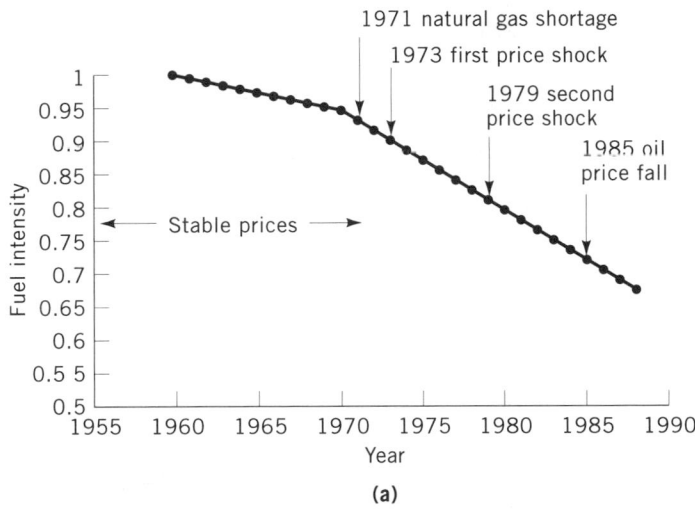

(a)

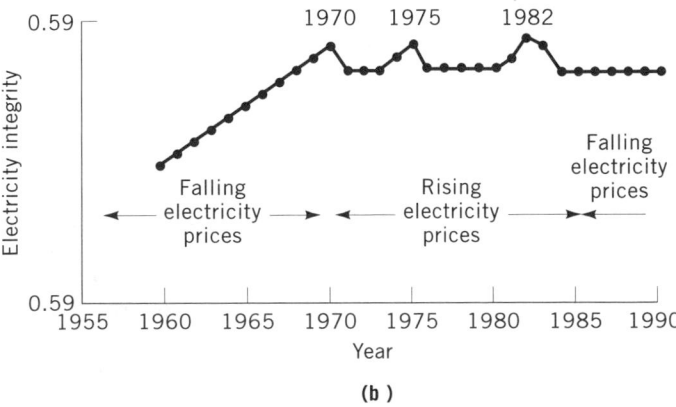

(b)

Figure 12. Fuel and electricity intensity histories (sketches).

12**a**) shows a slow decline from (before) 1958 to the early 1970s, when prices were constant or falling, and a rapid decline from the early 1970s, when the natural gas shortage and first oil price shock occurred. (Data have not been analyzed beyond 1985.) The typical pattern for electricity intensity (Fig. 12b) shows a rapid increase from (before) 1958 to the early 1970s, when prices were falling. Electricity intensities have been roughly constant since the early 1970s.

What is desired is a simple analytical representation of these histories, with parameters that are easily interpreted. This is usually achieved through regression analysis (32): the researcher decides (ideally, after qualitative examination of the data, thought about what affects energy intensity, and many trials with different analytic expressions) on an expression, usually a linear mathematical formula, and uses a statistical package to find the parameter values that minimize the sum of the squares of the differences between the formula and historical data. There are many problems with selection of data, poor data, selection of the independent variables, and their mathematical form. There is considerable variation among the results. One good review is available (49).

Three kinds of parameters are determined: energy price elasticities, trends with production or time, and capacity–use dependence. The last of these will be briefly discussed, followed by the two more important kinds of parameters. The capacity–use effect is seen in the sketch for electricity intensity for the recession years of 1970, 1975, and 1982 (Fig. 12**b**). Roughly speaking, it is as if the motors were all left running where there is a downturn in production, so there would be a sizable increase in electricity intensity. Smaller but similar effects are seen for fuel in some sectors. It is interesting that simplistic economic ideas suggest that in a downturn the least efficient facilities would be shut down so that efficiency would increase instead of decrease, as observed. In fact, many other considerations than simple cost are involved in industrial decision making, such as the role of particular products and plant location in marketing.

Own-price electricities are the best known parameters in this kind of analysis (32): how much does an energy intensity decline due to decisions made as a result of a price increase in that form of energy? The assumed relationship may be:

$$EI(t) = EI(t_o)\left(\frac{p}{p_o}\right)^A e^{Bt}$$

where A is the own-price elasticity and B is a time trend (alternatively formulated as a production dependence). One recent analysis shows that the long-term elasticity is low ($A = -0.1$ to -0.3) in sectors where energy use is intensive and high (-0.5 to -1.0) in sectors where energy use is not intensive. For example, for cement, both fuel and electricity intensities are estimated to be -0.2 (47). This means that a 10% increase in price would lead to a factor of $1.1^{-0.2} = 0.98$, or an eventual 2% decline in energy intensity. On the other hand, the long-term elasticity for electricity in general manufacturing is -0.5. Then a 10% price increase would lead to a 5% decline in electricity intensity. (The exponential formula itself should be used for large price increases, eg, if $p/p_o = 2$ and $A = -0.5$ then the electricity intensity is modeled to decline by a factor of $2^{-0.5} = 0.71$, or 29%.) In truly light manufacturing, especially the fast-growth high technology sectors mentioned above, the elasticity for electricity is about -1.0. For this sector, a doubling of prices would thus essentially lead to a factor of $2^{-1} = 0.50$, or a 50% decline in electricity intensity. These results for general and light manufacturing, excluding the materials manufacturing industries, are roughly consistent with the conservation supply curve of Figure 11.

The relative sizes of these elasticities are qualitatively consistent with the underlying physical opportunities. In energy-intensive sectors an effort has been made over the decades to make the process efficient, so elasticities are near 0. In nonintensive sectors, energy efficiency has had little or no priority and thermodynamics says that no energy use is required in principle. For these sectors the elasticities are near -1. Moreover, because electricity is a newer form of energy than fuel in most sectors, the long run opportunity for improving the efficiency of electricity use in existing applications is good. The elasticities for electricity tend to be more negative than those for fuel, ie, the long-term response to price increases is greater for electricity.

The autonomous trends, represented by the parameters B, were discussed above. In summary, the trend in aggregate electricity intensity, shown in Figure 7, for 1970–1985 is a combination of four effects: (1) a decline associated with the shift away from production in electricity-intensive industries and movement into production in high tech industries, (2) a decline associated with increasing electricity prices, (3) a decline due to efficiency improvement associated with general productivity improvement, and (4) a substantial increase associated with continuing electrification. (The last two are autonomous changes.) The net overall result was a gradual decline in the aggregate intensity for that period.

Policy Analysis

Barriers to Adoption of Efficiency Improvements. Analysts have identified a variety of barriers to adoption of cost-effective conservation measures (50–53). The industrial sector is similar in behavior to the residential and commercial sector but has much more expertise, is more cost conscious, has a tendency to follow "rational" models of investment evaluation, and has more constraints on timing (54). Barriers relevant to the industrial sector in terms of effects are shown below:

High implicit discount rate.
    Capital rationing, internal investments restricted.
    Capital rationing, market share or other strategic investments are favored.
    Limited access to capital, financial market failure.
    Split incentives, disconnect between investor and user.

Bias against small projects, poor evaluation techniques, or perceived riskiness of energy efficiency investments.
Limited application (low penetration).
    Unsuitably of some sites.
    Information gap, slow diffusion of knowledge, lack of skills at plant.
    New technology has not been demonstrated in a closely similar case.
Delay in adoption.
    Scheduling considerations.
    Investor's financial position.

The three most important barriers are discussed more fully.

Capital Funds Are Limited for Conservation Projects. More generally, funds are limited for small- and medium-size cost-cutting projects. Available anecdotal information strongly suggests that in U.S. industry in the 1970s and 1980s implicit discount rates were high, approximately 30 to 100% per year compared with real interest rates of perhaps 5% and returns on stock of perhaps 15%. A study by the Alliance to Save Energy found that this behavior is primarily associated with capital availability problems, in particular the practice at many firms of rationing capital for all projects except the large projects specifically analyzed in detail by top management (55,56). Such capital rationing is widely practiced to enable top management easily to control capital requirements and thus the corporation's external financial arrangements (57,58). Usually external capital markets do not deny a firm access to capital, but they operate in a way that inhibits raising funds for discretionary purposes.

Firms tend to rely on internal cash flow to finance the smaller discretionary investments like energy conservation projects. In principle, top management could follow textbook rules, encouraging divisions and plants to propose as many good projects as they can, with top management raising the funds for all those that are likely to return more than the cost of capital. Instead, in most firms capital is severely rationed to divisions and plants for the medium and small projects for which they have primary responsibility, regardless of the investment opportunities that might exist. Under capital rationing, top management reserves the often arduous task of raising outside funds for its own strategic projects. (It is often arduous because lenders or investors tend to want some control over the firm and management wants to keep as much control as possible.) Capital rationing is widely practiced by marginal firms with poor cash flow. It is also often practiced by firms with good cash flow where top management has a financial strategy that strongly limits the availability of capital for internal investments, and by many firms that have highly centralized management.

Because of capital rationing, funds are limited for discretionary small and medium projects. Under this limitation, little effort is made to identify energy conservation projects. One should not misunderstand the rate-of-return criterion met by projects undertaken in this context. It is not that all good projects with a better rate of return are done but that the projects being done have a return at least that high. If more good opportunities were identified, the minimum return on investment would tend to increase, because the total capital allocation is fixed.

Many economists object as a matter of principle to the notion that U.S. markets, in this case capital markets for firms and investment allocation within firms, do not perform close to perfection. In their view, if management practices capital rationing, there must be real costs at the firm that management is more or less properly taking into account. Accepting the spirit of these concerns for the moment, consider two reasons for reluctance to make small cost-reduction investments, like conservation projects.

Some Sites for Conservation Technology Are Unsuitable. Although a technology is, in principle, applicable at a site, the application at that site may be disqualified by decision makers at the firm for several reasons: the product being made there has poor market prospects, the production process is expected to be changed, the plant is likely to be closed or sold in the next few years, the equipment that would be retrofitted is expected to be replaced, or management does not fully trust the people responsible for the project at the site. A substantial fraction of nominally applicable sites will be in this category at any time. This phenomenon does not, however, explain the high implicit discount rates that are common at sites that are suitable.

Smaller Projects Are Discounted Because They Tend to Return Less Than Projected. The perceived riskiness is more speculative as a reason for limited investment in conservation projects. There are two arguments. One involves the risk that the physical performance of smaller projects may tend to be poorer than project designs predict. This is not a large concern for industrial energy projects and is not included in the cost assessment, because most projects are straightforward and design is usually carried out by technical experts. The other argument is that engineering and management costs are likely to be relatively large for smaller projects but not included in the cost assessment. For small projects (eg, under $100,000), this can be an important issue. For medium-size projects (costing one or several million dollars), this argument does not explain the observed implicit rates of 30% per year and more.

This has not been a full-blown review of the controversy between energy analysts and engineers who find that there are many cost-effective opportunities to reduce costs and their economist critics who believe that opportunities not undertaken cannot be cost effective (59). The broad practice of capital rationing for all locally controlled investment projects (ie, plant- and branch-level investment decisions) is powerful evidence that high implicit discount rates are, primarily, a market failure. However, the dispute can only be settled by extensive empirical work.

Policies to Improve Industrial Energy Efficiency. Public policies aimed at improving industrial energy efficiency, or reduced CO_2 emissions, can be categorized by government action or by their impact. Generic policies available to governments are regulation, fiscal incentives, information, and technology policies (60,61). It is, perhaps, most useful to categorize policies by type of impact. Government policies may affect energy prices, financial criteria for equipment investment, rate of adoption of energy efficient equipment, rate of technical innovation, and criteria for investments other than financial. Consider some of these policies (62).

Energy Taxes Including Revenue-neutral Schemes. Relative to other sectors, industry responds strongly to energy prices. But this response is nevertheless inhibited by two effects: stiff financial criteria, like a 2-yr payback requirement for small conservation investments at many firms, and the low priority given energy cost savings at most firms in industries with low energy intensity, where energy costs are only 3% of value added or less. Nevertheless, the response to energy price increases was substantial in the 1970s and early 1980s. If energy prices are to be raised by taxation, an important question is what would happen to the revenues. There are three principal options: (*1*) return the revenues to the firms in proportion to their prior energy intensities, such that the tax is revenue neutral (discussed below), (*2*) use part of the revenues to fund incentives and technology policies (discussed below), and (*3*) use the revenues for general government purposes.

The political difficulties of substantially raising energy prices, especially their uneven impact, must be carefully considered. There is no escaping the major negative impact on the energy supply industries. There may be other heavily affected sectors. In particular, substantial energy price increases could seriously affect energy-intensive manufacturing. Unless the revenues were returned as suggested in option 1, their product prices would have to increase. In itself, this would probably not have an important impact on overall sales, because competition between quite different products is not likely to be affected by modest price increases. For example, plastics based on petroleum did not appear to be adversely affected in their competition with wood-based plastics during the recent oil price shocks. However, the competition between domestic and foreign producers of the same energy-intensive product would be altered if foreign competitors were not subject to similar energy tax increases. A possible mitigating policy would be refunds (as suggested above) or tariffs and refunds for energy-intensive products at the border. If these corrections were applied, say, only to products whose incremental energy costs due to the taxation were over 5% of sale price, relatively few sectors would be affected. Unfortunately, the administration of such tariffs and refunds might be complex.

Energy or carbon taxes could be part of a revenue-neutral fee-refund scheme, in which the tax revenues from each sector could be refunded to firms in that sector on a different basis. For example, the tax might be based on carbon emissions while the refund might be in proportion to production. This approach has two important advantages. While providing the economic inducement to reduce emissions it does so in a way that does not penalize any manufacturing sector and does not create inflationary pressures. There are, however, serious disadvantages in implementation. The outstanding difficulty is caused by the heterogeneity of products and product processes. This means that the administration of refunds would become complex or iniquitous as the level and mix of production by any firm changed. Another disadvantage is that little of the carbon tax might be passed on to customers, so they would not get the full price signal.

Utility Regulation. A specific policy tool for encouraging manufacturers to invest in conservation in spite of their often high discount rates is utility rebates to pay part of the investment cost. The concept is that many conservation investments are more attractive to society than energy supply investments. With their longer time horizon it may be reasonable to encourage utilities to share in the conservation investments. One approach for motivating this is adjustable returns to the utility on its rate base, determined by an evaluation of utility performance in minimizing its customers' costs. If most utilities were motivated to do this rather than simply selling kilowatt hours, as at present, innovative and effective programs might result (63). Many states have instituted programs of this form but without effective motivation. In a few states, like California, New York, and Massachusetts, motivation has been created through financial rewards to utilities. It will be interesting to evaluate program success by sector and kinds of technology (64–66).

Energy Performance–based Tax Credits. Narrowly focused performance-based taxes and rebates to influence equipment purchases by manufacturers are probably not appropriate for industry, but tax credits for broadly defined conservation projects, or groups of projects, may be appropriate. Tax credits for conservation projects were tried (1980–1985), with a 10% credit for qualifying equipment specified on a short list (Energy Tax Act of 1978 and Crude Oil Windfall Profits Tax Act of 1980). A study based on interviews and analysis of detailed data from 15 large energy-intensive firms showed that the 10% credit and short list did not influence firms significantly in their investment behavior (55). However, it is plausible that a much larger credit and more generous qualifying criteria might get a strong response. This incentive approach would have the flaw of generating windfalls for firms that would invest even without the credit.

General Financial Policies to Reduce Discount Rates. Conservation investments, process-change investments, and private R&D and demonstration programs are believed to be strongly inhibited in the United States by the short-time horizons of industrial managers and financial markets. Striking differences in plant and equipment investment and improvements in energy and labor productivity between Japanese and U.S. manufacturers can be conveniently described in terms of a difference in time horizons. Much more controversial is any analysis of causes or of policies to ameliorate the situation. One dramatic analysis, created in large part by a corporate leader in energy conservation equipment manufacture, associates the short-term horizon with low net national savings due to government policies such as income taxation of capital gains, social security as a pay-as-you-go system, and federal deficits (67). The purpose here is not, however, to propose specific policies but to point out the potential importance of this policy area. For example, if firms' de facto capital recovery rates for conservation projects could in some way be reduced from 33 to 20% (payback increased

from 3 to 5 yr), then an energy price increase of 60% could motivate as much investment as a doubling of prices under present conditions.

Performance Standards for Equipment. The manufacturing sector is highly heterogeneous and most decisions are made by technically well-informed people (68). It is, therefore, not a sector where performance standards are generally appropriate. In the 1978 National Energy Conservation Policy Act energy-efficiency standards for motors and pumps were proposed and a study was mandated, but the concept was dropped. In the Energy Policy Act of 1992, efficiency standards for most general-purpose motors (1–200 hp) are included. While perhaps useful, such standards will not accomplish that much in the industrial sector. The real opportunity is in using motors more effectively in what they drive. This involves custom design. For lighting equipment, performance standards might, however, be effective. Lighting must be appropriate for production needs, but the energy efficiency of mass-produced lighting products (lamps, ballasts, fixtures) is of little or no concern to the plant engineer or manufacturer. Initial lighting-equipment performance standards are included in the Energy Policy Act of 1992.

Research and Development Programs. Many economists assume that because the U.S. R&D effort is large and generally successful it is basically sound, but there are many who criticize the design and priorities of the entire R&D system as it relates to "civilian" industries. The latter viewpoint can be summarized as follows: the basis of the problem is the underinvestment in research that characterizes the private sector, where a firm cannot expect to capture most of the benefits of research (because competitors can quickly copy successful ideas). The federal government's policy of leaving R&D to the private sector except in selected industries has meant massive research support in areas of military interest, and in certain other areas with historical precedent, but little research support in most civilian manufacturing sectors. Manufacturing processes for pulp and paper, inorganic chemicals, construction materials, and metals have not been fundamentally reexamined in light of the revolutions in basic science, modeling technology, and sensing and control technology that have occurred in recent decades. Engineering school education and research in these areas have largely withered.

In light of this analysis, a policy option is creation of strong R&D programs, including specialized research centers, in energy-intensive manufacturing processes and in other technology-based areas important to energy intensity reduction. This option differs from DOE conservation programs, which have selectively supported narrowly defined development efforts. The policy option suggested here involves creation of coherent programs of general research.

In creating such policies, one must be aware that R&D policies take a long time to have an impact. Many of the well-known fundamental process technologies were three or four decades or more in development and diffusion. However, the speed of creation and adoption may depend on the kind of technology. In particular, rapid development and adoption characterized many of the smaller-scale energy efficiency technologies of the 1970s–1980s. The timing of the introduction of technology for electricity conservation in automotive manufacturing (69) is listed below.

New technology (first marketed after 1970).
>   Microprocessor-controlled energy management.
>   Variable speed drive based on inexpensive power semiconductors.
>   Die cushioning.
>   High efficiency lamps and lighting systems.
>   Large microprocessor-controlled heating and ventilation systems.
>   Aircraft-derivative gas turbine cogeneration.

Old technology.
>   Isolation valving for segmenting compressed-air systems.
>   Small compressors for satellite compressed-air systems.
>   High efficiency motors.
>   Cog belts.

Demonstration Programs. The speed of diffusion of new technology depends on its demonstration, but the effectiveness of demonstration programs is controversial. The positive argument is that firms will delay use of a new technology, or new application of an existing technology, until they are assured by at least one demonstration, in the same manufacturing situation that concerns them, that production processes and product will be satisfactory. A typical Ford Motor Co. plant manager would like a technology to have been successfully demonstrated at another Ford plant. Demonstration in a GM plant might do, but the same technology in a different industry would not. Public subsidization of demonstrations and dissemination of technical information about them could accelerate the process. There are examples of successful programs (70). The negative argument is that the public sector tends to do a poor job in selecting technologies for demonstration and a poor job of managing its role in any demonstration. There are important examples of such failures (71).

Considerable experience has been gained about the design of demonstration programs in the past decade or so in the United States and Europe. Several characteristics of a successful program appear to be avoidance of large projects, a relatively small financial role for government, and a strong dissemination activity.

Goal Setting and Reporting of Performance. Goal setting (with the level of energy-intensity reduction being voluntary) data collection, and reporting of energy intensity improvement (through trade associations in each subsector of manufacturing) were mandated in the 1975 Energy Policy and Conservation Act. They were considered by many energy managers at firms to be remarkably effective, considering the mild nature of the legislation. The requirements expired in 1985. (The compilation and publication of the data by DOE was not considered effective.) The principal benefit from this policy was seen by energy managers as the creation of lines of responsibility within the firm for energy conservation and creation of energy consumption reporting systems within the firm. Similar requirements can be imagined for the area of recycled materials.

ENERGY EFFICIENCY OF TRANSPORTATION

Transportation is a critical sector for energy use because passenger travel is growing rapidly, because petroleum is the main energy resource used (requiring enormous imports in the United States and many other countries), and because it is the largest source for air pollution in most metropolitan areas. In this section the growth factors, especially in driving private motor vehicles, are considered. Then technical and economic aspects of improving the energy efficiency of motor vehicles are discussed. Some of the physical and technological issues will be explored in detail. Finally, a range of public policy issues is discussed. An up-to-date overview of U.S. transportation and energy issues is available (72).

Historical Statistics

The organization of the discussion is based on representing energy use as the sum of products of activity times intensity (eq. 1):

$$\text{Energy Consumption Rate} = \Sigma_i A_i \, (EI)_i$$

Transportation Activity. In the United States, passenger transportation dominates freight in importance from energy and environmental perspectives (Table 2). Passenger transportation averages 13,600 passenger mi/yr per capita or 37 mi/day (73). Of this, 85% is in motorized personal vehicles (cars as well as personal vans, utility vehicles like Jeeps, and pickup trucks when used like cars). This dominance of personal vehicles in per capita travel vehicles characterizes all industrialized countries, but the United States and Canada lead in the amount per capita of travel (36).

Total travel by personal motor vehicle has boomed in the United States since World War II, growing 4.7% per year from 1950 to 1970, and 3.2% per year from 1970 to 1990. As with industrial production, the 1950s and 1960s were decades of especially intense growth in physical activity.

Energy Use and Intensities, 1990. Table 2 shows energy, activity, and energy intensity in some detail for all of transportation in 1990. The main characteristic is the dominance of personal passenger vehicles. Personal passenger vehicles account for 52% of all transportation energy use and for 85% of all passenger miles (PM). (These data may underestimate the use of light trucks as personal passenger vehicles).

The personal vehicle is more energy intensive than the other forms of passenger transportation, but not by as much as many think. The average car is shown in Table 2 to have an energy intensity of 5980 Btu per mile (an in-use fuel economy of 20.9 mpg). An urban transit bus has an energy-intensity of 3740 Btu/PM. So a car with two

Table 2. Transportation Energy Use and Activity in the United States (1990)[a]

Use	Activity Billions[b]	Units[c]	Energy, quads[d]	Energy Intensity, 1000 Btu/activity unit
Passenger				
automobiles	1515	VM	9.07	5.98
personal trucks	296	VM	2.68	9.06
motorcycles	10	VM	0.02	2.50
buses[e]	118	PM	0.16	1.38
railroad[f]	25	PM	0.08	3.13[g]
air (domestic and international)	359	PM	1.93	5.00
Subtotal			*13.94*	
Freight				
intercity truck[h]	735	TM[i]	2.47	3.36
other truck	182	VM	2.40	13.2
railroad[j]	1034	TM[i]	0.43	0.41
water	816	TM[i]	0.32	0.40
domestic				
foreign	1.04	T[k]	0.92	
pipeline	280[j]	TM[i]	0.72	2.6[g]
natural gas				
oil and products	583	TM[i]	0.16	0.27
other			0.05	
Subtotal			*7.47*	
Miscellaneous				
recreation boating			0.25	
general aviation			0.13	
military			0.76	
Total			*22.5*	

[a] From Refs. 73 and 80.
[b] For kilometers, multiply by 1.609.
[c] VM, vehicle mile; TM, short ton mile; PM, passenger mile; T, short ton.
[d] For exajoules, multiply by 1.055.
[e] Includes intercity, urban transit, and school.
[f] Includes intercity and urban transit.
[g] Electrical energy accounted for in terms of primary fuel use.
[h] Rural to rural and intracity deliveries excluded; see text.
[i] For tonne kilometers, multiply by 1.460.
[j] Average trip length for gas assumed to be 620 mi.
[k] For metric tonnes, multiply by 0.907.

people, or a car with one person but twice the average fuel economy, not only goes where and when the driver wants but has lower energy intensity per passenger mile than an average urban bus. (The low energy intensity for buses in Table 2 is due to school buses and the shaky assumption that their average passenger load is 20. The average load of the urban transit bus is taken to be is 10.) The energy intensity of certificated air carriers is also not as great as one might, at first, think.

Freight energy use is also dominated by highway vehicles, but freight activity measured in ton miles (TM) is dominated by nonhighway modes. The nonhighway freight modes are much less energy intensive than heavy trucks. Note that gas pipelines are fairly energy intensive, however, because gas is much more difficult to pump than a liquid. There are at least three principal categories of freight: (1) light short-distance loads for which vehicle mile (VM) is the relevant measure of service; (2) long dis-

tance loads that are packaged or finished goods; and (3) bulk cargoes, like coal, ores, grains, and semifinished goods. The ton mile is a good measure for the latter two services. The main point is that competition or substitutability across these categories is usually not practical, thus the much lower energy intensity of waterborne freight than intercity truck does not indicate much of an opportunity for intermodal substitution. The rail freight energy intensity is an average of somewhat lower intensity bulk cargo shipping and somewhat higher intensity finished goods shipping. In the latter area there is substantial opportunity for substituting rail for long-distance trucking, with transfer from and to trucks at the ends. This is now going on, using containers, putting truck trailers on flat cars, or putting a new kind of trailer directly on the rails. One requirement for this substitution is that the railroads achieve modern scheduling goals, eg, for transferring loads between trains.

Table 3. Transportation Energy Use and Activity, 1970–1990

Use	Energy, quads		Activity Unit	Growth Rates, %/yr	
	1970	1990		Activity	Energy Intensity
Passenger					
automobiles	8.48	9.05	VM	2.5	−2.2
personal trucks	0.61	2.68	VM	9.4	−2.0
(combined)	(9.09)	(11.73)		(3.2)	(−1.9)
buses	0.11	0.16	PM	1.7	0.4
railroad	0.04	0.06	PM	1.3	0.0
air	1.36	2.19	PM	6.4	−4.0
Subtotal	*10.60*	*14.14*			
Freight					
light truck	0.94	1.48	VM	3.9	−1.7
other truck	1.53	3.39	TM	4.9	−0.9
railroad	0.50	0.43	TM	1.5	−2.3
water (dom.)	0.33	0.33	TM	1.6	−1.6
pipeline energy	0.99	0.87	TM	0.9	−1.5
Subtotal	*4.28*	*6.50*			
Total	*14.88*	*20.64*			

Trends in Energy Use and Their Decomposition. Many of the transportation activities have been tracked in consistent or nearly consistent data series since 1970 and before. Energy consumption and average growth rates for activity during the period 1970–1990 are shown in Table 3. There are special weaknesses underlying some of the numbers in this table: the division of light truck VM into personal passenger and freight, the assignment of ton miles to all other trucks, and the assignment of ton miles to natural gas pipelines. In addition, several minor categories have been omitted. Table 3 reveals the rapid growth of the light truck as a personal passenger vehicle. It also shows the growing importance of air travel as well as the relatively slow growth of some freight activities. In the latter connection, there has been a relative decline in bulk materials transportation that parallels that in manufacture of bulk materials discussed earlier.

Table 3 shows strong improvements, in particular (modal) energy intensities, with amazing progress by airlines (cutting the intensity to less than half in 20 yr) and major progress by railroads and in light vehicles. The highly organized commercial transportation sectors, airlines and railroads, were able to do a great deal. The airline story, including the large potential for continued gains has been told (74). The light vehicle story, especially future possibilities, is the main subject of this section.

The data in Table 3 have been set up to enable a Divisia decomposition of the sources of change in aggregate energy consumption, an analysis that does not require activities in different subsectors to be measured in the same units (see eq. 6). The average growth rate in energy use (approximately) equals the average growth in activity plus the average growth in energy intensity (both energy weighted).

The results of the analysis are summarized in Table 4. The behavior for transportation as a whole is essentially the same as that for personal passenger vehicles alone: growth in activity at an average 4%/yr and a rapid decline in energy intensity, such that energy use grew only 1.5%/yr in this period. This leads to an important prediction: the decline in energy intensity is expected to end because the legislation that caused the improvement in personal passenger vehicles has now worked its way through the economy. Further decline would require further legislation, much higher energy prices, or some other event. So transportation energy use will soon begin to grow much faster than it has in the last 20 yr.

The separate results for passenger and freight activity show what is not surprising to any observer of the U.S. scene: travel is increasing rapidly, but so is the energy efficiency with which it is provided. Freight activity has been increasing less rapidly, a characteristic of an affluent and mature society. At the same time, it has proven more difficult to improve the energy efficiency of freight services. The truck freight data, excluding light trucks, involves a major inconsistency. "Intercity freight" grew 78% from 1970 to 1990, while truck vehicle miles grew 141% (and in addition the trucks became larger). By using the intercity freight growth rate of 2.9% instead of the 4.9% growth rate based on vehicle miles and increased size, one obtains the results shown in parentheses in Table 4. These results conform even more strongly to the qualitative characteristics just mentioned.

If a Divisia decomposition is carried out for personal motor vehicles alone (cars and light trucks), because the units of activity are the same, $G(Q)$ and $G(S)$ can be cal-

Table 4. Divisia Analysis of Energy Used for Transportation 1970–1990, Growth Rates in Percent per Year

	Activity	Energy Intensity	Energy
Passenger	3.8	−2.3	1.5
Freight	3.4 (2.6)	−1.4 (−0.5)	2.1
Total	*3.7*	*−2.0*	*1.7*

culated (see eq. 2). From Table 3, it is found that the growth rate for overall activity $G(Q) = 3.2$, the growth rate for the shift toward light trucks $G(S) = 0.4$, and the real intensity growth rate $G(EI) = -2.2$ (all in percent per year). (While these results are roughly correct, they depend on an educated guess for the 1970 use of light trucks as personal passenger vehicles.)

To understand these results, it is necessary to probe them in more detail. What is responsible for the rapid growth in travel? What is responsible for the decline in energy intensity? In the next two sections these questions will be explored with respect to personal passenger vehicles.

Growth in Driving per Driver

Vehicle travel continues to grow in spite of arguments that saturation is imminent. Total vehicle miles per adult (ie, total vehicle miles divided by the population aged 16 and over) is found from a regression analysis to be almost proportional to real disposable income per capita, corrected by a small fuel price elasticity effect (of -0.1), indicating that a 10% increase in the fuel price induces a 1% decline in consumption (75). This short-term response to increased gasoline prices is well determined, because it is a direct feature of the price shocks of 1974, 1978–1982, and 1990 (76).

The agreement between travel and income may not be more than the similarity of two increasing time trends. In any case, the causes can be expressed in more useful terms. Demographics show that much of the growth in driving since the late 1960s is associated with women moving into the labor force, those women becoming drivers, and those women driving distances that are moving toward men's distances. In 1969, 39% of women over 15 years old were employed; in 1990, 54% were. In 1969 women with driver's licenses drove an average of 5400 mi/yr; in 1990 they drove 9500 mi/yr (77).

There is, in addition, an upward trend in the annual miles per driver by employed men. This distance has been growing 1.3%/yr (1969–1990), and continues to grow. Employed men in the 25–45 age group average about 20,000 mi/yr. At an estimated average 35 mph, this is 1.6 h/day. This large fraction of time is the basis for believing that saturation must soon set in. But when and at what level?

A hint of the future might be obtained from cross-sectional comparisons (from place to place), rather than from longitudinal variations (from year to year). For example, it is possible to examine the cross-sectional effect of income on driving (and on that basis predict the future assuming that per capita incomes will increase). Of course, poor people drive less, but at issue here is whether upper-middle-class people drive more than lower-middle-class people. Unfortunately, this kind of question is difficult to resolve. Many things change with income, like household size, local population densities, and regional patterns. Depending on what else is explicitly accounted for, income is or is not associated with increased driving. Better research is needed.

A different analysis is under way that may shed light on the issue: the dependence of driving per driver on the kind of community in which he or she lives. The principal variable of interest is population density. (Other variables of interest may be density of services, pedestrian friendliness like sidewalks, frequency and density of public transportation, and distance to major centers. In addition, there are variables like income, and vehicle and fuel costs.)

Vehicle miles per capita for metropolitan areas around the world, with gasoline consumption as surrogate, were studied (78). Vehicle miles per capita for communities within metropolitan areas of California have been examined (79). These studies are not definitive because one should question the quality of the data and their interpretation, in part because effects of income have not yet been carefully addressed. Nevertheless, there is a strong population density dependence of driving:

$$VMPC \approx \text{const } \rho^{-\alpha} \qquad (9)$$

where VMPC is vehicle miles per driver or vehicles miles per capita, and ρ is residential density (eg, people per square mile of developable land). One finds that α ranges from 0.5 to 1.0. This is an interesting line of research for two reasons. First, residential density should be expected to be a critical determinant of travel. People need access to work, services, social contacts, etc. Second, the choice of place to live, eg, urban versus rural, depends to some extent on public policies.

An interesting accompanying result of these studies is that people who drive less, travel much less. Each mile of travel on transit does not substitute for a mile of travel in a private vehicle or vice versa. Rather, if one lives in a denser community, one uses transit a little more and drives much less, with the result that 1 mi of travel on transit appears to substitute for, perhaps, 5 mi of travel by car or truck. This result is preliminary.

If one accepts, for the moment, this strong dependence on residential density in the United States, the important issue is: What motivates the growth of housing in low density communities? Is it primarily inherent human desires, as the real estate industry would suggest, or are there public policies that influence many people to live in rural or suburban areas, while nevertheless gaining a living in a city? Of course, some people strongly prefer to live in rural-like settings, but economic incentives are also powerful.

Public policies support investment in the land on which one's house stands, and newly developed areas are relatively cheap. Home mortgage interest is not taxed and infrastructure for real estate development is a prime focus of local government. Indeed, many U.S. local governments have been run by developers since they were founded. Such land investment has proven the best way to create a nestegg for those who are not in a business of their own or not specialists in speculation. Public policy also supports highway travel through diligent efforts to build and maintain safe high speed roads and to keep fuel prices low. Further exploration of the issue is beyond the scope of this article.

The description of vehicle miles traveled on the basis of population or of the number of drivers, as just presented, is in contrast to one based on the vehicles in use, an approach that has been widely used for analysis. The trouble with using miles per vehicle is that, in the United States, a fundamental shift in the use of private vehicles is now beginning to take place. The number of households with more vehicles than drivers is becoming large (77). This trend toward extra, probably special-use, vehicles may well continue as vehicles are kept in service longer and the adult population grows more slowly. For example, the median age of cars in use has increased 2 yr since the early 1970s (80). The growth in the number of vehicles and, especially, their use is thus difficult to analyze.

Choice of the variables on which the analysis is based is critical to the perspective created by the analysis as well as to the particular results. For example, the passenger transportation activity considered is usually passenger miles (mobility) or, as has been emphasized here, vehicle miles. With either of these variables there is a tendency to associate a substantial increase with progress. Certainly poorer Americans and people in poorer countries have less mobility. An alternative kind of activity variables is access, the number of trip destinations achieved. Because many people achieve destinations by walking in their neighborhoods or by short vehicular trips, there is a substantial difference between access and mobility or vehicle miles. In a course project at the University of Michigan (with a small sample size), the same amount of access per person-week for people in rural and urban samples was found, but twice as much travel was noted by the rural respondents. If data on access were available and the analysis of energy use for travel were carried out with access as the activity variable, one would probably find that fuel use per access has not been declining the way fuel use per vehicle mile has. This would create a different picture and, perhaps, different prognoses than the analysis presented here, with its emphasis on vehicles and only brief mention of community.

Fuel Economy Improvement: Technical Opportunities

Although this analysis will address a time period of about 15 yr, it is restricted to modifications of the present kind of gasoline-fueled vehicle (of today's size and power characteristics). This constraint is due to the need to be brief, not to lack of interest in alternative fuels, new propulsion systems, and alternative modes of transportation, all of which are important. Conventional vehicles are important because this country's gasoline-based personal transportation system involves an enormous investment in physical and human capital that will not be quickly replaced. Accordingly, moderate changes in light vehicles are an important means of addressing energy and environmental issues.

Fuel economy in miles per gallon is the number of miles of travel by a vehicle over a standardized sequence of speeds such that 1 gal of a standard fuel is consumed. Improvements in fuel economy are efficiency improvements as defined earlier (even though this generalized kind of efficiency does not include the concept of 100% efficiency).

Many believe that reducing vehicle size (and thus weight) is the best way to increase fuel economy. Reducing the maximum power per unit of vehicle weight can also make a major contribution. The subject of this analysis is yet a third kind of change: technological improvement without reducing vehicle size or performance. This change was behind most of the progress from 1975 to 1988. The industry is undergoing a remarkable period of increasing technical capabilities: electronic controls, new materials, and the capability, through computers, to design a car in detail without having to go through many stages of trial and error with real engines and vehicles are making it practical to do the things only dreamed of by earlier automotive engineers. This technological ferment can be sensed by reading papers of the Society of Automotive Engineers and attending its conferences.

The technologies for improving vehicle fuel economy fall into two categories. The first is load reduction, or decreasing the power required of the engine by reducing air drag, rolling resistance, weight, drivetrain friction, and vehicle accessory loads. The second is efficiency improvement of the engine, or increasing the effectiveness with which the energy in fuel is converted to useful work.

Load Reduction. In recent years the maximum power of new car engines has increased. The average new car power to weight ratio has risen from a low of 32 hp/1000 lb for the period of 1980–1982 to 43 hp/1000 lb in 1993, and acceleration times have fallen (81). This increase in power has been a useful marketing tool, as can be seen from the fact that many customers are sold the high power version of a model.

The power delivered to the drive wheels during typical patterns of driving is relatively low, however. High power is required only in unusual driving conditions, such as acceleration at speeds far over legal limits and climbing mountains at high speeds, situations that most drivers rarely encounter. Vehicles of average weight with modest engines, say 30 hp/1000 lb, can readily be used with good transmission management, to accelerate rapidly at moderate speeds, so acceleration at moderate speeds is not a rationale for high power.

In urban driving, a typical new U.S. car requires an average engine power output of 7 hp. This is low compared to engine capabilities that average 141 hp for 1993 models (81). At 7 hp, more fuel is being used merely to overcome the internal frictions of a typical engine than to provide the output power. The efficiency of an engine at part load is much lower than its best efficiency. This suggests the importance of engine downsizing coupled with load reduction and aggressive transmission management as key strategies for improving fuel economy.

For today's average new car, leaving the engine transmission unchanged, a 10% reduction in engine load results in a 4% reduction in fuel use in urban driving and a 5% reduction in highway driving (82). Technologies include reducing aerodynamic drag, reducing tire rolling resistance, and weight reduction at fixed vehicle size. Drivetrain efficiency can be improved through technologies such as torque converter lockup, electronically controlled standard gearing, and reducing transmission friction. Accessory loads (the largest of which is air-conditioning) can be cut by running accessories only when needed, improving component efficiencies, and reducing the need to run the accessories. Overall, there is a near-term potential for roughly a 15% reduction in weight and a 25% reduction in engine load, which would alone yield a 12% improvement in fuel economy (ie, without engine downsizing).

Types of Engines. There are two main kinds of internal combustion engines in use (83). In the spark ignition engine, combustion occurs by means of a flame front that proceeds from the spark through the mixture of vaporized fuel and the air. Typically the fuel:air ratio of the mixture is chemically correct, or stoichiometric, in that all the fuel present and all the oxygen in the air, could combine to form CO_2 and H_2O. (Ignition and smooth flame front propagation are challenging to achieve if the air:fuel ratio is more than about 1.5 times stoichiometric.) Variable (reduced) power output is achieved by reducing proportionately the fuel and the air admitted to the chamber. The amount of air is regulated by a throttle, which constricts the inlet to the intake manifold, thus creating a partial vacuum as each piston is pulled out during the intake stroke. The fuel is introduced through controlled injection. (Only about 0.017 as much fuel vapor as air is needed, by volume.) In cars sold in the United States, a catalytic converter oxidizes CO and hydrocarbons in the exhaust while simultaneously removing oxygen from (reducing) the NO_x. For this three-way balanced reaction to be achieved, the fuel:air mixture introduced to the engine must be stoichiometric.

The diesel engine uses fuel droplets and high compression. With the compression stroke, the temperature and pressure are high. After fuel injection, combustion spontaneously occurs on the surfaces of the droplets. The overall fuel:air ratio is typically lean (ie, excess air compared to stoichiometric). Variable power output is achieved by changing the amount of fuel injected, the amount of air admitted always being the same (no throttle). Very lean mixtures are satisfactory. With today's catalytic converters, NO_x in the diesel exhaust cannot be reduced, so it may be difficult to achieve low NO_x emissions in the regulatory test. In addition, soot or carbon can be emitted, especially when the fuel:air mixture approaches stoichiometric at high power output. New diesel truck engines are, however, meeting new emissions regulations. (Whether this means that soot will not be seen from new diesel trucks when they accelerate remains to be seen.) Finally, diesel engines are larger and heavier than spark ignition engines for the same power output. (They are also more robust.) These are not advantageous properties for most private passenger vehicles.

Thermal Efficiency. Engine efficiency is the product of two factors: thermal efficiency, expressing how much of the fuel energy is converted into work moving the pistons, and mechanical efficiency, the fraction of that work that is delivered by the engine to the vehicle (the rest going to overcome frictions in operating the engine). (Note: In many expositions the term *thermal efficiency* is used for the overall engine efficiency rather than the efficiency neglecting frictions as here.) Even the best practical combustion-based engines (boilers and steam turbines at electric power plants) have thermal efficiencies of only about 40%, and these engines are large, expensive, and stationary. About 50% efficiency is being achieved in new electric power plants with combined cycle technology, which in-

volves energy recovery from the exhaust gases after the main energy conversion. (These efficiencies are based on the higher heating value of fuel.)

The thermal efficiency of typical internal combustion engines is about 40%, relative to the lower heating value of the fuel (or 37% relative to the higher heating value). This could, perhaps, be increased to near 50% through several changes: increased compression ratio, lean burn (increased air:fuel ratio), recovery of work from the exhaust, faster combustion, effective control of working characteristics such as the air:fuel ratio for each cylinder and each cycle, and control of valve timing and enhancement of breathing so that intake and exhaust are optimized at each engine speed (84).

One of the most interesting and promising of these is lean burn spark ignition engines. Lean burn is advantageous in terms of efficiency because a gas of simple molecules when heated increases its pressure more than a gas of complex molecules like vaporized gasoline; with complex molecules much of the thermal energy is diverted into motions internal to the molecule. Radically increasing the air:fuel ratio by a factor of 1.5 (above the chemically correct stoichiometric value), eg, would nominally increase efficiency by 11%. Moreover, if the air:fuel ratio could be widely varied while still obtaining satisfactory combustion with spark ignition engines, this method could be partly substituted for the throttle to regulate engine power output. Significant additional mechanical efficiency benefits would result. But lean operation prevents the emission control system from reducing nitrogen oxides, so NO_x emissions from the engine might not be as low as one would hope. Moreover, lean mixtures can fail to ignite (misfire) or lead to incomplete combustion. One way to address this problem is to make the mixture near the spark plug richer (stratified charge). Several engine manufacturers are working to overcome these drawbacks, and Honda, Toyota, and Mitsubishi have first-generation lean burn engines in production cars. Moreover, a more radical approach to lean burn is said to be achieving success: the two-stroke engine with modern fuel injection and controls.

In summary, improving thermal efficiency by a factor of as much as 1.25, from roughly 40 to 50%, is a potentially important but difficult goal. One way to achieve it is to solve the environmental problems of the diesel and adopt modern direct-injection diesels such as those now in use in several European cars. Another way to approach it is to develop successful lean burn spark ignition engines. Still another way is to switch to a fuel with much simpler molecules and high octane (eg, hydrogen or methane), designing a high efficiency engine for that fuel. Achieving still greater improvements in thermal efficiency in internal combustion engines is likely to be impractical.

Mechanical Efficiency. The mechanical efficiency of typical U.S. cars averaged over urban and highway driving is about 43%. It is lower for high powered cars and higher for low powered cars. The mechanical efficiency is 0 when the engine provides no power output (an idling engine). Near wide-open throttle, the mechanical efficiency is about 80%. Unlike thermal efficiency, for which it is not practical to achieve efficiencies more than about 50%, it is practical to achieve mechanical efficiencies approaching 100%.

The following analysis rests on a simple approximation for fuel use as a sum of two terms: one to overcome engine friction, the other to provide output power (85). The validity of this approximation is exemplified by Figure 13, in which fuel energy converted per revolution is shown on the y-axis and energy output per revolution on the x-axis, at various engine speeds. All the operating points lie essentially on one straight line. Thus the rate of fuel use has the form

$$P_f = aN + P_b / n_t \qquad (10)$$

where N is the engine speed (revolutions per second) and P_b is the engine power output or load (in kw or hp). The constant n_t is the thermal efficiency discussed above, and a is the fuel energy per revolution needed to overcome engine frictions. There are three kinds of friction involved in operating the engine: the energy used for pumping, for overcoming rubbing friction, and for driving engine accessories. Pumping refers to moving the air and vaporized fuel into the cylinders and the combustion products out through the exhaust system. (For a given engine, a is essentially constant. Among engines, it is roughly proportional to engine displacement.)

The engine power output is the product of the mechanical and thermal efficiencies times the fuel energy input:

$$P_b = n_m \, n_t \, P_f \qquad (11)$$

Figure 14 illustrates the two efficiencies. The slope of the line $\Delta P_f / \Delta P_b$ is a measure of the reciprocal of the ther-

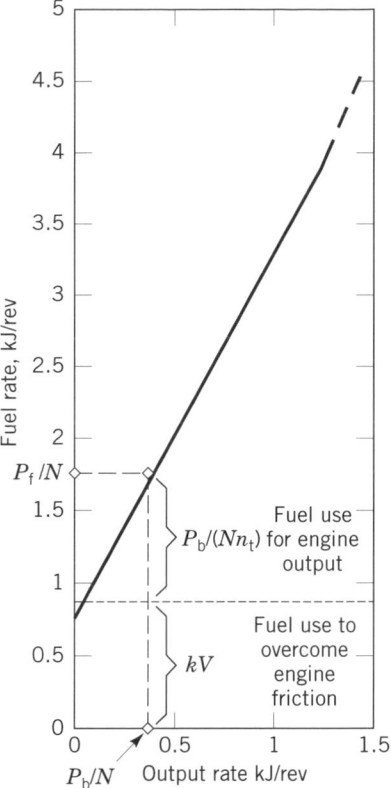

Figure 14. Thermal and mechanical efficiencies.

mal efficiency; here n_t is roughly 40%, independent of the load P_b. The mechanical efficiency varies strongly with the load. Solving equation 11 for n_m:

$$n_m = \frac{P_b}{n_t}/P_f = \frac{P_b/N}{a \, n_t + P_b/N} \qquad (12)$$

At the point shown in Figure 14, $P_b/N = 0.34$ kJ/rev, and $P_f/N = 1.75$ kJ/rev, so that $n_m = 0.49$, or 49%. This is a typical highway driving situation.

In a typical urban driving situation, ($P_b/N = 0.20$, $P_f/N = 1.40$ kJ/rev in Figure 14), the mechanical efficiency is 36%, but in urban driving the car also spends some time idling, where $n_m = 0$, and some at high power, where n_m is high. At the top of the solid line in Figure 14 ($P_b/N = 1.3$, $P_f/N = 4.2$ kJ/rev), the mechanical efficiency is 77%.

Five general kinds of technology for improving average mechanical efficiency are

- Aggressive transmission management (ATM) to reduce average engine speeds.
- Reduced displacement, or engine size, at constant maximum power.
- Reduced rubbing friction and more efficient engine accessories.
- Stop–start (idle–off).
- Reduced pumping (elimination of throttling).

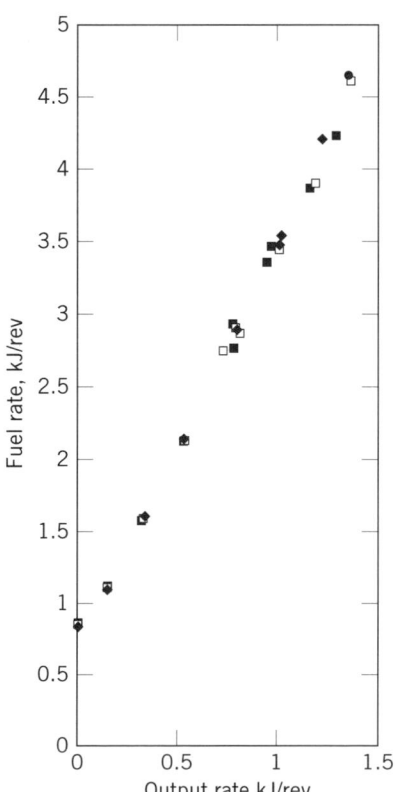

Figure 13. Fuel rate versus work output rate, Chevrolet V6 200CID 198 engine.

Technologies to progress in these five directions have been extensively discussed (76,86–89). It is of interest here to consider briefly the first two categories, ATM and engine downsizing. Not only are these important but they illustrate the rapid change in technology that is occurring and the possible conflict between some efficiency technologies and the product or service being offered. The tendency is to think of efficiency improvement as noninterfering. For example, additional wall insulation does not reduce the usefulness of a house, it may even expand the comfortable area. But some automotive efficiency improvements can reduce the amenities offered by a car.

ATM. Modifying the transmission to reduce engine speed at a given power output is the best established method for improving average mechanical efficiency (90). To implement it, more gears and lower gear ratios can be built into the transmission and, in driving, gears are shifted up as soon as feasible. A feel for aggressive transmission management can be obtained by driving a 1990–1991 Honda CRX HF with a shift indicator light on the dashboard. As the car accelerates, the upshift light comes on very soon. If one follows the shift light's suggestion, one shifts up at the much lower engine speeds than is typical.

A critical consideration for fuel economy is the span, or the ratio of the highest to the lowest gear ratio. Consider a standard manual transmission. In the lowest gear (highest gear ratio), clutch slip is involved in getting the car moving, but the gear ratio must still be high enough to enable the engine to begin to move a stopped car up a grade (91). Good fuel economy performance in highway driving requires a low gear ratio in the highest gear. But the large span desired will not be feasible unless the ratios of adjacent gears are close enough to make shifting convenient. This requires many gears. For fuel economy, six forward gears are preferable to five in a manual transmission. For automatic transmissions with fluid coupling, fewer gears are needed because the lowest gear provides, roughly speaking, the function of the two lowest gears in manual transmissions. Four-speed automatics are now being widely used and five-speeds are being discussed.

Aside from the issue of creating the transmission technology, manufacturers may be reluctant to reduce engine speeds for two reasons. First, the engine may not run smoothly at low speed. Second, relatively high power is not immediately available at low engine speeds.

A Smaller High Tech Engine. The most rapid change in vehicle technology in recent years has been in specific power, the ratio of maximum engine power to engine size, or displacement. The average specific power increased 3.3% per year from 1976 to 1993 (81). This trend is expected to continue. The technologies are more valves per cylinder, high tech valve cams, higher compression ratios, advanced fuel injection, sophisticated controls (eg, of ignition timing), and tuning of the intake and exhaust manifolds. Variable valve timing is beginning to come into play. More sophisticated controls are in the offing, enabling management of cycle to cycle and cylinder to cylinder variations.

If engine displacement is reduced in proportion to the increase in specific power, then maximum power can be maintained, while the friction parameter a is reduced es-

sentially in proportion to the displacement. With today's average new car, a 10% engine downsizing results in a 6.5% increase in efficiency (average over urban and highway driving cycles including the benefit of weight reduction).

Both engine downsizing and ATM reduce the available power at a typical engine speed. Consider the combined effect of these two technologies: In Figure 15 two overlapping engine maps are shown. Two variables fix the point at which any engine is operating at a given time. In Figure 15, engine speed and power output are the variables. A low engine speed (point A) and a higher engine speed (point A') with the same power level are shown. With an older design large engine, it is possible to move directly from point A to power at level B by opening the throttle (depressing the accelerator pedal). On the other hand, with a small high tech engine, high power at point B' is immediately available if the driver starts from the higher engine speed at A', but downshifting is necessary if the driver starts from A.

This process is familiar to drivers of many cars with a four-cylinder engine and an automatic transmission, in which downshift and engine speedup occur when the accelerator pedal is floored. The action of declutching, engine speedup, and reclutching takes time; with a well-designed system, however, it takes half a second or less. Although this delay is a minor loss of amenity, it is a loss of amenity. In this respect, the ATM and engine downsizing technology are not analogous to many other technological changes that can be implemented without loss of amenity.

The Potential for Improving Automotive Efficiency. An ambitious goal for improving the average mechanical efficiency, with technology already in production or in development, would be a factor of 1.4, from about 43 to

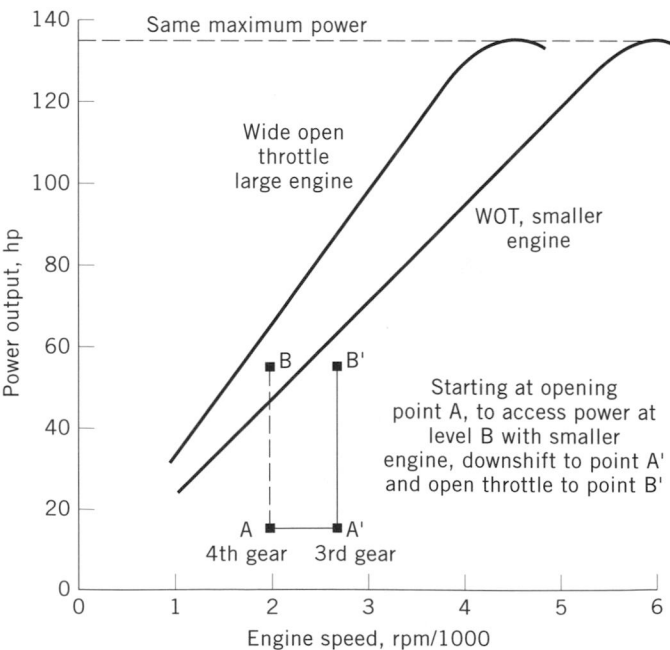

Figure 15. Downsized engine, transmission management.

about 60%, still much less than the 80% achieved in the best operating region of today's engines. This involves a combination of reductions: (1) reducing the three kinds of engine friction, (2) reducing the size of the engine while maintaining power capabilities, and (3) reducing the average engine speed. This combination is exemplified by examining the aN term of equation 10, which is responsible for all the mechanical inefficiency. (Note that $1 - n_m = aN/P_f$.) Because a is essentially proportional to the engine displacement V, $aN = kVN$. The three reductions just listed are in the three factors, k, V, and N, respectively.

Consider the following technological scenario. With an engine of today's general type, thermal efficiency is increased 3%. With a vehicle of today's general type, weight is reduced 15% and overall load is reduced 27%. The average mechanical efficiency is increased 28% by the following steps: k is reduced 15%, mostly through reduced pumping and accessory losses. V is reduced 15% because of a corresponding vehicle weight reduction and an additional 40% through continuing specific-power increases. N is maintained at its base value, in spite of the major downsizing of the engine, through ATM (Table 5).

The accounting scheme of Table 5 may be unfamiliar. The allocation of fuel use in the table is based on five energy sinks, the fuel used to enable the work done at the five sinks shown. Two other accounting schemes are common: (1) only the work done at the final sinks (including the transmission) is shown. For the car called AVCAR corresponding to Table 5, the energies in this scheme are total engine losses 3480 (of which 478 is at idle), trans-

mission 73, vehicle accessories 101, tires 231, air drag 220, and brakes 202, all in kJ/mi). (2) The engine frictions as well as the "lost work" associated with thermal inefficiency at the engine (as well as transmission loss) are allocated to the final sinks. For AVCAR in this scheme the energies are engine idle 478, vehicle accessories 467, tires 1190, air drag 1133, and brakes 1039.

In this scenario fuel economy is increased by 78% to 50 mpg. This goal has already been reached by two production cars, Geo Metro and Honda Civic VX. However, unlike the latter two cars, the car considered here is of average size for 1993 models and has an average power to weight ratio. The fuel use to overcome engine friction (top row of Table 5) is reduced a whopping 60%, but a good measure of the feasibility is the mechanical efficiency achieved (just under 60%), which is still low. The car described in Table 5 is hardly the ultimate that can be achieved with a fuel and internal-combustion (IC) based vehicle. The thermal efficiency was not pushed to a high value; neither a diesel engine or a lean burn spark ignition engine is considered. Nor was the load pushed to a really low value. In particular, radical new materials would enable weight reduction by perhaps 33%, instead of the 15% assumed. Moreover, a much narrower and smaller car for short trips and less carrying capacity, could enable even greater load reduction.

Another option, still based on the internal combustion engine, is the hybrid car, using energy storage, such as battery, flywheel, and/or electric capacitor, to enable the IC engine to operate only in the zone of its high mechanical efficiency. Efficient storage technology, which is reliable and of moderate cost, is needed. Such a car could achieve 80 mpg or perhaps much better (92).

The Economics of Auto Efficiency Improvement

The analysis of technologies available for improving conventional vehicles and their costs has been updated to include recent technological advances (93). The central result of this analysis is that roughly a 70% improvement in average new car fuel economy at fixed performance would be cost effective and could be implemented in a decade or so. The incremental retail cost of a new car, associated with these measures, is $770 (in 1993 dollars), an amount similar to the estimated cost of improvements to improve fuel economy at fixed performance made since the mid 1970s. The criteria behind this cost-effectiveness result are 5%/yr real discount rate, similar to actual auto loan rates, and 12-yr life, corresponding to vehicle life (rather than the period of initial ownership).

Examples of conservation supply curves for new cars show that the cost-effective improvements in vehicular efficiency (ie, in mpg) range from about 25 (93) to 70% (94), depending on the list of technologies, the capital recovery factor, and the price of gasoline used for comparison. And these examples are from two sources with a somewhat similar set of performance and cost estimates for each technology.

However, while the incremental retail cost implied by implementing many of the particular technologies is rather uncertain, the cost of a major group of the technologies is limited on the high side. The reason is that

Table 5. Specific Fuel Energy Consumption for Two Cars, kJ/mi[a]

	AVCAR'93	High MPG
Engine friction		
under power	1835	787
idle	478	228
Subtotal	*2313*	*1015*
Loads		
tires	619	371
air drag	590	444
brakes	541	432
vehicle access	243	158
Subtotal	*1993*	*1404*
Total fuel energy	*4307*	*2191*
Fuel economy, mpg	28.0	49.9
Vehicle Characteristics		
inertial weight, lb[b]	3234	2749
displacement, L[b]	2.77	1.40
N/v, rpm/mph[b]	34.0	34.0
engine friction k, kJ/rev · l	0.25	0.212
thermal efficiency η_t, %	41.5	42.7
rolling resistance C_R	0.010	0.008
air resistance $C_D A$, m²	0.663	0.530
transmission efficiency, ε	0.90	0.93

[a] EPA composite driving cycle, no adjustment.
[b] Sales-weighted averages from Ref. 81; the remainder of the AVCAR characteristics roughly describe modern cars.

most of them have been implemented in low priced production vehicles, without major impact on the vehicles' prices as illustrated by the Honda Civic VX (94). There simply is no room for these measures to add costs much greater than those estimated, if they are brought into production when models and components are being changed for other reasons.

Perhaps more important than these steady-state costs, are the up-front conversion costs. How much development and retooling is needed to incorporate these technologies in most cars, such that the manufacturing cost would be acceptable and the reliability, over millions of vehicles, would be excellent? Even though almost all the technologies mentioned are incorporated in some version of a production car, these up-front costs could be substantial for manufacturers who are not adept at changing the manufacturing technology or if the pace of change were rapid, not permitting the changes to be made when other large changes were made for other reasons.

These problems are particularly acute for engines, the site of most of the efficiency technologies. Engine manufacture is highly automated with the perhaps-surprising consequence that it is inflexible. The time between major changes in engines is long. Some Japanese manufacturers have, however, succeeded in moving to the next generation of engine manufacturing technology, where conversion of production lines for a new high tech engine is relatively easy. The U.S. manufacturers are catching up in this respect.

Policy Analysis

Selection of Policies for Discussion. The goal of a good transportation system with respect to daily travel is to provide convenient "access" at moderate cost, while maintaining or improving values such as safety and environmental quality. In a given situation, improving access may involve enabling people to travel farther, but it may instead involve reconfiguring land use patterns so that less travel is needed, eg, so the access can be achieved by walking.

Policy areas that bear directly on improved access are (95)

- Land use, including urban planning and community design.
- Public transport.
- Substitution of communication for transportation.
- Traffic and parking management, road controls, and road design.
- Driver behavior, including vehicle maintenance and driving style.
- Improvements in light vehicles.

Land use and public transportation policies are a principal focus of those interested in improved access (71,96,97). Their potential impact is suggested by the fact that per capita gasoline use in the Toronto metropolitan area is roughly half that in Houston, Phoenix, or Detroit (78). Yet Toronto is not that different. It is a relatively affluent, high quality of life North American metropolis.

The key characteristics of Toronto that appear to be responsible for its low gasoline use are high density and a well-developed public transportation system, both associated with regional government.

Substitutes for transportation, such as telecommunications, which enable some people to work and shop at home, and satellite at places of work that rely heavily on telecommunications also have significant potential. These are not primarily issues for public policy, but technology policies and regulation of communication systems are important to their success.

Traffic management in the form of speed limits, high occupancy vehicle lanes, and car pooling assistance have had modest success in reducing travel demand and fuel consumption (98). A component to increase the success of many of these programs is charging full cost for parking privileges (99). Road charges in congested areas have long been considered in Europe and Asia and have been successful in Singapore, where they have been combined with provision of extensive modern public transport (100).

Highway controls such as sophisticated signal management to encourage smooth traffic flow in congested areas can be effective. "Metering" lights, eg, to control the entrance of vehicles onto freeways are effective in enabling the main body of traffic to move smoothly (101). A new generation of highway information and controls, with many potential forms, is the subject of widespread research: intelligent vehicle and highway systems (IVHS) could, if they do not go far, turn out to be no more than an improved means for advertising local businesses, or they could, if more successful, be a large step toward automating traffic flow on high speed arteries. In the latter case the traffic smoothing capability (such that average speed is close to the maximum sought) can yield significant fuel economy benefits (102).

Driving behavior is also potentially important but difficult to influence. Proper vehicle maintenance, such as regular engine tuning and maintaining correct tire pressure, and driving style, ie, smooth flow instead of rapid stops, can contribute perhaps 10% to the fuel economy of the average car. Public education may be useful in this area.

While all these areas are important to efforts to reduce the environmental impact and energy consumption of light vehicles, in the interest of brevity the focus of this section is on policies that encourage technological improvements in light vehicles. Improving light vehicle technology will not be a sufficient means of resolving all the environmental and energy problems created by light vehicle use, but it may be the most important, because merely modifying conventional vehicles could cut fuel use by almost 50%.

Several aspects of improving vehicle fuel economy are of interest to policymakers:

- Modifications to conventional vehicles.
- Alternative fuels.
- Radical vehicle technology, such as the fuel cell vehicle or the very light, small commuter car.
- Interactions with vehicle safety and with emissions of pollutants.

The emphasis here is on the first topic. Consider four kinds of public policies: fuel taxes, standards for fuel economy, fees and rebates related to their fuel economy, and assistance for developing improved technology. To evaluate these policies, it is necessary first to consider briefly why the present market does not create an efficient vehicle fleet.

Barriers to Improved Fuel Economy. A significant barrier to fuel economy improvement was removed by standardized testing. Before the current fuel economy tests, the various claims that were made could not be evaluated by the public. The tests available are not highly accurate in mimicking actual driving; they apparently underestimate fuel use by roughly 20% (103,104). But they are accurate enough to order most vehicles correctly according to their actual fuel consumption in typical driving.

The major remaining barrier is the apparent low priority given to operating costs by new vehicle buyers. Among the total costs of owning and operating a car, fuel plays a small role, about 10%, and the net annual savings from buying a higher fuel economy car are small (76,105). Surveys suggest that new car buyers find fuel economy to be a secondary consideration. Many other attributes have higher priority: brand, safety, interior volume, trunk size, handling, price, acceleration, reliability, etc (106). Indeed, manufacturers have decided that fuel economy is of so little interest to buyers at this time that they offer it only as part of a package in bottom-of-the-market vehicles (such as the Geo Metro), making it impossible for buyers simply to choose added fuel economy at extra cost while preserving the other vehicle attributes in which they are interested. The Honda Civic VX is the only car for which a buyer can pay more to obtain the same vehicle performance and amenities but with a better fuel economy; it is not a best-seller.

This barrier at the point of vehicle purchase should not be misinterpreted as a barrier to policymaking. Many buyers of a new car are not much interested in the fuel economy of their new car, since both the personal savings and the societal benefits associated with their purchase of a single high fuel economy car would be small. The same people as citizens may well be interested in vehicle fuel economy. In other words, when everyone participates in a move toward higher fuel economy, it becomes much more interesting.

Fuel Pricing. There are at least two issues to discuss in terms of fuel pricing: rationales for taxing fuels and the likely consequences of increased fuel prices. The starting point is, of course, the damage caused by use of vehicles and the costs borne to avoid or mitigate such damage. Many of these costs are directly associated with energy use. Many have not been internalized, ie, the costs are not paid by people in relationship to the use of vehicles or to the energy used in operating the vehicles (107). The advantage of fuel taxes is the potential breadth of response by individuals, based on their individual interests. (Economists refer to this as an efficiency.) Thus, in response to a fuel price increase, an individual may change behavior in any of a wide variety of ways, from land use to amount of driving to choice of vehicle. And those choices depend on the values and circumstances of each individual.

Unfortunately, the effect of a price increase on gasoline use is only known in terms of the small short-term response discussed above. The longer-term response, involving choice of vehicle and residential community, eg, is not known with any reliability for today's conditions, although there have been regression analyses (108).

Experience overseas suggests that a large fuel tax (of $1.00–3.00/gal as in Europe and Japan) would not by itself have a dramatic effect on vehicle choice. It might have more effect on driving (36). Such a large tax would certainly not be politically feasible in the United States at present.

Regulatory Standards for Fuel Economy. The Motor Vehicle Information and Cost Savings Act of 1975 set corporate average fuel economy (CAFE) standards that required the fuel economy of new cars to increase from about 14 mpg in the early 1970s to 27.5 mpg by 1985. The act provided flexibility for manufacturers by applying the standard to the sales-weighted average for each corporation, instead of to each vehicle. Moreover, the secretary of transportation was given the discretion to set a lower standard, as was done for 1986 through 1989, on appeal from manufacturers. The discretion to set standards for light trucks was also left to the secretary of transportation.

In hearings on the 1975 Act, the manufacturers stated that the technology to achieve 27.5 mpg was not available on the proposed time scale and that the only way to achieve the standard would be by making the average car much smaller. They said it would "outlaw full-size sedans and station wagons" (Chrysler), "require all sub-compact vehicles" (Ford), and "restrict availability of 5 and 6 passenger cars regardless of consumer needs" (General Motors) (109). Indeed, there was some reduction in the ratio of maximum power to weight, although almost none in the interior volume, in the early 1980s (81). By the late 1980s, however, the manufacturers were achieving the mandated standards with vehicles of interior volume and maximum power equal to and higher than those of the early 1970s. The CAFE standards were thus an important example of successful "technology forcing" by regulation.

Auto manufacturers have argued that higher gasoline taxes drove the fuel economy improvement of the late 1970s and early 1980s. That is, car buyers demanded higher fuel economy, and the manufacturers responded. A detailed look at the data does not support this argument; the CAFE regulations were the critical factor (110). However, synergism between performance regulation and fuel prices was helpful and is an important lesson for policymaking.

General Motors and Ford have also argued that the CAFE formulation placed them at a disadvantage because their mix of vehicles includes large cars while those of Asian manufacturers do not. As a consequence, they argue, it has been easier and less expensive for the Asian manufacturers to meet the standards. Most of the recent legislation proposed in the U.S. Congress to strengthen the fuel economy standards addresses this problem by changing the basis of the standards so that each manufacturer is required to improve its fuel economy by the same percentage above its base year fuel economy. Size-weighted average fuel economies could also serve the same purpose.

Because substantially higher fuel economies are practical and cost-effective and because society has a major interest in reducing petroleum demand, it is not surprising that stronger regulatory standards for fuel economy are actively being considered in Congress. Senator Bryan sponsored a bill that would have required each manufacturer to increase its average fuel economy 40% above its 1988 level by 2001. On average, the bill would have required new cars to reach 40 mpg. It was supported by a majority of the Senate but failed to overcome a filibuster in late 1990. The bill was reintroduced in 1992 and again came close, but failed.

Automobile manufacturers strongly oppose the legislation and claim, as they did in 1975 before the first CAFE standards were passed, that it is not practical to improve fuel economy substantially except by moving, on the average, to much smaller cars. Manufacturers are stonewalling on this point. Other, more compelling, reasons for their opposition are

- Large tooling investments would be needed to make the changes, especially if a moderately rapid time table is required as has been proposed.
- The required rate of improvement in fuel economy might prevent manufacturers from fully exploiting sales opportunities for low fuel economy models already in production.
- High fuel economy standards might somewhat restrict designers' options in developing new vehicles and markets.

It is important to address such concerns by creating a schedule of strengthened standards, allowing adequate time for manufacturers to adjust, and by enacting policy packages (with components discussed elsewhere in this section) such that the burden of compliance would not fall entirely on the manufacturer. Policies in addition to fuel economy standards should be enacted that motivate buyers to select high fuel economy vehicles.

Incentives at New Vehicle Purchase. The gas guzzler tax, enacted as part of the Energy Tax Act of 1978, has tended to be overlooked as an effective policy tool for improving fuel economy. There is strong evidence that the gas guzzler tax played an important role in improving new car fuel economy, especially between 1983 and 1986 (76). In 1978, Congress also considered rebates for new cars with high fuel economy, but rejected the concept. Such a program of fees and rebates can be designed to be revenue neutral, such that total rebates roughly equal total fees (111).

Fees and rebates on the purchase of a new vehicle could be important tools to improve fuel economy and emissions (112). Given U.S. society's sensitivity to first cost, it is probably easier and more effective to adjust for market imperfections and influence new car fuel economy and emission levels at the point of capital equipment purchase than it is to adjust for imperfections in the course of operations with, eg, a gasoline tax.

Technology Push and Pull Policies. The policies discussed above indirectly encourage the creation of new technology to meet the changed economic conditions or regulatory constraints. Experience shows, however, that a more direct policy focus on new technology can be highly effective (113). Before considering the policies, briefly consider the possibilities for new technology.

New technology here means vehicles, and their energy supply systems, which could radically reduce energy requirements and emissions but which are not close to being in mass production. (This technology would go beyond that considered in constructing Table 5.) There are three potential types of vehicle technologies:

1. Vehicles with much higher fuel economy but still based on gasoline or diesel fuel and still serving four or more passengers with, roughly, today's driving capabilities.
2. Special-purpose vehicles requiring much less energy at the drive wheels, such as small commuter cars.
3. Alternative-fuel vehicles.

In group 1 would be hybrid vehicles, mentioned above, which use energy storage to enable the fuel engine to achieve much higher mechanical efficiency that at present. Another possibility would incorporate an advanced, direct-injection, diesel engine. This engine, now entering production in Europe, is about 25% more efficient than comparable conventional gasoline-powered engines. High fuel economy prototype vehicles incorporating advanced diesels have been built or partially developed by Volvo, Volkswagen, Renault, Peugeot, and Toyota, with in-use fuel economies estimated to be almost 70 mpg and higher (114). Diesel emissions problems would have to be solved.

In group 2, there are vehicles such as the proposed Lean Machine and the demonstration electric vehicle called Impact, both developed by General Motors. The Lean Machine is a two seater with the passenger seat behind the driver. Both the Lean Machine and the Impact are small, have little air and tire drag, and require very little power to be delivered to the wheels in typical driving. (The fact that the Impact is an electric vehicle is incidental to this discussion.) Both of these prototype vehicles happen to have rather high acceleration performance. It is not clear if that is an important attribute for marketing such a vehicle. Safety is a critical issue for such small vehicles. It may be important to consider separate lanes on high speed roadways.

In group 3 there is an enormous range of possibilities. Only two of the most exciting are mentioned here: hybrid-electric and fuel-cell vehicles. The hybrid has an internal-combustion engine and one or two kinds of storage that can be used to drive an electric motor. An attractive combination uses the engine only in a high efficiency and low emissions operating region. One of the storage devices would provide high power, eg, to enable a good 0–60 mph acceleration time. The other would be energy storage that could be used to balance energy flows and might also be used for all-electric operation in restricted urban zones. The hybrid could overcome the severe disabilities of electric vehicles: their short daily range, long battery recharge period, short battery life under vehicular conditions, and high cost.

The fuel cell is a kind of battery, but different from the common electrochemical cell, with the advantage of relying on a stored fluid fuel like methanol or hydrogen. The fuel cell converts the chemical energy of the fuel to electricity without combustion. Extremely little, if any, emissions are associated with fuel-cell operation, with the exception of water and carbon dioxide. Much higher efficiencies of conversion are possible than with the present kind of engine, because the second law does not limit the efficiency.

Let us briefly consider some technology policies. The U.S. government has been highly effective in encouraging new technology in some sectors like agriculture, commercial aircraft, and semiconductors. The tools used are technology push and technology pull. Technology push concerns the creation of technology: research, invention, development, and demonstration. This is not a linear sequence of activities, in which one follows the next but a complex interaction in which new technologies are created. Technology push policies involve government support for research, development, and demonstration (R,D&D) and government encouragement of private sector R,D&D through tax incentives, patent law, etc.

Technology pull concerns the demand for new technology, ie, demand for it after it reaches initial commercial status. It cannot be overemphasized that the existence of a likely market for a new or improved process or product strongly motivates development and production of new technologies, and the apparent absence of a market strongly inhibits them. Government policies can provide technology pull through government purchases and by encouraging the private sector's propensity to purchase new technology (115).

An example of technology push policy is government-supported research and development on generic technologies that could form the basis for many new product developments. Modest government involvement in transportation R&D is proving beneficial in electrochemistry (new and improved batteries), combustion (understanding of knock and soot formation), and ceramic insulation (for the combustion chamber). It would be valuable to continue support in these areas and greatly expand the government's efforts in, eg, engine friction and control approaches for hybrid-electric vehicles.

Another kind of technology push policy is the cooperative government–industry venture, closely managed by committees and involving substantial industry investment. An example is the Advanced Battery Consortium for electric vehicles. As with the Steel Initiative mentioned above, there are serious difficulties with this kind of approach because the firms involved are often mature, and although willing to join the venture, they may, in fact, be lukewarm about radical innovation.

An attractive example of a technology pull policy would be providing extra fuel economy credit to manufacturers who produce automobiles or light trucks that attain exceptionally high levels of fuel economy. Such a provision would reward manufacturers for aggressively introducing new technology, providing an incentive for manufacturers to take a substantial leap forward with fuel economy technologies, instead of taking small incremental steps. The incentive could be made especially strong for improving the fuel economy of mid-size and large cars.

CONCLUSIONS

The opportunity to improve the energy efficiency of society is excellent. Researchers have a long way to go before bumping up against limitations imposed by physical laws. If cost-effective investments were widely adopted, and people followed efficient practices, the resulting energy efficiency improvement would substantially ease environment problems, reduce capital requirements for energy, remove the petroleum security problems of the United States, and enable renewable energy supplies to be more easily implemented. (What is cost effective depends, however, on who makes the evaluation, especially on their time horizon or discount rate.)

Both the mix of activities and the efficiencies of the narrowly defined activities are involved in the overall energy efficiency of society. (Here activity refers, eg, to annual travel by different modes and annual use of various manufactured materials.) And the activity mix and real energy intensity can both be improved through changes in technology, behavior, and social institutions. In addition, new technologies are easing the way.

In the examination of industry and transportation, it was found that the opportunity for efficiency improvement may have different characteristics. The mix of activities in industry is tending toward higher efficiency, while the mix in transportation is tending toward lower energy efficiency. Meanwhile, the sectoral, or modal, energy intensities show great room for technological improvement. Moreover, large corporations (which predominate in industry, and as producers, and often operators, of transportation equipment) have proven highly capable of improving their own energy efficiency when motivated. Individuals and smaller firms have less technical capability and may be harder to motivate.

Many feel that market-based decision making is the only proper means for society's energy efficiency to be determined. This is, however, simplistic. One can ask, What market? Today's? There are three problems with today's energy-use markets. (1) Energy prices do not reflect many of the costs involved in providing energy, such as unregulated environmental damage, petroleum security expenditures, and supply subsidies. (2) Market failures are common in the design of products (relative to their energy performance), for which the cost of energy is a small part of the total. When the motivation is strong, striking design improvements can be made that are cost-effective for general application. For example, new high performance portable computers are designed to have much smaller energy consumption than personal computers plugged into the grid. With some government encouragement, these improvements, especially with respect to stand-by operation, are now beginning to be used in standard PCs. (3) Market failures are also common in consumer decisions relating to equipment purchase and its energy efficiency. There appears to be a threshold in relative cost below which individuals and firms do not give energy any attention, aside from complaining about the bills. The problems are especially acute where the energy-related decisions are made by a series of players.

Public policies can be appropriate tools for achieving the benefits of increased energy efficiency. Until recently, energy policies focused on expanding the supply of energy and encouraging the exploitation of natural resources more generally (5). The same general kinds of policies are gradually being adopted to encourage efficiency improvement. While this represents a significant shift in philosophy and does adversely impact particular sectors, such as energy supply, the policies need not be draconian or intervene in the details of everyday life.

Having said that, it must be added that while some kinds of energy efficiency improvement are proving relatively easy to achieve, other kinds of improvement are difficult to achieve. Progress will be difficult in several areas.

An important example of improvement that should continue to be relatively easy is enhanced performance of mass-produced equipment, such as automobiles, household appliances, lighting equipment and commercial building space conditioning equipment. That is not to say that these improvements are implemented (in the absence of high energy prices) without intervention. The gains have been and will be made with the help of regulatory performance standards; government support of research, development, and demonstration; and economic incentives such as rebate programs by regulated utilities. In many cases, effective public policy interventions have yet to be made.

Another area of relatively rapid improvement is in the mix of industrial activities. Consumption has gradually been shifting from massive products to high technology products, which are less energy intensive to produce and often less energy intensive to use. This development is a fundamental historic change, not the result of public policy intervention, although several policies nudge it along, especially policies to encourage high tech industries and policies to reduce the environmental impacts of industries that extract and process raw materials.

Other aspects of energy use are less encouraging: improving the efficiencies of custom processes and facilities, like most housing and industrial production processes, and moving the mix of travel activities toward more efficiency. These aspects of energy use will be hard to improve without the commitment of extraordinary skills and resources or large energy price increases.

The difficulty with custom-designed processes and facilities is that policies aimed at overcoming market failures of the kinds just mentioned tend to be cumbersome and ineffective. For example, government and utilities can and do offer energy audits to identify efficiency improvement opportunities. But the customer often does not know if the auditor and the vendor who would install the improvement really do good work. More important, energy has a low priority for most customers. Their attention is focused elsewhere. For example, the factories provided with recommendations by a (Department of Energy sponsored) Energy Audit and Diagnostic Center will usually only undertake projects that pay back in a year or less.

There are, nevertheless, possibilities for progress in this difficult area. The keys are found in new technologies and in changes that have multiple benefits rather than in narrowly focused energy efficiency improvement. For example, industrial processes can be changed with sensors and controls, so that the product is of higher quality, there are fewer rejects, maintenance requirements are anticipated, production becomes more flexible, and energy consumption is reduced. In contrast, pushing insulation and heat recovery to their simple cost-effective limits may prove to be inflexible and, not, in the end, be a sound way to use limited capital and engineering resources.

The pattern of increasing per capita auto and light truck travel is encouraged by land-use policies favoring this growth, and those policies will be difficult to change. Personal values, established self-government by towns, and tax and other long-established policies favoring low density land development in rural areas will resist change. Improvements in land-use patterns and associated travel needs require, at least, the political will to remove the incentives for development of rural lands. There are some growing nonenergy reasons for eliminating those incentives, such as environmental concerns and saturation in the devotion of land to the automobile. Because energy in itself is seldom a governing consideration, such mutual benefits are essential if energy efficiency is to be increased rapidly. To be improved rapidly, energy efficiency must be seen as part of economic efficiency and part of environmental quality.

BIBLIOGRAPHY

1. J. Priest, *Energy Principles, Problems, Alternatives,* 4th ed., Addison-Wesley Publishing Co., Inc., Reading, Mass., 1991.
2. American Institute of Physics, *Efficient Use of Energy, Part 1–A Physics Perspective, Conference Proceedings,* vol. 25, New York, 1975.
3. J. H. Keenan, *Thermodynamics,* John Wiley & Sons, Inc., New York, 1941.
4. R. U. Ayres, *Energy Efficiency in the U.S. Economy: A New Case for Conservation,* International Institute for Applied Systems Analysis, Laxenburg, Austria, 1989.
5. M. Ross and R. Williams, *Our Energy: Regaining Control,* McGraw-Hill Book Co., Inc. New York, 1981.
6. Scientific American, *Energy for Planet Earth,* W. H. Freeman Publishers, San Francisco, 1990.
7. C. D. Masters, D. H. Root, and E. D. Attanasi, *Ann. Rev. Energy* **15,** 23–51 (1990).
8. National Research Council, *Undiscovered Oil and Gas Resources: An Evaluation of the Department of Interior's 1989 Assessment Procedures,* National Academy Press, Washington, D.C., 1991.
9. World Resources Institute, *World Resources 1992–93,* Washington, D.C., 1992.
10. U.S. Congress, *U.S. Natural Gas Availability: Gas Supply Through the Year 2000,* Office of Technology Assessment, Washington, D.C., 1985.
11. R. Nehring, *Ann. Rev. Energy* **7,** 175–200 (1982).

12. G. W. Hinman in R. Howes and A. Fainberg, eds., *The Energy Sourcebook: A Guide to Technology Resources and Policy,* American Institute of Physics, New York, 1991, pp. 99–126.

13. T. B. Johansson, H. Kelly, A. K. N. Reddy, and R. H. Williams, *Renewable Energy: Sources for Fuels and Electricity,* Island Press, Washington, D.C., 1993.

14. U.S. Department of Commerce, *Survey of Current Business,* Washington, D.C. (monthly).

15. J. M. Hollander, ed., *The Energy Environment Connection,* Island Press, Washington, D.C., 1992.

16. T. E. Graedal and P. J. Crutzen, *Sci. Amer.* **261,** 58–68 (Sept. 1989).

17. S. Sillman, *Ann. Rev. Energy Environ.* **18,** 31–56 (1993).

18. V. A. Mohnen, *Sci. Amer.* **259,** 30–38 (Aug. 1988).

19. S. M. Schneider, *Sci. Amer.* **263,** 312–320 (1990).

20. A. Revkin, *Global Warming: Understanding the Forecast,* Abbeville Press, New York, 1992.

21. J. P. Peixoto and A. H. Oort, *Physics of Climate,* American Institute of Physics, New York, 1992.

22. U.S. Congress, *Changing by Degrees: Steps to Reduce Greenhouse Gases,* Office of Technology Assessment, Washington, D.C., 1991.

23. M. Grubb and co-workers, *Energy Policies and the Greenhouse Effect,* 2 vols., Dartmouth Publishing, Aldershot, UK, 1991.

24. V. Anderson, *Energy Efficiency Policies,* Routledge, London, 1993.

25. G. Boyd, J. McDonald, M. Ross, and D. Hanson, *Energy J.* **8**(2), 77–97 (1987).

26. W. Leontief, *Sci. Amer.,* **185** 15 (Oct. 1951).

27. U.S. Congress, *Energy Use and the U.S. Economy,* Office of Technology Assessment, Washington, D.C., 1990.

28. J. M. Roop, *The Trade Effects of Energy Use in the U.S. Economy: An Input-Output Analysis,* paper presented at the North American meeting of the International Association of Energy Economists, 1986.

29. Energy Information Administration, *International Energy Annual 1991,* Washington, D.C., 1992.

30. J. Goldemberg, T. B. Johansson, A. K. N. Reddy, and R. H. Williams, *Energy for a Sustainable World,* World Resources Institute, Washington, D.C., 1987.

31. A. J. Meier, J. Wright, and A. H. Rosenfeld, *Supplying Energy Through Greater Efficiency,* University of California Press, Berkeley, 1983.

32. P. G. LeBel, *Energy Economics and Technology,* Johns Hopkins University Press, Baltimore, Md., 1982.

33. U.S. Congress, *Building Energy Efficiency,* Office of Technology Assessment, Washington, D.C., 1992.

34. A. H. Rosenfeld and E. Ward, in Ref. 15, pp. 233–257.

35. U.S. Congress, *Industrial Energy Efficiency,* Office of Technology Assessment, Washington, D.C., 1993.

36. L. Schipper and S. Meyers, *Energy Efficiency and Human Activity,* Cambridge University Press, Cambridge, UK, 1992.

37. M. Ross, *Proc. Nat. Acad. Sci. U. S. A.* **89,** 827–831 (1992).

38. R. Herman, S. A. Ardekani, and J. H. Ausubel, in J. H. Ausubel, ed., *Technology and Environment,* National Academy Press, Washington, D.C., 1989, pp. 50–69.

39. R. C. Marlay, *Science* **226,** 1277–1283 (1984).

40. R. H. Williams, E. D. Larson, and M. H. Ross, *Ann. Rev. Energy* **12,** 19–144 (1987).

41. Jack Faucett Assoc., *National Energy Accounts 1958–1985,* Bethesda Md., 1989.

42. D. Steinmeyer in Ref. 15, pp. 319–343.

43. C. Berg, *Science* **199,** 608–614 (1978).

44. P. S. Schmidt, *Electricity and Industrial Productivity: A Technical and Economic Perspective,* Pergamon Press, Oxford, UK, 1984.

45. L. Argote and D. Epple, *Science* **247,** 920–924 (1990).

46. W. H. Joyce, in J. Tester, D. Wood, and N. Ferrari, eds., *Energy and the Environment in the 21st Century,* MIT Press, Cambridge, Mass., pp. 427–435, 1991.

47. M. H. Ross, P. Thimmapuram, R. E. Fisher, and W. Maciorowski, *Long-Term Industrial Energy Forecasting (LIEF) Model,* ANL/EAIS/TM-95, Argonne National Laboratory, Argonne, Ill., 1992.

48. S. R. Venkateswaran and H. E. Lowitt, *The U.S. Cement Industry: An Energy Perspective,* a report by Energetics Inc., Columbia, Md., 1988.

49. M. J. King, *Guide to the INDEPTH Level 1 Econometric Models: Final Report,* Electric Power Research Inst., Palo Alto, Calif., 1990.

50. R. Carlsmith, W. Chandler, J. McMahon, and D. Santini, *Energy Efficiency: How Far Can We Go?* ORNLTM-11441, Oak Ridge National Laboratory, Oak Ridge, Tenn., 1990.

51. J. G. Koomey, *Energy Efficiency Choices in New Office Buildings: An Investigation of Market Failures and Corrective Policies,* Ph.D. dissertation, University of California, Berkeley, 1990.

52. A. C. Fisher and M. H. Rothkopf, *Energy Policy* **17,** 397–406 (1989).

53. C. Blumstein, B. Krieg, L. Schipper, and C. York, *Energy* **5,** 355–371 (1980).

54. J. S. Peters, ACEEE, *Summer Study,* American Council for an Energy-Efficient Economy, Washington, D.C., 1988.

55. Alliance to Save Energy, *Industrial Investment in Energy Efficiency: Opportunities, Management Practices and Tax Incentives,* Washington, D.C., 1983.

56. M. Ross, *Financial Manage.,* 15–22 (Winter: 1986).

57. J. C. van Horne, *Financial Management and Policy,* Prentice-Hall, Inc., Englewood Cliffs, N.J., 1980.

58. R. Pike and B. Neale, *Corporate Finance and Investment: Decisions and Strategies,* Prentice-Hall, Inc., Englewood, N.J., 1993.

59. W. D. Montgomery in M. A. Kuliasha, A. Zucker, and K. J. Ballew, eds., *Technologies for a Greenhouse-Constrained Society,* Lewis Publishers, Boca Raton, Fla., 1992.

60. D. A. Lashof and D. Tirpak, *Policy Options for Stabilizing Global Climate,* Office of Policy, Planning and Evaluation, U.S. Environmental Protection Agency, Washington, D.C., 1989.

61. U.S. Department of Energy, *A Compendium of Options for Government Policy to Encourage Private Sector Responses to Potential Climate Change,* Washington, D.C., 1989.

62. W. H. Chandler, H. Geller, and M. Ledbetter, *Energy Efficiency: A New Agenda,* American Council for an Energy-Efficient Economy, Washington, D.C., 1988.

63. D. H. Moskovitz, *Ann. Rev. Energy* **15,** 399–421 (1990).

64. E. Hirst, *A Good Integrated Resource Plan: Guidelines for Electric Utilities and Regulators,* ORNL/CON-354, Oak Ridge National Laboratory, Oak Ridge, Tenn., 1992.

65. E. Hirst, *Electric-Utility DSM Program Costs and Effects: 1991 to 2001,* ORNL/CON-364, Oak Ridge National Laboratory, Oak Ridge, Tenn., 1993.

66. S. Nadel, *Ann. Rev. Energy Environ.* **17,** 507–536 (1992).

67. G. Hatsopoulos, P. Krugman, and L. Summers, *Science* **241,** 299–307 (1988).

68. H. S. Geller and S. M. Nadel, *Consensus National Efficiency Standards for Lamps, Motors, Showerheads, and Faucets, and Commercial HVAC Equipment,* American Council for an Energy-Efficient Economy, Washington, D.C., 1992.

69. A Price and M. Ross, *Electricity J.* **2,** 40–52 (July 1989).

70. H. Geller, J. Harris, M. D. Levine, and A. H. Rosenfeld, *Ann. Rev. Energy* **12,** 357–396 (1987).

71. J. Ahearne, *Why Federal Research Fails,* discussion paper EM 88-02, Resources for the Future, Washington, D.C., 1988.

72. D. Gordon, *Steering a New Course: Transportation, Energy and the Environment,* Union of Concerned Scientists, Boston, 1991.

73. S. C. Davis and S. G. Strang, *Transportation Energy Data Book: 13,* ORNL-6743, Oak Ridge National Laboratory, Oak Ridge, Tenn., 1993.

74. D. L. Greene, *Ann. Rev. Energy Environ.* **17,** 537–573 (1992).

75. M. Ross, *Ann. Rev. Energy* **14,** 131–171 (1989).

76. M. Ross and M. Ledbetter, in Ref. 15, pp. 258–318.

77. Federal Highway Administration, *1990 Nationwide Personal Transportation Survey, Summary of Travel Trends,* Washington, D.C., 1992.

78. P. G. Newman and J. R. Kenworthy, *Cities and Automobile Dependence: A Sourcebook,* Gower Technical, Aldershot, UK, 1988.

79. J. Holtzclaw, *Explaining Urban Density and Transit Impacts on Auto Use,* Testimony before the California Energy Commission, Docket No. 89-CR-90, Sacramento, April 23, 1990.

80. American Automobile Manufacturers Association, *Facts and Figures '93,* Washington, D.C., 1993.

81. J. D. Murrell, K. H. Hellman and R. M. Heavenrich, *Light Duty Automotive Technology and Fuel Economy Trends Through 1993,* U.S. Environmental Protection Agency, Office of Mobile Sources, Ann Arbor, Mich., 1993.

82. F. An and M. Ross, *A Model of Fuel Economy and Driving Patterns,* 930328, Society of Automotive Engineers, Warrendale, Pa., 1993.

83. R. Stone, *Introduction to Internal Combustion Engines,* Macmillan Publishers, London, 1985.

84. C. A. Amann, *The Automotive Engine–a Future Perspective,* 891666, Society of Automotive Engineers, Warrandale, Pa., 1989.

85. M. Ross and F. An, *The Use of Fuel by Spark Ignition Engines,* 930329, Society of Automotive Engineers, Warrendale, Pa., 1993.

86. Energy and Environmental Analysis Inc., *Analysis of the Capabilities of Domestic Auto-Manufacturers to Improve Corporate Average Fuel Economy,* U.S. Department of Energy, Arlington, Va., 1986.

87. U. Seiffert and P. Walzer, *Automobile Technology of the Future,* Society of Automotive Engineers, Warrendale Pa., 1991.

88. U.S. Congress, *Improving Automobile Fuel Economy: New Standards, New Approaches,* Report OTA-E-504, Office of Technology Assessment, Washington, D.C., 1991.

89. National Research Council, *Automotive Fuel Economy: How Far Should We Go?* National Academy Press, Washington, D.C., 1992.

90. K. C. Ludema in J. C. Hilliard and G. S. Springer, eds., *Fuel Economy in Road Vehicles Powered by Spark Ignition Engines,* Plenum Press, New York, 1984.

91. R. Stone, *Automobile Fuel Economy,* Macmillan Education, London, 1989.

92. A. B. Lovins, J. W. Barnett, and L. H. Lovins, *Supercars: the Coming Light Vehicle Revolution,* Rocky Mountain Institute, Snowmass, Colo., 1993.

93. J. DeCicco and M. Ross, *An Updated Assessment of the Near-Term Potential for Improving Automotive Fuel Economy,* American Council for an Energy-Efficient Economy, Washington, D.C., 1993.

94. J. G. Koomey, D. E. Schechter, and D. Gordon, *Cost Effectiveness of Fuel Economy Improvements in 1992 Honda Civic Hatchbacks,* Transportation Research Record 1416, National Research Council, Washington, D.C., 1993.

95. M. Ledbetter and M. Ross in J. Byrne and D. Rich, eds., *Energy and Environment, Energy Policy Studies,* Vol. 6, Transaction Publishers, New Brunswick, N.J., 1992, pp. 187–233.

96. B. S. Pushkarev and J. M. Zupan, *Public Transportation and Land Use Policy,* Indiana University Press, Bloomington, 1977.

97. R. W. Burchell and D. Listokin, eds., *Energy and Land Use,* Center for Urban Policy Research, Rutgers University, Piscataway, N.J., 1982.

98. M. Burke, *High Occupancy Vehicle Facilities: General Characteristics and Potential Fuel Savings,* American Council for an Energy-Efficient Economy, Washington, D.C., 1990.

99. M. Replogle, *U.S. Transportation Policy: Let's Make it Sustainable,* Institute for Transportation and Development Policy, Washington, D.C., 1990.

100. B. W. Ang in D. L. Bleviss and M. L. Birk, eds., *Driving New Directions: Transportation Experiences and Options in Developing Countries,* International Institute for Energy Conservation, Washington, D.C., 1991, pp. 41–51.

101. Institute of Transportation Engineers, *A Toolbox for Alleviating Traffic Congestion,* Washington, D.C., 1989.

102. F. An and M. Ross, *A Model of Fuel Economy with Applications to Driving Cycles and Traffic Management,* Transportation Research Record 1416, National Research Council, Washington, D.C., 1993.

103. K. Hellman and J. D. Murrell, *Development of Adjustment Factors for the EPA City and Highway MPG Values,* 8400496, Society of Automotive Engineers, Warrendale, Pa., 1984.

104. M. Mintz, A. Vyas, and L. A. Conley, *Differences Between EPA-test and In-use Fuel Economy: Are the Correction Factors Correct?* Transportation Research, Record 1416, Washington, D.C., 1993.

105. F. von Hippel and B. G. Levi, *Resources Conservation,* **10,** 103–124 (1987).

106. P. S. McCarthy and R. Tay, *Transportation Res.* **23A,** 367–375 (1989).

107. J. J. MacKenzie, R. C. Dower, and D. Chen, *The Going Rate: What it Really Costs to Drive,* World Resources Institute, Washington, D.C., 1992.

108. D. R. Bohi and M. B. Zimmerman, *Ann. Rev. Energy* **9,** 105–154 (1984).

109. Energy Conservation Coalition, *The Auto Industry on Fuel Efficiency: Yesterday and Today,* Washington, D.C., 1989.

110. D. L. Greene, *Energy J.* **11**(3), 37–57 (1990).

111. D. Gordon and L. Levenson, *J. Policy Anal. Manage.* **9,** 409–415 (1990).

112. J. DeCicco, H. Geller, and J. H. Morrill, *Feebates for Fuel Economy: Market Incentives for Encouraging Production and Sales of Efficient Vehicles,* American Council for an Energy-Efficient Economy, Washington, D.C., 1993.

113. R. R. Nelson, ed., *Government and Technical Progress,* Pergamon Press, New York, 1982.

114. D. Bleviss, *The New Oil Crisis and Fuel Economy Technologies: Preparing the Light Transportation Industry for the 1990s,* Quorum Press, New York, 1988.

115. M. Ross and R. H. Socolow, *Issues Sci. Technol.* **7,** 61–66 (Spring 1991).

ENERGY EFFICIENCY CALCULATIONS

WILLIAM KENNEY
Exxon Research and Engineering
Florham Park, New Jersey

Equipment and processes with high energy efficiency are generally valued. Indeed, consumer advocates have arranged for practical measures of efficiency (eg, cost/year to operate) to be attached to a number of household appliances and automobiles. Yet, the basis for such efficiency projections often remains vague. Invariably, the tests that provide the data for these efficiency claims follow some government protocol intended to match the patterns in which an "average user" would employ the appliance or automobile. Although these protocols remain as obscure as the habits of the "average user" to most consumers, the labels do provide some guidance to the purchaser of relatively standard devices about the energy consumption of the alternatives that are being considered. See also THERMODYNAMICS; ENERGY AUDITING; ENERGY CONVERSION FACTORS.

However, when specialized industrial equipment and processes are considered, many bases for efficiency calculations are possible. Not only is there a choice between a first law (of thermodynamics) and a second law basis, but assumptions are also made about ambient conditions, the boundaries of the system to be included, coolant temperatures (winter vs summer vs average), the purity of working fluids and lubricants, fouling, maintenance status, and so forth.

During the energy crisis of the late 1970s, Senator Kennedy argued that all processes and equipment should be evaluated by their second law efficiency, which is generally quite low relative to the more conventional first law approach. The implication was that because most industrial processes had low second law efficiencies, they could be markedly improved if only the industry would attend to the problem. There was, perhaps, even the inference that the government might not permit certain low efficiency processes or equipment to continue to operate. In that time frame, much work was done in the second law analysis of processes and equipment, and the government funded a number of research projects aimed at improving analytical techniques and actual processes and hardware in pursuit of higher efficiencies.

Some rejected the second law approach because in many cases, high second law efficiencies could only be achieved theoretically; practical processes and equipment to carry out the desired operations did not exist. For example, there were no high temperature fuel cells that would permit electricity generation directly from a combustion reaction, and no membranes existed that could separate the components of the flue gas so that they could be expanded in an engine to their partial pressure in the atmosphere and generate power in the process.

Despite the fog that surrounds the concept, analysts have found the concept of efficiency useful over the years. The comparison of the performance of a practical device or process to that of a well-defined standard has been of value in many applications. A low efficiency does highlight steps or systems that may be suitable for upgrade (1–3). In some cases, improvement may be inherently impossible, but these are often easily identified. For example, in a Carnot cycle, the efficiency of the engine depends only on the temperature difference between the source of heat and the sink to which waste heat is rejected. The best possible efficiency for the process is given by (4):

$$E = \frac{T - T_0}{T}$$

where all temperatures are absolute. For a heat engine operating between 1,000°F (1460°R) and 100°F (560°R) the maximum efficiency possible is

$$E = \frac{1460 - 560}{1460} = 0.614$$

Aspiring to an efficiency of 65% is futile. However, the best steam-power plant cycles have an efficiency of 35–40%, so some modest improvement may be possible.

Low temperature power cycles, such as those that operate on waste heat or on the temperature differences between the surface and the depths of the ocean, have much lower ideal efficiencies. Suppose the ocean surface is 60°F, with 30°F at depth. The maximum possible efficiency is

$$E = \frac{520 - 490}{520} = 0.057 \text{ or} < 6\%$$

Similarly, an engine working with waste heat at 200°F and rejecting heat at 70°F can aspire to an ideal efficiency of

$$E = \frac{660 - 530}{660} = 0.197 \text{ or} < 20\%$$

Thus, low temperature heat engines clearly have less potential to be efficient. If a practical system is to be developed, it must have other attributes that would compensate for the inherent efficiency limitations of the cycle (5).

In cases where lower efficiencies are not inherent, technology improvements might have significant impact. For example, improvements in the materials for and design of rotors in gas turbines have allowed the machines to run reliably at higher temperatures, resulting in higher efficiencies, ie, closer to the theoretical limitation set by combustion temperature.

CALCULATING EFFICIENCIES

First Law

The first law efficiency of a process can be captured by an enthalpy balance, that is, by simply calculating the fraction of the input energy which is captured in the process. For the boiler in the steam-power cycle shown in Figure 1, the fraction of the combustion heat absorbed in the water equals the first law efficiency of the boiler. For modern boilers, this is typically 85–90% and is controlled by reducing the stack temperature of the flue gas as low as practical (in regard to economics and corrosion). In the system shown, 12.549 MBtu/h are fired in the boiler; when the flue gas exits to the atmosphere at 200°F, its enthalpy has been reduced to 1.631 MBtu/h by exchange with the water and combustion air in the boiler. Thus, 87% of the fuel heat has been captured in the steam, ie, the first law efficiency of the boiler is 87%.

The first law efficiency of a heat engine, like the steam turbine, is virtually 100%. The enthalpy in the steam is reduced only by the amount of work extracted (and some very small amount for the friction in the machine). Similarly, the efficiency of a heat exchanger, like the condenser in the steam-power plant, is 100% in that all of the heat contained in the hot stream is captured in the cooling medium. Of course, this represents the heat-rejection step so that unless warming the cooling medium is a desired objective, this energy is lost.

If the overall efficiency of the steam-power process is considered to be based on only the amount of work produced, the first law efficiency is calculated simply by dividing the work produced by the heating value of the fuel burned. If the boiler first law efficiency is 87%, as previously discussed, this is calculated by

$$EFF = \frac{W}{Q_F} = \frac{3413}{12549} = 27.2\%$$

This equation ignores the work needed to pump the process water and the cooling medium, and the forced air fans for the boiler (if any). These factors could be significant.

If both power and process heat are desired, the potential for higher efficiency is presented. One approach to providing a combined load is shown in Figure 2. Two separate operations are shown: the steam-power cycle of Figure 1 and a separate boiler to provide steam for a process heat load. The useful products from this arrangement are the 3.413 MBtu/h of work and 17.645 MBtu/h of process heat (Q_P) the fuel fired increased to 32.830 MBtu/h. The first law efficiency is then given by

$$EFF = \frac{W + Q_p}{Q_F} = \frac{3.413 + 17.645}{32.830} = 64.1\%$$

As before, the heat rejected in the condenser and that lost in the flue gas is not counted as useful product from the process.

In a cogeneration configuration (see Fig. 3), both power and useful heat are produced in a single cycle. In many cases, the turbine exhaust steam is used to provide heat to some process or to commercial or residential buildings. Thus, a second component to the useful energy output is present and must be added to the numerator of the efficiency equation, resulting in a higher efficiency. This higher efficiency is, of course, the reason that legal incentives exist to use cogeneration.

From a practical standpoint, the base process conditions of the steam-power cycle must be changed to accommodate the dual role of providing both the power and the required process steam at 125 psia. Fundamentally, the exhaust pressure of the turbine must be raised to a level higher than 125 psia to provide a driving force to transfer heat to the process. Thus, 200 psia was chosen, and it was assumed that an additional heat exchanger would be provided so that the very pure water needed to produce the high pressure steam would not be contaminated by potential leakage in a process heat exchanger. This change in exhaust pressure means that the total steam flow through the turbine must be increased from 7.9 to 19.2×10^3 lb/hr (Kpph) because less work is extracted per pound of flow. However, only one boiler is required and total fuel fired is reduced to 24.21 MBtu/h because the heat rejected at the condenser is totally eliminated. Thus, the numerator in the efficiency equation remains the same, but the denominator is lower, yielding yet a higher

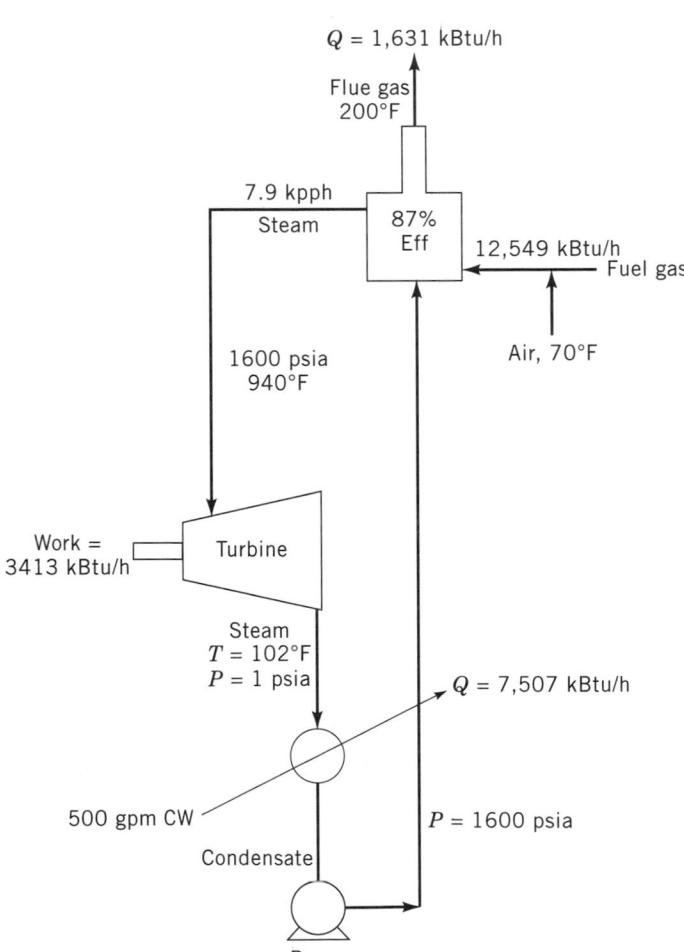

Figure 1. Heat balance for typical high pressure steam-power cycle showing fuel fired and heat rejected per megawatt hour power produced.

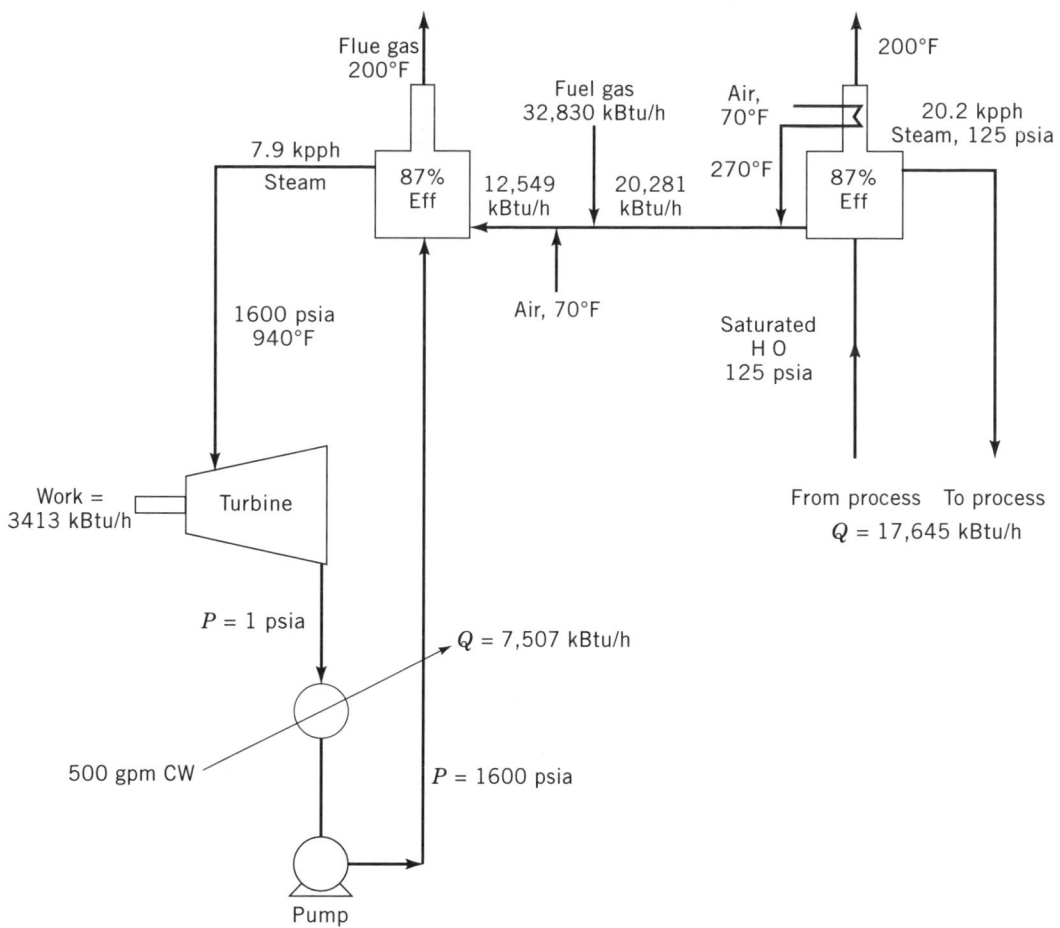

Figure 2. Heat balance for two separate systems producing 1 megawatt hour of power and a fixed amount of process heat. Heating requirements are additive.

first law efficiency:

$$\text{EFF} = \frac{W + Q_p}{Q_F} = \frac{3.413 + 17.645}{24.21} = 87\%$$

Not surprisingly, this equals the boiler efficiency, as there are no other losses taken into account.

Second Law

Calculating the second law efficiency of a process is another matter. Fundamentally, this efficiency is calculated from the availability (exergy) balance of the process. One approach is to take the ratio of the availability (exergy) in the products of the process to that input to the process. The appropriate boundaries need to be set up so that products and availability input are clearly defined. Thus, the efficiency is calculated as follows:

$$\text{EFF} = \frac{\text{Availability products}}{\text{Availability input}}$$

A second approach is to compare the availability (exergy, work) required or produced in the process to that of a reversible process producing the same product from the same inputs. For example, the Carnot cycle is a reversible

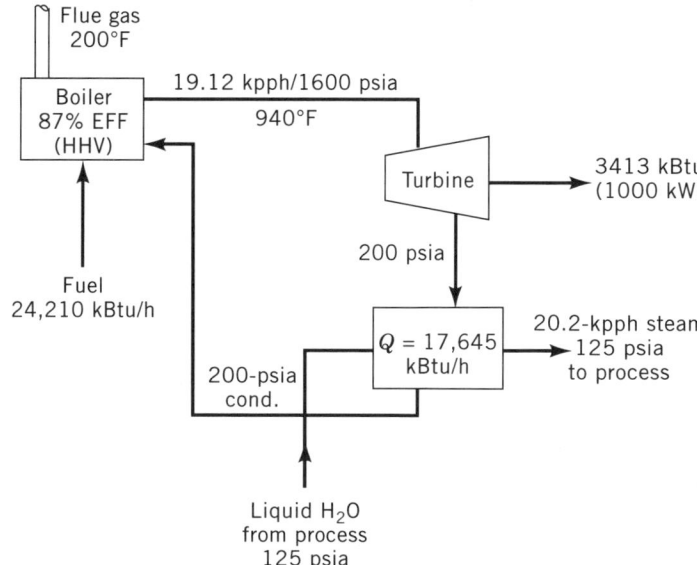

Figure 3. Balanced cogeneration cycle producing the same power and process heat demand as in Figure 2. Fuel fired is reduced because waste at condenser is eliminated, and turbine exhaust conditions and steam flow rate changed significantly.

process for a heat engine. Therefore, the amount of work produced from a given amount of heat at T_1 with rejection at T_0 in a real system might be compared to that which is theoretically available from a Carnot cycle operating between the same temperatures. In a process that requires work (availability) input, such as refrigeration, the reversible work required would be compared to that actually required to measure the efficiency. Thus, the efficiency for a work-producing process is

$$EFF = \frac{W_A}{W_I} = \frac{\Delta A_A}{\Delta A_I}$$

For a process that requires work, the ratio is inverted, ie,

$$EFF = \frac{W_I}{W_A} = \frac{\Delta A_I}{\Delta A_A}$$

Sign conventions must also be considered in these calculations. Work performed on the surroundings is considered positive and that done by the surroundings to the process is negative. Conversely, heat transferred into the process is positive and that rejected to the surroundings is negative.

In parallel with the development of availability balances, the work or availability lost or degraded in a process can be calculated directly from the total entropy change $(\Delta S)_T$ of the system:

$$W_L = \Delta A_L = T_0(\Delta S)_T$$

Even in reversible processes, an entropy change of the system will occur, causing some loss or degradation of availability. In a reversible process, the entropy change will be minimum, and the process will require minimal resources to implement; any real process will require more. As stated earlier, practical ways to approach reversible process steps may not exist.

To illustrate second law efficiencies, consider the processes in Figures 1–3. In Figure 1, the only product of value is considered to be the work generated. As work is 100% availability, both the first law and second law efficiencies are the same (neglecting the small difference between the availability and the fuel value for the fuel fired), ie,

$$EFF = \frac{3,413}{12,549} = 27.1\%$$

However, when analyzing a process that produces a product other than work (Figs. 2 and 3), a different result emerges. The availability of the steam heat produced is not the same as that of the product produced directly as work. (From a conceptual standpoint, its capacity to cause change is clearly less than that of the electricity.) Thus, the second law efficiency of the heat plus power process is significantly lower than the corresponding first law efficiency.

Table 1 lists the availability balances for the two processes and the resultant second law efficiency calculated by the ratio of product availability to that input. For sim-

Table 1. Lost Availability Comparison of Conventional and Cogeneration Steam-Power Systems

Parameter	Conventional System[a]	Cogeneration System[b]
Availability input, kBtu		
Fuel at 21,370 Btu/lb	32,830[c]	24,210[c]
Boiler feed water at 344°F and 125 psia	1,103	1,103
Total input	33,933	25,313
Availability output, kBtu		
Electricity	3,413	3,413
Steam	7,130	7,130
Total output	10,543	10,543
Second law efficiency, %	31.1[c]	41.7[c]
Total lost availability, kBtu	23,390	14,770

[a] See Figure 2.
[b] See Figure 3.
[c] Approximate figures; it has been demonstrated that the correct data should be 29.362 MBtu vs 32.830 and 21.640 MBtu/hr vs 24.210 for the conventional and cogeneration configurations, respectively; the corresponding efficiencies will be 34.6% and 46.3% (6).

plicity, it continues with the assumption that the fuel value corresponds to the input availability for the fuel and calculates the water and steam values from Keenen and Keyes, "Steam Tables." This simplified analysis is presented to reinforce the relationship between fuel fired and availability losses.

The results are clear; the useful output of both systems is identical, as is the input from the 125-psia boiler feed water. The difference between the two cycles rests with the fuel requirement. Because combustion is a particularly irreversible process, more lost work is associated with the conventional process, and the thermodynamic efficiency is lower, thus providing quantitative support for cogeneration in its simplest form.

The simple analysis of the two processes can be broken down to identify further the sources of lost work in each process. A more detailed analysis is given in Table 2, which was calculated by the more rigorous approach of Sussman. (6) Combustion losses are proportional to the amount of fuel consumed. Steam-generation losses are

Table 2. Sources of Availability Losses in Conventional and Cogeneration Steam-Power Heating Systems

	Lost Availability, kBtu	
Source	Conventional System[a]	Cogeneration System[b]
High pressure boiler		
Combustion	3686	7155
Steam generation	2604	3872
Low pressure boiler		
Combustion	5958	
Steam generation	6122	
Turbine (net)	1072	615
Heat exchange	419	606

[a] See Figure 2.
[b] See Figure 3.

really a function of the temperature-driving forces used in the boilers. The lower the steam pressure produced in a fired boiler, the greater the loss of available energy. Note that the work lost in heat exchange for the cogeneration cycle occurs in the 200-psia/125-psia boiler, and its contribution to the total lost work is only one-tenth that of steam generation in the fired 125-psia boiler. Turbine losses are lower because the exhaust pressure is higher in the cogeneration cycle.

In this simplified preview of the insight provided by thermodynamics, the first fundamental guidelines for the engineer dedicated to reducing a plant's energy bill emerge:

1. Minimize combustion.
2. Generate steam at maximum pressure and generate power from all low pressure steam demand (ie, maximize cogeneration).
3. Use steam for process heat at minimum temperature.

Note also the differences between the efficiency values and locations of losses between a first law viewpoint and the one advanced here. First law boiler efficiencies are listed at 87%, with the main waste occurring at the condenser. From the second law viewpoint, there is little lost in the condenser. Thus, a different energy conservation strategy is indicated by available energy analysis compared to simple energy accounting. The value of the second law approach lies not in the efficiency calculated, but rather in the analysis required to calculate the efficiency and in the basis the analysis provides for further identification of the process steps that offer opportunity for improvement.

As the total availability of the fuel in calculating the second law efficiencies of these processes has been considered, any differences that might have existed between the two approaches to calculating efficiencies mentioned previously are virtually eliminated. In an ideal combustion process, all of the availability liberated in the combustion reaction in the products would be recovered. In addition, the flue gases would be cooled to ambient temperatures (T_0) before leaving the system, and the reaction would take place in a reversible fuel cell.

However, this ideal process is far-removed from the process considered in the analysis. Yet, by considering all of the availability of the fuel as the basis for efficiency calculations, essentially the same result is obtained for efficiency. If only the availability of the flue gas in the boilers would be considered as input, a different (higher) result would be obtained because the availability loss of the combustion reaction would have been ignored. Thus, careful attention to the boundaries of the system being analyzed is important; identifying the original source of the availability driving the process is essential to correct analysis.

For the sake of comparison, the efficiency ranges of other processes that are of industrial and commercial interest are shown in Table 3. Those with high combustion components have the lowest second law efficiencies, whereas those driven by other sources of availability (eg, oxygen separation) give higher results. However, the data

Table 3. Typical Efficiencies

Process	Second Law Efficiency	First Law Efficiency
Ethylene manufacture (from ethane)	22–25%	~80%
Phthalic anhydride manufacture	22–28%	70–80%
Steam/power generation	33–38%	33–38%
Petroleum refining	~15%	~90%
Oxygen separation from air	27%	70–80%
High pressure steamboiler	50%	85–90%
Domestic water heater (fuel fired)	2–3%	40%
Residential heater (fuel fired)	10%	60%

sources used in Table 3 are disparate, so that no direct comparison is possible.

A BROADER VIEW OF EFFICIENCY

Two views of efficiency have been explored so far: the first law (Btu accounting approach) and the second law (energy quality/usefulness overlay). A third, more selective, approach to measuring efficiency appears to be growing. Yet unnamed, it will be referred here as "molecular efficiency" and defined as follows:

"The optimum management of all the molecules involved in a process system so that the desired product is produced with the minimum overall consumption of nature's resources."

The molecular considerations began with recycling scrap metal and has recently spread to recycling paper and container materials and the elimination of nondegradable packaging by some companies in the food-provider chain. These actions were driven by the growing problem in solid-waste disposal and by the realization that these materials, having served a useful purpose for society, had not yet exhausted that usefulness and could be used again.

A second force in molecular management has been SARA III, the Superfund reauthorization act in the United States, which requires manufacturing companies to report certain materials rejected to the environment in any form. These companies were thus forced to measure emissions, explore their sources, and understand how to manage better the materials in process to minimize the public outcry at the magnitude of the emissions reported. In addition, concepts such as "just-in-time" raw material supply offered economic advantages not only in working capital, but also in space utilization, work flow, and worker attitudes.

In many industries, the optimum use of feed and energy has become essential in a competitive world market to cut processing costs and to reduce waste handling costs, the administrative burdens of permits, tracking wastes, ensuring disposal methods and sites meet all present (and future) regulations, and dealing with public outrage. In addition, exposure to potential future liabilities is reduced.

Society has taken various initiatives to force a new definition of efficient operations upon those who would remain a viable enterprise. Thus, "pollution prevention," rather than clean-up at the end of the pipe, or no clean-up at all, has become a new measure of efficient operation. The evolution of automobile manufacturing in the United States provides an example of this process. Under the pressures for improved fuel economy, automobiles became smaller, lighter, driven by less powerful engines, and festooned with less chromeplated trim. All of this improved fuel economy, but reduced user functionality. In addition, less waste was produced and some wastes were eliminated altogether, eg, chromeplating bath chemicals.

As benefits from this "defunctionizing" phase approached the point of diminishing returns, newer technologies emerged. (That is, work began on the front end of the process to solve problems and restore functionality.) Better, longer-lasting lubricating oils were introduced to reduce waste; unleaded gasoline reduced emissions of a toxic substance and made possible the use of catalysts to reduce exhaust emissions further; and reliable, more economic fuel injection was developed to enhance fuel efficiency, thus restoring some of the lost performance encountered in the first phases.

Today, an even greater fundamental understanding of the manufacture and operation of an automobile is being used to maintain the car owner's independence while addressing environmental issues. Much of this comes from the availability of new materials, but some comes from further refinements in understanding combustion. New materials for bumpers and other car parts have restored some of the lost safety functionality formerly provided by larger cars, and have even enhanced the tolerance of some automobile parts. Anti-lock brakes and air bags are but a few of the newer technologies aimed at producing a safer vehicle; reformulated gasolines containing oxygenated compounds will further reduce emissions.

The common theme of these improvements is the application of new molecules to the problem and the better management and control of all molecules utilized in building and operating the product. This may well be the viewpoint required to be both competitive and acceptable to society in the future.

EFFICIENCY AND COST

The operating cost benefits of fundamental efficiency improvements are usually fairly obvious: lower energy cost, lower waste handling costs, lower administration costs, and lower exposures to future potential environmental liabilities. However, many believe that these benefits can only be achieved at higher investment. There are appreciable data (1) that demonstrate that truly fundamental improvements actually save investment as well as operating costs.

Often, cost-benefit analyses of what appear to be efficiency/investment trade-offs are flawed by either a too-narrow focus or inaccurate data. Utility cost figures do not necessarily represent the real cost of the availability supplied; a prime example of a mispriced utility is steam. Steam is usually priced on the basis of its enthalpy content, regardless of its pressure, and is often burdened with high fixed costs related to ensuring reliability of sup-

ply (such as 100% spare boiler capacity). This may have been a good decision when fuel cost $0.20/mBtu, but today, numerous implant steam pricing systems actually value low pressure steam more highly than high pressure steam, in spite of the fact that low pressure steam's capability to cause useful change in the plant is much lower than that of the high pressure steam.

Thermodynamicists will state that pricing steam (or any other utility) according to its availability (exergy) content is the only way to obtain correct analysis. However, pricing systems should be tested against their impact on decisions ie, do the prices established provide acceptable returns on energy efficiency investments? For example, if low pressure steam condenses at high enough temperatures to provide the needed process heat, but high pressure steam can do the same job in a smaller heat exchanger, then the price of low pressure steam must be sufficiently lower than that of the high pressure steam to give a good return on the larger investment for the heat exchanger. Of course, the low pressure steam cannot be produced by simply expanding high pressure steam through a valve for these values to be real; it must be produced by using the higher pressure material to produce work or do some other task that cannot be performed by the lower pressure steam. Otherwise, its capacity to cause change, or its availability content, is simply wasted.

A brief comparison of steam prices based on enthalpy content and availability content is often instructive. Assume that 1,500-psia steam is available at a cost of $10/1,000 lb, and that the plant also uses 140-psia and 15-psia steam. The prices developed for these three pressure levels by ratios of enthalpy and availability are shown in Table 4. This range of potential values would make it difficult to decide on any particular investment.

In many cases, using availability-based prices makes it difficult to justify using back-pressure turbines to produce the lower pressure steam. The decision will be based on the turbine efficiency, investment, and cost of supplying the required power by other means, eg, purchased electricity. Using enthalpy ratio prices, the credit for the exhaust steam would be much higher and turbines are thus easily justified.

The third test of correct steam pricing is whether the recovery of waste heat can be justified. Again, high values for low pressure steam make this easy to do. Unfortunately, many plants witnessed an overabundance of low pressure steam produced because of such pricing, steam that then had to be wasted somewhere in the system because it was in excess of plant needs.

To price utilities accurately, how the plant energy balances impact on the prime inputs of availability (usually fuel and electricity) must be known. If recovery of waste

Table 4. A Comparison of Steam Prices

| Pressure | Prices | |
	Enthalpy Ratio	Availability Ratio
1,500 psia	$10/1,000 lb	$10/1,000 lb
140 psia	8.34	4.77
15 psia	7.88	3.35

heat can be traced back to a reduction in fuel somewhere in the plant, its true value can be calculated. Similarly, if generating work can be traced back to an electricity or fuel saving, its true value can also be seen. The benefit part of the cost-benefit ratio can now be correctly measured.

Cost is often obscured by a too-narrow focus. Investments required outside the plant premises, to provide the increased demand (which must ultimately be paid for in higher prices or taxes), are often not factored into the equation. This became clear in global studies provided for the federal government in the late 1970s. If the circle of analysis is drawn wide enough, the true cost to society of providing new supply or recovering energy can be calculated. The barrier to this more global approach is the challenge of accounting for the benefits to a single enterprise. Consortia have been organized to clear these barriers, but this is often not a trivial problem.

Yet, there is hope for those who take a systems approach. Data has been presented for cases where both capital and operating costs were saved in new investments. For example, one case compared the investment and operating costs of using the overhead vapors from a higher temperature tower to reboil two lower temperature towers versus simply rejecting this heat to cooling water and reboiling the two towers with steam. Energy cost savings (reduced steam, slight increase in fuel consumption) were \$1.8 million/year. Increased investment costs for heat exchangers (pumps and piping) were \$2.3 million higher, but these were balanced by savings in cooling water system and incremental boiler capacity of \$3.5 million. (This analysis did not include any credits for reduced waste handling costs.) It is often true that investment in utility system capacity is greater than that in the process equipment to render it unnecessary, if the utility system investment can legitimately be drawn into the decision envelope.

Both costs and benefits must be assessed on the basis of the impact of any improvement on the entire plant energy system. The net savings on fuel and power usage, and the full impact on plant investment, should be considered in calculating cost-benefit ratios. At least for new investments, fundamental improvements in efficiency can offer savings in both investment and operating costs.

BIBLIOGRAPHY

1. W. F. Kenney, *Energy Conservation in the Process Industries,* Academic Press, Inc., Orlando, Fla., 1984.
2. E. P. Gyftopolous and T. F. Wiomer, *Potential Fuel Effectiveness in Industry,* Ballanger, Cambridge, Mass., 1974.
3. G. M. Reisad, "Available Energy Conversion and Utilization in the U.S.," *Journal of Engineering for Power* **97,** 429–434 (1975).
4. J. B. Fenn, *Engines, Energy and Entropy,* W. H. Freeman & Co., New York, 1982.
5. W. J. O'Brien, "Low Level Heat Recovery Technology," *Proceedings of the Fifth Industrial Energy Conservation Technology Conference,* Houston, Tex., 1982.
6. M. Sussman, *Availability (Energy) Analysis,* Milliken, Boston, 1980.

ENERGY MANAGEMENT, PRINCIPLES

Douglas A. Decker
Johnson Control, Inc.
Milwaukee, Wisconsin

HISTORICAL PERSPECTIVE

Viewed from a historical perspective, the statement "there is no time like the present" overwhelmingly applies to energy management: the systematic, ongoing science of favorably balancing an organization's energy needs and goals with the capabilities of people, technologies, and energy sources. At the risk of oversimplifying the concept, it is important to manage energy successfully because energy ultimately drives, in a modern, mechanized world, production aimed at meeting all three basic human needs: food, clothing, and shelter. The quality of human lives is thus defined by how, and how well, energy is used.

For those managing energy as it relates to buildings, the universe of energy management now presents a multitude of cost-effective and productive results. These results include improved management efficiencies, energy savings, operations and maintenance labor reduction, and the increased comfort and productivity of those using buildings. International events; national, state, and local governmental policies; public awareness; and levels of technology today offer those concerned with energy management the mix of incentives and resources that make results both reliable and rewarding. The incentives and resources range from utility rebates to enhanced understanding by those outside the energy management realm, to the greater societal benefits such as reduced greenhouse gases and less impact on the environment.

At one time energy managers faced a swinging pendulum of concern. As energy prices rose and fell, as supply and demand vacillated, and as governmental administrations came and went, energy managers faced varying and frustrating standards and levels of concern from officials, influentials, and the public at large. In fact, many of the 1970's energy czars in public and private institutions have long since been forgotten.

In addition to a historical perspective, this article discusses the organizational levels of energy management, the facilities management systems and services that help attain energy management goals, the economics of energy management, and the future of energy management. See also ENERGY AUDITING; ENERGY CONSERVATION.

Events and Trends

A quick review of the events and trends of the last few years reveals why the pendulum of concern may now have stopped swinging and may reflect a position that is best for continued advances in energy management. Among the top international events that have promoted energy management is the end of the Cold War. With fewer financial, governmental, and technological resources dedicated to strategic military needs, more resources are being devoted to tightening up the way strategic economic and domestic needs are managed. Indeed, the shift from military to domestic needs is complementary in terms of what constitutes national security, the capacity of the United States to protect and advance the interests of its people. The end of the Cold War cleared the national agenda for energy management. But the frequent and sudden shocks to world oil prices and supplies, emanating from the Middle East, should be given credit for putting energy management on the agenda in the first place. The oil crisis of 1973, the Iranian revolution of 1980, and the Persian Gulf War of 1991, typify events that heightened U.S. awareness of the need to have an energy strategy.

Sudden, dramatic changes in world oil prices harm the United States and other nations much more than a persistent but gradual rise in price. Even if the average price over the long term is the same, shocks do more harm. This stems not from how much oil is imported but from how oil dependent the U.S. economy is, its capacity for switching to alternative fuels, oil reserve stocks around the world, and the additional international production capacity that can be brought on line.

Leaders perceive achieving greater energy stability to be a principal part of enhancing America's security. They recognize that the United States must insulate itself from the shocks that emanate from the Persian Gulf, whose oil fields provide 25% of the oil the world presently consumes and nearly 66% of the world's proved oil reserves (1).

Data also show that conservation does not necessarily mean a decline in living standards, a position once held by many leaders. Since the oil shocks of the early 1970s, Americans have enjoyed a 35% rise in the gross national product without increasing their energy consumption (Fig. 1). The main reason is that the living standard–related services energy provides, such as comfort and mobility, are generated much more efficiently today than they were during the early 1970s. Similarly, by slowing the growth in demand for new energy capacity, it has been estimated that conservation could liberate 10% of U.S. industrial investment capital for other uses (2).

At national, state, local, and corporate levels the United States has faced disruptions, economic conditions, and budget constraints that have caused leaders to concentrate on the efficient use of funds, which equates to support of energy conservation and efficiency standards. In recent years a steady stream of congressional hearings and public forums involving leaders from the private and public sectors have led to several federal initiatives.

The National Energy Strategy, established by President George Bush in 1989, set the stage for Executive Order 12759, signed by Bush in 1991. Both advanced a commitment to improving energy use in federal buildings, facilities, and vehicles as part of a strategy to counter disruptions, economic conditions, and budget constraints. As part of this campaign to renew a federal energy conservation ethic, an industry coalition of trade organizations representing 4.7 million members collectively and individually supported signing of the Executive Order.

Executive Order 12759 led to the Energy Policy Act of 1992. Broadly stated, the purpose of the Energy Policy Act is to foster greater energy conservation in the traditional sectors of energy consumption: industrial, transportation, residential, and commercial. It requires states to incorporate energy efficiency standards in their commercial building codes that meet or exceed the 1992 voluntary standards established by the American Society of Heating, Refrigerating and Air-Conditioning Engineers (ASHRAE). The Energy Policy Act also requires the Department of Energy (DOE) to issue a new building code that contains minimum energy efficiency standards for new buildings.

Recent events and initiatives support the notion of sustained, politically bipartisan energy management awareness. In passing the Energy Policy Act, Congress has displayed leadership in promoting energy management. President Bill Clinton and Vice President Al Gore have track records of commitment to energy management and conservation. In his 1993 Earth Day address, Clinton announced initiatives to stimulate the energy technology industry, energy conservation, and a clean environment.

Two other Clinton initiatives indicate a move to a sustained energy policy for the country. The Climate Change Action Plan has a goal of reducing harmful greenhouse gases to 1990 levels by the year 2000 and creating jobs by promoting energy efficiency through public–private part-

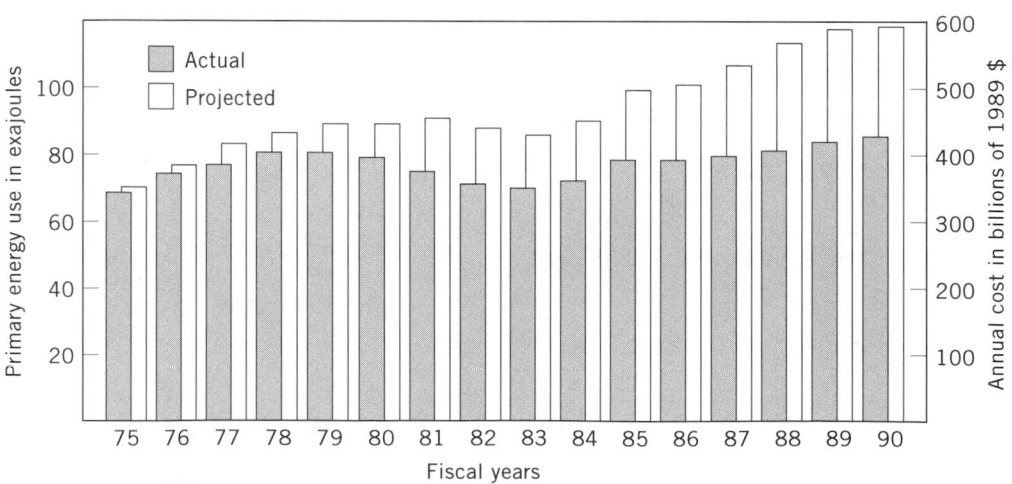

Figure 1. GNP projected energy use and GNP actual energy use. Data from the DOE and the Lawrence Berkeley Laboratory. ▨, Actual; □, projected.

nerships. Rather than relying exclusively on command-and-control mandates that tend to stifle innovation, the partnerships reflect the mutual responsibility of both the private sector and the government.

These investments are estimated to save over $60 billion in reduced costs between 1994 and 2000, with continued benefits of over $200 billion in energy savings between 2001 and 2010. The voluntary nature of the programs is important because the threat of additional regulations and the success of existing voluntary programs will induce companies to participate.

The EPA's Green Lights program is one of the successful voluntary initiatives promoted in the Climate Change Action Plan. Almost 1200 firms currently participate by upgrading lighting systems and reducing electricity use by nearly 200 million kW annually, at a savings of $17.7 million. Approximately 1 in every 20 commercial buildings in the United States is now involved in Green Lights.

The costs will fall and the benefits will rise as participants become more experienced with the implementation process of voluntary programs. In addition, businesses can show they are good citizens by saving energy, helping the environment, and contributing to the local economy by hiring local workers to implement voluntary programs.

While the Climate Change Action Plan addresses private industry, another Clinton initiative directs the federal government to reduce energy more than ever before. The Executive Order on Energy Efficiency and Water Conservation at Federal Facilities supercedes the Bush Executive Order. The Clinton mandate calls for a 30% reduction in energy use in federal buildings by 2005 from 1985 levels. The order includes increased uses of innovative financing, such as performance contracting and utility demand side management, to implement the measures. It calls for agencies and employees to be rewarded for excellence in energy efficiency, and it holds agencies and employees responsible for meeting the goals.

By signing this new Executive Order and announcing his intentions to hold the agencies accountable, Clinton showed his personal commitment to the issue. In addition, the very nature of the document will break down barriers and encourage participation by federal agencies. It will result in the federal government acting as a role model for the rest of the country and proving that energy efficiency is the best economic and environmental course of action an organization can take.

Sustained, active leadership as demonstrated by the Bush and Clinton administrations is essential to making energy management work. Without active leadership and funding, laws such as the Energy Policy Act become totally irrelevant. Sustained industry leadership, such as that which led to the Energy Policy Act, will ensure that legislation reflects the will and ability of the marketplace to implement energy strategies, avoiding controversy.

Incentives to Implement

There is growing recognition that investment in energy-efficient technologies creates jobs. For example, a study conducted for the DOE reported that for each $1 million of energy efficiency improvements made within local hospitals, employment would increase by 56.1 jobs over a 20-yr period (3).

Perhaps one of the strongest incentives to implement energy management programs and technologies will come from efforts by private industry to improve productivity, while also reducing energy costs. A joint study by the Center for Architectural Research and the Center for Services Research and Education at Rensselaer Polytechnic Institute studied the effect of environmentally responsive workstations (ERWs) on the productivity of office workers at the new West Bend (Wisconsin) Mutual Insurance Company headquarters (4). The study reported the combined effect of the new building with the ERWs was a 16% increase in employee productivity. The ability of employees to control their internal environment was credited with having a significant impact on this increase. The integrated design of ERWs allow people to control temperature, airflow, noise, and light within their work spaces. Equally important, energy-efficient building design, energy management control systems (EMCS), and the ERWs contributed to a reduction in utility costs per square foot of building space, from $0.18 per square foot to $0.11 per square foot.

ERWs represent one of the many off-the-shelf technologies available to help energy managers. For federal energy managers, the General Services Administration and Department of Defense have procurement schedules that include proven energy-efficient and energy management products. Overall, energy technology improves with each year, making efficiency easier and providing better paybacks. Corporations are joining in strategic partnerships to advance the compatibility, and hence cost-effectiveness, of their equipment. And creative financing methods allow public officials, corporations and commercial building owners to defer large capital outlays of funds for energy-efficient equipment.

Times have indeed changed, and for those concerned with energy management, there is no time like the present. An indepth understanding of this new and improved universe of energy management can lead the way to opportunity.

But has the pendulum of concern stopped swinging permanently? Will the mix of events, policies, incentives, and resources favor strong energy management in the future as much as it does now? The forces that influence energy management, including environmental issues, utility supply and demand, industry cooperation, ease of use, economic benefit, consumer focus, and political will, all will play a role in determining whether the pendulum has stopped and will forever favor a commitment to energy management.

THE LEVELS OF ENERGY MANAGEMENT

To manage energy at any level, with an enhanced likelihood of success, it is necessary to have knowledge of the major regulations, agencies, issues, resources, trends, and technologies as they affect each level of the energy management model or pyramid (Fig. 2). Such knowledge also permits energy managers to implement winning strategies that will satisfy criteria set by entities outside as well as inside their organizations.

This section considers the levels of energy management as they relate to public (government), commercial, and industrial buildings to be

1. Federal, including the executive and congressional branches of the U.S. government.
2. State and local government.
3. Utilities.
4. Private industry, professional, and academic.
5. Corporate.
6. Building and facility.

In a country based on a free market economy and individual rights, the role of the federal government in promoting energy management and efficiency will always be subject to debate. The traditional U.S. policy has been one of market reliance, allowing markets to determine prices, quantities, and technology choices. In certain cases and periods of history, markets cannot or do not work efficiently. During these periods government action must be taken to remove or overcome barriers to efficient market operation.

However, no one who seeks to improve this nation's energy efficiency for reasons of national security, international industrial competitiveness, or reducing federal energy outlays can doubt that federal, state, and local governments can play a positive role in fostering energy efficiency. There is also the compelling need to protect the environment.

The extraction, conversion, and consumption of energy creates more global environmental damage than almost any other human activity (5). The damage inflicted includes deforestation, nuclear waste and CO_2-induced climate change. Because research indicates that pollution has economic as well as social costs, there is a need for those who manage the economy and society to take the lead in protecting the environment. Energy efficiency initiatives are a required part of leadership at the government level as well as all other levels of energy management. The challenge for government is to develop programs, laws, and policies for promoting better energy management. Government initiatives can provide significant stimuli to the free market. Help is needed to overcome barriers such as short payback times required by investors, lack of investment capital, lack of information, and limited supply of and demand for energy-efficient products.

The case for government-supported research and development of energy-efficient technologies is based on evidence that projected returns on investments make support worthwhile. One study showed that increased use of renewable technologies and energy-efficient products and procedures can save American consumers and businesses hundreds of billions of dollars over the next 40 yr (Fig. 3). Depending on the mix of energy-efficient technologies, market penetration rates, and pollution emission objec-

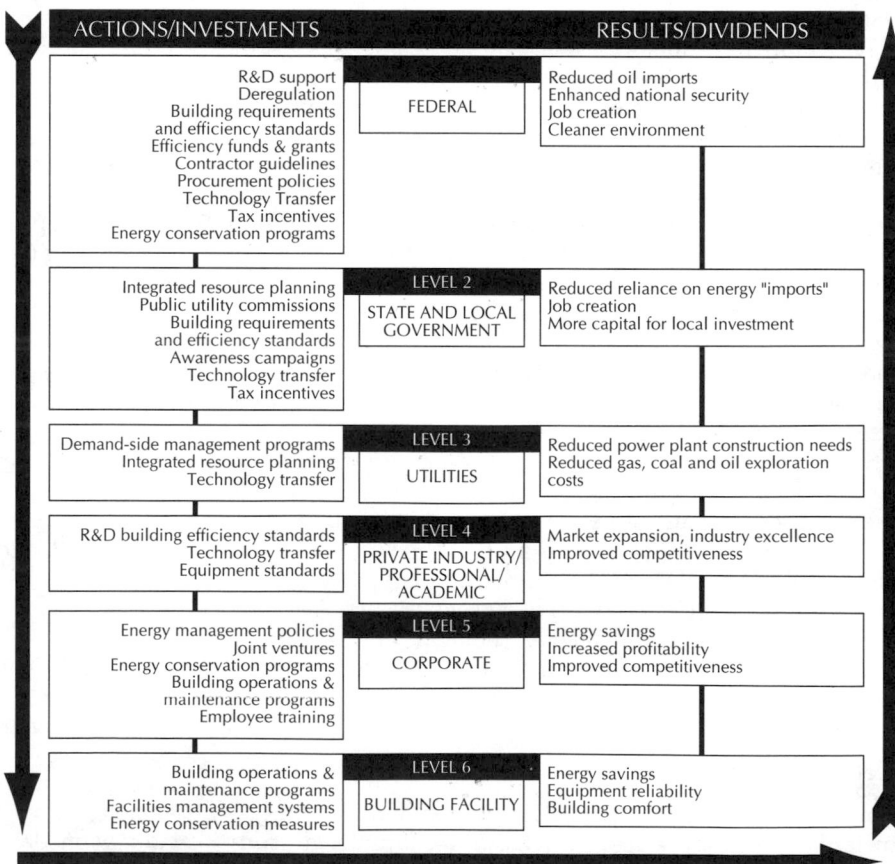

Figure 2. Levels of energy management.

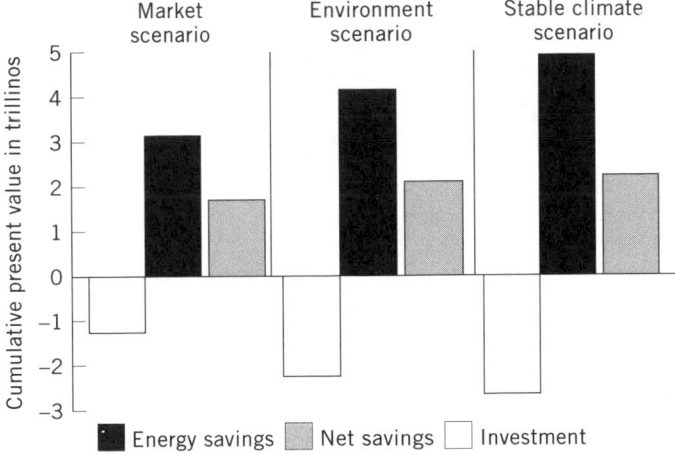

Figure 3. Analysis of energy savings and costs for three levels of investment in energy-efficient and renewable energy technologies, 1988–2030. The market scenario assumes moderate market penetration rates for the technologies. The environment scenario assumes more rapid market penetration rates and additional use of energy-efficient and renewable energy technologies. The stable climate scenario is based on use of a low and high cost mix of technologies. All scenarios lead to reductions in primary energy requirements and pollutant emissions from a reference scenario. The reference scenario was adapted from DOE projections, reflecting a current reliance on fossil fuels, nuclear power, and limited energy efficiency improvements. For additional discussion, see Ref. 6.

tives, the net savings estimated for a 40-yr period range from $1.9 trillion to $2.3 trillion (6).

Noting that residential and commercial buildings account for about 33% of U.S. energy consumption, at an annual cost of $170 billion, a separate government-authorized study projected significant savings using technologies that are already available. The 1992 study estimated that using commercially available, cost-effective technologies, building energy consumption could be reduced up to 33% by 2015, compared with a business-

as-usual projection that did not employ these technologies (7).

Federal Government

In response to overwhelming evidence, the trend is toward the federal government taking a leadership role in promoting energy management and stimulating the private sector. Just as the Clean Air Act saw government leading the way toward a cleaner environment by setting regulations for pollution control, the Energy Policy Act of 1992 positions the government in a leadership role for energy conservation and management. The Energy Policy Act affects every level of energy management, and energy managers in the public as well as private sector.

From a building technology perspective, the Energy Policy Act provides leadership by supporting the research and development of energy-efficient technologies. The act sets the stage for energy efficiency standards for buildings, lighting products, motors, and other equipment. Efficiency will be possible by allowing more competition and permitting the federal government to act as a consumer of new technologies. Reflecting increasingly higher federal energy costs, the lag in funding for energy conservation in federal agencies (Figs. 4 and 5), and a desire to lead by example, the Energy Policy Act makes the previously voluntary goal of reducing energy consumption per gross square foot in federal buildings by 10% before 1995 into a requirement and extends it to require a reduction of 20% by 2000.

The act also impacts the Federal Building Code. Under the new law the DOE must issue a new building code that contains minimum energy efficiency standards for new federal buildings. The standards must be at least equivalent to those of the 1992 Council of American Building Officials (CABO) and ASHRAE.

Increased competition will come in part from a deregulated electric utility industry. The Energy Policy Act opens up the electric utility industry to competition, which should bode well in the long term for price stability and energy efficiency incentives. At the heart of this

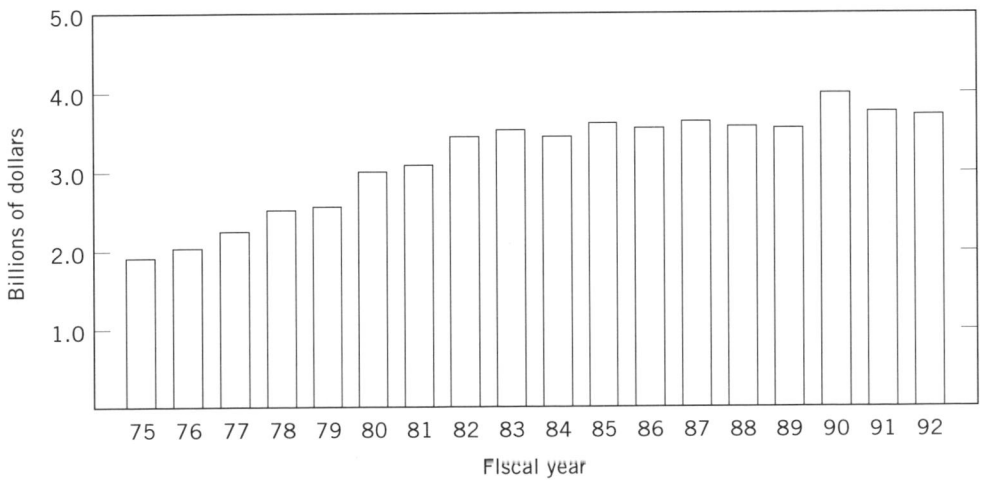

Figure 4. Federal energy costs for buildings and facilities, 1975–1992. Data from the DOE.

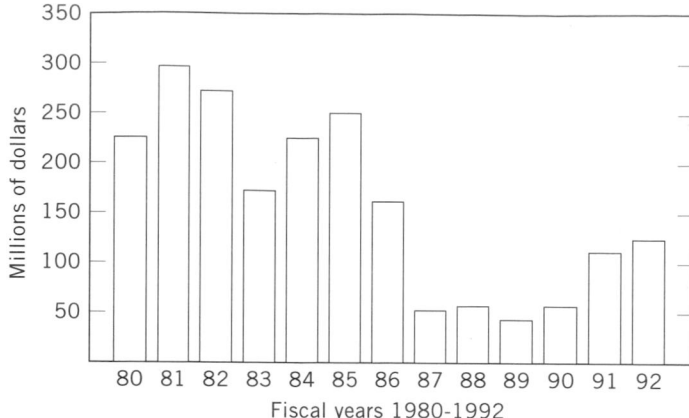

Figure 5. Federal building energy efficiency funding during fiscal years 1980–1992. Data from the DOE.

movement toward increased competition are provisions for exempt wholesale generators (EWGs). EWGs will be able to sell electricity to utilities and, according to the act, must have fair access to transmission lines.

Just as deregulation of the telephone industry led to restructuring along smaller, regional, and functional business lines, many experts believe this attempt to open up competition in the utility industry will spur development of separate power generation, transmission, and distribution companies.

The Energy Policy Act also focuses on commercial buildings in the private sector. The Energy Policy Act requires states "to incorporate energy efficiency standards in their commercial building codes that meet or exceed the 1992 voluntary standards established by ASHRAE." The standards are supported by the requirement that the DOE "provide incentive funding to states for implementation of this provision and to improve and implement their codes."

The purpose of the standard, officially known as ASHRAE/IES Standard 90.1, is to set minimum requirements for the energy-efficient design of new buildings and construction, to provide criteria for energy-efficient design and methodologies for measuring projects against these criteria, and to provide guidance in designing energy-efficient buildings and building systems (8). Although Standard 90.1 is not a code, its requirements can and have been adopted by governmental agencies empowered to enact codes through legislative or regulatory processes.

Standard 90.1 has a comparative standard in the federal government. The full reference for this standard is Code of Federal Regulations (CFR), Title 10, Part 435, Subpart A, *Performance Standards for New Commercial and Multi-Family High Rise Residential Buildings*, U.S. Department of Energy (DOE), January 1989. The DOE standard is intended to deal with private construction of federal buildings; Standard 90.1 encompasses all new construction, except low-rise residential, in all climates across the United States.

The act aims to enhance energy efficiency in process-oriented industries. The act requires the DOE to promote the use of energy-efficient technologies in certain manu-facturing industries, such as food and food products, lumber and wood products, and petroleum and coal products. There is a tax break for commercial and industrial buildings that comply. The Energy Policy Act excludes from gross income 40% the value of any subsidy provided by a public utility for the purchase and installation of an energy conservation measure in 1994, 50% in 1995, and 65% in 1996 and thereafter. This incentive only applies to amounts received after December 31, 1993.

For the energy manager looking to finance energy efficiency initiatives, today's market offers a number of options. The most widely used in recent history have been the incentives offered by utilities as part of demand side management programs. These incentives typically take the form of rebates tied to purchases of energy-efficient equipment. Energy managers of federal buildings should note that the Energy Policy Act of 1992 authorized federal agencies to participate in utility incentive programs. A total of 50% of the rebates or payments received from these programs are to be used for the purchase of additional energy efficiency measures.

Investment tax credits also have been used to offset the initial costs of energy-efficient equipment. Offered by states but not currently by the federal government, these tax credits complement utility-sponsored demand side management programs and help industries lower their tax liability as well as their future energy costs (9). States also have authorized bond issues to cover the cost of energy efficiency projects at state and municipal buildings.

Relatively new is the emergence of performance-based contracting as a financing method. In performance-based contracting, one or two suppliers plan and implement a building renovation in its entirety, acting as general contractors. Financing options are tied to performance guarantees, with the contractor taking as payments future energy savings. Performance-based contracting offers energy managers single-source convenience, control over expenditures, predictable costs, supplier accountability, guaranteed service delivery, and unlimited options. Usually these contracts include operations and maintenance services as a means of ensuring favorable results.

The federal government recognizes the importance of performance-based contracting. The Energy Policy Act of 1992 simplified the procedures federal agencies need to follow to enter into energy-saving performance contracts. Congress intended to simplify this form of financing as a way for the private sector to invest in federal energy projects (10). The Energy Policy Act offers an additional financing opportunity to federal energy managers via the Federal Energy Efficiency Fund. Under the Energy Policy Act, the secretary of energy may award monies from this fund to any federal agency for energy conservation investments. The fund is capitalized with $10 million in fiscal year 1994, $50 million in fiscal year 1995, and will be capitalized with "such sums as may be necessary" in the fiscal years that follow.

The Energy Policy Act also opens up the possibility for an energy conservation bank with deeper pockets than the Federal Energy Efficiency Fund. The Energy Policy Act directs top-level federal treasury, budget, and energy officials to conduct a detailed study of financing options for investments in energy conservation measures by federal agencies. The study must include a review of energy banks and revolving funds as long-term options.

It should be noted that funding as provided for in the Energy Policy Act is but a small piece of the total funding available to federal energy managers. The Department of Energy's annual report for fiscal year 1991 (11) lists 13 federal agencies that will spend a projected $124.5 million in energy conservation measures in fiscal year 1992, with the trend moving toward increased funding. These agencies, such as the General Services Administration, each allocate a portion of their operational budgets to energy conservation projects.

Without continued economic incentives to reduce energy consumption, energy efficiency goals would be tigers without teeth. Public policy and societal goals are not enough to initiate and sustain energy efficiency campaigns, especially in a free market economy. To advance and protect their efforts, energy managers must incorporate into their programs strong economic underpinnings in the form of realistic savings estimates, periodic review of results, appropriate tools and services, and a favorable financing package.

Refinements to the Energy Policy Act will be made as subsequent presidents and Congresses advance their interpretations of the nation's energy priorities. The federal government's ability to impact the energy manager with legislation and programs is shared by state and local governments.

State and Local Governments

State governments have a proven track record of energy management. In fact, the Energy Policy Act reflects legislation that has been successful at a state energy management level. Many states have effective programs in integrated resource planning (IRP), building energy codes, and equipment standards. All states and most U.S. territories, including American Samoa, Guam, and Puerto Rico, have energy offices or departments of state government responsible for coordinating energy management efforts. A state's public utility commission (PUC) regulates the supply of energy in its various forms to businesses and consumers. The PUCs and their federal counterpart, the FERC, also regulate access to transmission lines, an issue brought to the fore by the Energy Policy Act.

The approach that an individual state takes to energy efficiency is affected in part by the degree of dependence on energy "imported" into the state. Iowa, for example, is an agricultural state with relatively few resources of its own for energy production. More than 98% of its total energy is imported. The state thus wisely chooses to take a strong leadership position regarding energy management. Iowa is known for energy programs that apply to all traditional sectors of energy consumption, including industrial, transportation, residential, and commercial (12).

Individual states also vary on the emphasis they place on energy management within these sectors, a result of different levels of consumption and vulnerability arising from that sector (Fig. 6). In Nevada, the industrial and commercial sectors are largely untouched by state energy programs. This can probably be attributed to perceived need: Nevada's industrial sector accounts for 24% of state energy use compared with the national average of 36%. Nevada's commercial sector uses approximately 18% of the total energy consumed, in line with a national average of 16%. Demand by casinos for electricity is considerable, but casinos have been the target of past energy-saving programs and are considered efficient (14). By contrast, the transportation sector comprises 37% of total state energy consumption compared with a national average of 28%. Nevada is geographically the seventh largest state, and state officials assert that tourists driving from California to Nevada contribute to the state's poor ranking in transportation energy efficiency (15). Tourists notwithstanding, Nevada officials place considerable emphasis on leadership in the transportation area, running car-care clinics and designating October as car-care month to educate consumers about the environmental, safety and energy-savings benefits of a well-tuned, well-maintained car. Nevada employers may soon be required to participate in ride sharing and mass transportation efforts.

New York State considers all sectors and end uses and is the most energy-efficient state. The lowest energy use per capita can be attributed to the fact that a large percentage of the state's population lives in New York City

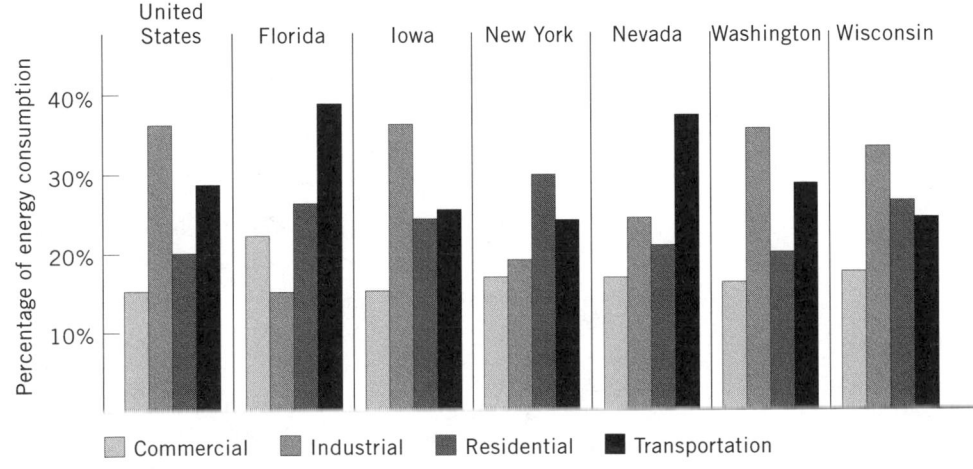

Figure 6. Six-state comparison of energy use by commercial, industrial, residential, and transportation users. Data from the DOE and Ref. 13.

where mass transit and multifamily dwellings are the norm.

Wisconsin has implemented an ambitious $50 million six-year energy conservation program that will significantly reduce energy consumption in state facilities. The program is designed to result in cost savings to taxpayers and to have a positive effect on the environment. Wisconsin is also responding to energy and environmental challenges by implementing a program called "2000 by 2000" (16). This program will place 2000 alternative-fueled vehicles on the road by the year 2000. This commitment is 33% of the state's fleet and represents three times the requirements of the federal government's Energy Policy Act. Wisconsin involves the private sector in its programs through coalitions and task forces formed to address specific issues.

To have an effective energy management plan on a state level, a majority of states use their authority to regulate utilities and thus manage energy supplies and prices. Integrated resource planning (IRP) is a cornerstone of most state energy plans (17). IRP requires utility companies to give as much consideration to conservation programs as they do to building new power plants. IRP simultaneously examines all energy-saving and energy-producing options to optimize resources and minimize total consumer costs, while considering environmental and health concerns (18).

In early 1994, all but seven states have either initiated or established IRP. The popularity does not translate to commonality, however. Nearly every state government,

public utility commission, or utility adds its own perspective to the process. IRP generally takes the following seven steps:

1. Development of a load forecast.
2. Inventory of existing resources.
3. Identification of future electricity needs that will not be met by existing resources.
4. Identification of potential resource options, including demand-side reduction programs.
5. Screening of options to identify those that are feasible and economic.
6. Performance of some form of uncertainty analysis.
7. Selection of a preferred mix of resources.

Figure 7 shows a model of the IRP process. State planners use the model as a tool to evaluate demand-side options such as conservation and load management, to compare these to supply-side options, and to structure an energy policy that incorporates environmental and social costs. The goal is to develop a long-term energy policy that acquires the most inexpensive resources first and internalizes social costs in the rate structure.

Most of the people involved in state utility issues who were interviewed for a study said that quantifying the environmental costs of electricity is likely to be the primary issue for utility planners in the next decade (19). Fostering energy efficiency is thus conducted with varying levels of intensity from state to state. States generally recognize

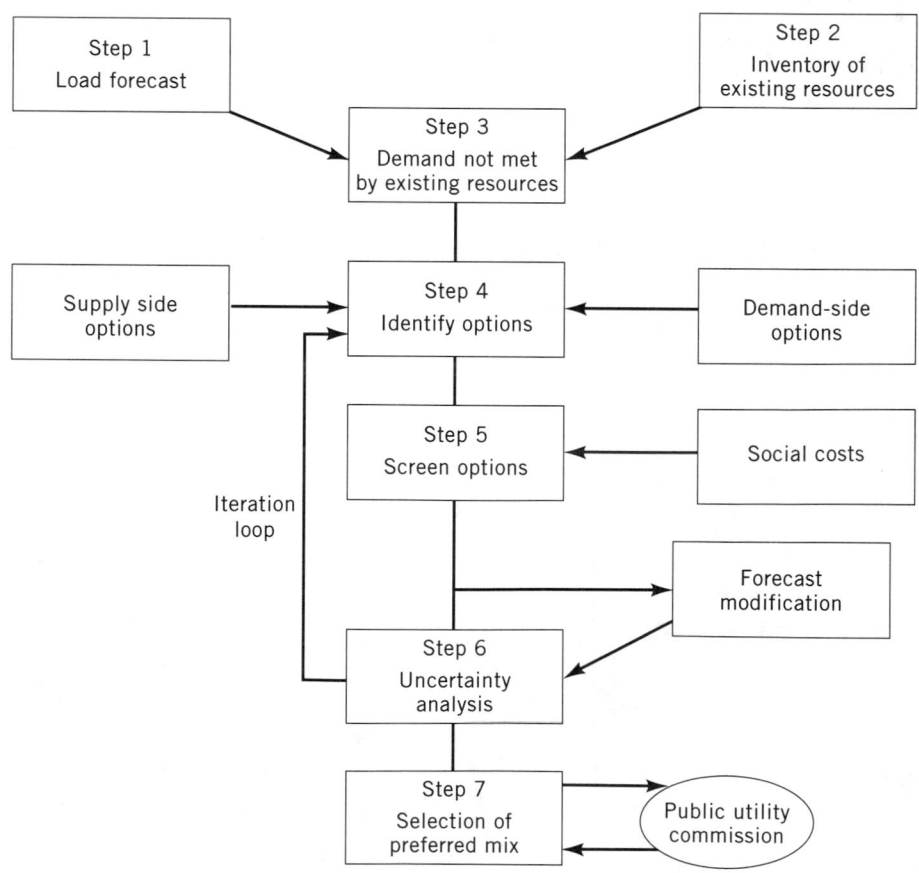

Figure 7. Schematic diagram of IRP process. Data from Ref. 12.

there are spinoff benefits from increased energy efficiency and conservation. Reducing expenditures to import energy means more money can be spent on goods and services to increase the state's sales revenues and employment levels. Savings put into financial institutions increase the supply of loanable funds and stimulate new business investments in the states. Decreasing dependence on imported energy is another spinoff benefit, as are improved air and water quality.

Utilities

The challenges that utilities and their customers have faced in the past stem, in part, from the Public Utility Holding Company Act of 1935 (PUHCA). Competition among wholesale electricity suppliers was not a practical possibility in 1935 (20). The utility structure imposed by PUHCA assumes that electricity will be generated by the local utility. Electric utilities vary widely in their ability to minimize power plant construction costs. PUHCA is thus a major obstacle to relying on the most efficient firms to build and operate the power plants that will be needed in the future.

Over the years the U.S. electricity supply system has been integrated into large regional networks. The Public Utility Regulatory Policies Act of 1978 led to limited competition among a small group of wholesale suppliers. The experience has led public policymakers to conclude that greater competition among wholesale suppliers is both feasible and likely to be beneficial, hence the Energy Policy Act's attempt to open up the nation's utilities to competition.

Demand side management (DSM) is a process used by utilities to plan, implement, and evaluate programs designed to influence the amount or timing of customer energy use. DSM programs affect the system energy and total capacity that a utility provides to meet demand. DSM uses four basic techniques to influence energy use. Peak clipping reduces system peak loads, valley filling builds off-peak loads through methods such as thermal energy storage, load shifting shifts load from on-peak to off-peak periods, and strategic conservation reduces end-use consumption. Gas utilities have recently begun to use DSM strategies as natural gas becomes more recognized as a bridging fuel to replace imported oil.

All utilities implementing DSM programs must consider the effect these programs have on sales and earnings. Investor-owned utilities in particular recognize that reduced sales mean reduced earnings. Pulling utilities in the other direction is the realization that without DSM programs, load growth could ensue that would necessitate the need for constructing new and expensive generating stations.

There are existing mechanisms that utilities and their regulators can use to deal with the conflict between shareholders and customers. These include command and control regulation; frequent rate cases, alternative rate designs that set energy and demand charges equal to short-term costs; compensation to the utility for the revenues lost because of its energy efficiency programs; and decoupling of electric revenues from sales (21).

A new method called statistical recoupling could help answer criticisms of other methods. Utilities would construct statistical models to estimate electricity use. The difference between estimated and actual electricity use would be used to adjust utility revenues. Refunds to customers or increases to retail electricity prices would take place in the following year.

Private Industry, Professional, and Academic

Professional associations, private industry groups, and academic organizations represent another level of energy management. One of the most influential is ASHRAE. ASHRAE Standard 90.1 represents the model used by many states and the federal government to develop regulations. The Council of American Building Officials (CABO), often works with ASHRAE. The Energy Policy Act of 1992 directs these two organizations to develop voluntary energy codes for buildings.

Other important groups are the Advanced Building Systems Integration Consortium (ABSIC), the Alliance to Save Energy (ASE), the American Institute of Architects (AIA), the Association of Energy Engineers (AEE), the Building Owners and Managers Association (BOMA), the Energy Efficient Building Association (EEBA), Intelligent Buildings Institute (IBI), the International Facility Management Association (IFMA), and the U.S. Energy Association (USEA). The University of California at Berkeley and its Lawrence Berkeley Laboratory is extremely active in researching energy-efficient technologies and transferring information to the public and private sectors. These organizations collect data on energy use, sponsor seminars and conferences, produce targeted publications, and track policy changes and technologies that affect their constituencies.

In many cases strategic partnerships are formed between industry associations or companies from private industry to develop codes, standards, and equipment solutions. Industry excellence, societal goals, and economic opportunity all drive these efforts.

Corporate

The corporate level of energy management represents the place where many people direct their energy efficiency initiatives. Together, the industrial processes and office buildings of the corporate world are big consumers of energy: the industrial and commercial building sectors combine to use 52% of the nation's energy annually (22).

Corporations should have an energy management policy as a matter of self-interest. Those that aggressively manage their energy use also reduce their exposure to potential environmental risks, cost increases, and future energy shortages. The energy management policies developed by companies range from formal policies that establish broad objectives to less formal policies that set reduction goals, and the time frames and means of achieving them (23).

Companies may employ design standards, operating standards and maintenance standards into their plans. Individual company cultures and management styles affect the nature of these plans. Energy managers in these organizations face moving targets that include process innovation, long-term quality planning, energy assessments of building and equipment purchases, employee or tenant awareness, and waste minimization and recovery.

A common denominator for the most successful programs appears to be assigning direct responsibility for implementing a company's energy policy to an energy manager. These energy managers should report to top management and also have the authority to make change. They work with division managers to develop and implement company policy in individual facilities. At their best, information and responsibility in these efforts flows up and down. Goals are set at the management level with input from the facility or building level, and accountability is at the lowest possible level.

Some of the strategies these energy managers use are monitoring energy usage and cost and establishing building performance criteria, purchasing reliable energy supplies, performing regular energy surveys of facilities and buildings, developing energy-efficient building design criteria, and providing technical assistance to facility and corporate staff. The use of facility management systems, another important strategy, will be discussed in detail below.

One of the most important aspects of energy management at this level is flexibility. A dramatic upswing in business may increase energy use, even though efficiency goals are being met. A drop in production may make it harder for some facilities to achieve goals because the plant is no longer operating 24 hours a day. The opportunity to reuse energy by-products is thus reduced.

Implementation of energy management policies pays off through even simple actions. At a General Motors plant, employees were given a monthly report that showed through computer-generated graphics the cost of neglecting to turn off lights and equipment. An awareness campaign resulted in 50% less energy wasted on lighting and 86% less energy wasted on major equipment, saving more than $309,500 a year (22).

Building Level

At a building level, energy managers, plant managers, and facility engineers implement policies developed in coordination with the corporate level. New building construction plans can be scrutinized with the goal of providing a comfortable environment while using energy cost effectively. Energy audits or detailed engineering studies can be done on existing buildings to determine if energy conservation measures, such as equipment retrofits, make operational and economic sense.

Building owners and managers need to recognize however, that interrelations and tradeoffs exist between energy and building environmental goals. For example, during nonoccupied periods opportunities exist for reductions of air flow that conserve energy and maintain air quality. During occupied periods, variable air volume systems provide a way to balance energy conservation and air quality (24).

Companies often put purchasing decisions on plant and equipment through energy assessments, requiring those making the decisions not only to consider the cost of the equipment but also to review how things can be done differently to save energy or reduce waste. Bringing employee talents and corporate resources together to focus on energy management can be as effective as purchasing expensive technologies. For commercial buildings, educating tenants about energy conservation measures can help keep building operating costs, and rents, down. Lighting in particular is an important area of concern.

Lighting typically comprises 30–50% of a building's electrical load. In the absence of computerized occupancy sensors and lighting controls, people should be encouraged to turn off lights that are not needed; lighting options such as retrofitting fixtures with optical reflectors and energy-efficient electronic ballasts can further reduce electrical load. By combining various lighting strategies, energy consumption savings can reach 40–60%.

Regulations, agencies, issues, resources, trends, and technologies, as they affect the above levels of energy management, will change as society's desire and ability to manipulate energy likewise changes. Strong technologies now exist, as well as an understanding of the long-term benefits of energy efficiency. At every level of energy management, there also is the will to improve our use of energy.

FACILITIES MANAGEMENT SYSTEMS AND SERVICES

Facilities management systems (FMS), also called energy management control systems (ECMS), represent technologies that act on all the goals of efficient energy management. Facilities management systems and related services, however, deliver benefits beyond energy savings. Broadly stated, facilities management systems, and related equipment and services create a comfortable, safe building environment; control energy and operating costs; and provide vital business information management to help optimize resource use. Equally important, FMS provide vital communications that are used to operate buildings and manage building activities more effectively.

One of the biggest challenges in the past has been how to optimize use of the many subsystems that must achieve building management objectives (25). These subsystems include heating, ventilation, and air-conditioning (HVAC); lighting; security; fire management; and power distribution.

Automation solutions previously involved master control rooms. The master control room attempted functionally to integrate building subsystems by placing all operator terminals in one control room. The building operator exercised the concept of "management by exception," seeking to manage variables so conditions stayed within predetermined parameters. The master control room was the place where optimizing strategies and the orchestration of schedules were conceived, tested, and tuned.

However, only a few high technology buildings have had the luxury of the master control room. More common has been the use of a nonintegrated control scheme. In most buildings, subsystems were managed passively or reactively. Analogue, digital, and quite often pneumatic displays were monitored by time-consuming visits to remote corners, rather than conveyed to a central point.

Temperatures and other variables relating to specific setpoints were regulated as opposed to optimized. Important operational data on systems were observed and forgotten instead of being tracked as trends and constantly processed for management-by-exception reporting.

More recently, facility managers have used direct digital controllers to enhance their control capabilities. Mounted on or near the equipment they control, these programmable devices can stand alone or send and receive data as part of a network. They do not depend on the network for basic control nor do they encumber the network by sending unwanted information. When asked or polled, they can respond fully.

Noncompatibility as a Hindrance

As various mechanical and electrical subsystems grew into the electronic age, each subsystem panel acquired information, but seldom would it capitalize on opportunities to distribute information to other panels. The existence of a direct communications port on both panels has not been enough to have one device talking to another across a network. The protocol or programming language used by a vendor for a given device has typically not been compatible with central system software or devices from other subsystems. There has thus been two levels of noncompatibility: peer-to-peer noncompatibility between individual manufacturers and noncompatibility between manufacturers. This incompatibility formed an impediment to centralized reporting and a more coordinated control scheme.

Fortunately, recent developments in building system integration and advances in personal computer technology have made possible the seamless integration of subsystems. An evolution occurred toward total integration by a single vendor who is able to supply compatible equipment for all subsystems. These network-based FMS are designed to be integrated and interactive on every level. Peer-to-peer networks and data highways provide infrastructure while personal computers and their graphic displays provide system operators with an improved means of knowing how their systems are operating. Operators have a number of ways of responding to conditions.

This desired flexibility is possible because current FMS are fully distributed systems built to achieve stand-alone control, supervisory control, and information management. Distributed systems offer many benefits. They ensure a more efficient system, because bottlenecks are eliminated and operators can make better use of processor power. Distributed systems offer a more reliable system through improved fault-tolerance; no single point of failure can bring down an entire system. The FMS are designed as scaleable systems, with building owners buying only what they need now and having the option of expanding later without the loss of original investments.

Connectivity at Work in FMS

Energy managers increasingly rely on the ability of devices employed in facilities management systems to exchange useful information, displaying connectivity with other devices and building management equipment. The basic architecture of these FMS consists of programmable control panels, called network control units (NCUs), and operator workstations that communicate with each other over a high-speed communication network or data highway (Fig. 8). NCUs directly control central plant equipment, while the management of smaller air handlers, heat pumps, lighting circuits, and other building subsystems is delegated to application specific controllers (ASCs).

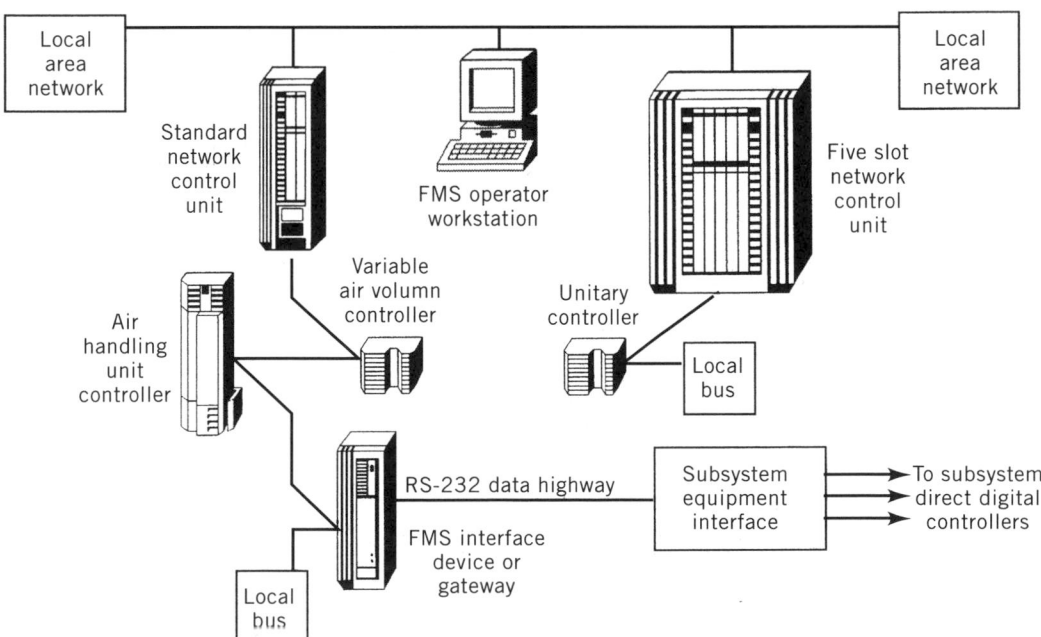

Figure 8. Basic architecture of a facilities management system. Data from Johnson Controls, Inc.

With connectivity, vendors work together on equipment and protocol standards to produce an open control system for equipment deployed in buildings. This results in devices being able to communicate with other devices across "transparent" barriers. Connectivity works because it understands the protocols of individual devices. Under connectivity, integration can be the function of interface units or gateways positioned between the controlled devices on a network's data highway. Alternatively, interface modules can be built into the hardware of central control equipment. Through short lead times for development of a common language or protocol, connectivity shortens technology curves, or the time it takes to introduce improvements to technology.

Connectivity also strives to use existing equipment and staff expertise. Existing HVAC equipment, for example, can be modified with sensors to make interactive communications with a broader control scheme possible. Because adaptations such as these do not represent wholesale changes, the current level of operator knowledge remains viable, and substantial retraining is not required.

Fostered by partnership, connectivity allows for even greater partnership by suppliers. Vendors of the different principal subsystems can come together to propose and provide packaged solutions that offer the best equipment and best technology, without the user being tied to a solution offered by a single supplier.

Guiding strategies for retrofits, renovations, and upgrades to FMS and new construction projects would include getting suppliers involved early in a team approach. With connectivity and technologies advancing at such a rapid rate, it is possible that building management personnel may not be aware of all the control synergies possible, including those linked with energy conservation measures. Power monitoring, lighting control and building access, for example, can easily be brought into the control system. Equipment suppliers are more likely to have at their fingertips information on "holistic" system possibilities, the possibilities for subsystems to interact optimally to form a well-tuned whole-building system.

Also, it is important to view building system needs from the perspectives of different types of building users. Office space has needs different from those of manufacturing areas, computer rooms, and controlled environments such as "clean rooms." Users or tenants that occupy multiple buildings may have an additional layer of needs. To bring suppliers and building users into the planning picture may well require a formal approach such as a building audit or survey. Factors such as the age of the existing FMS and how recently a study was conducted determine the level of complexity required by the audit or survey. The formal assessment will contain a technical proposal that lists options including energy conservation measures as well as the cost and benefits of those options.

To move beyond this stage toward implementation requires that the building owner or manager choose certain vendors and options over others. Adding connectivity potential to traditional vendor and project qualification criteria will ensure that a new or retrofitted FMS makes the most of current open system technologies.

A fully interactive building system would include physical and functional access to building operations data

from any system or subsystem within the building, effectively interfacing all controllers, processors, or data acquisition paths (26). The most frequently integrated, useful, and owner-demanded controls are listed below (see also Fig. 9).

Boiler controls
Fume hood controls
HVAC controls
Lighting control systems
Maintenance management control systems
Motor drive controls
Fire management systems
Power monitoring systems
Programmable logic controllers
Security management and access control systems
Uninterruptible power supply (UPS) systems

Energy management becomes much easier with an integrated control scheme. The integration of power monitoring and other subsystems with a centralized facilities management control system provides an example of the interprocess interaction that is possible from a single-seat user interface.

FMS as a Management Tool

Given a peak demand or time-of-day billing situation, it could become necessary to reduce or shift electrical use, because it saves operating dollars. An FMS can accomplish load reduction without a significant impact on occupant comfort. The lighting load could be decreased by using light sensors to monitor daylight levels, and adjust lighting conditions, provided sufficient daylight is available. Occupancy schedules, either programmed or monitored by the access control system, would automatically alter the load reduction strategy.

One of the strategies that can be employed is the use of ice storage for air-conditioning. This strategy uses electrical energy to prechill water for use during peak demand periods. This trend will continue because of financial incentives offered by utilities to adopt this form of load reduction. The FMS will be called on to optimize the start and stop time in accomplishing the above mission and providing documentation of these strategies.

Another off-shoot of these incentives will result in owners installing generators for off-peak demand and the use of load levelers. The generators will either be self-contained or use energy from a power plant's waste heat. In using waste heat, steam generated for heating purposes could be slightly increased and used to power steam turbines before channeling the energy for warming buildings.

Building owners will see value in generating their own peak energy and use alternative forms of energy production to minimize their peak demand. Included in these strategies may be operating an on-site peak demand generator and battery load-leveling.

Achieving effective energy management through use of integrated FMS requires properly trained operators working with the right equipment and over the long term,

	HVAC and controls	Fire	Lighting	Maintenance management
Fire	Integrate HVAC with fire alarm system for smoke control.			
Lighting	Use light switch to determine occupancy and start fan system.			Scheduled O & M program for maximum energy efficiency.
Security	Use occupancy sensor to start fan system.	Lights on in intrusion zone.	Lights on in intrusion zone.	
Electrical equipment and power monitoring	Optimize and monitor demand limit	Doors open for evacuation.	Schedule lighting	Monitor alarms from on-site or off-site. Monitor run time for lead/lag and maintenance purposes.
Chillers and boilers	Optimize	Shut down in case of fire.		
HVAC and controls				

Figure 9. Subsystem connectivity matrix. Data from Johnson Controls, Inc.

strong operations and maintenance (O&M) programs. These O&M programs ensure that equipment runs at peak energy efficiency. Scheduled servicing of equipment will increase the efficiency of machinery.

The FMS can monitor the operating efficiency of major energy-consuming equipment and alert management of the need for scheduling maintenance. Other advanced technologies that are employed include vibration analysis. This technique can predict failures before their occurrence. The procedures noted, all made possible by an FMS, can provide the building owner with the knowledge and power to perform predictive maintenance to save time, repair costs, and energy.

Planning as an Inclusive Process

Building experts acknowledge that one of the greatest challenges to greater efficiency is the fragmentation of the building industry (27). One consideration in the design of an FMS is to gather the views and needs of those who will operate the system once it is installed. The best systems have included facility managers in the system design process.

One trend that is occurring as a result of looking at the building holistically is a concept called the "smart building." According to the Intelligent Buildings Institute, a smart or intelligent building is one that provides a productive and cost-effective environment through optimization of its four basic components: structure, systems, services, and management as well as the interrelationships between them (28).

To develop a smart building, experts advocate an approach called participatory design. Specialists (including architects, engineers, facility managers, telecommunications experts, construction managers, owners, and even end-users) meet before a building is designed. The result of this meeting is a building plan, complete with facilities

management systems, that optimizes the four basic building components. Smart buildings incorporate flexibility into building design and subsystems to meet the changing needs of occupants and to adjust automatically to internal and external conditions.

To counter static, inflexible building environments, smart buildings may employ environmentally responsive workstations (ERWs). Specified in open office environments, ERWs allow workstation occupants to control, modulate, and maintain the environmental qualities at the workstation (4). ERWs are at their best if they are integrated with an environmentally responsive architecture. ERWs and their benefits will be discussed below.

Yet while the energy management related advantages of smart buildings will chiefly be available to those with a new building, significant benefits are within the grasp of energy managers overseeing any existing building. A participatory design approach can certainly apply to renovations and retrofits of existing buildings to achieve the optimal building environment. If the building is designed properly, an integrated FMS exists, and hence effective energy management is in place, the optimal building environment is attainable. Facilities management, HVAC, and lighting and power monitoring systems as well as building material technologies offer every energy manager a host of energy management tools.

ECONOMICS OF ENERGY CONSERVATION

Public policy, societal goals, and the information management capabilities of an FMS notwithstanding, the most compelling reason for many organizations to change their use of energy is the opportunity to make or save money. Organizations typically ask energy managers to develop a winning formula by which the economic benefits derived from energy management efforts outweigh their costs.

Organizations soon learn that a progressive energy management program can be a profit center that yields bottom-line results, and can be a safety net to protect an organization from future energy shocks (23). In some organizations, it is believed that to show results one must start by buying one's way to energy efficiency. In reality, economic benefits can begin by managing the way to savings. Organizations can start with a simple campaign to save energy through turning off unneeded lights and equipment. The energy savings that accrue can boost profits. For example, a company implements a plan that nets $10,000/yr. in energy savings. If that company has a modest pretax profit margin of 10%, the company would have needed to sell $100,000 in products to earn the same $10,000.

Pre-tax profits from reductions in energy use can also be demonstrated when considering energy cost as a percentage of revenue. For example, consider a company whose pretax profit is equal to 10% of revenues. If that company's energy bill is 15% of revenues and the company reduces energy use by 4%, the company's pretax profit would increase by 6% (Fig. 10).

Generally speaking, the lower the profit margin, the greater the impact energy savings will have on the bottom line. Energy managers at government agencies and not-for-profit organizations, while not concerned with profit margins, can appreciate the increased dollars available for organizational purposes, especially if those entities depend on outside sources of funding.

To achieve greater savings, organizations inside and outside of the private sector find they must, in effect, spend money to make money. Historically, this has been difficult for organizations for a number of reasons. Some organizations find it difficult to invest in energy-efficient technology when they are operating with tight budgets or close to their profit margins. In more secure companies, spending for energy efficiency competes with other capital investments, with research and development, and with sales and marketing needs. It may also be hard for organizations to justify investments in energy technologies at times when energy costs are not rising and the costs of new equipment are. Many organizations typically require equipment to pay for itself in 2–4 years. Choices for a given equipment upgrade or specification may include

energy-efficient equipment that has initial costs higher than conventional, less efficient technologies. A comparison of initial or first costs does not normally put energy-friendly technologies in more favorable and realistic contexts.

Evaluating Cost-effectiveness

There are several mathematical methods for evaluating the cost-effectiveness of equipment purchases or energy efficiency alternatives (7). Payback, the simplest method, divides the initial cost by the annual savings in dollars. The payback method is easy to understand but ignores the time value of money, the value of savings after the payback period, and the limited life of some measures.

Life cycle costing analyzes the aggregate cost of a capital outlay. Total aggregate costs would include the costs of energy consumed, maintenance, and impact on other systems. It demonstrates the real cost of a piece of equipment or building and includes not only the initial cost but also the total expense of operation over a useful life. In an energy efficiency context, life cycle costing enables energy managers to compare different procurement or specification alternatives to show the savings accrued over time.

There are two competing calculations used to attain life cycle savings estimates. Net present value translates all future costs and savings into their equivalent in today's dollars. If a piece of energy-efficient equipment is projected to have a useful life of 20 yr, the total savings would be

$$20 \text{ yr} \times \text{annual savings in dollars} = \text{total savings}$$
$$\text{total savings} - \text{initial costs} = \text{net present value}$$
$$\text{of savings}$$

A second life cycle calculation that is considered more realistic recognizes that a dollar received a year from now is worth less than a dollar received today. The theory is that a dollar received today can be put into an interest-earning account and thus grow in the year ahead. Future savings can be discounted by using percentages to reflect the time value of money:

$$\text{total savings with discounting} - \text{initial costs}$$
$$= \text{total savings in discounted dollars}$$

Another method is the cost of conserved energy (CCE). This method first measures how much one pays for each unit of energy saved, independent of fuel price. The CCE is then compared with the cost of the supplied energy it displaces to determine the merits of an energy conservation measure. The CCE is defined as the initial cost times the capital recovery factor (CRF), which converts an initial investment into an equivalent series of annual payments incorporating a discount rate, divided by the annual energy savings in energy units. The equation for Capital Recovery Factor is

$$i(1 + i)^n/(1 + i)^{n-1}$$

where i is the discount rate and n is the number of years.

$$\text{initial cost} \times \text{CRF/annual savings} = \text{CCE}$$

	Percent reduction in energy use				
	−2%	−4%	−6%	−8%	−10%
30%	+6%	+12%	+18%	+24%	+30%
15%	+3%	+6%	+9%	+12%	+15%
10%	+2%	+4%	+6%	+9%	+10%
5%	+1%	+2%	+3%	+4%	+5%

Energy costs as a percentage of revenue

Figure 10. Pretax profit from energy savings based on pretax profit equal to 10% of revenues. Data from the DOE and National Association of Manufacturers.

To complete the energy manager's understanding of these mathematical tools, a discussion of electricity conversion ratios is required. Analyses of energy use often require that different forms of energy such as natural gas, oil, and electricity be combined into one common measure, typically Btus. Converting electricity into Btus presents a problem because there is no one correct conversion method. The site conversion ratio converts 1 kW·h of electricity directly into heat or 3412 Btu. This ratio ignores that energy is used to produce that 1 kW·h of electricity. An alternative to the site conversion ratio is the primary conversion ratio of 10,240 Btu/kW·h, which includes the energy used to produce the electricity.

While the primary conversion ratio has drawbacks, among them not accounting for energy losses during transportation, it does allow for a more accurate comparison of the true energy savings resulting from increases in the efficiency of electricity use. Cents per kilowatt·hour calculated using the site conversion ratio are significantly higher than the price calculated using the primary conversion ratio, reflecting the omission of the true costs of producing electricity.

When reviewing savings estimates, especially those provided by outside vendors bidding on possible projects, energy managers should check to see that computations are based on consistent methods and should have the vendors make adjustments. Energy managers may want to use the organization's own internal methods and principles, possibly those used in its financial accounting, as benchmarks for computing potential energy savings. Benchmark methods can be stipulated up front as part of the project specification and design process.

When a commercial building is being designed, attention must be paid to the way in which it will function as a whole. Because initial costs of construction are comparable to energy costs over a building's life span, an increase of only a few percent in efficiency still translates into a considerable sum of money (27).

Computer Programs as Tools

Energy managers have in computers and software programs powerful tools for analyzing building performance, including a building's response to ECMs. One example of these computer programs is *DOE-2,* a building energy analysis program. *DOE-2* was originally developed by the DOE's Office of Building Technologies as an objective, public-domain tool to assist in evaluating building performance (29).

Used worldwide, the *DOE-2* model is an analytical tool, a design tool, a tool to evaluate individual technologies by climate zone, and a cost–benefit evaluation tool. *DOE-2* has been used in the development of building standards and as an educational tool for new engineers, designers and architects. An upgrade of *DOE-2* and an improved version called *DOE-3* will expand the simulation tools available to energy managers. *DOE-3* will be easier to use, provide built-in guidance, and allow models for new HVAC technologies to be quickly built up from component modules (30).

Building energy analysis programs typically perform the calculations such as life cycle costing. Many of these programs allow users quickly and completely to examine the impact of building standards including ASHRAE Standard 90.1. For the energy manager juggling a host of possible discount rates, energy rates, and financing levels, these computer programs represent a valuable tool.

A Dynamic Environment

A few cautionary notes on estimates of energy savings. Estimates of potential savings are important for planning, but the aim of energy management programs is to realize actual reductions in energy use and overall costs (31). Actual savings may be less than expected. This can be because the estimates were based on idealized engineering analyses that may be different from conditions in practice.

Production levels can effect actual savings, especially if production in an industrial setting is increased, thereby requiring more energy. Weather abnormalities and unforeseen occupancy changes can also effect actual savings. Building use in general is dynamic, with new equipment and load factors changing constantly.

Case Study: Improved Productivity

Relatively new to the list of benefits attributed to energy-efficient buildings and technologies, but with potentially significant impact on the justification of FMS, is improved productivity. As earlier noted, a 1991 study of the new corporate headquarters of the West Bend (Wisconsin) Mutual Insurance Company reported an increase in productivity of 16% over productivity in the old building vacated during the course of the research (4). The new building used smart building techniques, including participatory design and energy-efficient technologies. But the focal point of the research was ERWs. ERWs provide integrated heating, cooling, lighting, ventilation, and other environmental qualities directly to the occupants of workstations.

To measure the impact of ERWs themselves, several units were randomly turned off during a 2-week period. Researchers found that productivity, measured as the number of claim files processed by individual employees, decreased 13%. Overall, researchers estimated that the ERWs were responsible for an increase in productivity of approximately 2%. One of the insurance company's executives estimated that the productivity increases resulted in an annual savings of $260,000, with the system of ERWs paying for itself in less than 1 yr (32).

When reviewing ERWs, managers should keep in mind that increased productivity could be used to justify costs. An increase in productivity of a few percentage points could have a big impact on their payback, because the cost of labor is generally the single largest cost factor in the production of goods and services. An energy management program with strong economic underpinnings, when joined with high quality technologies and management practices, forms a comprehensive approach to meeting the challenges of conserving energy.

Case Study: Performance Contracting

There is perhaps no better example of an energy management program with strong economic underpinnings than one based on performance contracting. As noted earlier, under performance contracting a supplier will finance the project for the organization, with the contractor taking as payments future energy savings.

An example is the Pendleton Memorial Methodist Hospital in New Orleans. To help control climbing utility bills, Pendleton initiated a long-range, multimillion-dollar energy conservation program. Under the energy services agreement with the 317-bed hospital, the supplier, Johnson Controls, Inc., provided both the financing and retrofitting of the principal improvements to the HVAC and lighting systems. The monitoring and control of the HVAC and lighting systems, duties that once consumed physical plant staff time, are now aided by a facilities management system. Plant personnel use computer terminals to view dynamic graphs of system schematics and change equipment schedules from a central, remote location.

The improvements are projected to save approximately $2 million over the next 10 yr. The supplier will share in those savings as compensation for financing the program and monitoring the program's progress. Meanwhile, even with the scheduled addition of a 39,624 m², six-story building, the plant operations staff will be able to manage the total 152,400 m² complex without additional personnel.

Case Study: Connectivity

At a commercial building in Oak Ridge, Tennessee, a facilities management system equipped with an interface device extends the capabilities of the building's old and new HVAC equipment. A zone-control scheme and lack of control power once led to 8° to 10° temperature swings from the north to the south side of the 17,069 m² building.

Two separate suppliers teamed-up in cooperative planning and specifying efforts that incorporated a 9144 m² addition to the existing building. They offered the building owner off-the-shelf connectivity between factory-mounted smart devices on HVAC equipment and a facilities management system, with the FMS recognizing all the control points in the building.

The new, connectivity-enhanced equipment and control scheme allows staff to operate the building at maximum efficiency and to monitor or adjust equipment at a central location. Building tenants have enhanced comfort and no temperature swings, with the control system using 140 thermostats instead of the old system's 12. Overall, the building owner credits the cooperative efforts with increasing the value of the building by saving time and by creating a happy tenant.

THE FUTURE OF ENERGY MANAGEMENT

Priorities are changing. Governments and businesses know they must make better use of limited natural and economic resources (33). Will the pendulum of concern swing back toward apathy, or has it stopped permanently in favor of a more responsible use of energy? For energy managers who favor the latter, a discussion of the future of energy management is equal parts products and programs, obstacles, and an agenda for progress.

The degrees to which products and programs will be successful depend on ease of use or application. A product's application as a building management tool requires that facility management personnel and building users be readily able to use it and can readily equate its use with the target benefits of increased efficiency and comfort.

An initiative such as a program designed to achieve increased energy efficiency in an organization's buildings must also be readily understood by its participants. Users of federal buildings need to be shown that while giving them a more comfortable and productive place to work or learn, increased energy efficiency conserves the economic resources of the nation's taxpayers and the funds available to federal agencies. It must be understood by all that strong energy management enhances national security by making the United States more self-sufficient and less susceptible to international disruptions of price and supply.

Energy Efficiency and Labor Efficiency

Much of the above rationale translates to users of commercial and industrial buildings. Increased energy efficiency can help them be more comfortable, happy and productive. In any building, the costs of the occupants, ie, their wages and benefits, far outweigh any of the operating costs in a building. Worker productivity, therefore, should be a prime consideration as the future unfolds. Because labor costs comprise such a large component of production expense, productivity increases of even 1–2% provide quick returns on investment for the latest building technologies (34). Energy managers should seek to apply technologies that improve the productivity of workers. This could include such equipment as environmentally responsive workstations.

Increased energy efficiency, from a traditional perspective, conserves the organization's limited financial resources, allowing more to be returned to the organization through product research and development, employee compensation and benefits, additional jobs, and job security. The organization itself will be less susceptible to disruptions in energy prices and supply that could affect product cost and hence market competitiveness.

Future developments also will depend on successful measurement and management. Energy conservation programs and measures must have efficiency and payback targets; technologies and products must have efficiency standards. Equally important is the provision for evaluating the performance of programs and products. If the performance of initiatives cannot be measured in some fashion, they cannot be managed. Facility management and control systems provide this measurement and management function on an interactive, continuous process basis. Energy audits study and report on energy use in a building. Energy managers and their organizations can use the information provided by these systems to further optimize their strategies.

The participatory design of intelligent buildings has been discussed as a means of delivering results to new buildings. Owners of existing buildings should do more in the future to incorporate elements of the participatory design process into retrofit and renovation projects.

Money to Save Money

Product technologies hold the future of energy management. Goals in developing them should be to help employees work smarter, not harder. Facility management systems have been discussed as proven technologies; they need to be more universally applied and their capabilities more completely tapped.

Simple solutions, such as "lights off" campaigns, do not depend on sophisticated technologies or large-scale infusions of capital. But the biggest obstacle to energy efficiency is that it can take money to save money. Where possible national, state, and local government leaders must expand the monetary resources made available to energy managers. Loans, grants, appropriations, and municipal bonds form just a part of the picture.

Legislation establishing self-financing energy efficiency banks should be promulgated to provide additional sources of funds. Legislation and procedures that enable energy managers in the government sector to easily form public–private partnerships must likewise be advanced.

The trend is toward developing for utilities (and hence their customers) the type of marketplace that encourages competition and efficiency. This will conserve natural and economic resources, and as it has done in other deregulated industries, should stabilize energy prices. Regulations such as those promoting the use of statistical recoupling of revenues would give investor-owned utilities an incentive to run energy efficiency programs, avoiding the disincentive that an energy management program causes when it reduces earnings. Public utility commissions promote this concept.

Cultivating Quality

In the future, energy-conservation experience and management capabilities should count for more in the vendor selection process as should the vendor's willingness to submit proposals based on cooperative or progressive financing methods. Qualified energy management personnel should also be cultivated. Energy managers will require a career path, adequate compensation, and a recurrent training program. Incentives in the form of recognition and enhanced compensation should be expanded as a means of rewarding and retaining qualified personnel.

Organizations, particularly government agencies, must continue to expand their energy-consumption measurement, record-keeping, and analysis activities. It will be difficult to meet future goals without good statistics. If energy managers can appropriately measure energy use, energy managers will be able to develop strategies and purchase systems designed to meet their requirements. Energy efficiency standards for buildings and products will continue to be reviewed and refined. Energy efficiency standards must be reinforced by operating procedures that require documentation of building and energy management activities and a commitment by top management to meet these goals.

In the past, proprietary equipment protocols and computer languages have impeded interprocess interaction between computerized equipment. Attempts have been made by international and domestic organizations to develop standard protocols. In this way, users can take advantage of the best technologies available, regardless of the manufacturer, and be assured they are compatible with equipment already in place.

Compatibility is already achievable, however, using the various protocols developed by the private sector. Interface devices and strategic partnerships forged by forward-thinking suppliers already satisfy the requirements of many public, commercial and industrial buildings. More suppliers, especially those that refuse to open their equipment to a compatible or connective scheme, need to join this movement.

To enhance energy management in the future, maintenance procedures need to be improved. Managing the maintenance schedules of the building's systems is a consuming task, but a task that is well-suited to computerized facility management systems. These systems can help ensure that correct equipment maintenance is performed and documented. This will result in longer equipment life and minimal down-time.

Creating Jobs Through Energy Efficiency

In the years to come, energy efficiency improvements will lower energy bills and reduce the cost of energy services, cut oil imports, reduce pollutant emissions, and provide other benefits. The creation of jobs is an equally compelling reason to improve energy efficiency. By shifting economic activity away from energy supply and by saving consumers and businesses money that will be re-invested throughout the economy, energy efficiency improvements will result in a net increase in jobs and personal income. It also will make America more globally competitive.

The American Council for an Energy-Efficient Economy estimates that efficiency improvements consistent with an attainable 2.4% annual reduction in national energy intensity could create nearly 500,000 new jobs by the year 2000 and nearly 1.1 million new jobs by 2010 on a net basis (35). The ultimate goal must be to make energy efficiency a top priority in the minds of individual leaders, executives, operations and maintenance personnel, and building users. Progress has been made, but the momentum must be sustained. Certain obstacles remain, including inadequate funding, inconsistent commitment from top management, limited vendor qualifications, personnel limitations, deficient recordkeeping and maintenance procedures, and weak enforcement of building standards.

The movement toward the universal adoption of energy management philosophies must continue if the benefits of reduced energy costs and increased building comfort are to be obtained on a building or facility level. The more universal adoption of energy management philosophies also will ensure that the goals of enhanced organizational efficiency, national security, reduced pollution, and economic vitality are realized over the long term.

In this era of increased energy awareness, is it possible to say the pendulum of concern has stopped swinging? Given the factors that have affected the will and shortened the attention spans of participants at each level of energy management in the past, it must be concluded that the jury is still out. Perceived benefit and self-interest, as always, will determine the position of the pendulum at any period in history.

BIBLIOGRAPHY

1. U.S. Department of Commerce, *National Energy Strategy,* Springfield, Va., 1991.
2. A. H. Rosenfeld and D. Hafemeister, *Sci. Am.,* 78, (Apr. 1988).
3. S. Laitner, *Using Input-Output Analysis in Energy Policy Review,* Economic Research Associates, Eugene, Ore., 1993, p. 27.
4. W. Kroner, J. A. Stark-Martin, and T. Willemain, *Rensselaer's West Bend Mutual Study: Using Advanced Office Technology to Increase Productivity,* The Center for Architectural Research, Troy, N.Y., 1992.
5. Energy and Environment Division, Lawrence Berkeley Laboratory, *Center for Building Science,* FY 1992 Annual Report, MS 90-3058, University of California, Berkeley, 1993.
6. Alliance to Save Energy, American Council for an Energy-Efficient Economy, Natural Resources Defense Council, Union of Concerned Scientists, *America's Energy Choices,* the Union of Concerned Scientists, Cambridge, Mass., 1991.
7. U.S. Congress, Office of Technology Assessment, *Building Energy Efficiency,* OTA-E-518, U.S. Government Printing Office, Washington, D.C., 1992.
8. American Society of Heating, Refrigerating and Air-Conditioning Engineers, Inc., *ASHRAE / IES Standard 90.1-1989 User's Manual,* 1992.
9. Commonwealth of Massachusetts, Massachusetts Division of Energy Resources, *The Massachusetts Energy Plan,* 1993.
10. M. Ginsberg, *Energy User News,* 24, (June 1993).
11. U.S. Department of Energy, *Annual Report to Congress on Federal Government Energy Management and Conservation Programs, Fiscal Year 1991,* 1992.
12. National Conference of State Legislatures, *Energy Manage. Conservation,* 161, 1993.
13. U.S. Department of Energy, *Energy Manage. Conservation* (1988).
14. Ref. 12, p. 191.
15. Ref. 12, p. 190.
16. *State of Wisconsin, Governor Thompson's Alternative Fuels Task Force,* 1993.
17. National Conference of State Legislatures, *State Legislative Rep.* **17**(23), (1992).
18. Ref. 12, p. 37.
19. Ref. 12, p. 130.
20. U.S. Department of Commerce, National Technical Information Service, *National Energy Strategy: Powerful Ideas for America,* Springfield, Va., 1991.
21. *Energy Conservation Digest* **16**(8), 73 (1993).
22. Ref. 12, p. 138.
23. U.S. Department of Energy, National Association of Manufacturers, *Energy Efficiency: The Competitive Edge,* Washington D.C., 1991.
24. N. Camejo, paper presented at the 14th World Energy Engineering Congress, 1991.
25. J. Enright, *New for the Nineties: Connectivity in Building Systems,* 1993.
26. T. R. Weaver in *The Intelligent Building Sourcebook,* Fairmont Press, Lilburn, Ga., 1988, p. 16.
27. R. Bevington and A. H. Rosenfeld, *Sci. Am.,* 77 (Sept. 1990).
28. N. M. Post, *Eng. News Rec.* (May 17, 1993).
29. Alliance to Save Energy, American Council for an Energy-Efficient Economy, *Achieving Greater Energy Efficiency in Buildings: The Role of DOE's Office of Building Technologies,* Washington, D.C., 1992.
30. Energy and Environment Division, Lawrence Berkeley Laboratory, *Building Technologies Program 1991 Annual Report,* University of California, Berkeley, 1991.
31. U.S. Congress, Office of Technology Assessment, *Energy Efficiency in the Federal Government: Government by Good Example?,* OTA-E-492, U.S. Government Printing Office, Washington, D.C., 1991, p. 46.
32. P. E. Beck, *Consulting-Specifying Eng.,* 35 (Jan. 1993).
33. K. Rospond, *Service Reporter,* 27 (June 1993).
34. D. A. Decker, *Testimony before the U.S. Senate Committee on Governmental Affairs,* Feb. 18, 1991.
35. H. Geller, J. DeCicco, S. Laitner, *Energy Efficiency and Job Creation: The Employment and Income Benefits from Investing in Energy Conserving Technologies,* American Council for an Energy-Efficient Economy, Washington D.C., 1992.

ENERGY MANAGEMENT, PROCESS

DAN STEINMEYER
Monsanto Company
St. Louis, Missouri

In the chemical industry, which is inherently energy intensive, energy costs, including feedstock, average approximately 8% of the value added. For large-volume chemicals these costs represent a much higher fraction. For example, for nitrogen-based fertilizer the energy costs are approximately 70% of the value added.

Energy management includes energy conservation, but also encompasses utility system reliability; the intermesh of process design with utility systems; purchasing, including plant location for minimum energy cost; environmental impacts of energy use; tracking energy performance; and the optimization of energy against capital in equipment selection.

ENERGY AND THE CHEMICAL INDUSTRY

Table 1 is an estimate of energy usage by United States industry for 1988 (1). The chemical industry used 21% of the energy consumed by the U.S. industrial sector, and the other three related process industries, paper, petroleum, and primary metals, combined for an additional 50% of the industrial consumption.

Feedstocks

A separate breakdown between fuels and feedstocks for the chemical industry (2) shows that the quantity of hydrocarbons used directly for feedstock is about as great as that used for fuel (see FUELS, SYNTHETIC). Much of this feedstock is oxidized accompanied by the release of heat, and in many processes, by-product energy from feedstock oxidation dominates purchased fuel and electricity. A classic example is the manufacture of nitric acid (qv) [7697-37-2], HNO_3. Ammonia [7664-41-7], NH_3, is burned in air on a catalyst at a pressure around 1 MPa (10 atm) and a temperature above 900°C (3). As shown in Figure 1, the process is built around the heat and power recovery

Table 1. Energy Usage by United States Industry in 1988[a]

	Consumption, EJ[b,c]				
Energy Source	Chemicals	Petroleum Refining	Paper	Primary Metals	Other
Electricity					
Direct use	0.44	0.11	0.20	0.54	1.24
Loss[d]	0.88	0.22	0.40	1.08	2.48
Oil	0.15	0.11	0.20	0.06	0.34
Gas and LPG	3.11	0.78	0.45	0.79	2.10
Coal and coke	0.34		0.34	1.59	0.68
Other	0.58	5.65[e]	1.31[f]	0.05	0.51
Total	*5.50*	*6.87*	*2.90*	*4.11*	*7.35*

[a] Values include feedstocks.

[b] One EJ = 1×10^{18} Joules.

[c] To convert Joules to Btu, multiply by 0.95×10^{-3}.

[d] Assumes delivered electricity has one-third of the energy content of the fuel used to produce it; ie, 1 Joule of electrical supply to an industrial user represents 3 Joules of fuel consumed.

[e] For petroleum, this value represents primarily feedstock used in nonenergy products such as asphalt, as well as some feedstock used for petrochemicals.

[f] For paper, this value represents the fuel value of the biomass.

from this reaction, including the integration of the power recovery turbine and the air compressor. Another example is the reaction of propylene [9003-07-0], C_3H_6, and ammonia to make acrylonitrile [107-13-1], C_3H_3N. Here, two-thirds of the hydrogen is combusted just to satisfy stoichiometry. In addition, CO and CO_2 formation consumes about 15% of the feed propylene.

Fuels

Two-thirds of the fuel used by the United States chemical industry in 1988 was natural gas [8006-14-2], which is clean and easy to combust (see GAS, NATURAL). Although relatively inexpensive at the wellhead, natural gas is costly to transport. Hence the chemical industry is concentrated in regions where natural gas is produced, keeping the average price paid by the U.S. chemical industry for natural gas in 1988 to only 80% of the average U.S. industrial price (1). Similarly the movement of chemical commodity production to the Middle East is driven by the desire to obtain low cost natural gas.

Waste Fuel Utilization

It is always preferable to minimize or upgrade by-products for sale as chemicals, however when this is not feasi-

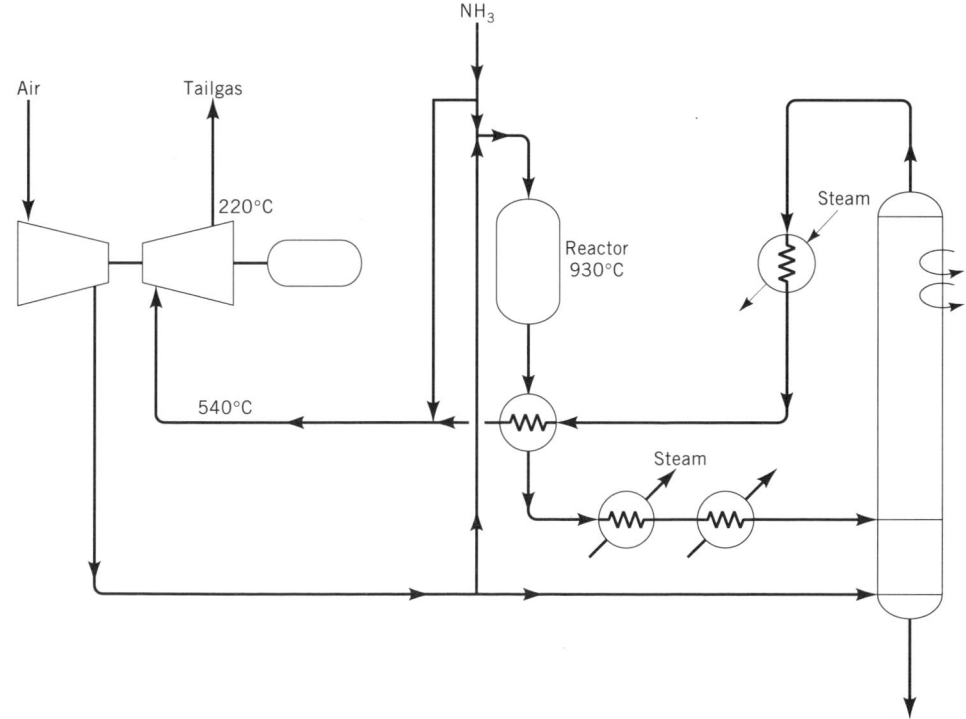

Figure 1. Schematic of nitric acid from ammonia showing integration of reactor heat recovery, power recovery from tailgas, and air compresssion (3).

ble the fuel value can still be recovered. Increased combustion of by-product gases and liquids was one of the principal components in the improvement in energy efficiency that occurred in the industry in the 1980s. An example of waste fuel utilization is the incineration of the off-gases from an acrylonitrile reactor followed by generation of high pressure steam, shown in Figure 2.

Electricity

Electricity, including the losses associated with production, represents 24% of the total energy used by the chemical industry (Table 1). On a cost basis, electricity represents a higher share at 29% of the energy bill including feedstocks. Increases in electrical costs have provided the driving force for increased cogeneration, ie, the recovery of power as a by-product of other process plant operations (see also POWER GENERATION). The historic cogeneration example is the steam turbine associated with the boiler plant. The relatively high cost of electricity has also led designers to focus on the efficiency of rotating equipment, and has motivated a closer look at how processes can be controlled to reduce power, using such innovations as variable frequency motor drives.

Energy Efficiency Improvements

Energy management is basically a game of economics played with a special set of technical rules. A saving of millions of kilowatt hours or gigaJoules is only communicable when converted into dollars, or into a ratio of dollars saved per dollar of incremental investment. The question of where in a process to focus on improvement is often answered by determining the biggest energy cost items in a plant.

Efficiency improvement is driven by two distinct forces: technological progress which is the long-term trend, and cost optimization which is the short-term response to price swings. The baseline, long-term trend has been in the range of 2 to 3% per year improvement. The forecast for the 1990s is 1 to 2% per year, reflecting a more mature industry. A 1 to 2% per year energy reduction still yields a large saving over time.

The U.S. chemical industry achieved an annual reduction of 4.2% in energy input per unit of output for the period 1975–1985 (2). This higher reduction resulted from cost optimization, the tradeoff of increased capital for reduced energy use, that was driven by energy prices (4). In contrast, from 1985 to 1990, the energy input per unit of

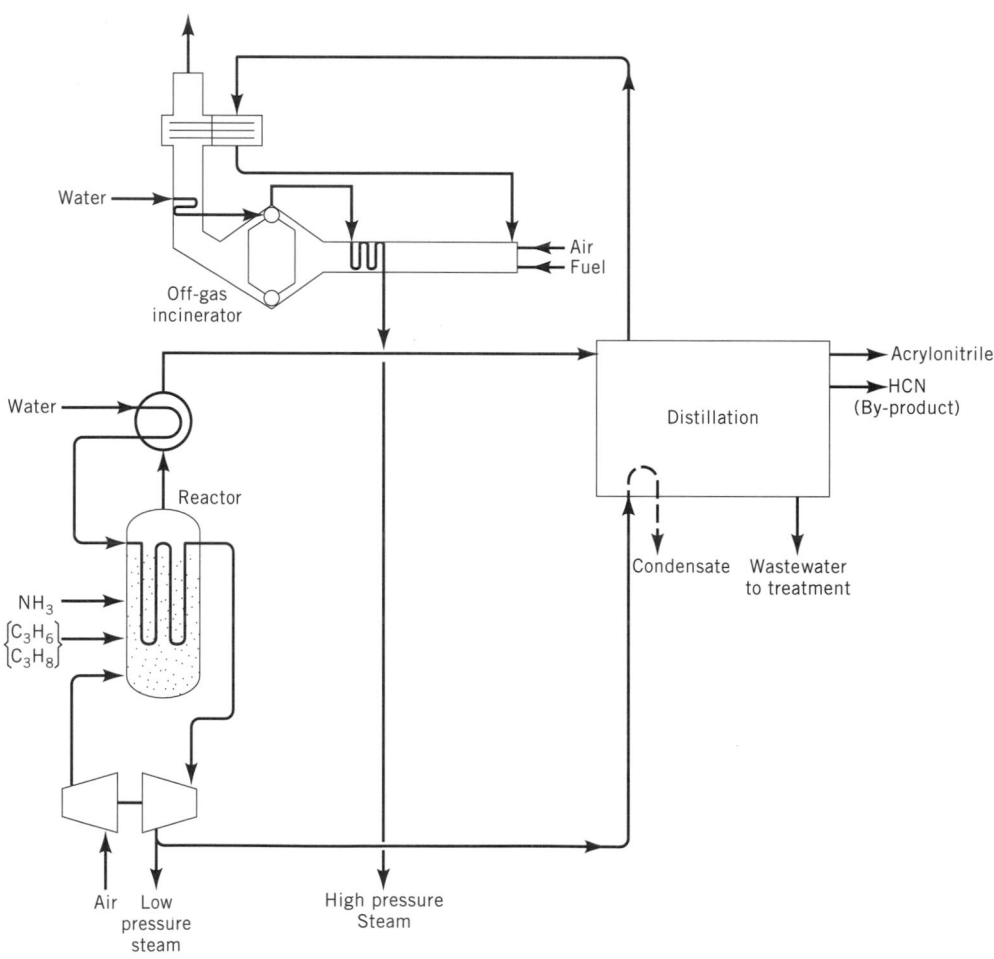

Figure 2. Acrylonitrile process showing integration of waste heat from reactor and off-gas incinerator. The reaction in the reactor is $C_3H_6 + NH_3 + 1.5\ O_2 \rightarrow C_3H_3N + 3\ H_2O +$ heat; in the off-gas incinerator, $C_3H_8 + 5\ O_2 \rightarrow 3\ CO_2 + 4\ H_2O +$ heat.

output has been almost flat (2) as a consequence of falling prices. The average price the U.S. chemical industry paid for natural gas fell by one-third between 1985 and 1988 (1,5).

Whereas energy conservation is an important component of cost reduction for the chemical industry, conservation is rarely the only driving force for technological change. Much of the increased energy efficiency comes as a by-product of changes made for other reasons such as higher quality, increased product yield, lower pollution, increased safety, and lower capital. For example, process energy integration in design, enabled by computer simulation, saves energy as well as capital; substitution of variable speed drives on motors for control valves saves energy as well as capital in the supply of power. One of the roles of energy management is to be sure energy use reduction is considered whenever processes are changed.

The refinement of processes such as occurred for production of low density polyethylene, where a process requiring over 100 MPa (1000 atm) was replaced by one taking place at 2 MPa (20 atm), may continue, but such refinement is expected to be less than in the 1970s and 1980s. The introduction of biotechnology-derived processes is expected to cause a shift to lower temperature and lower pressure processing in the chemical industry, but as of this writing the timing is unclear.

Energy and the Environment

The impact of energy usage on gaseous emissions has emerged as a primary environmental issue, and regulatory action has required emission reductions in NO_x and SO_2 (see AIR POLLUTION; AIR POLLUTION CONTROL METHODS). Control of NO_x is achieved by limiting the temperature of combustion and limiting excess oxygen. Control of SO_2 either requires changing fuel or flue gas scrubbing. Because the preferred fuel of the chemical industry is already predominantly low sulfur natural gas, the primary impact on the chemical industry of SO_2 regulation is expected to be to raise the price of electricity derived from coal.

Issues related to gases such as CO_2, which contribute to the global greenhouse effect, are also rising in importance (see ATMOSPHERIC MODELS). Energy conservation directly reduces CO_2 emissions. The elimination of fugitive hydrocarbon emissions as a result of improved maintenance procedures is also a tangible step that the industry is taking.

ENERGY TECHNOLOGY

Energy management requires the merging of such technologies as thermodynamics, process synthesis, heat transfer, combustion chemistry, and mechanical engineering.

Thermodynamics

The first law of thermodynamics, which states that energy can neither be created nor destroyed, dictates that the total energy entering an industrial plant equals the total of all of the energy that exits. Feedstock, fuel, and electricity count equally, and a plant should always be able to close its energy balance to within 10%. If the energy balance does not close, there probably is a big opportunity for saving.

The second law of thermodynamics focuses on the quality, or value, of energy. The measure of quality is the fraction of a given quantity of energy that can be converted to work. What is valued in energy purchased is the ability to do work. Electricity, for example, can be totally converted to work, whereas only a small fraction of the heat rejected to a cooling tower can make this transition. As a result, electricity is a much more valuable and more costly commodity.

Unlike the conservation guaranteed by the first law, the second law states that every operation involves some loss of work potential, or energy. The second law is a very powerful tool for process analysis, because this law tells what is theoretically possible, and pinpoints the quantitative loss in work potential at different points in a process.

Typically, the biggest loss that occurs in chemical processes is in the combustion step (6). One-third of the work potential of natural gas is lost when it is burned with unpreheated air. Figure 3 shows a conventional and a second law heat balance. The conventional analysis only

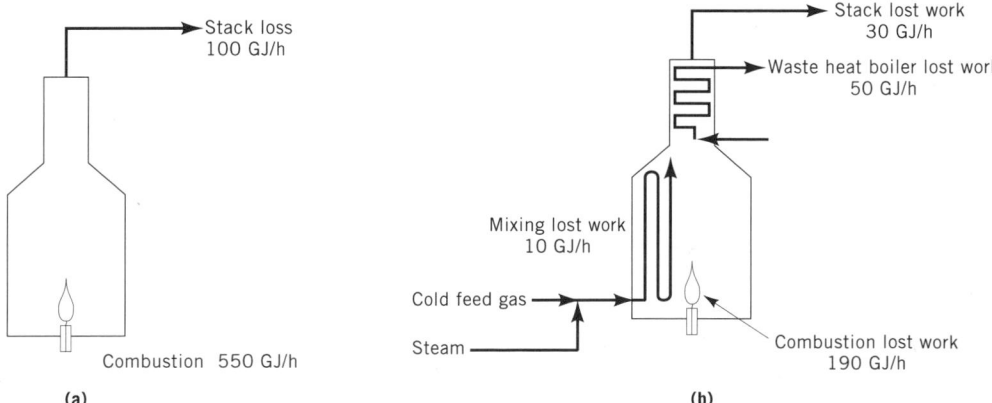

Figure 3. (**a**) A conventional heat balance, and (**b**) a balance employing the second law of thermodynamics. To convert J/h to Btu/h, multiply by 0.95×10^{-3}.

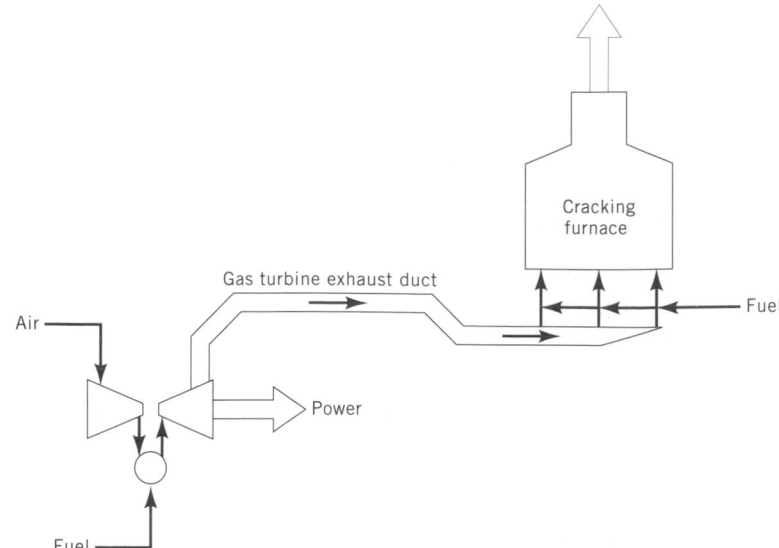

Figure 4. Use of gas turbine air preheat for ethylene cracking furnace. The gas turbine exhaust duct contains 17% oxygen at 400°C.

points to recovery of heat from the stack as an energy improvement. Second law analysis shows that other losses are much greater.

The second law can also suggest appropriate corrective action. For example, in combustion, preheating the air or firing at high pressure in a gas turbine, as is done for an ethylene cracking furnace, improves energy efficiency by reducing the lost work of combustion (Fig. 4).

Converting Heat to Work

There has been a historic bias in the chemical industry to think of energy use in terms of fuel and steam systems. A more fundamental approach is to minimize the input of work potential embedded in the fuel and feedstock, as well as work purchased directly as electricity. Steam is really just a medium of exchange, like money in an economy.

Waste heat tends to be very visible. It can be seen directly in steam plumes and easily measured by determining the temperature of the discharges. The loss of work potential or excess use of work is much harder to spot, but often is larger and more easily corrected. Small inefficient turbines, oversized pumps run on minimum bypasses, rewound motors having efficiency far below design, higher than optimum temperature differences that lead to below optimum power recovery, organics discharged in wastewater, and pressure drops taken across control valves are all examples of loss of work potential.

Economics of Energy Levels and Power Recovery

Cogeneration in a Steam System. The value of energy in a process stream can always be estimated from the theoretical work potential, ie, the determination of how much power can be obtained by running an ideal cycle between the actual temperature and the rejection temperature. However, in a steam system a more tangible approach is possible, because steam at high pressure can be let down

through a turbine for power. The shaft work developed by the turbine is sometimes referred to as by-product power, and the process is referred to as cogeneration.

The by-product power takes only 40% as much energy to produce as on-purpose firing for power only. A comparison of Figure 5b and 5d illustrates this point. By-product power also takes much less equipment, as can be seen in Figure 6. By-product power also avoids the losses and capital associated with use of electricity such as power lines, transformers, and motors. As a result, by-product power is much cheaper than power purchased as electricity.

There is, however, only a limited quantity of by-product power available, and for large process operations the demand for power is usually far greater than the simple steam cycle can produce. Many steam system design decisions fall back to the question of how to raise the ratio of by-product power to process heat. One simple approach is to limit the turbines that are used to extract power to large sizes, where high efficiency can be obtained.

Another way to raise the power/heat ratio is by raising the pressure of the steam system. An increase in pressure from 4.2 to 10.1 MPa (600 to 1500 psi) almost doubles the power associated with a given steam load. (Power/heat ratio increases from 0.12 to 0.20). This, however, comes at appreciable capital cost for alloy materials of construction in the boiler, piping, and turbines. It also requires a substantially higher cost treatment system for the boiler feedwater, and mandates a relatively high recovery factor for condensate.

By-product power does not give enough power to match the demand for many processes such as ammonia synthesis, and designs have historically incorporated condensing turbines for incremental power with heat rejection to cooling water. A more effective response is use of the gas turbine combined cycle shown by Figures 5c and 6c.

Gas Turbines and the Combined Cycle. The combined cycle first fires fuel into a gas turbine and greatly increases

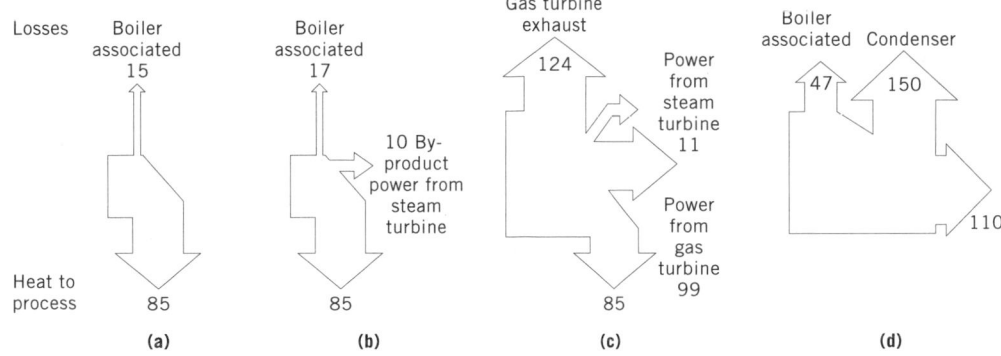

Figure 5. Relative energy flows showing power generation and heat losses in GJ/h for (**a**), boiler only; (**b**), boiler + steam turbine; (**c**), combined cycle employing gas turbine; and (**d**), condensing steam for power only. To convert J/h to Btu/h, multiply by 0.95 × 10⁻³.

Parameter	(a)	(b)	(c)	(d)
fuel	100	112	319	307
incremental fuel	0	12	219	307
power	0	10	110	110
power/incremental fuel		0.85	0.5	0.36
power/heat to process	0	0.12	1.29	

the power extracted per unit of steam produced. As the numbers in Figure 5 show, the simple steam turbine gives by-product power at the lowest incremental energy use. The combined cycle shown in Figure 5**c** is intermediate in energy per unit of power between Figure 5**b** and Figure 5**d**. The big advantage of the gas turbine in cogeneration is that it permits a much higher ratio of power to heat. This ratio, which is routinely >0.8, gets bigger as the unit size of the turbine increases (7). The ratio of power to heat is also larger for aero-derivative systems which are basically jet engines exhausting into power recovery turbines. Because the original design assumed only a gas cycle, ie, no waste heat boiler, the aero-derivative was designed for a higher compression ratio, resulting in a cooler (500°C) gas turbine exhaust. Cooler exhaust means less heat to surrender, and hence a higher power/heat ratio. The heavy-duty machines of comparable vintage run discharge temperatures about 100°C higher. The aero-derivatives are also lighter in weight, giving a second advantage for a process operation because the lightweight permits very fast plug-out/plug-in maintenance using spare rotating assemblies.

In contrast, the heavy-duty gas turbines, designed with capital cost per kilowatt as a key criterion, cost less for the same power output and come in much larger sizes than the aero-derivatives. This means that very large (>100 MW) installations invariably use the heavy-duty type. Heavy-duty turbines have benefited from such technology developments of the aero-derivatives as increased firing temperatures permitted by use of high temperature alloys (qv) and internal cooling of the blading. These developments have been a factor in continuing increased efficiency and increased output from a given frame size.

Gas turbine cogeneration is inherently relatively low in capital expenditures. The absence of heat exchange surface in the gas turbine part of the cycle provides the basic capital advantage, and standardized equipment, prepackaged as skid mounted components, adds to the capital advantage. When these factors are coupled with low priced natural gas, a situation results in which petrochemical plants have become exporters of cogenerated power to utilities. The gas turbine also has advantages that are firmly rooted in thermodynamics. It utilizes energy directly at a high temperature level, without large driving forces for pressure drop and temperature difference.

Most gas turbine applications in the chemical industry are tied to the steam cycle, but the turbines can be integrated anywhere in the process where there is a large requirement for fired fuel. An example is the use of the heat in the gas turbine exhaust as preheated air for ethylene cracking furnaces as shown in Figure 4 (8).

The combined cycle is also applicable to dedicated power production. When the steam from the waste heat boiler is fed to a condensing turbine, overall conversion efficiencies of fuel to electricity in excess of 50% can be achieved. A few public utility power plants use this cycle, but in general utilities have been slow to convert to gas turbines. Most electricity is generated by the cycle shown in Figures 5**d** and 6**d**.

Power Recovery in Other Systems. Steam is by far the biggest opportunity for power recovery from pressure letdown, but others such as tailgas expanders in nitric acid plants (Fig. 1) and on catalytic crackers, also exist. An example of power recovery in liquid systems, is the letdown of the high pressure, rich absorbent used for H_2S/CO_2 removal in NH_3 plants. Letdown can occur in a turbine directly coupled to the pump used to boost the lean absorbent back to the absorber pressure.

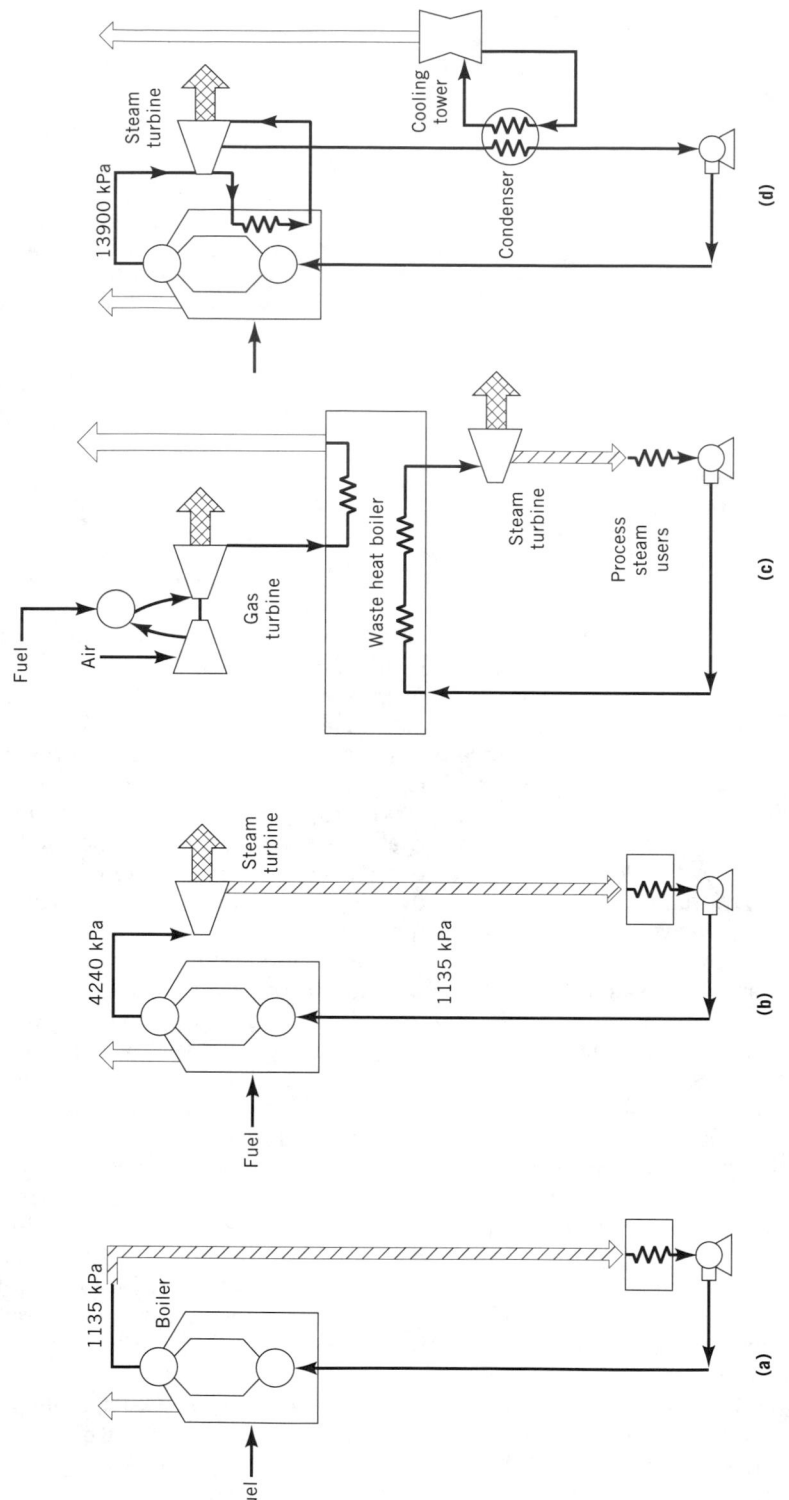

Figure 6. Schematics showing the principal equipment components for the energy systems shown in Figure 5. (**a**) through (**d**) correspond to Figure 5**a** through 5**d**, respectively. To convert kPa to psi, multiply by 0.145.

Heat Recovery, Energy Balances, and Heat-Exchange Networks

The goal of heat recovery is to be sure that energy does the maximum useful work as it cascades to ambient. An energy balance is a summary of all of the energy sources and all of the energy sinks for a unit operation, a process unit, or an entire manufacturing plant. Table 2 gives an energy balance for a simple propane-fired dryer. The energy balance is almost as important to understanding how a process works as the material balance. The energy balance is the basic tool for analyzing an operation for energy conservation opportunities. When incorporated into a computer program, the energy balance becomes the base for a model of the process. Operational changes, system configuration, and equipment alterations can be evaluated via the model.

The heat exchanger network analysis, sometimes called pinch technology, is a special kind of model that has been developed into a sophisticated way of attacking heat recovery problems. This type of analysis defines the optimum interchange between all the heat sinks and heat sources. It is similar to the concept used by furnace designers for many years for matching various heat sinks against the flue gas source. Using a temperature vs enthalpy plot, the analysis can be done manually or via computer. It can even be set up to adjust distillation sequence or operating pressures to improve energy efficiency and minimize capital (9).

Waste-Heat Boilers. In most chemical process plants, the steam system is the integrating energy system. Recovering waste heat by generating steam makes the heat usable in any part of the plant served by the steam system. Many waste-heat recovery boilers are unique and adapted to fit a particular process. There is a long history of process waste-heat boiler failure resulting from inadequate

Table 2. Product Dryer Analysis Heat Balance

Material	Mass, kg/h	Energy, MJ/h[a]
Inputs		
Fuel, C_3H_8	130	6,553
Air		
Combustion	6,817	106
Secondary	14,846	232
In-leakage	4,289	67
Water with product	1,731	354
Dry product solids	4,478	249
Totals in	*32,291*	*7,561*
Outputs		
Water vapor		
From product	1,445	3,808
From combustion of H_2	212	560
Air and combustion products	25,870	1,933
Dry products solids	4,478	458
Water with product out	286	146
Heat losses		656
Totals out	*32,291*	*7,561*

[a] To convert J/h to Btu/h, multiply by 0.95×10^{-3}.

attention to detail in design and the failure to maintain water quality (10). The high heat-transfer coefficients of boiling water are dependent upon clean surfaces. Designers should match the hardware as closely as possible to demonstrated designs, and operators should be sure that water treatment is monitored. Incinerators and gas turbines also involve heat recovery boilers. Here, a number of fairly standard designs have evolved (11).

Product-to-Feed Heat Interchange. Heat exchange is commonly used to cool the product of a thermal process by preheating the feed to that process, thus providing a natural stabilizing, feed forward type of process integration. Product-to-feed interchange is common on reactors as well as distillation trains.

Combustion Air Preheat. Flue gas to air exchange, a type of product-to-feed heat exchange, is extremely important because of the large loss associated with the combustion of unpreheated air. This exchange process has generated fairly unique types of hardware such as the Ljungstrom or rotary wheel regenerator shown in Figure 7a; the brick checkerwork regenerators used in metallurgical furnaces, hot oil, or hot water belts (also called "liquid runarounds") (Fig. 7b); and heat pipes (Fig. 8). Liquid runaround systems make it practical to use finned surface on both gas exchange surfaces. These are particularly useful for retrofit because of the ability to move the heat to physically separated units.

The heat pipe exchanger is a variant on the liquid runaround system where each tube (pipe) is sealed on both ends and filled with a vaporizing–condensing heat-transfer medium. At the hot end of the pipe, liquid is vaporized and moves to the cold end. At the cold end, vapor condenses and returns to the heat intake end. The flows are driven by gravity and capillary wicking. The heat pipe is particularly useful because it permits very compact, side by side ducting arrangements with countercurrent flow, as shown in Figure 8.

Boiler Economizers. Heat exchangers that use boiler flue gases to preheat the boiler feedwater are termed boiler economizers.

Heat Pumps. The use of heat pumps adds a compressor to boost the temperature level of rejected heat. It can be very effective in small plants having few opportunities for heat interchange. However, in large facilities a closer look usually shows an alternative for use of waste heat. The fuel/steam focus of energy use has led to application of heat pumps in applications where a broader examination might suggest a simpler system of heat recovery.

Heat Recovery Equipment

Factors that limit heat recovery applications are corrosion, fouling, safety, and cost of heat-exchange surface. Most heat interchange utilizes shell and tube-type units because of the rugged construction, ease of mechanical cleaning, and ease of fabrication in a variety of materials.

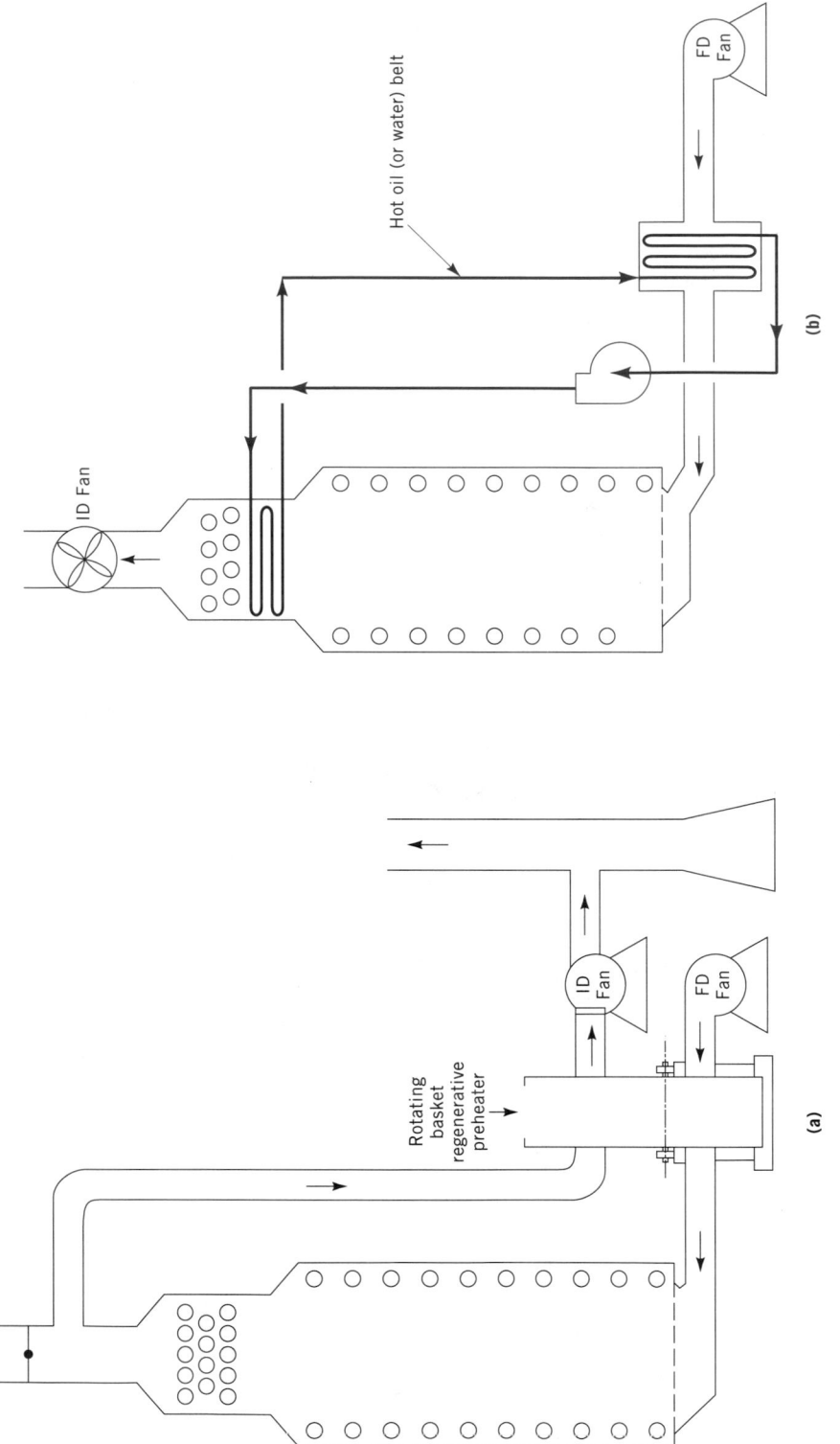

Figure 7. Air preheaters where ID = induced draft and FD = forced draft fan. (**a**) Rotating metal basket or Lungstrom regenerative preheater; and (**b**) hot oil or water belt (liquid runaround) used to move convection section heat to air preheater in furnace retrofit.

Hot oil (or water) belt

ID Fan

FD Fan

(b)

Rotating basket regenerative preheater

ID Fan

FD Fan

(a)

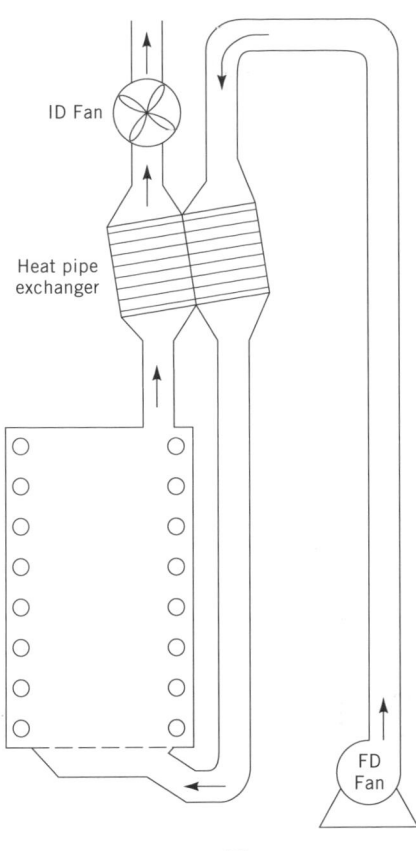

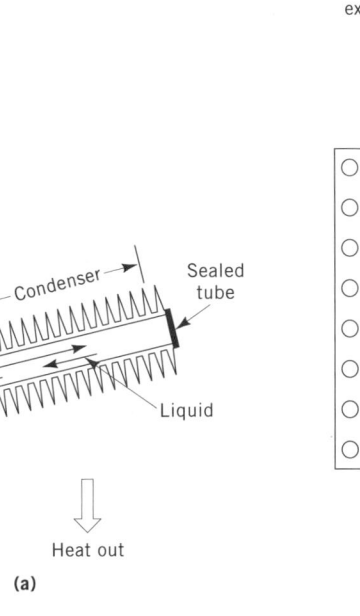

Figure 8. (**a**) Heat pipe showing the use of finned tubing for both heating and cooling; and (**b**), heat pipe exchanger air heater system. ID = induced draft and FD = forced draft fan.

However, there is a rich assortment of other heat exchangers. Examples found in chemical plants in special applications include the following.

Plate heat exchangers are made by sandwiching thin sheets of metal that have a corrugated pattern pressed into them. The corrugations provide mechanical support where the sheets contact each other, permitting compact, low cost construction and generating high turbulence and high heat-transfer coefficients. These plates are normally separated by gaskets, and are particularly useful in food processing, where cleaning is facilitated. However, the gaskets are the mechanical weak link, and limit the application in chemical plants to nontoxic, noninflammable fluids.

Brazed-fin aluminum cores are made of aluminum sheets having corrugated, cut layers sandwiched between flat sheets. The package is brazed together to form a very compact unit. The cut corrugated layers act as fins for both sides of the exchanger. The brazed aluminum construction needs to be protected from fire by special insulation or a coldbox filled with perlite. These units are used almost exclusively for very clean, cryogenic services such as air separation or hydrogen purification.

In spiral plate construction, two plates are welded together and rolled into a jelly-roll shape. The prime advantage is that there is a single flow passage. Any pluggage generates a high local pressure drop and tends to erode the deposit. Thus the unit is less subject to pluggage than in the other constructions having parallel flow paths.

UTILITY SYSTEMS

Steam

The steam system serves as the integrating energy system in most chemical process plants. Steam holds this unique position because it is an excellent heat-transfer medium over a wide range of temperatures. Water gives high heat-transfer coefficients whether in liquid phase, boiling, or in condensation. In addition, water is safe, nonpolluting, and if proper water treatment is maintained, noncorrosive to carbon steel.

Steam Balances. The steam balance is usually the most important plantwide energy balance. It shows each service requirement, including the use of steam as a working fluid to develop power.

There are, however, some limitations. A steam balance depicts the steam flows at a given point in time or as an average over some period. The steam flow on a cold weekday morning in winter is quite different from that on a Sunday in summer, and neither matches the annual average steam balance. Startup flows are also usually far different and merit their own special balance. It is also wise to prepare a balance for the beginning and the end of the cycle between unit shutdowns. For example, the power required by a turbine driving a compressor rises as the compressor efficiency falls, and process heating requirements rise as interchangers foul. Use of a computer-based model

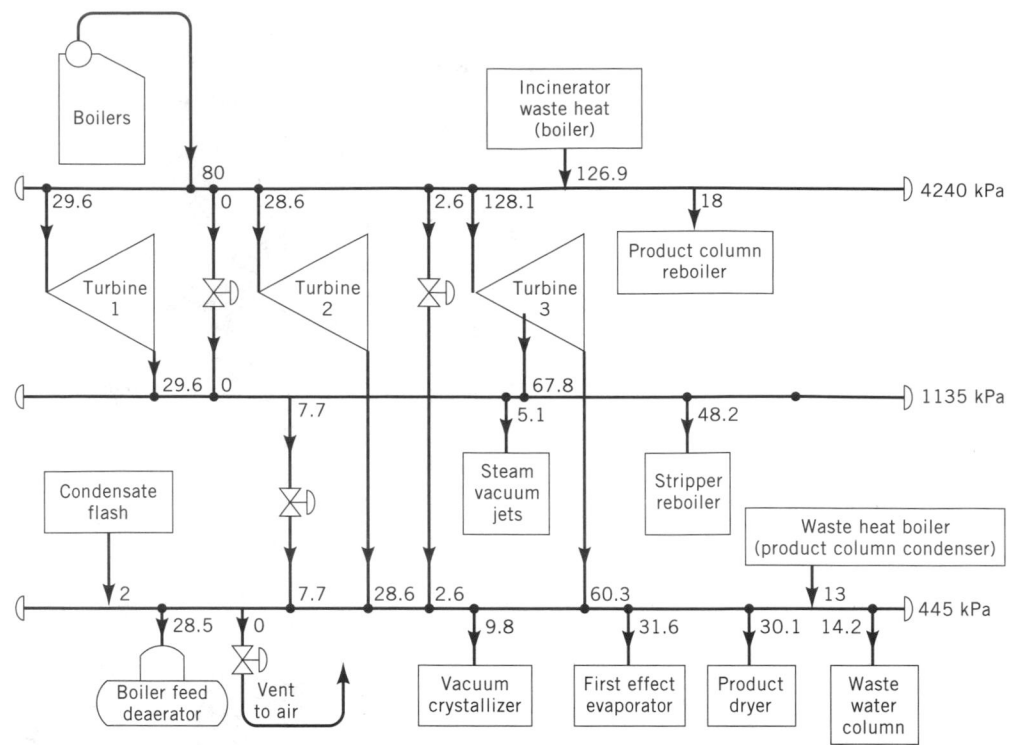

Figure 9. Schematic steam balance where the numbers represent steam flows in metric tons per hour. See Table 3.

permits easy adaptation to the calendar and onstream cycle.

There are two ways of presenting steam balance data, schematically or tabularly. For both presentation types, a balance is made at each pressure level. In a schematic balance, such as that shown in Figure 9, horizontal lines are drawn for each pressure. The steam-using equipment is shown between the lines, and individual flows are shown vertically. Table 3 contains the same data as shown in Figure 9. In both cases the steam balance has

Table 3. Steam Balance, Flows in Metric Tons per Hour[a]

Equipment	4240 kPa[b]		1135 kPa[b]		445 kPa[c]	
	Supply	Use	Supply	Use	Supply	Use
Boilers	80.0					
Incinerator waste heat	126.9					
Turbine 1		29.6	29.6			
Turbine 2		28.6			28.6	
Turbine 3		128.1	67.8		60.3	
Pressure reducing valve		0	0			
Pressure reducing valve		2.6			2.6	
Pressure reducing valve				7.7	7.7	
Product column reboiler		18.0				
Steam vacuum jets				5.1		
Stripper reboiler				48.2		
Finishing column reboiler				36.4		
Condensate flash					2.0	
Product column condenser						
Waste-heat boiler					13.0	
Vacuum crystallizer						9.8
1st effect evaporator						31.6
Product dryer						30.1
Wastewater column						14.2
Boiler feed deaerator						28.5
Total	*206.9*	*206.9*	*97.4*	*97.4*	*114.2*	*114.2*

[a] See Figure 9.
[b] To convert kPa to psi, multiply by 0.145.

been simplified to show only mass flows. A separate balance should be developed that identifies energy flows, including heat losses and power extraction from the turbines.

Steam Turbines. Historically, back-pressure steam turbines were used as drives throughout processes to increase reliability to cover electrical power failures. A typical turbine would be a single-stage 375 kW machine having throttle steam at 4240 kPa (600 psig) and exhaust at 1135 kPa (150 psig). The turbine would be controlled by a centrifugal fly-weight governor operating a single throttle valve. The efficiency would be about 40% when operated at rated conditions; ie, for the amount of steam passing through the turbine, it would develop 40% of the power that could be developed by an ideal turbine, expanding the steam isentropically. The efficiency was substantially lower when the machine was operated at part load, because a large portion of the pressure drop at part load was lost across the throttling valve, producing no work.

Because of increased emphasis on maximizing cogenerated power, newer plants are trying to utilize back-pressure turbines only in applications where efficiencies above 70% can be attained. This typically means limiting the applications to the large (>1000 kW) drives, and using small machines only where they are necessary for the safe shutdown of the unit. Multistage turbines are used even on the smaller loads.

Most large plants also have some condensing turbines to handle process and seasonal swings and provide some flexibility to the steam balance. For large (>15,000 kW) applications, condensing turbines can compete with purchased electricity. For small applications, power can be provided at much lower cost by motors. Condensing turbines generally have high reliability, and are also used where the costs of electrical power failure, in process downtime, are high. Public utility plants usually have condensing turbines at the bottom of a power cycle as shown by Figure 6**d**.

Condensate Return Systems. In a process plant, steam traps are used to drain and return condensate. Given proper application and continuous maintenance, these can operate with minimal steam leakage. Correct installation is also important (12).

For draining principal items of process equipment, level-controlled condensate chambers provide much better performance and reliability than steam traps. Usage is generally justified when condensate flow is greater than 4500 kg/h.

Electrical

The plant electrical system is sometimes more important than the steam system. The electrical system consists of the utility company's entry substation, any in-plant generating equipment, primary distribution feeders, secondary substations and transformers, final distribution cables, and various items of switch-gear, protective relays, motors starters, motors, lighting control panels, and capacitors to adjust power factor.

Electric Motors. Except for electrolytic, eg, chlorine production electric furnace processes, eg, phosphorus, typically 95% of the electricity used in a chemical plant is for electric motor drives. Induction electric motors in general use in chemical process plants range in efficiency from 90 to 95% depending primarily on size. The larger motors are more efficient. For any size, a range of efficiencies is available and high efficiency motors are somewhat more expensive than standard ones as shown in Figure 10. This price increment is normally justified in chemical process plants because of the high number of annual operating hours (13). For heating and ventilationg operations, the lower operating hours sometimes make the lower efficiency units the cost-effective choice.

Variable Frequency Drives. An important energy by-product of solid-state electronics is the relatively low cost variable speed drive. These electronic devices adjust the frequency of current to control motor speed such that a pump can be controlled directly to deliver the right flow without the need for a control valve and its inherent pressure drop. Figure 11 shows that at rated load the variable speed drive uses only about 70% as much power as a standard throttle control valve system, and at half load, it uses only about 25% as much power.

In addition to energy conservation, the variable speed drives offer better control because of a faster response, ie, reduced dead band. They are also sometimes chosen for safety reasons because of elimination of the control sta-

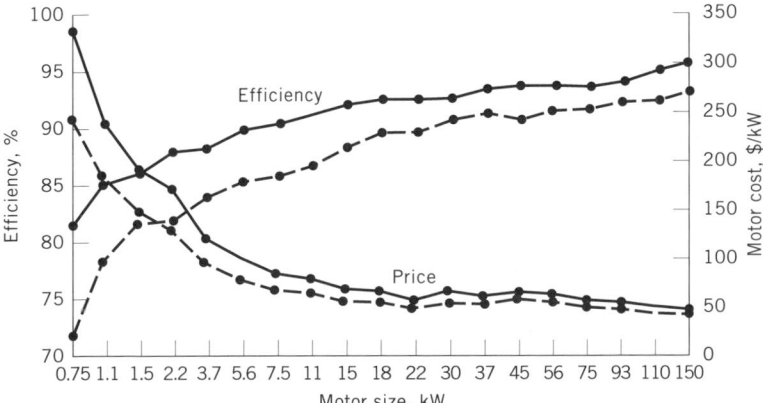

Figure 10. Full-load efficiencies and wholesale prices (in 1988 U.S. dollars) of (---) standard and (—) energy efficient (EEM) three-phase motors (13).

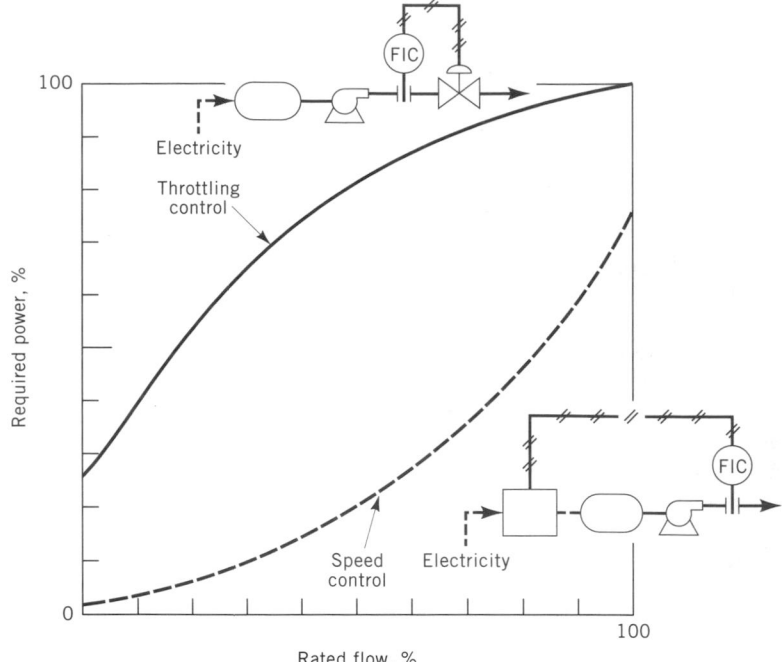

Figure 11. Power saving for variable speed drives. Power input for variable speed adjusts with flow to naturally match the frictional losses. FIC = flow indicating controller.

tion and accompanying valving. The capital saved by use of a smaller motor and elimination of the control valve partially compensates for the cost of the drive.

Lighting. High pressure sodium or metal halide lights offer significant savings over mercury vapor for process areas. Fluorescent lights have an even greater savings over incandescent lights for areas having low (mounting height <4 m) ceilings. Some other features that minimize electrical use in lighting are circuit arrangement so that unneeded lights can be turned off during non-use hours, and efficient reflectors and ballasts.

Other Energy Systems

Chemical plants usually require cooling water, compressed air, and fuel distribution systems. Sometimes also included are refrigeration, pressurized hot water, or specialized heat-transfer fluids such as Therminol liquid or condensing vapor. Each of these systems serves the process and reliability is the most important characteristic. Thus a project in any of them that achieves a 10% reduction in energy cost at the expense of a 1% loss of reliability loses money for the operation.

Cooling Water. The primary reliability concern is that water chemistry must be maintained in a low fouling, noncorroding regime. In addition, water flow velocity must be maintained above a certain threshold (ca 0.5 m/s in tubeside flow) to avoid fouling and corrosion.

The principal energy cost concern is avoidance of flows greatly in excess of need. Some cooling water systems are operated at such low temperature rises and high flows that pumping power cost equals the cost of the original heat input that it rejects. Care should be taken in design to assure that the system is balanced, and all heat exchangers utilize available pressure drop with the design flow. Every user should have a thermowell in the cooling

water outlet to monitor temperature rise, and enough flow and temperature measuring elements should be provided to check overall heat balances (see TEMPERATURE MEASUREMENT).

Compressed Air. Enough flow elements should be provided to permit accurate assessment of use by different operating departments within a plant. Compressed air is often unmetered; thus there is little motivation to reduce use. A large fraction is often lost through leakage at fittings.

A key question for energy reduction is whether a lower discharge pressure can be used. Dropping pressure from 790 kPa (100 psig) to 650 kPa (80 psig) reduces the required power 12%, and reduces the driving force for air leakage losses by 20%. Controls such as inlet guidevanes on the air compressor can be provided to trim pressure to the required level. Another action that lowers energy use is lowering inlet and interstage temperatures. A drop from 30 to 20°C typically reduces power by 3%.

Refrigeration. In processes such as olefin separations, the economic importance of refrigeration exceeds that of the steam system. Refrigeration is an extremely valuable utility because of the work required to raise the energy to ambient temperature. Its value goes up directly as the temperature gap between ambient and use level goes up. For example, whereas refrigeration at −25°C is worth approximately as much as heat at 250°C, refrigeration at −75°C is worth twice as much.

A series of energy audit questions asking what can be done to reduce the work of raising this heat to ambient, or whether refrigeration is really necessary, should be addressed. Energy costs are cut if cooling water can be used directly or used for part of the year or for part of the load. An energy savings can also be realized if a higher refrigeration temperature can be used. If the refrigeration need is above 5°C, and there is waste heat available above

90°C, an absorption refrigeration system can be used to supply refrigeration.

The optimization of heat-transfer surfaces also plays a role. At the optimum, the lifetime cost of a surface is approximately equal in value to the lifetime cost of power used to overcome the temperature differential in the condenser and evaporator. Additionally, condensation on insulation is a sign of questionable insulation (see INSULATION, THERMAL). Frost is a certain signal that insulation can be improved.

Condensing Organic Vapor. The eutectic mixture of diphenyl and diphenyl oxide is an excellent vapor medium for precise temperature control at temperatures higher than those practical using steam. This mixture can achieve 315°C while holding pressure at 304 kPa (3 atm) absolute. In contrast, steam would require 10.6 MPa (105 atm) pressure.

These systems, commercially known as Therminol VP or Dowtherm A, differ from steam in some key areas which can result in operating problems unless handled properly in design (14). The low pressure–high temperature operation means that the $\Delta T / \Delta P$ ratio at saturation is quite high; for example, at 315°C the ratio is 25 times that of steam. This means that a pressure drop that would be nominal in a steam system (10 kpa (0.1 atm)), can not be tolerated if precise temperature control is needed.

Another difference is that molecular weight is much higher than that of the common noncondensables, and hence the noncondensables arc harder to purge. In contrast, in a steam system almost all noncondensables are heavier than steam and tend to flush out with the condensate.

CAPITAL AND EQUIPMENT AREAS

Virtually all chemical processing is energy driven, but in separations such as distillation, drying, and evaporation, this is particularly clear. All three of these processes are simple thermal operations that involve separation through vaporization, and only a minor change in the chemical energy of the products. The capital related costs of those operations are typically three to five times as large on a lifetime basis as the energy costs. In almost all cases there is a balance between capital and energy costs, and typically one is traded against the other to achieve the lowest overall cost. Much can be said about this energy/capital trade based on first principles and engineering correlations (15).

Insulation

A surprisingly important capital element of energy management is insulation. On large projects the capital cost of insulation is in the same range as that for heat exchangers or distillation towers and trays. At the optimum insulation thickness, the lifetime value of the insulation approximates the lifetime value of the heat loss; ie, insulation is as costly as the heat loss that it prevents. Uninsulated flanges, when they exist, are a particularly severe loss point, and when flanges need to be opened periodically, insulation via removable blankets is usually justified.

Insulation provides other functions in addition to energy conservation. A key role for insulation is safety. It protects personnel from burns and minimizes hot surfaces that could ignite inflammables. It also protects equipment, piping, and contents in event of fire. Thus materials such as mineral wool are sometimes used despite relatively poor thermal qualities.

Corrosion under insulation is also a concern, particularly in refrigeration systems. The specification of the insulation system needs to include painting, vapor barriers, and external metal jackets (16).

Compressors

Compression equipment accounts for a large fraction of power use as well as a large fraction of installed capital. Usually the energy bill for a compressor is large enough to warrant a very visible monitor of the driver, such as a control room electric meter. Testing programs to ensure operation of the compressor and its driver at peak energy efficiency are also justified. Temperature rise across a compressor is a simple and effective way to monitor efficiency.

Compressors are relatively fragile, precision machines, and require more care in specification and maintenance than any other part of the process. The largest process compressors are axial flow machines, for example the air compressors for gas turbines. Centrifugal machines can also handle large volumes, for example the cracked gas compressors in ethylene plants. Centrifugal machines are a bit less efficient, but are more rugged and tolerant of fouling service. Smaller volume compressors are reciprocating or rotary designs.

Energy consumption is but one of the selection criteria. For example, a reciprocating compressor usually delivers 5 to 25% higher efficiency than a centrifugal machine. In the size range where a single unit compressor can handle the flow, this usually pays for increased maintenance, but it rarely justifies the increased capital of parallel units in competition with a larger single train centrifugal.

A compressor is typically a specially designed device, and comes with far less surplus capacity than other process components. As a result compressors merit great care in specification of flow, inlet pressure, and discharge pressure. Similarly, the control system and equipment need to be carefully matched to provide turndown with maximum efficiency.

Because of the large volumetric flow inherent in gases, the cost of power for incremental pressure drop is high. To a first approximation, incremental power is proportional to the volumetric flow of suction or discharge, multiplied by incremental pressure drop; hence the high importance of pressure drop otpimization for the associated piping and heat exchangers. Volumetric flow also varies with the absolute value of temperature; hence suction cooling is another area for optimization. The drive for lower temperatures also provides the incentive for adding compression stages, with intercooling between stages.

Pumps

Energy use for pumps can best be controlled by design for the proper flow and discharge pressure. Constant speed electric motor driven pumps, having a large margin of safety on flow, are particularly wasteful. One solution is use of the variable speed drive. Another solution in an operating unit is trimming of oversized impellers.

Vacuum Systems

The basic question in vacuum systems is what can be done to cut design inert loading. Historically, inert flows were whimsically overspecified, and the vacuum systems were run until mechanical deterioration brought the system capacity down to the actual inert loading. One factor driving toward greater use of vacuum pumps is the large reduction they achieve in effluent to be treated.

Boilers and Process Furnaces

Boilers and process fired heaters are the entry point for the energy released from burning fuel. Fuel combustion is irreversible, and fired heaters are typically the principal loss point for work potential (6). The high irreversibility results from taking the chemical energy of fuel and degrading it to heat. Air preheat cuts energy losses by cutting fuel firing and increasing the flame burst temperature.

A more obvious energy loss is the heat to the stack flue gases. The sensible heat losses can be minimized by reduced total air flow, ie, low excess air operation. Flue gas losses are also minimized by lowering the discharge temperature via increased heat recovery in economizers, air preheaters, etc. When fuels containing sulfur are burned, the final exit flue gas temperature is usually not permitted to go below about 100°C because of several problems relating to sulfuric acid corrosion. Special economizers having Teflon-coated tubes permit lower temperatures but are not commonly used.

Inadequate mixing of air and fuel can result in unburned combustibles in the flue gases. This results in energy losses, environmental problems, and damage from afterburning in the convection section. Heat leakage through refractories constitutes still another type of energy loss from combustion equipment. Because of the high temperature, the heat leakage is higher than on most equipment and can represent as much as 3% of fired fuel. On newer designs this value is typically closer to 1%.

Distillation

Reflux Rate. The optimum reflux rate for a distillation column depends on the value of energy, but is generally between 1.05 times and 1.25 times the reflux rate, which could be used with infinite trays. At this level, excess reflux is a secondary contributor to column inefficiency. However, when designing to this tolerance, correct vapor–liquid equilibrium data and adequate controls are essential.

Control. Energy savings for improved control are surprisingly high. A 2 to 20% savings from a series of control projects has been reported (17).

A reflux reduction of 15% is typical. Improved control achieves this by permitting a reduction in the margin of safety that the operators use to handle changes in feed conditions. The key element is the addition of feed-forward capability, which automatically handles changes

in feed flow and composition. One of the reasons for increased use of features such as feed-forward control is the reduced cost of computers and online analyzers.

Reboilers and Condensers. The real work used in a distillation varies with the temperature difference between the heating medium and the cooling medium. Part of this differential is the difference in boiling points between the overhead stream and the bottom stream. However, a large portion often results from the temperature difference used in the reboiler and condenser. The optimum differential is generally under 20°C, and if refrigeration is used, the optimum can be as small as 3°C. A signal that an excessive temperature difference may exist is a condenser or reboiler having a shell diameter less than one-third the diameter of the column.

There are a number of ways to provide the heating or cooling medium at temperatures closer to the optimum level. One is by use of double-effect distillation, which uses the overhead vapor from one column as the heat source for another column such that the second column's reboiler becomes the first column's condenser. This basically cuts the temperature differential in half, and shows up as an energy saving because external heat is supplied to only one of the units.

Column Pressure Drop. Another element that sets the temperature differential across the distillation is the pressure drop in the column and its auxiliaries. One of the more recent changes is the introduction of special structured packings (see DISTILLATION) which give extremely low (10% of an equivalent column with trays) pressure drop. This energy benefit can show up in an overhead temperature high enough to permit generation of by-product steam. It can also show up in a variety of other ways including lower bottoms temperature, yielding less fouling and product degradation to by-products, as in the styrene-ethylbenzene separation.

The Overall Process. Each distillation column should be examined in context with the rest of the process as well as by itself. For example, an opportunity for energy saving may be a reduction of process recycles via a purer overhead or bottoms stream. Lowering or raising column pressure to facilitate interchange with other parts of the process is another possible opportunity.

Drying

A typical dryer mass and energy balance (Fig. 9, Table 3) shows that the heat loss is 10% of the fuel input. Improving insulation is one of the simplest ways to reduce energy input. Another simple way to reduce energy input is improving the dewatering of the feed. There is a great difference in energy input for dewatering as compared to the subsequent drying step, as shown by Figure 12.

Some of the other energy conservation approaches applicable to dryers are interchange between the stack vapor and the incoming dryer air; recovering sensible heat from the product; use of waste heat from another operation for air preheat; and using less, but hotter drying air. This last is limited to nonheat-sensitive materials.

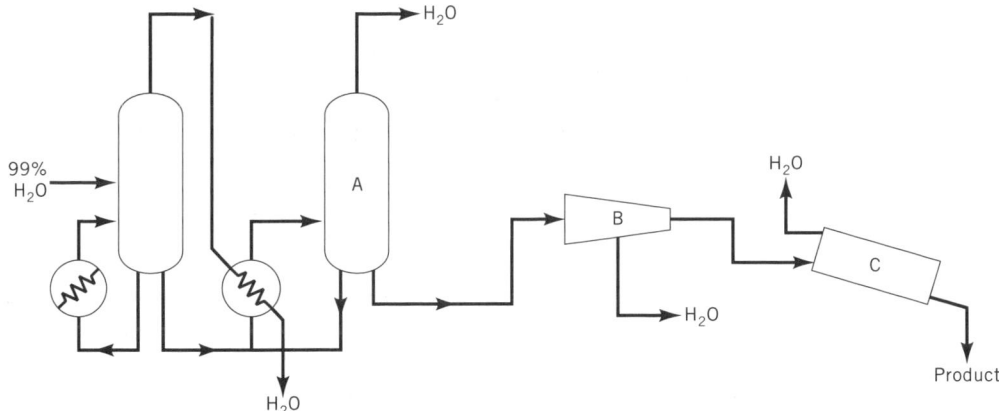

Figure 12. The relative energy input for removal of 1 kg of water relative to heat of vaporization is 0.7 in A, the double-effect crystallizer; 0.015 in B, the dewatering centrifuge; and 5 in C, the rotary dryer.

Evaporation. In most chemical industry evaporation systems, the objective is product recovery, although occasionally the objective is concentration of an organic waste from an aqueous solution, to facilitate incineration. Similar equipment is used extensively for desalination of salt or brackish water.

A single-effect evaporator produces slightly less than a kilogram of water vapor per kilogram of steam. By using the vapor produced by the first-effect as the heat source for a second-effect evaporator, steam use can be essentially halved. The performance can be improved almost in proportion to the number of effects employed. Six- and seven-effect evaporators are common in the wood pulp industry for concentration of black liquor. However, as the number of effects goes up, the temperature driving force is spread over the additional units and the capital cost increases almost in direct proportion to the number of effects. As a consequence, high alloy systems are often limited to single- or double-effect.

Much as reverse osmosis can compete with evaporation in desalination applications, osmosis should also be considered as an alternative for process evaporation. Reverse osmosis is particularly attractive where the inlet stream is greater than 99% water.

ENERGY ACCOUNTING AND IMPROVEMENT

Energy Accounting

Long-term costs at plant gate meters are costs that should be considered in project evaluations. The real benefit of a technical action can be quite different from the savings allocated by the accounting system. The benefit also depends on other changes that happen at the plant site. For example, the incremental value of steam may be near zero because of production from waste heat boilers and utility boilers operating at minimum load. Prudent planning credits a steam saving project based on the probable plant energy balance during the project's operation rather than on current allocated cost.

The incremental cost of electricity is usually dictated by contract. Incremental cost is normally less than the average cost because most utilities have descending rate scales for high volume industrial customers and also be-

cause there is a large fixed charged associated with past peak demand. An important factor in the control of operations is recognizing that the cost of electricity is typically one-third for peak demand and two-thirds for the actual incremental use. The demand charge is typically set by the peak usage in a month or 12 month period. The demand charge recognizes that a large component in the cost of electricity is the capital used by the public utility to generate and transmit power. Many chemical plants have load shedding systems to avoid setting a higher peak and increasing demand charges. Usually the shed load is for nonprocess applications such as battery charging of forklift trucks, but sometimes large users such as electric furnaces need to be curtailed as well.

Because it is so readily measurable, energy conservation offers a unique opportunity for continuous improvement. Energy conservation is something that individuals want to do, and given the opportunity to see ideas converted into measurable progress, personnel pursue conservation with enthusiasm, generating larger savings than deemed possible (18).

Measurements and Audits

The enabling element of continuous improvement is measurement. An old rule of thumb says that increased accuracy in measuring an energy use ultimately yields a reduction in use equal to 10% of the increased closure of the balance. A basic principle of economics is that any thing that is free is used in excess, ie, an unmetered electrical use is bigger than expected by at least 10%. Metering of the cost elements at each unit in a chemical plant provides effective accountability. Measurements should be linked via computer software to production as well as to weather to result in maximum feedback.

Reprinted from the *Kirk-Othmer Encyclopedia of Chemical Technology,* 4th ed., Vol. 9, John Wiley & Sons, New York, 1994.

BIBLIOGRAPHY

"Energy Management" in *ECT* 3rd ed., Vol. 9, pp. 21–45, by A. F. Waterland, Waterland, Viar & Associates, Inc.

1. *Manufacturing Energy Consumption Survey: 1988*, U.S. Department of Energy, Washington, D.C., 1991.
2. *U.S. Chemical Industry Statistical Handbook*, Chemical Manufacturers Association, Washington, D.C., 1991.
3. W. M. Weiss, personal communication, Monsanto Enviro-Chem Systems Inc., St. Louis, Mo., 1991.
4. M. H. Ross and D. E. Steinmeyer, *Sci. Am.* **163**, 88 (1990).
5. *Manufacturing Energy Consumption Survey: 1985*, U.S. Department of Energy, Washington, D.C., 1988.
6. W. F. Kenney, *Energy Conservation in the Process Industries*, Academic Press, Inc., New York, 1984.
7. J. M. Kovacek, *Cogeneration Application Considerations*, General Electric Co., Schenectady, N.Y., GER-3430B, 1991.
8. W. F. Kenney, "Combustion Air Preheat on Steam Cracker Furnaces," *Proceedings, 1983 Industrial Energy Conservation Technology Conference*, Texas Industrial Commission, p. 595.
9. R. Smith and B. Linnhoff, *Chem. Eng. Res. Des.* **66**, 195 (1988).
10. P. S. Gupton and A. S. Krisher, *Chem. Eng. Prog.* **69**(1), 47 (1973).
11. V. Ganapathy, *Waste Heat Boiler Deskbook*, Prentice Hall, Inc., Englewood Cliffs, N.J., 1991.
12. F. S. Pychewicz, "Steam Traps—The Oft Forgotten Energy Conservation Treasure," *Proceedings, 1985 Industrial Energy Conservation Technology Conference*, Texas Industrial Commission, Houston, Tex., p. 392.
13. E. D. Larson and L. J. Nilsson, *ASHRAE Trans.* **97**(2), 363 (1991).
14. D. R. Frikken, K. S. Rosenberg, and D. E. Steinmeyer, *Chem. Eng.*, 86 (June 9, 1975).
15. D. E. Steinmeyer, *CHEMTECH* 188 (Mar. 1982).
16. L. G. Britton and H. G. Clem, *Chem. Eng. Prog.* **87**(11), 87 (1991).
17. T. L. Tolliver, *Chem. Eng.*, 99 (Nov. 24, 1986).
18. K. Nelson, *Chem. Proc.*, 77 (Jan. 1989).

ENERGY POLICY PLANNING (UNITED STATES)

Amy R. Maron
William Lanouette
Gary R. Boss
General Accounting Office

For decades, the federal government has attempted to develop a national energy policy. However, until 1977 when the Department of Energy (DOE) was established, the federal government did not have a framework for consolidating its efforts on energy matters. By the end of the decade, the executive branch's authority over energy and natural resources had expanded, and analytic and forecasting capabilities had been introduced. In addition, the first national energy plan had been developed and a formalized biennial process for planning energy policy had been established under title VIII of the Department of Energy Organization Act (1). See also ENERGY TAXATION; ENVIRONMENTAL ECONOMICS.

OBJECTIVES OF THE NATIONAL ENERGY POLICY PLAN

The Department of Energy Organization Act, which created the Department of Energy, also provided a mechanism through which a coordinated national energy policy could be formulated and implemented. This mechanism, in title VIII of the act, requires that the President submit a National Energy Policy Plan (NEPP) to the Congress by April 1 every 2 years, beginning in 1979. Title VIII called for the use of more than 50 kinds of information in developing the plan. The 5 most prominent elements, which incorporate many of these 50 provisions, are

- Holding public hearings to seek the views of a range of interests, including regional, state, and local governments and the private sector.
- Establishing 5- and 10-year energy production, utilization, and conservation objectives that pay particular attention to the need for full employment, price stability, energy security, economic growth, environmental protection, and nuclear nonproliferation as well as special regional needs and the efficient utilization of public and private resources.
- Identifying strategies to be followed to achieve the objectives and outlining the appropriate policies and actions of the federal government to maximize private production and investment in each significant energy supply sector.
- Estimating the domestic and foreign energy supplies on which the United States will be expected to rely and evaluating current and foreseeable trends in the price, quality, management, and utilization of energy resources and the effects of those trends on the social, economic, environmental, and other requirements of the nation; and
- Submitting whatever data and analyses are necessary to support the objectives, resource needs, and policy recommendations of the plan.

By calling on administrations to set objectives; identify the strategies needed to achieve the objectives; project energy supply, demand, and prices; provide the data and analyses to support goals and strategies; and invite public input throughout, the Congress sought a comprehensive approach that uses the most timely and relevant information available. At the time DOE was created, the federal government was not required by law to develop a comprehensive energy policy, nor was much of the information needed to achieve such a task available. In total, there have been six submissions under title VIII: one by the Carter administration (in 1979), four by the Reagan administration (in 1981, 1983, 1985, and 1987), and one by the Bush administration in 1991.

In late 1992, the Congress enacted the Energy Policy Act of 1992 (P.L. 102-486). The act modified title VIII to require that future plans also include a "least-cost energy strategy." This new provision entails developing, among other things, inventories and estimates of the costs of energy resources, preparing a program to ensure adequate supplies of energy, and estimating the life-cycle costs of existing energy production facilities. The act also requires the Secretary of Energy to prepare a report that assesses, among other things, the feasibility of stabilizing and/or reducing greenhouse gases; the extent to which the

United States is responding, compared with other countries, to recommendations made by the National Academy of Sciences in its 1991 report *Policy Implications of Greenhouse Warming*; the feasibility of complying with recent international agreements on reducing greenhouse gases; and the potential impacts of policies needed to comply with these agreements. As part of the least-cost energy analysis, the Secretary is required to consider this new report in connection with the plan submitted under title VIII.

ENERGY PLANNING BEFORE 1977

Comprehensive national energy planning had been considered by the federal government for 40 years. However, it was the oil price and supply shocks of the 1970s that prompted the most concerted planning and government reorganization efforts regarding energy in this century— efforts designed largely to reduce the nation's dependence on imported oil through consideration of all energy sources and through conservation. These events eventually led to the title VIII provisions in the 1977 act.

As early as 1939, a report by an Energy Resources Committee, formed as part of the Temporary National Economic Committee, noted that all energy resources are closely interrelated and require the systematic attention of government (2). The report recommended that a "national energy resources policy" be prepared that would address these interrelationships rather than focusing on a fuel-by-fuel approach. Other comprehensive efforts followed, including those by the Department of the Interior and the National Security Resources Board from 1947 to 1949, the 1950–1952 Materials Policy Commission (known as the Paley Commission), the 1955 Cabinet Committee on Energy Supplies and Resources Policy, the U.S. Senate's 1961 National Fuels and Energy Study, and the Johnson administration's 1964 *Resource Policies for a Great Society: Report to the President by the Task Force on Natural Resources.*

Those who interacted with the federal government on energy matters found a lack of cohesion in the federal structure. For example, an oil executive said in 1972 that "the present dispersion of effort among some 61 different government agencies creates delays and confusion and will inevitably tend to accentuate whatever energy shortages may lie ahead." (3) While complaining about fragmentation and inefficiency, this executive acknowledged a strong energy policy role for the executive branch. In response to growing concerns about the dispersion of energy functions, President Nixon outlined a "comprehensive, integrated national energy policy" in April 1973 and in June established an Energy Policy Office to consolidate energy policy planning in the White House.

By the early 1970s, energy policy had crossed a watershed when, for the first time, supplies were inadequate to meet demand at prevailing energy prices. The role of the federal government shifted from support for energy-production industries to the regulation of increasingly (and temporarily) scarce and costly resources, particularly oil. By the spring of 1974, the Arab oil embargo that followed the November 1973 Arab-Israeli war had caused oil prices to climb to four times their level at the beginning of 1973. In early 1974, the Nixon administration launched "Project Independence," a research and development plan aimed at greater energy self-sufficiency for the nation.

During the same decade, increased concern about whether the nation would have adequate energy supplies at affordable prices to meet growing demand also prompted interest in employing economic and engineering analysis to forecast future energy needs and assist with policy decisions. In 1974, the Federal Energy Administration was created. Economic modelers at the Federal Energy Administration developed an 800-page draft *Project Independence Report*, using an analytical approach that "marked a turning point in the assessment of policy options" (4). Later that year, President Ford created an Energy Resources Council charged with developing a "single national energy policy and program." In November 1974, the final *Project Independence Report* was released.

In an April 1977 speech, President Carter declared the policies necessary for increasing energy self-sufficiency to be the "moral equivalent of war." Within three months of taking office, his administration had prepared the "National Energy Plan," (often referred to as "NEP I") (5). The plan was developed by a presidential team of economists, public administrators, and attorneys and was accompanied by the legislation to implement it.

In August of 1977, the Congress enacted the Department of Energy Organization Act, creating a single cabinet-level department responsible for overall coordination of energy programs. The act also established, in title VIII, a governmental planning function as a permanent mechanism for developing and implementing a national energy policy.

Under title VIII, the biennial energy plan is submitted by the President, while the Secretary of Energy is the principal adviser on formulating the plan and the principal agency head who defends the plan before the Congress. The Senate report explaining title VIII's planning provisions said that the plan should integrate the many disparate viewpoints within the executive branch. For all title VIII submissions, the Department of Energy has played a leading role as the integrator of views within the executive branch.

The 1977 National Energy Plan served as a kind of model for title VIII because it incorporated many of the criteria and exercises later codified in the statute. For example, it set the following goals for 1985: to reduce the annual growth rate of energy demand to below 2%, reduce gasoline consumption by 10%, cut U.S. oil imports from 16 to 6 million barrels a day, establish a 1-billion barrel Strategic Petroleum Reserve in the United States to protect against supply disruptions, increase coal production by two-thirds, insulate to minimum energy efficiency standards 90% of American homes and all new buildings, and use solar energy in more than 2.5 million homes.

This strategy reflected a time when many in the administration and the Congress believed that energy policy needed greater federal planning and control. Domestic oil production had been steadily declining from 1972 to 1976, while imports as a percent of consumption had risen from almost 29% to almost 42% during the same period. Two weeks before the 1977 plan was submitted to the Congress, the President had declared that the "energy crisis" was the "greatest challenge that our country will face during our lifetime." The administration's centralized approach to energy problems set the tone for the title VIII provisions that were subsequently enacted.

NATIONAL ENERGY PLANS HAVE VARIED SIGNIFICANTLY

The extent to which past energy plans conformed with the provisions of title VIII varied significantly from administration to administration. None of the energy plans submitted fully addressed the statute's provisions; the 1979 and 1991 plans represented the most thorough efforts to incorporate title VIII's provisions. Most plans contained strategies for achieving general goals, and all discussed energy trends, although none set 5- and 10-year objectives specified in the act. Furthermore, few plans offered analysis supporting their statements. Public hearings to solicit a range of input were held for all but one of the plans.

The primary factors that influenced how the administrations approached planning were (1) their differing views on the role of government in energy planning, supply, and price regulation and (2) developments in the nation's energy situation. Table 1 illustrates the wide variation in how the plans conformed with title VIII's provisions.

NO PLANS SET SPECIFIC 5- AND 10-YEAR OBJECTIVES

Although title VIII calls for 5- and 10-year energy production, utilization, and conservation objectives, no plans followed this approach. The only time such objectives were established was before the law was enacted: in the 1977 National Energy Plan. As noted earlier, this plan set numerous objectives to be achieved for 1985. Each subsequent plan instead developed its own unique approach to objective-setting.

1979: National Energy Plan II

The first plan submitted under title VIII's provisions, in 1979, did not specifically set 5- and 10-year objectives for energy production, utilization, and conservation. Instead, the plan, known as NEP II, reiterated the President's 1977 vision of a long-term energy transition, declaring broad objectives for the three periods of the transition: near-term (1979–1985), mid-term (1985-2000), and long-term (2000 and beyond). The plan's near-term objective was to reduce the nation's dependence on foreign oil and vulnerability to supply interruptions. The mid-term objectives were to (1) seek to keep imports sufficiently low to protect national security and extend the time it took before world oil demand reached the limits of production capacity and (2) develop the capability to use new, higher-priced technologies as world oil prices rose. The long-term objective was to use renewable and essentially inexhaustible sources of energy to sustain the economy. The plan

Table 1. National Energy Policy Plans, 1979–1991

Title of Plan, Date Submitted, and Administration Responsible	Conformity with Title VII's Provisions				
	5- and 10-Year Energy Objectives	Strategies to Achieve Objectives	Projections of Energy Prices and Supplies	Supporting Analyses	Public Hearings
National Energy Plan II (NEP II), May 1979: Carter administration	No	Strategies to achieve general objectives only	Yes	Partial	Yes
Securing America's Energy Future: The National Energy Policy Plan (NEP III), July 1981; Reagan administration	No	Strategies to achieve general objectives only	Yes	None	Yes
The National Energy Policy Plan, Oct. 1983; Reagan administration	No	Strategies to achieve general objectives only	Yes	None	Yes
The National Energy Policy Plan, Mar. 1986;[a] Reagan administration	No	Strategies to achieve general objectives only	Yes	Partial	Yes
Energy Security: A Report to the President of the United States, Mar. 1987; Reagan administration	No	Weighed pros and cons of alternative strategies	Yes	Partial	No
The National Energy Strategy, Feb. 1991; Bush administration	No	Strategies to achieve general objectives only	Yes	Partial	Yes

[a] The plan that was due in 1985 was not submitted until 1986. However, in this article this plan is referred to as the 1985 plan.
Note: No plan was submitted for the 1989 requirement. DOE officials told us that a summary of the public hearings prepared for the 1991 plan was used to meet the 1989 requirement and issued in April 1990.

mentioned specific times for achieving only a few objectives, such as reducing oil import demand by 5 percent of the consumption in International Energy Agency countries by the end of 1979 (the result of an agreement that year by the agency's governing board) and curbing the federal government's energy use by 5% in the year ending March 31, 1980.

By the time the 1979 plan was completed, major elements of the legislative package that accompanied the 1977 National Energy Plan had been enacted. As a result, few new strategies were announced in the 1979 plan; almost all of the specific strategies in the plan represented implementation of the newly authorized programs. The plan included a few new strategies. One strategy called for developing a 10-year energy conservation plan for federal buildings. Another strategy envisioned new legislation to phase out controls on domestic crude oil prices and establish a tax on the so-called windfall profits resulting from the decontrol of oil prices. The proceeds of this tax were to be assigned, in part, to subsidize mass transit and alternative energy technologies.

1981: Securing America's Energy Future: The National Energy Policy Plan

The 1981 plan, known as NEP III, did not include 5- and 10-year energy production, utilization, or conservation objectives. Instead, it offered "guiding principles" for energy policy that included both broad objectives and specific strategies such as

- Recognizing that even though efficient displacement of imported oil is an important objective, achieving a low level of U.S. oil imports at any cost is not a major criterion for the nation's energy security and economic health.
- Cooperating with international oil partners.
- Increasing oil stockpiles against potential disruptions in world markets.
- Eliminating controls or other impediments that could discourage the private sector from dealing with disruptions efficiently should they recur.
- Implementing the President's Economic Recovery Program.
- Refocusing federal spending for energy-related purposes to cases in which the private sector is unlikely to invest.

The 1981 plan also briefly discussed other forthcoming actions, chief of which was the implementation of the President's Economic Recovery Program. To address the special needs of the poor that result from higher energy prices, the plan counted on the Economic Recovery Program to deal directly with the problems of inflation and unemployment. The plan also described as forthcoming actions a review of regulations affecting the coal and electric utility industries and improvements to the licensing process for nuclear power plants.

1983: The National Energy Policy Plan

The 1983 plan also lacked 5- and 10-year energy objectives. Instead, it contained a single objective, two strategies for pursuing that objective, and a discussion of specific federal programs and actions determined by those strategies. The plan's objective was broadly to "foster an adequate supply of energy at reasonable costs." The strategies to meet these objectives were, according to the plan, to minimize federal control and involvement while maintaining public health and safety and environmental quality, and to promote a balanced, mixed energy resource system responsive to both domestic and international market forces that would also protect U.S. national security interests.

Describing the 1983 plan's approach to title VIII's provisions on 5- and 10-year objectives, a former DOE official who worked on the 1983 plan explained that the Department made a deliberate decision not to develop a "prescriptive set of domestic regulations," but instead structured the plan as a statement of policy, rather than as a plan. The administration, the official stated, believed that market forces rather than mandates would lead to socially desirable results for energy. As an example, he pointed to the administration's early opposition to setting minimum efficiency standards for appliances. Although appliance standards were eventually set by the Congress, the administration's approach of not setting general energy standards in the plan was not challenged and continued to be used in successive plans.

1985: The National Energy Policy Plan

The 1985 plan, like the three plans before it, did not set 5- and 10-year objectives and therefore did not set specific strategies to achieve objectives. However, continuing the trend begun in 1981, the plan discussed strategies to achieve the general objective of adequate supplies of energy at reasonable costs. These strategies were similar to those described in the 1983 plan: opening up additional offshore oil and gas fields; eliminating federal price controls on domestic natural gas production; redirecting research dollars to basic, rather than applied, science; and introducing legislation to reform licensing and regulation of nuclear power plants.

1987: Energy Security

The 1987 plan, entitled *Energy Security: A Report to the President of the United States*, also did not establish 5- and 10-year objectives. According to a former DOE official, because the report was prepared with the purpose of examining the domestic and international dimensions of energy security, it was not intended as a prescriptive document containing specific 5- and 10-year energy goals or strategies. The report did not purport to make any choices but rather to show what choices could be made, the former official said.

These choices were the strategies that could be used to reduce vulnerability to supply disruptions and enhance the domestic oil industry. For each strategy, the report weighed the pros and cons. However, no specific changes to existing law were proposed along with the report. The report raised, in general terms, the administration's belief that a revised regulatory framework would allow natural gas to compete more freely with other fuels, but it discussed few specific actions for making such revisions. The report also noted that DOE would pay increasing attention to electricity policy, although no specific steps were enumerated. The report included a general review of potential changes in federal requirements and regulations to increase development and improve the transportation of coal. More specific was a discussion of a four-part initiative for nuclear power, including licensing reform, federal-industry cooperation on reactor design, rate regulation, and creation of a repository for high-level waste disposal.

1991: National Energy Strategy

The Department of Energy had intended to develop a plan to meet the 1989 requirement. But the report submitted in April 1990 was an interim report on the 1991 plan, containing only a summary of public opinion but no proposed objectives and strategies. However, the 1991 title VIII submission, known as the National Energy Strategy (NES), had an overall objective that had been set by the President in July 1989: "achieving balance among our increasing need for energy at reasonable prices, our commitment to a safer, healthier environment, our determination to maintain an economy second to none, and our goal to reduce dependence by ourselves and our friends and allies on potentially unreliable energy suppliers." The plan also had more distinct objectives and approaches for energy production, utilization, and conservation than any plan before it. These goals included diversifying sources of oil supply outside the Persian Gulf, increasing energy efficiency, and enhancing environmental quality. Although few goals were identified with a specific date for their achievement, one objective was to have an operating nuclear fusion demonstration plant by about 2025 and an operating commercial plant by about 2040.

Although no 5- and 10-year timetables were included for most of its objectives, the NES gave more attention than any previous plan to identifying specific strategies that might achieve the plan's many distinct objectives. The plan included approaches for improving electricity generation and use in commercial and residential sectors, allowing private access to oil and gas on federal lands, and increasing the ability of natural gas to compete with other fuels. The plan also listed administrative actions and legislative proposals for a number of areas, and the administration submitted a major legislative package to the Congress shortly after the NES was released.

ALL PLANS DESCRIBED ENERGY TRENDS

As set forth in title VIII, all of the plans provided estimates of energy supply and demand for the major energy sources (oil, gas, coal, nuclear, and renewable energy) and descriptions of current and foreseeable trends in world oil prices. Most plans discussed, in varying depth, the effects of these trends on the economy. Only the 1991 plan discussed the impact of energy trends on the environment.

The 1979 plan's evaluation of energy trends was contained in three statistical appendixes published during the year after the plan was released. These appendixes predicted energy supply, demand, and prices to the year 2000. The appendixes also analyzed likely macroeconomic impacts of the President's April 1979 oil price decontrol and tax proposals (on inflation, employment, the trade balance, and household expenditures on energy in 10 regions of the country and for nine income-group levels). The appendixes also included detailed input-output model calculations showing the likely effects of the new proposals on 157 sectors of the economy.

A separate series of reports in support of the 1981 plan evaluated current and foreseeable energy trends and their effects on the economic health of the nation. These reports presented the macroeconomic impacts of energy prices for the period 1980–90; described energy and economic interaction in 1973–1980; analyzed the effects of energy price changes on various income groups; and provided estimates of energy prices, consumption, and supply. All projections assumed "the continuation of existing programs and policies." The reports stated that "*since the projections take into account only policies or programs that were in effect as of June 1981, the projections should not be viewed as a statement of Administration energy goals* [italics in original]. The plan projections are a starting point, or 'base case' for evaluating the potential impacts of new energy initiatives or developments."

The 1983 plan summarized projections of oil prices, energy consumption, energy production, and primary electricity inputs. These projections were presented in a technical report, *Energy Projections to the Year 2010*. The plan stated, however, that these projections "do not necessarily represent Administration policy or the beliefs of the President or the Secretary of Energy." A second technical report in support of the plan, *Energy Activity and Its Impact upon the Economy*, analyzed the macroeconomic effects of projected high and low energy prices using two different economic models.

The 1985 plan also contained forecasts for oil prices and energy consumption and production. A technical report, *National Energy Policy Plan Projections to 2010*, summarized expected world oil prices and U.S. energy supply and demand. Like the 1979 plan, the 1985 plan showed energy trends that might result if certain elements in the administration's energy policy were accepted. These elements included comprehensive decontrol of the natural gas market, increased rates of leasing federal lands for energy development, and a focus of federal research and development on long-term development rather than on subsidized commercialization.

The 1987 plan also provided an analysis of future energy trends, but it did not describe the potential effects of proposed administrative or legislative actions. Instead, it used two general scenarios. (1) A higher world oil price, indicating less dependence on imported oil, and (2) a lower world oil price, indicating greater dependence. These scenarios were used to generate projections of energy consumption, production, and imports for 1985–1995, by both world region and end-use sector. The 1987 report contained the most comprehensive worldwide projections to date; the trends were developed through an interagency process that included analysts from DOE and other federal agencies.

The 1990 *Interim Report* on the National Energy Strategy compared future energy production, consumption, and utilization trends from the Energy Information Administration and academic and interest groups. Then, in 1991, the NES presented DOE's projections of energy prices, supply, and demand for 1990–2030. This plan also described trends in environmental emissions from energy sources. In addition, it provided possible scenarios with and without many of the proposed strategies. A technical annex published in July 1991 explained some methodologies and assumptions used for these projections.

PLANS DID NOT ALWAYS PROVIDE DATA AND ANALYSIS

Title VIII also states that the plan should be accompanied by "whatever data and analysis are necessary to support the [plan's] objectives, resource needs, and policy recommendations." Title VIII did not specify what types of data and analysis should be submitted or at what time, but the Senate report on title VIII suggested that "the relevant data and analysis necessary to demonstrate the feasibility of achieving the plan's objectives and to support the estimates made in the plan" be included in a report "accompanying the plan." The types and timing of data and analysis included varied from plan to plan.

One approach to providing data and analysis to support the plan's objectives was to project future outcomes—such as impacts on the economy, energy supply and demand, and the environment—should the proposals in the plans be adopted. This approach was used in the 1979, 1985, and 1991 plans.

Neither the 1981 nor the 1983 plan included analysis to support its purposes. But as described earlier, by projecting possible results of the administration's proposals, the 1985 plan provided the data and analyses to support some of the plan's objectives (decontrolling natural gas, leasing of federal lands, refocusing federal research and development).

Although the 1987 submission discussed a wide range of theoretical strategies, an appendix to the report presented the data and analyses used to support only one proposal under consideration: a hypothetical oil import fee. The administration rejected this proposal because its analysis showed adverse macroeconomic impacts.

In 1991, the projections in the NES that reflected proposed administration strategies concerned (1) changes in the mix of fuels used in generating electricity or in transportation, (2) changes in oil import levels, (3) the impacts on electricity prices of clean coal technologies, and (4) changes in the levels of emissions and effluent from various pollutants. The plan presented more data and analyses to support its strategies than had any previous plan. Nevertheless, the NES lacked analytical support for several of its strategies. DOE has subsequently published four technical annexes containing analytical support for the proposals. In July 1991 testimony on the process DOE used in developing this plan (6), the U.S. General Accounting Office (GAO) stated that the administration had not published alternative analyses that it had examined in developing the plan's policy options, such as those examined at the request of the cabinet-level Economic Policy Council. At the time, the GAO stated that disclosure of all the relevant analyses conducted for the plan would have provided the Congress with better information to judge the relative merits of various energy policy proposals.

PUBLIC HEARINGS WERE USUALLY HELD

Another title VIII objective is active participation through public hearings. Except for the 1987 plan, each administration held public hearings to solicit input for its national energy plan. The public-hearings process for the 1991 National Energy Strategy, conducted over an 18-month period, was the most extensive effort in this regard.

For the 1979 plan, DOE conducted six seminars in Washington, D.C., to obtain the views of several principal constituency groups concerned with energy policy. Represented were energy producers and consumers, state and local government agencies, large and small businesses, large industrial energy users, environmental groups, and labor. These seminars were followed by public hearings in six cities: Boston, Dallas, Denver, Omaha, San Francisco, and Washington, D.C. Members of Congress, representatives of state and local governments, and environmental and other interest groups were also consulted. An appendix to the plan summarized the public participation.

For the 1981 plan, DOE held public hearings to solicit views of minority groups in Atlanta, Boston, Dallas, Denver, Kansas City, and San Francisco. The plan provided a supplemental report on these hearings. DOE sought comments for preparation of the 1983 plan in an invitation to hearings published in the *Federal Register*. Hearings were held in seven U.S. cities. The comments of the 136 persons who testified and the 111 who submitted written responses were summarized by general topic in the final plan. The plan reported that "energy security and emergency preparedness" were "mentioned frequently," while "some believe" in "free-market forces" to ensure fair and equitable distribution of energy supplies during an emergency. Similarly, the plan reported that while "many respondents expressed concern" about the role of nuclear energy in the nation's future and "questioned the continued funding" in light of "the Administration's commitment to free-market forces, others supported the development of nuclear power."

For the 1985 plan, DOE received 275 letters in response to a notice published in the *Federal Register* soliciting comments on energy policy. Public hearings were held in seven U.S. cities, and 124 people testified. In a two-page summary, DOE grouped public comments into five broad categories. Of these five categories, conservation was the issue most often cited, specifically, in support of continued federal funding of residential conservation and energy-efficiency programs and extension of energy tax credits. The summary stated that some speakers encouraged a "more market-based approach." Comments on fossil fuel and environmental issues were said to include support for offshore oil and gas leasing, full deregulation of natural gas, and increased research into nuclear waste disposal and acid rain. Nuclear energy prompted a range of suggestions as well, from streamlining licensing procedures to lessening federal promotion of nuclear power plants.

For the 1987 plan, DOE made a limited effort to obtain public participation. A *Federal Register* notice soliciting public comments resulted in 50 submissions from state and local government officials, trade associations and industry representatives, public interest groups, university and research organizations, and private citizens. A second *Federal Register* request for comments on a "Study of Crude Oil Production and Refining Capacity in the United States" produced 28 written comments. But no public hearings were held to provide input to the report.

Public hearings were a major feature of the process for developing the 1991 NES. In April 1990, a year after the 1989 plan was due, DOE released the *Interim Report* on the NES (a summary of these hearings). As noted earlier, the *Interim Report* contained no proposed objectives, strategies, or data and analysis to support such objectives.

For the NES, DOE held 18 sessions in 14 cities, more hearings than had been held for any previous plan. Over 499 witnesses from 43 states, two U.S. territories, and two Canadian provinces appeared, and DOE received more than 2,000 written submissions. When DOE released its *Interim Report* on April 2, 1990, it said the report was not intended as a first draft of the NES but as a step in building a national consensus.

ADMINISTRATIONS' VIEWS AND ENERGY CONDITIONS SHAPED PLANNING APPROACHES

The different approaches taken in the title VIII submissions prepared through 1991 largely reflected the specific views of each administration regarding the proper role for the federal government in energy policy. Changes in the nation's energy situation itself during the last 15 years have also influenced approaches to energy planning.

Stronger Federal Role Was Sought During the 1970s

A belief that a federal government plan could reduce dependence on foreign oil supplies and develop domestic alternatives shaped the national energy plans produced in 1977 and 1979. According to an energy analyst who witnessed much of the debate at the time, a prevailing view was that "if the government sat down and thought ahead, then we would have a better energy policy for the nation." Developing models which portrayed future scenarios and seeking public opinion, according to this expert, were also believed to contribute to a better energy policy.

The 1979 plan was issued about a month after the end of the Iranian oil embargo that had led to rapidly increasing world oil prices. Reducing dependence on foreign oil and the vulnerability to higher world oil prices were the principal energy concerns of the administration, and the 1979 plan described the economic, political, and strategic risks posed by continued dependence on foreign oil. At the same time, the plan predicted that U.S. dependence on imported oil would endure, that world oil prices would continue to rise, and that the nation would in turn continue to be vulnerable to oil price shocks or supply shortages.

During the 1970s, as a result of two big oil price shocks and a natural gas shortage, as well as a belief that energy supplies would be inadequate to meet growing demand, a crisis atmosphere shaped U.S. energy policies. These policies, which envisioned an activist government and national energy management, were designed primarily to insulate consumers from high world oil prices and increase energy self-sufficiency. As a result, by the end of the decade a number of demand-reducing (conservation) and supply-enhancing policies were put in place. Conservation measures included oil price controls, natural gas regulation, a tax on windfall profits, taxes on fuel-inefficient vehicles, tax credits for purchases of energy-saving equipment, and weatherization grants for low-income households, schools, and hospitals. To increase supplies of alternatives to imported oil, policies encouraged funding for renewable energy research and development, tax incentives for domestic production, and the development of synthetic fuels. The Department of Energy was created to manage many of these new programs and to fund research and development of renewable resources and conservation.

However, by mid-1979 some policies favoring more government involvement in energy began to be reversed. While the 1979 plan, published in May of that year, proposed continued federal support for alternatives to imported oil, it also proposed less federal intervention in some areas. The removal of federal price controls on domestic crude oil and implementation of the Natural Gas Policy Act of 1978, which deregulated some natural gas, were major features of the 1979 plan.

More Market-Based Approach Was Pursued in the 1980s

During the 1980s, the administration's views about the appropriate role of government in energy matters shifted further, and these views were reflected in each of the plans produced during that decade. The plans revealed the administration's opposition to setting energy goals with specific timetables, intervention in energy markets, and planning for different supply and demand conditions.

The 1981 plan departed from energy policies that had been in place since the 1973–1974 energy crises. The plan stated that "increased reliance on market decisions" rather than a "stubborn reliance on government dictates" was likely to lead to "the most appropriate energy policy." The plan presented a general set of guiding principles that were consistent with the administration's Economic Recovery Program—a plan for federal spending reductions, tax cuts, and regulatory reform. The American economy, not the government, according to these principles, would choose the appropriate energy consumption level for a strong, productive, and secure society in the year 2000. According to the plan, the best guarantee of maintaining a wholesome balance among competing interests in regard to energy lay in reversing policies that insert the government into the energy market and "allow[ing] the American people themselves to make free and fully informed choices."

Furthermore, on December 17, 1981, the President announced his intention to propose to the Congress a reorganization of federal energy programs. Federal efforts to finance the commercialization of energy technologies were to be greatly reduced. Concluding that a cabinet-level department was no longer necessary for managing energy matters, and that other highly critical energy functions (including DOE's nuclear weapons and basic research activities) could be more effectively carried out elsewhere in the government, the President proposed abolishing DOE and shifting many of its principal functions to other federal departments.

Beginning in early 1981 and continuing throughout most of the 1980s, a number of other developments had taken place in energy systems and markets as well as in government policies toward them. One of the most far-reaching of the government energy reforms was the suspension of oil price controls with deregulation of the oil industry in 1981. In addition, during this decade, federal funding for conservation programs and alternative energy resources was reduced and funds were redirected toward basic research. The government-sponsored synthetic fuels and breeder reactor projects were canceled, efficiency standards for automobiles were softened, natural gas prices were further decontrolled, and the windfall profits tax was eventually rescinded.

Federal policies to reduce government intervention in energy markets were further bolstered by simultaneous developments not necessarily related to the administration's energy policies. Oil prices had peaked in 1981, were fully decontrolled by September 30, 1981, and continued to fall until 1987. During the period 1981–86, the world price of oil had dropped from about $50 to about $17 a barrel, adjusted for inflation. From 1981 through 1985, electric utilities cut back on their petroleum consumption. Falling world oil prices between 1981 and 1986 also precipitated a drop in the value of U.S. production. As a result, domestic producers were receiving less value for each barrel of oil.

The 1983 plan attributed several of these developments to an improved national energy situation after 1981. As had the 1981 plan, it rejected the notion of government planning but did indicate that "protecting the environment, maintaining health and safety standards, and improving energy security are appropriate government responsibilities" and that "limited control and intervention may be required to reflect nonmonetary costs to society as a whole of energy production and use."

The 1985 plan, stating that "heavyhanded government planning has been abandoned," noted the success of the *"energy policy planning"* [italics in original] of recent years as "distinct from the earlier efforts at micromanaged energy planning." Repeating the goal of the administration's 1983 plan—that Americans should have an adequate supply of energy at a reasonable cost—the 1985 plan stated that progress had been made since 1981 through a climate of reduced government regulation, fewer controls, lower tax rates, and freer international trade in energy. According to the plan, "Market-oriented policies that build upon America's vast production and conservation resources and its technological genius, free of arbitrary regulation, provide the best hope of maintaining the momentum of energy progress of the past five years."

After the 1985 plan was issued, the U.S. energy situation took another turn, marked by steady increases of oil imports from countries in the politically unstable Middle East, continued plummeting of world oil prices, and a declining domestic oil industry. Moreover, in the spring of 1987, the United States and several of its allies had begun using the military to protect the safety of Kuwaiti oil tankers in the Persian Gulf in Operation Earnest Will.

The 1987 report *Energy Security* was issued in the wake of these events. In presenting the report, the Secretary of Energy stated that it had been written at the request of the President in response to his concern over declining domestic oil production and rising oil imports. When the report was published, imported oil accounted for over 40% of U.S. consumption. According to the report, worldwide oil price reductions, coupled with lower inflation and lower interest rates, had deeply affected the domestic oil industry. The net income of the 22 largest oil companies had been cut in half between 1985 and 1986, and exploration expenditures and active drilling by oil companies had declined by more than 40%. The most pressing question raised by the oil price collapse, according to the report, was what would happen if the United States and its principal allies and trading partners became much more dependent on oil supplies from the Persian Gulf region and from other member countries of the Organization of Petroleum Exporting Countries (OPEC). Because the administration was primarily concerned about rising oil imports and not about developing a national energy policy plan, the report made no specific proposals but rather weighed the costs and benefits of a range of policy options for meeting the nation's energy security objectives.

1991 Plan Responded to Concerns About Oil Vulnerability, the Economy, and the Environment

Concerns about the vulnerability of the United States to "sudden, dramatic changes in world oil prices" and about maintaining a "safer, healthier environment" and "an economy second to none" shaped the NES issued in 1991. Among the steps toward achieving these goals, the plan emphasized diversifying world oil supplies, increasing funding for renewable energy programs, extending renewable energy tax credits, and implementing the Clean Air Act Amendments of 1990.

Although oil imports as a percentage of U.S. consumption continued to rise in the latter half of the 1980s, concerns since then have been less about the level of these imports than about the vulnerability of the United States economy to oil price shocks and developing ways to mitigate the effects of such shocks. The NES stated that the United States would continue to be heavily dependent on oil imports into the next century, but that alternatives to Persian Gulf supplies in particular, such as oil from Western Hemisphere sources, domestic oil supplies, and alternative fuels, should be developed.

Several energy programs that had been scaled back during the 1980s were revived somewhat during 1991 and 1992. Between fiscal years 1980 and 1990, appropriations for DOE's renewable, fossil, nuclear, and conservation research and development programs fell by 83, 50, 68, and 34%, respectively. The administration's budgets in fiscal years 1991, 1992, and 1993 all proposed increases above the previous year's levels in these programs.

Despite these changes, the NES largely repeated a pattern of opposition to government energy planning, goal setting, and intervention in energy markets. In fact, as noted earlier, this document was not called a National Energy Policy Plan but rather a National Energy Strategy, although DOE officials told us it was submitted in response to title VIII. A DOE official in the Office of the Assistant Secretary for Domestic and International Energy Policy who worked on the plan stated that setting objectives or goals represented a kind of "command and control" approach to energy policy that the administration opposed. The NES reflected the administration's philosophy that, wherever possible, markets should determine energy prices, quantities, and technology choices, and that when markets fail to do so, government actions should be aimed at removing or overcoming barriers to efficient market operation. For example, the plan proposed removing regulations that prohibited energy exploration in the Arctic National Wildlife Refuge and the Outer Continental Shelf, further decontrolling natural gas, and accelerating the licensing of commercial nuclear power plants.

ENERGY PLANNING IS A CHALLENGING PROCESS

Developing consensus on energy policy can be a contentious activity, as administrations have learned in 15 years of responding to title VIII. Addressing title VIII's many provisions requires coming to grips with many complex issues, such as conflicting national goals, differing agency missions, and regional disparities in energy supplies. The difficulty of resolving these and other issues does not fully explain why plans have not addressed all of title VIII's provisions, but it does illustrate the challenge of completing the planning exercise within a short time.

Balancing Conflicting Goals and Values Is Difficult

Preparing energy plans requires coping with multiple and often conflicting goals and social values. Under title VIII, energy plans are to consider the "needs for full employment, price stability, energy security, economic growth, environmental protection, nuclear non-proliferation, special regional needs, and the efficient utilization of public and private resources." In the 1979 plan, the administration characterized these conflicts as "a complex tangle of sometimes competing national goals: market efficiency and greater production, equity among income classes and regions, environmental protection, national security, economic growth, and inflationary restraint. It will be difficult, and sometimes impossible, to reconcile all these goals." In 1983, the administration's commitment to securing an "adequate supply of energy at reasonable costs" signified a belief that Americans had abundant, affordable energy but that financial assistance to low-income Americans might be necessary. During the first of many hearings in 1991 on comprehensive energy legislation, the president of Cambridge Energy Research Associates and author, Daniel Yergin, observed that over the last 20 years energy policy has pursued three sometimes contradictory objectives—"cheap energy, secure energy, and clean energy" (7).

Conflicting goals also occur in other aspects of energy policy. Fundamental differences between the interests of consumers and producers blend with geographic disparities to make agreement on energy policy contentious. For example, most domestic oil is produced in the Pacific and West South Central states or offshore (in the Gulf of Mexico, in Alaska, or on the West Coast). But the Middle Atlantic, South Atlantic, and East North Central states, which produce almost no oil, account for 41% of U.S. consumption, giving rise to questions of fairness.

Energy Policy Involves Multiple Federal Agencies

Interagency conflicts arise from the multiple goals that need to be addressed by a comprehensive energy policy. For example, a Department of Energy or Department of the Interior program to encourage energy production may conflict with the Environmental Protection Agency's missions to protect the air, land, and water.

Furthermore, despite the consolidation of administrative and information-gathering functions within the new Department of Energy in 1977, important policy activities continue to take place elsewhere within the executive branch. At the time DOE was created, energy functions were spread among 20 executive departments and agencies, and more than 100 energy data programs were run by four separate entities. While some of this fragmentation has been corrected, jurisdiction over energy exploration on federal lands and the Outer Continental Shelf remains with the Department of the Interior, fuel-efficiency standards for motor vehicles are the responsibility of the Department of Transportation, and tax policy affecting both domestic and imported energy sources is initiated and regulated by the Department of the Treasury.

Cognizant of these diverse authorities, the drafters of title VIII placed the responsibility for developing plans not with the Secretary of Energy but with the President. Traditionally, however, the Secretary of Energy has managed the process of developing a plan.

Other National Issues Compete With Energy for Attention

A final reason for the difficulty of creating an energy plan is that energy policy is an adjunct to so many other policy matters. Energy serves as a medium or a catalyst to social, political, and economic activities and, as such, is inextricably linked to the complexity of the nation's heterogeneous society. Yet, unless policymakers perceive a crisis, such as after the first oil shock in 1973–1974, or when increased reliance on imported oil threatened the domestic industry in 1986–1987, or when some oil supplies were lost during the Persian Gulf War in 1990–1991, energy does not always receive consistent and focused attention.

A NATIONAL ENERGY PLANNING PROCESS IS USEFUL

Despite the inconsistencies in the approaches taken in past energy plans, experts we interviewed believed that the process of preparing energy plans is beneficial. Energy planning gives the federal government an opportunity to periodically assess long-term energy needs and develop a "base case" against which to weigh future decisions. It also provides a forum for competing interests to express their views and have them challenged.

GAO believes that title VIII provides a useful framework for achieving these planning benefits. However, the frequency with which plans must be submitted and the timing of those submissions under title VIII have not contributed to more effective planning. Once an administration prepares a comprehensive plan setting forth its goals and strategies, little is gained by repeating the full planning exercise two years later unless a new administration or energy developments warrant a new plan. It will be particularly difficult for incoming administrations to fully address all of title VIII's provisions, including additional provisions in the Energy Policy Act of 1992, by April of the first year in office.

EXPERTS SUPPORT THE NEED FOR ENERGY PLANNING

After 15 years of plans with varying approaches, there is no clear agreement among energy experts, agency officials, and congressional staff about the type of plan that may be the most useful to the Congress. Many believe the process itself is the most useful aspect of energy planning. A few including congressional staff, questioned whether the planning process should be continued, given that title VIII's objectives have been disregarded in the past and the Congress does not always find the plans useful.

Despite the inconsistencies of past plans, most of the experts and officials generally believed that a periodic evaluation of energy trends such as prices, supplies, and consumption is an important executive branch function, serving to focus attention and debate on key issues. Most who had an opinion on the question believed that requiring such as evaluation as often as every two years was probably not necessary. In addition, most favored the approach of setting some kind of energy goal or goals. But those interviewed did not all agree that the administration should set goals with specific dates for their achievement and accompanying strategies, as currently specified in title VIII but not included in the plans that have been submitted to date. This wide range of opinion may help explain why title VIII's provision for 5- and 10-year objectives has not been uniformly observed.

Opinions Differ on the Value of Objective Analysis Versus Policy-Oriented Plans

Some experts, including congressional staff, when asked what kind of national energy plan or report they believed would be most useful, stated that a "state of energy" report that also contained projections but that did not serve as a justification for current policy or make additional recommendations might be the most valuable. Such a report, they believed, would present a vision of the future under various energy price, supply, and demand scenarios. The information in such a report would be useful to a wide variety of interests, and those who chose to do so could challenge the data and assumptions. Although the 1987 report *Energy Security* did not address all of title VIII's provisions, it was regarded by some experts we interviewed as highly valuable because it objectively examined the future of the nation's energy needs.

On the other hand, some DOE officials and other experts believed that plans that contained policy statements and specific strategies, such as the 1991 plan, were more useful than those that contained only relatively neutral analyses of future scenarios. One congressional staff member believed that a plan should lay out all possible paths to reach the same goal. A few saw no value in continuing the planning exercise.

Because of the important role energy plays in the economy and national security, and its impact on the environment, the need to develop a national energy policy plan or energy "strategy" (as DOE called the 1991 title VIII submission) was a principal objective of the Department of Energy Organization Act in 1977. In a June 1990 report, GAO stated that DOE's effort to develop a national energy strategy in 1989 was a step in the right direction toward addressing the nation's future energy needs and the environmental and budgetary implications that should be considered when developing energy policies (8).

Experts Believe Goals Are Important to Energy Planning

From the first plan, administrations have chosen not to set 5- and 10-year objectives for energy production, utilization, and conservation. However, most plans did contain one or more goals without dates attached to them. Goals with specific target dates were only set in 1977, before title VIII was enacted. When asked whether objectives with target dates or some other kinds of quantifiable goals should be part of plans, those we interviewed did not agree on any single approach. But they all agreed that the nation needs at least "general energy goals." On the question of the level of specificity that such goals should include (eg, the percentage of oil imports in five years), concerns were raised that (1) the choice of specific goals runs the risk of choosing the wrong numbers, (2) failing to meet the goals does not necessarily indicate a lack of progress, (3) choosing specific goals sets up a debate over how much the nation would be willing to spend to reach them, and (4) the goals themselves become more important than the policies employed to achieve them. One argument in favor of specific goals was that since the administration currently has specific goals for education, nutrition, and health, why not for energy? Some of those we interviewed believed that certain environmental goals already imposed or that may be imposed, such as meeting fuel efficiency standards or national targets for carbon dioxide emissions by specific dates, could influence energy policy.

The Process Itself May Be the Most Valuable Part

Some experts noted that the real value of a plan is in the process that the administration must go through to prepare it. Soliciting a wide range of opinions, developing models, weighing alternative policies, setting goals, and choosing strategies may be more important than the final product itself, they argued. The exercise itself stimulates debate on the issues and serves as a forum for discussing alternatives and dealing with competing interests. "The point was that everything was heavily debated; that is exactly what you want in putting together an energy plan," stated a former DOE official in describing the approach taken to develop the 1987 plan.

FREQUENCY AND TIMING REQUIREMENTS DO NOT CONTRIBUTE TO BETTER PLANS

The requirement that plans be submitted biennially has contributed to the differing approaches and inconsistent adherence to title VIII's provisions. Incoming administrations have had difficulty meeting title VIII's provisions in their first year of office and have, in fact, chosen not to address all of title VIII's provisions every two years. In 1981, a new administration provided only a limited response to title VIII in a plan that set forth no specific energy policies or strategies but offered only "guiding principles." In addition, in responding to title VIII's provision for projections of energy prices and supplies, a supplement to the plan, *Energy Projections to the Year 2000*, pointed out that "since the projections take into account only policies or programs that were in effect as of June 1981, the projections should not be viewed as a statement of Administration energy goals. The plan's projections are a starting point, or base case, for evaluating the potential impacts of new energy initiatives or developments."

In response to the requirement that a plan be submitted in 1989, another new administration prepared only a summary of its public hearings and issued it a year late. The administration chose instead to work toward a comprehensive plan, the National Energy Strategy (NES), in time for the 1991 deadline. To explain the approach taken by this administration, one DOE official who worked on the NES said that in January 1989 when the new administration was still choosing staff, the Department was not ready to publish even a general energy policy statement. Yet the deadline for submission of the plan fell three months later. This official and others also acknowledged the time-consuming process of developing the NES. Because of the level of effort involved, DOE officials said that the process for meeting title VIII's requirement in 1993 would be a much more limited effort.

Plans submitted by the same administration every one to two years are not likely to vary much from one year to the next, according to some current and former DOE staff. Another DOE official stated that starting to prepare a new plan immediately after completing one leaves no time for retrospection. In addition, requiring an administration to submit a new plan every two years if no significant changes in the energy situation or in administration policies warrant a substantially revised plan could result in plans that are generally identical.

By the spring of 1993, with a National Energy Policy Plan due on April 1st, the incoming Clinton administration requested a one-year extension of the deadline. Since then, the Energy Department planned to issue a National Energy Strategy in 1995 that is based on the theme of "sustainable development."

CONCLUSIONS

Title VIII was developed at a time when the Congress believed that a biennial, step-by-step planning process in the executive branch would lead to a more efficient na-

tional energy policy. But administrations have not always closely followed the process of setting objectives, developing strategies, and projecting energy supply and demand from plan to plan. Each plan has reflected the current administration's philosophy toward energy and toward planning itself as well as the status of energy prices and markets. No plan has fully addressed all of title VIII's provisions.

Developing the required plans is difficult and sometimes contentious because of the time needed to address title VIII's provisions, the fact that energy policy cuts across many issues, and the large number of competing interests that must be considered. Thus, the current requirement that plans be submitted biennially, with the additional analysis the 1992 energy legislation requires, is unlikely to result in the comprehensive planning exercise the Congress intended in title VIII. Changing the frequency and timing to require that plans be submitted every four years, with the first deadline falling during the administration's second year in office, would allow each administration time to develop, and the Congress to review, a thorough energy policy statement.

GAO and most experts agree that there is value in periodically evaluating the nation's energy needs and developing a strategy to address these needs. The process of developing an energy plan serves as a forum for debating and discussing alternatives and for dealing with competing interests.

BIBLIOGRAPHY

1. *Department of Energy Organization Act, P.L. 95-91*, August 4, 1977.
2. C. D. Goodwin, "The Truman Administration: Toward a National Energy Policy," in C. D. Goodwin, ed., *Energy Policy in Perspective: Today's Problems, Yesterday's Solutions*, The Brookings Institution, Washington, D.C., 1981, pp. 6–7.
3. R. H. K. Vietor, Energy Policy in America Since 1945: A Study of Business-Government Relations, Cambridge University Press, Cambridge, UK, 1984, p. 320.
4. N. de Marchi, "Energy Policy Under Nixon: Mainly Putting Out Fires," in Ref. 2, p. 395.
5. *The National Energy Plan*, Executive Office of the President, Office of Energy Policy and Planning, Washington, D.C. April 29, 1977.
6. *Full Disclosure of National Energy Strategy Analyses Needed to Enhance Strategy's Credibility*, GAO/T-RCED-91-76, GAO, Washington, D.C., July 8, 1991.
7. *National Energy Strategy: A New Start*, statement of D. Yergin before the Subcommittee on Energy and Power, House Committee on Energy and Commerce, Feb. 20, 1991.
8. *Energy Policy: Developing Strategies for Energy Policies in the 1990s*, GAO/RCED-90-85, GAO, Washington, D.C., June 9, 1990.

Reading List

H. H. Landsberg, Ed., *Making National Energy Policy*, Resources for the Future, Washington, D.C., 1993.

A. R. Maron, W. J. Lanouette, G. R. Boss, *Energy Policy: Changes Needed to Make National Energy Planning More Useful*, U.S. General Accounting Office, Washington, D.C., GAO/RCED-93-29, April, 1993.

ENERGY TAXATION: AUTOMOBILE FUELS

David Gushee
Salvatore Lazzari
Congressional Research Service
Washington, D.C.

Over the past few years, increasing emphasis has been placed on the role that alternative motor vehicle fuels might play in reducing oil imports, improving urban air quality, and providing domestic jobs. During the past five years Congress has passed three bills designed to stimulate the technology, economics, and infrastructure needed for these fuels to compete with the incumbent fuels, gasoline and diesel.

The latest of these bills, the Energy Policy Act of 1992, includes tax incentives applied to alternative fuel vehicles and to refueling facilities for alternative fuels. However, no action has been taken to rationalize the disparate highway taxes applied to the different fuels that result from a series of independent actions taken over many decades.

Similarly, at the state level, motor vehicle fuel taxation has been driven by the twin needs of revenue generation and highway infrastructure development. Motor fuels are taxed primarily on a gallonage basis without regard to energy content. Until recently, and even then only in a few states, conscious stimulation of alternative motor fuels has not been a policy goal. In most of the states where stimulation has been a goal, ethanol has been the primary beneficiary, with its use mainly as an additive to gasoline rather than as an alternative fuel.

As a result, the different fuels have widely disparate tax rates. Thus, estimated prices at the retail service station pump or equivalent, particularly when corrected to a common energy content, are affected very differently by highway taxes. Compressed natural gas (CNG) and electricity are favored, electricity by not being viewed as a highway fuel and thus not carrying a highway tax burden at either the Federal or state level, and CNG, (a) by virtue of being taxed at the Federal level only at the newly-imposed additional deficit reduction rate, and (b) by being treated on an equivalent energy content basis by the states. Three others, propane, liquefied natural gas (LNG), and methanol, are significantly disadvantaged, because they have lower energy densities than gasoline (and the energy-equated CNG) and thus pay higher tax rates per unit of energy. Methanol is disadvantaged the most of any alternative fuel.

FEDERAL HIGHWAY TAXES ON MOTOR FUELS

Structure

The Internal Revenue Code (IRC) imposes excise taxes on a variety of motor fuels used in highway transportation, with the tax rates, varying by type of fuel and its use. The traditional fuels, gasoline and diesel, used in highway motor vehicles, are taxed at 18.4¢ per gallon and 24.4¢ per gallon, respectively (Internal §4081 and 4091).

These tax rates have several components: The 18.4¢ rate comprises an 11.5¢ highway trust fund rate (which finances the Federal Highway Trust Fund), a 6.8¢ deficit reduction rate (which is designated to the general fund for deficit reduction), and a 0.1¢ Leaking Underground Storage Tank (LUST) trust fund rate. (The LUST trust fund is a Federal program that finances the cost of cleaning up spills from underground fuel tanks). The 24.4¢ diesel tax rate comprises the 17.5¢ highway trust fund rate, the 6.8¢ deficit reduction rate, and the 0.1¢ LUST trust fund rate (§4091, and §4041(a)).

Many non-highway uses of motor fuels, such as farm uses or commercial uses in stationary motors and some highway uses of motor fuels such as uses by school districts or state and local governments, are tax-exempt. Some non-highway uses of motor fuels are, however, taxed at varying rates also. The 18.4¢ tax rate on gasoline and special motor fuels generally applies to fuels used in noncommercial motorboats. Fuel used for transportation on island waterways by commercial cargo vessels is taxed at 21.4¢ per gallon, rising to 24.4¢ by 1995 (§4042). Fuels used in noncommercial aviation are taxed at either 19.4¢ per gallon (in the case of gasoline, §4041(c), and §4081) or 21.9¢ per gallon for jet fuel (§4041(c) and §4091). Jet fuel used in commercial aviation only pays the 0.1¢ LUST fund tax. However, diesel used in trains is taxed at 6.9¢ per gallon, comprising the deficit reduction rate and the LUST fund rate only.

Liquid special motor fuels such as naphtha, benzene, benzol, casinghead gasoline, and natural gasoline–these are also known as gasoline substitutes–are subject to the 18.4¢ rate if the fuel is used in a highway vehicle for purposes that are not specifically tax-exempt (§4041(a)(2)). Liquefied petroleum gas (LPG) is taxed at 18.3¢ per gallon since it is the only gasoline substitute not subject to the 0.1¢ LUST tax. Where LPG is sold by weight, the equivalent of a gallon for purposes of computing the motor fuels tax is 4.25 pounds per gallon (see Rev. Rul. 71-464, 1971-2 CB 357).

Mixtures of motor fuels and biomass-derived alcohols (either methanol or ethanol) are partially tax-exempt, with the amount of the exemption depending upon the fraction of alcohol that is in the mixture and the type of alcohol. Gasohol mixtures–blends of gasoline and ethanol that are 10% ethanol–are taxed at 13.0¢ per gallon (the exemption is 5.4¢). Under the IRC, blends of gasoline with biomass-derived methanol would also qualify, but such blends are disqualified under the Clean Air Act because of the associated increase of emissions of ozone-forming pollutants. Mixtures that are 7.7% ethanol are taxed at 14.24¢ per gallon (the exemption is 4.16¢). And finally, mixtures that are 5.7% ethanol are taxed at 15.32¢ per gallon (the exemption is 3.08¢ per gallon). In all these cases, the exemption equates to 54 cents per gallon of ethanol used.

The exemption for alcohol fuels mixtures also applies to blends of diesel and biomass-derived alcohol and blends of a special motor fuel and biomass-derived alcohol. Alcohol blended with diesel–sometimes called "dieselhol"–is taxed at the rate of 19.0¢ (the exemption is 5.4¢, the same as for gasoline). Alcohol blended with one of the special motor fuels is assessed a 13.0¢ tax rate (the exemption is again 5.4¢).

It should be noted that in all these cases of alcohol blends, the alcohol must be at least 190 proof (95% pure alcohol) and the alcohol cannot be derived from petroleum, natural gas, or coal (including peat). In the case of the blended fuels, methanol produced from coal or natural gas did not originally qualify for the tax exemptions. This dispensation is because when the ethanol exemption was first introduced in 1977, the Congress believed that the cost of producing methanol from coal and natural gas was low enough without a Federal tax subsidy; whereas, the cost of producing methanol from wood and ethanol from grain was costly and did warrant a subsidy.

Exempt alcohols can currently only be derived from biomass. Thus, while both methanol and ethanol qualify for the exemption, methanol is in effect disqualified because it currently is mostly produced from natural gas. However, in the case of special motor fuels that are 85% alcohol (ethanol or methanol) derived from natural gas, there is a separate exemption of 7.0¢ per gallon (the tax rate is 11.4¢ per gallon (§4041(m)). This partial exemption has the effect for methanol of equalizing the Federal tax rate to that of gasoline on an energy-content basis.

Alcohol fuels mixtures that contain 85% alcohol made from biomass are taxed at varying rates. In the case of 85% ethanol blends, the tax rate is 12.35¢ consisting of a 5.5¢ highway trust fund rate (a 6.0¢ exemption), the 6.8¢ deficit reduction rate, and a 0.05¢ LUST fund rate. In the case of 85% methanol blends, the tax rate is 12.95¢ per gallon, consisting of a 6.1¢ highway trust fund rate (a 5.4¢ exemption), the 6.8¢ deficit reduction rate, and the 0.05¢ LUST fund rate. Note that the LUST fund rate on these 85% mixtures is one-half the rate that applies to all other taxable fuels (4041(b)(2)).

Finally, it is important to underscore the point that the highway motor fuels excise taxes historically applied to liquid fuels only; electricity and gaseous fuels, such as compressed natural gas, were not taxed. This situation is still true of electricity. Compressed natural gas, as a result of the Omnibus Budget Reconciliation Act of Intermodal Surface Transportation Efficiency Act of 1991 (P.L. 102-240). It should be noted, however, that the various motor fuels have always had expiration dates, which have always been extended prior to the actual expiration. The excise tax exemptions for alcohol fuels expire on September 30, 2000.

A summary of the tax rates on the various highway motor fuels is given in Table 1. The first row of numbers in Table 1 presents the tax rates on a per gallon basis, as discussed in the text. As these data clearly show, motor fuel tax rates display wide variation among different fuels.

The second row of numbers in Table 1 shows the energy content per gallon of each of the fuels in British thermal units (Btu). All of the alternative fuels listed have less energy per gallon than gasoline or diesel fuel. The amount of energy in a "gallon" of CNG is about 20 percent of the energy in a gallon of gasoline. The exact amount depends upon the temperature and pressure, which will vary depending on the fuel fill rate and the ambient temperature. Therefore, industry practice is to measure the amount of gas fed into the fuel tank, to calculate its energy content, and to divide the energy content by a factor to convert the energy to billable quantity (eg, therms, which equals 100,000 Btus). The number of therms equivalent to the energy content of a gallon of gasoline is defined by each State. The intent is to equate the energy content of a "gallon" of CNG to the energy content of a gallon of gasoline. Table 1, follows that intent. The higher energy density of gasoline and diesel fuel is one of their advantages compared to the alternatives, in that a given storage capacity for fuel leads to a longer driving range per tankful.

The third row of numbers in Table 1 shows the tax rates adjusted for the energy content compared to that of a gallon of gasoline. For example, since propane contains 85,000 Btu per gallon compared to gasoline's 115,000 Btu, 35% more of it is required to deliver the same quantity of energy as a gallon of gasoline. Thus, the "effective" excise tax rate is 24.9¢ per gallon, 35% more than the 18.4¢ per gallon tax. In Table 1, we have used lower heat values than those used by the Energy Information Administration and by some authors. The difference is whether water vapor is calculated as vapor (which results in lower heat values) or as liquid. The choice of one or the other does not affect the conclusions. Another cause of differences in heat values used in various sources is that all of these fuels are mixtures whose compositions vary from place to place and time to time.

Evolution of Federal Highway Excise Taxes on Motor Fuels

The present structure of Federal excise taxes on motor fuels evolved from three public policy concerns: (1) reve-

Table 1. Federal Highway Motor Fuels Taxes Adjusted for Energy Content

Gasoline	Diesel	Methanol (M100)	Ethanol (E100)	Natural Gas (CNG)	Natural Gas (LNG)	Propane (LPG)	Electricity
Tax per gallon, cents							
18.4[a]	24.4	11.4	13.0	5.6	18.4	18.3	None
Energy content per gallon (Btu)							
115,000	135,000	57,000	76,000	115,000	75,000	85,000	
Tax adjusted for energy content (cents per gallon of gasoline equivalent)[b]							
18.4	20.8	23.0	19.7	5.6	28.2	24.9	None

[a] In addition to the highway tax (18.3 cents/gal) and the LUST Fund tax (0.10 cents/gal), there are two additional taxes on imported petroleum products, including gasoline: the Hazardous Substance Trust Fund (0.23 cents/gal), and Oil Spill Liability Trust Fund (0.12 cents/gal). Domestically produced gasoline carries equivalent taxes, but they are imposed on the crude oil used in its manufacture.
[b] Obtained by multiplying a fuel's "unadjusted" tax by the ratio of per gallon energy content of gasoline to that of the particular fuel.

nue generation for budget deficit reduction; (2) revenue generation for highway infrastructure financing; and (3) energy policy considerations.

Deficit Reduction. Revenue for purposes of deficit reduction was the rationale for enacting the gasoline tax in 1932 and for several of the many subsequent increases in tax rates as well as extensions of the expiration dates. The Federal excise tax on gasoline was first enacted as part of the Revenue Act of 1932 (P.L. 154), although the first known Federal proposal to tax gasoline goes back to the Wilson Administration (1). The tax, which was initially 1¢ per gallon, was enacted as part of a program of tax increases designed to generate additional revenue to reduce budget deficits, which were looming due to the deepest and longest economic recession in U.S. history. Revenues from the tax were allocated to the general fund for deficit reduction.

Revenue generation for deficit reduction was also the underlying rationale for the gasoline tax rate increases of 1940, 1941, 1951, and 1954. The increases of 1951, which were part of the Revenue Act of 1951, raised the gasoline tax from 1.5¢ to 2¢ per gallon and introduced the tax on diesel fuel, also at the rate of 2¢ per gallon. Revenue generation to help finance the Korean War was an additional reason for these tax rate increases. Revenue generation for deficit reduction was part of the rationale for the tax rate increases of 1990. The Omnibus Budget and Reconciliation Act of 1990 (P.L. 101-508) raised the gasoline and diesel fuel taxes, which had increased to 9.1¢ and 15.1¢ per gallon respectively during the 1980s, by another 5¢ per gallon and authorized that revenues from 2.5¢ of the 5.0¢ increase would go toward deficit reduction rather than the highway trust fund. The 1990 law also allowed diesel fuel used in trains to be taxed at 2.6¢ per gallon, with 2.5¢ for deficit reduction and 0.1¢ for the LUST fund. Prior to the 1990 law, diesel used in trains was tax-exempt because it was a non-highway use.

Highway Finance. In 1956, the Highway Trust Fund was created under the Federal-Aid Highway Act of 1956 (P.L. 84-627). This act marked a fundamental change in Federal highway financing—from general revenues to motor fuels taxes. All gasoline tax revenues and most other highway user revenues went into that fund for highway construction finance. The purpose of the trust fund was to finance the cost of the interstate highway system. Thus, revenues to finance highway infrastructure rather than revenue generation for deficit reduction became the primary rationale underlying most of the increases in tax rates and expansion of the tax bases since then. From 1956 to 1982, there were two increases in tax rates, several extensions of expiration dates, and repeals of scheduled declines in tax rates. Each of these amendments was made to generate more money for the Highway Trust Fund programs.

Beginning in late 1982, another objective was added to the list. Rather than fiscal deficits or energy security, attention began to focus on the large portion of the roads and highways in this country that had fallen into disrepair, and on the unemployment rate, which had risen steeply as a result of the 1981–1982 economic recession. Between 1982 and 1990 there were four increases in the

motor fuels excise taxes. Title I of the Surface Transportation Assistance Act of 1982 (P.L. 97-424) boosted the motor fuel excise taxes by 5 cents per gallon (to 9 cents). The 1982 law also provided that 1¢ of the 5¢ increase would be allocated to a special mass transit account. The Tax Reform Act of 1984 (P.L. 98-369) increased the diesel fuel tax another 6¢ per gallon in return for a repeal of a scheduled boost in truck taxes based on vehicle weights. This made the tax on diesel fuel 15¢ per gallon. A 0.1¢ per gallon tax was added by the Superfund Amendments and Reauthorization Act of 1986 (P.L. 99-499) to pay for the cleaning up of leaking underground storage tanks.

Energy Policy Considerations. Beginning in the 1970s, energy policy considerations began to influence both the level of motor fuel taxation and, more importantly, the structure of tax rates. Reducing petroleum consumption and importation made it easier to support motor fuels excise tax increase proposals, and was the rationale for reducing the tax rates on alternative fuels, particularly alcohol fuels. Proposals to increase the Federal excise tax on gasoline became common during and after the 1973–1974 Arab oil embargo and subsequent rises in crude oil prices. Coming in the aftermath of the 1973–1974 oil shock, such proposals were intended largely to reduce consumption of motor fuels (by raising their prices), and thereby reduce oil imports. Perhaps the most ambitious of these proposals was that of Senator Henry Jackson, proposing to increase the tax by $1.00 per gallon.

The concept of taxing alternative fuels at lower rates, which began in the middle 1970s in response to the first oil shock, was actually realized in 1978 with the enactment of the Energy Tax Act (P.L. 95-618). The Energy Tax Act (ETA) also provided for the gas-guzzler tax, incentives for van pooling, and miscellaneous energy tax provisions. The underlying rationale for the ETA was the perceived failures in the energy markets in allocating resources efficiently and fairly, in coping with the 1973 oil embargo, and in adjusting to the sharp increases in energy prices, the shortages, and the associated adverse economic and social problems. Prior to this law, there were no special exemptions for highway use of alternative motor fuels.

The Federal exemption for alcohol fuels under the 1978 law was for the full amount of the gasoline tax: 4¢ per gallon. The Crude Oil Windfall Profits Tax (P.L. 96-223) extended the 4¢ exemption from October 1, 1984, to December 31, 1992. The Surface Transportation Assistance Act of 1982 (P.L. 97-424) raised the gasoline tax from 4¢ to 9¢ per gallon and also changed the exemption for gasohol from the complete 4¢ exemption to a partial 5¢ exemption (gasohol would be taxed at 4¢ per gallon instead of 9¢ per gallon). The Deficit Reduction Act of 1984 (P.L. 98-369) raised the diesel fuel tax from 9¢ to 15¢ per gallon as part of a compromise that also lowered the highway use taxes on trucks. The 1984 tax law also raised the gasohol exemption from 5¢ to 6¢ (i.e., it reduced the tax rate for gasohol from 4¢ to 3¢), and retained the 9¢ exemption for "neat" alcohol fuels, and provided that alcohol produced from natural gas would also qualify for the exemption.

The Tax Reform Act of 1986 (P.L. 99-514) reduced the excise tax exemption for 85 percent alcohol from 9¢ to 6¢

Table 2. State Tax Rates on Motor Fuels—As of October 1, 1993 (Cents per Gallon)

State	Gasoline	Diesel	Methanol	Ethanol	CNG	LNG	Propane	Electricity
Alabama	18	19	18	18	[c]	18	[c]	None
Alaska	8	8	None	None	[b]	[b]	None	None
Arizona	18	18	18	18	1	18	18	None
Arkansas	18.7	18.7	18.5	18.5	5	5	16.5[c]	None
California	17	17	8.5	8.5	7	6	[d]	None
Colorado	22	20.5	20.5	20.5	20.5[d]	20.5[c]	20.5[d]	None
Connecticut[a]	29	18	28	28	29	29	18	None
Delaware	22	19	19	19	19	19	19	None
Dist. of Columbia	20	20	20	20	20	20	20	None
Florida[a]	11.8	21	11.8	11.8	[c]	[c]	[c]	None
Georgia[a]	7.5	7.5	7.5	7.5	7.5	7.5	7.5	None
Hawaii[a]	16	16	16	16	[b]	[b]	11	None
Idaho	21	21	21	21	21[d]	21[d]	15.2[d]	None
Illinois[a]	19	21.5	19	19	19	19	19	None
Indiana[a]	15	16	15	15	[c]	[c]	[c]	None
Iowa[a]	20	22.5	20	20	18.4	20	20	None
Kansas[a]	18	20	20	None	17	17	17[d]	None
Kentucky[a]	15.4	12.4	15	15.4	15	15	15	None
Louisiana[a]	20	20	20	20	20[d]	20	20[d]	None
Maine	19	20	[b]	[b]	[b]	[b]	18	None
Maryland	23.5	24.75	23.5	23.5	23.5	23.5	23.5	None
Massachusetts[a]	21	21	21	21	21	21	9.6	None
Michigan	15	15	15	15	None	None	15	None
Minnesota[a]	20	20	20	18	20	20	20	None
Mississippi	18.2	18.2	18.2	18.2	18.2	18.2	17	None
Missouri	13.03	13	13	13.03	[c]	13	[c]	[c]
Montana[a]	24	24	24	24	7.49	7.49	[c]	None
Nebraska[a]	24.4	24.4	24.4	24.4	24.4	24.4	23.8	None
Nevada[a]	23	27	23	23	23	23	17	None
New Hampshire[a]	18.7	18.7	[c]	[c]	[c]	[c]	18	None
New Jersey[a]	10.5	13.5	10.5	10.5	5.25	5.25	5.25	None
New Mexico[a]	23	19	23	22	23[d]	[b]	18[d]	None
New York[a]	23.03	25.03	22.84	22.84	22.84	22.84	8	None
North Carolina[a]	22.3	22.3	22.3	22.3	25.6	22.3	22.3	None
North Dakota[a]	17	17	17	17	17	17	17	None
Ohio[a]	22	22	22	22	22	22	22	None
Oklahoma	17	14	17	17	[c]	[c]	[c]	None
Oregon[a]	24	24	24	24	24	24	24	None
Pennsylvania[a]	22.4	22.4	12	22.4	12	12	22.4	None
Rhode Island[a]	28	28	28	28	28	28	28	None
South Carolina	16	16	16	16	16	16	16	None
South Dakota[a]	18	18	16	16	18	18	16	None
Tennessee	21.4	17	21.4	21.4	13	14	14[d]	None
Texas	20	20	20	20	[c]	[c]	[c]	None

per gallon (for sales made beginning in 1987). The Technical and Miscellaneous Revenue Act of 1988 (P.L. 100-647) made minor liberalizations to the excise tax rules. Finally, the OBRA of 1990 reduced the alcohol fuels exemption to 5.4¢ per gallon. The 1990 OBRA also introduced a new tax credit for small ethanol producers (less than 15 millon gallons per year).

The Energy Policy Act of 1992 (P.L. 102-486) extended the gasohol excise tax exemption to gasohol that contains less than 10 percent alcohol. Two categories of gasohol mixtures were prescribed: mixtures containing 7.7-percent alcohol; and mixtures containing 5.7-percent alcohol. The exemption for 7.7 percent mixtures is 4.16¢ per gal-

lon; the exemption for 5.5 percent mixtures is 3.08¢ per gallon.

The Omnibus Budget Reconciliation Act of 1993, in addition to imposing a new tax on CNG, raised the tax rate on motor fuels used in highway transportation by 4.3¢ per gallon.

STATE HIGHWAY TAXES ON MOTOR FUELS

Structure

States tax motor fuels through a combination of excise taxes and other fees and taxes. State excise taxes are tab-

Table 2. *(Continued)*

State	Gasoline	Diesel	Methanol	Ethanol	CNG	LNG	Propane	Electricity
Utah	19	19	19	19	19	19	19[d]	None
Vermont[a]	16	17	None	None	None	None	[c]	None
Virginia[a]	17.5	16	17.5	17.5	10	10	10	10
Washington[a]	23	23	23	3.45	[c]	[c]	[c]	None
West Virginia[a]	25.35	25.35	25.35	25.35	25.35	25.35	25.35	None
Wisconsin[a]	23.2	23.2	23.2	23.2	23.2	23.2	23.2	None
Wyoming[a]	9	9	None	None	None	None	None	None

[a] The following states have special provisions:

Arkansas: Natural gas tax rate will increase as number of vehicles using it increases. Breakpoints are 1000, 1500, 2000, 2500, and 3000. At 3000, tax rate will be 16.5 cents per equivalent gallon.

California: Gasoline and diesel tax rates rise to 18 cents per gallon on 1/1/94.

Connecticut: Gasoline tax will go up one cent per gallon on 1/1/94. Propane was given the diesel tax rate on 10/1/93.

Florida: Tax rates are adjusted annually. For gasoline and gasohol, there is also a State Comprehensive Enhanced Transportation System Tax (SCETS) that varies by county from 0 to 4.2 cents per gallon. CNG, LNG, and LPG pay on a decal system; out of state vehicles pay 4 cents per gallon.

Hawaii: Ethanol is exempt from the State's general excise tax of 4%.

Illinois: Motor carriers pay an additional 5.9 cents per gallon on diesel.

Iowa: Ethanol produced in the State receives 20 cents per gallon subsidy.

Kansas: Ethanol is taxed as a motor fuel only when blended as gasohol. Trucks can buy pertrip permits in lieu of highway tax.

Kentucky: Tax rates are adjusted quarterly. A 2 percent surtax is imposed on gasoline and 4.7 percent on special fuels for any vehicle with 3 or more axles. There is an additional 2 cents per gallon on vehicles over 50,000 lbs.

Louisiana: There is a producer credit of $1.40 per gallon of ethanol.

Massachusetts: Tax rates are adjusted quarterly.

Minnesota: There is a credit of 20 cents per gallon of ethanol used to make gasohol. The LPG decal system has been repealed. Taxation by energy equivalence is likely in 1995. A temporary reduced tax rate for E85, CNG, LNG, and LPG is possible.

Missouri: The decal fee increases with vehicle weight. Three State agencies are considering highway tax changes to encourage alternative fuels.

Montana: There is an alcohol distillers' credit of 30 cents/gallon of ethanol produced in qualified facilities in the State.

Nebraska: Rates are adjusted quarterly. There is a producer incentive credit of 20 cents per gallon of ethanol produced in qualified facilities in the State.

Nevada: 125 cubic feet of natural gas or LPG is defined as equal to one gallon of gasoline and subject to the gasoline tax rate.

New Hampshire: Alternative fuel vehicles pay twice the usual registration fee in lieu of the gallonage tax.

New Jersey: There is a gross receipts tax of 4 cents per gallon on gasoline, diesel, and on-road propane. There is also a proposal to reduce the price of natural gas sold for motor vehicle use.

New York: Tax rates are adjusted annually. The rates include a State Petroleum Business Tax not applied to propane. There is a State sales tax of 4% and local sales taxes ranging from 0 to 4.5%.

North Carolina: Tax rates are adjusted semiannually. CNG is taxed at 22.3 cents per 100,000 Btu. The State is considering changing that to a higher energy content gallon equivalency, such as 115,000 or 120,000.

North Dakota: A special excise tax of 2 percent is imposed on all sales of LPG and diesel that are exempted from the gallonage tax if the fuel is sold for use in the State. There is a producer credit of 40 cents per gallon of agriculturally derived alcohol produced in the State and used to make gasohol.

Ohio: Dealers are refunded 15 cents per gallon of each qualified fuel (methanol or ethanol) blended with unleaded gasoline.

Oklahoma: The decal fee increases with increasing vehicle weight.

Oregon: 50% of local property taxes are waived for five years for an ethanol production plant.

Pennsylvania: Motor carriers pay an additional 6 cents per gallon.

Rhode Island: Tax rates are adjusted quarterly.

South Dakota: There is a credit at the rate of the gasoline tax to distributors blending ethanol with gasoline to produce ethanol. There is also a producer incentive payment of 20 cents per gallon of ethanol used in gasohol.

Vermont: Diesel vehicles over 10,000 lbs. pay 26 cents/gallon. Incentive proposals for AFVs have not **South Carolina:** Legislature to consider committee proposals to tax on energy equivalence, with reduced tax for clean-burning fuels, the reduction proportional to the air pollution benefit passed.

Virginia: Favorable tax rates for CNG, LNG, LPG, hydrogen, hythane, and electricity begin on 1/1/94 and run until 7/1/98.

Washington: Decal free for CNG, LNG, and LPG increases with vehicle weight. There is a credit of 60 percent of the gasoline tax for every gallon of alcohol used in gasohol, year-round for small producers (<10 million gallons per year) and for all but the 4 winter months for others.

West Virgina: Tax rates are adjusted annually.

Wisconsin: Tax rates are adjusted annually. The legislature has discussed waiving taxes on alternative fuels, but specific legislation has not moved forward.

Wyoming: All alternative fuels are subject to sales and use taxes. WDoT is studying alternative fuels tax policy.

[b] No tax because State has not addressed tax policy for this fuel.

[c] Registered vehicles using this fuel must pay an annual fee in lieu of the gallonage tax.

[d] Registered vehicles may pay an annual fee in lieu of the gallonage tax.

Sources: Federal Highway Administration. Office of Highway Information Management. Telephone calls to State energy and taxation departments.

ulated in Table 2. Some impose sales taxes on top of the Federal and State excise taxes; others use the pre-tax gasoline price as the base. Many states impose specific-purpose fees such as their own versions of the Leaking Underground Storage Tank tax, inspection fees, and the like. Fifteen states have authorized cities and counties to add their own charges. Communities in 11 of those states have chosen to do so.

Many states have tax policies for ethanol used in motor fuels and propane used as a motor fuel. With respect to ethanol, the special provisions take the form of a waiver of part of the highway tax or a credit per gallon of ethanol produced in the state and used in the motor fuel (as in gasohol, for example) or, in some cases, both. In most of these cases, however, the state does not have a specified policy for taxing ethanol as the primary component of the motor fuel as in, for example, E100 (100% ethanol) or E85 (85% ethanol and 15% gasoline).

With respect to propane, a number of states substitute a fee on vehicles (mandatory in some states, optional in others) based on vehicle type in place of the fuel tax. The fee in most cases comes out close to what the tax would have cost for a vehicle of average mileage. High–mileage vehicles end up paying a somewhat lower tax rate per gallon. These states also have a gallonage tax for out-of-state or nonfee-paying vehicles or, in some cases, require an LPG-fueled vehicle to buy a permit to operate in the state. Two states (Colorado and Texas), now have fee systems for natural gas-fueled vehicles.

Most states, however, do not have special policies for ethanol (as discussed above), methanol, natural gas (either compressed or liquefied), or electricity, since they are too new as motor fuels to have received specific attention. In most states, these new alternative fuels have been swept up under a general approach; if liquid, they are taxed on a volume-equivalent basis while, for compressed natural gas, they are taxed on some definition of energy equivalence to gasoline. In a few states, some alternative fuels are not taxed at all, either through lack of attention or because of special policy (particularly the case where a fuel-producing state seeks to develop motor fuel markets for its indigenous fuel). There has not been a standard definition of Btu equivalency be-tween CNG and gasoline. The equivalency depends on the gas composition, the pressure to which it is compressed, and the gasoline to which it is compared. The National Conference on Weights and Measures had recommended a standard Btu equivalency to gasoline for CNG of 1.14 therms (100 cubic feet at standard atmospheric conditions). The energy content of a therm varies over time by a few percent but averages about 100,000 Btu. This recommendation was recently changed, after a review of the issues, to a standard weight of 4.6601 lbs. of gas as the equivalent of one gallon of gasoline. The recommendation has not yet been adopted by states.

As the impact of the Energy Policy Act of 1992 is felt in the states (which must soon start to buy alternative fuel vehicles for their fleets), administrative and legislative attention to tax policy is increasing rapidly, with a number of states already having taken actions such as substituting up-front registration fees for natural gas (and sometimes propane) for the usual highway tax. Arkansas is the only state so far to introduce a reduced tax on an alternative fuel (compressed natural gas), phasing the tax back toward equivalence with gasoline as the number of natural gas vehicles increases. Several other states are considering similar approaches but are finding resistance from those sectors dependent on highway tax revenues.

Thirty-three states have sales taxes on motor fuels. They range from 2% to over 6%. They are applied differently from state to state–in some cases to the sales price of gasoline less all highway taxes, in other cases to the sales price including all highway taxes. Details are shown in Table 3.

Evolution of State Highway Excise Taxes on Motor Fuels

The present structure of state motor fuels taxes evolved predominantly from the need for revenues to finance highway construction, and secondarily for general revenue purposes, reasons generally true for each state. Energy policy considerations were not a factor in the evolution of the structure of motor fuels taxes in the states, although in recent years some states introduced special provisions for ethanol and propane, as was discussed above.

New York was the first state to require the licensing of automobiles. By 1917, every state had similar rules. Oregon enacted the first gasoline tax in 1919 at the rate of 1¢ per gallon thus initiating the policy of user financing of highway spending (2). By 1929, every state in the Union had a gasoline tax for highway finance, at rates ranging from 3¢ to 7¢ per gallon. Prior to 1919, all highway construction was financed from general tax revenues, generated primarily from the property tax.

Between 1919 and 1980, state gasoline taxes were changed infrequently, and usually .01¢ at a time. But, through most of this period, state gasoline tax collections grew enormously due to the growth in the demand for highway travel and in the number of automobiles. For instance, gasoline tax collections grew from $1 million in 1919 to $1,124 million in 1947. By 1948, motor fuels taxes yielded more revenue than any other state excise tax.

Lagging gasoline tax revenues due to energy conservation in response to the energy price shocks of the 1970s, combined with rising highway repair costs and demand for additional highways, created pressure to raise gasoline tax rates. As a result, during the 1980s most states increased taxes frequently and by relatively large amounts. Between 1980 and 1988, for example, there were 107 increases in gasoline taxes at the state level (3,4).

IMPACT OF FEDERAL AND STATE MOTOR FUEL TAXES

The combination of Federal and state motor fuel and sales taxes on gasoline adds up to 47 cents or more to the pre-tax price at the service station. For alternative fuels with lower energy densities than gasoline, the taxes, when applied per gallon of fuel unadjusted for energy content add an additional bite–up to 25 to 30 cents for methanol and up to 10 cents for propane. The precise amount extra depends on the state's gallonage tax and sales tax (if any), the fuel's energy content relative to gasoline, and the vehicle efficiency of fuel use compared to that for gasoline.

Electricity is such a special case that highway taxes, were they to be applied, would pay only a minor role compared to vehicle purchase price and battery cost and short life.

Since the Federal highway tax on compressed natural gas is only about 30% of that applied to gasoline, since state highway taxes on CNG are imposed in most cases on a Btu equivalency basis, and since the state highway taxes on the other alternative fuels are, in most cases, on a gallon basis rather than on energy equivalency, the net effect of the combined Federal and state taxes on the prices at the pump of alternative fuels compared to each other and compared to gasoline is very significant. Table

Table 3. State Sales Taxes on Motor Fuels

State	Percent	Remarks
Alabama	4	Applies to fuel not taxable under gallonage tax laws
Arizona	5	Applies to fuel not taxed under the motor fuel or use fuel taxes
Arkansas	4.5	Special fuel for municipal buses and gasoline are exempt
California	6	Applies to sales price including Federal and State motor fuel taxes
District of Columbia	6	Applies to fuel not taxable under gallonage tax laws
Georgia	4	Applies to sales price including Federal motor fuel tax
Hawaii	4	Applies to sales price excluding Federal and State motor fuel taxes. Alcohol fuels are exempt
Illinois	6.25	Applies to sales price excluding Federal and State motor fuel taxes. For gasohol, only 70 percent of the price is subject to sales tax
Indiana	5	Applies to sales price excluding Federal and State motor fuel taxes
Iowa	5	Fuel on which the gallonage tax was paid and not refunded is exempt. Gasohol is exempt
Kansas	4.9	Applies to fuels not taxable under the gallonage tax laws
Kentucky	6	Applies to sales price, exclusive of Federal tax, of fuels not taxable under gallonage tax laws
Louisiana	4	Fuels subject to gallonage tax are exempt. Gasohol is exempt if alcohol is produced in State
Maine	6	Applies to motor fuel not taxed at the maximum rate for highway use under the gallonage tax laws
Massachusetts	5	Applies to fuels not taxable under the gallonage tax laws
Michigan	4	Applies to sales price including Federal motor fuel tax except for certain multi passenger, for hire vehicles on scheduled routes
Minnesota	6	Applies to fuels not taxable under the gallonage tax laws
New Mexico	5	Applies to fuels not taxable under gallonage tax laws. Ethanol blends deductible under the gasoline tax laws are exempt
New York	4	Applies to sales price including Federal motor fuel tax
North Dakota	5	Applies to fuels not taxable under gallonage tax laws
Ohio	6	Applies to fuels not taxable under gallonage tax laws
Oklahoma	4.5	Applies to fuels not taxable under gallonage tax laws
Pennsylvania	6	Applies to fuels not taxable under gallonage tax laws
Rhode Island	7	Applies to sales price. Gasoline is exempt
South Carolina	5	Applies to aviation gasoline only
South Dakota	4	Applies to fuels not taxable under gallonage tax laws
Tennessee	4.5	Applies to aviation fuel only
Texas	6.25	Applies to fuels not taxed or exempted under other laws
Utah	5	Applies to fuels not taxable under gallonage tax laws
Virginia	2	Applies to retail sales within counties and cities with subway or bus systems owned and operated by transportation agencies
Washington	6.5	Applies to sales price excluding Federal and State gallonage taxes. Alcohol for use as a motor vehicle fuel is exempt
Wisconsin	6	Applies to fuels not taxable under gallonage tax laws
Wyoming	3	Applies to sales price of LPG. Gasoline and diesel subject to gallonage tax are exempt

Source: Federal Highway Administration. Office of Highway Information Management.

4 summarizes the estimated service station pump prices per gallon and per gasoline-equivalent gallon by state. This table shows clearly that the way highway taxes are applied narrows the difference between diesel fuel and gasoline, while among the alternative fuels it heavily favors CNG, disfavors LPG somewhat, disfavors LNG and ethanol considerably, and just about wipes out methanol in most states as a viable economic competitor.

The key factors driving these outcomes are the disparate Federal highway taxes and the strong tendency in the states to tax on a gallonage basis unadjusted for the fuels' energy contents (except for CNG and, in a few cases, LPG).

State-by-State Fuel Price Assumptions Underlying Table 4

To construct Table 4, Relative State-by-State Fuel Prices, the pretax prices at the service station pump must be estimated. For gasoline, an average price at the refinery rack was assumed to be $0.607 per gallon. An average combined distribution cost and service station markup of $0.209 per gallon was assumed. The pump pretax price would thus be $0.816. With a Federal gasoline tax of $0.184 per gallon, the final pump price, not counting any sales taxes or other local taxes, is $1.00 plus the state gasoline tax. Clearly, the assumptions include the goal of making the calculations as simple as possible, but, in addition, they are based on industry experience. The estimate is consistent with data over the past several years from the Energy Information Administration.

This gasoline price is a composite of regular and premium gasolines. Average premium pump price is about .15¢ higher; regular is about .05¢ lower.

For diesel fuel, the starting point is a recent estimate by the *Lundberg Letter* that the national average pump price was $1.22 per gallon. Subtracting the Federal tax of $0.244 and the median State tax of $0.19 per gallon (taken from the Federal Highway Administration report

Table 4. Relative State-by-State Fuel Prices at the Pump (October 1, 1993) (Cents per Gallon of Gasoline and Alternative Fuel Price Relative to Gasoline)[a]

State	Gasoline[b] $/Gal.(Index)	Diesel/ Gasoline Gal.	GGE[c]	Methanol/ Gasoline Gal.	GGE	CNG/ Gasoline[e] GGE	LNG/ Gasoline[f] Gal.	GGE	LPG/ Gasoline[g] Gal.	GGE
Alabama	$1.18(1.00)	1.05	0.90	0.81	1.41	0.78	0.87	1.33	0.82	1.12
Alaska	$1.08(1.00)	1.05	0.89	0.72	1.25	h	h	h	0.73	0.99
Arizona	$1.18(1.00)	1.04	0.89	0.81	1.41	0.59	0.87	1.33	0.82	1.12
Arkansas	$1.19(1.00)	1.04	0.89	0.81	1.41	0.62	0.75	1.15	0.81	1.09
California	$1.17(1.00)	1.05	0.89	0.73	1.28	0.65	0.77	1.18	0.73	0.99
Colorado	$1.22(1.00)	1.03	0.88	0.80	1.40	0.73	0.86	1.32	0.82	1.11
Connecticut	$1.29(1.00)	0.96	0.81	0.82	1.43	0.76	0.88	1.35	0.75	1.02
Delaware	$1.22(1.00)	1.02	0.87	0.79	1.38	0.72	0.85	1.30	0.81	1.09
Dist of Columbia	$1.20(1.00)	1.04	0.89	0.81	1.42	0.74	0.87	1.33	0.83	1.12
Florida	$1.12(1.00)	1.13	0.96	0.80	1.39	0.88	0.86	1.32	0.81	1.10
Georgia	$1.08(1.00)	1.05	0.89	0.79	1.38	0.71	0.85	1.31	0.81	1.09
Hawaii	$1.16(1.00)	1.05	0.89	0.80	1.41	h	h	h	0.78	1.05
Idaho	$1.21(1.00)	1.04	0.89	0.81	1.42	0.74	0.87	1.33	0.78	1.06
Illinois	$1.19(1.00)	1.07	0.91	0.81	1.42	0.80	0.87	1.33	0.83	1.12
Indiana	$1.15(1.00)	1.05	0.90	0.80	1.40	0.91	0.86	1.32	0.82	1.11
Iowa	$1.20(1.00)	1.07	0.91	0.81	1.42	0.97	0.87	1.33	0.83	1.12
Kansas	$1.18(1.00)	1.06	0.90	0.82	1.44	0.73	0.86	1.32	0.82	1.10
Kentucky	$1.15(1.00)	1.02	0.87	0.80	1.40	0.90	0.86	1.32	0.82	1.11
Louisiana	$1.20(1.00)	1.04	0.89	0.81	1.42	0.84	0.87	1.33	0.83	1.12
Maine	$1.19(1.00)	1.05	0.90	h	h	h	h	h	0.82	1.11
Maryland	$1.24(1.00)	1.05	0.90	0.82	1.43	0.75	0.87	1.34	0.83	1.13
Massachusetts	$1.21(1.00)	1.04	0.89	0.81	1.42	0.92	0.87	1.33	0.73	0.99
Michigan	$1.15(1.00)	1.05	0.89	0.80	1.40	0.77	0.73	1.12	0.82	1.11
Minnesota	$1.20(1.00)	1.04	0.89	0.81	1.42	0.87	0.87	1.33	0.83	1.12
Mississippi	$1.18(1.00)	1.04	0.89	0.81	1.41	0.74	0.87	1.33	0.81	1.10
Missouri	$1.13(1.00)	1.05	0.89	0.80	1.40	0.88	0.86	1.32	0.82	1.10
Montana	$1.24(1.00)	1.04	0.89	0.82	1.43	0.74	0.74	1.14	0.83	1.13
Nebraska	$1.24(1.00)	1.04	0.89	0.82	1.43	0.91	0.87	1.34	0.83	1.12
Nevada	$1.23(1.00)	1.08	0.92	0.82	1.43	0.75	0.87	1.34	0.78	1.06
New Hampshire	$1.19(1.00)	1.04	0.89	0.81	1.42	0.94	0.87	1.33	0.82	1.11
New Jersey	$1.11(1.00)	1.08	0.92	0.79	1.39	0.86	0.81	1.24	0.77	1.04
New Mexico	$1.23(1.00)	1.01	0.86	0.82	1.43	0.75	h	h	0.79	1.07
New York	$1.23(1.00)	1.06	0.90	0.81	1.42	0.75	0.87	1.34	0.71	0.96
North Carolina	$1.22(1.00)	1.04	0.89	0.81	1.43	0.92	0.87	1.34	0.83	1.12
North Dakota	$1.17(1.00)	1.05	0.89	0.81	1.41	0.85	0.87	1.33	0.82	1.11
Ohio	$1.21(1.00)	1.04	0.89	0.81	1.42	0.81	0.87	1.34	0.83	1.12
Oklahoma	$1.17(1.00)	1.02	0.87	0.81	1.41	0.73	0.87	1.33	0.82	1.11
Oregon	$1.24(1.00)	1.04	0.89	0.82	1.43	0.94	0.87	1.34	0.83	1.13
Pennsylvania	$1.22(1.00)	1.04	0.89	0.73	1.28	0.81	0.79	1.21	0.83	1.12
Rhode Island	$1.26(1.00)	1.04	0.89	0.82	1.44	0.80	0.88	1.35	0.84	1.13
South Carolina	$1.16(1.00)	1.05	0.89	0.80	1.41	0.73	0.86	1.33	0.82	1.11
South Dakota	$1.18(1.00)	1.04	0.89	0.79	1.38	0.93	0.87	1.33	0.81	1.09
Tennessee	$1.21(1.00)	1.01	0.86	0.81	1.42	0.89	0.81	1.24	0.77	1.04
Texas	$1.20(1.00)	1.04	0.89	0.81	1.42	0.88	0.87	1.33	0.83	1.12
Utah	$1.19(1.00)	1.04	0.89	0.81	1.42	0.74	0.87	1.33	0.83	1.12
Vermont	$1.16(1.00)	1.05	0.90	0.67	1.17	0.74	0.73	1.11	0.82	1.11
Virginia	$1.18(1.00)	1.03	0.88	0.81	1.41	0.88	0.80	1.23	0.76	1.03
Washington	$1.23(1.00)	1.04	0.89	0.82	1.43	0.85	0.87	1.34	0.83	1.13
West Virginia	$1.25(1.00)	1.04	0.89	0.82	1.43	0.75	0.87	1.34	0.83	1.13
Wisconsin	$1.23(1.00)	1.04	0.89	0.82	1.43	0.88	0.87	1.34	0.83	1.13
Wyoming	$1.09(1.00)	1.05	0.89	0.71	1.24	0.80	0.77	1.19	0.73	0.98

[a] Ref. 3. Estimated gasoline prices are presented in $/gallon. For each State, the estimated gasoline price becomes the base point against which the other fuels are compared. All other prices are relative to that State's gasoline price. For Alabama, for example, the diesel price of $1.24 is divided by the gasoline price of $1.18 to give the index of 1.05, or 105% of the gasoline price per gallon. The diesel price on an energy basis is only 90% of the gasoline price per Btu.

[b] Pretax gasoline pump price assumed to be $0.816 per gallon.

[c] GGE: Gallons of gasoline equivalent (adjusts price for energy content of the fuel). Pretax pump price for diesel assumed to be $0.809/gallon.

[d] Pretax methanol pump price assumed to be $0.659 per gallon. Methanol FFV assumed to be 12 percent more efficient on methanol than on gasoline, based on California data.

[e] Pretax CNG pump price assumed to be $0.69 per gallon at 8.7 gallons per 1000 cubic feet.

[f] Pretax LNG pump price assumed to be $0.659 per gallon.

[g] Pretax LPG pump price assumed to be $0.61 per gallon. In States with fees in lieu of gallonage taxes, an average effective tax rate has been estimated.

[h] Alaska and Hawaii have not addressed their tax policy toward CNG and LNG.

of October 1, 1993) gives an average pretax pump price of $0.786, or $1.03 plus the state tax.

Methanol price at the plant was assumed to be $0.45 per gallon which, with a distribution cost of $0.209 and a Federal tax of $0.114, brings the pump price to $0.773 plus the State tax. Methanol price is currently varying between $0.35 and $0.45 per gallon; a number of studies estimate that a price nearer to $0.45 per gallon is needed to attract project-based investment capital.

An average "city gate" price for natural gas of $2.75 per thousand cubic feet was assumed. This estimate is "soft" in that it is highly dependent on how far down the gas pipeline the take-off point is and whether the gas will be priced on an interruptible or noninterruptible (higher price) basis. The closer to the wellhead, the lower this price is likely to be and vice versa. The greater the volume of gas going to vehicles, the greater the likelihood that it will be priced on a noninterruptible basis.

The assumed "city gate" price for a natural gas equates to $0.315 per energy-equivalent gallon of gasoline. Adding $0.209 for local distribution and service station markup, plus an additional $0.11 per gallon for compression costs brings the pretax pump price to $0.634 per gallon. The Federal highway tax of 48.54 cents per million Btu (5.6 cents per gallon equivalent) brings the price at the pump to $0.69 per gallon of gasoline equivalent before State tax is added.

For liquefied natural gas, Gas Research Institute studies (5) on the economics of LNG as a vehicle fuel, considered a number of different potential ways to make and deliver LNG, and estimated delivered cost to be anywhere from $0.48 per gallon to $1.03 per gallon, depending on the scale of operation, the location, and the method of liquefaction. For this exercise, the case where LNG is imported and trucked 250 miles to fuel a medium duty fleet; has been picked; estimated delivered cost is $0.66 per gallon.

An LPG price at the refinery or gas separation plant of $0.40 per gallon was assumed. Adding the standard assumed $0.209 (adjusted to $0.21 for simplicity) for distribution and $0.183 for Federal highway tax brings the price to $0.79 plus the State tax. However, retail markups for LPG vary widely; the markup assumed here assumes dealer commitment to a vehicle fuel market.

Except for the methanol flexible fuel vehicle (FFV), no adjustment has been made for the potential improvements in energy efficiency which might be available from use of engines designed specifically to take advantage of the alternative fuels's characteristics. For the methanol FFV, an assumption of a 12% better efficiency on methanol than on gasoline has been made, based on operating experience gained in California on such vehicles over a decade. As a result, the multiplier from gallons of gasoline used is 1.76, instead of 2, the energy content ratio. Potential design efficiencies might reduce the energy equivalent pump price for alternative fuels, including a dedicated methanol vehicle on M100 (100% methanol), by as much as 20%. Such adjustments were not made for the other fuels because engines designed to take advantage of each fuel's characteristics are not yet available commercially.

The pump prices for each alternative fuel in each state calculated on the basis of these assumptions were then divided by the gasoline price to get a ratio of the alternative fuel price to the gasoline price, on both the volume and energy bases. The ratios show clearly, for example, that lower pump prices for methanol and LNG per gallon (ratios less than 1.0) translate into higher pump prices per unit of energy (ratios higher than 1.0).

States charging a fee for alternative fuel vehicles instead of a tax collected at the pump have in general set their fee at a level designed to generate about the same amount of revenue as a vehicle driven an average number of miles. Thus, in those states, the higher the mileage actually driven, the lower the imputed tax rate per gallon. For LPG, for example, in states where the ratio in energy terms is close to but higher than 1.0, vehicles driven more than the average 15,000 miles per year would become economic and, the more miles driven, the more economic they become.

CONCLUSIONS

All of these assumptions are challengeable for any specific location or specific set of circumstances with unique characteristics. Pump prices vary for a number of reasons not related to taxes. Nonetheless, the assumptions used generate a reasonable starting point from which to compare the impacts of the fuel energy density and other alternative fuel characteristics in combination with the highly variable Federal and state taxes.

An obvious conclusion is that compressed natural gas, which is taxed at an energy-equivalent rate, benefits the most among the alternative fuels from highway excise taxes, most of which are imposed on a gallonage basis. Methanol, as the contending liquid fuel with the lowest energy content, is the fuel subject to the greatest disadvantage.

The survey from which these numbers were generated showed also that states are beginning to think more explicitly about tax policies for the alternative fuels, particularly in light of the new mandates for their fleet vehicles in the Energy Policy Act and the mandates in the Clean Air Act Amendments of 1990 for cleaner vehicles and fuels.

However, states also report that the need for the revenue generated by highway taxes is just about as strong an imperative as is the need to provide incentives for alternative fuels. So far, the states are not of one mind on how to respond and will be actively considering their options over the next several years.

BIBLIOGRAPHY

1. Much of this historical information is taken from: U.S. Library of Congress, Congressional Research Service, *Federal Excise Tax on Gasoline and the Highway Trust Fund: A Short History*, Report No. 94-354-174 E, Washington, D.C., April 22, 1994.

2. A. M. Sharp and B. F. Sharp, *Public Finance: An Introduction to the Study of the Public Economy*, Business Publications, Inc., Austin, Texas, 1970, p. 377.

3. The Road Information Program, *1989 State Highway Funding Methods*, p. 20.

4. J. H. Bowman and J. L. Mikesell, "Recent Changes in State Gasoline Taxation: An Analysis of Structure and Rates," *National Tax J.* **36** (June 1983).

5. *Preliminary Assessment of LNG Vehicle Technology, Economics, and Safety Issues*, Revision 1, GRI 91/0347, Prepared for GRI by Acurex Environmental Systems Division, Jan. 10, 1992.

ENERGY TAXATION: SUBSIDIES FOR BIOMASS

SALVATORE LAZZARI
Congressional Research Service
Washington, D.C.

Biomass is broadly defined as any organic material or substance other than oil, natural gas, coal, or any product or byproduct of oil, natural gas, and coal. Biomass includes plants, wood, crops, plant and animal wastes, solid wastes including municipal and industrial waste, sewage, and sludge. Historically, the Federal tax system favored oil and gas, which qualified for two major tax subsidies, over energy from biomass and other alternative energy resources, which received no tax subsidies. Oil and gas companies could use percentage depletion (instead of cost depletion) for the recovery of the investment in a well, and could expense (deduct currently, rather than capitalize) intangible drilling costs (IDCs). This preferential treatment was curtailed during the 1970s and 1980s. Restrictions were imposed upon the two oil and gas tax subsidies, and several new oil taxes were introduced, all of which raised the industry's effective tax rates and narrowed its preferential treatment relative to investments in other industries. And a variety of tax incentives were introduced targeted specifically for the development of alternative energy resources. While these new tax incentives have been provided to a wide spectrum of renewable and nonconventional forms of energy resources (solar, wind, geothermal, synfuels, shale oil, coalbed methane and many others), biofuels (such as ethanol) and other types of biomass energy became the focus of virtually every energy tax legislation.

The Federal tax system has shifted away from oil and gas toward energy conservation and the development of alternative energy resources such as biomass. Combined with the two expired biomass tax incentives, the general investment tax incentives, and the spending programs and Federal research and development programs in effect at that time, these biomass tax incentives contributed to the enormous growth in the biomass industry over the last 20 years and they have helped to sustain it today.

This chapter examines the provisions of the Federal tax code that promote biomass energy and biofuels. The first section describes each incentive in detail, along with its limitations, legislative history and any expansion or liberalization adopted by the recently enacted Energy Policy Act of 1992. The second section describes the new biomass tax incentives enacted as part of the 1992 act. The third section compares the estimated tax benefits for biomass energy and other energy alternatives in relation to oil and gas, the benchmark energy resource. Current Federal energy tax policy clearly favors biomass and other alternatives to oil and gas.

This study examines only those tax provisions (incentives) specifically targeted to biomass. Oil and gas tax incentives and penalties will be discussed only to the extent that they are useful in placing the biomass tax incentives in perspective. General tax provisions of the corporate or individual income tax systems (such as accelerated depreciation, or tax rate structure) are not discussed because they are not likely to produce significant differential effects. However, under certain restrictive conditions, some types of biomass equipment may qualify for a slightly accelerated depreciation (1). Two biomass tax incentives—the 1978 business energy investment tax credit and the 1980 tax-exempt bond provisions for biomass facilities are not discussed since they have expired.

CURRENT FEDERAL TAX INCENTIVES FOR BIOMASS ENERGY

The current Federal tax code contains four incentives for biomass energy: the partial exemptions from the various motor fuels excise taxes, particularly the 5.4¢ per-gallon exemption for gasohol; the $5.35 per barrel alternative fuels production tax credit; the tax credits for blended and pure ethanol fuels; and the new 10¢ per gallon small ethanol-producer credit. Each of these tax incentives is provided for the conversion of biomass, broadly defined, into either a liquid or gaseous fuel or for the use of the biofuel; no incentives are currently provided for the direct combustion of biomass, such as wood. However, the Energy Policy Act of 1992 contains a tax incentive for biomass used directly to generate electricity. Moreover, in the case of biomass conversion, the tax incentives are for the conversion of biomass into an alcohol fuel, primarily ethanol.

Excise Tax Exemptions for Alcohol Fuels

Current Law. The most important tax incentives for biofuels are the exemptions for blends of ethanol and motor fuels, such as gasoline and diesel, from the various Federal motor fuels excise tax. The current Internal Revenue Code (IRC) imposes excise taxes on a variety of motor fuels used in highway transportation; the tax rates depend on the type of fuel and its use. Gasoline and special motor fuels (gasoline substitutes and additives such as liquified petroleum gas, naphtha, and benzene) are taxed at 18.4¢ per gallon (IRC §4041(a)(2), and §4081). Diesel fuel is taxed at 24.4¢ per gallon (§4091, and §4041(a)). Fuels used in noncommercial aviation are taxed at either 19.4¢ per gallon (in the case of gasoline, §4041(c), and §4081) or 21.9¢ per gallon for jet fuel (§4041(c), and §4091). Fuel used for transportation on inland waterways by commercial cargo vessels is taxed at 21.4¢ per gallon, rising to 24.4¢ by 1995 (§4042). The 18.4¢ tax also applies to gasoline and special motor fuels used in noncommercial motorboats.

In the first three of the above four excise taxes, mixtures of ethyl alcohol and the otherwise taxable fuel are partially tax-exempt. For example, ethanol blended either with gasoline ("gasohol") or one of the other qualifying special motor fuels is exempt from 5.4¢ of the 18.4¢ per gallon tax (ie, the tax rate is 13.0¢. Ethanol blended with diesel—sometimes called "dieselhol"—is exempt from 5.4¢ of the 24.4¢ tax (the tax rate is 19.0¢). Ethanol blended with either aviation gasoline or jet fuel also qualifies for a 5.4¢ exemption (making the taxes on the ethanol mixtures 14.0¢ and 16.5¢, respectively).

These tax exemptions apply to mixtures that are at least 10% ethanol. If the fuel contains at least 85% pure alcohol the exemption is 6.05¢. This tax structure would make the tax rates for mixtures with 85% alcohol as follows: 12.35¢ for gasoline and alcohol; 18.35¢ for diesel and alcohol; 13.35¢ for gasoline and alcohol used in noncommercial aviation; and 15.55¢ for jet fuel and alcohol.

The excise tax exemptions for alcohol fuels mixtures, particularly the 5.4¢ gasohol exemption, has been the single most important tax incentive for biofuels, causing (along with high oil prices and state tax exemptions) fuel ethanol production (produced mostly from corn) to increase from about 40 million gallons in 1978 to over 1 billion gallons in 1991.

The 5.4¢ per gallon exemption for gasohol represents 29% of the 18.4¢ gasoline tax. When the gasoline tax was 9.1¢ per gallon the gasohol tax exemption was raised to 6¢ per gallon (66% of the tax). Under the original 1978 statute, the exemption was 4¢ per gallon (100% of the tax). Thus in absolute terms, the gasohol tax exemption first increased from 4¢ to 6¢, then decreased to 5.4¢. In relative terms, however, the gasohol tax exemption has steadily decreased. The dieselhol tax exemption has decreased from 100% to 40%, and to 22% of the excise tax.

Limitations. It should be noted that, in the case of alcohol blends, the alcohol must be at least 190 proof (95% pure alcohol), the mixture must be at least 10% alcohol, and the alcohol cannot be derived from petroleum, natural gas, or coal (including peat). The latter limitation means that exempt alcohol is, generally, derived from biomass. The only exception to the latter rule is for special motor fuels that are 85% alcohol (ethanol or methanol) derived from natural gas, which is exempt for 7.0¢ of the 18.4¢ tax (the tax rate is 7.1¢ per gallon (§4041(m)).

In the case of the blended fuels, methanol produced from coal or natural gas does not qualify for the tax exemptions because in 1977, when the ethanol exemption was first introduced, Congress believed that the cost of producing methanol from coal and natural gas was low at that time and did not warrant a Federal tax subsidy; whereas, the cost of producing methanol from wood and ethanol gram grain was costly and did warrant a subsidy.

Finally, it is important to underscore the point that the highway motor fuels excise taxes historically applied to liquid fuels only; electricity and gaseous fuels, such as compressed natural gas, were tax-exempt. Compressed natural gas, as a result of the Omnibus Budget Reconciliation Act of 1993 (P.L. 93-66), is now taxed at 48.54¢ per million Btu, the equivalent of 5.6¢ per gallon of gasoline.

Legislative History. The excise tax exemptions for alcohol fuels originated with the Senate's version of the Energy Tax Act of 1978 (P.L. 95-618), which introduced the special tax provisions for biomass and other alternative energy resources, and signaled the new shift in Federal energy tax policy.

The other four components of Carter's National Energy Plan were: the Public Utilities and Regulatories Policies Act (P.L. 95-617); the National Energy Conservation Policy Act (P.L. 95-619); the Powerplant and Industrial Fuel Use Act (P.L. 95-620); and the Natural Gas Policy Act (P.L. 95-621). For biomass energy, the Energy Tax Act introduced the excise tax exemptions for alcohol fuels, the business energy investment tax credits, and the tax-exempt bond provisions to provide financing incentives. The Energy Tax Act also provided for the gas-guzzler tax, incentives for van pooling, and miscellaneous energy tax provisions.

The Tax Act was part of the Carter administration's National Energy Plan. However, tax provisions for conversion to alternative transportation fuels (gasohol) were not part of the original House bill (H.R. 8444, 95th Congress), which embodied the tax provisions of the National Energy Plan. The version of H.R. 8444 reported out of the Senate Finance Committee included the excise tax exemptions for gasohol fuels. The exemption was intended to induce a substitution of ethanol and methanol for gasoline used in transportation. The underlying rationale for the ETA was the perceived failures in the energy markets in allocating resources efficiently and fairly, in coping with the 1973 oil embargo, and in adjusting to the sharp increases in energy prices, the shortages, and the associated adverse economic and social problems. The belief was that the Federal Government had to influence resource allocations through tax incentives and other financial incentives. Each of these tax instruments was intended to contribute to the general goal of conserving conventional energy resources (primarily oil and gas), stimulate production of alternatives to oil and gas, reduce oil imports, and achieve energy security.

The original exemption was for the full amount of the gasoline tax: 4¢ per gallon. The Crude Oil Windfall Profits Tax (P.L. 96-223) extended the 4¢ exemption from October 1, 1984 to December 31, 1992. The Surface Transportation Assistance Act of 1982 (P.L. 97-424) raised the gasoline tax from 4¢ to 9¢ per gallon and also changed the exemption for gasohol from the complete 4¢ exemption to a partial 5¢ exemption (gasohol would be taxed at 4¢ per gallon instead of 9¢ per gallon). The Tax Reform Act of 1984 (P.L. 98-369) raised the diesel fuel tax from 9 to 15¢ per gallon as part of a compromise that also lowered the highway use taxes on trucks. The 1984 tax law also raised the gasohol exemption from 5 to 6¢ (ie, it reduced the tax rate for gasohol from 4 to 3¢), and retained the 9¢ exemption for "neat" alcohol fuels, and provided that alcohol produced from natural gas would also qualify for the exemption. The Tax Reform Act of 1986 (P.L. 99-514) reduced the excise tax exemption for 85% alcohol from 9¢ to 6¢ per gallon (for sales made beginning on 1987). The Technical and Miscellaneous Revenue Act of 1988 (P.L. 100-647) made minor liberalizations to the excise tax rules. Finally, the OBRA of 1990 reduced the exemption to 5.4¢ per gallon.

Amendments in the Energy Policy Act of 1992. P.L. 102-486 extends the gasohol excise tax exemption to gasohol that contains less than 10% alcohol. Two categories of gasohol mixtures are prescribed: mixtures containing 7.7% alcohol; and mixtures containing 5.7% alcohol. The exemption for 7.7% mixtures is 4.16¢ per gallon (the tax rate is 9.94¢); the exemption for 5.5% mixtures is 3.08¢ per gallon (the tax rate is 11.02¢).

The Alternative Fuels Production Tax Credit

The second major tax break for biomass fuels is the alternative fuels production tax credit, also known as the "§29 tax credit," named after the section of the IRC in which it resides.

Current Law. The alternative fuels production tax credit is a credit against the producer's income tax for the *production and sale* of fuels derived from a wide variety of alternative energy resources. Qualifying fuels are grouped into three categories: oil from shale or tar sands; liquid or gaseous synthetic fuels from coal, gas from coal seams (coalbed methane), tight sands, Devonian shale and geopressurized brine; and liquid or gaseous fuels from wood, agricultural byproducts, and other biomass. Certain types of alcohol fuels qualify for this credit including both ethanol and methanol, and alcohol produced from coal and lignite. Moreover, alcohol fuels produced from coal or lignite may be used as feedstocks, unlike other fuels, without invalidating the tax credit.

Two other types of biomass have qualified for the production tax credit under the provisions of the original 1980 law: processed wood fuel treated to increase the BTU content of the wood by at least 40%, and steam from agricultural products. These two types of biomass no longer qualify for the tax credit under original expiration dates. The credit for processed wood fuel expired on October 1, 1983, and the credit for steam from solid agricultural products expired on December 31, 1985. Solid agricultural products included only solid byproducts of farming or agriculture and excluded timber products or other forms of biomass. Thus waste wood used directly as a fuel did not qualify for the credit.

The production tax credit is $3.00 per barrel, in real terms, in barrels of oil or oil equivalent. The credit is linked with the BTU (British Thermal Unit) content of oil. Each 5.8 million BTUs of fuel, the energy content of one barrel of oil, qualifies for the $3.00 credit. The $3.00 amount is adjusted for inflation (using 1979 as the base year and the GNP deflator as the price index), which makes the current credit about $5.75 per barrel of oil equivalent (for qualifying gases the credit is about $1.00 per thousand cubic feet).

The availability of the credit is linked to the average wellhead price of domestic crude oil (called the reference price). When the reference price of oil is below $23.50 (in real terms), the tax credit becomes available; when the price of oil is between $23.50 and $29.50, the credit is phased out proportionately; when the price of oil is above $29.50, no credit is available. These trigger or threshold prices are also adjusted for inflation so that a comparison may be made with the reference price in nominal terms. At this writing, the phase-out range in current dollars is between $40 and $50 per barrel. With the market price of West Texas Intermediate crude oil (the reference price in nominal terms) at about $22 per barrel, well below the $40 ($23.50 times the inflation adjustment factor of 1.70), the credit is available. The Congress has believed that when oil prices are high, market incentives should suffice to stimulate production of alternative fuels.

Until recently, most of the current credit has gone for coalbed methane gas and very little for biomass. Coalbed methane is eligible for the production tax credit, and it has generated much of the revenue losses from the credit. Coalbed methane is a colorless and odorless natural gas that permeates coal seams, and is virtually identical to conventional natural gas. Under IRC§29, coalbed methane is treated as an unconventional gas because it resides in an unconventional location—coal beds—as opposed to conventional gas reservoirs. The combined effect of the $1.00 per MCF tax credit (which was at times 100% of natural gas prices) and declining production costs (due to technological advances in coalbed methane drilling and production techniques) was sufficient to offset the decline in oil and natural gas prices, and the resulting decline in domestic conventional natural gas production. Data show that production of coalbed methane has increased from 0.1 billion cubic feet in 1980 to over 300 billion cubic feet in 1991, a large part of it in response to the production tax credits, and virtually all of it at the expense of conventional gas production. The credit for coalbed methane benefits largely oil and gas producers, both independent producers and major integrated oil companies, and complicates calculation of the special tax benefits to biomass and other alternative energy resources in relationship to the net tax burdens on oil and gas.

Limitations. The production tax credit is available for fuels produced through December 31, 2002. However, the facilities must be placed-in-service (or from wells drilled) after 1979 and before 1993. To prevent "double dipping" the credit is reduced by any subsidized financing (grants, loans, tax-exempt financing) energy tax credits, including the enhanced oil recovery tax credit enacted as part of the OBRA 90. The credit is nonrefundable and is limited to the excess of a taxpayer's regular tax over several tax credits and the tentative minimum tax.

Legislative History. The production tax credit was introduced by the 1980 windfall profit tax (WPT) and has been amended several times. Most of the amendments have concerned the placed-in-service deadlines. The original 1980 WPT established a placed-in-service deadline of December 31, 1989 and was extended to December 31,1990 by the Technical and Miscellaneous Revenue Act of 1988, and to December 31, 1991 by OBRA of 1990, which also liberalized the treatment of tight-sands gas.

Amendments in the Energy Policy Act of 1992. H.R. 776 provides for a three-year extension of the placed-in-service rule, and an extension of the credit, for biomass and synthetic fuels only. Under these new rules, producers of gas from biomass or liquid, gaseous or solid synthetic fuels from coal or lignite, pursuant to a binding contract signed before January 1, 1996, would have until December 31, 1996 to build a facility and begin production. If these conditions are met, then production tax credits for these fuels would be available for another five years through December 31, 2007 (instead of December 31, 2002). These amendments mean that production of biomass and synfuels from new facilities or new wells will no longer qualify for the §29 credit if the new facilities are placed in service after 1997. For all other alternative fuels, production from facilities or wells placed in service after 1992, including coalbed methane wells, no longer qualify for the credit.

The Two Alcohol Fuels Tax Credits

In place of an exemption, the Federal tax code provides an alcohol fuels mixtures tax credit (the blender's credit), and a tax credit for straight alcohol fuel. Both credits are part of IRC §40.

Current Law. The blender's tax credit is 54¢ per gallon for alcohol that is at least 190 proof, and 40¢ per gallon for alcohol that is at least 150 proof but less than 190 proof. No credit is available for alcohol that is less than 150 proof. This credit is available to the blender only for use as a motor fuel in a trade or business whether produced and used by him or produced and sold by him. The straight alcohol fuels credit is 60¢ per gallon for alcohol that is at least 190 proof, and 40¢ per gallon for alcohol between 150 and 190 proof. This credit is typically available to the retail seller that dispenses it in the fuel tank of the buyer's vehicle. These credits have been available since October 1, 1980, and will; under current law, continue to be available through December 31, 2000.

The alcohol fuels tax credits apply to most types of ethanol (ie, alcohol derived from renewable energy resources such as vegetative matter, crops, and other biomass) and to methanol derived from wood. The alcohol cannot be derived from petroleum, natural gas, or coal (including peat). This rule effectively limits the tax credits to ethanol since most economically feasible methanol is derived primarily from natural gas and does not qualify for the credits. Currently, about 95% of current ethanol production is derived from corn. A 1990 IRS ruling has allowed mixtures of gasoline and ETBE—Ethyl Tertiary Butyl Ether–to qualify for the 54¢ blender's credit. ETBE is a compound that results from a chemical reaction between ethanol (which may be produced from renewable energy) and isobutylene. ETBE, which is no longer an alcohol after its chemical reaction, is being considered for use as a substitute for MTBE–Methyl Tertiary Butyl Ether—as the oxygenate in reformulated gasoline mentioned under the Clean Air Act of 1990 for use in designated ozone nonattainment areas. Allowing ETBE to qualify for the blender's tax credit is designed to stimulate the production of ethanol for use in reformulated gasoline and to reduce the production of MTBE, an alternative oxygenate made from natural gas. This strategy would increase the share of the U.S. corn crop allocated to ethanol production above the current 4–5%. It would also significantly increase Federal revenue losses from the alcohol fuels credits, which heretofore have been negligible due to blender's use of the exemption over the credit.

Limitations. There are several limitations to the two alcohol fuels tax credits: First, the alcohol cannot be derived from petroleum, coal, or natural gas. Thus, the alcohol must be derived from renewable energy resources such as vegetative matter, crops, wood, and other biomass to qualify for the credits. This limitation means that the credits are currently available only to ethanol, since most economically feasible methanol is made from natural gas, which does not qualify for the credits. Second, the alcohol

fuels tax credits are a component of the general business credit under IRC §38, which includes the jobs tax credit, research and development tax credit, low-income housing tax credit, and other credits. The general business tax credit is limited to the taxpayer's net income tax over the larger of either 25% net regular tax liability above $25,000 or the tentative minimum tax. Any unused general business credits may generally be carried forward 15 years or carried back three years.

Any taxpayer who claims the credit must also report it as gross income for the tax year in which the credit is earned (IRC §87). Fourth, the credits are not refundable; they may be used only against a positive tax liability; they are of no value of the producer has no tax liability. Finally, and more importantly, the credits are offset by the excise tax exemptions claimed on the same fuel. Typically, blenders prefer the excise tax exemption over the credit because the exemption is an immediate "up-front" tax benefit, which increases cash-flow, while the benefits from the tax credit must await the preparation of the tax returns. The limitations under IRC §38 also could reduce the value of the alcohol fuels tax credits.

Legislative History. The alcohol fuels tax credits were enacted as part of the 1980 WPT at the initial amounts of 40¢ and 30¢ per gallon for proofs greater than 190 and between 150 and 190 respectively, and raised by the Surface Transportation Assistance Act of 1982 to 50¢ and 37.5¢, respectively. The Tax Reform Act of 1984 raised the credits to 60¢ and 45¢. This law also introduced the provision that alcohol produced from peat, like coal, would not qualify for the credits.

Ethanol Tax Credit for Small Producers

An additional tax incentive for biofuels production is the small ethanol-producer credit introduced as part of OBRA 90.

Current Law. IRC §40 provides for an income tax credit of 10¢ per gallon ($4.20 per barrel) for up to 15 million gallons of annual ethanol production by a small ethanol producer. A "small producer" is defined as one with ethanol production capacity of less than 30 million gallons per year (about 2,000 barrels per day). This credit is available for alcohol produced by a small producer and sold to another person for blending into a qualified mixture in the buyer's trade or business, for use as a fuel in the buyer's trade or business or for sale at retail where such fuel is placed in the fuel tank of the retail customer. Casual off-farm production of ethanol does not qualify for this credit.

Limitations. The new ethanol tax credit is limited to small producers, and aggregation rules are provided to prevent the credit from benefiting producers wth a capacity in excess of the 30 million gallons per year limit, or from going to production in excess of the 15 million gallon limit.

Legislative History. The small-producers credit was added by the Omnibus Budget Reconciliation Act of 1990 (P.L. 101-508); it has not been amended.

New Biomass Tax Incentives in The Energy Policy Act of 1992

The end of 1992 has witnessed the enactment of broadly-based energy legislation—The Energy Policy Act of 1992—designed to resuscitate a Federal energy policy that many argued was missing. The law, P.L. 102-486, which some call the crowning achievement of the 102nd Congress, includes numerous provisions intended to reduce dependence on petroleum imports: a restructuring of the electric utility industry to enhance competition, promotion of alternative transportation fuels, provisions that mandate new and more stringent energy efficiency standards, and numerous other provisions. Title XIX of the law contains a variety of energy taxes and tax incentives. In addition to the liberalization of existing biomass tax incentives, as discussed above, H.R. 776 creates two new tax incentives for biomass: an income tax deduction for the costs of a vehicle that burns alcohol or some other clean-burning fuel; and an income tax credit for electricity generated from "closed-loop" biomass facilities.

Income Tax Deduction for Alcohol Fuel Vehicles and Alcohol Storage Facilities

Beginning on July 1, 1993, the purchaser of a new vehicle that burns either ethanol, methanol, ether, any combination of these, or some other clean-burning fuel, has qualified for a limited income tax deduction for the costs of the vehicle allocable to the engine and any collateral equipment used to store or deliver the fuel. The maximum deduction is $2,000 for cars, $5,000 for light trucks, and $50,000 for heavy trucks or buses. This deduction is also provided for the costs of retrofitting used non-qualifying vehicles to clean-fuel burning vehicles, and for the costs of equipment used to store or dispense the alcohol fuels and other clean-burning fuels into the tank of the clean-burning vehicle. Qualifying storage and dispensing equipment includes equipment used to compress natural gas into a usable fuel, provided that the equipment is located on the site that dispenses the fuel into the vehicle. The storage equipment deduction is limited to $100,000 per year. Alcohol fuels must contain 85% alcohol. Other clean burning fuels that qualify for the deduction are compressed natural gas, liquefied petroleum gas, and hydrogen.

Income Tax Credit for Closed-Loop Biomass Facilities

The second new tax incentive in P.L. 102-486 that would benefit biomass energy is the income tax credit for production of electricity from "closed-loop" biomass systems defined as systems that use renewable plant matter exclusively as an energy source to generate electricity. The credit, which also applies to electricity generated from wind energy systems, equals 1.5¢ per kilowatt hour of electricity and is phased out, proportionately, as the national average price of electricity produced from renewables rises from 8.0¢ to 11¢ per kilowatt hour. Both the credit and the phase-out limit are adjusted for inflation. It is important to note that biomass is narrowly defined for purposes of this new credit. It does not apply to municipal or agricultural waste or to scrap wood and other wastes. This credit will be part of the general business credit, and its limitations, which were discussed above. It will also be offset by any type of grant or subsidized financing, including tax-exempt bond financing.

THE MAGNITUDE OF BIOMASS ENERGY TAX BENEFITS IN RELATION TO OIL AND GAS

This section on the study estimates the tax benefits to biomass and other energy alternatives that result from the various tax incentives peculiar to that industry. These are compared to any tax benefits that might accrue to oil and gas, the benchmark energy resource, as a result of Federal tax incentives and taxes peculiar to that industry. For each industry, first we estimate aggregate tax benefits, then we estimate tax benefits per unit of output.

Aggregate tax benefits to each industry are measured as the sum of the losses in Federal tax revenue—also referred to as "tax expenditures"—that result from each of the various tax incentives peculiar to that industry. The source for these estimates is the annual tax expenditure study published by Congress's Joint Committee on Taxation, which shows these losses for each of the industry-specific tax incentives. It is assumed that these losses in Federal tax revenues are equivalent to the reduction in industry tax liabilities. Tax benefits per unit of output are measured as the total industry tax benefits divided by estimated output. The latter is an indicator of the effect of the tax benefits on relative prices among the fuels, which forms the basis of economic decisions. This particular framework of analysis abstracts from the general provisions of the income tax laws; it considers only those provisions peculiar or special to each industry. The comparison will show that biomass energy receives a fairly sizeable net tax subsidy compared with oil and gas, which is subject to a small, but positve, net tax.

Aggregate Tax Benefits

Table 1 shows the four tax incentives for biomass and other alternative energy resources and the corresponding annual revenue loss—reductions in industry tax liabilities—resulting solely from those incentives as calculated by the Joint Committee on Taxation. The table includes only the incentives prior to the enactment of P.L. 102-486.

House of Representatives Bill 776 contains tax relief for both the oil and gas industry as well as the biomass industry. In absolute dollar terms, the tax relief for the two industries over five years is about the same, reducing taxes by about $200 million annually. But, the tax relief for biomass and alternative energy resources is larger, in relation to industry size, than for oil and gas. Moreover, oil and gas would still be a net special taxpayer, while the net tax subsidy to alternative energy resources would increase. Thus after H.R. 776, the posture of energy tax policy is even more slanted in favor of biomass and other energy alternatives. These tax incentives resulting from H.R. 776 are excluded from our analysis because they are negligible for FY93, the period of analysis, and because we were unable to separate the tax decreases for biomass from the tax decreases for other qualifying fuels (2). The biomass and alternative energy industry tax liabilities will be approximately $1,250 million lower in FY93 as a result of the four tax incentives peculiar to that industry.

Table 1. Special Tax Incentives for Biomass and Alternative Energy Industry and Corresponding Reductions in Industry Tax Liabilities, Fiscal Year 1993[a]

Description of Tax Incentive	Principal Limitation	Expiration Date	Reduction in Tax Liabilities ($ millions)
	Special Tax Incentives (−)		
5.4¢ Excise tax exemption	Must be at least 10% alcohol	90-30-2000	−400
$5.35 Per barrel alternative fuels production tax credit	Market oil price below $50 per barrel	12-31-2000	−800
54¢ Per gallon alcohol fuels tax credits	Applies primarily to ethanol	12-31-2000	−50
10¢ Per gallon new ethanol producer credit	Up to 15 million gallons of output per year	12-31-2000	[b]
	Special Taxes (+)		
None			0
			Net Tax = −1,250

[a] Ref. 3
[b] The revenue loss for this provision is part of the revenue loss corresponding to the alcohol fuels tax credits, as shown in the third row of the table

In other words, biomass and other alternative energy resources will receive a net tax subsidy, ie, above and beyond the general provisions of the income tax laws. Note that, unlike oil and gas, there are no industry-specific Federal taxes targeted for this industry. It is difficult to determine the share of this $1,250 million that accrues specifically to biomass energy because the sum revenue loss corresponding to the alternative fuels tax credit is not disaggregated. Much of this benefit undoubtedly accrues to biomass, but a portion of it also accrues to such fuels

as coalbed methane, which some consider not to be an alternative fuel. If half of this tax benefit accrues to coalbed methane production, then the total tax benefit to alternative energy resources is reduced to $850 million for FY 1993.

In order to put these estimates in perspective, Table 2 shows Joint Committee of Taxation calculations of the various industry-specific Federal tax incentives and penalties imposed on oil and gas. Oil and gas is helped from four tax subsidies (including the two discussed in the in-

Table 2. Special Oil and Gas Tax Incentives and Taxes for and the Corresponding Reduction (−) and Increase (+) in Tax Liabilities[a]

Description of Tax Provision	Principal Limitation	Expiration Date	Revenue Effect, ($ Millions)
	Special Tax Subsidies (−)		
Expensing of intangible drilling costs			
Corporations must amortize 30% of costs	None	None	−200
	Available only for domestic properties		
Percentage depletion allowance[b]	Availabe to independents on 1,000 barrels per day	None	−100
Exempt from passive loss rules	None	None	[c]
15% Credit for enhanced oil recovery	Oil prices < $28	None	−50
	Special Taxes (+)		
9.7¢ Per barrel Superfund tax	None	12-31-95	+553
5.0¢ Per barrel oil spill tax	None	12-31-94	+285
Alternative minimum tax	None	None	[c]
			Net Tax = +488

[a] Refs. 3, 5.
[b] Includes special tax incentives for stripper oil and heavy oil enacted as part of OBRA 900.
[c] The revenue effects from this provision is part of the estimated revenue loss from the expensing and percentage depletion tax incentives.

troduction) and is penalized by three special taxes. The tax subsidies are: percentage depletion allowance, which is available only to independent producers, and only for limited quantities of output; expensing of domestic IDCs, which is available to all who drill for oil and gas, subject to limitations for corporate producers; exemption from the material participation requirements of the passive loss limitation rules; and a new 15% tax credit for enhanced oil recovery techniques. The tax penalties are: the alternative minimum tax imposed on oil and gas producers; a 9.7¢ tax per barrel on domestic and imported crude oil to finance Superfund; and a 5¢ tax per barrel on domestic and imported crude oil to finance the Oil Spill Liability Trust Fund.

The Comprehensive Environmental Response, Compensation, and Liability Act of 1980 (P.L. 96-510)–commonly known as Superfund–originally imposed a 0.79¢ excise tax on oil received by a domestic refinery. Under the Superfund Amendments and Reauthorization Act of 1986 (P.L. 99-499), this tax was increased to 8.2¢ per barrel of domestic oil and 11.7¢ per barrel of imported oil. This tax differential was ruled to be in violation of GATT–the General Agreement on Tariffs and Trade–to which the United States is a signatory and a 9.7¢ tax was imposed equally on domestic and imported oil as part of the Steel Trade Liberalization Program Implementation Act (P.L. 101-221). The 5¢ per barrel tax on crude oil for the ol spill liability trust fund was enacted under Omnibus Budget Reconciliation Act of 1989 (P.L. 101-239). This tax, also part of the 1980 Superfund law, was originally 1.3¢ per barrel, but no revenues were ever collected because the enabling legislation required to activate the tax under the 1908 law was not enacted. Another excise tax on domestic crude oil was introduced in the 1908 windfall profit tax but this tax was repealed in 1988 by the Omnibus Trade and Competitiveness Act (P.L. 100-418).

Table 2 shows that the four special tax subsidies for oil and gas development are projected to reduce industry tax liabilities by about $350 million in FY 1993 (net of the alternative minimum tax). If the estimated $400 million production tax credit for coalbed methane is scored as a tax subsidy for conventional fuels, then the oil and gas subsidy increases to $750 million for FY 1993. Table 2 also shows that there are two special excise taxes on oil and gas that increase the tax burdens on the oil and gas industry. These special taxes are projected to increase the tax burden on oil and gas by an estmated $838 million in FY 1993. The net effect of the special tax preferences and special taxes is a net special tax burden of +$488 million for FY 1993 ($838–350) without coalbed methane and +$88 million with coalbed methane.

The revenue loss from the expensing of IDCs shown in Table 2 has two important shortcomings: First, it excludes expensing of losses from abandoned properties because the Joint Tax Committee, which determines tax expenditure items and estimates the corresponding revenue losses, does not consider it a tax expenditure. Second, a revenue loss estimate for one year shows only a cash-flow effect, which is not a good indicator of the value of IDCs that are based on timing. Calculations that take these two shortcomings into account lower the next tax burden on the oil industry somewhat, but it is still a positive tax burden. These calculations do not affect the net tax subsidy to biomass and other energy alternatives, therefore do not affect the major conclusions of the report.

Tax Benefits Per Unit of Output

Estimates of aggregate tax benefits do not take into account the fact that economic decisions are based on relative prices per unit of output. To estimate the effect of the various tax incentives and taxes (in the case of oil and gas) on the relative prices of the two fuels we merely have to correct for the size difference between the two industries, which is significant. The U.S. oil and gas industry is ostensibly very large in comparison with the alternative energy industry, which was basically nonexistent until the 1970s. In 1990, for example, oil and gas output accounted for 53% of total domestic energy production, as compared with 9% for total renewable energy and 4.5% for biomass, broadly defined (4).

Size differences between the two industries are accounted for by computing the net tax subsidy per unit of output for both the biomass and alternative energy sources industry and the oil and gas industry. This requires an estimate of each industry's output, which is then projected for 1993. In the case of oil and gas, this is relatively easy undertaking because the data are readily available. The absence of output data makes it more difficult for biomass and other alternative fuels. As a result, output data must be imputed from the tax expenditure data presented in table 1. At $5.35 per barrel, the production tax credit is estimated to lose $800 million in FY 1993, implying industry output of 149.5 million barrels of oil equivalent of alternative fuels ($800 ÷ $5.35). The same approach is applied to each tax incentive item in Table 1, and the results are summed to determine total energy output for 1993. Each industry's total tax subsidy (for the alternative energy industry) or net penalty (for the oil and gas industry) is divided by each industry's total energy output. These estimates are made under two scenarios, depending upon whether coalbed methane is treated as an alternative or conventional fuel.

If coalbed methane is treated as an alternative fuel, the estimated Federal subsidy for alternative energy resources as a whole is about 72¢ per million BTUs or about $4.25 per barrel of oil equivalent. In contrast, oil and gas experiences a net tax burden estimated to be about 1.3¢ per million BTUs, which is equivalent to about 7.5¢ per barrel of oil. When coalbed methane is treated as a conventional fuel, the tax subsidy for alternative energy sources declines to about $3.00 per barrel and the net tax burden on oil and gas increases to about 2¢ per barrel. Thus, the net special tax burden on the oil and gas industry is small per unit of output, while the net tax subsidy for alternative energy resources is large per unit of output.

These estimates suggest that Federal tax incentives lower the market price of biomass energy by an estimated $3.00 to $4.25 per barrel of oil equivalent ($0.50 to $0.75 per MCF of gas) below what the market price would be without the incentives. Clearly, the Federal tax incentives help to narrow, but probably not completely eliminate the competitive disadvantage of biomass and other energy alternatives relative to conventional oil and gas. The estimates also indicate that Federal energy tax policy is, at this writing, slanted in favor of alternative energy resources, a policy posture that constitutes a reversal of the historical posture in favor of oil and gas.

BIBLIOGRAPHY

1. G. L. Middleton, Jr., Impact of Future Tax Incentive Legislation on the Development of Biomass Energy. *Biolog*, **7**, 7–11 (Feb./March 1990).
2. U.S. Congress House, Committee On Ways and Means, *Comprehensive National Energy Policy Act*, House Report No. 102-474, Part 6, U.S. Government Printing Office, Washington, D.C., May 5, 1992, p. 94.
3. U.S. Congress, Joint Tax Committee, *Estimates of Federal Tax Expenditures for Fiscal Years 1993–1997*, Joint Committee Print, Washington, D.C., April 24, 1992, p. 24.
4. U.S. Department of Energy, *Annual Energy Review*, DOE/ EIA-0384(91), Washington, D.C., June, 1992, pp. 241–257.
5. U.S. Congress, Joint Committee on Taxation, *Schedule of Present Federal Excise Taxes (As of January 1, 1992)*. Joint Committee Print No. JCS-7-92, 102d Cong., 2d sess. U.S. Government Printing Office, Washington, D.C., March 27, 1992, 35p.

ENERGY TAXATION: BIOMASS

SALVATORE LAZZARI
Congressional Research Service
Washington, D.C.

In general, the outlook for Federal energy tax policy is quite favorable for biomass. Biomass energy enjoys considerable support in the Congress, in the environmental community, and among farming interests. Biomass is a renewable energy resource with a huge resource base. Wood, an important energy resource in the underdeveloped world, is still used as a fuel in the United States. A variety of liquid and gaseous fuels can be produced from crops, crop wastes, residues, solid waste, municipal waste, and other plant and animal matter. Direct combustion of biomass can generate steam, heat, and electricity, in addition to ethanol, methanol, synthetic natural gas, and synthetic gasoline.

In the short-run, Federal energy tax policy is likely to witness frequent amendments of the type enacted in OBRA 1990, or recently in the Energy Policy Act of 1992. Federal energy tax policy will be activist and interventionist, as tax policy in general is likely to be. While both biomass and oil and gas are likely to be the focus of the energy tax laws, the thrust of this short-term policy will likely be to favor alternatives to oil and gas, especially biomass. The trend in energy tax policy that began in the 1970s is likely to continue. For example, we are likely to witness additional tax incentives for oil and gas, in the spirit of those recently enacted in OBRA 1990 (which introduced a new tax credit for enhanced oil recovery costs). These incentives proposals are motivated by the concern over declining domestic oil production, which peaked in 1970 and has since declined by an average of 4 percent per year, and rising oil imports which reached new records recently. These incentives are likely to focus on marginal, high cost oil such as from small wells or from enhanced oil recovery projects. However, any future tax breaks for oil are likely to be small due to the large deficits in the Federal budgets, and they are unlikely to have important implications for the development of biomass and other alternative energy resources, which is affected relatively more by the price of oil.

In the longer run, the outlook for biomass energy tax incentives will be shaped by three main issues: oil import dependence and energy security, which has driven energy policy for the last 20 years; air pollution problems (greenhouse effects and global warming, ozone); and Federal budget deficits. The first two issues suggest tax policy initiatives that promote alternative transportation fuels and alternative forms of energy such as biomass, which can improve the environment and reduce energy consumption and petroleum imports.

Large budget deficits, however, make it unlikely that significant tax incentives or subsidies will be enacted for either oil and gas or biomass. Rather, large budget deficits suggest that energy taxes on fossil fuels (particularly oil and coal) will be the preferred policy instrument, rather than tax subsidies, to achieve energy, environmental and fiscal policy objectives. Three types of energy tax options are likely to be seriously considered by Federal policymakers: gasoline taxes, oil import taxes, and broadly-based energy taxes such as carbon taxes. Each of these proposals would have important implications for the future of biomass.

HIGHER GASOLINE TAXES

The Federal excise tax on gasoline and special motor fuels is currently 18.4¢ per gallon; the rate on diesel is 24.4¢ per gallon. As recently as 1982 the tax rate on motor fuels was 4¢ per gallon. Increases in rates were enacted in 1982, 1984, and 1990. Including State and local gas taxes, the current tax rate on gasoline averages over 30¢ per gallon in the U.S. (about 30% of retail price), considerably lower than Western industrialized countries where gas taxes average about $3.50 per gallon (about 70% of price), and approach $4.00 per gallon (Italy). Low excise taxes and falling oil prices recently have reduced the real price of gasoline to the lowest level in about forty years. Proposals to increase gasoline excise taxes were common after the 1973–1974 oil embargo and subsequent increases in oil prices. Such proposals were motivated primarily as an effective method to conserve fuel and reduce imported oil. Gasoline tax hike proposals are perennial favorites in Washington as a relatively painless revenue raising and deficit reducing option. More recently proponents have mentioned gas tax hikes as an environmental policy, and as a way to finance public infrastructure investments. One popular, recent proposal would have hiked the gas tax by a total of 50¢ per gallon to be phased-in at 10¢ per year.

Higher gas taxes can simultaneously generate large revenues, reduce petroleum consumption and imports, and reduce vehicle emissions. It is estimated that revenues increase by more than $1 billion per year for every penny increase gas tax increase. An additional $200 million per year can be collected from each penny increment to the diesel fuel tax. A sizeable tax could be imposed and still leave gasoline prices relatively low. Nearly 2/3 of the 17 million barrels per day of petroleum is used in transportation, and nearly 2/3 of that—approximately the level of petroleum imports—is used in passenger cars. Emissions from mobile sources are a significant source of urban air pollution due to the large numbers of vehicles and the total miles driven.

In addition to the potential revenue, energy security, and environmental benefits, higher taxes on gasoline would be an important economic incentive for further development of biofuels, such as ethanol, methanol, and other alternative transportation fuels. A higher gas tax could render methanol from natural gas much more competitive, especially if the excise tax exemption limitations are repealed. If the increases are large enough ethanol from biomass might also be rendered competitive with gasoline and diesel. For a more comprehensive discussion of the pros and cons of a gas tax increase, see Ref. 1.

OIL IMPORT DEPENDENCE AND OIL IMPORT TAXES

Another energy tax option that would have significant economic benefits for biomass energy is the oil import tax or fee. As was mentioned above, there are presently two Federal excise taxes on imported and domestic oil that finance (in part) the Superfund and the oil spill trust fund. From 1981 to 1986, various oil import fee proposals were introduced primarily for the purpose of raising revenues and reducing large Federal budget deficits. The collapse of oil prices in 1986 generated much interest in an oil import tax as a way of helping the distressed U.S. oil industry. More recently, the oil import tax is also viewed as a method of lessening U.S. dependence upon foreign oil, thereby enhancing energy security and improving the balance of trade.

Federal policymakers are concerned that excessive dependence upon OPEC (Organization of Petroleum Exporting Countries) and other foreign oil producers entails a vulnerability to either supply disruptions or a sudden price spike. U.S. oil imports have been rising from about 8% of demand in 1947 to a record 54 percent of total oil demand, a total of 9.1 million barrels per day, in January 1990. Over 50% of oil imports come from OPEC. This is in sharp contrast to the 1930s, when the U.S. was the world's leading oil producer and exporter, accounting for about 2/3 of world oil output. Oil import dependence and vulnerability and energy security have been the single most important issue driving U.S. energy policy and energy tax policy over the last 20 years. This concern goes back to the two energy crises of the 1970s and was accentuated by the oil price shock of 1990–1991 caused by the conflict with Iraq.

Although many policies have been proposed to address the concern of energy security and rising dependence upon foreign oil, including tax incentives to expand oil and gas production and an increase in the gasoline excise tax, one prominent proposal is the imposition of a sizeable tax, usually of $5 or $10 per barrel, on imported crude oil and petroleum products. Taxes of these magnitudes would raise oil prices and promote the production of not only oil and gas and other fossil fuels, but also stimulate the commercialization of alternative energy technologies and the production of renewable and unconventional energy resources. In terms of its effects on the posture of energy tax policy, an oil import tax would, initially tend to favor oil and gas because the tax would raise oil prices and stimulate increased investment in oil and gas. In the long run, higher oil prices would lead to substitution of biomass and alternative energy resources for oil and gas. A sizeable oil import tax would, thus, be very beneficial in stimulating the development of biomass and other energy alternatives (2).

The oil import fee would also raise revenues, although its potential is significantly less than the gasoline tax. This is an important feature when $400 billion deficits are looming. The environmental consequences, however, are not all positive, which renders the oil import fee a less desirable option than the gasoline tax. Opponents of the fee also argue that it would increase inflation and unemployment in the economy, hurt lower income groups more than higher income groups, and discriminate against oil consuming States in the Northeast.

CARBON TAXES, BTU TAXES, AND OTHER ENERGY TAXES

A variety of broad-based energy taxes proposed in recent years to address the United States' energy, environmental, and fiscal problems would have significant economic effects on biomass energy. These include carbon taxes (excise taxes on fossil fuels based on their carbon content), BTU taxes (taxes based on the heat content of various fuels), ad-valorem taxes on energy (excises on the value of energy resources), and other broadly-based energy taxes. While the precise impacts on biomass and other energy alternatives varies, these proposals would, like the oil import tax, increase the price of oil (the benchmark energy resource) and promote production and use of energy from biomass (3).

BIBLIOGRAPHY

1. S. Lazzari, *Gasoline Excise Tax: Economic Impacts of an Increase*, Issue Brief No. 93028, Washington, Congressional Research Service, 1992, 13p.
2. S. Lazzari, *Taxation of Imported Oil*, Issue Brief No. 87189, Washington, Congressional Research Service, 1988. 14p.
3. S. Lazzari, *Btu Taxes, Carbon Taxes, and Other Energy Tax Options for Deficit Reduction*, CRS Report #90-384 E, 1990, 19 p.

ENVIRONMENTAL ANALYSIS

The results given by analytical methods are critical to making sound decisions about the changes in the environment and the sources of those changes. Sometimes what is required is only a qualitative determination of the elemental and molecular components of a selected specimen such as trace elements in soil. Other times the goal is the quantitative measurement of the fractional distribution of constituents in a gas such as the emissions in the tailpipe from an automobile. Sometimes it is also necessary to monitor a stream over an extended period of time. Information concerning analytical methods useful for specific purposes may be found in many articles dispersed alphabetically throughout the *Encyclopedia*. The follow articles are introductions to specific analytical methods of considerable importance for environmental analysis.

ENVIRONMENTAL ANALYSIS (CHEMICAL SENSOR APPLICATIONS)

Barry M. Wise
Jiri Janata
Pacific Northwest Laboratory
Richland, Washington

The latter part of the 20th Century has been dubbed the "information age." In fact, the information age really began with an explosion of data which resulted from the use of computers and data acquisition hardware. The increased availability of data has driven the need for advanced data processing algorithms that turn data into information. As data algorithms and computer hardware have advanced, the need for new sources of data has become apparent. The result is the self-stoking cycle of data and information that is occurring now. Nowhere is this more apparent than in the field of chemical sensors.

Chemical sensors provide data about the chemical state of our surroundings, such as the atmosphere of our habitats, about the course of dynamic chemical processes taking place in chemical manufacturing plants, etc. However, in most complex sensing situations, the simple functional relationship between the concentration of some species in the system and the raw electrical output from the sensor for that species is rarely adequate. Interferences from other species present in the system often result in biased measurements. The response of analytical chemists has been to collect more data on the system under observation and attempt to deconvolute the measurements.

Modern statistical techniques popularly known as chemometrics have been developed to overcome this measurement deconvolution problem. It has been amply demonstrated that multivariate measurements, combined with modern data and system modeling methods, can produce information well beyond the limits achievable with individual sensors (1). Thus, the advanced data processing techniques have become an integral part of the sensing process.

The driving force behind environmental science is the need to obtain a rational and quantitative assessment of the ecological impact of past and present human activities on the future prospects for our continued existence. The development of a dynamic ecological model is the subject of this emerging scientific discipline. Because many of the interactions between humans and environment are chemical in nature, chemical sensors are expected to play an important role in environmental science. Predictably,

they will be called "environmental sensors" very much like "biosensors" became a poorly defined but popular catchword for a group of sensors that had anything to do with "bio-" several years ago.

There are close to 1000 papers dealing with chemical sensors published annually, and the trend indicates further increase in this number (2–4). There are actually more sensor papers than species that are sensed. This paradox is partly due to the fact that most applications require a "customized" sensor with different attributes dictated by the application. A classic example of this situation is the sensing of hydrogen ion. Although the principle of operation of a pH sensor is the same, its implementation for pH measurement in different situations, for example soils, chemical process streams or *in vivo* measurement of pH in whole blood, are very different.

There have been several reviews on the general subject of environmental sensing published already (5–11) and the monitoring of effluent gases has been covered as a separate topic (12–15). In spite of the apparently large number of environmental sensors, there have been few successful implementations of chemical sensors in environmental applications. This is due largely to the problems created by non-specific sensors and the great number of potentially interfering species in most real world scenarios.

SCOPE OF ENVIRONMENTAL SENSING

Environmental problems range from those that make life unpleasant or difficult to those that could lead to a large alteration of the present ecosystem and ability of humans to survive. Most of the well-publicized environmental issues such as the effect of fluorocarbons on the ozone layer in the upper atmosphere, the effect of phosphates on natural water, adverse health effect of pesticides, etc, fall nearer the "make-life-difficult" end of the spectrum (16).

A good example for an environmental sensor in a make-life-difficult application is the monitoring of stack emissions from industrial processes. Large chemical and power plants can afford fairly expensive equipment for monitoring emissions (such as on-line spectrometers), but smaller plants will have to rely on less expensive sensing alternatives. It has also recently been found that in many urban areas, such as Los Angeles, the main contributors to air pollution are nonpoint sources such as dry cleaners and bakeries. Small businesses such as these cannot support expensive equipment for monitoring emissions, yet it is clear that if progress is to be made in improving air quality, these small sources must be monitored.

Another potentially huge market for environmental sensors is the detection of leaks of volatile organic compounds from valves and fittings in chemical plants. Current regulations require frequent testing of these systems, but it is entirely possible that continuous monitoring will eventually be required. Related areas include the monitoring of chemical storage tanks for fugitive emissions and the monitoring of oil rigs.

An example of a potentially more serious environmental issue arises from the prolonged and concentrated production of military materials, namely transuranium fuel for nuclear weapons. The United States Department of Energy production site at Hanford, Washington, is the largest U.S. environmental problem of this type. Specific examples of environmental problems at Hanford will be referred to throughout this article. Unlike other synthetic materials, such as synthetic polymers for consumer products, transuranics and fission products have been produced by nuclear reactions and are not subject to chemical degradation processes. The lifetimes of some transuranium elements (such as Pu^{239} and Am^{241}) and fission products (such at Tc^{99} and I^{129}) are very long, as measured on the historical time scale of the human species, and many of these materials are known to be chemically and genetically toxic. Release of large quantities of these materials into the biosphere could make life "more than difficult."

The entire spectrum of chemicals has been used during the large-scale production of nuclear materials at Hanford (17). These chemicals include halogenated hydrocarbons, heavy metals, nitrates, nitrites, ferrocyanides, toxic and flammable gases, a variety of complexing agents, etc. Any of those materials would feature prominently on a list of environmental contaminants. What makes them particularly serious, in the context of the nuclear materials production site, are the quantities involved and the fact that they are mixed in unknown ratios with each other and the radioactive waste. In addition, soil and groundwater contamination, both chemical and radiochemical, is a problem at Hanford. Several thousand wells at the Hanford site must be monitored for groundwater contaminants. These wells are checked at least yearly, with many of them requiring testing more often than that. Sensing technology that would eliminate sampling of the wells and transport of the samples to testing laboratories would greatly improve the efficiency of the monitoring effort. There are many other site specific needs within the DOE complex.

CHEMICAL INTELLIGENCE

Chemical sensors are only one component of *characterization*, which is defined as follows:

> Characterization is the act of developing a model or understanding of the chemical, physical, and spatial properties of a system.

In this context, characterization can be seen as chemical intelligence, which provides operational information about the chemical state of the system under consideration. From the analytical point of view data is obtained from: (*1*) off-line, *discontinuous*, batch chemical analysis; (*2*) on-line, *continuous*, *in situ* chemical sensors and *ex situ* sensor systems; and (*3*) transformed into information through application of advanced data processing. This scheme is diagrammed in Figure 1.

The validity of the information obtained by these two intertwined routes is critically dependent on the *sampling*. There is a problem with the meaning of the term "characterization" in systems as complicated as a waste

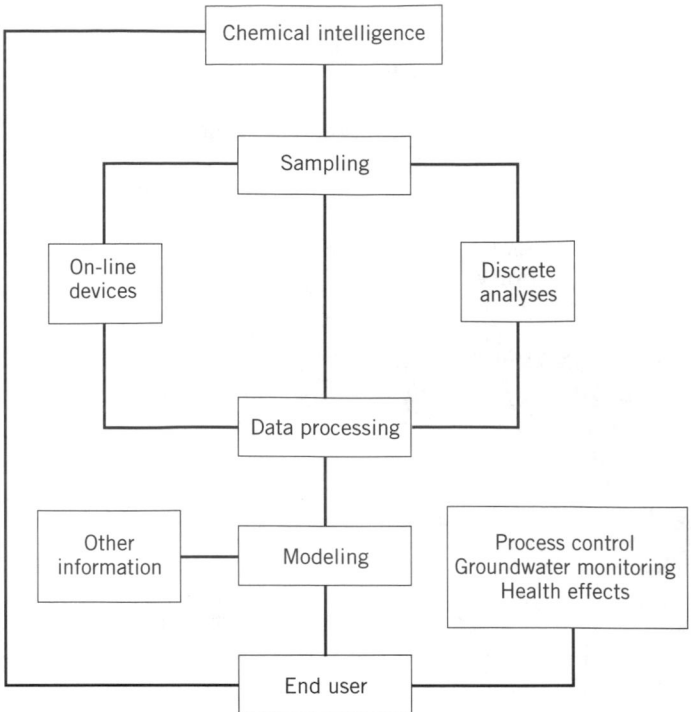

Figure 1. Chemical intelligence scheme.

tank or a human body in which several phases (or body compartments) coexist in a state of quasi-equilibrium. The reason is that a sample obtained from one phase is in a complex and often undefined relationship with respect to the other phases. The perturbation of such a system during any operation, such as the sampling event itself, may perturb any such relationship and render the result of the analysis invalid. The critical step in such a case is the separation of the phases and/or homogenization of the sample. Only then can a rational characterization of the original content of the object be contemplated. The situation is substantially similar in the health effects and in soils and groundwater areas. In all these cases, *the sampling space needs to be precisely delineated.*

On the other hand, local sampling or local sensing may have merit in obtaining chemical intelligence about critical localities within a complex system. For example, measurement of hydrogen-ion activity at the boundary between the organic liquid and aqueous phase adjacent to the container wall may be critically important in assessing the corrosion risk of that metal container. Similarly, accumulation of a heavy metal in certain organ tissues may be highly important for assessing the associated health risks.

The characterization sequence defined in chemical intelligence terms typically begins with:

1. Noninvasive physical diagnosis
2. Separation of macroscopic phases
3. Batch analysis and continuous sensing of the individual phases
4. Development and updating of a model of the system

The last step yields more detailed information about the chemical content of the sample. This information node represents the boundary condition on which the further treatment during the processing stage depends. From this information the strategy for development of a *model*, decisions about the raw data treatment, and deployment of additional chemical sensors/techniques must evolve.

Advantages and Limitations of Chemical Sensors

Chemical sensors are *complementary* to batch chemical analysis. Ordinary chemical sensors are generally not suitable for sustained, long-term quantitative monitoring because the baseline stability is rarely adequate. On the other hand they can economically provide valuable information about rapidly changing systems, something that batch techniques cannot do. If the required accuracy is traded for stability and selectivity, it is possible to design a sensor that performs according to its specification for a very long (months) period of time. Such sensors are often called *threshold detectors or alarms*. Many scenarios can be envisaged in which such devices would be useful in the framework of environmental monitoring. A better approach to solving the baseline stability problem may be design of higher order chemical sensors and sensor arrays, as will be shown later.

Single chemical sensors generally do not perform well in situations where many different chemical species are present, which is typically the case in environmental monitoring applications. This is the issue of chemical selectivity. Univariate sensors cannot reach the selectivity of modern analytical instruments (eg, of the hyphenated techniques such as gc-ms) in which the analytical separation step (eg, gc or hplc) precedes the actual quantification step. The trade off, *vis-a-vis* the sensor approach, is the speed of information acquisition, portability, remote operation, and economy. Again, it will be shown that higher performance can be expected from higher order sensors and sensor arrays.

A very unique aspect of chemical sensors is that some of them can be designed to respond to the change of *activity*, rather than concentration, of a given species. This is critically important in obtaining information about processes that are driven by thermodynamic, rather than by kinetic considerations. Thus, for example, interphasial partitioning equilibrium occurring in soil transport are driven by the activities rather than concentrations of the species involved.

Given the dangerous nature of some environmental remediation operations, remote sensing may be required. Another driving factor is the economy of obtaining information through chemical sensing as opposed to a discrete chemical analysis. These special considerations may affect the relative importance of the two routes of information acquisition shown in Figure 1. Multiple data inputs are then processed in order to formulate and refine the model of any given operation.

It is important first to identify the distinguishing features of environmental sensors by identifying the special requirements encountered in environmental sensing. In detection and determination of various species in the environment gaseous (eg, air), solid (eg, soils) and liquid (eg, groundwater) samples are encountered. The volume of the sample and the availability of the sampling space are not expected to be limiting factors. Therefore, the size of the individual chemical sensor is not a dominating issue. However, design and fabrication of multisensor arrays is important from the point of view of enhancement of chemical selectivity. Thus, microfabrication is expected to play an important role in development of environmental sensors. For environmental processing (ie, remediation) the speed of response may be the most important factor. On the other hand the detection limit may be most important in mapping the dynamics of contaminant spreading in soils, air, and groundwater. The attributes of chemical sensors required for different types of sensing situations are below.

Environmental	Health	Processing
Detection limit	Size	Speed
Selectivity	Safety	Safety
Stability	Selectivity	Robustness
Dynamic range	Detection limit	Stability

DEFINITION OF TERMS

In view of the existing apparent misconceptions about chemical sensors, it is pertinent to review the terminology and some essential features of these devices (18). Chemical sensors translate information about the concentration of chemical species in the sample into an electrical signal. Chemical sensors can be categorized by the four basic mechanisms by which this is done: thermal, mass, electrochemical, and optical. These devices operate in a *continuous* mode, in contrast to *chemical sensor systems*, such as flow-injection analysis or chromatography that requires manipulation of the sample and provides information in a *discrete* manner. Sensor systems can be automated and are expected to play an important role in environmental characterization. There is a low limit of concentration below which the sensor does not respond to a change of concentration of the detected species. This is the so-called *detection limit*. At the high end of concentration, the sensor again does not respond to the concentration change. This is called the *saturation limit*. The difference between these two limits is the *dynamic range*, which from the practical point of view is the usable concentration range of the sensor. Different sensors have different dynamic ranges. The slope of the signal vs. concentration dependence within the dynamic range is called the *sensitivity* of the sensor. Another important parameter is the *selectivity*, which is defined as the ability of the sensor to respond to the species of interest while being inert to other species. *Robustness* is a qualitative term that describes the ability of the sensor to maintain its essential performance characteristics under adverse operating conditions. Robustness is related to the operational lifetime of the sensor. Speed of response of the sensor is characterized by its time-constant, which for exponential processes is given as the time required for the signal to reach 63% of its final value. For sensors that are governed by more complicated response kinetics the response times of 95% or 90% are often used.

All of these parameters that characterize sensor performance are critically dependent on the conditions under which the sensor is used. This is why there are so many seemingly identical reports of applications of seemingly identical sensors. The key in a rational use of any sensor is to design it, package it, and evaluate its performance under the intended conditions of its application.

CHEMICAL SENSORS IN ENVIRONMENTAL REMEDIATION

It is useful to divide chemical environmental sensors according to their application. At the Hanford site, environmental remediation includes retrieval and processing of tank wastes and contaminated soils and requires monitoring of process input streams and process unit operations for control purposes and effluent monitoring. Also sensors are used for the monitoring of contaminants already released into the biosphere. Environmental monitoring would include mapping out of the movement of already released contaminants through soil and groundwater and the monitoring of wastes stored in underground tanks. The monitoring and remediation of contaminated industrial sites, such as toxic waste dumps, will require similar sensors.

Dynamic Models

A dynamic model relates the response of a system to a stimulus as a function of time. Dynamic models can be obtained directly from theoretical consideration of the governing physical laws of the system (such as material and energy balances) and other known or estimated system properties (mass and heat transfer coefficients, equilibrium constants, reaction mechanism, etc). Often, however, many of the fundamental properties of the system are unknown or not known accurately enough to allow development of an accurate dynamic model. In these cases it may be necessary to develop a model based entirely on data from controlled experiments (more desirable) or from observations of the system (less desirable). There is an extensive body of literature on the development of dynamic models. The foremost publication is probably *System Identification* (19). Recently, more work has been done on nonparametric model methods which require less *a priori* knowledge of the process dynamics (see, eg, ref. 20,21).

Advanced Process Control

The waste tanks at Hanford and many contaminated industrial sites contain an incredibly complex, heterogeneous mixtures. A rough idea about the number of possible sampling spaces can be obtained from the Gibbs phase rule which, loosely interpreted, says that number of phases will be comparable to the number of components (ie, chemicals). Even though some species are linked by equilibrium relationships and others may be accumulated only at interphases, the number of coexisting phases is expected to be large. This conclusion is supported by photographs of Hanford waste tank interiors. An example is

shown in Figure 2. From the sampling point of view, it can be immediately concluded that the condensed phases must be first separated before any meaningful information can be obtained. On the other hand, the composition of the head space in a waste tank or the vapors over contaminated soil can be obtained both by batch sampling and analysis and by direct sensing techniques.

Sensors used in environmental remediation will be required to meet rigorous safety and robustness requirements. For instance, sensors used in nuclear waste process could be exposed to very high radiation fields (~1000 rad/h). In other applications, potentially explosive gases exist in high concentrations and impose restrictions on the operational safety of electrical devices. It is expected that optical sensors will play a prominent role in these processing situations (22).

Even the simplest processing operation such as stirring or physical separation of the condensed phases (eg, filtration) must be done under controlled conditions in order to ensure the safety of the operation. Sensors for this application do not need to be particularly accurate, but they must be fast. Processing operations that result in some conversion of the raw waste materials can be very complex. A diagram of a more sophisticated operation such as the so called "clean option" process for the separation of Cs^+ from Hanford tank waste is shown in Figure 3. The nodes for possible deployment of sensors and/or discrete automatic assay have been added to probable locations in this diagram. They include sensors and sensing systems for ions and for neutral species in both liquid and gaseous phase.

The raw data obtained from this distributed sensor network are processed through the control system, which incorporates a dynamic model and is linked by the actua-

Figure 2. Photograph of the inside a double-shell tanks taken with a remotely controlled camera.

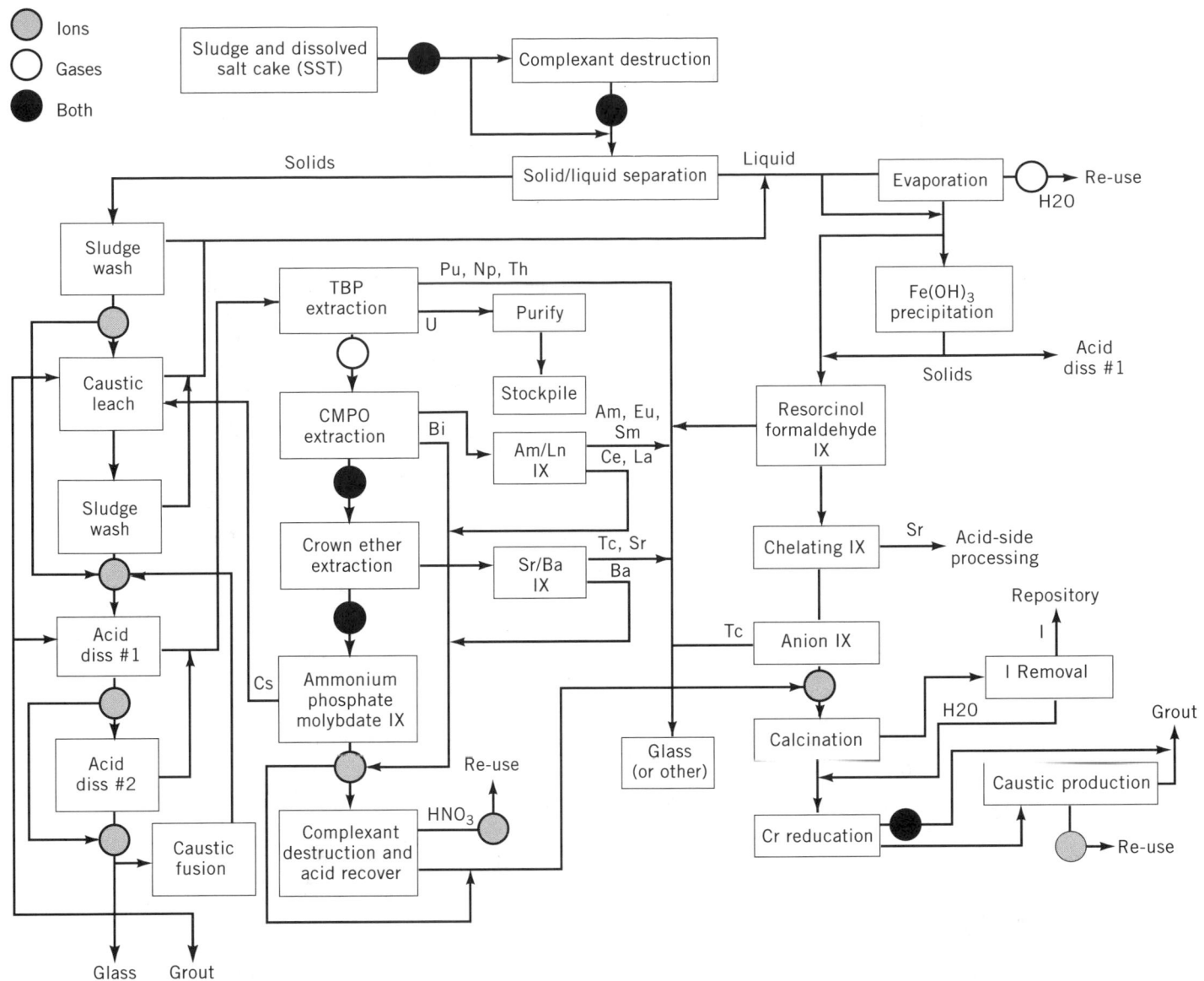

Figure 3. Flow-diagram of the proposed "clean option" process for extraction of Cs$^+$ with inserted "chemical intelligence" points.

tors back to the process. This feedback loop constitutes *advanced process control* (Fig. 4). Any operation in which macroscopic quantities of materials are manipulated must have advanced process control to operate the system safely and efficiently.

A dynamic model is an essential part of the control of a dynamic system. All control schemes use either an implicit or explicit model of the system. In classical PID control (proportional, integral, derivative), the implicit model is the response time and approximate order of the system. In more advanced control systems, an explicit model of the system is incorporated directly in the control loop (see Fig. 4a). In the figure the system dynamic model is placed in parallel with the system and given identical inputs (u). Thus the difference between the output of the model ($y^\wedge$) and the measured output of the system (y) form an estimate of any unmeasured disturbance to the system (v). The controller bases its actions on this disturbance esti-

mate and the desired output of the system, the setpoint. Modern control models consist, in part, of a mathematical inverse of the dynamic system model. In this sense the controller cancels the dynamics of the system and replaces them with the desired system dynamics. The more accurately the model represents the system, the more accurately the system can be controlled to conform to the setpoint.

The dynamic model is also used to decrease the uncertainty in the measurements. This problem was first solved by Kalman and the approach of optimally combining information on system inputs, system outputs and the dynamic system model to improve measurements is known as Kalman filtering (23). This procedure can also be used to monitor the health of the process sensors and determine if sensor failures have occurred. Thus, it can be seen that system models can improve the effective accuracy and robustness of the process sensors.

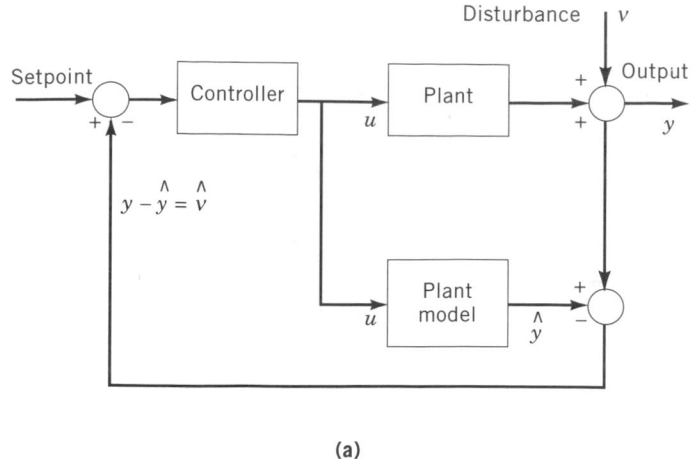

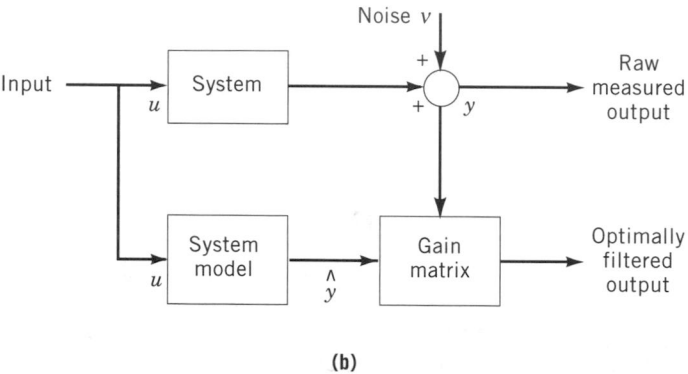

Figure 4. Schematic diagram of dynamic model used in (**a**) advanced process control loop and (**b**) for soil monitoring.

CHEMICAL SENSORS IN ENVIRONMENTAL MONITORING

The contamination of soils and groundwater has resulted from the discharge of the hazardous liquid wastes over the preceding half century. This was an accepted industrial practice over this period. At Hanford, soil and groundwater contamination was the result of nuclear materials production. For example, "inverse wells," where liquid waste was injected into the subterranean zone, were used for liquid waste disposal until 1980. More typical industrial processes would be the production of petrochemicals such as pesticides and plastics. Chemical sensors used in agriculture should be directly applicable to many of these sensing needs. However, there are special requirements, such as sensing of the organic liquids and vapors in the subterranean (vadose) zone, that require design of special class of sensors. Evaluation of the soil and groundwater contamination at Hanford has been prepared (24). Similar studies will be required, and in many cases have been prepared, for industrial sites.

In contrast to processing, changes of contaminant concentrations in soils and groundwater are slow. Moreover, the concentrations of the species to be monitored are often low and the important parameter is likely to be the *activity* rather than concentration. This consideration virtually eliminates optical sensing of ionic species as a viable pos-

sibility (25). The monitoring of the movement of contaminants through the soils and groundwater at Hanford is done primarily through a system of sampling wells from which samples of the liquid or suspension of soils can be obtained from various depths and at various time intervals (Figs. 5, 6). The samples are usually brought to the laboratory for analysis. Deployment in *in situ* chemical sensors in this situation is a principal economic incentive. Because the information must be obtained over long periods of time the calibration stability is of premium importance. This and the requirement of low detection limits makes the deployment of self-calibrating sensor systems and higher order sensors particularly attractive.

The pH of the soil, and its buffer and redox capacity play a dominating role in the dynamics of subterranean plumes of heavy metals and transuranium metals. Sensors and sensing systems for these applications should be at the top of the priority list.

Dynamic modeling and Kalman filtering may be applied to the problem of sensing changes in soils and groundwater in much the same way that it is applied in processing, with the exception that the feedback part of the loop no longer exists. This is shown in Figure 7b. Here the output from the model and raw measurement on the system are combined in an optimal fashion to achieve a better estimate. The general approach has been demonstrated by several researchers in analytical chemistry (23,26).

STRATEGIES FOR DEVELOPMENT OF ADVANCED CHEMICAL SENSORS

It is possible to classify sensing systems according to the dimensionality of the data they produce. The dimensionality of the data defines the *order* of the system. For instance, a pH electrode, which takes one datum per mea-

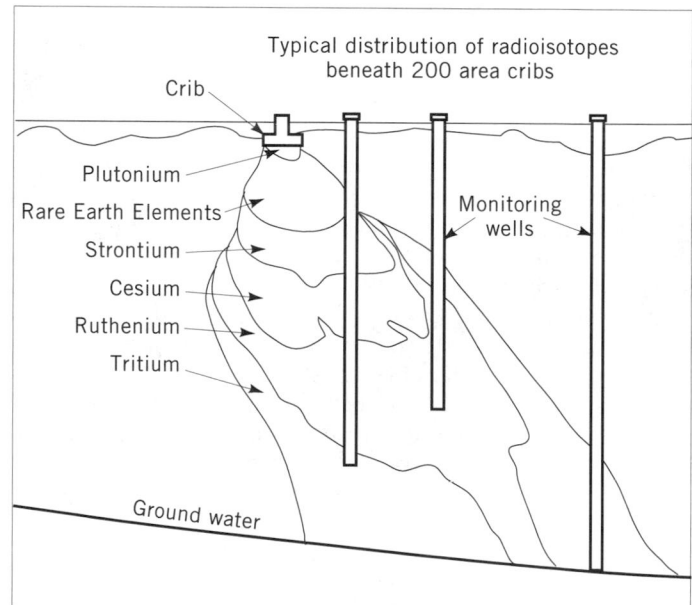

Figure 5. Discharge of the waste into a crib and ensuing contamination.

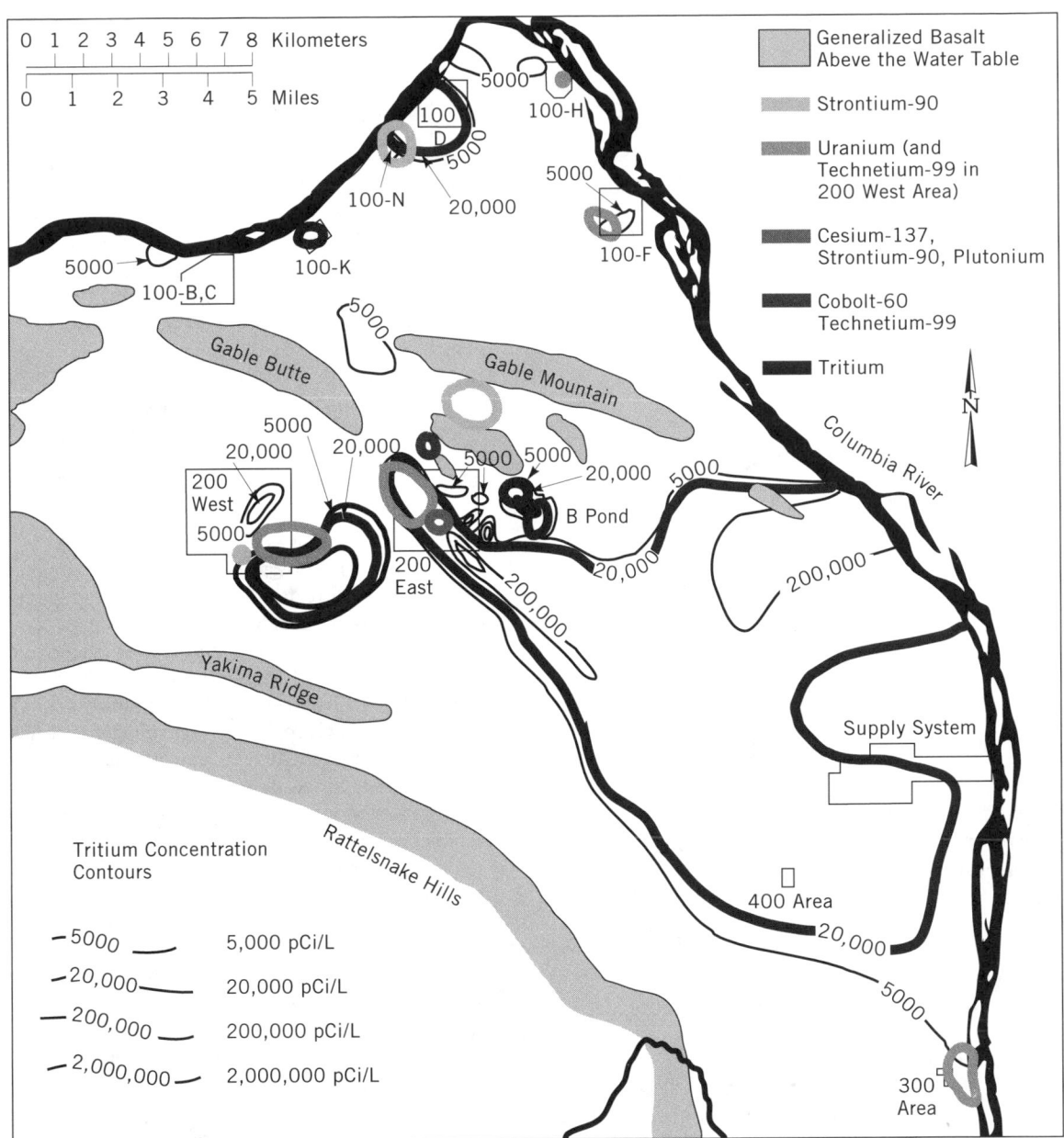

Figure 6. Distribution of various contaminants at the Hanford site and comparison to the Drinking Water Standard.

surement, is a zero-order instrument since a single point has dimension zero. A spectroscopic instrument, which measures absorbance as a function of wavelength, is a first-order instrument. Each measurement produces a vector of data. A vector, of course, is one-dimensional (a first-order tensor). There is a growing trend towards instruments that produce a matrix (second-order tensor) of data with each measurement. Typical examples are the hyphenated techniques, such as gc-ms, and many types of time resolved spectroscopy. In these techniques, measurements are made as a function of two variables to produce a matrix of measurements. For example, in gc-ms the quantity of ions produced is measured as a function of their atomic mass and the retention time on the gc.

It is impossible to detect the presence of an interferent with a zero-order instrument. There are no data with

which to cross reference the measurement. In a first-order instrument, it is generally possible to detect the presence of interferents; however, it is not generally possible to correct for the interferents and quantitate the species of interest correctly in their presence. With a second-order instrument, both detection of interferents and quantitation in their presence is possible.

The concept of order is relatively new to chemical sensors and sensing systems. Traditional chemical sensors provide output that is uniquely related to the concentration (or activity) of one species. They are economical and simple to operate. However, they require frequent calibration and any deviation from the calibration status between calibration steps is not detected. Another drawback of a zero-order sensor is that it relies solely on the specificity of the interaction of the detected species with the

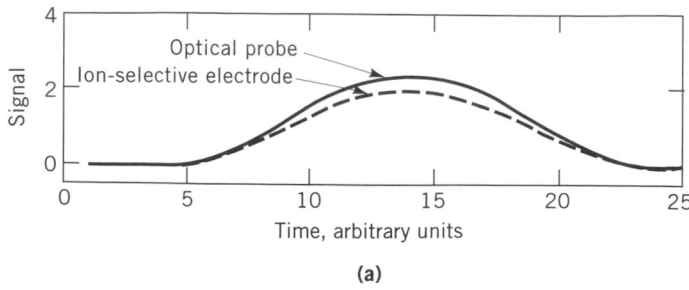

(a)

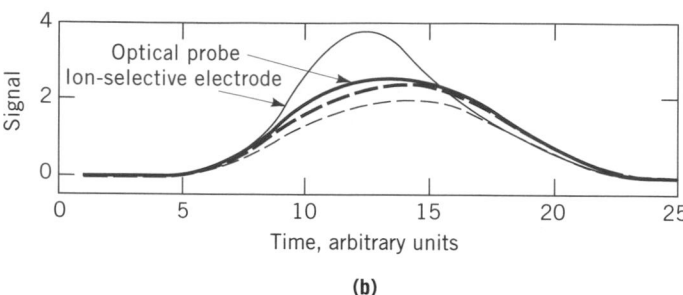

(b)

Figure 7. Response of second-order probe consisting of low resolution separation step followed by ion-selective electrode (light gray) and optic probe (solid black). (**a**) Response of the system to pure analyte (dashed lines). (**b**) Response of the system with interferent.

selective layer for the selectivity of its response. It is a unique and fortunate coincidence that a glass pH electrode has such an extraordinary selectivity and dynamic range for the most ubiquitous chemical species, the hydrogen ion. Ion selective electrodes, which represent an entire class of most developed chemical sensors, do not match the performance of a glass pH electrode for detection of other ions.

In the last decade individual zero-order sensors have been grouped to form first-order instruments. The output of these systems have been processed through advanced statistical and mathematical algorithms collectively known as chemometrics.

Chemometrics

Chemometrics can be broadly defined as *the science of relating measurements made on a chemical system to the state of the system via application of mathematical or statistical methods*. It is the name given to a collection of methods that have been found useful in chemical applications.

A principal area of interest in chemometrics is the general model identification problem. This problem can be divided into calibration of analytical instruments and sensors and identification of dynamic models of chemical systems.

Methods for calibration of zero-order instruments have existed since the time of Newton and Gauss. Typically, measurements are made with the instrument on several known samples. A simple bivariate model is then developed relating the instrument response to the property or concentration of interest. The model is often a simple

least squares fit of the response to the system property, though often a linearizing transformation must be used first. An example of this is the calibration of a pH electrode. The response of the instrument (in millivolts) is a nonlinear function of the concentration (activity) of hydrogen ion. Taking the log of the hydrogen ion concentration (ie, pH) linearizes the problem since the relationship between response and $\log(\text{activity of } H^+) == pH$ is linear. The final model consists of a slope, relating the sensitivity of the measurement to changes in pH and temperature and an intercept that gives the baseline offset, due to factors such as the reference electrode and the liquid junction that connects it to the sample.

In some cases the instrument response is a nonlinear function of the property of interest and, unlike pH measurement, it is not possible to obtain a linearizing relationship explicitly. In these cases a general nonlinear regression technique must be used to calibrate the instrument.

Once the sensor is calibrated, the model is used to relate instrument response to the property or concentration of interest in an unknown sample. This system works well until an unknown interferent that changes the sensitivity of the instrument to the property is introduced into the system. Unfortunately, it is impossible to detect the presence of such an interferent without some knowledge about the system other than the instrument responses.

Often, it is necessary to make measurements in systems where the concentration of more than one analyte or system property varies. This would not be a problem if it were possible to build sensors that were totally specific for the analyte or property of interest. In general, however, it is not possible to do this and instead we must rely upon collections of sensors that are only partially selective for the analyte of interest. Models are then developed (using any of the methods discussed below) that relate the responses on all of the sensor elements to the analyte or property of interest.

There are many possible calibration methods for these linear first order systems. Perhaps the oldest is the classical least squares (CLS) approach (also known as the K-matrix method) (1,26), which requires knowledge of the pure component response of each analyte to each sensor. In some cases, however, it is not possible to obtain pure component responses and another method must be employed. In these instances, techniques such as multiple linear regression (MLR), principal components regression (PCR) (27), and partial least squares (PLS) regression (28,29), which can all be unified under the technique continuum regression (CR) (20,30), and ridge regression (RR) (31) may be employed. In each case data is collected on the response of the sensors to known mixtures of analytes. It is critical that the mixtures "span the space" of expected concentrations during use of the measurement system. This includes species that may not be of interest analytically but whose concentration may vary during the course of using the sensing system.

The MLR approach (also known as ordinary least squares or OLS) fits a model to the data that minimizes the sum of squares error of the fit of the model. In this sense it provides an unbiased estimate of the property of interest. In systems where there are more sensors than independently varying analytes or if the response of the instrument to several analytes is identical, the MLR problem becomes "ill-conditioned." This "co-linearity" problem is due to the fact that if there are fewer independent analytes than measurements, some of the measurements are necessarily linear combinations of other measurements. In effect, there are more parameters in the model than independent variations in the data. In these cases the MLR model obtained may be a strong function of small changes in the calibration data caused by "noise" and may fit the calibration set quite well but not provide accurate estimates for new samples.

The PCR, PLS, and CR approaches solve the co-linearity problem by determining linear combinations of variables (sensor elements) or latent variables that are independent. In general, these methods will produce a smaller number of latent variables than original variables. This "reduced set" of latent variables are then regressed against the properties of interest. A cross-validation procedure is used to determine the optimum number of latent variables to retain for prediction of new samples.

Ridge regression solves the co-linearity problem in a different way, essentially by restricting the magnitude of the coefficients that relate the sensor response to the analyte concentration. Experience has shown that this method can be very effective in practice. However, ridge regression does not provide the useful diagnostics for detecting of outlier samples, i.e., samples that contain analytes not in the original calibration matrix.

An example of a multivariate calibration problem follows. Suppose that you have two sensors, one which measures the activity of Mg^{2+} in solution by ion-selective electrode and the other which measures concentration by a fiber optic probe. It is desirable to use both of these measurements to calculate $[Mg^{2+}]$, but in the example calibration set, only $[Mg^{2+}]$ was allowed to change.

If the calibration done for this system was attempted with MLR, it would be found that the response of the two sensors was co-linear (assuming that the log of the ion-selective electrode was used and that no nonlinear effects were observed) and a good solution would not be obtained. Using a CR method, it would be found that one linear combination of the response of both instruments would be used to form the calibration model. The CR model is essentially two models in one. The first model relates the response of the two probes to each other and the second model relates their collective response to $[Mg^{2+}]$. This is illustrated graphically in Figure 8, where the response of the ion-selective electrode and the fiber optic probe are plotted against each other. Note that a straight line fits through the data quite well. Any nonsystematic deviations from the line can be modeled as noise (random fluctuations of the output of the sensors). Systematic deviations from a line with a slope of 1 would be due to the fact that the response of the optical probe does not depend upon the activity coefficient of $[Mg^{2+}]$ while the ion-selective electrode does. The concentration of Mg^{2+} is estimated from the projection of the measured data onto the line. This is commonly known as a "score" value.

The CR techniques offer the advantage that they provide a model (the latent variables) that describes the normal variation in the response of the sensors (in linear al-

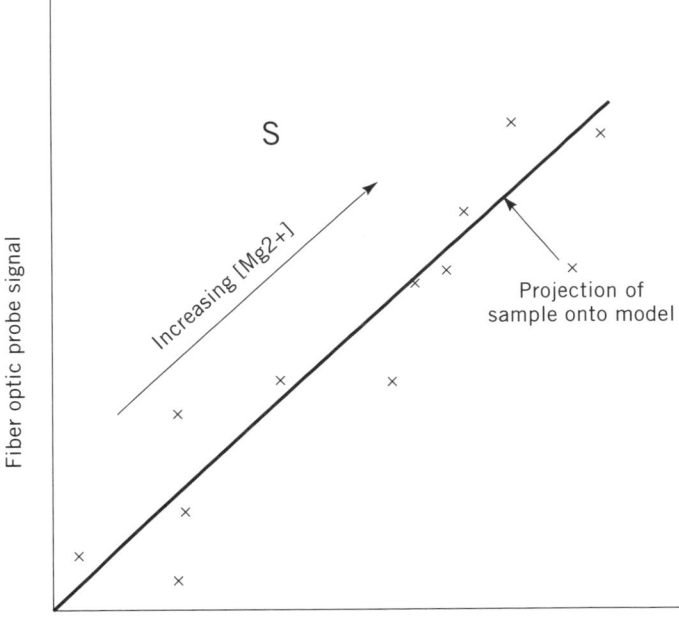

Figure 8. Relationship of linearized ion selective electrode signal to fiber optic probe signal under calibration conditions (xs) with an unusual sample (**S**).

gebra terms, the subspace spanned by the calibration set responses). When a new sample is analyzed, the difference between the new sample response and the normal subspace, or residuals, can be calculated (these are the deviations from the line in Figure 8). Statistical tests can be used to determine if the residuals are significant or within normal variation. It is in this way that interferents are detected. When the instrument responses do not "span the same space" as the calibration set, something new must have been added to the system, or the properties of the instrument itself must have changed.

Now suppose that an unknown change was made in the system (say, addition of another ion for which the ion selective electrode has some response), which affected the apparent activity of the Mg^{2+} relative to its concentration. When the CR model was applied, it would be found that the relationship between the measurements had changed. This is shown graphically in Figure 8, where the new data point is plotted. Note that the residual is very large, much larger than that associated with random fluctuations in the system. In this instance, it would be known that the model was no longer valid and that the estimates of $[Mg^{2+}]$ were no longer reliable.

Unfortunately, even though it is possible to detect the presence of an interferent, it is not possible to correct for it and do accurate quantitation on new samples. This is due to the fact that the exact nature of the change is impossible to quantify since it is not known how orthogonal the interferent is to normal changes in the system. In our example, deviation from the line relating the response of the two sensors could just as easily be caused by changes to either sensor. For instance, change of ionic strength could also account for such a shift.

Second order techniques can be classed as "bilinear" or

"non-bilinear." In a bilinear technique, the response of the instrument to a single pure analyte can be modeled as the (linear) outer product of two vectors (32–34). The resulting matrix of data will have a rank of one (rank = number of linearly independent columns or rows in the matrix). Many common hyphenated instruments involving a separation step are bilinear, including lc/uv, gc/ms, gc/ftir, and emission-excitation fluorescence. A nonbilinear instrument produces data matrices that have rank greater than one for pure analytes. Examples of this are ms/ms and 2d nmr, where the rank of pure analyte matrices may be as high as 20–30 (35,36).

As might be expected, different techniques have been developed to deal with calibration of second order instruments depending upon whether the data produced is bilinear, weakly nonbilinear or strongly nonbilinear. The Generalized Rank Annihilation Method (GRAM) works well for bilinear data. Several approaches have been suggested for nonbilinear data, including Nonbilinear Rank Annihilation (37) and Residual Bilinearization (38,39). Most recently, techniques such as Restricted Tucker Models have been developed to deal with cases where the data is only mildly nonbilinear (40).

GRAM can be used with bilinear data to quantitate for the species of interest in spite of the presence of interferents. The basic idea behind GRAM is that multiples of the pure analyte response are subtracted from the sample response and the rank of the response matrix is observed. When the multiple of the pure analyte response is correct, the rank decreases by 1. In the presence of interferents, the response matrix rank will increase, but GRAM will work in the same way since it looks for changes in the rank of the response matrix.

Moving back to our example, assume for a moment that we have combined our sensors for measurement of $[Mg^{2+}]$ with a low resolution separation step preceding the measurement. This would result in a "bilinear" second-order instrument. Suppose that the interferent is sufficiently different from the analyte of interest so that the separation step results in some separation (the interferent and analyte do not completely overlap). The response of the instrument to the pure analyte and to the mixture is now shown in Figure 7. The rank of the pure analyte matrix is 1, while that matrix with addition of interferent has a rank of 2. By subtracting multiples of the pure analyte response from the response of the mixture, a point can be found that reduces the matrix rank back to one. This point represents the correct amount of analyte present in the system.

As mentioned above, it is not always possible to identify a linearizing transformation that will allow one to use the linear techniques discussed above for calibration. There are a variety of approaches being used for such nonlinear problems. In the last several years, artificial neural networks (ANNs) have been applied to sensor calibration problems (41). Additional techniques include locally weighted regression (42), multivariate adaptive regression splines (43), and PLS techniques with nonlinear inner relationships (44,45).

In remote systems that may be typical of the monitoring of hazardous waste sites, the recalibration of an instrument is likely to become an issue. It would be desirable to transfer a calibration (mathematical model) from another instrument in a nonhazardous environment. This problem has been studied by several researchers and some methods are beginning to gain acceptance. Some spectroscopic instrument manufacturers are developing the concept of an *absolute virtual instrument* and relating the response of any particular instrument to this virtual response.

The basic idea behind calibration transfer is that it is possible to develop a "transform" between instruments based on the results of testing a few "well chosen" samples on both instruments. A full calibration model is then developed on one instrument, usually the laboratory (rather than field) instrument that is generally both easier to work with (get to) and of higher quality. The calibration for the second instrument is the transform that relates the instruments times to the calibration for the first instrument.

There are several methods for developing a transform between first order instruments. The piece-wise direct standardization (PDS) method has been shown to be quite effective for first-order instruments such as spectroscopic instruments and arrays of temperature sensors where it is expected that the correlation between responses on specific channels decreases with "distance" (46,47). With temperature measurement, this distance is the physical distance between sensors while for spectroscopic measurements it is the difference between wavelengths.

As might be expected, less work has been done to date on transforms for second-order instruments. The second-order problem is considerably more complex than the first-order problem. However, simultaneous standardization of both instrument dimensions has been demonstrated for simulated lc-uv data (48).

Microfabrication

Almost every type of chemical sensor can be made by using silicon-based microfabrication technology (49,50). In the context of the needs for environmental sensing the fabrication of multisensor arrays is particularly important. Multisensor chemiresistor arrays or multiple chemically sensitive field-effect transistors (CHEMFET) chips are the obvious example. Microfabrication can be extended to development of miniature sensing systems such as a potentiostatic electrochemical microcell (51). The requirements on lateral resolution of the fabrication process rarely exceed 5 μm; however, ability to deposit and geometrically pattern unusual combinations of materials is important. Because, during the course of its lifetime, the solid state device comes in an intimate contact with quite harsh environments new encapsulation procedures had to be developed (52). The microfabrication of environmental sensors is driven by the need for multisensor capability rather than by the size of the available sampling space.

Synergism Between Sensor and Remediation Process Development

As a general rule it can be stated that the interphasial interaction on which a macroscopic advanced process, eg, for environmental remediation, is based should also be the basis for development of an optimal chemical sensor. In other words, the development of advanced processes must go hand-in-hand with the development of chemical sensors and sensor systems for advanced process control. Thorough knowledge of interphasial physical chemistry is the mandatory common base for these two elements of environmental restoration.

For example, a chemical sensor for Cs^+ based on a conventional polymeric membrane/ionophore combination may not be usable in the high radiation fields encountered during the separation of the Cs^{137} isotope. Instead a new sensor based on radiation resistant Cs ion exchanger may have to be developed. Such a sensor may have a lower selectivity than the conventional polymer-based Cs ion-selective electrode but in terms of overall usability may be more appropriate. On the other hand, development of a new material for a Cs^+ sensor may lead to a new or improved material for the macroscopic conversion process. In any case, it is apparent that concurrent development of remediation processes and the required process sensors will result faster development and a higher likelihood of success.

CONCLUSIONS

A superficial examination of environmental sensing needs may lead to an erroneous conclusion that any of the existing, even commercially available, chemical sensors may be used and that no further sensor development is required. In exceptional circumstances this may be true; however, the state of characterization at Hanford and other DOE and industrial sites clearly shows that this is not the general case. The problem lies in the fact that the specifics of each sensing situation can degrade the performance characteristics of a commercially available sensor and in the extreme case make it unusable. This is often the result of unknown interferents in environmental samples that lead to biased sensor response. A careful examination and understanding of the operating principle of a given sensor under the application conditions is the least requirement for its rational deployment.

It is clear that there are many applications in environmental monitoring and remediation where no suitable sensor or sensor system exists and new devices need to be developed. It is the author's contention that first order sensors will become the norm for environmental applications and that the potential advantages of second order sensors will make them common. This will require continued development of both sensors, producing ever more data, and calibration and diagnostic methods, for extracting the required information. Thus, the field of environmental sensors will experience the data and information cycle characteristic of the "information age."

Acknowledgment

Help of Roy E. Gephart in compiling the Hanford inventory information is gratefully acknowledged.

BIBLIOGRAPHY

1. M. A. Sharaf, D. L. Illman, and B. R. Kowalski, *Chemometrics*, John Wiley & Sons, Inc., New York, 1986.

2. J. Janata, and A. Bezegh, *Anal. Chem.* **60,** 62R–74R (1988).

3. J. Janata, *Anal. Chem.* **62,** 33R–44R (1990).

4. J. Janata, *Anal. Chem.* **64,** 196R–219R (1992).

5. T. Hori, T.; ed., International Symposium on New Sensors and Methods for Environmental Characterization 1986, Kyoto, Japan *Pure Appl. Chem.* **59**(4) (1987).

6. R. B. Smart, *Hazard Assess. Chem.* **5,** 1–27 (1987).

7. J. D. R. Thomas, *Int. J. Environ. Anal. Chem.* **38**(2), 157–169 (1990).

8. R. Kalvoda, *Electroanalysis* (*N.Y.*), **2**(5), 341–346 (1990).

9. S. P. Banerjee, *J. Mines, Met. Fuels*, **37**(5), 178–185 (1989).

10. A. A. Tumanov and E. A. Korostyleva, *Zh. Anal. Khim.* **45**(7), 1304–1311 (1990).

11. I. Giannini, *Fis. Tecnol.* (*Bologna*) **12**(1), 23–34 (1989).

12. Chen, Aifan, Luo, Ruixian, Tan, Thiam Chye, Liu, Chung Chiun, *Sens. Actuators*, **19**(3), 237–248 (1989).

13. L. L. Altpeter, Jr., T. A. Williams, M. Rupich and H. Wise, *Oper. Sect. Proc. Am. Gas Assoc.*, Volume Date 1989, p. 144–147.

14. J. Riegel and K. H. Haerdtl, *Sens. Actuators, B*, **B1**(1–6), 54–57 (1990).

15. A. Accorsi, G. Delapierre, C. Vauchier, and D. Charlot, *Sens. Actuators, B*, **B4**(3–4), 539–543 (1991).

16. A. Gore, *Earth in the Balance*, Plume Publishers, 1993.

17. D. L. Illman, *Chemical and Engineering News*, 9–21 (June 21, 1993).

18. J. Janata, *Principles of Chemical Sensors*, Plenum, New York, 1989.

19. L. Ljung, *System Identification—Theory for the User*, Prentice-Hall International, London, 1989.

20. B. M. Wise and N. L. Ricker, *J. Chemometrics* **7,** 1–14 (1993).

21. B. M. Wise and N. L. Ricker, *Process Control and Quality*, **4,** 77–86 (1992).

22. M. A. Arnold, *Anal. Chem.* **64,** 1015A–1025A (1992).

23. R. K. Woodruff, R. W. Hanf, and R. E. Lundgren, *Hanford Site Environmental Report for Calendar Year 1991*.

24. J. Janata, *Anal. Chem.* **64,** 921A–927A (1992).

25. S. D. Brown, *Chemom. Intell. Lab. Syst.* **10,** 87–105 (1991).

26. C. L. Erickson, M. J. Lysaght, and J. B. Callis, *Analytical Chemistry*, **64,** 1155A–1163A (1992).

27. T. Naes and H. Martens, *J. Chemometrics* **2,** 155–167 (1988).

28. P. Geladi and B. R. Kowalski, *Anal. Chim. Acta* **185,** 1–18 (1986).

29. A. Lorber, L. E. Wangen, and B. R. Kowalski, *J. Chemometrics* **1,** 19–31 (1987).

30. M. Stone and R. J. Brooks, *J. R. Statist. Soc. B*, **52,** 337–369 (1990).

31. H. R. Draper and H. Smith, *Applied Regression Analysis*, 2nd ed. John Wiley & Sons, Inc., New York, 1981.

32. E. Sanchez and B. R. Kowalski, *J. Chemometrics* **2,** 247–263 (1988).

33. E. Sanchez and B. R. Kowalski, *J. Chemometrics* **2,** 265–280 (1988).

34. T. Hirschfeld, *Anal. Chem.* **52,** 297A–312A (1980).

35. D. E. Wilson, W. Lindberg, and B. R. Kowalski, *J. Am. Chem. Soc.* **111,** 3797–3804 (1989).

36. Y. Wang, O. Borgen, B. R. Kowalski, M. Gu, and F. Turecek, *J. Chemometrics*, in press.

37. B. E. Wilson, W. Lindberg, and B. R. Kowalski, *J. Am. Chem. Soc.* **111,** 3797–3804 (1989).

38. J. Ohman, P. Geladi, and S. Wold, *J. Chemometrics* **4,** 79–91 (1990).

39. J. Ohman, P. Geladi, and S. Wold, *J. Chemometrics* **4,** 135–148 (1990).

40. A. K. Smilde, Y. Wang, and B. R. Kowalski, *J. Chemometrics,* (Dec. 1992).

41. R. Long, H. T. Mayfield, M. V. Henley, and P. R. Kromann, *Analytical Chemistry* **63,** 1256–1261 (1991).

42. T. Naes, T. Isaksson, and B. Kowalski, *Analytical Chemistry,* **62,** 664–673 (1990).

43. J. H. Friedman, *Annals of Statistics,* **19,** 1–141 (1991).

44. S. Wold, N. Kettaneh-Wold, and B. Skagerberg, *Chemometrics and Intelligent Laboratory Systems* **7,** 53–65 (1989).

45. S. Wold, *Chemometrics and Intelligent Laboratory Systems* **14,** 71–84 (1992).

46. Y. Wang, D. J. Veltkamp, and B. R. Kowalski, *Analytical Chemistry,* **63,** 2750–2756 (1991).

47. Y. Wang, M. J. Lysaght, and B. R. Kowalski, *Analytical Chemistry,* **64,** 562–565 (1992).

48. Y. Wang and B. R. Kowalski, *Analytical Chemistry,* in press.

49. J. N. Zemel, *Rev. Sci. Instrum.* **61,** 1579–1606 (1990).

50. A. Manz, J. C. James, E. Verpoorte, H. Luedi, H. M. Widmer, and D. J. Harrison, *Trends Anal. Chem.* **10**(5), 144–149 (1991).

51. M. Koudelka, F. Rohner-Jeanrenaud, J. Terrattaz, E. Bobbioni-Harsch, N. F. DeRooij, B. Jeanrenaud, *Biosens. Bioelectron.* **6,** 31–36 (1991).

52. K. Domansky, J. Janata, M. Josowicz, and D. Petelenz, *Analyst,* **118,** 335–340 (1993).

ENVIRONMENTAL ANALYSIS (MASS SPECTROMETRY)

JOAN BURSEY
Radian Corporation
Research Triangle Park, North Carolina

Mass spectrometry pertains to the separation of charged particles (ions) according to their mass-to-charge ratio. The separated ions and their relative abundances are then indicative of the structure of the original molecule. The most common route to formation of the charged particles is by interaction of vaporized molecules with energetic electrons emitted from a filament, usually at an energy of 70 electron volts (eV), a process called electron ionization. Positively charged ions, as well as negatively charged ions and excited neutral species, are formed in the ion source of the mass spectrometer, and then electrically accelerated into a device to perform mass analysis. More than 99% of the sample molecules are removed from the ion source continuously by vacuum pumps, which maintain the mass spectrometer under vacuum. The separation of the charged particles is accomplished in numerous ways:

- Magnetic Sector Mass Analyzer. In a magnetic sector mass spectrometer, a magnetic field acts as a mass analyzer. The mass of the ions which can pass through the magnetic field at any given time depends upon the radius of ion path in the magnetic field, the strength of the magnetic field, and the potential with which the ion is accelerated out of the ion source in which it is formed.

- Quadrupole Mass Filter. In a quadrupole mass filter, ions from the ion source are electrically accelerated into a combination of radiofrequency (rf) and electric (dc) fields on four precisely-machined stainless steel rods, with opposite polarities on pairs of opposing rods. Ion paths oscillate according to a changing rf field. At any given time, only ions of a particular mass can achieve a stable orbit and arrive at the detector; all other ions are in unstable orbits and are discharged by impact with the rods.

- High Resolution Mass Spectrometer. The magnetic sector mass spectrometer and the quadrupole mass filter typically separate unit masses. Separating mass spectrometric peaks with a small mass difference (very much less than one mass unit difference) requires the use of high resolution mass spectrometry. Ions formed in the ion source of a mass spectrometer exhibit a wide range of energies. This spread in energies causes divergence in the orbits of these ions through a mass analyzer and, consequently, limits the ability of the instrumentation to achieve resolution of masses. If the ions which are formed are focused according to energies before they enter a magnetic field, a much higher level of resolution is attainable in the mass analysis. One common arrangement of double focusing instrumentation places an energy focus (electrostatic sector) prior to the magnetic sector. It is also possible to achieve high resolution in other configurations, such as a magnetic sector preceding an electrostatic sector.

- Time-of-Flight Mass Analyzer. In a time-of-flight mass analyzer, ions are alternately formed and accelerated; mass values of the ions are differentiated on the basis of the time of their arrival at the detector.

- Ion Trap Mass Spectrometer. In an ion trap mass spectrometer, ions are typically formed by interaction with a filament that emits electrons. The electron beam is gated so that packets of electrons are produced. When a packet of ions is formed, the ions and ionic decomposition products are trapped by a cyclotron radiofrequency voltage. When this trapping rf voltage is swept, ejection of ions from the trapping cell occurs, and the ejected ions are detected.

When vaporized molecules are bombarded with energetic electrons, the excess energy transferred to the vaporized molecules results not only in their ionization but also in their fragmentation: charged fragments of the original molecule are deflected through the appropriate field and separated according to their mass-to-charge ratios, to produce a characteristic pattern of mass and abundance. Extensive scholarly efforts have been devoted to the interpretation of the mechanism of the fragmentation which occurs. Common fragmentation features have been elucidated for families of molecules, and pattern recognition techniques have been developed to allow computerized identification of the molecule from its fragmentation pattern. The utility of the mass spectrometer in environmental applications stems from its ability to produce these characteristic fragmentation patterns, which allow the qualitative identification of a molecule as well as quantitative analysis as a function of the signal which is produced.

ENVIRONMENTAL ANALYSIS: APPLICATIONS

The mass spectrometer, as a scientific instrument, has been in use since the historic experiments of J. J. Thomson in 1910, when the isotopes of neon (^{20}Ne and ^{22}Ne) were discovered. F. W. Aston built the first mass spectrometer in 1919 to use velocity focusing. However, only since the coupling of the gas chromatograph and computer with the mass spectrometer in approximately the last twenty years has the application of the mass spectrometer to environmental analysis been extensive. The crux of environmental analysis is the ability to identify and perform quantitative analysis of trace quantities of a substance present in a complex matrix. Separation of the analyte of interest from the matrix is usually the most serious problem facing the environmental analyst. The mass spectrometer itself provides no separation of analytes from the matrix: analyte and matrix constituents are ionized together in the ion source of the mass spectrometer, and differentiation of the resulting mass of signals at a large number of masses is a challenge to the most powerful mass spectrometric system. The ability to perform separations, to differentiate the analyte from the environmental matrix in which it is found, is crucial to the ability to perform environmental analysis.

Gas Chromatograph/Mass Spectrometer (Gc/Ms)

The gas chromatograph is central to the utilization of the analytical capabilities of the mass spectrometer. The commercial availability of computerized gas chromatograph/mass spectrometer systems in the late 1970s led to the designation of mass spectrometry as the primary monitoring tool for trace organic compounds by the United States Environmental Protection Agency.

The gas chromatograph, first employed in 1905, is used to perform separations of organic compounds from complex matrices due to their differential solubilities in a liquid phase coated on particles in a column or chemically bonded in a thin film with the walls of a column through which a stream of inert gas, usually helium, is flowing. The basis for the separation is the partitioning of the sample in and out of the liquid phase. The gas chromatograph operates at a positive pressure of helium; most mass spectrometers require a vacuum of 10^{-5} to 10^{-6} torr for successful operation. The development of a molecular separator, a device fabricated usually of glass, has been required for the coupling of the two apparently disparate techniques: the pressurized gas stream exiting from the gas chromatograph enters the separator and, since the rapidly diffusing helium particles are pumped away at a greater rate than the heavier molecules of the separated sample components, an enrichment of the chromatographic effluent is achieved and the flow of helium is reduced to a level at which the vacuum system of the mass spectrometer can operate. In many cases, if a chromatographic column of very narrow internal diameter (0.25 mm to 0.32 mm) is used, the flow of helium required to operate the column is sufficiently low (1–2 mL/min) so that the gas chromatograph can be coupled directly to the mass spectrometer without the use of a separator; the pumping system of the mass spectrometer will accommodate this low flow of helium. Columns of these narrow diameters, usually made of fused silica capillary tubing, are a development of the last ten to fifteen years. Fused silica capillary columns are now in very common use in laboratories around the world.

With the routine coupling of the gas chromatograph and the mass spectrometer and with the computerization of the mass spectrometer, the application of the coupled technique of gas chromatography/mass spectrometry (gc/ms) to environmental analysis, as well as any other areas of analysis requiring separation of complex mixtures with qualitative and quantitative characterization of the components, could proceed. The use of a prior separation technique prior to the application of mass spectrometric techniques is essential because environmental matrices are typically very complex: the compounds of environmental interest are present as minor constituents in a medium such as sewage sludge or process waste which may contain hundreds of other components. The gas chromatograph can be coupled successfully to all types of mass analyzers.

Gc/Ms Applications

One of the extensive applications of gc/ms to environmental analysis is in the area of execution of numbered methods promulgated by the United States Environmental Protection Agency and other government agencies and scientific groups such as the American Society for Testing and Materials. In a numbered method, the entire procedure of taking a sample, preparing the sample for analysis, performing the analysis, interpreting and reporting the results, and ensuring the quality of the analysis is addressed in detail. In areas of concern such as the remediation of Superfund chemical dumpsites, for example, the ability of mass spectrometry to perform qualitative and quantitative analysis is essential in establishing the presence of toxic volatile and semivolatile organic compounds prior to remediation, and the absence of these compounds after remediation of the site has taken place. A chromatographic method, such as gas chromatography, is usually a very sensitive mode of analysis, but the result is nonspecific; that is, the gas chromatographic analyst can never be absolutely sure of the identification of the compounds which are observed, since retention time and detector response are the only parameters which the gas chromatograph can report. In addition, it is impossible to say with certainty in a chromatographic analysis, that no other compound is eluting at the same time as the compound of interest. When the unique ability of the mass spectrometer to produce a characteristic fragmentation pattern is coupled with the ability of the gas chromatograph to provide retention time information and resolve the compound of interest from its matrix, very sensitive (typically in the nanogram or picogram range) and highly specific analytical information is produced. Gc/ms, as applied in numbered methods, is used for qualitative and quantitative analysis of organic compounds in matrices such as drinking water, wastewater, groundwater, sewage sludge, soil, sediment, gaseous stationary source emissions, ambient air, indoor air, and ash. The numbered methods are modified and adapted to address matrices as diverse as still bottoms, incinerator flyash, process feeds, barrel coatings, printed circuit boards, wire coatings, cement, wood, wood smoke, upholstery, food, building materials, and a wide variety of complex liquid, solid, and gaseous matrices.

The challenge in the area of development of techniques and technology is always to make the analysis more sensitive and more specific. Computer models have been formulated to allow the assessment of the risk posed to populations by the presence of trace levels of organic constituents in media such as air, groundwater and drinking water. With levels of acceptable risk proceeding ever lower, it is essential that the analytical technology detect the presence of the compounds of interest at ever lower levels in the various media of environmental interest.

Toxic Dioxins: High Resolution Gc/Ms

Polychlorinated dibenzodioxins (PCDD) and dibenzofurans (PCDF) are among the most toxic synthetic organic chemicals known to humans. The entire group of dioxins and furans comprises 210 compounds, of which approximately one dozen compounds are considered to be very toxic and one compound, 2,3,7,8-tetrachloridibenzo-p-dioxin, is extremely toxic. Public attention has been focused on this group of compounds because of prominent incidents of environmental contamination which have occurred at locations such as Seveso, Italy; Times Beach, Missouri; and Love Canal, New York. PCDD and PCDF are formed as byproducts in the synthesis of herbicides and are formed in combustion processes (for example, in a municipal solid waste incinerator) by unknown mechanisms during the combustion of halogenated organic compounds such as polychlorinated biphenyls and chlorinated phenols. PCDD and PCDF have also been reported in vegetation, human and animal tissue, milk, blood, in discharges from pulp and paper mills, and in paper products.

Because of the toxicity of the polychlorinated dibenzodioxins and dibenzofurans and the high cost of dealing with their presence in the environment (the evacuation of Times Beach, for example), it is absolutely essential that identification of these compounds be correct beyond a reasonable doubt and that quantitative data be accurate. For these reasons, gas chromatography/mass spectrometry is used as the method of choice in the analysis of samples containing these compounds. However, the mass spectrometer found in the average laboratory is not adequate for the performance of this level of sensitive and specific analysis. A specialized mass spectrometer capable of high resolution of masses is required to perform the analysis at the required level of sensitivity and specificity.

High resolution gas chromatographic techniques are required for the separation of the 210 chemically similar compounds which make up this family. Additionally, sophisticated laboratory techniques for the extraction of these compounds from a complex environmental matrix such as flyash, with extensive compound purification techniques, are required for a successful analysis. With the coupling of the sensitivity and mass resolution of the instrumentation with the accurate masses of the molecule and its fragment ions and using complex computerized algorithms to relate accurate mass to the information yielded by the high resolution mass spectrometer, it is possible to obtain accurate mass measurement to allow for the qualitative identification and quantitative analysis of polychlorinated dibenzodioxins and dibenzofurans at the parts per trillion (10^{-12} grams/gram) level. The sensitivity is critical to be able to provide an accurate assessment of the risk posed by the presence of these chemicals. Accurate identification, beyond any reasonable doubt, is required as the basis of regulations or legal actions.

Applications of High Resolution Gc/Ms

The single most valuable piece of information which can be obtained from the mass spectrum is the molecular weight of an unknown compound. When the analyst has established the molecular weight of an unknown compound, potential atomic compositions can be suggested for the molecule and the fragmentation pattern observed in the mass spectrum can be interpreted to determine the arrangement of atoms in the original molecule. In many cases, the molecular weight information, coupled with the fragmentations which occur in the molecule under electron ionization, is sufficient to specify a structure for the molecule.

Frequently, the mass spectrum of a compound will identify that compound as a member of a family but cannot specify the particular isomer. Often the coupling of chromatographic retention times with the mass spectrum can specify the atomic composition, but a complete identification of the compound, including the exact position within the molecule of all of the functional groups, cannot be made. Sometimes the ionized molecular species formed under conditions of electron ionization is sufficiently energetic that extensive fragmentation occurs with formation of very stable fragments and no ion with a mass characteristic of the molecular weight is observed. High resolution mass spectrometry is extensively employed in environmental analysis for accurate mass identifications of unknown compounds, even when these compounds are present in complex matrices.

A mass spectrometric technique which can be quite useful in obtaining molecular weight information is chemical ionization. In chemical ionization mass spectrometry, the vaporized compound of interest enters the ion source as a minor constituent of a reagent gas consisting mostly of methane, isobutane, water, or ammonia, for example. The major constituent of the mixture (the reagent gas) is ionized and the compound of interest is ionized by ionic chemical reaction with the reagent gas. Reactions which typically occur, with judicious selection of the appropriate reagent gas, involve transfer of a proton or characteristic addition of some group such as a methyl group to the compound of interest. Because the ionization process occurs by chemical reaction, high levels of energy are not transferred to the compound of interest so minimal fragmentation occurs. The species which is detected in the mass spectrometer is typically the (M + H) ion, ie, the protonated molecule. Since the analyst is aware of the common reactions which can occur with the use of a given reagent gas, the mass of the charged adduct species can be related to the molecular weight of the original molecule to aid in the identification of an unknown compound.

If the chemical ionization reactions are performed in a high resolution mass spectrometer, accurate mass measurements can be used to determine the composition of an ion at a given mass. If chemical ionization techniques are combined with standard electron ionization techniques (not necessarily in the same analysis), both molecular weight information and fragmentation patterns can be used for the characterization of unknown compounds. Instrumentation which is commercially available can apply both electron ionization and chemical ionization techniques, nearly simultaneously.

Research/Methodologies: Lc/Ms Applications

Not all compounds of environmental interest are directly amenable to gas chromatographic techniques for separation of the compounds of interest from the environmental matrix in which they occur. Many compounds are thermally unstable and decompose upon vaporization or application of heat, while other compounds are insufficiently volatile for analysis by gas chromatography because they cannot be vaporized. In these situations, high performance liquid chromatography (hplc), a liquid, ambient temperature analog to gas chromatography, can be employed as a separation technique without the requirement of vaporization of the compound. Technological advances over the last 5–10 years have been applied to the coupling of liquid chromatography with mass spectrometry. Although several commercial lc/ms systems are available, many different interfacing methodologies are currently being used for coupling with the mass spectrometer, with variable levels of reproducibility and reliability. No single interface for lc/ms is universally accepted, and operating procedures are not standard.

The research field of lc/ms is moving in the direction of standardization of instrumentation and operating parameters. No numbered methods are presently available for application of lc/ms to the solution of environmental problems, but numbered methods in this area are being written and will be evaluated and adopted within the next few years. The lc/ms technique is essential in environmental analysis for qualitative and quantitative analysis of dyes and dyestuffs in media such as groundwater and wastewater. Dyes are typically large and very polar organic molecules. Their molecular weight and polar nature, for the most part, preclude their analysis by gas chromatography. The polar groups in the molecule in many cases make the dye molecules biologically active and either toxic or carcinogenic or both. The magnitude of the environmental problem afforded by the presence of aqueous waste containing dyes cannot be assessed accurately until successful analytical methods are available. Lc/ms techniques are being developed to address this area. Hplc/ms techniques have also been applied to the qualitative and quantitative analysis of pesticides and herbicides, especially the highly polar pesticides such as carbamates, which are not amenable to gas chromatographic techniques.

The problem of separating compounds of interest from complex environmental matrices so that qualitative and quantitative analytical mass spectrometric techniques can be applied extends into chromatographic research. It has been more than twenty years since the first reports on supercritical fluid chromatography (sfc), but coupling of supercritical chromatographic techniques with mass spectrometry is a development of the last ten years. The coupled technique, sfc/ms, is still far from a routine laboratory technique but the technology and areas of application are developing rapidly. The applications of lc/ms and sfc/ms overlap in many areas, but there are also many areas of application in which the two techniques complement each other.

Supercritical fluids exist at temperatures and pressures above the supercritical point of a compound. By controlling pressure and temperature, the physical properties of a supercritical fluid are variable between normal gas and nearly liquid. At each density, the solvent characteristics of the supercritical fluid are different. By judicious selection of the supercritical fluid and various materials which can be added to this fluid as modifiers, it is possible to perform selective extractions with a minimum of solvent. With the appropriate chromatographic equipment, it is possible to use supercritical fluid to perform chromatographic separations. Lower viscosities and higher diffusion coefficients observed in supercritical fluids (as compared to liquids) produce far higher chromatographic efficiency compared to hplc techniques.

Interfacing Options: Sfc/Ms

Since supercritical fluids are pressurized dense gases, there are numerous options for interfacing SFC to mass spectrometry. A direct fluid injection interface allows the supercritical fluid to expand into a region where ionization can occur. Since a gas is produced upon expansion of a supercritical fluid, either electron ionization or chemical ionization can occur upon expansion and ionization of the fluid. Various techniques which involve cooling and condensation of the supercritical fluid can be used, so that the interfaces which are effective for lc/ms are effective for sfc/ms. The development of the instrumentation is a prominent area of research. Some of the areas of application of sfc/ms include characterization of polymeric mixtures such as surfactants, and labile and nonvolatile compounds such as pesticides, herbicides, rodenticides, mycotoxins, polar substituted biphenyls, and alkaloids. Surfactants are components of laundry detergents and are widely dispersed in the environment. Analytical techniques which could be applied to their characterization and quantitative analysis have only recently become available. Azo dyes are widely used polar compounds which are difficult to separate and analyze by gas chromatographic techniques. Because of their highly polar nature, the compounds do not chromatograph well, if at all. Hplc/ms is successful in characterizing some of the members of the class, and sfc/ms is proving to have a broad application in this area as well.

Analysis of Semivolatile Organic Compounds: Gc/Ftir/Ms

Another combined technique with important advantages for the analysis of semivolatile or even relatively nonvolatile organic compounds is the combination of gas chromatography with Fourier transform infrared spectroscopy with mass spectrometry (gc/ftir/ms). At least one combined system is commercially available to allow the collection of ftir data simultaneously with MS information to provide complementary spectral information. The mass spectrometer used in this combination is a quadrupole mass filter. The ms provides molecular weight data, a characteristic fragmentation pattern, and isotopic cluster information. The ftir can distinguish isomers, provide frequency data for organic functional groups, and can provide absorption information for quantitative analysis, as a confirmation of the quantitative information available from the mass spectrometric measurements. The two techniques are complementary in that the ftir can be used to confirm the identifications made by the mass spectrometer. In some cases, when the mass spectrometer is able to characterize a compound as a member of a particular class, the ftir can even pinpoint the location of organic functional groups within the molecule. The characterization of specific isomers is frequently impossible for the mass spectrometer alone. In many cases, the ability of the ftir to characterize organic functional groups may yield sufficient information to assess biological hazard of constituents of the environment.

The application of the gc/ftir/ms systems to nonvolatile organic compounds has been slow to develop, although these compounds constitute the greater portion of the extractable portion of an environmental sample, and this nonvolatile fraction of an environmental sample frequently exhibits mutagenic properties. This nonvolatile material is difficult to analyze because nonvolatile materials do not vaporize readily and do not chromatograph using standard techniques. However, some nonvolatile materials can often be made amenable to chromatographic analysis by preparation of an appropriate derivative which will enhance the volatility of a nonvolatile compound. Compounds which are nonvolatile because of polar groups in the molecule can frequently be converted to an analyzable form by chemical reaction. Common examples of derivatization include preparation of esters of carboxylic acids and preparation of 2,4-dinitrophenylhydrazone derivatives of aldehydes and ketones, as well as ethers of alcohols.

Ms/Ms Technique

A technique known as mass spectrometry/mass spectrometry (ms/ms) often utilizes three quadrupole mass filters in tandem. The first quadrupole is used to perform a mass separation of a vaporized species. The second quadrupole serves as a reaction chamber, where a gas is introduced to collide with the energetic ions emerging from the first quadrupole. Highly specific reactions occur in this colli-

sion cell, and the products of these reactions are mass-analyzed in the final quadrupole. By utilizing characteristic reactions which can occur for a particular species, it is possible to characterize trace quantities of a given organic compound in a complex matrix without extensive purification of the compound or removal of the matrix. The ms/ms technique for mass spectrometric analysis can be combined with gas chromatographic separation, with liquid chromatographic separation, or with supercritical fluid chromatography. The hplc/ms/ms technique has been applied to the analysis of dioxins and furans without prior cleanup, to the analysis of carbamate and organophosphorus pesticides, to analysis of dyes, dye wastes, and dye-manufacturing intermediates and byproducts, and to manufacturing wastes. A mobile-van-based ms/ms system has been used for the direct analysis of gaseous mixtures in the field. The net effect of utilization of highly specific ionic reactions to identify compounds can be achieved by other combinations of mass spectrometric technique than three quadrupole analyzers; for example, the ion trap detector can trap ions and the occurrence of ionic chemical reactions can be observed, so the ion trap functions as an ms/ms system.

Icp/Ms Applications

The quadrupole mass filter has also been coupled to inductively coupled argon plasma as an ion source. The primary application of the coupled inductively coupled plasma/mass spectrometry (icp/ms) technique is to inorganic analysis, such as the determination of trace metals in various types of water such as groundwater, sea water, lake water, and drinking water. The technique is also applicable to the characterization of trace metals in soils and sediments, for as many as 22 elements. The combined technique of icp/ms can also be used to determine accurate elemental isotopic ratios. These accurate isotopic ratios may be used to pursue the source of metallic environmental contaminants so that regulatory action can be taken for remediation of the contamination. Different trace elements may be used to define regional sources for atmospheric aerosol pollution. Because toxicological assessment requires speciated information for trace metals rather than a value for total metal concentration, the next development in the area of icp/ms will probably be interfacing this technique with an amenable chromatographic separation technique such as ion chromatography or hplc.

Api: Advantages

It has been possible to form ions at atmospheric pressure and analyze these ions using mass spectrometric techniques for more than 30 years. The use of atmospheric pressure ionization (api) offers definite advantages for coupling of mass spectrometry with chromatographic techniques, since chromatographic techniques have a gaseous or liquid effluent that is at atmospheric pressure or above. Use of conventional mass spectrometric techniques requires that this pressurized flow be attenuated to accommodate the vacuum system of the mass spectrometer. In api, the ion source region is separated from the high-vacuum mass analyzer by a very small orifice. This orifice must be large enough to allow the entrance of sufficient ions to perform analysis, while keeping the pressure sufficiently low to allow efficient operation of the mass analyzer. Historically, quadrupole mass filters have been used for apims, although some use of magnetic sector mass spectrometers and ion trap mass spectrometers has been reported. Ions can be formed by interaction with a thin foil of radioactive material which emits energetic particles, or by interaction with a corona discharge needle (an electrical discharge at the tip of a needle held at high voltage). When the ions are formed, they undergo reaction with the air that is present as well as with other gaseous molecules. Thus, the ionic products that are ultimately observed are the products of chemical ionization reactions, and the ion chemistry that occurs can be very complex. Both positive and negative ions can be formed under api conditions. Formation of positive or negative species can be enhanced by judicious choice of reagent gases. Strongly electronegative compounds such as pesticides are more sensitive as negative ions. Since the ionization is mild, little fragmentation of the ions occurs and molecular weight information is readily available. However, to obtain structural information, the instrumentation can be modified for ms/ms capability. Chromatographic introduction systems such as gas chromatography, high performance liquid chromatography, and supercritical fluid chromatography can be utilized as sample introduction systems. Capillary electrophoresis, a simple high-efficiency separation for mixtures containing organic ions in solution, can also be used as a sample introduction system for apims, for environmental applications such as the characterization of sulfonylurea herbicides, sulfonated azo dyes, and phenoxyacetic acids.

FTICRMS: Technique for Mass Measurement

Fourier transform ion cyclotron resonance mass spectrometry (fticrms) is a technique for conversion of the mass-to-charge ratio of an ion to a cyclotron orbital frequency which is experimentally measurable. Since frequency can be measured more accurately than any other physical property, fticrms offers the potential for ultrahigh accuracy in mass measurement. Extensive instrument development research is presently being performed, with the coupling of numerous ionization techniques to produce ionized species of nonvolatile and high molecular weight compounds, which can then be subjected to accurate mass measurement of compounds of environmental interest using fticrms.

Surface Mass Spectrometry: Applications

Surface phenomena are an area of very high interest in environmental applications. Organic compounds can condense into solids from the gas phase, and this condensing organic material can interact with particulate matter, being deposited on the surface of the particle or becoming entrained within a porous particle. The environmental impact of the organic material is governed by the molecular structure of the uppermost molecular layers of the surface of a solid. Analytical elemental information, which can be obtained by digestion or extraction of the particulate matter with subsequent analysis for organic or inorganic analytes, provides characterization of the bulk composition of the particulate and is not sufficient to characterize the particulate surface. An analytical instrument with high sensitivity (these organic materials are present on the surface in a monolayer) and sufficient spatial resolution to provide detailed molecular information about the surface structures, is required.

Surface mass spectrometry shows high sensitivity and can provide detailed and specific molecular information at the surface of a particle: identification and quantitative analysis of elements, isotopes, and organic molecular species (nonvolatile and thermally labile compounds) can be performed. Two mass spectrometric techniques have been applied to the solution of this analytical problem: static secondary ion ms (sims) and laser secondary neutral ms (laser snms). In sims, surface species are desorbed by particle bombardment at an energy of thousands of electron volts. Adsorbed particles are ionized during their bombardment, and time-of-flight (tof) mass spectrometry offers a high transmission rate for these ions with detection of ions of all masses in parallel, although initial development of the field employed both magnetic sector mass spectrometers and quadrupole mass filters. With an unlimited mass range, high resolution and accurate determination of masses, and short time for analysis, tof–sims is a very powerful technique for the analysis of surfaces. Mass spectrometric surface characterization techniques have been applied to the characterization of polymers, and to identification and localization of contaminants on semiconductor surfaces (tof–sims can be used to determine the step in a manufacturing process where surface contamination occurs). Failure analysis, where failure in function can be characterized by a change in the molecular structure of the outer layers, is a prominent application of the tof–sims technique.

Laser Snms: Applications

Formation of positive ions from a surface using laser excitation (laser snms), in combination with the use of an ion microprobe, yields quantitative elemental mapping coupled with high sensitivity. Laser snms may be applied to elemental analysis as well as the characterization of molecular surfaces. Surfaces such as metals and ceramics are amenable to analysis, with flat or rough surfaces. Environmental applications are currently being reported in the literature, and technologies and techniques are under development in laboratories around the world.

Ims: Laboratory Techniques

Ion mobility spectrometry (ims) is a technique which has been in existence since the sixties, primarily for the study of ion-molecule reactions in the research laboratory and for qualitative trace analysis. Like many other laboratory techniques, ims is finding environmental applications in the continuous monitoring of gases such as ammonia, hydrogen chloride, hydrogen fluoride, chlorine, and chlorine dioxide. These gases are all toxic and corrosive, and are difficult to monitor by other techniques. In an ion mobility spectrometer, a diluted sample (diluted through a dilution probe from stationary source emissions or from chemical processes) is forced over a semi-permeable membrane. Purified air sweeps the sample into a reaction region, where the sample is ionized by a weak plasma formed by a small radioactive source (^{63}Ni). Ionized sample molecules drift through the cell under an electric field, and are periodically introduced into a drift tube by an electronic shutter grid. In the drift tube, the ions separate on the basis of charge, mass, and shape. The current at the detector is measured as a function of time, so the operation is equivalent to a time-of-flight mass spectrometer. Development of this type of system for use as a continuous emissions monitor is in its very early stages; its ultimate application in this area will depend upon its cost, ease of use, transportability, and maintenance.

Fab Analysis Technique

Fast atom bombardment analysis (fab) is a mass spectrometric ionization technique which improves the response of the ms system to solid nonvolatile materials. From its beginnings, mass spectrometry has focused on the analysis of samples in the vapor phase, an approach which usually requires heating samples in order to vaporize them. There has, therefore, always been some level of concern about the contribution of thermal processes to the mass spectral fragmentation which occurs. In the fast atom bombardment technique, a beam of fast argon atoms is directed to the surface of a target which is carrying the sample. The fast argon atoms are produced in a fast atom "gun": argon ions with an energy between 5 and 10 kV are generated by a gaseous discharge. The ions formed in this discharge pass through a collision chamber which is filled with neutral argon atoms. Collisions occur between the charged and neutral atoms. Charge exchange occurs, usually without loss of significant amounts of kinetic energy, and a beam of fast argon atoms is produced. When these fast argon atoms hit the surface of a sample, ions are generated and focussed into the analytical portion of the instrument. At present, the major applications of the fab technique have been large biomolecules, but use of the fab technique coupled with higher performance liquid chromatography will broaden the area of application to large thermally-labile molecules of environmental interest such as dyes and toxins for which there has been no viable analytical technique.

CONCLUSION

In most cases, the transition from research laboratory technique to commercial availability and hence to extensive application is quite short—sometimes a period of one to two years. In other cases, particularly coupled high performance liquid chromatography/mass spectrometry techniques, the sheer complexity of the coupled technologies makes widespread routine application slower. The instrumentation is costly, and a high level of expertise is required for effective utilization of the technique. The areas of application for hplc/ms have been far less urgent than the areas which required the development of gc/ms techniques and their nearly universal application. The current trends are to make the instrumentation simpler, more compact (and, therefore, less costly) and to make the instrumentation accessible for more laboratories in order to broaden the areas of application.

BIBLIOGRAPHY

J. J. Thomson, *Rays of Positive Electricity*, Longmans, Green, 1913.

D. H. Smith, ed., *Computer-Assisted Structure Elucidation*, American Chemical Society, Washington, D.C, 1977, p. 151.

W. H. McFadden, *Techniques of Combined Gas Chromatography/Mass Spectrometry*, Wiley, New York, 1973, p. 463.

P. H. Dawson, *Quadrupole Mass Spectrometry and its Applications*, Elsevier, Amsterdam, The Netherlands, 1976.

W. W. Lowrance, ed., *Public Health Risks of the Dioxins*, William Kaufmann, Inc., Los Altos, Calif., 1984.

M. Gough, *Dioxin, Agent Orange—the Facts*, Plenum Press, New York, 1986.

J. S. Stanley, T. M. Sack, *Protocol for the Analysis of 2,3,7,8-Tetrachlorodiobenzo-p-Dioxin by High Resolution Gas Chromatography/High Resolution Mass Spectrometry*, Jan. 1986, EPA 600/4-86-004.

J. M. L. Penninger, M. Radosz, M. A. McHugh, V. J. Krukonis, eds., *Supercritical Fluid Technology*, Elsevier, Amsterdam, The Netherlands, 1985.

F. W. McLafferty, ed., *Tandem Mass Spectrometry*, John Wiley & Sons, Inc., New York, 1983.

M. V. Buchanan, ed., *Fourier Transform Mass Spectrometry: Evolution, Innovation, and Applications*, American Chemical Society, Washington, D.C., 1987.

D. M. Lubman, ed., *Lasers in Mass Spectrometry*, Oxford University Press, New York, 1990.

D. P. Woodruff, T. A. Delchar, *Modern Techniques of Surface Science*, Cambridge University Press, Cambridge, UK, 1986.

A. S. Czanderna, D. M. Hercules, *Ion Spectroscopies for Surface Analysis*, Plenum Press, New York, 1991.

G. W. A. Milne, ed., *Mass Spectrometry: Techniques and Applications*, Wiley-Interscience, New York, 1971.

ENVIRONMENTAL ANALYSIS—MASS SPECTROMETRY, ADVANCES

RONALD A. HITES
School of Public and Environmental Affairs
and Department of Chemistry
Indiana University
Bloomington, Indiana

Innovations in mass spectrometry are being applied increasingly to environmental issues. The primary features of environmental mass spectrometry can be usefully compared with those of biomedical mass spectrometry, because these are the two largest areas of application for this analytical technique.

Environmental mass spectrometry tends to deal with anthropogenic compounds with molecular weights less than about 1000 daltons. On occasion, petroleum and the combustion of organic compounds are studied; but in both cases, there is an anthropogenic component. On the other hand, biomedical mass spectrometry tends to deal with natural products. Some of these are of modest molecular weight, but most have very high molecular weights and include biopolymers such as proteins and carbohydrates.

Environmental mass spectrometry tends to be highly quantitative. The utmost sensitivity is sought, and many laboratories routinely measure a few picograms of many compounds. Biomedical mass spectrometry, however, tends to be largely qualitative, and the materials studied have a high mass. For example, the sequence of amino acids in a protein might be the experimental goal; analysis of proteins of molecular weights of 150,000 or more is now routine (1).

Environmental mass spectrometry often uses official methods promulgated or approved by regulatory agencies, such as the Environmental Protection Agency (EPA) in the United States (2). Biomedical methods tend to use *ad hoc* procedures designed to solve the problem at hand. There is however, one area of commonality: environmental mass spectrometry of pesticide metabolites is similar to biomedical drug metabolite studies.

FIELDABLE INSTRUMENTS

A growing trend is the use of fieldable instruments that can be taken to the site of environmental contamination, for example, a hazardous waste landfill that is being cleaned up. Here, fieldable mass spectrometers are used to measure the quantity of specific compounds as the cleanup proceeds. Such an approach reduces the analytical costs associated with the cleanup through rapid turnaround and high throughput of samples. While instruments have been taken to the site in a truck, van, or recreational vehicle, an instrument that can be carried by a worker (Figure 1) has been developed by Urban and coworkers (3).

MEMBRANE MASS SPECTROMETRY

"Membrane mass spectrometry" is illustrated by the apparatus shown in Figure 2. The sample (water or air) pas-

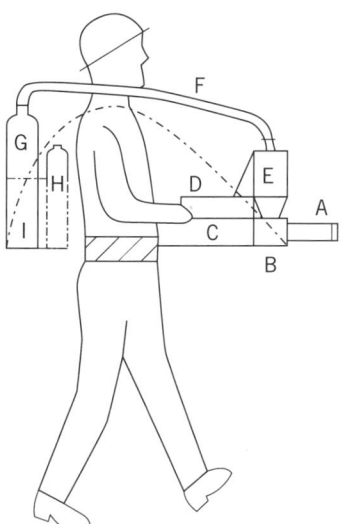

Figure 1. A fieldable mass spectrometer that can be carried by a single worker; from reference 3. Its total weight is about 35 kg. A, Inlet and gc column; B, mass spectrometer; C, electronics; D, computer; E, molecular drag pump; F, vacuum hose; G, vacuum reservoir; H, carrier gas; I, battery (24 VDC).

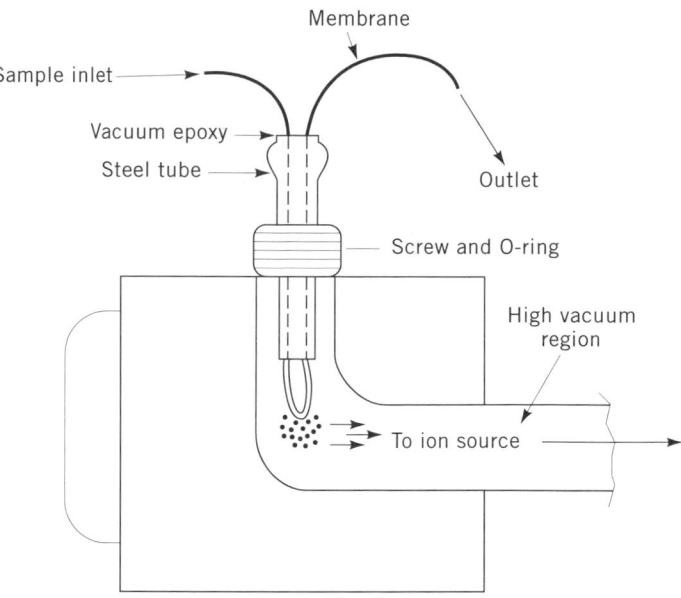

Figure 2. Principle of operation of membrane mass spectrometry; from reference 4.

ses through a tube made from a semipermeable polymer that is in the vacuum system of the ion source. The analytes of interest pass through the membrane and into the ion source. By using a triple-quadrupole mass spectrometer, considerable qualitative and quantitative information can be obtained without subjecting the sample to chromatography. This work has been popularized in the last several years in Cooks's laboratory (4), but important contributions have also been made by Enke and his students (5).

LIQUID CHROMATOGRAPHY MASS SPECTROMETRY

Considerable attention has also been given to liquid chromatographic mass spectrometry (LC/MS). Work by Behymer and colleagues (6), for example, has used particle beam LC/MS to develop an "official" EPA method for the analysis of benzidine. Unfortunately, developmental issues still exist for some environmental applications of LC/MS. For example, chromatographic resolution and sensitivity are often inadequate for some samples.

LC/MS is ideal for thermally unstable or nonvolatile compounds because these compounds cannot be analyzed by GC/MS. Unfortunately, many LC/MS methods heat the sample in the interface. This reduces the suitability of the techniques for just those compounds for which they should be most suited.

One exception, continuous flow fast-atom bombardment (CF-FAB), takes compounds from a flowing liquid stream directly into the ionic phase. Azo dyes have been analyzed by LC/CF-FAB using tandem mass spectrometry in our laboratory. Figure 3 (top) shows the liquid chromatogram of a mixture of five azo dyes, all of which are aromatic sulfonates. Note that the peak widths increase with time as a result of isocratic elution. Figure 3 (bottom) is the product ion spectra of peaks B and C, which

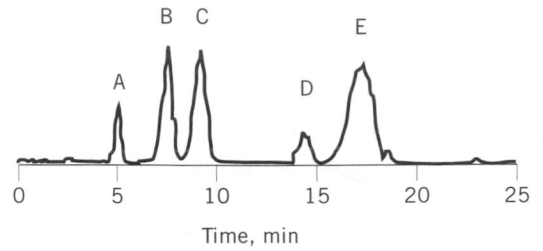

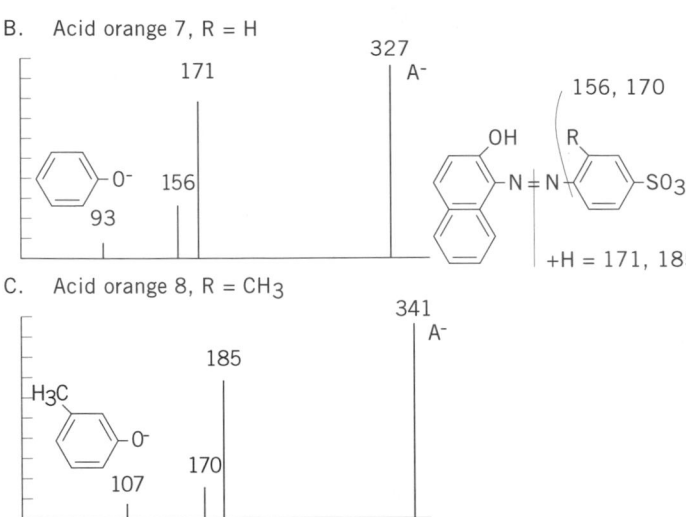

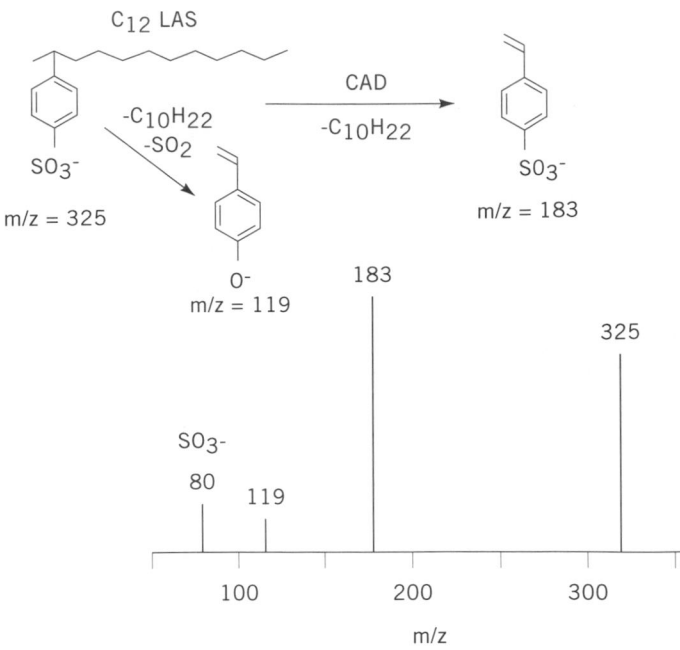

Figure 4. Product ion mass spectrum of a C_{12} linear alkyl benzene sulfonate and scheme showing the sources of the various ions.

Figure 3. Liquid chromatogram of a standard mixture of 5 azo dyes detected by continuous-flow fast-atom bombardment mass spectrometry (top) and daughter ion mass spectra of A^- from peaks *B* and *C* (bottom).

are the dyes Acid Orange 7 and 8, respectively. These compounds differ only by the presence of a methyl group *ortho* to the azo linkage; therefore, the spectra are simply offset by 14 daltons. Abundant ions at m/z 171 and 185 are due to cleavage of the azo group with a hydrogen rearrangement. The ions at m/z 93 and 107 are due to further rearrangement of the sulfonate group.

FLOW INJECTION MASS SPECTROMETRY

The cleanness and simplicity of the spectra in Figure 3 imply that in some cases, chromatographic separation before mass spectrometric analysis may not be necessary. If the chromatographic step could be avoided, the throughput would be increased by a factor of at least 5 to 10. This concept has been demonstrated for linear alkylbenzene sulfonates (LAS), widely used surfactants (7).

A typical LAS structure is shown in the upper left of Figure 4. The compound shown has 12 carbons in the side chain, commercial mixtures would have homologues with 10 to 13 carbons in the side chain. All LAS homologues show an abundant product ion at m/z 183, due to cleavage *beta* to the aromatic ring. Therefore, scanning the precursor ions of m/z 183 analyzes all the homologues in an LAS mixture. Branched alkylbenzene sulfonates can be distinguished from their linear cousins because the product ion at m/z 183 is shifted to m/z 197.

Because the different homologues give equal molar responses, the CF-FAB technique can be used for the quantitative analysis of LAS with only a simple calibration (7). For a sample containing 10 nanograms of LAS, the signal-to-noise ratio is about 5:1; this gives a sensitivity of about 6 picomoles.

The data obtained from an analysis of LAS samples in the input and primary treatment stages of a wastewater treatment plant and in the river 100 m downstream from the discharge point are shown in Figure 5. Concentrations of the 10- to 13-carbon LAS homologues decrease by factors of 10 to 80 between the influent and the river. This entire analysis, including calibration, took about one hour (the LASs were isolated by solid phase extraction), less time than going to the wastewater treatment plant to obtain the samples.

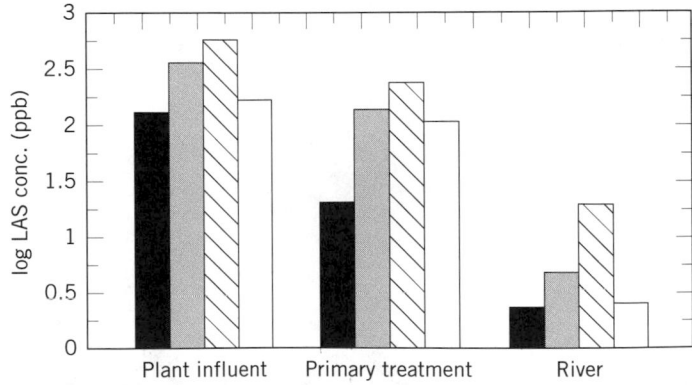

Figure 5. Concentrations (log scale) of C_{10} to C_{13} linear benzene sulfonates (LAS) in wastewater, in the treatment plant, and in the receiving river as measured by parent ion scans of m/z 183.

NEGATIVE IONIZATION MASS SPECTROMETRY

Electron capture (EC) mass spectrometry is a very important tool for environmental analysis (8). The sensitivity and selectivity of this technique make it particularly useful for analytes that contain electronegative functional groups or atoms. For example, many organochlorides, such as PCBs and chlordane, can be analyzed by electron capture mass spectrometry without extensive sample cleanup and with exquisite sensitivity. Figure 6 shows the electron impact (EI) mass spectrum of endosulfan (top) and the electron capture, mass spectrum of endosulfan (bottom) (8). This is probably the world's ugliest EI mass spectrum; there is no molecular ion, and very few structurally specific fragment ions are present. Furthermore, because of the large number of fragment ions, the *absolute* abundance of each ion is smaller than if most of the ions were at only a few m/z values. On the other hand, the EC mass spectrum shows an abundant molecular anion at m/z 404 and a relatively abundant ion because of the loss of a chlorine and the addition of a hydrogen at m/z 370. The electron capture mass spectrum is clearly more useful for the quantitation of this compound.

EC mass spectrometry is being used for the routine analysis of endosulfan in ambient air. An example of the data that can be obtained with this technique is shown in Figure 7, a plot of the natural logarithm of the concentra-

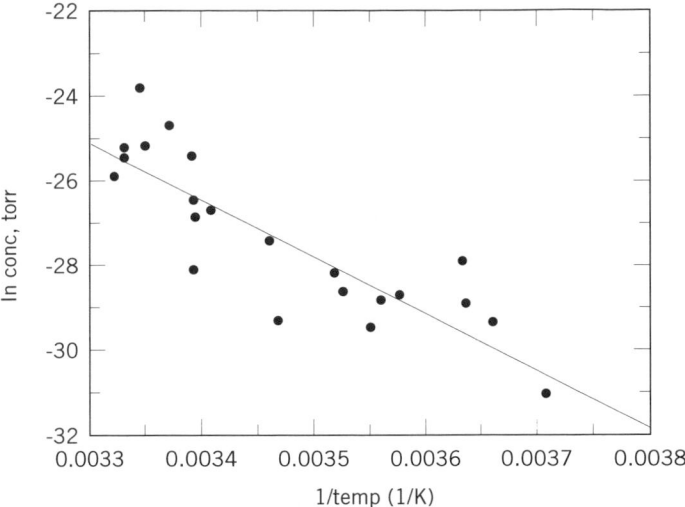

Figure 7. Ambient atmospheric concentrations of endosulfan measured in Bloomington, IN, plotted as a function of atmospheric temperature.

tion (in torr) versus the reciprocal of the absolute temperature. The slope of this line gives the heat a vaporization of endosulfan, 110 kJ/mole. Each air sample takes less than one hour to analyze becasue there is no sample cleanup.

DIOXINS

2,3,7,8-Tetrachlorodibenzo-*p*-dioxin alone has been responsible for the sale of more mass spectrometers than any other compound. Even today, 15 to 20 years after the first warnings about this compound were made public, debate about its toxic effects on humans continues. The literature on the use of mass spectrometry for the analysis of dioxins is so vast that one example suffices.

There are two major sources of dioxins and the related dibenzofurans: by-products in chlorinated aromatic compounds (9) and the combustion of municipal and chemical wastes (10). The latter was an important discovery. No longer could the simple presence of dioxin in a sample indict a chemical production facility. Indeed, it had been suggested that "dioxins have been with us since the advent of fire" (11).

Chlorinated dioxins and furans formed during combustion are emitted into the atmosphere. Depending on the ambient temperature, some of these compounds are adsorbed onto particles and some are in the vapor state. In either case, these compounds can travel through the atmosphere for considerable distances. While they are in the atmosphere, several things can happen to them.

The compounds can reequilibrate between the particle and vapor phases; this is a temperature-dependent process. They may also degrade by photo-oxidative or other chemical processes; the extent of this degradation depends on the physical state of the reactant. Eventually, the dioxins and furans leave the atmosphere by a number of routes. Particles with their load of absorbed compounds

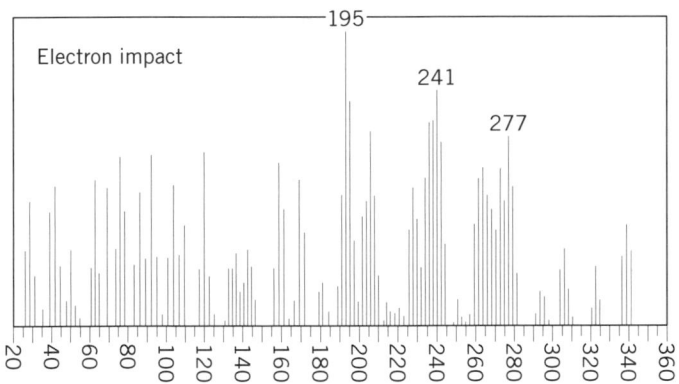

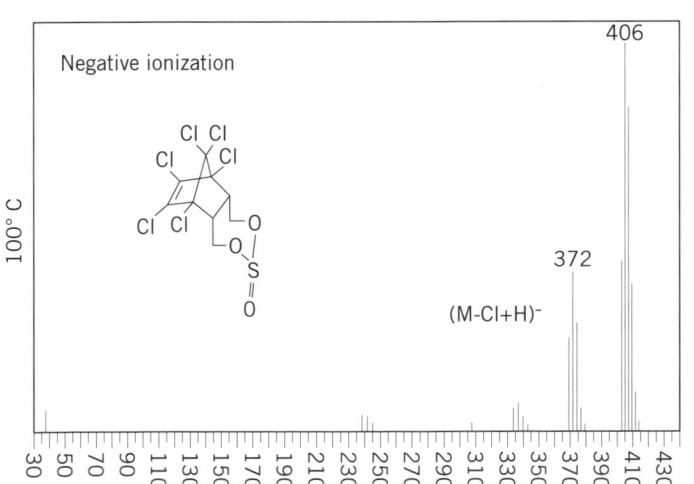

Figure 6. Electron impact mass spectrum (top) and electron capture, negative ion, mass spectrum (bottom) of endosulfan.

settle out of the air, scavenging both particle-bound and vapor-phase compounds. Dioxins and furans from industrial sources also enter the atmosphere; however, except for sporadic and localized events, these sources are minor.

Lake sediment preserves a record of atmospheric deposition because there is a rapid transport of material deposited on the top of a lake to its bottom and the regular accumulation of sediment at the bottom. Cores of sediment from the bottom of a lake, sliced into 0.5- to 1-cm layers, have been analyzed for the tetrachloro- through the octachlorodioxins and furans with isotope dilution and EC GC/MS at sensitivities of less than 100 attomoles (12). Radioisotopic analysis establishes when the sediment core was in contact with the atmosphere.

Figure 8 gives the concentrations of the dioxins and furans in a sediment core taken from Siskiwit Lake (on Isle Royale in Southern Lake Superior [13]) as a function of year of deposition. These data are typical for all the sediment cores that have been studied (12–15). Octachlorodioxin is always the most abundant of the compounds; the heptachlorodioxins and -furans are next in abundance. Although other chlorinated dioxins and furans are present, their concentrations are very small. Moreover, the concentrations of dioxins and furans have not been constant over the last century. The concentrations maximized about 1970, and they were at unmeasurable levels before 1930. This finding suggests that atmospheric dioxin and furan levels increased slowly starting about 1935 and have decreased since about 1970.

Sediment cores from the other Great Lakes and from three high-altitude lakes in Switzerland have also been analyzed (14,15). In every case, dioxins and furans were not present in the sediments before about 1935, and by implication, they were not present in the atmosphere then. This is true despite large differences in both the rates of sediment accumulation and the locations of the lakes. The overall average horizon date is 1938, a date well after the "advent of fire."

What happened in the mid- to late 1930s that led to the emission of dioxins? The products produced by the chemical industry changed about then. Before World War II (1939 to 1945), the chemical industry was commodity based, selling large amounts of inorganic products. During the war, organic products were introduced, and plastics became an important part of the chemical industry. Some of these products were organochlorine based; polyvinylchloride is but one example. As waste materials containing these chemicals were burned, dioxins and furans were produced and released into the atmosphere. These compounds eventually ended up in lake sediments.

Incidentally, coal combustion cannot account for the increase in dioxin and furan levels. Coal combustion has been almost constant since 1910; there has been no major shift either in amount burned or in combustion technology since 1935 (12,15). The 1970 maximum in concentration in most sediment cores suggests that emission-control devices, which were beginning to be widely installed at about this time, have been effective in removing dioxins, furans, and the more conventional air pollutants.

BIBLIOGRAPHY

1. M. M. Siegel, I. J. Hollander, M. Karas, A. Ingendoh, and F. Hillenkamp, *Proc. 38th ASMS Conf. on Mass Spectrom.*, 1990, 158.

2. R. A. Hites and W. L. Budde, *Environ. Sci. Technol.* **25**, 998 (1991).

3. D. T. Urban, N. S. Arnold, and H. L. C. Meuzelaar, *Proc. 38th ASMS Conf. on Mass Spectrom.*, 1990, 615.

4. J. S. Brodbelt and R. G. Cooks, *Anal. Chem.* **57**, 1153 (1985).

5. M. A. LaPack, J. C. Tou, and C. G. Enke, *Anal. Chem.* **62**, 1265 (1990).

6. T. D. Behymer, T. A. Bellar, and W. L. Budde, *Anal. Chem.*, **62**, 1686 (1990).

7. A. J. Borgerding and R. A. Hites, *Anal. Chem.* **64**, 1449 (1992).

8. T. W. Burgoyne and R. A. Hites, *Environ. Sci. Technol.* **27**, 910 (1993).

9. T. Pollock, *Dioxins and Furans: Questions and Answers*, Academy of Natural Sciences, Philadelphia, 1989.

10. K. Olie, P. L. Vermuelen, and O. Hutzinger, *Chemosphere* **6**, 455 (1977).

11. R. L. Rawls, *Chem. Engin. News* (Feb. 12, 1979), p. 23.

12. J. M. Czuczwa and R. A. Hites, *Environ. Sci. Technol.* **18**, 444 (1984).

13. J. M. Czuczwa, B. D. McVeety, and R. A. Hites, *Science* **226**, 568 (1984).

14. J. M. Czuczwa, F. Niessen, and R. A. Hites, *Chemosphere* **14**, 1175 (1985).

15. J. M. Czuczwa and R. A. Hites, *Environ. Sci. Technol.* **20**, 195 (1986).

ENVIRONMENTAL ANALYSIS (OPTICAL SPECTROSCOPY)

JAMES F. KELLY
ROBIN S. McDOWELL
Pacific Northwest Laboratory
Richland, Washington

Optical spectroscopy, broadly defined as spectroscopic measurements and instrumentation using the electromagnetic spectrum from the ultraviolet to the infrared

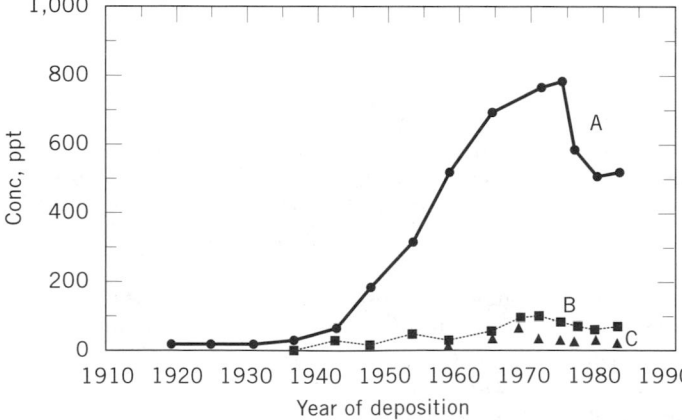

Figure 8. Concentrations of octa- (A) and heptachlorodioxins (B) and of heptachlorofurans (C) (in parts per trillion) versus year of deposition into Siskiwit Lake on Isle Royale.

and microwave regions, is a versatile and powerful analytical technique with many environmental applications. This article covers atmospheric analysis and pollution monitoring, in which the sample consists of an *in situ* volume of air in the atmosphere, and laboratory analysis, where samples of environmental interest are either removed to or constrained within a laboratory environment and analyzed there. The distinction is not rigid, eg, laser detection and ranging (lidar) involves sounding the atmosphere by both remote optical interrogation methods and mobile air samplers that employ laboratory instrumentation to effect *in situ* point-source analyses.

As in any analytical process, the object is to determine the composition of the sample (speciation) and to measure the amounts of different species present (quantification). Spectroscopic analysis can meet both of these demands and may offer advantages over other analytical methods. Spectroscopic techniques can identify and quantify species in a single measurement; they can detect a wide range of compounds and at the same time are highly specific, allowing the molecular identification of each species in multicomponent mixtures. They are quantitatively accurate, and they are noninvasive, may not require sample collection or pretreatment, do not introduce sample contamination, and can be remotely situated, requiring only optical access. They are capable of giving rapid results in real time, and they are easily adaptable to continuous long-term monitoring, including data logging and analysis. Environmental conditions such as temperature and pressure can be recovered from the data.

In spectroscopic analysis, species are identified by the positions and shapes of their absorption, scattering, or emission features, and quantified by the intensities of these features. Numerous applications of optical methods have been made to pollution and atmospheric monitoring; these rely on a few basic mechanisms of light–matter interaction. Absorption spectroscopy in the infrared (ir) or ultraviolet–visible (uv–vis) regions records energy depletion of transmitted radiation at characteristic resonance frequencies involving various rotational, vibrational, and/or electronic energy levels in molecules or atoms.

Scattering techniques monitor the change of a probe signal resulting from scattering by atmospheric species of interest. This can involve either elastic (energy-conserving) interactions, such as Rayleigh or Mie scattering, with the photons undergoing only a change in momentum, or the inelastic (energy-changing) Raman effect, in which a strong monochromatic probe beam is scattered with discrete changes in frequency. A spectroscopic analysis of the scattered Raman light reveals spectral shifts characteristic of the different species. Rayleigh scattering is much stronger (ca 10^3 times) than Raman scattering and occurs for all atoms, molecules, and small particles with diameters smaller than the wavelength of the probe light (eg, $\gtrsim 0.05\ \lambda$), while even stronger Mie scattering occurs when particulate sizes are $> 0.05\ \lambda$. These elastic scattering processes are a fundamental source of background scatter (noise) but can also serve as the source of a return signal for lidar sounding.

Fluorescence detection measures the emission of light from atoms or molecules that have been selectively excited to higher electronic levels by a spectrally intense light source and then decay to lower lying energy states. Such laser-induced fluorescence (lif) is commonly used to study transparent gases, liquids, and solids, while x-ray fluorescence is used extensively to analyze opaque or heterogeneous samples such as rock specimens.

Radiometry is the direct detection and spectroscopic analysis of radiation emitted naturally by the target molecules (ie, emission spectroscopy). This can involve thermal emissions of the atmosphere and of opaque surfaces or fluorescence from naturally induced excitation phenomena, such as airglow from the upper atmosphere due to solar wind and cosmic-ray bombardment.

Many novel interrogation schemes for each of these physical processes have been developed around the experimental details of available light sources, detectors, and spectral analyzers. The advent of powerful and broadly tunable laser sources has led to the ascendancy of the first three processes for active remote sensing of effluents and for ultrasensitive detection of trace samples under laboratory conditions. The laser's extremely high spectral intensity (photons per unit bandwidth) and spatial coherence (low divergence, allowing tight focusing) makes even weak scattering processes like the Raman effect useful over large distances (kilometers). These attributes have also been exploited to induce a variety of new nonlinear responses in media. New laser spectroscopies like multiphoton absorption (leading to fluorescence or ionization) have permitted the detection of single atoms in background gases at atmospheric pressure. Multiphoton scattering processes can be induced by powerful lasers, causing trace effluents to undergo stimulated emission (induced lasing) in highly preferred directions, thus enhancing the detection limits over spontaneous processes that scatter into many directions.

An extensive review of all such techniques is beyond the scope of this article. The conceptual issues of important analytical methodologies and instrumentation will be presented in the following sections. The reader is referred to more detailed treatments of the basic principles of spectroscopy, radiative transfer, and active laser-based remote sensing (1–3), passive remote atmospheric sounding and radiometry (4–6), conventional spectroscopic methodologies and basic atomic physics (7,8), and molecular spectroscopy and structure (9,10). Special treatments of infrared spectroscopy (11,12), laser principles and laser spectroscopy (13,14), and laboratory-based ultrasensitive spectroscopies (15,16) may be useful. Specific methodologies of optical applications to analytical chemistry are treated in the monograph series *Chemical Analysis* and new developments in lasers and their applications are reviewed periodically in the *Springer Series on Optical Sciences*.

SPECTROSCOPIC BACKGROUND

Electromagnetic radiation is characterized by its wavelength λ, frequency ν, or wave number $\bar{\nu}$, which are related by

$$\nu = c\,/\,\lambda,\ \bar{\nu} = 1/\lambda = \nu/c \qquad (1)$$

where c is the speed of light. The photon energy is $E = h\nu = hc\bar{\nu}$, where h is Planck's constant and hence is proportional to frequency (expressed in some multiple of cycles per second or Hertz) and wave number (usually given in cm^{-1}). The units for wavelength are commonly nm or μm (1 nm = 10 Å = 10^{-9}m; 1 μm = 10^{-6} m). The conversion of wave numbers to frequency is given directly as $\nu = c\bar{\nu}$; ie, 1 cm^{-1} = 3 × 10^{10} Hz = 30 GHz.

For convenience, the electromagnetic spectrum is divided into several energy regions characterized by the differing experimental techniques employed and the various nuclear, atomic, and molecular processes that can be studied. These are, in order of increasing energy (decreasing wavelength), with their approximate wavelength limits: radio waves (>30 cm), microwaves (1 mm–30 cm), far infrared (50–1000 μm), mid infrared (2.5–50 μm), near infrared (0.8–2.5 μm), visible (0.4–0.8 μm or 400–800 nm), near ultraviolet (180–400 nm), vacuum ultraviolet (10–180 nm), x rays (0.01–10 nm), and γ rays (<0.01 nm, or energy >0.1 MeV). The regions most useful for atmospheric and pollution monitoring are the mid infrared to near ultraviolet.

Molecular spectroscopy can be divided into the three broad areas: rotational, vibrational, and electronic transitions, named in order of increasing energy required for excitation. Rotational transitions occur in the far infrared and microwave regions (wavelengths longer than about 100 μm, except for a few light molecules such as H$_2$O that extend to shorter wavelengths). From about 15 μm (670 cm^{-1}) out to the extreme far infrared–microwave region near 1000 μm (1 mm or 10 cm^{-1}), there is strong and near-continuous absorption due to the rotational spectrum of water vapor, and atmospheric studies are difficult or impossible here. Still longer wavelengths (> 1 mm) comprise the various microwave and radar bands; these are useful for the detection and ranging of extended objects, ranging in size from raindrops to aircraft to large weather systems. Spectroscopy in the region covering 1 mm to 1 m has been applied successfully to the study of molecules in astronomical sources and interstellar dust clouds (radioastronomy), but this is not a useful region for general spectroscopic monitoring of the earth's atmosphere, because microwave absorption by molecules is extremely weak. However, spectroscopy in the region 260–280 GHz ($\sim$ 1.1 mm) has been used to quantify trace polar molecules such as ClO, O$_3$, HO$_2$, HCN, and N$_2$O at the ppt level in the stratosphere, using a ground-based millimeter-wave superheterodyne receiver (17). In such cases, the large fluence available from microwave sources, used in conjunction with coherent (heterodyne) detection, can overcome weak absorption.

Molecules vibrate at fundamental frequencies corresponding to electromagnetic radiation with wavelengths of approximately 2.5 μm (wavenumber 4000 cm^{-1}) to 100 μm (100 cm^{-1}) (the mid infrared), with some overtone and combination transitions at shorter wavelengths. Because this provides enough energy to excite rotational motions also, the infrared spectrum of a gas consists of rovibrational bands in which the basic vibrational transition is broadened by the superposition of numerous lower energy rotational transitions that occur simultaneously with the vibration. The vibrational frequencies depend on the masses of the atoms involved in the various motions and on the force constants and geometry of the bonds connecting them. At the same time, the shapes of the bands are determined by the rotational structure and hence by the molecular symmetry and moments of inertia. The rovibrational spectrum thus provides direct molecular structural information, resulting in high specificity: the vibrational spectrum of any molecule is unique (except those of optical isomers). Every molecule, except homonuclear diatomics such as O$_2$, N$_2$, and the halogens, has at least one vibrational absorption in the infrared.

Still shorter wavelength (higher energy) radiation will promote transitions between electronic orbitals in atoms and molecules. Outer bonding electrons can usually be excited by near ultraviolet or visible radiation. At higher energies, in the vacuum ultraviolet, innershell transitions occur, but this region is not useful for the present purposes because of strong absorption by atmospheric O$_2$ for $\lambda < 180$ nm. Deep innershell electronic transitions can be induced via x-ray excitation, and this provides a technique for laboratory elemental analysis. Electronic transitions in molecules will also include structure due to associated vibrational and rotational transitions (rovibronic band). Electronic transitions tend to have larger absorption cross-sections than rotational or vibrational transitions and hence higher sensitivity for analytical purposes, but spectral overlap and interferences are more likely to be problems.

LINEWIDTHS, LINE STRENGTHS, AND DETECTIVITY

The detectivity for a given species in optical spectroscopy is a product figure-of-merit of spectral resolution and sensitivity. Resolution, the ability to distinguish between transitions of nearly equal wavelength, is determined practically by the degree to which neighboring spectral features overlap. All spectral features have a narrow natural width, but line-broadening effects and instrumental limitations cause additional line broadening.

The sensitivity of an optical process is determined by the strength of the photon–molecule interaction, expressed usually as a frequency-dependent cross-section $\sigma(\nu)$ (cm^2) for absorption, or as a differential cross-section $d\sigma(\nu)/d\Omega$ (cm^2/steradian) for scattering (1). The latter measures the likelihood that active species scatter some portion of the incident laser fluence $\Phi(\nu)$ (photons/cm^2) into a viewing solid angle $\Delta\Omega$ (steradians) (Fig. 1). In most cases of practical significance for environmental monitoring, the frequency-dependent cross-sections can be written separately in the form

$$\sigma(\nu) = \sigma_0 \cdot \mathcal{L}(\nu - \nu_0, \Delta\nu), \qquad (2)$$

where σ_0 (or $d\sigma/d\Omega$) is the peak value for absorption (scattering) and $\mathcal{L}(\nu - \nu_0, \Delta\nu)$ is a symmetrical lineshape function that is parameterized by a center frequency ν_0 and linewidth $\Delta\nu$.

An important "semiclassical" measure of the frequency-integrated absorption cross-section (Ladenburg's formula) is

$$\int \sigma(\nu)d\nu = \pi r_c c f \equiv S \text{ (cm}^2\cdot\text{frequency)} \qquad (3)$$

where r_c is the classical electron radius (2.8 × 10^{-13} cm) and f is the oscillator strength for the radiative transition

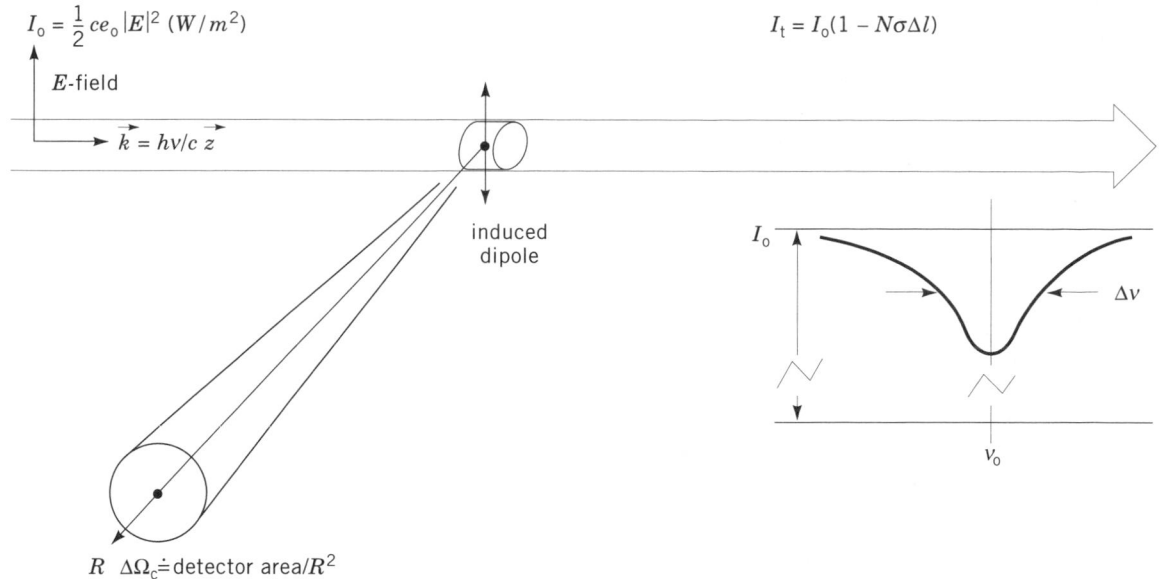

$$I_0 = \frac{1}{2} c e_0 |E|^2 \ (W/m^2)$$

$$I_t = I_0(1 - N\sigma \Delta l)$$

$$\vec{k} = h\nu/c \ \vec{z}$$

induced dipole

$$R \quad \Delta\Omega_c \doteq \text{detector area}/R^2$$

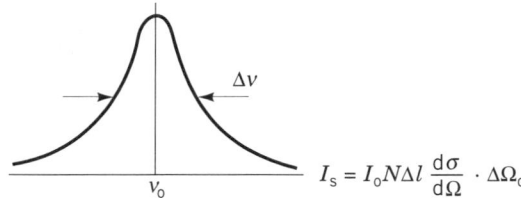

$$I_s = I_0 N \Delta l \frac{d\sigma}{d\Omega} \cdot \Delta\Omega_c$$

Figure 1. Schematic of an absorption–scattering process. An incident electromagnetic field of intensity I_0, with an associated electric field $\vec{E}$, induces dipole oscillation in the absorbers, resulting in light I_s scattered into a collection solid angle $\Delta\Omega_c$, and corresponding absorption resulting in transmitted intensity I_t.

(12,18). Similar expressions can be written for the differential fluorescence or scattering cross-sections, since they are interrelated processes (1). The f-values or line strengths S can be calculated from first principles of quantum mechanics but most often are obtained empirically. S is related to a matrix element d_{ij} connecting the initial and final states of a transition, $S \propto (d_{ij})^2$, where d_{ij} has typical dimensions of the atom ($\sim 10^{-8}$ cm). The maximum value of the line shape function $\mathfrak{L}(\nu - \nu_0, \Delta\nu)$ is inversely proportional to the line width, so the peak absorption (or angle-integrated scattering) cross-sections can be approximated as

$$\sigma_0 \doteq S / \Delta\nu \tag{4}$$

The peak absorption (scattering) cross-sections are thus a useful comparative measure of detectivity, because the latter is a product of the line strength and the practical line resolution. Table 1 lists typical peak values for absorption, fluorescence, and Rayleigh and Raman scattering. Four causes of line broadening are of importance in environmental analysis:

1. Natural broadening is due to the intrinsic lifetimes of the states involved in the transition. The line width (full width at half-maximum, fwhm) $\Delta\nu_N$ is $(2\pi\tau)^{-1}$, where τ is the natural lifetime for a transition to undergo spontaneous decay in the absence of external perturbations.

$\Delta\nu_N$ is $\leq$ 80 MHz for strong uv–visible transitions and $\sim$ 1 MHz for ir vibrational transitions. These line widths can be resolved by modern laser spectroscopies under laboratory conditions.

2. Collisional broadening results when an active absorber or scatterer experiences perturbations by other atoms or molecules. These collisions effectively reduce the natural lifetime of the optically active system, so the net broadening depends on a characteristic impact time τ_c:

$$\Delta\nu_C = \frac{1}{2\pi}\left(\frac{1}{\tau} + \frac{1}{\tau_c}\right) \tag{5}$$

3. Doppler broadening arises from the random thermal agitation of the active radiators, each of which, in its own rest frame, see the applied light field with a different frequency. When averaged over a Maxwellian velocity distribution, this yields a line width (fwhm)

$$\Delta\bar{\nu}(\text{cm}^{-1}) = 7.16 \times 10^{-7}\bar{\nu}_0\sqrt{T/A}, \tag{6}$$

where T is the sample temperature (K), A the molecular weight of the species (amu), and $\bar{\nu}_0$ is the transition energy (cm^{-1}). Typical Doppler line widths are $\sim$0.05 cm^{-1} for visible transitions ($\sim$500 nm) at room temperature, and are correspondingly larger (smaller) for uv (ir) transitions. Unless special nonlinear spectroscopies are ex-

Table 1. Typical Lineshape Parameters and Relationships

Parameter	Rotational Transitions	Vibrational Transitions	Electronic Transitions
Line frequency, $\bar{\nu}$	1–100 cm^{-1}	100–$4{,}000$ cm^{-1}	up to $50{,}000$ cm^{-1}
Natural line width, $\Delta\bar{\nu}_N$	$<10^{-11}$ cm^{-1}	$<10^{-7}$ cm^{-1}	$<3\times10^{-3}$ cm^{-1}
Doppler width at 300 K, $\Delta\bar{\nu}_D$	<0.0003 cm^{-1}	<0.01 cm^{-1}	0.01–0.2 cm^{-1}
Natural radiative lifetime, τ_N	$>10^{-1}$ s	$>2\times10^{-4}$ s	$>2\times10^{-9}$ s
Peak Doppler broadened absorption cross-section, σ_A	$<10^{-20}$ cm^2	$\leq10^{-18}$ cm^2	10^{-11}–10^{-16} cm^2 (atoms) $<10^{-17}$ cm^2 (molecules)
Peak Rayleigh differential scattering cross-section, $d\sigma/d\Omega$	Negligible	Negligible	$<2\times10^{-13}$ cm^2 sr^{-1} (atoms)[a] $<10^{-26}$ cm^2 sr^{-1} (molecules)[b]
Peak Raman differential scattering cross-section, $d\sigma/d\Omega$	$<10^{-27}$ cm^2 sr^{-1}	$<10^{-28}$ cm^2 sr^{-1}	$\sim10^{-24}$ cm^2 sr^{-1} (atoms)
Peak fluorescence differential scattering cross-section, $d\sigma/d\Omega$	Negligible	Negligible	$<5\times10^{-16}$ cm^2 sr^{-1} (atoms)[c] 10^{-20}–10^{-25} cm^2 sr^{-1} (molecules)

[a] Values are for resonant scattering by atomic vapors.

[b] Values are for nonresonant scattering of visible–near uv radiation by atmospheric gases; resonant Raman scattering approaches 10^{-24} cm^2 sr^{-1}.

[c] Values are for STP atmospheric conditions.

ploited, the Doppler broadening of a transition represents the fundamental lower limit for practical resolution.

Natural and collisional broadening are considered homogeneous processes, because all radiators experience the same local effects. Such lines have a Lorentzian shape, with $\Delta\nu(\text{fwhm}) = (2\pi\tau')^{-1}$, where τ' is the effective "lifetime" of a radiator's uninterrupted oscillation period. The homogeneous line profile is distorted by inhomogeneous broadening effects, of which the most important is the Doppler broadening, which has a Gaussian distribution. For gaseous samples at STP, absorptions and emissions actually have the Voigt profile, which is the mathematical combination (convolution) of the Lorentzian and Gaussian shapes.

4. Instrumental broadening occurs when the spectral resolution of the probe light or spectral analyzer are larger than the intrinsic broadening of the radiating species. The observed line shapes are the convolution of the instrumental line shape and the environmentally perturbed Voigt profile (Fig. 2) (8,13).

With inadequate resolution much of the information in a complex vapor-phase spectrum can be blended or lost, with consequent degradation of specificity and loss of quantitative accuracy. To make full use of the spectral information available, the instrumental resolution should be comparable with, or smaller than, the spectral width of the transitions observed. Under atmospheric conditions, collisional broadening is the dominant mechanism. There is usually a linear relationship between pressure and line width, with pressure-broadening coefficients typically in the range 0.01–0.5 cm^{-1} atm^{-1} for infrared rovibrational transitions. Ir spectrometers used in remote sensing will typically have 0.1-cm^{-1} resolution for scanning the troposphere (lower atmosphere), while analysis of the stratosphere (heights > 25 km) will often require < 0.01 cm^{-1}. Usually somewhat lower resolution will suffice for the uv–visible region because of the combined effects of collisional and Doppler broadening. For general analysis, 0.5–1 cm^{-1} may be adequate. Studies in liquid-phase solutions generally require even lower spectral resolution, because solvent effects significantly broaden molecular transitions to well over 1 cm^{-1}.

ATMOSPHERIC ANALYSIS

Two limiting cases define the range of analytical situations that might be encountered: (1) characterization of multicomponent mixtures, which might be a portion of the atmosphere itself or the emission plume from a stack or incinerator, for which numerous species are of concern and for which the nature of the pollutants might change with time, and (2) monitoring of emissions from particular sources, for which relatively few species, whose identity may already be known, must be quantified continuously, but there is less need for qualitative analysis.

Infrared Absorption Spectroscopy

The strong fundamental vibrational transitions of molecules fall in the mid infrared (2.5–100 μm) range and provide two types of specificity (12). At the shorter of these wavelengths ($\lambda \gtrsim 8$ μm) many chemical functional groups exhibit characteristic "group frequencies" that are relatively independent of the molecular environment and provide useful information on the chemical nature of the absorber. At longer wavelengths the frequencies are influenced more by the skeletal vibrations of the molecule: this is the "fingerprint region," where even similar molecules may have quite different spectra, allowing individual identification. Overtones and combinations of some of these vibrational modes appear at shorter wavelengths in the near infrared (nir) or visible but with intensities reduced by ca 100 for the first overtone and an additional factor of 10 for each successive higher overtone. Nir is useful for some industrial analyses (19), for which the potential loss of sensitivity may be compensated by the ability to analyze concentrated process streams (including aqueous systems) that would totally absorb mid-ir radiation, but most atmospheric monitoring occurs in the mid infrared.

Infrared methods are well established for qualitative and quantitative analyses of molecules and are noted for their high specificity and reasonably good quantitative accuracy (20). The pattern of the infrared absorption frequencies for vapor-phase molecules reflects the nature of the rotational and vibrational structure of a given molecule and so is unique for each species. While the rota-

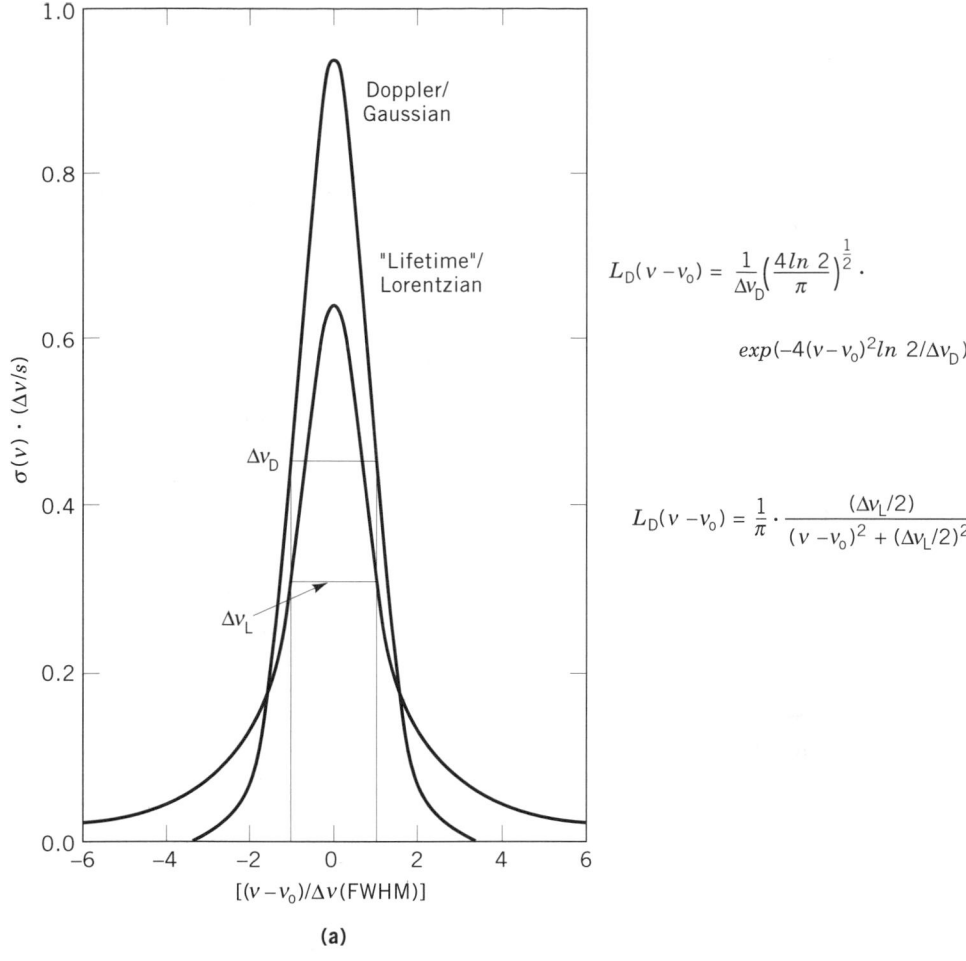

$$L_D(v-v_0) = \frac{1}{\Delta v_D}\left(\frac{4 ln\ 2}{\pi}\right)^{\frac{1}{2}} \cdot$$

$$exp(-4(v-v_0)^2 ln\ 2/\Delta v_D)$$

$$L_D(v-v_0) = \frac{1}{\pi} \cdot \frac{(\Delta v_L/2)}{(v-v_0)^2 + (\Delta v_L/2)^2}$$

(a)

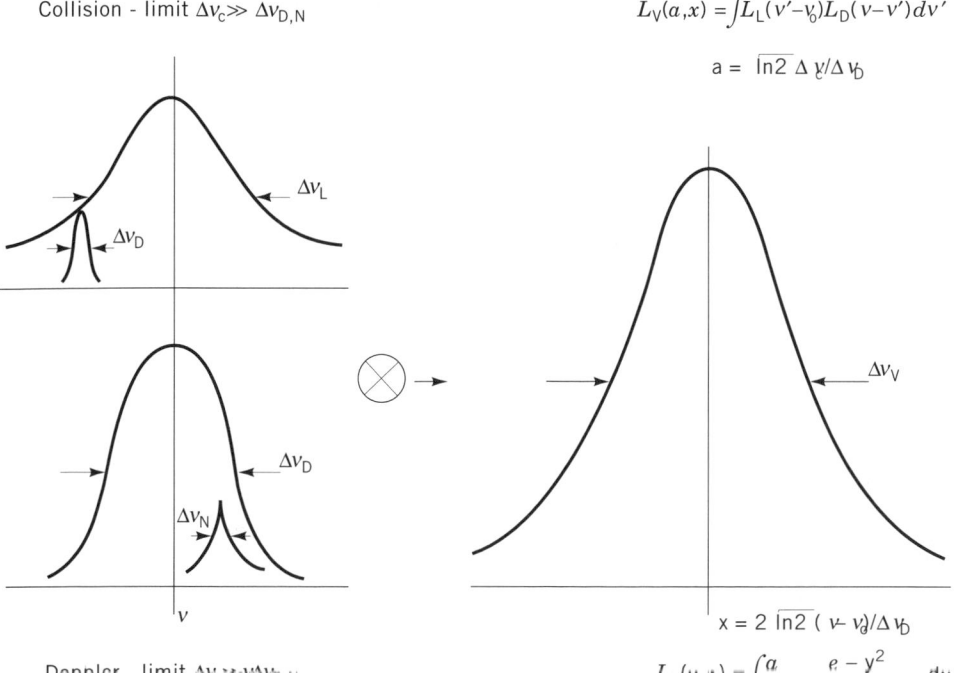

$$L_V(a,x) = \int L_L(v'-v_0)L_D(v-v')dv'$$

$$a = \sqrt{ln2}\ \Delta v/\Delta v_D$$

$$x = 2\sqrt{ln2}\ (v-v_0)/\Delta v_D$$

$$L_V(a,x) = \int \frac{a}{\pi} \cdot \frac{e^{-y^2}}{(x-y)^2 + a^2}\,dy$$

(b)

Figure 2. Comparison of line shape functions $\mathfrak{L}(v-v_0,\Delta v)$. (**a**) Normalized (integrated areas equal 1) absorption cross-sections are shown for the two natural limits of pure Doppler and lifetime broadening; under typical conditions, the line shape is a combination of two or more effects. (**b**) The convolution of a Lorentzian with a Gaussian line shape leads to the Voigt profile.

tional structure enhances specificity, there are some disadvantages to the resulting band broadening. The band intensity is spread over all the component rotational transitions, resulting in lower peak cross-sections for absorption (the analytical parameter that determines sensitivity), and the broadened bands are more likely to overlap and interfere. However, if the rotational structure can be resolved, valuable information about the temperature distributions of remote samples can be recovered. The high specificity of ir spectroscopy reduces difficulties caused by interferences, as when one is seeking a minor species present in a complex mixture. Atmospheric H_2O and CO_2 can interfere at some wavelengths (21), but for most molecules at least one absorption feature can be found that falls within any of several broad transmission windows. In process monitoring, commercial methods have been developed to remove water vapor from stack emissions for analytical purposes. Particulates smaller than the wavelength of the radiation ($< 1 \mu m$) have little effect on an infrared beam, but heavy loadings of larger particulates may present a problem.

Quantitative accuracy of most infrared methods can be characterized as adequate but not outstanding; typically, one might expect uncertainties of about 5%. In assessing quantitative accuracy it is assumed that the optical path length through the sample is known. This is true for laboratory analyses and can be achieved in stack configurations, but in the open atmosphere one obtains a column density averaged over the total path, which may not reflect the concentration of a "slug" of pollutant that occupies only a small volume in the sampled region. This may be acceptable when the location of the analyte is already known, as when sampling over a stack or a highway.

Detection limits depend strongly on the intrinsic band strengths and on the optical configuration employed (especially the path length) but are typically in the subparts-per-million range. For optically thin, scatter-free conditions, the intensity of radiation transmitted I_T is related to the incident intensity I_0 by the Beer-Bouguer-Lambert law,

$$I_T(\nu) = I_0(\nu)e^{-\sigma(\nu)Nl} \qquad (7)$$

where $\sigma(\nu)$ is the absorption cross-section and N is the density of absorbing species along the path length l. The fractional absorption is

$$\frac{I_0 - I_T}{I_0} = 1 - e^{-\sigma Nl} \approx \sigma Nl \qquad (8)$$

where the latter limit of weak absorption usually holds for atmospheric samples.

A typical peak absorption cross-section for an infrared transition is $\sigma = 10^{-18}$ cm^2. If the minimum detectable fractional absorption is assumed to be 0.01 and $l = 10$ m, then the minimum detectable density is $N = 10^{13}$ cm^{-3} or $(10^{13}$ mol cm$^{-3})/L \approx 400$ ppb, where L is the Loschmidt number (2.69×10^{19} mol cm^{-3} at STP). The relatively low sensitivity of conventional absorption spectroscopy stems from the fact that concentration measurements are made as the difference of two large numbers, I_0 and I_T. Novel techniques have been developed to improve sensitivity by increasing the effective integrated sample path length; improving the practical resolution of the absorbing features, so the measured peak absorption is as large as theoretically possible; and/or eliminating the background subtraction by "null-background" techniques.

Sampling. The two basic optical arrangements for real-time monitoring by absorption are point to point, for which the source and detector are separated and the atmospheric volume to be sampled lies between them, and single-ended remote sensing, in which a probe beam is returned from a retroreflecting target (a mirror or topographic feature) and traverses the sampled volume twice. Because sensitivity is proportional to the path length, mirrored multipass sample cells (White or Herriot cells) (22) are often used in the laboratory to obtain optical paths of 200 m or more with a small volume.

It has been assumed here that the optical path from source to sample to detector will be through the atmosphere. This is the usual arrangement, but it may be desirable in some cases to transmit the beam in part through fiber optics, which are cylindrical wave guides that channel electromagnetic radiation (23,24). This allows one safely to access a small sample region in remote or harsh environments far removed from the spectrometer. Silica fibers are available that transmit from 220 nm in the uv to 2.3 μm in the infrared with low losses; fibers of fluoride and chalcogenide glasses extend further into the ir, to about 5 μm, but have not found wide application because of their brittleness. An elaboration of fiberoptic sensors has given rise to optrodes, a specialized type of analytical instrumentation. Optrodes (25) have optical transducers mounted at the distal end of a fiber to monitor chemically selective changes in transmission or fluorescence, often employing an immobilized reagent on the surface of the fiber. Conventional ir absorption can be measured with a transmission-based optrode; the output of a low power, light-emitting diode is directed over a silica fiber, through the sample, and returned over another fiber for analysis. Such systems have been demonstrated for monitoring explosive gases (methane, propane) and NH_3, using near and mid ir absorption.

Numerous specialized sampling techniques are available (26). Planar wave guide cells have been developed that employ frustrated multiple total internal reflection, where the infrared beam is transmitted inside a thin, highly refractive crystal (ZnSe or Ge) and is absorbed by a sample in optical contact with the surface. Matrix isolation is a cryogenic sampling technique that freezes analytes onto windows, internal-reflection substrates, or mirrors for spectroscopic analysis. Cold metal mirrors offer the easiest sampling strategy, by which a thin film is condensed onto the mirror, from which the probe beam is reflected, traversing the sample twice (27). For atmospheric samples, CO_2 naturally present will condense as a matrix in which other minor species are trapped in individual lattice sites, so that Doppler and collisional broadening are eliminated, and because there is no molecular rotation in a solid, the resulting collapse of the rotational structure results in all the intensity of the band being concentrated in one sharp absorption line (ca 1 cm^{-1} wide), increasing sensitivity and reducing the likelihood of interferences. Detection limits are on the order of 10 μg for moderate absorbers, such as NO_x or CO_2, to 100 ng or less for a strong absorber such as a metal carbonyl. This approach involves some time delay, both for the deposition of the sample and for subsequent removal of the film from the substrate in preparation for the next analysis.

Nondispersive ir Gas Analyzers. A simple method of infrared analysis is to monitor the absorption of a specific band of a species by using two narrow bandpass filters to isolate the analytical frequency and a nearby reference frequency where there is no absorption (28). Such nondispersive infrared (ndir) spectroscopy can be an effective low cost method for gas analysis. Typically, radiation from a source is collimated, passed through the filters and sample cell, and focused on a detector. The resulting on- and off-resonance signals are processed. An analyzer can be tailored with the proper filter set for any species with a strong ir absorption that is reasonably free from interferences, and multichannel units can accommodate several filters to monitor different absorption frequencies and a background. Sensitivities are of the order of 0.1% to low ppm levels, depending on the species, and response is fast, providing continuous real-time analysis. Such ndir analyzers are currently marketed for CO, CO_2, CH_4, C_2H_6, freons, SF_6, and total hydrocarbons, the last monitoring a wavelength region around 3.4 μm, where most hydrocarbons exhibit absorption due to the C—H stretching vibration.

Correlation Spectrometers. In a correlation spectrometer, the beam, after traversing the sample, is alternated between two cells (6,29). One contains the analyte gas, so radiation passed by this cell is unaffected by concentration changes in the sample; the other cell is empty (or contains an interfering gas, if one is present). The fractional change between the two signals provides an output sensitive to the species sought and relatively independent of other disturbances. Correlation analyzers can be sensitive and specific and have an advantage over filter systems in that they are readily adaptable to different gases as conditions warrant, it being necessary only to fill the appropriate cell with the target gas. They lack flexibility because they can analyze for just one specific species, and interfering gases can sometimes cause difficulties.

Ir Spectrometers. An outstanding characteristic of rovibrational spectroscopy is its ability to analyze for a broad range of compounds, but this requires obtaining a significant portion of the infrared spectrum, say from 2 to 25 μm, with a recording spectrometer. From such a spectrum one can identify and quantify nearly all species present.

There are two basic methods for recording broad-band infrared spectra: dispersive and interferometric. In dispersive spectrometers a narrow beam defined by a mechanical entrance slit is dispersed by a prism, or (for better resolution) a diffraction grating, into its constituent wavelengths. Because photographic emulsions are not useful at wavelengths $\gtrsim 1.3$ μm, the dispersed radiation is swept slowly across an exit slit by the rotation of the prism or grating, and the signal is photoelectrically detected, amplified, and recorded to furnish a spectral plot of intensity versus wavelength. The optical path between the entrance and the exit slits serves to isolate a single narrow frequency or wavelength range out of the spectrum, and accordingly this portion of the instrument is called a monochromator; when the source and detector optics and amplifier and recording electronics are added, it becomes a recording spectrophotometer.

Commercial ir spectrophotometers usually have a double-beam configuration, in which a system of rotating mirrors switches the source beam alternately through the sample and through an equivalent reference path many times per second; these signals are compared at the detector, and their ratio provides a spectrum free from the influences of unwanted atmospheric absorption, variations in source intensity, etc. High performance dispersive spectrophotometers have for the most part been replaced by interferometric Fourier-transform spectrometers (discussed below), which offer many advantages in performance and flexibility. But inexpensive grating instruments will still play a large role in pollution monitoring, because multielement detector arrays can allow multiplexed detection of several absorption features simultaneously. Such polychromators have a solid-state optoelectronic array detector in the focal plane of a monochromator. These devices are discussed in more detail below; most are silicon based and usable only for $\lambda < 1.1$ μm, but commercially available Pt-Si devices now cover 1.5–5.0 μm, and research continues to extend their response to longer wavelengths, which will strongly impact ir instrumentation. InSb and HgCdTe arrays are also becoming available for still longer wavelengths (5.5 and 11 μm, respectively), but are very expensive. An array-equipped ir polychromator might provide spectral readouts for both speciation and quantification with much greater flexibility than a tunable laser and at much less expense than a Fourier-transform spectrometer.

Fourier-Transform Spectrometers. A Fourier-transform spectrometer (FTS) is actually a Michelson interferometer that records the interference produced between two alternate light paths reaching the same detector, as the optical delay (or phase retardation) of one of these paths is changed by the linear motion of a mirror (4,11). The resulting signal strength as a function of mirror travel is an interferogram, from which the desired spectrum (intensity as a function of wave number) can be obtained by the mathematical procedure known as a Fourier transform. Such instruments incorporate a dedicated computer to perform the transform using the FFT (fast Fourier transform, or Cooley-Tukey) algorithm. They are expensive and sophisticated devices, but they offer a high level of performance that makes them attractive for optical monitoring.

There are several reasons for the high performance of Fourier instruments compared with dispersive spectrometers. While a scanning spectrometer must record the spectrum sequentially, one spectral slit width at a time, an interferometer processes information from all frequencies simultaneously: the multiplex (or Fellgett) advantage. In this it has some of the characteristics of a spectrometer with a continuously recording array detector, but it offers much better resolution and broader spectral coverage. Interferometers also have a throughput (or Jacquinot) advantage, in that they accept a large solid angle of radiation and hence pass a much greater light flux than do spectrometers, which are limited by the necessity of defining the beam with narrow slits. These two advantages can be converted into orders-of-magnitude improvements in resolution, scan time, and/or signal:noise (S:N) ratio, the three mutually related instrumental parameters that are of the greatest practical interest to the analyst. The resolution of a FTS is approximately the reciprocal of the maximum optical path difference or of twice the maximum mirror travel. A resolution of 0.1 cm^{-1} thus requires a mirror travel of only 5 cm, and many commercial instruments offer at least this capability.

Many types of FTIR spectrometers are marketed, ranging from simple, rugged, low resolution instruments intended for basic analytical applications to expensive high performance instruments, offering high resolution and broad flexibility for research spectroscopy. A typical system for recording mid infrared spectra in the 2–25 μm region might include a silicon–carbide globar source, a germanium-coated KBr beam splitter, and a pyroelectric or photoconductive detector. Different spectral regions can be accessed with appropriate substitutions of source, beam splitter, and detector. The sample compartments of commercial instruments are designed for standard 10-cm gas absorption cells, but they can also accommodate multiple-reflection cells, and transfer optics are available or can be designed for other configurations that might be needed for pollution and atmospheric monitoring. FTIR spectrometers have now become an effective tool for quantifying pollutants in remote sensing spectroscopy and have allowed temperature determinations to within ±2 K from rotational analysis of spectra, as shown in Fig. 3. FTIR provides near-real-time results. The scan time depends to some extent on the mirror travel (ie, resolution) required, but its main determinant is the number of individual scans that are co-added to provide a final spectrum with an acceptable S:N level. A typical commercial FTIR can produce a spectrum from 10 co-added interferograms at a resolution of 1 cm^{-1} in ca 10 s.

Compact, robust FTIR instrumentation is currently marketed specifically for industrial analysis and on-line process monitoring. Permanently aligned, industrially hardened instruments are vibration and shock resistant and sealed against dust and moisture. Using sources that need no external cooling and room temperature detectors, they require no utilities other than a 120-V line. Units are available that can monitor a dozen or more selected gases with sensitivities of better than 1 ppm for most pollutants and with recording and automatic alarm capabilities.

Laser Sources. A (fixed-frequency) laser that happens to emit at a pollutant absorption frequency can provide a sensitive method of analysis (13,14). The most useful for this purpose is the CO_2 laser, which provides high power monochromatic emission in the region 8.7–11.8 μm; this falls in an atmospheric window and also corresponds to the "fingerprint" region in which many organic molecules have characteristic absorptions. The emission is not, unfortunately, continuously tunable, but $^{12}C^{16}O_2$ will lase on any of some 100 rovibrational lines, and its various isotopomers ($^{13}C^{16}O_2$, $^{16}O^{12}C^{18}O$, etc) provide many hundreds of additional lines. Analysis with line-tunable lasers offers little flexibility because the frequency must be selected for a given species, but it can provide high sensitivity in some cases. The detection limits depend on the offset or detuning between the analyte absorption feature and the laser emission line. Sensitivity varies from 3 ppb/km for ethylene (a strong absorber of a specific CO_2 line) to 3 ppm/km for SO_2. The CO_2 laser is particularly suited for long-range remote sensing because of its high efficiency and power.

Tunable ir Laser Sources. Fully tunable lasers have the advantage that the output frequency can be adjusted to correspond exactly to that needed for a specific analyte (30–32). The development of nearly monochromatic lasers that can be continuously frequency tuned throughout much of the ir region has revolutionized vibrational spectroscopy. Such a source provides in effect a compact high resolution spectrometer without the need for a bulky monochromator or for the computational complexities of

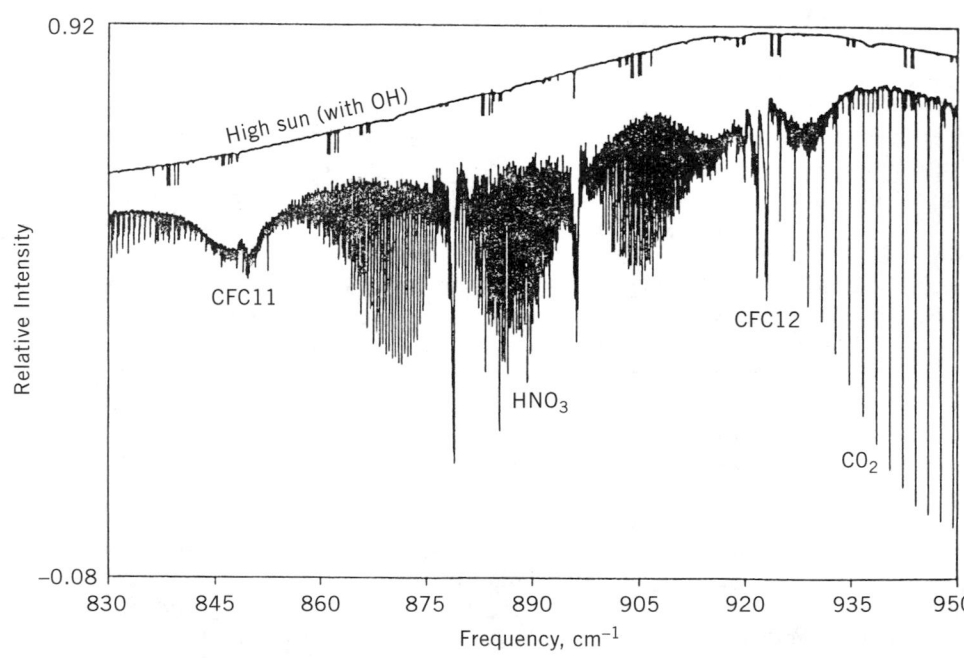

Figure 3. Real spectrum of the atmosphere at a tangent height of 17 km, taken with a remote sensing Fourier transform spectrometer (4).

Fourier-transform instrumentation. Several such systems have been developed, but semiconductor diode lasers are by far the most widely used. Tunable lead salt semiconductor diode lasers (TDLs) can be fabricated to cover any portion of the mid infrared from 3.3 to 28 μm. They can be tuned, by varying the temperature and/or injection current, quasi-continuously over frequency ranges of 50–100 cm^{-1} in continuous scans of over 1 cm^{-1}.

Commercial TDL systems (33) are modular, consisting of a cryogenic laser source assembly, a power and control module, collimating optics, a simple mode-selecting monochromator, and a Ge:Cu detector with lock-in amplifier. While the laser source assembly has usually included a relatively expensive closed-cycle helium refrigerator, recent improvements in manufacturing technology have produced high performance diodes that operate reliably above 77 K, allowing the use of simple liquid nitrogen Dewars with substantial reduction in cost and complexity.

The limited tuning ranges of TDLs make them unsuitable for general survey spectroscopy, but once a specific interference-free molecular absorption feature has been identified, their high resolution (ca 0.0003 cm^{-1}) and easy tunability are useful for monitoring such a feature. In addition to continuous tuning, they can be operated at discrete frequencies, tuning on and off a resonance many times per second to provide instant quantification. Industrial analytical applications to TDLs include stack-gas monitors for H_2SO_4 and other pollutants.

Null-Background Techniques. Conventional absorption spectroscopies measure small fractional changes of transmitted light (eq. 8), to determine concentrations. The uncertainties in obtaining small differences between large values of the incident and transmitted light limits measurable absorption values $N\sigma l$ to $\geq 10^{-3}$. A significant improvement can be achieved with null-background absorption methods that exploit the high spectral sensitivity and frequency stability of laser sources. These techniques have allowed sensitive determinations of $N\sigma l \sim 10^{-7}$ under laboratory conditions.

One of these techniques is harmonic or derivative spectroscopy (30), which involves modulating a relatively narrow-band light source, passing the modulated radiation through the sample, and then using notch filters and synchronously tuned amplification of the signal at the modulation frequency (or at one of its nth-order harmonics). Scanning the wavelength of the light source gives a signal that is proportional to the first (or nth-order) derivative of the absorption profile. Greater sensitivity results from elimination of the background and reduction of low frequency drifts in the light source, and the method discriminates against spurious signals that do not have a sharp wavelength dependence at the modulated frequency. The modulation can be effected by varying either the amplitude or the frequency of the source; the latter (fm modulation) offers the highest sensitivity, while amplitude modulation (am) adds additional frequency components that are harder to filter during postdetection processing. TDLs can be readily frequency modulated with high fidelity (little residual am) by adding a small ac modulation to their dc bias current. Balloon-borne TDL spectrometers have measured absorbances of $\sim 10^{-5}$ with modulation frequencies of ≤ 10 kHz (34), and such systems have

been proposed as *in situ* probes of planetary atmospheres (35).

Another method of eliminating background radiation is to measure secondary signals that are directly indicative of the energy depletion of the light. Examples include photoacoustic (pas) and photothermal spectroscopies, which measure the change of a thermodynamic parameter during absorption (15). When a molecular gas, liquid, or solid absorbs radiation, some of the energy is converted into kinetic motion and the sample is heated; this nearly adiabatic transfer induces pressure pulses that can be synchronously detected by a sensitive microphone (spectrophone). The principal applications of pas have been in solid sampling and in the analysis of gases at low concentrations. A closed gas cell is required (for maximum sensitivity, an acoustically resonant one), and the microphones used are very small, so pas is better suited for the analysis of small volumes of gas rather than for probing broad regions of the atmosphere. Gas cells that can accommodate flowing samples have been designed for atmospheric and pollution studies. An alternative technique, useful for localized gas concentrations such as might arise from a leak, is photoacoustic detection and ranging (padar) (36): a laser pulse tuned to an absorption line of the gas generates an acoustic signal that is detected with a parabolic microphone, with a range resolution of 1 cm out to 100 m. In photoacoustic deflection spectroscopy, the pressure changes are observed remotely by tracking the deflection of a probe laser; this technique has been exploited in studies of flame chemistry, where a spectrophone is not feasible because of background acoustic noise and thermal dissipation.

Examples of Infrared Analysis. Volatile organic compounds all have sufficiently distinctive ir spectra to be easily identified and quantified. Detection limits depend on the intrinsic strengths of the absorption bands used, but as a rough guide, detectability (ppm) is given approximately by (0.1 to 1)/(path length (m)). Ground-based solar ir spectra obtained at the South Pole have been used to monitor the column density of freon-12 and confirmed its secular increase (37). Ethylene (C_2H_4) has been analyzed at the 20-ppt level in air using pas at 10 μm (38); the system could detect the production of this gas, a plant growth regulator, by a single orchid flower. Ethylene emissions from a petrochemical plant have been followed at concentrations of some ppb and ranges up to 500 m with ir absorption using two CO_2 lasers and retroreflecting mirrors to return the signal (39).

Ultraviolet Absorption Spectroscopy

Ultraviolet absorption spectroscopy is similar to infrared, but it detects electronic rather than vibrational transitions, using shorter wavelength (higher energy) photons (40). The region of interest for the present purposes is 200–400 nm, extending for some molecules into the visible. Uv spectroscopy is a mature technique that has long been used for both qualitative and quantitative analysis.

The photon energies in this region can promote electronic transitions between outer (bonding) orbitals. This is also enough energy to excite vibrational and rotational transitions, so the electronic spectra of vapor phase molecules consist of highly structured rovibronic bands, offering the potential for high specificity. One potential problem with uv atmospheric monitoring is that this shorter wavelength radiation is much more susceptible to the effects of Rayleigh and particulate scattering and of turbulence than is the infrared. Also, spectroscopic congestion is a problem, with many strong transitions occurring in a relatively narrow wavelength band. Spectral overlap limits the use of uv spectroscopy for identifying complex mixtures of pollutants; most of the compounds sought will absorb in about the same spectral region, and so will tend to interfere. This limits the flexibility of the method, making uv less suitable than ir for general survey spectroscopy and long-range identification of unknown constituents. On the other hand, uv offers high sensitivity and excellent quantitative accuracy, for the greater uv absorption cross-sections (Table 1) and more efficient uv sources and detectors result in detection limits several orders of magnitude better than in the ir. Sampling methods for uv absorption spectroscopy are essentially the same as in the infrared; window materials are less of a problem, for quartz and glass are suitable. Uv atmospheric monitoring, like ir, provides an average column density of absorbing species.

Dispersive Spectrometers. Packaged uv spectrophotometers are manufactured similar to those widely used in the ir and typically operate in the uv–visible –near ir regions (say, 190 nm to 3 μm), with a tungsten–halogen or deuterium lamp source, holographic gratings, and for a detector either a photomultiplier tube (PMT) or (in the near infrared) a photoconductor such as PbS. Typical resolution may be 0.05 nm in the uv–vis and 0.5 cm^{-1} in the near infrared for a 0.2-m monochromator; it is proportionally better with larger instruments.

Simple monochromators are also widely used with solid-state imaging arrays (41), called optoelectronic imaging devices (OIDs) or optical multichannel analyzers (OMAs), which can be placed in the focal plane of a monochromator to record the spectrum nearly instantaneously. These are in effect the modern equivalent of the traditional photographic plate, with the advantage over an emulsion of rapid response with real-time results, instead of requiring long exposure times and subsequent development of the image. They also have a multiplex advantage in that for a given observation time, the signal:noise ratio is increased by a factor of $\sqrt{N}$ when N detectors are used or, for a given S:N, recording time is reduced by 1/N. An array detector consists of a set of photodiodes that respond to incident electromagnetic radiation, together with an integral electronic readout scheme that serially accesses the signal from each photodetector. The usual spectral response extends from ca 170 nm to ca 1 μm, and they are available in lengths of > 5000 pixels. Arrays with time-gated windows as short as 5 ns have also become useful in the spectroscopy of short-lived species and unstable chemical systems and in kinetics studies, for which they have obvious advantages over mechanically scanned spectrophotometers with single-channel detectors and are often superior to Fourier-transform spectroscopy. This speed is less of a necessity for pollution monitoring, for which time resolutions of the order of some seconds are perfectly adequate. Two-dimensional arrays, designed for image recording, also have spectroscopic applications in kinetics and chemical reaction monitoring, for which a temporally changing spectrum can be rastered across the second dimension of the arrray for time-resolved spectroscopy.

Fourier-Transform Spectrometers. Interferometry is difficult in the uv region because of much higher demands on optical alignment and mechanical stability imposed on the instrumentation by the shorter wavelength of the radiation. Most commercial Fourier-transform spectrometers have been designed for visible and (especially) infrared use. While in principle any interferometer can be operated in the uv with the proper choice of source, beam splitter, and detector, in practice it has proved difficult to obtain reliability from commercial instruments at wavelengths much shorter than the visible. Recently, some manufacturers have claimed satisfactory performance out to 55,000 cm^{-1} (182 nm), but this was achieved only with difficulty. This covers the region of potential usefulness for atmospheric monitoring, but there may be problems in operating such delicate instrumentation outside the laboratory in the open atmosphere and/or under conditions of vibrational and thermal stress.

Tunable Laser Sources. Tunable uv–vis lasers are better developed than comparable infrared systems (14). Dye lasers provide especially useful continuously tunable sources of near uv–visible–near ir light. The lasing action takes place in certain large organic dye molecules in solution, which are optically pumped to excited electronic states by more intense sources and then fluoresce with high quantum efficiency down to the ground state, emitting a broad band of longer wavelengths. The output frequency is selected with a dispersive optical cavity. Pulsed dye lasers (pumped with fixed-frequency Nd:YAG, excimer, or Cu-vapor lasers or by flash lamps) can provide high peak powers at repetition rates to 100 Hz and line widths of 0.1–1 cm^{-1}, which can be improved to < 0.01 cm^{-1} with an intracavity etalon. A given dye–solvent combination can typically be tuned continuously over a 40–80 nm range, and overall coverage with available dyes is from about 320 nm in the near uv to 1.2 μm in the near ir. With Ar$^+$ or Kr$^+$ pump lasers, cw operation is possible with output powers of 0.1–1 W. Multimode cw cavities can achieve line widths of a few GHz, while commercial ring-laser geometries improve the resolution to < 0.5 MHz over a 1-cm^{-1} continuous scan. Recently Ti:sapphire has been used as the gain medium for near-infrared cw ring lasers that can provide > 1-W output over much of the range 700–900 nm. Tunable solid-state optical parametric oscillators (OPOs) are being developed that tune continuously from 400 nm to 2 μm with a resolution of 0.1 cm^{-1}.

Tunable dye lasers are widely used in laboratory research on spectroscopy and photochemistry. For routine monitoring of pollutants they have serious drawbacks: they are inherently complex devices needing a separate pump laser (or at least a flash lamp), with demanding optical and alignment requirements (especially the synchronization of the tuning elements), and usually requiring a flowing dye system (for cw operation, a liquid dye jet) to dissipate heat generated by the pump. One could not reasonably consider the remote installation of a dye laser in an industrial environment, such as could be done with a tunable diode laser. But their high output power and tunability make them attractive as sources for lidar, Raman, and fluorescence analysis.

Examples of Ultraviolet Analysis. Uv absorption may be a quite satisfactory analytical technique when there are only a few species present. Volatile organics all have distinctive rovibronic bands in the near uv, which in the vapor phase exhibit detailed rovibronic structure and are in principle suitable for identification and quantification if interferences can be avoided. Perhaps the most suitable molecule for uv monitoring is ozone, which has strong bands in the near ultraviolet: the Huggins (300–370 nm) and Hartley (210–300 nm) bands. This pollutant has been monitored near 283 nm by differential optical absorption spectroscopy (doas), using a correlation technique in which the absorption spectrum is compared with that of a known amount of O_3 (42). In this case light from a Xe lamp was returned from a retroreflector 750 m distant; ozone in an urban environment was continuously monitored over 1 yr with this apparatus, with a sensitivity of about 3 $\mu g/m^3$ (2.5 ppb). In cases for which detecting O_3 is of paramount importance and few interfering species are present, simple uv systems like these are appropriate. Various pollutants have been recorded in the Black Forest at the ppb level over paths of ca 9 km, using retroreflectors to return the beam (43), including the OH radical using a dye laser source at 308 nm and O_3, NO_2, SO_2, and CH_2O with a Xe arc source and spectrometer. Other examples of uv–visible analyses of organics include formaldehyde in the ppb range by pas, using a pulsed laser at 303 nm (44) and formic and acetic acids by pas at 220 nm, with detection limits of ca 120 ppb (45).

Scattering Techniques

In lidar, a laser pulse is propagated coaxially along a telescope's field of view, and the return signal from atmospheric scattering is collected by the telescope for detection and in some cases spectral analysis (1). The azimuth and elevation of the scatterers is determined from the orientation of the telescope and the range from the time delay of the return signal. Several scattering processes can contribute to the signal: Rayleigh, elastic scattering (no change in frequency) from particles much smaller than the wavelength of the probe beam (atoms, molecules, fine dust), Mie (or Tyndall) elastic scattering from larger particulates or aerosols, and Raman inelastic scattering from molecules, with a frequency shift characteristic of the molecule. Fluorescence can be considered a natural limit of resonant Rayleigh or Raman scattering from excited electronic states of atoms and molecules, but this technique is sufficiently unique to justify its separate treatment below. An important modification of lidar is differential absorption lidar (dial), which uses two laser frequencies, one on and one off resonance, to monitor a specific molecular absorption feature; the backscatter can be either from a retroreflecting target or from aerosol and dust returns. These methods will all be referred to as lidar techniques, with the term *simple lidar ranging* used

to distinguish, when necessary, range and velocity determination by elastic scattering from the more sophisticated Raman, dial, and fluorescence methods.

The accurate and rapid three-dimensional spatial information provided about target species by laser scattering techniques makes these ideal for monitoring air mass movements and plume transport and for tracking aerosol and pollutant species (46). Such data are useful in meteorological site characterization and are vital for directing and evaluating emergency response to an inadvertent hazardous release. In addition to information on air mass position, Raman and fluorescence scattering both contain molecular spectral information that can identify the target species much as can absorption spectroscopy. These latter techniques, and dial, can also be used for quantification.

Lidar techniques can be very sensitive in some circumstances. Ground-based sensing of Na and Li atoms in the stratosphere (range 30–90 km) has been reported for concentrations of only a few atoms/cm^3. More typical detection ranges in the lower atmosphere are of the order of hundreds of meters and concentrations of ppm to ppb. Quantitative accuracy is difficult to assess, especially considering the variety of molecular processes subsumed under the term *lidar*, but generally is comparable with absorption spectroscopy: uncertainties of ca 5% can be expected for typical atmospheric conditions and perhaps 1% under optimum conditions. The sample in lidar and other scattering techniques is the column of air interrogated by the probe beam and is determined by the orientation of that beam. The return signal is proportional to the column density of the scattering molecules rather than to concentration at a given point. However, point concentrations can be estimated from signal return times, which can yield the range of the scatterers with resolutions on the order of a meter. Some further characteristics of the individual scattering methodologies are described below.

Elastic Scattering: Lidar. Lidar is a ranging technique that detects an elastically scattered return signal, predominantly from particulates and aerosols. The backscattered signal contains information on the distance of the scatterers (from the temporal delay of the return), their abundance (from the intensity), and in Doppler lidar (below), their velocity components along the line of sight (from the Doppler shift). No spectral information is present, so the method does not provide the identification or quantification of specific compounds. It is included here because of the ready adaptability of lidar instrumentation to such techniques as dial, Raman, and fluorescence and also because it is useful in problems closely related to pollution monitoring, such as tracking plume transport and dispersal.

Lidar range resolution depends on the laser pulse width and the detector gate width. As an example, temporal pulse lengths of 10 ns are readily available from excimer and solid-state lasers, corresponding to a spatial length of some 0.3 m. The beam spread for a laser of divergence of 0.5 mrad is 5 m at a range of 10 km, so spatial resolution of the order of a few meters is readily achievable at kilometer distances. Quantification is difficult, even for a single known species, because not all of the factors that affect the return signal intensity (laser power, fraction of signal returned, extinction coefficient, optical efficiency of the system, etc) are adequately known.

Aerosols and particulates are strong scatterers and are present in most plant and stack emission sources. Accordingly, the use of lidar for air mass and plume tracking is well established and has been proven in many field applications. Dense aerosols can be successfully located with a fast lidar search pattern and tracked over daylight ranges approaching 150 km under favorable conditions. Even faint plumes, orders of magnitude fainter than visible contrails, can be observed at 40 km or more in clear air, though a range limit of 10 km is more typical under usual atmospheric conditions near sea level. With lidar one can use time-dependent studies of aerosol transport to obtain wind velocity vectors that characterize actual wind fields and plume transport paths under various meteorological conditions and to study plume dispersal, stack effluent behavior near buildings, flow characteristics over complex terrain and the effects of terrain features, the influence of inversion and boundary layers, vortex and eddy structure, molecular flux gradients, etc. Deriving such information from meteorological sampling would require an expensive three-dimensional array of many sensors covering an extensive ground area to heights of hundreds of meters.

Dial. Two-frequency differential absorption lidar (47,48) combines lidar and absorption spectroscopy by using two probe frequencies, one at a molecular or atomic absorption feature and one at a nearby frequency where there is no absorption. The frequencies are alternated, generating a small differential absorption signal from the much stronger Raleigh or Mie scattering returns, which are relatively insensitive to small changes in frequency. Because one of the two probe frequencies provide essentially a reference signal, some of the quantification problems that arise with simple single-frequency lidar are overcome. One can thus obtain absolute range-resolved concentrations of specific molecules.

Dial has the disadvantage of requiring considerably greater experimental complexity than simple lidar systems. Either two different lasers must be used as the sources or a single laser must be tuned on and off resonance. Usually a tunable laser is chosen, for economy, but there are obviously stringent requirements on the tunability and monochromaticity of the laser selected, because it must be able to cycle repeatedly and rapidly to precise frequencies; missing the peak of an absorption feature by any significant fraction of a line width will greatly degrade quantitative accuracy. But with an appropriate tunable source, dial can provide species-specific concentration and range information that simple lidar cannot. Because the frequencies must be chosen carefully for any given species, dial does not offer the qualitative analytical capability and flexibility of spectroscopic methods.

If one is willing to sacrifice range information for sensitivity, dial can be performed with the signal returned by a reflecting target such as a mirror, topographic feature, or cloud. Because molecular scattering is not involved in the signal return, this is now essentially a single-ended remote sensing infrared absorption technique, giving path-averaged or column densities. Sensitivities for some molecules of the order of 1 to 10 ppb-km can be achieved, ie, a few parts per billion over ranges of several kilometers. In ranging experiments, concentration-path products of a few ppm-m can be measured out to ranges of 1–5 km.

Range-resolved dial measurements can provide three-dimensional maps of the measured species, but here the return signal is much smaller than that from a retroreflecting target. This requires either higher power lasers or sophisticated detection techniques such as photon counting or heterodyne detection. Most such work has been limited to relatively abundant pollutants such as O_3, SO_2, and NO_2 in urban and industrial sites. Detection limits in most cases are of the order of ppm. It should be noted that some innocuous compounds that have high sensitivity to dial, such as ethylene, simple freons, isopropanol, and SF_6, might be deliberately released as plume tracers.

Raman Scattering. Molecular Raman scattering consists of the inelastic scattering of an incident photon $\bar{\nu}_L$ to a new frequency $\bar{\nu}_R = \bar{\nu}_L \pm \bar{\nu}_{mol}$, where $\bar{\nu}_{mol}$ is the energy taken ($-$) or given up ($+$) by the molecule as a vibrational and/or a rotational transition (1,49,50). Typically the uv–vis laser frequency $\bar{\nu}_L$ is far detuned from any electronic or vibrational transition energy (ie, is nonresonant), in which case the differential scattering cross-section is

$$\frac{d\sigma}{d\Omega} \cong \left(\frac{2\pi}{\lambda}\right)^4 A_{if}^2 \sin^2 \theta \qquad (9)$$

where λ is the wavelength of the incident light and A_{if} (the polarizability matrix element connecting the initial and final molecular states) has values of $< 10^{-24}$ cm³ (comparable with the "volume" d^3 of the molecule). Typical nonresonant angle-integrated Raman cross-sections are of the order 10^{-27} cm² for uv–vis radiation and correspondingly smaller (by λ^{-4}) for longer wavelengths. If the frequency of the laser is within a few absorption line widths of an electronic transition, the resonant enhancements can be considerable ($\sim 10^3$). The selection rules for Raman scattering differ from those of absorption; in particular, rotational and vibrational transitions in homonuclear diatomics such as N_2 and O_2 appear strongly in the Raman effect. Because all molecules are Raman active, Raman lidar can provide a complete range-resolved spectroscopic fingerprint of each species in real time.

In Raman lidar, spectrometric detection is necessary to resolve the wavelength-shifted signal from the elastically scattered background. The Raman spectrum consists of a strong line at the exciting frequency, due to Rayleigh scattering, and much weaker (by factors of 10^3 or more) red-shifted lines at lower frequencies (Stokes lines) caused by inelastic scattering in which the molecules have gained energy from the photons, plus corresponding but weaker blue-shifted anti-Stokes lines at higher frequencies, indicating collisions in which the molecules have lost rovibrational energy to the photons. Usually a reasonably broad spectral region is detected that may include several Raman-shifted lines. The resulting molecular rovibrational spectra permit species identification and quantification.

The relative strengths of the Stokes and anti-Stokes lines can provide valuable additional information about the temperature of the sample.

Raman scattering has one major disadvantage, its inherent weakness. The process must, therefore, be excited by high energy pulsed lasers; it is most useful in the visible–uv region, where sensitive photomultiplier detectors can be used, and it is most applicable to abundant species or to short-range analysis. Atmospheric N_2 can be detected at ranges of tens of km, but most pollutants are limited to a hundred meters or so. Sensitivity is no better than 100 ppm, or even into the parts-per-thousand range, for most species. Some of these drawbacks can be overcome with nonlinear and coherent techniques, but these are most often used in laboratory bench studies. On the other hand, Raman is broadly applicable and highly specific and selective, because the return signal from every laser pulse contains spectral information on each molecular constituent present, and several components can be monitored simultaneously. Atmospheric water causes less interference than with infrared absorption spectroscopy, because water has a small Raman scattering cross-section. In contrast, many organic compounds are strong Raman scatterers. The laser wavelength used is often not critical because it does not have to be resonant with a molecular transition.

Laser Raman spectroscopy has been widely used for probing flame and combustion chemistry with a spatial resolution of better than 1 mm^3. In the case of atmospheric monitoring, an intense and dependable laser source is required, and there are advantages in using wavelengths of 200–370 nm, where sunlight is absorbed by stratospheric ozone, but oxygen absorption is not significant. Such solar-blind operation eliminates background solar and sky radiation, allowing the reliable detection of small Raman signals at any time of day or night. Typically, the returning Raman signal is detected as a function of time, and the signal intensity yields concentration versus range. Concentrations can be calibrated by ratioing the Raman signal from the monitored species against that from N_2, with separate filtered photomultiplier tubes providing the analyte and N_2 signals.

Special Raman Spectroscopies. Because of the weakness of the Raman effect, even with the advantages of laser excitation typically no more than 10^{-8} of the incident laser photons are converted to a usable Raman signal. In addition, collision-induced off-resonant fluorescence is often a problem, overwhelming the much weaker Raman scattering. Thus the sensitivity of conventional spontaneous Raman spectroscopy does not compare with that of absorption spectroscopy. This drawback can often be overcome with some clever alternative approaches.

In resonance Raman (RR) scattering, the probe frequency is tuned near an electronic absorption of the molecule under investigation, which increases the intensities of both fluorescence and Raman scattering by several orders of magnitude. This permits the highly selective enhancement of a particular species in a mixture, and the sensitivity can be increased to the point at which trace constituents can be routinely monitored. To date RR has been applied to a relatively few molecules with accessible electronic transitions such as O_2 and the halogens.

For the intense exciting beams provided by laser sources, the dipole moment induced in the molecule may vary nonlinearly with the electric field, giving rise to new nonlinear spectroscopic phenomena. As the irradiance of the source increases above the threshold for nonlinear interaction, multiphoton scattering and absorption processes arise. Because the scattered radiation is coherently driven by the pump laser, it emerges from the sample in a specific direction with a small solid angle of divergence. In contrast, noncoherent Raman signals are relatively isotropic and are emitted into a solid angle of 4π steradians. A coherent, directional signal can be collected efficiently and is much easier to separate from incoherent emissions such as fluorescence and Rayleigh scattering.

All scattering processes can be induced to stimulated emission at sufficiently high threshold intensities, but the substantial laser flux required makes these schemes impractical except for laboratory use. Stimulated Raman scattering (SRS) by atmospheric N_2 has the lowest threshold, by virtue of its high concentration in the atmosphere, with a threshold for gain of ~ 2 MW/cm^2.

Doppler lidar. If the signal return in any of these lidar processes is detected with sufficient spectral resolution to allow a measurement of the Doppler shift, then obviously one can determine the velocity component of the probed species along the line of sight. Such shifts are small ($\Delta\nu/\nu \approx 10^{-8}$ at 10 μm for a velocity of 1 m/s), but can be measured by optical heterodyne methods, using high resolution laser sources (bandwidth < 1 MHz) (51). Optical heterodyne detection involves the superposition of two signals in the same detector (photomixer) to produce a beat frequency at the difference between them. In the case of Doppler lidar, the return signal usually has a well-defined frequency offset and high spatial and temporal coherence that allows high quality heterodyne detection with a reference beam from the incident source laser. Spatial distortion and frequency broadening of the return signal by atmospheric inhomogeneities and turbulence, both in the transit path and by aerosol backscatter, will determine the ultimate sensitivities of heterodyne detection. Unless there are heavy winds, the line width of the return signal is typically less than 1 MHz for 10-μm radiation, which sets the limits for the laser stability needed. Effects of spatial distortion can be minimized by reducing the field of view of the receiving optics. Most such work has been carried out with CO_2 lasers at 10 μm. Doppler velocimetry has been done in conjunction with simple lidar ranging, dial, Raman, and fluorescence measurements. It can supply useful additional information on airmass movements but is not an analytical technique for monitoring pollutant species.

Instrumentation. All lidar methods share the basic requirement of an intense, nearly monochromatic exciting source, usually a laser (1–3). Given this source, lidar, dial, Raman, and lif can all be employed, depending on how the source is tuned and whether or not the return signal is spectroscopically analyzed. Because much of the instrumentation, including the expensive laser, monochromator, and detector system, can be shared by the various techniques, it makes sense to treat them together for this purpose.

The source is the most important feature of lidar instrumentation and must have sufficient output power at the appropriate wavelength to provide an adequate return signal; for dial and fluorescence lidar, tunability is also a requirement. In the infrared, laser pulse energies of 1–10 mJ are needed for detection of the return from a hard target at a range of a few km, and about 1 J for range-resolved returns from atmospheric aerosols. Heterodyne detection can reduce these requirements by factors of 10 to 100. Lower energy is also acceptable in the visible and uv, where better detectors are available and there is more atmospheric backscatter. These considerations have limited most experiments to certain lasers: discretely tunable systems such as rare gas–halide excimer (190–355 nm), ruby (694 nm), Nd:YAG (0.95–1.8 μm), CO (4.9–8.2 μm), and CO_2 (8.7–11.8 μm), and continuously tunable dye lasers (340–1150 nm) OPOs (400–2000 nm), and alexandrite lasers (700–800 nm).

Lidar systems are not commercially available as such and must be assembled from components. Many such systems have been described in the literature; typical is a portable lidar developed for spatial and temporal atmospheric monitoring (52). It consists of a laser source, a beam-steering theodolite, a 16-in receiving telescope coaxial with the laser, and a grating spectrometer with microchannel-plate signal intensifiers and a 1024-element silicon OMA detector, all mounted on a truck that carries its own 12-kW diesel generator and necessary utilities. The usual Raman-scatter laser is a high-power KrF excimer emitting at 248 nm; the detectors can be made solar blind so the system can operate day or night. To demonstrate the usefulness of remote sensing data for agricultural measurements, this system has been used to study water vapor concentrations and transport properties over a cotton field with a resolution of 1.5 m at ranges of 100–500 m, revealing fine-scale details of distribution, eddy structure, and vertical flux gradients and velocities. The same equipment can monitor a pollutant such as CCl_4 at the 0.1% level in air at distances over 1 km. Other portable lidar systems have been described in the literature (53,54).

Examples of Analysis by Scattering Methods.

Most pollutants thus far investigated by lidar techniques have been relatively simple species such as NO, O_3, SO_2, NO_2, and H_2S. There has, however, been some interesting work reported on organics. Methane leaks from underground pipelines and landfills were detected with a two-frequency HeNe laser at 2948 cm^{-1} using topographic-backscatter dial (55); detection limits were 3 ppm-m out to 90-m ranges. Ethylene (C_2H_4) has been measured over a refinery using a CO_2-laser dial system, providing a 3-D concentration map for a 70 × 400-m area with a sensitivity of about 10 ppb (56). Other pollutant analyses that have been carried out with lidar-related techniques include ammonia by dial (detection limit 5 ppb at ranges to 2.7 km) (57), SO_2 at high concentrations (0.1–20%) by cars (58), and hydrazine (N_2H_4) and methyl hydrazines by dial (40 ppb at ranges to 5 km) (59).

Laser-induced Fluorescence

The emission of uv, visible, or ir light as a result of a radiative transition from an excited to a lower state is broadly termed luminescence (1,46,60,61). Such processes are conventionally divided into short-lived (usually spin-allowed) emission or fluorescence, and long-lived (usually spin-forbidden) emission or phosphorescence. "Short-" and "long-lived" are defined relative to the natural lifetimes of allowed radiative transitions, which are of the order of nanoseconds for visible light. The resonant Raman scattering discussed above can be considered a variety of fluorescence, but there are differences that justify treating fluorescence as a separate technique and limiting scattering to only those processses that do not involve electronic excitation. The redistribution of radiation by fluorescence and scattering (both Rayleigh and resonant Raman) becomes subtle under near-resonant conditions, especially at high incident laser intensities and high pressures. Generally, fluorescence is considered to be the portion of incident laser radiation that undergoes spectral broadening as a result of the natural lifetime and collisional broadening of the excited state, whereas resonant Raman and Rayleigh scattering are considered to be two-photon scattering processes that are not so broadened.

Fluorescence (and resonant Raman) lidars require tunable pump lasers but offer greater sensitivity than spontaneous Raman lidar. In principle, any strong broad-band light source, such as xenon discharge lamps, can serve as an optical pumping source, but lasers are preferable because their high spectral intensity ensures that a greater fraction of absorbers can be selectively pumped into a desired excited state. Usually laser excitation provides sufficient discrimination of an analyte, but Rayleigh scattering by other species could mask the presence of a trace species. This is especially true if the temporal decay of the fluorescence cannot be studied. In such cases, a branch fluorescence to other states of the absorber can serve as an additional spectral discrimination of its presence. If there is just one well-isolated fluorescence transition, then it is economical to use interference band-pass filters to reject other emission frequencies. The analysis of complex effluent streams may be more effective with a polychromator.

The strength of a fluorescence signal is directly related to the absorption cross section σ_A:

$$\frac{d\delta}{d\Omega} = \frac{1}{4\pi} \sigma_A F. \tag{10}$$

The geometrical factor ($1/4\pi$) accounts for the isotropic reemission in all possible directions, and F represents a fluorescence yield factor. F seldom approaches unity because excited atoms often fluoresce to several different energy states, but the greatest yield reduction usually results from nonradiative quenching of the excited states before fluorescence ever occurs. At atmospheric pressures, the most important quenching occurs by collisional transfer to the ground or other states, which competes directly with the main emission. Quenching causes fluorescence intensities to depend strongly on pressure, temperature, and composition, and the resulting signal may be difficult to interpret quantitatively. Fluorescence is typically more intense in the uv than in the ir, due to both the greater uv absorption cross-sections and the longer ir radiative lifetimes, which permit more collisional deexcitation. Typically, $F < 10^{-3}$ for atomic and molecular transitions in the uv and is progressively smaller for longer-wavelength transitions.

Other types of luminescence than fluorescence might be useful for analysis in some situations. In chemiluminescence, eg, part of the energy of a chemical reaction is emitted as light. Methods of sensitive nitric oxide analysis (62–64) have been based on the detection of chemiluminescence generated by the reaction of NO with ozone, and ozone can be detected by the emission at 585 nm from its reaction with ethylene. Such techniques can be extremely sensitive, reaching ppb detection limits with relatively simple instrumentation. Chemiluminescent analysis tends to be highly specific, well suited for detecting a certain species but not for general monitoring of more than a few substances.

Photoluminescence has long been applied to both qualitative and quantitative analysis. Lif provides, much as does infrared absorption spectroscopy, "fingerprints" of different organic molecules, and these can be quantified by measuring fluorescence intensities. Emission methods such as lif frequently can offer much greater sensitivity than absorption analysis, because one is measuring small signals on a near-zero background rather than the small difference between two large signals (I and I_0), which is much more difficult. Selectivity can also be quite high in lif, because both the pump frequency and the fluorescence frequency can be individually chosen for optimum performance. Additional selectivity may be possible with measurements of fluorescence lifetimes and polarization behavior. There are, however, several factors that limit the applications of lif as a monitoring technique in the troposphere. Quenching of the excited states of species at atmospheric pressure has already been mentioned. Operation in the infrared beyond 1 μm is limited by the lack of sensitive photomultiplier detectors. And solar background radiation can interfere. Most applications reported to date have been to atomic species (Na, K, Li, Ca, Ca$^+$) and to radicals such as OH, CN, NH, and CH (63). Lif detection of trace contaminants in the stratosphere has been especially effective, because the collisional redistribution at pressures less than 30 Torr is small.

Some variations of lif should be mentioned that are applicable in special situations. Because molecular fluorescence extends over a broad wavelength region, it is of limited use for analyzing multicomponent systems. But by scanning both the excitation wavelength and the fluorescence detection wavelength synchronously with a fixed wavelength separation, the fluorescence can be reduced to a narrow signal only at the region of overlap between the excitation and fluorescence spectra. This technique, synchronous detection by lif (sdlif), allows the components in complex mixtures to be distinguished. A tunable laser is normally used, but similar results have been reported (at lower sensitivity) with a white-light source followed by a monochromator. Another technique is multiple-photon lif, in which the fluorescing state is pumped by a two-photon or higher process, using a source of longer wavelength than would be required for conventional lif. This results in a considerably weakened fluorescence signal but certain advantages accrue: because the fluorescence is blue shifted relative to the pump, noise sources can be discriminated against by using long-wavelength blocking filters and solar-blind PMTs; and the longer pump wavelength

reduces scattering problems and may excite fewer potentially interfering species.

Lif requires an intense exciting laser and equipment for spectroscopically and temporally analyzing the return fluorescence signal. It thus can be combined with the lidar group of techniques (simple lidar ranging, dial, and Raman scattering), and in fact most lidar installations for atmospheric monitoring are equipped also for laser fluorescence measurements. As with other scattering techniques, in lif the "sample" is the column of air interrogated by the probe beam and is determined by the orientation of that beam. The integrated return signal is proportional to the column density of the scattering molecules, but point concentrations can often be estimated by the analysis of signal return times.

Fluorescence optrodes have been developed using reagents immobilized on porous glass and polymer films. Here, as with ir absorption optrodes, the effective sample space is a small volume right at the tip of the optrode. Such devices have been demonstrated for the environmental monitoring of such gases as CO_2, H_2S, NH_3, I_2, and Cl_2.

Examples of Analysis by Lif. Despite its disadvantages, lif may be appropriate for certain molecules that are strong fluorescers even at atmospheric pressure. An important group of such compounds are the polycyclic aromatic hydrocarbons (PAHs) such as anthracene, pyrene, chrysene, and benzo-[α]-pyrene. These are potent carcinogens that are produced by combustion and appear especially in the exhaust of diesel engines; they may also occur in incinerator stack gases as products of incomplete combustion. Many of them are near-ideal fluorophors, with large fluorescence quantum yields and long fluorescence lifetimes. Detection limits of a few ppt in water solution have been reported using lif, and comparable sensitivities should be achievable in the vapor phase (66).

Another strong fluorescer is the tryptophan amino acid component in some bacteria, and this has been made the basis of an effective tracer technique for following air mass movements. Live spores of a harmless species such as *Bacillus globigii* are dispersed into an airflow and their fluorescence monitored by lidar. The spore plume can easily be detected at distances of several kilometers in minute releases.

Radiometry

Radiometry is the detection and measurement of radiant electromagnetic energy (4–6). The term properly includes spectroscopy, but it is convenient to consider separately under this heading the direct detection of molecular emissions, as distinguished from absorption and scattering techniques in which the sample is actively illuminated. Because this detection will include the spectroscopic analysis of the emitted radiation, radiometry as used here is synonymous with emission spectroscopy. Such detection of the weak emissions of remote thermal sources is the ultimate passive and noninvasive technique, requiring not even an optical probe of the sampled volume. At any temperature above absolute zero, some fraction of a sample of atoms or molecules will be in excited levels that have been populated by thermal energy; transitions from these states to the ground state will radiate this energy as thermal emission. The emission maximum for molecules at T = 300 K is near 10 μm, in the mid infrared. This radiation will be strongest at just those molecular frequencies at which absorption occurs, ie, at the energies of rovibrational transitions. Thermal emission, therefore, carries essentially the same qualitative and quantitative information as does an infrared absorption spectrum.

Thermal radiation is intrinsically weak, and in the presence of intense background radiation from sunlight, its direct spectroscopic detection and analysis is difficult. There is, however, a sensitive technique for detecting these weak thermal signals that has applications in pollution monitoring and atmospheric studies: laser heterodyne radiometry, in which one measures not the emitted frequency itself but the beat frequency between this and another, accurately known, frequency. The incident radiation is combined with the output of a coherent local oscillator (lo), usually a fixed-frequency laser, in a high speed photomixer, thus generating a difference frequency called the intermediate frequency (if), which is synchronously detected and amplified. The if preserves the spectral characteristics of the source but shifts this information into the radiofrequency region where sensitive radio detection techniques can be used, thus providing significantly higher sensitivities than can be obtained with background-limited incoherent detection. With a fixed-frequency lo, the spectral analysis is performed by a multichannel rf filter bank, with the channel width determining the resolution, which is typically better than 0.05 cm^{-1}. The tuning range is limited by the if bandwidth of the mixer; for the HgCdTe photodiodes used in the infrared this yields a spectrum in the region $\pm$0.08 cm^{-1} (2.4 GHz) around the lo frequency. The requirement of finding a molecular transition this close to a gas laser frequency is highly restrictive; the technique is not suitable for speciation but rather for monitoring one or a few specific molecular features. Tunable diode lasers now have sufficient power to serve as local oscillators, permitting continuous tunability out to 30 μm, but one is still limited with any single TDL to a tuning range of some 100 cm^{-1}.

Heterodyne radiometry provides excellent sensitivity, illustrated by the fact that it has been used (with appropriate receiving telescopes) for detection of various constituents of stellar and planetary atmospheres. Quantitative accuracy is generally good, but the signal strength depends on the total emissivity of all molecules radiating in the field of view at the detected frequency, and thus it yields an integrated column abundance.

For direct (nonheterodyne) detection of ir emission, any dispersive spectrometer or Fourier-transform interferometer can in principle be used, but in pollution monitoring, where the emitting molecules are generally at 300 K or colder, a dispersive spectrometer is unlikely to provide satisfactory S:N ratios for these weak signals. The multiplex and throughput advantages of FTIR instruments make them much better suited for this purpose. Recognizing this potential use, many commercial FTIR spectrometers have an input port that can accept the beam from a collecting telescope without disturbing the usual source optics. The *Voyager* interferometric spectrometer (iris) spectra from the outer planets and Titan illustrate that direct FTIR emission spectroscopy can offer effective analytical capabilities. In contrast, instrumentation for laser

heterodyne radiometry is fairly complex and must be assembled from individual components (67,68). The optical system normally includes a collecting telescope, a beam combiner, and a monochromator for coarse calibraton. Detecting and signal-processing electronics are necessarily elaborate: the signal and lo are combined in a wide-bandwidth photomixer–detector to generate the if, which is processed with a lock-in rf amplifier and signal averager. As an example of what can be accomplished in industrial monitoring, smokestack effluents have been analyzed by FTIR (range 74 m), using $H_2O:CO_2$ concentration ratios to discriminate between gas and oil combustion (69). Field-deployable spectrometers are available that achieve sensitivities of the order of ppm-m out to 1 km using retroreflectors.

LABORATORY LASER ANALYSIS OF ENVIRONMENTAL SAMPLES

The principles of optical detection in the laboratory are similar to those outlined for atmospheric analysis (14,16). However, several laser spectroscopies best exploit the unique resolution and/or high focused intensities that are achieved only in the laboratory. This section describes some very high resolution, or ultrasensitive, laser spectroscopies used primarily for laboratory analysis. Some of these techniques are now being adapted for field applications or as *in situ* sensors, but most are still fairly complex or cumbersome because of the lasers or sampling methods involved.

Absorption Methods

Absorption detectivity can be improved by increasing the spectral resolution, by using null-background techniques, and by increasing the integrated path density Nl of the sample by multipass or other methods, such as using long hollow-core wave guides.

Atomic and Molecular Beam Spectroscopy. In the absence of collisions, Doppler broadening $\Delta\nu_D$ usually overwhelms the radiative absorption line widths $\Delta\nu_N$ of optical transitions, unless the sample is extremely "cold" (70). Strong uv–vis transitions have natural widths of 10–50 MHz, while mid ir transitions have natural line widths of <1 MHz. Doppler broadening of room temperature uv–vis transitions is of order 1–3 GHz, so the effective peak absorption cross-sections measured by narrow-line lasers are $\sim(\Delta\nu_N/\Delta\nu_D)$ smaller than their theoretical maximum. Put differently, the absorbances measured with high resolution lasers (line widths $<\Delta\nu_N$) can be enhanced by $(\Delta\nu_D/\Delta\nu_N)$ if the Doppler broadening is suppressed. This can be accomplished in molecular beams by probing the beam transversely to its direction of flow; the frequency shift of the absorption becomes negligible when $\vec{k} \cdot \vec{u} \cong 0$, where $\vec{k}$ is the momentum of the radiation field ($\propto \bar{\nu}/c$) and $\vec{\mu}$ is the forward velocity of the molecular beam. In practice, slight divergences produce residual Doppler broadening, but the peak absorption cross-sections can be improved dramatically with this refinement.

For molecular analytes there is another reason to employ beam techniques. Under conditions of high pressure expansion into vacuum, the transverse motions of the effusive particles can be strongly ordered by collisional effects near the orifice of the jet (the regime of supersonic expansion). The rapid expansion also results in further cooling the thermal motions of the molecules, leading to a collapse of much of the rotational and vibrational structure into transitions originating from a few lowest lying states, and predominantly the ground state. The effective rovibrational energies of the molecules can be cooled to <10 K, although their kinetic energy of translation along the direction of the jet is quite high. The conditions downstream are also nearly collisionless. A supersonic molecular beam thus has the threefold benefit of greatly reducing environmental perturbations that degrade specificity; concentrating the molecules into a few quantum states, which enhances both specificity and sensitivity; and increasing the flux (particles passing through a unit area per second) of analyte probed by the laser beam(s). The disadvantage is that supersonic beams require large and expensive vacuum pumps. While much beam work has been primarily concerned with the basic science of molecular structure and aggregation, the methods developed may eventually find widespread use in environmental analysis. In addition to detecting light absorption by its direct attenuation, the absorption process has also been monitored in molecular beams by directly observing the increased energy of the beam with sensitive pyrometers or other calorimeters. Laboratory detection limits for strongly absorbing species are at the <10 ppb level using 1-mW lead salt diode lasers and can theoretically be improved by at least two orders of magnitude. Detection of atoms or molecules with uv–vis lasers can achieve higher sensitivities, but typically more sensitive fluorescence techniques are used. Fluorescence by ir transitions is usually undetectable for $\lambda > 3$ μm.

Nonlinear Saturated Absorption. Early laser researchers observed that gas-phase lasers exhibited a slight dip in their emitted gain at certain well-defined wavelengths within the Doppler-broadened gain profile of the laser's working medium (16). These "Lamb dips," named after their discoverer, result from intense nonlinear laser-matter interactions. A velocity subset of the randomly moving gas particles can undergo strong resonance with the narrow-line laser light, which suppresses their availability as emitters. This phenomenon, saturated absorption, permitted the first sub-Doppler line shapes to be resolved in static cells and is a useful method of studying absorption features at the highest possible resolution, without the need to collimate molecular beams. Saturated absorption offers high selectivity of detection, at higher sample densities, than in molecular beams. Gas pressures < pressures <10^2 Pa will typically produce collisional broadening comparable with or smaller than the natural radiative line widths in the visible range. The signal strength is the result of higher order nonlinear absorption cross-sections, which make the absolute quantification of species densities difficult to obtain.

Saturated absorption is used mainly as a sensitive null-background spectroscopy to measure precise line positions. Saturated absorption profiles are observed when a strong pump beam (intensity 10–100 mW/mm^2) and weaker probe beam (from the same laser) are made to overlap and counterpropagate through the sample. As the laser frequency is tuned, the probe beam measures pump-

induced changes of the absorbing medium's response (the nonlinear susceptibility). The advantage of saturated absorption over many other nonlinear optical techniques is that it measures resonant absorption positions precisely with modest cw laser powers (<100 mW), which minimizes light-field–induced level shifts and broadening. The technique is used most in the uv–vis, because of the availability of suitable lasers. Also, ir transitions are substantially weaker, and more powerful lasers that can saturate ir transitions, such as pulse lasers, often do not afford significant improvements in resolution. Furthermore, higher order multiphoton absorption or scattering processes are often induced with higher power lasers that offer new possibilities for sensitive or high resolution detection.

Doppler-free Multiphoton Absorption. A monochromatic light source of frequency ν_0 will be observed by a particle in thermal motion at a new apparent frequency, $\nu_0(1 + v_z/c)$, where v_z is the velocity component of the absorber along the direction of the photon (16). If the absorber could interact with two photons, one each from two counterpropagating beams from the same laser, then the transition frequency observed by the particle would be

$$\nu = \nu_0 \left(1 - \frac{v_z}{c}\right) + \nu_0 \left(1 + \frac{v_v}{c}\right) = 2\nu_0 \qquad (11)$$

The frequency of this sum is independent of the velocity of any one absorber. When simultaneous two-photon absorption is induced in this way, the Doppler width of an entire thermal collection of atoms can be suppressed. There is the possibility that two photons can be absorbed from one direction, but only a much smaller velocity subset of molecules participates, so the Doppler-free process results in a strong narrow absorption by all species, with peak amplitude $\sim(\Delta\nu_D/\Delta\nu_N)$ times larger than the broader Doppler-shifted two-photon absorptions from each beam. For weak, nearly forbidden transitions this can represent a 1000-fold improvement in selectivity and sensitivity. Although this is an important high resolution spectroscopic technique, the sensitivity is especially low for cw lasers, unless there is a near resonant intermediate state that can enhance the cross-section for a two-photon absorption. A two-photon absorption is analogous to the Raman process, discussed above, and distinct from a sequential two-step transition, because the former is not lifetime broadened by the intermediate state, while the latter experiences greater line broadening due to each resonant interaction. Multiple resonance experiments trade off some spectral selectivity but gain significant enhancement of sensitivity (>10^3).

Multiphoton Ionization. Photoionization of atoms or molecules can permit sensitive detection by absorption because the ionic signal is a clear signature, without background (14–16,71–73). The first demonstrations of multiphoton ionization (mpi) in the 1960s used the focused output of powerful lasers to produce plasma sparks in air; this is nonresonant mpi from the simultaneous absorption of several light quanta to energies much greater than that carried by a single photon. Mpi has been used in both gases and liquids and at interfaces, but it is most useful as a probe of low pressure gases or molecular beams, for which the charged photofragments can be extracted with high efficiency and analyzed by electron spectrometers or mass spectrometers. Mpi and its many variants are categorized here as null-background absorption techniques, but they are often considered as a distinct category of nonlinear optical spectroscopies, requiring initiation by an intense laser irradiance (typically >100 MW/cm²).

Resonant Ionization Spectroscopy. Multiphoton absorption through bound states is a potentially important spectroscopic probe of high lying levels and/or levels normally forbidden by the selection rules of one-photon absorption. The development of powerful, broadly tunable dye lasers made resonantly enhanced multiphoton absorption practical; the ionization effected by this process is an especially sensitive spectroscopic technique for detecting trace absorbers, because charged photofragments can easily be collected with nearly unit probability (unlike photons). Resonant ionization spectroscopy (ris) is sensitive enough to detect single atoms of an analyte like Cs, even in the presence of an atmosphere of noble gas (72). In this demonstration, the background buffer gas was used in a gas proportional counter (similar to a Geiger counter); the resonantly enhanced ionization of the trace Cs induced a strong electrical breakdown that greatly amplified the laser-induced signal.

Resonant Ionization Mass Spectrometry. The use of resonantly enhanced multiphoton ionization (rempi) for analytical spectroscopic detection of atoms and molecules became practical when electron–ion multipliers and mass spectrometers were used to detect the charged photofragments. Resonant ionization mass spectrometry (rims) is now a mature analytical methodology. The optimization of the laser ionization process has evolved to the point at which this is usually the most efficient method to induce ionization of trace analytes. Selective ionization by rempi is now used routinely to eliminate isobaric interferences and enhance isotopic selection by mass spectrometry. There are many technical trade-offs that must be considered in detail, but generally laser-induced ionization of the analyte can provide as much as 10^7–10^8 isotopic selectivity and accomplish efficient (~10%) ionization of the desired sample. The selectivity can be enhanced by factors of ~10^4 by subsequent mass spectrometry. Under laboratory conditions, cascading rempi with high resolution mass spectrometers has demonstrated isotopic selection greater than 10^{11} with sufficient throughput to analyze microgram samples in <10 h. Detection limits of 10^{-17} g have been achieved with moderately long-lived isotopes, representing the reliable analytic detection of less than 50,000 trace atoms. Isotopic detection of several hundred trace atoms like Pb, Cd, and Tl is theoretically possible (73).

Achieving these levels of trace detection depends on improving details of the ionization and detection of the analyte, but three basic strategies can be identified that are crucial to any ultrasensitive rims experiment: (1) availability of narrow line width lasers capable of resolving the natural line widths of the transitions being probed, (2) ionization induced by Doppler-free multiphoton absorption or multiple resonance schemes that permit the natural line width to be nearly resolved, and (3) the

use of atomic or molecular beam techniques to reduce the effective Doppler line widths of neighboring lines so that undesirable off-resonant absorption pathways are suppressed.

The high isotopic sensitivities noted above for elemental analyses circumvented another fundamental issue of resolution by using cw lasers. In the case of resonant transitions in atoms, the atomic transitions are strong, so spectrally intense lasers can often induce rapid stimulated absorption and stimulated emission of a transition, which leads to light-field–induced shifts and broadening of transitions. The rapid excitation–deexcitation (known as Rabi cycling) further shortens the effective lifetime of the excited atoms, so the "lifetime-limited" width is increased with greater light intensity, which degrades spectral resolution and reduces peak sensitivity. A transition is said to be saturated when the rate of Rabi cycling equals the natural transition rate for spontaneous emission by the radiator. Because the peak absorption cross-sections are greatest for transitions with natural line widths, it is always most effective to induce multiphoton transitions with high resolution lasers at powers just sufficient to begin saturating the desired transitions. Besides optimizing the desired excitation pathway for highest selectivity, this also reduces the off-resonant ionization paths, which are usually nonlinear in laser intensity I (eg, if n laser photons are needed to ionize an atom, then the nonresonant ionization signal scales as I^n). Detectivity, which is related to the ratio of resonant to nonresonant ionization, is thus optimized. Atomic transitions require a laser irradiance of ~100 mW/mm² to achieve saturation if their natural width can be resolved. Cw lasers can easily achieve this irradiance and spectral resolution, and they provide the added advantage of unity-duty-cycle interrogation of the sample. Cw-double resonance ionization mass spectrometry (drims) (73) has achieved detection limits approaching 10^{-18} g for ^{90}Sr and offers similar isotopic detection limits for other species like ^{210}Pb.

Rims is a particularly sensitive null-background absorption technique, but it is sometimes cumbersome to implement without pretreating samples for use in vacuum systems, and it suffers from limited throughput. Ris and Rims techniques require fairly low analyte densities, because a dense discharge tends to experience electron–ion recombination. This and other space-charge perturbations limit efficient extraction of the ions from the laser focal volume when the intensity of photofragments is >10^8 cm⁻³.

Photoacoustic Spectroscopy. A number of other null-background absorption techniques have been developed to detect small absorptions in gases, liquids, and solids, or at their interfaces, as well as in flame or plasma discharges (14,74,75). Many of these techniques are simpler than rims, and can achieve impressive detection limits with higher sampling throughput. These techniques can be categorized as calorimetric (the absorbed radiation induces thermal changes in the medium) or direct. Important examples of the former are photoacoustic and photothermal spectroscopy. Under optimal conditions, both these techniques can measure absorbances as low as 10^{-7}.

Photoacoustic methods rely on the efficient transformation of absorbed radiation into heat by various radiationless processes of relaxation, and the efficient conversion of local heating into pressure waves. Gas-phase samples are the least effective media for conversion, but the technology of gas microphonic cavities (spectrophones) is fairly well developed, so absorbances of $<10^{-6}$ in 1-cm path lengths can be achieved. Because the acoustic signal is proportional to the product of the source intensity and the energy absorption in the sample, high sensitivity can be achieved by increasing the power. Liquid or solid samples can be placed in direct contact with piezoelectric or pyroelectric transducers that convert pressure waves or temperature changes to an electrical signal. Properly coupling the sample to the transducer is important for obtaining high sensitivity. Usually liquids and solids offer good elastic coupling to piezoelectric crystals, and frequency responses to >1 GHz have been obtained. Absorbances in both aqueous and solid samples of ca 10^{-7} or slightly less have been measured.

The principal applications of pas have been in solid sampling and in the analysis of gases at low concentrations, but because the tuning range of the source lasers must be selected for the specific analytes wanted, it has limited flexibility. Within its limits, pas can provide high specificity and sensitivity. Pas detection of various volatile organic compounds at the ppb and sub-ppb levels has been demonstrated using line-tunable CO and CO_2 lasers. Such compounds as vinyl chloride and acrolein have been detected with limits of 1 to 3 ppb. The ultimate potential of pas may be in trace absorption studies of interfacial systems, such as thin films and coatings. In work on gas–solid absorption, pas has demonstrated a detection limit of 0.002 monolayers of species like SF_6 and NH_3 on silver substrates deposited on pyroelectric detectors (76).

Photothermal Methods. Photothermal methods include two related approaches, thermal lensing and photothermal deflection, which in principle can be as sensitive as photoacoustic detection and which are truly noncontact optical methods of interrogation. In both methods, the probe beam can pass through, be reflected from, or make glancing passes over a sample. Thermal lensing is primarily a low speed technique that monitors the thermal focusing or defocusing of a laser. The change of focusing of a pump laser beam can be studied directly, but the most sensitive methods use an amplified modulated cw laser periodically to heat a sample, while the transmission of a coaxial probe laser is synchronously monitored with a pinhole-detector combination. This configuration has been used to measure absorbances of $<10^{-7}$, but in general this method is limited by its geometry to the study of flat, homogeneous, and reasonably transparent samples and substrates. Photothermal deflection is more adaptable and is especially suited to analyzing thin films. It is useful for studying solid–liquid interfaces, with the flexibility to change quickly the position of measurement by adjusting the probe-pump beam overlap. Quantification by photothermal deflection is difficult, because one is studying sharp thermal gradients at the edge of the pump beam.

Optogalvanic Effect. Laser absorption in high temperature flames or gaseous discharges can change the steady-state rate of ionization; the resulting impedance (conductance) changes are sensed as changes in the steady-state discharge voltage established between two electrodes. When cw lasers are used, their amplitude is modulated and the ac response of the discharge is monitored synchronously, while transients generated by pulsed lasers are detected using gated electronics. The optogalvanic effect (oge) is becoming a useful method of diagnosing combustion and gaseous discharges, with excellent potential for trace element detection and chemical analysis. Ir lasers have been used to induce the oge in small molecules like NO, N_2O, CO_2, NH_3, SO_2, H_2S, and H_2O_2 (77), while an assortment of atomic and small molecular species have been studied with visible dye lasers. Molecular detection with visible laser sources is still effective, despite the fact that weak overtone transitions are probed. Uv and multiphoton absorption, which promotes transitions to high lying levels near the first ionization limit, appear to be the most desirable methods to effect strong oge signals. Uv absorption with the oge in aspirated flames and sputter discharge sources have shown potential for ultrasensitive detection of trace elements down to 10^{-15} g limits (78).

Frequency-modulated (fm) Spectroscopy. Fm absorption spectroscopy is analogous to the modulation detection techniques discussed earlier but was developed for use with well-stabilized single-mode lasers. It has emerged as an effective laboratory method to measure very weak absorption (and dispersive) spectral features. The technique combines high sensitivity and acquisition speed to achieve absorbance sensitivities $N\sigma l < 10^{-7}$ in a 1-s integration time, usng 1 mW of cw laser power; this corresponds to shot-noise–limited detection for laser light undergoing natural fluctuations of phase. Stabilized laser radiation exhibits well-defined statistical properties for a stable single-mode laser producing n photons the variance is given by the Poisson distribution of $(\Delta n)^2 = \overline{n}$. For a 1-mW single-mode laser light source, $n > 10^{15}$, so the fluctuations of a well-stabilized laser will produce shot noise of $\sqrt{n}/n \leq 10^{-7}$. Sensitivity can be improved by increasing the probe intensity, for it scales as $(I_0)^{1/2}$. Shot-noise–limited detection has been demonstrated with visible cw dye lasers (79) and lead salt laser diodes in the mid infrared (80). Fms has also been demonstrated with single-line sources, most notably with a CO_2 laser, as well as pulse lasers, and it has been successfully implemented after second harmonic conversion to the uv. Standard nanosecond pulse lasers of 5 ns duration should yield a single-shot absorption sensitivity of 2×10^{-3}, and this has been nearly achieved experimentally.

The technique involves frequency modulating a laser source at ω_0 to produce a carrier frequency with sidebands at $\omega_0 \pm m\omega_m$, where $m\omega_m$ is an integer multiple of the modulation frequency. For laser diodes, this can be effected by modulating the drive current with standard rf sources, whereas dye lasers and many other single-line sources can be frequency modulated using an external electrooptic modulator (eom). The technique of generating side bands on a laser field is based on the equivalence of phase modulation and frequency modulation given by

$$E(t) = E_0\sin(\omega_0 t + \eta\sin\omega_m t) = E_0 \sum J_n(\eta)\sin(\omega_0 \pm n\omega_m)t$$

(12)

Applying a sinusoidal change of phase $\eta\sin\omega_m t$ at modulation frequency ω_m on a monochromatic light field produces side bands. A variety of eom materials are available to frequency-modulate lasers at 0.42–5.2 μm (LiTaO$_3$ and LiNbO$_3$) and in the deeper ir (GaAs, 2.0–11 μm; CdTe, 1.0–25 μm).

The pure fm laser light is passed through a sample, where near-resonant absorption and/or dispersion leads to differential absorption (scattering) of the side bands. This produces a resultant am beat frequency, which can be coherently detected by standard phase-sensitive mixing technology. In principle, fm spectroscopy produces zero-background signal in the absence of an absorption process, so signal calibration is straightforward while yielding a high S:N ratio with low absorption. If the side band modulation ω_m is small compared with the spectral line width, the observed signature is a derivative of the line shape when the carrier frequency is tuned through a resonance (as in modulation spectroscopy). Greater S:N is achieved when the side band spacing is much larger than the spectral line width. This imposes a technical difficulty of requiring high speed detectors with bandwidths exceeding 3 GHz if Doppler- or collision-broadened lines are to be detected at the greatest sensitivity. This drawback can be eliminated by using a two-tone fm scheme, in which a pair of closely spaced side bands (with large average modulation frequency) are demodulated at their smaller relative separation. Two-tone demodulation yields second-derivative line shapes. The demodulation frequency is still kept fairly high (50–100 MHz), but it is now within limits imposed by standard detector technology. The advantage of using these high radio frequencies is that the signal is recovered in a higher frequency band where the noise of the laser amplitude fluctuations is usually negligible, so that the detection sensitivity is now limited by amplifier noise and/or quantum fluctuations. A balloon-borne *in situ* laser sensor (bliss) using fm absorption has demonstrated ~0.5 ppb sensitivities over 500-m path lengths for soundings made in the stratosphere above 20 km (32). Although impressive, this is still more than two orders of magnitude less sensitive than theoretical limits for fms. The bliss experiments were limited by using a small phase modulation index η and quite low modulation frequency (<2 kHz).

With a low power lead salt laser diode, one can typically detect molecular species at atmospheric pressures to the 0.5 ppb level over a single-pass 10-cm path length. The spectral resolution could be enhanced significantly (10- to 100-fold) by reducing the ambient pressure to eliminate collisional line broadening. Trace sensitivity would remain relatively invariant because the peak line strength of a molecular absorption remains fixed until either the Doppler or laser line width is achieved. This improvement in selectivity is especially important when heavier molecules are being probed. Fm detection has also been used with saturation absorption to achieve high resolution and sensitivity. Currently there is extensive work to develop "high temperature" ir laser diodes for thermoelectrically cooled operation near 250 K. Such devices in conjunction with fms detection could provide sub-ppb molecular analysis in small *in situ* sensor packages.

Laser-induced Fluorescence

Fluorescence detection is one of the most sensitive schemes for detecting uv–vis emission when environmen-

tal perturbations can be minimized so that scattering and quenching are negligible. Lif has demonstrated ultrasensitive elemental detection sensitivity (eg, <1 atom/mm^3) with excellent isotopic discrimination using atomic beam sources (66). These levels of sensitivity are generally obtained using resonance fluorescence of atomic systems having a dominant transition to the ground state with little branching to other levels. The Group IA and IIA metals are especially suited to ultrasensitive resonance lif, because their principal absorption lines are strong and have branching ratios to the ground state of >99%. Lif used in conjunction with beam sources is useful for spectroscopic analysis of atomic and molecular structure but less so for analytic quantification, because many collisional and radiative redistribution effects complicate the analysis. Under laboratory conditions, these secondary effects can be eliminated by design or by careful temporal analysis. In such cases, lif has provided useful information about collisional transfer rate coefficients, with applications to kinetic modeling of atmospheric chemistry and gaseous discharges.

Coherent Scattering Spectroscopies

As the irradiance of the source increases above the threshold for nonlinear interaction, multiphoton processes arise, in which two or more incident photons are effectively scattered (14,16,81). Figure 4 shows some examples of stimulated scattering processes that are induced by a strong pump laser of frequency ν_L, detuned near a resonance condition. The very intense light fields shift the atomic or molecular level positions slightly (ac Stark shifts) and induce new resonant conditions (denoted by the dashed lines). With sufficiently high intensities, two, three, or more photon scatterings can be induced to emit strongly in the direction of the incident pump beam. Stimulated Rayleigh scattering occurs at $\nu_R = \nu_L$, stimulated Raman scattering ν_S is induced to lower lying levels, and stimulated three- and four-photon scatterings are induced at frequencies ν_3 and ν_4 close to the pump laser frequency. Two laser photons can also create scattering at ν_{HRS} from higher excited states, known as hyper-Raman scattering, and stimulated emissions via these higher excited states can induce cascade emissions back to lower levels. Certain combinations of scatterings (four-wave mixing at ν_4 (ν_4')) can return the scatterers back to the ground state, to undergo continued cycling. These latter processes have the potential for significant amplification because the available scatterers are never depleted. The scattered radiation from these processes is coherent and directional: it emerges from the sample in a small solid angle and can be efficiently collected and separated from incoherent fluorescence and Raman scattering. Some of these, and other, coherent emission phenomena have laboratory applications for air monitoring.

Even for large detunings in molecules, with high excitation power coherent pumping can occur, leading to stimulated Raman scattering (SRS) at $\nu = \nu_L \pm \nu_{vib}$ (Fig. 5a). The amplified frequency occurs at the strongest Raman transition in the molecule, and the amplification process can increase the scattered signal by as much as 10^{13} over the spontaneous Raman effect. SRS is characterized by high conversion efficiency, resulting in high intensity and

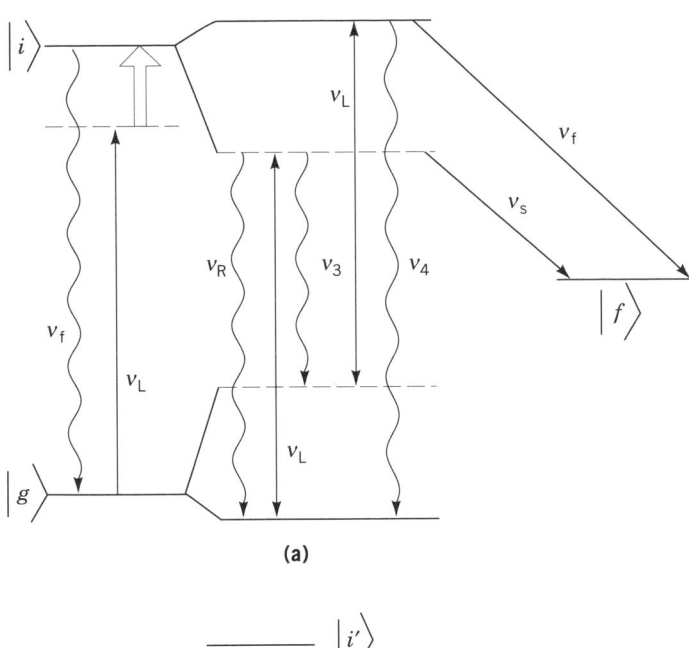

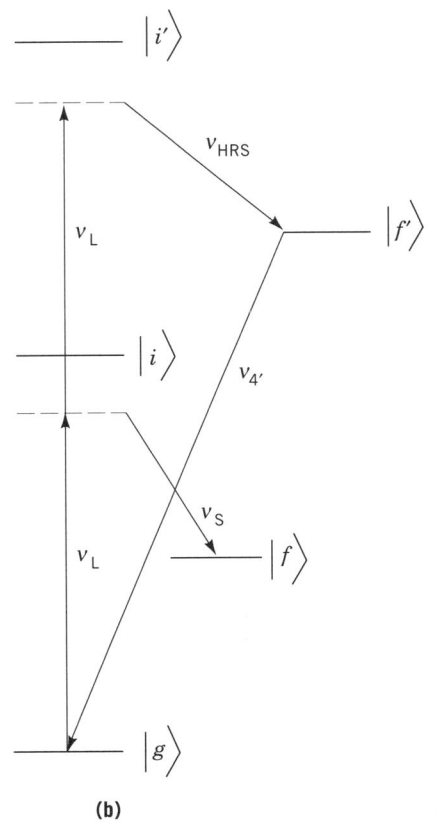

Figure 4. Various scattering–fluorescence processes can be induced in the presence of strong light fields. (**a**) A two-level radiator being pumped by a strong laser source at ν_L. The positions of the ground and intermediate levels are shifted and additional "dressed" levels are induced (denoted by dashes). These lead to Rayleigh scattering at ν_R, induced scattering at ν_3 and ν_4, and possible Stokes Raman emission at ν_S to lower lying levels. If collisions occur (denoted by the outline arrow), collision-induced fluorescence occurs at ν_F and $\nu_{F'}$. At sufficiently high intensities all scattering–fluorescence processes can be induced to create stimulated emission. (**b**) Possible two-photon scattering via higher excited states and possible hyper-Raman scattering to a different final state $|f'\rangle$. In some cases a radiative transition can occur between $|f'\rangle$ and $|g\rangle$, leading to four-wave mixing and efficient generation of ν_4'.

strong coherent forward scattering; at high enough intensities the SRS signals can also experience SRS to create new frequencies at $\nu = \nu_L \pm n\nu_{vib}$, where n is an integer.

While SRS is of limited use in analysis, some variations of it have more promise. Stimulated Raman gain spectroscopy (SRGS) can be considered as a stimulated gain or induced emission process at a Stokes frequency ν_S (see Fig. 5b). Two lasers are used, a pump ν_L and a tunable probe. The intensity of either of these beams is monitored as the probe laser is scanned. When the probe ν_P is scanned, the Raman (Stokes) frequency $\nu = \nu_L - \nu_{vib}$ is driven and ν_P gains in intensity at the expense of ν_L. The inverse of this process (Fig. 5c) might be called stimulated Raman loss spectroscopy but for historical reasons is generally termed inverse Raman scattering (IRS). Here there is stimulated loss at an anti-Stokes–shifted frequency, with gain on the pump beam ν_L: the probe is attenuated at frequencies corresponding to Raman-active vibrations.

Finally, perhaps the most useful stimulated technique is coherent anti-Stokes Raman spectroscopy (CARS). This employs two pump beams ν_1 and ν_2, interacting through the third-order nonlinear susceptibility of the medium to generate coherent anti-Stokes emission at $\nu_3 = 2\nu_1 - \nu_2$ in a four-photon mixing process (Fig. 5d). The laserlike beam of υ_3 is greatly enhanced when the frequency interval $\nu_1 - \nu_2$ is made equal to a Raman-active molecular infrared frequency by tuning the pump line ν_2. In addition to the high intensity of this scattering (signals stronger by factors of 10^5 to 10^{10} than spontaneous Raman), the anti-Stokes output is shifted to higher frequencies, so CARS spectra are free from Rayleigh interference from both lasers. (Coherent Stokes Raman scattering also occurs, but does not have this Rayleigh rejection advantage, offers no advantages over CARS, and is little used.) The laserlike CARS output beam emerges at a slightly different angle than the pump beams, and so it can be easily separated from interfering incoherent light and collected and detected with high efficiency. Because a single frequency is determined by the resonance conditions of the lasers, no monochromator is necessary, and the resolution depends only on the spectral purity of the exciting laser, which can reach 10^{-3} cm^{-1}. A significant disadvantage to CARS is the presence of a nonresonant background signal that can mask the weaker resonant signal; CARS detection limits are, therefore, 0.1% to as high as 1%. But like linear Raman scattering, CARS signals can be resonantly enhanced by tuning υ_1 near an electronic transition, improving detection limits by many orders of magnitude.

CARS has been widely used in combustion and flame diagnostics, for the identification and quantification of major species with high spatial resolution. Rovibrational bands of the molecule of interest can be scanned to obtain temperature information with an accuracy of $\pm 10°$ or better. Because the laser sources in CARS must be tuned so their frequency difference corresponds to a resonant frequency in a single molecule, this is not a technique that can monitor many species simultaneously. Beam alignment is critical for optimum results, for the phase-matching angle must be realized to within 0.1°. The theory of the process is complex, and not all factors contributing to signal intensity are fully understood; especially troublesome is the nonlinear dependence of signal intensity on species density. But adequate quantitative accuracy can

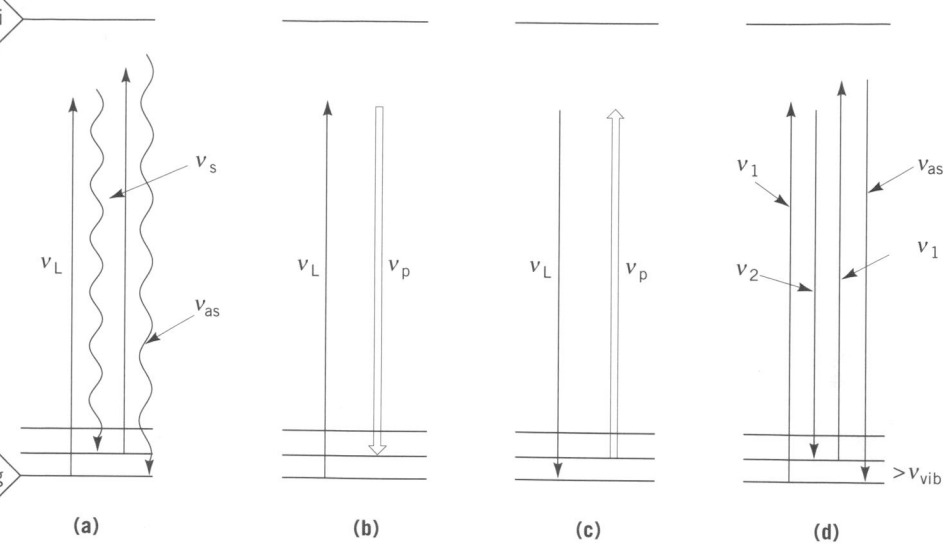

Figure 5. Possible molecular Raman scattering processes. (**a**) Spontaneous Raman scattering can occur from a ground or excited state, producing a Stokes or anti-Stokes emission at $\nu_L - \nu_{vib}$ and $\nu_L + \nu_{vib}$, respectively. If a second probe laser is tuned to $\nu_S = \nu_L - \nu_{vib}$, then (**b**) gain can occur at ν_P (SRGS) or, (**c**) if ν_P is strong, on ν_L (inverse Raman scattering). (**d**) If the intensity *and* momenta of the pump (ν_1) and probe (ν_2) lasers are matched, then strong gain can occur at the anti-Stokes transition (CARS).

be achieved by ratioing the scattered signal against that from atmospheric N_2. Other potential problems include signal degradation by scattering and turbulence and the experimental complexity inherent in using two lasers.

Degenerate Four-wave Mixing

Degenerate four-wave mixing (dfwm), a variation of the coherent Raman scattering process, has novel characteristics that offer great promise as a sensitive analytical technique. Dfwm uses a strong nonresonant pump and weaker probe beam from the same laser to induce a large backscattered coherent signal at the same laser frequency (82). All coherent scattering spectroscopies involve a nonlinear response by the scattering medium that is related to a third-order susceptibility, which is often small compared with the linear susceptibility for incoherent scattering or absorption. As such, intense laser irradiance (>1 MW/cm²) is typically needed to induce most scatterings, even near resonance. Dfwm is like CARS, except that all four legs of excitation–scattering are near resonance, so the induced signal is quite strong and requires irradiance 10^3-fold smaller than CARS. Also, dfwm has the unusual property of being self-corrective, unlike CARS, which requires rigorous alignments to achieve proper spatial overlap (phase matching). Dfwm is simple to align and has the proven unique capability of producing a highly directional return signal that can faithfully "unmap" the spatial optical distortions of inhomogeneous or turbulent media. This offers two benefits: it ensures excellent phase matching over longer lengths (and, therefore, greater signal enhancement) and the return signal exhibits negligible optical distortion. Finally, the process can employ a Doppler-free pump-probe scheme to induce the response, which makes it especially useful for studying high temperature combustion or plasma process streams. The sub-Doppler dfwm response probes a smaller velocity subset of the medium, which reduces sensitivity. Nevertheless, dfwm is almost the perfect noncontact sampling method, from simplicity of execution, imaging, and spectral sensitivity. It does have the drawback of nonlinear signal strength, but

progress has been made toward using it to quantify species concentration. Currently, dfwm has been used for high fidelity spatial mapping of analytes in flames and has shown excellent detection sensitivities for molecular species (for example, OH at concentrations of 10^{10} cm^{-3}), and it is even more sensitive to strong atomic transitions. Detection of excited-state Na atoms in the range of 10^8 cm^{-3} has been reported with an irradiance of ~1 kW/cm² (83). This corresponds to a total sensitivity of ~10^6 atoms in a 1-mm diameter by 1-cm long interaction volume.

FUTURE TRENDS IN LASER ANALYTICAL METHODS

An update of ultrasensitive laser spectroscopies is published every odd year as part of a series on laser-based ultrasensitive spectroscopy and detection, in the *Proceedings of the SPIE—The International Society for Optical Engineering*. Typically, three categories of ultrasensitive detection processes are reviewed: rims, single atom–molecule fluorescence detection, and advanced concepts in absorption measurements. Currently, the methodologies and principles used in laser-based analysis are those discussed above, but new techniques of laser atom or molecule deflection, cooling, and trapping are quickly emerging as potential new analytical techniques. Many of the laser spectroscopies that are now being exploited for analysis first appeared in basic research to make precise physical measurements. The applications to ultrasensitive analytical chemistry were often obvious, but generally occurred as secondary developments 10–15 yr later.

How the latest developments in laser spectroscopy may prove out as analytical methods can be guessed by realizing two important consequences of Ladenburg's formula (eq. 3): the peak absorption cross-section for any radiative process varies inversely with the resolved line width of the transition, and when the limit of the natural line width is achieved this peak absorption cross-section scales simply as $\sigma_0 = \lambda_0^2$, where λ_0 is the transition wavelength. In principle, then, *any* radiative transition could be induced to absorb strongly if the laser line width could be made to match the natural width, *and* if the environmental perturbations of the absorber can be reduced below the natural line width. Transitions with natural line widths of ≥ 1 MHz can be resolved with commercial lasers by Doppler-free techniques of laser and/or molecular beam spectroscopies. Much weaker transitions could ultimately be resolved if the atoms or molecules can be cooled and trapped so that residual Dopper and collisional broadening were reduced to limits approaching a few hundreds of Hertz. The straightforward detection of weak or nearly forbidden transitions is already under way in several laboratories, where ions or neutral atoms are trapped and cooled to ~1 μK; despite such low temperatures this still represents a Doppler line width in excess of 40 kHz in the visible. Currently tunable cw dye lasers can be stabilized to line widths ~1 Hz, while tunable laser diodes are now capable of stabilities ≤ 10 kHz. It can be anticipated that these ultrastable laser sources, used with atom or molecule laser cooling–trapping concepts, could become the absorption spectrometers for future chemical and elemental analysis.

BIBLIOGRAPHY

1. R. M. Measures, *Laser Remote Sensing*, John Wiley & Sons, Inc., New York, 1984.

2. R. M. Measures, ed., *Laser Remote Chemical Analysis*, John Wiley & Sons, Inc., New York, 1988.

3. E. D. Hinkley, ed., *Laser Monitoring of the Atmosphere*, Springer-Verlag, Berlin, 1976.

4. R. Beer, *Remote Sensing by Fourier Transform Spectrometry*, John Wiley & Sons, Inc., New York, 1992.

5. J. T. Houghton, F. W. Taylor, and C. D. Rodgers, *Remote Sounding of Atmospheres*, Cambridge University Press, Cambridge, 1984.

6. H. S. Chen, *Space Remote Sensing Systems*, Academic Press, Inc., Orlando, Fla., 1985.

7. A. Corney, *Atomic and Laser Spectroscopy*, Oxford University Press, 1977.

8. A. P. Thorne, *Spectrophysics*, 2nd ed., Chapman and Hall, London, 1988.

9. J. I. Steinfeld, *Molecules and Radiation*, 2nd ed., MIT Press, Cambridge, Mass., 1985.

10. C. H. Townes and A. L. Schawlow, *Microwave Spectroscopy*, Dover, New York, 1985.

11. P. R. Griffiths and J. A. de Haseth, *Fourier Transform Infrared Spectroscopy*, John Wiley & Sons, Inc., New York, 1986.

12. J. T. Houghton and S. D. Smith, *Infra-Red Physics*, Oxford University Press, Oxford, 1966.

13. P. W. Milonni and J. H. Eberly, *Lasers*, John Wiley & Sons, Inc., New York, 1988.

14. W. Demtröder, *Laser Spectroscopy*, Springer-Verlag, Berlin, 1982.

15. D. S. Kliger, ed., *Ultrasensitive Laser Spectroscopy*, Academic Press, Inc., New York, 1983.

16. M. D. Levenson and S. S. Kano, *Introduction to Nonlinear Laser Spectroscopy*, Academic Press, Inc., Boston, 1988.

17. A. Parrish, R. L. deZafra, P. M. Solomon, and J. W. Barrett, *Radio Sci.* **23**, 106–118 (1988).

18. A. C. G. Mitchell and M. W. Zemansky, *Resonance Radiation and Excited Atoms*, Cambridge University Press, Cambridge, 1961.

19. D. A. Burns and E. W. Ciurczak, eds., *Handbook of Near-Infrared Analysis*, Marcel Dekker, Inc., New York, 1992.

20. K. Narahari Rao and A. Weber, eds., *Spectroscopy of the Earth's Atmosphere and Interstellar Medium*, Academic Press, Inc., New York, 1992.

21. R. Beer in Ref. 2, pp. 85–162.

22. J. Altmann, R. Baumgart, and C. Weitkamp, *Appl. Optics* **20**, 995 (1981), and references therein.

23. M. J. Webb, *Spectroscopy* **4**(6), 26–34 (July–Aug. 1989).

24. U. Krull and R. S. Brown in Ref. 2, pp. 505–532.

25. S. M. Angel, *Spectroscopy* **2**(4), 38–48 (Apr. 1987).

26. P. B. Coleman, ed., *Practical Sampling Techniques for Infrared Analysis*, CRC Press, Boca Raton, Fla., 1993.

27. D. W. T. Griffith and G. Schuster, *J. Atmos. Chem.* **5**, 59–81 (1987).

28. D. W. Hill and T. Powell, *Nondispersive Infrared Gas Analysis in Science, Medicine, and Industry*, Hilger, London, 1968.

29. R. H. Wiens and H. H. Zwick in J. S. Mattson, H. B. Mack, Jr., and H. C. MacDonald Jr., eds., *Infrared, Correlation, and Fourier Transform Spectroscopy*, Marcel Dekker, Inc., New York, 1977.

30. C. Webster, R. Menzies, and E. D. Hinkley in Ref. 2, pp. 163–272.

31. R. Grisar, H. Preier, G. Schmidtke, and G. Restelli, eds., *Monitoring of Gaseous Pollutants by Tunable Diode Lasers*, D. Reidel, Dordrecht, The Netherlands, 1987.

32. R. S. McDowell, *Vibrational Spectra Struct.* **10**, 1–151 (1981).

33. W. Lo, ed., *Tunable Diode Laser Development and Spectroscopy Applications, Proc. SPIE* **438** (1983).

34. C. R. Webster and R. D. May, *J. Geophys. Res.* **92**, 11931 (1987).

35. C. R. Webster and co-workers, *Appl. Opt.* **29**, 907–917 (1990).

36. D. J. Brassington, *J. Phys.* **D15**, 219–228 (1982).

37. C. P. Rinsland, A. Goldman, F. J. Murcray, F. H. Murcray, D. G. Murcray, and J. S. Levine, *Appl. Opt.* **27**, 627–630 (1988).

38. F. J. M. Harren, J. Reuss, E. J. Woltering, and D. D. Bicanic, *Appl. Spectrosc.* **44**, 1360–1368 (1990).

39. H. Ahlberg, S. Lundqvist, and B. Olsson, *Appl. Opt.* **24**, 3924–3928 (1985).

40. R. E. Huffman, *Atmospheric Ultraviolet Remote Sensing*, Academic Press, Inc., San Diego, 1992.

41. J. V. Sweedler, R. D. Jalkian, and M. B. Denton, *Appl. Spectrosc.* **43**, 953–962 (1989).

42. H. Axelsson, H. Edner, B. Galle, P. Ragnarson, and M. Rudin, *Appl. Spectrosc.* **44**, 1654–1658 (1990).

43. U. Platt, M. Rateike, W. Junkermann, J. Rudolph, and D. H. Ehhalt, *J. Geophys. Res.* **93**, 5159 (1988).

44. M. Boutonnat, D. A. Gilmore, K. A. Keilbach, N. Oliphant, and C. H. Atkinson, *Appl. Spectrosc.* **42**, 1520–1524 (1988).

45. P. V. Cvijin, D. A. Gilmore, and G. H. Atkinson, *Appl. Spectrosc.* **42**, 770–774 (1988).

46. D. K. Killinger and N. Menyuk, *Science* **235**, 37–45 (1987).

47. K. Fredriksson, in Ref. 2, pp. 273–332.

48. K. A. Fredriksson, *Appl. Opt.* **24**, 3297–3304 (1985).

49. H. Inaba in Ref. 3, pp. 153–236.

50. J. G. Grasselli and B. J. Bulkin, eds., *Analytical Raman Spectroscopy*, John Wiley & Sons, Inc., New York, 1991.

51. R. T. Menzies in Ref. 3, pp. 297–353.

52. F. J. Barnes, R. R. Karl, K. E. Kunkel, and G. L. Stone, *Remote Sens. Environ.* **32,** 81–90 (1990).

53. J. D. Houston, S. Sizgoric, A. Ulitsky, and J. Banic, *Appl. Opt.* **25,** 2115–2121 (1986).

54. H. Edner, K. Fredriksson, A. Sunesson, S. Svanberg, L. Unéus, and W. Wendt, *Appl. Opt.* **26,** 4330–4338 (1987).

55. W. B. Grant, *Appl. Opt.* **25,** 709–719 (1986).

56. K. W. Rothe, *Radio Electron. Eng.* **50,** 567–574 (1980).

57. A. P. Force, D. K. Killinger, W. E. DeFeo, and N. Menyuk, *Appl. Opt.* **24,** 2837–2841 (1985).

58. M. Aldén and W. Wendt, *Appl. Spectrosc.* **42,** 1421–1427 (1988).

59. N. Menyuk, D. K. Killinger, and W. E. DeFeo, *Appl. Opt.* **21,** 2275–2286 (1982).

60. J. R. Lakowicz, *Principles of Fluorescence Spectroscopy*, Plenum Press, New York, 1983.

61. E. L. Wehry, ed., *Modern Fluorescence Spectroscopy*, Vol. 1 of *Modern Analytical Chemistry Series*; Plenum Press, New York, 1976.

62. B. A. Ridley and L. C. Howlett, *Rev. Sci. Instrum.* **45,** 742–746 (1974).

63. H. I. Schiff, D. Pepper, and B. D. Ridley, *J. Geophys. Res.* **84,** 7895–7897 (1979).

64. A. Torres, *J. Geophys. Res.* **90,** 12875–12880 (1985).

65. A. E. S. Green, ed., *The Middle Ultraviolet: Its Science and Technology*, John Wiley & Sons, Inc., New York, 1966.

66. J. H. Richardson, *Mod. Fluorescence Spectrosc.* **4,** 1–24 (1981).

67. R. T. Ku and D. L. Spears, *Opt. Lett.* **1,** 84–86 (1977).

68. D. Glenar, T. Kostiuk, D. E. Jennings, D. Buhl, and M. J. Mumma, *Appl. Opt.* **21,** 253–259 (1982).

69. R. C. Carlson, A. F. Hayden, and W. B. Telfair, *Appl. Opt.* **27,** 4952–4959 (1988).

70. G. Scoles, *Atomic and Molecular Beam Methods*, vol. 1, Oxford University Press, New York, 1988.

71. V. S. Letokhov, *Laser Photoionization Spectroscopy*, Academic Press, Inc., Orlando, Fla., 1987.

72. G. S. Hurst and co-workers, *Rev. Mod. Phys.* **51,** 767 (1979).

73. B. A. Bushaw, *Prog. Analyt. Spectrosc.* **12,** 247–276 (1989).

74. P. Hess and J. Pelzl, eds., *Photoacoustic and Photothermal Phenomena*, Springer-Verlag, Berlin, 1987.

75. V. P. Zharov and V. S. Letokhov, *Laser Optoacoustic Spectroscopy*, Springer-Verlag, Berlin, 1986.

76. H. Coufal, F. Trager, T. J. Chuang, and A. C. Tam, *Surface Sci.* **145,** L504 (1984).

77. R. E. Muenchausen et al., *Opt. Commun.* **48,** 317 (1984), and references therein.

78. B. W. Smith, G. A. Petrucci, R. G. Badini, and J. D. Winefordner, *Anal. Chem.* **65,** 118–122 (1993).

79. M. G. Gehrtz, G. C. Bjorklund, and E. A. Whittaker, *J. Opt. Soc. Am. B* **2,** 1510 (1985).

80. C. B. Carlisle, D. E. Cooper, and H. Preier, *Appl. Optics* **28,** 2567 (1990).

81. A. B. Harvey, *Chemical Applications of Nonlinear Raman Spectroscopy*, Academic Press, New York, 1981.

82. R. L. Farrow and D. J. Rakestraw, *Science* **257,** 1894–1900 (1992), and references therein.

83. P. Ewart and S. V. O'Leary, *J. Phys. B* **15,** 3669 (1982); **17,** 4609 (1984).

ENVIRONMENTAL ECONOMICS, SURVEY

J. Lon Carlson
Illinois State University
Normal, Illinois

The foundation for the study of environmental economics can be traced to the work of Pigou (1) and, in particular, his work on the theory of externalities. Systematic economic analyses of environmental problems started to appear in the 1960s (2,3). Since that time, a considerable amount of research has been devoted to numerous issues that arise within the context of environmental policymaking. An excellent review of the important advances that have occurred in the field over the last 25 yr is available (4). In addition to a large number of journal articles, monographs, reports, numerous books have also been written on the subject (5–11).

The purpose of this article is to highlight the principal features of the theory of environmental economics and the practical implications of that theory for the development of environmental policy. (See also ENERGY EFFICIENCY; ENERGY POLICY PLANNING; ENERGY TAXATION.)

ECONOMIC THEORY

The Standard Market Model

One of the fundamental premises of microeconomic analysis is that any action should be undertaken up to the point at which the marginal benefits gained equal the marginal costs incurred. Thus, eg, in the case of a marketed good, additional units of the good should be produced so long as consumers' willingness to pay for the last unit produced exceeds the marginal costs of production. This concept is illustrated in Figure 1, which depicts the

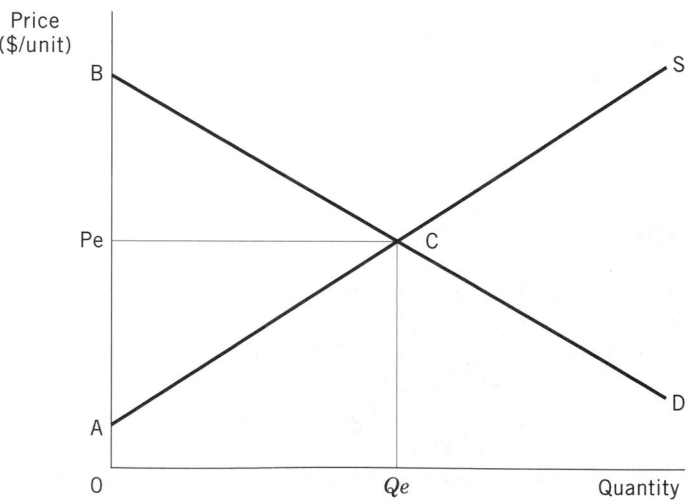

Figure 1. Supply and demand for a normal good. The equilibrium price and quantity of good X are P_e and Q_e. Total willingness to pay for Q_e units is $OBCQ_e$. Total expenditures equal OP_eCQ_e. and total consumer surplus is, therefore, P_eBC. Producer surplus is AP_eC. Thus the combined net benefits from the production and consumption of Q_e units are equal to the area ABC.

demand for and supply of good X. The market demand curve is the horizontal sum of the demand curves of the individual consumers of X. The demand curve for X, which is downward sloping to reflect the declining marginal benefits that are derived from the consumption of X, can also be thought of as a willingness-to-pay schedule. The market supply curve is the horizontal sum of the marginal cost curves of the firms that produce X. The supply curve is upward sloping reflecting the increasing opportunity costs of producing additional units of X. Opportunity costs are increasing as a result of diminishing returns in production. Discussion of the theory of supply and demand are available (12–15).

According to Figure 1, the equilibrium price and quantity of X are P_e and Q_e. Both consumers and producers benefit from the production of X. Consumers realize net benefits to the extent that total willingness to pay exceeds the total amount of money spent on X. In Figure 1, total willingness to pay for Q_e units of X is equal to the area $0BCQ_e$. However, because the equilibrium price is P_e per unit, consumers only spend $P_e \times Q_e$, or the area $0P_eCQ_e$ for the Q_e units. The difference between total willingness to pay and total expenditures, the area P_eBC, is called consumer surplus, and represents the net benefits to consumers. Producer surplus, which measures the difference between total revenues and total variable costs, represents the net benefit to the producers of X. In Figure 1, producer surplus is equal to the area AP_eC. Thus the combined benefits from the production and consumption of Q_e units are equal to the area ABC. Net benefits are maximized by producing Q_e. For output levels less than Q_e, net benefits would increase with additional production because marginal benefits exceed marginal costs. Production in excess of Q_e would cause net benefits to decline since marginal costs exceed marginal benefits. The concepts of consumer and producer surplus are developed within the framework of welfare economics. A number of excellent treatments of the theory of welfare economics are available (16–18).

Assuming that all of the costs and benefits associated with the production and consumption of X are incurred by either the producers or consumers of X, the equilibrium in Figure 1 is socially efficient. Moreover, in the course of determining the socially efficient level of output, the market has also determined the socially efficient allocation of inputs to the production of X. This is especially important in light of the constraint imposed by scarcity of resources. A number of conditions must be met for a privately determined equilibrium to be economically efficient from society's perspective, including full information, a high degree of competition on the part of both sellers and buyers, and a well-defined system of property rights. As is discussed below, in many situations one or more of these conditions is violated. The result is a socially inefficient equilibrium.

External Costs

In the preceding example it was assumed that all costs and benefits associated with the production and consumption of X were incurred by the producers and consumers of X. However, in many situations this assumption does not hold. Instead, some costs are also incurred by individuals who are neither producers nor consumers of the good in question. In the usual market setting, the producer does not have an incentive to consider such costs. Hence, they are external to the decision-making process. Such costs are referred to as negative externalities. External benefits are also possible. In this case, the marginal benefits curve, ie, the demand curve, understates the benefits derived from the consumption of each unit of the good.

A considerable amount of attention has been devoted to the concept of externalities and how to define them (analysts have identified a number of different types of externalities). For the purposes of this discussion, an externality can be defined as "the case where an action of one economic agent affects the utility or production possibilities of another in a way that is not reflected in the market place" (19). Pollution is an excellent example of a negative externality. The adverse effects of pollution, such as acid deposition or the risks associated with the land disposal of hazardous wastes, constitute costs to third parties.

The failure to consider external costs in the decision-making process results in the overallocation of resources to the production of certain goods. Figure 2 illustrates the situation in which the production of X also results in the production of a pollutant, eg, air emissions. Assume that the adverse effects of the air emissions impose costs on society that are not accounted for in the firm's decision-making process. The curve labeled MPC represents the firms' private, or internal, production costs. These costs include expenditures on labor and raw materials and any other out-of-pocket expenses. The curve labeled MSC represents the marginal social costs of production, which are equal to the sum of internal and external costs. Thus the vertical difference between MPC and MSC represents the marginal external costs resulting from the production of

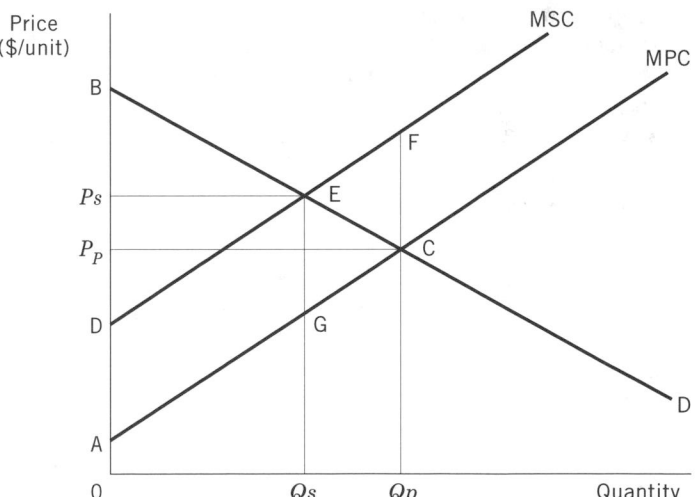

Figure 2. The effects of external costs, illustrating the situation in which production also results in external costs. The curve labeled MPC represents the firms' private, or internal, production costs; the curve labeled MSC represents the marginal social costs of production (the sum of internal and external costs). The socially efficient price and level of output are P_s and Q_s. The net benefits of producing Q_s units are equal to the area DBE.

each successive unit of X. For simplicity it has been assumed that the marginal external costs are constant. However, an assumption of increasing marginal external costs is also quite plausible. In this case, the MSC curve would be steeper than the MPC curve. In addition, it is assumed that there are no external benefits associated with the consumption of X. The socially efficient equilibrium is found by equating marginal benefits and marginal social costs. In Figure 2, the socially efficient price and level of output are P_s and Q_s. Comparing this equilibrium with the privately determined equilibrium (P_p, Q_p) confirms the earlier observation; failure to incorporate external costs into the decision-making process results in the overallocation of resources to the good in question. Failure to account for external benefits results in an underallocation of resources to the good in question; in this case, the privately determined equilibrium quantity would be too low. The net benefits of producing Q_s units are equal to the area DBE. For units of output to the right of Q_s, marginal social costs exceed the marginal benefit from each additional unit of output. Production of the privately determined equilibrium output would result in a loss of benefits equal to the area EFC.

Further consideration of Figure 2 reveals another important observation: in many cases, the optimal amount of pollution is not zero. Production of Q_s units results in external costs equal to the area $ADEG$; pollution is still being produced. However, according to the demand curve for X, the benefits from production of X exceed the costs, both internal and external, incurred. Thus external costs are still being borne by some segment of the population. Because the gainers (those benefitting from the production of X) could compensate the losers (those incurring external costs) and still be better off, the outcome is considered to be economically efficient. This condition is sometimes referred to as the Kaldor-Hicks compensation test (20). The fact that those experiencing the external costs are not compensated is immaterial to the question of efficiency. In fact, it has been shown that compensation of those experiencing external costs would result in inefficient behavior on their part unless the compensation is in the form of a lump-sum payment (21).

Internalizing Negative Externalities

The failure of firms and individuals to incorporate external costs into their decision-making processes results from a lack of incentives to do otherwise. In a market economy, the assumed primary objective of firms is the maximization of profits. A necessary condition for profit maximization is cost minimization. Thus firms have an incentive to manage their wastes and any resulting pollution at the lowest possible cost; as used here, pollution refers to any substance, energy form, or action that, when introduced into the natural environment, results in a lowering of the ambient quality level (22). To the extent that there are no restrictions, eg, a positive price, on using the atmosphere, waterways or land as a waste sink these options will be exploited. Failure to price the use of environmental media as waste sinks on the basis of the costs such uses impose on society results in the overallocation of these resources to such uses. Individual consumers are

confronted with a similar situation. While the objective is different (utility maximization as opposed to profit maximization) the outcome is the same. Consumption results in, among other things, the generation of wastes (packaging and so forth). To the extent that waste disposal services are underpriced, ie, price does not reflect both the internal and external costs of waste management, too much consumer waste will be generated and disposed of.

Environmental media (air, water, and land) will only be priced and used efficiently if property rights for these media are well defined. In fact, property rights to the air and water historically have been poorly defined or nonexistent. When property rights are not completely defined, the resulting market equilibrium is likely to be socially inefficient. However, the question of whether poorly defined property rights require some form of market intervention depends on the level of transactions costs incurred in resolving disputes involving external costs. As has been demonstrated (23), so long as transactions costs, ie, the costs of bringing together the affected parties and negotiating a settlement, are low, private negotiation will result in the efficient allocation of property rights. Regardless of how the property right in question is initially allocated, it will eventually go to the party that values it most highly. This conclusion has come to be known as the Coase Theorem. However, an equally important point made by Coase is that in the majority of cases involving disputes over property rights, transactions costs are *not* low. In such cases, private negotiation is not possible. Thus the initial allocation of the property right will have important implications for the resulting allocation of resources. In those cases where the resulting allocation is inefficient as a result of, say, failure to consider external costs, it may be necessary for the government to intervene in the market in an effort to move the affected parties toward the efficient solution.

There are a number of approaches that can be employed to address the problem of negative externalities, including reliance on legal remedies, direct regulation, and the use of market-based incentives (6,8,11). Legal remedies include the use of civil and criminal law to address disputes over property rights (4,8,11). Civil law, such as the use of a negligence standard, allows the plaintiff the opportunity to recover damages incurred as a result of some action by the defendant. Actions brought under a criminal standard provide for punitive measures in the event that the defendant is found to be at fault. Direct regulation involves the use of various standards to control the behavior of individuals. Direct regulation has been the chief means employed by the U.S. Environmental Protection Agency (EPA) to control the amount of pollution generated in the United States. Market-based incentives, including pollution charges (taxes) and permit systems, are also used to control the generation of pollution. As the name implies, these approaches rely on market forces, rather than direct government intervention, to determine who will ultimately be responsible for pollution control.

Public Goods

Thus far, the discussion has focused on the effects of externalities, such as pollution, on the allocation of resources. An equally important issue concerns the benefits that are derived from pollution control. To be specific, once pollution is controlled, everyone who was adversely affected by the pollution benefits. Moreover, the amount of benefit one person derives from pollution control does not affect the amount of benefits available to others. A single unit of pollution control can be "consumed" simultaneously by any number of individuals. In addition, individuals cannot be excluded from enjoyment of the benefits of pollution control. Goods that possess these properties, referred to respectively as indivisibility and nonexcludability, are called public goods. Public goods stand in contrast to private goods, which are both divisible and excludable. Examples of public goods include national defense, police and fire protection, national parks, and pollution control.

The fact that pollution control is a public good has important implications for the measurement of the benefits of pollution control and the manner in which pollution control will be achieved. With respect to benefits, because a public good is indivisible, the aggregate benefits from, or market demand for, each unit of the public good are calculated by summing the willingness to pay of all affected individuals *for each unit* of the good produced. This amounts to vertically summing the individual demand curves to derive the market demand curve for the public good. In the case of private goods, the market demand curve is calculated by horizontally summing the individual demand curves. This approach yields the total number of units that will be purchased at each price (ie, at each level of willingness to pay). Discussions of the derivation of the demand curves for private and public goods are available (8,12).

The practical importance of the distinction between the market demand for public goods and private goods can be illustrated with a simple example. Assume that five individuals each wish to purchase a unit of good X, and that the willingness to pay of each of the five individuals for one unit of X is $10. If X is a private good, five units of X must be produced to satisfy the demands of all five individuals. In addition, the marginal cost of the fifth unit of X cannot exceed $10 if the demand of all five individuals is to be met in an efficient manner. If the marginal cost of the fifth unit (or the first through fourth unit for that matter) exceeds $10, marginal cost exceeds marginal benefit ($10); production of the marginal unit is inefficient. In contrast, if X is a public good, as a result of the property of indivisibility a single unit of X provides marginal benefits of $10 to each individual. Collectively, they are willing to pay $50 for one unit. Thus marginal cost could be as much as $50 and production of the good would still be efficient.

Because public goods are indivisible, the marginal benefit from each unit of the good produced is potentially much greater than it would be if the good were a private good. However, the property of indivisibility, as well as nonexcludability, also prevents public goods from being produced in a market setting. Because public goods cannot be divided into discrete units for sale and individuals cannot be excluded from consuming the good unless they pay the market price, private entrepreneurs have no incentive to engage in the production of public goods. In-

stead, production of public goods must be undertaken by the government or, alternatively, the government must take actions to ensure that the public good is produced in the private sector. Thus the government has assumed the responsibility of providing for the nation's defense as well as police and fire protection. In the case of pollution control, the government has taken steps, primarily legislative and regulatory, designed to force firms to internalize the external costs of pollution and, in so doing, produce pollution control.

ANALYTICAL APPROACHES TO POLLUTION CONTROL POLICY

Depending on the needs and objectives of the decision maker, various analytical approaches can be employed to assess the effects of alternative policies designed to improve environmental quality. The most basic approach is referred to as impact analysis. The fact that it is the most basic approach should not be interpreted to imply that impact analysis is a simple, straightforward procedure. In many instances, as is described below, impact analyses can be quite complex and subject to considerable uncertainty.

Cost-effectiveness analysis considers the costs of alternative policy options and provides the decision maker with insights regarding the value of the resources that will be required to achieve a particular objective, such as a specified reduction in a particular type of air emissions. Benefit–cost analysis is intended to shed light on the relative value of the benefits and costs of policy alternatives. Depending on the decision rule that is used, benefit–cost analysis can assist in the identification of the economically efficient solution or those options for which net benefits are positive. Risk assessment is applicable to those situations in which policy is intended to influence the risks of certain adverse effects on humans or the environment.

Impact Analysis

The purpose of an impact analysis is to summarize the various impacts a proposed action is predicted to have on the affected environment. The affected environment can include both the natural environment and human activities. Thus impacts may include effects on natural resources, environmental quality, and economic activities. Environmental and economic impact assessments often are required at the federal and state levels. The National Environmental Policy Act (NEPA) of 1969 requires that an impact analysis be conducted for all major federal actions. The result of many of these analyses is an Environmental Impact Statement (EIS), which summarizes the predicted impacts of the proposed action and alternatives to the proposed action on the affected environment. The content of an EIS is dictated by regulations promulgated by the Council on Environmental quality (CEQ), which was created by NEPA and is responsible for, among other things, overseeing the EIS process (24). At the state level, similar requirements are often imposed on regulatory agencies. For example, in Illinois an Economic Impact Statement (Ecis) is required as part of the process of promulgating many environmental regulations. The determination of whether an Ecis is required in a particular situation is the responsibility of the Illinois Pollution Control Board.

Impact analyses, such as an EIS, provide decision makers with a large amount of information on the proposed action and its alternatives. However, because the impacts vary across the different affected resources, and some impacts are positive while others are negative, it is often difficult to determine, on net, whether the proposed action will be beneficial when viewed from society's perspective. As has been pointed out, "Current environmental impact statements are more sophisticated than their early predecessors. . . . Historically, however, the tendency had been to issue huge environmental impact statements that are virtually impossible to comprehend in their entirety" (25).

The EIS process has become more sophisticated over time. For example, the U.S. Bureau of Reclamation (BOR) has completed a draft of an EIS that examines the impacts of the operations of Glen Canyon Dam on the affected environment (26). This EIS employs state-of-the-art analytical techniques to analyze the impacts of a change in the operations of Glen Canyon Dam on such variables as water quality, sedimentation and river bed characteristics, and beaches and backwaters; biological resources including vegetation, wildlife and habitat potential, and fisheries; recreational resources including fishing, day rafting, white-water rafting, and camping; hydropower resources; sociocultural resources; economic resources; and aesthetic resources. With respect to hydropower and economic resources, sophisticated models were developed to analyze the impacts of a change in dam operations on the mix of resources (hydropower and fossil-fuel generation) that would be required to meet peak and off-peak energy demands and the implications for electrical power supplier in the region. In addition, these models were used to estimate the change in retail rates charged to various power customers in the affected region. The recreation analysis provided, among other things, estimates of the change in the value of recreational activities (angling and white-water boating) resulting from a change in streamflows below the dam. A separate component of the analysis is attempting to estimate the effects of a change in ambient environmental conditions, resulting from a change in dam operations, on the so-called nonuse values individuals attach to the affected environment. The term nonuse value refers to the amount an individual would be willing to pay for a change in environmental quality at a specific location, even though they do not intend to visit or otherwise use the site personally. As this brief discussion illustrates, impact analysis can be extremely complex, time-consuming, and costly. Thus far, the Glen Canyon EIS has required several years and tens of millions of dollars to complete.

Cost-effectiveness Analysis

Cost-effectiveness analysis involves comparison of the costs of alternative approaches to achieving a specific objective, eg, a reduction in pollution (27). Unlike benefit–cost analysis, which requires monetization of both the

benefits and costs of the proposed action, cost-effectiveness analysis does not require that benefits be expressed in monetary terms. Instead, benefits are expressed as the reduction in different types of damages or sources of damages, eg, decreased risk of mortality or decrease in the concentration of a pollutant in a body of water. Cost-effectiveness can be assessed by choosing a target level of effectiveness (or goal) for which the least-cost alternative is identified. Conversely, a fixed budget can be applied to alternative strategies to identify the alternative that yields the maximum effectiveness for the funds available. In either case, performing a cost-effectiveness analysis requires the implementation of three steps: (1) identification of the alternatives to be evaluated, (2) selection of the appropriate measure(s) of effectiveness, and (3) measurement of the costs incurred and the effectiveness of each alternative.

Identification of the preferred policy option depends on whether the effects of the proposed actions vary across the policy alternatives. If each of the policy alternatives has the same effect, ie, each alternative reduces pollution, and therefore damages, by the same amount, the decision rule is straightforward: select the policy alternative that achieves the desired reduction in pollution at least cost. The least-cost solution is determined by calculating the cost-effectiveness ratio, which is simply the total cost of an alternative divided by its total effectiveness, for each alternative. The costs of implementing the alternative actions are then compared to determine the least-cost strategy. While this approach is rather straightforward in the case in which each alternative yields the same level of effectiveness, it can produce ambiguous results when alternatives yield differing levels of effectiveness.

When the effectiveness of the alternatives varies, selection of the appropriate decision rule is more complex. For example, one could calculate the average cost per unit of effectiveness as is described above. However, the appropriate decision rule is not obvious. Consider, for example, the situation in which one option results in both a higher average cost and total amount of effectiveness than another option. In this case, it is not clear which option should be preferred. As an alternative, one could compare the marginal costs of pollution control across the various options. A frequently cited decision criterion is the equivalent marginal cost approach (28). The marginal cost of effectiveness is measured as the change in costs associated with a one unit change in the effectiveness of the action. In this approach alternatives that have the same marginal cost effectiveness are grouped together, and a predetermined selection process is then used to maximize the stated objective. For the marginal cost effectiveness criteria to be applicable, a number of conditions must be met, including continuity among alternatives, costs, and effectiveness measures.

As an alternative to the marginal cost approach, incremental cost-effectiveness is measured as the difference between the cost of an alternative and the cost of the next most stringent (or effective) alternative divided by the difference in effectiveness between the two alternatives (28). For example, assume two alternatives, A and B, are contiguous and A is more stringent than B, ie, A yields a greater level of effectiveness. In this case, incremental cost-effectiveness (IC-E) is measured as:

$$IC\text{-}E = \frac{C_A - C_B}{E_A - E_B}$$

The incremental cost effectiveness approach is most useful in the case of discrete alternatives and subalternatives. The alternatives are grouped according to equality of incremental cost-effectiveness. Once grouped in this way, a choice among alternatives can be made.

Finally, the decision rule could be based on minimization of control costs subject to some minimum threshold level of control. Alternatively, the threshold could be defined as the maximum cost that could be accepted (to be economic). In either case, the threshold value could be used as a decision point for choosing alternatives. Any alternative that results in a level of effectiveness or cost that falls short of or exceeds the corresponding threshold value would be eliminated from consideration.

Benefit-Cost Analysis

Simply put, the objective of a benefit-cost analysis is to compare the benefits and costs of a proposed policy action. More thorough treatments of the subject are available (29–31). Depending on the objectives of the decision maker, benefit–cost analysis can be more or less rigorous. For example, if the objective is simply to ensure that the net benefits of a proposed policy action are positive, the net present value criterion is sufficient (32). Under this decision rule, so long as the present value of the total benefits exceeds the present value of total costs, the proposed action is acceptable. This is equivalent to requiring that the ratio of total benefits to total costs (appropriately discounted) be greater than 1.

In the case in which the objective of the analysis is identification of the economically efficient policy action the decision rule is much more rigorous. Identification of the economically efficient solution requires knowledge of the marginal benefits and marginal costs of pollution control. This concept is illustrated in Figure 3, which depicts the marginal benefits and marginal costs associated with increasing levels of pollution control. The external costs of a particular pollutant, such as the sulfur emissions from a coal-fired generating unit are measured by the resulting damages imposed on society. As such, the benefits of successive reductions in the quantity of sulfur emissions are measured by society's valuation of the reduction in damages attributable to each unit of emissions reduction. Technically, this benefit is measured as the value of the reduction in the risk of environmental damage that results from the reduction in pollution. In Figure 3, pollution reduction is measured on the horizontal axis. The dollar value of the marginal costs and benefits associated with successive increments of pollution reduction is measured on the vertical axis.

The marginal social benefit (MSB) curve represents society's willingness to pay for each additional unit of pollution control. The area under the MSB curve to the left of any level of pollution reduction represents the total benefits of that level of pollution control. As illustrated in Figure 3, the marginal benefits of pollution control are as-

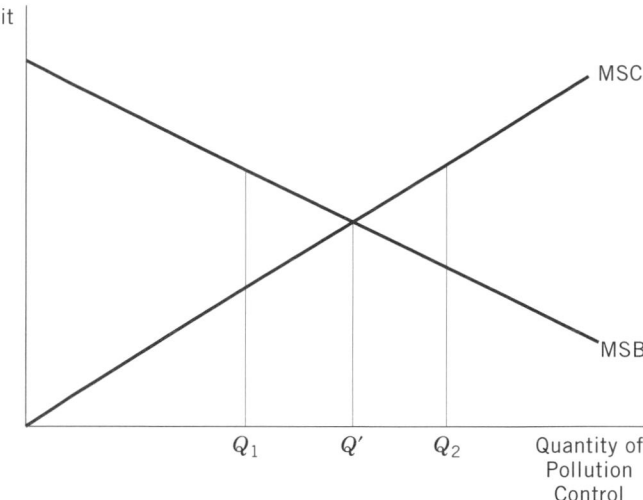

Figure 3. Determining the optimal level of pollution control, depicting the marginal benefits and marginal costs associated with increasing levels of pollution control. The marginal social benefit (MSB) curve represents society's willingness to pay for each additional unit of pollution control. The marginal social cost (MSC) curve represents the incremental cost of each additional unit of pollution control. The optimal level of pollution control is Q'. Any point other than Q' is inefficient. At levels such as Q_1, MSB > MSC: marginal net benefits are positive. At levels such as Q_2, MSB < MSC, and marginal net benefits are negative.

sumed to decrease as the total amount of pollution control increases. The marginal social cost (MSC) curve represents the incremental cost of each additional unit of pollution control. The area under the MSC curve to the left of any particular level of control represents the total cost of pollution control. As the total amount of pollution control increases, the marginal cost of each additional unit of pollution control is assumed to increase. This feature of pollution control costs, which has its origins in the law of diminishing returns, is regularly observed in actual situations involving pollution control (33).

The optimal level of pollution control is that level at which marginal benefits and marginal costs are equal (Q' in Fig. 3). Any point other than Q' is inefficient. At levels of control less than Q', such as Q_1, the marginal benefits of additional control exceed the marginal costs incurred. As such, the net benefits of additional pollution control would be positive. At levels of control in excess of Q', such as to Q_2, the marginal costs of pollution control exceed marginal benefits: marginal net benefits are negative. Another important point to emphasize is that the optimal level of pollution control is not the level at which total benefits equal total costs (8,11).

As an alternative to the model presented in Figure 3, the question of how much pollution to control can be approached from the perspective of cost minimization. Figure 4 depicts the marginal costs of pollution control and marginal damages resulting from each increment of pollution. The amount of pollution generated is measured on the horizontal axis, this contrasts with Figure 3, in which the amount of pollution controlled is measured on the horizontal axis. Marginal damage costs, depicted by the curve labeled MDC, are assumed to increase as the amount of pollution increases. (Marginal damage costs

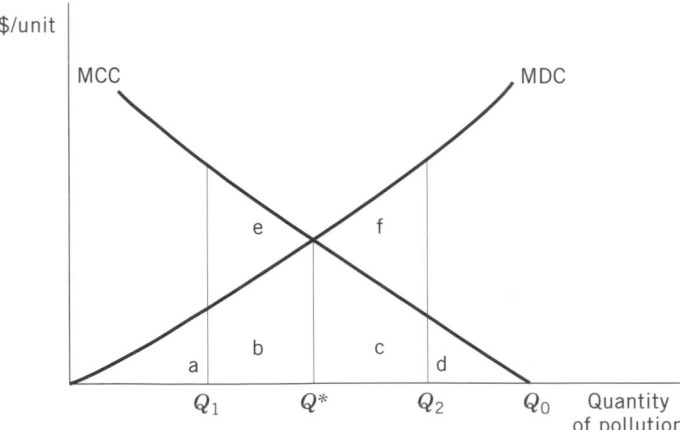

Figure 4. Marginal control costs versus marginal damage costs of pollution, depicting the marginal costs of pollution control (MCC) and marginal damage costs (MDC) resulting from each increment of pollution. In this example, the total costs of pollution are minimized by allowing Q^* units to be generated. The total costs attributable to Q^* units of pollution are equal to the area $a + b + c + d$, where $a + b$ represents total damage costs and $c + d$ represents total control costs. Restricting the level of pollution to something like Q_1 would result in MCC $>$ MDC; total costs would increase by the area e. In a similar manner allowing the level of pollution to exceed Q^*, eg, Q_2, MDC $>$ MCC and total costs would increase by the area f.

could also be assumed to be constant without changing the results of the analysis that follows.) Marginal control costs, depicted by the curve labeled MCC, decrease as the level of pollution increases, reflecting reduced levels of control. (Q_0 represents the amount of pollution that would be generated in the absence of market intervention.)

For the situation illustrated in Figure 4, the total costs associated with the pollutant in question would be minimized by allowing Q^* units to be generated. The total costs attributable to Q^* units of pollution, ie, the sum of damage costs and control costs, is equal to the area $a + b + c + d$, where $a + b$ represents total damage costs and $c + d$ represents total control costs. Restricting the level of pollution to something less than Q^*, such as Q_1, would result in marginal control costs that exceed marginal damage costs avoided. Thus relative to the cost minimizing solution, total costs would increase by the area e. In a similar manner if the level of pollution were to exceed Q^*, eg, Q_2, marginal damage costs would exceed marginal control costs and total costs would once again exceed the minimum, this time by the area f.

In fact, Figures 3 and 4 convey the same information with respect to the optimal amount of pollution, albeit from different perspectives. This is because the marginal benefits from pollution control are measured by the reduction in damages attributable to the pollution that is being controlled. Thus Q^* in Figure 4 is equal to the initial amount of pollution minus Q' in Figure 3, the optimal amount of pollution control.

A number of important points need to be borne in mind when discussing benefit–cost analysis. First, in many instances the costs and benefits of pollution control will be dispersed over many individuals and entities. Aggregating the gains and losses of all affected parties and then comparing these aggregate measures ignores the question

of distribution. However, recall the earlier discussion of the Kaldor-Hicks compensation test. According to this criterion, if the gainers can compensate the losers and still be better off, efficiency is increased. This approach allows decision makers to consider a wider range of alternatives than would be available if the rule were that no one could be made worse off. However, it does ignore questions of equity and fairness. To be specific, the compensation test does not require that compensation of the losers actually occur. Another important point concerns the various difficulties that are likely to be encountered when conducting a benefit–cost analysis.

Risk Analysis

Risk analysis (11) has been developed in recent years as an alternative approach to addressing certain environmental issues. Risk analysis is composed of two basic components: risk assessment and risk management. The concept of risk analysis can be illustrated by reference to the case of superfund sites. Superfund sites are locations at which hazardous materials or wastes have been deposited, and that have been designated for inclusion on the National Priorities List by the EPA. Superfund sites are of concern due to the risks they pose to the ambient environment. These risks include the potential for contamination of ground or surface waters, as well as the soil and atmosphere around the site. Contamination can then lead to adverse effects on ecosystems and human health and welfare.

Risk assessment is used to determine the types of adverse effects that could occur as a result of contamination or migration of materials away from the site and the probability of such occurrences. Thus, in the case of a superfund site such as an abandoned hazardous waste landfill, the risk assessment would entail identification of the types of wastes that were buried at the site, the possible pathways of migration of materials away from the site, the adverse effects that could result from exposure to such materials, and the probabilities of migration, exposure, and occurrence of adverse effects. Risk management focuses on the question of what constitutes the appropriate policy response to the risks that have been identified in the risk assessment. For example, it might be determined that disturbing the site would result in a greater probability of exposure than simply leaving the site intact and monitoring the area around the site. In addition, available evidence suggests that the values that society places on different types of risks can vary considerably. Thus willingness to pay, and therefore the benefits from risk mitigation, is likely to vary depending on the specifics of the situation. Studies of the public's attitudes toward different types of risks suggests that responses are influenced by a number of factors including whether the risk is voluntarily or involuntarily assumed and the specific types of adverse effects that could occur (34–36).

Comparison–Contrast of the Approaches

Going from impact analysis to cost-effectiveness analysis to benefit–cost analysis and risk analysis, the amount of information that can be made available to decision makers increases, at least in principle. However, the information and data requirements also increase. In effect, the outputs of an impact assessment serve as inputs to a cost-effectiveness analysis. In a similar fashion, the results of a cost-effectiveness analysis constitute information required to complete a benefit–cost analysis. In the case of impact analysis, the primary focus is on the effects the policy alternatives will have on the affected environment. Cost-effectiveness analysis goes beyond an impact analysis by estimating the opportunity costs, measured in dollars, of each of the policy alternatives. To complete a benefit–cost analysis it is also necessary to express, to the extent possible, the effects (benefits) of each alternative in monetary terms. Consequently, some discussion of cost and benefits estimation is in order. Brief discussions of the components of total value and the role of discounting are also presented.

Cost Estimation Techniques. The costs of a particular pollution control policy are measured by the value of the resources that must be used to achieve the policy's objectives and any net welfare losses in the affected markets (37). Economists refer to this concept as opportunity cost. Opportunity cost is measured as the value of the next best use of the resources in question. Consider, for example, a proposed regulation that would require sewage treatment plants to install an additional filtration device that would reduce the concentration of a particular chemical in the public's drinking water supply. In this case, the cost of the proposed regulation would depend on the cost of the filtration device, installation costs, and any operation and maintenance costs that would be incurred over the service life of the device as well as price and quantity adjustments in the affected markets. This example raises a number of important issues, including the methods used to obtain cost data and the use of market prices to approximate opportunity cost.

In practice, the most common approaches to cost estimation include the use of surveys, construction of an engineering process model (or series of models), or some combination of the two (38). Surveys are used to compile estimates of compliance costs from those entities, often firms in an industry, that would be subject to the proposed policy. Engineering process models entail construction of a mathematical model of a representative firm or set of firms in the affected industry. The model relates inputs and outputs and enables the analyst to estimate the costs of production for a given set of input prices and level of demand. Both approaches have their respective strengths and weaknesses. Surveys rely on input from those entities who are presumably in the best position to estimate compliance costs, ie, the affected firms. However, there is an incentive for affected firms to overstate expected compliance costs in an effort to influence the stringency of the final policy. Process models are able to circumvent this problem, but are limited since the models are for a "representative" firm (or firms) in the industry. Actual firms may differ significantly from the hypothetical average and therefore incur costs that differ significantly from those estimated using the process model. A detailed discussion of alternative cost estimation methods including various types of process models and econometric estimation techniques is presented in (39).

The extent to which pollution control expenditures based on market prices reflect the actual opportunity costs of a proposed policy is another important issue in cost estimation. If all of the markets for inputs and outputs in the economy were perfectly competitive, market prices would be an accurate measure of opportunity cost. However, in many of the markets in the U.S. economy, competition is limited. The extreme case is monopoly, in which there is a single seller of the good or input in question. When competition is constrained, for whatever reason, a likely outcome is that price will exceed marginal cost, which is the correct measure of opportunity cost, assuming marginal cost is determined in competitive markets). Hence, expenditures will tend to overstate the actual costs, ie, opportunity costs, of the proposed regulation. The difference between price and marginal cost, although paid by the purchaser of the good or input, represents a transfer of income and does not reflect the value of resources required to comply with the proposed policy. Hence, it is not properly considered as part of the costs of the proposed policy from society's perspective.

Benefits Estimation. Estimation of the benefits of a particular policy option is complicated by the fact that many of the effects of pollution control are not valued in a market setting. Consider, for example, the situation in which a pollutant adversely affects agriculture and water-related recreation. To the extent that the proposed policy reduces adverse effects on agriculture, such benefits could be measured by the market value of the increased production of the affected outputs. In contrast, water-related recreation is not "purchased" in the usual sense. Instead, recreators purchase a set of inputs, eg, boats, swimming gear, fishing tackle, and so forth, and combine these with a specific site to produce an output, water-related recreation. Thus what is of interest is the value attached to the quality of the recreation site. Because the price paid for access to a site is generally low or nonexistent, this price is not a good indicator of an individual's willingness to pay for access. Thus the value of the site is unknown. However, economists have developed a number of techniques, which can be classified as either indirect or direct, to assess such values. Indirect methods include the hedonic estimation technique and the travel cost method. The direct method consists of so-called contingent valuation.

Hedonic Estimation Technique. The hedonic estimation technique, first developed in the early 1970s (40,41), is based on the observation that in many situations, willingness to pay for a good depends on a number of characteristics of the good (4,9,42). Consider, for example, the case of housing. The price paid for a house is likely to depend on a number of factors, including overall size (square footage), the number rooms (baths, bedrooms, etc), and the location of the house (including quality of the neighborhood and schools). It is also reasonable to expect that the level of environmental quality could influence willingness to pay for a specific house. For example, two houses may be identical in all respects with the exception that one house is located next to a municipal waste landfill and the other is located in a quiet suburban neighborhood. In this case, it is reasonable to expect that the price of the latter would be higher than the price of the former. The objective of the hedonic estimation technique is to estimate that portion of the price that is attributable to the environmental characteristic of interest, eg, distance from a landfill. A good example of the application of the hedonic technique is presented in (43).

The hedonic technique has also been applied to the problem of risk valuation. Studies (44,45) have attempted to estimate that portion of the wage rate paid for certain jobs that represents compensation for an increased risk of death. The results of these "wage-risk premium" analyses have been used by some analysts to estimate the value of policies that would mitigate certain environmental risks. While this approach does shed some light on the potential magnitude of the benefits of such policies, it raises a number of issues including the difference between voluntarily and involuntarily assumed risks and the resulting value of a change in each type of risk.

In cases in which property values or wage estimates are available, the hedonic technique can prove useful in estimating willingness to pay for environmental improvements. However, this is also a main source of its limitations (46). To be specific, many instances arise in which the good to be valued is not related to property values or wages. Thus willingness to pay for environmental improvements must be estimated by alternative means.

Travel Cost Method. One particular case in which the hedonic technique often is not particularly useful is that of recreation values. Pollution can adversely affect the quality, and therefore value of recreation, especially water-based recreation. However, in many cases there is no fee for access to a recreation site, and in other cases, eg, national parks, there is a nominal fee for access; this price is far less than actual value of the experience to the recreator. Consequently, the hedonic technique cannot be used directly. An alternative approach (which was first proposed by Harold Hotelling in a letter to the director of the National Park Service in 1947 and since described formally by (47)), uses travel costs as a proxy for the price paid for access to a particular recreation site. The travel cost method has been used in a number of different studies to estimate the value of recreation (eg, angling) at numerous locations. Briefly, the demand for access to a recreational resource is estimated on the basis of the travel costs (both direct and implicit) incurred by individuals using the site. The area under the demand curve is then taken as a measure of the benefits derived from the site. Extended discussions of the travel cost method are available (4,9,42). Applications and assessments of the travel cost method can also be found (48–50).

Contingent Valuation. Contingent valuation entails the use of a survey instrument to generate individuals' estimates of their willingness to pay for the item in question, eg, an improvement in visibility at the Grand Canyon. The resulting estimates are then statistically analyzed to generate estimates of the affected population's aggregate willingness to pay for the proposed policy. Detailed discussions of the contingent valuation method have been presented (5,52). In many situations, contingent valuation constitutes the only method for estimating willingness to pay for a change in environmental quality. This is especially true in such cases as a change in the characteristics of a specific recreation site or the estimation of so-called nonuse values (which are discussed below).

The practice of contingent valuation has evolved considerably over the past 15 yr. Many of the changes have occurred in direct response to the numerous criticisms that have been leveled at previous studies. These criticisms include, but are not limited to, potential biases that may enter into the process (53). For example, starting point bias, or anchoring, occurs when the respondent's expressed willingness to pay is influenced by an initial value of willingness to pay suggested in the survey. To the extent that the respondent sees the potential to gain from a particular response, even though the response does not reflect his or her true willingness to pay, strategic bias can occur. Hypothetical bias refers to the possibility that the respondent may react to the hypothetical nature of the survey and in so doing respond in a manner that does not reflect his or her true willingness to pay for the change in question.

Each of the issues noted above, as well as a variety of others, has been the subject of careful analysis. And in many cases, significant progress has been made in overcoming such obstacles. However, in spite of the considerable advances that have made, the validity of the contingent valuation method is still a matter of some dispute. For example, the results of one recent study (54) suggest that survey respondents may have considerable difficulty isolating the specific good or environmental attribute that is being valued in the survey. While such results have in turn been challenged (55), other analysts have argued for significant revisions to the approach to enhance the validity of the resulting estimates of willingness to pay for changes in environmental quality (56).

Total Valuation: Use Values Versus Nonuse Values. In many cases, pollution control results in reduced damages, eg, adverse health effects and ecosystem damage, as well as an improvement in the quality of certain resources. To ensure that all of the benefits of a proposed policy are included, it is necessary to consider the change in the total value of the affected resources. The total value of an environmental attribute or natural resource is composed of two components: use value and nonuse value. In addition, the concept of option value has been developed to further distinguish the separate components that make up the total value of a resource.

Use value refers to the value of the direct (and sometimes indirect) uses to which a resource can be put. The concept of option value was first developed by Weisbrod (57). In its initial conception, option value referred to the value an individual attaches to the option to use a resource he or she is not currently using but may decide to use at some future point in time. Subsequent theoretical work has led to the rejection of this interpretation. According to Smith (58), option value is the difference between two different measures of use value in the presence of uncertainty and, as such, is not a separate component of the total value of a resource. Nonuse values include what have been termed existence value and bequest value. Existence value refers to the value an individual attaches to the existence or quality of a resource, even though he or she does not intend to ever use the resource personally. Bequest value refers to the value individuals may attach to the preservation of resources for future generations.

There is, in fact, some disagreement among economists regarding the legitimacy of nonuse value as a value distinct from use value. Some critics have argued that there is little or no reason to believe that existence values (or bequest values) are not already reflected in use values. It has been suggested that although existence values may be distinct from use values, they should nonetheless be excluded from benefit–cost analyses, because they reflect considerations other than the efficiency motive (59). More recently, it has been argued that nonuse values present a number of problems that raise serious questions about the legitimacy of their inclusion in the decision-making process (60). It is argued that the range of possible existence values may be limitless and that accurate estimation of existence values is extremely difficult, if not impossible. Other arguments against the consideration of nonuse values are available (61).

Proponents of the case for nonuse values have offered strong rebuttals to the arguments described above. For example, there is a well-developed theoretical basis for the consideration of non-use values (62,63). In addition, the fact that nonuse values are difficult to estimate is not, by itself, sufficient justification to exclude them from policy analyses (64). On the empirical side of the debate, studies suggest that nonuse values may account for a substantial portion of the total value of a resource. For example, a number of studies were reviewed that estimated use and nonuse values of particular resources (65). The authors concluded that "non-use benefits generally are at least half as great as recreational use benefits" (66). This suggests that, all else constant, failure to consider nonuse values could result in serious underestimation of the benefits of many proposed environmental policies.

The Question of Discounting. Because many of the costs and benefits of pollution control policies will occur in the future, discounting is also an important issue. To be specific, costs and benefits that occur at various times in the future must be adjusted so that all costs and benefits can be compared on an equal basis. The usual approach to this problem is to calculate the present value of all future costs and benefits and then aggregate the respective values for purposes of comparison. The immediate question that arises is what value of the discount rate to use. Not surprisingly, this issue has been the subject of considerable debate. Excellent discussions can be found in a number of references (31,67,68). It is sufficient here to note that this issue is far from settled. Discount rates used by various federal agencies in recent years have ranged from 10% by the Office of Management and Budget to 2% by the Congressional Budget Office (69).

ECONOMIC EFFICIENCY VERSUS COST EFFECTIVENESS: TRADE-OFFS IN POLICY MAKING

Standard Setting

As the discussion of benefit–cost analysis suggested, identification of the economically efficient level of pollution control is often difficult, if not impossible, to achieve in practice. This stems largely from the fact that the benefits of pollution control are often difficult to measure, let alone quantify in dollars. In addition, benefit–cost analysis is concerned only with the issue of efficiency. No consider-

ation is given to such factors as equity, eg, income redistribution, or the feasibility of the efficient solution. However, these are important concerns for policymakers. In light of these constraints, environmental policy in the United States has tended to rely on the use of specific standards to achieve improvements in environmental quality. The Clean Air Act, for example, requires that primary ambient air quality standards be set at a level that ensures protection of human health, without consideration of the costs that might be incurred in achieving the proposed standard. Although establishment of a standard circumvents the problem of identifying the efficient solution, ie, the optimal amount of pollution reduction, the question of how to achieve the standard remains. Any one (or combination) of a number of different policy options could be employed. However, from the perspective of efficiency, the objective is to identify the policy option that achieves the standard at least cost.

For a particular standard to be met at least cost, the marginal costs of compliance must be equal across all of the affected firms. To see this, assume that two different firms (A and B) are the only generators of a particular pollutant and that between them they currently generate a total of 50 units of pollution per time period. Assume also that the control authority (eg, EPA) has determined that the aggregate amount of the pollutant generated per time period should be limited to 34 units, ie, the total amount of pollution must be reduced by 16 units. Figure 5 depicts the marginal costs of pollution control incurred by each of the two firms. Marginal cost is measured on the vertical axis and pollution control is measured on the horizontal axis. The amount of pollution control undertaken by firm A is measured moving from left to right. The amount of pollution control by firm B is measured moving from right to left.

According to Figure 5, requiring firm A to control 5 units of pollution and firm B to control 11 units would result in the control of 16 units (the reduction necessary to meet the standard). With this allocation of pollution control marginal costs of control are equal across the two firms and total costs of control equal the area $a + b + c$. A deviation in either direction (eg, to an equal reduction of 8 units by each firm) would increase the marginal costs incurred by one firm by more than the reduction in marginal costs experienced by other firm; total costs increase. Allocating control responsibility equally across the two firms would result in total control costs of $a + b + c + d$, or a net increase of d. Thus the total costs of pollution control are minimized by allocating pollution control such that the marginal control costs of the affected firms are equal. The equal marginal cost rule for cost minimization extends to any number of firms.

Policy Options

Following one approach (70), it is useful to think of policy in terms of distinct components: goals, objectives, and instruments. In the case of pollution, the goal of the different policies that could be implemented might be described as "an improvement in the overall level of environmental quality." To this end, policies can be designed to achieve different objectives such as a reduction in the quantity or harmful characteristics of a particular water pollutant, or

Figure 5. Determining the cost-effective allocation of pollution control. The marginal costs of pollution control incurred by each of two firms are shown. The amount of pollution control undertaken by firm A is measured moving from left to right. The amount of control by firm B is measured moving from right to left. Assume that 16 units of pollution must be eliminated. Requiring firm A to control 5 units of pollution and firm B to control 11 units would result in the control of 16 units. With this allocation of pollution control marginal costs of control are equal across the two firms and total costs of control equal the area $a + b + c$. A deviation in either direction, eg, to an equal reduction of 8 units by each firm, would increase the marginal costs incurred by one firm by more than the reduction in marginal costs experienced by the other firm; total costs increase.

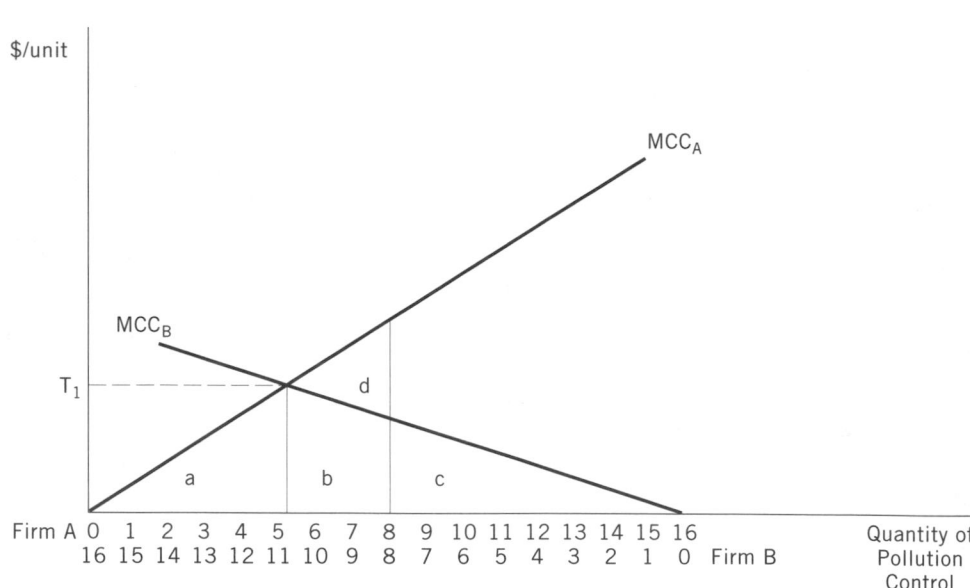

an increase in air quality. Instruments refer to the specific methods that are used to achieve the policy's objectives and, ultimately, the policy goals. In the case of pollution control policies, the primary instruments include direct regulation, charges, and permits. Legal remedies can also be used (4,8,11).

Direct Regulation. Historically, direct regulation has constituted the dominant approach to achieving improvements in environmental quality. Also referred to as command and control (CAC), direct regulation may, eg, require firms to use a particular production process or one that achieves a specified level of performance, reduce emissions or effluents to specified levels, or otherwise meet specific criteria. In addition, direct regulation generally requires that all affected firms meet the same criteria, eg, restrict emissions of a particular air pollutant below some threshold level.

One of the advantages of direct regulation is its relative simplicity. Once a standard has been established, each affected firm is responsible for meeting it. However, if the standard is applied uniformly across affected firms, it is almost certain that it will not be cost effective. The exception is the case in which all of the affected firms have identical marginal control costs. In this special case, a uniformly applied standard would result in marginal costs that are equal across the affected firms and the outcome would be cost effective. The upshot is that, in most cases, for direct regulation to be cost effective it is necessary to require the affected firms to control pollution by varying amounts. To be specific, policymakers would have to ascertain the marginal costs of pollution control of each of the affected firms and then impose a standard on each firm that achieves the cost-effective outcome. In many, if not virtually all, cases this approach would require a substantial amount of data that are in all likelihood not available to regulators. It is also reasonable to expect that this would involve the expenditure of significant resources

by the regulatory authority and be time-consuming as well.

Two additional points regarding the CAC approach are worth noting. The first concerns the distinction between design criteria and performance criteria. Design criteria specify technologies, inputs, chemical processes, and so forth that must be employed by affected firms. In contrast, performance criteria specify a necessary end result that can be achieved by any means, presently available or yet to be developed, available to the firm. Performance standards possess significant advantages relative to design criteria, especially when dealing with complex production processes. In particular, performance standards allow, and encourage, firms to innovate in the effort to achieve a particular standard at least cost.

A second point is that, relative to the incentives-based approaches discussed below, attempting to achieve a standard via direct regulation is likely to result in a greater amount of pollution control than attempts to achieve the same standard using an incentives-based mechanisms. Thus, *ceteris paribus*, the benefits associated with the use of CAC are likely to be larger as well. This fact must be considered when comparing the cost-effectiveness of alternative policy strategies.

Charges. Economists have identified alternatives to direct regulation that rely on market-based incentives (ie, prices) to control pollution. One approach forces firms to account for the external costs they generate by imposing a per unit charge (tax) on the firm equal to the external cost of the firm's production activities (1); more recent discussions of the use of charges to control pollution are available (7,8,71). In particular, a charge could be levied on each unit of pollution (emissions) produced by the firm. Such a charge affects the firm's decision-making process by altering the relative prices of the inputs to the production process. Referring to the previous example, by imposing a charge equal to T_1 (the level at which the firms' marginal control costs are equal) on each of the affected firms, pollution would be reduced by 16 units. To ensure that the standard is met in a cost-effective manner, the charge levied on each unit of pollution generated by a particular source should be a function of whether the pollutant is uniformly or nonuniformly mixed. A discussion of the difference between uniformly and nonuniformly mixed pollutants and how this issue affects the determination of the appropriate charge is available (72).

Firm A would reduce its pollution by 5 units because the marginal cost of control is less than the tax on the first 5 units of waste. However, paying the tax is cheaper than reducing waste beyond 5 units. Using the same logic, firm B would choose to reduce the amount of pollution it generates by 11 units. Thus the charge leads firms to undertake the cost-effective allocation of pollution control.

In order for the use of charges to result in the desired level of pollution control, the control authority must identify the appropriate level of the charge. To accomplish this, policymakers must have information on the control costs incurred by the affected firms to be able to predict the amount of pollution reduction a charge of a given magnitude is likely to induce. A charge that is set too high will result in overcontrol of pollution relative to the standard. If the charge is set too low, the opposite outcome will occur. An iterative process that entails adjusting the charge until the standard is met could be used to overcome this problem. However, this approach is politically unattractive and may have adverse incentives effects on firms who must periodically modify their pollution control strategies in response to changes in charge rates.

Permits. A second incentives-based approach to pollution control policy entails the creation of a market for permits to pollute. Description of this approach are available (3,7,8,73). A system of transferable permits entails the creation of a market in which a predetermined number of permits to emit a specific pollutant within a well-defined geographic region can be bought and sold. Like charges, the use of transferable permits results in a cost-effective allocation of pollution control across affected firms. Referring once again to the example illustrated in Figure 5, the regulatory authority would issue a total of 34 permits, assuming each permit represents one unit of pollution. Thus the two firms must reduce their emissions by a combined total of 16 units for the standard to be met. Regardless of the initial allocation of the permits, firms would find it in their interest to trade permits until the per unit price of the permits was equal to the marginal cost of control for both firms. The firms would in turn engage in the same amount of pollution reduction that they did under the charge scheme. As in the case of charges, to ensure that the standard is met in a cost-effective manner, the amount of pollution that each permit allows the holder to emit should be a function of whether the pollutant is uniformly or nonuniformly mixed (74).

Comparison of Alternative Instruments. The use of permits offers a number of attractive features relative to both direct regulation and charges. Compared with direct regulation, permits allow firms to achieve the target level of control in a cost-effective manner. In addition, permits (as well as charges) have the added advantage of offering firms a greater degree of flexibility with respect to how they will comply with the standard. Compared with charges, permits offer a greater degree of certainty with respect the amount of pollution control that will occur. In addition, in the case in which an industry is growing, permits ensure that the amount of pollution that is generated will remain fixed at the target level. In the case of charges, total pollution generated is likely to increase with an increase in the number of firms in the affected market. A potential drawback of permits is that there may be less incentive to reduce control costs than there would be under a charge system. Additional discussion of these issues can be found (4,75,76).

The potential cost savings from the use of incentives-based approaches has been the subject of numerous studies (73). The results of many early studies suggested that the use of charges or permits to achieve a specific standard could result in substantial cost savings relative to the use of direct regulation. For example, studies of various air pollution control policies have estimated that the use of direct regulation could result in total costs as much as 22 times greater than the costs that would be incurred under a system of permits (77). However, more recent work suggests that in many instances the predicted cost savings attributable to incentives-based systems, and in particular permits, may greatly exceed actual savings. As has been described (78), this outcome is the result of differences between the manner in which permit trading is assumed to occur (multilateral and simultaneously) and the actual trading process (which is more likely to be bilateral and sequential). Nonetheless, market-based incentives continue to be preferred to direct regulation in most settings on efficiency grounds.

CURRENT POLICY

The majority of pollution control laws that have been passed over time are directed at specific environmental media. The Clean Air Act and Water Pollution Control Act (and their respective amendments) govern emissions to the air and water ways. The Resource Conservation and Recovery Act, as amended, governs the land disposal of wastes and, in particular, hazardous wastes. Other laws, such as the Toxic Substances Control Act; the Comprehensive Environmental Response, Compensation, and Liability Act; the Federal Insecticide, Fungicide, and Rodenticide Act; the Safe Water Drinking Act; and their respective amendments are directed at specific sources of pollution. A complete discussion of the various aspects of this body of policy is obviously well beyond the scope of this article. The discussion that follows is intended only to highlight some of the major features of these laws and provide references to more complete discussions of environmental policy in the United States.

Air Pollution Control

At the federal level, air pollution control policy (8,10) dates back to the Air Pollution Control Act passed in 1955. Since that time, the federal approach to air pollution control responsibility has evolved considerably. Before 1970, policy relied heavily on the individual states to undertake steps to reduce the amount of air emissions. The role of the federal government was restricted primarily to funding research on the effects of air pollution and air pollution control strategies, and assistance to the individual states in support of their air pollution control efforts. However, once it became apparent that economic incentives and other factors were preventing states from assuming an active role, the federal government took the lead in developing policies designed to reduce the amount of air pollution in the United States.

The passage of the Clean Air Act amendments of 1970 marked a significant change in air pollution control policy, with the federal government assuming much more responsibility than it had previously. Three features of the 1970 amendments are particularly noteworthy. First, the amendments called on the administrator of the EPA to establish the National Ambient Air Quality Standards for a number of specific air pollutants. These standards were to be set so as to provide "an adequate margin of safety" without regard to the costs of achieving the standard. Second, uniform emissions standards were established for mobile sources of air emission, ie, cars. Third, in the case of new stationary sources of air emissions (eg, factories and electric utilities), standards were to be based on the best available technology that was deemed to be affordable. Thus Congress did allow for some consideration of costs in the setting of new source performance standard (NSPS). However, referring to the previous discussion of the conditions necessary for efficiency or cost-effectiveness, it is clear the requirements just described fail on both counts.

In spite of the limitations posed by the language in the Clean Air Act, the EPA has modified its air pollution control program over time in a number of innovative ways. Three innovations, which are parts of the EPA's emissions trading program and reflect a move toward greater use of economic incentives include the offset program, the bubble policy, and emission banking.

The offset program allows new air pollution sources in a particular geographic region to pay existing sources to reduce their emissions below the level required by the standard in lieu of installing control technology at the new plant so long as the total level of the pollutant does not increase (and, in fact decreases). Thus new sources are able to exploit the least-cost approach to pollution reduction in the geographic region so long as the air quality in the region is better after the new plant is established. The bubble policy allows a generator (or group of generators) to treat a number of closely situated sources, usually within a given production facility, as if they were encased in a giant bubble. The standard then applies to emissions coming out of the bubble. Firms are able to reduce emissions from those sources within the production facility that are the cheapest to control and to relax controls on the more costly sources, thus reducing total control costs. The emission banking program gives firms credits for reducing emissions below the emissions level allowed by the standard. These credits can then be saved for use at a later time or sold to another firm. This approach gives existing firms an incentive to adopt new, cost-saving pollution-control technologies.

The Clean Air Act Amendments of 1990. The Clean Air Act Amendments (CAAA) of 1990 contain five main sections that deal with the following topics: motor vehicles and fuels, urban air quality, air toxics, acid deposition, and the problems resulting from stratospheric ozone. In many respects, the CAAA are similar to previous legislation. For example, the use of CAC is still the principle policy instrument, and control of criteria pollutants is a major issue (as was the case with legislation passed in both 1970 and 1977). However, Title IV, which is directed at the control of SO_2 and NO_x emissions from coal-burning electric utilities, provides for the use of allowances (permits) to control SO_2 emissions. In addition, the CAAA identify 189 specific toxic air emissions for which the EPA is to set emissions standards. Discussions of the major features of the CAAA of 1990 and their possible impacts can be found (79,80).

Title IV: Allowances for SO₂ Emissions. The objective of Title IV of the CAAA is to reduce SO_2 emissions to roughly 50% of their current level, or 8.9 million tons per year. This objective is to be achieved, in large part, by restricting the total amount of SO_2 emissions from electric utilities. The requirements of Title IV will be implemented in two phases. Phase I, which runs from January 1, 1995, to December 31, 1999, will affect 110 utility plants. Phase II, which begins on January 1, 2000, will affect all of the remaining fossil-fuel burning plants. In each phase allowances, or permits, for SO_2 emissions will be allocated to each of the affected firms. Each allowance represents the right to emit one ton of SO_2 in a specified calendar year. Allowances can be used to offset current emissions, sold to other entities, or saved for use at a later date.

According to a recent report (80), many of the utilities affected during Phase I plan to comply with the new restrictions on SO_2 by using a number of approaches. The most common response appears to be the use of coal blending and fuel switching. Other approaches include the installation of flue gas desulfurization, reduced utilization, altered dispatch ordering, and purchase of allowances. Somewhat surprisingly, it appears that the use of allowances purchased from other plants will be relatively limited. Possible explanations from this include uncertainty with respect to the regulatory treatment of allowances and future regulatory actions regarding hazardous air pollutants and CO_2 emissions.

Water Pollution Control

Water pollution control policy at the federal level dates back to the Water Pollution Control Act of 1948. A concise summary of the history of water pollution control policy has been presented (81). Additional discussions of the economic aspects of water pollution can be found (4,8,11). Water pollution is attributable to a variety of sources. Rivers and lakes have been used as natural conduits for many types of commercial and industrial wastes. Agricultural practices have contributed to water pollution through runoff containing pesticides, herbicides, and fertilizers as well as groundwater contamination as water leaches down through the soil. Leachate contamination of groundwater has also been linked to various waste management practices, eg, landfills, and deep well injection. Sewage treatment plants can also contribute to water pollution depending on the level of biological oxygen demand present in the wastes that are ultimately released from the plant.

With respect to energy-related issues, water pollution problems and impacts on the quality of water resources are most obvious in the cases of acid deposition, nuclear power generation, and hydro power. One of the most common illustrations of the potential effects of acid deposition is the status of many of the lakes in the Adirondacks. Acid rain has been credited with rendering many of the lakes in this region and elsewhere unable to support aquatic life. While acid deposition has been traced to SO_2 emissions and is therefore addressed via air pollution policy, it nonetheless has significant adverse effects on water quality. The cooling requirements of nuclear-power and coal-fired plants can also have water-related impacts. For example, as a result of a significant difference in temperature, the discharge water from plants built next to a lake or river may affect the ecosystem of the receiving water body. Such concerns have had significant impacts on the siting and licensing process for power plants. In the case of hydropower generation, the construction of a dam and the resulting reservoir has various irreversible effects on the surrounding environment. While some of these impacts are viewed as costs (eg, loss of scenic vistas, range land, or other uses of the flooded area) many benefits also arise, such as the increased opportunities for water-based recreation.

Land Pollution

Compared with air and water pollution, land-based pollution received relatively little attention until the passage of the Resource Conservation and Recovery Act (RCRA) in 1976. The primary thrust of RCRA is to control the generation and disposal of hazardous wastes from "cradle to grave." In the time since RCRA was passed (and amended in 1984) the EPA and the states have developed a large body of regulations designed to reduce the risks posed by hazardous wastes.

Energy production facilities such as coal-fired and nuclear-powered electric plants pose special problems with respect to land-based waste management. In 1993, the EPA published its final regulatory determination for certain coal combustion wastes (eg, fly ash, bottom ash, slag, and FGD wastes) exempting the wastes from RCRA Subtitle C hazardous waste regulations. However, EPA retained the option of considering these wastes in its ongoing assessment of industrial nonhazardous wastes under Subtitle D. Consequently, these wastes could be subject to requirements that are more stringent than existing regulations under Subtitle D as a result of future actions by the agency. In addition, the determination did not address low volume, potentially hazardous "clean coal technology" wastes. Additional information must be acquired

before a determination can be made regarding these wastes. One possible outcome is that these wastes could be classified as hazardous in the future.

High level radioactive wastes, which are the by-product of nuclear-powered generation of electricity, are governed by the Nuclear Waste Policy Act (NWPA) of 1982. Although the NWPA requires the U.S. Department of Energy to develop an underground depository for high level wastes, the United States is still without a single repository for such wastes. Attempts to site such a facility, most recently in Nevada near Yucca Mountain, have met considerable opposition from the public and many local government officials. Low level radioactive wastes are also posing serious problems. At present, there are only three repositories for such wastes in the United States and these facilities are filling rapidly. Moreover, the prospects for developing additional sites look bleak (82).

Impacts of Pollution Control Policy

Viewed from an economic perspective, the primary interest is in the relative costs and benefits that have been realized over time. According to a study by the EPA (83) annualized environmental compliance costs in 1990 (measured in 1986 dollars) were approximately $28 billion for air pollution, $42.4 billion for water pollution, and $26.5 billion for land pollution. Other environmental statutes added another $1.6 billion to the total. However, these figures only reflect out-of-pocket expenditures and as such are likely to understate the true economic costs of pollution control. While estimation of the benefits that have accrued over time is even more difficult, the evidence that has been collected suggests that total air pollution control benefits exceeded total costs through 1981 (84). However, water pollution is another matter; available estimates suggest that total costs have exceeded total benefits (85). In both instances, there is considerable room for improvement vis-à-vis cost effectiveness.

CASE STUDY: THE USE OF ADDERS IN LEAST-COST UTILITY PLANNING

Over the past decade the concepts of least-cost utility planning and integrated resource planning (IRP) have become important components in the regulation of electric utilities. The objective of IRP is to identify the mix of energy sources and demand-side management (DSM) techniques that satisfies an expected level of demand and reliability requirements at a lower cost than any other possible combination of energy sources and DSM. As a rule, in those jurisdictions where IRP is used, each utility is required to formulate an integrated resource plan for meeting its obligation to serve.

As was discussed earlier, efficiency requires that the price of a good (in this case, electrical service) reflect all of the marginal costs, both internal and external, associated with the production of that good. Thus electricity produced by a coal-fired unit should be priced to include both the internal costs, eg, fuel and labor expenses, and the external costs resulting from, among other things, the

SO, NO_x, CO, CO_2, and particulate emissions from the unit. Assuming that output is priced on the basis of marginal costs, failure to include any external costs will result in a price for coal-fired electricity that is too low. To be specific, price will be less than full marginal costs of production. Moreover, the relative price of electricity produced by a coal-fired unit (ie, its price relative to the price of electricity produced by other types of generating units) may be understated, depending on how externalities attributable to other sources are treated. The result is that this source of electricity could make up too large a proportion of the utility's energy portfolio, ceteris paribus.

Numerous states have attempted to address the problem of externalities resulting from the production of electricity by requiring utilities to explicitly consider external costs in the process of planning capacity expansions. As of 1991, 19 states had implemented such requirements (86). Such costs have come to be referred to as "externality adders." In some cases, estimates of the external costs of different types of emissions, measured, eg, in dollars per kilowatt hour, are added to production costs to arrive at a total cost figure. In other cases, the cost of additional pollution control equipment is added to production costs to account for external costs. However, such efforts have been criticized by different agencies, including the Department of Energy (DOE) and the Federal Energy Regulatory Commission (FERC) (87), and numerous analysts (88,89). Three issues are addressed briefly here: (*1*) the validity of existing estimates of external costs of pollution and the methods used to estimate such costs, (*2*) the potential distortionary effects of adders, and (*3*) the implications of existing pollution-control policies for determining the correct value of adders.

Validity of Available Estimates. Questions regarding the validity of available estimates are certainly justified. Reviews of the literature on externalities associated with electricity production have pointed out that many of the existing estimates of external costs are relatively crude approximations that, in many cases, have been derived via flawed research methods (86,90,91). However, the fact that there are questions about how well existing estimates approximate actual external costs is not, in and of itself, sufficient support for the argument that until better estimates are forthcoming, public utility commissions (PUCs) should simply ignore external costs in the planning process. Rather, it emphasizes the need for the use of extreme care in how such estimates are used.

The question of whether existing estimates are a good approximation of actual costs notwithstanding, the different methods that have been used to generate estimates of external costs merit further consideration. In practice, three approaches have been used: mitigation or averting costs, control costs, and the damage function approach (86,90). Of these, mitigation costs and the damage function approach have the most to recommend them from a theoretical perspective. In addition, the damage function approach is the only one that attempts to relate costs directly to the damages attributable to the pollutant in question.

In some cases, the costs that would be incurred to mitigate or avert the potential damages attributable to a pol-
lutant have been used to measure damage costs. For example, potential damages to agricultural crops might be averted by increased use of fertilizer, water, or some other input (4). The value of the additional resources required to offset the potential damage constitutes the damage costs attributable to the pollutant, ie, the additional production costs constitute the damages incurred by farmers. However, in many cases, opportunities to offset fully the adverse effects of a pollutant are unavailable, and hence this approach cannot be used.

As a second approach, some analysts have suggested that pollution control costs already incurred provide insights to the marginal damage costs of certain pollutants (92). This assertion is based on the assumption that because policymakers have decided to require the existing level of control, it is reasonable to assume that the perceived marginal benefits of the existing level of control (ie, marginal damages avoided) are equal to or exceed the marginal costs of control, ie, policymakers are behaving in an efficiency enhancing manner. However, this argument has no sound basis on which to stand. In many, if not most, instances, pollution standards have been based on considerations other than economic efficiency. For example, according to the Clean Air Act, the standards for criteria pollutants are to be set so as to ensure the health and safety of the affected population. The act specifically states that control costs are not to be considered in setting these standards. A number of other environmental statutes also explicitly exclude the consideration of costs or benefits in the decision-making process. Thus there is no reason to believe that marginal control costs are in any way related to marginal damage costs.

The theoretically valid approach to estimation of the external costs of pollution is to rely on the use of damage functions and measures of the corresponding willingness to pay for a reduction in damage. Admittedly, this approach is anthropocentric, ie, assumes that all values should be based on the relationship between a specific damage and its effects on the welfare of humans. As such, pollution is only considered a problem to the extent that any damages from the pollution have economic value. Under this approach, the amount of damage and resulting value of that damage attributable to a particular pollutant are determined by the interaction of a number of factors.

As a practical matter, the damage function approach entails several complex steps, the first of which is to identify the specific types of damages that could be attributed to the pollutant in question. The second step is to estimate the relationship between the quantity of the pollutant and the resulting level of damages. In the third step, the relationship estimated in step two is used to estimate the level of damages associated with a given level of the pollutant. The fourth step entails assigning a monetary value to the damages calculated in step three. Each of these steps is likely to be data intensive. In addition, there is likely to be considerable uncertainty regarding the functional form of the damage function and the estimates of the monetary value of damages.

Data limitations (with respect to both quantity and quality) have the potential to affect each of the steps outlined above. For example, to be able to relate a pollutant to various damages, all of the potential damages that could be linked to the pollutant must be identified. However, in many cases, latency is a problem, ie, many adverse effects do not manifest themselves for many years. As such, there may be a considerable time lag between exposure to a pollutant and the occurrence of adverse effects. Limited data will also affect estimation of the damage function and, therefore, estimates of the amount of damage attributable to a given quantity, eg, concentration, of a pollutant. Finally, data limitations with respect to willingness to pay to avoid damages constrain the estimation of the monetary value of the damages.

Data limitations and other uncertainties notwithstanding, the damage function approach focuses directly on the link between pollutants and damages, measured in both physical and monetary terms. Thus, of the three approaches that have been discussed here, it is the most defensible from a theoretical perspective. In addition, it is the only method that is applicable, at least in theory, to all of the types of damages that might be linked to electricity production. It is for this reason that agencies such as the DOE and FERC have decided to rely on the damage function approach to estimate the external costs of electricity production (87).

Potential Distortionary Effects. A second issue concerns the distortionary effects that the use of externality adders might introduce in the IRP process (88). To be specific, external costs currently are considered only in the case of new capacity brought on line by independently owned utilities (IOUs), ie, utilities whose rates are governed by PUCs. Although the production activities of other sources of electrical power, such as cogeneration facilities and utilities located outside a particular PUC's jurisdiction, may also result in external costs, these sources are not subject to the same requirements. In the case of utilities located outside of the PUC's jurisdiction, the exception would be those instances in which the state where the "outside" utility is located also requires consideration of external costs. However, external costs per unit of output would have to be measured using the same method to ensure that no distortion in relative costs occurs.

Forcing IOUs to include external costs when determining the marginal cost of power could put them at a competitive disadvantage relative to alternative suppliers solely because the alternative suppliers do not have to factor external costs into their marginal cost calculations. The consequent distortion of relative costs could result in alternative suppliers producing a larger share of the total amount of electricity consumed in the region. If these alternative suppliers also produce greater amounts of pollution per unit of output than the regulated IOU, the result could be an increase in the total amount of air pollution in the region. Obviously, this is not the intended result of the use of externality adders.

Impact of the Current Regulatory Structure. The fact that electric utilities have already undertaken efforts to internalize at least some of the external costs resulting from the generation of electricity also has important implications for the socially efficient quantities of electricity and corresponding pollution. Over the past two decades elec-

tric utilities have undertaken a number of steps, including the installation of flue gas desulfurization (FGD) units, coal cleaning, and the adoption of newer "clean coal" or "green field" technologies that result in reduced emissions of SO_2, NO_x, particulates, and other air-borne pollutants. The effect of these efforts has been to internalize at least some amount of what were previously external costs. In the context of Figure 2, such efforts have shifted the MPC curve to the left, toward the MSC curve.

Pollution control efforts also have affected the marginal external costs attributable to electricity production. Referring again to Figure 2, the value of marginal external costs is based on preregulatory conditions; internal costs and a specific relationship between the firm's production function and the amount of pollution that is produced per unit of output in the absence of pollution control policies. The amount of pollution produced per unit of output and the value of the damages attributable to each unit of pollution combine to determine the value of external costs at each level of output. Depending on the actions taken by a utility in response to legislative and regulatory mandates directed at pollution reduction, marginal external costs per unit output are likely to have decreased (93). This would be the case if, eg, the types of pollutants produced are unchanged, but there is a reduction in the quantity of one or more of the pollutants produced per unit of output.

The preceding discussion suggests that marginal external costs and the amount of inefficiency attributable to a failure to account for such costs in the decision-making process in the electricity sector have both decreased as a result of regulations implemented to date. However, without knowing the actual value of the marginal external costs associated with electricity production and the marginal costs of pollution control it is not possible to determine whether the resulting level of pollution is efficient. Three outcomes are possible; the efficient level of pollution, undercontrol relative to the efficient level, and overcontrol relative to the efficient level. As has been demonstrated (89), and summarized below, the correct value of any adder that should be included in the calculation of the marginal social costs of electricity will depend on the type of policy instrument that has been used.

In the first case, assuming pollution control responsibility has been allocated across firms in a cost-effective manner (marginal control costs are equal across the affected firms) and the amount of control has been set at the efficient level, firms are producing the socially efficient output level. However, depending on the policy that was used to achieve this outcome, any remaining external costs may or may not be internalized, ie, they may or may not be included in the firm's decision-making process. If pollution control has been achieved via direct regulation, any remaining external costs should be added to private costs to ensure that total social costs are accounted for. However, if pollution control has been achieved via either charges or permits, the correct adder is zero. In the case of charges, the efficient output level will only occur if the charge is set equal to marginal external cost at the efficient level of output. Thus, if as has been assumed, the efficient solution has been achieved, it follows that the charge is an accurate measure of marginal damage costs.

Because the charge is paid for each unit of pollution emitted by the firm, external costs have been internalized. In the case of permits, so long as the rules governing the trading of permits reflect the characteristics of the pollutant being regulated, all external costs have been internalized as well; determination of the appropriate trading rules depends primarily on the characteristics of the pollutant in question (8). Incorporating adders in either of these latter two cases would result in double counting of external costs.

A second possibility is that the current level of pollution control is less than the efficient level. In this case, if current policy consists of direct regulation, the external cost attributable to any remaining pollution would once again constitute the correct value of the adder. In the case in which charges are being used to control pollution, the adder should be set equal to the difference between the current charge and actual external costs. Under a system of permits, and assuming the correct rules for trading are in place, the correct value for the adder is once again zero. In this latter case, the appropriate response to the undercontrol of pollution would be to decrease the number of permits to a level consistent with the economically efficient level of pollution control.

The third possibility is that there is excess pollution control. In this case, and assuming policy consists of direct regulation, the external cost attributable to any remaining pollution would once again constitute the correct value for the adder. While one might be tempted to argue for a negative adder to compensate for the overcontrol attributable to excessive restrictions on pollution, this would only introduce further distortions into the market. In the case in which charges are being used to control pollution, the adder should be set equal to the difference between the current charge and actual external costs; in the case of overcontrol this amount will be negative. Under a system of permits, and assuming the correct rules for trading, the correct value for the adder is once again zero. In this latter case, the appropriate response to the overcontrol of pollution would be to increase the number of permits to a level consistent with the economically efficient level of pollution control.

Presumably, the purpose of adders is to improve the overall efficiency of the electricity sector by accounting for external costs in the decision-making process. However, as the preceding discussion illustrates there are a number of important issues that must be addressed for adders to achieve that purpose. Failure to do so could result in even greater inefficiencies.

BIBLIOGRAPHY

1. A. C. Pigou, *The Economics of Welfare*, 4th ed., Macmillan, London, 1932.
2. A. V. Kneese and B. T. Bower, *Managing Water Quality: Economics, Technology, Institutions*, Johns Hopkins University Press for Resources for the Future, Inc., Baltimore, Md., 1968.
3. J. H. Dales, *Pollution, Property, and Prices*, University of Toronto Press, Toronto, Ont., 1968.
4. M. L. Cropper and W. E. Oates, *J. Econ. Lit.*, **30**(2), 675–740 (1992).
5. K. G. Maler, *Environmental Economics: A Theoretical Inquiry*, Johns Hopkins University Press for Resources for the Future, Inc., Baltimore, Md., 1974.
6. W. J. Baumol and W. E. Oates, *Economics, Environmental Policy, and Quality of Life*, Prentice-Hall, Inc., New York, 1979.
7. W. J. Baumol and W. E. Oates, *The Theory of Environmental Policy*, 2nd ed., Cambridge University Press, Cambridge, UK, 1988.
8. T. Tietenberg, *Environmental and Natural Resource Economics*, 3rd ed., HarperCollins, New York, 1992.
9. A. M. Freeman, *Measurement of Environmental and Resource Values: Theories and Methods*, Resources for the Future, Washington, D.C., 1993.
10. P. R. Portney, ed., *Public Policies for Environmental Protection*, Resources for the Future, Washington, D.C., 1990.
11. B. C. Field, *Environmental Economics: An Introduction*, McGraw-Hill Book Co., Inc., New York, 1994.
12. A. E. Dillingham, N. T. Skaggs, and J. L. Carlson, *Economics: Individual Choice and its Consequences*, Allyn & Bacon, Needham Heights, Mass., 1992.
13. P. A. Samuelson and W. D. Nordhaus, *Economics*, 14th ed., McGraw-Hill Book Co., Inc., New York, 1992.
14. W. Nicholson, *Macroeconomic Theory: Basic Principles and Extensions*, 2nd. ed., Dryden Press, Hinsdale, Ill., 1978.
15. E. Silberberg, *The Structure of Economics: A Mathematical Analysis*, McGraw-Hill Book Co., Inc., New York, 1978.
16. P.-O. Johansson, *An Introduction to Modern Welfare Economics*, Cambridge University Press, New York, 1991.
17. R. Boadway and N. Bruce, *Welfare Economics*, Basil Blackwell, Inc., Cambridge, Mass., 1984.
18. R. E. Just, D. L. Hueth, and A. Schmitz, *Applied Welfare Economics and Public Policy*, Prentice-Hall, Inc., Englewood Cliffs, N.J., 1982.
19. Ref. 18, p. 269.
20. Ref. 18, chapt. 3.
21. Ref. 7, chapt. 4.
22. Ref. 11, p. 28.
23. R. H. Coase, *J. Law Econ.* **3**, 1–44 (1960).
24. *Environmental Quality 1984*, Council on Environmental Quality, Washington, D.C., 1985.
25. Ref. 8, p. 95.
26. U.S. Bureau of Reclamation, *Operation of Glen Canyon Dam Colorado River Storage Project, Arizona*, Draft Environmental Impact Statement, Salt Lake City, 1993.
27. J. L. Carlson and co-workers, *An Economic Study on Proposed IPCB Regulation R86-9: Hazardous Waste Prohibitions*, ILENR/RE-EA-90/10, Illinois Department of Energy and Natural Resources, Springfield, 1990.
28. T. G. Walton and A. C. Basala, *Cost-Effectiveness Analysis and Environmental Quality Management*, paper presented at the 74th meeting of the Air Pollution Control Association, Philadelphia, 1981.
29. E. J. Mishan, *Cost-Benefit Analysis*, 2nd. ed., Praeger, New York, 1976.
30. R. Sugden and A. Williams, *The Principles of Practical Cost-Benefit Analysis*, Oxford University Press, Oxford, UK, 1986.
31. E. M. Gramlich, *Benefit-Cost Analysis of Government Programs*, Prentice-Hall Inc., Englewood Cliffs, N.J., 1981.
32. Ref. 8, chapt. 4.
33. Ref. 8, p. 365.
34. C. Starr, *Science* **165**, 1232–1238 (1969).

35. D. Kahneman and A. Tversky, *Econometrica* **47**(2), 263–291 (1979).
36. M. C. Weinstein and R. J. Quinn, *Natural Resource J.* **23**, 659–673 (1983).
37. Ref. 4, pp. 721–722.
38. Ref. 8, pp. 84–86.
39. J. B. Braden and co-workers, *Pollution Control Cost Analysis*, U.S. Environmental Protection Agency, Washington, D.C., 1985.
40. Z. Griliches, ed., *Price Indexes and Quality Change*, Harvard University Press, Cambridge, Mass., 1971.
41. S. Rosen, *J. Political Econ.* **82**, 34–55 (1974).
42. P.-O. Johansson, *The Economic Theory and Measurement of Environmental Benefits*, Cambridge University Press, Cambridge, UK, 1987.
43. D. S. Brookshire and co-workers in V. K. Smith, ed., *Advances in Applied Microeconomics*, Vol. 1, JAI Press, Greenwich, Conn., 1981.
44. R. Arnould and L. Nichols, *J. Political Econ.* **91**(2), 332–340 (1983).
45. W. K. Viscusi, *Rev. Econ. Statistics* **60**(3), 408–416 (1978).
46. Ref. 42, p. 111.
47. M. Clawson and J. L. Knetsch, *Economics of Outdoor Recreation*, Johns Hopkins University Press for Resources for the Future, Baltimore, Md., 1966.
48. W. J. Vaughn and C. S. Russell, *Land Econ.* **58**, 450–463 (1982).
49. N. E. Bockstael, K. E. McConnell, and I. E. Strand in J. B. Braden and C. D. Kolstad, eds., *Measuring the Demand for Environmental Quality*, North-Holland, Amsterdam, 1991.
50. V. K. Smith and Y. Kaoru, *Am. J. Agri. Econ.* **72**, 419–433 (1990).
51. R. C. Mitchell and R. T. Carson, *Using Surveys to Value Public Goods: The Contingent Valuation Method*, Resources for the Future, Washington, D.C., 1989.
52. R. G. Cummings, D. S. Brookshire, and W. D. Schultze, *Valuing Public Goods: The Contingent Valuation Method*, Rowman & Allanheld, Totwa, N.J., 1986.
53. Ref. 51, chapts. 5 and 11.
54. D. Kahneman and J. L. Knetsch, *J. Environ. Econ. Manage.* **22**(1), 57–70 (1992).
55. V. K. Smith, *J. Environ. Econ. Manage.* **22**(1), 71–89 (1992).
56. R. Gregory, S. Lichtenstein, and P. Slovic, *J. Risk Uncertainty* **7**, 177–197 (1993).
57. B. A. Weisbrod, *Q. J. Econ.* **78**, 471–477 (1964).
58. V. K. Smith, *South. Econ. J.* **54**, 19–26 (1987).
59. D. S. Brookshire, L. S. Eubanks, and C. S. Sorg, *Water Resources Res.* **22**, 1509–1518 (1986).
60. D. H. Rosenthal, D. H. and R. H. Nelson, *J. Policy Anal. Manage.* **11**(1), 116–122 (1992).
61. J. Quiggin, *J. Policy Anal. Manage.* **12**(1), 195–199 (1993).
62. A. Randall in Ref. 49.
63. R. C. Bishop and M. P. Welsh, *Land Econ.* **68**(4), 405–417 (1992).
64. R. Kopp, *J. Policy Anal. Manage.* **11**(1), 123–130 (1992).
65. A. Fisher, and R. Raucher in V. K. Smith, ed., *Advances in Applied Micro-Economics*, Vol. 3, JAI Press, Greewich, Conn., 1984, pp. 37–66.
66. Ref. 65, p. 60.
67. R. C. Lind, *J. Environ. Econ. Manage.* **18**(2), S-29–S-50 (1990).
68. W. R. Cline, *The Economics of Global Warming*, Institute for International Economics, Washington, D.C., 1992.
69. Ref. 11, p. 122.
70. A. Breton, *The Economic Theory of Representative Government*, Aldine, Chicago, 1974.
71. F. R. Anderson and co-workers, *Environmental Improvement Through Economic Incentives*, Johns Hopkins University Press, Baltimore, Md., 1977.
72. Ref. 8, pp. 372–380.
73. T. H. Tietenberg, *Emissions Trading: An Exercise in Reforming Pollution Policy*, Resources for the Future, Washington, D.C., 1985.
74. Ref. 8, pp. 375–382.
75. M. L. Weitzman, *Rev. Econ. Studies* **41**(4), 477–491 (1974).
76. P. Bohm and C. F. Russell, in A. V. Kneese and J. L. Sweeney, eds., *Handbook of Natural Resource and Energy and Economics*, Vol. 1, North-Holland, Amsterdam, 1985.
77. Ref. 10, pp. 70–74.
78. S. Atkinson and T. H. Tietenberg, *J. Environ. Econ. Manage.* **21**(1), 17–31 (1991).
79. Ref. 11, pp. 295–298.
80. Bailey and co-workers, *Examination of Utility Phase I Compliance Choices and State Reactions to Title IV of the Clean Air Act Amendments of 1990*, ANL/DIS/TM-2, Argonne National Laboratory, Argonne, Ill., 1993.
81. Ref. 10, chapt. 4.
82. Ref. 11, pp. 342–343.
83. Ref. 4, p. 712.
84. Ref. 10, p. 69.
85. Ref. 10, pp. 122–127.
86. R. L. Ottinger and co-workers, *Environmental Costs of Electricity*, Oceana Publications, New York, 1991.
87. Federal Energy Regulatory Commission Staff, *Report on Section 808, Renewable Energy and Energy Conservation Incentives of the Clean Air Act Amendments of 1990*, FERC, Washington, D.C., 1992.
88. P. L. Joskow, *Elec. J.* **5**, 53–67 (1992).
89. M. A. Freeman and co-workers, *Elec. J.* **5**, 18–25 (1992).
90. C. B. Szpunar and J. L. Gillette, *Environmental Externalities: Applying the Concept to Asian Coal-Based Power Generation*, ANL/EAIS/TM-90, Argonne National Laboratory, Argonne, Ill, 1992.
91. D. Pearce, C. Bann, and S. Georgiou, *The Social Cost of Fuel Cycles*, UK Department of Trade and Industry by the Centre for Social and Economic Research on the Global Environment, London, 1992.
92. P. Chernick and E. Caverhill, *Elec. J.* **4**(2), 46–53 (1991).
93. Ref. 18, pp. 274–275.

ENVIRONMENTAL CONSERVATION ORGANIZATIONS

JIMMY LANGMAN and HEATHER AUYANG
Earth Island Institute
San Francisco, California

Many consider environmental issues to be the most pressing problems confronting humanity in the 1990's. As public concern for the environment increases, so does the need for a comprehensive guide to organizations addressing ecological challenges.

Activists, research, scientific, and legal groups, all devoted to addressing and providing innovative solutions to various facets of the global environmental crisis are listed. Issue areas cover a wide range: from energy to marine protection to land conservation and restoration to population growth.

Fax numbers for many organizations, and in some cases computer mailbox addresses have been included. The Institute for Global Communications, 18 De Boom St., San Francisco, CA, 94107, (415) 442-0220; econet e-mail address igcoffice@igc.apc.org offers more information on computer networking.

Abalone Alliance
2940 16th Street, Room 310
San Francisco, CA 94103
P(415) 861-0592
E-Mail(EcoNet): abalone
Through raising public awareness about the dangers of nuclear power, the ultimate goal of the Alliance is a nuclear-free world.

Acid Rain Foundation, Inc.
1410 Varsity Dr.
Raleigh, NC 27606
P(919) 828-9443
Created to foster a greater understanding of air quality issues, including acid rain, air pollutants, and global climate change, and to help bring about solutions.

African Wildlife Foundation
1717 Massachusetts Avenue, NW
Washington, DC 20036
P(202) 265-8393
F(202) 265-2361
Finances and operates wildlife conservation projects in Africa.

Alaska Conservation Foundation
430 West 7th Ave., Suite 215
Anchorage, AK 99501
P(907) 276-1917
F(907) 274-4145
Provides supplemental financial support to Alaskan groups emphasizing activist projects with an immediate environmental focus.

Alaska Environmental Lobby, Inc.
P.O. Box 22151
Juneau, AK 99802
P(907) 463-3366
A parent organization for 19 Alaskan environmental groups which lobby the Alaskan legislature for better environmental laws.

Alliance for the Wild Rockies
Box 8731
Missoula, MT 59807
Dedicated to preserving wilderness in the Wild Rockies Bioregion; now has 39 member organizations which work at local, state, and national levels to educate the public.

Alternative Energy Resources Organization
44 N. Last Chance Gulch #9
Helena, MT 59874
P(406) 443-7272
Fosters exchange of information between U.S. and Canadian farmers and farm organizations concerning sustainable farming and agricultural management.

American Cetacean Society
P.O. Box 2639
San Pedro, CA 90731-0943
P(213) 548-6279
F(213) 548-6950
Dedicated to the conservation and protection of whales, dolphins, porpoises and the oceans they live in.

American Conservation Association, Inc.
30 Rockefeller Plaza, Rm. 5402
New York, NY 10112
P(212) 649-5600
Preserves and develops natural resources for public use.

American Farmland Trust
1920 N St., NW, Suite 400
Washington, DC 20036
P(202) 659-5170
Dedicated to preserving America's farmland, this organization purchases land and/or its development rights.

American Friends Service Committee
2160 Lake Street
San Francisco, CA 94121
P(415) 752-7766
This independent Quaker committee works through nonviolent means on environmental and social justice issues.

American Forestry Association
1516 P St., NW
Washington, DC 20005
P(202) 667-3300
Educates the public of the best ways to use natural resources.

American Littoral Society
Sandy Hook
Highlands, NJ 07732
P(201) 291-0055
A national organization dedicated to the study and conservation of coastal habitats and a publisher of scientific and popular materials.

American Oceans Campaign
725 Arizona Ave. #102
Santa Monica, CA 90401
P(301) 452-2206
The primary focus of this organization is to establish a National Oceans Policy which will end toxic and sewage dumping in the ocean and stop offshore drilling in sensitive areas.

American Rivers
801 Pennsylvania Ave., SE
Suite 303
Washington, DC 20003-2167
P(202) 547-6900
F(202) 543-6142
E-Mail(EcoNet): amrivers
Dedicated to preserving the rivers and landscapes of the United States.

American Water Resources Association
5410 Grosvenor Ln., Suite 220
Bethesda, MD 20814
P(301) 493-8600
This group works to advance the research, planning, and development of water resources in the United States.

American Wildlands
7500 E. Arapahoe Rd., #355
Englewood, CO 80112
P(303) 771-0380
The primary focus of this group is to preserve the timberlands, wildlands, and wildlife of the west.

Anglers for Clean Water, Inc.
P.O. Box 17900
Montgomery, AL 36141
P(205) 272-9530
Educates the public on the conditions of pollution nationwide and promotes clean streams, lakes, and rivers.

AT Work
300 Broadway, Suite 28
San Francisco, CA 94133
P(415) 788-3666
F(415) 788-7324
E-Mail(EcoNet): atwork
AT Work supports appropriate technology projects in the Third World with materials and skilled volunteers, focusing on the promotion of socially sustainable communities.

Atlantic Center for the Environment
39 S. Main St.
Ipswich, MA 01938
P(508) 356-0038
This organization promotes sustainable development in the Atlantic Region by developing leadership opportunities and providing technical assistance.

Atlantic Salmon Federation
P.O. Box 684
Ipswich, MA 01938
P(506) 529-8889
F(506) 529-4438
Dedicated to bringing the salmon back through stream restoration, public education, and hatcheries.

Better World Society
1100 Seventeenth St., NW Suite 502
Washington, DC 20036
P(202) 659-1833
F(202) 331-3779
Dedicated to developing programming which will foster awareness of global issues of sustainability and promote the benefits of preserving the world's rainforests.

California Action Network
P.O. Box 464
Davis, CA 95617
P(916) 756-8518
This organization focuses on the water uses of the agricultural industry in order to redefine the water allocation policy in California.

California Energy Commission
1516 9th Street, MS-25
Sacramento, CA 95814-5512
P(916) 324-3000
Forecasts energy uses, promotes energy conservation, develops renewable energy resources and alternative energy-generating technologies.

California Energy Extension Service
1400 Tenth St., Rm. 209
Sacramento, CA 95814
P(916) 323-4388
Works to lower energy use for small-scale consumers by energy management projects and demonstrations.

California Marine Mammal Center
Marin Headlands
Golden Gate National Recreation Area
Sausalito, CA 94965
P(415) 331-0161
This organization rescues injured and orphaned marine mammals from the California coast in hopes of returning them to the wild, and conducts research in marine mammal science and medicine.

California Trout
870 Market St., Suite 859
San Francisco, CA 94102
P(415) 392-8887
Dedicated to preserving trout fisheries and habitats in California.

California Wilderness Coalition
2655 Portage Bay East, Suite 5
Davis, CA 95616
P(916) 758-0380
E-mail(EcoNet): jeaton
This group promotes preservation of wildlands throughout California.

Californians Against Waste
909 12th St., Suite 201
Sacramento, CA 95814
P(916) 443-5422

This organization focuses on recycling programs and legislation across the country and publishes a quarterly newsletter, *Waste Watch.*

CARE
660 First Ave.
New York, NY 10016
P(212) 686-3110
F(212) 696-4005
CARE works to improve the livelihoods of the rural poor through sustainable resource management.

CEIP Fund
68 Harrison Ave.
Boston, MA 02111
P(617) 426-4375
F(617) 423-0998
An environmental career organization offering job and intern opportunities at local, state and federal government organizations, corporations and consulting firms, and non-profit organizations.

Catalina Conservancy
P.O. Box 2739
Avalon, CA 90704
P(310) 510-1421
Works to preserve the many native plants and animals of Catalina Island.

Cause for Concern
RD 1, Box 570
Stewartsville, NJ 08886
P(201) 479-6778
Cause for Concern works to increase consumer awareness and promote informed choices of environmentally sound products.

Center for Conservation Biology
Department of Biological Sciences
Stanford University
Stanford, CA 94305
P(415) 723-5924
F(415) 723-5920
E-Mail(EcoNet): conbio
Works on the development of conservation biology and its application to environmental problems.

Center for Economic Conversion
222 View St., Suite C
Mountain View, CA 94041
P(415) 968-8798
F(415) 968-1126
E-Mail(EcoNet): cec
CEC works for the conversion of military production towards projects that encourage environmental restoration.

Center for Environment, Commerce, and Energy
733 6th St., SE, Suite 1
Washington, DC 20003
P(202) 543-3939

The Center focuses on air quality and pollution, water resources and pollution, land use, and toxic substances.

Center for Environmental Education, Inc.
1725 DeSales St., NW
Washington, DC 20036
P(202) 429-5609
Dedicated to the conservation of marine species and their habitats.

Center for Marine Conservation, Inc.
1725 DeSales St., NW, Suite 500
Washington, DC 20036
P(202) 429-5609
F(202) 872-0619
Determines who are major manufacturers of pollution in marine habitats and seeks to bring action against them.

Center for Resource Economics
1718 Connecticut Ave., NW
Suite 300
Washington, DC 20009
P(202) 232-7933
This organization develops, publishes, and markets books concerning global environmental problems.

Central States Resource Center
809 S. Fifth
Champaign, IL 61820
P(217) 344-2371
Grass-roots organization researches hazardous waste, water and transportation issues.

Cetacean Society International
Box 9145
Wethersfield, CT 06109
P(203) 563-6444
Works to protect the oceans and marine inhabitants.

Children & The Environment: UNEP Youth Forum
United Nations
New York, NY 10017
P(212) 963-4931
United Nations committee which works to bring together school children active in environmental issues at an annual conference in New York.

Citizen Action
1300 Connecticut Ave., NW
Washington, DC 20036
P(202) 857-5153
Encourages citizens to increase involvement in environmental, social and economic decisions.

Citizens Against Nuclear Power & Weapons
53 W. Jackson, Suite 1306
Chicago, IL 60604
P(312) 786-9041
Opposed to nuclear power and weapons, this organization educates the public on alternatives.

Citizens Clearinghouse for Hazardous Waste
119 Rowell Court
Falls Church, VA 22046
P(703) 237-2249
Works to halt illegal dumping of hazardous wastes and promote recycling and other alternatives to landfills and incinerators.

Citizens Environmental Task Force
321 Calle Loma Norte
Santa Fe, NM 87501
P(505) 983-2894
F(505) 982-6412
Through education and lobbying this citizens group deals with the management of natural resources.

Citizens for a Better Environment
33 East Congress Parkway
Suite 523
Chicago, IL 60605
P(312) 939-1530
This organization fights environmental health threats.

Citizens for Alternatives to Chemical Contamination
9496 School St.
Lake, MI 48632
P(517) 544-3318
As well as functioning as a clearinghouse to alert citizens to potentially harmful substances, this organization also focuses on education.

Clean Sites, Inc.
1199 North Fairfax St., Suite 400
Alexandria, VA 22314
P(703) 683-8522
F(703) 548-8773
Works to accelerate cleanup of hazardous wastes by assisting cleanup and developing effective public policy.

Clean Water Action
1320 18th St., NW
Washington, DC 20036
P(202) 457-1286
This national organization works for safe water as well as protection of our nation's resources.

Climate Change & Energy Program
1616 P Street, NW
Washington, DC 20036
P(202) 332-0900
F(202) 332-0905
E-Mail(EcoNet)
This organization presses for reduction of carbon dioxide emissions from fossil fuels by working on better energy and transportation policies.

Coalition for New Budget Priorities
43 Samoset St.
Dorchestor, MA 02124
P(617) 727-4596
Demands to cut military spending and to use the money towards social and environmental projects.

Coast Alliance
235 Pennsylvania Ave., SE
Washington, DC 20003
P(202) 546-9554
Increase public awareness of the value of the coast and works to strengthen policies and programs to protect coastal ecosystems.

Coastal Conservation Association, Inc.
4801 Woodway, Suite 220 West
Houston, TX 77056
P(713) 626-4222
This association promotes and conserves marine and plant life along coastal areas of the United States.

Commonweal
P.O. Box 316, 451 Mesa Rd.
Bolinas, CA 94924
P(415) 868-0970
This nonprofit institute fosters projects serving humanity and the Earth.

Community Environmental Council
930 Miramonte Dr.
Santa Barbara, CA 93109
P(805) 963-0583
F(805) 962-9080
Through research and education, the Council promotes recycling of resources and sustainable development.

Concern, Inc.
1794 Columbia Rd., NW
Washington, DC 20009
P(202) 328-8160
F(202) 328-8161
Concern provides environmental information to communities in order to find solutions to problems that threaten public health.

Conservation Foundation
1250 24th St., NW
Washington, DC 20037
P(202) 293-4800
Promotes wise use of the earth's resources through research and public education.

Conservation Fund
1800 North Kent St., Suite 1120
Arlington, VA 22209
P(703) 525-6300
A national nonprofit organization searching for innovative ways to advance land and water conservation.

Conservation Law Foundation, Inc.
3 Joy St.
Boston, MA 02108
P(617) 742-2540
An environmental law firm that takes polluters to court, and works to pass legislation and to protect important natural resources.

Co-op America
2100 M St., NW, Suite 403
Washington, DC 20063
P(202) 872-5307 or (800) 424-2667
Provides alternatives for consumers to buy and invest
   in businesses that are concerned about the environment.

Council on Economic Priorities
30 Irving Place
New York, NY 10003
P(212) 420-1133 or (800) 822-6435
F(212) 420-0988
Promotes solutions to environmental problems and fos-
   ters arms control by encouraging corporate social re-
   sponsibility.

Council on Ocean Law
1709 New York Ave., NW, Suite 700
Washington, DC 20006
P(202) 347-3766
F(202) 638-0036
Furthers the laws governing ocean uses and preserves
   the ocean's abundance.

Cousteau Society, Inc., The
930 W. 21st St.
Norfolk, VA 23517
P(804) 627-1144
F(804) 627-7547
Telex: 6974570 COUSTEAUNFK
Dedicated to the study and exploration of the earth's
   oceans and to the education of the public on the pol-
   lution and development of the oceans.

Craighead Environmental Research Institute
Box 156
Moose, WY 83012
P(307) 733-3387
The Craigheads have been active in wildlife research
   from the migratory habits of caribou to the effects of
   environmental contamination on the raptor popu-
   lation.

Critical Mass Energy Project of Public Citizen
215 Pennsylvania Ave., SE
Washington, DC 20003
P(202) 546-4996
Opposes nuclear power and promotes safe alternatives
   through lobbying and litigation.

J.N. (Ding) Darling Foundation, Inc.
c/o J.M. Redman, Treasurer
P.O. Box 703
Des Moines, IA 50303-0703
P(515) 281-0812
Committed to conservation education by providing
   grants to students and initiating educational
   projects.

Declaration of Earth Ethics
700 East Daisy Lane

Milwaukee, WI 53217-3632
P(414) 351-2737
An urgent statement to establish world harmony in
   conjunction with the ecosystem.

Defenders of Wildlife
1244 19th St., NW
Washington, DC 20036
P(202) 659-9510
F(202) 833-3349
This group opposes any practice that harms wildlife di-
   versity by advocating governmental, citizen, and le-
   gal action.

Duck Unlimited, Inc.
One Waterfowl Way
Long Grove, IL 60047
P(708) 438-4300
Works to increase waterfowl and other wildlife popula-
   tion on the North American continent by restoration
   and management of wetland areas.

Earth First!
P.O. Box 5871
Tucson, AZ 85703
P(602) 622-1371
E-Mail(EcoNet): earthfirst
A nonviolent movement that works to preserve natural
   diversity as typified in wildness.

Earth Island Institute
300 Broadway, Suite 28
San Francisco, CA 94133
P(415) 788-3666
F(415) 788-7324
E-Mail(EcoNet): earthisland
Telex: 6502829302 MCI UW
Develops innovative projects for the conservation, pres-
   ervation, and restoration of the global environment.

Earth Right Institute
Gates-Briggs Building
Room 322
White River Junction, VT 05001
P(802) 295-7734
An educational and action center for environmental
   concerns.

EarthDance
P.O. Box 2155
Asheville, NC 28802
P(704) 252-8188
An educational project of the Youth Environmental
   Service Network that links young people from all
   over the world.

Earthsave Foundation
706 Frederick St.
Santa Cruz, CA 95062
P(408) 423-4069
F(408) 425-0255

EarthSave focuses on an ecologically sustainable future by providing education and developing leadership.

Earthwatch
P.O. Box 403
Mt. Auburn St.
Watertown, MA 02272
P(617) 926-8200
F(617) 926-8532
Telex: 5106006452
One of the largest private sponsors of research expeditions in the world, they provide funds to assist environmental scholars and scientists internationally.

Eco-Cycle
P.O. Box 4193
Boulder, CO 80306
P(303) 444-6634
Collects and recycles a portion of Boulder's solid waste to financial self-sufficiency in order to demonstrate the potential of recycling.

Eco-Home Network
4344 Russell Ave.
Los Angeles, CA 90027
P(213) 662-5207
Promotes and demonstrates urban ecological living.

Ecological Society of America
Center for Environmental Society
Arizona State University
Tempe, AZ 85287
P(602) 965-3000
Through published research the Center facilitates the dissemination of ecological principles.

EcoNet
3226 Sacramento St.
San Francisco, CA 94115
P(415) 923-0900
E-Mail(EcoNet)
An environmental electronic conference network.

Educational Communications, Inc.
P.O. Box 35473
Los Angeles, CA 90035
P(213) 559-9160
Works on media productions of all environmental issues to foster a better quality of life on earth.

Elmwood Institute
P.O. Box 5765
Berkeley, CA 94705
P(510) 845-4595
F(510) 845-1439
Founded by noted physicist Fritjof Capra, the Institute is known for its application of systemic thinking to a wide variety of contemporary issues.

Endangered Species UPDATE
School of Natural Resources

University of Michigan
Ann Arbor, MI 48109-1115
P(313) 763-3243
Magazine of data and articles that reports on endangered species.

Energy Conservation Coalition
1525 New Hampshire Ave., NW
Washington, DC 20036
P(202) 745-4874
This citizens' group is dedicated to finding and improving the efficiency of the nation's energy use.

Environmental Action Foundation, Inc.
46930 Carroll Ave
Tacoma Park, MD 20912
P(301) 891-1100
F(301) 891-2218
A combination of Environmental Action and the Environmental Task Force, the foundation lobbies the courts to make positive environmental changes.

Environmental Defense Center
906 Garden Street, Suite 2
Santa Barbara, CA 93101
P(805) 963-1622
F(809) 962-3162
E-mail(EcoNet): edc
A legal defense organization that counsels citizens on how to apply environmental laws.

Environmental Defense Fund, Inc.
257 Park Avenue South
New York, NY 10010
P(212) 505-2100
F(212) 505-2375
E-Mail(EcoNet): edf
This organization works to enforce laws preserving our natural resources, pursuing projects that fight against acid rain and the destruction of tropical rain forests.

Environmental Law Institute
1616 P St., NW, Suite 200
Washington, DC 20036
P(202) 328-5150
Works on research and education of environmental law and policy.

Environmental Protection Information Center, Inc.
P.O. Box 397
Garberville, CA 95440
P(707) 923-2931
Focuses on efforts to stop the clear-cutting of old growth forests.

Food First!/Institute for Food and Development Policy
145 Ninth Street
San Francisco, CA 94103
P(415) 864-8555
F(415) 864-3909

Encourages citizens to participate in finding solutions to critical social problems locally, nationally, and globally.

Forest Trust
P.O. Box 9238
Santa Fe, NM 87504-9238
P(505) 983-8992
Dedicated to improving forest ecosystems and resources through innovative land management and resource protection.

Fossil Fuel Policy Action Institute
P.O. Drawer 8558
Fredericksburg, VA 22404
P(703) 371-0222
Works to connect the U.S. internationally in a stronger environmental movement for projects such as tracking global warming.

Freshwater Foundation
2500 Shadywood Rd., Box 90
Navarre, MN 55392
P(612) 471-8407
F(612) 471-8142
Focuses on education and research of usable water to help people understand water issues and their environment.

Friends of the Earth
218 D St., SE
Washington, DC 20003
P(202) 544-2600
F(202) 543-4710
E-Mail(EcoNet): foedc
Telex: 650-192-5483
Fights to protect the earth from environmental disaster such as ozone depletion, global warming, and rainforest destruction.

Friends of the River, Inc.
Friends of the River Foundation
Bldg. C, Fort Mason Ctr.
San Francisco, CA 94123
P(415) 771-0400
This premier river preservation organization is dedicated to preserving over 100 natural river segments in California.

Fundamental Action to Conserve Energy
75 Day St.
Fitchburg, MA 01420
P(508) 345-5385
Through education, action, and a co-op store, this organization is dedicated to energy conservation, water conservation, and recycling.

Future Resources Associates, Inc.
2000 Center St., Suite 418
Berkeley, CA 94704
P(510) 644-2700
A consulting firm in energy and development issues.

Garden Club of America
598 Madison Ave.
New York, NY 10022
P(212) 753-8287
Dedicated to conservation and restoration of the environment, advocating wise land use and pollution control.

Global Action Plan
449 A Rt. 28A
West Hurley, NY 12491
P(914) 331-1312
F(914) 331-3241
Global Action Plan works on environmental restoration through prompting individual action.

Global Conservation, Protection, Restoration (CPR) Service
300 Broadway, Suite 28
San Francisco, CA 94133
P(415) 788-3666
F(415) 788-7324
E-mail(EcoNet): earthisland
Catalyses environmental restoration through education and advocacy, and by providing people with the opportunity to volunteer.

Global ReLeaf
1516 P St., NW
Washington, DC 20009
P(202) 667-3300
F(202) 667-7751
Organizes tree planting, providing technical assistance for new urban forestry programs.

Global Tomorrow Coalition, Inc.
1325 G St., NW
Suite 915
Washington, DC 20005-3104
P(202) 628-4016
F(202) 628-4018
E-Mail(EcoNet): gtc
Educates the public on social, environmental and economic issues to foster sustainable development in the U.S. and international communities.

Great Swamp Research Institute
Office of the College Dean
College of Natural Sciences and Math
Indiana University of Pennsylvania
305 Weyandt Hall
Indiana, PA 15705
P(412) 357-2609
Seeks innovative solutions to the increasing number of environmental problems.

Greenbelt Alliance
116 Montgomery, Suite 640
San Francisco, CA 94105
P(415) 543-4291
Works to protect the Greenbelt and promote better land development.

Greenhouse Crisis Foundation
1130 17th St., NW
Suite 630
Washington, DC 20036
P(202) 466-2823
F(202) 429-9602
E-Mail(EcoNet): gcf
Telex: 904059-WAS
Unites organizations to find solutions to the problem of
    global warming.

Greenpeace
1436 U St., NW
Washington, DC 20009
P(202) 462-1177
F(202) 462-4507
Telex: 89-2359
The largest environmental group in the world,
    Greenpeace was formed in 1971 and is dedicated to
    "protect all forms of life and to obstruct wrongs,
    without committing violence."

Human Environment Center
1001 Connecticut Ave., NW
Suite 827
Washington, DC 20036
P(202) 331-8387
The Human Environment Center encourages environ-
    mental organizations to work together and serves as
    a clearinghouse for youth conservation and service
    corps.

Humane Society of the United States
2100 L St., NW
Washington, DC 20037
P(202) 452-1100 or (800) 223-5400
F(202) 778-6132
Committed to the welfare of animals, this organization
    works to protect both domestic and wild species.

Inform
381 Park Ave. South
New York, NY 10016
P(212) 689-4040
F(212) 447-0689
A research, educational, and action organization that
    focuses on critical environmental issues.

Institute for Alternative Agriculture
9200 Edmonston Rd., Suite 117
Greenbelt, MD 20770
P(301) 441-8777
F(301) 220-0164
Encourages and facilitates environmentally sound and
    low cost farming methods.

Institute for Environmental Negotiation
Campbell Hall
Univ of Virginia
Charlottesville, VA 22903
P(804) 924-1970
Serves as a mediator between governments, busi-

nesses, and citizens to settle issues dealing with
    land use and other environmental issues.

Institute of Environmental Sciences
940 E Northwest Hwy.
Mount Prospect, IL 60056
P(312) 255-1561
Focuses on trying to understand the relationship be-
    tween nature and the impact of humanity in order
    to promote reliable and safe operations avoiding fur-
    ther contamination of the ecosystem.

International Ecology Society
1471 Barclay St.
St. Paul, MN 55106-1405
P(612) 774-4971
Committed to the correct treatment of animals and the
    protection of natural wildlands.

International Fund for Animal Welfare
P.O. Box 193
Yarmouth Port, MA 02675
P(508) 362-4944
Works to preserve wildlife and secure humane treat-
    ment of domestic animals.

International Rivers Network
1847 Berkeley Way
Berkeley, CA 94703
P(510) 848-1155
F(510) 848-1008
E-Mail(EcoNet): irn
Telex: 6503532706
Works on reclamation and restoration of rivers and in-
    forms the public concerning issues such as devel-
    opment.

Izzak Walton League of America, Inc., The
1401 Wilson Blvd., Level B
Arlington, VA 22209
P(703) 528-1818
Committed to educating the public in the conservation
    and restoration of soil, water and other natural re-
    sources.

Jackson Hole Alliance for Responsible Planning
Box 2728
Jackson, WY 83001
P(307) 733-9417
Publishes the *Alliance Newsletter* periodically and fo-
    cuses on maintaining high quality land usage in NW
    Wyoming.

John Muir Institute for Environmental Studies, Inc.
743 Wilson St.
Napa, CA 94559
P(707) 252-8333
Works to identify and study environmental problems.

Keep America Beautiful, Inc.
9 West Broad St.
Stamford, CT 06902
P(203) 323-8987

Through community level participation, this organization is dedicated to improving waste handling and recycling practices.

Land and Water Fund of the Rockies
1405 Arapahoe Ave.
Suite 200
Boulder, CO 80302
P(303) 444-1188
Develops legal strategies with grassroot groups working on pollution issues.

Land Institute
2440 E Water Well Rd.
Salina, KN 67401
P(913) 823-5376
Works on agricultural issues in order to sustain the health of the earth.

League for Coastal Protection
P.O. Box 421698
San Francisco, CA 94142-1698
P(415) 777-0221
Dedicated to protecting California's coastal waters, beaches, wetlands and wildlife.

League of Conservation Voters
1150 Connecticut Ave., NW
Suite 201
Washington, DC 20036
P(202) 785-8683
Works to help elect pro-environmental candidates to the U.S. House of Representatives and the Senate.

League To Save Lake Tahoe
P.O. Box 10110
S. Lake Tahoe, CA 95731
P(916) 541-5388
Non-profit organization committed to the preservation of the Lake Tahoe Basin.

Legal Environmental Assistance Foundation
203 North Gadsden St.
Suite 200
Tallahassee, FL 32301
P(904) 681-2591
This law foundation works to provide technical assistance in environmental and civil rights issues to grassroot organizations and citizens.

Local Environmental Action Group
717 1/2 Pujo St.
P.O. Box 3244
Lake Charles, LA 70601-4368
P(318) 474-6133
E-Mail(Environet)
This action group works to restore the ecosystem, from toxics and waste reduction to air quality issues.

Low Input Sustainable Agriculture
237 Hatchville Rd.
East Falmouth, MA 02536
P(508) 564-6301

Works on developing environmentally sound methods for food production, energy use and waste recycling.

Maine Organic Farmers and Gardeners Association
P.O. Box 2176
Augusta, ME 04338-2176
P(207) 622-3118
Committed to aiding farmers and gardeners grow organic food as well as helping to increase yield and public awareness about the benefits of organic food.

Marine Mammal Fund
Fort Mason, Center, Bldg. E
San Francisco, CA 94123
P(415) 921-3140
Projects and research to protect sea mammals including many films that show the tragic death of marine life.

Max McGraw Wildlife Foundation
P.O. Box 9
Dundee, IL 60118
P(312) 741-8000
The Foundation conducts research, management, and conservation education in wildlife and fisheries.

Midwest Consortium on Groundwater and Farm Chemicals
c/o Minnesota Project
2222 Elm St., SE
Minneapolis, MN 55414
P(612) 378-2142
Focuses on projects that protect groundwater from agricultural pesticides.

Mono Lake Committee
P.O. Box 29
Lee Vining, CA 93541
P(619) 647-6386 or 647-6596
This organization is dedicated to the preservation of Mono Lake.

National Academy of Sciences
2101 Constitution Ave., NW
Washington, DC 20037
P(202) 334-2644
F(202) 334-2614
The National Academy of Sciences studies issues such as environmental exposure, epidemiology and toxicology.

National Association for Environmental Education
P.O. Box 400
Troy, OH 45373
P(513) 649-3000
Works to aid individuals who work in environmental education, research, and service.

National Association for State River Conservation Programs
801 Pennsylvania Ave., SE
Suite 302

Washington, DC 20003
P(202) 543-2682
Encourages conservation and restoration of rivers and
their shore land environments as well as fostering a
forum to discuss river conservation.

National Association of Conservation Districts
509 Capitol Ct., NE
Washington, DC 20002
P(202) 547-6223
Works to promote conservation of natural resources.

National Audubon Society
950 Third Ave.
New York, NY 10022
P(212) 832-3200
F(212) 593-6254
A lobbying, litigation and citizens' action organization
that works to protect the world's ecosystem.

National Coalition Against the Misuse of Pesticides
530 7th St., SE
Washington, DC 20003
P(202) 543-5450
Focuses on the pesticide poisoning problem and promotes better alternative pest management solutions.

National Coalition for Marine Conservation
P.O. Box 23298
Savannah, GA 31403
P(912) 234-8062
This nonprofit, privately supported organization fo-
cuses on the conservation of ocean fish and their en-
vironment by increasing the public's awareness of
its responsibility for the natural world.

National Environmental Health Association
720 S. Colorado Blvd.
South Tower, 970
Denver, CO 80222
P(303) 756-9090
F(303) 691-9490
Supports people interested in environmental issues
and is the largest society of environmental health
practioners in the nation.

National Geographic Society
Colorado and M Sts., NW
Washington, DC 20036
P(202) 857-7000 or (800) 638-4077
F(202) 775-6141
Through exploration and research projects, the Na-
tional Geographic Society has increased the knowl-
edge of earth, sea, sky, and space.

National Parks and Conservation Association
1015 31st St., NW
Washington, DC 20007
P(202) 944-8530

This organization founded in 1919 acts to preserve our
national parks using research, wilderness preservation, and direct action.

National Recycling Coalition
1101 30th St., NW
Suite 305
Washington, DC 20007
P(202) 625-6406
Works to maximize recycling and conservation.

National Toxics Campaign
1168 Commonwealth Ave.
Boston, MA 02134
P(617) 232-0327
A coalition of citizens, consumer organizations, environmental groups, family farmers, and others working for solutions to the nation's toxic and environmental problems.

National Water Well Association
6375 Riverside Dr.
Dublin, OH 43017
P(614) 761-1711
F(614) 761-3446
Committed to the protection of ground water through
education and the publication of such water issues
as toxic substances and solid wastes.

National Wetlands Technical Council
1616 P St., NW Suite 200
Washington, DC 20036
P(202) 328-5150
Advises on wetlands policies and research priorities to
provide assistance to the nation's wetland conservation efforts.

National Wildflower Research Center
2600 FM 973 North
Austin, TX 78725
P(512) 929-3600
Dedicated to the restoration and conservation of native
plants, providing information through publications,
seminars, programs, and tours.

National Wildlife Federation
1400 Sixteenth St., NW
Washington, DC 20036-2266
P(202) 797-6800
E-Mail(EcoNet): nwfdc
With over 5 million members, this organization pro-
motes the wise use and proper management of natu-
ral resources upon which humanity depends.

National Wildlife Refuge Association
10824 Fox Hunt Ln.
Potomac, MD 20854
P(303) 249-8717
Dedicated to the protection of the National Wildlife
Refuge System by increasing public awareness and
appreciation of the system.

National Woodland Owners Association
374 Maple Ave., E., Suite 204
Vienna, VA 22180
P(703) 255-2700
A nationwide association of woodland owners that work together toward wise management of nonindustrial private forest lands.

Native Americans for a Clean Environment
P.O. Box 1671
Tahlequah, Oklahoma 74465
P(918) 458-4322
Focuses on nuclear industry issues as well as environmental issues in order to raise public awareness.

Native Forest Council
P.O. Box 2171
Eugene, OR 97402
P(503) 688-2600
F(503) 461-2156
Dedicated to preservation and protection of all remaining native forests in the United States.

Natural Areas Association
320 S. Third St.
Rockford, IL 61104
P(815) 964-6666
Works to identify, preserve and manage natural areas and diversity.

Natural Guard
125 W. 44th St., Suite 11E
New York, NY 10036
P(212) 704-0346
F(212) 869-3045
Acts to encourage young people, especially in inner cities, to address environmental issues.

Natural Land Institute
302 S Third St.
Rockford, IL 61104
P(815) 964-6666
Purchases or receives natural areas for preservation.

Natural Lands Trust
Hildacy Farm, 1031 Palmers Mill Road
Media, PA 19063
This organization concentrates on the protection and preservation of natural lands in the mid-Atlantic region.

Natural Resources Council of America
801 Pennsylvania Ave., SE
Suite 410
Washington, DC 20003
P(202) 547-7553
Concerned with proper management of natural resources in the public interest by providing citizens with policy information on conservation issues.

Natural Resources Defense Council
40 West 20th St.

New York, NY 10011
P(212) 727-2700
E-Mail(EcoNet): nrdc
This lobbying group works to protect endangered natural resources, combining legal action and a scientific approach.

Nature Conservancy
1815 North Lynn St.
Arlington, VA 22209
P(703) 841-5300
F(703) 841-1283
E-Mail(EcoNet): natconsv
In order to maintain genetic diversity among species, the Conservancy is committed to finding, maintaining, and protecting natural lands and ecosystems.

Negative Population Growth, Inc.
16 East 42nd St., Suite 1042
New York, NY 10017
P(201) 837-3555
As population growth is becoming a serious issue, this organization educates on the need to limit human population for sustainable development.

New Alchemy Institute
237 Hatchville Rd.
East Falmouth, MA 02536
P(508) 564-6301
The goal of this organization is to work towards a world which meets all basic human needs of food, water, and shelter.

New York Zoological Society
The Zoological Park
Bronx, NY 10460
P(212) 220-5100
Funds research and development of wildlife management, establishes parks and reserves, and works to increase the public understanding of zoology and the environment.

North American Lake Management Society
1000 Connecticut Ave., NW
Suite 300
Washington, DC 20036
P(202) 466-8550
Works to protect and restore lakes, reservoirs and their watersheds. This organization also sponsors annual conferences, and regional workshops as well as publications.

North American Wildlife Foundation
102 Wilmot Rd., Suite 410
Deerfield, IL 60015
P(708) 940-7776
F(708) 940-3739
Operates research programs concerned with wetland ecology and improving marsh management.

Northwest Coalition for Alternative Pesticides
P.O. Box 1393
Eugene, OR 97440
P(503) 344-5044
The main goal of the Northwest Coalition is to elimi-
nate the use of pesticides by education, watch-
dogging, and direct action.

Northwest Renewable Resources Center
1133 Dexter Horton Bldg.
710 Second Ave.
Seattle, WA 98104
P(206) 623-7361
Tries to offer alternatives on disputes dealing with natural resources issues.

Ocean Alliance
Building E
Fort Mason Center
San Francisco, CA 94123
P(415) 441-5970
Dedicated to preserve and protect the ocean through
conservation and marine education programs.

Oceanic Society
Executive Offices
1536 16th St., NW
Washington, DC 20036
P(202) 329-0098
Promotes understanding and stewardship of marine
and coastal environments as well as working to pro-
tect the oceans.

Oregon Natural Resources Council
Yeon Bldg., Suite 1050
522 Southwest Fifth Avenue
Portland, OR 97204
P(503) 223-9001
Concentrates on preserving wild forest lands by lob-
bying for better forest management.

Organization of Wildlife Planners
Box 7921
Madison, WI 53707
P(307) 777-7461
Focuses on constructing effective management systems
of resources.

Pacific Energy and Resource Center
Building 1055
Fort Cronkite
Sausalito, CA 94965
P(415) 332-8200
Provides policy research, professional consulting ser-
vices and natural resource education programs such
as exhibits and lectures.

Pacific Whale Foundation
Kealia Beach Plaza, Ste. 25
101 N. Kihei Rd.
Kihei, HI 96753
P(808) 879-8811

Works to conserve and research all marine mammals,
focusing primarily on Hawaiian humpback whales.

Partners for Livable Places
1429 21st Street, NW
Washington, D.C. 20036
P(202) 887-5990
Works to improve economic resources and the quality
of life of certain communities by preservation, conservation and cultural resources management.

Peregrine Fund, Inc., The
5666 West Flying Hawk Ln.
Boise, ID 83709
P(208) 362-3716
The Fund works on the preservation and restoration of
peregrine falcons.

Pesticide Action Network
P.O. Box 610
San Francisco, CA 94101
P(415) 771-2763
F(415) 541-9253
E-Mail(EcoNet): panna
Telex: 15683472 PANNA
Works to replace pesticides with sustainable agricul-
ture practices and technologies.

Planning and Conservation League Foundation
909 12th St., Suite 203
Sacramento, CA 95814
P(916) 444-8726
This organization dedicated to environmental education and research works on solving problems concerning energy, environmental research, water quality, and coastal use.

Population-Environmental Balance
1325 G St., NW Suite 1003
Washington, DC 20005
P(202) 879-3000
F(202) 879-3019
Works to educate the public on the relationship be-
tween population growth and the well-being of the
United States.

Prevention of Global Warming Project
26 Church St.
Cambridge, MA 02238
P(617) 547-5552
Seeks to involve scientists in the need to develop a na-
tional energy policy based on renewable energy.

Public Media Center
466 Green Street
San Francisco, CA 94133
P(415) 434-1403
PMC offers a service for environmental organizations
to create and run effective ads on TV, radio, and in
the written media.

Rainforest Action Network
301 Broadway, Suite A
San Francisco, CA 94133
P(415) 398-4404
F(415) 398-2732
E-Mail(EcoNet): rainforest
Telex: 151276475
This organization works to preserve the rainforests by drawing attention to the short term use of rainforests and by educating the public as to citizen actions in pressuring businesses to change.

Rainforest Alliance
270 Lafayette St., Suite 512
New York, NY 10012
P(212) 941-1900
F(212) 941-4986
This organization's goals are to preserve the rainforests and to create a way of utilizing the rainforests without further destruction.

Redwood Alliance
P.O. Box 293
Arcata, CA 95521
P(707) 822-7884
This organization promotes safe energy alternatives to nuclear power through education and lobbying.

Redwood Community Action Agency
904 G St.
Eureka, CA 95502
P(707) 445-0881
Among their many projects, this agency promotes the preservation of redwood forests, and the belief that improving the environment enhances the local economy.

Rene Dubos Center for Human Environments
100 East 85th St.
NY 10028
P(212) 249-7745
Works to create new environmental values and resolution of environmental conflicts through education, research, and publications.

Renew America
1400 16th St., NW Suite 710
Washington, DC 20036
P(202) 232-2252
Works to build a sustainable society by encouraging effective public policy.

Renewable Natural Resources Foundation
5430 Grosvenor Ln.
Bethesda, MD 20814
P(301) 493-9101
Renewable natural resources and public policy alternatives are the focuses of this organization.

Resource Renewal Institute
Ft. Cronkite, #1055

Sausalito, CA 94965
P(415) 332-8082
Creates long-term solutions to renewable resource problems by training professionals to implement changes and by setting up data banks.

Resources for the Future
1616 P St., NW
Washington, DC 20036
P(202) 328-5000
Develops research projects and education in the use of natural resources, quality of the environment, and conservation.

Restoring the Earth
1713C MLK Jr. Way
Berkeley, CA 94709
P(510) 843-2645
Develops creative solutions to environmental problems by means of ecological restoration.

Rocky Mountain Institute
1739 Snowmass Creek Road
Snowmass, CO 81654-9199
P(303) 927-3128
F(303) 927-4178
E-Mail(EcoNet): rmi
Promotes sustainable agriculture activities and ecologically sound and economically viable communities.

Sacramento River Preservation Trust
P.O. Box 5366
Chico, CA 95927
P(916) 345-4050
The purpose of this organization is to preserve, protect and improve the Sacramento River by educating the public

Save the Dolphin Project
Earth Island Institute
300 Broadway, Suite 28
San Francisco, CA 94133
P(415) 788-3666/800 3-DOLFIN
E-Mail(EcoNet): earthisland
This project is determined to eliminate the unnecessary slaughter of dolphins by promoting legal, political, and economic reforms.

Save the Redwoods League
114 Sansome St., Rm. 605
San Francisco, CA 94104
P(415) 362-2352
Hopes to rescue the redwoods by creating Redwood Parks and educating the public of the value of the Sequoias.

Sea Shepherd Conversation Society
P.O. Box 7000-S
Redondo Beach CA 90277
P(213) 394-3198

This is a direct, nonviolent action organization that focuses on illegal whaling, the dolphin slaughter, and drift netting.

Sea Turtle Restoration Project
300 Broadway, Suite 28
San Francisco, CA 94133
P(415) 788-3666
F(415) 788-7324
E-Mail(EcoNet): earthisland
Acts to preserve, restore and investigate threats to the world's endangered sea turtles

Sierra Club
730 Polk St.
San Francisco, CA 94109
P(415) 776-2211
F(415) 776-0350
With 57 chapters and 386 groups in North America, this organization's work includes legislation, litigation, public information, publishing, wilderness outings, and conferences.

Sierra Club Legal Defense Fund, Inc.
180 Montgomery St., Suite 1400
San Francisco, CA 94104
P(415) 627-6700
F(415) 627-6740
This public interest law firm brings lawsuits on behalf of environmentalists and citizens organizations to protect the environment.

Soil and Water Conservation Society
7515 NE Ankeny Rd.
Ankeny, IA 50021-9764
P(515) 289-2331
F(515) 289-1227
Through education this organization promotes good land and water use, and natural resources conservation.

Spill Control
400 Renaissance Center
10th Floor
Detroit, MI 48243-1895
P(313) 567-0500
Organized to combat pollution incidents by minimizing their effects.

Student Conservation Association, Inc.
P.O. Box 550
Charlestown, NH 03603
P(603) 826-4301
F(603) 826-7755
Publishes *Job Scan,* a monthly environmental job listing, and supports volunteers in conservation of public lands.

Student Environmental Action Coalition
P.O. Box 1168
Chapel Hill, NC 27514

P(919) 967-4600
F(919) 962-5604
E-Mail: SEAC UNC BITNET
The largest student environmental organization in the United States, SEAC was founded in 1989 to provide network services for campus environmental groups.

Surfrider Foundation
P.O. Box 230754
Encinitas, CA 92023
P(619) 792-9940
The Surfrider Foundation works towards enhancing the quality of water and beaches.

Thorne Ecological Institute
5398 Manhattan Cir.
Boulder, CO 80303
P(303) 499-3647
Brings together adults in order to educate them on the application of ecological principles to enhance the human and natural environment.

Threshold, Inc.
International Center for Environmental Renewal
Drawer CU
Bisbee, AZ 85603
P(602) 432-7353
Established in 1972, this organization works to develop ecologically sound alternatives in order to improve upon human understanding of the ecosystem.

Treepeople
12601 Mulholland Dr.
Beverly Hills, CA 90210
P(818) 753-4600
F(818) 753-4625
E-Mail(EcoNet): treepeople
The goal of this organization is to end global deforestation by taking surplus trees and sending them to needy areas for planting, and by encouraging citizens to plant trees.

Trout Unlimited
800 Follin Lane, Suite 250
Vienna, VA 22180-4906
P(703) 281-1100
F(703) 281-1825
Works to maintain trout fisheries by protecting and preserving rivers and streams.

Trust for Public Land
116 New Montgomery St.
4th Floor
San Francisco, CA 94105
P(415) 495-4014
F(415) 495-4103
A land conservation organization preserving land for present and future generations.

Union of Concerned Scientists
26 Church St.

Cambridge, MA 02238
P(617) 547-5552
F(617) 664-9405
Scientists and citizens concerned about the effects of technology on society, analyzing issues such as nuclear power and weapons.

United Nations Environment Programme
United Nations
Rm. DC2-0803
New York, NY 10017
P(212) 963-8138
Serves as an advocate for environmental management among U.N. agencies, monitoring and assessing issues of water, land and atmospheric areas.

US Public Interest Research Group
215 Pennsylvania Ave., SE
Washington, DC 20003
P(202) 546-9707
US PIRG engages in many issues including the environment, energy and government reform, and consumer protection.

Urban Creeks Council
2530 San Pablo Avenue
Berkeley, CA 94702
P(510) 540-6669
The Urban Creeks Council encourages the preservation and restoration of natural streams in urban or human-made environments.

Urban Ecology
P.O. Box 10144
Berkeley, CA 94709
P(510) 549-1724
Builds and rebuilds cities in a manner that makes them environmentally healthy and culturally viable.

Urban Habitat Program
300 Broadway, Suite 28
San Francisco, CA 94133-3312
P(415) 788-3666
F(415) 788-7324
EcoNet: earthisland
Develops multi-cultural, urban environmental leadership in the Bay Area.

Water Pollution Control Federation
601 Wythe St.
Alexandria, VA 22314-1994
P(703) 684-2400
F(703) 684-2492
Concerned with the collection and disposal of domestic and industrial wastewater.

Wetlands for Wildlife, Inc.
P.O. Box 344
West Bend, WI 53095
P(414) 628-0103

Promotes and participates in the preservation of wetlands and wildlife habitat in the United States.

Whale Center
3929 Piedmont Ave.
Oakland, CA 94611
P(415) 654-6621
Works to protect whales and their ocean habitat.

Wilderness Society
900 17th St., NW
Washington, DC 20006-2596
P(202) 833-2300
F(202) 429-3958
Dedicated to preserving the US's forests, parks, rivers and shorelands.

Wilderness Watch
P.O. Box 782
Sturgeon Bay, WI 54235
P(414) 743-1238
Based on scientific research, this organization is dedicated to sustained use of the United States's woodlands and waters.

Wildlife Conservation International
New York Zoological Society
185th St., and So. Blvd., Bldg. A
Bronx, NY 10460
P(212) 220-5155
Strives to understand the ecosystem in order to apply that research to conservation efforts.

Wildlife Forever
12301 Whitewater Dr., Suite 210
Minnetonka, MN 55343
P(612) 936-0605
Promotes the need for scientific wilderness management and preservation.

Wildlife Habitat Enhancement Council
1010 Wayne Ave., Suite 1240
Silver Springs, MD 20910
P(301) 588-8994
Works to bring together conservation and corporate communities to benefit environmental legislation.

Wildlife Information Center, Inc.
629 Green St.
Allentown, PA 18102
P(215) 434-1637
Researches wildlife conservation and educates the public through outreach programs.

Wildlife Management Institute
Suite 725
1101 14th St., NW
Washington, DC 20005
P(202) 371-1808
Works to improve professional management of natural resources to benefit all of society.

Wildlife Society
5410 Grosvenor Ln.
Bethesda, MD 20814
P(301) 897-9770
F(301) 530-2471
Organized in 1937, this organization engages in wildlife research, management, education, and administration to increase awareness and appreciation of wildlife values.

Windstar Foundation
2317 Snowmass Creek Rd.
Snowmass, CO 81654
P(303) 927-4777
F(303) 927-4779
Serves to inspire citizens towards direct action in achieving an environmentally sustainable future.

World Environment Center, Inc.
419 Park Ave., Suite 1404
New York, NY 10016
P(212) 683-4700
F(212) 683-5053
Serves to strengthen the bridge between industry and government to achieve environmentally beneficial goals.

World Forestry Center
4033 SW Canyon Rd.
Portland, OR 97221
P(503) 228-1367
A nonprofit organization dedicated to preserving the forests as well as other natural resources through school and community education and publications.

World Nature Association, Inc.
P.O. Box 673
Silver Spring, MD 20901
P(301) 593-2522
Funds small projects and scholarships for conservation and education projects in various parts of the world.

World Resources Institute
1709 New York Ave., NW
Washington, DC 20006
P(202) 638-6300
World Resources Institute provides environmental information to facilitate debate among people with differing perspectives.

World Wildlife Fund
1250 24th St., NW
Washington, DC 20037
P(202) 293-4800
F(202) 293-9211
Telex: 64505 PANDA
Works to protect the world's wildlife and biological diversity.

Worldwatch Institute
1776 Massachusetts Ave., NW

Washington, DC 20036
P(202) 452-1999
A research organization elucidating global problems and issues such as energy, population growth and migration, global economy, and the environment.

Zero Population Growth, Inc.
1400 16th St., NW Suite 320
Washington, DC 20036
P(202) 332-2200
Through educational material and lobbying this organization encourages public support for population stabilization.

EXHAUST GAS RECIRCULATION

RAMON ESPINO
Exxon Research and Engineering
Annandale, New Jersey

One of the first deliberate steps taken in engine design to reduce atmospheric pollution consisted of recirculating a fraction of the exhaust gas. The dilution of the air–fuel mixture with a relatively inert exhaust gas reduced the maximum combustion temperature significantly. Since the formation of NO in an internal combustion engine is strongly influenced by temperature, Exhaust Gas Recirculation (EGR) allowed NO_x emissions to decrease below the 4 g/mile level by the mid-1970s. (See also AIR POLLUTION: AUTOMOBILE; AUTOMOBILE EMISSIONS—CONTROL; AUTOMOTIVE ENGINES.

The first techniques used to recirculate exhaust gases to the combustion chamber accomplished the task, but at a price in terms of performance. One method was to increase the time overlap when both the exhaust valve and the inlet valve are open. This leads to a certain amount of exhaust combustion gas remaining in the combustion chamber. This method of lowering the combustion temperature and NO emissions should be called Exhaust Gas Retention instead of Exhaust Gas Recirculation. A true recirculation method was implemented in parallel consisting of a calibrated tube that bled a constant volume fraction of the high pressure exhaust gas into the lower pressure air–fuel mixture in the inlet manifold of the engine.

These two methods suffered from very poor idling and poor acceleration response at low speeds (when the amount of exhaust gas was relatively large compared with the fresh air–fuel mixture). Moreover, these techniques significantly reduced NO emissions for a level of about 6 g/mile prior to 1968 to almost 4 g/mile by 1973. However, the standards for NO emissions in the USA were below that level by 1974.

To meet the more stringent standards, most engine manufacturers developed variable-flow Exhaust Gas Recirculation (EGR) systems. They also went back to the normal valve overlap. The variable-flow EGR system substituted the fixed diameter pipe connecting the exhaust to the inlet manifold with a spring-loaded, vacuum-controlled, temperature compensated metering valve. The range of flow of the exhaust gas varies between zero when the car is idling and reaching a peak in flow when the engine is cruising in the 30–70 mph range. The valve also closes when the engine is operated at wide open throttle (WOT). The temperature control is used to shut off recirculation when the ambient temperature is below 14°C. With zero EGR during idling and wide open throttle, the problems of poor idling and slow acceleration were reduced. The new design increased the amount of exhaust gas at the conditions for peak NO formation and this enabled auto manufacturers to meet the 3g/mile standard for 1974/1975.

Exhaust gas recirculation has a negative impact on engine performance and fuel economy. Every percent of exhaust gas that is recirculated yields a corresponding reduction in fuel economy and acceleration times. When the variable flow EGR valves were introduced, there were also many concerns about their operability and driveability. The government required that these EGR systems be inspected every 12,000 miles. However, the design proved rugged and inspection schedules have been extended to 30–40,000 mile intervals. While a number of European cars still use EGR, most American and Japanese OEMs have moved from EGR to catalytic exhaust gas converters to reduce NO. Further restrictions in NO emissions standards, as well as better vehicle performance were the main drivers for this change.

BIBLIOGRAPHY

K. Owen and T. Coley, *Automotive Fuels Handbook,* Society of Automotive Engineers, Inc.

E. F. Obert, *Internal Combustion Engines an Air Pollution,* Harper and Row, Publishers.

Effect of Automotive Emissions Requirements on Gasoline Characteristics, ASTM Publication, STP 487.

EXHAUST CONTROL, AUTOMOTIVE

John J. Mooney
Engelhard Corporation
Iselin, New Jersey

Automobiles and trucks consume large amounts of gasoline, producing commensurately large amounts of gaseous exhaust consisting primarily of carbon dioxide [124-38-9], water, unburned hydrocarbons (qv), carbon monoxide [630-08-0] (qv), and oxides of nitrogen, NO_x. The latter three atmospheric pollutants have been regulated since the 1970s by the U.S. government and more stringently by the State of California (see Air pollution). Automobile companies have developed fuel metering and exhaust systems using the catalytic converter to meet emission regulations. Carbon dioxide emissions are indirectly controlled by corporate average fuel economy (CAFE) standards for passenger cars and small trucks.

Prior to emission control, passenger car and truck emissions together were the largest contributors to atmospheric pollution in the United States (1). By 1994, when the Tier I exhaust emission standards mandated by the Clean Air Act Amendments of 1990 take effect in the United States, the degree of cleanup for automobiles and small trucks must be 97.4% for hydrocarbon emissions, 96.0% for carbon monoxide, and 90% for oxides of nitrogen as compared to pre-control of exhaust emissions. For areas in the United States that do not reach minimum ambient air standards by the end of the 1990s, the U.S. Congress has set conditional Tier II standards for vehicles to take effect as early as the year 2003. Also, California, because of unusually severe atmospheric pollution conditions, has established the most stringent automobile exhaust regulations in the world to control exhaust emissions to the absolute minimum levels. These California standards, called Low Emission Vehicle (LEV) Standards, start to take effect in 1996. Other states are expected to adopt the California standards.

Key to achieving the degree of exhaust emission control by the early 1990s were the monolithic catalytic converter, the three-way catalyst, and the closed loop system based on the oxygen (qv) sensor (see Sensors). Initially, the catalytic converter contained an oxidation catalyst to control hydrocarbon (HC) and carbon monoxide emissions. Exhaust gas recirculation was used to control NO_x emissions. Subsequently, the three-way catalyst was developed to control HC, CO, and NO_x in a single catalyst. Over 200 million vehicles have been equipped with the catalytic converter which has been rated among the top 10 engineering breakthroughs of the twentieth century (2). Emission control is achieved without negatively affecting fuel economy or performance. The shift to alternative transportation fuels such as methanol, ethanol, natural gas, liquefied petroleum gas (LPG), and reformulated gasoline in accordance with the Clean Air Act and the National Energy Policy Act of 1992 (Gas, natural) is expected to produce further modifications to the catalytic converter.

Diesel engine emission control technology is not discussed herein. As of early 1993 only one passenger car manufacturer was marketing diesel fuel cars. Emission control technology for diesel engines, used in some light-duty trucks and in medium- and heavy-duty trucks, is evolving at a rapid pace. This technology includes engine modifications such as high pressure fuel injection, variable valve timing, and intercooled turbochargers. Catalytic aftertreatment that is quite different from that discussed herein has also been developed.

See also Air pollution: automobile; Automotive emissions; Automotive engines; Clean air act of 1990, mobile sources.

EMISSION REGULATIONS AND TESTING

In the United States, federal regulations require automobile manufacturers to certify that vehicles are in compliance with exhaust emission standards when tested under specific test procedures.

Clean Air Act Amendments

Exhaust emission standards were set by the Clean Air Act Amendments of 1970 requiring automobile manufacturers to control the amount of hydrocarbons, carbon monoxide, and oxides of nitrogen emitted from vehicles.

Table 1. Federal Exhaust Emission Standards for Conventionally Fueled Passenger Cars and Light Trucks[a], g/km

Vehicle and Year	Fleet, %	NMHC[b]	CO	NO$_x$	Particulates
Light-duty (0–2,700 kg gross vehicle weight rating) vehicles					
Passenger cars					
1991–1993	100	0.25[c]	2.1	0.6	0.12
1994	40	0.16[d]	2.1	0.2	0.05
1994[e]	40	0.19	2.6	0.4	0.06
1995	80	0.10[d]	2.1	0.2	0.05
1995[e]	80	0.19	2.6	0.4	0.06
1996–2003	100	0.10[d]	2.1	0.2	0.05
1996–2003[e]	100	0.19	2.6	0.4	0.06
2004–2006[e,f]		0.08	1.1	0.1	0.06
Trucks, loaded vehicle weight <1700 kg					
1991–1993	100	0.50[c]	6.2	0.7	0.08
1994	40	0.16[d]	2.1	0.2	0.08
1994[e]	40	0.19	2.6	0.4	
1995	80	0.16[d]	2.1	0.2	0.05[g]
1995[e]	80	0.19	2.6	0.4	0.06[g]
1996–2003	100	0.16[d]	2.1	0.2	0.05[h,i]
1996—2003[e]	100	0.19	2.6	0.4	0.06[h,i]
2004—2006[e,f]		0.08	1.1	0.1	0.6
Trucks, loaded vehicle weight 1700–2600 kg					
1991–1993[j]	100	0.50[c]	6.2	1.1	0.08
1994	40	0.20[d]	2.7	0.4	
1994[e]	40	0.25	3.4	0.60	0.08
1995	80	0.20[d]	2.7	0.4	0.05[g]
1995[e]	80	0.25	3.4	0.60	0.06[g,h]
1996–2003	100	0.20[d]	2.7	0.4	0.05[h,i]
1996–2003[e]	100	0.25	3.4	0.60	0.06
Light-duty (2,700–3,900 kg gross vehicle weight rating) trucks					
Trucks, 1,700–2,600 test weight[k]					
1991–1993[j]	100	0.50[c]	6.2	1.1	0.08
1994[j]	100	0.50[c]	6.2	1.1	0.08
1995[j]	100	0.50[c]	6.2	1.1	0.08
1996	50	0.20	2.7	0.4	
1996[j]	50	0.28[d]	4.0	0.61	0.06
1997–2003	100	0.20[d]	2.7	0.4	
1997–2003[e]	100	0.28	4.0	0.61	0.06
Trucks, >2,600 kg test weight[k]					
1991–1993[j]	100	0.50[c]	6.2	1.1	0.08
1994[j]	100	0.50[c]	6.2	1.1	0.08
1995	100	0.50[c]	6.2	1.1	0.08
1996[j]	50	0.24[d]	3.1	0.7	
1996[j]	50	0.35	4.5	0.95	0.07
1997–2003	100	0.24[d]	3.1	0.7	
1997–2003[e]	100	0.35	4.5	0.95	0.07

[a] The useful life of the emissions control system is expected to be five years or 80,000 km, unless otherwise noted.
[b] NMHC = nonmethane-reactive hydrocarbons.
[c] Total hydrocarbons.
[d] A set of intermediate in-use standards also applies during the phase-in period, 1994–1997 for passenger cars and small light-duty trucks, and 1996–1998 for larger light-duty trucks.
[e] Useful life is 10 years or 160,000 km; in-use compliance is seven years or 120,000 km.
[f] After an EPA/OTA study due June 1, 1997, the EPA shall by December 31, 1999, set standards more stringent than 1996 standards for passenger cars and certain light trucks, effective after January 1, 2003, but not later than model year 2006.
[g] Corresponds to 40% of fleet for particulate.
[h] Corresponds to 80% of fleet for particulate.
[i] 100% of fleet in 1997 and thereafter.
[j] Useful life is 11 years or 192,000 km; in-use compliance is seven years or 144,000 km starting 1994.
[k] Test weight = (gross vehicle weight rating + curb weight)/2.

The act was amended in 1990. Table 1 summarizes the emission standards for passenger cars and light trucks between the years 1991 and 2006. In 1994, emissions from vehicles come under more stringent control, and the useful life of the emission control system is extended from five years or 80,000 km to 10 years or 160,000 km. The regulations are set up so that after 80,000 km or five years of actual use, a slightly more relaxed standard is applicable for the next 80,000 km.

Test Procedure

To comply with emission standards, representative vehicles must be run for 80,000 km (Appendix IV of the Federal Test Procedure (FTP)) (3). The first 6,400 km are considered a break-in portion. Exhaust emissions are measured each 8,000 km between approximately 6,400 and 80,000 km of accumulation and a deterioration factor (DF) of emissions is calculated. A DF of 1.15 for HC indicates that HC emissions increased by 15% between 6,400 and 80,000 km, and were within the 80,000 km standard. This DF is applied to the 6,400 km emission test data points for all other model variations of the family of vehicles represented by the 80,000 km durability car.

The test for evaluating individual vehicle emissions, the FTP (4), specifies that a test vehicle be stored in an area where the ambient temperature is between 20 and 29°C for at least 12 hours immediately prior to the emission test. Then, the vehicle is placed on a chassis dynamometer which is calibrated for the vehicle weight and road load. The vehicle is started and driven for 41 min on a prescribed cycle of accelerations, cruises, decelerations, idles, a 10-min shutdown (called the hot soak), and a period of rerun.

During the FTP, all exhaust passes through a constant volume sampling system (CVS) which dilutes the exhaust with air so that the total flow of air plus exhaust is constant. A sample of the diluted gas is metered to three gas sample collection bags sequentially. The first, or cold start bag, contains the gas sample from the first 505 seconds of the test. The second bag is the gas sample of the hot transient portion of the test, between 506 and 1372 seconds, after which time the ignition is turned off for 10 minutes. The third bag gas sample is taken after the 10 min hot soak, from the point of ignition of the hot restart, for 505 seconds. Each bag is then measured for hydrocarbons, carbon monoxide, and oxides of nitrogen using, respectively, a flame-ionization detector (FID) HC analyzer, an infrared CO analyzer, and a chemiluminescent NO_x analyzer. A CO_2 analyzer (NDIR) is used to calculate dilution. The concentration measurements are converted into grams of each emission per unit of distance. The FTP prescribes weighting factors for each phase to give a composite value for the total test.

Typical HC and CO emissions from a vehicle undergoing the FTP are shown in Figure 1a. Figure 1b shows the inlet gas temperature to a typical underfloor catalytic converter. The HC and CO are very high during the first 100 seconds after the engine is started, and drop off precipitously after the engine is up to running temperature. From 75 to 85% of the HC and CO emissions pass through the catalytic converter during the time the converter is

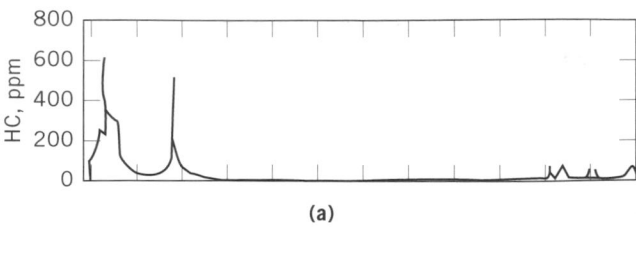

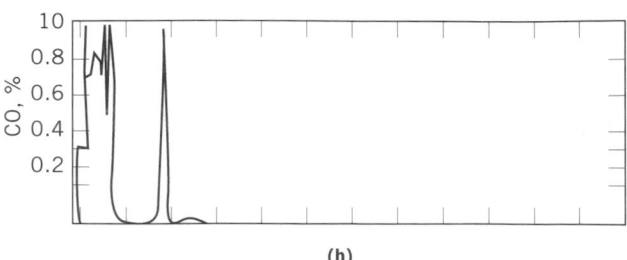

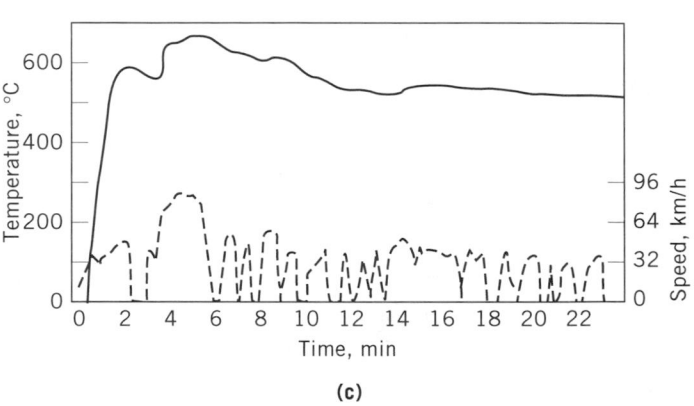

Figure 1. CO and hydrocarbon tailpipe emissions. Data from a test vehicle during a test cycle where the catalyst was mounted ~1.2 m from the exhaust part of the engine: (**a**) hydrocarbon and (**b**) CO tailpipe emissions; and (**c**), (—) catalyst temperature and (---) speed. As can be seen, the principal CO and hydrocarbon emissions occur during the cold start for this vehicle, ie, during the catalyst warmup period. When hot, the catalyst is very effective. In practice, one can expect between 60 and 90% of the engine CO and hydrocarbon emissions, as measured over the whole test cycle, to be removed by the catalyst after 50,000 miles of use (6).

being heated by the hot exhaust gases (5). Very little of these gases pass unconverted through a hot catalyst of a properly calibrated system.

EXHAUST GAS COMPOSITION

The exhaust composition from gasoline/air combustion is dependent on many factors. Total combustion in the engine is not possible even when excess oxygen is present (6). Formation of the air/fuel mixture as well as the design of the combustion chamber influence the combustion process, as do engine power and ignition system timing. However, for emission control, the main factor affecting the composition of the exhaust gas is the air/fuel mixture or ratio. The composition of exhaust varies according to the air/fuel ratio as shown in Figure 2 (7) for a standard

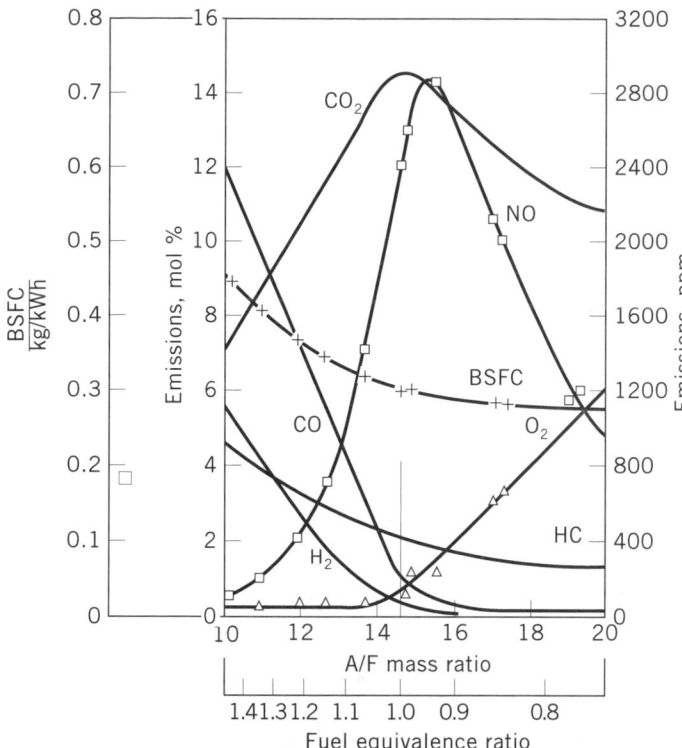

Figure 2. Effect of mixture strength on exhaust gas composition (dry basis) and brake specific fuel consumption (BSFC) for an unsupercharged automotive-type engine using indolene fuel, H/C = 1.86, where the ignition is tuned to achieve maximum best torque (MBT), the brake mean effective pressure (BNEP) is 386 kPa at 1200 rpm (7,8).

gasoline fuel where the hydrogen to carbon ratio (H/C) is 1.86. The exhaust gas composition changes as the H/C ratio changes (9), as shown in Figure 3 (8,10,11).

Unburned hydrocarbons in the exhaust originate primarily from crevices in the combustion chamber, such as gaps between the piston and cylinder wall, where the combustion flame cannot burn. The composition of unburned hydrocarbons is dictated primarily by the composition of the fuel (12). Carbon monoxide results from areas of insufficient oxygen. Oxides of nitrogen are produced in the high temperature zones during combustion by the reaction of nitrogen molecules and oxygen atoms thermally produced from oxygen and oxygen-containing species, according to the Zeldovich mechanism (6,13).

Hydrocarbons and carbon monoxide emissions can be minimized by lean air/fuel mixtures (Fig. 2), but lean air/fuel mixtures maximize NO_x emissions. Very lean mixtures (> 20 air/fuel) result in reduced CO and NO_x, but in increased HC emissions owing to unstable combustion. The turning point is known as the lean limit. Improvements in lean-burn engines extend the lean limit. Rich mixtures, which contain excess fuel and insufficient air, produce high HC and CO concentrations in the exhaust. Very rich mixtures are typically used for small air-cooled engines, needed because of the cooling effect of the gasoline as it vaporizes in the cylinder, where CO exhaust concentrations are 4 to 5% or more.

The best power is achieved slightly rich of stoichiometric air/fuel; the best fuel economy is achieved slightly lean of stoichiometric mixture.

Over 150 hydrogen and carbon species are present in the exhaust of a gasoline fueled engine (14–18) including methane, various paraffins, olefins, aldehydes, aromatics, and polycyclic hydrocarbons as well as unburned gasoline. Exhaust gas also contains sulfur dioxide from the combustion of sulfur contained in the gasoline. The average U.S. gasoline sulfur content is about 300 ppm, which results in about 20 ppm SO_2 in the exhaust. Gasoline sulfur has ranged between 50 and >1000 ppm in the United States. Reformulated gasolines are expected to have lower sulfur because of the greater processing by the petroleum (qv) refinery (11,12). Additionally, the State of California has specified a Phase II gasoline having a sulfur content between 30–40 ppm for the LEV program.

Exhaust gas also contains small amounts of hydrogen cyanide and ammonia depending on the air/fuel ratio.

EMISSION CONTROL SYSTEM

A typical 1993 model year automobile emission control system containing a multipoint fuel injection fuel metering system, which meters the fuel in response to a mea-

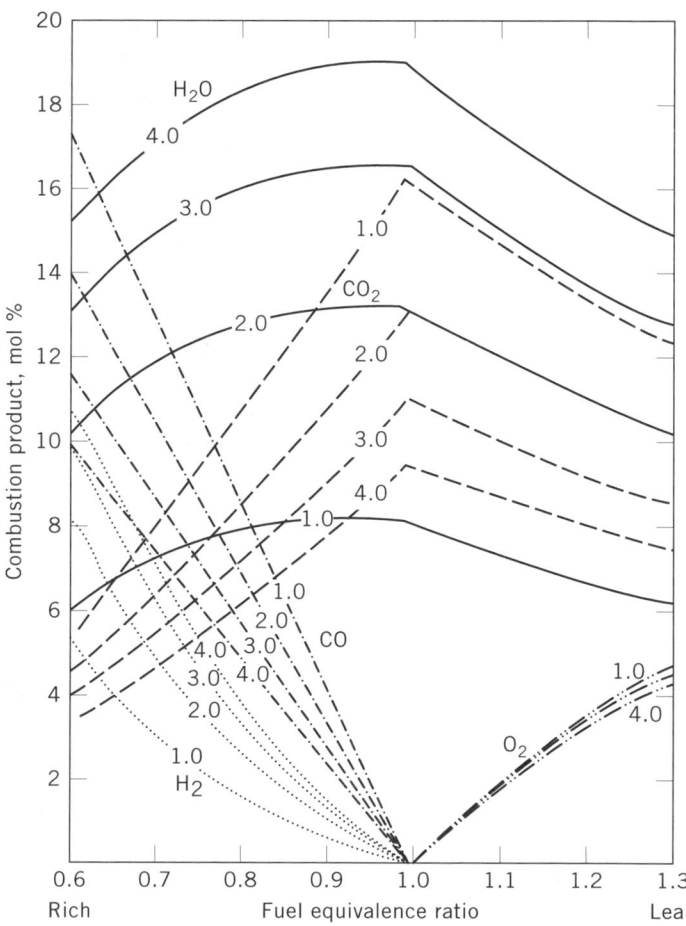

Figure 3. Theoretical mole percent of the principal combustion products of hydrocarbon fuels for fuel hydrogen:carbon ratios from 1, eg, C_6H_6, to 4, eg, CH_4, wet basis, where (—) represents H_2O; (– – –) CO_2; (–·–··–) CO; (–··–··–) O_2; and (········) H_2 (8,10,11).

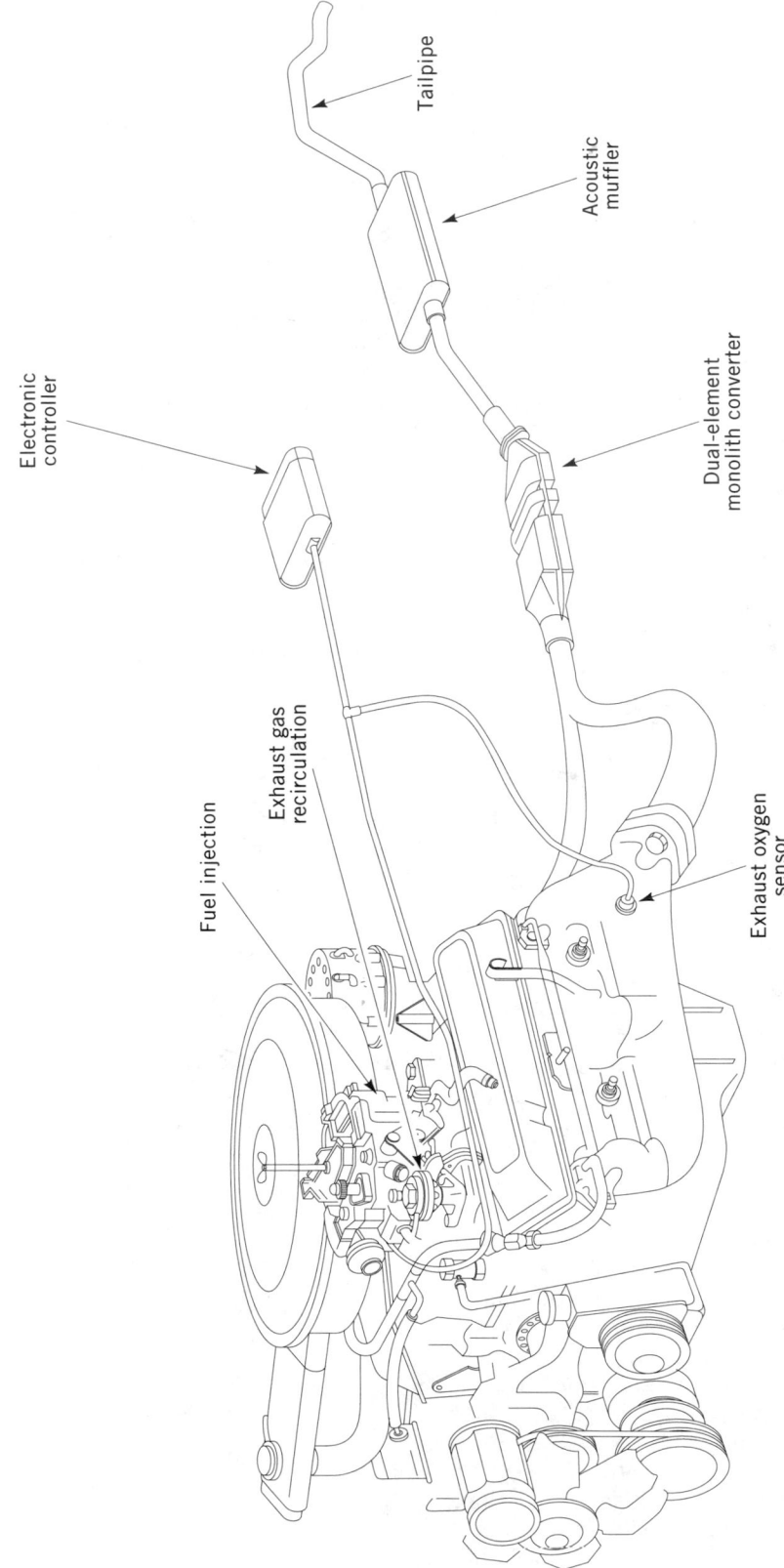

Figure 4. Closed-loop dual-catalyst system for emissions control using dual element mono-lith converter, which is three-way and oxidizing.

sured amount of air, is shown in Figure 4. Incorporated into the exhaust stream prior to the catalytic converter is an oxygen sensor which indicates whether the exhaust air/fuel mixture is rich or lean of the stoichiometric point (defined as neither air nor fuel in excess). The catalytic converter is located in the exhaust line leading from the exhaust manifold, upstream of the acoustic muffler. The signal generated by the oxygen sensor is sent to the computer controller as is a signal from an air flow measurement device. The computer controller regulates the fuel metered by the fuel injection system in response to the air measurement signal. Air flow varies in response to the throttle position and load (inlet vacuum). The oxygen sensor quickly detects any change in oxygen concentration in the exhaust and the controller adjusts for this change. Thus the air/fuel ratio is constantly being adjusted back and forth slightly rich and slightly lean of the stoichiometric mixture. The three-way conversion (TWC) catalyst therefore receives exhaust gas that reflects this constant change back and forth in air/fuel mixture and is designed to operate under those conditions to convert NO_x by reduction and HC and CO by oxidation of at least 80 to 90%.

Catalytic Converter

The converter consists of a catalytic unit contained in a metal canister which surrounds the fragile ceramic catalytic unit with a steel shell. The converter shell assembly is usually made from an iron/chrome Series 409 muffler-grade stainless steel that is resistant to internal and external oxidative corrosion. In between the steel shell and the exterior of the catalytic unit is a compliant layer that grips the catalytic unit with sufficient force to prevent movement of the catalytic unit within the canister, and which compensates for the differences in thermal expansion between the catalyst and the metal shell. Several types of compliant layers are used, and all have spring-like properties under compression that provide the necessary gripping force at all exhaust temperatures. Corrugated knitted wire mesh was the first successful compliant layer. As of this writing, a material based on vermiculite, which expands upon application of heat (about 300°C), is used, as is a wire mesh material wrapped several times around the catalytic unit. The compliant layer mounting system has proved to be durable for the life of the vehicle. The converter design, flow, and pressure drop characteristics are described in the literature (19–23).

The catalytic unit is designed to provide enough surface area so that all exhaust gases contact the catalyst surfaces as they pass from the engine to the tailpipe (24–27). In order to function quickly after the engine is started, the catalytic unit must rapidly heat up to operating temperature. It therefore must possess good heat-exchange properties to extract the necessary heat from the exhaust gas. Once the minimum catalytic operation temperature is reached, the catalytic unit is designed to maximize transfer of the pollutants from the exhaust gas to the surface of the catalytic unit. Heat transfer and mass transfer (qv) are driven by temperature difference and concentration difference, respectively. At operating temperatures above 300 or 350°C, the catalytic reactions are so fast that only the exterior surfaces of the catalyst are utilized for the catalytic function (28).

Automobile exhaust catalysts have been developed that maximize the catalyst surface area available to the flowing exhaust gas without incurring excessive pressure drop. Two types have been extensively studied: the monolithic honeycomb type and the pellet type.

Use of the pelletted converter, developed and used by General Motors starting in 1975, has declined since 1980. The advantage of the pelletted converter, which consists of a packed bed of small spherical beads about 3 mm in diameter, is that the pellets were less costly to manufacture than the monolithic honeycomb. Disadvantages were the pelletted converter had 2 to 3 times more weight and volume, took longer to heat up, and was more susceptible to attrition and loss of catalyst in use. The monolithic honeycomb can be mounted in any orientation, whereas the pelletted converter had to be downflow. Additionally, the pressure drop of the monolithic honeycomb is one-half to one-quarter that of a similar function pelletted converter.

Pelletted converters are used by General Motors for heavy-duty gasoline fueled trucks because the pellets have better high temperature physical characteristics and the converter has been redesigned to minimize pressure drop for this application (19). However, these are expected to be replaced by monolithic converters. In Japan, taxicabs have used pelletted converters and LPG (propane) fuel since the 1960s. The catalyst in these converters is changed once per year. It is thought easier to change pellets than to change a monolithic honeycomb catalyst. Although this practice of changing catalyst remains, the monolithic honeycomb converter remains active and fully functional for the life of the vehicle, especially when operated on a clean fuel such as LPG.

Catalytic Unit

The catalytic unit consists of an activated coating layer spread uniformly on a monolithic substrate. The catalyst predominantly used in the United States and Canada is known as the three-way conversion (TWC) catalyst, because it destroys all three types of regulated pollutants: HC, CO, and NO_x. Between 1975 and the early 1980s, an oxidation catalyst was used. Its use declined with the development of the TWC catalyst. The TWC catalytic efficiency is shown in Figure 5. At temperatures of >300°C a TWC destroys HC, CO, and NO_x effectively when the air/fuel mixture is close to stoichiometric, as shown. Other conventions for describing the air/fuel mixture are use of the Greek letter lambda (λ) where $\lambda = 1.0$ is the stoichiometric mixture, REDOX ratio, and equivalence ratio. When the air/fuel mixture is near the stoichiometric point, an optimum value of conversion for all three components is achieved. The principal chemical reactions of a TWC catalyst are

Oxidation reactions

$$H_2 + \tfrac{1}{2}O_2 \rightarrow H_2O$$
$$CO + \tfrac{1}{2}O_2 \rightarrow CO_2$$
$$C_mH_n + (m + n/4)O_2 \rightarrow m\,CO_2 + n/2\,H_2O$$

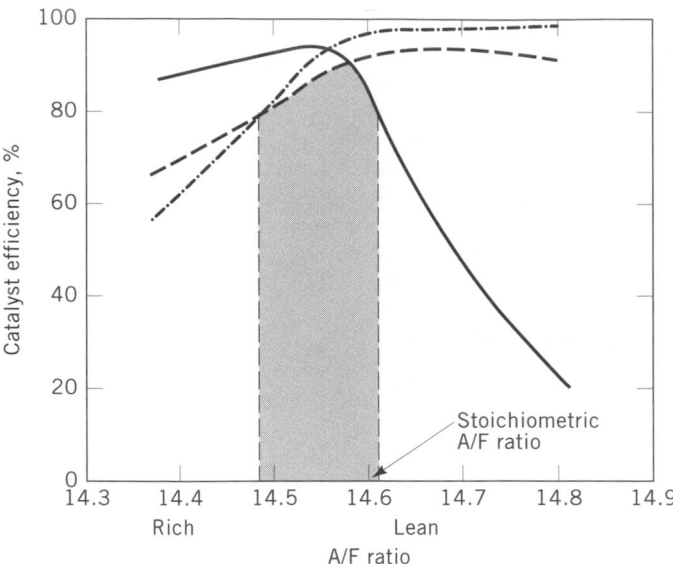

Figure 5. Conversion of (—) NO_x, (–·–·) CO, and (– –) hydrocarbons for a TWC as a function of the air/fuel (A/F) ratio. The shaded area shows the A/F ratio window where the TWC catalyst functions at 80% efficiency at 400°C.

TWC reactions at stoichiometric A/F mixture

$$CO + NO \rightarrow \tfrac{1}{2} N_2 + CO_2$$

$$C_mH_n + 2(m + n/4)\, NO \rightarrow (m + n/4)\, N_2 \\ + n/2\, H_2O + m\, CO_2$$

$$H_2 + NO \rightarrow \tfrac{1}{2} N_2 + H_2O$$

$$CO + H_2O \rightarrow CO_2 + H_2 \quad \text{water gas shift}$$

$$C_mH_n + m\, H_2O \rightarrow m\, CO + \left(m + \frac{n}{2}\right) H_2 \quad \text{steam reforming}$$

Other NO-reduction reactions

$$2\, NO + 5\, CO + 3\, H_2O \rightarrow 2\, NH_3 + 5\, CO_2$$

$$2\, NO + 5\, H_2 \rightarrow 2\, NH_3 + 2\, H_2O$$

$$NO + \text{hydrocarbons} \rightarrow N_2 + H_2O + CO_2 + CO + NH_3$$

$$\left.\begin{array}{l} 2\, NO + CO \rightarrow N_2O + CO_2 \\ 2\, NO + H_2 \rightarrow N_2O + H_2O \end{array}\right\}$$

at 200°C (below exhaust gas temperature)

Fuel sulfur reactions

$$S + O_2 \rightarrow SO_2$$

$$SO_2 + \tfrac{1}{2} O_2 \xrightarrow[\text{Pt}]{} SO_3$$

$$3\, SO_3 + Al_2O_3 \underset{>600°C}{\rightleftharpoons} Al_2(SO_4)_3$$

$$Al_2(SO_4)_3 + 12\, H_2 \xrightarrow[\text{PM catalyst}]{} 3\, H_2S + Al_2O_3 + 9\, H_2O$$

$$SO_2 + 3\, H_2 \xrightarrow[\text{PM catalyst}]{} H_2S + 2\, H_2O$$

where PM is precious metal (6,29).

Activated Coating. The activated coating layer of a TWC is applied to the high geometric surface area monolithic honeycomb body or substrate. Precious metals are dispersed within this catalytic layer, which contains particles as small as 10 μm. The thickness of the layer is between 25 and 100 μm. The honeycomb substrate typically has a geometric surface area of 27 cm²/cm³ so that a typical vehicle catalyst volume of 1.6 L would contain 4.6 m² of geometric surface area. The activated coating layer increases the total surface to 7000 m², ie, more than three orders of magnitude. The total surface area consists of the combined internal and external surface of all the minute particles contained in the activated coating layer. Total surface area is measured by the classic Brunauer-Emmet-Teller (BET) technique, which involves N_2 adsorption at low temperature (30).

The small (10 μm) coating particles are typically aluminum oxide [1344-28-1], Al_2O_3. These particles can have BET surface areas of 100 to 300 m²/g. The thermal and physical properties of alumina crystalline phases vary according to the starting phase (aluminum hydroxide or hydrate) and thermal treatment.

Alumina is used because it is relatively inert and provides the high surface area needed to efficiently disperse the expensive active catalytic components. However, no one alumina phase possesses the thermal, physical, and chemical properties ideal for the perfect activated coating layer. A great deal of research has been carried out in search of modifications that can make one or more of the alumina crystalline phases more suitable. For instance, components such as ceria, baria, lanthana, or zirconia are added to enhance the thermal characteristics of the alumina. Figure 6 shows the thermal performance of an alumina-activated coating material.

An unstabilized high surface area alumina sinters severely upon exposure to temperatures over 900°C. Sintering is a process by which the small internal pores in the particles coalesce and lose large fractions of the total surface area. This process is to be avoided because it occludes some of the precious metal catalyst sites. The network of small pores and passages for gas transfer collapses and restricts free gas exchange into and out of the activated catalyst layer resulting in thermal deactivation of the catalyst.

The activated coating layer must possess two additional properties. It must adhere tenaciously to the monolithic honeycomb surface under conditions of rapid ther-

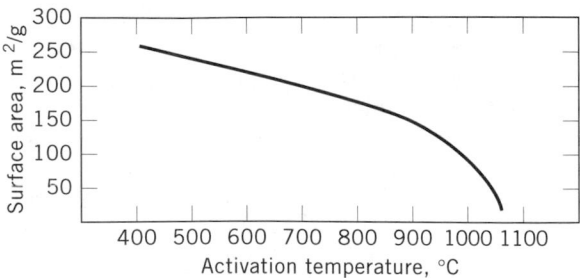

Figure 6. Surface area stability of Pural SB. The activation time was 3 h.

mal changes, high flow, and moisture condensation, evaporation, or freezing. It must have an open porous structure to permit easy gas passage into the coating layer and back into the main exhaust stream. It must maintain this porous structure even after exposure to temperatures exceeding 900°C.

Precious Metal Catalysts. Precious metals are deposited throughout the TWC-activated coating layer. Rhodium plays an important role in the reduction of NO_x, and is combined with platinum and/or palladium for the oxidation of HC and CO. Only a small amount of these expensive materials is used (31) The metals are dispersed on the high surface area particles as precious metal solutions, and then reduced to small metal crystals by various techniques. Catalytic reactions occur on the precious metal surfaces. Whereas metal within the crystal cannot directly participate in the catalytic process, it can play a role when surface metal oxides are influenced through strong metal to support reactions (SMSI) (32,33). Some exhaust gas reactions, for instance the oxidation of alkanes, require larger Pt crystals than other reactions, such as the oxidation of CO (34).

The small precious metal crystals can exist as metal crystallites or as metal oxides, both of which are catalytic (31). Rhodium oxide has a tendency to react with alumina to form a solid solution (35). To minimize this reaction, zirconia is used with the alumina (36). Publications regarding the TWC function of precious metals abound (37–42).

Catalytic Support Body Monolithic Honeycomb Unit

The terms substrate and brick are also used to describe the high geometric surface area material upon which the active coating material is placed. Monolithic honeycomb catalytic support material comes in both ceramic and metallic form. Both are used in automobile catalysts and each possesses unique properties. A common property is a high geometric surface area which is inert and does not react with the catalytic layer.

Ceramic. The ceramic substrate is made from a mixture of silicon dioxide, talc, and kaolin to make the compound cordierite [12182-53-5]. Cordierite possesses a very low coefficient of thermal expansion and is thermal-shock resistant. The manufacturing process involves extruding the starting mixture (which is mixed with water and kneaded into a sort of dough) through a complex die to form the honeycomb structure. The extruded piece is dried and fired in a kiln to form the cordierite. The outside or circumferential dimension is formed by the die, and the length is cut later with a ceramic saw.

The extrusion of the honeycomb was a significant advance aiding the adoption of the monolithic honeycomb catalyst by the automobile industry as the preferred means of emission control (43). Coupled with this was the development of the mounting system (44) securing the fragile ceramic substrate-based catalytic unit in a protective canister. Some of the physical properties considered when selecting a substrate for a catalytic converter are given in Table 2. The most common cell structure used is 62 cells/cm^2 with a 0.152 mm thick wall. This cell density is about optimal with respect to pressure drop, geometric surface area, physical strength, and general ruggedness (24–27).

Metallic. Metallic substrates are also used as a support for the activated coating layer. A class of metal alloys containing Fe, Cr, and Al, when stabilized with Y or Ce, have excellent oxidation resistance at the extreme temperatures found in automobile exhaust (45). Melting temperatures are about in the same range as that of the ceramic cordierite. A feature of these alloys is that upon heating in air, an aluminum oxide surface layer is formed. The yttria or ceria is present to stabilize the aluminum oxide surface and resist the spalling of the alumina which would otherwise occur. The ceria-stabilized alloy develops a whisker-like surface. The catalytic activated coating layer has been found to adhere very well to these surfaces.

The metallic substrate is designed to provide a cell density similar to that of the ceramic counterpart, and can be constructed in versions of 46, 62, or 93 cells/cm^2. However, the wall thickness can be thinner, ie, 0.05 to 0.15 mm thick (46). Additionally, 0.05 mm thick wall material can be constructed in versions of 124 to 248 cells/cm^2. The unique advantage that a metal substrate has

Table 2. Physical Properties of Catalytic Converter Substrates

Parameter	Cordierite Value		
Cell structure			
Wall thickness, mm	0.152	0.203	0.304
cells/cm^2	62	46	46
Pressure drop, %	100	90	120
Geometric surface area, cm^2/cm^3	27	24	22
Isostatic strength,[a] N/m^2	>103	>103	>206
Bulk density, g/cm^3	0.4	0.4	0.5
Cross-sectional shape			

[a] To convert N/m^2 (=Pa) to kgf/cm^2, mulitply by 1.02 × 10^{-5}.

over its ceramic counterpart is that either the same geometric surface can be made into a smaller volume, or that for the same volume and geometric surface area, there is a lower pressure drop (47). The 124 to 248 cells/m^2 metal versions have an additional potential performance advantage over ceramic because of large increases in geometric surface area. Metallic substrate catalysts are used by Porsche for high performance vehicles and by Chrysler for the Viper model (45,48). In both of these cases, the lower exhaust gas pressure drop results in increased horsepower and performance. In the future, metallic supports may be used for small catalysts located close to the exhaust manifold in order to heat up faster in emission control systems calibrated to meet the strict California Low Emission Standards.

Mass Transfer

Exhaust gas catalytic treatment depends on the efficient contact of the exhaust gas and the catalyst. During the initial seconds after start of the engine, hot gases from the exhaust valve of the engine pass through the exhaust manifold and encounter the catalytic converter. Turbulent flow conditions (Reynolds numbers above 2000) exist in response to the exhaust stroke of each cylinder (about 6 to 25 times per second) times the number of cylinders. However, laminar flow conditions are reached a short (~ 0.6 cm) distance after entering the cell passages of the honeycomb (5,49–52).

The process of catalyst heating and initiation of the catalytic function is shown in Figure 7, where there are three or four distinct regions. Depending on the location of the catalytic unit in the exhaust system and the thermal mass present prior to the catalytic unit, it can take from 30 to 120 seconds for the catalyst unit to reach the catalyst ignition temperature of approximately 250–300°C (Region I). At temperatures above this ignition point, the activity of the catalyst increases rapidly with temperature (Region II). Some catalyst reaches a point where the sharp rise with temperature abruptly takes on a mild positive slope (Region III). Then, a point is reached at which catalytic performance improves only slightly to an increase in temperature (Region IV).

In Region I (below the ignition point) there is no catalytic activity. Within Region II, the specific activity of the catalyst is the rate-limiting step; this is called the kinetically controlled region. Highly active catalysts have a lower ignition point and exhibit large increases in catalytic performance associated with small increases in temperature. The behavior of a catalyst in Region II is important to selecting an auto emission catalyst because catalyst light-off is a prime factor in achieving adequate emission control early in the FTP test.

Region III is not present in all catalysts. It depends on the porous structure design of the catalyst and is often only found in a used catalyst. If present, the rate-limiting step is called pore diffusion control. The volume of the catalytic unit dictates the catalyst performance in Regions III and IV. Catalysts clogged with masking poisons such as lube oil ash would exhibit Region III behavior. Region IV is a mass-transfer limited region, ie, catalytic reactions occur so fast that the rate-limiting factor is getting the reactants to the surface of the catalyst. Mathematical models describing the entire process are found in several references (5,49–52).

Catalyst Function

Automobile exhaust catalysts are perfect examples of materials that accelerate a chemical reaction but are not consumed. Reactions are completed on the catalyst surface and the products leave. Thus the catalyst performs its function over and over again. The catalyst also permits reactions to occur at considerably lower temperatures. For instance, CO reacts with oxygen above 700°C at a substantial rate. An automobile exhaust catalyst enables the reaction to occur at a temperature of about 250°C and at a much faster rate and in a smaller reactor volume. This is also the case for the combustion of hydrocarbons.

Concerning the reduction of NO$_x$, automobile three-way catalysts exhibit a property called selectivity. Catalyst selectivity occurs when several reactions are thermodynamically possible but one reaction proceeds at a faster rate than another. In the case of a TWC catalyst, CO, HC, and H$_2$ are all potential reductants of NO. On the other hand, O$_2$ is present, which oxidizes the CO, HC, and H$_2$. If these oxidation reactions are too rapid, no reductant is available to convert NO. Using modern TWC catalysts, however, NO reduction is fast enough that it is substantially completed before the reductants are consumed by O$_2$.

Two classes of metals have been examined for potential use as catalytic materials for automobile exhaust control. These consist of some of the transitional base metal series, for instance, cobalt, copper, chromium, nickel, manganese, and vanadium; and the precious metal series consisting of platinum [7440-06-4], Pt; palladium [7440-05-3], Pd; rhodium [7440-16-6], Rh; iridium, [7439-88-5], Ir; and ruthenium [7440-18-8], Ru. Specific catalyst activities are shown in Table 3.

The precious metals possess much higher specific catalytic activity than do the base metals. In addition, base metal catalysts sinter upon exposure to the exhaust gas temperatures found in engine exhaust, thereby losing the catalytic performance needed for low temperature operation. Also, the base metals deactivate because of reactions

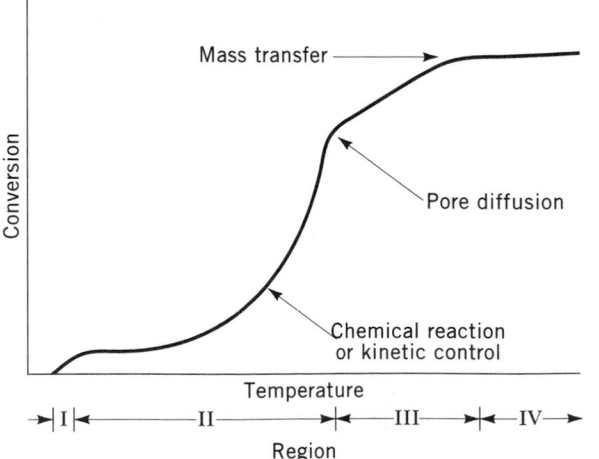

Figure 7. Conversion vs temperature.

Table 3. Specific Reaction Rates

| Catalyst | $CO + O_2^a$ | | $C_2H_4 + O_2^b$ | |
	Temperature, °C	R^c	Temperature, °C	R^c
Pd (wire)	300	500	300	100
Pt (wire)	300	100	300	10
Au (wire)	300	15	300	0.03
Co_3O_4	200	25 + 5	300	0.33
CuO	200	11	300	0.6
$CuCr_2O_4$	200	5	300	0.8
$LaCoO_3$	200	2.3	400	0.53
$BaCoO_3$	300	5.3	400	0.1
SnO_2	300	5.2	400	0.4
MnO_2	300	3.4	300	0.04
$LaMnO_3$	300	2	400	0.3
$La_{0.5}Sr_{0.5}MnO_3$	300	1.2	400	0.26
$La_{0.5}Pb_{0.3}MnO_3$	300	0.5	400	<0.05
Fe_2O_3	300	0.4	400	0.06
$FeCr_2O_4$	300	0.33		
Cr_2O_3	300	0.03	300	0.006
NiO	300	1.5		
ZrO_2	300	0.013	400	0.002

^{a} 1% O_2, 1% CO, 0% H_2O.
^{b} 1% O_2, 0.1% C_2H_2, ~0.1% H_2O.
^{c} Units in terms of CO_2 formation mL/(min·m²).

with sulfur compounds at the low temperature end of auto exhaust. As a result, a base metal automobile exhaust catalyst would need to be considerably larger than a precious metal one and, even if a large bed were used, it would not heat up quickly enough to achieve the catalytic performance demanded of the emission control systems (6).

Catalyst function in an exhaust gas stream can be understood by examining the catalyst performance curves shown in Figures 5 and 8. Figure 8 shows catalyst function at a specific set of flow conditions as the catalyst inlet temperature is slowly increased. This test, conducted in a laboratory reactor, uses a gaseous mix to simulate the composition of automobile exhaust. As the temperature increases from ambient to about 200°C, there is no apparent action on the part of the catalyst to consume any of the reactants. The carbon monoxide present strongly chemisorbs on the surface of the catalyst, and prevents oxygen access. As the inlet gas temperature approaches 200°C, the CO bonds to the metal surface are relaxed. Oxygen molecules are now able to chemisorb, and catalytic ignition occurs.

Beyond the catalytic ignition point there is a rapid increase in catalytic performance with small increases in temperature. A measure of catalyst performance has been the temperature at which 50% conversion of reactant is achieved. For carbon monoxide this is often referred to as T_{50} CO. The catalyst light-off property is important for exhaust emission control because the catalyst light-off must occur reliably every time the engine is started, even after extreme in-use engine operating conditions.

A representation of a catalytic reaction is shown in Figure 9. A gaseous pollutant such as CO diffuses from the main gas stream into the porous catalyst matrix. There is a net CO flow from the main flow stream to the catalytic surface because CO is at higher concentration in the ex-

haust stream, and it is attracted to a precious metal site that is temporarily unoccupied. For the same reasons, oxygen molecules are attracted to the site. The oxygen molecule is chemisorbed onto the precious metal surface where oxygen molecule bond stretching occurs, leading to dissociation to oxygen atoms that bond to the catalyst surface. The rate at which the reaction proceeds depends strongly on the temperature at the catalyst surface. The reaction of carbon monoxide and oxygen proceeds to form CO_2, which desorbs and reenters the exhaust. There is less CO_2 in the flowing exhaust stream and more at the surface so there is a net flow of CO_2 away from the catalyst surface.

The oxidation of HC and CO must proceed in balance with the reduction of NO_x by CO, HC, or H_2. For the NO_x removal reaction, a reductant is required. First NO is ad-

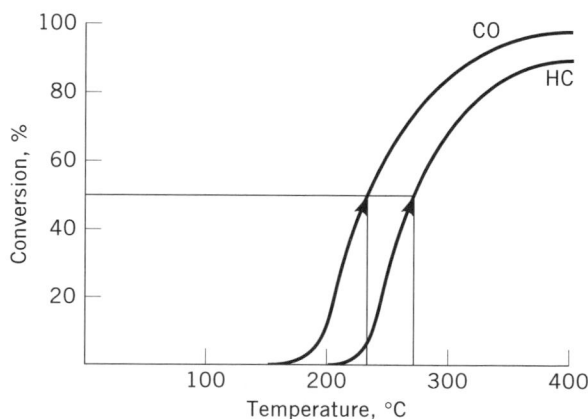

Figure 8. Effect of temperature on catalyst performance using synthetic exhaust: 0.45% CO, 12% CO_2, 500 ppm NO, 0.15% H_2, 3% O_2, 200 ppm C_3H_6, 10% H_2O, and remainder N_2. The T_{50} for CO and HC are 230°C and 270°C, respectively. See text.

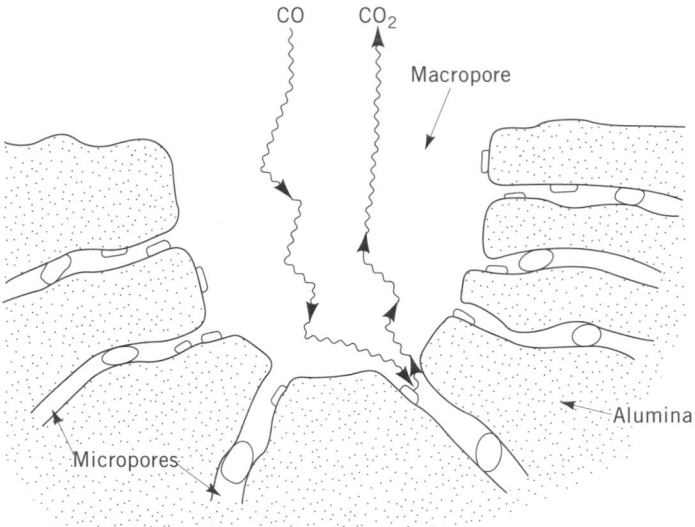

Figure 9. Catalyst pore and reaction. The CO diffuses into a precious metal site ; reacts with O_2; and leaves as CO_2.

sorbed on the catalyst surface and dissociates forming N_2 which leaves the surface, but the O atoms remain. CO is required to remove the O atoms to complete the reaction cycle (53).

A TWC catalyst must be able to partition enough CO present in the exhaust for each of these reactions and provide a surface that has preference for NO adsorption. Rhodium has a slight preference for NO adsorption rather than O_2 adsorption; Pt prefers O_2. Rh also does not catalyze the unwanted NH_3 reaction as does Pt, and Rh is more sinter-resistant than Pt (6). However, the concentrations of O_2 and NO have to be balanced for the preferred maximum reduction of NO and oxidation of CO. This occurs at approximately the stoichiometric point with just enough oxidants (O_2 and NO_x) and reductants (CO, HC, and H_2). If the mixture is too rich there is not enough O_2 and no matter how active the catalyst, some CO and HC is not converted. If the mixture is too lean, there is too

much O_2 and the NO cannot effectively compete for the catalyst sites (53–58).

In an actual exhaust system controlled by the signal of the oxygen sensor, stoichiometry is never maintained, rather, it cycles periodically rich and lean one to three times per second, ie, one-half of the time there is too much oxygen and one-half of the time there is too little. Incorporation of cerium oxide or other oxygen storage components solves this problem. The ceria adsorbs O_2 that would otherwise escape during the lean half cycle, and during the rich half cycle the CO reacts with the adsorbed O_2 (32,44,59–63). The TWC catalyst effectiveness is dependent on the use of Rh to reduce NO_x and upon the proper balance of reactants; the oxygen sensor plays the key role in maintaining this proper balance.

Catalyst Durability

Automobile catalysts last for the life of the vehicle and still function well at the time the vehicle is scrapped. However, there is potential for decline in total catalytic performance from exposure to very high temperatures, accumulation of catalyst poisons, or loss of the active layer (29,64–68).

High Temperatures. Exposure to temperatures above 1000°C can cause the active layer surfaces to sinter, collapse, agglomerate, shrink, and possibly exfoliate. Collapse of the surfaces occludes precious metal sites, preventing free passage to these sites. Also, the small precious metal crystals can unite and grow into larger crystals, thereby diminishing the total precious metal surface and thus the total number of active surface sites. If catalyst temperatures exceed 1200°C, the substrate softens and perhaps shrinks. At about 1465°C, the substrate melts and is destroyed. A thermally damaged pore is shown in Figure 10a.

High catalyst temperatures are caused by two types of factors. The first are engine related and can be corrected by calibration. That is, ignition timing and the air/fuel mixture, as well as the engine load, govern the exhaust gas temperature and thus the catalyst inlet gas tempera-

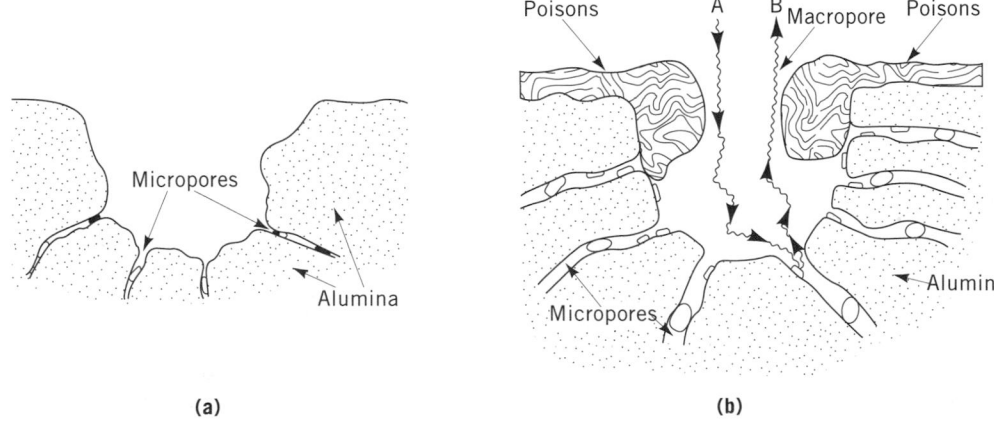

(a)   (b)

Figure 10. Catalyst macropores showing □ noble metal sites and (**a**) narrowed micropores after exposure to high temperatures where □ represents thermally damaged noble metal sites; and (**b**) pore mouth plugging from poisons where A, if allowed, diffuses in to be converted to B.

ture. The catalyst consumes unburned exhaust components, and most of the corresponding reactions are exothermic, resulting in increased catalyst and exhaust gas temperatures (33). For instance, the combustion of 1% CO would result in a 87°C adiabatic temperature rise of the exhaust gas (33); the combustion of 1000 ppm C_6H_6 would yield a 103°C temperature rise.

The second type is the quality of combustion. A misfire or partial misfire of the air/fuel charge in the combustion chamber causes the unburned air and fuel mixture to pass to the catalyst, where combustion occurs. The temperature rise from combustion in the catalyst is considerable. Two misfiring cylinders in a four-cylinder engine could cause the catalyst to approach the melting temperature of the substrate (1465°C) (69).

There are many reasons for misfire. The principal causes are interruption of ignition energy caused by faulty or wet ignition wiring and either lean or rich noncombustible air/fuel mixtures. Electronic ignition, and the development of very durable ignition wiring, as well as highly improved fuel metering systems, have greatly lessened the occurrence of misfire. Unleaded fuel does not clog, corrode, or wear spark plugs, extending spark plug life. Thus the ignition and fuel metering systems are vastly improved over pre-1975, pre-catalyst equipped vehicles. Catalysts have been designed to resist thermal degradation by the incorporation of base metal oxide stabilizers (70).

Poisons and Inhibitors. Catalyst poisons and inhibitors can come from the fuel, the lube oil, from engine wear and corrosion products, and from air ingestion. There are two types of catalyst poisons: one poisons active sites, the other is a masking agent.

The main catalyst site poison for many years was tetraethyllead [78-00-2], $C_4H_{12}Pb$, even after use of unleaded gasoline. Not only is lead a catalyst poison, but automotive source lead is also a health hazard (66). The source of this lead came from manufacture and distribution of leaded and unleaded gasoline in common transport equipment and storage facilities (67). In the early 1990s, so little leaded gasoline was being distributed that Pb contamination was approaching zero (<0.26 ppm/L).

The mechanism of poisoning automobile exhaust catalysts has been identified (71). Upon combustion in the cylinder tetraethyllead (TEL) produces lead oxide which would accumulate in the combustion chamber except that ethylene dibromide [106-93-4] or other similar halide compounds were added to the gasoline along with TEL to form volatile lead halide compounds. Thus lead deposits in the cylinder and on the spark plugs are minimized. Volatile lead halides (bromides or chlorides) would then exit the combustion chamber, and such volatile compounds would diffuse to catalyst surfaces by the same mechanisms as do carbon monoxide compounds. When adsorbed on the precious metal catalyst site, lead halide renders the catalytic site inactive.

Lead compounds were not found on the surrounding activated coating layer, rather only associated with the precious metal. The Pt sites are less poisoned by lead than are Pd or Rh sites because the Pt sites are protected by the sulfur in the fuel. Fuel sulfur is converted to SO_2

in the combustion process, and Pt easily oxidizes SO_2 to SO_3 on the catalyst site. The SO_3 reacts with the lead compounds to form $PbSO_4$, which then moves off the catalyst site so that lead sulfate is not a severe catalyst poison. Neither Pd nor Rh is as active for the SO_2 to SO_3 reaction, and therefore do not enjoy the same protection as Pt.

Sulfur oxides resulting from fuel sulfur combustion often inhibit catalyst performance in Regions II, III, and a portion of Region IV (see Fig. 7) depending on the precious metals employed in the catalyst and on the air/fuel ratio. Monolithic catalysts generally recover performance when lower sulfur gasoline is used so the inhibition is temporary. Pd is more susceptible than Rh or Pt. The last is the most resistant. Pd-containing catalysts located in hotter exhaust stream locations, ie, close to the exhaust manifold, function with little sulfur inhibition (72–74).

Fuel sulfur is also responsible for a phenomena known as storage and release of sulfur compounds. Sulfur oxides (SO_2, SO_3) easily react with ceria, an oxygen storage compound incorporated into most TWC catalysts, and also with alumina. When the air/fuel mixture temporarily goes rich and the catalyst temperature is in a certain range, the stored sulfur is released as H_2S yielding a rotten egg odor to the exhaust. A small amount of nickel oxide incorporated into the TWC removes the H_2S and releases it later as SO_2 (75–79).

Masking agents also deactivate catalysts and are loosely called poisons. Small droplets of lubricating oil, for instance, are emitted from the engine and deposit on the surfaces of the catalyst. Lubricating oil contains a small amount of inorganic elements such as zinc, phosphorus, calcium, and barium (see LUBRICATION AND LUBRICANTS). When the organic fractions of lube oil are combusted on the surface of the catalyst, these inorganic fractions can remain on the catalyst, usually as oxides. Zinc oxide and phosphorus pentoxide have been found in prodigious amounts on catalyst surfaces. A zinc pyrophosphate glaze has been found to form on the catalyst surface after exposure to certain temperatures (80–82). This glaze completely masks large areas of the catalyst surfaces, preventing passage of the gases into the catalyst porous structure. Materials such as calcium inhibit the formation of this glaze.

Silicone residue introduced to gasoline with toluene plugged catalysts on vehicles (83). Also a manganese-based octane improver known as MMT has been shown to clog catalyst surfaces (84).

The accumulation of matter on the surface of the catalyst restricts gas passage into the catalyst by a mechanism known as pore mouth plugging (see Fig. 10**b**). It takes only a small amount of material on the surface of the catalyst to restrict the free passage of gases into and out of the active porous catalyst layer.

Activated Layer Loss. Loss of the catalytic layer is the third method of deactivation. Attrition, erosion, or loss of adhesion and exfoliation of the active catalytic layer all result in loss of catalyst performance. The monolithic honeycomb catalyst is designed to be resistant to all of these mechanisms. There is some erosion of the inlet edge of the cells at the entrance to the monolithic honeycomb, but this loss is minor. The pelletted catalyst is more susceptible to attrition losses because the pellets in the catalytic bed rub against each other. Improvements in the design of the pelletted converter, the surface hardness of the pellets, and the depth of the active layer of the pellets also minimize loss of catalyst performance from attrition in that converter.

OXYGEN SENSOR AND THE CLOSED LOOP FUEL METERING SYSTEM

The first commercial application of the TWC catalyst and closed loop fuel metering system using an oxygen sensor came with the introduction of the 1977 Volvo for the California market (85–88). This catalyst was developed by Engelhard. Other car companies introduced the system the following year, and in the 1990s almost 100% of U.S. and Canada passenger cars and light-duty vehicles utilize it (89). The fully developed system is described in the literature (90).

The function of the oxygen sensor and the closed loop fuel metering system is to maintain the air and fuel mixture at the stoichiometric condition as it passes into the engine for combustion; ie, there should be no excess air or excess fuel. The main purpose is to permit the TWC catalyst to operate effectively to control HC, CO, and NO_x emissions. The oxygen sensor is located in the exhaust system ahead of the catalyst so that it is exposed to the exhaust of all cylinders (see Fig. 4). The sensor analyzes the combustion event after it happens. Therefore, the system is sometimes called a closed loop feedback system. There is an inherent time delay in such a system and thus the system is constantly correcting the air/fuel mixture cycles around the stoichiometric control point rather than maintaining a desired air/fuel mixture.

Oxygen Sensor

The oxygen sensor is also known as the lambda sonde or lambda sensor, from the greek letter used to denote the air/fuel ratio, and as an exhaust gas oxygen sensor (EGO). The lambda sensor responds in about 3 ms to a change in exhaust gas composition, yielding a voltage signal that is monitored by the computer control unit. The signal variation is from 50 mV at $\lambda = 1.05$ (lean) to 900 mV at $\lambda = 0.99$ (rich) as shown in Figure 11. The voltage changes sharply in the immediate vicinity of the stoichiometric ratio. A schematic of the lambda sensor is shown in Figure 12**a**. The sensor consists of a ceramic porous solid electrolyte that is ionically conductive at operating temperatures. The outside of the ceramic is coated with platinum electrodes: one electrode exposed to the exhaust gas, the second to outside air. Between these electrodes is a zirconia solid electrolyte. The voltage generated depends on the oxygen concentration difference between each electrode. The exhaust platinum electrode is a catalyst at the exhaust gas surface and equilibrates the mixture, consuming, by catalytic reaction, any unreacted oxygen and CO, HC, and H_2, yielding a net amount of oxygen present. Near the stoichiometric point, ie, $\lambda = 1.0$, the difference between atmospheric O_2 and exhaust O_2, and thus the voltage generated, approach a maximum. Thus

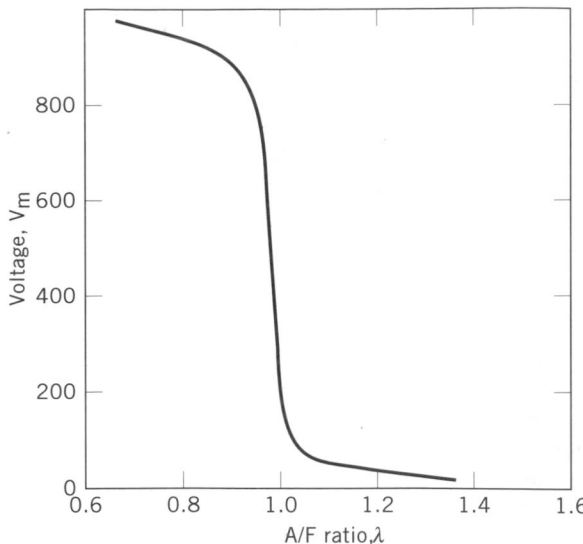

Figure 11. The A/F ratio and the lambda sensor's voltage signal. Courtesy of Robert Bosch.

the sensor signal is used to detect the stoichiometric point (91–95). In order for the sensor to function properly, it needs to be heated to above 350°C. To accommodate the need for quick catalyst heatup, a heated version of the oxygen sensor is used (96–99) (see Fig. 12**b**).

Computer Controller

The computer controller takes inputs of speed, load, and temperature to assist the engine in cold starting and to select the optimal air/fuel trim adjustment for optimal power, fuel economy, and emissions control. The control system uses the sensor signal as a switching device using the logic "rich! go lean!" or "lean! go rich!" Sophisticated adjustment logic is incorporated into the control unit to correct the air/fuel mixture in anticipation of change.

The oxygen sensor closed loop system automatically compensates for changes in fuel content or air density. For instance, the stoichiometric air/fuel mixture is maintained even when the vehicle climbs from sea level to high altitudes where the air density is lower.

The oxygen sensor is, however, fooled by certain gases. For instance, methanol combustion results in higher H_2 content in the exhaust, and because H_2 has a much higher rate of diffusion than other exhaust gases, a higher concentration is registered at the exhaust side electrode of the sensor than is actually present in the exhaust, causing a richer signal. The computer controller would then adjust the air/fuel to a leaner setting. This factor has to be compensated for when using fuels such as methanol.

Closed loop control has been designed for both carburetors and fuel injection metering systems. The latter are used in almost all 1990 models. Two types of fuel metering exist: a single fuel injector to serve all cylinders, called single-point fuel injection; and fuel injectors for each cylinder, called multipoint fuel injection. The multipoint fuel injection systems may be continuous or individual electrically activated. An electronic fuel injection valve is located in the inlet manifold just ahead of each

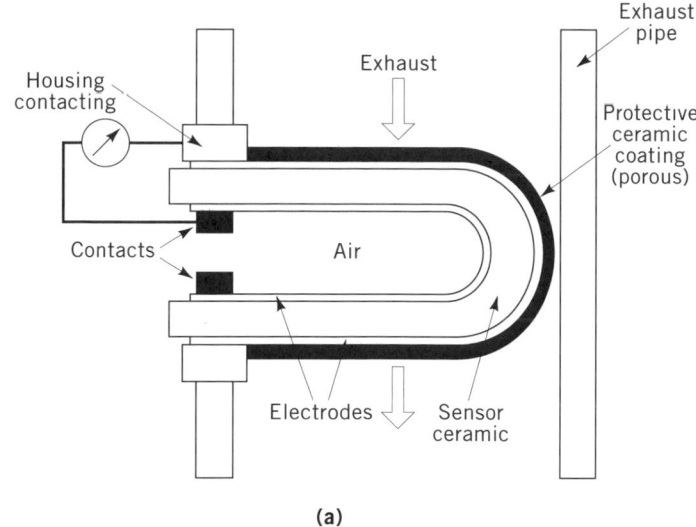

(a)

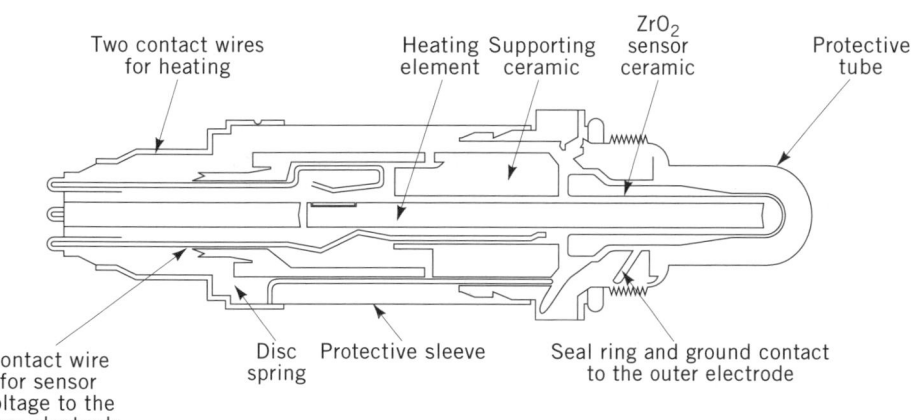

(b)

Figure 12. Schematic of lambda sensor (**a**) in exhaust pipe and (**b**) internal components. Courtesy of Robert Bosch.

inlet valve. All valves are connected in parallel and open for a calibrated time as called for by the computer controller. The injected fuel is swept into the cylinder along with the air. The latest development is sequential fuel injection that meters fuel to each cylinder according to the firing order.

All closed loop control systems must measure the amount of air needed under all conditions of engine demand. Air measurement is most often done using a hot wire anemometer, usually referred to as a mass air meter (99,100).

The performance of the catalytic converter is affected by the conditions of air/fuel control provided by the fuel metering system. A slowly responding fuel metering system can dramatically decrease the conversion efficiency of the converter compared to a fast response multipoint fuel injection system.

On-Board Diagnostics

State of California regulations require that vehicle engines and exhaust emission control systems be monitored by an on-board system to assure continued functional performance. The program is called OBD-II, and requires that engine misfire, the catalytic converter, and the evap-

orative emission control system be monitored (101). The U.S. EPA is expected to adopt a similar regulation.

One system for measuring catalyst failure is based on two oxygen sensors, one located in the normal control location, the other downstream of the catalyst (102,103). The second O_2 sensor indicates relative catalyst performance by measuring the ability to respond to a change in air/fuel mixture. Other techniques using temperatures sensors have also been described (104–107). Whereas the dual O_2 sensor method is likely to be used initially, a criticism of the two O_2 sensors system has been reported (44) showing that properly functioning catalysts would be detected as a failure by the method.

Oxidation Catalyst

An oxidation catalyst requires air to oxidize unburned hydrocarbons and carbon monoxide. Air is provided with an engine driven air pump or with a pulse air device. Oxidation catalysts were used in 1975 through 1981 models but thereafter declined in popularity. Oxidation catalysts may be used in the future for lean burn engines and two stroke engines.

The oxidation catalyst (OC) operates according to the same principles described for a TWC catalyst except that the catalyst only oxides HC, CO, and H_2. It does not reduce NO_x emissions because it operates in excess O_2 environments. One concern regarding oxidation catalysts was the ability to oxidize sulfur dioxide to sulfur trioxide, because the latter then reacts with water to form a sulfuric acid mist which is emitted from the tailpipe. The SO_2 emitted has the same ultimate fate in that SO_2 is oxidized in the atmosphere to SO_3 which then dissolves in water droplets as sulfuric acid.

Some OCs were of the monolithic honeycomb type, but all General Motors cars used pelletted OCs. For a period in the late 1970s and throughout the 1980s, both TWC and OC were used in a dual-bed catalyst. Oxygen needed for OC performance was provided by an engine driven air pump or a reed valve (pulse air valve) positioned in the exhaust pipe or manifold. In the case of dual-bed catalysts, air was injected between the TWC and OC. OCs utilize Pt and/or Pd but not Rh as a rule. The OC is subjected to the same exhaust environments as the TWC, except the air/fuel ratio is different (108,109).

Exhaust Gas Recirculation

In one method of NO_x emission control, exhaust gas is fed back into the inlet manifold and mixed with the fuel and inlet air. The resultant mixture upon combustion in the cylinder results in lower peak combustion temperature and lower NO_x formation because the reaction of $N_2 + O_2 \rightarrow NO_x$ is strongly dependent on the combustion flame temperature (99,109–112). The degree of NO_x depression is dependent on the amount of exhaust gas recirculation (EGR) as shown in Figure 13. EGR provides a diluent gas having high molecular weight and CO_2 which absorbs heat. Also, EGR affects the flame speed of the mixture, and thus provides a certain antiknock quality to the combustion process. The impact of EGR on engine parameters has been detailed (113).

EGR can seriously degrade engine performance, especially at idle, under load at low speed, and during cold start. Control of the amount of EGR during these phases can be accomplished by the same electronic computer controller used in the closed loop oxygen sensor TWC system. Thus the desired NO_x reduction is achieved while at the same time retaining good driveability.

OTHER EMISSIONS CONTROL

Evaporative Emission

Fumes emitted from stored fuel or fuel left in the fuel delivery system are also regulated by U.S. EPA standards. Gasoline consists of a variety of hydrocarbons ranging from high volatility butane (C-4) to lower volatility C-8 to C-10 hydrocarbons. The high volatility HCs are necessary for cold start, and are especially necessary for temperatures below which choking is needed to start the engine. Stored fuel and fuel left in the fuel system evaporates into the atmosphere.

Evaporative emissions sources are as follows. (1) Diurnal evaporative emissions are those that occur as ambient temperature fluctuates between daily high and low. The actual quantities of loss depend on the composition of the

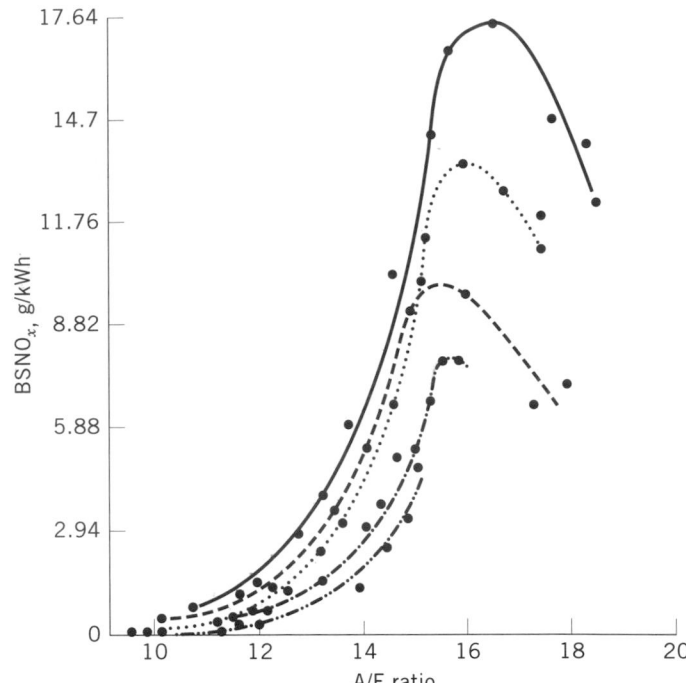

Figure 13. Effect of EGR on brake specific NO_x ($BSNO_x$) production where (—) represents no EGR, (– – –) no EGR and a 20° retard, (······) 5% EGR, (·–·–·) 10% EGR, and (–··–··) 15% EGR.

fuel and the daily temperatures. High concentrations of high volatility C-4 result in high vapor pressure and high evaporative emissions. (2) When the engine is turned off the engine and exhaust system heat up the fuel lines. This is known as hot soak, and fuel left in the fuel metering system or inlet manifold evaporates. (3) Evaporation also results from running loss, which occurs during operation at unusual conditions such as at low speeds on a hot day when the rate of vapor generation exceeds the rate that the engine can consume the vapor. (4) Evaporative loss source occurs during fueling as the liquid fuel displaces the fuel tank vapor.

Fuel vapor pressure, or Reid vapor pressure (RVP) is predominantly influenced by the amount of C-4 in the fuel. California, other of the United States, and the EPA have or are restricting fuel RVP for both seasonal and areas (latitude). These restrictions greatly reduce all gasoline evaporative losses.

The EPA regulation prohibits more than 2 grams per test to escape into the atmosphere (114). The test consists of a diurnal cycle of 1 hour where the temperature of the fuel is raised from 15.6 to 28.9°C during a 17 mile run on a chassis dynamometer. An immediate hot soak in a shed enclosure follows the dynamometer run.

The common method of controlling evaporative emission is an activated charcoal canister that connects to the intake manifold. When the engine is shut off, the valve permits hydrocarbon fumes to be absorbed and stored by the activated charcoal. When the engine is operating, the stored hydrocarbons are purged back into the manifold where they enter the combustion process. A new California regulation also controls running loss emissions of hydrocarbons to 0.03 g/km. There is discussion as to whether control of fueling emissions should be on-board the vehicle or incorporated into the fueling station system (115–118). As of 1993 the control is in the fueling station system.

Crankcase Emissions

Exhaust gases are also found in the crankcase. The principal source of these exhaust gases results from what is known as blow-by or gases from the combustion chamber, past the piston rings to the crankcase, or from the intake and exhaust valve mechanisms. As the engine wears, blow-by increases. Control systems are required to feed these gases back into the inlet manifold so that the hydrocarbons and carbon monoxide contained are consumed in the combustion process. The control device used is called a positive crankcase ventilation (PCV) valve. This device or one of similar function is required by law (119–121). Formerly, these gases were ventilated to the atmosphere through the engine breather tube. The test to measure crankcase emissions is defined in Reference 121.

ALTERNATIVE FUELS

Under the National Energy Policy Act of 1992 nonpetroleum-based transportation fuels are to be introduced in the United States. Such fuels include natural gas (see GAS, NATURAL), liquefied petroleum gas (qv) (LPG), methanol (qv), ethanol (qv), and hydrogen (qv), although hydrogen fuels are not expected to be a factor until after the year 2000 (see also ALCOHOL FUELS; HYDROGEN ENERGY).

Natural Gas

Natural gas, an abundant fuel resource in the United States, has sufficient reserves to fuel over 10×10^6 U.S. vehicles per year for the next 50 years (122). Natural gas is used in two forms as a transportation fuel compressed or liquefied at low temperatures. Tanks for the storage of compressed natural gas are heavy and larger in volume than for liquid fuels. However, the added cost is offset by an expected lower pump price compared to gasoline (123). Whereas the lack of public natural gas fueling stations and other factors make natural gas more attractive for fleet vehicles in the United States, small compressors for overnight natural gas refueling at household sites have been developed. Natural gas has a high antiknock quality, and engines designed for this fuel can have a higher compression ratio, yielding higher power and fuel economy. A disadvantage of compressed natural gas is that the driving range for a fuel load is usually 50% of that of gasoline. Work on an adsorptive system of natural gas storage at reduced storage pressures has been announced that should reduce compressed tank costs and may be able to improve range per fueling (124). Dual fuel systems have been developed in which the engine can run on either natural gas or gasoline.

Emission control for natural gas fueled engines consists of either the same principal components as used for gasoline or those designed for lean combustion. No evaporative emission control is required.

The catalysts for natural gas run engines are designed differently, however, because a primary portion of the exhaust from natural gas combustion is methane, and the catalysts are based on Pd rather than Pt. Spark ignited engines calibrated for natural gas usually are calibrated at stoichiometric conditions. Diesel engines re-engineered for natural gas are usually spark ignited and calibrated very lean. Pd/Rh is used for TWC catalysts for the stoichiometric engine; Pd oxidation catalysts are used for the lean re-engineered diesel engine. Both the U.S. EPA and the California standards exclude methane in the calculation of hydrocarbon emissions because methane has low reactivity in the atmosphere and has low ozone (qv) formation potential (125), and natural sources produce large quantities of methane. Methane, however, is a strong global warming gas and is regulated by a total hydrocarbon standard. The symmetrical methane molecule is the most refractory of the hydrocarbon family, and resists reactions on traditional auto exhaust catalysts at low temperatures. Pd catalysts need 450°C to oxidize methane but are nevertheless much better than all other precious metal catalysts. Methane is a poor reductant of NO_x (126,127). More recently, a Pd-based catalyst having poor dispersion (low metal surface area) was found to be superior to highly dispersed Pd for these reactions (128). Pd suffers temporary deactivation from SO_2 present in the exhaust. Natural gas contains very little sulfur. A small amount of mercaptan is added to give a characteristic odor to the fuel so that leaks in the distribution system can be detected. Pd-based catalysts therefore maintain performance much better than when used with gasoline (129–131).

LPG

LPG could be a principal alternative transportation fuel if its other uses were displaced by natural gas. A significant number of LPG fueling stations are located throughout the United States. LPG is a liquid fuel and does not suffer the same driving range problem as natural gas. Because LPG vapor pressure is high, the storage tank has to withstand 2800 kPa (400 psi).

The emission control system for LPG is the same as is used for gasoline fueled engines with the exception of the fuel metering system. No evaporative emission system is required. Both Pt–Rh and Pd–Rh catalysts are good for emission control of LPG fuel exhaust. Pt provides the lowest light off temperature for C_3H_8. The sulfur content of LPG is also very low so that Pd catalysts perform very well.

Methanol

Methanol is a liquid fuel made from natural gas, but can also be made from coal (qv), and the U.S. has huge coal resources. Engines designed for 100% methanol usually take advantage of the properties of methanol to provide high power. Methanol has high antiknock properties and can be used in high compression engines. Although methanol has a lower energy content, it has a higher latent heat of vaporization than gasoline. Evaporation in the inlet manifold cools and densifies the air/fuel charge to the engine, thus increasing the energy charge to the engine.

Theoretically, therefore, the engine could be smaller and still provide the same power as a gasoline fueled engine. The disadvantage of 100% methanol engines is cold starting of the engine. Methanol has very low vapor pressure at cold temperatures, so there is insufficient vapor phase to provide a combustible mixture with a choked amount of air to start the engine.

Addition of 15% gasoline to methanol to produce M85 fuel is an alternative. At temperatures above −6.7°C, reliable ignition of M85 fuel occurs because the gasoline provides the vapor phase necessary for ignition under choked condition.

Engines are also designed to use either gasoline or methanol and any mixture thereof (132–136). Such a system utilizes the same fuel storage system, and is called a flexible fueled vehicle (FFV). The closed loop oxygen sensor and TWC catalyst system is perfect for the flexible fueled vehicle. Optimal emissions control requires a fuel sensor to detect the ratio of each fuel being metered at any time and to correct total fuel flow.

The principal hydrocarbon emissions from a methanol fueled car are methanol and formaldehyde. Formaldehyde is a carcinogen and regulated to very low levels by California and the U.S. EPA (0.015 g/m for light-duty vehicles (LDV)) and standards going to 0.008 g/m for California ultra low emission vehicle (ULEV) standards. Uncontrolled formaldehyde emissions from a methanol fueled light-duty vehicle are on the order of 0.60 g/m. Most TWC catalysts destroy formaldehyde effectively after operating temperature is reached, ie, about 1 to 2 minutes after the engine is started. However, during the 1 to 2 minutes that the catalyst is heating, about 0.05 to 0.1 g/m of formaldehyde is passed unreacted through the catalyst. The emission control system for either 100% methanol or for M85 must have catalysts located as close as possible to the manifold to heat up faster. Specially designed Pd–Rh TWC catalysts and close coupled manifold catalysts are being utilized for FFV to minimize formaldehyde emissions as well as to meet the other emission standards (136). An electrically heated catalyst system obtained very low formaldehyde emissions from a methanol fueled vehicle (137).

FUTURE ENGINES AND EMISSION CONTROL SYSTEMS

Two engines are under development as of this writing: the two-stroke engine and the lean burn engine. The driving forces behind this development are fuel economy and global warming.

The National Highway and Safety Administration regulates vehicle fuel consumption through the Corporate Average Fuel Economy (CAFE). When vehicle manufacturers submit vehicles for certification, fuel consumption values must be given for a specified city and highway driving cycle (138). The average must be equal to or more than 11.6 km/L (27.5 mpg). Carbon dioxide is a principal global warming gas and conservation actions, such as increased CAFE, are thought to be a measure to stave off the buildup of CO_2 in the atmosphere.

Whereas automobile companies continue to improve the four-stroke engine, making it more efficient and more powerful to run vehicles that are smaller and lighter and have sufficient space, the two-stroke and the lean burn engines are also being studied.

Two-Stroke Engine

The two-stroke engine has the potential of delivering the same power as the four-stroke engine, but has fewer moving parts, lower weight, and a smaller engine volume, ie, it would be ideal for smaller, lighter, more fuel efficient vehicles having smaller engine compartments. Baseline emissions of two-stroke engines are very high in hydrocarbons. For example, a four-stroke engine may have 250 ppm hydrocarbons in its exhaust; the two-stroke engine could have 5000 ppm or greater. Designs that use inlet valves, fuel injection, and air compression have reduced baseline emissions, but have added complexity to the original simple two-stroke engine (139–147).

The operating air/fuel mixture of the two-stroke engine designs range from 1.3 to 2.0 stoichiometric. This lean mixture plus the characteristic internal exhaust gas recirculation lowers the peak combustion temperatures and results in low NO_x formation.

Emission control systems for two-stroke engines depend heavily on an efficient oxidation catalyst. These may be based on Pt and/or Pd. Higher lube oil consumption characteristics of two-stroke engines may result in modification to the lube oil or require the development of oxidation catalysts more resistant to lube oil ash compounds.

Lean Burn Engine

An engine calibrated at >18/1 A/F ratio, ie, very lean, encounters what is known as the lean limit (148). At this point, ignition of the air/fuel mixture becomes more difficult to achieve and subsequent propagation of the flame more difficult to sustain. Engineering solutions to both problems are diametrically opposed. At A/F ratios even more lean, combustion within the cylinder deteriorates and hydrocarbon emissions increase. Another result is loss of power. However, the production of NO_x is decreased at leaner A/F mixtures, and more importantly, thermal efficiency is improved and specific fuel consumption decreases. If engine designs could push back the lean limit, a practical lean burn engine may be a reality (149). To overcome the poor power and performance of a lean burn engine, a partial lean burn system was designed. The engine operated lean during idle and low cruise engine operating conditions. When acceleration was required, the engine adjusted to the stoichiometric air/fuel ratio. Toyota marketed a lean burn engine in Japan in the late 1980s and for a period in Europe without much customer acceptance (150). More recently, since 1990, several automobile manufacturers, such as Honda, Mitsubishi, Nissan, and Hyundai, have announced the development of the lean burn engine (151,152). All of these later developments used the Ford partial lean and partial stoichiometric calibration. Emission control for the lean burn/stoichiometric engine requires a TWC catalyst that can operate as a TWC catalyst at stoichiometric air/fuel mixtures and as an oxidation catalyst when the engine is operated lean.

Work on the development of the full time lean burn engine continues with efforts to push back the lean burn limit. A problem recognized by the developers is that although low basic engine NO_x emissions arc possible, it is not yet possible to meet the NO_x emissions required by California and the Tier II emission levels of the U.S. Clean Air Act of 1990 (see Table 1).

There has been a growing demand for a lean NO_x catalyst in order to decrease the relatively low NO_x emission of the lean burn engine sufficiently to meet the future standards. Lean NO_x catalysts have been developed based on zeolites. Cu-promoted ZSM-5 zeolite has shown ability to reduce NO_x in an exhaust having excess oxygen at an efficiency of 30 to 50% (153). Durability is not proven. Research has revealed that certain hydrocarbons are preferred for the reduction of NO_x, and that CO and H_2 apparently do not reduce NO_x over such lean NO_x catalysts (154). Considerable effort is being expended to develop a practical lean NO_x catalyst system (155–159).

Emission Control Technologies

The California low emission vehicle (LEV) standards has spawned investigations into new technologies and methods for further reducing automobile exhaust emissions. The target is to reduce emissions, especially HC emissions, which occur during the two minutes after a vehicle has been started (53). It is estimated that 70 to 80% of nonmethane HCs that escape conversion by the catalytic converter do so during this time before the catalyst is fully functional.

Technologies being investigated include improved catalysts that can be located closer to the exhaust manifold (high temperature resistant), reduced thermal mass in the exhaust system, use of alternative fuels, use of air pumps to assist catalyst light off, and engine adjustments that yield higher exhaust temperatures for this period. Three new technologies are undergoing intense development.

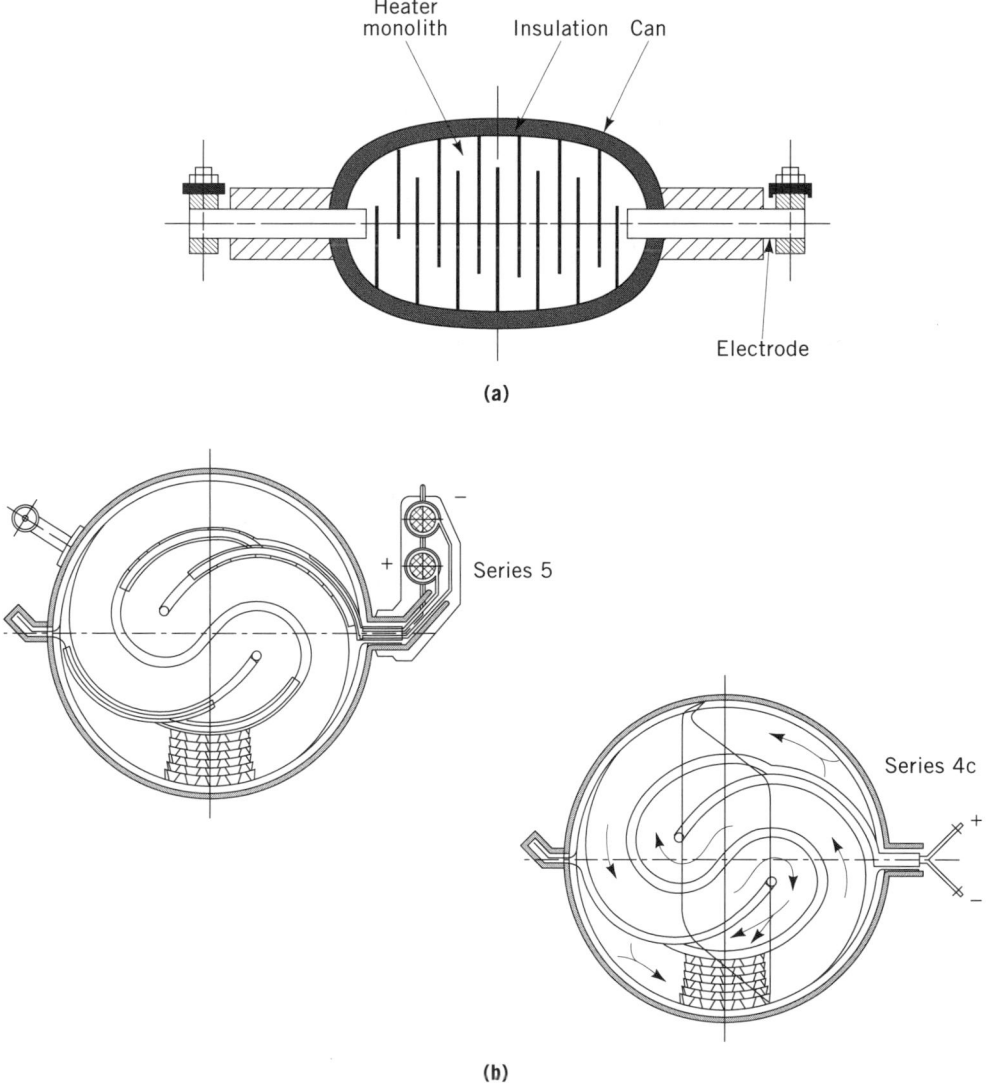

Figure 14. Cross-sectional schematics of electrically heated catalyst (EHC) for emission control (**a**) extruded sintered metal powder EHC (160); (**b**) two sintered metal foil EHCs. Courtesy of Emitec GmbH.

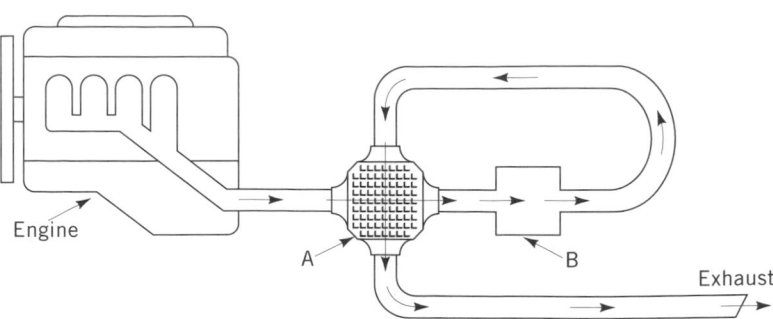

Figure 15. Low hydrocarbon emission control system utilizing a cross-flow heat exchanger TWC catalyst, A, and a zeolite-based hydrocarbon absorber system. Cold start HCs are absorbed by the hydrocarbon trap, B, until the cross-flow heat exchanger catalyst is hot enough to destroy the HCs that subsequently are eluted from the HC trap.

Electrically Heated Catalyst. An electrically heated catalyst (EHC) is an electrical resistive metallic foil or matrix which, like a toaster, will heat upon the application of voltage. Two basic types are being developed. One is based on a metal foil; the other is based on sintered metal powder. Figure 14a shows a cross-sectional view of a metal foil type, and Figure 14b shows a cross-sectional view of the sintered metal powder type. A voltage, typically 12 V or 24 V supplied from the automobile battery, is applied across the electrodes for about 15 to 20 seconds, in order to heat the foil or metal matrix from ambient temperature to 300–600°C in this time period. The metal foil or metal matrix surfaces are covered with a catalytic layer as described for the monolithic converter. Thus, when the metal surface achieves 300°C, catalyst light-off occurs. The system usually employs an air pump so that there is sufficient oxygen present to consume significant quantities of CO and HCs, resulting in an additional temperature rise.

The main converter, which is located downstream of the EHC, heats to functional temperature much more quickly because of catalytic combustion of exhaust gases that would otherwise pass unconverted through the catalyst during the cold start period. The EHC theoretical power required for a reference case (161) was 1600 watts to heat an EHC to 400°C in 15 s in order to initiate the catalytic reactions and obtain the resultant exotherm of the chemical energy contained in the exhaust. Demonstrations have been made of energy requirements of 15–20 Wh and 2 to 3 kW of power (160,161). Such systems have achieved nonmethane HC emissions below the California ULEV standard of 0.025 g/km. The principal issues of the EHC are system durability, battery life, system complexity, and cost (137,162–168).

Hydrocarbon Trap System. The concept of a hydrocarbon trap or adsorber system is based on molecular sieve hydrocarbon adsorber systems. The temperatures at which hydrocarbon adsorption takes place exist in the auto engine exhaust system during the period of cold start of an automobile when the catalytic control system has not yet reached functional temperature. Zeolites have been reportedly useful for hydrocarbon adsorption (53,169). Zeolites desorb hydrocarbons at temperatures of 400°C, ie, once the catalytic control system is functional. Therefore, hydrocarbons adsorbed by the zeolite can also be desorbed then oxidized by a catalyst. Methods to accomplish cold start hydrocarbon adsorption, heatup of the main catalyst, and desorption have been identified. Some of these systems use exhaust pipe valves to divert the exhaust gases to the hydrocarbon trap for the low temperature portion, and by-pass the gases around the trap after the main catalyst has heated up. One device that uses a heat exchanger is shown in Figure 15 (44). The Si–Al ratio in the zeolite is important, and by lowering the alumina content, the zeolite is rendered more hydrophobic and more able to adsorb hydrocarbons. Adsorption of 50 to 60% of the cold start hydrocarbons have been demonstrated (44,169–171). Zeolites do not efficiently adsorb C-2 to C-4 hydrocarbons.

Exhaust Gas Igniter. Exhaust gases during cold start have high concentrations of CO and HCs because of air restriction caused by choking. Using a severe choke strategy and addition of air by a pump the exhaust gas/air mixture is flammable. The Ford Exhaust Gas Igniter incorporates a spark igniter in the exhaust line to ignite and sustain the flame. The flame heats up the main catalyst in 15 to 20 seconds, and in the process consumes those hydrocarbons and carbon monoxide that would otherwise pass through the cold catalyst (172,173). The system uses energy available in the exhaust gases during the initial seconds of engine starting. On the other hand, the choked condition is extreme, and there are associated engine concerns.

Reprinted from *Kirk-Othmer Encyclopedia of Chemical Technology*, 4th ed., Vol. 9, John Wiley & Sons, Inc., New York, 1994.

BIBLIOGRAPHY

1. M. P. Walsh, *Plat. Met. Rev.* **30**(3), 106–115 (1986).

2. C. Csere, *Car Driver,* 60–64 (Jan. 1988).

3. *Code of Federal Regulations,* Title 40, Part 86, Appendix IV, AMA Mileage Accumulation Route, Washington, D.C., July 1992, p. 951.

4. *1991 SAE Handbook,* Vol. 3. Society of Automotive Engineers, Warrendale, Pa., 1991, Chapt. 25.

5. A. L. Boehman, S. Niksa, and R. J. Moffatt, *A Comparison of Rate Laws for CO Oxidation Over Pt on Alumina,* SAE 930252, Society of Automotive Engineers, Warrendale, Pa., 1993.

6. J. T. Kummer, *Prog. Ener. Combus. Sci.* **6**, 177–199 (1981).

7. B. A. D'Alleva, *Procedure and Charts for Estimating Exhaust Gas Quantities and Composition,* General Motors Laboratory Report 372, Warren, Mich., 1960.

8. N. A. Heinen and D. J. Patterson, *Combustion Engine Economy, Emissions, and Controls,* University of Michigan, Ann Arbor, Mich., Chapt. 5, 1992.

9. L. Eltinge, *Fuel–Air Ratio and Distribution from Exhaust Gas Composition,* SAE 680114, Society of Automotive Engineers, Warrendale, Pa., 1968.

10. J. B. Edwards, *Combustion: The Formation and Emissions of Trace Species,* Ann Arbor Science Publishers, Ann Arbor, Mich., 1974, p. 16.

11. J. D. Benson and co-workers, *Effects of Gasoline Sulfur on Mass Exhaust Emissions–Auto/Oil Air Quality Improvement Research Program,* SAE 912323, Society of Automotive Engineers, Warrendale, Pa., 1991.

12. *Auto/Oil Air Quality Improvement Research Program,* SAE SP-920, Society of Automotive Engineers, Warrendale, Pa., 1992, 16 pp.

13. P. Blumberg and J. T. Kummer, *Combust. Sci. Technol.* **4,** 73 (1971).

14. N. Pelz and co-workers, *The Composition of Gasoline Engine Hydrocarbon Emissions–An Evaluation of Catalytic and Fuel Effects,* SAE 902074, Society of Automotive Engineers, Warrendale, Pa., 1990.

15. P. R. Shore, D. T. Humphries, and O. Hadded, *Speciated Hydrocarbon Emissions from Aromatic, Olefinic and Parafinic Model Fuels,* SAE 930373, Society of Automotive Engineers, Warrendale, Pa., 1993.

16. G. J. den Otter, R. E. Malpas, and T. D. B. Morgan, *Effect of Gasoline Reformulation on Exhaust Emissions in Current European Vehicles,* SAE 930372, Society of Automotive Engineers, Warrendale, Pa., 1993.

17. J. Laurikko and N-O. Nyland, *Regulated and Unregulated Emissions from Catalyst Vehicles at Low Ambient Temperatures,* SAE 930946, Society of Automotive Engineers, Warrendale, Pa., 1993. Good reference for low ambient temperature emissions.

18. P. R. Shore and R. S. de Vries, *On-Line Hydrocarbon Speciation Using FTIR and CI-MS,* SAE 922246, Society of Automotive Engineers, Warrendale, Pa., 1992.

19. D. W. Wendland and W. R. Matthes, *Visualization of Automotive Catalytic Converter Internal Flows,* SAE 861554, 1986; D. W. Wendland, P. L. Sorrell, and J. E. Kreucher, *Sources of Monolithic Catalytic Converter Pressure Loss,* SAE 912372, Society of Automotive Engineers, Warrendale, Pa., 1991.

20. H. Weltans, H. Bressler, and P. Krause, *Influence of Catalytic Converters on Acoustics of Exhaust Systems for European Cars,* SAE 910836, Society of Automotive Engineers, Warrendale, Pa., 1991.

21. D. W. Wendland, W. R. Matthes, and P. L. Sorrell, *Effect of Header Truncation on Monolith Converter Emission Control Performance,* SAE 922340, Society of Automotive Engineers, Warrendale, Pa., 1992.

22. H. Weltans and co-workers, *Optimization of Catalytic Converter Gas Flow Distribution by CFD Prediction,* SAE 930780, Society of Automotive Engineers, Warrendale, Pa., 1993.

23. I. Sword and co-workers, SAE 930943, Society of Automotive Engineers, Warrendale, Pa., 1993.

24. J. Howitt, *Thin Wall Ceramics as Monolithic Catalyst Supports,* SAE 910611, Society of Automotive Engineers, Warrendale, Pa., 1991.

25. L. S. Socha, J. P. Day, and E. M. Barnett, *Impact of Catalytic Support Design Parameters on FTP Emissions,* SAE 892041, Society of Automotive Engineers, Warrendale, Pa., 1989.

26. H. Yamamoto and co-workers, *Warm-Up Characteristics of Thin Wall Honeycomb Catalysts,* SAE 910611, Society of Automotive Engineers, Warrendale, Pa., 1991.

27. J. P. Day and L. S. Socha, Jr., *The Design of Automotive Catalyst Supports for Improved Pressure Drop and Conversion Efficiency,* SAE 910371, Society of Automotive Engineers, Warrendale, Pa., 1991.

28. M. Luoma, P. Lappi, and R. Lylykangas, *Evaluation of High Cell Density Z-Flow Catalyst,* SAE 930940, Society of Automotive Engineers, Warrendale, Pa., 1993. Good reference for mass-transfer limited model reactions.

29. K. Otto, W. B. Williamson, and H. Gandhi, *Ceram. Eng. Sci. Proc.* **2**(6), (May/June, 1981). Good review of various catalyst deactivation processes.

30. R. J. Farrauto and M. C. Hodson, "Catalyst Characterization," *Encyclopedia of Physical Science and Technology,* Vol. 2, Academic Press, Inc., New York, 1987.

31. J. T. Kummer, "The Use of Noble Metals in Automobile Exhaust Catalyst," *Chicago Symposium,* 1985.

32. J. E. Kubsh, J. S. Rieck, and N. D. Spencer, in A. Crucq, Ed., *Catalysis and Automotive Pollution Control II,* Elsevier, New York, 1991, p. 125.

33. J. K. Hockmuth and co-workers, *Hydrocarbon Traps for Controlling Cold Start Emissions,* SAE 930739, Society of Automotive Engineers, Warrendale, Pa., 1993.

34. Y. T. Yu Yao, *Indust. Eng. Chem. Prod. Res. Dev.* **19,** 293 (1980).

35. H. C. Yao, H. K. Stephen, and H. S. Gandhi, *J. Catal.* **61,** 547 (1980).

36. H. C. Yao, S. Japar, and M. Shelef, *J. Catal.* **50,** 407 (1977).

37. S. H. Oh, P. J. Mitchel, and R. M. Siewert, *J. Catal.* **132,** 287 (1991).

38. R. F. Hicks, C. Rigano, and B. Pang, *Catal. Lett.* **6,** 271 (1990).

39. T. Yamada, K. Kayano, and M. Funibike, *The Effectiveness of Pd for Converting Hydrocarbons in TWC Catalysts,* SAE 930253, Society of Automotive Engineers, Warrendale, Pa., 1993.

40. J. C. Dettling and W. K. Lui, *A Non-Rhodium Three-Way Catalyst for Automotive Applications,* SAE 920094, Society of Automotive Engineers, Warrendale, Pa., 1992.

41. C. N. Montreiul, S. D. Williams, and A. A. Adanczyk, *Modeling Current Generation Catalytic Converters: Laboratory Experiments and Kinetic Parameter Optimization–Steady State Kinetics,* SAE 920096, Society of Automotive Engineers, Warrendale, Pa., 1992.

42. M. A. Harkonen and P. Talvitie, *Optimization of Metallic TWC Behavior and Precious Metal Costs,* SAE 920395, Society of Automotive Engineers, Warrendale, Pa., 1992.

43. J. T. Kummer, *J. Phys. Chem.* **90,** 4747 (1986).

44. G. B. Fisher and co-workers, *The Role of Ceria in Automotive Exhaust Catalysis and OBD-II Catalyst Monitoring,* SAE 931034, Society of Automotive Engineers, Warrendale, Pa., 1993.

45. S. Pelters, F. W. Kaiser, and W. Maus, *The Development and Application of Metal Supported Catalyst for Porsche's 911 Carrera 4,* SAE 890488, Society of Automotive Engineers, Warrendale, Pa., 1989.

46. F. W. Kaiser and S. Pelters, *Comparison of Metal Supported Catalysts with Different Cell Geometries,* SAE 910837, Society of Automotive Engineers, Warrendale, Pa., 1991.

47. P. Oser, *Novel AutoCatalyst Concepts and Strategies for the Future with Emphasis on Metal Supports,* SAE 880319, Society of Automotive Engineers, Warrendale, Pa., 1988.

48. K. Nishizawa and co-workers, *Metal Supported Automotive Catalysts for Use in Europe,* SAE 880188, Society of Automotive Engineers, Warrendale, Pa., 1988.

49. S. H. Oh and D. L. Van Ostrom, *A Three Dimensional Model for the Analysis of Transient Thermal and Conversional Characteristics of Monolithic Catalytic Converters,* SAE 880282, Society of Automotive Engineers, Warrendale, Pa., 1988.

50. J. C. W. Kuo, C. R. Morgan, and H. G. Lasson, *Mathematical Modelling of CO and HC Catalytic Converter Systems,* SAE 710289, Society of Automotive Engineers, Warrendale, Pa., 1971.

51. C. N. Montreuil, S. D. Williams, and A. A. Adanczyk, *Modeling Current Generation Catalytic Converters; Laboratory Experiments and Kinetic Parameter Optimization–Steady State Kinetics,* SAE 920096, Society of Automotive Engineers, Warrendale, Pa., 1992.

52. K. S. Creamer and J. H. Sanders, *Evaluation of a Catalytic Converter for a 3.73 kW Natural Gas Engine,* SAE 930221, Society of Automotive Engineers, Warrendale, Pa., 1993.

53. S. Tauster, Engelhard Corp., Iselin, N.J., 1992.

54. C. M. Friend, *Sci. Am.* **268**(4), 74–79 (Apr. 1993).

55. M. Golze and co-workers, *Phys. Rev. Lett.* **53**(8), 850–853 (1993).

56. R. J. Maddix, *Science* **V233,** 1159–1166 (Sept. 12, 1986).

57. G. Fisher and co-workers, "Mechanism of the Nitric Oxide–Carbon Monoxide–Oxygen Reaction Over a Single Crystal Rhodium Catalyst," in M. J. Philips and M. Ternan, eds., *Proceedings of the 9th International Congress on Catalysis,* Vol. 3, *Characterization and Metal Catalysts,* Chemical Institute of Canada, Ottawa, 1988.

58. J. T. Yates, Jr., *Chem. Eng. News.* **70**(13), 22–35 (Mar. 30, 1992).

59. J. C. Summers and S. A. Ansen, *J. Catal.* **58,** 131 (1979).

60. H. C. Yao and Y. F. Yu Yao, *J. Catal.* **86,** 254 (1984).

61. K. Ihara, H. Marakami, and K. Ohkubo, *Improvement of Three-Way Catalyst Performance by Optimizing Ceria Impregnation,* SAE 902168, Society of Automotive Engineers, Warrendale, Pa., 1990.

62. A. F. Duvell, R. R. Rojaram, H. A. Shaw, and T. J. Truex, in A. Crucq, ed., in Ref. 32, p. 139.

63. R. Hicks and co-workers, *J. Catal.* **122,** 296–306 (1990).

64. H. S. Gandhi, W. B. Williamson, and J. L. Bomback, *Appl. Catal.* **3,** 79–88 (1982).

65. E. Jobson and co-workers, *Deterioration of Three-Way Automotive Catalysts, Part I–Steady State and Transient Emission of Aged Catalyst,* SAE 930937, Society of Automotive Engineers, Warrendale, Pa., 1993.

66. S. Lundgren and E. Jobson, *Deterioration of Three-Way Automotive Catalysts, Part II–Oxygen Storage Capacity at Exhaust Conditions,* SAE 930944, Society of Automotive Engineers, Warrendale, Pa., 1993.

67. L. A. Carol, N. E. Newman, and G. S. Mann, *High Temperature Deactivation of Three-Way Catalyst,* SAE 892040, Society of Automotive Engineers, Warrendale, Pa., 1989.

68. N. Miyoshi and co-workers, *Development of Thermal Resistant Three-Way Catalyst,* SAE 891970, Society of Automotive Engineers, Warrendale, Pa., 1989.

69. C. D. Tyree, *Emission Level in Catalyst Temperature as a Function of Ignition-Induced Misfire,* SAE 920298, Society of Automotive Engineers, Warrendale, Pa., 1992.

70. N. Miyoshi and co-workers, *Development of Thermal Resistant Three-Way Catalysts,* SAE 891970, Society of Automotive Engineers, Warrendale, Pa., 1989.

71. M. P. Walsh, *The Advantages of Removing Lead from Gasoline and Using Catalytic Converters to Control Vehicle Exhaust Pollution,* sponsored by Corning Glass Works, 1983.

72. Needleman and co-workers, *New Engl. J. Med.* **300**(13), (Mar. 29, 1979).

73. A. M. Hochhauser and co-workers, *Effects of Gasoline Sulfur Level on Mass Exhaust Emissions–Auto/Oil Air Quality Improvement Research Program,* SAE 912323, Society of Automotive Engineers, Warrendale, Pa., 1991.

74. J. C. Summers and co-workers, SAE 920558, Society of Automotive Engineers, Warrendale, Pa., 1992.

75. U.S. Pat. 4,552,733 (Nov. 12, 1985), C. Thompson, J. Mooney, C. D. Keith, and W. Mannion (to Engelhard Corp.)

76. I. Gottberg, E. Hogberg, and K. Weber, *Sulfur Storage and Hydrogen Sulphide Release From a Three-Way Catalyst Equipped Car,* SAE 890491, Society of Automotive Engineers, Warrendale, Pa., 1989.

77. J. C. Dettling and co-workers, *Control of H2S Emissions from High-Tech TWC Converters,* SAE 900506, Society of Automotive Engineers, Warrendale, Pa., 1990.

78. S. E. Golunski and S. A. Roth, *Catal. Today* **9,** 109 (1991).

79. U.S. Pat. 5196390 (Mar. 23, 1993), S. Tauster, J. Dettling, and J. Mooney (to Engelhard Corp).

80. W. B. Williamson and co-workers, *Catalyst Deactivation Due to Glaze Formation from Oil-Derived Phosphorus and Zinc,* SAE 841406, Society of Automotive Engineers, Warrendale, Pa., 1984.

81. K. Inoue, T. Kurahashi, and T. Negishi, SAE 92064, Society of Automotive Engineers, Warrendale, Pa., 1992.

82. W. B. Williamson and co-workers, *Appl. Catal.* **15,** 277–292 (1985).

83. H. S. Gandhi and co-workers, "Affinity of Lead for Moble Metals on Different Supports," *Surface Interface Analy.* **6**(4), (1984).

84. W. B. Williamson, H. S. Gandhi, and E. E. Weaver, *Effects of Fuel Additive MMT on Contaminent Retention and Catalyst Performance,* SAE 821193, Society of Automotive Engineers, Warrendale, Pa., 1982.

85. G. T. Engh and S. Wallman, *Development of the Volvo Lambda-Sond System,* SAE 770295, Society of Automotive Engineers, Warrendale, Pa., 1977.

86. J. J. Mooney, C. E. Thompson and J. C. Dettling, *Three-Way Conversion Catalysts–Part of the New Emission Control System,* SAE 770365, Society of Automotive Engineers, Warrendale, Pa., 1977.

87. E. Koberstein, *Characterization of Multifunctional Catalysts for Automotive Exhaust Purifications,* SAE 770366, Society of Automotive Engineers, Warrendale, Pa., 1977.

88. P. Oser, *Catalyst Systems with and Emphasis on Three-Way Conversion and Novel Concepts,* SAE 790306, Society of Automotive Engineers, Warrendale, Pa., 1979.

89. K. Oblander, J. Abthoff, and H. D. Schuster. *Der Drieweg-katalysator–eine Abgasreinigungstechologie fur Kraftfahrzenge mit Ottomotoren (Three-Way Catalyst–An Exhaust Cleaning Technology for Automobiles with Internal Combustion Engines),* VDI-Berichte 630, VDI-Verlag, Dusseldorf, Germany, 1984.

90. J. Abthoff and co-workers, *MTZ Motortechnische Zeitschrift* **53,** 11 (1992).

91. E. Logothetis, *Sci. Technol. Zirconia,* **3,** 388 (1981); R. E. Hetrich, W. A. Fate, and W. C. Vassell, *Oxygen Sensing by Electrochemical Pumping,* SAE 8104333, Society of Automotive Engineers, Warrendale, Pa., 1981.

92. W. J. Fleming, *Zirconia Oxygen Sensor—An Equivalent Circuit Model,* SAE 800020, Society of Automotive Engineers, Warrendale, Pa., 1980.

93. E. Hamann, H. Manger, and L. Stencke, *Lambda Sensor with Y2O3–Stabilized ZrO2–Ceramic for Application in Automotive Emission Control Systems,* SAE 770401, Society of Automotive Engineers, Warrendale, Pa., 1977.

94. J. W. Butler and co-workers, *Fast Response Zirconia Sensor–Based Instrument for Measurement of the Air/Fuel Ratio of Combustion Exhaust,* SAE 840061, Society of Automotive Engineers, Warrendale, Pa., 1984.

95. I. Gorille, N. Rittmansberger, and P. Werner, *Bosch Electronic Fuel Injection with Closed Loop Control,* SAE 750368, Society of Automotive Engineers, Warrendale, Pa., 1975.

96. E. Hendricks, T. Vesterholm, and S. C. Sorenson, *Non Linear Closed Loop SI Engine Control Observers,* SAE 920237, Society of Automotive Engineers, Warrendale, Pa., 1992.

97. A. J. Beumont, A. D. Noble, and A. Scariobrick, "Adaptive Transient Air Fuel Ratio Control to Minimize Gasoline Engine Emissions," *Fisita Congress,* London, 1992.

98. M. J. Anderson, *A Feedback A/F Control System for Low Emission Vehicles,* SAE 930388, Society of Automotive Engineers, Warrendale, Pa., 1993.

99. *Automotive Handbook,* 2nd ed., Robert Bosch GmbH, Stuttgart, Germany, 1986, p. 442.

100. C. O. Probst, *Bosch Fuel Injection & Engine Management,* Robert Bentley, Cambridge, Mass., 1989.

101. Technical Support Document *Revisions to the Malfunction and Diagnostic System Requirements Applicable to 1994 and Later New California Passenger Cars, Light-Duty Trucks, and Medium Duty Vehicles with Feedback Fuel Control Systems (OBD-II),* California Air Resources Board, Sacramento, Sept. 14, 1989.

102. J. W. Koupal, M. A. Sabourin, and W. V. Clemmens, *Detection of Catalyst Failure On-Vehicle Using the Dual Oxygen Sensor Method,* SAE 91061, Society of Automotive Engineers, Warrendale, Pa., 1991.

103. W. Clemmens, M. Sabourin, and T. Rao, *Detection of Catalyst Failure Using On-Board Diagnostics,* SAE 900062, Society of Automotive Engineers, Warrendale, Pa., 1992.

104. S. H. Oh, *Thermal Response of a Monolith Catalyst Converter During Sustained Misfiring: A Computational Study,* SAE 881591, Society of Automotive Engineers, Warrendale, Pa., 1988.

105. W. Cai, Novel Sensors for On Vehicle Measurements of Emissions, Ph.D. dissertation University of Cambridge, Mass., 1992.

106. W. Cai and N. Collings, *A Catalytic Oxidation Sensor for the On-Board Detection of Misfire and Catalyst Efficiency,* SAE 922248, Society of Automotive Engineers, Warrendale, Pa., 1992.

107. N. Collings and co-workers, *A Linear Catalyst Temperature Sensor for Exhaust Gas Ignition (EGI) and On Board Diagnostics of Misfire and Catalyst Efficiency,* SAE 930938, Society of Automotive Engineers, Warrendale, Pa., 1993.

108. D. D. Bech and co-workers, *The Performance of Pd, Pt and Pd–Pt Catalysts in Lean Exhaust,* SAE 930084, Society of Automotive Engineers, Warrendale, Pa., 1993.

109. H. E. Jaaskelainen and J. S. Wallace, *Performance and Emissions of a Natural Gas-Fueled 16-Valve DOHC Four-Cylinder Engine,* SAE 930380, Society of Automotive Engineers, Warrendale, Pa., 1993.

110. A. A. Quader, *Why Intake Charge Dilution Decreases Nitric Oxide Emissions from Spark Ignited Engines,* SAE 710009, Society of Automotive Engineers, Warrendale, Pa., 1971.

111. J. A. Harrington and R. C. Shishu, *Zirconia Oxygen Sensor–An Equivalent Circuit Model,* SAE 730476, Society of Automotive Engineers, Warrendale, Pa., 1973.

112. E. A. Mayer, *Electro-Pneumatic Control Valve for EGR/ATC Actuation,* SAE 810464, Society of Automotive Engineers, Warrendale, Pa., 1981.

113. J. J. Gumbleton and co-workers, *Optimizing Engine Parameters with Exhaust Gas Recirculation,* SAE 740104, Society of Automotive Engineers, Warrendale, Pa., 1974.

114. Ref. 4, Chapt. 25.89, SAE J171.

115. H. M. Haskey, W. R. Cadman, and T. F. Liberty, *The Development of a Real-Time Evaporative Emission Test,* SAE 901110, Society of Automotive Engineers, Warrendale, Pa., 1990.

116. L. L. Lave, W. E. Wecher, W. S. Reis, and D. A. Ross, *Environ. Sci. Technol.* **24,** 8 (1990).

117. *The Effects of Temperature and Fuel Volatility on Vehicle Evaporative Emissions,* Report No 90/1, Concawe, Brussels, Belgium, 1990.

118. T. Cam, K. Cullen, and S. L. Baldus, *Running Loss Temperature Profile,* SAE 930078, Society of Automotive Engineers, Warrendale, Pa., 1993; H. M. Haskew, W. R. Cadman, and T. F. Liberty, *Evaporative Emissions Under Real-Time Conditions,* SAE 891121, Society of Automotive Engineers, Warrendale, Pa., 1989.

119. C. B. Tracy and W. W. Frank, *Fuels, Lubricants, and Positive Crankcase Ventilation System,* SAE PT112, 451, Society of Automotive Engineers, Warrendale, Pa., 1963–1966.

120. R. J. Templin, *Discussion of Reference 2,* SAE TP-6, 249, Society of Automotive Engineers, Warrendale, Pa., 1964.

121. Ref. 4, Chapt. 25.68, SAE J900.

122. A. Unich, R. M. Bada, and K. W. Lyons, *Natural Gas: A Promising Fuel for I.C. Engines,* SAE 930929, Society of Automotive Engineers, Warrendale, Pa., 1993. An extensive reference source.

123. American Gas Association (AGA), "Projected Natural Gas Demand From Vehicles Under the Mobile Sources Provisions of the Clean Air Act Ammendments," *AGA Energy Analysis EA 1990–1991,* Chicago, 1991.

124. M. Samsa, *Potential for Compressed Natural Gas Vehicles in Centrally-Fueled Automobile, Truck and Bus Fleet Application,* Gas Research Institute, Chicago, 1991, pp. 44–61.

125. Technical staff report on Reactivity Adjustment Factors (RAF), California Air Resources Board, 1993.

126. J. J. White, *Low Emission Catalysts for Natural Gas Engines,* GRI Report 91/0214, Gas Research Institute, Chicago, Southwest Research Institute, San Antonio, SwRI 3178–22, 1991.

127. J. Klimstra, *Catalytic Converters for Natural Gas Engines–A Measurement and Control Problem,* SAE 872165, Society of Automotive Engineers, Warrendale, Pa., 1987.

128. R. Hicks and co-workers, *Structure Sensitivity of Methane Oxidation over Platinum and Palladium; J. Catal.,* 280–306 (1990).

129. W. M. Burkmyre, W. E. Liss, and M. Church, *Natural Gas Converter Performance and Durability,* SAE 930222, Society of Automotive Engineers, Warrendale, Pa., 1993.

130. S. Subramanian, R. J. Kudla, and M. S. Chattha, *Treatment of Natural Gas Vehicle Exhaust,* SAE 930223, Society of Automotive Engineers, Warrendale, Pa., 1993.

131. J. E. Sinor and B. K. Bailey, *Current and Potential Future Performance of Ethanol Fuels,* SAE 930376, Society of Automotive Engineers, Warrendale, Pa., 1993.

132. J. J. Mooney, J. G. Hansel, and K. R. Burns, *Three-Way Conversion Catalysts on Vehicles Fueled with Ethanol–Gasoline Mexuture,* SAE 790428, Society of Automotive Engineers, Warrendale, Pa., 1979.

133. R. K. Pfefly, University of Santa Clara, personal communication, 1979.

134. C. L. Mynng and co-workers, *Research and Development of Hyundai Flexible Fuel Vehicles (FFVs),* SAE 930330, Society of Automotive Engineers, Warrendale, Pa., 1993.

135. T. Suga and Y. Hamazaki, *Development of Honda Flexible Fuel Vehicle,* SAE 922276, Society of Automotive Engineers, Warrendale, Pa., 1992.

136. J. K. Hochmuth and J. J. Mooney, *Catalytic Control of Emissions from M-85 Fueled Vehicles,* SAE 930219, Society of Automotive Engineers, Warrendale, Pa., 1993.

137. K. H. Hellman, G. K. Piotrowski, and R. M. Schaefer, *Start Catalyst Systems Employing Heated Catalyst Technology for Control of Emissions from Methanol-Fueled Vehicles,* SAE 93082, Society of Automotive Engineers, Warrendale, Pa., 1993.

138. Ref. 4, Chapt. 24.373.

139. R. Douglas and G. P. Blair, *Fuel Injection of a Two-Stroke Cycle Spark Ignition Engine,* SAE 820952, Society of Automotive Engineers, Warrendale, Pa., 1982.

140. D. Plohberger and co-workers, *Development of a Fuel-Injected Two-Stroke Gasoline Engine,* SAE 880170, Society of Automotive Engineers, Warrendale, Pa., 1988.

141. M. Nuti, *A Variable Timing Electronically Controlled High Pressure Injection System for 2S S.I. Engines,* SAE 9009799, Society of Automotive Engineers, Warrendale, Pa., 1990.

142. G. E. Hundleby, *Development of a Poppet-Valved Two-Stroke Engine–The Flagship Concept,* SAE 900802, Society of Automotive Engineers, Warrendale, Pa., 1990.

143. G. P. Blair and co-workers, "The Reduction of Emissions and Fuel Consumption by Direct Air-Assisted Fuel Injection into a Two-Stroke Engine," *The 4th Graz Two-Wheeler Symposium,* Technical University, Graz, Austria, 1991.

144. J. Stokes and co-workers, *Development Experience of a Popped-Valved Two-Stroke Flagship Engine,* SAE 920778, Society of Automotive Engineers, Warrendale, Pa., 1992.

145. H. H. Huang and co-workers, *Improvement of Irregular Combustion of Two-Stroke Engine by Skip Injection Control,* SAE 922310, Society of Automotive Engineers, Warrendale, Pa., 1992.

146. C. Csere, *Car Driver,* 87–95, (June 1991).

147. *Auto. Eng.* **99**(7), 11–14 (July 1991).

148. F. Markus, *Car Driver,* 72–75 (Feb. 1992).

149. K. Oblander, J. Abthoff, and L. Fricker, "From Engine Testbench to Vehicle–An Approach to Lean Burn by Dual Ignition," *I. Mech E. C80/79,* (1979).

150. S. Matsushita and co-workers, *Development of the Toyota Lean Combustion System,* SAE 850044, Society of Automotive Engineers, Warrendale, Pa., 1985.

151. I-Y Ohm and co-workers, *Development of HMC Axially Stratified Lean Combustion Engine,* SAE 930879, Society of Automotive Engineers, Warrendale, Pa., 1993.

152. T. Inoue and co-workers, *Toyota Lean Combustion System– the Third Generation System,* SAE 930873, Society of Automotive Engineers, Warrendale, Pa., 1993.

153. M. Iwamoto and H. Hamada, *Catal. Today* **10,** 57 (1991).

154. B. H. Engler and co-workers, *Catalytic Reaction of NO_x with Hydrocarbons under Lean Diesel Exhaust Gas Conditions,* SAE 930735, Society of Automotive Engineers, Warrendale, Pa., 1993.

155. W. Held and co-workers, *Catalytic NO_x Reduction in Net Oxidizing Exhaust Gas,* SAE 900496, Society of Automotive Engineers, Warrendale, Pa., 1990.

156. T. J. Truex, R. A. Searles, and D. C. Sun. *Plat. Met. Rev.* **36**(1), 2–11 (Jan. 1992). A good NO_x reduction background reference.

157. G. Muramatsu and co-workers, *Catalytic Reduction of NO_x in Diesel Exhaust,* SAE 930135, Society of Automotive Engineers, Warrendale, Pa., 1993.

158. M. J. Heimrich and M. L. DeViney, *Lean NO_x Catalyst Evaluation and Characterization,* SAE 930736, Society of Automotive Engineers, Warrendale, Pa., 1993.

159. D. R. Monroe and co-workers, *Evaluation of a Cu–Zeolite Catalyst to Remove NO_x from Lean Exhaust,* SAE 930737, Society of Automotive Engineers, Warrendale, Pa., 1993.

160. L. S. Socha, D. F. Thompson, and P. S. Weber. *Reduced Energy and Power Consumption for Electrically Heated Extruded Metal Converters,* SAE 930383, Society of Automotive Engineers, Warrendale, Pa., 1993.

161. F. W. Kaiser and co-workers, *Optimization of an Electrically-Heated Catalytic Converter System–Calculations and Application,* SAE 930384, Society of Automotive Engineers, Warrendale, Pa., 1993.

162. M. J. Heimrich, S. Albu, and J. Osborne, *Electrically Heated System Conversion on Two Current Technology Vehicles,* SAE 910612, Society of Automotive Engineers, Warrendale, Pa., 1991.

163. I. Gottberg and co-workers, *New Potential Exhaust Gas Aftertreatment Technologies for Clean Car Legislation,* SAE 910849, Society of Automotive Engineers, Warrendale, Pa., 1991.

164. K. H. Hellman, C. K. Piotroski, and R. M. Shaefer, *Evaluation of Different Resistively Heated Technologies,* SAE 912382, Society of Automotive Engineers, Warrendale, Pa., 1991.

165. R. J. Hurley and co-workers, *Experiences with Electrically Heated Catalysts,* SAE 912384, Society of Automotive Engineers, Warrendale, Pa., 1991.

166. J. E. Kubsh and P. W. Lissuik, *Vehicle Emission Performance with an Electrically Heated Converter System,* SAE 912385, Society of Automotive Engineers, Warrendale, Pa., 1991.

167. L. S. Socha and D. F. Thompson, *Electrically Heated Extruded Metal Converter for Low Emission Vehicles,* SAE 920093, Society of Automotive Engineers, Warrendale, Pa., 1992.

168. O. Haddad and co-workers. *The Achievement of ULEV Emission Standards for Large High Performance Vehicles,* SAE 930389, Society of Automotive Engineers, Warrendale, Pa., 1993.

169. EP 424966 (1990), T. Minami and T. Nagase.

170. B. H. Engler and co-workers, *Reduction of Exhaust Gas Emission by Using Hydrocarbon Adsorber Systems,* SAE 930738, Society of Automotive Engineers, Warrendale, Pa., 1993.

171. M. J. Heimrich, L. R. Smith, and J. Kitowski, *Cold-Start Hydrocarbon Collection for Advanced Exhaust Emission Control,* SAE 920847, Society of Automotive Engineers, Warrendale, Pa., 1992.

172. T. Ma, N. Collings, and T. Hands, *Exhaust Gas Ignition (EGI)–A New Concept for Rapid Light-Off of Automotive*

Exhaust Catalyst, SAE 920400, Society of Automotive Engineers, Warrendale, Pa., 1992.

173. N. Collings and co-workers, *A Linear Catalyst Temperature Sensor for Exhaust Gas Ignition (EGI) and In-Board Diagnostics of Misfire and Catalyst Efficiency,* SAE 930938, Society of Automotive Engineers, Warrendale, Pa., 1993.

General References

Alcoa Alumina Handbook

R. B. Bird and co-workers, *Transport Phenomena,* John Wiley & Sons, Inc., New York, 1965.

Code of Federal Regulations, Title 40, Part 86, Washington, D.C., July 1, 1992.

K. C. Taylor, in J. R. Anderson and J. R. Boudart, eds., *Catalysis Science and Technology,* Berlin, 1984, Chapt. 2, pp. 119–170.

G. C. Bond and G. Webb, eds., "Catalysis," *Royal Soc. Chem., London,* **6,** (1984).

D. D. Eley, H. Pines, and P. B. Wiez, eds., *Advances in Catalysis,* Vol. 33, Academic Press, Inc., New York, 1985.

H. Heinman and J. J. Carberry, eds., *Catalysis Reviews–Science and Engineering,* Vol. 26, Marcel-Dekker, New York, 1984.

C. N. Satterfield, *Heterogeous Catalysis in Practice,* 1st ed., McGraw-Hill Book Co., Inc., New York, 1980.

J. R. Mondt, *J. Eng. Gas Turbines Power* **109**(2), 200–206 (1987).

C. D. Falk and J. J. Mooney, *Three-Way Conversion Catalysts–Effect of Closed Loop Feedback Control and Other Parameters on Catalyst Efficiency,* SAE 800462, Society of Automotive Engineers, Warrendale, PA., 1980.

M. Luomo, P. Loppi, and R. Lylykaugas, *Evolution of High Cell Density Z-Flow Catalyst,* SAE 930940, Society of Automotive Engineers, Warrendale, Pa., 1993.

EXHAUST CONTROL, INDUSTRIAL

Ronald L. Berglund
The M. W. Kellogg Company
Houston, Texas

Limits for exhaust emissions from industry, transportation, power generation (qv), and other sources are increasingly legislated (see also EXHAUST CONTROL, AUTOMOTIVE) (1,2). One of the principal factors driving research and development in the petroleum (qv) and chemical processing industries in the 1990s is control of industrial exhaust releases. Much of the growth of environmental control technology is expected to come from new or improved products that reduce such air pollutants as carbon monoxide [630-08-0], (CO), volatile organic compounds (VOCs), nitrogen oxides (NO$_x$), or other hazardous air pollutants (see AIR POLLUTION). The mandates set forth in the 1990 amendments to the Clean Air Act (CAAA) (see CLEAN AIR ACT OF 1990) push pollution control methodology well beyond what, as of this writing, is in general practice, stimulating research in many areas associated with exhaust system control (see AIR POLLUTION CONTROL METHODS). In all, these amendments set specific limits for 189 air toxics, as well as control limits for VOCs, nitrogen oxides, and the so-called criteria pollutants. An estimated 40,000 facilities, including establishments as diverse as bakeries and chemical plants are affected by the CAAA (3).

There are 10 potential sources of industrial exhaust pollutants which may be generated in a production facility (4): (*1*) unreacted raw materials; (*2*) impurities in the reactants; (*3*) undesirable by-products; (*4*) spent auxiliary materials such as catalysts, oils, solvents, etc; (*5*) off-spec product; (*6*) maintenance, ie, wastes and materials; (*7*) exhausts generated during start-up and shutdown; (*8*) exhausts generated from process upsets and spills; (*9*) exhausts generated from product and waste handling, sampling, storage, and treatment; and (*10*) fugitive sources.

Exhaust streams generally fall into two general categories, intrinsic and extrinsic. The intrinsic wastes represent impurities present in the reactants, by-products, co-products, and residues as well as spent materials used as part of the process, ie, sources (*1*)–(*5*). These materials must be removed from the system if the process is to continue to operate safely. Extrinsic wastes are generated during operation of the unit, but are more functional in nature. These are generic to the process industries overall and not necessarily inherent to a specific process configuration, ie, sources (*6*)–(*10*). Waste generation may occur as a result of unit upsets, selection of auxiliary equipment, fugitive leaks, process shutdown, sample collection and handling, solvent selection, or waste handling practices.

CONTROL STRATEGY EVALUATION

There are two broad strategies for reducing volatile organic compound (VOC) emissions from a production facility: (*1*) altering the design, operation, maintenance, or manufacturing strategy so as to reduce the quantity or toxicity of air emissions produced, or (*2*) installing aftertreatment controls to destroy the pollutants in the generated air emission stream (5). Whether the exhaust stream contains a specific hazardous air pollutant, a VOC, a nitrogen oxide, or carbon monoxide, the best way to control the pollutant is to prevent its formation in the first place. Many technologies are being developed that seek to minimize the generation of undesirable by-products by modifying specific process materials or operating conditions. Whereas process economics or product quality may restrict the general applicability of these approaches, an increased understanding of the mechanisms and conditions by which a pollutant is created is leading to significant breakthroughs in burner design and operation (for nitrogen oxide control), equipment design, maintenance, and operation for fugitive and vent VOC emission control, and product and waste storage and handling design and operation (for VOC emission control).

Once an undesirable material is created, the most widely used approach to exhaust emission control is the application of add-on control devices (6). For organic vapors, these devices can be one of two types, combustion or capture. Applicable combustion devices include thermal incinerators (qv), ie, rotary kilns, liquid injection combustors, fixed hearths, and fluidized-bed combustors; catalytic oxidation devices; flares; or boilers/process heat-

Table 1. Emission Control Technologies

Technology	Reduction Effectiveness	Recovery	Waste Generation	Advantages	Disadvantages
Activated carbon adsorption	90–98%	Chemical recovery possible with regeneration	Spent carbon or regenerant	Good for wide variety of VOCs	Carbon replacement, regeneration costs, potential for bed fires
Adsorption in wet scrubbers	75–90%+	Chemical recovery possible through decanting/distillation	Spent solvent or regenerant	Simple operation	Not efficient at low concentration
Vapor condensation	50–80%	Chemical recovery possible through decanting/treatment	Liquid wastes, needs off-gas treatment	Simple operation, effective for high VOC concentration	Low removals applicability limits to some VOCs, high power costs
Thermal oxidation	99%	Heat recovery	NO_x generation, CO_2 generation	Handles any VOC concentration	High operating costs, capital costs, temperatures, and maintenance
Catalytic oxidation	95–99%	Heat recovery	Spent catalyst regeneration acids and alkalines	Simple systems, lower T than thermal economical operation	Fouling of catalysts, temperature limits

ers. Primary applicable capture devices include condensers, adsorbers, and absorbers, although such techniques as precipitation and membrane filtration are finding increased application. A comparison of the primary control alternatives is shown in Table 1.

The most desirable of the control alternatives is capture of the emitted materials followed by recycle back into the process. However, the removal efficiencies of the capture techniques generally depend strongly on the physical and chemical characteristics of the exhaust gas and the pollutants considered. Combustion devices are the more commonly applied control devices, because these are capable of a high level of removal efficiencies, ie, destruction for a variety of chemical compounds under a range of conditions. Although installation of emission control devices requires capital expenditures, these may generate useful materials and be net consumers or producers of energy. The selection of an emission control technology is affected by nine interrelated parameters: (1) temperature, T, of the inlet stream to be treated; (2) residence time; (3) process exhaust flow rate; (4) auxiliary fuel needs; (5) optimum energy use; (6) primary chemical composition of exhaust stream; (7) regulations governing destruction requirements; (8) the gas stream's explosive properties or heat of combustion; and (9) impurities in the gas stream. A process flow diagram for the consideration of these parameters is shown in Figure 1. Given the many factors involved, an economic analysis is often needed to select the best control option for a given application.

Capture devices are discussed extensively elsewhere (see AIR POLLUTION CONTROL METHOD). Oxidation devices are either thermal units that use heat alone or catalytic units in which the exhaust gas is passed over a catalyst usually at an elevated temperature. These latter speed oxidation and are able to operate at temperatures well below those of thermal systems.

OXIDIZATION DEVICES

Thermal Oxidation

Thermal oxidation is one of the best known methods for industrial waste gas disposal. Unlike capture methods, eg, carbon adsorption, thermal oxidation is an ultimate disposal method destroying the objectionable combustible compounds in the waste gas rather than collecting them. There is no solvent or adsorbent of which to dispose or regenerate. On the other hand, there is no product to recover. A primary advantage of thermal oxidation is that virtually any gaseous organic stream can be safety and cleanly incinerated, provided proper engineering design is used (7).

A thermal oxidizer is a chemical reactor in which the reaction is activated by heat and is characterized by a specific rate of reactant consumption. There are at least two chemical reactants, an oxidizing agent and a reducing agent. The rate of reaction is related both to the nature and to the concentration of reactants, and to the conditions of activation, ie, the temperature (activation), turbulence (mixing of reactants), and time of interaction.

Thermal oxidation relies on a homogenous gas-phase reaction condition. Exhaust emissions from industrial sources usually contain organic compounds (the reducing agents) well mixed with oxygen (the oxidizing agent). Imparting the necessary, uniform temperature for reaction within this mixture is of primary importance in the design of the oxidizer and related equipment. Proper activation requires establishing the minimum required temperature (650–800°C) for an adequate time (0.1–0.3 s). General design consideration is given to minimizing heat input and reactor size under the constraints of time, turbulence, and temperature.

Thermal oxidization devices are widely used, and gen-

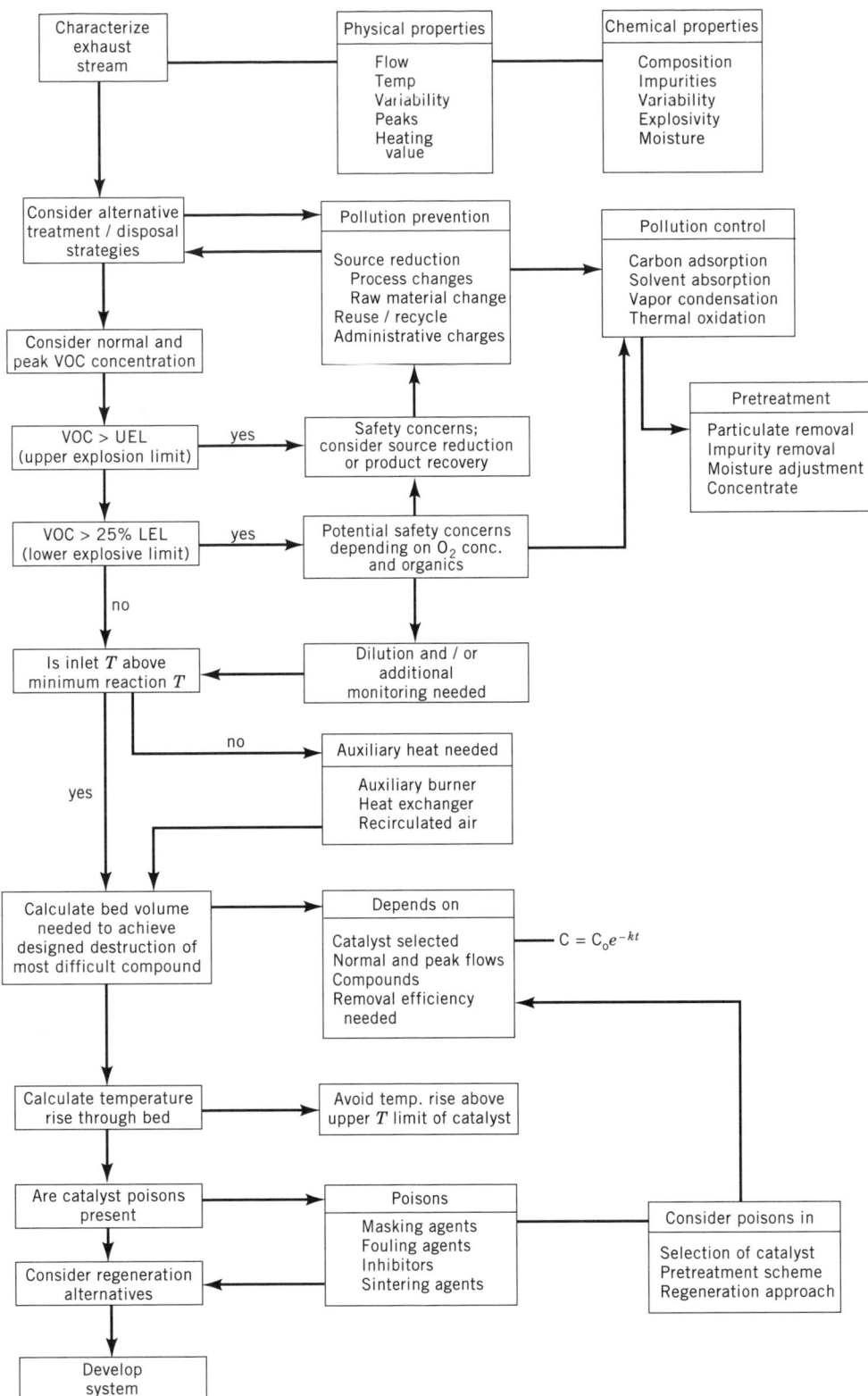

Figure 1. Process flow diagram for the selection of an exhaust control system.

erally provide a high degree of assurance that the process oxidizes the material in the exhaust gas. The high temperature operation causes other problems, however, especially compared to alternatives such as catalytic oxidation. The thermal oxidation devices often incur higher fuel costs because of the higher temperatures necessary, and require exotic high temperature materials, because the high temperatures entailed can bring about serious mechanical design problems as a result of operating in the temperature range in which metal creep takes place

(1,8). In addition, equipment durability is reduced by the extent of thermal cycling. Thermal oxidation systems are susceptible to thermal stress effects which result in distortion of the ductwork and heat-exchange surfaces, creating the potential for cracks and leaks. A further consequence of these high temperatures is that a thermal oxidizer may produce nitrogen oxides (NO_x), and sometimes yield undesirable by-products such as dioxins from chlorinated materials (5).

Some of the problems associated with thermal oxidizers have been attributed to the necessary coupling of the mixing, the reaction chemistry, and the heat release in the burning zone of the system. These limitations can reportedly be avoided by using a packed-bed flameless thermal oxidizer which is under development. This system relies on radiant heat from a large heat sink to raise the temperature of the exhaust gas to its ignition temperature. The heat sink, a ceramic matrix, is preheated radiatively using an electric preheating element, or a natural gas preheater, prior to introducing the exhaust gas. Because the system temperature is reasonably constant, NO_x generation within the flame is minimized. High (99.99%) VOC reductions at low contact times were reported (9). However, as with all thermal oxidation systems, this system is most effective for higher concentration exhaust streams, and requires significant auxiliary heat to treat low concentration streams.

Catalytic Oxidization

A principal technology for control of exhaust gas pollutants is the catalyzed conversion of these substances into innocuous chemical species, such as water and carbon dioxide. This is typically a thermally activated process commonly called catalytic oxidation, and is a proven method for reducing VOC concentrations to the levels mandated by the CAAA. Catalytic oxidation is also used for treatment of industrial exhausts containing halogenated compounds.

As an exhaust control technology, catalytic oxidation enjoys some significant advantages over thermal oxidation. The former often occurs at temperatures that are less than half those required for the latter, consequently saving fuel and maintenance costs. Lower temperatures allow use of exhaust stream heat exchangers of a low grade stainless steel rather than the expensive high temperature alloy steels. Furthermore, these lower temperatures tend to avoid the emissions problems arising from the thermal oxidation processes (10,11).

Critical factors that need to be considered when selecting an oxidation system include (12) (1) waste stream heating value and explosive properties. Low heating values resulting from low VOC concentration make catalytic systems more attractive, because low concentrations increase fuel usage in thermal systems; (2) waste gas components that might affect catalyst performance. Catalyst formulations have overcome many problems owing to contaminants, and a guard bed can be used in catalytic systems to protect the catalyst; (3) the type of fuel available and optimum energy use. Natural gas and no. 2 fuel oil can work well in catalytic systems, although sulfur in the fuel oil may be a problem in some applications (13). Other

fuels should be evaluated on a case-by-case basis; and (4) space and weight limitations on the control technology. Catalysts are favored for small, light systems.

There are situations where thermal oxidation may be preferred over catalytic oxidation: for exhaust streams that contain significant amounts of catalyst poisons and/or fouling agents, thermal oxidation may be the only technically feasible control; where extremely high VOC destruction efficiencies of difficult to control VOC species are required, thermal oxidation may attain higher performance; and for relatively rich VOC waste gas streams, ie, having ≥20–25% lower explosive limit (LEL), the gas stream's explosive properties and the potential for catalyst overheating may require the addition of dilution air to the waste gas stream (12).

Whereas the catalytic converter has been used in automobiles to control air pollutants only since 1975 (5), catalytic oxidation of industrial exhaust emissions began in the late 1940s, and is a reasonably mature technology (14). Initially it was used only in circumstances where an extremely serious odor problem was associated with an industrial system, or where the concentration of organic solvents in the gases to be discharged to the air was high enough that these could be burned and the heat utilized in the process (8). By the mid-1950s there were several dozen catalytic incinerators in California, primarily in Los Angeles county, the first sizable area within the United States to experience a serious air pollution problem. Early applications of this technology involved some serious odor, eye irritation, or visible organic emission problems resulting from halogen poisoning and catalyst fouling (15).

The chemical industry was the first to utilize catalytic oxidation extensively for emission control, building units capable of treating up to 50 m³/s (100,000 scfm) of exhaust gas containing VOCs. Catalytic systems accounted for roughly one-fourth of the $200 million market for VOC control systems in 1992, and over one thousand catalytic oxidation devices were in place by the end of that year (5).

Catalysts. A catalyst has been defined as a substance that increases the rate at which a chemical reaction approaches equilibrium without becoming permanently involved in the reaction (16). Thus a catalyst accelerates the kinetics of the reaction by lowering the reaction's activation energy (5), ie, by introducing a less difficult path for the reactants to follow. For VOC oxidation, a catalyst decreases the temperature, or time required for oxidation, and hence also decreases the capital, maintenance, and operating costs of the system.

A key feature of a catalyst is that the catalytic material is not consumed by the chemical oxidation reactions, rather it remains unaltered by the reactions which occur generally on its surface and thus remains available for an infinite number of successive oxidation reactions.

Many chemical elements exhibit catalytic activity (5) which, within limits, is inversely related to the strength of chemisorption of the VOCs and oxygen, provided that adsorption is sufficiently strong to achieve a high surface coverage (17). If the chemisorption is too strong, the catalyst is quickly deactivated as the active sites become irre-

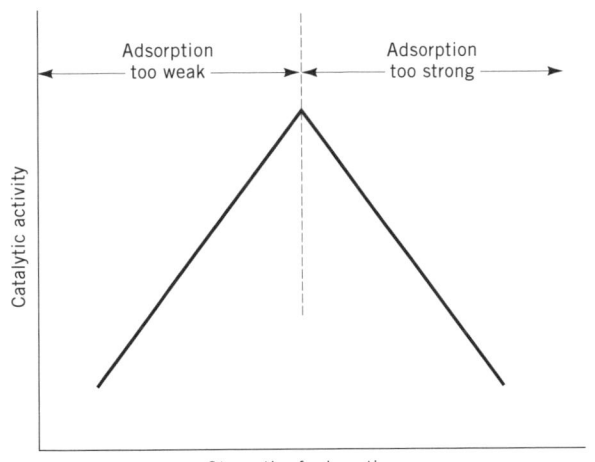

Figure 2. Catalytic activity as a function of adsorption strength (17). Courtesy of Oxford University Press.

versibly covered. If the chemisorption is too weak, only a small fraction of the surface is covered and the activity is very low (17) (Fig. 2).

Catalysts vary both in terms of compositional material and physical structure (18). The catalyst basically consists of the catalyst itself, which is a finely divided metal (14,17,19); a high surface area carrier; and a support structure. Three types of conventional metal catalysts are used for oxidation reactions: single- or mixed-metal oxides, noble (precious) metals, or a combination of the two (19).

The precious metal or metal oxide imparts high intrinsic activity, the carrier provides a stable, high surface area for catalyst dispersion, and the mechanical support gives a high geometric surface area for physical support and engineering design features (20). Only the correct combination of these components provides suitable performance and long catalyst life of a properly designed catalytic system (21).

Metal Oxides. The metal oxides are defined as oxides of the metals occurring in Groups 3–12 (IIIB to IIB) of the Periodic Table. These oxides, characterized by high electron mobility and the positive oxidation state of the metal, are generally less active as catalysts than are the supported nobel metals, but the oxides are somewhat more resistant to poisoning. The most active single-metal oxide catalysts for complete oxidation of a variety of oxidation reactions are usually found to be the oxides of the first-row transition metals, V, Cr, Mn, Fe, Co, Ni, and Cu.

Noble Metals. Noble or precious metals, ie, Pt, Pd, Ag, and Au, are frequently alloyed with the closely related metals, Ru, Rh, Os, and Ir. These are usually supported on a metal oxide such as α-alumina, α-Al_2O_3, or silica, SiO_2. The most frequently used precious metal components are platinum [7440-06-4], Pt, palladium [7440-05-3], Pd, and rhodium [7440-16-6], Rh. The precious metals are more commonly used because of the ability to operate at lower temperatures. As a general rule, platinum is more active for the oxidation of paraffinic hydrocarbons;

palladium is more active for the oxidation of unsaturated hydrocarbons and CO (19).

Each precious metal or base metal oxide has unique characteristics, and the correct metal or combination of metals must be selected for each exhaust control application. The metal loading of the supported metal oxide catalysts is typically much greater than for nobel metals, because of the lower inherent activity per exposed atom of catalyst. This higher overall metal loading, however, can make the system more tolerant of catalyst poisons. Some compounds can quickly poison the limited sites available on the noble metal catalysts (19).

Carrier. The metal catalyst is generally dispersed on a high surface area carrier, ie, the carrier is given a washcoat of catalyst, such that very small (2–3 nm dia) precious metal crystallites are widely dispersed over the surface area, serving two basic functions. It maximizes the use of the costly precious metal, and provides a large surface area thereby increasing gas contact and associated catalytic reactions (18).

Proper selection of the carrier is critical (20). For example, in most cases, the carrier of a precious metal catalyst is a high surface area alumina having an effective surface area in the order of 120 m^2/g of material. Alumina is often used because of its unique phase transformation properties. Various phases of the aluminum hydroxides exist as a function of temperature and starting phase. For a catalyst to be stable, the correct starting phase of alumina must be selected for the projected commercial operating temperature. Otherwise the alumina may undergo a transition during operation resulting in a carrier of less surface area, and hence less catalytic activity (1).

Efforts to redesign catalyst formulations involve both catalyst and washcoat. One thrust of this research is to manipulate the catalytic surface so that it can handle larger quantities of catalyst poisons and to incorporate more catalytic sites, redistributed within the washcoat to make them more accessible to exhaust molecules. Altering the composition of the alumina washcoat by including various nonprecious metal oxides, such as oxides of barium, cerium, and lanthanum as stabilizers, is being looked at to promote catalyst activity before sintering and stabilize precious metal dispersion (18). Reformulation efforts are aided by use of computer controls and cleaner reactants and continuous monitoring, all of which help make exhaust composition more predictable (1).

The Support Structure. After the catalytic element is placed on the high surface area carrier, it is deposited on a mechanical support structure which determines the form of the catalyst. The support structure may have many forms, such as spheres, pellets, woven mesh, screen, honeycomb, or other ceramic matrix structures designed to maximize catalyst surface area (6). Some of the advantages of these different supports are listed in Table 2.

The pelleted and honeycomb support structures are most widely used; honeycombs are the most commonly employed. Pelleted structures are generally spherical beads or cylinders having diameters ranging from 0.16–0.64 cm. The pellets are assembled into a packed bed containing large numbers of these pellets through which the

Table 2. Conventional Catalyst Bed Geometries[a]

Geometry	Advantages	Disadvantages
Metal ribbons	Low pressure drop; high surface-to-volume	Less active than ceramic-supported catalysts
Spherical pellets	Can be used in both fixed and fluidized beds	High pressure drop; attrition problem
Ceramic rods	Low pressure drop	Low surface-to-volume ratio
Ceramic honeycomb	Low pressure drop; high surface-to-volume ratio	May have nonuniform catalyst coating
Metal honeycomb	Low pressure drop; high surface-to-volume ratio; high mechanical strength	Less active than ceramic-supported catalysts

[a] Ref. 14.

exhaust passes. The honeycomb supports are monolithic structures having numerous parallel channels through which the exhaust passes, the channel sizes ranging from about 8 to 50 cells/cm² of catalyst frontal area. Each cell has a width opening ranging from 0.29 to 0.13 cm, respectively. Some commercial honeycombs are available from 1.6 to 100 cells/cm² (20). The shape in the individual honeycomb channel is unlimited, eg, circle, square, triangle, etc.

Although more expensive to fabricate than the pelleted catalyst, and usually more difficult to replace or regenerate, the honeycomb catalyst is more widely used because it affords lower pressure losses from gas flow; it is less likely to collect particulates (fixed-bed) or has no losses of catalyst through attrition, compared to fluidized-bed; and it allows a more versatile catalyst bed design (18), having a well-defined flow pattern (no channeling) and a reactor that can be oriented in any direction.

The honeycomb structure is either fabricated of ceramic or stainless steel. The high surface area carrier and catalytic precious metal crystallites are coated onto the walls of the channels in the honeycomb. The honeycomb catalyst blocks generally range from 15 to 30 cm square at depths from 5 to 10 cm. These blocks are packed into larger modules containing many catalyst blocks. Flow through a honeycomb catalyst structure is shown in Figure 3 (18,22). A typical 30 cells/cm² honeycomb structure has about 4600 m² of geometric wall area per cubic meter of catalyst volume (18). The actual shape of the individual

honeycomb channel is unlimited, eg, it may be circular, square, triangular, etc. In addition, the channel density can be varied. Commercial honeycombs are available that range from 1.5 to 100 cells/cm² (10 to 600 cells/in.² (20).

Only by using the carrier can the catalyst be sufficiently active because the majority of applications require 10 to 100 m²/g of surface area (20). Surface areas for a typical monolith support structure and a carrier are given (20).

Catalyst	BET surface area, m²/g
geometric smooth monolith wall	0.003
natural porosity of monolith wall	0.3
carrier on monolith wall	20

Mechanistic Models. A general theory of the mechanism for the complete heterogeneous catalytic oxidation of low molecular weight vapors at trace concentrations in air does not exist. As with many catalytic reactions, however, certain observations have led to a general hypothesis (17).

The overall process of any catalytic reaction is a combination of mass-transfer (qv), describing transport of reactants and products to and from the interior of a solid catalyst, and chemical reaction kinetics, describing chemical reaction sequences on the catalyst surface. The most cost-effective catalytic oxidation systems require use of a solid catalyst material having a high specific surface area, ie, high surface area per net weight of catalyst. The pres-

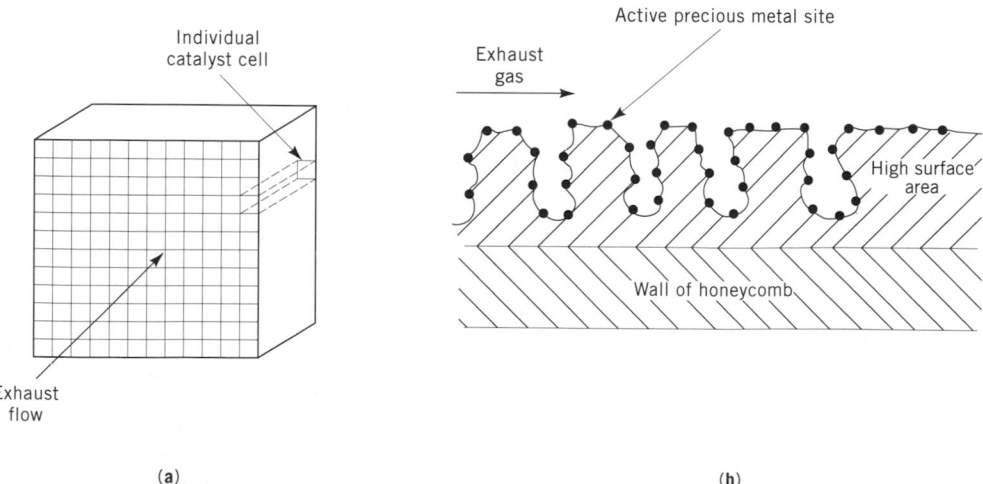

Figure 3. Schematic of the flow through (**a**) a honeycomb catalyst structure and (**b**) a cross section of a honeycomb channel.

ence of many small pores necessarily introduces pore transport diffusion resistance as a factor in the overall, or global, kinetics. The overall process consists of (17,23): (1) transport of reactants from the bulk fluid through the gas film boundary layer to the surface of the catalytic particle; (2) transport of reactants into the catalyst particle by diffusion through the catalyst pores; (3) chemisorption of at least one reactant on the catalyst surface; (4) chemical reaction between chemisorbed species or between a chemisorbed species and a physisorbed or fluid-phase reactant; (5) desorption of reaction products from the catalyst surface; (6) diffusive transport of products through the catalyst pores to the surface of the catalyst particle; and (7) mass transfer of products through the exterior gas film to the bulk fluid.

In principle, any of these steps or some combination can be rate controlling. In practice, temperature plays a primary role in determining the rate-controlling stage. Any comprehensive analysis of actual catalytic oxidation systems of practical interest must include a quantitative understanding of the relative effects of mass transfer (steps 1,2,6,7) and surface reaction (steps 3,4,5). The temperature relationship of these two mechanisms is shown in Figure 4 (17,18,20). As a catalyst is heated, conversion of the pollutant is negligible until a critical temperature is reached, then the rate of conversion increases rapidly with rising temperature. This is referred to as the kinetically limited region. Conversion increases in this region because catalytic reaction rates increase with temperature, until the catalyst's normal operating temperature is achieved. Then the conversion rate increases only slightly with further temperature rise in the mass-transfer limited region. At some advanced temperature, the conditions reach a point where thermal oxidation begins to play a role, and the rate of conversion again increases rapidly.

In the mass-transfer limited region, conversion is most commonly increased by using more catalyst volume or by increasing cell density, which increases the catalytic wall

area per volume of catalyst. When the temperature reaches a point where thermal oxidation begins to play a role, catalyst deactivation may become a concern.

Reaction Rate. The kinetics for a single catalytic reaction can be modeled as

$$-r_m = k(T)f(C)n$$

where $-r_m$ is the rate of the main reaction; $k(T)$ is the rate constant, a function of temperature, T; $f(C)$ is the function of reactant and product concentration, C; and n is the effectiveness factor, which accounts for pore-diffusional resistance (24). The form of the terms $k(T)$ and $f(C)$ depends on the kinetic model for the system. Kinetic models for the catalytic oxidation can either be empirical or mechanistic.

Empirical Models. In the case of an empirical equation, the model is a power law rate equation that expresses the rate as a product of a rate constant and the reactant concentrations raised to a power (17), such as

$$r_m = kC_1^a C_2^b$$

where r_m is the reaction rate; k is the rate constant; C_1 is the concentration of reactant 1; C_2 is the concentration of reactant 2; and a and b are empirically determined reaction orders.

For combustion of simple hydrocarbons, the oxidation reactions appear to follow classical first-order reaction kinetics sufficiently closely that practical designs can be established by application of the empirical theory (8). For example, the general reaction for a hydrocarbon:

$$C_xH_y + (x + y/2)\, O_2 \rightarrow x\, CO_2 + (y/2)\, H_2O$$

can be represented by the rate equation

$$r_m = (dC/dt) = -kC$$

where C = hydrocarbon concentration, r_m = rate of change of contaminant concentration, t = time, and k = reaction rate constant, which must be determined experimentally from the burning of various organic materials. The pattern of variation with t is predictable from kinetic theory and follows the Arrhenius equation,

$$k = A \exp(-\Delta E/RT)$$

where A is the Arrhenius collision constant, ΔE the activation energy, and R is the universal gas constant. A catalyst increases the rate of reaction by adsorbing gas molecules on catalytically active sites. The catalyst may function simply to bring about a higher concentration of reactive materials at the surface than is present in the bulk gas phase, which has the effect of increasing the collision constant, A, or the catalyst may modify a molecule of adsorbed gas by adding or removing an electron or by physically opening a bond. This has the effect of decreasing the activation energy, ΔE, in the Arrhenius equation. In either circumstance, it is necessary for the reactive ma-

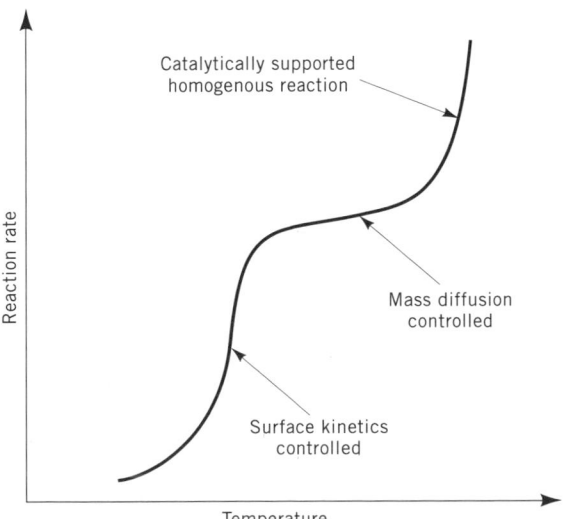

Figure 4. Reaction rate profile as a function of temperature (20).

terials to reach the active catalyst surface by diffusion through the gas phase, and for the reaction products to leave the surface. For the conditions encountered in most hydrocarbon emission control applications, the oxygen partial pressure is much larger than the organic reactant partial pressure, and can be treated as a constant.

Mechanistic kinetic expressions are often used to represent the rate data obtained in laboratory studies, and to explain quantitatively the effects observed in the field. Several types of mechanisms have been proposed. These differ primarily in complexity, and on whether the mechanism assumes that one compound that is adsorbed on the catalyst surface reacts with the other compound in the gas phase, eg, the Eley-Rideal mechanism (23); or that both compounds are adsorbed on the catalyst surface before they react, eg, the Langmuir-Hinshelwood mechanism (25).

The volatile organic compounds on the list of hazardous air pollutants under the CAAA have been classified into four main categories: (1) pure hydrocarbons (qv), (2) halogenated hydrocarbons, (3) nitrogenated hydrocarbons, and (4) oxygenated hydrocarbons. The compounds in these groups are characterized by the following oxidation reactions (26):

Hydrocarbons

$$C_6H_6 + 9\,O_2 \rightarrow 6\,CO_2 + 6\,H_2O$$

Halogenated hydrocarbons

$$CCl_4 + O_2 + 2\,H_2O \rightarrow CO_2 + 4\,HCl + O_2$$

Nitrogenated hydrocarbons

$$2\,HCN + 3\,O_2 \rightarrow 2\,CO_2 + 2\,H_2O + 2\,N_2$$

Oxygenated hydrocarbons

$$C_4H_8O + 5\tfrac{1}{2}\,O_2 \rightarrow 4\,CO_2 + 4\,H_2O$$

Temperature reaction rate profiles for representative compounds are available (21,26). Particularly important are the operating temperatures required before destruction is initiated. Chemical reactivity by compound class from high to low is (27) alcohols > cellosolves/dioxane > aldehydes > aromatics > ketones > acetates > alkanes > chlorinated hydrocarbons. In general, within a class the higher the molecular weight, the higher the relative destructibility. All of these compound classes, except chlorinated hydrocarbons, can be destroyed with 98–99% efficiency at sufficiently low space velocities and/or high enough inlet temperatures (28). Table 3 (22) presents oxidation temperatures for a number of hydrocarbons.

Historically, the destruction efficiency for chlorinated hydrocarbons is quite low. In addition, tests conducted after the chlorinated hydrocarbon is treated show that the catalyst is partially deactivated. More recent advancements in catalyst technology have resulted in the development of a number of catalysts and catalytic systems capable of handling most chlorinated hydrocarbons under a variety of conditions (19).

Table 3. Ignition Temperatures for 90% Conversion[a]

Component	Temp, °C
Hydrogen	93
Acetylene	177
Carbon monoxide	218
Cyclohexanone	218
Propylene	232
Toluene	232
2-Propanol	260
Ethylene	260
Benzene	260
Xylene	260
Ethanol	260
Methyl ethyl ketone	274
Ethyl acetate	288
Cyclohexane	288
n-Hexane	316
Methyl isobutyl ketone	316
Propane	399
Methane	427

[a] Ref. 22.

Mixture Effects. Care must be taken in determining the oxidation kinetics for a mixture of chemicals (29). In principle, given one set of conditions and a two-component mixture, the overall conversion of one component A may be controlled by mass transfer to the catalyst surface and the conversion of another component B by surface-reaction kinetics. Of course, the controlling regime (mass transfer or reaction) can change with temperature. Thus for two independent parallel reactions the combined effect of diffusional and reaction rate resistances can have a considerable influence on the relative rate of the two reactions. Additionally, a third, fourth, or nth component can conceivably affect the other components by, for instance, competing more successfully for active surface sites than B while simultaneously influencing the mass transfer of A. Thus even for a simple two- or three-component mixture, interpretation of observed results can be difficult. Extrapolation of mixture behavior from single-component data is ill-advised.

In a mixture of n-hexane and benzene (29), the deep catalytic oxidation rates of benzene and n-hexane in the binary mixture are lower than when these compounds are singly present. The kinetics of the individual compounds can be adequately represented by the Mars-VanKrevelen mechanism. This model needs refinements to predict the kinetics for the mixture.

One important consideration in any catalyst oxidation process for a complex mixture in the exhaust stream is the possible formation of hazardous incomplete oxidation products. Whereas the concentration in the effluent may be reduced to acceptable levels by mild basic aqueous scrubbing or additional vent gas treatment, studying the kinetics of the mixture and optimizing the destruction cycle can drastically reduce the potential for such emissions.

Design and Operation. The destruction efficiency of a catalytic oxidation system is determined by the system design. It is impossible to predict *a priori* the temperature and residence time needed to obtain a given level of conversion of a mixture in a catalytic oxidation system. Control efficiency is determined by process characteristics such as concentration of VOCs emitted, flow rate, process fluctuations that may occur in flow rate, temperature, concentrations of other materials in the process stream, and the governing permit regulation, such as the mass-emission limit. Design and operational characteristics that can affect the destruction efficiency include inlet temperature to the catalyst bed, volume of catalyst, and quantity and type of noble metal or metal oxide used.

Catalytic oxidation systems are normally designed for destruction efficiencies that range from 90 to 98% (27). In the early 1980s, typical design requirements were for 90% or higher VOC conversions. More recently, however, an increasing number of applications require 95 to 98% conversions to meet the more stringent emission standards (20).

Operational Considerations. The performance of catalytic incinerators (28) is affected by catalyst inlet temperature, space velocity, superficial gas velocity (at the catalyst inlet), bed geometry, species present and concentration, mixture composition, and waste contaminants. Catalyst inlet temperatures strongly affect destruction efficiency. Mixture compositions, air-to-gas (fuel) ratio, space velocity, and inlet concentration all show marginal or statistically insignificant effects (30).

Operating Temperature. The operating temperature needed to achieve a particular VOC destruction efficiency depends primarily on the species of pollutants contained in the waste stream, the concentration of the pollutants, and the catalyst type (14). One of the most important factors is the hydrocarbon species. Each has a catalytic initiation temperature which is also dependent on the type of catalyst used (14).

For a given inlet temperature, the quantity of supplied heat may be provided by (6) the heat supplied from the combustion of supplemental fuel, the sensible heat contained in the emission stream as it enters the catalytic system, and the sensible heat gained by the emission stream through heat exchange with hot flue gas (6). Three types of systems for catalytic oxidation of VOCs are shown in Figure 5 (11). The simplest (Fig. 5**a**) uses a direct contact open flame to preheat the gas stream upstream of the catalyst. The second (Fig. 5**b**) involves only a catalyst bed over which the gas stream passes, usually after some indirect preheating. The third (Fig. 5**c**) involves more extensive indirect preheating and heat exchange. The difference in the three configurations is the method for preheating the gas.

There are two general temperature policies: increasing the temperature over time to compensate for loss of catalyst activity, or operating at the maximum allowable temperature. These temperature approaches tend to maximize destruction, yet may also lead to loss of product

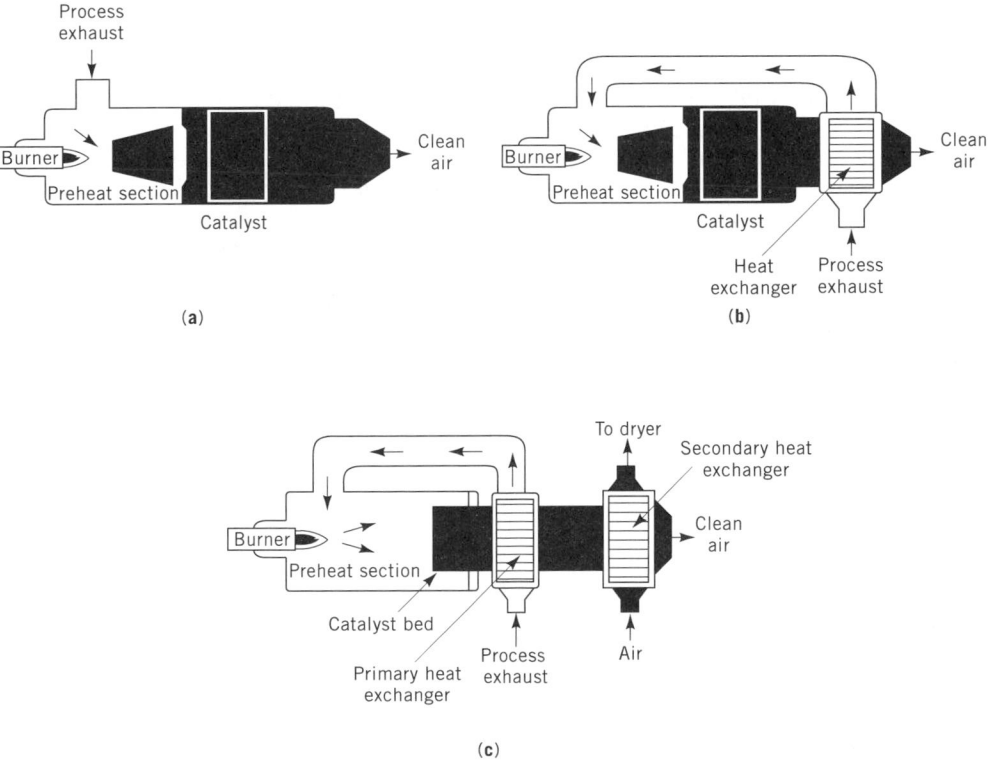

Figure 5. Catalytic system designs (11) of (**a**) basic VOC catalytic converter containing a preheater section, a reactor housing the catalyst, and essential controls, ducting, instrumentation, and other elements; (**b**) a heat exchanger using the cleaned air exiting the reactor to raise the temperature of the incoming process exhaust; and (**c**) extracting additional heat from the exit gases by a secondary heat exchanger.

selectivity. Selectivity typically decreases with increasing temperature; faster deactivation; and increased costs for reactor materials, fabrication, and temperature controls.

Reactor Design. The catalytic reactor is designed to be operated in the mass-transfer controlled catalytic region. The prime design parameter is the geometric surface area. The honeycomb catalyst shows substantial advantage over other forms because of the high geometric surface areas obtainable with low pressure drop (20).

Catalyst Selection. The choice of catalyst is one of the most important design decisions. Selection is usually based on activity, selectivity, stability, mechanical strength, and cost (31). Stability and mechanical strength, which make for steady, long-term performance, are the key characteristics. The basic strategy in process design is to minimize catalyst deactivation, while optimizing pollutant destruction.

Both catalyst space velocity and bed geometry play a role. The gas hourly space velocity (GHSV) is used to relate the volumetric flow rate to the catalyst volume. GHSV has units of inverse hour and is defined as the volume flow rate per catalyst volume.

The size of the catalyst bed depends mainly on the degree of VOC reduction required (14). VOC destruction efficiencies up to 95% can usually be attained using reasonable space velocities (14). However, the low GHSVs, and subsequently high catalyst volumes required to achieve extremely high (eg, 99%) conversions, can sometimes make catalytic oxidation uneconomical. Conventional bed geometries may be found in the literature (14).

Process Conditions. To effectively design a catalytic control system, the Manufacturers of Emissions Controls Association recommends the following data be obtained (5): list of all VOCs present and range of concentration of each, flow rate of exhaust and expected variability, oxygen concentration in exhaust and expected variability, temperature of exhaust and expected variability, static pressure, potential uses for heat recovery, particular performance criteria and/or regulations to be met, capture efficiency, ie, fraction of all organic vapors generated by the processes that are directed to the control device, presence of hydrocarbon aerosols in the effluent exhaust, identity and quantity of all inorganic and organic particulate, amount of noncombustibles, presence of possible catalyst deactivators, and anticipated start-up/shutdown frequency of the system.

Pilot Studies. Applications requiring the reduction of VOC emissions have increased dramatically. On-site pilot tests are beneficial in providing useful information regarding VOC emission reduction applications. Information that can be obtained includes optimum catalyst operating conditions, the presence of contaminants in the gas stream, and the effects of these contaminants.

Catalyst Inhibition. A number of potential applications for catalytic oxidation of organic materials have resulted in serious odor or eye irritation, or visible emission problems (8). Some of these failures are a result of fouling of the catalyst surface. Others occur because materials such as halogens in the gas stream interfere with or suppress the activity of the catalyst, or because the substances react with the precious metals, rendering them permanently inactive. Finally, all catalysts eventually deteriorate by aging or thermal processes (8).

Many of the exhaust streams that must be purified contain significant amounts of halogenated organics, such as polychlorinated ethanes and ethylenes vented in the manufacture of vinyl chloride monomer or released in usage solvents (32). However, the catalysts used in the conventional catalytic oxidation are severely inhibited by the halogen atoms in these compounds (32). Other trace contaminants of concern in air streams may include phosphorus, nitrogen, and sulfur-containing compounds. Whereas gases containing chlorine, sulfur, and other atoms can deactivate supported noble metal catalysts such as platinum, chlorinated VOC can be treated by certain supported metal oxide catalysts (7).

The four basic mechanisms of catalyst decay are shown in Figure 6 (5,18,24). These are fouling or masking, poisoning, thermal degradation through aging or sintering, and loss of catalyst material through formation and escape of vapors. Poisoning and vapor transport are basically chemical phenomena, whereas fouling is mechanical. Table 4 lists substances that inhibit catalyst activity (5).

Masking or Fouling. Masking or fouling is a physical deposition of species from the fluid phase onto the catalyst surface (Fig. 6a) that results in blockage of reaction sites or pores (24). Masking or fouling is caused by a gradual accumulation of noncombusted, solid material that mechanically coats the catalyst's surface and prevents or slows down the diffusion of reactants to the catalyst.

Typical masking or fouling agents include (5,8,21,24) airborne dust or dirt; metal oxides formed from materials in the process, such as silicon dioxide ash remaining when silicone compounds are oxidized; aggregate compound formation on the catalyst surface, ie, phosphorus for lubricating oils; corrosion products from the duct system; and organic char or tars formation from incomplete combustion products, often caused by too low a reactor operating temperature.

Low levels of particulates or potential poisons can sometimes be tolerated without a dramatic decrease in performance. Generally it has been recommended that the maximum particulate concentration not exceed 115 mg/m^2 and that the maximum poison concentration not exceed 25 ppm (14). In addition, every effort should be made to avoid flow over a cold catalyst bed for any extended period of time, as a process stream containing volatile organics may condense on a cold catalyst bed (5).

Combustible masking materials such as organic char may be partially or completely removed by periodic elevations of the catalyst bed temperature. Noncombustible masking materials may be removed by air lancing or aqueous washing generally with a leaching solution (20,21).

Poisons. Halogens, sulfur dioxide [7446-09-5], SO_2, nitrogen dioxide [10102-44-0], NO_2, and numerous other materials act as catalyst suppressants for precious-metal

Table 4. Substances that Inhibit Catalyst Activity[a]

Type of Inhibitor	Effect	Examples	Regeneration
Fast-acting inhibitors	Reduction of catalyst activity at rate depending on concentration and temperature	Phosphorus, bismuth, lead, arsenic, antimony, mercury	Catalyst regeneration is sometimes difficult or impossible
Slow-acting inhibitors	Reduction of catalyst activity; higher concentrations than those of fast-acting catalyst inhibitors may be tolerated	Iron, tin, silicon	Catalyst regeneration remains difficult or impossible
Reversible inhibitors/maskers	Surface coating of catalyst active area; rate also dependent on concentration and temperature	Sulfur, halogens, silicon, zinc, phosphorus	Regeneration is possible
Surface maskers	Surface coating of catalyst active areas	Organic solids	Removed by increasing catalyst temperature or by acid and alkaline washing
Surface eroders and maskers	Surface coating of catalyst active area, or erosion of catalyst surface; both result in loss of catalyst activity; rate dependent on particle size, grain loading, and gas stream velocity	Inert particulates	Surface coating is easily removed by washing
Thermal degradation and sintering	Loss of catalyst surface area because of catalyst dispersion and crystal growth, or catalyst support collapse through sintering	Higher temperatures for extended time, temperature excursions, hot spots in bed	Regeneration generally very difficult; best avoided by operating in optimum temperature range and avoiding temperature excursions
Vapor transport and attrition	Loss of catalytic material through formation of metal carbonyl oxides, sulfides, and halides, or surface shear effects resulting from exhaust gas velocity, particulates, or thermal shock	CO, NO, hydrogen sulfide, halogens, and particulates	Must replace lost catalytic material; vaporization generally not a factor; attrition particularly important in fluidized beds

[a] Refs. 5 and 22.

oxidation catalysts. These compounds tend to adsorb strongly on the catalytic surface, preventing the reactants from doing so. The strength of adsorption is ordinarily such that the suppressant materials can gradually be stripped off after there is no longer a concentration of suppressant materials in the gas stream passing through the catalyst (8). In other cases, the adsorption is irreversible. A poison blocks the catalytic sites, and may also induce changes in the surface to result in formation of compounds (24). Active precious-metal sites become inactive, reducing catalyst performance (see Fig. 6b).

At low (>450°C) temperatures, the presence of these materials, particularly the oxides, leads to simple masking or fouling. In some cases, a catalyst that shows reduced activity believed to be from poisoning may simply be masked, and activity can be rejuvenated by cleaning with aqueous leaching solutions (21).

Poisoning is operationally defined. Often catalysts believed to be permanently poisoned can be regenerated (5). A species may be a poison in some reactions, but not in others, depending on its adsorption strength relative to that of other species competing for catalytic sites (24), and the temperature of the system. Catalysis poisons have been classified according to chemical species, types of reactions poisoned, and selectivity for active catalyst sites (24).

Group 14 and 15 (VA and VIA) elements act as poisons. The interaction depends on the poison's oxidation state and chemical structure. For sulfur, the order of decreasing poisoning activity is $H_2S > SO_2 > SO_4^{2-}$. Adsorption studies indicate that H_2S adsorbs strongly and dissociates on nickel surfaces. The sulfur adsorbs essentially irreversibly and over most of the catalyst–metal surface. It has been observed that SO_2 and SO_3 also poison catalysts differently (13); SO_2 selectively adsorbs on Pt or Pd in an oxidation catalyst, whereas SO_3 reacts with the Al_2O_3 carrier, forming $Al_2(SO_4)_3$, which destroys the structure of the catalyst. The latter can be prevented by using a more inert support such as SiO_2 or TiO_2. The former requires a change in operating conditions such as a higher temperature. When No. 2 fuel oil is used in some gas turbines, the sulfur compound in the fuel oil can be converted to SO_2 at levels of 40 to 150 ppm in the exhaust. For such applications, the presence of 100 to 200 ppm SO_2 can require 150 to 200°C higher temperature for the catalyst to give the same CO conversion as without SO_2.

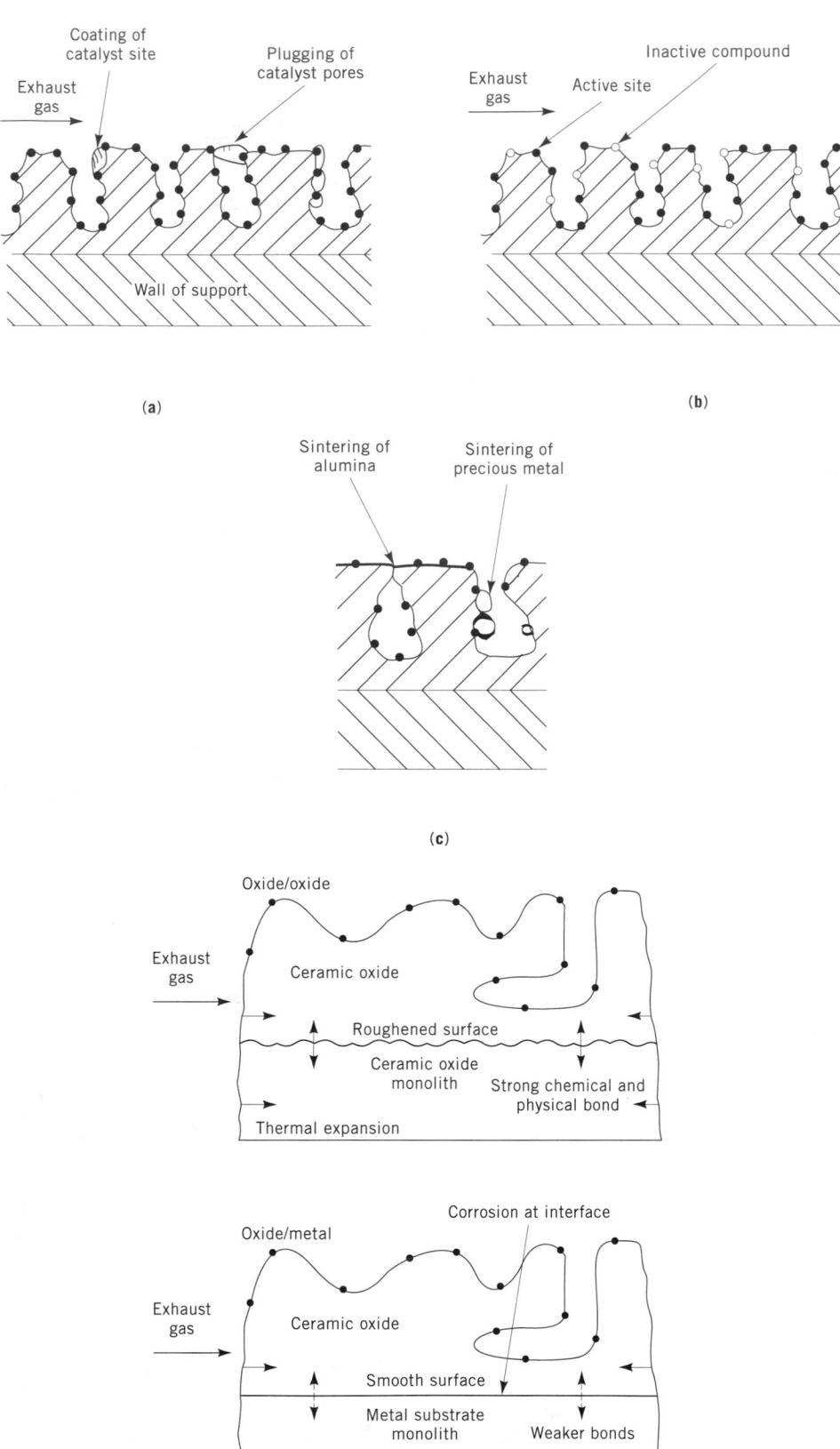

Figure 6. Catalyst inhibition mechanisms where (●) are active catalyst sites; ▨, the catalyst carrier; and ▨ the catalytic support: (**a**) masking of catalyst; (**b**) poisoning of catalyst; (**c**) thermal aging of catalyst; and (**d**) attrition of ceramic oxide metal substrate monolith system, which causes the loss of active catalytic material resulting in less catalyst in the reactor unit and eventual loss in performance.

Toxic heavy metals and ions, eg, Pb, Hg, Bi, Sn, Zn, Cd, Cu, and Fe, may form alloys with catalytic metals (24). Materials such as metallic lead, zinc, and arsenic react irreversibly with precious metals and make the surface unavailable for catalytic reactions. Poisoning by heavy metals ordinarily destroys the activity of a precious-metal catalyst (8).

Molecules having unsaturated bonds, eg, CO, NO, HCN, and benzene, may chemisorb through multiple bonds (24).

Catalysts having improved poison resistance have been developed. Catalysts are available that can destroy chlorine-, fluorine-, or bromine-containing organic compounds (5).

Thermal Degradation and Sintering. Thermally induced deactivation of catalysts may result from redispersion, ie, loss of catalytic surface area because of crystal growth in the catalyst phase (21,24,33) or from sintering, ie, loss of catalyst-support area because of support collapse (18). Sintering processes generally take place at high (>500°C) temperatures and are generally accelerated by the presence of water vapor (see Fig. 6c). Another thermal effect is the transformation of catalytic phases to noncatalytic ones, eg, the reaction of nickel and alumina to form nickel aluminate (24). Each catalyst has a recommended temperature window of operation. At temperatures above this window (usually ≥760°C), sintering can occur.

Loss of Catalyst by Vapor Transport. The direct volatilization of catalytic metals is generally not a factor in catalytic processes, but catalytic metal can be lost through formation of metal carbonyl oxides, sulfides, and halides in environments containing CO, NO, O_2 and H_2S, and halogens (24).

The ceramic oxide carrier is bonded to the monolith by both chemical and physical means. The bonding differs for a ceramic monolith and a metallic monolith. Attrition is a physical loss of the carrier from the monolith from the surface shear effects caused by the exhaust gas, a sudden start-up or shutdown causing a thermal shock as a result of different coefficients of thermal expansion at the boundary between the carrier and the monolith, physical vibration of the catalyzed honeycomb, or abrasion from particulates in the exhaust air (21) (see Fig. 6d).

Avoiding Catalyst Deactivation. Catalyst deactivation is more easily prevented than cured. Poisoning by impurities may be prevented by removing impurities from the reactants. Carbon deposition and coking may be prevented by minimizing formation of precursors and manipulating mass-transfer regimes so as to minimize the carbon's or coke's effect on activity. Most sintering is irreversible, or reversible only with great difficulty, so it is important to choose reaction, ie, lower temperatures, that do not sinter the catalyst. Additionally, when process upsets that could release inhibitors or cause small fluctuations in the heating value of the oxidizer are highly probable a thermal system is favored over a catalytic one (7).

Except for No. 2, fuel oil should not be considered as auxiliary fuel when using a catalytic system because of the sulfur and vanadium the fuel oil may contain (7). In some cases even the sulfur in No. 2 fuel oil can present a problem. Galvanized metal should not be used in process ovens or ductwork because zinc is a catalyst poison.

Proper system design and catalyst maintenance are key to minimizing deactivation and providing long-term catalyst service. For example, control of air dilution, use of temperature control loops, and use of catalysts having high intrinsic thermal stability can provide necessary protection against high temperature damage caused by reaction exotherms and from operational upsets (20).

Experimental Evaluation. Often the deactivation kinetics for a catalytic oxidation system can be evaluated in a series of laboratory studies (24). Reactors should be gradientless with respect to reactant poison concentration and temperature. Heat- and mass-transfer effects should be avoided because these disguise the intrinsic kinetics. Experiments should be designed to study one deactivation process at a time, and accelerated targets must be representative of the process. Deactivation can be accelerated by using smaller amounts of catalyst, operating at higher temperatures or different pressures, at greater residence times, or at different gas compositions.

Whereas changing catalyst volume or residence time rarely yields complications, changing temperature or pressure could introduce sintering. The properties of the catalyst should be measured both before and after deactivation and inlet and outlet streams should be analyzed by chromatography (qv) or spectrometry.

Around 1972, it was reasoned that the problem of catalyst deactivation could not always be entirely eliminated, but that continuous replacement of a portion of the catalyst bed during normal operation would allow continuing operation at high efficiency even in the presence of poisoning agents. Hence the fluidized bed was born (8). In some applications, fluidized-bed oxidation processes overcame poisoning, masking, and thermal aging. A process in which performance depends on the continuous attrition of the external surface of the catalyst particles, however, has many unattractive features, including the effort required for trapping, collecting, and disposing of the fine particulate released from the reactor (32).

At least one printer using catalytic oxidation has experienced relatively rapid catalyst deactivation, requiring replacement after useful lives as short as 3–6 months, when producing high quality printed matter using some of the most desirable lithographic printing plates and materials (34). It was observed that the use of phosphorus additives caused rapid deactivation of the conventional catalyst used to destroy the hydrocarbons in the solvent-laden air (SLA) discharged from the press dryer. A precious-metal catalyst containing platinum and palladium was being used in this application. It had replaced an earlier base metal catalyst, which showed rapid deactivation as a result of sulfur in the SLA, presumably introduced in the fuel used to fire the heatset dryer. It was found that the P concentration in the SLA might be as high as 0.16 ppm. The phosphorus concentration on the deactivated catalyst was found to be 1.4% of the catalyst. The printer was urged by the supplier to find and eliminate the cause of the phosphorus contamination of the waste gas entering the catalyst bed. However, after it was determined that use of phosphorus-containing additives was crucial to many of the high quality printing jobs, attention was directed at the catalyst bed itself.

A catalyst with a substantially improved resistance to poisoning by phosphorus in catalytic oxidation applications was developed. In part, the catalyst in this program permitted printers to use lithographic technology without paying an unreasonable cost in terms of frequent replacement of oxidation catalysts.

Catalyst Reactivation. Some catalytic systems are reported to have operated continuously for more than 10 years with little or no loss in control efficiency (5). In most processes catalysts inevitably lose activity, and when the activity has declined to a critical level, the catalyst needs to be discarded or regenerated (24). Regeneration is only possible when the deactivation is reversible by chemical washing or heat treatment or oxidation (20,24).

Thermal Treatment. A thermal treatment for catalyst regeneration is usually effective when deactivation is a result of coking or masking of the catalyst surface. Thermal treatment can usually be done on-site, by elevating the temperature of the catalyst bed by 50 to 100°C above the normal operating point and running at this oxidizing condition for a specified limited period of time (20). The elevated temperature vaporizes or oxidizes the organic compounds or char that may be masking the catalyst surface.

Physical Treatment. If inspection of the catalyst indicates deposits again, or if an excessive pressure drop across the catalyst is noted, then the catalytic bed may be lanced, on-site, using compressed air or water until the deposits are removed. Abrasion by contact with excessively high pressure from the compressed air should be avoided (20). If this treatment is combined with heating, hot spots or overtemperatures that could further deactivate the catalyst should be avoided (24). In many cases, periodic maintenance, removing the catalyst bed and blowing or washing off residues, has restored catalyst to original or near-original activity levels (5).

Chemical Treatment. The most involved regeneration technique is chemical treatment (20) which often follows thermal or physical treatment, after the char and particulate matter has been removed. Acid solution soaks, glacial acetic acid, and oxalic acid are often used. The bed is then rinsed with water, lanced with air, and dried in air. More involved is use of an alkaline solution such as potassium hydroxide, or the combination of acid washes and alkaline washes. The most complex treatment is a combination of water, alkaline, and acid washes followed by air lancing and drying. The catalyst should not be appreciably degraded by the particular chemical treatment used.

Analyses of a catalyst used in a process involving cleaning products and pigments and achieving a hydrocarbon destruction capacity of only 13% showed deposition of P, Sn, Pb, and Na contaminants (20). Initial acid treatment increased the hydrocarbon destruction capacity from 13 to 63%. Alkaline treatment increased the capacity to 90% of that new.

EXHAUST CONTROL TECHNOLOGIES

In addition to VOCs, specific industrial exhaust control technologies are available for nitrogen oxides, NO_x, carbon monoxide, CO, halogenated hydrocarbon, and sulfur and sulfur oxides, SO_x.

Nitrogen Oxides

Annual releases of nitrogen oxides (NO_x) into the atmosphere amounted to ca 550×10^6 t in the early 1990s. A number of states, in addition to California, regulate NO_x emissions (35). The production of nitrogen oxides can be controlled to some degree by reducing formation in the combustion system. The rate of NO_x formation for any given fuel and combuster design are controlled by the local oxygen concentration, temperature, and time history of the combustion products. Techniques employed to reduce NO_x formation are collectively referred to as combustion controls and U.S. power plants have shown that furnace modifications can be a cost-effective approach to reducing NO_x emissions. Combustion control technologies include operational modifications, such as low excess air, biased firing, and burners-out-of-service, which can achieve 20–30% NO_x reduction; and equipment modifications such as low NO_x burners, overfire air, and reburning, which can achieve 40–60% reduction (36). As of this writing, approximately 600 boilers having 10,000 MW of capacity use combustion modifications to comply with the New Source Performance Standards (NSPS) for NO_x emissions (37).

When NO_x destruction efficiencies approaching 90% are required, some form of post-combustion technology applied downstream of the combustion zone is needed to reduce the NO_x formed during the combustion process. Three post-combustion NO_x control technologies are utilized: selective catalytic reduction (SCR); nonselective catalytic reduction (NSCR); and selective noncatalytic reduction (SNCR).

Selective Catalytic Reduction. Selective catalytic reduction (SCR) is widely used in Japan and Europe to control NO_x emissions (1). SCR converts the NO_x in an oxygen-containing exhaust stream to molecular N_2 and H_2O using ammonia as the reducing agent in the presence of a catalyst. NO_x removals of 90% are achievable. The primary variable is temperature, which depends on catalyst type (38). The principal components of an SCR system include the catalyst, the SCR reactor, the ammonia injection grid (AIG), the ammonia–air dilution system, the ammonia storage–vaporization system, the ammonia addition control system, and a continuous emissions monitoring system (39).

The AIG is used to uniformly inject diluted ammonia into the reactor. Uniform mixing of the ammonia into the flue gas is necessary to maintain catalyst performance at its highest level and to minimize ammonia leakage (ammonia slip) past the catalyst.

The ammonia–air dilution system dilutes the vaporized ammonia by a factor of 20 to 25 with air for better admixing through the AIG and to prevent explosive ammonia–air mixtures. Once the catalyst volume is selected, the NO_x removal is set by the NH_3/NO_x mole ratio at the inlet of the SCR system (39).

SCR was first developed in the United States in the late 1950s, targeted at nitric acid tail-gas exhausts, using precious-metal catalysts. In the mid-1970s, SCR entered widespread commercial use in Japan using base metal catalysts. The first SCR systems in Germany started up in 1986 (39), and German utilities are committed to installing SCR systems on the majority of oil- and coal-fired boilers to achieve between 60 and 90% NO_x reductions (39). By the end of 1992, there were about 120 SCR plants in service in Germany alone (40). A limited amount of experience has been documented in the United States (41,42), although commercial service began in 1985 and is expected to increase (43).

Performance criteria for SCR are analogous to those for other catalytic oxidation systems: NO_x conversion, pressure drop, catalyst/system life, cost, and minimum SO_2 oxidations to SO_3. An optimum SCR catalyst is one that meets both the pressure drop and NO_x conversion targets with the minimum catalyst volume. Because of the interrelationship between cell density, pressure drop, and catalyst volume, a wide range of optional catalyst cell densities are needed for optimizing SCR system performance.

Reactions. The SCR process is termed selective because the ammonia reacts selectively with NO_x at temperatures $> 232°C$ in the presence of excess oxygen (44). The optimum temperature range for the SCR catalyst is determined by balancing the needs of the redox reactions.

SCR reactions

$$4\,NO + 4\,NH_3 + O_2 \rightarrow 4\,N_2 + 6\,H_2O$$
$$2\,NO_2 + 4\,NH_3 + O_2 \rightarrow 3\,N_2 + 6\,H_2O$$

The NO reduction is the most important because NO_2 accounts for only 5–10% of the NO_x in most exhaust gases.

Ammonia oxidation reactions

$$4\,NH_3 + 5\,O_2 \rightarrow 4\,NO + 6\,H_2O$$
$$4\,NH_3 + 3\,O_2 \rightarrow 4\,N_2 + 6\,H_2O$$

When sulfur dioxide is also present there are important side reactions in which SO_2 is oxidized to SO_3. The main side reaction in the SCR catalyst is the conversion of SO_2 to SO_3, thus facilitating the reaction above. The SO_3 in turn reacts with ammonia to form ammonium sulfates.

$$NH_3 + SO_3 + H_2O \rightarrow NH_4HSO_4$$
$$2\,NH_3 + SO_3 + H_2O \rightarrow (NH_4)_2SO_4$$

The formation of ammonium bisulfate is strongly temperature dependent. Formation is favored at the lower temperatures. The temperature at which ammonium bisulfate is not formed depends strongly on the SO_3 concentration in the exhaust gas. The temperature needed to minimize bisulfate formation has been reported to increase by about 15°C (around about 350°C) when the SO_3 concentration increases from 5 to 15 ppm (23). The formation of the bisulfate is reversible, ie, if the temperature is raised to 20°C above the minimum temperature, the reaction is shifted to result in the decomposition of the bisul-

fate formed. When chlorides are present, ammonium chlorides can be formed:

$$NH_3 + HCl \rightarrow NH_4Cl$$

When sulfuric acid is present, ammonium bisulfate can be formed:

$$NH_3 + H_2SO_4 \rightarrow NH_4HSO_4$$

These various reactions should be minimized to avoid plugging the catalyst and to prevent fouling of the downstream air preheaters, when these components condense from the gas at the lower temperatures.

The SCR Process. The first step in the SCR reaction is the adsorption of the ammonia on the catalyst. SCR catalysts can adsorb considerable amounts of ammonia (45). However, the adsorption must be selective and high enough to yield reasonable cycle times for typical industrial catalyst loadings, ie, uptakes in excess of 0.1% by weight. The rate of adsorption must be comparable to the rate of reaction to ensure that suitable fronts are formed. The rate of desorption must be slow. Ideally the adsorption isotherm is rectangular. For optimum performance, the reaction must be irreversible and free of side reactions.

It has been suggested that the first step, weak coadsorption of NO and O_2 on a reduced vanadium site, may represent the slow step in the mechanism. Subsequent formation of a N_2O_3-like intermediate could be a fast step, because it is known that in the gas phase the equilibrium of $NO–NO_2$ and N_2O_3 is established within microseconds (35).

At low temperatures the SCR reactions dominate and nitrogen oxide conversion increases with increasing temperature. But as temperature increases, the ammonia oxidation reactions become relatively more important. As the temperature increases further, the destruction of ammonia and generation of nitrogen oxides by the oxidation reactions causes overall nitrogen oxide conversion to reach a plateau then decreases with increasing temperatures. Examples are shown in Figure 7 (44).

In the SCR process, ammonia, usually diluted with air or steam, is injected through a grid system into the flue/exhaust stream upstream of a catalyst bed (37). The effectiveness of the SCR process is also dependent on the NH_3 to NO_x ratio. The ammonia injection rate and distribution must be controlled to yield an approximately 1:1 molar ratio. At a given temperature and space velocity, as the molar ratio increases to approximately 1:1, the NO_x reduction increases. At operations above 1:1, however, the amount of ammonia passing through the system increases (38). This ammonia slip can be caused by catalyst deterioration, by poor velocity distribution, or inhomogeneous ammonia distribution in the bed.

Types of SCR Catalysts. The catalysts used in the SCR were initially formed into spherical shapes that were placed either in fixed-bed reactors for clean gas applications or moving-bed reactors where dust was present. The moving-bed reactors added complexity to the design and in some applications resulted in unacceptable catalyst

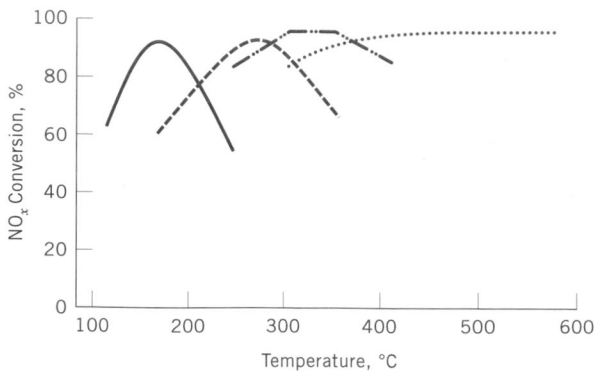

Figure 7. NO$_x$ reduction vs temperature for SCR catalysts: (—), (———), precious metal; (–··–), base metal; and (········) zeolites (41).

abrasion. As of 1993 most SCR catalysts are either supported on a ceramic or metallic honeycomb or are directly extruded as a honeycomb (1). A typical honeycomb block has face dimensions of 150 by 150 mm and can be as long as one meter. The number of cells per block varies from 20 by 20 up to 45 by 45 (39).

No SCR catalyst can operate economically over the whole temperature range possible for combustion systems. As a result, three general classes of catalysts have evolved for commercial SCR systems (44): precious-metal catalysts for operation at low temperatures, base metals for operation at medium temperatures, and zeolites for operation at higher temperatures.

The precious-metal platinum catalysts were primarily developed in the 1960s for operation at temperatures between about 200 and 300°C (1,38,44). However, because of sensitivity to poisons, these catalysts are unsuitable for many combustion applications. Variations in sulfur levels of as little as 0.4 ppm can shift the catalyst required temperature window completely out of a system's operating temperature range (44). Additionally, operation with liquid fuels is further complicated by the potential for deposition of ammonium sulfate salts within the pores of the catalyst (44). These low temperature catalysts exhibit NO$_x$ conversion that rises with increasing temperature, then rapidly drops off, as oxidation of ammonia to nitrogen oxides begins to dominate the reaction (see Fig. 7).

The most popular SCR catalyst formulations are those that were developed in Japan in the late 1970s comprised of base metal oxides such as vanadium pentoxide [1314-62-1], V$_2$O$_5$, supported on titanium dioxide [13463-67-7], TiO$_2$ (1). As for low temperature catalysts, NO$_x$ conversion rises with increasing temperatures to a plateau and then falls as ammonia oxidation begins to dominate the SCR reaction. However, peak conversion occurs in the temperature range between 300 and 450°C, and the fall-off in NO$_x$ conversion is more gradual than for low temperature catalysis (44).

A family of zeolite catalysts has been developed, and is being increasingly used in the United States in SCR applications. Zeolites which can function at higher temperatures than the conventional catalysts, are claimed to be effective over the range of 300 to 600°C, having an optimum temperature range from 360 to 580°C (37,38). However, ammonia oxidation to NO$_x$ begins around 450°C and

is predominant at temperatures in excess of 500°C. Zeolites suffer the same performance and potential damage problems as conventional catalysts when used outside the optimum temperature range. In particular, at around 550°C the zeolite structure may be irreversibly degraded because of loss of pore density. Zeolite catalysts have not been continuously operated commercially at temperatures above 500° C (37).

Using zeolite catalysts, the NO$_x$ reduction takes place inside a molecular sieve ceramic body rather than on the surface of a metallic catalyst. This difference is reported to reduce the effect of particulates, soot, SO$_2$/SO$_3$ conversions, heavy metals, etc, which poison, plug, and mask metal catalysts. Zeolites have been in use in Europe since the mid-1980s and there are approximately 100 installations on stream. Process applications range from use of natural gas to coal as fuel. Typically, nitrogen oxide levels are reduced 80 to 90% (37).

Catalyst Selection. For an SCR application, catalyst selection depends largely on the temperature of the flue gas being treated. A given catalyst exhibits optimum performance within a temperature range of about 30 to 50°C. Below this optimum temperature range, the catalyst activity is greatly reduced, allowing unreacted ammonia to slip through. Above this range, ammonia begins to be oxidized to form additional NO$_x$. Operations having adequate temperature controls are important, as are uniform flue gas temperatures (37,38).

Problems. A number of difficulties in utilizing SCR operations have been identified. Problems in European installations are of particular interest because SCR systems are subjected to conditions not experienced in Japan, but also encountered in the United States. Difficulties include matching the NH$_3$ injection pattern to the nonuniform flow of NO$_x$ in the ductwork ahead of the SCR rector; inability to optimize the NH$_3$ injection rate by feedback control of slip ammonia for lack of a reliable NH$_3$ monitor; erosion and plugging on units retrofitted to boilers that fire high ash coal; catalyst deactivation caused by arsenic poisoning on wet-bottom units that recycle flyash; process control under load swings; and spent catalyst disposal. For medium to high sulfur coals, the potential exists for accelerated catalyst deactivation caused by sulfur poisoning and contamination by trace metals in flyash, and deposit buildup and corrosion of the air heater.

For boilers, SCR DENO$_x$ plants can be installed at the exhaust exit just before the air preheater. These are called high dust plants (17 g dust per cubic meter) (23) because the flue gas still contains volatile trace elements as well as flyash particles. When the SCR system is installed after the flue gas desulfurization (FGD) system, it is called a low dust or tail-end plant. The high dust plant has the advantages of not requiring any additional energy because of the high temperatures present. The low dust plant requires a regenerative heat exchanger, but requires less catalyst because (40): there is less dust to contaminate the bed, potential catalyst poisons are removed from the flue gas by the FGD system, and a high activity catalyst can be used when only low concentrations of poisons such as SO$_2$ remain after the flue gas system.

Other problems that can be associated with the high

dust plant can include alkali deterioration from sodium or potassium in the stack gas deposition on the bed, calcium deposition, when calcium in the flue gas reacts with sulfur trioxide, or formation and deposition of ammonium bisulfate. In addition, plugging of the air preheater as well as contamination of flyash and FGD wastewater discharges by ammonia are avoided if the SCR system is located after the FGD (23).

A significant problem area for initial SCR systems has been the continuous emission monitoring (CEM) systems. In power plants, all sites equipped with CEM systems report the highest failure frequency. The CEM systems are the most labor intensive component, requiring as much as full-time attention from one technician. At one power plant CEM systems were responsible for 100% of 73 reported SCR system shutdowns (38). As CEM systems improve, these concerns may disappear.

Nonselective Catalytic Reduction. Hydrocarbons, hydrogen, or carbon monoxide can be used as reducing agents for NO_x in applications where the exhaust oxygen concentration is low, as it is in fuel rich-burn reciprocating engines, where it is less than 1%, and in nitric acid plants, when it is from 2 to 3%. This approach is called nonselective catalytic reduction (NSCR). In some applications, the oxygen must be removed from the feed stream prior to the catalyst (35). An oxygen sensor in the exhaust stream signals the air–fuel delivery system to adjust the air–fuel ratio so it is just slightly fuel-rich, having enough reducing agent present to react with all the oxygen and nitrogen oxides (1).

Nonselective catalytic reduction systems are often referred to as three-way conversions. These systems reduce NO_x, unburned hydrocarbon, and CO simultaneously. In the presence of the catalyst, the NO_x are reduced by the CO resulting in N_2 and CO_2 (37). A mixture of platinum and rhodium has been generally used to promote this reaction (37). It has also been reported that a catalyst using palladium has been used in this application (1). The catalyst operation temperature limits are 350 to 800°C, and 425 to 650°C are the most desirable. Temperatures above 800°C result in catalyst sintering (37). Automotive exhaust control systems are generally NSCR systems, often shortened to NCR.

Typically NO_x conversion ranges from 80 to 95% and there are corresponding decreases in CO and hydrocarbon concentrations. Potential problems associated with NSCR applications include catalyst poisoning by oil additives, such as phosphorus and zinc, and inadequate control systems (37).

Carbon Monoxide

Carbon monoxide is emitted by gas turbine power plants, reciprocating engines, and coal-fired boilers and heaters. CO can be controlled by a precious-metal oxidation catalyst on a ceramic or metal honeycomb. The catalyst promotes reaction of the gas with oxygen to form CO_2 at efficiencies that can exceed 95%. CO oxidation catalyst technology is broadening to applications requiring better catalyst durability, such as the combustion of heavy oil, coal (qv), municipal solid waste (qv), and wood (qv). Research is underway to help cope with particulates and

contaminants, such as flyash and lubricating oil, in gases generated by these fuels (1).

CO conversion is a function of both temperature and catalyst volume, and increases rapidly beginning at just under 100°C until it reaches a plateau at about 150°C. But, unlike NO_x catalysts, above 150°C there is little benefit to further increasing the temperature (44). Above 150°C, the CO conversion is controlled by the bulk phase gas mass transfer of CO to the honeycomb surface. That is, the catalyst is highly active, and its intrinsic CO removal rate is exceedingly greater than the actual gas transport rate (21). When the activity falls to such an extent that the conversion is no longer controlled by gas mass transfer, a decline of CO conversion occurs, and a suitable regeneration technique is needed (21).

It has been reported that below about 370°C, sulfur oxides reversibly inhibit CO conversion activity. This inhibition is greater at lower temperatures. CO conversion activity returns to normal shortly after removal of the sulfur from the exhaust (44). Above about 315°C, sulfur oxides react with the high surface area oxides to disperse the precious-metal catalytic agents and irreversibly poison CO conversion activity.

Catalyst contamination from sources such as turbine lubricant and boiler feed water additives is usually much more severe than deactivation by sulfur compounds in the turbine exhaust. Catalyst formulation can be adjusted to improve poison tolerance, but no catalyst is immune to a contaminant that coats its surface and prevents access of CO to the active sites. Between 1986 and 1990 over 25 commercial CO oxidation catalyst systems operated on gas turbine cogeneration systems, meeting both CO conversion (40 to 90%) and pressure drop requirements.

Halogenated Hydrocarbons

Destruction of halogenated hydrocarbons presents unique challenges to a catalytic oxidation system (45–51). The first step in any control strategy for halogenated hydrocarbons is recovery and recycling (45). However, even upon full implementation of economic recovery steps, significant halocarbon emissions can remain. In other cases, halogenated hydrocarbons are present as impurities in exhaust streams (45). Impurity sources are often intermittent and dispersed.

The principal advantage of a catalytic oxidation system for halogenated hydrocarbons is in operating cost savings. Catalytically stabilized combusters improve the incineration conditions, but still must employ very high temperatures as compared to VOC combustors; eg, carbon tetrachloride [56-23-5], CCl_4, has a 40-fold lower heat of combustion than a typical organic vapor such as toluene [108-88-3], thus CCl_4 requires much more supplemental fuel to burn than do typical organics (45). Alternatively, the low temperature catalytic oxidation process is typically designed for a maximum adiabatic temperature rise of only 200°C. This would correspond to only about 1500 ppm of an organic compound in the exhaust stream. But, with the lower heat of combustion, up to 40,000 ppm of carbon tetrachloride could be treated in the same temperature rise, or with less dilution air.

By-Product Formation. The presence of halogenated hydrocarbons dramatically increases the yield of aldehydes from the oxidation process (45). For example, in the partial oxidation of methane on a PdO sponge catalyst (19), methylene chloride, CH_2Cl_2, was added in pulses to the inlet gas, which also contained the methane. Methane oxidation was strongly inhibited and formaldehyde was formed. The formaldehyde production continued even after the methylene chloride addition was stopped, suggesting a strong interaction of chlorine with the catalyst. However, pulses of pure CH_4 plus oxygen gradually restored the original activity to the catalyst, indicating that the effect of this interaction was reversible.

Catalyst Deactivation. Catalyst deactivation (45) by halogen degradation is a very difficult problem particularly for platinum (PGM) catalysts, which make up about 75% of the catalysts used for VOC destruction (10). The problem may well lie with the catalyst carrier or washcoat. Alumina, for example, a common washcoat, can react with a chlorinated hydrocarbon in a gas stream to form aluminum chloride which can then interact with the metal. Fluid-bed reactors have been used to offset catalyst deactivation but these are large and costly (45).

Catalytic Reaction. The desired reaction of the chlorine group on a chlorinated hydrocarbon is

$$RCl + O_2 \rightarrow CO_2 + HCl + R'$$

It is important to produce HCl rather than elemental chlorine, Cl_2, because HCl can be easily scrubbed out of the exhaust stream, whereas Cl_2 is very difficult to scrub from the reactor off-gas. If the halogenated hydrocarbon is deficient in hydrogen relative to that needed to produce HCl, low levels of water vapor may be needed in the entering stream (45) and an optional water injector may be utilized. For example, trichloroethylene [79-01-6], C_2HCl_3, and carbon tetrachloride require some water vapor as a source of hydrogen (45).

$$C_2HCl_3 + H_2O + 1.5\,O_2 \rightarrow 2\,CO_2 + 3\,HCl$$

Groundwater contaminated with chlorinated hydrocarbons is being remediated by a conventional air stripper or a rotary stripper, producing an air stream containing the halogenated hydrocarbon vapors and saturated with water vapor (45), which is then passed through a catalyst bed.

At least two catalytic processes have been used to purify halogenated streams. Both utilize fluidized beds of probably nonnoble metal catalyst particles. One has been estimated to oxidize > 9000 t/yr of chlorinated wastes from a vinyl chloride monomer plant (45). Several companies have commercialized catalysts which are reported to resist deactivation from a wider range of halogens. These newer catalysts may allow the required operating temperatures to be reduced, and still convert over 95% of the halocarbon, such as trichlorethylene, from an exhaust stream. Conversions of C-1 chlorocarbons utilizing an Englehardt HDC catalyst are shown in Figure 8. For this system, as the number of chlorine atoms increases, the temperatures required for destruction decreases.

USES

Catalytic oxidation of exhaust streams is increasingly used in those industries involved in the following (13,15,17):

Surface Coatings

Aerospace
Automobile
Auto refinishing
Can coating
Coil coating
Fabric coating
Large appliances
Marine vessels
Metal furniture
Paper coating
Plastic parts coating
Wire coating and enameling
Wood furniture

Printing Inks

Flexographic
Lithographic
Rotogravure
Screen printing

Solvent Usage

Adhesives
Disk manufacture
Dry cleaning
Fiber glass manufacture
Food tobacco manufacture
Metal cleaning
Pharmaceutical
Photo finishing labs
Semiconductor manufacture

Chemical and Petroleum Processes

Cumene manufacture
Ethylene oxide manufacture
Acrylonitrile manufacture
Caprolactam manufacture
Maleic anhydride manufacture
Monomer venting
Phthalic anhydride manufacture
Paint and ink manufacture
Petroleum product refining
Petroleum marketing
Resin manufacture
Textile processing

Industrial/Commercial Processes

Aircraft manufacture
Asceptic packaging
Asphalt blowing
Automotive parts manufacture
Breweries/wineries
Carbon fiber manufacture
Catalyst regeneration
Coffee roasting
Commercial charbroiling
Electronics manufacture
Film coating
Filter paper processing
Food deep frying
Gas purification
Glove manufacture
Hospital sterilizers
Peanut and coffee roasting
Plywood manufacture
Rubber processing
Spray painting
Tire manufacture
Wood treating

Engines

Diesel engines
Lean burn internal combustion
Natural gas compressors
Oil field steam generation
Rich burn internal combustion
Gas turbine power generation

Cross Media Transfer

Air stripping
Groundwater cleanup
Soil remediation (landfills)
Hazardous waste treatment
Odor removal from sewage gases

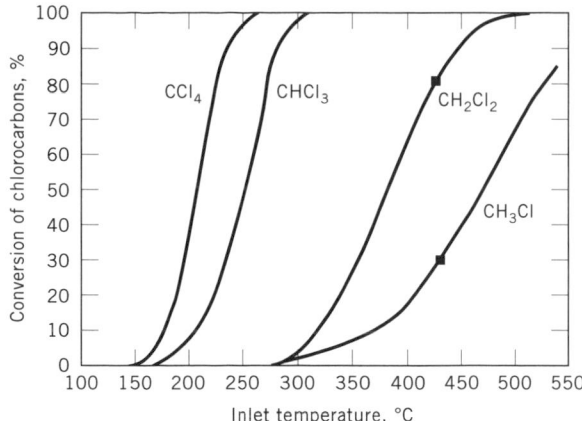

Figure 8. Destruction of C-1 chlorocarbons over HDC in the presence of 1.5% H_2O in air at 15,000 h^{-1} GHSV at STP. Chlorocarbon concentrations are CCl_4, 900 ppmv; $CHCl_3$, 500 ppmv; CH_2Cl_2, 800 ppmv; and CH_3Cl, 600 ppmv.

The most important factors affecting performance are operating temperature, surface velocity, contaminant concentration and composition, catalyst properties, and the presence or absence of poisons or inhibitors.

Air Stripping of Groundwater

Treatment of exhaust streams from the air stripping of contaminated groundwater is a particular challenge, because the emissions from air stripping units may consist of a complex mixture of both fuel and solvent fractions (6). The catalytic oxidation of any given compound is generally negatively impacted by the presence of others in mixtures, and higher catalyst bed operating temperatures are required to achieve adequate destruction.

Some catalysts exposed to air stripping off-gas were subject to deactivation. However, using a catalytic oxidizer at a U.S. Coast Guard facility (Traverse City, Mich.) for the destruction of benzene, toluene, and xylene stripped from the groundwater, the catalytic oxidization unit operated at 260 to 315°C, and was able to achieve 90% destruction efficiency.

Printing and Graphic Arts

In the graphic arts industry, the catalyst in the oxidizer needs to be monitored regularly because it is susceptible to contamination by phosphorus from fountain solutions, silica from silicone gloss enhancer sprays, and chlorides from chlorinated solvents or blanket wash solutions. Phosphorus and silica accumulate most rapidly on the leading edge of the catalyst bed, deactivating the catalyst by masking the precious metals. In a fluidized-bed configuration, the catalyst surface is continually renewed by abrasion and the problem of masking the catalyst surface with silicones is avoided.

Chemical Processing

Terephthalic Acid Production. The control of exhaust from production of pure terephthalic acid (PTA) has been a challenge (see CARBOXYLIC ACIDS) (52). Eight million metric tons of PTA are produced annually worldwide for use primarily in high grade polyester fiber production. Based on a *para*-xylene feedstock, vent gases from the process contain such by-products as methyl acetate, organic acids, and often methyl bromide. These exhausts have been estimated to total 34,000 m^3 worldwide. Historically, the presence of the methyl bromide limited the use of fixed-bed catalytic oxidation as a control technology using precious-metal catalysts. Thus base metal catalysts in fluidized-bed reactors have been the primary catalytic technology of choice. In this application, the continuous abrasion of the outer layer of the catalyst particle exposes a fresh surface of unpoisoned material to the reactants, allowing the catalyst to effectively treat the exhaust stream.

In the late 1980s, however, the discovery of a noble metal catalyst that could tolerate and destroy halogenated hydrocarbons such as methyl bromide in a fixed-bed system was reported (52,53). The products of the reaction were water, carbon dioxide, hydrogen bromide, and bromine. Generally, a scrubber would be needed to prevent downstream equipment corrosion. However, if the focus of the control is the VOCs and the CO rather than the methyl bromide, a modified catalyst formulation can be used that is able to tolerate the methyl bromide, but not destroy it. In this case the methyl bromide passes through the bed unaffected, and designing the system to avoid downstream effects is not necessary. Destruction efficiencies of hydrocarbons and CO of better than 95% have been reported, and methyl bromide destructions between 0 and 85% (52).

Latex Monomer Production. ARI Technologies, Inc. has introduced a catalyst system which, it is claimed, can operate at an average bed temperature of 370°C while achieving conversion efficiency in excess of 99.99% on exhaust streams from latex monomer production (see LATEX TECHNOLOGY).

Acrylonitrile Manufacture. In the manufacture of acrylonitrile (qv), off-gases containing from 1–3% of CO plus various hydrocarbons are emitted. Catalytic beds of platinum-group metals are used to reduce the regulated compounds to acceptable levels. Close attention to bed design is required to prevent the formation of appreciable quantities of NO_x caused by the fixation of combustion–air nitrogen. Some NO_x also is produced from fuel nitrogen by oxidation. Because of the high thermal energy content of the off-gases, considerable heat recovery is possible in abating acrylonitrile plant emissions.

Vinyl Monomer Manufacturing. Process vent gases containing small quantities of halogenated hydrocarbons and substantial quantities of nonhalogenated hydrocarbons have been successfully reduced to comply with regulatory objectives in large-scale laboratory–pilot catalytic fume abaters having satisfactory long-term catalyst performance. The design freedoms offered by precious metals on ceramic honeycomb support catalysts have been demonstrated in equipment that utilizes the heat energy resulting from the substantial exotherm of the nonhalogenated hydrocarbon oxidation to preheat the exhaust gases. Fuel consumption is thereby minimized.

Coatings Industries

Surface coating processes (qv) produce similar air pollution problems in a number of different industries.

Can Manufacturing. An internal coating is necessary to protect the purity and flavor of can contents for beverages or any edible product that might react with the container metal. Both the exterior decorative and interior sanitary coatings are applied to the metal surface by rolls or spray guns using a solvent vehicle. Catalytic oxidation systems are used by the principal can manufacturers to treat coatings exhaust streams. The can manufacturers' industry is estimated to utilize more catalysts than any of the other surface coating industries.

A large number of diverse solvents are used in exterior and interior coatings in plants for manufacturing both three- and two-piece cans. Most of the organic solvents are found in the cure-oven exhausts at concentrations of 2–16% of the lower explosive limit (LEL). The oven exhaust volumes are usually 1–35 m³/s. When burned, these concentrations of combustibles provide an exotherm of 30 to 220°C. The heat that is released is used for preheating the incoming effluent and/or heating the cure oven by recycling the hot, cleaned gases to the supply blowers or by heating makeup air by heat exchange. A few plants use the heat of the cleaned exhaust to produce hot water for the two-piece can line washers, hot air for dry-off ovens, or building space heating. For example, one large can company utilizes the heat energy contained in the stream, leaving some of their catalytic fume abaters to supply all the heat energy required by the oven's heating zones, which have no burners. The fuel energy supplied to the catalytic fume abater is less than would be needed to heat the oven if the solvent fumes were exhausted directly to the atmosphere without use of the fume abater. The exhaust rate of the oven is adjusted to maintain a solvent concentration of at least 8% of the LEL, equivalent to a 110°C temperature differential.

The various reaction rate properties of the different solvents influence the design of a catalytic reactor. For example, for a specific catalyst bed design, an effluent stream containing a preponderance of monohydric alcohols, aromatic hydrocarbons, or propylene requires a lower catalyst operating temperature than that required for solvents such as isophorone and short-chain acetates.

Design considerations and costs of the catalyst, hardware, and a fume control system are directly proportional to the oven exhaust volume. The size of the catalyst bed often ranges from 1.0 m³ at 0°C and 101 kPa per 1000 m³/min of exhaust, to 2 m³ for 1000 m³/min of exhaust. Catalyst performance at a number of can plant installations has been enhanced by proper maintenance. Annual analytical measurements show reduction of solvent hydrocarbons to be in excess of 90% for 3–6 years, the equivalent of 12,000 to 30,000 operating hours. When propane was the only available fuel, the catalyst cost was recovered by fuel savings (vs thermal incineration prior to the catalyst retrofit) in two to three months. In numerous cases the fuel savings paid for the catalyst in 6 to 12 months.

Can manufacturers often regenerate the catalyst beds on an annual or biannual basis during a weekend downtime. Both air lancing and an aqueous bath are utilized to remove noncombustible particulates that mask the active sites. Frequently, condensed organic material on the catalyst is removed by short-term (4–6 h) heating excursions to 370 or 430°C; the organic matter is removed much like a self-cleaning oven. The gaseous and organic smoke, which is usually evolved from the first few cm of catalyst bed depth, is oxidized in the latter part of the bed. If allowed to operate too long at temperatures that promote condensation, high boiling organic compounds, the subsequent carbon char that is formed, may require temperatures of 480 to 540°C to convert the carbon to carbon monoxide for subsequent oxidation. The higher temperatures required for burn-offs should be approached in small (0–30°C) increments to bring about slow evolution and partial oxidation; this prevents autogenous combustion of local high concentrations of combustible material.

Day-to-day operating techniques that are employed by one large can manufacturer and are intended to prevent organic condensation are dictated by the use of a low cost, well-established, sanitary coating for beer and beverage three-piece cans. Polybutadiene and other sanitary coatings may have volatile resin monomers entrained in the oven atmosphere as a result of rapid evaporation of solvent before polymerization takes place. A short (4 to 6 h) heating excursion up to a catalyst inlet temperature of 400°C after use of the coating usually burns off any condensed organic materials. It has become standard practice in some plants to turn the catalytic afterburner up to 370°C for these coatings vs the normal 315°C operating temperature for vinyls, acrylics, etc.

Coil Coating. Coil coating is the prefinishing of many sheet metal items with protective and decorative coatings that are applied by roll coating on one or both sides of a fast-moving metal strip. The metal strip (from 13 mm to 1.7 m in width) unwinding from a coil travels at rates of 30–150 m/min through the coating applicator rolls and bake ovens. It is rewound into a coil for transport to a forming operation for products that are to be used in cans, appliances, industrial and residential siding, shelving, cars, gutters, downspouts, etc. The source of hydrocarbon emissions in coil coating is the coating application area and the cure-oven exhaust. The coatings include primers, finishes, and metal protective (5 μm) films or backers.

The increasing use of siliconized coatings for weather durability caused severe masking problems for the all-metal, filter mesh-like catalyst elements available in the 1970s. Interest in catalytic afterburners increased when dispersed-phase precious metal–alumina-on-ceramic honeycomb catalysts offered economically attractive results.

The hot (260–370°C) oven exhaust can be oxidized catalytically without preheat, but when the coater-area exhaust (at room temperature) is combined with the oven exhaust, a preheat burner becomes necessary. The greatest energy savings potential having the least capital investment is obtained by recycling a portion of the hot, cleaned exhaust to the oven. This principle has been demonstrated at a number of can manufacturing plants and at least four coil coating facilities. One operator preheats oven make-up air which has been taken from the oven cooler section by means of a heat exchanger; whereas others recycle directly to the oven. In one case, the heat energy for the dry-off oven is supplied by the catalytic incinerator exhaust (482°C) remaining after supplying most of the heat energy to operate the four zones in the oven. The concentration of solvents in the exhaust is about 12% LEL (167°C temperature differential). The net fuel energy consumption is about 20% of that required to fire the paint, bake, and dry-off ovens without fume control.

Coil coaters operate equipment continuously and, in most cases, operate catalytic fume abaters 6000–7000 h/yr. Under these conditions the anticipated catalyst life is years, with an annual aqueous solution cleaning. However, the catalyst may last no more than two years if frequent maintenance is needed, such as in-place air lancing every 60 to 90 days to remove noncombustible particulates. Frequent maintenance may be needed if coatings such as siliconized polyester (15–40% silicones) comprise 30% of the coatings put through the system.

Filter Paper Processing. In the fabrication of fuel oil and air filters for vehicles such as motorcycles and diesel locomotives, heat processing of the filter paper is required to cure the resin (usually phenolic) with which the paper (qv) is impregnated. The cure-oven exhaust, which contains water vapor, alcohols, and dimers and trimers of phenol, produces a typical blue haze aerosol having a pungent odor. The concentration of organic substances in the exhaust is usually rather low.

The paper-impregnation drying oven exhausts contain high concentrations (10–20% LEL) of alcohols and some resin monomer. Vinyl resins and melamine resins, which sometimes also contain organic phosphate fire retardants, may be used for air filters. The organic phosphates could shorten catalyst life depending on the mechanism of reduction of catalyst activity. Mild acid leaching removes iron and phosphorus from partially deactivated catalyst and has restored activity in at least one known case.

Catalysis is utilized in the majority of new paper filter cure ovens as part of the oven recirculation/burner system which is designed to keep the oven interior free of condensed resins and provide an exhaust without opacity or odor. The application of catalytic fume control to the exhaust of paper-impregnation dryers permits a net fuel saving by oxidation of easy-to-burn methyl or isopropyl alcohol, or both, at adequate concentrations to achieve a 110–220°C exotherm.

BIBLIOGRAPHY

1. R. J. Farrauto, R. M. Heck, and B. K. Speronnelo, *C & EN*, 34–44 (Sept. 7, 1992).

2. C. C. Lewis, *CPI Purch.* 29–33 (Aug. 1991).

3. J. C. Summers, J. E. Sawyer, and A. C. Frost "The 1990 Clean Air Act and Catalytic Emission Control Technology for Stationary Sources," in R. G. Silver, J. E. Sawyer, and J. C. Summers, eds., *Catalytic Control of Air Pollution: Mobile and Stationary Sources,* ACS Symposium Series 495, 1992.

4. R. L. Berglund and C. T. Lawson, "Pollution Prevention in the Chemical Process Industries," *Chem. Eng.* (Sept. 1991).

5. *Catalytic Control of VOC Emissions, A Cost Effective Viable Technology for Industrial, Commercial and Waste Processing Facilities,* Manufacturers of Emission Controls Association, Washington, D.C., 1992.

6. M. Kosusko and C. M. Nunez, "Destruction of Volatile Organic Compounds Using Catalytic Oxidation," *JAWMA* **40**(2) (Feb. 1990).

7. D. R. van der Vaart, W. M. Vatvuk, and A. H. Wehe *JAWMA* **41**(1), 92–98 (Jan. 1991).

8. L. C. Hardison and E. J. Dowd, *Chem Eng Prog.* 31–35 (Aug. 1977).

9. R. J. Martin, R. E. Smyth, and J. T. Schofield, "Elimination of Petroleum Industry Air Toxic Emissions with a Flameless Thermal Oxidizer," *Petro-Safe '93,* Houston, Tex., Jan. 29, 1993.

10. G. Parkinson, *Chem Eng.* 37–43 (July, 1991).

11. A. F. Hodel, *Chem Proc.,* 88–90 (June 1992).

12. R. Yarrington and L. Morris, "The VOC-Incinerator Option," *Indust. Finish.* (Mar. 1992).

13. J. Chen, R. M. Heck, and R. J. Farraoto *Catal. Today* **11,** 517–545 (1992).

14. M. S. Jennings, N. E. Krohn, and R. S. Berry, *Control of Industrial VOC Emissions by Catalytic Incineration,* Vol. 1, U.S. Environmental Protection Agency, Research Triangle Park, N.C., July 1984.

15. E. J. Dowd, W. M. Sheffer, and G. E. Addison, "A Historical Perspective on the Future of Catalytic Oxidation of VOCs," paper 92-109.03, *85th Annual Meeting of Air and Waste Management Association,* Kansas City, Mo., June 21–26, 1993.

16. D. M. VanBenshchoten, "On-Site Pilot Testing Demonstrates Catalytic Emission Control Technology for New VOC Applications," paper 92-109.01, presented at *85th Annual AWMA Meeting & Exhibition,* Kansas City, Mo., June 21–26, 1992.

17. J. J. Spivey, *Ind. Eng. Chem. Res.* **26,** 2165–2180 (1987).

18. K. R. Bruns, "Use of Catalysts for VOC Control," presented at the *New England Environmental Expo,* Boston, Mass., Apr. 10–12, 1990.

19. J. J. Spivey and J. B. Butt, *Catal. Today* **11,** 465–500 (1992).

20. R. M. Heck, M. Durilla, A. G. Bouney, and J. M. Chen "Air Pollution Control–Ten Years Operating Experience with Commercial Catalyst Regeneration," paper presented at the *81st APCA Annual Meeting & Exhibition,* Dallas, Tex., June 19, 1988.

21. R. M. Heck, J. M. Chen, and M. F. Collins "Oxidation Catalyst for Cogeneration Applications–Regeneration of Commercial Catalyst," paper 90-105.1, presented at the *83rd Annual AWMA Meeting & Exhibition,* Pittsburgh, Pa., June 24–29, 1990.

22. R. E. Kenson, "Control of Volatile Organic Emissions," Bulletin 1015, Series 1000, Met-Pro Corp., 1981.

23. W. L. Prins and Z. L. Nuninga, *Catal. Today* **16,** 187–105 (1993).

24. C. H. Bartholomew, *Chem. Eng.,* 96–119 (Nov. 12, 1984).

25. A. C. Frost and co-workers, *Environ. Sci. Technol.* **25**(12), 2065–2070 (1991).

26. D. Ciccilella and B. Holt *Environ. Protec.,* 41–47 (Sept. 1992).

27. K. J. Herbert, "Catalysts for Volatile Organic Control in the 1990's," presented at the *1990 Incineration Conference,* San Diego, Calif., May 14–18, 1990.

28. M. A. Palazzolo, J. I. Steinmetz, D. L. Lewis, and J. F. Beltz "Parametric Evaluation of VOC/HAP Destruction via Catalytic Incineration," U.S. Environmental Protection Agency, Report no. EPA/600/S2-85/041, Research Triangle Park, N.C., July 1985.

29. S. K. Gangwal, M. E. Mulling, J. J. Spivey, and P. R. Caffrey, *Appl. Catal.* **36**, 231–247 (1988).

30. M. A. Palazzolo, and C. L. Jamgonhian "Destruction of Chlorinated Hydrocarbons by Catalytic Oxidation," U.S. Environmental Protection Agency, Report EPA/600/82-86/079, Research Triangle Park, N.C., Jan. 1987.

31. Difford and Spensor, *Chem. Eng. Prog.* **71**, 31 (1975).

32. G. R. Lester, "Catalytic Destruction of Hazardous Halogenated Organic Chemicals," presented at the *82nd Annual AWMA Meeting & Exhibition*, Anaheim, Calif., June 25–30, 1992.

33. S. E. Wanke and P. C. Flynn, *Cat. Rev. Sci. Eng.* **12**, 93 (1975).

34. G. R. Lester, and J. C. Summers "Poison-Resistant Catalyst for Purification of Web Offset Press Exhaust," presented at the *Air Pollution Control Association, 81st Annual Meeting*, Dallas, Tex., June 19–24, 1988.

35. F. Luck and J. Roiron, *Catal. Today* **4**, 205–218 (1989).

36. E. Cichanowicz, *Power Eng.*, 36–38 (Aug. 1988).

37. L. M. Campbell, D. K. Stone, and G. S. Shareef, *Sourcebook: NOₓ Control Technology Data,* EPA Report NO. EPA600/S2-91/029., Washington, D.C., Aug. 1991.

38. G. S. Shareef, D. K. Stone, K. R. Ferry, K. L. Johnson, and K. S. Locke "Selective Catalytic Reduction NOₓ Control for Small Natural Gas-Fired Prime Movers," paper 92-136.06, in Ref. 16.

39. J. R. Donnelly and B. Brown "Joy/Kawasaki Selective Reduction DE-NOX Technology," paper No 89-96B.6, in Ref. 32.

40. H. Gutberlet and B. Schallert, *Catal. Today* **16**, 207–236 (1993).

41. R. D. Walloch, *Oil Gas J.,* 39–41 (June 13, 1988).

42. R. Craig, G. Robinson, and P. Hatfield, "Performance of High Temperature SCR Catalyst System at Unocal Science and Technology Division," paper 92-109-08, in Ref. 15.

43. D. L. Champagne, *Gas Turb. World,* 20–23 (Nov.–Dec. 1987).

44. B. K. Speronello, J. M. Chen, and R. M. Heck "Family of Versatile Catalyst Technologies for NOₓ and CO Removal in Co-Generation," paper 92-109.06, in Ref. 16.

45. D. W. Agar and W. Ruppel, "Extended Reactor Concept for Dynamic DeNOₓ Design," *Chem. Eng. Sci.* **43**(8), (1988); J. R. Kittrell, C. W. Quinian, and J. W. Eldridge *J. Air Waste Manage. Assoc.* **41**(8), 1129–1133 (Aug. 1991).

46. T. D. Hylton, *Environ. Prog.* **11**(1), 54–57 (Feb. 1992).

47. Y. Wang, H. Shaw, and R. J. Farrauto, "Catalytic Oxidation of Trace Concentrations of Trichloroethylene over 1.5% Platinum on alpha-Alumina," in Ref. 3.

48. T. C. Yu, H. Shaw, and R. J. Farrauto, "Catalytic Oxidation of Trichloroethylene over PdO Catalyst on Alpha Al₂O₃," in Ref. 3.

49. J. L. Lin and B. E. Bent, "Thermal Decomposition of Halogenated Hydrocarbons on a Cu (111) Surface," in Ref. 3.

50. S. L. Hung and L. D. Pfefferie *Environ. Sci. Technol.* **23**(9), 1085–1091 (1989).

51. M. Mirghanbari, D. J. Muno, and J. A. Bacchetti, *Chem. Proc.* **45**, 14 (Dec. 1982).

52. T. G. Otchy and K. J. Herbert "First Large Scale Catalytic Oxidation System for PTA Plant CO and VOC Abatement," paper presented at the *85th Annual Meeting & Exhibition, AWMA,* June 1992, p. 2026.

53. K. J. Herbert, "Catalytic Oxidation of the Vent Gas from a PTA Plant," 1992.

EXTRA HEAVY OILS

JAMES SPEIGHT
Western Research Institute
Laramie, Wyoming

Petroleum and the equivalent term "crude oil" cover a vast assortment of materials consisting of gaseous, liquid, and solid hydrocarbon-type chemical compounds that occur in sedimentary deposits throughout the world (1).

The constituents of petroleum vary widely in specific gravity and viscosity. Metal-containing constituents, notably those compounds that contain vanadium and nickel, usually occur in the more viscous crude oils in amounts up to several thousand parts per million and can affect the processing of these feedstocks.

See also PETROLEUM MARKETS; PETROLEUM REFINING.

DEFINITIONS

When petroleum occurs in a reservoir that allows the crude material to be recovered by pumping operations as a free-flowing dark to light colored liquid, it is often referred to as "conventional" petroleum.

Heavy oils are the other "types" of petroleum that are different from conventional petroleum insofar as they are much more difficult to recover from the subsurface reservoir. The definition of heavy oils is usually based on the API gravity or viscosity, and the definition is quite arbitrary although there have been attempts to rationalize the definition based upon viscosity, API gravity, and density.

For many years, petroleum and heavy oils were very generally defined in terms of physical properties. For example, heavy oils were considered to be those petroleum-type materials that had gravity somewhat less than 20° API, with the heavy oils falling into the API gravity range 10–15° (eg, Cold Lake heavy crude oil = 12° API), and extra heavy oils, such as tar sand bitumens, falling into the 5–10° API range (eg, Athabasca bitumen = 8° API). Residua would vary depending upon the temperature at which distillation was terminated, but usually vacuum residua were in the range 2–8° API.

This has led to the development of a more formal method of classification which depends upon gravity and viscosity (Table 1 and Fig. 1). This system affords a better classification of petroleum, heavy oils and bitumen; the scale can also be used for residua or other heavy feedstocks.

Heavy oils have a much higher viscosity (and lower API gravity) than conventional petroleum, and primary recovery of these petroleum types usually requires thermal stimulation of the reservoir. The generic term "heavy oil" is often applied to a petroleum that has an API gravity of less than 20 degrees and usually, but not always, a sulfur content higher than 2% by weight. Furthermore, in contrast to conventional crude oils, heavy oils are darker in color and may even be black.

The term "heavy oil" has also been arbitrarily used to describe both the heavy oils that require thermal stimulation of recovery from the reservoir and the bitumen in

Table 1. Crude Oil Classification Using Specific Gravity, API Gravity, and Viscosity

Type of Crude	Characteristics
1. Conventional or "light" crude oil	Density–gravity range less than 934 kg/g³ (>20° API)
2. "Heavy" crude oil	Density–gravity range from 1000 kg/m³ to more than 934 kg/m³ (10° API to <20° API)
	Maximum viscosity of 10,000 mPa.s (cp)
3. "Extra-heavy" crude oil; may also include atmospheric residue (bp >340°C)	Density–gravity greater than 1000 kg/m³ (<10° API)
	Maximum viscosity of 10,000 mPa.s (cp)
4. Tar sand bitumen or natural asphalt; may also include vacuum residua (bp >510°C)	Viscosity greater than 10,000 mPa.s (cp)
	Density–gravity greater than 1000 kg/m³ (<10° API)

bituminous sand (tar sand) formations from which the heavy bituminous material is recovered by a mining operation.

Extra heavy oils are materials that occur in the near-solid state and are almost incapable of free flow under ambient conditions. Therefore, it is more appropriate that native asphalt (often referred to as "bitumen") that occurs in various locations throughout the world (2) be also included in the extra heavy oil definition. Bitumen includes a wide variety of reddish brown to black materials of semisolid, viscous to brittle character that can exist in nature with no mineral impurity or with mineral matter contents that exceed 50% by weight. The bitumen is frequently found filling pores and crevices of sandstones, limestones, or argillaceous sediments, in which case the organic and associated mineral matrix is known as rock asphalt.

The expression "tar sand" is commonly used to describe these sandstone reservoirs that are impregnated with a heavy, viscous material that cannot be retrieved through a well by conventional oil production techniques. However, the term "tar sand" is actually a misnomer; more correctly, the name tar is usually applied to the heavy product remaining after the destructive distillation of coal or other organic matter.

It is incorrect to refer to native bituminous materials as tar or pitch. Although the word tar is descriptive of the

black, heavy bituminous material, it is best to avoid its use with respect to natural materials, and to restrict the meaning to the volatile or near-volatile products produced in the destructive distillation of such organic substances as coal. In the simplest sense, pitch is the distillation residue of the various types of tars.

Thus, alternative names, such as bituminous sand or oil sand, are gradually finding usage, with the former name (bituminous sand) more technically correct. The term "oil sand" is also used in the same way as the term "tar sand" and these terms are used interchangeably throughout this text.

Other materials, albeit not naturally-occurring but not a chemically-altered material, that fit into the category of extra heavy oils are the residua (singular: residuum; often shortened to "resid").

A residuum is the nonvolatile material obtained from petroleum after nondestructive distillation has removed all the volatile materials. The temperature of the distillation is usually maintained below 350°C, since the rate of thermal decomposition of petroleum constituents is substantial above 350°C. They are obtained by distillation of a crude oil under atmospheric pressure (atmospheric residuum) or under reduced pressure (vacuum residuum) (Fig. 2).

Residua are black, viscous materials which may be liquid at room temperature (generally atmospheric residua) or almost solid (generally vacuum residua), depending upon the nature of the crude oil. The differences between a parent petroleum and the residua are due to the relative amounts of various constituents present, which are removed or remain by virtue of their relative volatility.

When a residuum is obtained from a crude oil and thermal decomposition has commenced, it is more usual to refer to this product as pitch. Being a thermally altered material, pitch is not classed as an extra heavy oil.

Asphalt prepared from petroleum has also often been referred to as extra heavy oil because the product closely resembles native asphalt. However, in many cases the manufactured asphalt is substantially different from the bitumen. It is recommended that the manufactured asphalt not be included in the category of extra heavy oils.

When the asphalt is produced simply by distillation of an asphaltic crude, the product can be referred to as residual, or straight run, petroleum asphalt. It is actually a residuum. If the asphalt is prepared by solvent extraction

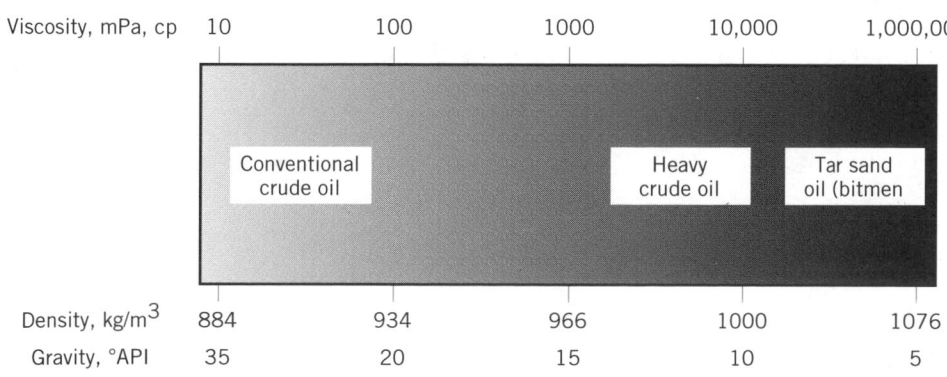

Figure 1. General classification of crude oil by viscosity, density, and API gravity.

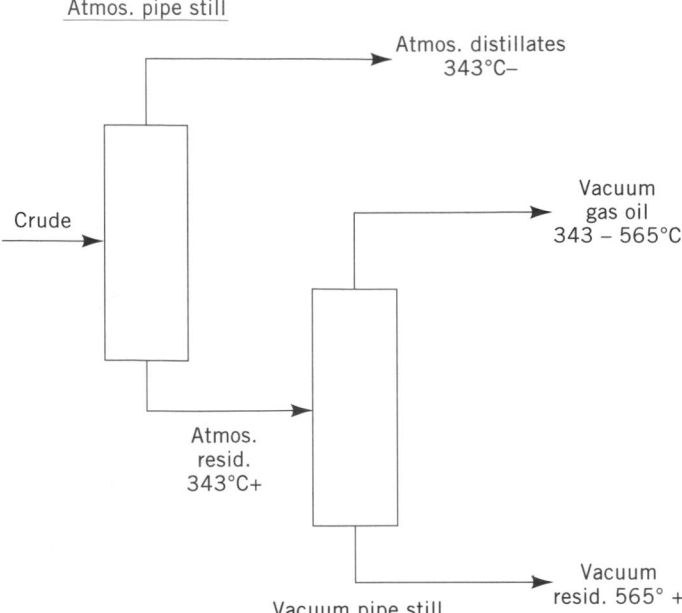

Atmos. pipe still

Atmos. distillates
343°C−

Crude

Vacuum
gas oil
343 − 565°C

Atmos.
resid.
343°C+

Vacuum
resid. 565° +

Vacuum pipe still

Figure 2. Production of residua during refinery distillation.

of residua or by light hydrocarbon (propane) precipitation, or if blown or otherwise treated, the term should be modified accordingly to qualify the product (eg, propane asphalt).

Tars are the result of the destructive distillation (at 450–1290°C) of many bituminous or other organic materials and are brown to black, oily, viscous liquids. In spite of the similarity of tars to many of the heavy and extra heavy oils, tar is most commonly produced from bituminous coal and is generally understood to refer to the coal product, although it is advisable to specify coal tar if there is the possibility of ambiguity.

PROPERTIES

Density (specific gravity) has been, since the early years of the industry, the principal and often the only specification of crude oil feedstocks and products, and was taken as an index of the proportion of gasoline and, particularly, kerosene present.

As long as only one kind of petroleum was in use the relations were approximately true, but as crude oils having other properties were discovered, and came into use, the significance of density measurements disappeared. However, crude oils of particular types are still rated by gravity. The use of density values has been advocated for application based on the American Petroleum Institute (API) gravity (Fig. 1).

The chemical composition of extra heavy oil is complex. Physical methods of fractionation indicate high proportions of asphaltenes and resins, even in amounts up to 50% (or higher). In addition, the presence of ash-forming metallic constituents, including such organometallic compounds as those of vanadium and nickel, is also a distinguishing feature of the extra heavy oils.

The properties of extra heavy oils can be summarized quite conveniently. These feedstocks (1) are usually nonvolatile below 200°C, (2) have an API gravity less than 10° and a high viscosity, (3) contain high proportions of asphaltenes and resins, (4) contain high proportions, often more than 2% w/w, of sulfur as organically-bound sulfur, and (5) contain high proportions, several thousand parts per million, of metallic ash-forming constituents.

BIBLIOGRAPHY

1. J. G. Speight, *The Chemistry and Technology of Petroleum,* Second Edition, Marcel Dekker Inc., New York, 1991.
2. J. G. Speight, *In Fuel Science and Technology Handbook,* Marcel Dekker Inc., New York, 1990.

F

FISCHER-TROPSCH PROCESSES AND PRODUCTS

Rocco A. Fiato
Exxon Research and Engineering Co.
Florham Park, New Jersey

HISTORY

The conversion of carbon monoxide and hydrogen to methane over nickel or cobalt catalysts was first reported in 1902 (1). A decade later, Badische Anilin und Soda Fabrik (BASF) was granted patents on metal-oxide catalysts for the conversion of carbon monoxide and hydrogen to a mixture of hydrocarbons and oxygenates at elevated pressures. In the 1920s, Fischer and Tropsch (2) reported on their catalyst development efforts, which included the production of oxygenates over alkali-promoted iron catalysts as well as the synthesis of higher hydrocarbons at atmospheric pressure over nickel- or cobalt-based catalysts. During this period, researchers at BASF and I. G. Farben obtained a number of patents on methanol synthesis over zinc oxide–based catalysts. By 1928 they developed a better understanding of catalyst promoter effects and had identified a series of alkali-ZnO-Cr$_2$O$_3$ catalysts for the selective production of higher alcohols.

Ni-ThO$_2$-kieselguhr and Co-ThO$_2$-kieselguhr hydrocarbon synthesis catalysts were not fully developed until the early 1930s. Early pilot plant tests indicated that the cobalt-based catalysts were more selective to higher hydrocarbons. Realizing the importance of these findings, extensive pilot-plant tests were conducted that were critical to the commercial development of this catalyst system. Studies in the mid-1930s demonstrated the utility of cobalt catalysts at medium pressures. During operations at 5.1×10^5 to 20.3×10^5 Pa, it was found that increased yields of more paraffinic and higher molecular weight products were possible, that catalyst life was extended, and that the need for periodic hydrotreating to restore activity was avoided.

Low and medium pressure cobalt-based catalyst technology was commercialized from 1935 to 1939 in nine commercial plants with total design capacity of ca 671,328 t/yr of liquid fuels. The largest facility at Schwarzheide with a capacity of ca. 190,572 t/yr was based on the normal low pressure process. Four other plants were based on the low pressure process: two on the medium pressure process, and two had both low and medium pressure operating systems. Product distributions for these processes are summarized in Table 1 (3).

The discovery of iron-based catalysts for medium pressure synthesis occurred in 1937, and alkali promoters were shown to increase the yield of higher hydrocarbons over these catalysts. The development of iron-based catalysts as substitutes for conventional cobalt catalysts was pursued from 1939 to 1944. Other studies focused on more effective reactor systems including fixed beds with hot gas or liquid recycle and three-phase slurry reactors for improved heat transfer.

In 1938, the utility of ruthenium-based catalysts for the production of high molecular weight waxes at elevated pressures (10–100 MPa) was demonstrated. Shortly thereafter, the isosynthesis process was developed for production of low molecular weight isoparaffins at high pressures and temperatures over thorium oxide and mixtures of aluminum and zinc oxides.

Roelen, who had made principal contributions to the development of the low-pressure Fischer-Tropsch (F-T) process, discovered the OXO process for olefin hydrocarbonylation to higher alcohols in 1938. This reaction sequence, which is the likely source of aldehyde and alcohol by-products in F-T synthesis, has become a cornerstone in the petrochemical industry for production of plasticizer alcohols and surfactants.

Hydrocarbon Research, Inc., initiated efforts to construct a large-scale natural gas–based Fischer-Tropsch synthesis facility in Brownsville, Texas, in the early 1950s. Startup difficulties with its fluid-bed F-T synthesis reactor led to design changes that delayed full-scale operations until 1953. Although technically successful, this operation was not economically viable at the high feed gas costs experienced in 1953 and was immediately shut down.

Demonstration of three-phase slurry reactor technology also occurred in this period, with construction of the 10 TPD Rheinpreussen-Koppers plant in Homberg-Niederhein, Germany (Table 2) (4,5). This facility was successfully operated, and its superior temperature control capabilities was demonstrated over a wide range of operating conditions tailored for low, medium, or high molecular weight product formation (Table 3).

Since 1955 when the first South African plant was constructed, there have been a number of commercial-scale operations initiated (6). A more detailed description of these and related demonstration scale activities is provided below. Finally, an overview of significant development programs on use of synthesis gas to produce hydrocarbon fuels, oxygenates, or higher alcohols is provided.

See also Natural gas; Manufactured gas; Petroleum markets.

PROCESS CHEMISTRY

A wide range of carbonaceous feedstocks can be used to generate synthesis gas for the F-T process. These include natural gas, coal, refinery coke, heavy oil residues, and biomass. These feedstocks are differentiated by their H:C content, the types of technology used to convert them to synthesis gas, and the H$_2$:CO ratio that is inherent in the most efficient and lowest cost synthesis gas generation route.

The overall reaction for Fischer-Tropsch synthesis from carbon monoxide and hydrogen generates a range of par-

Table 1. Products from Normal and Medium Pressure Synthesis[a]

Operating Mode[b]	Normal	Medium Pressure
Pressure, kPa	101	707
Temperature, °C	180–195	175–195
C_5+ products, wt % selectivity		
Gasoline, <185°C	56	35
Diesel, 185°–320°C	33	35
Oil, 320°+ C	11	30

[a] From Ref. 3.
[b] Conditions: 2:1 H_2:CO, ca 20% inerts, two-stage/no recycle.

affin, olefin, and oxygenated products with the average molecular weight distribution varying as a function of catalyst, synthesis gas feed composition, and operating conditions. In the absence of water–gas shift activity, the synthesis of higher hydrocarbons has a 2:1 H_2:CO consumption ratio, with water formed as the primary oxygen-containing by-product:

$$2\ H_2 + CO \rightarrow -CH_2- + H_2O \qquad (1)$$

Depending on the catalyst and reaction conditions employed, this may be accompanied by varying levels of the water–gas shift reaction:

$$H_2O + CO \rightleftarrows CO_2 + H_2 \qquad (2)$$

Table 2. Rheinpreussen Pilot-Plant Operating Conditions[a,b]

Effective Reactor Space	
Volume suspension including dispersed gas	10,000 L
Catalyst	880 kg# Fe
Synthesis pressure	1.2 MPa
Synthesis gas, CO:H_2	1.5
Synthesis gas amount	2,700 Nm^3/h
Linear velocity of the compressed gas at operating temperature in relation to the free reactor cross section	9.5 cm/s
Total amount of CO + H_2 used	2,300 Nm^3/h
per m^3 reactor space	230 Nm^3/h
per kg Fe	2.6 Nm^3/h
Average synthesis temperature	268 °C
CO conversion	91%
CO + H_2 conversion	89%

Synthesis Products Related to CO-H_2 Employed Hydrocarbons	
C_1+	178 g/Nm^3
$C_1 + C_2$	12 g/Nm^3
C_3	166 g/Nm^3
Oxygenates in synthesis water	3 g/Nm^3
Space-time yield of products: C_3+ including oxygenated products in 24 h	930 kg/m^3 reactor space

[a] From Refs. 3 and 4.
[b] Operating data and results of the one-step liquid-phase eynthesis on precipitated iron-catalyst with single pass of gas. Mode of operation aimed at gasoline production.

Table 3. Rheinpreussen Pilot-Plant Selectivity in Gasoline Production Operating Mode[a]

Factor	g/Nm^3 CO + H_2	Weight Percent of Total Production, C_1+	Percent Olefin
Methane + ethane	5.7	3.2	0
Ethylene	6.3	3.6	100
C_3	40.3	22.6	75–85
C_4	9.1	5.1	70–80
Fractionation 40°–180°C	95.5	53.6	70
Fractionation 180°–220°C	7.1	4.0	48
Fractionation 220°–320°C	10.7	6.0	37
Fractionation >320°C	3.3	1.9	7
Total	178.0	100.0	

[a] From Refs. 3 and 4.

Consequently, in cases in which Fischer-Tropsch synthesis occurs with stoichiometric shift activity, the H_2:CO consumption ratio drops to 0.5:1, and CO_2 is generated as the primary oxygen-containing by-product:

$$H_2 + 2\ CO \rightarrow -CH_2- + CO^2 \qquad (3)$$

Cobalt- and ruthenium-based catalysts typically operate with a 2:1 consumption ratio. Consumption ratios higher than 2:1 are possible when light paraffins are formed, ie, C_nH_{2n+2} where $n < 4$, due to their hydrogen rich stoichiometry. Iron-based catalysts tend to operate at higher temperatures that are more conducive to water–gas shift activity and have consumption ratios approaching 0.5:1.

Anderson-Schulz-Flory Reaction Kinetics and Mechanism

The production of hydrocarbons on these catalyst systems is believed to involve formation of CH_x intermediates for hydrocarbon chain initiation and growth, generation of surface intermediates of varying chain length, and termination steps, resulting in the formation of paraffin, olefin, or oxygenated products. In the early 1950s, a general polymerization-termination kinetic model was developed to account for this reaction sequence (7). Kinetics for these individual reactions steps are largely determined by the nature of the catalyst and operating conditions (8).

The process has been interpreted as involving a polymerization sequence in which a single carbon atom containing initiator is formed on the catalyst surface, and chain growth occurs via addition of single carbon (eg, CH_x) atom intermediates. Growing hydrocarbon intermediates can continue to participate in the polymerization reaction or can terminate to olefins or paraffins as primary reaction products. A similar sequence involving CO insertion to form surface acyl-type intermediates generates aldehydes or carboxylic acids that can undergo further reaction to form alcohols and carboxylic esters.

The Anderson-Schulz-Flory (ASF) polymerization model assumes that the entire sequence can be characterized by a single value parameter alpha (α), the relative probability of chain growth, which is independent of the chain length of a growing intermediate (8). According to the ASF model, there should be a linear relationship between the logarithm of the mole fraction of a given product versus carbon number with the slope reflecting the α

of that system. In practice, however, deviations from a linear ASF distribution are frequently observed, and α is found to increase with carbon number.

Several recent studies have shown ASF distributions with a dual α are possible, ie, where α has a lower value up to C_n and a higher value for C_{n+} products. A number of theories have been advanced to explain these deviations. These distributions may originate from catalysts containing two inherently different sites, ie, one with a low and one with a high probability of chain growth (9–12). Alternatively, the reincorporation of primary olefin products to form intermediates that reenter the polymerization sequence, grow and ultimately desorb as higher molecular weight products may also explain the double α type distributions (13–16).

Periodically, studies have been conducted to develop catalysts that generate products with a truncated ASF distribution, ie, a decreasing α with increasing carbon number, by which higher yields of intermediate range hydrocarbons could be realized. In theory, these systems would overcome the selectivity limits that are associated with the normal ASF polymerization sequence by controlling chain growth pathways or via shape selectivity to limit transition state geometry. In the majority of cases reported, however, non-ASF distributions can be traced to experimental errors or to the effect of secondary reactions such as dehydrocyclization or hydroisomerization–cracking that transform the primary ASF product distribution into lower molecular weight aromatic or isoparaffinic homologs.

COMMERCIAL FISCHER-TROPSCH PROCESSES

Commercial Operations at Sasol

In 1955, the South African Coal, Oil, and Gas Corp. (Sasol) initiated commercial operations at Sasolburg for the conversion of coal to liquid fuels and petrochemicals (17,18). The facility (Sasol One) is equipped with Lurgi gasifier technology for synthesis gas generation from coal and POX technology for generating synthesis gas from by-product methane. The F-T synthesis step is achieved in ARGE fixed-bed and Kellogg-Synthol circulating fluid-bed (CFB) reactors. The original Kellogg design has been improved over the years, and the Sasol Two and Sasol Three units constructed at Secunda in the early 1980s have incorporated many design and process flow scheme changes to maximize system performance. The newer CFB reactors have nearly three times the capacity of those at Sasol One. A total of 16 of these reactors are used at the Secunda facility to produce more than 11,000,000 B of fuels per year.

Principal products from the SASOL plants include paraffin waxes, from the ARGE fixed-bed reactor, as well as a wide range of liquid fuels and oxygenated petrochemicals. Efforts to improve the F-T reactor technology are continuing, with recent emphasis on demonstration of fixed-fluid-bed reactors for high temperature F-T and slurry bubble-column reactor systems for low temperature F-T operation.

The ARGE fixed-bed reactor system (Fig. 1) operates with pelleted catalyst packed inside the tubes of a tube-

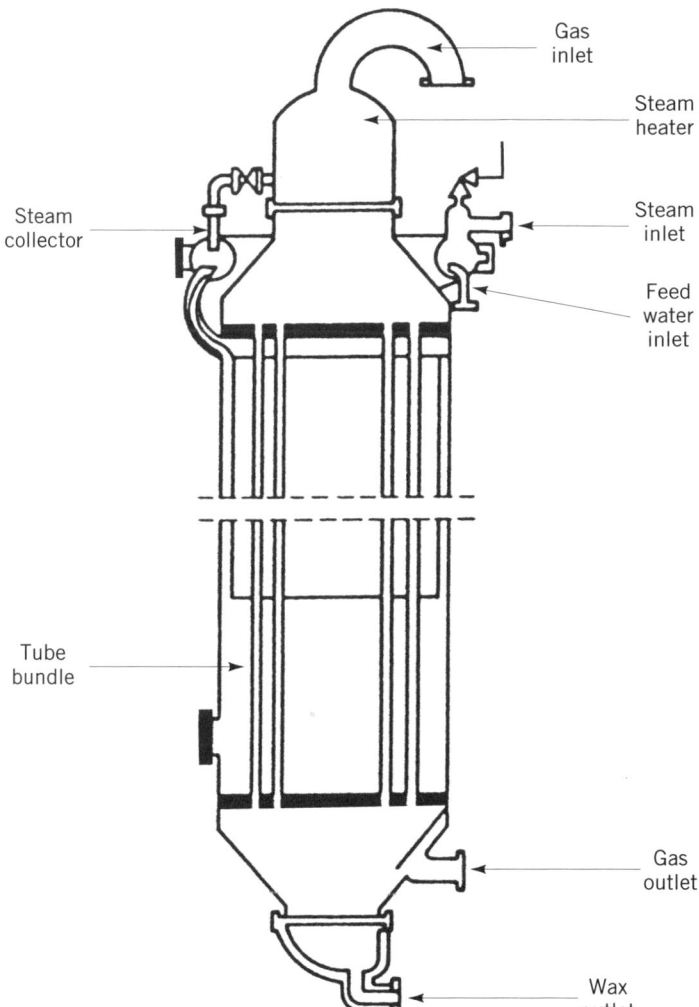

Figure 1. Fixed-bed Arge reactor.

and-shell heat exchanger. Each vessel contains about 40 m^3 of 2- to 5-mm catalyst pellets housed in 2050 tubes (5 cm ID $\times$ 12 m long). Heat removal from the fixed bed is achieved by passing boiler feed water through the shell side of the system. The operating temperature maintained by controlling the pressure of the produced 1.6+ MPa steam. Additional temperature control is possible by gas recycle through external exchangers.

The fixed-bed process employs a precipitated four-component iron-based catalyst (Fe:Si:Cu:K) tailored to operate at relatively low temperatures that favor hydrocarbon wax formation. Long catalyst life (>300 days) is achieved by careful control of feed gas purity and limiting peak operating temperatures to prevent carbon buildup and hydrothermal sintering. Representative operating data are shown in Table 4 and product distribution data for the ARGE reactor are shown in Table 5. The Kellogg-Synthol circulating fluid-bed (CFB) reactors for Sasol One were designed for medium pressure operation (2.2 MPa) at elevated (320°–359°C) temperatures. This operating mode is tailored to produce gasoline range hydrocarbons as the principal product (Table 5).

As shown in Figure 2, circulating catalyst is introduced into fresh feed gas via a standpipe equipped with slide

Table 4. Data for Arge Fixed-bed Synthesis Reactors[a]

Thickness of catalyst bed	46 cm
Height of catalyst bed	12 m
Operating pressure	2.0–3.0 MPa
Operating temperature	220°–260°C
Cooling surface	230 m^2/Nm^3 converted CO + H_2
Fresh gas charge	500–700 vol/vol·h
Daily production per cubic meter of catalyst (one step)	1250 kg C_2 +

[a] From Ref. 3.

valves. The mixed catalyst and feed gas are then transported throughout the horizontal inlet line into a vertical section that houses a 2.3-m-dia reactor containing two sections of heat-transfer coils that operate with a circulating oil coolant. The catalyst and products are transported overhead through the top bend to a hopper where they are separated through internal cyclones.

The Synthol CFB process employs a fused iron catalyst prepared from iron mill-scale in a 1500°C electric arc. A single charge at Sasol One consists of about 109 t of < 70-μm-size crushed catalyst, which is activated by hydrogen pretreatment. This material is circulated through the reaction zone at a rate of 7258 t/h during normal operations. Useful catalyst lifetimes are on the order of 40–80 days are reported before a full turnaround is required. The units at Sasol Two and Sasol Three are equipped with continuous catalyst addition capabilities that have likely increased their turnaround time.

The Synthol CFB system uses hydrogen-rich synthesis gas, ie, H_2:CO > 2.7:1 feed ratio, which is well in excess of the 2:1 stoichiometry for F-T synthesis, with no water–gas shift. Surplus hydrogen not only serves to prevent excessive catalyst carburization and but also induces reverse shift to convert CO_2 to CO for subsequent F-T synthesis. In addition, use of high H_2:CO ratios together with

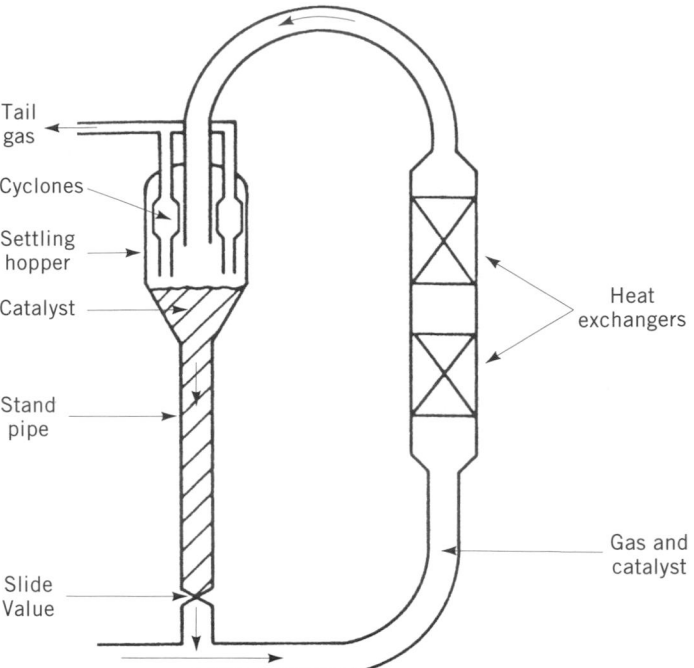

Figure 2. Circulating fluidized-bed synthol reactor.

high operating temperatures favors production of easily vaporized low molecular weight products. The formation of high molecular weight hydrocarbons is to be avoided, because they can lead to catalyst particle sticking and loss of fluidization properties. Tables 6 and 7 provide additional information on the product slate from the Synthol and ARGE reaction systems.

The successful commercialization of a new fixed-fluid-bed (FFB) reactor was recently announced (20). This system employs high efficiency cyclones or porous metal filters to maintain well-fluidized catalyst within the reaction zone. The FFB design eliminates all the equipment and operating difficulties associated with catalyst circulation and gas recompression. The system accommodates a higher density of heat-transfer coils and needs much less catalyst inventory than the CFB. Sasol claims that a full-scale FFB will be able to produce over 16.5 KBD of hydrocarbon product. Moreover, capital costs are estimated to be only 50% of those for a similar capacity CFB, and these compact and less complex reactors are more easily maintained. The overall properties of this system should result in increased thermal efficiency. The overall efficiency of a conceptual natural gas to liquid hydrocarbons process is

Table 5. Comparison of Arge and Synthol Process Selectivity[a]

Factor	Arge	Synthol
Temperature, average, °C	232	330
Selectivities		
CH_4	5.0	10.0
C_2H_4	0.2	4.0
C_2H_6	2.4	6.0
C_3H_6	2.0	12.0
C_3H_8	2.8	2.0
C_4H_8	3.0	8.0
C_4H_{10}	2.2	1.0
C_5+	3.5	8.0
Gasoline C_5–C_{12}	19.0	31.0
Diesel C_{13}–C_{18}	15.0	5.0
Heavy oil and wax		
C_{19}–C_{21}	6.0	1.0
C_{22}–C_{30}	17.0	3.0
C_{31}	18.0	2.0
nonacid chemicals	3.5	6.0
acids	0.4	1.0
	100.0	100.0
Tail gas and fresh feed	0.53	0.33

[a] From Ref. 19.

Table 6. C_{12}–C_{18} Product Distribution from Sasol[a]

Distribution of C_{12} to C_{18} Fraction	Synthol Circulating Fluid Bed	ARGE Fixed Bed
Percent olefins	73	26
Percent paraffins	10	66
Percent aromatics	10	0
Percent oxygenates	7	8
Percent straight chain	60	93

[a] From Refs. 7 and 8.

Table 7. Composition of F-T Process Water[a]

Type	Compound	Synthol	Arge
Nonacids (ref. pbw, ex. H_2O)			
	CH_3OH	1	25
	C_2H_5OH	55	50
	C_3H_7OH	16	11
	C_4H_9OH	7	6
	C_5+ alcohols	2	3
	aldehydes	5	2
	acetone	10	2
	higher ketones	4	1
Acids (ref. pbw, ex. H_2O)			
	acetic	70	
	propionic	16	
	butyric	9	

[a] From Refs. 7 and 8.

estimated at 64.7% for the high pressure FFB system versus 61.9% for the normal pressure CFB.

Commercial Operations at Mossel Bay

The first natural gas to fuels project (Mossgas) in South Africa was commissioned in 1992 in Mossel Bay (21). This facility uses Lurgi autothermal reforming for syngas generation followed by Synthol CFB reactors for Fischer-Tropsch synthesis. The raw F-T products are separated and further processed to gasoline and diesel range hydrocarbons by a combination of upgrading steps. Key processing steps include C_1 isomerization and alkylation, C_3/C_4 olefin plus C_5+ condensate oligomerization to distillate, naphtha hydrotreating to remove oxygenates followed by separation with C_5/C_6 isomerization and C_7+ aromatization to gasoline. Although this project is based on proven technology, it clearly demonstrates the versatility of F-T synthesis to operate with widely varying feedstocks and to produce useful intermediates for a range of fuels synthesis schemes.

SHELL MIDDLE DISTILLATE SYNTHESIS (SMDS)

In 1989, Shell announced plans to construct a plant in Bintulu, Malaysia, for conversion of natural gas to more transportable hydrocarbon liquids. The first Shell Middle Distillate Synthesis (SMDS) plant will generate nearly 12,000 B/day of liquids from 30.5 million m^3/day of natural gas. Shell, together with Mitsubishi, Petronas, and the Malaysian government are jointly sponsoring this project, which is scheduled to start commercial production in 1993 (22–25).

The SMDS process differs from that at Mossgas in that it uses natural gas to produce a heavy distillate-range product rather than gasoline. The overall SMDS flow scheme includes the Shell Gasification Process (SGP) to generate synthesis gas via noncatalytic POX of natural gas, heavy paraffin synthesis (HPS) via low temperature Fischer-Tropsch catalysis, and heavy paraffin conversion (HPC) via selective hydrocracking and isomerization of the C_5+ F-T hydrocarbons.

The overall process is designed to maximize yield of paraffin wax in the F-T synthesis step, which is then sub-jected to selective hydrocracking and hydroisomerization to kerosene and gasoil. This approach produces higher yields of liquid products than are generated with single-step F-T synthesis, where operating with a low ASF alpha would simultaneously produce high yields of C_4- by-products. Additional control over the relative yield of kerosene and gasoil is possible by varying the operating conditions in the C_5+ hydrocracking step.

Shell Gasification Process (SGP) for Synthesis Gas Generation

The production of synthesis gas via noncatalytic partial oxidation of methane with oxygen can theoretically produce a 2:1 H_2:CO synthesis gas, which is near the ideal 2.1:1 stoichiometry in F-T synthesis with no water–gas shift. In practice, however, the SGP step produces a hydrogen-lean synthesis gas with < 1.7:1 ratio. This requires supplemental hydrogen to match the 2.1:1 F-T stoichiometry. Shell conducts steam reforming of the C_1/C_4 process tail gas in the Hydrogen Manufacturing Unit (HMU) to produce the needed hydrogen for the HPS and HPC process steps.

SGP can be operated at temperatures from 1300° to 1500°C and pressures up to 7 MPa. Shell claims overall carbon efficiencies of about 95% with less than 1% leak of unreacted methane and less than 4% total CO_2 plus N_2 and H_2O. A specially designed waste-heat exchanger is employed to cool the 1300°C process gas to about 340°C and simultaneously generate high pressure steam for use in the air separation plant. The high pressure capability of SGP allows HPS to be conducted at preferred operating pressures of > 3 MPa. The Malaysian facility will reportedly employ type 1200 SGP units, each with a syngas production capacity of 1.2 million Nm^3/day.

The effective use of light hydrocarbon tail gas in HMU and the recovery of process heat for O_2 production are essential to achieving high overall thermal efficiency during synthesis gas generation. Shell has discussed the fact that about 50% of the capital investment is tied up in the syngas generation step and that ongoing research should lead to additional cost savings.

Shell Heavy Paraffin Synthesis (HPS) via Fischer-Tropsch Synthesis

Shell has discussed a wide range of reactor systems that were considered in the development of the SMDS. The types of reactors considered included circulating fluidized bed, ebullating fluidized bed, tube and shell fixed bed, and three-phase slurry bed. Shell has discussed the relative limitations of these systems, which include minimum volatility requirements for maintaining fluid-bed hydrodynamics, intraparticle diffusion limitations and effects on C_5+ selectivity in large-particle fixed or ebullating-bed catalysts, and external gas-to-liquid mass-transfer limitations and associated productivity constraints in three-phase slurry reactors.

Shell eliminated several of these potential reactor options due to the high productivity wax synthesis performance targets that had been established for SMDS. Fluid-bed reactors were eliminated from consideration due to the high molecular weight nature of raw SMDS wax and the inability efficiently to separate this from HPS catalyst. Similarly, diffusion constraints associated with the ebullating or slurry bed reactors and potential scaleup difficulties with the latter also eliminated them from consideration. Shell concluded that the potential problems with intraparticle diffusion limitations in the fixed-bed could be overcome by controlling catalyst size and shape, and through use of active metals in an eggshell distribution. The fixed-bed reactor was ultimately selected for use in its Malaysian project.

The tube-and-shell reactors employed in the Malaysian plant are similar to those used in Sasol's fixed-bed ARGE process. Recent reports indicate that four tube-and-shell reactors are being employed in the 12 KBD plant, with each vessel estimated to have 25,000 catalyst-filled tubes. Relatively large catalyst pellets of >1 mm dia. are used to minimize the pressure drop across the reactor's tubes.

Shell has obtained numerous patents on cobalt-based F-T catalysts for the selective production of C_5+ hydrocarbons from synthesis gas (26–28). Methods for producing catalyst pellets with the active metal in an eggshell distribution have been described. This approach eliminates metal from the diffusion-limited core of the catalyst pellet where, because of the high diffusivity of H_2, high $H_2:CO$ ratios would otherwise lead to lower ASF alphas and possible hydrogenolytic routes that favor C_4- formation. As a consequence, the eggshell catalysts produce high (> 88%) yields of C_5+ paraffins at the elevated temperatures (> 220°C) of their HPS step.

Shell Heavy Paraffin Conversion (HPC) via Selective Hydrocracking

The raw C_5+ stream from the HPS step is finally converted to finished products over a selective hydrocracking catalyst with mild isomerization activity. This relative distribution of naphtha, kerosene, and gasoil products can be varied, depending on the severity of the hydrocracking step. Shell has disclosed operating modes that favor kerosene or gasoil as the major product (Table 8).

The relative distribution of products from the HPC step is similar to that reported for highly selective hydrocrack-

ing of pure linear paraffin feeds (29–30). Shell has developed specially tailored catalysts that facilitate internal versus terminal C-C bond scission. This results in minimal C_1/C_3 formation with the remaining products generated in nearly equimolar quantities. Shell's reported strategy for upgrading F-T liquids in its MDS process involves treatment of the entire C_5+ stream over a Pt/SiO_2-Al_2O_3 catalyst. It has also developed a wide range of Pt- and Ni-based catalysts on traditional metal oxide- and zeolite-based supports for selective conversion of heavy hydrocarbons to jet and diesel.

Shell patents disclose an HPC step that employs a bifunctional catalyst, eg, Pt/SiO_2-Al_2O_3, operated at 13+ MPa, 340°–370°C, 0.5–1.25 LHSV with an $H_2:oil$ feed ratio of 2000 Nl/l. The company has recently disclosed a series of Pt and Ni catalysts with beta-zeolite or Y-zeolite supports that could be employed at lower operating pressures (22–25).

In the latest discussions of the MDS process, Shell highlights its ability to exploit carbon number reactivity differences in the C_5+ stream by preferentially converting higher molecular weight paraffins to jet and diesel range products. The use of a bifunctional hydrocracking catalysts that facilitate C-C bond breaking via beta-hydride scission in nonclassical carbenium ion intermediates, rather than terminal C-C cracking, significantly reduces the yield of C_1 and C_2 by-products. Moreover, because this mode of cracking is more facile with secondary and tertiary intermediates, the cracked C_5- products have a high iso:normal ratios.

A recent example of the integrated CO hydrogenation and paraffin upgrading includes F-T synthesis over Co-Zr/SiO$_2$ at 220°C with an $H_2:CO$ ratio of 1.8 at 3 MPa and 500 v/v/h; paraffin upgrading over 0.8% Pt/SiO_2-Al_2O_3 at 345°C with an $H_2:oil$ ratio of 2000 Nl/l at 13.5 MPa and 1.25 LHSV. Product distributions from F-T synthesis and single pass C_5+ upgrading are shown in Table 9. In a typical example, the company first subjects the C_5+ stream to mild Co-Mo/Al$_2$O$_3$-catalyzed hydrotreatment to remove oxygenates and unsaturates before conducting the hydro-isomerization–hydrocracking step over Pt/SiO_2-Al_2O_3. In the latter, it indicated that recycle of the 360°C+ fraction and/or higher operating temperatures give higher yields of jet and naphtha range products, although no details were given. Shell also has shown that naphtha range products from C_{16}/C_{17} hydrocracking have a similar distribution to that reported by Weitkamp over Pt/Ca-Y zeolite, suggesting that a high degree of control over catalyst

Table 8. SMDS Products and Properties[a]

Upgrading Severity	Medium	Heavy
Selectivity, wt %		
Gas naphtha	15	25
Kerosene	25	50
Gas oil	60	25
Major Product		
Property	Gas Oil	Kerosene
boiling range	250°–360°C	150°–250°C
pour	−10°C	
freeze		−47°C
blending cetane	75	
smoke point		>100 mm

[a] From Refs. 22–28.

Table 9. Shell HPC Upgrading—Representative Product Distributions[a]

Weight Percent	F-T Synthesis	C_5 + Upgrading
150°C-	4.2	6.9
150°–250°C	18.7	22.3
250°–360°C	24.9	44.7
360°–400°C	10.4	11.0
400°C oil	3.1	10.6
400°C paraffins	38.7	4.5

[a] From Refs. 29 and 30.

metal and acid functionality has been achieved. It recently reported that further upgrading of the naphtha (relatively low iso:normal ratio) will be required to produce high octane mogas.

Potential environmental aspects of the SMDS products have also been highlighted. The absence of sulfur, nitrogen, and aromatics results in improved emissions properties for the jet and diesel fuels, a property that is common to SMDS and related F-T based paraffinic fuels (31). In contrast to the gasoline range hydrocarbons produced from Sasol's high temperature CFB operation, SMDS generates a linear paraffinic naphtha stream. Although useful as a chemical feedstock, this material would require further upgrading to achieve the octane necessary for gasoline.

Shell has reported overall 88–90% selectivity to C_5+ hydrocarbons in its process with the kerosene and gasoil having properties summarized below. It has also claimed an overall thermal efficiency of 63% based on the heating values (LHV) of the gaseous feed and distillate products.

EXXON ADVANCED GAS CONVERSION TECHNOLOGY

In 1992, Exxon announced that it was nearing completion on a program to demonstrate advanced technology for the conversion of natural gas to premium quality liquid fuel and petrochemical feedstocks. Research on this new process, Advanced Gas Conversion for the 21st Century (AGC-21), was initiated in the early 1980s. AGC-21 involves three principal process steps: synthesis gas generation, hydrocarbon synthesis, and product upgrading (32,33).

The synthesis gas generation step converts natural gas, oxygen and steam to a $2:1$ $H_2:CO$ ratio. The synthesis gas generator incorporates a novel fluid-bed reactor system in which partial oxidation and steam-reforming reactions are carried out in a single large reactor containing fluidized catalyst particles. The product gases pass through a gas–solids separation and waste heat recovery system, yielding a clean syngas of the desired composition. This approach was chosen to build on Exxon's experience in fluid catalytic cracking, Flexicoking and other large-scale fluid-bed processes. The fluid-bed synthesis gas generation (FBSG) unit provides significant thermal efficiency and economy of scale benefits versus conventional technology options, including those that involve combined POX and steam reforming.

The Hydrocarbon Synthesis (HCS) step involves new slurry-bed reactor technology, which is a significant departure from more conventional fixed or circulating fluid bed systems. In this step, proprietary high performance catalysts were developed for use in a multiphase slurry reactor. Highly efficient heat removal and scaleup of multiphase fluid dynamics were studied during the process demonstration program.

The product from HCS is finally subjected to a hydroisomerization upgrading step similar to fixed-bed hydroprocessing. This step yields products that are free of sulfur, nitrogen, nickel, vanadium, and aromatics. A wide range of product options are available from premium quality jet and diesel to specialty lubes and waxes. As shown in Table 10, the relative yield of jet and diesel versus catalytic cracker feed can be varied by controlling the severity of the upgrading step. At lower severity, the upgrading step will produce a maximum yield of feed for refinery cat cracking and lubes production. At higher severity, the yield of cat feed boiling range material can be eliminated to give 70% of the product in the jet and diesel range. Chemical intermediates can also be manufactured for production of solvents, alcohols, and high performance polymers.

The unique catalytic reactor technology associated with the FBSG and HCS process steps was demonstrated in process development units (PDU) located at the Exxon Research and Development Laboratories (ERDL) in Baton Rouge, Louisiana. This facility was sized to provide all of the process, design, and equipment performance data needed for scaleup to full commercial size. This includes testing of tonnage quantities of catalysts prepared by commercial vendors in large-scale equipment, using commercial catalyst production techniques. The PDUs were sized sufficiently large to match the flow regimes expected in commercial-scale reactors. Key parts of the equipment were sized to match all the critical temperatures, pressures, velocities, and hydrodynamic parameters needed to ensure scaleup with confidence.

PDU Construction and Operation

The FBSG unit was constructed during 1989. The HCS unit was constructed during 1990, while the FBSG unit was in operation.

FBSG Unit. The FBSG reactor vessel is 1.52 m in diameter with an internal refractory lining. It has a close-coupled external single cyclone, also refractory lined. Supporting facilities include a feed gas sulfur removal system, a preheat furnace, waste heat recovery, and heat exchange and gas cleanup systems. Steam for process feed is generated on site. The natural gas feed comes from several gas fields in southern Louisiana. It is treated in a gas plant located across from the refinery. The gas plant uses amine treating plus cryogenic separation of light gases. Methane content varies from 98% to as low as 90%. Sulfur content is normally less than 1 ppm. Oxygen comes into the facility by commercial pipeline that supplies oxygen to many plants along the Mississippi River.

The FBSG process uses a proprietary catalyst system, with numerous patents issued and applications still pending. A variety of catalysts, prepared by commercial cata-

Table 10. AGC-21 Process—Relative Product Distribution[a]

Product	Maximum Cat Feed	Maximum Diesel and Jet
Naphtha	15	30
Diesel and jet	50	70
Cat cracking feed	35	0

[a] From Refs. 32 and 33.

lyst manufacturers in large quantity, were tested. The FBSG development included testing several oxygen and natural gas distributor arrangements. Tests were also conducted on a variety of materials of construction for key components in the reactor and heat recovery systems.

After an initial year of testing, the FBSG PDU then supplied synthesis gas to the HCS PDU on a sustained basis, with both units operating successfully in fully integrated fashion. FBSG operations were conducted from 1990 through 1992, covering a 3-yr span, with some periods of downtime for turnarounds, during which inspections were conducted and some modifications were implemented.

HCS Unit. The HCS slurry reactor is 1.22 m in diameter and contains an internal heat transfer coil, which uses boiling water in the tubes to remove the heat generated by the Fischer-Tropsch reaction. A proprietary product recovery–catalyst separation technology was developed. Its operation has been excellent, and it has played a key role in the development of the slurry reactor approach.

The HCS slurry catalyst is a proprietary formulation that is covered by a number of patents. Extensive exploratory and predevelopment work was carried out at both the Exxon Corporate Research Laboratories in Clinton, New Jersey, and at ERDL to develop the preferred system now used. Catalysts tested in the HCS PDU were produced by commercial catalyst manufacturers in large quantities to meet detailed specifications.

A wide range of operating conditions were explored during 2 yr of operation of the HCS PDU. Tests also covered modifications to the internal heat transfer coil and operations with different reactor heights. Run lengths of 2–3 months of round-the-clock operation were typical. Throughputs equivalent to about 31.8 m³/day were achieved. Operation of the integrated FBSG–HCS PDUs was generally regarded as smooth and quite robust to upsets.

RESEARCH AND DEVELOPMENT PROGRAMS FOR SYNTHESIS GAS CONVERSION

Oil shortages in the early 1970s promoted a resurgence of research activity on Fischer-Tropsch and related chemistries for synthesis gas conversion to hydrocarbon fuels and fuel oxygenates. Large R&D programs on new catalysts and reactor technology for Fischer-Tropsch synthesis were undertaken by Gulf-Badger from 1970 to 1984; by the U.S. Department of Energy with Mobil from 1980 to 1984; by Air Products from 1980 to the present; and by an international Norwegian-sponsored effort among Statoil, Sintef, and Pittsburgh-based GTO from 1986 to the present. Related technology for methanol synthesis and conversion to gasoline was commercialized by Mobil in 1986. During the 1980s new catalytic processes for the direct conversion of synthesis gas to C_2+ alcohols were independently developed by ENI, IFP in collaboration with Idemitsu, and Lurgi. Continued impetus for this work has been provided by the recent commercialization of SMDS in Malaysia and Sasol CFB technology in Mossel Bay, South Africa.

Gulf-Badger

The Gulf-Badger (G-B) program for converting natural gas to hydrocarbon fuels originally focused on conventional steam reforming of methane for synthesis gas production followed by fixed-bed or fluid-bed F-T synthesis to a distillate range product (34). This effort continued until the mid-1980s when Gulf was purchased by Chevron. Many of the catalyst and process patents later acquired by Shell are clearly relevant to the SMDS process.

Gulf researchers worked on a range of Fischer-Tropsch catalysts with improved distillate productivity. At that time, they were at the leading edge of research on supported cobalt catalyst technology for normal paraffin synthesis from 2:1 H_2:CO. They developed improved rare earth modifiers and Group VIII metal promoter formulations as well as highly effective reduction–oxidation pretreatment procedures that achieved state-of-the-art performance.

G-B conducted fixed-bed pilot reactor studies in a 2.54-cm-dia tubular reactor consisting of two 6.1-m sections connected in series. They reported performance of 40–60% CO conversion per pass with >70% C_5+ selectivity for operations at 200°–255°C, 1.6 MPa, and H_2:CO of 1.5–2.0 at a feed rate of 500–1000 v/v/h. Promising results were also obtained in fluid-bed pilot reactor studies in which 70–75% C_5+ selectivity was achieved at >40% conversion per pass at 200–400 kPa with a 2:1 H_2:CO ratio.

A full-scale demonstration of the fixed-bed process was undertaken in a ca 35 B/D demonstration plant fed by 1.0×10^4 m³/d of natural gas feed. This unit included natural gas treatment before the steam reformer, which included recycle CO_2 for controlling the H_2:CO ratio. After compression and CO_2 removal, the synthesis gas was passed through a membrane separator to adjust the ratio to the desired 1.5–2.0:1 H_2:CO ratio. A bundle of 15.24-m-long reactor tubes was used to simulate a segment of a commercial tube and shell reactor. Although details of this operation were not published, unofficial reports indicate that the demonstration was successful.

U.S. Department of Energy

The Department of Energy (DOE) has sponsored research on indirect liquefaction of coal via F-T synthesis for many years. A number of industrial and academic laboratories are still involved in studies of catalyst preparation and characterization, reactor modeling and catalytic process development. In the early 1980s, Air Products and Mobil independently conducted studies on catalyst and reactor technology for slurry-phase F-T synthesis. Air Products examined cobalt carbonyl based catalysts for direct production of liquid fuels. Mobil focused on iron-based F-T synthesis with hydrogen lean (H_2:CO < 0.8) synthesis gas followed by ZSM-5 catalyzed conversion of the volatile F-T olefin–alcohol products to gasoline.

Air Products

A significant research effort was undertaken in the 1980s on the preparation, characterization, and performance of cobalt carbonyl–based catalysts for application to slurry-phase F-T synthesis (35). A principal objective of this study was to identify key properties that distinguish the carbonyl-based catalysts for more conventional metal salt–based analogues. The study was aimed at maximizing the yield of liquid fuels from the F-T step while minimizing the formation of methane and heavy wax. The influence of alkaline earth and early transition metal promoters was examined.

Laboratory-scale fixed-bed and continuous-stirred-tank reactor (CSTR) tests were conducted on catalysts with Al_2O_3, SiO_2, or MgO containing supports, Ti, or Zr promoters and 3–10 wt % cobalt loading levels (36). In general, the highest activities were found with SiO_2 as the support. The highest overall productivity to liquid fuels was achieved with $Zr(OPr)_4$ promoted Co/SiO_2 (Table 11). An extended 4400-hr CSTR test was conducted with 4.4% Co/7.6% Zr/SiO_2 at temperatures from 220° to 280°C and H_2:CO feed ratios from 1:1 to 2:1 at feed rates from 1.8 to 2.0 Nl/g catalyst:h at 2.07 MPa. The catalyst gradually lost activity with an approximate half-life of about 100 days.

C_5+ selectivity was typically above 70 wt % with methane selectivity at less than 15 wt %. The cobalt-based catalysts exhibited little water–gas shift activity and exhibited a 2:1 H_2:CO feed gas consumption ratio. Unsuccessful attempts were reported to make the cobalt-based system useful with CO rich feed by use of water–gas shift cocatalysts. The overall laboratory program was completed in the mid-1980s, and no further work was reported.

In 1992, Air Products initiated tests of conventional iron-based catalysts in its Laporte bubble-column slurry reactor. Preliminary results were encouraging with a 20+ day run reported. Aside from problems with catalyst–product separation, performance was similar to that achieved in the Rheinpreussen system.

Mobil

In 1980, Mobil initiated a program on slurry phase F-T synthesis. It focused on a two-stage process that involved slurry phase F-T synthesis to olefin or alcohol products followed by zeolite-catalyzed fixed-bed condensation of the F-T products to high quality gasoline (37). A 5-cm-dia by 7.62-m-long bubble-column F-T reactor in series with two parallel fixed-bed condensation reactors. Their program involved design and construction of an integrated pilot plant for each process step, evaluation and parametric studies of iron-based F-T catalysts, gasoline characterization, and conceptual design and process economics evaluations of possible commercial projects.

Table 11. Effect of Cobalt Source on Slurry Reactor Performance[a]

Cobalt Source	Bulk Activity, mol/h·kg	Selectivity, wt %[b]			
		C_1	C_{2-4}	C_{5-23}	$C_{24}+$
$Co_2(CO)_8$	35	7.9	13.7	69.3	9.1
$Co(NO_3)_2 \cdot 6 H_2O$	16	10.9	6.5	54.6	28.0

[a] From Ref. 35.
[b] Conditions: 2.1 MPa, 1:1 H_2:CO ratio, SHSV 2.0 nL/h·g catalyst, 240°C.

The bubble-column system used three-component iron-based catalysts, eg, $Fe/Cu/K_2CO_3$, that were activated by H_2/CO induction. Hydrodynamic studies in hot and cold mock-up reactors and fluid dynamic model calculations were used to establish preferred F-T operating conditions. The unit was initially operated with internal filters for catalyst product separation. This filter system was later replaced with on-line catalyst settling vessels.

A number of runs up to 86 days long were conducted. The unit was operated at temperatures ranging from 240° to 280°C and pressures from 1.13 to 2.52 MPa. Synthesis gas conversions up to 85–91% were sustained for periods as long as 60 days, and total productivities of ≥800 g hydrocarbon/1 g iron were observed. Selectivities to C_1/C_2 as low as 1.7% were observed with maximum reactor wax yields up to 80% with a nominal 0.7 feed ratio.

The second stage of the ZSM-5 zeolite-catalyzed conversion of F-T products to gasoline was conducted at temperatures ranging from 288° to 466°C. The raw product had R+O octane numbers of 82 to 98, depending on the conditions of the final conversion step. The catalyst, which did not require frequent regeneration, was fully reactivated after two such treatments.

Mobil successfully completed this DOE-sponsored program in the mid-1980s and did not pursue this effort to commercialization. Economics studies conducted in 1983 suggested that additional process improvements were likely required for the F-T based process to be economic.

In a related area, Mobil did commercialize new technology for converting methanol to gasoline (MTG) (38–39). A 14.5 KBD unit was constructed in New Zealand with commercial startup in late 1985. Mobile has acknowledged, that this technically successful operation is not yet economically attractive due to the relatively low price of crude oil.

Statoil

Statoil initiated studies on natural gas conversion to liquid fuels in the late 1980s. Its gas to middle distillates (GMD) project includes efforts from contract research institutes Sintef and the Centre for Industrial Research in Norway as well as the U.S. firm GTO Inc. Their process scheme involves autothermic-type reforming to synthesis gas, slurry phase F-T synthesis over cobalt-based catalysts to produce paraffin wax, and selective hydrocracking of the wax to distillate range liquids.

Statoil has considered alternative synthesis gas generation technologies and for CO hydrogenation schemes to methanol have concluded that the highest efficiency and lowest cost routes involve POX integrated with some degree of steam reforming. Its studies indicate that, relative to steam reforming (SR), noncatalytic partial oxidation (POX), and autothermic reforming (AR), a combined process (steam + autothermic) will provide 3–5% higher thermal efficiency and be 5–10% less costly than individual process routes. It also notes that more tightly integrated schemes such as Uhde's combined autothermal reforming (CAR) and ICI's gas heated reforming (GHR) offer additional efficiency and cost credits (40).

The development of viable slurry phase F-T technology is the focal point of Statoil's research program (41). It believes that maintaining satisfactory catalyst performance and achieving continuous product extraction are the two main challenges in developing economically viable technology. It has also noted the benefits of slurry, which include significantly better heat-transfer control and heat use over a fixed bed, the ability to achieve continuous catalyst recovery and regeneration, and the potential to operate with high conversion per pass without encountering hot spots that plague fixed-bed systems.

Statoil has developed a range of laboratory- and pilot-scale reactor facilities to explore slurry F-T synthesis. Catalyst testing and development are conducted in small-scale fixed-bed, CSTR, and 2.54-cm-dia bubble-column reactors. Larger-scale pilot-plant tests are performed in a 5-cm-dia bubble-column reactor, and hydrodynamics tests in a 25.4-cm-dia hot-flow unit. It is reportedly considering further scaleup to a 35 BD bubble-column facility.

A large part of the company's effort involves development of supported cobalt catalysts tailored for use under slurry F-T process conditions. Statoil has developed a family of highly active alumina-supported catalysts containing rare earth and rhenium or precious metal promoters (Pt, Rh, Ir). High selectivity to paraffinic liquids with less than 10% CH_4 and less than 3% CO_2 have been reported in laboratory-scale fixed-bed tests at 101 kPa, 195°–205°C, at feed rates of 1680 $cm^3/g \cdot h$. Hydrocarbon productivities of 0.6–1.4 g C^6+/g catalyst·h were reported for the 2.54-cm-dia bubble-column reactor system in 1000+ h tests at 225°C and 3 MPa with 6.8 wt % initial catalyst loading (42). Although it is reportedly operational, operating data from the 5-cm-dia pilot plant has not yet been published.

HIGHER ALCOHOL SYNTHESIS

A range of programs were initiated in the 1980s on the selective production of higher alcohols from synthesis gas. The most significant programs were conducted by ENI-Snamprogetti, Lurgi, and IFP in collaboration with Idemitsu-Kosan, DOW/Union Carbide, the U.S. Department of Energy, the Japanese C1 Group, and several universities contract research groups (43). Catalyst for these studies were based on modifications to systems designed for conventional Fischer-Tropsch synthesis, methanol synthesis, olefin hydrocarbonylation, or alcohol homologation chemistry. Although several programs were pursued through relatively large demonstration-scale testing, none has achieved commercial status. Operating conditions employed in large-scale tests at ENI, Lurgi, and IFP-Idemitsu are summarized in Table 12.

ENI-Snamprogetti

In the early 1980s, Snamprogetti, in collaboration with ENICHEM and Haldor Topsoe, conducted laboratory and pilot-plant testing of modified methanol catalysts for higher alcohol synthesis (44). These studies ultimately led to the development of alkali modified Zn–Cr catalysts that exhibited reasonably good selectivity to C_2+ alcohols at 350°–420°C and 10–16 MPa of 0.5–3.0:1 $H_2:CO$ at space velocities of 3000+ h^{-1} in adiabatic fixed-bed reac-

Table 12. Comparative Operating Characteristics of Various Mixed Alcohol Processes[a]

Process	Snamprogetti MAS	lurgl OCTAMIX	IFP Substifuel
Thermal efficiency, %	56	60–62	<50
Catalyst	Zn/Cr/K	Cu/ZnO/Al	Cu/Co/Al
Operating temperature, °C	350–420	285–300	260–320
Operating pressure, MPa	10–16	6–9	6–10
CO, conversion per pass, %	14	13–23	12–15
Liquid product selectivity, %	90	95	70–75
Alcohol productivity, kg/L catalyst/h	0.23	0.46	0.10–0.15

[a] From Ref. 43.

tors (Table 13). This technology for methanol and higher alcohol synthesis, the "MAS" process, was demonstrated in a retrofit high pressure methanol plant in Pisticci, Italy. Product recovery was complicated by the fact that raw alcohol contains large quantities of water. A three-step recovery sequence, including azeotropic distillation of higher alcohols, was successfully demonstrated to produce alcohol streams with less than 0.1 wt % water. By 1986, over 2 yr of tests at a 15 KT/Y production rate had been completed and the technology was claimed to be commercially ready.

The MAS alcohols were trial marketed as a gasoline additive by AGIP under the brand name Super E. Although the alcohols were shown to be viable additives in the limited test data available from those trials, this approach has not been implemented commercially.

Lurgi

In the early 1980s Lurgi, in collaboration with Sud-Chemie, began tests to identify promising catalysts for production of higher alcohols. These efforts focused on electronic and structural promoters for catalysts that operated under the low pressure and temperature conditions of Lurgi's methanol process (Table 12). Catalysts that were successfully operated in 12,000-h fixed-bed pilot-plant tests at ca 1 TPD capacity were selected for further study in Lurgi's 20 BPD process demonstration unit, which was commissioned in 1989 (45,46).

Table 13. Higher Alcohol Synthesis—Product Distributions[a]

Process, Alcohol Distribution, wt %	Snamprogetti MAS	Lurgl OCTAMIX	IFP Substifuel
C_1	70.0	64.5	57.5
C_2	2.5	11.5	28.5
C_3	3.4	5.2	7.1
C_4	12.5	7.4	2.8
C_5+	9.5	7.4	2.5
Other oxygenates	2.0	3.2	0.6

[a] From Ref. 43.

Cu/Zn/Al catalysts, which appear to be structurally modified versions of Lurgi's low pressure methanol synthesis formulations, are operated in their low pressure tube-and-shell reactor system at 250°–300°C and 5–10 MPa (Table 12). The stoichiometric feed ratio for the Octamix process is about 1:1 H_2:CO, because the alcohol formation sequence is accompanied by near equilibrium levels of water–gas shift activity. This simplifies the alcohol recovery step, since by-product water (<1.5 wt % in the crude product) from CO hydrogenation is converted to CO_2, which is more easily separated from the alcohol product.

The relative yield of C_1/C_2+ alcohols may be varied by operating at different temperatures and H_2:CO feed ratios (Tables 13 and 14). In addition, the level of C_2+ alcohols can also be increased by recycling methanol and volatile products to the primary synthesis step.

By 1990, Lurgi had declared this process ready for commercial application, and Texas Methanol Corp. had been given rights to market the technology in the United States for plants up to 200 TPD capacity. Texas Methanol obtained necessary waivers for use of the mixed alcohol blend as a gasoline additive. The higher alcohols are good cosolvents and help increase water tolerance of methanol-containing gasoline blends. Widespread use of the higher alcohol additives has been delayed, however, due to increased use of more fungible MTBE and related ether additives.

IFP–Idemitsu-Kosan

IFP initiated studies on improved catalysts for higher alcohol synthesis in the late 1970s. Shortly thereafter, they initiated joint studies with Idemitsu-Kosan under sponsorship of the Japanese Research Association for Petroleum Alternatives Development (RAPAD) and the Ministry of International Trade and Industry (MITI). In the mid-1980s, IFP's copper-modified cobalt and Idemitsu's nickel-promoted cobalt catalysts were evaluated in a 20 BPD PDU facility located in Chiba, Japan. The initial se-

ries of PDU tests employed the first generation Cu-Co-Al catalyst system with 2:1 H_2:CO ratio at pressures from 6 to 10 MPa and temperatures from 260° to 320°C with CO_2 recycled back to the synthesis step (Tables 12 and 13). C_2+ alcohol selectivities of 40–50 wt % dropped to about 35% after 100 h on-stream time. After unit modifications to reduce the CO_2 content of the recycle stream, the Ni-Co catalyst appeared to be more stable with 35–40 wt % selectivity to C_2+ alcohols for maintained for 300+ h at 6–7 MPa.

Purified alcohols from the PDU tests were further evaluated in car fleet tests and found to have good octane-blending properties and to function as a cosolvent for methanol. The methanol higher alcohol blends still appear to suffer from relatively low water-tolerance and phase-separation problems that, together with their relatively high projected manufacturing cost, are some of the factors limiting their ability to compete with ether-based additives.

BIBLIOGRAPHY

1. P. Sabatier and J. D. Senderens, *C. R. Hebd. Seance Acad. Sci.* **134**, 514 (1902).
2. F. Fischer and H. Tropsch, *Brennstoff-Chem.* **4**, 276 (1923).
3. C. D. Frohning, W. Rottig, and F. Schnur in J. Falbe, ed., *Chemierohstoffe aus Kohle,* Georg Thieme Verlag, Stuttgart, 1977.
4. H. Kolbel, P. Ackermann, and F. Engelhardt, *Erdoel Kohle* **9**, 303 (1956).
5. H. Kolbel and P. Ackermann, *Chem. Ing. Tech.* **28**, 381 (1956).
6. M. E. Dry and J. C. Hoogendoorn, *Catal. Rev. Sci. Eng.* **23**(1–2), 266–278 (1981).
7. R. B. Anderson, in P. H. Emmett, ed., *Catalysis, IV,* 2nd ed., Van Nostrand & Rheinhold Co., Inc., New York (1961).
8. R. B. Anderson, *The Fischer-Tropsch Synthesis,* Academic Press, Orlando, Fla., 1984.
9. L.-M. Tau, H. A. Dabbagh, S. Bau, and B. H. Davis, *Catalysis Lett.* **7**, 127–140 (1990).
10. L.-M. Tau, H. A. Dabbagh, and B. H. Davis, *Energy Fuels* **4**, 94–99 (1990).
11. R. Snel and R. L. Espinoza, *J. Mol. Catal.* **43**, 237–247 (1987).
12. K. Krishna and A. T. Bell, *Catalysis Lett.* **14**, 305–313 (1992).
13. C. J. Kim, U.S. Pat. 4,547,525 (Oct. 15, 1985), C. J. Kim (to Exxon Research and Engineering Co.).
14. E. Iglesia, S. C. Reyes, and R. J. Madon, *J. Catal.* **129**, 238 (1991).
15. R. J. Madon, S. C. Reyes, and E. Iglesia, *J. Phys. Chem.* **95**, 7795 (1991).
16. R. A. Dictor and A. T. Bell, *Ind. Eng. Chem. Process Des. Dev.* **22**, 678 (1983).
17. M. E. Dry, *Catalysis Lett.* **7**, 241–252 (1990) and references therein.
18. M. E. Dry, paper presented at the 191st ACS National Meeting, 1987.
19. J. C. Hoogendoorn, *Conversion of Coal into Fuels and Chemicals in South Africa,* paper presented at the 3rd International Coal Conference, Sydney, Australia, Oct. 6–8, 1976.

Table 14. Lurgl Octamix—Product Distributions[a]

Operating Conditions			
H_2-CO feed ratio	1.0	1.0	0.95
Pressure, MPa	7	7	10
Temperature, °C	270	270	275
Recycle tail gas and fresh, w/w	—	0.3	—

Product Selectivity, wt %			
Alcohols			
C_1	59.7	59.6	49.8
C_2	7.4	15.1	9.3
C_3	3.7	6.0	4.7
C_4	8.2	6.6	10.2
C_5	3.6	3.2	4.5
C_6+	6.8	4.9	8.4
Other oxygenates	10.2	4.1	12.7
Hydrocarbons	0.1	0.1	0.1
Water	0.3	0.4	0.3

[a] From Refs. 45, 46.

20. B. Jager, M. E. Dry, T. Shingles, and A. P. Steynberg, *Catalysis Lett.* **7**, 293–302 (1990).

21. S. T. J. Van Rensburg, *Chem. SA.*, 41–42 (Feb. 1990).

22. J. Ansorge and A. Hoek, paper presented at the ACS National Meeting, San Francisco, Apr. 5–10, 1992.

23. J. Eilers, S. A. Posthuma, and S. T. Sie, *Catalysis Lett.* **7**, 253–270 (1990).

24. G. A. Bekker, *Petromin*, 52–60 (Aug. 1990).

25. S. T. Sie, J. Eilers, and J. K. Minderhout, *9th Int. Cong. Catal.* **2**, 743–750 (1988).

26. U.S. Pat. 4,686,238 (Aug. 11, 1987), D. Bode and S. T. Sie (to Shell).

27. U.S. Pat. 4,729,981 (Mar. 8, 1988), T. P. Kobulinski and co-workers (to Shell).

28. Ger. Pat. 2,130,133 A (May 31, 1984), A. Hoek and co-workers (to Shell).

29. J. Weitkamp in J. W. Ward, ed., *ACS Symposium Series*, Vol. 20, 1975, pp. 1–27.

30. U.S. Pat. 4,385,193 (May 24, 1983), H. M. J. Bijwaard, M. A. M. Boersma, and S. T. Sie (to Shell).

31. J. E. Sinor Consultants, *Clean Fuels Rep.* **2**(4), 165–166, (Sept. 1990).

32. J. M. Hochman, G. C. Lahn, R. F. Bauman and B. Eisenberg, *Development of Advanced Gas Conversion Technology*, paper presented at the European Applied Research Conference on Natural Gas, Trondheim, May 1–3, 1992.

33. B. Eisenberg, G. C. Lahn, R. A. Fiato, and G. R. Say, paper presented at Alternate Energy '93, Colorado Springs, Apr. 27–30, 1993.

34. A. H. Singleton and S. Regier, *Hydrocarbon Processing* **62**(5), 71 (1983).

35. H. P. Withers, K. F. Eliezer, and J. W. Mitchell, *Novel Fischer-Tropsch Slurry Catalysts and Process Concepts for Selective Transportation Fuel Production, Final Report*, DOE/PETC Contract DE-AC22-84PC70030, 1987.

36. H. P. Withers, K. F. Eliezer, and J. W. Mitchell, *Ind. Eng. Chem. Res.* **29**, 1807–1814 (1990).

37. J. C. W. Kuo and co-workers, *Slurry Fischer-Tropsch / Mobil Two-Stage Process of Converting Syngas to High Octane Gasoline*, DOE/PC/3022-10, DE84004411, 1983.

38. C. D. Chang, *Catal. Today* **13**(1), 103–111 (1992).

39. C. D. Chang, *Stud. Surf. Sci. Catal.* **61**, 393–404 (1991).

40. A. Solbakken in A. Holmen, ed., *Natural Gas Conversion*, Elsevier, Amsterdam, The Netherlands, 1991.

41. P. Roterud, E. Ritter and A. Solbakken, paper presented at the SPUNG Gas Utilization Seminar, Trondheim, Sept. 26, 1989.

42. U.S. Pat. 4,857,559 (Aug. 15, 1989), S. Eri and J. Goodwin (to Gas-To-Oil, Inc.).

43. A. H. El Sawy, *Evaluation of Mixed Alcohol Production Processes and Catalysts*, DOE Report DE900100325, Contract DE-AC04-76DP00789, Apr. 1990.

44. A. Paggini and D. Sanfilippo, *MAS Process: From Research to Commercialization*, paper presented at the AlChE Spring National Meeting, New Orleans, Apr. 6–10, 1986.

45. E. Supp, *Octamix, A Fuel Methanol that Needs No Cosolvent*, paper presented at the London Conference on the European Fuel Oxygenates Market, July 1986.

46. H. Goehna and P. Koenig, *The Octamix Process*, paper presented at the U.S. DOE Indirect Liquefaction Contractor's Meeting, Pittsburgh, Nov. 13–15, 1989.

FLAMMABLE LIMITS

RAMON ESPINO
Exxon Research and Engineering
Annandale, New Jersey

The concentration of a combustible substance in air has to be above a certain minimum level for the combustion process associated with flame propagation to occur. There is also a maximum concentration point above which flammable combustion will not occur. These two concentration levels define the flammable limits of a substance in air. The minimum and maximum concentration vary according to the chemical composition of the combustible material. Most normal paraffinic hydrocarbons have lower limits around 1 volume % in air and maximum limits around 3 volume %. Olefinic compounds have higher combustion limits around 10% by volume and acetylenes up to 80% by volume. See also AUTOMOTIVE ENGINES; HAZARD ANALYSIS OF ENERGY FACILITIES.

The lower combustion limit corresponds to the minimum concentration of the combustible component in air that is required for combustion to occur rapidly and essentially to the complete conversion of the burnable substance into water and carbon oxides. This is also referred at the minimum concentration required to propagate the flame. Above this concentration the energy released during the combustion is sufficient to cause the surrounding molecules to oxidize (burn). As the concentration of the combustible increases so does the amount of energy released. The maximum energy release occurs at the stoichiometric ratio of the combustible and air. This ratio defines the amount of combustible and oxygen that must exist for complete conversion of the two reactants into combustion products, in most cases carbon oxides and water. If the combustible material contains nitrogen or sulfur atoms, these are converted to their corresponding oxides.

If the concentration of the combustible substance in air is above the stoichiometric ratio, the combustion process will leave a portion of the combustible material unreacted and the energy release per volume of mixture will be less, too. Further increases in the concentration of the combustible will leave more unreacted material and less energy released. The upper flammable limit defines the concentration of combustible in air where the energy release per volume of mixture is not sufficient to maintain the combustion process. In other words, there is insufficient energy for the flame to propagate. The flammable limits of combustible materials are also influenced by pressure and the concentration of oxygen. Increasing the absolute pressure of the system decreases the lower flammable limit while increasing the oxygen concentration raises the upper flammable limit. The flammable limits of many compounds have been measured experimentally, but for many others, they are estimated by comparison with values for similar compounds. This is particularly true for homologous series. For example, the flammable limits have been measured for ethylene, propylene, 1-butene and 1-pentene, but for most other olefins only have estimates of the lower limit are known.

Table 1 lists the flammable limits for some important

Table 1. Flammability Limits[a]

	Air		Oxygen	
	Lower	Upper	Lower	Upper
Methane	5.0	15.0	5.1	61.0
Ethylene	2.7	36.0	2.9	80.0
Gasoline	1.3	7.1	1.3	—
Kerosene	0.7	5.0	0.7	—
Ethyl ether	1.9	36.0	2.0	82.0
Methanol	6.7	36.0	6.7	93.0
Ethanol	3.3	19.0	3.3	—
Benzene	1.3	7.9	1.3	30.0
Toluene	1.2	7.1	1.2	—
Ammonia	15.0	28.0	15.0	79.0
Hydrogen	4.0	75.0	4.0	95.0

[a] Ref. 1.

compounds at atmospheric pressure, both in air and pure oxygen. Note that the concentrations are given in terms of volume % of the combustible in air or in oxygen. The values mentioned before were in volume percent of combustible material in air.

Knowledge of the flammable limits of a combustible substance is critical from a safety point of view. Many household fires occur when liquids or solids wet with combustibles are left in confined spaces. The concentration of the combustible material in air can reach the flammable range in the confined space, and a spark or an open flame trigger the combustion process which can rapidly turn into a major fire.

Knowledge of the flammability limit is important in the storage of the hydrocarbons in tanks. The vapor space of the gasoline tank of a car is well above the upper flammability limit and so are the storage tanks in refineries and chemical plants. Of course, tanks can rupture or overflow, resulting in mixtures of hydrocarbon and air within the flammability limits. Therefore, it is basic common sense to avoid open flames and electrical sparks around flammable materials. If motors are to be used, they should be "explosion proof," meaning that the area where a spark may occur is isolated from hydrocarbon vapors.

Knowledge of the combustion limits of materials is also critical in the controlled oxidation of hydrocarbons, ammonia, and many other combustibles. For example, butane can be safely oxidized to maleic anhydride, with air and a suitable catalyst, if the concentration of butane is maintained below its lower flammable limit. One can also operate safely if the concentration of butane is maintained above the upper flammability limit. The same principles apply to the oxidation of ethylene to ethylene oxide, ammonia to nitrogen monoxide, benzene to maleic anhydride, etc.

BIBLIOGRAPHY

1. *Fire Protection Handbook*, 15th ed.

Reading List

D. W. Bartknecht, *Dust Explosions*, Springer-Verlag, New York, 1989.

Flammable and Combustible Liquids Code Handbook, M. F. Henry, ed., National Fire Protection Association.
Physical Constants of Hydrocarbons C_1 to C_{10}, ASTM Special Technical Publication No. 109A.

FLASH POINT

RAMON ESPINO
Exxon Research and Engineering
Annandale, New Jersey

The temperature at which the vapor concentration of a combustible material reaches the lower flammability limit is called the flash point of the liquid. The flash point of a liquid varies with the absolute pressure of the system, the oxygen content of the atmosphere and the composition of the liquid. If the pressure and oxygen content are higher than atmospheric, the flash point of the liquid is decreased. The flash point of a liquid can be measured by various methods. The most commonly used is the Tagliabue (TAG) method recommended by the National Fire Protection Association.

All flammable liquids are combustibles, but not all combustibles are flammable liquids. The critical property that differentiates one from the other is the "volatility" of the liquid. Liquids that have relatively low boiling points, certainly those boiling below 37.8°C, are flammable. The other extreme is of course heating oil, which certainly is combustible but not flammable. Lubricating oil, such as that used in automobiles, is another commonplace substance that is combustible but not flammable. The Occupational Safety and Hazard Agency (OSHA) has established definitions for flammable liquids and combustible liquids. Flammable liquids are called Class I liquids and they are subdivided into the following groups:

IA - liquids with flash points below 22.8°C and boiling points below 37.8°C

IB - liquids with flash points below 22.8°C and boiling points above 37.8°C

IC - liquids with flash points below 37.8°C

Class II materials are not flammable but combustible. In this category are liquids with flash points below 60°C. Liquids with flash points above 60°C are in Class III. Those with flash points below 93°C are in classification IIIA and those with flash points above 93°C are classified as IIIB. Examples of materials in all those categories are given in Table 1.

It is clear that many flammable and combustible compounds are widely used in daily life since they serve important functions (lubrication, provide energy for home heating or transportation). However, we must be very conscious of the hazard associated with their use since they can ignite, burn or even explode. One should never leave gasoline containers open to the atmosphere, particularly in enclosed spaces like garages. It is also not recommended to leave materials soaked with paint or lacquer and enamel thinners in confined spaces, such as closets or basements.

Table 1. Flash Points and Boiling Points of Various Compounds[a]

Compound	Flash Point °C	Boiling Point °C	OSHA Class
Pentane	−40	36.6	IA
Ethyl ether	−40	35	IA
Gasoline	−43	38–200	IB
Benzene	−11.1	80	IB
Lacquer, enamel thinners	−6.7 to −4.4	20–40	IB
Most alcohols	−10 to −16	65–70	IB
Xylenes	27.2 to 30	138–143	IC
Turpentine	35	above 90	IC
Kerosene	37.8	150–300	II
Oil paint and varnish cleaners	40 to 60	Not measured	II
Varsol	41.1	157–201	II
Heavy duel oil	65.6	Not measured	IIIA
Lubricating oils	121–245	Not measured	IIIB

[a] Ref. 1.

BIBLIOGRAPHY

1. *Fire Protection Handbook*, 15th ed., National Fire Protection Assoc.

Reading List

R. Stephenson, *Flash Point of Organic and Organometallic Compounds*, Elsevier, New York, 1987.

M. F. Henry, ed., *Flammable and Combustible Liquids Code Handbook*, National Fire Protection Association.

Fire Protection for Laboratories Using Chemicals, NFPA 45, National Fire Protection Association.

FOREST RESOURCES

JOHN S. SPENCER, JR.
USDA Forest Service
St. Paul, Minnesota

The world's forests have often presented a dilemma to those who lived nearby. Forests have provided food, shelter, and raw material, while hindering agriculture, national growth, and expansion. Shifting agriculturists cleared the forest to grow crops, and forest clearing expanded as sedentary agriculture became a way of life. Area of this old-growth forest continues to diminish today as previously inaccessible areas are opened to utilization.

Today's global forests, which may cover from one-half to two-thirds of the forested area of preagricultural times, provide a broad array of commodities, amenities, and environmental services. Fuelwood, wildlife habitat, pasture for livestock, industrial forest products, recreation, soil moisture retention, production of atmospheric oxygen, climate regulation, a source of new agricultural or grazing land, and spiritual renewal are a few examples. Currently, a growing world population, an increasingly industrialized world economy with accelerating energy needs, and rising material expectations of the people combine to generate unparalleled pressures on the global forest re-

source. Today, the clash between the philosophies of the environmentalist and those of the forest commodity user is more pronounced than ever as people are becoming more aware of the disparate benefits that flow from the forest, as well as the consequences of realizing those benefits. Society's goal, which is both elusive and difficult to achieve, is to find ways to use the forest bounty that are environmentally sound, socially acceptable, and economically rewarding.

See also ACID RAIN; CARBON CYCLE; CARBON STORAGE IN FORESTS; FUELS FROM BIOMASS.

HISTORICAL PERSPECTIVE

A large accessible supply of wood, the principal building material and fuel of past societies, was essential to the flowering of civilizations. History abounds with examples of societies that have grown rapidly because of an abundance of wood and that have collapsed after exhausting their forests. In 2700 BC the Sumerians in Mesopotamia thrived in the lower reaches of the Tigris and Euphrates Rivers of present-day Iraq. But by 2000 BC the Sumerian empire had collapsed largely because of the progressive decline of barley yields caused by salinization of soils, triggered by the clearing of the forests in the watersheds of the two rivers, exposing salt-rich sedimentary rock on the denuded slopes (1).

The extensive forests surrounding Athens probably provided the material needed to build the Athenian fleet that defeated the Persian navy in 480 BC, thrusting Athens into its Golden Age. By the time of the defeat in 404 BC of Athens by Sparta in the Peloponnesian War, the forests had been denuded and much of the topsoil had washed away, prompting Aristotle to recommend that laws be adopted to protect forests and to regulate their use.

The Macedonians translated their wealth in forests to economic, political, and military power, and in the process conquered much of what was then the civilized world under Alexander the Great, from 334 BC to 323 BC.

The Romans responded to their wood scarcity by seizing new forests in Iberia, Gaul, Britain, and North Africa, transferring Rome's problems to the provinces. Rome financed the growth of its empire largely from silver mined and smelted in Spain, using local wood for fuel. During the 400 years they operated, the furnaces consumed an estimated 500 million trees, deforesting more than 1.8 million hectares (7,000 square miles) of the Spanish landscape (1). When silver production declined, not because the ore supply was exhausted but because fuel was inaccessible, emperors were forced to debase the coinage progressively, until by the end of the third century AD the public and the government had little confidence in the almost-silverless metal.

The scarcity of wood in England from centuries of iron- and glass-making, ship-building, and domestic heating and cooking generated much of England's interest in timber-rich North America in the seventeenth and eighteenth centuries, especially in the tall white pines of New England prized for masts on Royal Navy ships.

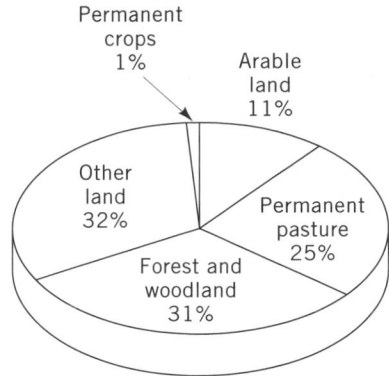

Figure 1. The world's land area by type of land use (3).

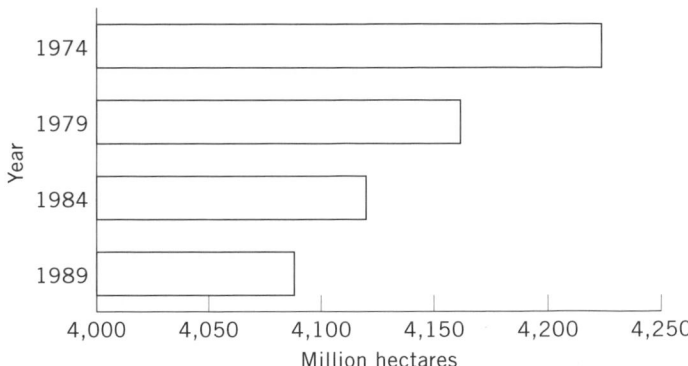

Figure 2. Area of global forest and woodland, 1974, 1979, 1984, and 1989 (3).

As in centuries past, the distribution and condition of the world's forests continue to be altered by human activities, generated by increasing demands for space, food, fiber, and energy, and fueled by a population that grows by nearly 88 million people annually (2).

DISTRIBUTION AND CLASSIFICATION OF GLOBAL FORESTS

Distribution

The United Nations Food and Agriculture Organization (FAO) estimates the world's land area as of 1989 to be 13,076 million hectares (32,311 million acres; hectares may be converted to acres by multiplying by 2.471) (3), of which forest and woodland account for 31% (Fig. 1).

Forests of the world range in composition and structure from the closed, old-growth Douglas-fir (*Pseudotsuga menziesii*) forests of the mountains of Oregon and Washington to the open, dry woodland of the plains of Africa. Closed forests are those in which the canopy allows little light to fall to the ground, and open forests are those in which the canopy has openings that permit the ground to receive some light. The FAO defines forest and woodland as land under natural or planted stands of trees, whether productive or not, including land from which forests have been cleared but which will be reforested in the foreseeable future. The FAO estimates the world's area of forest and woodland as of 1989 to be 4,087 million hectares (10,098 million acres) (Table 1); approximately the same size as the total combined areas of North, Central, and South America (3).

The global area of forest and woodland continues to shrink as it is used for other purposes. Area declined from 4,224 to 4,087 million hectares from 1974 to 1989 (Fig. 2), a loss of 137 million hectares (3.3%). In general, however, temperate forest area increased, and tropical forest area declined. Temperate forests are generally located on the globe between the treeless tundra region near the poles, and the Tropics of Cancer and Capricorn; tropical forests are generally located within or near the Tropics of Cancer and Capricorn. A much larger area of temperate forest is planted annually than tropical forest, and tropical forests are being converted at a much faster rate than temperate forests. Also, some nonforest lands, such as cropland or pasture, may revert naturally to forest if abandoned, a situation that occurs more frequently in the temperate forest region, because there is less competition there for agricultural land than in the tropics.

Between 1974 and 1989, the rate of decline of forest and woodland area in Oceania was greater than that of any other region (16.8%). Oceania is made up of a wide group of islands in the Pacific Ocean, including Australia, New Zealand, American Samoa, French Polynesia, Cook Islands, Fiji, Guam, Midway Islands, Papua New Guinea, Solomon Islands, Wake Island, and others. The rate of

Table 1. Area of Forest and Woodland in the World by Region, 1974 and 1989[a]

Region	Area of Forest and Woodland[b]		Difference
	1974	1989	
Former Soviet Union	920,000	946,000	+ 26,000
South America	954,520	891,338	(−) 63,182
North and Central America	707,609	716,175	+ 8,566
Africa	722,661	683,574	(−) 39,087
Asia	576,939	535,398	(−) 41,541
Oceania	189,005	157,245	(−) 31,760
Europe	153,639	156,964	+ 3,325
Total	*4,224,373*	*4,086,694*	*(−)137,679*

[a] Reference 3.
[b] In thousand hectares.

Table 2. Area of Forest and Woodland by Major Forested Country, 1989

Country	Area[a]
Former Soviet Union	946
Brazil	553
Canada	358
United States	294
Zaire	175
China	126
Indonesia	113
Australia	106
Peru	69
India	67
Argentina	59
Bolivia	56
Angola	53
Colombia	51
Sudan	45
Mexico	43
Tanzania	41
Papua New Guinea	38
Central African Republic	36
Myanmar	32
Venezuela	30
Zambia	29
Sweden	28

[a] In million hectares.

decline was next largest in Asia (7.2%), followed by South America (6.6%), and Africa (5.4%). However, the largest area of forest and woodland converted to other uses was in South America (63 million hectares), led by Brazil (35 million hectares).

The former Soviet Union ranked first in 1989 among individual countries in area of forest and woodland (946 million hectares), 71% greater than second-place Brazil (Table 2).

Closed forests, defined by FAO as land with a forest cover and with tree crowns covering more than 20% of the land area, make up between 2,500 and 3,000 million hectares, or about two-thirds of the total forest and woodland area (4). The remainder is open woodland, discontinuous forest stands with tree-crown cover of 10–20%, perhaps interspersed with grass or brush cover.

Classification of Forests

Because of the diverse nature of forests, several systems for classifying them have been developed. One system, devised by the FAO (5), includes the five groups discussed below.

Cool Coniferous Forests. These boreal forests are found in the northern latitudes of North America, Scandinavia, and the former Soviet Union, as well as at high elevations in temperate zones. Principal tree species, ie, spruce (*Picea* spp), fir (*Abies* spp), larch (*Larix* spp), aspen (*Populus* spp), and birch (*Betula* spp), are small and growth is slow because of the short growing season and cool temperatures.

Temperate Mixed Forests. This mix of forest ecosystems is found between the cool coniferous forests and the tropical forests of the northern hemisphere, and south of the tropical forests of the southern hemisphere. The range includes southern Canada, the United States, Europe (except Scandinavia), much of the former Soviet Union, China, Japan, parts of East Asia, Chile, Argentina, New Zealand, and Australia.

Tropical Moist Evergreen Forests. These tropical rainforests are found where annual precipitation is more than 2,000 mm and evenly distributed throughout the year. They are generally located in the Amazon Basin, northern South America, eastern coastal Central America, southern Mexico, equatorial Africa, western coastal India, Southeast Asia, and northeast coastal Australia. Tropical rainforests include the most biologically diverse ecosystems on earth, containing nearly half the world's known plant and animal species (6). As elevation increases, rainforests grade into cloud forests, where trees are shorter and less diverse.

Tropical Moist Deciduous Forests. These tropical forests are located where annual precipitation is 1,000 to 2,000 mm, and where a dry season occurs for at least one month each year. They are generally found in central South America, western coastal Central America, south-central Africa, India, and Southeast Asia. Commercially valuable teak (*Tectona* spp) and Philippine mahogany (*Shorea* spp) are found in these forests in Asia.

Dry Forests. These forests are located in both temperate and tropical zones where annual precipitation is less than 1,000 mm. They range from closed forests to open woodland, thornlands, shrublands, savannahs, and other sparse woody vegetation. These forests generally occur in the southwestern United States, Mexico, South and Central America, the Mediterranean Basin, India, Australia, and over much of sub-Saharan Africa. In the arid western United States, pinyon pine (*Pinus edulis*)–juniper (*Juniperus* spp) stands are representative of dry forests.

Limiting Factors of Forests

Climate and geographical location determine distribution and type of forests. Temperature is the limiting factor at higher latitudes and elevations. Generally speaking, forests do not grow poleward beyond the isotherm of 10°C (50°F) average temperature, eg, northern Norway and Siberia, during the growing season (7). However, in areas close to the sea they may extend from this isotherm toward the pole, and in continental areas they may contract toward the equator. The amount and seasonal distribution of precipitation determine the nature of forests at midlatitudes. Tropical rainforests exist where rainfall is greater than 2,000 mm (about 80 in.) annually, eg, Brazil's Amazon Basin, while open woodland, scrub, and savannah exist where rainfall is less than 1,000 mm annually, eg, plains of New Mexico. Little woodland is found where rainfall amounts to less than 400 mm/yr, eg, Sahara Desert. Other factors such as soil type and drainage, slope, and aspect modify the local situation further to produce the mosaic of diverse forests found throughout the world.

IMPORTANCE OF FORESTS

Ecological Significance of Forests

Much more remains to be learned about the many complex interactions in forests between trees, other plants, animals, soil, and the environment, as well as about the role of forests in maintaining the health of the planet. However, we do know that forest ecosystems provide a host of environmental services including maintaining biological diversity, providing wildlife habitat, cycling nutrients, producing oxygen, affecting regional rainfall patterns, and sequestering carbon in the global carbon pool. They also regulate streamflow, reduce flooding, store water, moderate wind erosion, and reclaim degraded land.

Disturbance of forest ecosystems in a particular location may result in important changes in other ecosystems that may be separated by great distances. For example, removal of Central and South American tropical forests used as wintering grounds by many species of neotropical migratory songbirds, along with fragmentation of the forest habitat in their summer range and along their migratory routes, are thought to be the primary reasons for the decline in the numbers of these birds.

Forest decline, ie, needle loss, tree growth reduction, and tree death, has been observed over broad areas in Europe and North America, and is thought to be caused by acid rain or airborne pollutants. However, at this time ongoing research has documented only that ozone has caused decline in the forest health of ponderosa pine (*Pinus ponderosa*) in southern California, but can not confirm pollutant-caused decline elsewhere (8). Ozone is produced in the atmosphere when nitric oxide (NO) is created by the combustion of fossil fuels and emitted as exhaust, such as from automobiles. The nitric oxide combines with oxidants in the atmosphere to form nitrogen dioxide (NO_2), which, in the presence of sunlight, breaks down into free oxygen atoms (O) and nitric oxide (NO). The oxygen atoms then combine with oxygen molecules (O_2) in the atmosphere to produce ozone (O_3).

Many pollutants may simply stress trees, predisposing them to natural causes of death such as drought, insects, and disease. However, continuing research may reveal a causal effect between other pollutants and forest decline. Already there is a growing body of evidence to link the decline of red spruce (*Picea rubens*) in the northern Appalachian Mountains of the United States to acid deposition. Over the past three decades, red spruce in the higher elevations of the region have exhibited reduced rates of growth and poor crown condition. In some areas, more than 50% of the red spruce trees have died. This area of forest decline is unique because of its high degree of exposure to acidic depositions due to the frequent immersion in clouds, which are more polluted than precipitation. Acidic deposition has been linked to increased frequency and severity of winter injury to red spruce foilage, and to reduction of nutrients in the soil that are essential to tree growth (9).

Scientists at the U.S. Department of Agriculture (USDA) Forest Service's Northeastern Forest Experiment Station in Durham, New Hampshire, have found that calcium is being depleted from mountain forest soils in parts of the eastern United States by leaching (probably induced by acid deposition) and by frequent whole-tree clearcutting of forests (10). At the present rate of loss, 50% of the total soil and biomass calcium could be removed within 120 yr, a level that would result in serious calcium deficiency and subsequent reduced tree growth. As a result of these findings, timber harvest practices in the eastern United States now leave more large, woody residue on the site so the calcium and other nutrients within it will be returned to the soil as the residue decomposes.

Industrial emissions can spread far beyond the point of origin and deposit pollutants which build up in forest ecosystems with no present visible adverse effects. USDA Forest Service scientists at the North Central Forest Experiment Station Laboratory in Grand Rapids, Minnesota, demonstrated a relationship between emissions from fossil fuel combustion and the acidity of precipitation, as evidenced by near background levels of acidic deposition in northwestern Minnesota, increasing to high levels in southeastern Michigan (11). These emissions probably originated in places like the industrialized Ohio River valley, and were carried by winds to the Lake States. Studies of lakes along this deposition gradient found evidence of water chemistry changes related to acidic deposition. Investigations of soil and tree tissue along the gradient showed that the amount of sulfate deposited in acid deposition is related to the amount of sulfur in soil and trees, suggesting that effects of acid deposition may be cumulative and that continued monitoring is advisable. No apparent effect of elevated sulfur levels on tree growth has been observed to date, regardless of location on the gradient. However, long-term effects of this and other buildups of potential pollutants are unknown.

Volume of Timber in Forests

The diversity of forests translates into widely different volumes of standing timber. It is estimated (12) that average volumes per hectare range from 50 cubic meters in dry savannah woodlands to 150 cubic meters in temperate forests, and to 350 cubic meters in tropical rainforests. These averages translate into an estimate of total standing volume for global closed forests of 396 billion cubic meters, half of which is in tropical rainforests. By comparison, there were an estimated 23 billion cubic meters (816 billion cubic feet) in live trees on timberland in the United States in 1987 (13). Some of the global volume is in tree species considered unsuitable for industrial forest products, is in forests legally protected from logging, or is located where exploitation is precluded by inaccessibility. The area of exploitable closed forests is estimated to be about 2,150 million hectares with around 300 billion cubic meters of standing volume (14). Exploitable forests are those on which there are no legal, economic, or technical restrictions on timber harvesting.

Production of Wood from Forests

The FAO estimates that total production (removals) of roundwood from global forests in 1991 was 3,429 million cubic meters (Table 3), or about 1.2% of the exploitable inventory (15). Roundwood is defined as wood in the

Table 3. Volume of Global Roundwood Production by Region, 1991[a]

Region	Roundwood Production[b]	Total, %
Asia	1,086.1	31.7
North and Central America	737.4	21.5
Africa	527.2	15.4
Former Soviet Union	355.4	10.3
South America	345.4	10.1
Europe	335.5	9.8
Oceania	42.4	1.2
Total	*3,429.4*	*100.0*

[a] Reference 15.
[b] Million cu. meters.

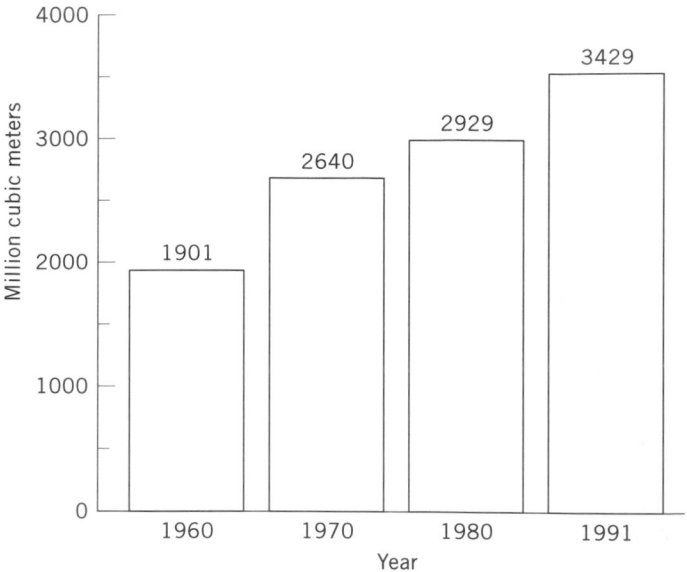

Figure 3. Volume of global roundwood production 1960, 1970, 1980, and 1991 (15).

rough, from logs, bolts, or other round sections cut from trees. Asia produced the largest share of the world's output (32%). The United States led all individual countries with 495.8 million cubic meters, followed by the former Soviet Union (355.4 million), China (282.3 million), India (279.8 million), Brazil (264.6 million), Canada (178.0 million), and Indonesia (173.0 million).

The 1991 production of 3,429 million cubic meters was 80% higher than the 1960 production of 1,901 million cubic meters (Fig. 3). Coniferous (softwood) species accounted for about 39% of the total production, and deciduous (hardwood) species accounted for 61%. Fifty-three percent of the volume produced was used for fuelwood and charcoal, and 47% was industrial roundwood such as saw logs, veneer logs, and pulpwood. Developed countries produced 76% of the industrial roundwood; developing countries produced 87% of the fuelwood and charcoal, still the principal source of energy for three-fourths of the population of the developing world.

Wood: A Vital but Dwindling Energy Source in Developing Countries

The silent energy crisis, unknown by most people in industrialized countries, and unaffected by oil price fluctuations, is the daily struggle by more than one third of the world's population to find wood to cook their meals. An Indian official enunciated the looming fuelwood crisis in many countries when he said, "Even if we somehow grow enough food for our people in the year 2000, how in the world will they cook it?"

About half the world's roundwood production in 1991 was used to produce energy (15). Wood is the primary source of energy for poorer urban households and for the vast majority of rural households in developing countries, where other fuels are economically beyond the reach of most people. The developed countries consume wood primarily for industrial purposes (lumber, plywood, paper, etc), but the developing world consumes wood primarily for energy production. Eighty percent of the total roundwood consumption in developing countries in 1986 was for fuelwood (16). Fuelwood, which may be converted to charcoal, is usually consumed in-country for domestic heating and cooking, and to a lesser extent, for small industrial enterprises, such as baking, pottery making, and coffee and tobacco drying. Urban and rural fuelwood needs differ, as illustrated in Africa (south of the Sahara) where 67 to 75% of urban domestic energy use is from fuelwood, compared to 90 to 98% of rural energy use (17).

An estimated 1,688.7 million cubic meters of fuelwood were produced from global forests in 1991 (15), as shown in Table 4. The 1991 volume was 23% greater than the 1980 volume (1,373.3 million cubic meters). Eighty-seven percent of the volume was from hardwood species, and

Table 4. Volume of Global Fuelwood Production by Region and by Softwood and Hardwood Production, 1991[a,b]

Region	Softwood Production	Hardwood Production	Total Fuelwood Production
Asia	99.0	713.7	812.7
Africa	6.5	395.4	401.9
South America	16.3	180.4	196.7
North and Central America	29.9	108.7	138.6
Former Soviet Union	52.7	28.4	81.1
Europe	15.3	33.9	49.2
Oceania	0.6	7.9	8.5
Total	*220.3*	*1,468.4*	*1,688.7*

[a] Reference 15.
[b] In million cubic meters.

the remaining 13% was from softwood species. Asia produced 48% of the fuelwood, followed by Africa (24%) and South America (12%). India led all individual countries in fuelwood production with 243 million cubic meters (14% of the world's production), followed by China (192 million), Brazil (154 million), Indonesia (143 million), and Nigeria (94 million).

Dead trees and branches have traditionally supplied much of the fuelwood requirement in many countries, but increasingly demand must be satisfied by cutting live trees. The pressure to use livewood is exacerbated by rapid population growth, urbanization, or restriction of access for wood gathering, as parts of the forest that were formerly common property become privatized or nationalized (18). The consequences of such pressure on a limited forest resource is often degradation of the resource and fuelwood scarcity. There are now shortages of fuelwood in 57 developing countries. Of the 2,000 million people there who depended on fuelwood in 1980, approximately 100 million experienced acute fuelwood scarcity, and 1,050 million people lived in areas of increasing fuelwood scarcity and met their fuelwood needs at the expense of depleting existing wood resources. By the year 2000, the number of people in situations of acute shortages will increase to 2,400 million unless action is taken (19).

In many cases, local populations do not perceive an impending fuelwood shortage, and are unaware that by depleting their forest capital, they are destroying their future source of fuelwood and other forest resources. However, trees represent a capital stock that appreciates through growth, and can regenerate and maintain itself as long as timber harvesting and other removals do not exceed the level of regeneration and net growth.

This potential fuelwood crisis takes an increasing human toll; in parts of Tanzania, villagers had to walk 10 km for firewood in 1983, compared to only 1 km twenty years earlier (20). In parts of East Africa, a family may spend up to 40% of its income to purchase fuelwood (21). In parts of West Africa, people now have only one instead of the customary two cooked meals a day (6). The fuelwood shortage also has a profound effect on soils and crop yields because people are forced to burn animal dung and crop residues that once were plowed under to increase soil fertility. However, Dewees (22) argues that the projected shortage may be overstated because it does not take into account practices that could moderate future consumption, such as fuel sharing, shared cooking, and increased labor use for fuel collection.

National energy planners in many developing countries neglected fuelwood in energy development programs because of their presumption that family incomes would increase as a result of national development programs, allowing households to substitute oil, liquid petroleum gas, kerosene, electricity, or coal for wood. This has not occurred because of the still prohibitive costs of nonwood energy sources for most people, and the outlook is for fuelwood to continue to be a principal energy source in the foreseeable future for most developing countries (17). Improvement of the fuelwood situation requires a two-pronged approach: (1) an increase in fuelwood production, and (2) better fuelwood conservation and conversion efficiency. It has been suggested (23) that the integration of trees into agricultural systems is the most worthwhile method of increasing fuelwood production (see SOCIAL FORESTRY). Widespread use of more efficient cookstoves, better use of waste wood, and improved methods of charcoal production will stretch existing fuelwood stocks.

Nontimber and Amenity Uses of the Forest

Besides timber products, forests provide nuts, fruits, herbs, medicinal plants, pharmaceuticals, resins, gums, oils, forage, commercial flowers, spices, syrups, thatch, and rattan. Additionally, forests protect soil and watersheds; provide areas for ecosystem research; furnish opportunities for recreation and spiritual renewal; and inspire literature, music, religion, and art.

TEMPERATE FORESTS

Depending on their location on the globe, forests can be classed as either temperate or tropical. The temperate forest zone generally occurs between the treeless tundra region and the frost-free tropical forest region in both the northern (north temperate zone) and southern (south temperate zone) hemispheres. The north temperate zone includes the boreal forest, often called "taiga," located closest to the tundra in the northern hemisphere, and composed of stands of spruces, firs, larches, birches, and aspens. The southern hemisphere contains nothing comparable to the boreal forest. The remainder of the two forest zones contains stands of other frost-hardy coniferous, or deciduous, or mixed coniferous–deciduous trees, such as the longleaf (*Pinus palustris*)–slash pine (*Pinus ellottii*) stands of the southeastern United States, or the maple (*Acer* spp)–beech (*Fagus grandifolia*)–birch (*Betula* spp) stands of the Lake States and northeastern United States. Tree species found naturally in the northern hemisphere are related to but different from those in the southern hemisphere. Temperate zone forests occur largely in developed countries, where they are intensively used and generally receive a higher level of forest management than tropical forests.

In 1990, the area of forest land (land with tree crown cover of more than 20% of the area) in the developed temperate zone was estimated to be 1,432 million hectares (24), or 27% of the land area of the zone. The developed temperate zone includes all the countries in Europe (including Cyprus, Israel, Turkey, and the former Soviet Union), Canada, the United States, Japan, Australia, and New Zealand. The proportion of forest to total land area ranges from 66% in Japan and Finland to 5% in Israel and Australia. No estimates of recent change in area of temperate forest land are available. However, the FAO estimates that the area of forest and other wooded land (includes forest, open woodland and scrub, and shrub and brushland) increased by 2.0 million hectares between 1980 and 1990, probably from new plantations and from other land reverting naturally back to forest. An average of 63% of the temperate forest area is exploitable. The exploitable share by country ranges from 100% for Belgium, Czeck and Slovak Federal Republic, Denmark, and the United Kingdom, to 28% for New Zealand. Eighty-one percent of the temperate forest is publicly owned, ranging from 100% in the former Soviet Union to an average of 62% in North America and 49% in Europe.

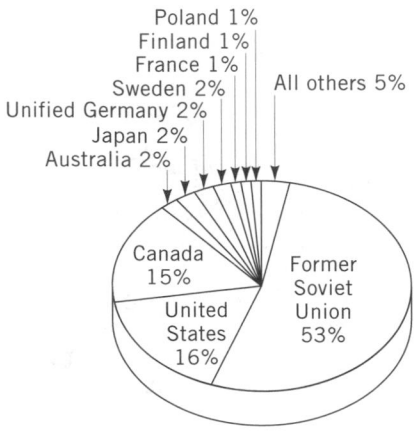

Figure 4. Percent of total volume of standing, live (growing-stock) trees in global temperate forests, by country, 1980s (24).

The volume of standing, live (growing-stock) trees in global temperate forests in the 1980s amounted to 159 billion cubic meters, about three-quarters of which is in softwood species and one-quarter of which is in hardwood species (24). More than half of the standing volume is in the former Soviet Union (Fig. 4).

TROPICAL FORESTS

The global tropical forest situation is of particular importance because of the alarming rate at which forests are being cleared. Between 1980 and 1990, global tropical forest area declined from 1,910 to 1,756 million hectares, an 8% loss (25). This amounts to about 29 hectares or 72 acres each minute. This rapid rate of increase has gener-

ated increasing concerns about the environmental consequences of the loss of these forests; the economic impact of the forest resource, especially to developing countries; and the social effects of this deforestation on indigenous persons. The term tropical forest may be misleading in that it implies a forest in a moist tropical environment. In fact, tropical forests grow on both humid and semiarid lands within or near the Tropics of Cancer and Capricorn in Asia, the Pacific region, Africa, Latin America, and the Caribbean region. The FAO classifies tropical forests into six primary ecological zones, as shown in Table 5.

The 1990 area of tropical forests (1,756 million hectares) roughly equals the total area of South America. The largest tropical forest area is in Latin America and the Caribbean region (918 million hectares), followed by Africa (528 million hectares), and Asia and the Pacific region (310 million hectares), as shown in Table 6.

The greatest area loss was in the Latin America and Caribbean region where 74.1 million hectares (7.5% of the region's total) were deforested between 1980 and 1990. The FAO defines deforestation as a change of land use with depletion of tree crown cover to less than 10%. However, the largest proportional loss occurred in the Asia and Pacific region where the 39.0 million hectares deforested amounted to 11.2% of the region's total. The average of 15.4 million hectares (an area the size of the state of Georgia in the United States) of tropical forest cleared annually during 1981–1990, is 36% higher than the average of 11.3 million hectares cleared annually during 1976–1980 (27).

In addition to this deforestation, a large but unestimated area of tropical forest is degraded, changed from closed to open forest by human activity or by natural phenomena (insects, disease, wind, fire, or flooding), but with

Table 5. Characteristics of Tropical Forests by Ecological Zone[a]

Ecological Zone	Type of Forest	Elevation, m[b]	Mean Annual Rainfall, mm	Region
Tropical rainforest	Wet and very moist evergreen and semievergreen forests	<800	>2,000	Brazil, Ecuador, Colombia, Central Africa, Indonesia, Malaysia
Moist deciduous forest	Moist semideciduous and deciduous forest, woodlands, and tree savanna	<800	1,000–2,000	Brazil, Bolivia, Paraguay, Central and West Africa, India, Bangladesh
Dry deciduous forest	Drought-deciduous and evergreen forest, woodlands, and tree savanna	<800	500–1,000	Brazil, Bolivia, Paraguay, eastern and tropical southern and Sub-sahelian Africa, Sudan, India, Thailand, Laos, Myanmar
Very dry forest	Discontinuous thickets, tree/shrub savanna	<800	200–500	Namibia, Zimbabwe, Kenya, Sudan, Ethiopia, Madagascar, Sub-sahelian Africa, northwestern India, Pakistan
Desert	Tree/shrub steppe	<800	<200	Sub-sahelian Africa, Namibia, Pakistan
Hill and montane forest	Premontane to alpine forest	>800 (upland)	Includes wet, moist, and dry conditions	Bolivia, Peru, Ethiopia, northern Pakistan, Nepal, Bhutan, Mexico

[a] Reference 26.
[b] Lowland unless indicated.

Table 6. Area of Global Tropical Forests by Region, 1980 and 1990[a]

Region	Tropical Forest Area, 1980[b]	Tropical Forest Area, 1990[b]	Area Deforested Annually, 1981–1990[b]	Annual Rate of Deforestation, 1981–1990, %
Latin America and Caribbean	992.2	918.1	7.4	0.8
Central America and Mexico	79.2	68.1	1.1	1.5
Caribbean	48.3	47.1	0.1	0.3
Tropical South America	864.6	802.9	6.2	0.7
Asia and Pacific	349.6	310.6	3.9	1.2
South Asia	69.4	63.9	0.6	0.8
Continental Southeast Asia	88.4	75.2	1.3	1.6
Insular Southeast Asia	154.7	135.4	1.9	1.3
Pacific	37.1	36.0	0.1	0.3
Africa	568.6	527.6	4.1	0.7
West Sahelian Africa	43.7	40.8	0.3	0.7
East Sahelian Africa	71.4	65.5	0.6	0.9
West Africa	61.5	55.6	0.6	1.0
Central Africa	215.5	204.1	1.1	0.5
Tropical Southern Africa	159.3	145.9	1.3	0.9
Insular Africa	17.1	15.8	0.1	0.8
Total[c]	*1,910.4*	*1,756.3*	*15.4*	*0.8*

[a] Reference 25.
[b] Million hectares.
[c] Totals may not add due to rounding.

enough live tree crown cover left per hectare to be classified as forest. Degraded forests contain a reduced volume of timber and biomass, may contain fewer commercially desirable tree species, exhibit a loss of biodiversity, and may reflect damage to soils. Many of these degraded lands will be further cleared and converted to other uses in the future. Clearly, there is cause for concern over the declining area of tropical forests, and the ecological impact of their conversion (28).

Although efforts to manage tropical forests are increasing, the results are not encouraging. The FAO defines forest management as inventorying the site and stand, preparing a management plan, and applying silvicultural treatments and logging controls in a forest area set aside to be managed for future production of forestry goods and services. The FAO and International Timber Trade Organization (ITTO) studied the status of forest management in the tropics and reported that South Asia, principally India, accounted for the largest area under management in 1980: 59 million hectares covered by working plans, or 3.1% of the 1980 global tropical forest area (25). However, much of that area is being degraded slowly by intense grazing of livestock and unauthorized fuelwood removals by local people. Outside of South Asia, the tropical forest area under sustained yield management was less than 1 million hectares.

The FAO estimates that about three-fourths of the tropical forest area in 1990 was in two zones, the tropical rainforest and moist deciduous forest zones (25). The largest area deforested annually between 1981 and 1990 was in the moist deciduous forest (6.1 million hectares) (Table 7). Annual deforestation rates are highest in the

Table 7. Area of Global Tropical Forest in 1990, Area Deforested Annually Between 1981 and 1990, and Annual Rates of Deforestation, by Ecological Zone[a]

Ecological Zone	Area, 1990[b]	Area Deforested Annually, 1981–1990[b]	Annual Rate of Deforestation, 1981–1990, %
Tropical rainforest	718.3	4.6	0.6
Moist deciduous forest	587.3	6.1	1.0
Dry and very dry deciduous forest	238.3	2.2	0.9
Hill and montane forest	204.3	2.5	1.1
Desert and alpine zone	8.1	0.1	1.0
Total[c]	*1,756.3*	*15.4*	*0.8*

[a] Reference 25.
[b] Million hectares.
[c] Totals may not add due to rounding.

hill and montane forest (1.1%). The hill and montane forest is perhaps most threatened because of the relatively small area of forest remaining, a population density that is higher than the average for tropical forests, and the high deforestation rate.

DEFORESTATION

Causes of Deforestation

The forest area in some developed regions has expanded (Table 1) as economic development has encouraged the reversion of agricultural lands to forest. However, in developing countries, the trend is toward deforestation, particularly in tropical forests. Although broad issues of poverty, rapidly increasing population pressures, unequal political power, lack of opportunities to make a living, landlessness, and inadequate knowledge and means to exploit the tropical forest without destroying it are at the root of deforestation, there are more specific causes.

Shifting cultivation (also called slash and burn agriculture), practiced by landless indigenous people who clear trees to grow subsistence crops, is the principal cause of deforestation in the tropics, accounting for 70, 50, and 35%, respectively, in Africa, Asia, and tropical America (29). Farmers must move to new sites after several years because most tropical forest soils are of low productivity; the already limited nutrients in these soils are lost to leaching and erosion, rather than being quickly recycled into plants as they had been under a forest cover. The abandoned patches, called forest fallows, are being created at the rate of more than 5 million hectares per year, but may revert back to forest if left undisturbed. However, because rising populations and the ensuing competition for land are forcing farmers to return to these fallows at increasingly shorter intervals, little of this land is allowed to revert to forest. About 500 million people (nearly 10% of the world population) and 240 million hectares of closed forest are involved in shifting cultivation, which is increasing at an average annual rate of 1.25% (30). Improvement of the efficiency of agricultural production in many developing countries is a prerequisite for controlling tropical deforestation (31).

Deforestation also occurs because conversion of forest to pasture for domestic animals is widely practiced, stimulated by a growing network of roads, and, in some places, by government incentives. Fuelwood gathering may be an important deforestation agent in dry forests.

Commercial logging may not be a primary cause of deforestation in the tropics (except in parts of West Africa) because the number of trees left after logging may be sufficient to classify the site as forested. However, it is often a secondary cause because new logging roads permit shifting cultivators and fuelwood gatherers to gain access to logged areas and fell the remaining trees. When logging is performed poorly, it results in a degraded forest.

Finally, expansion of agribusinesses that grow oil palm, rubber, and fruit trees, and ornamental plants has resulted in forests being cleared, and government-sponsored programs that resettle landless farmers on forested sites have contributed to deforestation around the world.

Consequences of Deforestation

A large effect of deforestation, although incompletely understood, is its contributions to global warming by releasing carbon dioxide and other greenhouse gases into the atmosphere from the cutting and burning of trees.

Other consequences of deforestation are better known, including loss of industrial timber and nontimber products, and loss of long-term forest productivity on the site. In many places the lack of fuelwood following deforestation challenges local people, especially where fuelwood had already been scarce. Forest fragmentation, the reduction of a large block of forest to many smaller tracts, promotes loss of biodiversity because some species of plants and animals require large continuous areas of similar habitat to survive. Agriculture may be negatively impacted if deforestation causes soil loss or compaction, or sedimentation of irrigation systems. Human life and downstream structures may be endangered by floods that may be intensified by clearing forests on upstream watersheds. Species of plants and animals, which may occupy narrow ecological niches and whose potential value to humans is unknown, may be eliminated: it is estimated (32) that two-thirds (3.3 million) of all species of organisms on earth live in the tropics, and that two-thirds (2.2 million) of these are unique to tropical forests. Indigenous people may be forced into a new way of life for which they are unprepared. Encroaching deserts may follow deforestation in dry forest areas.

REVERSING DEFORESTATION

Forest Plantations

Estimates of the area of global forest plantations (temperate and tropical) differ, but center around a total of 100 million hectares as of 1985, or about 3.5% of the closed forest area. It is estimated (33) that the four countries or regions with the largest plantation areas (the former Soviet Union, western Europe, China, and the United States) account for 62% of the total (Table 8).

An estimated 60% of the area of global plantations is devoted to industrial timber production, and the remaining 40% is devoted to nonindustrial uses, such as fuelwood production and environmental protection (29). Most of the plantation area is outside the tropical zone, yet most of the deforestation occurs in the tropics. Although the area planted is increasing at an accelerating rate, global deforestation surpasses it by about 5 million hectares annually. Most plantations contain only one or several tree species, and cannot replace a natural forest in terms of biological diversity. Nevertheless, plantations play a crucial role in slowing down the loss of natural forests by supplying large quantities of industrial timber from small, highly productive areas. For example, in Latin America, industrial plantations made up less than 1% of the forest area, but accounted for 30% of industrial wood production during the 1980s. This proportion is expected to rise to 50% by the year 2,000 (34).

Tropical forest plantations in 90 countries as of 1990 amounted to 43.9 million hectares (25). Based on an average tree survival rate of 70% from 56 plantation invento-

Table 8. Area of the World's Forest Plantations by Countries or Regions, ca 1985[a]

Country or Region	Plantation Area[b]	Percent
Developed countries		
Former Soviet Union	21,900	22.8
Western Europe	13,000	13.6
United States	12,100	12.6
Japan	9,600	10.0
Canada	1,500	1.6
New Zealand	1,100	1.2
Australia	800	0.8
Subtotal	*60,000*	*62.6*
Developing countries		
China	12,700	13.2
Brazil	6,100	6.4
India	3,100	3.2
Indonesia	2,600	2.7
Republic of Korea	2,000	2.1
Chile	1,200	1.3
Argentina	800	0.8
Others	7,400	7.7
Subtotal	*35,900*	*37.4*
Total	*95,900*	*100.0*

[a] Reference 33.
[b] In thousand hectares.

ries in 18 tropical countries, the FAO estimates the net area of successful plantations to be 30.7 million hectares. Between 1981 and 1990, the annual deforested area (15.4 million hectares) was about 8.5 times greater than the annual net planted area (1.8 million hectares). Five countries accounted for 85% of the gross plantation area: India (18.9 million hectares), Indonesia (8.8 million hectares), Brazil (7.0 million hectares), Vietnam (2.1 million hectares), and Thailand (0.8 million hectares). About 60% of the total reported plantation area was established during the period 1981 to 1990. The principal tree species planted were pines (*Pinus* spp.) and teak (*Tectona* spp.), planted mostly for industrial timber production; eucalypts (*Eucalyptus* spp.), planted for both industrial and nonindustrial purposes; and acacias (*Acacia* spp.), used primarily for fuelwood and gum arabic production.

Agroforestry: A Positive Tool

Agroforestry, the deliberate integration of trees, shrubs, and other woody perennials on the same land with agricultural crops, livestock, or both, recognizes the symbiotic relationship between trees and agriculture that has existed in most countries for centuries. It offers a practical way to intensify land use and to reduce the economic and social pressures for clearing additional closed forests. Growing annual crops in rows between nitrogen-fixing trees or shrubs, and growing canopy trees for plantation crops, eg, coffee or cacao, are examples of this strategy, which represents part of a solution to the growing fuelwood shortage.

Conservation agroforestry, the practice of working trees into agricultural and community ecosystems for hu-

man, as well as economic and environmental, benefits, is a variation of traditional agroforestry that is gaining acceptance in many parts of the semiarid world. Examples of these practices include planting windbreaks, living fences, livestock havens, and riparian buffer strips, and establishing wildlife habitat.

Social Forestry

A modification of agroforestry is social forestry, a broad range of tree- or forest-related activities that rural people and community groups in developing countries undertake to provide products for their own use and to generate local income (17). Examples are: farmers growing wood to sell or use for firewood; individuals or communities earning income from the gathering, processing, and sale of minor forest products (fruits, nuts, herbs, basketry materials, etc); or governments or other groups planting trees on public lands to meet local village needs. A basic premise is to change land use in such ways that people get what they need on a sustainable basis from a relatively fixed, or even shrinking, land base. Another premise is the critical importance of widespread local participation, possible only when people have the ability to cooperate, the knowledge of what is required, the proper mix of incentives to stimulate them, and the institutions to support and sustain them. Solutions to many environmental problems ultimately reside with each land user and his/her land use practices (35).

Social forestry programs are helping to ameliorate some of the important issues facing developing countries, such as energy shortages, food shortages, and unemployment. Fuelwood production is one of the key objectives of social forestry because of seriously declining fuelwood availability and the closely linked environmental instability. Other objectives are to reverse declining agricultural productivity associated with poor land use, deforestation, erosion, and declining water supplies, and to provide employment opportunities and income from forestry and related processing activities. To stimulate the practice of social forestry in developing countries, the World Bank made loans totaling $1.3 billion during the decade 1977 to 1986 (17).

SUMMARY

Forests are a renewable resource, and are part of a complex network of natural systems that affect life on the globe. When forests are misused, the consequences may be felt in places far removed from the offending act. Increasingly, international cooperation, public education, political will, personal incentives, and institutional structures are needed to manage the world's forests in harmony with the environment and with economic development.

BIBLIOGRAPHY

1. J. Perlin, *A Forest Journey: The Role of Wood in the Development of Civilization*, W. W. Norton and Co., New York, 1989.
2. World Resources Institute, *World Resources 1990–91*, Oxford University Press, New York, 1990.

3. Food and Agriculture Organization of the United Nations (FAO), *Production Yearbook, 1990*, Vol. 44, FAO Statistics Series No. 99, Rome, 1991.

4. A. Mather, *Global Forest Resources*, Belhaven Press, London, 1990.

5. A. Sommer, *Unasylva* **28**(112/113), 5–25 (1976).

6. J. Spears and E. Ayensu, in R. Repetto, ed., *The Global Possible*, Yale University Press, New Haven, 1985, pp. 299–335.

7. H. Gleason and A. Cronquist, *The Natural Geography of Plants*, Columbia University Press, New York, 1964.

8. P. Miller, in R. Olson, D. Binkley, and M. Bohm, eds., *The Response of Western Forests to Air Pollution*, Springer-Verlag, New York, 1992; *Ecological Studies* **97**, 461–500 (1992).

9. C. Eagar and M. Adams, eds., *Ecology and Decline of Red Spruce in the Eastern United States*, Springer-Verlag, New York, 1992.

10. C. Federer and co-workers, *Env. Manage.* **13**, 593–601 (1989).

11. E. Verry and A. Harris, *Water Resources Res.* **24**, 481–492 (1988).

12. R. Persson, *World Forest Resources: Review of the World's Forest Resources in the Early 1970's*, Research Note 17, Royal College of Forestry, Stockholm, 1974.

13. K. Waddell, D. Oswald, and D. Powell, *Forest Statistics of the United States, 1987*, U.S. Department of Agriculture, Forest Service, Resource Bulletin PNW-RB-168, Portland, Oreg., 1989.

14. C. Binkley and D. Dykstra, in M. Kallin, D. Dykstra, and C. Binkley, eds., *The Global Forest Sector: An Analytical Perspective*, John Wiley & Sons, Inc., New York, 1987, pp. 508–533.

15. Food and Agriculture Organization of the United Nations (FAO), *Yearbook of Forest Products, 1991*, FAO Forestry Series No. 26, FAO Statistics Series No. 110, Rome, 1993.

16. Food and Agriculture Organization of the United Nations (FAO), *Forest Products: World Outlook Projections*, FAO Forestry Paper 84, Rome, 1988.

17. H. Gregersen, S. Draper, and D. Elz, eds., *People and Trees—The Role of Social Forestry in Sustainable Development*, Economic Development Institute of the World Bank, Washington, D.C., 1989.

18. G. Goodman, *Ambio* **16**, 111–119 (1987).

19. Food and Agriculture Organization of the United Nations (FAO), *Fuelwood Supplies in Developing Countries*, FAO, Rome, 1983.

20. M. Fergus, *Geographical J.* **142**, 29–38 (1983).

21. E. Mnzava, *Unasylva* **33**(131), 24–29 (1981).

22. P. Dewees, *World Dev.* **17**(8), 1159–1172 (1989).

23. C. Bailly and co-workers, *Bois For. Trop.* **197**, 23–43 (1982).

24. Food and Agriculture Organization of the United Nations (FAO), *The Forest Resources of the Temperate Zones*, Vol. 1, The UN-ECE/FAO 1990 Forest Resource Assessment, ECE/TIM/62, New York, 1992.

25. Food and Agriculture Organization of the United Nations (FAO), *Summary of the Final Report of the Forest Resources Assessment 1990 for the Tropical World*, (Report prepared for the United Nations Eleventh Session of the Committee on Forestry), Rome, 1993.

26. Personal communication, Dr. K. D. Singh, Project Coordinator, Forest Resources Assessment 1990 Project, FAO, Rome.

27. Food and Agriculture Organization of the United Nations (FAO), *Second Interim Report of the State of Tropical Forests*, FAO, Forest Resources Assessment 1990 Project, Hand-out

at 10th World Forestry Congress, Paris, Sept. 1991, Rome, 1991.

28. B. Freezailah, "The Evolving Role of International Organizations," in *Proceedings of Society of American Foresters 1992 National Convention*, Oct. 25–28, 1992, Richmond, Va., 1992, pp. 399–403.

29. J. Lanly, *Tropical Forest Resources*, Food and Agriculture Organization of the United Nations, FAO Forestry Paper 30, Rome, 1982.

30. J. Lanly, *Unasylva* **37**(147), 17–21 (1985).

31. A. Lugo, "Tropical Forest Management—Time to Do Something About It," in Ref. 28, pp. 382–391.

32. N. Myers, in R. Hoage, ed., *Animal Extinctions: What Everyone Should Know*, Smithsonian Institution Press, Washington, D.C., 1985.

33. J. Laarman and R. Sedjo, *Global Forests: Issues for Six Billion People*, McGraw-Hill, Inc., New York, 1992.

34. S. McGaughey and H. Gregersen, *Forest-Based Development in Latin America*, Inter-American Development Bank, Washington, D.C., 1983.

35. L. Brown, *The Woodlands Forum* **3**(2), 1–4 (1986).

FORESTRY, SUSTAINABLE

DAVID N. WEAR
USDA Forest Service
Research Triangle Park, North Carolina

Sustainability has become the primary criterion for evaluating human endeavors that involve natural resources and influence the environment. Defined in such a way, and given the ecological maxim that all things are connected, sustainability applies to nearly every human endeavor. Because forestry represents a globally important, perhaps the greatest, direct interface between human endeavor and natural process, its practice has an especially important influence on the function of ecological systems and the quality of the environment. Sustainable forestry is now seen by many as the ultimate objective for the practice and profession of forestry.

See also AGRICULTURE AND ENERGY; BIODIVERSITY MAINTENANCE; CARBON CYCLE; CARBON STORAGE IN FORESTS.

The word sustainable implies a stasis. That is, it suggests a process that can be continued into perpetuity. Beyond some consensus on the definition of sustainability at this broad level, that activities undertaken today should not compromise the opportunities left for tomorrow, there is very little agreement on how sustainability should be measured and implemented in the practice of forestry.

There are several reasons for disparate definitions. One relates to differences in basic values that define "land ethics" (1). For example, deep differences in land ethics may be rooted in the difference between modern rural and urban cultures. Clearly, not all people value the same quantities and qualities in forests and nature. However, a great deal of dissonance can perhaps be traced to contradictions in the root concept of sustainability. Stasis is perhaps not an especially useful concept for guiding the management of essentially dynamic ecological systems in a world of expanding populations and changing climate.

Consequently, practical guiding principles for forest management have yet to result from the discourse on sustainability. This, in fact, may be an unreasonable expectation. At this time it is perhaps most accurate to view sustainability as a platform upon which many public concerns regarding the practice of resource management and its consequences for ecosystem health are debated and, one may hope, eventually sorted out.

Resource management issues have, of course, been individually debated in several forums. By and large, however, the debates have been undertaken in the context of protest, advocacy, and litigation. In contrast, sustainability seems to be emerging as a focal point for discussions between researchers in several disciplines, policy makers, and advocates as well, on how to construct management objectives and approaches that reconcile the direct use of resources with ecological health and environmental quality (2–4). The task is enormous. Viewed in this way, defining sustainability emerges from the fray of resource management and environmental debates as an attempt at synthesis; as an attempt to recognize the tradeoffs and interactions between human endeavors and consumption and resulting environmental health. Defining sustainability therefore holds promise as a positive rather than a reactionary endeavor that has gathered considerable attention both in the United States and globally.

For sustainability to move from a platform for discussion and synthesis to a means to guiding resource management towards large-system goals, several challenges must be surmounted. These range from defining the structure and operation of forest-ecological systems to fashioning instruments and institutions that can effectively address resource and environmental goals at appropriate scales. Of course, these issues hinge on being able to settle on some notion of what it is that should be sustained.

FORESTRY AND CONCEPTS OF PRODUCTIVITY

Forestry throughout the world has historically involved some element of social crusade. Not surprisingly, these social concerns have typically related to protecting and enhancing the productivity of forest lands. Modern concerns, especially those brought under the discussion of sustainability, can be easily seen as extensions of historical movements to conserve or protect natural resources.

In the United States (of which the forestry history serves as the vehicle for discussing the evolution of resource management in this article), forestry started in the late 1800s as a public response to the private exploitation of resources. The rhetoric of the times saw severe timber shortages resulting from the expansive harvesting that moved from region to region in the U.S. The expanding country would be crippled by a shortage of its most important raw material.

Conservation, broadly defined as wise use of renewable resources, was the public sector's response to rapidly increasing timber consumption and deforestation at the start of industrial expansion in the United States. Between 1850 and 1909, per capita consumption of wood products increased four-fold (5). This expansion derived from vast demands for wood to construct the railroads, to build cities and ships, and to smelt iron (6).

The strong demand for wood products was fed by a lumber industry in the United States that moved from the Northeast to the lake states and eventually to the Pacific Northwest and the South. The scale of forest clearing and lumber production was enormous and impressive. However, it was not the scale of production alone that suggested impending timber famine. It was also the prevalence of fraud and theft in the procurement of timber from public lands that fueled the perception of resource use gone out of control (7).

The setting in which forestry developed in the U.S. strongly parallels conditions now motivating conservation activities in developing parts of the world. In both cases, rapidly increasing populations coupled with uncertain frontier land ownership have resulted in rapid deforestation. Typically, other pressures for land use or improved transportation have resulted in forests becoming more accessible and therefore more marketable. Also, the demand for agricultural land has typically fueled deforestation. Both then and now, public intervention to control forest exploitation has emerged as the solution to resource degradation.

In the late nineteenth century, the private timber industry was seen as failing to account for the long-term economic interests of the developing U.S. The response was a conservation movement based on wise use and reforestation. As a result of these efforts, a system of, first, federal forest reserves and then, national forests was established. Presidential authority to establish forest reserves from the public domain was set out by the Creative Act of 1891. The national forests were established with the Organic Act of 1897 and the Weeks Law of 1911 allowed for the purchase of land and the establishment of national forests in the eastern United States. At roughly the same time, forestry attained credibility as a profession and the first forestry schools and the two principal professional forestry associations were established in the United States (8).

The young profession of U.S. forestry sought to bring the forests of the country under a scientific model of management, and its earliest textbooks reflect this emphasis on forest management as an engineering exercise. At root, however, forestry was based on a social vision. Indicative of this view, the Society of American Foresters in 1917 defined forestry as "the science and art of managing forests in *continuity* for forest purposes" (9). This and other textbooks of the day clearly establish perpetual production as the goal of forestry practice. In the same text, forest management is described as being motivated by ". . . the ultimate aim of securing a sustained yield" (10).

Sustained yield, perhaps a primitive form of sustainability, has provided the motivating concept as well as the guiding ideology for forestry throughout its history. Its value in the latter role has only recently been challenged. The context of the rapid deforestation experienced in the 1890s and the early part of the twentieth century, easily justified a public forestry that sought maximum output to ward off national timber shortages. That is, the government intervened in private resource use on the grounds that private production could not provide for the future (11).

Sustained yield provided, then, the initial impetus for

the development of forestry as a profession in the U.S. Social movements in general are most effectively motivated by policy capsules that summarize the good qualities of highly complex social phenomena (12). These encapsulations often take on a life of their own, forming the ideology of a profession. As such, they are discarded only with great trepidation and debate.

The shortcomings of a forest policy and practice guided solely by the goal of maximum perpetual yields of timber were evident early on. They related to two arguments. One was that many goods and services in addition to timber were valued outputs from forests. These included water and forage as well as recreation and wildlife habitats. The other was that society was not necessarily served best by producing more timber at any cost. Because it focused on a single measure of production, the sustained yield model of forest management could not accommodate value tradeoffs.

The history of policy and legislation directed at the management of national forests can be seen as gradually expanding the spectrum of forest products and services. The progression of events that expanded the agenda of public forestry from one focused on timber and watershed protection commenced early in the century and led up to the Multiple Use Sustained Yield Act (MUSYA) of 1960 (13). The MUSYA clearly defines a broad agenda for forestry in the United States: "It is the policy of the Congress that the national forests are established and shall be administered for outdoor recreation, range, timber, watershed, and wildlife and fish purposes."

While the MUSYA came only in 1960, the concerns for the nontimber benefits derived from forests were clear from the start. The enabling legislation for the National Forests emphasized the protection of waterways in addition to sustained timber production. Wildlife, recreation and wilderness considerations developed in the U.S. Forest Service beginning in the 1920s (13).

Public forests have been seen as progressively more than just trees. Since 1960, a progression of forest management and environmental protection laws have codified the multiple concerns and procedures for managing the government's forest assets. Public lands do not, however, constitute the entirety of forests in the United States. It is important to remember that, from the Multiple-Use Sustained-Yield Act on, public and private visions of forestry have essentially diverged. This is not necessarily a distinct dichotomy. Some public holdings, for example State trust lands in the western states, are managed for maximum long-run profit, though constrained to a sustained yield style of management. In addition, environmental and species protection laws clearly have important influence on private land management.

As the complement of objectives has expanded, sustained yield has become a broader and more ambiguous encapsulation of the goals of private and public forestry. Because it encapsulates so much, it can easily be interpreted in many different ways. The private sector may view sustained yield as the physical timber production model emphasized by the forestry profession in its early years. The public sector is perhaps more likely to view it as a part of multiple-use sustained yield as defined by MUSY. It is quite possible and perhaps likely that debates over what constitutes the difference between the tradition of sustained yield and the innovation of sustainability is based on these different, though rarely discussed, definitions of the root concept.

Multiple Use did not, however, supplant sustained yield as the guiding principle for public forestry. Rather as the name Multiple-Use Sustained-Yield indicates, the concept was simply extended to the full range of uses listed in the Act. While this may appear egalitarian, a simple extension to all relevant resources is simply not possible. This is because once more than one resource service is addressed, it becomes impossible to unambiguously maximize their production. That is, as long as the multiple resources are not perfectly complementary (and they clearly are not), producing more of one often implies producing less of another. Multiple use therefore raises the issue of tradeoffs.

However, the MUSY did not explicitly define how tradeoffs should be resolved. At the time, it can be argued that the greater availability of forests (especially on a per capita basis) limited the extent of conflicts over the use of public forests. That is, the tradeoffs were not, initially, severe because there was more than enough room to accommodate most uses. However, the most modern phase of public forestry is marked by the process of sorting out tradeoffs that have become progressively more severe. Driven primarily by the National Forest Management Act of 1976, National Forest managers sought to define resource tradeoffs on public lands through the 1980s. The platform for this exercise was a massive planning exercise guided by the NFMA with public involvement and impact analysis guided by the National Environmental Policy Act of 1969 (NEPA).

Sorting out tradeoffs between dissimilar goods requires boiling the arguments down to a common currency. The analyses undertaken in the 1980s were expansive and involved scientific management tools in the form of linear programming models and cost benefit analysis, as well as massive efforts to collect and utilize public opinion to define the relative values of resource uses. The result was a complex public forest management process that matched an equally complex set of resource management concerns.

The outcome of public forest planning and management in the United States, rather than improving consensus over appropriate forest usage, fomented an intense level of debate. Hundreds of administrative appeals and law suits resulted. Some were not satisfied with the outcome of the process; their relative values did not reflect the perceived majority view. Others took issue with the structure of the process, that is, the casting of the problem as a scientific exercise left out important elements that could not be modeled. It is also likely that the practice of placing dollar values on all resources, for example wilderness preservation, offended many at philosophical levels. Another argument directly relevant to the discussion of sustainability is that the process focused on many parts of resource management issues but failed to address the integrity of overarching ecological systems.

In sum, public forestry in the United States has become an increasingly complex and an increasingly divisive issue. As forests and their services become more scarce the issues become increasingly difficult. Not surprisingly, public forestry is where much of the discussions of sustainable forestry has focused. Sustainable forestry can be viewed as an outgrowth of the public discourse over national forests that started in the 1970s, reached a crescendo in the 1980s, and continues to the present.

MODERN ORIGINS OF SUSTAINABILITY

Concerns for sustainability are, however, much broader than forest management issues. Rather, "sustainability" or "sustainable development" has become an important criterion for judging all human activities that impinge on ecological systems. The modern focus on sustainability is easily traced to the Brundtland Commission Report (World Commission on Environment and Development 1987), which emphasizes that environmental health is and will continue to be a necessary condition for economic prosperity. It also makes an explicit link between poverty and disparate distributions of wealth and the state of this ecological-economic system. Gro Harlem Brundtland, the commission's chair and the report's namesake, views "alleviating poverty (as) priority number one . . ." in successfully achieving sustainability (14).

Sustainability is therefore an extremely broad concept, encapsulating ecological, physical, economic, and social aspects of resource use and management. It can indeed be argued that the concept is too broad to be productive, especially if it spawns endless debates on defining what sustainability means. Perhaps, however, debate over what is and who affects sustainability is productive in itself. If the linkages between ecological health, economic prosperity and the human condition are critical, then discourse between ecologists, economists, anthropologists, policy makers, etc, is a necessary first step in formulating these ideas and then organizing human activities towards desired ends. Sustainability has begun to stimulate discussions between disparate disciplines and spawn interdisciplinary investigations (2,4).

While sustainability may not enjoy ubiquitous definition, it does command near-ubiquitous attention. For example, sustainable development was the general theme of the 1992 Earth Summit. It encapsulates a number of concepts regarding the connection between humankind and its environment and serves as a focal point in a world of disparate ideas, models, and policies. Perhaps the concepts of sustainable development can focus disparate scientific disciplines on these clearly important issues in the way that George Marshall seemed to organize disparate viewpoints to address the complex problems of post-war Europe. It can be argued that, in both cases, simplified statements catalyzed action where careful analysis of complicated issues could (and can) not. It may therefore be more useful to view sustainability as useful for organizing ideas about the impacts of human activity and growth on the functioning of the biosphere, than as any operable set of physical standards or guidelines at this point.

If sustainability is about the interactions between resource use, environmental health, and the human condition, then forestry is clearly a complete microcosm of the sustainability debate. Forestry is the most land-extensive human endeavor so that forestry practice may have enormous influence on the function of ecological systems. For example, commercial tree planting can replace old and mixed species forests with vast monocultures of a single species and, in the process, completely alter the matrix of sub-canopy flora and fauna as well. Impacts can range from small local to large regional scales.

Several other elements of sustainable development are also found in the context of forestry issues. The time required to regrow forests to commercial or old-growth conditions may be an impediment to regenerating land and has been raised as a point of concern for centuries. That is, popular sentiment has held that the very nature of the exercise of growing trees precludes investment in sustainable forestry. Accordingly forestry decisions are often seen as favoring the present over future generations. Intergenerational equity is at the crux of sustainability issues (15).

The pattern of forest allocations within a particular generation also define a set of issues within forestry. As a primary resource, wood production has a larger than average variation. Resulting boom and bust patterns of resource use typically lead to expanding and contracting economies and considerable dislocation in rural areas. Public forest policy is often motivated by a desire to stabilize the rural communities built upon wood products (16). In this sense forest policy is often concerned with equity and distribution of income in poorer rural areas (17).

Another area where forests (but not necessarily forestry) are central to sustainable development is forest clearing and land conversions. Clearing aboriginal forests is often the first step in transforming a place dominated by nature and ecological structure to one that is dominated by humankind and its cultural structures. While, in an increasingly global and crowded community, it may be counter productive to separate humankind from "natural" systems (2), forest harvesting remains a symbol of man's conquest of nature.

As with most symbolic meaning, the issues are actually much more subtle. Not all forest harvesting is an act of deforestation. Rather, it may also be the first step in reforestation, either active tree planting or a more passive natural regeneration. Furthermore, the practice of forestry (forest growing as well as harvesting) may in fact be the only practical alternative to less desirable alternatives such as an erosive agricultural use. In comparison, the active practice of forestry may prevent soil erosion and other ancillary problems found in the agricultural uses of marginal lands. In such a case, forestry plays a positive role in sustaining soil and land productivity.

Forestry seems to capture or is central to several elements of the broader sustainable development debate. It defines a direct interaction between human consumption and ecological function and directly involves intergenerational tradeoffs. Sustainable forestry is, accordingly, a complex issue, effectively encapsulating ecological, intergenerational, equity, and productivity concerns. Structuring these multiple issues into coherent resource management strategies and policies remains the essential challenge in developing an operational definition of sustainable forestry.

ESTABLISHING SUSTAINABILITY GOALS

The first step in establishing a practice of sustainable forestry is defining a set of objectives. Concerns for environmental quality as well as human welfare constantly raise questions about tradeoffs. It is clear that sustainability is not an absolute concept. That is, there appears to be no single "ecologically correct" approach to managing ecosystems. Rather, and especially due to an expanding human population, priorities and objectives must be carefully articulated. Without objectives, sustainable policy and management decisions are essentially arbitrary and their effectiveness cannot be measured. However, while crucial, defining these goals is far from straightforward.

There are many different perspectives in the discussion of sustainable forestry. For example, there is the obvious split between an industrial or agricultural view and the natural system view of forests. In policy discussions, there is also a parallel dichotomy between how the social sciences and the natural sciences view sustainability issues. Both perspectives recognize that ecological health is important and that sustainable forestry requires a long-term view. Differences arise over how to view human use of ecological assets, the role of ecosystem health in human welfare, and the emphasis placed on human endeavors in global systems.

The differences between economic and ecological perspectives define the salient features of defining sustainability goals. While generally aimed at the same issues it is important to recognize that the tension between economic and ecological perspectives may often result from confusing the ends and means of sustainable forestry.

The social sciences, quite naturally, put the human condition at the center of the discourse over resource use and ecological health. While anthropology, economics, and sociology may measure welfare in decidedly different terms, ultimately they measure the health of ecosystems in terms of observed and expected effects on human welfare. In addition, the social sciences place great emphasis on the ability of humanity to adapt to changing conditions and to improve its lot in the face of impending resource scarcity.

Economics in particular views the human community as an adaptive organism and actively engages human progress and adaptability as important elements in global systems. Notably, the twin engines of substitution and technology have guided humanity away from resource scarcity and has, at least through the modern age, vastly increased the productivity of the earth's reources in terms of their consumable products. Assuming that degraded ecological systems define a kind of scarcity that can be signaled through resource markets, these signals might similarly motivate innovations that would mitigate ecological damage and improve the human condition. These signals take the form of changes in price and the correct signal is a price that reflects the true cost of a resource, including the costs of ecological damages.

At the root of this line of argument is the assumption that shifting ecological conditions can be recognized and signaled in the process of economic enterprise. There are several reasons why this may be a weak assumption. One is simply that many environmental qualities are not traded in markets. Examples include classical externalities of production, such as air and water pollution. Another reason is that a considerable lag may occur between human actions and ecological consequences. This is often the case with species extinction arising from habitat destruction, where populations may fragment and gradually decline before being recognized as threatened. In such cases, impacts are cumulative, with effects being felt only several years or decades after initial causes.

The argument that technological innovation and substitution will always alleviate problems is a somewhat polar case. Economics, in the main, recognizes several factors that may justify a society's intervention in the free operation of markets. All of these relate to the welfare of citizens and include the fair pricing of goods, protection from crime, and providing a quality environment. In fact, recent government initiatives in the United States have created markets for environmental goods such as air quality. By placing these goods in the care of markets (i.e., internalizing them), these policies attempt to allow scarcity to be recognized by producers and consumers and thereby be alleviated by substitution and technological innovation.

There are perhaps more fundamental objections to the hope that economic man's present and future ingenuity can alleviate ecological problems. One is that the world of the future will be a new and wholly unprecedented condition due mainly to expanding human populations (18). Simply put, the problems of the world are becoming progressively larger and they may not be effectively anticipated by the marketplace. Another argument is that human ingenuity necessarily approaches a point of diminishing returns. That is, while the processing of materials through industry and agriculture has become more and more efficient, all processes are eventually bounded by the laws of thermodynamics (19). Of course, it is not clear just how close the laws of thermodynamics are to becoming binding constraints on human enterprize.

It may also be argued that ecological health is too broad and intricate to be understood by the firms and consumers who interact in markets. That is, the integrity, quality, and interactions of complex ecosystems is what matters, not their mere abundance. If this is the case, then it follows that market mechanisms that have historically recognized scarcity and averted resource famine are simply not designed to respond to changes to ecological systems. Furthermore, and at a most fundamental level, ecological health may therefore depend on a concept that is incongruous with the competitive market place. It may depend on the orchestration of resource management among different and otherwise competing landowners and firms.

This notion of an essential complexity that cannot be recognized or protected by a collection of freely-acting economic agents, is a central element of the natural science perspective on sustainability. This perspective places the condition of ecological systems rather than human condition at the center of the discourse on sustainability. This focus leads to rules for managing ecosystems in ways that preserve their natural complexity and information, generally without consideration of tradeoffs or cost. Typically, the perspective defines desired ecosystem conditions without explicit consideration of the role of humankind in ecosystems (2).

Of course how to measure ecosystem health remains an unanswered question. Because these systems represent complex and dynamic assemblages of organisms and processes, defining essential linkages is at present impossible. In light of this uncertainty, various rules for managing ecosystems have been proposed. For example, Franklin (20) suggests that two principles guide ecosystem management. One reflects a concern for the essential genetic information contained in an ecosystem: "prevent . . . the degradation of the productive capacity of our lands and waters—no net loss of productivity." The other reflects concerns for process: "prevent . . . the loss of genetic diversity, including species—no loss of genetic potential" (20).

Concern for preserving biological information or biodiversity is an ubiquitous element of contemporary conservation. The modern rate of species loss is estimated to be several times higher than any recorded by history or fossils. The cause of species loss is easily traced to the impacts of development and other human activities which, as a rule, reduce the complexity of ecosystems.

The erosion of biodiversity may have severe consequences. One consequence is related to the loss of biological information (in the form of survival strategies encoded in DNA) that has evolved over the millennia. Extinction erases the biochemical recipes of species which may have human benefit in future application.

However, more severe consequences may relate, not to the specific information encoded within any particular species, but to the interactions among different species. Organisms do not interact with their physical environment in isolation. Rather organisms are connected, for example, in the transfer of energy from one trophic level to another, or as vectors in the reproduction of other organisms. Commensal as well as parasitic relationships define the overall organization of an ecosystem and its cycling of materials.

These interactions and a complex coevolution of species suggests that, while species have a range of adaptability, their survival and persistence are inextricably tied to other species. The biological parable for this type of dependence is the interacting population dynamics of predator and prey, for example, between foxes and rabbits. Fluctuations in predator or prey populations are constantly moderated by starvation or procreation. Loss of the predator or the prey eliminates a critical regulatory process, thereby severely affecting the remaining species and its habitat. However, many important ecological interactions are decidedly more subtle, involving multiple interactions between and among flora and fauna. Damage to the specific complexes of interacting species that defines habitat is the most common cause of extinction. Productive intervention in the demise of endangered species is, accordingly, very difficult.

Followed to conclusion, this line of argument also suggests that loss of a species may eventually lead to collapse of larger systems of interacting species. System disfunction clearly depends, however, on the amplitude of other species adaptability. The human population, while capable of massive system intervention and modification is, in the end, tied to other organisms as a part of ecological systems. Accordingly, species persistence may have direct bearing on human prospects.

Because of these important relationships, biodiversity has been proposed as the key index for measuring the sustainability of human endeavors. Concern for extinction in particular, has become a critical element in checking development in the U.S. as well as in other countries. The U.S. Endangered Species Act of 1973 lays out strict regulations for the protection of species that become threatened by development or other causes. In fact, some argue that the Endangered Species Act places a nearly absolute value on the preservation of individual species once they are declared endangered. Evidence for this valuation can be found in recent actions taken to protect several species, including the Northern Spotted Owl in the Pacific Northwest. The human cost of protecting owl habitat on commercial timber lands has been very high.

However, the process of monitoring and protecting individual species that are on the brink of extinction may have only limited effectiveness. Because there is often a lag between the loss of important habitat and eventual species endangerment, the ESA may signal a problem only after it is insurmountable (21). Furthermore, a general demise of an ecosystem's function and health, for example, severe forest decline observed in the Blue Mountains of Oregon, may not be signaled by an endangered species. In such cases, the consequences of changes may be very severe at a local scale while not immediately threatening the persistence of species at a larger scale.

In addition, species persistence can only be monitored for those species that have been identified. While a large share of the vertebrates and vascular plants have been cataloged, these groups make up a very small share of the world's biota, and the proportion of species actually cataloged is placed somewhere between 2 and 15 percent (22). Accordingly, the extinction of many species may simply go unnoticed.

An index of biodiversity, based on the number of species that occur in an area, can be an essentially arbitrary and even misleading indicator of ecological function when temporal species dynamics are considered (23). That is, species may flourish or decline depending on the stage of a forest's development. Furthermore, common biodiversity indices are not weighted to place emphasis on critical or scarce species. Instead all species are treated equally. At present the information required to understand the role of species and therein identify critical species is not known.

One way of summarizing the differences between ecological and economic perspectives on sustainability then is the relative emphasis placed on ecological health and human welfare. Economics views sustainable development in general as an issue of intergenerational fairness. That is, it is a matter of arranging natural resource use and investment so that future living standards are protected (24). If markets work well and therefore prices reflect the true quality and costs of the resources we use, then sustainability can readily be achieved in a market economy. Substitutions and technical progress will steer humanity away from resource famine, and investment will transfer the necessary wealth to posterity. Of course, markets don't always work well and the major role of sustainability policy is to correct for these problems and to adjust prices so that they reflect true scarcities.

In countries where markets are not the primary mechanism for resource allocation or where markets are not mature, the issues are more basic. In such cases, sustainability policy would still aim to get prices right, but creating and stabilizing the market foundations must necessarily come first. Chief among these foundations is secure ownership of property, including land. Without a reasonable assurance of long-term tenure, there can be no incentive for long-term resource management. As a result the dual effects of over-exploitation and under-investment short change future generations. Regardless of the setting, the economic version of sustainability goals is establishing the setting in which well-informed decisions about resource use can be made.

The ecological perspective can be summarized as a concern for the maintenance of ecological systems. Its tacit assumptions are that human progress has steadily eroded the diversity of life and the integrity of ecosystem processes and that the implication of this erosion is a potentially catastrophic eventual shift in the structure of ecosystems at local and global scales. While admittedly uncertain, the implication of a rapidly unravelling system of fundamental energy transfer and resource production is a greatly diminished human prospect.

The ecological perspective on sustainability takes issue with the economic perspective at two basic levels. One is that the erosion of biological diversity (biological capital) has progressed at an increasingly rapid rate over the last century and that markets have not recognized the phenomenon, mainly because there is no linear connection between this scarcity and human welfare. Rather, the connection is subtle and ultimately cumulative. If the effects of an erosion of biodiversity are irreversible, and felt only after several decades, then we cannot rely on markets to account for these values.

The economist might reply that the market mechanism should not be abandoned because of incomplete information. Instead, resource prices should be adjusted to reflect these important values. However, it might be countered that, in fact, the market mechanism is fundamentally flawed for protecting ecological functions. Due to the spatial nature of important ecological processes, individual landowners and resource managers cannot effect sustainable forest ecosystems while pursuing independent ends. As discussed above, this suggests that ecosystem maintenance in the presence of resource management may require an orchestration of management efforts across owners.

At the same time, the ecological perspective is generally not set within a social context. Protecting ecosystem health and ecological functions is only one of several important social goals and cannot be accomplished without cost. Without comparing the costs of achieving these goals with the derived benefits as well as with other social objectives, calls to protect ecosystems imply either arbitrary valuation or an absolute value for ecosystem health. One implication is that ecosystems hold intrinsic value that overwhelms all other social concerns.

However, the outcome of undisclosed value comparisons does not define the motivations behind them. There are at least two ways to interpret values in this context. One is what Toman (15) calls "ecological centrisism," where nature is accorded values and rights. The other is that an undisclosed tradeoff analysis places high priority on ecological functions. It may be that the consequences of ecological failure are so severe that they overwhelm all foreseeable costs of ecological protection. This latter perspective might be labeled ecological utilitarianism.

Are there ways to reconcile these disparate views on the need for and the structure of sustainability goals? Perhaps there is, if one takes the perspective of ecological utilitarianism, rather than ecological centrisism. With the latter, the prospects for consensus are limited simply because a conflict with absolute values cannot be resolved. However, if the perspective is utilitarian, then differences in opinions regarding sustainability goals are the result of different perspectives on physical outcomes and risk. These differences can be ameliorated over time with better information on outcomes. That is, science can play a strong role in the latter situation, but not the former.

Taking this perspective allows us to reconcile these perspectives in a hierarchial approach to defining sustainability goals. At one level, the goals of ecological sustainability need to be prioritized with other important social goals such as improving education, fighting crime, and reducing poverty. At another level, science needs to define the connections between "states of nature" and different levels of social welfare. That is, science can inform the process of social goal setting with information on feasible states of nature and their anticipated outcomes. Clearly, this area of knowledge is fraught with uncertainty, so that relationships between nature and welfare need to be couched in the language of risk analysis. Communicating the degree of uncertainty is as important as communicating the state of knowledge.

The process of sorting out sustainability goals in a social context begins with defining the relative values social systems place on natural systems as well as an understanding (albeit necessarily limited) of the connections between natural productivity and human prosperity. This stage of the process seeks an understanding of how much society is willing to pay for ecological health, in direct terms of forgone consumption and in terms of transfers of natural wealth to future generations.

Ultimately this stage of the process can only be conducted as a political exercise where the discourse is played out in the context of an economy and other social concerns. Clearly the context is crucial and the ranking of ecological goals will necessarily differ between countries. For example, France, Pakistan, Ukraine, and Brazil will likely place different relative values on alleviating poverty, improving transportation infrastructure, and ecological sustainability.

EFFECTING SUSTAINABLE FORESTRY

The previous section focused on formulating and articulating sustainable forestry goals. This section focuses on the subsequent issue of designing plausible approaches to moving the practice of forestry towards these goals. The following discusses three aspects of implementing action to achieve sustainability. One is the collection of physical management strategies that might be applied. These necessarily depend on the scale of application and on site specific features of the forest. Another aspect is the influence of human institutions on sustainability. The practicality of any management strategy depends crucially on the institutional setting within which it is applied. For example, the institutions that define land ownership and taxation can strongly influence resource management decisions. The final aspect of implementing sustainable forestry is a program of research that addresses ecological uncertainty and monitors the effectiveness of different forest management strategies.

Management Approaches

Regardless of their form, goals for sustainable forestry can only be implemented through modifications in the practice of forestry. Forest management will need to address two kinds of questions. One is where will forestry be practiced and, conversely, where will natural processes be left to operate without significant human intervention. The other is how will forestry (the practice of harvesting and regenerating trees) be practiced.

Allocating forest lands to various management emphases has long been a part of public forestry in the United States. While, historically, public forestry has emphasized producing the largest number of multiple use services from each forest stand, it is clear that not all of these services are complimentary. Accordingly, in most cases multiple use can only be accomplished by zoning forest lands to different uses (25). For example, different areas may be zoned to emphasize timber, forage, wildlife, or recreation values.

Another important forest land allocation mechanism in the United States is wilderness designation. Since passage of the Wilderness Act in 1973 (and the subsequent Eastern Wilderness Act of 1975), the setting aside of large blocks of large, essentially undisturbed, natural areas has been a major focus of conservation efforts. The Acts set aside several wilderness areas but also set out a mechanism for adding to this initial constellation. The process requires Congressional action to designate an area as wilderness and fully engages public debate over land use and wilderness values.

While the processes of forest zoning on National Forests and Wilderness allocation have long histories, the process has been motivated by a set of goals that are distinct from ecological sustainability. As discussed above, there are important dissimilarities between a multiple-use agenda and one that focuses on biodiversity and ecological processes. In addition, wilderness designation has been motivated not by relative ecologic merits but primarily by the lack of human disturbance or presence and the potential for wilderness recreation. Accordingly, the present system of wilderness areas are generally remote and mountainous. These physical barriers to access are, in fact, what prevented their development.

The present collection of preserved areas in the U.S.,

wilderness areas along with national parks and wildlife preserves, does not fully represent the spectrum of ecological conditions. Scarce and therefore important ecological conditions are significantly missing from this network of preserves. These and other problems of designing reserve systems to effectively protect diversity and ecological function have been the focus of the relatively new field of conservation biology (26).

Among these concerns is how to define the minimum size of an effective reserve. In effect, there are certain economies of scale associated with ecosystems. Species persistence relates not only to the quantity of available habitat but to a complement of different habitats and their interconnections. In addition, the shape of the reserve is critical because of the phenomenon known as edge effect (27). At the interface between a forest and another kind of land cover, for example, pasture, environmental conditions are different from conditions in the interior of the forest, and therefore support a different complement of species. Therefore, the shape and size of the reserve interact to define the relative extent of edge and interior habitat conditions. Relatively large reserve areas may be shaped in ways that cause disturbance-related edge habitat to dominate.

Edge effects and the importance of connections between various habitats suggests that the condition of lands immediately adjacent to reserves can play a crucial role in their function. Because the condition of adjacent areas defines edge effects as well as potential barriers or corridors, for example, for seed dispersal and animal migration, their management is a critical element in defining the effectiveness of reserves. Coordinating management on adjacent lands is a challenge especially in a multiple owner landscape.

The phenomenon of spatial juxtaposition raises a complex set of issues for sustainable forestry. Ecological functions are rarely determined by conditions found within a small area. Rather they are connected across the landscape among areas. One example of these spatial connections is a large mammal that requires foraging habitat as well as hiding and resting cover. Another is the gravitational definition of water courses that connects upstream activities to downstream impacts.

These spatial connections define an "economy of configuration" (28) in addition to an economy of scale in the function of ecological reserves. That is, the ability of reserves to protect biodiversity or ecological function depends crucially on where they are located in absolute terms and in reference to other landscape conditions. In addition, these conditions are dynamic—forests may be harvested and then regrown—so that controlling for ecological function may become very complex.

One element of the problem is simply defining where high-valued (ie, scarce) ecological conditions exist. One way of prioritizing areas is to order ecosystems by their species diversity. An example of an ecosystem type with an exceptionally high diversity of species is wetlands, especially riparian corridors. These zones of gradation between water courses and upland habitats are extremely dynamic and diverse (29). They define portions of the landscape that, because of their flood-related disturbance regimes, natural variability, and connections across land forms, serve as species reservoirs in their respective ecosystems. Accordingly they may deserve special management emphasis.

At a larger scale, areas of ecological importance can be identified by mapping the distribution of endangered species. Areas that have relatively large concentrations of threatened and endangered species may indicate where ecological functions are imperilled. Accordingly, the ecological returns to preserves may be highest in these areas.

In general, however, ecology has yet to provide a comprehensive scheme for classifying ecosystems at any scale (30). Instead the focus of classification and systematics has been on the identification and classification of individual species. Without a classification of ecosystem types, however, it is difficult to judge indeed where ecological conditions are scarce or threatened.

While one aspect of a physical strategy for sustainable forestry focuses on where to undertake forestry, the other focuses on how to manage forests. The application of silviculture ultimately defines the condition of forests under management. Because managed forests represent the largest share of forested lands, at least in the temperate zones, the practice of silviculture plays a critical role in defining the ecological condition and biodiversity provided by the entire landscape.

Managed forest lands differ from unmanaged lands in several ways. In general, however, forest management tends to simplify the spatial distribution and the internal structure of forest stands (31). This reduction in complexity results from "streamlining" stands to efficiently produce high yields of commercial timber products.

At temperate latitudes, the typical transition to a managed forest involves harvesting a mixed-species and mixed-age forest and replacing it with a single, fast growing species. This species is generally adapted to early seral conditions and quickly occupies the site and grows to maturity, in a financial rather than a biological sense. Timber harvesting is then associated with a regeneration method that establishes a new forest. The regeneration method is the crucial activity defining the structure of the subsequent forest.

Timber harvesting combined with regeneration methods defines what Smith (32) calls a silvicultural system. These systems lead to either an even-aged or an uneven-aged forest stand. Even aged stands can be established by cutting the entire stand (clearcutting) and then either planting seedlings or spreading seed across the site. Two other approaches allow for natural regeneration in connection with a harvest. One leaves a few scattered trees to naturally seed the site (seed-tree methods). The other removes trees in two or three stages so that natural regeneration can take place under the shelter of partial forest cover (shelterwood methods). An uneven-aged forest can be maintained by constantly removing a small proportion of the forest in a way that promotes regeneration (33). Another type of forest stand is generated by vegetative reproduction or coppice regeneration. These forests are regrown from root stock and are generally used to produce fuelwood or other forms of biomass fuel stocks.

The choice of silvicultural methods, therefore, has direct implications for the age and species composition of forest stands. Even-aged approaches yield forest stands of the same age and species. However, and depending on the growth period and the amount of effort expended on controlling vegetation, an understory of competing tree species will also emerge. These are the natural successors of the early seral species emphasized by even-aged management. In contrast, uneven-aged management through selection silvicultural systems, generally yields a highly diverse stand in terms of both species and age composition.

The majority of forest management utilizes even-aged silvilculture. The result in some regions is a great reduction in the species composition of managed forests. Furthermore, the genetic diversity of the regenerated species may often be quite limited. This is because tree nurseries and seed orchards breed trees to emphasize certain productivities and disease resistance. Quite often this can result in a small number of superior trees parenting vast areas of forest. As a result, the adaptive amplitude of a commercial species can be greatly reduced from its natural range.

Because commercial forests are harvested much earlier than natural disturbance would ordinarily replace forests, the average age of managed forests is much lower than typically found in a natural setting. This obviously skews habitat conditions to favor species complexes in early seral stages and away from later seral stages. The endangerment of the Red-Cockaded woodpecker and Northern spotted Owl for example, reflect substantial reductions of old-forest habitats in the southern and Pacific Northwestern parts of the United States, respectively. Preserving what remains of old-growth conditions has been a focus of efforts to preserve regional biodiversity.

Active forest management is not, however, the only major vector of change in forested systems. In addition to forestry, other uses of land may have both direct and indirect effects. For example, in some rural areas—notably the Southern Appalachians—low density residential development continues to spread from central urban areas and across large areas of forest. Agricultural demands for land have also played an historically important role in modifying the landscape and clearing forests for crop and pasture uses continues to be behind the majority of global deforestation activities.

These activities reduce habitat by removing forests. However, they also have indirect impacts on remaining forests. One effect is that the remaining forest patches are smaller and increasingly isolated from one another. In addition, the interspersion of land uses also increases the occurrence of edge habitat within forest patches. This may completely change the character of remaining forests in a developed area. Accordingly, measuring the amount of forest disturbance may greatly understate the area of forest that is actually impacted by development.

Another vector of change in forest systems is human effects on natural disturbance regimes. The most effective example of this is the campaign against forest fires waged over the last century. This campaign has been enormously successful and has greatly reduced the area of forest burned by natural and human-caused fires (34). Protecting large areas of forests in the temperate zones from fire has also resulted in eliminating the chief natural agent of material cycling in forests. As a result, the composition of forests has shifted away from fire adapted species. In addition, some larger-scale effects have also occurred. Among these are an increased severity and extent of insect epidemics in the U.S. (eg, southern pine beetles and mountain pine beetles, gypsy moth, etc). Also, because fuel loads in forest stands are no longer regulated by frequent and low intensity natural fires, fuel builds up in forests over time and eventually promotes large, high intensity fires (35). The fires in and around Yellowstone National Park during the summer of 1989 is one example of this meta-phenomenon.

All of these factors change and, in one way or another, decrease the complexity of a forested landscape. Accordingly, the choice of silvicultural systems and forest management strategies can have a great impact on the ecological function and biodiversity of forests. There are ways to balance the production of valuable services and products from forests while maintaining more of their ecological integrity.

A critical factor in determining ecosystem level sustainability is not just the quantity of certain kinds of habitat but their spatial juxtaposition. While not a traditional focus of forest management, spatial arrangement of harvest units is increasingly emphasized in public forestry (36). Careful selection of harvest patches through time allows important habitat connections to be maintained while providing a continuous flow of timber products (37,38). In addition, the size of cutting units is variable and can be adjusted to an appropriate scale for a given ecosystem type and set of objectives. For example, smaller units spread disturbance more widely across a landscape. However, depending on the intensity of harvesting, smaller cutting units may also lead to smaller forest patches, greatly increasing edge effects.

Streams and riparian corridors define especially important connectors in a landscape (29), and the effects of forestry activities, such as sedimentation, water flow, and debris in streams, may have critical impacts on watershed integrity (29). Because of their central role in ecological health, riparian corridors and wetlands have been a focus of regulation. Some argue that maintaining ecosystem health will require even more careful management of these critical landscape elements.

The techniques of harvesting and regeneration (ie, silvicultural systems) within cutting units may have the greatest impact on the species and age structure of forest patches. The key to managing for these conditions depends on understanding the dynamic processes of regeneration, growth, and species succession particular to individual ecosystem types. New silvicultural systems may be designed to mimic natural disturbance regimes (39) thereby maintaining more ecological structure and biodiversity.

Shifting from even-aged to uneven-aged silvicultural systems is a direct means of increasing species diversity within a forest patch. However, this less intensive harvest approach requires accessing forest stands much more frequently and accordingly, increasing a different kind of human intervention. The more frequent presence of people and machinery may render areas ineffective as habitat for certain animal species, especially mammals. In effect, while less intensive, uneven-aged management spreads annual disturbance regimes out across a much larger area.

Other choices in the design of forest management strategies may also hold crucial influence over the structure and function of large forested landscapes. For example, techniques used to prepare a site for regeneration will have a critical influence on the nutrients left on a site following harvest. Rather than clear harvested sites—often an aesthetically desirable approach—more stems and other waste can be left to decompose on the site. Decisions regarding site preparation need to be balanced against resulting fuel loads and fire hazards as well, but they afford a means to increasing forest material cycling and could play a role in protecting and extending soil productivity, a critical element of sustainable forestry.

Certain engineering aspects of forest management plans may also have a critical role in defining the composition and health of regenerated forests. These include decisions regarding the type and the location of forest roads. Road design is perhaps the most important variable in determining stream sedimentation (40) and roads can serve as corridors for the spread of exotic weed species. It may be possible to apply low-impact temporary roads where highly engineered permanent roads have been constructed in the past. In addition, the equipment used to harvest trees, move logs to loading areas, and then transport logs to market will also have residual impacts of ecosystems through, for example, soil compaction. Alternative harvesting machinery and skidders may be applied to mitigate these impacts where important.

Institutions

The preceding discussion emphasizes a combination of site and system level management approaches to sustainable forestry. Implementing these types of approaches may be feasible in the setting of public land management where large areas are under a common management agency, though this may not be as straightforward as it appears. However, system-level planning is extraordinarily difficult in the mixed private ownership setting of most forest areas. The institutional setting of forest management, including land ownership and regulations, defines the crucial context for designing forest policies to accomplish sustainable forestry.

An institution is any rule or organization which governs the behavior of humans. In the context of forest management, relevant institutions include the structures of land ownership (public and private), the size and distribution of land holdings, and policy instruments, such as forest management laws and property taxes, which influence land management decisions. Because human behavior, expressed through land use decisions, is the dominant cause of landscape change, institutions define crucial control mechanisms for achieving goals otherwise unaddressed by human activities.

There are two broad institutional contexts for land management in the United States. One is private ownership with extensive property rights held by autonomous landowners. Under the right conditions, enlightened self-interest should guide landowners to allocate land to highest-valued uses and, in the process, to effectively produce marketable goods. However, there are goods and services which do not transact in markets but which may be of considerable value to society, including the ecological services discussed above. The production of nonmarket goods is a primary rationale for public ownership of forest land, the other major institutional context of land management (11). In theory, public land management aims to provide all important goods and services including those left unaddressed by markets.

Of course, public ownership is not the only mechanism for providing nonmarket goods. The actions of individual land owners might be directed towards producing other benefits by altering their incentives and selectively restricting property rights. An example of the former would be a severance tax on forest products, and an example of the latter would be forest practices regulations. These types of policy tools have a cost side, including the costs of administration, but also the cost of foregone market benefits. Balancing these regulatory costs against public benefits is a critical part of policy design. However, using these types of policy tools to address economies of configuration may be very costly.

The costs of public land management and of central planning in general may also be high, and this has been highlighted by recent international events. The private or market model of resource allocation seems to have enjoyed near global vindication and a great run of international favor. The fall of Soviet-style socialism, the Reagan Revolution, and the Sagebrush Rebellion all involve passing the control of resources from the public to the private sector. Even environmental advocates have adopted market arguments for regulating air quality with tradeable pollution permits and managing wilderness through private vendors, in effect, creating markets for environmental goods.

The collapse of centrally planned economies and the development of environmental markets reflect, on the one hand, the cumbersome and costly approach of central planning, and on the other, the efficiency of market allocation through competition. Where markets are well-structured or where they can be effectively created, there seems to be no more efficient mechanism for allocating resources. However, not all problems can be solved with market tools and economies of configuration seem to define problems which may not be shaped into a well-structured market. There seems to be little hope of organizing the actions of neighboring landowners through regulated competition alone. Rather, organizing landscape patterns and spatial processes seems to call for directly orchestrating the decisions of landowners.

Standard policy tools are applied without spatial discrimination across land owners and landscapes. They apply to all lands. However, in many cases critical habitat or corridors may lie on only a small portion of the relevant landscape. In effect, the policy may regulate and therefore penalize land management on all acres to influence production on a few, imposing considerable costs. Furthermore, it is unclear that these policy tools could be designed to influence these types of issues at all. A severance tax might, for example, encourage forest owners to grow trees longer so that, on average, more old-forest habitat is available but if harvesting occurs in large blocks then this may not address the configuration of habitats at the right scale.

One approach to directly shaping landscapes might call for land management zoning. Of course, severe land use regulations amount to de facto public ownership. That is, regulation can remove nearly all of the relevant property rights, often without compensation. In this kind of regulatory environment uncertainty might be high, encouraging accelerated resource extraction while policies and regulations are being developed. It may well be better to condemn and purchase the critical properties outright. As far as orchestrated action is concerned, public control over designated critical habitat seems to have an advantage over other regulatory mechanisms because it allows specific important areas to be targeted.

Sustainability seems to require an approach which can orchestrate land conditions across landscapes. The policy question, as usual, is whether the medicine is worse than the illness. While some type of land management coordination seems called for, central planning of any sort has its own kinds of costs, related mainly to the burdens of collecting and processing massive amounts of information and enforcing complicated plans. On the other hand, marginal regulation may be costly in two ways. It may be ineffective for targeting highly important habitats and it imposes costs on the management of all lands in the process. Marginal regulation may be more benign in an ideological sense because it works through markets but at the same time it may be very expensive.

At present, efforts to implement ecologically sustainable forestry have focused on the public lands, especially the national forests (41). These entities have a comparative advantage in this regard because they control large, though not necessarily contiguous, areas of forested land. Even so, managing for ecological conditions will require expanding the focus of national forest management to consider the context of neighboring forest lands and connections at a larger scale (42). In effect, public lands become control mechanisms in a larger dynamic landscape.

Shifting management of the national forests to focus on ecological health may be a cost-effective approach to sustainable forestry. Allowing public lands to specialize in this area, might reduce the need to impose onerous regulation on private forests which efficiently produce commercial timber products.

However, this kind of specialization (ecological services on public lands, commercial products on private lands) raises a crucial question for sustainability. Is the present arrangement of public lands, defined in another time with different objectives, sufficient to achieve sustainability goals? If not, then what lands should be public and what lands should be private? An inventory of critical habitat would be a first step towards evaluating the potential impacts of this kind of approach.

The preceding discussion has focused on difficult system-level approaches to sustainable forestry. At the level of forest management practices implementation may be much more straightforward. However, these approaches necessarily address different aspects of sustainable forestry. There is a long history of requiring "Best Management Practices" (BMP's) and other regulatory stipulations for the practice of forestry (43). These regulations have been applied at state and local levels and address everything from adequate forest regeneration to environmental protection.

BMP's could obviously be extended to provide for certain kinds of management in special areas. For example, they could, as they have in the past, target wetlands and riparian corridors. In addition, they could be targeted to other scarce and important types of habitat. These types of regulatory tools hold the same kinds of enforcement costs, but not the same extent as the management orchestration discussed above. Rather, enforcement can easily be implemented through, for example, a harvest permitting process.

Research

The third element of implementing sustainable forestry is a research program aimed at generating new ecological and economic insights into forest management. In addition, research will need to focus on monitoring the impacts of forest management efforts. The preceding discussion mentioned several areas where knowledge deficits exist. These include the following:

1. Classification of ecosystem types and definition of relatively scarce types.

2. Understanding the relative valuation of ecological systems.

3. Establishing the relationships between ecological conditions and human welfare.

4. Defining the effects of various silvicultural practices on forest stand dynamics and resulting biodiversity.

5. Designing institutional structures to affect sustainable forestry in a cost effective manner.

BIBLIOGRAPHY

1. A. Leopold, *A Sand County Almanac*, Oxford University Press, New York, 1949, 226 pp.

2. J. Lubchenco and co-workers, *Ecology* **72**(2), 371–412 (1991).

3. M. A. Harwell and co-workers, "Ecological Sustainability and Human Institutions: Case Studies of Three Biosphere Reserves," *Research Proposal submitted to the U.S. Man and the Biosphere Program*, May 1991.

4. R. G. Lee and co-workers, "Land Use Patterns in the Olympic and Southern Appalachian Biosphere Reserves: Implications for Long-Term Sustainable Development and Environmental Vitality," *Research Proposal submitted to U.S. Man and the Biosphere Program*, May 15, 1990.

5. T. R. Cox, R. S. Maxwell, P. D. Thomas, and J. J. Malone, *This Well-Wooded Land: Americans and Their Forests from Colonial Times to the Present*, University of Nebraska Press, Lincoln, 1985, 325 p.

6. Ref. 5, pp. 111–132.

7. G. Pinchot, *Breaking New Ground*, reprinted by Island Press, Washington, D.C., 1947, pp. 79–86.

8. Ref. 7, pp. 147–153.

9. A. B. Recknagel and J. Bentley, *Forest Management*. John Wiley & Sons, Inc., London, 1919, 269 pp.

10. Ref. 9, p. 124.

11. J. V. Krutilla and J. A. Haigh, *Environmental Law* **8**, 373–415 (1978).

12. A. Downs, *Inside Bureaucracy*, Little, Brown, and Company, Boston, 1967, 292 pp.

13. C. F. Wilkenson and H. M. Anderson, *Land and Resource Planning in the National Forests*, Island Press, Washington, D.C., 1987, 396 p.

14. Technology Review, "The Road from Rio: An interview with Gro Harlem Brundtland," *MIT Technology Review* **96**(3), 60–65 (1993).

15. M. A. Towman, *Resources* 3–6, (Winter 1992).

16. C. H. Schallou and R. M. Alston, *Environmental Law* **17**(3), 429–482 (1987).

17. D. N. Wear and W. F. Hyde, *J. of Business Administration* **21**, 297–314 (1992).

18. P. R. Ehrlich and A. H. Ehrlich, "Humanity at the crossroads" in H. E. Daly, ed., *Economics, Ecology, and Ethics: Essays Toward a Steady-State Economy*, W. H. Freeman and Co., New York, 1973, pp. 38–43.

19. H. E. Daly, *J. of Pol. Econ.* **76**(3), 392–406 (1968).

20. J. F. Franklin, "The fundamentals of ecosystem management with applications in the Pacific Northwest," in G. H. Aplet, ed., *Defining Sustainable Forestry*, Island Press, 1993, pp. 127–144.

21. D. J. Rohlf, *Conservation Biology* **5**(3), 273–282 (1991).

22. P. H. Raven and E. O. Wilson, *Science* **258**, 1099–1100 (Nov. 13, 1992).

23. J. F. Franklin, *Ecol. Applications* **3**(2), 202–205 (1993).

24. R. Solow, *An Almost Practical Step Toward Sustainability*, Resources for the Future, 1992, 22 pp.

25. S. E. Daniels, *Environ. Law* **17**(3), 483–506 (1987).

26. J. E. Rodiek and E. G. Bolen, eds., *Wildlife and Habitats in Maanaged Landscapes*, Island Press, Washington, D.C., 1991, 220 pp.

27. D. A. Saunders, R. J. Hobbs, and C. R. Margules, *Conserv. Biol.* **5**(1), 18–32 (1991).

28. D. N. Wear, "Forest Management, Institutions, and Ecological Sustainability," *Proceedings of the Appalachian Society of American Foresters Meeting*, Asheville, N.C., 1992, pp. 12–16.

29. R. J. Naiman, H. DeCamps, and M. Pollock, *Ecol. Applications* **3**(2), 209–212 (1993).

30. G. H. Orians, *Ecol. Applications* **3**(2), 209–212 (1993).

31. A. J. Hansen, T. A. Spies, F. J. Swamson, and J. L. Ohmann, *BioScience* **41**(6), 382–392 (1991).

32. D. M. Smith, *The Practice of Silviculture*, John Wiley & Sons, Inc., New York, 1962, 578 pp.

33. Ref. 31, pp. 353–358 for extensive discussion of these techniques.

34. S. F. Arno, *J. Forestry* **78**(8), 460–465 (1980).

35. M. A. Marsden, "Modeling the Effect of Wildfire Frequency on Forest Structure and Succession in the Northern Rocky Mountains," *J. Environ. Manag.* **16**, 45–62 (1983).

36. J. P. Roise, *Forest Science* **36**, 487–501 (1990).

37. J. G. Hof and L. A. Joyce, *Forest Science* **38**, 489–508 (1992).

38. J. Sessions, *Forest Science* **38**(1), 203–207 (1992).

39. J. F. Franklin and R. T. T. Forman, *Landscape Ecology* **1**, 5–18 (1987).

40. W. F. Megahan and W. J. Kidd, *J. Forestry* **70**, 136–141 (1972).

41. D. J. Brooks and G. E. Grant, *J. Forestry* **90**(1), 25–28 (1992).

42. S. K. Swallow, and D. N. Wear, *J. Environ. Econ. Manag.* **25**(2), 103–120 (1993).

43. F. W. Cubbage, J. O'Laughlin, and C. S. Bullock III, *Forest Resource Policy*, John Wiley & Sons, New York, 1993, 562 pp.

Reading List

R. W. Fri, "Sustainable Development: Can We Put These Principals into Practice?" *J. Forestry* **89**(7), 24–26 (1991).

R. P. Gale and S. M. Cordray, "What Should Forests Sustain? Eight Answers." *J. Forestry* **89**(5), 31–36 (1991).

R. J. Kopp, "The Role of Natural Assets in Economic Development," *Resources* 7–10 (Winter 1992).

R. G. Lee, R. Flamm, M. G. Turner, C. Bledsoe, P. Chandler, D. De Faerrari, R. Gottfried, R. J. Naiman, N. Schumaker, and D. Wear, "Integrating Sustainable Development and Environmental Vitality," in R. J. Naiman, ed., *Watershed Management: Balancing Sustainability and Environmental Change*, Springer-Verlag, New York, 1992, pp. 497–518.

R. G. Lee, "Ecologically Effective Social Organization as a Requirement for Sustaining Watershed Ecosystems," in R. J. Naiman, ed., *Watershed Management: Balancing Sustainability and Environmental Change*, Springer-Verlag, New York, 1992, pp. 73–90.

D. Ludwig, R. Hilborn, and C. Walters, "Uncertainty, Resource Exploitation, and Conservation: Lessons from History," *Ecolog. Applications* **3**(4), 547–549 (1993).

G. H. Orians, "Ecological Concepts of Sustainability," *Environment* **32**(9), 10–39 (1990).

J. R. Probst and T. R. Crow, "Integrating Biological Diversity and Resource Management," *J. Forestry* **89**(2), 12–17 (1991).

G. C. Ray and J. F. Grassle, "Marine Biological Diversity." *BioScience* **41**(7), 453–457 (1991).

H. Salwasser, "Sustainability as a Conservation Paradigm." *Conserv. Biol.* **4**(3), 213–216 (1990).

M. G. Turner, V. H. Dale, and R. H. Gardner, "Predicting Across Scales: Theory Development and Testing," *Landscape Ecology* **3**(3/4), 245–252 (1989).

FUEL CELLS

JOHN APPLEBY
Texas A&M University
College Station, Texas

DIRECT ENERGY CONVERSION

Fuel cells (1) are devices which directly convert the Gibbs energy ("free energy") of fuel oxidation into work in the form of direct current electricity. They are devices operating at constant temperature with a controlled combination of reactants, unlike the uncontrolled process which takes place when the fuel is burned to produce heat. Work is a directed flow of energy in the form of particles in motion, whereas heat is a random movement of particles in all directions. From the second law of thermodynamics, only part of energy in the random form of heat can be transformed into the directed form of work after the heat is put through a suitable thermodynamic cycle in a specially designed machine. Whenever heat is converted into work by any thermodynamic device, it is taken up at a high temperature T_2 from a heat source and is rejected at a lower temperature T_1 to a heat sink. It is then subject to the efficiency limitations of the Carnot theorem, so that the maximum efficiency for the conversion of heat into work is $(T_2 - T_1)/T_2$ in a reversible thermodynamic cycle. The earliest thermal machines were steam piston engines, operating on the Rankine cycle, which were followed by similar turbine engines. Others are hot-gas piston engines, operating on the Stirling cycle, internal combustion piston engines, operating on the Otto and diesel cycles, and gas turbines, operating on the Brayton cycle (2).

A number of different devices are said to offer direct energy conversion. These include thermoelectric and thermionic converters and photovoltaic cells. The first two convert heat, and the third photons, into direct current electricity. Thermoelectric and thermionic converters are not strictly direct energy conversion devices. They are simply thermal engines which convert heat into electrical work using no moving parts, compared with the devices listed above which first produce mechanical work using moving parts, followed by electricity using a rotating or linear generator.

By definition, a direct energy conversion device is one which can directly convert work in one form of Gibbs energy into another. A photovoltaic cell accepts Gibbs energy in the form of directed beams of photons and converts it to electrical work in the form of a directed stream of electrons under isothermal conditions. It consumes no heat which is converted into work. A fuel cell, a specialized form of a battery, directly converts the Gibbs energy of a chemical reaction into a directed stream of electrons, again under isothermal conditions. Again, it consumes no heat which is converted into work. Both photovoltaic cells and batteries or fuel cells are true direct energy conversion devices, which are not subject to the Carnot limitation, since Gibbs energy can in principle be converted at 100% efficiency. In practice, efficiencies will be less than this value, and the systems will reject heat. A brief introduction to the thermodynamics of heat engines and electrochemical direct energy conversion devices follows.

See also BATTERIES; HYBRID VEHICLES; ELECTRIC POWER.

THERMODYNAMICS OF ENERGY CONVERSION

The first law of thermodynamics states that energy is conserved in a system of constant mass. If the energy of the system changes, it must do so by a combination of exchanging heat with its surroundings and performing work on its surroundings or having work performed on it. If dE is an infinitesimally small change in internal energy, q is the amount of heat absorbed by the system, and w is the amount of work performed by the system. If we consider systems at constant pressure P, then $w = P\,dV$, where dV is the change in volume of the system, doing work by moving against the pressure. Thus, from the first law,

$$dE = q - P\,dV \qquad (1)$$

The second law of thermodynamics connects the change in the amount of heat and the absolute temperature of the system (T in kelvins, K) by the change in state of disorder or entropy, dS, such that $dS = q/T$. Thus, combining the first and second laws,

$$dE = T\, dS - P\, dV \qquad (2)$$

If we define the heat content of the system, H, as the sum of the energy E and the energy required to occupy the volume of the system at constant pressure, PV, then

$$dH = dE + P\, dV + V\, dP = q + V\, dP \qquad (3)$$

Thus, for a system at constant pressure, $dH = q$. The "bound" energy of the system at constant pressure, the part associated with its finite temperature, is equal to $T\, dS$. Thus the change in "free" energy at constant pressure, which may be converted into work, is

$$dG = dH - T\, dS = q - T\, dS \qquad (4)$$

We can change the infinitesimal changes to gross changes at constant pressure and temperature by integration, writing Δ instead of d. Thus

$$\Delta G = \Delta H - T\, \Delta S = Q - T\, \Delta S \qquad (5)$$

where Q is the gross change in heat absorbed by the system. In any system at constant temperature and pressure, a change in ΔH, or a change in Q, can produce a maximum amount of useful work equal to ΔG, the Gibbs energy of reaction.

When a constant-pressure system consists of chemical equivalents of two substances which interact in an exothermic chemical reaction, the system acquires a quantity of energy equal to the heat or enthalpy of reaction per equivalent, ΔH. For exothermic reactions, ΔH is a negative quantity representing the loss of energy from the interacting atoms. If we assume that the process is spontaneous, so that all of the energy appears as heat and none as work, i.e., the system remains at constant volume V_1, then the temperature of the system will be raised from its initial value T_1 to T_2, and its pressure from P_1 to P_2. If the system is then allowed to expand isothermally to pressure P_1 and volume V_2, it will do work on its surroundings. It is then allowed to cool down at constant volume to its original temperature T_1, rejecting heat at that temperature to its surroundings. It will then be at a pressure P_3, where $P_3 < P_1$. Following this, it contracts to its original volume V_1 and pressure P_1 at constant temperature T_1. During this part of the cycle, the surroundings do work on the system. The net work w performed by the system is given by the heat in at T_2 minus the heat rejected at T_1, i.e., the difference in area between the isotherms at T_2 and T_1, respectively. These are given by $\int V\, dP$ integrated between the appropriate limits. From the ideal-gas law, for all states, $PV = nRT$, where n is the number of equivalents per mole and R is the gas constant. Thus, w and q' are given by the expressions

$$w = nR(T_2 - T_1)\ln(V_2/V_1) \qquad q' = nR(T_2)\ln(V_2/V_1) \qquad (6)$$

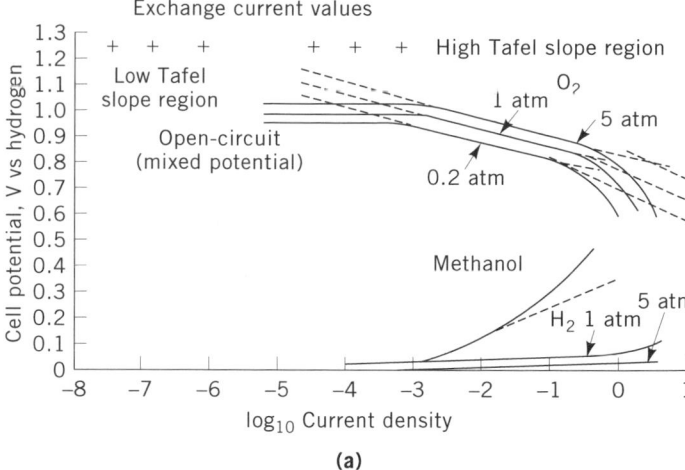

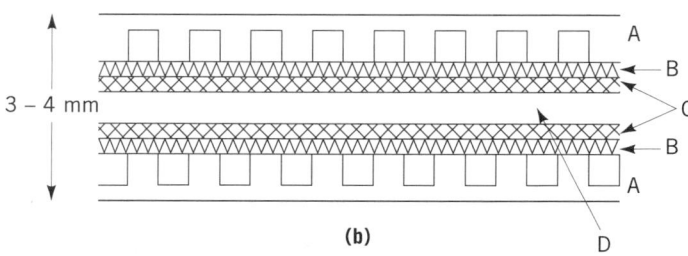

Figure 1. Hydrogen and oxygen polarizations at 70°C at 1 atm and 5 atm pressure in an acid fuel cell with a proton exchange membrane (PEM) electrolyte (a stable organic perfluorinated polymer sulfonic acid polymer), showing deviations from oxygen reduction Tafel slope due to diffusion limitations at high current density. Polarization of methanol is illustrated. Tafel lines in the low- and high-slope regions are shown dashed. A schematic fuel cell cross section is also shown.

Thus ε, the theoretical thermal efficiency for converting heat into work for this particular cycle, is given by

$$\varepsilon = (T_2 - T_1)/T_2 \qquad (7)$$

ie, Carnot's theorem. The cycle considered above is the Stirling cycle (2). Carnot's theorem was devised for a cycle in which the system accepts heat in a compression at constant entropy from T_1 to T_2. It then expands isothermally at T_2, then expands at constant entropy while its temperature changes from T_2 to T_1, during which it rejects heat to the sink. This is followed by isothermal compression and then constant-entropy compression at constant entropy while it receives heat to increase its temperature from T_1 to T_2, thus completing the cycle. In principle, an ideal cycle has two isothermal stages and two at constant volume (the Stirling cycle), with two either at constant entropy (the Carnot cycle) or at constant pressure (the Ericsson cycle). All show the same relationship for the maximum efficiency (2).

For an ideal gaseous chemical fuel and oxidant, we can arrive at the maximum Carnot efficiency as follows. We assume that the combination acts as its own working fluid for one complete thermodynamic cycle. We also assume that the specific heat at constant pressure is independent of temperature in the ideal system; then it follows that ΔH and ΔS are temperature independent. Since no work is done during the initial reaction, the change in ΔG is zero, and from Eq. (5), $T_2 \Delta S = \Delta H$, where ΔS is the entropy of reaction. At T_1, the Gibbs energy of reaction is

$$\Delta G_{T_1} = \Delta H - T_1 \Delta S \qquad (8)$$

Thus, since $T_2 = \Delta H / \Delta S$ and $T_1 = (\Delta H - \Delta G_{T_1})/\Delta S$, the maximum efficiency of a thermal machine which uses the heat from this chemical process, rejecting it at temperature T_1, is

$$(T_2 - T_1)/T_2 = \Delta G_{T_1}/\Delta H \qquad (9)$$

Thus, a thermal engine operating on the heat from a chemical reaction having a heat source at the maximum spontaneous temperature that the system can reach and rejecting energy at a temperature T_1 would have a maximum efficiency equal to $\Delta G_{T_1} / \Delta H$, where ΔG_{T_1} and ΔH are the Gibbs energy and enthalpy of reaction at T_1. Operation of a heat engine using hydrogen as the fuel and oxygen as the oxidant, producing liquid water as the product at 25°C (298 K) would require a source temperature T_2 approaching 2000°C (2273 K). Materials considerations make this impossible, to which must be added the limitations of heat transfer to the working fluid and the nonideality of both the materials used and of the thermal cycle itself. Practical machines, e.g., gas turbines, are restricted to source temperatures of 1250°C (1523 K). In a combined steam cycle application, their sink temperatures are about 40°C (313 K). Thus, such a system has a theoretical maximum efficiency of about 79%. The maximum practical efficiency resulting from the nonideality of the thermodynamic cycle and practical losses is about two-thirds of the theoretical value, or 52%. Because the fuel used in a gas turbine is burned to give gaseous water among the products, the efficiency is usually based on the so-called lower heating value (LHV) of the fuel, the enthalpy or heat of combustion to give gaseous water. The higher heating value (HHV) is that in which the water is condensed. The values at the boiling point of water differ by the latent heat of vaporization. A condensing steam engine or combined cycle, or a direct energy conversion device operating below the boiling point of water, operates on the HHV of the fuel used.

Electrochemical Energy Conversion

An electrochemical battery converts the Gibbs energy in the form of "electrochemical free energy" of pairs of chemical compounds to Gibbs energy in the form of electrical work as the compounds combine under controlled conditions (3). A familiar battery is the primary zinc–manganese dioxide system, used in many familiar devices. Zinc is an oxidizable metal, which forms a partially soluble oxide or hydroxide in alkaline solutions such as potassium hydroxide. Manganese dioxide is a relatively strong oxidizing agent. If powdered manganese dioxide and zinc are intimately mixed and heated, zinc oxide and a lower oxide of manganese will be the products, and heat will be produced corresponding to $-\Delta H$, the enthalpy of reaction. This could be transmitted to a suitable thermodynamic working fluid, e.g., a gas in a Stirling cycle engine, and mechanical work will be produced at an efficiency subject to the Carnot limit.

An intimate mixture of zinc and manganese dioxide can only produce heat when reaction occurs. However, a layer of zinc placed parallel to a layer of manganese dioxide, both separated by a layer of an electrolyte such as potassium hydroxide, can react to produce Gibbs energy in the form of direct current (dc) electricity. If the zinc and manganese dioxide layers make good electronic contact with conducting wires and an electric motor, a voltage is developed between the two electrodes which drives the electric motor. This voltage represents electrical work which is opposite equal in magnitude to $-\Delta G$, the Gibbs energy of reaction. The reaction at the zinc and manganese dioxide electrodes will be

$$Zn + 2\,OH^- \rightarrow ZnO + H_2O + 2\,e^- \qquad (10)$$

and

$$2\,MnO_2 + 2\,H_2O + 2\,e^- \rightarrow 2\,MnOOH + 2\,OH^- \qquad (11)$$

respectively. The overall reaction is

$$Zn + H_2O + 2\,MnO_2 \rightarrow ZnO + 2\,MnOOH \qquad (12)$$

The electrons flow from the zinc anode, which lies at a negative potential to the positive MnO_2 cathode. In the external circuit, the electrons do work by passing through the motor. The overall circuit is completed by a corresponding transfer of OH^- ions from the cathode to the anode through the electrolyte, which is not an electronic conductor. The above are the initial reactions of the system. As time progresses, the first electron transfer reaction of Mn^{4+} in MnO_2 to Mn^{3+} will be exhausted, and it will be replaced by the transfer from Mn^{3+} in MnOOH to Mn^{3+} in $Mn(OH)_2$. These reactions are quite rapid, and they are sufficiently thermodynamically reversible to develop the theoretical voltage or cell potential difference corresponding to $-\Delta G$ for the relevant reactant and product compositions and/or concentrations at low overall currents, where infrared (IR) drops in the external electronic circuit and in the internal ionic circuit are minimized. Overall, the enthalpy of reaction of the zinc and MnO_2 is partly converted into electrical work and partly dissipated as entropic heat loss according to the general equation combining the first and second laws:

$$\Delta H = \Delta G_T + T \Delta S \qquad (13)$$

where T is the temperature of operation of the battery, at which ΔG_T is the Gibbs energy of reaction.

To extract the Gibbs energy of reaction of two chemical compounds as work in the form of electrical energy, the compounds must react with each other isothermally. They must react separately, ie, unmixed. They must react via a common medium. The common medium in the above example is KOH solution, which also serves as a reservoir for the water required as a reactant at the MnO_2 surface. Because it is ionically conducting, the KOH solution allows the OH^- ion produced in the reduction of MnO_2 to flow to the reacting zinc, where it is consumed, so that the total amount of OH^- ion in the system remains constant as a function of time. The zinc and MnO_2 will only react if an electronic pathway exists between them, so that electrons can be transferred from the oxidizing zinc to the reducing MnO_2. This occurs via the electronically conducting wire joining the two reactants. Extraction of the Gibbs energy of reaction therefore can only be undertaken if both the reactants are separated by an ionically conducting medium which conducts largely via an ion which is produced in the electrochemical reaction at one of the electrode materials and consumed at the other. It also requires simultaneous collection of the electrons at the anodic reactant and means of conducting them to the cathodic reactant. Finally, it requires separation of the pathways for both ionic and electronic conduction. If the ionically conducting medium between the electrodes is also an electronic conductor, no external work will be done, and ΔG will appear not as work, but as heat resulting from an irreversible change in entropy, ΔS_{irrev}. Exactly the same will happen if the zinc and MnO_2 reactants are allowed to touch, which will allow electronic flow between the electrodes. In both these cases, the electronic pathway will be short-circuited, and the zinc anode and manganese dioxide cathode will no longer have a potential difference between them corresponding to the value of ΔG but will show zero potential difference, so that all the enthalpy of reaction is dissipated as heat. The reaction conditions then are totally irreversible, as they are in a normal thermal chemical reaction. Under these conditions

$$\Delta H = T\,\Delta S_{irrev} + T\,\Delta S \qquad (14)$$

The complete separation of charges is provided by the totally ionic nature of the electrolyte, as is its ability to sustain a potential difference between the electrodes corresponding to the Gibbs energy of reaction. This is because a double layer of positive and negative ions in the electrolyte at the surface of each electrode acts as a capacitor which can sustain an electric field equal to about 10^8 V/cm at potentials within the window of electrochemical stability of the electrolyte, beyond which it is itself oxidized or reduced. This surface capacitor is known as the Helmholtz double layer.

Technical energy conversion devices (i.e., electrochemical cells or batteries) use an electrolyte and external electronic conductors with the highest possible ionic conductivity to minimize losses under net current flow. The reacting electrodes themselves (in the above example, zinc and MnO_2) must also be made as electronically conducting as possible. This would be particularly true of the MnO_2 electrode, which would be mixed with a stable electronic conductor such as graphite or conducting carbon. To provide a total system with the maximum Gibbs energy conversion efficiency, an efficient electromagnetic device (ie, an electric motor) is required to convert the directed flow of electrons under the potential difference between the electrodes into mechanical work.

Biological Direct Energy Conversion

In principle, it is also possible to directly short out the electronic pathway between the electrodes and substitute an ionic (electrolytic) pathway at open circuit. If this contains a hypothetical device which can convert an ionic potential or concentration gradient into work at high efficiency, then the overall effect is similar to the operation of a battery. A satisfactory method of converting an ionic gradient into work has not been devised. Nature has opted to use the ionic gradient approach to create useful work in the natural equivalent of the battery (or more accurately fuel cells), the mitochondrion, the direct energy converter in the cells of living matter and the basis of the respiratory cycle (4). These have a shorted electronic circuit between the anodic and cathodic electrode sites and use the ionic gradient created by a semiopen proton circuit across a biological membrane for the synthesis of adenosine triphosphate (ATP) from adenosine diphosphate (ADP) and phosphate ion. The Gibbs energy thus stored in the ATP can be used throughout the organism. The mitochondria are in effect batteries which oxidize reduced fuel ultimately derived from glucose, ie, nicotinamide adenine dinucleotide phosphate (NADPH), to its oxidized form NADP at the anode (5). This is recycled elsewhere. The mitochondrion reduces molecular oxygen at its cathodic sites, ie,

$$4\,NADPH \rightarrow 4\,NADP + 4\,H^+ + 4\,e^- \quad (anode) \qquad (15)$$

$$O_2 + 4\,H_2O + 4\,e^- \rightarrow 4\,OH^- \quad (cathode) \qquad (16)$$

$$4\,NADPH + O_2 \rightarrow 4\,NADP + 2\,H_2O \quad (overall) \qquad (17)$$

The H^+ ions created in the anodic reaction in an essentially neutral medium, together with the OH^- ions created in the cathode process, result in the pH gradient which allows the synthesis of ATP.

FUEL CELLS

The above brief discussion of the operation of primary batteries and natural mitochondria allows an introduction to the concept of the fuel cell. Primary batteries contain electrode materials which are consumable. As the materials are oxidized at the anode and reduced at the cathode, the electrode structure degrades and finally collapses, and the battery will cease to operate. A fuel cell is a primary battery with invariant electrode structures and reaction sites, arranged in such a way that an oxidizable material (a fuel) can be fed continuously as required to the anode. Similarly, a reducible material (an oxidant) can be continuously fed to the cathode. Finally, provision is needed to remove the reaction product as it is formed. Thus, a fuel cell is a primary battery which is capable of continued operation as long as an oxidizable fuel and an oxidant are fed to it. The fuel plays the part of metallic zinc, and the oxidant the part of MnO_2 in the primary battery discussed above.

Fuel cells are generally required to operate on common hydrogen-containing fuels, eg, hydrocarbons and their derivatives, and hydrogen itself. For reasons to be explained, the fuel which is actually electrochemically consumed in high-performance fuel cells is gaseous hydrogen, although there are certain exceptions. Because of its availability, the oxidant is almost invariably molecular oxygen. For terrestrial purposes, this is obtained from air without separation, whereas for use in space cryogenic oxygen is normally used. Systems operating on halogens as oxidants are infrequently encountered, but they are sometimes referred to as fuel cells. Electrochemical cells operating on an oxidizing metal at the anode (eg, zinc, aluminum, sodium, lithium) with an oxygen cathode are also sometimes referred to as fuel cells, but they are more accurately metal–air batteries. They are usually primary (not-rechargeable) systems, with the exception of those with zinc anodes, which can be recharged in certain aqueous electrolytes.

FUEL CELL ELECTROLYTES

In a hydrogen–oxygen fuel cell, possible electrode reactions might be

$$2\,H_2 \rightarrow 4\,H^+ + 4\,e^- \quad \text{(anode)} \tag{18}$$

$$O_2 + 4\,H_2O + 4\,e^- \rightarrow 4\,OH^- \quad \text{(cathode)} \tag{19}$$

$$2\,H_2 + O_2 \rightarrow 2\,H_2O \quad \text{(overall)} \tag{20}$$

It can be seen that these are the same as those in a mitochondrion, except for the lack of the NADP hydrogen carrier. However, the mitochondrion operates in a dilute physiological electrolyte with a mean pH which is close to neutral. This contains very small residual amounts of H^+ or OH^- ions, which are overwhelmed by their rate of production during the reaction. It also operates by the use of a special semipermeable membrane to maintain a pH gradient to do chemical work. The electronic circuit is short circuited through the membrane via nature's equivalent of copper wire, a series of successive reduction–oxidation (redox) couples embedded in the membrane which permit the passage of electrons from one to the other to complete the circuit.

In contrast, a man-made fuel cell must perform all work in the external electronic circuit, and hence it must minimize the voltage losses in the ionic circuit. Above all, it must avoid any concentration gradients for ions formed (or consumed) in the electrode processes, since these result in an irreversible change in the effective Gibbs energy of reaction equal to $RT \ln(C_1/C_2)$, where C_1/C_2 is the concentration ratio of the H^+ or OH^- ions (6). In electrical terms, this represents an opposing voltage equal to $RT/nF \ln(C_1/C_2)$ per mole, or $RT/F \ln(C_1/C_2)$ per equivalent, where n is the change on the ion and F is the faraday (96,485 in coulombs per equivalent or in joules per electron-volt, J/eV). To avoid this loss, the fuel cell must use an electrolyte whose majority conductor is the ion produced in the reaction at one electrode and consumed at the other. An electrolyte for a high-power-density aqueous fuel cell must therefore be a concentrated solution containing this ion. Whereas mitochondria function on the pH change due to the concentration gradients which easily build up in dilute neutral solutions, a practical aque-

ous-electrolyte fuel cell operating at economically effective current densities requires a strongly acid or strongly alkaline solution to function well. Under these conditions, the cell reactions will be

$$2\,H_2 \rightarrow 4\,H^+ + 4\,e^- \quad \text{(anode, acid electrolyte)} \tag{21}$$

$$O_2 + 4\,H^+ + 4\,e^- \rightarrow 2\,H_2O$$
$$\text{(cathode, acid electrolyte)} \tag{22}$$

$$2\,H_2 + 4\,OH^- + 4\,e^- \rightarrow 2\,H_2O$$
$$\text{(anode, alkaline electrolyte)} \tag{23}$$

$$O_2 + 4\,H_2O + 4\,e^- \rightarrow 4\,OH^-$$
$$\text{(cathode, alkaline electrolyte)} \tag{24}$$

In molten salt electrolytes, the rule that the majority charge carrier should be an ion produced at one electrode and consumed at the other requires the ions to be H^+ (in a molten proton conductor), OH^- (in a molten hydroxide), or O^{2-} (in a molten oxide). The latter would be produced in a simple four-electron charge transfer process in the reduction of molecular oxygen. Molten proton conductors such as acid salts, eg, sodium bisulfate, are corrosive and often not stable, so they have not been widely examined as electrolytes. Molten alkaline hydroxides are also corrosive, and suitable molten oxides only exist at temperatures which are beyond the capabilities of economic fuel cell construction materials. If CO_2 is present in the fuel gas stream, molten hydroxides will react with it to form carbonates, which would normally be solids at the cell operating temperatures. In consequence, the only molten salt electrolyte which has found application is a mixture of molten alkali carbonates, usually Li–Na or Li–K. A mixture is used to lower the melting point and to give improved ion transport and other physicochemical properties. In the molten carbonate fuel cell (MCFC), oxide ion (O^{2-}) carries the current from anode to cathode in the form of a carrier ion, CO_3^{2-}. This is formed by supplying CO_2 along with O_2 as a reactant at the cathode:

$$O_2 + 2\,CO_2 + 4\,e^- \rightarrow 2\,CO_3^{2-} \tag{25}$$

The reaction with hydrogen at the anode is

$$2\,H_2 + 2\,CO_3^{2-} \rightarrow 2\,H_2O + 2\,CO_2 + 4\,e^- \tag{26}$$

The CO_2 in the anode exit gas stream is then collected and recycled to the cathode gas. In this way, the overall cell process is reaction 20. The use of CO_2 as a depolarizer in this manner eliminates the possibility of concentration gradients. In principle, other oxyanions could be used in this way, but they involve the use of nongaseous or low-volatility oxides (eg, SiO_2, B_2O_3, or P_2O_5) or oxides with limited stability (eg, SO_3, which is reduced under anodic conditions). Thus, carbonates appear to be the only choice for a system based on the use of hydrogen-containing fuels and oxygen in molten electrolytic media. In addition, both carbonates and CO_2 show good stability in regard to oxidation and reduction and are generally nonaggressive in regard to corrosion behavior.

While no molten oxides are feasible as electrolytes, at sufficiently high temperatures certain ionically doped oxides become effective O^{2-} ion conductors. Thus, they may be used as electrolytes in the solid oxide fuel cell (SOFC) operating on hydrogen-containing fuel and oxygen.

As already stated, a primary dc generator operating on hydrogen with chlorine or bromine as oxidant is also a fuel cell. Such systems have not been widely used, since the halogens are not generally available as oxidants and they are also not easy to store and handle. However, with chlorine as oxidant, the anodic reaction would be (20), or

$$H_2 + 2\ Cl^- + 2\ e^- \rightarrow 2\ HCl \tag{27}$$

The corresponding cathode reactions would be

$$Cl_2 + 2\ H^+ + 2\ e^- \rightarrow 2\ H_2O \tag{28}$$

or

$$Cl_2 \rightarrow 2\ Cl^- \tag{29}$$

$$H_2 + Cl_2 \rightarrow 2\ HCl \tag{30}$$

(overall). Such cells will require either an acid electrolyte, a concentrated chloride solution, or a molten chloride salt as the electrolyte.

FUEL CELL THERMODYNAMICS

Reversible Thermodynamics

From Eq. 8, it is evident that the maximum amount of work which can be obtained from the isothermal electrochemical oxidation of a fuel at temperature T is ΔG_T. If the entropy of reaction, ΔS, is positive, then $-\Delta G$ will be greater than $-\Delta H$, i.e., the electrochemical cell operating under quasi-reversible thermodynamic conditions will absorb enthalpy equal to $T\ \Delta S$ from its surroundings; ie, it will operate endothermically, thus making up the difference between the available enthalpy of reaction and the reversible electrical work done (5). In addition, $-\Delta G$ will then increase as T increases. In general, oxidation reactions in which ΔS is significantly positive involve a net increase in the number of molecules in the gas phase on going from reactants to products. Thus, while methane and ethylene oxidation, ie,

$$CH_4 + 2O_2 \rightarrow CO_2 + H_2O$$
$$C_2H_4 + 3\ O_2 \rightarrow 2\ CO_2 + 2\ H_2O \tag{31}$$

have ΔS values which are close to zero if the product is gaseous water, those for higher paraffins and olefins will be positive, eg,

$$C_2H_6 + 3.5\ O_2 \rightarrow 2\ CO_2 + 3\ H_2O$$
$$C_3H_6 + 4.5\ O_2 \rightarrow 3\ CO_2 + 3\ H_2O \tag{32}$$

Similarly, ΔS for the oxidation of gaseous methanol and higher saturated alcohols to gaseous water will also be positive:

$$CH_3OH + 1.5\ O_2 \rightarrow CO_2 + 2\ H_2O \tag{33}$$

as is the oxidation of gaseous ammonia to gaseous water:

$$2\ NH_3 + 1.5\ O_2 \rightarrow N_2 + 3\ H_2O \tag{34}$$

Another often-quoted example is the oxidation of solid carbon to CO:

$$C + 0.5\ O_2 \rightarrow CO \tag{35}$$

In all of the above, a reversible fuel cell operating with the above reactant and product states could in principle have a thermal efficiency greater than 100%. If liquid water is the product in reactions (31)–(34), then ΔS will be negative and the corresponding thermal efficiencies will be less than 100%. For example, for gaseous methanol oxidation, 2.5 molecules of gaseous reactants yield 1 molecule of gaseous products. If the methanol reactant is liquid, 1.5 molecules of reactant yield 1 molecule of products. Here ΔS will still be negative, though less so.

Irreversible Effects

Unfortunately, all of the above reversible thermodynamic considerations are only of academic interest, since they imply zero net rate of reaction. A fuel cell is intended to convert a fuel at a net rate, not produce reversible work at zero rate of working. A fuel cell operating with reactants and products whose rates of reaction on appropriate catalytic surfaces are sufficiently high to prevent interference from any competing electrochemical processes should give the reversible thermodynamic potential ($-\Delta G$) for the process. Indeed, this was the classical method for measuring the values of ΔG and ΔS (ie, $-\partial \Delta G / \partial T$) of reaction, the results being by independent measurement of ΔH by calorimetry. However, under conditions of net reaction rate, the measured values of the cell potential are always less than the reversible value; ie, some of the available Gibbs energy is converted into $T\ \Delta S_{irrev}$, as in Eq. 14.

FUEL CELL KINETICS

Overpotential

The deviations from the reversible thermodynamic potential values due to the effect of reaction rate are usually called overpotentials or overvoltages. They are conceptually explained, at least in outline, by the theory of irreversible thermodynamics of Prigogine (7), using the so-called Marcellin-DeDonder equation (8), as applied to electrochemical processes by Van Rysselburgh (9). In short, this states that to induce a net rate, a reaction must be driven by part of the (reversible) Gibbs energy of reaction, known as the (irreversible) reaction affinity ($-\Delta A$). This appears as irreversible (ie, irrecoverable) enthalpy, $T\ \Delta S_{irrev}$. At least for relatively small displacements from thermodynamic equilibrium, ie, for low net rates of reaction, DeDonder considered affinity to be proportional to net rate. Measurements of the rates of electrochemical processes are unique in having been able to determine the form of the affinity–rate relationship to be determined over a wider affinity (ie, electrochemical overpotential) range. This is illustrated by Tafel's equation (10), which was first demonstrated for a typical electrochemical reaction (cathodic hydrogen evolution from acid electrolytes). It can be written as

$$V = a + b \ln j_c = a + b \ln nFk \tag{36}$$

where V is the electrode potential, j_c is the cathodic reaction rate expressed as a current density (A/cm^2), k is the rate (expressed in molecules/cm^2 s), n is the number of electrons successively transferred per reacting molecule, and a and b are constants for a given reaction and substrate at a given temperature. The constant b is in most cases proportional to T and is equal to $-RT/\alpha_c F$, where α_c, the cathodic transfer coefficient (11), is a constant which depends on the reaction mechanism (12–16). The negative sign indicates that the cathodic current density increases as V decreases. In most cases, transfer coefficients lie between 0.5 and 1.5. In the anodic direction (hydrogen oxidation), the corresponding equation has constants a' and b', where b' $RT/\alpha'F$. For cases which are commonly encountered,

$$\alpha_c = n' + \beta(n_{\rm rds}) \quad \text{and} \quad \alpha_a = n'' + (1 - \beta)(n_{\rm rds}) \tag{37}$$

where α_a is the corresponding transfer coefficient; β is a dimensionless quantity called the symmetry factor, which is usually close to 0.5; n' is the number of electrons which are transferred in the reaction sequence before each single rate-determining step (rds) in the cathodic direction, in which $n_{\rm rds}$ electrons are transferred; and n'' is the number similarly transferred in the anodic direction. Since only one electron is permitted to be transformed at a time, $n_{\rm rds}$ can only be 1 (for an electrochemical rds) or zero (for a chemical rds). Thus, $\alpha_c + \alpha_a = n$, where n is the number of electrons transferred in the overall process per unit rds (12,13). If one unit rds occurs per molecule of reactant, then n is identical with n in Eq. 36. If ν rate-determining step units occur for each overall reaction, it is equal to n/ν when n is the value in Eq. 36. The value of ν is known as the stoichiometric number (12–14).

The net electrode reaction is the sum of the anodic (positive current) and cathodic (negative current) processes. Equation (36) therefore may be written as

$$\begin{aligned} j = j_a - j_c &= [\exp(a/b)\exp - (V/b)] \\ &\quad - [\exp(a'/b')\exp - (V/b')] \end{aligned} \tag{38}$$

where j_a is the anodic current density and j is the net (experimental) current density. By definition, at the reversible potential, $j = 0$ and $j_a = j_c$. Thus, the current for the anodic and cathodic processes at the reversible potential is equal to $\exp[(a - a')/(b' - b)]$ or $\exp[\alpha_c\alpha_a(a - a')/n]$. This can be conveniently written as j_0, the current at equilibrium or the exchange current (17). Thus, if η is the overpotential measured from the equilibrium potential, the net current at any overpotential η will be given by

$$j = j_0[\exp(\alpha_a F\eta/RT) - \exp - (\alpha_c F\eta/RT)] \tag{39}$$

This is known as the Butler–Volmer equation (11,18). Expanding the exponentials for small values of η, this gives

$$j = j_0[(\alpha_a + \alpha_c)F\eta]/RT = j_0(nF\eta/RT) \tag{40}$$

This expression (19) will still be accurate to within 2% for $\eta \approx 0.5\alpha_c F\eta/RT \approx 0.5\alpha_a F\eta/RT$. This may be true for η of ± 0.06 V at 298 K (25°C) or ± 0.2 V at 923 K (650°C). Not too far from thermodynamic equilibrium, Eq. 40 shows a linear relationship between reaction rate and irreversible affinity, which is equal to $\pm F\eta/n$. This is in agreement with the classical irreversible thermodynamic treatment of Prigogine (7), which shows this to be a special case of a more generalized concept, in which the rate away from equilibrium is related exponentially to the affinities (8,9). In Eq. (39), the notional irreversible reaction affinities A_a (anodic) and A_c (cathodic) are given by $-\alpha_a F\eta/n$ and $\alpha_c F\eta/n$, respectively. The $d\eta/d \ln j$ relationships for the individual anodic and cathodic reaction in Eq. 39, ie, $RT/\alpha_a F$ and $-RT/\alpha_c F$, are known as the Tafel slopes. They are conventionally expressed in $\log_{10}$ terms in units of millivolts per decade of current density.

This short account gives the more important characteristics of electrochemical reaction rate theory, ie, current density–overpotential relationships, so that some of the operational characteristics of fuel cells can be appreciated. Readers are referred to Refs. 13 and 14 for further details. The displacement of the net reactions from equilibrium depends of the ratio j/j_0. For a rapid reaction (i.e., one with a low enthalpy of activation, say 20 kJ/mol) on an effective electrocatalyst, j_0 may be 10^{-3} A/cm^{-2} at 298 K (25°C) and 3×10^{-3} A/cm^{-2} at 342 K (75°C). This is approximately the case of the hydrogen oxidation reaction in aqueous media on platinum electrocatalysts. To maximize the current density which can be obtained, fuel cell electrodes consist of high-surface-area electrocatalyst dispersed to give the most effective area of contact. An electrode containing platinum electrocatalyst dispersed in 10–20% by weight on high-surface-area carbon is typical, using platinum with a specific surface area of 100 m^2/g, indicating a particle radius of about 1.5 nm; ie, about 50% of the platinum atoms are on the surface of the particles. If the catalyst loading is 0.1 mg/cm^{-2}, each geometrical square centimeter of electrode therefore contains 100 cm^2 of catalyst. If all of the catalyst is available, then the effective value of j_0 is 3×10^{-1} A/cm^{-2}. The hydrogen electrode reaction is quite complex close to equilibrium with alternative parallel mechanisms. These are reviewed in Refs. 15, 16, and 19. However, the maximum displacement from the equilibrium potential will be given by the values $\alpha_a = 0.5$ and $\alpha_c = 0.5$, which represent values for mechanisms which predominate away from equilibrium. Thus, if j is to be 0.5 A/cm^{-2}, $\eta \approx 0.075$ V, ie, the electrode will function with a displacement of only 75 mV from the equilibrium potential. The existence of the alternative mechanisms makes the actual value even less than this.

CARBON-CONTAINING FUELS

The high reactivity, ie, exchange current, of hydrogen is unfortunately the exception, rather than the rule. Other hydrogen-containing fuels, with the exception of hydrazine and hydroxylamine, without exception show much lower rates of reaction (21). Equation 39 shows that when η exceeds approximately $1.1RT/\alpha F$, where α is the transfer coefficient for the reaction in the desired direction, the back reaction becomes negligible, and the current density depends on the exponential of the overpotential. Thermodynamic data for different fuels are listed in Ref. 1 (p. 18). The standard Gibbs energy for the two-electron oxidation of gaseous hydrogen by gaseous oxygen to liquid water is 237.3 kJ/mol, so that the difference between the reversible electrode potentials of a standard oxygen electrode and a standard hydrogen electrode in the same aqueous electrolyte at 298 K will be 1.229 V. The corresponding value for the four-electron hydrazine oxidation process, ie,

$$N_2H_4 + 2\,O_2 \rightarrow N_2 + 2\,H_2O \qquad (41)$$

is 602.4 kJ/mol; ie, the reversible hydrazine electrode will lie at a potential -1.525 V below that of a hypothetical reversible oxygen electrode, ie, -0.226 V more cathodic of the reversible hydrogen electrode. Thus on an active electrocatalyst such as platinum, hydrazine will decompose water, and 14.8% of the Gibbs energy of hydrazine oxidation will be irrecoverably lost as heat ($T\,\Delta S_{irrev}$). This fact, together with the cost and dangerous nature of hydrazine, has not made it a practical fuel, although its use was attempted in the 1960s for military applications such as for torpedo propulsion, often with specialized liquid oxidants such as concentrated hydrogen peroxide. Ammonia is much less reactive than hydrazine and hydrogen, and if it is available for use in fuel cells, it is best decomposed to hydrogen by thermal cracking. We should note that undecomposed ammonia reacts with acid electrolytes, making its use undesirable in acid fuel cells.

Carbon-containing fuels with the highest electrochemical activity are those which are partially oxidized and contain one carbon atom, ie, formic acid (HCOOH, two-electron oxidation, -1.480 V), formaldehyde (HCHO, four-electron oxidation, -1.350 V), and methanol (CH$_3$OH, six-electron oxidation, -1.214 V). The voltages are the corresponding potential s vs. the hypothetical reversible oxygen electrode. They are respectively -0.251 V, -0.121 V, and $+0.015$ V vs. the reversible hydrogen electrode in the same electrolyte. Like hydrazine, formic acid in particular should be capable of decomposing water with the production of hydrogen on a catalytic surface. The kinetics of this reaction are very slow. Even though formic acid is normally regarded as a relatively strong reducing agent, its electrocatalytic activity is low, as is that of formaldehyde. Neither material has been seriously considered as a fuel, since their degree of oxidation is high. Therefore their available enthalpy of combustion per gram is low. For the series gaseous methane (eight electrons), liquid methanol (six electrons), gaseous formaldehyde (four electrons), and solid formic acid (two electrons), the enthalpy of combustion per equivalent (HHV) is in the order 1.0, 1.09, 1.26, and 1.21. With allowance for the latent heat of the initial states, it can be seen that the available enthalpy of combustion of successive hydrogens increases as oxidation becomes more extensive. However, the enthalpies of combustion per unit mass fall dramatically in the ratio 1:0.41:0.34:0.11. For this and toxicological reasons, only methanol has been seriously considered as a fuel.

The electrochemical oxidation of methanol and other oxidized carbon species in aqueous media has been extensively studied (21–23). The most effective electrocatalysts are platinum alloys, e.g., those with tin and ruthenium (22,23). The first step in the electrochemical oxidation of organic compounds is adsorption, which takes place with partial dissociation of the C—H bond. This is associated with the presence of platinum in the electrocatalyst. This step is followed by oxidation of the adsorbed hydrogen. Partial oxidation of the carbon then occurs, via OH species derived from partial oxidation of water molecules, which are adsorbed on the electrocatalyst surface. The OH species are associated with alloying metals which are more oxidizable than platinum. We should note that water is required as a reactant for the oxidation process for methanol or any other carbonaceous compound. Finally, an adsorbed CO group remains, which acts as a catalyst poison (24). This group is identical with that form if carbon monoxide is present during the electrooxidation of gaseous hydrogen. Thus, the reaction is self-poisoning, and the CO group can only be entirely eliminated by exposing the electrocatalyst to high positive potentials of about -0.4 V vs. the hypothetical reversible oxygen potential, or $+0.83$ V vs. hydrogen. If liquid methanol in water is used as feedstock, the effective surface concentration of reactants is about 10^3 times higher than that of gaseous hydrogen. Even under these conditions, the oxidation of methanol has a j_0 value about 10^{-4} times less than that of gaseous hydrogen oxidation. Thus, methanol oxidation will reach the same rates as hydrogen oxidation close to the reversible hydrogen potential only at about $+0.5$ V on the same potential scale.

For hydrocarbons and compounds with fully reduced carbons, poisoning effects are worse, and reaction rates are orders of magnitude less (24). For example, ethanol is much less active than methanol, but it is still much more reactive than ethane. Ethylene is also more reactive than ethane (24). In contrast, methane is totally inert even at 200°C in aqueous media (concentrated phosphoric acid), where other hydrocarbons and carbon compounds show slight activity provided that very high (and totally uneconomical) catalyst loadings are used to give highest redundancy against the effects of poisoning. As a result, direct hydrocarbon fuel cells have remained curiosities. Direct methanol cells are still of some interest, but their low cell potential (ie, low thermal efficiency), generally low current density, requirement for high catalyst loading, and the general tendency for unreacted methanol to diffuse through the electrolyte and affect the performance of the oxygen cathode have limited their application.

Since reaction rates increase with temperature, the question arises as to whether yet higher temperatures will give useful reaction rates for carbonaceous fuels. With appropriate catalysts, they react with steam to produce H$_2$ and CO/CO$_2$ mixtures, depending on the temperature, CO$_2$ being favored at low temperatures. At temperatures of 200°C and above, methanol will steam-reform readily via this reaction. Hydrocarbons fed directly to high-temperature fuel cells (HTFCs), eg, the molten carbonate (MCFC) system at 650°C, may show some tendency to oxidize, but they also decompose by cracking, producing carbon deposits. One effective way of eliminating cracking is the injection of steam with the hydrocarbon to shift the equilibrium

$$C + H_2O \rightarrow CO + H_2 \qquad (42)$$

toward the right. This results in an environment in which the hydrocarbon itself is likely to steam-reform, eg, for methane,

$$CH_4 + H_2O \rightarrow CO + 3\,H_2 \qquad (43)$$

Like reaction (42), this is endothermic. The hydrogen formed is oxidized much more rapidly than either methane and CO, so it is rapidly consumed to produce more steam. The CO undergoes the rapid exothermic water–gas shift reaction with excess steam to produce more hydrogen:

$$CO + H_2O \rightarrow CO_2 + H_2 \qquad (44)$$

Thus, at first sight methane is being oxidized directly, where it is in effect being decomposed to CO_2 and H_2 in the gas phase; only the latter is oxidized at the electrode surfaces. For methane, the steam-reforming reaction normally takes place at high rates only at about 800°C on suitable catalysts. However, at 650°C, with the same catalysts, its rates are sufficient to maintain the reactions on fuel cell electrodes. Methanol requires a much lower reforming temperature (about 300°C to obtain rates which are sufficiently high for industrial processes). In contrast, ethanol, or any other compound containing fully reduced carbons, requires high temperatures for steam reforming.

HYDROGEN–OXYGEN FUEL CELL

It is apparent from the above discussion that the only practical high-performance fuel for use in fuel cells is hydrogen, which shows oxidation behavior on suitable catalysts which is sufficiently rapid to be close to being thermodynamically reversible, even at ambient temperatures. If common fuels, eg, natural gas, other hydrocarbons, or even coal, are to be employed with fuel cells, they must first be converted into hydrogen or hydrogen-rich gases. This is addressed below.

At ambient temperatures, the highest rates for electrochemical hydrogen oxidation are on platinum and palladium (15). These catalysts may be supported on high-surface-area carbon blacks (25,26), which may be graphitized for the highest corrosion resistance (27,28). Typical catalyst powders have 10% by weight of platinum on carbons with a specific surface area of about 250 m^2/g. The corresponding specific surface area of the platinum catalyst itself is about 100 m^2/g. A typical loading at the hydrogen anode is 0.25 mg/cm^{-2}, resulting in an effect catalyst area of 250 cm^2 per geometric square centimeter (28).

Platinum catalysts of this type are generally used in aqueous media and in the phosphoric acid fuel cell operating at up to approximately 200°C. As temperature rises, the nature of the electrocatalyst becomes less critical, since reaction rates rise with temperature. The usable specific surface areas of suitable catalysts are also lower as the operating temperature increases, because of the increased rate of sintering. In fuel cells operating at about 650°C in molten carbonate electrolyte, sintered nickel anodes containing a sintering and creep inhibitor are used, with a specific area of about 1 m^2/g. At 1000°C, nickel cermets are used as anodes with solid oxide electrolytes.

So far, we have said nothing about the reactivity of oxygen in fuel cells. Both hydrogen and oxygen molecules (dihydrogen and dioxygen) have two atoms. However, the two hydrogen atoms are joined by a single two-electron bond. When this is adsorbed on a suitable catalyst (preferably platinum or palladium, somewhat less effectively nickel) for which the Gibbs energy of adsorption is near zero or slightly negative, it will adsorb with electron transfer to two adjacent catalyst surface atoms. This results in dissociation of the bond, which can be followed by rapid electron transfer to give protons. Three possible rate-determining steps can occur in the anodic direction (with corresponding steps for hydrogen evolution):

$$H_2 \rightarrow 2H_{ads} \qquad (45)$$

$$H_{ads} \rightarrow H^+ \qquad (46)$$

$$H_2 \rightarrow H^+ + H_2 \qquad (47)$$

Reaction (45) in the anodic direction is often called the combination or Tafel reaction, (46) the discharge or Volmer reaction, and (47) the ion plus atom or Heyrovsky reaction, after their proposers. The first is normally rate-determining near-equilibrium on platinum-group metals, whereas the others are important at high overpotentials and on less reactive surfaces (15,20).

Other single-bonded molecules, eg, difluorine or dichlorine, also have high electrode reactivity, for the same reasons as hydrogen. However, dioxygen is much less reactive. Its Periodic Table neighbor, dinitrogen, shows scarcely any reactivity at all. Dioxygen has a double (four-electron) bond, whereas dinitrogen has a triple (six-electron) bond. This accounts in part for the increasing stability of the molecules. Since electrons must be transferred one at a time in a reaction sequence, dihydrogen only requires two electronic steps for oxidation, giving two protons. However, dioxygen requires four separate electronic steps to complete its reduction. If water is to be the final product, this requires four protons also, either from an acid electrolyte or from water molecules. This leads to a complex and less probable (ie, slower) reaction sequence for dioxygen reduction, in which many parallel pathways are possible (15,29). Reduction of dinitrogen to ammonia would require an even more complex and less probable sequence. As a result, the electrochemical reduction of dinitrogen is so slow that it does happen to any significant extent at ambient pressures over the whole of the fuel cell operating temperature range.

Dioxygen reduction on effective catalysts, e.g., platinum, does not have excessively high activation energies, but it seems to require specialized or favorable sites which must occur only in small numbers on the electrocatalyst surface (29). Since electrocatalysts such as platinum show reaction with water, ie, coverage with adsorbed OH, which favors the oxidation of hydrocarbon residues discussed in the previous section, sites showing the most ready adsorption are occupied (24), reducing the chances of dioxygen adsorption and reduction.

As a result, dioxygen reduction is a slow process close to ambient temperatures, with complex kinetics, and often shows two Tafel slopes corresponding to a change of mechanism or a change in oxygen radical adsorption conditions. Near ambient temperatures in acid electrolyte, the low-overpotential slope is equal to RT/F (about 60 mV/pdecade, $\alpha_c = 1.0$), with j_0 equal to approximately 10^{-10} A/cm^{-2} (30). Crossover to a higher slope ($2RT/F$, about 120 mV/decade, $\alpha_c = 0.5$) occurs at about 0.8 V vs. hydrogen, i.e., about -0.43 V from the hypothetical reversible oxygen potential, with an extrapolated j_0 equal to about 5×10^{-10} A/cm^{-2} (29). The reversible oxygen potential was called "hypothetical" above because it is never attained. At about 0.95 V vs. hydrogen, the cathodic process encounters equal and opposite anodic processes such as oxidation of the electrocatalyst or its substrate (31) and/or the oxidation of impurities (32). In consequence, in aqueous medium the open-circuit potential of practical oxygen electrodes normally lies about 0.25 V below the theoretical value under these conditions.

In the absence of other limitations, the kinetic rates of both the hydrogen and oxygen electrode processes are proportional to the concentrations of reacting molecules, i.e., to the gas partial pressures. The electrodes used in practical fuel cells are termed *three-phase-boundary* or *gaseous diffusion* electrodes. They have porous structures to maximize the area of contact between the gas, electronically conducting catalyst, and ionic phases and to maximize gaseous diffusion. Even so, local depletion of a reactant may occur at high current densities due to the inability of gaseous diffusion to keep up with reaction rate. Product accumulation (eg, water vapor at the oxygen cathode in acid media) will also dilute the reactant. This results in deviations from the Tafel line at high current densities. These are more marked for oxygen than for hydrogen, because of the lower diffusivity of the heavier molecule. Figure 1 shows typical hydrogen and oxygen polarizations for platinum electrodes in a stable organic perfluorinated polymer sulfonic acid electrolyte in a proton exchange membrane fuel cell (PEMFC, see below) at 70°C with hydrogen and oxygen at atmospheric pressure and 5 atm absolute (atma) pressure. A typical polarization plot for methanol oxidation is also included. We should note that the deviation from the Tafel line for oxygen reduction due to diffusion occurs before the higher Tafel slope is observed. The j_0 values are those for high-surface-area area electrodes. The figure includes a schematic cross section of the fuel cell.

In general, the low rates of dioxygen reduction require heavier catalyst loadings than those at the anode to obtain effective performance. Acid fuel cells for terrestrial applications normally use cathodes with loadings in the range 0.5–0.75 mg/cm^{-2}, although efforts are being made to reduce these values to facilitate a wider range of applications. The carbon-supported platinum catalysts are similar to those used at the anode, generally using graphitized carbons because of the aggressive conditions at the cathode, especially in phosphoric acid fuel cells operating in the 200°C range (28). It has been discovered that a number of platinum alloys with a base metal as a minority component are more active than pure platinum (28,33–35) by up to about 50 mV at constant current density, implying a three-fold increase in j_0 on these materials compared with the value on pure platinum. It is assumed that these are now being employed by developers.

Due to the low rate of oxygen reduction, the fuel cell whose performance is illustrated in Figure 1 can only produce about 0.8 V at usable current densities (ca. 0.5 A/cm^{-2} with air at 5 atma or oxygen at 1 atma). Under these conditions, the Gibbs energy efficiency for hydrogen consumed in the cell is $V/(-\Delta G/nF)$, where V is the cell voltage, ie, 0.8 V/1.23, or 65.0%. The corresponding thermal efficiency, $V/(-\Delta H/nF)$, is 0.8/1.254, or 63.7% (LHV), or 0.8/1.48, or 54.0% (HHV). For methanol at the same current density, the cell voltage is 0.45. The corresponding efficiencies are (Gibbs energy) 0.45/1.216, or 37.0% (thermal HHV), 0.45/1.255, or 35.9% (thermal LHV), or 0.45/1.103, or 40.8%.

The lack of reactivity of dioxygen at ambient temperatures must have been an important factor in allowing complex living matter to develop, after nature learned to produce the gas via photosynthesis from sunlight. This reaction, and the corresponding reduction reaction in mitochondria, required the development of active biochemical catalysts. If oxygen had been as reactive as, eg, chlorine, life as we know it would not have been possible. However, dioxygen becomes very reactive at high temperatures, at red heat and beyond. In high-temperature fuel cells, the oxygen electrode may show very rapid kinetics, similar to those for hydrogen at ambient temperatures. Unfortunately, the consequent reduction in overpotential is offset by the change in ΔG with temperature according to Eq. 8. This is illustrated in Figure 2. As a result of this effect, the voltage developed by a fuel cell at high temperatures is not very different from that in aqueous electrolytes, in spite of the vastly improved kinetics for dioxygen reduction. However, noble metal catalysts are not required in high-temperature cells, which in any case possess other advantages which are considered in a later section.

So far, we have assumed that gaseous hydrogen and oxygen reactants and product water can be made to react, virtually to completion, under standard thermodynamic conditions (with regard to the phase and the pressure, if not necessarily to the temperature). This is indeed possible if the fuel cell operates below the boiling point of water, so that liquid water is the product, which can be readily separated from the reactants. Under these conditions, pure hydrogen and oxygen at 1 atm pressure can be supplied "dead-headed" to the anode and cathode of the cell. This implies that each anode and cathode is a closed-ended chamber into which the gases are supplied at constant pressure until they are consumed. Under these conditions, the cell operates on the thermal energy represented by the HHV of combustion of hydrogen. However, terrestrial fuel cells normally operate on air rather than on pure oxygen. In this case, air is allowed to enter into the cathode chamber and flows toward the cathode exit, oxygen being consumed along the way. The oxygen therefore becomes progressively more dilute, and the current density becomes lower as the air traverses the cell. Increasing the utilization of the oxygen in the air results in diminishing returns, but clearly, the mean partial pressure of oxygen is considerably less than that in the standard state at 1 atm.

In an acid fuel cell, conduction is via H$^+$ ions produced at the anode, which react to produce water at the cathode.

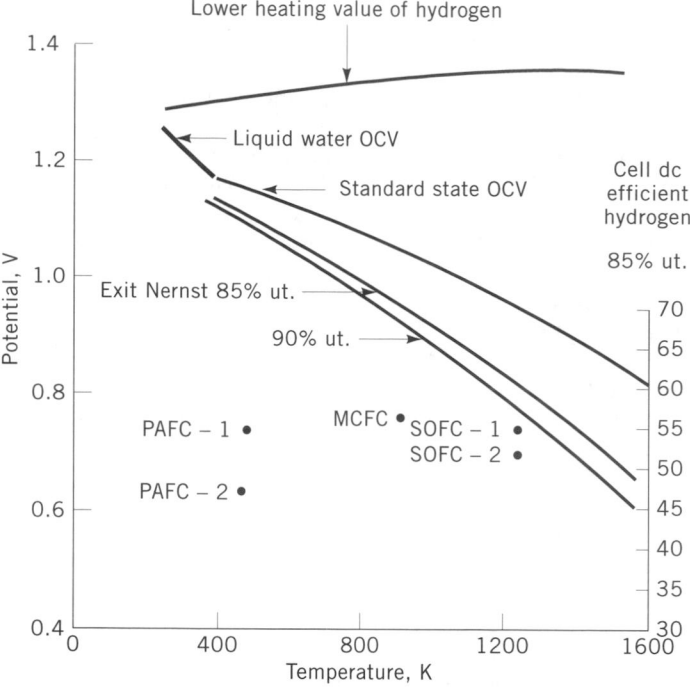

Figure 2. The change in $-\Delta G°$ for hydrogen oxidation to liquid and gaseous water with temperature (T, K) in electron-volts (eV, volts per equivalent, ie, reversible electrochemical cell voltages, E); $E = -\Delta G°/nF$, where F is the Faraday (96,485 coulombs per equivalent, or J/eV), and $n = 2$ equivalents per mole for hydrogen oxidation to water. Calculated from the expression $\Delta G°$ (J/mol) $= -240,200 + 3.93T \ln T + 0.0069\,T^2 - 1.55 \times 10^{-6}T^3 + 16.4T$. (Cf. G. Lewis and M. Randall, *Thermodynamics and Free Energy of Chemical Substances*, McGraw-Hill, New York, 1923.) The heat of reaction to gaseous water ($-\Delta H$ in eV) is superimposed. The exit Nernst plots show the highest theoretical cell voltage which may be obtained at 85 and 90% hydrogen utilization respectively. Scales at right are the cell dc HHV efficiency (for hydrogen, equal to 85% of LHV to gaseous water product). Operational fuel cell data points (all on reformed natural gas): PAFC-1: proposed International Fuel Cells (IFC) pressurized PAFC (1985) at 0.216 A/cm²; PAFC-2: IFC 200 kW nonpressurized PAFC at 0.325 A/cm²; MCFC: state-of-the-art (1990) at 0.16 A/cm²; SOFC-1, -2; Westinghouse 1993 at 0.15 and 0.30 A/cm².

If the cell operates at temperatures sufficient to produce water vapor, this further dilutes the cathode air, more so toward the cathode chamber exit, further reducing the partial pressure of oxygen. In general, a maximum oxygen utilization from air is usually 85%, which also determines the partial pressure of water vapor in the cell, and hence the electrolyte concentrations, which may have important implications for the ionic conductivity of the electrolyte. The cathode air is almost always run on a one-pass basis without feedback to economize on the work of circulation. In an alkaline fuel cell (AFC), ionic conduction is via OH^- ions produced at the cathode, which react with hydrogen, producing electrons and water at the anode. Hence, if pure hydrogen is used as fuel, it is diluted by water vapor as it passes through the anode chamber. It is not possible to consume the hydrogen, since if this were to be so, both the current density and the reversible hydrogen potential

vs. that of the oxygen electrode (ie, the cell potential) would go to zero. In consequence, the hydrogen is generally operated in a feedback loop, in which a condenser continuously removes water vapor, discharging the latent heat of evaporation. In such cases, the hydrogen fuel may be operated at 30% utilization per pass, depending on the temperature. Aerospace AFCs, such as the United Technologies 12-kW PC-17C units in the Space Shuttle *Orbiter*, operate on this principle using pure oxygen as oxidant at 4 atm pressure to increase performance and thus render the system more compact (36).

For terrestrial applications, pure hydrogen is generally not available as the fuel. Fuel cells are supplied with manufactured fuel, consisting of various mixtures of H_2 and CO and/or CO_2. Just as it is not feasible to separate pure oxygen from air, extraction of pure hydrogen from such mixtures is neither energetically nor financially economical. These gas mixtures are not appropriate for use with AFCs, since the CO_2 present will neutralize the alkaline hydroxide electrolyte. The aqueous carbonates which form precipitate in the cell and the electrolytes, have poor conductivity characteristics, and permit the formation of OH^- concentration gradients which seriously reduce performance. Pure hydrogen (and indeed chemically scrubbed air to eliminate CO_2) is required for use with terrestrial AFCs. In spite of their potential advantages (see below), these considerations have severely limited their ordinary use. Mixtures of H_2 and CO_2, with CO reduced by the water–gas shift reaction [reaction (44)] to levels which avoid anode poisoning (see above), are the norm for acid fuel cells. As these mixtures react, the hydrogen becomes progressively more dilute toward the anode exit. All of the hydrogen cannot be used, since the current density, as well as the cell voltage, would then go to zero. The practical utilization represents what is economically feasible. Since the anode exit gas, even at 80% utilization, has some heating value which can be used for other purposes, such as to provide heat for steam reforming, reaction (43) in an external reactor. Thus a figure of 80% is often used as a practical utilization level. Any further increase in utilization would seriously affect cell voltage.

In contrast to acid–electrolyte fuel cells, the HTFCs can handle CO, since it is spontaneously converted inside the anode chamber of the cell to H_2 and CO_2, via the water–gas shift process, reaction (44). We have already remarked that the HTFCs (MCFC and SOFC) rely on O^{2-} transport from the cathode to the anode, either alone or via CO_3^{2-} as a carrier ion. Thus, water vapor is produced at the anode as H_2 is oxidized, which drives the water–gas shift reaction, consuming any CO which is present until equilibrium is reached. As hydrogen is consumed across the face of the fuel cell anode, the fuel gas becomes more and more dilute, resulting in a fall in current density. Under these conditions, it is not possible to drive the reaction to zero hydrogen content in the fuel gas stream. As before the hydrogen utilization is determined by economic considerations. In the SOFC, it is usually limited to 85%, and it may be lower in the MCFC. If internal steam reforming of the fuel [eg, methane (natural gas), reaction (43)] is used with fuel cell stack waste heat, then the heating value of the gas in the anode exit stream is wasted, although it might serve for cogeneration purposes.

HISTORY OF FUEL CELL DEVELOPMENT

The battery was invented by Volta in 1800 using the zinc–silver oxide couple. It was immediately used to show that electricity could decompose many compounds by electrolysis, which, eg, led to the discovery of sodium and potassium by Davy in 1807. The realization that water electrolysis might be reversible was first realized by Grove, a British jurist and gentleman scientist, in 1839, with the invention of what he called the "gaseous voltaic battery." Each cell consisted of two platinum wires platinized with platinum black dipping into sulfuric acid electrolyte. One had a meniscus in contact with hydrogen gas, the other with oxygen. He could draw a small current from the cell at a voltage of about 0.5–0.6 V. Four such cells in series could be seen to decompose water. A series of papers describing Grove's fuel cells appeared in the *Philosophical Magazine* and the *Proceedings of the Royal Society* over the period 1839–1845 (37). Grove realized that an effective fuel cell would require what he described as a "notable surface of action," ie, the largest possible interface per unit area between the gas phase, the electrolyte, and the electrically conducting active surface, which would later be called the electrocatalyst. In later editions (38) of his book *The Correlation of Physical Forces*, first published in 1846, he described fuel cells operating fuels other than hydrogen. In 1889, the German-born scientists Mond and Langer constructed a fuel cell prototype in England which resembled many modern designs (39). It consisted of two sheets of perforated gold leaf covered with platinum black, both in contact with a thin stable porous diaphragm or matrix containing the electrolyte. When one electrode contacted hydrogen and the other air at ordinary temperatures, they were able to obtain 0.73 V at a current density of 3.5 A/cm². They were the first to realize the high efficiency of direct electrochemical energy conversion, stating that "this gives a useful effect of nearly 50% of the total energy contained in the hydrogen absorbed in the battery." This was before the work of Gibbs was generally known, and Mond and Langer meant 50% of the enthalpy (HHV) of combustion of hydrogen. The standard enthalpy of combustion of hydrogen to liquid water is 286.0 kJ/mol, or 143 kJ/equivalent. In voltage terms, this is equal to 143/F or 1.482 eV. Thus, the HHV thermal efficiency of Mond and Langer's cell (the ratio of the work done to the enthalpy of combustion in electronvolts) was 49.3%.

In his discourse on the foundation of the Bunsengesellschaft in 1894 (40), Ostwald looked forward to a future of electrochemical, rather than thermal, combustion, which would be more efficient and much cleaner than the coalfired steam engine of his day. However, the competition was the internal combustion engine operating on oil distillates, which took over the world for all but very large steam plants. Even so, work was going on at about the same time on the direct use of coal in high-temperature fuel cells, eg, that of Jacques and of Baur and co-workers. In later work, Baur was the first to make use of molten carbonate as a high-temperature electrolyte.

Grove had remarked on the difficulty of maintaining what he called "a notable surface of action" in his electrodes. Those of Mond and Langer became wetted by the electrolyte an the product water formed as time progress, so that the surface available for gaseous reaction was reduced. Wetting of high-temperature electrodes by molten carbonate electrolyte was controlled by Baur and co-workers by using capillary action. Their matrix contained a finer mean porosity than the electrodes, so that the latter did not flood (41).

An excellent review of early fuel cell work in given in Ref. 21. Starting in 1933, Bacon in England used concentrated KOH electrolyte in pressurized cells at 200°C. This work continued (with a break during World War II) until the late 1950s (42). Schmid had previously used a fine-pore platinum black electrode on a coarse-pore graphite layer in 1923 in an attempt to control wetting (43). The electrodes eventually used by Bacon consisted of fine-pore sintered nickel toward the gas side, with a coarse-pore support toward the electrolyte, which was circulated to remove heat. Pure hydrogen and oxygen were used as fuel and oxidant at 45 bars pressure. Bacon had demonstrated 1.0 A/cm² at 0.8 V under these conditions by the late 1950s (42). Justi and Winsel developed dual-porosity nickel electrodes in parallel withh Bacon's later work, but their structures were intended for use in more dilute aqueous KOH at low temperatures and pressure. The lower reactivity of the gases under these conditions was offset by the use of high-surface-area Raney nickel in the active layer (44). Another approach to maintaining a high-surface-area three-dimensional microscale meniscus in the electrode for aqueous electrolytes is the incorporation of nonwetting materials in the electrodes for meniscus control. Heise and Schumacher (45) attempted to use paraffin wax for this purpose in 1932. After polytetrafluoroethylene (PTFE, Teflon, E.I. DuPont de Nemours) became available in the 1950s, it began to be used for this purpose, apparently first by workers at the laboratories of General Electric, Schenectady, NY, and at Union Carbide Corporation, Parma, OH (21). It is now used in virtually all fuel cell electrodes for use in aqueous media to control internal wetting of electrodes and to maximize the total area of triple contact between gas, electrolyte, and electronically conducting catalyst per unit of geometrical area (the "three-phase boundary").

Acid fuel cells demonstrated during the later 1950s, eg, at the General Electric Research and Development Laboratory, used high-surface-area Teflon-bonded platinum black electrodes and sulfuric acid as the electrolyte (21). The aim was to develop systems which would operate directly on hydrocarbon fuels, which could not be used in alkaline fuel cells because of the problem of carbonate formation in the electrolyte. It soon became apparent that high temperatures in excess of 150°C would be required for effective operation, even with electrodes containing 40 mg/cm² of pure platinum black. Sulfuric acid could only be operated up to about 80–90°C, since it was reduced (at least by hydrogen) at the anode. A search for other possible acids showed that only phosphoric acid had both the required stability and lack of volatility to operate at these temperatures. Orthophosphoric acid is not normally considered to be a strong acid, but at temperatures beyond the boiling point of water it becomes successively dehydrated, forming pyrophosphate polymer chains. Sulfuric acid (a relatively strong acid) shows the same property, though in a less degree, since it can dimerize to form pyrosulfuric acid. Their structures are as follows:

$$[(OH)_3P{=}O]_n$$
$$\rightarrow -O-PO(OH)-O-PO(OH)-O-PO(OH)- \quad (48)$$

$$(OH)_2S({=}O)_2 \rightarrow HO-S({=}O)_2-O-S({=}O)_2-OH \quad (49)$$

The corresponding Group IV acid, silicic acid, is not only weaker than phosphoric acid, but its valence state means that corresponding polymers are totally deprotonated. The protons can ionize in both the Group V and Group VI pyroacids, but the fact that phosphoric acid is a relatively weak acid with a proton on each phosphate group gives it unique properties. Not only can it form stable polyanions of the type $-P({=}O)_2^--O-$, but each of these groups also has the capability of forming cations with the $-P(OH)_2^+-O-$ structure. Pyrosulfuric acid is not only unstable at the hydrogen anode, but also has a much lower probability of forming corresponding cations. That the cation, anion, and neutral groups in pyrophosphoric acid have an approximately equal probability of formation and that protons can transfer down the chain via a hopping mechanism from one phosphate group to the next give pyrophosphoric acid an ionic conductivity which is apparently unique. Other aqueous acids, including sulfuric and pyrosulfuric, have protons which become immobile as water is lost, so that their monohydrates (in which the proton is present as H_3O^+) form ion pairs with their anions, giving very low conductivity. When phosphoric acid was first used as a high-temperature acidic electrolyte, its unique proton conductivity properties were accepted, but the fact that other potentially stable acids might not show similar properties was overlooked. At constant water activity, the conductivity of phosphoric acid improved with temperature, whereas that of true aqueous acids decreases. In effect, the phosphoric acid fuel cell (PAFC) operating in the temperature range in which it shows advantageous properties and performance (ie, 150–210°C) is not strictly an aqueous fuel cell, but a molten acid salt system (46).

Even at temperatures up to the maximum available, the performance of the PAFC as a direct converter of hydrocarbons was shown to be disappointingly low. It showed uneconomic performance even with methanol fuel. However, it could operate at temperatures well beyond the boiling point of water, and therefore its waste heat could be used to raise steam for a reformer and shift converter, which would increase the efficiency of steam reforming of natural gas and light hydrocarbons. In addition, the platinum anode of an acid fuel cell operating at 90°C is poisoned by the presence of more than a few ppm of CO, whereas at 200°C, a PAFC anode will tolerate up to 1–2% of CO in the anode inlet gas from a shift con-

verter. These two features attracted the engineers at Pratt and Whitney Aircraft Division of United Aircraft to the PAFC system for use with reformed hydrocarbon fuels in the 1960s (1,28). Even though the Bacon-type AFC would operate at similar temperatures, it required total purification of both hydrogen and air streams to eliminate CO_2. The first in particular was considered to be economically and energetically impossible.

The challenge presented by the PAFC was that of the high cost of materials. Pourbaix's Atlas of Electrochemical Equilibria at 25°C (47), the standard compilation on the thermodynamic stability of the elements in noncomplexing aqueous solutions at 25°C as a function of potential and pH, shows that only niobium, tantalum, gold, and platinum and certain of its relatives were thermodynamically stable under PAFC cathode operating conditions, even at ordinary temperatures. By the early 1970s, it had been shown that various carbons were *kinetically*, if not *thermodynamically*, stable at the cathode and could show reasonable lifetimes (27,28). After graphitization, stabilities were even better. Careful work showed that corrosion was dependent on the partial pressure of water vapor and temperature, which allowed definitions of the temperature and pressure ranges in which stable operation could be expected. This led to the development of low-loading platinum electrodes supported on carbon and graphite structural elements, which could be fabricated at low cost in mass production. Since acid electrolyte fuel cells require platinum metal catalysts, an early goal was to obtain a performance equal to that of the best high-loading platinum systems with low-loading electrodes. During the early 1970s, a great deal of effort went into the development of carbon-supported catalysts, which allowed stable surface surface areas (in square meters per gram) about five times higher than those of pure platinum black (about 20 m^2/g). At the same time, it was clear that platinum was not being effectively used in high-loading electrodes, so effforts where made to increase the amount in contact with the gas phase and the electrolyte in thinner electrodes. As a result, platinum loadings were lowered by more than one order of magnitude and platinum utilizations were increased to over 50% (28). Performance increased to as much as 325 mA/cm^2 and 0.62 V at atmospheric pressure in the most recent PAFC systems (28).

In parallel to work on the direct-hydrocarbon PAFC system, the General Electric Company developed another acid system for application as a power source for the *Gemini* space capsule, in which its product water could also be used for drinking. The system had to be simple and light, because of the weight constraints of the small space capsule. The system was called the Solid Polymer Electrolyte system (SPE) by its developer and used a solid sulfonic acid electrolyte based on a nonfluorinated polystyrene divinyl benzene copolymer backbone ion-exchange membrane (21,48). Water dissolved in the material, but it was not itself soluble in water, so that product water was rejected automatically in pure form. It was completed by platinum black electrodes applied to the membrane and bonded under pressure. An early demonstrator made a suborbital flight on October 30, 1960, as the first fuel cell in space. The *Gemini* contract was awarded in early 1962, and the result was a-1 kW (nominal) pure hydrogen–oxygen sytem operating at 2 atm and 50°C at 0.78 V, but only 37 mA/cm². The system weighed 31 kg/kW, corresponding to 27 L/kW, and equal to 9 kg/m² of active area. In contrast, the same statistics for the Pratt & Whitney 1.5-kW, 115-kg PC-3A Apollo AFC prototype of the same year were 77 kg/kW, 100 L/kW, and 92 kg/m² of active area (36). The United Technologies 12-kW PC-17C Space Shuttle *Orbiter* system of 1974 represented a great improvement at 5.3 kg/kW (of which 55% were the cell stacks), 12.8 L/kW, and 26.0 kg/m² (36). In spite of its poor performance, the *Gemini* system remains one of the lightest ever constructed in terms of kilograms per square meter of active area.

The early material used in the *Gemini* fuel cell was much less stable than the later fluorinated materials typified by Nafion, a product of E. I. DuPont de Nemours and Company developed as a sodium-conducting membrane to produce pure NaOH for the chlor-alkali industry, replacing the mercury cell, which posed environmental problems. Fluorination made Nafion not only more stable than sulfonated polystyrene divinyl benzene sulfonic acid, but it also had much lower ionic resistance and higher performance (48). The fluorinated backbone of Nafion contains the same elements as a combination of the Teflons TFE (polytetrafluoroethylene), FEP (perfluoroethylene-perfluoropropylene copolymer), and PFA (perfluoroalkoxy), ie,

$$-(CF_2CF_2)_n-CF(CF_2-)-O-[CF_2-CF(CF_3)-O-]_m$$
$$(CF_2)_xSO_3H \quad (50)$$

where n is normally 2–3, $m = 1$, and $x = 2$. The material is a relatively dilute acid, with about the same conductivity as 1 M sulfuric acid (48). It behaves as a typical aqueous acid from the viewpoint of conductivity (46), so that its use is limited to temperatures below the boiling point of water. Work at GE showed cells using Nafion to be capable of attaining 10 times the current density of the *Gemini* fuel cell under comparable conditions, provided that the cathodes were improved by adding Teflon to provide a surface to avoid flooding by liquid product water (48). When GE sold its Nafion-based technology to the Hamilton Standard Division of United Technologies Corporation in 1984, the successor company retained all rights to the SPE trademark. In consequence, the technology is now known elsewhere as proton exchange membrane (or polymer electrolyte membrane), PEM.

Inspired by work by Greger (49) and Gorin (50), which was founded on experiments by Baur in the early 1920s (41), development of today's MCFC was started in the Netherlands by Broers and Ketelaar in the 1950s (51). The laboratory cells made by Broers and Ketelaar used sintered nickel anodes similar to those of Bacon's AFC, though with only one sintered layer. Broers and co-workers used various alkaline carbonate eutectics as electrolytes, usually retained by a magnesia matrix. Different cathodes were used, including silver, copper, and lithium-doped nickel oxide prepared in situ from sintered nickel structures similar to the anode. Only nickel oxide was shown to be sufficiently stable. Finally, lithium aluminate was adopted as the matrix material in late work by Broers and by the Institute of Gas Technology (IGT) in

Chicago in the late 1960s. Early work at IGT showed that stainless steel was sufficiently stable at the cathode side as a structural component, although it required cladding with nickel to avoid corrosion in the humid atmosphere at the cathode outlet. It also required protection from corrosion at the edges because of the presence of an oxidizing atmosphere outside and a reducing atmosphere inside, in the present of electrolyte. This required the use of aluminizing. The electrolyte was standardized as 62 mol % Li_2CO_3–38 mol % K_2CO_3 eutectic operating at a mean temperature of 650°C (52).

In work started in the late 1950s at Westinghouse, Weissbart and Ruka (53) were inspired by the operation of the zirconia-based oxide-ion-conducting high-temperature light source, the "Nernst Glower," invented in 1900 (54). The geometry of the system has seen changes to improve assembly (1,55), but the basis is a thin layer of yttria-doped zirconia oxide-ion-conducting electrolyte operating at 1000°C, which is also used in the oxygen sensors of spark ignition engines with catalytic converters. The anode is a nickel cermet, and stable conducting oxides with compatible coefficients of thermal expansion are used for the cathode. Construction and materials in recent systems are discussed in a later section.

FUEL CELL CLASSIFICATION

Today's fuel cells are descended from the above systems. They may be classified in various ways, ie, by their temperature of operation (high-, medium-, and low-temperature fuel cells) or by the type of electrolyte they use. The latter classification is perhaps the most convenient and logical, since the electrolyte used determines the operating temperature range, the operating characteristics, the materials which can be used, the type of catalysts required, and the overall performance.

AFCs are confined to using KOH electrolytes. The main reason for this choice is the higher solubility of potassium carbonate than sodium carbonate. Any CO_2 entering an AFC is converted to carbonate, so that both fuel and oxidant supplies must be as free as possible from this contaminant. The fuel must be pure hydrogen and the oxidant pure oxygen or CO_2-scrubbed air. Hydrogen manufactured from natural gas for aerospace applications always contains traces of CO_2, so some accumulation of carbonate in the electrolyte is inevitable. The use of KOH instead of NaOH reduces the possibility of precipitation of carbonate deposits in the electrodes. Even though lithiated sintered nickel oxide cathodes had a marginal lifetime for the 1968 and later Apollo missions (ie, approximately 15 days at 260°C), other materials considerations have now reduced the operating temperature of the AFC to about 115°C. Systems using practical materials are obliged to operate at 80°C or less. The lower limit of operating temperature is determined by the freezing point of the electrolyte. A cell designed to operate at 65°C will generally use 34 wt % or approximately 8 N KOH as the electrolyte, and it will be able to operate down to perhaps −30°C (34 wt %), although its low-temperature performance is limited by increasing electrolyte viscosity and reduced ionic conductivity.

The only true acid fuel cell electrolytes in practical use today are the perfluorinated proton exchange membrane materials such as Nafion, a somewhat similar material from the Dow Chemical Company in which $m = 0$ and $x = 2$ in formula (50), and the material known as Aciplex-S from the Asahi Chemical Industry Company, in which $m = 0-2$ and $x = 2-5$. The materials are made using slightly different chemistry and have slightly different properties, with equivalent weights of 1100, 800, and 1000, respectively. Typical electrolyte film thicknesses are in the $100-125$-μm range, resulting in cells with resistances of about 0.1 Ω-cm^{-2} when operating at temperatures below the boiling point of water under the particular imposed pressure conditions. Such cells operate at about $65-70°C$ under atmospheric pressure conditions and up to slightly over $100°C$ under pressurized conditions. In all cases, it must be noted that the oxidant (dioxygen) and reaction product (water molecules) compete for space. Insufficient water vapor pressure (whether from product water or from external humidification) means low ionic conductivity of the electrolyte. In contrast, high water vapor pressure will reduce dioxygen partial pressure, giving poor cathode performance. This gives an upper limit to the operating temperature of the system of about $105°C$ at a pressure of 5 atma. It can operate down to slightly below $0°C$. The acid electrolyte system is CO_2 rejecting and can therefore be used in principle with H_2/CO_2 fuel gas mixtures. However, the anode is sensitive to traces of CO and is poisoned by amounts exceeding a few ppm, even at its maximum operating temperature. Any excess anode catalyst exposed to the gas phase is sufficiently active to allow the reverse water–gas shift reaction to go to equilibrium, causing poisoning by initially CO-free H_2/CO_2 fuel gas mixtures. It was not clear at the time of writing if this problem was insuperable or whether elimination of excess catalyst would make it tolerable. If it is a problem, the PEMFC, like the AFC, will require pure hydrogen fuel.

We have already seen that the PAFC uses a self-ionizing oxyacid as its electrolyte. This does not require water for conduction, so it can operate at temperatures which exceed the boiling point of water. Systems designed for operation at $190-200°C$ use an electrolyte concentration corresponding to 97 wt % orthophosphoric acid, which is in equilibrium with the typically partial pressure of product water vapor in this temperature range. This electrolyte will freeze at about $+40°C$. The lower limit of operating temperature is determined by performance considerations, and it is usually about $150°C$ for use with H_2/CO_2 fuel gas mixtures which have been water–gas shifted to give acceptable CO contents. At lower temperatures, its CO tolerance is reduced. The upper limit of operating temperature is determined by materials considerations at the cathode, involving corrosion of the graphitized carbon cathode support and slow sintering and area loss of the platinum catalyst. The latter shows a voltage decay which is proportional to log time (28). Support corrosion seems to result in a linear voltage decay with time, and since it is an anodic electrochemical process, its rate increases exponentially as the cell potential rises, according to the Tafel equation (Eq. 36). As a result,

the cell must be maintained below a predetermined voltage representing the onset of serious corrosion and must never be allowed to reach the open circuit voltage under air, especially when operating under pressure. This is achieved by the use of nitrogen blanketing at the cathodes on shutdown. Corrosion of the graphite intercell separator plate (bipolar plate) can also be a problem, especially if there is an ionic contact so that a galvanic corrosion cell can be set up driven by the potential difference between sites which cathodically reduce oxygen and anodic corrosion sites (28).

Like the PAFC, both the MCFC and SOFC have upper operating temperature which is limited by materials considerations. In the case of the MCFC, the maximum temperature point in the cell (about $675°C$) is determined by corrosion of components in the presence of molten carbonate electrolyte. The lowest temperature point (perhaps $625°C$) is determined by performance considerations. In the SOFC, the upper temperature limit of the cell (about $1050°C$) is determined by interdiffusion of solid-state components, particularly manganese into zirconia, which result in a loss in desired properties. The lower temperature limit is determined by performance considerations, which are largely due to increased electronic and/or ionic resistance of cell and intercell components. Today's minimum operating temperature with standard components is about $925°C$. Attempts are being made to reduce this to allow the use of steel peripheral components (heat exchangers) instead of those constructed from superalloys of ceramics.

Thus, fuel cells can operate from below ambient temperature to about $115°C$, which is followed by a gap to another operating range from 150 to $200°C$. Followed by other gaps, there are two other operating ranges from 625 to $675°C$ and from 925 to $1050°C$. These ranges are determined by the nature and properties of the only available electrolytes.

ENERGY CONVERSION EFFICIENCY OF FUEL CELLS

It is clear from the discussion given earlier that the maximum nominal thermal efficiency of a fuel cell operating at temperature T will be given by $\Delta G_{T1}/\Delta H°$, where ΔG_{T1} is the Gibbs energy of reaction at T_1 and where $-\Delta H°$ is the enthalpy of combustion of the hydrogen used in the fuel cell, which is normally defined under standard conditions. In practice, this can only be achieved in an ideal fuel cell operating under conditions which are truly thermodynamically reversible, ie, at zero net forward rate. A real fuel cell must consume fuel at a positive rate, and since fuel is valuable, it must consume as much of it as possible. A hypothetical cell with infinitely fast anode and cathode kinetics producing a combustion product which may be continuous separated and removed from the reacting anode and cathode gas streams would indeed have an efficiency given by the above expression. An example would be a cell producing liquid water as a product, in which the correct stoichiometric quantities of hydrogen and oxygen at 1 standard atmosphere pressure are continuously supplied to the anode and cathode, respectively.

In practice, all fuel cells operating under temperature conditions where liquid water is a product 1 atm pressure are limited by kinetic considerations, so the example is only of theoretical interest.

Practical fuel cells are required to operate at the highest possible power densities, which means at the highest temperatures their materials allow. This almost invariably means that product water is eliminated in vapor form, although certain low-temperature alkaline fuel cell and acid (PEM systems) do produce product water. In most cell designs, water vapor will progressively dilute either the anode or the cathode reactant gas. This depends on whether the cell operates by conduction of a proton carrier ion (H_3O^+) or an oixde carrier ion (eg, OH^-, O^{2-}, or CO_3^{2-}). In the former case, product water vapor is formed on the cathode side and, in the latter case, at the anode side. In aqueous systems, the electrolyte has a composition which is not far from being in equilibrium with the product water vapor, so both anode and cathode side become diluted. The molten salt (ie, molten carbonate) and solid oxide systems produce all their water vapor at the anode side.

In principle, the appropriate reactant gas or gases can be bled from the operating cell and cooled, and the water can be condensed as a liquid after a single pass of a volume of reactant gas which is much larger than the stoichiometric requirement. The reactant gases may then be recycled to the cell after heat exchange. This will ensure that the reactant gas remains at a high concentration everywhere in the cell. Fuel cell developers are usually reluctant to do this, because of the high cost of heat exchange equipment and because of the loss of total system efficiency due to the pumping and circulation losses and requirement to make up for overall heat losses. In real systems, especially those operating on practical fuel and oxidant mixtures, a line must be drawn between efficiency and capital cost. As a result, one or both of the reactant gases will normally be allowed to become progressively more dilute as it traverses the cell from inlet to exit, as the concentration of product water vapor rises.

If pure hydrogen and oxygen are used in an acid fuel cell, eg, a PEM system, water is produced on the cathode side. The water vapor will be in equilibrium with the water activity in the PEM electrolyte, which will in turn humidify the hydrogen at the anode. The system must include a provision to remove product water. In principle, water can be collected at either the anode or the cathode, but collection is more efficient at the cathode, where the product is formed. Hence, the hydrogen anode may be operated dead headed, and the oxygen at the cathode may be converted to water until its partial pressure becomes too low to give satisfactory performance, after which water is removed in a condenser or by pressurizing, and the oxygen is recycled. Pure hydrogen–oxygen systems are generally for space applications and are normally operated pressurized to give maximum performance. Under these conditions, pressurization at temperatures of about 90°C will result in the idealized situation of formation of liquid water at the cathode. This must be removed from the cathode itself, to avoid blockage of the reaction surface by a liquid film. Wicking may be used to redirect the

product water. This approach was used in the General Electric Company's early PEM fuel cell for the *Gemini* space capsule in 1965. In a recent development, United Technologies uses a microporous membrane in the cathode chamber opposite and parallel to the plane of the electrode. Pure water is circulated on the opposite side of this membrane to cool the cell, and the system is designed so that the cathode gas is at a higher pressure than the circulating water, so that product water is forced through the membrane and eliminated from the cell. Hence, both pure hydrogen and oxygen can be used dead headed (Fig. 3). In practice, the reactant gases are never absolutely pure, and inert impurities, generally nitrogen and argon, will build up in the anode and/or cathode chambers. Hence, a purge is required from time to time to remove impurities or a small amount of gas may be bled off at the dead-headed end of a large cell.

In alkaline space cells, typified by the United Technologies PC-17C used in the Space Shuttle, water vapor is produced at the anode side in a matrix cell. This dilutes the KOH electrolyte, which must be provided with an expansion volume in the form of an electrolyte reservoir plate (ERP) which is an electronically conducting sintered nickel structure to carry the bipolar current flow. It is in

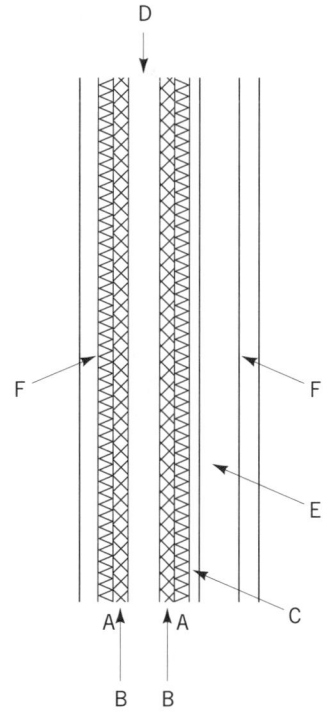

Figure 3. Schematic of recent IFC aerospace PEMFC operated dead headed on pure hydrogen and oxygen which uses a microporous membrane in the cathode chamber opposite and parallel to the plane of the electrode. Pure water is circulated on the opposite side of this membrane to cool the cell, and the system is designed so that the cathode gas is at a higher pressure than the circulating water, so that product water is forced through the membrane and eliminated from the cell. A, A, porous conducting reactant flow field; B, B, active electrode layers; C, water-permeable membrane for liquid water product removal on pressurized cathode side; E, coding water flow; F, F, bipolar plates.

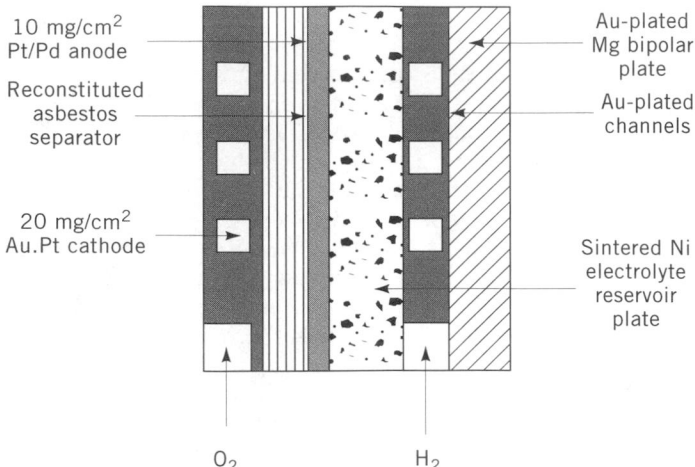

10 mg/cm² Pt/Pd anode

Reconstituted asbestos separator

20 mg/cm² Au.Pt cathode

Au-plated Mg bipolar plate

Au-plated channels

Sintered Ni electrolyte reservoir plate

O₂   H₂

Figure 4. Schematic of United Technologies (now IFC) PC-17C pure hydrogen–oxygen (4 atma) AFC used in the Space Shuttle, in which water vapor is produced at the anode side in a cross-flow matrix cell. The KOH electrolyte is provided with an expansion volume in contact with the anode in the form of an electrolyte reservoir plate (ERP), an electronically conducting sintered nickel structure to carry the bipolar current flow. It is arranged so that it does not interfere with the flow of hydrogen, which passes through a feedback loop for product water and heat removal. The oxygen cathode is operated dead headed. The schematic does not include two-piece dielectric liquid cooling plates.

contact with the anode (Fig. 4), arranged so that it does not interfere with the flow of hydrogen. The hydrogen is partially converted in the cell, to about 50% utilization, and the mixture of hydrogen and water vapor is recycled in a feedback loop which contains a condenser for product water removal. The oxygen supply is operated dead headed.

Whereas space cells must carry and use all of their reactant gases, for terrestrial use, the oxidant is normally air, which is available in any required quantity subject to the energy required to circulate and/or to compress it, if applicable. Oxygen separation or concentration from air has been dismissed as impractical, since it requires too much energy. Thus, air is used directly at the fuel cell cathode. In acid electrolyte cells, eg, the PEM, product water vapor is formed in the air as it traverses the cell, as oxygen is consumed. The major design consideration is how much air must be circulated, ie, the oxygen utilization in the cell. If ambient-pressure air is used, an oxygen utilization of 75% will result in an oxygen partial pressure of only 0.0435 atm at the cathode exit, where the water vapor pressure will be 0.26 atm, if the temperature is high enough to sustain this value. This oxygen partial pressure will normally be too low to sustain sufficiently high reaction rates close to the cathode exit. In contrast, if the oxygen utilization is 25%, the volume of air required will be three times higher, and the water vapor pressure at the exit will be only 0.095 atm. This will almost certainly result in drying out of the membrane. Humidification of the incoming air will then be required. As a compromise, 50% utilization is normally the aim, which will result in exit oxygen and water vapor partial pressures of 0.091 and 0.182 atm, respectively. The situation at an

acid fuel cell anode in which hydrogen is supplied as part of a gas mixture with inert components, eg, from a reformer and shift converter, is similar. Using natural gas as the fuel with a steam-to-carbon ratio of 3, followed by two-stage water–gas shifting will result in a mixture with the composition $4H_2 + H_2O + CO_2$, along with small amounts of CO (about 1.5% for feedstock for the PAFC). Initially, the mixture contains about 67% of hydrogen. At 80% utilization, the hydrogen content is reduced to 13.3%, ie, a H_2–CO ratio of about 9:1. Under these conditions, the operation of the hydrogen anode begins to show a reduction in performance due to the start of the effects of poisoning by CO, so this is about the practical limit for the anode exit gas composition.

Since the cathodes of aqueous cells are entirely limited by kinetics, the low concentration of oxygen in air near the exit is the controlling factor, since the presence of product water has no effect on reaction rate in a kinetically controlled situation. However, in an alkaline cell in which water is produced at the anode, the system may operate close to thermodynamic equilibrium due to the rapid hydrogen oxidation kinetics. Hence, as the reaction is driven toward higher hydrogen utilizations and rising product water partial pressures, from the Nernst equation the reversible cell potential will be effectively reduced by the amount.

$$\Delta V = RT/2F(\ln p_{H_2}/p_{H_2O^\circ} - \ln p_{H_2}/p_{H_2O}) \tag{51}$$

as the reactant traverses the cell. In this expression, $p_{H_2^\circ}$ and $p_{H_2O^\circ}$ are the partial pressures of hydrogen and water vapor at the anode inlet and p_{H_2} and p_{H_2O} are the corresponding values at a given point in the cell. It is clear from material balance considerations that

$$p_{H_2} = p_{H_2^\circ} - (p_{H_2O} - p_{H_2O^\circ}) \quad \text{and} \quad p_{H_2O} = p_{H_2O^\circ} + (p_{H_2^\circ} - p_{H_2}) \tag{52}$$

In high-temperature fuel cells (MCFCs and SOFCs) both electrodes show the reversible potential under zero-current conditions. The revesible or Nernst potential at the cell anode exit then takes on a particular significance which is easiest to appreciate for parallel-flow (coflow or counterflow) systems. It is the highest potential the cell can attain under the particular conditions of fuel utilization assumed at the exit, when no driving force for reaction will be present and the current density will have fallen to zero. Neglecting any small Nernst effects at the cathode, the effect of utilizations of 50% ($p_{H_2} = p_{H_2O} = 0.5$ atm), 75% ($p_{H_2} = 0.25$ atm, $p_{H_2O} = 0.75$ atm), 85% ($p_{H_2} = 0.15$ atm, $p_{H_2O} = 0.85$ atm), and 90% ($p_{H_2} = 0.1$ atm, $p_{H_2O} = 0.9$ atm) at an anode operating on pure hydrogen is shown as a function of temperature in Figure 2. The maximum operating voltage of the cell shows significant losses as operating temperatures increase. This is a major practical problem of the SOFC, in which the anode and cathode polarizations may be negligible and the IR reduced to acceptable values yet may operate at lower cell voltages than the MCFC, in which the anode and cathode polarizations are certainly not negligible. Thus, the MCFC is the most efficient fuel cell to date.

DESIGN CONSIDERATIONS

Fuels and Fuel Clean-up

Fuels Available. The anodic reactant in a fuel cell should be hydrogen. However, common fuels are hydrocarbons and their derivatives, which must therefore be converted to either a hydrogen-rich gas or pure hydrogen. The most important fuel considered is as-delivered natural gas, which is generally of lower cost than other hydrocarbons and derivatives. For certain applications, the use of methanol, ethanol, kerosene, jet fuel, diesel fuel, military combined jet–diesel fuel (JP-8), gasoline, liquid propane gas (LPG), naphtha, or refinery off-gas may be desirable. The use of coal-derived synthesis gas is a special case, which is discussed separately. Finally, biomass gasified by fermentation, by steam reforming, or by oxidative gasification is possible.

Steam Reforming. Steam reforming is the normal first stage of conversion of carbonaceous fuels, although partial oxidation with injected steam (adiabatic reforming), which may use a less efficient but much more compact reactor, may be considered for special applications. Since both steam reforming and adiabatic reforming are catalytic processes, they are not tolerant of catalyst poisons, particularly sulfur in quantities above a fraction of 1 ppm. In consequence, the first stage of fuel treatment is sulfur removal, usually via hydrodesulfurization. This involves injection of hydrogen into the high-temperature fuel stream to convert organic sulfur compounds to H_2S, which is absorbed by an active bed, eg, zinc oxide. The absorbant bed is generally expendable.

Because of the difficulty of handling sulfur, feedstock with the lowest possible sulfur content is desirable. Natural gas is desulfurized before it is exported via the interstate pipeline system, but it has a minimum of 1–2 ppm of a mercaptan/thioether added as an odorant before distribution. Methanol and ethanol are sold sulfur free. Before the 1990 Clean Air Act, gasoline in the United States was required to have a sulfur content below 0.1 wt % (1000 ppm) but was usually lower. For diesel fuel, the U.S. military specification since 1965 (MIL-T-5624G, November 5, 1965) have been a maximum of 0.5% (5000 ppm), but the fuel used was usually much better, except in certain overseas locations. Until the late 1980s, diesel fuel in the United States typically contained 1500–3000 ppm of sulfur. The 1990 Clean Air Act specifies "base" or "baseline" gasoline for vehicle emissions testing variously at 316 and 339 ppm sulfur. The 1990 Clean Air Reauthorization Act set an upper limit of 1000 ppm sulfur on diesel fuel for vehicle certification until the 1993 model year, with a maximum level of 500 ppm for all fuel sold after October 1, 1993. The EEC requires that the sulfur content of diesel motor fuel should be reduced to 200 ppm by 1996. Jet fuel is specified to have less than 4000 ppm, but the current amounts in JP-4, JP-5, JP-8, and Jet-A are closer to 300 ppm. In consequence, natural gas and sometimes methanol are the most desirable fuels. The heavier hydrocarbons require higher steam–carbon ratios for reforming than natural gas to prevent cracking and should be completely volatilized before entering the reactor. The

U.S. military has expressed interest in small fuel cells with fuel processors using JP-8 feedstock for mobile battlefield power, but this is currently beyond the state-of-the-art. Power on remote Japanese islands may be provided in the future using PAFC units operating on desulfurized kerosene using specially designed reformers (56).

A further possibility is the use of methanol for this application, which is unique in allowing effective reforming at low temperature (ca 300°C, rather than over 700°C). Methanol is produced from reformate (synthesis gas) in a catalytic process and thus is sulfur free as delivered. At the low temperature required for methanol reforming, the reaction essentially goes to completion (ie, to CO_2 + $4H_2$), so that no shift converters are needed to reduce CO to levels which do not poison the PAFC anode. This advantage may make methanol attractive for use with fuel cells for use in electric vehicles.

Integrated Reforming. A PAFC operating at 190–200°C can supply the excess steam required for efficient reforming via waste heat from the cell stack. In general, steam-to-carbon ratios of 2.5–3.5 are used, compared with a theoretical value of 1 for reaction (43), to drive the reaction as far as possible to equilibrium and improve kinetics. The anode off-gas is used in the reformer burner to provide the enthalpy of reforming. The availability of free excess steam reduces the energy requirement for reforming by 11–14%, giving a substantial gain in overall system efficiency. The reforming process is followed by two-stage shift conversion [reaction (44)] at high and low temperatures to reduce CO levels to acceptable values for the PAFC operating at 190–200°C, ie, 1.5% CO in the incoming fuel gas, i.e., a ratio of H_2 to CO of about 25 at the anode exit at 80% hydrogen utilization in the fuel cell. The overall HHV efficiency of a large pressurized 11-MW PAFC unit was estimated at 42% overall (net ac), or 44.3% (gross d-c power output).

Hydrogen Fuel. The HHV efficiency of the fuel cell operating on hydrogen at 0.73 V is 49.3%. AFCs require pure hydrogen as a fuel. As was already stated, recent work has shown that PEMs may not operate effectively on reformate containing a few ppm of CO produced by partial oxidation following low-temperature methanol reforming or two-stage water–gas shift after natural gas reforming. State-of-the-art hydrogen production from natural gas might use an advanced reformer, perhaps based on the integrated concept developed by Rolls-Royce and Associates (57). Operating as an autothermal reformer, this has a natural gas to reformate HHV efficiency of 91% (85.7% LHV). We can assume an advanced reformer with a conventional burner with an integrated shift converter with the same efficiency [eg, in a system of flat-plate type (56) to produce an 80% H_2–20% CO_2 mixture]. The latest data for the work requirements to separate CO_2 using pressure-swing adsorption indicate a maximum of 0.35 kW·h/H-m^3 (56). If this electric power is produced from natural gas at a modest 38% LHV efficiency, this will produce hydrogen from natural gas at 78.7% LHV efficiency. Assuming 5% compression loss for hydrogen storage or transportation, a hydrogen-fueled AFC or PEM operating at 0.73 V could operate from natural gas at 43.5% LHV efficiency (39.2% HHV), based on natural gas input with advanced processing and hydrogen separation. This may provide an effective way of reducing greenhouse gas emissions in the transportation field in the future.

Coal Gas. This is a special case. As produced in a gasifier, a partial oxidation device with steam injection to perform what is in effect noncatalytic high-temperature adiabatic reforming, the resulting gas contains variable amounts of H_2, CO, and steam, with some CO_2 and CH_4, together with N_2 if air is the reactant instead of oxygen separated via a liquid air plant. The exact gas composition depends on the gasifier type, i.e., on the reaction temperature and chemistry, which depends on whether air of oxygen is used and on the oxygen–carbon and steam-to-carbon ratio. The available gasification reactors have been reviewed. Depending on the coal used, the resulting gas may contain considerable amounts of H_2S and COS, which may be removed via a number of chemical or physical processes. Most require that the synthesis gas should be cooled to ambient temperature, e.g., the chemical Stretford process, which uses sodium carbonate as a stripping agent, with a regenerative step to produce H_2S, 50% of which is then combusted with air to SO_2. The resulting SO_2 and the remaining H_2S are then reacted in the gas-phases Claus process to produce sulfur and water. In the Selexol process, the synthesis gas is chilled to near 0°C, and the sulfur products are removed via an organic solvent, which also removes some of the CO_2. These processes require multiple heat exchangers, which add to the capital cost of the plant. In consequence, high-temperature desulfurization processes, typically using metal oxides which may be regenerated, are being sought if the gas is to be used in a high-temperature fuel cell.

In general, systems producing clean coal gas from coal may be up to about 75% efficient, depending on the gasifier used. A gasifier operating under strongly oxidizing conditions will produce a gas which is CO rich, whereas one operating under more reducing conditions will produce a gas which is relatively richer in methane. The former has a higher heating value, but it must be water–gas shifted to give hydrogen for use in the fuel cell. Since this is an exothermic reaction, energy is lost in the transformation. However, if the gas is relatively rich in methane, the latter may be transformed by steam reforming in an endothermic reaction using high-temperature fuel cell waste heat to give hydrogen, which has a higher heating value per equivalent than methane, so that an upgrading of the gas stream occurs. If the gases are directly burnt in a gas turbine, a gas of higher heating value per equivalent, containing a larger percentage of CO, will be desirable. However, if a fuel cell is to be used, especially one including internal reforming, then a methane-rich stream will lead to a higher system efficiency. Thus the overall system efficiency will rely on an understanding of the thermodynamics and kinetics of the gasifier and the fuel cell itself.

The simplest assumption would be one in which pure hydrogen is produced from coal after high- and low-temperature water–gas shift and separation steps. The total efficiency of this process is minimum of about 67% (HHV)

and a maximum of about 75%, depending on the gasifier used. If the hydrogen derived from coal gas is then used in an (unspecified) fuel cell system operating at 0.73 V at 100% utilization, in which no steam recovery for gasification occurs, then the overall system HHV efficiency will be given by multiplying the gasification efficiency by 0.73/1.48, ie, an unimpressive 33–37%. If clean synthesis gas produced at 80% efficiency can be directly used in the fuel cell at 85% utilization, the corresponding efficiency at the same cell voltage will be $0.8 \times 0.85 \times 0.73/1.48$, i.e., 33.5%. However, if the anode exit gas stream and fuel cell waste heat can be used in the process, e.g., for the production of steam and for preheating, then the efficiency may rise to 42%. Efficient integration of the total system can yield even higher values.

The objective of fuel-cell-combined cycle-gasifier processes is to recover as much waste heat as possible from each element of the system. An oxygen-blown gasifier which yields a medium-Btu gas containing no nitrogen is most advantageous. It will generally be pressurized and will yield a clean low-sulfur pressurized gas stream after clean-up. Such a gas stream will most advantageously be used in a pressurized fuel cell, preferably operating at high temperature. The thermodynamics will favor the MCFC, especially if internal reforming is used, so that a reducing gas from a gasifier with low oxygen demand can be used to improve overall efficiency. In a scheme in which the fuel cell is the major power-producing element, anode exit gas may be used to operate a small gas turbine, which will pressurize cathode air and produce a small amount of shaft power. Waste heat is used to raise steam for gasification. All remaining waste heat from the system is collected and used in a steam turbine bottoming cycle. Such a system, operating at 0.8 V at 85% utilization, may have an HHV efficiency of 50%. However, its initial capital cost and the cost of replacing fuel cell stacks at 5-year intervals may be such that it would be more economical to regard a smaller MCFC as a "topping cycle" for an integrated gasifier gas turbine combined cycle (IGGCC) plant, in which a larger turbine burns some of the synthesis gas and the anode exit fuel. The efficiency of such a plant will be lower (about 45% based on the coal HHV), but lower capital and O&M cost may mean less expensive electricity. A simplified plant design is shown in Figure 5. Their low emissions, combined with their relatively high efficiency, may make them attractive in the next century, either with MCFC or SOFC systems.

State-of-the-Art Fuel Cell Generators

On-Site PAFC Systems. The PAFC represents the most developed integrated fuel cell generation technology, with the start of system design for operating on natural gas fuel using an integrated fuel processing dating from the American Gas Association TARGET (Team to Advance Research in Gas Energy Technology) Program, conducted at the Pratt and Whitney Division of United Technologies Corporation, in 1967. This was intended to supply individual households with 12.5-kW combined heat and power units, making them independent of the electricity distribution network. The cost of delivered gas was then one-sixth of the cost of delivered electricity. In many parts of

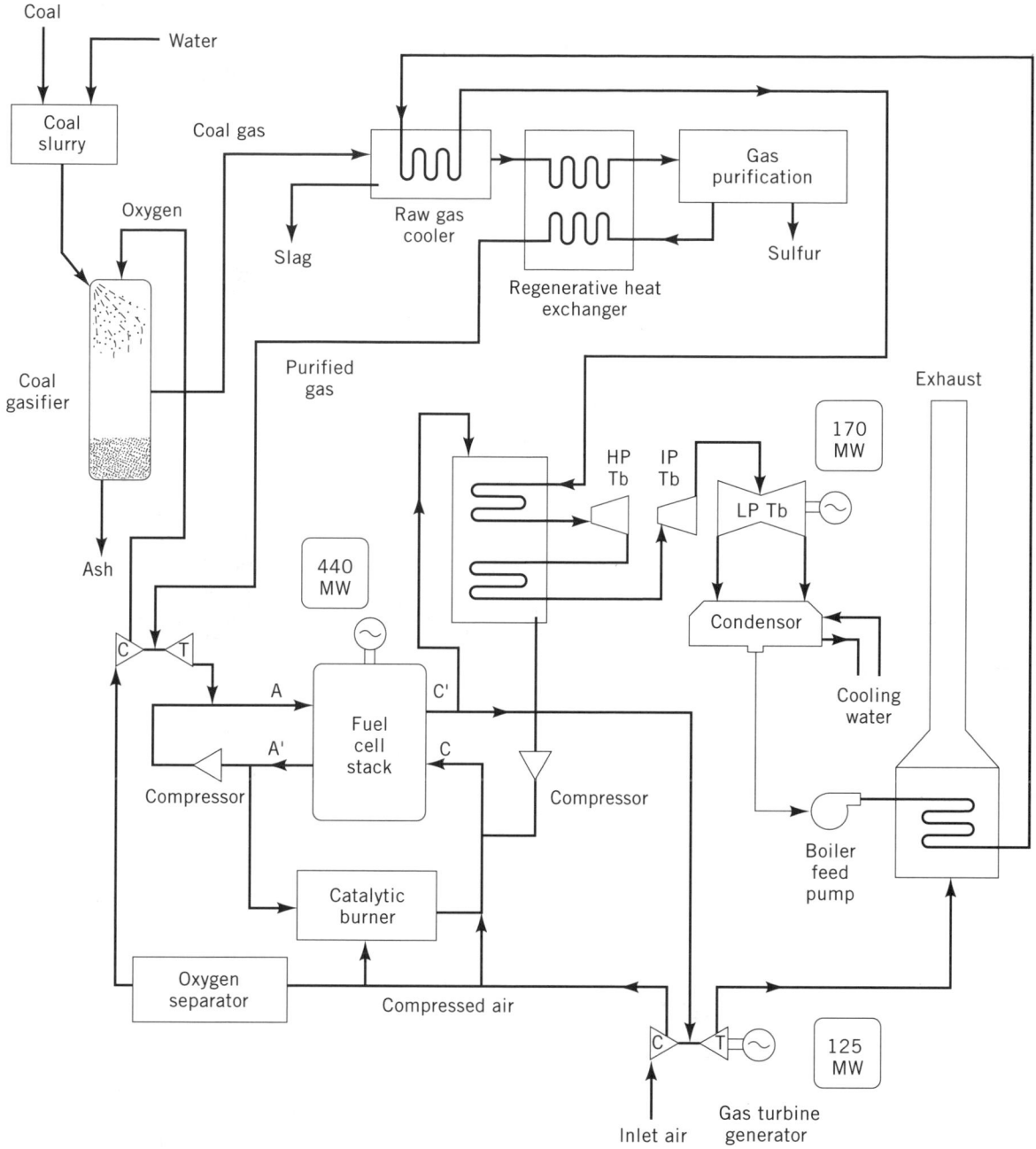

Figure 5. A 735-MW plant design (General Electric Company, 1980) for a MCFC "topping cycle" for an integrated gasifier gas turbine combined cycle (IGCC) plant with approximately 50% efficiency; A, anode gas entry; A', anode gas exit; C, cathode gas entry; C', cathode gas exit. In this design, 440 MW is from the fuel cell, 125 MW from the gas turbine, and 170 MW from the steam bottoming cycle. In other proposed designs, some coal gas is burned directly in a turbine without first passing through the fuel cell anode. This will have a lower efficiency (45–50%) and may have a lower capital and O&M cost, therefore providing a lower cost of electricity.

the United States today, delivered electricity may cost 5.5¢/kW·h, or $16.12/MMBtu; whereas delivered gas may cost $5.50/MMBtu, about one-third of the cost of electricity. Even at the cost ratio of gas to electricity in the late 1960s, it became evident that the small-atmospheric-pressure 12.5-kW TARGET unit would not be economical, at least under the then-prevalent materials and economics situation, so that scale-up would be desirable. A number

of units were however given proof-of-principle testing during the 1970s, and they showed successful operation, although their materials were high cost, e.g., pure platinum electrodes in high loadings per square centimeter of electrode area were used. They showed the principle of integrating PAFC cell stacks to the fuel processor so that steam for reforming could be raised, increasing overall system efficiency.

The first scale-up was to 40 kW (electrical), in the proposed PC-18 on-site intended for what was variously called dual energy use (DEUS), combined heat and power (CHP), or cogeneration, which made its debut for experimental operation in the mid-1970s. When the program was completed in the late 1980s, a further increase in scale to a 200-kW (electrical) cogeneration unit was considered feasible to provide electric power and heat for an industrial plant, service industry, or neighborhood applications. The 40-kW unit consisted of a natural gas desulfurizer, a fuel processing unit producing a mixture of hydrogen and CO_2 from natural gas by steam reforming, and water–gas shift conversion to remove CO. This was integrated with the cell stack, so that waste heat supplied steam for reforming. The dc power produced by the fuel cell stack was converted into utility-quality AC power via an electronic inverter.

The scaled-up unit (the PC-25) was planned to produce equal amounts of electricity and heat in the form of hot water plus some steam. It operated at atmospheric pressure at about 325 mA/cm^2 at 0.62 V at a net HHV electrical efficiency of about 36%. The PC-25 was simpler than the PC-18 and was designed for less rapid load-following rates. It was also more reliable and was able to operate for longer periods without attention, with a service call every 6 months. It was controlled by microprocessors and was turned on, off, or to whatever load was required by remote dispatching. Its noise level was 55 dB at the fence, and its specified NO_x (as NO_2) emissions were to be less than 0.02 lb/MMBtu (HHV) (8.6 g/GJ or 86 g/MW·h). This would have corresponded to 26 ppmv in the raw exhaust gas. The maximum total hydrocarbon emissions were required to be the same as those for NO_2 on a gram-per-gigajoule basis, with CO emissions not more than 10 times higher. In practice, its measured emissions were 0.45 ppmv of NO_2, 0.04 ppmv nonmethane hydrocarbons (NMHCs), and 1.4 ppmv CO (15% oxygen, dry basis). These correspond to the negligible levels of 2.4, 4.6, and 0.04 g/MW·h, respectively (56). It was stated that the burner exhaust was below the NMHC background value for air in the Los Angeles Basin. These emissions should be compared with 0.45 lb/MMBtu (1940 g/MW·h as NO_2 at 36% efficiency), which will be required by the 1990 Clean Air Act for low-NO_x burners in coal plants, whose average emissions in 1990 were 2500 g/MW·h. The corresponding average values in oil-fired and natural-gas-fired steam plants were 1220 g/MW·h and 820 g/MW·h, respectively. For small gas turbines (25% LHV efficiency, 150 ppmv at 15% oxygen, dry basis), they were 4450 g/MW·h, and for large gas turbines (1990 technology, 75 ppmv, specified as for small turbines), they were 1670 g/MW·h. Values of 360 g/MW·h could be attained in an advanced gas turbine combined cycle with the latest combustor technology (52% LHV efficiency, 25 ppmv, specified as before). According to the turbine developers, advanced combustors could deliver one-third of these ppmv values, ie, 120 g/MW·h. The SO_x and particulate emissions of the fuel cell were undetectable, and it required no make-up water. With these specifications, it could be located anywhere, eg, in subdivisions, in shopping malls, or in basements or on roofs of large downtown buildings. It could potentially be delivered to a prepared

concrete pad, and after connecting utilities and check-out, it could be operating in 9 days.

In 1992, it was available from the manufacturing arm (ONSI) of a joint venture called International Fuel Cells (IFC), which was organized in 1985 by the Power Systems Division of United Technologies Corporation, a corporate successor of a subsidiary of Pratt & Whitney, with Toshiba Corporation as a minority stockholder. One feature of UTC's stack was the use of purified water circulating in parallel-flow copper (in the PC-18) or later serpentine stainless steel tubes (in the PC-25) with appropriate Teflon protection in graphite cooling plates between groups of five cells to perform cooling via flash evaporation and to provide steam for reforming. Its cost in 1993 depended on the package offered but was a minimum of about \$3500/kW (electric). Whether this will be competitive will depend on the niche markets which can be identified for cogeneration applications. IFC was examining methods of cost reduction, such as the use of 1990 technology electronics in the inverter, rather than those of 1985, as well as general weight reduction of the system from 80,000 lb in the prototype to 60,000 lb in production versions (i.e, a cost of \$11.60/lb), with a target of 40,000 lb. Approximately 60 PC-25 units were operating or on order in mid-1993 for installation at sites in the United States, Japan, Europe, and Korea.

By 1980, other developers were entering the field. These included Engelhard Industries, a major catalyst supplier, and Energy Research Corporation (ERC) in the United States. The beginnings of efforts in Japan by Toshiba, Hitachi, Fuji Electric, and Mitsubishi Electric took place at the same time. Westinghouse in the United States and Sanyo in Japan took licenses to ERC's stack technology. This used air-cooled cooling plates and did not include an acid reservoir but relied on various means of electrolyte make-up after specified periods of operation. All of ERC's licensees used a solid ribbed bipolar plate for reactant gas circulation. Steam was raised for reforming in this system via a heat exchanger. Engelhard's cooling system used cooling plates incorporating tubes and a flow of dielectric oil, which again raised steam via a heat exchanger. This system was licensed to Fuji Electric in the late 1980s. Both ERC and Engelhard cells used ribbed electrode structure for gas circulation. Engelhard's cells used a unique bipolar plate construction, consisting of porous ribbed graphite on each side with solid flat internal components (the ABA plate). This could be designed to have an electrolyte reservoir. Reportedly, IFC now uses a similar concept to the Engelhard bipolar plate of ABA structure in the PC-25. This structure incorporates an electrolyte reservoir and is called "Configuration B."

The Fuji Electric Company in Japan developed on-site PAFC units in sizes starting at 50 kW (the FP-50), which showed an LHV efficiency of 41% in a version operating on natural gas and 44% in a 200-kW version operating on methanol for use as a stationary cogeneration unit on the 300 inhabited remote islands off the Japanese mainland. The heat recovery efficiency on these units was about 35%, and measured NO_2 emissions were 2 ppmv (7% oxygen, dry basis), ie, 4 g/MW·h (56).

Mitsubishi Electric had a similar unit to the PC-25 operating on natural gas, producing 200 kW, and weighing

25 metric tons. Its LHV efficiency was 40%, and it produced less that 1 ppmv NO_2. The prototype of this unit operated successfully for over 13,000 h at the Plaza Hotel, Osaka, during 1990 and 1991, with very little performance degradation. Toshiba is manufacturing a modified version of the PC-25 for use in Japan, where IFC does not compete with Toshiba products, although it does elsewhere. Finally, Japanese government agencies are expecting Hitachi to again become involved in PAFC technology, following its extensive work in the early 1980s.

It was expected that, by 1995, Japanese electric utilities would introduce about 30 PAFC units in the 50–200-kW class, as a preliminary to building up about 1000 MW of capacity by the year 2000. Japanese gas utilities were expected to introduce similar amounts. Demonstrations of PAFCs in Japan (whether of Japanese or overseas technology) were to be one-third financed by the Japanese government during the early 1990s as an aid to commercialization. In the decade from 1983 to 1993, over 100 PAFC units of different sizes had been operated in Japan.

PAFC Electric Utility Units. During the late 1970s and early 1980s, the Power Systems Division of United Technologies Corporation (UTC) actively pursued the development of electric utility power generation equipment, which should cost less than on-site cogeneration units and make a profit of electric sales alone. In 1973–1975, a 27-MW unit was proposed in the form of truck-transportable, factory-manufactured modules. By 1978, a 1-MW brassboard unit had been tested reasonably successfully at developer's facility at South Windsor, CT. This unit was pressurized, to increase cell voltage, and also operated at slightly higher voltage. This could in principle give increased system efficiency, but the cell bipolar plates used plastics which did not prove satisfactory because of corrosion, especially pressurized cells at higher temperatures (190°C) and voltages with higher water vapor partial pressures. As a result, it was determined that bipolar plates and catalyst supports should be made from pure graphite. At approximately the same time (ca. 1978), it was realized that evaporation of the phosphoric acid electrolyte as P_2O_5 hydrates was sufficient to warrant inclusion of a porous graphite substrate, which was ribbed to allow passage of gases, whose capillary forces would allow storage of sufficient electrolyte for 40,000 operating hours at 205°C under pressurized water vapor conditions. This was to become known as the "ribbed substrate stack." This used a flat graphite bipolar plate with the ribbed substrate at each electrode, to allow for appropriate reactant gas circulation in large cells, pressurized or not. The ribbed substrate was partially wet-proofed with PTFE when the electrodes were sintered to aid electrolyte retention and to allow open porosity for the diffusion of gaseous reactants and product water between the electrode surfaces and the gas channels.

The first demonstration of a 4.5-MW (AC) unit was to have taken place in New York City by approximately 1978. This unit used preribbed substrate (but all-graphite) stack technology with 0.35-m^2 cells. The fuel processor and dc–ac inverter of the unit were successfully tested on natural gas and light naphtha, but by the time initial problems had been worked out and the necessary fire department regulations complied with in 1982–1983, the older stack technology had passed its storage lifetime. Retrofitting with new stacks was not deemed to be cost effective. In consequence, the unit never produced electricity, but it did prove that the computerized remote-dispatch fuel processor would operate. It also showed that a fuel cell would be the only new generator which would comply with New York City safety codes, having demonstrated that the fuel processor, coded as a "refinery" was safe and nonpolluting. In any case, by 1983, a similar unit was successfully operating at the Tokyo Electric Power Company (TEPCO) at Goi, near Tokyo, Japan. It used "ribbed-substate" stacks and performed according to specifications on natural gas at 37% HHV efficiency at 0.65 V and 4.8 atma operating pressure at 190°C.

During the 1980s, a new design with 0.93-m^2 cells was proposed by UTC as a generator for urban sites with negligible emissions, no cooling water requirements, and an imperceptible noise level of 55 dBA at the fence. This would have been an 11-MW unit with a lower power footprint than that of the 4.5-MW unit, to ease maintenance problems. The unit was designated PC-23 and would have used eighteen 650-kW (dc) stacks. After certain utility interest in the United States, it failed to attract orders because of low energy costs, its high capital cost, lack of utility financing, and not least, utilities that had specific requirements which could not be met by a standardized design. For example, the City of Palo Alto, CA, realized that it could not afford the cost of an urban site, in spite of the advantages of operating a dispersed generator on essentially free land-fill gas, which would have to be disposed of anyway. In another case, Consolidated Edison Company in Manhattan found itself in a similar predicament, despite the advantages of a pollution-free dispersed generator in this location, where the cost of bringing in new power lines is very high; much larger units that 11 MW were required, designed with multilevel subsystems to conserve building area.

TEPCO ordered an 11-mW unit using 18 IFC stacks of the PC-23 type with chemical engineering and on-site construction by Toshiba for installation at the old 4.5-MW demonstrator site at Goi. This was specified to operate at 42% net AC HHV efficiency on natural gas (at 0.73 V at 215 mA/cm^2 rated end-of-life performance at 8.2 atma operating pressure at 205°C). The unit was required to have a cold-start-up time of 6 h and a hot-start-up time of 2.5 h. This operated starting in 1991. In late 1992, it was announced that an operational simplification compared with previous PAFCs did not prove successful, and had damaged some stacks by corrosion. The simplification involved using spent reformer burner gas which contained small amounts of oxygen and a large amount of steam as the blanket gas around the stacks instead of dry nitrogen. The affected stacks were removed, nitrogen was used as a blanket gas, and the unit has operated at reduced power according to specifications. Its NO_2 emissions were measured as 1 ppmv and were always less than 3 ppmv (7% oxygen, dry basis, ie, 1.8–5 g/MW·h). Other similar units of improved type may be operated in Japan and Europe in the future. Two of the same PC-23 stacks have operated since late 1993 in a 1-MW pressurized unit in Milan, Italy (Progetto Volta), with a Haldor-Topsøe reformer and chemical engineering by Kinetics Technology International (KTI).

In Japan, 1-MW demonstrator PAFCs operating under pressurized conditions were ordered in the early 1980s (starting in 1981) under the Moonlight Program with NEDO (now the National Energy and Industrial Development Organization) acting as manager. Operation was to be at 0.7 V at 220 mA/cm^2 and 190°C at 4 atma and 205°C at 7–8 atma. Toshiba and Hitachi were selected to develop the higher pressure and temperature technology, with Fuji Electric and MELCO as contractors for the lower temperature and pressure unit. The prime contractors were to develop both stacks and balance-of-plant. The first unit was installed at the Chita plant of the Chubu Electric Power Company, near Nagoya. It had two Toshiba and two Hitachi stacks, a Toshiba reformer, and a Hitachi turbocompressor. The other unit was located at a Kansai Electric Power Company plant in Osaka. It had two Fuji Electric and two MELCO stacks. After many start-up difficulties, both units operated in 1987–1988, the first for 1018 h and the second for 2045 h.

As a result of this experience, a pre-prototype 5-MW pressurized unit was ordered from Fuji Electric, which has been installed at the Kansai Electric Power Company near Osaka and is expected to operate in mid-1994.

PAFC Issues. Since PAFCs use an acid electrolyte, it is inevitable that platinum catalysts must be used. The key to reducing platinum loadings to economic levels is that as much as possible of the total platinum surface area can be used effectively. If the platinum particles on graphitized high-surface-area carbon have an initial specific area of about 100 m^2/g, which will eventually result in 50 m^2/g as aging occurs according to a logarithmic time scale, then 5.5 g platinum/cm^2 resulted in satisfactory performance. Of this two-thirds was at the cathode (0.5 mg/cm^2) and one-third at the anode (0.25 mg/cm^2). More recently, 0.1 mg/cm^2 has been used at the anode (58). At July 1993 platinum prices, these loadings represent about $50–65/kW for the noble metal catalyst. In consequence, some developers may use somewhat higher cathode loadings to increase the system efficiency and stack cost effectiveness. Certain base-metal platinum catalysts, eg, Pt–30% Cr, appear to result in stable materials with little or no base-metal corrosion and 50% greater activity at constant cathode potential. Since cathodes generally cannot double current density beyond the state-of-the-art levels available today because of diffusion limitations, the increase in performance of the alloy catalysts is better taken in the form of an increased cell voltage, eg, by 50 mV. This underscores the fact that today's electrodes are essentially current density limited for any given set of reactant partial pressures. Under present operating conditions, the remaining components of the PAFC cells are sufficiently stable to allow 40,000 h operation. Some corrosion of the graphitized high-surface-area carbon catalyst support occurs at the cathode, which contributes to platinum catalyst area loss. The matrix material which immobilizes the electrolyte between the electrodes by absorption and serves as a bubble-pressure barrier is a layer of fine-particle-size silicon carbide about 0.25 mm thick bonded with some PTFE. It generally shows high stability. The graphite parts show good stability provided that they are not exposed to conditions where galvanic corrosion might occur. This may happen, e.g., if the exterior surfaces of bipolar plates are exposed to an oxygen atmosphere in the presence of electrolyte films, so that the graphite corrodes (cf. the pressurized 11-MW unit at Goi, above).

There are three economic issues with the electric utility PAFC, namely efficiency, durability, and capital cost. All of the above are involved in determining the cost of electricity, the ultimate economic criterion. The efficiency of a pressurized PAFC operating today on natural gas is 42% (HHV) or 46.2% (LHV). It can be produced in units as small as a few megawatts, is an excellent environmental neighbor, and can be dispersed within the grid to avoid the requirements for new rights-of-way, which is an increasingly difficult problem for utilities in urban areas. Dispersed PAFCs incorporating an electronic inverter have capacitive and inductive qualities which help stabilize the network. All these allow site-specific credits for electric utilities, which may in some cases be considerable.

However, the PAFC today is expensive. The unpressurized PC-25 200-kW cogeneration unit in limited-series production costs approximately $3500/kW. The next prototype multimegawatt electric utility unit is likely to cost $10,000/kW. In mass production, costs will certainly fall on a learning curve (28). In the early to mid-1980s it was expected that by 1993 eight 10-MW units per year would be installed at a cost (adjusted for inflation, 1993 dollars) of about $2500/kW. This would be close to the break-even point for a cogeneration unit but twice the affordable cost of a stand-alone electric utility generator (28). These figures still stand today. At $2500/kW, the PC-25 should start to break even in some cogeneration applications. At $1500/kW, it will be widely used. Similarly, a multimegawatt plant may break even in a stand-alone electric utility application if its installed cost is $1200/kW. In niche markets, the allowable cost might be higher.

Today, we are a long way from these costs, and little or no learning is taking place on how to further reduce costs and improve performance of pressurized systems because no large units are being bought. The 200-kW units are still really only technology demonstrators, although the first improved PC-25C units with increased stack power density and 33% less weight and volume than previous models will be offered at $3000/kW starting in 1995 (56). The fuel cell stack is not the problem. Its materials are affordable, and in mass production its cost will not be a large fraction of the total. The inverter, which cost $3000/kW (1993 dollars) represented a cost problem in the early 1980s. However, its cost was predicted to fall to $75/kW (in 1978 dollars, ie, $150/kW in 1993 dollars) by 1990 as progress was made in power electronics. Today its cost in 1993 dollars is $75/kW. The problem is the cost of the fuel processor and other subsystems, which require redesign and simplification to reduce weight, increase performance by better integration, and lower cost. The systems are extremely complex and contain up to 13 costly heat exchangers for efficient energy integration. System weight must also be further reduced. The weight of electrochemically active stack components is only 2.8 kg/kW, whereas the present Japanese and U.S. complete on-site systems weigh 125–135 kg/kW (59), although the PC-25C is projected to offer 90 kg/kW (56). The 11-mW Goi PAFC unit, with its 61-ton, 6-m-tall reformer, weighs 74 kg/kW without foundations and building structures.

The lifetime of the stack is another issue. Its design life is 40,000 operating hours for decay corresponding to the nameplate performance. While this is 5 years, the life of the balance of the plant will be 30 years, so six sets of stacks will be required during the system lifetime. This will clearly further increase operating costs compared with conventional generating equipment. Even at projected cost levels, stack replacement will cost more than the total initial plant cost over the system lifetime. Stack life must be continuously improved, even at some loss of efficiency after long operating times. However, in 1992 it was reported that stack power densities had been increased by 33% at constant voltage and initial decay had been reduced to 2 mV/1000 h for the first 10,000 h (60). This seems to be satisfactory, since performance decay tends to level out with time (28).

An electric utility might prefer to incrementally add capacity as it is required, rather than invest in large multigigawatt central stations. However, central steam plants are also not favored over natural-gas-powered gas turbine combined (steam) cycle plants, which are available in sizes greater than 100+ MW in efficiencies up to 52% (LHV) today. Such a plant might cost only $850/kW installed, is reliable, and has low pollution compared with conventional generating capacity. Its emissions are well within the requirements of the 1990 Clean Air Act. The emissions of the PAFC may be 100 times less (in grams per megawatt-hour), but this is perceived as really unnecessary. In consequence, it is difficult to see how larger electric utility PAFCs are going to find a niche in the marketplace. However, the on-site cogeneration PAFC unit seems assured of a place.

MCFCs. During the 1980s, the MCFC went from laboratory single cells to small stacks (52). Much of the development work during this period was conducted at United Technologies (International Fuel Cells), the Institute of Gas Technology (IGT) in Chicago, and Energy Research Corporation, Danbury, CT (ERC). Scale-up of UTC stacks with external manifolding containing a porous gasket material the height of the stack showed eventual failure (at about 1000–4000 h) due to loss of electrolyte via the gasket. This was traced to an electroosmotic mechanism, which was alleviated by modifications to the gasket or, in another approach, by the use of internal manifolding in the stack (1). During the same period, nickel anode creep was reduced to acceptable limits by the use of nickel alloys, first with chromium, then with aluminum (1).

Other problems which required solution included corrosion of stainless steel cathode current collectors and the cathode sides of stainless steel bipolar plates, which were rendered immune from corrosion by wet anode exit gas by nickel cladding on the anode side. To avoid galvanic corrosion around the edges associated with omnipresent carbonate electrolyte films, the sides of the plates were aluminized. This places an electronically insulating layer in the galvanic pathway, eliminating corrosion of this type. A life-limiting factor involves slow dissolution of the lithiated nickel oxide cathode, which allows migration of Ni^{2+} dissolved in the electrolyte and precipitated as nickel nodules close to the anode. This may eventually cause shorting of the cell (52). The rate of nickel oxide dissolution was proportional to CO_2 partial pressure, so that it was estimated that the lifetime of pressurized cells operating on a cathode gas mixture containing 30% CO_2 at 5 atma pressure would be less then 10,000 h, although 40,000 h might be attained at atmospheric pressure (1). The major aim of the U.S. Department of Energy Development Program managed from Morgantown, WV, was the development of pressurized cells to operate on coal gas in large integrated gasifier combined-cycle systems. An active effort was therefore started at Argonne National Laboratory and elsewhere to find a stable substitute cathode (1).

Active support in the United States included emphasis on simple, inexpensive natural gas systems supported by the Electric Power Research Institute on behalf of the electric utility industry. System designs in the mid-1980s by ERC using internal reforming showed the possibility of attaining 50% HHV efficiency in atmospheric pressure systems using natural gas fuel. These used a minimum of heat exchangers with internal reforming (1).

At about the same time, programs were started in Japan to develop the "second generation" MCFC to follow the PAFC. The developers selected by NEDO were Hitachi and Ishikawajima-Harima Heavy Industries (IHI). The latter used technology developed at the IGT. Other programs were at Toshiba, which eventually licensed IFC technology, and at MELCO, which had licensed ERC technology. By the late 1980s, 1- and 10-kW stacks had been tested and cells had been scaled up to 1 m². Both of NEDO's developers were using internally manifold stacks, which were in one single piece at IHI and in four separate cells per 1-m² plate in a "window pane" arrangement at Hitachi. Testing of 1-m² components in Japan is continuing.

In the United States, the major developers supported by the U.S. DOE are (1993) MC-Power Corporation, a company using IGT technology, and ERC. The emphasis of the first is on pressurized (3-atma) 1-m² technology with internal manifolding, using external reforming of natural gas with integrated flat-plate reformers. Their system contains four heat exchangers, which may be used in part to supply cogenerated steam equal to 40% of electrical output. ERC concentrates on externally manifolded systems with internal reforming of natural gas using indirect internal reforming (IIR) plates within the stack (one per six cells) which remove sensible heat, with cells scaled up to 0.56 m². The program at IFC uses externally manifolded 0.74-m² cells in natural gas systems incorporating sensible heat reforming in the anode feedback loop. the large stacks are designed to be capable of thermal cycling, including cool-down to ambient temperature.

Emphasis in all cases is on simplification. This is best illustrated by developments at ERC. To avoid the use of multiple heat exchangers and condensers, anode exit gas is burned with excess air, and the hot, raw product is used as cathode gas. The system contains only three heat exchangers integrated in a single unit for normal operation, for hydrodesulfurization of natural gas, for natural gas preheat, and for superheating steam. An auxiliary heat exchanger and burner are used for start-up. To supply sufficient heat to the incoming cathode gas stream, fuel utilization is limited to 75%. No feedback loops are used, and cooling is via internal reforming (80% within the IIR plates, 20% in the fuel cell anodes) combined with sensible heat removal from process gases. Steam reforming in this system uses a steam–carbon ratio of 2.5, so that the effective composition of the anode inlet stream for each mole of methane is 1 mol of CO_2, 4 mol of H_2, and 0.5 mol of H_2O. At 75% utilization, the anode exit stream becomes 4 mol of CO_2, 1 mol of H_2, and 3.5 mol of H_2O. This is combusted with 3.75 mol of oxygen in air, to give a cathode inlet gas composition of 4 mol of CO_2, 3.75 mol of O_2, and 17 mol of n_2 and 4.5 mol of H_2O. Thus, the cathode feedstock consists of 13.7% CO_2 and 12.8% O_2, which is directly used in a single pass through the cell at utilizations of 75% for CO_2 and 40% for O_2. For each mole of methane oxidized at the anode, 3 mol of hydrogen is oxidized at 75% fuel utilization; therefore 3 mol of CO_2 and 1.5 mol of CO_2 are combined at the cathode to produce CO_3^{2-} ion, which is transferred to the anode. Thus, the cathode exit gas composition corresponds to 4.0% CO_2 and 9.0% O_2, and the mean composition is 9.3% CO_2 and 11.1% O_2.

Even with these very dilute gas compositions, the developers have succeeded in operating atmospheric pressure cells with quite acceptable polarization slopes, resulting in a performance at ERC of 0.76 V at 160 mA/cm^2 and 0.715 V at 200 mA/cm^2 (Fig. 6). One advantage of operating on dilute CO_2 mixtures is that the lifetime of conventional lithiated nickel oxide cathodes may be extended to 100,000 h or more, which would mean that cathode dissolution would no longer limit life. At 0.76 V and 75% fuel utilization, the gross d-c system efficiency will be 49.4% (HHV). The net a-c efficiency is expected to

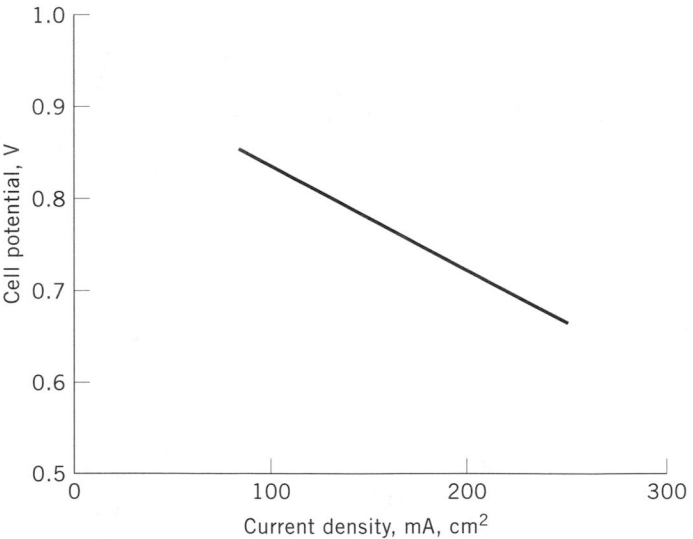

Figure 6. Performance curve of Energy Research Corporation MCFC operating at atmospheric pressure at 75% reformate utilization with dilute CO_2–O_2 cathode gas mixture (mean composition about 0.1 atm for each reactant) to simplify the system and give lowest cost.

be about 46.5% HHV or 51.6% LHV. This is competitive with the best 100+ MW combined cycles can be achieved in units as small as 1 MW or less. While present systems cost several thousand dollars per kilowatt, developers expect to achieve $1300/kW in limited-series product. These prices should be attractive for on-site cogeneration. Stack costs of $300/kW are anticipated, based on average materials costs of about $3/lb, i.e., $120/kW.

At the time of this writing plans were for one 2-MW demonstrator at Santa Clara, CA, using ERC indirect internal reforming technology. This is expected to complete acceptance testing in late 1994–early 1995. Two 250-kW demonstrations of MC-Power technology would be conducted, at the UNOCAL Science and Technology Center in Brea, CA, in early 1994 and at the Kaiser Permanante Hospital in San Diego in 1995. The latter would be a prepackaged plant. Plans for multimegawatt systems will follow. Siting considerations caused the hospital unit to be abandoned in early 1994, and it will be replaced by a unit at Miramar Naval Air Station, San Diego.

Raw MCFC exhaust emissions have been measured by IHI. Nitrogen oxides were below the limit of detection, CO was 10–45 ppmv (40–175 g/MW·h), and methane was 150 ppmv (340 g/MW·h). These emissions were from the reformer burner. In the Santa Clara plant, all gas will exhaust through the fuel cell cathode after combustion, and emissions are expected to be below the limits of detection.

SOFC. The most successful SOFC technology to date has been the tubular technology at Westinghouse (Fig. 7). This was developed as a system with a porous nonconducting calcia-stabilized zirconia-supported tube 1.56 cm in diameter on which was deposited a layer of strontium-doped lanthanum manganite, which was then sintered to become the cathode. This was followed by a masking operation, after which a strip of magnesium-doped lanthanum chromite was deposited via a process called electrochemical vapor deposition (EVD). In this, the tube is exposed on the outside to the mixed chlorides in the vapor phase, and hydrogen and steam are supplied on the inside. Since the material is electronically conductive with slight ionic conductivity, it continues to grow even when densified, via hydrogen oxidation and chloride reduction with conductivity via the dense ceramic layer. This layer is called the interconnect and is stable under both oxidizing and reducing conditions. The interconnect is then masked, and a 250-μm layer of dense yttria-stabilized zirconia electrolyte is deposited, again by EVD. The anode (nickel powder) is then deposited as a slurry. It is dried and then turned into a cermet impregnated with yttria-stabilized zirconia electrolyte, again by EVD. The final step is nickel plating of the interconnect. The complete tubes are up to 1 m in length with an active area of about 450 cm^2. There are plans for further scale-up. The tubes are connected from the interconnect of one of the external nickel cermet anode of the next by nickel felts which sinter in place in situ at the operating temperature of 1000°C. Preheated air (four to six times stoichiometric for the cathode process and cooling) is supplied to the interior of the tubular cells, and reformate is supplied to the bottom of the tubes at the outside. The system is typically operated at 85%

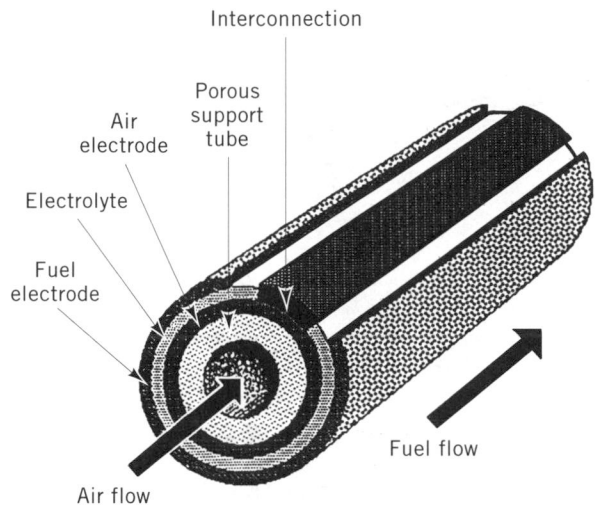

Interconnection

Porous
support
tube

Air
electrode

Electrolyte

Fuel
electrode

Air flow

Fuel flow

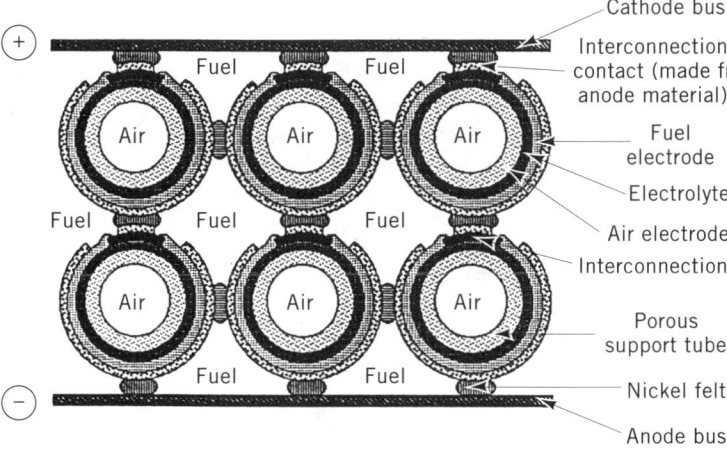

Cathode bus

Interconnection
contact (made fr
anode material)

Fuel
electrode

Electrolyte

Air electrode

Interconnection

Porous
support tube

Nickel felt

Anode bus

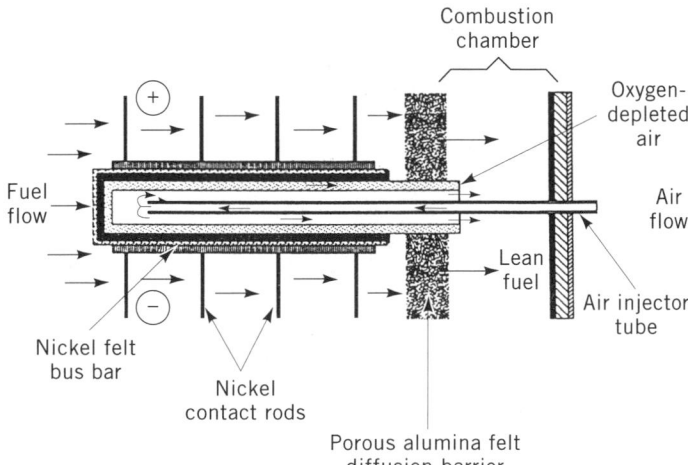

Combustion
chamber

Oxygen-
depleted
air

Air
flow

Air injector
tube

Lean
fuel

Fuel
flow

Nickel felt
bus bar

Nickel
contact rods

Porous alumina felt
diffusion barrier

Figure 7. Schematics of Westinghouse tubular SOFC technology. Top: tubular cell. Center: cross section of bundle of tubular cells (1.56 cm outside diameter). Bottom: side view of cell in bundle, showing manifolding.

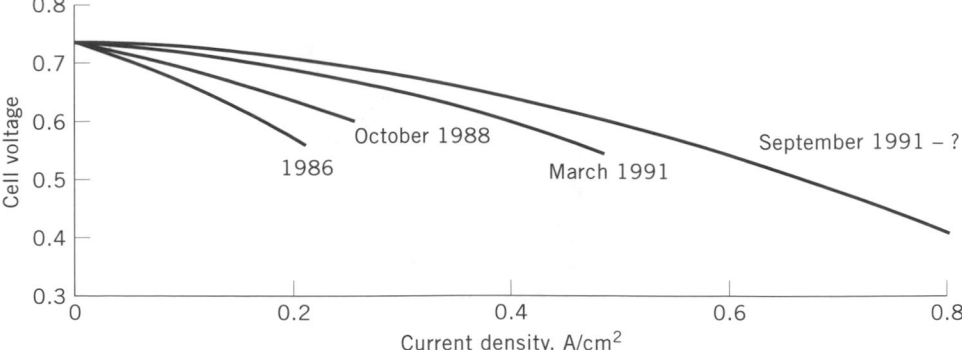

Figure 8. Performance progress of Westinghouse tubular SOFC technology to date. Latest data (September 1991 to 1994) use air electrode support tube. Conditions: 1000°C, natural gas reformate, 85% fuel utilization.

fuel utilization. Spent fuel and air are allowed to mix and burn in a ceramic plenum chamber at the top of the cells, which acts as an air preheater. Air enters the plenum chamber at 600°C from a stainless steel heat exchanger. Reforming of fuel is accomplished in a reforming preheater.

Since 1991, the calcia-stabilized zirconia support tube has been replaced by a tube made of air electrode material, which serves as the cathode. This has greatly improved the peripheral electronic conductivity of the system, which is required for current collection between tubes. The performance is now excellent (Fig. 8), although it is naturally limited by the thermodynamic considerations discussed in Figure 2. The waste heat from the system may be used for cogeneration or in a bottoming cycle. In 1994, Westinghouse was scaling demonstration 25-kW units up to 100 kW using 1-m-length tubes. Two 25-kW units with five hundred seventy-six 50-cm cells arranged in 18 cell bundles, each in strings of 192 cells, had been tested in Japan by gas industry and electric industry consortia. Good performance was obtained, and individual bundles had performed well to 6000–10,000 h. Emissions were measured on smaller 3-kW Westinghouse systems at Osaka Gas and Tokyo Gas in the early 1990s. They varied from 0.3 to 1.3 ppmv NO_2 at six times stoichiometric air flow, or about 7–30 g/MW·h.

Worldwide interest in the SOFC has grown enormously since 1990, since it is perceived as a simple system which will use what are expected to be inexpensive ceramic materials. Unlike molten carbonate, it requires no electrolyte management techniques and should show good durability. Emphasis is on small planar systems, operating at temperatures less than 1000°C to avoid the use of ceramic heat exchangers (cf. the Westinghouse system). These may also allow the use of metal bipolar plates between cells. Such approaches may require alternative ceramics (61).

AFCs. Modifications of the Bacon cell with sintered nickel electrodes operating at up to 260°C, as in the Project Apollo fuel cell, have been abandoned due to the short lifetime of nickel cathodes at high temperatures, which were required so that performance would not be significant by operating at less than 4 bars pressure rather than 45 atma (36). This change was necessary to reduce the mass of the pressure containment. Stable alternative materials, eg, rhodium, are not cost-effective. For simplicity,

a stationary (although free) electrolyte was used in the Apollo fuel cell, rather than the circulating system favored by Bacon. In the Space Shuttle fuel cell system, whose design was frozen in the 1970s, an immobilized stationary electrolyte was used. This contained a purified crysotile mat as a matrix in each cell. This set a limit to the cell operating temperature at about 80°C. In consequence, Teflon-bonded noble metal electrodes were used (gold with 20% platinum at the cathode in 20 mg/cm² total loading on a gold-plated nickel screen, platinum with 20% palladium at the anode in 10 mg/cm² loading on a silver-plated nickel screen). The system operated at 4 atma on pure hydrogen and oxygen and could give a nominal performance of 0.86 V at 470 mA/cm² and 0.8 V at 750 mA/cm². Stacks consisted of 32 cells each with 465 cm² active area. Each Space Shuttle unit in its final form had three parallel-connected stacks, each capable of 6 kW at nominal power, derated to give the normal power requirement of 12 kW. The Space Shuttle contained three such units, weighing about 120 kg. The AFC uses PC-17C Space Shuttle fuel cell cathodes consisting of 10 mg/cm² of 10 wt % platinum and 90 wt % gold black on a gold-plated nickel screen. Modifications of these cells using advanced materials operating at high temperatures and pressures with thin components will offer up to 6 A/cm² at 0.8 V. They were developed for applications under the Strategic Defense Initiative. An example of the performance of such a cell is shown in Figure 9.

Clearly, electrodes of the type used in the Space Shuttle fuel cell are unaffordable for common applications, and they are normally replaced by low-loading carbon-supported platinum electrodes. However, little recent work has been carried out to optimize these for common applications, because of the requirement for pure hydrogen fuel and CO_2-purified air for the AFC. Much of the available technology is that of the 1960s, but the system does have two attractive features. The first is that it will normally operate at a potential 50–100 mV higher than an acid fuel cell (eg, the PEM) with similar electrodes, because of its lower oxygen electrode Tafel slope. The second is that it can used less expensive materials. Carbons are attacked in alkaline media at the oxygen electrode above about 80°C, but a wide range of metals are passive under these conditions. A number of alternative macrocyclic electrocatalysts exist as alternatives to platinum which perform equally well (36). They include carbon-supported cobalt oxides associated with surface nitrogen, which are

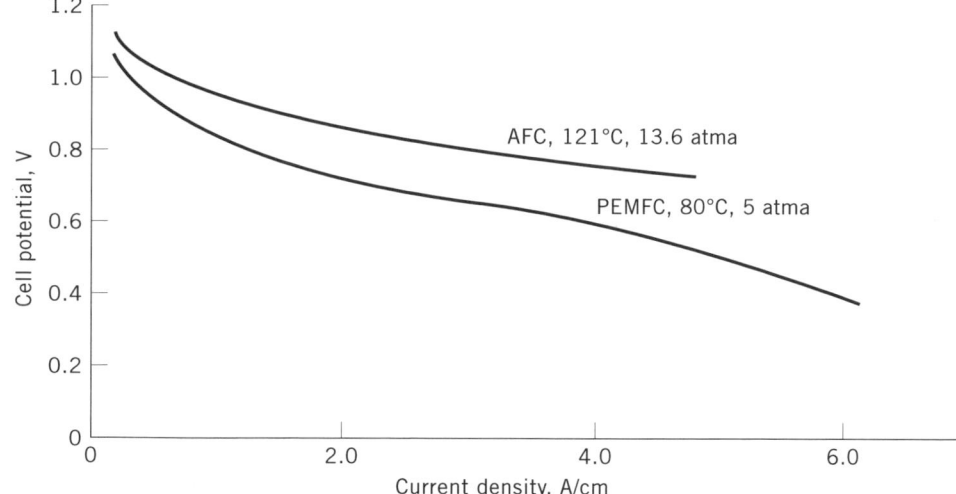

Figure 9. Performance curves of PEM (with Dow Chemical Company membrane) and AFC systems with high noble metal catalyst loadings for specialized military or aerospace applications on pure hydrogen and oxygen (PEMFC, 4 mg/cm² platinum; AFC, Space Shuttle cathodes with 20 mg/cm² 80/20 gold–platinum on a gold-plated nickel screen).

formed by the pyrolysis of cobalt macrocyclics such as the tetraazaannulene or the tetramethoxyphenylporphyrin, which is related in structure to the phthalocyanines and hemoglobin (36). The AFC with simple materials requires a further examination for use with hydrogen-fueled electric vehicles.

The PEMFC. For specialized military or aerospace applications, this may be operated under pressure on pure hydrogen and oxygen, with 4 mg/cm² platinum black electrodes. An example of the performance of such a system is shown in Figure 9, which illustrates the lower performance of the cathode in acid electrolyte compared with alkali, which results from the lower Tafel slope. The PEMFC system is also attractive for terrestrial applications, particularly for electric vehicles. During the late 1980s, the system did not show much promise for this application because of the poor interface formed between typical electrodes, which required high platinum loadings.

This problem is discussed in Ref. 48. Since 1986, remarkable progress in platinum utilization has occurred, which is illustrated in Figure 10. This plot of the logarithm of a performance parameter as a function of time is a good example of a learning curve. The kinetics of the oxygen electrode in the PEM are more than an order of magnitude more rapid in the PEM at 80°C than in the PAFC at 190°C. Today, the best results observed so far (November 1993, Fig. 11) are about a factor of 2 higher than those for the PAFC (about 800 mA/cm² at 0.6 V, compared with 350 mA/cm², with hydrogen and air at atmospheric pressure), but these results can be obtained with platinum loadings which are 10 times lower (0.05 mg/cm² Pt at the cathode, vs. 0.5 mg/cm²). With an anode with only 0.025 mg/cm² of Pt, this indicates a total platinum catalyst requirement of only 0.15 g per peak kilowatt. A small car can therefore be powered by 3 g of platinum costing $40, similar to the amount of noble metal in a catalytic converter.

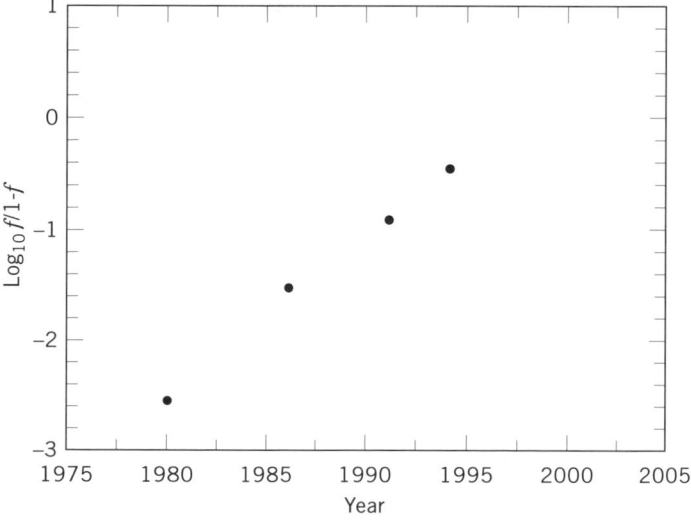

Figure 10. Logarithm of $f/(1-f)$ as a function of year, where f is platinum utilization in the useful cell voltage range for hydrogen–air PEMFCs [log $f/(1-f)$ = 0 indicates 50% utilization, expected about 1997).

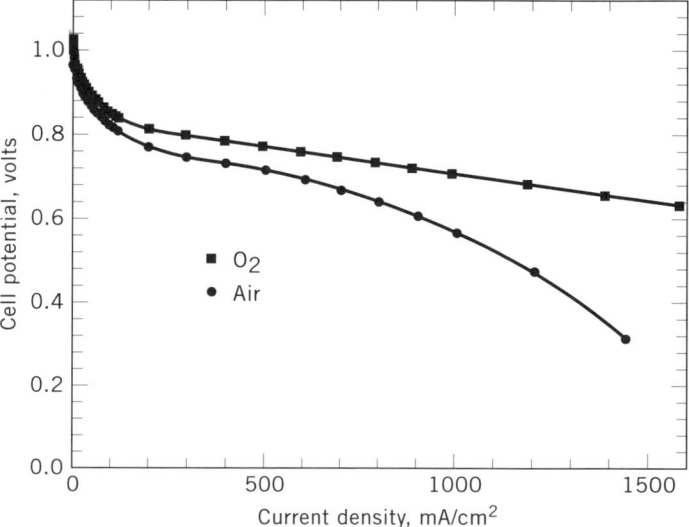

Figure 11. Best cell performance data to November 1993 on PEMFCs with ultra-low platinum loading, Asahi Chemical Industry Company Aciplex-S electrolyte membrane, hydrogen–air, 1 atma.

A small 1000-kg subcompact commuter car requires about 0.15 kW/km for 95 kph cruise, or 0.1 kW·h/kg for urban driving. Based on an averaged figure, a 300-km range will require 37.5 kW·h. This could be achieved with 1000 mol of hydrogen, or slightly more than 100 L of compressed gas at 200 atma. How this hydrogen might be stored (as compressed gas, metal hydride, liquid hydrogen, or some other form) may be the basis of the future transportation economy.

BIBLIOGRAPHY

1. A. J. Appleby and F. R. Foulkes, *Fuel Cell Handbook*, Van Nostrand Reinhold, New York, 1989.

2. K. Wark, *Thermodynamics*, McGraw-Hill, New York, 1983; A. W. Culp, *Principles of Energy Conversion*, McGraw-Hill, New York, 1979.

3. J. O'M. Bockris and A. K. N. Reddy, *Modern Electrochemistry*, Plenum, New York, 1970.

4. F. M. Harold, *The Vital Force: A Study of Bioenergetics*, Freeman, New York, 1986.

5. J. O'M. Bockris and S. Srinivasan, *Fuel Cells: Their Electrochemistry*, McGraw-Hill, New York, 1969.

6. E. A. Moelwyn-Hughes, *Physical Chemistry*, Pergamon, New York, 1961.

7. I. Prigogine, *Introduction àla Thermodynamique des Processus Irreversibles*, Gallimard, Paris, 1971.

8. P. Van Rysselberghe, *J. Chem. Phys.* 29, 640–642 (1958).

9. P. van Rysselberghe, in J. O'M. Bockris, ed., *Modern Aspects of Electrochemistry*, vol. 4, Plenum, New York, 1966, pp. 1–46.

10. J. Tafel, *Zeitschrift für Physikalische Chemie* 50, 641–712 (1905).

11. J. A. V. Butler, *Transactions of the Faraday Society* 19, 729–734, 734–739 (1924); T. Erdey-Gruz and M. Volmer, *Zeitschrift fuer Physikalische Chemie* 150A, 203–213 (1930).

12. R. Parsons, *Transactions of the Faraday Society* 47, 1332–1344 (1951).

13. T. P. Hoare, in *Proceedings of the 8th Meeting CITCE, Madrid, 1956*, Butterworth, London, 1958, pp. 439–444.

14. H. Mauser, *Zeitschrift fuer Electrochemie* 62, 419–425 (1958).

15. A. J. Appleby, "Electrocatalysis," in B. E. Conway and J. O'M. Bockris, eds., *Modern Aspects of Electrochemistry*, vol. 9, Plenum, New York, 1974, pp. 369–478.

16. A. J. Appleby, "Electrocatalysis," in B. E. Conway, J. O'M. Bockris, E. Yeager, S. U. M. Khan, and R. E. White, eds., *Comprehensive Treatise of Electrochemistry*, vol. 7, Plenum, New York, 1983, pp. 173–239.

17. J. A. V. Butler, *Proceedings of the Royal Society of London* 157A, 423–433 (1936).

18. P. Delahay, *Double Layer and Electrode Kinetics*, Interscience, New York, 1965.

19. J. A. V. Butler, *Transactions of the Faraday Society* 28, 379–382 (1932).

20. R. Parsons, *Transactions of the Faraday Society* 54, 1053–1063 (1958).

21. H. A. Leibhavsky and E. J. Cairns, *Fuel Cells and Fuel Batteries*, Wiley, New York, 1968.

22. B. D. McNicol, *Specialist Reports on Catalysis*, vol. 2, The Chemical Society, London, 1979.

23. B. D. McNicol, R. T. Short, and A. G. Chapman, *Journal of the Chemical Society, Faraday Transcripts 1* 72, 2735–2743 (1976); *Journal of Applied Electrochemistry* 6, 221–227 (1976); M. R. Andrew, J. S. Drury, B. D. McNichol, C. Pinnington, and R. T. Short, *Journal of Applied Electrochemistry* 6, 99–106 (1976).

24. H. Dahms and J. O'M. Bockris, *Journal of the Electrochemical Society* 111, 728–736 (1964); A. T. Kuhn, H. Wroblowa, and J. O'M. Bockris, *Transactions of the Faraday Society* 63, 1458–1467 (1967).

25. U.S. Pats. 3,992,331 and 3,992,512 (November 16, 1976), 4,044,193 (August 23, 1977), 4,059,541 (November 22, 1977), and 4,082,695 (April 4, 1978), H. G. Petrow and R. J. Allen (to Prototech., Inc.).

26. U.S. Pats. 4,136,056 and 4,137,373 (January 23, 1979), V. L. Jalan and C. L. Bushnell (to United Technologies, Inc.).

27. A. J. Appleby, in S. Sarangapani, J. R. Akridge, and B. Schumm, eds., *The Electrochemistry of Carbon*, The Electrochemical Society, Princeton, N.J., 1964, pp. 251–272.

28. A. J. Appleby, *Energy* 11, 13–94 (1986).

29. A. J. Appleby, *Journal of Electroanalytical Chemistry* 357, 117–179 (1993).

30. A. Damjanovic and V. Brusic, *Electrochim. Acta* 12, 615–628 (1967).

31. A. J. Appleby, *Journal of Electroanalytical Interfacial Electrochemistry* 35, 193–207 (1972).

32. J. O'M. Bockris and A. K. S. Huq, *Proceedings of the Royal Society of London* A237, 277–296 (1956).

33. U.S. Pats. 4,186,410 (January 29, 1980), V. M. Jalan and D. A. Landsman; 4,192,907 (March 11, 1980), V. M. Jalan, D. A. Landsman, and D. M. Lee; 4,202,934 (May 13, 1980), V. M. Jalan (all to United Technologies, Inc.).

34. P. N. Ross, "Oxygen Reduction on Supported Pt Alloys and Intermetallic Compounds in Phosphoric Acid," EPRI-EM-1553, Electric Power Research Institute, Palo Alto, Ca., 1980.

35. V. M. Jalan and E. J. Taylor, *Journal of the Electrochemical Society* 130, 2299–2302 (1983).

36. J. O'M Bockris and A. J. Appleby, *Energy* 11, 95–135 (1986).

37. W. R. Grove, *Philosophical Magazine, Series 3* 14, 127–130 (1839); 21, 417–420 (1843); *Proceedings of the Royal Society of London* 4, 463 (1843); 5, 557–559 (1845).

38. W. R. Grove, *The Correlation of Physical Forces*, 6th. ed., Longmans Green, London, 1874.

39. L. Mond and C. Langer, *Proceedings of the Royal Society of London* 46, 296–304 (1889).

40. W. Ostwald, *Zeitschrift fuer Elektrochemie* 1, 122–125 (1894).

41. E. Baur, W. D. Treadwell, and G. Trumpler, *Zeitschrift fuer Electrochemie* 27, 199–208 (1921).

42. A. M. Adams, F. T. Bacon, and R. G. H. Watson, in W. Mitchell, Jr., ed., *Fuel Cells*, Academic, New York, 1963, pp. 129–192.

43. A. Schmid, *Die Diffusionsgaselektrode*, Enke, Stuttgart, 1923.

44. E. W. Justi and A. W. Winsel, *Cold Combustion Fuel Cells*, Steiner, Wiesbaden, 1962.

45. G. W. Heise and E. A. Schumacher, *Transactions of the Electrochemical Society* 52, 383–391 (1932).

46. A. J. Appleby, O. A. Velev, J-G. LeHelloco, A. Parthasarthy, S. Srinvasan, D. D. DesMarteau, M. S. Gillette, and J. K. Ghosh, *Journal of the Electrochemical Society* 140, 109–111 (1993).

47. M. Pourbaix, *Atlas of Electrochemical Equilibria at 25°C*, National Association of Corrosion Engineers, Houston, Tex., 1966.

48. A. J. Appleby and E. B. Yeager, *Energy* **11**, 137–152 (1986).

49. U.S. Pat. 2,175,523 (October 10, 1939), H. H. Greger (to H. H. Greger).

50. U.S. Pat. 2,581,651 (January 8, 1952), 2,654,661 (October 6, 1953), 2,654,662 (October 6, 1953), E. Gorin (to Pittsburgh Consolidated Coal Company).

51. G. H. J. Broers and J. A. A. Ketelaar, in G. H. Young, eds., *Fuel Cells*, vol. 1, Rheinhold, New York, 1960, pp. 78–93.

52. R. Selman, *Energy* **11**, 153–208 (1986).

53. J. Weissbart and R. Ruka, *Journal of the Electrochemical Society* **109**, 723–726 (1962).

54. W. Nernst and W. Wild, *Zeitschrift fuer Elektrochemie* **7**, 373–380 (1900).

55. J. T. Brown, *Energy* **11**, 209–229 (1986).

56. A. J. Appleby, *Energy* **19** (in press).

57. J. P. Shoesmith, R. D. Collins, M. K. Oakley, and R. D. Stevenson, *Journal of Power Sources* **49**, 129–142 (1994).

58. J. H. Hirschenhofer, D. B. Stauffer, and R. R. Engleman, *Fuel Cells, A Handbook (Revision 3)*, U.S. Department of Energy, Office of Fossil Energy, Morgantown Energy Technology Center, Morgantown, W. Va., 1994.

59. A. J. Appleby, *Journal of Power Sources* **37**, 223–239 (1992).

60. "Advanced Water-Cooling PAFC Development, Final Report," Report No. DE/MC/24221-3130, U.S. DOE Contract No. DE-AC21-88MC24221, International Fuel Cells, South Windsor, Conn., September 1992.

61. B. H. C. Steele, *Journal of Power Sources* **49**, 1–14 (1994).

FUEL RESOURCES

David A. Tillman
Jeffrey Warshauer
David E. Prinzing
Foster Wheeler Environmental Coop.
Sacramento, California

The wheel is considered to be the greatest invention and fire the greatest discovery of all time. Together, the invention of the wheel and the discovery of fire as a useful force have led to the application of energy. From the invention of the wheel has come such innovations as steam and combustion turbines, rotors and stators used in electricity generation, diesel and Otto-cycle engines for transportation systems, and windmills, water wheels, and hydroelectric turbines. Similarly, the harnessing of fire has led to the use of various materials as fuels: coal (qv), lignite (see LIGNITE AND BROWN COAL), petroleum (qv), natural gas (see GAS, NATURAL), tar sands (qv), oil shale (qv), peat, wood (qv), and the biofuels (see FUELS FROM BIOMASS), organic wastes (see FUELS FROM WASTE), uranium and nuclear power, wind, falling water (for hydroelectric power), geothermal steam and hot water (see GEOTHERMAL ENERGY), sunlight, ocean thermal gradients, and the range of conversion products including both electricity and synthesis gas from coal (see FUELS, SYNTHETIC) have been used. These fuels are used both to power the wheel-related inventions and to supply energy for process applications: iron- and steelmaking, nonferrous metal smelting and refining, process heat and steam for pulp and paper

operations, process energy for chemicals manufacture, etc. Harnessed fuels supply the needs of commercial and residential users as well.

Evaluations of fuel resources or total fuel supply focus on critical economic and environmental issues as well as existence. These issues include availability, utilization patterns, environmental consequences, and related economic considerations (See also ENERGY CONSUMPTION IN THE UNITED STATES; RENEWABLE RESOURCES).

HISTORICAL PATTERNS IN FUEL UTILIZATION

Preindustrial society relied primarily on wood, other biomass, and falling water for energy. These energy sources provided carbon for steelmaking, heat for domestic and commercial purposes, energy for modest shaft power applications, eg, grinding of grain, and fuel for transportation on riverboats and early railroads. These fuels were readily available and could be gathered up or otherwise harnessed with little capital investment and scant attention to technology. U.S. energy consumption by fuel source from 1870 to 1990 is shown in Table 1.

Industrialization in the United States and northern Europe demanded significant sources of carbon for steelmaking, fuels for pumping water from mines, and energy for manufacturing processes. As the process of industrialization gained momentum manufacturing shifted away from optimal sites along rivers and connected regional economies with transcontinental railroads. Industrialization created a national economy, along with strong regional economies, through the use of energy for manufacturing and transportation systems, and coal was the fuel of choice (Table 1). With the advent of industrialization also came the shift in agriculture toward development of mechanized equipment and chemical fertilizers, and petroleum became the dominant fuel. The emergence of pipelines (qv), has enabled natural gas, then complemented oil, to be the desired form of energy. Fuel selection factors, in all cases, include availability, energy density (J/kg or J/m^3), energy transportability, fuel cost, and fuel reliability.

Fuel Production and Consumption Since 1970

In recent hisotry, both technological and political forces have influenced fuel consumption in the United States and throughout the world. Events such as the oil embargo of 1973, the political unrest in the Middle East, and the collapse of the Union of the Soviet Socialist Republics, have caused disruptions and shifts in petroleum supply systems. The emergence of the North Sea oil field, the construction of the Alaska Pipeline bringing North Slope, Alaska crude to refineries, and other technical developments have also occurred. Most recently, environmental concerns have influenced the selection of fuels, such as the potential to form air pollutants such as particulates NO_x, SO_2, and most recently air toxics (3,7–10) (see AIR POLLUTION; AIR POLLUTION CONTROL METHODS). Moreover, there has been the passage of numerous energy and environmental laws within the United States. Legislation has included the Clean Air Act amendments of 1990, and the National Energy Policy Act of 1992. These laws comple-

Table 1. U.S. Energy Consumption by Source from 1870–1990, Exajoules (EJ)[a,b]

Year	Wood and Biomass	Coal	Petroleum	Natural Gas	Hydroelectric	Nuclear[c]	Other[d]	Total
1870	3.1	1.1						4.1
1880	3.1	2.1	0.1			qc		5.2
1890	2.6	4.3	0.2	0.3				7.5
1900	2.1	7.2	0.2	0.3	0.3			10.1
1910	2.0	13.4	1.1	0.5	0.5			17.5
1920	1.7	16.4	2.7	0.8	0.8			22.4
1930	1.6	14.4	5.7	2.1	0.8			24.5
1940	1.5	13.2	7.9	2.9	1.0			26.3
1950	1.3	13.6	14.2	6.5	1.5			37.1
1960	0.3	10.7	21.2	13.4	1.8			47.4
1970	1.1	13.4	28.9	23.2	2.9	0.2	0.01	69.6
1980	2.53	16.27	36.08	21.51	3.29	2.89	0.08	80.13
1985	2.6	18.44	32.62	18.81	3.54	4.38	0.21	78.01
1986		18.21	33.97	17.63	3.58	4.72	0.23	78.33
1987		19.00	34.68	18.72	3.24	5.18	0.26	81.08[e]
1988		19.89	36.10	19.57	2.79	5.97	0.30	84.61
1989		19.98	36.09	20.45	3.04	5.99	0.26	85.81
1990	3.3	20.17	35.40	20.36	3.11	6.50	0.22	85.76

[a] References 1–6.
[b] To convert EJ to Btu, multiply by 9.48×10^{14}.
[c] Nuclear energy is that generated by electric utilities.
[d] Other includes net imports of coal coke and electricity produced from wood, waste, wind, photovoltaic, and solar thermal sources connected to electric utility distribution systems. It does not include consumption of wood energy (other than consumed by electric utility industry).
[e] An estimated additional 2.5 EJ of wood energy was consumed for residential heating and light industry.

ment the move toward energy conservation, and the emphasis on materials recycling (qv) for resource management. Further, actions by local and state regulatory agencies in the 1990s, including Public Utility Commissions, have further increased the complexity of fuel supply in the United States.

Trends in commercial fuel, eg, fossil fuel, hydroelectric power, nuclear power, production and consumption in the United States and in the Organization of Economic Cooperation and Development (OECD) countries, are shown in Tables 2 and 3. These trends indicate (6,13); (1) a significant resurgence in the production and use of coal throughout the U.S. economy; (2) a continued decline in

the domestic U.S. production of crude oil and natural gas leading to increased imports of these hydrocarbons (qv); and (3) a continued trend of energy conservation, expressed in terms of energy consumed per dollar of gross domestic product.

U.S. ENERGY PRODUCTION, CONSUMPTION, AND AVAILABILITY

Production and consumption of commercially available fossil fuel, nuclear power, and hydroelectric power in the United States for the year 1992 is shown in Table 2 (12).

Table 2. U.S. Energy Production and Consumption, 1982–1992, EJ[a,b]

Fuel Source	1982	1984	1986	1988	1990	1992
Consumption						
Petroleum	31.90	32.76	33.97	36.11	35.40	35.31
Dry Natural gas	19.52	19.53	17.63	19.57	20.36	21.44
Coal	16.17	18.01	18.21	19.88	20.15	19.96
Hydroelectric	3.77	4.01	3.64	2.81	3.11	2.94
Nuclear	3.30	3.75	4.72	5.97	6.50	7.02
Total	*74.66*	*78.06*	*78.17*	*84.34*	*85.52*	*86.67*
Production						
Crude oil	19.32	19.89	19.39	18.23	16.43	16.02
Natural gas liquids	2.31	2.40	2.27	2.38	2.39	2.49
Dry natural gas	19.26	18.92	17.38	18.48	19.37	19.27
Coal	19.66	20.8	20.58	21.88	23.70	22.75
Hydroelectric	3.45	3.57	3.24	2.46	3.09	2.65
Nuclear	3.30	3.75	4.72	5.97	6.50	7.02
Total	*67.30*	*69.33*	*67.58*	*69.40*	*71.38*	*70.20*

[a] References 11, 12.
[b] To convert EJ to Btu, multiply by 9.48×10^{14}.

Table 3. Total Final Consumption per Gross Domestic Product OECD Countries, 1973–1989[a,b]

Country	1973	1979	1987	1988	1989
OECD North America	0.45	0.41	0.32	0.32	0.31
Canada	0.55	0.52	0.41	0.41	0.41
United States	0.44	0.40	0.31	0.31	0.31
OECD Pacific	0.30	0.26	0.20	0.20	0.20
Australia	0.36	0.35	0.31	0.31	0.31
Japan	0.30	0.25	0.18	0.19	0.18
New Zealand	0.33	0.35	0.39	0.40	0.41
OECD Europe	0.38	0.35	0.30	0.29	0.28
Austria	0.35	0.34	0.30	0.29	0.28
Belgium	0.58	0.51	0.40	0.39	0.38
Denmark	0.36	0.33	0.23	0.23	0.22
Finland	0.50	0.44	0.38	0.36	0.36
France	0.36	0.31	0.26	0.25	0.24
Germany	0.39	0.36	0.30	0.29	0.28
Greece	0.38	0.38	0.40	0.40	0.41
Iceland	0.48	0.35	0.29	0.31	0.33
Ireland	0.45	0.43	0.38	0.36	0.35
Italy	0.34	0.29	0.25	0.24	0.24
Luxembourg	1.43	1.18	0.78	0.76	0.78
Netherlands	0.49	0.47	0.41	0.38	0.37
Norway	0.39	0.36	0.31	0.30	0.29
Portugal	0.39	0.43	0.46	0.49	0.49
Spain	0.31	0.35	0.29	0.30	0.29
Sweden	0.44	0.41	0.32	0.32	0.30
Switzerland	0.21	0.21	0.20	0.19	0.19
Turkey	0.65	0.62	0.61	0.60	0.61
United Kingdom	0.40	0.36	0.30	0.29	0.28
Total OECD	*0.41*	*0.37*	*0.29*	*0.29*	*0.28*

[a] Reference 13.
[b] Ratio of total final consumption of energy to gross domestic product (GDP). Measured in metric tons of oil equivalent per $1000 of GDP at 1985 prices and exchange rates; changes in ratios over time reflect the combined effects of efficiency improvements, structural changes, and fuel substitution.

Table 4. Electricity Supply and Disposition, 1990[a]

Supply and Disposition	Quantity, kW·h × 10⁹	Percent of Total
Fuel type for electric utilities generation		
Coal	1560	55.6
Petroleum	117	4.2
Natural gas	264	9.4
Nuclear power	577	20.5
Pumped storage hydroelectric	−2	
Renewable sources/other[b]	293	10.4
Total	*2808*	*100*
imports	2	
Fuel type for nonutilities[c] generation		
Coal	33	15
Petroleum	5	2.3
Natural gas	100	45.9
Renewable sources/other[b,d]	80	36.7
Total	*218*	*100*
Sales to utilities	106	
Generation for own use	111	
Electricity sales by sector		
Residential	924	34.1
Commercial/other	843	31.0
Industrial	946	34.9
Total	*2713*	*100*

[a] Reference 14.
[b] Renewable sources/other includes hydroelectric, geothermal, wood, wood waste, municipal solid waste, other biomass, and solar and wind power.
[c] Nonutilities includes cogenerators, small power producers, and all other sources, except electric utilities which produce electricity for self-use or for delivery to the grid. The generation values for nonutilities represent gross generation rather than net generation (net of station use).
[d] Includes waste heat, blast furnace gas, and coke oven gas.

Coal production is most significant followed by natural gas and petroleum. Electricity generation and utilization patterns are shown in Table 4. Coal is overwhelmingly the most significant energy source used to generate electricity.

The data presented in Tables 2 and 4 focus on commercially traded sources of energy. During the period 1970–1990, increased emphasis was placed on renewable energy resources (qv), including wood and wood waste; municipal solid waste and refuse-derived fuel; other sources of biomass and waste, eg, agricultural crop wastes, tire-derived fuels, and selected hazardous wastes burned as fuel substitutes in cement kilns; wind and solar energy; geothermal steam and hot water; andd other unconventional energy sources. Estimates of the contribution of these energy sources vary. As of this writing biofuel utilization in the United States runs about 3.7 EJ/yr (3.5 × 10¹⁵ Btu/yr) in support of process energy needs for industry, cogeneration facilities, and small stand-alone power plants (5), and geothermal energy is about 0.21 EJ/yr (0.2 × 10¹⁵ Btu/yr) (6).

Coal Availability and Utilization

There are vast reserves of coal (qv) and lignite (see LIGNITE AND BROWN COAL) in the United States (see Table 5). The total reserve base exceeds 425 billion metric tons

Table 5. U.S. Coal Reserves by State, 1990, EJ[a,b,c,d]

State	Reserves	State	Reserves
Alabama	114	Montana	2,848
Alaska	146	New Mexico	106
Arizona	6		
Arkansas	10	North Dakota	229
Colorado	403	Ohio	438
		Oklahoma	38
Illinois	1,857	Pennsylvania	691
Indiana	241	South Dakota	9
Iowa	52	Tennessee	20
Kansas	23	Texas	316
Kentucky	697	Utah	146
Louisiana	12	Virginia	62
Maryland	18	Washington	34
Michigan	3	West Virginia	880
Missouri	143	Wyoming	1,614
Total	*11,155*		

[a] References 6 and 15.
[b] Reserve data is based on demonstrated reserve base. Minable reserves differ from these figures.
[c] Georgia, Idaho, North Carolina, and Oregon also have some reserves.
[d] To convert EJ to Btu, multiply 9.48 × 10¹⁴.

Table 6. Largest Coal-Producing States in 1990, EJ[a,b]

State	1990 Production	Rank
Wyoming	4.37	1
Kentucky	4.11	2
West Virginia	4.02	3
Pennsylvania	1.67	4
Illinois	1.43	5
Texas	1.32	6
Virginia	1.11	7
Montana	0.89	8
Indiana	0.85	9
Ohio	0.84	10
Total	*20.63*	

[a] Reference 6.

[b] To convert EJ to Btu, multiply by 9.48×10^{14}.

equivalent to 11,200 EJ (10.6×10^{18} Btu) and is distributed throughout 32 states. This reserve base has increased by 8.3% since the 1970s despite the high levels of fuel production (6). Total U.S. recoverable reserves exceed 240 billion metric tons or 6100 EJ (5.8×10^{18} Btu) and are distributed among three geographic areas: the Appalachian, Interior, and Western coal producing regions. Of these, the western region contains 53.6% of the recoverable reserves, the interior region 25.8%, and the Appalachian region 20.6%. Reserves can also be evaluated in terms of the sulfur (qv) content of the coal. The sulfur is important owing to environmental considerations. Of the recoverable reserves, 34.3% contains <0.6% sulfur, 33.9% contains 0.61–1.67% sulfur, and 31.8% contains >1.68% sulfur.

Coal production and consumption in the 1990s reflects the shift toward the use of western, lower sulfur coal. In 1970, West Virginia, Kentucky, and Pennsylvania ranked 1–3 in coal production, respectively. In 1990, Wyoming, Kentucky, and West Virginia held those ranks, and Texas and Montana entered the top 10 coal producers. Whereas Appalachia remained the most significant energy production region, the western coal producing states surpassed the interior states in solid fossil fuel production (see Table 6). The average coal heating value reflected the shift from Appalachia and the interior to the west, declining from 25.8×10^6 J/kg (11.1×10^3 Btu/lb) in 1973 to 24.8×10^6 J/kg (10.7×10^3 Btu/lb) in 1980 and 24.3×10^6 J/kg (10.4×10^3 Btu/lb) in 1990. The shift in coal production toward western coal deposits also reflects the shift in coal utilization patterns (see Table 7). Electric utilities are in-

creasing coal consumption on both absolute and percentage bases, whereas coke plants, other industrial operations, and residential and commercial coal users are decreasing use of this solid fossil fuel.

Environmental considerations also were reflected in coal production and consumption statistics, including regional production patterns and economic sector utilization characteristics. Average coal sulfur content, as produced, declined from 2.3% in 1973 to 1.6% in 1980 and 1.3% in 1990. Coal ash content declined similarly, from 13.1% in 1973 to 11.1% in 1980 and 9.9% in 1990. These numbers clearly reflect a trend toward utilization of coal that produces less SO_2 and less flyash to capture. Emissions from coal in 1990s were 14×10^6 t/yr of SO_2 and 450×10^3 t/yr of particulates generated by coal combustion at electric utilities. The total coal combustion emissions from all sources were only slightly higher than the emissions from electric utility coal utilization (6).

Oil and Natural Gas Availability and Utilization

U.S. resources and reserves of petroleum (qv) and natural gas (see GAS, NATURAL), including natural gas liquids (NGL) are limited. As of January 1, 1992, U.S. reserves of petroleum were some 151 EJ (24.7×10^9 bbl) and U.S. reserves of natural gas were 182 EJ (17.3×10^{16} Btu) (11). Since 1976, the United States has experienced a significant decline in oil reserves. In 1976, proven petroleum reserves totaled 205 EJ (33.5×10^9 bbl). Between 1976 and 1993, some 210 EJ (3.4×10^{10} bbl) were added to the reserves, and 263.5 EJ (4.31×10^{10} bbl) were produced, yielding a net reserve loss of 53.8 EJ (8.8×10^9 bbl) (14). Similarly from 1976 to 1992, there was a net reserve loss of 44.5 EJ (4.22×10^{16} Btu) of dry natural gas (16).

As shown in Table 8, U.S. distribution of oil and natural gas reserves is centered in Alaska, California, Texas, Oklahoma, Louisiana, and the U.S. outer-continental shelf. Alaska reserves include both the Prudhoe Bay deposits and the Cook Inlet fields. California deposits include those in Santa Barbara, the Wilmington Field, the Elk Hills Naval Petroleum Reserve No. 1 at Bakersfield, and other offshore oil deposits. The Yates Field, Austin Chalk formation, and Permian Basin are among the producing sources of petroleum and natural gas in Texas.

The decrease in petroleum and natural gas reserves has encouraged interest in and discovery and development of unconventional sources of these hydrocarbons. Principal alternatives to conventional petroleum reserves include oil shale (qv) and tar sands (qv). Oil shale re-

Table 7. U.S. Coal Consumption by Sector, 1970–1990, EJ[a,b]

Year	Electric Utilities	Industrial Coke Plants	Industrial Other	Residential and Commercial	Total Consumption
1970	7.60	2.29	2.15	0.38	11.42
1975	9.64	1.98	1.51	0.22	13.35
1980	13.51	1.58	1.43	0.15	16.67
1985	16.47	0.98	1.79	0.19	19.43
1990	18.36	0.94	1.81	0.16	21.27

[a] Reference 6.

[b] To convert EJ to Btu, multiply by 948×10^{14}.

Table 8. Crude Oil and Natural Gas Proved Reserves, EJ[a,b,c]

State	Oil Proved Reserves	Gas Proved Reserve[d]
Alaska	37.22	10.22
California	25.80	3.19
Louisiana	4.15	11.79
Oklahoma	4.28	15.83
Texas	41.59	39.87
Wyoming	4.63	11.02
Federal offshore	16.03	31.02
Total	*133.72*	*122.94*

[a] Reference 16.
[b] To convert EJ to Btu, multiply by 9.48×10^{14}.
[c] As of Dec. 31, 1991.
[d] Gas reserves equal dry natural gas plus natural gas liquids.

serves in the United States are estimated at 20,000 EJ (19.4×10^{18} Btu) and estimates of tar sands and oil sands reserves are on the order of 11 EJ (10×10^{15} Btu) (see TAR SANDS). Of particular interest are the McKittrick, Fellows, and Taft quadrangles of California, the Asphalt Ridge area of Utah, the Asphalt, Kentucky area, and related geographic regions.

The unconventional reserves of natural gas occur principally in the form of recoverable methane from coal beds. As of 1991, reserves of coal bed methane totaled 8.6 EJ (8.2×10^{15} Btu), principally in the states of Alabama, Colorado, and New Mexico (16).

Domestic petroleum, natural gas, and natural gas liquids production has declined at a rate commensurate with the decrease in reserves (see Table 2). Consequently, the reserves/production ratio, expressed in years, remained relatively constant from about 1970 through 1992, at 9–11 years (16). Much of the production in the early 1990s is the result of enhanced oil recovery techniques: water flooding, steam flooding, CO_2 injection, and natural gas reinjection.

Whereas the use of petroleum and natural gas is significant in the electricity generating sector, this usage declined from 1970 to 1990, in part owing to the 1977 Fuel Use Act (see Table 9). The legislation of the 1990s and the growth of independent power producers (IPP) generating electricity for utilities in combined cycle combustion turbine (CCCT) facilities, may mean a reversal in the trend for oil and natural gas utilization for power generation (qv). In any event, total U.S. oil and gas consumption (Table 1) remains high, and these are the fuels of choice for residential, commercial, industrial, and transportation applications.

Table 9. Fuel for Electric Utility Generation of Electricity 1970–1990, kW·h $\times 10^{9}$[a]

Year	Petroleum	Gas-Fired	Internal Combustion and Gas Turbine	Total
1970	174	361	22	557
1975	273	288	28	589
1980	238	326	28	592
1985	97	279	16	392
1990	113	246	22	381

[a] Reference 6.

Other Fuel Availability and Utilization

As shown in Table 2, nuclear, hydroelectric, and geothermal resources now contribute some 9.8 EJ (9.3×10^{15} Btu) annually to the U.S. economy. Of these energy sources, nuclear power is the dominant force having over 70% of the total. U.S. nuclear power production continued to increase through 1990, but nuclear electricity generation may have peaked at the 6.5 EJ for political and social reasons. Hydroelectric power generation remains relatively stable. There are annual variations in supply which depend on local weather, eg, rainfall, snowpack, and regional economic conditions. Geothermal energy (qv) has been developed to only a modest extent.

Biomass and waste fuels contributed some 3.7 EJ to the economy (see FUELS FROM BIOMASS; FUELS FROM WASTE). These fuels include wood and wood waste; spent pulping liquor at pulp and paper mills; agricultural materials such as rice hulls, bagasse, cotton gin trash, coffee grounds, and a variety of manures. When wood waste and numerous other forms of biomass are added to municipal solid waste (MSW), refuse-derived fuel (RDF), methane recovered from landfills and sewage treatment plants, and special industrial and municipal wastes such as tire-derived fuel, these together contribute about 5 EJ (4.7×10^{15} Btu) to the U.S. economy (17). Of these fuels, wood and the biofuels are typically employed in industrial settings either to generate process steam (qv) or to cogenerate electricity and process steam. Some condensing power plants have been built by such utilities as Washington Water Power, Burlington Electric, and several IPP firms. There are some 1500 MW$_e$ of electricity generating capacity based on wood and the biofuels in existence as of 1993.

MSW incinerators (qv) are typically designed to reduce the volume of solid waste and to generate electricity in condensing power stations. Incineration of unprocessed municipal waste alone recovers energy from about 34,500 t/d or 109 million metric tons of MSW annually in some 74 incinerators throughout the United States. This represents 1.1 EJ (1.05×10^{15} Btu) of energy recovered annually (18). Additionally there are some 20 RDF facilities processing from 200 to 2000 t/d of MSW into a more refined fuel (19). Representative projects are shown in Table 10.

Other sources of energy worth noting are the extensive wind farms, solar projects, and related engineering unconventional technologies. These renewable resources provide only small quantities of energy to the U.S. economy as of this writing.

Trends in Energy Technology and Future Fuel Consumption

Increased economic activity usually means an increase in energy consumption particularly for generation of electricity, manufacturing of products, transportation, and residential and commercial applications. Regulatory and political requirements associated with energy supply and utilization require increasing attention to environmental concerns in order to ensure reliable energy availability without undue environmental degradation. Thus attention is being paid to increasing the efficiency of fuel utilization as well as to reducing the formation of airborne emissions ranging from particulates NO_x and SO_2

Table 10. Municipal Waste to Energy Projects[a]

Location	Capacity, t/d
Mass burn	
Hillsborough County, Fla.	1100
Pinnelas County, Fla.	2700
Tampa, Fla.	910
Baltimore, Md.	2000
North Andover, Mass.	1350
Saugus, Mass.	1350
Peekskill, N.Y.	2000
Tulsa, Okla.	1000
Marion County, Oreg.	500
Nashville, Tenn.	1000
Refuse-derived fuel	
Akron, Ohio	1000
Duluth, Minn.	360
Niagara Falls, N.Y.	1800
Dade County, Fla.	2700
Columbus, Ohio	1800
Hartford, Conn.	1800

[a] References 18, 19.

to the management of air toxics such as HCl and trace metals.

Coal. Technologies traditionally deployed for coal utilization include using pulverized coal (PC), cyclone, and stoker-fired boilers. For PC boilers, technologies being deployed or developed include the use of micrometer-sized coal, staged fuel-staged air low NO_x burners, limestone injection multistage burners (LIMB), reburning for NO_x control, and advanced techniques for overfire air management. Cyclone-fired boilers also are capitalizing on reburning technologies and air management techniques. Further, both PC and cyclone-fired boilers are utilizing cofiring techniques, blending nitrogen and sulfur-free biomass fuels, and low sulfur tire-derived fuels with the coal for both cost control and emissions reduction (17,20).

Advanced coal utilization technologies include the development of bubbling, circulating, and pressurized fluidized-bed combustion for electricity generation and process energy production (see COAL CONVERSION PROCESSES; FLUIDIZATION). Since the early 1980s, over 250 fluidized beds have been installed that have capacities ranging from 25 GJ/h to 1 TJ/h. Whereas fluidized beds are fired using every solid fuel available, these beds are predominantly used for coal combustion. Projects include the Shawnee #10 boiler of TVA, the Black Dog project of Northern States Power, and the Colorado-Ute circulating fluidized-bed project. Advanced coal utilization technologies also include integrated gasification combined cycle (IGCC) systems where coal is gasified and the low or medium heat-value gas is then utilized in a combustion turbine and the exhaust is ducted to a heat recovery steam generator (HRSG). The initial demonstration of this technology was at the Cool Water project near Barstow, California. More recent projects are under development in Polk and Martin counties in Florida, and at the Tracy Station of Sierra Pacific Power Co. in Nevada. Post combustion technologies including alternative acid gas scrubbing technologies,

urea or ammonia injection for NO_x control, and combinations of air pollution control systems are also emerging to support coal utilization.

Petroleum and Natural Gas. The dominant technologies under development for oil and gas include advances in combustion turbine design. These include applying aircraft engine technology to power generation (qv), and placing emphasis on higher temperatures and increased efficiencies in the combustion turbine. Other technologies of significance include dry low NO_x combustion systems employing principles of staged fuel-staged air, and various catalytic and noncatalytic ammonia injection systems in the post-combustion environment.

Other Fuels. The emerging technologies for unconventional and renewable fuels somewhat mirror those associated with coal: cofiring of biofuels and coal in PC and cyclone boilers, fluidized-bed combustion systems fed with biofuel or a blend of various solid fuels, combustion air management, and gasification–combustion systems. Additionally, technologies associated with post-combustion pollution control for the alternative solid fuels are similar to those developed for coal utilization. At the same time, significant advances have been made in such alternative technologies as fuel cells (qv), photovoltaic cells (qv), wind energy, and the broad range of renewable resources. All of these technologies, combined with those for the dominant fossil fuels, are designed to promote fuel supply and utilization within the framework of an environmentally conscious society.

WORLD FUEL RESERVES, PRODUCTION, AND CONSUMPTION

Energy reserves, production, and consumption in the world economy are given in Tables 11, 12, and 13, respectively. As these tables indicate the overwhelming sources of petroleum reserves, and supoply are in the Middle East. Other significant sources of reserves include Russia, the North Sea, North American countries, and parts of southeast Asia. There are also significant concentrations of coal reserves in Russia and China. The dominant coal-producing countries include China and the United States, plus Poland, South Africa, Australia, India, Germany, and the United Kingdom. China is the single largest coal producer and consumer, utilizing over 22 EJ/yr (21×10^{15} Btu/yr) of this solid fossil fuel.

In addition to the significant consumption of coal and lignite, petroleum, and natural gas, several countries utilize modest quantities of alternative fossil fuels. Canada obtains some of its energy from the Athabasca tar sands development (the Great Canadian Oil Sands Project). Oil shale is burned at two 1600 MW power plants in Estonia for electricity generation. World reserves of tar sands total some 6400 EJ (6.1×10^{18} Btu), and world reserves of oil shale total some 20,400 EJ (19.3×10^{18} Btu).

Renewable and unconventional energy sources are used more extensively in other parts of the world than in the United States. Tables 12 and 13 document the significance of hydroelectric power throughout industrialized and developing economies. Biofuels are also a significant

Table 11. World Fossil Fuel Reserves, EJ[a,b]

Region	Coal Reserves 1991[c]		Petroleum Reserves 1992[d]		Natural Gas Reserves 1992[d]	
North America	6,570	(6,565)	499	(500)	346	(344)
Central and South America	254	(254)	419	(447)	172	(193)
Western Europe	2,964	(2,578)	90	(136)	187	(222)
Eastern Europe and former USSR	7,818	(8,252)	358	(378)	1,818	(1,929)
Middle East	5	(5)	4,048	(3,651)	1,360	(1,387)
Africa	1,623	(1,623)	370	(462)	320	(344)
Far East and Ocenia	7,950	(7,942)	270	(345)	309	(401)
Total	*27,185*	*(27,221)*	*6,054*	*(5,918)*	*4,512*	*(4,820)*

[a] Refs. 11 and 12.
[b] To convert EJ to Btu, multiply by 9.48×10^{14}.
[c] Data from the World Energy Council. Values in parentheses are from British Petroleum.
[d] Data from *Oil and Gas Journal*. Data in parentheses are from *World Oil*.

Table 12. World Energy Production 1991, EJ[a,b]

Region	Crude Oil	Natural Gas Liquids	Dry Natural Gas	Coal	Hydroelectric	Nuclear	Total
North America	26.22	3.75	24.64	24.73	6.64	7.94	102.28
Central and South America	10.56	0.31	2.56	0.82	3.73	0.11	14.90
Western Europe	9.52	0.46	8.21	9.60	5.06	8.10	68.64
Eastern Europe and former USSR	22.86	0.89	29.52	20.97	2.67	2.93	74.12
Middle East	36.83	1.64	4.59	0.03	0.14	0	12.53
Africa	15.04	0.43	2.81	4.64	0.48	0.10	10.49
Far East and Oceania	14.85	0.33	6.29	36.38	4.81	3.21	82.97
Total	*135.88*	*7.81*	*78.61*	*97.18*	*23.51*	*22.40*	*365.40*

[a] Ref. 11.
[b] To convert EJ to Btu, multiply by 9.48×10^{14}.

Table 13. World Energy Consumption, 1991, EJ[a,b]

Region	Petroleum	Natural Gas	Coal	Hydroelectric	Nuclear	Total
North America	41.95	24.58	21.17	6.64	7.94	102.28
Central and South America	7.88	2.46	0.72	3.73	0.11	14.90
Western Europe	29.67	12.29	13.70	4.98	8.00	68.64
Eastern Europe and former USSR	20.44	27.63	20.27	2.75	3.03	74.12
Middle East	7.78	4.45	0.16	0.14	0.00	12.53
Africa	4.76	1.62	3.53	0.48	0.10	10.49
Far East and Oceania	31.19	6.42	37.34	4.81	3.21	82.97
Total	*143.67*	*79.44*	*96.91*	*23.51*	*22.40*	*365.93*

[a] Ref. 11.
[b] To convert EJ to Btu, multiply by 9.48×10^{14}.

Table 14. World Net Electricity Consumption, 1982–1991, kW·h $\times 10^9$ [a]

Region	1982	1983	1984	1985	1986	1987	1988	1989	1990	1991
North America[b]	390.1	406.3	435.9	458.1	477	499.1	524.2	541.6	543.2	552.5
United States	2,086.4	2,151.0	2,285.8	2,324.0	2,368.8	2,457.3	2,578.1	2,646.8	2,712.6	2,759.3
Central and South America	309.9	333.4	355.6	371.1	391.1	403.7	421.7	427.8	437.0	446.9
Western Europe	1,725.1	1,783.4	1,872.2	1,962.6	2,001.7	2,064.9	2,069.4	2,150.8	2,188.9	2,217.1
Eastern Europe and former USSR	1,554.4	1,610.8	1,694.8	1,748.9	1,727.0	1,787.5	1,825.7	1,853.9	1,838.8	1,751.0
Middle East	111.4	121.7	131.3	143.5	149.4	158.3	169.7	180.2	195.6	185.6
Africa	182.9	202.0	206.5	217.7	248.6	254.7	270.5	280.1	282.0	281.0
Far East and Ocenia	1,300.3	1,387.0	1,477.8	1,568.9	1,654.4	1,783.0	1,904.8	2,047.9	2,199.1	2,288.5
World total	*7,661.6*	*7,995.6*	*8,461.0*	*8,794.9*	*9,018.1*	*9,408.5*	*9,764.1*	*10,128.4*	*10,397.2*	*10,481.9*

[a] Ref. 11.
[b] Excluding the United States.

Table 15. Projections of World and U.S. Energy Consumption, 1995–2005, EJ[a,b]

Energy Source	1995			2000			2005		
	Base	Low	High	Base	Low	High	Base	Low	High
			World projection[c]						
Oil	152	141	157	159	147	170	171	152	190
Gas	82	80	83	92	89	94	107	100	112
Coal	106	102	108	112	107	116	129	118	136
Nuclear	24	23	24	26	25	26	31	27	32
Other	31	31	32	35	34	36	42	39	46
Total	*395*	*377*	*404*	*424*	*402*	*442*	*480*	*436*	*516*
			U.S. projection						
Oil	38	37	38	40	39	41	42	40	44
Gas	23	23	23	25	24	25	26	25	27
Coal	21	21	21	21	21	22	22	22	23
Nuclear	7	7	7	7	7	7	7	7	8
Other	8	8	8	9	9	10	10	10	10
Total	*97*	*96*	*97*	*102*	*100*	*105*	*107*	*104*	*112*

[a] Ref. 14.
[b] To convert EJ to Btu, multiply by 9.48×10^{14}.
[c] World consumption totals also include the United States.

contributor to certain economies, with proportional contributions as follows: Kenya, 75%; India, 50%; China, 33%; Brazil, 25%; and Scandanavia, 10% (5,21). Peat is a significant source of energy for Russia, Finland, and Ireland.

World electricity generation and consumption, shown in Table 14 increased from 1982 to 1991, and is expected to continue to do so. Further, the industrialized economies are focusing on issues of energy conservation, materials conservation through recycling, and environmental protection. Given the world trends in fuels availability and consumption, projections of energy production and consumpton have been made, as shown in Table 15. These projections, to the year 2005, reflect the emphases on fuel availability, energy economics, and environmental awareness of a world community. Further, they reflect the trend toward increased technology development, leading to economically and environmentally sound energy utilization.

Reprinted from *Kirk-Othmer Encyclopedia of Chemical Technology*, 4th ed., vol. 11, John Wiley & Sons, Inc., New York, 1994.

BIBLIOGRAPHY

1. H. Enzer, W. Dupree, and S. Miller, *Energy Perspectives: A Presentation of Major Energy Related Data*, U.S. Department of the Interior, Washington, D.C., 1975.
2. H. H. Lansberg, L. L. Fishman, and J. L. Fisher, *Resources in America's Future*, Johns Hopkins University Press, Baltimore, Md., 1963.
3. H. C. Hottel, and J. B. Howard, *New Energy Technology: Some Facts and Assessments*, MIT Press, Cambridge, Mass., 1971.
4. *A National Plan for Energy Research, Development, and Demonstration: Creating Energy Choices for the Future*, Energy Research and Development Administration, Washington, D.C., 1976.
5. D. A. Tillman, *The Combustion of Solid Fuels and Wastes*, Academic Press, Inc., San Diego, Calif., 1991.
6. *The U.S. Coal Industry, 1970–1990: Two Decades of Change*, Energy Information Agency, Washington, D.C., 1992.
7. F. W. Brownell, *Clean Air Handbook*, Government Institutes, Inc., Rockville, Md., 1993.
8. S. Bruchey, *Growth of the Modern American Economy*, Dodd, Mead, New York, 1975.
9. N. Rosenberg, *Technology and American Economic Growth*, Harper and Row, New York, 1972.
10. H. M. Jones, *The Age of Energy*, Viking Press, New York, 1970.
11. *International Energy Annual 1991*, Energy Information Agency, Washington, D.C., 1992.
12. *EIA's Annual Energy Review 1992*, Energy Information Agency, Washington, D.C., 1993.
13. *Energy Policies of IEA Countries, 1990 Review*, International Energy Agency, Washington, D.C., 1991.
14. *Annual Energy Outlook, 1993*, Energy Information Agency, Washington, D.C., 1993.
15. *Keystone Coal Industry Manual*, Maclean Hunter Publishing Co., Chicago, 1992.
16. *U.S. Crude Oil, Natural Gas, and Natural Gas Liquids Reserves*, Energy Information Agency, Washington, D.C., 1992.
17. D. Tillman, E. Hughes, and B. Gold, "Cofiring of Biofuels in Coal Fired Boilers: Results of Case Study Analysis," *Proceedings First Biomass Conference of the Americas: Energy, Environment, Agriculture, and Industry*, Burlington, Vt., 1993.
18. D. Tillman, A. Rossi, and K. Vick, *Incineration of Municipal and Hazardous Solid Wastes*, Academic Press, Inc., San Diego, Calif., 1989.
19. J. L. Smith, *Early and Current Systems Utilizing Refuse Derived Fuels*, Combustion Engineering Co., Windsor, Conn., 1986.
20. V. Nast, G. Eirschele, and W. Hutchinson, "TDF Co-firing Experience in a Cyclone Boiler," *Proceedings, Strategic Benefits of Biomass and Waste Fuels Conference*, EPRI, Washington, D.C., 1993.
21. D. O. Hall and R. P. Overend, eds., *Biomass, Regenerable Energy*, John Wiley & Sons, Inc., New York, 1987.

FUELS FROM BIOMASS

Donald L. Klass
Entech International, Inc.
Barrington, Illinois

The contribution of biomass energy to U.S. energy consumption in the late 1970s was over 1.8×10^{15} kJ/yr [850,000 barrels of oil equivalent per day (BOE/d)] or more than 2% total energy consumption (1). By 1987, biomass energy had increased to approximately 3.1×10^{15} kJ/yr (1,400,000 BOE/d, 3.0×10^{15} Btu/yr), or 3.7% of U.S. primary consumption (2). Projections indicate that by the year 2000, the biomass energy contribution will increase to about 4.2×10^5 kJ/yr (1,900,000 BOE/d), ie, over 4% of total U.S. primary energy consumption (2). Land- and water-based vegetation, organic wastes, and photosynthetic organisms are categorized as biomass and are nonfossil, renewable carbon resources from which energy, eg, heat, steam, and electric power, and solid, liquid, and gaseous fuels, ie, biofuels, can be produced and utilized as fossil fuel substitutes.

See also Alcohol fuels; Commercial availability of energy technology; Forest resources; Forestry, sustainable; Manufactured gas; Energy taxation, biomass.

Renewable carbon resources is a misnomer because the earth's carbon is in a perpetual state of flux. Carbon is not consumed such that it is no longer available in any form. Reversible and irreversible chemical reactions occur in such a manner that the carbon cycle makes all forms of carbon, including fossil resources, renewable. It is simply a matter of time that makes one carbon form more renewable than another. If it is presumed that replacement does in fact occur, natural processes eventually will replenish depleted petroleum or natural gas deposits in several million years. Fixed carbon-containing materials that renew themselves often enough to make them continuously available in large quantities are needed to maintain and supplement energy supplies; biomass is a principal source of such carbon.

The capture of solar energy as fixed carbon in biomass via photosynthesis is the initial step in the growth of biomass. It is depicted by the equation

$$CO_2 + H_2O + light \xrightarrow{\text{chlorophyll}} (CH_2O) + O_2$$

Carbohydrate, represented by the building block CH_2O, is the primary organic product. For each gram mole of carbon fixed, about 470 kJ (112 kcal) is absorbed. Oxygen liberated in the process comes exclusively from the water, according to radioactive tracer experiments. There are many unanswered questions regarding the detailed molecular mechanisms of photosynthesis, but the prerequisites for plant biomass production are well established; ie, carbon dioxide, light in the visible region of the electromagnetic spectrum, the sensitizing catalyst chlorophyll, and a living plant are essential. The upper limit of the capture efficiency of incident solar radiation in biomass is estimated to range from about 8% to as high as 15%; in most situations it is generally in the range 1% or less (3).

The primary features of biomass-to-energy technology as a source of synthetic fuels are illustrated in Figure 1. Conventionally, biomass is harvested for feed, food, and

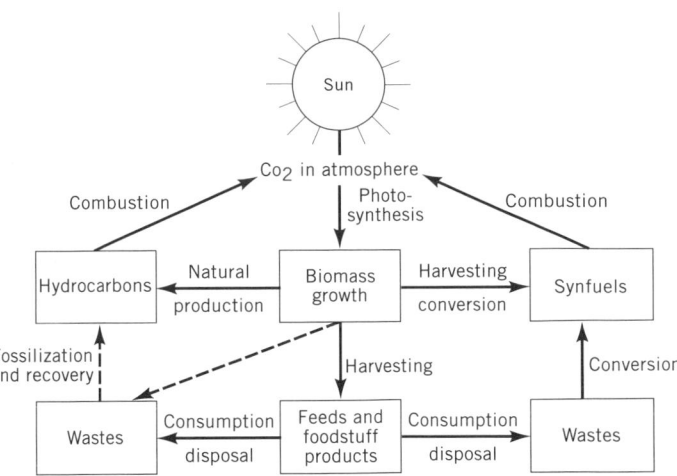

Figure 1. Biomass-to-energy technology.

materials-of-construction applications, or is left in the growth areas where natural decomposition occurs. Decomposing biomass, and waste products from the harvesting and processing of biomass, disposed of on or in land, in theory can be partially recovered after a long period of time as fossil fuels. This is indicated by the dashed lines in Figure 1. Alternatively, biomass, and any wastes that result from its processing or consumption, can be converted directly into synthetic organic fuels if suitable conversion processes are available. The energy content of biomass also can be diverted to direct heating applications by combustion. Certain species of biomass can be grown, eg, the rubber tree (*Hevea braziliensis*), in which high energy hydrocarbons are formed within the biomass by natural biochemical mechanisms. The biomass serves the dual role of a carbon-fixing mechanism and a continuous source of hydrocarbons without being consumed in the process. Other plants, such as the guayule bush, also produce hydrocarbons but must be harvested to recover them. Thus, conceptually, there are several different pathways by which energy products and synthetic fuels might be manufactured (Fig. 1).

Fixed carbon supplies also can be developed from renewable carbon sources by the conversion of carbon dioxide outside the biomass species into synthetic fuels and organic intermediates. The ambient air, which in 1992 contained an average of 350 ppm by volume of carbon dioxide, the dissolved carbon dioxide and carbonates in the oceans, and the earth's large terrestrial carbonate deposits serve as renewable carbon sources. However, because carbon dioxide is the final oxidation state of fixed carbon, it contains no chemical energy and energy must be supplied in a reduction step. A convenient method of supplying the required energy and reducing the oxidation state is to reduce carbon dioxide with elemental hydrogen. The end product can be, eg, methane (CH_4), the dominant component of natural gas.

$$CO_2 + 4H_2 \rightarrow CH_4 + 2H_2O$$

With all components in the ideal gas state, the standard enthalpy of the process is exothermic by -165 kJ (-39.4 kcal) per gram mole of methane formed. Biomass can serve as the original source of hydrogen, which then effectively acts as an energy carrier from the biomass to the carbon dioxide, to produce substitute (or synthetic) natural gas (SNG).

Distribution of Carbon

Estimation of the amount of biomass carbon on the earth's surface is a problem in global statistical analysis. Although reasonable projections have been made using the best available data, maps, surveys, and a host of assumptions, the validity of the results is impossible to support with hard data because of the nature of the problem. Nevertheless, such analyses must be performed to assess the feasibility of biomass energy systems and the gross types of biomass available for energy applications.

The results of one such study are summarized in Table 1 (4). Each ecosystem on the earth is considered in terms of area, mean net carbon production per year, and standing biomass carbon; ie, carbon contained in biomass on the earth's surface and not including carbon stored in biomass underground. Forest biomass, produced on only 9.5% of the earth's surface, contributes more than any other source to the total net carbon fixed on earth. Marine sources of net fixed carbon also are high because of the large area of earth occupied by water. However, the high turnover rates of carbon in the marine environment result in relatively small steady-state quantities of standing car-

bon. The low turnover rates of forest biomass make it the largest contributor to standing carbon reserves. Forests produce about 43% of the net carbon fixed each year and contain over 89% of the standing biomass carbon of the earth; tropical forests are the largest sources of these carbon reserves. Temperate deciduous and evergreen forests also are large sources of biomass carbon, followed by the savanna and grasslands. Cultivated land is one of the smaller producers of fixed carbon and is only about 9% of the total terrestrial area of the earth.

Human activity, particularly in the developing world, continues to make it more difficult to sustain the world's biomass growth areas. It has been estimated that tropical forests are disappearing at a rate of tens of thousands of hm^2 per year. Satellite imaging and field surveys show that Brazil alone has a deforestation rate of approximately 8×10^6 hm^2/yr (5). At a mean net carbon yield for tropical rain forests of 9.90 t/hm^2·yr (4) (4.42 short ton/acre·yr), this rate of deforestation corresponds to a loss of 79.2×10^6 t/yr of net biomass carbon productivity.

The remaining carbon transport mechanisms on earth are primarily physical mechanisms, such as the solution of carbonate sediments in the sea and the release of dissolved carbon dioxide to the atmosphere by the hydrosphere (6). The great bulk of carbon, however, is contained in the lithosphere as carbonates in rock. These carbon deposits contain little or no stored chemical energy, although some high temperature deposits could pro-

Table 1. Estimate of Net Photosynthetic Production of Dry Biomass Carbon and Standing Biomass Carbon for World Biosphere[a,b]

Ecosystem	Area, 10^6 km^{2c}	Mean Net Carbon Production t/(hm^2·yr)^c	Mean Net Carbon Production 10^9 t/yr^c	Standing Biomass Carbon t/hm^2	Standing Biomass Carbon 10^9 t
Tropical rain forest	17.0	9.90	16.83	202.5	344
Boreal forest	12.0	3.60	4.32	90.0	108
Tropical seasonal forest	7.5	7.20	5.40	157.5	118
Temperate deciduous forest	7.0	5.40	3.78	135.0	95
Temperate evergreen forest	5.0	5.85	2.93	157.5	79
Total	*48.5*		*33.26*		*744*
Extreme desert-rock, sand, ice	24.0	0.01	0.02	0.1	0.2
Desert and semidesert scrub	18.0	0.41	0.74	3.2	5.8
Savanna	15.0	4.05	6.08	18.0	27.0
Cultivated land	14.0	2.93	4.10	4.5	6.3
Temperate grassland	9.0	2.70	2.43	7.2	6.5
Woodland and shrubland	8.5	3.15	2.68	27.0	23.0
Tundra and alpine	8.0	0.63	0.50	2.7	2.2
Swamp and marsh	2.0	13.50	2.70	67.5	14.0
Lake and stream	2.0	1.80	0.36	0.1	0.02
Total	*100.5*		*19.61*		*85*
Total continental	*149.0*		*52.87*		*829*
Open ocean	332.0	0.56	18.59	0.1	3.3
Continental shelf	36.6	1.62	4.31	0.004	0.1
Estuaries excluding marsh	1.4	6.75	0.95	4.5	0.6
Algae beds and reefs	0.6	11.25	0.68	9.0	0.5
Upwelling zones	0.4	2.25	0.09	0.9	0.04
Total marine	*361.0*		*24.62*		*4.5*
Grand total	*510.0*		*77.49*		*833.5*

[a] Ref. 4.
[b] Dry biomass is assumed to contain 45% carbon.
[c] 1 km^2 = 1 $\times$ 10^6 m^2 (0.3861 sq. mi); to convert t/(hm^2·yr) to short ton/(acre·yr), divide by 2.24.

vide considerable thermal energy, and all of the energy for a synfuel system must be supplied by a second raw material, such as elemental hydrogen. These carbon deposits consist of lithospheric sediments and atmospheric and hydrospheric carbon dioxide. Together, these carbon sources comprise 99.9% of the total carbon estimated to exist on the earth. Fossil fuel deposits are only about 0.05% of the total, and the nonfossil energy-containing deposits make up the remainder, about 0.02%.

Biomass carbon is thus a very small, but important, fraction of the total carbon inventory on earth. It helps maintain the delicate balance among the atmosphere, hydrosphere, and biosphere necessary to support all life forms, and serves as a perpetual source of food and materials. Biomass carbon also has served as a primary energy source for the industrialized nations of the world; it continues to do so for developing countries. Biomass carbon may again become a dominant source of energy products throughout the world because of fossil fuel depletion and environmental problems, eg, the effect that large-scale fossil fuel combustion is believed by many to have on atmospheric carbon dioxide build-up (7). The utilization of biomass carbon as a primary energy source does not add any new carbon dioxide to the atmosphere; it is simply recycled between the surface of the earth and the air over a period of time that is extremely short compared to the recycling time of fossil-derived carbon dioxide.

ENERGY POTENTIAL

The percentage of energy demand that could be satisfied by particular nonfossil energy resources can be estimated by examination of the potential amounts of energy and biofuels that can be produced from renewable carbon resources and comparison of these amounts with fossil fuel demands.

The average daily incident solar radiation, or insolation, that strikes the earth's surface worldwide is about 220 W/m^2 (1675 Btu/ft^2). The annual insolation on 0.01% of the earth's surface is approximately equal to all energy consumed (ca 1992) by humans in one year, ie, 321×10^{18} J (305×10^{15} Btu). In the United States, the world's largest energy consumer, annual energy consumption is equivalent (1992) to the insolation on about 0.1 to 0.2% of its total surface.

Based on the state of technology in the early 1990s, the most widespread and practical mechanism for capture of this energy is biomass formation. The energy content of standing biomass carbon, ie, the above-ground biomass reservoir that in theory could be harvested and used as an energy resource (Table 1) is about 110 times the world's annual energy consumption (8). Using a nominal biomass heating value of 16×10^9 J/dry t (13.8×10^6 Btu/short ton), the solar energy trapped in 17.9×10^9 t of biomass, or about 8×10^9 t of biomass carbon, would be equivalent to the world's fossil fuel consumption in 1990 of 286×10^{18} J. It is estimated that 77×10^9 t of carbon, or 171×10^9 t of biomass equivalent, most of it wild and not controlled, is fixed on the earth each year. Biomass should therefore be considered as a raw material for conversion to large supplies of renewable substitute fossil fuels. Under controlled conditions dedicated biomass crops could be grown specifically for energy applications.

A realistic assessment of biomass as an energy resource is made by calculating average surface areas needed to produce sufficient biomass at different annual yields to meet certain percentages of fuel demand for a particular country (Table 2). These required areas are then compared with surface areas available. The conditions of biomass production and conversion used in Table 2 are either within the range of 1993 technology and agricultural practice, or are believed to be attainable in the future.

Figure 2 shows the three yield levels in Table 2 together with the percentage of the U.S. area needed to supply SNG from biomass for any selected gas demand. Although relatively large areas are required, the use of

Table 2. Potential Substitute Natural Gas in United States from Biomass at Different Crop Yields

Demand, %[a]	Average Area Required, km^{2b,c,d}		
	25 t/(hm^2·yr)	50 t/(hm^2·yr)	100 t/(hm^2·yr)
1.58	20,400	10,200	5,100
10	129,000	64,500	32,300
50	645,500	323,000	161,000
100	1,291,000	645,500	323,000

[a] United States demand estimated to be 244×10^8 GJ or 653×10^9 m^3 (231×10^{11} standard cubic feet at 15.5°C, 101.5 kPa (60°F, 30.00 in Hg) dry (SCF)). A percentage of 1.58 is equal to a daily production of 28.3×10^6 m^3 at normal conditions (1×10^9 SCF) of SNG.
[b] Biomass, whether trees, plants, grasses, algae, or water plants, has a heating value of 15.1×10^9 J/dry t, and is converted in integrated biomass planting, harvesting, and conversion systems to SNG at an overall thermal efficiency of 50%.
[c] 1 km^2 = 0.3861 sq. mi.
[d] Yields expressed as dry t.

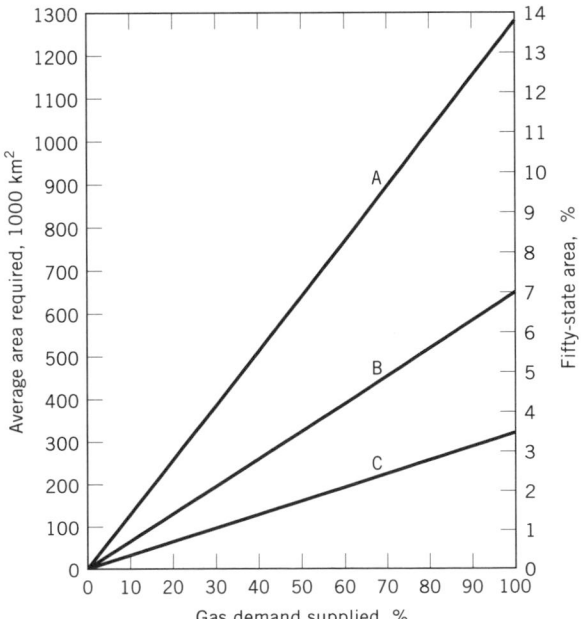

Figure 2. Average area required and percent of U.S. area vs gas demand, %, for a total gas demand of 24.4 EJ/yr. A, 25 t/hm^2·yr; B, 50 t/hm^2·yr; C, 100 t/hm^2·yr (Table 2). To convert t/hm^2·yr to short ton/acre·yr, divide by 2.24.

Table 3. Land and Water Areas of United States[a]

Area Classification	10^6 hm^{2b}	%
Nonfederal land		
Forest[c]	179.41	18.8
Rangeland[d]	178.66	18.7
Other land[e]	279.09	29.2
Transition land[f]	14.41	1.5
	651.57	*68.2*
Federal land		
Forest[c]	102.14	10.7
Rangeland[d]	133.10	13.9
Other land[e]	25.70	2.7
	260.94	*27.3*
Water		
Inland water	24.75	2.6
Other water	19.28	2.0
	44.03	*4.6*
Total land and water	956.55	

[a] Ref. 9. Data for forest, rangeland, and other land as of 1982; data on inland water as of 1980; data on other water as of 1970.

[b] 1 hm^2 = 2.471 acre.

[c] At least 10% stocked by trees of any size, or formerly having such tree cover and not currently developed for nonforest use.

[d] Climax vegetation is predominantly grasses, grass-like plants, forbs, and shrubs suitable for grazing and browsing.

[e] Includes crop and pasture land and farmsteads, strip mines, permanent snow and ice, and land that does not fit into any other land cover.

[f] Forest land that carries grasses or forage plants used for grazing as the predominant vegetation.

land- or freshwater-based biomass for energy applications is still practical. The area distribution pattern of the United States (Table 3) shows selected areas or combinations of areas that might be utilized for biomass energy applications (9), ie, areas not used for productive purposes. It is possible that biomass for both energy and foodstuffs, or energy and forest products applications, can be grown simultaneously or sequentially in ways that would benefit both. Relatively small portions of the bordering oceans also might supply needed biomass growth areas, ie, marine plants would be grown and harvested. The steady-state carbon supplies in marine ecosystems can conceivably be increased under controlled conditions over current low levels by means of marine biomass energy plantations in areas of the ocean dedicated to this objective.

Waste biomass is another large renewable carbon resource. It consists of a wide range of materials and includes municipal solid wastes (MSW), municipal sewage, industrial wastes, animal manures, agricultural crop and forestry residues, landscaping and tree clippings and trash, and dead biomass that results from nature's life cycles. Several of these wastes can cause serious health or environmental problems if not disposed of properly. Some wastes, such as MSW, can be considered a source of recyclables such as metals and glass in addition to energy. Waste biomass is a potential energy resource in the same manner as virgin biomass.

To assess the potential availability and impact of energy from wastes on energy demand, the energy contents and availabilities of different types of wastes generated must be considered. For example, in the United States an average of 2.3 kg of MSW/d is discarded per person. From an energy standpoint, one t of MSW has an as-received energy content of about 10.5×10^9 joule (10 million Btu), so about 2.2 EJ/yr of energy potential resides in the MSW discarded in the United States.

The amount of energy that can actually be recovered from a given waste and utilized depends on the waste type. The amount of available MSW is larger than the total amount of available agricultural wastes even though much larger quantities of agricultural wastes are generated. A larger percentage of MSW is collected for centralized disposal than the corresponding amounts of agricultural wastes, most of which are left in the fields where generated; the collection costs would be prohibitive for most agricultural wastes.

Several studies estimate the potential of available virgin and waste biomass as energy resources (Table 4) (10). In Table 4, the projected potential of the recoverable materials is about 25% of the theoretical maximum; woody biomass is about 70% of the total recoverable potential. These estimates of biomass energy potential are based on existing, sustainable biomass production and do not include new, dedicated biomass energy plantations that might be developed.

U.S. Market Penetration

Table 5 shows U.S. consumption of biomass energy in 1990 and projected consumption for 2000 (10,11). The projected consumption for 2000 is about 50% greater than the consumption of biomass energy in 1990.

A projection of biomass energy consumption in the United States for the years 2000, 2010, 2020, and 2030 is shown in Table 6 by end-use sector (12). This analysis is based on a National Premiums Scenario which assumes that specific market incentives are applied to all new renewable energy technology deployment. The scenario depends on the enactment of federal legislation equivalent to a fossil fuel consumption tax. Any incentives over and above those in place (ca 1992) for use of renewable energy

Table 4. Potential U.S. Biomass Energy Available in 2000, EJ[a,b]

Energy Source	Estimated Recoverable	Theoretical Maximum
Wood and wood wastes	11.0	26.4
Municipal solid wastes		
Combustion	1.9	2.1
Landfill methane	0.2	1.1
Herbaceous biomass and agricultural residues	1.1	15.8
Aquatic biomass	0.8	8.1
Industrial solid wastes	0.2	2.2
Sewage methane	0.1	0.2
Manure methane	0.05	0.9
Miscellaneous wastes	0.05	1.1
Total	*15.4*	*57.9*

[a] Ref. 10. 1 EJ = 0.9488 $\times$ 10^{15} Btu.

[b] Gross heating value of biomass or methane. Conversion of biomass or methane to another biofuel requires that the process conversion efficiency be used to reduce the potential energy available. These figures do not include additional biomass from dedicated energy plantations.

Table 5. U.S. Consumption of Biomass Energy, EJ[a]

Resource	1990,[b] EJ	2000,[c] EJ
Wood and wood wastes		
Industrial sector	1.646	2.2
Residential sector	0.828	1.1
Commercial sector	0.023	0.04
Utilities	0.013	0.01
Subtotal (wood)	*2.510*	3.35
Municipal solid wastes	0.304	0.63
Agricultural and industrial wastes	0.040	0.08
Methane		
Landfill gas	0.033	0.100
Digester gas	0.003	0.004
Thermal gasification	0.001	0.002
subtotal (methane)	*0.037*	*0.106*
Transportation fuels		
Ethanol	0.063[d]	0.1[d]
Other biofuels	0	0.1
Subtotal (transportation fuels)	*0.063*	*0.2*
Total all resources	*2.954*	*4.37*
Primary energy consumption, %	3.3	4.8

[a] 1 EJ = 0.9488 × 10^{15} Btu. To convert from EJ to barrels of oil equivalent per day (BOE/d), multiply by 448,200.
[b] Refs. 10 and 11. Total energy consumption including biomass energy estimated to be 88.426 EJ in 1990.
[c] Ref. 10. Assumes noncrisis conditions, tax incentives and PURPA in place continued to 2000, no legislative mandates to embark on an off-oil campaign, and total consumption of 91.7 EJ in 2000.
[d] Domestic consumption only.

will have a significant impact on biomass energy consumption.

The market penetration of synthetic fuels from biomass and wastes in the United States depends on several basic factors, eg, demand, price, performance, competitive feedstock uses, government incentives, whether established fuel is replaced by a chemically identical fuel or a different product, and cost and availability of other fuels such as oil and natural gas. Detailed analyses have been performed to predict the market penetration of biomass energy well into the twenty-first century. A range of from

Table 6. Projected Biomass Energy Consumption in the United States from 2000 to 2030,[a] EJ[b]

End Use Sector	2000	2010	2020	2030
Industry[c]	2.85	3.53	4.00	4.48
Electricity[d]	3.18	4.41	4.95	5.48
Buildings[e]	1.05	1.53	1.90	2.28
Liquid fuels[f]	0.33	1.00	1.58	2.95
Total	*7.41*	*10.47*	*12.43*	*15.19*

[a] Ref. 12.
[b] 1 EJ = 0.9488 × 10^{15} Btu. Assumes market incentives of 2 ¢/kWh on fossil fuel-based electricity generation, $2.00/10^6 Btu on direct coal and petroleum consumption, and $1.00/10^6 Btu on direct natural gas consumption.
[c] Combustion of wood and wood wastes.
[d] Electric power derived from present (ca 1992) technology via the combustion of wood and wood wastes, MSW, agricultural wastes, landfill and digester gas, and advanced digestion and turbine technology.
[e] Biomass combustion in wood stoves.
[f] Ethanol from grains, and ethanol, methanol, and gasoline from energy crops.

3 to about 21 EJ seems to characterize the results of most of these studies.

U.S. capacity for producing biofuels manufactured by biological or thermal conversion of biomass must be dramatically increased to approach the potential contributions based on biomass availability. For example, an incremental EJ per year of methane requires about 210 times the biological methane production capacity that now exists, and an incremental EJ per year of fuel ethanol requires about 14 times existing ethanol fermentation plant capacity. The long lead times necessary to design and construct large biomass conversion plants makes it unlikely that sufficient capacity can be placed on-line before 2000 to satisfy EJ blocks of energy demand. However, plant capacities can be rapidly increased if a concerted effort is made by government and private sectors.

Projections of market penetrations and contributions to primary consumption of energy from biomass are subject to much criticism and contain significant errors. However, even though these projections may be incorrect, they are necessary to assess the future role and impact of renewable energy resources, and to help in deciding whether a potential renewable energy resource should be developed.

Global Market Penetration. The consumption of all energy resources worldwide in 1990, according to the United Nations, is presented in Table 7 (8). Detailed and time-consuming analysis is necessary to assure validity of the results for an energy resource that is as widespread, dispersed, and disaggregated as biomass, and many nations of the world do not require the archiving of historical energy production and consumption data. Table 7 indicates that biomass energy is a significant source of energy in the developing regions of the world; Africa, 36.8% of total energy consumed; South America, 23.7%; and Asia, 10.9%. It is a small energy resource in the industrialized areas relative to fossil fuels. The markets for biomass energy as replacements and substitutes for fossil fuels are large and have only been developed to a limited extent. As fossil fuels are either phased out because of environmental issues or become less available because of depletion, biomass energy is expected to acquire an increasingly larger share of the organic fuels market.

CHEMICAL CHARACTERISTICS OF BIOMASS

The chemical characteristics of biomass vary over a broad range because of the many different types of species. Table 8 compares the typical analyses and energy contents of land- and water-based biomass, ie, wood, grass, kelp, and water hyacinth, and waste biomass, ie, manure, urban refuse, and primary sewage sludge, with those of cellulose, peat, and bituminous coal. Pure cellulose, a representative primary photosynthetic product, has a carbon content of 44.4%. Most of the renewable carbon sources listed in Table 8 have carbon contents near this value. When adjusted for moisture and ash contents, it is seen that with the exception of the sludge sample, the carbon contents are slightly higher than that of cellulose, but span a relatively narrow range.

Table 7. Global Energy Consumption in 1990, EJ[a]

Region	Fossil Fuels[b]			Electricity[c]	Biomass[d]	Total[e]
	Solids	Liquids	Gases			
Africa	2.96	3.36	1.55	0.18	4.68	12.73
N. America	21.55	38.48	22.13	4.69	1.77[f]	88.62
S. America	0.68	4.66	2.09	1.29	2.71	11.43
Asia	35.52	27.58	8.38	2.57	8.89	82.94
Europe[g]	35.18	40.90	37.16	6.25	1.29	120.85
Oceania	1.64	1.70	0.85	0.14	0.19	4.53
Total world	*97.52*	*116.68*	*72.18*	*15.13*	*19.53*	*321.10*

[a] Ref. 8.

[b] Solids are hard coal, lignite, peat, and oil shale. Liquids are crude petroleum and natural gas liquids. Gases are natural gas.

[c] Includes hydro, nuclear, and geothermal sources, but not fossil fuel-based electricity, which is included in fossil fuels.

[d] Includes fuelwood, charcoal bagasse, and animal, crop, pulp, paper, and municipal solid wastes, but does not include derived biofuels.

[e] Sums of individual figures may not equal totals because of rounding.

[f] Less than Table 5 value of 2.954 EJ for the United States because does not include biofuels from biomass.

[g] Includes former Soviet Union.

The organic components that make up biomass depend on the species. Alpha-cellulose [9004-34-6], or cellulose as it is more generally known, is the chief structural element and a principal constituent of many biomass types. In trees, eg, the concentration of cellulose is about 40 to 50% of the dry weight; materials such as lignin and compounds related to cellulose, such as hemicelluloses, comprise most of the remaining organic components. However, cellulose is not always the dominant component in the carbohydrate fraction of biomass. For example, it is a minor component in giant brown kelp; mannitol [87-78-5], a hexahydric alcohol that can be formed by reduction of the aldehyde group of D-glucose to a methylol group, and alginic acid [9005-32-7], a polymer of mannuronic and glucuronic acids, are the primary carbohydrates.

Fat and protein content of plant biomass are much less on a percentage basis than the carbohydrate components. Fatty constituents are usually present at the lowest con-centration; the protein fraction is much higher in concentration but lower than that of carbohydrate. Crude protein values can be approximated by multiplying the organic nitrogen analyses by 6.25. The average weight percentage of nitrogen in pure dry protein is about 16%, although the protein content of each biomass species can best be determined by amino acid assay. The calculated crude protein values of the biomass species listed in Table 8 range from a low of about 0% for pine wood, to a high of about 30% for Kentucky Blue Grass. For grasses, the protein content is strongly dependent on growing procedures used before harvest, particularly fertilization methods. However, some biomass species, such as legumes, fix nitrogen from the ambient atmosphere and often contain high protein concentrations.

The energy content of biomass is a very important parameter from the standpoint of conversion to energy products and synfuels. The different components in biomass

Table 8. Composition and Heating Value of Biomass, Wastes, Peat, and Coal

Analysis	Pure Cellulose	Pine Wood	Kentucky Bluegrass	Giant Brown Kelp[a]	Feedlot Manure	Urban Refuse[b]	Primary Sewage Sludge	Reed Sedge Peat	Illinois Bituminous Coal
Elemental, wt %									
C	44.44	51.8	45.8	27.65	35.1	41.2	43.75	52.8	69.0
H	6.22	6.3	5.9	3.73	5.3	5.5	6.24	5.45	5.4
O	49.34	41.3	29.6	28.16	33.2	38.7	19.35	31.24	14.3
N		0.1	4.8	1.22	2.5	0.5	3.16	2.54	1.6
S		0.0	0.4	0.34	0.4	0.2	0.97	0.23	1.0
C (MAF)[c]	44.44	52.1	52.9	45.3	45.9	47.9	59.5	57.2	75.6
Proximate, wt %									
Moisture		5–50	10–70	85–95	20–70	18.4	90–97	84.0	7.3
Volatile matter		99.5	86.5	61.1	76.5	86.1	73.47	92.26	91.3
Ash		0.5	13.5	38.9	23.5	13.9	26.53	7.74	8.7
High heating value, MJ/kg[d]									
Dry	17.51	21.24	18.73	10.01	13.37	12.67[e]	19.86	20.79	28.28
MAF[c]	17.51	21.35	21.65	16.38	17.48		27.03	22.53	30.97
Carbon	39.40	41.00	40.90	36.20	38.09		45.39	39.38	40.99

[a] *Macrocystis pyrifera.*

[b] Combustible fraction.

[c] Moisture and ash free.

[d] To convert MJ/kg to Btu/lb, multiply by 430.

[e] As received with metals.

Table 9. Fuel Values of Biomass Components[a,b]

Component	Carbon, %	MJ/kg[c,d]
Monosaccharides	40	15.6
Disaccharides	42	16.7
Polysaccharides	44	17.5
Lignin	63	25.1
Crude protein	53	24.0
Fat	75	39.8
Carbohydrate	41–44	16.7–17.7
Crude fiber[e]	47–50	18.8–19.8

[a] Ref. 13. Approximate values.
[b] Product water in liquid state.
[c] Dry.
[d] To convert MJ/kg to Btu/lb, multiply by 430.
[e] Contains ca 15–30% lignin.

have different heats of combustion because of different chemical structures and carbon content. Table 9 lists heating values for each of the classes of organic compounds. The more reduced the state of carbon in each class, the higher the heating value. As carbon content increases and degree of oxygenation is reduced, the structures become more hydrocarbon-like and heating value increases. Fatty components thus have the highest heating values of the components in Table 9. Cellulose, the dominant component in most biomass, has a high heating value of 17.51 MJ/kg (7533 Btu/lb).

Typical low heating values of selected biomass are given in Table 10. The water-based algae, *Chlorella*, has a higher energy content value than woody and fibrous materials because of its higher lipid or protein contents. Oils derived form plant seeds are much higher in energy content and approach the heating value of paraffinic hydrocarbons. High concentrations of inorganic components in a given biomass species greatly affect its heating value

Table 10. Low Heating Values[a] **of Biomass and Fossil Materials**[b]

Material	MJ/kg[c]
Wood	
Pine	21.03
Beech	20.07
Birch	20.03
Oak	19.20
Oak bark	20.36
Bamboo	19.23
Fiber	
Coconut shells	20.21
Buckwheat hulls	19.63
Bagasse	19.25
Green algae	
Chlorella	26.98
Seed oils	
Cottonseed	39.77
Rapeseed	39.77
Linseed	39.50
Amorphous carbon	33.8
Paraffinic hydrocarbon	43.3
Crude oil	48.2

[a] Product water in vapor state.
[b] Refs. 14 and 15.
[c] Dry; to convert MJ/kg to Btu/lb, multiply by 430.

because inorganic materials generally do not contribute to heat of combustion, eg, giant brown kelp, which leaves an ash residue equivalent to about 40 wt % of the dry weight, has a high heating value on a dry basis of about 10 MJ/kg, and on a dry, ash-free basis, the heating value is about 16 MJ/kg (Table 8).

BIOMASS CONVERSION

Various processes can be used to produce energy or gaseous, liquid, and solid fuels from biomass and wastes. In addition, chemicals can be produced by a wide range of processing techniques. The following list summarizes the principal feed, process, and product variables considered in developing a synfuel-from-biomass process.

Feeds	Conversion Processes	Products
Land-based	Separation,	Energy
trees	combustion,	thermal, steam,
plants	pyrolysis,	electric
grasses	hydrogenation,	Solid fuels
Water-	anaerobic	char,
based	fermentation,	combustibles
single-	aerobic	Gaseous fuels
cell algae	fermentation,	methane (SNG),
multicell	biophotolysis,	hydrogen, low
algae	partial oxidation,	and medium
water	steam reforming,	thermal-value
plants	chemical	gas, light
	hydrolysis,	hydrocarbons
	enzyme	Liquid fuels
	hydrolysis, other	methanol,
	chemical	ethanol, higher
	conversions,	hydrocarbons,
	natural processes	oils
		Chemicals

There are many interacting parameters and possible feedstock–process–product combinations, but all are not feasible from a practical standpoint; eg, the separation of small amounts of metals present in biomass and the direct combustion of high moisture content algae are technically possible, but energetically unfavorable.

Moisture content of the biomass chosen is especially important in the selection of a suitable conversion process. The giant brown kelp, *Macrocystis pyrifera*, contains as high as 95% intracellular water, so thermal gasification techniques such as pyrolysis and hydrogasification cannot be used directly without first drying the algae. Anaerobic digestion methods are preferred because the water does not need to be removed. Wood, on the other hand, can often be processed by several different thermal conversion techniques without drying. Figure 3 illustrates the effects of thermal drying on biomass used for synfuel production as SNG. A large portion of a feed's equivalent energy content can be expended for drying, so the properties of the feed must be considered carefully in relation to the conversion process.

Table 11 lists the important feed characteristics to be examined when developing a successful conversion pro-

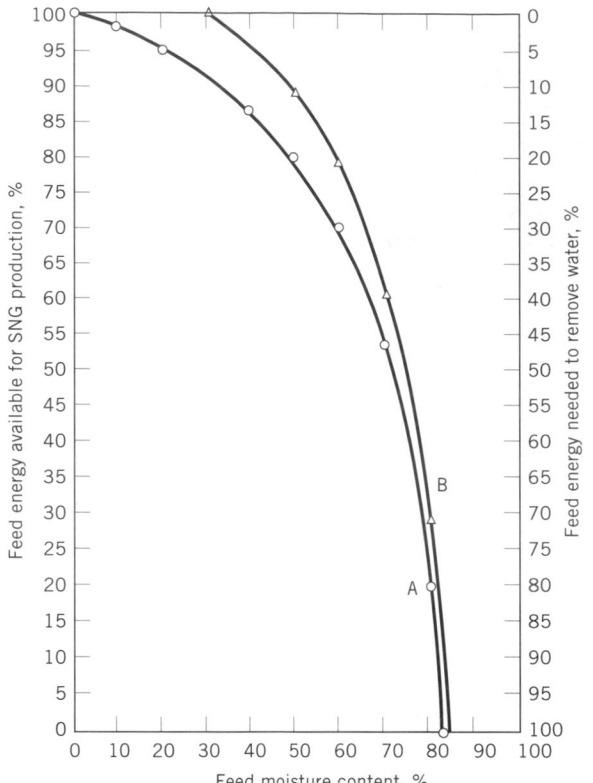

Figure 3. Effect of feed moisture content on energy available for synfuel production. Assumes feed has a heating value of 11.63 MJ/kg (5000 Btu/lb) dry. A, 0% moisture in dried feed; B, 30% moisture in dried feed. For example, reduction of an initial moisture content of 70 wt % by thermal drying to 30% moisture content requires the equivalent of 37% feed energy content and leaves 63% feed energy available for SNG production.

cess for a specific biomass feedstock. A particular process also may have specific requirements within a given process type; eg, anaerobic digestion and alcoholic fermentation are both biological conversion processes, but animal manure, which has a relatively high biodegradability, is not equally applicable as a feedstock for both processes. The degree of complexity of the process design also affects practical utility of the conversion process. Some processes, such as combustion, are simple in design, whereas others, such as alcoholic fermentation, consist of several different unit operations and are complex. Capital and operating

costs are dictated by the particular process design, the logistics of raw material supply, plant size, and operating conditions. Generally, the more complex the process, the higher the costs.

The need to meet environmental regulations can affect processing costs. Undesirable air emissions may have to be eliminated and liquid effluents and solid residues treated and disposed of by incineration or/and landfilling. It is possible for biomass conversion processes that utilize waste feedstocks to combine waste disposal and treatment with energy and/or biofuel production so that credits can be taken for negative feedstock costs and tipping or receiving fees.

The primary types of conversion processes for biomass can be divided into four groups, ie, physical, biological–biochemical, thermal, and chemical.

Physical Processes

Particle Size Reduction. Changes in the physical characteristics of a biomass feedstock often are required before it can be used as a fuel. Particle size reduction (qv) is performed to prepare the material for direct fuel use, for fabrication into fuel pellets, or for a conversion process. Particle size of the biomass also is reduced to reduce its storage volume, to transport the material as a slurry or pneumatically, or to facilitate separation of the components.

The ultimate particle size required depends on the conversion process used, eg, for thermal gasification processes the particle size of the material converted can influence the rate at which the gasification process occurs. Biological processes, such as anaerobic digestion to produce methane, also are affected by the size of the particle; the smaller the particle, the higher the reaction rate because more surface area is exposed to the organisms. Particle size reduction consists of one or more unit operations that make up the front end of the total processing system. Two basic types of machines are in commercial use (ca 1992) for particle size reduction, ie, wet shredders and dry shredders. Wet shredders utilize a hydropulping mechanism in which a high speed cutting blade pulverizes a water slurry of the feed over a perforated plate. The pulped material passes through the plate and the nonpulping materials are ejected. The two most common types of dry shredders in commercial use are the vertical and horizontal shaft hammermills. Rotating metal hammers on a

Table 11. Biomass Feedstock Characteristics that Affect Suitability of Conversion Process

Feedstock Characteristic	Process Type			
	Physical	Thermal	Biological	Chemical
Water content	+	+	+	+
Energy content	+	+		+
Chemical composition		+	+	+
Bulk component analysis	+	+	+	+
Size distribution	+	+	+	+
Noncombustibles	+	+	+	
Biodegradability			+	
Carbon reactivity		+		+
Organism content/type			+	

shaft reduce the particle size of the feed material until the particles are small enough to drop through the grate openings. Particle size reduction units such as agricultural choppers and tree chippers are usually hammermills or are equipped with knife blades that reduce particle size by a cutting or shearing action. Maintenance costs for dry shredders generally are higher than those for wet shredders.

Separation. It may be desirable to separate the feedstock into two or more components for different applications. Examples include separation of agricultural biomass into foodstuffs and residues that may serve as fuel or as a raw material for synfuel manufacture, separation of forest biomass into the darker bark-containing fractions and the pulpable components, separation of marine biomass to isolate various chemicals, and separation of urban refuse into the combustible fraction, ie, refuse-derived fuel (RDF), and metals and glass for recycling. Common operations such as screening, air classification, magnetic separation, extraction, distillation, filtration, and crystallization often are used as well as industry-specific methods characteristic of farming, forest products, and specialized industries.

Drying. Drying refers to the vaporization of all or part of the water in the feedstock. In cases where the biomass or waste is thermally processed directly for energy recovery, it may be necessary to partially dry the raw feed before conversion; otherwise, more energy might be consumed to operate the process than that produced in the form of recovered energy or fuels. Open-air solar drying is perhaps the cheapest drying method if it can be used. Raw materials that are not sufficiently stable to be dried by solar methods can be dried more rapidly using spray driers, drum driers, and convection ovens. For large-scale applications, forced-air-type furnaces and driers designed to use stack gases are more efficient. Special driers such as those that use powdered feeds, and hot metal balls separated from the feed for reheating, also have been successful.

Fabrication. Processes for fabricating solid fuel pellets from a variety of feedstocks, particularly RDF, wood, and wood and agricultural residues, have been developed. The pellets are manufactured by extrusion and other techniques and, in some cases, a binding agent such as a thermoplastic resin is incorporated during fabrication. The fabricated products are reported to be more uniform in combustion characteristics than the raw biomass. Depending on the composition of the additives in the pelletized fuel, the heat of combustion can be higher or lower than that of the unpelletized material.

Biological–Biochemical Processes

Fermentation is a biological process in which a water slurry or solution of raw material interacts with microorganisms and is enzymatically converted to other products. Biomass can be subjected to fermentation conditions to form a variety of products. Two of the most common fermentation processes yield methane and ethanol. Biochem-

ical processes include those that occur naturally within the biomass.

Anaerobic Digestion. Methane can be produced from water slurries of biomass by anaerobic digestion in the presence of mixed populations of anaerobes. This process has been used for many years to stabilize municipal sewage sludges for purposes of disposal. Presuming the biomass is all cellulose, the chemistry can be represented in simplified form as follows:

$$(C_6H_{10}H_6)_x + x\,H_2O \xrightarrow{\text{hydrolysis}} x\,C_6H_{12}O_6$$

$$x\,C_6H_{12}O_6 \xrightarrow{\text{acidification}} 3x\,CH_3COOH$$

$$3x\,CH_3COOH \xrightarrow{\text{methanation}} 3x\,CH_4 + 3x\,CO_2$$

Complex organic compounds are first converted in the water slurry to lower molecular weight soluble products, primarily carboxylic acids, by the acidogenic bacteria present in the digester. Methanogenic bacteria then convert these intermediates to a medium heat value (MHV) gas which has heating values ranging from about 19.6 to 29.4 MJ/m^3 at normal conditions [500 to 750 Btu/SCF, dry at 60°F, 30 in. Hg (15.5°C, 101.5 kPa)]. The principal components in the gas are methane and carbon dioxide (Fig. 4). Residual ungasified solids which contain more nitrogen, phosphorus, and potassium than the feed solids also are formed. In some systems, these solids have application as animal feeds and fertilizers. The conventional high-rate digestion process is conducted under nonsterile conditions in large, mixed, anaerobic fermentation vessels at near-ambient pressures, temperatures of about 35°C (mesophilic range) or 55°C (thermophilic range), and reactor residence times of 10 to 20 days. The pH is maintained in the range 6.8 to 7.2. The raw digester gas has been used for many years as a fuel to heat the digesters and for steam and electric power production. It also can be upgraded to SNG by removing the carbon dioxide by means of adsorption or acid-gas scrubbing processes.

The overall thermal efficiency of the anaerobic digestion of biomass to methane is a function of the process design and the raw material characteristics. Without pretreatment of the feed or the recycled ungasified solids to increase biodegradability, the feed can generally be gasified at overall thermal efficiencies ranging from about 30 to 60%. Yields of methane can range up to 0.30 m^3 at normal conditions/kg (5 SCF/lb) of volatile solids (VS) added to the digesters. A typical biological gasification plant can contain hydrolysis units, anaerobic digesters, gas cleanup and dehydration units, and liquid effluent treatment units (Fig. 5). Advanced digester designs under development include two-phase, plug-flow, packed-bed, fluidized-bed, and sludge blanket digesters (16).

Typical methane yields and volatile solids reductions observed under standard high-rate conditions are shown in Table 12. Longer detention times will increase the values of these parameters, eg, a methane yield of 0.284 m^3 at normal conditions/kg VS added (4.79 SCF/lb VS added) and volatile solids reduction of 53.9% for giant brown kelp at a detention time of 18 days instead of the correspond-

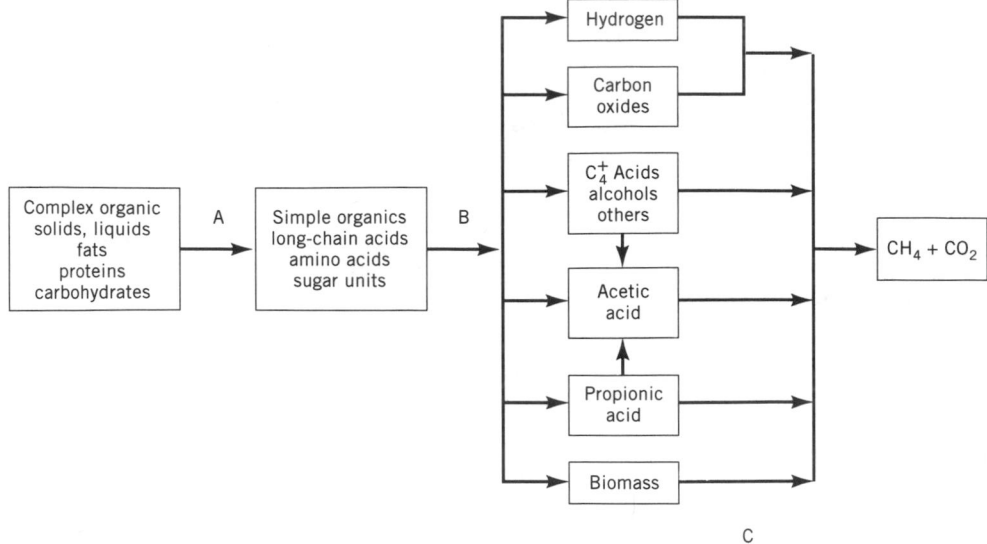

Figure 4. Microbial phases in anaerobic digestion: A, Hydrolysis; B, acidification; C, methane fermentation.

ing values of 0.229 and 43.7 at 12 days under standard high rate conditions. However, improvements might be desirable in the reverse direction, ie, at shorter detention times.

Alcoholic Fermentation. Certain types of starchy biomass such as corn and high sugar crops are readily converted to ethanol under anaerobic fermentation conditions in the presence of specific yeasts (*Saccharomyces cerevisiae*) and other organisms (Fig. 6). However, alcoholic fermentation of other types of biomass, such as wood and municipal wastes that contain high concentrations of cellulose, can be performed in high yield only after the

cellulosics are converted to sugar concentrates by acid- or enzyme-catalyzed hydrolysis:

$$(C_6H_{10}O_5) + x\,H_2O \xrightarrow{\text{hydrolysis}} x\,C_6H_{12}O_6$$

$$x\,C_6H_{12}O_6 \xrightarrow{\text{fermentation}} 2x\,C_2H_5OH + 2x\,CO_2$$

Advanced processes for conversion of cellulosics are being developed (17). A commercial alcohol fermentation plant for biomass would include units to shred and separate the combustible fermentable organic fraction from the nonfermentable components, hydrolysis units to produce glucose concentrates if the feed were high in low de-

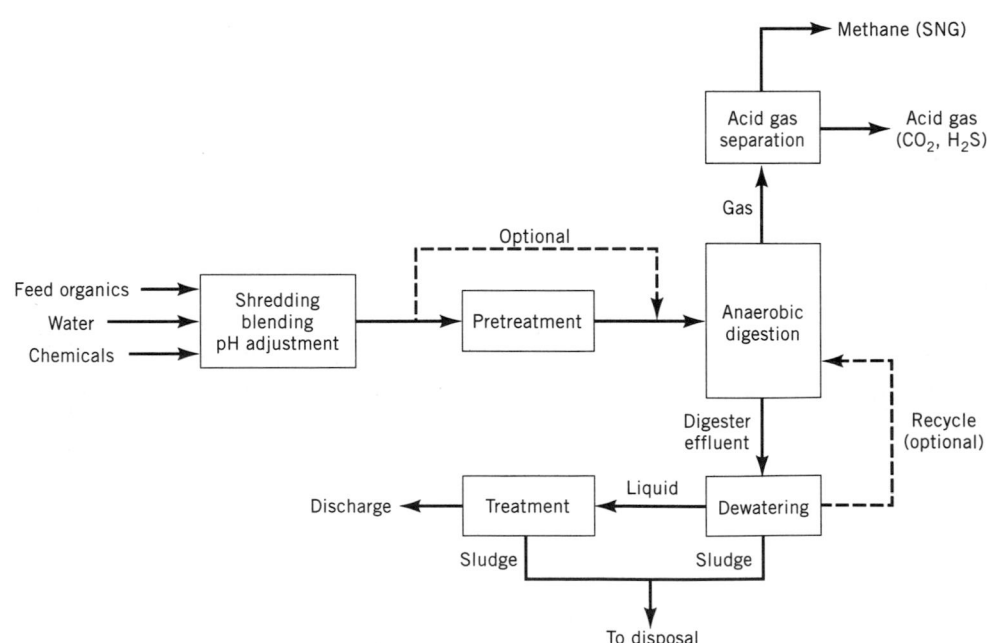

Figure 5. Methane production by anaerobic digestion of biomass.

Table 12. Comparison of Methane Fermentation Performance Under High Rate Mesophillic Conditions[a,b]

Component or Measure of Performance	Primary Sewage Sludge	Primary Activated Sludge	RDF– Sludge Blend	Biomass– Waste Blend	Coastal Bermuda Grass	Kentucky Bluegrass	Giant Brown Kelp	Water Hyacinth
Carbon, wt % (dry)	43.7	41.8	42.1	43.1	47.1	46.2	26.0	41.0
Nitrogen, wt % (dry)	4.02	4.32	1.91	1.64	1.96	4.3	2.55	1.96
Phosphorus, wt % (dry)	0.59	1.30	0.81	0.43	0.24		0.48	0.46
Ash, wt % of total solids	26.5	23.5	8.4	17.2	5.05	10.5	45.8	22.7
Volatile matter, wt % of total solids	73.5	76.5	91.6	82.8	95.0	89.8	54.2	77.3
Heating value, MJ/kg (dry)	19.86	18.31	17.20	20.92	19.04	19.19	10.26	16.02
C/N ratio	10.9	9.7	22.0	26.3	24.0	10.7	10.2	20.9
C/P ratio	74.1	32.2	52.0	100	196		54.2	89.1
Gas production rate, volume(n)/liquid volume-day	0.74	0.84	0.59	0.52	0.56	0.52	0.62	0.47
Methane in gas, mol %	68.5	65.5	60.0	62.0	55.9	60.4	58.4	62.8
Methane yield, $m^3(n)$/kg VS added	0.313	0.327	0.210	0.201	0.208	0.150	0.229	0.185
Volatile solids reduction, %	41.5	49.0	36.7	33.3	37.5	25.1	43.7	29.8
Substrate energy in gas, %	46.2	54.4	39.7	38.3	41.2	27.6	49.1	35.7

[a] Ref. 16.
[b] Daily feeding, continuous mixing, 35°C, pH 6.7 to 7.2, 12-day hydraulic retention time, 1.6 kg volatile solids/m^3·d except for kelp, which was 2.1 kg VS/m^3·d. All biomass substrates 1.2 mm or less in size.

gradability cellulose components, fermenters, distillation towers, and dehydration units (Fig. 7). Under conventional conditions, the degradable organics are converted in the fermenters, at high efficiencies and residence times of 1 to 2 days, to a beer that contains about 10% ethanol. This broth is heated to remove the product alcohol as overhead by distillation, and the resulting 50 to 55% alcohol distillate is distilled again to yield 95% alcohol and by-product aldehydes and fusel oil. Bottoms from the beer still contain low volatility components from the fermentation called stillage, which is often processed further to yield high protein animal feeds.

The thermal efficiency of ethanol production from fermentable sugars is high, but the overall thermal efficiency of the process is low because of the many energy-consuming steps, the nonfermentable fraction in biomass, and the by-products formed. Alcohol yields are about 40 to 50 wt % of the weight of the fermentable fraction in the feed. Substantial improvements in the overall thermal efficiency of alcoholic fermentation are possible by improving the thermal efficiencies of the auxiliary unit operations. Alcoholic fermentation under reduced pressure, systems in which polysaccharide hydrolysis and fermentation occur together, and improved heat exchange and alcohol-drying processes are under development to provide improved process performance.

Biophotolysis. The decomposition of water (splitting) to hydrogen and oxygen using the radiant energy of visible light and the photosynthetic apparatus of green plants, certain bacteria, and blue-green algae, is called biophotolysis. The concept has been studied in the laboratory, but has not been developed to the point where a practical process exists. Basically, biophotolysis involves the oxidation

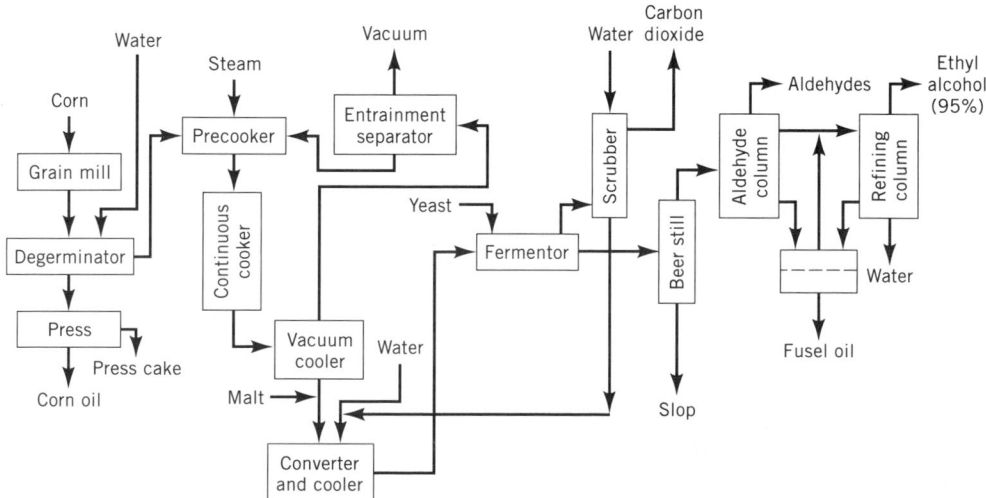

Figure 6. Flow scheme for manufacture of ethyl alcohol from corn.

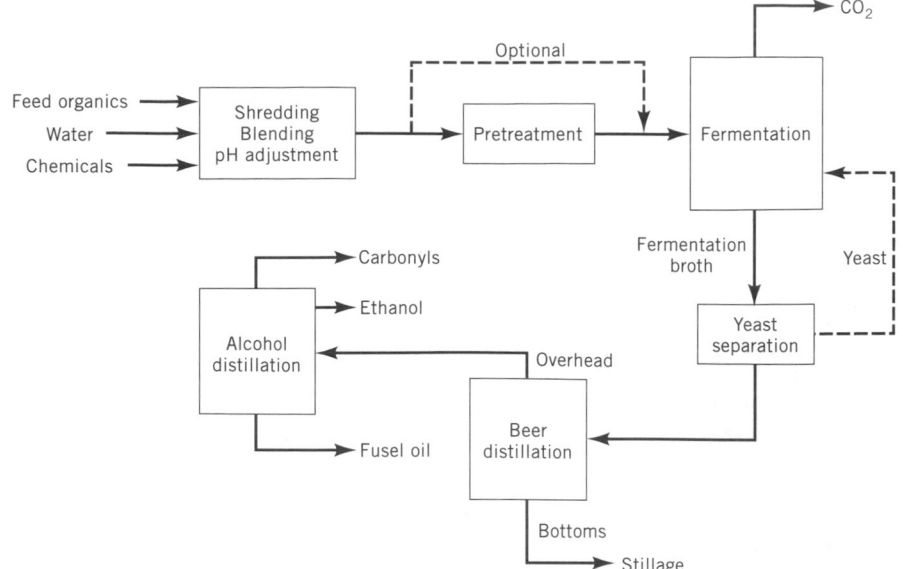

Figure 7. Ethanol production by alcoholic fermentation.

of water to liberate molecular oxygen and electrons which are raised from the level of the water-oxygen couple, eg, +0.8 V, to 0.0 V by Photosystem II.

$$H_2O \xrightarrow{\text{light}} \tfrac{1}{2}O_2 + 2\,H^+ + 2\,e^-$$

The electrons undergo the equivalent of a partial oxidation process in a dark reaction to a positive potential of +0.4 V, and Photosystem I then raises the potential of the electrons to as high as −0.7 V. Under normal photosynthesis conditions, these electrons reduce tryphosphopyridine-nucleotide (TPN) to TPNH, which reduces carbon dioxide to organic plant material. In the biophotolysis of water, these electrons are diverted from carbon dioxide to a microbial hydrogenase for reduction of protons to hydrogen:

$$2\,H^+ + 2\,e^- \rightarrow H_2$$

Thus, the overall chemistry is simply the photolysis of water to hydrogen and oxygen.

Synthetic water-splitting membranes that contain the biochemical and other catalysts necessary to form hydrogen also are under development. These membranes mimic natural photosynthesis except that the electrons are directed to form hydrogen. Several sensitizers and catalysts are needed to complete the cycle, but progress is being made. Various single-stage schemes, in which hydrogen and oxygen are produced separately, have been studied, and the thermodynamic feasibility of the chemistry has been experimentally demonstrated.

The upper limit of efficiency of the biophotolysis of water has been projected to be 3% for well-controlled systems. This limits the capital cost of useful systems to low cost materials and designs. But the concept of water biophotolysis to afford a continuous, renewable source of hydrogen is quite attractive, and may one day lead to practical hydrogen-generating systems.

Natural Processes. Hydrocarbon production in land-based biomass by natural chemical mechanisms is a well-known phenomenon. Commercial production of natural rubber, the highly stereospecific polymer cis-1,4-polyisoprene [9003-31-0], is an established technology. Natural rubber has a mol wt range between about 500,000 and 2,000,000 and is tapped as a latex from the hevea rubber tree (Hevea braziliensis). The desert shrub guayule, which grows in the southwestern United States and in northern Mexico, is another biomass species studied as a source of natural rubber almost identical to hevea rubber (18). The idea of growing guayule and extracting the rubber latex from the whole plant was tested in full-scale plantations during the rubber shortage in World War II and found technically feasible. Terpene extraction from pine trees and other biomass species is also established technology.

Many plants native to North America, or that can be grown there, have been tested as sources of oils (triglycerides) and hydrocarbons (19,20). The objectives of this work have been to identify those biomass species that produce hydrocarbons, especially those of lower molecular weight than natural rubber, so that they would be more amenable to standard petroleum refining methods; to characterize hydrocarbon yields and those of other organic compounds; and to learn what controls the structure and molecular weight of hydrocarbons within the plant so that genetic manipulation or other biomass modifications can be applied to control hydrocarbon structures. Some efforts have concentrated on desert plants that might be grown in arid or semiarid areas without competition from biomass grown for foodstuffs. Other work has been aimed at perennial species adapted to wide areas of North America. Several biomass species have been found to contain oils and/or hydrocarbons (Table 13). It is apparent that oil or hydrocarbon formation is not limited to any one family or type of biomass. Interestingly, some species in the Euphorbiaceae family, which includes Hevea braziliensis, form hydrocarbons having molecular weights considerably less than that of natural rubber at

Table 13. Oil- and Hydrocarbon-Producing Biomass Species Potentially Suitable for North America[a]

Family	Genus and Species	Common Name
Aceraceae	*Acer saccharinum*	Silver maple
Anacardiaceae	*Rhus glabra*	Smooth sumac
Asclepiadiaceae	*Asclepias incarnata*	Swamp milkweed
	sublata	Desert milkweed
	syriaca	Common milkweed
	Cryptostegia grandiflora	Madagascar rubber vine
Buxaceae	*Simmondsia chinensis*	Jojoba
Caesalpiniaceae	*Copaifera langsdorfi*	Copaiba
	multijuga	
Caprifoliaceae	*Lonicera tartarica*	Red tarterium honeysuckle
	Sambucus canadensis	Common elder
	Symphoricarpos orbiculatus	Corral berry
Companulaceae	*Companula americana*	Tall bellflower
Compositae	*Ambrosia trifida*	Giant ragweed
	Cacalia atriplicifolia	Pale Indian plantain
	Chrysathamnus nauseosus	Rabbitbrush
	Circsium discolor	Field thistle
	Eupathorium altissimum	Tall boneset
	Parthenium argentatum	Guayule
	Silphium integrifolium	Rosin weed
	laciniatum	Compass plant
	terbinthinaceum	Prairie dock
	Solidago graminifolia	Grass-leaved goldenrod
	leavenworthii	Edison's goldenrod
	rigida	Stiff goldenrod
	Sonchus arvensis	Sow thistle
	Vernonia fasciculata	Ironweed
Curcurbitaceae	*Cucurbita foetidissima*	Buffalo gourd
Euphorbiaceae	*Euphorbia denta*	
	lathyris	Mole plant, gopher plant
	pulcherima	Poinsetta
	tirucalli	African milk bush
Gramineae	*Agropyron repens*	Quack grass
	Elymus canadensis	Wild rye
	Phalaris canariensis	Canary grass
Labiatae	*Pycnanthemum incanum*	Western mountain mint
	Teucrium canadensis	American germander
Lauraceae	*Sassafras albidium*	Sassafras
Rhamnaceae	*Ceanothus americanus*	New Jersey tea
Rosaceae	*Prunus americanus*	Wild plum
Phytolaccaceae	*Phytolacea americana*	Pokeweed

[a] Ref. 20.

yields as high as 10 wt % of the plant. This corresponds to hydrocarbon yields of about 3.97 $m^3/hm^2\cdot yr$ (25 bbl/$hm^2\cdot yr$).

Figure 8 illustrates one of the processing schemes used for separating various components in a hydrocarbon-containing plant. Acetone extraction removes the polyphenols, glycerides, and sterols, and benzene extraction removes the hydrocarbons. If the biomass species in question contain low concentrations of the nonhydrocarbon components, exclusive of the carbohydrate and protein fractions, direct extraction of the hydrocarbons with benzene or a similar solvent might be preferred.

The principal steps in the mechanism of polyisoprene formation in plants are known and should help to improve the natural production of hydrocarbons. Mevalonic acid, a key intermediate derived from plant carbohydrate via acetylcoenzyme A, is transformed into isopentenyl pyrophosphate (IPP) via phosphorylation, dehydration, and decarboxylation (see ALKALOIDS). IPP then rearranges to dimethylallyl pyrophosphate (DMAPP). DMAPP and IPP react with each other, releasing pyrophosphate to form another allyl pyrophosphate containing 10 carbon atoms. The chain can successively build up by five-carbon units to yield polyisoprenes by head-to-tail condensations; alternatively, tail-to-tail condensations of two C_{15} units can yield squalene, a precursor of sterols. Similar condensation of two C_{20} units yields phytoene, a precursor of carotenoids. This information is expected to help in the development of genetic methods to control the hydrocarbon structures and yields.

Other sources of natural oils are the oilseed crops, many of which have been used in nonfuel applications,

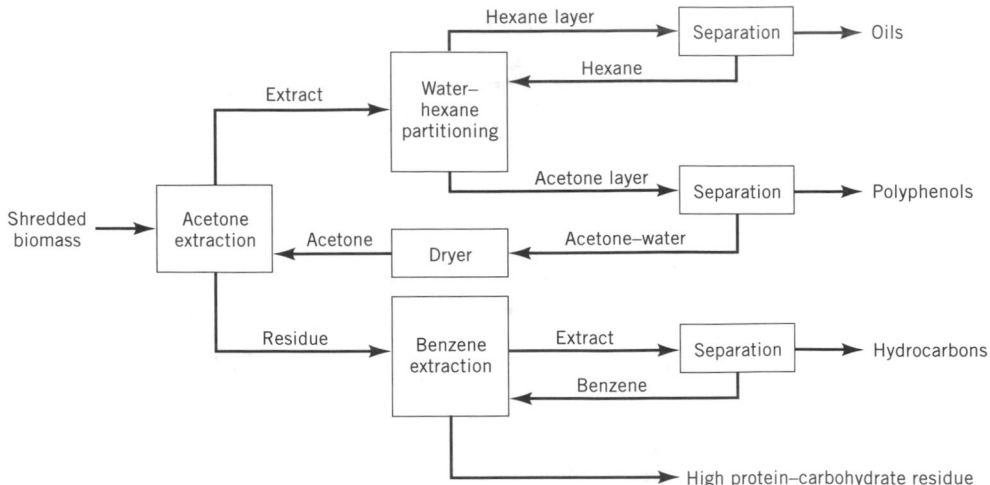

Figure 8. Operations for processing hydrocarbon-containing biomass.

and microalgae (21). Oils from these types of biomass are largely triglycerides and typically contain three long-chain primary fatty acids, each bound to one of the carbon atoms of glycerol via an ester linkage. The viscosity and other properties of these oils vary with the degree of saturation of the fatty acid; the more paraffinic oils have higher viscosities and melting points. The oils can be upgraded to diesel fuels or gasoline plus diesel fuels by transesterification or by catalytic cracking or hydrocracking. Transesterification with methanol or ethanol yields monoesters in the C_{15-20} range depending on the oil source (22). The monoesters have viscosity and volatility characteristics similar to conventional diesel fuels. Carbon build-up and crankcase oil contamination in diesel engines vary with the degree of saturation and with the service characteristics of the diesel engine. Catalytic cracking of the oils over shape-selective zeolites gives substantial yields of aromatic-rich gasoline-range liquids (23). Catalytic hydrocracking of similar vegetable oils is reported to yield diesel fuel additives (24) or high quality gasolines (25).

Average and potential yields for seed oils are shown in Table 14 (26). If the seed oil is the only product for which revenue is realized, even the high potential yields are insufficient to justify seed oil use as fuel at 1992 petroleum prices, ie, about $20–22 per 159 L (42-gal bbl) of crude oil. Some studies indicate that small-scale transesterification facilities operated as farm cooperatives in the United States can produce biodiesel fuels from seed oils at a profit provided advantage is taken of the Minor Oilseed Provision of the 1990 Farm Bill (27). In this option, the farmer grows rapeseed, for example, on land that is removed from production of a crop such as corn, wheat, cotton, and soybeans. The 1990 Farm Bill permits a farmer to harvest and sell minor oilseed crops grown on this set-aside land without losing his program participation payment. In effect, the farmer is paid land rent by the Government, but can still produce minor oilseed crops.

Work on microalgae has focused on the growth of these organisms under conditions that promote lipid, ie, algal oil, formation. This eliminates the high cost of cell harvest because the lipids often can be separated by simple flotation or extraction (28). Lipid yields greater than 50% of the cell dry weight have been reported when the organisms are grown under nitrogen-limited conditions (29). The oils are high in triglycerides and can be transesterified to form biodiesel fuels in the same manner as seed oils (30). However, the estimated production costs of these fuels still appear to be too high to compete with petroleum-based diesel fuels (28,30).

Thermal Processes

Thermal processes for the production of energy and fuels from biomass and wastes usually involve irreversible chemical reactions, heat, and the transfer of chemical energy from reactants to products. The two largest classes of thermal processes are combustion and pyrolysis. A third class of processes can be described either as a combination of combustion and pyrolysis reactions, or as a thermochemical process in which conversion of the feed is facilitated by a reactant such as water or hydrogen. For convenience, these processes are grouped together as miscellaneous processes. Gaseous, liquid, and solid fuels can be produced by pyrolysis processes and several processes in the miscellaneous category.

Combustion. Complete combustion, eg, incineration, direct firing, burning, is the rapid chemical reaction of the feed and oxygen to form carbon dioxide, water, and heat. The heat released is a function of the enthalpy of combustion of the biomass. Agricultural products, such as bagasse generated in sugarcane plantations, forestry residues, wood chips, RDF, and even raw garbage, have been used as fuels in combustion systems for many years. The recovered heat has been used for steam production, electric power production in a steam-electric plant, and drying.

Many types of combustion equipment are available commercially. The basic differences in various units reside mainly in the design of the combustion chambers, the operating temperature, and the heat transfer mechanism.

Table 14. Commercial Yields of Oilseeds and Seed Oils in the United States[a,b]

| | Seed Yield, kg/hm² | | Seed Oil Yield | | | |
| | | | Average | | Potential | |
Species	Average	Potential	kg/hm²	L/hm²	kg/hm²	L/hm²
Castorbean[c]						
(*Ricinus communis*)	950	3,810	428	449	1,504	1,590
Chinese						
tallow tree						
(*Sapium sebiferum*)	(12,553)[d]		(5,548)[d]			(6,270)[d]
Cotton[c]						
(*Gossypium hirsutum*)	887	1,910	142	150	343	370
Crambe[c]						
(*Crambe abssinica*)	1,121	2,350	392	421	824	940
Corn (high oil)						
(*Zea mays*)		5,940			596	650
Flax						
(*Linum usitatis-simum*)	795	1,790	284	309	758	840
Peanut						
(*Arachis hypogaea*)	2,378	5,160	754	814	1,634	1,780
Safflower						
(*Carthamus tinctorius*)	1,676	2,470	553	599	888	940
Soybean						
(*Glycine max*)	1,980	3,360	354	383	591	650
Sunflower						
(*Helianthus annuus*)	1,325	2,470	530	571	986	1,030
Winter rape						
(*Brassica napus*)		2,690			1,074	1,220

[a] Ref. 26.
[b] Growth conditions are dryland unless otherwise noted.
[c] Irrigated.
[d] Not an average yield from several sources; it is one reported yield equivalent to 6,270 L/hm² of oil plus tallow. It is believed that yield would be substantially less than this in a managed dense stand, but still higher than that of conventional oilseed crops.

Refractory-lined furnaces operating at about 1000°C were standard until the introduction in early 1990s of water-wall incinerators. Ash buildup occurs rapidly in refractory-lined furnaces, and excess air must be introduced to limit the wall temperature. The water-wall incinerator has combustion chamber walls containing banks of tubes through which water is circulated, thereby reducing the amount of cooling air needed. Heat is transferred directly to the tubes to produce steam.

Another type of combustion unit operates at about 1600°C to produce a molten slag which forms a granular frit on quenching rather than the usual ash. The higher operating temperature is obtained by preheating the combustion air or by burning auxiliary fuel.

Fluidized-bed combustion represents still another approach. In these systems, air is dispersed through an orifice plate at the bottom of the combustion unit. The dispersed air passes through a bed of sand or residual inorganic particles recovered from combustion causing the effective volume of the bed to increase and the bed to become fluidized. The feed is fed to this rapidly mixed bed, where flameless combustion occurs at about 650°C. This temperature is substantially below flame temperature and because of the lower heat input requirements, many high moisture feeds can be combusted without supplemental fuel. Many other furnace variations, such as stationary and rotating shaft furnaces, suspension firing systems, and stationary and moving grates, are in com-

mercial use or available for biomass and waste combustion applications.

The specific design most appropriate for biomass, waste combustion, and energy recovery depends on the kinds, amounts, and characteristics of the feed; the ultimate energy form desired, eg, heat, steam, electric; the relationship of the system to other units in the plant, independent or integrated; whether recycling or co-combustion is practiced; the disposal method for residues; and environmental factors.

Pyrolysis. Pyrolysis, eg, retorting, destructive distillation, carbonization, is the thermal decomposition of an organic material in the absence of oxygen. For biomass and wastes, pyrolysis generally starts at temperatures near 300 to 375°C. Chars, organic liquids, gases, and water are formed in varying amounts, depending particularly on the feed composition, heating rate, pyrolysis temperature, and residence time in the pyrolysis reactor. Higher temperatures and longer residence times promote gas production, while higher liquid and char yields result from lower temperatures and shorter residence times. No matter what the pyrolysis conditions are, with the exception of extremely high temperatures, the product mixture has a complex composition and selectivity for specific products is low even with a single feed component.

Depending on the pyrolysis temperature, the char fraction contains inorganic materials ashed to varying degrees, any unconverted organic solids, and fixed carbon residues produced on thermal decomposition of the organics. The liquid fraction contains a complex mixture of organic chemicals having much lower average molecular weights than the feed. For highly cellulosic feeds, the liquid fraction will usually contain acids, alcohols, aldehydes, ketones, esters, heterocyclic derivatives, and phenolic compounds. The pyrolysis gas is a low heat value (LHV) gas having a heating value of 3.9 to 15.7 MJ/m³ at normal conditions (100 to 400 Btu/SCF). It contains carbon dioxide, carbon monoxide, methane, hydrogen, ethane, ethylene, minor amounts of higher gaseous organics, and water vapor. It is immediately apparent that if pure pyrolysis products are desired, product separation is a significant problem or further processing to refine the products is necessary.

Pyrolysis processes may be endothermic or exothermic, depending on the temperature of the reacting system. For most biomass feeds containing highly oxygenated cellulosic fractions as the principal components, pyrolysis is endothermic at low temperatures and exothermic at high temperatures. Energy to drive the process often is obtained from a portion of the feed or the pyrolysis products such as the char. At low temperatures, pyrolysis generally is reaction-rate controlled; at high temperatures, the process becomes mass-transfer controlled. The experimental data in Tables 15 and 16 show how temperature affects product yields and gas and char compositions with the combustible fraction of municipal solid waste (31). Gas yield increases as the temperature is increased from 500 to 900°C. Although the heating value of the product gas remains about the same, significant increases in hydrogen concentration and energy yield in the gas occur with increasing temperature. Substantial decreases occur in the carbon dioxide concentration over the same temperature range. Also, as the temperature increases from 500 to 900°C, the char yields decrease along with the volatile matter concentration, but the energy value of the char does not undergo similar changes.

Pyrolysis reactor designs are as varied as combustion unit designs. They include fixed beds, moving beds, suspended beds, fluidized beds, entrained feed solids reactors, stationary vertical shaft reactors, inclined rotating kilns, horizontal shaft kilns, high temperature (1000 to 3000°C) electrically heated reactors with gas-blanketed walls, single and multihearth reactors, and a host of other designs. One of the more innovative pyrolysis processes in development for gas production is a fluidized two-bed system (32–34). This system uses two fluidized-bed reactors containing sand as a heat transfer medium. Combus-

Table 16. Char and Gas Composition from Pyrolysis of Municipal Solid Waste Organics[a]

Component	Temperature, °C			
	500	650	800	900
Gas				
Carbon dioxide, mol %	44.8	31.8	20.6	18.3
Carbon monoxide, mol %	33.5	30.5	34.1	35.3
Methane, mol %	12.4	15.9	13.7	10.5
Hydrogen, mol %	5.56	16.6	28.6	32.5
Ethane, mol %	3.03	3.06	0.77	1.07
Ethylene, mol %	0.45	2.18	2.24	2.43
High heat value (HHV), MJ/m³[b]	12.3	15.8	15.4	15.1
Char				
Volatile matter, %	21.8	15.1	8.13	8.30
Fixed carbon, %	70.5	70.7	79.1	77.2
Ash, %	7.71	14.3	12.8	14.5
High heat value (HHV), MJ/kg[c]	28.1	28.6	26.7	26.5

[a] Ref. 31.
[b] At normal conditions. To convert MJ/m³ to Btu/SCF, multiply by 25.45.
[c] To convert MJ/kg to Btu/lb, multiply by 430.

tion of char from the pyrolysis reactor takes place within the combustion reactor. The heat released supplies the energy for pyrolysis of the combustible fraction in the pyrolysis reactor. Heat transfer is accomplished by sand flow from the combustion reactor at 950°C to the pyrolysis reactor at 800°C and return of the sand to the combustion reactor (Fig. 9). This configuration separates the combustion and pyrolysis reactions and keeps the nitrogen in air separated from the pyrolysis gas. It yields a pyrolysis gas that can be readily upgraded to SNG by shifting, scrubbing, and methanating without regard to nitrogen separation. The initial pyrolysis gas from high cellulose feeds contains about 37 mol % hydrogen, 35 mol % carbon monoxide, 16 mol % carbon dioxide, and 11 mol % methane. This is an LHV gas having a high heating value of about 13.5 MJ/m³ at normal conditions (344 Btu/SCF). The projected gas yields are about 667 m³ at normal conditions (17,000 SCF) of pyrolysis gas and about 196 m³ at normal conditions (5000 SCF) of methane per dry ton of feed (32).

An example of a liquid fuel production system under development in a short-residence time pyrolysis process is shown in Figure 10 (35). In this process, high cellulose RDF is dried in a rotary kiln to about 4 wt % moisture content, and finely divided to a particle size of which 80% is smaller than 14 mesh (1200 μm). The feed, about 0.23

Table 15. Product Yields from Pyrolysis of Municipal Solid Waste Organics[a]

Temperature, °C	Products, wt %			Gas Yield	
	Gases	Liquids	Char	Combustibles,[b] m³/kg	Combustibles, MJ/kg
500	12.3	61.1	24.7	0.114	1.39
650	18.6	59.2	21.8	0.166	2.63
800	23.7	59.7	17.2	0.216	3.33
900	24.4	58.7	17.7	0.202	3.05

[a] Ref. 31.
[b] At normal conditions.

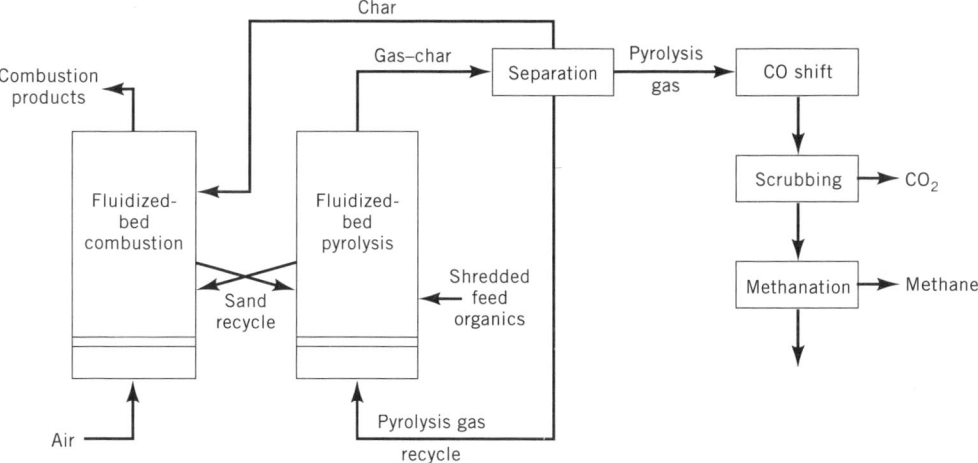

Figure 9. Methane production by pyrolysis using sand and char recycle in fluidized two-bed system.

kg of recycled char preheated to 760°C per kilogram of this finely divided material, is rapidly passed through the pyrolysis reactor. The raw product mixture, which consists of product gas, the char fed to the reactor, and new char formed on pyrolysis, leaves the reactor at about 510°C. Separation of the gas from the char and rapid quenching to about 80°C yields the liquid fuel. The remaining gas goes through a series of cleanup steps for in-plant use. Part of the gas is used as an oxygen-free solids transport medium and part of it as fuel. The raw product yields are about 10 wt % water, 20 wt % char, 30 wt % gas, and 40 wt % liquid fuel. The product char has a heating value of about 20.9 MJ/kg (9000 Btu/lb), contains about 30 wt % ash, and is produced at an overall yield of about 7.5 wt % of the dry feed. The corresponding values

of the liquid fuel are about 24.4 MJ/kg (10,500 Btu/lb), 0.2 to 0.4% ash, and 22.5 wt % of dry feed as received (approximately 1 bbl/short ton of raw refuse). This product has been proposed for use as a heating oil; its properties are compared with a typical No. 6 fuel oil in Table 17. It is apparent that some differences exist, but successful combustion trials in a utility boiler with the liquid fuel have been performed.

A report on the continuous flash pyrolysis of biomass at atmospheric pressure to produce liquids indicates that pyrolysis temperatures must be optimized to maximize liquid yields (36). It has been found that a sharp maximum in the liquid yields vs temperature curves exist and that the yields drop off sharply on both sides of this maximum. Pure cellulose has been found to have an optimum

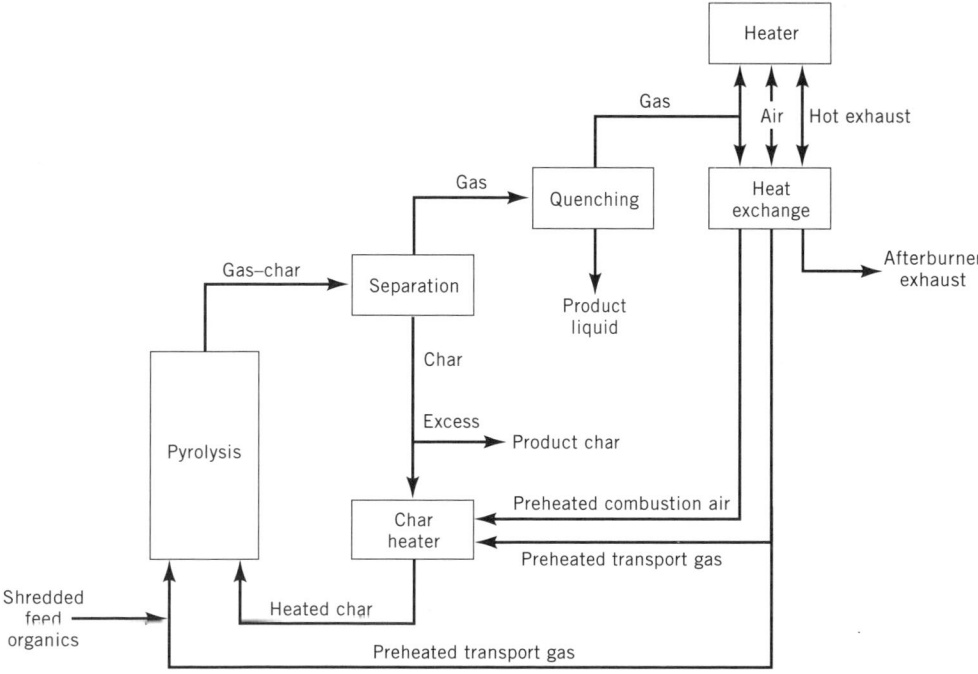

Figure 10. Liquid-fuel production by flash pyrolysis using char recycle.

Table 17. Properties and Analysis of Liquid Fuel and No. 6 Fuel Oil[a]

Properties	Liquid Fuel	No. 6 Fuel Oil
Heating value, MJ/kg[b]	24.6	42.3
Density, g/cm³	1.3	0.98
Pour point, °C	32[c]	15–30
Flash point, °C	56[c]	65
Viscosity at 87.8°C (SUs)	1150[c]	90–250
Pumping temperature, °C	71	46
Atomization temperature, °C	116	104
Analysis,[d] wt %		
Carbon	57.5	85.7
Hydrogen	7.6	10.5
Sulfur	0.1–0.3	0.5–3.5
Chlorine	0.3	
Nitrogen	0.9	2.0
Oxygen	33.4	
Ash	0.2–0.4	0.5

[a] Ref. 35. Liquid fuel produced by flash pyrolysis using char recycle (Fig. 10).
[b] To convert MJ/kg to Btu/lb, multiply by 430.
[c] Containing 14% water as produced.
[d] Dry basis.

temperature for liquids at 500°C, while the wheat straw and wood species tested have optimum temperatures at 600°C and 500°C, respectively. Organic liquid yields were of the order of 65 wt % of the dry biomass fed, but contained relatively large quantities of organic acids.

Miscellaneous Thermal Processes. Many thermal conversion processes can be classified as partial oxidation processes in which the biomass or waste is supplied with less than the stoichiometric amount of oxygen needed for complete combustion. Under these conditions, LHV gases similar to pyrolysis gases are formed that can contain high concentrations of hydrogen and carbon monoxide. Such gaseous mixtures are termed synthesis gases and can be converted to a large number of chemicals and synthetic fuels by established processes (Fig. 11). In some partial oxidation processes, the various chemical reactions may occur simultaneously in the same reactor zone. In others, the reactor may be divided into a combustion zone, which supplies the heat to promote pyrolysis in a second zone, and perhaps a third zone for drying, the overall result of which is partial oxidation. Both air and pure oxygen have been utilized for such systems.

In one system, the three-zoned vertical shaft reactor furnace (Fig. 12), coarsely shredded feed is fed to the top of the furnace. As it descends through the first zone, the charge is dried by the ascending hot gases, which are also partially cleaned by the feed. The gas is reduced in temperature from about 315°C to the range of 40 to 200°C. The dried feed then enters the pyrolysis zone in which the temperature ranges from 315 to 1000°C. The resulting char and ash then descend to the hearth zone, where the char is partially oxidized with pure oxygen. Slagging temperatures near 1650°C occur in this zone and the resulting molten slag of metal oxides forms a liquid pool at the bottom of the hearth. Continuous withdrawal of the pool and quenching forms a sterile granular frit. The product gas is processed to remove fly ash and liquids, which are recycled to the reactor. A typical gas analysis is 40 mol % carbon monoxide, 20 mol % hydrogen, 23 mol % carbon dioxide, 5 mol % methane, and 5 mol % C_2. This gas has an HHV of about 14.5 MJ/m³ at normal conditions (370 Btu/SCF).

An example of partial oxidation in which air is supplied without zone separation in the gasifier is the molten salt process (39). In this process, the shredded biomass or waste and air are continuously introduced beneath the surface of a sodium carbonate-containing melt which is maintained at about 1000°C. The resulting gas passes through the melt. Acid gases are absorbed by the alkaline salts and the ash is also retained in the melt. The melt is continuously withdrawn for processing to remove the ash and returned to the gasifier. No tars or liquid products are formed in this rather simple process. The heating value of the gases produced depends on the amount of air supplied, and is essentially independent of the type of feed organics. The greater the deficiency of air needed to achieve complete combustion, the higher the fuel value of the product gas. Thus with about 20, 50, and 75% of the theoretical air, the respective high heating values (HHVs) of the gas are about 9.0, 4.3, and 2.2 MJ/m³ at normal conditions (230, 110, and 55 Btu/SCF).

Steam also is blended with air in some gasification units to promote the overall process via the endothermic steam–carbon reaction to form carbon monoxide and hydrogen. This was common practice at the turn of the nineteenth century, when so-called producer gasifiers were employed to manufacture LHV gas from different types of biomass and wastes. The producer gas from biomass and wastes had heating values around 5.9 MJ/m³ at normal conditions (150 Btu/SCF), and the energy yields as gas ranged up to about 70% of the energy contained in the feed. Many gasifier designs were offered for the manufacture of producer gas from biomass and wastes; several types of units are still available for purchase (ca 1992). Thousands of producer gasifiers operating on air and wood were used during World War II, particularly in Sweden, to power automobiles, trucks, and buses. The engines needed only slight modification to operate on LHV producer gas.

Hydrogenation. Another approach to the production of energy products from biomass and wastes is based on hydrogenation. Hydrogen, which can be either generated from the feed or the conversion products, or obtained from an independent source, reacts directly with the feed organics or intermediate process streams at elevated pressures and temperatures to yield substitute fuels. In the-

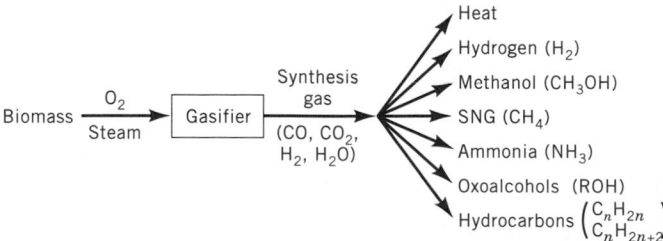

Figure 11. Applications of synthesis gas from biomass (37).

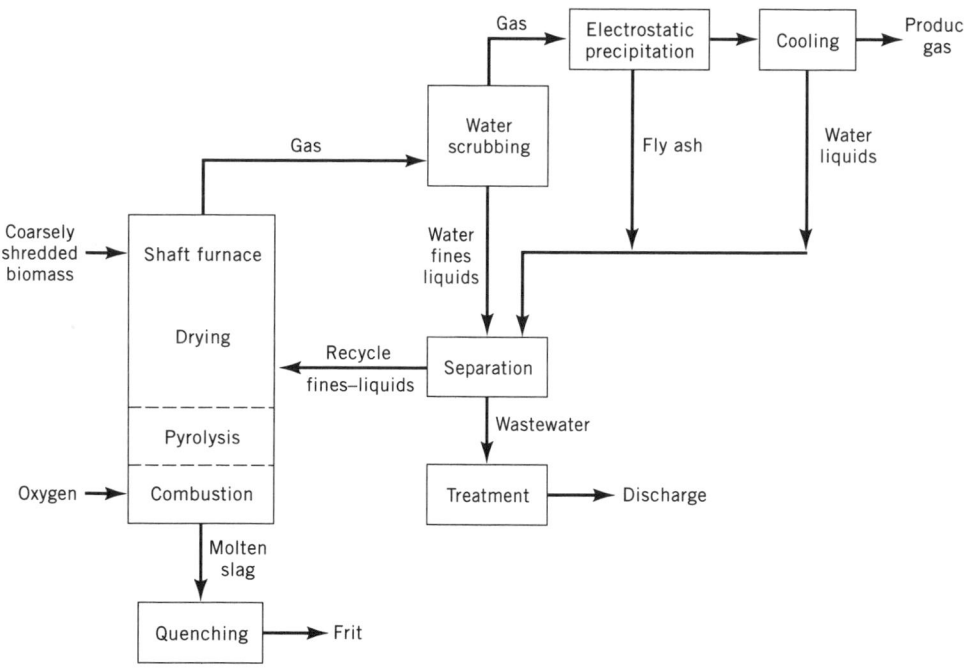

Figure 12. Production of synthesis gas in three-zone shaft reactor furnace (38).

ory, highly oxygenated feeds should be capable of reduction to liquid and gaseous fuels at any level between the initial oxidation state of the feed and methane.

$$R(OH)_x + y\,H_2 \rightarrow RH_y(OH)_{x-y} + y\,H_2O$$

$$R - R' + H_2 \rightarrow RH + R'H$$

For a cellulosic material containing hydroxyl groups, the reactions might consist of dehydroxylation and depolymerization by hydrogenolysis, during which there is a transition from solid to liquid to gas.

Most of the work on hydrogenation has been concentrated on hydrogasification to produce methane as the final product. One route to methane involves the sequential production of synthesis gas and then methanation of the carbon monoxide with hydrogen to yield methane. The routes shown in Figure 13 involve the direct reaction of the feed with hydrogen (40). In this process, shredded feed is converted with hydrogen-containing gas to a gas containing relatively high methane concentrations in the first-stage reactor. The product char from the first stage is used in a second-stage reactor to generate the hydrogen-rich synthesis gas for the first stage. From experi-

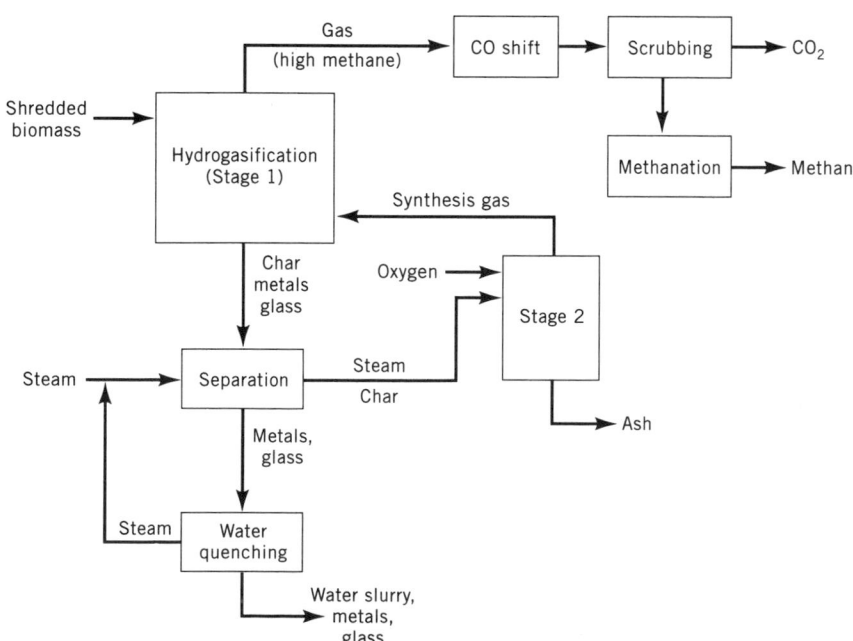

Figure 13. Methane production by hydrogasification.

Table 18. Gas Composition and Yield from Integrated Hydrogasification Process at Stage 1 [a]

Product	Free Fall	Moving Bed
Composition, mol %		
Carbon monoxide	45.9	51.9
Hydrogen	31.9	13.3
Methane	10.4	17.2
Carbon dioxide	10.1	16.1
Ethane	1.2	1.1
Benzene	0.5	0.4
Yield, m³/kg[b] dry feed	1.1	0.95
Fraction of total methane produced in stage 1 after methanation	0.26	0.52

[a] Fig. 13, estimated.

[b] To convert m³/kg to ft³/lb, multiply by 16.0.

mental results with the first-stage hydrogasifier operated in the free-fall and moving-bed modes at 1.72 MPa and 870°C with pure hydrogen, calculations shown in Table 18 were made to estimate the composition and yield of the high methane gas produced when the first stage is integrated with an entrained char gasifier as the second stage. Although the methane content of the raw product gas is projected to be higher in the moving-bed reactor than in the falling-bed reactor, gas from the first stage must still be adjusted in H_2/CO ratio in a shift converter, scrubbed to remove carbon dioxide, and methanated to obtain SNG.

Another hydrogenation process utilizes internally generated hydrogen for hydroconversion in a single-stage, noncatalytic, fluidized-bed reactor (41). Biomass is converted in the reactor, which is operated at about 2.1 kPa, 800°C, and residence times of a few minutes with steam-oxygen injection. About 95% carbon conversion is anticipated to produce a medium heat value (MHV) gas which is subjected to the shift reaction, scrubbing, and methanation to form SNG. The cold gas thermal efficiencies are estimated to be about 60%.

Another advancement involves low temperature catalytic gasification of 2 to 10% aqueous biomass slurries or solutions that range from dilute organics in wastewater to waste sludges from food processing (42). The estimated residence time in the metallic catalyst bed is less than 10 min at 360°C and 20,635 kPa (3000 psi) at liquid hourly space velocities of 1.8 to 4.6 L of feedstock/L of catalyst/hr depending on the feedstock. The product fuel gas contains 45–70 vol % methane, 25–50 vol % carbon dioxide, and less than 5% hydrogen with as much as 2% ethane. The by-product water stream carries residual organics from 40 to 500 ppm COD. The product gas is MHV gas produced directly in contrast to MHV gas-phase processes that require either oxygen in place of air or a two-bed reaction system to keep the nitrogen in air separated from the fuel gas product.

Studies on the gasification of wood in the presence of steam and hydrogen show that steam gasification proceeds at a much higher rate than hydrogasification (43). Carbon conversions 30 to 40% higher than those achieved with hydrogen can be achieved with steam at comparable residence times. Steam/wood weight ratios up to 0.45 promoted increased carbon conversion, but had little effect

on methane concentration. Other experiments show that potassium carbonate-catalyzed steam gasification of wood in combination with commercial methanation and cracking catalysts can yield gas mixtures containing essentially equal volumes of methane and carbon dioxide at steam/wood weight ratios below 0.25, with atmospheric pressure and temperatures near 700°C (44). Other catalyst combinations produced high yields of product gas containing about 2:1 hydrogen/carbon monoxide and little methane at steam/wood weight ratios of about 0.75 and a temperature of 750°C. Typical results for both of these studies are shown in Table 19. The steam/wood ratios and the catalysts used can have significant effects on the product gas compositions. The composition of the product gas also can be manipulated depending on whether a synthesis gas or a fuel gas is desired.

Direct hydroliquefaction of biomass or wastes can be achieved by direct hydrogenation of wood chips on treatment at 10,132 kPa and 340 to 350°C with water and Raney nickel catalyst (45). The wood is completely converted to an oily liquid, methane, and other hydrocarbon gases. Batch reaction times of 4 hours give oil yields of about 35 wt % of the feed; the oil contains about 12 wt % oxygen and has a heating value of about 37.2 MJ/kg (16,000 Btu/lb). Distillation yields a significant fraction that boils in the same range as diesel fuel and is completely miscible with it.

A catalytic liquefaction process for heavy liquids production reacts biomass in a water solution of sodium carbonate and carbon monoxide gas at elevated temperature and pressure to form a heavy liquid fuel (46). Biomass and the combustible fraction of wastes have been converted at weight yields of 40 to 60% at temperatures of 250 to 425°C and pressures of 10 to 28 MPa. Lower viscosity products are generally obtained at higher reaction temperatures and solid or semisolid products are obtained when the reaction temperature is below 300°C. However, the high nitrogen and oxygen contents and the boiling characteristics and high viscosity range of the liquid products make it difficult to classify them as synthetic crude

Table 19. Product Gases from Steam Gasification of Wood

Gas or Parameter	Value			
Gas composition, mol %				
H_2	0	53	29	50
CH_4	52	4	15	17
CO_2	48	12	17	11
CO	0	30	34	17
Reactor temperature, °C	740	750	696	762
Pressure, kPa (gauge)	0	0	129	159
Primary catalyst	K_2CO_3	K_2CO_3	none	wood ash
Secondary catalyst	Ni:SiAl	SiAl	none	none
Steam/wood weight ratio	0.25	0.75	0.24	0.56
Carbon conversion to gas, %	64	77	68	52
Feed energy in gas, %	76	78		
Heating value of gas,[a] MJ/m³	20.6	12.1	16.6	17.7
Reference	44[b]	44[b]	45[c]	45[c]

[a] At normal conditions.

[b] Laboratory results with unspecified wood.

[c] Process development unit (PDU) results with unspecified hardwood.

oils. Conventional refining methods could not be used to upgrade this kind of material to standard petroleum derivatives. The original process consisted of a sequence of steps: drying and grinding wood chips to a fine powder, mixing the powder with recycled product oil (30% powder to 70% oil), blending the mixture with water containing sodium carbonate, and treatment of the slurry with synthesis gas at about 27,579 kPa (4000 psi) and 370°C. The modified process consists of partially hydrolyzing the wood in slightly acid water and treating the water slurry containing dissolved sugars and about 20% solids with synthesis gas and sodium carbonate at 27,579 kPa and 370°C on a once-through basis. The resulting oil product yield is about 1 bbl/400 kg (158.9 L/400 kg) of chips and is approximately equivalent to No. 6 grade boiler fuel. It contains about 50% phenolics, 18% high boiling alcohols, 18% hydrocarbons, and 10% water.

Study of the mechanism of this complex reduction-liquefaction suggests that part of the mechanism involves formate production from carbonate, dehydration of the vicinal hydroxyl groups in the cellulosic feed to carbonyl compounds via enols, reduction of the carbonyl group to an alcohol by formate and water, and regeneration of formate (46). In view of the complex nature of the reactants and products, it is likely that a complete understanding of all of the chemical reactions that occur will not be developed. However, the liquefaction mechanism probably involves catalytic hydrogenation because carbon monoxide would be expected to form at least some hydrogen by the water-gas shift reaction.

Chemical Processes

Biological–biochemical and thermal conversion processes are chemical processes, too, but a few specific chemical processes are mentioned separately because they are directed more to conventional chemical processing and production. These processes have been grouped together as chemical processes.

Chemicals have long been manufactured from biomass, especially wood (silvichemicals), by many different fermentation and thermochemical methods. For example, continuous pyrolysis of wood was used by the Ford Motor

Table 20. Product Yields from Wood Pyrolysis[a,b]

Product	Yield per t Dry Wood
Gas,[c] m^3	156
Charcoal, kg	300
Ethyl acetate, L	61.1
Creosote oil, L	13.6
Methanol, L	13.0
Ethyl formate, L	5.3
Methyl acetate, L	3.9
Methyl ethyl ketone, L	2.7
Other ketones, L	0.9
Allyl alcohol, L	0.2
Soluble tar, L	91.8
Pitch, kg	33.0

[a] Ref. 47.
[b] Feed: 70% maple, 25% birch, 5% ash, elm, and oak; av temperature, 515°C.
[c] CO_2, 37.9 mol %; CO, 23.4 mol %; CH_4, 16.8 mol %; N_2, 16.0 mol %; O_2, 2.4 mol %; H_2, 2.2 mol %; hydrocarbons, 1.2 mol %. To convert m^3 to ft^3, multiply by 35.3.

Co. in 1929 for the manufacture of various chemicals (Table 20) (47). Wood alcohol (methanol) was manufactured on a large scale by destructive distillation of wood for many years until the 1930s and early 1940s, when the economics became more favorable for methanol manufacture from fossil fuel-derived synthesis gas.

In the production of chemicals from biomass, wood is still the raw material of choice for the manufacture of certain chemicals, although many of them cannot compete with fossil-based products. The chemistry of silvichemical production is related directly to the chemical composition of trees, ie, 50% cellulose, 25% hemicelluloses, and 25% lignins. However, specialty chemicals are often manufactured from nonwoody biomass because they occur naturally in certain plant species or can easily be derived from these plants. Examples are the alginic acids from *Macrocystis pyrifera*, ie, giant brown kelp, and physiologically active alkaloids from particular plants. Ethanol has been manufactured for chemical applications from starchy biomass by fermentation for many years. Figure 14 lists some of the more important primary biomass-derived

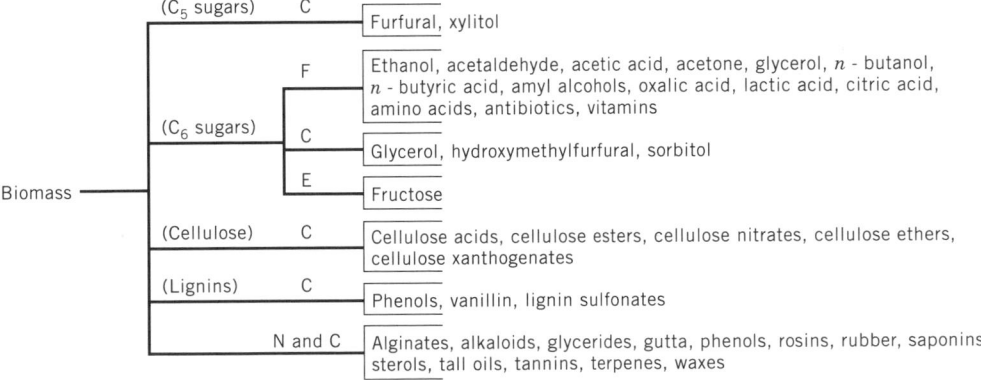

Figure 14. Primary biomass-derived chemicals. Dominant processing methods are chemical (C), fermentation (F), enzymic (E), and natural (N) processes; products in parentheses represents intermediates.

chemicals, the principal intermediates, and the dominant processing methods used. All chemicals listed in Figure 14 are either manufactured commercially in 1992 by the indicated routes or were manufactured in the past. Secondary processing of these primary chemicals would appear to make it possible to manufacture almost all heavy and fine organics produced from fossil raw materials, eg, ethanol (qv) can be converted to ethylene, acetaldehyde, and acetic acid, which can be converted to other organic chemicals by established routes.

The availability of C_6 sugars such as glucose is an important factor in the development of a biomass chemicals industry as alluded to in Figure 14. Unfortunately, most biomass is higher in cellulose than C_6 sugars. Cellulose, because of its relatively low biodegradability, cannot easily be converted to fermentation alcohol and other products without first liberating the monosaccharides. Trees and fibrous biomass, such as plant stalks and reedy plants, contain cellulose in partially crystalline form sometimes complexed with other materials. The low degradability of wood cellulose, which can exist as lignin-cellulose complexes, is attributed to these factors. Such forms of cellulose can be degraded to glucose concentrates by several hydrolytic methods, the most common of which is hydrolysis with dilute sulfuric acid. At the temperatures necessary to form glucose, a large portion of the product is ordinarily converted to by-products such as hydroxymethylfurfural. Glucose yields are usually near 50% of theory because of by-product formation. Enzyme-catalyzed hydrolysis of cellulose has afforded much higher yields of glucose, but the particle size of the cellulosic material subjected to hydrolysis must be reduced to facilitate depolymerization. The cost of particle size reduction tends to outweigh the advantages of higher glucose yields. One approach to improving yield of monosaccharides is first to dissolve the cellulose in a solvent, separate the insolubles from the cellulose solution, and then subject the solution to hydrolysis conditions or precipitate the cellulose before hydrolysis. This method destroys crystallinity and any cellulose complex that may be present, and makes it possible to achieve high yields of glucose even with highly fibrous biomass. Special solvents such as Cadoxen [14874-24-9] and 65% sulfuric acid have been suggested for this application (48). Another approach involves the use of a blocking agent, acetone, which temporarily forms a cyclic ketal (acetonide) with vicinal diols to protect the sugars from degradation under hydrolysis conditions (49). The process has been reported to afford very high yields of sugars and to be equally applicable to high fiber biomass residues.

As mentioned in the biological–biochemical section, another approach to improve alcoholic fermentation com-

bines saccharification and fermentation, ie, simultaneous saccharification and fermentation (SSF). Enzyme-catalyzed cellulose hydrolysis and fermentation to alcohol takes place in the same vessel in the presence of enzyme and yeast (50). Reduced fermenter pressures and enzyme and yeast recycling result in 70 to 80% ethanol yields. These process modifications, coupled with more energy-efficient distillation and heat exchanger improvements, are projected to make fermentation ethanol from low value biomass competitive with industrial ethanol (51).

Multiproduct processes for biomass chemical plants in which the operating conditions can be manipulated to vary the product distribution as a function of demand or other factors, or in which an optimum mix of products is chosen based on feedstock characteristics, appear to have some merit even though no full-scale systems have yet been commercialized. The process depicted in Figure 15 illustrates how such a plant might function. Mild acid hydrolysis of the hemicelluloses in wood affords either a predominantly xylose or mannose solution, depending on the type of wood feed, and a cellulose–lignin residue. Strong acid treatment of this residue yields a glucose solution, which can be combined with mannose for alcoholic fermentation, and a lignin residue. Phenols can be made from this residue by hydrogenation, and furfural can be made from xylose by strong acid treatment. The products are thus ethanol, furfural, and phenols.

There are many different routes to organic chemicals from biomass because of its high polysaccharide content and reactivity. The practical value of the conversion processes selected for commercial use with biomass will depend strongly on the availability and price of the same chemicals produced from petroleum and natural gas.

BIOMASS PRODUCTION

The manufacture of synfuels and energy products from biomass requires that suitable quantities of biomass be grown, harvested, and transported to the conversion plant site. Many variables must be considered when selecting the proper species or mixture of species for operation of a system: growth cycle; fertilization; insolation; temperature; precipitation; propagation and planting procedures; soil and water needs; harvesting methods; disease resistance; growth area competition from biomass for food, feed, and fiber; growth area availability; possibilities for simultaneous or sequential growth of biomass for synfuels and foodstuff or other applications; and nutrient depletion. At least 250,000 botanical species, of which only about 300 are cash crops, are known in the world. A relatively small number are, and will be, used as biomass

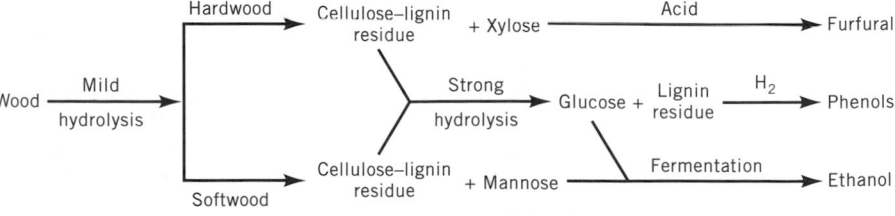

Figure 15. Furfural, ethanol, and phenols production from wood in a multiproduct process biomass chemical plant (52). Wood is a ca 50% cellulose, 25% lignin, and 25% hemicellulose.

feedstocks for the manufacture of synfuels and energy products.

In the ideal case, biomass chosen for energy applications should be high yield, low cash-value species that have short growth cycles and grow well in the area and climate chosen for the biomass energy system. Fertilization requirements should be low and possibly nil if the species selected fix ambient nitrogen, thereby minimizing the amount of external nutrients that must be supplied to the growth areas. In areas having low annual rainfall, the species grown should have low water needs and be able to efficiently utilize available precipitation. For land-based biomass, the requirements should be such that the crops can grow well on low grade soils and do not need the best classes of agricultural land. After harvesting, growth should commence again without the need for replanting. Surprisingly, several biomass species meet many of these idealized characteristics and appear to be quite suitable for energy applications. There are a number of important factors that relate to biomass production for energy applications.

Photosynthesis

The basic biochemical pathways in ambient carbon dioxide fixation involve decomposition of water to form oxygen, protons, and electrons; transport of these electrons to a higher energy level via Photosystems I, II, and several electron transfer agents; concomitant generation of reduced nicotinamide adenine dinucleotide [53-57-6] (NADPH) and adenosine triphosphate [56-65-5] (ATP); and reductive assimilation of carbon dioxide to carbohydrate. The initial process is believed to be the absorption of light by chlorophyll [1406-65-1], which promotes decomposition of water. The ejected electrons are accepted by ferredoxin [9080-02-8] (Fd), a nonheme iron protein. The reduced Fd initiates a series of electron transfers to generate ATP from adenosine diphosphate [58-64-0] (ADP), inorganic phosphate, and NADPH. For each of the two light reactions, one photon is required to transfer each electron; a total of eight photons is thus required to fix one molecule of carbon dioxide. Assuming that the carbon dioxide is in the gaseous phase and that the initial product is glucose, the standard Gibbs free energy change at 25°C is +0.48 MJ(+114 kcal) per mole of carbon dioxide assimilated and the corresponding enthalpy change is +0.47 MJ (+112 kcal).

The maximum efficiency with which photosynthesis can occur has been estimated by several methods. The upper limit has been projected to range from about 8 to 15%, depending on the assumptions made; ie, the maximum amount of solar energy trapped as chemical energy in the biomass is 8 to 15% of the energy of the incident solar radiation. The rationale in support of this efficiency limitation helps to point out some aspects of biomass production as they relate to energy applications.

The relationship of the energy and wavelength of a photon is energy $= \hbar c/\lambda$ where $\hbar$ is Planck's constant, 6.624×10^{-34}; c is velocity of light; and λ is wavelength. Assume that the wavelength of the light absorbed is 575×10^{-9} m and is equivalent to the light absorbed between the blue (400×10^{-9} m) and red (700×10^{-9} m)

ends of the visible spectrum. This assumption has been made by several investigators for green plants to calculate the upper limit of photosynthesis efficiency. The energy absorbed in the fixation of one mole of carbon dioxide, which requires 8 photons/molecule, is then given by

energy absorbed
$$= [(6.624 \times 10^{-34})(3.00 \times 10^8)/(575 \times 10^{-9})]$$
$$\times 8 \times 6.024 \times 10^{23}$$
$$= 1.67 \text{ MJ}$$

Since 0.47 MJ of solar energy is trapped as chemical energy in this process, the maximum efficiency for total white-light absorption is 28.1%. Further adjustments are usually made to account for the percentages of photosynthetically active radiation in white light, the light that can actually be absorbed, and respiration. The amount of photosynthetically active radiation in solar radiation that reaches the earth is estimated to be about 43%. The fraction of the incident light absorbed is a function of many factors, such as leaf size, canopy shape, and reflectance of the plant; it is estimated to have an upper limit of 80%. This effectively corresponds to the utilization of eight photons out of every 10 in the active incident radiation. The third factor results from biomass respiration. A portion of the stored energy is used by the plant, the amount of which depends on the properties of the biomass species and the environment. For purposes of calculation, assume that about 25% of the trapped solar energy is used by the plant, thereby resulting in an upper limit for retention of the nonrespired energy of 75%. The upper limit for the efficiency of photosynthetic fixation of biomass can now be estimated to be 7.2%, ie, $0.281 \times 0.43 \times 0.80 \times 0.75$. For the case where little or no energy is lost by respiration, the upper limit is estimated to be 9.7%, ie, $0.281 \times 0.43 \times 0.80$. The low efficiency limit might correspond to land-based biomass, while the higher efficiency limit might be closer to water-based biomass such as unicellular algae. These figures can be transformed into dry biomass yields by assuming that all of the fixed carbon dioxide is contained in the biomass as cellulose, $(C_6H_{10}O_5)_x$, from the equation

$$Y = \frac{CIE}{F}$$

where Y is yield of dry biomass, t/ha·yr; C is a constant, 3.1536; I is average insolation, W/m²; E is solar energy capture efficiency, %; and F is energy content of dry biomass, MJ/kg. Thus for high cellulose dry biomass, an average insolation of 184 W/m² (1404 Btu/ft²·d), which is the average insolation for the continental United States, a solar energy capture efficiency of 7.2%, and a high heat of combustion of 17.51 MJ/kg for cellulose, the yield of dry biomass is 239 t/hm²·yr (107 short tons/acre·yr). The corresponding value for an energy capture efficiency of 9.7% is 321 t/hm²·yr (143 short t/acre·yr). These yields of organic matter can be viewed as an approximation of the theoretical yield limits for land- and water-based biomass. Some estimates of maximum yield are higher and some are lower than these figures, depending on the values used for I, E, and F, but they serve as a guideline to indicate the highest yields that a biomass production system could be expected to achieve under normal environmental conditions. Unfortunately, real biomass yields rarely approach these limits. Sugarcane, for example, which is one of the high yielding species of biomass, typically produces total dry plant matter at yields of about 80 t/hm²·yr (36 short tons/acre·yr).

Yield is plotted against solar energy capture efficiency in Figure 16 for insolation values of 150 and 250 W/m², which spans the range commonly encountered in the United States, and for biomass energy values of 12 and 19 MJ/kg (dry). The higher the efficiency of photosynthesis, the higher the biomass yield. For a given solar energy capture efficiency and incident solar radiation, the yield is projected to be lower at the higher biomass energy values, ie, curves A and C, curves B and D. From an energy production standpoint, this means that a higher energy content biomass could be harvested at lower yield levels and still compete with higher yielding but lower energy content biomass species. It is also apparent that for a given solar energy capture efficiency, yields similar to those obtained with higher energy content species should be possible with a lower energy content species even when it is grown at a lower insolation, eg, curves B and C. Finally, at the solar energy capture efficiency usually encountered in the field, about 1% or less, the spread in

yields is much less than at the higher energy capture efficiencies. It is important to emphasize that this interpretation of biomass yield as functions of insolation, energy content, and energy capture, although based on sound principles, is still a theoretical analysis of living systems. Because of the many uncontrollable factors in the field, such as the changes that occur in climate and seasonal changes in biomass composition, departures from the norm can be expected.

The previous discussion of photosynthesis has concentrated on the gross features of ambient carbon dioxide fixation in biomass. However, the biochemical pathways involved in the conversion of carbon dioxide to carbohydrate play an important role in understanding the molecular events of biomass growth. Three different biochemical energy transfer pathways occur during carbon dioxide fixation (53–55). One pathway, the Calvin three-carbon cycle, involves an initial three-carbon intermediate of phosphoglyceric acid [82-11-1] (PGA). This cycle, often referred to as the reductive pentose phosphate cycle, is used by autotrophic photochemolithotrophic bacteria, algae, and green plants. For every three molecules of carbon dioxide converted to glucose in a dark reaction, nine ATP, six NADPH, and 12 Fd molecules are required. Six molecules of PGA are formed from three molecules each of carbon dioxide and ribulose-1,5-diphosphate (RuDP) in the chloroplasts. After these carboxylation reactions, a reductive phase occurs in which six molecules of PGA are successively transformed into six molecules of diphosphoglyceric acid (DPGA) and then six molecules of 3-phosphoglyceraldehyde, a triose phosphate. Five molecules of the triose phosphate are then used to regenerate three molecules of RuDP, which initiates the cycle again. The other triose phosphate molecule is used to generate glucose-6-phosphate and energy and electron carriers. Plant biomass species that use the Calvin cycle are called C_3 plants; it is common in many fruits, legumes, grains, and vegetables. C_3 plants usually exhibit low rates of photosynthesis at light saturation, low light saturation points, sensitivity to oxygen concentration, rapid photorespiration, and high carbon dioxide compensation points, ie, about 50 ppm. The carbon dioxide compensation point is the carbon dioxide concentration in the surrounding environment below which more carbon dioxide is respired by the plant than is photosynthetically fixed. Typical C_3 biomass species are peas, sugar beet, spinach, alfalfa, *Chlorella*, *Eucalyptus*, potato, soybean, tobacco, oats, barley, wheat, tall fescue, sunflower, rice, and cotton.

The second pathway is called the C_4 cycle because the carbon dioxide is initially fixed as the four-carbon dicarboxylic acids, malic or aspartic acids. Phosphoenol pyruvate (PEP) reacts with one molecule of carbon dioxide to form oxaloacetate (OAA) in the mesophyll of the biomass, and then malate or aspartate is formed. The C_4 acid is transported to the bundle sheath cells, where decarboxylation occurs to regenerate pyruvate, which is returned to the mesophyll cells to initiate another cycle. The carbon dioxide liberated in the bundle sheath cells enters the C_3 cycle in the usual manner. Thus no net carbon dioxide is fixed in the portion of the C_4 cycle, and it is the combination with the C_3 cycle which ultimately results in carbon dioxide fixation. The subtle differences between the C_4

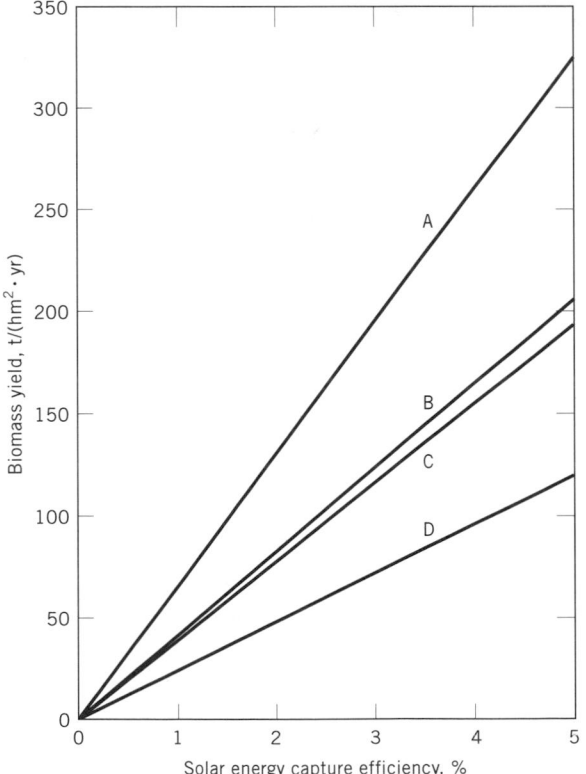

Figure 16. Effect of solar energy capture efficiency on biomass yield. A, insolation value (I) of 250 W/m², and biomass energy value (F) of 12 MJ/kg (dry); B, I of 250 W/m², F of 19 MJ/kg; C, I of 150 W/m², F of 12 MJ/kg; D, I of 150 W/m², F of 19 MJ/kg. To convert W/m² to Btu/(ft²·d), multiply by 7.616; to convert MJ/kg to Btu/lb, multiply by 430.

and C_3 cycles are believed responsible for the wide variations in biomass properties. In contrast to C_3 biomass, C_4 biomass is usually produced at higher yields and has higher rates of photosynthesis, high light saturation points, insensitivity to oxygen concentrations below 21 mol %, low levels of respiration, low carbon dioxide compensation points, and greater efficiency of water usage. C_4 biomass often occurs in areas of high insolation, hot daytime temperatures, and seasonal dry periods. Typical C_4 biomass includes important crops such as corn, sugarcane, and sorghum, and forage species and tropical grasses such as Bermuda Grass. Even crabgrass is a C_4 biomass. At least 100 genera in 10 plant families are known to exhibit the C_4 cycle.

The third pathway is called crassulacean acid metabolism (CAM). CAM refers to the capacity of chloroplast-containing tissues to fix carbon dioxide in the dark via phosphoenolpyruvate carboxylase leading to the synthesis of free malic acid. The mechanism involves the β-carboxylation of PEP by this enzyme and the subsequent reduction of OAA by malate dehydrogenase. CAM has been documented in at least 18 families, including the family Crassulaceae, and 109 genera of the Angiospermae. Biomass species in the CAM category are typically adapted to arid environments, have low photosynthesis rates, and have high water usage efficiencies. Examples are cactus plants and the succulents, such as pineapple. The information developed to date on CAM biomass indicates that CAM has evolved so that initial carbon dioxide fixation can take place in the dark with much less water loss than the fully light-dependent C_3 and C_4 pathways. CAM biomass also conserves carbon by recycling endogenously formed carbon dioxide. Several CAM species show temperature optima in the range 12 to 17°C for carbon dioxide fixation in the dark. The stomates in CAM plants open at night to allow entry of carbon dioxide and then close by day to minimize water loss. The carboxylic acids formed in the dark are converted to carbohydrates when the radiant energy is available during the day. Relatively few CAM plants have been exploited commercially.

Significant differences in net photosynthetic assimilation of carbon dioxide are apparent between C_3, C_4, and CAM biomass species. One of the principal reasons for the generally lower yields of C_3 biomass is its higher rate of photorespiration; if the photorespiration rate could be reduced, the net yield of biomass would increase. Considerable research is in progress (ca 1992) to achieve this rate reduction by chemical and genetic methods, but as yet, only limited yield improvements have been made. Such an achievement with C_3 biomass would be expected to be very beneficial for foodstuff production and biomass energy applications.

The specific carbon dioxide-fixing mechanism used by a plant will affect the efficiency of photosynthesis, so from an energy utilization standpoint, it is desirable to choose plants that exhibit high photosynthesis rates to maximize the yields of biomass in the shortest possible time. There are numerous factors that affect the efficiency of photosynthesis other than the carbon dioxide-fixing mechanism, eg, insolation; amounts of available water, nutrients, and carbon dioxide; temperature; and transmission, reflection, and biochemical energy losses within or near

the plant. For lower plants such as the green algae, many of these parameters are under human control. For conventional biomass growth subjected to the natural elements, it is not feasible to control all of them.

Climate and Environmental Factors

The biomass species selected for energy applications and the climate must be compatible to facilitate operation of fuel farms. The three primary climatic parameters that have the most influence on the productivity of an indigenous or transplanted species are insolation, rainfall, and temperature. Natural fluctuations in these factors remove them from human control, but the information compiled over the years in meteorological records and from agricultural practice supplies a valuable data bank from which to develop biomass energy applications. Ambient carbon dioxide concentration and the availability of nutrients are also important factors in biomass production.

Insolation. The intensity of the incident solar radiation at the earth's surface is a key factor in photosynthesis; natural biomass growth will not take place without solar energy. Insolation varies with location and is high in the tropics and near the equator. The approximate changes with latitude are illustrated in Table 21. At a given latitude, the incident energy is not constant and often exhibits large changes over relatively short distances. A more quantitative summary of insolation values over the continental United States is shown in Table 22. To place the amount of energy that strikes the earth in the proper perspective, the annual insolation on about 0.1 to 0.2% of the surface of the continental United States is equivalent to all the energy consumed by the United States in one year. The production figures shown in Table 23 represent annual yields obtained under good growth conditions (13,14,53,59,60). The estimated solar energy capture efficiencies (SECD) for the biomass listed assumes that all organic matter is cellulose. These are only rough approximations and most are probably too high, but they indicate that C_4 plants are usually better photosynthesizers than C_3 plants, and that high insolation alone does not always correlate with high biomass yield and capture efficiency. Although there are a few exceptions in Table 23, there appears to be a trend in this direction.

Precipitation. Precipitation as rain, snow, sleet, or hail is governed by movement of air and is generally abundant

Table 21. Insolation at Various Latitudes for Clear Atmospheres[a]

Location	Latitude	Insolation, W/m²[b]		
		Maximum	Minimum	Average[c]
Equator	0°	315	236	263
Tropics	23.5°	341	171	263
Mid-earth	45°	355	70.9	210
Polar circle	66.5°	328	0	158

[a] Ref. 56.
[b] To convert W/m² to Btu/ft²·d), multiply by 7.616.
[c] Yearly total divided by 365.

Table 22. Daily Solar Radiation in the United States[a]

| Location | Total Daily Insolation, W/m²[b,c] | | | | |
	January	April	July	October	Annual[c]
Tucson, Arizona	146	289	288	208	229
Fresno, California	93.2	290	338	187	229
Lakeland, Florida	135	260	247	189	210
Indianapolis, Indiana	90.0	188	242	120	157
Lake Charles, Louisiana	109	215	236	175	191
Saint Cloud, Minnesota	75.9	178	275	104	157
Glasgow, Montana	72.0	190	299	118	175
Ely, Nevada	108	257	288	176	210
Oklahoma City, Oklahoma	80.8	212	264	155	183
San Antonio, Texas	113	198	286	182	199
Burlington, Vermont	76.3	182	208	99.7	146
Sterling, Virginia	90.9	173	233	113	159
Seattle–Tacoma, Washington	36.5	179	276	98.1	151

[a] Ref. 57.
[b] To convert W/m² to Btu/(ft²·d), multiply by 7.616.
[c] Average.

wherever air currents are predominately upward. The greatest precipitation should therefore occur near the equator. The average annual rainfall in the United States is about 79 cm.

The moisture needs of aquatic biomass presumably are met in full because growth occurs in liquid water, but the growth of land biomass often can be water-limited. Requirements for good growth of many biomass species have been found to be in the range of 50 to 76 cm of annual rainfall (63). Some crops, such as wheat, exhibit good growth with much less water, but they are in the minority. Without irrigation, water is supplied during the growing season by the water in the soil at the beginning of the season and by rainfall. Figure 17 depicts the normal precipitation recorded in the 48-state area during the normal growing season, April to September. This type of information and the established requirements for the growth of land-based biomass can be used to divide the

Table 23. Annual Production of Dry Matter and Solar Energy Capture Efficiency (SECD)[a]

Location	Biomass Community	Productivity, t/(hm²·yr)[b]	Insolation, W/m²[c]	SECD,[d] %
Sweden	Enthrophic lake angiosperm	7.2	106	0.38
Denmark	Phytoplankton	8.6	133	0.36
Mississippi	Water hyacinth	11.0–33.0	194	0.31–0.94
Minnesota	Maize	24.0	169	0.79
New Zealand	Temperate grassland	29.1	159	1.02
West Indies	Tropical marine angiosperm	30.3	212	0.79
Nova Scotia	Sublittoral seaweed	32.1	133	1.34
Georgia	Subtropical saltmarsh	32.1	194	0.92
England	Coniferous forest, 0–21 yr	34.1	106	1.79
Israel	Maize	34.1	239	0.79
New South Wales	Rice	35.0	186	1.04
Congo	Tree plantation	36.1	212	0.95
Holland	Maize, rye, two harvests	37.0	106	1.94
Marshall Islands	Green algae	39.0	212	1.02
FRG	Temperate reedswamp	46.0	133	1.92
Puerto Rico	*Panicum maximum*	48.9	212	1.28
California	Algae, sewage pond	49.3–74.2	218	1.26–1.89
Colombia	Pangola grass	50.2	186	1.50
West Indies	Tropical forest, mixed ages	59.0	212	1.55
Hawaii	Sugarcane	74.9	186	2.24
Puerto Rico	*Pennisetum purpurcum*	84.5	212	2.21
Java	Sugarcane	86.8	186	2.59
Puerto Rico	Napier grass	106	212	2.78
Thailand	Green Algae	164	186	4.90

[a] Refs. 13, 14, 53, 59, 60.
[b] To convert t/(hm²·yr) to short ton/(acre·yr), divide by 2.24.
[c] Average. To convert W/m² to Btu/(ft²·d), multiply by 7.616.
[d] Approximate estimates of solar energy capture efficiencies; probably too high.

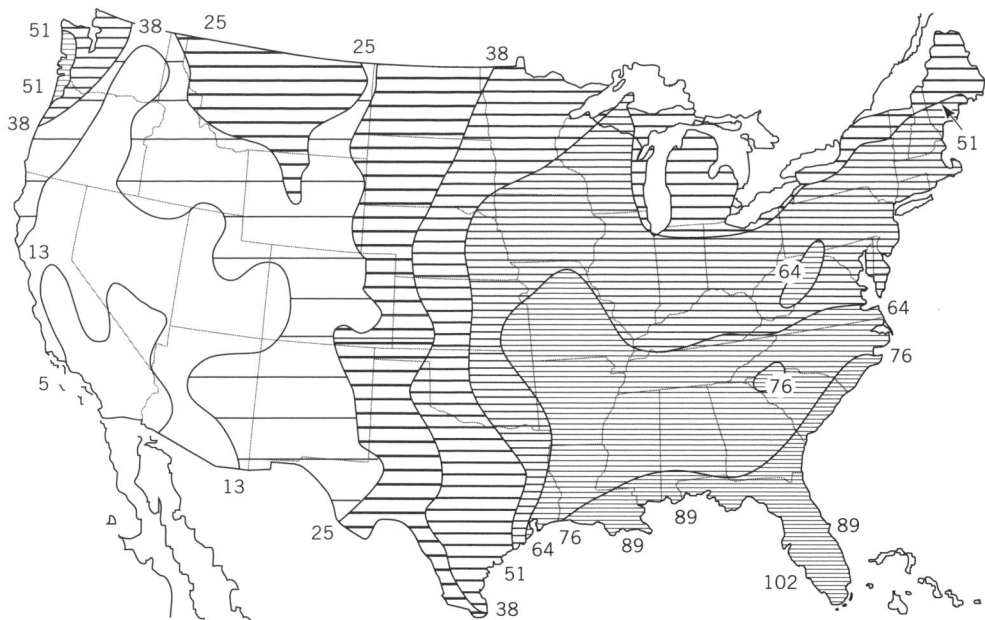

Figure 17. Normal precipitation in cm during the growing season, April to September, in the United States (61).

United States into precipitation regions. Regions more productive for biomass generally correlate with precipitation regions. It should be realized, however, that rainfall alone is not quantitatively related to productivity of land biomass because of the differences in soil characteristics, water evaporation rates, and infiltration. Also, certain areas that have low rainfall can be made productive through irrigation. Finally, some areas of the country that vary widely in precipitation as a function of time, such as many Western states, will produce moderate biomass yields, and often sufficient yields of cash crops without irrigation, to justify commercial production.

Temperature. Most biomass species grow well in the United States at temperatures between 15.6 and 32.3°C. Typical examples are corn, kenaf, and napier grass. Tropical grasses and certain warm-season biomass have optimum growth temperatures in the range 35 to 40°C, but the minimum growth temperature is still near 15°C (64). Cool-weather biomass such as wheat may show favorable growth below 15°C, and certain marine biomass such as the giant brown kelp only survive in water at temperatures below 20 to 22°C (65). The growing season is longer in the southern portion of the United States; in some areas such as Hawaii, the Gulf states, southern California, and the southeastern Atlantic states, the temperature is usually conducive to biomass growth for most of the year (61,62).

The effect of temperature fluctuations on net carbon dioxide uptake is illustrated by the curves in Figure 18. As the temperature increases, net photosynthesis increases for cotton and sorghum to a maximum value and then rapidly declines. Ideally, the biomass species grown in an area should have a maximum rate of net photosynthesis as close as possible to the average temperature during the growing season in that area.

Ambient Carbon Dioxide Concentration. Many studies have been performed which show that higher concentrations of carbon dioxide than are normally present in air will promote more carbon fixation and increase biomass yields. In confined, environmentally controlled enclosures

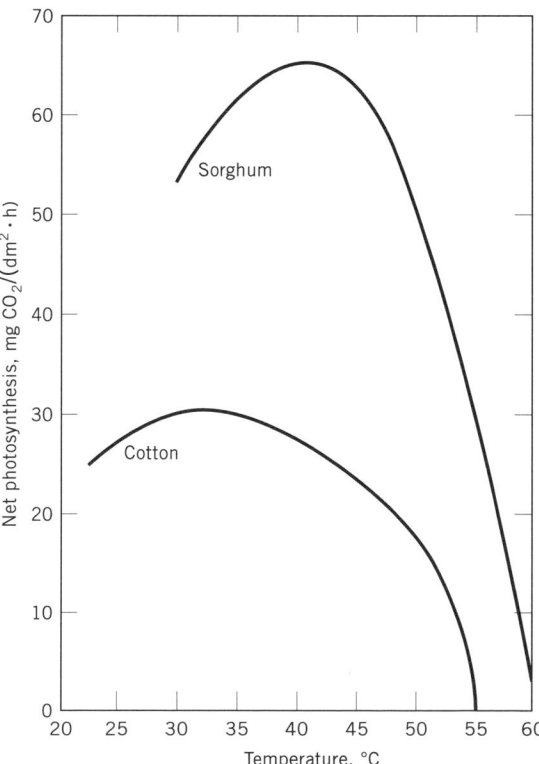

Figure 18. Effect of temperature on net photosynthesis for sorghum and cotton leaves. To convert mg/(dm²·h) to lb/(ft²·h), multiply by 2.373×10^{-3}.

such as hothouses, carbon dioxide-enriched air can be used to stimulate growth. This is not practical in large-scale open systems such as those envisaged for biomass energy farms. For aquatic biomass production, carbon dioxide enrichment of the water phase may be an attractive method of promoting biomass growth if carbon dioxide concentration is a limiting factor; the growth of biomass often occurs by uptake of carbon dioxide from both the air and liquid phase near the surface.

For some high growth-rate biomass species, the carbon dioxide concentration in the air among the leaves of the plant often is considerably less than that in the surrounding atmosphere. Photosynthesis may be limited by the carbon dioxide concentrations under these conditions when wind velocities are low and insolation is high.

Nutrients. All living biomass requires nutrients other than carbon, hydrogen, and oxygen to synthesize cellular material. Principal nutrients are nitrogen, phosphorus, and potassium; other nutrients required in lesser amounts are sulfur, sodium, magnesium, calcium, iron, manganese, cobalt, copper, zinc, and molybdenum. The last five nutrients, as well as a few others not listed, are sometimes referred to as micronutrients because only trace quantities are needed to stimulate growth. For land-based biomass, these elements are usually supplied by the soil, so the nutrients are depleted if they are not replaced through fertilization. Some biomass, such as the legumes, are able to meet all or part of their nitrogen requirements through fixation of ambient nitrogen. Water-based biomass such as marine kelp use the natural nutrients in ocean waters. Freshwater biomass such as water hyacinth is often grown in water enriched with nutrients in the form of wastewater. The growth of the plant is stimulated, and at the same time the influent waste is stabilized because its components are taken up by the plant as nutrients. So-called luxuriant growth of water hyacinth on sewage, in which more than the needed nutrients are removed from the waste, can be used as a substitute wastewater treatment method (28).

Whole plants typically contain 2 wt % N, 1 wt % K, and 0.5 wt % P, so at a yield of 20 t/hm^2·yr (8.9 short ton/acre·yr), harvesting of the whole plant without return of any of the plant parts to the soil corresponds to the annual removal of 400 kg N, 200 kg K, and 100 kg P per hm^2. This illustrates the importance of fertilization, especially of these macronutrients, to maintain fertility of the soil. Biomass growth is often nutrient-limited and yield correlates with fertilizer dose rates. Average nitrogen fertilizer applications for production of wheat, rice, potato, and brussel sprouts are about 73, 134, 148, and 180 kg/hm^2, respectively, in the United States (67). Estimates of balanced fertilizers needed to produce various land biomass species are shown in Table 24. Note that alfalfa does not require added nitrogen because of its nitrogen-fixing ability. It is estimated that this legume can fix from about 130 to 600 kg of elemental nitrogen per hm^2 annually (68).

Normal weathering processes that occur in nutritious soils release nutrients, but they are not available at rates that promote maximum biomass yields. Fertilization is usually necessary to maximize yields. Since nitrogenous fertilizers are largely manufactured from fossil fuels,

Table 24. Fertilizer Requirements of Biomass[a,b], kg

Biomass	N	P$_2$O$_5$	K$_2$O	CaO
Alfalfa	0.0	12.3	34.0	20.7
Corn	11.8	5.7	10.0	0.0
Kenaf	13.9	5.0	10.0	16.1
Napier grass	9.6	9.3	15.8	8.5
Slash pine[c]	3.8	0.9	1.6	2.3
Potato	16.8	5.3	28.3	0.0
Sugar beet	18.0	5.4	31.2	6.1
Sycamore	7.3	2.8	4.7	0.0
Wheat	12.9	5.3	8.4	0.0

[a] Ref. 62.
[b] Estimated per whole dry plant.
[c] Five years.

mainly natural gas, and since fertilizer needs are usually the most energy intensive of all inputs in a biomass production system, careful analysis of the integrated biomass production-conversion system is needed to ensure that net energy production is positive. Trade-offs between synfuel outputs, nonsolar energy inputs, and biomass yields are required to operate a system that produces only energy products.

Land Availability

The availability of sufficient land suitable for production of land-based biomass can be estimated for the United States by several techniques. One method relies on the land capabilities classification scheme developed by the U.S. Department of Agriculture (69), in which land is divided into eight classes. Classes I to III are suited for cultivation of many kinds of crops; Class IV is suited only for limited production; and Classes V to VIII are useful only for permanent vegetation such as grasses and trees. The U.S. Department of Agriculture surveyed nonfederal land usage for 1987 in terms of these classifications (70). Out of 568 million hm^2, which corresponds to 60% of the 50-state area, 43% of the land (246.3 × 10^6 hm^2) was in Classes I to III, 13% (75.59 × 10^6 hm^2) in Class IV; and 43% (246.3 × 10^6 hm^2) in Classes V to VIII. The actual usage of this land at the time of the survey is shown in Table 25 (70). Table 25 illustrates that of all the land judged suitable for cultivation in Classes I to III, only about one-half of it is actually used as cropland (70), and that the combined areas of pasture, range, and forest lands is about 66% of the total nonfederal lands. There is ample opportunity to produce biomass for energy applications on nonfederal land not used for foodstuffs production. Large areas of land in Classes V to VIII not suited for cultivation, and sizable areas in Classes I to IV not being used for crop production, also would appear to be available for biomass energy applications; land used for crop production could be considered for simultaneous or sequential growth of biomass for foodstuffs and energy. Portions of federally owned lands, which are not included in the survey, might also be dedicated to biomass energy applications. Careful design and management of land-based biomass production areas could result in improvement or upgrading of lands to higher land capability classifications.

Table 25. Nonfederal Rural Land Use in United States by Type, 1987,[a] 10^6 hm^2

Land Class	Cropland[b]	Pastureland[c]	Rangeland[d]	Forest[e]	Minor	Total
I	11.58	0.82	0.17	0.66	0.23	13.47
II	77.31	12.81	6.58	18.00	2.66	117.36
III	54.28	16.00	18.76	24.26	2.23	115.53
IV	18.60	10.30	21.62	23.56	1.51	75.59
V	1.16	1.86	1.99	7.52	1.03	13.55
VI	6.56	6.84	53.34	37.34	2.63	106.71
VII	1.59	3.91	58.40	46.88	4.08	114.86
VIII	0.035	0.067	1.68	1.40	7.81	11.00
other[f]	0	0	0	0	2.08	2.08
Total	*171.12*	*52.60*	*162.56*	*159.62*	*24.25*	*570.15*
%	30.01	9.22	28.51	28.00	4.25	

[a] Ref. 70. Totals may not be precise summations due to rounding.
[b] Land used for production of crops for harvest alone or in rotation with grasses and legumes.
[c] Land used for the production of adapted, introduced, or native species in a pure stand, grass mixture, or a grass–legume mixture.
[d] Land on which the vegetation is predominantly grasses, grass-like plants, forbs, or shrubs suitable for grazing or browsing.
[e] Land that is at least 10% stocked by forest trees of any size or formerly having had such tree cover and not currently developed for nonforest use.
[f] Land, such as farmsteads, stripmines, quarries, and other lands, that do not fit into any other land class category.

Water Availability

The production of marine biomass in the ocean, even on the largest scale envisaged for energy applications, would require only a very small fraction of the available ocean areas. The U.S. Navy has estimated that a square area 753 km on each edge off the coast of California may be sufficient to produce enough giant brown kelp for conversion to methane to supply all of the nation's natural gas needs. This large area is very small when compared with the total area of the Pacific Ocean. Also, the benefits to other marine life from a large kelp plantation have been well documented (65). Any conflicts that might arise would be concerned primarily with ocean traffic. With the proper plantation design for marine biomass and precautionary measures to warn approaching ships, it is expected that marine biomass growth could be sustained over long periods.

Freshwater biomass in theory can be grown on the 20 million hm^2 of fresh water in the United States. However, several difficulties mitigate against large-scale freshwater biomass energy systems. About 80% of the fresh water in the United States is located in the northern states, while several of the freshwater biomass species considered for energy applications require a warm climate such as that found in Gulf states. The freshwater areas suitable for biomass production in the southern states, however, are much smaller than those in the North, and the density of usage is higher in southern inland waters. Overall, these characteristics make small-scale aquatic biomass production systems more feasible for energy applications. It may be advisable in the future to examine the possibility of constructing large artificial lakes for this purpose; this does not seem practical in the early 1990s.

Land-Based Biomass

Much effort to evaluate land-based biomass energy applications has been expended. This work aims at selecting high yield biomass species, characterizing physical and chemical properties, defining growth requirements, and rating energy use potential. Several species have been proposed specifically for energy usage, while others have been recommended for multiple uses, one of which is as an energy resource. The latter case is exemplified by sugarcane; bagasse, the fibrous material remaining after sugar extraction, is used in several sugar factories as a boiler fuel. Most land-based biomass plantations operated for energy production or synfuel manufacture also will yield products for nonenergy markets. Land-based biomass for energy production can be divided into forest biomass, grasses, and cultivated plants.

Forest Biomass. About one-third of the world's land area is forestland. Broadleaved evergreen trees are a dominant species in tropical rain forests near the equator (71). In the northern hemisphere, stands of coniferous softwood trees such as spruce, fir, and larch dominate in the boreal forests at the higher latitudes, while both the broadleaved deciduous hardwoods such as oak, beach, and maple, and the conifers such as pine and fir, are found in the middle latitudes. Silviculture, ie, the growth of trees, is practiced by five basic methods: exploitative, conventional extensive, conventional intensive, naturalistic, and short-rotation (71). The exploitative method harvests trees without regard to regeneration. The conventional extensive method harvests mature trees so that natural regeneration is encouraged. Conventional intensive silviculture grows and harvests commercial tree species in essentially pure stands such as Douglas fir and pine on tree farms. The naturalistic method has been defined as the growth of selected mixed tree species, including hardwoods, in which the species are selected to match the ecology of the site. The last method, short-rotation silviculture, ie, short-rotation intensive culture (SRIC) or short-rotation woody crops (SRWC), has been suggested as the most suitable method for energy applications. In this technique, trees that grow quickly are harvested every few years, in contrast to once every 20 or more years. Fast-growing trees such as cottonwood, red alder, and aspen are intensively cultivated and mechanically harvested every 3 to 6 years when they are 3 to 6 m high and only a few centimeters in diameter. The young trees are converted into chips for further processing or direct fuel use and the small remaining stems or stumps form new sprouts by coppice growth and are intensively cultivated again. SRWC production affords dry yields of several tons of biomass per hm^2 annually without large energy inputs for fertilization, irrigation, cultivation, and harvesting.

Historically, trees have been important resources and still serve as significant energy resources in many developing countries. Several studies of temperate forests indicate productivities from about 9 to 28 t/hm^2·yr, while the corresponding yields of tropical forests are higher, ranging from about 20 to 50 t/hm^2·yr (72). These yields are obtained using conventional forestry methods over long periods of time, ie, 20 to 50 years or more. Productivity is initially low in a new forest, slowly increases for about the first 20 years, and then begins to decline. Coniferous forests will grow even in the winter months if the temperatures are not too low; they do not exhibit the yield fluctuations characteristic of deciduous forests.

One of the tree species studied in great detail as a renewable energy source is the *Eucalyptus* (73), an evergreen tree which belongs to the myrtle family, *Myrtaceae*. There are approximately 450 to 700 identifiable species of *Eucalyptus*. The *Eucalyptus* is a rapidly growing tree native to Australia and is a prime candidate for energy use because it reaches a size suitable for harvesting in about seven years. Several species have the ability to coppice, ie, resprout, after harvesting; as many as four harvests can be obtained from a single stump before replanting is necessary. In several South American countries, *Eucalyptus* trees are converted to charcoal and used as fuel. *Eucalyptus* wood has also been used to power integrated sawmill, wood distillation, and charcoal–iron plants in Western Australia. Several large areas of marginal land in the United States may be suitable for establishing *Eucalyptus* energy farms. These areas are in the western and central regions of California and the southeastern United States.

Various species and hybrids of the genus *Populus* are some of the more promising candidates for SRWC growth and harvesting as an energy resource (74). The group has long been cultivated in Europe and more recently in North America. Poplar hybrids are easily developed and the resulting progeny are propagated vegetatively using stem cuttings. Short-rotation growth of poplar hybrids has been reported by several investigators to afford yields of biomass that range as high as 112–202 green t/hm^2·yr (50–90 green short ton/acre·yr) (75). These results were reported with very high density plantings and selected clones; this type of tree growth has been termed woodgrass in which the tree crop is harvested several times each growing season in the same manner as perennial grasses. However, there is some dispute regarding the benefits of woodgrass growth vs SRWC growth (76).

It can be concluded from other studies that deciduous trees are preferred over conifers for the production of biofuels (77). Several species can be started readily from clones, resprout copiously and vigorously from their stumps at least five or six times without loss of vigor, and exhibit rapid initial growth. They also can be grown on sites with slopes as steep as 25%, where precipitation is 50 cm or more per year. It has been estimated that yields between about 18 and 22 dry t/hm^2·yr are possible on a sustained basis almost anywhere in the Eastern and Central time zones in the United States from deciduous trees grown in dense plantings. A representative list of deciduous trees judged to have desirable growth characteristics for methane plantations, and shown to grow satisfactorily at high planting densities on short and repeated harvest cycles, is available (77).

Grasses. Grasses are very abundant forms of biomass (78). About 400 genera and 6000 species are distributed all over the world and grow in all land habitats capable of supporting higher forms of plant life. Grasslands cover over one-half the continental United States; about two-thirds of this land is privately owned. Grass, as a family Gramineae, includes the great fruit crops, wheat, rice, corn, sugarcane, sorghum, millet, barley, and oats. Grass also includes the many species of sod crops that provide forage or pasturage for all types of farm animals. In the concept of grassland agriculture, grass also includes grass-related species such as the legumes family, ie, the clovers, alfalfas, and many others. Grasses are grown as farm crops, for decorative purposes, for preserving the balance of productive capacity of lands by crop rotation, for controlling erosion on sloping lands, for the protection of water sheds, and for the stabilization of arid areas. Many advances in grassland agriculture have been made since the 1940s through breeding and the use of improved species of grass, alone or in seeding mixtures; cultural practices, including amending the soil to promote herbage growth best suited for specific purposes; and the adoption of better harvesting and storage techniques. Until the mid-1980s, very little of this effort had been directed to energy applications. A few examples of energy applications of grasses can be found, ie, as the combustion of bagasse for steam and electric power, but many other opportunities exist that have not been developed.

Perennial grasses have been suggested as candidate raw materials for conversion to synfuels (77). Most perennial grasses can be grown vegetatively, and they reestablish themselves rapidly after harvesting. Also, more than one harvest can usually be obtained per year. Warm-season grasses are preferred over cool-season grasses because their growth increases rather than declines as the temperature rises to its maximum in the summer months. In certain areas, rainfall is adequate to permit harvesting every 3 to 4 weeks from late February into November, and yields between about 18 and 24 t/hm^2·yr of dry grasses may be obtainable in managed grasslands. Table 26 lists promising warm-season grasses proposed for conversion to synfuels.

Experimental work has shown that cool-season grasses such as Kentucky bluegrass and warm-season grasses such as Coastal Bermuda grass (*Cynodon dactylon*) can be converted to methane by conventional anaerobic digestion techniques (79,80). The compositions of some grasses indicate that fertilizatioin procedures can incorporate certain nutrients into the harvested grass so that they can be converted by biological means without the use of excessive chemical additions to the conversion units (80).

Sugarcane is used commercially as a combination food and fuel crop. A great deal of information has been com-

Table 26. Warm-Season Grass Species for Methane Plantations[a,b]

Species	Localities[c]	Comments
Perennial sorghums and their hybrids	Plains, South	Sudan grasses, Johnson grasses, and other warm-season hybrids are promising for localities with alkaline soils; several harvests per year
Bermuda grasses[d]	South and South Central states	Most promising of all warm-season grasses, especially for localities with acid soils; can be harvested several times per year
Related to sugarcane[e]	Louisiana and Florida	Limited suitable sites
Related to bamboo[f]	South Central United States	
Bahia grasses	Florida and southern coastal plains	Competes with Bermuda grasses when fertilized; effect on overall yield is in dispute

[a] Ref. 77.
[b] High annual yields in the range of 18–22 dry t/(hm$^2 \cdot$yr) unless otherwise noted.
[c] Regions in which species grow naturally, have been successfully introduced, or have been tested extensively.
[d] Coastal, midland, and Suwanne grasses.
[e] Very high annual yield up to 45 dry (t/(hm$^2 \cdot$yr) in specially suitable sites.
[f] Untested annual yield.

piled about sugarcane, and it might well be used as a model for other biomass energy systems. It grows rapidly, produces high yields, the fibrous bagasse is used as boiler fuel, and cane-derived ethanol is used as a motor fuel in gasoline blends, ie, gasohol. Sugarcane plantations and the associated sugar processing and ethanol plants are in reality biomass fuel farms. About one-half of the organic material in sugarcane is sugar and the other half is fiber. Dry cane yields per year have been reported to range as high as 80 to 85 t/hm^2 (36 to 38 short ton/acre). Normal cultivation of sugarcane provides dry annual sugarcane yields of about 50 to 59 t/hm^2 (22 to 26 short ton/acre) (13).

Other productive grasses given serious consideration as raw materials for the production of energy and synfuels include sorghum and their highbreds. Tropical grasses are very productive and normally yield 50 to 60 t of organic matter per hm^2 annually on good sites (13). The tropical fodder grass *Digitaria decumbens* has been grown at yields of organic matter as high as 85 t/hm$^2 \cdot$yr (38 short ton/acre·yr) (13).

There are many grasses and related plants that can be considered for energy applications because they have the desirable characteristics needed for land-based biomass energy systems.

Other Cultivated Crops. Other high yielding land biomass species have been proposed as renewable energy sources (81). Promising species are kenaf, *Hibiscus cannabinus*, an annual plant reproducing by seed only; sunflower, *Helianthus annuus L.*, which is an annual oil seed crop grown in several parts of North America; and a few others, such as the polyisoprene-containing plant species described previously. Kenaf is highly fibrous and exhibits rapid growth, high yields, and high cellulose content. It is a potential pulp crop and is several times more productive than the traditional pulpwood trees. Maximum economic growth usually occurs in less than 6 months, and consequently two croppings may be possible in certain regions of the United States. Without irrigation, heights of 4 to 5 m are average in Florida and Louisiana; 6-m plants have

been observed under near-optimum growing conditions. Yields as high as 45 t/hm$^2 \cdot$yr have been observed on experimental test plots in Florida, and it has been suggested that similar yields could be achieved in the Southwest with irrigation.

The sunflower is a prime candidate for biomass energy applications because of its rapid growth, wide adaptability, drought tolerance, short growing season, massive vegetative production, and adaptability to root harvesting. Dry yields have been projected to be as high as 34 t/hm^2 per growing season.

Water-Based Biomass

The average net annual productivities of dry organic matter on good growth sites for land- and water-based biomass are shown in Table 27. With the exception of phytoplankton, which generally has lower net productivities, aquatic biomass seems to exhibit higher net organic yields than land biomass. Water-based biomass considered to be the most suitable for energy applications include the unicellular and multicellular algae and water plants.

Algae. Unicellular algae, eg, the species *Chlorella* and *Scenedesmus*, have been produced by continuous processes in outdoor light at high photosynthesis efficiencies. *Chlorella*, for example, has been produced at a rate as high as 1.1 dry t/hm$^2 \cdot$day (82); this corresponds to an annual rate of 401 t/hm$^2 \cdot$yr. These figures are probably in error, but there is no theoretical reason why yields cannot achieve these high values because the process of producing algae can be almost totally controlled.

Algae production is not composed only of surface growth. Algae are produced as slurries in lakes and ponds, so the depth of the biomass-producing area as well as plant yield per unit volume of water are important parameters. The nutrients for algae production can be supplied by sewage and other wastewaters. It should be pointed out that most unicellular algae are grown in fresh water, which limits their energy applications to small-

Table 27. Annual Biomass Yields on Fertile Sites[a,b]

Dry Organics, t/(hm²·yr)	Climate	Ecosystem Type	Remarks
1	Arid	Desert	Better yield if hot and irrigated
2		Ocean phytoplankton	
2	Temperate	Lake phytoplankton	Little influence by humans
3		Coastal phytoplankton	Probably higher in some polluted estuaries
6	Temperate	Polluted lake phytoplankton	In agricultural and sewage runoffs
6	Temperate	Freshwater submerged macrophytes	
12	Temperate	Deciduous forest	
17	Tropical	Freshwater submerged macrophytes	
20	Temperate	Terrestrial herbs	Possibly higher yields if grazed
22	Temperate	Agriculture, annuals	
28	Temperate	Coniferous forests	
29	Temperate	Marine submerged macrophytes	
30	Temperate	Agriculture, perennials	
30		Salt marsh	
30	Tropical	Agriculture, annuals	Including perennials in continental climates
35	Tropical	Marine submerged macrophytes	Including coral reefs
38	Temperate	Reedswamp	
40	Subtropical	Cultivated algae	Better yield if CO_2 supplied
50	Tropical	Rain forest	
75	Tropical	Agriculture, perennials, reedswamp	

[a] Ref. 13.
[b] Average net values.

scale algae farms. The high water content of unicellular algae also limits the conversion processes to biological methods.

Macroscopic multicellular algae, or seaweeds, have been considered as renewable energy resources. Candidates include the giant brown kelp *Macrocystis pyrifera* (83–85), the red algae *Rhodophyta*, and the floating algae *Sargassum*. Giant brown kelp has been studied in detail and harvested commercially off the California coast for many years (65). Because of its high potassium content, it was used as a commercial source of potash during World War I; in the 1990s, organic gums and thickening agents and alginic acid derivatives are manufactured from it. Laminaria seaweed is harvested off the East Coast for the manufacture of alginic acid derivatives. In tropical seas not cooled by upwelled water, species of the *Sargassum* variety of algae may be suitable as renewable energy resources. Several species of *Sargassum* grow naturally around reefs surrounding the Hawaiian Islands. However, only a small amount of research has been done on *Sargassum* and little detailed information is available about this algae. A considerable amount of data on yields and growth requirements is available, however, on the *Macrocystis* and *Laminaria* varieties. The high water content of macroscopic algae suggests that biological conversion processes, rather than thermochemical conversion processes, should be used for synfuel manufacture. The manufacture of coproducts from macroscopic algae, such as polysaccharide derivatives, along with synfuel may make it feasible to use thermochemical processing techniques on intermediate process streams.

Water Plants. The productivity of some salt marshes is similar to that of seaweeds. *Spartina alterniflora* has been grown at net annual dry yields of about 33 t/hm²·yr (14.7 short ton/acre·yr), including underground material, on optimum sites (13). Other emergent communities in brackish water, including mangrove swamps, appear to have annual organic productivities of up to 35 t/hm²·yr (15.6 short ton/acre·yr) (13); insufficient information is available to judge their value in biomass energy systems. Freshwater swamps may be highly productive and offer opportunities for energy production. Both the reed *Arundo donax* and bulrush *Scirpus lacustris* appear to produce 57 to 59 t/hm²·yr (25.4–26.3 short ton/acre·yr) yields (13); if these could be sustained, they will be suitable candidates for biomass energy applications.

A strong candidate for energy applications is the water hyacinth, *Eichhornia crassipes* (3,86). This species of aquatic biomass is highly productive, grows in warm climates, and has submerged roots and aerial leaves like reedswamp plants. It is estimated that water hyacinth could be produced at rates up to about 150 t/hm²·yr (67 short ton/acre·yr) if the plants are grown in a good climate, the young plants always predominated, and the water surface was always completely covered (13). Some evidence has been obtained to support this growth rate (87,88). If it can be sustained on a steady-state basis, water hyacinth may be one of the best candidates as a nonfossil carbon source for synfuels manufacture. It has no competitive uses (ca 1992) and is considered to be an undesirable species on inland waterways. Many attempts have been made to rid navigable streams in Florida of water hyacinth without success; the plant is a very hardy, disease-resistant species (89).

SYSTEMS ANALYSIS

The overall design of an integrated biomass-to-synfuel system is very important to its successful operation. The system is large and requires coordination of many different operations, such as planting, growing, harvesting, transporting, and converting biomass to gaseous and liquid synfuels. The detailed design of a biomass-to-synfuel system depends on several parameters, such as the type, size, number, and location of the biomass growth and processing areas. In the ideal case, synfuel production plants are located in or near the biomass growth areas to minimize cost of transporting the harvested biomass to the plant. All nonfuel effluents are recycled to the growth areas as shown in Figure 19. This type of synfuel plantation, if developed, would be equivalent to an isolated system with inputs of solar radiation, air, carbon dioxide, and minimal water, and one output, synfuel. The nutrients are kept within the ideal system so that the addition of external fertilizers and chemicals is not necessary. Also, environmental and disposal problems are minimized.

Various modifications of the idealized design in Figure 19 can be conceived for large-scale usage. One modification might consist of the addition of wastewater influent into the biomass growth area and the growth of water hyacinth in the Southern United States for two purposes, ie, the treatment of wastewater by luxuriant uptake of nutrients and the conversion of water hyacinth to synfuels. In this case, inorganic material would build up in the biomass growth area so that the residual material from the conversion plant could be partially removed or bled from the system as the synfuel is produced. This product might be considered to be a coproduct along with the synfuel.

Alternatively, short-rotation hybrid poplar and selected grasses can be multicropped on an energy plantation in the U.S. Northwest and harvested for conversion to liquid transportation fuels and cogenerated power for on-site use in a centrally located conversion plant. The salable products are liquid biofuels and surplus steam and electric power. This type of design may be especially useful for larger land-based systems.

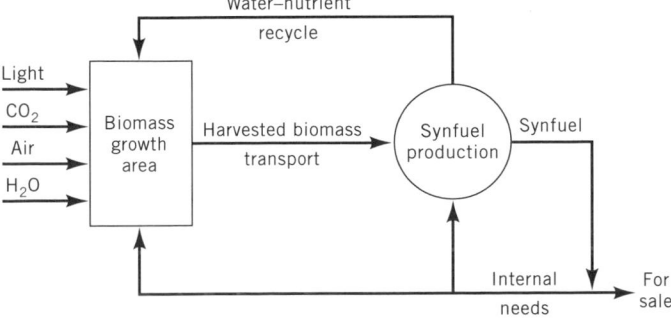

Figure 19. Idealized biomass-to-synfuel plantation system.

Another possibility, especially for small-scale farm use, is integration of agricultural crop, farm animal, and biofuel production into one system, eg, a farmer in the Midwest United States might grow corn as feedstock for a farm cooperative fuel ethanol plant. The residual distillers dried grains from the plant is used as hog feed, and the hog manure is used to generate medium heating value (MHV) fuel gas by anaerobic digestion. The fuel gas is used as plant fuel and the residual ungasified solids, which are high in nitrogen, potassium, and phosphorus, are recycled to the fields as fertilizer to grow more corn. The salable products are ethanol and hogs; the residuals are kept within the system.

Still another possibility is a marine biomass plantation such as that envisaged for giant brown kelp grown off the California Coast and conversion of the kelp to methane in a system similar to that shown in Figure 19. The location of the SNG plant could be either on a floating platform near the kelp growth area or located on shore, in which case the biomass or fuel transport requirements would be different.

Many different biomass energy system configurations are possible. As the technology is refined and developed to the point where commercialization activities are well under way, optimum designs will evolve. A great deal of attention has been given to the cost factors in the operation of biomass energy systems for the production of energy and biofuels. Of equal importance is the net energy production efficiency of the total system.

Economics

The practical value of biomass energy ultimately depends on the end-user costs of salable energy and biofuels. Consequently, many economic analyses have been performed on biomass production, conversion, and integrated biofuels systems. Conflicts abound when attempts are made to compare results developed by two or more groups for the same biofuels because methodologies are not the same. Technical assumptions made by each group are sometimes so different that valid comparisons cannot be made even when the same economic ground rules are employed. Comparative analyses, especially for hypothetical processes conducted by an individual or group of individuals working together, should be more indicative of the economic performance and ranking of groups of biofuels systems.

Several important generalizations can be made. The first is that fossil fuel prices are primary competition for biomass energy. Table 28 summarizes 1990 U.S. tabulations of average, consumption-weighted, delivered fossil fuel prices by end-use sector (90). The delivered price of a given fossil fuel is not the same to each end user; ie, the residential sector normally pays more for fuels than the other sectors, and large end users pay less.

In the context of biomass energy costs, dry, woody, and fibrous biomass species have an energy content of approximately 20 MJ/kg (8600 Btu/lb) or 20 GJ/t (17.2 MBtu/short ton). If such types of biomass were available at delivered costs of \$1.00/GJ (\$1.054/MBtu), or \$20.00/dry t (\$18.14/dry short ton), biomass on a strict energy content basis without conversion to biofuels would cost less than

Table 28. U.S. Delivered Fossil Fuel Prices to End Users, 1990, $/GJ[a]

Fossil Fuel	Residential	Commercial	Industrial	Transportation[b]	Utility Electricity[c]
Coal	2.87	1.52	1.60		1.38
Natural gas	5.34	4.45	2.79		2.20
Petroleum	8.43	5.65	5.32	7.90	3.24
LPG	10.38	8.17	5.12	8.03	
Kerosine	8.41	6.40	6.25		
Distillate fuel	7.60	5.79	5.39	8.03	
Motor gasoline		8.68	8.68	8.65	
Residual fuel		3.25	2.94	2.83	

[a] Ref. 90. All figures are consumption-weighted averages for all states in nominal dollars and include taxes.

[b] Aviation gasoline is delivered at $8.84/GJ; jet fuel at $5.39/GJ.

[c] Heavy oil, ie, grade nos. 4, 5, and 6, and residual fuel oils; light oils, ie, no. 2 heating oil, kerosine, and jet fuel; and petroleum coke are delivered at $3.13/GJ, $5.33/GJ, and $0.79/GJ, respectively.

most of the delivered fossil fuels listed in Table 28. The U.S. Department of Energy has set cost goals of delivered biomass energy drops at $1.90–2.13/GJ ($2.00–2.25/MBtu) (91) and $0.18/L ($0.67/gal) for fuel ethanol from biomass without subsidies in 2000, $0.22 to $0.26/L ($0.85 to $1.00/gal) by the year 2007 for biocrude-derived gasoline, $3.32/GJ ($3.50/MBtu) for methane from the anaerobic digestion of biomass by the year 2000, and 4.5 cents/kWh for electricity from biomass by the late 1990s (92).

An economic analysis of the delivered costs of biomass energy in 1990 dollars has been performed (ca 1992) for herbaceous and woody biomass for different regions of the United States (91). The analysis was done for each decade from 1990 to 2030 for Class I and II lands; results for biomass grown on Class II lands for the years 1990 and 2030 are shown in Table 29. Estimates of the total pro-

duction costs for biomass were calculated with discounted cash flow models, one for the herbaceous crops switchgrass, napier grass, and sorghum, and one for the short-rotation production of sycamore and hybrid poplar trees. The delivered costs are shown in Table 29 in 1990 $/dry t and 1990 $/MJ and are tabulated by region and biomass species. The yield figures for 1990 were obtained from the literature and the projected yields for 2030 were assumed achievable from continued research. The annual, dry biomass yields per unit area have a great influence on the final estimated costs. This analysis indicates that the lowest cost energy crop of those chosen may be different for different regions of the country. A few of the biomass-region combinations appear to come close to providing delivered biomass energy near the U.S. Department of Energy cost goal. Realizing that there are many differences in the methodologies and assumptions used to compile the 1990

Table 29. Projected Delivered Costs for Candidate Biomass Energy Crops in 1990 and 2030[a,b]

Region and Species	1990 Assumed Yield[c]	1990 Delivered Cost $/t	1990 Delivered Cost $/GJ	2030 Assumed Yield[c]	2030 Delivered Cost $/t	2030 Delivered Cost $/GJ
Great Lakes						
Switchgrass	7.6	104.07	5.26	15.5	61.32	3.60
Energy sorghum	15.5	62.56	3.17	30.9	36.79	2.16
Hybrid poplar	10.1	113.79	5.76	15.9	72.82	4.29
Southeast						
Switchgrass	7.6	105.89	5.36	17.3	52.91	3.11
Napier grass	13.9	63.72	3.22	30.9	33.31	1.96
Sycamore	8.1	88.61	4.49	14.3	53.19	3.13
Great Plains						
Switchgrass	5.4	74.32	3.77	10.3	44.05	2.59
Energy sorghum	6.3	91.73	4.65	13.7	48.07	2.83
Northeast						
Hybrid poplar	8.1	105.26	5.33	11.9	71.69	4.26
Pacific Northwest						
Hybrid poplar	15.5	66.69	3.56	23.8	44.73	2.63

[a] Ref. 91. Discounted cash-flow models account for use of capital, working capital, income taxes, time value of money, and operating expenses. Real after-tax return assumed to be 12.0%. Short-rotation model used for sycamore and poplar. Herbaceous model used for other species. Costs in 1990 dollars. Dry tons.

[b] Yields in 1990 obtained from literature on Class II lands. Average total field yields are for entire region on prime to good soil, less harvesting and storage losses. Yields in 2030 assumed to be attained through research and genetic improvements. Short-rotation woody crops (hybrid poplar and sycamore) grown on six-year rotations on six independent plots. Net income is negative for first five years for each plot.

[c] Yield in t/hm$^2 \cdot$hr.

costs for delivered fossil fuels in Table 28 and delivered biomass energy in Table 29, it is evident that many of the biomass energy costs are competitive with those of fossil fuels in several end-use sectors, even without incorporating yield improvements that are expected to evolve from continued research on biomass energy crops.

It is essential to recognize several other factors, in addition to the cost of virgin biomass and its conversion to biofuels, when considering whether the costs of biomass energy are competitive with the costs of other energy resources and fuels. Some potential biomass energy feedstocks have negative values; ie, waste biomass of several types such as municipal biosolids, municipal solid wastes, and certain industrial and commercial wastes must be disposed of at additional cost by environmentally acceptable methods. Many generators of these wastes will pay a service company for removing and disposing of the wastes, and many of the generators will undertake the task on their own. These kinds of feedstocks often provide an additional economic benefit and revenue stream that can help support commercial use of biomass energy.

Another factor is the potential economic benefit that may be realized due to possible future environmental regulations from utilizing both waste and virgin biomass as energy resources. Carbon taxes imposed on the use of fossil fuels in the United States to help reduce undesirable automobile and power plant emissions to the atmosphere would provide additional economic incentives to stimulate development of new biomass energy systems. Certain tax credits and subsidies are already available for commercial use of specific types of biomass energy systems (93).

Energetics

The net energy production efficiency of an integrated biomass energy system is extremely important to its development and practical use. The ultimate goal is to design and operate environmentally acceptable systems to produce new supplies of salable energy whether they be low heat value gas, substitute natural gas, substitute gasolines or diesel fuels, methanol, ethanol, hydrogen, or electric power from biomass at the lowest possible cost and energy consumption. It is necessary to quantify how much energy is expended and how much salable energy is produced in each fully integrated system. An energy budget similar to an economic budget should be prepared because the capital, operating, and salable energy cost projections and the conversion process efficiency are insufficient to choose and design the best systems. These values do not necessarily correlate with net energy production. Also, the capital energy investment consumed during construction of the system should be recovered during its operation. Comparative analyses of similar systems for production of synthetic liquid and gaseous fuels from the same feedstock or of different systems that yield the same fuels from different biomass should be performed by consideration of the economics and the net energetics.

One method of analyzing net energetics of a biomass energy system is to let E_f, E_x, and E_p represent the energy content of the dry biomass feed, E_f, the sum of the external nonsolar energy inputs into the total system, E_x, and the energy content of the salable fuel products, E_p. The ratio $(E_p - E_x)/E_x$, which can be termed the net energy production ratio, indicates how much more, or less, salable fuel energy is produced than that consumed in the integrated system if the external energy consumed is replaced and it is assumed that the biomass feed energy is zero. This is a reasonable assumption because the energy value of biomass is derived essentially 100% from solar radiation. Net energy production ratios greater than zero indicate that an amount of energy equivalent to the sum of the external nonsolar energy inputs and an additional energy increment of salable fuel are produced; the larger the ratio, the larger the increment. The ratio $100E_p/(E_f + E_x)$ is the overall fuel production efficiency of the system or simply the energy output divided by the gross energy input. Finally, another useful ratio is the value of net energy output divided by the gross energy out, $100(E_p - E_x)/E_p$. This ratio expresses the percentage of the energy output that is new energy added to the economy.

A simple model for biomass energy production, excluding conversion to synfuels, is illustrated in Table 30 (94) which presents the results of an analysis of a short-rotation tree plantation that produces dry biomass at a yield of 11.2 t/hm²·yr. The principal sources of energy consumption are fertilization, which includes the energy cost of ammonia production, and biomass transportation. The important result of this study is that the net energy production efficiency is high and that about 14 times more energy is produced in the form of woody biomass than the external nonsolar energy inputs needed to operate the system.

A mixed biomass plantation, including species of short-rotation hardwoods, sunflower, and kenaf, is projected to produce dry biomass at a yield of 67.3 t/hm²·yr (95).

Table 30. Net Energy Analysis of Short-Rotation Wood Biomass Production[a,b]

Operation	Energy Consumed per Dry Wood Produced, MJ/t[c]	Total Consumption, %
Cultivation and planting	9.3	0.8
Fertilization	604	52.1
Harvesting	87	7.5
Transport to trucks	115	9.9
Load trucks	55	4.7
80.5-km transport to user	221	19.0
Unload	nil	
Auxiliary	69	5.9
Total	*1160*	*99.9*
Biomass energy produced, GJ/(hm²·yr)[d]	196[e]	
Energy production efficiency, %	93.4[f]	
Net energy production ratio	14.1[g]	

[a] Ref. 94.
[b] Yield is 11.2 t/(hm²·yr) dry; 20-yr planting cycle; fertilization is 224 kg N/(hm²·yr).
[c] To convert MJ/t to Btu/1000 lb, multiply by 430.
[d] To convert GJ to 10⁶ Btu, divide by 1.054.
[e] Assumes heating value of biomass is 17.5 MJ/dry kg.
[f] (Net energy output ÷ gross energy output) × 100, ie, $100(E_p - E_x)/E_p$.
[g] Net energy output ÷ energy consumed, ie, $(E_p - E_x)/E_x$.

Table 31. Net Energy Production Ratios for Ethanol Production from Corn in Integrated System[a]

Salable Energy Products, E_p	Nonfeed Energy Inputs, E_x	N[b]
Alcohol	Corn production, fermentation, bottoms drying	−0.65
Alcohol	Corn production, fermentation	−0.51
Alcohol, chemicals	Corn production, fermentation	−0.50
Alcohol, chemicals, cattle feed	Corn production, fermentation, bottoms drying	−0.44
Alcohol, chemicals, cattle feed	Corn production, fermentation, bottoms drying, 50% residuals[c]	−0.10
Alcohol, chemicals, cattle feed	Corn production, fermentation, bottoms drying, 75% residuals[c]	0.29
Alcohol	Corn production, fermentation, 75% residuals[c]	1.43
Alcohol, chemicals	Corn production, fermentation, 75% residuals[c]	1.47

[a] Ref. 96.
[b] $N = (E_p - E_x)/E_x$.
[c] Percent of cobs and stalks collected and used as fuel within system to replace fossil fuel inputs.

Again, fertilization consumes the largest amount of external energy, but the energy cost for irrigation is also high. The net energy production efficiency is high; about 20 times more energy is produced as biomass energy than the external nonsolar energy inputs.

These systems analyses suggest that biomass plantations can be designed to operate at high net energy efficiencies, and that further improvements might best be incorporated in the systems by concentrating on fertilization. The use of nitrogen-fixing biomass and the recycling of nutrients from the conversion facilities may offer additional benefits.

For the total integrated biomass production–conversion system, the arithmetic product of the efficiencies of biomass production and conversion is the efficiency of the overall system. An overall conversion efficiency near 45% would thus be produced by integrating the biomass plantation illustrated in Table 30 with a conversion process that operated at an overall efficiency of 50%. Every operation in the series is thus equally important.

These simplified treatments correspond to net energy analyses using the First Law of Thermodynamics. Some energy analysts feel that only an analysis based on the Second Law can provide the ultimate answers in terms of

where more available energy, in the thermodynamic sense, can be found to permit efficiency maximization. Others believe that the conventional energy balance is optimal because it is more realistic and easier to use. Indeed, for integrated synfuel production systems, entropic losses may not always be definable for all segments of the system, and a rigorous Second Law analysis may not be possible.

Location of the system boundaries also is important in the net energy analysis of integrated biomass energy systems. Thus tractors may be used to plant and harvest biomass. The fuel requirements of the tractors are certainly part of E_x, but is the energy expended in manufacturing the tractors also part of E_x? Some analysts believe that a complete study should trace all materials of construction and fossil fuels used back to their original locations in the ground.

Another important factor in net energy analysis concerns energy credits taken for by-products; they have an important effect and can determine whether the net energy production efficiency is positive or negative. An example of this effect is shown in Table 31 (96), which was derived from the integrated alcohol-from-corn production system illustrated in Figure 20. The boundary of the sys-

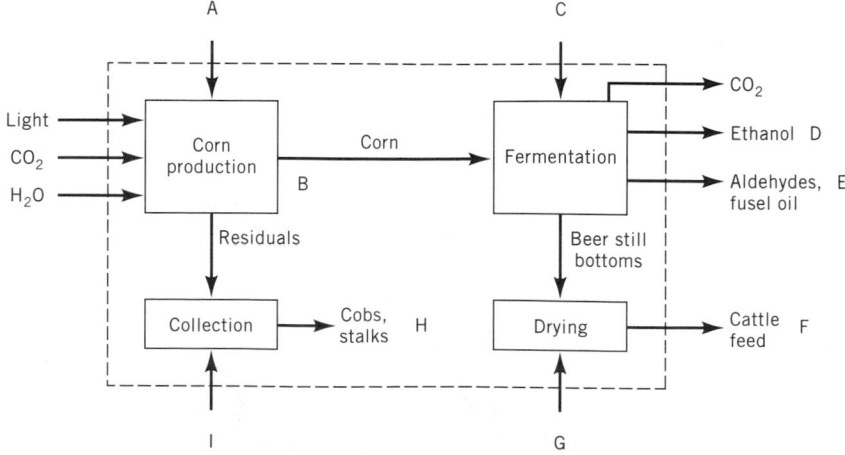

Figure 20. Energy inputs and outputs to manufacture 3.785 L of anhydrous ethanol from corn. (———) denotes system boundary. All KJ figures are lower heating value (LHV). A, 48,469 KJ; B, 128,314 KJ; C, 113,748 KJ; D, 79,682 KJ; E, 1,138 KJ; F, 47,483 KJ; G, 66,634 KJ; H, 174,768 KJ; I, 1,644 KJ. To convert KJ to Btu, divide by 1.054.

tem depicted in Figure 20 circumscribes all the operations necessary to grow and harvest the corn, to collect residual cobs and stalks if they are used as fuel, to operate the fermentation plant for the production of anhydrous alcohol and by-product chemicals, and to dry the stillage to produce distillers' dried grains plus solubles for sale as cattle feed. The capital energy investment in the system is not included within the boundary.

Various net energy production ratios were calculated as shown in Table 31. It is apparent that the ratio can be either positive or negative, depending on whether credit is taken for the by-product chemicals and cattle feed, whether the energy for drying the stillage is included as an input to the system, and whether a portion of the residual corn cobs and stalks is collected and used as fuel within the system to replace fossil fuel inputs. To permit comparisons, net energy analyses must be clearly specified as to all details. On the whole, net energy production in a modern corn-to-ethanol plant would appear to be borderline if petroleum fuels comprise a significant part of the nonfeed energy inputs. However, to improve net energy production, these fuels can be replaced by fuels generated within the system or by renewable fuels, and credit can be taken for the by-products. Another route is to use more of the corn plant or the stillage as feedstock to the fermentation plant. The use of the grain alone as feedstock places a severe limitation on net energy production. Some of the information available on designs of corn wet-milling plants in which corn oil and other products are produced indicate that integration with an alcohol plant may be more efficient than a conventional corn alcohol plant.

COMMERCIAL USE OF BIOMASS ENERGY IN THE UNITED STATES

Relatively few biomass energy technologies are in commercial use in 1992. With two possible exceptions, ie, a few small tree plantations, and fuel ethanol from corn, sugarcane, and sugar beet, there is no biomass species grown in the United States specifically for conversion to biofuels. But conversion processes in commercial use in the early 1990s still span the basic technologies of combustion, gasification, and liquefaction. These include combustion of wood, wood wastes, and forestry residues; combustion of agricultural residues such as rice husks and bagasse; combustion of MSW and RDF for simultaneous disposal and energy recovery; biological gasification of animal manures in farm- and feedlot-scale anaerobic digestion systems for simultaneous waste disposal and production of biogas as well as upgraded solids for feed, fertilizer, and animal bedding; biological gasification of municipal wastewater sewage by anaerobic digestion for simultaneous waste disposal and biogas production; biological gasification of MSW in sanitary landfills and recovery of biogas for fuel, which also mitigates environmental and safety problems caused by gas migration in the landfill; thermal gasification of biomass for LHV gas production for on-site use; and alcoholic fermentation of starchy and sugar crops for fuel ethanol for use as a fuel extender and octane enhancer in motor gasoline blends.

Most of these commercial processes have been in use for many years. Some were greatly improved in the early 1990s, such as alcoholic fermentation that has undergone large process steam requirement reductions, thereby increasing net energy production efficiencies of fuel ethanol. Other commercial processes are relatively new, such as two-phase anaerobic digestion that permits higher plant capacities at low capital costs and the production of higher methane-content biogas at higher rates.

Inventories of commercial usage in the United States are available (97,98) Table 32 offers a summary (2). Although most of the data available does not refer to a specific time or year, wood use as a fuel in the industrial and residential sectors is responsible for the largest portion of biofuels consumption in the United States. Those states that have large forest products industries are the principal wood energy users. Similarly, states in the Corn Belt are the largest fuel ethanol producers. With few exceptions, those states having the most populated cities tend to process more municipal solid wastes by simultaneous disposal-energy recovery technologies. The biomass energy industry covers the entire nation; not one state is

Table 32. Biofuels Utilization and Production and Biomass-Fueled Electric Power Plant Capacities in the United States[a]

Utilization	United States, Total	Energy Equivalents,[b] 10^{12} kJ/yr
Wood fuel utilized		
Commercial, 10^6 t/yr	83	1,553
Residential, 10^3 m³/yr	198,496	1,154
Agricultural wastes utilized as fuels, 10^6 t/yr	9.5	142
Fuel ethanol produced, 10^6 L/yr	2,542	53.5
Biogas produced, 10^6 m³/yr	9.3	0.2
MSW converted to energy,[c] 10^6 t/yr	14.0	113
Biomass-fueled electric plant capacity, MW	5,154	

[a] Ref. 2. The indicated biofuels consumption and capacity figures are the estimated values for various time periods since 1985 and do not refer to a specific year.

[b] The energy equivalents were calculated by the author using the following assumptions for Btu values: HHV of commercial wood fuel is 16×10^6 Btu/t, HHV of residential wood fuel is 20×10^6 Btu/cord, HHV of agricultural waste is 12.9×10^6 Btu/t, HHV of ethanol is 75,500 Btu/gal, HHV of biogas is 500 Btu/cf, HHV of MSW is 7×10^6 Btu/t, HHV of 1.00 barrel of oil equivalent (BOE) is 5.8×10^6 Btu.

[c] Ref. 98. These values are from 1987 and 1988. Energy products are steam and/or electric power.

devoid of commercial biofuels production or utilization. The practical limitations to transport distance of some biomass such as wood, and the requirement for nearby or local processing, correlates with the concentration of biomass energy-processing facilities by state.

Combustion

The Public Utility Regulatory Policies Act of 1978 (PURPA), which provides benefits to cogenerators and small power producers, has stimulated commercialization of biomass combustion for electric power production. To be eligible for benefits under PURPA, small power production systems are limited to 80 MW and must receive 75% or more of their total energy input from renewables. Cogeneration systems do not have these limitations. Emissions limitations must be taken into consideration in developing commercial biofuels combustion projects. The regulations apply in a rather complex fashion for boilers burning either wood or municipal solid wastes alone or in combination with fossil fuels (99); even wood-burning stoves must meet national standards. With few exceptions, all stoves built after July 1988 must comply with a strict set of environmental regulations established by the U.S. Environmental Protection Agency and some states (100).

Wood Fuels

Wood fuels are the largest contributor to biomass energy usage. Approximately 88% of the total is attributed to wood energy consumption. Distribution of the annual consumption of wood fuel by region and sector from 1980 to 1984 is available (101); the South and the industrial sector are the largest wood energy consumers. Table 33 presents the composition of wood fuel by resource type from 1972 to 1987 (102). The largest concentration of wood fuel usage is in the lumber and wood products industry and the pulpwood and paper industry. It has been reported that the pulp and paper industry meets about 70% of its own fuel needs with wood energy (103). Over the period covered in Table 33, the pulp and paper industry utilized about one-half of total wood energy consumption, primarily as black liquor. This lignin-containing material is a by-product of the pulping process and is not wood as such. It is noteworthy that commercial, utility, and other indus-trial usage of wood fuels is a very small part of total consumption, and that residential usage more than doubled over the time period 1972–1984, whereas total wood energy usage increased about 61%.

The increase in residential fuelwood consumption over this period parallels the sharp increase in costs of oil, natural gas, and electricity and can be tracked by the number of wood-burning stoves in homes. Between 1950 and 1973, the estimated number of stoves dropped from 7.3 million to 2.6 million, but grew to an estimated 11 to 14 million in 1981 (104). The trend in the wood-burning stove inventory suggests that a four-fold increase in residential fuelwood use may have occurred during the 1970s and 1980s, which can be explained in part by heavy rural wood burning (105). However, since attaining about a one EJ level of consumption in 1984, there has been little gain in total residential wood energy use (106). Steady gains in industrial use have been counterbalanced by a decrease in residential use (106).

The data in Table 33 indicate that there is a small but relatively steady increase in total wood energy usage; that the industries having captive sources of wood or wood-derived products consume the bulk of wood energy, although at a generally flat rate; that there appears to be much opportunity for growth in wood energy consumption in the utility and commercial sectors; and that residential fuelwood usage has shown the largest incremental growth until the late 1980s.

Municipal Solid Waste

In the early 1990s, the need to dispose of municipal solid waste (MSW) in U.S. cities has created a biofuels industry because there is little or no other recourse (107). Landfills and garbage dumps are being phased out in many communities. Combustion of MSW, ie, mass-burn systems, and RDF, ie, refuse-derived fuel, has become an established waste disposal–energy recovery industry.

In May 1988 the United States had 105 operating MSW-to-energy plants, 29 plants under construction, 61 plants in the advanced planning stage, and 5 plants temporarily shutdown (98). About one-half of the 105 operating plants had been placed in operation since 1985, and about 80% of all operating plants use mass-burn, modular technology. Mass-burn, waterwall designs predominated for plants under construction or in the ad-

Table 33. Wood Energy Use, 10^{12} kJ

Sector	1972	1978	1982	1986	1987
Lumber and wood products industry	373.1	444.7	362.5	474.3	484.8
Pulp and paper industry					
As hog fuel	473.2	104.0	178.1	268.7	271.9
As bark	94.5	99.3	116.9	139.1	142.3
As black liquor	737.8	854.8	811.5	949.6	1001.3
Other industry	38.9	47.4	45.3	51.6	51.6
Residential	420.5	689.3	985.4	927.5	885.3
Commercial	9.5	15.8	23.2	23.1	23.2
Utilities	3.2	1.1	3.2	9.4	9.5
Total	2150.7	2256.7	2526.7	2843.3	2869.9

[a] Refs. 2 and 102.

vanced planning stage. The sum total design capacity of these MSW-to-energy plants and corresponded to about 30% of the total MSW generated in the United States. A 1988 inventory of 25 MSW-to-energy plants that have been permanently shutdown shows the technologies for these plants consist of both mass-burn and RDF combustion systems, as well as a few pyrolysis plants. They have a combined total design capacity of 13,500 t/d, and most were built in the 1970s (98). The majority of these plants had operating difficulties caused by equipment and environmental concerns as well as cost factors. Problems that can cause permanent shutdown do not appear to be prevalent in more modern designs, presumably because combustion and waste-processing equipment have been improved.

According to the Solid Waste Association of North America, by 1992 there tense 137 municipal solid waste-to-energy plants operating in 36 states. They process about 16% of the 185 million tons of solid waste generated in the United States, and produce the equivalent of more than 2,300 MW of electricity. Nearly 100 other waste-to-energy projects are in various stages of planning or implementation. The U.S. Environmental Protection Agency has estimated that there will be more than 300 waste-to-energy plants in operation in the United States by 2000; these will process about 25% of U.S. municipal solid waste.

Power Production

PURPA has stimulated commercial use of biomass combustion for electric power production. It has prompted many companies to build and operate small power plants fueled with fossil and nonfossil energy resources. The power is used on-site in many cases, and the surplus is injected into the grid for sale to the utility at avoided cost. The number of U.S. filings as of December 31, 1987, submitted to the Federal Energy Regulatory Commission (FERC) for biomass-fueled and all cogeneration and small power production facilities illustrates the phenomenal growth of this industry from fiscal 1980 to 1988 (Table 34) (108,109). Of the total number of 1730 cogeneration filings that have been qualified by FERC as eligible for PURPA benefits, 138 (8.0% of total) are biomass-fueled and have a capacity of about 2699 MW, ie, 6.0% of total. Similarly, of the 1987 small power production filings that have been qualified by FERC, 468 (23.6% of total) are biomass-fueled and have a capacity of about 6260 MW, ie, 36.8% of total. The number of filings qualified by FERC and the total electric capacity for each biomass type are: 231 and 3703.8 MW of wood wastes, 140 and 654.951 MW for biogas, 131 and 2546.186 MW for MSW, and 108 and 1719.01 MW for agricultural wastes (109).

Data on the installed electric generating capacity and generation in 1986 by nonutility and utility power producers as of December 31, 1986 are available (110,111). Nonutility generators accounted for 3.45% of total U.S. capacity (25.3×10^3 MW) (73.3×10^4 MW) in 1986 and generated 4.31% (112×10^6 MWh) of the power produced in the same year. Biomass-fueled electric capacity and generation was 19.2% (4.9×10^3 MW) and 21.2% (23.7×10^6 MWh) respectively, of total nonutility capacity and generation. Biomass-fueled capacity experienced a 16% increase in 1986 over 1985, the same as natural gas, but it was not possible to determine the percentage of the total power production that was sold to the electric utilities and used on-site. Total production should be substantially more than the excess sold to the electric utilities. Overall, the chemical, paper, and lumber industries accounted for over one-half of the total nonutility capacity in 1986, and three states accounted for 45% of total nonutility generation, ie, Texas, 26% of total; California, 12% of total; and Louisiana, 7% of total. There were 2449 nonutility producers with operating facilities in 1986, a 15.8% increase over 1985, 75% of whose capacity was interconnected to electric utility systems.

Installed nonutility electric generation capacity and generation for each biofuel is presented in Table 35 for

Table 34. Biomass-fueled Cogeneration and Small Power-production Capacities and Facilities, kW

Facility	1980	1985	1988	Biomass Total	Qualified Total All[c]
			Cogeneration		
New	400(1)	383,003 (23)	13,685 (3)	1,617,390 (122)	41,947,273 (1,705)
Existing		115,000 (2)		570,497 (18)	3,341,629 (78)
Both[d]	161,000 (1)	88,000 (1)		615,000 (0)	1,784,543 (33)
Total	161,400 (2)	586,003 (26)	13,685 (3)	2,802,887 (149)	47,073,445 (1,816)
Qualified[c]	161,400 (2)	585,503 (24)	725 (2)	2,698,802 (138)	44,943,616 (1,730)
			Small power production		
New	50 (1)	1,078,690 (112)	339,611 (19)	6,219,125 (487)	16,869,289 (1,964)
Existing		−800 (2)		116,925 (9)	319,099 (62)
Both[d]				11,500 (2)	136,485 (15)
Total	50 (1)	1,077,890 (114)	339,611 (19)	6,347,550 (498)	17,324,873 (2,041)
Qualified[c]	50 (1)	1,075,490 (110)	339,611 (19)	6,260,309 (468)	17,006,761 (1,987)

[a] Refs. 108 and 109. Filings under Public Utilities Regulation Policies Act of 1978. Totals are for years 1980–1988.

[b] Number of facilities in parentheses.

[c] Qualiied for PURPA benefits, ie, only owners or operators of facilities who claim qualifying status for PURPA benefits would make filings, some filings are submitted after the facilities begin operation, FERC does not review notices of qualifying status and has not completed review of all listed applications for certification, and the data provided in the filings are not verified by FERC inspection of the facilities.

[d] Combination of existing and new incremental capacity.

Table 35. Installed Nonutility Electricity Generation Capacity and Generation by Biofuel, 1986[a,b]

Biofuel	Cogeneration		Small Power Producers		Total	
	Capacity, MW	Generation, MWh	Capacity, MW	Generation, MWh	Capacity, MW	Generation, MWh
Wood	3,119.8	16,650,778	624.7	2,403,718	3,744.5	19,054,496
Agricultural wastes	252.0	1,022,573	68.2	310,387	320.2	1,322,960
Municipal solid waste	75.4	217,599	463.9	2,198,941	539.3	2,416,540
Landfill gas	0	0	184.6	622,031	184.6	622,031
Digester gas	21.2	117,146	49.6	186,750	70.8	303,896
Total	*3,468.4*	*18,008,096*	*1,391.0*	*5,721,827*	*4,859.4*	*23,729,923*

[a] Ref. 110.

[b] Total number of facilities reported in early 1985 was 111 in operation, 50 under construction, and 72 in the planning stages (112, 113).

1986. Wood was the largest contributor in 1986 to total capacity and generation, followed in decreasing order by MSW, agricultural wastes, landfill gas, and digester gas. The incremental changes in capacity and generation between 1985 and 1986 were 11.5% and 1.7% for wood, 7.8% and 15.9% for agricultural wastes, 30.4% and 9.2% for MSW, 184% and 207% for landfill gas, and 7.4% and 27.0% for digester gas. All incremental changes were positive.

Utility production of biomass-fueled electric power is much less than nonutility production. In early 1985, there were only 18 facilities having a total capacity of 245 MW, ie, nine fueled with wood (180.7 MW), five fueled with MSW (33.8 MW), two fueled with agricultural residues (22.5 MW), and two fueled with digester gas (8 MW) (112,113). The largest was a 50-MW plant in Burlington, Vermont (114).

Gasification

Conversion of biomass to gaseous fuels can be accomplished by several methods; only two are used by the biomass energy industry (ca 1992). One is thermal gasification in which LHV gas, ie, producer gas, is produced. The other process is anaerobic digestion, which yields an MHV gas.

Thermal Gasification. A survey of commercial gasifiers in use and under construction in 1983 indicates that the commercialization rate was low, but that several process developers and vendors have installed 30 to 35 operating systems (115). Feed rates of biomass range from 0.1 to 13.6 t/h, LHV gas output ranges from 1.1 to 211 billion J/h (1.0 to 200 million Btu/h) for a wide range of reactor configurations, and the gas is used for several different applications. Several large U.S. plants have been shutdown because of operating difficulties, eg, plants in Baltimore, Maryland; Orlando, Florida; and El Cajon, California.

By 1993, many U.S. gasifier vendors had gone out of business or were focusing marketing activities overseas or on other conversion technologies, particularly combustion for power generation, in states where combined federal and state incentives make economic factors attractive. Some existing gasification installations also have been shutdown and placed in a stand-by mode until natural gas prices make biomass gasification competitive again.

A survey of commercial thermal gasification in the United States shows that few gasifiers have been installed since 1984 (115). Most units in use are retrofitted to small boilers, dryers, and kilns. The majority of existing units operate at 0.14 to 1.0 t/h of wood wastes on updraft moving grates. The results of this survey are summarized in Table 36. Assuming all 35 of these units are operated continuously, extremely unlikely, the maximum amount of LHV gas that can be produced is about 0.003 to 0.006 EJ/yr (222–445 m³/d).

Biological Gasification. Several surveys have been performed on the use of anaerobic digestion for biogas production and waste stabilization. In 1984, 84 farm-scale and industrial anaerobic digestion facilities, exclusive of municipal wastewater treatment plants, were identified in 32 states and were estimated to have a total reactor volume of 90,600 m³ (117). Individual digesters were found to range in size up to 13,000 m³, but the majority had volumes in the range 249 to 750 m³. A comprehensive inventory was conducted in 1985 (118). Exclusive of municipal wastewater treatment, 96 farm-scale and industrial anaerobic digestion systems were found to have been built between 1972 and mid-1985 in 30 states; in Puerto Rico, 2 units. Eighty-seven of the systems had digestion

Table 36. Commercial Thermochemical Gasifiers, June 1988[a]

Size, 10⁶ kJ/h[b]	Number of Gasifiers	Feedstock	Gas Use
Updraft reactor			
6.0	2	Corn cobs	Corn dryer
4.9–25.0	4	Wood	Space or process heat
0.94–25.9	14	Wood	Dry kiln, space heat
12.0–69.9	3	Wood	Brick kilns
Downdraft reactor			
0.2–6.0	5	Wood, peach pits	Greenhouse
10.0	2	Wood	Power boiler
Fluid-bed reactor			
25.0	2	Rice hulls	Process heat
82.2	1	Wood	Power boiler
124.4	2	Wood	Clay dryers

[a] Ref. 116. All units are LHV gasifiers.

[b] To convert 10⁶ kJ/h to 10⁶ Btu/h, divide by 1.054.

volumes of 100 m³ or more; 60 were operational, 7 were temporarily shutdown, and others were in various stages of design or development. Forty-four of the operational or shutdown systems, ie, 43.900 m³, were used for digesting animal manures, 35 for dairy or beef cattle manure, and the remainder for swine or poultry manure. Fourteen of the facilities, ie, 107,900 m³, provided wastewater cleanup services to agricultural product processing plants, breweries, and related food production facilities. The designs that were used extensively for the farm-scale digesters were plug flow and stirred tank configurations and unheated lagoons. Only a few additional commercial digestion systems have been installed since this inventory was completed.

An estimate of the potential methane production possible from existing (ca 1992) municipal wastewater treatment plants that produce and use biogas as a fuel and from the farm-scale and industrial anaerobic digestion plants identified in the inventory is presented in Table 37. The maximum amount of methane that could be produced from this commercial anaerobic digestion capacity under conventional operating conditions is about 0.005 EJ/yr (2,400 BOE/d).

Gas production is considerably greater from commercial landfill methane recovery systems. In 1987, 94 plants (50 operational, 44 scheduled) had an estimated design production of 1.2×10^6 m³/d and an estimated actual production of 0.314×10^6 m³/d, or 114×10^6 m³/yr; estimated electric capacity was 231.2 MW (120–123).

The initial biogas recovered is an MHV gas and is often upgraded to high heat value (HHV) gas when used for blending with natural gas supplies. The annual production of HHV gas in 1987, produced by 11 HHV gasification facilities, was 116×10^6 m³ of pipeline-quality gas, ie, 0.004 EJ (121). This is an increase from the 1980 production of 11.3×10^6 m³. Another 38 landfill gas recovery plants produced an estimated 218×10^6 m³ of MHV gas, ie, 0.005 EJ. Additions to production can be expected because of landfill recovery sites that have been identified as suitable for methane recovery. In 1988, there were 51 sites in preliminary evaluation and 42 sites were proposed as potential sites (121).

Liquefaction

Since the 1970s attempts have been made to commercialize biomass pyrolysis for combined waste disposal–liquid fuels production. None of these plants were in use in 1992 because of operating difficulties and economic factors;

only one type of biomass liquefaction process, alcoholic fermentation for ethanol, is used commercially for the production of liquid fuels.

Fermentation ethanol, primarily from corn, but also from sugarcane, sugar beet, or derivatives, has shown extraordinarily high production rate increases since 1979 when it was reintroduced in the United States as a blending component in motor fuels. In 1979, 24 operating plants, with a design capacity of 151×10^6 L/yr, produced 75.7×10^6 L/yr; 35 additional plants were planned. By 1988, the number of plants had increased to 55, with a design capacity of 3743×10^6 L/yr and an actual production of 3160×10^6 L/yr; 70 additional plants were shut down, with total unused design capacity of 1400×10^6 L/yr (124). Tax incentives provided by the federal and state governments coupled with generally high gasoline prices, low corn prices, and the phase-out of leaded fuels, have helped establish the fuel ethanol industry.

Since most fuel ethanol manufactured in the United States is made from corn, its price plays a crucial role in determining the competitive position of ethanol in an open market. With corn priced at about $2.50/bu, the embedded feedstock cost of product ethanol is about $0.14–0.23/L ($0.52–0.87 gal), depending on overall yield and by-products ignored (125). Fuel ethanol plants may have contingency plans to close if corn prices rise to a certain level, eg, $3.50/bu or above (126).

A listing of fuel ethanol plants in operation, with total anhydrous capacity of 3743.2×10^6 L/yr, is available (124). Leading producers include Archer Daniels Midland, with over one-half of all domestically produced fuel ethanol from four locations: Pekin Energy Co., Pekin, Illinois; South Point Ethanol, South Point, Ohio; and New Energy Company of Indiana, South Bend. Several plants terminated operations even when the price of corn was in the $2.00/bu range. Continuous operation at a profit is difficult to sustain when the selling price of fuel ethanol must remain competitive with gasoline prices and alternative octane-enhancing methods. Without tax incentives, it is doubtful that fuel ethanol producers, particularly those who operate smaller plants and use older technologies, will be able to survive during times of high corn prices and low crude oil prices.

Biomass Production

In 1992, there was no biomass species grown and harvested in the United States specifically for conversion to biofuels, with the possible exceptions of feedstocks for fuel

Table 37. Potential Methane Production from Commercial U.S. Anaerobic Digestion Systems

Number of Plants	Feedstock	Estimated Digester Volume, 10^6 m³	Estimated Methane Production Potential[a]	
			10^6 m³/d	EJ/yr
209[b]	Municipal wastewater	0.213	0.208	0.0028
44	Animal manures	0.044	0.043	0.0006
14	Industrial wastes	0.108	0.105	0.0015
Total			0.356	0.0049

[a] Calculated assuming 65 vol % of methane in product gas and 1.5 vol gas/culture vol·d.
[b] Ref. 119. These are treatment plant unit processes, not individual digesters, that produce and use digester gas; the flow capacity is 14.2×10^3 m³/d.
[c] Calculated assuming 15-d hydraulic retention time (HRT).

ethanol and a few tree plantations. This is understandable from an economic standpoint. For example, the average natural gas price in the United States in 1991 at the point of production, not end use cost, was estimated to be $1.51/BJ ($1.59/MBtu) (U.S. Energy Information Agency, Washington, D.C.). For biomass to compete on an equivalent basis, it must be grown, harvested, and gasified to produce methane at an average cost of $1.51/GJ ($1.59/MBtu). Assuming an unrealistic gasification cost of zero, the maximum biomass cost that is acceptable under this condition is $29.73/dry ton. At an optimistic yield of 4.45 dry t/hm^2·yr (10 dry short ton/acre·yr) a biomass energy crop producer for a gasification plant will realize not more than $667.60/hm^2·yr ($270.30/acre·yr), a marginal amount to permit a net return on an energy crop without other incentives. This simplistic calculation emphasizes the effect of depressed fossil fuel prices on biomass energy crops. Negative feedstock costs, ie, wastes, substantial by-product credits, captive uses, and/or tax incentives, are needed to justify energy crop production on strict economic grounds.

Most of the commercial tree plantations that produce wood for captive use as a raw material in manufacturing operations use a portion as fuel. Examples of short-rotation plantations are listed in Table 38 (127). Paper companies in the southeastern United States are reported to have short-rotation plantings also, eg, Weyerhaeuser, James River Corp., Buckeye Cellulose, and Lykes Brothers, but the intensity of maintenance is not known (127).

The advances in biomass growth technologies developed in the United States for agricultural crops, trees, and aquatic species, and that are commercial and being improved further through research, are available for growth of biomass energy crops. Multicropping designs and multiple-use crops will be the most likely candidates for biomass energy when conditions warrant commercial plantations.

ECONOMIC AND LEGISLATIVE IMPACTS

An interminable number of studies have been performed to predict future energy consumption patterns, resources, imports, and prices. If the predictions of higher oil prices had been accurate in the late 1970s, or if the oil price had stabilized at its peak in 1981, the biomass energy industry would have exhibited much greater growth than it has (128).

Biofuels usage has slowly increased since the mid-1980s because of environmental problems, eg, MSW disposal; favorable legislation, eg, tax incentives and PURPA; and combinations of both, eg, oxygenated transportation fuels. Although environmental problems continue to increase, many tax incentives for alternative renewable energy resources have been reduced or eliminated (129). Commercialization of biomass energy in 1992 is driven by waste disposal, alternative fuels and environmental issues, and the available incentives for PURPA power plants.

Capacity Limitations and Biofuels Markets. Large biofuels markets exist (130–133), eg, production of fermentation ethanol for use as a gasoline extender (see ALCOHOL FUELS). Even with existing (1987) and planned additions to ethanol plant capacities, less than 10% of gasoline sales could be satisfied with ethanol–gasoline blends of 10 vol % ethanol; the maximum volumetric displacement of gasoline possible is about 1%. The same condition applies to methanol and alcohol derivatives, ie, methyl t-butyl ether [1634-04-4] and ethyl-t-butyl ether.

In 1987, taxable motor gasoline sales were 415.89 × 10^9 L (109.88 × 10^9 gal) (131). In the same year, the methanol nameplate capacity was 5.30 × 10^9 L (1.40 × 10^9 gal) and actual production was 4.12 × 10^9 L (1.09 × 10^9 gal); synthetic ethanol capacity was 0.80 × 10^9 L (0.21 × 10^9 gal) and 0.30 × 10^9 L (0.08 × 10^9) was actually produced (133,134); fermentation ethanol capacity was 3.62 × 10^9 L (0.957 × 10^9 gal) and actual production for blending with gasoline was 2.84 × 10^9 L (0.750 × 10^9 gal) (124).

Only a small portion of motor fuel needs could be satisfied if truly large-scale alcohol–gasoline blending or fuel switching occurred via transition to fuel-flexible vehicles and ultimately to neat alcohol-fueled vehicles (132).

Capacities for producing virtually all biofuels manufactured by biological or thermal conversion of biomass must be dramatically increased to approach their potential contribution to primary energy demand (Table 4). As already pointed out, an incremental EJ per year of biogas requires about 210 times the existing digestion capacity, including wastewater treatment plants, whereas an incremental EJ

Table 38. Commercial Tree Production for Energy Use,[a] 1988

Company	Area, hm^2	Species	Rotation, yr	Comments
Simpson Timber Co.[b]	283	Eucalyptus		
West Vaco[c]	6475	Cottonwood, Sycamore	10	Primarily for pulp with some to fuel paper mills
Packaging Corp. of America[d]	1214	Hybrid poplar		
Hagerstown[e]	202	Hybrid poplar		Wastewater disposal site, energy use of wood planned
Reynolds Metals Co.[f]	91	Hybrid poplar	6	Captive energy use of wood planned
Union Corp. of North Carolina	8903	Sweetgum	10	Captive for pulp with some to fuel paper mills
James River Corp. of Nevada	2975	Hybrid poplar	6	Captive for fiber and fuel for paper mills, larger plantings are planned

[a] Ref. 127.   [b] California.   [c] Kentucky.
[d] Michigan   [e] Maryland.   [f] New York.

per year of ethanol requires about 14 times the existing fermentation plant capacity. Thus, biofuels cannot be expected to satisfy large EJ markets in the short- to midterm. Since most nonwaste-derived biofuels are not economically competitive with fossil fuels in the early 1990s, large additions to plant capacity will not occur except in those cases where environmental concerns or legislative incentives are governing factors.

Investment Opportunities and Capital Requirements. Despite some of the temporary economic problems that confront the biomass energy industry in the early 1990s, several business opportunities are being developed at rapid rates. These projects are distributed across the nation and include landfill gas recovery plants, MSW-to-energy systems, and nonutility power generation that qualifies under PURPA. Conventional combustion technology is utilized in the majority of plants; gasification seems to have been largely ignored and should offer several advantages (112). A production tax credit equivalent to $0.48/m³ ($3.00/BOE) indexed to inflation and linked to the price of oil is available; it amounted to about $0.71/GJ ($0.75/MBtu) of product gas in mid-1985 (112,129), and can have a significant beneficial impact on the profitability of a biofuels project. The lower the cost of oil, the greater the credit. Taking the most optimistic view of the language in the law, wastes are included in the definition of biomass, so it appears the production tax credit is applicable to all of the above projects, not just those based on wood and other nonwaste biomass.

The Tax Reform Act of 1986 has resulted in a transition away from capital supplied by individuals for the financing of biofuels projects toward conventional financing and greater use of institutional capital sources (128,134). The capital requirements can be large, eg, $20.7 billion in 10 years to complete 240 early and advanced-planned MSW-to-energy plant projects (98), and $35 million per plant for 300–400 small plants (112). In the United States, however, there are also biofuels opportunities that do not involve such large capital needs. Numerous landfill gas recovery plants have been installed for well under $30 million each; most have a capital cost of $5 million or less, depending on scale and end use, although the capital cost of one of the largest landfill gas recovery projects is in the $20 million range (115).

The Energy Policy Act of 1992 (H.R. 776) has liberalized the rules concerning biofuels and provided tax incentives for increased usage. Many states also have gasohol fuel tax exemptions in place, and some have enacted legislation that requires use of oxygenated fuels under certain conditions. Most of these laws impact favorably on biofuels usage.

Many energy analysts believe that it is only a matter of time before petroleum prices, the economic parameter that influences almost all other energy prices, begin to return to market prices of at least $3.97/m³ ($25/bbl). It is widely believed that the gas bubble, which provided excess gas deliverability in the 1980s, will decline in the 1990s. Thus, energy prices are expected to rise again under any scenario. If petroleum prices stabilize at or continue to increase to levels over $3.97/m³ ($25/bbl) it is expected that this, along with environmental issues, will provide the market forces that will increase biomass energy usage.

RESEARCH

A large variety of biomass feedstock developments and advanced conversion processes for the production of energy, fuels, and chemicals are in the research stage in the United States. Many other countries are also developing biomass energy in the laboratory and in the field. The research is aimed at reducing the cost of biomass and increasing the efficiency of production of the final products, eg, new fuels, substitute fossil fuels, and energy, so that biofuels can compete with other energy resources, especially fossil fuels.

Feedstock Development

Most of the research in process in the United States in the early 1990s on the selection of suitable biomass species for energy applications is limited to laboratory studies and small-scale test plots. Many of the research programs on feedstock development were started in the 1970s or early 1980s.

Herbaceous Biomass. Considerable research has been conducted to screen and select nonwoody herbaceous plants as candidates for biomass plants that are unexplored in the continental United States; other research has concentrated on cash crops such as sugarcane and sweet sorghum; and still other research has emphasized tropical grasses. In the late 1970s, a comprehensive screening study of the United States generated a list of 280 promising candidates from which up to 20 species were recommended for field experiments in each region of the country (135). The four highest-yielding species recommended for further tests in each region are listed in Table 39 (135). Since many of the plants in the original list of 280 species had not been grown for commercial use, the production costs were estimated as shown in Table 40 for the various classes of herbaceous species and used in conjunction with yield and other data to develop the recommendations in Table 39.

A large number of small-scale field tests on potential herbaceous energy crops have been carried out. The productivity ranges for some of the most important species for the midwestern and southeastern United States are shown in Table 41 (136). The results of this research helped to establish a strategy that these crops should be primarily grasses and legumes produced using management systems similar to those used for conventional forage crops. It was concluded from this work that the ideal selection of herbaceous energy crops for these areas would consist of at least one annual species, one warm-season perennial species, one cool-season perennial species, and one legume. Production rates, cost estimates, and environmental considerations indicate that perennial species will be preferred to annual species on many sites, but annuals may be more important in crop rotations.

In greenhouse, small-plot, and field-scale tests conducted to screen tropical grasses as energy crops, three

Table 39. Reported Maximum Productivities for Recommended Herbaceous Plants[a]

U.S. Region[b]	Species	Yield, dry t/hm² · yr
Southeastern prairie delta and coast	Kenaf	29.1
	Napier grass	28.5
	Bermuda grass	26.9
	Forage sorghum	26.9
General farm and North Atlantic	Kenaf	18.6
	Sorghum hybrid	18.4
	Bermuda grass	15.9
	Smooth bromegrass	13.9
Central	Forage sorghum	25.6
	Hybrid sorghum	19.1
	Reed canary grass	17.0
	Tall fescue	15.7
Lake states and Northeast	Jerusalem artichoke	32.1
	Sunflower	20.0
	Reed canary grass	13.7
	Common milkweed	12.3
Central and southwestern plains and plateaus	Kenaf	33.0
	Colorado River hemp	25.1
	Switchgrass	22.4
	Sunn hemp	21.3
Northern and western great plains	Jerusalem artichoke	32.1
	Sunchoke	28.5
	Sunflower	19.7
	Milkvetch	16.1
Western range	Alfalfa	17.9
	Blue panic grass	17.9
	Cane bluestem	10.8
	Buffalo gourd	10.1
Northwestern/ Rocky Mountain	Milkvetch	12.1
	Kochia	11.0
	Russian thistle	10.1
	Alfalfa	8.1
California subtropical	Sudan grass	35.9
	Sudan–sorghum hybrid	31.6
	Forage sorghum	28.9
	Alfalfa	19.1

[a] Ref. 135.
[b] As defined by U.S. Dept. of Agriculture, Agriculture Handbook 296, Mar. 1972; excludes Alaska and Hawaii.

categories have emerged, based on the time required to maximize dry-matter yields: short-rotation species (2–3 months), intermediate-rotation species (4–6 months), and long-rotation species (12–18 months) (137). A sorghum–sudan grass hybrid (Sordan 70A), the forage grass napier grass, and sugarcane are outstanding candidates in these categories. Minimum-tillage grasses that produce moderate yields with little attention are wild *Saccharum* clones, and Johnson grass in a fourth category. The maximum yield observed was 61.6 dry t/hm²·yr for sugarcane propagated at narrow row centers over 12 months. The estimated maximum yield is of the order of 112 dry t/hm²·yr using new generations of sugarcane and the propagation of ratoon, ie, regrowth, plants for several years after a given crop is planted.

Overall, research on the development of herbaceous energy crops shows that a broad range of plant species may ultimately be prime energy crops.

Short-Rotation Woody Crops. Research to develop trees as energy crops via short-rotation intensive culture (SRIC) made significant progress in the 1980s. Projections indicate that yields of organic matter can be substantially increased by coppicing techniques and genetic improvements. Advanced designs of whole-tree harvesters, logging residue collection and chipping units, and rapid planting machinery have progressed to the point where prototype units are being evaluated in the field. It is expected that several of these devices will be manufactured for commercial use. Some tree species being targeted for research are red alder, black cottonwood, Douglas fir, and ponderosa pine in the Northwest; *Eucalyptus*, mesquite, Chinese tallow, and the leucaena in the West and Southwest; sycamore, eastern cottonwood, black locust, catalpa, sugar maple, poplar, and conifers in the Midwest; sycamore, sweetgum, European black alder, and loblolly pine in the Southwest; and sycamore, poplar, willow, and sugar maple in the East. Generally, tree growth in test plots is studied in terms of soil type and the requirements for site preparation, planting density, irrigation, fertilization, weed control, disease control, and nutrients. Harvesting methods are also important, especially in the case of coppice growth for SRIC hardwoods. Although tree species native to the region are usually included in the experimental design, non-native and hybrid species are often tested too. Advanced biotechnological methods and techniques, such as tissue culture propagation, genetic transformation, and somaclonal variation, are being used in research to clonally propagate individual genotypes and to regenerate genetically modified species.

After an intensive 10-year research effort, short-rotation woody crop yields in the United States, based on data accumulated to 1992, were projected to be 9, 9, 11, 17, and 17 dry t/hm²·yr in the Northeast, South/Southeast,

Table 40. Production Costs for Annual Herbaceous Plants[a]

Plant Groups	Model Crop Used	Whole Plant Yield, dry t/(hm² · yr)	Cost, $/t
Tall grasses	Corn	17.3	19.1
Short grasses	Wheat	9.9	17.2
Tall broadleaves	Sunflower	15.0	12.7
Short broadleaves	Sugar beet	13.9	77.1
Legumes	Alfalfa[b]	13.7	20.9
Tubers	Potatoes	9.2	136

[a] Ref. 135. Average cost.
[b] Is a perennial.

Table 41. Productivity Rates for Reproductive Herbaceous Biomass species in Southeast and Midwest, dry t/hm$^2 \cdot$yr^a

Biomass Type and Species	Southeast	Midwest
Annuals		
Warm-season		
Sorghums^b	0.2–19.0	1.9–29.1
Cool-season		
Winter rye^c	0.0–7.2	2.4–6.1
Perennials^d		
Warm-season		
Switchgrass^c	2.9–14.0	2.5–13.4
Weeping lovegrass^c	5.4–13.7	
Napier grass–energycane^b	20.4–28.3	
Cool-season		
Reed canary grass^c		2.7–10.8
Legumes		
Alfalfa		1.6–17.4
Flatpea	2.1–12.9	3.9–10.2
Sericea lespedeza	1.8–11.1	

^{a} Ref. 136. Figures are average annual productivities.
^{b} Thick-stemmed grass.
^{c} Thin-stemmed grass.
^{d} Productivity rates after 1–2 yr establishment period.

Midwest/Lake, Northwest, and Subtropics, respectively (138). The corresponding research goals are 15, 18, 20, 30, and 30 dry t/hm$^2 \cdot$yr. Hybrid poplar, which can grow in many parts of the United States, and *Eucalyptus*, which is limited to Hawaii, Florida, southern Texas, and part of California, have shown the greatest potential thus far for attaining exceptionally fast growth rates (138). Both have achieved yields in the range of 20 to 43 t/hm$^2 \cdot$yr in experimental trials with selected clones. Research indicates other promising species to be black locust, sycamore, sweetgum, and silver maple.

Hybridizing techniques seem to be leading to super trees that have short growth cycles and yield larger quantities of woody biomass. Fast-growing clones are being developed for energy farms in which the trees are ready for harvest in as little as 10 years and yield up to 30 m^3/hm$^2 \cdot$yr. Genetic and environmental manipulation has also led to valuable techniques for the fast growth of saplings in artificial light and with controlled atmospheres, humidity, and nutrition. The growth of infant trees in a few months is equivalent to what can be obtained in several years by conventional techniques.

Chemical injections into pine trees have been reported to have stimulatory effects on the natural production of resins and terpenes and may result in high yields of these valuable chemicals. Combined oleoresin–timber production in mixed stands of pine and timber trees is under development, and it appears that when short-rotation forestry is used, the yields of energy products and timber can be substantially higher than the yields from separate operations.

One of the largest research projects on SRIC trees in the Western World, the Large European Bioenergy Project (LEBEN), was reported to be scheduled for initiation in the Abruzzo Region of Italy in the mid-1980s and to be established near the end of that decade (139,140). This project integrates SRIC tree production, agricultural energy crops and residues, and biomass conversion to fuels and energy. About 400,000 t/yr of biomass consisting of 260,000 t/yr of woody biomass from 700 hm^2, and 120,000 t/yr of agricultural residues from 700 hm^2 of vineyards and olive and fruit orchards, will be used. Later, 110,000 t/yr of energy crops from 1050 hm^2 will be utilized. The energy products include liquid fuels, ie, biomass-derived oil and charcoal, 200 million kWh/yr of electric power, and waste heat for injection into the regional agro-forestry and industrial sectors. This project is still in the start-up stage.

In the Amazon jungle of Brazil, perhaps the largest SRIC tree plantation in the world is being integrated with energy, pulp, and chemical production facilities (139). Fast-growing Caribbean pine, *Eucalyptus*, and *Gmelina arborea* grown on 51,400 hm^2 are converted to these products. Although gmelina failed on some of the planted sites, it is doing well on about one-third of the planted sites that have the best soil conditions. The other tree species are apparently being grown successfully on the other two-thirds of the sites with sandy and transition soils. It has been reported that the Brazilian Government and industry have taken control of the project from non-Brazilian interests.

Aquatic Biomass. Aquatic biomass, particularly micro- and macroalgae, are more efficient at converting incident solar radiation to chemical energy than are most other biomass species. For this reason, and the fact that most aquatic plants do not have commercial markets, research was performed in the late 1970s and 1980s to evaluate several species as energy crops. The overall goals of the research have generally been directed to either biomass production, often with simultaneous waste treatment, for subsequent conversion to fuels by fermentation, or to species that contain valuable products. The aquatics studied and their main applications are microalgae for liquid fuels, the macrophyte water hyacinth for wastewater treatment and conversion to methane, and marine macroalgae for specialty chemicals or conversion to methane.

Research in the United States on microalgae focuses on the growth of these organisms under conditions that promote lipid formation. This eliminates the high cost of cell harvest because the lipids often can be separated by simple flotation or extraction. The United States Department of Energy research program on microalgae in the 1980s was one of the largest of its kind. It consisted of several projects and emphasized the isolation and characterization of the organisms and the development of microalgae that afford high oil yields. The research included projects on siting studies; collection, screening, and characterization of microalgae; growth of certain species in laboratory and small-scale production systems; exploration of innovative approaches to microalgae production; and innovative methods for increasing oil formation. Some microalgae, such as *Botryococcus braunii*, have been reported to produce lipid yields that are 40–50% of the dry cell weight under nitrogen-limited conditions (29). However, in other research, *B. braunii* has been reported to yield 20–52% of the dry cell weight as liquid hydrocarbons (139).

Conversion

Combustion. Biomass combustion accounted for about 4% of total U.S. energy consumption in 1992, primarily in the industrial, residential, and utility sectors. Electric power capacity fueled by biomass grew from 200 MW in the early 1980s to about 6000 MW in 1992. The direct combustion of biomass for heat, steam, and power has been, and is expected to continue to be, the principal end use of biomass energy. Conventional biomass-fired technology uses a variety of combustion equipment designs that are usually capable of burning a wet, nonhomogeneous fuel with large variations in moisture content and particle size (141). Spreader stoker-fired boilers have evolved from the designs of the past to systems which include several designs for controlled fuel distribution and automatic ash removal. Research on biomass combustion has focused on improvements of existing systems with respect to ease of operation, increased efficiency, and lower capital and operating costs; emission controls and abatement; and development of new technologies to permit utilization of solid biomass fuels in a wider range of applications (142). Some of the biomass combustion research developments since the early 1930s include whole-tree burning technologies (143), cyclonic incineration of waste biomass (144), direct wood-fired gas turbines (145), improved combustion cycles for biomass (141), fluid-bed biomass combustion (146), pulverized biomass combustion (147), catalytic wood-burning stoves (148), cofiring of biomass and fossil fuels to reduce emissions (149), and control of biomass combustion to reduce emissions (150). Even though the burning of biomass is one of the oldest energy-producing methods used, research continues to make significant advancements in the art and science of biomass combustion. Recent U.S. legislation concerned with air quality and waste biomass disposal has a significant impact on the direction of ongoing research to develop advanced biomass combustion systems (151).

Anaerobic Digestion (Methane Fermentation). A large amount of research was performed on the anaerobic digestion of biomass in the 1970s and 1980s to develop biological gasification processes that are capable of producing methane (16). Basic research on methane fermentation provides a better understanding of the kinetics and mechanisms of biomass conversion under anaerobic conditions; improvements in digestion efficiencies in terms of methane yield and volatile solids reduction have been slow to evolve from this knowledge. A large number of agricultural residues, animal wastes, and biomass species have been evaluted as potential feedstocks for methane production in laboratory digesters and small-scale digestion facilities. Considerable laboratory work has also been done to develop pre- and post-digestion treatments that improve biodegradability. A plateau of about 50–60% volatile solids destruction efficiencies and energy recoveries in the product gas seems to exist for most methane fermentation systems.

Research in the early 1990s has addressed several potentially beneficial methods of improving the process, eg, the two-phase digestion in which the acetogenic and methanogenic phases are physically separated. Practical implementation of two-phase digestion is achieved by control of the hydraulic retention times in the acid and methane reactors or reaction zones. This process configuration provides several advantages over conventional high rate digestion such as enhanced stability, an optimum environment for acetogenic and methanogenic bacteria, substantial increases in throughput rates for given size reactors, increased gas and methane production rates, and higher methane content in the product gas (16,152). The process has been scaled-up for treatment of municipal biosolids, ie, the Acimet Process (153), and has been applied to industrial wastes (16). From a practical standpoint, two-phase anaerobic digestion of biomass is capable of retrofit to existing digestion systems of any design and is projected to be capable of doubling plant capacity at about 50% of the capital cost of a grassroots plant. A significant market is anticipated for this advanced technology, particularly for wastewater treatment.

Some of the other research studies have addressed topics such as high solids biomass digestion (154), the utilization of superthermophilic organisms (155), advanced reactor designs (156), landfill gas enhancement (157), and microbiology of the mixed cultures involved in methane fermentation (158).

Thermochemical Gasification. Extensive research and pilot studies have been carried out since 1970 to develop thermochemical processes for biomass conversion to energy and fuels. Basic studies on the effects of various operating conditions and reactor configurations have been performed in the laboratory and at the process development unit (PDU) and pilot scales on steam, steam-air, air-blown, and oxygen-blown gasification, and on hydrogasification. Other research has also been done on the rapid pyrolysis of biomass which, in addition to gaseous products, yields coproduct liquids and solids.

Over one million air-blown gasifiers were built during World War II to manufacture LHV gas to power vehicles and to generate steam and electric power. Units are available in a variety of designs, some of which have been retrofitted to gas-fired furnaces. Although some research is in progress to refine air-blown wood gasifiers in North America, particularly portable units, most of the research has been conducted in Europe. The Swedish automobile manufacturers, Volvo and Saab, have ongoing programs to develop a standard gasifier design suitable for mass production.

Research on thermochemical biomass gasification in North America has tended to concentrate on MHV gas production, scale-up of the advanced process concepts that have been evaluated at the PDU scale, and the problems that need to be solved to permit large-scale thermochemical biomass gasifiers to be operated in a reliable fashion for power production, especially advanced power cycles. Many different reactor designs have been evaluated under a wide range of operating conditions. Exemplary advanced gasifiers and gasification systems, some of which are in the scale-up stage, include IGT's single-stage, pressurized, fluid-bed gasifier; the National Renewable Energy Laboratory's (NREL) pressurized, fixed-bed, downdraft gasifier; the fire tube-heated, fluid-bed system of the

University of Missouri-Rolla; the indirectly heated, fluid-bed, dual reactor system of Battelle Columbus Laboratory; the pulse-enhanced, indirectly heated, fluid-bed gasifier of M.T.C.I.; and the catalytic, pressurized, gasification system for wet biomass of Pacific Northwest Laboratory.

An example of a scale-up project is the pressurized, fluid-bed Renugas™ plant in Hawaii (159). The gasifier is designed for 63.5 t/d of sugarcane bagasse. In addition to bagasse, other feedstocks such as wood, waste biomass, and RDF may be evaluated. The demonstration provides process informaton for both air- and oxygen-blown gasification at low and high pressures. Renugas will be evaluated for both fuel gas and synthesis gas production, and for electric power production with advanced power generation schemes.

Although only limited research has been carried out on small-scale, LHV, producer gasifiers for biomass, significant design advancements have been made even though they have been used for over 100 years. One development is the open-top, stratified, downdraft gasifier in which air is drawn in through successive reaction strata (160). The unit is simple to operate, inexpensive, and can be close-coupled to an engine–generator set without complex gas-cleaning equipment. The gasifier dimensions are sized to deliver gas to the engine based on its fuel-rate requirements, and no controls are needed.

Emissions from the thermochemical gasification of biomass can affect the operation of advanced power cycles and include particulates, alkalies, oils and tars, and heavy metals. One of the high priority research efforts is to develop hot-gas cleanup methods that will permit biomass gasification to supply suitable fuel gas for these cycles (161). Some of the other research needs that have been identified include versatile feed-handling systems for a wide variety of biomass feedstocks; biomass feeding systems for high pressure gasifiers; determination of the effects of additives including catalysts for minimizing tar production and capturing contaminants; and suitable ash disposal and wastewater treatment technologies (162).

Liquefaction. Figure 21 outlines most of the biomass liquefaction methods under development. There are essentially three basic types of biomass liquefaction technologies, ie, fermentation, natural, and thermochemical processes.

Much research has been conducted since the early 1970s to improve the alcohol fermentation process, ie, the technology by which biomass is converted to fuel ethanol (96,112,125). A large portion of this research has been focused on development of the process so that it will be suitable for conversion of low grade lignocellulosics. This type of biomass generally contains about 50 wt % celluloses, 25 wt % hemicelluloses, and 25 wt % lignins. The hexose sugars obtained on hydrolysis of celluloses are converted by conventional alcohol-forming yeasts to ethanol. The pentose sugars require other specialized organisms for conversion. The lignins are essentially inert.

Research has focused on minimizing the energy inputs for the distillation steps needed to produce 190 and 200 proof ethanol; on the fermentation process itself to increase ethanol yields and reduce fermentation times; on the pretreatment and hydrolysis processes needed to af-

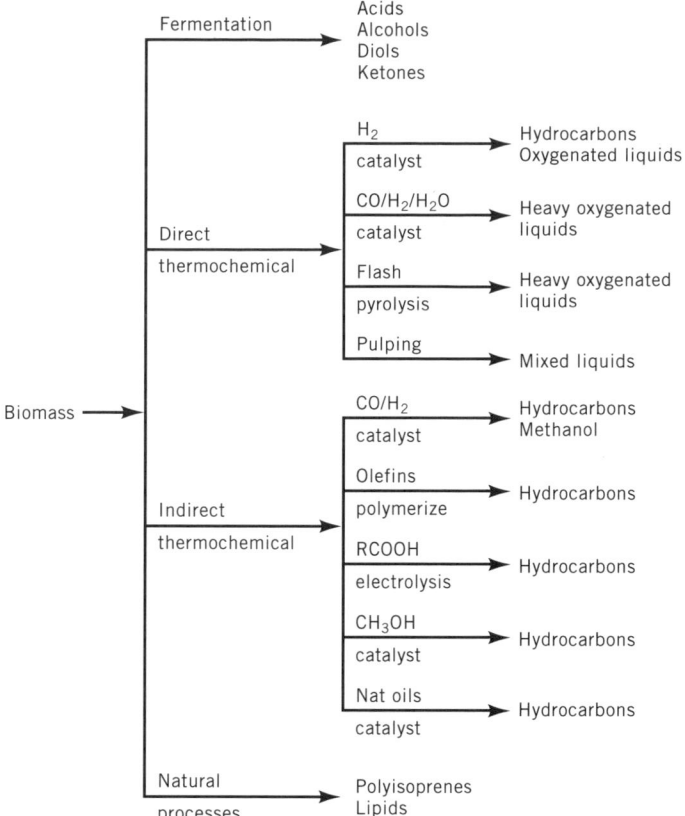

Figure 21. Biomass liquefaction routes under development (ca 1992).

ford high yields of sugars and to make low grade cellulosics suitable feedstocks for fermentation; on development of organisms that ferment pentoses separately, and pentoses and hexoses together; on development of advanced processes that permit simultaneous saccharification and fermentation (SSF); and on development of immobilized organisms and genetically engineered bacteria that afford much shorter fermentation times and avoid yeast recycling.

Several advancements have been developed and incorporated into grain fermentation processes to improve operating efficiencies and reduce energy consumption. Ethanol yields and production rates have improved slightly, but not significantly. In contrast, research on the development of low grade biomass fermentation processes is on the verge of process demonstrations with a variety of feedstocks such as waste paper, agricultural residues, wood biomass, and refuse-derived fuel.

Much of the research to apply ethanol fermentation to low grade biomass is funded by the U.S. Department of Energy. The goal is to produce fuel ethanol at a cost of $0.145/L, or $6.90/GJ, by 1995 (166).

Oilseed and Vegetable Oil Fuels. Limited research has continued on the utilization of seed and vegetable oils as motor fuels, particularly as substitute diesel fuels and diesel fuel extenders (164). Work has focused on studies of the yields and properties of oils from oilseed and vegetable oil crops, the performance of neat oils and oil–diesel fuel blends as fuels for compression ignition engines, improvement of the transesterification process and the fuel characteristics of the resulting esters as diesel fuels, upgrading vegetable oils to gasolines and diesel fuels by hydrocracking processes, and field tests of the liquid fuels made from seed and vegetable oils in trucks and buses. Although several operating problems have been observed, such as lubricating oil deterioration and crystal formation in cooler weather even with some of the lower viscosity seed oil esters, significant advances have been made. Esters of selected vegetable oils are very promising candidates for both indirect and direct fuel-injected engines.

The cost of seed and vegetable oil fuels is still not competitive with petroleum-based diesel fuels. The cost of esterification alone can add up to 50% to the cost of the fuel, depending on the size of the processing operation and the market value of by-product glycerol. In 1992 the cost of oilseed and vegetable oil diesel fuels was about twice that of conventional diesel fuels. One approach to elimination of the cost differential is to use waste vegetable oils from large-scale restaurant operations as the feedstock for transesterification plants (165). This offers the possibility of taking credit for waste oil disposal, ie, the analogue of a tipping fee in the solid waste disposal field, and of recycling of oils as fuel. Another approach is to utilize tax credit legislation available in the United States to farmers who produce minor oilseed crops in place of principal commodity crops.

Thermochemical Liquefaction. Most of the research done since 1970 on the direct thermochemical liquefaction of biomass has been concentrated on the use of various pyrolytic techniques for the production of liquid fuels and fuel components (96,112,125,166,167). Some of the techniques investigated are entrained-flow pyrolysis, vacuum pyrolysis, rapid and flash pyrolysis, ultrafast pyrolysis in vortex reactors, fluid-bed pyrolysis, low temperature pyrolysis at long reaction times, and updraft fixed-bed pyrolysis. Other research has been done to develop low cost, upgrading methods to convert the complex mixtures formed on pyrolysis of biomass to high quality transportation fuels, and to study liquefaction at high pressures via solvolysis, steam–water treatment, catalytic hydrotreatment, and noncatalytic and catalytic treatment in aqueous systems.

Essentially all of these conversion processes are technically feasible and can be used to convert biomass to a wide range of liquid products. Unfortunately, because of the complex composition of biomass and the chemistry of direct thermal cracking of biomass, complex product mixtures are always formed. Selectivities for individual products are low; this is sometimes advantageous. Because of the oxygenated nature of biomass, higher yields of certain oxygenated products can be obtained. This offers the possibility of producing specific organic liquids that have higher intrinsic value as chemicals rather than as fuels. A few research efforts are cited here to illustrate the versatility of direct liquefaction processes for biomass.

The principal products in conventional, long-term pyrolysis of biomass at about 400°C or lower are char, gas, organic liquids, and water, as shown in Table 20. Research has shown that fast pyrolysis of biomass at 475–525°C and vapor residence times of a few seconds or less can maximize organic liquid yields. Wood and grasses yield 55–65 wt % and 40–65 wt % of the dry biomass as organic liquids, respectively (168). Products from the fast pyrolysis of wood, for example, contain significant amounts of low molecular weight oxygenated compounds such as hydroxyacetaldehyde, acetaldehyde, formic acid, acetic acid, and glyoxal. Fast pyrolysis of waste lignocellulosics such as newsprint or pulp mill sludges also affords similar liquid products. Pretreatment of the wood before pyrolysis gives dramatic changes in product selectivity.

Ultrafast pyrolysis in the vortex reactor is capable of pyrolyzing biomass at high heat-transfer rates on the reactor wall by ablation and has been found to be useful for a variety of biomass and waste feedstocks. In this reactor, biomass particles are entrained tangentially at high velocities by a carrier gas into the vortex reactor, which causes the biomass particles to be preferentially heated relative to the carrier gas and the pyrolysis vapors (169). Products recovered from this innovative reactor have been demonstrated to be about 55 wt % organic liquids, 14 wt % gases, 13 wt % char, and 12 wt % water (94% closure). The pyrolysis vapors can be condensed to a low viscosity liquid, thought to be suitable for combustion in furnaces and turbines, or can be cracked to form about 15 wt % C_2+ hydrocarbons. Zeolites have been found to catalyze the conversion of the pyrolysis vapors to gasoline range hydrocarbons in yields that approach the theoretical upper limit as determined from stoichiometry (170). The energy conversion efficiency was about 45% for C_2–C_8 range hydrocarbons and 55% for C_2+ hydrocarbons. The ablative fast pyrolysis process has been scaled up to a unit that converts 11,400 t/yr of waste sawdust into 5.3 million L/yr of fuel oil and 1,720 t/yr of charcoal (171). This relatively small plant is expected to have a pretax revenue of $263,000 presuming the feedstock cost is zero, and the fuel oil and char can be sold for $4.74/GJ and $72.60/t, respectively. The capital cost of the plant was $850,000.

Prospects

Despite the slow development of renewable biomass as a primary source of energy, the large research effort in progress on feedstock production and conversion is expected to lead to greater commercialization of advanced energy and organic chemical processes based on biomass. Small-scale systems for the individual farmer are being designed and marketed to make it possible to install and operate complete on-site total energy packages that will supply all of the farm's energy requirements. These systems will be fueled with captive sources of biomass and wastes generated on the farm. It is likely that large building complexes such as schools, apartments, shopping malls, and theme parks in urban areas will be able to incorporate similar systems using captive wastes and delivered biomass. Individual, small-scale, farmers' cooperative and industrial-scale fuel ethanol plants will continue to be built and operated as long as government tax incentives are provided. Tax subsidies for fuel ethanol are expected to become unnecessary as the technologies for use of low grade lignocellulosic feedstocks for fuel ethanol are perfected. New, larger scale, biomass-fueled and waste-to-energy power plants, especially those that incorporate cogeneration, will continue to show modest growth as the technology advances and the disposal of waste biomass in an environmentally acceptable manner is implemented. The development of improved methane fermentation processes for waste biomass such as municipal biosolids, industrial wastes from food-processing and beverage alcohol plants, and refuse-derived fuel, is expected to result in more efficient waste treatment and disposal and increased methane recovery and utilization.

Fossil fuels are still sufficiently low in cost to make the economics of large-scale production of substitute transportation fuels, fuel gases, and fuel oils from biomass borderline or unattractive if the biomass systems are used only to produce energy. Large-scale integrated biomass energy plantations are therefore not expected to be constructed and operated until some time during the first or second quarter of the twenty-first century. Biomass grown strictly as profitable energy crops is expected to occur in that time frame as fossil fuels are phased out or their prices increase because of shortages or additional taxes.

Growing environmental concerns and federal and state environmental regulations are expected to be the driving force behind increased usage of biomass energy. Carbon taxes applied to fossil fuel usage, especially for vehicles and utility power plants, are expected to provide very strong incentives to convert to renewable biomass energy resources for both mobile and stationary applications.

Reprinted from *Kirk-Othmer Encyclopedia of Chemical Technology*, 4th ed., vol. 12, John Wiley & Sons, Inc., New York, 1995.

BIBLIOGRAPHY

1. D. L. Klass, in D. L. Klass and J. W. Weatherly III, eds., *Energy from Biomass and Wastes IV*, IGT, Chicago, 1980, pp. 1–41.
2. D. L. Klass, in D. L. Klass, ed., *Energy from Biomass and Wastes XIII*, IGT, Chicago, 1990, pp. 1–46.
3. D. L. Klass, *Chemtech* **4**(3), 161 (1974).
4. R. H. Whittaker and G. E. Likens, in H. Leith and R. H. Whittaker, eds., *Primary Productivity of the Biosphere*, Springer Verlag, New York, 1975.
5. R. Repetto, *Sci. Am.* **262**(4), 36 (1990).
6. B. Bolin, "The Carbon Cycle," in *The Biosphere*, W. H. Freeman and Co., San Francisco, Calif., 1970.
7. J. T. Houghton, G. J. Jenkins, and J. J. Ephraums, eds., *Climate Change: The IPCC Scientific Assessment*, Cambridge University Press, Cambridge, 1990, 365 pp; D. L. Klass, *Energy & Environment* **3**(2), 109 (1992); D. L. Klass, *Energy Policy* **21**(11), 1076 (1993).
8. United Nations, *1990 Energy Statistics Yearbook*, Department of Economic and Social Development, New York, 1992.
9. *RPA Assessment of the Forest and Rangeland Situation in the U.S., 1989*, No. 26, USDA, Forest Service Washington, D.C., Oct. 1989.
10. D. L. Klass, *Chemtech* **20**(12), 720 (1990).
11. U.S. Department of Energy, *Estimates of U.S. Biofuels Consumption 1990*, DOE/EIA-0548(90), Energy Information Administration, Washington, D.C., Oct. 1991.
12. U.S. Department of Energy, *The Potential of Renewable Energy, An Interlaboratory White Paper*, SERI/TP-260-3674, DE90000322, Office of Policy, Planning and Analysis, Washington, D.C., Mar. 1990.
13. D. F. Westlake, *Biol. Rev.* **38**, 385 (1963).
14. J. S. Burlew, ed., *Algae Culture From Laboratory to Pilot Plant*, Publication 600, Carnegie Institute of Washington, Washington, D.C., 1953, pp. 55–62.
15. C. D. Hodgman, ed., *Handbook of Chemistry and Physics*, 31st ed., Chemical Rubber Publishing Co., Cleveland, Ohio, 1949, p. 1537.
16. D. L. Klass, *Science* **223**, 1021 (1984).
17. L. R. Lynd and co-workers, *Science* **251**, 1318 (1991).
18. *Guayule: An Alternative Source of Natural Rubber*, National Academy of Sciences, Washington, D.C., 1977.
19. R. A. Buchannan and F. O. Otey, "Multi-Use Oil- and Hydrocarbon-Producing Crops in Adaptive Systems for Food, Material, and Energy Production," paper presented at *19th Annual Meeting, Society for Economic Botany*, St. Louis, Mo., June 11–14, 1978.
20. M. Calvin, *Chemtech* **7**(6), 353 (1977); *Bioscience* **29**, 533 (1979); *Die Naturwissenschaften* **67**, 525 (1980); E. K. Nemethy, J. W. Otvos, and M. Calvin, in D. L. Klass and G. H. Emert, eds., *Fuels from Biomass and Wastes*, Ann Arbor Science Publishers, Ann Arbor, Mich., 1981; J. D. Johnson and C. W. Hinman, *Science* **208**, 460 (1980).
21. D. K. Schmalzer and co-workers, *Biocrude Suitabilities for Petroleum Refineries*, ANL/CNSV-69, Argonne National Laboratory, Argonne, Ill., June 1988.
22. K. R. Kaufman, in E. D. Schultz and R. P. Morgan, eds., *Fuels and Chemicals from Oil Seeds: Technology and Policy Options*, Westview Press, Boulder, Colo., 1982, pp. 143–174.
23. P. B. Weisz and J. F. Marshall, *Science* **206**, 257 (1979); R. M. Furrer and N. N. Bakshi, in Ref. 2, pp. 897–914.
24. M. Stumborg and co-workers, in D. L. Klass, ed., *Energy from Biomass and Wastes XVI*, IGT, Chicago, 1993, pp. 721–738.
25. E. S. Olson and R. K. Sharma, in Ref. 24, pp. 739–751.
26. E. S. Lipinsky and co-workers, in Ref. 22, pp. 205–223.
27. E. E. Gavett and D. VanDyne, in Ref. 24, pp. 709–719.
28. D. L. Klass, in D. L. Klass, ed., *Energy from Biomass and Wastes IX*, IGT, Chicago, 1985, pp. 1–83.
29. D. M. Tillett and J. R. Benemann, in D. L. Klass, ed., *Energy from Biomass and Wastes XI*, IGT, Chicago, 1988, pp. 771–786.
30. A. M. Hill and D. A. Feinberg, *Fuel Products from Microalgae*, SERI/TP-231-2348, Solar Energy Research Institute, Golden, Colo., 1984.
31. D. A. Hoffman and R. A. Fitz, *Environ. Sci. Technol.* **2**(11), 1023 (1968).
32. S. B. Alpert and co-workers, *Pyrolysis of Solid Wastes: A Technical and Economic Assessment*, NTIS PB 218-231, SRI, Menlo Park, Calif., Sept. 1972.
33. M. A. Paisley, H. F. Feldmann, and H. R. Appelbaum, in D. L. Klass and H. H. Elliott, eds., *Energy from Biomass and Wastes VIII*, IGT, Chicago, 1984, pp. 675–696.
34. R. C. Bailie, "Results from Commercial-Demonstration Pyrolysis Facilities (35–45 tons/day refuse) Extended to Producing Synfuels from Biomass," in D. L. Klass and J. W. Weatherly III, eds., *Energy from Biomass and Wastes V*, IGT, Chicago, 1981, pp. 549–569.
35. G. T. Preston, "Resource Recovery and Flash Pyrolysis of Municipal Refuse," in F. Ekman, ed., *Clean Fuels from Bio-*

mass, Sewage, Urban Refuse, Agricultural Wastes, IGT, Chicago, 1976, pp. 89–114.

36. D. S. Scott and J. Piskorz, "Continuous Flash Pyrolysis of Wood for Production of Liquid Fuels," in D. L. Klass and H. H. Elliott, eds., *Energy from Biomass and Wastes VII*, IGT, Chicago, 1983, pp. 1123–1146.

37. D. L. Klass, "Wastes and Biomass as Energy Resources: An Overview," in ref. 35, pp. 21–58.

38. T. F. Fisher, M. L. Kasbohm, and J. R. Rivero, "The Purox system," in ref. 35, pp. 447–459.

39. S. J. Yosim and K. M. Barclay, *Preprints of Papers, 171st National Meeting ACS, Div. of Fuel Chem. 21* (1), 73 (Apr. 5–9, 1976).

40. H. F. Feldmann and co-workers, *Hydrocarbon process. 55* (11), 201 (1976).

41. S. P. Babu, D. Q. Tran, and S. P. Singh, "Noncatalytic Fluidized-Bed Hydroconversion of Biomass to Substitute Natural Gas," in ref. 1, pp. 369–385.

42. D. C. Elliott and co-workers, "Low-Temperature, Catalytic Gasification of Wastes for Simultaneous Disposal and Energy Recovery," in D. L. Klass, ed., *Energy from Biomass and Wastes XV*, IGT, Chicago, 1991, pp. 1013–1021.

43. H. F. Feldmann and co-workers, "Gasification of Forest Residues," in D. L. Klass, ed., *Biomass as a Nonfossil Fuel Source*, ACS Symposium Series 144, American Chemical Society, Washington, D.C., 1980, pp. 351–375.

44. L. K. Mudge and co-workers, "Catalytic Gasification of Biomass," in *3rd Annual Biomass Energy Systems Conference Proceedings The National Biomass Program*, SERI/TP-33-285, Solar Energy Research Institute, Golden, Colo., 1979, pp. 351–357.

45. D. G. B. Boocock and D. Mackay, "The Production of Liquid Hydrocarbons by Wood Hydrogenation," in ref. 1, pp. 765–777.

46. H. R. Appel and co-workers, "Conversion of Cellulosic Wastes to Oil," U.S. Bur. of Mines Rep. Invest. *8013* (1975).

47. E. R. Riegel, *Industrial Chemistry,* 2nd ed., The Chemical Catalog Co., New York, 1933, Chapt. 16, p. 257.

48. M. R. Ladisch, C. M. Ladisch, and G. T. Tsao, *Science 201* 743, (1978).

49. L. Paszner and co-workers, "Two-Stage, Continuous Hydrolysis of Wood by the Acid Catalyzed Organosolv Saccharification (ACOS) Process," in ref. 24, in press.

50. W. H. Hoge, U.S. Patent 4,009,075, assigned to BioIndustries, Inc., Feb. 22, 1977.

51. L. R. Lynd and co-workers, *Science 251,* 1318 (1991).

52. R. Katzen Associates, *Chemicals from Wood Wastes,* U.S. Department of Agriculture Forest Products Laboratory, Madison, Wisc., Dec. 14, 1975.

53. R. S. Loomis, W. A. Williams, and A. E. Hall, *Ann. Rev. Plant Physiol. 22* 431 (1971).

54. E. I. Rabinowitch, *Photosynthesis*, Vols. 1–2 Interscience, New York, 1956.

55. C. B. Osmond, *Ann. Rev. Plant Physiol. 29* 379 (1978).

56. B. J. Brinkworth, *Solar Energy for Man,* John Wiley & Sons, Inc., New York, 1973.

57. U.S. Dept. of Commerce, *Climatological Data, National Summary*, Vol. 21, Nos. 1–12, U.S. Government Printing Office, Washington, D.C., 1970.

58. H. J. Critchfield, *General Climatology*, 3rd ed., Prentice-Hall, Inc., Englewood Cliffs, N.J., 1974, p. 22.

59. R. S. Loomis and W. A. Williams, *Crop Sci. 3,* 67 (1963).

60. T. R. Schneider, *Energy Convers. 13*, 77 (1973).

61. S. S. Visher, *Climatic Atlas of the United States*, Harvard University Press, Cambridge, Mass., 1954.

62. *Statistical Abstracts of the United States*, U.S. Department of Commerce, U.S. Government Printing Office, Washington, D.C., 1976.

63. W. L. Roller and co-workers, *Grown Organic Matter as a Fuel Raw Material Source, NASA Report CR-2608*, Ohio Agricultural Research and Development Center, Washington, D.C., Oct. 1975.

64. M. M. Ludlow and G. L. Wilson, *J. Aust. Inst. Agric. Sci. 36*, 43 (Mar. 1970).

65. W. J. North, ed., *The Biology of Giant Kelp Beds (Macrocystis) in California*, Cramer, Lehre, FRG, 1971, p. 12.

66. T. A. El-Sharkawy and J. D. Hesketh, *Crop Sci. 4*, 514 (1964).

67. J. Krummel, in Ref. 35, pp. 359–370.

68. H. J. Evans and L. E. Barber, *Science 197*, 332 (1977).

69. *Land Capability Classification, Agricultural Handbook 210*, U.S. Department of Agriculture, Soil Conservation Service, Washington, D.C., 1966, 21 pp.

70. *Summary Report 1987 National Resources Inventory*, No. 790, U.S. Department of Agriculture, Soil Conservation Service, Washington, D.C., Dec. 1989, 37 pp.

71. S. H. Spurr, *Sci. Am. 240*, 76 (1979).

72. A. A. Nichiporovich, *Photosynthesis of Productive Systems*, Israel Program for Scientific Translations, Jerusalem, Israel, 1967.

73. E. O. Mariani, in D. L. Klass and W. W. Waterman, eds., *Energy from Biomass and Wastes*, IGT, Chicago, 1978, pp. 29–38.

74. R. L. Sajdak and co-workers, in Ref. 43, pp. 21–48.

75. J. C. Dula, in Ref. 33, pp. 193–207.

76. L. L. Wright and co-workers, in D. L. Klass, ed., *Energy from Biomass and Waste XII*, IGT, Chicago, 1989, pp. 261–274.

77. *Solar SNG, Final Report American Gas Association Project IU-114-1*, Prepared by InterTechnology Corp., American Gas Association, Washington, D.C., Oct. 1975.

78. *Grass: The Yearbook of Agriculture 1948*, U.S. Department of Agriculture, U.S. Government Printing Office, Washington, D.C., 1948.

79. D. L. Klass, S. Ghosh, and J. R. Conrad, in Ref. 35, pp. 229–252.

80. D. L. Klass and S. Ghosh, in Ref. 43, pp. 229–249.

81. J. A. Alich, Jr., and R. E. Inman, *Effective Utilization of Solar Energy to Produce Clean Fuel, Grant No. GI 38723*, Final Report for National Science Foundation, Stanford Research Institute, Palo Alto, Calif., June 1974.

82. R. Retovsky, *Continuous Cultifation of Algae, Theoretical and Methodological Bases of Continuous Culture of Microorganisms*, Academic Press, Inc., New York, 1966.

83. D. L. Klass and S. Ghosh, in W. W. Waterman, ed., *Clean Fuels from Biomass and Wastes*, IGT, Chicago, 1977, pp. 323–351.

84. D. L. Klass, S. Ghosh, and D. P. Chynoweth, *Process Biochem. 14,* 18 (1979).

85. D. P. Chynoweth, D. L. Klass, and S. Ghosh, in Ref. 73, pp. 229–251.

86. D. L. Klass and S. Ghosh, in D. L. Klass and G. H. Emert, eds., *Fuels from Biomass and Wastes*, Ann Arbor Science Publishers, Ann Arbor, Mich., 1981, pp. 129–149.

87. M. G. McGarry, *Process Biochem. 6,* 50 (1971).

88. J. L. Yount and R. A. Grossman, *J. Water Pollut. Control Fed.* **42,** 173 (1970).

89. E. S. Dell Fosse, in Ref. 83, pp. 73–99.

90. *State Energy Price and Expenditure Report 1990*, DOE/EIA-0376(90), U.S. Department of Energy, Energy Information Administration, Washington, D.C., Sept. 1992.

91. M. D. Fraser, in Ref. 24, pp. 295–330.

92. R. F. Moorer, D. K. Walter, and S. Gronich, in Ref. 24, pp. 139–153.

93. S. Lazzari, in Ref. 24, pp. 275–294.

94. N. Smith and T. J. Corcoran, *Preprints of Papers Presented at 171st National Meeting, ACS, Fuel Chemistry Division, Symposium on Net Energetics of Integrated Synfuel Systems* **21**(2), 9 (Apr. 1976).

95. R. E. Inman, in Ref. 94, pp. 21–27.

96. D. L. Klass, *Energy Topics*, 1 (Apr. 14, 1980).

97. National Wood Energy Association, *NWEA State Biomass Statistical Directory*, Arlington, Va., 1988.

98. E. Berenyi and R. Gould, *1988–1989 Resource Recovery Yearbook*, Governmental Advisory Associates, Inc., New York, 1988, 718 pp.

99. R. M. Dykes, in Ref. 76, pp. 379–397.

100. *Stove and Fireplace Catalog IX*, Consolidated Dutchwest, Plymouth, Mass., 1988, 67 pp.

101. *Estimates of U.S. Wood Energy Consumption 1980–1983*, DOE/EIA-0341(83), U.S. Department of Energy, Energy Information Administration, Washington, D.C., Nov. 1984.

102. J. W. Koning, Jr. and K. E. Skog, in D. L. Klass, ed., *Energy from Biomass and Wastes X*, IGT, Chicago, 1986, pp. 1309–1322; J. C. Nicolello, *U.S. Pulp and Paper Industry's Energy Use-Calendar Year 1986*, New York, Apr. 20, 1987; J. C. Nicello, *U.S. Pulp and Paper Industry's Energy Use-Calendar Year 1987*, New York, May 17, 1988; *Annual Energy Review 1987*, DOE/EIA-0384(87). U.S. Department of Energy, Energy Information Addministration, Washington, D.C., 1988.

103. National Wood Energy Association, *Wood Energy, America's Renewable Resource*, Arlington, Va., 1988, 2 pp.

104. *Past, Present, and Future Trends in the U.S. Forest Sector: 1952–2040, Review Draft*, U.S. Department of Agriculture, Forest Service, Washington, D.C., June 1988.

105. J. I. Zerbe, *Forum for Applied Research and Public Policy*, 38–47 Winter (1988).

106. Table 1, in Ref. 11.

107. D. L. Klass and C. T. Sen, *Chem. Eng. Prog.* **83**(7), 46 (1987).

108. *The Qualifying Facilities Report*, Federal Energy Regulatory Commission, Washington, D.C., Jan. 11, 1988.

109. J. L. Easterly, personal communication, Meridian Corp., Alexandria, Va., Aug. 16, 1988.

110. D. A. Flint and C. Norris, *1986 Capacity and Generation of Non-Utility Sources of Energy*, Edison Electric Institute, Washington, D.C., July 1988.

111. B. DeCampo, D. A. Flint, and C. Norris, *Electric Perspectives*, 22 (Summer 1988).

112. D. L. Klass, *Resources and Conservation* **15,** 7 (1987).

113. J. L. Easterly, S. Lees, and B. Detwiler, *Electric Power from Biofuels: Planned and Existing Projects in the U.S.*, rev. Jan. 1985, DOE/CE/307841/1, U.S. Department of Energy, Washington, D.C., Aug. 1985.

114. C. Tewksbury, in D. L. Klass, ed., *Energy from Biomass and Wastes X*, IGT, Chicago, 1987, pp. 555–578.

115. D. L. Klass, *Resources and Conservation* **11,** 157 (1985).

116. T. R. Miles and T. R. Miles, Jr., *Biomass* **18,** 163 (1989).

117. R. L. Wentworth, "Anaerobic Digestion in North America," paper presented at *Symposium Anaerobic Digestion and Carbohydrate Hydrolysis of Waste, sponsored by Commission of the European Communities*, Luxembourg, May 8–10, 1984.

118. J. H. Ashworth, Y. M. Bihun, and M. Lazarus, *Universe of U.S. Commercial-Scale Anaerobic Digesters: Results of SERI/ARD Data Collection*, Solar Energy Research Institute, Golden, Colo., May 30, 1985; J. H. Ashworth, *Problems With Installed Commercial Anaerobic Digesters in the United States: Results of Site Visits*, Rev. ed., Solar Energy Research Institute, Golden, Colo., Nov. 6, 1985.

119. *1984 Needs Survey Report to Congress: Planned and Existing Projects in the U.S.*, rev. Jan. 1985, DOE/CE/30784/1, U.S. Department of Energy, Washington, D.C., Aug. 1985.

120. Table 19, in Ref. 28.

121. S. Doelph, *Gas Energy Review* **16**(1), 14 (1988).

122. "Landfill Gas Summary Update," *Waste Age* **19**(3), 167 (1988).

123. "Resource Recovery Activities," *City Currents* **6**(4), 1 (1987).

124. F. L. Potter, personal communication, Information Resources, Inc., Washington, D.C., Aug. 11, 1988.

125. D. L. Klass, *Energy Topics*, 1 (Aug. 1, 1983).

126. *Alc. Update*, Aug. 8, 1988; *Alc. Wk.* **9**(32) (Aug. 8, 1988).

127. L. L. Wright, personal communication, Oak Ridge National Laboratory, Oak Ridge, Tenn., July 1988.

128. S. Fenn, *Institutional Investment in Renewable Energy Technologies*, Renewable Energy Institute, Washington, D.C., Feb.. 1987, 50 pp.

129. S. Lazzari, *A History of Federal Tax Policy: Conventional as Compared to Renewable and Nonconventional Energy Resources*, 88-455E, The Library of Congress, Congressional Research Service, Washington, D.C., June 7, 1988.

130. W. A. Rains, paper presented at *1982 Annual Meeting, National Petroleum Refiners Association*, San Antonio, Tex., Mar. 21–23, 1983.

131. W. R. Keene, Lundberg Survey, Inc., N. Hollywood, Calif., Aug. 16, 1988.

132. *Chem. & Eng. News* **66**(25), 40 (June 20, 1988).

133. *Assessment of Costs and Benefits of Flexible and Alternative Fuel Use in the U.S. Transportation Sector, Progress Report One*, DOE/PE-0080, U.S. Department of Energy, Washington, D.C., Jan. 1988.

134. B. Paul, *WSJ LXIX* (217), Sec. 2, 19 (Aug. 19, 1988).

135. K. A. Saterson and M. W. Luppold, *3rd Annual Biomass Energy Systems Conference Proceedings*, SERI/TP-33-285, U.S. Department of Energy, Golden, Colo., June 5–7, 1979, pp. 245–254.

136. J. H. Cushman and A. F. Turhollow, in D. L. Klass, ed., *Energy from Biomass and Wastes XIV*, IGT, Chicago, 1991, pp. 465–480.

137. A. G. Alexander, in Ref. 136, pp. 367–374.

138. L. L. Wright, in L. L. Wright and W. G. Hohenstein, eds., *Biomass Energy Production in the United States: Situation and Outlook*, Oak Ridge National Laboratory, Oak Ridge, Tenn., Aug. 1992, Chapt. 2.

139. D. L. Klass, in Ref. 114, pp. 13–113.

140. G. Grassi, in Ref. 114, pp. 1545–1562.

141. A. Ismail and R. Quick, in Ref. 42, pp. 1063–1100.

142. J. E. Robert and E. N. Hogan, in Ref. 42, pp. 1245–1265.

143. L. D. Ostlie and T. E. Drennen, in Ref. 76, pp. 621–650.

144. A. Rehmat and M. Khinkis, in Ref. 42, pp. 1111–1139.

145. J. T. Hamrick, in Ref. 114, pp. 517–528.

146. M. L. Murphy, in Ref. 29, pp. 371–380; in Ref. 42, pp. 1167–1179; S. C. Bhattacharya and W. Wu, in Ref. 76, pp. 591–601.

147. J. F. L. Lincoln and T. C. Litchney, in Ref. 29, pp. 357–369.

148. S. G. Barnett and S. J. Morgan, in Ref. 136, pp. 191–236.

149. G. A. Norton and A. D. Levine, in Ref. 2, pp. 513–527.

150. C. Tewksbury, in Ref. 42, pp. 95–127.

151. S. M. Turner and D. A. Rowley, in Ref. 42, pp. 43–63; D. R. Patrick, in Ref. 42, pp. 65–72; R. N. Sampson, in Ref. 42, pp. 159–186.

152. U.S. Pat. 4,022,665 (May 10, 1977), S. Ghosh and D. L. Klass (IGT); U.S. Pat. 4,318,993 (Mar. 9, 1982), S. Ghosh and D. L. Klass (IGT); T. L. Miller and S. Ghosh, in Ref. 136, pp. 869–876.

153. *Acimet Technical Briefing*, Illinois Department of Energy and Natural Resources and the DuPage Group, Woodridge, Ill, Oct. 7, 1992.

154. W. J. Jewell and co-workers, in Ref. 28, pp. 669–693; R. Legrand and W. J. Jewell, in Ref. 114, pp. 1077–1095; B. De Wilde and L. De Baere, in Ref. 136, pp. 915–929; C. J. Rivard, in Ref. 24, pp. 1025–1041.

155. J. W. Deming, in Ref. 114, pp. 1097–1111.

156. R. W. Meyer and W. R. Guthrie, in Ref. 28, pp. 857–872; S. R. Harper, G. E. Valentine, and C. C. Ross, in Ref. 29, pp. 637–664; L. M. Safley and P. D. Lusk, in Ref. 136, pp. 955–980; R. R. Dague and S. Sung, in Ref. 24, pp. 1001–1023.

157. J. J. Walsh and co-workers, in Ref. 114, pp. 1115–1125; P. Fletcher, in Ref. 76, pp. 1001–1027.

158. A. J. L. Macario and co-workers, in Ref. 114, pp. 1009–1020; S. J. Schropp and co-workers, in Ref. 114, pp. 1035–1043; D. P. Chynoweth and co-workers, in Ref. 76, pp. 965–981; V. Chitra and K. Ramasamy, in Ref. 24, pp. 1043–1060.

159. A. R. Trenka and co-workers, in Ref. 42, pp. 1051–1061.

160. H. LaFontaine, in Ref. 29, pp. 561–575.

161. *Summary Report for Hot-Gas Cleanup, Compiled by Institute of Gas Technology*, for International Energy Agency, IGT, Chicago, Dec. 1991, 50 pp.

162. *Research Needs for Thermal Gasification of Biomass, Compiled by Studsvik AB Thermal Processes*, for International Energy Agency, IGT, Chicago, Mar. 1992, 4 pp.

163. *Conservation and Renewable Energy Technologies for Transportation*, DOE/CH 10093-84, U.S. Department of Energy, Washington, D.C., Nov. 1990, 20 pp.

164. G. R. Quick, in G. Robbelen, R. K. Downey, and A. Ashri, eds., *Oil Crops of the World*, McGraw-Hill Publishing Co., New York, 1989, pp. 118–131.

165. T. B. Reed, M. S. Graboski, and S. Gaver, in Ref. 42, pp. 907–914.

166. A. V. Bridgwater and G. Grassi, eds., *Biomass Pyrolysis Liquids Upgrading and Utilization*, Elsevier Applied Science, New York, 1991, 377 pp.

167. D. C. Elliott and co-workers, *Energy & Fuels* **5**, 399 (1991).

168. D. S. Scott, J. Piskorz and D. Radlein, in Ref. 24, pp. 797–809.

169. J. Diebold, in Ref. 166, pp. 341–350.

170. J. Diebold and co-workers, in E. Hogan and co-workers, eds., *Biomass Thermal Processing*, The Chameleon Press Ltd., London, 1992, pp. 101–108.

171. D. A. Johnson, G. R. Tomberlin, and W. A. Ayres, in Ref. 42, pp. 915–925.

General References

D. L. Klass, ed., *Energy from Biomass and Wastes*, Vols. 4–16, IGT, Chicago, 1980–1993.

D. L. Klass, ed., *Biomass as a Nonfossil Fuel Source*, ACS Symposium Series 144, American Chemical Society, Washington, D.C., 1981, 564 pp.

D. L. Klass and G. H. Emert, eds., *Fuels from Biomass and Wastes*, Ann Arbor Science Publishers, Inc., Ann Arbor, Mich., 1981, 592 pp.

J. L. Jones and S. B. Radding, eds., *Thermal Conversion of Solid Wastes and Biomass*, ACS Symposium Series 130, American Chemical Society, Washington, D.C., 1980, 747 pp.

S. S. Sofer and O. R. Zaborsky, eds., *Biomass Conversion Processes for Energy and Fuels*, Plenum Press, New York, 1981, 420 pp.

E. Hogan and co-workers, eds., *Biomass Thermal Processing*, The Chameleon Press Limited, London, 1992, 255 pp.

A. V. Bridgwater and G. Grassi, eds., *Biomass Pyrolysis Liquids Upgrading and Utilization*, Elsevier Applied Science, London, 1991, 377 pp.

A. V. Bridgwater, ed., *Thermochemical Processing of Biomass*, Butterworths, London, 1984, 344 pp.

Biomass, International Directory of Companies, Processes & Equipment, Macmillan Publishers, Ltd., New York, 1986, 243 pp.

D. L. Klass, ed., *A Directory of U.S. Renewable Energy Technology Vendors, Biomass, Photovoltaics, Solar Thermal, Wind*, Biomass Energy Research Association, Washington, D.C., 1990, for U.S. Agency for International Development, 74 pp.

P. F. Bente, Jr., ed., *Bio-Energy Directory*, The Bio-Energy Council, Washington, D.C., 1980, 768 pp.

Biofuels Technical Information Guide, SERI/SP-220-3366, Solar Energy Research Institute, Golden, Colo., Apr. 1989, 198 pp.

A Guide to Federal Programs in Biomass Energy, Meridian Corporation and Science Applications International Corp., Washington, D.C., Sept. 1984, 157 pp.

W. H. Smith, ed., *Biomass Energy Development*, Plenum Press, New York, 1986, 668 pp.

First Biomass Conference of the Americas: Energy, Environment, Agriculture, and Industry, NREL/CP-200-5768, DE930/0050, Proceedings Vols. I–III, National Renewable Energy Laboratory, Golden, Colo., 1993, 1942 pp.

FUELS FROM WASTE

DAVID A. TILLMAN
Foster Wheeler Environmental Corporation
Sacramento, California

A significant number and variety of organic wastes are combusted in energy recovery systems including municipal solid waste (MSW), various forms of refuse-derived fuel (RDF) produced from MSW, and municipal sewage sludge; bark and other wood wastes from sawmills and other forest industry operations; spent pulping liquor from chemical pulp mills such as kraft and sulfite mills; wastewater treatment solids (WTS) or sludges from pulp and paper operations; agribusiness wastes including bagasse from sugar-refining operations, rice hulls, orchard and vineyard prunings, cotton gin trash, and a host of other food and fiber-producing operations; manure from feedlots and dairy cattle, chickens, and other agricultural animals; methane-rich gases generated from municipal-waste landfills; industrial trash and specific wastes such as demolition debris, tire derived fuel (TDF), broken pallets, unrecyclable paper wastes, and related materials; off-gases from pulp mills and chemical manufacturers; incinerable hazardous wastes generated regularly as a function of production processes, eg, spent solvents, or found on Superfund sites targeted for clean-up; and a broad range of other specific specialty wastes. The practice of incinerating these materials has become increasingly prevalent (ca 1990) in order to accomplish disposal in a cost-effective, environmentally sensitive manner. The combustion of such wastes already contributes some 5 EJ (5×10^{15} Btu) to the U.S. economy and over 15 EJ ($>14 \times 10^{15}$ Btu) to the economies of the industrialized world (1). Combustion of such wastes reduces the volume of material which must be disposed of in a landfill, reduces the airborne emissions resulting from plant operations and landfill operations, and permits some economic benefit through energy recovery.

The technologies used to combust wastes depend on the form and location of components to be burned. Typically solid wastes are burned, alone or in combination, and both with and without supplementary fossil fuels. Solid wastes can be burned in mass-burn or pile-burning systems such as hearth furnaces, spreader–stokers, ashing and slagging rotary kilns, or fluidized beds. The choice of combustion technology depends on the degree of waste preparation which is practical; the availability of existing combustion systems, eg, a spreader–stoker for hog fuel utilization, adapted to the cofiring of hog fuel and WTS; and the type of energy recovery contemplated. Energy recovery from the solid wastes can be accomplished in the form of medium or high pressure steam, eg, 4.5–8.6 MPa/672–783 K) suitable for cogeneration or condensing power generation purposes; low pressure steam, eg, 314–1030 kPa, saturated, suitable for process purposes; or the direct production of process heat in the form of heated air or hot combustion products. Energy recovery from gaseous wastes can be accomplished through electricity generation from gas-fired boilers, combustion turbines, or internal combustion engines. Alternatively, these gaseous fuels can be used to generate process heat in conventional fashion.

The success of waste-to-energy programs using municipal wastes and biowastes reduces the volume of material being interred in the ground in landfills. This action also changes the character of materials being landfilled, reducing the organic content with its associated generation of methane gas, and leachates with their significant concentrations of organic compounds. Waste-to-energy, applied to municipal and biomass wastes, can simultaneously provide renewable energy while addressing environmental issues.

Critical concerns associated with energy generation from wastes include fuel composition characteristics; combustion characteristics; formation and control of airborne emissions including both criteria pollutants and air toxics, eg, trace metals; and the characteristics of bottom and flyash generated from waste combustion. These issues are particularly important given the U.S. Clean Air Act Amendments of 1990, the Resource Recovery and Conser-

vation Act (RRCA), and related state and regional regulations. Further, these issues are of critical importance given the capital intensive nature of organic waste-to-energy systems.

See also COMMERCIAL AVAILABILITY OF ENERGY TECHNOLOGY; ELECTRIC POWER GENERATION; EXHAUST CONTROL, INDUSTRIAL; INCINERATION; WASTE-TO-ENERGY.

FUEL CHARACTERISTICS OF ORGANIC WASTES

Fuel characteristics of organic wastes include physical characteristics such as state, specific gravity, bulk density, porosity, and void volume, and related thermal properties; traditional chemical analyses such as proximate and ultimate analyses, including chlorine; calorific content; elemental analyses of the ash, including trace metal contents, base/acid, slagging, and fouling ratios of the various ash products; and certain chemical structural analyses such as aromaticity. These characteristics are governed by the sources of waste-based fuels. They determine the performance of materials in fuel preparation systems such as particle size reduction and drying systems, and also govern the combustion characteristics of the various wastes being burned.

Sources of Waste-Based Fuels

The general architecture of waste-based fuels is a function of waste origination. MSW characteristics are governed by the product composition of the waste stream, as shown in Table 1. The composition of RDF is governed by the processing technologies used to generate the fuel. RDF production technologies involve, at a minimum, coarse shredding of the MSW stream, followed by magnetic separation of ferrous metals. Primary separation techniques for concentration of combustibles involve trommels, air classifiers, or eccentric screens. Trommels have become the most popular separation systems; their overall separation efficiency can be as high as 98.5% (Table 2). Process flow sheets using trommel separation of MSW follow the pattern shown in Figure 1. The composition of MSW, and RDF, ultimately is a function not only of the general composition of the waste stream and the RDF production technology, but also of community and industrial recycling programs. Such programs are acceler-

Table 1. Product Composition for Municipal Solid Waste,[a] Wt %

Product	1990[b]	2000[c]
Paper and paperboard	38.3	41.0
Yard waste	17.0	15.3
Food waste	7.7	6.8
Plastics	8.3	9.8
Wood	3.7	3.8
Textiles	2.2	2.2
Rubber and leather	2.5	2.4
Glass	8.8	7.6
Metals	9.4	9.0
Miscellaneous	2.1	2.1

[a] Ref. 2.
[b] Approximate.
[c] Estimated.

Table 2. MSW Separation Efficiencies for Trommels as a Function of Waste Component,[a] Wt %

Waste component	Separation efficiency
Paper, plastic	61.1–69.4
Other combustibles[b]	74.6–86.8
Ferrous metal	61.6–80.1
Aluminum	76.7–93.6
Glass, stones, and other	96.6–100
Fines	97.0–98.0
Overall efficiency	*81.0–98.5*

[a] Ref. 3.
[b] For example, wood.

ating and will influence the amount and relative concentration of paper, plastic, aluminum, and other commodities in the waste stream.

The basic architecture of wood-waste fuels is governed by sawmill or plywood mill configuration, and the consequent blend of bark, trim ends, sawdust, planer shavings, and related residuals. All chippable wastes typically are directed to pulp chips. Planer shavings and some sawdust may be directed to alternative products including oriented strand board (OSB), particleboard, animal bedding, a range of other materials applications, and fuel. The characteristics of pulp-mill wastes, eg, bark, WTS, and spent pulping liquor, also are determined by the production processes. The characteristics of wastes from food processing, eg, bagasse, rice hulls, peach pits, cotton gin trash, etc, also are governed by the basic product manufacturing technology and its efficiency of separation.

Physical Properties

Physical properties of waste as fuels are defined in accordance with the specific materials under consideration. The greatest degree of definition exists for wood and related biofuels. The least degree of definition exists for MSW, related RDF products, and the broad array of hazardous wastes. Table 3 compares the physical property data of some representative combustible wastes with the traditional fossil fuel bituminous coal. The solid organic wastes typically have specific gravities or bulk densities much lower than those associated with coal and lignite.

Specific gravity is the most critical of the characteristics in Table 3. It is governed by ash content of the material, is the primary determinant of bulk density, along with particle size and shape, and is related to specific heat and other thermal properties. Specific gravity governs the porosity or fractional void volume of the waste material, ie,

$$FVV = (1 - SG)/1.5 \qquad (1)$$

where FVV is fractional void volume, SG is specific gravity, and the value, 1.5, is the approximate specific gravity of the cell wall in wood fiber (5). Specific gravity and moisture content (MC) together determine thermal conductivity characteristics, k, of cellulosic waste-based fuels:

$$k_{MC<30\%} = SG (1.39 + 0.028 \times MC) + 0.165 \qquad (2)$$

and

$$k_{MC>30\%} = SG (1.39 + 0.028 \times MC) + 0.165 \qquad (3)$$

Specific gravity is directly related to the bulk density of waste fuels prepared in a variety of ways. Solid oven-dry (OD) wood, for example, has a typical bulk density of 48.1 kg/m³ (30 lb/ft³). In coarse hogged form, eg, <1.9-cm minor dimension, this bulk density declines to about 24 kg/m³ (15 lb/ft³). In pulverized form, at a particle size <0.16 cm, this bulk density declines to 16–19 kg/m³ (10–2 lb/ft³). Similar relationships hold for municipal waste, agricultural wastes, and related fuels.

Chemical Composition

Chemical compositional data include proximate and ultimate analyses, measures of aromaticity and reactivity, elemental composition of ash, and trace metal compositions of fuel and ash. All of these characteristics impact the combustion processes associated with wastes as fuels. Table 4 presents an analysis of a variety of wood-waste fuels; these energy sources have modest energy contents.

The analysis of solid fuels (Table 4) contains the bases for calculating reactivities, ie, volatile:fixed carbon ratios, volatile carbon:total carbon ratios, hydrogen:carbon and oxygen:carbon ratios, and aromaticity, which is estimated

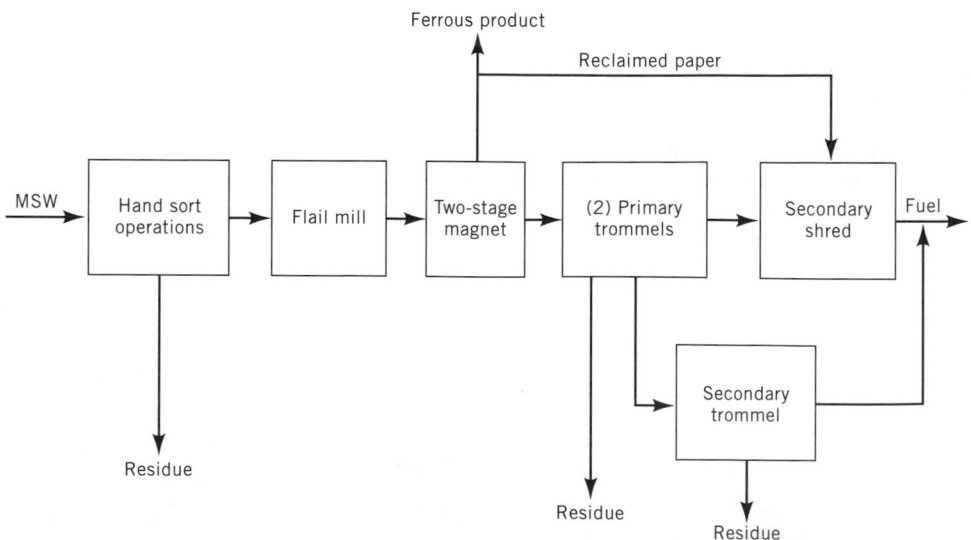

Figure 1. Simplified schematic flow sheet for the production of a moderate RDF, using trommel separation.

Table 3. Physical Properties of Waste-Based Fuels[a]

Fuel	Specific Gravity[b]	Bulk Density,[c] kg/m³	Moisture Content, wt %
Municipal waste		160–320	25–35
Waste paper	1.2–1.4	80–160	15–25
Waste wood	0.37–0.65	100–320	5–15[d]
			40–65[e]
Bagasse			50–55
Rice hulls			7–10
Orchard and vineyard prunings	0.45–0.55		20–40
Bituminous coal	1.12–1.35	672–1393	3.5–5.0

[a] Refs. 1 and 4.
[b] Ovendry.
[c] To convert kg/m² to lb/ft³, divide by 16.01.
[d] Dry waste.
[e] Wet waste.

from the chemical components of the waste stream. The typical source of aromatic carbon in waste fuels, such as municipal solid waste and wood waste, is lignin (qv) found in either groundwood-based papers or wood products. Lignin [9005-53-2] has a typical empirical formula (1) of $C_9H_{10}(OCH_3)_{0.9-1.7}$, and a higher heating value (HHV) of 26.7 MJ/kg (11,500 Btu/lb), resulting from its basic building blocks of phenyl propane units. These basic building blocks contain aromatic structures. Other sources of aromatic structures in waste fuels include plastic polymers in the waste stream. The aromaticity of a solid fuel can be estimated as a function of H:C atomic ratio (6):

H:C atomic ratio	Carbons as aromatic carbons, %
1.5	0.0
1.2	3.0
0.7	9.0
0.5	16.0

Typically, 40–50% of the carbon atoms in lignite are in aromatic structures while 60–70% of the carbon atoms in Illinois bituminous coal are in aromatic structures (7,8). By all of these measures, waste fuels are significantly more reactive than coal, peat, and other combustible solids.

The chemical analysis of waste fuels also demonstrates that the wood-based fuels contain virtually no sulfur and little nitrogen. Unless the hog fuel contains bark from logs previously stored in salt-water, the chlorine content is very modest to nonmeasurable.

Municipal waste contains, nominally, about 0.5% nitrogen and 0.5% chlorine, the latter coming largely from plastics. Municipal waste also contains moderate amounts of sulfur. The actual composition of MSW, or RDF generated from MSW, is a function of the relative percentages of various components in the waste stream as shown in Table 5. Use of these wastes provides a means for reducing acid gas emissions from energy generation.

Ash Characteristics. The elemental ash composition of biomass waste and municipal solid waste differs dramatically from that of coal (qv). Wood wastes have ash compo-

Table 4. Analysis of Wood-Based Fuels,[a] Wt %

Wood material	Volatile Matter	Fixed Carbon	Ash	C	H	O	N	S	HHV,[b] MJ/kg
Big leaf maple	87.9	11.5	0.6	49.9	6.1	43.3	0.14	0.03	16.9
Douglas fir	87.3	12.6	0.1	50.6	6.2	43.0	0.06	0.02	18.3
Douglas fir bark	73.6	25.9	0.5	54.1	6.1	38.8	0.17	[c]	19.7
Oak									
Black[d]	85.6	13.0	1.4	49.0	6.0	43.5	0.15	0.02	16.8
Tan[e]	87.1	12.4	0.5	48.3	6.1	45.0	0.03	0.03	17.2
Oak bark			5.3	49.7	5.4	39.3	0.2	0.1	17.5
Pine bark			2.9	53.4	5.6	37.9	0.1	0.1	18.4
Pitch pine			1.1	59.0	7.2	32.7			21.8
Popular			0.7	51.6	6.3	41.5			17.2
Red alder	87.1	12.5	0.4	49.6	6.1	43.8	0.13	0.07	17.3
Red alder bark	77.3	19.7	3.0	50.9	5.5	40.7	0.39		17.2
Western hemlock[f]	87.0	12.7	0.3	50.4	5.8	41.4	0.1	0.1	19.8

[a] Ref. 4.
[b] Higher heating value (OD basis); to convert MJ/g to Btu/lb, multiply by 430.3.
[c] Trace amounts.
[d] Black oak bark has 81.0 wt % volatile matter, 16.9 wt % fixed carbon, and 2.1 wt % ash.
[e] Tan oak bark has 76.3 wt % voltatile matter, 20.8 wt % fixed carbon, and 2.9 wt % ash.
[f] Western hemlock bark has 73.9 wt % volatile matter, 24.3 % fixed carbon, and 0.8 wt % ash.

Table 5. Components of Municipal Solid Waste[a]

Material	Components, wt %								
	Carbon	Hydrogen	Oxygen	Nitrogen	Chlorine	Sulfur	Moisture	Ash	HHV, MJ/kg[b]
Corrugated paper	36.79	5.08	35.41	0.11	0.12	0.23	20.0	2.26	13.0
Newsprint	36.62	4.66	31.76	0.11	0.11	0.19	25.0	1.55	13.0
Magazine stock	32.93	4.64	32.85	0.11	0.13	0.21	16.0	13.13	11.4
Other paper	32.41	4.51	29.91	0.31	0.61	0.19	23.0	9.06	11.5
Plastics	56.43	7.79	8.05	0.85	3.00	0.29	15.0	8.59	24.2
Rubber and leather	43.09	5.37	11.57	1.34	4.97	1.17	10.0	22.49	17.6
Wood	41.20	5.03	34.55	0.24	0.09	0.07	16.0	2.82	14.5
Textiles	37.23	5.02	27.11	3.11	0.27	0.28	25.0	1.98	13.8
Yard waste	23.29	2.93	17.54	0.89	0.13	0.15	45.0	10.07	8.37
Food waste	17.93	2.55	12.85	1.13	0.38	0.06	60.0	5.10	6.82
Fines[c]	15.03	1.91	12.15	0.50	0.36	0.15	25.0	44.90	5.41

[a] Ref. 9.
[b] To convert MJ/kg to Btu/lb, multiply by 430.3.
[c] Smaller than 2.54 cm (1 in.).

sitions that are quite alkaline (Table 6) and that have consequent low ash fusion temperatures (Table 7). When firing solid wastes with coal or lignite, the potential exists to have eutectic mixtures formed by the two ash products.

The Clean Air Act of 1990 has made trace metal content in fuels and wastes the final ash-related compositional characteristic of significance. Considerable attention is paid (ca 1993) to emissions of such metals as arsenic, cadmium, chromium, lead, mercury, silver, and zinc. The concentration of these metals in both grate ash and flyash is of significance as a result of federal and state requirements; of particular importance is the mobility of metals. This mobility, and the consequent toxicity of the ash product, is determined by the Toxic Characteristic Leaching Procedure (tclp) test. Tables 8–10 present trace metal contents for wood wastes and agricultural wastes, municipal waste, and refuse-derived fuel, respectively. In Table 8, the specific concentration of various components in the RDF governs the expected average concentration of trace metals.

Biofuels, ie, wood and agricultural waste, are relatively low in metal contents, and typically have a lower metals content when compared to coals being burned for energy generation. However, municipal waste and its derivative fuels (RDF) can be quite high in trace metals. The RDF production process removes approximately 67% of the incoming metals content, but significant quantities of components such as lead and cadmium remain in some compositions of RDF. These metals must be controlled for safe energy generation from such combustible materials. The wood waste in RDF is typically not the same as the wood waste from forest products manufacture. Commonly the wood in RDF is treated with compounds, eg, copper chromium arsenate (CCA), which make it more suitable in outdoor service, such as in deck construction. Such wood treating adds trace metals to the fuel feed (13).

COMBUSTION OF SOLID WASTE-BASED FUELS

It is useful to examine the combustion process applied to solid wastes as fuels and sources of energy. All solid wastes are quite variable in composition, moisture content, and heating value. Consequently, they typically are burned in systems such as grate-fired furnaces or fluidized-bed boilers where significant fuel variability can be tolerated.

Combustion characteristics of consequence include the overall mechanism of solid waste combustion, factors governing rates of waste fuels combustion, temperatures as-

Table 6. Elemental Analysis of Wood Waste Ash[a]

Compound	CAS Registry Number	Source, wt%		
		Pine Bark	Oak Bark	Spruce Bark
SiO_2	[14808-60-7]	39.0	11.1	32.0
Fe_2O_3	[1309-37-1]	3.0	3.3	6.4
TiO_2	[13463-67-7]	0.2	0.1	0.8
Al_2O_3	[1344-28-1]	14.0	0.1	11.0
MnO_4	[12502-70-4]	b	b	1.5
CaO	[1305-78-8]	25.5	64.5	25.3
MgO	[1309-48-4]	6.5	1.2	4.1
Na_2O	[12401-86-4]	1.3	8.9	8.0
K_2O	[12136-45-7]	6.0	0.2	2.4
SO_3	[7446-11-9]	0.3	2.0	2.1
Cl	[7782-50-5]	b	b	b

[a] Ref. 10.
[b] Trace amounts.

Table 7. Ash Fusion Temperatures for Some Wood Waste Ash,[a] K

Wood Species	Initial		Softening		Fluid	
	Oxidizing	Reducing	Oxidizing	Reducing	Oxidizing	Reducing
Tan oak	1663	1650	1713	1711	1730	1727
Pine bark	1483	1467	1522	1500	1561	1540
Oak bark	1744	1750	1772	1766	1783	1777

[a] Refs. 4 and 10.

sociated with waste oxidation, and pollution-formation mechanisms.

Mechanisms and Rates of Combustion

All solid fuels and wastes burn according to a general global mechanism (Fig. 2). The solid particle is first heated. Following heating, the particle dries as the moisture bound in the pore structure and on the surface of the particle evaporates. Only after moisture evolution does pyrolysis initiate to any great extent. The pyrolysis process is followed by char oxidation, which completes the process.

The rate of solid waste combustion is controlled by diffusion, rather than by reaction kinetics. In general, the time required for combustion of a single particle of waste (1) can be expressed as:

$$T_b = T_h + T_d + T_p + T_{co} \qquad (4)$$

where T_b is time for complete particle burnout, T_h is time for initial particle heatup, T_d is time required for particle drying, T_p is time required to pyrolyze the particle into volatiles and char, and T_{co} is time for char oxidation. The first two terms, initial heating plus drying, $T_h + T_d$, can be taken as the drying step. This time component is governed by the temperature of the environment, the particle size, the moisture content of the particle, and the porosity of the particle. The term T_p is strictly governed by heat transfer through the particle (14,15). The rate of pyrolysis is governed by the heat capacity of the solid waste, its porosity, and its thermal conductivity. The T_{co} term is mass-transfer-limited, with diffusion of oxygen to the surface of the char particle being rate-limiting. Of these steps, either drying or char oxidation may be rate-limiting, as shown in Figure 3, depending on the moisture content of the solid waste.

Temperatures Associated with Combustion

The temperatures achieved by solid waste combustion are typically lower than those associated with fossil fuel oxidation, and are governed by the following general equation (1):

$$T_{f, \text{solid waste}} = [695 - 10.1\,\mathrm{MC}_t + 1734(1/\mathrm{SR}) + 0.61(A - 298)]\,\mathrm{K} \qquad (5)$$

Where T_f is flame temperature is K; MC_t is moisture content of the waste, expressed on a total weight basis; SR is defined as stoichiometric ratio or moles O_2 available/moles O_2 required for complete oxidation of the carbon, hydrogen, and sulfur in the fuel, ie, $1/\mathrm{SR}$ = equivalence ratio; and A is temperature of the combustion air, expressed in K. In English units, this equation is as follows:

$$T_{f, \text{solid waste}} = [3870 - 15.6\,\mathrm{MC}_t - 130.4\,\mathrm{EO}_2 + 0.59(A - 77)]\,°\mathrm{F} \qquad (6)$$

Table 8. Trace Metal Concentrations[a] in Ash from Agricultural Biofuels and Wood-Fired Boilers, mg/kg

Metal	Agricultural Biofuel			Wood-Fired Boilers[b]
	Cotton Gin Trash	Orchard Prunings	Vineyard Prunings	
Barium	120	220	41	130
Silver	<0.08	<0.08	<0.08	<0.08
Arsenic	12	5.5	3.4	3.0–6.3
Beryllium	0.1	0.1	0.06	0.1
Cadmium	1.1	0.36	0.39	1.5–16
Cobalt	14	9.0	2.8	8.5–20
Chromium	20	12	11	16.8–25
Copper	23	14	31	40–76.9
Mercury	<0.05	<0.05	<0.05	<0.05–<0.5
Molybdenum	16	2	2	3.0–14
Nickel	4.6	5.8	4.4	11–50
Lead	21	22	55	38–70
Antimony	10	10	10	10
Selenium	<0.2	<0.2	<0.2	5.0
Vanadium	20	12	11	26–27
Zinc	87	190	40	130–560
Thallium	15	10	2	6.5

[a] Ref. 1.
[b] Range of concentrations from various locations.

Table 9. Trace Metals in Municipal Solid Waste and Solid Waste Ash,[a] ppmw

Metal	Solid Waste[b,c]	Solid Waste Ash
Arsenic	0.73–12.5	2.9–50
Barium	19.8–675	79–2,700
Beryllium	ND–0.6	ND–2.4
Boron	6–43.5	24–174
Cadmium	0.05–25	0.18–100
Chromium	3.0–375	12–1,500
Cobalt	0.43–22.75	1.7–91
Copper	10–1,475	40–5,900
Lead	7.75–9.15	31–36,600
Magnesium	175–4,000	700–16,000
Molybdenum	0.6–72.5	2.4–290
Manganese	3.5–782.5	14–3,130
Mercury	ND–4.38	0.05–17.5
Nickel	3.25–3,228	13–12,910
Selenium	0.03–12.5	0.10–50
Strontium	3.0–160	12–640
Zinc	23–11,500	92–46,000

[a] Range of concentration. Ref. 11.
[b] Based on ash measurements. Imputed to waste.
[c] ND = nondetectable.

where EO_2 is excess oxygen in the stack gas, ie, total, not dry, basis.

Whereas solid wastes can achieve significant flame temperatures, they are substantially below those associated with fossil fuels. These differences are largely caused by the lower calorific value; the chemical composition, eg, oxygen content of the waste; and the higher moisture and ash contents commonly associated with various solid wastes. Typically for grate-fired systems the use of wastes as fuel requires maintaining temperatures in excess of 1256 K (1800°F) and for residence times exceeding 2 s in order to ensure complete combustion and minimize dioxin and furan formation. As shown from equation 5, such temperatures are readily achieved under most conditions.

Given the mechanisms and temperatures, waste combustion systems typically employ higher percentages of excess air, and typically also have lower cross-sectional and volumetric heat release rates than those associated with fossil fuels. Representative combustion conditions are shown in Table 11 for wet wood waste, with 50–60% moisture total basis, municipal solid waste, and RDF.

Formation of Airborne Emissions

Airborne emissions are formed from combustion of waste fuels as a function of certain physical and chemical reactions and mechanisms. In grate-fired systems, particulate emissions result from particles being swept through the furnace and boiler in the gaseous combustion products, and from incomplete oxidation of the solid particles, with consequent char carryover. If pile burning is used, eg, the mass burn units employed for unprocessed MSW, typically only 20–25% of the unburned solids and inerts exit the combustion system as flyash. If spreader–stoker technologies are employed, between 75 and 90% of the unburned solids and inerts may exit the combustion system in the form of flyash.

Sulfur dioxide [7446-09-5] is formed as a result of sulfur oxidation, and hydrogen chloride is formed when chlo-

Table 10. Analysis of Refuse-Derived Fuel[a]

Parameter	Glossy Paper	Nonglossy Paper	Cardboard	Film Plastics	Rubber, Leather, and Hard Plastics	Wood and Textiles	Other Organics	Total RDF
				Trace metals, mg/kg fuel				
Arsenic	3.1	3.3	3.5	2.7	2.5	5.2	4.6	4.0
Barium	285.1	78.9	48.7	186.5	724.3	96.7	210.0	173.2
Beryllium	1.1	1.3	1.2	0.5	0.4	1.5	1.5	1.3
Cadmium	1.1	1.3	3.8	7.7	17.3	3.0	3.1	3.4
Chromium	23.8	37.3	23.2	69.4	95.9	34.8	44.5	42.7
Copper	74.8	40.3	27.0	2740.7	12.1	202.3	61.4	220.1
Lead	88.4	621.2	66.2	836.6	668.1	747.6	475.1	495.5
Manganese	61.2	137.6	101.1	311.8	83.1	183.9	367.3	260.4
Mercury	0.3	0.7	0.4	1.0	0.4	0.9	1.2	0.9
Nickel	10.4	15.5	25.5	45.6	170.4	27.4	17.7	24.4
Selenium	3.1	2.9	3.3	2.1	2.0	3.3	3.0	2.9
Strontium	62.4	73.2	47.8	88.5	88.6	198.9	474.8	283.2
Zinc	164.5	227.6	161.4	482.2	2494.5	449.4	360.0	380.2
				Ultimate analysis, wt %				
Carbon	43.4	47.3	49.6	59.8	53.8	50.1	34.6	41.1
Hydrogen	5.3	6.1	6.4	8.2	8.9	6.0	4.3	5.3
Oxygen	27.5	32.0	35.7	13.8	23.3	31.5	41.1	35.2
Nitrogen	0.62	1.58	0.72	1.01	0.83	1.07	1.07	1.12
Sulfur	0.25	0.25	0.24	0.56	0.57	0.28	0.38	0.34
Chlorine	0.04	0.04	0.05	0.10	0.05	0.05	0.01	0.07
Ash	23.0	12.7	7.4	16.6	12.5	11.0	18.3	16.5
Higher heating value, MJ/kg[b]	14.7	19.7	18.5	31.0	25.4	21.0	16.5	18.7

[a] Fuel produced in Tacoma, Wash. Values on ovendry (OD) fuel basis. Ref. 12.
[b] To convert MJ/kg to Btu/lb, multiply by 430.3.

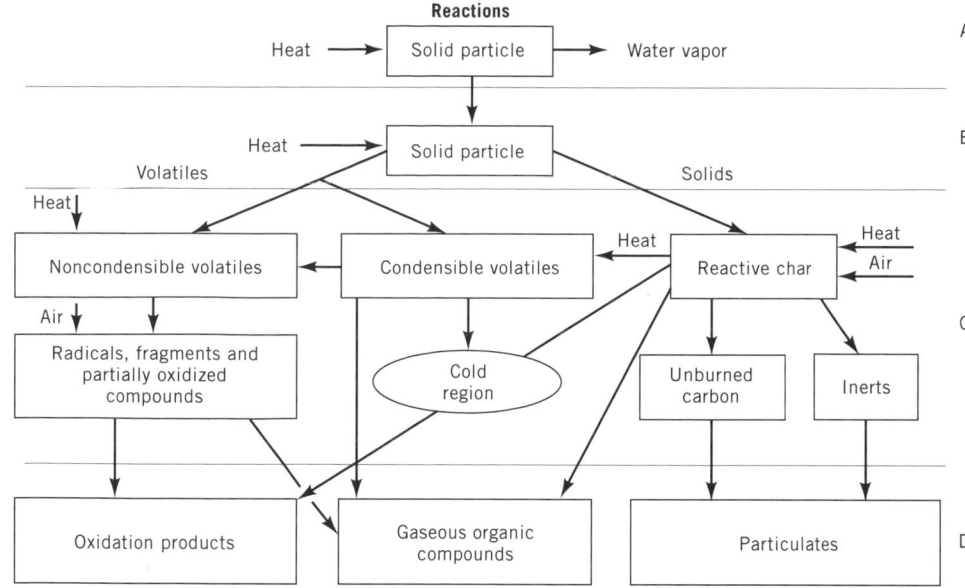

Figure 2. Overall schematic of solid fuel combustion (1). Reaction sequence is A, heating and drying; B, solid particle pyrolysis; C, oxidation; and D, post-combustion. In the oxidation sequence, left and center comprise the gas-phase region, right is the gas–solids region. Noncondensible volatiles include CO, CO_2, CH_4, NH_3, H_2O; condensible volatiles are C-6–C-20 compounds; oxidation products are CO_2, H_2O, O_2, N_2, NO_1; gaseous organic compounds are CO, hydrocarbons, and polyaromatic hydrocarbons (PAHs); and particulates are inerts, condensation products, and solid carbon products.

rides from plastics compete with oxygen as an oxidant for hydrogen. Typically the sulfur is considered to react completely to form SO_2, and the chlorine is treated as the preferred oxidant for hydrogen. In practice, however, significant fractions of sulfur do not oxidize completely, and at high temperatures some of the chlorine atoms may not form HCl.

Nitrogen oxide, NO_x, formation results from conversion of nitrogen in the fuel to NO, since combustion temperatures are below those typically associated with thermal NO_x formation, eg, 1483°C as the threshold for thermal NO_x has been documented (17,18). The conversion of fuel nitrogen to NO_x typically proceeds along the pathways of nitrogen volatilization followed by oxidation of the nitrogen volatiles in the presence of excess oxygen. In the absence of available oxygen, the nitrogen volatiles react with each other to form N_2. Conversion of nitrogen from waste fuels into NO_x is typically 15–25% of the fuel nitrogen converted, depending on combustion technology and firing conditions.

Formation of emissions from fluidized-bed combustion is considerably different from that associated with grate-fired systems. Flyash generation is a design parameter, and typically >90% of all solids are removed from the system as flyash. SO_2 and HCl are controlled by reactions with calcium in the bed, where the limestone fed to the bed first calcines to CaO and CO_2, and then the lime reacts with sulfur dioxide and oxygen, or with hydrogen chloride, to form calcium sulfate and calcium chloride, respectively. SO_2 and HCl capture rates of 70–90% are readily achieved with fluidized beds. The limestone in the bed plus the very low combustion temperatures inhibit conversion of fuel N to NO_x.

Trace metal emissions from waste combustion are a function of metal content in the feed, combustion temperatures, and the percentage of ash exiting the combustion

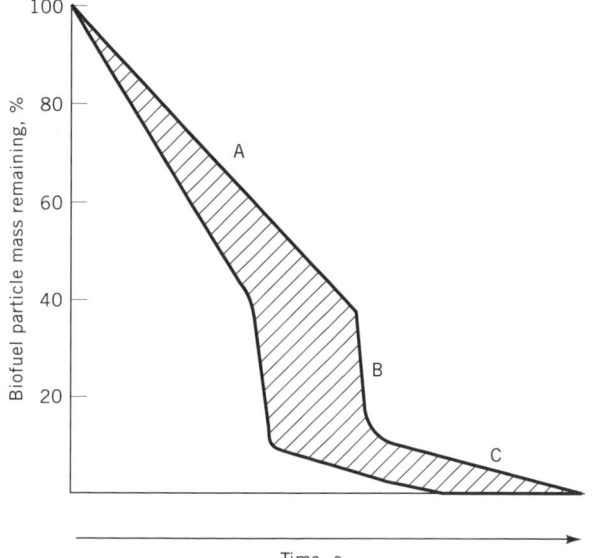

Figure 3. Schematic of the sequential nature of waste fuel combustion (1). A, particle heating and drying; B, solid particle pyrolysis; and C, char oxidation. A and C may be rate-limiting.

Table 11. Combustion Conditions for Conventional Waste Fuel Boilers[a]

Parameter	Wood[b]	MSW	RDF
Maximum fuel moisture, %	55–60	30–40	20–35
Grate fuel feed rate, kg/m²h[c]	1000–1500		
Grate heat release, GJ/m²h[d]	8.5–11.35	3.4	5.7–8.5
Volumetric heat release, MJ/m³h[e]	480–560	335–410	450–480
Stoichiometric ratio	1.25–1.5	1.8–2.0	1.6–1.8
Excess air, %	25–50	80–100	60–80
Representative overfire air, %	20–40	30–40	30–40

[a] Refs. 1 and 16.
[b] Wet wood waste, 50–60% moisture total basis.
[c] To convert kg/(m²h) to lb/(ft²h), multiply by 0.204.
[d] To convert GJ/(m²h) to Btu/(ft²h), multiply by 8.8×10^4.
[e] To convert MJ/(m³h) to Btu/(ft³h), multiply by 26.9.

chamber as flyash. They are also a function of the temperatures in the air pollution control system, eg, the precipitator, baghouse, or scrubber, and the consequent degree to which these metals undergo homogeneous nucleation and become a fine flume, eg, submicron particles in the flyash, or undergo heterogeneous condensation on existing flyash particles. For some metals, such as arsenic and lead, emissions are also a function of the combustion system and the presence of lime. In fluidized beds, it has been shown that the arsenic and lead are captured and stabilized by the presence of reactive lime (1,19).

Dioxin [828-00-2] and furan [110-00-9] emissions are the final pollutants of consideration and are of most concern for combustors using MSW, RDF, or hazardous waste. Dioxins are formed, at some concentration level ranging from inconsequential to problematical, whenever aromatic compounds and trace quantities of chlorine are present in the boiler feed. Several mechanisms have been postulated for dioxin and furan emission formation including the passage of such molecules, unreacted, through the furnace; the formation of dioxins and furans from such precursers as lignins and trace concentrations of chlorine; and the formation of dioxins in post-combustion reactions in the economizer section at temperatures of about 550–700 K (2,20,21). Of these, the post-combustion mechanism is shown to dominate. However, the impact of this mechanism can be minimized by maintaining temperatures in excess of 983°C for 2 s, while ensuring complete mixing of fuels and oxidants and ensuring the absence of cool zones in the furnace. A general equation for approximating dioxin and furan emissions is as follows (22):

$$D + F_{\mu g/Nm^3} = 0.0376(EA) - 3.305 \qquad (7)$$

where $D + F$ is dioxins plus furans in the gaseous combustion products, corrected to 12% CO_2, and EA is the percentage of excess air above about 70%. Dioxin emissions are a problem more for MSW burners than other types of waste fuel systems, largely as a result of chlorine in the waste feed. Dioxin emissions typically are minimized in fluidized-bed combustion as a consequence of the solids mixing and solids turbulence in the bed.

APPLICATIONS OF FUELS FROM WASTE

Because fuels from combustible organic wastes have long been economic in specific industries such as pulp mills, sawmills, sugar mills or factories, and other biomass processing operations, and because municipal waste-to-energy is becoming increasingly cost effective, these systems are continuing to be installed. The typical industrial system is used either to generate process steam or to generate both electricity and steam in a cogeneration application. Typically these applications involve power boilers which are the essential source of process energy in pulp mills and food processing operations. Typical larger installations generate some 200–300 t/h of steam used to generate 25–35 MW plus process heat. As stand-alone electricity generating stations, these units are capable of

50–60 MW, with a typical thermal efficiency of 65–75%. Thermal efficiency depends on moisture and ash content of the feed waste, consequent firing conditions employed, and extent of heat-transfer surface available for combustion air heating as well as steam generation.

Since the early 1980s, there have been several stand-alone power plants built to fire biomass wastes including such materials as wood waste, rice hulls, and vineyard prunings; these facilities typically generate 20–50 MW_e for sale to electric utilities. MSW and RDF are typically consumed in condensing power plants generating 15–50 MW_e while reducing the volume of solids to be landfilled. These units have thermal efficiencies comparable to the large power boilers of the pulp and paper industry, depending again on waste fuel condition and firing strategy.

There has been increased interest in firing wood waste as a supplement to coal in either pulverized coal (PC) or cyclone boilers at 1–5% of heat input. This application has been demonstrated by such electric utilities as Santee-Cooper, Tennessee Valley Authority, Georgia Power, Delmarva, and Northern States Power. Cofiring wood waste with coal in higher percentages, eg, 10–15% of heat input, in PC and cyclone boilers is being carefully considered by the Electric Power Research Institute (EPRI) and Tennessee Valley Authority (TVA). This practice may have the potential to maximize the thermal efficiency of waste fuel combustion. If this practice becomes widespread, it will offer another avenue for use of fuels from waste.

Reprinted from *Kirk-Othmer Encyclopedia of Chemical Technology*, 4th ed., vol. 12, John Wiley & Sons, Inc., New York, 1995.

BIBLIOGRAPHY

1. D. A. Tillman, *The Combustion of Solid Fuels and Wastes*, Academic Press, San Diego, Calif., 1991.

2. W. R. Seeker, W. S. Lanier, and M. P. Heap, *Municipal Waste Combustion Study: Combustion Control of MSW Combustors to Minimize Emissions of Trace Organics*, EER Corporation, Irvine, Calif., 1987.

3. J. Barton, *Evaluation of Trommels for Waste to Energy Plants, Phase I*, National Center for Resource Recovery, Washington, D.C., 1982.

4. A. J. Rossi, in D. A. Tillman and E. Jahn, eds., *Progress in Biomass Conversion*, Vol. 5, Academic Press, New York, 1984, pp. 69–99.

5. U.S. Forest Service, *Wood Handbook: Wood as an Engineering Material*, U.S. Government Printing Office, Washington, D.C., 1974.

6. F. Shafizadeh and Y. Sekuguchi, *Carbon* **21**(5), 511–516 (1983).

7. K. E. Chung and I. B. Goldberg, in *Proceedings of the 12th Annual EPRI Contractors' Conference on Fuel Science and Conversion*, EPRI, Palo Alto, Calif., 1988.

8. K. E. Chung, I. B. Goldberg, and J. J. Ratto, *Chemical Structure and Liquefaction Reactivity of Coal*, EPRI, Palo Alto, Calif., 1987.

9. R. E. Kaiser, "Physical-Chemical Character of Municipal Refuse," *Proceedings of the 1975 International Symposium on Energy Recovery from Refuse*, University of Louisville, Louisville, Ky., 1975.

10. *Steam: Its Generation and Use*, 40th ed., Babcock & Wilcox, Barberton, Ohio, 1991.

11. A. .M. Ujihara and M. Gough, *Managing Ash From Municipal Waste Incinerators*, Resources for the Future, Washington, D.C., 1989.

12. D. A. Tillman and C. Leone, "Control of Trace Metals in Flyash at the Tacoma, Washington Multifuels Incinerator," *Proceedings of the American Flame Research Committee Fall International Symposium*, San Francisco, 1990.

13. D. A. Tillman, "The Fate of Arsenic at the Tacoma Steam Plant #2," paper presented at *1992 Fall International Symposium*, American Flame Research Committee, Boston, Mass., 1992.

14. M. Hertzberg, I. A. Zlochower, and J. Edwards, *Coal Particle Pyrolysis Temperatures and Mechanisms*, RI 9169, U.S. Department of the Interior, Bureau of Mines, Washington, D.C., 1988.

15. M. Hertzberg, I. Zlochower, R. Conti, and K. Cashdollar, *Am. Chem. Soc.* **32**(3), 24–41 (1987).

16. D. A. Tillman, A. J. Rossi, and K. M. Vick, *Incineration of Municipal and Hazardous Solid Wastes*, Academic Press, San Diego, Calif., 1989.

17. D. W. Pershing and J. Wendt, *Proceedings of the 16th International Symposium*, The Combustion Institute, Pittsburgh, Pa., 1976.

18. D. W. Pershing and co-workers, *Proceedings of the 17th International Symposium*, The Combustion Institute, Pittsburgh, Pa., 1978.

19. T. C. Ho and co-workers, "Metal Capture During Fluidized Bed Incineration of Wastes Contaminated with Lead Chloride," presented at the *Second International Congress on Toxic Combustion By-Products: Formation and Control*, Salt Lake City, Utah, Mar. 26–29, 1991.

20. R. G. Barton, W. O. Clark, W. S. Lanier, and W. R. Seeker, "Dioxin Emissions During Waste Incineration," presented at *Spring Meeting, Western States Section of the Combustion Institute*, Salt Lake City, Utah, 1988.

21. *National Incinerator Testing and Evaluation Program: Mass Burning Incinerator Technology*, Vol. II, Lavalin, Inc. Quebec City, Quebec, Canada, 1987.

22. M. Beychok, *Atmos. Envir.* **21**(1), 29–36 (1987).

FUELS, SYNTHETIC, GASEOUS

James G. Speight
Western Reserve Institute
Laramie, Wyoming

Substitute or synthetic natural gas (SNG) has been known for several centuries. When SNG was first discovered, natural gas was largely unknown as a fuel and was more a religious phenomenon (see GAS, NATURAL) (1). Coal was the first significant source of substitute natural gas and in the early stages of SNG production the product was more commonly known under variations of the name coal gas (2,3). Whereas coal continues to be a principal source of substitute natural gas (4) a more recently recognized source is petroleum (5). See also MANUFACTURED GAS; NATURAL GAS.

GAS FROM COAL

Coal can be converted to gas by several routes (2,6–11), but often a particular process is a combination of options chosen on the basis of the product desired, ie, low, medium, or high heat-value gas. In a very general sense, coal gas is the term applied to the mixture of gaseous constituents that are produced during the thermal decomposition of coal at temperatures in excess of 500°C (>930°F), often in the absence of oxygen (air) (3). A solid residue (coke, char), tars, and other liquids are also produced in the process:

$$C_{coal} + heat \rightarrow C_{char} + tar/liquid + CO + CO_2 + H_2$$

The tars and other liquids (liquor) are removed by condensation leaving principally hydrogen, carbon monoxide, and carbon dioxide in the gas phase. This gaseous product also contains low boiling hydrocarbons (qv), sulfur-containing gases, and nitrogen-containing gases including ammonia (qv) and hydrogen cyanide. The solid residue is then treated under a variety of conditions to produce other fuels which vary from a purified char to different types of gaseous mixtures. The amounts of gas, coke, tar, and other liquid products vary according to the method used for the carbonization (especially the retort configuration), and process temperature, as well as the nature (rank) of the coal (3,11).

The recorded chronology of the coal-to-gas conversion technology began in 1670 when a cleryman, John Clayton, in Wakefield, Yorkshire, produced in the laboratory a luminous gas by destructive distillation of coal (12). At the same time, experiments were also underway elsewhere to carbonize coal to produce coke, but the process was not practical on any significant scale until 1730 (12). In 1792, coal was distilled in an iron retort by a Scottish engineer, who used the by-product gas to illuminate his home (13).

The conversion of coal to gas on an industrial scale dates to the early nineteenth century (14). The gas, often referred to as manufactured gas, was produced in coke ovens or similar types of retorts by simply heating coal to vaporize the volatile constituents. Estimates based on modern data indicate that the gas mixture probably contained hydrogen (qv) (ca 50%), methane (ca 30%), carbon monoxide and carbon dioxide (ca 15%), and some inert material, such as nitrogen (qv), from which a heating value of approximately 20.5 MJ/m³ (550 Btu/ft³) can be estimated (6).

Blue gas, or blue-water gas, so-called because of the color of the flame upon burning (10), was discovered in 1780 when steam was passed over incandescent carbon (qv), and the blue-water gas process was developed over the period 1859–1875. Successful commercial application of the process came about in 1875 with the introduction of the carburetted gas jet. The heating value of the gas was low, ca 10.2 MJ/m³ (275 Btu/ft³), and on occasion oil was added to the gas to enhance the heating value. The new product was given the name carburetted water gas and the technique satisfied part of the original aim by adding luminosity to gas lights (10).

Coke-oven gas is a by-product fuel gas derived from coking coals by the process of carbonization. The first by-

Table 1. Analyses of Fuel Gases

Type of Fuel Gas	Gas Composition, vol %							Heat Value,[a] MJ/m³
	CO	CO₂	H₂	N₂	O₂	CH₄	Illuminants	
Blast-furnace	27.5	10.0	3.0	58.0	1.0	0.5		3.8
Producer (bituminous)	27.0	4.5	14.0	50.9	0.6	3.0		5.6
Blue-water	42.8	3.0	49.9	3.3	0.5	0.5		11.5
Carburetted water	33.4	3.9	34.6	7.9	0.9	10.4	8.9[b]	20.0
Retort coal	8.6	1.5	52.5	3.5	0.3	31.4	2.2[c]	21.5
Coke-oven	6.3	1.8	53.0	3.4	0.2	1.6	3.7[d]	21.9
Natural								36.1
Mid-continent		0.8		3.2		96.0		
Pennsylvania				1.1		67.6	31.3[e]	46.0

[a] To convert MJ/m³ to Btu/ft³, multiply by 26.86.
[b] 6.7 vol % C_2H_4 plus 2.2 vol % C_6H_6.
[c] 1.1 vol % each of C_2H_4 and C_6H_6.
[d] 2.7 vol % C_2H_4 plus 1.0 vol % C_6H_6.
[e] 31.3 vol % C_2H_6.

product coke ovens were constructed in France in 1856. Since then they have gradually replaced the old and primitive method of beehive coking for the production of metallurgical coke. Coke-oven gas is produced in an analogous manner to retort coal gas, with operating conditions, mainly temperature, set for maximum carbon yield. The resulting gas is, consequently, poor in illuminants, but excellent as a fuel. Typical analyses and heat content of common fuel gases vary (Table 1) and depend on the source as well as the method of production.

In Germany, large-scale production of synthetic fuels from coal began in 1910 and necessitated the conversion of coal to carbon monoxide and hydrogen.

Water gas reaction

$$C_{coal} + H_2O \rightarrow CO + H_2$$

The mixture of carbon monoxide and hydrogen is enriched with hydrogen from the water gas catalytic (Bosch) process, ie, water gas shift reaction, and passed over a cobalt–thoria catalyst to form straight-chain, ie, linear, paraffins, olefins, and alcohols in what is known as the Fisher-Tropsch synthesis.

$$n\,CO + (2n + 1)H_2 \xrightarrow[\text{catalyst}]{\text{cobalt}} C_nH_{2n+2} + n\,H_2O$$

$$2n\,CO + (n + 1)H_2 \xrightarrow[\text{catalyst}]{\text{iron}} C_nH_{2n+2} + n\,CO_2$$

$$n\,CO + 2n\,H_2 \xrightarrow[\text{catalyst}]{\text{cobalt}} C_nH_{2n} + n\,H_2O$$

$$2n\,CO + n\,H_2 \xrightarrow[\text{catalyst}]{\text{iron}} C_nH_{2n} + n\,CO_2$$

$$n\,CO + 2n\,H_2 \xrightarrow[\text{catalyst}]{\text{cobalt}} C_nH_{2n+1}OH + (n - 1)H_2O$$

$$(2n - 1)CO + (n + 1)H_2 \xrightarrow[\text{catalyst}]{\text{iron}} C_nH_{2n+1}OH + (n - 1)CO_2$$

In Sasolburg, South Africa, a commercial plant using the Fischer-Tropsch process was completed in 1950 and began producing a variety of liquid fuels and chemicals. The facility has been expanded to produce a considerable portion of South Africa's energy requirements (15,16).

In 1948, the first demonstration of suspension gasification was successfully completed by Koppers, Inc. The product gas was of 11.2 MJ/m³ (300 Btu/ft³) calorific value and consisted primarily of a mixture of hydrogen and oxides of carbon. In the United States, so-called second-generation coal gasification processes came into being as a result of the recognized need to develop reliable, domestic energy sources to replace the rapidly diminishing supply of conventional fuels (3,9). More recently, the biological conversion of coal and synthesis gas (carbon monoxide–hydrogen mixtures) into liquid fuels by methanogenic bacteria has received some attention (17–19).

Gas Products

The originally designated names of the gaseous mixtures are used herein, with the understanding that since their introduction there may be differences in means of production and in the make-up of the gaseous products. Properties of fuel gases are available (20). There are standard tests to determine properties and character of gaseous mixtures. These tests are accepted by the American Society for Testing and Materials (ASTM), by the British Standards Institution (BSI), by the Institute of Petroleum (IP), and by the International Standards Organization (ISO) (1,10,20).

Low Heat-Value Gas. Low heat-value (low Btu) gas (7) consists of a mixture of carbon monoxide and hydrogen and has a heating value of less than 11 MJ/m³ (300 Btu/ft³), but more often in the range 3.3–5.6 MJ/m³ (90–150 Btu/ft³). The gas is formed by partial combustion of coal with air, usually in the presence of steam (7).

$$2\,C_{coal} + O_2 \rightarrow 2\,CO$$

$$C_{coal} + H_2O \rightarrow CO + H_2$$

$$CO + H_2O \rightarrow CO_2 + H_2$$

This gas is of interest to industry as a fuel gas or even, on occasion, as a raw material from which ammonia and other compounds may be synthesized.

The first gas producer making low heat-value gas was built in 1832. (The product was a combustible carbon monoxide–hydrogen mixture containing ca 50 vol % nitrogen). The open-hearth or Siemens-Martin process, built in 1861 for pig iron refining, increased low heat-value gas use. The use of producer gas as a fuel for heating furnaces continued to increase until the turn of the century when natural gas began to supplant manufactured fuel gas.

The combustible components of the gas are carbon monoxide and hydrogen, but combustion (heat) value varies because of dilution with carbon dioxide and with nitrogen. The gas has a low flame temperature unless the combustion air is strongly preheated. Its use has been limited essentially to steel (qv) mills, where it is produced as a by-product of blast furnaces. A common choice of equipment for the smaller gas producers is the Wellman-Galusha unit because of its long history of successful operation (21).

Medium Heat-Value Gas. Medium heat-value (medium Btu) gas (6,7) has a heating value between 9 and 26 MJ/m^3 (250 and 700 Btu/ft^3). At the lower end of this range, the gas is produced like low heat-value gas, with the notable exception that an air separation plant is added and relatively pure oxygen (qv) is used instead of air to partially oxidize the coal. This eliminates the potential for nitrogen in the product and increases the heating value of the product to 10.6 MJ/m^3 (285 Btu/ft^3). Medium heat-value gas consists of a mixture of methane, carbon monoxide, hydrogen, and various other gases and is suitable as a fuel for industrial consumers.

High Heat-Value Gas. High heat-value (high Btu) gas (7) has a heating value usually in excess of 33.5 MJ/m^3 (900 Btu/ft^3). This is the gaseous fuel that is often referred to as substitute or synthetic natural gas (SNG), or pipeline-quality gas. It consists predominantly of methane and is compatible with natural gas insofar as it may be mixed with, or substituted for, natural gas.

Any of the medium heat-value gases that consist of carbon monoxide and hydrogen (often called synthesis gas) can be converted to high heat-value gas by methanation (22), a low temperature catalytic process that combines carbon monoxide and hydrogen to form methane and water.

$$CO + 3 H_2 \rightarrow CH_4 + H_2O$$

Prior to methanation, the gas product from the gasifier must be thoroughly purified, especially from sulfur compounds the precursors of which are widespread throughout coal (23). Moreover, the composition of the gas must be adjusted, if required, to contain three parts hydrogen to one part carbon monoxide to fit the stoichiometry of methane production. This is accomplished by application of a catalytic water gas shift reaction.

$$CO + H_2O \rightleftarrows CO_2 + H_2$$

The ratio of hydrogen to carbon monoxide is controlled by shifting only part of the gas stream. After the shift, the carbon dioxide, which is formed in the gasifier and in the water gas reaction, and the sulfur compounds formed during gasification, are removed from the gas.

The processes that have been developed for the production of synthetic natural gas are often configured to produce as much methane in the gasification step as possible thereby minimizing the need for a methanation step. In addition, methane formation is highly exothermic which contributes to process efficiency by the production of heat in the gasifier, where the heat can be used for the endothermic steam–carbon reaction to produce carbon monoxide and hydrogen.

$$C + H_2O \rightarrow CO + H_2$$

Carbonization

Next to combustion, carbonization represents one of the largest uses of coal (2,24–26). Carbonization is essentially a process for the production of a carbonaceous residue by thermal decomposition, accompanied by simultaneous removal of distillate, of organic substances.

$$C_{organic} \rightarrow C_{coke/char/carbon} + liquids + gases$$

This process may also be referred to as destructive distillation. It has been applied to a whole range of organic materials, more particularly to natural products such as wood (qv), sugar (qv), and vegetable matter to produce charcoal (see FUELS FROM BIOMASS). However, in the present context, coal usually yields coke, which is physically dissimilar from charcoal and appears with the more familiar honeycomb-type structure (27).

The original process of heating coal in rounded heaps, the hearth process, remained the principal method of coke production for over a century, although an improved oven in the form of a beehive was developed in the Durham-Newcastle area of England in about 1759 (2,26,28). These processes lacked the capability to collect the volatile products, both liquids and gases. It was not until the midnineteenth century, with the introduction of indirectly heated slot ovens, that it became possible to collect the liquid and gaseous products for further use.

Coal carbonization processes are generally defined according to process operating temperature. Terms are defined in Table 2.

Low Temperature Carbonization. Low temperature carbonization, when the process does not exceed 700°C, was mainly developed as a process to supply town gas for lighting purposes as well as to provide a smokeless (devolatilized) solid fuel for domestic consumption (30). However, the process by-products (tars) were also found to be valuable insofar as they served as feedstocks for an emerging chemical industry and were also converted to gasolines, heating oils, and lubricants (31).

Coals preferred for the low temperature carbonization were usually lignites or subbituminous, as well as high volatile bituminous, coals (see LIGNITE AND BROWN COAL).

Table 2. Coal Carbonization Methods[a]

Carbonization Process	Final Temperature, °C	Products	Processes
Low temperature	500–700	Reactive coke and high tar yield	Rexco (700°C) made in cylindrical vertical retorts; coalite (650°C) made in vertical tubes
Medium temperature	700–900	Reactive coke with high gas yield, or domestic briquettes	Town gas and gas coke (obsolete); phurnacite, low volatile steam coal, pitch-bound briquettes carbonized at 800°C
High temperature	900–1050	Hard, unreactive coal for metallurgical use	Foundry coke (900°C); blast-furnace coke (950–1050°)

[a] Ref. 29.

These yield porous solid products over the temperature range 600–700°C. Certain of the higher rank (caking) coals were less suitable for the process, unless steps were taken to destroy the caking properties, because of the tendency of these higher rank coals to adhere to the walls of the carbonization chamber.

The options for efficient low temperature carbonization of coal include vertical and horizontal retorts which have been used for batch and continuous processes. In addition, stationary and revolving horizontal retorts have also been operated successfully, and there are also several process options employing fluidized or gas-entrained coal. Coke production from batch-type carbonization of coal has been supplanted by a variety of continuous retorting processes which allow much greater throughput rates than were previously possible. These processes employ rectangular or cylindrical vessels of sufficient height to carbonize the coal while it travels from the top of the vessel to the bottom and usually employ the principle of heating the coal by means of a countercurrent flow of hot combustion gas. Most notable of these types of carbonizers are the Lurgi-Spulgas retort and the Koppers continuous steaming oven (2).

High Temperature Carbonization. When heated at temperatures in excess of 700°C (1290°F), low temperature chars lose their reactivity through devolatilization and also suffer a decrease in porosity. High temperature carbonization, at temperatures >900°C, is, therefore, employed for the production of coke (27). As for the low temperature processes, the tars produced in high temperature ovens are also sources of chemicals and chemical intermediates (32).

A newer concept has been developed that is given the name mild gasification (33). It is not a gasification process in the true sense of the word. The process temperature is some several hundred degrees lower, hence the term mild, than the usual gasification process temperature and the objective is not to produce a gaseous fuel but to produce a high value char (carbon) and liquid products. Gas is produced, but to a lesser extent.

Documented efforts at cokemaking date from 1584 (34), and have seen various adaptations of conventional wood-charring methods to the production of coke including the eventual evolution of the beehive oven, which by the mid-nineteenth century had become the most common vessel for the coking of coal (2). The heat for the process was supplied by burning the volatile matter released from the coal and, consequently, the carbonization would progress from the top of the bed to the base of the bed and the coke was retrieved from the side of the oven at process completion.

Some beehive ovens, having various improvements and additions of waste heat boilers, thereby allowing heat recovery from the combustion products, may still be in operation. Generally, however, the beehive oven has been replaced by wall-heated, horizontal chamber, ie, slot, ovens in which higher temperatures can be achieved as well as a better control over the quality of the coke. Modern slot-type coke ovens are approximately 15 m long, approximately 6 m high, and the width is chosen to suit the carbonization behavior of the coal to be processed. For example, the most common widths are ca 0.5 m, but some ovens may be as narrow as 0.3 m, or as wide as 0.6 m.

Several (usually 20 or more, alternating with similar cells that contain heating ducts) of these chambers are constructed in the form of a battery over a common firing system through which the hot combustion gas is conveyed to the ducts. The flat roof of the battery acts as the surface for a mobile charging car from which the coal (25–40 t) enters each oven through three openings along the top. The coke product is pushed from the rear of the oven through the opened front section onto a quenching platform or into rail cars that then move the coke through water sprays. The gas and tar by-products of the process are collected for further processing or for on-site use as fuel.

Most modern coke ovens operate on a regenerative heating cycle in order to obtain as much surplus gas as possible for use on the works, or for sale. If coke-oven gas is used for heating the ovens, the majority of the gas is

Table 3. Gas Composition Requirements for Substitute Natural Gas and for Power Generation

Characteristic	Product Requirement	
	Synthetic Natural Gas	Power Generation
Methane content	High, less synthesis required	Low, probably means no tars
H_2/CO ratio	High, less shifting required	Low, CO more efficient fuel
Moisture content	High, steam required for shift	Low, lower condensate treatment costs
Outlet temperature	Low, maximizes methane minimizes sensible heat loss leads to high cold gas efficiency	High, precludes tar formation provides for steam generation reduces cold gas efficiency
Gasifier oxidant	O_2 only, cost of N_2 removal excessive	Air or O_2, low heating gas value acceptable fuel

surplus to requirements. If producer gas is used for heating, much of the coke-oven gas is surplus.

The main difference between gas works and coke oven practice is that, in a gas works, maximum gas yield is a primary consideration whereas in the coke works the quality of the coke is the first consideration. These effects are obtained by choice of a coal feedstock that is suitable to the task. For example, use of lower volatile coals in coke ovens, compared to coals used in gas works, produces lower yields of gas when operating at the same temperatures. In addition, the choice of heating (carbonizing) conditions and the type of retort also play a principal role (10,35).

Gasification

The gasification of coal is essentially the conversion of coal by any one of a variety of processes to produce combustible gases (7,8,11,22,36–38). Primary gasification is the thermal decomposition of coal to produce mixtures containing various proportions of carbon monoxide, carbon dioxide, hydrogen, water, methane, hydrogen sulfide, and nitrogen, as well as products such as tar, oils, and phenols. A solid char product may also be produced, and often represents the bulk of the weight of the original coal.

Secondary gasification involves gasification of the char from the primary gasifier, usually by reaction of the hot char and water vapor to produce carbon monoxide and hydrogen.

$$C_{char} + H_2O \rightarrow CO + H_2$$

The gaseous product from a gasifier generally contains large amounts of carbon monoxide and hydrogen, plus lesser amounts of other gases and may be of low, medium, or high heat value depending on the defined use (Table 3) (39,40).

The importance of coal gasification as a means of producing fuel gas(es) for industrial use cannot be underplayed. But coal gasification systems also have undesirable features. A range of undesirable products are also produced which must be removed before the products are used to provide fuel and/or to generate electric power (22,41).

Chemistry. Coal gasification involves the thermal decomposition of coal and the reaction of the carbon in the coal, and other pyrolysis products with oxygen, water, and hydrogen to produce fuel gases such as methane by internal hydrogen shifts

$$C_{coal} + H_{coal} \rightarrow CH_4$$

or through the agency of added (external) hydrogen

$$C_{coal} + 2 H_2 \rightarrow CH_4$$

although the reactions are more numerous and more complex as can be seen in Table 4.

If air is used as a combustant, the product gas contains nitrogen and, depending on design characteristics, has a

Table 4. Gasification Reactions

Reaction	ΔH, kJ[a]
Gasification zone, 595–1205°C	
$C + CO_2 \rightarrow 2\ CO$	9.65[b]
$CO + H_2O \rightarrow CO_2 + H_2$	−1.93[c]
$C + H_2O \rightarrow CO + H_2$	7.76[d]
$C + 2 H_2 \rightarrow CH_4$	−5.21
$C + 2 H_2O \rightarrow 2 H_2 + CO_2$	5.95
Combustion zone, >1205°C	
$2 C + O_2 \rightarrow 2\ CO$	−22.6
$C + O_2 \rightarrow CO_2$	−6.74
Methanation reactions	
$CO + 3 H_2 \rightarrow CH_4 + H_2O$	−12.9
$CO_2 + 4 H_2 \rightarrow CH_4 + 2 H_2O$	−11.0
$2 C + 2 H_2O \rightarrow CH_4 + CO_2$	0.62
Other reactions	
$2 H_2 + O_2 \rightarrow 2 H_2O$	
$CO + 2 H_2 \rightarrow CH_3OH$	
$2 C + H_2 \rightarrow C_2H_2$	
$CH_4 + 2 H_2O \rightarrow CO_2 + 4 H_2$	

[a] To convert kJ to kcal, divide by 4.184
[b] Boudouard reaction.
[c] Water gas shift reaction.
[d] Water gas reaction.

heating value of approximately $5.6–11.2 \times 10^3$ MJ/m^3 (150–300 Btu/ft^3). The use of pure oxygen, although expensive, results in a product gas having a heating value of 11–15 MJ/m^3 (300–400 Btu/ft^3) and carbon dioxide and hydrogen sulfide as by-products.

If a high heat-value gas (33.5–37.3 MJ/m^3 (900–1000 Btu/ft^3)) ie, SNG, is the desired product, efforts must be made to increase the methane content.

Shift conversion reaction

$$CO + H_2O \rightarrow CO_2 + H_2$$
$$CO + 3 H_2 \rightarrow CH_4 + H_2O$$
$$2 CO + 2 H_2 \rightarrow CH_4 + CO_2$$
$$CO + 4 H_2 \rightarrow CH_4 + 2 H_2O$$

The gasification is performed using oxygen and steam (qv), usually at elevated pressures. The steam–oxygen ratio along with reaction temperature and pressure determine the equilibrium gas composition. The reaction rates for these reactions are relatively slow and heats of formation are negative. Catalysts may be necessary for complete reaction (2,3,24,42,43).

Process Parameters. The most notable effects in gasifiers are those of pressure (Fig. 1) and coal character. Some initial processing of the coal feedstock may be required. The type and degree of pretreatment is a function of the process and/or the type of coal.

Depending on the type of coal being processed and the analysis of the gas product desired, some or all of the following processing steps may be required: (*1*) pretreatment of the coal (if caking is a problem); (*2*) primary gasification of the coal; (*3*) secondary gasification of the carbonaceous residue from the primary gasifier; (*4*) removal of carbon dioxide, hydrogen sulfide, and other acid gases; (*5*) shift conversion for adjustment of the carbon monoxide–hydrogen mole ratio to the desired ratio; and (*6*) catalytic methanation of the carbon monoxide–hydrogen mixture to form methane. If high heat-value gas is desired, all of these processing steps are required because coal gasifiers do not yield methane in the concentrations required.

An example of application of a pretreatment option occurs when the coal displays caking or agglomerating characteristics. Such coals are usually not amenable to gasification processes employing fluidized-bed or moving-bed reactors. The pretreatment involves a mild oxidation treatment, usually consisting of low temperature heating of the coal in the presence of air or oxygen. This destroys the caking characteristics of coals.

Gasification technologies for the production of high heat-value gas do not all depend entirely on catalytic methanation, that is, the direct addition of hydrogen to coal under pressure to form methane.

The hydrogen-rich gas for hydrogasification can be manufactured from steam by using the char that leaves the hydrogasifier. Appreciable quantities of methane are formed directly in the primary gasifier and the heat released by methane formation is at a sufficiently high temperature to be used in the steam–carbon reaction to produce hydrogen so that less oxygen is used to produce heat for the steam–carbon reaction. Hence, less heat is lost in the low temperature methanation step, thereby leading to a higher overall process efficiency.

There are three fundamental reactor types for gasification processes: (*1*) a gasifier reactor, (*2*) a devolatilizer, and (*3*) a hydrogasifier (Fig. 2). The choice of a particular design is available for each type, eg, whether or not two stages should be involved depending on the ultimate product gas desired. Reactors may also be designed to operate over a range of pressure from atmospheric to high pressure.

Gasification processes have been classified on the basis of the heat-value of the produced gas. It is also possible to classify gasification processes according to the type of reactor and whether or not the system reacts under pressure. Additionally, gasification processes can be segregated according to the bed types, which differ in the ability to accept and use caking coals. Thus gasification processes can be divided into four categories based on reactor (bed) configuration: (*1*) fixed bed, (*2*) moving bed, (*3*) fluidized bed, and (*4*) entrained bed.

In a fixed-bed process the coal is supported by a grate and combustion gases, ie, steam, air, oxygen, etc, pass through the supported coal whereupon the hot produced

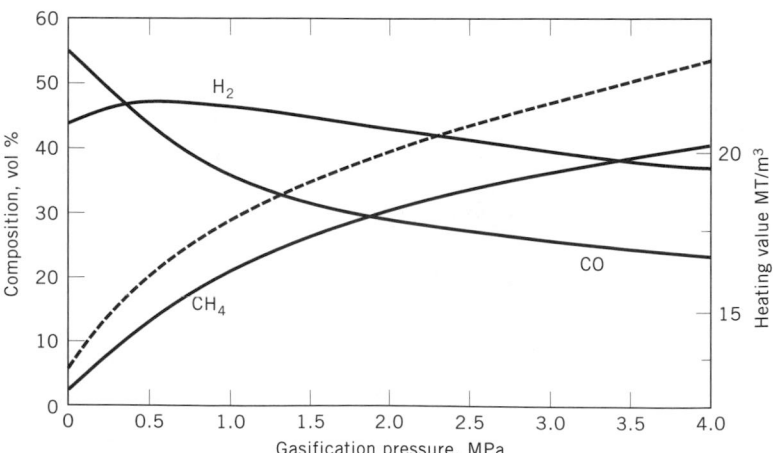

Figure 1. Variation of (—) gas composition and (---) heating value with gasifier pressure. To convert MJ/m^3 to Btu/ft^3, multiply by 26.86. To convert MPa to psi, multiply by 145.

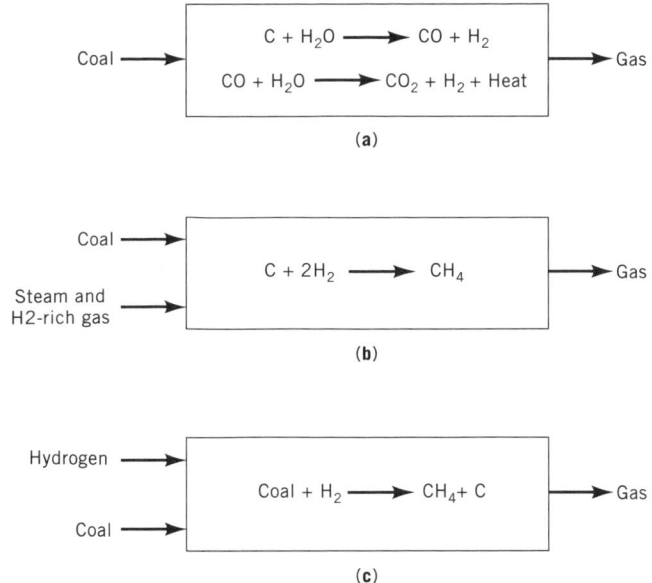

Figure 2. Chemistry of (**a**) gasifier; (**b**) hydrogasifier; and (**c**) devolatization processes. The gaseous product of (**a**) is of low heating value; that of (**b**) and (**c**) is of intermediate heating value.

gases exit from the top of the reactor. Heat is supplied internally or from an outside source, but caking coals cannot be used in an unmodified fixed-bed reactor. In the moving-bed system (Fig. 3), coal is fed to the top of the bed and ash leaves the bottom with the product gases being produced in the hot zone just prior to being released from the bed.

The fluidized-bed system (Fig. 3) uses finely sized coal particles and the bed exhibits liquid-like characteristics when a gas flows upward through the bed. Gas flowing through the coal produces turbulent lifting and separation of particles and the result is an expanded bed having

greater coal surface area to promote the chemical reaction. These systems, however, have only a limited ability to handle caking coals.

An entrainment system (Fig. 3) uses finely sized coal particles blown into the gas steam prior to entry into the reactor and combustion occurs with the coal particles suspended in the gas phase; the entrained system is suitable for both caking and noncaking coals.

Gasifier Options. The standard Wellman-Galusha unit, used for noncaking coals, is an atmospheric pressure, air-blown gasifier (Fig. 4) fed by a two-compartment lockhopper system. The upper coal-storage compartment feeds coal intermittently into the feeding compartment, which then feeds coal into the gasifier section almost continuously except for brief periods when the feeding compartment is being loaded from storage. In small units up to 1.5 m internal diameter ash is removed by shaker grates. In larger units, ash is removed continuously from the bottom of the gasifier into an ash-hopper section by a revolving grate.

The grate is constructed of flat, circular steel plates set one above the other with edges overlapping. The grate, revolving eccentrically within the gasifier, causes ash to fall from the coal bed as the space between the grate and the shell increases, and then pushes the ash down into the ash-hopper as the space decreases. The smaller size units are brick-lined, although the larger sizes are unlined and water-jacketed. Combustion air, provided by a fan, passes over the warm (82°C) jacket water causing the water to vaporize to provide the necessary steam for gasification. Gas leaves from above the fuel bed at 428–538°C for bituminous coal.

On the other hand, the agitator type of Wellman-Galusha unit, used for gasification of any type of coal, uses a slowly revolving horizontal arm which also spirals vertically below the surface of the fuel bed to retard channeling and maintain a uniform bed. Use of the agitator

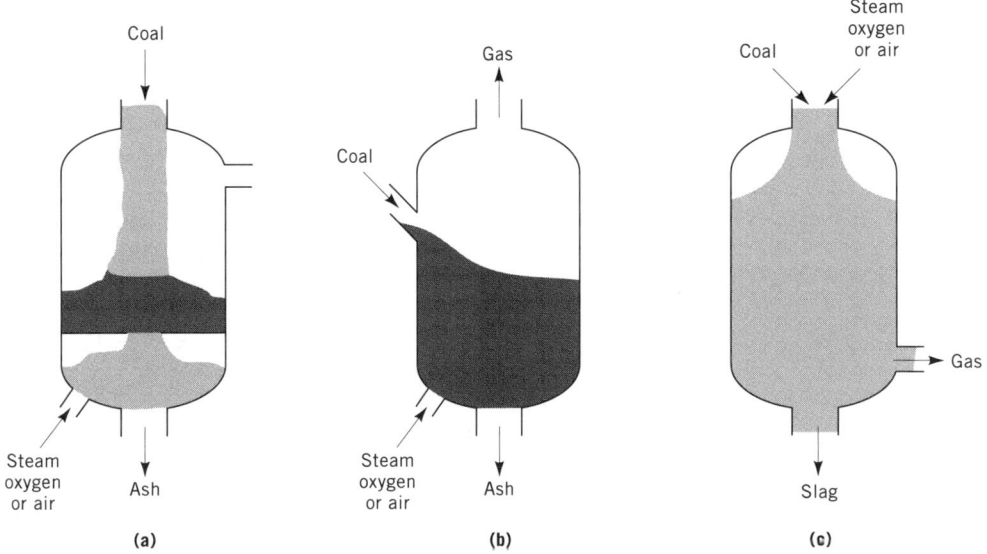

Figure 3. Gasifier systems: (**a**), moving bed (dry ash); (**b**), fluidized bed; and (**c**), entrained flow.

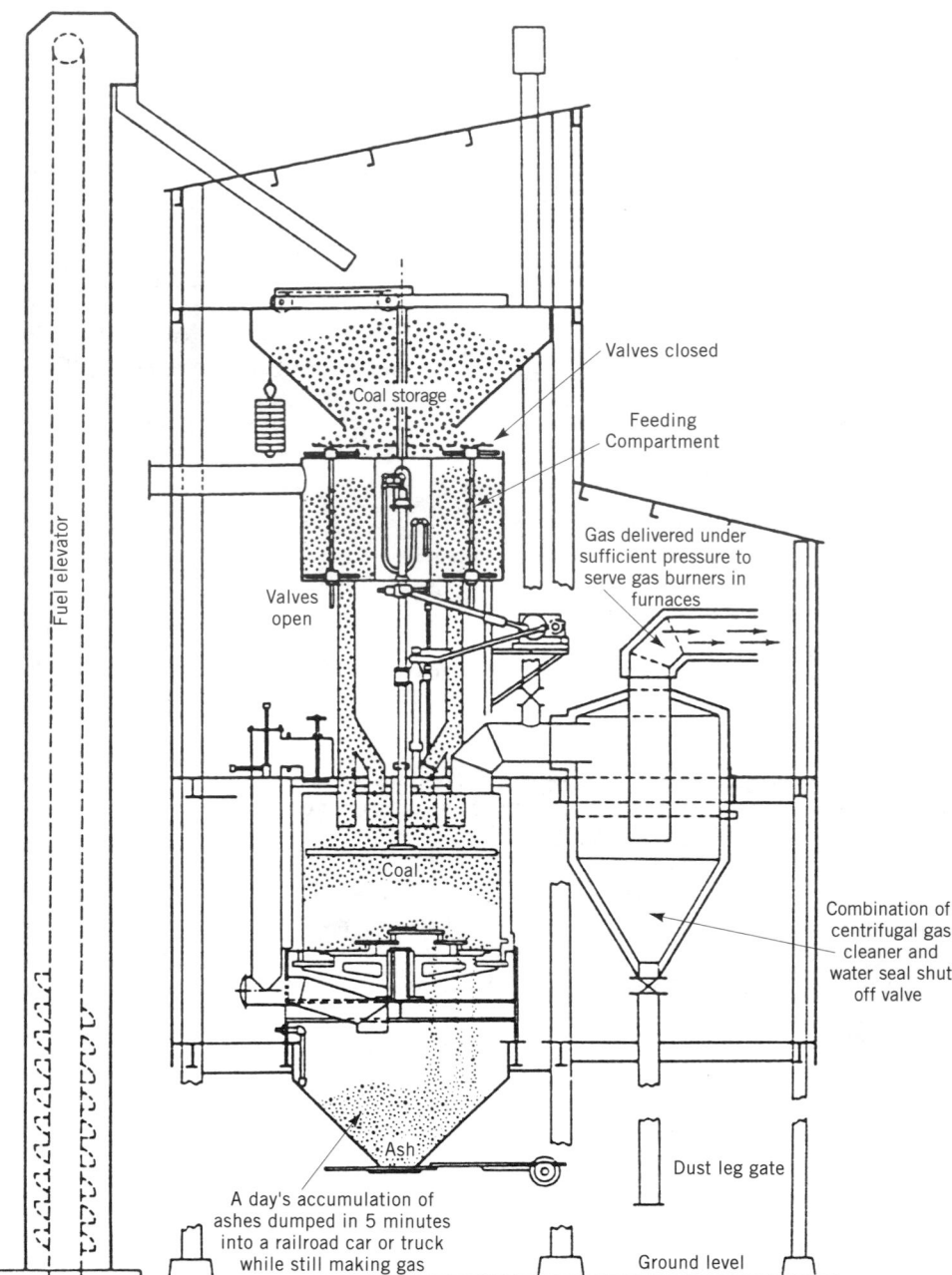

Valves closed

Feeding Compartment

Coal storage

Gas delivered under sufficient pressure to serve gas burners in furnaces

Valves open

Fuel elevator

Coal

Combination of centrifugal gas cleaner and water seal shut off valve

Ash

Dust leg gate

A day's accumulation of ashes dumped in 5 minutes into a railroad car or truck while still making gas

Ground level

Figure 4. A Wellman-Galusha agitator-type gas producer.

not only allows operation with caking coals but also can increase the capacity of the gasifier by about 25% for use with other coals.

The Winkler gasifier (Fig. 5) is an example of a medium heat-value gas producer which, when oxygen is employed, yields a gas product composed mainly of carbon monoxide and hydrogen (43).

In the process, finely crushed coal is gasified at atmospheric pressure in a fluidized state; oxygen and steam are introduced at the base of the gasifier. The coal is fed by lockhoppers and a screw feeder into the bottom of the fuel bed. Sintered ash particles settle on a grate, where they are cooled by the incoming oxygen and steam; a rotating, cooled rabble moves the ash toward a discharge port. The ash is then conveyed pneumatically to a disposal hopper.

The gas, along with entrained ash and char particles, which are subjected to further gasification in the large space above the fluid bed, exit the gasifier at 954–1010°C. The hot gas is passed through a waste-heat boiler to recover the sensible heat, and then through a dry cyclone. Solid particles are removed in both units. The gas is further cooled and cleaned by wet scrubbing, and if required, an electrostatic precipitator is included in the gas-treatment stream.

The Koppers-Totzek process is a second example of a process for the production of a medium heat-value gas (44,45). Whereas the Winkler process employs a fluidized bed, the Koppers-Totzek process uses an entrained flow system. In the Koppers-Totzek process, dried and pulverized coal is conveyed continuously by a screw into a mixing nozzle. From there a high velocity stream of steam

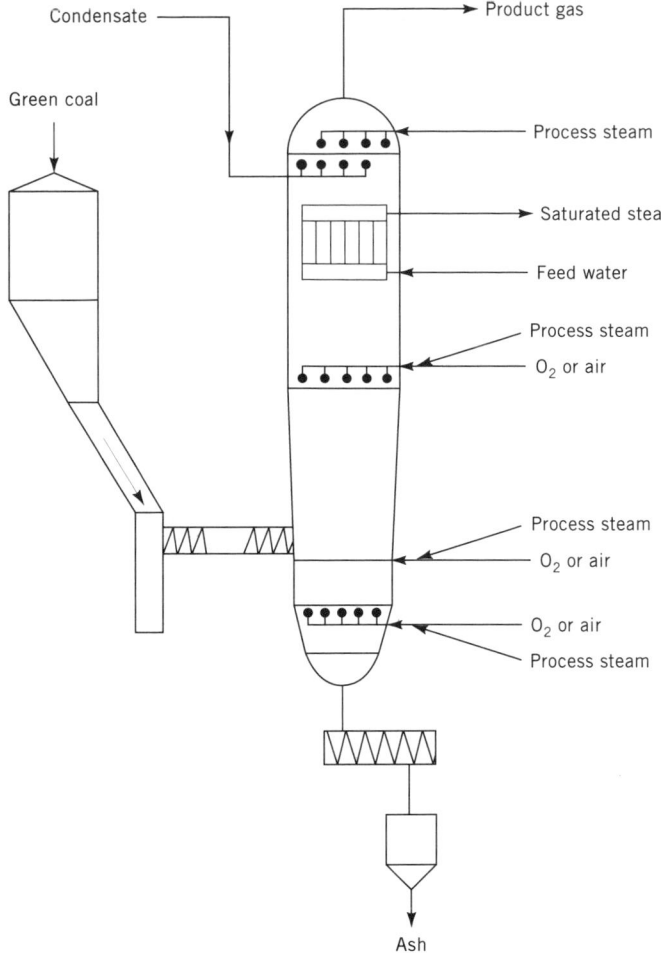

Figure 5. The Winkler gas producer.

and oxygen entrains the coal into a gasifier. The gasifier (Fig. 6) is a cylindrical vesel with a refractory-lining that is designed to conduct a selected amount of heat to a surrounding water jacket in which low pressure process steam is generated. The lining is thin (about 5 cm) and

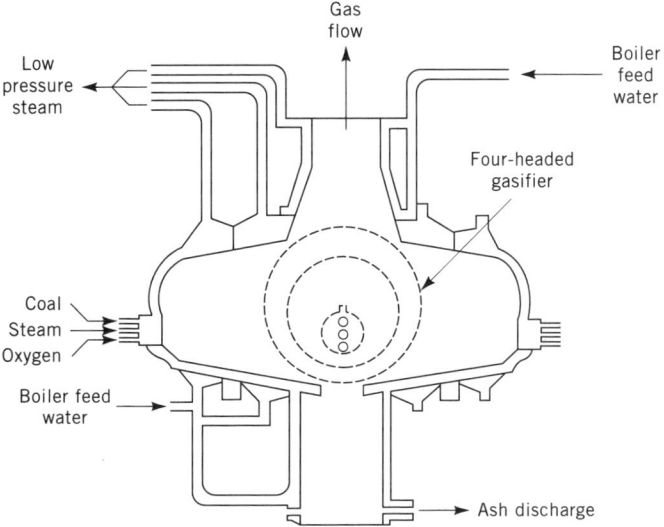

Figure 6. The Koppers-Totzek gas producer.

made of a high alumina cast material. In a two-headed gasifier two burner heads are placed 180° apart at either end of the vessel. Four burner heads, 90° apart, are used in a four-headed gasifier. The largest gasifiers are 3–4 m diameter at the middle, tapering to 2–3 m at the burner ends and are about 19 m long. The reactor volume is about 30 m³ for the two-headed design, and 64 m³ for the larger, four-headed models.

The process is carried at moderate (slightly above atmospheric) pressures, but at very high temperatures that reach a maximum of 1900°C. Even though the reaction time is short (0.6–0.8 s) the high temperature prevents the occurrence of any condensable hydrocarbons, phenols, and/or tar in the product gas. The absence of liquid simplifies the subsequent gas clean-up steps.

Normally ca 50% of the coal ash is removed from the bottom of the gasifier as a quenched slag. The balance is carried overhead in the gas as droplets which are solidified when the gas is cooled with a water spray. A fluxing agent is added, if required, to the coal to lower the ash fusion temperature and increase the molten slag viscosity.

Conversion of carbon in the coal to gas is very high. With low rank coal, such as lignite and subbituminous coal, conversion may border on 100%, and for highly volatile A coals, it is on the order of 90–95%. Unconverted carbon appears mainly in the overhead material. Sulfur removal is facilitated in the process because typically 90% of it appears in the gas as hydrogen sulfide, H_2S, and 10% as carbonyl sulfide, COS; carbon disulfide, CS_2, and/or methyl thiol, CH_3SH, are not usually formed.

The production of synthetic natural gas can be achieved by use of the Lurgi gasifier (Fig. 7), which is similar in principle to the Wellman-Galusha unit and is designed for operation at pressures up to 3.1 MPa (450 psi) (46). Three distinct reaction zones are identifiable in a pressurized (1.9–2.9 MPa (280–425 psi)) Lurgi reactor: (1) the drying/devolatilization and pyrolysis zone, 370–595°C, nearest the coal feed end, commences the process by converting the coal to a reactive char; (2) the gasification zone (595–1205°C); and (3) the combustion zone (>1205°C), nearest the discharge end, which provides the heat requirements for the endothermic steam–carbon reaction. Equations and reaction enthalpies for the last two zones are given in Table 4.

The operating conditions in the gasifier (temperature and pressure) and the reaction kinetics (residence time, concentration of the constituents, and rate constants) determine the extent of conversion or approach to equilibrium.

The coal is fed through a lockhopper mounted on the top of the gasifier where a rotating distributor provides uniform coal feed across the bed. When processing caking coals, blades attached to the distributor rotate within the bed to break up agglomerates. A revolving grate supports the bed at the bottom and serves as a distributor for steam and oxygen. Solid residue is removed at the bottom of the gasifier through an ash lockhopper. The entire gasifier vessel is surrounded by a water jacket in which high pressure steam is generated.

Crude gas leaves from the top of the gasifier at 288–593°C depending on the type of coal used. The composi-

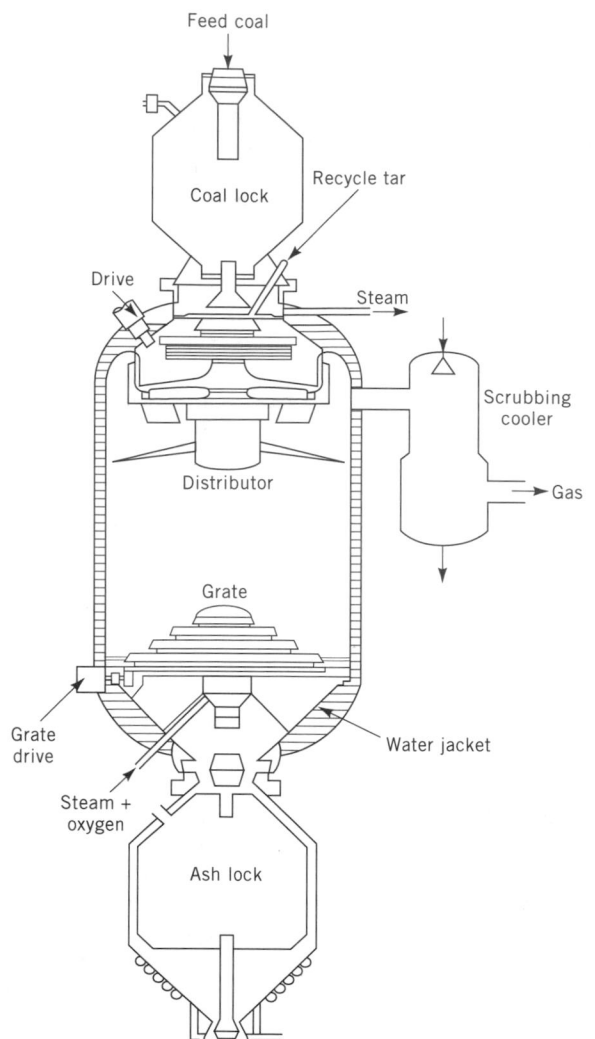

Figure 7. The pressurized Lurgi gas producer.

tion of gas also depends on the type of coal and is notable for the relatively high methane content when contrasted to gases produced at lower pressures or higher temperatures. These gas products can be used as produced for electric power production or can be treated to remove carbon dioxide and hydrocarbons to provide synthesis gas for ammonia, methanol, and synthetic oil production. The gas is made suitable for methanation, to produce synthetic natural gas, by a partial shift and carbon dioxide and sulfur removal.

As in most of the high heat-value processes, the raw gas is in the medium heat-value gas range and can be employed directly in that form. Removing the carbon dioxide raises the heating value, but not enough to render the product worthwhile as a high heat-value gas without methanation.

Methanation of the clean desulfurized main gas (less than 1 ppm total sulfur) is accomplished in the presence of a nickel catalyst at temperatures from 260–400°C and pressure range of 2–2.8 MPa (300–400 psi). Equations and reaction enthalpies are given in Table 4.

Hydrogenation of the oxides of carbon to methane according to the above reactions is sometimes referred to as

the Sabatier reactions. Because of the high exothermicity of the methanization reactions, adequate and precise cooling is necessary in order to avoid catalyst deactivation, sintering, and carbon deposition by thermal cracking.

Catalytic methanation processes include (1) fixed or fluidized catalyst-bed reactors where temperature rise is controlled by heat exchange or by direct cooling using product gas recycle; (2) through wall-cooled reactor where temperature is controlled by heat removal through the walls of catalyst-filled tubes; (3) tube-wall reactors where a nickel–aluminum alloy is flame-sprayed and treated to form a Raney-nickel catalyst bonded to the reactor tube heat-exchange surface; and (4) slurry or liquid-phase (oil) methanation.

To enable interchangeability of the SNG with natural gas, on a calorific, flame, and toxicity basis, the synthetically produced gas consists of a minimum of 89 vol % methane, a maximum of 0.1% carbon monoxide, and up to 10% hydrogen. The specified minimum acceptable gross heating value is approximately 34.6 MJ/m³ (930 Btu/ft³).

In a combined power cycle operation, clean (sulfur- and particulate-free) gas is burned with air in the combustor at elevated pressure. The gas is either low or medium heat-value, depending on the method of gasification.

The hot gases from the combustor, temperature controlled to 980°C by excess air, are expanded through the gas turbine, driving the air compressor and generating electricity. Sensible heat in the gas turbine exhaust is recovered in a waste heat boiler by generating steam for additional electrical power production.

The use of hot gas clean-up methods to remove the sulfur and particulates from the gasified fuel increases turbine performance by a few percentage points over the cold clean-up systems. Hot gas clean-up permits use of the sensible heat and enables retention of the carbon dioxide and water vapor in the gasified fuel, thus enhancing turbine performance. Further, additional power may be generated, prior to combustion with air, in an expansion turbine as the hot product fuel gas is expanded to optimum pressure level for the combined cycle. Future advances in gas turbine technology (turbine inlet temperature above 1650°C) are expected to boost the overall combined cycle efficiency substantially.

More recently, advanced generation gasifiers have been under development, and commercialization of some of the systems has become a reality (36,41). In these newer developments, the emphasis has shifted to a greater throughput, relevant to the older gasifiers, and also to high carbon conversion levels and, thus, higher efficiency units.

For example, the Texaco entrained system features coal–water slurry feeding a pressurized oxygen-blown gasifier with a quench zone for slag cooling (Fig. 8). In fact, the coal is partially oxidized to provide the heat for the gasification reactions. The Dow gasifier also utilizes a coal–water slurry fed system whereas the Shell gasifier features a dry-feed entrained gasification system which operates at elevated temperature and pressure. The Kellogg Rust Westinghouse system and the Institute of Gas Technology U-Gas system are representative of ash agglomerating fluidized-bed systems.

In response to the disadvantage that the dry ash Lurgi

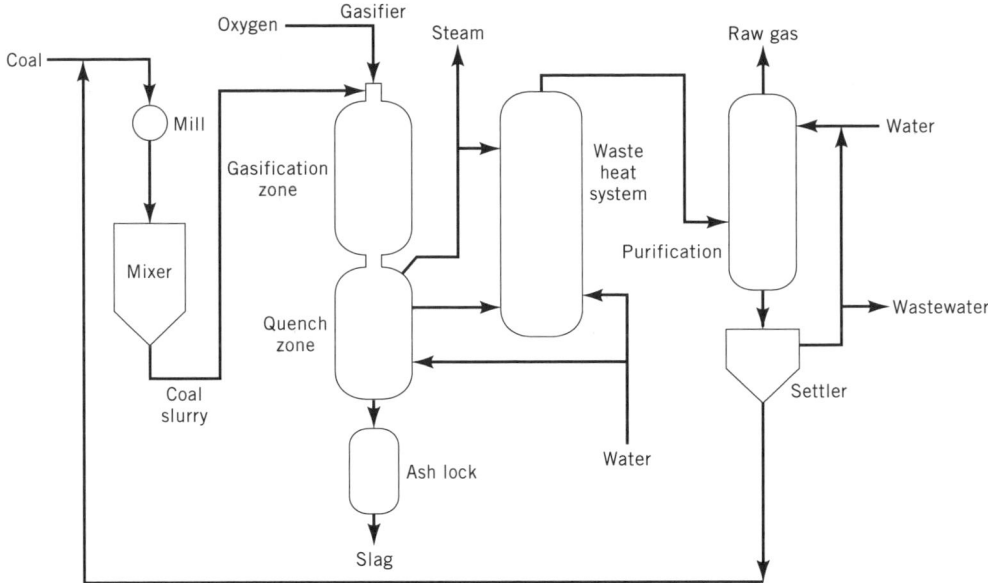

Figure 8. The Texaco gasification process.

gasifier requires that temperatures have to be below the ash melting point to prevent clinkering, improvements have been sought in the unit; as a result the British Gas-Lurgi GmbH gasifier came into being. This unit is basically similar to the dry ash Lurgi unit insofar as the top of the unit is identical but the bottom has been modified to include a slag quench vessel (Fig. 9). Thus the ash melts at the high temperatures in the combustion zone (up to 2000°C) and forms a slag that runs into the quench chamber, which is in reality a water bath where the slag forms granules of solid ash. Temperatures and reaction rates are high in the gasification zone so that coal residence time is markedly educed over that of the dry ash unit.

In summary, these second-generation gasifiers offer promise for the future in terms of increased efficiency as well as for use of other feedstocks, such as biomass. The older, first-generation gasifiers, however, continue to be used.

Combustion

Coal combustion, not being in the strictest sense a process for the generation of gaseous synfuels, is nevertheless an important use of coal as a source of gaseous fuels. Coal combustion, an old art and probably the oldest known use of this fossil fuel, is an accumulation of complex chemical and physical phenomena. The complexity of coal itself and the variable process parameters all contribute to the overall process (8,10,47–50).

There are two principal methods of coal combustion: fixed-bed combustion and combustion in suspension. The first fixed beds, eg, open fires, fireplaces, and domestic stoves, were simple in principle. Suspension burning of coal began in the early 1900s with the development of pulverized coal-fired systems, and by the 1920s these systems were in widespread use. Spreader stokers, which were developed in the 1930s, combined both principles by providing for the smaller particles of coal to be burned in

suspension and larger particles to be burned on a grate (10).

A significant issue in combustors in the mid-1990s is the performance of the process in an environmentally acceptable manner through the use of either low sulfur coal or post-combustion clean-up of the flue gases. Thus there is a marked trend to more efficient methods of coal combustion and, in fact, a combustion system that is able to accept coal without the necessity of a post-combustion treatment or without emitting objectionable amounts of sulfur oxides, nitrogen oxides, and particulates is very desirable (51,52).

The parameters of rank and moisture content are regarded as determining factors in combustibility as it relates to both heating value and ease of reaction as well as to the generation of pollutants (48). Thus, whereas the lower rank coals may appear to be more reactive than higher rank coals, though exhibiting a lower heat-value and thereby implying that rank does not affect combustibility, environmental constraints arise through the occurrence of heteroatoms, ie, noncarbon atoms such as nitrogen and sulfur, in the coal. At the same time, anthracites, which have a low volatile matter content, are generally more difficult to burn than bituminous coals.

Chemistry. In direct combustion coal is burned to convert the chemical energy of the coal into thermal energy, ie, the carbon and hydrogen in the coal are oxidized into carbon dioxide and water.

$$2\ H_{coal} + O_2 \rightarrow H_2O$$

After burning, the sensible heat in the products of combustion can then be converted into steam that can be used for external work or can be converted directly into energy to drive a shaft, eg, in a gas turbine. In fact, the combustion process actually represents a means of achieving the complete oxidation of coal.

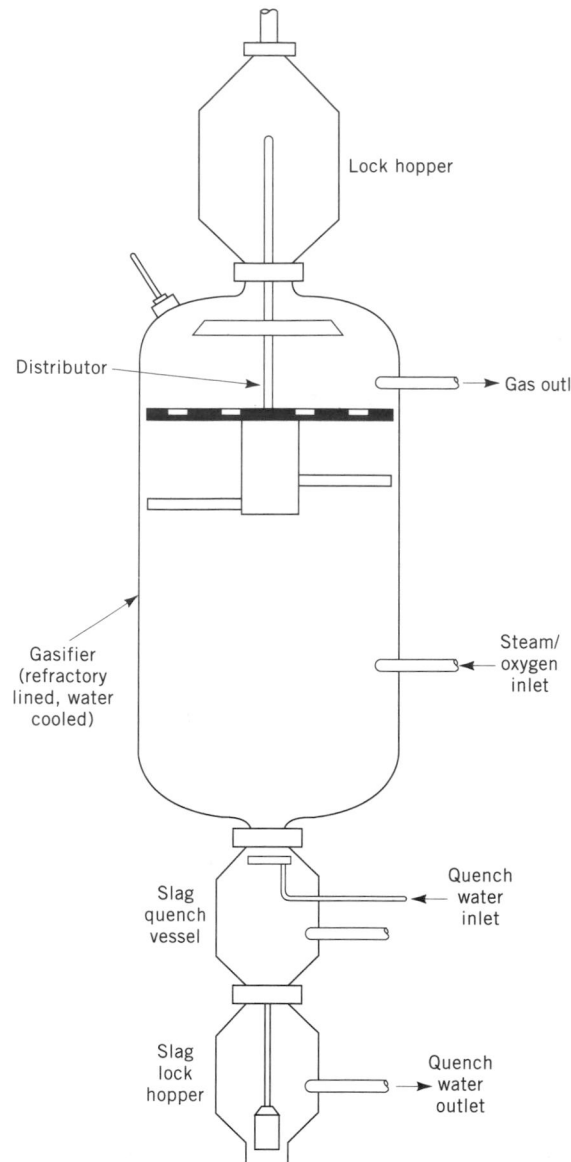

Figure 9. The British Gas-Lurgi slagging gasifier.

The combustion of coal may be simply represented as the staged oxidation of coal carbon to carbon dioxide.

$$2\,C_{coal} + O_2 \rightarrow 2\,CO$$
$$2\,CO + O_2 \rightarrow 2\,CO_2$$

with any reactions of the hydrogen in the coal being considered to be of secondary importance. Other types of combustion systems may be rate-controlled as a result of the onset of the Boudouard reaction (see Table 4).

The complex nature of coal as a molecular entity (2,3,24,25,35,37,53) has resulted in the chemical explanations of coal combustion being confined to the carbon in the system. The hydrogen and other elements have received much less attention but the system is extremely complex and the heteroatoms, eg, nitrogen, oxygen, and sulfur, exert an influence on the combustion. It is this latter that influences environmental aspects.

For example, the conversion of nitrogen and sulfur, during coal combustion, to the respective oxides during combustion cannot be ignored:

$$S_{coal} + O_2 \rightarrow SO_2$$
$$2\,SO_2 + O_2 \rightarrow 2\,SO_3$$
$$2\,N_{coal} + O_2 \rightarrow 2\,NO$$
$$2\,NO + O_2 \rightarrow 2\,NO_2$$
$$N_{coal} + O_2 \rightarrow NO_2$$

The sulfur and nitrogen oxides that escape into the atmosphere can be converted to acids by reaction with moisture in the atmosphere (see also AIR POLLUTION; AIR QUALITY MODELING.)

Sulfurous acid	$SO_2 + H_2O \rightarrow H_2SO_3$
Sulfuric acid	$2\,SO_2 + O_2 \rightarrow 2\,SO_3$
	$SO_3 + H_2O \rightarrow H_2SO_4$
Nitrous acid	$NO + H_2O \rightarrow H_2NO_2$
Nitric acid	$2\,NO + O_2 \rightarrow 2\,NO_2$
	$NO_2 + H_2O \rightarrow HNO_3$

Combustion Systems. Combustion systems vary in nature depending on the nature of the feedstock and the air needed for the combustion process (54). However, the two principal types of coal-burning systems are usually referred to as layer and chambered. The former refers to fixed beds; the latter is more specifically for pulverized fuel.

Fixed or Slowly Moving Beds. For fuel-bed burning on a grate, a distillation effect occurs. The result is that liquid components which are formed volatilize before combustion temperatures are reached; cracking may also occur. The ignition of coal in a bed is almost entirely by radiation from hot refractory arches and from the flame burning of volatiles. In fixed beds, the radiant heat above the bed can only penetrate a short distance into the bed.

Consequently, convective heat transfer determines the intensity of warming up and ignition. In addition, convective heat transfer also plays an important part in the overall flame-to-surface transmission. The reaction of gases is greatly accelerated by contact with hot surfaces and, whereas the reaction away from the walls may proceed slowly, reaction at the surface proceeds much more rapidly.

Fluidized Beds. Fluidized-bed combustion occurs in expanded beds (Fig. 3) and at relatively lower (925°C) temperatures; high convective transfer rates exist resulting from the bed motion. Fluidized systems can operate under substantial pressures thereby allowing more efficient gas clean-up. Fluidized-bed combustion is a means for providing high heat-transfer rates, controlling sulfur, and reducing nitrogen oxide emissions from the low temperatures in the combustion zone.

There are, however, problems associated with pollution control. Whereas the sulfur may be removed downstream using suitable ancillary controls, the sulfur may also be captured in the bed, thereby adding to the separations and recycle problems. Capture during combustion, however, is recognized as the ideal and is a source of optimism for fluidized combustion.

A fluidized bed is an excellent medium for contacting gases with solids, and this can be exploited in a combustor because sulfur dioxide emissions can be reduced by adding limestone, $CaCO_3$, or dolomite, $CaCO_3 \cdot MgCO_3$, to the bed.

$$2 SO_2 + O_2 \rightarrow 2 SO_3$$
$$SO_3 + CaCO_3 \rightarrow CaSO_4 + CO_2$$

or

$$2 SO_2 + O_2 + 2 CaCO_3 \rightarrow 2 CaSO_4 + 2 CO_2$$

The spent sorbent from fluidized-bed combustion may be taken directly to disposal and is much easier than the disposal of salts produced by wet limestone scrubbing. Alternatively, the spent sorbent may be regenerated using synthesis gas, CO/H_2.

$$CaSO_4 + H_2 \rightarrow CaO + H_2O + SO_2$$
$$CaSO_4 + CO \rightarrow CaO + CO_2 + SO_2$$

The calcium oxide product is supplemented with fresh limestone and returned to the fluidized bed. Two undesirable side reactions can occur in the regeneration of spent lime leading to the production of calcium sulfide:

$$CaSO_4 + 4 H_2 \rightarrow CaS + 4 H_2O$$
$$CaSO_4 + 4 CO \rightarrow CaS + 4 CO_2$$

which results in the recirculation of sulfur to the bed.

Entrained Systems. In entrained systems, fine grinding and increased retention times intensify combustion but the temperature of the carrier and degree of dispersion are also important. In practice, the coal is introduced at high velocities which may be greater than 30 m/s and involve expansion from a jet to the combustion chamber.

Types of entrained systems include cyclone furnaces, which have been used for various coals. Other systems have been developed and utilized for the injection of coal–oil slurries into blast furnaces or for the burning of coal–water slurries. The cyclone furnace, developed in the 1940s to burn coal having low ash-fusion temperatures, is a horizontally inclined, water-cooled, tubular furnace in which crushed coal is burned with air entering the furnace tangentially. Temperatures may be on the order of 1700°C and the ash in the coal is converted to a molten slag that is removed from the base of the unit. Coal fines burn in suspension; the larger pieces are captured by the molten slag and burn rapidly.

GAS FROM OTHER FOSSIL FUELS

As of this writing natural gas is a plentiful resource, and there has been a marked tendency not to use other fossil fuels as SNG sources. However, petroleum and oil shale (qv) have been the subject of extensive research efforts. These represent other sources of gaseous fuels and are worthy of mention here.

Petroleum

Thermal cracking (pyrolysis) of petroleum or fractions thereof was an important method for producing gas in the years following its use for increasing the heat content of water gas. Many water gas sets operations were converted into oil-gasification units (55). Some of these have been used for baseload city gas supply, but most find use for peak-load situations in the winter.

In the 1940s, the hydrogasification of oil was investigated as a follow-up to the work on the hydrogasification of coal (56,57). In the ensuing years, further work was carried out as a supplement to the work on thermal cracking (58,59), and during the early 1960s it became evident that light distillates having end boiling points <182°C and containing no sulfur could be catalytically reformed by an autothermic process to pure methane (60). This method was extensively used in England until natural (North Sea) gas came into use.

Prior to the discovery of plentiful supplies of natural gas, and depending on the definition of the resources (1), there were plans to accommodate any shortfalls in gas supply from solid fossil fuels and from gaseous resources by the conversion of hydrocarbon (petroleum) liquids to lower molecular weight gaseous products.

$$CH_{petroleum} \rightarrow CH_4$$

Thermal Cracking. In addition to the gases obtained by distillation of crude petroleum, further highly volatile products result from the subsequent processing of naphtha and middle distillate to produce gasoline, as well as from hydrodesulfurization processes involving treatment of naphthas, distillates, and residual fuels (5,61) and from the coking or similar thermal treatment of vacuum gas oils and residual fuel oils (5).

The chemistry of the oil-to-gas conversion has been established for several decades and can be described in general terms although the primary and secondary reactions can be truly complex (5). The composition of the gases produced from a wide variety of feedstocks depends not only on the severity of cracking but often to an equal or lesser extent on the feedstock type (5,62,63). In general terms, gas heating values are on the order of 30–50 MJ/m³ (950–1350 Btu/ft³).

Catalytic Processes. A second group of refining operations which contribute to gas production are the catalytic cracking processes, such as fluid-bed catalytic cracking, and other variants, in which heavy gas oils are converted into gas, naphthas, fuel oil, and coke (5).

The catalysts promote steam reforming reactions that lead to a product gas containing more hydrogen and carbon monoxide and fewer unsaturated hydrocarbon products than the gas product from a noncatalytic process (5). Cracking severities are higher than those from thermal cracking, and the resulting gas is more suitable for use as a medium heat-value gas than the rich gas produced by straight thermal cracking. The catalyst also influences the reactions rates in the thermal cracking reactions, which can lead to higher gas yields and lower tar and carbon yields (5).

The basic chemical premise involved in making synthetic natural gas from heavier feedstocks is the addition of hydrogen to the oil:

$$CH_3(CH_2)_nCH_3 + (n + 1)H_2 \rightarrow (n + 2)CH_4$$

In general terms, as the molecular weight of the feedstock is increased, similar operating conditions of hydrogasification lead to decreasing hydrocarbon gas yields, increasing yields of aromatic liquids, with carbon also appearing as a product.

The principal secondary variable that influences yields of gaseous products from petroleum feedstocks of various types is the aromatic content of the feedstock. For example, a feedstock of a given H/C (C/H) ratio that contains a large proportion of aromatic species is more likely to produce a larger proportion of liquid products and elemental carbon than a feedstock that is predominantly paraffinic (5).

Another option for processing crude oils which are too heavy to be hydrogasified directly involves first hydrocracking the crude oil to yield a low boiling product suitable for gas production and a high boiling product suitable for hydrogen production by partial oxidation (57). Alternatively, it may be acceptable for carbon deposition to occur during hydrogasification which can then be used for heat or for hydrogen production (64,65).

Partial Oxidation. It is often desirable to augment the supply of naturally occurring or by-product gaseous fuels or to produce gaseous fuels of well-defined composition and combustion characteristics (5). This is particularly true in areas where the refinery fuel (natural gas) is in poor supply and/or where the manufacture of fuel gases, originally from coal and more recently from petroleum, has become well established.

Almost all petroleum fractions can be converted into gaseous fuels, although conversion processes for the heavier fractions require more elaborate technology to achieve the necessary purity and uniformity of the manufactured gas stream (5). In addition, the thermal yield from the gasification of heavier feedstocks is invariably lower than that of gasifying light naphtha or liquefied petroleum gas(es) because, in addition to the production of hydrogen, carbon monoxide, and gaseous hydrocarbons, heavy feedstocks also yield some tar and coke.

As in the case of coal, synthetic natural gas can be produced from heavy oil by partially oxidizing the oil to a mixture of carbon monoxide and hydrogen

$$2 CH_{petroleum} + O_2 \rightarrow 2 CO + H_2$$

which is methanated catalytically to produce methane of any required purity. The initial partial oxidation step consists of the reaction of the feedstock with a quantity of oxygen insufficient to burn it completely, making a mixture consisting of carbon monoxide, carbon dioxide, hydrogen, and steam. Success in partially oxidizing heavy feed-

stocks depends mainly on details of the burner design (66). The ratio of hydrogen to carbon monoxide in the product gas is a function of reaction temperature and stoichiometry and can be adjusted, if desired, by varying the ratio of carrier steam to oil fed to the unit.

To make synthetic natural gas by partial oxidation, virtually all of the methane in the product gas must be produced by catalytic methanation of carbon monoxide and hydrogen. The feed is mixed with recycled carbon and fed, together with steam and oxygen, to a reactor in which partial combustion takes place. Heat from the reaction gasifies the rest of the feed, and by-product coke is formed. The heavier feedstocks tend to produce more carbon than can be consumed through recycling; thus some must be withdrawn. After carbon separation by water scrubbing, the synthesis gas that is available can be converted into hydrogen or into synthetic natural gas by methanation.

Steam Reforming. When relatively light feedstocks, eg, naphthas having ca 180°C end boiling point and limited aromatic content, are available, high nickel content catalysts can be used to simultaneously conduct a variety of near-autothermic reactions. This results in the essentially complete conversions of the feedstocks to methane:

$$CH_3(CH_2)_3CH_3 + 2 H_2O \rightarrow 4 CH_4 + CO_2$$

Because of limitations on the activity of practical catalysts, this reaction must be carried out in stages, the first of which is carried out at 425–480°C and 1.5–2.9 MPa (200–400 psi) and amounts approximately to the following reaction:

$$C_5H_{12} + 3 H_2O \rightarrow 3 CH_4 + CO_2 + 3 H_2 + CO$$

In ensuing catalytic stages, usually termed hydrogasification and methanation (not to be confused with the noncatalytic, direct hydrogasification processes described above), the remaining carbon monoxide and hydrogen are reacted to produce additional methane.

Oil Shale

Oil shale (qv) is a sedimentary rock that contains organic matter, referred to as kerogen, and another natural resource of some consequence that could be exploited as a source of synthetic natural gas (67–69). However, as of this writing, oil shale has found little use as a source of substitute natural gas.

Biomass

Biomass is simply defined for these purposes as any organic waste material, such as agricultural residues, animal manure, forestry residues, municipal waste, and sewage, which originated from a living organism (70–74).

Biomass is another material that can produce a mixture of carbonaceous solid and liquid products as well as gas:

$$C_{organic} \rightarrow C_{coke/char/carbon} + liquids + gases$$

Whereas biomass has not received the same attention as coal as a source of gaseous fuels, questions about the security of fossil energy supplies related to the availability of natural and substitute gas have led to a search for more reliable and less expensive energy sources (75). Biomass resources are variable, but it has been estimated that substantial amounts (up to 20×10^6 mJ (20×10^{15} Btu)) of energy, representing ca 19% of the annual energy consumption in the United States. In addition, environmental issues associated with the use of coal have led some energy producers to question the use of large central energy generating plants. However, biomass may be a gaseous fuel source whose time is approaching (see FUELS FROM BIOMASS).

The means by which synthetic gaseous fuels could be produced from a variety of biomass sources are variable and many of the known gasification technologies can be applied to the problem (70,71,76–82). For example, the Lurgi circulatory fluidized-bed gasifier is available for the production of gaseous products from biomass feedstocks as well as from coal (83,84).

GAS TREATING

The reducing conditions in gasification reactors effect the conversion of the sulfur and nitrogen in the feed coal to hydrogen sulfide, H_2S, and ammonia, NH_3. Some carbonyl sulfide, COS, carbon disulfide, CS_2, mercaptans, RSH, and hydrogen cyanide, HCN, are also formed in the gasifier. These compounds, along with carbon dioxide, are removed simultaneously, either selectively or nonselectively, from the gas stream in the clean-up stages of the process using commercially available physical or chemical solvents and scrubbing agents (1,5,85–88).

Solvents used for hydrogen sulfide absorption include aqueous solutions of ethanolamine (monoethanolamine, MEA), diethanolamine (DEA), and diisopropanolamine (DIPA) among others:

$$2\,RNH_2 + H_2S \rightarrow (RNH_3)_2S$$

$$(RNH_3)_2S + H_2S \rightarrow 2\,RNH_3HS$$

$$2\,RNH_2 + CO_2 + H_2O \rightarrow (RNH_3)_2CO_3$$

$$(RNH_3)_2CO_3 + CO_2 + H_2O \rightarrow 2\,RNH_3HCO_3$$

$$2\,RNH_2 + CO_2 \rightarrow RNHCOONH_3R$$

These solvents differ in volatility and selectivity for the removal of H_2S, mercaptans, and CO_2 from gases of different composition. Other alkaline solvents used for the absorption of acidic components in gases include potassium carbonate, K_2CO_3, solutions combined with a variety of activators and solubilizers to improve gas–liquid contacting.

Whereas most alkaline solvent absorption processes result in gases of acceptable purity for most purposes, it is often essential to remove the last traces of residual sulfur compounds from gas streams. This is in addition to ensuring product purity such as the removal of water, higher hydrocarbons, and dissolved elemental sulfur from liquefied petroleum gas. Removal can be accomplished by passing the gas over a bed of molecular sieves, synthetic zeolites commercially available in several proprietary forms. Impurities are retained by the packed bed, and when the latter is saturated it can be regenerated by passing hot clean gas or hot nitrogen, generally in a reverse direction.

By-product water formed in the methanation reactions is condensed by either refrigeration or compression and cooling. The remaining product gas, principally methane, is compressed to desired pipeline pressures of 3.4–6.9 MPa (500–1000 psi). Final traces of water are absorbed on solica gel or molecular sieves, or removed by a drying agent such as sulfuric acid, H_2SO_4. Other desiccants may be used, such as activated alumina, diethylene glycol, or concentrated solutions of calcium chloride.

Reprinted from *Kirk-Othmer Encyclopedia of Technology,* 4th ed. Vol. 12, John Wiley & Sons, Inc. New York, 1995

BIBLIOGRAPHY

1. J. G. Speight, ed., *Fuel Science and Technology Handbook,* Marcel Dekker, Inc., New York, 1990.
2. J. G. Speight, *The Chemistry and Technology of Coal,* Marcel Dekker, Inc., New York, 1983.
3. R. A. Hessley, in Ref. 1, pp. 645–734.
4. E. J. Parente and A. Thumann, eds., *The Emerging Synthetic Fuel Industry,* Fairmont Press, Atlanta, Ga., 1981.
5. J. G. Speight, *The Chemistry and Technology of Petroleum,* 2nd ed., Marcel Dekker, Inc., New York, 1991.
6. A. Kasem, *Three Clean Fuels from Coal: Technology and Economics,* Marcel Dekker, Inc., New York, 1979.
7. L. L. Anderson and D. A. Tillman, *Synthetic Fuels from Coal,* John Wiley & Sons, Inc., New York, 1979.
8. A. D. Dainton, in G. J. Pitt and G. R. Millward, eds., *Coal and Modern Coal Processing: An Introduction,* Academic Press, Inc., New York, 1979.
9. D. M. Considine, *Energy Technology Handbook,* McGraw-Hill Book Co., Inc., New York, 1977.
10. A. Francis and M. C. Peters, *Fuels and Fuel Technology,* Pergamon Press, Inc., New York, 1980.
11. J. L. Johnson, in M. A. Elliott, ed., *Chemistry of Coal Utilization,* 2nd suppl. vol. John Wiley & Sons, Inc., New York, 1981, Chapt. 23.
12. A. Elton, in C. Singer and co-workers, eds., *A History of Technology,* Vol. 4, Oxford University Press, Oxford, U.K., 1958, Chapt. 9.
13. L. Shnidman, in H. H. Lowry, ed., *Chemistry of Coal Utilization,* John Wiley & Sons, Inc., New York, 1945, Chapt. 30.
14. C. M. Jarvis, in Ref. 12, Vol. 5, Chapt. 10.
15. P. F. Mako and W. A. Samuel, in R. A. Meyers, ed., *Handbook of Synfuels Technology,* McGraw-Hill Book Co., Inc., New York, 1984, Chapt. 2–1.
16. *Oil Gas J.* **90**(3), 53 (1992).
17. K. T. Klasson and co-workers, C. Akin and J. Smith, eds., *Gas, Oil, Coal, and Environmental Biotechnology II,* Institute of Gas Technology, Chicago, 1990, p. 408.
18. B. D. Faison, *Crit. Revs. Biotechnol.* **11,** 347 (1991).
19. S. R. Bull, *Energy Sources* **13,** 433 (1991).
20. R. C. Reid, J. M. Prausnitz, and T. K. Sherwood, *The Properties of Gases and Liquids,* McGraw-Hill Book Co., Inc., New York, 1977.
21. *Wellman Galusha Gas Producers,* research bulletin no. 576A, McDowell-Wellman Engineering Co., Cleveland, Ohio, May 1976.
22. R. F. Probstein and R. E. Hicks, *Synthetic Fuels,* pH Press, Cambridge, Mass., 1990.

23. J. S. Sinninghe Damste and J. W. de Leeuw, *Fuel Processing Technol.* **30,** 109 (1992).

24. N. Berkowitz, *Introduction to Coal Technology,* Academic Press, Inc., New York, 1979.

25. R. K. Hessley, J. W. Reasoner, and J. T. Riley, *Coal Science,* John Wiley & Sons, Inc., New York, 1986.

26. M. O. Holowaty and co-workers, in R. A. Meyers, ed., *Coal Handbook,* Marcel Dekker, Inc., New York, 1981, Chapt. 9.

27. W. Eisenhut, in Ref. 11, Chapt. 14.

28. F. W. Gibbs, in Ref. 12, Vol. 3, Chapt. 25.

29. G. J. Pitt and G. R. Millward, eds., in Ref. 8, p. 52.

30. L. Seglin and S. A. Bresler, in Ref. 11, Chapt. 13.

31. E. Aristoff, R. W. Rieve, and H. Shalit, in Ref. 11, Chapt. 16.

32. D. McNeil, in Ref. 11, Chapt. 17.

33. C. Y. Cha and co-workers, *Report No. DOE/MC/24268-2700 (DE89000967),* United States Department of Energy, Washington, D.C., 1988.

34. F. M. Fess, *History of Coke Oven Technology,* Gluckauf Verlag, Essen, Germany, 1957.

35. M. A. Elliott, ed., in Ref. 11.

36. D. Hebden and H. J. F. Stroud, in Ref. 11, Chapt. 24.

37. R. A. Meyers, ed., in Ref. 15.

38. D. R. Simbeck, R. L. Dickenson, and A. J. Moll, *Energy Prog.* **2,** 42 (1982).

39. W. W. Bodle and J. Huebler, in Ref. 26, Chapt. 10.

40. L. K. Rath and J. R. Longanbach, *Energy Sources* **13,** 443 (1991).

41. S. Alpert and M. J. Gluckman, *Ann. Rev. Energy* **11,** 315 (1986).

42. R. A. Meyers, ed., in Ref. 26.

43. J. H. Martin, I. N. Banchik, and T. K. Suhramaniam, *Report of the Committee on Production of Manufactured Gases,* report no. 1GU/B-76, London, 1976.

44. J. F. Farnsworth, *Ind. Heat.* **41**(11), 38 (1974).

45. J. F. Farnsworth, *Proceedings Coal Gas Fundamentals Symposium,* Institute of Gas Technology, Chicago, 1979.

46. J. C. Hoogendoorn, *Proceedings of the Ninth Pipeline Gas Symposium,* Chicago, Oct. 31–Nov. 2, 1977.

47. R. Essenhigh, in Ref. 11, Chapt. 19.

48. M. A. Field and co-workers, *Combustion of Pulverized Coal,* British Coal Utilization Research Association, Leatherhead, Surrey, U.K., 1967.

49. A. Levy and co-workers, in Ref. 26, Chapt. 8.

50. N. Chigier, *Combustion Measurements,* Hemisphere Publishing Corp., New York, 1991.

51. United States Congress, *Public Law 101-549, An Act to Amend the Clean Air Act to Provide for Attainment and Maintenance of Health Protective National Ambient Air Quality Standards, and for Other Purposes,* Nov. 15, 1990.

52. United States Department of Energy, *Clean Coal Technology Demonstration Program,* DOE/FE-0219P, U.S. Dept. of Energy, Washington, D.C., Feb. 1991.

53. J. E. Funk, in J. A. Kent, ed., *Riegel's Handbook of Industrial Chemistry,* Van Nostrand Reinhold Co., New York, 1983, Chapt. 3.

54. F. J. Ceely and E. L. Daman, in Ref. 26, Chapt. 20.

55. J. M. Reid, *Proceedings SNG Symposium 1,* Institute of Gas Technology, Chicago, May 12–16, 1973.

56. F. J. Dent, *Gas J.* **288**(12), 600, 606, 610 (1956).

57. F. J. Dent, *Gas World,* **144**(11), 1078, 1080 (1956).

58. H. R. Linden and E. S. Pettyjohn, *Research Bulletin No. 12,* Institute of Gas Technology, Chicago, 1952.

59. G. B. Schultz and H. R. Linden, *Research Bulletin No. 29,* Institute of Gas Technology, Chicago, 1960.

60. F. J. Dent, *Proceedings of the Ninth International Gas Conference,* The Hague (Scheveningen), the Netherlands, Sept. 1–4, 1964.

61. J. G. Speight, *The Desulfurization of Heavy Oils and Residua,* Marcel Dekker, Inc., New York, 1981.

62. B. B. Bennett, *J. Inst. Fuel* **35**(8), 338 (1962).

63. H. R. Linden and M. A. Elliot, *Am. Gas J.* **186**(2), 22 (1959).

64. *Oil Gas J.* **71**(7), 36, 37 (1973).

65. *Oil Gas J.* **71**(15), 32, 33 (1973).

66. C. J. Kuhre and C. J. Shearer, *Oil Gas J.* **71**(36), 85 (1971).

67. P. Nowacki, *Oil Shale Technical Data Handbook,* Noyes Data Corp., Park Ridge, N.J., 1981.

68. V. D. Allred, ed., *Oil Shale Processing Technology,* Center for Professional Advancement, East Brunswick, N.J., 1982.

69. C. S. Scouten, in Ref. 1.

70. J. S. Robinson, *Fuels from Biomass: Technology and Feasibility,* Noyes Data Corp., Park Ridge, N.J., 1980.

71. J. L. Jones and S. B. Radding, eds., *Thermal Conversion of Solid Wastes and Biomass, Symposium Series No. 130,* American Chemical Society, Washington, D.C., 1980.

72. M. P. Kannan and G. N. Richard, *Fuel* **69,** 747 (1990).

73. L. Jimenez, J. L. Bonilla, and J. L. Ferrer, *Fuel* **70,** 223 (1991).

74. L. Jimenez and F. Gonzalez, *Fuel* **70,** 947 (1991).

75. *Energy World* **145**(3), 11 (1987).

76. M. P. Sharma and B. Prasad, *Energy Management (New Delhi)* **10**(4), 297 (1986).

77. A. A. C. M. Beenackers and W. P. M. van Swaaij, *Thermochemical Processing of Biomass,* Butterworths, London, 1984, p. 91.

78. G. J. Esplin, D. P. C. Fung, and C. C. Hsu, *Can. J. Chem. Eng.* **64,** 651 (1986).

79. S. Gaur and co-workers, *Fuel Sci. Technol. Int.* **10,** 1461 (1992).

80. M. A. McMahon and co-workers, *Preprints, Div. Fuel Chem.* **36**(4), 1670 (1991).

81. K. Dura-Swamy and co-workers, *Preprints, Div. Fuel Chem.* **36**(4), 1677 (1991).

82. J. T. Hamrick, *Preprints, Div. Fuel Chem.* **36**(4), 1986 (1991).

83. R. Reimert and co-workers, *Bioenergy 84,* Vol. 3, Elsevier Applied Science Publishers, London, p. 102.

FUELS, SYNTHETIC, LIQUID

Scott Han
Clarence D. Chang
Mobil Research and Development Corporation
Princeton, New Jersey

The creation of liquids to be used as fuels from sources other than natural crude petroleum (qv) broadly defines synthetic liquid fuels. Hence, fuel liquids prepared from naturally occurring bitumen deposits qualify as synthetics, even though these sources are natural liquids. Synthetic liquid fuels have characteristics approaching those of the liquid fuels in commerce, specifically gasoline, kerosene, jet fuel, and fuel oil (see Aircraft fuels; Coal liquefaction; Diesel fuel; Middle distillate; Transportation fuels, automotive). For much of the twentieth century, the synthetic fuels emphasis was on liquid products derived from coal (qv) upgrading or by extraction or hydrogenation of organic matter in coke liquids, coal tars, tar sands (qv), or bitumen deposits. More recently, however, much of the direction involving synthetic fuels technology has changed. There are two reasons.

The potential of natural gas, which typically has 85–95% methane, has been recognized as a plentiful and clean alternative feedstock to crude oil (see GAS, NATURAL). Estimates (1–3) place worldwide natural gas reserves at ca 1×10^{14} m^3 (3.5×10^{15} ft^3) corresponding to the energy equivalent of ca 1×10^{11} m^3 (637×10^9 bbl) of oil. As of this writing, the rate of discovery of proven natural gas reserves is increasing faster than the rate of natural gas production. Many of the large natural gas deposits are located in areas where abundant crude oil resources lie such as in the Middle East and Russia. However, huge reserves of natural gas are also found in many other regions of the world, providing oil-deficient countries access to a plentiful energy source. The gas is frequently located in remote areas far from centers of consumption, and pipeline costs can account for as much as one-third of the total natural gas cost (1) (see PIPELINES). Thus tremendous strategic and economic incentives exist for on-site gas conversion to liquids.

In general, the proven technology to upgrade methane is via steam reforming to produce synthesis gas, CO + H$_2$. Such a gas mixture is clean and when converted to liquids produces fuels substantially free of heteroatoms such as sulfur and nitrogen. Two commercial units utilizing the synthesis gas from natural gas technology in combination with novel downstream conversion processes have been commercialized.

The direct methane conversion technology, which has received the most research attention, involves the oxidative coupling of methane to produce higher hydrocarbons (qv) such as ethylene (qv). These olefinic products may be upgraded to liquid fuels via catalytic oligomerization processes.

A second trend in synthetic fuels is increased attention to oxygenates as alternative fuels (4) as a result of the growing environmental concern about burning fossil-based fuels. The environmental impact of the oxygenates, such as methanol (qv), ethanol (qv), and methyl *tert*-butyl ether (MTBE) is still under debate, but these alternative liquid fuels are gaining new prominence. The U.S. Alternative Fuels Act of 1988, and the endorsement of oxygenate fuels that act contains, clearly underscore the idea that economics is no longer the sole consideration with regard to alternative fuels production (5).

Despite reduced prominence, coal technology is well positioned to provide synthetic fuels for the future. World petroleum and natural gas production are expected ultimately to level off and then decline. Coal gasification to synthesis gas is utilized to synthesize liquid fuels in much the same manner as natural gas steam reforming technology. Although as of this writing world activity in coal liquefaction technology is minimal, the extensive development and detailed demonstration of processes for converting coal to liquid fuels should serve as solid foundation for the synthetic fuel needs of the future.

Coal, tar, and heavy oil fuel reserves are widely distributed throughout the world. In the Western hemisphere, Canada has large tar sand, bitumen (very heavy crude oil), and coal deposits. The United States has very large reserves of coal and shale. Coal comprises ca 85% of the U.S. recoverable fossil energy reserves (6). Venezuela has an enormous bitumen deposit and Brazil has significant oil shale (qv) reserves. Coal is also found in Brazil, Colombia, Mexico, and Peru. Worldwide, the total resource base of these reserves is immense and may constitute >90% of the hydrocarbon resources in place.

The driving force behind the production of combustible liquids before 1900 was the search for low cost lighting. Gas produced during coal distillation was used to light homes at the end of the eighteenth century (7). Large-scale use of coal, which began in England in the nineteenth century, led to significant reductions in the costs of hydrocarbon liquids. The production of coal tar, and the separation therefrom of various coal liquids concomitant to the production of illuminating gas, probably predates production from the coking operations associated with iron (qv) production. The coal tars produced in gas works may have been the first synthetic liquids turned to fuel use in quantity.

Proof of the existence of benzene in the light oil derived from coal tar (8) first established coal tar and coal as chemical raw materials. Soon thereafter the separation of coal-tar light oil into substantially pure fractions produced a number of the aromatic components now known to be present in significant quantities in petroleum-derived liquid fuels. Indeed, these separation procedures were for the recovery of benzene–toluene–xylene and related substances, ie, benzol or motor benzol, from coke-oven operations (8).

By the middle of the nineteenth century it was realized, both in England and in the United States, that kerosene, or coal oil, distilled from coal, could produce a luminous combustion flame. Commercialization was rapid. By the time of the U.S. Civil War, ~87 m^3/yr (23,000 gal/yr) of lamp oil was produced in the United States from the distillation not only of coal, but also of oil shale and natural bitumen. In 1859, high gravity, low sulfur crude oil was discovered in the United States. This produced high quality kerosene with minimal processing, and the world's first oil boom erupted. Until the end of the nineteenth century, kerosene was the only substance of value extracted from natural crude oil. It cost too much for heating purposes, but was used widely for lighting until replaced by electricity. Refiners slowly learned to use the residues from kerosene production, and as the market for lamp oil collapsed, gasoline increased in value. The widespread use of liquids as fuels dates from that time.

Liquid fuels possess inherent advantageous characteristics in terms of being more readily stored, transported, and metered than gases, solids, or tars. Liquid fuels also are generally easy to process or clean by chemical and catalytic means. The energy densities of clean hydrocarbon liquids may be very high relative to gas, solid, or semisolid fuel substances. Moreover, liquid fuels are the most compatible with the twentieth century world fuel infrastructure because most fuel-powered conveyances are designed to function only with relatively clean, low viscosity liquids. In general, liquid hydrocarbon fuels possess an intermediate hydrogen-to-carbon content. Production of synthetic fuels from alternative feedstocks to natural petroleum crude oil is based on adjusting the hydrogen-to-carbon ratio to the desired intermediate level.

There is an inherent economic penalty associated with producing liquids from either natural gas or solid coal feedstock. Synthetic liquid fuels technologies are generally not economically competitive with crude oil processing in the absence of extraneous influences such as price supports or regulations.

INDIRECT LIQUEFACTION/CONVERSION TO LIQUID FUELS

Indirect liquefaction of coal and conversion of natural gas to synthetic liquid fuels is defined by technology that involves an intermediate step to generate synthesis gas, $CO + H_2$. The main reactions involved in the generation of synthesis gas are the coal gasification reactions:

Combustion

$$C + O_2 \rightleftharpoons CO_2$$
$$\Delta H_{298\,K} = -394 \text{ kJ/mol } (-94.2 \text{ kcal/mol}) \quad (1)$$

Gasification

$$C + 1/2\, O_2 \rightleftharpoons CO$$
$$\Delta H_{298\,K} = -111 \text{ kJ/mol } (-26.5 \text{ kcal/mol}) \quad (2)$$

$$C + H_2O\,(g) \rightleftharpoons CO + H_2$$
$$\Delta H_{298\,K} = +131 \text{ kJ/mol } (31.3 \text{ kcal/mol}) \quad (3)$$

$$C + CO_2 \rightleftharpoons 2\,CO$$
$$\Delta H_{298\,K} = +172 \text{ kJ/mol } (41.1 \text{ kcal/mol}) \quad (4)$$

Water gas shift

$$CO + H_2O\,(g) \rightleftharpoons CO_2 + H_2$$
$$\Delta H_{298\,K} = -41 \text{ kJ/mol } (-9.8 \text{ kcal/mol}) \quad (5)$$

the methane steam reforming reactions:

Partial oxidation

$$CH_4 + 1/2\, O_2 \rightleftharpoons CO + 2\,H_2$$
$$\Delta H_{298\,K} = -36 \text{ kJ/mol } (-8.6 \text{ kcal/mol}) \quad (6)$$

Reforming

$$CH_4 + H_2O\,(g) \rightleftharpoons CO + 3\,H_2$$
$$\Delta H_{298\,K} = 206 \text{ kJ/mol } (49.2 \text{ kcal/mol}) \quad (7)$$

and the water gas shift reaction (eq. 5), used to increase the H_2/CO ratio of the product synthesis gas.

Coal gasification technology dates to the early nineteenth century but has been largely replaced by natural gas and oil. A more hydrogen-rich synthesis gas is produced at a lower capital investment. Steam reforming of natural gas is applied widely on an industrial scale (9,10) and in particular for the production of hydrogen (qv).

The conversion of coal and natural gas to liquid fuels via indirect technology can be achieved by the routes shown in Figure 1. Two pathways from synthesis gas can be taken. Both have been commercialized. One pathway involves coupling with Fischer-Tropsch technology to produce fuel range hydrocarbons directly or upon further processing. Using coal feedstock, this route has been commercialized in South Africa since the 1950s and a process using natural gas was commercialized in Malaysia by Shell Oil Co. in 1993. An alternative route relies on the production of methanol from synthesis gas and subsequent transformation of the methanol to fuels using zeolite catalyst technology introduced by Mobil Oil Corp. This route was commercialized using indigenous natural gas in New Zealand in 1985.

Coal Upgrading via Fischer-Tropsch

The synthesis of methane by the catalytic reduction of carbon monoxide and hydrogen over nickel and cobalt catalysts at atmospheric pressure was reported in 1902 (11).

$$CO + 3\,H_2 \rightarrow CH_4 + H_2O\,(l)$$
$$\Delta H_{298\,K} = -250 \text{ kJ/mol } (-59.8 \text{ kcal/mol}) \quad (8)$$
$$2\,CO + 2\,H_2 \rightarrow CH_4 + CO_2$$
$$\Delta H_{298\,K} = -247 \text{ kJ/mol } (59.0 \text{ kcal/mol}) \quad (9)$$

In the early 1920s Badische Anilin- und Soda-Fabrik announced the specific catalytic conversion of carbon monox-

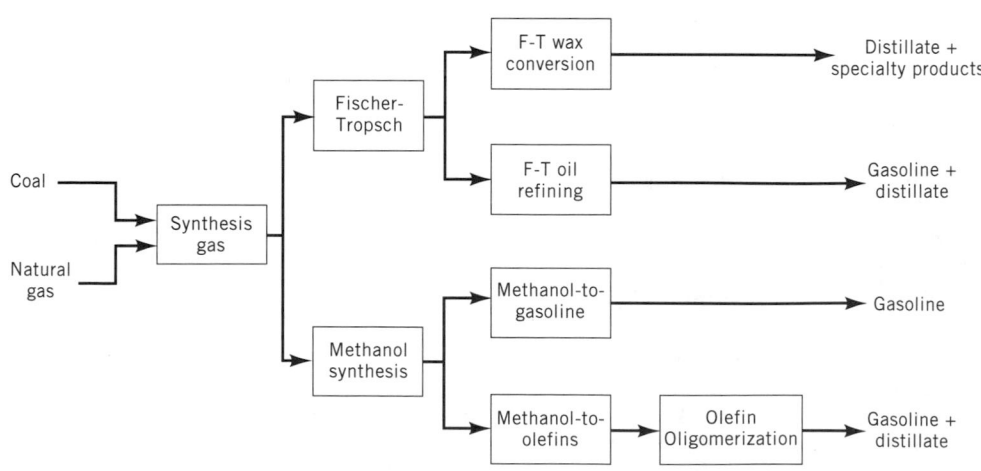

Figure 1. Routes to liquid fuels from natural gas and coal via synthesis gas. F-T is the Fischer-Tropsch process.

ide and hydrogen at 20–30 MPa (200–300 atm) and 300–400°C to methanol (12,13), a process subsequently widely industrialized. At the same time Fischer and Tropsch announced the Synthine process (14,15), in which an iron catalyst effects the reaction of carbon monoxide and hydrogen to produce a mixture of alcohols, aldehydes (qv), ketones (qv), and fatty acids at atmospheric pressure.

In the classical normal pressure synthesis (16), higher hydrocarbons are produced by net reactions similar to those observed in the early 1900s, but at temperatures below the level at which methane is formed:

$$n\ CO + 2n\ H_2 \rightarrow C_nH_{2n} + n\ H_2O + \text{heat} \qquad (10)$$

$$2n\ CO + n\ H_2 \rightarrow C_nH_{2n} + n\ CO_2 + \text{heat} \qquad (11)$$

In the mid-1930s improvements in catalysts and techniques (17–19) culminated in the licensing of the process to Ruhrchemie to produce liquid hydrocarbons and paraffin waxes using precipitated cobalt-on-kieselguhr catalysts. Subsequently, a medium pressure synthesis was developed (20) at 0.5–2 MPa (5–20 atm) using dispersed cobalt catalysts which improved hydrocarbon yields by 10–15%. The yields of paraffin wax, in particular, could be increased to 45% of the total liquid product. Hydrotreating of catalyst (required in the normal pressure process) was avoided, and catalyst life was extended (21–23). There is a marked influence of pressure on product yields. Beyond the optimum pressure of about 2 MPa (20 atm), paraffin yield decreases. Little change is found in the gasoline and gas oil yields, however.

Furthermore it was discovered that reasonable yields could be obtained using precipitated iron catalysts at 1–3 MPa (10–30 atm), and that very high melting waxes could be synthesized at 10–100 MPa (100–1000 atm) over ruthenium catalysts. At the same time a related process, the oxo synthesis, was announced (24). Early in World War II, the iso-synthesis process was developed for the production of low molecular weight isoparaffins at high temperatures and pressures over thoria and mixtures of alumina and zinc oxide (25–28). In the early 1960s polymethylenes were synthesized using activated ruthenium catalysts at high pressures.

Industrial operation of the Fischer-Tropsch synthesis involved five steps: (1) synthesis gas manufacture; (2) gas purification by removal of water and dust, and hydrogen sulfide and organic sulfur compounds; (3) synthesis of hydrocarbons; (4) condensation of liquid products and recovery of gasoline from product gas; and (5) fractionation of synthetic products. Only the synthesis reactor and its method of operation were unique to the process. For low pressure synthesis the reactor incorporated elaborate bundles of cooling tubes immersed in the catalyst, whereby circulating water removed the heat of reaction, limiting the conversion to methane which produced high temperatures. In the pressure process, bundles of concentric tubes, with catalyst arranged in the annuli, through and around which cooling water flowed, served as conversion units. In both systems, the conversion units each contained about 10 m³ (ca 350 ft³) of catalyst, and were rated at a capacity of ca 4.8 m³ (30 bbl) of liquid product per day.

During World War II, nine commercial plants were operated in Germany, five using the normal pressure synthesis, two the medium pressure process, and two having converters of both types. The largest plants had capacities of ca 400 m³/d (2500 bbl/d) of liquid products. Cobalt catalysts were used exclusively.

Development Outside Germany. In the late 1930s experimental work in England (29–31) led to the erection of large pilot facilities for Fischer-Tropsch studies (32). In France, a commercial facility near Calais produced ca 150 m³ (940 bbl) of liquid hydrocarbons per day. In Japan, two full-scale plants were also operated under Ruhrchemie license. Combined capacity was ca 400 m³ (2500 bbl) of liquids per day.

In the mid-1930s Universal Oil Products reported (33,34) that gasoline of improved quality could be produced by cracking the high boiling fractions of Fischer liquids, and a consortium, the Hydrocarbon Synthesis, Inc., entered into an agreement with Ruhrchemie to license the Fischer synthesis outside Germany.

In 1955 the South African Coal, Oil, and Gas Corp. (Sasol) commercialized the production of liquid fuels utilizing Fischer-Tropsch technology (35). This Sasol One complex has evolved into the streaming of second-generation plants, known as Sasol Two and Three. The Sasol One process, shown in Figure 2a (36), combines fixed-bed Ruhrchemie-Lurgi Arge reactor units with fluidized-bed Synthol process technology (37). For Sasol One, 16,000 t/d of coal is crushed and gasified with steam and oxygen. After a number of gas purification steps in which by-products and gas impurities are removed, the pure gas is processed in both fixed- and fluidized-bed units simultaneously. Table 1 gives product selectivity comparisons of fixed-bed and Synthol operations. Conversion to hydrocarbons is higher in the Synthol unit and the H_2/CO ratio is also higher. Because the fixed-bed Arge reactor favors the formation of straight-chain paraffins, there is greater production of diesel and wax fractions than the Synthol unit. The Arge reactor products have lower gasoline octane number but higher diesel cetane number relative to Synthol. The high wax production using the Arge reactor was disadvantageous at the time owing to market limitations of wax fuels. Sasol One produced a vast array of chemical and fuel products, including gasoline at 1.5×10^6 t/yr.

Table 1. Product Selectivities for Commercial Fixed-Bed and Synthol Units[a]

Product	Fixed-bed	Synthol
Methane	2.0	10
Ethylene	0.1	4
Ethane	1.8	4
Propylene	2.7	12
Propane	1.7	2
Butenes	3.1	9
Butanes	1.9	2
C_5 and higher	83.5	51
Soluble chemicals	3.0	5
Water-soluble acids	0.2	1

[a] Ref. 36.

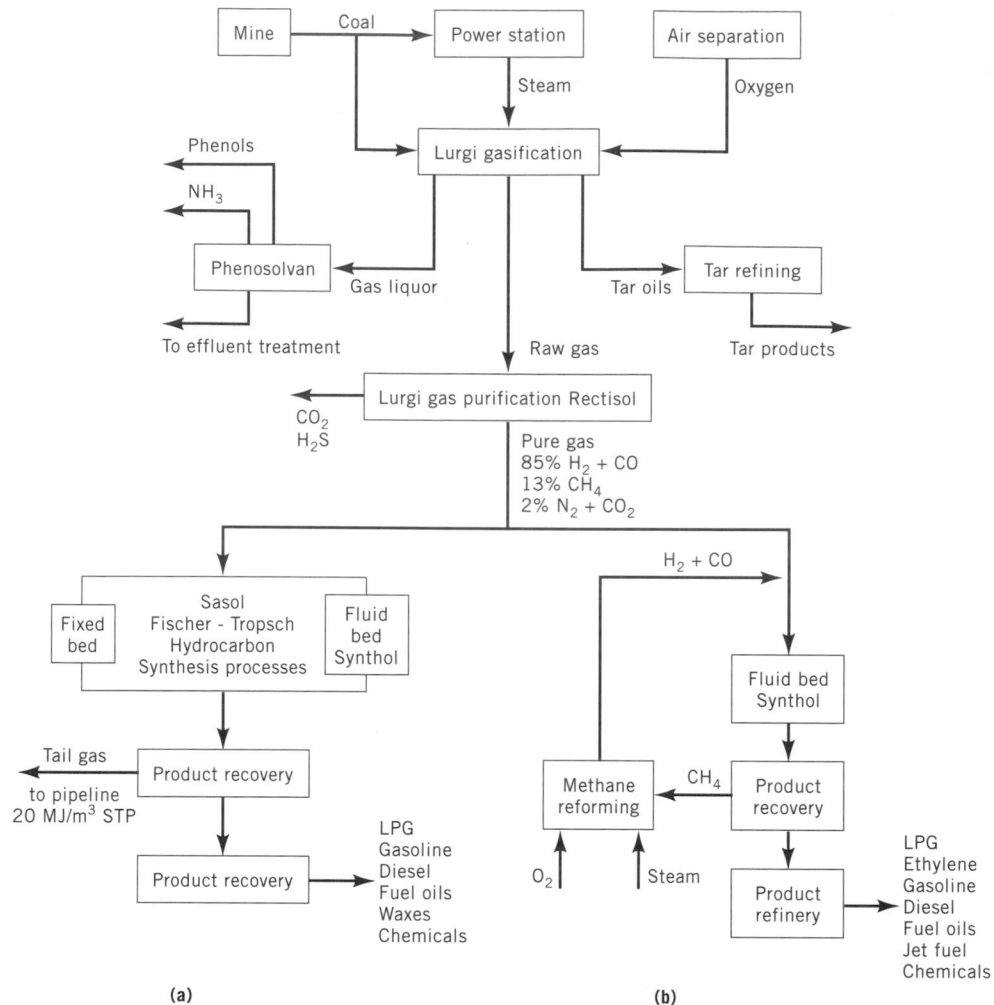

Figure 2. Process flow sheet of (**a**) Sasol One and (**b**) the Sasol synfuel process for Sasol Two and Three (36). LPG is liquefied petroleum gas; other terms are defined in the text. To convert J to cal, divide by 4.184.

The 1973 oil crisis resulted in the Sasol Two unit which started up in early 1980 followed by the nearly identical Sasol Three unit two years later. Figure 2**b** gives the schematic flow diagrams for the Sasol Two and Three processes. Sasol Two uses 36 Lurgi gasifiers in parallel to process ca 31,000 t/d of sized coal. By-product effluents and gas impurities are removed in Rectisol (sulfur compounds and CO_2 removal), Phenosolvan (oxygen compounds and ammonia removal), and tar separation units. Synthol fluid-bed units were employed because of the product distribution and ease of design scale-up. Approximately 80,000 t/d of coal are needed for the two plants. Composition and manufacturing information for Sasol Fischer-Tropsch catalysts are trade secrets, but the catalyst is widely accepted as being an alkali-metal promoted iron-based material.

More recently, Sasol commercialized a new type of fluidized-bed reactor and was also operating a higher pressure commercial fixed-bed reactor (38). In 1989, a commercial scale fixed fluid-bed reactor was commissioned having a capacity similar to existing commercial reactors at Sasol One (39). This effort is aimed at expanded production of higher value chemicals, in particular waxes (qv) and linear olefins.

Properties. Fischer-Tropsch liquid obtained using cobalt catalysts is roughly equivalent to a very paraffinic natural petroleum oil but is not so complex a mixture. Straight-chain, saturated aliphatic molecules predominate but monoolefins may be present in an appreciable concentration. Alcohols, fatty acids, and other oxygenated compounds may represent less than 1% of the total liquid product. The normal pressure synthesis yields ca 60% gasoline, 30% gas oil, and 10% paraffin (mp 20–100°C). The medium pressure synthesis yields 35% gasoline, 35% gas oil, and 30% paraffin. The octane rating of the gasoline is too low for direct use as motor fuel (40).

Most of the German gasoline production was blended into motor fuels using benzene derived from coking. The gas oil could be used directly as a superior diesel fuel; some was also used in soap manufacture. The paraffin (referred to as gatsch) was used primarily for the synthesis of fatty acids and hard soaps. The propane and butane gases were also used as motor fuels. Some propylene and butylenes were polymerized over phosphoric acid to high octane gasoline, and some olefins to lubricating oils. Typical values for the composition of the technical-scale reaction products of the normal and medium pressure synthesis are available (41).

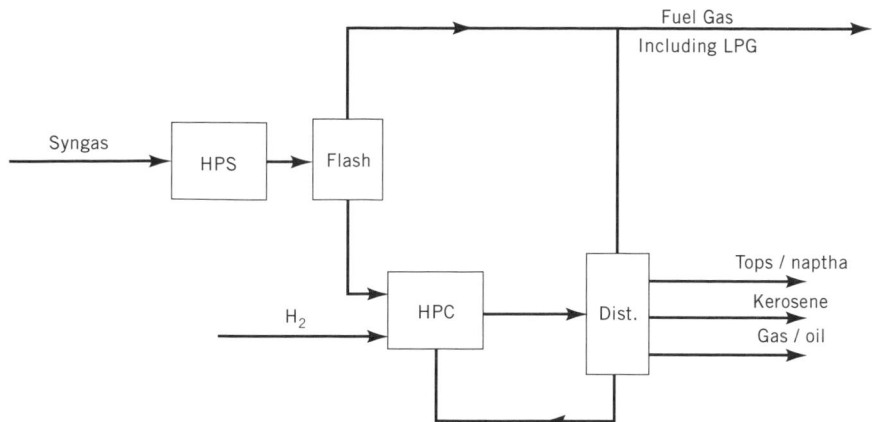

Figure 3. The Shell middle distillate synthesis (SMDS) process. HPS = heavy paraffin synthesis. HPC = heavy paraffin conversion.

Natural Gas Upgrading via Fischer-Tropsch

In the United States, as in other countries, scarcities from World War II revived interest in the synthesis of fuel substances. A study of the economics of Fischer synthesis led to the conclusion that the large-scale production of gasoline from natural gas offered hope for commercial utility. In the Hydrocol process (Hydrocarbon Research, Inc.) natural gas was treated with high purity oxygen to produce the synthesis gas which was converted in fluidized beds of iron catalysts (42).

Shell Middle Distillate Synthesis. The Shell middle distillate synthesis (SMDS) process developed by Shell Oil Co., uses remote natural gas as the feedstock (43–45). A simplified flow scheme is given in Figure 3. This two-step process involves Fischer-Tropsch synthesis of paraffinic wax called the heavy paraffin synthesis (HPS). The wax is subsequently hydrocracked and hydroisomerized to yield a middle distillate boiling range product in the heavy paraffin conversion (HPC). In the HPS stage, wax is maximized by using a proprietary catalyst having high selectivity toward heavier products and by the use of a tubular, fixed-bed Arge-type reactor. The HPC stage employs a commercial hydrocracking catalyst in a trickle flow reactor. The effect of hydrocracking light paraffins is shown in Figure 4. The HPC step allows for production of narrow range hydrocarbons not possible with conventional Fischer-Tropsch technology.

After years of bench-scale and pilot-plant studies, construction was begun on a gca 1600 m³/d (10,000 bbl/d) unit in Sarawak, Malaysia, by Shell in a joint venture with Mitsubishi and the Malaysian government. Plant commissioning was in early 1993 at a capital investment of ca $600–700 × 10⁶. The plant uses natural gas from offshore fields and is located adjacent to the existing Malaysian Liquefied Natural Gas (LNG) plant. The production of liquid transportation fuels via SMDS cannot compete economically with fuels derived from crude oil. However, economics vary greatly with site location, and subsidies from the Malaysian government, eg, reduced natural gas cost, brought this plant to commercialization. In addition, premium selling prices for the high quality products made from SMDS are a primary influence on commercialization potential (44).

A similar process to SMDS using an improved catalyst is under development by Norway's state oil company, den norske state olijeselskap AS (Statoil) (46). High synthesis gas conversion per pass and high selectivity to wax are claimed. The process has been studied in bubble columns and a demonstration plant is planned.

Properties. Shell's two-step SMDS technology allows for process flexibility and varied product slates. The liquid product obtained consists of naphtha, kerosene, and gas oil in ratios from 15:25:60 to 25:50:25, depending on process conditions. Of particular note are the high quality gas oil and kerosene. Table 2 gives SMDS product qualities for these fractions.

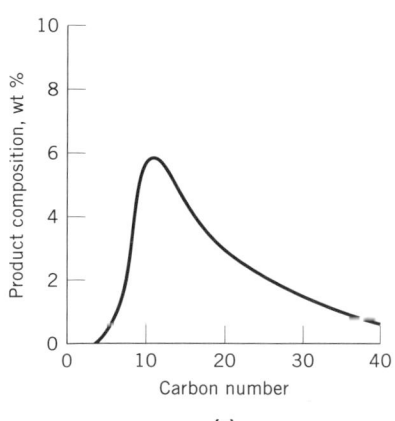

(a)

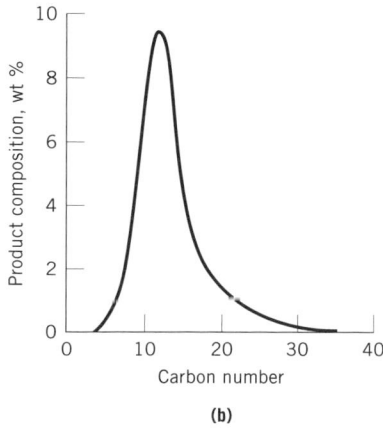

(b)

Figure 4. Product compositions as a function of carbon number for the Shell middle distillate synthesis process: (**a**) the Fischer-Tropsch product following HPS, and (**b**) the final hydrocracking product following HPC. See text (45).

Table 2. SMDS Product Quality[a,b]

Parameter	SMDS	Specification
Gas oil		
Cetane number	70	40 to 50
Cloud point, °C	−10	−10 to + 20
Kerosene		
Smoke point, mm	45	19
Freezing point, °C	−47	−47

[a] Ref. 44.

[b] The tops/naphtha fraction is similar to straight-run material.

The products manufactured are predominantly paraffinic, free from sulfur, nitrogen, and other impurities, and have excellent combustion properties. The very high cetane number and smoke point indicate clean-burning hydrocarbon liquids having reduced harmful exhaust emissions. SMDS has also been proposed to produce chemical intermediates, paraffinic solvents, and extra high viscosity index (XHVI) lubeoils (see LUBRICATION AND LUBRICANTS) (44).

Liquid Fuels via Methanol Synthesis and Conversion

Methanol is produced catalytically from synthesis gas. By-products such as ethers, formates, and higher hydrocarbons are formed in side reactions and are found in the crude methanol product. Whereas for many years methanol was produced from coal, after World War II low cost natural gas and light petroleum fractions replaced coal as the feedstock.

Methanol-to-Gasoline. The most significant development in synthetic fuels technology since the discovery of the Fischer-Tropsch process is the Mobil methanol-to-gasoline (MTG) process (47–49). Methanol is efficiently transformed into C_2–C_{10} hydrocarbons in a reaction catalyzed by the synthetic zeolite ZSM-5 (50–52). The MTG reaction path is presented in Figure 5 (47). The reaction

sequence can be summarized as

$$n/2 \ [2 \ CH_3OH \rightleftharpoons CH_3OCH_3 + H_2O] \rightarrow C_nH_{2n} \rightarrow n[CH_2] \tag{12}$$

where $[CH_2]$ represents an average paraffin–aromatic mixture.

How the initial C—C bond is formed from the C_1 progenitor is unknown and much debated (48,53–55). Light olefins are key intermediates in the reaction sequence. These undergo further transformation, ultimately forming aromatics and light paraffins. Table 3 lists a typical MTG product distribution. A unique characteristic of these products is an abrupt termination in carbon number at around C_{10}. This is a consequence of molecular shape-selectivity (56–58), a property of ZSM-5. The composition and properties of the C_5+ fraction are those of a typical premium aromatic gasoline. Interestingly, C_{10} also is the end point of conventional gasoline.

The MTG process was developed for synfuel production in response to the 1973 oil crisis and the steep rise in crude prices that followed. Because methanol can be made from any gasifiable carbonaceous source, including coal, natural gas, and biomass, the MTG process provided a new alternative to petroleum for liquid fuels production. New Zealand, heavily dependent on foreign oil imports, utilizes the MTG process to convert vast offshore reserves of natural gas to gasoline (59).

Two versions of the MTG process, one using a fixed bed, the other a fluid bed, have been developed. The fixed-bed process was selected for installation in the New Zealand gas-to-gasoline (GTG) complex, situated on the North Island between the villages of Waitara and Motonui on the Tasman seacoast (60). A simplified block flow diagram of the complex is shown in Figure 6 (61). The plant processes over 3.7×10^6 m³/d (130×10^6 SCF/d) of gas from the offshore Maui field supplemented by gas from the Kapuni field, first to methanol, and thence to 2.3×10^3 m³/d (14,500 bbl/d) of gasoline. Methanol feed to the MTG section is synthesized using the ICI low pressure process (62) in two trains, each with a capacity of 2200 t/d.

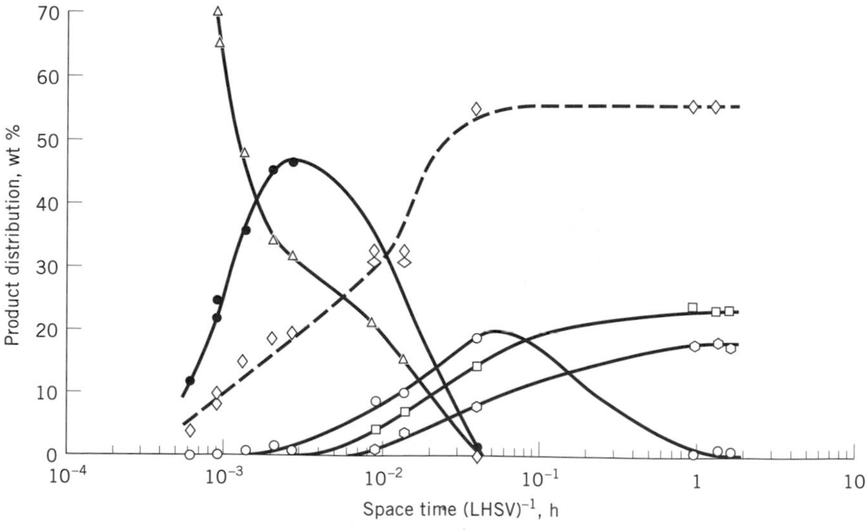

Figure 5. Methanol-to-hydrocarbons reaction path at 371°C △—△— methanol; ●—●— = dimethyl ether; where (◇—◇—) is water; (□—□—) paraffins (and C_6+ olefins); (○—○—) aromatics; and (○—○—) C_2—C_5 olefins. LHSV = Liquid hourly space velocity.

Table 3. MTG Product Distribution[a,b]

Hydrocarbon	Distribution, wt %
Methane	1.0
Ethane	0.6
Ethylene	0.5
Propane	16.2
Propylene	1.0
i-Butane	18.7
n-Butane	5.6
Butenes	1.3
i-Pentane	7.8
n-Pentane	1.3
Pentenes	0.5
C_6+ Aliphatics	4.3
Benzene	1.7
Toluene	10.5
Ethylbenzene	0.8
Xylenes	17.2
C_9 Aromatics	7.5
C_{10} Aromatics	3.3
C_{11}+ Aromatics	0.2

[a] Reaction conditions of 371°C and LHSV of 1.0 h^{-1}.
[b] 100% conversion.

Table 4. MTG Gasoline Quality[a]

Parameter	Average	Range
Density at 15°C, kg/m^3	730	728–733
Reid vapor pressure, kPa[b]	86.2	83.4–91.0
Octane number		
Research	92.2	92.0–92.5
Motor	82.6	82.2–83.0
Durene content, wt %	2.0	1.74–2.29
Induction period, min	325	260–370
Distillation, % evaporation		
at 70°C	31.5	29.5–34.5
at 100°C	53.2	51.5–55.5
at 180°C	94.9	94.0–96.5
Distillation end point, °C	240.5	196–209

[a] Ref. 67
[b] To convert kPa to psia, multiply by 0.145.

and nitrogen levels in the gasoline are virtually nil. The MTG process produces ca 3–7 wt % durene [95-93-2] (1,2,4,5-tetra-methylbenzene) but the level is reduced to ca 2 wt % in the finished gasoline product by hydrodealkylation of the durene in a separate catalytic reactor.

Methanol-to-Olefins and Olefins-to-Gasoline-and-Distillate. Because the MTG process produces primarily gasoline, a variation of that process has been developed which allows for production of gasoline and distillate fuel (68). This process integrates two known technologies, methanol-to-olefins (MTO) and Mobil olefins-to-gasoline-and-distillate (MOGD). The MTO/MOGD process schematic is shown in Figure 9. The combined process produces gasoline and distillate in various proportions and, if needed, olefinic by-products.

In the MTO process, methanol is converted over ZSM-5 giving high (up to ca 80 wt % hydrocarbons) olefin yields and low ethylene and light saturate yields. The low ethylene yields are desirable in achieving high distillate yields using MOGD. Figure 5 shows that the production of olefins rather than gasoline from methanol is governed by the kinetics of methanol conversion over ZSM-5 catalyst (69). Generally, catalyst and process variables which increase methanol conversion decrease olefins yield.

The MTO process employs a turbulent fluid-bed reactor system and typical conversions exceed 99.9%. The coked catalyst is continuously withdrawn from the reactor and burned in a regenerator. Coke yield and catalyst circulation are an order of magnitude lower than in fluid catalytic cracking (FCC). The MTO process was first scaled up in a 0.64 m^3/d (4 bbl/d) pilot plant and a successful 15.9 m^3/d (100 bbl/d) demonstration plant was operated in Germany with U.S. and German government support.

The MOGD process oligomerizes light olefins to gasoline and distillate products over ZSM-5 zeolite catalyst. Gasoline and distillate selectivity is >95% of the feed olefins and gasoline/distillate product ratios can vary, depending on process conditions, from 0.2 to >100. High octane MTO gasoline is separated before the MOGD section and blended with MOGD gasoline. Some MOGD gasoline may be recycled. The distillate product requires hydrofinishing. Generally, the process scheme uses four fixed-bed reactors, three on-line and one in regeneration. A large-

A flow diagram of the MTG section is shown in Figure 7. Methanol feed, vaporized by heat exchange with reactor effluent gases, is converted in a first-stage reactor containing an alumina catalyst to an equilibrium mixture of methanol, dimethyl ether (DME), and water. This is combined with recycle light gas, which serves to remove reaction heat from the highly exothermic MTG reaction, and enters the reactors containing ZSM-5 catalyst. As indicated in Figure 7, five parallel swing reactors are used. Four reactors are on feed while the fifth is under regeneration. The multiple-bed configuration is used to minimize pressure drop as well as to control product selectivity. Reaction conditions are 360–415°C, 2.17 × 10^3 kPa (315 psia), and 9/1 recycle/fresh feed ratio. The overall thermal efficiency of the plant is ca 53%.

A fluid-bed version of the MTG process has been developed (60,63–65) and demonstrated in semiworks scale of 15.9 m^3/d (100 bbl/d), but has not been commercialized to date (ca 1993). Heat management of the exothermic MTG reaction is greatly facilitated by use of fluid-bed reactors. The turbulent bed, with its excellent heat-transfer characteristics, ensures isothermality through the reaction zone and permits steam generation by direct exchange with steam coils in the bed. A schematic diagram appears in Figure 8. The reactor system consists of three principal parts: the reactor, the catalyst regenerator, and an external catalyst cooler. The reactor is also equipped with internal heat-exchanger tubes. Methanol is converted in a single pass at 380–430°C, 276–414 kPa (40–60 psia). The methanol feed rate is 500–1050 kg/h. The fluid-bed demonstration was carried out in 1982–1983 (66).

Properties. Table 4 contains typical gasoline quality data from the New Zealand plant (67). MTG gasoline typically contains 60 vol % saturates, ie, paraffins and naphthenes; 10 vol % olefins; and 30 vol % aromatics. Sulfur

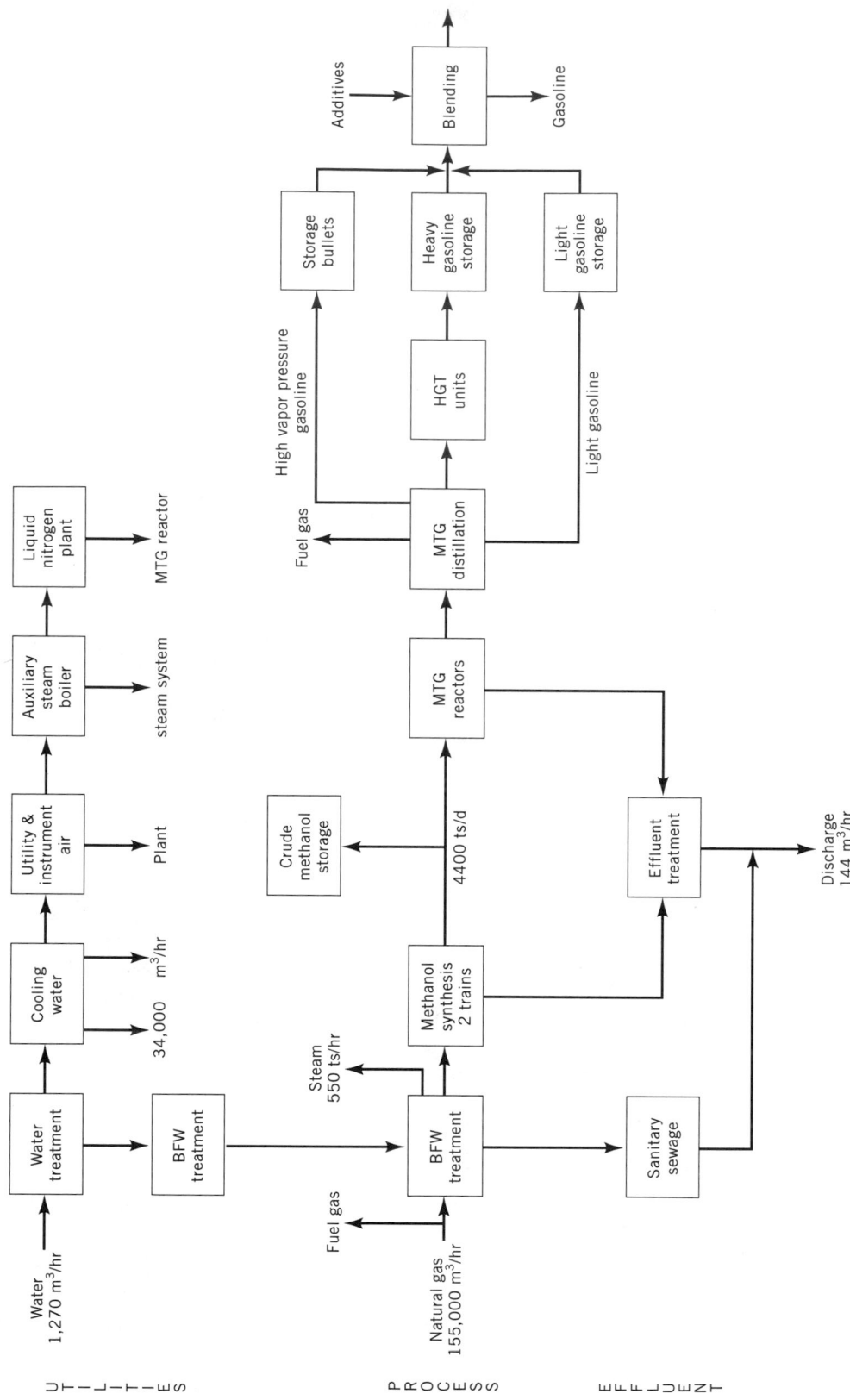

Figure 6. Simplified block flow diagram for the New Zealand gas-to-gasoline (GTG) plant (61). To convert m³/h to gal/min, multiply by 4.40. HGT = heavy gasoline treatment facility; MTG = methanol-to-gasoline; BFW = boiler feed water.

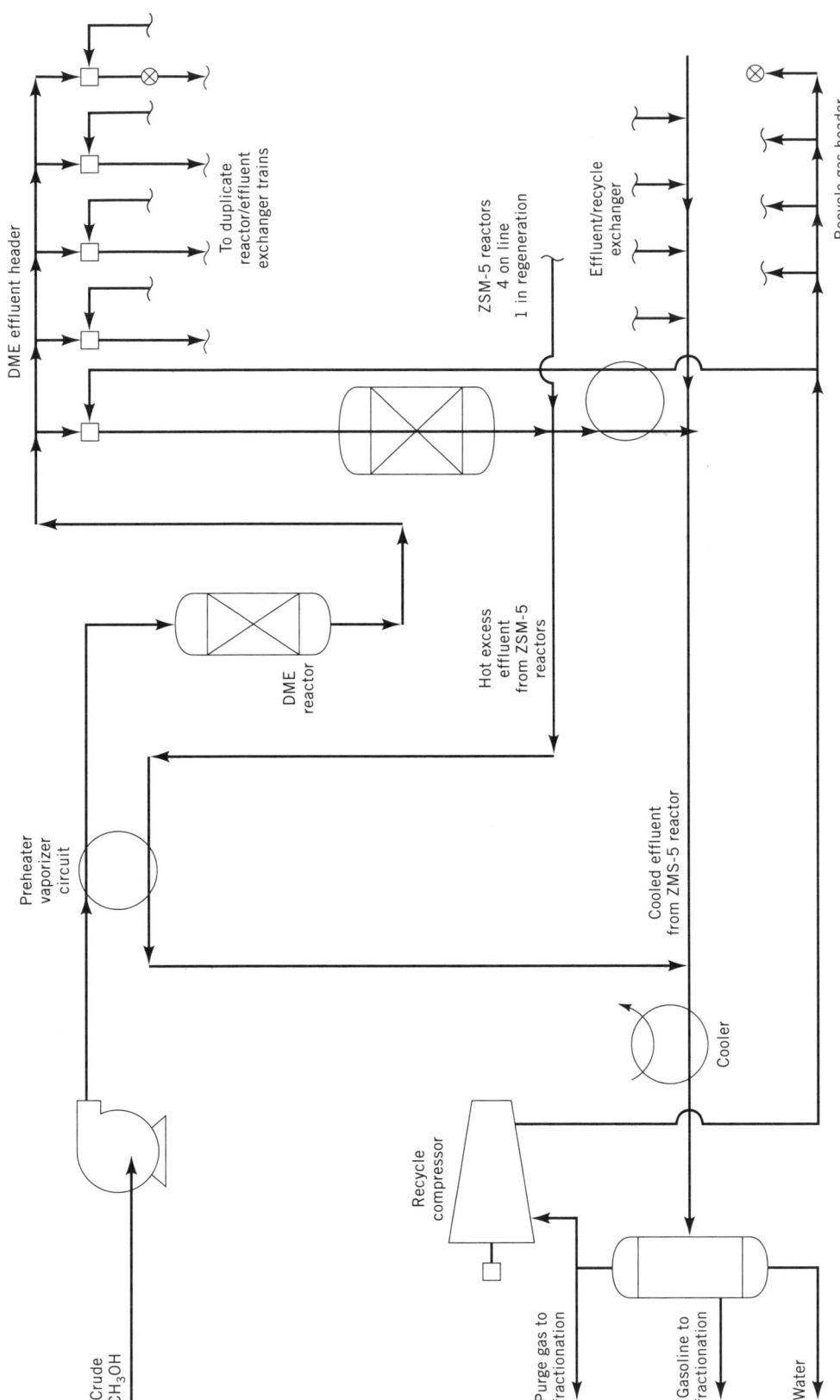

Figure 7. A methanol-to-gasoline (MTG) fixed-bed process flow diagram. DME = dimethyl ether.

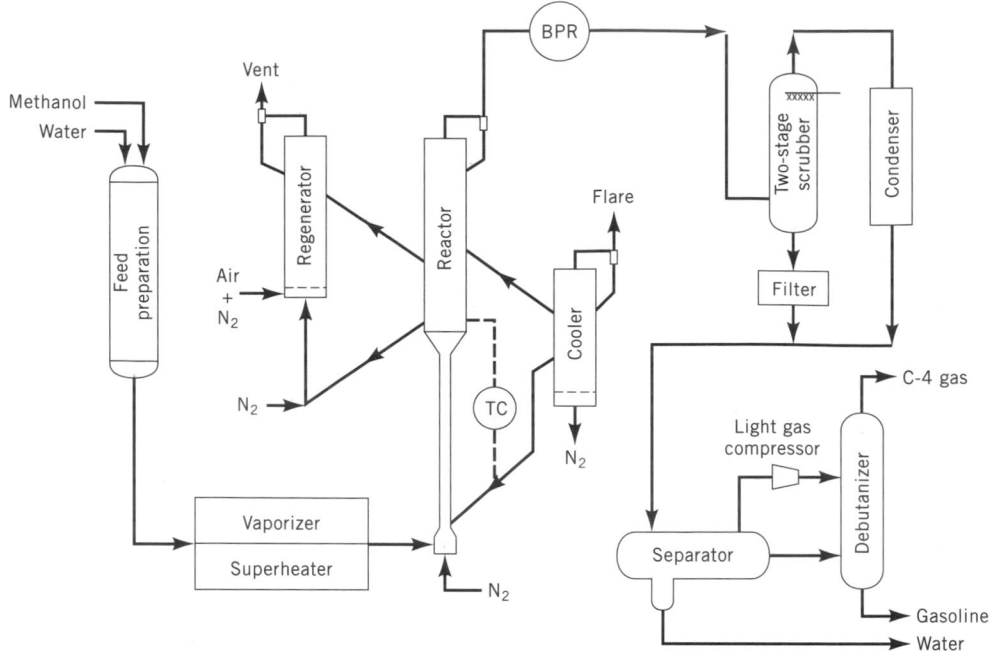

Figure 8. Fluid-bed MTG demonstration plant schematic diagram. BPR = Back pressure regulator; TC = temperature controller.

scale MOGD refinery test run was conducted by Mobil in 1981.

Properties. The gasoline product from the integrated MTO/MOGD process is predominately olefinic and aromatic. The gasoline quality (ca 89 octane) is comparable to FCC gasoline. Typical distillate product properties are given in Table 5. After hydrofinishing, the distillate product is mostly isoparaffinic and has high cetane index, low

pour point, and negligible sulfur content. MOGD diesel fuel has somewhat lower density than typical conventional fuels (0.8 vs 0.86). Low aromatics levels contribute to a stable jet fuel with very little smoke emission during combustion. MOGD diesel and jet fuels meet or exceed all conventional specifications.

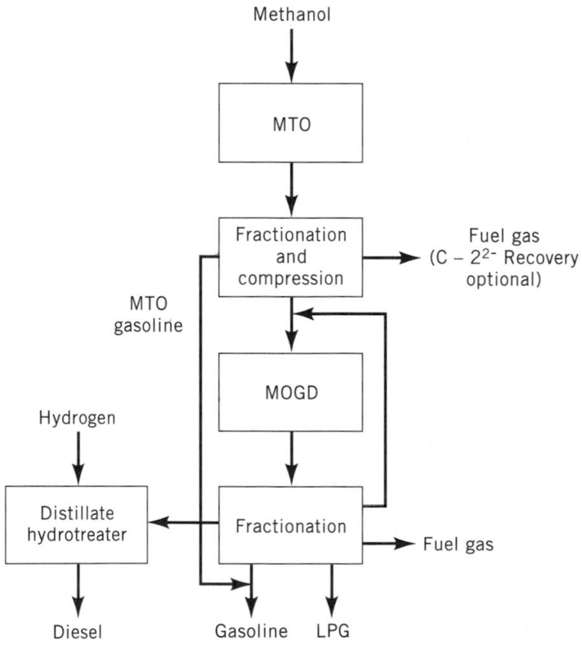

DIRECT CONVERSION OF NATURAL GAS TO LIQUID FUELS

The capital costs associated with indirect natural gas upgrading technology are high, thus research and development has focused on direct conversion of natural gas to liquid fuels. Direct conversion is defined as upgrading methane to the desired liquid fuels products while bypassing the synthesis gas step, ie, direct transformation to oxygenates or higher hydrocarbons. Direct upgrading routes which have been extensively studied include direct partial oxidation to oxygenates, oxidative coupling to higher hydrocarbons, and pyrolysis to higher hydrocarbons. Owing to the inert nature of methane, the technol-

Figure 9. Methanol-to-olefins (MTO) and Mobil olefins-to-gasoline (MOGD) and distillate process schematic.

Table 5. MTO/MOGD Distillate Properties

Parameter	Total Distillate	Jet Fuel	Diesel Fuel
Quantity, vol %	100	30	70
Density, g/mL	0.792	0.774	0.800
Pour point, °C	−50		−30
Freeze point, °C	−60	−60	
Flash point, °C	60	50	100
Cetane number	50		52
Smoke point, mm	25	25	
Aromatics, vol %	4	5	
Sulfur, ppm	50		

ogy is limited by the yields of desired products which in turn affects the process economics. Only one direct oxidative methane conversion process has been commercialized. A plant at Copsa Mica (Romania) in the 1940s (70) produced formaldehyde directly from methane and air by partial oxidation. This plant is no longer in operation. Plants to produce acetylene from methane by high temperature pyrolysis routes have been commercialized.

Generally, the most developed processes involve oxidative coupling of methane to higher hydrocarbons. Oxidative coupling converts methane to ethane and ethylene by

$$2\ CH_4 + 1/2\ O_2 \rightarrow H_3CCH_3 + H_2O \qquad (13)$$

$$H_3CCH_3 + 1/2\ O_2 \rightarrow H_2C{=}H_2 + H_2O \qquad (14)$$

The process can be operated in two modes: co-fed and redox. The co-fed mode employs addition of O_2 to the methane/natural gas feed and subsequent conversion over a metal oxide catalyst. The redox mode requires the oxidant to be from the lattice oxygen of a reducible metal oxide in the reactor bed. After methane oxidation has consumed nearly all the lattice oxygen, the reduced metal oxide is reoxidized using an air stream. Both methods have processing advantages and disadvantages. In all cases, however, the process is run to maximize production of the more desired ethylene product.

Direct conversion of natural gas to liquids has been actively researched. Process economics are highly variable and it is unclear whether direct natural gas conversion technologies are competitive with the established indirect processes. Some emerging technologies in this area are presented herein.

ARCO Gas-to-Gasoline Process

A two-step process using oxidative coupling to upgrade natural gas to liquid fuels has been proposed by ARCO (Atlantic Richfield Co.) (71,72). A simplified process scheme is given in Figure 10. Methane is passed through a redox-mode oxidative coupling reactor which generates $C_2 +$ hydrocarbons such as ethylene. The olefinic products are then oligomerized over a zeolite catalyst in a second reactor to produce gasoline and distillate. Unreacted methane is recycled. ARCO claims 25% conversion of

methane with 75% $C_2 +$ selectivity (ethylene to ethane ratio up to 10) in the oxidative coupling first stage and 95% ethylene conversion with 70% selectivity to gasoline and distillate in the olefin oligomerization second stage. This technology has been developed through bench-scale and pilot-plant stages.

OXCO Process

The OXCO process for upgrading natural gas has been proposed by the Commonwealth for Scientific and Industrial Research Organization (CSIRO) in Australia (73,74). This process involves $C_2 +$ pyrolysis and oxidative coupling of natural gas in a two-stage reactor; the entire concept is shown schematically in Figure 11. The methane in natural gas is separated and oxidatively coupled in a fluidized-bed reactor operating in co-fed mode to produce ethylene and ethane. The higher alkanes from the natural gas as well as the product ethane from the first stage are injected into an oxygen-free pyrolysis stage to make additional ethylene. The heat from the coupling reactor is used in the pyrolysis reaction. The overall carbon conversion to unsaturates plus CO_2 per pass is 30% with an overall carbon selectivity to unsaturates of 86%. The ethylene may be subsequently upgraded by oligomerization to liquid fuels. This process, which produces higher yields of ethylene and has a more favorable heat balance than conventional oxidative coupling technology, has been demonstrated in 30- and 60-mm fluidized-bed reactors.

Properties. Liquid fuels derived from oxidative coupling/olefin oligomerization processes would be expected to have properties similar to those derived from olefin oligomerization pathways such as MTO/MOGD.

OXYGENATE FUELS

Alcohols and ethers, especially methanol, ethanol, and methyl *tert*-butyl ether [1634-04-4] (MTBE), have been widely used separately or in blends with gasolines (reformulated gasoline) and other hydrocarbons to fuel internal combustion engines. Fuel properties of key oxygenates are presented in Table 6 (5). These compounds, as a class, may be considered to be partially oxidized, ie, each has a

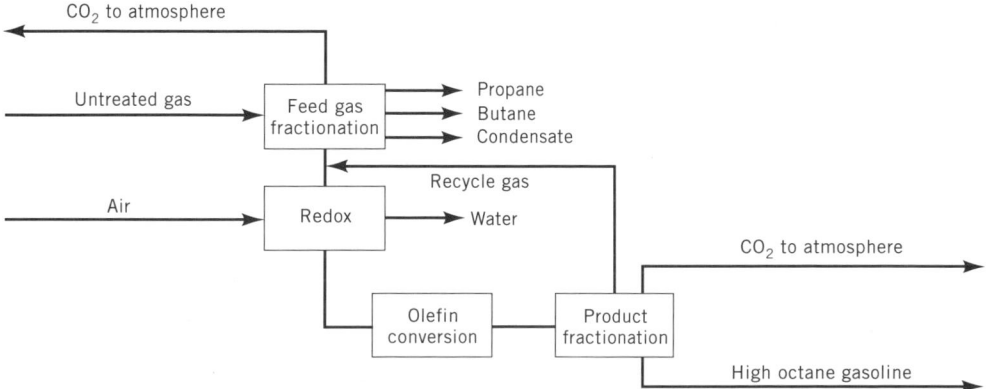

Figure 10. Simplified flow diagram depicting the ARCO gas-to-gasoline process for a conceptual gasoline production plant (72).

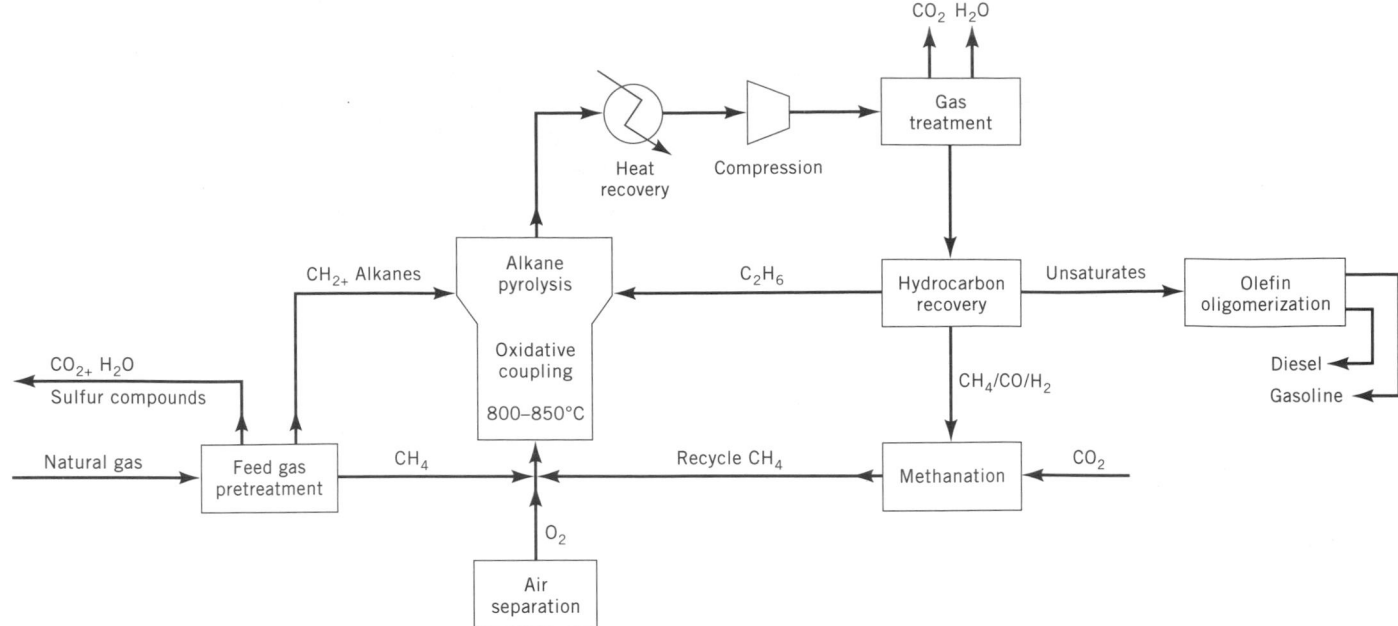

Figure 11. The OXCO process for natural gas conversion (74).

mole of oxidized hydrogen. They differ from the hydrocarbons that make up gasoline principally in lower heating values and in higher vaporization heat requirements. This constitutes a serious disadvantage to the substitution of oxygenates, especially lower alcohols, for motor gasoline. For example, the heating value of methanol is about half that of gasoline on an equivalent volume basis. Other properties which greatly influence the potential of oxygenates as fuels include octane performance, solubility in gasoline, effect on gasoline vapor pressure, sensitivity to water, and evaporative/exhaust emissions. Oxygenate fuels tests are often debated because the tests employed were developed for conventional gasolines.

The addition of small percentages of oxygenates to gasoline can produce large gains in octane. Thus, as blending components in gasoline, oxygenates improve octane quality. As neat fuels for spark-ignition engines, octane values

for oxygenates are not useful in determining knock-limited compression ratios for vehicles because of the lean carburetor settings relative to gasoline. Neither do these values represent the octane performance of oxygenates when blended with gasoline.

In part because neat alcohols are insufficiently volatile to enable a cold engine to start, even at moderate temperatures, the use of neat alcohols for automotive motor fuel is problematic (see ALCOHOL FUELS). Manufacturers have, however, reported that alcohol-powered cars, after being started and warmed up, can have the same or better driveability as gasoline cars (75–77). As for gasoline vehicles, port fuel injector fouling has occurred in some methanol vehicles and has affected driveability and emissions. Other problems related to high alcohol content gasoline in conventional engines include vapor lock and corrosion. Flexible-fuel vehicles (FFV), which can operate on either

Table 6. Fuel Oxygenates Properties[a]

Oxygenate	Blending Octane, 1/2(RON + MON)[b]	Heat of Combustion, MJ/L[c]	Specific Gravity	Boiling Point, °C
Methanol	101	18.0	0.79	64.6
Ethanol	101	21.3	0.79	78.5
2-Propanol	106	26.4	0.79	82.4
2-Butanol	99	28.3	0.80	99.5
tert-Butyl alcohol	100	28.1	0.80	82.6
MTBE[d]	108	30.2	0.75	55.4
ETBE[e]	111	32.5	0.74	72.8
TAME[f]	102	31.2	0.77	86.3
Gasoline	87	34.8	0.74	

[a] Ref. 5.
[b] RON = research octane number; MON = motor octane number.
[c] To convert MJ/L to Btu/gal, multiply by 3589.
[d] MTBE = methyl tert-butyl ether.
[e] ETBE = ethyl tert-butyl ether.
[f] TAME = tert-amyl methyl ether.

neat methanol or gasoline, or mixtures thereof, are being evaluated.

Gasoline blends containing oxygenates change the emissions characteristics of a motor vehicle designed for gasoline. Oxygenates and oxygenate-blends approved for use by the U.S. government are expected to have desirable emissions features as automotive fuels, and governmental environmental mandates and regulations have necessitated increased examination and implementation of oxygenates as fuels. As of this writing, however, no process can produce alcohols or ethers at equivalent or lower cost per volume than gasolines derived from natural petroleum.

Methanol

Methanol (qv) production in the 1990s is dominated by reaction of synthesis gas produced from natural gas. The economics of producing methanol as fuel are highly variable and site-specific. The natural gas feedstock has a broad range of values depending on location. Delivered costs for methanol are probably double that for gasoline from petroleum. Impurities, including water, vary according to the synthesis process employed. The term methyl fuel describes certain methanol products that may also contain significant quantities of water and higher alcohols as well as other oxygenated compounds. Water removal from such mixtures by distillative processes is generally complicated because of the formation of azeotropes. Methanol is used to produce MTBE, another oxygenate fuel.

Methanol as a fuel has been proposed in various ratios with gasoline. In gasoline formulations having relatively low methanol content, eg, M3 (3% methanol) and M15 (15% methanol), solubilizers are used and stringently dry conditions must be maintained. High methanol content fuels, M85 (85% methanol) and M100 (neat methanol), have special engine requirements. The use of high methanol content fuels is limited by methanol cost plus poor compatibility with the existing gasoline infrastructure.

Methanol is more soluble in aromatic than paraffinic hydrocarbons. Thus varying gasoline compositions can affect fuel blends. At room temperature, the solubility of methanol in gasoline is very limited in the presence of water. Generally, cosolvents are added to methanol–gasoline blends to enhance water tolerance. Methanol is practically insoluble in diesel fuel.

Concerns about using methanol–gasoline blend fuels include problems with vapor lock, cold start, and warmup. Oxygenate–gasoline blends, and in particular those containing methanol, have unusual volatility characteristics and cannot be accurately characterized using test methods developed for gasoline. Vapor pressure is an important volatility parameter that is adversely affected by the addition of methanol. Gasoline blends containing oxygenates form nonideal solutions with varying characteristics. In general, methanol–gasoline blend fuels exhibit increases in Reid vapor pressure (RVP) over that of gasoline itself. This effect contributes to vapor lock and evaporative emissions. The data available for assessing methanol's impact on exhaust emissions, and consequently on air quality, are limited. Formaldehyde (qv) has been reported in the exhausts of cars fueled with straight methanol or methanol–gasoline blends (see also EXHAUST CONTROL, AUTOMOTIVE).

The use of methanol as a motor fuel has been discussed since the 1920s. Straight methanol has long been a preferred fuel for racing engines because of the much higher compression ratios at which methanol may be combusted relative to gasolines. This is translatable for racing purposes at equivalent power outputs to engines of considerably reduced weight. However, fuel consumptions are roughly three times that of gasoline on a km/L (mi/gal) basis, and extremely high emissions of unburned fuel and carbon monoxide can result (78).

In Germany in the early 1950s, a 50:50 mixture of methanol and 2-propanol was blended with gasoline, first at a level of 7.5% and later at 1.5% (79). Complaints about stalling, power loss, and phase separation caused the ratio to be changed to 60:40::methanol:2-propanol but this apparently aggravated the problems. The practice was discontinued in 1970 when a tax was placed on alcohol.

In the United States, the Clean Air Act of 1970 imposed limitations on composition of new fuels, and as such methanol-containing fuels were required to obtain Environmental Protection Agency (EPA) waivers. Upon enactment of the Clean Air Act Amendments of 1977, EPA set for waiver unleaded fuels containing 2 wt % maximum oxygenates excluding methanol (0.3 vol % maximum). Questions regarding methanol's influence on emissions, water separation, and fuel system components were raised (80).

In 1979 Sun Oil Co. was granted a waiver for a gasoline blend containing 2.75 vol % methanol and an equal volume of tert-butyl alcohol (TBA) (2 wt % total oxygen). Cosolvents such as TBA were shown to reduce adverse effects of methanol on volatility and water tolerance. ARCO obtained EPA waiver in 1981 for a 3.5 wt % oxygen fuel blend containing Oxinol, also comprising equal parts of methanol and TBA. In 1985 a waiver was granted to Du Pont, Inc. for a gasoline blend containing 5 vol % maximum methanol with at least 2.5 vol % higher alcohol cosolvents. The waiver incorporated a water tolerance or phase separation requirement (81).

The most extensive worldwide program on methanol blend gasoline was in Italy where from 1982 to 1987 a 1.9×10^4 m³/yr (5×10^6 gal/yr) plant produced a mixture containing 69% methanol. The balance contained higher alcohols. This mixture was blended into gasoline at the 4.3% level and marketed successfully as a premium gasoline known as Super E (82).

Methanol, a clean burning fuel relative to conventional industrial fuels other than natural gas, can be used advantageously in stationary turbines and boilers because of its low flame luminosity and combustion temperature. Low NO_x emissions and virtually no sulfur or particulate emissions have been observed (83). Methanol is also considered for dual fuel (methanol plus oil or natural gas) combustion power boilers (84) as well as to fuel gas turbines in combined methanol/electric power production plants using coal gasification (85).

Owing to its properties, methanol is not recommended for aircraft or marine fuel uses. Methanol cannot be used in conventional diesel-powered vehicles without modifications to the fuel system and engine. Simple methanol–diesel blends are not possible because of insolubility. Heavy-duty diesel engines have been adapted to use neat methanol by many U.S. manufacturers, and several are being used in field demonstrations (82) (see ALCOHOL FUELS).

Ethanol

Ethanol (qv) is produced both from ethylene (qv) derived from the cracking of petroleum fractions and by the fermentation of sugars derived from grains or other biomass (see SUGAR). Many of its relevant properties are similar to those of methanol. Although ethanol may be a more desirable fuel or fuel component than methanol, its significantly higher cost (volume basis) may outweigh these advantages. Broad implementation of ethanol-containing fuels would require government action, eg, in the form of subsidies to farmers and fuel waivers.

The term gasohol has come into wide usage to identify, generally, a blend of gasoline and ethanol, with the latter derived from grain. The term may also be applied to blends of methanol or other alcohols in gasolines or other hydrocarbons, without regard to sources of components.

Brazil's Alcohol Program.

In Brazil, the enactment of legislation in 1931 made ethanol addition to gasoline compulsory at a level of 5% (86). Excess molasses and sugar were converted to alcohol in distilleries attached to sugar mills as a means to stabilize sugar prices. Production of fuel ethanol in the 1990s is mostly from biomass.

Starting in the city of Sao Paulo in 1977, and extending to the entire state of Sao Paulo in 1978, a gasohol incorporating 20% ethanol was mandated. Brazil's National Alcohol Program (Proalcool) set an initial goal of providing the 20% fuel mixture nationwide by 1980–1981 and a system of special tax, warranty, and price considerations were enacted to advance the aims of Proalcool.

For a considerable period, >90% of the new cars in Brazil operated on E96 fuel, or a mixture of 96% ethanol and 4% water (82). The engines have high compression ratios (ca 12:1) to utilize the high knock resistance of ethanol and deliver optimum fuel economy. In 1989 more than one-third of Brazil's 10 million automobiles operated on 96% ethanol/4% water fuel. The remainder ran on gasoline blends containing up to 20% ethanol (5).

Gasohol in the United States. Over 90% of the fuel ethanol in the United States is produced from corn. Typically, 0.035 m³ (1 bushel) of corn yields 9.5 L (2.5 gal) of ethanol. Ethanol is produced by either dry or wet milling (87). Selection of the process depends on market demand for the by-products of the two processes. More than two-thirds of the ethanol in the United States is produced by wet milling. Depending on the process used, the full cost of ethanol after by-product credits has been estimated to be between $0.25–0.53/L ($1–2/gal) for new plants (88). Feedstock costs are a significant factor in the production of fuel ethanol. A change in corn price of $0.29/m³ ($1.00/bushel) affects the costs of ethanol by $0.08/L ($0.30/gal).

Ethanol can also be produced from cellulose or biomass such as wood (qv), corn stover, and municipal solid wastes (see ENERGY FROM BIOMASS; FUELS FROM WASTE). Each of these resources has inherent technical or economic problems. The Tennessee Valley Authority (TVA) is operating a 2 t/d pilot plant on converting cellulose to ethanol.

After the oil embargo in 1973, gasohol use was stimulated by tax incentives. An application for EPA waiver of gasohol fuels (up to 10 vol % ethanol) was granted in 1979. From 1981 to 1983 the California Energy Commission field tested alcohol-powered cars equipped with a gasoline-assist starting system, ie, having an onboard auxiliary supply of volatile fuel for cold start tests. In 1989 about 8% of U.S. gasoline contained 10% ethanol plus a corrosion inhibitor (82). As of this writing, government waivers of RVP standards for gasohol fuels are being considered (89).

Methyl t-Butyl Ether

MTBE is produced by reaction of isobutene and methanol on acid ion-exchange resins. The supply of isobutene, obtained from hydrocarbon cracking units or by dehydration of tert-butyl alcohol, is limited relative to that of methanol. The cost to produce MTBE from by-product isobutene has been estimated to be between $0.13 to $0.16/L ($0.50–0.60/gal) (90). Direct production of isobutene by dehydrogenation of isobutane or isomerization of mixed butenes are expensive processes that have seen less commercial use in the United States.

More than 95% of MTBE produced worldwide is used to blend with gasoline. In 1987 U.S. production of MTBE exceeded 3.8×10^6 m³/yr (1×10^9 gal/yr) (82). The worldwide capacity for MTBE is increasing, especially in the United States and Europe, and has been projected to exceed production for years to come.

MTBE's gain in prominence as a fuel-blend component is a result of inherent technical advantages over other oxygenates, especially the lower alcohols. MTBE has a high blending octane number (Table 6) although this number varies somewhat with gasoline composition. The low vapor pressure relative to the lower alcohols results in no increase in RVP for MTBE-gasoline blends and consequently better evaporative emission and vapor lock characteristics. No phase separation occurs in blends with other fuels. MTBE, in blends of <20 vol % with gasoline, does not deleteriously affect other fuel or driving characteristics such as cold start, fuel consumption, and engine materials compatibility.

MTBE has been used in motor fuels in Europe since the early 1970s and is undergoing rapid growth, particularly in the United States. MTBE-blended gasoline containing up to 11 vol % MTBE received EPA waiver in 1981. Later legislation increased the MTBE waiver up to 15 vol %. In 1987–1988 Colorado began mandating use of winter oxygenate-based fuels in the Denver region. About 90% of the fuel in this period used a gasoline blend containing 8 vol % MTBE and in 1988–1989 the fuel was required to contain at least 2% oxygen (11 vol % MTBE). Based on the success of this program and EPA assessments that CO reductions of 10–20% over the next decade were possible with oxygenate-blend fuels, numerous state governments enacted legislation requiring the use of these fuels in winter and in cities having high ozone (smog) concentrations. The Clean Air Act Amendments of 1990 have mandated the use of reformulated gasolines, especially in serious ozone problem areas, by 1995.

The effectiveness of MTBE, however, is under discussion (91). Based on Denver, Colorado vehicle emissions data from 1981 to 1991 and theoretical models, Colorado scientists have claimed that the use of MTBE-blended fuels had no statistically significant effect on atmospheric CO levels, but increased pollutants such as formaldehyde. A drop in CO levels in Denver during this time period was attributed to fleet turnover of older, more polluting cars being replaced by newer cars having cleaner burning engines. In addition, health problems associated with direct exposure to MTBE in Fairbanks, Alaska has resulted in EPA exemption of the oxygenated fuel requirement in that area (91).

DIRECT LIQUEFACTION OF COAL

Direct liquefaction, the production of liquids from feed coal in a single processing scheme without a synthesis gas intermediate step, includes two routes for the upgrading of coal: hydrogenation and pyrolysis. In hydrogenation, the conversion of coal to liquids having higher hydrogen-to-carbon ratio involves the addition of hydrogen. Generally, the additional hydrogen required is added either from molecular hydrogen or from a hydrogen-donor solvent such as tetralin. Processes classified under pyrolysis are those which produce liquids by removal of carbon. This occurs when coal is thermally processed under inert or reducing atmospheric conditions. The use of hydrogen in a pyrolytic process to increase yields of distillate products is known as hydropyrolysis. Coal carbonization to produce metallurgical coke involves much the same chemistry as pyrolysis.

Coal and Coal-Tar Hydrogenation

If paraffinic and olefinic liquids are extracted from solid fuel substances, the hydrogen content of the residual material is reduced even further, and the residues become more refractory. The yields of liquids so derivable are generally low, even when a significant fraction of the hydrogen is extractable. Thus production of fuel liquids from nonliquid fuel substances such as coal and coal tars may be enhanced only by the introduction of additional hydrogen in a synthesis process. The principal differences in the processes are from the modes in which hydrogen is introduced and the catalysts used.

Hydrogenation of coal and other carbonaceous matter using high pressure hydrogen has been patented (92), and subsequently the Nobel Prize in chemistry was won for this accomplishment. By 1992, a 1 t/d plant was operating and using hydrogen at 10 MPa (100 atm) and 400°C to treat brown coal tar to give a liquid that comprised 25 wt % gasoline boiling at 75–210°C and 40 wt % middle oil, 210–300°C (see LIGNITE AND BROWN COAL). The pitch residue had a specific gravity of 1.04, and a solidifi-

cation point of 15°C. The degree of liquefaction was shown to increase with decrease in oil rank (93). Liquid products were of low quality, being high in oxygen, nitrogen, and sulfur content, owing to low hydrogenation rates and polymerization of primary products (94).

In 1935 an ICI coal hydrogenation plant at Billingham, U.K., produced ca 136,000 t/yr motor fuel from bituminous coal and coal tar. By 1936, 272,000 t/yr of motor fuel were produced by improved hydrogenation of brown coal and coal tar at a facility constructed at Leuna and some 363,000 t/yr was being produced in three other German plants (95). Two years later the total German output from these facilities was ca 1.4×10^6 t/yr (96). The number of coal hydrogenation plants in Germany increased during World War II to 12, with total capacity of about 4×10^6 t/yr (100,000 bbl/d) of aviation and motor gasolines.

Experimental plants for hydrogenating coal or coal tar were operated in Japan, France, Canada, and in the United States before or during World War II. Much of that technology has remained proprietary. In general, coal-in-oil slurries containing iodine or stannous oxalate catalyst were subjected to liquid-phase hydrogenation at pressures of 25–70 MPa (250–700 atm). Liquids produced were fractionated, and the middle oils were then subjected to vapor-phase hydrogenation over molybdenum-, cobalt-, or tungsten sulfide-on-alumina catalysts (97). About 1 t of crude motor fuel was recovered from 4.5 t of coal, from which all necessary hydrogen and power requirements for the production were also obtained.

Developments in the United States. A large number of proprietary coal hydrogenation process variants have been proposed. Much of the technology originally directed to the catalytic hydrogenation of coals and coal tars in Germany has been applied to the hydrorefining of petroleum fractions, but U.S. commercial interest in coal hydrogenation was offset by the relative abundance of domestic petroleum up to World War II.

The huge demand for liquid fuels during World War II prompted the passage of the Synthetic Liquid Fuels Act of 1944. There were various programs relating to demonstration plants to produce liquid fuels from coal, oil shale, and other substances, including agricultural and forestry products. The Bureau of Mines had begun work on coal liquefaction in 1936, at which time a 45 kg/d experimental coal hydrogenation unit was constructed (98). The expanded program, after 1944, culminated in the construction and operation of a 45 t/d coal hydrogenation demonstration plant at Louisiana, Missouri, in 1949 (99), where a variety of problems and processing variations were investigated (100,101). Cost studies (102) showed that production of gasoline from coal hydrogenation could not compete with using natural petroleum as a gasoline source. The demonstration plant operations were terminated in 1953.

Work on coal hydrogenation continued by the Bureau of Mines on a laboratory scale (103–105). In one of these variants (106) coal-oil pastes admixed with catalyst in tubular reactors were hydrogenated at high pressure and low residence times to give improved yields of liquid products. The original thrust of the work was to hydrodesulfurize coal economically to produce environmentally acceptable boiler fuel (107). In the mid-1970s, a process sponsored by the Bureau of Mines named Synthoil (108) was developed, but the efforts were terminated by 1978 owing to limited catalyst lifetimes.

H-Coal Process. The H-coal process (Hydrocarbon Research, Inc., HRI, subsidiary of Dynalectron Corp.), for the conversion of coal to liquid products (109), is an application of HRI's ebullated-bed technology for the conversion of heavy oil residues into lighter fractions. Coal is dried, pulverized, and slurried with coal-derived oil (110). The coal-oil slurry is charged continuously with hydrogen to a reactor of unique design (111) containing a bed of ebullated catalyst, where the coal is hydrogenated and converted to liquid and gaseous products. The liquid product is a synthetic crude oil that can be converted to gasoline or heating oil by conventional refining processes. Alternatively, under milder operating conditions, a clean fuel gas and low sulfur fuel oils may be produced. The relative yields of these products depend on the desired sulfur level in the heavy fuel oil. In general, reaction products are separated by fractionation and absorption (qv). Unreacted coal may be fed into a fluid coker that produces gas, gas oil, and dry char. The coker gas oil, along with gas oils separated from the main reactor effluent, may be subjected to hydrocracking for conversion to lighter products.

In 1976, Ashland Oil (Ashland Synthetic Fuels, Inc.) was awarded the prime contract to construct a 540 t/d H-coal pilot plant adjacent to its refinery at Catlettsburg, Kentucky, by an industry–government underwriting consortium. Construction was completed in 1980 (112). The pilot-plant operation ended in early 1983.

Properties. The properties of naphtha, gas oil, and H-oil products from an H-coal operation are given in Table 7. These analyses are for liquids produced from the syncrude operating mode. Whereas these liquids are very low in sulfur compared with typical petroleum fractions, they are high in oxygen and nitrogen levels. No residual oil products (bp >540°C) are formed.

Solvent-Refined Coal Process. In the 1920s the anthracene oil fraction recovered from pyrolysis, or coking, of coal was utilized to extract 35–40% of bituminous coals at low pressures for the purpose of manufacturing low cost newspaper inks (113). Tetralin was found to have higher solvent power for coals, and the I. G. Farben Pott-Broche process (114) was developed, wherein a mixture of cresol and tetralin was used to dissolve ca 75% of brown coals at 13.8 MPa (2000 psi) and 427°C. The extract was filtered, and the filtrate vacuum distilled. The overhead was distilled a second time at atmospheric pressure to separate solvent, which was recycled to extraction, and a heavier liquid, which was sent to hydrogenation. The bottoms product from vacuum distillation, or solvent-extracted coal, was carbonized to produce electrode carbon. Filter cake from the filters was coked in rotary kilns for tar and oil recovery. A variety of liquid products were obtained from the solvent extraction-hydrogenation system (113). A similar process was employed in Japan during World War II to produce electrode coke, asphalt (qv), and carbonized fuel briquettes (115).

In the United States there was little interest in solvent processing of coals. A method to reduce the sulfur content of coal extracts by heating with sodium hydroxide and zinc oxide was, however, patented in 1940 (116). In the 1960s the technical feasibility of a coal deashing process was studied (117), and a pilot plant able to process ca 45 t/d was completed in late 1974 (118).

A flow diagram of the solvent-refined coal or SRC process is shown in Figure 12. Coal is pulverized and mixed with a solvent to form a slurry containing 25–35 wt % coal. The slurry is pressurized to ca 7 MPa (1000 psig), mixed with hydrogen, and heated to ca 425°C. The solution reactions are completed in ca 20 min and the reaction product flashed to separate gases. The liquid is filtered to remove the mineral residue (ash and undissolved coal) and fractionated to recover the solvent, which is recycled.

The liquid remaining after the solvent has been recovered is a heavy residual fuel called solvent-refined coal, containing less than 0.8 wt % sulfur and 0.1 wt % ash. It melts at ca 177°C and has a heating value of ca 37 MJ/kg (16,000 Btu/lb), regardless of the quality of the coal feedstock. The activity of the solvent is apparently more important than the action of gaseous hydrogen in this type of uncatalyzed hydrogenation. Research has been directed to the use of petroleum-derived aromatic oils as start-up solvents (118).

Table 7. Properties of Syncrude from H-Coal Process[a]

Property	Initial to 190°C	190–343°C	343–524°C	Total
		Boiling Range		
Specific gravity, (°API)[b]	0.767 (53.0)	0.915 (23.2)	1.05 (3.5)	0.863 (32.4)
Vol % on total	40.0	54.2	5.8	100.0
Analysis, wt %				
Carbon	84.5	88.8	89.4	87.3
Hydrogen	13.6	11.0	10.2	11.9
Oxygen	1.7			0.6
Nitrogen	0.1	1.0	0.1	0.1
Sulfur	0.1	0.1	0.3	0.1
Total	*100*	*100*	*100*	*100*

[a] Ref. 111.

[b] $°API = \dfrac{141.5}{\rho} - 131.5$ where ρ is specific gravity.

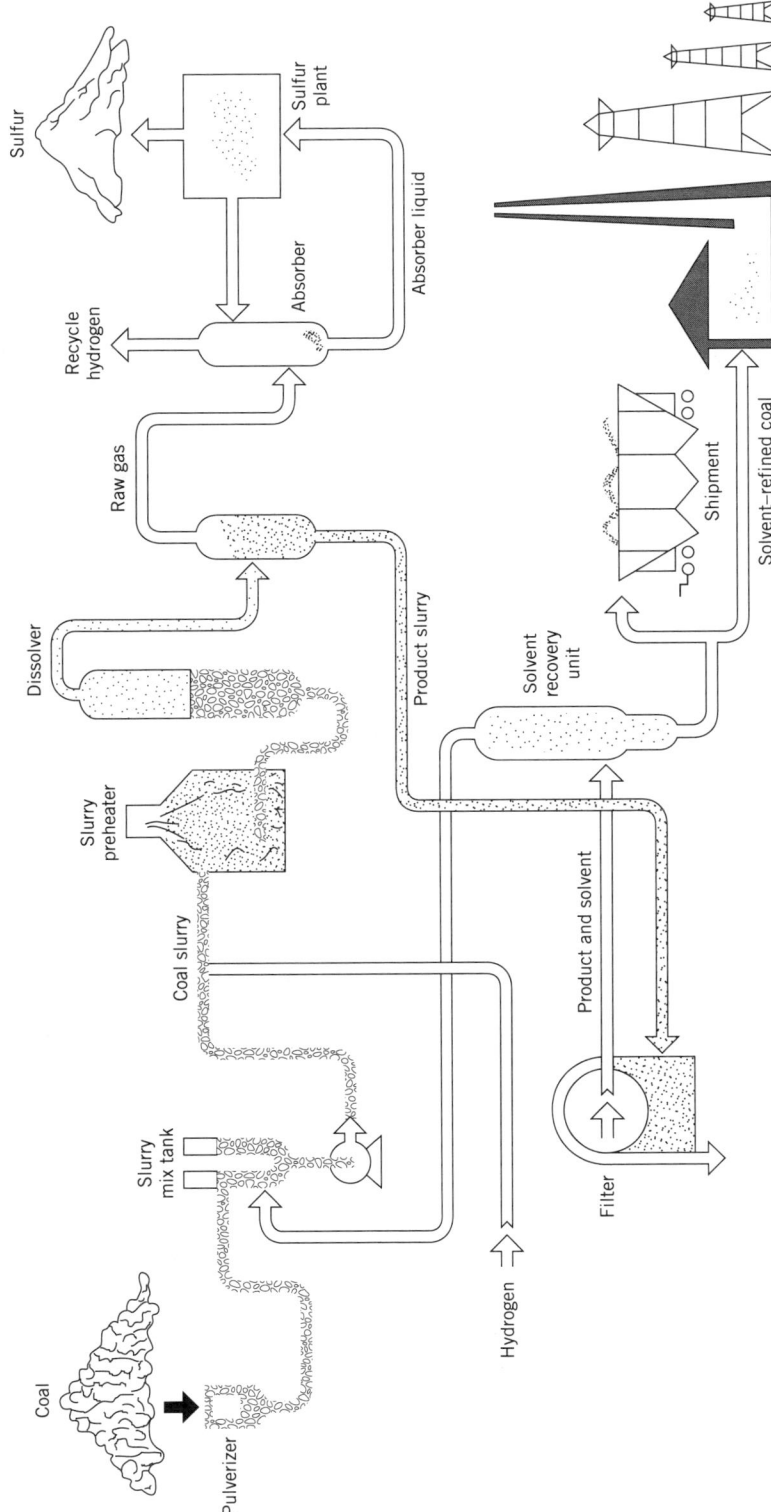

Figure 12. Solvent-refined coal process (119).

In the early 1970s production of low sulfur, ashless (solid) boiler fuel was the preferred commercial application (119). This basic process (SRC-I) yielded small amounts of liquid oil products with additional processing. Liquid output was significantly increased by the coal-oil-gas (COG) refinery concept (120–122) which incorporated high degrees of hydroconversion and hydrotreating. A SCR-II process has been developed, in which hydrocracking occurs in the solution (hydrogenation) vessels (123). A low viscosity fuel oil is the primary distillate product in this case, although naphtha and LPG are also recovered.

Two pilot plants have been built and operated to demonstrate the feasibility of the SRC process. These included a 6 t/d plant at Wilsonville, Alabama (*vide infra*) and a 50 t/d plant at Ft. Lewis, Washington which was operated from 1974 to 1981.

In an effort to obtain higher value products from SRC processes, a hydrocracking step was added to convert resid to distillate liquids. The addition of a hydrocracker to the SRC-I process was called nonintegrated two-stage liquefaction (NTSL). The NTSL process was essentially two separate processes in series: coal liquefaction and resid upgrading. NTSL processes were inefficient owing to the inherent limitations of the SRC-I process and the high hydrocracker severities required.

Properties. The properties of the liquid fuel oil produced by the SRC-II process are influenced by the particular processing configuration. However, in general, it is an oil boiling between 177 and 487°C, having a specific gravity of 0.99–1.00, and a viscosity at 38°C of 40 SUs (123). Pipeline gas, propane and butane (LPG), and naphtha are also recovered from an SRC-II complex.

Exxon Donor Solvent-Coal Liquefaction Process. The EDS process from Exxon is a hydrogenation process using a donor solvent for the direct conversion of a broad range of coals to liquid hydrocarbons (124). In the process sequence, shown in Figure 13 the feed coal is crushed, dried, and slurried with hydrogenated recycle solvent (the donor solvent) and fed to the reactor with hydrogen. The reactor is an upward plug-flow design operating at 430–480°C and at ca 14 MPa (2000 psi) total pressure.

The reactor effluent is separated by conventional distillation into recycle solvent, light gases, C_4 to 537°C bp distillate, and a heavy vacuum bottoms stream containing unconverted coal and ash. The recycle solvent is hydrogenated in a separate reactor and sent back to the liquefaction reactor.

The heavy vacuum bottoms stream is fed to a Flexicoking unit. This is a commercial (125,126) petroleum process that employs circulating fluidized beds at low (0.3 MPa (50 psi)) pressures and intermediate temperatures, ie, 480–650°C in the coker and 815–980°C in the gasifier, to produce high yields of liquids or gases from organic material present in the feed. Residual carbon is rejected with the ash from the gasifier fluidized bed. The total liquid product is a blend of streams from liquefaction and the Flexicoker.

The EDS process was developed starting from 1976 in a 10-year joint undertaking between DOE and private industry (127). Under the direction of Exxon Co. USA, a 250 t/d pilot plant was operated at Baytown, Texas. Operation of this unit began in 1980 and was completed by late 1982.

Properties. Pilot-unit data indicate the EDS process may accommodate a wide variety of coal types. Overall process yields from bituminous, subbituminous, and lignite coals, which include liquids from both liquefaction and Flexicoking, are shown in Figure 14. The liquids produced have higher nitrogen contents than are found in similar petroleum fractions. Sulfur contents reflect the sulfur levels of the starting coals: ca 4.0 wt % sulfur in the dry bituminous coal; 0.5 wt % in the subbituminous; and 1.2 wt % sulfur in the dry lignite.

Table 8 shows that the naphthas produced by the EDS process have higher concentrations of cycloparaffins and phenols than do petroleum-derived naphthas, whereas the normal paraffins are present in much lower concen-

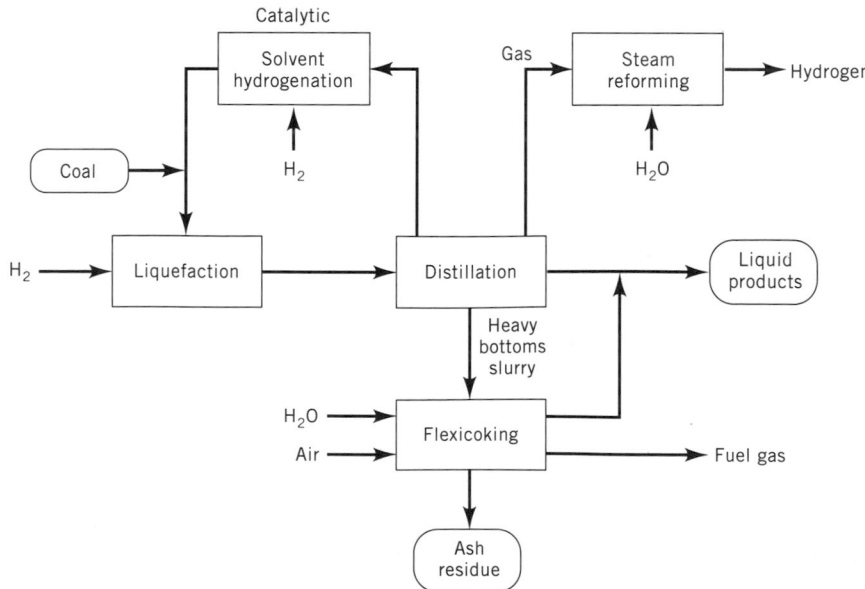

Figure 13. Exxon donor solvent process (124).

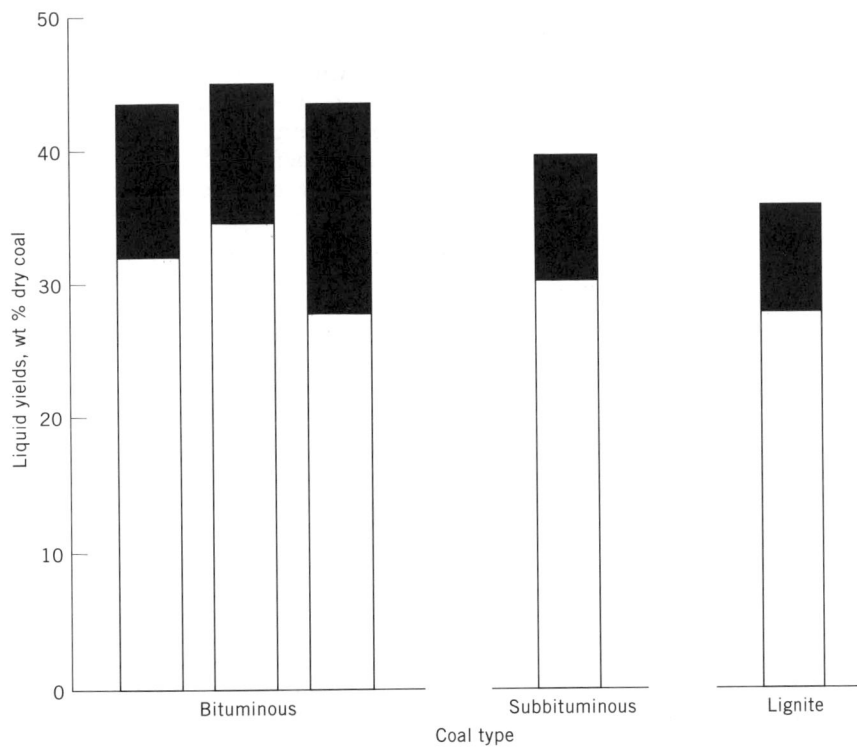

Figure 14. Preferred liquefaction-coking liquid yields in the EDS process for various coals where ■ represents Flexicoking liquids and, □ liquefaction liquids (124). A, Ireland (West Virginia); B, Monterey (Illinois); C, Burning Star (Illinois); D, Wyodak (Wyoming); and E, Big Brown (Texas).

trations. The sulfur and nitrogen concentrations in coal naphthas are high compared to those in petroleum naphthas.

Gas oil fractions (204–565°C) from coal liquefaction show even greater differences in composition compared to petroleum-derived counterparts than do the naphtha fractions (128). The coal-gas oils consist mostly of aromatics (60%), polar heteroaromatics (25%), asphaltenes (8–15%), and saturated compounds (<10%). Petroleum-gas oils, on the other hand, contain more than 50% saturated compounds, less than 5% polar heteroaromatics, and no as-

phaltenes. Furthermore, the aromatics of petroleum-gas oils have longer side chains.

Coal Liquefaction at Wilsonville. Starting in 1974 the Advanced Coal Liquefaction R&D Facility at Wilsonville, Alabama operated a 6 t/d pilot plant and studied various coal liquefaction processing schemes. The facility, cosponsored by the DOE, the Electric Power Research Institute (EPRI) and Amoco Oil Co, was shut down in early 1992.

Initial operation at the Wilsonville pilot plant was in SRC-I mode and later evolved into a two-stage process

Table 8. Composition of Naphthas[a] from Various Sources[b]

Component, wt %	EDS Coal Liquefaction		Petroleum Naphthas	
	Illinois Coal	Wyodak Coal	Cycloparaffinic[c]	Paraffinic[d]
Saturated compounds	69.9	60.8	77.9	81.3
Paraffins	13.4	19.4	36.0	58.2
Cycloparaffins	56.5	41.4	41.9	23.1
Olefins[e]	0.5	3.7		
Aromatics	17.0	28.6	22.1	18.7
Benzenes	11.7	25.1	20.8	17.2
Indanes, tetralins	5.1	3.5	1.0	1.2
Indenes	0.2	0.0		
Naphthalenes	0.01	0.0	0.3	0.3
Phenols	12.6	6.9	Traces	Traces
Total	*100.0*	*100.0*	*100.0*	*100.0*
Sulfur	0.57	0.10	0.035	0.049
Nitrogen	0.15	0.18	0.0001	0.0001
Oxygen	1.60	2.49		

[a] C$_5$ to 204°C bp.
[b] Ref. 128.
[c] Prudhoe Bay.
[d] Arab Light.
[e] Values are approximate.

(129) by operation in NTSL mode. NTSL limitations described previously combined with high hydrogen consumptions resulted in subsequent focus on a staged integrated approach, which was to be the basis for all further studies at Wilsonville.

The integrated two-stage process (ITSL) combined short contact time liquefaction in one reactor with ebullated-bed hydrocracking in a second stage (130). The short contact time conditions permitted better hydrogen transfer from the solvent rather than from the gas phase. The hydrocracking step operated at lower severity resulting in lowered gas make and improved hydrogen efficiency. Recycle solvent was generated from the hydrocracked distillates and coupled the two reaction stages. Results of ITSL processing of Illinois No. 6 coal at Wilsonville are given in Table 9. Distillate yields and coal throughput for ITSL were higher than those obtained by NTSL.

Further developments of the ITSL process resulted in incremental gains in distillate yields (131). Reconfigured integrated two-stage liquefaction (RITSL) involved placing the solvent deasher after the hydrocracker thus producing a recycle solvent consisting of deashed resid and distillate. This resulted in reduction of feed to the deasher and reduced organic rejection. Close coupled integrated two-stage liquefaction (CC-ITSL) linked the two reactors and removed several operations between the two stages. A deleterious effect of these two processing modes was increased hydrogen consumption over ITSL.

From 1985 to 1992, development activity at Wilsonville was on catalytic two-stage liquefaction (CTSL). CTSL, initiated by HRI (132), consists of catalytic processing in two ebullated-bed reactors which lower reaction temperatures and increase distillate yields, up to 78% yield. CTSL results from Wilsonville for Illinois No. 6 coal are also given in Table 9. Distillate yields were shown to be significantly higher for CTSL over ITSL; however, hydrogen consumption was somewhat increased.

Properties. CTSL distillates have qualities comparable to or better than No. 2 fuel oil and have good hydrogen content and low heteroatom contents. Distillates having a higher boiling point distribution from Wilsonville CTSL operation (131) showed 26.8°API gravity with heteroatom levels of 0.11 wt % sulfur, <1 wt % oxygen, and 0.16 wt % nitrogen.

Coal Pyrolysis

Pyrolysis is the destructive distillation of coal in the absence of oxygen typically at temperatures between 400

Table 9. Wilsonville Plant Operating Conditions and Yields for ITSL and CTSL Modes[a]

Parameter	Mode of Operation[b]	
	ITSL	CTSL
Operating conditions		
Run number	7242BC; 243JK/244B	253A
Catalyst	Shell 32M	Shell 317
First stage[c]		
Average reactor temperature, °C	460; 432	432
Space velocity	690; 450[d]	4.8[e]
Pressure, MPa[f]	17; 10–17	17.9
Second stage		
Average reactor temperature, °C	382	404
Space velocity, feed/catalyst[e]	1.0	4.3
Catalyst age, resid/catalyst	278–441; 380–850	100–250
Yields[g]		
C_1–C_3 gas	4; 6	6
C_4 + distillate	54; 59	70
Resid	8; 6	−1
Hydrogen consumption	4.9; 5.1	6.8
Other		
Hydrogen efficiency, C_4+ distillate/H_2 consumed	11, 11.5	10.3
Distillate Selectivity, C_1–C_3/C_4 + distillate	0.07; 0.10	0.08
Energy Content of feed coal reject to ash concentrate, %	24; 20–23	20

[a] Feed is Illinois No. 6 coal.
[b] CTSL = catalytic two-stage liquefaction; ITSL = integrated two stage liquefaction.
[c] First stage is thermal for ITSL.
[d] Value given is coal space velocity at temp >371°C in kg/m³.
[e] Value given is in h⁻¹.
[f] To convert MPa to psia, multiply by 145.
[g] Wt % on a moisture- and ash-free (MAF) coal basis.

and 500°C (133). As the temperature of carbonaceous matter is increased, decomposition ultimately occurs. Melting and dehydration may also occur. Coals exhibit more or less definite decomposition temperatures, as indicated by melting and rapid evolution of volatile components, including potential fuel liquids, during destructive distillation (134). Table 10 summarizes an extensive survey of North American coals subjected to laboratory pyrolysis. The yields of light oils so derived average no more than ca 8.3 L/t (2 gal/short ton), and tar yields of ca 125 L/t (30 gal/short ton) are optimum for high volatile bituminous coals (135).

Coal pyrolysis has been studied at both reduced and elevated pressures (136), and in the presence of a variety of agents and atmospheres (137). Although important to the study of coal structure and reactions, coal pyrolysis, as a means to generate liquids, has proved to have limited commercial value.

COED Process. Sponsored by the Office of Coal Research of the U.S. Department of the Interior, the COED process was developed by FMC Corp. as Project Char-Oil-Energy Development (COED) through 1975 (138–140). Bench-scale experiments led the way to construction in 1965 of a process development unit employing multistage, fluidized-bed pyrolysis to process 45 kg/h (141). Correlated studies included hydrotreating of COED oil (142), high temperature hydrodesulfurization of COED char, and investigations of char-oil and char-water slurry pipelining economics (143). A pilot plant capable of processing up to 33 t/d and hydrotreating 4.7 m³/d (30 bbl/d) was started up in 1970 (144), and was operated successfully for a number of years (145).

The COED concept (139), designed to recover liquid, gaseous, and solid fuel components, consists of four stages. Heat is generated by the reaction of oxygen with a portion of the char in the last pyrolysis stage and is also introduced by the air combustion of gas to dry feed coal. The number of stages in the pyrolysis, and the operating temperatures in each, may be varied to accommodate high volatile bituminous and subbituminous feed coals with widely ranging caking or agglomerating properties.

Oil condensed from the released volatiles from the second stage is filtered and catalytically hydrotreated at high pressure to produce a synthetic crude oil. Medium heat-content gas produced after the removal of H_2S and CO_2 is suitable as clean fuel. The pyrolysis gas produced, however, is insufficient to provide the fuel requirement for the total plant. Residual char, 50–60% of the feed coal, has a heating value and sulfur content about the same as feed coal, and its utilization may thus largely dictate process utility.

Properties. The properties of char products from two possible coal feeds, a low sulfur Western coal, and a high sulfur Midwestern coal, are shown in Table 11. The char derived from the low sulfur Western coal may be directly suitable as plant fuel, with only minor addition of clean process gas to stabilize its combustion. Flue gas desulfurization may not be required. Flue gas from the combustion of the char derived from the high sulfur Illinois coal, however, requires desulfurization before it may be discharged into the atmosphere.

Typical COED syncrude properties are shown in Table 12. The properties of the oil products depend heavily on the severity of hydroprocessing. The degree of severity also markedly affects costs associated with hydrogen production and compression. Syncrudes derived from Western coals have much higher paraffin and lower aromatic content than those produced from Illinois coal. In general, properties of COED products have been found compatible with expected industrial requirements.

Occidental Petroleum Coal Conversion Process. Garrett R&D Co. (now the Occidental Research Co.) developed the Oxy Coal Conversion process based on mathematical simulation for heating coal particles in the pyrolysis unit. It was estimated that coal particles of 100-mm diameter could be heated throughout their volumes to decomposition temperature (450–540°C) within 0.1 s. A large pilot facility was constructed at LaVerne, California, in 1971. This unit was reported to operate successfully at feed rates up to 136 kg/h (3.2 t/d).

Hot product char carries heat into the entrained bed to obtain the high heat-transfer rates required. Feed coal

Table 10. Average Yields and Range of Yields of Fischer Assay of Various Coals[a,b]

Rank of coal	Coke, % Average	Coke, % Range	Tar, L/t[c] Average	Tar, L/t[c] Range	Light oil, L/t[c] Average	Light oil, L/t[c] Range	Gas, m³/t[d] Average	Gas, m³/t[d] Range	Water, % Average	Water, % Range
Semianthracite			3.2		0.14					
Low volatile bituminous	89.7	85.8–93.3	39.6	29.0–58.4	4.69	3.36–7.41	59.8	54.4–66.6	3.2	1.1–6.6
Medium voltaile bituminous	83.3	77.4–90.4	86.9	44.6–117.8	7.68	4.92–10.58	66.0	47.3–76.2	4.1	2.8–7.0
High volatile A bituminous	75.5	68.8–81.4	142.1	105.3–187.2	10.53	6.81–15.09	67.0	57.5–80.2	6.0	3.0–9.2
High voltaile B bituminous	70.4	66.0–73.2	139.4	111.8–198.3	10.03	7.13–15.82	68.3	56.4–82.3	11.1	10.2–13.1
High volatile C bituminous	67.1	65.4–68.6	124.2	85.1–178.5	8.65	5.93–12.47	61.2	53.0–70.4	15.9	12.0–19.1
High volatile C bituminous or subbituminous A	59.1		94.3	84.6–112.2	7.59	6.26–8.88	90.4		23.4	
Subbituminous A			81.9	81.0–82.8	6.21	6.12–6.26				
Subbituminous B	57.6	54.8–59.9	70.8	60.7–76.8	6.12	5.24–7.13	90.4	62.2–93.8	27.8	23.3–30.4
Lignite	36.5		69.9	30.8–124.2	5.47	2.90–8.69	71.4		44.0	
Cannel	58.8	44.1–69.0	338.1	247.0–498.2	23.28	16.84–34.13	61.5	51.0–72.1	3.7	2.0–4.8

[a] Ref. 135.
[b] As-received basis; maximum temperature, 500°C.
[c] To convert L/t to gal/short ton, divide by 4.6.
[d] To convert m³/t to ft³/short ton, multiply by 29.4.

Table 11. Properties of COED Char Product[a]

Property	Utah Coal	Illinois No.6
Volatile matter, wt %	6.1	2.7
Fixed carbon, wt %	80.2	77.0
Ash, wt %	13.7	20.3
Higher heating value, MJ/kg,[b] dry	28.6	25.6
Elemental analysis, wt %, dry		
Carbon	81.5	73.4
Hydrogen	1.3	0.8
Nitrogen	1.5	1.0
Sulfur	0.5	3.4
Oxygen	1.5	1.0
Chlorine	0.006	0.1
Iron[c]	0.28	

[a] Ref. 139.

[b] To convert MJ/kg to Btu/lb, multiply by 430.

[c] Included in ash above.

must be dried and pulverized. A portion of the char recovered from the reactor product stream is cooled and discharged as product. The remainder is reheated to 650–870°C in a char heater blown with air. Gases from the reactor are cooled and scrubbed free of product tar. Hydrogen sulfide is removed from the gas, and a portion is recycled to serve as the entrainment medium.

Table 12. Typical COED Syncrude Properties[a]

Property	Utah A-Seam	Illinois No. 6 Seam
Specific gravity, (°API)[b]	0.934 (20)	0.929 (22)
Pour point, °C	16	−18
Flash point, closed cup, °C	24	16
Viscosity, at 38°C, mm²/s (= cSt)	8	5
Ash, wt %	0.01	0.01
Moisture, wt %	0.1	0.1
Metals, ppm	10	10
Elemental analysis, wt %		
C	87.2	87.1
H	11.0	10.9
N	0.2	0.3
O	1.4	1.6
S	0.1	0.1
ASTM distillation initial bp, °C	138	88
10%	221	134
30%	277	199
50%	349	270
70%	416	316
90%	493	362
End point (95%)	510	397
Hydrocarbon type analysis, liquid vol %		
Paraffins		10.4
Olefins	23.7	0
Naphthenes	0	41.4
Aromatics	42.2	48.2
	34.1	

[a] Ref. 139.

[b] $°API = \dfrac{141.5}{\rho} - 131.5$ where ρ is specific gravity.

Properties. A high volatile western Kentucky bituminous coal, the tar yield of which by Fischer assay was ca 16%, gave a tar yield of ca 26% at a pyrolysis temperature of 537°C (146–148). Tar yield peaked at ca 35% at 577°C and dropped off to 22% at 617°C. The char heating value is essentially equal to that of the starting coal, and the tar has a lower hydrogen content than other pyrolysis tars. The product char is not suitable for direct combustion because of its 2.6% sulfur content.

The TOSCOAL Process. The Oil Shale Corp. (TOSCO) piloted the low temperature carbonization of Wyoming subbituminous coals over a two-year period in its 23 t/d pilot plant at Rocky Falls, Colorado (149). The principal objective was the upgrading of the heating value in order to reduce transportation costs on a heating value basis. Hence, the solid char product from the process represented 50 wt % of the starting coal but had 80% of its heating value.

Furthermore, 60–100 L (14–24 gal) oil, having sulfur content below 0.4 wt %, could be recovered per metric ton coal from pyrolysis at 427–517°C. The recovered oil was suitable as low sulfur fuel. Figure 15 is a flow sheet of the Rocky Flats pilot plant. Coal is fed from hoppers to a dilute-phase, fluid-bed preheater and transported to a pyrolysis drum, where it is contacted by hot ceramic balls. Pyrolysis drum effluent is passed over a trommel screen that permits char product to fall through. Product char is thereafter cooled and sent to storage. The ceramic balls are recycled and pyrolysis vapors are condensed and fractionated.

Properties. Results for the operation using subbituminous coal from the Wyodad mine near Gillette, Wyoming, are shown in Table 13. Char yields decreased with increasing temperature, and oil yields increased. The Fischer assay laboratory method closely approximated the yields and product assays that were obtained with the TOSCOAL process.

The volatiles contents of product chars decreased from ca 25–16% with temperature. Char (lower) heating values, on the other hand, increased from ca 26.75 MJ/kg (11,500 Btu/lb) to 29.5 MJ/kg (12,700 Btu/lb) with temperature. Chars in this range of heating values are suitable for boiler fuel application and the low sulfur content (about equal to that of the starting coal) permits direct combustion. These char products, however, are pyrophoric and require special handling in storage and transportation systems.

Properties of the tar oil products are given in Table 14. The oils change only slightly with change in the retorting temperature; sulfur levels are low. The fraction boiling up to 230°C contains 65 wt % of phenols, cresols, and cresylic acids.

TOSCO tar oils have high viscosity and may not be transported by conventional pipelines. Heating values of product gas on a dry, acid gas-free basis are in the natural gas range if butanes and heavier components are included.

Coal Carbonization. In the by-product recovery of a modern coke oven, coal tar is removed first by cooling the

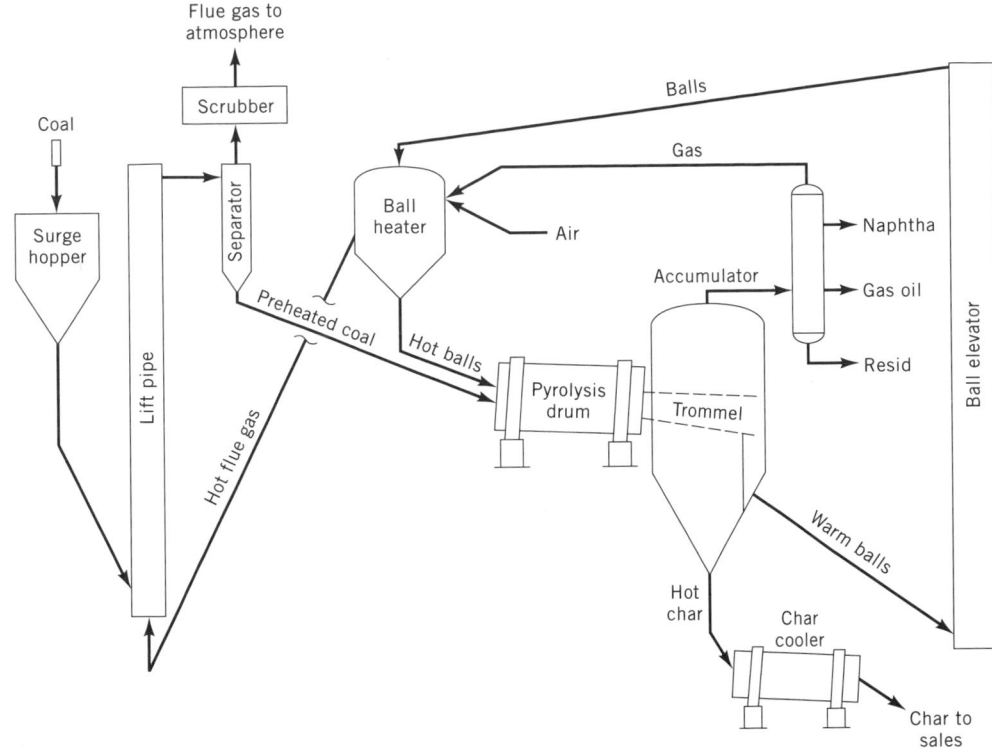

Figure 15. TOSCOAL process (149).

gases emanating, and light oil is removed last by scrubbing the gas with solvents. Other products, including ammonia, phenols, pyridine, or naphthalene, may be recovered between these operations. The constituents of coal tar, light oil, and gas usually overlap considerably, ie, the fractional condensation does not effectively separate individual components. Assuming the lowest boiling coal tar constituent to be benzene (bp 80.09°C), and the highest boiling to be naphthalene, the overlapping compositions of gas, light oil, and tar may be as shown graphically in Figure 16. Many chemical compounds have been identified (8) in these substances. Included are most of the significant constituents of petroleum-derived fuel liquids, although only a few components are present in sufficient quantity to make commercial recovery feasible.

The precise compositions of the light oil and coal tar recovered from coke-oven gas is a distinct function of the

design of the recovery system, as well as of the properties of the starting coal. In general, 12.5–16.7 L/t (3–4 gal/m light oil per short ton) of coal carbonized is recovered from high temperature coke-oven operations. Light oil may contain 55–70% benzene, 12–20% toluene, and 4–7% xylene. Unrecovered light oil appearing in the effluent coal gas may comprise ca 1 vol % and contribute ca 5% of the gas's heating value. Refining of light oil consists mainly of sulfuric acid washing, followed by fractional distillation.

Large-scale recovery of light oil was commercialized in England, Germany, and the United States toward the end of the nineteenth century (151). Industrial coal-tar production dates from the earliest operation of coal-gas facilities. The principal bulk commodities derived from coal tar are wood-preserving oils, road tars, industrial pitches, and coke. Naphthalene is obtained from tar oils by crystallization, tar acids are derived by extraction of tar oils with caustic, and tar bases by extraction with sulfuric acid. Coal tars generally contain less than 1% benzene and toluene, and may contain up to 1% xylene. The total U.S. production of BTX from coke-oven operations is insignificant compared to petroleum product consumptions.

OTHER PROCESSES

Shale Oil

In the United States, shale oil, or oil derivable from oil shale, represents the largest potential source of liquid hydrocarbons that can be readily processed to fuel liquids similar to those derived from natural petroleum. Some countries produce liquid fuels from oil shale. There is no such industry in the United States although more than 50 companies were producing oil from coal and shale in

Table 13. TOSCOAL Retorting of Wyodak Coal[a,b]

Component	Retort Temperature		
	427°C	482°C	521°C
Char	524.5	505.8	484.4
Gas, C_3 and lighter	59.5	78.4	63.0
Oil, C_4 and heavier	57.0	71.5	93.1
Water[c]	351.0	351.0	351.0
Total	*992.0*	*1006.7*	*991.5*
Recovery, %	*99.2*	*100.7*	*99.1*

[a] Ref. 150.
[b] Product yields, kg/t, of as-mined coal.
[c] Value assumed from Fischer assay and moisture content. The addition of steam to the process prevented accurate measurement of water produced in retorting.

Table 14. TOSCOAL Oil Properties[a]

Properties	Retort temperature		
	427°C[b]	482°C	521°C
Analysis, wt %			
Carbon	81.4	80.7	80.9
Hydrogen	9.3	9.1	8.7
Oxygen	8.3	9.4	9.3
Nitrogen	0.48	0.7	0.7
Sulfur	0.43	0.2	0.2
Chlorine	0.0	0.0	0.0
Ash	0.0	0.2	0.1
Total	*99.91*	*100.3*	*99.9*
Heating values			
Gross, MJ/kg[c]	38.59	37.72	37.13
Net, MJ/kg[c]	36.61	35.75	35.26
Specific gravity, (°API)[d]		1.040 (4.5)	1.061 (1.9)
Primary oil	1.015 (7.9)	0.985 (12.1)	1.027 (6.2)
Calculated, with C₄ and heavier components of gas added	0.978 (13.2)		
Pour point, °C	32	38	35
Conradson carbon, wt %	7.6	9.9	11.4
Distillation,[e] °C			199
2.5 vol %	212	216	235
20 vol %	302	288	385
50 vol %	407	413	
Viscosity, SUs			
at 82°C	122	123	128
at 90°C	63	66	69

[a] Ref. 150.
[b] Feed coal was different from that used for 482 and 521°C.
[c] To convert MJ/kg to Btu/lb, multiply by 430.
[d] $°API = \dfrac{141.5}{\rho} - 131.5$ where ρ is specific gravity.
[e] Combination of true boiling point and D1160 distillations.

the United States in 1860 (152,153), and after the oil embargo of 1973 several companies reactivated shale-oil process development programs (154,155). Petroleum supply and price stability has since severely curtailed shale oil development. In addition, complex environmental issues (156) further prohibit demonstration of commercial designs.

Heavy Oil

The definitions used to distinguish among naturally occurring heavy petroleum oils, bitumens, asphalts, and tars, are subject to broad variations. More than 10% of the world's current crude oil production has an API gravity below 20°, or a specific gravity greater than $0.934\frac{15}{15}$.

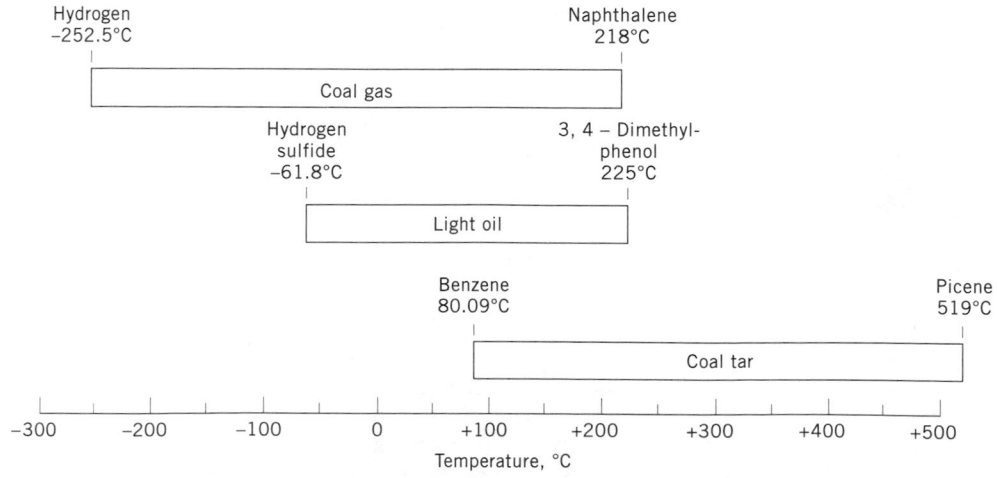

Figure 16. Boiling ranges of carbonization products (8).

Oils having sp $\text{gr}^{15}_{15} > 0.904-0.934$ (20–25°API) are considered heavy oils in most classifications. However, Safaniyah crude oil produced in Saudi Arabia having sp gr^{15}_{15} of 0.893 (27°API) carries the designation Arabian heavy, and in petroleum parlance is generally referred to as heavy crude oil. Yet its production method does not differ from that of Arabian medium or Arabian light crude oils.

Energy in the form of injected water or CO_2 may be supplied to increase the rate of production of light crude oils. Application of heat to the reservoirs, eg, using hot water, steam, heated CO_2, fireflood, or *in situ* combustion, however, is generally associated with the production of heavier, viscid crudes.

Heavy crude oil is widely distributed, and it is difficult to estimate reserves separate from normal crude oil reserves or from tar sands deposits. Estimates of petroleum reserves frequently include a large heavy oil component, which can only be produced at significantly higher cost than light oil.

Most heavy oil production is concentrated in California, Canada, and Venezuela. There is significant production of heavy oil in California from the Kern River field near Bakersfield and in Canada from the Cold Lake deposit in Alberta. Production generally involves steam drives, or the injection of steam into reservoirs through special wells in prescribed sequences. Oil–water mixtures are recovered, and often separated water is treated and reinjected.

Heavy oil may be upgraded through two main routes: coking and hydroprocessing. Virtually all established upgrading schemes involve some variant of those two routes. The challenges in upgrading and refining are from the low hydrogen content and specific gravity and high sulfur, nitrogen, and metals content of the heavy oil.

Tar Sands

Tar sands (qv) are considered to be sedimentary rocks having natural porosity where the pore volume is occupied by viscous, petroleum-like hydrocarbons. The terms oil sands, rock asphalts, asphaltic sandstones, and malthas or malthites have all been applied to the same resource. The hydrocarbon component of tar sands is properly termed bitumen.

Distinctions between tar sands' bitumens and heavy oils are based largely on differences in viscosities. The bitumen in oil sand has a specific gravity of less than 0.986 g/mL (12°API), and thus oil sands may be regarded as a source of extremely heavy crude oil. Whereas heavy oils might be produced by the same techniques used for the lighter crude oils, the bitumens in tar sands are too viscous for these techniques. Consequently these oil-bearing stones have to be mined and specially processed to recover contained hydrocarbon.

Tar sands have been reported on every continent except Australia and Antarctica. The best known deposits are the Athabasca of Canada, where almost 60,000 km^2 in northeastern Alberta is underlain with an estimated 138×10^9 m^3 (870×10^9 bbl) of recoverable bitumen (157). The Alberta deposits may contain up to 215×10^9 m^3 (1350×10^9 bbl) of bitumen reserves. Venezuela may have the largest accumulations in the world; the Orinoco heavy-oil belt has been estimated by some (157) to contain as high as 636×10^9 m^3 (4000×10^9 bbl). The Olenek reserves in the former USSR may contain ca 95×10^9 m^3 (600×10^9 bbl). The United States is estimated to have deposits of about 4.5×10^9 m^3 (28×10^9 bbl).

The Great Canadian Oil Sands, Ltd. (GCO) (Sun Oil Co.) has been operating a plant at Fort McMurray, Alberta, Canada, since 1967. Initially, some 8050 t/d (55,000 bbl/d) of synthetic crude oil was produced from coking (158) with the project expanding to 9220 t/d (63,000 bbl/d). Since 1978, Syncrude Canada has been producing ca 22,000 m^3/d (140,000 bbl) synthetic crude oil by fluid coking from their plant at Cold Lake, Alberta, Canada (159) with expansion planned for ca 35,000 m^3/d (225,000 bbl/d).

ECONOMIC ASPECTS OF SYNTHETIC FUELS

As of this writing, processes for production of synthetic liquid fuels by upgrading natural gas, coal, or heavy oil are generally not directly competitive with crude oil upgrading (160). The key controlling factors in the economics are crude oil price and availability. Many economic analyses for synthetic liquid fuels give a crude oil price target whereupon the alternative technology becomes attractive, but these studies sometimes neglect the fact that the natural gas, coal, and heavy oil prices often track those of crude oil. In addition, conversion of a refractory gas (methane) or solid (coal) to liquids is a greater technical challenge than that of processing crude oil. Thus there are processing cost penalties which inevitably exist even after considerable technological development. Nevertheless, synthetic fuels technology is projected to play a primary role in providing liquid fuels once crude oil depletion is of concern. Economic competitiveness plays a reduced role in commercialization only when environmental legislation mandates the use of certain fuels such as oxygenates.

The commercialization potential of synthetic fuels technology relies on site-specific economic and political factors. This complex network of factors may include capital costs, crude oil price, product yields and value, government subsidies, strategic impact, alternative uses for the feed, and environmental and geographical constraints. Whereas no direct coal liquefaction process has gone to commercial stage, technologies involving indirect conversion of natural gas or coal have been commercialized. In all cases, special conditions allowed the technology to progress. In the Sasol project at South Africa, coal upgrading was possible due to factors such as no indigenous petroleum, minimal environmental standards, and cheap labor (160). The New Zealand GTG process became economically feasible owing to high oil prices, abundance of indigenous natural gas, and government commitment to energy self-sufficiency (59). Government support and long-term strategic benefits were also keys to Shell's SMDS project in Malaysia.

At 1994 crude oil prices of ca \$94–125/m^3 (ca \$15–20/bbl), conversion of natural gas to liquid fuels exists only in unique situations. For natural gas upgrading via New Zealand-type technology, economics by Mobil for a 1987 plant start-up on the U.S. Gulf Coast (161) indicated an investment of 895×10^6 was required for a 2.3×10^3 m^3/d (14,600 bbl/d) gasoline production unit. Thus this and other natural gas-to-fuels processes are highly capital-intensive and capital recovery remains the dominant factor even with incremental advances in conventional technology. This is especially the case using indirect upgrading of natural gas because the cost of synthesis gas manufacture may account for more than 50% of the total process capital cost (44). An analysis by Shell of the SMDS process published in 1988 showed capital expense for a 1600 m^3/d (10,000 bbl/d) plant to be 300×10^6 for a developed site in an industrialized country and 600×10^6 for a developing site in a developing country (44). Direct upgrading of natural gas to gasoline and distillate by oxidative coupling plus olefin oligomerization has been evaluated to be ca 10% costlier in capital than upgrading via indirect technologies (162).

Natural gas upgrading economics may be affected by additional factors. The increasing use of compressed natural gas (CNG) directly as fuel in vehicles provides an alternative market which affects both gas price and values (see GAS, NATURAL). The hostility of the remote site environment where the natural gas is located may contribute to additional costs, eg, offshore sites require platforms and submarine pipelines.

The economic feasibility of coal upgrading, and in particular direct coal liquefaction, are closely tied to crude oil price and capital expense. H-coal technology was evaluated as a base case in 1981 and the results showed economic feasibility was possible only at crude oil price of $630/m^3 ($100/bbl) or greater (160). A more recent analysis by HRI on coal/oil coprocessing technology indicated the required light crude price was $138–182/m^3 ($22–29/bbl) for economic feasibility (163). The cost of capital could add over $60/m^3 ($10/bbl) to the cost of products. A study of EDS, H-coal, ITSL, and two-stage Wilsonville systems showed capital costs for a 30,000 t/d plant processing Illinois No. 6 coal to run between $4100–$4700 \times 10^6$ (131). Required break-even selling prices for products from these technologies ranged from $226–314/m^3 ($36–50/bbl). The two-stage system was the most economical at $229.94/m^3 ($36.56/bbl). An evaluation of coprocessing Cold Lake vacuum bottoms using Alberta subbituminous coal in a 3200 m^3/d (20,000 bbl/d) synthetic crude oil production unit indicated a selling price of $189–220/m^3 ($30–35/bbl) was necessary for the process to be competitive (164). In general, the economics of direct coal liquefaction depend more on the high cost of liquefaction rather than the cost for upgrading product coal liquids (165).

Factors which may affect the cost of coal upgrading are environmental considerations such as toxicity, hazardous waste disposal, and carcinogenic properties (131). These and other environmental problems from process streams, untreated wastewaters, and raw products would figure significantly into the cost of commercialization.

Reprinted from *Kirk-Othmer Encyclopedia of Chemical Technology*, 4th ed., Vol. 12, John Wiley & Sons, Inc., New York, 1995.

BIBLIOGRAPHY

1. H. Mimoun, *New J. Chem.* **11**, 4 (1987).
2. U. Preuss and M. Baerns, *Chem. Eng. Technol.* **10**, 297 (1987).
3. "Liquid Fuels From Natural Gas," *Petrole Informations,* API 34-5250, 96 (May 1987).
4. *Chem. Eng. News,* 25 (Aug. 14, 1989).
5. E. E. Ecklund and G. A. Mills, *Chemtech,* 549 (Sept. 1989).
6. L. Haar, in W. P. Earley and J. W. Weatherly, eds., *Advances in Coal Utilization Technology IV,* Institutes of Gas Technology, Chicago, 1981, pp. 787–952.
7. L. Shnidman, in H. H. Lowry, ed., *Chemistry of Coal Utilization,* Vol. 2, John Wiley & Sons, Inc., New York, 1945, pp. 1252–1286.
8. W. L. Glowacki, in Ref. 7, pp. 1136–1231; E. O. Rhodes, in Ref. 7, pp. 1136–1231.
9. J. R. Rostrup-Nielsen, *Steam Reforming Catalysts,* Teknisk Forlag A/S, Copenhagen, 1975.
10. *Catalyst Handbook,* Wolfe Scientific Texts, London, 1970.
11. P. Sabatier, *Catalysis, Then and Now,* Part II, Franklin Publishing Co., Englewood, N.J., 1965.
12. Fr. Pat. 571,356 (May 16, 1924), (to Badische Anilin- und Soda-Fabrik).
13. Fr. Pat. 580,905 (Nov. 19, 1924), (to Badische Anilin- und Soda-Fabrik).
14. F. Fischer and H. Tropsch, *Ber.* **56**, 2428 (1923).
15. F. Fischer, *Die Umwandlung der Kohle in Ole,* Borntraeger, Berlin, 1923, p. 320.
16. F. Fischer and H. Tropsch, *Ber.* **59**, 830, 832, 923 (1926).
17. F. Fischer, *Brennstoff-Chem.* **11**, 492 (1930).
18. F. Fischer and H. Koch, *Brennstoff-Chem.* **13**, 61 (1932).
19. F. Fischer and K. Meyer, *Brennstoff-Chem.* **12**, 225 (1931).
20. F. Fischer and H. Pichler, *Brennstoff-Chem.* **20**, 41, 221 (1939).
21. K. Fischer, *Comparison of I. G. Work on Fischer Synthesis, Technical Oil Mission Report, Reel 13,* Library of Congress, Washington, D.C., July 1941.
22. H. Pichler, *Medium Pressure Synthesis on Iron Catalyst, (Pat. Appl), Technical Oil Mission Report, Reel 100,* Library of Congress, Washington, D.C., 1937–1943.
23. H. Pichler, *Medium Pressure Synthesis on Iron Catalyst, Technical Oil Mission Report, Reel 101,* Library of Congress, Washington, D.C., June 1940.
24. U.S. Pat. 2,327,066 (Aug. 17, 1943). O. Roelen.
25. F. Fischer, *Ole Kohle* **39**, 517 (1943).
26. H. H. Storch, N. Golumbic, and R. B. Anderson, *The Fischer-Tropsch and Related Synthesis,* John Wiley & Sons, Inc., New York, 1951.
27. W. G. Frankenburg, V. I. Komarewsky, and E. D. Rideal, *Advances in Catalysis,* Vol I., Academic Press, Inc., New York, 1948, pp. 115–156.
28. H. H. Storch, in Ref. 7, p. 1797.
29. O. C. Elvins and A. W. Nash, *Fuel* **5**, 263 (1926).
30. O. C. Elvins, *J. Soc. Chem. Ind. (London)* **56**, 473T (1927).
31. A. Erdeley and A. W. Nash, *J. Soc. Chem. Ind. (London)* **47**, 219T (1928).
32. W. W. Myddleton, *Chim. Ind.* **37**, 863 (1937); *J. Inst. Fuel* **11**, 477 (1938); *Colliery Guardian* **157**, 286 (1938).
33. G. Egloff, *Brennst.-Chem.* **18**, 11 (1937).

34. G. Egloff, E. F. Nelson, and J. C. Morrell, *Ind. Eng. Chem.* **29,** 555 (1937).

35. F. Mako and W. A. Samuel, in R. A. Meyers, ed., *Handbook of Synfuels Technology,* McGraw-Hill, Inc., New York, 1984, pp. 2-5–2-43.

36. J. C. Hoogendoorn, *Phil. Trans. R. Soc. Lond. A* **300,** 99 (1981).

37. W. B. Johnson, *Pet. Ref.* **35** (Dec. 1956).

38. M. E. Dry, "Fischer-Tropsch Synthesis Over Iron Catalysts," paper presented at *1990 Spring AlChE National Meeting,* Orlando, Fla., Mar. 18–22, 1990.

39. B. Jager and co-workers, in Ref. 38.

40. A. E. Sands, H. W. Wainwright, and L. D. Schmidt, *Ind. Eng. Chem.* **40,** 607 (1948); A. E. Sands and L. D. Schmidt, *Ind. Eng. Chem.* **42,** 2277 (1950).

41. H. Pichler, *Technical Oil Mission Report, Reel 259,* Library of Congress, Washington, D.C., 1947, frames 467–654.

42. P. C. Keith, *Oil Gas J.* **345**(6), 102 (1946).

43. *Oil Gas J.,* 74 (Feb. 17, 1986).

44. M. J. v. d. Burgt and co-workers, in D. M. Bibby and co-workers, eds., *Methane Conversion,* Elsevier Science, Inc., New York, 1988, pp. 473–482.

45. I. E. Maxwell and J. E. Naber, *Catal. Lett.* **12,** 105 (1992).

46. P. T. Roterud and co-workers, in Ref. 38.

47. C. D. Chang and A. J. Silvestri, *J. Catal.* **47,** 249 (1977).

48. C. D. Chang, *Catal. Rev.-Sci. Eng.* **25,** 1 (1983).

49. C. D. Chang and A. J. Silvestri, *Chemtech* **17,** 624 (1987).

50. U.S. Pat. 3,702,886 (1972), R. J. Argauer and G. R. Landolt (to Mobil Oil Corp.).

51. G. T. Kokotailo and co-workers, *Nature* **272,** 437 (1978).

52. D. H. Olson, G. T. Kokotailo, and S. L. Lawton, *J. Phys. Chem.* **85,** 2238 (1981).

53. C. D. Chang, in Ref. 44, pp. 127–143.

54. G. J. Hutchings and R. Hunter, *Catal. Today* **6,** 279 (1990).

55. F. Bauer, *ZfI-Mitt.* **156,** 31 (1990).

56. P. B. Weisz and V. J. Frilette, *J. Phys. Chem.* **64,** 382 (1960).

57. S. M. Csicsery, *ACS Monograph* **171,** 680 (1976).

58. N. Y. Chen, W. E. Garwood, and F. G. Dwyer, *Shape Selective Catalysis in Industrial Applications,* Marcel Dekker, Inc., New York, 1989.

59. C. J. Maiden, in Ref. 44, pp. 1–16.

60. J. E. Penick, W. Lee, and J. Maziuk, *ACS Symp. Ser.* **226,** 19 (1983).

61. J. Z. Bem, in Ref. 44, pp. 663–678.

62. P. L. Rogerson, in Ref. 35, pp. 2-45–2-73.

63. A. Y. Kam, M. Schreiner, and S. Yurchak, in Ref. 35, pp. 2-75–2-111.

64. D. Liederman and co-workers, *Ind. Eng. Chem. Proc. Des. Devel.* **17,** 340 (1978).

65. H. R. Grimmer, N. Thiagarajian, and E. Nitschke, in Ref. 44, pp. 273–291.

66. K. H. Keim and co-workers, *Erdol. Erdgas, Kohle* **103,** 82 (1987).

67. K. G. Allum and A. R. Williams, in Ref. 44, pp. 691–711.

68. A. A. Avidan, in Ref. 44, pp. 307–323.

69. C. D. Chang, *Catal. Rev.-Sci. Eng.* **26** (3&4), 323 (1984).

70. M. M. Holm and E. H. Reichl, *Fiat Report No. 1085,* Office of Military Government for Germany (U.S.), Mar. 31, 1947.

71. J. A. Sofranko, "Gas to Gasoline: The ARCO GTG Process," paper presented at *Bicentenary Catalysis Meeting,* Sydney, Australia, Sept. 1988.

72. J. A. Sofranko and J. C. Jubin, "Natural Gas to Gasoline: The ARCO GTG Process," paper presented at *International Chemical Congress of Pacific Basin Societies,* Honolulu, Hawaii, Dec. 1989.

73. J. H. Edwards, K. T. Do, and R. J. Tyler, in Ref. 72.

74. J. H. Edwards, K. T. Do, and R. J. Tyler, in E. E. Wolf, ed., *Methane Conversion by Oxidative Processes,* Van Nostrand Reinhold, New York, 1992, pp. 429–462.

75. R. J. Nichols, "Applications of Alternative Fuels," *SAE Special Publication SP-531,* Society of Automotive Engineers, Warrendale, Pa., Nov. 1982.

76. R. A. Potter, "Neat Methanol Fuel Injection Fleet Alternative Fuels Study," paper presented at *Fourth Washington Conference on Alcohol,* Washington, D.C., Nov. 1984.

77. N. D. Brinkman, *Ener. Res.* **3,** 243 (1979).

78. T. Powell, "Racing Experiences with Methanol and Ethanol Based Motor-Fuel Blends," paper 750124, *Automotive Engineering Congress and Exposition,* Society of Automotive Engineers, Detroit, Mich., Feb. 1975.

79. American Petroleum Institute, Task Force EF-18 of the Committee on Mobile Source Emissions, *Alcohols–A Technical Assessment of Their Application as Fuels, Publication No. 4261,* API, New York, July 1976.

80. U.S. Environmental Protection Agency, *Fed. Reg.* **46**(144) (July 28, 1981).

81. U.S. Environmental Protection Agency, *Fed. Reg.* **50**(12), 2615 (Jan. 17, 1985).

82. G. A. Mills and E. E. Ecklund, *Chemtech,* 626 (Oct. 1989).

83. KVB, Inc., *KVB Report Number 72-804830-1998,* Vol. 2, California Energy Commission, Irvine, Calif., Mar. 1985, pp. 1–2.

84. A. J. Weir and co-workers, "Methanol Dual-Fuel Combustion," paper presented at *1987 Joint Symposium on Stationary Combustion NO_x Control,* New Orleans, La., Mar. 23–26, 1987.

85. S. B. Alpert and D. F. Spencer, *Methanol and Liquid Fuels from Coal–Recent Advances,* Electric Power Research Institute, Palo Alto, Calif., 1987.

86. V. Yand and S. C. Trindade, *Chem. Eng. Prog.,* 11 (Apr. 1979).

87. *Alcohols: Economics and Future U.S. Gasoline Markets,* Information Resources, Inc., Washington, D.C., 1984.

88. H. L. Muller and S. P. Ho, "Economics and Energy Balance of Ethanol as Motor Fuel," paper presented at *1986 Spring AIChE National Meeting,* New Orleans, La., Apr. 1986.

89. *Chem. Eng. News,* 7 (Nov. 2, 1992).

90. ARCO Chemical Co., *Testimony to the Colorado Air Quality Control Commission on Proposed Regulation No. 13 (Oxygenate Mandate Program),* Denver, Colo., June 4, 1987.

91. *Chem. Eng. News,* 28 (Apr. 12, 1993).

92. U.S. Pat. 1,251,954 (Jan. 1, 1918), F. Bergius and J. Billwiller.

93. F. Fischer and H. Tropsch, *Ges. Abhandl. Kenntis Kohle* **2,** 154 (1918).

94. F. Bergius, *Pet. Z.* **22,** 1275 (1926).

95. *Gas World* **104,** 421 (1936).

96. *Pet. Times* **42,** 641 (1939).

97. H. H. Storch and co-workers, *U.S. Bur. Mines. Tech. Pap.* **622** (1941).

98. A. C. Fieldner and co-workers, *U.S. Bur Mines Tech. Pap.* **666** (1944).

99. M. L. Kastens and co-workers, *Ind. Eng. Chem.* **41,** 870 (1949).

100. J. L. Wiley and H. C. Anderson, *U.S. Bur. Mines Bull.* **485,** I (1950), II (1951), III (1952).

101. C. C. Chaffee and L. L. Hirst, *Ind. Eng. Chem.* **45,** 822 (1953).

102. Bituminous Coal Staff, *U.S. Bur. Mines Rep. Invest.* **5506** (1959).

103. E. L. Clark and co-workers, *Ind. Eng. Chem.* **42,** 861 (1950).

104. L. L. Newman and A. P. Pipilen, *Gas Age* **119**(10), 16 (1957); **119**(11), 18 (1957).

105. U.S. Pat. 2,860,101 (Nov. 11, 1958), M. G. Pelipetz (to the United States of America).

106. S. Akhtar, S. Friedman, and P. M. Yavorsky, *U.S. Bur. Mines Tech. Prog. Rep.* **35** (1971).

107. S. Akhtar and co-workers, "Process for Hydrodesulfurization of Coal," paper presented at *71st National AIChE Meeting,* Dallas, Tex., Feb. 20, 1972.

108. B. Linville and J. D. Spencer, *U.S. Bur. Mines Inf. Cir.* **8612** (1973).

109. U.S. Pat. 3,321,393 (May 23, 1967), S. C. Schuman, R. H. Wolk, and M. C. Chervenak (to Hydrocarbon Research, Inc.).

110. *Coal Age,* 101 (May 1976).

111. G. A. Johnson and co-workers, "Present Status of the H-Coal Process," paper 30, *IGT Coal Symposium,* Chicago, 1973.

112. J. E. Papso, in Ref. 35, pp. 1-47–1-63.

113. "High Pressure Hydrogenation at Ludwigshafen-Heidelberg," Vol. IA, *FIAT Final Report No. 1317, ATI No. 92,762,* Central Air Documents Office, Dayton, Ohio, 1951.

114. H. H. Lowry and H. J. Rose, *U.S. Bur. Mines Inf. Cir.* **7420** (1947).

115. A. Baba and co-workers, *Rep. Resources Res. Inst. Jpn.,* (22) (1955).

116. U.S. Pat. 2,221,866 (Nov. 19, 1940), H. Dreyfus.

117. D. L. Kloepper and co-workers, *Solvent Processing of Coal to Produce a De-ashed Product,* Contract 14-01-0001-275, OCR Report No. 9, U.S. Government Printing Office, Washington, D.C., 1965.

118. V. L. Brant and B. K. Schmid, *Chem. Eng. Prog.* **68**(12), 55 (1969).

119. B. K. Schmid and W. C. Bull, "Production of Ashless, Low-Sulfur Boiler Fuels From Coal," paper presented at *ACS Division of Fuel Chem. Symposium on Pollution Control,* New York, Sept. 12, 1971.

120. *Demonstration Plant, Clean Boiler Fuels From Coal,* OCR R&D report no. 82, Interim report no. 1, Vols, 1–3, Ralph M. Parsons Co., Los Angeles, Calif., 1973–1975.

121. M. E. Frank and B. K. Schmid, "Economic Evaluation and Process Design of a Coal–Oil–Gas (COG) Refinery," paper presented at *Symposium on Conceptual Plants for the Production of Synthetic Fuels From Coal, AIChE 65th Annual Meeting,* New York, Nov. 26, 1972.

122. U.S. Pat. 3,341,447 (Sept. 12, 1967), W. C. Bull and co-workers (to the United States of America and Gulf Oil Corp.).

123. B. K. Schmid and D. M. Jackson, "The SRC-II Process," paper presented at *Third Annual International Conference on Coal Gasification and Liquefaction,* University of Pittsburgh, Aug. 3–5, 1976; D. M. Jackson and B. K. Schmid, "Production of Distillate Fuels by SRC-II," paper presented at *ACS Div. of Ind. and Eng. Chem. Symposium,* Colorado Springs, Col., Feb. 12, 1979.

124. W. R. Epperly and J. W. Taunton, "Exxon Donor Solvent Coal Liquefaction Process Development," paper presented at *Coal Dilemma II ACS Meeting,* Colorado Springs, Colo., Feb. 12, 1979.

125. D. E. Blaser and A. M. Edelman, "Flexicoking for Improved Utilization of Hydrocarbon Resources," paper presented at *API 43rd Mid-Year Meeting,* Toronto, Canada, May 8, 1978.

126. S. F. Massenzio, in Ref. 35, pp. 6-3–6-18.

127. T. A. Cavanaugh, W. R. Epperly and D. T. Wade, in Ref. 35, pp. 1-3–1-46.

128. L. E. Swabb, Jr., G. K. Vick, and T. Aczel, "The Liquefaction of Solid Carbonaceous Materials," paper presented at *The World Conference on Future Sources of Organic Raw Materials,* Toronto, Can., July 10, 1978.

129. E. L. Huffman, *Proceedings of the Third Annual International Conference on Coal Gasification and Liquefaction,* Pittsburgh, Pa., 1976.

130. H. D. Schindler, J. M. Chen, and J. D. Potts, *Final Technical Report on DOE Contract No. DE-AC22-79ET14804,* Department of Energy, Washington, D.C., 1983.

131. H. D. Schindler, *Final Technical Report on DOE Contract No. D-AC01-87ER30110,* Vol. 2, Department of Energy, Washington, D.C., 1989.

132. A. G. Comolli and co-workers, *Proceedings of the DOE Direct Liquefaction Contractors' Review Meeting,* Pittsburgh, Pa., 1986.

133. M. G. Thomas, in B. R. Cooper and W. A. Ellingson, eds., *The Science and Technology of Coal and Coal Liquefaction,* Plenum Press, New York, 1984, pp. 231–261.

134. M. J. Burges and R. V. Wheeler, *Fuel* **5,** 65 (1926).

135. W. A. Selvig and W. H. Ode, *U.S. Bur. Mines. Bull.* **571,** (1957).

136. H. C. Howard, in Ref. 7, Vol. 1, pp. 761–773.

137. *Ibid.,* Suppl. Vol., pp. 340–394.

138. J. F. Jones and co-workers, *Chem. Eng. Prog.* **62**(2), 73 (1966).

139. J. A. Hamshar, H. D. Terzian, and L. J. Scotti, "Clean Fuels From Coal by the COED Process," paper presented at *EPA Symposium on Environmental Aspects of Fuel Conversion Technology,* St. Louis, Mo., May 1974.

140. C. D. Kalfadelis and E. M. Magee, *Evaluation of Pollution Control in Fossil Fuel Conversion Processes, Liquefaction, Section 1, COED Process,* EPA-650/2-74-009e, Environmental Protection Agency, Washington, D.C., 1975.

141. R. T. Eddinger and co-workers, *Char Oil Energy Development, Office of Coal Research R&D Report No. 11,* Vol. 1 (PB 169,562) and Vol. 2 (PB 169,563), U.S. Government Printing Office, Washington, D.C., Mar. 1966.

142. J. F. Jones and co-workers, *Char Oil Energy Development, Office of Coal Research R&D Report No. 11,* Vol. 1 (PB 173,916) and Vol. 2 (PB 173,917), U.S. Government Printing Office, Washington, D.C., Feb. 1967.

143. M. E. Sacks and co-workers, *Char Oil Energy Development, Office of Coal Research Report 56, Interim Report No. 2,* GPO Cat. No. 163.10:56/Int.2, U.S. Government Printing Office, Washington, D.C., Jan. 1971.

144. J. F. Jones and co-workers, *Char Oil Energy Development, Office of Coal Research R&D Report No. 56, Final Report,* GPO Cat. No. 163.10:56, U.S. Government Printing Office, Washington, D.C., May 1972.

145. J. F. Jones and co-workers, *Char Oil Energy Development, Office of Coal Research R&D Report No. 73, Interim Report No. 1,* GPO Cat. No. 163.10:73/Int 1, U.S. Government Printing Office, Washington, D.C., Dec. 1972.

146. A. Sass, "The Garrett Research and Development Company Process for the Conversion of Coal into Liquid Fuels," paper presented at *65th Annual AIChE Meeting,* New York, Nov. 29, 1972.

147. *Oil Gas J.,* 78 (Aug. 26, 1974).

148. A. Sass, *Chem. Eng. Prog.* **70**(1), 72 (1974).

149. F. B. Carlson and co-workers, *Chem. Eng. Prog.* **69**(3), 50 (1973).

150. F. B. Carlson, L. H. Yardumian, and M. T. Atwood, "The TOSCOAL Process for Low Temperature Pyrolysis of Coal," paper presented at *American Institute of Mining, Metallugical, and Petroleum Engineers,* San Francisco, Calif., Feb. 22, 1972, and to *American Institute of Chemical Engineers,* New York, Nov. 29, 1972.

151. W. Tiddy and M. J. Miller, *Am. Gas J.* **153**(3), 7 (1940).

152. M. J. Gavin, *U.S. Bur. Mines Bull.* **210**, (1922).

153. M. J. Gavin and J. S. Desmond, *U.S. Bur. Mines Bull.* **315** (1930).

154. H. Shaw, C. D. Kalfadelis, and C. E. Jahnig, *Evaluation of Methods to Produce Aviation Turbine Fuels From Synthetic Crude Oils-Phase I, Technical Report AFAPL-TR-75-10,* Vol. I, Air Force Aero Propulsion Laboratory, Wright-Patterson Air Force Base, Dayton, Ohio, Mar. 1975.

155. C. D. Kalfadelis, *Evaluation of Methods to Produce Aviation Turbine Fuels From Synthetic Crude Oils-Phase II, Technical Report AFAPL-TR-75-10,* Vol. II, Air Force Aero Propulsion Laboratory, Wright-Patterson Air Force Base, Dayton, Ohio, May 1976.

156. Colony Development Operation, *An Environmental Impact Analysis for a Shale Oil Complex at Parachute Creek, Colorado,* Vols. 1–3, Denver, Colo., 1974.

157. H. L. Erskine, in Ref. 35, pp. 5-3–5-32.

158. R. D. Hynphreys, F. K. Spragins, and D. R. Craig, "Oil Sands–Canada's First Answer to the Energy Shortage," *Proceedings 9th World Petroleum Congress,* Tokyo, Japan, May 11, 1975, Vol. 5, p. 17.

159. C. W. Bowman, R. S. Phillips, and L. R. Turner, in Ref. 35, pp. 5-33–5-79.

160. M. Crow and co-workers, *Synthetic Fuel Technology Development in the United States–A Retrospective Assessment,* Praeger Publishing, New York, 1988.

161. S. Yurchak and S. S. Wong, "Mobil Methanol Conversion Technology," *Proceedings IGT Asian Natural Gas Seminar,* Singapore, 1992, pp. 593–618.

162. J. L. Matherne and G. L. Culp, in Ref. 74, pp. 463–482.

163. J. E. Duddy, S. B. Panvelker, and G. A. Popper, "Commercial Economics of HRI Coal/Oil Co-Processing Technology," paper presented at *1990 Summer AIChE National Meeting,* San Diego, Ca., 1990.

164. M. Ikura and J. F. Kelly, "A Techno-Economic Evaluation of CANMET Coprocessing Technology," *Proceedings Annual International Pittsburgh Coal Conference,* 1990, pp. 719–728.

165. J. G. Sikonia, B. R. Shah, and M. A. Ulowetz, "Technical and Economic Assessment of Petroleum, Heavy Oil, Shale Oil and Coal Liquid Refining," paper presented at *Synfuels' 3rd Worldwide Symposium,* Washington, D.C., Nov. 1–3, 1983.

General Reference

American Petroleum Institute, *Alcohols and Ethers-A Technical Assessment of Their Application as Fuels and Fuel Components,* API Publication 4261, American Petroleum Institute, Washington, D.C., July 1988.

FUSION ENERGY

WILLIAM R. ELLIS
Raytheon Engineers & Constructors
New York, New York

As far as is known, nuclear fusion, which drives the stars, including the sun, is the primary source of energy in the universe. The process of nuclear fusion releases enormous amounts of energy. It occurs when the nuclei of lighter elements, such as hydrogen, are fused together at extremely high temperatures and pressure to form heavier elements, such as helium. Whereas practical methods for harnessing fusion reactions and realizing the potential of this energy source have been sought since the 1950s, achieving the benefits of power from fusion has proved to be a difficult, long-term challenge. See also ELECTRIC POWER GENERATION; COLD FUSION; NUCLEAR POWER.

Fusion is widely held to be the ultimate resource for the world's long-term energy needs. The fuel reserves for fusion are virtually limitless and available to all countries. Fusion fuels can be extracted from water. Additionally, fusion promises to be an energy source which is potentially safe and environmentally benign. Radiological and proliferation hazards are much smaller than for fission power plants. The atmospheric impact is negligible compared to fossil fuels, and adverse impacts on the Earth's ecological and geophysical processes are smaller than for large-scale renewable energy sources (see also FUELS, SYNTHETIC; GAS, NATURAL). The economics and costs of fusion power plants are still being studied, but appear comparable to those for other medium- and long-term energy sources. The tantalizing promise of affordable essentially unlimited supplies of clean, safe energy, free of political boundaries, has motivated a worldwide research effort to develop this energy resource.

The nuclear burning mechanism of the Sun was elucidated in the 1930s (1). In a complex sequence of reactions starting with hydrogen, atomic nuclei are fused to form heavier species. Because of a mass deficit, Δm, exhibited by the reaction products, large amounts of energy, E, are released, as dictated by the well-known Einstein equivalence $E = \Delta mc^2$ where c is the speed of light. Large-scale fusion energy production was demonstrated dramatically on earth in the early 1950s with the explosion of thermonuclear fusion, ie, hydrogen, bombs. These weapons used the heat of nuclear fission (atomic bombs) to cause the fusion of deuterium [16873-17-9], D, and tritium [15086-10-9], ^{3}H or T. Subsequently, an international research effort was undertaken to harness this awesome power on a controllable scale for peaceful purposes. Several impressive advances in the 1980s and early 1990s have led to a well-founded feeling of optimism that fusion energy should become a practical energy source during the early twenty-first century.

In order to effect a fusion reaction between two atomic nuclei, it is necessary that these nuclei be brought together closely enough to experience an attractive nuclear force. All nuclei are positively charged and repel one another via Coulomb's law, the electrostatic law of the repulsion of like charges. This electrostatic barrier can be overcome by imparting sufficient kinetic energy to the reacting species so that the nuclei can approach closely enough together that quantum mechanical tunneling can occur. The repulsive forces increase rapidly with the magnitude of the nuclear charge; therefore, nuclear fusion research has concentrated on the lightest elements and the isotopes having the lowest atomic numbers.

The reactions of deuterium, tritium, and helium-3 [14762-55-1], ^{3}He, having nuclear charge of 1, 1, and 2, respectively, are the easiest to initiate. These have the highest fusion reaction cross-sections and the lowest reactant energies.

DEUTERIUM–TRITIUM FUSION

The D–T reaction involving the two heavy isotopes of hydrogen

$$D + T \rightarrow \,^4He + n$$

is especially attractive to fusion scientists because of its relative ease of ignition. The products of this reaction are an alpha particle, ie, the helium nucleus, ^{4}He, and a free neutron, n, carrying kinetic energies of 3.5 and 14.1 MeV, respectively. In an electric power-generating facility the neutrons would be absorbed in a blanket surrounding the fusion region, and the kinetic energy converted into heat. Conventional power conversion systems could then be used to transform this heat into electrical energy. Fusion reactions are extremely energetic, and yields are measured in units of millions of electron volts, MeV (1 MeV = 1.6×10^{-13} J).

Another set of reactions of practical interest involves only deuterons. The D–D reaction can proceed along either of two pathways with roughly equal probabilities:

$$D + D \rightarrow T + p + 4.0 \text{ MeV}$$

or

$$D + D \rightarrow \,^3He + n + 3.2 \text{ MeV}$$

Finally, the D–^{3}He reaction

$$D + \,^3He \rightarrow \,^4He + p$$

is noteworthy not only because of its high (18.3 MeV) energy release, but also because the reaction products are both charged particles, offering the possibility of high efficiency, direct energy conversion. Direct energy conversion would involve the extraction of the positively charged ions and negatively charged electrons from the reaction region directly onto collection electrodes having a potential difference of the same order of magnitude as the mean kinetic energy of the charged particles. A variation of these above reaction schemes is the catalyzed D–D reaction wherein the external feedstock is deuterium but the ^{3}He and T produced in the D–D reactions are recycled and burned *in situ* to enhance the net energy yield.

Because of its relatively high reactivity (2), the D–T fusion-fuel cycle is very likely to be employed in the first generation of fusion reactors. This implies the use of a neutron absorbing blanket and thermal (Carnot) conversion efficiencies. Deuterium, also known as heavy hydro-

gen, occurs naturally in the ratio of 1:6700 relative to ordinary hydrogen; 30,000 kg water contains one kilogram of deuterium. The separation of deuterium from water is a relatively simple and inexpensive process. Tritium, on the other hand, is a radioactive isotope of hydrogen found in nature only in trace amounts and has a half-life of only 12.36 yr. The initial inventory of tritium for a power-producing D–T fusion power plant is a few kilograms and could be supplied, for example, from heavy-water fission reactors where it is produced as a by-product. Further tritium needs can be met by breeding additional tritium in the fusion reactor itself, by absorbing the fusion-produced neutrons in a blanket of lithium and exploiting the reaction

$$n + \,^6Li \rightarrow T + \,^4He + 4.8 \text{ MeV}$$

^{6}Li accounts for about 7.5% of natural lithium and is abundantly available in the earth's crust and oceans. Detailed fusion-blanket designs incorporate additional isotopes, such as ^{7}Li and ^{9}Be, which provide neutron-multiplying reactions, to compensate for the leakage and absorption losses of neutrons. A D–T fusion reactor, then, is in reality a consumer of deuterium and lithium. The estimated reserves of lithium should prove sufficient for at least several hundred years of D–T fusion power plant operation, even allowing for a significant increase in the world demand for energy (3). A common fusion evolution scenario relies on D–T fusion to fulfill the energy needs until the more difficult fuel cycle involving pure deuterium can be implemented. Then deuterium alone would be the fuel for the fusion energy economy. Because each liter of seawater contains enough deuterium to supply the energy equivalent of 300 L of gasoline, long-term energy needs would be assured.

Although the D–T reaction is the easiest route to fusion power production, it is no easy task to meet the conditions required to produce net fusion energy. Relative kinetic energies between the deuterons and tritons of 10 keV or more are necessary for practical energy generation, corresponding to relative particle velocities on the order of 10^6 m/s. Fusion-produced neutrons have, in fact, been created by impinging a beam of accelerated deuterons onto a solid target containing tritium. Unfortunately, a fusion reactor cannot be built around this concept because most of the incident beam energy is dissipated nonproductively through scattering and collision events in the target, and only a relatively small number of energy-producing fusion reactions occur. Other approaches, involving colliding beams of particles, have been proposed, but such schemes are inherently of very low power density and are not likely to yield practical energy sources.

PLASMA CONDITIONS REQUIRED FOR NET ENERGY RELEASE

The most promising approach to attaining significant reaction rates is to heat the reacting species to a high temperature, thereby imparting large kinetic energies to the nuclei in the form of thermal motions. By doing so, the particles, eg, deuterons and tritons, may scatter among themselves many times before undergoing fusion reactions, without losing significant energy from the system. At any given temperature, a system of particles in thermal equilibrium is characterized by a Maxwellian distribution of kinetic energies. The particles at the high energy end of this distribution account for most of the fusion reactions in fusion experiments.

The fusion fuel, when undergoing thermonuclear reactions, exists as an ionized gas called a plasma. In physics, the plasma state usually means a high temperature gas of net electrical neutrality consisting of free electrons and ions exhibiting collective behavior. The collection of charged particles exhibits characteristics of an electrically conducting fluid that can interact with electromagnetic fields. As such, its physical behavior is much more complex than that of an ordinary gas, and plasma confinement can be disrupted or reduced by many different kinds of plasma instabilities and other loss mechanisms. There exists a large literature and a number of outstanding books on plasma physics, such as Reference 4.

In a plasma undergoing fusion reactions, the reactivity, and thus the fusion-power output rate, increases with increasing temperature. However, over a wide range of temperatures, as the temperature of the plasma is raised, the radiation losses are also increased, primarily because of bremsstrahlung, or continuum, ie, braking, radiation from the electrons. For any fusion-fuel system there exists a unique temperature at which the fusion power production is precisely balanced by the radiation losses. This temperature is called the ideal ignition temperature, and equals about 50 million K (5 keV) for a D–T plasma (1 keV = 11.6×10^6 K). For a D–D plasma, this temperature is considerably higher, about 400×10^6 K (40 keV), a fact which considerably increases the difficulty of using pure deuterium fuel. Furthermore, a fusion system must be operated above the ideal ignition temperature for net power production, typically by a factor of 2–5.

Besides having to satisfy a minimum temperature requirement, the plasma must be sufficiently dense and contained for a long enough time to yield net power. If the plasma burns above the ideal ignition temperature for some time period, τ, the fusion energy released must at least equal the energy required to heat the plasma to that temperature plus the energy radiated during that period. It can be shown that this condition is met by requiring that the product of the plasma density, n, and confinement time, τ, exceed a characteristic value which depends only on the temperature. The minimum value of the product $n\tau$ represents the least stringent condition for the plasma to be a net producer of fusion energy. For D–T plasmas, this minimum occurs at a temperature of about 100×10^6 K, for which $n\tau \sim 10^{20}$ s/m^3. This minimum $n\tau$ product is called the Lawson criterion (5). For D–D, the minimum $n\tau$ product is about 10^{22} s/m^3 at a higher temperature, again indicating that a pure deuterium system requires a higher quality of confinement. A commonly used measure of the quality of plasma confinement is given by the triple product of the plasma density, n, ion temperature T_i, and energy confinement time, τ, usually expressed in units of keV·s/m^3. A primary goal of fusion research is to achieve $n\tau T_i$ values of $\sim 10^{22}$ keV·s/m^3, as required for a D–T reactor. Experiments as of this writing

(1993) have reached a value of 1.1×10^{21} keV·s/m^3 in the JT-60 tokamak in Japan (6).

Plasmas at fusion temperatures cannot be kept in ordinary containers because the energetic ions and electrons would rapidly collide with the walls and dissipate their energy. A significant loss mechanism results from enhanced radiation by the electrons in the presence of impurity ions sputtered off the container walls by the plasma. Therefore, some method must be found to contain the plasma at elevated temperatures without using material containers.

Once a fusion reaction has begun in a confined plasma, it is planned to sustain it by using the hot, charged-particle reaction products, eg, alpha particles in the case of D–T fusion, to heat other, colder fuel particles to the reaction temperature. If no additional external heat input is required to sustain the reaction, the plasma is said to have reached the ignition condition. Achieving ignition is another primary goal of fusion research.

PATHS TO FUSION POWER

Two diverse technical approaches to fusion power, magnetic confinement fusion, also known as magnetic fusion energy (MFE) and inertial confinement fusion, also known as inertial fusion energy (IFE) are being pursued worldwide. These form the basis of a large number of fusion research programs. Magnetic confinement techniques, studied since the 1950s, are based on the principle that charged particles such as electrons and ions, ie, deuterons and tritons, tend to be bound to magnetic lines of force. Thus the essence of the magnetic confinement approach is to trap a hot plasma in a suitably chosen magnetic field configuration for a long enough time to achieve a net energy release, which typically requires an energy confinement time of about one second. In the alternative IFE approach, fusion conditions are achieved by heating and compressing small amounts of fuel ions, contained in capsules, to the ignition condition by means of tightly focused energetic beams of charged particles or photons. In this case the confinement time can be much shorter, typically less than a millionth of a second.

Magnetic Confinement

In magnetic confinement, strong magnetic fields are used to confine the plasma. Electrons and ions in magnetic fields spiral in circles around the field lines but translate freely along the direction of the magnetic field. Thus the magnetic field of a long solenoid, for example, confines the plasma in two directions but does not prevent the particles from streaming from either end of the system. Furthermore, collisions between particles displace them from one field line to another, producing a net diffusion of plasma across the field toward the walls of the container. By employing more complex magnetic field configurations, fusion researchers have made significant progress toward solving the problem of magnetic confinement of plasmas by substantially reducing plasma losses.

One of the earliest configurations studied was the simple magnetic mirror. A simple mirror system is depicted in Figure 1. Particles gyrating about the field lines move freely along these lines until they enter regions of increased field strength at either end of the device. Conservation of angular momentum considerations dictate that, as the particles approach the end regions, they gyrate more energetically about the field lines and slow down in the direction of motion along the lines. Ultimately, their kinetic energy is completely converted into gyration energy, at which point the particles are reflected from these mirror points and return to the central, weaker field region. Particles having motion exactly along the axis of the device are not reflected and are lost through the ends. Although ingenious attempts have been made to reduce end losses from mirror machines and to make them stable against magnetohydrodynamic (MHD) and other instabilities, all single-cell mirror reactor designs have suffered from a high recirculating power fraction, ie, a large fraction of the output power has to be used to operate the reactor itself. In single-cell mirror machines these losses are fundamentally too high. The machines are referred to as too lossy, and the amount of injected power required to maintain the plasma, usually in the form of high energy neutral beams, has been too large to be practical.

A more advanced mirror approach involving multicells, called the tandem mirror, has been studied as a means to overcome the leakage problem. One way to view the tandem mirror is as a long uniform magnetic solenoid with two single-cell mirrors installed at the ends to electrostatically plug the device. Plasma end losses are impeded by electrostatic potentials developed by the plasma as the electrons and ions attempt to leave the device at different rates.

Another mirror variation is the field-reversed mirror configuration, in which the diamagnetic nature of the plasma is exploited to cause the interior magnetic field lines within the central region of a single-cell mirror to reverse and close on themselves. The plasma current responsible for this field modification is at right angles to the original field lines.

The problem with all the mirror approaches is that none has achieved the degree of confinement quality that the closed systems have. Closed systems are characterized by magnetic field lines that close on themselves so that charged particles following the field lines remain confined within the system.

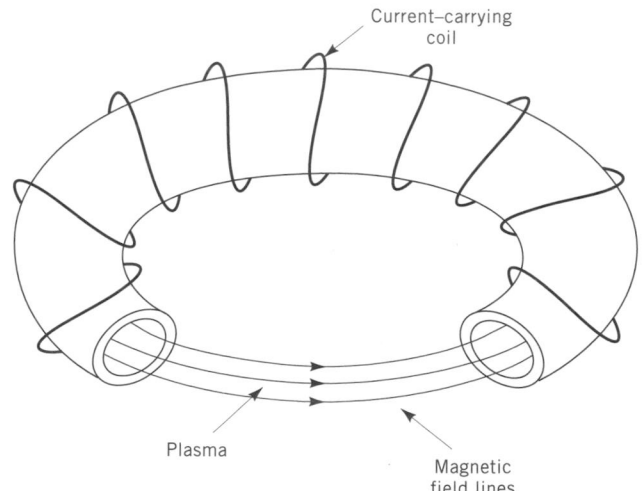

Figure 2. Cutaway view of a simple toroidal field configuration.

The simplest way of producing a closed configuration is to employ a torus or doughnut-shaped container having current-carrying coils wrapped around the minor diameter as shown in Figure 2. In this geometry, the magnetic lines of force are circles that traverse the torus and provide endless paths for the plasma ions and electrons to spiral about. Unfortunately, such a simple toroidal configuration is well known to have very poor confinement properties, because the magnetic field strength is not constant across the plasma. Instead, it is stronger at the inner wall and weaker at the outer wall of the toroidal chamber. As a result, the positive ions and electrons drift in opposite directions across the field lines and establish an electric field within the plasma. This electric field, coupled with the applied magnetic field, then causes the plasma as a whole to move to the container wall and dissipate its energy (2).

This deleterious effect can be obviated by introducing additional components of magnetic field, causing the field lines to circumscribe the torus without ever closing on themselves. The net magnetic field is then composed of a major, or toroidal, field component produced by the current coils, plus a smaller poloidal component which gives the desired twist to the lines. Particle drifts weaken or nullify the harmful electrical field and the plasma no longer tends to move to the walls.

Several geometries for producing the required poloidal magnetic field component have been studied. The class of plasma devices called tokamaks generates the poloidal (ie, around the minor circumference) field component from a toroidal (ie, around the major circumference) current in the plasma itself, either induced by a pulse from an external transformer or driven by external current-drive systems. The basic components of a tokamak are shown in Figure 3. External current-drive systems, such as high energy neutral beam injection or radio-frequency (r-f) current drive, impart a net toroidal momentum to one of the charged species, ions or electrons. A toroidal system related to the tokamak, which has a higher plasma energy density, is called the reversed field pinch (RFP). The RFP is an inherently pulsed, or batch-burn device.

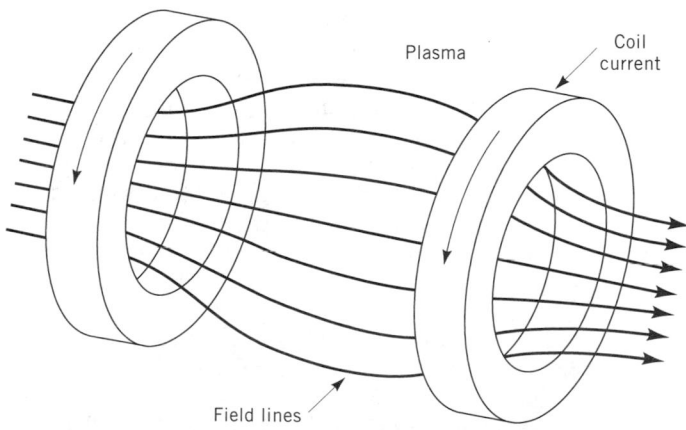

Figure 1. Simple magnetic mirror open configuration.

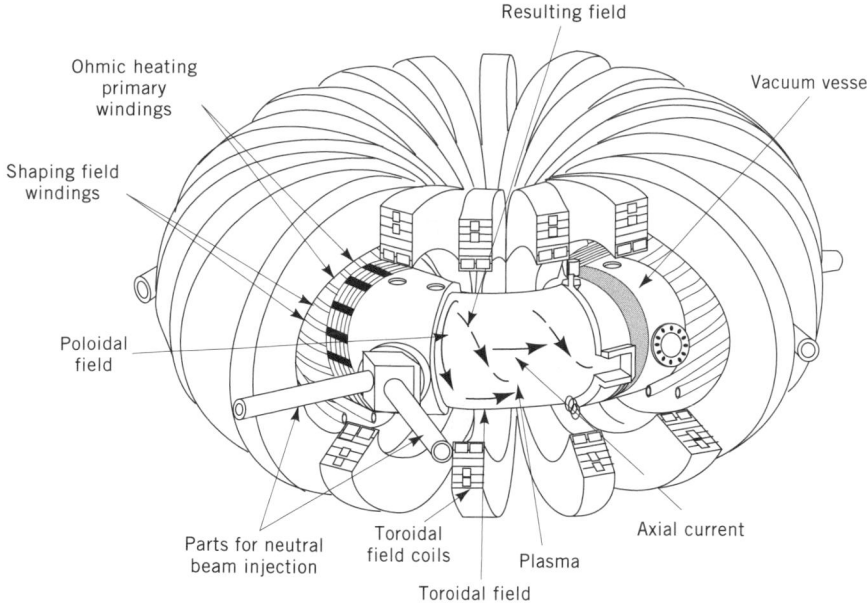

Figure 3. The tokamak fusion approach.

The poloidal field component can also be created externally by using current-carrying coils that wind helically around the outside surface of the torus. Such devices, called stellerators or torsatrons, have the advantage of not requiring a net toroidal current. The helical field windings, however, make these machines mechanically and magnetically more complex than the tokamak.

Tokamak. The design concept that has come the closest by far to achieving energy breakeven conditions is the tokamak. Invented in the 1950s by the Russian physicists Andrei Sakharov and Igor E. Tamm (7), the tokamak derives its name from the Russian acronym for toroidal magnetic chamber. Technical progress in tokamaks was dramatic in the late 1980s and early 1990s (8,9). Central ion temperatures of 400×10^6 K have been reached, and energy confinement times have increased from 0.02 to about 1.4 seconds for strongly heated plasmas. The result has been $n\tau T_i$ triple products of about 10^{21} keV·s/m³, compared to a value of $\sim 10^{22}$ keV·s/m³ required for a steady state D–T power plant.

Other important parameters have also shown dramatic gains. The normalized plasma pressure which is usually called beta, the plasma pressure divided by the confining magnetic field pressure, has been increased fourfold, to about 10%. This value is actually higher than that needed in a reactor. Bootstrap currents have been measured for the first time in several experiments. Bootstrap current is the name given to a toroidal current, theoretically predicted to arise spontaneously in tokamaks under near-reactor conditions. These in principle can eliminate much of the need for external current drive. Bootstrap currents open the possibility of a self-sustained steady-state tokamak power plant. Results from the large Japanese tokamak JT-60 are particularly interesting in this regard, where up to 80% of the 500,000 A of total plasma current is attributed to the bootstrap effect (8). Table 1 summarizes some of the progress made in the parameters of interest for magnetic fusion since 1971.

Some of the tokamaks in operation around the world, on which the data in Table 1 were obtained are

Designation	Tokamak	Location
ALC-A	Alcator-A	Plasma Fusion Center, Massachusetts Institute of Technology (MIT), Cambridge, Mass.

Table 1. Tokamak Plasma Parameters

Parameter	Achieved 1971	Achieved 1981	Achieved 1991	Required for Steady-State D–T power plant
Central ion temperature, T_i, keV	0.5	7	35	30
Central electron temperature, T_e, keV	1.5	3.5	15	30
Energy confinement time, τ, s	0.007	0.02	1.4	3
Triple product, $n\tau T_i$, keV · s/m³	1.5×10^{17}	5.5×10^{18}	9×10^{20}	7×10^{21}
Normalized plasma pressure (β), %	0.1	3	11	5
Fusion reactivity, reactions per second				
D–D		3×10^{14}	1×10^{17}	
D–T			6×10^{17}	10^{21}

Designation	Tokamak	Location
ALC-C	Alcator-C	Plasma Fusion Center, MIT
ASDEX	Axially Symmetric Divertor Experiment	Max Planck Institute for Plasma Physics, Garching, Germany
ATC	Adiabatic Toroidal Compressor	Princeton Plasma Physics Laboratory (PPPL), Princeton, N.J.
C-MOD	ALC-C Modified	MIT
DIII	Doublet III	General Atomics, San Diego, Calif.
DIII-D	Doublet III-D	General Atomics, San Diego, Calif.
ISX-B	Impurity Studies Experiment B	Oak Ridge National Laboratory (ORNL), Oak Ridge, Tenn.
JET	Joint European Torus	Abingdon, England
JFT-2M		Japan Atomic Energy Research Institute, Tokai, Japan
JT-60		Japan Atomic Energy Research Institute, Naka, Japan
ORMAK	Oak Ridge Tokamak	ORNL
PDX	Princeton Divertor Experiment	PPPL
PLT	Princeton Large Torus	PPPL
T-3	Tokamak-3	Kurchatov Institute, Moscow
T-10	Tokamak-10	Kurchatov Institute, Moscow
T-15	Tokamak-15	Kurchatov Institute, Moscow
TFR	Tokamak Fontenay-aux-Roses	Centre d'Etudes Nucleaire, Fontenay-aux-Roses, France
TFTR	Tokamak Fusion Test Reactor	PPPL

Additionally, two other tokamaks, the International Thermonuclear Experimental Reactor (ITER) for which the location is under negotiation, and the Tokamak Physics Experiment at PPPL, Princeton, New Jersey, are proposed. The most impressive advances have been obtained on the three biggest tokamaks, TFTR, JET, and JT-60, which are located in the United States, Europe, and Japan, respectively. As of this writing fusion energy development in the United States is dependent on federal funding (10–12).

Until 1992, tokamak experiments were performed using deuterium or hydrogen only. The use of radioactive tritium greatly complicates the operation of experimental facilities, impeding the pace of research. Certain experiments, however, such as those directly involving D–T fusion, cannot be done without the use of tritium. A European research team in 1992 produced nearly 2 million watts of fusion power for about one second in the JET device, and opened the modern frontier of D–T fusion experiments (13). Only about half of the JET fusion energy release came from fusion in the thermal plasma, at temperatures of 15–20 keV. The other half came from fusion of the injected tritium beams striking the deuterium in the plasma. The ratio of tritium to deuterium was about 2% in JET. If a 50:50 mixture of tritium and deuterium had been used instead, an amount of fusion energy would have been released roughly equal to the energy required to heat and sustain the plasma, giving an energy gain, Q, of about unity. In December 1993, scientists at the Princeton Plasma Physics Laboratory initiated a series of experiments on the Tokamak Fusion Test Reactor (TFTR), introducing D–T fuel into the machine and producing over 6 MW of fusion power. For the first time in a tokamak experiment an approximately 50:50 mixture of deuterium and tritium was used as the fusion fuel. Preliminary analysis of the first 100 experimental runs indicated that the confinement in a D–T fuel mixture was better than in a pure deuterium plasma, the ion and electron temperatures were higher, and the plasma stored energy longer. No enhanced loss of alpha particles (the product of D–T fusion reactions) was observed as the fusion power was increased. These results are encouraging for tokamak-based power generation.

International Thermonuclear Experimental Reactor. One of the largest obstacles to the development of fusion power has been that high powered, and correspondingly expensive, research facilities are needed at each step of the reactor development path. ITER (pronounced "eater") is a project supported by the United States, Japan, the European community, and Russia, wherein each party contributes equally to the effort and shares equally in the results (9). The main reason for making the ITER an international effort is cost sharing. The project is managed under the auspices of the International Atomic Energy Agency (IAEA), and the design is based on the tokamak concept. The central purpose of the ITER is to demonstrate the scientific and technological feasibility of fusion power by achieving, for the first time, controlled ignition and extended burn in a D–T plasma. ITER is expected to accomplish this by demonstrating technologies essential to a reactor in an integr ꞈd system, and by integrated steady-state testing of the high heat-flux and nuclear components (9).

A conceptual design of ITER, done in 1988–1990 by an international team (14), utilizes superconducting magnets. The heating and current drive are provided by a combination of 1.3 MeV negative-ion neutral beams, lower-hybrid frequency rf, and electron-cyclotron frequency rf. The negatively charged beams of deuterons or tritons are to be accelerated to 1.3 MeV, neutralized, and then injected, unperturbed by the confining magnetic field, into the plasma. The design is based on a conservative assessment of physics knowledge and allows for operational and experimental flexibility. The conceptual design calls for a plasma major radius of 6 m, plasma minor radius of 2.1 m, plasma current of 22 MA, magnetic field of 4.85 T, average neutron wall loading of about 1 MW/ m², and fusion power of about 1 GW thermal.

The second phase of ITER, the engineering design activity (EDA), was begun in 1992 and is scheduled to be completed in 1998. The ITER engineering design is being conducted at three co-centers: San Diego, California; Garching, Germany; and Naka, Japan. At these co-centers, multinational teams focus on developing a mature design in sufficient detail to allow the construction of the machine, with industrial vendors able to bid on the fabrication and installation of ITER systems. The first ITER plasma could be made as early as 2005. D–T operation could begin a few years later.

Inertial Confinement

Because the maximum plasma density that can be confined is determined by the field strength of available magnets, MFE plasmas at power plant conditions are very diffuse. Typical plasma densities are on the order of one hundred-thousandth that of air at STP. The Lawson criterion is met by confining the plasma energy for periods of about one second. A totally different approach to controlled fusion attempts to create a much denser reacting plasma which, therefore, needs to be confined for a correspondingly shorter time. This is the basis of inertial fusion energy (IFE). In the IFE approach, small capsules or pellets containing fusion fuel are compressed to extremely high densities by intense, focused beams of photons or energetic charged particles as shown in Figure 4. Because of the substantially higher densities involved, the confinement times for IFE can be much shorter. In fact, no external means are required to effect the confinement; the inertia of the fuel mass is sufficient for net energy release to occur before the fuel flies apart. Typical burn times and fuel densities are 10^{-10} s and 10^{31}–10^{32} ions/m³, respectively. These densities correspond to a few hundred to a few thousand times that of ordinary condensed solids. IFE fusion produces the equivalent of small thermonuclear explosions in the target chamber. An IFE power plant design, therefore, must deal with very different physics and technology issues than an MFE power plant, although some requirements, such as tritium breeding, are common to both. Some of the challenges facing IFE power plants include the highly pulsed nature of the burn, the high rate at which the targets must be made and transported to the beam focus, and the interface between the driver beams and the reactor chamber (15).

Drivers. In inertial fusion the fuel is compressed and heated using driver beams. Achieving ignition requires a large amount of energy to be precisely controlled and delivered to the fuel target in a very short time, and the target must be capable of absorbing this energy efficiently. To produce net energy, the IFE system must have gain, ie, more energy output than was used to make, compress, and heat the fuel. Driver efficiency and capsule design and fabrication are therefore important issues for an IFE reactor (16).

The necessary energy can be delivered to the fuel by a variety of possible drivers. The four types of drivers receiving the most research attention are solid state lasers,

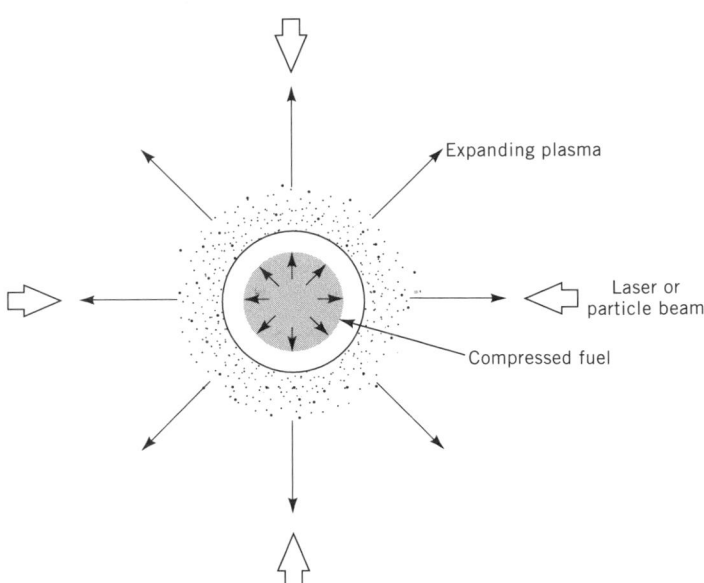

Figure 4. IFE capsule compression.

KrF lasers, light-ion accelerators, and heavy ion accelerators. The leading driver for target physics experiments worldwide is the solid-state laser, and in particular the Nd:glass laser. The reason is that the irradiances required for IFE are in the 10^{18}–10^{19} W/m² range (17). The Nd:glass laser was the first driver which could produce these large power densities on target and it has remained in the forefront because of its high performance, reliable technology, and relative ease of maintenance. Low efficiencies and pulse rates have traditionally eliminated Nd:glass lasers from serious consideration in IFE reactor designs. However, new Nd:glass technology, replacing flash lamp pumping with higher efficiency diode pumping and utilizing crystalline disks and gas cooling, could change this view. Higher driver efficiencies are achievable in KrF lasers and particle beam accelerators. Particle beams have thus far had difficulty in achieving the low divergences and small focal spots required for IFE experiments, a technical area where lasers have a natural advantage. In IFE power plants, however, focal spots as large as 1 cm are permitted, and it appears that both light ion and heavy ion drivers could meet this requirement.

Targets. Two types of IFE targets have been investigated, known as direct and indirect drive targets. Direct-drive targets absorb the energy of the driver directly onto the fuel capsule, whereas indirect-drive targets use a cavity, called a hohlraum, to convert the driver energy to x-rays which are then absorbed by the fuel capsule. This latter method can tolerate greater inhomogeneities in driver illumination, albeit at the expense of the efficient delivery of energy to the capsule.

The extremely high peak power densities available in particle beams and lasers can heat the small amounts of matter in the fuel capsules to the temperatures required for fusion. In order to attain such temperatures, however, the mass of the fuel capsules must be kept quite low. As a result, the capsules are quite small. Typical dimensions are less than 1 mm. Fuel capsules in power plants could be larger (up to 1 cm) because of the increased driver energies available.

Laser Fusion. The largest and most powerful operating laser in the world is the NOVA 10-beam Nd:glass laser facility at the Lawrence Livermore National Laboratory in California. NOVA can deliver up to 40 kJ of 351-nm light in a 1-ns pulse onto the target. NOVA is primarily used for indirect-drive experiments. Other large Nd:glass laser facilities include the GEKKO XII laser at Osaka University in Japan, and the OMEGA laser at the Laboratory for Laser Energetics at the University of Rochester (Rochester, New York). The latter is used primarily for direct-drive experiments.

The krypton-fluoride (KrF) laser, which uses a gaseous lasing medium, can in principle operate at much higher pulse repetition rates and efficiencies than solid-state Nd:glass lasers. Moreover, the shorter (250 nm) wavelength and broad bandwidth, both of which improve coupling to the target, provide additional advantages. However, the use of KrF lasers is complicated by the long pulse length, which, for the 1 ns time scales of IFE, has to be shortened by a factor of about 100. At least two schemes to do this have been proposed and demonstrated (15). In one method, angular multiplexing, many short, low power pulses are sent sequentially through the laser power amplifier stage for the entire duration of the pumping pulse, each at a different angle. After traversing paths of different optical length, these pulses are recombined at the target into a single high amplitude short pulse. In the second method, a long pulse is extracted and subsequently shortened in a Raman scattering cell filled with, for example, SF_6 gas (see INFRARED TECHNOLOGY AND RAMAN SPECTROSCOPY). Through Raman backscattering, the pulse can be shortened by the desired factor of 100. KrF laser technology is not as well developed as the technology for Nd:glass lasers, however, and no KrF lasers have been constructed which are as powerful as NOVA. The efficiency of KrF lasers may also fall a little short of that needed for a power plant.

Particle Beam Fusion. Advances in pulsed power technology have enabled large quantities of electrical energy to be generated in short pulses with relatively high efficiency and low cost. In a light-ion particle accelerator, an initial electrical pulse of the required energy is progressively shortened through a series of pulse forming steps to be delivered with an amplitude of several tens of megavolts to a diode which emits and accelerates the selected ions, eg, lithium, across a short gap to converge on the fuel capsule. The light-ion Particle Beam Fusion Accelerator II (PBFA II) at Sandia National Laboratory in New Mexico is the most energetic particle beam driver, delivering up to 1 MJ on target. However, obtaining good beam divergence has been a challenge.

To survive the effects of the target explosion, the diode must be located at least several meters away from the target. The diode on PBFA II is only about 15 cm from the target. Long-lived, reliable diodes having 10 Hz repetition rates and beam-transport systems several orders of magnitude longer than those in use as of this writing, are required to make a light-ion beam power plant feasible (15).

The Fusion Policy Advisory Committee of the Department of Energy has identified the heavy-ion accelerator as the leading candidate for an IFE power plant driver (16). The reasons include ruggedness, reliability, high pulse-rate capabilities, and potential for high efficiency. There are two different technologies being developed for heavy-ion accelerators: induction acceleration and radio frequency (rf) acceleration. The induction accelerator approach is pursued mainly in the United States, at the Lawrence Berkeley Laboratory. The rf accelerator approach is pursued primarily in Europe and Japan (15). The same types of heavy ions can be utilized in both approaches; typically cesium, bismuth, or xenon are chosen. To obtain the required $10^{18}–10^{19}$ W/m^2 on target for a power plant, using targets of 1 cm^2 size and accelerator energies limited to 5 GeV (to provide the requisite stopping distance inside the target fuel), particle beam currents of around 100,000 A are required. These currents are quite large compared to traditional high energy physics accelerators, and in experiments where high currents have been achieved, the beam divergence has been unsatisfactorily large.

ENVIRONMENTAL AND SAFETY ASPECTS

Fusion power plants are expected to be relatively benign environmentally when compared to other sources of power. A 1989 National Research Council report cites the environmental issue as a persuasive reason for pursuing the fusion energy option (10). A general environmental advantage of nuclear power plants whether fission or fusion, compared to fossil fuel plants, is the minimization of mining requirements and no emission of noxious effluents. A further advantage of fusion, relative to fission, is the absence of meltdown dangers and avoidance of long-lived radioactive wastes (see NUCLEAR REACTORS). An accidental runaway reaction cannot occur in a fusion reactor for two reasons. First, the amount of deuterium and tritium in the reactor at any given time is small, and any uncontrolled burning quickly consumes all the available fuel and extinguishes itself. Second, a neutron chain reaction of the fission-reactor type is impossible in fusion, because fusion reaction rates are not sustained by neutrons.

The fusion of deuterium and tritium produces only energetic neutrons and alpha particles (helium nuclei), which are not themselves radioactive. The 14-MeV neutrons are absorbed in a blanket surrounding the reacting plasma, and the only unavoidable ash of the D–T reaction is ordinary helium gas. The main concern about radiation comes from a secondary process, namely activation of the reactor components by the fusion neutrons. The secondary nuclear reactions which result from the energetic neutrons depend on the materials selected for the reactor blanket and support structure (18). The materials aspects of fusion reactors have been reviewed (19), and the calculated decay of radioactivity following shutdown of D–T fusion reactors constructed of various materials is shown in Figure 5, together with that of a fission reactor (8,18). If advanced structural materials such as silicon carbide,

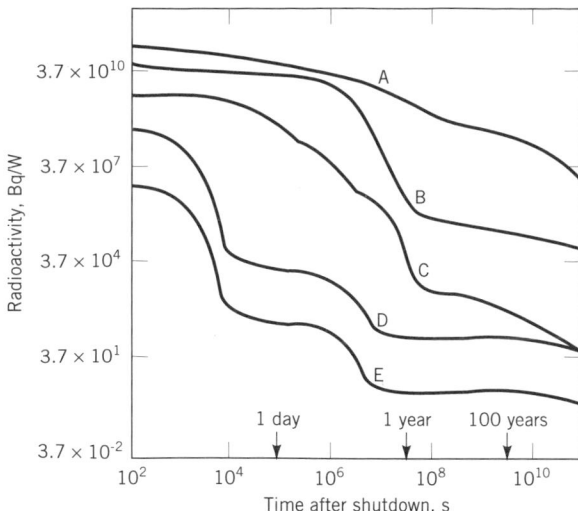

Figure 5. Radioactivity after shutdown per watt of thermal power for A, a liquid-metal fast breeder reactor, and for a D–T fusion reactor made of various structural materials: B, HT-9 ferritic steel; C, V-15Cr-5Ti vanadium–chromium–titanium alloy; and D, silicon carbide, SiC. There is the million-fold advantage of SiC over steel a day after shutdown. The radioactivity level after shutdown is also given for E, a SiC fusion reactor using the neutron reduced D–^{3}He fuel cycle.

SiC, can be used, fusion power plants are expected to reduce the amount of radioactive waste by six orders of magnitude or more.

A D–T fusion power plant is expected to have a tritium inventory of a few kilograms. Tritium is a relatively short-lived (12.36 year half-life) and benign (beta emitter) radioactive material, and represents a radiological hazard many orders of magnitude less than does the fuel inventory in a fission reactor. Clearly, however, fusion reactors must be designed to preclude the accidental release of tritium or any other volatile radioactive material. There is no need to have fissile materials present in a fusion reactor, and relatively simple inspection techniques should suffice to prevent any clandestine breeding of fissile materials, eg, for potential weapons diversion.

FUTURE DEVELOPMENTS AND APPLICATIONS

The goal of fusion development is central station electrical power generation. Using the D–T fuel cycle, power would be extracted from the thermalization of the neutron kinetic energy deposited in the blanket. Pulsed systems such as inertial fusion require storage techniques to provide a continuous output of electrical power. In some cases, this storage medium may be simply the thermal blanket surrounding the reaction chamber. In MFE, significant technological challenges include the development of large superconducting magnets, efficient current drive systems, and adequate diverter plates and plasma facing components to handle the high particle and radiation heat loads. Provisions must also be made for the replacement and maintenance of components by remote handling techniques.

Potential fusion applications other than electricity production have received some study. For example, radiation and high temperature heat from a fusion power plant could be used to produce hydrogen by the electrolysis or radiolysis of water, which could be employed in the synthesis of portable chemical fuels for transportation or industrial use. The transmutation of radioactive actinide wastes from fission reactors may also be feasible. This idea would utilize the neutrons from a fusion power plant to convert hazardous isotopes into more benign and easier-to-handle species. The practicality of these concepts requires further analysis.

Fusion energy research is also the primary avenue for the development of plasma physics as a scientific discipline. The technologies and the science of plasmas developed en route to fusion power are already important in other applications and fields of science.

COLD FUSION

In the spring of 1989, it was announced that electrochemists at the University of Utah had produced a sustained nuclear fusion reaction at room temperature, using simple equipment available in any high school laboratory. The process, referred to as cold fusion, consists of loading deuterium into pieces of palladium metal by electrolysis of heavy water, D_2O, thereby developing a sufficiently large density of deuterium nuclei in the metal lattice to cause fusion between these nuclei to occur. These results have proven extremely difficult to confirm (20,21). Neutrons usually have not been detected in cold fusion experiments, so that the D–D fusion reaction familiar to nuclear physicists does not seem to be the explanation for the experimental results, which typically involve the release of heat and sometimes gamma rays.

Room temperature fusion reactions, albeit low probability ones, are not a new concept, having been postulated in 1948 and verified experimentally in 1956 (22), in a form of fusion known as muon-catalyzed fusion. Since the 1989 announcement, however, international scientific skepticism has grown to the point that cold fusion is not considered a serious subject by most scientists. Follow-on experiments, conducted in many prestigious laboratories, have failed to confirm the claims, and although some unexplained and intellectually interesting phenomena have been recorded, the results have remained irreproducible and, thus far, not accepted by the scientific community.

Reprinted from *Kirk-Othmer Encyclopedia of Chemical Technology*, 4th ed., Vol. 12, John Wiley & Sons, Inc., New York, 1995.

BIBLIOGRAPHY

1. H. A. Bethe, *Phys. Rev.* **55**, 103 (1939).
2. S. Glasstone and R. H. Lovberg, *Controlled Thermonuclear Reactions,* D. Van Nostrand, New York, 1960.
3. R. A. Gross, *Fusion Energy,* John Wiley & Sons, Inc., New York, 1984.
4. L. Spitzer, Jr., *Physics and Fully Ionized Gases,* 2nd rev. ed., John Wiley & Sons, Inc., New York, 1962.

5. J. D. Lawson, *Proc. Phys. Soc.* **B70,** 6(1957).

6. O. Naito and co-workers, *Plasma Phys. Control. Fusion* **35,** B215–B222 (1993); T. Kondo and co-workers, "High Performance and Current Drive Experiments in JA-ERI Tokamak-60 Upgrade," *Phys. Plasmas* (in print) (1994); H. Ninomiya and co-workers, "Recent Progress and Future Prospect of the JT-60 Program," in the *Proceedings of the 15th Symposium on Fusion Engineering,* Hyannis, Mass., 1993.

7. I. E. Tamm and A. D. Sakharov, in M. A. Leontovich, ed., *Plasma Physics and the Problem of Controlled Thermonuclear Reactions,* Vol. 1, Pergamon Press, New York, 1961.

8. H. P. Furth, *Science* **249,** 1522 (Sept. 1990); J. G. Gordey, R. J. Goldston, and R. R. Parker, *Phys. Today,* 22 (Jan. 1992).

9. R. W. Conn and co-workers, *Sci. Am.* **266,** 103 (Apr. 1992).

10. *Pacing the U.S. Magnetic Fusion Program,* National Academy Press, Washington, D.C., 1989.

11. *Fusion Policy Advisory Committee Final Report, DOE / S-0081,* Department of Energy, Washington, D.C., Sept. 1990.

12. *National Energy Strategy, First Edition 1991 / 1992,* Department of Energy, Washington, D.C., Feb. 1991.

13. The JET Team, *Nuc. Fusion* **32,** 187 (1992).

14. International Atomic Energy Agency, *ITER Conceptual Design Report,* IAEA, Vienna, 1991.

15. W. J. Hogan, R. Bangerter, and G. L. Kulcinski, *Phys. Today,* 42 (Sept. 1992).

16. *Review of the Department of Energy's Inertial Confinement Fusion Programs,* National Academy Press, Washington, D.C., Sept. 1990.

17. J. D. Lindl, R. L. McCrory, and E. M. Campbell, *Phys. Today,* 32 (Sept. 1992).

18. R. W. Conn and co-workers, *Nucl. Fusion* **30,** 1919 (1990).

19. J. P. Holdren and co-workers, *Report of the Senior Committee on Environmental, Safety, and Economic Aspects of Magnetic Fusion Energy,* report UCRL-53766, Lawrence Livermore National Laboratory, Livermore, Calif., Sept. 25, 1989.

20. *Cold Fusion Research, DOE Report S-0073,* U.S. Dept. of Energy, Washington, D.C., Nov. 1989.

21. F. Close, *Too Hot to Handle: The Role for Cold Fusion,* Princeton University Press, Princeton, N.J., 1991.

22. B. V. Lewenstein and co-workers, *Forum for Applied Research and Public Policy,* **7**(4), 67–107 (Winter 1992).